ENCYCLOPEDIA
OF
FOOD
MICROBIOLOGY

Copyright © 2000 by ACADEMIC PRESS

The following articles are US government works in the public domain and are not subject to copyright: **Aspergillus**: Introduction; **Aspergillus**: *Aspergillus flavus*; *Cyclospora*; Helminths **and Nematodes**

Academic Press
A Harcourt Science and Technology Company
24–28 Oval Road, London NW1 7DX, UK
http://www.hbuk.co.uk/ap/

Academic Press
525 B Street, Suite 1900, San Diego, California 92101-4495, USA
http://www.apnet.com

ISBN 0-12-227070-3

A catalogue for this encyclopedia is available from the British Library

Library of Congress Catalog Card Number: 98-87954

Access for a limited period to an on-line version of the Encyclopedia of Food Microbiology is included in the purchase price of the print edition. This on-line version has been uniquely and persistently identified by the Digital Object Identifier (DOI)

10.1006/rwfm.2000

By following the link

http://dx.doi.org/10.1006/rwfm.2000

from any Web Browser, buyers of the Encyclopedia of Food Microbiology will find instructions on how to register for access.

Typeset by Selwood Systems, Midsomer Norton, Bath
Printed in Great Britain by The Bath Press, Bath

00 01 02 03 04 05 BP 9 8 7 6 5 4 3 2 1

ENCYCLOPEDIA
OF
FOOD
MICROBIOLOGY

Editor-in-Chief

RICHARD K. ROBINSON

Editors

CARL A. BATT

PRADIP D. PATEL

ACADEMIC PRESS

A Harcourt Science and Technology Company

**San Diego   San Francisco   New York   Boston
London   Sydney   Tokyo**

EDITORIAL ADVISORY BOARD

FOREWORD

Public concern about food safety has never been greater. In part this is due to the ever increasing demand from consumers for higher and higher standards. But new food-borne pathogens like *E. coli* O157 have emerged in recent years to become important public health problems, and changes in production and manufacturing sometimes reopen doors of opportunity for old ones. A powerful reminder that food scientists have much unfinished business to attend to is provided by the succession of food scares that generate strong stories for the media.

Experience tells us that science must underpin all approaches to food safety, whether through the application and implementation of well-tried approaches or the development of new or improved methods. Microbiologists have had a central role in this since the high quality work of pioneers like van Ermengem on botulism and Gaffky on typhoid more than a century ago. The large amount of important data that has accumulated since then joins with the current rapid rates of technological and scientific advance to make the need for a structured and authoritative source of information a very pressing one. It is provided by this encyclopedia.

These are exciting times for food microbiologists. Expectations are high that as scientists we will soon provide answers to the many problems still posed by microbes – from spoilage to food poisoning. Approaches like HACCP are making everyone think hard about how best to apply the data we have to develop better ways for reducing and eliminating food-borne pathogens. The pace of scientific developments continues to accelerate and more and better methods are available for the detection and enumeration of microbes than ever before. The microbes themselves continue to evolve and so present moving targets. The solid foundation presented by the mass of information in this encyclopedia provides the launching pad and guide for meeting these challenges.

It could be said that a penalty of working in food microbiology is that because the subject is broad-ranging, mature and dynamic, its practitioners, teachers and students have to know about many things in breadth and depth. For most of us, of course, this is not a penalty but an attractive bonus because of its intellectual challenge. I am particularly pleased to be associated with the encyclopedia because it will help us all to meet this test with confidence. I wish it every success.

Professor H Pennington
Department of Medical Microbiology
University of Aberdeen

PREFACE

Although food microbiology and food safety have, in recent times, become major concerns for governments around the world, equally important is the fact that, without yeasts and bacteria, popular meals like bread and cheese would not exist. Consequently, a knowledge of the relationship between foodstuffs and the activities of bacteria, yeasts and mycelial fungi has become a top priority for everyone associated with food and its production. Farmers have concerns related to produce harvesting and storage, food processors have to generate wholesome retail products that are both free from pathogenic organisms and have a satisfactory shelf life and, last but not least, food handlers and consumers need to be aware of the procedures necessary to ensure that food is safely prepared and stored.

In order for these disparate groups to operate successfully, accurate and objective information about the microbiology of foods is essential, and this encyclopedia seeks to provide a source of such information. In some areas, introductory articles are provided to guide readers who may be less familiar with the subject but, in general, superficiality has been avoided. Thus, the coverage has been developed to include details of all the important groups of bacteria, fungi, viruses and parasites, the various methods that can be employed for their detection in foods, the factors that govern the behaviour of the same organisms, together with an analysis of likely outcomes of microbial growth/metabolism in terms of disease and/or spoilage. A further series of articles describes the contribution of microorganisms to industrial fermentations, to traditional food fermentations from the Middle or Far East, as well as during the production of the fermented foods like bread, cheese or yoghurt that are so familiar in industrialized societies. The division of these topics into 358 articles of approximately 4000 words, has meant that the contributing authors have been able to handle their specialist subject(s) in real depth.

Obviously, another group of editors might have approached the project in a different manner, but we feel confident that this encyclopedia will provide readers at all levels of expertise with the data being sought. A point enhanced, perhaps, by the inclusion at the end of each article of a list for further reading, comprising a selection of review articles and key research papers that should encourage further exploration of any selected topic. If this confidence is borne out in practice, then the efforts of the contributors, the members of the Editorial Board and the editorial team from Academic Press will be well rewarded, for raising the scientific profile of food microbiology is long overdue.

R.K. Robinson, C.A. Batt, P.D. Patel
Editors

INTRODUCTION

The advent of antibiotics gave the general public, and many professional microbiologists as well, the feeling that bacterial diseases were under control, and the elimination of smallpox and the control of polio suggested that even viruses posed few problems. However, this complacency has received a nasty jolt over the last decade, and the emergence of HIV and multiple-drug-resistant strains of bacteria has become a major concern for the medical profession. The food industry has been similarly shaken by the appearance of new, and potentially fatal, strains of *Escherichia coli*, a species that for over 100 years was regarded as little more than a nuisance. Equally unexpected was the devastating impact of BSE, and fresh reports of the activities of so-called 'emerging food-borne pathogens' are appearing with alarming regularity.

In some cases, it has been possible to understand, with the advantages of hindsight, why a particular species of bacterium, fungus or protozoan has become a major risk to human health while, on other occasions, the vagaries of nature have left the 'experts' totally bemused. However, even in these latter situations, control over the threat posed to food supplies has to be instituted, but the ability of the food industry, in conjunction with Public Health and other bodies, to develop effective responses can only be as good as the scientific knowledge available. In the case of food microbiology, this background has to be derived from a wide range of sources. Thus, agricultural practices may alter the biochemistry of a crop and, perhaps, its microflora as well; the microflora of any given foodstuff and/or processing facility will have specific characteristics that need to be understood before control is possible; techniques must be available to monitor a retail food for microorganisms that would pose a risk to the consumer. As the procedures necessary to monitor these various facets become ever more sophisticated, so fewer microbiologists can claim total competence, and the need for a specialist source of outside knowledge increases.

It is this latter need that the *Encyclopedia of Food Microbiology* seeks to satisfy for, within this work, a busy microbiologist can find details of all the important genera of food-borne bacteria and fungi, how the same genera may react in different foods and under different environmental conditions, and how to detect the growth and/or metabolism of the same organisms in foods using classical or modern techniques. In order to place this information into a broader context, the reader can explore the latest advice concerning food standards/specifications, or the role of monitoring systems like HACCP in achieving product targets for specific microorganisms; potential concerns over viruses and protozoa are also evaluated in the light of current knowledge. Readers interested in fermented foods will find the pertinent information in a similarly accessible form; indeed, purchasers of the print version of the encyclopedia will be entitled to register for access to the on-line version as well. This form allows the user the benefit of extensive hypertext linking and advanced search tools, adding value to the encyclopedia as a reference source, teaching aid and text for general interest.

It is inevitable, of course, that short articles written to a tight deadline may have omissions, but it is to be hoped that such faults arc minimal and, in any event, more than compensated for through the careful selections of further reading. If this optimism is justified, then the major credit rests with the authors of each article. They are all recognized as experts in their fields, and their willing participation has been much appreciated by the editors. The role of the Editorial Advisory Board merits a special mention as well, for their constructive criticisms of the list of articles, their suggestions for authors and their expert refereeing of the manuscripts has provided a solid foundation for the entire enterprise.

However, the finest manuscripts are of little value to the scientific community until they have been published, and the editorial team at Academic Press – Carey Chapman (Editor-in-Chief), Tina Holland (Associate Editor), Nick Fallon (Commissioning Editor), Laura O'Neill (Editorial Assistant), Tamsin Cousins (Production Project Manager), Richard Willis (Freelance Project Manager), Emma Parkinson (Electronic Publishing Developer), Peter Lord (Publishing Services Manager), Emma Krikler (Picture Researcher) – have been outstanding in their support of the project. Obviously, each member of the team has made an important contribution, but it must be recorded that the role of Tina Holland has been absolutely invaluable. Thus, not only has Tina co-ordinated the numerous inputs from the editors, referees and authors, but even found time to help the editors with the location of authors; the editors acknowledge this unstinting assistance with much gratitude.

R.K. Robinson, C.A. Batt, P.D. Patel
Editors

GUIDE TO USE OF THE ENCYCLOPEDIA

Structure of the Encyclopedia

The material in the Encyclopedia is arranged as a series of entries in alphabetical order. Some entries comprise a single article, whilst entries on more diverse subjects consist of several articles that deal with various aspects of the topic. In the latter case the articles are arranged in a logical sequence within an entry.

To help you realize the full potential of the material in the Encyclopedia we have provided three features to help you find the topic of your choice.

1. Contents Lists

Your first point of reference will probably be the contents list. The complete contents list appearing in each volume will provide you with both the volume number and the page number of the entry. On the opening page of an entry a contents list is provided so that the full details of the articles within the entry are immediately available.

Alternatively you may choose to browse through a volume using the alphabetical order of the entries as your guide. To assist you in identifying your location within the Encyclopedia a running headline indicates the current entry and the current article within that entry.

You will find 'dummy entries' where obvious synonyms exist for entries or where we have grouped together related topics. Dummy entries appear in both the contents list and the body of the text. For example, a dummy entry appears for Butter which directs you to Milk and Milk Products: Microbiology of Cream and Butter, where the material is located.

Example

If you were attempting to locate material on Dairy Products via the contents list.

DAIRY PRODUCTS *see BRUCELLA*: Problems with Dairy Products; CHEESE: In the Market Place; Microbiology of Cheese-making and Maturation; Mould-ripened Varieties; Role of Specific Groups of Bacteria; Microflora of White-brined Cheeses; FERMENTED MILKS: Yoghurt; Products from Northern Europe; Products of Eastern Europe and Asia; PROBIOTIC BACTERIA: Detection and Estimation in Fermented and Non-fermented Dairy Products

At the appropriate location in the contents list, the page numbers for articles under *Brucella*, etc. are given.

If you were trying to locate the material by browsing through the text and you looked up Dairy Products then the following information would be provided.

> **DAIRY PRODUCTS** *see BRUCELLA*: Problems with Dairy Products; **CHEESE**: In the Market Place; Microbiology of Cheese-making and Maturation; Mould-ripened Varieties; Role of Specific Groups of Bacteria; Microflora of White-brined Cheeses; **FERMENTED MILKS**: Yoghurt; Products from Northern Europe; Products of Eastern Europe and Asia; **PROBIOTIC BACTERIA**: Detection and Estimation in Fermented and Non-fermented Dairy Products.

Alternatively, if you were looking up *Brucella* the following information would be provided.

BRUCELLA

Contents
Characteristics
Problems with Dairy Products

2. Cross References

All of the articles in the Encyclopedia have an extensive list of cross references which appear at the end of each article, e.g.:

ATP BIOLUMINESCENCE/Application in Dairy Industry.
See also: **Acetobacter. ATP Bioluminescence**: Application in Meat Industry; Application in Hygiene Monitoring; Application in Beverage Microbiology. **Bacteriophage-based Techniques for Detection of Food-borne Pathogens. Biophysical Techniques for Enhancing Microbiological Analysis**: Future Developments. **Electrical Techniques**: Food Spoilage Flora and Total Viable Count (TVC). **Immuno-magnetic Particle-based Techniques**: Overview. **Rapid Methods for Food Hygiene Inspection. Total Viable Counts**: Pour Plate Technique; Spread Plate Technique; Specific Techniques; MPN; Metabolic Activity Tests; Microscopy. **Ultrasonic Imaging**: Non-destructive Methods to Detect Sterility of Aseptic Packages. **Ultrasonic Standing Waves**.

3. Index

The index will provide you with the volume number and page number of where the material is to be located, and the index entries differentiate between material that is a whole article, is part of an article or is data presented in a table. On the opening page of the index, detailed notes are provided.

4. Colour Plates

The colour figures for each volume have been grouped together in a plate section. The location of this section is cited both in the contents list and before the *See also* list of the pertinent articles.

5. Contributors

A full list of contributors appears at the beginning of each volume.

CONTRIBUTORS

Lahsen Ababouch
Department of Food Microbiology and Quality Control
Institut Agronomique et Vétérinaire Hassan II
Rabat
Morocco

D Abramson
Agriculture & Agri-Food Canada
Cereal Research Centre
195 Dafoe Road
Winnipeg
Manitoba
R3T 2M9
Canada

Ann M Adams
Seafood Products Research Center
US Food and Drug Administration
PO Box 3012
22201 23rd Drive SE
Bothell
WA 98041–3012
USA

Martin R Adams
School of Biological Sciences
University of Surrey
Guildford
GU2 5XH
UK

G E Age
PO Box 553
Wageningen
The Netherlands

M Ahmed
Food Control Laboratory
PO Box 7463
Dubai
United Arab Emirates

Imad Ali Ahmed
Central Food Control Laboratory, Ajman Municipality
PO Box 3717
Ajman
UAE

Peter Ahnert
Department of Biochemistry
Ohio State University
Columbus
OH 43210
USA

William R Aimutis
Land O' Lakes Inc.
PO Box 674101
St Paul
Minnesota
55164–0101
USA

J H Al-Jedah
Central Laboratories
Ministry of Public Health
Qatar

Cameron Alexander
Macromolecular Science Department
Institute of Food Research,
Reading Laboratory
Earley Gate
Whiteknights Road
Reading
RG6 6BZ
UK

Marcos Alguacil
Departmento de Genética
Facultad de Ciencias, Universidad de Málaga
Spain

M Z Ali
Central Laboratories
Ministry of Public Health
Qatar

M D Alur
Food Technology Division
Bhabha Atomic Research Centre
Mumbai 400085
India

R Miguel Amaguaña
US Food and Drug Administration
Washington, DC
USA

Vilma Moratade de Ambrosini
Centro de Referencia para Lactobacilos and Universidad
Nacional de Tucumán
Casilla de Correo 211
(4000)-Tucuman
Argentina

Wallace H Andrews
US Food and Drug Administration
Washington, DC 20204
USA

Dilip K Arora
Department of Botany
Banaras Hindu University
Varanasi 221 005
India

B Austin
Department of Biological Sciences
Heriot-Watt University
Riccarton
Edinburgh EH14 4AS
Scotland, UK

Aslan Azizi
Iranian Agricultural Engineering Research Institute
Agricultural Research Organization
Evin Tehran
Iran

S De Baets
Laboratory of Industrial Microbiology and Biocatalysis
Department of Biochemical and Microbial Technology
Faculty of Agricultural and Applied Biological Sciences
University of Gent
Coupure links 653
B-9000
Gent
Belgium

Les Baillie
Biomedical Sciences
DERA
CBD Porton Down
Salisbury
Wiltshire
UK

Gustavo V Barbosa-Cánovas
Biological Systems Engineering
Washington State University
Pullman
Washington 99164–6120
USA

J Baranyi
Institute for Food Research
Reading
UK

Eduardo Bárzana
Departamento de Alimentos y Biotecnología
Facultad de Química
Universidad Nacional Autónoma de México
Mexico City 04510
Mexico

Carl A Batt
Department of Food Science
Cornell University
Ithaca
NY 14853
USA

Derrick A Bautista
Saskatchewan Food Product Innovation Program
Department of Applied Microbiology and Food Science
University of Saskatchewan
Canada

S H Beattie
Hannah Research Institute
Ayr KA6 5HL
UK

H Beck
Department for Health Service
South Bavaria
Veterinärstrasse 2
85764 Oberschleissheim
Germany

Reginald Bennett
FDA
Center for Food Safety and Applied Nutrition
Washington, DC
USA

Marjon H J Bennik
Agrotechnological Research Institution (ATO-DLO)
Bornsesteeg 59
6709 PD
Wageningen
The Netherlands

Merlin S Bergdoll (dec)
Food Research Institute
University of Wisconsin-Madison
Madison, WI
USA

R G Berger
Food Chemistry
University of Hannover
Germany

K Berghof
BioteCon Gesellschaft für Biotechnologische
Entwicklung und Consulting
Hermannswerder Haus 17
14473 Potsdam
Germany

P A Bertram-Drogatz
Mediport VC Management GmbH
Wiesenweg 10
12247 Berlin
Germany

Gail D Betts
Campden and Chorleywood Food Research Association
Chipping Campden
Gloucestershire
GL55 6LD
UK

R R Beumer
Wageningen Agricultural University
Laboratory of Food Microbiology
Bomenweg 2
NL 6703 HD Wageningen
The Netherlands

Rijkelt R Beumer
Wageningen University and Research Centre
Department of Food Technology and Nutritional Sciences
Bomenweg 2
NL 6703 HD Wageningen
The Netherlands

Saumya Bhaduri
Microbial Food Safety Research Unit
Eastern Regional Research Center
Agricultural Research Service
US Department of Agriculture
600 East Mermaid Lane
Wyndmoor
PA 19038
USA

Deepak Bhatnagar
Southern Regional Research Center
Agricultural Research Service
US Department of Agriculture
LA
USA

J R Bickert
Halosource Corporation
First Avenue South
Seattle
WA 98104
USA

Hanno Biebl
GBF – National Research Centre for Biotechnology
Braunschweig
Germany

Clive de W Blackburn
Microbiology Unit
Unilever Research Colworth
Colworth House
Sharnbrook
Bedford
UK

I S Blair
Food Studies Research Unit
University of Ulster at Jordanstown
Shore Road
Newtownabbey
Co. Antrim
Northern Ireland
BT37 9QB

G Blank
Department of Food Science
University of Manitoba
Winnipeg
MB
Canada

Hans P Blaschek
Department of Food Science and Human Nutrition
University of Illinois
488 Animal Science Lab
1207 West Gregory Drive
Urbana
IL 61801
USA

D Blivet
AFSSA
Ploufragan
France

R G Board
South Lodge
Northleigh
Bradford-on-Avon
Wiltshire
UK

Enne de Boer
Inspectorate for Health Protection
PO Box 9012
7200 GN Zutphen
The Netherlands

Christine Bonaparte
Department of Dairy Research and Bacteriology
Agricultural University
Gregor Mendel-Str. 33
A-1180 Vienna
Austria

Kathryn J Boor
Department of Food Science
Cornell University
Ithaca
NY 14853
USA

A Botha
Department of Microbiology
University of Stellenbosch
Stellenbosch 7600
South Africa

W Richard Bowen
Biochemical Engineering Group
Centre for Complex Fluids Processing
University of Wales Swansea
Singleton Park
Swansea
SA2 8PP
UK

Catherine Bowles
Leatherhead Food Research Association
Leatherhead
Surrey
UK

Patrick Boyaval
INRA
Laboratoire de Recherches de Technologie Laitière
65 rue de Saint-Brieuc
35042
Rennes Cedex
France

F Bozoğlu
Department of Food Engineering
Middle East Technical University
Ankara
Turkey

Astrid Brandis-Heep
Philipps Universität
Fachbereich Biologie
Laboratorium für Mikrobiologie
D-35032 Marburg
Germany

Susan Brewer
Department of Food Science and Human Nutrition
University of Illinois
Urbana
Illinois
USA

Aaron L Brody
Rubbright Brody Inc.
PO Box 956187
Duluth
Georgia
30095–9504
USA

Bruce E Brown
B. E. Brown Associates
328 Stone Quarry Priv.
Ottawa
Ontario
K1K 3Y2
Canada

G Bruggeman
Laboratory of Industrial Microbiology and Biocatalysis
Department of Biochemical and Microbial Technology
Faculty of Agricultural and Applied Biological Sciences
University of Gent
Coupure links 653
B-9000
Gent
Belgium

Andreas Bubert
Department for Microbiology
Theodor-Boveri Institute for Biosciences
University of Würzburg
Am Hubland
97074 Würzburg
Germany

Ken Buckle
Department of Food Science and Technology
The University of New South Wales
Sydney
Australia

Lloyd B Bullerman
Department of Food Science and Technology
University of Nebraska
PO Box 830919
Lincoln
NE 68583–0919
USA

Justino Burgos
Food Technology Section
Department of Animal Production and Food Science
University of Zaragoza
Spain

Frank F Busta
Department of Food Science and Nutrition
University of Minnesota
St Paul
Minnesota 55108
USA

Daniel Cabral
Departmento de Ciencias Biológicas
Facultad de Ciencias Exactas y Naturales
Pabellon II 4to piso – Ciudad Universitaria
1428 Buenos Aires
Argentina

María Luisa Calderón-Miranda
Biological Systems Engineering
Washington State University
Pullman
Washington 99164–6120
USA

Geoffrey Campbell-Platt
Gyosei Liaison Office
Gyosei College
London Road
Reading
Berks RG1 5AQ
UK

Iain Campbell
International Centre for Brewing and Distilling
Heriot-Watt University
Edinburgh
EH14 4AS
Scotland

Frédéric Carlin
Institut National de la Recherche Agronomique
Unité de Technologie des Produits Végétaux
Site Agroparc
84914
Avignon
Cedex 9
France

Brigitte Carpentier
National Veterinary and Food Research Centre
22 rue Pierre Curie
F-94709
Maisons-Alfort Cedex
France

Maria da Glória S Carvalho
Departamento de Microbiologia Médica
Instituto de Microbiologia
Universidade Federal do Rio de Janeiro
Rio de Janeiro 21941
Brazil

O Cerf
Alfort Veterinary School
7 Avenue du Général de Gaulle
F-94704
Maisons-Alfort Cedex
France

Lourdes Pérez Chabela
Universidad Autónoma Metropolitana-Iztapalapa
Mexico
Apartado Postal 55–535
CP 09340 Mexico DF
Mexico

Perng-Kuang Chang
Southern Regional Research Center
Agricultural Research Service
US Department of Agriculture
LA
USA

E A Charter
Canadian Inovatech Inc.
31212 Peardonville Road
Abbotsford
BC V2T 6K8
Canada

Parimal Chattopadhyay
Department of Food Technology and Biochemical
Engineering
Jadavpur University
Calcutta-700 032
India

Yusuf Chisti
Department of Chemical Engineering
University of Almería
E-04071 Almería
Spain

Thomas E Cleveland
Southern Regional Research Center
Agricultural Research Service
US Department of Agriculture
LA
USA

Dean O Cliver
University of California, Davis, School of Veterinary
Medicine
Department of Population Health and Reproduction
One Shields Avenue
Davis
California 95616–8743
USA

T E Cloete
Department of Microbiology and Plant Pathology
Faculty of Biological and Agricultural Sciences
University of Pretoria
Pretoria 0002
South Africa

Roland Cocker
Cocker Consulting
Bergeendlaan 16
1343 AR Almere
The Netherlands

Timothy M Cogan
Dairy Products Research Centre
Teagasc
Fermoy
Ireland

David Collins-Thompson
Nestlé Research and Development Center
210 Housatonic Avenue
New Milford
Connecticut
USA

Janet E L Corry
Division of Food Animal Science
Department of Clinical Veterinary Science
University of Bristol
Langford
Bristol
BS40 5DT
UK

Aldo Corsetti
Institute of Dairy Microbiology
Faculty of Agriculture of Perugia
06126 S. Costanzo
Perugia
Italy

Polly D Courtney
Department of Food Science and Technology
Ohio State University
2121 Fyffe Road
Columbus
OH 43210
USA

M A Cousin
Department of Food Science
Purdue University
West Lafayette
Indiana
47907–1160
USA

N D Cowell
Elstead
Godalming
Surrey
GU8 6HT
UK

Julian Cox
Department of Food Science and Technology
The University of New South Wales
Sydney
Australia

C Gerald Crawford
US Department of Agriculture
Agricultural Research Service
Eastern Regional Research Center
600 E. Mermaid Lane
Wyndmoor
PA 19038
USA

Theresa L Cromeans
Department of Environmental Sciences and Engineering
School of Public Health
University of North Carolina
North Carolina
USA

Kofitsyo S Cudjoe
Department of Pharmacology
Microbiology and Food Hygiene
Norwegian College of Veterinary Medicine
PO Box 8146 Dep
0033 Oslo
Norway

David Cunliffe
Macromolecular Science Department
Institute of Food Research
Reading Laboratory
Earley Gate, Whiteknights Road
Reading
RG6 6BZ
UK

Ladislav Čurda
Department of Dairy and Fat Technology
Prague Institute of Chemical Technology
Czech Republic

G J Curiel
Unilever Research Vlaardingen
PO Box 114
3130 AC Vlaardingen
The Netherlands

G D W Curtis
Bacteriology Department
John Radcliffe Hospital
Oxford
UK

Michael K Dahl
Department of Microbiology
University of Erlangen
Staudtstrasse 5
91058 Erlangen
Germany

Crispin R Dass
The Heart Research Institute Ltd
145 Missenden Road
Camperdown
Sydney
NSW 2050
Australia

E Alison Davies
Technical Services & Research Department
Aplin & Barrett Ltd (Cultor Food Science)
15 North Street
Beaminster
Dorset
DT8 3DZ
UK

Brian P F Day
Campden and Chorleywood Food Research Association
Chipping Campden
Gloucestershire
GL55 6LD
UK

J M Debevere
Laboratory of Food Microbiology and Food Preservation
Faculty of Agricultural and Applied Biological Sciences
University of Ghent
Coupure Links 654
9000 Ghent
Belgium

Joss Delves-Broughton
Technical Services and Research Department
Aplin & Barrett Ltd (Cultor Food Science)
15 North Street
Beaminster
Dorset
DT8 3DZ
UK

Stephen P Denyer
Department of Pharmacy
The University of Brighton
Cockcroft Building
Moulescoomb
Brighton
BN2 4GJ
UK

P M Desmarchelier
Food Safety and Quality
Food Science Australia
PO Box 3312
Tingalpo DC
Queensland 4173
Australia

Janice Dewar
CSIR Food Science and Technology
PO Box 395
Pretoria 001
South Africa

Vinod K Dhir
Biotec Laboratories Ltd
32 Anson Road
Martlesham Heath
Ipswich
Suffolk
IP5 3RD
UK

M W Dick
Department of Botany
University of Reading
Reading
RG6 6AU
UK

Vivian M Dillon
Department of Biology and Biochemistry
University of Bath
Bath
UK

Eleftherios H Drosinos
Department of Food Science and Technology
Laboratory of Microbiology and Biotechnology of Foods
Agricultural University of Athens
Iera Odos 75
Athens
Greece

F M Dugan
USDA–ARS Western Regional Plant Introduction Station
Washington State University
Washington
USA

B Egan
Marine Biological and Chemical Consultants Ltd
Bangor
UK

H M J van Elijk
Unilever Research Vlaardingen
PO Box 114
3130 AC Vlaardingen
The Netherlands

Hartmut Eisgruber
Institute for Hygiene and Technology of Foods of Animal
Origin, Veterinary Faculty
Ludwig-Maximilians University
80539 Munich
Germany

Phyllis Entis
QA Life Sciences, Inc.
6645 Nancy Ridge Drive
San Diego
CA 92121
USA

John P Erickson
Microbiology – Research and Development
Bestfoods Technical Center
Somerset
New Jersey
USA

Douglas E Eveleigh
Department of Microbiology
Rutgers University
Cook College
76 Lipman Drive
New Brunswick
NJ 08901-8525
USA

Richard R Facklam
Streptococcus Laboratory
Respiratory Diseases Branch
Division of Bacterial and Mycotic Diseases
Centres for Disease Control and Prevention
Mail Stop CO-2
Atlanta
GA 30333
USA

M Fandke
BioteCon Gesellschaft für Biotechnologische
Entwicklung und Consulting
Hermannswerder Haus 17
14473 Potsdam
Germany

Nana Y Farkye
Dairy Products Technology Center
Dairy Science Department
California Polytechnic State University
San Luis Obispo
CA 93407
USA

Manuel Fidalgo
Departmento de Genética
Facultad de Ciencias, Universidad de Málaga
Spain

Christopher W Fisher
Department of Food Science and Human Nutrition
University of Illinois
Urbana
IL 61801
USA

G H Fleet
CRC for Food Industry Innovation
Department of Food Science and Technology
The University of New South Wales
Sydney
New South Wales 2052
Australia

Harry J Flint
Rowett Research Institute
Greenburn Road
Bucksburn
Aberdeen
UK

Samuel Formal
Department of Microbiology and Immunology
Uniformed Services University of the Health Sciences
F Edward Hébert School of Medicine
4301 Jones Bridge Road
Bethesda
MD 20814
USA

Pina M Fratamico
US Department of Agriculture
Agricultural Research Service
Eastern Regional Research Center
600 E. Mermaid Lane
Wyndmoor
PA 19038
USA

Colin Fricker
Thames Water Utilities
Manor Farm Road
Reading
RG2 0JN
UK

Daniel Y C Fung
Department of Animal Sciences and Industry
Kansas State University
Manhattan
Kansas 66506
USA

H Ray Gamble
United States Department of Agriculture
Agricultural Research Service
Parasite Biology and Epidemiology Laboratory
Building 1040, Room 103, BARC-East
Beltsville
MD 20705
USA

Indrawati Gandjar
Department of Biology
Faculty of Science and Mathematics
University of Indonesia
Jakarta
Indonesia

Mariano García-Garibay
Departamento de Biotechnología
Universidad Autónoma Metropolitana
Iztapalapa, Apartado Postal 55–535
Mexico City 09340
Mexico

María-Luisa García-López
Department of Food Hygiene and Food Technology
University of León
24071-León
Spain

S K Garg
Department of Microbiology
Dr Ram Manohar Lohia Avadh University
Faizabad 224 001
India

A Gasch
BioteCon Gesellschaft für Biotechnologische
Entwicklung und Consulting
Hermannswerder Haus 17
14473 Potsdam
Germany

Michel Gautier
Ecole Nationale Supérieure d'Agronomie
Institut National de la Recherche Agronomique
65 rue de SrBrieuc
35042
Rennes cédex
France

Gerd Gellissen
Rhein Biotech GmbH
EichsFelder Str. 11
40595 Düsseldorf
Germany

N P Ghildyal
Fermentation Technology and Bioengineering
Department
Central Food Technological Research Institute
Mysore 570013
India

M Gibert
Institut Pasteur
Unité Interactions Bactéries Cellules
28 rue du Dr Roux
75724 Paris
Cedex 15
France

Glenn R Gibson
Microbiology Department
Institute of Food Research
Reading
UK

M C te Giffel
Wageningen Agricultural University
Laboratory of Food Microbiology
Bomenweg 2
NL 6703 HD Wageningen
The Netherlands

A Gilmour
Food Science Division (Food Microbiology)
Department of Agriculture for Northern Ireland
Agriculture and Food Science Centre
Newforge Lane
Belfast
BT9 5PX
Northern Ireland, UK

Giorgio Giraffa
Istituto Sperimentale Lattiero Caseario
Via A. Lombardo
11 – 26900 Lodi
Italy

R W A Girdwood
Scottish Parasite Diagnostic Laboratory
Stobhill Hospital
Glasgow
GL21 3UW
UK

Andrew D Goater
Institute of Molecular and Biomolecular Electronics
University of Wales
Dean St
Bangor
Gwynedd
LL57 1UT
UK

Marco Gobbetti
Instituto di Produzioni e Preparazioni Alimentari
Facoltà di Agraria di Foggia
Via Napoli 25
71100 Foggia
Italy

Millicent C Goldschmidt
Department of Basic Sciences
Dental Branch
The University of Texas Health Center at Houston
6516 John Freeman Avenue
Houston
Texas 77030
USA

Lorena Gómez-Ruiz
Departamento de Biotechnología
Universidad Autónoma Metropolitana
Iztapalapa, Apartado Postal 55–535
Mexico City 09340
Mexico

Katsuya Gomi
Division of Life Science
Graduate School of Agricultural Science
Tohoku University
Japan

M Marcela Góngora-Nieto
Biological Systems Engineering
Washington State University
Pullman
Washington 99164–6120
USA

S Gonzalez
Universidad Nacional de Tucumán, Argentina
Cerela–Conicet
San Miguel de Tucumán
Argentina

Silvia N Gonzalez
Centro de Referencia para Lactobacilos (Cerela) and
Universidad Nacional de Tucumán
Chacabuco 145 (4000)
Tucumán
Argentina

Leon G M Gorris
Unit Microbiology and Preservation
Unilever Research Vlaardingen
PO Box 114
3130 AC Vlaardingen
The Netherlands

Grahame W Gould
17 Dove Road
Bedford
MK41 7AA
UK

M K Gowthaman
Fermentation Technology and Bioengineering
Department
Mysore 570013
India

Lone Gram
Danish Institute for Fisheries Research
Department of Seafood Research
Technical University of Denmark Bldg 221
DK-2800 Lyngby
Denmark

AGE Griffioen
Stichting EFFI
PO Box 553 Wageningen
The Netherlands

Mansel W Griffiths
Department of Food Science
University of Guelph
Guelph
Ontario
N1G 2W1
Canada

C Grönewald
BioteCon Gesellschaft für Biotechnologische
Entwicklung und Consulting
Hermannswerder Haus 17
14473 Potsdam
Germany

Isabel Guerrero
Universidad Autónoma Metropolitana-Iztapalapa
Mexico
Apartado Postal 55–535
CP 09340 Mexico DF
Mexico

G C Gürakan
Middle East Technical University
Ankara
Turkey

Carlos Horacio Gusils
Centro de Referencia para Lactobacilos and Universidad
Nacional de Tucumán
Casilla de Correo 211
(4000)-Tucuman
Argentina

Thomas S Hammack
US Food and Drug Administration
Washington, DC 20204
USA

S A S Hanna,
48 Kensington Street
Newton
MA 02460
USA

Karen M J Hansen
Saskatchewan Food Product Innovation Program
University of Saskatchewan
Saskatoon
SK
S7N 5A8
Canada

J Harvey
Food Science Division (Food Microbiology)
Department of Agriculture for Northern Ireland
Agriculture and Food Science Centre
Newforge Lane
Belfast
BT9 5PX
Northern Ireland, UK

Wilma C Hazeleger
Wageningen University and Research Centre
Department of Food Technology and Nutritional Sciences
Bomenweg 2
NL 6703 HD Wageningen
The Netherlands

G M Heard
CRC for Food Industry Innovation
Department of Food Science and Technology
University of New South Wales
Sydney
New South Wales 2052
Australia

Nidal Hilal
Biochemical Engineering Group
Centre for Complex Fluids Processing
Department of Chemical and Biological Process Engineering
University of Wales Swansea
Singleton Park
Swansea SA2 8PP
UK

G Hildebrandt
Institute for Food Hygiene
Free University of Berlin
Germany

Colin Hill
Department of Microbiology and National Food Biotechnology Centre
University College
Cork
Ireland

A D Hitchins
Center for Food Safety and Applied Nutrition
Food and Drug Administration
Washington, DC
USA

Jill E Hobbs
George Morris Centre
345, 2116 27th Avenue NE
Calgary
Alberta
T2E 7A6
Canada

Ailsa D Hocking,
CSIRO Food Science Australia
Riverside Corporate Park
North Ryde
New South Wales 2113
Australia

Cornelis P Hollenberg
Institut für Microbiology
Heinrich-Heine-Universität Düsseldorf
40225 Düsseldorf
Germany

Richard A Holley
Department of Food Science
University of Manitoba
Winnipeg
Manitoba
R3T 2N2
Canada

Wilhelm H Holzapfel
Institute of Hygiene and Toxicology
Federal Research Centre for Nutrition
Bundesforschungsanstalt
Haid-und-Neu-Str. 9
D-7613 Karlsruhe
Germany

Rolf K Hommel
Cell Technologie Leipzig
Fontanestr. 21
Leipzig
D-04289
Germany

Dallas G Hoover
Department of Animal and Food Sciences
University of Delaware
Newark
DE 19717–1303
USA

Thomas W Huber
Medical Microbiology and Immunology
Texas A&M College of Medicine
Temple
Texas
USA

Robert Hutkins
Department of Food Science and Technology
University of Nebraska
338 FIC
Lincoln
NE 68583–0919
USA

Cheng-An Hwang
Nestlé Research and Development Center
210 Housatonic Avenue
New Milford
Connecticut
USA

John J Iandolo
Department of Microbiology and Immunology
University of Oklahoma Health Sciences Center
Oklahoma City
OK 73190
USA

Y Iimura
Department of Applied Chemistry and Biotechnology
Yamanashi University
Kofu
Japan

Charlotte Nexmann Jacobsen
Department of Dairy and Food Research
Royal Veterinary and Agricultural University
Rolighedsvej 3,0
1958 Frederiksberg C
Denmark

Mogens Jakobsen
Department of Dairy and Food Research
Royal Veterinary and Agricultural University
Rolighedsvej 3,0
1958 Frederiksberg C
Denmark

Dieter Jahn
Institute for Organic Chemistry and Biochemistry
Albert Ludwigs University Freiburg
Albertstr. 21
79104 Freiburg
Germany

B Jarvis
Ross Biosciences Ltd
Daubies Farm
Upton Bishop
Ross-on-Wye
Herefordshire
HR9 7UR
UK

Ian Jenson
Gist-brocades Australia Pty, Ltd
Moorebank
NSW
Australia

Juan Jimenez
Departmento de Genética
Facultad de Ciencias, Universidad de Málaga
Spain

Karen C Jinneman
Department of Veterinary Science and Microbiology
University of Arizona
Tucson
AZ 85721
USA

Juan Jofre
Department of Microbiology
University of Barcelona
Spain

Eric Johansen
Department of Genetics and Microbiology
Chr. Hansen A/S
10–12 Bøge Allé
DK-2970
Hørsholm
Denmark

Nick Johns
Independent Research Consultant
15 Collingwood Close
Steepletower
Hethersett
Norwich NR9 3QE
UK

Eric A Johnson
Department of Food Microbiology
Food Research Institute, University of Wisconsin
Madison
WI
USA

Clifford H Johnson
US Environmental Protection Agency
Cincinatti
Ohio
USA

Rafael Jordano
Department of Food Science and Technology
Campus Rabanales, University of Córdoba
E-14071 Córdoba
Spain

Richard Joseph
Department of Food Microbiology
Central Food Technological Research Institute
Mysore
570 013
India

Vinod K Joshi
Department of Post-harvest Technology
Dr YSP University of Horticulture and Foresty
Nauni
Solan-173 230
India

Vijay K Juneja
United States Department of Agriculture
Eastern Regional Research Center
600 East Mermaid Lane
Wyndmoor
Pennsylvania
USA

G Kalantzopoulos
Department of Food Science and Technology
Agricultural University of Athens
Greece

Chitkala Kalidas
Field of Microbiology
Department of Food Science
Cornell University
Ithaca NY 14853
USA

A Kambamanoli-Dimou
Department of Animal Production
Technological Education Institute
Larissa
Greece

Peter Kämpfer
Institut für Angewandte Mikrobiologie
Justus-Liebig-Universität Giessen
Senckenbergstr. 3
D-35390 Giessen
Germany

N G Karanth
Fermentation Technology and Bioengineering
Department
Mysore 570013
India

Embit Kartadarma
Department of Food Science and Technology
The University of New South Wales
Sydney
Australia

K L Kauppi
University of Minnesota
Department of Food Science and Nutrition
St Paul
USA

C A Kaysner
US Food and Drug Administration
22201 23rd Drive SE
Bothell
Washington 98021
USA

William A Kerr
Department of Economics
University of Calgary
2500 University Drive NW
Calgary
Alberta
T2N 1N4
Canada

Tajalli Keshavarz
Department of Biotechnology
University of Westminster
115 New Cavendish Street
London
W1M 8JS
UK

George G Khachatourians
Department of Applied Microbiology and Food Science
University of Saskatchewan
Saskatoon
Canada

W Kim
Department of Microbiology
University of Georgia
Athens
Georgia
USA

P M Kirk
CABI Bioscience UK Centre (Egham)
Bakeham Lane
Egham
Surrey
TW20 9TY

Todd R Klaenhammer
Departments of Food Science and Microbiology
Southeast Dairy Foods Research Center
Box 7624
North Carolina State University
Raleigh
NC 27695–7624
USA

Hans-Peter Kleber
Institut für Biochemie
Fakultöt für Biowissenschaften
Pharmazie und Psychologie
Universität Leipzig
Talstr. 33
Leipzig
D-04103
Germany

Thomas J Klem
Department of Food Science
Cornell University
USA

Wolfgang Kneifel
Department of Dairy Research and Bacteriology
Agricultural University
Gregor Mendel-Str. 33
A-1180 Vienna
Austria

Barb Kohn
VICAM LP
313 Pleasant Street
Watertown
MA 02172
USA

C Koob
BioteCon Gesellschaft für Biotechnologische
Entwicklung und Consulting
Hermannswerder Haus 17
14473 Potsdam
Germany

P Kotzekidou
Department of Food Science and Technology
Faculty of Agriculture
Aristotle University of Thessaloniki
PB 250
GR 540 06
Thessaloniki
Greece

K Krist
Meat and Livestock Australia
Sydney
Australia

Pushpa R Kulkarni
University Department of Chemical Technology
University of Mumbai
Matunga
Mumbai 400 019
India

Madhu Kulshreshtha
Division of Plant Pathology
Indian Agricultural Research Institute
New Delhi 11012
India

Susumu Kumagai
Department of Biomedical Food Research
National Institute of Infectious Diseases
Toyama 1–23–1
Shinjuku-ku
Tokyo 162–8640
Japan

G Lagarde
Inovatech Europe B.V.
Landbouwweg
The Netherlands

Keith A Lampel
US Food and Drug Administration
Center for Food Safety and Applied Nutrition HFS-327
200 C St SW
Washington
DC 20204
USA

S Leaper
Campden and Chorleywood Food Research Association
Chipping Campden
Gloucestershire
GL55 6LD
UK

J David Legan
Microbiology Department
Nabisco Research
PO Box 1944
DeForest Avenue
East Hanover
NJ 017871
USA

J J Leisner
Department of Veterinary Microbiology
Royal Veterinary and Agricultural University
Stigbøjlen 4
DK-1870 Frederiksberg C
Denmark

H L M Lelieveld
Unilever Research Vlaardingen
PO Box 114
3130 AC Vlaardingen
The Netherlands

D F Lewis
Food Systems Division
SAC
Auchincruive
Ayr KA6 5HW
Scotland
UK

M J Lewis
Department of Food Science and Technology
University of Reading
UK

E Litopoulou-Tzanetaki
Department of Food Science, Faculty of Agriculture
Aristotle University of Thessaloniki
54006
Thessaloniki
Greece

Aline Lonvaud-Funel
Faculty of Œnology
University Victor Segalen Bordeaux 2
351, Cours de la Libération
33405 Talence Cedex
France

S E Lopez
Departamento de Ciencias Biológicas
Facultad de Ciencias Exactas y Naturales
Pabellon II 4to piso – Ciudad Universitaria
1428 Buenos Aires
Argentina

G Love
Centre for Electron Optical Studies
University of Bath
Claverton Down
Bath
BA2 7AY
UK

Robert W Lovitt
Biochemical Engineering Group
Centre for Complex Fluids Processing
Department of Chemical and Biological Process
Engineering
University of Wales Swansea
Singleton Park
Swansea
SA2 8PP
UK

Majella Maher
National Diagnostics Centre
National University of Ireland
Galway
Ireland

R H Madden
Food Microbiology
Food Science Department
Department of Agriculture for Northern Ireland and
Queen's University of Belfast
Newforge Lane
Belfast
BT9 5PX
Northern Ireland

T Mahmutoğlu
TATKO TAS
Gayrettepe
Istanbul
Turkey

K A Malik
Chairman
Pakistan Agricultural Research Council
Islamabad
Pakistan

Miguel Prieto Maradona
Department of Food Hygiene and Food Technology
University of León
24071-León
Spain

Scott E Martin
Department of Food Science and Human Nutrition
University of Illinois
486 Animal Sciences Laboratory
1207 West Gregory Drive
Urbana
IL 61801
USA

L Martínková
Laboratory of Biotransformation
Institute of Microbiology
Academy of Sciences of the Czech Republic
Prague
Czech Republic

Tina Mattila-Sandholm
VTT Biotechnology and Food Research
Tietotie 2
Espoo
PO Box 1501
FIN-02044 VTT
Finland

D A McDowell
Food Studies Research Unit
University of Ulster at Jordanstown
Shore Road
Newtownabbey
Co. Antrim
Northern Ireland
BT37 9QB

Denise N McKenna
Microbiology – Research and Development
Bestfoods Technical Center
Somerset
New Jersey
USA

M A S McMahon
Food Studies Research Unit
University of Ulster at Jordanstown
Shore Road
Newtownabbey
Co. Antrim
Northern Ireland
BT37 9QB

T A McMeekin
School of Agricultural Science
University of Tasmania
Hobart
Australia

Luis M Medina
Department of Food Science and Technology
Campus Rabanales
University of Córdoba
E-14071 Córdoba
Spain

Aubrey F Mendonca
Iowa State University
Department of Food Science and Human Nutrition
Ames
Iowa
USA

James W Messer
US Environmental Protection Agency
Cincinnati
Ohio
USA

M C Misra
Fermentation Technology and Bioengineering
Department
Central Food Technological Research Institute
Mysore 570013
India

Vikram V Mistry
Dairy Science Department
South Dakota State University
Brookings
South Dakota 57007
USA

D R Modi
Department of Microbiology
Dr Ram Manohar Lohia Avadh University
Faizabad 224 001
India

Richard J Mole
Biotec Laboratories Ltd.
32 Anson Road
Martlesham Heath
Ipswich
Suffolk
IP5 3RD
UK

M C Montel
Station de Recherches sur la Viande
INRA
63122 Saint Genès Champanelle
France

M Moresi
Istituto di Tecnologie Agroalimentari
Università della Tuscia
Via S C de Lellis
01100 Viterbo
Italy

André Morin
Imperial Tobacco Limited
3810 rue St-Antoine
Montreal
Quebec H4C 1B5
Canada

Maurice O Moss
School of Biological Sciences, University of Surrey
Guildford
GU2 5XH
UK

M A Mostert
Unilever Research Vlaardingen
PO Box 114
3130 AC Vlaardingen
The Netherlands

Donald Muir
Hannah Research Institute
Ayr
KA6 5HL
Scotland, UK

Maite Muniesa
Department of Microbiology
University of Barcelona
Spain

E A Murano
Center for Food Safety and Department of Animal
Science
Texas A&M University
Texas
USA

M J Murphy
CBD Porton Down
Salisbury
SP4 0JQ
UK

K Darwin Murrell
Agricultural Research Service
US Department of Agriculture
Beltsville
Maryland 20705
USA

C K K Nair
Radiation Biology Division
Bhabha Atomic Research Centre
Mumbai 400 085
India

Motoi Nakao
Horiba Ltd
Miyanohigashimachi
Kisshoin
Minami-ku
Kyoto
Japan
601–8510

A W Nichol
Charles Sturt University
NSW
Australia

D S Nichols
School of Agricultural Science
University of Tasmania
Hobart
Australia

Poonam Nigam
Biotechnology Research Group
School of Applied Biological and Chemical Sciences
University of Ulster
Coleraine BT52 1SA
UK

M de Nijs
TNO Nutrition and Food Research Institute
Division of Microbiology and Quality Management
PO Box 360
3700 AJ Zeist
The Netherlands

S H W Notermans
TNO Nutrition and Food Research Institute
PO Box 360
3700 AJ Zeist
The Netherlands

Martha Nuñez
Centro de Referencia par Lactobacilos (Cerela)
Chacabuco 145 (4000)
Tucumán
Argentina

George-John E Nychas
Department of Food Science and Technology
Laboratory of Microbiology and Biotechnology of Foods
Agricultural University of Athens
Iera Odos 75
Athens
11855
Greece

R E O'Connor-Shaw
Food Microbiology Consultant
Birkdale
Queensland
Australia

Louise O'Connor
National Diagnostics Centre
National University of Ireland
Galway
Ireland

Triona O'Keeffe
Department of Microbiology and National Food
Biotechnology Centre
University College
Cork
Ireland

Rachel M Oakley
United Biscuits (UK Ltd)
High Wycombe
Buckinghamshire
HP12 4JX
UK

Yuji Oda
Department of Applied Biological Science
Fukuyama University
Fukuyama
Hiroshima 729–0292
Japan

Lucy J Ogbadu
Department of Biological Sciences
Benue State University
Makurdi
Nigeria

Guillermo Oliver
Centro Referencia para Lactobacilos and Universidad
Nacional de Tucumán
Casilla de Correo 211
(4000)-Tucuman
Argentina

Ynes R Ortega
Seafood Products Research Center
US Food and Drug Administration
PO Box 3012
22201 23rd Drive SE
Bothell
WA 98041–3012
USA

Andrés Otero
Department of Food Hygiene and Food Technology
University of León
24071-León
Spain

Kozo Ouchi
Kyowa Hakko Kogyo Co. Ltd
1–6–1 Ohtemachi
Chiyoda-ku
Tokyo 100–8185
Japan

Barbaros H Özer
Department of Food Science and Technology
Faculty of Agriculture
University of Harran
63040
Şanlıurfa
Turkey

Dilek Özer
GAP Regional Development Administration
Şanlıurfa
Turkey

J Palacios
Universidad Nacional de Tucumán, Argentina
Cerela-Conicet
San Miguel de Tucumán
Argentina

Ashok Pandey
Laboratorio de Processos Biotecnologicos
Universidade Federal do Parana
Departmento de Engenharia Quimica
CEP 81531-970 Curitiba-PR
Brazil

Photis Papademas
Department of Food Science and Technology
University of Reading
Whiteknights
Reading
Berkshire
RG6 6AP
UK

A Pardigol
BioteCon Gesellschaft für Biotechnologische
Entwicklung und Consulting
Hermannswerder Haus 17
14473 Potsdam
Germany

E Parente
Dipartimento di Biologia, Difesa e Biotecnologie Agro-
Forestali
Università della Basilicata
Via N Sauro 85
85100 Potenza
Italy

Zahida Parveen
University of Huddersfield
Department of Chemical and Biological Sciences
Queensgate
Huddersfield
HD1 3DH
UK

P Patáková
Faculty of Food and Biochemical Technology
Institute of Chemical Technology
Prague
Czech Republic

Pradip Patel
Science and Technology Group
Leatherhead Food Research Association
Randalls Road
Leatherhead
Surrey
KT22 7RY
UK

Margaret Patterson
Food Science Division
Department of Agriculture for Northern Ireland and The
Queen's University of Belfast
Agriculture and Food Science Centre
Newforge Lane
Belfast
BT9 5PX
UK

P A Pawar
Fermentation Technology and Bioengineering
Department
Central Food Technological Research Institute
Mysore 570013
India

Janet B Payeur
National Veterinary Services Laboratories
Veterinary Services
Animal and Plant Health Inspection Service
Department of Agriculture
1800 Dayton Road
Ames
IA 50010
USA

Gary A Payne
Department of Plant Pathology
North Carolina State University
Raleigh
North Carolina
USA

Ron Pethig
Institute of Molecular and Biomolecular Electronics
University of Wales
Dean St
Bangor
Gwynedd
LL57 1UT
UK

L Petit
Unité Interactions Bactéries Cellules
Institut Pasteur
28 rue du Dr Roux
75724 Paris
Cedex 15
France

William A Petri Jr
Department of Medicine, Division of Infectious Diseases
University of Virginia Health Sciences Center
MR4, Room 2115, 300 Park Place
Charlottesville
VA 22908
USA

M R A Pillai
Isotope Division
Bhabha Atomic Rsearch Centre
Mumbai 400 085
India

D W Pimbley
Leatherhead Food Research Association
Randalls Road
Leatherhead
Surrey
KT22 7RY
UK

J I Pitt
CSIRO Food Science Australia
Riverside Corporate Park
North Ryde
New South Wales 2113
Australia

M R Popoff
Institut Pasteur
Unité Interactions Bactéries Cellules
28 rue du Dr Roux
75724 Paris
Cedex 15
France

U J Potter
Centre for Electron Optical Studies
University of Bath
Claverton Down
Bath
BA2 7AY
UK

B Pourkomailian
Department of Food Safety and Preservation
Leatherhead Food RA
Randalls Road
Surrey
UK

K Prabhakar
Department of Meat Science and Technology
College of Veterinary Science
Tirupati 517 502
India

W Praphailong
National Center for Genetic Engineering and
Biotechnology
Rajdhevee
Bangkok
Thailand

M S Prasad
Fermentation Technology and Bioengineering
Department
Mysore 570013
India

J. C du Preez
Department of Microbiology and Biochemistry
University of the Orange Free State
PO Box 339
Bloemfontein 9300
South Africa

Barry H Pyle
Montana State University
Bozeman
Montana
USA

Laura Raaska
VTT Biotechnology and Food Research
PO Box 1501
FIN-02044 VTT
Finland

Moshe Raccach
Food Science Program
School of Agribusiness and Resource Management
Arizona State University East
Mesa
Arizona 85206–0180
USA

Fatemeh Rafii,
Division of Microbiology
National Center for Toxicological Research, US FDA
Jefferson
AR
USA

M I Rajoka,
National Institute for Biotechnology and Genetic
Engineering (NIBGE)
PO Box 577
Faisalabad
Pakistan

Javier Raso
Biological Systems Engineering
Washington State University
Pullman
Washington 99164-6120
USA

K S Reddy
Department of Meat Science and Technology
College of Veterinary Science
Tirupati 517 502
India

S M Reddy
Department of Botany
Kakatiya University
Warangal
506 009
India

Wim Reybroeck
Department for Animal Product Quality and
Transformation Technology
Agricultural Research Centre CLO-Ghent
Melle
Belgium

V G Reyes
Food Science Australia
Private Bag 16
Sneydes Road
Werribee
Victoria
VIC 3030
Australia

E W Rice
US Environmental Protection Agency
Cincinnati
Ohio 45268
USA

Jouko Ridell
Department of Food and Environmental Hygiene, Faculty
of Veterinary Medicine
University of Helsinki
Finland

R K Robinson
Department of Food Science
University of Reading
Whiteknights
Reading
Berkshire RG6 6AP
UK

Hubert Roginski
Gilbert Chandler College
The University of Melbourne
Sneydes Road
Werribee
Victoria
3030
Australia

Alexandra Rompf
Institute for Organic Chemistry and Biochemistry
Albert Ludwigs University Freiburg
Albertstr. 21
79104 Freiburg
Germany

T Ross
School of Agricultural Science
University of Tasmania
Hobart
Australia

T Roukas
Department of Food Science and Technology
Aristotle University of Thessaloniki
Greece

M T Rowe
Food Microbiology
Food Science Department
Department of Agriculture for Northern Ireland and
Queen's University of Belfast
Newforge Lane
Belfast
BT9 5PX
Northern Ireland

W Michael Russell
Departments of Food Science and Microbiology
Southeast Dairy Foods Research Center
Box 7624
North Carolina State University
Raleigh
NC 27695–7624
USA

G Salvat
AFSSA
Ploufragan
France

R Sandhir
Department of Biochemistry
Dr Ram Manohar Lohia Avadh University
Faizabad 224 001
India

Robi C Sandlin
Department of Microbiology and Immunology
Uniformed Services University of the Health Sciences
F Edward Hébert School of Medicine
4301 Jones Bridge Road
Bethesda
MD 20814
USA

Jesús-Angel Santos
Department of Food Hygiene and Food Technology
University of León
24071-León
Spain

A K Sarbhoy
Division of Plant Pathology
Indian Agricultural Research Institute
New Delhi 110012
India

David Sartory
Severn Trent Water
Shrewsbury
UK

Joanna M Schaenman
Department of Medicine
Division of Infectious Diseases
University of Virginia Health Sciences Center
MR4, Room 2115, 300 Park Place
Charlottesville
VA 22908
USA

Barbara Schalch
Institute of Hygiene and Technology of Food of Animal Origin
Ludwig-Maximilians-University Munich
Veterinary Faculty
Veterinärstr. 13
81369 Munich
Germany

P Scheu
BioteCon Gesellschaft für Biotechnologische Entwicklung und Consulting
Hermannswerder Haus 17
14473 Potsdam
Germany

Bernard W Senior
Department of Medical Microbiology
University of Dundee Medical School
Ninewells Hospital
Dundee
DD1 9SY
UK

Gilbert Shama
Department of Chemical Engineering
Loughborough University
UK

Arun Sharma
Food Technology Division
Bhabha Atomic Research Centre
Mumbai 400 085
India

M Shin
Faculty of Pharmaceutical Sciences
Kobe Gakuin University
Kobe
Japan

J Silva
Universidad Nacional de Tucumán, Argentina
Cerela–Conicet
San Miguel de Tucumán
Argentina

Dalel Singh
Microbiology Department
CCS Haryana Agricultural University
Hisar
125 004
India

Rekha S Singhal
University Department of Chemical Technology
University of Mumbai
Matunga
Mumbai 400 019
India

Emanuele Smacchi
Institute of Industrie Agranie (Microbiologia)
Faculty of Agriculture of Perugia 06126 S. Constanzo
Perguia
Italy

Christopher A Smart
Macromolecular Science Department
Institute of Food Research
Reading Laboratory
Earley Gate
Whiteknights Road
Reading R66 6BZ
UK

H V Smith
Scottish Parasite Diagnostic Laboratory
Stobhill Hospital
Glasgow
G21 3UW
Scotland, UK

O Peter Snyder
Hospitality Institute of Technology and Management
670 Transfer Road
Suite 21A
St Paul
MN 55114
USA

Mark D Sobsey
Department of Environmental Sciences and Engineering
School of Public Health
University of North Carolina
North Carolina
USA

Carlos R Soccol
Laboratorio de Processos Biotecnologicos
Departamento de Engenharia Quimica
Universidade Federal do Parana
CEP 81531–970
Curitiba-PR
Brazil

M El Soda
Department of Dairy Science and Technology
Faculty of Agriculture
Alexandria University
Alexandria
Egypt

R A Somerville
Neuropathogenesis Unit
Institute for Animal Health
West Mains Road
Edinburgh
EH9 3JF
UK

N H C Sparks
Department of Biochemistry and Nutrition
Scottish Agricultural College
Auchincruive
Ayr
Scotland

M Van Speybroeck
Laboratory of Industrial Microbiology and Biocatalysis
Department of Biochemical and Microbial Technology
Faculty of Agricultural and Applied Biological Sciences
University of Gent
Coupure links 653
B-9000
Gent
Belgium

D J Squirrell
CBD Porton Down
Salisbury
SP4 0JQ
UK

E Stackebrandt
DSMZ – German Collection of Microorganisms and Cell
Cultures
Brunswick
Germany
Deutsche Sammlung von Mikroorganisem und
Mascheroder
Weg 1 B
38124, Braunschweig
Germany

Jacques Stark
Gist-brocades Food Specialties
R&D
Delft
The Netherlands

Colin S Stewart
Rowett Research Institute
Greenburn Road
Bucksburn
Aberdeen
UK

G G Stewart
International Centre for Brewing and Distilling
Heriot-Watt University
Riccarton
Edinburgh
Scotland
EH14 4AS
UK

Gordon S A B Stewart (dec)
Department of Pharmaceutical Sciences
The University of Nottingham
University Park
Nottingham
NG7 2RD
UK

Duncan E S Stewart-Tull
University of Glasgow
Glasgow
G12 8QQ
UK

A Stolle
Institute of Hygiene and Technology of Food of Animal
Origin
Ludwig-Maximilians-University Munich
Veterinary Faculty
Veterinärstr. 13
81369
Munich
Germany

Liz Straszynski
Alcontrol Laboratories
Bradford
UK

M Stratford
Microbiology Section
Unilever Research
Colworth House
Sharnbrook
Bedfordshire
MK44 1LQ
UK

M Surekha
Department of Botany
Kakatiya University
Warangal
506 009
India

B C Sutton
Apple Tree Cottage
Blackheath
Wenhaston
Suffolk
IP19 9HD
UK

Barry G Swanson
Food Science and Human Nutrition
Washington State University
Pullman
Washington 99164–6376
USA

Jyoti Prakash Tamang
Microbiology Research Laboratory
Department of Botany
Sikkim Government College
Gangtok
Sikkim 737 102
India

A Y Tamime
Scottish Agricultural College
Auchincruive
Ayr
UK

J S Tang
American Type Culture Collection
10801 University Blvd
Manassas
VA 20110-2209
USA

Chrysoula C Tassou
National Agricultural Research Foundation
Institute of Technology of Agricultural Products
S. Venizelou 1
Lycovrisi 14123
Athens
Greece

S R Tatini
University of Minnesota
Department of Food Science and Nutrition
1334 Eckles Ave
St Paul
MN 55108
USA

D M Taylor
Neuropathogenesis Unit
Institute for Animal Health
West Mains Road
Edinburgh
EH9 3JF
UK

John R N Taylor
Cereal Foods Research Unit
Department of Food Science
University of Pretoria
Pretoria 0002
South Africa

Lúcia Martins Teixeira
Departamento de Microbiologia Médica
Instituto de Microbiologia
Universidade Federal do Rio de Janeiro
Rio de Janeiro 21941
Brazil

Paula C M Teixeira
Escola Superior de Biotecnologia
Rua Dr António Benardino de Almeida
4200 Porto
Portugal

J Theron
Department of Microbiology and Plant Pathology
Faculty of Biological and Agricultural Sciences
University of Pretoria
Pretoria 0002
South Africa

Linda V Thomas
Aplin & Barrett Ltd
15 North Street
Beaminster
Dorset
DT8 3DZ
UK

Angus Thompson
Technical Centre
Scottish Courage Brewing Ltd
Sugarhouse Close
160 Canongate
Edinburgh
EH8 8DD
UK

Ulf Thrane
c/o Eastern Cereal and Oilseed Research Centre
K.W. Neatby Building, FM 1006,
Agriculture and Agri-Food Canada
Ottowa
Ontario K1A 0C6
Canada

Mary Lou Tortorello
National Center for Food Safety and Technology
US Food and Drug Administration
6502 South Archer Road
Summit-Argo
Illinois 60501
USA

Hau-Yang Tsen
Department of Food Science
National Chung Hsing University
Taichung
Taiwan
Republic of China

Nezihe Tunail
Department of Food Engineering
Faculty of Agriculture
University of Ankara
Dişkapì
Ankara
Turkey

D R Twiddy
Consultant Microbiologist
27 Guildford Road
Horsham
West Sussex
RH12 1LU
UK

N Tzanetakis
Department of Food Science
Faculty of Agriculture
Aristotle University of Thessaloniki
54006
Thessaloniki
Greece

C Umezawa
Faculty of Pharmaceutical Sciences
Kobe Gakuin University
Kobe
Japan

F Untermann
Institute for Food Safety and Hygiene
University of Zurich
Switzerland

Matthias Upmann,
Institute of Meat Hygiene
Meat Technology and Food Science
Veterinary University of Vienna
Veterinärplatz 1
A-1210 Vienna
Austria

Tümer Uraz
Ankara University
Faculty of Agriculture
Department of Dairy Technology
Ankara
Turkey

M R Uyttendaele
Laboratory of Food Microbiology and Food Preservation
Faculty of Agricultural and Applied Biological Sciences
University of Ghent
Coupure Links 654
9000 Ghent
Belgium

E J Vandamme
Laboratory of Industrial Microbiology and Biocatalysis
Department of Biochemical and Microbial Technology
Faculty of Agricultural and Applied Biological Sciences
University of Gent
Coupure links 653
B-9000
Gent
Belgium

P T Vanhooren
Laboratory of Industrial Microbiology and Biocatalysis
Department of Biochemical and Microbial Technology
Faculty of Agricultural and Applied Biological Sciences
University of Gent
Coupure links 653
B-9000
Gent
Belgium

L Le Vay
School of Ocean Sciences
University of Wales
Bangor
UK

P H In't Veld
National Institute of Public Health and the Environment
Microbiological Laboratory for Health Protection
PO Box 1
3720 BA Bilthoven
The Netherlands

Kasthuri Venkateswaran
Jet Propulsion Laboratory
National Aeronautics and Space Administration
Planetary Protection and Exobiology, M/S 89–2, 4800
Oakgrove Dr.
Pasadena
CA 91109
USA

V Venugopal
Food Technology Division
Bhabha Atomic Research Centre
Mumbai 400 085
India

Christine Vernozy-Rozand
Food Research Unit National Veterinary School
Lyon
France Ecole Nationale Véténaire de Lyon
France

B C Viljoen
Department of Microbiology and Biochemistry
University of the Orange Free State
Bloemfontein
South Africa

Birte Fonnesbech Vogel
Danish Institute for Fisheries Research
Department of Seafood Research
Technical University of Denmark Bldg 221
DK-2800 Lyngby
Denmark

Philip A Voysey
Microbiology Department
Campden and Chorleywood Food Research Association
Chipping Campden
Gloucestershire
GL55 6LD
UK

Martin Wagner
Institute for Milk Hygiene
Milk Technology and Food Science
University for Veterinary Medicine
Veterinärplatz 1
1210 Vienna
Austria

Graeme M Walker
Reader of Biotechnology
Division of Biological Sciences
School of Science and Engineering
University of Abertay Dundee
Dundee
DD1 1HG
Scotland

P Wareing
Natural Resources Institute
Chatham Maritime
Kent
ME4 4TB
UK

John Watkins
CREH Analytical
Leeds
UK

Ian A Watson
University of Glasgow
Glasgow
G12 8QQ
UK

Bart Weimer
Center for Microbe Detection and Physiology
Utah State University
Nutrition and Food Sciences
Logan
UT 84322–8700
USA

Irene V Wesley
Enteric Diseases and Food Safety Research
USDA, ARS, National Animal Disease Center
Ames IA 50010
USA

W B Whitman
Department of Microbiology
University of Georgia
Athens
Georgia
USA

Martin Wiedmann
Department of Food Science
Cornell University
Ithaca
NY 14853
USA

R C Wigley
Boghall House
Linlithgow
West Lothian
EH49 7LR
Scotland

R Andrew Wilbey
Department of Food Science
University of Reading
Whiteknights
Reading
UK

F Wilborn
BioteCon Gesellschaft für Biotechnologische
Entwicklung und Consulting
Hermannswerder Haus 17
14473 Potsdam
Germany

A G Williams
Hannah Research Institute
Ayr
KA6 5HL
UK

Alan Williams
Campden and Chorleywood Food Research Association
Chipping Campden
Gloucestershire GL55 6LD
UK

J F Williams
Department of Microbiology
Michigan State University
East Lansing
MI 48824
USA

Michael G Williams
3M Center
260–6B-01
St Paul
MN55144–1000
USA

Caroline L Willis
Public Health Laboratory Service
Southampton,
UK

F Y K Wong
Food Science Australia
Cannon Hill
Queensland
Australia

Brian J B Wood
Reader in Applied Microbiology
Dept. of Bioscience and Biotechnology
University of Strathclyde
Royal College Building
George Street
Glasgow
G1 1XW
Scotland

S D Worley
Department of Chemistry
Auburn University
Auburn
AL 36849
US

Atte von Wright
Department of Biochemistry and Biotechnology
University of Kuopio
PO Box 1627
FIN-70211 Kuopio
Finland

Chris J Wright
Biochemical Engineering Group
Centre for Complex Fluids Processing
Department of Chemical and Biological Process
Engineering
University of Wales Swansea
Singleton Park
Swansea
SA2 8PP
UK

Peter Wyn-Jones
Sunderland University
UK

Hideshi Yanase
Department of Biotechnology
Faculty of Engineering
Tottori University
4–101 Koyama-cho-minami
Tottori
Tottori 680–0945
Japan

Yeehn Yeeh
Institute of Basic Science
Inje University
Obang-dong
Kimhae 621–749
South Korea

Seyhun Yurdugül
Middle East Technical University
Department of Biochemistry
Ankara
Turkey

Klaus-Jürgen Zaadhof
Institute for Hygiene and Technology of Foods of Animal
Origin
Veterinary Faculty
Ludwig-Maximilians University
80539 Munich
Germany

Gerald Zirnstein
Centers for Disease Control
GA
USA

Cynthia Zook
Department of Food Science and
University of Minnesota
St Paul
MN 55108
USA

CONTENTS

VOLUME 1

E

VOLUME 2

F

VOLUME 3

N

O

P

Q

Fatty Acids *see* **Fermentation (Industrial)**: Production of Oils and Fatty Acids.

FERMENTATION (INDUSTRIAL)

Contents

Basic Considerations

Yusuf Chisti, Department of Chemical Engineering, University of Almería, Spain

Introduction

Fermentation processes utilize microorganisms to convert solid or liquid substrates into various products. The substrates used vary widely, any material that supports microbial growth being a potential substrate. Similarly, fermentation-derived products show tremendous variety. Commonly consumed fermented products include bread, cheese, sausage, pickled vegetables, cocoa, beer, wine, citric acid, glutamic acid and soy sauce.

Types of Fermentation

Most commercially useful fermentations may be classified as either solid-state or submerged cultures. In solid-state fermentations, the microorganisms grow on a moist solid with little or no 'free' water, although capillary water may be present. Examples of this type of fermentation are seen in mushroom cultivation, bread-making and the processing of cocoa, and in the manufacture of some traditional foods, e.g. miso (soy paste), saké, soy sauce, tempeh (soybean cake) and gari (cassava), which are now produced in large industrial operations. Submerged fermentations may use a dissolved substrate, e.g. sugar solution, or a solid substrate, suspended in a large amount of water to form a slurry. Submerged fermentations are used for pickling vegetables, producing yoghurt, brewing beer and producing wine and soy sauce.

Solid-state and submerged fermentations may each be subdivided – into oxygen-requiring aerobic processes, and anaerobic processes that must be conducted in the absence of oxygen. Examples of aerobic fermentations include submerged-culture citric acid production by *Aspergillus niger* and solid-state koji fermentations (used in the production of soy sauce). Fermented meat products such as bologna sausage (polony), dry sausage, pepperoni and salami are produced by solid-state anaerobic fermentations utilizing acid-forming bacteria, particularly *Lactobacillus*, *Pediococcus* and *Micrococcus* species. A submerged-culture anaerobic fermentation occurs in yoghurt-making.

Fermentations may require only a single species of microorganism to effect the desired chemical change. In this case the substrate may be sterilized, to kill unwanted species prior to inoculation with the desired microorganism. However, most food fermentations are non-sterile. Typically fermentations used in food processing require the participation of several microbial species, acting simultaneously and/or sequentially, to give a product with the desired properties, including appearance, aroma, texture and taste. In non-sterile fermentations, the culture environment may be tailored specifically to favour the desired microorganisms. For example, the salt content may be high, the pH may be low, or the water activity may be reduced by additives such as salt or sugar.

Factors Influencing Fermentations

A fermentation is influenced by numerous factors, including temperature, pH, nature and composition of the medium, dissolved O_2, dissolved CO_2, operational system (e.g. batch, fed-batch, continuous), feeding with precursors, mixing (cycling through varying environments), and shear rates in the fermenter. Variations in these factors may affect: the rate of fermentation; the product spectrum and yield; the organoleptic properties of the product (appearance, taste, smell and texture); the generation of toxins; nutritional quality; and other physico-chemical properties.

The formulation of the fermentation medium affects the yield, rate and product profile. The medium must provide the necessary amounts of carbon, nitrogen, trace elements and micronutrients (e.g. vitamins). Specific types of carbon and nitrogen sources may be required, and the carbon:nitrogen ratio may have to be controlled. An understanding of fermentation biochemistry is essential for developing a medium with an appropriate formulation. Concentrations of certain nutrients may have to be varied in a specific way during a fermentation to achieve the desired result. Some trace elements may have to be avoided – for example, minute amounts of iron reduce yields in citric acid production by *Aspergillus niger*. Additional factors, such as cost, availability, and batch-to-batch variability also affect the choice of medium.

Submerged Fermentations

Fermentation Systems

Industrial fermentations may be carried out either batchwise, as fed-batch operations, or as continuous cultures (**Fig. 1**). Batch and fed-batch operations are quite common, continuous fermentations being relatively rare. For example, continuous brewing is used commercially, but most beer breweries use batch processes.

In batch processing, a batch of culture medium in a fermenter is inoculated with a microorganism (the 'starter culture'). The fermentation proceeds for a certain duration (the 'fermentation time' or 'batch time'), and the product is harvested. Batch fermentations typically extend over 4–5 days, but some traditional food fermentations may last months. In fed-batch fermentations, sterile culture medium is added either continuously or periodically to the inoculated

fermentation batch. The volume of the fermenting broth increases with each addition of the medium, and the fermenter is harvested after the batch time.

In continuous fermentations, sterile medium is fed continuously into a fermenter and the fermented product is continuously withdrawn, so the fermentation volume remains unchanged. Typically, continuous fermentations are started as batch cultures and feeding begins after the microbial population has reached a certain concentration. In some continuous fermentations, a small part of the harvested culture may be recycled, to continuously inoculate the sterile feed medium entering the fermenter (Fig. 1(D)). Whether continuous inoculation is necessary depends on the type of mixing in the fermenter. 'Plug flow' fermentation devices (Fig. 1(D)), such as long tubes that do not allow back mixing, must be inoculated continuously. Elements of fluid moving along in a plug flow device behave like tiny batch fermenters. Hence, true batch fermentation processes are relatively easily transformed into continuous operations in plug flow fermenters, especially if pH control and aeration are not required. Continuous cultures are particularly susceptible to microbial contamination, but in some cases the fermentation conditions may be selected (e.g. low pH, high alcohol or salt content) to favour the desired microorganisms compared to potential contaminants.

In a 'well-mixed' continuous fermenter (Fig. 1(C)), the feed rate of the medium should be such that the dilution rate, i.e. the ratio of the volumetric feed rate to the constant culture volume, remains less than the maximum specific growth rate of the microorganism in the particular medium and at the particular fermentation conditions. If the dilution rate exceeds the maximum specific growth rate, the microorganism will be washed out of the fermenter.

Industrial fermentations are mostly batch operations. Typically, a pure starter culture (or seed), maintained under carefully controlled conditions, is used to inoculate sterile Petri dishes or liquid medium in the shake flasks. After sufficient growth, the pre-culture is used to inoculate the 'seed' fermenter. Because industrial fermentations tend to be large (typically 150–250 m^3), the inoculum is built up through several successively larger stages, to 5–10% of the working volume of the production fermenter. A culture in rapid exponential growth is normally used for inoculation. Slower-growing microorganisms require larger inocula, to reduce the total duration of the fermentation. An excessively long fermentation time (or batch time) reduces productivity (amount of product produced per unit time per unit volume of fermenter), and increases costs. Sometimes inoculation spores,

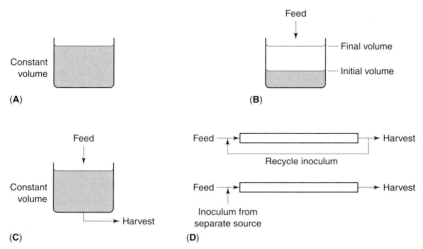

Figure 1 Fermentation methodologies. (**A**) Batch fermentation. (**B**) Fed-batch culture. (**C**) Continuous-flow well-mixed fermentation. (**D**) Continuous plug flow fermentation, with and without recycling of inoculum.

produced as seeds, are blown directly into large fermentation vessels with the ingoing air.

Microbial Growth

Microbial growth in a newly inoculated batch fermenter typically follows the pattern shown in **Figure 2**. Initially, in the lag phase, the cell concentration does not increase very much. The length of the lag phase depends on the growth history of the inoculum, the composition of the medium, and the amount of culture used for inoculation. An excessively long lag phase ties up the fermenter unproductively – hence the duration of the lag phase should be minimized. Short lag phases occur when: the composition of the medium and the environmental conditions in the seed culture and the production vessel are identical (hence less time is needed for adaptation); the dilution shock is small (i.e. a large amount of inoculum is used); and the cells in the inoculum are in the late exponential phase of growth. The lag phase is essentially an adaptation period in a new environment. The lag phase is followed by exponential growth, during which the cell mass increases exponentially. Eventually, as the

nutrients are exhausted and inhibitory products of metabolism build up, the culture enters a stationary phase. Ultimately, starvation causes cell death and lysis, and hence the biomass concentration declines.

Exponential growth can be described by **Equation 1**:

$$\frac{dX}{dt} = \mu X - k_d X \qquad \text{(Equation 1)}$$

where: X is the biomass concentration at time t; μ is the specific growth rate (i.e. growth rate per unit cell mass); and k_d is the specific death rate. During exponential growth, the specific death rate is negligible and Equation 1 reduces to **Equation 2**:

$$\frac{dX}{dt} = \mu X \qquad \text{(Equation 2)}$$

For a cell mass concentration X_0 at the beginning of the exponential growth (X_0 usually equalling the concentration of inoculum in the fermenter), and taking the time at which exponential growth commences as zero, Equation 2 can be integrated to produce **Equation 3**:

$$\ln \frac{X}{X_0} = \mu t \qquad \text{(Equation 3)}$$

Using Equation 3, the biomass doubling time, t_d, can be derived (**Equation 4**):

$$t_d = \frac{\ln 2}{\mu} \qquad \text{(Equation 4)}$$

Doubling times typically range over 45–160 min. Bacteria generally grow faster than yeasts, and yeasts multiply faster than moulds. The maximum biomass concentration in submerged microbial fermentations is typically 40–50 kg m^{-3}.

The specific growth rate μ depends on the concentration S of the growth-limiting substrate, until

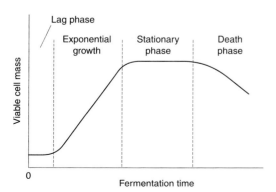

Figure 2 Typical growth profile of microorganisms in a submerged culture.

the concentration is increased to a non-limiting level and μ attains its maximum value μ_{max}. The dependence of the growth rate on substrate concentration typically follows Monod kinetics. Thus the specific growth rate is given as **Equation 5**:

$$\mu = \mu_{max} \frac{S}{k_s + S} \qquad \text{(Equation 5)}$$

where k_s is the saturation constant. Numerically, k_s is the concentration of the growth-limiting substrate when the specific growth rate is half its maximum value.

An excessively high substrate concentration may also limit growth, for instance by lowering water activity. Moreover, certain substrates inhibit product formation, and in yet other cases, a fermentation product may inhibit biomass growth. For example, ethanol produced in the fermentation of sugar by yeast can be inhibitory to cells. Multiple lag phases (or diauxic growth) are sometimes seen when two or more growth-supporting substrates are available. As the preferentially-utilized substrate is exhausted, the cells enter a lag phase while the biochemical machinery needed for metabolizing the second substrate is developed. Growth then resumes. Details of the kinetics of continuous culture, fed-batch fermentation, product formation and more complex phenomena, such as the inhibition of growth by substrates and products, are given in the references listed under Further Reading.

Aeration and Oxygen Demand

Submerged cultures are most commonly aerated by bubbling with sterile air. Typically, in small fermenters, the maximum aeration rate does not exceed 1 volume of air per unit volume of culture broth. In large bubble columns and stirred vessels, the maximum superficial aeration velocity tends to be $< 0.1 \text{ m s}^{-1}$. Superficial aeration velocity is the volume flow rate of air divided by the cross-sectional area of fermenter. Significantly higher aeration rates are achievable in airlift fermenters. In these, aeration gas is forced through perforated plates, perforated pipes or single-hole spargers located near the bottom of the fermenter. Because O_2 is only slightly soluble in aqueous culture broths, even a short interruption of aeration results in the available O_2 becoming quickly exhausted, causing irreversible damage to the culture. Thus uninterrupted aeration is necessary. Prior to use for aeration, any suspended particles, microorganisms and spores in the gas are removed by filtering through microporous membrane filters.

The O_2 requirements of a fermentation depend on the microbial species, the concentration of cells, and the type of substrate. O_2 supply must at least equal O_2 demand, or the fermentation will be O_2-limited. O_2 demand is especially difficult to meet in viscous fermentation broths and in broths containing a large concentration of O_2-consuming cells. As a general guide, the capability of a fermenter in terms of O_2 supply depends on the aeration rate, the agitation intensity and the properties of the culture broth. In large fermenters, supplying O_2 becomes difficult when demand exceeds $4\text{--}5 \text{ kg m}^{-3} \text{ h}^{-1}$.

At concentrations of dissolved O_2 below a critical level, the amount of O_2 limits microbial growth. The critical dissolved O_2 level depends on the microorganism, the culture temperature and the substrate being oxidized. The higher the critical dissolved O_2 value, the greater the likelihood that O_2 transfer will become limiting. Under typical culture conditions, fungi such as *Penicillium chrysogenum* and *Aspergillus oryzae* have a critical dissolved O_2 value of about $3.2 \times 10^{-4} \text{ kg m}^{-3}$. For bakers' yeast and *Escherichia coli*, the critical dissolved O_2 values are $6.4 \times 10^{-5} \text{ kg m}^{-3}$ and $12.8 \times 10^{-5} \text{ kg m}^{-3}$ respectively.

The aeration of fermentation broths generates foam. Typically, 20–30% of the fermenter volume must be left empty to accommodate the foam and allow for gas disengagement. In addition, mechanical 'foam breakers' and chemical antifoaming agents are commonly used. Typical antifoams are silicone oils, vegetable oils and substances based on low-molecular-weight polypropylene glycol or polyethylene glycol. Emulsified antifoams are more effective, because they disperse better in the fermenter. Antifoams are added in response to signals from a foam sensor. The excessive use of antifoams may interfere with some downstream separations, such as membrane filtrations – hydrophobic silicone antifoams are particularly troublesome.

Heat Generation and Removal

All fermentations generate heat. In submerged cultures, typically $3\text{--}15 \text{ kW m}^{-3}$ comes from microbial activity. In addition, mechanical agitation of the broth produces up to 15 kW m^{-3}. Consequently, a fermenter must be cooled to prevent a rise in temperature and damage to the culture. Heat removal tends to be difficult, because typically the temperature of the cooling water is only a few degrees lower than that of the fermentation broth. Therefore industrial fermentations are commonly limited by their heat-transfer capability. The ability to remove heat depends on: the surface area available for heat exchange; the temperature difference between the broth and the cooling water; the properties of the broth and the coolant; and the turbulence in these fluids. The geometry of the fermenter determines the surface area that can be provided for heat exchange. Heat generation due to metabolism depends on the rate of O_2 consumption, and heat removal in large vessels becomes difficult as the rate of O_2 consumption approaches $5 \, kg \, m^{-3} \, h^{-1}$.

A fermenter must provide for heat transfer during sterilization and subsequent cooling, as well as removing metabolic heat. Liquid medium, or a slurry, for a batch fermentation may be sterilized using batch or continuous processes. In batch processes, the medium or some of its components and the fermenter itself are commonly sterilized together in a single step, by heating the medium inside the fermenter. Steam may be injected directly into the medium, or heating may take place through the fermenter wall.

Heating to high temperatures (typically 121°C) during sterilization often leads to undesirable reactions between components of the medium. Such reactions reduce the yield, by destroying nutrients or by generating compounds which inhibit growth. This thermal damage can be prevented or reduced by sterilizing only certain components of the medium in the fermenter and adding other, separately-sterilized components, later. Sugars and nitrogen-containing components are often sterilized separately. Dissolved nutrients that are especially susceptible to thermal degradation may be sterilized by passage through hydrophilic polymer filters, which retain particles of $0.45 \, \mu m$ or more. Even finer filters (e.g. retaining particles of $0.2 \, \mu m$) are also available.

The heating and cooling of a large fermentation batch takes time, and ties up a fermenter unproductively. In addition, the longer a medium remains at a high temperature, the greater the thermal degradation or loss of nutrients. Therefore, continuous sterilization of the culture medium en route to a presterilized fermenter is preferable, even for batch fermentations. Continuous sterilization is rapid and it limits nutrient loss – however, the initial capital expense is greater, because a separate sterilizer is necessary.

Photosynthetic Microorganisms

Photosynthetic cultures of microalgae and cyanobacteria require light and CO_2 as nutrients. Microalgae such as *Chlorella* and the cyanobacterium *Spirulina* are produced commercially as health foods in Asia. Algae are also cultivated as aquaculture feeds for shellfish.

Typically, open ponds or shallow channels are used for the outdoor photosynthetic culture of microalgae. Culture may be limited by the availability of light, but under intense sunlight, photoinhibition limits productivity. Temperature variations also affect performance.

More controlled production is achieved in outdoor tubular photobioreactors, bubble columns and airlift systems. Tubular bioreactors use a 'solar receiver', consisting of either a continuous tube looped into several U-shapes to fit a compact area, or several parallel tubes connected to common headers at either end. The continuous looped-tube arrangement is less adaptable, because the length of the tube cannot exceed a certain value: photosynthetically-produced O_2 builds up along the tube, and high levels of dissolved O_2 inhibit photosynthesis. The parallel-tube arrangement can be readily scaled up by increasing the number of tubes. Typically, the tubes are 0.05–0.08 m in diameter and the continuous-run length of any tube does not exceed 50 m. However, greater lengths may be feasible, depending on the flow velocity in the tube. The tubular solar receivers may be mounted horizontally, or horizontal tubes may be stacked in a ladder configuration, forming the rungs of the ladder. The latter arrangement reduces the area of land required.

The culture is circulated through the tubes by an airlift pump or other suitable low-shear mechanism. The maximum flow rate is limited by the tolerance of the algae to hydrodynamic stress. The flow velocity is usually $0.3–0.5 \, m \, s^{-1}$. The tube diameter is limited by the need to achieve adequate penetration of light. This declines as the cell concentration increases, due to self-shading. Closed, temperature-controlled outdoor tubular systems attain significantly higher productivity than open channels. The protein content of the algal biomass, and the adequacy of the development of colour (chlorophyll) affect the acceptability of the product.

Among other types of culture system, airlift devices tend to perform better than bubble columns because only part of the airlift system is aerated and hence the penetration of light is less affected by air bubbles. Conventional external-loop airlift devices may not be suitable because of the relatively high hydrodynamic shear rates they generate. However, concentric-tube airlift devices, with gas forced into the draft tube (zone of poor light penetration), are likely to perform well. Also, split-cylinder types of airlift system may be suitable. However, the volume of the aerated zone in any airlift device for microalgal culture should not exceed approximately 40% of the total volume of the circulating zones. This way the light blocking effect of bubbles remains confined to a small zone.

Submerged-culture Fermenters

Types The major types of submerged-culture bioreactor are:

- stirred-tank fermenter
- bubble column
- airlift fermenter

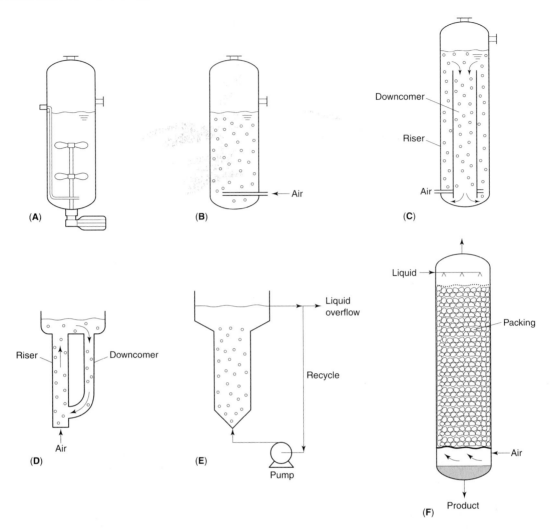

Figure 3 Types of submerged-culture fermenter. (**A**) Stirred-tank fermenter. (**B**) Bubble column. (**C**) Internal-loop airlift fermenter. (**D**) External-loop airlift fermenter. (**E**) Fluidized-bed fermenter. (**F**) Trickle-bed fermenter.

- fluidized-bed fermenter
- trickle-bed fermenter.

These are shown in **Figure 3**.

Stirred-tank Fermenter (See Fig. 3(A).) This is a cylindrical vessel with a working height-to-diameter ratio (aspect ratio) of 3–4. A central shaft supports three to four impellers, placed about 1 impeller-diameter apart. Various types of impeller, that direct the flow axially (parallel to the shaft) or radially (outwards from the shaft) may be used (**Fig. 4**). Sometimes axial- and radial-flow impellers are used on the same shaft. The vessel is provided with four equally spaced vertical baffles, that extend from near the walls into the vessel. Typically, the baffle width is 8–10% of the vessel diameter.

Bubble Column (See Fig. 3(B).) This is a cylindrical vessel with a working aspect ratio of 4–6. It is sparged at the bottom, and the compressed gas provides agitation. Although simple, it is not widely used because

of its poor performance relative to other systems. It is not suitable for very viscous broths or those containing large amounts of solids.

Airlift Fermenters (See Figs. 3(C) and 3(D).) These come in internal-loop and external-loop designs. In the internal-loop design, the aerated riser and the unaerated downcomer are contained in the same shell. In the external-loop configuration, the riser and the downcomer are separate tubes that are linked near the top and the bottom. Liquid circulates between the riser (upward flow) and the downcomer (downward flow). The working aspect ratio of airlift fermenters is 6 or greater. Generally, these are very capable fermenters, except for handling the most viscous broths. Their ability to suspend solids and transfer O_2 and heat is good. The hydrodynamic shear is low. The external-loop design is relatively little-used in industry.

Fluidized-bed Fermenter (See Fig. 3(E).) These are

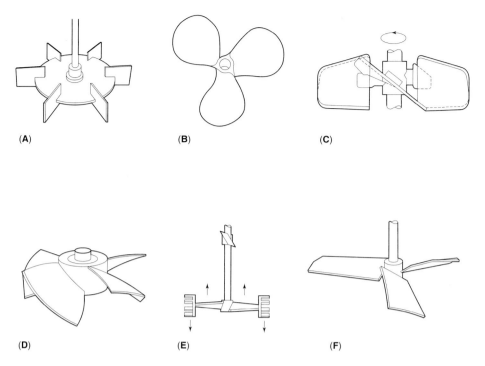

Figure 4 Impellers for stirred-tank fermenters. (**A**) Rushton disc turbine (radial flow). (**B**) Marine propeller (axial flow). (**C**) Lightnin' hydrofoil (axial flow). (**D**) Prochem hydrofoil (axial flow). (**E**) Intermig (axial flow). (**F**) Chemineer hydrofoil (axial flow).

similar to bubble columns with an expanded cross section near the top. Fresh or recirculated liquid is continuously pumped into the bottom of the vessel, at a velocity that is sufficient to fluidize the solids or maintain them in suspension. These fermenters need an external pump. The expanded top section slows the local velocity of the upward flow, such that the solids are not washed out of the bioreactor.

Trickle-bed Fermenter (See Fig. 3(F).) These consist of a cylindrical vessel packed with support material (e.g. woodchips, rocks, plastic structures). The support has large open spaces, for the flow of liquid and gas and the growth of microorganisms on the solid support. A liquid nutrient broth is sprayed onto the top of the support material, and trickles down the bed. Air may flow up the bed, countercurrent to the liquid flow. These fermenters are used in vinegar production, as well as in other processes. They are suitable for liquids with low viscosity and few suspended solids.

Design Irrespective of their configuration, industrial bioreactors for sterile operations are designed as pressure vessels, capable of being sterilized *in situ* with saturated steam at a minimum guage pressure of 0.11 MPa. Typically, the bioreactor is designed for a maximum allowable working pressure of 0.28–0.31 MPa (guage) and a temperature of 150–180°C. The vessels are designed to withstand a full vacuum. Modern commercial fermenters are predominantly made of stainless steel. Type 316L stainless steel is preferred, but the less expensive Type 304L (or 304) may be used in less corrosive situations. Fermenters are typically designed with clean-in-place capability.

A typical submerged-culture vessel has the features shown in **Figure 5**. Sight glasses in the side and top of the vessel allow for easy viewing. The top sight glass can be cleaned during fermentation, using a short-duration spray of sterile water derived from condensed steam. An external lamp is provided, to light the vessel through the sight glass or a separate window. The vessel has ports for sensors of pH, temperature and dissolved O_2. A steam-sterilizable sampling valve is provided. Connections for the introduction of acid and alkali (for pH control), antifoam agents, substrate and inoculum are located above the liquid level in the bioreactor vessel. Additional ports on the top support a foam-sensing electrode, a pressure sensor and sometimes other instruments. Filter-sterilized gas for aeration is supplied through a submerged sparger. Sometimes CO_2 or ammonia may be added to the aeration gas, for pH control.

A harvest valve is located at the lowest point on the fermenter. A mechanical agitator, entering from either the top or the bottom, may be used. The agitator shaft supports one or more impellers, of various designs (Fig. 4). A high-speed mechanical foam breaker may be provided at the top of the vessel, and waste gas may exit through the foam breaker. Commonly, the exhaust gas line also has a heat exchanger, to condense and return water in the gas to the fermenter. The top

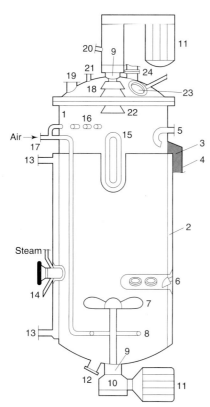

Figure 4 A typical submerged-culture fermenter. (**1**) Reactor vessel. (**2**) Jacket. (**3**) Insulation. (**4**) Protective shroud. (**5**) Inoculum connection. (**6**) Ports for sensors of pH, temperature and dissolved O_2. (**7**) Agitator. (**8**) Gas sparger. (**9**) Mechanical seal. (**10**) Reducing gearbox. (**11**) Motor. (**12**) Harvest nozzle. (**13**) Jacket connections. (**14**) Sample valve with steam connection. (**15**) Sight glass. (**16**) Connections for acids, alkalis and antifoam agents. (**17**) Air inlet. (**18**) Removable top. (**19**) Medium feed nozzle. (**20**) Air exhaust nozzle (connects to condenser, not shown). (**21**) Instrumentation ports for foam sensor, pressure gauge and other devices. (**22**) Centrifugal foam breaker. (**23**) Sight glass with light (not shown) and steam connection. (**24**) Rupture disc nozzle. Vertical baffles are not shown. Baffles are mounted on brackets attached to the wall. A small clearance remains between the wall and the closest vertical edge of the baffle.

of the fermenter is either removable or provided with a manhole. A port on the top supports a 'rupture disc' that is piped to a drain. The disc is intended to protect the vessel in the event of a pressure build-up. The fermentation vessel is jacketed for heat exchange, and the jacket may be covered with fibreglass insulation and a protective metal shroud. Additional surfaces for heat exchange, typically coils, may be located inside the vessel.

The equipment for fermenting slurries containing undissolved solid substrates is identical to that used in submerged-culture processes. Commonly-used slurry fermenters include stirred tanks, bubble columns, and airlift vessels.

Selection Considerations in selecting industrial fermenters are:

1. Nature of substrate solid, liquid, suspended slurry, water-immiscible oils).
2. Flow behaviour (rheology), broth viscosity and type of fluid (e.g. Newtonian, viscoelastic, pseudoplastic, Bingham plastic).
3. Nature and amount of suspended solids in broth.
4. Whether fermentation is aerobic or anaerobic, and O_2 demand.
5. Mixing requirements.
6. Heat-transfer needs.
7. Shear tolerance of microorganism, substrate and product.
8. Sterility requirements.
9. Process kinetics, batch or continuous operation, single-stage or multistage fermentation.
10. Desired process flexibility.
11. Capital and operational costs.
12. Local technological capability and potential for technology transfer.

Solid-state Fermentations

Substrate Characteristics

Water Activity Typically, solid-state fermentations are carried out with little or no free water. Excessive moisture tends to aggregate the substrate particles, and hence aeration is made difficult. For example steamed rice, a common substrate, becomes sticky when the moisture level exceeds 30–35% w/w. Percentage moisture by itself is unreliable for predicting growth: for a given microorganism growing on different substrates, the optimum moisture level may differ widely. This water activity correlates with microbial growth. The water activity of the substrate is the ratio of the vapour pressure of water in the substrate to the saturated vapour pressure of pure water at the temperature of the substrate. Water activity equals 1/100th of the relative humidity (RH%) of the air in equilibrium with the substrate. Typically, water activities of < 0.9 do not support bacterial growth, but yeasts and fungi can grow at water activities as low as 0.7. Thus the low-moisture environment of many solid-state fermentations favours yeasts and fungi.

The water activity depends on the concentrations of dissolved solutes, and so sometimes salts, sugars or other solutes are added to alter the water activity. Different additives may influence the fermentation differently, even though the change in water activity produced may be the same. Furthermore, the fermentation process itself leads to changes in the water activity, as products are formed and the substrate is hydrolysed, e.g. the oxidation of carbohydrates produces water. During fermentation, the water activity is controlled by aeration with humidified air and, sometimes, by intermittent spraying with water.

Particle Size The size of substrate particles affects the extent and the rate of microbial colonization, air penetration and CO_2 removal, as well as the downstream extraction and handling characteristics. Small particles, with high surface-to-volume ratios, are preferred because they present a relatively large surface for microbial action. However, particles that are too small and shapes that pack together tightly (e.g. flat flakes, cubes) are undesirable because close packing reduces the interparticle voids that are essential for aeration. Similarly, too many fine particles in a batch of larger particles will fill up the voids.

Substrate pH The pH is not normally controlled in solid-state fermentations, but initial adjustments may be made during the preparation of the substrate. The buffering capacity of many substrates effectively checks large changes in pH during fermentation. This is particularly true of protein-rich substrates, especially if deamination of the protein is minimal. Some pH stability can be obtained by using a combination of urea and ammonium sulphate as the nitrogen source in the substrate. In the absence of other contributing nitrogen sources, an equimolar combination of ammonium sulphate and urea is expected to yield the greatest pH stability.

Aeration and Agitation

Aeration plays an important role in removing CO_2 and controlling temperature and moisture. In some cases, an increased concentration of CO_2 may be severely inhibitory, while an increase in the partial pressure of O_2 may improve productivity. Deep layers and heaps of substrate may require forced aeration and agitation. Forced aeration rates vary widely, a typical range being $(0.05-0.2) \times 10^{-3} \, \text{m}^3 \, \text{kg}^{-1} \, \text{min}^{-1}$. Occasional turning and mixing improve O_2 transfer and reduce compaction and mycelial binding of the substrate particles. However, excessive agitation is undesirable because continuous agitation damages the surface hyphae – although mixing suppresses sporulation, which is often unwanted. The frequency of agitation may be purely experience-based, as in the occasional turning of a fermenting heap of cocoa beans, or it may be adjusted in response to a temperature controller.

Heat Transfer

The biomass levels in solid-state fermentations, at $10-30 \, \text{kg m}^{-3}$, are lower than those in submerged cultures. However, because there is little water, the heat generated per unit of fermenting mass tends to be much greater in solid-state fermentations than in submerged cultures, and again because there is little water to absorb the heat, the temperature can rise rapidly. The cumulative metabolic heat generation in fermentations producing koji, for the manufacture of a variety of products, has been noted at $419-2387 \, \text{kJ}$ per kilogram solids. Higher values, up to $13\,398 \, \text{kJ} \, \text{kg}^{-1}$, have been observed during composting. Peak heat generation rates in koji processes lie in the range $71-159 \, \text{kJ} \, \text{kg}^{-1} \text{h}^{-1}$ but average rates are more moderate, at $25-67 \, \text{kJ} \, \text{kg}^{-1} \text{h}^{-1}$. The peak rate of production of metabolic heat during the fermentation of readily oxidized substrates, such as starch, can be much greater than that associated with typical koji processes.

The substrate temperature is controlled mostly through evaporative cooling – hence drier air provides a better cooling effect. The intermittent spraying of cool water is sometimes necessary to prevent dehydration of the substrate. The air temperature and humidity are also controlled. Occasionally, the substrate-containing metal trays may also be cooled (by circulating a coolant), even though most substrates are relatively dry and porous, and hence are poor conductors. The intermittent agitation of substrate heaps further aids heat removal. However, despite much effort, temperature gradients in the substrate do occur, particularly during peak microbial growth.

Koji Fermentations

Koji fermentations are widely practised, typical examples of solid-state fermentations. Koji comprises soybeans or grain on which mould is growing, and has been used in oriental food preparation for thousands of years. Koji is a source of fungal enzymes, which digest proteins, carbohydrates and lipids into nutrients which are used by other microorganisms in subsequent fermentations. Koji is available in many varieties, which differ in terms of the mould, the substrate, the method of preparation and the stage of harvest. The production of soy sauce, miso and saké involves koji fermentation. Koji technology is also employed in the production of citric acid in Japan. The production of soy sauce (shoyu in Japanese) koji is detailed below, as an example of a typical industrial solid-substrate fermentation.

The koji for soy sauce is made from soybeans and wheat. Soybeans, or defatted soybean flakes or grits are moistened and cooked (e.g. for 0.25 min or less, at about 170°C) in continuous pressure cookers. The cooked beans are mixed with roasted, cracked wheat, the ratio of wheat to beans varying with the variety of shoyu. The mixed substrate is inoculated with a pure culture of *Aspergillus oryzae* (or *A. sojae*), the fungal spore density at inoculation being about 2.5×10^8 spores per kilogram of wet solids. After a 3-day fermentation, the substrate mass becomes green-yellow because of sporulation. The koji is then harvested, for use in a second submerged fermentation step. Koji production is highly automated and continuous – processes producing up to $4150 \, kg \, h^{-1}$ of koji have been described. Similar large-scale operations are used to produce koji for miso and saké in Japan.

Solid-state Fermenters

Solid-state fermentation devices vary in technical sophistication, from very primitive banana-leaf wrappings, bamboo baskets and substrate heaps to the highly automated machines used mainly in Japan. Some 'less sophisticated' fermentation systems, e.g. the fermentation of cocoa beans in heaps, are quite effective at large-scale processing. Also, some of the continuous, highly mechanized processes for the fermentation of soy sauce, that are successful in Japan, are not suitable for less highly developed locations in Asia. Thus, fermentation practice must be tailored to local conditions.

The use of pressure vessels is not the norm for solid-state fermentation. The commonly used devices are:

- tray fermenter
- static-bed fermenter
- tunnel fermenter
- rotary disc fermenter
- rotary drum fermenter
- agitated-tank fermenter
- continuous screw fermenter.

These are described below. Large concrete or brick fermentation chambers, or koji rooms, may be lined with steel, typically Type 304 stainless steel. For more corrosion-resistant construction, Type 304L and 316L stainless steels are used.

Tray Fermenter This is a simple type of fermenter, widely used in small- and medium-scale koji operations in Asia (see **Fig. 6**). The trays are made of wood, metal or plastic, and often have a perforated or wire-mesh base to achieve improved aeration. The substrate is fermented in shallow ($\leqslant 0.15 \, m$ deep) layers. The trays may be covered with cheesecloth to reduce contamination, but processing is non-sterile. Single or stacked trays may be located in chambers in which the temperature and humidity are controlled, or simply in ventilated areas. Inoculation and occasional mixing are done manually, although the handling, filling, emptying and washing of trays may be automated. Despite some automation, tray fermenters are labour-intensive, and require a large production area. Hence the potential for scaling up production is limited.

Static-bed Fermenter This is an adaptation of the tray fermenter (**Fig. 7**). It employs a single, larger and deeper, static bed of substrate located in an insulated chamber. O_2 is supplied by forced aeration through the bed of substrate.

Tunnel Fermenter This is an adaptation of the static-bed device (**Fig. 8**). Typically, the bed of solids is quite long but no deeper than 0.5 m. Fermentation using this equipment may be highly automated, by way of mechanisms for mixing, inoculation, continuous feeding and harvest of the substrate.

Rotary Disc Fermenter The rotary disc fermenter consists of upper and lower chambers, each with a circular perforated disc to support the bed of substrate

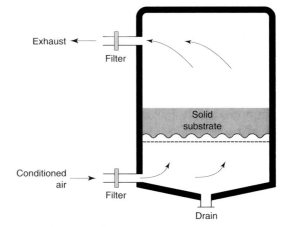

Figure 7 Static-bed fermenter.

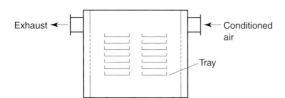

Figure 6 Tray fermenter.

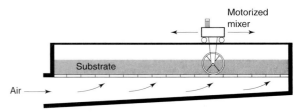

Figure 8 Tunnel fermenter.

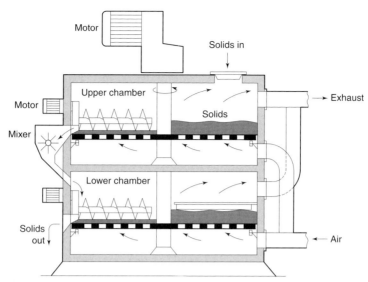

Figure 9 Rotary disc fermenter.

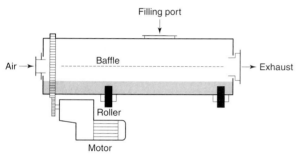

Figure 10 Rotary drum fermenter.

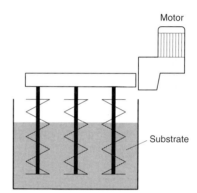

Figure 11 Agitated-tank fermenter.

(**Fig. 9**). A common central shaft rotates the discs. Inoculated substrate is introduced into the upper chamber, and slowly moved to the transfer screw. The upper screw transfers the partly fermented solids through a mixer to the lower chamber, where further fermentation occurs. The fermented substrate is harvested using the lower transfer screw. Both chambers are aerated with humidified, temperature-controlled air. Rotary disc fermenters are used in large-scale koji production in Japan.

Rotary Drum Fermenter The cylindrical drum of the rotary drum fermenter is supported on rollers, and rotated at 1–5 r.p.m. around the long axis (**Fig. 10**). Rotation may be intermittent, and the speed may vary with the fermentation stage. Straight or curved baffles inside the drum aid in the tumbling of the substrate, hence improving aeration and temperature control. Sometimes the drum can be inclined, causing the substrate to move from the higher inlet end to the lower outlet during rotation. Aeration occurs through coaxial inlet and exhaust nozzles.

Agitated-tank Fermenter In this type of fermenter, either one or more helical-screw agitators are mounted in cylindrical or rectangular tanks, to agitate the fermented substrate (**Fig. 11**). Sometimes, the screws extend into the tank from mobile trolleys, that ride on horizontal rails located above the tank. Another stirred-tank configuration is the paddle fermenter. This is similar to the rotary drum device, but the drum is stationary and periodic mixing is achieved by motor-driven paddles supported on a concentric shaft.

Continuous Screw Fermenter In this type of fermenter, sterilized, cooled and inoculated substrate is fed in through the inlet of the non-aerated chamber (**Fig. 12**). The solids are moved towards the harvest port by the screw, and the speed of rotation and the length of the screw control the fermentation time. This type of fermenter is suitable for continuous anaerobic or microaerophilic fermentations.

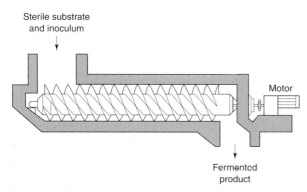

Sterile substrate
and inoculum

Motor

Fermented
product

Figure 12 Continuous screw fermenter.

Safe Fermentation Practice

The microorganisms used in certain industrial fermentations are potentially harmful. Certain strains have caused fatal infections in immunocompromised individuals, and rare cases of fatal disease in previously healthy adults have also been reported. Microbial spores and fermentation products, as well as microbes, have been implicated in occupational diseases. Most physiologically active fermentation products are potentially disruptive to health, and certain products are highly toxic. The product spectrum of a given microorganism often depends on the fermentation conditions. Under certain environmental conditions, some organisms, e.g. *Aspergillus flavus* and *A. oryzae*, are known to produce lethal toxins, and specific strains of the blue-veined cheese mould *Penicillium roqueforti* also produce mycotoxins under narrowly defined environmental conditions. Poor operational practice and failings in process and plant design can increase the risks. The safety aspects of industrial fermentations are considered in some of the literature cited under Further Reading. Consumer safety, product quality and the cleanliness of a fermentation product should be ensured by compliance with Good Manufacturing Practices (GMP).

See also: **Fermentation (Industrial)**: Basic Considerations; Control of Fermentation Conditions; Recovery of Metabolites; Production of Xanthan Gum; Production of Organic Acids; Production of Oils and Fatty Acids; Colours/Flavours Derived by Fermentation. **Fermented Foods**: Origins and Applications; Fermented Vegetable Products; Fermented Meat Products; Fermentations of the Far East; Beverages from Sorghum and Millet; Fermented Fish Products. **Fermented Milks**: Range of Products; Yoghurt; Products from Northern Europe; Products of Eastern Europe and Asia.

Further Reading

Chisti Y (1989) *Airlift Bioreactors*. London: Elsevier Applied Science Publishers.

Chisti Y (1992) Build better industrial bioreactors. *Chem. Eng. Prog.* 88(1): 55–58.

Chisti Y (1992) Assure bioreactor sterility. *Chem. Eng. Prog.* 88(9): 80–85.

Chisti Y (1998) Biosafety. In: Subramanian G (ed.) *Bioseparation and Bioprocessing*. Vol. 2, p. 379. New York: Wiley-VCH.

Chisti Y (1999) Solid substrate fermentations, enzyme production, food enrichment. In: Flickinger MC and Drew SW (eds) *Encyclopedia of Bioprocess Technology*. Vol. 5, p. 2446. New York: John Wiley.

Chisti Y and Moo-Young M (1991) Fermentation technology, bioprocessing, scale-up and manufacture. In: Moses V and Cape RE (eds) *Biotechnology: The Science and the Business*. P. 167. New York: Harwood Academic Publishers.

Chisti Y and Moo-Young M (1994) Clean-in-place systems for industrial bioreactors: Design, validation and operation. *J. Ind. Microbiol.* 13: 201–207.

Crueger W and Crueger A (1990) *Biotechnology: A Textbook of Industrial Microbiology*, 2nd edn. Madison: Science Tech Publishers.

Doran PM (1995) *Bioprocess Engineering Principles*. London: Academic Press.

Hambleton P, Melling J and Salusbury TT (eds) (1994) *Biosafety in Industrial Biotechnology*. London: Chapman & Hall.

Steinkraus KH (ed.) (1989) *Industrialization of Indigenous Fermented Foods*. New York: Marcel Dekker.

Wang DIC, Cooney CL, Demain AL et al (1979) *Fermentation and Enzyme Technology*. New York: John Wiley.

Ward OP (1989) *Fermentation Biotechnology*. Stony Stratford: Open University Press.

Media for Industrial Fermentations

Graeme M Walker, School of Molecular and Life Sciences, University of Abertay Dundee, UK

Introduction

The production of foods and beverages from fermentable carbon sources by microorganisms represents the oldest and most economically significant of all biotechnologies. A wide array of plant- and animal-based complex media for the industrial cultivation of bacteria, fungi and yeasts are employed in the food industry (**Table 1**).

The composition of a fermentation medium in terms of nutrients, their bioavailability, and the

Table 1   Fermentation media for food microorganisms

Media	Microorganisms	Products
Barley malt wort	Yeasts (*Saccharomyces* spp.)	Ale, Scotch malt whisky, spent yeast
Barley malt plus unmalted cereals (e.g. rye, wheat, maize, sorghum)	Yeasts (*Saccharomyces* spp.)	Lager, Scotch grain whisky (yielding blended Scotch on mixing with malt whisky), Bourbon whiskey, neutral spirits (e.g. gin, vodka, liqueurs)
Ethanol	*Acetobacter* spp.	Vinegars (e.g. from beer, wine, cider)
Rice	*Aspergillus oryzae*, yeasts	Pachwai, saké, sochu, arrack, binuburan
Potatoes, artichokes, *Agave* spp., sweet potatoes	Yeasts	Aquavit, vodka, pulque, tequila, awamori
Cane and beet molasses	Yeasts, fungi, bacteria	Yeast biomass for baking, brewing, wine-making and distilling Yeast extracts/enzymes Rum, citric acid, glutamic acid
Wine must, fruit juices, honey	Yeasts, lactic acid bacteria	Wine, Cognac, Armagnac, brandy, grappa, kirsch, slivovitz, cider, perry, mead
Milk, cheese whey	Lactic acid bacteria, yeasts, fungi	Bacterial and fungal starter cultures for dairy produce Cheese, yoghurts, buttermilk, sour cream, acidophilus milk, koumiss, taette, kefir Lactic acid Ethanol (for potable spirits, cream liqueurs) Spent yeast/food yeast
Starch hydrolysates, glucose syrups	Fungi, yeasts, bacteria	Mycoprotein Fermented beverages Microbial proteases, lipases, carbohydrases, organic acids
Water, CO_2 and sunlight	*Chlorella*, *Scenedesmus*, *Spirulina*	Food source Protein/vitamin supplements
Solid substrate media		
Soya, wheat	*Aspergillus oryzae*, yeasts, lactic acid bacteria	Soy sauce (shoyu), tofu, tempeh, miso
Wheat flour	Yeasts, lactic acid bacteria	Bread, sourdough breads, rye breads, pumpernickel
Peanut presscake	*Neurospora sitophila*	Ontijom
Meat, fish	Lactic acid bacteria, fungi	Sausages (*Pediococcus cerevisiae*), fish sauces (halophilic *Bacillus* spp.), cured hams (*Aspergillus* and *Penicillium* spp.)
Plants, vegetables	Lactic acid and other bacteria	Sauerkraut (cabbage), pickles (e.g. cucumber), olives, tea, cocoa, coffee (pectinolytic *Bacillus* and *Erwinia* spp.)
Straw, manure, sawdust,	*Pleurotus* spp.	Oyster mushroom
Oak wood	*Lentinula edodes*	Shii-take mushroom
	Volvariella volvacea	Chinese (padi-straw) mushroom
	Agaricus bisporus	Button mushroom
Milk curd	*Penicillium* spp.	Mould-ripened cheeses
	Propionibacterium spp.	Swiss-type cheeses
Wheat bran	Fungi (e.g. *Aspergillus niger*)	Food-processing enzymes
Tea leaves	*Acetobacter xylinum*, *Schizosaccharomyces pombe*	Teekwass

absence of potentially toxic or inhibitory constituents, are crucially important for the metabolism and growth of food microorganisms. The cost of the medium is also important – raw materials account for a significant proportion (generally over 50%) of the overall costs of production of a fermented food. Historically, the choice of media for large-scale fermentations has been based on price and availability rather than on microbial physiology.

Microorganisms require carbon, nitrogen, inorganic ions, growth factors, energy and water in order to metabolize and grow. The sources of major, minor and trace nutrients in media commonly employed in industrial food fermentations are described below.

Sources of Utilizable Carbon

All food microorganisms are chemoorganotrophs, with the exception of some photosynthetic microalgae – that is, they obtain their energy and carbon by metabolizing organic substrates. These include: carbon biopolymers (e.g. starch, pectin); hexose and pentose monosaccharides (e.g. glucose, xylose); disaccharides (e.g. sucrose, maltose); trisaccharides (e.g. maltotriose); oligosaccharides (e.g. maltodextrins); alcohols (e.g. ethanol); polyols (e.g. glycerol); organic acids (e.g. lactic acid, acetic acid); fatty acids; amino acids; peptides; and polypeptides. All biosynthetically produced organic compounds of plant or animal

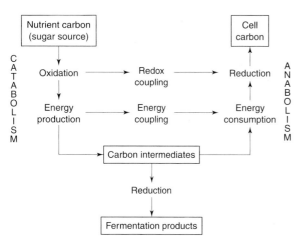

Figure 1 Microbial metabolism of sugars.

origin have the potential to serve as substrates for microbial fermentation, but the capability of microorganisms of using particular carbon sources varies between genera and species. Sugars represent the main fermentable carbon sources for food microorganisms. The main catabolic and anabolic fates of sugars in microbial metabolism are outlined in **Figure 1**.

A small proportion of the carbon assimilated by chemoorganotrophic bacteria, fungi and yeasts may be in the form of CO_2. This is 'fixed' using anaplerotic enzymes, such as phosphoenolpyruvate carboxykinase and pyruvate carboxylase.

Although glucose is commonly used as the sole carbon and energy source for the growth of food microorganisms in the laboratory, it is not generally freely available in industrial fermentation media. In these media, the more common carbon sources are maltose, sucrose, fructose, xylose and lactose – indeed, glucose frequently exhibits a repressive effect on the assimilation of other sugars by microorganisms, e.g. the yeast *Saccharomyces cerevisiae* during the fermentation of the complex sugar mixtures found in malt wort and cereal dough. Molasses, corn steep liquor and sulphite waste liquor are complex carbon sources which also supply necessary nitrogenous and mineral nutrients (**Table 2**).

Molasses, derived from the refining of sugar cane or sugar beet, is a globally employed fermentation substrate. Its composition varies with the specific production process and the geographical location. Different names are applied to molasses, depending on the mode of sugar production from which it was recovered. Thus, blackstrap molasses is the residual liquor following the crystallization of sucrose from sugar cane; beet molasses is generated similarly from sugar beet; refinery molasses differs from blackstrap molasses only in that it is the residual mother liquor which accumulates in the refining of crude sucrose by recrystallization; high-test molasses contains much of the original sugar of cane juice, which has been partially hydrolysed (inverted) to glucose and fructose; and hydrol is molasses resulting from the production of crystalline glucose from corn starch. **Table 3** provides more quantitative information on the composition of cane and beet molasses.

Corn steep liquor is the water extract (concentrated to 50% solids) which results from the steeping of maize during the production of corn starch, gluten and other corn products. It contains high levels of lactic acid, resulting from the growth of lactic acid bacteria and fungi. Corn steep liquor is thus a natural product of fermentation. Sulphite waste liquor is the spent liquor from the paper-pulping industry, and remains after wood is digested to cellulose pulp by calcium bisulphite under heat and pressure. Sulphite waste liquor contains 10% solids, of which 20% comprises sugars (hexoses and pentoses). Sulphite waste liquor cannot be fermented directly – the free SO_2 or sulphurous acid must first be removed, by steam stripping or precipitation with lime.

Molasses, corn steep liquor and sulphite waste liquor generally represent complete nutritional sources for the growth of microorganisms, but certain components may be limiting, unavailable, inhibitory or toxic. For example, molasses for the propagation of bakers' yeast or potable alcohol fermentations needs to be supplemented with assimilable nitrogen and phosphorus sources (e.g. in the form of diammonium hydrogen phosphate).

Sources of Utilizable Nitrogen

The nitrogen in microbial growth media serves an anabolic role in the biosynthesis of structural proteins and functional enzymes and nucleic acids. Some nitrogen sources, notably amino acids, may be catabolized immediately on entry into the cell, and the products may be important in determining the flavour of certain foods (e.g. higher alcohols and diacetyl in fermented beverages). Food microorganisms are non-diazotrophic (i.e. cannot fix atmospheric N_2, and therefore require a supply of either organic or inorganic nitrogen sources (**Table 4**).

Simple inorganic nitrogen sources such as gaseous ammonia or ammonium salts are widely utilized. Ammonium sulphate and diammonium hydrogen phosphate are also useful as sources of assimilable sulphur and phosphorus, respectively. Nitrate and urea may be employed as nitrogen sources: the former may be reduced to ammonia by many bacteria, fungi and yeasts, using the assimilatory enzyme nitrate reductase; urea may be used as an inexpensive nitrogen source in certain industrial fermentation media, e.g. molasses. However, urea is not recommended for

Table 2 Principal constituents of selected media used in food fermentations

	Molasses	Beer wort	Wine must	Cheese whey	Corn steep liquor
Carbon source	Sucrose Fructose Glucose Raffinose	Maltose Sucrose Fructose Glucose Maltotriose Maltodextrins	Glucose Fructose Sucrose Pentoses	Lactose	Glucose Other residual sugars

Nitrogen source	Protein
	Other nitrogenous
	  compounds
		Amino acids
		Ammonium ions
		Amino nitrogen
		  compounds
			Amino acids
			Amino nitrogen
			  compounds
				Globulin
				Albumin
				Amino and urea
				  nitrogen
				  compounds
					Amino acids
					Proteins

Minerals	Phosphorus Potassium Magnesium Sulphur	Phosphorus Potassium Magnesium Sulphur	Phosphorus Potassium Magnesium Sulphur (sulphite often present)	Phosphorus Potassium Magnesium Sulphur	Phosphorus Potassium Magnesium Sulphur

Vitamins	Range present, but
	  biotin may be
	  deficient in beet
	  molasses
		Range present, but biotin
		  may occasionally be
		  deficient
			Range present	Range present	Biotin
					Pyridoxine
					Thiamin

Trace elements	Range present, but
	  manganese (Mn^{++})
	  may be limiting
		Range present, but
		  Zinc (ZN^{++}) may be
		  limiting
			Range present	Iron
				Zinc
				Manganese
				Calcium
				Copper
					Range present

Other components	Unfermentable
	  sugars
	Organic acids
	Waxes
	Pigments
	Silica
	Pesticide residues
	Caramelized
	  compounds
	Betaine
		Maltodextrins not
		  fermented by yeasts
		Pyrazines
		Hop compounds
			Pentose sugars not
			  fermented by
			  yeasts
			Tartaric and malic
			  acids
			Decanoic and
			  octanoic acids
				Lipids
				NaCl
				Lactic and citric
				  acids
					High levels of
					  lactic acid
					  present
					Fat
					Fibre

the production of potable spirit beverages by fermentation by yeasts, owing to the possibility of the formation of carcinogenic ethylcarbamate during the distillation process.

Complex, organic forms of nitrogen are found in various types of hydrolysed plant protein material (**Table 5**). For example, corn steep liquor, casein hydrolysate, soybean meal, barley malt and yeast extract provide mixtures of peptides and amino acids that invariably support higher rates of growth and fermentation than those achieved using inorganic nitrogen sources. Peptones are protein hydrolysates derived from meat, casein, gelatin, keratin, peanuts, soybean meal, cottonseeds and sunflower seeds, but are relatively expensive sources of nitrogen for industrial applications. The individual amino acids present in complex mixtures may be assimilated sequentially by food microorganisms, but the presence of ammonium ions may inhibit amino acid uptake. In the case of some microbes, the provision of amino acids in the form of peptides may result in better growth than the provision of the same amino acids in free form.

Ammonia may be assimilated by either the glutamate dehydrogenase pathway or the glutamine synthetase–glutamate synthase pathway. In the former pathway, the reductive amination of α-ketoglutarate forms L-glutamate, while in the latter pathway, L-glutamate is aminated by glutamine synthetase to form L-glutamine. One of the amide groups of the L-glutamine is then transferred to α-ketoglutarate by glutamate synthase, yielding two molecules of L-glutamate. The precise pathway adopted depends on the microbial species and the concentrations of available ammonia and intracellular amino acids.

Table 3 Composition of cane and beet molasses

Main constituent	Components	Typical composition[a] (% weight except for vitamins, mg kg^{-1})	
		Cane molasses	Beet molasses
pH		5–6	7–9
Sugars	Total sugars	50–65	49–58
	Sucrose	30–40	47–55
	Invert sugar	10–25	0.2–2.0
	Unfermentable sugars	3–5	1.0
Other carbon compounds	Gums, starch pentosans, hexitols, organic acids, waxes	10–15	10–20
Nitrogenous compounds	Crude protein, amino acids and other nitrogenous compounds	3.0	8–12
Minerals	Phosphorus	0.10	0.02
	Potassium	3.0	5.0
	Sulphur	0.55	0.33
	Magnesium	0.35	0.12
	Calcium	0.74	0.23
	Sodium	0.25	0.5
	Ash	9.0	5.0
Vitamins	Thiamin	1.8	1.3
	Riboflavin	2.5	0.40
	Pyridoxine	5	5
	Nicotinic acid	200	50
	Pantothenic acid	60	100
	Folic acid	0.04	0.20
	Biotin	1.2	0.05
	Choline	750	500
	Inositol	6000	8000

[a]Figures quoted are representative of typical molasses – composition will vary depending on country of origin, method of production, etc.

Table 4 Commonly employed nitrogen sources in food fermentation media

Organic N sources	Inorganic N sources
Corn steep liquor	$(NH_4)_2SO_4$
Casein hydrolysate	NH_4Cl
Soybean meal	NH_3
Yeast extract	$(NH_4)_2HPO_4$
Barley malt	$(NH_4)_2PO_4$
Dried distillers' solubles	NH_4NO_3
Pharmamedia (cottonseed flour)	Urea NH_4OH
Corn gluten meal	
Linseed meal	
Rice and wheat meal	

Sources of Inorganic Ions

Around 8% of the dry weight of microbial cells comprises inorganic ions. The 'bulk' or macronutrients are required in millimolar concentrations and are nitrogen, phosphorus, sulphur, potassium and magnesium. Micronutrients (or trace elements) are required in micromolar or less concentrations and play specific metabolic roles. They include sodium, calcium, chlorine, iron, cobalt, zinc, molydenum, copper, manganese, nickel and selenium. Several metal ions may be toxic to microorganisms at con-

centrations of around 100 µmol l^{-1}, e.g. silver, arsenic, barium, caesium, cadmium, mercury, lithium and lead. **Table 6** summarizes the major requirements of microorganisms in terms of inorganic ions.

Most of the complex fermentation media used in industry, and the water used to dilute such media, contain adequate levels of inorganic ions for microbial growth, so supplementation with minerals is not usually necessary. The media for fermentations usually contain around 70–90% water, which acts as the solvent for the nutrients contained in the media and also supplies trace metals. The ionic content of the water used is very influential in determining product quality, e.g. the flavour of beer.

Occasionally, certain metals are present in concentrations which are sub-optimal for efficient fermentation. Also, the bioavailability of metal ions may be compromised as a result of sterilization, precipitation, chelation or binding to inert surfaces. The separate sterilization of metal ion supplements may then be necessary to counteract such losses (see below). Some metals may interact antagonistically, e.g. the inhibition of essential magnesium-dependent cellular functions by calcium. In contrast, however,

Table 5   Nitrogenous components of selected fermentation media

Medium	Dry matter (%)	Total protein (%)	Individual amino acids (%)												
			Arg	Cys	Gly	His	Ile	Leu	Lys	Met	Phe	Thr	Trp	Tyr	Val
Corn steep liquor	50	24	0.4	0.5	1.1	0.3	0.9	0.1	0.2	0.5	0.3	–	–	0.1	0.5
Dried distillers solubles	92	26	1.0	0.6	1.1	0.7	1.6	2.1	0.9	0.6	1.5	1.0	1.0	0.7	1.5
Pharmamedia (cottonseed flour)	99	59	12.3	1.5	3.8	3.0	3.3	6.1	4.5	1.5	5.9	3.3	0.95	3.4	4.6
Soybean meal	90	45	3.2	0.7	2.9	1.1	2.3	3.4	3.0	0.7	2.1	1.9	0.6	1.7	2.4
Wheat flour	90	13	0.8	0.2	–	0.3	0.6	1.0	0.5	0.2	0.7	0.4	0.2	0.5	0.6
Whey powder	95	12	0.4	0.4	0.7	0.2	0.7	1.2	1.0	0.4	0.5	0.6	0.2	0.5	0.6
Linseed meal	92	36	2.5	0.6	0.2	0.5	1.3	2.1	1.0	0.8	1.8	1.4	0.7	1.7	1.8
Brewers' yeast	95	43	2.2	0.6	3.4	1.3	2.7	3.3	3.4	1.0	1.8	2.5	0.8	1.9	2.4
Yeast autolysate	70	55	2.1	0.3	1.6	0.9	2.0	2.9	3.2	0.5	1.6	1.9	0.8	–	2.3
Casein hydrolysate	97	15	0.5	0.07	0.9	0.5	1.1	2.9	1.3	1.1	0.9	1.3	0.01	0.5	1.7

Table 6   Inorganic ion requirements of microorganisms

Element	Common source	Typical concentrations needed for growth	Cellular functions
Macronutrients			
Phosphorus and nitrogen	$(NH_4)_2HPO_4$	$10\ mmol\ l^{-1}$	Energy transduction; nucleic acid and membrane structure
Potassium	KCl	$5\ mmol\ l^{-1}$	Ionic balance; enzyme activity
Magnesium and sulphur	$MgSO_4.7H_2O$	$2\ mmol\ l^{-1}$	Transphosphorylase activity; cell and organelle structure (Mg); sulphydryl amino acids and vitamins (S)
Micronutrients			
Calcium	$CaCl_2$	$< \mu mol\ l^{-1}$	Possible second messenger in signal transduction; bacterial sporulation
Copper	$CuSO_4.5H_2O$	$1\ \mu mol\ l^{-1}$	Redox pigments
Iron	$FeCl_3.H_2O$	$2\ \mu mol\ l^{-1}$	Haem proteins, e.g. cytochromes
Manganese	$MnSO_4.H_2O$	$2\ \mu mol\ l^{-1}$	Enzyme activity
Zinc	$ZnCl_2$	$5\ \mu mol\ l^{-1}$	Alcohol dehydrogenase activity
Nickel	$NiCl_2$	$5\ \mu mol\ l^{-1}$	Urease activity
Molybdenum	Na_2MoO_4	$0.1\ \mu mol\ l^{-1}$	Nitrate metabolism; vitamin B_{12}
Toxic ions			
Heavy metals, e.g. cadmium, lead		$> 100\ \mu mol\ l^{-1}$	Toxic

some metals may act synergistically, e.g. magnesium and cobalt in fermentations involving bacterial glucose isomerase. The media constituents and water may contain inhibitory or toxic ions. The levels of potentially toxic ions can be limited by chelating agents naturally present in the medium (e.g. citric acid, polyphosphates); by chelating agents added as supplements (e.g. EDTA); or by ion-exchange pretreatments.

By controlling the availability of some metal ions, it is possible to control the progress of certain food fermentations. For example, low levels of manganese (of the order of parts per billion) must be carefully maintained in citric acid fermentations using *Aspergillus niger*, because manganese deficiency is a prerequisite for the overproduction of citric acid. In contrast, manganese is an important activator of lactate dehydrogenase in the production of lactic acid by homofermentative species of *Lactobacillus*. In the production of ethanol by *Saccharomyces cerevisiae*, it is crucially important to maintain high levels of bioavailable magnesium to ensure maximal fermentation performance.

Major requirements for magnesium, phosphorus and sulphur can be met by the supplementation of crude media with appropriate salts (e.g. magnesium

Table 7 Vitamin content of selected fermentation media

Medium	Vitamins (mg kg^{-1})						
	Biotin	Choline	Niacin	Pantothenate	Pyridoxine	Riboflavin	Thiamin
Dried distillers' solubles	2.9	4400	110	20		15	5.5
Blackstrap molasses	1.2	750	200	60	5	2.5	1.8
Pharmamedia	1.5	3270	83	12	16	4.8	4.0
Whey powder		2420	11	48	2.9	20	4.0
Brewers' yeast		4840	498	121	50	35	75
Wheat flour		880	62	13		1.1	5.1
Soybean meal		2673	26	15		3.3	
Corn steep liquor	0.88				19		0.88
Barley malt			50	8.6		2.9	3.7

sulphate). It is not possible to generalize the ionic requirements of food microorganisms, because of differences between strains, chelation by different media and ionic interactions. However, it is possible to optimize the metal ion concentrations in individual media by using elemental mass balances, programmed search techniques and surface-response statistical modelling.

Sources of Growth Factors

Growth factors are organic compounds which are required in very low concentrations and which perform specific catalytic or structural roles in microbial physiology. They include vitamins, purines and pyrimidines, nucleotides and nucleosides, amino acids, fatty acids, sterols and polyamines. An auxotroph is a microorganism that is unable to synthesize one or more essential growth factors, and will not grow in fermentation media lacking them. For example, *Saccharomyces cerevisiae* is auxotrophic for ergosterol and oleic acid when propagated under strictly anaerobic conditions, because O_2 is required for the biosynthesis of the sterols and unsaturated fatty acids which are essential for the development of the cell membrane.

A relative growth factor requirement is revealed when the addition of growth factors stimulates microbial growth. Microorganisms differ greatly in their requirements for growth factors. *Lactobacillus* species are particularly fastidious, requiring a range of growth factors. Many microorganisms require vitamins in the fermentation medium, at micromolar levels. These include biotin (which serves as a cofactor in carboxylase-mediated reactions); pantothenic acid (a component of coenzyme A, which is involved in acetylation reactions); nicotinic acid (in the form of nicotinamide, which is involved in redox reactions); and thiamin (as thiamin pyrophosphate, which is involved in decarboxylation reactions). **Table 7** lists the vitamin content of certain fermentation media.

Most complex nutritional substrates provide the vitamins necessary for microbial fermentations, although some media may be limiting in certain vitamins. For example, beet molasses is generally deficient in biotin, and cane blackstrap molasses may occasionally be deficient in pantothenic acid and inositol. Mixtures of beet and cane molasses are therefore used, to ensure adequate levels of vitamins for the optimal growth of bakers' yeast. Yeast extracts are rich sources of vitamins for use in industrial fermentations; more expensive sources include soy flour, malt sprouts and malt extract. Some fermentations benefit from the addition of commercially available media supplements or 'foods'. For example, yeast foods based on mixtures of yeast extract, ammonium phosphate and minerals (e.g. magnesium and zinc) may be employed in alcohol fermentations, to ensure consistent yeast activity.

Design and Preparation of Food Fermentation Media

Media Design

Several important criteria must be considered in the design and preparation of media for food fermentations. These are:

1. Media supply: cost effectiveness (raw materials, transport, storage); consistency and reliability of supply; nutritional variability; world political situations; alternative carbon and nitrogen sources.
2. Media type: liquid or solid; complex, defined or semi-synthetic; nutrient-limited; used for propagation or fermentation; balanced or unbalanced.
3. Media properties: foaming characteristics; colour/pigmentation; heat-labile components; toxic components; buffering capacity; viscosity; particulate nature; biochemical oxygen demand loading; control of redox potential; ionic interactions; microbiological stability in storage.
4. Media treatment necessary: sterile, pasteurized or non-sterile; separate treatments for heat-labile components; pre-treatments (e.g. centrifugation, acidification, ion-exchange, clarification, pre-

hydrolysis); ease of product recovery; effluent treatment.

In most cases, complex, inexpensive and readily available agriculturally derived media are employed. Such media are, however, notoriously variable from batch to batch in terms of nutritional consistency. For example, the composition of molasses varies in terms of sugar and inorganic ions according to the country of origin and the production processes. Similarly, if corn steep liquor is intended to supply a particular amino acid or growth factor at critically low levels, its concentration in each batch of media should be monitored. The maintenance of reproducible fermentations using complex, undefined and variable media is thus fraught with difficulties.

The design of chemically defined, synthetic media allows the nutritional needs of the microorganisms to be addressed, and this can improve control over fermentation performance. For example, defined media can be designed so as to limit the availability of carbon, nitrogen, phosphorus, metal ions or growth factors during fermentation, and such limitation may cause a shift in the balance between growth of the microorganism and the production of desired metabolites. Defined media are more expensive than complex media, but in certain cases may be preferred (e.g. in mycoprotein production).

Semi-synthetic media can also be employed. These are mixtures of defined chemicals and non-defined complex nutrient substrates. For example, a typical medium for a lactic acid bacterial fermentation may comprise: glucose (as the carbon and energy source); diammonium hydrogen phosphate (as the nitrogen and phosphorus source); calcium carbonate (to neutralize lactic acid); and malt sprouts (as the source of growth factors and trace elements).

The desired levels of particular nutrients in the medium depend on whether the desired product is cellular biomass or primary metabolites. For example, in molasses used as an industrial medium for the production of bakers' yeast, the levels of assimilable sugar must be kept low, by controlled incremental nutrient-delivery regimes, to prevent the repression of respiration by glucose. However, molasses can also be used for the production of potable ethanol, and in this case the sugar levels are kept high, in order to support fermentative metabolism.

The method of preparation of fermentation media is very influential in determining choice. The following are important considerations in the preparation of media in bulk:

1. Composition of ingredients; quality/impurities; carbon : nitrogen ratio; batch-to-batch variability; bioavailability of metal ions and growth factors.

Table 8 Wet-heat sterilization conditions

Temperature (°C)	Time (min)	Pressure (kPa)
121	15	103.4
126	10	137.8
134	3	206.7
140	0.67	261.8

2. Order of solution or suspension of ingredients; pH adjustments needed before and after sterilization; effects of sterilization on minerals, salt precipitation, etc.
3. Changes in the medium before inoculation, due to temperature, aeration, agitation, presence of antifoams, etc.

The most important criterion in the choice of media for industrial food fermentations is cost. The cost of the production media virtually dictates the selling price of a particular commodity. World politics may affect the price and availability of fermentation substrates, so it is advisable always to have alternative substrates to hand. The costs of media pre-treatment (e.g. ion-exchange, acid/enzymic hydrolysis, pH control, antifoams) are also significant, as are the product recovery and effluent treatment costs. Such costs can, however, be reduced by judicious approaches to media design and preparation.

Media Sterilization

The prevention of microbial contamination is fundamental to many industrial food fermentation processes. Media sterilization is the destruction or removal of all forms of microbial life from the aqueous feedstock. In industrial fermentations, components such as vessels, pipework, media, inlet air and exhaust gases are frequently sterilized by a combination of wet-heat and filtration methods. Wet-heat methods are less expensive and more effective than dry-heat methods, and so are commonly employed in fermentation industries to destroy unwanted microorganisms. The wet-heat sterilization conditions typically used to kill all microorganisms, including bacterial spores, and are listed in **Table 8**. These conditions may be achieved in an autoclave in an atmosphere of saturated steam.

However, most heat treatments of industrial fermentation media are designed to selectively kill only those microorganisms of particular concern. Pasteurization is the method commonly employed for destroying frequently encountered pathogenic bacteria. **Table 9** gives the pasteurization conditions used for common food fermentation media.

Strategies for the bulk sterilization of fermentation media include *in situ* steam injection of a full charge

Table 9 Pasteurization conditions for some food fermentation media

Medium	Food product	Typical treatment
Molasses	Citric acid	100°C briefly, then acidified to pH 2.5
Molasses	Bakers' yeast	Preheated to 70°C, then flash sterilized at 136°C for 15–30 s
Malt wort	Beer	90–100°C for 1 h, in the presence of hops
Milk	Yoghurt (skimmed milk) Cheese (full-fat milk)	High temperature short time method (flash pasteurization), e.g. 72°C for 15 s; 88°C for 1 s; 90°C for 0.5 s; or 96°C for 0.05 s Low temperature long time method (batch pasteurization), e.g. 63°C for 30 min

Table 10 Problems and solutions relating to media sterilization

Problem	Solutions
Sugar caramelization	Sterilize sugars separately, and add aseptically
Metal precipitation	Sterilize phosphate source and metal salts separately
Unsuccessful sterilization of particulate and viscous media	Ensure sufficient agitation in fermenter to achieve heat transfer
Very high initial bioburden	Good housekeeping, cleaning in place, elimination of residues, sterilization of pipework dead legs

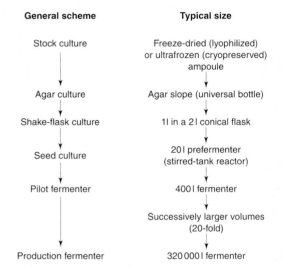

Figure 2 Inoculum preparation.

of non-sterile medium in the fermenter, or steam conduction through attemperation jackets in agitated fermenters. Alternatively, the media and vessels may be sterilized separately prior to fermentation. Antifoam agents, especially those which are oil-based, are often difficult to sterilize. Inert, silicone-based antifoams may be used which, although expensive, are nontoxic towards microorganisms.

The loss of available carbon and the build-up of potentially toxic or inhibitory compounds may occur during heat treatments employed to sterilize or pasteurize growth media. For example, the excessive heating of molasses may generate undesired caramelization products, following the Maillard reaction between reducing sugars and the free amino groups in proteins. Heat may also destroy vitamins and other growth factors essential for microbial growth. Some of the problems that may be encountered during media sterilization, and their possible avoidance measures, are listed in **Table 10**.

Fermenter inlet and exhaust gases are generally sterilized by filtration, using either depth filters (e.g. comprising porous ceramic, granular carbon, glass-fibre or synthetic membranes) or filtration cartridges (microfilters, or microporous membrane sheets or mats).

Inocula Preparation

The development of microbial inocula for food fermentations is important, in order to provide sufficient

amounts of viable and vital biomass to carry out large-scale production effectively. Most food fermentations, with the notable exception of traditional brewing employ specifically grown inocula which are discarded at the end of the fermentation process. This avoids genetic instability and microbial contamination, and ensures the provision of high-viability cultures. The final inoculation levels are high – often 10–20% – because this reduces the fermentation time and suppresses the growth of contaminants.

Figure 2 outlines a general scheme for the multi-stage build-up of inoculum biomass, from the laboratory stock culture to the amount needed to inoculate the industrial production fermenter. The total amount of biomass produced depends on the properties of the medium rather than the inoculum size, but larger inocula enable maximum growth to be achieved more rapidly. With regard to the media used for the development of inocula, transfers of inocula during multistage build-up should be made between identical or similar media. This ensures that production stage growth is simply an extension of the prior seed stage growth. If the utilization of a particular component of a production medium requires the microbial cells to become enzymatically adapted, then this substrate should be included in the medium used for the development of the inoculum, to prevent de-adaptation and the prolongation of lag phases during growth.

During inoculum development, cellular biomass is required rather than fermentation products, therefore the medium and the conditions must be properly balanced, to encourage respiratory growth and discourage fermentative metabolism. For example, in the brewing industry, the propagation of seed yeast should be conducted aerobically with sugar limitation (preferably in fed-batch mode), to ensure that sufficient conditioned biomass is produced prior to the commencement of anaerobic alcoholic fermentation.

See also:
Fermentation (Industrial): Basic Considerations. **Lactobacillus**: Introduction. **Saccharomyces**: Saccharomyces cerevisiae.

Further Reading

Harlander SK and Labuza TP (1986) *Biotechnology in Food Processing.* New Jersey: Noyes Publications.

Jackson AT (1990) *Process Engineering in Biotechnology.* Milton Keynes: Open University Press.

Jones GA and Jones AM (1995) Microbial growth and physiology. In: Hui YH and Khachatourians GG (eds) *Food Biotechnology. Microorganisms.* P. 47. New York: VCH Publishers.

McNeill B and Harvey LM (eds) (1990) *Fermentation. A Practical Approach.* Oxford: IRL Press.

Miller TL and Churchill BW (1986) Substrates for large-scale fermentations. In: Demain AL and Solomon NA (eds) *Manual of Industrial Microbiology and Biotechnology.* P. 122. Washington: American Society for Microbiology.

Präve P, Faust U, Sittig U and Sukatsch DA (1987) *Fundamentals of Biotechnology.* Weinheim: VCH Publishers.

Rogers PL and Fleet GH (eds) (1989) *Biotechnology and the Food Industry.* Amsterdam: Gordon & Breach.

Stanbury PF (1992) *Principles of Fermentation Technology,* 2nd edn. Oxford: Pergamon Press.

Stowell JD, Beardsmore AJ, Keevil CW and Woodward JR (1987) *Carbon Substrates in Biotechnology.* Oxford: IRL Press.

Vogel HC (ed.) (1983) *Fermentation and Biochemical Engineering Handbook.* New Jersey: Noyes Publications.

Control of Fermentation Conditions

Tajalli Keshavarz, Department of Biotechnology, University of Westminster, London, UK

Introduction

The ultimate aim of an industrial fermentation process is to produce the desired bioproduct as fast as possible and in the greatest possible quantities, in the simplest and cheapest possible way. In practice, compromises have to be made, in order to provide the optimal conditions for the production of a particular product.

The physiology of the microorganisms and the relevant metabolic pathways must be well-understood, and the needs of the microorganisms, e.g. in terms of nutrients and air (in the case of aerobic cultures) must be satisfied in the best possible way. These needs often change as the microbial biomass increases in concentration and the environmental conditions (e.g. nutrient composition, temperature, pH) are altered. For example, the needs for carbon and molecular O_2 sources may be different at different stages of batch fermentations. Most industrial fermentation processes are fed-batch processes, in which nutrients are added to the fermenter, either intermittently or continuously. To ensure that the requirements of the culture are met as the fermentation progresses, the environmental conditions, including the concentration of nutrients, must be controlled. Continuous manual control is not feasible in practice – it would be expensive, impractical and at best inefficient. In order to maintain the optimal conditions for fermentation, there is a need for robust automatic control.

Control Systems

Control systems for bioprocesses normally comprise the following elements:

- monitoring and measuring devices
- controllers
- operators.

The bioprocess itself is also considered as a part of the control system which is often called a 'control loop' (**Fig. 1**).

An automatic control loop deals with the environmental conditions and all other aspects of a culture external to the microorganisms. Controlling the environment, i.e. the fermentation conditions, elicits the best response from a microorganism in terms of the desired product(s).

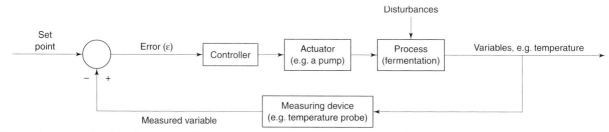

Figure 1 Conventional feedback control loop.

Measuring Equipment

The role of a sensor (measuring element) is to recognize a process variable and in response to produce a signal, which is sent to a controller. Examples of variables (properties) are temperature, pH and dissolved oxygen tension (DOT), and in order to control as many fermentation variables as possible, a wide range of sensors is needed. Some of the fermentation parameters that can be measured by sensors at present are:

1. Physical: temperature; pressure; gas flow rate; liquid inlet and outlet flow rates; culture level; culture volume; culture weight; culture viscosity; colour; agitation power; agitation speed; foaming; gas hold-up.
2. Chemical: dissolved O_2; dissolved CO_2; redox potential; general gas analysis; pH; nutrients; intermediates; products; conductivity; ionic strength.
3. Biochemical: carbohydrates; total proteins; vitamins; nucleic acids; ATP ADP AMP; NAD/NADH; enzymes; amino acids; cell mass composition.
4. Biological: total cell count; viable cell count; biomass concentration; morphology; cell size and age; doubling time; contamination.

Some sensors, e.g. temperature and pH probes, have been available for a long time, but there has been considerable progress in the design and construction of novel sensors over recent years.

Measuring and analytical equipment may be classified in different ways. In relation to fermenters, they may be categorized as 'on-line', 'at-line', or 'off-line'. In relation to the characteristics of the culture, they may be categorized in terms of their use for physical, chemical, biochemical or biological measurements. Fast-response sensors are needed to facilitate the efficient control of a bioprocess: the time taken to measure a variable should be compatible with the time taken for it to change.

Physical fermentation variables, such as temperature, foaming and the flow rate of gases and media (in fed-batch and continuous fermentations) are measured on-line. Some chemical variables, e.g. pH and dissolved O_2 and CO_2, are also measured on-line, i.e. in real time. On-line measurement is not available for all variables (e.g. biomass composition and some ingredients of the broth), but in some cases, analysis can be made at-line. In these cases, the property is measured using analytical equipment linked to the fermenter, without the need for an operator to remove samples from the fermenter for later off-line analysis. Examples of techniques which can be used at-line include: high-pressure liquid chromatography (HPLC); nuclear magnetic resonance (NMR); flow cytometry; fluorometry; and image analysis. At-line analysis does not, however, reflect real-time events in the fermenter.

The off-line measurement of fermentation parameters that cannot be measured on-line or at-line is routine in laboratories. A range of chemical and biochemical assays have traditionally been employed.

Table 1 lists examples of on-line and at-line measurements for monitoring fermentations, together with a brief description of the related equipment. Steam-sterilizable sampling equipment (autosamplers) is needed for the analysis of fermentation culture broth and cell characteristics at-line. The major problems in designing efficient and reliable autosamplers have been: the presence of gas bubbles and solid particles; the blockage of filters (separating cells from the broth); and the clogging of the connection lines between the fermenter and the sampler over long periods. A good autosampler should be stable and durable and should have a low dead volume to avoid losing large volumes of culture through sampling. This is particularly important if the total culture volume is small, the duration of fermentation is long, and frequent sampling is required. **Figure 2** shows a fermentation system incorporating at-line analysis. As the design of autosamplers has improved, it has become possible to measure more fermentation culture parameters at-line. For example, it is now possible to obtain automatic at-line assays of glucose, lactose, ammonia, urea, phosphate, sulphate, organic compounds and penicillin. The use of non-sterilizable biosensors (see below) at-line yields additional information about fermentations.

Table 1 Examples of on-line and at-line measurements for monitoring fermentations

Fermentation property	On-line measurement equipment	At-line measurement equipment	Brief description of measuring element
Temperature	Thermistor	–	Exhibits a large change in resistance with a small change in temperature; stable; cheap
pH	pH probe	–	Steam-sterilizable combined reference electrode, made of silver/silver chloride plus potassium chloride
Pressure (head space)	Pressure gauge	–	E.g. Bourdon gauge
Stirrer rotation speed (r.p.m.)	Tachometer	–	Uses of electromagnetic induction, light sensing or magnetic force
Dissolved O_2 tension	O_2 electrode	–	Steam-sterilizable polarographic electrode, with silver anode and platinum or gold cathode; measures partial pressure of O_2
Gas-phase O_2, CO_2 and other gases and volatiles	Process mass spectrometer	–	
Ionic composition	–	Range of ion-specific electrodes	Electrodes are not steam-sterilizable; response time varies from 10 s to a few minutes; ions measured include ammonium, calcium, potassium, magnesium, phosphate (PO_4^{3-}), sulphate ion (SO_4^{2-})
Media and culture ingredients; biomass; products	Optical-fibre sensor	–	NIR (near infrared) absorbance (460–1200 nm) utilized for rapid analysis
Substrates and products	–	Biosensors	Non-steam-sterilizable electrodes; enzymes or microbial cells are immobilized on a membrane, and a change in pH or Po_2 is detected

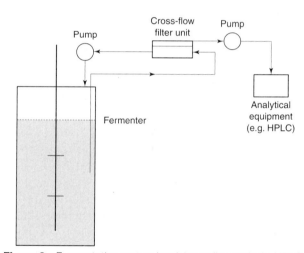

Figure 2 Fermentation system involving at-line analysis. HPLC, high-pressure liquid chromatography.

Biomass Concentration

Biomass is one of the most important fermentation variables which needs to be controlled, and many attempts have been made at designing equipment for its real-time measurement. Traditionally, biomass has been measured off-line: usually, a culture sample is taken and either its turbidity (in the case of bacteria) or dry weight (in the case of fungi) is measured. However, off-line data cannot contribute efficiently to fermentation control, and in recent years *in situ*

methods of biomass estimation have been introduced. These methods may be classified as optical, calorimetric, acoustic, fluorimetric or capacitance-based.

The turbidity of a bacterial culture can be measured using a steam-sterilizable flow cell linked to a computer. As cell numbers and the turbidity increase, however, corrections must be made to the readings. In addition, the wall of the flow cell must be cleaned frequently, to minimize adherent microbial growth.

The calorimetric method of biomass estimation measures the heat produced by metabolically active cells, but has not been widely used. The acoustic method employs the relationship between the resonant frequency of a liquid and its specific gravity, but can yield erroneous results due to the presence of CO_2 microbubbles – the amount of CO_2 depends on not only the concentration of the microbial cells but also their metabolic activity. In addition, the presence of suspended solid particles affects the results.

The fluorimetric method utilizes the excitation of NADH/NADPH by ultraviolet light: a detector measures the fluorescence. However, the fluorescent-active ingredients of the medium and the metabolic state of the cells may interfere with the results. The combined detection of infrared radiation and culture fluorescence could provide a measurement of the total and viable biomass concentration.

The concentration of cells can be measured on-line using a laser flow cytometer (although these are very expensive) or a Coulter counter. Recently, a steam-sterilizable *in situ* probe for the measurement of culture capacitance has been introduced. Several reports suggest that the results are reliable when the probe is used for bacterial or fungal cultures grown in defined or complex media, with different modes of fermentation and at different scales. These studies have shown a linear correlation between off-line biomass measurements and on-line capacitance values.

Biosensors

The development of biosensors has enabled the provision of more comprehensive data from fermentation cultures. Biosensors use a biological sensing device together with a transducer, to produce an electrical signal from a biological change. The operation of a biosensor is shown in **Figure 3**.

The use of enzymes in biosensors has enabled the selective monitoring of fermentation cultures. One of the limitations of biosensors is that they cannot be steam-sterilized and lack durability. Even if steam-sterilizable biosensors were developed, their repeated sterilization, for use in sequential fermentations, would shorten their life. Other important criteria in the design of biosensors are sensitivity, stability, linearity and reproducibility of response. Physical and chemical interference from the culture broth can cause problems in the performance of biosensors.

When it is not possible to use a biosensor in a fermenter for on-line measurement, it may be used at-line. Some of the biosensors used for the monitoring of fermentation variables at-line are:

1. Glucose: glucose oxidase immobilized on an oxygen electrode, in a mixture with bovine serum albumin and glutaraldehyde: measures glucose concentration in the range $0.2–2$ mmol l^{-1}.
2. Urea: comprises a pH electrode and a urease immobilized membrane; effective life span is around 20 days for detection of urea in the concentration range of about $17–170$ mmol l^{-1}.
3. Alcohol: immobilized cell membrane of *Gluconobacter oxydans* in calcium alginate containing PQQ (pyrrolo-quinoline quinone), coated with a

nitrocellulose layer; ethanol concentrations of up to 20 mg l^{-1} can be detected.
4. Integrated multi-biosensor: a variety is available, using different enzyme-immobilized membranes: e.g. electrodes for the simultaneous measurement of glucose and galactose, or of potassium, sodium and calcium ions.

Indirect Measurement

If the direct measurement of a fermentation variable is not feasible, indirect methods may be possible. A remarkable example is the use of process mass spectrometry for the analysis of fermentation exhaust gases, volatile materials and light organic acids. The process mass spectrometer measures O_2 and CO_2 on-line, and is a sensitive method for the early detection of contamination, because each culture has its own specific respiration profile. The O_2 uptake rate, CO_2 evolution rate and respiratory quotient are important physiological parameters which can be monitored by the process mass spectrometer. Using this information, it is possible to estimate the volumetric O_2 mass transfer coefficient (K_La), the biomass concentration, the substrate utilization rate and the specific growth rate. The on-line data available through mass spectrometry, together with that obtained through at-line liquid-phase analysis, facilitates the use of sophisticated process controls.

Neural Networks

In spite of recent advances in the development of sensors for the on-line measurement of process variables, many fermentation parameters, e.g. metabolite concentrations, cannot be monitored on-line. In cases in which some of these variables may have significant roles in process optimization, artificial neural networks can be powerful tools in process control. Neural networks work on the principle of learning from previous experiments. A neural network comprises non-linear interconnected processing units. Three layers (input, hidden, and output) with different connecting weights (strength) are linked, and the strengths can be adjusted to fit a particular case and produce the desired output. This process of adjustment is called 'training' of the network. In fermentation processes, on-line, at-line, and off-line measurements of environmental and state-variables, e.g. temperature, pH, DOT and concentration of components of the medium, can be used as input values for a neural network. The output parameters of the fermentation process, e.g. concentrations of biomass and of measurable metabolites, can be used to train the network. The aim is to minimize the difference between the desired output and the predicted output.

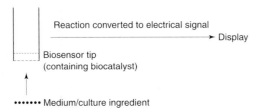

Reaction converted to electrical signal
→ Display

Biosensor tip
(containing biocatalyst)

······ Medium/culture ingredient

Figure 3 Simplified operation of a biosensor.

Table 2 Controllers and their characteristics

Controller	Characteristics
On/off	Used with on/off actuators, e.g. a valve or a pump
	There is a delay in response
	There is oscillation around the set point
	Not suitable for processes with large, abrupt changes
Proportional	The controller output is proportional to the error (the difference between the desired value and the measured value)
	The oscillation dampens quickly
	A new equilibrium is created, which is different from the original one (the difference is called the offset)
	The controller may be set at high or low sensitivity (low or high proportional band)
	At high sensitivity, there is high oscillation (similar to the on/off controller) and small offset
	At low sensitivity, the oscillation is reduced but there is higher offset
Integral	The controller output is an integral of the error
	Early response is slow
	The deviation from the set point is high
	The system settles down with no offset
Derivative	The controller output is the derivative of the error
	Provides the ability to control the extent and rate of the oscillations of a controlled system
	If there is no change in the error, then there will be no control action
	There is a fast damping action against the error
PID	Combination of proportional, integral and derivative actions. The control action is proportional, and is the derivative of the error. The relative weightings can be adjusted to the requirements of the process
	There is no offset
	Care should be taken in tuning of the controller, otherwise unwanted fluctuations will adversely affect the process

Control Systems

Feedback Control

In a simple control loop with feedback control, the controller receives a signal from the measuring element, compares it with a set point (desired value) and then responds with a control action, calculated according to an internal algorithm. The controller may be pneumatic, electronic or computerized (digital). The use of computers with advanced programs has introduced the possibility of the efficient control of fermentation by mimicking PID control (see below) and by the simultaneous handling of several control loops. The controller should be robust, should produce fast and reliable responses and should match the requirements of the fermentation system. A conventional feedback control system is shown in Figure 1.

In general, four types of controller can be identified, based on the control action: on/off or two-position; proportional; integral; and derivative. A fifth type combines proportional, integral and derivative (PID) control actions in a single system (**Table 2**). The on/off controller is the simplest type, the most complex being the PID controller. The more complex controllers offer potential for improved control of fermentation, but care must be taken – a poorly tuned PID controller can cause fermentation failure.

More complex control systems than feedback systems include cascade control, feed-forward control, and adaptive control (see below).

Cascade Control

When two processes to be controlled are closely linked, it is conventional to apply feedback control to each. It is possible to cascade the control loops, as shown in **Figure 4**. This is usually done when a fast process, which is subject to disturbances, is linked to a slower process. An example is the control of the culture temperature in a jacketed fermenter together with the control of the water temperature in the fermenter jacket.

Feed-forward Control

If the control element (the actuator) is slow, but the disturbances are large and change rapidly, then the disturbances should be compensated by using feed-forward control. The degree of compensation must be calculated or estimated, and synchronized with the disturbance.

Adaptive Control

Fermentation is a non-linear process, which varies with time. An efficient controller for such a process should adapt to the changing process characteristics. When the differences between the outcomes of conventional control and the desired outcomes are significant, an adaptive control system might be beneficial. In such a system, the controller learns about the process as it controls the process. Adaptive control is particularly useful for batch fermentations, although its successful application to fed-batch fermentations has been reported.

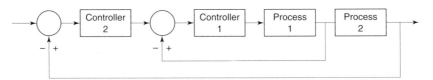

Figure 4 Cascade control loop for the control of the temperature of the water in a fermenter jacket (process 1) and that of the culture in the fermenter (process 2).

Computer Control

The falling cost of computers has facilitated their use in small- and large-scale fermentations. Computers can be linked to measuring equipment, to enhance data acquisition; data analysis and fermentation control and optimization. Computers may also be used to model fermentation processes, as a basis for improving process control.

Data acquisition systems comprise both hardware and software. On-line and at-line process data (from the measuring equipment), and also off-line data, can be stored in the system for analysis.

The data from various items of measuring equipment can be compared, combined and analysed using mathematical programs. The indirect measurement of fermentation variables is also possible, and provides valuable information about the growth of a culture. Depending on the software, different types of control may be applied to a fermentation process. These types may be categorized as follows:

1. Simple: control of valves and pumps, feeding of medium, removal of culture, based on a pre-set time.
2. Direct digital control: control of fermentation parameters (acting as a conventional controller).
3. Digital set-point control: computer acts a data logging system, and also changes the set points when necessary.
4. Complex: on-line data handling and analysis; provision of at-line and off-line data for comprehensive data accumulation and documentation; calculations for more complex control algorithms.
5. Advanced: use of on-line measuring equipment only, together with mathematical models and trained neural network systems, for process optimization.

The simplest type of use of computers in fermentation control is in operating valves, pumps, etc. Batching of the medium and the time-dependent addition or removal of nutrients to or from the culture (in fed-batch or continuous cultures) fall into this category.

More complex control involves the use of feedback algorithms, for the simultaneous control of fermentation variables such as temperature and pH. In such control, which is often called 'direct digital con-trol', a computer replaces the conventional analogue controller, and sensors are linked to the computer through an interface, which converts the continuous analogue signal from the sensors to a digital signal appropriate for the computer. The interface is usually in the form of an analogue-to-digital converter. There is a trend towards the provision of measuring equipment which emits a digital signal: a binary coded digital signal from a sensor can be used as an input for the computer. The computer compares the digital signal with the set point and generates a digital output signal, which is translated into an analogue signal to an actuator.

Computers provide fermentation process control with a high degree of flexibility and precision. However, if a computer is used as the sole controller, minor failure may cause serious problems, and so appropriate backup is necessary. This difficulty can be avoided by using computers in conjunction with conventional electrical controllers, the latter providing conventional control, while the computer plays a 'supervisory' role. The computer logs and stores data from the measuring equipment, and changes the set points when necessary. This type of system is usually called 'digital set-point control' or 'supervisory set-point control'.

The use of computers in fermentation control has expanded significantly as they have become increasingly cheap and powerful. Advanced control strategies, relying on sophisticated programs, can now be applied. The availability of more on-line and at-line measured data, together with the use of neural networks, facilitate improved control and hence optimized production (see **Fig. 5**).

Solid-state Fermentations

The control of fermentation is not always easy, one of the main reasons being that the accurate on-line measurement of environmental and state variables is not always possible. A range of on-line and at-line measuring equipment is available for submerged fermentations, but the parameters associated with solid-state fermentations are less easily measured. A major problem arises because cell biomass is grown on a solid medium and remains attached to it. In addition, until recently the design of bioreactors for solid-state fermentations tended to be basic. New designs aim to

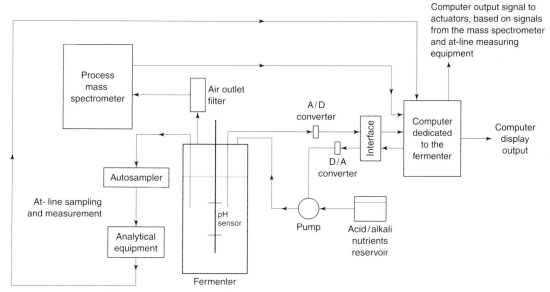

Figure 5 Computer control of fermentation. A/D, analogue-to-digital; D/A, digital-to-analogue.

achieve good mixing and heat transfer, and improved monitoring and control.

Two of the variables which must be controlled are heat generation and moisture, and on-line sensors are available for these parameters. The aeration of a moist solid medium in an aerobic solid-state fermentation is another important factor influencing productivity, and can be controlled: the on-line measurement and monitoring of O_2 and CO_2 in the exhaust gases is possible when a process mass spectrometer is employed and infrared CO_2 analysers and paramagnetic O_2 analysers carry out delayed monitoring of these gases. The indirect estimation of biomass and specific growth rate has become possible through the measurement of the rates of O_2 uptake and CO_2 evolution. The monitoring and control of the pH of solid-state fermentations, however, is not easy.

Submerged Fermentations

Most industrial fermentations are aerobic. In an aerobic batch culture, as the cells grow, the overall culture requirement for O_2 increases. Although some variables (e.g. temperature) are best maintained at a constant value throughout a batch fermentation, other variables, including O_2 concentration, must be controlled at a changing level, if optimal productivity is desired. In this particular case, the automatic control of the speed of the stirrer can maintain the DOT above a desired value. In some batch fermentations, such as those for the production of secondary metabolites, a fast initial growth rate, for biomass production, and slower growth subsequently, are needed to optimize the process in terms of productivity. The environmental variables, as well as the

state variables, can be controlled using on-line measuring equipment linked to a computer, thus achieving optimized batch fermentation.

In fed-batch fermentations, extended and improved productivity can be achieved by the continuous or intermittent addition to the culture of one or more of the ingredients of the medium, particularly if knowledge-based strategies for feeding the culture are adopted. Computer control, using appropriate algorithms, is necessary for the implementation of an accurate and efficient feeding strategy. For the more complex fed-batch fermentations such as 'feed and bleed' systems, and for continuous culture, computers can provide accurate control of the inlet and outlet flows. Time-based changes to a fermentation, such as the addition of certain ingredients at particular times and the routine sampling of the culture for at-line analysis, are made possible by computerization.

See also: **Fermentation (Industrial)**: Basic Considerations; Media for Industrial Fermentations; Recovery of Metabolites.

Further Reading

Atkinson B and Mavituna F (1991) *Biochemical Engineering and Biotechnology Handbook*. New York: Macmillan.

Austin GD, Watson RWJ and D'Amore T (1994) Studies of on-line viable yeast biomass with a capacitance biomass monitor. *Biotechnol. Bioeng.* 43: 337–341.

Bailey JE and Ollis DF (1986) *Biochemical Engineering Fundamentals*. New York: McGraw-Hill.

Bucke C and Chaplin MF (1990) *Enzyme Technology*. Cambridge: Cambridge University Press.

Cass AEG (ed.) (1990) *Biosensors: A Practical Approach.* Oxford: Oxford University Press.

Doran PM (1995) *Bioprocess Engineering Principles.* London: Academic Press.

Fehrenbach R, Comberbach M and Petre JO (1992) On-line biomass monitoring by capacitance measurement. *J. Biotechnol.* 23: 303–344.

Fish NM, Fox RI and Thornhill NF (eds) (1989) *Computer Applications in Fermentation Technology: Modelling and Control of Biotechnological Processes.* London: Elsevier Applied Science.

Karube I and Yokoyama K (1991) Biosensors. In: Cheremisinoff PN and Ferrante LM (eds) *Biotechnology Current Progress.* P. 1. Lancaster: Technomic Publishing.

Leigh JR (1983) *Essentials of Nonlinear Control Theory.* London: Peter Peregrinus.

Leigh JR (ed.) (1987) *Modelling and Control of Fermentation Processes.* London: Peter Peregrinus.

Leigh JR (1987) *Applied Control Theory.* London: Peter Peregrinus.

Lonsane BK, Ghildyal NP, Budiatman S and Ramakrishna SV (1985) Engineering aspects of solid-state fermentation. *Enzyme Microb. Technol.* 7: 258–365.

Mutharasan R and Fletcher FA (1998) On-line monitoring of intracellular properties and its use in bioreactor operation. In: Galindo E and Ramisez OT (eds) *Advances in Bioprocess Engineering II.* P. 53. London: Kluwer Academic Publishers.

Nigam P and Singh D (1994) Solid-state (substrate) fermentation systems and their applications in biotechnology. *J. Basic Microbiol.* 34(6): 405–423.

Präve P, Faust U, Sittig W and Sukatsch DA (eds) (1987) *Fundamentals of Biotechnology.* Weinheim: VCH.

Royce PN (1993) A discussion of recent development in fermentation monitoring and control from a practical perspective. *Critical Reviews in Biotechnology* 13(2): 117–149.

Saucedo-Castaneda G, Trejo-Hernandez MR, Lonsane BK et al (1994) On-line automated monitoring and control system for CO_2 and O_2 in aerobic and anaerobic solid-state fermentations. *Process Biochemistry* 29: 13–24.

Schugerl K (1997) *Bioprocess Monitoring.* New York: John Wiley.

Schugerl K, Lorenz A, Lubbert J et al (1986) Pros and cons: on-line versus off-line analysis of fermentations. *TIBTECH*: 4: 11–15.

Stanbury PF, Whitaker A and Hall SJ (1995) *Principles of Fermentation Technology.* Oxford: Pergamon Press.

Treskatis SK, Olgerdinger H and Gilles ED (1997) Morphological characterization of filamentous microorganisms in submerged cultures by on-line digital image analysis and pattern recognition. *Biotechnol. Bioeng.* 53: 191–201.

Recovery of Metabolites

P A Pawar, M C Misra, N P Ghildyal and **N G Karanth**, Fermentation Technology and Bioengineering Department, Central Food Technological Research Institute, Mysore, India

Introduction

The separation of desired products from a fermentation broth, and their purification to the required level, are crucial and challenging steps in industrial fermentation. Product recovery usually accounts for a significant proportion of the product cost, and may account for 20–60% of the total manufacturing cost. The choice of recovery process depends on the location of the product (intracellular or extracellular); its concentration; the physical and chemical properties of the metabolite and the culture broth; the intended use of the product; the purity required; impurities present in the broth; and the market price of the final product. In the case of recombinant products, recovery could represent the major manufacturing cost.

The typical sequence of product recovery operations begins with the separation of insoluble components, e.g. whole microbial cells and cell debris. This may involve centrifugation and filtration. The desired product is then isolated from the broth and concentrated. This may involve one or more of: extraction and adsorption by solvents; elution; precipitation; ion exchange; gel filtration; and membrane techniques. The techniques of isolation and concentration are described below.

Solvent Extraction

Solvent extraction generally means either the separation of components based on differences between their solubility in two phases, or alternatively solid–liquid extraction (leaching). Liquid–liquid extraction involves intimate contact between the culture medium and a suitable solvent, in which one or more of the desired components is more soluble. This is followed by physical separation of the two phases, by settling or centrifugation. The solvent-rich solution containing the extracted component, and the residue, which contains less solvent, are called the 'extract' and the 'raffinate' respectively. The solvent from the extract is often recovered by distillation.

Extraction may comprise a single stage or a number of stages, or may be a continuous process. The flow of liquids is generally countercurrent, although other types of flow are possible. Continuous countercurrent differential-type contactors remove solvent from the extract and then reflux the residue, to achieve improved separation.

The choice of solvent depends mainly on cost, toxicity and partition (or distribution) coefficient (i.e. the ratio of the concentrations of the component in the extract and the raffinate). The partition coefficient can be altered by modifying the temperature or pH, or by the addition of salts.

In solid–liquid extraction, the solid (biomass) is brought into intimate contact with the solvent. The residual biomass is then separated, and the product is recovered from the solution. Countercurrent, co-current and other flow schemes are used.

A multistage countercurrent scheme is shown in **Figure 1**. If there are n mixer/separators connected in series, the raffinate passes from vessel 1 to vessel n and the product-enriched solvent flows in the opposite direction, from vessel n to vessel 1.

Applications

The recovery of lactic acid, acetic acid and β-carotene involves solvent extraction.

In the recovery of lactic acid, bacterial proteins, calcium and heavy metals are first removed from the fermented broth. The lactic acid is then extracted in aqueous solution by using isopropyl ether in countercurrent flow. After further extraction using distilled water in countercurrent flow, the resulting aqueous extract is decolorized and subjected to ion-exchange, and is then concentrated by evaporation to food-grade lactic acid.

In the recovery of acetic acid from submerged vinegar fermentations containing ethanol, the acetic acid is extracted using ethyl acetate and is recovered by distillation.

In the recovery of β-carotene from *Bacillus trispora* cells, the mycelium is first dehydrated and then treated with methylene chloride, which extracts the β-carotene. The extract is concentrated by evaporation at low temperature, and pure β-carotene is crystallized from solvents such as acetone and chloroform.

The extraction of enzymes from mouldy bran using water, the extraction of fat from *Rhodotorula* cells using organic solvents such as hexane, are examples of solid–liquid extraction.

Table 1 Composition and surface area of some commercial adsorbents

Adsorbent (commercial name)	Composition	Surface area ($m^2 g^{-1}$)
Nuchar CEE	Carbon	740
Duolite S-30	Phenyl formaldehyde	128
ES-40	Styrene-DVB	110
ICN Alumina	Alumina	155–220
A-305 CS Alumina	Alumina	325
Matrex Silica	Silica	500–600
Amberlite XAD-2	Styrene-DVB	300
Pittsburg PCC SGL	Carbon	1000–1200

DVB, divinyl benzene.

Adsorption and Elution

In adsorption chromatography, molecules of solute are physically adsorbed onto the surface of the adsorbent, by van der Waals' and hydrogen-bonding associations. The solute is then eluted from the adsorbent by a pure solvent, e.g. chloroform, hexane, ethyl ether, or by a mixture of solvents. This separation process is based on the partition of substances between the (polar) adsorbent column material and the (non-polar) solvent. This technique is generally used to separate non-polar molecules.

The usefulness of an adsorbent in a recovery process depends on: its composition; the presence and type of functional groups at its surface; its porosity and surface area; the degree of its polarity; and its relative hydrophobic/hydrophilic properties. Most adsorbents used in industrial purification processes have a surface area greater than $100 \, m^2 g^{-1}$ and particles in the size range 150–1500 μm.

The most useful adsorbents for the recovery of fermentation products are activated carbon, oxides of silicon, aluminium and cross-linked organic polymers. Some of the important commercial adsorbents and their properties are listed in **Table 1**.

Column Operation

Fixed-bed column operations are most commonly used for industrial adsorption. The column holds the adsorbent particles, and the fluid containing the desired solute is passed through the column, as shown in **Figure 2**. Either pressure or gravity can be used as the force driving the flow of fluid downwards. Initially, most of the solute is adsorbed, so its

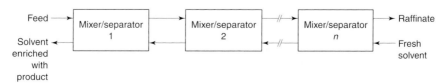

Figure 1 Countercurrent multistage extraction system.

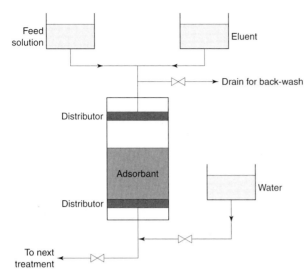

Figure 2 Operation of a fixed-bed adsorption column.

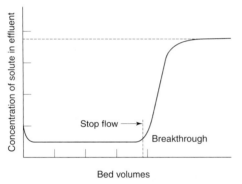

Figure 3 Concentration of adsorbed solute in column effluent, during column loading.

concentration in the effluent is low. As the operation progresses, the concentration of solute in the effluent rises – slowly at first, then abruptly. When this abrupt rise of the breakthrough occurs, the flow is stopped (**Fig. 3**). The unadsorbed impurities are then removed from the bed by washing with water. The treated fluid is collected at the bottom of the column, and discharged to the next adsorption column or treatment unit. The adsorbed solute is then selectively eluted, using an appropriate eluent. The elution curve is shown in **Figure 4**. Generally, the adsorbent is regenerated and used again. Granular activated carbon is often used as the adsorbent, and is regenerated by thermal treatment, which oxidizes the adsorbed organic impurities.

Adsorption Isotherm

When an adsorbate and an adsorbent are in equilibrium, the distribution of the solute between the solid and fluid phases is constant, and no further net adsorption occurs. Adsorption equilibrium data are available as adsorption isotherms, which give the concentration of the solute in the adsorbed phase as a

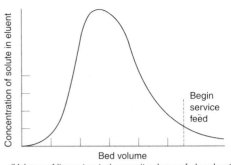

Figure 4 Elution curve, showing concentration of adsorbed solute eluted as a function of eluent volume.

function of its concentration in the unadsorbed phase, at a given temperature. Adsorption isotherms are useful for selecting the most appropriate adsorbent. They also play a crucial role in predicting the performance of adsorption systems.

Several types of isotherm have been developed to describe adsorption relationships. A commonly used one is Freundlich's adsorption isotherm for liquid–solid systems (**Equation 1**).

$$q = Ky^n \qquad \text{(Equation 1)}$$

where q is the amount of solute adsorbed and y is the equilibrium concentration of adsorbate; K and n are empirical constants, determined experimentally. If the value of n is less than 1, this indicates favourable adsorption.

Applications

Adsorption is used to remove unwanted molecules, e.g. coloured impurities, from fermentation products, e.g. lactic and citric acid fermented broths. Adsorption is also used to increase the concentration of a fermentation product when it has been eluted from the adsorbent. For example, vitamin B_{12} is recovered by passing the eluent through an adsorption column packed with commercial cross-linked styrene adsorbents, from which it is eluted, in concentrated form, with methanol.

Precipitation

The recovery of metabolites from fermented broth by precipitation involves adjusting the dielectric constant, ionic strength, temperature and pH of the system. The desired molecules are made insoluble, and the crude precipitated concentrate is separated for further processing. On a large scale, precipitation is achieved by either salting-out, solvents or polyelectrolytes. Batch or continuous operations are used for precipitation.

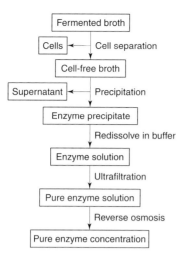

Figure 5 Process for the recovery of α-amylase from fermented broth.

Salting-out

This is the precipitation technique most widely used for the concentration and fractionation of enzymes and proteins. The addition of salts creates an imbalance between electrostatic forces (tending to keep proteins in solution) and hydrophobic forces (tending to cause agglomeration of proteins), and results in precipitation.

The salting-out of proteins is empirically described by **Equation 2**:

$$\log S = \log S_0 + kC \qquad \text{(Equation 2)}$$

where S and S_0 are the solubility of protein in the presence and absence of salts respectively; k is a constant, which depends on the proteins, the salts used and the temperature; and c is the concentration of the salt(s).

A wide range of neutral salts (e.g. citrates, phosphates, sulphates) are effective in the precipitation of proteins, but ammonium sulphate is the common choice because of its high solubility, relatively low cost and, in some cases, its ability to stabilize enzymes. The precipitated proteins are usually contaminated with residual salts, which are removed by diafiltration or gel filtration (see below).

The applicability of salting-out depends on the characteristics of the proteins; the selection of the salt and its concentration; the method of contact; and the economics. A process for the recovery of α-amylases from fermented broth is shown in **Figure 5**.

Organic Solvents

The addition of organic solvents reduces the dielectric constant of a protein solution, making the protein molecules less polar in character, enhancing the protein–protein interaction, and thus leading to agglomeration and the precipitation of the proteins.

The solubility of a protein is related to the dielectric constant of the system at its isoelectric point, and is given by **Equation 3**:

$$\log S = K/D^2 + \log S_0 \qquad \text{(Equation 3)}$$

where S and S_0 are the solubility of the protein in the presence and absence of the solvent respectively; D is the dielectric constant of the organic solvent–water mixture; and K is a constant dependent on the dielectric constant of the original aqueous solution.

Ethanol, methanol, isopropanol and acetone are the solvents usually used for large-scale precipitation. The entire operation is carried out in flameproof conditions and at a temperature below 4°C. The solvents are recovered by distillation. A serious disadvantage of using solvents is the requirement for fireproof motors and switches and other specialized equipment, which increases the capital cost of the plant.

Organic solvents are also used for recovering metabolites other than proteins, e.g. the recovery of xanthan gum from fermented broth.

Polyelectrolytes

Polyelectrolytes, such as polyacrylic acid, polyethyleneimine, carboxymethylcellulose and polyphosphates, have been used on an industrial scale for the separation and purification of enzymes from fermented broth. The mechanism of precipitation involves partly salting-out and partly reduction of the water of solvation. Usually, low concentrations of polyelectrolytes (0.05–0.10%) bring about precipitation, but the disadvantage is the high cost of the polyelectrolytes.

This method of recovering metabolites has been applied successfully on an industrial scale for the purification of amyloglucosidases and microbial rennet from fermented broth.

Ion Exchange and Gel Filtration

In many fermentation processes, ion exchange and gel filtration are used to isolate and purify relatively low concentrations of metabolites.

Ion Exchange

In this process, ions that are electrochemically bound to an insoluble and chemically inert matrix are reversibly replaced by ions in solution. This is described by **Equation 4**:

$$R^+A^- + B^- = R^+B^- + A^- \qquad \text{(Equation 4)}$$

where R^+A^- is an anion exchanger in the A^- form and B^- is an anion in solution. Similarly, a cation exchanger possesses negatively charged groups that reversibly bind cations.

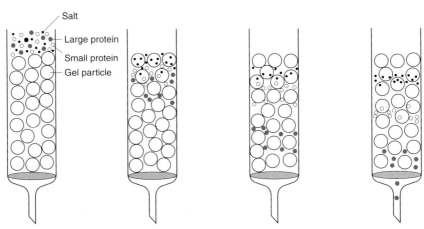

Figure 6 Progressive separation of molecules of different sizes by gel filtration.

Metabolites with both positive and negative charges can bind to either the cation or the anion exchanger, depending on their net charge. The affinity with which a metabolite binds to a given ion exchanger depends on the identities and concentrations of the other ions in solution, because of the competition between ions for the binding sites of the ion exchanger. The binding affinities of metabolites bearing acidic or basic groups are also highly dependent on pH, because of the variation of their net charges with pH.

Gel Filtration

Gel filtration is also known as size-exclusion chromatography or molecular-sieve chromatography. In this process, separation is based on the differing ability (due to differing molecular size) of molecules in the sample to enter the pores of the gel filtration medium. The stationary phase in this technique consists of beads of a hydrated, sponge-like material which has pores of molecular dimensions and with a narrow range of sizes. When an aqueous solution, containing molecules of various sizes, is passed through a column containing such 'molecular sieves', molecules that are larger than the pores of the filtration medium move quickly through the column. Smaller molecules enter the pores of the gel and move slowly through the column (**Fig. 6**). The molecules are eluted in order of decreasing molecular size. The molecular mass of the smallest molecule unable to penetrate the pores of a given gel is said to be the 'exclusion limit' of the gel.

Matrix

The most important consideration in designing a large-scale purification process concerns the type of matrix. A matrix should be hydrophilic, macroporous, rigid, spherical, chemically stable, inert, robust and reusable. It should also be readily available and cheap.

Table 2 Ion exchange matrices commonly used in biomolecule purification

Matrix	Type	Ionizable group
Polystyrene	Basic	Trimethyl benzyl ammonium $-CH_2N^+(CH_3)_3$
	Acidic	Sulphonated $-SO_3H$
DEAE-cellulose	Basic	Diethyl aminoethyl $-CH_2CH_2N(C_2H_5)_2$
CM-cellulose	Acidic	Carboxymethyl $-CH_2COOH$
P-cellulose	Strongly and weakly acidic	$-OPO_3H_2$
DEAE-Sephadex	Basic cross-linked dextran gel	$-CH_2CH_2N(C_2H_5)_2$
CM-Sephadex	Acidic cross-linked dextran gel	$-CH_2COOH$

Ion exchangers consist of a support matrix, to which charged groups are covalently attached. The chemical nature of the charged groups determines the types of ion that bind to the ion exchanger, and the strength with which they bind. Support matrices used in ion exchangers for metabolite purification include styrene, cellulose, agarose, cross-linked polyacrylamide and polydextran-based gels. **Table 2** lists some commercially available ion exchangers used for the purification of biomolecules. The fractionation ranges of several gels that are commonly used for separating biological molecules are listed in **Table 3**.

Equipment for Large-scale Chromatography

The most common method of chromatography used in batch processing is the packed-bed operation, involving an ion-exchange or gel chromatography column. The dimensions of large-scale columns vary, being 5–500 cm in diameter and 15–3000 cm in height, depending on the type of resin and the appli-

Table 3   Commonly used gel filtration matrices

Matrix	Type	Fractionation range (kDa)
Sephadex G-10	Dextran	0.05–0.7
Sephadex G-50	Dextran	1–30
Sephadex G-200	Dextran	5–600
Bio-Gel p-2	Polyacrylamide	0.1–1.8
Bio-Gel p-100	Polyacrylamide	5–100
Bio-Gel p-300	Polyacrylamide	60–400
Sepharose 6B	Agarose	10–4000
Sepharose 4B	Agarose	60–20 000
Sepharose 2B	Agarose	70–40 000

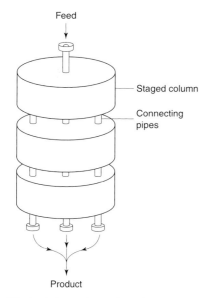

Figure 7   Stacked chromatography column.

cation. A fundamental requirement of all column designs is that they must allow for the uniform and efficient packing of the chromatographic medium. The pores in the medium support must be sufficiently small to retain the medium without clogging. The sample to be purified must be uniformly distributed over the chromatographic bed, and this becomes critical as the diameter of the column is increased.

When the gel medium is relatively hard and strong, it can be accommodated in a single column. However, the softer gel media used for protein purification have low mechanical strengths, and so the bed must be broken into several columns, each with a diameter in the range 15–45 cm. These column sections are arranged in stacks, to minimize loss of pressure and compression stresses in each section (**Fig. 7**). Scaling up is achieved by increasing the column diameter.

Chromatographic Column Operation

The general operating procedures for ion exchange and gel filtration are:

1. Soak the resin before adding it to the column, to allow it to reach its hydrated volume, and then load the column.
2. Wash the resin with distilled water, and allow the resin to settle.
3. Rinse the column with distilled water or buffer until the pH of the effluent equals that of the water or buffer.
4. Start the treatment cycle, and continue it until the metabolite concentration in the effluent reaches that in the feed solution (the breakthrough point).
5. Backwash the resin with distilled water, to 50–100% bed expansion, for 5–10 min.
6. Elute the metabolite, by gradient elution.
7. Regenerate the resin, at a flow rate that allows at least 45 min contact time.
8. Rinse the column with distilled water or buffer, until the pH of the effluent has reached that of the water or buffer.

Applications

Industrially, lysine is separated from fermented broth by ion exchange. The acidity of the fermented broth is adjusted to a pH value of 2 with hydrochloric acid, and then passed through a column of strongly acidic cation-exchange resin, in the NH_4^+ form. Dilute aqueous ammonia is used to elute the lysine from the resin.

Gel filtration is very effective for removing salts, and for removing solvents from mixtures containing biological products. It is particularly useful in the final purification step, for producing enzymes of high purity. Both ion exchange and gel filtration are valuable tools for the purification of proteins of high value on a large scale. They have been used in the production of insulin, interferons, hormones and analytical and medicinal enzymes.

Membrane Separation

A membrane filter is a barrier which is capable of redistributing the components in a fluid stream, using a pressure differential as the driving force, and without involving a change of phase. Depending on the size of the membrane pores, molecular sieving of the components takes place when a pressure difference (1–100 bar) is applied.

Depending on the pore size, the process is classified as microfiltration (MF), ultrafiltration (UF) or reverse osmosis (RO) (**Fig. 8**). MF and UF membranes have pore diameters in the ranges 0.2–10 μm and 0.01–2.0 μm respectively. UF membranes are generally described in terms of cut-off values relating to molecular sizes, and are available in the range 1–500 kDa. RO membranes are nonporous, and allow ionic

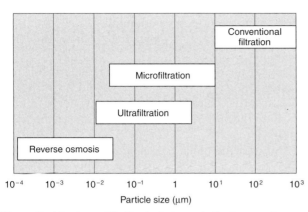

Figure 8 Membrane filtration techniques, indicating membrane pore size.

solutes, typically of molecular size less than 0.001 μm, to be separated. MF and UF are more widely used than RO in primary recovery stages.

An important disadvantage of conventional filtration is that resistance to the flow of filtrate increases as deposits build up on the filter. This is partly overcome by 'tangential-flow filtration' (TFF), in which the medium containing the desired product is made to flow across the functional surface of a membrane. A decrease in pressure across the membrane drives the fluid through the separation barrier. Particles that cannot pass through the membrane are swept away by the incoming fluid, reducing the accumulation of particles on the functional surface of the membrane. The configuration of a typical TFF system is shown schematically in **Figure 9**. The filtrate, or permeate, passes through the membrane and the retentate (also known as the 'concentrate' or the 'reject'), is returned to the reservoir that holds the sample solution.

The rate of filtration is described by the term 'flux', which is defined as the volume of filtrate per unit time per unit surface area of membrane. An effective filtration process has a constant, high flux for a long period, although there is usually some reduction in the flux with time, due to concentration polarization and fouling of the membrane. For process optimization, the appropriate combination of membrane pore size, recirculation rate, membrane surface area,

operating pressure and feed concentration must be used.

Membranes

There are two types of membrane: symmetric (isotropic), the applications of which are limited to MF; and asymmetric (anisotropic), which can be used for MF, UF and RO. Asymmetric membranes have a thin (0.2 μm), dense layer or skin on top and a spongy support layer (about 100 μm thick) underneath. **Table 4** lists some common membranes and their properties.

Equipment for Membrane Separation

There are three basic types of membrane filtration device (**Fig. 10**), as well as some variations on these configurations. The plate-and-frame and the spiral systems use a combination of sheet membranes and separator screens: in the spiral system, the membranes and screens are in a cylindrical cartridge, and in the plate-and-frame system, they are between two plates. Both these devices can withstand pressures of up to 1480 kPa. The hollow-fibre system comprises a cluster of fibres inside a cartridge. The characteristics of the separator screens and the internal diameter of the fibres control the flow, and hence the final concentration of the filtered material. Operating pressure is an important factor: it depends on the design of the

(A) Plate-and-frame device

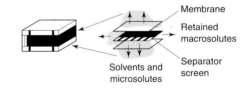

(B) Spiral system

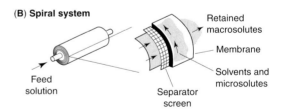

(C) Hollow-fibre system

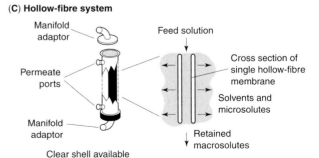

Figure 10 Devices for membrane filtration. (**A**) Plate-and-frame device. (**B**) Spiral system. (**C**) Hollow-fibre system.

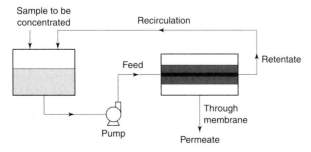

Figure 9 Batch tangential-flow filtration process.

Table 4 Common membrane filters, used for microfiltration (MF), ultrafiltration (UF) or reverse osmosis (RO)

Material	Application	pH range	Approximate maximum temperature (°C)
Cellulose acetate	MF, UF, RO	3.5–10	75
Mixed cellulose esters	MF, UF, RO	4–8.5	120
Polysulphone	MF, UF, RO	1–14	130
Polyester	MF, UF	NA	150
Polyamide	MF, UF, RO	2–12	NA
Polytetrafluoroethylene (PTFE)	MF	1–14	140
Polypropylene	MF	1–14	130
Ceramic	MF	1–13	140

NA, not available.

device rather than on the membrane, and is about 270–400 kPa.

A variation on these devices is a tubular module comprising a stainless-steel shell containing several perforated stainless-steel tubes, each containing a membrane. Such modules have relatively large, open feed channels, with internal diameters of 12–25 mm and lengths of 0.6–6.5 m. The feed solution flows in a turbulent manner inside each tube, the inner wall of which contains the membrane. The permeate passes through the membrane and the support material, and is collected in the shell.

Other Membrane Separation Techniques

Other techniques which are available include:

1. Electrodialysis: this employs semipermeable ion-exchange membranes that are impervious to water. The separation is electrically driven, in contrast to membrane filtration, which is pressure-driven.
2. Diafiltration is simply an alternative method of using an ultrafilter. It can be used to transfer a macromolecule or protein from one solvent to another – in effect, washing one solvent out by continuously adding another. This technique is frequently used in the preparation of feed solutions for chromatography, for which the product must always be in a buffered solution.
3. Nanofiltration is a process which can be considered as an extension of reverse osmosis. It differs from RO in that the membranes allow ionic species to pass through, but retain uncharged molecules of molecular mass in excess of 200 Da.
4. Pervaporation is a technique in which a liquid feed mixture is separated by partial vaporization through a nonporous selectively permeable membrane.

Applications

Membrane separation is used in the recovery of cells from fermented broth, e.g. recombinant *Escherichia coli* cells and *Zymomonas mobilis* cells are recovered by TFF. This technique is also used for the separation of cell debris, e.g. the separation of *Saccharomyces cerevisiae* cell debris after the cells have been disrupted in order to recover invertase. TFF is also used for the concentration or diafiltration of a product stream after an intermediate purification step such as chromatography. For example, in the concentration and purification of α-amylases from fermented broth, TFF is used to remove residual substrate and salts, and to separate lactase by diafiltration from an aqueous suspension containing ammonium sulphate.

Extraction of Intracellular Products

Developments in recombinant DNA technology have made it possible to produce an increased number of bioactive compounds, such as human and bovine interferons, insulin, growth hormones and many enzymes, which are synthesized intracellularly. The recovery of these compounds demands the disintegration of the cells without affecting their biological activity. The efficacy of disintegration depends on: the type of cells; the growth conditions used in the microbial culture; the composition of the cell wall; temperature; the shear forces applied; the nature of the intracellular products; and the conditions of storage.

Many techniques can be used in the laboratory, but only a few are suitable for large-scale operations. These include: physico-mechanical methods, such as liquid or solid shearing, agitation with abrasives, osmotic shock or freeze-thawing; chemical methods, such as the use of detergents or acid/alkali treatment; and the enzymatic method.

Physico-mechanical Methods

Liquid shear has been widely used in large-scale enzyme recovery. The high-pressure homogenizer used in the food industry is very effective at disrupting bacterial and yeast cells (**Fig. 11**). It consists of a high-pressure positive displacement pump, with an adjustable valve and a restricted orifice. During operation, the cell suspension is drawn through a one-way valve and pushes against the operative valve, which is set at the selected operating pressure. The cells then pass through a narrow channel between the valve and the impact ring, and at the exit of the narrow orifice there is a sudden fall in pressure, which causes cell disintegration. It may be necessary to recycle the slurry through the homogenizer a number of times. The degree of disintegration depends on the homogenizer pressure; the valve design; the number of passes through the valve assembly; the temperature of operation; and the cell concentration.

Cell disruption in a high-pressure homogenizer follows first-order kinetics, described by **Equation 5**:

$$\ln\left\{ R_m / (R_m - R) \right\} = KNP^n \qquad \text{(Equation 5)}$$

where R_m is the maximum obtainable protein release after N is the number of passes through the valve; R is the concentration of proteins released at time t; K is the first-order rate constant (s^{-1}); P is the operating pressure; and n is the pressure exponent. The value of n is a measure of the resistance of a microorganism to disruption.

The liquid shear disruption method has been used for the recovery of intracellular products from *Saccharomyces cerevisiae*, *Escherichia coli*, *Candida lipolytica*, *C. utilis*, *Pseudomonas aeruginosa*, *P. putida* and *Aspergillus niger*. A typical flow sheet for the recovery of intracellular enzymes is shown in **Figure 12**.

Agitation with Abrasives

Cell disruption can be brought about by a high-speed bead mill, containing a series of rotating discs and a charge of glass ballotini (**Fig. 13**). The cell suspension is agitated at high speed. The mill chamber is almost full of grinding beads during operation, the optimum concentration of ballotini usually being 70–90% of the volume of the chamber. The optimal bead size

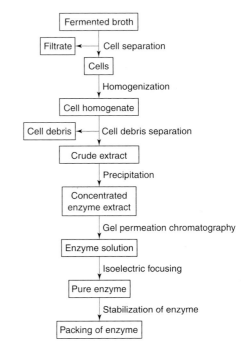

Figure 12 Recovery of intracellular enzymes.

depends on the density and viscosity of the feed, and is usually in the range 0.2–0.5 mm for bacteria and 0.4–0.7 mm for yeasts. Cell disintegration depends on the bead loading, the design of the agitator and its speed. The rotating discs accelerate the beads in a radial direction, and they form streaming layers, with different velocities. In this way, shearing forces are created, leading to cell disruption. The frequency and strength of collisions during milling also contribute to cell disruption.

High-speed bead mills have been used to disrupt the cells of several yeasts, bacteria and fungi.

Chemical Methods

Chemicals, including acids, alkalis, surfactants and solvents, have been used for cell lysis. For example, the selective extraction of cholesterol oxidase from *Nocardia rhedocrous*, by cell permeabilization using the surfactant Triton X-100, has been reported. Toluene has been used for the recovery of proteins from the cells of *Escherichia coli* and yeasts, due to its ability to act on the inner membrane phospholipids, thus dissolving the membrane. Phenol has been used to extract tRNA from *E. coli*. Other chemicals, including cholate and sodium dodecylsulphate, act on the cytoplasmic and outer cell membranes. Chemicals such as acids and alkalis are not selective, and tend to damage sensitive proteins and enzymes along with the cell wall.

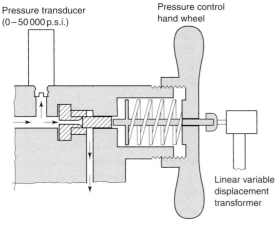

Figure 11 High-pressure homogenizer valve assembly. Shaded areas are stainless steel; hatched area is the Stellite valve mechanism.

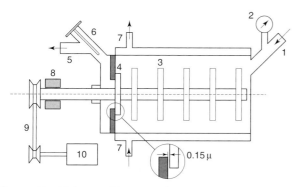

Figure 13 High-speed bead mill (Dyno mill). (1) Inlet for suspension. (2) Manometer. (3) Rotating disc. (4) Slit for separation of glass beads from the suspension. (5) Outlet for suspension. (6) Thermometer. (7) Cooling water/inlet and outlet. (8) Bearings. (9) Variable drive. (10) Drive motor.

Enzymatic Method

Lytic enzymes, such as lysozyme, Zymolase (a commercial enzyme) and lysostaphin, have been used in the laboratory owing to their specificity for the structural elements of the cell wall. Lysozyme is obtained commercially from chicken egg-whites. It catalyses the hydrolysis of $\beta(1\rightarrow4)$-glucosidic bonds in the peptidoglycan layer of bacterial cell walls. As a result, the internal osmotic pressure of the cell ruptures the periplasmic membrane, releasing the intracellular products into the surrounding medium. Lysozyme has been used for the lysis of *Pseudomonas putida*, to obtain alkaline hydroxylase, and also to release invertase from *Saccharomyces cerevisiae*. Polyhydroxyalkonoates have been obtained from *Alcaligenes eutrophus* by using a cocktail of hydrolytic enzymes, comprising lysozyme, phospholipase, lecithinase, proteinase and alcalase.

See also: **Fermentation (Industrial)**: Basic Considerations; Media for Industrial Fermentations; Control of Fermentation Conditions.

Further Reading

Applegate LE (1984) Membrane Separation Processes. *Chem. Eng.* June 11, pp. 64–89.

Avshalom M (ed.) (1988) *Advances In Biotechnological Processes*. Vol. 8. New York: Alan R Liss.

Bell DJ, Hoare M and Dunnil P (1983) *Adv. Biochem. Eng..* P. 26. Berlin: Springer.

Belter PA, Cusseler EL and Hu WS (1988) *Bioseparations: Downstream Processing for Biotechnology*. New York: John Wiley.

Bjurstrom EE (1985) *Biotechnology. Chem. Eng.* February 18, pp. 126–158.

Caughlin RT and Thomson AR (1983) *Protein Recovery and Purification*. Oxfordshire: Harwell.

Cheryan M (1986) *Ultrafiltration Handbook*. Lancaster: Technomics.

Darbyshire J (1981) Large scale enzyme extraction and recovery. In: Wiseman A (ed.) *Topics in Enzyme and Fermentation Biotechnology*. Vol. 5, p. 164. New York: John Wiley.

Dechow FJ (1989) *Separation and Purification Techniques in Biotechnology*. New Jersey: Noyes Publications.

King CJ (1971) *Separation Processes*. New York: McGraw-Hill.

Moo-Young M (ed.) (1985) *Comprehensive Biotechnology*. Vol. 2. Oxford: Pergamon Press.

Moshe DW and Marcu D (1988) Disintegration of Microorganisms. In: Avshalom M (ed.) *Downstream Processes: Equipment and Techniques, Advances in Biotechnological Processes*. Vol. 8. New York: Alan R Liss.

Rehm HJ and Reed G (1983) *Biotechnology*. Vol. 3. Weinheim: Verlag Chemie.

Scopes RK (1982) *Protein Purification Principles and Practice*. Berlin: Springer.

Stanbury PF and Whitaker A (eds) (1984) The recovery and purification of fermentation products. In: *Principles of Fermentation Technology*, Oxford: Pergamon Press.

Walter V (1982) Isolation of hydrophilic fermentation products by adsorption chromatography. *J. Chem. Tech. Biotechnol* 32: 109–118.

Wang DIC, Cooney CL, Demain AL et al (1979) Enzyme Isolation. In: *Fermentation and Enzyme Technology*. New York: John Wiley.

Yusuf C and Moo-Young M (1986) Disruption of microbial cells for intracellular products. *Enzyme Microb. Technol.* 8: 194–204.

Production of Xanthan Gum

M K Gowthaman, **M S Prasad** and **N G Karanth**, Central Food Technological Research Institute, Mysore, India

Introduction

Xanthan gum is the most versatile microbial exopolysaccharide. It is synthesized by the bacterium *Xanthomonas campestris*. Xanthan gum has many applications, reflecting its physico-chemical properties.

In the 1950s, a strain of *X. campestris* isolated in the Northern Regional Research Laboratory of the US Department of Agriculture was found to produce a polysaccharide of potential commercial importance: this was denoted as B-1459, the culture of *X. campestris* which produced it being NRRL B-1459. The chemical composition of the polysaccharide, whether produced using cabbage extract or using a synthetic medium, was identified in terms of physico-chemical characteristics and structure.

Structure

The primary structure of xanthan gum is shown in **Figure 1**. Each xanthan gum repeat unit contains five sugar residues: two β-D-glucosyl residues, two D-mannose residues and one D-glucuronic acid residue. The possession of trisaccharide side-chains on alternating glucosyl residues distinguishes xanthan gum from cellulose: the O-3 position of alternating glucosyl residues carries a trisaccharide side-chain comprising one D-glucuronic acid residue and two D-mannose residues. Approximately half of the terminal D-mannose residues bear a pyruvic acid moiety. At least some of the proximal D-mannose residues carry acetyl groups. The glucuronic acid residues usually occur as mixed calcium, sodium and potassium salts. The structural details of xanthan gum differ with different strains of *X. campestris*.

The rheological behaviour of xanthan gum is strongly influenced by molecular weight. The molecular weight of xanthan gum, as determined by electron microscopy, light scattering, viscometry and ultracentrifugation, is estimated to be about 2×10^{6}. This corresponds to approximately 2000 repeat units per polymer molecule.

Electron microscopy has revealed the helical nature of the xanthan gum molecule, and X-ray diffraction studies on oriented xanthan gum fibres have identified the molecular conformation as a right-handed, five-fold helix. This conformation is believed to be responsible for a number of the physico-chemical properties of xanthan gum in solution.

Properties

Xanthan gum is a white to cream-coloured free-flowing powder. It is soluble in both cold and hot water, but is insoluble in most organic solvents.

Rheology

Its industrial importance is based on its rheology in water-based systems. Xanthan gum solutions are characterized by high levels of pseudoplasticity, i.e. the apparent viscosity decreases with increasing shearing force but is almost instantaneously regained with a decrease in shear. This feature is attributed to the formation of complex aggregates, involving hydrogen bonds and polymer entanglement. The high viscosity with low shearing forces is due to the immobile molecules being entangled in a highly ordered network. This property makes xanthan gum a very effective thickener, stabilizer or suspension medium. The disaggregation of the network, and the alignment of individual polymer molecules in the direction of a shearing force, lead to loss of viscosity. When the shearing force decreases, the aggregates reform. The changes in viscosity are instantaneous, and no hysteresis (lagging) is evident.

Unlike other commercial polysaccharides, xanthan gum has a well-defined yield value, which relates to its abilities to stabilize emulsions and act as a suspension medium. The rigid helical conformation of xanthan gum results in its viscosity being relatively insensitive to differences in ionic strength and pH. The protection of the backbone of the molecule by the side-chains results in the superior stability of xanthan gum, com-

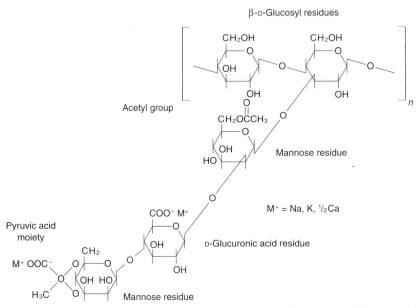

Figure 1 Primary structure of xanthan gum, showing one repeat unit comprising five sugar residues: two glucosyl, two mannose and one glucuronic acid.

pared with other polysaccharides, when exposed to acids, alkalis and enzymes. The pyruvate content of xanthan gum influences its viscosity in salt solutions.

Comparison with Other Polysaccharides

Xanthan gum solutions have the following advantages over other polysaccharides:

1. High viscosity at low concentrations: for example, a solution with a concentration of 1% appears almost gel-like at rest, yet pours readily and has a very low resistance to mixing and pumping. The reduction in viscosity with increasing shear is important for the pourability of suspensions and emulsions, and hence for the efficacy of xanthan gum as a processing aid.
2. High resistance to pH variations in the range of 2–12 make xanthan well-suited to foods. Excellent stability is shown at low pH over long periods of time.
3. High resistance to temperature variations, even in the presence of acids and salts. Excellent freeze–thaw ability, with practically no syneresis. Viscosity is not affected by temperatures in the range 0–100°C, and even after heat treatments such as pasteurization and sterilization, viscosity is recovered after cooling. The pyruvate content of xanthan gum influences its thermal stability.
4. Direct solubility of xanthan gum in 5% acetic acid, 5% sulphuric acid, 5% nitric acid, 25% phosphoric acid and 5% sodium hydroxide render it appropriate for many applications.
5. Reasonable stability for several months at ambient temperature.
6. Compatibility with most of the commercially available thickeners, including cellulose derivatives, starch, pectin, gelatin, dextrin, alginate and carrageenan. Xanthan gum shows a synergistic increase in viscosity with galactomannan, i.e. the observed viscosity is higher than the sum of the viscosities of the individual gums. By blending different gums with xanthan gum in different proportions, very specific and defined characteristics can be obtained, e.g. viscosity, pseudoplasticity, 'mouth feel'.

Reaction with Chemicals and Biochemicals

Strong oxidizing agents, e.g. persulphates, peroxides and hypochlorites, degrade xanthan gum. Reducing agents do not generally affect its stability. Methanol, ethanol, isopropanol and acetone have no effect on aqueous solutions of xanthan gum up to a certain concentration, after which precipitation of the gum occurs. Xanthan gum is compatible with non-ionic surfactants in concentrations up to 20%. Anionic and amphoteric surfactants tend to salt out xanthan gum at a concentration of 15%, but this depends on the presence of acids, bases and salts. Most commercial enzymes, including α-amylase, amyloglucosidase, cellulases, pectinases and proteases, cannot degrade xanthan gum.

Toxicity

Toxicological studies on the suitability of xanthan gum for food use have indicated its safety. In 1969, the US Food and Drug Administration (FDA) permitted the use of xanthan gum in food products without any quantity limitations, and xanthan gum has also been approved for food use by many other countries, including the European Community.

Applications

Xanthan is used mainly as a suspension agent; for viscosity control; for gelation and for flocculation, in both food and non-food applications.

Food Applications

These can be categorized as follows:

- thickening agent: in sauces and syrups
- gelling agent: in milk-based desserts, confectionery, jellies, pie fillings and pastry fillings
- colloid: as a stabilizer in ice cream, salad dressings and fruit drinks
- synergistic gel formation: in synthetic meat gels
- for the retardation of ice or sugar crystal formation: in ice cream, ice lollies, etc.
- for providing stability at low pH: in salad dressings.

Non-food Applications

The applications are:

- suspension: agricultural chemicals, biocides, fungicides and pesticides; e.g. for ingredients of glazes in the ceramic industry; emulsion and water-based paints; polishes for leather; pigments and emulsion inks for textiles and paper
- viscosity control: in oil drilling, to suspend rock cuttings from oil wells; in abrasives
- gelling: in explosives
- flocculation: in ore extraction, water clarification
- mobility control: to enhance oil recovery.

Microbial Production of Xanthan Gum

The major operations involved in the microbial production of xanthan are:

- organism and inoculum preparation
- media preparation

- fermentation
- downstream processing.

Organism and Inoculum Preparation

Xanthan gum is produced by the bacterium *Xanthomonas campestris*, maintained on a sucrose–tryptone–yeast-extract agar (STYA) slant. An inoculum of 5–10% is needed for optimum polysaccharide production. The inocula are prepared by transferring cells from an STYA slant (at 28°C) to a tube containing 7 ml STYA medium (pH 7.0), and incubating at 28°C and 160 r.p.m. for 24 h. The inoculum is then transferred to 250 ml flasks and, 24 h later, to the laboratory fermenters.

Media

The components of the media used in industry are mainly inexpensive and complex, being natural raw materials. Tap water is generally used for dilution. The carbon source may be glucose, sucrose or starch, in the concentration range 1–5%. Concentrations higher than 5% tend to inhibit both growth and xanthan gum production. Acid whey from cottage cheese manufacture is another effective source of carbon.

The cost of media sterilization can be reduced if the media components are sterilized in separate streams. This simplifies the procedure, and also allows greater flexibility, through the use of modern process monitoring and control equipment. In xanthan gum production, continuous sterilization does not offer much economic advantage because the volumes of media handled are relatively small. In addition, suspended components such as soya protein may not be easily sterilized and may cause fouling of the surfaces at which heat exchange occurs. The separate sterilization of carbohydrates and the nitrogen source(s) also allows the rate of addition to be varied during the process, and in the case of carbohydrates may avoid caramelization.

Fermentation

Successful xanthan gum fermentation requires a clear understanding of the microbial environment and the kinetics involved, with particular reference to the high viscosity of the fermentation broth. The dynamic rheology of the broth has a profound influence on the bioreaction rates, power consumption, heat and mass transfer, and mixing. The kinetics are influenced significantly by spatial variations in the concentrations of substrate, biomass and the polymer itself, which are attributable to the rheology of the polymer and its production.

Kinetics Xanthan gum is generally considered as a secondary metabolite (i.e. not associated with growth), produced when a carbohydrate source is present in excess. However, the specific growth rate of the organism is a major determinant of the rate of production of xanthan gum. The organism is able to assimilate substrates and synthesize intermediates for polymer formation at sustained rates, in spite of adverse environmental conditions. In continuous cultures, the overall rate of xanthan gum production by *X. campestris* is almost constant over the dilution rate range of 0.05–0.20 h^{-1}, with the amount of xanthan gum produced per unit of cell mass increasing with decreasing growth rate.

Effect of Substrates In general, media with a high carbon-to-nitrogen ratio are preferred for xanthan gum production. The conversion of glucose to xanthan gum gives a theoretical yield of about 85%, and conversions of 70–80% of the glucose consumed are common in well-run fermentations. However, yields of 50–60% are reasonable taking into account the cells, other organic material and inorganic salts which are co-precipitated during the separation and recovery of the product. Sucrose appears to be a better substrate than glucose and also other saccharides, including dextrins, sorbose, galactose, rhamnose, mannose, maltose, trehalose, cellobiose, lactose, ribose and arabinose. A concentration of 4% sucrose has been found to be optimal.

Nitrogen is the next limiting nutrient after carbon, the carbon-to-nitrogen ratio being critical. High nitrogen levels in the early stages of fermentation support rapid cell growth, but during the later stages the nitrogen levels are allowed to drop. This saves raw materials, and also yields a purer product. A variety of nitrogen sources can be used for xanthan gum production, including dried distillers' solubles, urea, the juice produced as a by-product of kenaf plants, peptone, meat peptone, soya peptone, ammonium nitrate and corn steep liquor. Limiting the nitrogen levels helps to inhibit the growth of the bacteria, and to stimulate xanthan gum synthesis.

Conversions of over 70% are achieved when 4% sucrose is supplemented with organic acids or glutamate. Also, xanthan gum titres up to 3.5% can be achieved. A viscosity of 15 000 mPa s can be obtained with 4% sucrose and 1% succinate. The addition of pyruvate, succinate or α-ketoglutarate stimulates higher yields from 4% sucrose, but has less effect on yields from glucose. Also, sucrose results in a higher specific xanthan gum production than does glucose. The addition of citric acid improves xanthan gum productivity, and the addition of corn steep liquor at 1 g l^{-1} increases the yield and viscosity of xanthan gum produced by cells grown in sucrose. Corn steep liquor also shortens the cultivation time and promotes better sugar utilization.

Effect of Operational Parameters The optimum temperature for xanthan gum production is reported as 28°C, and that for the growth of *X. campestris* is 24–30°C. In a continuous culture, under controlled and nitrogen-limited steady-state conditions, the cell concentration remains constant at 20–37.5°C. However, the conversion of glucose to polymer and the culture viscosity are strongly affected by temperature, the maximum values being attained at 30°C.

Normally, the pH tends to decrease during fermentation from its initial value of 7.0. A lack of pH control and poor medium buffering lead to a sharp drop in pH, and the growth of *X. campestris* and xanthan gum formation cease when the pH falls to 5.5. The optimum pH value for xanthan gum production is 7.0. Potassium hydroxide solution is generally used for the control of pH in the fermentation: it serves as both a titrant and a potassium source, and although it is more expensive than sodium hydroxide, it is preferred because it yields xanthan gum containing the desirable K^+ forms of mannose and glucuronic acid.

Dissolved oxygen is another parameter of critical importance. Oxygen is required for the synthesis of components of the polymer, as well as for the oxidation of the reduced pyridine nucleotides.

Impeller Design and Oxygen Transfer

The high apparent viscosities built up during the course of fermentation, and the thixotropic nature of the broth, result in effective O_2 transfer being critical for successful fermentations with high yields. Because of the lack of homogeneity in the broth as the fermentation proceeds, viscosity gradients build up and promote the channelling of air up the centre of the fermentation vessel. Hence impeller design is highly critical for the creation of sufficient turbulence, bulk mixing and O_2 transfer.

Higher average shear rates are obtained with extended, large-diameter impellers, such as anchor stirrers and helical ribbon impellers, than with small-diameter turbine impellers running at higher speeds but with the same power input. Double helical ribbon impellers running at a moderate speed are usually suitable, although they are not very effective at micromixing, which is essential for fast cell growth and substrate conversion. Two impellers driven at different speeds, one to effect circulation and the other to disperse O_2, are suggested for the achievement of better micromixing. The viscous, pseudoplastic broth preserves the fine division of air bubbles, preventing the recoalescence which occurs in some other aerobic fermentations. Double helical impeller and combination of disc turbine with helical are special impellers with large profiled blades, designed to meet the needs of processes such as xanthan gum fermentation, which demand high levels of gas dispersion, bulk blending and mass transfer. The manufacturers call these impellers 'high-efficiency hydrofoils', and claim that they can disperse about 85% more air than a disc turbine with the same power input.

Heat Transfer The dissipation of heat during fermentation is essential, not only for the removal of the metabolic heat generated during growth and the synthesis of xanthan gum, but also for cooling the sterilized medium. Xanthan gum fermentation generates much less energy than that of antibiotics, and so the fermenters can be as large as $50 \, m^3$ without needing internal cooling coils – sufficient cooling being achieved through the vessel walls or external cooling coils. However, in fermenters larger than $50 \, m^3$, heat transfer becomes a problem because the large internal cooling coils which are necessary are a hindrance to the free flow of the broth.

Rheology The feature of xanthan gum fermentation that distinguishes it from other microbial processes is the remarkable rheological behaviour exhibited by the fermentation broth, which poses distinctive problems in xanthan gum manufacture. The fermentation broth is characterized by high viscosity at low concentrations, because the polymer is continuous with the water phase. In contrast, in antibiotic fermentations the viscosity is due to the mycelia of the fungal culture, which are discontinuous with the water phase.

Xanthan gum broths exhibit significant thixotropy, and also the Weissenberg effect – i.e. the broth climbs the impeller shaft. This effect is due to viscoelastic forces. Viscoelasticity may be caused by high concentrations of calcium in the medium.

Manufacturing Process Xanthan gum production is illustrated in **Figure 2**. The fermentation begins with the transfer of inoculum from slants of *X. campestris* to Erlenmeyer flasks. After shaking for 24 h, the flask contents, now an actively growing culture, are transferred to a laboratory fermenter. Here, the culture inoculum is grown for another 24 h. The inoculum is then propagated in increasingly larger fermenters prior to addition to the main fermenter, the capacity of which may be of the order of $100 \, m^3$. During each stage of the inoculum development, aseptic conditions are maintained and the pH, temperature and carbon and nitrogen concentrations are strictly controlled.

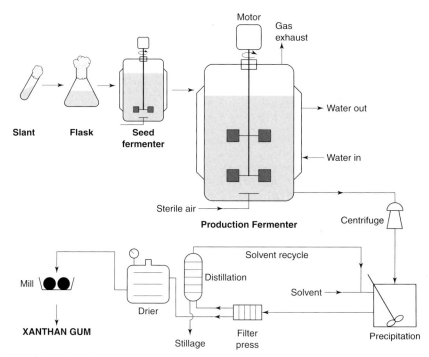

Figure 2 Industrial production of xanthan gum.

The fermentation is carried out over 72–96 h, and the viscosity and biomass are monitored constantly. At the end of the fermentation, the broth is pasteurized to kill all the cells. Downstream processing follows.

Downstream Processing

The recovery, concentration and purification of xanthan gum constitute a significant fraction of the total production cost. The concentration of the polymer in the final broth is about 15–30 kg m^{-3}. The aims of the recovery operation are:

- to obtain xanthan gum in a form which is solid; microbiologically stable; easy to handle, transport and store; and which can be readily redissolved or diluted for any application
- purification: to reduce the level of non-polymer solids, and to improve the functional performance, colour, odour and other qualities of the xanthan gum
- to deactivate undesirable enzymes, e.g. cellulases, pectinases
- to modify the chemical properties of the polymer for special applications.

The removal of cells is required mainly for food applications. In order to remove cells by conventional centrifugation, filtration and flocculation, the broth has to be diluted several times to obtain a suitable viscosity. Alternatively, cell removal can be achieved by treating the broth with proteolytic and lytic enzymes, which break down the cells into molecules of low molecular weight.

After the removal of cells, the xanthan gum is separated from the solvent water. This is accomplished very effectively by precipitation with either isopropanol or ethanol. The choice depends on costs, practicability and the final specification of the product, isopropanol being widely favoured. The volume of alcohol required for xanthan gum precipitation is dependent on the concentrations of certain salts, but is independent of xanthan gum concentration. The addition of an electrolyte, usually potassium chloride (1%), lowers the isopropanol requirement by about 30%. Hence, increasing the concentration of product during fermentation, and increasing the salt content of the broth before precipitation, considerably decrease the amount of alcohol required. If the cells have not been removed earlier, they are precipitated along with the xanthan gum.

The next stage in processing is drying, which is made easier by first removing water from the wet precipitate (which may contain the bulk of the microbial cells), by pressing or centrifugation. Forced-air driers, vacuum driers (continuous or batch), drum driers and spray-driers are commonly used. The correct drying conditions are essential if chemical degradation and excessive changes in the colour or solubility of the product are to be avoided. Rapid drying of the polymer results in case hardening, which imparts poor hydratability to the product, i.e. the dispersed product takes longer to reach its final viscosity. The drying conditions also affect its dispersibility. Spray-drying can be used to process the broth directly, for applications that do not require a cell-free product.

The dried precipitate is finally milled, to obtain a product of mesh size 40–200. The milled product contains no viable *X. campestris* cells, in order to conform to the FDA criteria for food-grade products. Great care during milling is required, to avoid excessive heating, which can lead to the darkening or degradation of the polymer. Owing to the hygroscopic nature of the milled product, containers with a low permeability to water are used for packaging. The absorption of moisture can cause clumping during redissolution and, in some cases, hydrolytic degradation of the product.

Economics

The current cost of production of xanthan gum is higher than that of the traditional polysaccharides, e.g. corn starch, cellulose-derived products and plant gums, which dominate the market. Several means of reducing the overall costs have been suggested, including:

- reuse of medium: the residue from the distillation column contains glucose and other carbohydrates, nitrogenous materials and minerals, and could be reused as a growth medium if supplemented appropriately
- recovery of heat: the large amounts of heat lost during distillation may be transferred via heat exchangers to the incoming material
- meticulous control of solvent losses, due to vapour leakage from equipment, spillage, inadequate cooling of solvent, etc: such losses constitute a major proportion of total costs.

The applications of microbial xanthan gum are steadily increasing, and global demand is likewise expected to increase. Microbial xanthan gum has several competitive advantages over alternative products, e.g. plant-derived gums, owing to its diverse and unique properties. The higher demand foreseen in the future should help to lower the costs of production through the introduction of economies of scale.

*See also: **Xanthomonas**.*

Further Reading

Cottrell IW and Kang KS (1978) *Dev. Ind. Microbiol.* 17: 117–131.

Rocks JK (1971) Xanthan gum. *Journal of Food Technolology.* Vol. 25: I: 476–480.

Sandford PA (1979) Extracellular Microbial Poly-

saccharide. In: Stuarter Tipson R and Horton D (eds). *Adv. Carbohydr. Chem. Biochem.* 36: 265–313. New York: Academic Press.

Slodki ME and Cadmus ML (1979) Production of Microbial Polysaccharide. In: Perlman D (ed.) *Adv. Appl. Microbiol.* Vol. 23: 19–24. New York: Academic Press.

Vincent A (1985) In: Wiseman A (ed.) *Topics in Enzyme and Fermentation Technology Production.* Vol. 10, p. 109. Chichester: Ellis Horwood.

Production of Organic Acids

M Moresi, Istituto di Tecnologie Agroalimentari, Università della Tuscia, Viterbo, Italy

E Parente, Dipartimento di Biologia, Difesa e Biotecnologie Agro-Forestali, Università della Basilicata, Potenza, Italy

Introduction

The organic acids commonly used as food acidulants include, in order of decreasing importance, citric, acetic, lactic, tartaric, malic, gluconic, propionic and fumaric acids (**Table 1**). They are amongst the most versatile ingredients in the food and beverage industry, because they are soluble and hydroscopic, and also have the abilities of buffering and chelation. They are synthesized commercially either chemically or biotechnologically (by fermentation or using enzymes), or are extracted from the residues remaining after wine-making, by leaching as shown in Table 1.

In terms of volume, citric and acetic acids together account for around three-quarters of food acidulant usage, with inorganic phosphoric acid next in order of importance. Citric acid has the widest range of applications, the uses of phosphoric and acetic acids being limited, almost exclusively to the soft drink cola in the case of phosphoric acid, and to vinegar, sauces and condiments in the case of acetic acid.

This article covers only the production of citric, gluconic, lactic and propionic acids, being all those produced by fermentation.

Citric Acid

Citric acid (2-hydroxy-1,2,3-propane-tricarboxylic acid: $C_6H_8O_7$) is widely distributed in natural raw materials (e.g. lime, lemon and raspberry). It is commercially available in the monohydrated form (mol. mass 210.13 Da, relative density 1.542 at 20°C, and heat of combustion 1962 kJ mol^{-1} at 25°C). It is quite a strong tricarboxylic acid ($K_1 = 7.45 \times 10^{-4}$; $K_2 = 1.73 \times 10^{-5}$; $K_3 = 4.02 \times 10^{-7}$; all at 25°C where the subscript i refers to the 1st, 2nd or 3rd dissociation

Table 1 Food acidulants: Chemical formulas, molecular weight, dissociation constants at 25°C, world output, and production methods.

Acidulant	Chemical formula	Mol. wt	pK_i	World output (metric tonnes)	Production Methods [1,2]
Gluconic acid (and δ-glucono-lactone)	COOH | H–C–OH | HO–CH | H–C–OH | H–C–OH | CH₂–OH	196.16	3.70 (25°C)	50 000	F 67% C 33%
Citric acid	CH₂–COOH | HO–C–COOH | CH₂–COOH	192.13	6.40; 4.76; 3.13 (25°C)	840 000	F 100%
Malic acid	CH₂–COOH | HO–CH–COOH	134.09	5.11; 3.40 (25°C)	30 000	C 70% E 30%
Tartaric acid	HO–CH–COOH | HO–CH–COOH	150.09	4.34; 2.98 (25°C)	30 000	L 100%
Fumaric acid	HC–COOH ‖ HC–COOH	116.07	4.44; 3.03 (18°C)	20 000	C 100%
Lactic acid	CH₃ | HO–CH | COOH	90.08	3.08 (100°C)	35 000	F 50% C 50%
Propionic acid	CH₃CH₂COOH	74.08	4.87 (25°C)	50 000	C 100%
Acetic acid	CH₃ | COOH	60.65	4.75 (25°C)	120 000	F 100%

[1] Percentage of total production for food uses.
[2] F, fermentation; C, chemical synthesis; E, enzymatic synthesis; L, leaching.

step), and is soluble in water, with a pleasant, acid taste.

Citric acid was first isolated in 1784 by Scheele, who precipitated it as calcium citrate by adding calcium hydroxide to lemon juice. Prior to 1920, citric acid was almost exclusively produced in Sicily from lemons: the company Arenella (Palermo) was the world's leading producer of citric acid, operating as a monopoly until the advent of the citric acid fermentation technique in Belgium in 1919 and in the US in 1923.

Synthesis of Citric Acid

Citric acid is produced by several moulds (*Penicillium* spp., *Aspergillus niger*, *A. wentii*, *Trichoderma viride*), yeasts (e.g. *Yarrowia lipolytica*, *Candida guilliermondii*) and bacteria (e.g. *Arthrobacter*) from a variety of substrates (glucose, sucrose, *n*-alkanes). However, industrial processes have been developed only for the production of citric acid from sugars (glucose and sucrose) by *A. niger* and from sugars and *n*-alkanes by yeasts. Industrial strains which produce citric acid are not freely available, but a few can be obtained from international culture collections. These include: *A. niger* NRRL 2270, NRRL 599, ATCC 11414 and ATCC 9142; and *Y. lipolytica* ATCC 20346, ATCC 20390, NRRL Y-7576 and NRRL Y-1095.

The metabolic pathways involved in the over-production of citric acid by *A. niger* are shown in **Figure 1**. Essential for overproduction are a high flux of metabolites through the glycolysis (5–10 mmol min^{-1} · mg protein^{-1}), a block of the reactions of the tricarboxylic acid (TCA) cycle that degrade citrate, and an anaplerotic sequence, to replenish the oxaloacetate (OAA) used for the synthesis of citrate. Key regulatory enzymes in citric acid synthesis are phosphofructokinase (PFK1), pyruvate carboxylase (PC), citrate synthase (CS) and α-ketoglutarate dehydrogenase (KDH). PFK1 is inhibited by citrate, but this inhibition is counteracted by the presence of high levels of NH$_4^+$ and by the accumulation of fructose-2,6-biphosphate (FBP). FBP, which lowers the Michaelis-Menten constant (K_m) of PFK1, counteracts its inhibition by citrate and inhibits gluconeogenesis, thus increasing the flux of carbon through glycolysis during acidogenesis. 6-Phospho-fructo-2-kinase (PFK2), the enzyme responsible for FBP production, is poorly regulated and its activity is only influenced by its substrates. The regulation of pyruvate kinase (PK) does not seem to play a role in the overproduction of citric acid. CS activity in *A. niger* is regulated by the level of OAA, which is produced in the anaplerotic reaction catalysed by pyruvate carboxylase (PC), a cytosolic enzyme induced by high sugar concentrations. Malate is produced from OAA by cytosolic malate dehydrogenase, and is involved in the antiport of citrate from the mitochondrion: inside the mitochondrion, it is oxidized back to OAA. In normal conditions (i.e. with no citrate accumulation), citrate is catabolized in the mitochondrion by the action of aconitase (ACT), NADP$^+$-specific isocitrate dehydrogenase (IDH) and KDH. The levels of citrate, OAA and NADH observed during acidogenesis result in the inhibition of KDH and NADP$^+$-specific IDH.

Although the regulation of enzyme activity is critical in controlling the metabolite flux toward citric acid overproduction, the most important steps in controlling the rate of the pathway are glucose transport and hexokinase (HK) activity. A low-affinity glucose carrier ($K_m = 3.67$ mmol), which is induced by high sugar concentration, is active during citrate overproduction and is only partially inhibited by low pH and high citrate concentration. The improvement of strains of microorganisms for citric acid production has traditionally been achieved by mutagenesis and screening, but it has been postulated that the overexpression of HK and the glucose transport systems would result in the maximum possible increases in citric acid production. This strategy appears to be viable, the overexpression (20–30-fold) of PK and PFK1 already having been achieved in *A. niger*.

Respiratory-chain enzymes have also been found to play a role in citric acid overproduction. An alternative pathway, sensitive to salicyl hydroxamic acid (SHAM), is apparently used during acidogenesis to reoxidize the NADH produced during glycolysis. This malfunction of the normal respiratory chain is due to the diminished activity of NADH–UQ-oxidoreductase and other respiratory-chain enzymes.

The metabolic changes necessary for the overproduction of citric acid by *A. niger* are induced by high sugar concentrations and manganese deficiency, although other factors (including the concentrations of phosphate, nitrogen and trace metals, a low pH, and a high dissolved O$_2$ concentration) are important. Very low concentrations of manganese (< 10 mg m^{-3}) are critical: they result in decreased activity of the pentose phosphate pathway; increased glycolytic flux; high intracellular NH$_4^+$ concentration; increased turnover of nucleic acids and proteins; changes in membrane lipid composition; alteration of the composition of the cell wall; and, hence, morphological changes.

The overproduction of citric acid by yeasts is relatively insensitive to trace metal concentration, but is triggered by the limitation of nutrients (nitrogen, sulphur, phosphorus or magnesium) coupled to a high rate of glucose utilization. This results in a high ATP/AMP ratio and, in turn, in the inactivation of NAD$^+$-specific IDH. The main anaplerotic reactions are the synthesis of OAA from pyruvate, catalysed by PC during its production from glucose, and the glyoxylate cycle during growth on *n*-alkanes. The accumulation of high levels of isocitrate (10–50% of the citrate produced), with respect to those predicted by the equilibrium constant of aconitase, is probably due to the high permeability of yeast mitochondria to isocitrate compared to citrate. Low cytoplasmic levels of citrate may be responsible for the reduced inhibition of glycolysis through negative feedback. Improvement of yeast strains for citric acid production is focused on obtaining strains with reduced levels of activity of isocitrate dehydrogenase and aconitase.

Methods of Manufacture of Citric Acid

Citric acid production currently uses submerged or surface fermentation processes, with beet molasses or glucose syrup as the main raw material and *A. niger* as the fermenting organism. The economics of the two processes are generally acknowledged to be com-

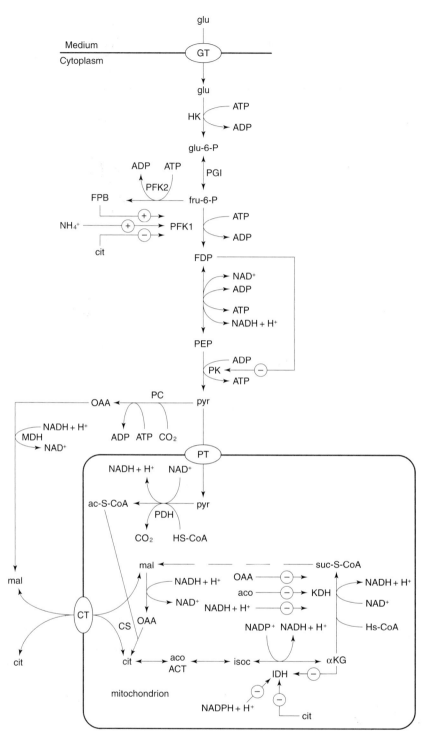

Figure 1 Metabolic pathway for citric acid overproduction by *Aspergillus niger*. Only relevant enzyme activities, substrates, products and effectors (−: inhibitor; +: activator) are shown. Enzymes and transport systems: GT, low-affinity glucose transport system; HK, hexokinase; PGI, phosphoglucose isomerase; PFK1, phosphofructokinase; PFK2, 6-phosphofructo-2-kinase; PK, pyruvate kinase; PC, pyruvate carboxylase; MDH, malate dehydrogenase; PT, pyruvate transport system; CT, citrate transport system; PDH, pyruvate dehydrogenase; CS, citrate synthase; ACT, aconitase; IDH, isocitrate dehydrogenase; KDH, α-ketoglutarate dehydrogenase. Substrates and products: glu, glucose; glu-6-P, glucose-6-phosphate; fru-6-P, fructose-6-phosphate; FDP, fructose-1,6-bisphosphate; FBP, fructose-2,6-bisphosphate; PEP, phosphoenolpyruvate; pyr, pyruvate; OAA, oxaloacetate; mal, malate; cit, citrate; ac-S-CoA, acetyl-coenzyme A; aco, *cis*-aconitate; isoc, isocitrate; αKG, α-ketoglutarate; suc-S-CoA, succinyl-coenzyme A.

parable, but submerged fermentation has been selected by the world's largest citric acid manufacturers (i.e. Bayer, ADM, Jungbunzlauer and Cargill) – probably because of the lower capital and labour costs. Surface fermentation is still used by Citrique Belge (Belgium) and Cerestar (Italy), because of its ease of process control and lower energy requirements.

Submerged fermentation is carried out in either stirred-tank reactors of capacity 150–200 m^3, or bubble columns of capacity 300–500 m^3 (or up to 1000 m^3, as claimed by some manufacturers). The main advantages of submerged as opposed to surface fermentation are asepsis during fermentation; automatic control of inoculation and fermentation procedures; shorter fermentation times; and greater yields of product.

Interest in citric acid production by yeasts has led to their use in submerged fermentations with sugar- or hydrocarbon-based media, to overcome the main disadvantages of fermentations using moulds, i.e. high sensitivity to trace metals and low production rates. However, no yeast-based process is known to be operating currently – the plant at Southport (NC, US) was shut down in 1982 after operating for 3–4 years, and the Liquichimica plant (Saline, Italy) never came into operation.

Production Media

As indicated in the section on synthesis of citric acid, its overproduction is obtained in media containing high concentrations of simple sugars (molasses or glucose syrup), in which mycelial growth is restricted by nutrient (phosphorus, manganese, iron or zinc) limitation. The typical composition of production media is given in **Table 2**. Nitrogen is usually added as ammonium nitrate or sulphate. Metals are removed by the pre-treatment of raw materials, especially molasses, by means of cation-exchange resins and/or

Table 2 Composition of production media used in the laboratory and industrially for the production of citric acid by *Aspergillus niger*

Component	Range of values	Typical values
Sucrose or glucose	125–225 kg m^{-3}	180 kg m^{-3}
NH$_4$NO$_3$ (or other NH$_4^+$ salt)	0.5–3.5 kg m^{-3}	1.5 kg m^{-3}
KH$_2$PO$_4$	0.5–2 kg m^{-3}	0.5 kg m^{-3}
MgSO$_4$.7H$_2$O	0.1–2.0 kg m^{-3}	0.25 kg m^{-3}
Fe^{++}	2–1300 mg m^{-3}	<200 mg m^{-3}
Zn^{++}	0–2900 mg m^{-3}	200–1500* mg m^{-3}
Cu^{++}	1–10 200 mg m^{-3}	200–1500* mg m^{-3}
Mn^{++}	0–46 mg m^{-3}	<2 mg m^{-3}

* The upper value is used to overcome the detrimental effects of iron and manganese on the mycelial structure and to restore proper pellet morphology

the addition of potassium hexacyanoferrate (HCF). The optimal iron concentration seems to depend on the fungal strain, but iron levels of 200 mg m^{-3} have been found to inhibit citrate production. The inhibitory effect of Fe^{++} can be counterbalanced by the addition of copper and zinc salts, either during the development of the inoculum or during the early growth of the mycelium in the production medium. The manganese concentration must be kept as low as possible (<10 mg m^{-3} in low-phosphate media). Some ingredients (including: 3–6% w/v methanol; 0.1–0.5% w/v corn, peanut and olive oils and 0.025–0.5% w/v starch) have been claimed to enhance the yield of citric acid.

The media are sterilized batchwise at 121°C for 15–30 min in the laboratory or on a pilot scale. Industrially, sterilization is continuous, using a three-stage plate heat exchanger.

Fermentation Process

Inoculation is generally carried out by aseptic transfer of conidia from a stock culture to agar media (in shallow trays, Petri dishes, slants). After 3–6 days at 30°C the conidia are harvested and inoculated into starch-rich media, to yield up to 10^{11} spores per cubic centimetre. The resulting culture may be directly transferred into fermenters of 10–20 m^3 capacity, to produce a pellet-type inoculum consisting of 1–5 × 10^5 pellets dm^{-3}, each pellet having a diameter of about 0.1–0.2 mm. The pellets are used as inoculum (5–10% v/v) for the industrial-scale production medium.

The fungus may develop different morphological forms, shown in **Figure 2**. The formation of a loose mycelium with long unbranched hyphae must be avoided, because it results in an enormous increase in the apparent viscosity of the culture broth. This limits the effective O$_2$ transfer rate, which results in little or no citric acid production. Small spherical, dense pellets (Fig. 2A, C) with short stubby hyphae (Fig. 2B) are generally regarded as the morphological form for producing optimal yields of citric acid. Frequent observation of pellet morphology during the early stages of the fermentation, using a microscope, allows the proliferation of hyphae to be controlled, in the case of adverse developments (Fig. 2D), by the addition of appropriate amounts of inhibiting compounds, such as HCF, zinc sulphate or copper sulphate. In stirred fermenters, the higher the impeller speed, the easier the formation of small pellets.

Fermentation is exothermic, and the temperature must be kept in the range 28–35°C by cooling. Assuming that the overall heat transfer coefficient is of the order of 500 kJ m^{-2} h^{-1}, and that the effective temperature difference between the fermentation medium and the cooling water is of the order of 5°C, the heat

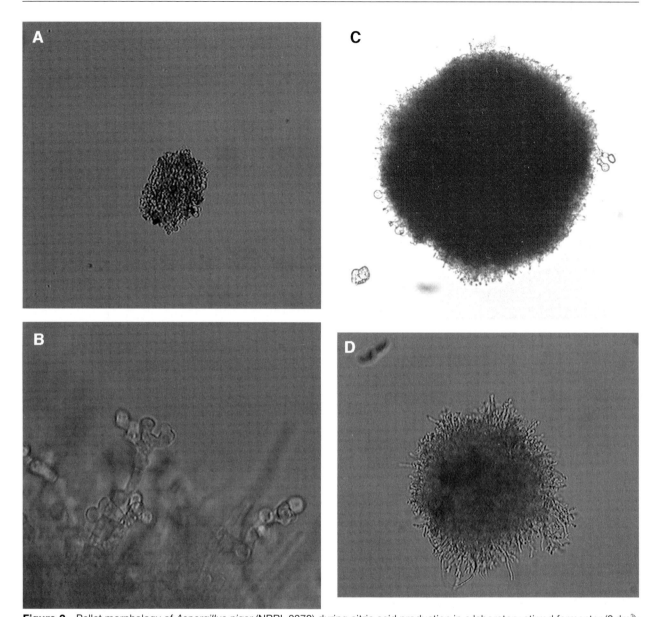

Figure 2 Pellet morphology of *Aspergillus niger* (NRRL 2270) during citric acid production in a laboratory stirred fermenter (2 dm^3); (**A**) Young pellet (× 100); (**B**) Stubby, bulbous hyphae with frequent branching, characteristic of citric acid production (× 400); (**C**) Compact pellet (× 40); (**D**) Degenerating pellet, with pointed, unbranched hyphae protruding from the pellet core (× 100).

transfer surface required to keep the fermentation temperature constant is about 3.2 m^2 per cubic metre of fermentation medium. Therefore, cooling by external jackets is only feasible if the internal diameter of the fermenter is not greater than 1 m.

A low pH and high dissolved O_2 concentration are essential for citric acid production. Initially, a decrease in pH occurs due to the uptake of ammonium. However, extreme pH values (<1.6) limit productivity, and once production of citric acid has started the pH is controlled at 2.2–2.6 by the addition of alkali (NH$_3$). pH should be < 3 to avoid the fungus producing oxalic and gluconic acids. The typical productivity of citric acid on an industrial scale (1–

1.5 kg m^{-3} h^{-1}) involves microbial O_2 demand rates of 0.3–0.5 kg m^{-3} h^{-1}. An adequate supply of O_2 is ensured by forcing 0.1–0.4 volumes of air per volume of medium per minute (vv^{-1} min^{-1}) through the sparger section of the fermenter at pressures of not <0.3–0.4 MPa, while the pressure at the top of the tank varies in the range 0.25–0.35 or 0.12–0.15 MPa, depending on the type of fermenter used (stirred or airlift). Foaming is controlled by adding specific antifoam agents. A temporary interruption of the air supply during fermentation does not seem to affect the performance of the culture, provided that the dissolved O_2 level is greater than 20% of the saturation value. Dissolved O_2 values approaching 0, for

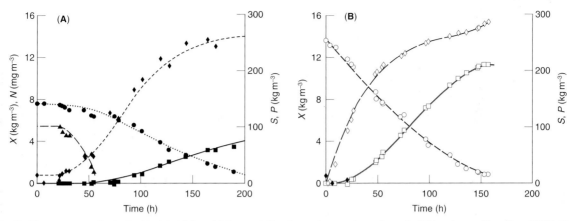

Figure 3 Time course of a typical batch citric acid fermentation from glucose-based media (**A**) by *Aspergillus niger* NRRL 2270 in a 2 dm³ stirred fermenter (closed symbols) and (**B**) by an industrial strain in a 400 m³ bubble column fermenter (open symbols). Concentrations of mycelial biomass (*X*: ◆, ◇), glucose (*S*: ●, ○), citric acid (*P*: ■, □) and ammoniac nitrogen (*N*: △) versus time. Data from the industrial-scale trial were provided by Dr A. Trunfio, Palcitric, Calitri, Italy, while the laboratory-scale trial was performed by the authors at the University of Basilicata, Potenza, Italy.

as long as 85 min followed by restoration of the air supply, are unable to inhibit mycelial growth and citric acid production permanently. However, they do reduce the product yield coefficient by up to 20%.

Fig. 3 shows the overall evolution of a typical batch fermentation of citric acid from glucose-based media by *A. niger* (NRRL 2270) in a stirred fermenter (2 dm³) (Fig. 3A) and by a mutant strain of *A. niger* in a bubble column (400 m³) (Fig. 3B). Two distinct phases are generally acknowledged. During the primary growth phase (the trophophase), no acid production occurs, and during the idiophase, secondary metabolism is dominant and acid production, with very little growth, is observed.

Table 3 shows the simplified overall stoichiometric reactions occurring during the trophophase and idiophase of the fermentation examined. In particular, it was assumed that during the trophophase the microorganism (represented by a formula based on elemental analysis: $CH_nO_pN_q$) replicates itself at the expense of a generic carbon source in the presence of ammonia as the only nitrogen source, whereas during the idiophase it undergoes further growth while decreasing progressively its intracellular nitrogen content and excreting citric acid in a medium practically devoid of any nitrogen source.

Assuming that no carbon atoms of sugar are converted into biomass, carbon dioxide or other by-products as shown by reaction (2) (i.e. when *y* and *d* = 0), the theoretical molar yield (*z*) of citric acid would be 1 or 2 if glucose or sucrose is used. This would be equivalent to 1.17 (or 1.23) kg of citric acid monohydrate per kilogram of glucose (or sucrose) consumed. In practice, the industrial yields range from 57 to 77% of the above theoretical value, the smaller figure

generally being associated with the surface fermentation technique.

Citric acid fermentation may be mathematically described by means of the set of kinetic equations shown in Table 3.

In accordance with the Herbert–Pirt maintenance concept, both product formation (r_p) and substrate consumption (r_s) rates may be linearly related to cell growth rate (r_x) and cell concentration (*X*). In this way, the well known Luedeking–Piret kinetics for product formation is regarded as a special case: in fact, the first term in equation 10 may be described as the product formation rate in association with the mycelial growth rate, while the second term is the non-growth-associated product formation rate. In both the fermentation trials shown in Figure 3, citric acid fermentation may be classified as the mixed-growth associated product formation type.

Microscopic description of this fermentation using *A. niger* pellets also has to account for oxygen diffusion phenomena from the bulk of the fermenting medium to the pellet surface and through the porous structure of the pellet itself. The low effective diffusivity of oxygen within the pellet limits mycelial activity to a peripheral spherical shell only; oxygen penetration depth ranges from 110 to 300 μm in 2 mm pellets.

Recovery and Purification Processes

The citrate-rich broths resulting from either surface or submerged-culture fermentations undergo similar downstream processing. The separation of citric acid may be carried out by one of three methods: direct crystallization, following concentration of the filtered liquor; precipitation as calcium citrate tetrahydrate;

Table 3 Citric acid fermentation: overall stoichiometric reactions, kinetic equations and instantaneous concentrations of mycelia, product, sugar and nitrogen sources

	Equation or reaction		Number
Trophophase reaction	$C_6H_{12}O_6 + a\ NH_3 + b\ O_2 \rightarrow y\ CH_nO_pN_q + d\ CO_2 + e\ H_2O$ glucose $\qquad\qquad\qquad$ mycelium		(1)
Idiophase reaction	$C_6H_{12}O_6 + b\ O_2 \rightarrow y\ CH_nO_pN_q + z\ C_6H_8O_7 + d\ CO_2 + e\ H_2O$ glucose $\qquad$ mycelium $\quad$ citric acid		(2)
	$r_X = dX/dt = \mu_X\ X;$		(3)
	$r_N = dN/dt = \begin{cases} -Y_{N/X}\ (dX/dt) \\ 0 \end{cases}$ for $\begin{cases} N \geq N_{lim} \\ N < N_{lim} \end{cases}$		(4) (5)
Kinetic equations	$\mu = \begin{cases} 0 \\ \mu_M \\ \mu_M\ (1 - X/X_M) \end{cases}$ for $\begin{cases} t \leq t_0 \\ t \leq t_{lim} \\ t > t_{lim} \end{cases}$		(6) (7) (8)
	$r_P = dP/dt = \begin{cases} 0 \\ Y_{P/X}\ (dX/dt) + m_P\ X; \end{cases}$ for $\begin{cases} t \leq t_{lim} \\ t > t_{lim} \end{cases}$		(9) (10)
	$r_S = -dS/dt = \quad Y_{S/X}\ (dX/dt) + m_S\ X;$		(11)
	$X = \begin{cases} X_0 \\ X_0\ \exp[\mu_M\ (t-t_0)] \\ \dfrac{X_M}{1 + \left(\dfrac{X_M}{X_0} - 1\right) e^{-\mu_M\ (t-t_{lim})}} \end{cases}$ for $\begin{cases} t \leq t_0 \\ t \leq t_{lim} \\ \\ t > t_{lim} \end{cases}$		(12) (13) (14)
Integral solutions of the differential kinetic equations	$N = \begin{cases} N_0 \\ N_0 - Y_{N/X}\ (X - X_0) \\ N_{lim} \end{cases}$ for $\begin{cases} t \leq t_0 \\ t \leq t_{lim} \\ t > t_{lim} \end{cases}$		(15) (16) (17)
	$P = \quad P_0 + m_P\ A(t) + Y_{P/X}\ (X - X_0)$		(18)
	$S = \quad S_0 - [m_S\ A(t) + Y_{S/X}\ (X - X_0)]$		(19)
	$A(t) = \begin{cases} 0 \\ \dfrac{X_M}{\mu_M} \ln\left\{1 \dfrac{X_M}{\mu_M}[1 - e^{\mu_M\ (t-t_{lim})}]\right\} \end{cases}$ for $\begin{cases} t \leq t_{lim} \\ t > t_{lim} \end{cases}$		(20) (21)

Nomenclature: $A(t)$, cumulative non-growth contribution to product formation; b, y, z, d, e, stoichiometric coefficients; m_P (m_S), specific rate of product formation (or substrate consumption) at zero cell growth rate; N_{lim}, critical concentration of nitrogen at the onset of citric acid production; P, citrate concentration; r_i, conversion rate of any reagent or product; t_{lim}, overall duration of the citrate lag phase; t_0, overall duration of the cell lag phase; S, substrate concentration; X, mycelium concentration; X_M, maximum mycelium concentration; μ, specific cell growth rate; μ_M, maximum specific cell growth rate; $Y_{N/X}$, $Y_{P/X}$ and $Y_{S/X}$, yield factors for ammoniac nitrogen, citrate and substrate on unit cell biomass. Subscripts: lim, referred to limiting concentration of the nitrogen source; N, nitrogen; P, citric acid; S, glucose; X, mycelium; 0, referred to the initial value.

or liquid extraction. Impurities present in molasses preclude direct crystallization unless highly refined raw materials, such as sucrose syrups or crystals, are used. Liquid extraction is used by only one company (Bayer), the precipitation process predominating.

A simplified process flow sheet of citric acid production from glucose syrup is shown in **Figure 4**.

Mycelia and suspended particles are separated under vacuum by continuous-belt filters. The precipitation of citric acid, as calcium citrate, is achieved by the addition of calcium hydroxide (lime) to the filtrate. The temperature at which this is performed is critical: amorphous tricalcium citrate tetrahydrate is generally obtained at around 70°C, but crystalline dicalcium acid citrate at 90°C. The removal of oxalic acid is unnecessary in the case of submerged-culture

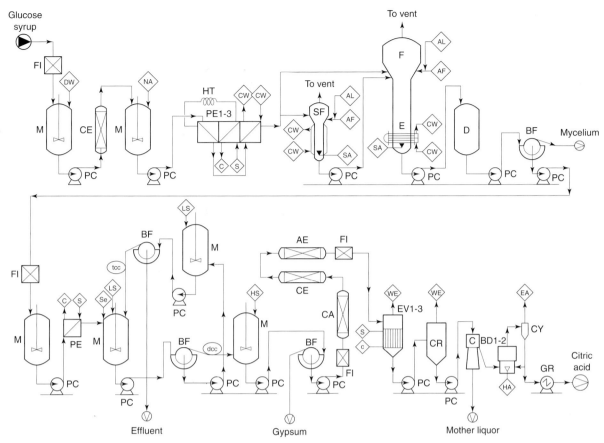

Figure 4 Process flow sheet of a typical citric acid fermentation from glucose-based media by *Aspergillus niger*. Equipment and utility identification items: AE, anion exchanger; AF, antifoam agent; AL, alkaline reagent; BD, fluidized bed drier; BF, vacuum belt filter; C, centrifuge; c, condensate; CA, activated carbon adsorber; CE, cation exchanger; CR, vacuum crystallizer; cw, cooling water; CY, cyclone; D, holding tank; dcc, dicalcium citrate; DW, dimineralized water; E, heat exchanger; EA, exhausted air; EV, evaporator; F, production bubble fermenter; FI, sterile pressure filter; GR, grinder; HA, hot air; HS, sulphuric acid; HT, holding tube; LS, lime slurry; M, mixer; NA, nutrients and additives; PC, centrifugal pump; PE, plate heat exchanger; S, low-pressure steam; SA, sterile compressed air; Se, dicalcium citrate seed; SF, seed bubble fermenter; tcc, tricalcium citrate; WE, water evaporated.

fermentations because of the stricter pH control. The residual citrate in the filtrate is precipitated as tricalcium citrate by the further addition of lime, until the pH reaches 5.8. An amorphous cluster of particles is recovered using another continuous-belt filter, and then recycled back to the liming step. In this way, the clusters dissolve in the citric acid-rich solution and converts into citric acid and calcium ions. The filtrate is discarded. The precipitation of dicalcium acid citrate or tricalcium citrate asks for one or one and a half mol of Ca(OH)$_2$ per mol of citric acid. Therefore the former uses one-third less lime, and consequently less sulphuric acid in the subsequent regeneration of citric acid (see below). Dicalcium acid citrate is also more easily filtered and washed, owing to its crystalline structure, but such a precipitation procedure asks for a high seed ratio of 10–25% to avoid the formation of tricalcium citrate.

The precipitate is washed to remove the impurities adsorbed on to it, which arise from the raw carbon source in the form of residual sugars and contaminants,

and from the autolysis of the fungus in the form of soluble proteins. The washed crystals and 98% w/w sulphuric acid are simultaneously, but separately, fed to a mixer containing a 40% citric acid solution at pH 0.5–0.6. This reaction results in the formation of citric acid and a precipitate of calcium sulphate dihydrate (gypsum). The final steps in refining the filtrate are the adsorption of coloured substances by activated carbon (the decolorization step) and the adsorption of residual calcium sulphate and iron and nickel salts by strong cation-exchange and weak anion-exchange columns (the demineralization step).

The resulting solution (containing 250–280 kg m^{-3} anhydrous citric acid) is concentrated using multiple-effect evaporators to about 700 kg m^{-3}, then fed into a vacuum crystallizer, operating at a temperature of either 35°C or 62°C, i.e. below or above the transition temperature (36.6°C) between the monohydrate and anhydrous forms, depending on the form being manufactured. Crystals of citric acid are separated by centrifugation and dehydrated in a two-stage fluidized bed drier, the first stage using hot air at 90°C and the second stage using air at 20°C and relative humidity 30–40% because the crystals are hygroscopic.

About 20% of the mother liquor is diluted with equipment cleansing water, decolorized, and fed back to the liming stage, while the remainder is decolorized and demineralized before being recycled to the crystallization unit. By following this procedure, the citric acid crystals need no additional purification in order to meet the specifications of the US Pharmacopeia (USP) or the Food and Chemical Codex (FCC).

In the liquid extraction process, the citric acid is extracted from the fermentation broth with a mixture of trilaurylamine, n-octanol and C_{10}- or C_{11}-isoparaffin. The extract is then heated and washed with water in countercurrent, yielding an aqueous citric acid solution. This is passed through a granular activated-carbon column, before undergoing the evaporation and crystallization steps described above.

Future Developments in Citric Acid Production

The production of citric acid has not received much attention in the form of modern methods of molecular biology, presumably because it is considered a mature area. The improvement of strains of *A. niger* used to be done by mutagenesis and selection, but metabolic engineering is now possible, in the light of increased knowledge about the regulation of acidogenesis.

The replacement of batch fermentations with semicontinuous processes would increase volumetric productivity and reduce specific production costs, but is presently precluded because of the deterioration of the mycelial structure, the mechanism of which is unknown. The effect of ammonium ions on prevention of inhibition of PFK_1 by citrate was discussed in the section on bysynthesis. Moreover, when the level of intracellular NH_4^+ falls below 1 mmol per g of dry cell, citric acid production stops and morphological degeneration, sporulation, as well as production of pigments, may result. This may be prevented by adding low levels of ammonium ions ($30 \, g \, m^{-3}$, equivalent to 2 mmol of intracellular NH_4^+ per g of dry cell) during fed-batch fermentations. The confirmation of such findings, at least on a pilot-fermenter scale and under prolonged operating conditions, could radically modify the current citric acid production technology. Similarly, the possibility of keeping microbial cells active and controlling their growth and production processes for several weeks or months, by immobilization within organic or inorganic matrices, represents a further challenge.

The traditional recovery processes give rise to several problems associated with the disposal of liquid effluents (their chemical oxygen demand being about $20 \, kg \, m^{-3}$) and solid by-products (i.e. about 0.15 kg of dried mycelium and 2 kg of gypsum per kilogram of anhydrous citric acid). Several alternative processes have been suggested to minimize the overall environmental impact of this recovery. The replacement of molasses with raw or hydrolysed starch-based materials, or raw sucrose-based materials, would simplify only the downstream processing. However, citric acid could be separated in a single step if tricalcium citrate were recovered from clarified, decolorized fermentation broths by electrodialysis, and citric acid were adsorbed on to weakly basic anionic-exchange resins or zeolites, using simulated moving-bed chromatographic technology, followed by desorption with water or dilute acidic solutions, or the use of liquid membranes. This approach would also avoid the formation of solid wastes, and hence the need for their for disposal.

Gluconic Acid

D-Gluconic acid (1,2,3,4,5-pentahydroxypentane-1-carboxylic acid: $C_6H_{12}O_7$) is an oxidation product of D-glucose. In aqueous solution it tends to esterify, leading to a complex equilibrium between gluconic acid and its two lactones, 1,5-lactone (D-glucono-δ-lactone) and 1,4-lactone (D-glucono-γ-lactone).

D-Gluconic acid is available commercially as a 50% aqueous solution (density $1230 \, kg \, m^{-3}$ at 20°C; pH 1.82). The acid and its derivatives are commonly applied in the pharmaceutical, food, feed and chemical industries, because of their extremely low toxicity and their ability to form water-soluble complexes with metallic ions (e.g. Ca^{++}, Fe^{+++}), especially in the presence of 5–10% sodium hydroxide. Sodium gluconate is the main industrial product. It is used as a sequestering agent (e.g. in bottle washing, cleaning the surface of metals and rust removal), and also to plasticize and retard the curing process of cement mixes. Calcium and iron gluconates are used to treat diseases caused by deficiencies of the corresponding cations (e.g. osteoporosis and anaemia). D-Glucono-δ-lactone is used as a latent acid, in baking powder for use in dry cake mixes, in meat processing and in 'instant' chemically leavened bread mixes. D-Glucono-γ-lactone is only made in small quantities, as a speciality.

Synthesis of Gluconic Acid

The conversion of glucose to gluconic acid is a simple oxidation process and chemical methods have been used in manufacturing, but fermentation is the current method of choice.

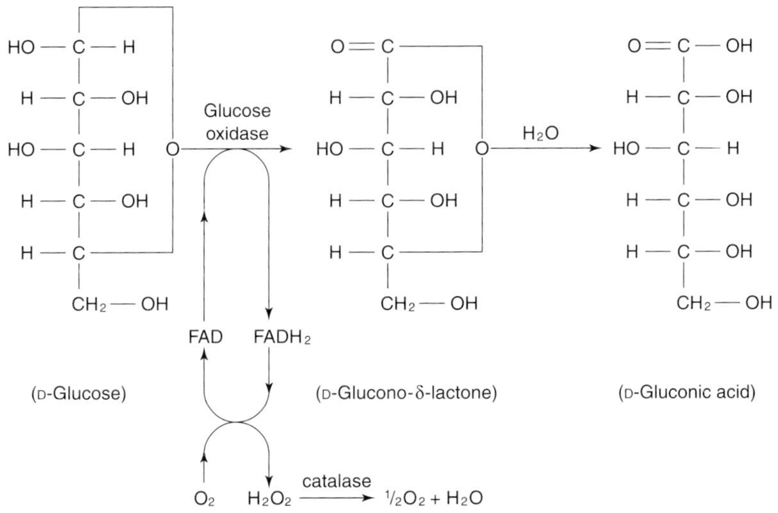

Figure 5 Synthesis of gluconic acid by *Aspergillus niger*. (FAD and FADH₂ are the oxidized and reduced forms respectively of flavin adenine dinucleotide.)

Calcium gluconate was first isolated in 1880 from a fermentation of glucose by a strain of *Mycoderma aceti*, in presence of calcium carbonate. The industrial-scale production of gluconic acid was started in 1923 by Chas. Pfizer & Co., New York. Subsequently, research at the US Department of Agriculture led to the semi-continuous production of sodium gluconate from glucose using *A. niger* NRRL 67. Several filamentous fungi of the genera *Penicillium* and *Aspergillus*, as well as bacteria (e.g. *Gluconobacter suboxydans*, *Pseudomonas ovalis*, *Gluconobacter* spp.), produce gluconic acid from glucose-based media. However, industrial processes have only been developed for the production of gluconic acid from glucose syrups using *Aspergillus niger*. The industrial strains are not freely available, but a few gluconic acid-producing strains (*A. niger* NRRL 3 and NRRL 67, and *Penicillium variabile* P 16 (University of Tuscia)) can be obtained from international culture collections. *Penicillium* species generally produce less gluconate than *Aspergillus*, but they have the advantage of excreting glucose oxidase (an important by-product) into the medium. This makes the enzyme recovery easier.

The formation of gluconic acid by *A. niger* is outlined in **Figure 5**. The enzyme glucose oxidase, that contains two flavin adenine dinucleotide (FAD) moieties, is responsible for removing two hydrogen atoms from glucose, yielding FADH₂ and glucono-δ-lactone. This is hydrolysed, to a variable extent, to gluconic acid. The FADH₂ reacts with O₂ to form hydrogen peroxide, which is converted into O₂ and water by the enzyme catalase. Both glucose oxidase and catalase are constitutive endoenzymes in *A. niger*.

Table 4 Composition of the production media used in the production of gluconic acid by *Aspergillus niger*

Component	Vegetative seed culture media	Gluconic acid production media
Glucose	40 kg m⁻³	120–350 kg m⁻³
NH₄NO₃ (or other NH₄⁺ salt)	2.4 kg m⁻³	0.4–0.5 kg m⁻³
KH₂PO₄	1.5 kg m⁻³	0.1–0.3 kg m⁻³
MgSO₄.7H₂O	3.57 kg m⁻³	0.1–0.3 kg m⁻³
Agar	1 kg m⁻³	0 kg m⁻³
Yeast extract	1 kg m⁻³	0 kg m⁻³
Corn steep liquor	0 kg m⁻³	0.2–0.4 kg m⁻³
ZnSO₄.7H₂O	100 mg m⁻³	0 mg m⁻³
CuSO₄.5H₂O	20 mg m⁻³	0 mg m⁻³
FeCl₃.6H₂O	300 mg m⁻³	0 mg m⁻³
MnSO₄.7H₂O	0 mg m⁻³	30 mg m⁻³
Initial pH	6.5	6.0

Manufacture of Gluconic Acid

Only the production of sodium gluconate by batch submerged fermentation from glucose syrups, using *A. niger* is described in this article. Glucose syrups of 70°Brix strength are generally used as the carbon sources in the preparation of the fermentation medium. **Table 4** lists the typical composition of the seed-culture and production media used in the laboratory and in industrial-scale trials.

The pH is first adjusted to 4.5 with sulphuric acid, and then the medium is sterilized at 121°C for 15–30 min, cooled to 33°C, and charged into the fermentation vessel. The pH is then raised to 6–6.5 with sodium hydroxide. A 2–5% inoculum is generally used. For the development of the inoculum, conidia are recovered from the stock agar slants and inoculated into vegetative seed culture medium (10⁶ conidia per cubic centimetre). After incubation at 30°C for

15–24 h, pellet-like mycelia are obtained and used to inoculate the seed fermenters at a concentration of 20–50 pellets per cubic centimetre.

Fermentation is carried out with continuous automatic control of: the sterile air flow rate (1.0–1.5 vv^{-1} min^{-1}), the temperature (33°C); the pressure on the top of the tank (2–3 bar); the pH (5.5–6.5); and the foam level. The pH is controlled by the addition of 30–50% NaOH solution, to neutralize the gluconic acid formed. The fermentation is complete within about 30 h, with yields of 0.97–1 kg of gluconic acid per kilogram of glucose consumed (the theoretical yield being 1.09 kg kg^{-1}) and gluconate productivities of 9–13 kg m^{-3} h^{-1}.

In fed-batch fermentations, the mycelium can be reused up to five times without loss of gluconate productivity, provided that the activity of glucose oxidase and the levels of micronutrients (i.e. iron and manganese) are kept under control. The recovery of the mycelium can be performed by allowing the mycelium to form a sediment at the bottom of the vessel in the absence of agitation and aeration, by aseptic centrifugation or by cross-flow microfiltration. The stepwise addition of glucose may increase the gluconate concentration in the fermented broth to 580 kg m^{-3}.

At the end of fermentation, the mycelium is removed using continuous vacuum belt filters and may be used as a source of glucose oxidase or disposed via incineration. The clarified broth, generally containing ca. 300 kg m^{-3} sodium gluconate, is filtered, decolorized using a granular activated-carbon column, concentrated under vacuum to 45–50% total solids, neutralized to pH 7.5 with NaOH, and then spray- or drum-dried.

If 50% gluconic acid is required, the concentrated liquor may be passed through a cation-exchanger to remove Na$^+$ ions. Further crystallization at 30–70°C or >70°C allows crystals of the δ-lactone or γ-lactone to be precipitated, respectively.

Lactic Acid

Lactic acid (2-hydroxypropionic acid: $C_3H_6O_3$) is currently produced by chemical synthesis or fermentation (Table 1). Of the two enantiomers, L-(+) and D-(–)-lactic acid, only the L-(+) isomer is used by human metabolism and, because of the slight toxicity of the D-(–) isomer, it is preferred for food use. Nevertheless, synthetic racemic (DL) lactic acid is the primary commercial form. The free acid is used as an acidulant/preservative in several food products (cheese, meat, jellies, beer); ammonium lactate is used as a source of non-protein nitrogen in feeds; sodium and calcium stearoyl lactylates are used as emulsifiers and dough conditioners. Interest in lactic acid production has

recently increased because of its use in the synthesis of biodegradable, biocompatible plastics and coatings.

Organisms/Metabolic Pathways

All lactic acid bacteria produce lactic acid by fermentation of sugars, but only homofermentative species are used.

Thermophilic lactobacilli (*Lactobacillus delbrueckii* subsp. *delbrueckii*, *L. delbrueckii* subsp. *bulgaricus*, *L. helveticus*) tolerate higher concentrations of lactate and higher temperatures (48–52°C) than mesophilic strains, involving higher productivity and yields and reduced contamination risks. However, they produce D-(–) or DL-lactic acid (some industrial strains which have been claimed to be *L. delbrueckii* produce L-(+)-lactic acid) and may be less suitable for food, feed or biomedical applications. Lactococci, mesophilic lactobacilli (*L. casei* subsp. *casei*, *L. amylophilus*) and thermophilic streptococci (*Streptococcus thermophilus*) have lower optimal temperatures and/or reduced acid tolerances, but they may be desirable for other reasons (production of pure L-(+)-lactic acid, hydrolysis of starch)

Homofermentative lactic acid bacteria ferment hexoses via the glycolytic pathway. Pyruvate is reduced to lactate by stereospecific lactate dehydrogenases (L-LDH and/or D-LDH). LDH is allosteric (activators : FDP and Mn^{++}) in lactococci and non-allosteric in homofermentative lactobacilli. Undissociated lactic acid acts as a non-competitive inhibitor for growth and lactic acid production by diffusing through the membrane and decreasing intracellular pH: pH control during fermentation reduces the inhibition, but the maximum lactic acid concentration achievable is usually smaller than 150 kg m^{-3}. Concomitant substrate and product inhibition has been reported for *L. delbrueckii*.

Lactic acid bacteria are fastidious microorganisms and require fermentation media supplemented with peptides (those made of 5–12 amino acids are most effective) and growth factors. *Bacillus* spp. can be used in place of lactic acid bacteria because of their reduced growth requirements and their capability in utilizing inorganic nitrogen sources. Even the mould *Rhizopus oryzae* may be used, but it also produces ethanol, resulting in lactic acid yields lower than 75%.

Substrate/Production/Recovery

Lactic acid can be produced from a variety of raw substrates (whey and whey permeate, beet and cane molasses, corn starch hydrolysates, wood hydrolysates). The liquor resulting from corn wet milling is the carbon source used by the largest manufacturers, such as ADM and Purac Biochem.

Some species (*L. amylophilus*) can hydrolyse starch,

Table 1 Commercial applications of fats and oils

(A) Products (edible)	Oil source
Margarine	Soybean oil, groundnut oil,
Cooking fat	cottonseed oil, sunflower
Cooking oils	oil, rapeseed oil, sesame oil,
Salad oils/mayonnaise/table oils	palm oil, some fish oils,
Ice cream	olive oil, castor oil,
Confectionery	lard and tallow
Pharmaceuticals	Coconut oil, palm kernel oil, castor oil

(B) Products (non-edible)	Oil source
Detergents and surfactants	Palm kernel, coconut oil
Soaps, metallic soaps, synthetic waxes	Palm oil
Paints and coatings	Linseed oil, tung oil, soybean oil, sunflower oil
Varnishes and lacquers	Linseed oil, tung oil
Inks	Various, mainly castor oil
Plastics and additives	Various, mainly soybean oil
Lubricants and cutting oils	Castor oil, coconut oil
Wood dressings, polishes	Tung oil
Leather dressing	Fish oils
Metal industry	Palm oil and tallow
Agrochemicals, long-chain quaternary compounds as herbicides, insecticides and fungicides	Various, mainly soybean oil
Evaporation retardants	Fatty alcohols from any source
Fabric softeners	Tallow

Table 2 Fatty acyl composition of lipids extracted from oleaginous yeasts of commercial interest

Strain	Relative % w/w of fatty acyl composition						Oil coefficient g oil accumulated / g lactose consumed
	C14:0	C16:0	C18:0	C18:1	C18:2	C18:3	
Candida 107	0.8	27.6	7.8	40.6	16.5		0.149
Apiotrichum curvatum ATCC 20509	0.4	29–32	9–11	47–51	7–7	0.5	0.202
Cryptococcus albidus IBFM Y-229	0.2	19.6	11.4	59.4	6.1	2.3	0.146

Metabolic Pathway

The ultimate precursor for the biosynthesis of saturated fatty acids is acetyl-CoA, which is derived from carbohydrate or amino acid sources. **Figure 1** shows how the intermediary metabolism is linked to fatty acid biosynthesis in oleaginous microorganisms.

The fatty acid synthesis is catalysed by a group of seven proteins – the fatty acid synthetases complex – in the cytosol. The usual end product is palmitic acid, which is the precursor of the other long-chain, saturated and unsaturated, fatty acids found in most microorganisms. Acetyl-CoA supplies only one of the eight acetyl units needed for the biosynthesis of palmitic acid ($CH_3(CH_2)_{14}COOH$); the other seven are provided in the form of malonyl-CoA.

The overall reaction is:

$$\text{Acetyl-CoA} + 7 \text{ malonyl-CoA} + 14\text{NADPH} + 14\text{H}^+ \rightarrow$$
$$\text{palmitic acid} + 7\text{CO}_2 + 8\text{CoA} + 14\text{NADP}^+ + 6\text{H}_2\text{O}$$

Figure 2 shows the steps involved in the biosynthesis of a triglyceride.

A correlation has been observed between possession of the enzyme ATP:citrate lyase and the ability of a yeast to accumulate more than 20% of its biomass as lipid. The significance of the enzyme is that it serves to produce the substrate for fatty acid biosynthesis, acetyl-CoA, from citrate:

$$\text{Citrate} + \text{ATP} + \text{CoA} \rightarrow$$
$$\text{acetyl-CoA} + \text{oxaloacetate} + \text{ADP} + \text{P}_i$$

Acetyl-CoA cannot be produced in the cytoplasm from pyruvate (this reaction proceeds in the mitochondria). Oleaginous yeasts and, probably, other oleaginous eukaryotic microorganisms accumulate citrate in the mitochondria which is then transported into the cytoplasm and there cleaved by ATP:citrate lyase.

As prokaryotic microorganisms do not have the compartmentalization of the mitochondrion to separate acetyl-CoA formation from acetyl-CoA utilization for fatty acid biosynthesis, there is no need for ATP:citrate lyase. Thus the absence of this enzyme in oleaginous bacteria has no biochemical significance.

Table 3 Analysis of bacterial cell mass for fat and total lipid composition

Species	Fat (%) dry weight	Total lipid (%) dry weight
Agrobacterium tumefaciens		
Grown on glycerol	1.8	2.1
Grown on glucose	0.9	6.1
Bacillus megaterium		21.0
B. subtilis	2.0	6.1
Bordetella pertussis	22.0	24.0
Brucella abortus	0.9	6.1
B. melitensis	4.8	5.3
B. suis	2.4	5.6
Corynebacterium diphtheria	6.3	8.8
Escherichia coli	4.0–5.0	4.0–6.0
Lactobacillus acidophilus	4.8	7.0
L. arabinosus		0.5
L. casei		0.6
Malleomyces mallei	5.0–7.0	6.0–8.0
Mycobacterium avium	2.2	15.2
M. phlei	3.0	7.0
M. tuberculosis var. bovis	3.3	14.3
M. tuberculosis var. humanis Ra.	5.6	16.1
M. tuberculosis var. humanis Rv.		21.1
Salmonella paratyphi C	2.8	3.6
S. typhosa	1.3	1.5

Patterns of Lipid Accumulation in Fermentations

Many eukaryotic microorganisms increase their lipid content if they are depleted of a nutrient provided that the supply of carbon to the cell stays plentiful. The course of lipid accumulation follows a biphasic pattern in batch cultures (**Fig. 3A**). In the first phase, when all nutrients are present in excess, there is a period of balanced growth during which the lipid content of cells stays approximately constant. This ends when a nutrient – usually nitrogen – is exhausted. There then follows an interim period of about 6 h during which the nitrogen pool (mainly amino acids) within the cells becomes diminished. When the cell is completely devoid of any further utilizable nitrogen, protein and nucleic acid biosynthesis ceases although the existing cell machinery continues to take up glucose and metabolize it into lipid. There is therefore a continued build-up of lipid in the second phase without much increase in total cell population.

With some of the slower-growing moulds, a different pattern has occasionally been reported where the rate of lipid accumulation appears to coincide

Table 4 Fermentative production of fats by yeasts

Yeast species	Substrate used in fermentation	Fat content (% dry wt of biomass)	Fat coefficient (g fat produced per 100 g substrate consumed)
Candida sp. no. 107	Glucose	42	22.5
	n-Alkanes	15–37	25
C. guilliermondii	n-Alkanes	30	
C. intermedia	n-Alkanes	20	
C. tropicalis	n-Alkanes	32	
Cryptococcus terricolous	Glucose	55–65	21
Hansenula anomala	Glucose	17	
H. ciferrii	Molasses	22	
H. saturnus	Molasses	20	
	Glucose	28	8
Lipomyces sp.	Glucose	67	20
	Xylose	48	17
	Various wastes and molasses	66	≤24
L. lipofer	Glucose	38	
	Peat moss hydrolysate	48	
L. starkeyi	Lactose	31	10
	Glucose	31–38	9–15
Rhodotorula glutinis		30–35	9–15
R. gracilis	Molasses	40	44 (short time only)
	Glucose	64	
	Sugar-cane syrup	67	21
	Glucose	64	15 overall; 44 (short time only)
	Ethanol	62	15
	Synthetic ethanol	60	14
	Glucose	66	17
	Alkanes	32	
R. longissima		20	

Table 5 Fermentative production of fats by fungi

Strain	Substrate used in fermentation	Fat content (% dry wt of biomass)	Fat coefficient (g fat produced per 100 g substrate consumed)
Aspergillus fischeri	Sucrose	32–53	12–20
A. fumigatus	Maltose and other sources	20	
A. nidulans	Glucose	27	9
	Glucose	15	7
A. ochraceus	Sucrose	48	13
A. terreus	Sucrose	51–57	10–13
	Starch	18–24	6
A. ustus	Lactose	36	12.7
Chaetomium globosum	Glucose	54	
Cladosporium fulvum	Sucrose	22–24	7
C. herbarum	Sucrose	20–29	7–11
Giberella fujkuroi (Fusarium moniliforme)	Glucose	45	7.8
Malbranchea pulchella	Glucose	27	
Mortierella vinacea	Acetate	28	
	Glucose	66	18
	Maltose	34	
Mucor miehei	Glucose	24	
M. pusillus	Glucose	26	
Myrothecium sp.	Not given	30	
Penicillium funiculosum	n-Alkanes	22	
P. gladioli	Sucrose	32	5.7
P. javanicum	Glucose	39	9
P. lilacinum	Date extract	23	
	Sucrose	35	25
P. soppi	n-Alkanes, sucrose	11–25	
	molasses	19	
P. spinulosum	Molasses, sucrose	25–64	6–16
Pythium irregulare	Glucose	30–42	
P. ultimum	Glucose	48	
Rhizopus sp.	Glucose	27	
R. arrhizus	Glucose, maltose	20	
Stibella thermophila	Glucose	38	

Table 6 Intracellular contents of AMP and ATP under steady conditions in continuous culture in oleaginous yeast Candida 107, and non-oleaginous yeast Candida utilis

Condition of growth	Dilution rate (h^{-1})	Candida 107 (nmol mg^{-1} cell wt)			Candida utilis (nmol mg^{-1} cell wt)		
		AMP	ATP	AMP/ATP	AMP	ATP	AMP/ATP
N-limiting (lipogenic)	0.05	1.6	2.2	0.73	3.5	0.7	5.0
	0.085	0.5	1.2	0.42	3.8	.0.9	4.2
C-limiting	0.05	9.0	0.5	18.0	3.9	1.6	2.4
	0.085	9.6	0.5	19.0	6.2	1.2	5.1

with the growth rate (Fig. 3B). In continuous culture (Fig. 3C), lipid accumulation can be achieved under nitrogen-limited conditions and at a slow dilution rate (specific growth rate) to allow the organism time to assimilate the carbon which is available to it within the fermenter. The specific rate of lipid biosynthesis is expressed as grams of lipid synthesized per gram of lipid-free cell weight per hour. The rate of lipid synthesis stays approximately constant but assumes a greater proportion of the cell's activity as the growth rate declines.

Substrates for Oleaginous Fermentation

Hydrocarbons

Production of lipids from hydrocarbons has been considered for technical uses. Hydrocarbons, and alkanes in particular, have the advantage over other substrates in that they can predetermine the chain length of the ensuing fatty acids found in the extracted lipids (**Fig. 4**). This may be a considerable advantage if a lipid with particular fatty acid substituents is wanted for any reason. Hydrocarbons, in general, also lead to

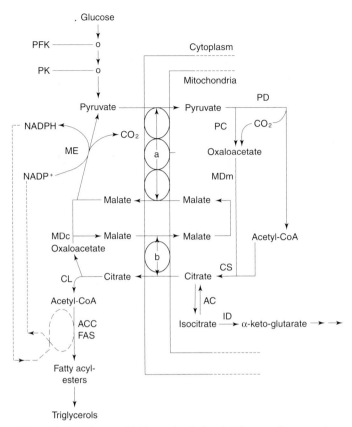

Figure 1 Intermediary metabolism linked to fatty acid biosynthesis in oleaginous microorganisms. Enzymes: ACC, acetyl-CoA carboxylase; AC, aconitase; CL, ATP : citrate lyase; CS, citrate synthase: FAS, fatty acid synthetase complex; ID, isocitrate dehydrogenase (inoperative due to lack of AMP as cofactor); MDc, malate dehydrogenase (cytosolic); MDm, malate dehydrogenase (mitochondrial); ME, malic enzyme; PC, pyruvate carboxylase; PD, pyruvate dehydrogenase; PFK, phosphofructokinase; PK, pyruvate kinase. Mitochondrial transport processes: a, interlinked pyruvate–malatetranslocase systems; b, citrate–malate translocase.

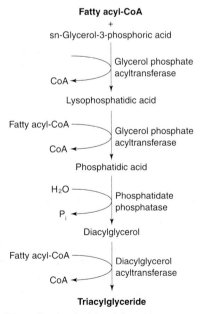

Figure 2 Biosynthesis of a triacylglyceride.

greater production of lipid, as a percentage of the cell biomass, than do carbohydrates. This may again be of advantage where a product such as wax may be wanted but is normally only found as a small percentage of the total biomass.

In addition to being useful for the production of specific fatty acids which are then recoverable as triacylglycerols or phospholipids, hydrocarbons can lead to the production of both ω and ω-1 hydroxy fatty acids and dicarboxylic acids.

Fatty Acids, Soapstocks and Oils

Desirable lipids can be produced by cultivating appropriate yeasts, usually of the genus *Candida*, *Torulopsis* and *Trichosporon*, though *Saccharomyces cerevisiae* and *C. utilis* have also been used on a mixed carbon source which includes a fatty acid or material containing a fatty acid. The fatty acids or oils may be up to $20 \, \mathrm{g \, l^{-1}}$ in the growth medium and, like alkanes, these then lead to high lipid contents: up to 65% and 67%. High relative percentages of stearic acid (if stearic acid had been included in the medium) may also be achieved in the yeast oil. This can then lead

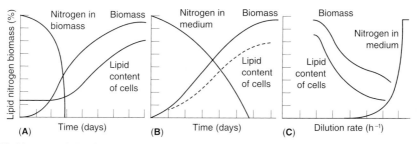

Figure 3 Patterns of lipid accumulation in eukaryotic microorganisms: (**A**) in a typical oleaginous organism growth in batch culture; (**B**) in an organism in which lipid accumulation parallels the growth rate; (**C**) in an oleaginous organism grown in continuous culture.

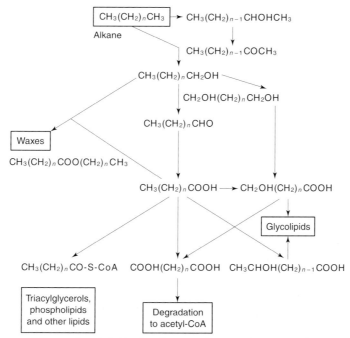

Figure 4 Pathways of alkane oxidation and assimilation in microorganisms. $n = 8\text{--}16$, especially 11–14.

to a yeast lipid which has some of the characteristics of cocoa butter. The possibilities of upgrading the cheap vegetable oils, such as palm oil, to more expensive materials have been considered. Such a process relies upon the various lipases of the organism carrying out the initial hydrolysis; the ensuing fatty acids are then incorporated directly into new triacylglycerols in the same manner as occurs when alkanes are used as substrate. Lipases may be isolated from microorganisms and, as immobilized preparations, then used to carry out transesterification reactions either between two different oils or between an oil and a fatty acid to produce high value-added commodities such as cocoa butter.

Other Substrates

These include various starchy crops and wastes, molasses, whey, peat hydrolysates and ethanol. The use of hydrolysed cellulose, which includes a wide range of materials such as peat, is no different from using carbohydrate. The utilization of pentoses arising from the hydrolysis of hemicelluloses is not detrimental for the formation of lipids and thus these materials could be considered to be convenient cheap substrate. The emphasis of all work with fermentation substrates focuses primarily on availability and cheapness. Very inexpensive, or even negative-cost, substrates available throughout the year are needed for the production of low-cost lipids. In several countries such a substrate could be whey. Lactose-utilizing oleaginous yeasts are well known. The best organism for this work is considered to be *C. curvata*. This process would obviously be worth consideration for whey disposal.

Microbial Production Systems

To produce high yields of oils, oleaginous fungi are grown in a medium containing a carbon source, but one in which another nutrient (usually nitrogen) is limiting. Exhaustion of the limiting nutrient brings to an end protein and nucleic acid synthesis by the fungus, while fats continue to be synthesized. Fat-producing fermentations can involve batch or con-

Table 7 Composition of some fermentation media for lipid biosynthesis

Component	Amount (g l⁻¹)
Cultivation of *Aspergillus fischeri*	
Commercial glucose	200
Ammonium nitrate	10
Potassium dihydrogen phosphate	6.8
Magnesium sulphate, 7H₂O	5.0
Ferric chloride, 6H₂O	0.16
Zinc sulphate, 7H₂O	0.05
Cultivation of *Aspergillus nidulans*	
Sucrose	170
Ammonium nitrate	3
Potassium sulphate	0.22
Zinc sulphate, 7H₂O	0.05
Sodium dihydrogen phosphate, 2H₂O	7.3
Magnesium sulphate, 7H₂O	5
Cultivation of *Penicillium javanicum*	
Glucose	200
Ammonium nitrate	2.25–3.4
Potassium dihydrogen phosphate	0.3–1.2
Magnesium sulphate, 7H₂O	0.25
Cultivation of *Rhodotorula gracilis*	
Glucose	30–40
Potassium nitrate	1.42
Diammonium hydrogen phosphate	0.33
Magnesium sulphate, 7H₂O	1.00
Corn steep liquor (50% dry wt)	0.05 (optional)
Airflow rate, 1 vvm; temperature, 28–29°C, pH 5.5–6.0	

tinuous culture. Yield efficiencies for oil production in general tend to be low, around 22–24%, so the optimum content of oil produced is around 40%. A wide variety of microorganisms produce oils, but sound economics dictate that preference is given to organisms that produce both high-quality oils and a single-cell protein, thereby giving the process an edge over alternative sources of oils.

Both yeasts and filamentous fungi have been evaluated for their ability to produce oils (Tables 4 and 5). The lipids extracted from yeasts consist primarily of triacylglycerols, which are also the main component of plant oils. The major fatty acids found in yeasts, in order of abundance, are oleic, palmitic, linolenic and stearic acids – a complement of fatty acids similar to that found in plants. Oleaginous fungi produce a more diverse range of lipid types and fatty acids. Some moulds contain high proportions of shorter-chain fatty acids (C12 and C14), while others contain high levels of mainly polyunsaturated acids (C18.2–C18.3).

Fermentation Media for Lipid Biosynthesis

Although the biosynthesis of lipids occurs throughout the growth of microorganisms, the accumulation of fat only begins when a nutrient other than the carbon source is exhausted. It follows that fat accumulation usually occurs at the end of growth, therefore the medium should be formulated to sustain rapid and extensive growth initially. The carbon source can be glucose, sucrose or other carbohydrates, ethanol or *n*-alkanes. The nutrient that becomes exhausted from the medium is usually the nitrogen source. Depletion of other nutrients (e.g. phosphate, sulphate and iron) can also induce fat formation in some microbial species. The optimum medium formulations for culture of some microorganisms are given in **Table 7**; glucose in *Rhodotorula gracilis* culture can be replaced by ethanol (2% vv⁻¹) or *n*-alkanes (20 g l⁻¹). **Tables 8–9** show the role of medium composition in fat biosynthesis.

Fermentative Production Processes

Most microorganisms start accumulating fat after the initial growth phase, hence batch culture is usually preferred. Two-stage continuous fermentation can be used to produce fat; the first stage is microbial growth and the second is fat accumulation.

Lipid accumulation is favoured by oleaginous microorganisms growing in a medium with a high carbon to nitrogen ratio, usually 50:1. In a batch culture, the organism grows until the nitrogen is consumed but thereafter continues to take up the excess carbon and convert this to lipid. Thus a biphasic growth pattern can be envisaged. With some of the slower-growing moulds, the rate of lipid accumulation appears to coincide with the growth rate. Lipid content of the cells increases at the same rate as growth proceeds.

In continuous culture, lipid accumulation is achieved by growing oleaginous microorganisms under nitrogen-limiting conditions at a dilution rate (specific growth rate) of about 30% of maximum. The build-up of lipid is dependent upon the correct balance being achieved between growth rate and the specific rate of lipid biosynthesis so that the optimum amount

Table 8 Effect of inorganic salt supplementation on fat biosynthesis in moulds grown on whey

Supplement	Mycelium (g dry wt)	Fat (%)	Fat coefficient
Aspergillus ustus			
None	0.73	14.0	4.8
0.2 mol l⁻¹ potassium sulphate + 0.1 mol l⁻¹ ammonium nitrate	0.81	20.0	3.8
Penicillium frequentans			
None	0.78	13.1	3.3
0.2 mol l⁻¹ potassium sulphate + 0.1 mol l⁻¹ ammonium nitrate	1.05	24.3	8.2

Table 9 Effect of nitrogen sources on fat biosynthesis by *Rhodotorula gracilis*

Nitrogen source	Aeration	Sugar consumed (%)	Fat content of medium (g l^{-1})	Yeast content (g l^{-1})	Fat content of yeast (%)	Fat coefficient[a]
Ammonium	−	78.9	3.7	9.0	47	12.3
sulphate	+	64.4	3.8	4.0	52	15.3
Asparagine	−	59.5	5.0	7.7	62	14.0
	+	80.4	7.5	11.7	72	18.4
Aspartic acid	−	69.9	5.8	8.9	65	15.8
	+	81.9	9.5	12.7	74	20.9
Urea	−	66.4	6.1	9.7	64	15.9
	+	71.7	6.3	9.6	67	16.2
Uric acid	−	79.2	5.1	7.9	64	11.9
	+	79.1	6.8	10.4	67	16.0

[a]g fat produced per 100 g substrate consumed.

of carbon can be diverted into lipid and the minimum into other cell components.

The efficiency of lipid accumulation in continuous culture is often the same as or better than in batch cultures where the same organism has been studied under both conditions. With *Candida* sp. 107, *R. glutinis*, *R. gracilis* and *C. curvata*, lipid yields of 17–22% have been obtained under both conditions of growth. A conversion of carbohydrate to lipid of 20% would appear to be near to a possible practical limit as the theoretical maximum is about 33 g triacylglycerol from 100 g glucose, assuming that all the carbon of the medium is converted into lipid without synthesis of any other cell component.

Factors Influencing Lipid Biosynthesis

Factors which affect the composition of the lipid within a cell also cause a rise or fall in the total amount of lipid within a cell. Any change in the growth condition of a microorganism can bring about a change in lipid composition in a batch culture. It is often stated that temperature affects the degree of unsaturation of the fatty acyl groups of the lipid; however, it must be borne in mind that lowering the temperature will slow the growth rate of the organism and simultaneously increase the amount of oxygen dissolved in the medium. Changes in both these conditions could then influence the metabolic status of the cells, resulting in the pH of the culture falling (or rising) and then the temperature effect cannot be interpreted. Therefore a chemostat where each growth condition can be changed *independently* of all the others is used.

Growth Rate

The influence of the growth rate on lipid accumulation in oleaginous microorganisms is a major determinant of the amount of lipid built up within the cells. Individual components of the phospholipids are subject to

striking alterations in relative proportions at different growth rates.

Substrate Concentration

Glucose has been the usual substrate whose concentration has been varied. Glucose, metabolized through the Embden–Meyerhof pathway, poses the most problems for eukaryotic microorganisms which can be divided into two groups according to whether they are Crabtree-positive or Crabtree-negative. In the former group, an increase in the concentration of glucose (or any other carbohydrate metabolized via the glycolytic pathway) results in repression of the synthesis of oxidative (respiratory) enzymes and is manifested by decreased oxygen uptake coupled with accumulation of a metabolically reduced intermediate such as ethanol. *S. cerevisiae* is a typical Crabtree-positive yeast; an increase in glucose concentration therefore brings about increased ethanol production even under aerobic conditions. The metabolic changes are probably associated with a decrease in mitochondrial components, which leads to a general decrease in total fatty acids, glycerophospholipids and sterol esters. In Crabtree-negative yeast, such as *C. utilis* and *S. fragilis*, there is an increase in lipid accumulated as the glucose concentration is increased. The increase in lipid is mainly triacylglycerol and, even though the total lipid content of such cells may still be only 20%, this serves to illustrate that modest lipid accumulation can still be achieved, even in non-oleaginous species.

Growth Substrate

Alkanes and other hydrocarbons profoundly influence lipid composition. Other growth substrates can change the amount of lipid accumulated as well as the fatty acids to be found within the lipid.

Temperature

The degree of unsaturation of the fatty acids of biological systems increases as the growth temperature is decreased in both prokaryotic and eukaryotic microorganisms. *S. cerevisiae* shows an increased content of triacylglycerols and glycerophospholipids at lower temperatures. The membrane lipids of thermophilic bacteria, capable of growth at 95°C and survival at perhaps 105°C, are considerably different from mesophilic organisms.

Oxygen

In eukaryotes, oxygen is required for the conversion of stearic acid to oleic acid and thence to linoleic and linolenic acids. Whilst there is little evidence to show that depriving a culture of oxygen leads to a decline in the amount of polyunsaturated fatty acids being produced, this need not always be the case as the outcome of oxygen deprivation will depend on the relative affinities of the fatty acid desaturase and, probably, cytochrome oxidase.

pH and Salinity

It is only at the extremes of pH and salinity that effects on lipid composition become strikingly evident. These are genotypic rather than phenotypic changes. Ordinarily the membranes of microorganisms are not capable of modification to allow them to survive in such environments and consequently the changes seen in these organisms tend to be minimal.

Other Factors

The influence of limiting the amount of nitrogen available to a culture can serve to increase the amount of lipid accumulated in oleaginous microorganisms. Limitations of other nutrients such as Mg, PO_4 may, or may not, bring about similar increases in lipid accumulation but they may bring about many changes in lipid composition independently of their effect on lipid accumulation. For example, phosphate limitation of growth of *Pseudomonas diminuta* results in the partial replacement of acidic phospholipids by acidic glycolipids. With *S. cerevisiae* sterol esters and triaclglycerol decline without any significant change occurring in the amounts of phosphilipids. Inositol deficiency with certain strains of *S. cerevisiae* and *S. carlsbergensis* leads to an increase in cell lipid. This effect is brought about by an increase in the activity of acetyl-CoA carboxylase. Thiamine deficiency in *S. carlsbergensis* produces a decline in content of all lipids: sterol esters, acylglycerols and all glycerophospholipids.

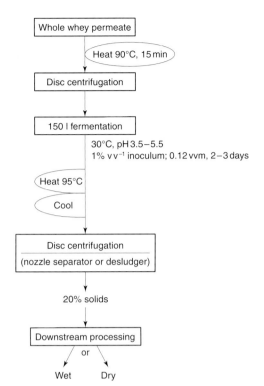

Figure 5 Pilot-scale production of yeast-oil fermentation.

Recovery and Purification

At pilot-scale level, in the development of an oil process by the fermentation of an oleaginous yeast *Apiotrichum curvatum* ATCC 20509 at a dairy factory, the best recovery method was to dry the concentrated yeast cream and extract the oil into hexane using an agitated bead mill. The initial scheme of operations to produce a kilogram of yeast oil is shown in **Figure 5**.

Oil Extraction

With oleaginous yeast breakdown of cell walls is necessary before solvents can efficiently remove the oil. Cell walls can be ruptured by autolysis, enzyme hydrolysis, acid or alkaline treatment and mechanical disintegration. Oil can be extracted either in wet downstream processing (**Fig. 6**) using ethanol : hexane and methanol : benzene with wet cells or in dry downstream processing (**Fig. 7**) drying a known quantity of washed yeast cells at 70°C for 24 h and extracting oil from the dry cell pellet with ethanol : hexane (1 : 1 v/v) using a high-speed disperser.

Oil Refining

Yeast oil can be refined using standard edible oil technology: acid degumming with phosphoric acid; alkali refining; bleaching; deodorization. Ethanol : hexane extracted oils must be alkali-refined to

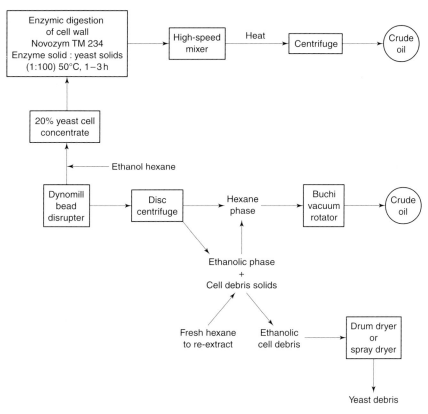

Figure 6 Pilot-scale production of yeast oil: wet downstream processing.

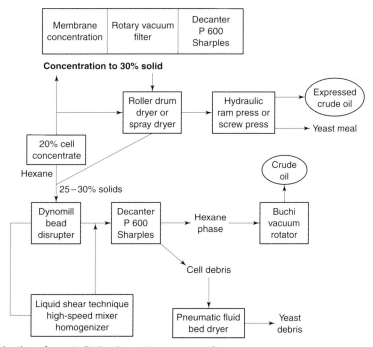

Figure 7 Pilot-scale production of yeast oil: dry downstream processing.

remove all the gums, soaps and surfactant material. Hexane-extracted oils are of higher quality and have the potential for economic refining. Oleochemical data before and after refining are presented in **Tables 10 and 11** with refined bleached deodorized (RBD) specification. Conditions of oil extraction are important with regard to refining; hexane extraction on dried yeast is the appropriate method.

Table 10 Oleochemical data of crude degummed yeast oil extracted from pilot-scale fermentation

Extraction method	Phosphorus content (p.p.m.)	Free fatty acid as oleic (%)	Peroxide value (mmol l^{-1} kg^{-1}	Colour lovibond 5 1/4"
Wet yeast (ethanol/hexane)	27	0.50	4.0	1.2R 12Y
Wet yeast (hexane)	18	0.51	3.0	1.2R 12Y
Dry yeast (ethanol/hexane)	24	1.35		2.5R 25Y
Dry yeast (hexane)	6	0.09	3.9	2.5R 25Y

Proposed specification for refined bleached deodorized yeast oil

Moisture and impurities	0.1% max
Lovibond colour (5 1/4")	30Y 30R max
Free fatty acid (as oleic)	0.1% max
Melting point (Barnicoat drop pt)	20–23°C
Iodine value	50–55
Specific gravity at 25°C	0.91–0.92
Peroxide value	4.0 mmol l^{-1} kg^{-1} max
Flavour	Free from foreign, rancid odours or flavours

Table 11 Oleochemical data of refined bleached deodorized yeast oil extracted from pilot scale fermentation

Extraction method	Phosphorus content (p.p.m.)	Iodine value	Peroxide value (mmol l^{-1} kg^{-1})	% Free fatty as oleic	Lovibond colour 5 1/4"	Melting point
Wet yeast (ethanol/hexane)	4	55	0.8	0.07	1.2R 12Y	21.4
Wet yeast (hexane)	6	57	0.4	0.08	1.2R 12Y	20.4
Dry yeast (ethanol/hexane)	9		0.2	0.15	1.5R 15Y	
Dry yeast (hexane)	0.2	51.5	1.0	0.10	2.5R 25Y	22.3

Commercial Importance

Microbial lipids are produced as part of a defatting process of yeasts grown on hydrocarbons and, although these lipids have interesting properties, there seems to be little likelihood of similar materials being deliberately produced elsewhere for reasons of cost.

The prospects for microbial oils lie in three possible areas: as substitutes for high-value plant oils; as novel materials which are unavailable from other sources; and as a saleable end product from waste processing. Microbial lipids which are novel materials do exist but similar materials can often be produced chemically, and this usually means cheaper.

Certain fungi also contain unusual oils and have been evaluated as a source of dietary essential fatty acids. *Mucor javanicus* and *M. isabellana* are used to produce polyunsaturated fatty acids of dietary importance, such as γ-linoleic acid (the main component of evening primrose oil, recommended to help women suffering from premenstrual tension). Large-scale production of this fatty acid is achieved using potato-paste dextrose as the substrate. Icosapentaenoic acid, which principally occurs in fish oils, can also be produced by a strain of *Mortierella alpina* in yields of up to 20% of the total fatty acid content.

Yeasts also have the potential to act as a commercial source of cocoa butter-like fats. Cocoa butter is probably the most expensive of all bulk oils and fats. The yeast *C. curvata* can be grown on wastes like whey, and by using the lactose component this produces a potentially useful triglyceride oil. By modifying the process, this yeast can be induced to make large amounts of stearic acid, a useful alternative to cocoa butter for use in the production of cosmetics and confectionery. Any increase in the price of cocoa butter is likely to make fatty acid production by yeast an economically viable proposition.

Crude oil has direct commercial applications in soap-making, animal feedstuffs and in textile lubricants. Refining is required to stabilize and purify the oil for wider uses in cosmetics, creams and lotions, food lubricants, food texture modifiers and blending oils.

Defatted microbial biomass, like its oilseed meal counterpart, may be sold as animal fodder. Single cell oil (SCO) has the advantage over single cell protein as SCO can be used for technical purposes without expensive toxicological trials.

See also: **Candida**: Introduction. **Mucor. Saccharomyces**: *Saccharomyces cerevisiae*; *Saccharomyces Brewer's Yeast.* **Torulopsis. Yeasts**: Production and Commercial Uses.

Further Reading

Erwin JD (1973) *Lipid and Biomembranes of Eucaryotic Microorganisms*. London: Academic Press.

Finnerty WR and Makula PA (1975) Microbial lipid metabolism. *Crit. Rev. Microbiol.* 4: 1–40.

Gunstone FD, Harwood JL and Padley FB (1994) *The Lipid Handbook*, 2nd edn. London: Chapman & Hall.

Gurr MI and James AT (1975) *Lipid Biochemistry: An Introduction*. London: Chapman & Hall.

Moo-Young M (ed.) (1985) *Comprehensive Biotechnology* Vol. 1. Oxford: Pergamon Press.

Morton RS (ed.) (1988) *Singe Cell Oil*. Sussex: Longman Scientific & Technical.

Weete JD (1980) *Lipid Biochemistry of Fungi and other Organisms*. New York: Plenum Press.

Colours/Flavours Derived by Fermentation

R G Berger, Food Chemistry, University of Hannover, Germany

The important nutritional, hygienic and technical characteristics of a food cannot be reliably assessed without analytical instrumentation. Colour and flavour (which includes odour, taste and texture in scientific Anglo-Saxon usage) are the signals that are immediately perceived by the optical and chemical senses of humans, and these attributes determine whether a certain food is appealing. Attractive food colours and flavours are usually translated into increased consumption, and this is a fundamental behavioural response rather than a simple emotional decision. However, colours and flavours are often sensitive to heat, oxygen, light and acid and, thus, changed or lost during processing and storage. Natural colourants and flavours, mainly derived from plants, and chemosynthetic compounds are used by the food industry to replenish and sometimes raise the genuine stock. Once the strong interaction of optical and chemical perception was recognized,

multifunctional food additives were developed. A beverage base, for example, may contain *Citrus* flavour, carotenoid preparation, and possibly a polysaccharide thickener and emulsifier. In a diversifying market of nutraceuticals, health food and foods for special age groups, alternative sources of natural food ingredients are of interest. Biotechnological routes to natural colours and (volatile) flavours will be discussed in this article.

Colours by Fermentation

Only a few *artificial* colourants still enjoy approval for food use by the authorities in both the US and Europe. Pigments that are chemosynthetic counterparts of regular food components are referred to as *nature identical*. About 15 of the latter compounds have been permitted, some with restrictions of application, by the Council of the EU. By analogy with flavour legislation, colours extracted from food or other biological sources are sometimes referred to as *natural*. Some representatives of this group, for example anatto, paprika and grape skin extract, are in fact exempt from certification in the US; in Europe, however, colour-enriched concentrates must comply with food additive legislation. Unprocessed fruit juice, beetroot juice or spinach are generally classified as food ingredients, not additives. Only a small proportion of naturally occurring food pigments is used in industry (**Table 1**). Of these, β-carotene, riboflavin and the *Monascus* pigments are commercially produced and some other colourants appear to be amenable to a biotechnological approach in the future.

Carotenes and Xanthophylis

Tetraterpene hydrocarbons and oxygenated compounds derived thereof have multiple physiological functions: filtering and harvesting light, transmitting energy, phototaxis, as attractant, provitamin, antioxidant and flavour precursor, and occur widely in nature. Photosynthetic and heterotrophic micro-

Table 1 Natural food colourants

Chemical group	No. of compounds identified	Colour	Occurrence	Known biotechnical producer
Flavonoids	600	Yellow to black	Leaves, flowers, fruits	Plant cell culture
Carotenoids/xanthophylls	570	Yellow to red	Leaves, seeds, fruits	Bacteria, yeast, fungi, algae
Quinones	400	Yellow to black	Flowers, nuts, berries	*Xenorhabdus, Fusarium*
Anthocyanins	120	Orange to red/violet	Leaves, fruits	Plant cell culture
Betaxanthins/betacyans	70	Yellow/red	Centrospermae	Plant cell culture
Chlorophylls	25	Green to olive	Leaves	Plant cell culture
Xanthones	20	Yellow	Fruits	None
Haems	10	Red to brown	Animals	None

Table 2 Microbial producers of carotenoids

Bacteria	*Agrobacterium, Brevibacterium, Erwinia, Erythrobacter, Rhodobacter, Streptomyces, Thermus*
Yeasts	*Cryptococcus, Phaffia, Rhodotorula, Sporobolomyces*
Fungi	*Blakeslea, Neurospora, Phycomyces*
Algae	*Dunaliella* and *Haematococcus*

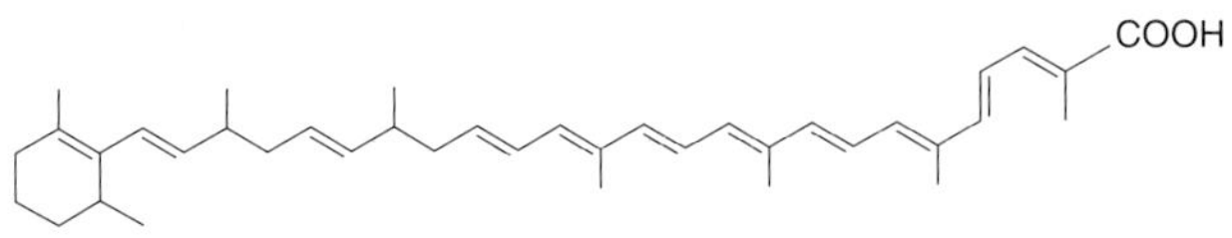

Figure 1 β-Torularhodin, a norcarotenoid from *Rhodotorula* yeast.

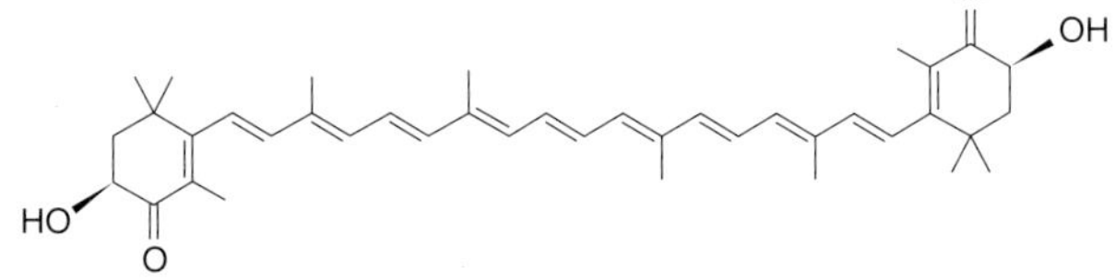

Figure 2 Astaxanthin, a marine xanthophyll.

organisms accumulate such pigments (**Table 2**). Microbial carotenoids are often structurally related but not identical to food constituents (**Fig. 1**).

Bacterial gene clusters coding for carotenoid synthesis were identified. At least nine genes of *Rhodobacter capsulatus*, located on a plasmid called pRS4021 and 46 kb in size, were required to synthesize neurosporene, the immediate precursor of lycopene (tomato colour). Seven genes sized 12.4 kb from *Erwinia herbicola* were cloned into *Escherichia coli* to produce food-related carotenoids. Expression in the heterologous host was successful, and a total pigment yield of 290 μg g^{-1} dry weight was found: this was several times more than in the original producer cells. The option selectively to mutate or switch off single genes to accumulate a certain carotenoid as a dead end product and to grow *E. coli* in proven high-density batch or fed-batch systems is luring.

Astaxanthin (**Fig. 2**), the principal xanthophyll of shrimp, crab and lobster (also deposited via the food chain in flamingo feathers), was accumulated in the heterobasidiomycetous yeast *Phaffia rhodozyma*. Wild strains grew on renewable resources, such as whey hydrolysate or molasses, and produced < 400 μg g^{-1} wet weight of total carotenoids. Nutritional supplements like yeast extract, or general terpenoid precursors, such as acetic acid or mevalonic acid, stimulated carotenoid accumulation, but, for various reasons, were excluded from practical application. Addition of 0.2% v/v ethanol increased the specific rate of carotenoid formation, particularly in overproducing mutants. This effect was attributed to

an induction of HMG-CoA reductase, a key enzyme in isoprene synthesis, combined with nonspecific activation of oxidative metabolism. Maximum carotenoid yields of 2.8 mg g^{-1} wet weight were reported.

Fungal producers were characterized by slow-growth, highly viscous and costly nutrient media and remarkable β-carotene yields of up to 3 g l^{-1}. Special inducers like β-ionone were needed for best yields. The carotenoids were solvent-extracted, washed and purified by crystallization. Batch cultivation of *Phycomyces blakeslea* or *Blakeslea trispora* on plant or animal lipids may take 1 week.

Industrial-scale production of β-carotene by *Dunaliella salina* and *D. bardawil* is carried out in the US, Australia and Israel. These algae are unicellular, obligate photoautotrophic, halotolerant and may contain more than 10% w/w β-carotene on a dry-weight basis. Apart from high photon flux, various modes of growth limitation appeared to stimulate carotene formation. Open sea-water ponds and more sophisticated thin-layer membrane reactors were operated. The halotolerant nature of the algae suppressed contaminants and predators in open culture systems. Two-stage cultivation was suggested to grow the cells rapidly under low-salinity conditions, and then to induce carotene synthesis in a high-salinity high-light environment. Similar physiology was found with *Haematococcus pluvialis*, a microalgal producer of asthaxanthin.

Riboflavin

Microbial overproducers of riboflavin (vitamin B$_2$) were mainly found among yeasts and fungi. Among others, processes based on the use of *Candida flareri*, *C. guilliermundii*, *Debaromyces subglobosus*, *Hansenula polymorpha*, *Saccharomyces* and *Torulopsis xylinus* have been patented. Productivities of 0.2 g l^{-1} h^{-1} were observed. Fungal bioprocesses using *Ashbya gossypi* and *Eremothecium ashbyii* and molasses, soybean or corn preparations and cod-liver oil resulted in the stable production of several grams per litre. The phosphate repression of riboflavin synthesis in *E. ashbyii* has been well described. The mechanism involved a suppression of flavin mononucleotide (FMN) hydrolase and stimulation of flavokinase, resulting in increased conversion of riboflavin into FMN.

The bacterial riboflavin operon was introduced into a *Bacillus* strain which, thereby, gained 10-fold productivity as compared to the original strain. It is to be expected that future industrial producers will be bacterial.

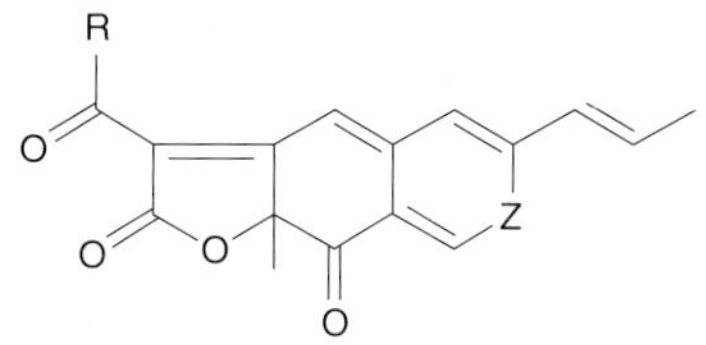

Figure 3 Pyranocoumarin pigments of *Monascus* fungi. R, pentyl or heptyl; Z, oxygen or secondary nitrogen.

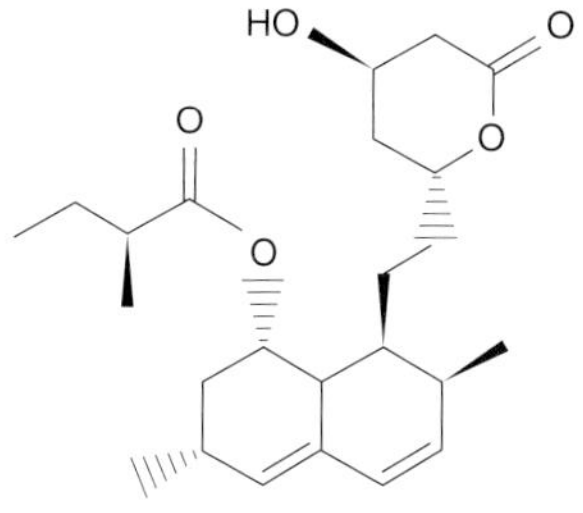

Figure 4 Lovastatin, an antibiotic from *Monascus ruber*.

Monascus Pigments

When selected strains of various species of the fungal genus *Monascus* (preferably *M. anka* and *M. purpureus*) are grown for 3 weeks on steamed rice, a coloured product (ang-kak, beni-koji) results that is powdered and used for colouring other oriental food items, such as fish, soy cheese, meat and wine. The pigments are yellow to purple, chemically stable, but not permitted as food additives in the western world (**Fig. 3**). The intracellular pigments are extracted from the mycelium using ethanol and can then be further reacted with ammonia or amino group-bearing food constituents, such as nucleotides or amino acids. The resulting conversion of the pyran ring into a (substituted) 1,4-dihydropyridine ring affects colour and improves water solubility. Inherent problems are the concurrent generation of antibiotics, such as citrinin or lovastatin (a potent inhibitor of HMG-CoA reductase; **Fig. 4**) by some *Monascus* spp. and the transformation of the solid-state process into a submerged variant. Protoplast fusion between *Aspergillus oryzae* and *M. anka* generated fusants with faster growth, but lower pigment content. Roller bottle cultivation and the use of a macroporous resin (XAD-7) have been applied with more success. Mutants with increased pigment content and release and restricted antibiotic production hold promise for future genetic engineering approaches.

Other Colourants

Indigo Introduction of an oxygen atom at position 3 of the indole molecule yields indoxyl which is rapidly oxidized to indigo. The disulphonic acid derivative (FD&C Blue 2, Indigotin) is approved for food use in the US and in Europe. Xylene oxidases and other enzymes have been shown to catalyse the required

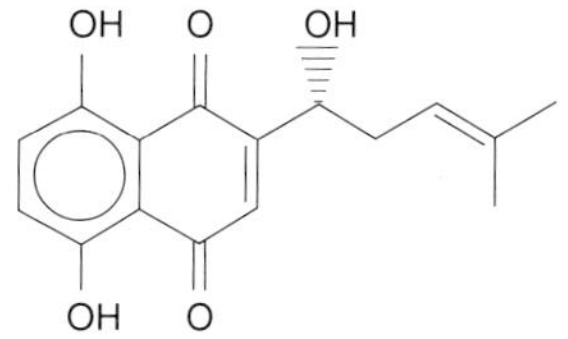

Figure 5 Shikonin, an antimycotic colourant from *Lithospermum erythrorhizon* plant cell culture.

oxidation reaction. First recombinant strains of *E. coli*-bearing plasmids with oxidase genes have become known.

Chromoproteins Cyanobacteriae, red algae, and *Cryptophyceae* contain blue, blue-red and blue-green chromophores (phycocyanins, phycoerythrins and phycoerythrocyanins) attached to a peptide backbone by a thioether bridge. These well water-soluble acyclic tetrapyrroles close the chlorophyll light absorption gap and function as a nitrogen depot. Cyanobacteria of the genus *Spirulina* are cultured in Mexico, southern California and Israel, but are marketed rather as alternative food than processed into colourant preparations. According to the physiological functions, high yields of phycobiliproteins have been achieved at high light and nitrogen (as urea or amino acids) levels. Photoinhibition must be avoided by an appropriate photoperiod, agitation and pond depth. *Spirulina* thrives in an alkaline environment which provides a natural protection mechanism against competitors and favours the dissolution of CO_2, the sole C-source, into the medium.

Quinones Laccaic acid, its 5-hydroxy derivative (kermes acid), and the 7-glucosyl derivative of kermes acid (carminic acid, cochineal) constitute a small group of anthraquinone pigments produced by scale insects. Cochineal is permitted for food use and imparts a stable, strong red colour. Anthraquinone and naphthoquinone structures have been isolated from both bacteria and fungi (for example, from *Trichoderma* and *Fusarium*), but thorough screenings have not become known. A patent was granted for a heat-stable mixture of *Monascus* pigments and laccaic acid.

Lithospermum erythrorhizon, an endangered Japanese plant species, produces the red naphthoquinone shikonin (**Fig. 5**). Starting in 1974, high-yielding plant cell mutants were visually selected and callus cultures raised from overproducing protoplasts. Concerted improvements have increased product yields of suspended plant cells to a level 10 times higher than in the wild plant. This pioneering work has led to one of the few industrial-scale bioprocesses with higher plant cells and is operated in a two-stage process in 200/750 l agitated propagation/production reactors.

Anthocyanins and Betalains Higher contents of these plant colours have been reported for plant cell cultures. Grape cell cultures accumulated anthocyanins up to 16% dry weight intracellularly. Some 30 plant cell cultures have been shown to date to accumulate significant levels of flavonoids that typically occurred in the mother plant from which they were established.

Both the yellow-orange betaxanthins and the red-violet betacyanins were produced in cell cultures of *Amaranthus*, *Beta*, *Chenopodium*, *Phytolacca* and *Portulaca*. Light, tyrosine feeding, and phytoeffector ratios were regarded as crucial. In contrast to most other pigments, betacyanin production in *Phytolacca americana* (pokeweek or pokeberry) correlated positively with cell proliferation. Maximum concentrations of 1% dry weight were found, and first patents have been filed.

In view of some inherent drawbacks, such as long doubling time, metabolic rigidity, non-growth-associated production and intracellular storage of secondary compounds, it has been argued whether plant cell culture would become a viable industrial option at all. Future introduction and expression of genes of higher plants in microbial hosts, however, is an intriguing perspective. On demonstration of an active gene in an in vitro plant system, the respective nucleic acid sequence can be isolated from a uniform, ever-growing source.

Commercial Importance

Factors which limit the use of natural colours are poor processing stability, low tinctorial strength and high price. Nature identical colours are one to two orders of magnitude cheaper and often serve the purpose just as well. It is with this group of products that fermentation colourants have to compete. Looking at the identical chemical structures of natural, nature identical and fermented colourants, the same technical properties must be assumed; however, fermentation products may be more difficult to purify. About half of the world's demand of riboflavin, for example, is produced by fermentation, but it is used as a feed supplement, not in human nutrition.

The world market of natural colourants (including caramel colour) may be in the range of $250 million, with the US accounting for about half of the market. The largest product group are the carotenoids with a share of about $50 million and an annual production of 40–50 tons. At a price of $1000 per kilogram or more, bioprocess-derived colourants are only competitive in niche markets. As production cost for low-volume products (which are mainly labour and investment) is more or less fixed, biocolourants cannot compete with compounds that are abundant in cultivable plants or relatively economical to synthesize.

Flavours by Fermentation

Unlike colours, the subdivision of volatile flavours into natural, nature identical (EC only) and artificial is based on legal definitions: the EC Guidelines 88/388/EWG, 91/71/EWG, 91/72/EWG and the US Code of Federal Regulations. Fermentation flavours are natural if obtained from natural substrates, and this status is emphasized on the product label or list of ingredients. The International Organization of the Flavour Industry (IOFI) and the Flavor Extract Manufacturers Association (FEMA) are continuously dealing with manufacturing details and product safety. Microbial flavours can occur along *de novo* syntheses, or in bioconversions upon adding a suitable precursor compound to the biocatalyst. Probably more than 1000 strains were reported to generate aroma compounds (**Table 3**). According to estimations, some 50–100 microbial flavours are commercially available.

Vanillin, Benzaldehyde, Aromatics

Of the 12 000 or so tons of vanillin consumed annually, less than 1% is extracted from the vanilla bean. The difference in price of nature identical ($12 per kilogram) and natural vanillin (< $4000 per kilogram) created a lot of discussion among biotechnologists. Processes based on various precursors and oxidase enzymes, bacteria (*Enterobacter*, *Klebsiella*, *Serratia*), fungi (Table 3), and plant cell culture were developed. Two commercial processes exist: pressure hydrolysis of curcumin from *Curcuma* roots and degradation of natural ferulic acid by *Pseudonocardia*. The latter patent claims yields $> 10 \, g \, l^{-1}$ and the sales price is in the range of $1000 per kilogram.

The liberation of benzaldehyde, the second most important flavour compound after vanillin, from its cyanogenic glycoside precursor in apricot pits raises the problem of concurrent formation of equimolar amounts of hydrocyanic acid. White rot fungi, such as *Bjerkandera*, *Dichomitus* and *Ischnoderma*, converted L-phenylalanine to benzaldehyde. In studies using deutero-labelled precursor, fungal pathways to benzaldehyde and 3-phenylpropanol (flowery, rose-like) were elucidated. *In situ* recovery of volatiles by fixed-bed adsorbent columns or immobilization of *Bjerkandera* on polyurethane foam cubes increased the yields to $> 1 \, g \, l^{-1}$. Benzaldehyde removal by pervaporation was also reported. More 'natural' benzaldehyde is currently sold rather than distilled from fruit pits, but it is not known if the difference originates from a bioprocess or from 'grey-zone'-chem-

Table 3 Microbially generated character impact components of food

Microorganism	Compounds (odour)
Bacteria	
Brevibacterium, Micrococcaceae sp.	Thioesters (smear cheese)
Lactobacillus lactis, *Streptococcus diacetylactis*, *Leuconostoc citrovorum*	Diacetyl (butter)
Acetobacter aceti, *Gluconobacter oxydans*, *Propionibacterium* sp., *Clostridium* sp.	Short-chain fatty acids (cheese)
Streptomyces citreus	Geosmin (beetroot)
Bacillus sp., *Pseudomonas* sp., *Streptomyces* sp.	Pyrazines (bakery, roasted notes)
Bacillus cereus	2-Acetyl-1-pyrroline (bread, popcorn)
Enterobacteriaceae	Nootkatone (grapefruit)
Yeast	
Candida sp., *Pichia* sp., *Saccharomyces* sp., *Saccharomycopsis*, *Rhodotorula glutinis*, *Yarrowia lipolytica*	C_4 to C_{12}-Alkanolides (milk fat, coconut, peach), carboxylic acid esters (fruity, spicy notes)
Kluyveromyces sp.	Phenylethanol and phenylethyl esters (honey, flowery)
Kluyveromyces lactis	Citronellol, geraniol, linalool (citrus)
Zygosaccharomyces rouxii	Furaneol (strawberry jam, caramel)
Fungi	
Bjerkandera adusta	Benzaldehyde (cherry)
Ischnoderma benzoinum	4-Methoxy benzaldehyde (aniseed)
Pycnoporous cinnabarinus	Vanillin (vanilla pod), methyl anthranilate (orange blossom)
Nidula niveo-tomentosa	4-(4-Hydroxyphenyl)-2-butanone (raspberry)
Phellinus sp.	Methyl salicylate (wintergreen)
Lentinus edodes	Lenthionin (shiitake)
Agaricus bisporus, *Grifola frondosa*, *Lentinus edodes*, *Morchella* sp., *Pleurotus pulmonarius*, *Rhizopogon rubescens*	C_8-Compounds (mushroom)
Pleurotus euosmus	Coumarins (woodruff)
Aspergillus and *Penicillium* sp.	Methyl ketones (blue cheese)
Botryodipoldia theobromae, *Gibberella fujikurio*	Jasmonates

istry. Advance techniques of isotopic pattern analysis and, where applicable, chiral analysis have been used to prove authenticity.

Cinnamic acid esters (exotic fruit), cinnamaldehyde (cinnamon), eugenol (clove), methyl benzoate (dry fruit) and benzyl acetate (jasmine) were among the volatiles found in submerged cultures of basidiomycetes. Supplementation of shikimate intermediates, lignin, or even lipids raised concentrations to several $100 \, mg \, l^{-1}$. These higher fungi should preferably be considered for the generation of aromatic metabolites.

Esters

Aliphatic and terpenic esters are key to most fruit flavours. The required alkyl and acyl moieties are produced by many microorganisms by oxidative shortening of fatty acids and partial reduction of the degradation products, or by a *Strecker* degradation of free amino acids, or by conversion of terpenoid precursors. Acyltransferases then attach the activated moieties onto the most abundant alcohols. As a result, ethyl esters of acetate and ethyl methylbutanoates dominate in yeast fermentations. *Pseudomonas fragi* converted milk fatty acids into ethyl butanoate and other esters, yielding off flavour milk products. *Clostridium* strains oxidized *n*-butanol to the cor-

responding acid and formed the symmetric ester. *Brevibacterium linens* and further bacterial strains have transferred methane thiol on to straight and branched chain fatty acids, as they occur in the strong-smelling coat of certain soft cheeses.

The observation that resting or even dried cells of *Rhizomucor* or *Rhizopus* were capable of forming esters led to the development of a lipase-mediated reverse hydrolysis technique: immobilized lipase or esterase introduced to the substrates in a microaqueous environment, in which the alcohol could be the reaction solvent. Ester formation largely depended on substrate partitioning in this heterogeneous system and was either achieved through direct condensation, through transesterification of two ester substrates, or through alcoholysis of one ester substrate. Terpenoid substrates with their multiple double bond particularly benefited from the mild reaction conditions. Genetically engineered *Candida* strains were the preferred producers for these lipases. As a consequence of the microaqueous conditions, thermal inactivation of the enzyme was slow at reaction temperatures < 100°C. Packed bed and membrane reactors were used, including an exotic gas-phase reactor for highly volatile esters.

Nature identical L-menthol, one of the bulk chemicals of the flavour industry (4000 tons per year), is in part produced by kinetic resolution of racemic methyl esters. Esterases from *Saccharomyces* or *Trichoderma* or from bacteria stereoselectively cleave the L-methyl ester. The remaining enantiomer ester with undesired sensory properties is, upon racemization, recycled into the process.

Lactones

4- and 5-alkanolides (γ- and δ-lactones) with 8–12 carbons impart fruity, nutty, and fatty aroma attributes to food. Generated by intramolecular condensation of oxo or hydroxy acid precursors delivered by the peroxysomal β-oxidation of fatty acids, many microorganisms formed and some accumulated odorous lactones. Reduction of the oxo function or introduction of the hydroxy function proceeded stereospecifically, resulting in high optical purities of the lactone products. For example, *Candida lipolytica* formed (4S)-dodecalactone, and *Pichia ohmeri* gave the opposite enantiomer. This is important to note, because different enantiomers possess different sensory characters and intensities (odour detection thresholds).

The hydroxy fatty acid precursor could be derived from a microbial hydroxylation, for example involving lipoxygenase, or could be of plant origin. Two decades after the first description of the conversion of ricinoleic acid (12-hydroxy oleic) to 4-hydroxydecanoic acid by *Candida*, this process was patented in 1985; castor oil that contains about 90% w/v ricinoleic acid was used as the sole carbon source in the process that yielded 5 g of 4-decanolide per litre. The initial price of $20 000 per kilogram had signalling character and started a race for better biocatalysts and precursors. More recently, genetic engineering has found target-oriented applications: after it was found, for example, that good growth of *Yarrowia lipolytica* was at the expense of 4-decanolide yield, a uracil auxotrophic strain was created. Productivity of the auxotrophic cells in the uracil-free medium was 10 to 20-fold higher. Today, about 30 processes have been published for various microbial lactones, and some appear to be operated on an industrial scale.

Terpenoids

The accumulation of volatile oligo-isoprenoids is not a plant domain. Small amounts of geosmin and bornanes were formed by *Actinomyces*, cyanobacteria and by some asco- and basidiomycetes. Farnesol appeared to leak from *Saccharomyces* cells in unfavourable cultivation conditions. Some members of the Ascomycete genus *Ceratocystis* are still unsurpassed concerning diversity and quantity of monoterpenes accumulated.

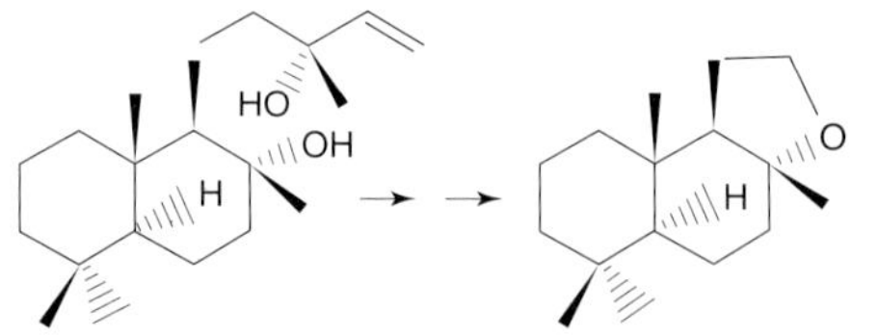

Figure 6 Microbial conversion of the plant diterpene sclareol into ambrox, a fragrance with woody odour.

Soil bacteria and filamentous fungi (*Aspergillus, Botrytis, Cladosporium, Fusarium, Rhizomucor, Penicillium, Trichoderma*) were particularly suited to transform many of the > 400 plant monoterpenes into potent flavours. Typical conversion substrates were the abundant hydrocarbons pinene, limonene and myrcene. However, no commercial process is running, because both precursor and product may be chemically and biochemically unstable; they are not well water-soluble, they are cytotoxic; and they are lost during an aerated process due to volatility.

More successful was the regioselective hydroxylation of patchouli alcohol to the 10-hydroxy derivative by a soil bacterium. The yield was in the range of 1 g l^{-1} which allowed subsequent chemical conversion to nor-patchoulol, the key compound of patchouli oil. Similarly, ambrox and sclareolide were generated from sclareol by *Cryptococcus* for perfumery applications (**Fig. 6**).

Other Commercial Flavours

Aliphatics Based on traditional fermentations, processes using bacteria, deuteromycetes or isolated enzyme mixtures were developed to convert milk fractions or other food or waste-stream materials into products rich in volatile fatty acids, 2-alkanones, butanedione, ethanal and other aldehydes, and alkanols. The resulting flavour mixtures usually represented typical fermentation flavours, such as cheese, soy sauce, or fermented beverage flavour. In contrast to the traditional fermentations, however, optimized conditions and repeated process cycles or multistage fermentation yielded strong flavour bases in a short time.

Mushroom Aroma Fungal lipoxygenases stereospecifically catalysed the peroxydation of unsaturated C_{18}-fatty acids; cleavage of the hydroperoxy products by a lyase generated aliphatic C_8-carbonyls and alcohols with typical mushroom flavour. Maximum yields were in the range of 50 mg g^{-1} dry weight. An economical evaluation showed that the cost of producing mushroom flavour from submerged mycelia was about the same as from the fruiting body, but the submerged culture was faster.

Pyrazines Thermally treated foods accumulate these heterocycles from aminoketone precursors. Cold formation in the gram per litre range occurred with bacteria. The ring nitrogen was obviously derived from amino acids, as many of the pyrazines still bear respective alkyl side chains in the α-position. Solid-state cultivation and a fixed-bed column reactor were used.

Raspberry Ketone 4-(4'-Hydroxyphenyl)-butan-2-one, an impact of raspberry flavour, represents all current problems and perspectives of a microbial flavour speciality. At concentrations < 1 mg kg^{-1} in the berry, economic isolation from the plant material was not feasible. A price of about $10 000 per kilogram was demanded for the same flavour product hydrolysed from a birch bark glycoside, betuloside. A microbial producer of this rare compound has also become known (Table 3), and precursor feeding increased the product level significantly. However, a bioreaction period of 2 weeks and special reactor requirements prevented industrial use.

Commercial Importance

A peak concentration of 1 g l^{-1} and a 5-year amortization period with annual depreciation was assumed to calculate break-even points for a bioprocess flavour. With annual sales of 1000 kg the cost price would be in the range of $1200 per kilogram; with an annual usage of 10 000 kg this would drop to about $300 per kilogram. Indeed, prices from $200 to $2000 for character impact components appear to be competitive. Natural bitter almond oil or some *Citrus* oils, depending on changing harvest, may typically sell in a range from $100 to $500.

Genetic modification of starter strains and other bacteria and yeasts has been undertaken to control butanedione and fermentation alcohol synthesis. While the single gene approach still dominates, the methodical limitations of transferring an entire pathway into a heterologous host are now being discussed with increasing optimism.

See also: **Bacillus**: Bacillus cereus. **Brevibacterium**. **Candida**: Candida lipolytica. **Cheese**: Mould-ripened Varieties. **Escherichia Coli**: Escherichia Coli. **Fermentation (Industrial)**: Recovery of Metabolites. **Genetic Engineering**: Modification of Yeast and Moulds; Modification of Bacteria. **Metabolic Pathways**: Lipid Metabolism; Production of Secondary Metabolites – Fungi; Production of Secondary Metabolites – Bacteria. **Monascus. Rhodotorula. Saccharomyces**: Saccharomyces cerevisiae. **Starter Cultures**: Uses in the Food Industry. **Streptomyces**.

Further Reading

Abraham B, Onken J, Reil G and Berger RG (1997) Strategies toward an efficient biotechnology of aromas. In: Kruse H-P and Rothe M (eds) *Flavour Perception/Aroma Evaluation*. P. 357. Bergholz-Rehbrücke: University of Potsdam.

Anke T (1997) (ed.) *Fungal Biotechnology*. London: Chapman & Hall.

Bajaj YPS (1988) *Biotechnology in Agriculture and Forestry*. Berlin: Springer.

Berger RG (1995) *Aroma Biotechnology*. Berlin: Springer.

Berger RG (1996) (ed.) *Biotechnology of Aroma Compounds. Advances in Biochemical Engineering/Biotechnology*. Vol. 55. Berlin: Springer.

Endress R (1994) *Plant Cell Biotechnology*. Berlin: Springer.

Gabelman A (1994) (ed.) *Bioprocess Production of Flavor, Fragrance, and Color Ingredients*. New York: Wiley.

Goodwin TW (1988) (ed.) *Plant Pigments*. London: Academic Press.

Lee BH (1996) (ed.) *Fundamentals of Food Biotechnology*. Weinheim: VCH.

Patterson RLS, Charlwood BV, MacLeod G and Williams AA (1992) (eds) *Bioformation of Flavours*. Cambridge: Royal Chemistry Society.

Takeoka GR, Teranishi R, Williams PJ and Kobajashi A (1996) (eds) *Biotechnology for Improved Foods and Flavors*. Washington, DC: American Chemistry Society.

FERMENTED FOODS

Contents

Origins and Applications

Geoffrey Campbell-Platt, University of Reading, UK

Introduction

Fermented foods are foods that have been subjected to the action of microorganisms or enzymes, in order to bring about desirable changes. Fermented foods originated several thousand years ago, in different parts of the world, when microorganisms were introduced incidentally into local foods. These microorganisms caused changes which helped to preserve the food or improved its appearance or flavour (**Table 1**). Fermented foods now play a major role in diets worldwide.

History, Development and Uses of Fermented Foods

The earliest peoples were generally nomadic hunters and gatherers; only when they settled did cultivation and agriculture develop.

The Middle East

It is believed that one of the first transitions to organized food cultivation and production occurred in the Middle East, in the Valleys of the rivers Tigris and Euphrates, more than 10 millennia ago. It was also in the Middle East that the discovery was made that keeping milk in animal skins produced pleasant fermented milk drinks and yoghurts, and that as fluid was lost and the milk coagulated, cheese could be produced by fermentation.

Asia and Oceania

In China, the major staple food crops cultivated around 10 000–8000 BC were rice and soybeans. Rice was processed into and preserved as fermented cereal porridges such as lao-chao, and rice wine (saké). As soybean cultivation and processing spread to Korea, Japan and Indonesia, soybeans were made into soy sauce, miso (soy paste) and tempeh. In Southeast Asia, fermented fish sauces, such as nuoc mam, and fermented fish pastes, such as bagoong, were protein-rich, flavoursome components of the diet.

A Chinese concept originally, now adopted throughout the world, is that a balanced meal should consist of rice or another basic staple, 'fan', plus 't'sai' – originally meaning vegetables or other accompanying food. T'sai added flavour, interest, fibre and many essential micronutrients.

In Korea, kimchi remains the traditional pickled vegetable food, made from a range of vegetables during the summer and consumed during the long, hard winter. Kimchi sometimes also contains fish, and is still regarded as an essential component of every Korean meal. Much kimchi is still produced domestically in Korea, where the estimated annual production is over 1 million tonnes.

In Papua New Guinea, the starchy plants taro or cocoyam were fermented with coconut into a gruel known as sapal.

Africa

In Africa, south of the Sahara, the staple foods were starch crops. Cassava dough was fermented in West Africa into gari, kokonte or lafun, and agbelima. The main cereal crops of Africa were sorghum, millet and later maize, and were used to make a variety of foods:

Table 1 History and origins of some fermented foods

Food	Approximate year of introduction	Region
Mushrooms	4000 BC	China
Soy sauce	3000 BC	China, Korea, Japan
Wine	3000 BC	North Africa, Europe
Fermented milk	3000 BC	Middle East
Cheese	2000 BC	Middle East
Beer	2000 BC	North Africa, China
Bread	1500 BC	Egypt, Europe
Sourdough bread	1000 BC	Europe
Fish sauce	1000 BC	Southeast Asia, North Africa
Pickled vegetables	1000 BC	China, Europe
Tea	200 BC	China

maheu, a sour fermented beverage, in Southern Africa; uji or togwa (porridge) in East Africa; and kenkey (dumpling) in West Africa.

In North Africa, the cereal teff, *Eragrotis tef*, was made into the flat bread njera in Ethiopia. The Egyptians discovered that the residue left after fermenting barley to make beer was rich in yeast, which could help to produce the raised or leavened bread which later came to dominate European diets.

Along the North African coast, the Romans planted grapes. These are rich in sugars and highly perishable, but after fermentation into alcoholic wine, kept well. Further south, palm wine was produced from the oil palm *Elaeis guineensis*. Palm wine, alcoholic beer from cereals and wine from grapes, along with fermented milks, were valuable beverages, being safe and nutritious at a time when water supplies were unreliable and of doubtful safety.

Europe

Throughout Europe, meat was fermented by lactic acid bacteria and micrococci, sometimes with fungi, to produce a range of fermented meats including salami and country ham. In these products, the activity of microbial lipase leads to the production of carbonyl compounds, including aldehydes and ketones, which are important flavour components.

Many fermented wheat and rye breads were produced in Europe, as well as a range of fermented cereal beverages, including the ales and beers of northern Europe and the lagers of central and eastern Europe. Apples were also fermented, into cider, and grapes were fermented to make red and white wines. Many of these fermented beverages were fortified, by either distillation, producing whisky from beer and brandy from wine, or fortification – the addition of brandy to wine to give sherry or port.

A wide range of cheeses was developed in Europe. The majority were made from cow's milk, but some were made from sheep's or goat's milk. The cheeses were pressed, ripened and matured for different periods, giving a wide range of types, from mild soft cheeses, such as cottage cheese, to strongly flavoured mould-ripened cheeses, such as Camembert and blue Stilton.

The Americas

Iquncq, dried cider duck, was produced by the Inuits of northern Canada, while in Peru and Brazil flat sheets of charqui were made from llama, alpaca, beef and sheep. As people migrated, their skills in producing fermented foods spread. Scandinavians took to North America their all-important sourdough starters, used to produce a range of sourdough breads. The practice of producing beers, lagers, wines, cheeses

Table 2 Worldwide production of some fermented foods

Food	Quantity (t)	Beverage	Quantity (hl)
Cheese	15 million	Beer	1000 million
Yoghurt	3 million	Wine	350 million
Mushrooms	1.5 million		
Fish sauce	300 000		
Dried stockfish	250 000		

Table 3 Individual consumption of some fermented foods: average per person per year

Food	Country	Annual consumption
Beer (l)	Germany	130
Wine (l)	Italy, Portugal,	90
	Argentina	70
Yoghurt (l)	Finland	40
	Netherlands	25
Kimchi (kg)	Korea	22
Tempeh (kg)	Indonesia	18
Soy sauce (l)	Japan	10
Cheese (kg)	UK	10
Miso (kg)	Japan	7

and fermented meats was also introduced to North and South America by immigrants. Nowadays, wines are produced in Argentina, Chile and California, and a wide range of cheeses is made in Latin America.

Current Production

The production and consumption of fermented foods (**Table 2**, **Table 3**) has now spread to such an extent that Argentina is a major producer and consumer of salami; West Africa regards French bread as its staple food; Australia produces, and helps the French to produce, world-class wines; New Zealand is a major producer of Cheddar; the Indian fermented legume and cereal products idli and dosa are enjoyed in Indian restaurants everywhere; and a second major soy sauce factory has been opened in the USA. The beans which are fermented to make cocoa and coffee are produced as far apart as Indonesia, Malaysia, Kenya, Ghana, Ecuador and Brazil, and coffee and chocolate are enjoyed everywhere.

Benefits and Importance of Fermented Foods

The term 'biological ennoblement' has been used to describe the nutritional benefits of fermented foods (**Table 4**). Fermented foods comprise about one-third of world consumption, and 20–40% (by weight) of individual diets. The three main groups of fermented foods are cereal products, beverages and dairy products.

Fermentation considerably increases variety in the human diet. Many interesting cuisines depend on

Table 4 Benefits of fermentation

Benefit	Raw material	Fermented food
Preservation		
	Milk	Yoghurt, cheese
Enhancement of safety		
Acid production	Fruit	Vinegar
Acid and alcohol	Barley	Beer
production	Grapes	Wine
Production of bacteriocins	Meat	Salami
Removal of toxic	Cassava	Gari, polvilho
components		azedo
	Soybean	Soy sauce
Enhancement of		
nutritional value		
Improved digestibility	Wheat	Bread
Retention of	Leafy	Kimchi
micronutrients	vegetables	
Increased fibre content	Coconut	Nata de coco
Synthesis of probiotic	Milk	Bifidus milk,
compounds		Yakult
Improvement of flavour	Coffee beans	Coffee
	Grapes	Wine

regional fermented food products for their individuality, and nowadays international cuisines are popular far distant from their origins.

Many staple food crops cannot be readily or safely consumed prior to processing. Raw wheat, barley and maize are not very appetizing, but the products of their fermentation – breads, porridges, dumplings and beers – are much more digestible and acceptable. People who cannot tolerate milk find that fermented milks, yoghurt and cheeses are more acceptable.

Important flavour compounds, including diacetyl and acetaldehyde, are produced by lactic acid bacteria during the fermentation of milk. Grapes are pleasant to eat raw, but are highly perishable. Their fermentation produces a range of wines with different characteristics, depending on the variety of the grape. Wines are much safer to drink than contaminated water. Raw legume beans contain lectins, which are toxic; cooking and fermentation produce a wide range of edible legume products.

Isoflavones, from soybeans, appear to be beneficial. Soy sauce is an essential component of many Eastern Asian dishes – more than 1 billion litres of soy sauce are produced every year in Japan alone. Fish sauces, which preserve the amino acids of highly perishable fish, have a similar role.

Several fermented products are particularly associated with festivals and celebrations. Lao-chao is the sweet–sour fermented rice produced in China for celebrations, and in Japan a special rice wine containing fermented plums, umeboshi saké, may be produced for New Year. Sparkling wines, such as champagne, are used at occasions such as European weddings, for toasts. In contrast, many fermented foods are regarded as essential in daily diets, for example, bread.

Applications of Fermentation

As one of the oldest forms of food preservation, fermentation has played a key role in enabling people to survive periods of food shortage. Fermentation was the earliest form of food biotechnology.

Fermented foods are the major group of 'functional foods', which provide extra benefits to our diet beyond those expected from the major nutrients present. Fermentation may result in particular desirable nutrients becoming more readily available, or in the amount of less desirable or toxic components being minimized. Knowledge of the mechanisms by which specific microorganisms and their enzymes change foods during fermentation facilitates control of the fermentation process. The traditional art of making a particular product has been progressively influenced by scientific knowledge, but still has an important role in combination with it. For example, some of the character of traditional fermented foods may be lost by using commercially produced starter cultures which comprise the major fermentative microorganisms but not all those involved in naturally occurring cultures – this may result in the absence of key flavour compounds.

Further research is needed into many aspects of the fermentation of foods. For example, the roles of bacteriocins, antimicrobial compounds produced by lactic acid bacteria, the predominant microorganisms in food fermentations, are incompletely understood. The significance of low levels of metabolites in generating important flavour compounds is appreciated, but knowledge about the associated chemical interactions is incomplete.

It is known that diet is important for human well-being, as well as being contributory or preventative in the development of a range of diseases. It is believed that several fermented foods help to improve the quality of human life and also delay or prevent the onset of some diseases, including cancer and some degenerative diseases. The importance of the polyphenolic antioxidants in tea and red wines has been recognized, and the role of certain lactic acid bacteria and bifidobacteria as probiotics in fermented milks is being researched intensively. Rapidly increasing knowledge about genes provides the possibility of 'improving' food fermentations, for example by using modified strains of starter cultures to increase the reliability of fermentations. Genetically modified organisms, and the enzymes they produce offer great potential for commercial gain, but their development and use must be carefully controlled and considered –

there is a risk that the natural, healthy image of fermented foods could be lost, and hence consumer demand eroded.

See also: **Bacteriocins**: Potential in Food Preservation. ***Bifidobacterium***. **Bread**: Sourdough Bread. **Cheese**: In the Market Place; Microbiology of Cheese-making and Maturation. **Cocoa and Coffee Fermentations**. **Fermented Foods**: Fermented Fish Products. Fermented Vegetable Products; Fermented Meat Products; Fermentations of the Far East; Beverages from Sorghum and Millet. **Fermented Milks**: Range of Products; Yoghurt; Products from Northern Europe; Products of Eastern Europe and Asia. **Lager**. **Probiotic Bacteria**: Detection and Estimation in Fermented and Non-fermented Dairy Products. **Wines**: The Malolactic Fermentation.

Further Reading

Campbell-Platt G (1987) *Fermented Foods of the World. A Dictionary and Guide*. London: Butterworth.

Campbell-Platt G and Cook PE (1989) *Fungi in the Production of Foods and Food Ingredients*. Journal of Applied Bacteriology Symposium Suppl. 117S.

Nout MJR and Motarjemi Y (eds) (1992) Fermented food safety. *Food Control* 8: 221–339.

Steinkraus KH (ed.) (1995) *Handbook of Indigenous Fermented Foods*, 2nd edn. New York: Marcel Dekker.

Wood BJB (ed.) (1998) *Microbiology of Fermented Foods* (2nd edn). London: Blackie.

Fermented Vegetable Products

Guillermo Oliver and **Martha Nuñez**, Centro de Referencia par Lactobacilos (Cerela), Tucumán, Argentina

Silvia Gonzalez, Centro de Referencia para Lactobacilos (Cerela) and Universidad Nacional de Tucumán, Argentina

Introduction

Fermented foods are an essential part of the diet of people in many countries, bread being one of the oldest traditional fermented products. Only a few traditional fermentations are considered here, mainly to illustrate the complexity of biochemical, sensory and nutritional changes that can result from microbial activity in raw materials. The fermentation of vegetables can be affected by several groups of microorganisms. Once harvested, fruits and vegetables are perishable due to the activity of bacteria, fungi, moulds, yeasts and enzymes. When acting on organic matter, these microorganisms and enzymes split their constituents, and turn them into simple compounds.

Destroying these agents or controlling their activity help to keep fruits and vegetables in good condition. Creation of unfavourable conditions, such as humidity and temperature, extends the duration of storage for a certain period of time without changes in the nutritional value.

In vegetable processing, salting and fermentation are related to each other. There is an unspecified number of fermented vegetables available in the market, however, at present, only cabbage for sauerkraut, cucumbers for pickles and olives are of real economic importance. The fermentation of vegetables is a complex network of interactive microbiological, biochemical, enzymatic and physico-chemical reactions. Lactic acid fermentation is a valuable tool for the production of a wide range of vegetable products. At present the naturally available strains of lactic acid bacteria are preferred (respecting consumer opinion) over genetically modified microorganisms. Factors which influence the process, interactions and quality attributes are shown in **Figure 1**. As vegetables have widely different properties the technologies applied to fermentation are different.

Sauerkraut

Acid (sour) cabbage (sauerkraut) is one of the most-consumed fermented vegetables in central and southern Europe and in the United States. In other countries consumption is less and industrial production is restricted. A number of microorganisms play a key role in the transformation of fresh shredded cabbage into the fermented product called sauerkraut. Under proper conditions the fermentation of shredded cabbage is a spontaneous process in which bacteria convert sugars present in cabbage into acids, alcohol

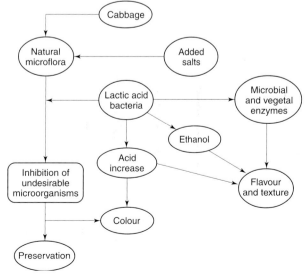

Figure 1 Factors involved in vegetal fermentation.

and carbonic anhydride. Simultaneously, cabbage proteins undergo biochemical changes. The major end products of this fermentation are lactic acid and, to a lesser amount, other acids. Lactic acid bacteria are the most important microorganisms responsible for cabbage fermentation. The organoleptic and nutritional characteristics of sauerkraut are: low caloric power, high fibre content, which improves intestinal transit, absence of lipids and high level of essential micronutrients.

Method of Manufacture

The technique used by industry is empirical and is based on a natural fermentation produced by microorganisms present in raw material. In recent years, artificial inoculation with selected strains and anaerobic conditions have been used in order to gain uniformity of the nutritional and sensory quality of fermented vegetables, to improve them, and to reduce the duration of the process. Several researchers have applied controlled fermentation on cucumber, and others have studied improvements in the fermentation process.

After removal of the outer leaves, green leaves and the core, the cabbage is washed and sliced into shreds as fine as 2.5 cm of thickness. This mass is placed in a fermenter in 2 kg layers alternating with salt layers, the first and last layers being salt. The salt added is 2.5% of the weight of the cabbage to enhance the release of tissue fluids during fermentation. In order to create anaerobic conditions the layers are pressed mechanically, to remove the cabbage juice which contains fermentable sugars and other nutrients required for microbial activity. Anaerobic conditions promote the growth of lactic acid bacteria. This process takes about 31 days during which the average temperature is 14°C (range 18–9.2°C, daily average oscillation 9.2°C).

The progress of fermentation is monitored every 2–4 days, by measuring the ascorbic acid and sugar content of the solid part, and the pH, acidity and chlorine content of the fermentation brine. Total microbiological and viability counts are determined and species are isolated and identified.

Microbiology of the Fermentation Process

Several hours after the cabbage has been placed in the fermenters, gas-forming microorganisms (*Leuconostoc mesenteroides*) initiate acid production. When an acidity of 0.25–0.3% is reached (calculated as lactic acid), the growth of cocci begins to slow and they gradually die. However, their enzymes are released by autolysis and continue to be active in the fermentation process. By the time the acidity has reached 0.7–1.0% most of the cocci have disappeared.

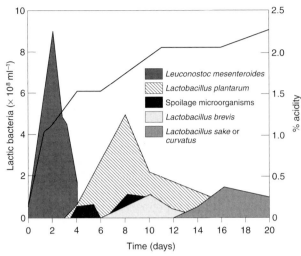

Figure 2 Microbiology of the fermentation process.

However, *Lactobacillus plantarum* and *Lactobacillus brevis* are active and increase the acidity to 1.5–2.0%, despite the inhibitory effect of the salt and low temperature. Finally, *Lactobacillus sake* and *Lactobacillus curvatus* complete the fermentation and produce an acidity of 2.0–2.5%.

The end products formed during the fermentation of the cabbage contain significant amounts of lactic acid (non-volatile) and small quantities of acetic acid and propionic acid (volatiles), a mixture of gases, CO_2 being the most important, small quantities of alcohol, and a mixture of aromatic esters. Acids associating with alcohols form esters which give flavour to high quality sauerkraut.

The acids produced by the bacteria during the fermentation process help to control the process by inhibiting putrefying organisms, and thus avoiding unfavourable changes in the sauerkraut. The characteristics and the quality of the product depend on the acidity and the esters formed during the fermentation process.

On completion of the process only traces of sugar remain, whereas lactic acid contents are high. The amounts of acetic acid and alcohol are products of *Leuconostoc mesenteroides*, the first species in the series. Lactic acid is produced by homofermentative non-gas-forming lactobacillus species. Any modification in the sequential order of desirable bacterial species causes a change of flavour and quality of the product. The sequence and quantity of bacteria involved in the fermentation process are shown in **Figure 2**.

Use of Starters

Starters obtained from vegetable matter, using techniques applied in fermented dairy products, are being increasingly used in sauerkraut production in order to acquire a uniform product of good quality. Furthermore, by using starter cultures, putrescine and histamine (biogenic amines) formation is inhibited.

Addition of *Leuconostoc mesenteroides*, which is predominant during the early stages of spontaneous fermentation, contributes to the good flavour of the end product. However, it alters the natural sequence of the species involved. Addition of *Lactobacillus curvatus* and *Lactobacillus sake* induces a higher acetic acid and lower lactic acid production than in a normal fermentation and the fermentation cannot be completed. When *Lactobacillus plantarum* is used as a starter the fermentation is again not completed, it has a bitter flavour and is more easily affected by yeasts.

The juice of a good fermentation can be used as a starter for other fermentations. The quality of the product depends on the microorganisms present and on the acidity, which should not be too high.

Defects and Spoilage of Sauerkraut

Spoilage of sauerkraut is mainly due to aerobic soil microorganisms. These do not act on sugars, but on proteins and, therefore, produce undesirable changes. Fortunately, these microorganisms are rapidly inhibited by the fermentation. Yeasts are not affected by acidity, and thus cause spoilage. Fungi need oxygen and cannot develop in closed tanks.

Ropy Sauerkraut Spoilage is caused by bacteria, fungi and yeasts, occurring due to lack of hygiene during harvest and processing of the cabbage. Bacteriological quality is defined by absence of coliforms, *Salmonella* and *Staphylococcus*. Yeasts are present at low concentration but these microorganisms do not grow during fermentation.

Mucilaginous Sauerkraut This is not very common. Flavour and odour are not seriously affected. It is caused by a slimy material produced by bacteria such as *Pediococcus acidilactici*. High temperatures favour the development of the latter.

Softening of Sauerkraut This is caused by a number of conditions, such as high oxygen levels (air), too little salt (below 0.8%) and temperature variations.

Darkening of Sauerkraut This is one of the most common defects caused by growth of spoilage organisms. Uniform distribution of the sauerkraut brine suppresses the growth of undesirable organisms that produce darkening.

Pink Sauerkraut This is caused by the growth of certain yeast strains which are able to produce an intense red pigment (e.g. *Torula glutinis*).

Poor Flavour of Sauerkraut This is caused by inhibition of heterofermentative lactic acid bacteria (lack of acetic acid and other suitable fermentative by-products).

Cucumber

Pickled gherkins are very popular and are mostly consumed without any other vegetables. Cucumbers for pickling must be harvested while still immature. They should not be crushed or bruised and should have a stalk of 1–2 cm. There should not be more than 6–8 h between harvesting and brining because spoilage is fast.

Selection

To obtain a uniform product a previous selection removing all crushed, unsound, decomposed, defective and too mature cucumbers should be carried out. Sorting is followed by size grading.

Brining

A wooden, clay or glass tank is filled with 10–15% salt brine to cover the bottom to a depth of 15 cm. Then, previously washed cucumbers are put into the container and the container is filled with brine to cover the cucumbers completely.

Cucumbers are submerged into the brine with a white, perforated wooden cover. Cucumbers consist of about 90% of water which causes the salt concentration of the brine to be lowered during fermentation. Brine should be kept at 40° salinometer to prevent the activity of putrefying bacteria. However, the salt concentration should not be higher as this would cause inhibition of the activity of lactic acid bacteria, producers of lactic acid. After 5–6 weeks, the brine is increased to 60° salinometer by adding salt. During fermentation, cucumbers change colour from bright green to olive green by action of acids on chlorophyll. The flesh of the pickles turns an attractive translucent white when the curing process finishes by expulsion of air from the tissue.

Fermentation Process

Initiating cultures are composed of a normal mixed flora of the gherkins. Fermentation is initiated by the following lactic acid bacteria, cited in order of importance: *Leuconostoc mesenteroides*, *Enterococcus faecalis*, *Pediococcus cerevisiae*, *Lactobacillus brevis* and *Lactobacillus plantarum*. Of these, *Pediococcus* and *Lactobacillus plantarum* are important, whereas *Lactobacillus brevis* is undesirable because of its ability to produce gas. *Lactobacillus plantarum* is the indispensable species.

During the initial fermentation stage, a great many unrelated bacteria, yeasts and fungi can be isolated. All these microorganisms are widespread in nature and, at the beginning of the fermentation, far outnumber the desirable lactic acid bacteria in uncontrolled fermentations. Therefore, the primary fermentation stage is the most important phase of the pickling process. During this period the number of lactic acid bacteria and the fermentative and oxidative yeasts increases rapidly, whereas the undesirable flora decreases rapidly and may even disappear. At the same time, there is a decrease in pH of the brine and an increase in total acidity. During an intermediate stage, when the brine is stabilized, a mixture of low acid-tolerant species of *Leuconostoc* and high acid-tolerant species of *Lactobacillus* and *Pediococcus* control coliform bacteria, which disappear completely after 10–14 days. During this period there is still a low number of yeasts while there is a further increase in total acidity.

Pediococcus cerevisiae, *Lactobacillus brevis* and *Lactobacillus plantarum* are responsible for the final fermentation stage and the completion of the lactic acid build-up. The remaining oxidative yeasts are held without further development by anaerobiosis.

Defects and Spoilage of Cucumber Pickles

Swelling This serious problem in cucumber fermentation generally increases with larger cucumbers, high fermentation temperatures and/or an augmentation of carbon dioxide in the brine. The last effect can be controlled by purging the brine with nitrogen to remove the carbon dioxide or by moderated agitation of the brine. Furthermore, homofermentative lactic acid bacteria can ferment malic acid with production of lactate and carbon dioxide which contributes to bloater damage in cucumber fermentation. Mutants of *Lactobacillus plantarum* lacking the ability to produce carbon dioxide by malolactic fermentation have been isolated in experimental laboratories, but commercial exploitation of these strains was not described.

Soft or Shrivelled Cucumber Pickles These can be the result of high acid formation by *Leuconostoc mesenteroides*. Organic acids used as source of energy by *Mycoderma* organisms may be troublesome because pH variation alters the normal fermentation sequence.

Fermentation of Olives

Olives (*Olea europaea*) are produced in great amounts in the Mediterranean area, California, South America and Australia among others. This fruit is processed for oil production (about 85%) and olives for table use (15%). World production of green olives (Spanish-type) amounts to 350 000 tons a year; approximately 150 000 tons are produced in Spain.

Green Olives

Industrial processing of green olives starts with treatment with diluted NaOH (2%, w/v) for 6 h to remove oleuropein, a bitter glucoside, which is degraded into glucose and aglycone. The latter are then further decomposed into simple molecules, such as elenoic acid and β-3,4-dihydroxy-phenylethyl. The olives are then washed with water for 8 h to remove excess alkali and then submerged in a 10% NaCl solution for lactic fermentation.

Sugars, vitamins and amino acids pass, by osmosis, from the olives into the brine solution, gradually converting it into a suitable medium for the growth of microorganisms which produce olive lactic fermentation. After 6–7 months, the olives have the particular characteristics of the end product.

Recent innovations of the traditional process for producing green olives for table use have helped to minimize the waste water volume and keep the characteristics of the end product. One of the most seriously affected characteristics is colour, an important quality attribute. Therefore, different pigments have been studied, and so far 12 compounds have been isolated, including chlorophylls and carotenoids.

From a microbiological viewpoint, the olive lactic fermentation is a highly complex process in which a number of microorganism strains are involved successively.

There are three stages in a normal olive fermentation (1) initial; (2) intermediate; and (3) the final stage, each of which has different microbiological characteristics. The initial stage is the most important in the process of olive curing. If the fermentation develops normally potential spoilage bacteria are rapidly eliminated principally due to developing lactic acid bacteria and their acid production. These belong to the genera *Streptococcus*, *Pediococcus* and *Leuconostoc* and to a minor extent *Lactobacillus*. If the fermentation follows a different pattern, spoilage microorganisms become predominant. Generally, these are Gram-negative bacteria, which, if not controlled, produce gas and are very proteolytic. They belong to the genera *Aerobacter*, *Escherichia*, *Aeromonas* and *Desulfovibrio*. The latter produces sulphur compounds during olive fermentation in brine. Gram-positive bacteria can also be a problem; *Clostridium* spp. form butyric acid and gas, and *Bacillus* spp. produce softening of the product. During the second stage of a normal fermentation lactic acid bacteria are predominant. Among these there are less acid-tolerant species, belonging to the genus *Leuconostoc*, and highly tolerant *Lactobacillus* species. During this stage *Leuconostoc* species are predominant. If the fermentation is normal, Gram-negative bacteria disappear completely after 12–15 days.

This stage takes two to three months. Initially *Leuconostoc* species are predominant, declining rapidly in number by the end, when they are replaced by more acid-tolerant *Lactobacillus* species. During the final stage, which lasts about 25 days and sometimes even less, olive fermentation is dominated by highly acid-tolerant *Lactobacillus* species. A small number of yeasts may remain, but generally they do not cause any problem as long as the fermenters are maintained under anaerobic conditions. The total acidity of the fermentation lowers the pH to 4.0 and sometimes even to 3.8. If these values are not reached, the olives may suffer spoilage known as zapatera, an abnormality produced by butyric fermentation that causes unpleasant odours. Propionic bacteria appear first, causing an increase in pH, and as a result the growth of *Clostridium* species. The latter are known to cause zapatera spoilage. The unpleasant odour has been attributed to the formation of cyclohexanecarboxylic acid.

During the fermentation process, lactic acid species appear which are predominant in different phases of the three stages mentioned before. *Leuconostoc mesenteroides* dominates the final phase of the first stage and the early phase of the intermediate stage. Lower numbers of *Leuconostoc dextranicum* are found. *Lactobacillus plantarum* dominates the final phase of the intermediate stage and the whole of the final stage. *Lactobacillus brevis* is the most numerous of the heterofermentative species and is found together with *Lactobacillus plantarum*. *Pediococcus cerevisiae* is found together with *Leuconostoc*.

It has been suggested that pure cultures of lactic acid bacteria could be used for curing of green olives. However, the use of pure cultures requires heat treatment of the fruits in order to eliminate natural microorganisms present in the fruits, which could interfere with the pure cultures and deviate the fermentation. Although there have been numerous experimental studies, some with good results, particularly with the use of *Lactobacillus plantarum*, the application of scientific findings on an industrial scale has been very limited.

Dark Olives

The industrial process for ripe olive production consists of three successive treatments with NaOH solutions of 1.5, 1.0 and 1.0% (w/v) which allows penetration to a depth of only 1 mm into the flesh through the skin and to the pit. During the intervals between each lye treatment, the fruits are suspended in water and air is bubbled. This results in a progressive darkening of the olives. The brown colour which develops during the oxidation process is due to polymer formation from orthodiphenols, hydroxytyrosol and caffeic acid. The oxidation rate increases as the pH becomes more alkaline. The colour obtained is not stable and to prevent this deterioration a solution of ferrous gluconate or lactate is added. These are the only additives for fixing the dark colour, that are authorized by the COI/CODEX International Regulation for Table Olives. The iron in addition to its effect on colour, has a role in the ripe olive texture similarly to other divalent cations such as calcium.

Safety and Nutritional Aspects of Fermented Products

Food microbiologists and sanitarians are interested in the microorganisms that produce toxic substances during fermentation or storage of fermented foods principally indigenous fermented foods. Although mycotogenic and pathogenic microorganisms can grow in unacceptable fermented or stored foods, traditional fermented products can be considered microbiologically safe. The nitrate content of fermented vegetables can vary depending on agricultural factors and fermentation conditions. Microorganisms convert nitrate into nitrite which in acidic conditions reacts with amines or amides to form nitrosamines which are carcinogenic. Reduction of the initial nitrate concentration can improve the safety of products. In this way several microorganisms, including lactic acid bacteria, are very useful. Some *Lactobacillus plantarum*, *Lactobacillus pentosus* and *Leuconostoc mesenteroides* strains contain nitrite reductase. However, at low pH values nitrate or nitrite reduction is not complete, so that the full beneficial effect of this enzyme is not seen in the fermented vegetable process. The quality and quantity of proteins is similar in raw and fermented vegetables, but the digestibility may be improved by the fermentation process. Regular consumers of traditional fermented products have recognized their positive effect.

See also: **Enterococcus. Lactobacillus**: Introduction; *Lactobacillus brevis*. **Leuconostoc**. **Preservatives**: Traditional Preservatives – Sodium Chloride. **Starter Cultures**: Uses in the Food Industry.

Further Reading

Buckenlüskes HJ (1993) Selection criteria for lactic acid bacteria to be used as starter cultures for various food commodities. *FEMS Microbiol. Rev.* 12: 253–272.

Beuchat LR (1987) *Traditional Fermented Food Products in Food and Beverage Micology*, 2nd edn. New York: Van Nostrand Reinhold.

Campbell-Platt G (1987) *Fermented Foods of the World: a Dictionary and Guide*. London: Butterworth.

Christ C, Lebault JM and Noel C (1980) *Preparation of Sauerkraut*. US Patent 4: 241–361. Washington DC.

Etchells JL, Fleming HP and Bell TB (1975) Factors Influencing the Growth of Lactic Acid Bacteria during the Fermentation of Brined Cucumbers. In: Carr et al (eds) *Lactic Acid Bacteria in Beverages and Food*. New York: Academic Press.

Fernández MJ, de Castro R, Garrido A et al (1985) In: *Biotecnología de la Aceituna de mesa*. CSIC: Madrid.

Fleming HP (1982) Fermented Vegetables. In: Rose AH (ed.) *Economic Microbiology Fermented Food*. New York: Academic Press.

Fleming HP, Daeschell MA, McFeeters RF and Pierson MD (1989) Butyric acid spoilage of fermented cucumbers. *J. Food Sci.* 54: 636–639.

Garrido A, García P, Brenes M and Romero C (1995) Iron content and colour of ripe olives. *Die Nahrung* 39: 67–76.

Jehano D and Le Guern J (1996) Les Légumes Fermentés. In: Bourgeois CM and Larpent JP (eds) *Microbiologie Alimentaire*. Vol. 2. *Aliments Fermentés et Fermentations Alimentaires*, 2nd edn. Paris: Tec & Doc.

Leclair J (1996) La fermentation de la choucroute. In: Bourgeois CM and Larpent JP (eds) *Microbiologie Alimentaire*. Vol. 2. *Ailments Fermentés et Fermentations Alimentaires*, 2nd edn. Paris: Tec & Doc.

Levin RE and Vaughn RH (1966) *Desulfovibrio destuarii* the causative agent of hydrogen sulfide spoilage of fermenting olive brines. *J. Food Sci.* 31: 768–772.

McDonald LC, Shieh DH, Fleming HP, McFeeters RF and Thompson RL (1994) Evaluation of malolactic fermentation. *Food Microbiol.* 10: 489–499.

Míngueus-Mosguera MI, Gandel-Rojas B, Montaño-Asquerino A and Garrido-Fernández J (1991) Determination of chlorophylls and carotenoids by high-performance liquid chromatography during olive lactic fermentation. *J. Chromatogr.* 585: 259–266.

Mountney GJ and Gould WA (1988) *Fermented Foods in Practical Food Microbiology and Technology*, 3rd edn. New York: Van Nostrand Reinhold.

Pelagatti O, Balloni W and Materassi R (1975) Controllo della stabilita biologica delle olive de mensa preparate con fermentazione lattiche controllate. *Ann. Iost. Sper. Elaiot.* V: 209–224.

Reddy NR and Pierson MD (1994) Reduction in antinutritional and toxic components in plants foods by fermentation. *Food Res. Int.* 27: 281–290.

Ruskin FR (1992) *Applications of Biotechnology to Traditional Fermented Foods*. Washington, DC: National Academy Press.

Fermented Meat Products

M C Montel, Station de Recherches sur la Viande, INRA, Saint Genès Champanelle, France

Introduction

For many centuries, traditional, empirical methods of improving the shelf life of meat have relied on salting or drying (with or without chopping), or smoking. Some methods involve microorganisms, which in addition to increasing the shelf life affect the texture and flavour of the final meat product, so that they are distinctly different from those of the raw material. Nowadays such products, called fermented meat products, are represented mainly by fermented sausages. The role of microorganisms in fresh sausages, cooked sausages and dry-cured ham is limited, and they are not considered in detail in this article.

The manufacturing methods of fermented meat products have changed and diversified over time, shown by the large variety of names given to these products. Individual regions tend to manufacture products with distinctive characteristics, and with names often related to their geographical origin. Different varieties of sausage are defined according to their formulation, their area of production and their physical and chemical characteristics. As the role of microorganisms in the development of the wholesomeness and organoleptic qualities of the product has become better understood, the manufacture of fermented meat products has become more reliable. Commercially made products have become increasingly safer, with increasingly standardized characteristics. Further progress is needed, to ensure that the distinctive characteristics of typical regional products can be retained at the same time.

Manufacture of Fermented Sausages

Fermented sausages are classified as semidry or dry sausages. They are generally prepared by mincing and mixing lean and fat meat with additives (e.g. nitrate, nitrite, NaCl, ascorbate) and seasonings (e.g. sugar, garlic, pepper). The proportion of meat and fat, the size of the pieces of meat, the type of meat used (e.g. pork, beef, veal) and the kind and the concentration of additives all contribute to the definition of the product. The meat mixture is always stuffed into casings, which can be natural (i.e. gut) or artificial (e.g. collagen) and which are available in a range of diameters. After stuffing, the products are fermented and air-dried. Fermentation and drying, which constitute ripening (the period between preparation and consumption), play an important role in the development of the characteristics of the final products.

The natural climatic conditions which used to create products typical of a particular area have now been replaced by controlled conditions. Nowadays, the desired characteristics are obtained by matching the extrinsic parameters, i.e. temperature, humidity and air velocity, to the intrinsic parameters, such as fat content, sodium content, degree of comminution and the nature and diameter of the casings. All these parameters vary considerably between different products, and are taken into account in the classification of semidry or dry sausages – which, unfortunately, is very difficult to simplify (**Table 1**). Semidry sausages are produced in the US by rapid fermentation at an elevated temperature, either without drying or with a short period of drying followed by cooking (at 60–68°C) at the end of fermentation. In Europe, the production processes used in the Mediterranean countries are generally quite different from those used in northern Europe. Traditional Mediterranean products have a slow ripening process, which allows the development of moulds and yeasts on the surface. In northern and central Europe, fermentation is generally combined with smoking, preventing the development of yeasts and moulds, and the drying period is shorter.

Microbiology of Fermentation

The raw meat mixture is naturally contaminated with many kinds of microorganisms. The initial microflora is complex, being derived from the raw meat, the natural casings, the working environment and the workers. It may include *Lactobacillus*, *Carnobacterium*, *Micrococcus*, *Staphylococcus*, *Pseudomonas*, *Acinetobacter*, *Enterococcus*, *Arthrobacter*, *Corynebacterium*, *Brochothrix* and *Listeria*, and also Enterobacteriaceae, yeasts and moulds.

Spoilage Bacteria

Gram-negative aerobic microorganisms (e.g. *Pseudomonas*, *Acinetobacter*) may be involved in the spoilage of the products, due to their proteolytic activity and/or the catabolism of sulphur-containing amino acids, which cause defects in texture and flavour. However, these disappear through the fermentation period. *Brochothrix thermosphacta*, a spoilage bacterium which produces cheesy odours, can grow to concentrations of 10^4 cfu g^{-1} during the fermentation period, but its concentration decreases during the ripening period. *Enterococcus* can reach concentrations of 10^4 cfu g^{-1} by the end of ripening in certain Mediterranean dry sausages.

Pathogenic Bacteria

The growth of many pathogenic bacteria, including *Clostridium*, *Staphylococcus aureus*, *Listeria* and Enterobacteriaceae (including *Salmonella*) is easily controlled if good practice is applied during the processing of the sausages (see below).

Desirable Flora

At the end of the fermentation period, lactic acid bacteria are generally the dominant bacterial flora

(see **Fig. 1**). The species *Lactobacillus curvatus*, *L. sakei*, *L. plantarum*, *L. viridescens*, *Carnobacterium divergens*, *C. piscicola* and *Leuconostoc* are present naturally, but *Pediococcus* is only found when inoculated as a starter culture. Their count generally exceeds 10^6 cfu g^{-1} and remains at this level during the whole ripening period. *Carnobacterium* is present during the fermentation period, but disappears afterwards. In dry sausages with a pH above 5, Gram-positive, catalase-positive cocci (including *Kocuria*, formerly *Micrococcus*, but mainly *Staphylococcus*) can constitute part of the dominant microflora. The species of *Staphylococcus* found, in addition to *S. aureus*, may include *S. warneri*, *S. saprophyticus*, *S. sciuri*, *S. cohnii*, *S. epidermis*, *S. xylosus* and *S. carnosus*.

During drying, yeasts (*Debaryomyces*, *Candida*, *Kluyveromyces*, *Hansenula*) and moulds (*Penicillium*, *Aspergillus*, *Mucor*, *Cladosporium*) are commonly found growing on the casings of sausages, mainly those made in southern Europe, which are ripened over a long time.

The desired organoleptic qualities (e.g. appearance, texture, flavour) of the products are associated with the dominance of lactic acid bacteria, which contribute to the inhibition of the spoilage and pathogenic bacteria. Certain species of coagulase-negative staphylococci (e.g. *S. carnosus*, *S. xylosus*) are also needed, to contribute to the development of colour and flavour. If the appropriate conditions are ensured, these bacteria grow naturally. However, to ensure reliability of the quality of fermented meat products, starter cultures are increasingly being used.

Starter Cultures for Fermented Sausages

The lactic acid bacteria commonly used in starter cultures belong to the species *Lactobacillus sakei*, *L. curvatus*, *L. plantarum*, *Pediococcus acidilactici* and *P. pentosaceus*. Starter cultures may also include the

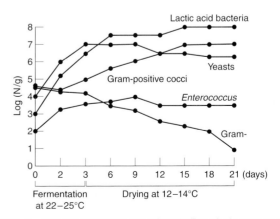

Figure 1 Growth of internal microflora in French dry sausage.

Gram-positive, catalase-positive cocci *Kocuria varians* (formerly *Micrococcus varians*), *Staphylococcus carnosus* subspp. *carnosus* and *utilis* and *S. xylosus*, but their inoculation is not systematic in northern European products and they are not used at all in American-type products. The total initial concentration of the starter bacteria is 10^6 cfu g^{-1}. Yeasts (*Debaryomyces hansenii, Candida famata*) and moulds (*Penicillium nalviogense, P. chrysogenum* or *P. camemberti*) are added to the surface of the casing.

Suitable strains must grow well in the physical and chemical conditions encountered in the sausages. They must be harmless to human health and non-pathogenic, and must not produce toxins or amines. Their biochemical activity must be able to control acidification, must ensure the stability of the colour of the product, and must affect its flavour positively (**Table 2**).

The fermented meat products most commonly prepared with starter cultures are sausages, but microbial activity is also used for increasing the shelf life, accelerating the ripening and improving the flavour of dry-cured hams and bacons. The microorganisms are inoculated by brine-salting or by the injection of brine. The same strains of bacteria are used as in starter cultures for dry sausage, but *Vibrio* species can also be inoculated.

Health-related Aspects of Fermented Meat Products

The ease of digestibility, reduction of blood cholesterol level, and probiotic effects claimed for some other fermented products has not been clearly demonstrated for fermented meat products. For example, it is not obvious that the beneficial effect of *Lactobacillus casei, L. acidophilus* or *Bifidobacterium* in dairy products would be shown in meat products, even in the case of species able to grow in meat products. The products of the oxidation of lipids – lipid peroxides and oxidized cholesterol – may represent a health risk, but this has not been evaluated for dry sausages. Safety and quality, with regard to pathogenic poisoning and bacterial spoilage, are best guaranteed by the use of starter cultures and by appropriate process controls.

Inhibition of Undesirable Bacteria

Many interacting intrinsic and extrinsic factors influence the development and stability of the microbial flora during ripening. Drying at temperatures below 25°C and the addition of nitrite, nitrate or sodium salts, either alone or in combination, inhibit *Salmonella* and *Clostridium*. The reduction of the redox potential during fermentation suppresses the aero-

philic bacteria and favours the desirable lactic acid bacteria. The reduction of water activity (to 0.96–0.86) during drying also stabilizes the product, by controlling undesirable bacteria such as *Clostridium* and *Staphylococcus aureus*. Sausages that have been ripened and dried over longer periods are safer with respect to enterohaemorrhagic *Escherichia coli* than those ripened for shorter periods, because the decreased a_w has a lethal effect on this bacterium.

Lactic acid bacteria are favoured by these conditions, and contribute significantly to the microbial safety of the product. Further, they produce inhibitory substances, including organic acids and bacteriocins, and compete for nutrients. Thus carbohydrate catabolism during fermentation is crucial. It is attributed mainly to *Lactobacillus* and *Pediococcus*, which can ferment glucose homofermentatively, producing mainly D- and L-lactate. Fermentation by heterofermentative species of *Lactobacillus* or *Leuconostoc* can also occur, in which case acetate, CO_2 and ethanol are produced. The lactic and acetic acids prevent the growth of undesirable bacteria. In the absence of O_2, non-dissociated lactic acid is able to inhibit *Brochothrix thermosphacta*. Acid production and the corresponding fall in pH are influenced by many factors, e.g. type and concentrations of sugars added, microorganisms present, temperature during fermentation and the diameter of the casings. A decreased pH contributes to the safety of the products: many pathogenic and spoilage microorganisms are acid-sensitive and do not grow at pH values below 5.5. However, enterohaemorrhagic *E. coli* has a high acid tolerance, and the most effective method of destroying it is by heating (at 65°C for 4 min).

Listeria can grow at pH values greater than 5, in a temperature range of 3–35°C, with 10% NaCl or in the presence of nitrite, but can be inhibited by the bacteriocins produced by lactic acid bacteria. However, the effect of bacteriocins is more obvious in vitro than in fermented meat products. The bacteriocins sakacin A, P and K and curvacin A, active against *L. monocytogenes* and *Enterococcus*, have been reported as being produced by strains of *Lactobacillus sakei* and *L. curvatus* respectively. Pediocins produced by *P. acidilactici* have also been considered as antilisterial agents.

Given good industrial practice, dry sausages have rarely been involved in salmonellosis or listeriosis.

Toxins

The formation of botulinum toxin in meat is prevented by the sensitivity of *Clostridium botulinum* to NaCl, nitrite, $a_w < 0.97$, pH and temperature, which also inhibit the growth of this bacterium. The production of enterotoxins by *Staphylococcus aureus* or

Table 1 Classification of semidry and dry sausages

	Semidry sausage	Dry sausage	
		Northern European type	Southern European type
Examples	Summer sausages	German salamis	Italian salamis (Milanese, Calebrese)
	German cervelat	Danish salamis	Saucissons secs
	Bologna sausages		Spanish chorizos
Raw mixture			
Meat : fat	Pork or beef, lean and fat	Lean pork : lean beef : fat pork (1 : 1 : 1)	Lean pork : fat pork (2 : 1)
Sugars (e.g. glucose, lactose, sucrose)	0.3–1.5%	0.3–0.8%	0–0.4%
Nitrate	–	–	< 300 p.p.m.
Nitrite	0–150 p.p.m	20–200 p.p.m	20–200 p.p.m
NaCl	2–2.5%	2–2.5%	2–2.5%
Seasoning (e.g. pepper, garlic, cardamom)	++	++	++
Starter cultures (10^6 cfu g^{-1})	Yes	Yes	Yes
Lactobacillus sakei, L. curvatus, L. plantarum	+	++	+++
Pediococcus acidilactici, P. pentosaceus	+++	+++	+
Staphylococcus carnosus, S. xylosus	–	++ or –	++
Kocuria varians	–	++ or –	+
Penicillium chrysogenum, P. nalviogense	+	–	++
Debaryomyces hansenii, Candida lipolytica	+	–	++
Fermentation period			
Time/temperature/relative humidity	15–20 h/27–41°C/90% (US)	18–48 h/20–30°C/58–95%	Day 1/22–24°C/94–96%
	18–48 h/20–32°C/85–95% (Germany)		Day 2/20–22°C/90–92%
			Day 3/18–20°C/85–88%
			or
			2–3 days or 22–25°C/ 90–95°C
Drying period			
Time/temperature/relative humidity	2–3 days/10°C/68–72% (US)	1–3 weeks/12–15°C/ 75–80%	4–6 weeks/12–15°C/ 75–78%
	10–25 days (Germany)		or
			8–14 weeks (traditional)
Method of production			
Smoking	Yes	Yes	No
Cooking	Yes	No	No
Characteristics of product			
Final pH	4.4–5	4.6–5.1	5.1–5.5
Final a_w	0.93–0.98	0.92–0.94	0.85–0.86
Water content	40–50%	30–40%	20–30%
Moisture/protein ratio (w/w)	2.3–3.7	2–2.3	1.6–1.9

+, occasionally used
++, frequently used
+++, regularly used
–, not used

by certain strains of coagulase-negative staphylococci (e.g. *S. warneri*, *S. saprophyticus*, *S. xylosus*) may be a problem, even in the absence of the growth of these bacteria. However, the production of enterotoxins by *S. aureus* can be inhibited by a high concentration of NaCl combined with low a_w and low pH, under anaerobic conditions. To minimize risk, great care must be taken to avoid enterotoxin-producing staphylococci when selecting strains for starter cultures.

The highly toxinogenic mould *Aspergillus flavus*, which can produce aflatoxins, is occasionally found on the surface of ripened salamis. Its growth, and toxin formation, are prevented by storage at low temperature and low humidity (75–90%). *Penicillium* species are predominant, and are capable of syn-

Table 2 Biochemical activity of microorganisms involved in the manufacture of fermented meat products

Metabolic process	Product	Tissue and microbial enzymes involved	Effect on qualities of the product	Strength of effect
Degradation of carbohydrates	Lactate and acetate (causing a decrease in pH)	Enzymes from lactic acid bacteria	Inhibition of undesirable bacteria	+++
			Development of texture and colour	++
	Diacetyl, acetoin	Enzymes from *Staphylococcus*	Development of flavour	+ (or –, in excess)
Degradation of proteins	Peptides	Tissue proteinases	Development of texture	– (in excess)
Degradation of peptides	Amino acids	Tissue peptidase, microbial peptidases	Development of flavour	++
Amino acid catabolism	Amines	Bacterial enzymes from contaminating flora	Reduces safety	––
	Aromatic compounds	Strecker reaction bacterial enzymes from *Staphylococcus* or lactic acid bacteria	Development of flavour (branched-chain amino acids)	++
			Development of flavour (sulphur-containing amino acids)	
Degradation of lipids	Fatty acids	Tissue lipase (mainly), *Staphylococcus* lipases (very minor contribution)	Development of flavour	++ (or –, in excess)
Oxidation of fatty acids	Aldehydes	Chemical reactions	Development of flavour	–
		Antioxidant enzymes from *Staphylococcus* and lactic acid bacteria	Development of flavour and colour	++
	Ketones	Tissue or fungal enzymes	Development of flavour	++
Reduction of nitrate		Nitrate reductase from *Staphylococcus*	Development of colour and flavour (limit oxidation)	+++
Reduction of nitrite		Nitrite reductase from lactic acid bacteria	Increases safety	++

thesizing mycotoxins. The growth of undesirable moulds can be prevented by the inoculation of suitable strains of *P. chrysogenum* or *P. nalviogense*, which do not produce mycotoxins. The growth of some undesirable moulds can also be inhibited by treatment with potassium sorbate or pimaricin.

Amines

Given good industrial practice, the consumption of fermented meat products does not represent a risk to the consumer, even if biogenic amines are formed during fermentation or ripening. The most toxic amine is histamine, but its content in dry sausages is low ($< 10\,\mu\mathrm{g\,g}^{-1}$) and at this concentration it does not induce cutaneous reactions, oedema or heart palpitations. The biogenic amines which are most frequently detected in fermented meat products are tyramine (1–56 mg per 100 g dry matter) and putrescine (1–19 mg per 100 g dry matter), but phenylethylamine, cadaverine, tryptamine, spermine and spermidine are also found. Tyramine, phenylethylamine and tryptamine can induce allergic reactions. They are not dangerous to healthy people, but can be more so in people who are immunodepressed or being treated with drugs.

As the factors involved in amine production become better understood, it will be controlled more easily. Microbial enzymes play an important role in amine production. For example, histamine can be produced by Enterobacteriaceae; and *Enterococcus* species, *Carnobacterium divergens*, *C. piscicola* and some strains of *Lactobacillus curvatus* are able to decarboxylate tyrosine, forming tyramine, and phenylalanine, forming phenylethylamine. Putrescine can be produced by *Pseudomonas* and Enterobacteriaceae. The production of amines in fermented meat products can be kept under control by minimizing microbial contamination of the raw materials and by selecting starter cultures which do not produce amines. Production can be further limited by controlling proteolysis, and hence restricting the availability of amino acids, by decreasing the pH (to < 5) and by lowering the temperature. The combination of nitrite with putrescine and cadaverine may result in the formation of nitrosamines. These are detected sporadically in dry sausages, but always at very low concentrations – the processing conditions are not favourable, and the free nitrite content is generally low because nitrites bind with meat pigment proteins or are converted into

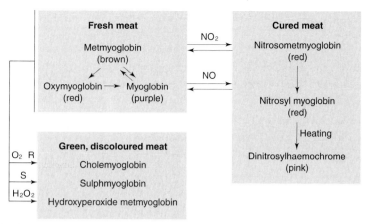

Figure 2 Development of colour in dry sausage.

other compounds. Nitrosamines are most commonly found in fried bacon and cooked sausages.

Development of Texture

Acidification is very important in the development of product texture. At pH values below 5.4, the meat proteins are close to their isoelectric point: thus they are denatured and coagulated, and their water-binding capacity decreases. These changes induce the release of water molecules, and so contribute to moisture loss and accelerate drying.

Development of Colour

The characteristic red colour of sausages results from the formation of either nitrosylmyoglobin or the stabilized chromogen nitrosohaemochrome, which are produced by the reaction between nitric oxide or nitrogen dioxide with myoglobin or metmyoglobin (**Fig. 2**). Curing is initiated by the reduction of nitrate (the curing agent) to nitrite, involving the nitrate reductase of staphylococci and *Kocuria*. Meat can be cured by the direct addition of nitrite. The nitrite is reduced to nitric oxide or nitrogen dioxide by the haem-independent nitrite reductase of *Lactobacillus plantarum*, *L. sakei* or *L. curvatus*. These species accelerate nitric oxide production by lowering the pH and by creating a reducing atmosphere. Nitrite is also degraded to ammonia, by the haem-dependent nitrite reductase of *L. plantarum* or *L. pentosus*.

Colour defects, in the form of greenish discoloration, can be produced by the reaction between H_2O_2 and myoglobin, to produce hydroxyperoxide metmyoglobin. H_2O_2 can be produced by bacteria found in sausages, but is generally broken down by the catalase of staphylococci or the pseudocatalase of *L. sakei*. Sulphydryl-producing bacteria can also cause a green discoloration.

Development of Flavour

The fermentation of meat products provides flavour, which is distinctly different from that of the raw material, and each variety of fermented product has its own characteristic taste and aroma. Consumer preferences regarding flavour vary – for example, the strong aroma of mould-ripened salami is liked by Mediterranean consumers, but is rejected by northern European and US consumers, who prefer products with a more acid taste and aroma. Chemically, flavour results from a subtle balance between different volatile and non-volatile compounds: some of the volatile compounds have been identified by the use of gas chromatography and mass spectrometry. Some volatile compounds are present in or arise from the ingredients added to the sausage mixture, e.g. seasonings, sugar and spices – for example, most of the terpenes originate from pepper, and sulphur compounds are present in garlic and onions. Aromatic hydrocarbons arise from smoking agents, and also from enzymatic reactions occurring simultaneously during fermentation and drying, which involve microbial and tissue enzymes as well as purely chemical reactions. The degradation of carbohydrates, lipids, proteins, fatty acids and amino acids yields a variety of compounds which contribute to the characteristic flavour of dry sausage.

Acids and Other Products of Carbohydrate Catabolism

Acid production and acidification play an important part in flavour development. Products with an acid taste are characterized by a low pH and a high content of D-lactate and L-lactate ($> 10 \, \mathrm{mg \, g^{-1}}$). It has been suggested that a high D-lactate concentration ($> 5 \, \mathrm{mg \, g^{-1}}$) is responsible for the acid taste; it is unclear whether other acids (e.g. acetic or formic acids or short-chain fatty acids) also contribute. The pH value is an important factor in controlling the microbial and tissue reactions involved in the development of flavour.

Products with a dominant 'dry sausage' aroma are inoculated with *Lactobacillus sakei*, *Pediococcus acidilactici* or *P. pentosaceus* in combination with *Staphylococcus carnosus* or *S. xylosus*. This results in a low desorption of diacetyl and acetoin – high desorption is undesirable, as it leads to a buttery or dairy aroma. These compounds may be produced following the inoculation of *S. saprophyticus* or *S. warneri*, which are able to catabolize pyruvate into diacetyl and acetoin.

Amino acids

The peptide and amino acid content of fermented meat products increases slightly during ripening, to an extent depending on the particular amino acid. Hence the final amino acid composition differs from product to product. The increased amino acid content is attributed to proteolysis and the activities of peptidases.

Initially, proteins are degraded into peptides, by endogenous enzymes (cathepsins D, B, H and L) – the degradation of the meat proteins actin, myosin and troponin is unaffected by the presence of bacteria. Most strains of lactic acid bacteria and catalase-positive, Gram-positive cocci have weak proteolytic activity. However, lactic acid bacteria also influence protein degradation by causing a decrease in pH, which results in reduced protein solubility and the increased activity of tissue cathepsin D. The further degradation of peptides into amino acids can be attributed to both tissue and microbial enzymes, as demonstrated by the lower amino acid concentration in sausages containing antibiotics compared with that in sausages containing bacteria. Peptidase activity has been found in lactic acid bacteria and in strains of *Staphylococcus* and *Kocuria*, but it is low at pH values encountered in meat products (i.e. < 6).

The contribution of amino acids and peptides to the flavour of these products is not well understood. Some amino acids and peptides have their own flavour, and have been identified in fermented meat products. However, the flavour of the final product is not influenced by changes in its amino acid profile, as achieved by adding different starter cultures, or by increasing the amino acid concentration by adding endogenous proteinase, even if maturation is accelerated. Further, a strong 'dry sausage' aroma can be obtained without increasing endogenous proteolysis. The amino acid content of dry sausages depends on the rates of production and catabolism.

Aromatic Compounds from Amino Acids

The degradation of amino acids into volatile compounds apparently plays an important role in the characteristic flavour of dry sausages. Dimethyl disulphide and carbon disulphide from methionine must be avoided, because they give putrid odours to both fresh and processed meats. In contrast, important components include aldehydes (e.g. 2- or 3-methyl butanal, 2-methyl propanal and benzaldehyde); alcohols (e.g. 2- or 3-methyl butanol); and acids (e.g. 2- or 3-methyl butanoic acid, 2-methyl propionic acid and phenyl acetic acid), resulting from the degradation of branched-chain amino acids (including leucine, isoleucine and valine) or phenylalanine. These have interesting aromatic properties (**Table 3**).

In dry sausage, *Staphylococcus* seems to play a more important role than *Lactobacillus sakei*, *Pediococcus acidilactici* or *P. pentosaceus* in the production of these volatile compounds. Dry sausage inoculated with *S. carnosus* contains higher quantities of 2- or 3-methyl butanal and 3-methyl butanoic acid than that inoculated with *S. warneri* and *S. saprophyticus*. The catabolism of branched-chain amino acids into aromatic compounds may involve the Strecker reaction, and may also involve bacteria. The bacterial degradation of leucine seems to involve chain reactions, beginning with transamination to produce α-ketoisocaproic acid. This is converted by non-oxidative decarboxylation into 3-methyl butanal, which is then oxidized to acids or reduced to alcohols. 3-Methylbutanoic acid may also be formed directly from α-ketoisocaproic acid. Leucine can be degraded in this way by *S. xylosus*, *S. carnosus*, *S. warneri* and *S. saprophyticus*, which produce mainly 3-methyl butanoic acid with smaller quantities of 3-methyl butanal and 3-methyl butanol. Leucine is broken down less actively by *L. sakei*, *L. curvatus*, *L. plantarum* and *P. acidilactici*. *Carnobacterium piscicola* produced a high amount of 3-methylbutanol.

Fatty Acids

Throughout ripening, the concentrations of free fatty acids and diglycerides increase whereas those of triglycerides decrease. The degradation of phospholipids occurs only in the later stages of ripening. Most of the lipolysis (60–100%) in fermented sausages is caused by lipases and phospholipases from pork fat, even if lipolytic microorganisms are present. *Lactobacillus* species and *Staphylococcus carnosus* are weakly lipolytic, whereas *S. saprophyticus*, *S. warneri* and some strains of *S. xylosus* have pronounced lipolytic activity in pork fat. Lipase activity has also been shown in yeasts and moulds. However, the inhibition of the bacterial flora, by adding antibiotics, does not

Table 3 Volatile compounds related to flavour of dry sausages

Volatile compounds	Odour	Threshold value		Origin
		In water	In oil	
Alkanes, alkenes				
pentane, heptane, octane, decane, dodecane	Odourless			Lipid oxidation
1-octene, 2-octene, 4-octene, 1-heptene	Odourless			Lipid oxidation
Phenolic compounds				
phenol, 4-methylphenol, 2-methoxy phenol	Smoked			Smoke
Terpenes				
α-pipene, β-pipene terpinolene, terpenoid, δ3-carene salinene	Curry, spice			Pepper
limonene	Lemon	0.2		Pepper
Furans				
2-methylfuran, propylfuran	Odourless			Lipid oxidation
Sulphur compounds				
dimethyl disulphide, methanediol, dimethyl trisulphide	Cabbage			
3-ethylthio-1-propene, 1-methylthio-1-propene, allyl-1-thiol, allyl methylsulphide	Garlic			Garlic
Aldehydes				
pentanal	Woody, fruity			Lipid oxidation
hexanal	Green leaves, fatty	0.015	0.3–0.6	Lipid oxidation
heptanal	Fatty, heavy	0.003		Lipid oxidation
nonanal	Fatty, waxy	0.045	0.32	Lipid oxidation
2-methyl butanal	Slightly fruity	0.003		Isoleucine catabolism
3-methyl butanal	Slightly fruity, malty	0.0002		Leucine catabolism
2-methyl propanal	Malty, buttery			Valine catabolism
benzaldehyde	Almond, bitter			Phenylalanine catabolism
phenylacetaldehyde	Harsh, hawthorn			Phenylalanine catabolism
Alcohols				
ethanol	Ether	0.8–100		Carbohydrate catabolism
2-methyl propanol	Vinegar, pungent			Valine catabolism
1,3-butanediol	Fat			Carbohydrate catabolism
2,3-butanediol	Buttery			
hexanol	Fruity	2.5		Lipid oxidation
1-octen-3-ol	Mushroom	0.001	0.0075–0.01	Lipid oxidation
phenylethanol	Floral	0.2		Phenylalanine catabolism
1-pentanol	Fruity	4		Lipid oxidation
3-methyl butanol	Fruity	4.75		Leucine catabolism
Ketones				
4-methyl-2-pentanone	Grass			Lipid oxidation
3-methyl-2-pentanone	Fruity	22		Lipid oxidation
2-pentanone	Wine, fruity			Lipid oxidation
2-heptanone	Blue cheese	0.05		Lipid oxidation
2-octanone	Fruity	0.15–1		Lipid oxidation
2-nonanone	Fruity	30		Lipid oxidation
2–3-pentanedione	Butter	0.04		Carbohydrate catabolism
2–3-butanedione (diacetyl)	Fat, butter			Carbohydrate catabolism
3-OH-2-butanone (acetoin)	Dairy products			Carbohydrate catabolism

Table 3—continued   Volatile compounds related to flavour of dry sausages

Volatile compounds	Odour	Threshold value		Origin
		In water	In oil	
Carboxylic acids				
acetic acid	Vinegar, pungent	22–54		Carbohydrate catabolism
3-methyl butanoic acid	Sweat socks	0.07		Leucine catabolism
2-methyl butanoic acid	Sweat socks			Isoleucine catabolism
2-methyl propanoic acid	Rancid butter, acid			Valine catabolism
butanoic, pentanoic acid	Blue cheese			Lipid oxidation
hexanoic acid	Cheesy	5–51		Lipid oxidation
Esters				
methyl acetate	Sweet, ether			Carbohydrate catabolism
ethyl acetate	Sweet, pineapple	5		Fermentation/ lipid oxidation
ethyl propanoate	Fruity, green	0.009		Fermentation/ lipid oxidation
ethyl butanoate	Pineapple			Fermentation/lipid oxidation
ethyl 3-methylbutanoate	Fruity			Fermentation/ leucine catabolism
ethyl 2-methylbutanoate	Acid, fruity			Fermentation/ isoleucine catabolism

reduce the free fatty acid content of dry sausage and the inoculation of highly lipolytic strains into dry sausage with a low pH causes only a very slight increase in free fatty acid content. In contrast, the inoculation of *Staphylococcus* species alone, without lactic acid bacteria, increases the free fatty acid content fivefold in comparison with that of sterile controls. Thus the inhibition of bacterial lipolytic activity in dry sausage can be explained by its low pH value (< 5.5). Low temperature and the presence of NaCl are also not conducive to bacterial lipolytic activity – indeed, most staphylococcal lipases have maximum activity at pH values over 7.5, and 37°C and without NaCl.

The role of fatty acids in product flavour has not yet been demonstrated. Short-chain fatty acids have sour tastes, but this characteristic diminishes with increasing chain length. The addition of exogenous lipases to dry sausage significantly and rapidly increases free fatty acid concentration, reducing the time needed for maturation but not reliably improving the flavour. Moreover, pronounced lipolytic activity does not seem to be necessary for the development of an intense 'dry sausage' aroma – very weakly lipolytic strains of *S. carnosus* or *S. xylosus* are capable of generating a strong aroma in a model of dry sausage.

Aromatic Compounds from Fatty Acids

Fatty acids can be oxidized into alkanes, alkenes, aldehydes, alcohols, ketones and acids by chemical or enzymatic reactions. Alkanes and alkenes are odourless, but the other compounds have very low sensory thresholds. Thus even if their content in sausages is very low, of the order of parts per million, they have a real effect on the flavour. Moreover, in sausages without spices they represent 60% of the volatile fraction. The compounds largely involved in the aroma of dry sausage are 2-methyl ketones (2-pentanone, 2-hexanone, 2-heptanone). A predominance of alkanals (hexanal, nonanal) is undesirable, causing a rancid aroma.

Chemical oxidation is dependent on the presence of oxidizing agents (e.g. iron, haem compounds, sodium salts) and antioxidants (e.g. nitrites, spices, SO_2). *Staphylococcus* and lactic acid bacteria can control oxidation. *Staphylococcus* and *Kocuria* (formerly *Micrococcus*), *Lactobacillus sakei*, *L. plantarum* and certain yeasts and moulds can break down H_2O_2 by their catalase or pseudocatalase activity. The nitrate reductase activity of bacteria (see above) can limit oxidation, because nitrites are strong antioxidants.

Enzymatic oxidation (β-oxidation) breaks down saturated fatty acids, with the formation of acetyl-CoA and propionyl-CoA, and can result in the liberation of the corresponding fatty acids. Methyl ketones or secondary alcohols can be produced by the β-oxidation accomplished by some fungi.

Esters

Ethyl esters are present in fermented meat products, and their fruity aroma contributes to that of the products. They result from esterification, due to either chemical reactions or microbial esterase activity (e.g. *Staphylococcus* esterase).

See also: **Bifidobacterium**. **Clostridium**: Introduction; *Clostridium botulinum*. **Dried Foods**. **Escherichia coli**: *Escherichia coli*. **Fermented Foods**: Fermentations of the Far East. **Lactobacillus**: Introduction; *Lactobacillus casei*. **Preservatives**: Traditional Preservatives – Sodium Chloride; Permitted Preservatives – Nitrate and Nitrite. **Salmonella**: Introduction. **Spoilage Problems**: Problems caused by Bacteria. **Starter Cultures**: Uses in the Food Industry.

Further Reading

Campbell-Platt G and Cook PE (eds) (1995) *Fermented Meats*. London: Blackie.

Hammes WP and Knauf HJ (1994) Starters in the processing of meat products. *Meat Science* 36: 155–168.

Johansson G, Berdagué JL, Larsson M, Tran N and Borch E (1994) Lipolysis, proteolysis and formation of volatile components during ripening of a fermented sausage with *Pediococcus pentosaceus* and *Staphylococcus xylosus* as starter cultures. *Meat Science* 38: 203–218.

Ladisko D and Lougovois V (1990) Lipid oxidation in muscle foods: a review. *Food Chemistry* 35: 295–314.

Lücke FK (1985) Microbiology of fermented foods. In: Wood BJB (ed.) *Fermented Sausages*. P. 41. Amsterdam: Elsevier.

Montel MC, Reitz J, Talon R, Berdagué JL and Rousset-Akrim S (1996) Biochemical activities of Micrococcaceae and their effects on the aromatic profiles and odours of a dry sausage model. *Food Microbiology* 13: 489–499.

Nychas GJE and Arkoudelos JS (1990) Staphylococci: their role in fermented sausages. *Journal of Applied Bacteriology* (Symposium Supplement) 167–188.

Ten Brink B, Damink C, Joosten HM and Huis int Veld JHJ (1990) Occurrence and formation of biologically active amines in foods. *International Journal of Food Microbiology* 11: 73–84.

Fermented Fish Products

J H Al-Jedah and **M Z Ali**, Central Laboratories, Ministry of Public Health, Qatar

Fish is the major source of animal protein in many rural and urban communities in the Far and Near East, and to a lesser extent in some African and European societies. Due to the perishable nature of fresh fish, various fish preservation techniques have been developed through history. Both inland and marine fish produce suffers from wide seasonal variation in catch volume. To maintain a stable fish supply during times of scarcity, drying, salting and fermentation were widely practised at cottage-level processing. Such preservation techniques could be used alone or in combinations. Moreover, for the salvage of a small fish harvest which is otherwise wasted, these techniques are employed on the shores immediately after catching and sorting. Fermentation combined with drying and/or salting is probably more appealing to fishing communities, providing a mélange of flavoured products throughout the world. Asia can be singled out as one of the richest geographical areas in fish fermentation, with a prominent culture of processing.

Fish sauce and paste are the two main products of fish fermentation. To classify all fermented fish products under these two categories is somewhat difficult, in view of the lack of standardization in cottage-level fish fermentation throughout the world. Fish sauce is a somewhat thinner product, with additives such as spices, fruits and cereal adjuncts. Fish paste is a thicker product with whole fermented fish, undergoing minimal straining during the course of the fermentation. Various examples of fermented fish products throughout Asia, the Middle East and Europe are covered, together with characterization of the finished products. Due to the relatively high salt content and the flavour, some fermented fish products are not used on their own, but as a condiment or as flavouring agents added to bland rice and cereal staples. The end product of sauce or paste is a brown liquid resembling soy sauce.

The term fermented fish is an ambiguous term, in view of the large number of cottage products. As an example, over 60 different types are available in Southeast Asia. The term fermented fish product can be defined as a type of processed fish, or part thereof, obtained by maceration or hydrolysis of chemical constituents – especially proteins – through microbial and natural enzymes in the process, normally in the presence of salt. Sometimes spices and cereal adjuncts are added to the fermenting fish liquor. Due to the subsistence margins of production, it is difficult to estimate the volumes of the product even in national statistics. They are mainly produced in villages at cottage level.

The Nature of Fermented Fish Products

The overall processing steps of fish paste and sauce are outlined in **Figure 1**. After the fish have been caught, small fish such as sardines are either dried or used fresh for fermentation (**Fig. 2**). These products undergo spontaneous fermentation with or without de-gutting or de-boning, in salt liquors. Fish paste is kept at ambient temperature for 1 month to several

Fish (fresh or dry)
↓
Cleaning of fresh or dry fish
↓
Addition of salt in large containers
↓
Thick fish paste product
↓
Maceration, mixing and dilution in water
↓
Addition of ground spices, cereal adjuncts
or fruit and filling in containers
↓
Fermentation at room temperature
↓
Ready-to-serve sauce (2 weeks to 1 year)

Figure 1   The overall processing of fish paste and sauce.

Figure 2   Dried sardines.

Figure 3   Fish paste.

months (**Fig. 3**). Various bacterial and natural enzymes proceed to solubilize fish proteins at this stage. In fact, various bacterial strains can be isolated during this phase, while the fungal and yeast loads tend to diminish throughout fermentation. Unlike conventional carbohydrate fermentations, fish is a protein-rich substrate, with minimal sugar content. Therefore, the fermentation process of fish mainly addresses the protein fraction, and the lipid fatty components of the raw material.

The processing of fish at the end of this phase produces a coarse, thick brown liquor, often referred to as fish paste. The paste is optionally processed further, resulting in a thinner product, referred to as a sauce (**Fig. 4**). For sauce production, the thick paste is macerated, mashed and mixed with other additives, according to the nature of the sauce and its country of origin. These additives include spices such as cumin, fennel seeds, pepper, cinnamon, ginger, coriander and thyme. Rice, barley or wheat adjuncts are added in some cases, probably to boost the activity of the lactic acid microflora. The fish paste with additives is left to ferment for a period of 2 weeks up to a number of years, and consumed when needed as a sauce.

The proper fermentation of fish sauce involves lactic acid bacteria such as *Lactococcus* spp., *Lactobacillus brevis* and *Pediococcus* spp. Other bacterial species have often been encountered at early stages of fish paste and sauce fermentation and decrease in number through the process. Such strains include *Bacillus* spp., *Micrococcus* spp. and *Pseudomonas* spp. The latter group are contaminants of the raw fish, where lactic acid bacteria tend to dominate at the end of the fermentation process. Extremely halophilic bacterial strains were isolated in Thai fish sauce (nam pla) and caught in coastal areas. In protein hydrolysation an array of bacterial enzymes and natural fish gut enzymes are responsible for breakdown during the solubilization process. In some cases, enzymes from plant and animal sources such as papain, ficin, bromelain or trypsin and chymosin have been used in an attempt to accelerate the rate of proteolysis in various fish fermentation procedures in Asia. The results have yet to be confirmed.

The flavour of fish paste and sauce is attributed to three groups:

1. Volatile components of fatty acids, methyl ketones, aldehydes and esters. Free fatty acids were reported to increase by the end of fermentation, at the expense of triglycerides. Aerobic fermentation with an ample supply of oxygen tends to result in a richer profile of volatile fatty acids. This type is often referred to as cheesy flavour, probably generated by lactic acid bacteria. However, there is evidence for deamination processes during the relatively long fermentation time that leads to conversion of the carbon skeleton of amino acids into the equivalent volatile fatty acids, such as the degradation of isoleucine into acetic acid. As an example, the common volatile fractions that contribute to the flavour of Thai fish sauce are acetic, propionic, isobutyric and isovaleric acids.
2. A meaty flavour is evident, but not fully explored due to the inherent complexities of the process.
3. Ammonia and amines such as trimethyl amine impart a characteristic flavour in fish fermentation.

Importance of Fermented Fish in Coastal Areas

Although fish is a good source of protein, it is a highly perishable commodity. It offers microorganisms good nutrient availability and suitable water activity (a_w) and moderate pH. Also, most fish caught in nets are too small to be sold commercially or are not generally consumed as fresh fish. The lack of or high cost of modern facilities and technologies of chilling, freezing and canning in developing countries and the very short life of fish have motivated a search for low-cost methods of preservation of large quantities of fish: such techniques include drying, salting and fermentation. Fermentation of fish is the main method used in coastal areas especially in Southeast Asia, in order to extend the shelf life of the fish, as well as to provide people with animal protein, since fish seems to be the main source of protein in the area.

The shelf life of fish in developed countries is greater, due to the preservation technologies. In the developing world, where these technologies are not available or too expensive, other methods are used to

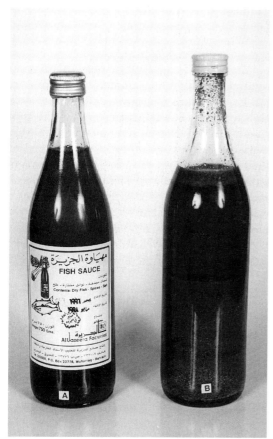

Figure 4 Fish sauce (Mehiawuh) (**A**) Factory-made (**B**) Home-made.

considered to be strong proteolytic spoilage groups, giving an objectionable red colour to the product.

The manufacturing procedures of fermented fish include packing all or part of the fish with salt (20–50%) in earthenware pots or jars which are tightly sealed. These pots are buried in the ground or exposed to the sun depending on whether the climate is rainy or sunny. Fermentable carbohydrate sources such as cooked rice and molasses may be added to provide the lactic acid bacteria with the energy to assist in the production of lactic acid. This plays a substantial role in fermentation and inhibition of the growth of spoilage microorganisms.

High salt concentrations are used in fermented fish products to combat microbial spoilage. Therefore, in the presence of high salt content, proteolytic enzymes become more important in the hydrolysation of fish and protein solubilization. Such proteolytic activity is not necessarily vigorous and a fairly long time is needed to accomplish the desired characteristic in the final products. Such enzyme activity is controlled by the fish species, the body part used in the process and the fishing season as well as the interacting microbial species during processing. Pelagic types of fish have higher proteolytic activity than other ground fish

Table 1 Marine and freshwater fish and shellfish mainly used in fermentation processes

Fish	Common name
Marine fish	
Rastrelliger spp.	Mackerel
Stolephorus indicus	Anchovy
Engraulis spp.	Anchovy
Stolephorus commersonii	Anchovy
Sardinella longiceps	Herring
Sardinella perforata	Deep-bodied herring
Sardinella fimbriata	Fimbriated herring
Leiognathus equullus	Slip-mouth
Decapterus macrosoma	Round scad
Freshwater fish	
Ophicephalus striatus	Mudfish
Trichogaster	Gourami
Anabas testudineus	Climbing perch
Clarias spp.	Catfish
Cirrhinus	Carp
Mugil cephalus	Bulti
Shellfish	
Ostrea/Crassosrea	Oyster
Atya spp.	Shrimp
Mytilus	Mussel
Ommastrephes spp.	Squid

preserve fish; the methods depend on factors such as the weather and energy available. Salting, drying and fermentation are mainly used in these areas. In Southeast Asia, where the climate is rainy, fermentation is one of the best methods used to preserve fish and extend its shelf life.

Examples of Fermented Fish Products

Fish sauce and fish paste are the most widely known and consumed fermented fish products. Different types of sea and freshwater fish can be used in the fermentation process (**Table 1**); however, small fish is most frequently used.

Fermented fish is produced in different parts of the world, and is the main meal in some areas.

Asian Fermented Fish Products

Large quantities of small fish or shrimps are fermented to produce fish or shrimp sauce and paste in Southeast Asia.

Fish Sauce Fish sauce is a brown liquid with a characteristic meaty flavour and aroma. It is mainly used as a condiment to flavour rice and other cereal dishes. Different types of fish sauce are well known in Southeast Asia and Japan.

Nuoc-mam is a fish sauce from Vietnam. It is a clear brown liquid with a distinguishing odour and flavour, rich in salt and soluble nitrogen compounds. Nouc-mam is prepared from different types of fish,

such as *Engraulis* spp., *Stolephorus* spp., *Clupea* spp. and others. It only takes a few months to produce nuoc-mam from small fish, although it may take 12–18 months when using large fish. Nuoc-mam is manufactured by mixing the fish with salt and pressing it by hand; the mixture is kept in an earthenware pot, which is then tightly sealed and buried in the ground. This method is different from the commercial practice.

Patis is a fish sauce from the Philippines. It is manufactured in a similar way to nuoc-mam, although the procedure is less complicated and shorter. It is usually prepared from fermented sardines, anchovies and shrimps. A ratio of one part salt to 3.5–4 parts of fish gives a product of 20–25% salt. It is reported that temperature up to 45°C at the first stage of fermentation (1–2 weeks) increases the rate of protein dehydration.

Nam-pla is the local name of a fish sauce which is widely processed in Thailand. Sea fish such as *Stolephorus* spp. and *Rastrelliger* spp. as well as freshwater fish such as *Cirrhinus* are used in the production of Nampla. Again, the manufacturing process is similar to that of nuoc-mam, but usually it is somewhat simpler. It takes about 6 months, and 2–3 years for better quality.

Budu is a Malaysian fish sauce, which is another brown liquid. Tamarind and palm sap sugar may be added to the mixture of fish and salt in earthenware jars.

Ketjapakan is a fish sauce produced in Borneo from *Stolephorus* spp., *Clupea* and from freshwater species of *Puntires* and *Osteochilus* in a similar method to nuoc-mam.

In some places in Southeast Asia the liquors obtained from salting fish in different ways are boiled and concentrated. These products are usually of poorer quality than normal fish sauce, for example, tuk-trey in Cambodia and petis in Indonesia.

Shottsuru is a product manufactured in Japan. This sauce is produced from sardines, anchovies, cuttle fish, herring, fish waste materials and molasses. Soybean may be added to shottsuru to convert it to shoyu.

Fish Paste Fish pastes are eaten almost everywhere in Southeast Asia, generally as a condiment for rice dishes (**Table 2**). They are more important nutritionally than fish sauce. There are two types of fish pastes: first fish and salt mixtures, and second, fish, salt and carbohydrate products.

Bagoong is a fish paste from the Philippines. It is prepared from *Stolephorus* spp., *Sardinella* spp. and *Decapterus* spp. A species of small shrimps, e.g. *Atya* sp., is also used. The fish are first cleaned, then mixed with 20–25% salt and kept in clay vats until the liquor

Table 2 Asian fish sauces and pastes

Country	Fish sauce	Fish paste
Burma	Ngapi	Ngapi
Japan	Shottsuru	Shiokara
Indonesia	Ketjap-ikan	Trassi, Trassi-ikan
Laos	Nam-pla	Padec
Cambodia	Tuk-trey	Prahoc
Malaysia	Budu	Belachan
Vietnam	Nuoc-mam	Man-ca (fish), mam-tom (shrimp)
Philippines	Patis	Bagoong
Thailand	Nam-pla	Kapi

becomes ready for consumption. To speed up the manufacture of bagoong, it can be stored at relatively high temperatures of about 45°C. In this type of fermented fish most of the protein breakdown is accomplished by the fish enzymes. According to the Philippine Pure Food and Drug Law, bagoong should contain 40% solids, 12.5% protein and 20–25% sodium chloride.

Balao-balao is another fish paste from the Philippines. It is generally prepared by mixing cooked rice, whole raw shrimps and salt (20% of the shrimp weight). The mixture is then allowed to ferment for several days. It is eaten either as a sauce or as a main dish after it is sautéed with garlic and onions.

Prahoc is a product of Cambodia. This is a fish paste corresponding to bagoong. Fish are beheaded, scaled, gutted, thoroughly washed, then drained for 24 h. The following day, fish are mixed with salt, dried in the sun, then powdered into a paste, and placed in open jars in the sun. The pickle which appears on the top is removed everyday and consumed. This phase of processing may take about 1 month. When no further pickle forms, the finished prahoc is ready to eat. It is mainly used in the preparation of soups, which play an important role in the Cambodian diet. Prahoc contains 37.8 g of nitrogen per litre, 22.4 g of which is soluble. In Cambodia, Vietnam and Laos fish pastes are also prepared by adding cooked glutinous rice, roasted rice, rice bran and other cereal products.

Padec is a fish paste from Laos prepared with salt and rice bran.

Man-ca-loc, Man-ca-sal, Man-ca-tre, Man-ca-no, and Man-ca-linh are different types of fish pastes consumed in Vietnam.

Trassi is another fish paste from Indonesia. It is manufactured in large quantities in high season. It can be prepared from fish as well as from shrimps. The method of manufacture is different from other fish paste methods. The paste is exposed to the sun in thin layers rather than put in deep containers. Probably selecting for aerobic fermentation results in more volatile constituents of the finished product.

Sida is a fish paste from Eastern India and Pakistan. It is prepared from small fish, mostly *Barbus* spp. It is another thick brownish liquor.

European Fermented Fish Products

Some fermented fish products are manufactured and consumed in Europe. Scandinavia seems to be the main producer of this type of food. Sunstromming is made in Sweden, and rakefisk in Norway. These are made from whole herring and trout. The fishes are immersed in brine for 1–2 days, eviscerated, retaining the roe or milt and packed in barrels with fresh brine. The final product is repacked in cans after fermentation. These products are usually consumed on special occasions.

Tidbits is another Scandinavian product. It is canned or bottled with vinegar, sugar and spices after maturation and filleting. In France, anchovies are prepared by salting the fish of the species *Engraulis encrasicholus*. The product is made from beheaded and gutted fish layered with salt in barrels. The mixture is weighed down with wooden or metal objects to keep the fish well pressed and to squeeze out the pickle. The fish mature at ambient temperature for 6–7 months.

In southern France pissala is a fish sauce prepared from small fish of the *Engraulis* sp., *Aphya* sp., and *Gobius* sp.

The Middle East and Africa

It is sometimes forgotten that the countries of the Middle East are surrounded by sea and crossed by rivers and that, hundreds of years ago, catching fish was the full-time job of the people in the region. Since modern methods of preservation were unknown, salting/drying and fermentation were the only ways to preserve fish. Consequently, many types of fermented fish were prepared in the Middle East (**Table 3**). Faseikh is a famous fermented fish processed in Egypt and Sudan from different types of fish. Both small and large *Tilapia* spp. are the main raw fish used in the processing.

Dried fish that undergoes some degree of fermentation is very common in Africa. As an example, kejeik is a fermented dried fish product common to Sudan and central Africa. The product is powdered and added to stew and relishes and thickened with okra after boiling.

Nutritional Considerations

Fish hydrolysate in the form of paste or sauce constitutes an important component of the daily diet of the communities of the Far and Near East (**Table 4**). Since these two products are used as condiments or flavouring agents in principle, no one eats a large daily

Table 3 The main types of fermented fish products found in the Middle East

Name of food	Species of fish	Country	Comments
Faseikh	Bouri, solti Tilapia spp., *Mugil cephalus*	Egypt	Usually eaten in public gardens during the festivities of Easter known as sham al-naseem
Turkeen	*Tilapia* spp. and other spp.	Sudan	Fermented fish sauce, used in stew preparation
Mindishi	Unknown	Sudan	Stew preparation
Salted sardines	*Sardinella* spp.	Gulf states	Side fish
Kejeik	Unknown	Sudan	Dried fish used in stew preparation
Tareeh	*Sardinella* spp.	Gulf states	Fermented fish paste, eaten with local bread
Katheef	Mixed	Gulf states	Fermented whole fish
Awal	Mixed	Gulf states	Dried, salted, fermented fish used in seasons of scarcity
Maleh	Mixed	Gulf states	Fermented whole or half fish in brine, used in stew
Sehnaor/metoot	*Sardinella* spp.	Gulf states	Dried and ground fish used for flavouring foods such as rice

Table 4 Chemical composition of some fermented fish products

Product	Country	Moisture (%)	Protein (%)	Fat (%)	NaCl (%)	Ash (%)
Katsuo shiokora	Japan	65.0	2.3	ND	14.8	ND
Makassar	Philippines	65.8	15.0	0.4	ND	16.9
Funasushi	Japan	72.5	38.7	21.7	ND	7.6
Mam caloc	Vietnam	45.6	22.6	1.1	5.8	11.5
Tareeh	Bahrain	58.5	14.9	1.8	ND	23.2
Mehiawa	Qatar	75.6	4.9	3.5	3.3	4.1
Turkeen	Sudan	71.5	14.9	5.8	7.5	3.6
Faseikh	Egypt	65.5	20.5	6.3	15.7	20.4

ND, not determined.

portion, but almost everyone eats a little bit every day. It is this continuous consumption that improves the daily intake of the relevant nutrients, such as protein and amino acids, to such populations. Although it is a side dish in some cases, fish sauce is present in almost all meals in Asian communities every day.

On hydrolysis during fermentation, fish protein has an amino acid composition that supplements the bland cereal diets of Asia. Particular emphasis is directed towards the high content of essential amino acids of fermented fish proteins such as lysine and methionine, which are always limiting factors in cereal diets. However, some amino acids such as arginine, tyrosine and isoleucine may be susceptible to degradation by the fermenting lactic acid bacteria, which contribute to the average amino nitrogen loss during fermentation of traditional fish products. It is obvious that the high salt content of fermented fish might limit the daily intake of protein from such products. The development of products with relatively lower salt content may serve as a good source of proteins in the low season of fishing.

Fat is another nutrient found in fermented fish products. Total fat ranges between 1 and 3% in most fish sauces. Fish paste such as funasushi (Japan) and

faseikh (Egypt) could supply a good source of fat. The addition of salt and spices contributes to the mineral composition of fermented fish. Fish fermentation may contribute to the mineral and vitamin nutrition in such communities.

The Risks

Despite the advantages of fermentation, there are possible risks from the consumption of fermented fish. The production of fermented fish often involves the use of salt (sodium chloride) at high concentrations (10–30%). Therefore, the continuous consumption of such foods could be one of the main factors leading to hypertension. Moreover, there is evidence to show that fish fermentation may result in elevated amounts of histamine and tyramine: both are regarded as allergy-inducing substances. In contrast, the use of contaminated raw materials in the production of fermented fish, inadequate storage of these products, and poor handling may lead to the growth and development of certain bacteria, such as *Clostridium botulinum* or *Staphylococcus aureus*, which produce bacterial toxins. Consumption of such foods may lead to outbreaks of food poisoning. In spite of the high concentration of salt and low pH in fermented fish, some food-poisoning cases have been reported. An outbreak of type E botulism among Eskimos from consumption of utjak and muktak (fermented flipper) was described. In addition, 10 outbreaks of food poisoning resulting from the consumption of izushi have been reported.

Fermented fish are consumed without cooking or heating, and may be prone to cross contamination due to poor handling. The transfer of *Salmonella* or faecal coliforms from handlers may be equally serious with food at pH above 4.5. Since fermented fish is often prepared in poor conditions with little attention to hygiene, this fermented fish product may be considered a potential vehicle for the transmission of food-borne diseases. There is little evidence that fish fermentation may render the product safe from *Listeria monocytogenes* and *Aeromonas hydrophila*.

See also: **Dried Foods**. **Fish**: Spoilage of Fish. **Preservatives**: Traditional Preservatives – Sodium Chloride.

Further Reading

Amano K (1962) The influence of fermentation on the nutritive value of special fermented fish products of South-east Asia. In: Hun E and Kvuzer R (eds) *Fish in Nutrition*. P. 180. London: Fishing Newa Books.

Campbell-Platt G (1985) *Fermented Foods of the World, A Dictionary and Guide*. London: Butterworth.

Davis HK (193) Fish spoilage. In: Macrae R, Robinson RK and Sadler MJ (eds) *Encyclopaedia of Food Science, Food Technology and Nutrition*. Vol. 3, p. 1872. London: Academic Press.

Dirar HA (1993) *The Indigenous Fermented Foods of the Sudan, A Study in African Food and Nutrition*. UK: CAB International.

Gieger E and Borgstrom G (1962) Fish protein-nutritive aspects. In: Borgstrom G (ed.) *Fish as Food*. P. 29. USA: Academic Press.

Ko SD (1982) Fermented foods of Indonesia except those based on soybeans. In: Moo-Young M (ed.) *Advances in Biotechnology*. Vol. 2, p. 523. New York: Pergamon Press.

Ko SD (1982) Indigenous fermented foods. In: Rose AH (ed.) *Fermented Foods*. Vol. 7, p. 16. London: Academic Press.

Mackie IM, Hardy R and Hobbs G (1971) *Fermented Fish Products*. Rome, Italy: FAO Fisheries.

Mizutani T, Kinizuk T, Rudde K and Ishige N (1992) Chemical components of fermented fish products. *Journal of Food Composition and Analysis* 5: 152–159.

Musaiger AO, Al-Mohizea IS, Al-Kanhal MA and Al Jedah JH (1990) Chemical and amino acid composition of four traditional foods consumed in the Arab Gulf states. *Food Chemistry* 36: 181–189.

Saisithi P, Kasemsran B, Liston J and Dollar AE (1966) Microbiology and chemistry of fermented fish. *Journal of Food Science* 31: 105–110.

Van Veen AG (1962) Fermented and dried seafoods products in South-east Asia. In: Borgstrom G (ed.) *Fish as Food*. Vol. 3, p. 227. New York: Academic Press.

Whitaker JR (1978) Biochemical changes occurring during the fermentation of high protein foods. *Journal of Food Technology* 32: 175–180.

Beverages from Sorghum and Millet

Janice Dewar, CSIR Food Science and Technology, Pretoria, South Africa

John R N Taylor, Cereal Foods Research Unit, Department of Food Science, University of Pretoria, South Africa

Throughout sub-Saharan Africa, the indigenous cereal grains – sorghum, *Sorghum bicolor* (L.) Moench; pearl millet, *Pennisetum glaucum* (L.) R.Br.; and finger millet, *Eleucine coracana* (L.) Gaertn. – have been used since prehistoric times to prepare traditional fermented beverages (**Fig. 1**). Today, these beverages remain as popular as ever, and their industrial manufacture has become increasingly common. Although the character of the industrially produced beverages is substantially the same as the traditional products, in general high proportions of maize, *Zea mays* (L.) – even maize alone – are used in their manufacture.

The fermented beverages may be alcoholic, or substantially nonalcoholic. The alcoholic beverages have many names, including sorghum beer (South Africa), opaque beer (Zimbabwe), Chibuku (throughout southern and eastern Africa) and pito (west Africa). Chibuku is a commercial brand name: the word is from the African 'mine' language Fanagalo and can be translated as 'according to the recipe book'. In this article the term 'sorghum beer' is used generically for the southern and eastern African product. Sorghum beer is characterized by its opacity due to the presence of semi-suspended particles of cereal, starch and yeast. The starch gives it a viscous consistency. The beer is sour in taste owing to the presence of lactic acid, and is consumed in an active state of fermentation. In contrast, pito is clear or somewhat cloudy and is not particularly sour in taste, but like sorghum beer it is consumed while it is actively fermenting.

The substantially nonalcoholic fermented beverages have been less well documented but all involve some form of lactic acid fermentation. Products include kunun-zaki, made from millet (Nigeria); hulu mur, from sorghum malt and sorghum flour (Sudan); and motoho oa mabele, from sorghum meal (South Africa). The best-known is mageu (mahewu or

Figure 1 Sorghum and millet growing together in a field in the Northern Province of South Africa (courtesy of Beryl Fabian, CSIR, South Africa).

magou) from southern Africa, which is a lactic acid-fermented gruel, widely produced industrially using almost exclusively maize.

Traditional Methods of Making Sorghum Beer

Sorghum beer is intrinsically linked with African culture, traditionally being brewed by the female members of the family for use at ceremonies including weddings, funerals and other social gatherings.

The basic sorghum beer process involves the following steps: souring (lactic acid fermentation), cooking (starch gelatinization), mashing (thinning and conversion of gelatinized starches to sugars), straining (grain separation) and alcoholic fermentation. It should be noted that under home brewing conditions it is very difficult to maintain the brews at the optimal temperatures for each of the steps. This leads to an overlap of the processes across different steps, so that two or more brewing steps may take place at the same time.

Malt is the vital ingredient in brewing. In essence, malting of any grain simply involves germinating the grain for several days to produce enzymes (specifically amylase enzymes) not present in the grain, then drying and milling it. In the brewing of sorghum beer, the malt amylase enzymes hydrolyse grain starches to sugars. The sugars are fermented to lactic acid by lactobacilli, and during alcoholic fermentation, yeasts convert the sugars to ethanol and carbon dioxide, producing beer.

In traditional home brewing, the grain is malted by first soaking it for about a day, either in large, water-filled clay vessels or in woven grass baskets placed in a nearby stream. After soaking, the steep water is drained and the damp grain allowed to germinate for 2–4 days depending upon the ambient conditions. Germination is conducted in small baskets or bags which allow the entry of air. At the end of germination, the germinated grain is spread out in a thin layer on the ground and sun-dried. Once dry, the malt is ground, commonly by hand on a stone.

The home-malted grain is then ready for use in the brewing process. A small portion of the ground malt is mixed with hot water – either slurried with the water and then cooked, or added to previously heated water. The warm mixture is then covered and allowed to stand overnight to sour. On the following day, the soured wort is diluted with additional water and more starchy (adjunct) material added; traditionally, this adjunct material is malt, or (more commonly) whole, coarsely ground sorghum or millet.

The entire mixture is then boiled (2–6 h), and then allowed to cool and thicken. The next day, a second batch of malt is added to the now cooled brew. The amylase enzymes in the malt liquefy the brew by converting the starch in the mash to sugars. Depending upon the taste of the brew and the experience of the brewer, the brew may be recooked, cooled and more malt added. The wild yeasts in the cooled mixture then utilize the sugars in the wort to produce ethanol and CO_2, giving a bubbling, effervescing liquid. The brew is allowed to ferment for a few days. When the brew is considered to be acceptable it is strained, for example by passing it through a braided grass strainer or similar equipment. The beer is then put in clay drinking vessels, and served with great ceremony to friends and family members. **Figure 2** shows a woman serving out beer with a ladle made from a gourd.

Home brewing today is done with recipes and procedures passed from generation to generation. Over the years these have been improved through the knowledge gained from experience and experimentation. Modern home brewers may make their own malt, or they may obtain it from large commercial maltsters who produce malt of more consistent quality. Over the years, milled maize has tended to replace malted

Figure 2 Serving traditionally made sorghum beer from clay pots using a gourd ladle (courtesy of Beryl Fabian, CSIR, South Africa).

Figure 4 Modern industrial sorghum beer brewery (courtesy of Traditional Beer Investments of Johannesburg, South Africa).

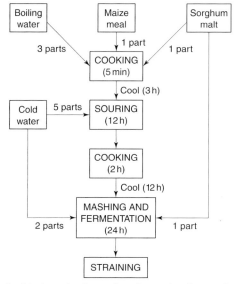

Figure 3 Modern-day home-brewing recipe for sorghum beer (courtesy of Nola, Randfontein, South Africa).

or unmalted sorghum or millet as the starchy adjunct. However, the procedures followed are essentially the same as those of ancient times. **Figure 3** shows a modern-day sorghum beer home-brewing recipe.

Industralization of the Brewing Process

Industrial sorghum beer brewing was initiated by municipal governments in South Africa and Zimbabwe. However, the adaptation of the traditional craft to factory scale was hampered by technical difficulties. In 1954, the Council for Scientific and Industrial Research (CSIR) in South Africa was com-

missioned to undertake applied research on behalf of the municipal factory sorghum beer brewers.

The spectacular increase in scale from home brewing (approximately 50–200 litre brews) to factory brewing (approximately 15 000–27 000 litre brews) has been made possible largely by the development and institution of carefully specified raw material standards and the use of specialized equipment. This brewing equipment allows rapid heating and cooling, enabling the key brewing steps to be split into separate operations and conducted under optimal conditions of temperature (**Fig. 4**).

Malting

The first major problem encountered in scaling-up sorghum beer production related to the quality of the sorghum malt. At that time, virtually nothing was known about the science and technology of sorghum malting. The CSIR researchers developed a sensitive method of determining the diastatic power (i.e. the joint level of α- and β-amylase activity) or sorghum malts, and this is still the most important quality criterion for sorghum malt.

Sorghum malt of high and consistent quality is required when it is used as an ingredient in industrial brewing. Today, most of the sorghum malt used in factory brewing is produced by pneumatic malting and is referred to as 'conversion' or 'industrial' malt. Sorghum grain is germinated indoors under conditions of controlled temperature, moisture and aeration to produce malt to a diastatic power specification of 28–35 sorghum diastatic units (SDU) per gram and a moisture content of approximately 10%. Another type of malt used in industrial brewing, referred to as 'souring' malt, serves as a source of lactobacilli as well as a substrate for lactic fermentation. Souring malt is generally produced by

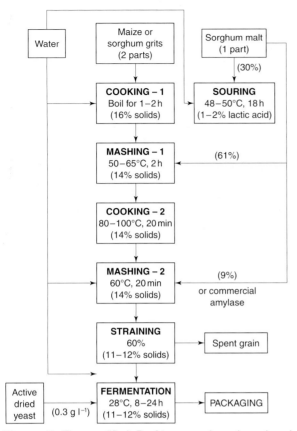

Figure 5 The modified Reef-type sorghum beer brewing process.

germinating sorghum grain in the open air on a concrete floor (floor malting).

The malting process can be physically split into three unit operations: steeping, germination and drying. Sorghum malt quality increases with steeping time (16–40 h) and the optimum steeping and germination temperature is between 25°C and 30°C. Within limits, the more water taken up by the grain during malting, the higher is the resulting malt quality. The malt is generally dried at temperatures no greater than 50°C until the moisture content of the malt is approximately 10%. Drying at this low temperature helps conserve the amylase activity of the malt, which is low in comparison with barley malt.

Modified Reef-type Process

A number of different industrial brewing processes are used in southern Africa, mainly because of differences in the taste preferences of the customers. The most common is the modified Reef-type (double cook) beer process (**Fig. 5**), originally developed in the gold-mining region of South Africa known as the 'Reef', and later modified to include a second cooking stage.

Souring In southern Africa, the lactic acid fermentation or souring is an essential part of the brewing process. The lactic acid sour lowers the pH of the beer to below 4.0, which helps to prevent the complete hydrolysis of starch into sugars, slows the rate of microbial spoilage and inhibits the growth of pathogenic organisms. The sour also contributes to the indigenous beer's characteristic sour flavour.

Some brewers add industrially produced lactic acid during the brewing process. In South Africa, however, most brewers ferment their own lactic acid sour. Most commonly this is done by slurrying a mixture of sorghum malt and water, inoculating it with a starter culture from a previous highly viable lactic acid bacteria sour (approximately 10% by volume) and heating it to about 48°C for 8–18 h, until the sour reaches 1–2% lactic acid. Unlike home brewing, the temperature of souring (in the range 48–50°C) is strictly controlled under industrial brewing conditions.

Cooking After the lactic acid fermentation the sour is diluted with water, then brewers' grits (refined maize or sorghum meal) are added as a starchy adjunct and the mixture is boiled for approximately 1–2 h under atmospheric conditions, or for less time under pressurized conditions. Industrial brewers have tended to abandon the use of whole unrefined grain as adjunct, owing to the high fat content of the grain which causes rancidity problems when stored in a milled form. The high fat content also negatively affects the beer foam. Refined, low-fat maize grits have become abundantly available and are considered to be a less variable raw material.

The purpose of cooking is to gelatinize the grain starches and render them more easily hydrolysed by the malt amylase enzymes later during mashing. After cooking, the boiled wort is cooled rapidly to approximately 60°C by circulating cool water through an internal coil. As a consequence of starch gelatinization, the mash is quite viscous at the end of this step.

Mashing Once the mash has been cooled to approximately 60°C, additional malt ('conversion' malt) is added. A small amount of conversion malt is sometimes added to the mash while it is cooling, to accelerate the conversion of starch to sugar. As a consequence of the malt α-amylases converting the gelatinized starch to dextrins, the mash is rapidly liquefied. The low pH of the mash (about 4.0) is below the optimum for sorghum α- and β-amylase activity, which limits starch hydrolysis, preventing the production of too thin a beer. Some industrial brewers supplement or substitute their costly conversion malt with less expensive commercially produced amylase enzymes (α-amylase and/or amyloglucosidase), to facilitate the thinning of the mash and fermentable sugar production. Under industrial conditions, mashing is carried out for about 2 h at constant temperature (50–65°C).

Second Cooking and Mashing The gelatinization temperature of sorghum starch is 68–72°C. Therefore, under the conditions of conversion at approximately 60°C in the previous step, the sorghum malt starch is not gelatinized. In the modified Reef-type brewing process the mash is recooked (80–100°C for approximately 20 min) at the end of mashing. This second cooking stage pasteurizes the mash and also gelatinizes the sorghum malt starch, leaving virtually no ungelatinized starch in the mash. Thereafter, the mash is cooled to 60°C and a second short massing period (approximately 20 min) is carried out either with a small amount of conversion malt or commercially produced amylase. The objective of this second conversion (mashing) stage is to reduce the viscosity in the mash to the desired level by the thinning action of the amylase enzymes; the second cooking and mashing steps effectively increase the amount of soluble solids in the mash.

Straining Straining removes the coarse cereal particles from the wort, and is carried out directly after mashing and prior to fermentation. Industrial brewers strain their mash at 60°C by the use of equipment such as centrifugal decanters or vibrating screens. Before straining there are approximately 14% solids in the wort and after straining approximately 11–12% solids remain.

Fermentation Plate or tubular heat exchangers are used to cool the strained wort to 28°C as quickly as possible. The wort is transferred to stainless steel vessels and alcoholic fermentation is brought about by pitching with commercially produced active dried yeast ($0.3 \, \mathrm{g} \, \mathrm{l}^{-1}$), rather than relying on the growth of wild yeasts in the wort as is the case in traditional home brewing. The mixture is then allowed to ferment for 8–24 h.

Packaging The beer is packaged in large bulk containers (100–500 l) for sale in beer halls. Alternatively, the beer is packaged in vented milk-type cartons or plastic containers (0.5–25 l) for sale in the retail market.

Unlike clear lager-type beer, sorghum beer is not filtered; indeed, not only is the yeast not removed, the beer is actually sold and consumed in a state of active fermentation. When ready for consumption the beer has an alcohol content of approximately 3% by weight and the shelf life of the product is approximately 5–10 days after straining.

Role of Microflora

Microflora of Alcoholic Beverages

The microbiology of sorghum beer has three components: the alcoholic fermentation, the lactic acid fermentation and beer spoilage.

Alcoholic Fermentation Because sorghum beer wort is not boiled, up to 20 different species of yeasts (wild types) have been found in beers that have not been pitched with yeast. Since the species match those of the sorghum malt, it is presumed that the malt is the primary inoculum in such beers. The yeast species found include a number of *Candida* species and *Torulopsis holmii*, although *Saccharomyces cerevisiae* usually predominates (**Table 1**). In view of this and because of its superior fermentative ability, it is probable that *S. cerevisiae* is the only organism of significance.

Industrially brewed sorghum beer initially relied on a natural, uninoculated fermentation. However, to give a more predictable fermentation, pitching with yeast was introduced in the 1950s. Professor J P van der Walt, of the CSIR in South Africa, isolated a number of strains of *S. cerevisiae* with superior fermentation properties from sorghum beers. Two of these strains, Y2 and Y48, have been produced by yeast companies in southern Africa for many years. They are distributed in an active dried form and are used by breweries throughout the region. The yeasts are top-fermenting strains which ferment rapidly at high temperatures (20–25°C). In contrast to conventional beer yeasts, they are non-flocculating strains; flocculation is not desired as it would cause the beer to sediment. The yeast is a major component of the foam on sorghum beer, which is more strictly speaking a 'yeast head'. Foam is an important criterion of beer quality and recent examination of Professor Van der Walt's original isolates has revealed that some of the strains have superior foaming characteristics.

Pito beer similarly has been found to contain a variety of yeast species, including *Candida* species, *Kluyveromyces* strains and possibly *Geotrichum candidum*, but as with sorghum beer, *S. cerevisiae* is probably the most important. In dolo, a beer similar to pito from Burkina Faso and Togo, *Candida krusei* and *S. cerevisiae* appear to be the most important yeast species.

Lactic Acid Fermentation The organisms involved depend to a great extent on whether the temperature of the lactic fermentation is controlled. With fer-

Table 1 Yeast flora of three sorghum beers

Species	Concentration (cells ml^{-1})		
	Beer 1	Beer 2	Beer 3
Candida fabianii		$1–2 \times 10^5$	
Candida guillermondi (var. guillermondii)		1×10^6	
Candida krusei	$1–2 \times 10^2$	$1–4 \times 10^4$	$1–2 \times 10^6$
Candida lusitaniae		$6–11 \times 10^4$	
Candida macedoniensis (asporogenus form of Kluyveromyces marxianus)	1×10^2	$1–5 \times 10^3$	$1–2 \times 10^6$
Candida pseudotropicalis	$1–2 \times 10^3$		
Candida robusta (asporogenus form of Saccharomyces cerevisiae)	$5–6 \times 10^2$	$5–6 \times 10^7$	$8–9 \times 10^7$
Candida rugosa		3×10^4	
Candida tropicalis	$2–3 \times 10^3$		
Candida valida	$1–2 \times 10^2$		
Kluyveromyces wikenii (asporogenus form)	$1–2 \times 10^2$		
Saccharomyces cerevisiae	$4–7 \times 10^7$	$5–8 \times 10^7$	1×10^8
Saccharomyces fibuligera	$1–3 \times 10^6$		
Torulopsis candida		3×10^3	
Torulopsis holmii		$5–7 \times 10^6$	
Trichosporon capitatum	$4–6 \times 10^2$		

From Novellie and De Schaepdrijver (1986) with kind permission of Elsevier Science, Amsterdam, the Netherlands.

mentations in the lower temperature range (up to 37°C) a wide range of organisms will grow, including mesophilic lactobacilli, pediococci, streptococci and even yeasts. The lower temperature also favours the growth of the heterofermentative lactic acid bacteria, *Betabacterium* spp. and *Leuconostoc* spp., which produce ethanol, acetic acid and formic acid in addition to lactic acid. The taste imparted by these and other by-products of heterofermentative fermentation are considered undesirable in some commercial sorghum beers, notably in South Africa. As a consequence, the lactic acid fermentation, known as souring, may be carried out as a separate step in the brewing process, under conditions of carefully controlled elevated temperature (48–50°C). These conditions favour the growth of thermophilic, homofermentative *Lactobacillus* spp., notably *L. delbrueckii*, otherwise known as *L. leichmannii*. Not only does this species produce solely lactic acid (which gives the beer a characteristic clean, sharp taste), unlike the mesophilic lactic bacteria it produces only the laevorotatory isomer which may be nutritionally preferable. The sour, a slurry containing 8–10% sorghum malt, is generally inoculated by a system of back slopping, although some breweries may use freeze-dried cultures of lactobacilli. The particular strains used apparently impart desirable flavours to the beer.

Beer Spoilage   Since sorghum beer is consumed in an active state of fermentation, the major problem is spoilage of the beer before it ferments out. Spoilage after the beer has fermented out is of less concern because once alcoholic fermentation finishes the beer becomes flat, the particles settle out of suspension and

it is no longer palatable. The main spoilage organisms during fermentation are mesophilic lactic bacteria. Homofermentative types such as *Lactobacillus planatarum*, *Pediococcus* and *Streptobacterium* spp. are responsible for 'over-souring' of the beer, as well as the production of volatile off flavours. However, it should be pointed out that, for some consumers, the sourer the beer the better. Possibly of more concern are the heterofermentative lactic acid bacteria such as *Lactobacillus fermentum*, *Leuconostoc* and *Betabacterium* spp, which produce acetic acid and can make the beer unacceptable. Also, if the desired yeast fails to grow properly, wild yeasts may spoil the beer through the production of undesirable flavours and pellicle formation (**Table 2**).

Under aerobic conditions, which occur when the beer has fermented out or has failed to ferment fully, acetic acid bacteria oxidize the ethanol in the beer to acetic acid. *Zymomonas lindneri* may ferment sugars to produce off flavours such as acetaldehyde and hydrogen sulphide.

Microflora of Nonalcoholic Beverages

Different types of starter cultures are used in commercial mageu manufacture. These include a multistrain, homofermentative thermophilic *Lactobacillus* culture producing D,L-lactic acid; a thermophilic culture (the identity of which is a commercial secret), which is probably a single strain and heterofermentative; and mesophilic heterofermentative cultures. When mageu is produced by spontaneous fermentation at ambient temperature, *Leuconostoc mesenteroides* and *Lactobacillus brevis* are found to

Table 2   Microorganisms associated with sorghum beer spoilage

Microorganism	Chemicals produced	Odours	Other effects
Bacteria			
Acetobacter	Acetic acid	Vinegar	Pellicle
Pediococcus	Lactic acid		Ropiness
Lactobacillus (homofermentative)	Lactic acid		Ropiness
Lactobacillus (heterofermentative)	Lactic acid and acetic acid	Vinegar	Ropiness and turbidity
Leuconostoc (heterofermentative)	Diacetyl	Butter/honey	Ropiness and turbidity
	2,3-Butadione	Sweet	
Zymomonas	Ethanol		
	Carbon dioxide		
	Acetaldehyde	Rotten apples	
	Hydrogen sulphide	Rotten eggs	
Obesumbacterium	Dimethyl sulphide	Parsnip, cooked cabbage, mouldy	
Yeasts			
Candida		Fruity	Pellicle
Pichia			Pellicle
Rhodotorula			Red coloration
Hansenula			Pellicle
Saccharomyces (wild strains)	Diacetyl	Phenolic Butter/honey	Superattenuation

From Haggblade and Holzapfel (1989), with kind permission of Marcel Dekker, Inc.

be the major organisms. The most important spoilage microorganisms of mageu are yeasts, with *Pichia* spp. dominating. Spoilage by *Acetobacter* spp. can also occur.

In the Nigerian millet beverage kunun-zaki, the *Lactobacillus fermentum* and *L. leichmannii* predominate at the end of fermentation. However, during the fermentation other dominant microorganisms include *Bacillus subtilis*, *Enterobacter aerogenes* and *E. cloaceae*. The presence of these and of potentially mycotoxic moulds (*Aspergillus* spp.) have given rise to concern about the biological safety of the home-prepared product.

Nutritional Aspects

The nutritional value of fermented beverages is influenced to a great extent by two factors, neither associated with fermentation. These are the concentration of cereal solids, and the degree to which the cereal has been refined. The total solids content of commercial mageu is between 5% and 8%, rather lower than the sugar content of many carbonated soft drinks. Higher concentrations are not possible, as the starch would increase the viscosity of the product to that of a porridge. Commercially manufactured beverages are increasingly made with cereal endosperm meal rather than partially processed or milled whole grain. Research has shown that brewing sorghum beer with maize grits (as is now common practice) instead of whole milled sorghum as adjunct substantially reduces its vitamin B and mineral content.

Advantages of Lactic Acid Fermentation

Many nutritional advantages have been claimed for the lactic acid fermentation of cereal beverages and porridges. These include improved starch digestibility, increased protein quantity, improved protein quality and digestibility, increased vitamin levels, decreased antinutritional factors such as tannins and phytate, improved mineral availability, production of bacteriocins and pre-biotics, and improved palatability and acceptability. Unfortunately, there are few firm data on the extent of these nutritional improvements. Probably the most unequivocal advantage of lactic acid fermentation is the reduction in pH. Sorghum beer and mageu have a pH of around 3.5. At this low pH pathogenic bacteria such as *Salmonella typhimurium* are destroyed, heat-resistant bacterial spores do not germinate and the rate of growth of spoilage bacteria is much reduced. Because many households in Africa still do not have access to potable water and few have refrigerators, these beverages are a valuable safe source of nutrition.

Sorghum Beer

Disregarding the question of whether the consumption of any alcoholic beverage is nutritionally desirable, the nutritive value of sorghum beer is here compared with that of conventional lager beer. There is little evidence, certainly with respect to commercial sorghum beer, to support the often-quoted statement that sorghum beer is more of a food than an alcoholic drink (**Table 3**). However, sorghum beer has some obvious and not-so-obvious nutritional advantages.

Table 3 Nutrient composition and contribution to recommended daily allowance (RDA) of commercial sorghum beer and lager beer

Nutrient (mg l^{-1} unless stated otherwise)	Sorghum beer	Lager beer	Contribution to RDA of 1 litre	
			Sorghum beer (%)	Lager beer (%)
Alcohol (% w/w)	2.50	3.59	–	–
Energy (kJ l^{-1})	1651	1605	15	14
Protein (g l^{-1})	5.4	2.8	10	5
Thiamine	0.24	0.03	17	2
Riboflavin	0.39	0.19	24	12
Nicotinic acid	2.93	4.72	16	26
Cu	0.21	0.09	7	3
Fe	1.90	0.07	19	0.7
Zn	1.40	0.10	9	0.7
Mn	1.60	0.16	40	4
Ca	48	44	6	6
Mg	137	83	39	24
K	312	357	8	9
Na	20	69	1	3
P	245	246	31	31

Reprinted from *SA Food Review* supplement, Feb/Mar 1988, p. 83, with kind permission of National Publishing (Pty) Ltd, Cape Town, South Africa.

Although the energy content of the two beers is similar, sorghum beer has less ethanol and substantially more complex carbohydrates. The protein content of sorghum beer is much higher, and much of this is yeast protein which has a better essential amino acid content than cereal protein. The higher thiamine and riboflavin levels are also due to the presence of yeast. The lower nicotinic acid level is due to the use of a high proportion of maize in the grist. The high levels of iron and some other minerals in sorghum beer are the consequence of leaching out of vessels, because of the low pH of the beer. In small-scale brewing where mild steel (as opposed to stainless steel) vessels are commonly used, excessively high levels of iron have been found in the beer. What is not shown in Table 3 is that sorghum beer – probably uniquely among beers – contains measurable quantities of both insoluble and soluble dietary fibre.

Other Fermentations Using Sorghum and Millets

Because sorghum and millets are uniquely suited to cultivation in the semiarid tropics, there is considerable interest in Africa and India in using them as a replacement for imported barley in the production of lagers, stouts and distilled beverages.

In 1988 the government of Nigeria banned the importation of barley malt. Today in Nigeria, lager beers are produced using unmalted sorghum grain as the sole cereal source. Commercial amylases and proteases are used to produce sugars and free amino nitrogen from the sorghum grain starch and protein. In addition, research and development is being undertaken worldwide to produce clear lager-type beers using malted sorghum, which would in itself provide the necessary hydrolytic enzymes. A number of technical problems have been encountered, including low yields of poorly fermentable extract and poor wort separation. However, good progress is being made in resolving these problems.

See also: **Candida**: Introduction. **Lactobacillus**: *Lactobacillus brevis*. **Packaging of Foods**: Packaging of Solids and Liquids. **Saccharomyces**: *Saccharomyces cerevisiae*. **Spoilage Problems**: Problems caused by Bacteria.

Further Reading

Daiber KH and Taylor JRN (1995) Opaque beers. In: Dendy DAV (ed.) *Sorghum and Millets: Chemistry and Technology*. P. 299. St Paul: American Association of Cereal Chemists.

Haggblade S and Holzapfel WH (1989) Industrialization of Africa's indigenous beer brewing. In: Steinkraus KH (ed.) *Industralization of Indigenous Fermented Foods*. P. 191. New York: Marcel Dekker.

Hallgren L (1995) Lager beers from sorghum. In: Dendy DAV (ed.) *Sorghum and Millets: Chemistry and Technology*. P. 283. St Paul: American Association of Cereal Chemists.

Holzapfel WH (1989) Industrialization of mageu fermentation in South Africa. In: Steinkraus KH (ed.) *Industrialization of Indigenous Fermented Foods*. P. 285. New York: Marcel Dekker.

Murty DS and Kumar KA (1995) Traditional uses of sorghum and millets. In: Dendy DAV (ed.) *Sorghum and Millets: Chemistry and Technology*. P. 185. St Paul: American Association of Cereal Chemists.

Novellie L (1986) Sorghum beer and related fermentations of Southern Africa. In: Hesseltine CW and Wang HL (eds) *Indigenous Fermented Food of Non-Western Origin*. P. 219. Berlin: Cramer.

Novellie L and De Schaepdrijver P (1986) Modern developments in traditional African beers. In: Adams MR (ed.) *Progress in Industrial Microbiology*. Vol. 23, p. 73. Amsterdam: Elsevier Science.

Palmer GH, Etokakpan OV and Igyor MA (1989) Review: sorghum as brewing material. *MIRCEN Journal* 5: 265–275.

Rooney LW, Kirleis AW and Murty DC (1986) Traditional foods from sorghum: their production, evaluation, and nutritional value. In: Pomeranz Y (ed.) *Advances in Cereal Science and Technology*. Vol. 8, p. 317. St Paul: American Association of Cereal Chemists.

South African Bureau of Standards (1970) Standard test method for the determination of the diastatic power of malts prepared from kaffircorn (sorghum) including bird-proof varieties, and from millet. SABS Method 235. Pretoria: South African Bureau of Standards.

South African Bureau of Standards (1979) Standard specification for: the production of mageu. SABS 1199–1979. Pretoria: South African Bureau of Standards.

Tomkins A, Alnwick D and Haggerty P (1988) Fermented foods for improving child feeding in eastern and southern Africa: a review. In: Alnwick D, Moses S and Schmidt OG (eds) *Improving Young Child Feeding in Eastern and Southern Africa: Household-Level Food Technology*. P. 136. Ottawa: International Research Development Centre.

Van Heerden IV (1989) Sorghum beer – a decade of nutritional research. *Proceeding of the Second Scientific and Technical Convention*. P. 293. Johannesburg: The Institute of Brewing Central and Southern African Section.

World Health Organization (1996) *Fermentation: Assessment and Research*. Report of a Joint FAO/WHO Workshop on fermentation as a household technology to improve food safety. Food Safety Unit, Division of Food and Nutrition. Geneva: WHO.

Fermentations of the Far East

Indrawati Gandjar, Department of Biology, Faculty of Science and Mathematics, University of Indonesia, Jakarta, Indonesia

Introduction

The application of microorganisms to human foods has been known for thousands of years, for example the leavening of bread and the production of beer, wine, saké, vinegar and cider. Various fermented foods originated in the Far East, and their production has been passed on from generation to generation. The producer always saves a small amount of the fermented product and uses it again for the next batch. In this article the most important fermented foods of Southeast Asia are described, including tempeh, tape, nata de coco, soy sauce and fish paste.

Table 1 Tempeh products of Indonesia. From Steinkraus (1996).

Substrate	Local name
Soybean (*Glycine max*)	Tempeh kedelai
Velvet bean (*Mucuna pruriens*)	Tempeh benguk
Jack bean (*Canavalia ensiformis*)	Tempeh koro pedang
Winged bean (*Psophocarpus tetragonolobus*)	Tempeh kecipir
Pigeon pea bean (*Cajanus cajan*)	Tempeh gude
Wild tamarind (*Leucaena leucocephala*)	Tempeh lamtoro
Peanut presscake	Tempeh bungkil kacang
Coconut presscake	Tempeh bongkrek
Solid residue from tofu manufacture	Tempeh gembus

Tempeh

Tempeh is a common term for a solid fermented food overgrown by the mycelia of a mould; it is compact like a cake, and can be cut in thin slices. The moulds involved are the genus *Rhizopus* (*Rhizopus oryzae*, *R. arrhizus*, *R. oligosporus* or *R. chinensis*, or a combination of them). The substrate can be soybeans, velvet beans, pigeon pea beans, peanut presscake, coconut presscake, seeds of wild tamarind (*Leucaena leucocephala*) and solid tofu wastes. Tempeh is a traditional fermented food from Indonesia and has its origin on the island of Java. The name of the tempeh product depends on the substrate used (**Table 1**).

Soybean Tempeh

Soybeam tempeh or tempeh kedelai is prepared as follows: soybeans (*Glycine max*) are washed, soaked overnight, hulled, washed, boiled or steamed until partially cooked, cooled, inoculated with the traditional tempeh inoculum (*usar* or *laru*), wrapped in banana leaves or in perforated polyvinyl bags, and incubated for 24 h at 30°C until the cotyledons of the soybeans are completely covered by mycelium. There are a number of variations in the preparation of tempeh.

Soybean tempeh has a pleasant, nutty aroma and taste when deep-fried. It can also be mixed with vegetables or meat and stir-fried or cooked in coconut milk. Tempeh is a very nutritious food and is consumed by all strata of the Indonesian population. It is produced commercially in the Netherlands, the USA, Canada, Malaysia and Singapore.

Nutritive Value Soybean tempeh is one of the best plant protein sources and is often used as a substitute for meat in the diet. The protein content of tempeh is over 40% (dry basis); the average protein efficiency ratio (PER) is 2.4 compared with 2.5 for casein; the biological value (BV) of tempeh is 58.7 compared with 80 for meat; the net protein utilization (NPU) is 56 compared with 65 for chicken meat; and the

Table 2 Nutrient content of tempeh gembus. From Steinkraus (1996).

Nutrient (per 100 g)	Steamed tofu residue	Tempeh gembus
Moisture (g)	82.6	84.9
Ash (g)	0.5	0.7
Fat (g)	2.7	2.1
Protein (g)	4.0	4.0
Carbohydrate (g)	10.1	8.4
Fibre (g)	29.6	30.9
Calcium (mg)	204.0	226.0
Iron (mg)	1.3	1.4
Thiamin (mg)	0.1	0.1
Vitamin C (mg)	0	0
Carotene (IU)	10.3	22.0

digestibility of soybean tempeh is 86.1%. Tempeh made by traditional processing contains vitamin B_{12} produced by bacteria; tempeh prepared using a pure culture of *Rhizopus* sp. does not. Some bacteria produce little or no vitamin B_{12}, whereas others produce relatively large quantities. Surveys for B_{12}-producing microorganisms have shown that fungi are not B_{12} producers, and this is true of *R. oligosporus*. When tempeh is made with only this species, no vitamin B_{12} is found in the product. Soybean tempeh contains isoflavone compounds (daidzein, glycetein, isogenestein), has an antioxidative activity, has the capacity to reduce levels of total cholesterol and low-density lipoprotein cholesterol, increases the bioavailability of the elements Zn, Fe, Mn, Ca and P, and has antidiarrhoeal properties. Soybean tempeh has never been reported to be toxic.

Tempeh Gembus

This fermented product is made from the residue of tofu manufacture, with *Rhizopus* moulds as the major microorganisms. A short report on this product appeared for the first time in 1972. This kind of tempeh was unknown before the Second World War. Around 1943, when food became scarce, people began to use the residue of the tofu factories as food. The residue may be mixed with grated coconut and palm sugar, steamed, and consumed as a snack. The nutrient content is listed in **Table 2**.

To prepare tempeh gembus, the residue is pressed to remove the excess water and steamed for 30–45 min; it is then cooled to 30°C, inoculated with ragi tempeh, wrapped in banana leaves and incubated for 24–48 h. In a good tempeh gembus, the residue should be completely bound together and covered by mycelia of *Rhizopus*. The product is soft, like a sponge, but can be cut in slices. It is consumed as a side dish with rice, in the same way as soybean tempeh.

Tempeh Enthu

Tempeh enthu is prepared from the solid refuse of tofu processing, mixed with coconut presscake or with grated coconut and inoculated with the traditional tempeh inoculum, which contains *Rhizopus* moulds. The next steps are the same as for tempeh gembus. The product is not as soft as tempeh gembus, and has a distinctive taste and aroma. No report of toxicity has been published, and there has been no research on this kind of tempeh.

Tempeh Bungkil Kacang

For the substrate peanut presscake is used, which is soaked overnight to remove any oil, then steamed and cooled; no starch is mixed in, unlike the production of red (orange) oncom. It is inoculated with ragi tempeh, moulded and placed on bamboo trays, covered with banana leaves, and incubated 24–48 h until the whole substrate is a compact mass owing to the growth of *Rhizopus* mould. The end product is dark grey or brown in colour. This kind of tempeh is also called 'black oncom' because it is prepared from the same substrate as red oncom. It is eaten fried or stirred with vegetables and chillies. The peanut presscake often contains aflatoxin B_1, but in tempeh production the aflatoxin is reduced by 70% by the mould.

Non-soybean Legume Tempeh

Other minor tropical legume seeds can also be fermented into tempeh. In rural areas on the island of Java these tempehs are a valuable dietary protein source (**Table 3**).

Tempeh benguk or velvet bean tempeh is made from the seeds of the legume plant *Mucuna pruriens*; tempeh koro pedang or jack bean tempeh is made from the seeds of *Canavalia ensiformis*; tempeh kecipir or winged bean tempeh is made from the seeds of *Psophocarpus tetragonolobus*; tempeh gude or pigeon pea tempeh is made from the seeds of *Cajanus cajan*; tempeh lamtoro or wild tamarind tempeh is made from the seeds of *Leucaena leucocephala*.

The methods of production of all these non-soybean tempehs are similar. The legume seeds are washed, soaked overnight, boiled in plenty of water and hulled; the cotyledons are then soaked in water again to remove any toxic substances present in the seeds, for 24–72 h (the soaking water is changed twice daily). Large cotyledons are chopped up in preparation for the next stage of washing, steaming and cooling. The seeds are then inoculated with ragi tempeh, wrapped in banana leaves, and incubated 24–48 h at 30°C.

Analysis of the nutrient and amino acid content of some of these tempehs has been carried out. The net protein utilization for velvet bean tempeh and winged

Table 3 Nutrient content of some legume seeds and the tempeh products. From Steinkraus (1996).

| Legume | Local name | Substrate | Nutrient (per 100 g) | | | | | | | | | | | |
			Moisture (%)	Ash (g)	Fat (g)	Protein (g)	Carbo-hydrate (by diff.) (g)	Fibre (g)	Ca (mg)	Fe (mg)	P (mg)	Caro-tene (total) (μg)	Vit. C (mg)	Vit. B₁ (mg)
Velvet bean	Koro	Seeds	11.0	3.3	3.8	28.7	50.7	1.7	138	7.6	445	0	0.0	0.30
(*Mucuna pruriens*)	benguk	Tempeh	59.2	0.7	2.2	13.4	23.5	2.0	108	3.8	53	0	0	0.40
Jack bean (*Canavalia ensiformis*)	Koro pedang	Seeds	10.8	3.0	4.1	20.5	61.6	7.0	150	6.2	272	0	1.0	0.29
	Kaos bakol	Tempeh	69.4	0.8	1.3	7.0	21.5	1.5	42	1.7	90	0	0	0.08
Pigeon pea (*Cajanus cajan*)	Kacang kayu	Seeds	16.1	4.2	1.0	20.7	59.0	4.6	146	4.7	445	0	0	0.30
	Kacang hiris	Tempeh	71.9	1.4	0.3	7.0	17.3	1.5	36	1.4	144	0	0	0.06
Winged bean (*Psophocarpus tetragonolobus*)	Kecipir	Seeds	10.4	4.2	16.9	34.4	34.1	10.7	468	6.8	182	0	0	0.30
	Botor Jaat	Tempeh	58.2	1.4	10.0	17.5	12.9	2.9	186	2.2	177	0	0	0.20
Broad bean (*Vicia faba*)	Kacang babi	Seeds	11.6	3.5	3.2	30.4	51.3	4.5	178	6.8	521	87	0	0.23
	Kacang-oncet	Tempeh	64.0	1.2	0.8	12.5	21.5	1.8	68	2.6	182	20	0	0.10

bean tempeh is 48–55% (average 51%), while the NPU for soybean tempeh is 50–60%. This indicates that velvet bean tempeh and winged bean tempeh can be used as a protein source as an alternative to soybean tempeh.

Oncom

There are two popular kinds of oncom. The first is oncom tahoo, which is a fermented product made from the solid residue of tofu manufacture with the mould *Neurospora intermedia* or *N. crassa* (red oncom). The second is oncom kacang, which is made from peanut presscake. Both products have an orange colour due to the thick layer of orange conidia produced by the mould. The production is as follows.

Oncom Tahoo

The solid residue of tofu production is pressed to remove excess tofu milk, steamed, cooled, mixed with tapioca (1% w/v) and moulded on bamboo trays. It is sprinkled with oncom from a previous batch, covered with banana leaves and incubated 48–72 h at 28–30°C. The compact orange cake obtained can be cut in slices. Oncom tahoo is eaten fried or mixed with vegetables and stirred.

Oncom Kacang

The peanut presscake is soaked overnight to remove remnants of oil, pressed, mixed with tapioca (onggok), steamed, cooled and moulded on bamboo trays. It is sprinkled with oncom kacang from a previous batch, covered with banana leaves (crushed to improve aeration) and incubated. The cake is turned

over after 48 h to allow the underside to be colonized by the mould. After 72 h the fermentation is finished, and a firm cake covered by a thick layer of orange conidia is obtained. Oncom kacang is consumed as a snack (slices are dipped in rice flour and deep-fried) or as a side dish with rice.

Tape

Tape is a sweet-sour alcoholic fermented food which can be prepared from rice, glutinous rice or cassava. Tape made from glutinous rice is more popular than that made from rice. Tape is very popular in Southeast Asia. The local names are *tape ketan* (Indonesia), *tapai* (Malaysia, Singapore, Brunei), *basi* (Philippines) and *khao-mak* (Thailand).

Tape Ketan

Glutinous rice is washed, soaked for 2 h or boiled in water until half-cooked, and then steamed until cooked. After cooling to room temperature it is sprinkled with powdered tape inoculum and mixed well. The mixture is then placed in glass jars or polyvinyl bags (10 cm × 20 cm), or packed in banana leaves (about 100 g) and incubated for 2–3 days at room temperature (about 30°C). The fermented product is a sweet-sour juicy mass with a pleasant alcoholic flavour and aroma. The tape ketan is consumed without cooking or pasteurization, and will become gradually sour if the fermentation time is prolonged. The microorganisms involved and are the mould *Amylomyces rouxii* and the yeast *Endomycopsis burtonii*.

Tape Nasi

Tape nasi is made from rice; otherwise the procedure is the same as for tape ketan. The product is often somewhat softer than tape ketan. Tape nasi is known as *tapai* or as *samsul* in Malaysia, as *binubudan* or *basi* in the Philippines, as *chao* in Cambodia and as *lao-chao* in China.

Brem

Brem Wine

Brem wine is a fermented drink, known as brem Bali in Indonesia. Black glutinous rice (a variety of glutinous rice) is used for this because of its colour. The rice is made into tape ketan which has a characteristic, dark purple-red colour. After an incubation time of 5 days liquid is formed and is separated from the substrate through a 'dripping sieve'. This liquid is aged for several months at 25°C, resulting in a brown wine containing 6–13% alcohol.

Brem Cake

There are two types of brem cake: brem madiun and brem wonogiri. Brem madiun is a solid product, yellowish-white in colour with a sweet-sour taste. It is used not only as a snack, but also as a traditional medicine to stimulate the blood circulation and to cure acne. The cake is made from tape ketan which is 5–8 days old. The tape is pressed to separate the juice, which is then boiled with stirring. The resultant thick mass is poured onto wooden tables covered with plastic sheets, cooled (8–12 h) and cut into squares of 5–7 cm, 0.5 cm thick.

Brem wonogiri is white in colour, softer and more sweet-sour than brem madiun, and is sold in circular blocks 5 cm in diameter. It is made from tape ketan which is 3–4 days old. The tape is pressed, and the juice obtained is boiled until a white mass rises, then poured and moulded on bamboo trays which are covered with banana leaves. The cake is stored on bamboo racks for 24 h, then sub-dried for another 24 h.

Soy Sauce

Soy sauce is a popular seasoning agent in the Far East, and is consumed by all strata of society. It is known in Malaysia and Singapore as *kicap*, in Indonesia as *kecap*, in Thailand as *see-in*, in the Philippines as *toyo*, in China as *in-yu*, as *kajang* in Korea, and as *shoyu* in Japan.

Indonesian Soy Sauce

Preparation The processing of Indonesia soy sauce (kecap) is as follows: whole black soybeans are cleaned, soaked overnight, boiled until soft, drained, cooled, placed on trays or bamboo baskets, and inoculated with tempeh inoculum. The whole mass is then covered with banana leaves, jute sacks or polyethylene sheets, and incubated for 4–7 days at room temperature (30°C). Sometimes cassava starch is added to the substrate before inoculation. The caking substrate or tempeh is crumbled, sun-dried, and the seed hulls and formed spores are blown away by mouth. The resulting mass is then put into earthenware or ceramic jars which contain 20–30% salt solution, and the mouths of the jars are covered with a thin cloth. For the next 1–4 months, the jars are exposed daily to sunlight with the cloth removed. The brine is then boiled and filtered. Palm sugar and spices are added to the filtrate, which is then boiled again with continuous stirring, and re-filtered. When cool, the kecap is ready for bottling. In Indonesia there are two kinds of kecap, depending on the amount of palm sugar added: *kecap manis* (sweetened soy sauce) and *kecap asin* (salty soy sauce).

Microorganisms The inoculum contains *Rhizopus* spp. and very often also *Aspergillus* sp. if starch is added to the beans. During the brine fermentation halophilic bacteria and yeasts are the active microorganisms. The exact species are not known.

Nutritive Value Owing to its high salt content, which limits its consumption, kecap cannot be considered as a protein source. The amino acids and peptides contribute more to the flavour and aroma of the kecap. In Indonesia it is believed that a good-quality kecap mixed with *Curcuma xanthorhiza* tuber can relieve stomach or liver disorders, and experience proves that kecap mixed with lime is a good cough medicine.

Malaysian Soy Sauce

The soy sauce produced in Malaysia (kicap) is of the Chinese or tamari type; it may be thin (*san-chan* or *pak-yan*) or thick (*sai-yan* or *lak-yan*). The preparation method is very similar to the preparation of Indonesian kecap, but instead of cassava starch, wheat flour is used. Incubation takes 6–7 days at room temperature, and incubation in the brine solution (16–23% salt, two parts brine to one part soya/wheat) in the sun is 2–6 months. The microorganisms active in the solid substrate fermentation are *Aspergillus oryzae*, *A. soyae*, *A. niger*, *A. flavus*, *Rhizopus* spp. and *Penicillium* spp. No toxicity from kicap after consumption has been reported. Locally produced kicap has a short shelf life.

Another type of soy sauce is made from a mixture of soybean and fish. The soybeans are cleaned, soaked, boiled until cooked, drained and cooled. The substrate is then inoculated with *Aspergillus oryzae* for 2 days at 30°C. The fermented mass is ground, and then mixed with deboned fish and 10% salt, followed by incubation at 55°C for 3 weeks. A salt brine of 15% is then added and the mass is again incubated for 5 weeks at 37°C. The liquid part (fish soy sauce) is separated from the residue, pasteurized and bottled.

Philippine Fish Sauce

Philippine fish sauce (patis) is a salty amber-yellow liquid prepared from fish or shrimp, and is very common in the Philippines. It is similar to Malaysian *kicap ikan* or Vietnamese *unoc-mam*. To prepare patis, the fish or shrimp are placed in jars and sprinkled with salt. The supernatant liquid is filtered and bottled. Although patis is very nutritious its consumption is limited by its high salt content. Patis is now manufactured on an industrial scale.

Taoco

Taoco is very popular in the Far East. This fermented product is a yellow-brown, salty soybean paste with the consistency of porridge, in which the cotyledons are still visible. Taoco is closely related to *taosi* (in the Philippines), *tao-chieo* (in Thailand) and *doenjang* (in Korea).

Indonesian Taoco

Preparation   The preparation methods are more or less similar in the various countries. Dry soybeans are washed, soaked overnight (or boiled in water for 1 h until half-cooked) and hulled. The cotyledons are boiled until cooked, drained and cooled. Roasted rice flour or tapioca starch is then mixed with the cotyledons. Tempeh inoculum is sprinkled on and mixed well with the cotyledon mass, which is then placed on trays, covered with banana leaves and incubated for 2–3 days at 28–32°C until the cotyledons are soft and covered by mycelium. The mass is sun-dried, then soaked in a brine solution (18–20% w/v) in a container or jar without a lid. A thin cloth is placed over the mouth of the jar. Incubation is carried out in sunlight for 3–4 weeks. The taoco is cooked and either bottled or placed in polyvinyl bags; it is then ready for consumption. Sometimes palm sugar is added to the taoco before it is cooked.

Microorganisms   The microorganisms in the fermented mass are *Aspergillus oryzae*, *Rhizopus oryzae* or *R. oligosporus*, or a combination of them. In the palm brine solution the microorganisms are *Hansenula* sp. and *Zygosaccharomyces soya* (yeasts), and *Lactobacillus delbrueckii* (bacterium).

Nutritive Value   Liquid taoco (not sun-dried) contains 10% protein, 5% fat and 24% carbohydrate. Taoco is not a protein source. Intake of this food is limited by its high salt content.

Philippine Taoco

Philippine taoco or tausi is prepared from soybeans mixed with roasted wheat flour and inoculated with *Aspergillus oryzae*. The inoculated substrate is placed on bamboo trays, covered with banana leaves and incubated for 2–3 days. The soybean cake is placed in jars and soaked in 18% (w/v) salt brine; it is then heated to boiling, after which it is ready for consumption.

Trasi

Trasi is a dark-brown paste made from fish (*trasi ikan*) or from very small shrimp (*trasi udang*) and used as a condiment in many side dishes in nearly every Indonesian household. Trasi is never consumed raw. It is usually roasted, then ground with red chilli, and is then called *sambal trasi*. Trasi is very similar to Malaysian *belachan*, to *bagoong*, *lamayo*, *burong hipong*, *dinailan* of the Philippines, to *kung-chom* and *kapi* of Thailand, to *mam-tom* of Vietnam, to *mam-ruoc* and *prahoc* of Cambodia, to *ngapi* of Myanmar, and to *padec* of Laos.

Trasi is produced by mixing very small shrimp (*Schizopodes* sp. or *Mytis* sp.) or small fish (*Stolephorus* sp.) with 15% salt. They are spread out on trays and sun-dried (1–3 days) until the moisture content drops to 50%. During this period halophilic bacteria (*Bacillus brevis*, *B. pumilus*, *B. megaterium*, *B. coagulans*, *B. subtilis*, *Micrococcus kristinae*) will degrade the macronutrients of the mass, and a very strong aroma develops. The mass is then pounded thoroughly until a thick paste is obtained, which is further sun-dried and moulded in small packages for sale at traditional markets, or in larger parcels (100–500 g). The chemical composition is moisture 40%, protein 30 g, fat 3.5 g, carbohydrate (by difference) 3.5 g, ash 20 g per 100 g.

Dage

Seeds of *Pangium edule* or *Pithecolobium lobatum* are washed, cooked, drained, cooled and placed in earthenware jars. The jars are buried for several days, during which a spontaneous bacterial (lactic acid bacteria) fermentation takes place. The fermented product has a very strong acid smell. Sometimes the seeds of the rubber plant *Hevea brasiliensis* or seeds of *Canavalia ensiformis* or *Mucuna pruriens* var. *utilis* are used. This fermented product is only popular in West Java and is consumed as a side dish with rice.

Dadih

Dadih is a fermented food prepared from buffalo milk (West Sumatra) or goat milk (Sumba, Sumbawa islands). The product is white and has the consistency of tofu or young cheese. The buffalo milk is poured into the hollow of bamboo stems, and the opening is covered with banana leaves. Incubation is carried out at room temperature (30°C) for 24–48 h, sometimes 72 h. The buffalo milk coagulates into a solid mass or soft cake which can be shaken out of the bamboo stem and cut into smaller parts. Dadih is consumed with rice as a side dish (in soup, or stirred with vegetables). The microorganisms involved are *Lactobacillus* spp. and *Candida* sp.

Indonesian Palm Wine

Palm wine (tuak) is less popular in Indonesia than in tropical Africa and South America. In certain areas of Indonesia the inflorescence of *Cocos nucifera*, *Arenga pinnata* or *Arenga obtusifolia* is cut off and the clear, brown sap which drips out is collected in bamboo stems. The open end of the stem is then covered with banana leaves. The fresh sap becomes turbid and bubbles of CO_2 arise after several hours due to a spontaneous natural fermentation. The microorganism involved is usually *Zymomonas mobilis*, and other lactic acid and acetic acid bacteria also may be present. The alcohol content of this traditional beverage is 2–3%. The tuak is transported in the bamboo stems to the market and sold by the glass (200 ml). The empty bamboo stems are not washed, but used again for the next fermentation batch.

Traditional Vinegar

In many villages throughout Southeast Asia vinegar is a home-made product.

Philippine Vinegar

In the Philippines vinegar is made from pineapple, palm juice, sugar cane or coconut milk. The juice is first fermented with *Saccharomyces cerevisiae* (approximately 7 days) to produce alcohol and then continued with *Acetobacter aceti* (approximately 1 month) to produce the vinegar, which contains around 3% acetic acid.

Thai Vinegar

Thai vinegar is made by the fermentation of coconut water, sugared coconut juice, pineapple, banana, glutinous rice or rice wine. The preparation is very similar to Philippine vinegar. However, the production of vinegar, from rice wine, the most popular substrate, takes 3–8 months.

Nata

Nata is a thick, white insoluble layer of polysaccharides (cellulose) and cells of the bacterium *Acetobacter*; it is very popular not only in the Philippines but also throughout Southeast Asia. Nata can be produced from pineapple juice (*nata de pina*) or from coconut milk (*nata de coco*).

Nata is prepared by sieving coconut milk or pineapple juice through a cloth, and then adding sugar. The mixture is boiled, then cooled, and glacial acetic acid is added. It is then inoculated with a 'mother of vinegar' or a previous nata batch. After 2 weeks a thick, white layer is formed on the surface of the liquid substrate. The layer is washed, boiled and candied, and is ready for consumption. On an industrial scale, a pure inoculum of *Acetobacter aceti* subsp. *xylinum* is used, and addition of 10% sucrose and $(NH_4)_2PO_4$ or $(NH_4)_2SO_4$ gives the highest yield.

Indonesian Tempoyak

Tempoyak is a fermented food made from the pulp of the durian fruit (*Durio zibethinus*). It is creamy, like butter, yellow-white in colour, and has a strong smell of hydrogen sulphide. Tempoyak is popular in Malaysia and in Indonesia, chiefly on the islands of Sumatra and Kalimentan, where good-quality durian fruits are available in large amounts.

Tempoyak is prepared by placing the durian pulp in jars, to which salt is added and carefully mixed. The jars are then tightly closed and kept at room temperature for 4–7 days, during which period spontaneous fermentation takes place. The finished tempoyak is sour and salty, sour being the dominant taste. In certain parts of Sumatra the salted pulp is placed in bamboo stems which are then covered with banana leaves, and after the tempoyak is taken out the bamboo stems are used for the next batch. Tempoyak is sometimes used to improve the appetite of convalescent patients.

Pakasam

Pakasam is a fermented fish food which is popular in Malaysia and in Indonesia on the islands of Sumatra, Kalimantan and Sulawesi. For its preparation, freshwater or marine fish, shrimp or cuttlefish can be used. The fish are scaled and gutted, washed, rubbed with salt, placed on perforated bamboo trays, covered with banana leaves, and incubated for 2 days at room temperature (first fermentation period). During this incubation period spontaneous fermentation takes place, most probably by lactic acid bacteria and yeasts. Because of the added salt, water will drip from the tissues of the fish. The fish are then placed in jars, mixed with cooked rice, and more salt is added. The jars are tightly closed to obtain an anaerobic environment (second fermentation period). After 2 weeks the pakasam is ready for consumption. The shelf life of this product is 1–2 years.

Indonesian pakasam is similar to Malaysian *pekasam*. In the Malaysian product roasted rice is used instead of boiled rice. In Thailand roasted rice is also used and the first fermentation period lasts sometimes for 3 months and the second fermentation period for 3 months to 1 year. The product is called *plara*.

Indonesian Ragi

Ragi means 'starter' or 'inoculum'. The accompanying word indicates its use: for example, ragi tape is an inoculum for the production of tape; ragi tempeh is an inoculum to produce tempeh.

Ragi Tape (Yeast Cakes)

This inoculum is similar to Thailand *luk-paeng*, Chinese *peh-chu*, Korean *nooruk*, Philippines *bubod* or *bubud levadura*.

Preparation Rice flour is mixed with powdered spices: garlic (*Allium sativum* Linn.), rhizome of 'langkuas' or 'laos' (*Apina galanga* Sw.), white pepper (*Piper nigrum* Linn.), red chillies (*Capsicum frutescens* Linn.) and rhizome of ginger. Some cane sugar and powdered ragi from a previous batch is added; a thick dough is made by adding water, and the mass is moulded into small, flattened balls about 3 cm in diameter. The cakes are placed on bamboo trays, covered with rice straw and incubated for 2–3 days at 30°C. During this period the microorganisms in the dough grow and reproduce. The addition of spices helps to inhibit unwanted microorganisms. The cakes are then sun-dried and are ready to be used as inoculum for the production of rice tape (*tape nasi*), glutinous rice tape (*tape ketan*), cassava tape, brem cake and brem wine. The microorganisms found in various kinds of ragi sold in the markets always include at least one mould (*Rhizopus* spp., *Mucor* sp., *Amylomyces rouxii*) and one or more yeasts (*Saccharomyces cerevisiae*, *Endomycopsis burtonii*, *Hansenula anomala*). The moulds possess strong amylolytic activities. There are reports that *Hansenula* sp. and *Endomycopsis* sp. contribute a pleasant aroma to the tape, while *Saccharomyces cerevisiae* is the main alcohol producer.

Ragi Tempeh

Ragi tempeh contains predominantly spores of the mould *Rhizopus*. In the villages ragi tempeh is also called *usar* or *laru*.

Traditional Ragi Tempeh Cooked soybean cotyledons are spread as a single layer on the undersurface of a *Hibiscus similis* or *Tectona grandis* leaf. Another leaf is placed on top to cover the cotyledons. This process is repeated until a pile of ten leaves is obtained. The piles are tied with string and wrapped in banana leaves, placed on bamboo trays and incubated at room temperature for 24–48 h or until the cotyledons are completely covered by mycelia and a heavy black sporulation is visible. The leaves are then air-dried. For use as an inoculum, one leaf is pulverized and mixed with 1 kg of cooked soybean cotyledons.

Industrial Ragi Tempeh Cooked rice is inoculated with spores of *Rhizopus oligosporus* or a mixture of *R. oryzae* and *R. oligosporus*, placed on stainless steel trays and incubated at 30°C until full sporulation is achieved. The mouldy rice is ground into a powder. To make tempeh, 1 g of the powder is mixed with 1 kg of cooked cotyledons.

See also: **Acetobacter**. **Aspergillus**: Introduction; *Aspergillus oryzae*; *Aspergillus flavus*. **Bacillus**: Introduction; *Bacillus subtilis*. **Fermented Milks**: Products of Eastern Europe and Asia. **Hansenula**. **Saccharomyces**: *Saccharomyces cerevisiae*. **Vinegar**. **Zygosaccharomyces**.

Further Reading

Campbell-Platt G (1987) *Fermented Foods of the World – A Dictionary and Guide*. London: Butterworth.

Hesseltine CW and Wang HL (1986) *Indigenous Fermented Foods of Non-Western Origin*. Mycologia Memoir 11. Berlin: Cramer.

Steinkraus KH (ed.) (1996) *Handbook of Indigenous Fermented Foods*, 2nd edn. New York: Marcel Dekker.

Sudarmadji S, Suparmo and Rahardjo (1997) Reinventing the hidden miracle of tempe. *Proceedings of the International Tempe Symposium*, 13–15 July, Bali.

Surono IS and Hosono A (1994) Microflora and their enzyme profile in 'terasi' starter. *Biosci. Biotech. Biochem.* 58 (6): 1167–1169.

FERMENTED MILKS

Contents
Range of Products
Yoghurt
Products from Northern Europe
Products of Eastern Europe and Asia

Range of Products

E Litopoulou-Tzanetaki and **N Tzanetakis**,
Department of Food Science, Faculty of Agriculture,
Aristotle University of Thessaloniki, 54006,
Thessaloniki, Greece

Introduction

Fermented milks have been developed as a means of preserving milk against spoilage. Their evolution through the ages has progressed from home manufacture, using a small portion of a previous batch as starter, to large-scale production, in which selected starters and automatic processes are used. Most fermented milks are made from cow's milk but sheep, goat, buffalo and horse milks are also used. The consumption of these products continues to increase, and there are good incentives to expand the range and quality of fermented milks, by using new isolates of fermentation microorganisms from the native flora of traditional products (**Table 1**).

Benefits to health and nutrition have been attributed to fermented milks. In today's world, consumers demand products of high quality in terms of nutritional, biological and dietetic value. Starter cultures play a central role in the achievement of these qualities.

Classification of Fermented Milks

There are some common steps for the production of all fermented milks (**Fig. 1**), but the fine details of manufacture differ from product to product. The bacteria used in individual fermentations (**Table 2**) are nowadays available as starter cultures, although for some traditional products (e.g. koumiss) non-defined starters are still used. The organisms in the starter culture determine the type of fermentation, and the resulting products have individual characteristics, which derive from the metabolic activities of the starter bacteria.

Fermented milks may be classified by starter culture and the type of fermentation involved, as follows:

- lactic fermentation – mesophilic or thermophilic
- fermentation by intestinal bacteria
- yeast–lactic fermentation
- mould–lactic fermentation.

Mesophilic Lactic Fermentations

Starters

Mesophilic starters belong to the genera *Lactococcus*, *Leuconostoc* or *Pediococcus*, and mesophilic cultures contain organisms which produce acid and flavour. The main acid-producer is *Lactococcus lactis* subsp. *cremoris*; *L. lactis* subsp. *lactis* is used to a lesser extent. The flavour-producers are *L. lactis* subsp. *lactis* biovar *diacetylactis* and leuconostocs.

The application of DNA and RNA hybridization techniques recently resulted in the reclassification of several of the lactic acid bacteria. Of the five species of *Lactococcus* which are now recognized, only *L. lactis* is used in dairy technology. Leuconostocs are still classified by phenotypic criteria (**Fig. 2**). However, recent studies on phylogenetic relationships suggest three well-distinguished groups, possibly three new genera, within the *Leuconostoc* group. The only species of *Pediococcus* used in fermented milks is *P. pentosaceus*.

Lactococci produce 0.5–0.7% L(+)-Lactate from lactose in milk. The leuconostocs in mesophilic starter cultures produce mainly D(–)-Lactate. Lactose is transported into cells by the phosphoenolpyruvate (PEP) system (also known as the phosphoenolpyruvate-dependent phosphotransferase system (PTS)) as lactose phosphate. This is hydrolyzed to glucose and galactose 6-phosphate by a phospho-β-galactosidase (β-P-gal). In leuconostocs, lactose is hydrolyzed directly by a permease and a β-galactosidase (β-gal), producing galactose and glucose. In either organism, the glucose moiety is catabolized by the Embden–Meyerhof–Parnas (EMP) pathway, the galactose 6-phosphate by the D-tagatose 6-phosphate pathway and the galactose by the Leloir pathway.

The leuconostocs and *L. lactis* biovar *diacetylactis* metabolize citrate to diacetyl, acetoin, 2,3-butylene glycol and CO_2. Citrate is transported into the cells by a citrate permease, which is plasmid-encoded and easily lost.

Table 1 Traditional fermented milks and derived products, made in various parts of the world

Product	Country	Type of milk	Microflora
Aoules	Algeria	Goat, ewe	
Arkhi	Mongolia	Mare	
Ayran	Turkey	Ewe, goat, cow	Yoghurt microorganisms
Bjaslag	Mongolia		Yoghurt microorganisms
Brano mliako	Bulgaria	Ewe	Yoghurt microorganisms
Bulgarian milk	Bulgaria	Ewe, cow	*Lactobacillus bulgaricus*
Chakka	India	Mixed buffalo and cow	*Lactococcus lactis* subsp. *lactis*
Chal	Turkmenistan	Camel	Thermophilic lactobacilli and streptococci
Churpi	Nepal	Buffalo, goat	
Dahi	India	Buffalo, goat	*Lactococcus lactis* subspp. *lactis*, *cremoris*, biovar *diacetylactis*, *Leuconostoc* spp., yoghurt microorganisms
Dough	Iran, Afghanistan	Ewe	Yoghurt microorganisms
Ergo	Ethiopia	Cow	Lactobacilli, (mainly *L. plantarum*), streptococci
Grurovina	Former Yugoslavia	Mixed ewe and cow	Yoghurt microorganisms (possibly)
Iria ri matti	Kenya	Cow	*Streptococcus thermophilus*
Jamid	Jordan	Goat	
Jub-jub	Lebanon	All types	Yoghurt microorganisms
Karmidinka	Poland	Cow	Yoghurt microorganisms
Kashk	Iran	Ewe, mixed ewe and cow	Yoghurt microorganisms
Katyk	Uzbekistan	Buffalo	*Streptococcus thermophilus*, *Thermobacterium* spp.
Kefir	Former Soviet Union	Ewe, cow	*Lactococcus lactis* subspp. *lactis*, *cremoris*, *Leuconostoc*, *Acetobacter aceti*, yeasts (lactose-positive or -negative)
Kisle mliake	Bulgaria	Mixed ewe and cow	Yoghurt microorganisms
Koumiss	Former Soviet Union	Mare	*Lactobacillus delbrueckii* subsp. *bulgaricus*, yeasts
Kurunga	Former Soviet Union	Cow	*Lactococcus lactis* subspp. *lactis*, *cremoris*, biovar *diacetylactis*, *Streptococcus thermophilus*, *Lactobacillus delbrueckii* subsp. *bulgaricus*, *L. acidophilus*, lactose-positive yeasts
Krut (kurt)	Afghanistan	Cow, ewe, goat	Lactococci, *Streptococcus thermophilus*, *Lactobacillus delbrueckii* subsp. *bulgaricus*
Laban (leben)	Middle East	All types	*Streptococcus thermophilus*, *Lactobacillus delbrueckii* subsp. *bulgaricus*, *L. acidophilus*, *Leuconostoc lactis*, *Kluyveromyces marxianus* subsp. *marxianus*, *Saccharomyces cerevisiae*
Laban khad	Egypt	All types	Mesophilic (buttermilk) microorganisms
Laban rayeb	Egypt		*Lactococcus lactis*, *Kluveromyces marxianus* subsp. *marxianus*, *Lactobacillus* casei
Laban zeer	Egypt	All types	*Lactococcus lactis*, *Leuconostoc* spp., *Lactobacillus casei*, *L. plantarum*, *L. brevis*
Labneh	Middle East	All types	Yoghurt microorganisms
Labnech anbaris	Middle East	All types	Yoghurt microorganisms
Lassi	India	Buffalo, goat	*Lactococcus lactis* subspp. *lactis*, *cremoris*, biovar *diacetylactis*, *Leuconostoc* spp., yoghurt starter bacteria
Lactofil	Sweden	Cow	*Lactococcus lactis* subspp. *lactis*, *cremoris*, biovar *diacetylactis*, *Leuconostoc atrovorum*
Langfil	Sweden	Cow	Butter starter bacteria
Matsun	Armenia	All types	Thermophilic bacteria (cocci and rods)
Maziva iala	Kenya	Cow	*Lactococcus lactis*, *Leuconostoc mesenteroides* subsp. *cremoris*
Nono	Nigeria	Goat	*Lactobacillus delbrueckii* subsp. *bulgaricus*, *L. helveticus*, *L. plantarum*, *Lactococcus lactis* subsp. *cremoris*
Prostokvasha	Former Soviet Union	Cow, buffalo	Mesophilic lactic acid bacteria
Roba	Iraq, Sudan, Egypt	Cow, buffalo, goat	Lactococci, lactobacilli, *Mycoderma*
Syuzma	Azerbaijan	Cow	Thermophilic cocci, *Lactobacillus delbrueckii* subsp. *bulgaricus*
Shubat	Kazakhstan	Camel	*Lactobacillus delbrueckii* subsp. *bulgaricus*, yeasts
Skyr	Iceland	Ewe, cow	Yoghurt bacteria, yeasts
Sooms tej	Hungary	Ewe	Lactic fermentation

Table 1 Continued

Srikhand	India	Buffalo, goat	Dahi microorganisms
Syuzma	Azerbaijan	Cow	Thermophilic cocci, *Lactobacillus delbrueckii* subsp. *bulgaricus*
Taettemelk	Finland		*Lactococcus lactis* subspp. *cremoris*, biovar *diacetylactis*, *Leuconostoc* spp.
Takammart	Algeria	Goat, ewe	
Tarag	Mongolia	All types	
Torba	Turkey	Mixed cow and ewe	Yoghurt microorganisms
Tulum yoghurt	Turkey	Cow, ewe, goat	Yoghurt microorganisms
Viili	Finland	Cow	*Lactococcus lactis* subspp. *lactis*, *cremoris*, biovar *diacetylactis*, *Leuconostoc mesenteroides* subsp. *dextranicum*, *Geotrichum candidum*
Xynogalo	Greece	Ewe	*Lactococcus lactis* subspp. *lactis*, *cremoris*, *Lactobacillus plantarum*, *L. maltomicus*, *L. casei*, *Leuconostoc lactis*, *L. mesenteroides*, *L. paramesenteroides*, *Enterococcus faecalis*, *E. faecium*, *E. durans*
Yiaourti	Greece	Ewe, goat, cow	*Streptococcus thermophilus*, *Lactobacillus delbrueckii* subsp. *bulgaricus*, *L. paracasei*, *Leuconostoc* spp., pediococci, enterococci
Ymer	Denmark	Cow	*Lactococcus lactis* subspp. *lactis*, *cremoris*, biovar *diacetylactis*, *Leuconostoc citrovorum*
Zabady	Egypt	All types	Yoghurt microorganisms
Zhentitsa	Carpathian	Ewe	Yoghurt microorganisms
Zimne sour milk	Former Yugoslavia	Ewe	Yoghurt microorganisms

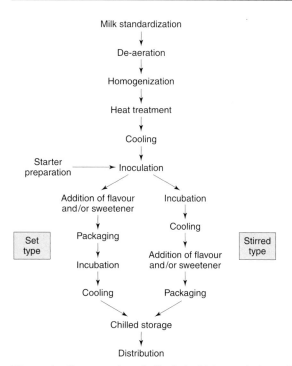

Figure 1 Common steps in the industrial manufacture of fermented milks.

The lactic acid bacteria degrade caseins. Cell wall and endocellular proteinases produce large peptides, which are degraded by exopeptidases and endopeptidases to small peptides. These are further degraded by dipeptidases, tripeptidases, prolidase or proline iminopeptidase. There is also evidence for the presence of intracellular peptide hydrolase systems in leuconostocs. Esterases from lactococci and *Leuconostoc mesenteroides* degrade short-chain fatty acids preferentially.

Mesophilic Fermented Milks

Buttermilk   This is made from either skimmed milk or milk with a fat content of 0.5–3.0%. The milk is fermented by a starter composed of *Lactococcus lactis* and *Leuconostoc mesenteroides* subsp. *cremoris* (inoculum 1–2%) at 20°C for 16–20 h. After fermentation, the lactic acid content of the milk is 0.8–0.9%. The fermented milk has to be cooled immediately, to avoid decreases in the diacetyl concentration. Natural or true buttermilk is the product that remains after churning ripened cream into butter. In Greece and other parts of the world, it is either consumed as it is or used to make pies. It is also used to feed animals.

Ymer and Similar Products   Ymer is a Danish traditional fermented milk. It is made from heat-treated cow milk by fermentation with a cream starter culture at 20–22°C, until a pH of 4.6 is attained. The product is then cut and about 50% of the whey is removed by indirect heating with water. The milk is then homogenized, cooled and packed. Lactafil is a traditional Swedish fermented milk, similar to ymer. Filmjolk is a Swedish product, made from cow milk which has been partially skimmed (3% fat) and heat-treated (90–91°C for 3 min), and a cream starter culture containing lactococci and leuconostocs in the ratio

Table 2 Bacteria associated with the manufacture of fermented milks, and their main function in milk

Bacteria	Examples of fermented milks	Main products of bacteria
Lactococci		
L. lactis subsp. lactis	Cultured buttermilk, kefir	Lactic acid
L. lactis subsp. cremoris	Cultured buttermilk, kefir	Lactic acid
L. lactis biovar diacetylactis	Cultured buttermilk, kefir	Diacetyl + lactic acid
Streptococci		
S. thermophilus	Yoghurt	Lactic acid + acetaldehyde
Leuconostocs		
L. mesenteroides subsp. mesenteroides	Kefir	Diacetyl + lactic acid
L. mesenteroides subsp. cremoris	Kefir	Diacetyl + lactic acid
L. mesenteroides subsp. dextranicum	Kefir	Diacetyl + lactic acid
Pediococci		
P. acidilactici	Biokys	Lactic acid
Lactobacilli		
L. delbrueckii subsp. delbrueckii	Lactic drinks	Lactic acid
L. delbrueckii subsp. lactis	Lactic drinks	Lactic acid
L. delbrueckii subsp. bulgaricus	Yoghurt, Bulgarian buttermilk	Acetaldehyde + lactic acid
L. helveticus	Kefir, koumiss	Lactic acid
L. acidophilus	Acidophilus milk, kefir	Lactic acid
L. paracasei subsp. paracasei	Lactic drinks	Lactic acid
L. rhamnosus	Kefir	Lactic acid
L. plantarum	Kefir	Lactic acid
L. kefir	Kefir	Lactic acid
L. kefiranofasciens	Kefir	Lactic acid
L. brevis	Kefir	Lactic acid + CO_2
L. fermentum	Kefir	Lactic acid + CO_2
Bifidobacteria		
B. adolescentis	Fermented milks	Lactic and acetic acid
B. bifidum	Yoghurt-like products	
B. breve		
B. infantis		
B. longum		

85 : 15. The milk is cooled to 20–21°C and is then fermented by the starter (inoculum 2%) for 20–24 h.

Nordic Ropy Milks These are traditional products of Norway, Sweden and neighbouring countries, but ropy milk is the generic name for any fermented milk made with mesophilic cocci which produce slime. Traditionally, in addition to the uncharacterized starter culture, the leaves of *Pinguicula vulgaris* and *Drosera* spp. are also added to the milk. The lactic microflora includes slime-producing strains of *Lactococcus lactis* subspp. *lactis* and *cremoris*. The grasses may introduce *Alcaligenes viscosus*, a bacterium which also produces slime.

Thermophilic Lactic Fermentations

Starters

The thermophilic lactic acid bacteria involved in the production of yoghurt and yoghurt-like products are *Streptococcus thermophilus* and *Lactobacillus delbrueckii* subsp. *bulgaricus*. These two microorganisms produce factors which stimulate each other's growth: this interaction is favourable to both but not obligatory, and is termed 'protocooperation'.

The taxonomic positions of thermophilic lactic acid bacteria have undergone various reclassifications. *S. thermophilus* is now included in the group 'other streptococci' of the genus *Streptococcus*, and *L. delbrueckii* subsp. *bulgaricus* is classified in Group I of the obligatory homofermentative lactobacilli. However, phylogenetic analysis by the rRNA sequencing technique suggests that the internal structure of *Lactobacillus* does not correlate with its metabolic traits and grouping. Although the phylogenetically instinctive *L. delbrueckii* group contains obligatory homofermentative lactobacilli, it does not contain them all. *L. fermentum* is also a peripheral member of this group.

Lactose is transferred into the bacterial cells by a permease system and subsequently split by a β-galactosidase into glucose and galactose. Galactose is normally released to the extracellular medium, although in some strains it can be metabolized by the Leloir pathway when lactose is limited. *S. thermophilus* produces 0.6–0.8% L(+)-Lactate and *L.*

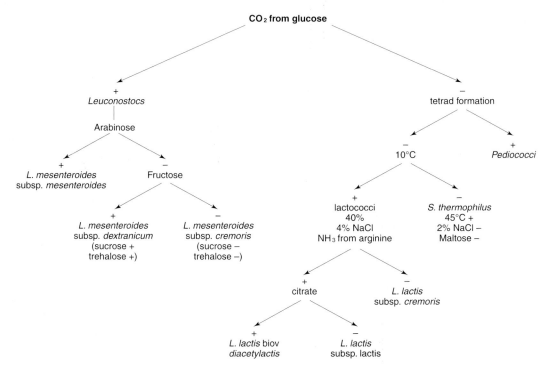

Figure 2 Differentiation of lactococci and leuconostocs associated with fermented milks, and of *Streptococcus thermophilus*.

Table 3 Proteolytic enzymes of thermophilic lactic acid bacteria used in fermented milks

Microorganism	Enzyme	Substrate
Streptococcus thermophilus	Proteinase	Casein
	Metalloaminopeptidases	Dipeptides
	X-propyl dipeptidyl aminopeptidase	N-terminal X-propyl and X-alanyl residues
	Leu-aminopeptidase	Leu-p-NA
	Dipeptidase	Dipeptides
	Endopeptidase	Glucagon and insulin-β chain
Lactobacillus delbrueckii subsp. bulgaricus	Proteinase	Casein
	Proteinase	Whey proteins
	Aminopeptidase N	Oligopeptides
	Aminopeptidase C	Oligopeptides
	Dipeptidase	Dipeptides
	X-propyl dipeptidyl aminopeptidase	N-terminal X-propyl and X-alanyl residues
	Proline iminopeptidase (cell wall, intracellular)	Pro-X dipeptides and Pro-Gly-Gly, mainly

delbrueckii subsp. *bulgaricus* produces 1.7–1.8% D(+)-Lactate. The typical aroma and flavour of plain yoghurt is particularly associated with acetaldehyde, which is produced by both bacteria, either from lactose via pyruvate or, in the case of *Lactobacillus*, by cleavage of threonine to glycine and acetaldehyde. Acetone, acetoin, diacetyl and ethanol may also be formed.

Both organisms produce extra- and intracellular proteinases. They also produce peptidases (**Table 3**) and cause free amino acids to accumulate in the milk. During yoghurt manufacture and storage, lipolysis may occur because of bacterial esterase activity.

However the lipolytic activities of the thermophilic starter organisms are generally low, so any volatile acids produced may derive from the hydrolysis of compounds other than lipids.

Thermophilic Fermented Milks

Yoghurt This has been developed in eastern Mediterranean countries over thousands of years. A quantity of the product from previous batches is still used to start fermentation in homes, small shops and creameries.

Yoghurt produced over the past few decades in the factories of Europe and other parts of the world results from the fermentation of cow milk by a defined microflora, which consists of *Streptococcus thermophilus* and *Lactobacillus delbrueckii* subsp. *bulgaricus*. The fermentation takes place at 42°C for 3 h, either in retailable containers or in bulk, producing 'set' or 'stirred' yoghurt respectively. The final product contains 1.2–1.4% lactic acid. Goat's milk can also be used to make yoghurt. Sheep's milk is traditionally used to make 'yiaourti' in Greece, using a small quantity of a previous batch as starter. The lactic microflora which develops during the fermentation consists of *S. thermophilus* and *L. delbrueckii* subsp. *bulgaricus*, but enterococci, pediococci, leuconostocs and *L. paracasei* may also be present. 'Zabady' (or 'laban zabady') is the traditional type of yoghurt made in Egypt from buffalo or cow milk, or a mixture of the two. Its lactic microflora consists, predominantly, of *S. thermophilus* and *L. delbrueckii* subsp. *bulgaricus*, but *L. casei*, *L. fermentum*, *L. viridescens*, *L. helveticus* and *Enterococcus faecalis* may also be found. 'Dahi' is made in India from buffalo, cow or mixed milk. The starter culture is composed of mesophilic cocci, with leuconostocs or yoghurt starter bacteria.

Bulgarian Buttermilk This is produced in Bulgaria from cow, sheep or goat milk, which is pasteurized, cooled, inoculated with *L. delbrueckii* subsp. *bulgaricus* (inoculum 2–5%) and fermented for 5 h at 42°C.

Other Products Products made with a starter culture of *S. thermophilus* alone (inoculum 5%) are 'kehran', 'karan' or 'heran' in Siberia, and 'lapte-akru' in Romania. For these products, heat-treated milk is fermented for 3–5 h at 40°C. 'Katyk' is made in Kazakhstan using a culture consisting of equal proportions of *S. thermophilus*, *Lactococcus lactis* biovar *diacetylactis* and *Lactobacillus helveticus* or *L. delbrueckii* subsp. *bulgaricus*.

Milks Fermented with Selected Intestinal Bacteria

Starters

The intestinal bacterial strains used in starters are lactobacilli, bifidobacteria (**Table 4**) and enterococci (*E. faecalis* and *E. faecium*), called probiotics.

DNA–DNA homology studies of *Lactobacillus acidophilus* strains identified groups and subgroups, which could represent six species. Strains belonging to group A-1 are suggested for dietary preparations. *L. acidophilus* and bifidobacteria may prevent the growth and development of many gastrointestinal organisms, and milk fermented by these organisms may exert therapeutic and/or prophylactic properties

for the consumer. A successful probiotic must fulfil the following criteria:

- human origin
- stability in gastric acid
- stability in bile
- antagonism against carcinogenic and pathogenic bacteria
- production of antibacterial substances
- adherence to human intestinal cells
- colonization of the human intestinal tract
- good growth in vitro
- safety in food and clinical use
- clinically validated and documented health effects.

L. acidophilus metabolizes carbohydrates by homolactic fermentation, as described for thermophilic lactic acid bacteria. In mixed starters, lactic acid production by *L. acidophilus* stimulates *Bifidobacterium bifidum* to produce acid. In addition, the products of proteolysis, resulting from the peptidolytic activity of *L. acidophilus*, stimulate acid production by *B. bifidum* and *B. longum*. Bifidobacteria metabolize carbohydrates by heterolactic fermentation, degrading hexoses through the phosphoketolase pathway to form L(+)-Lactate and acetate from glucose. They do not exhibit extracellular caseinolytic activity, but cause amino acids to accumulate in milk due to aminopeptidase and carboxypeptidase activities. It is likely that bifidobacteria have no lipolytic activity.

Therapeutic Products

Fermented milks made with *Lactobacillus acidophilus* and/or bifidobacteria are listed in **Table 5**.

Acidophilus Milk This is a non-traditional product, made from cow milk by fermentation with *Lactobacillus acidophilus* isolates from the faeces of healthy humans. Appropriate strains have the following characteristics: good growth in milk, to 10^8 cfu g^{-1} or ml^{-1}; production of L(+)-Lactic acid; reduction of milk pH $\leqslant 4.7$ in $\leqslant 20$ h with an inoculum of 10%; and little proteolytic activity. In the production of acidophilus milk, heat-treated (95°C) homogenized milk is cooled to 37°C and inoculated with 2–5% of the starter. *L. acidophilus* is inhibited by lactic acid concentrations of around 0.6%: therefore it is preferable to cease incubation when the lactic acid concentration approaches 0.65%. The retail product should contain 5×10^8 cfu ml^{-1} *L. acidophilus*.

Gefilac, Gefilus These are the trade names of fermented milks made in Finland using Lactobacillus GG (*L. rhamnosus*, ATCC 53103), a strain of human origin which fulfils the requirements for a probiotic microorganism.

Table 4 Differential characteristics of selected species of bifidobacteria, and their distribution in the intestines of humans and animals

Characteristic	B. adolescentis	B. bifidum	B. breve	B. infantis	B. longum
Arabinose	+	−	−	−	+
Gluconate	+	−	−	−	−
Maltose	+	−	+	+	+
Salicin	+	−	+	−	−
Infant human		+	+	+	+
Adult human		+			+
Monkey	+				+
Dog	+				
Pig	+				

Table 5 Fermented milks made using *Lactobacillus acidophilus* and/or bifidobacteria

Product	Country of origin	Starter culture
AB milk products	Denmark	*L. acidophilus, B. bifidum*
A38 fermented milk	Denmark	*L. acidophilus*, mesophilic lactic culture
Acidophilus milk	Many countries	*L. acidophilus*
Acidophilus yoghurt	Many countries	*L. acidophilus*, yoghurt starter bacteria
Acidophilus bifidus yoghurt	Germany	*L. acidophilus, B. bifidum* (or *B. longum*), yoghurt starter bacteria
ACO-yoghurt	Switzerland	*L. acidophilus*, yoghurt starter bacteria
Arla acidophilus	Norway	*L. acidophilus*
BA	France	*B. longum*, yoghurt starter bacteria
Bifidus milk	Germany	*B. bifidum* (or *B. longum*)
Bifidus yoghurt	Many countries	*B. bifidum* (or *B. longum*), yoghurt starter bacteria
Bifighurt	Germany	*B. longum*
Bifilact	Former Soviet Union	*Lactobacillus* spp., *Bifidobacterium* spp.
Biobest	Germany	Bifidobacteria, yoghurt starter bacteria
Biogarde	Germany	*L. acidophilus, B. bifidum, Streptococcus thermophilus*
Bioghurt	Germany	*L. acidophilus, B. bifidum, Streptococcus thermophilus*
Biokys	Former Czechoslovakia	*B. bifidum, L. acidophilus, Pediococcus acidilactici*
Biolat	Greece	*L. acidophilus*, bifidobacteria, yoghurt starter bacteria
Biomild	Germany	*L. acidophilus, Bifidobacterium* spp.
Cultura	Denmark	*L. acidophilus, B. bifidum*
Diphilus milk	France	*L. acidophilus, B. bifidum*
Kefir	Many countries	*L. acidophilus*, lactic acid bacteria, yeasts
Mil-Mil E	Japan	*L. acidophilus, B. bifidum, B. breve*
Miru-Miru	Japan	*L. acidophilus, B. breve, L. casei*
Ofilus	France	*Streptococcus thermophilus* (or *Lactococcus lactis* subsp. *cremoris*), *L. acidophilus, B. bifidum*
Progurt	Chile	*Lactococcus lactis* subsp. *cremoris*, biovar *diacetylactis*, *L. acidophilus, B. bifidum*
Smetana	Eastern Europe	*L. acidophilus, Lactococcus lactis* biovar *diacetylactis*
Sweet acidophilus bifidus milk	Japan	*L. acidophilus, B. longum*
Sweet bifidus milk	Japan, Germany	*Bifidobacterium* spp.

Yakult This is a product of Japan and the Far East, made using *L. paracasei* subsp. *paracasei* (strain Shirota), which behaves in the human gut in a way similar to *L. acidophilus* and bifidobacteria. The concentration of viable cells of *L. paracasei* subsp. *paracasei* in the product is $> 10^8$ per millilitre.

Bioghurt, Biogarde and Bifighurt These are the trade names of fermented milks made by the following starters: Bioghurt – *Streptococcus thermophilus* and *L. acidophilus*; Biogarde – *S. thermophilus*, *L. acidophilus* and *Bifidobacterium bifidum*; Bifighurt – *B. bifidum*. The microorganisms are initially grown as monocultures. By using inocula in the proportion of 10–20% the acidification time is shortened, and a product with abundant cells of *L. acidophilus* (10^7–10^8 ml^{-1}) and/or *B. bifidum* (10^6–10^7 ml^{-1}) is obtained. Incubation is terminated at pH 4.9–5.0, so that the retail product has a pH > 4.6, which allows the prolonged survival of the bacteria.

AB and Similar Milk Products AB milk products are produced when *L. acidophilus* and *B. bifidum*, grown separately, are used to ferment heat-treated cow milk (250 g *L. acidophilus* + 100 g *B. bifidum* per 100 l milk). The sour milks resemble yoghurt in terms of consistency and flavour. 'Mil-Mil' is the trade name of a product similar to AB fermented milk, and is made using *B. bifidum*, *B. breve* and *L. acidophilus*.

'Biokys' is the trade name of a product of the former Czechoslovakia, made from cow milk. It is a mixture of two fermented milks. Nine parts of heat-treated milk are inoculated with 1% of a starter containing *B. bifidum* and *Pediococcus acidilactici*, and are incubated at 37°C; one part of milk is inoculated with 1% of a starter of *L. acidophilus* and is incubated at 30°C. After fermentation, the two products are mixed and cooled.

Bifidus Milk This is made from heat-treated cow milk inoculated with *B. bifidum* and incubated at 37°–42°C. The final product contains 10^8–10^9 per millilitre bifidobacteria.

Acidophilus or bifidus yoghurts are produced either by a mixed fermentation of the two cultures, or by mixing yoghurt and a separately fermented acidophilus or bifidus milk together in a desired ratio.

Yeast–Lactic Fermentations

Starters

Kefir grains are white or slightly yellow and incorporate a microflora comprising lactic acid bacteria, acetic acid bacteria, yeasts, the mould *Geotrichum candidum* and various contaminants. The indigenous microflora of kefir grains is variable. The lactic microflora of 'grain starter' consists of *Lactococcus lactis* subspp. *lactis* and *cremoris*, homofermentative and heterofermentative lactobacilli (15 species), *Leuconostoc mesenteroides* subsp. *dextranicum* and *Streptococcus thermophilus*. Yeasts of the genera *Saccharomyces*, *Kluyveromyces*, *Candida*, *Mycotorula*, *Torulopsis*, *Cryptococcus*, *Torulaspora*, *Pichia*, and the acetic acid bacteria *Acetobacter aceti* and *A. racens* may also be found. Koumiss microflora is composed of lactobacilli, mainly *L. delbrueckii* subsp. *bulgaricus*, and yeasts (*Saccharomyces lactis*, *Saccharomyces cartilaginosus*, *Torula koumiss* and *Mycoderma* spp.).

Lactose can be utilized by homofermentative lactic acid bacteria, by the glycolytic and D-tagatose 6-phosphate pathways, and by heterofermentative lactic acid bacteria, by the phosphoketolase and Leloir pathways. It can also be used by lactose-fermenting yeasts, in alcoholic fermentations. Citrate may also be fermented. The most important volatile components are acetaldehyde, propionaldehyde, acetone, ethanol, 2-butanone, *n*-propyl alcohol, diacetyl and amyl alcohol. Proteolytic activities are exhibited, mainly by yeasts and acetic acid bacteria.

Products

Kefir This is a traditional product of the Caucasus, made from all types of milk. Traditionally, fresh milk was fermented in goatskin bags and hung in the house during the winter and outside during the summer. During incubation, microorganisms from grain are shed into the milk and ferment it. For commercial-scale fermentation, heat-treated (95°C, 10–15 min) skimmed milk is fermented at 18–22°C using kefir grains, which are then sieved out. The milk is then used as an inoculum, at 3–5%, for heat-treated skimmed milk which is incubated at 20–22°C for 10–12 h. The milk is then cooled to 8°C and ripened for another 12 h. Kefir cultures and beverages of high quality have the microbial composition summarized in **Table 6**. The final product has 1.0% titratable acidity, and contains 0.01–0.1% ethanol and small amounts of CO_2.

Koumiss This is a traditional product of Mongolia, Tibet and Russia, which is made from mare's milk. Koumiss is traditionally produced from raw milk and a portion of a previous batch. For industrial production, the milk is pasteurized (90–92°C, 5 min) and cooled to 26–28°C, and the starter is added in the proportion 10–30%.

Skyr This is a traditional product of Iceland, made from skimmed ewe's milk. The milk is boiled, cooled to 40°C and inoculated with product from a previous batch, which has been diluted with water. Rennet is also added. After fermentation the product is strained. The skyr microflora consists of *Streptococcus thermophilus*, *Lactobacillus delbrueckii* subsp. *bulgaricus*, *L. jugurti*, *L. helveticus* and lactose-fermenting yeasts.

Mould–Lactic Fermentations

The Finnish product 'vilii' is made by fermenting milk with *Lactococcus lactis* subsp. *lactis* biovar *diacetylactis*, *Leuconostoc mesenteroides* subsp. *cremoris* and the mould *Geotrichum candidum*. Low-fat (about 2.5%) or whole-fat (3.0%) milk is heat-treated at 85–90°C, cooled to 20°C and inoculated (inoculum 3–4%) with the starter. Fermentation is ended when the concentration of lactic acid reaches 0.9%. During incubation the fat rises to the surface. The mould, which develops on the floating fat, gives the product a velvet-like appearance.

Table 6 Viable counts of kefir grains, starter and beverage

Microbial group	Kefir grains (cfu g^{-1})	Kefir starter (cfu ml^{-1})	Kefir (cfu ml^{-1})
Lactococci	10^6 (cremoris)	10^8–10^9	10^9
Leuconostocs	10^6	10^7–10^8	10^7–10^8
Thermophilic lactobacilli	10^8	10^6	10^7–10^8
Mesophilic lactobacilli	10^6–10^9	10^2–10^3	
Acetic acid bacteria	10^8	10^5–10^6	10^4–10^5
Yeasts	10^6–10^8	10^5–10^6	10^4–10^5

Derived Products

Strained Yoghurt This is manufactured in the Middle East and neighbouring countries by straining natural yoghurt, usually using a cloth bag, over periods of several hours. It is named differently in different countries, e.g. 'strangisto yiaourti' in Greece, 'torba' in Turkey, 'labnah' in Saudi Arabia and 'labneh' in Lebanon. Strained yoghurt is used in Greece to make 'tzatziki', a food consisting of strained yoghurt with added cucumber, garlic, anise, vinegar, olive oil and salt. 'Kurut' is made in Turkey by pressing torba to remove more whey, adding 5% salt and drying it for up to 10 days. A similar product in Iran is called 'kashk'. 'Labneh anbaris' is made in some Middle East countries from labneh, by adding salt, forming it into balls and placing the balls in the sun for partial drying. The balls are then packed into jars and covered with oil.

Paskitan and Other Yoghurt-related Products 'Paskitan' is a home-made yoghurt-related product, made by Greeks originating from the Black Sea area. The yoghurt is made from cow milk by slow acidification over about a week, and the butterfat is removed by churning in a wooden churn (a 'xylag'). The remaining product is called 'tan' and is either consumed as it is or heated, without boiling, to coagulate the proteins. It is then cooled and strained in a cloth bag. The product, paskitan, is either consumed as it is or is lightly salted. It has a low pH (3.15–3.94) and usually contains yeasts (10^5–10^7 cfu g^{-1}) and psychrotrophic bacteria (10^4–10^7 cfu g^{-1}), but coliforms, staphylococci and enterococci are rarely found. Dried balls of paskitan, which are heavily salted, are called 'tsiortania'. They are used in dishes with macaroni or other foods.

'Churra', a product of Nepal and 'chokeret', a product of Turkey, are made like paskitan. Turkish nomads call the buttermilk remaining after making butter from yoghurt 'ayran'. After heating ayran, the coagulate formed is placed in cloth bags to drain and the final product is called chokeret. Another product, also called 'ayran', is made in Turkey by mixing 1% brine with an equal amount or twice the amount of yoghurt. 'Dough' is made in Iran from yoghurt diluted with an equal quantity of water and flavoured with 1% salt, mint, and other herbs.

Jamid, Aoules and Chakka The Bedouin in Jordan make jamid by transferring goat's milk, which has been naturally fermented in sheepskin bags, into a cloth bag to remove the whey. Salt is added to the strained fermented milk, which is then formed into balls and dried in the sun. From a similar fermented milk in Algeria, butterfat is removed, the remaining buttermilk is heated and the curd is formed into flat discs which are dried in the sun. The dried product is called 'aoules'. 'Chakka' is a traditional fermented milk of India made using different starters with buffalo and other types of milk from which the whey has been drained. After manufacture, chakka is used to make 'shirkhard', by mixing it with cream, sugar and cardamom.

Products containing Wheat 'Sweet trahanas' is made from sheep's milk, which is boiled, mixed with parboiled wheat and left to dry. 'Sour trahanas' is made from traditional yoghurt or soured ewe's milk, which is mixed with flour and left to dry. The dried mass is then ground into small particles. The lactic microflora of acid trahanas made from naturally acidified milk is composed of *Lactobacillus plantarum*, *L. brevis*, *Enterococcus faecalis*, *Leuconostoc mesenteroides*, *L. paramesenteroides*, *Pediococcus acidilactici* and *P. pentosaceus*. There is evidence that trahanas is an ancient Greek product, called 'amis' by the ancient Greeks.

A product known as 'kishk' in Egypt and 'kichkin' in Lebanon is made from 'laban khad', a fermented milk made by natural souring in skin bags, or 'laban zeer', which is laban khad that has been stored in earthenware pots. Three or more parts of laban khad or laban zeer are mixed with one part of wheat flour or parboiled wheat (bulgur) or cooked wheat (belila). The mixture is boiled and dried. 'Kuskuk' is made in Iraq by mixing one part of dried parboiled whole wheat and two parts of yoghurt, and allowing the mixture to ferment for a week. The curd from an equal volume of milk is then added, and the mixture is left for another 4–5 days for further fermentation. The product is then dried and pulverized. For the production of 'lebnye' in Syria, rice or wheat is cooked with fermented milk. Thyme and other herbs are added and the product is formed into balls, packed into stone jars and stored under olive oil.

Aims and Benefits of Fermented Milks

Since the beginning of the twentieth century, scientists have agreed that human wellbeing is dependent on a well-functioning intestinal flora – if disturbed in any way, the intestinal flora must quickly be restored and normalized. This re-establishment can be accelerated by the administration of lactobacilli, such as *L. acidophilus*, and/or bifidobacteria. Dietary variables that affect the intestinal microflora by acting as substrates are called prebiotics, and probiotic lactic acid bacteria. In addition to the nutritional and physiological value of the constituents of fermented milks, they have also significant biological effects on the consumer.

The sugars contained in fermented milks are broken down in the mammalian gut to monosaccharides. Some of these are consumed by the bacteria in the small intestine. However most monosaccharide is absorbed into the blood and serves as an energy source for the tissues. Lactose stimulates gastrointestinal activity and increases the absorption of phosphorus and calcium. Galactose contributes to the synthesis of nervous tissue in the early years of life, and increases the absorption of fats by the individual.

Both the optical isomers of lactic acid, which are amongst the end products of lactose metabolism by microorganisms, are absorbed from the intestinal tract. However only L(+)-Lactic acid is metabolized by humans – the D(−)-Lactic acid is almost entirely excreted in the urine. Lactic acid stimulates gastric secretion and speeds up the transport of gastric contents into the intestinal tract. The digestion of proteins in facilitated, and harmful bacteria are suppressed.

Fermented milks contain essential amino acids, which are necessary for the synthesis of proteins. Proteins are digested by enzymes in the stomach, duodenum and the rest of the small intestine and converted to peptides and amino acids. Amino acids and small peptides are absorbed into the body. Milk fat is of high value as an energy source and it supplies essential fatty acids and fat-soluble vitamins.

Fermented milks are also a source of minerals, including calcium, phosphorus, magnesium and iron. The minerals are absorbed in the stomach and in the small and large intestines. Calcium, phosphorus and magnesium contribute to the formation of bones and teeth, and iron, phosphorus, potassium, chlorine and iodine contribute to the formation of muscle and skin.

Minerals also regulate some biological functions. Fermented milks are an excellent source of vitamins, which are absorbed in the small intestine.

Promotion of Growth and Digestion Bifidobacteria are thought to promote metabolism and prevent loss of amino acids by suppressing the growth of putrefactive bacteria. Bifidobacterial substances such as lactulose, added to the diet of formula-fed infants, increased nitrogen retention and weight gain. *Lactobacillus acidophilus* produces lactic acid, H_2O_2 and antibiotics, breaks down bile acids and creates an environment for the efficient utilization of nutrients such as protein, calcium, iron and phosphorus.

Experiments on rats showed that feeding with yoghurt had a weight-promoting effect. It seems that the growth-promoting effect is greater with higher counts of lactic acid bacteria and fat content. The substance promoting a gain in body weight is, however, found only in the cells of *Streptococcus thermophilus*. The denaturation or partial predigestion of protein during yoghurt production increases the digestibility of milk proteins and hence their nutritional value.

Improved Vitamin Metabolism Bacteria such as bifidobacteria produce vitamins B_1, B_2, B_6 and B_{12}, nicotinic acid and folic acid in the intestine of healthy humans, and these can be utilized by the host. Intestinal bacteria also suppress thiaminolytic bacteria (e.g. *Bacillus thiaminolyticus*), which can break down vitamin B_1 and cause vitamin B_1 deficiency.

Increased Mineral Absorption The consumption of fermented milks may result in increased mineral absorption. Yoghurt is considered beneficial in treating geriatric osteoporosis because it is a good source of calcium, but results on the benefits of yoghurt consumption are conflicting. Nevertheless, there remains the possibility that fermented milks may help to lower the gastric pH of elderly people, and so increase mineral solubility and bioavailability.

Detoxification In vitro studies on selected strains of probiotic lactic acid bacteria demonstrated the ability of some strains to degrade aflatoxins. The use of these bacteria in the food industry for detoxification purposes looks promising as a strategy for reducing the risks from these extremely toxic and carcinogenic compounds.

Stress Situations Everyday stress situations can cause a decrease in the number of lactobacilli in the gastrointestinal flora, and alterations in its composition. The daily consumption of foods containing lactobacilli can help individuals to recover their normal gut flora.

Improvement of Lactose Utilization Lactose 'maldigestion' can be improved by the consumption of fermented milks, such as yoghurt and acidophilus products. The enhanced absorption of lactose after consuming yoghurt is a result of the intra-intestinal digestion of lactose by the β-galactosidase released by yoghurt bacteria – β-galactosidase survives the passage through the stomach of a lactase-deficient human, and hence acts as a substitute for the endogenous lactase. When *Lactobacillus acidophilus* colonizes the intestine, lactose utilization is improved significantly. Bile salts also increase the ability of these cells to utilize lactose. The response of individuals to fermented milks varies, from a marginal improvement of lactose digestion with *Bifidobacterium bifidum* to almost complete lactose digestion with *Lactobacillus bulgaricus*.

Control of Serum Cholesterol and of Blood Pressure The intestinal microflora may interfere with the absorption of cholesterol from the intestine. This effect seems to be restricted to thermophilic lactic acid bacteria. Those *Lactobacillus acidophilus* strains which are able to deconjugate bile salts may act on cholesterol in the intestine, with the effect of reducing its absorption from the intestinal tract. Yoghurt also may lower the cholesterol levels of humans.

It is possible that some peptides released by the activity of lactic acid bacteria on milk casein may inhibit enzymes (carboxypeptidases) catalysing the formation of substances that raise blood pressure. Thus, hypertension may be decreased.

See also: **Bifidobacterium**. **Enterococcus**. **Fermented Foods**: Origins and Applications; Fermentations of the Far East. **Fermented Milks**: Yoghurt; Products from Northern Europe; Products of Eastern Europe and Asia. **Geotrichum**. **Kluyveromyces**. **Lactobacillus**: Introduction. **Lactococcus**: Introduction. **Leuconostoc**. **Microflora of the Intestine**: The Natural Microflora of Humans. **Milk and Milk Products**: Microbiology of Liquid Milk. **Pediococcus**. **Probiotic Bacteria**: Detection and Estimation in Fermented and Non-fermented Dairy Products. **Saccharomyces**: Introduction. **Starter Cultures**: Uses in the Food Industry; Importance of Selected Genera. **Streptococcus**: Introduction; *Streptococcus thermophilus*. **Yeasts**: Production and Commercial Uses.

Further Reading

Chandan RC (ed) (1989) *Yoghurt: Nutritional and Health Properties*. Virginia: National Yoghurt Association.

Gilliland SE (1989) Acidophilus milk products: A review of potential benefits to consumers. *Journal of Dairy Science* 72: 2483–2494.

International Dairy Federation (1988) *Fermented milks – Science and Technology*. Brussels: IDF.

Kosikowski F and Mistry VV (eds) (1997) *Cheese and Fermented Milk Foods*, 3rd edn. Vol. I. Connecticut: Westport.

Kurman JA, Rasic TL and Krger M (1992) *Encyclopedia of Fermented Fresh Milk Products*. New York: AVI.

Law BA (ed) (1997) *Microbiology and Biochemistry of Cheese and Fermented Milk*, 2nd edn. London: Blackie Academic and Professional.

Nakasawa Y and Hosono A (eds) (1992) *Functions of Fermented Milk. Challenges for the Health Sciences*. London: Elsevier Applied Science.

Renner E (1991) *Cultured Dairy Products in Human Nutrition*. Bulletin 255. Brussels: IDF.

Robinson RK (ed) (1991) *Therapeutic Properties of Fermented Milks*. London: Elsevier Applied Science.

Salminen S and von Wright A (eds) (1993) *Lactic Acid Bacteria*. New York: Marcel Dekker.

Tamime AV and Robinson RK (1988) Fermented milks and their future trends. Part II. Technological aspects. *Journal of Dairy Research* 55: 281–307.

Tamime AV, Marshall VME and Robinson RK (1995) Microbiological and technological aspects of milks fermented by bifidobacteria. *Journal of Dairy Research* 62: 151–187.

Yoghurt

R K Robinson, University of Reading, UK

Yoghurt and similar fermented milks have been produced in the warmer regions around the Mediterranean for centuries, but the more widespread popularity of these products throughout Europe, North America and elsewhere is comparatively recent. The latter markets are dominated by two types of retail product. One variant, 'natural set yoghurt', has a firm, gel-like structure and a clean, mildly acidic and slightly aromatic flavour. The other, 'stirred yoghurt', has the consistency of double cream and its background flavour is usually modified by the addition of fruit and/or flavours and sugar (sucrose). Despite the apparently contrasting nature of the end products, the manufacturing procedures for both variants have much in common.

Table 1 Outline of the important stages in the manufacture of natural set or stirred fruit yoghurt

Processes	Materials	Comments
Standardization of fat: addition of skim-milk powder or vacuum/membrane concentration	Full-cream or skimmed milk at 14–16% SNF, and milk fat (0.1–4.5%)	Sucrose (7–10%) and stabilizers may be added to the base for stirred fruit yoghurt
Homogenization at 13.79 MPa and 50–55°C	Process milk	Reduces fat globules to < 2.0 μm and improves texture of end product
Heat treatment at 80–85°C for 30 min or 90–95°C for 5–10 min	Process milk	Reduces bacterial load and O_2 content of milk. Denatures whey proteins, which interact with κ-casein and improve texture of end product
Cooling to 30°C or 42°C and inoculation with culture	Inoculated milk	Set yoghurt is packaged at this point
Incubation for 16 h at 27–30°C or 3.5–4.5 h at 42°C	Milk coagulated by lactic acid, and flavour/texture compounds released by culture	Incubation rooms for set yoghurt or in-tank incubation for stirred yoghurt
Cooling: to 2–4°C for set yoghurt in cartons; to 15–20°C for stirred yoghurt in tanks	Fruit purée (10–15% addition rate) or fruit flavours for stirred varieties	The coagulum must be handled carefully to avoid damage to the warm gel
Packaging of stirred yoghurt and cooling to 2–4°C	Retail products: natural set yoghurt and stirred fruit yoghurt	Normally 120–150 g individual cartons or 500 g family packs

Method of Manufacture

The raw material for yoghurt production is usually fresh cow's milk, although the milk from other mammals, including the sheep, camel and buffalo, is equally suitable for fermentation. Goat's milk can also be employed, but due to the low level of α_{s1}-casein, the coagulum formed during fermentation is soft and the end product may lack the attractive 'mouth feel' of other yoghurts.

The content of milk fat in yoghurt can range from 0.4 to 4.5%, according to taste and/or market demand, but the critical feature of the milk in the context of yoghurt production is the level of 'solids-non-fat' (SNF). In cow's milk, the SNF level is 8.5–9.0%, of which around 4.5% is lactose, 3.3% protein (2.6% casein and 0.7% whey proteins) and 0.7% mineral salts. Each of these components is vital for the production of a satisfactory yoghurt. The lactose is an energy source for the starter bacteria (see below) and the protein, together with minerals (e.g. calcium and phosphorus), gives rise to the basic structure of the gel. However, the levels of protein in liquid milk are not sufficiently high to produce a satisfactory end product, and so the first step in manufacture is to raise the SNF content (see **Table 1**).

Heat Treatment

Once the desired level of SNF has been achieved, the milk is homogenized (often regarded as an optional step) then heat-treated. The latter stage, at temperature well above those for normal pasteurization, involves either passing the milk through a plate heat-

exchanger with a holding tube of sufficient capacity to raise the temperature to 90–95°C for 5–10 min, or heating the milk in a process vessel to 80–85°C, with a holding time of 30 min. The choice of treatment is simply a reflection of the sophistication of the available plant, but the step is essential to give a yoghurt with the desired textural properties. In particular, heating/holding alters the physico-chemical properties of the caseins and denatures the whey proteins. As a result, β-lactoglobulin may become attached to the κ-casein, so improving the texture of set yoghurt or the viscosity of stirred yoghurt, and the partial breakdown of other whey proteins gives rise to products that stimulate the activity of the starter culture.

Microbiology of the Process

After heat treatment, the milk is normally cooled to 42°C, prior to inoculation with a culture composed of equal numbers of *Streptococcus thermophilus* and *Lactobacillus delbrueckii* subsp. *bulgaricus* (usually known as *L. bulgaricus*). A prolonged period of incubation at 27–30°C can be employed if an overnight fermentation is preferred, but because mesophilic contaminants can grow readily at that temperature, most companies opt for 42°C. The technology for handling the cultures can vary, but whatever the means of inoculation, it is essential that both species are present (see **Fig. 1**). Indeed, retail cartons should, by convention of by law, only be labelled as 'yoghurt' if *L. bulgaricus* has been used for the fermentation. Terms such as 'mild yoghurt' and 'bio-yoghurt' are permitted in some countries, to describe a product that is

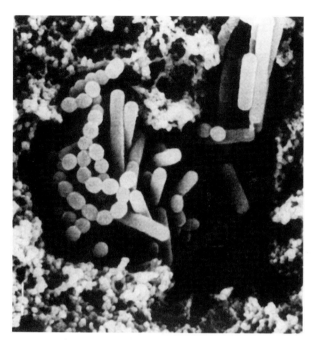

Figure 1 A scanning electron micrograph of natural yoghurt, showing the long chains of cocci (*Streptococcus thermophilus*) alongside shorter chains of rods (*Lactobacillus delbrueckii* subsp. *bulgaricus*). The matrix of coagulated milk proteins is also clearly visible. (Courtesy of Bottazzi-Bianchi, Institute of Microbiology, UCSC, Piacenza, Italy.)

yoghurt-like in texture, but they imply that the culture may be different from that used for normal yoghurt.

Streptococcus thermophilus is a Gram-positive bacterium with spherical or ovoid cells of 0.7–0.9 µm diameter, which is a natural inhabitant of raw milk in many parts of the world. In milk it occurs in long chains of 10–20 cells. It ferments lactose homofermentatively, to give L(+)-Lactic acid as the principal product. Glucose, fructose and manrose can also be metabolized, but the fermentation of galactose, maltose and sucrose is strain-specific. Although the taxonomic characteristics of both *S. thermophilus* and *L. bulgaricus* are well established, numerous strains of each species have evolved, or been selected in the laboratory – hence the loss or gain of alleles for specific aspects of metabolic performance is not uncommon.

The principal sugar in the yoghurt base, lactose, is actively transported across the cell membrane of *S. thermophilus*, by means of a membrane-located enzyme, galactoside permease. Inside the cell, the enzyme β-glactosidase hydrolyses the lactose to glucose and galactose. The glucose is metabolized to pyruvate via the Embden–Meyerhof–Parnas (EMP) pathway, and lactic dehydrogenase converts the pyruvate to lactic acid. In most strains of *S. thermophilus*,

the galactose and lactic acid leave the cell and accumulate in the medium, but some strains possess a galactokinase, that converts the galactose to galactose 1-phosphate. This is converted via the Leloir pathway to glucose 1-phosphate, which is metabolized in the usual manner. Other strains have the ability to form galactose 6-phosphate, which may then be transformed via the tagatose 6-phosphate pathway to glyceraldehyde 3-phosphate and on to pyruvate. However, the evidence for this latter pathway in *S. thermophilus* is disputed by some authorities.

The accumulation of galactose in yoghurt suggests that pathways other than the EMP pathway are suppressed in the presence of glucose derived from the hydrolysis of lactose. This view is supported by the fact that the uptake of galactose, involving a specific galactose permease, and its subsequent metabolism can only be induced under conditions of lactose deficiency. Indeed, lactose is always the preferred substrate – even glucose and fructose are absorbed slowly in comparison.

Despite its protein-rich habitat, *S. thermophilus* displays limited proteolytic ability. Its source of nitrogen is, at least initially, free amino acids, occurring naturally in the milk or released during the heat treatment. However, particular free amino acids, including glutamic acid, histidine, cysteine, methionine, valine and leucine, are not present in milk at levels sufficient to support the extensive growth of *S. thermophilus*. As a result, the maximum increase in cell numbers is possible only through the absorption of short-chain peptides released by *L. bulgaricus*, and their hydrolysis to the constituent amino acids.

The optimum temperature for the growth of *S. thermophilus* is 37°C, but it can be important that certain metabolic pathways, e.g. those for the synthesis of polysaccharides or acetaldehyde, have different optima. The growth of *S. thermophilus* ceases at 10°C.

Lactobacillus bulgaricus is also Gram-positive, but it occurs in milk as chains of three to four short rods, each 0.5–0.8 µm × 2.0–9.0 µm, with rounded ends. Its basic metabolism is homofermentative, to give D(–)-Lactic acid to a level of 1.7–2.1% in milk. This acid tolerance contrasts with that of *S. thermophilus*, whose metabolism is normally inhibited by levels of lactic acid of > 1.0% (in the range of pH 4.3–4.5). *L. bulgaricus* can, like *S. thermophilus*, utilize lactose, fructose and glucose, and some strains can utilize galactose.

In contrast to *S. thermophilus*, *L. bulgaricus* can hydrolyse casein, especially β-casein, by means of a wall-bound proteinase, to release polypeptides. However, the peptidase activity of *L. bulgaricus* is limited and it may use some of the free amino acids released by *S. thermophilus*, which can readily hydrolyse peptides.

The optimum growth temperature for *L. bulgaricus* is 45°C. Thus the temperature of 42°C selected for commercial production is an effective compromise between the growth optima of the two species.

If the yoghurt milk is examined microscopically immediately after inoculation, the balance between the two genera is usually one chain of *S. thermophilus* to one chain of *L. bulgaricus*. However, a differential colony count shows that *S. thermophilus* accounts for some 75–85% of the colony forming units (cfu). This contrast reflects the difference in chain length, and implies that the initial stages of the fermentation are dominated by *S. thermophilus*.

The use of these two organisms is partly historical in origin – they have frequently been isolated from natural yoghurt made by the indigenous tribes of the Middle East, where the high ambient temperature has led to the selection of thermophilic microfloras in fermented dairy products. There are, however, good reasons for continuing with the tradition, because when growing in milk, the two organisms interact synergistically. This protocooperation is based on two facts. Firstly, *S. thermophilus* grows more rapidly than *L. bulgaricus*, releasing lactic acid and CO_2 from the breakdown of urea in the milk by urease and usually also formic acid (up to $40\,\mu g\,ml^{-1}$), all of which stimulate the growth and metabolism of *L. bulgaricus*. Second, *L. bulgaricus* hydrolyses casein which, together with peptidase activity (predominantly originating from *S. thermophilus*) releases amino acids that are essential for the further development of both species.

The end result of this protocooperation is that both species actively metabolize lactose to lactic acid, so that the fermentation is complete within 3–4 h. In contrast, the same number of cells of one species alone might take 12–16 h to produce the same level of acidity.

In addition, the metabolites liberated by the two species give the yoghurt a flavour that is distinctly different from that of any other fermented milk. Acetaldehyde, at levels of up to $40\,mg\,kg^{-1}$ is the main component of the flavour profile. The major pathway for the production of acetaldehyde by *L. bulgaricus*, and to a lesser extent by *S. thermophilus*, is the conversion of threonine to glycine by threonine aldolase (Equation 1):

$$CH_3.CH(OH).CH(NH_2).COOH$$
Threonine

(Equation 1)

$$\longrightarrow CH_2.(NH_2)COOH + CH_3.CHO$$
Glycine   Acetaldehyde

In the absence of alcohol dehydrogenase in either *S. thermophilus* or *L. bulgaricus*, the acetaldehyde accumulates, to a level dependent on the strain involved. Other metabolic pathways, such as the transformation of pyruvate by α-carboxylase, may also be involved in causing a build-up of acetaldehyde. The rate of formation of acetaldehyde by *S. thermophilus* is temperature-dependent, the activity of threonine aldolase decreasing significantly as the temperature of incubation is raised above 30°C. The comparable enzyme in *L. bulgaricus* is unaffected by temperature, and hence this organism is probably the main source of acetaldehyde in commercially made yoghurt. Other compounds produced by the starter organisms, such as free fatty acids, amino acids, acetone, diacetyl, ketoacids and hydroxy-acids, also contribute to the final flavour. However, their importance in terms of perceived taste and aroma is poorly understood.

Finally, some strains of both species can produce appreciable levels of extracellular polysaccharides, such as glucans or polymers involving glucose, galactose and rhamnose as the constituent sugars. These metabolites considerably enhance the viscosity of the end product, because while some of the polysaccharide forms a layer over the bacterial cells, the remainder forms a network that binds the cells and the casein together as a viscous mass. The increased viscosity leads to increased consumer appeal, and so manufacturers of commercial starter cultures offer a range of starters, which differ with respect to polysaccharide synthesis. Strains may differ with respect to both the type of polysaccharide formed and its quantity. In the example shown in **Figure 2**, *L. bulgaricus* synthesizes a gum-like material while *S. thermophilus* produces a slimy glucan-like polymer. Blending *L. bulgaricus* with a non-polysaccharide-producing yoghurt culture (SC) increases the viscosity of the yoghurt in line with the rate of addition. In contrast, the polysaccharide from *S. thermophilus* enhances the viscosity of the yoghurt at low concentrations but, as its level increases along with the raised inoculation rate, its 'fluid' nature causes no increase in the viscosity of the yoghurt. Other strains of *S. thermophilus* possess a glycohydrolase capable of hydrolysing polysaccharides, which influence viscosity.

After inoculation, the milk follows one of the routes illustrated in **Figure 3**. As the fermentation begins, the population of *S. thermophilus* increases rapidly until it accounts for > 90% of the bacterial cells present. However, over the next 2 h the synergistic influence of the streptococci encourages more rapid growth and metabolism of *L. bulgaricus*. In addition, the growth of *S. thermophilus* is progressively

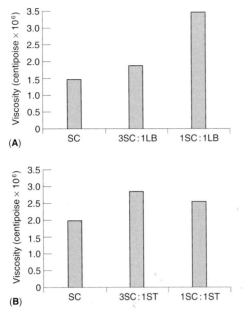

Figure 2 The effect on the viscosity of yoghurt of adding increasing volumes of either (**A**) *L. bulgaricus* (LB), a 'gum' producer, or (**B**) *S. thermophilus* (ST), a 'slime producer, to a standard, non-polysaccharide-producing starter culture (SC).

product may develop an excessively sharp, sour taste and the protein gel may begin to shrink, causing whey to separate as a discrete layer on the surface of the yoghurt. This free whey can be stirred back into the body of the product, but in set yoghurts, at least, its presence must be regarded as a fault.

Inhibitors of Starter Activity

The production and/or employment of starter cultures in the commercial production of yoghurt can be hindered by the microbiological problems detailed below.

Contamination

The starter culture may become contaminated by adventitious bacteria, yeasts or moulds or, possibly, bacteriophages. Problems from these sources should be containable with proper attention to hygiene but, because the risk is always present, many yoghurt manufacturers are turning to direct-to-vat cultures. These concentrated cultures are added directly to the process milk, minimizing the possibility of contamination.

Phages which infect *S. thermophilus* and *L. bulgaricus* have been isolated, but their presence in yoghurt is not as serious as in mesophilic fermentations. This contrast appears to arise because cheese milk is stirred continuously during acidification, and hence any phages released by the lysis of bacterial cells become distributed throughout the vat. Also, the undisturbed yoghurt milk coagulates quite rapidly, so that any proliferation of phages remains localized.

Changes in Culture Activity

The activity of the culture, e.g. rate of acid production or level of aroma/flavour compounds secreted, may change as a consequence of routine subculturing. The

inhibited by the accumulation of lactic acid, while the activity of *L. bulgaricus* is stimulated. After 4 h, the balance between the populations is restored to that in the freshly inoculated milk. The end result is that when the fermentation is complete, i.e. the acidity of the milk has risen to 1.2–1.4% lactic acid (pH around 4.2–4.3), the population of each starter organism may well exceed 2×10^7 cells ml^{-1}.

At an acidity of 1.2–1.4% lactic acid, which is probably the level preferred by most consumers, the milk proteins have coagulated to form a firm gel and the product must be cooled to avoid over-acidification. If this control is not exercised, then the

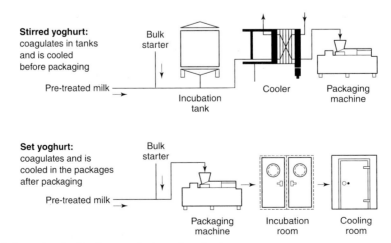

Figure 3 The principal stages involved in the manufacture of yoghurt.

reasons are not clear. The loss of plasmids encoding characteristics of technological importance (e.g. proteolytic activity, antibiotic resistance, secretion of glycoproteins) is well-documented in connection with the cultivation of *Lactococcus* species, but neither *L. bulgaricus* nor *S. thermophilus* appear to possess any plasmids of metabolic significance.

Inhibitory Substances in Milk

S. thermophilus is especially sensitive to antibiotics such as penicillin, streptomycin, neomycin and ampicillin, which are widely used to control mastitis. When grown alone, contamination by amounts of penicillin as low as 0.004 international units (iu) can inhibit cell wall development. Strains of *L. bulgaricus* tend to be more tolerant (threshold 0.02 iu of penicillin). Even when the two organisms are growing together under optimum conditions, 0.01 iu of penicillin can delay fermentation.

In addition, sanitizing agents employed to clean a plant, e.g. chlorine $(100\,mg\,l^{-1})$ or iodophors $(60\,mg\,l^{-1})$, can cause inhibition of mixed cultures. Hence the screening of bulk milk for microbiocidal agents is essential.

Strain Incompatibility

Incompatibility between strains of *L. bulgaricus* and *S. thermophilus* can almost completely prevent protocooperation, with the results that coagulation times are increased by several hours and the organoleptic quality of the end product may be extremely poor. In practice, therefore, the manufacturers of starter cultures take great care to select and blend only compatible strains, and manufacturers of yoghurt must avoid mixing cultures from different sources.

Final Processing

Yoghurt coagulated in retail cartons can be cooled by blowing cold air through the incubation room. Cooling of the base for stirred yoghurt requires either the circulation of chilled water (2°C) through the jacket/in-tank cooling system of the vessel, or pumping of the warm yoghurt (42°C) through a plate or tubular cooler. In the latter case, great care is necessary to avoid physical damage to the coagulum.

Once the yoghurt has been cooled to around 20°C, the metabolic activity of the culture almost ceases and the product is handled as described in Table 1.

Microbiological Quality

The severe heat treatment received by the process milk, together with the low pH of the final product, make yoghurt extremely safe in terms of public health because none of the recognized pathogens can survive or grow at 4.3 pH. For example, the spores of *Bacillus cereus* will not germinate at low pH, and organisms of concern in soft cheeses, e.g. *Listeria monocytogenes*, will be inactivated long before the yoghurt reaches the consumer. In addition, there is good evidence that some metabolites of yoghurt organisms can actively depress the viability of many enteric pathogens such as *Campyiobacter*, *Escherichia* and *Salmonella* species. Hydrogen peroxide is one such metabolite released by *L. bulgaricus*, some strains of which are reported to secrete an antibiotic called 'bulgarican'. *S. thermophilus* may also release a compound of low molecular weight, with bactericidal properties. However, the activity of these bacteriocins is strictly limited and in practice they do little more than reinforce the effects of acidity.

Spoilage can occur through the activities of acid-tolerant yeasts, or occasionally moulds. Widely distributed yeasts, like *Candida* or *Saccharomyces* species, can be associated with gas formation and/or 'doming' of the cartons of fruit yoghurts. The natural fruit sugars provide an abundance of substrates for fermentation, but because doming indicates a yeast count in the region of $1 \times 10^5\,cfu\,g^{-1}$ of yoghurt, it is more likely to be the result of contamination, by either 'dirty' plant or unpasteurized fruit. In the absence of severe temperature abuse, e.g. storing the yoghurt at 10–15°C, casual, airborne yeasts rarely proliferate sufficiently to cause problems.

In natural yoghurt, the principal sugar available is lactose and, because few yeasts can ferment it, the major concerns are species like *Kluyveromyces marxianus* var. *lactis* or *K. marxianus* var. *marxianus*. Both of these lactose-utilizing species grow readily on poorly cleaned surfaces, and hence high standards of hygiene are essential if contamination after heat treatment is to be avoided.

In cold northern climates, cases of yeast spoilage should be infrequent but it is nevertheless generally recommended that yoghurt for the retail market should have a yeast count of $< 10\,cfu\,g^{-1}$. The same figure applies for moulds – genera including *Mucur*, *Rhizopus*, *Penicillium* and *Aspergillus* can grow readily at the yoghurt/air interface of an undisturbed carton. In contrast to yeasts, just one spore of a fungus can spoil a carton of yoghurt by growing over the surface of the product. Hence protection of filling lines from airborne contamination is essential.

In warmer regions, where active spoilage might be

expected, either the sell-by date of the product is limited to 4–5 days post-production, compared to the 2–3 weeks typical of the UK. Alternatively, if regulations permit, sorbic acid is added (usually as potassium sorbate) at a level of up to 300 mg kg^{-1}. This preservative is extremely effective against yeasts, but has little effect on *L. bulgaricus* or *S. thermophilus*. This is important, because yoghurt, by definition, must contain an 'abundant population of viable bacteria of starter origin'. If any processing technique interferes with this population, the consumer must be informed on the packaging. For example, yoghurt that has been subjected to heat treatment after fermentation must be designated as 'heat-treated' or 'pasteurized' yoghurt.

Excessive acidity may result from continued starter activity during prolonged storage at temperature > 5°C, e.g. the acid-tolerant *L. bulgaricus* can generate lactic acid to levels of 1.7% or even above, depending on the strain. Such levels are too high for the palates of most consumers, and it is this post-production acidification that tends to determine the shelf life of commercially produced yoghurt. Some protection can be obtained by lowering the level of *L. bulgaricus* in the original inoculum, but this can reduce the degree of synergism and such products tend to lack the characteristic flavour of yoghurt.

Nutritional Consideration

The raised SNF of yoghurt milk improves the composition of yoghurt compared to milk, e.g. total protein 5.0% compared to 3.4%, and calcium 180 mg ml^{-1} compared to 120 mg ml^{-1}. In addition, the availability of these nutrients is enhanced as a result of the fermentation. For example, digestion of the milk protein of yoghurt occurs much more rapidly than when the same quantity is consumed in liquid milk and the acidity of yoghurt alters the bioavailability of minerals, e.g. calcium and zinc. Some changes in the levels of water-soluble vitamins have been recorded following fermentation (see **Table 2**). However, the extent of any increase or decrease is rather variable. This reflects differences in synthesis and utilization between the strains of *L. bulgaricus* and *S. thermophilus* comprising the starter cultures available on the market.

As already mentioned, all yoghurts should contain high counts of starter bacteria. Most authorities believe that their presence is beneficial to the consumer, although no strains of *S. thermophilus* survive either the low pH of the stomach (pH around 2.0) or the action of bile salts in the small intestine, and few strains of *L. bulgaricus* fare any better. However, as

Table 2 Some typical water-soluble vitamin contents of skim milk and low-fat yoghurt

Vitamin	Milk	Yoghurt
Thiamin, B$_1$ (µg)	42	40
Riboflavin, B$_2$ (µg)	180	200
Pyridoxine, B$_6$ (µg)	42	46
Cobalamin, B$_{12}$ (µg)	0.4	0.2
Folic acid (µg)	0.3	4.1
Nicotinic acid (µg)	480	125
Pantothenic acid (µg)	370	380
Biotin (µg)	1.6	2.6
Choline (mg)	4.8	0.6

Values are per 100 g.

these stresses cause the starter bacteria to autolyse, lactase (β-galactosidase) is released into the intestine, with the result that most of the residual lactose from the yoghurt is hydrolysed before the intestinal contents reach the colon. This enables lactose maldigesters (e.g. adults whose ileum has lost the facility to secrete lactase) to consume yoghurt with ease, whereas the same quantity of milk would cause severe abdominal discomfort due to CO_2 produced by the fermentation of lactose by bacteria in the colon.

In addition, there is evidence that the regular ingestion of yoghurt starter bacteria assists in the maintenance of healthy populations of lactobacilli in the distal region of the small intestine and bifidobacteria in the colon. The precise mechanism for this is not clear, but it is probably related to the release of vitamins or other growth factors from the degenerating cells of the starter culture.

It is relevant that *S. thermophilus* produces the L(+)-isomer of lactic acid because this form is completely metabolized by humans, with the release of energy. Ingestion of the D(−)-isomer secreted by *L. bulgaricus* can, especially in young children, give rise to metabolic disturbances. However, because early acidification is dominated by the activity of *S. thermophilus*, the lactic acid in normal yoghurt is mainly the L(+)-isomer.

See also: **Hazard Appraisal (HACCP)**: The Overall Concept. **Heat Treatment of Foods**: Principles of Pasteurization. *Lactobacillus*: Introduction; *Lactobacillus bulgaricus*. **Milk and Milk Products**: Microbiology of Liquid Milk. **Preservatives**: Permitted Preservatives – Sorbic Acid. **Spoilage Problems**: Problems caused by Bacteria. **Starter Cultures**: Uses in the Food Industry. *Streptococcus*: *Streptococcus thermophilus*. **Yeasts**: Production and Commercial Uses.

Further Reading

Chandan RC (ed) (1989) *Yogurt: Nutritional Health Properties*. Virginia: National Yogurt Association.

Gilliland SE (ed) (1985) *Bacterial Starter Cultures for Foods*. Boca Raton: CRC Press.

International Dairy Federation (1988) *Fermented Milks – Science and Technology*. Bulletin 227. Brussels: International Dairy Federation.

Marshall VM (1987) Fermented milks and their future trends. Part 1. Microbial aspects. *Journal of Dairy Research* 54: 559–574.

Rasic JL and Kurmann JA (1978) *Yoghurt – Scientific Grounds, Technology, Manufacture and Preparations*. Copenhagen: Technical Dairy Publishing House.

Robinson RK (ed) (1990) *Dairy Microbiology*, 2nd edn. London: Chapman & Hall.

Tamime AY and Robinson RK (1985) *Yoghurt – Science and Technology*. Oxford: Pergamon Press.

Tamime AY and Robinson RK (1988) Fermented milks and their future trends. Part II. Technological aspects. *Journal of Dairy Research* 55: 281–307.

Zourari A, Accolas JP and Desmazeaud MJ (1992) *Lait* 72: 1–34.

Products from Northern Europe

Hubert Roginski, Gilbert Chandler College, The University of Melbourne, Victoria, Australia

Introduction

Fermented milks have always played an important role in the nutrition of people inhabiting northern Europe. In Scandinavia in particular, the consumption of fermented milks has a long tradition. Historically, fermented milks, often referred to as sour milks, were made and consumed at home. Most of these products have a characteristic thick consistency and ropiness, and could be kept for weeks or even months in a cool room. In warmer parts of the world, ropiness of milk is treated with suspicion – with good reason, because it may be caused by organisms other than lactic acid bacteria.

The home production of fermented milks is now much less common in Scandinavia, but many types of commercially manufactured products have the same attributes as their home-made precursors. Consumption statistics provide ample evidence that traditional Nordic fermented milks continue to enjoy strong popularity. In 1993, fermented milks other than yoghurt accounted for 74% of the total fermented milk consumption in Sweden, 64% in Finland, 62% in each of Norway and Iceland, and 45% in Denmark.

Interest in Nordic fermented milks has recently been stimulated by the important prophylactic and therapeutic effects associated with the starter flora and its metabolites, observed in experiments on animals.

General Features

Nordic fermented milks constitute a group of products distinctly different from fermented milks made elsewhere – primarily due to their unique physical properties, characterized by high viscosity and ropiness. In the cool climate of northern Europe, home-made fermented milks are the result of the spontaneous growth of the mesophilic flora present in raw milk. Low temperatures seem to be critical for the domination of the milk environment by encapsulated strains of lactococci. Lactobacilli are only rarely isolated from spontaneously fermented milk in Scandinavia. Moulds and lactose-fermenting yeasts will grow well in acidified milk, and are therefore found in some types of Nordic fermented milks, influencing their properties.

Other traditional fermented products popular in Scandinavia include cultured buttermilks and cultured creams, which have a characteristic aroma and flavour, derived primarily from diacetyl and related volatile compounds. Also popular are concentrated milks, made using membrane-concentrated milk or by the removal of whey from the fresh coagulum.

The properties of Nordic fermented milks place them at the other end of the fermented milk spectrum from acid (Bulgarian) buttermilk, which has a lower viscosity but a much higher acidity, and a flavour resembling that of yoghurt.

Microorganisms Present

Diverse cultures are used commercially in Scandinavia for products of various types, and have been derived from traditional products. In many of these cultures, the bacteria responsible for ropiness are indispensable. However, the presence of non-ropy organisms is also essential in many cultures, to ensure the formation of metabolites imparting typical organoleptic properties to the product.

Overall, the following groups of microorganisms have been isolated from various types of traditional Nordic fermented milks:

- mesophilic bacteria: *Lactococcus lactis* subspp. *cremoris*, *lactis* and *lactis* biovar *diacetylactis*; *Leuconostoc mesenteroides* subspp. *cremoris* and *dextranicum*
- thermophilic bacteria: *Streptococcus thermophilus*, *Lactobacillus delbrueckii* subsp. *bulgaricus* and *L. helveticus*

- lactose-fermenting yeasts
- the mould *Geotrichum candidum*.

Forms of lactococci classified as 'intermediate' between *Lactococcus lactis* subspp. *cremoris* and *lactis* (on the basis of their physiological characteristics) have also been found in some ropy milks.

The total numbers of viable bacteria in these products can be very high. For example, concentrations of up to 10^{10} cfu g^{-1} have been reported in the Finnish ropy milk called 'pitkäpiimä'. This is a higher concentration of live bacteria than that in some starter cultures used in the commercial production of fermented milks.

Traditional Nordic fermented milks owe their characteristic high viscosity and ropiness to the vigorous growth of capsule-forming lactococci, mainly *Lactococcus lactis* subsp. *cremoris*. Encapsulated strains of *L. lactis* subspp. *lactis* and *lactis* biovar *diacetylactis* and *Leuconostoc* species have also been isolated from some products. However, many strains of these species which clearly do not produce capsules are also found in Nordic fermented milks.

Glycocalyx

Terms like 'slime', 'capsule' and 'mucoidal (Muc$^+$) phenotype' are used interchangeably in the literature, usually without ascertaining whether the material in question is capsule-type or slime-type. 'Capsule' is defined as a compact layer of polysaccharide and 'slime' as a diffuse layer of polysaccharide, both exterior to the cell wall. As a rule, the capsule of lactococci is tightly bound to the cell that produced it. In cases where the physical nature of the material has not been studied, it would be more appropriate to refer to 'glycocalyx', a more general term which describes the polysaccharide-containing material lying outside the cell wall, although this term is seldom used.

Figure 1 shows an encapsulated strain of *Lactococcus lactis* subsp. *cremoris*. In some places, the capsular material clearly does not follow the chain of lactococci. It is possible, however, that detachment from the cells occurred during the preparation of the slide. Slime seems to be produced in particular abundance during the late exponential and early stationary phases of growth.

Normally, both encapsulated and non-encapsulated forms of the same species can be isolated from a product. As an example, the types and numbers of bacteria isolated from pitkäpiimä are shown in **Table 1**. *L. lactis* subsp. *cremoris* was the organism most often encountered in this product. It is worth noting that encapsulated *L. lactis* subsp. *lactis* biovar *diacetylactis* was found only in samples of ropy milk which

Figure 1 *Lactococcus lactis* subsp. *cremoris*, strain Va, after 6 h of growth. Preparation stained with methylene blue, magnification × 1350. (From Forsén (1966)).

did not contain slime-forming cells of *L. lactis* subsp. *cremoris*. The growth of slime-forming lactococci isolated from pitkäpiimä depended on the presence of nicotinic acid and riboflavin in the growth medium. It has also been observed that strains of both *L. lactis* subspp. *cremoris* and *lactis* biovar *diacetylactis* found in pitkäpiimä were more likely to grow at 40°C than were the stains of these subspecies isolated from commercial starter cultures for fermented milks. The O$_2$ demand and CO$_2$ production of slime-forming cultures of *L. lactis* subspp. *cremoris* and *lactis* from pitkäpiimä has been observed, at 19°C, to be two to three times higher than those of the strains which do not produce slime.

Characteristics of Capsular Material

Studies of ropy fermented milk using a scanning electron microscope revealed that the polysaccharide slime material remains associated with the cells which produced it, while also attached to casein micelle clusters, forming a network responsible for the ropy consistency of the product. Other milk proteins may also be involved. In addition, proteins of bacterial origin may be present in slime. However, the results of slime composition studies are affected, to some extent, by the techniques of isolation and purification of the slime. In one study, the slime produced by a strain of *L. lactis* subsp. *cremoris* isolated from a Swedish ropy milk was found to contain 47% protein, 20% methyl pentoses, approximately 10% hexose-like sugars and almost 3% of sialic acid. The composition of a phosphorus-containing polysaccharide produced by *L. lactis* subsp. *cremoris* isolated from a sample of the Finnish fermented milk viili is shown in **Table 2**. Rhamnose, glucose and galactose were present in this phosphopolysaccharide at the molar ratio of 1 : 1.45 : 1.75. The purification procedure (preparative SDS-PAGE) removed protein, glycerol, hexosamine, sialic acid and uronic acids from the crude slime material. The phosphate group was attached to the β-D-galactopyranosyl residue. The molecular weight of this phosphopolysaccharide was estimated at 1.7×10^6 by gel permeation chromat-

Table 1   Types of bacteria identified in pitkäpiimä. The numbers of slime-forming strains are shown in parentheses (from Forsén (1966))

Ropy milk culture	Number of characterized colonies	Leuconostoc spp.	Lactococcus lactis subsp. lactis	Lactococcus lactis subsp. lactis biovar diacetylactis	Lactococcus lactis subsp. cremoris	Streptococcus viridans group[a]
H$_1$	29 (4)	9	2 (2)	–	18 (2)	–
H$_4$	16 (5)	3	6 (1)	2	5 (4)	–
H$_5$	20 (2)	6	3	4 (2)	7	–
VA	42 (10)	5 (5)	7 (4)	30 (1)	–	–
Jo	13 (3)	–	7 (1)	–	2 (2)	4
In	17 (4)	2	3 (1)	4	8 (3)	–
Total	137 (28)	(5)	(9)	(3)	(11)	
Percentage of slime-forming strains	100%	18%	32%	11%	39%	

[a] The *Streptococcus viridans* physiological group includes *Streptococcus thermophilus*, but the result reported above does not imply the presence of *S. thermophilus* in pitkäpiimä.

Table 2   Composition of the slime material and phosphopolysaccharide excreted by *Lactococcus lactis* subsp. *cremoris* SBT 0495 (from Nakajima et al (1990))

	Content (%)	
	Crude slime material	Purified polysaccharide
Total carbohydrate	42.3	78.9 (90.5)[a]
Rhamnose	–	21.7
Glucose	–	31.4
Galactose	–	38.0
Methyl pentose	20.6	21.4
Hexosamine	4.8	ND[b]
Sialic acid	0.8	ND[b]
Uronic acid	1.9	ND[b]
Glycerol	–	ND[b]
Phosphorus	1.8	3.1
Protein	21.2	ND[b]

[a] Value after hydrofluoric acid cleavage.
[b] Not detected.

ography. The phosphopolysaccharide has a negative charge: it is, therefore, possible that it could form complexes with basic proteins found on the cell surface of lactococci, due to electrostatic interactions.

Determinants of Slime Production

Plasmid DNA   Observation of the loss of mucoidal (Muc$^+$) phenotypes at higher incubation temperatures led to studies which have demonstrated the involvement of plasmid DNA in slime production by lactococci. Plasmids of different sizes have been demonstrated to code for slime production. The ropy phenotype of strains of *L. lactis* subsp. *cremoris* was found to be associated with a 17 MDa plasmid in Swedish strains and with a 30 MDa plasmid in Finnish strains. This indicates that distinctly different plasmids coding for the ability to produce slime are present in separate populations of starter bacteria. An 18.5 MDa plasmid from a strain of a different origin was also found to be associated with a (Muc$^+$) phenotype. Plasmids on which slime synthesis is encoded can be transferred to strains with no slime-forming ability by a number of currently available techniques, including electroporation and co-transfer involving a bigger plasmid, on which an ability to ferment lactose is encoded.

Process of Slime Production

Proteins of the cell surface are thought to play a role in slime production by lactococci. This view is supported by the observation that two cell-surface proteins of mol. wt 26 000 and 42 000, present only in slime-forming cells, disappeared when the ability to produce slime was lost after incubation at 30°C. The synthesis of these proteins is probably plasmid-encoded.

Experiments demonstrating the presence of lipoteichoic acid (LTA) antigen on the cell surface of *L. lactis* subsp. *cremoris* led to a hypothesis linking LTA with the formation of the capsule or slime. LTA, which was also detected in the growth medium in early stationary phase and during storage at refrigeration temperature, may form associations with proteins and polysaccharides of the slime material. In addition, LTA present on the cell surface inhibits autolytic enzymes of the bacterial cell wall, which has obvious advantages for the shelf life of the product. LTA also interferes with cell division: 20% of the cells of one particular slime-forming strain were found to have an abnormal morphology due to disturbances in cell division, attributed to LTA.

Even in the presence of an energy source, bacterial growth may be affected by limiting factor(s), such as an unfavourable physical or chemical environment or the exhaustion of an essential nutrient. It has been suggested that the continuing catabolism of energy substrates can then provide an impetus for the production of polymers, such as the slime material, as either energy reserves or waste products.

The variable rate at which lactococci produce slime is a well-known phenomenon, especially during incubation at higher temperatures, e.g. 30°C. In addition, the serial transfer of a ropy strain, even at lower temperatures which normally favour the production of slime material, often leads to the loss of this trait. Sometimes, this unpredictability of slime formation causes difficulties in the commercial manufacture of ropy fermented milks.

Although the loss of the ropiness trait from starter strains has been observed in the commercial production of ropy milks, a similar problem has not been reported in the traditional manufacture of these products, in which a mixture of undefined organisms is used. A mixed population of lactococci and leuconostocs is likely to have effective mechanisms for retention in the culture of the plasmid(s) on which the ability to produce capsular material is encoded, especially if other properties important for the survival of the bacterial population are encoded on the same plasmid(s). The data reported in Table 1 lend support to this concept.

Bacteriophages

Attack by bacteriophages can cause problems in the manufacture of ropy fermented milks. Thirteen phages of different morphology have been found in an extensive study of 90 viili samples from 20 dairies in Finland. The main morphological type was characterized by isometric heads with long, non-contractile tails. There were differences in the tail sizes and in the presence or absence of a collar, a baseplate, and a tail fibre. One phage had a very long head with a short tail. The host organisms for these phages included strains of *L. lactis* subspp. *cremoris* and *lactis* biovar *diacetylactis* and *Leuconostoc mesenteroides* subsp. *cremoris*. Some phages differed only in their host specificity.

A phage of non-encapsulated *Lactococcus lactis* subsp. *cremoris* isolated from viili was shown to dissolve the capsules of some strains, including a strain of *L. lactis* subsp. *lactis*. The capsules disappeared soon after infection, and maturing phages were seen inside the cells 2 h after infection, appearing as tightly packed bundles. The phage progeny particles were held together, probably by their tail structures, even after cell lysis, 5 h after infection. The phage titre declined rapidly after heating for 15 min at 50–70°C, but particles capable of plaque formation were detected even after heating for 5 min at 100°C.

Other Starter Organisms

Important microorganisms, other than lactococci, isolated from Nordic fermented milks are:

- *Lactobacillus helveticus*. It is significant that the optimum growth temperature of a strain isolated from the Norwegian and Finnish versions of tätmjölk was 10°C lower than that reported for other strains of this species. The thermophilic *L. helveticus* has also been isolated from the Icelandic fermented milk 'skyr'.
- *Streptococcus thermophilus* and *Lactobacillus delbrueckii* subsp. *bulgaricus*, found in skyr.
- The yeasts *Candida kefir* and its teleomorph *Kluyveromyces marxianus*, as well as *Torulopsis holmii* and its teleomorph *Saccharomyces exiguus*. These have been isolated from various types of Nordic fermented milks. The long shelf life of ropy milk has been attributed in part to the presence of yeasts, which are inhibitory against a number of potential mould contaminants. The inhibitory compound was inactivated by heat.
- The white mould *Geotrichum candidum*. This is found in viili, tette and other products. *G. candidum* isolated from a sample of ropy milk and added to a ropy culture (*Lactococcus lactis* subsp. *cremoris*), which was previously not associated with this mould, stimulated the growth of the culture in milk. A cell-free filtrate of the mould's growth medium (sterile whey) had a similar effect.

Products and Processes

In the domestic production of fermented milks, no heat treatment was applied to the milk prior to fermentation, and the milk was inoculated with the residue remaining in the fermentation vessel from the previous batch. In this way, a characteristic flora dominated by lactococci became established in the vessel. Nowadays, in the commercial production of Nordic fermented milks, defined strains of starter microorganisms are used in a carefully controlled manner. The type of starter flora has a distinct influence on the properties and shelf life of these products.

Scandinavian fermented milks fall into six major classes, as shown in **Table 3**. Not all products in each row are identical to each other; the table is provided as a guide.

Långfil

'Långfil' or 'tätmjölk' has a mild, sour taste and a ropy, dough-like consistency, produced by encapsulated variants of lactococci. Occasionally, lactobacilli are also found in this product. Traditionally, the leaves of plants such as butterwort (*Pinguicula*

Table 3 The major classes of Scandinavian fermented milks[a,b] (adapted from Bertelsen (1983))

Sweden	Norway	Denmark	Finland
Långfil	Tettemelk	–	Viili
Filmjölk	Kulturmelk	Tykmælk	Talouspiimä
Lättfil	Skummet kulturmelk	–	Rasvatonpiimä
Kärnmjölk	Kjernemelk	Kærnemælk	Kirnupiimä
Gräddfil	Rømme	Crème fraîche	Kermapiimä
Lactofil	–	Ymer	Kokkeli

[a] Each row represents a separate class of products.
[b] There are also many special product names used by commercial manufacturers of fermented milks.

vulgaris) and sundew (*Drosera* species, especially *Drosera rotundifolia*) were added to the milk before leaving it to sour. They seem to exert a rennet-like action. No relationship has been established between the bacterial flora of butterwort or other plants and the bacterial composition of lågfil. Traditionally, lågfil cultures were preserved for later use by simple techniques, such as dipping a piece of cloth (in Sweden) or birch twigs (in Norway) in the finished product and then allowing them to dry. A new batch was started by immersing the twigs, or placing the dried cloth, in fresh milk.

'Viili' contains various lactose-fermenting and *Geotrichum candidum* yeasts in addition to lactococci and leuconostocs. Viili made in Western and northern Finland is ropy, while viili from the eastern part of the country lacks ropiness. The cream layer is usually covered with the mould. Viili is made from non-homogenized milk and is eaten with a spoon.

'Pitkäpiima', which is always ropy, is used as a drink. In domestic production, the cream layer, removed from the pitkäpiima after fermentation, was used to make butter. 'Villipiimä', from which the cream is not removed, also serves as a drink.

The following steps are involved in the commercial production of viili:

- standardization of the fat content, to a minimum level of 3.9%
- heat treatment at 89–90°C for 15–30 min or 92–96°C for 4–5 min
- cooling to 18–20°C
- addition of 4–8% of starter culture
- incubation for 18–20 h, to pH 4.6
- cooling to the storage temperature, 8–10°C.

The final pH, after cooling, is 4.3–4.4 and the shelf life is between 10 and 15 days.

Cultured Milk

'Filmjölk', sometimes known as 'kulturmjölk' (cultured milk), is a popular Swedish fermented milk characterized by a typical flavour and aroma (derived primarily from diacetyl) and a fairly high viscosity, with a fat content of 3.0%. Filmjölk, which is used as a drink, was developed in the early 1930s. The starter cultures contain the acid-producer *Lactococcus lactis* subsp. *lactis* and the flavour and aroma producers *L. lactis* subsp. *lactis* biovar *diacetylactis* and *Leuconostoc mesenteroides* subsp. *cremoris*. All these strains are propagated together.

In the production of filmjölk:

- standardized milk is often de-aerated at 78°C, to prevent or to alleviate defects such as syneresis of the coagulum, granulation, lumpiness and low viscosity
- the de-aerated milk is homogenized at 10–20 MPa and 70°C, and then heated at 95°C for 2–6 min
- the milk is cooled to 17–24°C, and inoculated with 1% starter culture
- after 17–24 h, coagulum forms at pH 4.6
- the product is cooled to 8°C and packaged into retail containers.

Normally, a shelf life of 10 days at 8°C is achieved.

Filmjölk with 0.5% fat is called 'lättfil'. The same product is known as 'skummet kulturmelk' in Norway.

'Tykmælk', a characteristic Danish fermented milk, is produced commercially in a very similar manner to the Swedish filmjölk. Its fat content is 3.5%.

Buttermilk

Buttermilk ('kärnmjölk', 'kjernemelk', 'kærnemælk', 'kirnupiimä') of traditional type is made by churning cultured cream which has a relatively low fat content. In this manner, a small amount of cultured butter and a relatively high volume of buttermilk are obtained. A product known as 'cultured buttermilk' is made by the fermentation of skim milk or low-fat milk.

Starter cultures contain, in addition to the bacteria used for filmjölk manufacture, *Lactococcus lactis* subsp. *cremoris*. Slime-producing strains are included in the starter cultures for some types of cultured buttermilk made in Finland. The production process involves:

- heat treatment at 85–90°C for 20–30 min, or 92–96°C for 4–5 min
- cooling to 20–23°C
- incubation with 1–4% starter culture
- 15–20 h fermentation
- cooling to 5–10°C, during which the pH drops to 4.4–4.45.

The flavour and aroma of cultured buttermilk are very similar to those of filmjölk. The shelf life varies from 7 to 12 days.

Cultured Creams

'Gräddfil', 'rømme', 'crème fraîche', and 'kermapiimä' are cultured creams made by the fermentation of heat-treated cream with the same cultures that are used in the manufacture of cultured buttermilk. The legal requirements regarding the fat content of cultured creams and production practices vary between Scandinavian countries. For example in Denmark, production involves:

- the cream being standardized at either 9%, 18%, 38% or 50% of fat, then homogenized
- heat treatment at 90°C for 5 min
- cooling to the incubation temperature of 20–27°C
- inoculation with 2% starter culture
- 16–20 h incubation in the tank
- cooling to 5°C, during which the pH of the final product falls to 4.4.

In two types of Swedish gräddfil, the legally required fat content is 12% and 34.5%; Norwegian rømme contains 20% or 35% fat; and in Finland there are three types of cultured cream, at > 12%, 35–40% and 42% fat.

Concentrated Fermented Milks

These include 'ymer', lactofil', 'kokkeli' and 'skyr'.

Ymer is a Danish fermented milk, which contains at least 11% non-fat milk solids (including 5–6% protein) and 3.5% fat. It is usually produced from the ultrafiltered retentate, which contains about 15% total solids. Production involves the following:

- blending with cream and homogenization
- heat treatment and cooling to 20–27°C
- inoculation with 4% starter culture containing *Lactococcus lactis* subspp. *lactis* and *cremoris* and *Leuconostoc mesenteroides* subsp. *cremoris*
- fermentation until the pH reaches 4.5 (after 16–20 h)
- stirring, cooling to 5°C and storage for 24 h
- stirring again and packaging.

The cooling is often carried out in two stages, first to 14°C and then to 5°C.

In the traditional process, fermentation of the skim milk is followed by cutting of the coagulum and drainage of the whey at 40°C. Alternatively, a quarg separator can be used to remove the whey. Pasteurized cream is then blended with the skim product, and this is followed by homogenization, cooling to 12–14°C and packaging. The shelf life of ymer is 20 days at 5°C.

Lactofil, produced in Sweden, is similar to ymer. The fat content of lactofil is 5% and the starter culture is similar to those used in the manufacture of filmjölk and kärnmjölk.

Kokkeli, traditionally made in eastern Finland, is prepared by warming spontaneously soured milk in the oven and removing the separated whey.

Skyr, made in Iceland, is manufactured from skim milk. Unlike all other traditional Scandinavian fermented milks, it is fermented by thermophilic flora. The milk is then heated, to facilitate the syneresis of the protein coagulum. Whey separation is achieved by either straining through linen bags or using quarg separators. Ultrafiltration is now also used to recover whey proteins, which are added to the skyr before packaging. Microorganisms isolated from skyr include thermophilic lactic acid bacteria (*Streptococcus thermophilus*, *Lactobacillus delbrueckii* subsp. *bulgaricus* and *L. helveticus*) and lactose-fermenting yeasts. Typical skyr contains 17.5% of total solids, including > 13% protein and 0.4% fat.

Other Fermented Milks

Many types of fermented milks are produced in Scandinavia, and some do not fit into the classification outlined above, e.g. products which are mixtures of ropy milks with acidophilus milk.

New Developments The Scandinavian dairy industry is constantly looking for ways of adding new products to the established range. For example, a new type of sour milk has been developed in Finland, employing a combination of *Propionibacterium freudenreichii* subsp. *shermanii* and *Lactobacillus acidophilus*. Both of these are known to inhibit some of the common pathogenic bacteria in vitro.

Product Composition and Shelf Life

Composition

The composition of Nordic fermented milks in terms of total solids and fat content is determined by law. Often, the minimum numbers of viable microorganisms are also prescribed by the food standards.

Starter activity in milk leads to the production of lactic acid, as well as flavour and aroma compounds. For example, *Lactococcus lactis* subsp. *lactis* biovar *diacetylactis* metabolizes the citrate present in milk to diacetyl, acetoin, 2,3-butylene glycol and CO_2. Citrate is also metabolized by leuconostocs, mainly to diacetyl and acetoin (at pH $\leqslant$ 5.5), or to acetate (at pH > 5.5).

Differences in vitamin content have been observed between fermented milks and raw milk. For example, viili and cultured buttermilk (containing 1.9% fat) were found to contain more folate than raw milk, by 48% and 65% respectively. However, their vitamin C content was lower than that of milk, by 58% and

24% respectively, and their riboflavin content was lower by 12% and 11% respectively. The concentration of vitamin B_{12} in both products was also lower than in milk, by > 20%.

Shelf Life

Very high numbers of viable bacteria have been observed in some spontaneously fermented milks, which may be due to the reduced rate of cell autolysis (discussed above). The shelf life of traditional ropy fermented milks is exceptionally long, mainly due to the presence of capsular material, which improves the rheological behaviour of these products. A balance of 60–70% of ropy to 30–40% of non-ropy cells has been claimed to ensure that the product has desirable rheological characteristics. These proportions would depend on the amount and properties of the viscous material produced. Commercially made products are also noted for their long shelf life.

Even if some pre-existing defect diminishes the water-holding capacity of the protein matrix, e.g. if proteolysis were allowed to proceed unhindered in raw milk, the presence of capsular material would mask the syneresis of the protein coagulum. Features such as the cream layer and the presence of the mould *Geotrichum candidum* on the surface of some products provide added protection from spoilage organisms, thus further extending the shelf life.

Health-related Effects

Considerable progress has been made in demonstrating certain beneficial effects of Nordic fermented milks in animals. However, unequivocal experimental and/or epidemiological evidence needs to be gathered in order to substantiate claims of similar effects in humans.

Effects on Immunity

Fermented milks have been claimed to counter some of the detrimental effects of the gradual progression of humans, over the past few centuries, from a low-fat, high-fibre diet to a modern, more hygienic one, dominated by highly processed and refined foods rich in protein. It has been suggested that fermented milks may play an important immunomodulating role, for example they stimulate the functions of gut-associated lymphoreticular tissue (GALT), and this effect has been attributed mainly to antigenic structures of the surface of lactococci. In particular, *L. lactis* subsp. *cremoris* isolated from viili has been shown, in studies with human lymphocyte cultures, to stimulate the secretion of immunoglobulins, primarily those of the IgM class. In addition, T lymphocytes showed considerable proliferation in response to the same strain.

A substance active as a mitogen of murine B cells was purified from the slime produced by another strain of *L. lactis* subsp. *cremoris*, also isolated from viili. The mitogenic compound was a phospho-polysaccharide, containing rhamnose, glucose, galactose and phosphorus. Its activity was higher than that of unpurified slime. A significant induction of the cytotoxicity of peritoneal murine macrophages against sarcoma cells, by the capsular material of the same strain of *L. lactis* subsp. *cremoris*, has been demonstrated in vivo. A single intraperitoneal injection of the freeze-dried cells of the same strain retarded the growth of ascitic and solid sarcomas in mice. However, the same preparation did not show any direct cytotoxic activity against the same sarcoma cells in vitro, which suggests that the anti-tumour effect of this organism is through the enhancement of the cytotoxicity of the host's macrophages. The exact mechanism of this enhancement is not known, but the slime has been observed to increase glucose consumption in vitro by intraperitoneal macrophages.

Freeze-dried preparations of viili, långfil and ropy yoghurt, used as intraperitoneal injections daily for 9 days after the tumour inoculation, also significantly retarded the growth of murine solid sarcomas in vivo. The maximum anti-tumour effect was induced by a dose of $10 \, \text{mg kg}^{-1}$ of the viili preparation, $50 \, \text{mg kg}^{-1}$ of the långfil preparation and $100 \, \text{mg kg}^{-1}$ of the ropy yoghurt preparation. Thus the effect of both viili and långfil was clearly stronger than that of ropy yoghurt. All three preparations significantly enhanced the delayed cutaneous hypersensitivity (DCH) response to oxazolone, which was depressed in tumour-bearing mice. The anti-tumour effect of these ropy milks is thought to be mediated by the immune responses associated with host's macrophages and/or T cells.

Antimutagenic Activity

Strains of *L. lactis* subsp. *cremoris* isolated from viili, either ropy or non-ropy, reduced the mutagenicity of nitrosated beef extract by 40%, as determined by the Ames test, using *Salmonella typhimurium* as the test organism.

Lowering of Serum Cholesterol

In experiments on rats receiving diets containing viili, non-ropy fermented milk or acidified skim milk, the serum cholesterol level of rats on a viili-containing diet was the lowest of the three groups. Their ratio of high-density lipoprotein (HDL) cholesterol to total cholesterol was the highest of the three groups. The mechanism of this cholesterol-lowering effect is unknown.

Antibacterial Effects

The antagonistic effects of lactic acid bacteria against common pathogens and spoilage bacteria have been well-demonstrated in vitro. For example, lactococci and their capsular material have been shown to inhibit the growth of *Staphylococcus aureus*, *Escherichia coli* and some clostridia. Some of these lactococci are components of the starters for Nordic fermented milks.

See also: **Fermented Milks**: Range of Products. **Geotrichum**. **Kluyveromyces**. **Lactobacillus**: *Lactobacillus bulgaricus*; *Lactobacillus acidophilus*. **Lactococcus**: *Lactococcus lactis* Sub-species *lactis* and *cremoris*. **Leuconostoc**. **Probiotic Bacteria**: Detection and Estimation in Fermented and Non-fermented Dairy Products. **Propionibacterium**. **Streptococcus**: *Streptococcus thermophilus*.

Further Reading

Alm L and Larsson I (1983) Från forntid till framtid. Den nordiska tätmjölken – en produkt med gamla anor. *Nordisk Mejeriindustri* 10: 396–399.

Bertelsen E (1983) Kulturmjölksprodukter i Norden. *Nordisk Mejeriindustri* 10: 386–390.

Forsén R (1966) Die Langmilch (Pitkäpiimä). *Meijeritieteellinen Aikakauskirja* 26: 1–76.

Forsén R (1989) Characterization of antigenic surface structures of lactococci and their possible immunobiological effects. In: *Fermented Milks and Health*. Workshop Proceedings, Arnhem. Ede: NIZO.

Forsén R, Niskasaari K, Tasanen L and Nurmiaho-Lassila E-L (1989) Studies on slimy lactic acid fermentation: detection of lipoteichoic acid containing membrane antigens of *Lactococcus lactis* ssp. *cremoris* strains by crossed immunoelectrophoresis. *Netherlands Milk and Dairy Journal* 43: 383–393.

Gudmundsson B (1987) Skyr. *Scandinavian Dairy Industry* 4: 240–242.

IDF (1988) Fermented milks – science and technology. *International Dairy Federation Bulletin* 227.

IDF (1995) Consumption statistics for milk and milk products – 1993. *International Dairy Federation Bulletin* 301.

Kitazawa H, Toba T, Itoh T, Kumano N and Adachi S (1990) Antitumor activity of ropy sour milks in murine solid tumor. *Japanese Journal of Zootechnical Science* 61(11): 1033–1039.

Kitazawa H, Itoh T and Yamaguchi T (1991) Induction of macrophage cytotoxicity by slime products produced by encapsulated *Lactococcus lactis* ssp. *cremoris*. *Animal Science and Technology* 62(9): 861–866.

Kitazawa H, Toba T, Itoh T et al (1991) Antitumoral activity of slime-forming, encapsulated *Lactococcus lactis* ssp. *cremoris* isolated from Scandinavian ropy sour milk, 'viili'. *Animal Science and Technology* 62(3): 277–283.

Kitazawa H, Yamaguchi T, Miura M, Saito T and Itoh T (1993) B-cell mitogen produced by slime-forming, encapsulated *Lactococcus lactis* ssp. *cremoris* isolated from ropy sour milk, viili. *Journal of Dairy Science* 76: 1514–1519.

Kontusaari S and Forsén R (1989) Finnish fermented milk 'viili': involvement of two cell surface proteins in production of slime by *Streptococcus lactis* ssp. *cremoris*. *Journal of Dairy Science* 71: 3197–3202.

Laukkanen M, Antila P and Antila V (1988) The water-soluble vitamin contents of Finnish liquid milk products. *Meijeritieteellinen Aikakauskirja* 46(1): 7–24.

Macura D and Townsley PM (1984) Scandinavian ropy milk – identification and characterization of endogenous ropy lactic streptococci and their extracellular excretion. *Journal of Dairy Science* 67: 735–744.

Mantere-Alhonen S and Mäkinen E (1987) A new type of sour milk with propionibacteria. *Meijeritieteellinen Aikakauskirja* 45(1): 49–61.

Nakajima H (1995) Characteristics of fermented milk produced by slime-forming *Lactococcus lactis* subsp. *cremoris*. *Snow Brand R&D Reports* 104: 97–169.

Nakajima H, Toyoda S, Toba T et al (1990) A novel phosphopolysaccharide from slime-forming *Lactococcus lactis* subspecies *cremoris* SBT 0495. *Journal of Dairy Science* 73(6): 1472–1477.

Neve H, Geis A and Teuber M (1988) Plasmid-encoded functions of ropy lactic acid streptococcal strains from Scandinavian fermented milk. *Biochimie* 70: 437–442.

Robinson RK and Tamime AY (1990) Microbiology of fermented milks. In: Robinson RK (ed.) *Dairy Microbiology*, 2nd edn. Vol. 2, p. 291. London and New York: Elsevier Applied Science.

Saxelin M-L, Nurmiaho-Lassila E-L, Merilainen VT and Forsén R (1986) Ultrastructure and host specificity of bacteriophages of *Streptococcus cremoris*, *Streptococcus lactis* subsp. *diacetylactis*, and *Leuconostoc cremoris* from Finnish fermented milk 'viili'. *Applied and Environmental Microbiology* 52(4) 771–777.

Sundman V (1953) On the microbiology of Finnish ropy sour milk. *Thirteenth International Dairy Congress, The Hague* 3: 1420–1427.

Products of Eastern Europe and Asia

Dilek Özer, GAP Regional Development Administration, Şanlıurfa, Turkey

Barbaros H Özer, Department of Food Science and Technology, The University of Harran, Şanlıurfa, Turkey

Introduction

Fermented milks represent an increasing and very important percentage of retail milk products. Worldwide, around 400 different names are applied to fermented milk products, made by traditional or

Table 1   Fermented milk products of eastern Europe, the Balkans, the Middle East and central Asia

Product	Origin	Type of starter	Gas and alcohol production
Yoghurt	Turkey, Bulgaria	Thermophilic LAB	–
Dahi	India	Thermophilic LAB	–
Diluted yoghurt			
Ayran	Turkey	Thermophilic LAB	–
Doogh	Iran	Thermophilic LAB	–
Dehydrated yoghurt			
Kashk	Iran	Thermophilic LAB	–
Kaskg	Iran	Thermophilic LAB	–
Jub-jub	Lebanon	Thermophilic LAB	–
Kurut	Turkey	Thermophilic LAB	–
Concentrated yoghurt			
Labneh-lebneh	Lebanon and Arab countries	Thermophilic LAB	–
Tan or than	Armenia	Thermophilic LAB	–
Torba	Turkey	Thermophilic LAB	–
Tuzlu (salted) yoghurt	Turkey	Thermophilic LAB	–
Roba, rob	Iraq	Thermophilic LAB	–
Madzoon	Armenia	Thermophilic LAB	–
Tiaourti	Greece	Thermophilic LAB	–
Tarho	Hungary	Thermophilic LAB	–
Koumiss	Central Asia	Mesophilic or mixed LAB	+
Leben, laban	Iraq, Lebanon	Mesophilic or mixed LAB	
Kefir	Caucasus, former Soviet Union countries	LAB, acetic acid bacteria	
Miscellaneous			
Surk	Turkey	Thermophilic LAB, moulds	–
Tarhana	Turkey	Thermophilic LAB	–

LAB: lactic acid bacteria.

industrialized processes. There is a close relationship between each type of fermented milk and the region where it was first manufactured. For instance, in the subtropical conditions of the Middle East, products fermented by mesophilic lactic acid bacteria predominate.

In the area known as eastern Europe, including the Balkans and central Asia, which have long, eminent and well-kept traditions, fermented milks are widely consumed. Numerous fermented milk products are manufactured in these regions, but few are of commercial significance. The main fermented milk products of eastern Europe, the Balkans, the Middle East and central Asia are illustrated in **Table 1**. With the exception of normal and concentrated yoghurt, only kefir and, to a lesser extent, koumiss, have economic importance. In addition, 'tarhana,' a fermented product consisting of yoghurt, tomatoes and red peppers, is becoming a popular traditional product in the Balkans and central Europe. This article deals particularly with the technology, microbiology and nutritive benefits of kefir, koumiss and tarhana.

Kefir

Origins

Kefir is a fermented milk product which has been

consumed for thousands of years. It was first produced in the Caucasus Mountains, where the drink was fermented naturally in bags made of animal hides. In the latter part of the nineteenth century, production spread to eastern and central Europe and thence to other parts of the world. Kefir is now produced commercially in large quantities in the countries of the former Soviet Union, and in appreciable quantities in Poland, Germany, Sweden, Romania and other countries. It is also still produced traditionally, under a variety of names including kephir, kefer, kiaphur, knapon, kepi and kippi.

Production

Cow's and goat's milk are usually preferred for the manufacture of kefir, and either whole or skimmed milk, or a mixture, is used. The composition of the product is determined primarily by the raw material and the microflora of kefir grains (see below), and is subject to regional variations. Kefir is a self-carbonated, fermented beverage containing about 0.8–1% lactic acid and 1–2% alcohol. During fermentation, lactose-forming yeasts produce alcohol and CO_2 and lactic acid bacteria convert lactose to lactic acid. Some proteolysis occurs in the milk, and a yeasty aroma develops. The flavour of kefir is mildly

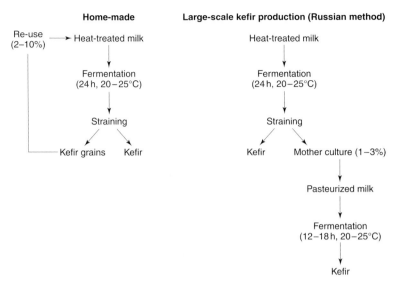

Figure 1   Home-made and large-scale (Russian method) kefir production.

alcoholic, yeasty and sour, with a tangy effervescence, and is related to the composition of the kefir grains.

Several methods exist for the manufacture of kefir, and modern techniques give a product with the same characteristics as those of traditional kefir. In the traditional method, pasteurized milk is inoculated with kefir grains, which are composed of lactic acid bacteria, yeasts, streptococci and acetic acid bacteria in various proportions. The usual initial heat treatment is at 90°C for 30 min. The inoculum consists of a 'mother culture' (see below) (2–10%, w/v). Fermentation is achieved at 20–25°C over about 24 h. The kefir grains are then removed, by straining. The filtrate is refrigerated overnight and the beverage, which contains live microorganisms from the grains, is ready for consumption (**Fig. 1**). Alternatively, a two-step fermentation (known as the Russian method) is applied, to stimulate the activity of the microorganisms and accelerate the changes in the milk. In the first step, the milk is inoculated with kefir grains (2–3%, w/v), to prepare a mother culture. Fermentation takes about 24 h at 20–25°C. The grains are removed and the filtrate, which is the mother culture, is added to pasteurized milk (1–3%, v/v). The second fermentation lasts 12–18 h at 20–25°C (Fig. 1).

Many problems are associated with traditional kefir production, and this has led to more modern methods. The traditional method allows the production of only small volumes of kefir, and involves several steps. Also, the production of CO_2 by yeasts often leads to 'blown' containers, which are mistakenly judged by the consumer to be 'spoiled'. The shelf life of traditional kefir is as short as 2–3 days. In order to resolve these difficulties and to facilitate large-scale pro-

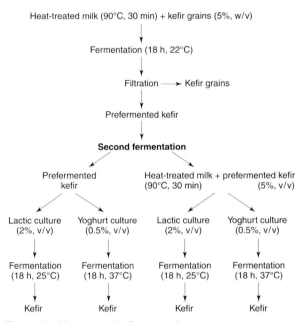

Figure 2   Modernized kefir production.

duction, the use of standard starter cultures, consisting mainly of streptococci, has gained popularity. Intensive studies have been carried out at the National Institute for Research in Dairying (NIRD) in Reading (UK), involving the comparison of a number of methods (**Fig. 2**). Kefir production can involve either a single-stage or a double-stage fermentation, the latter involving the production of a mother-like starter culture followed by fermentation with an additional starter culture. The second starter is chosen so as to include either aroma-producing or acid-producing bacteria, depending on the required characteristics of

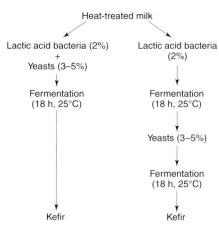

Figure 3 Kefir production using pure culture.

the kefir. The starter for lactic cultures is a blend of *Leuconostoc cremoris* and *Lactococcus lactis* subsp. *lactis* biovar *diacetylactis*, while that for yoghurt cultures is a blend of *Streptococcus thermophilus* and *Lactobacillus delbrueckii* subsp. *bulgaricus*.

Pure Cultures

In recent years, attention has focused on producing kefir from pure, defined cultures. These enable better control of the microorganisms involved, greater ease of production and more consistent quality. In addition, the shelf life of the kefir can be extended to 10–15 days at 4°C and its modification and improvement, e.g. in terms of health-related and nutritional aspects, are facilitated.

There are two basic procedures for producing kefir using pure cultures isolated from kefir grains. In the first, lactic acid bacteria and yeasts are added to heat-treated milk. In the second, the heat-treated milk is first fermented with lactic acid bacteria and the yeasts are added prior to the second fermentation (**Fig. 3**).

The main problem when using pure cultures is finding the balance between the bacteria and yeast strains which creates a product with the characteristic properties of traditional kefir, including both the organoleptic qualities and the health benefits. In practice this is rather difficult, the proportions of microorganisms in kefir grains themselves being influenced by many factors. The optimal rate of inoculation is in the range 1 : 30–1 : 50 (culture : milk). High inoculation rates of kefir grains (1 : 10–1 : 20) shorten the fermentation, but the growth rates of the yeasts and of both homo- and heterofermentative streptococci decrease compared with growth at lower inoculation rates (e.g. 1 : 50). The balance of microorganisms in kefir grains is also affected by agitation: frequent agitation during the fermentation may cause the numbers of bacteria and yeasts to increase. However, frequent washing of the grains with water leads to a

rapid decrease in the number of microorganisms; also, the fermentation takes longer and the taste and consistency of the end product become nonrepresentative of kefir. Thus the microflora of kefir can be controlled by altering the conditions of fermentation.

The following procedures should ensure a product of high quality and, to some extent, standardization:

- The milk should be changed at the same time every day.
- The ratio of grains to milk should be in the range 1 : 30–1 : 50.
- The milk should be heat-treated (90–95°C for 5–15 min, or ultra-high temperature).
- The milk should be fermented at 18–20°C.
- During incubation, the fermenting milk should be stirred two to three times.

Another critical step in ensuring the quality of kefir is post-fermentation cooling. Cooling must take place slowly (over 10–12 h) for the pronounced taste and aroma of kefir to be acquired, so that the necessary accumulation of homofermentative lactic streptococci and yeasts can occur.

Leakage and blowing of containers, as a result of excess CO_2 production, are major problems in the kefir industry. To avoid them, non-lactose-fermenting yeasts should be used – in which case sucrose supplementation is recommended.

Microbiology of Kefir Grains

Kefir grains are gelatinous granules, some 2–15 mm in diameter, consisting of a mixture of microorganisms grouped in a highly organized manner. Kefir grains have the chemical composition 89–90% water, 0.2% lipid, 3% protein, 6% sugar (mainly polysaccharide) and 0.7% ash. They should be stored wet at 4°C, or should be dried at room temperature for 36–48 h. Dried kefir grains retain their activity for 12–18 months, whereas wet grains retain their activity for only 8–10 days. Storage at –20°C is another effective method of preserving kefir grains.

The bacterial and yeast cells in the kefir grains are embedded in a slimy polysaccharide material named 'kefiran'. The bacteria responsible for its production have not yet been identified. Early studies revealed that *Lactobacillus brevis* was able to produce kefiran, but later a homofermentative lactobacillus, which was described as an atypical streptobacterium, was found to be responsible. Subsequently, a capsule-forming homofermentative bacterium, *Lactobacillus kefiranofaciens*, was isolated from kefir grains. The appearance of kefir grains is similar to that of the tiny florets of cauliflower. The proportions of bacteria and yeasts vary, depending on the source of the kefir

Table 2   Species of lactobacilli, leuconostocs, lactococci and yeasts isolated from different kefir grains

Leuconostocs	Lactococci	Lactobacilli	Yeasts
L. dextranicum	L. lactis subsp. lactis biovar	L. caucasicus	Kluyveromyces lactis
L. mesenteroides	diacetylactis	L. brevis	Klyveromyces marxianus var. marxianus
L. kefir	L. lactis subsp. cremoris	L. kefir	Kluyveromyces fragilis
	L. filant	L. casei	Torula (Candida) kefir
		L. plantarum	Candida holmii
		L. acidophilus	Saccharomyces cerevisiae
		L. kefiranofaciens	Zygosaccharomyces florentinus
		L. cellobiosus	Torulaspora delbrueckii
		L. helveticus subsp. jugurti	Saccharomyces exiguus
		L. lactis subsp. lactis	Candida pseudotropicalis
		L. rhamnosus	Saccharomyces globus
		L. fermentum	Saccharomyces dairensis
		L. paracasei subsp. paracasei	Saccharomyces unispores
		L. paracasei subsp. tolerans	Mycotorula kefyr
		L. parakefir	Mycotorula lactis
		L. viridescens	Candida friedrichii
		L. delbrueckii subsp.	Candida valida
		bulgaricus	

grains. The microorganisms present are listed in **Table 2.**

Nothing is yet known about the mechanisms of grain formation: attempts at making kefir grains from pure or crude cultures have not been successful. The outer layer of a kefir grain is dominated by rod-shaped lactic acid bacteria, while yeasts predominate in the core. The intermediate zones contain a balance between bacteria and yeasts, which changes progressively according to distance from the core.

Mesophilic homofermentative lactic streptococci (*Lactococcus lactis* subspp. *lactis* and *cremoris*) are the most active components of kefir grains, and they cause a rapid increase in acidity during the first hours of fermentation. *Lactobacillus brevis*, which is always heterofermentative, and the facultatively heterofermentative *Lactobacillus casei* subsp. *rhamnosus* are common in kefir starters. The number of mesophilic lactobacilli in kefir starters is not $> 10^2$–10^3 cfu ml^{-1}, and these organisms do not have any significant role in determining the quality of product or the formation of the grain. Mesophilic heterofermentative lactic streptococci (*Leuconostoc mesenteroides* and *L. mesenteroides* subsp. *dextranicum*) are mainly responsible for the development of the characteristic taste and aroma of kefir, and they may cause gas formation in association with yeasts. Yeasts have a symbiotic relationship with some of the other microorganisms in kefir, and contribute to the formation of its specific taste and aroma and to the release of CO_2.

Metabolites of Microorganisms

Lactic acid and lactate are the main metabolites of fermentation. The kefir microflora is dominated by mesophilic homofermentative lactic streptococci, and

so about 10 times more L(+)-lactic acid is formed than D(−)-lactic acid. CO_2 is produced as a result of the activity of yeasts and heterofermentative lactic streptococci. CO_2 plays an important role in the organoleptic characteristics of kefir, giving it the slight effervescence that is highly appreciated by consumers.

Alcohol, mainly ethanol, is another metabolite synthesized by the yeasts, and also by the lactic streptococci if they are not inhibited by the high acidity and the anaerobic conditions resulting from tight packing. The alcohol content of kefir varies, depending on its microbial composition and the conditions in which it was produced, but generally it is 1–2% (v/v). Ethanol continues to be produced when the growth of leuconostocs and yeasts has ceased.

Some vitamins are synthesized by both lactic acid bacteria and yeasts. While the levels of some vitamins increase, others are utilized by the microflora. Orotic acid, which is known to cause fat accumulation in the liver, is lost during the kefir fermentation; this might have a hypocholestaemic effect in humans. The B vitamins and vitamin P (riboflavenoid) group vitamins are accumulated in kefir.

Some components of the kefir microflora, particularly lactose-fermenting yeasts (*Torulaspora* spp.), have bacteriostatic and bactericidal effects on coliforms, but non-lactose-fermenting yeasts are ineffective against them. The total antibiotic activity of the kefir microflora is more effective than that of its individual components.

Kefiran is perhaps the most important metabolite in kefir, because it acts like glue and keeps the kefir grains intact. Kefiran consists of glucose and galactose in equal proportions, and is present in the capsular material of some large, rod-like bacteria, especially lactobacilli. Some 30–35% of the wet weight of kefir grains may be attributed to kefiran.

Nutritional and Health Benefits

In common with many other fermented foods, kefir has a remarkably high nutritional value and therapeutic effect. Its gross composition and calorific value are similar to those of the original milk, except that a quarter of the lactose is converted to lactic acid. The accumulation of free amino acids, as a result of proteolysis, contributes to the nutritive value of kefir, as does the formation of B and P group vitamins. CO_2 produced by the yeasts stimulates the appetite and aids digestion.

Koumiss

Koumiss is a drink with ancient origins, and is common in eastern Europe and central Asia. It is traditionally produced from mare's milk by a combined fermentation to lactic acid and alcohol, and its highly nutritive and curative characteristics are well known.

Production

Traditionally, koumiss was made by using part of the product of a previous day to seed freshly drawn mare's milk (usually unpasteurized) in 'saba' or 'turdusk', made from smoked horsehide. Fermentation takes some 3–8 h. Production ceased at the end of lactation, in late autumn, and the koumiss starter was put into a glass bottle which was sealed tightly and stored in a cool dark place until early summer, when production began again. The starter microorganisms were reactivated by keeping the koumiss starter at room temperature for about 24 h and then mixing with fresh milk three or four times. Alternatively, koumiss was mixed with cow's milk in the middle of winter and kept at room temperature. In early summer, the mixture was left for 4–5 days at 22–25°C, until gas formed, and it was then used as a starter.

The koumiss starter can also be preserved by drying. Before use, the dried starter (3–4 tablespoonfuls) is added to 5 l of fresh mare's milk and then left at room temperature for about 2–3 days. This fermented milk is blended with 6–7 l of fresh milk and further fermented.

In the event of problems with a starter, fermented wheat can be used as an inoculum or, alternatively, cooked and cooled millet blended with small amounts of honey and brewer's yeast can be used. This mixture is left at 35–40°C until it starts to swell, and is then wrapped in cheesecloth. The bag of starter is immersed in fresh mare's milk and when the acidity of the milk has increased sufficiently, the bag is

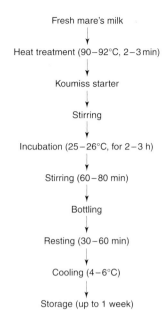

Fresh mare's milk
↓
Heat treatment (90–92°C, 2–3 min)
↓
Koumiss starter
↓
Stirring
↓
Incubation (25–26°C, for 2–3 h)
↓
Stirring (60–80 min)
↓
Bottling
↓
Resting (30–60 min)
↓
Cooling (4–6°C)
↓
Storage (up to 1 week)

Figure 4 Industrialized koumiss production, using mare's milk.

removed and the milk is added to fresh mare's milk. After the second fermentation, the fermented milk is used as normal starter.

Heat treatment is not essential as far as the physical quality of the product is concerned because koumiss is a liquid. However, heat treatment does provide a suitable environment for the growth of the starter culture. The industrial production of koumiss using mare's and cow's milks is outlined in **Figures 4** and **5** respectively.

Microorganisms

Pure cultures have been used in the manufacture of koumiss on an industrial scale since the 1960s. The microorganisms present in koumiss are listed in **Table 3**. Koumiss may have viable cell counts of 5×10^7 cfu ml^{-1} and 1–2×10^7 cfu ml^{-1} of bacteria and yeasts respectively.

During storage, the populations of bacteria and yeasts decline gradually due to the accumulation of lactic acid and ethanol. The yeasts can be divided into two groups: lactose-fermenting (e.g. *Saccharomyces lactis*, *Torula koumiss*) and non-lactose-fermenting (e.g. *Mycoderma* spp., *Saccharomyces cartilaginosus*). Streptococci do not have a significant role in the formation of the characteristic aroma or flavour of koumiss, because they are inhibited by lactic acid. *Acetobacter* spp. are also of only minor importance.

Composition

Mare's milk has lower levels of fat, protein, ash and total solids than cow's, goat's and sheep's milk. Koumiss made from mare's milk is sweeter than that

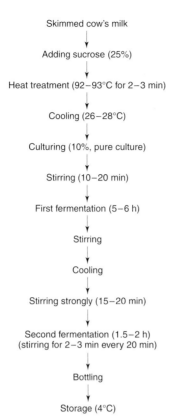

Skimmed cow's milk

↓

Adding sucrose (25%)

↓

Heat treatment (92–93°C for 2–3 min)

↓

Cooling (26–28°C)

↓

Culturing (10%, pure culture)

↓

Stirring (10–20 min)

↓

First fermentation (5–6 h)

↓

Stirring

↓

Cooling

↓

Stirring strongly (15–20 min)

↓

Second fermentation (1.5–2 h)
(stirring for 2–3 min every 20 min)

↓

Bottling

↓

Storage (4°C)

Figure 5 Industrialized koumiss production, using cow's milk.

1.8%), ethanol (0.6–2.5%) and CO_2 (0.5–0.9%). As in the case of other fermented dairy products, the precise characteristics of koumiss are determined by the starter microorganisms.

Health Benefits

It has long been speculated that koumiss can cure many illnesses, including tuberculosis, disorders of the stomach and colon, and hepatitis. In addition, the product shows antibiotic effects in vitro against *Escherichia coli*, *Staphylococcus aureus* and species of *Mycobacterium*, *Bacillus*, *Serratia* and *Shigella*.

Tarhana

Tarhana is a fermented food resulting from the combined fermentation of yoghurt with cracked wheat or flour. Tarhana is a traditional Turkish food, which is widely known and consumed in the Balkans and the Middle East under different names, including 'trahana' (Bulgaria), 'trahanas' (Greece), 'taron' (Macedonia), 'tarhonya' (Hungary), 'kisk' (Iraq), 'kishk' (Egypt, Syria and Lebanon) and 'goce' (Turkmenistan). Tarhana is used for making soup, and has a considerable share of the ready-soup market in the Balkans and the Middle East.

Table 3 Microorganisms present in koumiss starter

Lactic acid bacteria	Yeasts	Milk type
Lb. delbrueckii ssp. bulgaricus	Saccharomyces lactis	Cow, mare
Lb. casei	Pichia ssp.	Mare
Str. lactis subsp. lactis	Rhodotorula ssp.	Mare
Lb. lactis ssp. lactis	Torula lactis	Mare
Lb. leichmanii	Saccharomyces lactis	Mare
Lb. delbrueckii ssp. lactis	(Lactose-fermenting)	
	Mycoderma ssp.	Mare
	(Non-lactose-fermenting)	
	Saccharomyces cartilaginosus	Mare
	(Non-lactose-fermenting)	
	Torula koumiss	Mare
	(Lactose-fermenting)	
Lactic streptococci	Kluyveromyces lactis or Kluyveromyces fragilis	Cow
Lb. acidophilus	Kluyveromyces marxianus var. marxianus	Cow

made from cow's milk. It is milky-green in colour, light and fizzy, and has a sharp alcoholic and acidic taste. The digestibility of mare's milk is good, because it contains a high level of whey protein. The action of chymosin on this results in a soft coagulum containing appreciable amounts of the products of proteolysis, such as peptones, polypeptides and amino acids. Koumiss contains about 90% moisture, 2.1% protein (1.2% casein, 0.9% whey proteins), 6.4% lactose, 1.8% fat and 0.3% ash, as well as the main metabolites of fermentation, which are lactic acid (0.7–

Production

The method of manufacture of tarhana does not vary significantly from one region to another, although slight variations in the microbiological and compositional characteristics may occur depending on the locality and its traditions.

Production is based on a yoghurt fermentation, but concentrated yoghurt or sour milk can also be used. Usually, equal quantities of yoghurt and cracked wheat or flour are used. *Saccharomyces cerevisiae* is

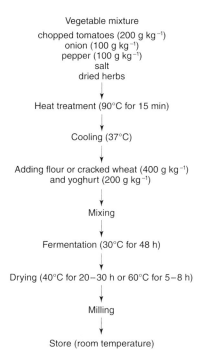

Vegetable mixture
chopped tomatoes (200 g kg^{-1})
onion (100 g kg^{-1})
pepper (100 g kg^{-1})
salt
dried herbs

↓

Heat treatment (90°C for 15 min)

↓

Cooling (37°C)

↓

Adding flour or cracked wheat (400 g kg^{-1})
and yoghurt (200 g kg^{-1})

↓

Mixing

↓

Fermentation (30°C for 48 h)

↓

Drying (40°C for 20–30 h or 60°C for 5–8 h)

↓

Milling

↓

Store (room temperature)

Figure 6 Tarhana production (Turkish-style).

generally added to the mixture, to give the characteristic aroma of tarhana. *Lactobacillus delbrueckii* subsp. *bulgaricus* and *Streptococcus thermophilus*, from the yoghurt, are responsible for the production of lactic acid and the yeast produces ethanol and CO_2. *Lactobacillus casei*, *L. plantarum* and *L. brevis* can also be added in order to improve the aroma and flavour.

The production of tarhana is outlined in **Figure 6**. A mixture of chopped vegetables, including tomatoes, onions, peppers, salt and dried herbs (mainly mint), is heated at 90°C for 15 min, and then cooled to 37°C. Cracked wheat or flour and yoghurt (or a suitable replacement) are added, and the mixture is fermented at 30°C for 48 h. After fermentation, the mixture is spread onto a large cloth (traditional method) or a pulsating tray (industrialized method), and is dried at 40°C for 20–30 h (traditional method) or 60°C for 5–8 h (industrialized method). The dried mixture is milled and stored at room temperature.

Composition

According to Turkish standards, tarhana must contain protein (min. 14%), moisture (max. 10%) and salt (max. 5%). A low level of lactose and a high level of hydrolysed starch facilitate the digestion of the product. It is rich in protein, calcium, iron and zinc. Unless dried in the sun, it is an important source of group B vitamins – direct sunlight causes the loss of riboflavin, which can be avoided by drying on pulsating trays.

See also: **Fermented Foods**: Origins and Applications. **Fermented Milks**: Yoghurt. **Heat Treatment of Foods**: Principles of Pasteurization. *Leuconostoc*. **Starter Cultures**: Uses in the Food Industry.

Further Reading

Bottazi V and Bianchi F (1980) A note on scanning electron microscopy of micro-organisms associated with the kefir granule. *Journal of Applied Bacteriology* 48: 265–268.

Duitschaever CL, Kemp N and Snith AK (1988) Microscopic studies of the microflora of kefir grains and of kefir made by different methods. *Milchwissenschaft* 43(8): 479–481.

Koroleva NS (1991) Products prepared with lactic acid bacteria and yeasts. In: Robinson RK (ed.) *Therapeutic Properties of Fermented Milk Products*. P. 159. Elsevier Science.

Marshall VM, Cole W and Brooker BE (1984) Observations on the structure of kefir grains and the distribution of the microflora. *Journal of Applied Bacteriology* 57: 491–497.

Marshall V and Cole A (1985) Methods for making kefir and fermented milks based on kefir. *Journal of Dairy Research* 52: 451–456.

Merin U and Rosenthal I (1986) Production of kefir from UHT milk. *Milchwissenschaft* 41(7): 395–396.

Tamime AY and Marshall VME (1997) Microbiology and Technology of Fermented Milks. In: Law BA (ed.) *Microbiology and Biochemistry of Cheese and Fermented Milk*, 2nd edn. P. 57. Blackie Academic & Professional.

Temiz A and Pirkul T (1991) Farkli bilesimlerde uretilen tarhanalarin kimyasal ve duyusal ozellikleri Chemical and organoleptic properties of tarhana produced of different compositions. *Gida* 16(1), 7–13.

Yaygin H (1992) *Kimiz ve Ozellikleri [Koumiss and its Properties]*. P. 69. Antaya: Yeni Matbaa.

Yokoi H, Watanabe T, Fujii Y, Toba T and Adachi S (1990) Isolation and characterisation of polysaccharide-producing bacteria from kefir grains. *Journal of Dairy Science* 73: 1684–1689.

Filtration *see* **Physical Removal of Microfloras**: Filtration.

FISH

Contents
Catching and Handling
Spoilage of Fish

Catching and Handling

Parimal Chattopadhyay, Department of Food Technology and Biochemical Engineering, Jadavpur University, Calcutta, India

Anatomy and Physiology of Fish

Fishes have certain common characteristics – for example backbones and gills – and all of them are cold-blooded. Most of them also have fins along the back and under the tail, two pairs of fins and a large vertical tail fin.

Cod may be considered as a typical commercial fish (**Fig. 1**). It is torpedo-shaped and is covered with a transparent, slimy skin below which lie row upon row of scales from head to tail. It has three vertical fins along the back, the dorsal fins, and two beneath the tail behind the vent, the ventral fins. In addition it has a pair of pectoral fins and a pair of pelvic fins; these fins act as stabilizers and brakes. The tail usually propels the fish. A fish possesses six senses: apart from the usual hearing, sight, smell, taste and touch, it has a series of delicate and sensitive nerve endings in the skin, situated mainly along the lateral line, which enable it to detect small water currents and changes in water pressure. It can detect small ripples in the water due to movement of other fish. Some species of fish have sensitive organs in the skin and on the fins by which they can 'taste' or 'smell' objects without eating them. Cod and similar species have a specially sensitive 'beard' or barbel, a smelling organ.

Bony fishes have a characteristic gill cover or operculum on each side of the body. This acts as a non-return valve. When the fish breathes in, the gill cover is closed against the body so that water enters only through the mouth. When the fish breathes out, the mouth is closed and water passes over the gills and out behind the gill covers. Sharks, dogfish, skates and rays which do not have bony skeletons (the cartilaginous fishes) differ slightly from the bony fishes such as cod and herring; they breathe in water through a special hole just behind each eye instead of through the mouth, and breathe it out through a separate series of gill slits, usually five, which lie on each side of the head.

A fish usually swallows food without chewing it, although some fish have teeth which are used for breaking up lumps of food. The food passes straight into the stomach. Fish can survive periods of starvation lasting many months; in some seas they are forced to fast either because food is not available or because they cannot hunt during the long Arctic night. Some species do not eat when they are preparing to spawn. Starvation is one possible cause of 'soft' fish, which is difficult to hang up for smoking or marketing fresh.

The stomach wall of fish contains microscopic glands that secrete digestive enzymes as soon as food is eaten. This is the reason why feed herring rapidly becomes soft and broken after death. These enzymes will digest any protein with which they come into contact, and after the death of the fish this includes the stomach and intestines themselves. Enzymes are produced in the microscopic glands in the lining of the stomach and intestine and pyloric caeca in the bony fish (**Fig. 2**). The latter are not found in cartilaginous fish. The bile, which is produced in the liver, enters the intestine just behind the stomach through a

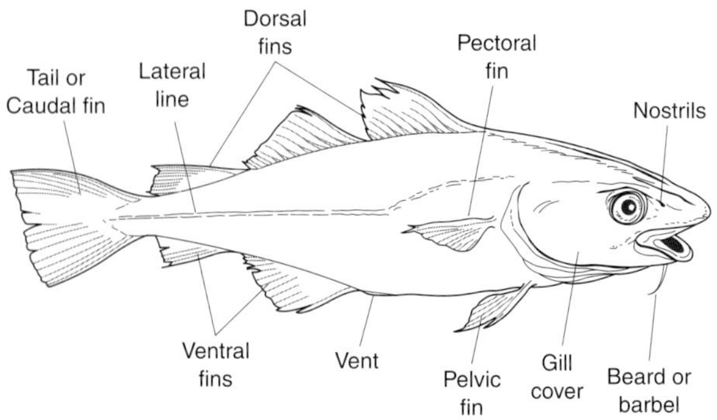

Figure 1 External anatomy of the cod.

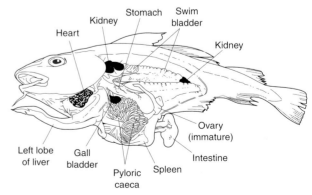

Figure 2 Dissection of a cod.

fine tube. A duct enters near this point from the pancreas, which produces digestive juices. The entire intestine of all fish is short. The main purpose of the gut is to digest food and absorb it through its walls into the body. The other main organs of fish are the liver, kidney, swim bladder and reproductive organs.

Fish are sensitive to temperature. The body temperature of fish is not controlled and it is, therefore, very close to that of the surrounding water. There is a range of temperatures in which any particular species of fish can live. The behaviour of fish is conditioned by the desire to find food and to reproduce, although other factors such as water temperature also have significant effect on fish behaviour.

A broad distinction is usually made between two types of fish, *pelagic* and *demersal*. Pelagic fish, such as herring, sprats and mackerel, are those that usually find their food (e.g. plankton) in the surface layers of the sea. Demersal fish are those such as cod, haddock and flatfish that lie on or near the bottom. Pelagic fish may become demersal during part of their life cycle; herring, for example, may be trawled in certain areas during part of the year. Demersal fish may become pelagic; dogfish, when herring is plentiful, may cause serious damage to driftnets. Fishing methods depend almost entirely upon the habits of the fish to be captured; the drift net and ring net are as unsuitable for fish of demersal habit as the bottom trawl is for fish shoaling on the surface. Successful fishing requires knowledge of how and where particular fish are likely to congregate at various times.

Composition of Body Tissues

The body tissues of fish include skin, flesh and bone. Skin consists mainly of water (about 80%) and protein (about 16%). Bone contains mineral matter, mostly calcium phosphate which amounts to about 14% of the total bone material; the rest is mainly water (about 75%) and protein (about 9%). The flesh is the most important tissue. It is made up chiefly of muscle fibres held together by connective tissue. These cells are surrounded by extracellular fluid. The flesh also contains blood vessels and nerve fibres. The fat content of fish varies with the season; however, in healthy fish flesh, fat and water together amount to about 80%. This value does not vary much. The protein content of healthy fish flesh is about 16–18%. Under conditions of prolonged partial starvation, protein content is depleted and the flesh may contain well over 80% water and only 3% protein.

Nitrogenous bases such as trimethylamine oxide and urea are plentiful in shark, dogfish, rays and skates. These are colourless compounds and without smell. When bacterial spoilage occurs, bacterial enzymes convert trimethylamine oxide into tri-

methylamine and ammonia from urea. Freshwater fish contain less trimethylamine oxide than marine fish. A little trimethylamine can produce strong odour. Free amino acids are involved in the development of brown colour and off flavour in dehydrated and canned fish due to the reaction of certain types of sugar with an amino group. The principal free sugar of fresh fish is glucose. The main significance of glucose in fish to the processor is in regard to the browning reactions just mentioned with free amino acids. Another sugar called ribose occurs in live fish attached to a complex nitrogenous substance and to phosphate; after death autolysis frees the sugar, which is a reactive browning agent. Fish flesh contains metals, e.g. potassium, sodium, calcium, magnesium, iron, copper, manganese, zinc and cobalt, and nonmetals, e.g. phosphorus, sulphur, chlorine and iodine. Fish contains almost all the vitamins necessary in the human diet; in particular it is a good source of vitamins A, D, B_1, B_2 and B_{12}. The composition of the edible flesh of various fishes is given in **Table 1**.

Microbiology of Fish

It is known that muscle tissues are sterile in healthy fish, while large populations of bacteria are present on the external surfaces, gills and intestines. There may be as much as 10^2–10^6 bacteria per square centimetre of skin surface; similarly, gill counts and intestinal counts may range from a few to 10^8 per gram, depending on whether or not the fish were feeding. During spoilage skin counts increase to 10^7 per square centimetre or more, and gill counts also increase. The microflora on fish are dominated by the Gram-negative bacteria, i.e. *Achromobacter*, *Flavobacterium*, *Pseudomonas* and sometimes *Vibrio*. However, some Gram-positive bacteria are also found (**Table 2**). The Gram-negative bacteria dominate after a few days in ice, but if fish are held at high temperature a mixed flora results. Fish spoilage bacteria reduce trimethylamine oxide to trimethylamine in the presence of triamine oxidase. Fish bacteria are mostly psychrophilic, growing at between 0°C and about 30°C, with some growing at temperatures as low as −7.5°C. Fish bacteria are sensitive to low pH. Most bacteria would not grow below pH 6.0 and this is one reason for the stable microbial population found during rigor mortis when the pH of flesh is in the range 6.2 to 6.5.

The organisms that dominate the spoilage microflora are now known to be *Pseudomonas* and *Alteromonas*, and their characteristics can be related to the spoilage process. Fish from warm water carry greater numbers of bacteria than cold-water fish and yield higher microbial counts when incubation tem-

Table 1 Analysis of the edible portion of raw fish. (Source: Wheaton and Lawson, 1985)

Common name	Scientific name	Content per 100 g raw fish					
		Protein (g)	Fat (g)	Moisture (g)	Ash (g)	Carbohydrate (g)	Cholesterol (mg)
Carp, Indian	*Labeo* sp.	14.3–19.1	0.5–24.5	72.5–82.1	0.9–1.4	0.3–0.4	–
Carp	*Cirrhinus* sp.	18.1–19.6	0.2–4.0	75.0–79.8	1.0–1.6	0.6–2.0	
Catfish (freshwater)	*Ictaluridae* spp.	15.4–22.8	0.3–11.0	68.0–82.6	0.9–1.7		–
Catfish (marine)	*Arlidae* spp.	12.7–21.2	0.2–2.9	75.1–81.1	0.9–1.6	0.4–0.6	–
Cod (Atlantic)	*Gadus morhua*	16.5–20.7	0.1–0.8	78.2–82.6	1.0–1.2	–	36.1–40.8
Eel (freshwater)	*Anguillidae* spp.	18.0	12.7–21.5	62.2–70.1	1.3	–	
Haddock	*Melanogramus aeglefinus*	15.4–19.6	0.1–1.2	79.1–81.7	1.0–1.2	–	
Halibut (Atlantic)	*Hippoglossus hippoglossus*	12.6–20.1	0.7–5.2	76.5–82.9	1.1	–	
Herring (Atlantic)	*Clupea harengus*	15.2–21.9	2.4–29.1	52.6–78.0	1.7	–	
Mackerel (Atlantic)	*Scomber scombrus*	15.1–23.1	0.7–24.0	49.3–78.6	1.0–3.0	–	80
Prawn	*Miscellaneous species*	8.9–23.2	0.3–3.1	67.5–80.6	1.6–5.2	–	
Salmon (pink)	*Oncorhynchus gorbuscha*	17.2–20.6	2.0–9.4	69.0–78.2	1.1–1.4	–	–
Sardine	*Sardinella* sp.	19.0	3.7	77.1	2.6	–	–
Shark	Mixed species	14.9–27.1	0.1–2.9	72.0–76.9	1.0–2.0	–	–

Table 2 Flora of marine fish by percentage of isolations in various generic groups (Source: Connell, 1980a)

Fish	Micrococcus (%)	Achromobacter (%)	Flavobacterium (%)	Pseudomonas (%)	Bacillus (%)	Misc. (%)
Haddock	4	23	8	22	24	18
Halibut	16	34	30	–	–	20
Herring	24	43	13	11	–	9
Cod	14	48	25	5	–	8
Salmon	13	54	5	8	2	19
Porgy	53	21	7	6	–	13
Skate	3	19	9	65	–	4
Lemon sole	1	22	5	69	–	3
Shrimp (pond culture)	–	2	–	3	–	83 (coryneform)

peratures are 35–37°C. This indicates that the microorganisms on fish in warm waters are more mesophilic. Similar variations were observed in the microbial populations of shrimp which are fished in all oceans of the world and taken from both cold and warm waters. Counts of shrimp and other bottom-dwelling creatures may be difficult to evaluate because they tend to be contaminated with sediment material. Some of the data available on bacterial population numbers of fish are shown in **Table 3**. The skin of cold-water fishes has a major population of Gram-negative bacteria, whereas the skin of truly warm-water fishes contains a majority of Gram-positive bacteria. Cold-water marine fish carry mainly *Moraxella*, *Acinetobacter*, *Pseudomonas*, *Flavobacterium* and *Vibrio*, while warm-water species carry mostly micrococci, coryneforms and *Bacillus*.

The bacteria in fish intestines vary depending on the food consumed but normally include *Vibrio*, *Achromobacter*, *Pseudomonas* and *Aeromonas* and small

Table 3 Bacterial counts on fish (Source: Connell, 1980a)

Fish	Count × 10^{-3}
Cod	47
Pollock	52
Whiting	77
Channel catfish	69–198 000
Hake	0.15
Rockfish	0.14
English sole	0.06
North Sea fish	0.1–100
Indian sardine	10–10 000
Flatfish	10
Mullet	10
Shrimp (cold-water)	0.53–169
Shrimp (tropical)	1–10 000
Shrimp (pond)	1.5–13

numbers of *Clostridium*. Freshwater fish may show slightly lower skin and gill counts than marine fish. The kinds of bacteria found on freshwater fish vary according to the microflora of the water; dominant

Table 4 Bacterial flora of fresh North Sea fish (1932–1970) (Source: Connell, 1980a)

1932		1960		1970	
Genus	Proportion (%)	Genus	Proportion (%)	Genus	Proportion (%)
Pseudomonas	5.0	Pseudomonas		Pseudomonas	
Achromobacter	56.0	Group I	–	Group I	–
Flavobacterium	11.0	Group II	16.0	Group II	13.0
Micrococcus	22.5	Group III	–	Group III	9.0
Luminous bacteria	1.0	Achromobacter	23.0	Moraxella	40.0
Others	4.5	Flavobacterium/Cytophaga	27.0	Acinetobacter	1.0
		Micrococcus	4.0	Flavobacterium/Cytophaga	10.0
		Luminous bacteria	1.0	Micrococcus	1.0
		Coryneforms	18.0	Luminous Vibrio spp.	< 1.0
		Vibrio/Aeromonas	1.0	Photobacterium	< 1.0
		Others	10.0	Arthrobacter	18.0
				Vibrio	< 1.0
				Aeromonas	< 1.0
				Others	3.0

genera are *Pseudomonas*, *Cytophaga*, *Aeromonas* and the coryneform groups, and warm-water freshwater fish may often carry *Salmonella*. *Clostridium botulinum* may be found in fish captured where sediments are contaminated with this organism. Although a few yeasts such as *Rhodotorula*, *Candida* and *Torulopsis* have been found in freshwater fish, moulds are rarely found in fish.

Flora of Newly Caught Fish

Effect of Environment and Species Different species of fish, such as cod, haddock, sole, skate and herring, caught in the North Sea at approximately the same season have very similar floras; but fish caught in different environments have different floras which reflect the flora of the water in which the fish are caught. This was also demonstrated by an easily identifiable organism – *Clostridium botulinum* type E – which was detected in the sea mud of Skagerrak and the Baltic. *Vibrio parahaemolyticus* was isolated from seawater fish and shellfish in Southeast Asia, in the east and west coasts of America, the Mediterranean and from fish landed in Baltic ports in Germany. The bacterial flora of the environment in which the fish is caught play a major role in determining the flora of the newly caught fish. The bacterial flora of North Sea fish as influenced by the season and environment are given in **Table 4**.

Effect of Handling

So far we have considered the flora of fish caught by line or net; however, in commercial practice the situation is very different. Fish are landed on the deck of the fishing vessel, where they may be trodden on; they are gutted and thus often contaminated with the gut contents which contain large numbers of bacteria; and are then washed in seawater and packed below in ice or frozen on board ship, where they may remain

for up to 17 days. On landing, the fish are laid out in market boxes or kits whose surfaces may carry a heavy bacterial load, and they are then removed to be processed, filleted and smoked before being despatched to the retailers. During these events, depending on the hygiene of handling, the acquired load of mesophilic contaminants may be considerable. The initial flora changes considerably during storage in ice, the *Pseudomonas* groups gradually predominating. On landing, the fish pick up organisms from boxes and surroundings. The bacterium *Erysipelothrix rhusiopathiae* is the cause of erysipeloid in fish handlers during the warm periods of the year all over the world. This organism has never been found in newly caught fish, but it regularly occurs on market fish, on fish boxes and on market floors in the summer months. The organism is a soil saprophyte which grows well in fish slime. Other indications of the effect of handling are the numbers and types of indicator organisms on fish and fishery products (**Table 5**).

Role of Handling on Bacterial Spoilage

Bacterial spoilage of fish proceeds even at 0–4°C, but can easily be prevented by storage at temperatures below 10°C. Immediately after death a complex series of enzymic changes take place in muscle, involving ATP and anserine. The flesh remains sterile (or nearly so) for 3–4 days at 0°C. It is believed that bacteria penetrate the gill tissue and proceed along the vascular system, particularly along the caudal vein through the kidney, and after some days into the flesh or through the intestines into the body cavity and belly walls, or through the skin into the flesh. Bacteria are confined to the surface layers before spoilage begins. The penetration into the tissues or along the blood vessels takes place with the progress of spoilage. This case is different in whole or eviscerated fish held at low temperature. In fillets, the bacterial penetration is

Table 5 Total counts and indicator organisms in some UK fishery products[a]

Product	No. of samples tested	Mean colony count per gram at 37°C	No. of coliforms MPN per 10 g	No. or % positive for S. aureus
Fish skin (direct from sea)	12	< 90	9.1	50 (2 +Ve)
Retail fillets	6	264 000	–	250 (1 +Ve)
Fish cakes	12	197 000	200 (42 E. coli)	42% in 1 g
Kippers	17	7 450 000	23 (6/gm+Ve)	0
Suggested standard		Not exceeding 10^5 per g at 35°C	Not exceeding 200 per g or 100 E. coli per g	Not exceeding 100 per g

MPN, most probable number; *E. coli*, *Escherichia coli*; *S. aureus*, *Staphylococcus aureus*.
[a]Source: Shewan J M (1971) *J. Appl. Bact.* 34(2), 299

more rapid but most activity occurs at the surfaces. This is a function of the oxidative nature of bacteria involved. There is a shift in bacterial types during the storage period. *Pseudomonas* become more dominant and *Moraxella*, *Achromobacter* and *Flavobacterium* persist at a decreasing level. The sulphur-containing compounds are most important as spoilage odour components, and *Pseudomonas fluorescens*, *P. pero-lens*, *P. putida* and *P. putrefaciens* have been shown to produce them from sulphur-containing amino acids.

Only two species of pathogenic bacteria occur on fish: *Clostridium botulinum* type E and *Vibrio para-haemolyticus*. *Clostridium botulinum* type E is found in marine and lake sediments and in fish intestine; it does not grow or produce toxin in living fish but is carried passively, and only becomes a hazard in mishandled processed products. *Vibrio para-haemolyticus* is found in the marine environment and on fish and shellfish when temperatures are high enough (above 15°C). The microorganism is distributed widely in inshore marine areas and can be readily isolated from marine animals in warm-water areas in summer months. However, not all *V. par-ahaemolyticus* strains are pathogenic and the level of natural occurrence rarely approaches infective numbers; food poisoning from this organism usually involves mishandling of seafood products. The vibrio is very sensitive to heat above 48°C and to cold, particularly 0–5°C. Most outbreaks of food poisoning derive from consumption of raw fish and also from eating shrimp recontaminated after cooking and held at temperatures allowing rapid growth. Generally *V. parahaemolyticus* grows very rapidly under favourable conditions.

Other potentially pathogenic bacteria associated with fish and shellfish include *Clostridium per-fringens*, *Staphylococcus*, *Salmonella*, *Shigella*, *Vibrio cholerae* and other vibrios. These organisms are derived by contamination from terrestrial sources. *Vibrio cholerae* can be very persistent in inshore marine environments. *Salmonella* may also persist in fish for long periods. *Salmonella typhimurium* per-sisted in warm-water marine fish and freshwater species for 30 days. *Salmonella* was detected in catfish from a retail market due to indigenous contamination. *Clostridium perfringens* was found in a number of fishes owing to contamination with sewage, which is the main source of this organism.

Bacteria occurring on fresh tuna can cause a public health problem; these organisms actively decarb-oxylate histidine to histamine and this is associated with scombroid poisoning, a condition that affects people consuming tuna, mackerel or related fishes containing more than 100 mg histamine per 100 mg of tissue. Bacteria implicated include *Morganella mor-ganii*, *Klebsiella pneumoniae* and *Hafnia alvei*. The occurrence of these pathogenic bacteria is important to the fish processor. Some microorganisms are commonly used by regulatory agencies as indices of haz-ardous conditions; they include coliforms and faecal bacteria such as *Escherichia coli*, *Staphylococcus* and sometimes enterococci. Such organisms should not be present on fresh-caught fish. Low levels of these organisms are found on iced fish when they are unloaded from the boats, and proper handling is required to maintain these low levels. In general, fresh and frozen fish should have less than 199 faecal bacteria per gram, less than 100 coagulase-negative staphylococci per gram and a total count no higher than 10^6 per gram. Changes in proportions or total numbers of indicator organisms are important for assessment of bacteriological effectiveness of pro-cessing and handling procedures.

Effect of Processing

The numbers and types of bacteria on fish are affected by simple primary processing operations (**Tables 6 and 7**). Since bacteria are mainly confined to the skin, gills and intestine of freshly caught fish, it might be expected that evisceration, beheading, filleting and skinning would greatly reduce the bacterial count on the final product. This depends on avoidance of cross-contamination and the addition of extraneous bac-teria from the environment. In practice, even with the

Table 6 Effect of processing on bacterial counts in shrimp[b]

	Hand peeled	Machine peeled
Initial	10^4	10^3
Brining	10^4	
Storage		10^{8a}
Blanching		10^2
Peeling	10^3	10^3
Cooking		< 10
Grading or packing	10^4	10^4

[a]From a storage hopper.
[b](Source: Connell, 1980a)

Table 7 Effect of processing on fish microflora (Source: Connell, 1980a)

	Newly caught fish	Auction	Fillets	Retail fillets
Bacterial count	8.4×10^3 per cm^2	7×10^4 per cm^2	7×10^5 per g	8.6×10^5 per g
		Percentage distribution		
Pseudomonas	18	15	6	11
Moraxella	8	15	6	7
Other Gram-negative bacteria	8	6	2	12
Micrococcus	49	41	76	45
Coryneforms	12	19	7	22
Other Gram-positive bacteria	4	2	–	–

Table 8 Effect of freezing on microflora of fish (ocean perch)[a]

	Fresh caught	Frozen
Bacterial count (per g)	4.9×10^5	8.3×10^4
Coliforms (per g)	0.6	0
Pseudomonas (%)	22.9	26.9
Achromobacter (%)	28.4	22
Flavobacterium (%)	16.0	12.1
Gram-positive bacilli (%)	27.2	24.8
Micrococcus (%)	4.3	1.5

[a](Source: Connell, 1980a)

use of chlorinated process water, machine processing, antiseptic dips for hands and knives and other precautions, fillets, steaks and other products from fresh fish processors usually carry bacterial counts of 10^3–10^5 per gram of fillet flesh, though occasionally lower counts are achieved and at other times much higher counts occur.

The major change qualitatively from primary processing is usually an increase in the relative proportion of Gram-positive bacteria and the appearance of bacteria associated with humans including some staphylococci and enteric bacteria. The processing of shrimp has been studied in detail. Cooked products usually show lower bacterial counts than uncooked products but subsequent operations cause an increase in count (Table 6). Shrimp on world markets typically have an average bacterial count of 10^6 per gram. Processing may or may not bring about major changes in composition of the microflora of shrimp.

Freezing The effect of freezing on the bacterial population of fish is difficult to predict. In general, there is some reduction in counts and the numbers continue to fall during storage in the frozen state (**Tables 8–10**). Gram-negative bacteria are more sensitive to freezing than Gram-positive bacteria, and bacterial spores are highly resistant. *Salmonella* and other members of the Enterobacteriaceae are among the more sensitive bacteria but there are great variations in the response of these organisms when present on fish. This variability is due to the range of different freezing processes for foods. Even in the case of cold-sensitive microorganisms such as *Vibrio parahaemolyticus* there may be some survival after freezing. From a practical standpoint, the conditions of freezing and subsequent cold storage most desirable for quality, i.e. rapid freezing and low non-fluctuating temperature during storage, are most protective of the bacteria present. Freezing simply preserves the bacterial status quo of the product. It is an effective method of halting bacterial action. Although a few microorganisms have been reported to be able to grow at $-7.5°C$, in practice there is no significant bacterial activity below $-5°C$ in seafood.

Canning Canned seafood falls into two categories from a bacteriological point of view: fully processed commercially sterile products, and semiconserved products. The fully processed products include canned tuna, salmon, shrimp, crab, sardines, and other fish, fish balls, etc. The heating process applied to these products is designed to destroy pathogenic bacteria and normal numbers of other organisms. Spore-formers if present in the unprocessed material in excessive numbers, or gaining entry into the can after processing through improper seaming or contaminated cooling water, may create problems. Flat sour spoilage due to *Bacillus stearothermophilus* which survive processing and multiply during slow cooling or storage at high temperature (45°C or above) can be a problem. Due to improper processing and infection from a leaky can, swollen or blown can may be caused by *Clostridium sporogenes*. Stability

Table 9 Effect of processing on bacteria of public health significance in precooked frozen seafoods

	Plate count (per g)		Coliforms	E. coli	Coagulase-positive Staphylococcus
	20°C	35°C	MPN	gm^{-1}	% + ve
Frozen blocks	10^4–10^5	10^4	< 10–10^2	< 10	64
Cut, battered and breaded	10^5–10^6	10^5	10–10^3	< 10–10^2	73
Precooked	10^4–10^5	10^3–10^4	< 10–10^1	0	52

No *Salmonella* were isolated from 293 samples tested.

Table 10 Microbial quality of frozen breaded seafood products (Source: Connell, 1980a)

	Percentage of samples containing					
	Plate count $< 10^6$ per g		E. coli < 3 per g		S. aureus < 100 per g	
	USA	Canada	USA	Canada	USA	Canada
Fish sticks	99.8		96.1		100	
Other breaded fish		100		100		100
Fish cakes	99.4		90.0		99.7	
Scallops	100		98.6		100	
Fish portion	99.2		97.9		100	
Breaded raw shrimps	82.2	7.3	98.8	99	99.7	83.3
Precooked shrimps	–	97.4		100		–

E. coli, *Escherichia coli*; *S. aureus*, *Staphylococcus aureus*.

of semiconserved seafood products is maintained by low pH, salt water activity (a_w) control, specific acid or other preservative, anaerobic condition and refrigeration. Failure to maintain the above conditions may permit growth of acidophilic bacteria, yeast or mould and sometimes dangerous bacteria. Botulism due to canned fish products is rare.

Salting and Drying   Preservation of fish and shrimp by drying, salting or both is widely practised throughout the world. The principal effect on microorganisms is due to the lowering of a_w, though sodium chloride itself in higher concentrations may be lethal for some bacteria and yeast owing to its osmotic effects. However, sensitive bacteria such as *Salmonella* which contaminate a dried product may persist on it for sometime.

Microbiological changes occur mainly during early stages of salting and drying ($a_w > 0.90$). The final population is dominated by micrococci and Gram-positive rods. Dried fish, shrimp and other seafood products are easily contaminated by mould spores which grow if the product becomes slightly moist. Mycotoxigenic moulds have been identified on such products. Small pelagic fishes are caught in enormous quantities by purse seine nets and undergo enzymatic liquefaction when held unrefrigerated in vessel holds or in outside holding bins at tropical temperatures. The major problem of product contamination has been *Salmonella*. Fish meal prepared from these fishes is therefore blamed for dissemination of *Salmonella* serotype throughout the world. Raw fish are con-

taminated in boats and holding areas of the plants. Boats and machinery should be completely cleaned with chlorinated water.

Smoking   Large quantities of fish are still treated by the smoking process, in which preservation is achieved by drying. In hot smoking the internal temperature normally exceeds 60°C, while in cold smoking it rarely exceeds 35°C. The initial brining process brings about a change in the original microflora. Generally, smoking shifts the balance of microflora from Gram-negative to Gram-positive. Coryneform bacteria, micrococci and *Bacillus* are the dominant forms present. The internal temperature reached by the fish during smoking controls the type of microflora dominating. As there is a risk of botulism due to the growth of *Clostridium botulinum* type E in smoked fish, the US Food and Drug Authority promulgated regulations for processes which would exclude any danger of botulism from this source.

Hard smoked and essentially dried fish products are normally spoiled by moulds.

Fermentation   There are many different fermented food products in Southeast Asia, notably fish sauce which is a digested fish product of brine and enzyme. Mixed fermentations of fish or shellfish and vegetable or cereals are used to prepare products such as *i-sushi*. Here lactic acid bacteria and moulds are involved. Lactic acid bacteria produce antibacterial substances which stabilize the product. Mould-fermented seafood products are popular in Japan; for example, *kojizuki* is prepared by adding *koji* to salted fish. Koji is prepared by growing *Aspergillus oryzae* on steamed rice. The mould provides proteinases and other enzymes required for flavour and texture change. Another Japanese mould product is *katsuobushi*, which is used as soup stock. The moulds grown on the partly dried product are *Aspergillus* and *Penicillium*. Their growth assists drying, reduces fat content and improves flavour. Products of fish ensilage in Europe mainly involve a lactic fermentation. Lactic starter cultures such as *Lactobacillus plantarum*, *Pediococcus* and others are added to the mixture of fish and carbohydrate source (cereal, cassava or molasses) and controlled digestion is allowed to proceed. This microbial process is very effective in utilizing waste fish for the production of a high-quality animal food.

Irradiation The most noticeable effect of radiation at the lower pasteurizing level (10–20 cGy) is to bring about a shift in the apparent spoilage flora from *Pseudomonas* to *Achromobacter* and Gram-positive bacteria and yeasts. At a slightly higher dose level (> 30 cGy), yeasts and Gram-positive bacteria dominate.

See also: **Clostridium**: *Clostridium botulinum*. **Dried Foods**. **Fermented Foods**: Fermented Fish Products. **Fish**: Spoilage of Fish. ***Flavobacterium***. **Freezing of Foods**: Growth and Survival of Microorganisms. **Heat Treatment of Foods**: Principles of Pasteurization. **Preservatives**: Traditional Preservatives – Sodium Chloride; Wood Smoke. ***Pseudomonas***: Introduction. ***Salmonella***: Introduction. **Spoilage Problems**: Problems caused by Bacteria. ***Vibrio***: Introduction, including *Vibrio vulnificus*, and *Vibrio parahaemolyticus*.

Further Reading

Aitken A, Mackie IM, Merrit JH and Windsor ML (1982) *Fish Handling and Processing*, 2nd edn. London: HMSO.

Borgstrom G (ed.) (1961–5) *Fish as Food*. Vols. I–IV. New York: Academic Press.

Burgess GHO, Cutting CL, Lovern JA and Waterman JJ (eds) (1965) *Fish Handling and Processing*. London: HMSO.

Connell JJ (ed.) (1980a) *Advances in Fish Science and Technology*. Farnham: Fishing News Books.

Connell JJ (1980b) *Control of Fish Quality*, 2nd edn. Farnham: Fishing News Books.

Kreuzer R (ed.) (1971) *Fish Inspection and Quality Control*. Farnham: Fishing News Books.

Love RM (1970) *The Chemical Biology of Fishes*. London: Academic Press.

Regenstein JM and Regenstein CE (1991) *Introduction to Fish Technology*. New York: Van Nostrand Reinhold.

Wheaton FW and Lawson TB (1985) *Processing Aquatic Food Products*. New York: John Wiley.

Spoilage of Fish

J J Leisner, Department of Veterinary Microbiology, Royal Veterinary and Agricultural University, Frederiksberg C, Denmark

L Gram, Danish Institute for Fisheries Research, Department of Seafood Research, Danish Technical University, Lyngby, Denmark

Introduction

The high degree of perishability of fish has limited its consumption in a fresh state to areas close to its capture. Traditional curing techniques based on combinations of salting, drying and smoking as well as more recent improvements in food technology have extended the shelf life of fish and fish products, so that seafoods may play an important role in nutrition for a wider range of human populations. The great diversity of fish products (**Table 1**) combined with a great variation in raw material and processing parameters used throughout the industry establish a wide range of products with different spoilage problems.

Spoilage is defined here as microbial or chemical changes that cause sensory changes to a degree that the food becomes unacceptable to the consumer. Toxin formation and toxicity due to pathogenic microorganisms, although unacceptable, is not dealt with.

Role of Autolysis, Lipid Oxidation and Microorganisms

Several different processes may produce undesirable sensory changes in fish products; these may be:

- autolytic (enzymatic) processes
- chemical processes
- microbiological processes.

These processes may result in visual changes (e.g.

Table 1 Types of fish products

Type of product	Important spoilage organisms
Chilled, stored aerobically	*Shewanella putrefaciens*, *Pseudomonas* spp.
Chilled, vacuum-packed	*Photobacterium phosphoreum*, *S. putrefaciens*, LAB
Chilled, MAP	*P. phosphoreum*, LAB
Lightly preserved fish products	LAB, Enterobacteriaceae, *P. phosphoreum*
Highly salted	Halophilic bacteria, moulds
Fermented	Moulds, LAB
Heat treated	Gram-positive spore-formers

LAB, lactic acid bacteria; MAP, modified-atmosphere packed.

discoloration, slime formation), changes in texture, gas formation, or, most commonly, unpleasant changes in odour and flavour of the product.

Autolytic Changes

It is generally accepted that the early decrease in the sensory quality of some fish products, notably chill-stored fresh fish, is caused by autolytic processes involving enzymes indigenous to the fish muscle. The autolytic processes post mortem result not only in rigor mortis but also in a well-defined degradation of nucleotides. Adenosine triphosphate (ATP) is catabolized by a series of dephosphorylation and deamination reactions to inosine monophosphate (IMP) which may be further degraded to hypoxanthine and ribose (**Fig. 1**). The initial degradative changes typically take place within days after catch and do not normally result in off odours or flavours. However, particularly in Japan, the nucleotide breakdown is used as a quality index. As fish vary in the rate of the different steps, the 'K value' (giving the ratio between hypoxanthine plus inosine and the total amount of ATP-related compounds) has been introduced as quality index. The degradation of nucleotides, particularly the formation of hypoxanthine, can also be caused by bacteria that degrade IMP. Thus the production of hypoxanthine in fresh, lean fish is primarily caused by growth of Gram-negative spoilage bacteria.

Many fish, mostly marine species, are rich in a nitrogenous compound, trimethylamine oxide (TMAO), which, during storage of fish products at above freezing temperatures, plays an important role in the bacterial spoilage process (see below). This compound may be autolytically degraded to dimethylamine (DMA) and formaldehyde, and this process is involved in the sensory changes occurring during frozen storage of fish where 'soapy' or 'cardboard-like' off odours and flavours develop.

In whole, uneviscerated fish, enzymes from the digestive tract may play an important role in tissue degradation and may result in bursting of the belly. This is typical of fatty pelagic fish like herring and mackerel and occurs in periods of high feeding activity.

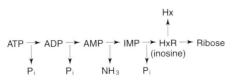

Figure 1 Degradation of ATP. IMP, inosine monophosphate; Hx, hypoxanthine.

Lipid Oxidation and Hydrolysis

Fish may from a technological point be grouped as either lean (e.g. cod, plaice) or fatty (e.g. herring, salmon). Fatty fish species accumulate lipid in the muscle and the lipid fraction is rich in unsaturated fatty acids, particularly the n-3 polyunsaturated fatty acids. If left exposed to air, these fish may develop serious quality defects due to changes of the lipid fraction. The most important changes taking place in the lipid fraction are rancidity caused by non-microbial processes, either due to autoxidation, which is a chemical reaction involving oxygen and unsaturated lipid, or due to autolytic hydrolytic activity. Both processes cause the development of off flavours and off odours characterized as rancid. For this reason, most fatty fish which have been processed to give an extended shelf life (e.g. by smoking) are distributed and sold as vacuum-packed products.

Microorganisms: the Specific Spoilage Organism Concept

Although autolytic and chemical processes may cause sensory changes leading to spoilage it is well established that microbial growth and activity are the main reason for the development of off odours and flavours rendering nonfrozen fish products unacceptable or spoiled.

Microorganisms on Newly Caught Fish The muscle and internal organs of healthy, freshly caught fish are usually sterile but the outer and inner surfaces of the live fish (skin, gills and alimentary tract) all carry substantial numbers of bacteria. Reported numbers on the skin have ranged from 10^2 colony forming units (cfu) per square centimetre to 10^7 cfu cm^{-2}, and from 10^3 cfu g^{-1} to 10^9 cfu g^{-1} in the gills and the gut. The microflora of the gut is highly variable from fish to fish, and some fish contain a high proportion (up to 90%) of unculturable bacteria that may be visualized by DNA or RNA staining.

Fish are poikilothermic (cold-blooded) animals and their microflora is therefore a reflection of the environment in which the fish is caught. Thus, the microflora on temperate-water fish is dominated by psychrotrophic Gram-negative bacteria of the genera *Pseudomonas*, *Psychrobacter*, *Acinetobacter*, *Shewanella*, *Flavobacterium* and the families *Vibrionaceae* and *Aeromonadaceae*, but some Gram-positive bacteria such as *Bacillus*, *Micrococcus*, *Clostridium*, *Corynebacterium* and lactic acid bacteria can also be found in varying proportions. The microflora on fish from tropical waters is composed of similar types of organisms although the proportion of Gram-positive bacteria and Enterobacteriaceae tends to be slightly higher. Fish may be caught in fresh or marine waters and this also influences the composition of the microflora. Vibrionaceae (*Vibrio* spp. and *Photobacterium* spp.) and *Shewanella* are typical of the marine environment whereas *Aeromonas* spp. are freshwater bacteria.

The Spoilage Microflora: Specific Spoilage Organisms   After catch, the microflora will change as a consequence of the contamination and, most important, the preservation processes used during production. The physical (e.g. atmosphere, temperature) and chemical (e.g. preservatives) conditions as well as interactions between the microorganisms will cause a selection of organisms capable of growing under the defined conditions. In general, the more severe the conditions, the fewer species will be able to grow. Thus only few halophilic bacteria will grow in highly salted products with 20–30% NaCl in the water phase, whereas a multitude of bacterial species may grow in lightly preserved fish products such as cold-smoked products. During spoiling a microflora typical of the product develops, termed the 'spoilage microflora', or the 'spoilage association'. Despite the large variation in initial microflora of the fresh fish and the many different parameters used for preservation, a remarkable consistency exists in terms of species growing in the different products. Of the different microbial species developing, only one or a few will be responsible for the production of the off odours and off flavours characterizing the spoilage. These species are called the specific spoilage organisms.

Fish as Substrate for Microbial Growth and Metabolism

Fish and fish products are, in general, excellent substrates for microbial growth. Like other food raw materials, unprocessed fish contains large quantities of water and is rich in non-protein nitrogen (NPN) such as free amino acids. Trimethylamine oxide is a part of the NPN fraction and is accumulated in many marine fish species and also in some freshwater species. Fish accumulate virtually no carbohydrate in the muscle (typically less than 0.5%) and very little lactic acid is produced post mortem. As the buffering capacity of fish muscle is high the pH rarely falls below 6.0, allowing many acid-sensitive bacteria to grow.

The major reservoir of substrates for bacterial metabolic activities important for spoilage is the water-soluble fraction, including TMAO, sulphur-containing amino acids and nucleotides (e.g. IMP and inosine) (**Table 2**). From these substrates a range of volatile compounds of importance for spoilage are produced, including TMA, sulphides, ammonia, ketones and aldehydes.

Reduction of TMAO to Trimethylamine

Trimethylamine oxide does not serve as a substrate for bacterial catabolism but is instead important as an alternative electron acceptor enabling some bacteria to exhibit rapid growth under anaerobic conditions. The product of this reaction is trimethylamine (TMA) which is an important component of the odour of stored fish, giving the typical fishy smell. Substrates (electron donors) for respiration include lactate and several of the free amino acids. The presence of TMAO contributes to a relatively high redox potential in the flesh since the E_h of the TMAO/TMA couple is +19 mV. The presence of TMAO has been suggested as an extra hurdle against anaerobic bacteria in salted fish.

Many Gram-negative bacteria growing in fish and fish products, e.g. *Shewanella putrefaciens*, *Photobacterium phosphoreum*, *Aeromonas* spp. and *Enterobacteriaceae*, are able to use TMAO as an electron acceptor.

Degradation of Amino Acids

Many bacteria causing spoilage produce one or several volatile sulphides. Very unpleasant putrid odours are caused by the production of H_2S from the sulphur-containing amino acid L-cysteine by *Shewanella putrefaciens* and some Vibrionaceae and by the production of methylmercaptan (CH_3SH) and dimethyl sulphide ($(CH_3)_2S$) from methionine by *S. putrefaciens*. *Pseudomonas* spp. are not typical H_2S producers but produce some of the other volatile sulphur compounds. A number of lactic acid bacteria isolated as part of the spoilage association of lightly preserved fish are capable of producing H_2S. Taurine, which is also sulphur-containing, occurs as free amino acid in very high concentrations in fish muscle and disappears from the fish flesh during storage, but this is because of leakage rather than bacterial attack. *Pseudomonas* spp. produce, apart from the sulphides, a number of volatile aldehydes, ketones and ethyl esters. Production of ethyl esters may be responsible for sweet, fruity odours and are typical products of the breakdown of amino acids.

Breakdown of Proteins

Although several of the spoilage bacteria important in fish and fish products exhibit extensive proteolytic potential (e.g. *Shewanella putrefaciens*, *Pseudomonas* spp.) it appears that turnover of the protein fraction is not of major importance in spoilage of fresh fish.

Breakdown of Nucleotides

Hypoxanthine, which may cause a bitter off flavour in fish, can be formed by degradation of nucleotides.

As described above, this process can be autolytic but microbial activity is involved to a larger extent. Several spoilage bacteria produce hypoxanthine from inosine or IMP, including *Pseudomonas* spp., *Shewanella putrefaciens* and *Photobacterium phosphoreum*.

Breakdown of Carbohydrate

The carbohydrate content of fresh fish is very low and accumulation of metabolic products from this substrate will only be of significance in fish products to which carbohydrates are added, e.g. sugar-salted fish. The most important product from carbohydrate catabolism is CO_2 which can be produced by the mixed acid pathway by Enterobacteriaceae, or by the phosphogluconate pathway by lactic acid bacteria. This CO_2 may result in swelling or 'blowing' of products. Polymers produced from carbohydrates by lactic acid bacteria may generate spoilage through slime formation. In addition, production of acid by the catabolism of carbohydrates by lactic acid bacteria may result in spoilage due to souring, as has been reported for sugar-salted fish containing high numbers of lactic acid bacteria.

Breakdown of Lipids

It is known that *Pseudomonas* spp. (for example) may produce lipases that are able to break down milk fat and thereby release fatty acids. The significance of this finding for fish products is, however, not known.

Spoilage of Different Types of Fish Products

Various methods of preservation with effects on water activity (a_w), pH, temperature or atmosphere may have a great effect on the microbial flora of fish and the corresponding spoilage pattern. Thus, Gram-positive bacteria will in general be more resistant to freezing and thawing, decreases in a_w and pH, and to conditions with low oxygen tension and no alternative electron acceptor present (e.g. TMAO). The spoilage patterns of such products may differ significantly from that observed for fresh, chilled fish.

Frozen Fish

Gram-negative bacteria are in general more sensitive to the effects of freezing and thawing than Gram-positive bacteria. However, at below freezing temperatures, bacteria play no role in the spoilage process. Instead, autolytic changes involving DMA and formaldehyde production are important in the production of typical spoilage off odours and off flavours.

Fresh Fish

Generally, the quality deterioration of fresh fish is characterized by an initial loss of fresh fish flavour which is species-specific but may in general be described as 'sweet' and 'seaweedy'. After a period where the odour and flavour are described as neutral or nonspecific, the first indications of off odours and flavours are detectable. These will progressively become more pronounced and eventually the fish is spoiled or putrid. The time to spoilage depends mainly on storage temperature and fish species. The initial quality loss in fish is caused by autolytic changes and is unrelated to microbiological activity. Of particular importance in this respect is the degradation of nucleotides (ATP-related compounds) as described above. The off odours and flavours developing depend on the fish species, the origin of the fish and the atmospheric conditions of storage. The spoilage of marine temperate-water fish stored aerobically is characterized by development of offensive fishy, rotten, H_2S off odours and flavours. This sensory impression is distinctly different for some tropical fish and freshwater fish, where fruity, sulphydryl off odours and flavours are more typical. The predominating bacteria of fish caught or harvested in temperate as well as in subtropical or tropical waters are, under aerobic iced storage, *Pseudomonas* spp. and *Shewanella putrefaciens* (**Table 3**). At ambient temperature (25°C), the microflora is dominated by mesophilic Vibrionaceae and, particularly if the fish are caught in polluted waters, mesophilic Enterobacteriaceae.

Shewanella putrefaciens is the specific spoilage bacteria of marine temperate-water fish stored aerobically in ice (**Fig. 2**), whereas *Pseudomonas* spp. are the specific spoilers of ice-stored tropical freshwater fish and are also, together with *S. putrefaciens*, spoilers of marine tropical fish stored on ice. At ambient temperature, motile aeromonads are the specific spoilers of aerobically stored freshwater fish.

In vacuum-packed ice-stored fish from temperate marine waters the specific spoilage organisms are *S. putrefaciens* or *Photobacterium phosphoreum*. The latter organism produces TMA in the same levels per gram of cell material as *S. putrefaciens* but does not cause such foul off odours, probably because it does not produce volatile sulphides. Differences in initial numbers of *S. putrefaciens* and *P. phosphoreum* probably determine which of the two becomes the most important spoilage organism of vacuum-packed fish from temperate marine waters. It is unlikely that *P. phosphoreum* plays a major role in the spoilage of freshwater fish as it requires NaCl. It has been reported that Gram-positive bacteria (lactic acid

Table 2 Substrate used and typical spoilage compounds produced by bacteria during storage of fresh and packed fish. From Gram and Huss (1999)

Substrate:	TMAO	Cysteine	Methionine	Other amino acids	IMP, inosine	Carbohydrates, lactate	Product examples
				Production (+) of spoilage compounds			
Compounds:	TMA	H_2S	CH_3SH, $(CH_3)_2S$	Ketones, esters, aldehydes, NH_3	Hypoxanthine	Acids	
Spoilage bacteria							
Shewanella putrefaciens	+	+	+	?	+	+	Iced marine fish
Pseudomonas sp.	–	–	+	+	+	?	Iced freshwater fish
Photobacterium phosphoreum	+	–	–	?	+	?	CO_2-packed fish
Vibrionaceae	+	+	?	?	?	?	Ambient stored fresh fish
Enterobacteriaceae	+	(+)	?	+	?	+	Lightly preserved fish
Lactic acid bacteria	–	(+)	?	+	?	+	Lightly preserved fish
Yeast	–	–	–	+	?	+	Sugar-salted fish
Anaerobic rods	–	–	?	+	?	?	Sous-vide fish

Table 3 Specific spoilage bacteria of fresh and packed fish stored chilled (< 4°C) or in ice. From Gram and Huss (1999)

Storage conditions	Specific spoilage organisms of fresh, chilled fish depending on source of fish			
	Temperate waters		Tropical waters	
	Marine	Fresh	Marine	Fresh
Aerobic	Shewanella putrefaciens	Pseudomonas sp.[a]	S. putrefaciens Pseudomonas sp.	Pseudomonas sp.
Anaerobic	S. putrefaciens Photobacterium phosphoreum	Gram-positive bacteria Lactic acid bacteria	Lactic acid bacteria Others?	Lactic acid bacteria?
CO_2 (20–70%)	P. phosphoreum	Lactic acid bacteria	Lactic acid bacteria TMAO-reducing bacteria	Lactic acid bacteria? TMAO-reducing bacteria

[a] Assumed to be the most likely spoilage bacteria as typical marine bacteria are not present.

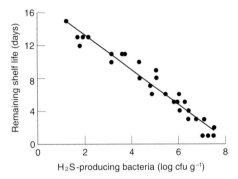

Figure 2 Correlation between remaining shelf life of iced cod and numbers of H₂S-producing bacteria (*Shewanella putrefaciens*). From Gram and Huss (1999).

bacteria) predominate in vacuum-packed trout after 4 weeks of storage on ice.

Carbon dioxide packing of marine fish from temperate waters inhibits the development of the respiratory organisms such as *Pseudomonas* and *S. putrefaciens* and their numbers rarely exceed 10^5–10^6 cfu g^{-1}. The availability of TMAO as an alternative electron acceptor mean that the dramatic extensions of shelf life seen with meat are not found for fish packed in a CO_2-containing atmosphere. This is because the development of TMA is similar to or delayed by only a few days compared with vacuum-packed storage, but is increased compared with aerobic storage. The responsible spoilage organism is the CO_2-tolerant *Photobacterium phosphoreum* (see Table 3) which can grow to levels of 10^7–10^8 cfu g^{-1} in CO_2-packed fresh fish products.

Carbon dioxide and vacuum packing of fish caught in freshwater or warmer waters where the heat-sensitive, NaCl-requiring *P. phosphoreum* is probably not as common, result in diminished TMA production. The microflora becomes dominated by various Gram-positive organisms, mainly lactic acid bacteria. However, as TMA can be detected later in the storage, TMAO-reducing organisms must be present at some level.

A special problem is encountered with elasmobranch fishes such as dogfish and sharks. These species contain high levels of urea, and bacterial urease activity in the product may generate ammonia, causing a pungent odour.

Lightly Preserved Fish Products

Lightly preserved products include fish preserved by low levels of salt (< 6% NaCl in the water phase), pH above 5, storage at chill temperatures (< 5°C) and, for some products, addition of preservatives (sorbate, benzoate, NO₂ or smoke). This is a group of high-value delicatessen products – cold-smoked, sugar-salted ('gravad') or marinated fish – that are typically consumed as ready-to-eat products with no heat treatment.

The spoilage of these products is complex and not well understood. Studies on cold-smoked salmon have shown that bacterial activity is the cause of spoilage (defined as unpleasant off odours and off flavours) although autolytic enzymes caused texture changes (softening). The off odours and flavours developing are variously described as putrid, cabbage-like, sour, bitter, fruity or sweet, and the normal shelf life of this product also varied considerably from about 3 weeks to 8 weeks for vacuum-packed, cold-smoked salmon (4–5% NaCl in water phase, pH 6.3–6.4) stored at 5°C.

The microflora developing in this type of product is dominated by lactic acid bacteria (LAB) which are often present at high levels (10^7–10^8 cfu g^{-1}) for several weeks before the products become spoiled (**Fig. 3**). Species of LAB frequently isolated include *Carnobacterium* spp., *Leuconostoc* spp., *Lactobacillus plantarum*, *Lactobacillus sake* and *Lactobacillus curvatus*. It has been observed that various isolates of LAB are able to produce some of the off odours ('sour', 'cabbage-like', 'sulphurous') associated with spoilage of cold-smoked salmon. It has also been demonstrated that several isolates of LAB from

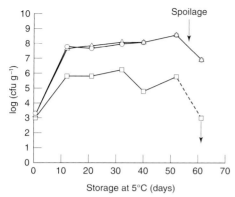

Figure 3 Changes in bacterial counts of aerobic, psychrotrophic bacteria (circles), lactic acid bacteria (triangles) and Enterobacteriaceae (squares) during storage of vacuum-packed, cold-smoked salmon (NaCl 4.6% w/w in water phase) at 5°C. From Gram and Huss (1999).

pickled fish, including isolates of *Carnobacterium* spp., were able to produce H_2S from cysteine, and in another study it was found that a strain of *L. sake* produced H_2S during growth on cold-smoked salmon. However, the direct association between LAB and the spoilage and shelf life of these products has not been established.

Besides LAB, the microflora of lightly preserved fish products may also contain psychrotrophic members of the Enterobacteriaceae (*Serratia liquefaciens*, *Enterobacter agglomerans*, etc.). Depending on the processes involved, *Photobacterium phosphoreum* may sometimes grow in these products. Lactic acid bacteria are not capable of reducing TMAO, and as TMA in some storage trials may be detected in (for example) cold-smoked salmon, Enterobacteriaceae and/or *P. phosphoreum* may be causing this change. In addition *Brochothrix thermosphacta*, which is a known spoilage bacteria of some meat products, may sporadically be isolated from lightly preserved fish products.

Salt-cured Products

Two basic techniques may be used in the salting of fish, resulting in dry-salted and wet-salted (or pickled) fish products. Only nonfatty fish are used for dry salting. There are two types of spoilage of this product. One is growth of extremely halophilic bacteria which cause a red discoloration on the surface of the fish ('pink fish'). This group of bacteria, which includes such genera as *Halobacterium* and *Halococcus*, may cause proteolytic spoilage. The other type of spoilage, known as 'dun', is caused by highly osmophilic moulds (*Sporendonema* and *Oospora*).

Wet salting or barrel salting is used for fatty fish species such as herring and anchovy. The fish are mixed with salt and kept in a closed container. The water phase salt in barrel-salted herring is typically

15–20%. Three types of spoilage are known for this product. The most common type is characterized by the presence of sour, sour/sweet and putrid off odours and flavours. This type of spoilage is caused by growth of a Gram-negative, halophilic, obligate anaerobic rod (up to 10^6–10^7 cfu g^{-1}). The growth of this organism is not possible until the general microflora has reduced all the TMAO, causing the E_h to drop to negative values. This may take more than 1 year at chill storage (2–4°C) as the general flora may consist of low levels (10^3–10^5 cfu g^{-1}) of mainly Gram-negative halophilic rods able to reduce TMAO. The second type of spoilage is characterized by the development of fruity off odours and is caused by growth to levels of 10^5 cfu g^{-1} of osmotolerant yeast species. Finally, the term 'ropiness' or 'ropy brine' is used to describe the type of spoilage in which the brine becomes highly viscous or slimy; this type is caused by a Gram-negative, halophilic, aerobic non-motile rod-shaped (*Moraxella*-like) bacterium.

Fermented Products

Fermented products may contain low concentrations of salt (maximum 8–10% NaCl) and carbohydrates. The fermentation process is often a lactic acid fermentation. Rice has to be added as source of fermentable carbohydrate although only a few LAB are amylolytic. Fermentation is therefore likely to depend on other carbohydrate sources such as molasses or garlic. Knowledge of the spoilage processes of this type of products is very limited, but excessive lactic acid souring (by LAB) and mould growth have been identified as causes of spoilage.

High-salt 'fermented' fish products such as fish sauce and paste are really autolysed products and the spoilage pattern of these products is not described here.

Heat-treated Products

Many types of seafood products receive a heat treatment as part of their processing. Products receiving only a mild treatment and distributed at chill temperatures (refrigerated processed foods with extended durability, REPFED) are particularly likely to spoil because of microbial action. Not surprisingly, it has been reported that Gram-positive spore-formers may spoil these products; for example, sous-vide-packed cod stored at 5°C was reported to have spoiled owing to growth of a Gram-positive spore-forming bacteria producing extremely obnoxious and putrid off odours.

Hot-smoked fish (receiving treatment temperatures of approximately 65°C) will, if packed aerobically, spoil owing to growth of moulds and yeast. Depending on the water activity, pseudomonads also may grow. If the fish is vacuum-packed, little change is seen in the microflora and the count remains low at approximately 10^3 cfu g^{-1} for weeks.

Gram-positive spore-formers may also be the causative spoilage organisms of canned seafoods, especially in low acid (pH > 4.5) foods that have not received adequate heat treatment. Spore-forming anaerobic *Clostridium* spp. produce gas from either carbohydrates or amino acids during growth, causing cans to swell. Anaerobic *Bacillus* spp. may break down carbohydrates in order to produce acid but not gas, thereby giving rise to the type of spoilage designated 'flat sour', which describes the characteristics of the can as well as the food.

See also: **Fermented Foods**: Fermented Fish Products. ***Pseudomonas***: Introduction. **Shewenella**.

Further Reading

Dalgaard P, Gram L and Huss HH (1993) Spoilage and shelf-life of cod fillets packed in vacuum or modified atmospheres. *International Journal of Food Microbiology* 19: 283–294.

Gram L (1992) Evaluation of the bacteriological quality of seafood. *International Journal of Food Microbiology* 16: 25–39.

Gram L and Huss HH (1996) Microbial spoilage of fish and fish products. *International Journal of Food Microbiology* 33: 121–137.

Gram L and Huss HH (1999) Microbiology of fish and fish products. In: Lund B, Baird-Parker AC and Gould CW (eds) *Microbiology of Foods*. Ch. 23. London: Chapman & Hall (in press).

Herbert RA, Hendrie MS, Gibson DM and Shewan JM (1971) Bacteria active in the spoilage of certain seafoods. *Journal of Applied Bacteriology* 34: 31–50.

Hobbs G and Hodgkiss W (1982) The bacteriology of fish handling and processing. In: Davies R (ed.) *Developments in Food Microbiology*. Vol. 1, p. 71. London: Applied Science.

Huss HH (1995) *Fresh Fish – Quality and Quality Changes*. FAO Fisheries Technological Papers 348. Rome: FAO.

Liston J (1980) Fish and shellfish and their products. In: Silliker JH, Eliott RP, Baird-Parker AC et al (eds) *Microbial Ecology of Foods*. Vol. II, *Food Commodities*. P. 567. New York: Academic Press.

FLAVOBACTERIUM

María-Luisa García-López, **Jesús-Angel Santos** and **Andrés Otero**, Department of Food Hygiene and Food Technology, University of León, Spain

Introduction

Numerous reports have illustrated the importance of yellow-pigmented Gram-negative bacteria, termed flavobacteria, in foods, water, aquatic environments and clinical specimens. However, recent taxonomic studies based on molecular techniques have made it clear that most of these 'Flavobacterium' strains should have been assigned to different (or even new) genera such as *Bergeyella*, *Cytophaga*, *Empedobacter*, *Sphingobacterium*, *Weeksella*, *Chryseobacterium* or *Myroides*. Misidentification of 'flavobacteria' from foods and other sources is not surprising because taxonomic heterogeneity and general uncertainty have characterized the genus *Flavobacterium* from its inception. At the present time, it is extremely difficult to determine whether 'authentic' flavobacteria are of significance in foods. Nevertheless as misnamed 'Flavobacterium' species are widely recognized as food contaminants, an exhaustive review of the published literature has been carried out with the purpose of providing information on the current taxonomic status of this genus and the proposed methods for isolation and identification. This article also covers the prevalence and spoilage activity of misnamed flavobacteria in foods, their applications and pathogenicity.

Flavobacterium and Related Genera

The genus *Flavobacterium* (*flavus*, yellow; *bakterion*, small rod) was first described in 1923 to include bacteria forming yellow- or orange-pigmented colonies on culture media. Because of this poor definition, numerous species, not always genetically related, were assigned to the genus, leading to a great taxonomic heterogeneity. In *Bergey's Manual of Systematic Bacteriology*, it was restricted to Gram-negative, non-spore-forming, aerobic, typically pigmented, non-motile and non-gliding rods. *Flavobacterium aquatile* was retained as the type species, although it is represented by a single strain, which is not the strain that was described originally.

Flavobacteria are closely related to the genera *Cytophaga* and *Flexibacter* and in *Bergey's Manual of Systematic Bacteriology* they all were placed in the Order Cytophagales, with two different families: Cytophagaceae (including the genera *Cytophaga*, *Capnocytophaga*, *Flexithrix* and *Sporocytophaga* with *Flexibacter*, *Microscilla* and *Chitinocyto-*

phaga being considered as related genera) and Flavobacteriaceae (the only genus being *Flavobacterium*). The families Bacteriodaceae and Spirosomaceae, though not included in the order, are closely related.

Further approaches, based on nucleic acid homologies, have shown that the genus *Flavobacterium* belongs to one of the 10 phylogenetic branches of Eubacteria, called the *Cytophaga–Flavobacterium–Bacteroides* group (or rRNA superfamily V), which has two subdivisions: the *Bacteroides* and *Flavobacterium* groups. Recent studies clarified the structure of the second subdivision, which includes three well-defined clusters: (I) the *Sphingobacterium* cluster (containing sphingophospholipids), (II) the *Cytophaga/Flexibacter* cluster, and (III) the family Flavobacteriaceae.

The members of the family Flavobacteriaceae are Gram-negative rods, are non-flagellated, non-motile or motile by gliding and non-spore-forming. Colonies are non-pigmented or pigmented by carotenoid and/or flexirubin types of pigments. Growth is aerobic or microaerobic to anaerobic, chemoorganotrophic, and menaquinone 6 is the only respiratory quinone. Cellulose is not decomposed. The G+C content is in the range 32–37 mol%. Sphingophospholipids are absent. The type genus is *Flavobacterium* and other genera included are *Bergeyella*, *Capnocytophaga*, *Myroides*, *Ornithobacterium*, *Riemerella* and *Weeksella* (all of which are able to grow at 37°C), as well as *Chryseobacterium*, *Empedobacter* and the taxa belonging to the [*Flexibacter*] *maritimus* rRNA branch.

The emended genus *Flavobacterium* is now limited to the type species, *F. aquatile*, and related species, the main characteristics being the yellow pigmentation of colonies (carotenoids and/or flexirubins), the gliding motility and aerobic growth (*F. hydatis* and *F. succinicans* can grow anaerobically under certain conditions).

The differentiation of the genus *Flavobacterium* from some related genera is summarized in **Table 1**.

Flavobacterium Species

The emended description of the genus *Flavobacterium* is based on the single type strain of *F. aquatile*. Thus, several species previously considered as belonging to the genus, have been reclassified and placed in different or new genera, and new species have been added to *Flavobacterium*. At this time, the genus *Flavobacterium* contains 10 species: *F. aquatile*, *F. branchiophilum*, *F. columnare*, *F. flevense*, *F. hydatis*, *F. johnsoniae*, *F. pectinovorum*, *F. psychrophilum*, *F. saccharophilum* and *F. succinicans*. Only *F. aquatile* and *F. branchiophilum* retain their previous taxonomic position, whereas all other species were relocated from their former genera (*Cytophaga* and *Flexibacter*).

Methods of Detection and Enumeration in Foods

Flavobacteria are not difficult to isolate and do not require enrichment procedures. Standard nutrient media allow growth of flavobacteria and can be used in general studies. Flavobacteria have been isolated from different foods using tryptone soya agar (TSA) and plate count agar (PCA) incubated either at 30°C/48 h or 7°C/10 days. A similar type of medium has been proposed for the isolation of yellow-pigmented bacteria from foods (w/v): beef extract (1%), peptone (1%), NaCl (0.5%) and agar (1.5%), with incubation at a temperature similar to the environment (i.e. 20–25°C) for about 4 days. Standard media such as TSA or PCA slants can be used to maintain the strains, but it is said that media with a low nutrient concentration (such as PMYA II (w/v): peptonized milk, 0.1%; yeast extract, 0.02%; sodium acetate, 0.002%; agar, 1.5%) are better suited for the maintenance of flavobacteria and also for the observation of gliding motility and colony swarming (spreading growth).

Several meat spoilage bacteria (including flavobacteria) can be detected with a polymerase chain reaction (PCR) assay (and gel electrophoresis detection of PCR products) in which universal primers

Table 1 Main phenotypic characteristics for the differentiation of *Flavobacterium* and related genera

	Flavobacterium	Chryseobacterium	Cytophaga	Empedobacter	Flexibacter	Myroides
Pigmented colonies	+	+	+	+	+	+
Gliding motility	+[a]	−	+	−	+	−
Catalase production	+	+	+	+	−	+
Degradation of cellulose	−	−	+	−	−	−
Growth at 37°C	−	+[b]	−	+[b]	−	+
Acid production from glucose	v	+[b]	+	+[b]	+	−

Symbols: +, positive reaction; −, negative reaction; v, variable within and between species.
[a]Gliding motility was observed in *F. aquatile* under certain conditions and not observed in *F. branchiophilum*.
[b]With some exceptions.

Table 2 Differentiation of species of *Flavobacterium* through phenotypic features

	Gliding motility	Flexirubin pigments	Degradation of gelatin	Degradation of casein	Degradation of chitin	β-Galactosidase activity	Sensitivity to O/129[a]	Oxidase activity
F. aquatile	+	–	+	+	–	v	–	+
F. branchiophilum	–	–	+	+	–	+	+	+
F. columnare	+	+	+	+	–	–	+	+
F. flevense	+	–	–	–	–	+	+	+
F. hydatis	+	+	+	+	(+)	+	–	v
F. johnsoniae	+	+	+	+	+	+	–	+
F. pectinovorum	+	+	+	+	+	+	+	+
F. psychrophilum	(+)	+	+	+	–	–	+	v
F. saccharophilum	+	+	+	+	–	+	+	–
F. succinicans	+	–	+	+	–	+	+	+

Symbols: +, positive; –, negative; (+), weakly positive; v, variable among strains.
[a]O/129, 2,4-diamino-6,7-diisopropylpteridine phosphate.

based on conserved 23S rDNA sequence of *Pseudomonas aeruginosa* are used.

Characterization and Identification

Phenotypic Characteristics

The *Flavobacterium* species are Gram-negative, aerobic rods. They produce acid aerobically from carbohydrates, with the exception of *F. columnare* and *F. psychrophilum*. Most species are able to grow in rich media (TSA), producing circular colonies that are convex with entire or wavy edges. *F. flevense* and *F. saccharophilum* are agarolytic, and the colonies are sunken into the agar medium. The optimum temperature for most species is 20–30°C (15–18°C for *F. psychrophilum*).

A large number of phenotypic tests useful in differentiating the 10 species included in the emended genus *Flavobacterium* have been developed. Some of the most discriminant tests are shown in **Table 2**.

Tests included in Table 2 can be performed by conventional methods. The observation of gliding motility can be done by direct microscopic observation (taking the cells from young cultures and from the edge of the colony) or by observation of spread of the colonies, an important criterion being the use of low nutrient concentration media. The production of flexirubin type pigments can be revealed by a reversible colour shift that occurs when the plate is flooded with 20% (w/v) KOH.

Rapid commercial systems, such as API 20NE, Vitek GNI, Biolog GN or BBL Cristal E/NF, include some former flavobacterial species in their databases, and can be used to carry out some differentiation tests. A table of API ZYM profiles for the 10 recognized species of *Flavobacterium* is found in Bernardet et al (1996).

There are several studies using a numerical taxonomy approach to classify strains of *Flavobacterium* isolated from dairy sources but the attribution of

the strains is not accurate in relation to the current taxonomy of *Flavobacterium* (the results of this work are discussed in the milk and dairy products section).

Chemotaxonomy

The predominant fatty acid components of the valid species of *Flavobacterium* are 15 : 0, 15 : 0 iso, 15 : 1 iso G (unknown position of the double bond), 15 : 0 iso 3OH, 16 : 0 iso 3OH, 17 : 0 iso 3OH and a summed featured of 15 : 0 iso 2OH, 16 : 1 ω 7c and 16 : 1 ω 7t. There are some differences in the fatty acid profiles of the species that can be of value for differentiating several species (e.g. 15 : 0 anteiso is almost absent in *F. hydatis* and only traces of 17 : 1 iso ω 9c are detected in *F. branchiophilum*).

The analysis of whole-cell protein profiles combined with a computer-assisted comparison of the patterns has been used for differentiating *Flavobacterium* species, but in some cases (*F. succinicans* and *F. branchiophilum*) the protein profiles are different within the strains tested.

Genetic Taxonomy

The current classification of flavobacteria is mainly based on data of DNA–DNA hybridization, DNA–rRNA hybridization and 16S rRNA oligonucleotide cataloguing and sequencing. However, no specific nucleic acid probes are available for the rapid detection and identification of flavobacteria.

Flavobacteria in Foods: Prevalence and Spoilage Potential

The primary habitats of the emended genus *Flavobacterium* are soil and fresh water, where they decompose organic matter. However, pigmented, strictly aerobic (non-fermentative), Gram-negative, nonmotile rods, which have been identified as *Flavobacterium* spp. or have been termed flavobacteria, have consistently been found among the bacteria isolated from fresh and marine waters, ocean sediments, sewage, food environments and foods, including fish and fish products, crustaceans and molluscs, meat and poultry, bulk raw milk and certain dairy products, shell eggs and egg products. These organisms also occur in chlorinated cooling water of canning plants and in canned products.

Although a large number of publications report the presence of flavobacteria in a variety of foods, conclusive data on the prevalence of *Flavobacterium* strains (genus and species) and their possible role in spoilage are not easy to obtain. This is mainly due to (1) the long and complex history of the classification, description and nomenclature of the genus *Flavobacterium* and allied genera and (2) the fact that most workers did not assign the strains to species and, when they did, a considerable proportion of the strains have been misnamed. Furthermore, many of the isolates quoted in the literature have been reclassified continuously.

In spite of this, since flavobacteria are widely accepted as common contaminants of proteinaceous foods and water, it is of interest to review the available literature on the incidence and spoilage potential of flavobacteria in foods, water and wastes from the food industries.

Fish and Fish Products, Crustaceans and Molluscs

The microflora of freshly caught finfish of marine origin captured in temperate water (halibut, rock sole, cod, pollock, sardines, Cape hake) commonly includes flavobacteria, but the specific spoilage bacteria, when stored in ice or in refrigerated sea water, are species of *Pseudomonas* and *Shewanella putrefaciens*. *Flavobacterium* spp. also occur in significant numbers in fresh and frozen fillets (of most of the above mentioned fish species and blue grenadier and herring), minced fish, surimi products, newly caught wild freshwater fish (salmon, brown trout, pike, *Clarias*) and freshly harvested pond-reared fish (tilapia, salmon, rainbow trout, striped bass). Again, upon refrigerated storage under aerobic conditions, other Gram-negative organisms, particularly *Pseudomonas*, ultimately spoil these products.

Although the initial microflora of raw crustaceans and shellfish invariably reflects the quality of the water from which the animals were harvested, *Flavobacterium* spp. are among the predominant groups of bacteria isolated from shrimps, crabs and crab meat, oysters and other bivalve molluscs. This is not surprising as many surveys have reported the presence of these microorganisms in marine habitats. In addition, they seem to be a part of the resident bacterial flora of shellfish and they also occur in the haemolymph of

normal crabs. It is generally assumed that flavobacteria are not likely to play a role in the deterioration of this type of seafood; however, a few papers have reported that strains that originated from live crustaceans became predominant by growth during refrigerated storage of crab meat and fresh and commercial shrimps.

Reptiles

The very limited information on the bacterial flora of reptiles used as a food resource shows that tail meat from captive Nile crocodiles and alligator meat have a significant proportion of flavobacteria, some of which were identified as *F. breve* (*Empedobacter brevis*) and *F. indologenes* (*Chryseobacterium indologenes*).

Meat and Meat Products

Flavobacteria are likely to be present among the initial microflora of red meat, their incidence and levels being usually much lower that those reported for seafood and freshwater fish. These bacteria appear to be more prevalent on pig carcasses taken from the processing line, but the spoilage flora of pork stored aerobically under chilled conditions is dominated by other Gram-negative bacteria (e.g. *Pseudomonas*). Significant numbers of flavobacteria have been detected in conventionally prepared curing solutions ('*F. diffusum*') and in thawed ground beef (*Flavobacterium* spp.).

The feathers and feet of live birds (chickens and turkeys) carry *Flavobacterium* spp. which originated from the rearing environment. This and their widespread occurrence in the processing plant, water supplies, chill tanks and soiled surfaces of equipment is reflected in the prevalence of flavobacteria among the main groups of microorganisms recovered from poultry carcasses immediately after processing. The spoilage flora that develops on whole carcasses and cut-up parts held in air at refrigeration temperatures is highly dependent on the combined effect of factors such as storage temperature, the inherent pH of poultry meat and the initial flora. Despite the relative importance of flavobacteria as a major initial contaminant of fresh poultry carcasses, they are not implicated in spoilage, mainly due to their lack of competitive growth.

Selected strains of presumptive flavobacteria have been tested to evaluate their spoilage activity on sterile bovine adipose tissue and chicken breast. Strains of *Flavobacterium* were capable of substantial growth on beef fat, the samples inoculated developing a fruit-like odour after 8–12 days storage at 2°C. In chicken breast meat inoculated with strains of *Flavobacterium* spp., the proportion of lower fatty acids (C_{14}–C_{17} plus isostearic acid and stearic acid) tended to increase with storage time. Ethanol and dimethylsulphide were produced by one strain of *Flavobacterium* (designated as *Flavobacterium* sp. 44) grown for 5 days at 10°C on irradition-sterilized chicken breast.

Milk and Dairy Products

Flavobacteria appear to be more significant in milk and dairy products than in other proteinaceous foods. These bacteria are accepted as a part of the psychrotrophic population of refrigerated bulk cows' milk where they can represent up to 20% of the aerobic psychrotrophic flora. A lower occurrence has also been reported in other milks, especially in goats' and ewes' milk. Because strains of flavobacteria can produce heat-resistant proteolytic, and often lipolytic, enzymes they have been implicated in a variety of defects such as development of off-flavours in pasteurized milk, surface taint and off-odours in butter, thinning in creamed rice, and spoilage of milk-based canned products. Flavobacteria have also been isolated from spoiled cream and they reduce the yield of Cheddar cheese. The fact that a significant percentage of flavobacteria strains from foods are not only without standing in the nomenclature but are also possibly not even members of the genus *Flavobacterium* is clearly illustrated by the literature dedicated to the identification of pigmented Gram-negative rods in milk and dairy products. The most comprehensive studies on *Flavobacterium–Cytophaga* isolates from dairy sources are those undertaken by Jooste and colleagues who, on the basis of numerical taxonomy, grouped 189 isolates in nine clusters. Three of the clusters, representing 18.7% of the isolates, were assigned to *F. balustinum*, *F. breve* and *F. multivorum*, now reclassified as *Chryseobacterium balustinum*, *Empedobacter brevis* and *Sphingobacterium multivorum*, respectively. A further two clusters were equated with *Flavobacterium* sp. group IIb (46.3% of isolates) and *Flavobacterium* sp. L 16/1 (22.7% of isolates). The second edition of the *Prokaryotes* includes the following species within group IIb: *F. gleum* (reclassified as *Chryseobacterium gleum*), *F. indologenes* (reclassified as *C. indologenes*), *F. indoltheticum* (reclassified as *C. indoltheticum*) and *F. meningosepticum* (reclassified as *C. meningosepticum*). Later, Botha and colleagues identified the 49 isolates of *Flavobacterium* sp. L16/1 as *Weeksella* spp., *Empedobacter brevis* and *Flavobacterium odoratum* (reclassified as *Myroides odoratus*). More recently, pigmented Gram-negative rods encountered during the routine analysis of milk products were accommodated in the genera *Weeksella* and *Bergeyella*. Other flavobacterial strains associated with dairy products have been reported as '*F. maloloris*' (butter taints, off-odours in skim milk and able to produce isovaleric acid), *F. aquatile*, *F. devorans* (reclassified as *Sphingomonas paucimobilis*), '*F. lutensis*' (producers of themoresistant lipases) and *F. suaveolens* (reclassified as *Microbacterium esteraromaticum*).

Eggs and Egg Products

Flavobacterium strains may be present on shell eggs from domestic chickens contaminated in the production environment and, occasionally, may act as primary invaders. Since they are often detected in rotten eggs, it has been assumed that they are capable of growth in eggs once the eggs' defences have been breached by primary invaders. It has also been suggested that flavobacteria are able to use the metal ions sequestered by the siderophores produced by the primary invaders. Defects produced by *Flavobacterium* strains isolated from liquid eggs and then inoculated into sterile liquid whole egg included coagulation, faecal odour and other off-odours. The spoilage activity of one *Flavobacterium* strain isolated from green rotten eggs was similar to that shown by other egg spoilage organisms (mainly *Pseudomonas*). Changes observed in inoculated egg white were less intense than those in yolk which developed a fishy, sulphuric or ammoniacal odour, as well as considerable levels of trimethylamine, H_2S and total volatile bases.

Water

Flavobacteria are likely to be present in water for drinking. Deterioration of the microbiological quality of city water, associated with these or other Gram-negative bacteria, depends on the presence of metabolizable organic compounds, as well as reduced levels or the absence of residual chlorine.

Secondary contamination, which may occur from a variety of sources (breakage or leaks, non-potable water introduced through cross-contamination, back siphoning), is another important factor. These organisms are also recovered from drinking water produced by domestic water filtration units. The autochthonous microflora of bottled water includes flavobacteria characterized by their low nutritional requirements. Finally, these organisms may be present in water used in food factories for a number of purposes.

Applications

No applications for strains belonging to the emended genus *Flavobacterium* are yet known. However, suitable strains termed *Flavobacterium* sp. or assigned to species as yet not properly described are of interest for their applied aspects. Thus, *F. arborescens* (reclassified as *Microbacterium arborescens*) is capable of producing large amounts of glucose isomerase, which is then used in the commercial production of high fructose corn syrup. Strains identified as *F. multivorum* (reclassified as *Sphingobacterium multivorum*) contained a β-fructosidase that was active in both inulin and sucrose. Fructose syrups obtained from inulin can compete as sweeteners with the high fructose corn syrups. *Flavobacterium* proteases produced by *F. balustinum* (*Chryseobacterium balustinum*) and other flavobacterial strains have been used for accelerating the flavour development of cheese. The *Flavobacterium* system was also suggested as a useful model for studying synthesis of extracellular proteins by immobilized cells. '*F. aurantiacum*' has been successfully used to remove aflatoxin B_1 from test substrates and foods (milk, vegetable oils, corn, peanuts, peanut butter and peanut milk). This bacterium metabolizes the toxin to water-soluble products and carbon dioxide. *Flavobacterium* sp. strain K39 produced chitinase. Chitinase-producing bacteria can inhibit fungal growth, e.g. that of plant-pathogenic fungi. Biodegradation of gamma-irradiated lignocellulosic materials (woods, barks and needles) by a *Flavobacterium* strain isolated from soil has been described. Sugars released by xylanolytic bacteria can be fermented to fuel, chemicals and foods. A cell-bound cyclodextrin-degrading enzyme from *Flavobacterium* strains (designated as *Flavobacterium* sp. and isolated from wheat bran and brook water) has been purified and characterized. The enzyme hydrolyses maltooligosaccharides and cyclodextrins to glucose, maltose and maltotriose. The mode of action of this enzyme on pullulan has also been described. α-Amylases produced by *Flavobacterium* spp. have proved useful in eliminating beta-cyclodextrins residues in egg yolk treated to remove cholesterol. Growth measurements with *Flavobacterium* sp. strain S12 (isolated from tap water) gives information about the concentration of maltose and starch-like compounds in drinking water, where these compounds may promote undesirable bacterial regrowth in distribution systems. *Flavobacterium* spp. are also useful in the treatment of dairy wastes. Their growth in the aerobic zone of facultative ponds degrades organic matter and produces CO_2 and the macronutrients needed by the algae. They are also among the major bacterial groups found in activated sludge and are active in biological filters, such as trickling filters.

Pathogenicity

Disease in Animals

At least, three of the species described included in the emended genus *Flavobacterium* (*F. branchiophilum*, *F. columnare* and *F. psychrophilum*) are pathogenic to fish and supposedly flavobacterial strains have also been associated with bovine mastitis. *Weeksella zoohelcum* (synonymous with *Flavobacterium* group IIj) is commonly found in dogs.

Disease in Humans

Strains belonging to species previously placed within the genus *Flavobacterium*, but nowadays reclassified and transferred to new or different genera, such as *Chryseobacterium* (*C. indologenes*, *C. gleum*, *C. balustinum* and *C. meningosepticum*), *Sphingobacterium* (*S. multivorum*, *S. spiritivorum*, *S. thalpophilum* and *S. mizutaii*), *Empedobacter* (*E. brevis*), *Myroides* (*M. odoratus*) and *Weeksella* (*W. virosa* and *W. zoohelcum*), have been recovered from the hospital environment and from clinical specimens (infected wounds, spinal fluid, urine, blood, faeces, etc.). Isolates most often incriminated as opportunistic pathogens are those belonging to the former *Flavobacterium* species group IIb, especially *Chryseobacterium meningosepticum*. Although the frequency of occurrence of 'pathogenic flavobacteria' is usually low (< 1%) and their virulence is considered 'low-grade' or doubtful, they can be of concern because of their resistance to many antimicrobial agents.

Among the supposed flavobacteria associated with food poisoning is one reported enterotoxigenic strain of *M. odoratus* and strains of *Flavobacterium farinofermentans* (*Burkholderia cocovenenans farinofermentans*) the latter being able to produce a heat-resistant, low-molecular-weight exotoxin (flavotoxin A) which produces toxic symptoms and pathological alterations in animal models.

See also: **Bacteria**: Classification of the Bacteria – Phylogenetic Approach. **Cheese**: In the Market Place. **Eggs**: Microbiology of Fresh Eggs. **Fish**: Spoilage of Fish. **Meat and Poultry**: Spoilage of Cooked Meats and Meat Products. **Milk and Milk Products**: Microbiology of Liquid Milk. **Shellfish (Molluscs and Crustacea)**: Contamination and Spoilage.

Further Reading

Bernardet JF, Segers P, Vancanneyt M et al (1996) Cutting a Gordian knot: Emended classification and description of the genus *Flavobacterium*, emended description of the family *Flavobacteriaceae*, and proposal and *Flavobacterium hydatis* nom. nov. (basonym, *Cytophaga aquatilis* Strohl and Tait 1978). *Int. J. Syst. Bacteriol.* 46: 128–148.

Botha WC, Jooste PJ and Britz TJ (1989) The taxonomic relationship of certain environmental flavobacteria to the genus *Weeksella*. *J. Appl. Bacteriol.* 67: 551–559.

Botha WC, Jooste PJ and Hugo CJ (1998) The incidence of *Weeksella-* and *Bergeyella*-like bacteria in the food environment. *J. Appl. Microbiol.* 84: 349–356.

Freeman LR, Silverman GJ, Angelini P, Merritt C and Esselen WB (1976) Volatiles produced by micro-organisms isolated from refrigerated chicken at spoilage. *Appl. Environ. Microbiol.* 32: 926–932.

Holmes B (1992) In: Balows A, Trüper HG, Dworkin M, Harder W and Schleifer K-H (eds) *The Prokaryotes. A Handbook on Habitat, Isolation and Identification of Bacteria*, 2nd edn, p. 3220. New York: Springer-Verlag.

International Commission on Microbiological Specifications for Foods (ICMSF) (1998) *Microorganisms in Foods 6. Microbial Ecology of Food Commodities.* London: Blackie Academic & Professional.

Jooste PJ, Britz TJ and De Haast J (1985) A numerical taxonomic study of *Flavobacterium-Cytophaga* strains from dairy sources. *J. Appl. Bacteriol.* 59: 311–323.

Line JE and Brackett RE (1995) Role of toxin concentration and second carbon source in microbial transformation of aflatoxin B1 by *Flavobacterium aurantiacum. J. Food Protect.* 58: 91–94.

Vancanneyt M, Segers P, Torck U et al (1996) Reclassification of *Flavobacterium odoratum* (Stutzer 1929) strains to a new genus, *Myroides odoratus* comb. nov. and *Myroides odoratimimus* sp. nov. *Int. J. Syst. Bacteriol.* 46: 926–932.

Flavours *see* **Fermentation (Industrial)**: Production of Colours/Flavours.

Flours *see* **Spoilage of Plant Products**: Cereals and Cereal Flours.

FLOW CYTOMETRY

Charlotte Nexmann Jacobsen and **Mogens Jakobsen**, Department of Dairy and Food Research, Royal Veterinary and Agricultural University, Frederiksberg, Denmark

Flow cytometry is a rapid technique originally developed for analysing mammalian cells. The technique permits analysis of single cells in a suspension, for example for DNA content, immunogenic properties, protein content, intracellular pH and enzyme activities. Despite the many advantages of flow cytometry, implementation of the technique in microbiological laboratories is still limited. In the following, development of flow cytometry is reviewed with special emphasis on its application in microbiology. The techniques and the possible methods of labelling bacteria and yeasts for flow cytometric studies are described, a short presentation of the commercially available instruments is given and future applications in food microbiology are discussed.

The Development of Flow Cytometry in Microbiology

During the last three decades flow cytometry has been widely accepted as a reliable and rapid technique for the study of mammalian cells. In 1965, the first efforts were made to replace autoradiography with flow cytometry when measuring the nucleic acid content of mammalian cells. The technique was improved for cell cycle analysis of human cells throughout the next decade. Flow cytometry proved particularly useful for analysis of human cells. The big difference in size and thereby DNA content between mammalian cells and yeasts and even more between mammalian cells and bacteria, caused a very slow development in the field of microbiology. In 1974 it was shown that *Escherichia coli* could be detected by flow cytometry using light scatter. Studies on yeasts were intensified, with work on protein determinations and evaluation of the purity of the cultures of *Saccharomyces cerevisiae* using antibodies. DNA determinations were performed for bacteria, yeasts and moulds. Despite this it was still difficult to study bacteria with the laser-based flow cytometers available.

In the late 1970s a new type of flow cytometer was introduced, based on a laminar flow of cells passing the focused light beam from a mercury arc lamp. By using this instrument it was possible to measure the DNA distributions of cultures of *E. coli* in exponential growth. Meanwhile, improved optics of the laser-based flow cytometers made these instruments satisfactory for analysis of bacteria too.

Since then, use of flow cytometry for microbiology has expanded. Specific detection of bacteria by flow cytometry using immunofluorescence in combination with nucleic acid staining has been described, determination of the guanine plus cytosine content of bacterial DNA by specific staining of AT- and GC-rich regions has been reported and cell viability of different bacterial cultures has been determined using dyes that are indicators of the membrane potential. The performance of membrane-potential stains has been compared with fluorogenic esters for viability assessment of various bacteria, indicating that no universal stain exists. Elsewhere fluorescein-di-β-D-galacto-pyranoside was found to be a useful substrate for flow cytometric determination of the β-galactosidase activity in bacteria. Light scatter determinations for differential counting of bacteria have also been reported.

Flow cytometric analysis of yeasts, has been used to study the viability and vitality during cider fermentation, for assessment of the fermentation efficiency, to discriminate between wild and lager yeasts and for on-line prediction of the shelf-life of beverages. Combined flow cytometric measurements of the membrane potential and intracellular pH of *S. cerevisiae* has also been reported together with measurements of the protein distributions of populations of *S. cerevisiae*. The protein content was correlated to the actual growth rate.

The applications of flow cytometry in microbiology are widespread and still expanding. With the opportunity for multiparameter analysis of single cells, flow cytometry rapidly provides the user with information not available with other techniques. The technique is therefore of great interest to microbiologists.

Flow Cytometric Analysis

Principles

Flow cytometric studies of bacteria and yeasts are based on staining with a fluorochrome. The stained cells in suspension are injected into the centre of a fast-moving carrier fluid called sheath fluid (**Fig. 1**). Caused by hydrodynamic focusing, the cells are centred in the middle and arrive one by one at the measuring area of the flow chamber where a light source of a defined wavelength is focused on the flow cell. The stained cells are excited at a particular wavelength and emit light at a longer wavelength. The individual cells traverse the focus of the light beam very rapidly, minimizing the risk of two cells being calculated as one. The emitted light is separated from the exciting light by optical filters and mirrors permitting only light of a certain wavelength to reach the photomultiplier, which measures the emitted light dividing it into channels according to the fluorescence intensity. The analysis is often presented as a histogram showing the number of cells versus fluorescence intensity. A narrow histogram indicates that the population is homogenous with regard to the parameter analysed.

Multiparameter Analysis

Most flow cytometers are able to analyse multiple parameters on each cell simultaneously, making it

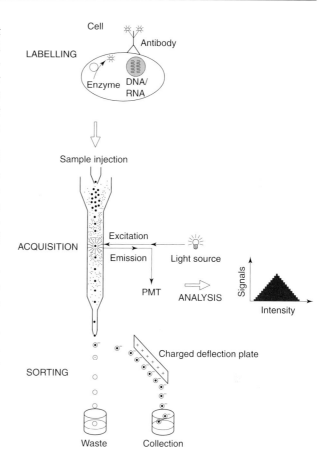

Figure 1 The cells are stained and injected into a narrow channel. Then, by hydrodynamic focusing, the cells are centred in the middle and arrive one by one at the flow chamber, where a light source of a defined wavelength is focused on the flow cell. The cells are excited at a particular wavelength and emit light at a longer wavelength. The cells pass through the flow chamber and leave the chamber as droplets. A voltage is applied to the droplets with the bacteria to be sorted, leaving them with a charge. The droplets are passed through an electrostatic field, where they are deflected according to their charge and collected.

easier to identify a certain cell population within a heterogenous population (Fig. 1). The number of parameters depends on the light source(s), scatter functions and number of photomultipliers. As an example, cells can be labelled with different fluorochromes like fluorescein isothiocyanate (FITC)-conjugated antibodies for specific cell surface antigens and propidium iodide for determination of DNA content, and the result correlated to the cell size and structure measured by forward and wide-angle scatter. Depending on the particular instrument, up to eight different parameters can be measured on the same cell, with the various signals being separated by filters and registered by different photomultipliers.

Light Scattering

Measurements of light scattered at small angles in the forward direction (forward scatter) as well as in larger angles (wide-angle scatter) to the light source, are offered by many flow cytometers. It is a simple and rapid measurement, that can be performed without any sample preparations. When light strikes a particle it is either absorbed or deflected to an extent depending on the size and the shape of the particle. It has been shown that the light from a laser beam deflected in small angles in the forward direction, is proportional to the volume of the particle, whereas the light scattered in larger angles measures a combination of size and internal structure due to scattering of internal structure such as mitochondria or lysosomes. The angles at which the scattered light is measured vary from instrument to instrument, but forward scatter is typically measured somewhere between 2 and 15° and wide angle scatter between 15 and 90°. In an arc lamp-based flow cytometer the light beam is very broad compared to the distinct light of a laser. With a very wide illumination field covering 90° to both sides in the forward direction, it is difficult to measure light scatter within the same narrow limits as a laser-based flow cytometer. However, it has been shown that an arc lamp-based flow cytometer measuring forward scatter between 2° and 90°, and wide-angle scatter between 20° and 80° is just as efficient for discriminating granulocytes and bacteria by cell size and granularity, as a laser-based instrument. Since the intensity of the light scattered decreases very rapidly with increasing scattering angle, the light scattered at the smallest angles will dominate and the light scattered at larger angles will be negligible in comparison.

When working with microorganisms one should be aware that the cell morphology varies according to the phase of growth as well as the growth conditions. Discrimination of bacterial species on the basis of their forward and wide-angle scattering characteristics has been reported, but it has been seen that the intensity of forward light scatter does not necessarily correlate with the cell size. In the same study bacteria with distinct internal or surface structures could generally be discriminated by the wide-angle scatter intensities, but two species with different internal structure were found to have similar wide-angle scatter characteristics. When combining the scatter measurements with staining of DNA, the individual species were discriminated more efficiently. In another study of recombinant *S. cerevisiae*, the correlation between forward light scattering and cell size was interrupted when the cells were dividing.

In conclusion measurements of light scattering properties is a rapid method to gain information of cell size and structure, but the measurements are not absolute and should be combined with another parameter such as DNA staining.

Figure 2 (**A**) One of the smallest and simplest flow cytometers (the Microcyte, Optoflow) and (**B**) one of the biggest and more advanced flow cytometers commercially available (the FACS Vantage, Becton Dickinson).

Cell Sorting

In a flow cytometer equipped with a cell sorter it is possible to separate and collect cells of interest (Fig. 1). Flow sorting is a combination of one or more of the flow cytometric parameters measured on cells and physical separation of exactly those cells having a specific characteristic. The more parameters used to describe the cells to be sorted, the more specific and effective is the resultant sorting. The principle of flow sorting is based on the deflected droplet method. Originally the cells were sorted according to their volume, but now the cells can be sorted on behalf of any of the parameters, offered by flow cytometry. The cells pass through the flow chamber where they are analysed and leave the chamber as droplets. A voltage is applied to the droplets containing the bacteria to be sorted, leaving them with a positive or negative charge. The droplets are passed through an electrostatic field, where they are deflected according to their charge and collected. Uncharged droplets pass through the electrostatic field to the waste container. Cell sorting has been used to separate *Staphylococcus aureus* from *E. coli* with final cell purities of 95% or greater. Starved cells of *Micrococcus luteus* were sorted into groups of viable, dormant and dead cells according to their ability to bind the dye rhodamine 123, whereas the β-galactosidase activity was used

Table 1 Special characteristics of various dyes for fluorescence staining of microorganisms

Category of staining	Dye	Excitation maximum (nm)	Emission maximum[a] (nm)
DNA staining	Ethidium bromide	510	595
	Propidium iodide	536	623
	Hoechst 33342	340	450
	DAPI	350	470
	Chromomycin	440	555
	Mithramycin	440	575
Protein/antibody	Fluorescein isothiocyanate	494	520
	R-phycoerythrin	480, 546, 565	578
Membrane potential	Rhodamine 123	507	529
	Dihexylocarbocyanine	482	500
	Oxonol dye	493	516
Enzyme activity	Fluorescein derivatives	494	520

[a]Values according to Haugland (1996) except for chromomycin A_3 and mithramycin which are from Shapiro (1995).

for sorting viable bacteria and yeasts with very high efficiency.

Cell sorting seems to be an effective method of separating cells of interest occurring at low levels in heterogeneous microbial populations.

Staining Reagents: Their Modes of Action and Applications

Characteristics of Staining Reagents

When staining cells for flow cytometric analysis, many compounds are available, such as reagents for staining specific nucleic acids, proteins, antibodies and for determination of membrane potential and enzymatic activity (**Table 1**).

The fluorochrome that is chosen should have excitation maxima at a wavelength compatible with the light source. In addition a large difference between the excitation maximum and the emission maximum to facilitate the separation of signals and minimize overlap, is important.

Staining of Nucleic Acids

As mentioned previously the first use of flow cytometry in biology was for staining of cellular DNA in mammalian cells. Generally the reagents for staining DNA bind stoichiometrically to nucleic acids, making them suitable for quantitative DNA measurements. For staining of DNA, it is necessary for the cells to be fixed or permeabilized for the dye to enter. Analysis of cellular DNA content has been successfully performed for yeasts as well as bacteria. For bacteria with their low content of DNA a sensitive instrument is necessary. Another problem is the large RNA/DNA ratio, making it necessary to treat both bacteria and yeasts with RNase before staining.

Ethidium bromide and propidium iodide are both dyes that stain double-stranded DNA and RNA. The dyes have excitation maxima in the UV range (320 and 342 nm) as well as in the blue–green spectral area at 510 and 518 nm, respectively, and emission maxima at 595 and 623 nm, respectively. Ethidium bromide staining in combination with forward scatter measurements was found to be a reliable and rapid method for detecting and counting bacteria. Propidium iodide has mainly been used in combination with immunofluorescent techniques for specific detection of bacteria in mixed populations.

Chromomycin A_3 and mithramycin are antibiotics binding preferentially to the GC-rich regions of DNA. They have excitation maxima at 440 nm and emission maxima at 555 and 575 nm, respectively. Unfortunately the low fluorescence intensity of mithramycin restricts its use. However, when used in combination with ethidium bromide a higher fluorescence is obtained. The stained cells are excited by a mercury arc lamp or a laser at 436–458 nm, where ethidium bromide has a very low absorption. It is excited by fluorescence resonance energy transfer from mithramycin molecules bound to the DNA, resulting in an increase of the fluorescence intensity by a factor of two.

This double staining in combination with light scattering has been used in numerous studies for characterization and differentiation of bacteria during growth on the basis of DNA and protein contents. Chromomycin A_3 has mainly been used in combination with the bis-benzimidazole Hoechst 33342, that binds preferentially to AT-rich regions of DNA. By combining these two dyes it is possible to differentiate bacterial species with different fractions of G+C base composition at a level of 4% difference. Chromomycin A_3 is excited at 457 nm laser and Hoechst 33342 at 340 nm, thus two light sources are needed to perform this study. Hoechst 33342 was found to stain bacteria with a higher fluorescence intensity to another AT-specific DNA dye, 4'-6'-diamino-2'-phenylindole (DAPI), making it easier to distinguish bacteria from the background fluorescence. DAPI staining has been used in combination with scatter measurements for detection of a *Pseudomonas* species in marine and fresh water samples and for quantification of a mixture of bacteria in samples from aquatic ecosystems. Recently two nucleic acid binding dyes hexidium iodide (HI) and SYTO 13 (both from Molecular Probes Inc.) were used in combination as a fluorescent Gram stain for flow cytometry and epifluorescence microscopy. Hi penetrated only the Gram-positive bacteria staining them red–orange, whereas SYTO 13 stained both Gram-positive and Gram-negative bacteria green. The green fluorescence was simultaneously quenched by the previously HI stained Gram-positive bacteria rendering the bacteria red–orange.

Staining of Cellular Protein

Staining of cellular protein can be used to obtain information about growth and metabolism of cell populations. The reagent of choice for this purpose is fluorescein isothiocyanate (FITC), which is excited by a laser at 488 nm and emits very strongly at 525 nm (green light). Protein staining is often used in combination with staining of DNA. For this purpose FITC could be combined with propidium iodide emitting in the red channel at 623 nm (red light). A strong correlation exists between measurements of cell size and protein content in mammalian cells and yeasts like *S. cerevisiae*. With bacteria it was possible to divide 19 different species into three groups after staining with FITC in combination with propidium iodide. Further it has been found, that the amount of gramicidin S produced by *Bacillus brevis* corresponded directly to the fluorescence intensity of cells after staining with FITC, making it possible to sort *B. brevis* populations on the basis of this.

Immunofluorescence Staining

For immunofluorescence labelling of bacteria, specific antibodies against bacterial cell surface determinants are required. If such antibodies are available, the next step is the binding of the fluorochrome. As for protein labelling, the most popular fluorescent label is FITC. If more than one epitope is analysed, R-phycoerythrin is the second most appropriate choice for labelling the antibodies. Both FITC and R-phycoerythrin is excited at 488 nm with emission maxima easily distinguishable at 520 and 578 nm, respectively (Table 1).

Detection of antigenic properties using immunofluorescence is routinely used with mammalian cells. Flow cytometric applications for detection of pathogens in foods or raw milk have been reported using FITC-labelled antibodies such as *Listeria monocytogenes* in raw milk. *Salmonella typhimurium* has been detected down to 1 cell ml^{-1} in eggs and 10–20 cells ml^{-1} in milk by flow cytometry using fluorescein-labelled monoclonal antibodies following a 6 h nonselective enrichment, and recently *Escherichia coli* O157 was detected in a model system using a polyclonal anti-*E. coli* O157 antibody making it possible to discriminate between *E. coli* O157 and 12 other strains of *E. coli* with a detection limit down to 10^3 cells ml^{-1}. *Staphylococcus aureus* has been detected in cheese, ham, milk and water buffalo milk using fluorescein-labelled anti-protein A antibodies. Immunofluorescent detection of pathogenic bacteria seems to be a promising technique though not applicable on all species at present.

Assessment of Cell Viability

Assessment of the physiological state of bacteria is an important criterion when evaluating the response to outside stress factors (e.g. starvation or heat treatment). It is generally accepted that bacteria should be into three groups according to their physiological condition: (1) viable cells with the ability to form a colony on an agar plate; (2) dead cells unable to form a colony under any circumstances; and (3) viable but non-culturable cells being microorganisms that are viable but only capable of forming a colony after a preliminary resuscitation step. The latter group could be further divided into groups showing various degrees of cell damage. Viable but non-culturable cells could be of particular concern, when pathogenic bacteria remain undetected by conventional methods, but are still capable of growth and virulence. Flow cytometric assessment of the viability of a microorganism has mainly been based on the measurements of the membrane potential or intracellular enzyme activity.

Measurements of Membrane Potential The membrane potential is the electrical potential caused by the existence of an ionic concentration gradient across the cell membrane. The distribution of a charged ion across the membrane and the membrane potential of the cell follows the Nernst equation:

$$E_{in} - E_{out} = -2.3(\mathbf{R}T/zF) \log (a_{in}/a_{out})$$

where E is the electropotential inside (in) and outside (out) and a is the unbound concentration of the particular ion. $\mathbf{R}$ is the Reynolds constant, T the absolute temperature, F the Faraday constant and z the charge of the iron. A cationic dye such as rhodamine 123 or the cyanine dye $DiOC_6(3)$, will accumulate inside a cell maintaining a membrane potential, whereas anionic dyes, such as the oxonols, will only accumulate inside a cell that has lost its membrane potential. After equilibrium of a cation dye has been established, depolarization of the cells will result in efflux of the dye from cells, whereas hyperpolarization will cause enhanced uptake of the dye. The magnitude of the membrane potential in polarized cells determines the fluorescence response according to the accumulation of fluorochrome. The ionophore valinomycin can be used to hyperpolarize cells since it facilitates potassium flux across the membrane whereas gramicidin increases both sodium and potassium transport across the membrane with resultant depolarization of the cells. In this way it was shown how *Staphylococcus aureus*, under normal conditions, was stained and detected by flow cytometry using rhodamine 123 and $DiOC_6(3)$, but also how the fluorescence of the cells was reduced to the level of unstained cells upon treatment with gramicidin, indicating that staining was dependent on the membrane potential of the cell. Similar experiments with *Micrococcus luteus* confirmed this dependence and showed how it is possible to distinguish viable, viable but nonculturable and dead cells from each other by flow cytometry. It was shown that viable but non-culturable cells were much more common than dead cells in starving cultures. The potential of using rhodamine 123 for assessment of viability has also been shown for Gram-negative bacteria. Rhodamine 123 is not a universal stain, however; it failed to stain *L. monocytogenes* as well as *Bacillus subtilis*, *Aeromonas salmonicida* and *Pseudomonas fluorescens*.

As mentioned before, the lipophilic anionic oxonol dyes, such as bis-(1,3-dibutylbarbiturate) trimethine oxonol, $(DiBaC_4(3))$, stains by the opposite mechanism to rhodamine 123 and $DiOC_6(3)$ and accumulates in depolarized cells only. A reduced membrane potential in the bacterial cells promotes the accumulation of $DiBaC_4(3)$ making it suitable for assessing sensitivity towards antibiotics such as gentamicin or monitoring of starvation of *Escherichia coli*. Discrimination of viability among starved bacterial populations of *E. coli* and *Salmonella typhimurium* was successfully performed using both rhodamine 123 and $DiBaC_4(3)$. The DNA stain, propidium iodide, was also included, but this failed to provide acceptable discrimination of the subpopulations arising during starvation. A comparison between rhodamine 123, carbocyanine, $DiBaC_4(3)$ and calcofluor white (CFW), showed that $DiBaC_4(3)$ and CFW were most suitable for distinguishing viable and non-viable populations of *E. coli*, *S. typhimurium* and *S. aureus*.

Measurements of Intracellular Enzyme Activity In addition to membrane potential sensitive dyes for estimation of viability, the enzyme activity of the cell has frequently been measured. There is a broad range of dyes that can be used, for example the esters of fluorescein, such as fluorescein diacetate (FDA), carboxyfluorescein diacetate (CFDA) and 2',7'-bis-(2-carboxyethyl)5(6)-carboxyfluorescein-acetoxymethylester (BCECF-AM). In principle, they are all colourless dye conjugates, that are cleaved enzymatically inside a cell, to release a fluorescent molecule. This molecule is retained in living cells, whereas dead cells will leak the fluorescent molecule. The intracellular accumulation and efflux of fluorescein in living cells has been reported to be dependent on changes in cellular metabolism, membrane potential and membrane integrity. To minimize efflux, the temperature needs to be kept low after staining. The passive efflux is dependent on the polarity of the compound. Since CFDA has a higher charge than FDA, the intracellular retention of fluorescein is higher. Because of its negative charge of four or five, BCECF-AM is retained even more. There seems to be an energy-driven pump in *Lactococcus lactis*, that, independent of the proton motive force, causes an efflux of BCECF-AM. In *Saccharomyces cerevisiae* the efflux seems to take place in an energy-dependent manner as well, probably via a secondary transport system. Because of its pK_a near neutral, BCECF-AM is also convenient for measuring the intracellular pH. For estimation of intracellular pH carboxyfluorescein succinimidyl ester (CFSE) has been used with very reliable results, since leakage of the fluorescent dye is minimal.

The application of viability markers in industrial quality assurance has been commercialized by Chemunex (Maisons-Alfort, France), detecting yeasts in fruit juices (see Appendix) and bacteria in frozen vegetables, thereby predicting the shelf life of the product.

A comparative flow cytometric study of five different viability stains for *Listeria monocytogenes*, showed that CFDA and Chemchrome B (Chemunex) were the most suitable stains throughout growth, producing distinct fluorescence histograms, with detection limits for *L. monocytogenes* of approx. 10^4 cells ml^{-1}. Rhodamine 123 and BCECF-AM stained cells with a very low fluorescence intensity (FI) whereas the viability kit BacLight (Molecular Probes Europe BV, Leiden, The Netherlands) gave high and reproducible stainings in the lag phase and at stationary growth but not during the exponential growth phase.

Flow cytometric analyses for assessing the fermentative ability of various yeasts by combined analyses of the cell cycle phases and the physiological state using FDA and propidium iodide has been described. Combined with protein staining, these parameters were found suitable for evaluation of yeast quality and prediction of its performance. FDA staining has also been used to study the growth kinetics of

Debaromyces hansenii under different combinations of temperature, pH and NaCl concentrations with reliable detection to 10^2 cells ml^{-1}.

Antibiotic susceptibility testing is another successful application of flow cytometry using viability markers. It has been used to study the antibiotic susceptibility of yeasts and bacteria, e.g. *Mycobacteria*, *Staphylococcus aureus* and *Escherichia coli* with results obtained within 2–5 h. The same technique could be useful when testing for bacteriocin production or activity.

The viability of microorganisms is an important area of study and flow cytometry is able to provide scientists with new information in this area. Unfortunately no universal staining method exists, thus the choice of staining reagent depends on the particular microorganism being studied.

Flow Cytometers

As shown in **Table 2**, a wide range of flow cytometers meeting different needs, are commercially available. One of the major differences between the instruments concerns the use of a mercury arc lamp or the more expensive laser system. Comparison of the two illumination systems concluded that the differences between the two systems are in the light intensity, the range of excitation wavelength and the price. The laser system produces light with high intensity at a well-defined wavelength, whereas the mercury arc lamp emits light at several wavelengths, e.g. 365, 405, 436, 546 and 578 nm. For small objects with weak fluorescence, as is often the case with bacteria, it is often advantageous to use the laser lamp. The ability to

measure various parameters determined by the number of light sources and photomultipliers, is another major difference. In addition some of the flow cytometers have the ability to sort the fluorescent cells from the sample. Automation is possible in many cases. Recently a small and simple portable flow cytometer, based on a diode-laser emitting light at 635 nm has been developed (Microcyte) primarily for use in routine microbiological analysis. In general advanced flow cytometers offering many parameters are expensive, whereas more simple types offering only a few parameters are less expensive, but still adequate for many purposes.

Future use of Flow Cytometry in Food Microbiology

For many purposes flow cytometry seems to be superior to the conventional methods for microbiological analysis. It represents a rapid technique based on the separate analysis of thousands of individual cells in a short time, providing additional information on live and dead cells and their physiology.

A disadvantage of the technique, is the need for a homogenous cell suspension free of particles disturbing the flow or blocking the narrow tubings. Furthermore, the detection limit of the instrument, often necessitates preenrichment of the sample being investigated, thereby making quantitative detection of a particular organism difficult.

Compared to other rapid techniques, such as enzyme linked immunosorbent assay and polymerase chain reaction, flow cytometry is a very sensitive technique and it can be made more sensitive if combined

Table 2 Characteristics of commercial flow cytometers

Company	Instrument	Light source	Optical parameters (PMT)	Sorting
Becton Dickinson				
	FACSCalibur	2 lasers	4 fluorescence and 2 scatter	–/+
	FACS Vantage*	3 lasers	5 fluorescence and 2 scatter	+
BIO-RAD				
	Bryte HS	1 lamp	3 fluorescence and 2 scatter	–
CHEMUNEX				
	DCount	1 laser	2 fluorescence	–
Coulter				
	EPICS XL	1 laser	4 fluorescence and 2 scatter	–
	EPICS ELITE	3 lasers	5 fluorescence and 2 scatter	+
Optoflow				
	Microcyte[a]	1 laserdiode	1 fluorescence and 1 scatter	–
Ortho				
	Cytoron Absolute	1 laser	3 fluorescence and 2 scatter	–
Partec				
	CA-IV	1 laser	2 fluorescence and 2 scatter	–
	PAS	1 laser and 1 lamp	4 fluorescence and 2 scatter	–/+
	PAS III	2 lasers and 1 lamp	6 fluorescence and 2 scatter	–/+

PMT, Photomultiplier; laser; argon ion laser; lamp, mercury arc lamp.
[a] See Figure 2.

with other techniques such as immunomagnetic separation.

For estimation of the physiological state of a culture by measuring the membrane potential or enzymatic activity, the technique seems superior to other techniques and could be used for routine screening of the viability and vitality of starter cultures before inoculation or for testing of the antimicrobial effect of bacteriocins and other compounds, as performed for antibiotics today. Until now flow cytometric analysis of microorganisms has mainly been used as a research tool, but a rising popularity for routine applications is predicted, together with the development of automated, low-cost flow cytometers.

See also: **ATP Bioluminescence**: Application in Beverage Microbiology. ***Bacillus***: Detection by Classical Cultural Techniques. **Direct (and Indirect) Conductimetric/Impedimetric Techniques**: Foodborne pathogens. **Hydrophobic Grid Membrane Filter Techniques (HGMF)**. **Immunomagnetic Particle-based Techniques**: Overview. ***Listeria***: Detection by Classical Cultural Techniques; Detection by Commercial Enzyme Immunoassays; Detection by Colorimetric DNA Hybridization; *Listeria monocytogenes* – Detection by Chemiluminescent DNA Hybridization; *Listeria monocytogenes* – Detection using NASBA (an Isothermic Nucleic Acid Amplification System). **Molecular Biology – in Microbiological Analysis. PCR-based Commercial Tests for Pathogens. Petrifilm – An Enhanced Cultural Technique. Reference Materials**. ***Salmonella***: Detection by Classical Cultural Techniques; Detection by Latex Agglutination Techniques; Detection by Enzyme Immunoassays; Detection by Colorimetric DNA Hybridization; Detection by Immunomagnetic Particle-based Assays. **Sampling Regimes & Statistical Evaluation of Microbiological Results**. ***Staphylococcus***: Detection by Cultural and Modern Techniques. **Total Viable Counts**: Microscopy.

Further Reading

Allman R (1993) Flow Cytometric Analysis of Heterogenous Bacterial Populations. In: Lloyd D (ed.) *Flow Cytometry in Microbiology*, p. 27. London: Springer-Verlag.

Al-Rubeai M and Emery AN (eds) (1996) *Flow Cytometry Applications in Cell Culture*. New York: Marcel Dekker.

Diaper JP and Edwards C (1994) The use of fluorogenic esters to detect viable bacteria by flow cytometry. *Journal of Applied Bacteriology* 77: 221–228.

Diaper JP, Tither K and Edwards C (1992) Rapid assessment of bacterial viability by flow cytometry. *Applied Microbiology and Biotechnology* 38: 268–272.

Edwards C, Porter J and West M (1996) Fluorescent probes for measuring physiological fitness of yeast. *Fermentation* 9: 288–293.

Haugland RP (1996) *Handbook of Fluorescent Probes and Chemicals*, 6th edn. Eugene, US: Molecular Probes.

Hutter KJ and Eipel HE (1978) Flow cytometric determinations of cellular substances in algae, bacteria, moulds and yeasts. *Antonie van Leeuwenhoek* 44: 269–282.

Lloyd D (1993) Flow cytometry: a technique waiting for microbiologists. In: Lloyd D (ed.) *Flow Cytometry in Microbiology*, p. 1. London: Springer-Verlag.

Macey MG (ed.) (1994) *Flow Cytometry – Clinical Applications*. Oxford: Black Scientific Applications.

Melamed MR, Lindbo T and Mendelsohn ML (eds) (1991) *Flow Cytometry and Sorting*. New York: Wiley-Liss.

Ormerod MG, (1994) (ed.) *Flow Cytometry. A Practical Approach*, 2nd edn. Oxford: IRL Press.

Shapiro HM (1995) *Practical Flow Cytometry* 3rd edn. New York: Wiley-Liss.

Van Dilla MA, Langlois RG, Pinkel D, Yajko D and Hadley WK (1983) Bacterial characterization by flow cytometry. *Science* 220: 620–621.

Appendix

CHEMFLOW 3: CHEMPREP Yeast Detection in Fruit Juices

1. Principle

Samples are treated with reagents which fluorescently label the viable microorganisms which are present. Only viable organisms are capable of enzymatically cleaving the non-fluorescent marker staining the fluorescent end products. The samples are then injected into a flow channel in which the microorganisms are aligned and focused into a laminar fluid stream. As the labelled cells pass the laser beam, they fluoresce and are individually detected and counted by sensitive photomultipliers.

2.0 Precautions

2.1 Between each batch of samples clean the ChemFlow 3 system with a diluted bleach solution (see 4.6).

2.2 At the end of each day, complete the system with the diluted bleach solution.

2.3 Do not mix different batches of a reagent.

2.4 Do not inject alcohol.

2.5 Do not re-use the consumables.

3.0 Apparatus

3.1 The ChemFlow 3 ChemPrep T (code: 100-A0007–01) is available from Chemunex. This analytical system should be used in accordance with the instructions in the user manual.

3.2 A centrifuge unit (code: 100-B0106–01) is available from Chemunex.

3.3 An incubator set at 30°C.

4.0   Reagents

Chemunex reagents required to perform the assay:

4.1   ChemSol S3 (code: 100-R3082–01). The reagent is used as a buffer and allows the laminar flow necessary in the channel for the analysis.

4.2   Standard D (code: 100-R5083–01). The reagent is used as a calibrator (fluorescent beads) and should be stored at 4–8°C.

4.3   ChemSol B5 (code: 100-R2022–01). The reagent is used as a labelling buffer and stored at room temperature before opening and at 4–8°C after opening.

4.4   ChemChrome V2 (code: 100-R1002–01). The reagent is the marker and should be stored at 4–8°C.

4.5   CSB (code: 100-R4081–01). The reagent is used to minimize background fluorescence and should be stored at room temperature before opening and at 4–8°C after opening.

General reagents:

4.6   The cleaning solution used is a diluted bleach solution in distilled water: a 6° Chl solution hypochlorite aqueous solution (approximately 3% active chloride).

4.7   Distilled water.

5.0   Materials

Chemunex consumable items required to perform the assay:

5.1   3 ml tubes (code: 100-C1002–01) – clean

5.2   35 ml tubes (code: 100-C1004–01) – clean

5.3   1 ml tubes (code: 100-C1011–01) – clean

5.4   Stoppers 35 ml (code: 100-C1008–01) – clean

5.5   ChemFilter 20 (code: 100-C2002–01) – clean

5.6   Blue Gilson pipette tips (code: 100-C3023–01) – clean

5.7   Needle (code: 100-C3002–01) – clean

General material:

5.8   50 ml tube holder

5.9   3 ml dispenser

5.10   Disposable serological pipettes (5 ml)

5.11   Gilson type pipette (P1000)

5.12   Paper tissues

5.13   Aluminium foil

6.0   Method.

6.1   Incubation of product: Incubate the product for 24 h at 25–30°C (incubator).

6.2   Solubilization

6.2.1   Vigorously shake the product

6.2.2   Sample 1 ml of fruit juice and dispense into a 35 ml tube

6.2.3   Add, with the bottle-top dispenser provided, 6 ml of ChemSol S3

6.2.4   Cap the tube with the 35 ml stoppers and shake for a few seconds using a Vortex mixer.

6.2.5   Pour the solubilized sample into a ChemFilter 20.

6.3   Centrifugation

6.3.1   Centrifuge 5 min at 1500 g (2800 r.p.m. using the Chemunex centrifuge)

6.3.2   Remove the paper filter of each tube and discard the supernatant until the last drop

6.3.3   It is important to ensure that any residue is completely removed from the tube walls. Wipe it away if necessary.

6.3.4   The pellet should not remain more than 30 min before staining.

6.4   Reagents preparation

6.4.1   Place the ChemSol S3 bottles for the ChemFlow 3 and for the automated system.

6.4.2   Place the ChemSol B5 as described in the user manual.

6.4.3   Place the ChemChrome V2 on the refrigerated rack.

6.4.4   Place the CSB solution on the automate rack.

6.4.5   Place the Standard D solution on the automate rack.

6.5   Labelling of samples

6.5.1   Place the 35 ml tubes containing the pellets obtained after step no. 6.3 on the automate rack.

6.5.2   Prior to use, the ChemFlow 3 should be tested using the Standard D control. Acceptance ranges are supplied on the ChemFlow 3 user manual document.

6.5.3   Initiate the program using the ChemPrep T software.

6.5.4   The ChemPrep T; delivers 3 ml of ChemSol B5 to each pellet; homogenizes the sample; transfers 1.1 ml of this suspension into a 1 ml tube; takes 11 μl of ChemChrome V2; homogenizes the sample; incubates at 40°C for 5 min, takes 55 μl of CSB; mixes in a 3 ml tube; injects in the ChemFlow 3 cytometer.

6.6   ChemFlow 3 analysis; 100 μl of labelled suspension is analysed in 30 s.

6.7   Result

The result is displayed on the screen of the ChemFlow 3 and transferred to the PC.

The final result displayed on PC is a number of counts per millilitre of product. The result is positive if GC is more than the threshold value which is defined during the validation period on site.

FOOD POISONING OUTBREAKS

S Notermans, TNO Nutrition and Food Research Institute, AJ Zeist, The Netherlands

Introduction

Food is thought to play a major role in the transmission of microorganisms causing infectious diseases, such as salmonellosis, campylobacteriosis, hepatitis A and listeriosis, or intoxications such as botulism and staphylococcal food poisoning. Food may also cause disease if it contains toxic chemicals – either occurring naturally (e.g. cyanogenic glycosides in cassava), or resulting from contamination with chemicals (e.g. toxic metals and ciguatoxin, the toxin of the dinoflagellate *Gambierdiscus toxicus*, which accumulates in the fish which feed on it, causing food poisoning when the fish is consumed). Damage to health may also occur if food contains physical agents such as glass splinters.

Food-borne illnesses are among the most widespread diseases of the contemporary world. In most cases the clinical picture is mild and self-limiting, but because of their high frequency of occurrence, the socio-economic impact is very significant. Some food-borne illnesses, however, show a very severe clinical picture and death rates are substantial. The average rate of death due to infections caused by the shiga-like toxin produced by *Escherichia coli* may be as high as 10%. The death rate due to listeriosis may even reach 30%.

Food products can become contaminated during various stages of production. The chain of events involved in primary production, harvesting, processing, distribution and final preparation is quite long, and there are many opportunities for the food to become contaminated. In addition, microorganisms such as bacteria and moulds may proliferate if the food is stored under conditions favourable for their growth.

Epidemiology and Control

In the case of most food-borne infections, vaccines are not available – but physical barriers can substantially lower the risk of transmission. The control of food-borne diseases, therefore, depends on understanding their mechanism of transmission well enough to prevent it. Detailed investigation of the contamination of foods often reveals specific points at which the safety of the food was compromised. For example, many cases of food-borne disease result from a sequence of several events that compromised good hygienic practices. A knowledge of factors contributing to food-borne disease facilitates the identification of specific control measures, and hence the prevention of such disease.

In addition, knowledge of the characteristics of food-borne organisms enables the identification of suitable conditions for food processing and preparation. If these conditions are well-selected, the growth of microorganisms can be inhibited or the organisms can be destroyed, e.g. by heat treatment.

Recent analyses of outbreaks of food-borne disease reveal that its epidemiology is changing. In many large outbreaks, only a small proportion of exposed consumers have become diseased. Large geographical regions are involved, some outbreaks spreading worldwide. Finally, the pathogens involved are mostly new and many of them, including strains of *Salmonella enteritidis*, *Campylobacter* and shiga-toxin-producing *Escherichia coli*, have reservoirs in healthy food animals – from which they spread to an increasing variety of foods.

Developments such as increasing world food trade, new production and processing technologies, increasing mass catering and changing eating habits make food-borne disease an evolving public health challenge. Thus information about the diseases is important for many reasons, some of which are discussed below.

Causative Organisms

During the last two decades many new food-borne pathogens have emerged or re-emerged. Examples are *Campylobacter jejuni*, *Cryptosporidium cayetanensis*, *Cyclospora*, *Listeria monocytogenes*, Norwalk-like viruses, *Salmonella enteritidis*, multi-antibiotic-resistant *S. typhimurium* DT104, *Vibrio vulnificus* and *Escherichia coli* O157, which produces a shiga-like toxin. These pathogens share a number of characteristics: virtually all are zoonotic, i.e. they have an animal reservoir, from which they spread to humans; they rarely cause illness in the infected animal host; and they rapidly spread globally. A knowledge of the aetiology of the related diseases and the characteristics of the microorganisms, such as growth limits and virulence, provides a basis for their control.

Identification

Clinically, food-borne disease is usually mild and self-limiting – therefore, rapid diagnosis is not often urgent in terms of protecting the victim. However, the rapid identification of the causes of so-called diffuse and widespread outbreaks of food-borne diseases is the subject of increasing interest in terms of public health. Due to the mass production of food and worldwide trade, outbreaks of food-borne disease involving many countries are becoming more common.

The foods involved usually have only low-level contamination, but are widely distributed. Outbreaks are often detected only because of a fortuitous concentration of cases in one location, or because laboratory-based typing of strains collected over a wide area identifies a diffuse surge in one subtype. In such situations, the coordinated efforts of teams in several districts and countries can pinpoint and sometimes eradicate the outbreak. An example of a large, multistage outbreak occurred in the USA in 1994. The outbreak was attributed to contamination by *S. enteritidis* of ice cream products manufactured during late August and September 1994. The contamination was the result of a basic failure: pasteurized ice cream mix was transported in tankers that had previously contained raw eggs. Although the contamination caused an estimated 250 000 illnesses, it was detected only when vigorous routine surveillance identified a surge in reported infections with *S. enteritidis* in one area of southern Minnesota. It was estimated that the concentration of *S. enteritidis* in the ice cream was 0.004–0.46 cells per gram and the attack rate was approximately 6.6%. Another outbreak of salmonellosis occurred in Germany in 1994, this time caused by paprika-powdered potato chips. The concentration of *Salmonella* spp. was 0.04–0.45 cells per gram. About one in 10 000 exposed persons contracted salmonellosis.

Food Safety

The general perception that food-borne diseases are benign generates little incentive for the investigation, reporting and monitoring of their incidence. In turn, this lack of information perpetuates the lack of appreciation of the significance of food-borne diseases for public health. Consequently, low priority and few resources tend to be assigned to food safety and a decreasing amount of information becomes available. Indeed, at present information about the incidence of food-borne disease is published by only a few countries. As a consequence, legislative action remains limited.

The situation may, however, improve drastically now that the member states of the World Trade Organization have agreed on the free trade of safe food. In addition, they have agreed that risk analysis is necessary for the assessment of safety. In consequence, a consensus is developing in support of a worldwide food safety policy, initiated by the WHO/FAO Codex Alimentarius Commission. Food safety objectives can be derived from the policy by risk analysis, and this approach should underpin a basis on which scientists, managers and consumers can communicate openly.

Although the general perception of the consumer is that 'food has to be safe', it is increasingly recognized that absolutely safe food does not exist. In addition, the degree of safety has economic consequences and costly preventive measures will only be taken if there is a clear need. Questions about the levels of safety that are necessary and realistic must be solved through communication between food producers and consumers. Such communication may be enhanced if clear information about health risks is available.

The production of safe food is largely dependent on adherence to food safety objectives, such as end product specification. Food safety objectives are based on preventive measures such as the use of safe raw food materials, good manufacturing practices and procedures with hazard analysis critical control points. The success of these preventive measures will be reflected in the incidence of food-borne disease.

Investigation of Food-borne Disease

Public health epidemiologists have five principal tools for monitoring and investigating food-borne illness:

- surveillance of laboratory reports of food-borne infections in humans
- monitoring of contamination of food and animals by specific pathogens
- investigation of intensive outbreaks
- collection of reports on outbreaks, at regional, national and global levels
- studies of sporadic infections.

In addition, records such as registrations of death, hospital discharges and notifications of disease can be used.

Surveillance of Food-borne Infections

Clinical laboratories routinely identify pathogenic organisms that may be food-borne, by testing clinical specimens such as blood or stool from patients. The regular reporting of the isolation of specific pathogens provides an important source of surveillance data. However, laboratory-based surveillance is dependent on an infrastructure of competent laboratories that provide routine diagnostic services, and often it also requires a central reference laboratory, that can confirm the identity of unusual isolates and provide quality assurance. Follow-up studies of cases identified through laboratory diagnostics provide additional epidemiological data, including information

Monitoring of Food and Animals

Monitoring farm animals for pathogenic micro-organisms can provide scientifically sound and statistically valid information on the occurrence and distribution of these agents, and any significant trends. The typing of the strains isolated from animals and human patients proves the origin of the causative organisms.

Microbiological surveys of foods can help in the definition of the risks of exposure to potential pathogens. The risks can then be taken into account in the control measures designed to reduce the contamination of food by pathogenic microorganisms. In the absence of epidemiological data on human illness, data on food microbiology may be the only available indicator of the risk of food-borne disease.

Investigation of Outbreaks of Food-borne Disease

A food-borne outbreak is generally defined as an incident in which two or more individuals experience a similar illness after indigestion of the same food or meal. A sporadic case involves a patient who has not knowingly been exposed to a similarly ill person. Sporadic cases are far more common than outbreaks, but are less easy to investigate and characterize. Consequently, in most countries it is only outbreaks of food-borne disease that are investigated and reported.

The most informative investigations of outbreaks start with the collection of data on the exposure of defined groups of patients (cases) and of healthy controls who have had a similar opportunity to become ill. **Table 1** summarizes the findings of such a case–control study into an outbreak of food-borne disease, which was known to be caused by *Salmonella enterica* in cheese made with unpasteurized goat's milk. The cases were defined as residents of France with a positive culture for *S. paratyphi*-B between 1 August and 30 November 1993, and with symptoms of gastroenteritis or septicaemia. For each case, a control matched in terms of age, gender and city of residence was selected from the telephone directory (those with

diarrhoea in the previous 3 months were excluded). The data summarized in Table 1 were collected by telephone interview. On the basis of their analysis, the conclusion of the study was that people who had eaten cheese of brand A had a twelve-fold increased risk of illness. Case–control studies are carried out to identify the factors involved in the aetiology of a disease, and to quantify risk factors – for example in Norway, to investigate sporadic *Campylobacter* infections, and also to investigate *Yersinia enterocolitica* infections. However, the risk factors identified may have only regional value.

Another type of study is a cohort study. This compares the incidence of illness amongst individuals who have or have not eaten certain food items during the same event. The ratio of the 'attack rates' is expressed as 'a relative risk'. Cohort studies are used less often than case–control studies, but constitute a convenient and powerful tool when those at risk can easily be listed, for example the passengers on a cruise or the guests at a party. The findings of a cohort study are presented in **Table 2**. This study investigated a large outbreak of gastroenteritis, caused by the diarrhoeal form of the toxin produced by *Bacillus cereus*. The cohort comprised all those attending a university field day, and the cases were defined as those with diarrhoea (three or more loose stools in a 24-hour period) within 5 days of the field day. Data were collected by means of a postal questionnaire. Analysis of the data showed that 26% of those who had eaten pork became ill, and that those who had eaten pork were five times more likely to become ill than those not eating pork.

Once a food source is implicated, further investigations, into its mode of preparation and the sources of the raw ingredients, may be warranted in order to identify how it became contaminated. Intensive laboratory investigations into leftovers, raw ingredients, etc. may identify the causative agent and clarify the mode of contamination. In this way it was shown that iceberg lettuce imported from Spain caused a large outbreak of *Shigella sonnei* infections in Norway and other European countries.

Table 1 Investigation of an outbreak of food poisoning, using a case–control study (reproduced with permission from Desenclos et al, *British Medical Journal* 312: 161–167, 1996)

Type of goat's-milk cheese	Cases	Controls	Matched odds ratio
Brand A	32	10	12
Brand unknown	9	8	6
Brand other than A	5	10	1.7
None	13	30	1 (reference)
Total	59	58	

Table 2 Investigation of an outbreak of food poisoning, using a cohort study (reproduced with permission from Luby et al, *Journal of Infectious Disease*, 167: 1452–1455, 1993)

Consumption of pork	Total cohort	Number of cases	Attack rate (%)	Relative risk
Yes	523	137	26	5.4 (1.4–21.0)
No	41	2	5	

Surveillance of Reports on Food-borne Disease Outbreaks

The value of data on outbreaks of food-borne disease depends heavily on the quality of the original investigations, including the use of standard methods for implicating aetiological agents and food vehicles, and the constancy of reporting. The collection and summary of data occasionally lead to discoveries such as the linkage of outbreaks of *Salmonella enteritidis* infections with eggs. More typically, well-selected data provide useful information about the spectrum of outbreaks associated with a particular aetiological agent, a particular type of food or a particular setting. New trends, such as the observed change in the epidemiology of food-borne outbreaks, can be revealed if data from different countries are put together. However, few papers provide detailed information on the number of organisms present and the response of individuals to them.

Studies of Sporadic Cases

Sporadic cases of food-borne infections are far more frequent than cases associated with identified outbreaks, and the preventive lessons learned from investigations of outbreaks may not always apply to sporadic cases. For example, outbreaks of *Campylobacter jejuni* infections in the USA are typically caused by the consumption of raw milk and untreated surface water. Eating poultry results in sporadic cases and were identified as a dominant source of campylobacteriosis after case–control studies were conducted.

During the 1990s, investigators demonstrated that *Listeria monocytogenes* can be found in a wide variety of processed foods, and that considerable growth of the organism may occur during prolonged refrigerated storage. However, only after case–control studies of sporadic cases were foods such as cheese, undercooked chicken and hot dogs identified as agents which transmit listeriosis. This finding resulted in educational campaigns and legislation, and more recent surveillance has indicated a decrease in the incidence of human cases of listeriosis.

Sentinel and Population Studies

Sentinel and population studies provide quantitative information about the incidence of food-borne pathogens in a certain region. The lack of essential information about the incidence of food-borne diseases led to the organization of a sentinel system in the Netherlands, in which 42 general practitioners participated. They reported all those patients with acute gastroenteritis who exhibited symptoms similar to a previously established case definition, and all of these patients were asked to fill in a questionnaire and to

Table 3 Investigation of gastroenteritis, using sentinel and population studies (reproduced with permission from Notermans S and Borgdorff MW, (1997))

Causative microorganism	Proportion of cases in sentinel study (%)	Incidence of gastroenteritis in population study (Cases per 1000 individuals per year)
Campylobacter	12–15	18–23
Salmonella	4–5	6–11
Shigella	0–3	
Pathogenic Escherichia coli	3	
Clostridium perfringens	3	
Rotavirus	6	

send a faecal sample to a regional laboratory for examination. The results are summarized in **Table 3**. *Campylobacter* species were the most frequently found organisms in faecal samples of the patients. The sentinel study revealed that in a year, 15 out of 1000 individuals seek medical attention for complaints of acute gastroenteritis. However, the true incidence must be much higher because not all such patients seek medical attention. Therefore, a population study was carried out at the same time at community level, to assess the true number of individuals with gastroenteritis. From these data, the true incidence of campylobacteriosis and salmonellosis was calculated. Sentinel and population studies thus provide information about both the incidence of disease and the proportion of patients who consult general practitioners about gastroenteritis.

Reporting of Food-borne Disease

Most countries have systems for reporting notifiable diseases, but few have programmes for the surveillance of food-borne disease. Hence little is known about such disease on a worldwide basis. Annual reports are generated by only a few countries, including England and Wales, Canada, Japan and the United States. However, during the 1990s many European countries began to produce reports on food-borne disease, under the auspices of the WHO surveillance programme for the control of food-borne infections and intoxications in Europe. A few other countries are also attempting to develop programmes for reporting food-borne disease, but are hampered by lack of resources.

Reports of outbreaks of food-borne disease are produced all over the world by scientists, who carry out, for example, case–control studies and sentinel and population studies. Although their work is usually done on an ad hoc basis, it is of paramount value in yielding information about new and emerging food-borne diseases.

Causes of Food Poisoning Outbreaks

Annual reports of countries generally provide only superficial information about food poisoning outbreaks, such as the types of organisms and foods involved, general factors contributing to the occurrence of disease and trends associated with the emergence of causative organisms.

Table 4 lists the most predominant organisms involved in outbreaks of food-borne disease. It incorporates the most recent reported data gathered in association with outbreaks in England and Wales, the Netherlands and the USA. The results presented are not representative for all food-borne complaints, because single cases, which represent the majority, are not included. In most reports only five to seven types of organism are mentioned, whereas more than 50 hazardous food-borne organisms are known.

Table 5 lists some food vehicles associated with food poisoning outbreaks. The information presented provides only a rough indication, and may be subject to much bias. In fact, all types of foods are involved in food-borne disease, although foods of animal origin are the most predominant.

The most important factors contributing to outbreaks of food-borne disease, as identified by the WHO Centre for Control of Foodborne Infections and Intoxications in Europe, include:

- poor general hygiene
- consumption of raw ingredients
- use of contaminated ingredients
- contamination by infected persons
- cross-contamination
- use of contaminated equipment
- failures in processing
- preparation too early in advance
- inadequate heating

- inadequate hot holding
- inadequate refrigeration
- too long a storage time
- contamination during final preparation
- inadequate heating before reuse.

These factors are useful pointers for improving general hygiene, and the majority of them are within the control of the consumer.

Trends become evident in the case of only some organisms. A particularly good example of such an organism is *Salmonella*, because many countries have well-developed systems for the surveillance of *Salmonella* serotypes, isolated from patients seeking medical attention.

Reliability of Information

Annual reports on food-borne disease provide a poor reflection of reality, showing only the tip of the iceberg. Normally only outbreaks (in which two or more persons are involved) are recorded, and so no information about sporadic cases is available. A sentinel and population study carried out in the Netherlands indicated that annually about 95 000–165 000

Table 5 Types of food involved in reported food-borne diseases of known aetiology, in order of frequency (reproduced with permission from Notermans S and Borgdorff MW (1997))

England and Wales 1992–1993	The Netherlands 1991–1994	Spain 1990–1992
Poultry meat	Chinese food	Eggs and mayonnaise
Eggs	Poultry and eggs	Bakery products
Pork and ham	Meat	Meat and poultry
Meats and pies	Foreign foods	Fish and shellfish
Dairy products	Dairy products	Cheese
Mixed foods	Bakery products	Milk
Bakery products	Fish and shellfish	Canned products
Rice	Fruit and vegetables	
Beef		
Lamb and mutton		
Puddings		
Vegetables		

Table 4 Microorganisms reported most frequently as the cause of food-borne and water-borne disease during the 1990s (Reproduced with permission from Notermans S and Borgdorff MW (1997))

Microorganism	Incidence of specific microorganism, as a percentage of total		
	England and Wales, 1992–1993	The Netherlands 1991–1994 (n = 148)	USA 1989–1992 (n = 657)
Salmonella	68	25	70
Clostridium perfringens	17	15	6
Escherichia coli	0.2 (Type O157)	14	2
Bacillus cereus	3	19	3
Staphylococcus aureus	2	5	6
Campylobacter	2	2	3
Viruses	6 (Small round structured viruses)	2	
Shigella	4		3
Clostridium botulinum			6

persons suffered from salmonellosis, and 270 000–345 000 suffered from campylobacteriosis (incidence rates 6–11 and 18–23 cases per 1000 individuals per year, respectively). These infections are considered to be mainly transmitted by food. In the Dutch annual overview of food-borne disease, only 1279 cases (1264 in outbreaks and 15 sporadic cases) were reported in the period covered by the sentinel and population studies, a causative agent being found in 17% of these cases. *Salmonella* and *Campylobacter* accounted for 27% and 10% respectively of the cases with known aetiology, i.e. 58 persons were diagnosed as suffering from salmonellosis and 21 from campylobacteriosis. Hence the annual report mentioned < 0.1% of the true frequency of disorders caused by food. Since sporadic cases are known to be strongly under-represented, published statistics must be treated with care – particularly if they are to be used as a basis for risk management. Nevertheless, reports on food-borne diseases are of the utmost importance in stimulating safe food production.

See also: **Biochemical and Modern Identification Techniques**: Food Poisoning Organisms. **Campylobacter**: Introduction. **Cryptosporidium**. **Cyclospora**. **Eggs**: Microbiology of Fresh Eggs. **Escherichia coli**: Escherichia coli; Detection of Enterotoxins of *E. coli*; *Escherichia coli 0157:H7*. **Hazard Appraisal (HACCP)**: The Overall Concept; Critical Control Points; Involvement of Regulatory Bodies. **Ice Cream**. **International Control of Microbiology**. **Listeria**: Listeria monocytogenes. **National Legislation, Guidelines & Standards Governing Microbiology**: European Union. **Process Hygiene**: Involvement of Regulatory Bodies. **Salmonella**: Introduction; *Salmonella enteritidis*.

Further Reading

Borgdorff MW and Motarjemi Y (1997) Surveillance of food-borne diseases: What are the options? WHO/FSF/FOS/97.3

Notermans S and Borgdorff M (1997) A global perspective of foodborne disease. *Journal of Food Protection* 60: 1395–1399.

World Health Organization (1997) Food safety and foodborne disease. In: *World Health Statistics Quarterly*. Vol. 50, no. 1/2.

Food Preservation *see* **Bacteriocins**: Potential in Food Preservation; **Heat Treatment of Foods**; **High Pressure Treatment of Foods**; **Lasers**: Inactivation Techniques; **Microbiology of Sous-Vide Products**; **Minimal Methods of Processing**; Electroporation – Pulsed Electric Fields; **Ultrasonic Standing Waves**; **Ultra-Violet Light**

FREEZING OF FOODS

Contents
Damage to Microbial Cells
Growth and Survival of Microorganisms

Damage to Microbial Cells

Rekha S Singhal and **Pushpa R Kulkarni**, University Department of Chemical Technology, University of Mumbai, Matunga, Mumbai 400 019, India

Thermophilic and mesophilic organisms cease to grow at temperatures above 0°C, but psychrophilic and psychrotrophic organisms can grow – albeit slowly – at temperatures below the freezing points of foods. *Bacillus psychrophilus*, for instance grows at – 5°C to – 7°C with a generation time of 204 h. Other psychrophilic bacteria can also grow at temperatures as low as – 7°C, while some moulds can apparently grow at – 10°C. Moulds such as *Fusarium sporotrichoides* can produce toxin at temperatures as low as – 4°C while cold-tolerant pathogenic bacteria such as *Yersinia enterocolitica* can grow at temperatures as low as – 2°C. Viruses are thought to become more stable on refrigeration and freezing.

Nature of Freezing Processes

Freezing commences in foods at – 1 to – 3°C, depending on the concentrations of solutes in the aqueous phase. As the temperature is reduced below the point where freezing starts, progressively larger fractions of the water are frozen. For instance, in meat a small fraction of water in meat is not frozen at – 50°C, while in fruits and vegetables the corresponding temperature is – 16 to – 20°C. The available water in

Table 1 Water activity (a_w) of meat at various subfreezing temperatures

Temperature (°C)	a_w
25	0.993
–1	0.990
–3	0.971
–5	0.953
–7	0.934
–9	0.916
–11	0.899
–13	0.881
–15	0.864
–l17	0.847
–19	0.831
–21	0.815
–25	0.784
–30	0.746

Table 2 Minimum water activity (a_w) permitting microbial growth

a_w	Microorganisms
0.98	*Clostridium botulinum* type C, few strains of *Pseudomonas*
0.97	*C. botulinum* type E
0.96	*Flavobacterium, Klebsiella, Shigella*; few strains of *Lactobacillus, Pseudomonas*
0.95	*Alcaligenes, Bacillus, Citrobacter, Clostridium botulinum* types A and B, *C. perfringens, Enterobacter, Escherichia, Proteus, Pseudomonas, Salmonella, Serratia* and *Vibrio*
0.94	*Lactobacillus, Microbacterium, Pediococcus*; few strains of *Streptococcus* and *Vibrio*
0.93	Few strains of *Lactobacillus, Streptococcus, Rhizopus, Mucor*
0.92	*Rhodotorula* and *Pichia*
0.91	*Corynebacterium*, anaerobic *Staphylococcus* and few strains of *Streptococcus*
0.90	Few strains of *Lactobacillus* and *Vibrio, Micrococcus, Pediococcus, Hansenula, Saccharomyces*
0.88	*Candida, Debaryomyces, Torulopsis, Cladosporium*
0.87	Few strains of *Debaryomyces*
0.86	Aerobic *Staphylococcus, Paecilomyces*
0.80	Few strains of *Saccharomyces, Aspergillus, Pencillium*
0.75	Halophilic bacteria, few strains of *Aspergillus*
0.70	*Eurotium, Chrysosporium*
0.62	Few strains of *Saccharomyces* and *Eurotium, Monascus*

foods is indicated by the term water activity (a_w) and this is defined as the ratio of vapour pressure of the water present in the food to the vapour pressure of pure water at the same temperature. Ice formation reduces the water that is potentially available for growth of microorganisms. This reduces a_w considerably, as can clearly be seen from **Table 1**, which shows the a_w of meat at various subfreezing temperatures. The minimal a_w permitting the growth of microorganisms is as shown in **Table 2**. As the temperature falls from 0°C, a series of ice–solute mixtures

or eutectics is formed, accompanied by an increasing concentration of dissolved solids in the unfrozen water. This concentration of solutes continues until the eutectic point is reached, when the remaining solution solidifies. The lowest temperature at which the solution remains liquid is referred to as the eutectic temperature. This value is – 21.8°C for sodium chloride, at which the concentration of the solution is about 5 mol l^{-1}. There is a strong interaction between low temperature and low a_w, which work synergistically for the cessation or inhibition of microbial growth.

The concept of freezing rate has been introduced, since it is meaningless to compare freezing times for products of vastly different sizes. Rate of freezing can be expressed in several ways. The most commonly used method is the change of temperature per unit time. Rates of freezing may be classified as slow when the substrate cools at rates < 10°C per hour, moderate when the substrate cools at rates between 10 and 50°C per hour, and rapid when the rate of cooling is > 50°C per hour. Most commercial freezing of food occurs at moderate rates. However, this drop varies from the surface of the food to the centre, making this approach less useful for characterization of freezing processes. Another approach to expressing freezing rate is in terms of average velocity of advancement of ice front from the surface to the thermal centre, and this is defined as ratio between the minimum distance from the surface to the thermal centre and the freezing time. Freezing rate according to this method is expressed as cm h^{-1}, and is classified as ultrarapid (over 10), rapid (1–10), normal (0.3–1.0), slow (0.1–0.3) and very slow (< 0.1).

The viability of the organisms is enhanced at rapid rates of freezing in the range of 1–10°C per minute. This is mainly because of lower contact time of the susceptible organisms with the harmful high solute concentration in the unfrozen water. Within the ultrarapid range of cooling at 10–100°C per minute, viability of organisms decreases. This is believed to be due to the formation of internal ice crystals which destroy the cell membrane. At extremely fast freezing rates ranging from 100 to 10 000°C per minute, such as those encountered in liquid nitrogen freezing, ice crystal formation is reduced and an improvement in bacterial viability is observed.

Freezing begins from the exterior of the food surface and progresses towards the interior. The freezing rate is a function of the size and surface area of the unit being frozen, its thermal conductivity and the temperature gradient within the food. Weight loss during freezing, termed desiccation, is inevitable due to evaporation or sublimation. It is undesirable, since excessive desiccation results in the removal of ice from the surface layers. This allows free access to oxygen which initiates and accelerates oxidative changes, causing irreversible changes in the taste, texture and appearance of the product surface. This is called freezer burn and is especially seen in foods of animal origin. Desiccation occurs when the surface is not cooled quickly – a condition that favours sublimation. Fast freezing lowers the surface temperature of the product quickly, resulting in negligible sublimation. Desiccation is hence common in foods subjected to slow freezing compared to rapid or ultrarapid freezing. The microorganisms distributed within the food will also be subjected to different freezing rates, and those on exposed surfaces will be exposed to desiccation similar to that of the food product.

Microbial cells behave like solute molecules during freezing and become partitioned and concentrated in the unfrozen part of the solution as ice crystals form. This can alter the pH, in some foods by as much as 2 pH units. Biological macromolecules such as DNA and proteins may be denatured by the increased ionic strength of the unfrozen phase.

Freezing affects the intracellular water and the water surrounding the microorganisms. Slow freezing rates cause formation of large ice crystals which move slowly through the food being frozen and thereby concentrate the solutes and microorganisms in the unfrozen phase ahead of them. Thus, microorganisms are exposed to high solute concentrations, and water is transferred from the interior of the cells to the medium by osmosis, leading to the plasmolysis of the cells. Temperature fluctuations during storage affect the movement of solutes, the growth of ice crystals, and also the loss of water from the product by sublimation. All these factors collectively damage the cells. On the other hand, rapid freezing causes numerous small ice crystals to be formed both intra- and extracellularly without the concentration of solutes that occur in slow freezing. Microbes therefore retain their normal size, although they do show membrane disintegrity as a manifestation of the formation of ice crystals. Lesions in the cell membrane reduce the ability of the microbial cell to retain low molecular solutes and maintain its internal environment. The effects of this damage are only visible after thawing and when the conditions are made suitable for the initiation of growth. Commercial freezing shows damages typical of both rapid and slow rates of freezing, although the cells are less damaged than those subjected to slow or rapid freezing; there is also less evidence of membrane damage and dehydration of microbial cells.

Freezing results in loss of cytoplasmic gases such as oxygen and carbon dioxide. A loss of oxygen to aerobic cells suppresses respiratory reactions.

Effect on Microflora

The cold sensitivity of a microbe in a food depends on the type, species, strain and physiological condition of the organism, the rate of freezing and the composition of the food. The lethal effects of freezing on the microbial cells is evident from the fact that supercooling and freezing at the same temperature have been shown to give 97% and 2% survival of cells, respectively. In general, microorganisms that are resistant to heat, chemical treatments or radiation are also resistant to freezing, and vice versa. However, exceptions to this general rule have been encountered. Cells of *Escherichia coli* that have grown under nitrogen limitation are more resistant to freezing, suggesting the formation of polyglucose of glycogen-like reserve material which can work as cryoprotectants by virtue of their ability to strengthen the cell envelope or outer membrane. Cryoprotectants are solutes that protect a material from freezing damage. The mechanism of cryoprotection may be through dilution of harmful solutes. Cryoprotectants help to minimize the translocation of components through damaged membranes, and thereby allow time for repair mechanisms to function.

Generally, bacteria in the exponential phase are more sensitive to freeze–thaw stress than those in the stationary phase. In the case of *Pseudomonas*, cells in the exponential phase are more sensitive than stationary-phase cells to freezing at cooling rates of the order of 10°C per minute while the reverse is found for cooling rates of further than 100°C per minute. An interesting observation has been reported: when exponential-phase competitor cells are introduced at a level of 10^8 cfu ml^{-1}, exponential-phase cells of *Salmonella typhimurium* (at 10^5 cells per millilitre) gain a level of resistance that is commensurate with that of the cells in the stationary phase. The genus of the competitor cells is unimportant, but there is an absolute requirement of the cells to be alive. Similarly, cells of *S. typhimurium* that have grown at 25°C are more resistant to freezing than those grown at 37°C. In the pH range found in foods, the percentage of microorganisms that survive freezing is usually greatest at neutral or slightly alkaline pH values. Similarly, addition of 50% sucrose to orange juice has been shown to increase the yeast cells that survive freezing from below 0.1 to 54%. Other food ingredients also alter the survival of microorganisms to freezing.

Thawing is more injurious to microorganisms than freezing, and the effects vary according to the species. This is particularly true at rapid freezing rates. Even simple thawing of a frozen microbial population without intervening storage causes slight to moderate reduction in number of live organisms. Many times it

Table 3 Number of viable microorganisms and percentage of total count in beef before and after freezing at $-30°C$

	Before freezing	After freezing
Microorganisms per gram	385 000	77 000
% of total count:		
Gram-positive bacteria:		
Micrococcus	5	2
Leuconostoc		
Lactobacillus	5	
Bacillus		2
Gram-negative bacteria:		
Pseudomonas	75	22
Vibrio + Aeromonas	6	7
Achromobacter	4	1
Yeasts		

has been observed that repeated freeze–thaw cycles are more lethal than continuous frozen storage.

In general, bacterial spores are unaffected by freezing. Clostridial spores, for instance, can survive almost 20 years of frozen storage. This has been attributed to the relatively dehydrated state of the spore protoplast with much of its water bound within the expanded cortex. Gram-positive cocci such as micrococci and streptococci and rods are generally more resistant to freezing than Gram-negative bacteria such as *Escherichia*, *Salmonella*, *Serratia*, *Pseudomonas*, *Moraxella* and *Vibrio*. An exception is *Clostridium perfringens*, which is very sensitive to frozen storage. Gram-positive cocci are usually more resistant than Gram-positive rods. This has been observed in yoghurts, where lactobacilli suffer a 4–6 log reduction in numbers during frozen storage and handling before inoculation in milk. The presence of mucoprotein complexes and diaminopimelic acid in the cell walls of Gram-positive cells seem to protect the membrane proteins against denaturation. This is seen from **Table 3**, which shows the number of viable organisms and percentage of total count in beef before and after freezing at $-30°C$. The spores and vegetative cells of yeasts and moulds survive freezing and frozen storage quite well. Frozen bread dough is a common food where the quality depends on the microbial survival. The survival of yeast cells is excellent if the dough is frozen immediately after the yeast is added, but poor if the yeast cells remain in the dough for even 1 h before the dough is frozen. The delay allows budding of the yeast cells – a state which increases their sensitivity to freezing. Hence, thawing of frozen bread dough during transportation and storage, to a temperature permitting yeast growth, can result in severe damage to the yeast during subsequent refreezing.

Survival rates of the order of 75% have been observed with air-dried conidia of *Aspergillus flavus* which have been frozen at $-73°C$ and thawed rapidly. Organisms that are sensitive to freezing are free-living amoebae, ciliated protozoa and nematodes.

Freeze-injury in microorganisms is caused by a sudden drop in temperature, ice formation and increased solute concentration. It causes loss of viability, leakage of the cellular materials, increased sensitivity to surfactants and other compounds, increased nutritional needs, extended lag and increased sensitivity to radiation. Death in frozen *E. coli*, for instance, is proportional to the amount of nucleic acid lost through leakage, and also to the protein-like substances and ATP. The sensitivity to surfactants and other inhibitory compounds like antibiotics is thought to be due to the impairment by freezing of barriers which normally protect the cells. Repair of the injured cells is possible in the presence of nutrients such as trypticase soy broth, potassium dihydrogen phosphate, pyruvate and ATP.

Pathogenicity is affected differently in different organisms due to freezing and thawing. While there is no loss in pathogenicity of *Salmonella gallinarium*, repeated freeze–thawing of *Treponema pallidum* was found to result in complete loss of pathogenicity. Other functions such as respiration are also affected in different ways depending on the organisms. While repeated freeze–thawing initially increased the respiratory activity of *E. coli* for the first several cycles, it was totally lost at the 40th cycle. Other studies point towards the temperature of freezing as altering the respiration rate of organisms. Freezing also induces reduction in phosphorylation activity, probably by some alteration in the electron transport particles within the membrane structure.

Importance of Freezing Rate and Final Temperature

Rate of cooling has a profound effect on the susceptibility of microorganisms to survive. Membrane lipids and membrane-associated processes such as active transport, respiration and growth are vital for the organism to survive. Viability of fungal spores such as *Alternaria*, *Fusarium*, *Penicillium* and *Aspergillus flavus* has been shown to be higher in meat samples that have been thawed rapidly. The introduction of liquid freezing has raised a question of survival of microorganisms around $-196°C$. A mixed aerobic population has been reportedly recovered from vegetables frozen at $-196°C$, and the same is the case with pathogenic bacteria and fungi. It has been concluded that microorganisms withstand the temperature of liquid nitrogen at least as well as – and perhaps better than – they withstand customary freezer temperatures.

The rate of thawing also influences the number of organisms surviving the freeze–thaw cycle: higher recoveries are recorded with faster thawing. Rate of thawing has little effect on microbial cells that have been frozen at low cooling rates ($< 100°C$ per minute). However, at rapid and ultrarapid cooling rates, survival is greater at rapid warming rates ($> 500°C$ per minute) than after slow thawing ($< 12°C$ per minute). This is attributed to ice crystal growth at slow rates of thawing which is prevented or minimized at rapid thawing rates.

The microflora of a thawed frozen food is entirely different from that of a fresh counterpart.

Nature of Sublethal Damage

The rate of freezing, storage temperature and temperature fluctuations during storage influence the extent of sublethal injury and death. Most work in scientific literature concentrates on cells which have been subjected to freeze–thaw cycles. Higher subfreezing temperatures (-2 to $-5°C$) are generally more injurious to cells than lower temperatures ($-10°C$). Low pH accelerates death and increases sublethal injury, while common food components such as sugars, amino acids, peptides and glycerol protect microorganisms from frozen damage (cryoprotectants). Such compounds are formed in the microenvironment of the microbial cells as leakage products due to damage to the integrity and permeability of the cell membrane. If a large number of microorganisms are present, the concentration of such compounds may be sufficient to provide some degree of cryoprotection. Salt concentration influences osmotic dehydration, and can be critical by increasing damage during slow freezing. The sensitivity of the injured cells to environmental stress such as low pH is greatly increased.

Sublethal injury of the microbial cells encountered during freezing is reversible, and can regain the characteristics of normal cells within several hours if suitable temperature and nutrients are made available. Gram-negative bacteria lose their ability to use inorganic nitrogen after freezing and need peptides as the starting material for protein synthesis, since they cannot utilize amino acids directly. However, under stress conditions or in the presence of inhibitory agents, such sublethally injured cells lose viability. These indicate that the genetic material of the cell is not permanently altered, and the consequences of freeze–injury are not transmitted at cell division.

Various mechanisms have been proposed to explain how organisms survive freezing and frozen storage. Among the many mechanisms, the following are exemplary:

1. Certain organisms such as *Vibrio parahaemolyticus* are halophilic and able to survive at low temperatures due to the high concentration of sodium chloride in the foods preserved under the said condition. Many bacteria, especially of marine origin, are stabilized in the presence of cations because they are able to maintain the integrity of the cell wall, which in turn stabilizes the cell envelope.
2. Inhibition of protein synthesis resulting from low temperature exposure of an organism has been reported by several workers.

Recovery from Sublethal Damage

Recovery of the injured cells from sublethal damage occurs during the lag phase of the organism, when there is an intense metabolic activity involving the synthesis of the RNA and ATP, and also reorganization of the membrane components such as lipopolysaccharides. All these phenomena take place without accompanying growth. Since a wide range of damage is caused by freezing and various conditions are needed for repair, a general agreement on optimum conditions for recovery of organisms on particular foods is not very clear. This is further complicated by the non-uniform distribution of the microorganisms in the food, a wide variation of the spread in counts (i.e. standard deviation), and possible differences in the behaviour of natural contaminants compared to artificially contaminated ones. The importance of energy synthesis in repair mechanisms was demonstrated by the observation that 2,4-dinitrophenol, an uncoupler of oxidative phosphorylation, inhibited repair or injury.

Examination of a number of factors, including different types of resuscitation media such as peptone buffer diluent, mineral-modified glutamate, trypticase soy agar and non-resuscitation PCA, on the recovery of natural microflora from frozen minced beef indicated a wide range of recovery from $+178\%$ to -50%. Conclusions on the best possible method with respect to recovery of injured cells are therefore difficult unless the design of the experiments is done carefully, taking into account the wide variation. Correlations between the physiological basis of injury and of repair are not available. Unless such data are available, design of recovery media for injured cells is difficult.

See also: **Aspergillus**: Aspergillus flavus. **Clostridium**: Clostridium perfringens. **Escherichia coli**: Escherichia coli. **Freezing of Foods**: Growth and Survival of Microorganisms. **Pseudomonas**: Introduction. **Salmonella**: Introduction. **Vibrio**: Introduction, including *Vibrio vulnificus*, and *Vibrio parahaemolyticus*.

Further Reading

Brown MH (1991) Microbiological Aspects of Frozen Foods. In: Bald WB (ed.) *Food Freezing: Today and Tomorrow*. London: Springer-Verlag.

Davies R and Obafemi A (1985) Response of Microorganisms to Freeze–thaw Stress. In: Robinson RK (ed.) *Microbiology of Frozen Foods*. London: Elsevier Applied Science.

Dodd CER, Sharman RL, Bloomfield SF, Booth IR and Stewart GSAB (1997) Inimical processes: bacterial self-destruction and sub-lethal injury. *Trends in Food Science and Technology* 8: 238–241.

Frazier WC (1967) Preservation by Use of Low Temperature. In: *Food Microbiology*, 2nd edn. New York: McGraw-Hill.

Gray RJH and Sorhaug T (1983) Response, Regulation and Utilization of Microbial Activities at Low Temperatures. In: Rose AH (ed.) *Economic Microbiology*. vol. 8. *Food Microbiology*. London: Academic Press.

Hayes PR (1985) Food Spoilage. In: *Food Microbiology and Hygiene*. London: Elsevier Applied Science.

Jay JM (1978) Food Preservation by the Use of Low Temperatures. In: *Modern Food Microbiology*. New York: D. Van Nostrand.

Michener HD and Elliott RP (1969) Microbiological Conditions Affecting Frozen Food Quality. In: Van Arsdel WB, Copley MJ and Olson RL (eds.) *Quality and Stability of Frozen Foods Time–Temperature Tolerance and its Significance*. New York: John Wiley.

Reid DS (1994) Basic Physical Phenomenon in the Freezing and Thawing of Plant and Animal tissues. In: Mallett CP (ed.) *Frozen Food Technology*. London: Blackie Academic & Professional.

Rosset R (1982) Chilling, Freezing and Thawing. In: Brown MH (ed.) *Meat Microbiology*. London: Applied Science.

Growth and Survival of Microorganisms

Parimal Chattopadhyay, Department of Food Technology and Biochemical Engineering, Jadavpur University, India

The low temperature preservation of foods is based on the principle that the activities of food-borne microorganisms can be slowed down at temperatures around freezing point and stopped at temperatures below freezing point. The metabolic reactions of microorganisms are catalysed by enzymes and its rate is dependent on temperature. The rate of reaction is increased with increase in temperature. The temperature coefficient (Q_{10}) value varies between 1.5 and 2.5. The temperature is always related to the relative humidity of the reaction environment and therefore sub-freezing temperatures also affect the

Table 1 Bacterial genera containing species or strains known to grow at or below 7°C

Gram negatives	Relative numbers[a]	Gram positives	Relative numbers[a]
Acinetobacter	XX	Bacillus	XX
Aeromonas	XX	Brevibacterium	X
Alcaligenes	X	Brochothrix	XXX
Alteromonas	XX	Carnobacterium	XXX
Cedecea	X	Clostridium	XX
Chromobacterium	X	Corynebacterium	X
Citrobacter	X	Deinococcus	X
Enterobacter	XX	Enterococcus	XXX
Erwinia	XX	Kurthia	X
Escherichia	X	Lactococcus	XX
Flavobacterium	XX	Lactobacillus	XX
Halobacterium	X	Leuconostoc	X
Hafnia	XX	Listeria	XX
Klebsiella	X	Micrococcus	XX
Moraxella	XX	Pediococcus	X
Morganella	X	Propionibacterium	X
Photobacterium	X	Vagococcus	XX
Pantoea	XX		
Providencia	X		
Pseudomonas	XXX		
Psychrobacter	XX		
Salmonella	X		
Serratia	XX		
Shewanella	XXX		
Vibrio	XXX		
Yersinia	XX		

[a]Relative importance and dominance as psychrotorophs: X, minor; XX, intermediate; XXX, very significant.

relative humidity and the pH of the system for microbial growth.

Temperature and Microbial Growth

Bacterial species and strains that can grow at or below 7°C are mainly Gram-negative with some extent Gram-positive microorganisms (**Table 1**). The lowest recorded temperature of growth for a microorganism in food is –34°C, yeasts and moulds are more likely to grow at temperatures below 0°C compared to bacteria (**Table 2**). This is due to the fact that fungi can grow under conditions of lower water activity (a_w). Bacteria are reported to grow at –20°C and around –12°C. Foods which support microbial growth at subzero temperatures include fruit juice concentrates, bacon, ice cream and certain fruits. These food products contain cryoprotectants that depress the freezing point of water.

Prefreezing Operations and Microbial Growth

Before freezing vegetables, the prefreezing operations include selecting, sorting, washing, blanching and packing. The spoiled foods are rejected before freez-

Table 2 Minimum reported growth temperatures of some food-borne microbial species and strains that grow at or below 7°C (Source: Roberts et al (1981))

Species/Strains	°C	Comments
Pink yeast	−34	
Pink yeasts (2)	−18	
Unspecified moulds	−12	
Vibrio spp.	−5	True psychrophiles
Yersinia enterocolitica	−2	
Unspecified coliforms	−2	
Brochothrix thermosphacta	−0.8	Within 7 days; 4°C for 10 days
Aeromonas hydrophila	−0.5	
Enterococcus spp.	0	Various species/strains
Leuconostoc carnosum	1.0	
L. gelidum	1.0	
Listeria monocytogenes	1.0	
Leuconostoc sp.	2.0	Within 12 days
L. sakelcurvatus	2.0	Within 12 days; 4°C in 10 days
Clostridium botulinum B, E, F	3.3	
Pantoea agglomerans	4.0	
Salmonella panama	4.0	In 4 weeks
Serratia liquefaciens	4.0	
Vibrio parahaemolyticus	5.0	
Salmonella heidelberg	5.3	
Pediococcus sp.	6.0	Weak growth in 8 days
Lactobacillus brevis	6.0	In 8 days
L. viridescens	6.0	In 8 days
Salmonella typhimurium	6.2	
Staphylococcus aureus	6.7	
Klebsiella pneumoniae	7.0	
Bacillus spp.	7.0	165 of 520 species/strains
Salmonella spp.	7.0	65 of 109, within 4 weeks

ing. Other foods like meats, seafoods, poultry should be as good as fresh.

The blanching process is carried out by immersing foods in hot water or by open steam. Apart from reducing the number of microorganisms on the foods, the blanching process helps to fix the green colour of vegetables, inactivates enzymes which may cause undesirable changes in foods during frozen storage and removes entrapped air in plant tissues which may create problems during freezing. The method of blanching depends on the type of food, size of packs etc. It is possible to reduce microbial loads by as much as 99% by blanching, if the operation is carried out carefully. Bacterial spores should not be allowed to recontaminate the food. Milk pasteurization temperatures (63°C for 30 min) destroy most vegetative bacterial cells which are responsible for spoilage of vegetables and blanching reduces the vegetative cells.

Effect of Freezing on Microbial Growth

There are three types of freezing process: (1) ultra-rapid freezing in which the temperature of the food is lowered to about −20°C within 5–10 min; (2) quick or fast freezing where the time requirement is 30 min for same temperature reduction; and (3) slow freezing which takes 3–72 h to achieve the same food temperature. As the freezing starts from the surface of the food material, the rate of freezing is expressed as the rate of movement of this freezing front towards the geometric centre of the food material. The freezing rate is $2 \, mm \, h^{-1}$ for slow freezing, $5–30 \, mm \, h^{-1}$ for quick or fast freezing and $50–1000 \, mm \, h^{-1}$ for ultra-rapid freezing.

Slow freezing favours the formation of large extracellular ice crystals and quick freezing favours the formation of small extracellular ice crystals. Crystal growth is one of the factors that affects the storage life of certain frozen foods. During storage these ice crystals grow in size and cause cell damage by disrupting membranes, cell walls and internal structures, so that when thawed the product is different from the original in texture and flavour. Food materials may be viewed as dilute biological systems whose freezing point varies from −1 to −3°C depending on the nature and concentration of solute present (**Fig. 1**). At the freezing point or on further cooling below the freezing point, ice crystals begin to separate or, in the absence of nucleation, the liquid becomes supercooled. As soon as ice begins to form, the dissolved solutes are concentrated in the remaining liquid. As the temperature is further reduced and more water is converted into ice the solute concentration rises gradually in the unfrozen water portion with a corresponding decrease in freezing point and water activity (a_w). The vapour pressures of water and ice at various temperatures and water activity are reported in **Table 3**. This phenomenon continues as the temperature is lowered, until the eutectic point is reached and the remaining solution then solidifies. Microbial cells, suspended in aqueous solutions during freezing, behave like solute molecules and become partitioned and concentrated in the unfrozen portion of the solution as ice crystals form. They are thus exposed to the effects of concentrated solute and to the results of localized ice crystal growth. The depression of freezing point of the remaining unfrozen water, the increased solute concentration and the progressive lowering of water activity (a_w) have increasingly deleterious effects on the microbial population. Therefore, organisms which are capable of growth in foods at subzero temperature must also tolerate lowered a_w. A small percentage of water remains unfrozen at temperatures well below −100°C, however, for practical

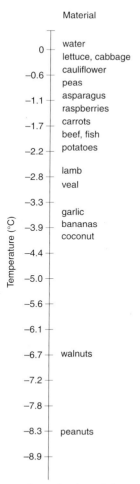

Figure 1   Freezing points of selected foods (from Desrosier (1977)).

Table 3   Vapour pressures of water and ice at various temperatures (Source: Jay (1992))

°C	Liquid water (mmHg)	Ice (mmHg)	$a_w = P_{ice}/P_{water}$
0	4.579	4.579	1.00
−5	3.163	3.013	0.983
−10	2.149	1.950	0.907
−15	1.436	1.241	0.864
−20	0.943	0.776	0.823
−25	0.607	0.476	0.784
−30	0.383	0.286	0.75
−40	0.142	0.097	0.68
−50	0.048	0.030	0.62

positive rods and cocci are more resistant than Gram-negative bacteria. The viability of organisms is enhanced as the freezing rate is increased (**Fig. 2**). This increase in survival may be due to the diminishing time that the susceptible organism is in contact with harmful high solute concentrations in the unfrozen water (curve a). When freezing is more rapid, viability decreases due to the formation of internal ice crystals causing destruction of the cell membranes (curve b). With extremely fast freezing rates, for example, using liquid nitrogen, ice crystal formation is reduced and is replaced by 'vitrification' (curve c). When foods are frozen commercially the bacterial viability will be predominantly as in curve a. There are certain substances such as glucose, milk solids, fats and sodium glutamate which are known to be 'protective' and improved viabilities are obtained in their presence. The mechanism of action of these cryoprotectants which prevent freezing damage to microbial cells are yet to be confirmed. The effects of freezing several species of *Salmonella* to −25.5°C and holding for up to 270 days are presented in **Table 4**. Although there was a significant reduction in viable numbers over the 270 days of storage with most species, in no case did all cells die.

Survival of Microorganisms at Low Temperature

A number of psychrophilic and psychrotrophic microorganisms can grow at temperatures between −5 and −7°C (e.g. *Bacillus psychrophilus*) but rarely at temperatures below −10°C. Such organisms have a long generation time (8–9 days) and therefore cannot proliferate during the freezing process which is much faster (10–90 min). Psychrophiles are different in sensitivity to cold shock from thermophiles and mesophiles. However, cold shock depends on the magnitude of the temperature differential rather than the exact low temperature required for growth.

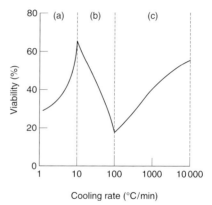

Figure 2   Effect of freezing on viability of typical Gram-negative rod (from Hayes (1992)).

purposes the 'freezable' water in meat and fish is totally frozen at −50 to −70°C and in fruits and vegetables at −16 to −20°C.

Although some microorganisms are killed by freezing, approximately 50% may survive depending on the type of organism, the rate of freezing and the composition of substrate being frozen. Bacterial spores are unaffected by freezing and in general Gram-

Table 4 Survival of pure cultures of enteric organisms in chicken chow mein at −25.5°C (Source: Jay (1992))

Organism	Bacterial count (10^5/g) after storage for (days)								
	0	2	5	9	14	28	50	92	270
Salmonella newington	7.5	56.0	27.0	21.7	11.1	11.1	3.2	5.0	2.2
S. typhimurium	167.0	245.0	134.0	118.0	11.0	95.5	31.0	90.0	34.0
S. typhi	128.5	45.5	21.8	17.3	10.6	4.5	2.6	2.3	0.86
S. gallinarum	68.5	87.0	45.0	36.5	29.0	17.9	14.9	8.3	4.8
S. anatum	100.0	79.0	55.0	52.5	33.5	29.4	22.6	16.2	4.2
S. paratyphi B	23.0	205.0	118.0	93.0	92.0	42.8	23.3	38.8	19.0

During freezing, the temperature continues to drop and a_w of the environment is also reduced; this allows selective growth of microorganism. Organisms that are resistant to dehydration, like yeast and moulds, grow in such environments rather than bacteria. Most spores can survive all conditions of freezing and thawing. Bacterial endospores are extremely resistant to freezing and storage at subzero temperature and 90% are reported to survive. This may be due to the fact that in the dehydrated spore protoplast much of the water is bound in an unfreezable state within the expanded cortex. Fungal spores are also resistant to freezing and it has been reported that 75% of air-dry conidia of *Aspergillus flavus* cooled rapidly to −73°C and thawed rapidly, survived. However, only 3.2% of spores survived if they were suspended in water before freezing and <1% survived if the frozen spores were thawed slowly. A few vegetative bacterial cells are insensitive to freezing. Some Gram-positive staphylococci, micrococci and streptococci are relatively resistant with survival levels of 50% or more. Organisms which are very sensitive to the effects of freezing include the free-living amoebae, ciliated protozoa and nematodes. Storage of frozen foods at −10 to −20°C for a few days is lethal to *Toxoplasma gondii*, *Entamoeba* and trypanosomes although cooling rates are critical and lethality seems to be associated with the formation of intracellular ice at rates of 3°C per minute or above. The majority of microorganisms resist the immediate effects of freezing but are sensitive to frozen storage. Generally, most Gram-positive organisms, including *Bacillus*, *Clostridium*, *Corynebacterium*, *Lactobacillus*, *Microbacterium*, *Micrococcus*, *Staphylococcus* and *Streptococcus*, together with some yeasts are relatively resistant to freezing, although some such as *Cl. perfringens* are very sensitive to frozen storage. Gram-negative organisms, such as *Escherichia*, *Salmonella*, *Serratia*, *Pseudomonas*, *Acinetobacter*, *Moraxella* and *Vibrio* are more sensitive to both freezing and frozen storage and their survival depends on a number of variables such as cooling velocity, temperature, cell concentration, storage time and thawing conditions. For example, raw meat contained 15% Gram-positive and 85%

Gram-negative bacteria before freezing at −30°C. After freezing the total viable count decreased from 385 000 to 77 000 per gram and the proportion of Gram-positive and Gram-negative was 70% and 30%, respectively. Exponential phase cells of *Salmonella typhimurium* frozen in distilled water at −30°C by immersion in liquid freon for 44 min, showed less than 0.1% survival after rapid thawing. The previous nutritional conditions of growth have an effect on the sensitivity of microbial cells to freezing. Exponential-phase cells of *S. typhimurium* grown in tryptone–soy broth were 100 times more sensitive to freezing at −30°C than cells grown to exponential phase in a minimal medium (salt–glucose).

Age and Growth Rate

It has been shown that exponential-phase cells of *Salmonella typhimurium* are much more sensitive to freeze–thaw stress than stationary-phase cells. However, this is not a generally recognized phenomenon. Sensitivity of *Pseudomonas* to freezing, at different phases of growth is dependent on freezing rate. At a lower cooling rate ($\leqslant 10°\text{C min}^{-1}$) exponential-phase cells were more sensitive, but at a higher rate ($\geqslant 100°\text{C min}^{-1}$) stationary-phase cells were more sensitive. Cells of *S. typhimurium* grown at 25°C are more resistant to freezing stress than cells grown at 30°C. This may be due to changes in membrane properties as a function of growth temperature or it may reflect the different growth rates at mid-exponential phase, which are 0.25 h^{-1} and 0.17 h^{-1} at 37°C and 25°C, respectively.

Protection Offered by Food Components

One important factor governing the survival of microorganisms in frozen foods is the protective effects of specific components of food such as proteins, peptides, sugars and fats as well as other substances. There are also constituents of foods which enhance freezing injury of cells. Due to the presence of protective substances freezing may have only minor destructive effects on the original microflora of some foods.

An increase in freezing resistance in yeast cells was observed in solutions of potassium chloride, potassium nitrate and sodium chloride, whereas solutions of lithium, calcium and magnesium salts decrease resistance. Thus, monovalent cations, such as sodium and potassium, tend to stabilize cells against freezing damage to a greater extent than anions, such as chloride and nitrate. Sodium chloride is widely used for the preservation of foods. Despite the stabilizing effects of low concentrations of sodium chloride, at the concentrations found in foods, the sensitivity of cells to freezing increases with increases in salt concentration. Survival is enhanced when glucose, sucrose, erythritol, diglycol or polyethylene glycol are present in the medium in which microorganisms are frozen. The protective activities of these substances are not due to their penetrating into the cells. Instead, the interference of these substances with intracellular or extracellular freezing is responsible for the protective action. Other substances which offer protection against freezing and thawing injury to microorganisms are proteins and protein-related compounds. Milk provides protection to microorganisms during freezing and thawing due to its colloidal character, but it is not clear how colloids stabilize living systems against freezing injury. Low-molecular-weight compounds like amino acids and peptides may play a role in protection.

In a living system, protein is denatured when cells are injured. At low temperatures, freezing rupture of intramolecular bonds in the protein molecule is the cause of denaturation. This leads to the loss of activities of some enzymes. Hydrogen bonds are important for the stability of protein molecules. Therefore, compounds which stabilize such bonds protect against enzyme inactivation. Amino acids and related low-molecular-weight substances are among these protective substances. They display their protective activity at the metabolic level.

Sodium glutamate and aspartate protected tubercle bacilli during freeze-drying. Substances which are capable of forming hydrogen bonds are involved in the recovery of microorganisms from heat injury. Chemical compounds resembling glutamic acid in chemical structure prevent the death of bacteria from freezing. Considerable protection is provided by aspartic acid, malic acid, cysteic acid, pyrrolidone carboxylic acid, α-aminopimelic acid, acetylglycine, DL-threonine and DL-allothreonine. Among the effective compounds, there are some, such as α-aminopimelic acid, cysteic and pyrrolidone carboxylic acids which are not metabolized by most bacteria. Compounds like glutamic acid and asparagine are readily metabolized but have no protective effect, but the molecules remain protective if the $-NH_2$ group is replaced by some other polar group such as NH, –OH or O. The effectiveness of these compounds for protecting against freezing damage is the presence of a –COOH group at alpha and omega positions and an electronegative group (e.g. $-NH_2$) on the alpha carbon.

Mortality of *Escherichia coli* is diminished if the cells are frozen in lysates of *E. coli*, indicating that constituents of the cells exhibit a protective activity. However, the concentration of the products of lysis does not affect the inactivation rate, but the presence of spent growth medium, i.e. a filtrate of a stationary culture, in the freezing medium protects the cells even in high dilution.

Food materials contain a great variety of substances which have a protective activity. For example, in meat and processed meat preparations, protein and protein-related compounds may stabilize the cellular proteins by some sort of physicochemical activity besides allowing repair of metabolic injury by furnishing essential metabolites. Competition between different effects is involved when microorganisms are exposed to freezing in complex environments, such as food materials.

See also: **Bacteria**: The Bacterial Cell. *Escherichia coli*: *Escherichia coli*. **Freezing of Foods**: Damage to Microbial Cells. *Salmonella*: Introduction.

Further Reading

Brown AD (1990) *Microbial Water Stress*. Chichester: Wiley.

Desrosier NW (1997) *The Technology of Food Preservation*, 4th edn. USA: AVI Publishing.

Fennema OR, Powrie WD and Marth EH (1973) *Low Temperature Preservation of Foods and Living Matter*. New York: Marcel Dekker.

Hawthorn H and Rolfe EJ (eds) (1968) *Low Temperature Biology of Food Stuffs*. Oxford: Pergamon Press.

Hayes PR (1992) *Food Microbiology and Hygiene*, 2nd edn. London: Elsevier Applied Science Publishers.

International Institute of Refrigeration (1972) *Recommendations for the Processing and Handling of Frozen Foods*, 2nd edn. Paris: International Institute of Refrigeration.

Jay JM (1992) *Modern Food Microbiology*, 4th edn. New York: Von Nostrand Reinhold.

Mazur P (1970) Cryobiology: the freezing of biological systems. *Science* 168: 939.

Mrak EM and Stewart GF (eds) (1955) *Advances in Food Research*, vol. VI. New York: Academic Press.

Roberts TA, Hobbs G, Christian JHB and Skovgaard N (eds) (1981) *Psychrotrophic Microorganisms in Spoilage and Pathogenicity*. New York: Academic Press.

Robinson RK (ed.) (1985) *Microbiology of Frozen Foods*. London: Elsevier Applied Science Publishers.

FUNGI

Contents

The Fungal Hypha

J Silva and **S Gonzalez**, Universidad Nacional de Tucumán, Argentina

J Palacios and **G Oliver**, CERELA-CONICET, San Miguel de Tucumán, Argentina

Copyright © 1999 Academic Press

Introduction

The word 'fungus' is used to cover a huge range of shapes and types of cellular, coenocytic, spherical, filamentous, simple, complex, mobile, immobile, parasitic, symbiotic, saprophytic, microscopic and macroscopic organisms, which makes any attempt at definition difficult. Fungi are commonly defined as 'chlorophyll-lacking eukaryotes, and hence heterotrophic, with the following characteristics: uni- or multinucleate, nutrient absorption, typically chitinous cell walls, meiosis takes place within a zygote, and lysin synthesis takes place via adipicamine acid'.

The number of fungi known to date is large (over 600 000 species), but many people think they may be as numerous as flowering plants (over 1 000 000 species). Fungi are of ancient lineage and there is fossil evidence of their existence in the Precambrian and Devonian eras.

Owing to their morphological variability fungi show great differences in size, structure and metabolic activity, forming different types of colonies and complicated fruiting bodies. The latter possess complicated production, propagation and dispersion mechanisms.

The Fungal Cell

Somatic Structures

Fungal cells are larger than their bacterial counterparts, but generally smaller than animal and plant cells. However, their cellular organization does not differ greatly from other eukaryotic cells, with the possession of a true nucleus and internal cell structures that are more complex than prokaryotic cells. The use of electron microscopy for the systematic study of fungal cells has revealed that their ultrastructure is similar to that of plant cells.

The cytoplasm is bounded by a plasmic membrane and consists of the usual organelles and inclusions, such as mitochondria, endoplasmic reticulum, ribosomes, vacuoles, vesicles, microtubules, crystals and polysaccharides. Golgi bodies or dictyosomes do not always appear in all fungi in their typical shape, but vary according to the different types of fungi (**Fig. 1.**)

Fungal cells are of two basic morphological types: hyphae (multicellular filamentous fungi) or yeasts (unicellular fungi). Some fungi can possess a mixture of these two structures, depending on prevalent growth conditions; this is known as dimorphism and is a particular characteristic of some pathogenic fungi.

Multicellular Fungi   The thallus of filamentous fungi typically consists of microscopic filaments which branch out in all directions, thus colonizing the substrate that serves as food. They can grow over or into the substrate. Each of these filaments is referred to as a hypha. The hypha comprises a thin, transparent tubular wall, whose interior is full of or covered with protoplasm of a different density. The mycelium is a

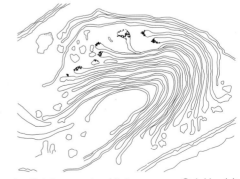

Figure 1   Golgi apparatus (dictyosome or Golgi body).

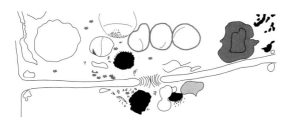

Figure 2 Septum with central pore.

structure composed of hyphae which form a weft or tissue, and varies according to its function.

The protoplasm held within the hyphae is interrupted at regular intervals by cross-walls called septa, which divide each hypha into sections or cells. In the more elementary filamentous fungi septa are only formed at the base of reproduction organs; rapidly growing hyphae are coenocytic, meaning that they are aseptate. When hyphae grow older septa are formed at several places. As one part of the hypha dies and the protoplasm draws back to the growing tip a septum is formed which separates the dead from the living section. The essential character of the coenocytic condition is that during growth nuclear division occurs without formation of new cells, leading to the development of a large mass of cytoplasm containing many nuclei.

Two types of septa are distinguished: primary septa are formed in connection with nuclear division and they remain between the two newly formed nuclei; adventitious septa are formed independently from nuclear division and they are related to changes in protoplasm concentration, as the protoplasm moves from one part of the hypha to another.

Septa vary in complexity according to their structure. All types seem to be formed by centrifugal growth, from the hypha wall towards the internal part. In some septa growth continues until the septum has been converted into a continuous plate. In others the septum remains incomplete, leaving a central pore which is often blocked. In the more complex fungi septa possess a central formation which consists of a barrel-shaped dilatation, flanked by a perforated membrane. This formation is called a dolipore septum (**Fig. 2**). Dolipore septa are lined by a membrane structure, the parenthesome or septal pore cap.

In fungi containing perforated septa protoplasts on either side of the septum are connected through the pore itself. These pores normally are large enough to allow nuclei and other organelles to pass through, so that nuclear movement is not necessarily impeded in septate fungi.

Each of the cells of septate hypha can contain one or more nuclei. The number of nuclei is characteristic of each fungus group, but most of them have multinucleate cells.

Unicellular Fungi Yeasts are predominantly unicellular fungi that generally reproduce by fission or budding (some people consider fission as a wide-based gemmation). They can be spherical, oval, elliptic, or elongated and cylindrical. Sometimes they form filamentous structures. Their size varies from 3 μm to 10 μm in width and 5 μm to 30 μm in length. Yeast species differ in the manner of budding (apical, lateral, bipolar or multipolar), and in some other morphological and physiological details.

After budding (an essential characteristic of yeasts), some species, which are generally unicellular, remain attached, forming a pseudomycelium which possesses a tiny micropore and minute filaments at the joining unions. These filaments (observed by scanning electron microscopy) provide the force that impedes separation of the buds. Sometimes, depending on the substrate, some yeasts can eventually form true mycelium like filamentous fungi, but their initial development always starts from budding yeasts.

The Cell Wall

The cell wall consists of several rigid layers, containing fibrils that are variably arranged. These fibrils maintain a characteristic cellular morphology, allowing interactions between fungi and the environment, other cells or the host. The cell wall is composed of polysaccharides (80–90%), glycoproteins (protein–polysaccharide complex), lipids and other components in smaller quantities. Insoluble polysaccharides such as cellulose, chitin, and α- and β-glucans make up the rigid matrix responsible for cell wall resistance.

Cell walls of all true fungi contain a minor percentage of chitin (glucosamine polymer), together with an amorphous matrix of hetero- and homopolysaccharides often attached to proteins, the latter providing adherence. The proteins are part of the extracellular enzymes. The cells walls of hyphae are 0.5 μm to 1.00 mm in diameter but at the apex the wall is thinner and simpler; it seems to have an inner layer of chitin and protein and an outer layer of protein. Probably more layer or wall materials are added behind the growth apex and they may contribute to the cell wall endurance when it matures. The lipid content may contribute to the surface properties (elasticity, sensitivity) and help prevent desiccation of the spores.

Pigments such as melanin may be incorporated into the cell wall or they can constitute an outer layer. Probably these pigments help to defend protoplasm against the hazardous effects of ultraviolet radiation; they could also protect from other organisms' lytic enzymes.

Chemical analysis of fungal cell walls enables the

subdivision of fungi into distinct evolutionary groups, and has improved present classification systems.

Filamentous basidiomycetes, ascomycetes and deuteromycetes possess chitin–β-glucans in their cell walls. Ascomycete and deuteromycete yeasts, on the other hand, have mannan–β-glucan, while basidiomycete yeasts have chitin–mannan. Zygomycetes contain chitin–chitosan, oomycetes cellulose–β-glucan, and hyphochytridiomycetes cellulose–chitin.

Composition of the cell wall of many fungus species is not always the same in all circumstances. Some substances appear in young hyphae and can nearly completely disappear in older structures; or other substances can precipitate, hiding the presence of the initial constituents, and making their detection more difficult. Moreover, it has been demonstrated that external factors such as culture medium composition, pH and temperature profoundly influence the chemical structure of fungal cell walls.

Fungal Growth

Most fungi grow at temperatures between 0°C and 30°C, but optimum temperatures vary between 20°C and 30°C. There exist several thermophilic species (e.g. fungi of the genus *Aspergillus* grow at temperatures close to 50°C), whereas others are psychrophilic, growing at relatively low temperatures (below 10°C). The ability of fungi to withstand extremely low temperatures when in a state of dormancy allows fungus cultures to be stored in liquid nitrogen at −196°C for prolonged periods.

Fungi prefer acid media for growth; pH 6 is most suitable for the majority of the species studied.

A low quantity of light, although not necessary for growth, is essential for sporulation of many species. Division into zones in certain species that have sporulation and non-sporulation zones seems to be induced by alternating light and dark periods. Although the process whereby light activates sporulation in fungi has not yet been identified, it has been hypothesized that light stops hyphal growth, initiating a cascade of processes leading to sporulation. It is commonly known that light is a key factor in dissemination of spores, because sporophores of many fungi demonstrate positive phototropism, so that spores are released towards the light.

On adequate substrates fungal hyphae can continue growing indefinitely. In nature fungal colonies have been observed that are hundreds of years old, as in the case of *Armillaria bulbosa* (Basidiomycetes), which was spread over 15 hectares (about 40 acres) in a forest in Michigan, USA. It had an estimated weight of 10 tonnes and was approximately 1500 years old.

The mycelium (a network of hyphae which make

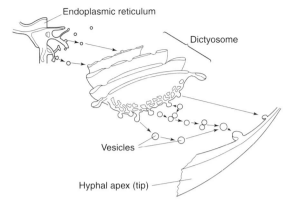

Figure 3 Sequence of growth at the hyphal tip. After Grove et al (1970).

up the fungal thallus) usually originates from a short germinal tube, which arises from a germinating spore. The mycelium has a tendency to grow uniformly in all directions from a central point, thus forming a spherical colony. A true sphere is rarely formed in nature, but fungi are able to produce a rapid growth response if substrate conditions and external factors are favourable. On solid synthetic media fungi tend to form circular colonies.

A hypha only grows at the apical end, and numerous experiments have affirmed this. Results have revealed that the hypha tip lacks almost all cellular inclusions and has a great number of cytoplasmic vesicles related to growth of this tip. In fungi with septate hyphae a refractive body (apical body) which can be stained densely is found near the hypha tip. Electron microscopy studies of active growing hypha tips led to the development of the vesicular hypothesis of hyphal growth (**Fig. 3**). These studies established that membrane material from the endoplasmic reticulum is transported to dictyosome cisterns by way of vesicles, which in turn are then incorporated by the dictyosomes. This is a transformation process in which endoplasmic reticulum-type membranes are converted into plasmic-type membranes. Thereafter the dictyosomes form secretory vesicles which migrate towards the hypha tip where they fuse with the plasmic membrane, releasing the contents which remain incorporated in the cell wall.

The role played by the internal pressure which hyphae must endure should be added to this process. This pressure is produced and maintained by cytoplasmic vacuoles, which are found at some distance from the hyphal tip. The importance of this pressure is now recognized and in some fungi it exceeds 150 atmospheres (15 MPa). It also contributes to stretching of the cell wall in the hypha tip. The vesicles found in the apical zone contain the necessary substances (polysaccharides, proteins, enzymes and salts)

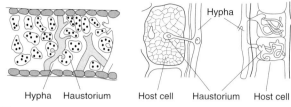

Figure 4 Different forms of haustoria.

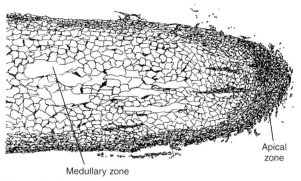

Figure 5 Longitudinal section through a rhizomorph.

both for synthesis and lysis processes of the fungal cell wall, which all in all lead to hyphal growth.

Special Vegetative Structures

Stolons or Runners These are hyphae destined for dissemination of the species on the substrate, forming an extended aerial structure of mycelium which allows the fungus to advance rapidly over the medium in all directions. The stolons are unbranched aerial hyphae which grow in a straight or arched manner over a long distance and connect groups of rhizoids. They are characteristic of the genera *Absidia* and *Rhyzopus* (Zygomycotina subdivision).

Rhizoids and Haustoria These are lateral outgrowths of intracellular hyphae specially modified for absorption of nutrients. These root-like hyphae, which enlarge the absorption surface for food substances, are called *rhizoids* in saprophytic fungi, and *haustoria* in parasites.

Parasitic fungi penetrate with haustoria into the host cells through little pores that the fungus has previously made in the cell wall. Haustoria are variously shaped, being knob-shaped in *Albugo candida*, large and irregularly swollen in *Peronospora parasitica* and branched in *Puccinia menthae* (**Fig. 4**).

Appressoria Modifications of hyphal structure and organization occur in relation to special functions. These are attachment elements, formed by modified or specialized hyphae which act as adhesion or anchorage organs. Appressoria are localized swellings of the tip of germ tubes or older hyphae that develop in response to contact with the host. They originate in infection hyphae and penetrate into the host epidermal cells (generally of plant origin). An appressorium sometimes has the same shape as a rhizoid or haustorium, but it differs from these, because it is provided with a mechanical adherence, as contrasted with opposite forces, thus fixing the mycelium to the substrate.

Rhizomorphs These structures are made up of bundles of hyphae of large diameter, forming complex networks. The outer cells develop a solid, thick cortex. The inner cells form a central meristem or medulla and are specialized for nutrient storage and transport.

The extreme part of a rhizomorph has a root-like structure, which enables longitudinal growth (**Fig. 5**). Rhizomorphs can reach a considerable length and are very common in wood-destructive fungi of the Basidiomycotina subdivision. They are able to grow through unusual substrates such as concrete walls and to advance several metres in a longitudinal direction.

See color Plates 14, 15, 16, 17 and 18.

See also: **Fungi**: The Fungal Hypha; Overview of the Classification of the Fungi; Classification of the Peronosporomycetes; Classification of the Zygomycetes; Classification of the Eukaryotic Ascomycetes; Classification of the Hemiascomycetes; Classification of the Deuteromycetes.

Further Reading

Bartnicki-Garcia S (1970) Cell wall composition and other biochemical markers in fungal phylogeny. In: Hardborne JB (ed.) *Phytochemical Phylogeny*. P. 81. New York: Academic Press.

Bartnicki-Garcia S (1973) Fundamental aspects of hyphal morphogenesis. In: Asworth JM and Smith JE (eds) *Microbial Differentiation*. P 245. Cambridge: Cambridge University Press.

Brock FD and Madigan MT (1991) *Biology of Microorganisms*, 6th edn. Englewood Cliffs: Prentice Hall.

Cooke RC and Whipps JM (1993) *Ecophysiology of Fungi*. Oxford: Blackwell.

Grove SN and Bracker CE (1970) Protoplasmic organization of hyphal tips among fungi: vesicles and Spitzenkörper. *Journal of Bacteriology*. 104: 989–1009.

Grove SN, Bracker CE and Morre DJ (1970) An ultrastructural basis for hyphal tip growth in *Pythium ultimum*. *American Journal of Botany*. 57: 245–266.

Kwon-Chung J and Bennet JL (1992) *Medical Mycology*. Philadelphia: Lea & Febiger.

Larpent JP and Larpent-Gourgaud M (1985) *Éléments de Microbiologie*. Paris: Herman.

Wainright PO, Hinkle G, Sogin ML and Stickel SK (1993) Monophyletic origin of the Metazoa: an evolutionary link with the Fungi. *Science* 260: 340–342.

Wesseels JGH (1994) Developmental regulation of fungal cell wall formation. *Annual Review of Phytopathology*. 32: 413–437.

Food-borne Fungi – Estimation by Classical Cultural Techniques

A K Sarbhoy and **Madhu Kulshreshtha**, Division of Plant Pathology, Indian Agricultural Research Institute, New Delhi, India

Many agricultural products can be turned into good foreign exchange earners. With about 70% of the population engaged in agriculture and with diverse geographic conditions, India promises to emerge in the foreseeable future as a major caterer for world needs in the processed food segment.

India's share in the world production of fruits is 7.7%. In the case of Indian vegetables, production is 45 million tonnes. At present, India's share in world exports has been small (1% of fruits and vegetables, 3.03% of fish and 0.15% of meat products). Horticultural crops (fruits and vegetables) are grown in about 7% of all the cropped area in the country. Mango, citrus, banana, guava and apple account for about 75–80% of production.

In these commodities 26% of losses occur due to diseases, including food-borne diseases. In fungi the food is accumulated as glycogen, fats and oils and derive nourishment from the living and dead organisms. Enzymes are also known to play an important role in converting various complex carbohydrates into simpler sugars which are ultimately utilized by the fungus. The mycelium may have the power to absorb the food and this is relegated to special modification of mycelial structures such as rhizoids, knobs and button-like processes. These are commonly known as haustoria and they enter the cells of the host plants through lenticels, stomata and clevices and derive food for different fungi.

Importance in Food Types/Categories

Cooking processes involve the interplay of chemical and physical reactions and also encompass biological reactions. Those commonly observed in everyday life are fermentation processes brought about by the activities of yeasts and bacteria. These are only part of the total microbial activity in food. Moulds are the major group which causes food-borne diseases and spoil many food commodities.

Moulds

Moulds may be white, grey, blue, green or orange. Moulds have a brush-like structure. They do not require as much moisture for growth as yeasts and bacteria and mostly do not require temperatures above average room temperature. They look fuzzy and like cotton wool and generally are unfit to eat. They are involved in spoilage of many foods; some are useful in the production of certain fermented foods. Some cheeses are known by their mouldy flavour.

Yeasts

Like moulds, yeasts may be useful or harmful in foods. Yeasts are unicellular fungi which vary in form from spherical to cylindrical. They reproduce by budding. Yeast is extensively used in making bread and certain other baked products. The most noticeable effect of yeast is the production of carbon dioxide, which expands the dough and makes the final product light and porous. The source of the CO_2 is sugar, which may be added when the dough is mixed or produced from the starch hydrolysis made possible by the enzyme diastase in flour.

Fermentation not only results in dough but also renders the gluten of the flour more elastic when combined with a liquid. Lactic and acetic acids form during fermentation. The increased acidity changes some of the insoluble proteins into soluble forms. Because the substances fermented are monosaccharides, starch, sucrose and maltose must be broken down into simple sugars (glucose and fructose) before they can be fermented by the action of the yeast. Yeast is capable of producing the enzymes sucrase and maltase, which bring about the splitting of sucrose and maltose into monosaccharides.

Ragi-like starter cultures were obtained from China, India, Sikkim, Indonesia (Java and Bali), Malaysia, Nepal, Philippines and Taiwan. These starters have different names in each country (murcha, bubood, Chinese yeast). The microorganisms involved in this process are Mucorales (*Rhizopus*, *Mucor*, etc.), yeasts and bacteria. *Amylomyces rouxis* was reported to survive for 5 years at room temperature and retained its amylolytic activity. Chlamydospore germination was first observed in *Amylomyces*; these are assumed to be the spores that remain alive when cultures are dried in ragi and similar starters. The remarkable thing about the starters was the consistent occurrence of the filamentous yeast, with the exception of *Amylomyces* from Nepal. Determining the number of colonies of *Amylomyces* is difficult because they resemble *Rhizopus* and spread rapidly in dilution plates.

Mycotoxins

In recent years much emphasis has been given to mycotoxins in human food and animal feed (**Table 1**). If consumed in large amounts they cause damage

Table 1 Mycotoxins involved in human and animal mycotoxicoses

Mycotoxins	Fungus	Biosynthetic origin	Found in:	Animals affected
Aflatoxins	Aspergillus flavus A. parasiticus	Polyketide	Soybeans, sorghum, cotton seeds, rice, corn nuts	All species
Citrinin	Penicillium	Polyketide	Cereal crops, grains	Pigs, guinea pigs, rats, rabbits, field mice
Cyclopiazonic	Penicillium cyclopium Asp. vesicolor	Amino acids	Processed foods, corn, peanuts	Experimental animals, dogs
Fumonisin	Fusarium monoliforme	Amino acids	Corn	Horses
Rubratoxin	Penicillium rubrum	Tricarboxylic acid cycle	Corn	Pigs, dogs, mice, rabbits, guinea pigs
Trichothecene	Stachybotrys atra, Myrothecium	Terpene	Cereals, crops, grain feeds	All species
Zearalenone	Fusarium roseum	Polyketide	Cereal grain	Pigs, cattle, poultry

to various organs such as the liver, kidney and gastrointestinal tract of animals and humans. These effects can accumulate when staple foods contaminated with mycotoxins producing fungi are consumed. The amount ingested over time will be considerable under normal conditions of harvest and storage. The risk of mycotoxins in a staple food such as rice, wheat and maize is much less than when stored under abnormal conditions, e.g. in cyclones, floods and unpredictable rain.

We know in great detail about the commodities attacked by the moulds which elaborate the mycotoxins, the carcinogenic effect of mycotoxins, the biosynthesis of aflatoxins, effects of mycotoxins on animals and to a certain extent their effect on human beings and the suppression of the immune system.

Principle and Characteristic Appearance of Yeasts on Cultural Media

Yeasts are of cosmopolitan origin. They can be isolated from the soil, excreta, milk, the vegetative parts of plants and from other habitats. Yeasts are commonly known as sugar fungi or Saccharomycetes.

The term yeast refers to Ascomycetes which possess a predominantly unicellular thallus and reproduce by budding, fission or both. Ascospores are produced in a naked ascus, originating from either a zygote or parthenogenetically by a single somatic cell.

Yeast cells vary in shape with the species. They are globose to ovoid, elongated or rectangular, sometimes adhere in chains, forming a pseudomycelium but do not represent the true mycelium. They are colourless individually, but when grown on artificial solid media they produce colonies (Czapek solution agar: Difco and BBL) which are white, cream or a brownish-white.

Yeasts are unicellular organisms and possess a definite cell wall made of chitin with a minute nucleus. Reproduction by multipolar budding, isogamous or heterogamous conjugation may or may not precede ascus formation. There are 1–4 spores per ascus. They are usually round or ovoid and seldom other shapes. Spores may conjugate before starting growth, vigorously fermenting glucose and usually other sugars, but nitrates are not utilized.

The order Saccharomycetales constitutes two recognized genera consisting of 16 species. The mycelium is well developed, but lacks a polysaccharide sheath. The septa contain clusters of minute pores, fragmenting to produce thallic conidia; asci are formed by the fusion of gametangia from adjacent cells or separate mycelia. They are usually elongated, or ellipsoid, and rarely ornamented. They usually have a mucous sheath and do not stain blue with iodine.

For a long time, the morphology of ballistoconidia has been considered an important taxonomic criterion for yeast classification. Recent extensive isolation studies revealed that three kinds of conidia (ballistoconidia, budding cells and stalked conidia) are produced by these yeasts, although productivity of the conidia is not stable. Therefore, the conidium ontogeny cannot always be reliable in the process of ballistoconidium-forming yeasts. Blastospores with a narrow base and arthrospores are known in yeasts. In this genus sporulation is multipolar all over the surface. In *Schizosaccharomyces* multiplication takes place by fission. Blastospores with a broad base followed by formation of a septum are typical in genera such as *Nadosma* or *Saccharomycoides*. In this case formation of the daughter cells takes place only at the end of the mother cell (bipolar) and can repeat itself by meristematic growth of the apical region. A number of daughter cells can be produced in basipetal succession as blown-out ends or by proliferation through the scars of the previous daughter cells. In some cases a collarette-like sheath may surround the scar. Nowadays, cell components such as proteins and nucleic acids are considered to be important criteria for estimation of relationships among taxa. The phylogenetic tree based on the partial sequences suggested that ballistoconidium-forming yeasts consisted of two branches and these corresponded to the existence of xylose as the cellular monosaccharide component.

Dimorphism – yeast form (YF) and filamentous pseudohyphal (PH) cells – is well known in *Saccharomyces cerevisiae*. Many laboratory strains of *Saccharomyces* contain a single mutation that blocks pseudohyphal growth, but wild strains are dimorphic.

Principle and Characteristic Appearance of Moulds on Cultural Media

Most moulds of importance in foods are now placed in the Fungi Imperfecti classification. Their reproductive cycles are fully characterized. Some of these teleomorphic states are placed under the Ascomycetous group of fungi.

Some important moulds involved in food-borne disease and their descriptions are given below. Some species produce white spot on refrigerated beef, whilst *Thamnidium* can also be found on refrigerated meats and decaying eggs. *Trichothecium* is represented as pink growth on foods such as fruits and vegetables. Other mould types in spoilage include *Byssochlamys* (decaying grapes), *Colletotrichium* (common contaminant of fruits and vegetables), *Candida*, *Debaryomyces*, *Hansenula* (cured meats, sausage, pickles and wines) and *Kloeckera* (off flavour and turbidity in wines). *Torula* can be found on refrigerated foods while *Trichosporon* is associated with meats and beers.

Aspergillus

One group of authors recognized 78 species of *Aspergillus*, while another numbered 132 species and nine varieties. The black *Aspergillus* are commonly called black moulds. Characteristics used to separate species include the colony morphology and conidium ontogeny when grown on Czapek Dox agar.

Vegetative mycelium consists of septate branching hyphae (colourless or bright-coloured). The conidial apparatus is developed as conidiophores and the heads form specialized, enlarged, thick-walled hyphal cells (the foot cells) producing conidiophores (stalks) as branches approximately perpendicular to the long axis of the foot cell and usually to the surface of the substrata. Conidiophores are non-septate or septate, usually enlarging upwards and broadening into turbinate, elliptical, hemispherical or globose fertile vesicles bearing cells or sterigmata, either parallel and clustered in terminal groups or radiating from the entire surface. The sterigmata are either in one series or as a double series with each primary sterigmatum bearing a cluster of two to several sterigmata at the apex. The conidia vary greatly in colour, size, shape and markings, they are produced successively from the tips of the sterigmata in basipetal succession (not produced by budding) and form unbranched chains arranged so that conidial heads are globose, radiate or compactly columnar.

Penicillium

The genus *Penicillium* is widespread and omnipresent. It is represented in all material habitats, giving the substrate a green or blue mouldy appearance. The generic name *Penicillium* has been applied to three species: *P. glaucum*, *P. candidum* and *P. expansum*.

The characteristics used to separate species include colony morphology when grown on Czapek Dox agar and malt extract agar (incubated at 25°C for 7–14 days), conidial colour, exudate and soluble pigment, reverse colour, growth and colony texture and medium buckling. In addition to the above media temperature, growth was compared on glycerol nitrate agar and at temperatures ranging from 5°C to 37°C.

Basically the *Penicillium* has a supporting stipe which leads to the final verticil of phialides directly (monoverticillate), through a prior verticil of short branches – metulae (biverticillate) or through a complex system of branches involving rami and ramuli (terveticillate). The shape of the phialides is of taxonomical importance as ampulliform and mamalliform are found in association with mono- and terverticillate penicilli, while the acerose type is consistent with the biverticillate penicilli. On the basis of penicillus morphology, the genus has been divided into four subgenera:

1. *Aspergilloides*: Penicilli monoverticillate, phialides ampulliform with sometimes one or two intercalary metulae
2. *Furcatum*: Penicilli biverticillate or irregularly monoverticillate with ampulliform phialides
3. *Penicillium*: Penicilli biverticillate or terverticillate (complex) with ampulliform phialides
4. *Biverticillum*: Penicilli strictly biverticillate, seldom branched, with acerose phialides.

Myrothecium

This genus comprises 13 species, including *Myrothecium inundatum*, *M. roridum* and *M. verrucuria*. It produces toxins known as trichothecenes (verrucarin A).

M. roridum and *M. verrucuria* are strong cellulose decomposers. On culture plates, they appear as white colonies with minute dark green droplets of sporodochia spread all over. Fruiting bodies are either synnema or sporodochia formed from closely compacted conidiophores, arising from a stroma bearing a mass of slimy green to black spores which become hard on drying. Hyaline or darkened setae are present or absent; the setae arise from the basal stroma.

Conidiophores are macronematous (morphologically different from vegetative hyphae), closely packed to form sporodochia and branched repeatedly. The ultimate branches bear conidiogenous cells (phialides) in whorls; they are hyaline, olivaceous or slightly darkened.

Conidiogenous cells are monophialidic and discrete, cylindrical, elevated or subulate, hyaline or darkened at apex. Conidia are aggregated in dark green or black slimy masses; they are simple, cylindrical, navicular, limoniform or broadly ellipsoidal. They often have a projecting truncate base; they are hyaline to pale olive, smooth or straightly marked and non-septate.

Alternaria

The host-specific toxins produced by *Helminthosporium*, *Tricothecium* and *Alternaria* include HC (*Helminthosporium carborum* – HC toxin), HV (*Helminthosporium victoriae* – Victoria), HS (*Helminthosporium sativum* – HS toxin), T (*Trichothecea*), AK (*Alternaria kikuchiana* – AK toxin). *A. alternata formae specialis* is the only species involved in toxin production.

Colonies are grey, blackish brown or black and hyphae are colourless. Conidiophores are macronematous, mononematous, simple or irregularly or loosely branched, pale brown, solitary or in groups. Conidiogenous cells are integrated, terminal or intercalary, mono- or polytrectic, sympodial and cicatrized. Conidia are formed in long, often branched chains with a short conical or cylindrical beak, not more than one-third of the length of conidium. Conidia are pale to mid golden brown, smooth or verruculose with up to eight transverse septa and several longitudinal septa, 20–63 (37) × 9–18 (13) μm; beak pale; 2–5 μm thick.

Rhizopus

This is differentiated on the basis of stolons and sporangiospores which arise above the rhizoids. Thirteen species of *Rhizopus* have been recognized on the basis of biochemical and morphological characters. In the *R. stolonifer* group there are only two species, both with several varieties. Numerous older taxa are treated as synonyms of *R. oryzae*. The colony can be differentiated when grown at 20 and 22°C for 7 days.

Rhizopus is readily distinguished from *Mucor* by the presence of stolons (runners), often several centimetres long, and of tufts of rhizoids (root-like hyphae) emerging from the points where the stolons touch the medium. The *Rhizopus* growth is also more vigorous

and it is grey in colour. Twenty-two species have been described, but some of these are rare, and a number of others are best regarded as strains of *R. stolonifer*.

Mucor

In 1873 it was shown that alcohol is produced by a pin mould, *Mucor mucedo* (later identified as *M. racemosus*), *M. rouxii*.

Mucor racemosus Fresenius is probably the most widely distributed of all species of *Mucor*. It is found on almost every kind of damp material. Colonies are grey or brownish-grey, loose in texture and normally less than 1 cm high; sporangiophores are simple at first, later becoming branched, with the branches irregularly arranged and unequal in length. Sporangia are globose and unequal in size but mostly small (20–70 μm diameter), with walls which break in pieces when handled. Columellae are spherical or ovate, with a collarette (a portion of the broken sporangial wall left *in situ*). Spores are mostly elliptical and 6–10 × 5–8 μm. The most characteristic feature of the species is the abundant production of chlamydospores, which are formed in aerial hyphae, along the sporangiophores, and even in the columellae. They are of diverse shapes, colourless or yellow, smooth and about 20 μm in diameter. This species can grow in submerged culture and, like yeast, produces alcohol.

Absidia

Absidia van Tieghem differs in several respects from *Rhizopus*. The rhizoids and stolons are not so clearly differentiated; the sporangiophores arise from the stolons and not from the points of attachment of the rhizoids; the sporangia are relatively small and pear-shaped and the most characteristic feature of all is that there is a well-marked apophysis – a funnel-shaped base to the sporangium, where the walls of the sporangium and columella are united. The zygospores, when present, are surrounded by coarse hairy outgrowths from one or both suspensors. This feature serves to place at once the homothallic species of the genus. Seventeen species were described; later the genus was divided into 29 species and two varieties.

Some species commonly occur in the soil and sometimes appear in mixed cultures from mouldy material, or as aerial contaminants.

Thamnidium Link ex. Wallroth

The main sporangiophores bear lateral clusters of sporangioles as well as large terminal sporangia. The sporangioles resemble miniature sporangia, containing approximately 24 spores and are formed on a richly branched outgrowth from near the base of the sporangiophore.

T. elegans Link has been reported most frequently

on culture media. It is greyish, granular and at first only a few millimetres high. Growth consists chiefly of sporangioles, later producing sporangiophores up to 1 cm or more long, bearing large terminal sporangia and sporangioles in dense clusters, borne on the ends of much branched hyphae spores, 4–9 × 2.3–2.5 µm. Asci develop rapidly on compact, without any trace of peridium. Each ascus contains eight ovate spores, 6–6.5 × 4.3–4.5 µm.

Cladosporium Link

Colonies on natural substratum and in culture are effuse. They are pale greyish blue, olive and velvety, reverse on glucose agar greenish black. Conidiophores are macronematous and micronematous, flexuous geniculate, nodose, septate, branched, smooth and brown. Ramoconidia are smooth with 0–2 septa 27–33 × 4.5–6.0 µm. Conidia are catenate mostly non-septate, long, cylindrical, obclavate, ellipsoidal 6–18 × 3–7.5 µm in diameter with protuberant scars present.

Cephalosporium Corda

Conidiophores lacking sporogenous cells and phial-ides arising directly and singly from the vegetative hyphae or from funiculose strands. Sporogenous cells are hyaline, tapering, producing phialospore in balls or rarely, in fragile chains at the apex; phialospores contain hyaline and are globose to ovoid or short, cylindrical and non-septate.

Geotrichum Link

This genus has conidiophores lacking vegetative hyphae and hyaline or lightly pigmented conidia. Arthrospores are produced by basipetal septation and fragmentation of the vegetative hyphae. They are very variable in width and length, non-septate, hyaline or sub-hyaline, cylindrical in shape with truncated ends. The end walls are sometimes slightly convex. In some species arthrospores round off to form chains of globose to subglobose spores.

Helminthosporium Link ex. Fries

Colonies on PDA (Potato Dextrose Agar) and PCA (Potato Carrot Agar) are effuse, dark and hairy. Mycelium is immersed in the substratum. Stromata are usually present. Conidiophores are often fasciculate, erect and brown to dark. Conidia develop laterally, often in verticils, through pores beneath the septa, while the tip of the conidiophores is actively growing. Growth of the conidiophore ceases with the formation of the terminal conidia. Conidia are formed singly, subhyaline to brown; they are usually obclavate, pseudoseptate, frequently with a dark brown to black protruding scar at the base.

Monilia Pers

In this genus the branching chains of hyaline blastospores with a bead-like appearance are distinctive.

Conidiophores are little differentiated from the vegetative hyphae. They are erect, simple or dichotomously or irregularly branched, hyaline and septate. Conidia blastospores are produced in apical succession by budding to give branching chains. Conidia is hyaline to sub-hyaline, continuous globose to ovoid. Often there are isthmus-like connectives between spores.

Sporotrichum Link

Non-septate, rough-walled aleuriospores borne laterally or terminally on undistinguished sporogenous cells characterize the common species of Sporotrichum.

Hyphae are broad, septate and hyaline to lightly pigmented. Conidiophores are little differentiated from the vegetative hyphae. Conidia are solitary, with broad attachment to the parental hypha. They are borne terminally or laterally on the sporogenous cells. They are non-septate, hyaline to lightly pigmented, usually rough-walled, seceding by rupture of the subtending sporogenous hypha. They are globose or, more often, ovoid to clavate and rounded at the distal end. They are truncated and often with an annular frill at the base.

Candida Berkhout

This is a yeast-like organism with a strong mycelial or pseudomycelial growth and multilateral budding, typical of Candida.

Cells are of varying shape. Reproduction is multilateral budding. Chlamydospores are sometimes present. Pseudomycelium is more or less abundantly developed. True mycelium may also occur. Blastospores are attached to the pseudomycelium in a way typical of the species. In liquid media there is submerged growth, often ring formation and a pellicle; oxidative assimilation is present and, in many species, there is a strong fermentative dissimilation.

Colletotrichum Corda

In culture, colonies of Colletotrichum may have sparse setae and produce pinkish, water-soaked colonies. The fruiting body is an acervulus; conidiophores are produced in a dense, even stand on a thin or well-developed stroma. Conidiophores are simple, short, hyaline, producing abundant phialophores. They are produced in mucus, ovoid, non-septate, short cylindrical, falcate or crescent-shaped, hyaline. They are pinkish in masses, frequently producing dark setae; they are stout, septate and darkly pigmented, acutely pointed at the apex.

Torulopsis Berlese

Cells are round or oval or, very rarely, somewhat elongated. Reproduction is by multilateral budding. Cells are only exceptionally capsulate. There is no reproduction of a starch-like compound, as in the genus *Cryptococcus*. Pseudomycelium is lacking or very primitive. There is a sediment and ring, but rarely a pellicle. It is produced in liquid media. The pellicle, if present, is never dry, dull or creeping. Fermentative ability is usually present and seldom absent. No red or yellow caratenoid pigments are produced.

Trichosporon Behrand

Trichosporon includes the asporogenous yeasts which reproduce by forming both blastospores and arthrospores.

Pseudomycelium and true mycelium develop abundantly; both blastospores and arthrospores are produced. Chlamydospores sometimes occur. It is mostly oxidative, and rarely fermentative.

Trichothecium Link

Colonies white to pale red or pink on PDA

Conidiophores are erect or sub-erect, produced singly or in groups. They are simple or sparingly branched, hyaline and septate. Conidia are produced in short fragile chains in basipetal succession, held together by mucus. Conidia are large, two-celled and hyaline with a well-marked truncate attachment point.

The short chains of two-celled conidia at the apex of a hyaline, simple conidiophore are diagnostic for the common *T. roseum*.

Rhodotorula Harrison

Cells are round, oval or elongated and reproduce by multilateral budding. They sometimes produce pseudomycelium. Distinct red or yellow pigments are produced, and they do not ferment.

This genus includes red or yellow yeasts which are microscopically similar to *Cryptococcus*.

Fusarium Link

There is a fruit body, when present; a sporodochium is sometimes lacking and sporogenous cells arise directly from the vegetative hyphae or from conidiophores. Conidiophores are solitary and simple or aggregated with complex branching. Ultimate branches terminate in sporogenous cells. Phialides taper distally, sometimes with an apical collarette. Phialospores are frequently of two types – there are large macrospores which have one to several septa, forming elongate phragmospores, hyaline, cylindrical or curved. They are frequently boat-shaped, with a well-marked foot cell at the attachment end of the spore. They are produced in mucus and in sliming down to form globoid heads or spore masses. Microspores are smaller and non-septate or single-septate; they are ovoid to short-cylindrical, gathering in short chains or, more commonly, in spore balls.

F. moniliforme and *F. graminearum* are the conidial state of *Gibberella fujikori* and *G. zeae* respectively (Hyopcreales).

Kloeckera Jane

Cells are ovate to lemon-shaped; reproduction is by bipolar budding. There are eight species recognized.

K. apiculata (Ress emend. Klocker) Jane is the imperfect stage of *Hanseniaspora valbyensis* and is found on grapes and associated with the flor of sherry. Cells are lemon-shaped or ovate, 5–8 (10) × 2–4.5 µm in liquid media; there is a sediment and thin surface ring. The formation of pseudomycelium is rare; only glucose, laevulose and mannose are fermented.

Debaryomyces Lodder and v. Rij nom conserv. (A. de Bary)

Cells are round or short and oval, propagating by multipolar budding, producing in liquid media a dry dull pellicle. Ascus formation is almost invariably preceded by conjugation, usually heterogamous. A cell conjugates with its own bud. Spores are round, usually with an oil-drop in the middle and sometimes finely warted.

Hansenula H. and P. Sydow

They grow as dry, wrinkled pellicles on the surface of the liquid and their metabolism of sugars is more oxidative than fermentative.

Cells are of various shapes – round, ovate and elongated cylindrical with a strong tendency to form pseudomycelium. Ascospores are hemispherical, hat-shaped or angular. Species of *Hansenula* ferment vigorously and can utilize nitrates.

Media Formulations

Media formulations are listed in **Table 2**.

Table 2 Media formulations

Czapek Dox
NaNo$_3$ 3 g, K$_2$HPO$_4$ 1 g, MgSO$_4$ 7H$_2$O 0.5 g, KCl 0.5 g, FeSO$_4$7H$_2$O 0.01 g, sucrose 30 g, agar-agar 15 g
Potato dextrose agar
Potato 250 g, dextrose 20 g, agar-agar 20 g, DW 1000 ml
Potato carrot agar
Peeled carrots 250 g, peeled potato 250 g, agar-agar 20 g, DW 1000 ml
Glucose
KNO$_3$ 2 g, KH$_2$PO$_4$ 1 g, MgSO$_4$7H$_2$O 0.5 g, glucose 50 g, agar-agar 20 g, DW 1000 ml

DW, distilled water

See also: **Mycotoxins**: Detection and Analysis by Classical Techniques; Immunological Techniques for Detection and Analysis. **Sampling Regimes and Statistical Evaluation of Microbiological Results**.

Further Reading

Booth C (1959) *Studies of Pyrenomycetes. IV. Nectria (Part I).* Mycology paper no. 73. p. 1. Kew, UK: Commonwealth Mycological Institute.

Calmette A (1892) Contribution à l'etude des ferments de l'amidon, la levûre chinoise. *Ann. Inst. Pasteur* 6: 604–620.

Fitz A (1873) Uber alkoholische Gahrung durch *Mucor mucedo. Ber. Dtsch. Chem. Ges.* 6: 48–58.

Hesseltine CW and Anderson A (1956) The genus *Thamnidium* and a study of the formation of its zygospores. *Am. J. Bot.* 43: 696–703.

Lender A (1908) Les Mucorinées de la Suisse. *Materiaux pour la flore cryptogamique Suisse* 3: 1–180.

Lodder J and Kreger van Rij NJW (1952) *The Yeasts. A Taxonomic Study.* P. 455. Amsterdam: North Holland Publishing.

Peckham GC (1974) *Foundation of Food Preparation* P. 51. New York: Macmillan.

Pitt JI (1979) The genus *Penicillium* and its teleomorphic states *Eupenicillium* and *Talaromyces*. P. 1. New York: Academic Press.

Raper KB and Fennell DI (1965) *The genus* Aspergillus. P. 1. Baltimore: Williams & Wilkins.

Raper KB, Thom C and Fennell DI (1949) *A Manual of the* Pencillia. P. 1. Baltimore: Williams & Wilkins.

Thom C (1930) *The* Penicillia. P. 1. Baltimore: Williams & Wilkins.

Overview of Classification of the Fungi

B C Sutton, Blackheath, Suffolk, UK, formerly of the International Mycological Institute, Egham, UK

Organisms are named and classified for purposes of reference and data retrieval. A classificatory scheme should reflect natural relationships, but scientists may attach varying degrees of importance to the criteria available. Thus not all authorities necessarily accept the same scheme. For organisms traditionally included in the 'fungi', a number of changes have been introduced to the study of their systematics. These include recognition of the artificiality of the three- (or even five-) kingdom classification system for living things; polyphyly of the 'fungi'; development and acceptance of data analysis techniques of phylogenetic systematics; and the inception, development and application of molecular techniques. These have resulted in phylogenetic classifications identified by monophyletic groups which contain an ancestor and all its descendants, i.e. they are based on evolutionary relationships.

The kingdom *Fungi*, as now generally accepted, includes four phyla: Chytridiomycota, Zygomycota, Ascomycota and Basidiomycota. Several other organisms formerly included in the 'fungi' are now treated in separate kingdoms and phyla. These are the kingdoms Stramenopila with the phyla Oomycota, Hyphochytriomycota and Labyrinthulomycota. The phyla Plasmodiophoromycota, Dictyosteliomycota, Acrasiomycota and Myxomycota are not a monophyletic group and are placed as protists in recent classificatory schemes. These stramenopiles and protists and the Chytridiomycota are of no significance in food microbiology, although some are important in the context of primary food production as plant pathogens. Of the true fungi comparatively few genera and species are significant as food sources, in food production, or as spoilage organisms.

Features Used to Define a True Fungus

A combination of features is used to define true fungi, including morphology, ultrastructure, chemistry, nutritional mode and latterly DNA profiles. Fungi are eukaryotic (with one or more nuclei in their cells bounded by a nuclear membrane and with paired DNA-containing chromosomes) and, unlike plants and algae, lack plastids. They are unicellular or filamentous and consist of multicellular coenocytic haploid hyphae which are hetero- or homokaryotic. Their somatic structures, with few exceptions, show comparatively little differentiation and practically no division of labour. Cell walls contain chitin and β-glucans. Subcellular organelles include mitochondria with flattened cristae, Golgi bodies or individual cisternae, and peroxisomes which are nearly always present. Cells are mostly non-flagellate; when present, flagella always lack mastigonemes. Reproduction is sexual (meiotic) or asexual (mitotic) by formation of spores, the diploid phase being generally short-lived. There is no amoeboid pseudopodial phase, although some have motile reproductive cells. True fungi are saprobic, parasitic or mutualistic with absorptive (osmotrophic), never phagotrophic, nutrition. Primary carbohydrate storage is by glycogen rather than starch.

Features Used in Classification

The four phyla accepted in the Fungi are monophyletic, so criteria used within these taxa for separating subordinate ranks – classes (-mycetes), orders (-ales), families (-aceae) and genera – are different for each phylum. Ascomycetes especially, and to a lesser extent basidiomycetes and other groups, have both meiotic (teleomorphic) and mitotic (anamorphic) reproductive states. In some, these states have become separated and remain uncorrelated. Evolution may possibly have taken place in either state. The result is that huge numbers of mitotic fungi (approximately 15 000 species) have no known teleomorphs, so separate artificial classifications have been constructed. Currently no taxon above the rank of genus is recognized for mitotic fungi. Those for which correlations have been proved are included in the appropriate classification of the teleomorph. For Chytridiomycota the ranks of class, order and family are used; in the Ascomycota only order and family are used; and in the Basidiomycota the ranks of class, subclass, order and family are used.

Phylum Chytridiomycota

Definition same as for the class.

Class Chytridiomycetes

Thallus coenocytic, holocarpic or eucarpic, monocentric, polycentric or mycelial. Cell walls chitinous at least in the hyphal stages. Mitochondrial cristae flat. Zoospores posteriorly monoflagellate or rarely polyflagellate, lacking mastigonemes or scales, but with an unique flagellar 'root' system and rumposomes. Aquatic saprobes or parasites on decaying and living organic material in freshwater soils; some are marine. Others are obligately anaerobic in guts of herbivores. None is of significance in food microbiology.

Phylum Zygomycota

Definition same as for the class Zygomycetes.

Class Zygomycetes

Mycelium coenocytic, walls chitinous. Zygospore (a thick-walled resting spore) produced in a zygosporangium after fusion of two gametangia. Mitotic reproduction by sporangiospores. No flagellate cells or centrioles present. Nutrition saprobic to facultative weak parasitic. Known from soil, dung, fruits and flowers, stored grain and fleshy plant organs, mushrooms, invertebrates, vertebrates and humans, and significant as spoilage and primary fermentative organisms.

Order Mucorales Asexual reproduction by multispored, few-spored or single-spored sporangia (sporangiola). Sexual reproduction by zygospores.

Family Mucoraceae Sporangia columellate, specialized sporangia absent. Zygospores smooth to warty, borne on opposed tongs-like or apposed, naked or appendaged suspensors. Polyphyletic.

- *Absidia* – pear-shaped sporangia produced in partial whorls at intervals along stolon-like branches. Sporangiospores subglobose to ellipsoid. Branches produce rhizoids at intervals but not opposite the sporangiophores. Zygospores surrounded by curved, unbranched suspensor appendages which may arise from one or both suspensors. Distributed worldwide in soil, stored grain, decaying vegetables and fruit, air, compost, animals and humans.
- *Actinomucor* – stolons, rhizoids and non-apophysate sporangia present. Sporangia hyaline to faintly coloured, globose, formed on repeatedly branched sporangiophores. Sporangiospores smooth, globose to irregular. Used as inoculum in production of sufu (traditional oriental food from fermented soya milk).
- *Mucor* – sporangia non-apophysate, globose and formed on branched and unbranched sporangiophores, from mycelium lacking stolons and rhizoids. Sporangiospores variable in shape. Important widely distributed food spoilage organisms, especially found in soil, dung, hay, stored grain, fruit, vegetables, milk, animals and humans. Also used in fermentation processes for sufu production.
- *Rhizopus* – rhizoids form at the base of sporangiophores which may grow in clusters. Habit stoloniferous, an aerial hypha grows out and where it touches the substratum rhizoids and sporangiophores are formed. Sporangia apophysate with sporangiospores irregularly angled and often striate. Worldwide in distribution but especially in tropical and subtropical areas, from soils, grain, water, vegetables and fruit. Used as a fermenting agent in production of tempeh from soya bean, and tempeh-bongrek from manioc. Also reported to produce toxic metabolites, but there is no strong evidence for its implication in mycotoxicoses.

Family Thamnidiaceae Sporangia diffluent, columellate. Sporangiola few to single-spored, persistent-walled and columellate, borne on same or separate, morphologically identical sporangiophores, or sporangiola only present. Zygospores warty, borne on opposed suspensors. Polyphyletic.

- *Thamnidium* – large, terminal columellate sporangia produced with dichotomous lateral branches bearing fewer-spored, non-columellate sporangiola. Sporangiola may also be borne on separate branch systems. Zygospores warty, borne on opposed suspensors. Cosmopolitan, commonly from dung, also soil, occasionally occurs as a food contaminant.

Family Cunninghamellaceae Sporangia absent, sporangiola single-spored. Zygospores warty, borne on opposed suspensors.

- *Cunninghamella* – sporangia absent, asexual conidia (sporangiola) are hyaline and borne singly on globose vesicles on branched or unbranched conidiophores. Zygospores are of the *Mucor* type, warty and borne on opposed suspensors. From soil, air, fruit, and occasionally as a food contaminant.

Family Syncephalastraceae Merosporangia present. Zygospores warty, borne on opposed suspensors.

- *Syncephalastrum* – aerial branches terminate in club-shaped or spherical vesicles which are multinucleate, budding over their surface to form cylindrical outgrowths, the merosporangial primordia. Cytoplasm cleaves into a single row of 5–10 sporangiospores and with shrinking of the sporangial wall the spores appear in chains. Zygospores resemble those of Mucoraceae. From tropical and subtropical regions from soil, dung and grain, and as a food contaminant.

Phylum Ascomycota

The ascus is diagnostic (a sac-like cell in which after karyogamy and meiosis generally eight ascospores are produced by free cell formation). Structure and presence of lamellated hyphal walls with a thin, electron-dense outer layer and relatively thick, electron-transparent inner layer is typical. Molecular sequence data are important in characterizing the phylum, in terms not only of the teleomorph (meiotic) states but also of the anamorph (mitotic) states. The current trend is for anamorphic (asexual, mitotic or mitosporic) states which have been correlated with teleomorphs (especially relevant in the ascomycetes) to be incorporated into teleomorph classifications. Those species that have not been correlated are still referred to as the mitosporic fungi (deuteromycetes) although molecular techniques are now making it increasingly feasible to predict where one of the group is likely to belong in teleomorphic classifications. This phylum contains most of the food spoilage organisms, many primary fermentative species, and a few edible ones. The significance of teleomorphs in food spoilage is minimal, most damage and problems being the result of activities of the anamorphs.

Classes

None.

Order Cyttariales Ascomata apothecial, formed within conspicuous, brightly coloured, fleshy compound stromata, spherical, but with a wide opening; interascal tissue of filiform paraphyses. Asci develop synchronously or sequentially, cylindrical, with active discharge, apex with an I+ ring, opening irregularly. Ascospores more or less ellipsoidal, aseptate, pale grey, thin-walled, without a sheath.

Family Cyttariaceae As for order.

- *Cyttaria* – characters as for order. Biotrophic on *Nothofagus* spp., often gall-forming; edible, sold in southern Chile.

Order Dothideales Ascomata very varied, apothecial, perithecial or cleistothecial, formed as lysigenous locules within stromatic tissue. Hymenium sometimes gelatinous and blueing in iodine. Interascal tissue of branched or anastomosed paraphysoids or pseudoparaphyses, initially attached at apex and base. Asci cylindrical, thick-walled, fissitunicate, rarely with apical structures. Ascospores nearly always septate, longitudinally asymmetrical, constricted at the primary septum, sometimes muriform, hyaline or brown, unornamented.

Family Diademaceae Ascomata globose to ellipsoid-elongate, typically opening by a disc-like operculum, but sometimes by a lysigenous pore or slit. Peridium of large, thick-walled pseudoparenchyma. Interascal tissue of pseudoparaphyses. Asci cylindrical, fissitunicate. Ascospores large, brown, muriform, usually radially asymmetrical.

- *Clathrospora* (anamorph *Alternaria*) – teleomorph rarely seen in context of food microbiology. *Alternaria* black or grey mycelium in culture. Conidiophores solitary, brown, simple or branched. Conidia dry, in long, often branched chains, obclavate, obpiriform, ovoid or ellipsoid with a short conical or cylindrical beak, and several transverse and longitudinal eusepta, pale to medium brown, smooth or verrucose, formed from preformed pores (loci) in association with sympodial growth of conidiogenous cell. Distribution worldwide, commonly saprobic on plant materials, foodstuffs, soil, textiles. Produces mycotoxins.

Family Leptosphaeriaceae Ascomata perithecial, often conical or papillate, immersed or erumpent, sometimes aggregated into small stromata. Ostiole periphysate. Peridium black, well-developed, sometimes thicker at the base, of thick-walled pseudoparenchyma. Interascal tissue of cellular pseudoparaphyses. Asci cylindrical, narrow, thin-walled, fissitunicate. Ascospores hyaline to brown, transversely septate, sometimes elongated and sheathed.

● *Leptosphaeria* (anamorph *Alternaria*) – see *Clathrospora*

Family Mycosphaerellaceae Ascomata small, immersed, often aggregated or on a weakly developed stroma, black, papillate. Ostiole lysigenous. Peridium thin, of pseudoparenchyma. Interascal tissue absent. Asci ovoid to saccate, fissitunicate. Ascospores usually hyaline, transversely septate, unsheathed.

● *Mycosphaerella* (anamorph *Cladosporium*) – teleomorph rarely seen in context of food microbiology. *Cladosporium* state olivaceous, grey-olive to blackish-brown mycelium in culture. Conidiophores solitary, brown, unbranched except towards the apices. Conidia dry, in branched chains, ellipsoid, fusiform, ovoid, subglobose, aseptate or with several transverse eusepta, pale to dark olivaceous brown, smooth, verruculose or echinulate, with a distinct scar at the base and several in the apical region if forming chains, arising from cicatrized loci produced synchronously, sympodially or irregularly by the conidiogenous cell. Distributed worldwide, commonly airborne, ubiquitous as saprobes and primary plant pathogens, also from soil, foodstuffs. Mycotoxins such as the dihydroisocoumarin cladosporin and the monomeric anthraquinone emodin are produced, but there is no evidence so far for significance in mycotoxicoses.

Order Eurotiales Ascomata small, cleistothecial, usually solitary. Peridium thin, membranous, often brightly coloured, varied in structure and rarely acellular and cyst-like. Interascal tissue absent. Asci clavate or saccate, thin-walled, evanescent, sometimes in chains. Ascospores varied, small, septate, often ornamented and with equatorial thickening, without a sheath.

Family Monascaceae Ascomata small, cleistothecial, globose, thin-walled. Peridium of flattened hyphae. Interascal tissue absent. Asci evanescent at early stage. Ascospores hyaline, aseptate, ellipsoidal, thick-walled.

● *Monascus* (anamorph *Basipetospora*) – mycelium brownish, grey-brown in the centre, with brick-red pigmentation on aot agar. Ascomata borne terminally on a long hypha, subhyaline to brown-red. Asci globose to subglobose. Ascospores yellowish, ovate-ellipsoid, with a hyaline wall, smooth. *Basipetospora* conidia borne in basipetal succession in chains on solitary, septate, erect conidiophores, retrogressively delimited, ovate to piriform, hyaline, aseptate, thin-walled, base truncate. Distributed worldwide, from soil, silage, dried foods, rice, oat seeds, soya, sorghum, tobacco, also used in fermentation of angkak to produce pigment for food products of fish, soya beans and some alcoholic beverages.

Family Trichocomaceae Ascomatal walls varied, pseudoparenchymatous or hyphal, sometimes thick and sclerotioid, usually bright-coloured. Ascogonia often coiled. Interascal tissue absent. Asci small, more or less globose, often formed in chains. Ascospores hyaline, usually bivalvate and ornamented.

● *Byssochlamys* (anamorph *Paecilomyces*) – mycelium white, becoming yellow-brown with development of anamorph. Ascomata discrete, confluent. Wall absent or minimal, of a loose weft of hyaline, thin, twisted hyphae. Ascogonia coiled around swollen antheridia. Asci globose to subglobose, eight-spored, stalked. Ascospores ellipsoid, smooth, pale yellow. *Paecilomyces* conidia in dry chains from phialidic conidiogenous cells on solitary, septate, erect conidiophores. Phialides in groups of two to five, cylindrical at the base, long-necked, on short supporting cells. Conidia cylindrical, hyaline, aseptate, with flattened ends, yellow. Distributed worldwide, in soil, bottled and tinned fruit, pasteurized food, airtight stored cereals; also produces mycotoxins.

● *Emericella* (anamorph *Aspergillus*) – ascomata surrounded by thick-walled hülle cells. Ascospores lenticular, smooth, with two equatorial crests, usually red to purple. *Aspergillus* conidia in dry chains, forming dark yellow-green columns from solitary, erect, aseptate, brown, smooth conidiophores. Phialides borne on supporting cells on swollen apices of conidiophores, short-necked. Conidia rough-walled, globose. Distributed worldwide, especially in soil, potatoes, grain, citrus, stored cereals, cotton. Mycotoxins formed. *Aspergillus* is used in the fermentation industry for production of vitamins, enzymes, organic acids, antibiotics, soy sauce, miso and saki.

● *Emericellopsis* (anamorph *Acremonium*) – ascomatal wall of several layers of hyaline, flattened cells. Ascospores ellipsoid with an initially smooth, wide, gelatinous layer collapsing to form three to six longitudinal wings. *Acremonium* conidia in chains collapsing into wet masses, from solitary, erect, aseptate or septate, simple or sparingly branched, smooth or slightly rough, hyaline or pale brown conidiophores. Phialides terminal, cylindrical. Conidia hyaline or sometimes pale brown, globose, subglobose, ellipsoid or fusiform, aseptate or sometimes septate. Distributed worldwide, from soil, cultivated fields, mud sediments, plant remains, hay, apples, pears. *Acremonium* also produces the weak trichothecene crotocin, and the metabolite cerulenin which enhances aflatoxin biosynthesis.

- *Eurotium* (anamorph *Aspergillus*) – ascomata mostly yellow with cellular smooth wall, one cell thick. Ascospores lenticular, smooth or roughened, furrowed or with equatorial crests. *Aspergillus* conidia in dry chains, forming grey to olive-green heads from solitary, erect, aseptate or septate, smooth or rough conidiophores. Phialides arising directly from swollen apex of the conidiophores, radiating, very short-necked. Conidia echinulate, globose, subglobose, ovate or ellipsoid, sometimes with both ends flattened. Distributed worldwide though predominant in tropical to subtropical areas, from soil, stored or decaying grain and food products, fruit, fruit juice, peas, milled rice, nuts, dried food products, spices, meat products. *Aspergillus* states produce a range of mycotoxins, but *Eurotium* states are aerophilic and the toxins they produce have not been studied extensively.
- *Neosartorya* (anamorph *Aspergillus*) – ascomatal wall of flattened pseudoparenchyma or hyphae, several cells thick. Ascospores biconvex, hyaline, with two equatorial crests and irregular surface ridges. *Aspergillus* conidia in dry chains forming olive-grey columns from solitary, erect, aseptate or septate, smooth conidiophores. Phialides formed directly on the swollen apex of the conidiophore, radiating, short-necked. Conidia slightly roughened, globose to subglobose or ellipsoid. Distributed worldwide, from soil, rice, cotton, potatoes, groundnuts, leather, paper products. Produce mycotoxins.
- *Talaromyces* (anamorph *Penicillium*) – ascomata soft, covered by and made of a few layers of well-developed networks of yellow pigmented hyphae, often heavily encrusted. Asci in chains. Ascospores yellow, sometimes red, broadly ellipsoid, sometimes thick-walled and spinulose. *Penicillium* conidia in dry chains from solitary, erect, branched, septate, smooth or rough conidiophores. Phialides from an apical branching system consisting of branch cells and supporting cells to the phialides (biverticillate), long, lageniform with a short,

narrow neck. Conidia subglobose, cylindrical, ellipsoid or fusiform, smooth or finely spinulose, hyaline, brown, brown-green or pale green. Distributed worldwide, from soil, organic substrates, rape, cotton, pears, wheat, barley, milled rice, pecan nuts, bagasse, often in tropical fruit juices, some species heat-resistant. *Penicillium* is used in production of antibiotics, enzymes, and manufacture and fermentation of cheeses, sugar, juice, brewing, and organic acids. *Penicillium* spp. are amongst the most toxigenic of moulds.

- *Eupenicillium* (anamorph *Penicillium*) – stromata sclerotioid, non-ostiolate, with walls of thick-walled pseudoparenchyma. Asci single or in chains. Ascospores aseptate, lenticular, spinulose, sometimes reticulate, with or without distinct equatorial ridges. *Penicillium* conidia in dry chains from solitary, erect, branched, septate, smooth or rough conidiophores. Phialides formed directly from conidiophore apices (monoverticillate) or as in species with *Talaromyces* teleomorphs, long or broadly lageniform, with a short, narrow neck. Conidia smooth, ovoid, subglobose, piriform or ellipsoid, hyaline. Distribution worldwide, from soil, groundnuts, fruit cake, canned fruit, corn, oranges. *Penicillium* species produce mycotoxins.
- *Thermoascus* (anamorph *Paecilomyces*) – thermophilic. Ascomata red, with a distinct pseudoparenchymatous wall several cells thick. Ascospores aseptate, echinulate, thick-walled, hyaline. For conidia see *Byssochlamys*.

Order Hypocreales Ascomata perithecial, rarely cleistothecial, sometimes in or on a stroma, more or less globose, sometimes ornamented, rarely setose. Ostiole periphysate. Peridium and stromatal tissues fleshy, usually brightly coloured. Interascal tissues of apical paraphyses, often evanescent. Asci cylindrical, thin-walled, sometimes with a small apical ring or a conspicuous apical cap, not blueing in iodine. Ascospores varied, hyaline or pale brown, usually septate, sometimes muriform, sometimes elongated and fragmenting, without a sheath.

Family Hypocreaceae Ascomata perithecial, rarely cleistothecial, sometimes either in or on a stroma, more or less globose, sometimes ornamented, rarely setose. Ostiole periphysate. Peridium and stromatal tissues fleshy, usually bright-coloured. Interascal tissue of apical paraphyses, often evanescent. Asci cylindrical, thin-walled, sometimes with an apical ring or conspicuous apical cap, not blueing in iodine. Ascospores varied, hyaline or pale brown, usually septate, sometimes muriform, sometimes elongated and fragmenting, lacking a sheath.

- *Gibberella* (anamorph *Fusarium*) – perithecia more or less superficial, often gregarious, with or without a basal stroma, fleshy, dark blue to violet. Paraphyses absent. Asci cylindrical, unitunicate, with an undifferentiated apex. Ascospores ellipsoid to fusiform, mostly triseptate, hyaline to subhyaline. *Fusarium* macroconidia in slimy yellow, brown, pink, red, violet or lilac masses, chains or dry masses from branched or unbranched, procumbent or erect, hyaline, smooth, septate conidiophores in sporodochial conidiomata. Phialides produced from apices of conidiophores or branches, slender or tapered, with one or sometimes several conidiogenous loci. Macroconidia hyaline, single- to multiseptate, fusiform to sickle-shaped, mostly with an elongated apex and a pedicellate basal cell. Microconidia usually aseptate, piriform, fusiform or ovoid, straight or curved, nearly always formed on aerial mycelium. Chlamydospores present or absent, intercalary, solitary, in chains or clusters, formed in hyphae or conidia. Distribution worldwide, from soil, aquatic and semiaquatic environments, stored grain and natural products. Mainly the *Fusarium* states are potent producers of mycotoxins in food.
- *Hypocrea* (anamorph *Trichoderma*) – teleomorph rarely produced in culture or seen in context of food microbiology. *Trichoderma* conidia in dry, powdery green, to yellow masses, from solitary, repeatedly branched, erect, hyaline, smooth, septate conidiophores which may end in sterile appendages. Phialides apical and lateral, often irregularly bent, flask-shaped with a short neck. Conidia aseptate, hyaline or usually green, smooth or roughened, globose, subglobose, ellipsoid, oblong or piriform. Distribution worldwide, from soil, stored grain, groundnuts, tomatoes, sweet potatoes, citrus fruit, pecan nuts. *Trichoderma* states also produce mycotoxins.
- *Nectria* (anamorphs *Acremonium, Fusarium*) – perithecia more or less superficial, often gregarious, with or without a basal stroma, fleshy, cream, orange, red, purple or violet, setose or glabrous. Paraphyses absent. Asci cylindrical, unitunicate, with an undifferentiated apex. Ascospores ellipsoid or naviculate with rounded ends, equally two-celled, smooth, minutely spinulose, or striate, hyaline, yellow or brown. Conidia (*Fusarium* state) as for *Gibberella*; *Acremonium* state as for *Emericellopsis*.

Order Leotiales (formerly known as Helotiales) Stromata usually absent, if present sclerotial. Ascomata apothecial, usually small, often brightly coloured, sessile or stipitate, sometimes with con-

spicuous hairs. Interascal tissue of simple paraphyses, variously shaped, apices sometimes swollen. Asci usually thin-walled, without separable wall layers, with an apical pore surrounded by an I+ or I– ring, apical apparatus variable. Ascospores usually small, simple or transversely septate, mostly hyaline, usually not quite longitudinally symmetrical, often smooth.

Family Dermateaceae Stroma absent. Ascomata small, flat or concave, usually sessile, grey-brown or black, occasionally immersed in plant tissue, then with a specialized opening mechanism, margin well-defined and often downy but without distinct hairs. Excipulum of brown, thin- or thick-walled isodiametric cells. Interascal tissue of simple paraphyses. Asci usually with a well-developed I+ or I– ring. Ascospores small, hyaline, septate or aseptate, often elongated. Two genera, *Mollisia* and *Pyrenopeziza*, are reported with *Phialophora*-like anamorphs. See *Phialophora*.

Family Sclerotiniaceae Stromata present. Ascomata apothecial, often long-stalked, usually brown, cupulate, lacking hairs, stalk often darker. Interascal tissue of simple paraphyses. Asci usually with an I+ apical ring. Ascospores large or small, ellipsoid, usually aseptate, hyaline or pale brown, often longitudinally symmetrical.

- *Botryotinia* (anamorph *Botrytis*) – teleomorph rarely seen in culture or encountered in context of food microbiology. *Botrytis* conidia formed in dry, powdery, grey masses from erect, brown, smooth, septate, solitary, hygroscopic conidiophores. Conidiogenous cells produced terminally on an apical head of small alternate branches, swollen, with many denticulate conidiogenous loci each forming a single conidium simultaneously. Conidia aseptate, rarely one or two septate, pale brown, globose, ovate or ellipsoid, smooth, hydrophobic. Microconidial state (*Myrioconium*) sometimes formed, sporodochial, phialidic with small, globose or subglobose, hyaline conidia. Sclerotia large, cortex black to brown with a white medulla, flattened to pulvinate, rounded to ellipsoid, smooth or wrinkled. Distribution worldwide, more commonly in humid temperate and subtropical regions, from soils both dry and aquatic, stored and in transit fruit and vegetables, fruit and leaf rots of strawberry, grape, cabbage, lettuce, neck rot in onions and shallots.

Order Microascales Stromata absent. Ascomata solitary, perithecial or cleistothecial, usually black, thin-walled, sometimes with well-developed smooth setae. Interascal tissue absent or rarely of undifferentiated hyphae. Centrum absent. Asci formed in chains, mostly globose, unstalked, very thin-walled, evanescent, eight-spored. Ascospores yellow or reddish brown, aseptate, often curved and with very inconspicuous germ pores, without a sheath.

Family Microascaceae Ascomatal wall entirely of black, small-celled pseudoparenchyma, perithecial or cleistothecial. Interascal hyphae absent. Ascospores smooth.

- *Microascus* (anamorph *Scopulariopsis)* – mycelium pigmented. Ascomata superficial or partly immersed, carbonaceous, glabrous or setose, with a well-differentiated, sometimes cylindrical ostiole. Asci ovate, evanescent. Ascospores extruded in cirrhi, small, aseptate, asymmetrical, often reniform, heart-shaped or triangular, dextrinoid when young, with a small germination pore. *Scopulariopsis* conidia in white to shades of brown, dry powdery masses from erect penicillately or verticillately branched, hyaline to pale brown, smooth, septate, solitary conidiophores. Conidiogenous cells terminal, cylindrical, repeatedly forming basipetal chains of conidia from percurrently proliferating loci giving rise to apical annellations. Conidia aseptate, hyaline or brown, globose, ovate or mitriform, with a truncate base, smooth or ornamented. Distribution worldwide, from soil, grain, fruit, soya beans, groundnuts, milled rice, and animal products such as meat, eggs, cheese, milk, butter.

Order Pezizales Stroma absent. Ascomata apothecial or cleistothecial, rarely absent, often large, discoid, cupulate or globose, sometimes stalked, often brightly coloured. Excipulum usually thick-walled, fleshy or membranous, of thin-walled pseudoparenchyma cells. Interascal tissue of simple or moniliform paraphyses, often pigmented and swollen apically, absent in cleistothecial taxa. Asci elongated, persistent, thin-walled, usually with no apical thickening, opening by a circular pore or vertical split, wall sometimes blueing in iodine, globose and in cleistothecial taxa indehiscent. Ascospores ellipsoidal, aseptate, hyaline to strongly pigmented, often ornamented, usually without a sheath. This order contains a number of families with genera in which species are edible, though not generally commercially produced. It includes the morels (family Morchellaceae, genus *Morchella*) and subterranean truffles, of which there are many species (family Helvellaceae, genus *Hydnotrya*; family Terfeziaceae, genus *Terfezia*; family Tuberaceae, genus *Tuber*).

Order Saccharomycetales Mycelium absent or poorly developed, when present septa with minute pores rather than a single simple pore. Vegetative cells proliferating by budding or fission. Walls usually lacking chitin except around bud scars, sometimes with I+ gel. Ascomata absent. Asci single or in chains, sometimes not differentiated morphologically from vegetative cells, usually at least eventually evanescent. Ascospores varied in shape, sometimes with equatorial or asymmetric thickenings.

Family Dipodascaceae Mycelium well-developed, lacking a polysaccharide sheath, septa with clusters of minute pores, fragmenting to form thallic conidia. Asci form by fusion of gametangia from adjacent cells or separate mycelia, usually elongated, more or less persistent, single- to multispored. Ascospores usually ellipsoid, rarely ornamented, usually with a mucous sheath, not blueing in iodine.

- *Galactomyces* (anamorph *Geotrichum*) – supporting hyphae of gametangia profusely septate. Gametangia on opposite sides of hyphal septa, globose to clavate, fusing at the apices to form the ascus. Asci subhyaline, subglobose to broadly ellipsoid, with one or two ascospores. Ascospores broadly ellipsoid, pale yellow-brown, with an echinate inner wall and an irregular exosporium wall, often with a hyaline equatorial furrow. *Geotrichum* conidia formed in white, smooth, often butyrous colonies from aerial, erect or decumbent hyaline mycelium functioning conidiogenously. Mycelium dichotomously branched at advancing edge. Conidiogenesis thallic. Conidia hyaline, aseptate, smooth, cylindrical, doliiform or ellipsoid. Distributed worldwide, from soil, water, air, cereals, grapes, citrus, bananas, tomatoes, cucumber, frozen fruit cakes, milk and milk products; also used with bacteria in fermentation of manioc to produce gari in West Africa.

Family Saccharomycetaceae Mycelium more or less absent. Vegetative cells reproducing by multilateral budding, more or less ellipsoid, without mucus. Asci morphologically similar to vegetative cells, not in well-developed chains, more or less globose, thinwalled, one to four spored, evanescent or persistent. Ascospores usually spherical, often ornamented with equatorial ridges. Fermentation positive. Coenzyme system usually Q-6. Many linked to *Candida* anamorphs which are polyphyletic.

- *Debaryomyces* (anamorph *Candida*) – colonies not pigmented or mucoid. Septate hyphae absent but septate expanding hyphae predominant in anamorph, yeast cells usually haploid. Asci usually persistent, one to four spored. Ascospores pitted or with blunt ridges, thin-walled, hyaline, spherical. *Candida* conidia produced by budding which is multilateral or acropetal leaving conidiogenous cells with or without broad denticles or scars, hyaline, aseptate, base attenuated, rounded. Fermentation present or nearly absent, nitrate not assimilated, acid production absent or weak, growth at 25°C. Common food-borne species.

- *Dekkera* (anamorph *Brettanomyces*) – colonies not pigmented or mucoid. Cells often cylindrical. Asci formed on hyphae, thin-walled, not crowned with an apical cell. Ascospores less than 5 μm diameter, galeate, smooth, usually hyaline. *Brettanomyces* conidia hyaline, with an attenuated rounded base, formed by multilateral or acropetal budding from conidiogenous cells with or without broad denticles or scars. Conidiophores absent. Strong acid production in glucose-containing media, fermentation present or nearly absent. Spoilage organism in beverages such as mineral waters and nonalcoholic drinks, and other old or spoilt beers, ciders and wines.

- *Issatchenkia* (anamorph *Candida*) – yeast cells often elongated, usually haploid, forming a pseudomycelium. Septate hyphae absent. Asci usually spherical, one to four spored, usually persistent. Ascospores thick-walled, spherical, hyaline, rather large, often irregular or with a sheath. Often heterothallic. Fermentation absent or weak, nitrate not assimilated, growth at 25°C. For conidia see *Debaryomyces*. Common food spoilage species.

- *Kluyveromyces* (anamorph *Candida*) – septate hyphae absent. Asci deliquescent. Ascospores reniform, oblate or nearly spherical, smooth, easily liberated from the ascus. Growth at 25°C. For conidia see *Debaryomyces*. Spoilage organisms in dairy products.

- *Pichia* (anamorph *Candida*) – asci without a tube-like base, formed on hyphae, thin-walled, not crowned with an apical cell, one to four spored. Ascospores less than 5 μm diameter, galeate (hat-shaped), smooth, usually hyaline. Nitrate not assimilated, acid production absent or weak. For conidia see *Debaryomyces*. Spoilage in tanning fluids, wine, soft drinks, beer, fermented vegetable, olive and pickle brines; preservative resistant.

- *Saccharomyces* (anamorph *Candida*) – septate hyphae absent. Diploid yeast cells become one to four spored asci, usually persistent. Ascospores thin-walled, smooth, spherical, usually hyaline. Fermentation strong, nitrate not assimilated, growth at 25°C. For conidia see *Debaryomyces*. Includes brewers' yeast used in beer, bread-making

and other fermentations; also a spoilage organism in fruit products such as juices, pulps and wines.

- *Torulaspora* (anamorph *Candida*) – septate hyphae absent, yeast cells usually haploid. Asci usually persistent. Ascospores verrucose or nearly smooth, thin-walled, hyaline, spherical. Fermentation present, nitrate not assimilated, growth at 25°C. For conidia see *Debaryomyces*. Food spoilage organisms e.g. in fruit juices, soft drinks and olive brines; preservative resistant.

- *Yarrowia* (anamorph *Candida*) – asci free or on septate hyphae. Ascospores irregular in size and shape, often hemiellipsoid or navicular, verrucose. Heterothallic, lipolytic. For conidia see *Debaryomyces*. Lipid food spoilage organisms, including margarine, olives, corn oil processing.

- *Zygosaccharomyces* – septate hyphae absent, yeast cells usually ellipsoid and haploid. Asci formed by two conjugating yeast cells, usually persistent, with one to four spherical ascospores. Ascospores usually smooth, rarely verrucose, usually thin-walled, spherical, hyaline. Fermentation present, nitrate not assimilated, osmotolerant, growth at 25°C. Spoilage organisms in high-sugar products such as wines, acid foods, fruit juices, soft drinks, honey, syrups, jelly; preservative resistant. Used in miso and soy sauce fermentations.

Order Sordariales Ascomata very rarely stromatic, perithecial or cleistothecial, thin- or thick-walled. Interascal tissue very inconspicuous or lacking. Asci cylindrical or clavate, persistent or evanescent, not fissitunicate. Ascospores mostly with at least one dark cell, with germ pore, often with a gelatinous evanescent sheath or appendages.

Family Coniochaetaceae Ascomata usually perithecial, solitary or aggregated, sometimes on a poorly developed subiculum. Interascal tissue inconspicuous, of paraphyses. Asci usually cylindrical, often with a small apical ring, sometimes I+ blue. Ascospores aseptate, dark brown, with a germ slit, sheath lacking. The genus *Coniochaeta* is reported with a *Phialophora*-like anamorph. See *Phialophora*.

Family Sordariaceae Ascomata dark, usually thick-walled and ostiolate. Interascal tissue of thin-walled undifferentiated cells, inconspicuous and often evanescent. Asci cylindrical, with a thickened I– ring. Ascospores brown, simple or very rarely septate, sometimes ornamented, often with a gelatinous sheath but lacking caudae.

- *Neurospora* (anamorph *Chrysonilia*) – mycelium extremely fast-growing, initially colourless, later pink to orange. Perithecia separate, piriform, short-necked, glabrous. Ascospores dark brown with nerve-like ribs ornamenting the outer wall. *Chrysonilia* conidia aseptate, ellipsoid or more or less cylindrical, globose, subglobose or irregular, hyaline, smooth, formed in dry chains with connectives from ascending to erect, smooth, septate, much-branched conidiophores. Widespread, especially Europe, USA and Asia, from bread (red bread mould) and related products, silage, meat, and transported and stored fruit; used in production of oncom mera by fermentation of soya bean products.

Phylum Basidiomycota

The basidium is diagnostic (a cell on the outside of which after karyogamy and meiosis generally four basidiospores are produced). Clamp connections sometimes formed in maintenance of dikaryotic condition. Septa of the dolipore type. Walls double-layered, lamellate and electron-opaque by electron microscopy. Molecular sequence data important in separations within the phylum.

Class Basidiomycetes

Definition same as for phylum.

Subclass Phragmobasidiomycetidae Metabasidium is divided by primary septa, usually cruciate or horizontal. One order, Auriculariales, containing edible species, some commercially produced.

Subclass Holobasidiomycetidae Metabasidium not divided by primary septa but may sometimes become adventitiously septate. Several orders, Agaricales, Boletales, Cantharellales, Cortinariales, Fistulinales, Gomphales, Hericiales, Lycoperdales, Poriales, Russulales, Thelephorales, containing edible species, some commercially produced (**Table 1**).

Mitosporic Fungi

The 'mitosporic fungi' (asexual, anamorphic, imperfect, conidial, deuteromycete fungi) are an artificial group without a formal nomenclature above the generic level, comprising the mitotic states (anamorphs) of the meiotic ascomycetes and basidiomycetes (teleomorphs) and mitotic fungi that have not been correlated with any meiotic states. They are characterized by the formation of conidia as a result of presumed mitosis. Separation of genera is primarily by mode of conidiogenesis and growth of the conidiogenous cell, with morphology of conidiomata, conidia and conidiophores as subsidiary criteria.

- *Acremonium* – see *Emericellopsis* and *Nectria*, but

Table 1 Orders, families and genera of basidiomycetes in which edible species have been reported

Order	Family	Genera
Auriculariales	Auriculariaceae	*Auricularia*
Agaricales	Agaricaceae	*Agaricus, Chamaemyces, Leucoagaricus, Macrolepiota*
	Amanitaceae	*Amanita, Limacella, Termitomyces*
	Bolbitiaceae	*Agrocybe*
	Coprinaceae	*Coprinus, Psathyrella*
	Entolomataceae	*Rhodocybe*
	Hygrophoraceae	*Camarophyllus, Hygrocybe, Hygrophorus*
	Pluteaceae	*Pluteus, Volvariella*
	Strophariaceae	*Kuehneromyces, Panaeolus, Pholiota, Psilocybe, Stropharia*
	Tricholomataceae	*Armillaria, Calocybe, Clitocybe, Collybia, Flammulina, Laccaria, Lentinula, Lyophyllum, Marasmius, Melanoleuca, Mycena, Oudemansiella, Pseudoclitocybe, Strobilurus, Tricholoma, Tricholomopsis*
Boletales	Boletaceae	*Boletus, Leccinum, Suillus*
	Gomphidiaceae	*Chroogomphus, Gomphidius*
	Gyrodontaceae	*Gyroporus*
	Hygrophoropsidaceae	*Hygrophoropsis*
	Paxillaceae	*Paxillus*
	Strobilomycetaceae	*Chalciporus*
	Xerocomaceae	*Phylloporus, Xerocomus*
Cantharellales	Cantharellaceae	*Cantharellus*
	Clavariadelphaceae	*Clavariadelphus*
	Craterellaceae	*Craterellus*
	Hydnaceae	*Hydnum*
	Sparassidaceae	*Sparassis*
Cortinariales	Cortinariaceae	*Cortinarius, Phaeolepiota, Rozites*
Fistulinales	Fistulinaceae	*Fistulina*
Gomphales	Ramariaceae	*Ramaria*
Hericiales	Hericiaceae	*Hericium*
	Lentinellaceae	*Lentinellus*
Lycoperdales	Lycoperdaceae	*Lycoperdon*
Poriales	Coriolaceae	*Grifola, Poria*
	Lentinaceae	*Lentinus, Pleurotus*
	Polyporaceae	*Polyporus*
Russulales	Russulaceae	*Lactarius, Russula*
Thelephorales	Thelephoraceae	*Sarcodon*

many species of polyphyletic ascomycete affinity have no known teleomorph.

- *Alternaria* – see *Clathrospora* and *Leptosphaeria*,

but many species of undoubted ascomycete affinity have no known teleomorph.

- *Aspergillus* – see *Emericella*, *Eurotium* and *Neoasartorya*, but many species of ascomycete affinity have no known teleomorph.
- *Aureobasidium* – colonies covered by slimy, yellow, cream, pink, brown or black masses of spores. Aerial mycelium scanty, immersed mycelium often dark brown. Conidiogenous cells undifferentiated, procumbent, intercalary or on short lateral branches. Conidia produced synchronously on multiple loci in dense groups on short scars or denticles, hyaline, smooth, with a truncate base. Distribution worldwide, saprobic, from soil, leaf surfaces, cereal seed, on flour, tomato, pecan nuts, fruit, fruit drinks.
- *Basipetospora* – see *Monascus*.
- *Botrytis* – see *Botryotinia*, but many species have not been linked to teleomorphs.
- *Brettanomyces* – see *Dekkera*. There are a number of species not linked to teleomorphs.
- *Candida* – see *Debaryomyces*, *Issatchenkia*, *Kluyveromyces*, *Pichia*, *Saccharomyces*, *Torulaspora* and *Yarrowia*, but many species of polyphyletic ascomycete affinity have not been linked with teleomorphs.
- *Chrysonilia* – see *Neurospora*.
- *Cladosporium* – see *Mycosphaerella*, but many species have not been linked to teleomorphs.
- *Epicoccum* – colonies fluffy, yellow, orange, red, brown or green. Conidiophores formed in black sporodochial conidiomata, closely branched, compact and dense. Conidiogenous cells pale brown, smooth or verrucose, integrated, terminal, determinate, cylindrical. Conidia solitary, dry, subspherical to piriform, dark golden-brown, often with a pale, protuberant basal cell, muriform, rough, opaque. Distribution worldwide, from soil, cereal seed, beans, mouldy paper, textiles.
- *Fusarium* – see *Nectria* and *Gibberella*, but many species have no known teleomorph.
- *Geotrichum* – see *Galactomyces*, but many species of polyphyletic ascomycete affinity have not been linked to teleomorphs.
- *Moniliella* – colonies acidophilic, restricted, smooth, velvety or cerebriform, cream then pale olivaceous or black-brown. Cells often budding to produce a pseudomycelium. Conidiophores undifferentiated, hyaline, smooth, repent. Conidia formed in acropetal chains from individual (conidiogenous) cells of the mycelium, hyaline, smooth, aseptate, ellipsoid. Thallic conidia also formed by fragmentation of hyphae, becoming thick-walled and brown. From Europe and USA,

occurring in pickles and vinegar, fruit juices, syrups and sauces.

- *Paecilomyces* – see *Byssochlamys* and *Thermoascus*, but many species of ascomycete affinity have no known teleomorph.
- *Penicillium* – see *Talaromyces* and *Eupenicillium*, but many species of ascomycete affinity have no known teleomorph.
- *Phialophora* – colonies slow-growing, olivaceous black, sometimes pink or brown. Conidiophores erect, hyaline or pale brown, branched or reduced to simple hyphae. Conidiogenous cells clustered or single, phialidic, lageniform or cylindrical, with a distinct darker collarette. Conidia formed in slimy heads or in chains, aseptate, globose to ellipsoid or curved, mostly hyaline, smooth. See *Coniochaeta*, *Mollisia*, *Pyrenopeziza*, with *Phialophora*-like anamorphs, also linked with *Geaumannomyces*, but several species with no known teleomorph. Worldwide in distribution but most common on decaying wood, wood pulp, secondarily soil-borne, from water, fermented corn dough, foodstuffs, butter, wheat.
- *Phoma* – colonies comparatively fast-growing, grey, olivaceous, brown, fluffy. Conidiomata pycnidial, black-brown, ostiolate, sometimes setose. Conidiophores absent. Conidiogenous cells ampulliform to doliiform, hyaline, smooth, phialidic. Conidia hyaline, smooth, aseptate or sometimes septate, ellipsoid, ovate, cylindrical. Dark-brown unicellular or multicellular chlamydospores sometimes formed. Some teleomorphs in Pleosporaceae (*Pleospora*), but most species with no known teleomorphs. Distribution worldwide, from soil, butter, rice grain, cement, litter, paint, wool and paper; also produces mycotoxins.
- *Rhodotorula* – colonies pink, with carotenoid pigment soluble in organic solvents, mycelium and/or pseudomycelium formed, cells usually small and narrow. Conidia spherical, ovate or clavate, with a narrow or rather broad base, budding. Sometimes assimilates nitrate, but fermentation absent. Teliospores absent, but basidium-like structures in some species indicate basidiomycete affinity with *Rhodosporidium* (Sporidiobolaceae). From wood, involved in spoilage of dairy products, fresh fruit, vegetables and seafoods, especially refrigerated foods.
- *Scopulariopsis* – see *Microascus*, but many species of ascomycete affinity have no known teleomorph.
- *Stachybotrys* – colonies black to black-green, powdery. Conidiophores erect, separate, simple or branched, septate, becoming brown and rough at the apex. Conidiogenous cells grouped at the conidiophore apex, phialidic, obovate, ellipsoid,

clavate or broadly fusiform, becoming olivaceous, with a small locus and no collarette. Conidia in large, slimy black heads, ellipsoid, reniform or subglobose, hyaline, grey, green, dark brown or black, sometimes striate, coarsely rough or warted, aseptate. Distribution worldwide, from soil, paper, cereal seed, textiles. Trichothecene mycotoxins produced such as satratoxin but its toxicity is unknown.

- *Trichoderma* – see *Hypocrea*, but many species of ascomycete affinity have no known teleomorph.
- *Trichosporon* – colonies slow-growing, white to cream, butyrous, smooth or wrinkled. Mycelium repent, hyaline. Conidiophores absent. Conidia of two types: (1) thallic, formed by fragmentation of the mycelium, cylindrical to ellipsoid; (2) blastic, formed in clusters near the ends of the thallic conidia or by budding of the lateral branches of the mycelium, subglobose, with a narrow distinct scar. Distribution worldwide, from humans and animals, saprobic in soil, fresh and sea water, plant material, fermented corn dough.
- *Trichothecium* – colonies powdery, pink. Conidiophores erect, separate, simple, unbranched, septate near the base, rough, apical cell functioning conidiogenously. Conidia formed in retrogressively delimited basipetal chains, appearance zigzagged, hyaline, smooth, one-septate, ellipsoid or piriform, thick-walled, with an obliquely truncate scar. Distribution worldwide, from soil, water, decaying plant material, leaf litter, cereal seed, pecan nuts, stored apples, fruit juices, foodstuffs especially flour products; also a potent producer of trichothecene mycotoxins, but significance to human health is unknown.
- *Ulocladium* – colonies black to olivaceous black. Conidiophores erect, separate, simple or branched, septate, smooth, straight, flexuous, often geniculate, geniculations associated with preformed loci (pores). Conidia dry, solitary or in short chains, obovoid to short ellipsoid, with several transverse and londitudinal or oblique eusepta, medium brown to olivaceous, smooth or verrucose, base conical, apex broadly rounded and becoming conidiogenous. Not uncommon, widely distributed, from soil, water, dung, paint, grasses, fibres, wood, paper, corn, seeds.
- *Verticillium* – colonies cottony, white to pale yellow, sometimes becoming black due to resting mycelium. Conidiophores erect, separate, septate, smooth, hyaline, simple, unbranched or branched. Conidiogenous cells solitary or produced in verticillate divergent whorls, long lageniform to aculeate, hyaline, phialidic. Conidia form in droplets at the apices of conidiogenous cells, hyaline, aseptate,

smooth, ellipsoid to cylindrical. Hyaline multicellular chlamydospores and microsclerotia sometimes formed. Distribution worldwide, commonly causing plant wilt diseases, from soil, paper, insects, seeds, bakers' yeast, potato tubers, commercially grown fungi; also forms mycotoxins.

- *Wallemia* – colonies xerophilic, restricted, fan-like or stellate, powdery, orange brown to black brown. Conidiophores erect, separate, cylindrical, smooth, pale brown. Conidiogenous cells apical, long lageniform to cylindrical, finally verrucose, forming a phialidic aperture without collarette from which a short chain of four thallic conidia is formed. Conidia initially cuboid, later globose, pale brown, finely warted. Distributed worldwide, from dry foodstuffs such as jams, marzipan, dates, bread, cake, salted fish, bacon, salted beans, milk, fruit, soil, air, hay, textiles.

Conclusion

It is not the purpose of this article to provide an extensive review of all the fungi involved in food microbiology. Plant pathogens play a significant preproduction role but are beyond the scope of this volume. Many spoilage organisms are opportunistic and the numbers potentially capable of causing problems are enormous, so only the most common have been mentioned. There are comparatively few commercially produced fungi, and these are greatly outnumbered by those which are edible but occur only in natural habitats. They are not pertinent to mainstream food microbiology.

See also: **Alternaria**. **Aspergillus**: Introduction. **Aureobasidium**. **Botrytis**. **Brettanomyces**. **Byssochlamys**. **Candida**: Introduction. **Debaryomyces**. **Fusarium**. **Geotrichum**. **Kluyveromyces**. **Monascus**. **Mucor**. **Mycotoxins**: Classification; Occurrence. **Penicillium**: Introduction. **Pichia**. **Rhodotorula**. **Saccharomyces**: Introduction. **Trichoderma**. **Trichothecium**. **Zygosaccharomyces**.

Further Reading

Alexopoulos CJ, Mims CW and Blackwell M (1996) *Introductory Mycology*, 4th edn. New York: John Wiley.

Barr DJS (1992) Evolution and kingdoms of organisms from the perspective of a mycologist. *Mycologia* 84: 1–11.

Betina V (1989) *Mycotoxins: Chemical, Biological and Environmental Aspects*. Bioactive Molecules 9. Amsterdam: Elsevier.

Beuchat LR (ed.) (1987) *Food and Beverage Mycology*, 2nd edn. New York: Van Nostrand Reinhold.

Carmichael JW, Kendrick WB, Conners IL and Sigler L (1980) *Genera of Hyphomycetes*. Edmonton: University of Alberta Press.

Davenport RR (1981) Yeasts and yeast-like organisms. In: Onions AHS, Allsopp D and Eggins HOW (eds) *Smith's Introduction to Industrial Mycology*, 7th edn. London: Edward Arnold.

Domsch KH, Gams W and Anderson TH (1980) *Compendium of Soil Fungi*. London: Academic Press.

Hawksworth DL (1994) (ed.) *Ascomycete Systematics: Problems and Perspectives in the Nineties*. NATO ASI Series A: Life Sciences 269. New York: Plenum Press.

Hawksworth DL, Kirk PM, Sutton BC and Pegler DN (1995) *Ainsworth & Bisby's Dictionary of the Fungi*, 8th edn. Wallingford: CAB International.

Kendrick WB (1979) *The Whole Fungus*. Ottawa: National Museums of Canada.

Pitt JI (1979) *The Genus Penicillium and its Teleomorphic states of Eupenicillium and Talaromyces*. London: Academic Press.

Rayner ADM, Brasier CM and Moore D (eds) (1987) *Evolutionary Biology of the Fungi*. Cambridge: Cambridge University Press.

Reynolds DR and Taylor JW (eds) (1993) *The Fungal Holomorph: Mitotic, Meiotic and Pleomorphic Speciation in Fungal Systematics*. Wallingford: CAB International.

Samson RA and Pitt JI (eds) (1990) *Modern Concepts in Penicillium and Aspergillus Classification*. NATO ASI Series A: Life Sciences 185. New York: Plenum Press.

Samson RA and van Reenen-Hoekstra ES (1988) *Introduction to Food-borne Fungi*, 3rd edn. Baarn: Centraalbureau voor Schimmelcultures.

Smith JE and Moss MO (1985) *Mycotoxins: Formation, Analysis and Significance*. Chichester: John Wiley.

Sutton BC (1980) *The Coelomycetes*. Kew: Commonwealth Mycological Institute.

Talbot PHB (1971) *Principles of Fungal Taxonomy*. London: Macmillan.

Classification of the Peronosporomycetes

M W Dick, Department of Botany, University of Reading, UK

Introduction

Physiologically and morphologically, as obligately osmotrophic heterotrophs, the Peronosporomycetes are 'fungi'. However, they are phylogenetically separate from the Mycota. The biflagellate, anisokont but non-straminipilous Plasmodiophorales and the uniflagellate Chytridiomycetes are likewise unrelated. The Chytridiomycetes may be an early offshoot from the phylogenetic line leading to the non-flagellate Mycota.

The Peronosporomycetes include the most numerous, most important and earliest known (with mid-eighteenth century reports for *Saprolegnia* on fish) water moulds. Study of the Peronosporomycetes has received attention since the 1840s, because of the sociohistoric significance of late blight of potato (*Phytophthora infestans*) and downy mildew of vines (*Plasmopara viticola*). Some of the most damaging groups of pathogens of food crops are Peronosporomycetes.

Many of the parasitic species, other than the root pathogens, have restricted host ranges; most are obligate parasites not available in axenic culture. The downy mildews (Peronosporales on advanced dicotyledons and Sclerosporales on panicoid grasses) are leaf and stem parasites; nematodes and rotifers are parasitized by the Myzocytiopsidales; arthropods by the Saprolegniales and Salilagenidiales; vertebrates by the Saprolegniales and Pythiales; and other Peronosporomycetes by related fungi.

Most species of Peronosporomycetes are freshwater or terrestrial; very few are strictly aquatic, but many are characteristic of wet marginal sites, or from seasonally or intermittently waterlogged soil. *Aqualinderella fermentans* is the only obligate anaerobe. In terrestrial and freshwater ecosystems the saprobic Peronosporomycetes have a major ecological role in degradation and recycling, as deduced from estimates of activity and biomass production from spore population sizes. Many of the saprobic and facultatively parasitic species are abundant, with worldwide distributions. A few taxa are confined to the pantropics or to a continental landmass, but strictly psychrophilic or thermophilic species have not been identified. Saprobic taxa survive in estuarine conditions, but such habitats may not be their primary niche: a few parasitic Peronosporomycetes are oligohaline or marine.

The Peronosporomycetes contains at least 900 and perhaps as many as 1500 species, depending on the species concepts used for the obligate parasites of angiosperms. The principal families in terms of numbers of species, frequency of isolation, and economic importance are the Peronosporaceae, Pythiaceae, Sclerosporaceae and Saprolegniaceae.

Class Diagnosis

The Peronosporomycetes are straminipilous fungi: fungi possessing (or evolved from organisms that once possessed) a biflagellate zoosporic phase in which the flagella are anisokont and heterokont, the anteriorly directed flagellum bearing two rows of tubular tripartite hairs (the straminipilous flagellum). Other straminipiles are photosynthetic (diatoms, brown algae, etc.) and the kingdom also includes further heterotrophs such as the uniflagellate straminipilous fungi (Hyphochytriomycetes), the Labyrinthista and the Lagenismatales.

Correlating characters are: walls primarily composed of β-1,3- and β-1,6-glucans; α,ε-diaminopimelic acid (DAP) lysine synthesis; a haplomitotic B ploidy cycle (mitosis confined to the diploid phase); oogamy; and mitochondria with tubular cristae.

A closed cruciform meiosis in which the nuclear membrane remains intact until the second telophase and the metaphase I and metaphase II spindle poles are in the same plane is characteristic of the class. These fungi are unique in possessing synchronous multiple meioses in a coenocyte with a haplomitotic B ploidy cycle (**Fig. 1**). The nature of sexual reproduction provides several of the primary criteria for classification within the Peronosporomycetes.

Features of the Class

Commercial Importance

Economically, the most important members of the Peronosporomycetes are the phytoparasites, particularly the root-rotting fungi and the downy mildews (**Table 1**). In many cases the hosts are crop plants such as sunflower, lettuce, cucurbits, vines, corn and millet, and the pathogens are important causes of crop failure. The recognition of the tightly circumscribed temperature and humidity optima for leaf infection of *Solanum* by zoospores of *Phytophthora infestans* has enabled sophisticated forecasting procedures to be developed. Crop losses due to *Pythium* are probably more considerable than has been recognized.

By far the most noteworthy arthropod parasites are *Aphanomyces* (Saprolegniales) on freshwater crustacea and *Salilagenidium* and *Halodaphnea* on marine crustacea. The introduction and spread of *Aphanomyces astaci* (Krebspest disease) in Europe has eliminated entire populations of the European crayfish from many river systems, from which recovery is improbable. Mariculture of prawns and shrimps in Asian coastal waters is subject to epidemics caused by species of the Salilagenidales. The disease of salmonid fish, ulcerative dermal necrosis, occasionally reaches epidemic levels, and continues to be under investigation because of its incidence in fish farming.

Although there are several examples of insect parasitism, *Lagenidium*, endoparasitic in mosquitoes, has been targeted for biological control of mosquito populations.

No member of the Peronosporomycetes is known as a source for any economically important product, although carbohydrate polymer production from cell wall material has potential from taxa with rapid hyphal growth rates and ease of culture.

Biochemistry

Saprobic species have long been studied because of their rapid biomass increase on simple media. Restricted availability of combined forms of nitrogen and sulphur for nutrition are of particular interest since they appear to provide phylogenetic markers of genetic deletions unlikely to be restored. The biochemistry of respiration has received attention because of facultatively or obligately fermentative abilities.

Nucleic Acids Genomic and mitochondrial DNA are now being used in discussions of relatedness at family, genus and species levels. The DNA G+C ratios vary widely in the Peronosporomycetes. Both the position (within the NTS of the rDNA repeat unit) and the occurrence of inverted copies of the 5S rDNA unit show variation within the subclasses. The mitochondrial genome length is between 36.4 kb and 73.0 kb and the presence of an inverted repeat (about 10–30 kb) has been established for some species.

Lysine Synthesis The diagnostic DAP lysine synthesis pathway is thought to have evolved before the α-aminoadipic acid (AAA) pathway (which is correlated with the presence of mitochondria with flat, plate-like cristae and often with chitinous cell walls). Unlike the AAA pathway, the DAP pathway may be associated with a range of mitochondrial types.

Cell Wall Materials The amount of wall fibrillar material is lower than in plants, and cellulose (β-1,4-glucan) tends to be masked by larger amounts of β-1,3- and β-1,6-glucans. Chlor-zinc iodide histochemistry, while an unreliable indicator for cellulose, remains the only way of indicating the wall chemistry of endobiotic parasites. Glucosamine occurs in the Peronosporomycetidae and Saprolegniomycetidae and the presence of chitin has been confirmed for Saprolegniaceae. Hydroxyproline-rich protein (HRP) is present in greater amounts in the Pythiaceae than in the Saprolegniaceae.

Sterol Metabolism: Secondary Metabolites Sterols have been implicated in various functions associated with sexual reproduction, including induction of sexuality (with indications that there may be substitution of analogue induction); directional growth of the gametangial axes; localized stimulation of wall softening at the point of contact between gametangia; and effect on meiosis.

Possibly all taxa of the Peronosporales are depend-

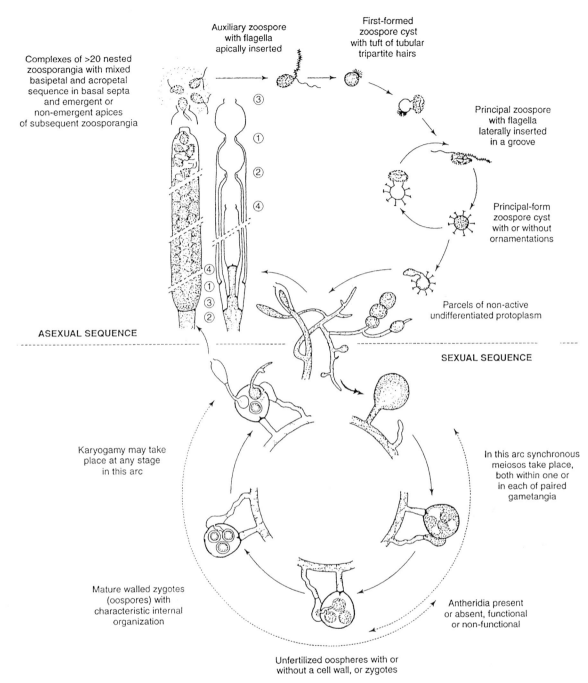

Auxiliary zoospore
with flagella
apically inserted

First-formed
zoospore cyst
with tuft of tubular
tripartite hairs

Complexes of >20 nested
zoosporangia with mixed
basipetal and acropetal
sequence in basal septa
and emergent or
non-emergent apices
of subsequent zoosporangia

Principal zoospore
with flagella
laterally inserted
in a groove

Principal-form
zoospore cyst
with or without
ornamentations

Parcels of non-active
undifferentiated protoplasm

ASEXUAL SEQUENCE

SEXUAL SEQUENCE

Karyogamy may take
place at any stage
in this arc

In this arc synchronous
meiosos take place,
both within one or
in each of paired
gametangia

Mature walled zygotes
(oospores) with
characteristic internal
organization

Antheridia present
or absent, functional
or non-functional

Unfertilized oospheres with or
without a cell wall, or zygotes

Figure 1   Illustrated and annotated life-history of *Saprolegnia*.

ent on exogenous sources of sterols. Evidence of partial dependence on exogenous sterol precursors exists for the Pythiales. The sterol requirements for sexual reproduction in *Pythium* and *Phytophthora* have been given most prominence, but the detailed requirements for exogenous sterols differ between genera, as does the ability to utilize sterols with certain substituents. Loss of sterol anabolic pathways occurs at subgeneric levels, as may be inferred from sexual reproductive capacity within the Pythiaceae and *Achlya* species.

Some sterol biosynthetic pathways have been regarded as being of a fundamental evolutionary significance equivalent to that of lysine synthesis. Information of phylogenetic value comes from cycloartenol and lanosterol synthesis. Lanosterol is formed from squalene oxide cyclization via cycloartenol in photosynthetic lineages, but directly in non-photosynthetic lineages, including the Peronosporomycetes.

Secondary metabolites may also be linked with steroid metabolism. For most of the Albuginaceae and

Table 1 Representative food and crop pathogens placed in the Peronosporomycetes

Food/crop	Genus/family	Disease	Geographic/cultural restriction/importance	Genus of pathogen
Potato	Solanum (Solanaceae)[a]	Late blight[b,c]		Phytophthora
Tobacco	Nicotiana (Solanaceae)[a]	Black shank	North America	Phytophthora
Tobacco	Nicotiana (Solanaceae)[a]	Downy mildew		Peronospora
Sunflower	Helianthus (Asteraceae)[a]	Downy mildew, stunting		Peronospora
Lettuce	Lactuca (Asteraceae)[a]	Leaf rot[b]		Bremia
Cucumber	Cucumis (Cucurbitaceae)[a]	Downy mildew	Northern glasshouse crops	Pseudoperonospora
Cucumber	Cucumis (Cucurbitaceae)[a]	Fruit rot	North America	Phytophthora
Melon	Cucumis (Cucurbitaceae)[a]	Root rot	Middle east	Pythium
Cocoa	Theobroma (Sturculiaceae)[a]	Black pod	Africa, South America	Phytophthora
Cocoa	Theobroma (Sturculiaceae)[a]	Pod rot	Africa	Trachysphaera
Carrot	Daucus (Apiaceae)[a]	Cavity root spot	Australia, UK	Pythium
Grapes	Vitis (Vitaceae)	Downy mildew		Plasmopara
Avocado	Persea (Lauraceae)	Root/collar rot[c]	America	Phytophthora
Pea	Pisum (Fabaceae)	Root rot	North America	Aphanomyces
Apple	Malus (Rosaceae)	Crown/collar rot		Pythium
Strawberry	Fragaria (Rosaceae)	Red core/stele[b]	New Zealand	Phytophthora
Raspberry	Rubus (Rosaceae)	Root rot[b]		Phytophthora
Onion	Allium (Liliaceae s.l.)	Downy mildew		Peronospora
Millet	Pennisetum (Panicoideae)	Downy mildew	India	Sclerospora
Sorghum	Sorghum (Panicoideae)	Downy mildew	India, Africa	Sclerospora
Maize	Zea (Panicoideae)	Downy mildew		Sclerospora
Sugar cane	Saccharum (Panicoideae)	Unstable ratooning	Queensland	Pachymetra
Trout		Ulcerative dermal necrosis		Saprolegnia
European crayfish		Krebspest		Aphanomyces

[a]Tubiflorous dicotyledons.
[b]The most important disease of the crop.
[c]Disease economically limiting in some regions.

the Peronosporaceae, hosts are found in the highly evolved, sympetalous Asteridae and three other unrelated and less highly evolved groups, the Rosidae, the Caryophyllidae and part of the Dilleniidae. These superorders are noted for the production of secondary metabolites which are frequently either toxic to other organisms (saponins and alkaloids) or oily (essential oils and mustard oils). Food plants are commonly from these superorders because their secondary metabolites confer palatability. The Sclerosporales have a totally different host preference in the Poaceae (Panicoideae), possibly related to C_4 photosynthesis and sulphated flavonoid production.

Polyene antibiotics, effective on Mycota, are ineffective on the Peronosporomycetes: since these antibiotics are thought to function by acting on membrane-bound sterols, fundamental differences between the membrane-bound sterols of the two groups of fungi could be inferred.

General Morphology: Characters Used in Taxonomy

The Assimilative System

Thallus form in the Peronosporomycetes is diverse, ranging from a mycelium of hyphae (analogous to hyphae of the Mycota, with tip growth) to allantoid or ellipsoid (holocarpic) cells, or monocentric and eucarpic thalli having an assimilative system composed of branched rhizoids. Obligate parasites may be entirely confined within a single host protoplast (endobiotic), intracellular (some hyphae invading the protoplasts of a host thallus), or intercellular with specialized side branches (haustoria) that penetrate the cell walls, but not the protoplasts, of the host cells.

In mycelial forms, hyphae vary in diameter from 1.0–3.5 μm (Pythiogeton, Verrucalvus) up to 150 μm (after intussusception, in older hyphae of Achlya). Generalized intussusception of wall material occurs in the monocentric thalli of Rhipidiaceae. Septa are normally present to delimit reproductive structures and exclusion septa sometimes develop in old mycelia of Pythium; plugs of wall material may replace septa at reproductive junctions in some families. 'Cellulin' granules (granules containing chitin) are found only in the Leptomitales. Rhizoid development may be more frequent in the class than is generally accepted. After initial plasmodial development, a walled thallus, which continues expanding, soon becomes apparent in Olpidiopsis.

The protoplasm is coenocytic and in wide hyphae bidirectional cytoplasmic streaming along cytoplasmic strands is usually seen. During active vegetative growth the nuclear cycle is short (about 36–76 min in Saprolegniaceae and 75–155 min in Pythiaceae). Linear growth rates on agar are very variable.

Vegetative Ultrastructure

Mitochondria with tubular cristae are conspicuous. Dictyosomes are also well developed. The two most abundant remaining vesicular systems are those of the lipid vesicles and the dense body vesicles (DBVs). The latter possess an electron-opaque core or inclusion surrounded by a more electron-lucent zone, with or without myelin-like configurations; they may have a role in vegetative vacuolation, and may be essential for the mobilization of the vast reserves of the mostly phosphorylated β-1,3-glucans; they produce β-1,3-glucans for zoosporangial evacuation and probably ooplasm cleavage in oosporogenesis; they have been implicated in oogonial wall formation of some Saprolegniaceae, and they coalesce to form the ooplast in the mature oospores of all Peronosporomycetes. The importance of DBVs may be related to the phosphate/polyphosphate storage differences between Mycota and the Peronosporomycetes. The Peronosporomycetes also have a highly developed complex of extrusomes associated with encystment and germination.

Asexual Reproduction

Asexual reproduction shows considerable adaptive diversity. Normally, a sporangium is delimited and differentiated from the eucarpic vegetative system. The nuclei of the sporangial protoplast do not normally undergo further division in the sporangium (exceptions include *Phytophthora* and the Peronosporaceae). Cleavage of zoospore initials occurs either within the zoosporangium (intrasporangial zoosporogenesis) or after discharge of the sporangial protoplast (extrasporangial zoosporogenesis).

Discharge of intrasporangial zoospores is achieved by imbibition of water through the sporangial cell wall, due to secretion of osmotically active β-1,3-glucans from the zoospore initials and other residues. *Pythium* is characterized by the extrusion of uncleaved multinucleate protoplasm into a homohylic vesicle formed simultaneously with discharge. In *Phytophthora* (and *Pseudoperonospora* of the Peronosporaceae) cleavage takes place within a persistent zoosporangial plasma membrane (the plasma-membranic vesicle). In some Peronosporomycetes asexual propagation is by means of hyphal bodies.

Sporangial regeneration (or hyphal regrowth) may occur by internal renewal (through the sporangial septum), by a lateral branch (cymose renewal), by basipetal development, or by limited internal renewal so that the successive sporangial septa are formed at approximately the same point on the axis (percurrent development, *Albugo*). The conidio-sporangiophore may be swollen (*Basidiophora*, *Sclerospora*) with dichotomous (*Peronospora*), pseudodichotomous (*Sclerospora*) or more irregular branching (*Plasmopara*).

The Zoospore

The asymmetric shape of the principal-form zoospore, reniform with a ventral groove (**Fig. 2**), is due to the microtubular cytoskeletal array from the microtubule organizing centre at the flagellar bases (**Fig. 3**). Substantial differences in zoospore volume (**Table 2**) may influence zoospore shape and ultrastructural complexity.

Flagella of different lengths, whether of identical morphology or with non-heterokont differences in morphology, are termed 'anisokont'. The term 'heterokont' is restricted to the possession of two different kinds of flagellum: the straminipilous flagellum and a posteriorly directed, unornamented flagellum, with or without a fibrillar surface coat. The straminipilous flagellum pulls the zoospore through the water because its hydrodynamic thrust is reversed by the two rows of stiff, tubular tripartite (flagellar) hairs (TTHs) held in the plane of the quasi-sinusoidal beat. Each TTH has a tubular shaft, a solid, tapered point of attachment, and two distal, diverging hairs, one long and one short. Most TTHs have a shaft length of 1.0–2.0 μm irrespective of the length of the flagellum or the volume of the zoospore (see Fig. 2).

Sometimes there is a sequence of two or more zoosporic phases (polyplanetism). Dimorphism of the zoospore occurs in some taxa, with an initial zoospore with sub-apically inserted flagella (the auxiliary zoospore). The cysts formed from each kind of zoospore are also often morphologically distinct with different kinds of ornamentation: the principal-form zoospores of some species of *Saprolegnia* have distinctive split-ended hairs (boat-hook hairs) (**Fig. 4**). The shedding or retraction of flagella at encystment is also a taxonomic variable.

The flagellar base is composed of the kinetosome and its two groups of attached roots of microtubules. Independent variation in the ultrastructure of each root occurs between taxa, and sometimes within a species. The ultrastructure of the transitional zone between the kinetosome and the flagellar axoneme also differs within the class (**Fig. 5**).

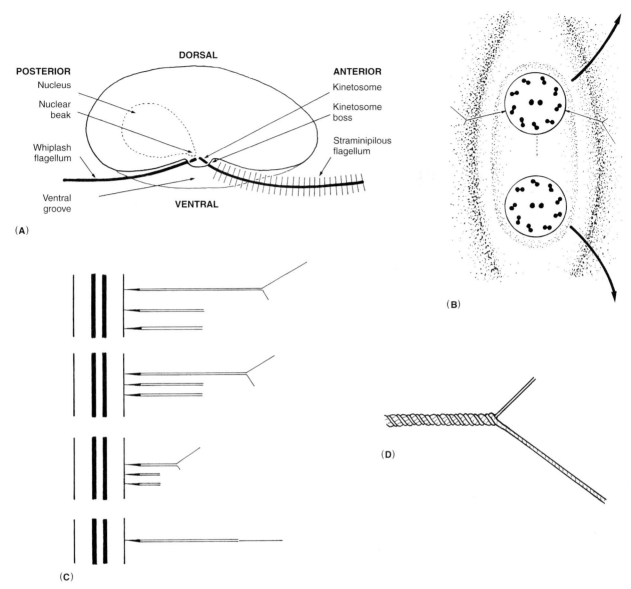

Figure 2 (**A**) Principal-form zoospore. (**B**) Arrangement of flagella in the ventral groove, direction of flagella indicated by large arrows, flagellar cross-sections with 9 + 2 axomes and tubular tripartite hair attachment (to upper, anterior flagellum) indicated. (**C**) Variation in lengths and densities of tubular tripartite hairs (straminipilous hairs). (**D**) Distal structure of a tubular tripartite hair.

Sexual Reproduction

Gametangia may be developed terminally, subterminally, in an intercalary position on main or branch hyphae, as terminal or lateral appendages to a non-mycelial thallus, or from the entire thallus. Gametangia are coenocytic meiogametangia, in which synchronous meioses occur. The numbers of nuclei entering the oogonial initials are greater than the numbers of mature uninucleate haploid female gametes, which in turn are more numerous than the numbers of zygotes in the oogonium. A comparable reduction in the number of nuclei occurs in the antheridium. In paired gametangia the meioses are also either simultaneous, or nearly so, between the two

protoplasts. After meiosis the contents of the receptor gametangium (oogonium) become separated as one or several uninucleate and initially unwalled gametes, but there is no differentiation of the male gametangial contents.

The evolution of the Peronosporomycetes is based on vegetative diploidy, and thus differs from that of other fungi. The occurrence of multiple synchronous meioses in a gametangium makes it possible for karyogamy to take place between two haploid nuclei from adjacent meioses in the same gametangium (automictic sexual reproduction). Sexual reproduction can thus occur without a separate male gametangium or antheridium. In the Myzocytiopsidales

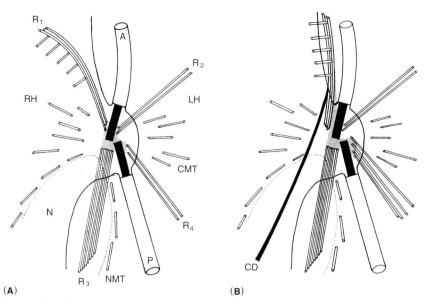

(A) **(B)**

Figure 3 Flagellar bases with skeletal roots of principal-form zoospores. (**A**) Saprolegniomycetidae. (**B**) Peronosporomycetidae. A, anterior flagellum; P, posterior flagellum; RH, right-hand side of zoospore; LH, left-hand side of zoospore; R_1, R_2, roots to the anterior flagellum; CD, R_1 cord; R_3, R_4, roots to the posterior flagellum; CMT, cytoplasmic microtubules; NMT, nuclear-associated microtubules; N, nucleus.

Table 2 Calculations of mean zoospore/zoospore cyst volume for a miscellaneous range of 20 taxa to give an indication of intergeneric variation, based on published cyst diameters – spheres, $\frac{4}{3}\Pi r^3$ – or mean zoospore dimensions assuming an ellipsoidal body – $\frac{4}{3}\Pi r_1 r_2^2$. Most members of the Peronosporomycetidae and the Saprolegniomycetidae have zoospore cyst diameters of 8–12 μm (268–905 μm^3); interspecific differences have taxonomic value; differences may reflect different life-history strategies.

Species	Estimated volume (μm^3)
Saprolegnia anisospora	4905 (large cysts)
	1200 (small cysts)
Pythiogeton autossytum	1767
Pythium anandrum	1288
Verrucalvus flavofaciens	755
Saprolegnia ferax	668
Phytophthora cinnamomi	606
Pythium aquatile	434
Salilagenidium callinectes	382
Lagenidium giganteum	350
Myzocytiopsis lenticularis	260
Lagena radicicola	256
Leptolegniella exoospora	243
Brevilegniella keratinophila	143
Aphanomyces amphigynus	113
Pythium angustatum	113
Halodaphnea parasitica	87
Gracea gracilis	33
Basidiophora entospora	29
Olpidiopsis brevispinosa	29
Olpidiopsis saprolegniae	14

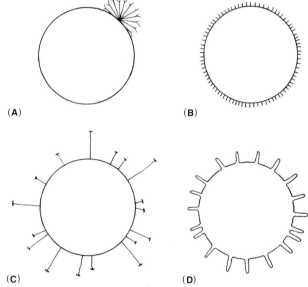

(A) **(B)**

(C) **(D)**

Fig. 4 Zoospore cyst ornamentations. (**A**) Auxiliary zoospore cyst with tuft of tubular tripartite hairs (*Saprolegnia*). (**B**) Principal-form zoospore cyst with short, simple hairs (*Phytophthora*). (**C**) Principal-form zoospore cyst with scattered 'boat-hook' hairs (*Saprolegnia*). (**D**) Principal-form zoospore cyst with papillae (*Leptomitus*).

and a few Pythiales the thallus becomes septate and adjacent segments assume the function of gametangia, whether automictic, or by homothallic pairing of equal-sized segments. In the Olpidiopsidales a contiguous but independent, smaller thallus (<10% of the receptor gametangium volume), the companion cell, functions as the heterothallic donor.

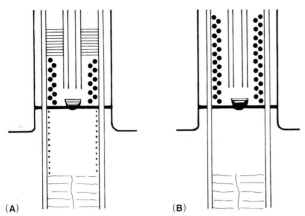

Figure 5 Transitional zones of (**A**) Saprolegniomycetidae and (**B**) Peronosporomycetidae.

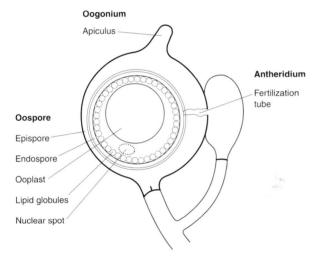

Figure 6 A typical oogonium and oospore.

Oogonium
Apiculus

Antheridium
Fertilization tube

Oospore
Epispore
Endospore
Ooplast
Lipid globules
Nuclear spot

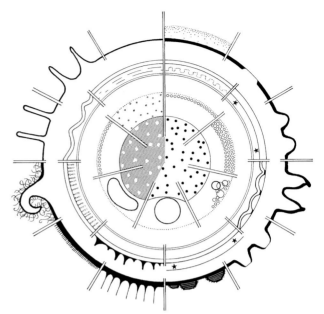

Figure 7 The biodiversity of oogonial and oosporic wall structure and oosporic cytoplasmic organization. The diagram is arranged as three partially exploded wheels of sectors: the outermost sector provides examples of oogonial wall structure; the middle sector displays the complex layers of the oospore wall (the usually thick endospore wall is resorbed on germination; the epispore wall is usually thin, but may become convoluted; the exospore wall (of periplasmic derivation) is blocked in); the innermost sector with the dotted line indicating the plasma membrane, indicates the cytoplasmic reorganization in the oospore (zygote). The central ooplast is diagrammed to indicate the phase reversal between a solid ooplast (hatched) with translucent zones and a fluid ooplast with granules in Brownian motion; the outer zone displays the differential coalescence of lipids. *indicates a fluid layer.

The majority of species of most of the genera are homothallic; heterothallism may be secondarily derived. The classic studies of heterothallism in *Achlya*, *Phytophthora* and *Pythium* have led to some understanding of the mating systems, morphogenesis of directional growth and penetration, and the identification of a C_{29} steroid sex hormone, antheridiol. Relative sexuality in the Peronosporales is thought to be a result of lethals. Estimates of chromosome numbers may be difficult to establish because of the possibilities of autopolyploidy, polysomy and chromosome polymorphy.

Karyogamy is presumed to occur in the oosphere after fertilization, or perhaps in the haploid coenocyte of the oogonium (see Fig. 1). The cytoplasmic reorganization of the oospore is indicative that functional or non-functional meioses precede oospore formation.

Many species have oogonia with a smooth, more or less spherical outline (**Fig. 6**), but many others are ornamented. Oogonial form depends on three criteria: initial expansion, secondary primordial initiation and wall deposition. The sequential or simultaneous expression of these criteria results in different kinds of oogonial ornamentation. No single group within the Peronosporomycetes displays all the morphological diversity, morphogenetic patterns and wall layering that can be found in the class (**Fig. 7**).

The morphology and morphogenesis of the antheridium are simpler than those of the oogonium. When the antheridium is of regular occurrence, the mode of application of the antheridium to the oogonium can be diagnostic. Antheridial origin can either be characteristic for a species, or highly variable within a species. Amphigynous antheridial development, in which the oogonial initial grows through the preformed antheridial initial, is characteristic of some species of *Phytophthora*. When an antheridium is present, and fertilization occurs, there is the injection of a small part of the antheridial protoplasm into the oogonium through a fertilization tube. Gametangial copulation (Myzocytiopsidales) occurs when the two gametic protoplasts condense to a common pore in the contiguous walls of the gametangia.

Table 3 Oospore dimensions for a miscellaneous range of 20 taxa to give an indication of intergeneric variation. Note that there is approximately a hundredfold difference in oospore volume from the top to the bottom of this list, with a wide spread for all orders and some genera. Overall diameters are used because most of the oospore wall thickness is due to endospore material which is resorbed on germination. Differences presumably reflect inoculum potential in relation to life-history strategies

Species	Oospore diameter (μm)	Oospore volume (μm)
Calyptralegnia achlyoides	51 (80)	69 456
Achlya megasperma	48	57 905
Aqualinderella fermentans	46	50 965
Pythium polymastum	44	44 602
Basidiophora entospora	42	38 792
Phytophthora megasperma	41	36 086
Verrucalvus flavofaciens	41	36 086
Peronospora media	38	28 730
Pachymetra chaunorhiza	34	20 579
Phytophthora infestans	30	14 137
Phytophthora citricola	22	5 575
Aphanomyces euteiches	22	5 575
Pythium acanthicum	21	4 849
Pythium ultimum var. ultimum	18	3 053
Myzocytiopsis lenticularis	18	3 053
Geolegnia inflata	14	1 436
Pythium parvum	13	1 150

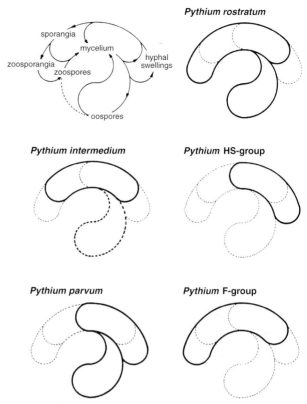

Figure 8 Congenitally foreshortened life-histories within the genus *Pythium*.

The morphogenesis of the oosphere is of major phylogenetic significance. The first distinction is between centripetal and centrifugal oosporogenesis; second is the extent of exclusion of part of the oogonioplasm for the oosphere(s) as periplasm. In the Peronosporales and Rhipidiales this periplasm may be substantial, persistent and sometimes initially nucleate. When the entire oogonial cavity is occupied by the oospore, the oospore is plerotic: very few taxa are truly plerotic. In *Pythium* the concept of an aplerotic index has been proposed to aid taxonomic assessment of species differences.

Within the oospore, there are two complementary processes of vesicle coalescence which proceed simultaneously. The DBVs gradually coalesce to form a large, single membrane-bound structure, the ooplast. At the same time there may be a variable degree of coalescence of lipid globules. The mature oospore of some species contains a single ooplast and a single lipid globule. Mean oospore size is very variable (**Table 3**).

The ooplast has different characteristics within the class. In most of the Saprolegniales the ooplast is fluid, with Brownian movement of granules, but in the Pythiales and Leptomitales the ooplast appears to be homogeneous.

The evolutionary strategies of straminipilous fungi and mycote fungi are similar in that they depend on large populations with generation times that are ephemeral in the context of the generation times of the macrobiota of the ecosystem (**Fig. 8**). Diploidy has resulted in the oospore population functioning in ecological or population genetics in a way analogous to that of heterokaryotic anamorph spore populations in other fungi; the oospore is not always a survival spore.

Characteristics of the Subclasses and the More Important Orders

Gross thallus structure has always been the prime criterion for separating the orders of the Peronosporomycetes (**Table 4**), but reliance on thallus morphology also resulted in many of the smaller and apparently simpler holocarpic species being unwarrantably grouped together.

Peronosporomycetidae

Mycelial fungi with centripetal oosporogenesis.

Peronosporales Oosporogenesis centripetal with persistent periplasm; conidio-sporangiophores well differentiated, persistent (and therefore well preserved in herbarium material). Mostly stem and leaf parasites of dicotyledons.

Table 4 Hierarchical classification of the class Peronosporomycetes

Kingdom **Straminipila** (subkingdom **Chromophyta**)
  Phylum **Heterokonta** (other phyla omitted from consideration)
    Class **Labyrinthista** (omitted from consideration)
      Order **Labyrinthulales**
      Order **Thraustochytriales**
    Subphylum **Peronosporomycotina**
      Class **Peronosporomycetes**
        Subclass **Peronosporomycetidae**
          Order **Peronosporales**
          Order **Pythiales**
            Genus *incertae sedis: Lagena*
        Subclass **Rhipidiomycetidae**
          Order **Rhipidiales**
        Subclass **Saprolegniomycetidae**
          Order **Saprolegniales**
          Order **Sclerosporales**
          Order **Leptomitales**
          Order **Salilagenidiales**
          Order **Olpidiopsidales**
            Genus *incertae sedis: Gracea*
          Order **Myzocytiopsidales**
            Genus *incertae sedis: Crypticola*
      Class **Hyphochytriomycetes** (omitted from consideration)
        Subclass **Hyphochytriomycetidae**
          Order **Hyphochytriales**
      Order *incertae sedis* in the subphylum
        Lagenismatales (omitted from consideration)

- Peronosporaceae: mycelium intercellular with large, lobate haustoria; conidio-sporangiophore frequently dichotomously branched, sporangia or conidio-sporangia borne on pedicels (downy mildews).
- Albuginaceae: mycelium intercellular with small spherical haustoria; conidio-sporangiophore percurrent with chains of zoosporangia (white blister rusts).

Pythiales Oosporogenesis centrifugal or centripetal with insignificant periplasm; conidio-sporangiophores rarely differentiated. Mostly root parasites of a wide range of vascular plants, or saprobes, rarely parasites of animals.

- Pythiaceae: hyphae of uniform diameter, usually 6–10 (15) μm in diameter; zoosporogenesis intrasporangial with a plasma-membranic vesicle or extrasporangial in a homohylic vesicle; antheridial gametogenesis following oogonial morphogenesis; oogonia modally uniovulate; oogonial wall usually thin; oospores never strictly plerotic; oospore with

a hyaline ooplast, lipid coalescence minimal; able to utilize SO_4^{2-}; variable ability to utilize different inorganic nitrogen sources; basal chromosome number $x = 5$.
- Pythiogetonaceae: hyphae very slender (usually < 5 μm in diameter), perhaps rhizoidal, with or without inflated sinuses, probably with anaerobic propensities; zoosporogenesis extrasporangial in a detachable homohylic vesicle; oogonial and antheridial morphogenesis nearly simultaneous; oogonium uniovulate; oospore plerotic, endospore wall very thick. Saprobic in submerged plant debris.

Saprolegniomycetidae

Mycelial or blastic fungi with centrifugal oosporogenesis.

Saprolegniales, Saprolegniaceae *sensu lato* Hyphae often stout with diameter increasing with age (up to 150 μm in diameter), most about 20 μm in diameter; zoosporogenesis intrasporangial; oosporogenesis centrifugal; antheridial gametogenesis following oogonial gametogenesis; oogonia often pluriovulate, oospores usually aplerotic; oospore with a fluid and granular ooplast, lipid coalescence variable (limited or to a single eccentric globule); unable to utilize SO_4^{2-} or NO_3^-; basal chromosome number $x = 3$. Saprobes, or parasites of animals, rarely phytoparasitic.

Sclerosporales Hyphae of extremely narrow diameter (< 5 μm); zoosporogenesis intrasporangial; oogonial wall thick, often verrucate; oospores plerotic or nearly so; oospore with a homogeneous ooplast, lipid coalescence limited. Parasites of Poaceae.

- Sclerosporaceae: haustoria present; zoosporangia or conidia borne on inflated sporangio-conidiophores; sporangio-conidiophores pseudo-dichotomously branched, evanescent, and therefore seldom well preserved in herbarium material. Graminicolous downy mildews.
- Verrucalvaceae: haustoria not known; zoosporangia without sporangiophores. Graminicolous root parasites.

Rhipidiomycetidae

Monocentric and polar fungi with centripetal oosporogenesis.

Rhipidiales, Rhipidiaceae Thallus with an inflated basal cell and rhizoids; thalloid segments separated by short, narrow, thick-walled isthmuses; zoosporogenesis intrasporangial; zoosporangia with a single exit tube and sometimes an evanescent protoplasma-membranic vesicle; zoospores of the principal form; gametangia differentiated into oogonia and antheridia; oosporogenesis periplasmic with nucleated periplasm; oogonia uniovulate; facultatively or obligately fermentative; requirement for organic nitrogen. Saprobic.

Other Orders

Leptomitales Thallus blastic, rarely allantoid or coralloid to pseudomycelial; zoosporogenesis usually intrasporangial; zoospores (or zoospore cysts) medium-large (> 175 μm³ volume) or large (> 300 μm³ volume); sexual reproduction homothallic, automictic or heterothallic; donor gametangia (when present) smaller than the oogonium; gametangia frequently not morphologically differentiated; oosporogenesis without nucleated periplasm. Freshwater or terrestrial families.

Salilagenidiales Thallus pseudomycelial, coralloid, allantoid or olpidioid, cytoplasm often with prominent granulation; zoosporogenesis intrasporangial or with a precipitative vesicle; zoospores (or zoospore cysts) medium-small (> 60 μm³ volume) or large (> 300 μm³ volume); gametangia frequently not morphologically differentiated; oosporogenesis without nucleated periplasm; oospore with a multilayered wall and granular ooplast, lipid droplets condensed to varying degrees, zoospores with lateral flagellar insertion at least in the second phase in polyplanetic species. Marine families.

Myzocytiopsidales (*incertae sedis*), **Myzocytiopsidaceae** Thallus initially tubular, sometimes branched, becoming septate and often disarticulating; sexual reproduction by copulation between contiguous segments.

Olpidiopsidales (*incertae sedis*), **Olpidiopsidaceae** Thallus more or less spherical, olpidioid, never septate; sexual reproduction by fusion of two disparately sized thalli.

Lagenaceae (*incertae sedis*) Thallus allantoid, nonpolar and without rhizoids; oosporogenesis with nucleated periplasm; presumed to be automictic.

Conclusion

A generic root for the class name, Peronosporomycetes, is preferred. The more familiar term 'Oomycetes', with a wider compass than the class Peronosporomycetes, is taxonomically imprecise, involving argument regarding the definitions of oogamy and the use of oogamy in a taxonomic context. The class almost certainly stems from an exclusively freshwater provenance. Geo-phytogeography indicates that the *Phytophthora*–Peronosporales evolutionary line must have been established long before the late Cretaceous. The evolution of the Peronosporomycetes from photosynthetic ancestors is unlikely because of the direct synthesis of lanosterol.

Members of the Peronosporomycetes are primarily distinguished from other fungi by their biflagellate zoospore with straminipilous ornamentation on the anteriorly directed flagellum. The Peronosporomycetes is one of three commonly accepted classes of straminipilous heterotrophs, two of which are fungal (the Labyrinthista is excluded). The separation of the small anamorphic class Hypochytriomycetes from the Peronosporomycetes may be less justified.

Within the Peronosporomycetes three subclasses have been recognized. The Peronosporomycetidae is species-rich (60% of the total number of species in the class) but restricted to four groups of phytopathogens and a few other parasites and saprobes. The Peronosporomycetidae is derived, with biochemical diversity largely confined to recognition systems. The Saprolegniomycetidae shows the greatest diversity in thallus morphology, physiology, cytology, ecology and parasitism. The Rhipidiomycetidae perhaps represents an early, monocentric, offshoot from the Peronosporomycetidae–Peronosporales line.

The development of antifungal agents effective on these fungi should lead to fundamental investigations of sterol metabolism and host/parasite recognition systems.

See also: **Fish**: Spoilage of Fish. **Fungi**: Overview of Classification of the Fungi. **Spoilage Problems**: Problems Caused by Fungi.

Further Reading

Alderman DJ (1986) Fungi as pathogens of non insect invertebrates. In: Samson RA, Vlak JM and Peters D (eds) *Fundamental and Applied Aspects of Invertebrate Pathology*. P. 354. Wageningen: Foundation of the Fourth International Colloquim of Invertebrate Pathology.

Alexopoulos CJ, Mims CW and Blackwell M (1996) *Introductory Mycology*, 4th edn. New York: Wiley.

Dick MW (1990) Phylum Oomycota. In: Margulis L, Corliss JO, Melkonian M and Chapman D (eds) *Handbook of Protoctista*. P. 661. Boston: Jones & Bartlett.

Dick MW (1992) Patterns of phenology in populations of zoosporic fungi. In: Carroll G and Wicklow D (eds) *The Fungal Community, Its Organization and Role in the Ecosystem*, 2nd edn. P. 355. New York: Marcel Dekker.

Dick MW (1995) Sexual reproduction in the Peronosporomycetes. *Canadian Journal of Botany* (Supplement 1, Sections E–H) 73: S712–S724.

Dick MW (1998) Peronosporomycetes. In: McLaughlin D and McLaughlin E (eds) *The Mycota.* Vol. 7 (in press). Berlin: Springer.

Erwin DC and Ribeiro OK (1996) Phytophthora *Diseases Worldwide.* St Paul: APS Press.

Fuller MS and Jaworski A (eds) (1987) *Zoosporic Fungi in Teaching and Research.* Athens, Ga: Southeastern Publishing.

Griffith JM, Davis AJ and Grant BR (1992) Target sites of fungicides to control Oomycetes. In: Köllered W (ed.) *Target Sites of Fungicide Action.* P. 69. Boca Raton: Chemical Rubber Co.

Hawksworth DL, Kirk PM, Sutton BC and Pegler DN (1995) *Ainsworth & Bisby's Dictionary of the Fungi,* 8th edn. Wallingford: CAB International.

Lucas JA, Shattock RC, Shaw DS and Cooke LR (eds) (1991) *Phytophthora.* Cambridge: Cambridge University Press.

Nes WD (1987) Biosynthesis and requirement for sterols in the growth and reproduction of Oomycetes. In: Fuller G and Nes WD (eds) *Ecology and Metabolism of Plant Lipids.* P. 304. Washington: American Chemical Society.

Sparrow FK (1960) *Aquatic Phycomycetes,* 2nd edn. Ann Arbor: University of Michigan Press.

Spencer DM (ed.) (1981) *The Downy Mildews.* London: Academic Press.

Classification of Zygomycetes

P M Kirk, CABI Bioscience UK Centre (Egham), Surrey

The Zygomycetes is the larger of the two classes of the phylum Zygomycota. The second class, the Trichomycetes, contains a group of organisms which are united mainly because of the ecological niche they inhabit; they are typically endocommensals, particularly being found in the digestive tract of the aquatic larvae of a number of insect groups. Others are found in other arthropod groups, including crustaceans and diplopods. One of the four orders shows some relationship with the Kickxellales (see below); the evidence comes from both ultrastructural and serological data. The other orders show increasing distance from the rest of the Zygomycota: one order has already been recognized as likely to prove to be totally unrelated. Lack of additional evidence, particularly from molecular data, prevents a reappraisal of their systematic position.

The fungi classified in the Zygomycetes are mainly saprobes, chiefly found in the soil, or parasites, especially of arthropods, and many are amongst the most widely distributed of the fungi, occurring in both temperate and tropical areas and both polar regions. The class is a relatively small one, with some 125 genera and 870 species compared to, for example, the Basidiomycetes which has nearly 14 000 species and

the Lecanorales (and order of Ascomycota) with 7000 species.

Members of the Zygomycetes impact variously on human society. For example, species of *Mucor* and *Rhizopus* cause decay of foodstuffs; these and the thermophilic species of *Absidia* and *Rhizomucor* are involved in storage problems usually associated with grain or other cereals. These thermophilic species and some species of *Cunninghamella, Mortierella* and *Saksenaea* are often found associated with medical and veterinary problems. A few species, *Choanephora cucurbitarum* and *Gilbertella persicaria*, are facultative pathogens of plants, but these have become less important as a result of better harvesting techniques.

The class is characterized by an anamorph producing aplanate mitospores, sporangiospores, of endogenous origin. The sporangiospores are formed within sporangia borne by branched or unbranched aerial sporangiophores. The columella is a typically globose or pyriform structure which projects into the sporangium, delimiting the area containing the sporangiospores from the rest of the sporangiophore. The anamorph is very variable in form, ranging from large multispored (> 1000 spores) sporangia borne on tall (> 5 cm) sporangiophores to single-spored sporangia (often termed sporangiola), which resemble conidia and are often referred to by this term, borne on simple, unspecialized sporangiophores. The single-spored sporangiolum may, in addition, be found associated with rather complex sporangiophores, the structure of which appears to have been mainly driven by evolution of mechanisms for spore liberation. The sporangiospore is presumed to be a spore of dispersal and colonization of suitable substrata. The multispored condition has been presumed to be the most primitive and the single-spored condition the more advanced, but there is no clear evidence for this.

The sexual stage, the teleomorph, is the zygospore, from which the names of the phylum and class are derived. The zygospore is formed as a result of hyphal conjugation. The modification of hyphae to form zygophores, in the case of heterothallic species, has been shown to be in response to chemical stimulants, which have been referred to as sex hormones and are based on β-carotene. The zygophores, often also termed progametangia, grow towards one another and as they make contact gametangia are delimited terminally by the production of a septum. The wall at the point of contact dissolves and plasmogamy (mixing of protoplasm) occurs followed by karyogamy (fusion of nuclei). In what have been referred to as the primitive representatives of the class, the zygospore is typically thick-walled and the wall develops brown pigmentation (melanin-like) and ornamentation in the form of warts or spines. In the presumed advanced forms the zygospore is typically thin-walled and the wall is lightly pigmented or not pigmented and smooth or only slightly ornamented.

Intermediate forms have varying degrees of wall thickening, pigmentation and ornamentation. The supporting hyphae (the remains of the zygophores after delimitation of the gametangia) are usually referred to as suspensors. These may be either opposed (on opposite sides of the zygospore) or apposed (lying almost parallel) or, rarely, tongs-like. The suspensors are sometimes ornamented with hyphal-like outgrowth. Zygospores are unknown in many species of Zygomycetes, which are otherwise included in the class because of the characters of their anamorph. The heterothallic condition is predominant in species where zygospores are known; the homothallic condition is somewhat restricted in occurrence. The zygospore has evolved as a spore of survival.

The Zygomycetes is presently classified into seven orders: Dimargaritales, Endogonales, Entomophthorales, Glomales, Kickxellales, Mucorales, and Zoopagales.

Dimargaritales

The Dimargaritales are a small (15 species) order of obligate mycoparasites of Mucorales (rarely *Chaetomium* spp.; Ascomycota: Sordariales) which are coprophilous in habitat. The single family, Dimargaritaceae, was formerly classified in the Mucorales along with Kickxellaceae and Piptocephalidaceae in a series termed the merosporangiferous mucorales, with Syncephalastraceae as the basal family of a presumed evolutionary series. Later evidence, including ultrastructural evidence, suggested that these families were only distantly related within the Zygomycetes and were better classified as separate orders; similarities were presumed to be due to convergent evolution, amongst other factors.

The Dimargaritales, with the type genus *Dimargaris* (etymology: two pearls, after the form of the septal plugs, particularly in the sporophores) and two other genera (*Dispira* and *Tieghemiomyces*), form asexual spores in pairs in a more or less cylindrical sporangium. The spore mass, borne on an often complex sporophore, may remain dry (dry-spored) or become enveloped in a liquid droplet (wet-spored); the latter is the more common. The germinating spores, along with many other mycoparasitic species in the Zygomycetes, are chemotropic and grow towards the hyphae of a suitable host where they form an appresorium and haustoria. Zygospores are thick-walled, faintly pigmented, smooth or with simple ornamentation and are formed by relatively unspecialized hyphae (zygophores).

Endogonales

The Endogonales, with the single family Endogonaceae of 21 species, are an order of mainly mycorrhizal fungi, in addition to some saprobes. They form ectomycorrhiza, the endomycorrhizal species having recently been transferred to the Glomales. They form mainly hypogeous sporocarp, containing only zygospores or chlamydospores.

Entomophthorales

The Entomophthorales are an order of mainly entomogenous fungi. They are usually associated with Diptera, although some are pathogens of other arthropods, nematodes and tardigrades, amongst others.

The typical life cycle of the entomopathogenic species of Entomophthorales involves the invasion of the host by germ hyphae produced by adhesive spores (termed conidia) which are airborne. The fungus invades the abdomen of the host and, following its death, sporophores are produced, typically between the individual segments of the abdomen, where a new crop of forcibly discharged spores are produced. Resting bodies are often produced within the host and the primary conidia also have the ability to produce typically smaller secondary conidia, usually of the same form or as morphologically distinct microconidia.

The Entomophthorales are divided into six families: Ancylistaceae, Basidiobolaceae, Completoriaceae, Entomophthoraceae, Neozygitaceae and Meristacraceae. However, recent molecular evidence has suggested that the Basidiobolaceae, with a single genus containing four species, is more closely related to the Chytridiales. The most common and widespread species, *Basidiobolus ranarum* (commonly found on frog dung), appears to be a saprophyte but one of the other species is known to cause a subcutaneous phycomycosis in humans. Recent reports in the press of a presumed member of the Chytridiales causing a fatal disease in tree frogs may provide further evidence for this relationship. In fact, a second genus of Basidiobolaceae is known but apparently is still undescribed; it was referred to as causing fatal mycoses in snakes. It is suggested that species of *Basidiobolus* are chytrids which have lost their flagella: there is additional ultrastructural evidence to support this in the form of what have been interpreted as structures in the spores which resemble those associated with flagella in true chytrids.

The remaining five families are quite clearly related

Table 1 Families in the order Entomophthorales

Family	Number of species
Ancylistaceae	32
Basidiobolaceae	4
Completoriaceae	1
Entomophthoraceae	131
Meristacraceae	6
Neozygitaceae	10

Table 2 Genera in the family Entomophthoracae

Genus	Number of species
Batkoa	4
Entomophaga	9
Entomophthora	11
Erynia	12
Eryniopsis	5
Furia	12
Massospora	11
Pandora	16
Strongwellsea	2
Tarichium	29
Zoophthora	20

to the Mucorales, considered the most primitive order of the Zygomycetes. Most fungi classified in these five families are obligate parasites and, therefore, it is only possible to grow them in pure culture with complex media often containing natural products. Even under these conditions it is unlikely that growth will be typical and certainly sporulation will rarely be present. The exceptions to this general character are species of *Conidiobolus* (Ancylistaceae) which are saprophytes from the soil and occur widely. They are frequently isolated from soil and are easy to grow in culture.

The classification of the Entomophthorales into families is based on the following characters: several aspects of nuclear cytology, mode of formation and germination of resting spores, nature of vegetative growth and development. The primary characters on which genera are based involve aspects of the primary conidia and are as follows: overall conidial and papillar morphology, nuclear numbers, unitunicate or bitunicate state of wall, mode of discharge, morphology of conidiogenous cells and conidiophores. Secondary characters include presence or absence and form of rhizoids and/or cystidia, the types of secondary conidia formed and characters of the resting spores, vegetative cells and general pathobiology. Species are separated on the basis of differences in shape and size of the characters used to circumscribe genera.

Within the Entomophthoraceae, by far the largest family in terms of number of species recognized, are found what could be considered to be the typical entomophthoralean species (**Table 1**). Here are found the species of the previously broadly circumscribed and heterogeneous genera *Entomophthora* (previously known more widely as *Empusa*). However, *Entomophthora* in a modern restricted sense is not now the largest genus of Entomophthoraceae (**Table 2**).

Order Kickxellales

The Kickxellales are a comparatively small order of rarely encountered fungi. They are mainly saprobes from soil or coprophilous, rarely as mycoparasites.

The single family, Kickxellaceae, contains eight genera and about 21 species.

Asexual reproduction is by unispored sporangiola borne on special fertile branches on often complex sporangiophores. At maturity the spores are either dry or, more frequently, are borne in a liquid droplet. Sexual reproduction is by relatively unspecialized non-pigmented, non-ornamented, often thin-walled zygospores.

The Kickxellales are of widespread occurrence but are relatively under-recorded and so their true distribution, like that for many of the fungi, is unclear. Some of the structurally more complex species appear to have a tropical or subtropical distribution, apparently favouring somewhat dry climates rather than the wet tropics, whilst others have a worldwide distribution.

Order Mucorales

The Mucorales is by far the largest order of Zygomycetes, with about 300 species, and contains some of the best-known representatives of the class. In addition, the genus *Mucor*, on which the order is based, is one amongst the small number of generic names which were the first to be applied to fungi; the generic name *Mucor* dates from P. Micheli (1729). The pin-moulds that in earlier times were often seen associated with mouldy bread are members of this order.

Asexual reproduction is by multispored or few-spored or one-spored sporangia (sporangiola). At maturity the sporangia are usually diffluent, forming liquid droplets, in which the sporangiospores are held. Some species are dry-spored at maturity and in one genus, *Pilobolus*, the entire sporangium is forcibly discharged. Sexual reproduction is by thick-walled, pigmented and typically ornamented zygospores. The Mucorales are cosmopolitan, having a worldwide distribution; they are amongst the most widespread of fungi, occurring in all land masses. Most are saprobes, especially in the soil, where they quickly colonize

Table 3 Families in the order Mucorales

Family	Number of species
Chaetocladiaceae	7
Choanephoraceae	5
Cunninghamellaceae	7
Gilbertellaceae	1
Mortierellaceae	106
Mucoraceae	121
Mycotyphaceae	6
Phycomycetaceae	3
Pilobolaceae	13
Radiomycetaceae	4
Saksenaeaceae	1
Syncephalastraceae	2
Thamnidiaceae	22

suitable substrata and rapidly break down simple carbohydrates. A few, species of *Chaetocladium* and *Parasitella*, are mycoparasites of other members of the Mucorales. Others, species of Choanephoraceae and *Gilbertella*, are facultative parasites of plants, often causing rot in flowers or fruit; a few (e.g. *Absidia corymbifera*, *Rhizomucor pusillus* and *Rhizopus microsporus*) are facultative parasites of animals (including humans).

The order is divided into 13 families (**Table 3**) and, although this classification is practical for identification purposes, it is not particularly natural: a number of the families differ from one another by few characters. Some families, for example the Choanephoraceae and Pilobolaceae, are what appear to be natural groupings as they not only have distinct anamorph characters but their zygospores are also diagnostic.

Recent evidence suggests that the Mortierellaceae is only distantly related to the other families and it is likely to be removed and placed in a new order.

Chaetocladiaceae

This is a small family of widespread distribution, particularly in the northern hemisphere. Sporangiophores terminate in sterile spines and bear unispored, pedicellate sporangiola on vesicular swellings (multispored sporangia are absent). The zygospores are rough-walled and the suspensors are opposed.

Choanephoraceae

This is a small, apparently natural family with distribution restricted to the tropics or subtropics. Sporangia and sporangiola are borne on separate and distinct sporangiophores. The sporangiospores are unique within the Mucorales as they possess hair-like appendages, in combination with a pigmented and differentially thickened (appearing striate under the microscope) wall. The zygospores are smooth but have an internal striate appearance and are borne on tongs-like suspensors.

Cunninghamellaceae

This family has a wide distribution; it is commoner in warm climates. Sporangiola are uni-spored (sporangia are absent) and the zygospores are warty and borne on opposed suspensors.

Gilbertellaceae

This monotypic family is only found in the tropics and subtropics; the single species has been associated with a dry rot in peaches. The multispored sporangia form sporangiospores with hair-like appendages. The zygospore wall is roughened and the suspensors are opposed.

Mortierellaceae

This large family has a widespread distribution, occurring mainly as saprobes in the soil. Sporangia and sporangiola typically have only a rudimentary columella or the columella is absent. The zygospores are smooth or angular and are borne on opposed suspensors. They are unknown in most species but where they have been observed they are either surrounded by a dense mat of hyphae or are naked – the norm for the Mucorales.

Mucoraceae

This is the type family of the genus and considered the most primitive. The sporangia are columellate and specialized sporangiola are absent. Zygospores are smooth to warty and are borne on opposed and tongs-like, or apposed, naked or appendaged suspensors. The family is clearly polyphyletic.

Mycotyphaceae

This small family has only recently been established, based on the unique mode of dehiscence of the sporangiola (sporangia are absent) and an unusual ability to form a yeast-like phase either in response to anaerobic conditions or due to high sugar concentrations. The six species are presently only known from a few localities but show a preference for warm, relatively dry climates.

Phycomycetaceae

This family contains a single genus, *Phycomyces*, characterized by tall, unbranched sporangiophores bearing large sporangia and by zygospores with coiled, tongs-like suspensors bearing branched appendages.

Pilobolaceae

This family, of mainly coprophilous species, contains three genera which can easily be arranged in a tentative evolutionary series, from a presumed primitive form through an intermediate form to an advanced form, exhibiting adaptations for spore liberation. In the advanced state spore liberation is by forcible discharge of the entire sporangium. There are distinct patterns in distribution; the primitive genus is restricted to temperate areas, the genus showing intermediate characters is tropical, whilst the advanced genus is worldwide. The multispored sporangia are columellate (the sporangium wall is black and persistent) and the zygospores are smooth and borne on tongs-like or apposed suspensors. This combination of characters leads to the speculation that this family is one of the more natural within the Mucorales.

Radiomycetaceae

This family has two genera, each with two species, characterized by sporangiola borne on complex ampullae, simple or branched, often stoloniferous sporangiophores and an absence of sporangia. Zygospores are smooth and are borne on apposed, appendaged suspensors.

Saksenaeaceae

The Saksenaeaceae is a monotypic family of unknown affinities within the Mucorales. The single species forms lageniform, columellate sporangia on unbranched sporangiophores arising from rhizoids. The zygospores are unknown.

Syncephalastraceae

This small family, containing a single genus, is defined by its unique (for the Mucorales) linear sporangia, termed merosporangia. These are superficially similar to the merosporangia found in the Piptocephalidaceae (Zoopagales) but are distinct when ultrastructural evidence is examined. The zygospores are warty and are borne on opposed suspensors; they are remarkably *Mucor*-like.

Thamnidiaceae

The Thamnidiaceae share many characteristics with the Mucoraceae and both may be seen as a single family containing a single series of genera, where there are no distinct discontinuities. The series ranges from large multispored sporangia at one end to single-spored sporangiola at the other. That the two families continue to be recognized is for purely practical reasons; the distinguishing character is the type of sporangia and/or sporangiola. The Thamnidiaceae

possess sporangiola (sometimes acolumellate) with persistent walls and, in some genera, columellate sporangia with a diffluent wall borne on the same or separate but morphologically identical sporangiophores. The Mucoraceae, on the other hand, do not form sporangiola, even though in some genera small sporangia, usually with persistent walls, are formed in ageing colonies. Zygospores are warty and are borne on opposed suspensors; they are largely identical to the zygospores found in the Mucoraceae. The family is considered polyphyletic.

Order Zoopagales

This order of five families and some 160 species is relatively unknown in terms of its frequency and distribution. Asexual reproduction is by spores, which have been termed conidia or merosporangia, and sexual reproduction is by zygospores. They appear to be cosmopolitan as haustorial parasites of fungi (mycoparasites), nematodes, amoebae and other small terrestrial animals.

Cochlonemataceae

This family of 29 species of ectoparasites or endoparasites is characterized by fertile hyphae only appearing outside the host. The zygospores are warty and are borne on spirally twisted suspensors.

Helicocephalidaceae

The sporophores in this small family of only eight species are unbranched and arise from rhizoids; the sporangiospores are large and pigmented. Zygospores are unknown. These fungi are obligate parasites of nematodes, rotifers and their eggs.

Piptocephalidaceae

The Piptocephalidaceae, with three genera containing 51 species, were originally classified in the Mucorales, along with members of the Dimargaritales and Kickxellales, in a presumed evolutionary series termed the merosporangiferous mucorales. The sporangiospores are borne in rod-like merosporangia on branched (*Kukuhaea* and *Piptocephalis*) or unbranched (*Syncephalis*) aerial sporangiophores. The zygospores are warty and are borne on tongs-like suspensors, which are often spirally wound, and with globose outgrowths (especially in *Syncephalis*). The species in this family are obligate mycoparasites of Mucorales (rarely, *Penicillium* and *Wynnea*).

Sigmoideomycetaceae

The dichotomously branched fertile hyphae of this small family of seven species are septate and the small, more or less hyaline spores are borne on fertile vesicles. Zygospores are unknown. They are parasitic on other fungi.

Zoopagaceae

This is the largest family of Zoopagales with some 63 species disposed in six genera. They are of cosmopolitan distribution and are predacious parasites of nematodes, amoebae and other small terrestrial animals. The spores are typically in chains borne on simple sporophores. Zygospores are warty and are borne on spirally twisted suspensors.

Order Glomales

The Glomales was recently segregated from the Endogonales for mycorrhizal species which form endomycorrhizal associations with plants. Two suborders are recognized: Glomineae (with two families: Acaulosporaceae, 37 species, and Glomaceae, 86 species) for the vesicular-arbuscular mycorrhiza (VAM)-forming species and Gigasporineae (single family of 37 species, Gigasporaceae) for the arbuscular mycorrhiza-forming species.

See also: **Fungi**: Overview of the Classification of the Fungi. *Mucor*.

Further Reading

Baxazy S (1993) *Flora of Poland* 24. *Entomophthorales*. Polish Academy of Sciences: Krakow.

Benjamin RK (1959) The merosporangiferous *Mucorales*. *Aliso* 4: 321–433.

Benjamin RK (1979) Zygomycetes and their Spores. In: Kendrick WB (ed.) *The Whole Fungus*. Vol. 2, p. 573. National Museums of Canada, Ottawa.

Duddington CL (1973) Zoopagales. In: Ainsworth GC, Sparrow FK and Sussman H (eds) *The Fungi*. Vol. 4B, p. 231. Academic Press: New York and London.

Halliday T (1998) A declining Amphibian conundrum. *Nature* 394: 418–419.

Hawksworth DL, Kirk PM, Sutton BC and Pegler DN (1995) *Ainsworth & Bisby's Dictionary of the Fungi*. CAB International, Wallingford.

Hesseltine CW (1965) A millennium of fungi, food, and fermentation. *Mycologia* 57: 149–197.

Hesseltine CW (1991) Zygomycetes in food fermentations. *Mycologist* 5: 162–169.

Humber RA (1989) Synopsis of a revised classification for the Entomophthorales (Zygomycotina). *Mycotaxon* 34: 441–460.

Morton JB and Benny GL (1990) Revised classification of arbuscular mycorrhizal fungi (Zygomycetes): a new order, Glomales, two new suborders, Glomineae and Gigasporineae, and two new families, Acaulosporaceae

and Gigasporaceae, with an emendation of Glomaceae. *Mycotaxon* 37: 471–491.

Zycha H, Siepmann P and Linemann G (1969) *Mucorales*. Cramer, Lehre.

Classification of the Eukaryotic Ascomycetes

M A Cousin, Department of Food Science, Purdue University, West Lafayette, USA

Introduction

The Ascomycotina or ascomycetes as they are commonly called, produce asci containing ascospores or sexual spores and many also produce asexual spores, like conidia; therefore, the terms 'teleomorph', 'anamorph', and 'holomorph' refer to the different reproductive states of the fungi. Most ascomycetous moulds are in the class Plectomycetes and order Eurotiales, whereas most ascomycetous yeasts are in the class Hemiascomycetes and the order Endomycetales. The most important aspects of ascomycetous moulds are the production of heat-resistant ascospores and their growth in food with low water activities of 0.61–0.80. Ascomycetous yeasts have some species that grow at similar low water activities in high-sugar foods, are resistant to chemical preservatives, and are used in producing fermented foods, especially from cereal grains. Some ascomycetous moulds and yeasts are involved in spoilage of many different types of foods.

Defining Features of the Ascomycetes

The Ascomycotina are characterized by the production of an ascus, which is a thin-walled structure containing usually 8, but sometimes as few as 1–4 or as many as 12–70, ascospores or sexual spores. The ascus is inside a fruiting body called an ascoma or ascocarp. At maturity, the ascus bursts, releasing the ascospores into the environment. Mycelia are septate and branched with one or more nuclei per cell.

Many fungi have the genetic ability to produce both sexual spores (ascospores) and asexual spores (conidia). When a fungus produces ascospores, it is termed a teleomorph and is named as a genus of the Ascomycotina. When the fungus produces conidia, it is termed an anamorph and is named as a genus of the Deuteromycotina or Fungi Imperfecti (**Fig. 1**). The term 'holomorph' refers to the fungus in all its different states, and the earliest name referring to the teleomorphic state should be used. There is still much confusion about the names of teleomorphs,

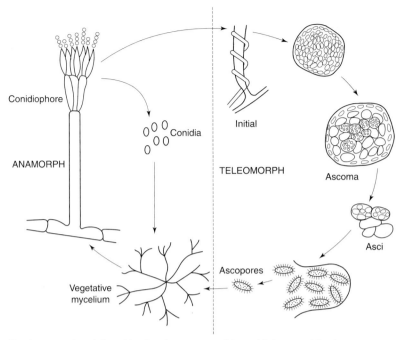

Figure 1   Ascomycete–Deuteromycete relationship showing anamorphic and teleomorphic stages.

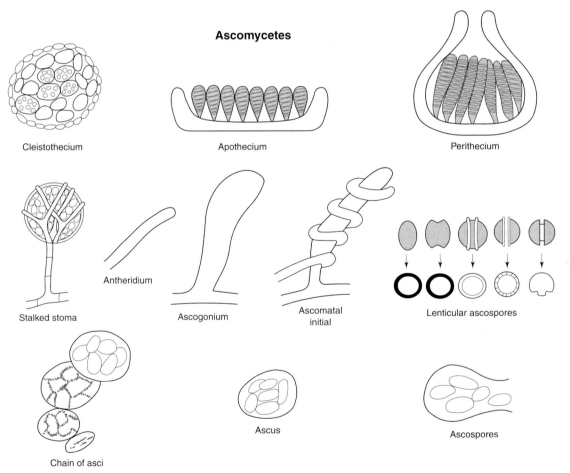

Figure 2   Structures for sexual propagation in Ascomycetes.

Table 1 Characteristics of some food-borne mould genera that produce ascospores

Yeast	Anamorph	Asci	Ascoma	Ascospores
Byssochlamys	Paecilomyces	Globose to subglobose	No fruiting bodies, white hyphae	Eight, ellipsoidal, thick and smooth-walled, pale yellow
Chaetomium	Botryotrichum	Cylindrical or clavate	Perithecia, brown with hairs, rhizoids	Subspherical to ellipsoidal
Emericella	Aspergillus	Globose to subglobose	Red cleistothecia with Hülle cells	Eight, ellipsoidal, lens-shaped with two longitudinal ridges, red-purple
Eupenicillium	Penicillium	Globose to ellipsoidal	Red, yellow, brown cleistothecia	Eight, ellipsoidal with spiny walls and two longitudinal ridges or furrows, pale yellow
Eurotium	Aspergillus	Globose to subglobose	Bright yellow cleistothecia	Ellipsoidal, lens-shaped with rough or smooth walls, with/without ridges or furrows, yellow
Monascus	Basipetospora	Globose to subglobose	Brown cleistothecia on stalk	Oval to ellipsoidal with smooth walls, yellow to colourless
Neosartorya	Aspergillus	Globose to flat	White cleistothecia and wall of flat cells attached to hyphae	Eight ascospores, ellipsoidal with two longitudinal ridges, colourless
Talaromyces	Penicillium, Geosmithia	Ellipsoidal to globose	Yellow or white gymnothecia	Eight ascospores, ellipsoidal with thick walls and spiny appearance, yellow
Xeromyces[a]		Thin-walled evanescent	White, brown cleistothecia on stalk	Two ascospores, D-shaped, smooth-walled

[a]No anamorph identified.

anamorphs and holomorphs. Earlier names given to anamorphs that are really holomorphs may still be in common usage. This can create confusion about the sexual nature of some fungi; therefore, some literature citations, especially older ones, can present misinformation about the fungi that produce ascospores. Also, since many of the ascomycetes produce conidia and other asexual spores similar to those produced by the deuteromycetes, the anamorphs of species that belong to *Aspergillus* and *Penicillium* have been mistakenly identified as producing ascospores.

General Features of Ascomycetes

The genera and species of ascomycetes are usually identified microscopically by the shape and size of the ascus and ascospores (**Fig. 2**). Generally, the ascospores have thick walls, are refractive and have different degrees of ornamentation, such as ridges, furrows, and rough to spiny walls (**Tables 1 and 2**). The ascoma can occur by itself or be in or on a vegetative hyphal mass. The ascoma can be flask-shaped (perithecium), cup-shaped (apothecium) or spherical (cleistothecium). Some asci are produced in cushion-like structures termed 'pseudothecia' or arise on masses of vegetative hyphae (stromata). The ascomycetous yeasts are in the order Endomycetales where the asci are formed from zygotes without the fruiting bodies. The shape of asci is used to place fungi into

systematic divisions. In the ascus, two haploid nuclei generally fuse to make a diploid which undergoes meiosis to form four haploid nuclei; these then divide by mitosis resulting in eight nuclei, which form the basis for most ascospore development. The number of ascospores is characteristic for both the genus and species. Ascospores are liberated from the ascus in the moulds; however, many yeasts do not liberate them but instead the ascospores germinate within the ascus and then penetrate the ascus wall (Table 2).

It generally takes asci, which are inside macroscopic structures such as cleistothecia or gymnothecia, over 10 days to mature at 25°C in ascomycetous moulds. Hence, these moulds grow more slowly than other moulds in foods and some may not produce ascospores for a long time, especially species of *Eupenicillium*. Hence the detection and identification of ascomycetous moulds will take more time than for other common moulds found in foods.

Basis for Division of Ascomycetes

Fungal nomenclature follows the rules and recommendations of the International Code of Botanical Nomenclature. The latest adoption is the Tokyo Code from the Fifteenth International Botanical Congress held in Yokohama, Japan in 1993. A *taxon* consists of the primary ranks of domain, kingdom, division or phylum, class, order, family, genus and species (in

Table 2 Characteristics of some yeast genera that produce ascospores

Yeast	Anamorph	Asci	Ascospores
Debaryomyces[a]	Candida	Conjugated	One to four, spherical to oval, warty or ridged walls, not liberated
Hanseniaspora[b]	Kloeckera	Unconjugated	One to four, hat-shaped or spherical with or without ledge and smooth or warty, not liberated
Kluyveromyces[a]	Candida	Conjugated	One to four (up to 70) ellipsoidal, reniform or crescentiform, liberated
Pichia[a]	Candida	Unconjugated	One to four (up to eight) spherical to oval, hat- or saturn-shaped, smooth, usually liberated
Saccharomyces[a]	Candida	Unconjugated	One to four (up to 12) spherical or oval, smooth or warty, not liberated
Saccharomycodes[b,c]		Unconjugated	One to four, spherical or oval, smooth, not liberated
Schizosaccharomyces[c,d]		Conjugated	Two to eight, spherical, ellipsoidal, reniform, smooth or warty, liberated
Torulaspora[a]	Candida	Conjugated	One to four, spherical, ellipsoidal, smooth or warty, not liberated
Yarrowia[a]	Candida	Unconjugated	One to four, spherical, oval, hat-, saturn- or walnut-shaped
Zygosaccharomyces[a]	Candida	Conjugated	One to four, spherical, ellipsoidal, not liberated

[a] Multilateral budding.
[b] Bipolar budding.
[c] No anamorph identified.
[d] Fission.

descending order). In the domain Eukaryota, the kingdom Fungi, the division Ascomycota and subdivision Ascomycotina, there were six classes (no longer recognized): Discomycetes (asci are in an open cup-shaped structure called an apothecium), Hemiascomycetes (yeasts in the family Saccharomycetaceae have asci that are not in ascocarps), Laboulbeniomycetes (ectoparasites of insects that can only be seen with a lens), Loculoascomycetes (double-walled asci are in a pseudothecium that is similar to a perithecium), Plectomycetes (asci are in an enclosed spherical cleistothecium) and Pyrenomycetes (asci are in a flask-shaped perithecium). The Ascomycotina is the largest subdivision of fungi with over 42 000 known species. A book based on the First International Workshop on Ascomycete Systematics presents discussion of the changing ideas of the ascomycetes based on the ascoma, ascus, ascospore and septal structures; secondary metabolites; molecular biology; ecological and population biology; and cladistics. The publication *Systema Ascomycetum* provides additional information on the ascomycetes, including the annually revised *Outlines in Systema Ascomycetum*.

The order Eurotiales *Byssochlamys*, *Emericella*, *Eupenicillium*, *Eurotium*, *Monascus*, *Neosartorya*, *Talaromyces* and *Xeromyces* (**Figures 3–7**) contains most of the important food-borne Ascomycotina. There are four other genera of ascomycetes that are found to a limited extent in foods. *Gibberella*, the teleomorph of *Fusarium* species, is not associated with foods; however, it is usually more of a concern with grains in the field. *Claviceps purpurea* has produced

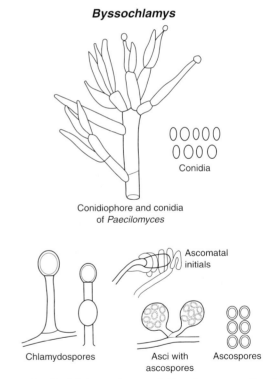

Figure 3 Conidial structures of *Paecilomyces* with chlamydospores, asci and ascospores of *Byssochlamys* sp.

ergot in rye. *Neurospora* species, teleomorphs of *Chrysonilia* species, are isolated from foods occasionally. *Chaetomium* species can be found in some foods, especially in tropical countries (see Table 1). *Sclerotinia* species, teleomorphs of *Botrytis* that cause soft rot of vegetables and *Morchella* or morel mushrooms are also important.

Eurotium

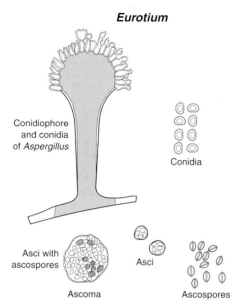

Conidiophore and conidia of *Aspergillus*

Conidia

Asci with ascospores

Ascoma

Asci

Ascospores

Figure 4 Conidial structures of *Aspergillus* with asci, ascoma and ascospores of *Eurotium* sp.

Neosartorya

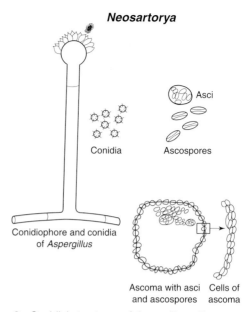

Asci

Conidia

Ascospores

Conidiophore and conidia of *Aspergillus*

Ascoma with asci and ascospores

Cells of ascoma

Figure 6 Conidial structures of *Aspergillus* with ascoma, asci and ascospores of *Neosartorya* sp.

Monascus

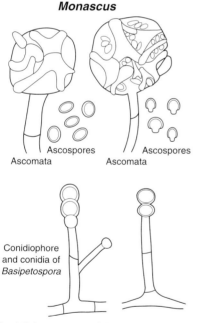

Ascospores

Ascomata

Ascospores

Ascomata

Conidiophore and conidia of *Basipetospora*

Figure 5 Conidial structures of *Basipetospora* with ascomata and ascospores of *Monascus* sp.

Talaromyces

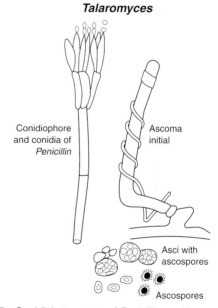

Conidiophore and conidia of *Penicillin*

Ascoma initial

Asci with ascospores

Ascospores

Figure 7 Conidial structures of *Penicillium* with ascoma, asci and ascospores of *Talaromyces* sp.

Commercial Importance of Ascomycetes

Yeasts in the order Saccharomycetales include several genera that are important in food, such as: *Debaryomyces*, *Hanseniaspora*, *Kluyveromyces*, *Pichia*, *Saccharomyces*, *Saccharomycodes*, *Schizosaccharomyces*, *Torulaspora*, *Yarrowia*, and *Zygosaccharomyces*. These yeasts are subdivided by the way in which they undergo vegetative reproduction, namely fission, bipolar budding or multilateral budding (Table 2).

Some ascomycetes are responsible for food spoilage; others are used in fermentations, especially species of yeasts (**Tables 3 and 4**). The major fermentative yeasts are strains of *Saccharomyces cerevisiae* used to make bread, beer, saké, wine and many other fermented foods. *Zygosaccharomyces rouxii* can be isolated from fermenting soy sauce. An ascomycetous mould genus is used for fermentation; *Monascus pilosus* and *M. purpureus* are used for making rice wine and

Table 3   Importance of the major ascomycetous moulds in foods

Mould genus	Species	Importance in foods
Byssochlamys	fulva, nivea	Heat-resistant spores in fruit products; B. nivea not common
Chaetomium	brasiliense, funicola	Isolated from cereal grains, legumes, nuts in tropical countries
Chaetomium	globosum	Isolated from wheat, cereal grains, legumes, nuts
Emericella	nidulans	Isolated from cereal grains, legumes, meats, nuts, produces mycotoxins
Eurotium	amstelodami, chevalieri, repens, rubrum	Xerophiles isolated from many foods: cereal grains, cheese, dried fruits, nuts
Monascus	ruber	Spoilage of high-moisture prunes; isolated from dried fish and other foods
Monascus	pilosus, purpureus	Fermented rice wine and kaoliang brandy
Neosartorya	fischeri	Produces heat-resistant ascospores in acid heat-processed fruit products
Talaromyces	macrosporus, bacillisporus	Produces heat-resistant ascospores in fruit products
Talaromyces	flavus, wortmannii	Isolated from cereals, nuts, meats, but no reports of spoilage of heat-treated foods
Xeromyces	bisporus	Xerophile that grows down to 0.61 a_w; isolated from liquorice, dried fruits

Table 4   Importance of some ascomycetous yeasts in foods

Yeast genus	Species	Importance
Debaryomyces	hansenii	$D_{110°C}$ 1.3 min; is salt-tolerant, film-former in brines, spoils orange juice, yoghurt
Hanseniaspora	guilliermondii, uvarum	Isolated from fruits, vegetables, wine, brined foods, soft drinks
Kluyveromyces	marxianus	Isolated from beer, dairy products, molasses, sugar cane, wine
Pichia	anomola, fermentans, membranaefaciens	Isolated from fruit juices, soft drinks, wine, beer, confections, dried fruits, mayonnaise, salad dressings; some species preservative resistant
Saccharomyces	cerevisiae	Fermentation yeast (bread, beer, wine, saké, whisky, cocoa, etc.)
Saccharomyces	exiguus	Sourdough bread (acid-tolerant)
Saccharomycoides	ludwigii	Spoils cider, isolated from beer and wine
Schizosaccharomyces	pombe	Xerotolerant, reduces malic acid in wine, preservative resistant
Torulaspora	delbrueckii	Used in fermented breads, isolated from many foods (dairy, fruits, high-sugar foods, vegetables)
Yarrowia	lipolytica	Isolated from refrigerated foods (dairy, meat, salads, seafoods)
Zygosaccharomyces	bailii	Spoils salad dressings, mayonnaise, ice cream mix, wine; preservative resistant
Zygosaccharomyces	rouxii	Xerotolerant to a_w 0.62, preservative resistant, isolated from fermented foods (soy sauce, cocoa, pickles)

kaoling brandy, respectively. Other ascomycetes are occasionally used in fermentations (Tables 3 and 4).

Generally, there are few ascospores present in most foods; therefore, it will take time to detect their presence. The major problem with ascospores in foods has been their resistance to heat, which allows them to survive the thermal treatments of pasteurization, canning or aseptic processing. This has been a particular concern with heat-processed fruit juices and products with fruit purée or pieces. Various D and z values have been reported. Values of $D_{90°C}$ of 1–12 min and a z value of 6–7°C have been recorded for *Byssochlamys fulva*, slightly higher than for *B. nivea* with a $D_{88°C}$ of 0.75–0.8 min and z of 4.0–6.1°C. For *Talaromyces macrosporus* values for $D_{90°C}$ of 2–7 min and a z value of 10.3°C were recorded, and for *Neosartorya fischeri*, $D_{88°C}$ was 1.2–16.2 min and z was 5.6°C. The ascospores of these moulds have been particularly troublesome in fruit products. In fact,

Byssochlamys species rarely are isolated from non-thermally processed spoiled acid foods. Yeast cells as well as mould hyphae and conidia are not as resistant to heat as ascospores. Mould ascospores are more resistant to heat than yeast ascospores.

Most of the xerophilic fungi belong to the *Ascomycotina* or closely related deuteromycetes. The most xerotolerant mould is *Xeromyces bisporus* that grows down to a water activity (a_w) of 0.61 in dried fruits, liquorice, fruit cakes and cookies with fruit fillings; however, it takes about 120 days to germinate at the a_w of 0.61. *Zygosaccharomyces rouxii* is the most xerotolerant yeast growing in high-sugar foods (sugar syrups, fruit concentrates, cake icings, confections, jams, chocolate sauces) with a_w minimum of 0.62. Other xerophiles are species of *Eurotium* (Table 3), which have been isolated from many different types of foods, such as dried fruits, stored cereals, dried meat and fish, and nuts. Since foods that have low water activities are not optimal for microbial growth, xerophiles generally will take months to become evident in foods.

Several ascomycetous yeasts are resistant to common chemical preservatives used in foods, especially benzoate and sorbate. *Pichia membranaefaciens* grows in foods with 1% acetic acid and up to 1500 mg kg^{-1} sodium benzoate at pH 4.0. *Schizosaccharomyces pombe* is resistant to sulphur dioxide at pH 3.0–3.5 and has been isolated from foods with a sulphur dioxide content of 120–250 mg kg^{-1}; it also grows in 600 mg l^{-1} of benzoic acid. *Zygosaccharomyces bailii* is resistant to several chemical preservatives, such as acetic, benzoic, propionic and sorbic acids and sulphur dioxide at levels of 400–800 mg l^{-1}. These yeasts have caused spoilage in foods preserved with vinegar, salad dressings and mayonnaise, sugar syrups, soft and sports drinks, and fruit products.

See also: **Preservatives**: Classification and Properties. **Spoilage Problems**: Problems Caused by Fungi. **Yeasts**: Production and Commercial Uses.

Further Reading

Barnett JA, Payne RW and Yarrow D (1990) *Yeasts. Characteristics and Identification*, 2nd edn. Cambridge: Cambridge University Press.

Deak T and Beuchat LR (1996) *Handbook of Food Spoilage Yeasts*. Boca Raton: CRC Press.

Greuter W, Barrie FR, Burdet HM et al (eds) (1994) *International Code of Botanical Nomenclature (Tokyo Code)* Königstein: Koeltz Scientific.

Hawksworth DL (ed.) (1994) *Ascomycete Systematics. Problems and Perspectives in the Nineties*. New York: Plenum Press.

Hawksworth DL, Kirk PM, Sutton BC and Pegler DN (1995) *Ainsworth & Bisby's Dictionary of the Fungi*, 8th edn. Wallingford: CAB International.

Ingold CT and Hudson HJ (1993) *The Biology of Fungi*, 6th edn. New York: Chapman & Hall.

Jay JM (1996) *Modern Food Microbiology*, 5th edn. New York: Chapman & Hall.

King AD, Pitt JI, Beuchat LR and Corry JEL (eds) (1986) *Methods for the Mycological Examination of Food*. New York: Plenum Press.

Kreger-van Rij NJW (1984) *The Yeasts. A Taxonomic Study*, 3rd edn. New York: Elsevier.

Pitt JI and Hocking AD (1997) *Fungi and Food Spoilage*, 2nd edn. New York: Blackie.

Samson RA, Hoekstra ES, Frisvad JC and Filtenborg O (1995) *Introduction to Food-borne Fungi*. Baarn: Centraalbureau voor Schimmelcultures.

Classification of the Deuteromycetes

B C Sutton, Blackheath, Suffolk, UK

Defining Features of the 'Class'

The deuteromycetes is an artificial grouping in that the phylogenetic relationships among taxa are mostly unknown or not apparent. They are the mitotic states of meiotic groups such as the basidiomycetes and especially the ascomycetes, or have evolved from them. A very small number of taxa have been correlated with meiotic states but the majority have not. Thus there is a residual body of taxa which cannot easily be incorporated into the classifications for meiotic fungi. This situation will become less problematic as the results of molecular characterization and their application to fungal systematics become more widespread. For the moment, however, there is a serious lack of information about DNA-based typification, and apart from classifications for ascomycetes and basidiomycetes there is still no taxonomic system to cope with these other fungi. Over the last 200 years separate classifications have been developed for mitotic fungi and until recently these have arisen independently from classifications for ascomycetes and basidiomycetes. Several names, both formal (nomenclatural) and informal (colloquial), have been used in the past for groups of mitotic fungi. These include Deuteromycotina, Deuteromycetes, Fungi Imperfecti, asexual fungi, conidial fungi, and anamorphic fungi. The most recent suggestion, accepted in the *Dictionary of the Fungi* (8th edn), is 'mitosporic fungi'. Colloquial names such as this and others have no nomenclatural standing. Neither these nor the formal names that have been proposed are in any way equivalent to the names used for taxonomic categories accepted in basidiomycete and ascomycete systematics that are governed by the *International Code of Botanical Nomenclature*. Although many class, subclass, order, suborder and family names have been used in the group in the past none of these is currently accepted, and if they are used at all it is in an informal manner. Even the use of generic and specific names (which are allowed by the Code) must be with qualification, for they are also not equivalent to those employed in ascomycetes and basidiomycetes. Sometimes they are referred to as form genera and form species. Despite this there is still a need for a framework on which to hang the information used in identification of mitosporic fungi. Until the 1950s taxa were differentiated primarily by the nature of the fruiting structures and conidial morphology.

However, since that time the systematics of the group has largely depended on aspects of the conidiogenous processes exhibited by the fungi involved.

Discussions concerning the classification of, or information frameworks for, mitosporic fungi ignore the ultimate aim to do away with the group and incorporate its members into the classifications for ascomycetes and basidiomycetes. The use of DNA technology is the key to accomplishing this.

The group is characterized by the absence of teleomorphic (meiotic) states. It is heterogeneous, i.e. polyphyletic. Reproduction is commonly by spores (conidia) produced mitotically (asexually) from conidiogenous cells which are sometimes free (as in yeasts) or more commonly formed on separate supporting hyphae (conidiophores) and/or cells which may be produced in or on organized fruiting structures (conidiomata). Taxa are separated by differences in conidiogenous events and the structures involved, conidiomatal form and development, conidial morphology, colony characteristics and the presence and nature of vegetative structures.

Organizational Framework for Mitosporic Fungi

The group is traditionally separated into two classes: Hyphomycetes, in which conidia are formed on separate hyphae or aggregations of hyphae, and Coelomycetes, where conidia are formed in closed or partly closed fruiting structures (conidiomata). However, these distinctions are now only used informally for convenience. The most recently accepted organizational framework for mitosporic fungi rests on the definition of genera by the events surrounding conidiogenesis. All other characters are of secondary importance, although some are used to corroborate or endorse the separations indicated by conidiogenesis. Differences in conidiogenous events are used to distinguish taxa, and in addition to this role the framework is helpful in that it is also a descriptive tool, i.e. precise words are used to describe the individual developmental stages. Previous attempts to use conidiogenesis as the basis for classification of the group proved to be unworkable because of their inability to cope with taxa that failed to fit into the recognized pigeonholes. The advantage of the newer system is that it provides the rationale to deal with all combinations of events, so problematic fungi such as *Trichothecium*, *Basipetospora* and *Wallemia* can be dealt with like any other within the framework.

The foundation of the framework is the recognition of a number of basic processes surrounding conidiogenesis. These are the ways in which walls are laid down in hyphae (apical-, diffuse-, and ring-wall

building), conidial initiation, conidial secession, conidial maturation, collarette production, and conidiogenous cell proliferation. When the system was first introduced in the *Dictionary of the Fungi* (7th edn), descriptive terminology was given for 13 genera (*Pseudallescheria*, *Cladosporium*, *Tritirachium*, *Trichoderma*, *Scopulariopsis*, *Spadicoides*, *Geotrichum*, *Pseudospiropes*, *Aspergillus*, *Trichothecium*, *Cladobotryum*, *Arthrinium* and *Basipetospora*). Since that time the descriptors have been widened to include 43 different combinations of events and these have been published in the eighth edition of the *Dictionary of the Fungi*. A few more have since been recognized and there is little doubt that more are likely to be identified in the future.

Genera Involved in Food Microbiology

In the following generic descriptions conidiogenesis is indicated by the conidiogenous event number listed and illustrated in the *Dictionary of the Fungi* (8th edn).

- *Acremonium* (teleomorphs *Emericellopsis*, *Nectria*) – conidia in chains collapsing into wet masses, from solitary, erect, aseptate or septate, simple or sparingly branched, smooth or slightly rough, hyaline or pale brown conidiophores. Conidiogenous cells (phialides) terminal, cylindrical. Conidiogenesis – *event 15, Fig. 25*. Conidia hyaline or sometimes pale brown, globose, subglobose, ellipsoid or fusiform, aseptate or sometimes septate. Distributed worldwide, from soil, cultivated fields, mud sediments, plant remains, hay, apples, pears, also produces mycotoxins.

- *Alternaria* (teleomorphs *Clathrospora*, *Leptosphaeria*) – teleomorph rarely seen in context of food microbiology. Colonies of black or grey mycelium. Conidiophores solitary, brown, simple or branched, showing sympodial growth in association with conidium production. Conidiogenesis – *event 26, Fig. 25*. Conidia dry, in long, often branched chains, obclavate, obpiriform, ovoid or ellipsoid with a short conical or cylindrical beak, and several transverse and longitudinal eusepta, pale to medium brown, smooth or verrucose. Distributed worldwide, some parasitic, commonly saprobic on plant materials, foodstuffs, soil, textiles. Produces mycotoxins.

- *Aspergillus* (teleomorph *Emericella*) – conidia in dry chains, forming dark yellow-green columns from solitary, erect, aseptate, brown, smooth conidiophores. Conidiogenous cells (phialides) borne on supporting cells on swollen apices of conidiophores, short-necked. Conidiogenesis – *event 32, Fig. 26*. Conidia rough-walled, globose. Distributed worldwide, especially from soil, potatoes, grain, citrus, stored cereals, cotton. Mycotoxins formed, also used in fermentation industry for production of vitamins, enzymes, organic acids, antibiotics, soy sauce, miso and saki.

- *Aspergillus* (teleomorph *Eurotium*) – forming grey to olive-green heads from septate or aseptate, smooth or rough conidiophores. Conidiogenous cells (phialides) arising directly from swollen apex of the conidiophores, radiating, very short-necked. Conidiogenesis – *event 32, Fig. 26*. Conidia echinulate, globose, subglobose, ovate or ellipsoid, sometimes with both ends flattened. Distributed worldwide, though predominant in tropical to subtropical areas, from soil, stored or decaying grain and food products, fruit, fruit juice, peas, milled rice, nuts, dried food products, spices, meat products. Also produces a range of mycotoxins.

- *Aspergillus* (teleomorph *Neosartorya*) – conidia in dry chains forming olive-grey columns from solitary, erect, aseptate or septate, smooth conidiophores. Conidiogenous cells (phialides) formed directly on the swollen apex of the conidiophore, radiating, short-necked. Conidiogenesis – *event 32, Fig. 26*. Conidia slightly roughened, globose to subglobose or ellipsoid. Distributed worldwide, from soil, rice, cotton, potatoes, groundnuts, leather, paper products; also produces mycotoxins.

- *Aureobasidium* – colonies covered by slimy yellow, cream, pink, brown or black masses of spores. Aerial mycelium scanty, immersed mycelium often dark brown. Conidiogenous cells undifferentiated, procumbent, intercalary or on short lateral branches. Conidiogenesis – *event 16, Fig. 25*. Conidia produced synchronously on multiple loci in dense groups on short scars or denticles, hyaline, smooth, with a truncate base. Distributed worldwide, saprobic, from soil, leaf surfaces, cereal seed, on flour, tomato, pecan nuts, fruit, fruit drinks.

- *Basipetospora* (teleomorph *Monascus*) – mycelium brownish, grey-brown in the centre, with brick-red pigmentation on oat agar. Conidiogenesis – *event 36, Fig. 26*. Conidia borne in basipetal succession in chains on solitary, septate, erect conidiophores, retrogressively delimited, ovate to piriform, hyaline, aseptate, thin-walled, base truncate. Distributed worldwide, from soil, silage, dried foods, rice, oat, seeds, soya, sorghum, tobacco, also used in fermentation of angkak to produce pigment for food products of fish, soya beans and some alcoholic beverages.

- *Botrytis* (teleomorph *Botryotinia*) – conidia formed in dry, powdery grey masses from erect, brown, smooth, septate, solitary, hygroscopic conidiophores. Conidiogenous cells produced ter-

minally on an apical head of small alternate branches, swollen, with many denticulate conidiogenous loci each forming a single conidium simultaneously. Conidiogenesis – *event 6, Fig. 24*. Conidia aseptate, rarely 1–2 septate, pale brown, globose, ovate or ellipsoid, smooth, hydrophobic. Microconidial state (*Myrioconium*) sometimes formed, sporodochial, phialidic (*event 15*), with small globose or subglobose hyaline conidia. Sclerotia large, cortex black to brown with a white medulla, flattened to pulvinate, rounded to ellipsoid, smooth or wrinkled. Distributed worldwide, but more commonly in humid temperate and subtropical regions, from soils both dry and aquatic, stored and in transit fruit and vegetables, causing fruit and leaf rots of strawberry, grape, cabbage, lettuce, neck rot in onions and shallots.

- *Brettanomyces* (teleomorph *Dekkera*) – conidia hyaline, with an attenuated rounded base, formed by multilateral or acropetal budding from conidiogenous cells with or without broad denticles or scars. Conidiophores absent. Conidiogenesis – *event 3, Fig. 24*. Strong acid production in glucose-containing media, fermentation present or nearly absent. Spoilage organisms in beverages such as mineral waters and nonalcoholic drinks, and lambic and other old or spoilt beers, ciders and wines.

- *Candida* (teleomorphs *Debaryomyces, Issatschenkia, Kluyveromyces, Pichia, Saccharomyces Torulaspora, Yarrowia*) – conidia produced by budding which is multilateral or acropetal leaving conidiogenous cells with or without denticles or scars, hyaline, aseptate, base attenuated, rounded. Conidiogenesis – *event 3, Fig. 24*. Fermentation present or nearly absent, nitrate not assimilated, acid production absent or weak, growth at 25°C. Common food-borne species.

- *Chrysonilia* (teleomorph *Neurospora*) – conidia aseptate, ellipsoid or more or less cylindrical, globose, subglobose or irregular, hyaline, smooth, formed in dry chains with connectives from ascending to erect, smooth, septate, much-branched conidiophores. Conidiogenesis – *event 38, Fig. 26*. Widespread, especially Europe, USA and Asia, from bread (red bread mould) and related products, silage, meat, and transported and stored fruit, used in the production of oncom mera by fermentation of soya bean products.

- *Cladosporium* (teleomorph *Mycosphaerella*) – teleomorph rarely seen in context of food microbiology. Colonies of olivaceous, grey-olive to blackish-brown mycelium. Conidiophores solitary, brown, unbranched except towards the apices. Conidiogenesis – *event 3, Fig. 24*. Conidia dry, in branched chains, ellipsoid, fusiform, ovoid, subglobose, aseptate or with several transverse eusepta, pale to dark olivaceous brown, smooth, verruculose or echinulate, with a distinct scar at the base and several in the apical region if forming chains, formed from cicatrized loci produced synchronously, sympodially or irregularly by the conidiogenous cell. Distributed worldwide, commonly airborne, ubiquitous as saprobes and primary plant pathogens, also from soil, foodstuffs. Produces mycotoxins.

- *Epicoccum* – colonies fluffy, yellow, orange, red, brown or green. Conidiophores formed in black sporodochial conidiomata, closely branched, compact and dense. Conidiogenous cells pale brown, smooth or verrucose, integrated, terminal, determinate, cylindrical. Conidiogenesis – *event 1, Fig. 24*. Conidia solitary, dry, subspherical to piriform, dark golden-brown, often with a pale protuberant basal cell, muriform, rough, opaque. Distributed worldwide, from soil, cereal seed, beans, mouldy paper, textiles.

- *Fusarium* (teleomorphs *Gibberella*, *Nectria*) – macroconidia in slimy yellow, brown, pink, red, violet or lilac masses, chains or dry masses from branched or unbranched, procumbent or erect, hyaline, smooth, septate conidiophores in sporodochial conidiomata. Conidiogenous cells (phialides) produced from apices of conidiophores or branches, slender or tapered, with one or sometimes several conidiogenous loci. Conidiogenesis – *event 15, Fig. 25*. Macroconidia hyaline, one or many septate, fusiform to sickle-shaped, mostly with an elongated apex and a pedicellate basal cell. Microconidia usually aseptate, piriform, fusiform or ovoid, straight or curved, nearly always formed on aerial mycelium. Chlamydospores present or absent, intercalary, solitary, in chains or clusters, formed in hyphae or conidia. Distributed worldwide, from soil, aquatic and semiaquatic environments, stored grain and natural products. Potent producers of mycotoxins.

- *Geotrichum* (teleomorph *Galactomyces*) – conidia formed in white, smooth, often butyrous colonies from aerial, erect or decumbent mycelium functioning conidiogenously. Mycelium dichotomously branched at the advancing edge. Conidiogenesis – *event 39, Fig. 26*. Conidia hyaline, aseptate, smooth, cylindrical, doliiform or ellipsoid. Distributed worldwide, from soil, air, water, cereals, grapes, citrus, bananas, tomatoes, cucumber, frozen fruit cakes, milk and milk products, also used with bacteria in fermentation of manioc to produce gari in West Africa.

- *Moniliella* – colonies acidophilic, restricted, smooth, velvety or cerebriform, cream then pale olivaceous or black-brown, cells often budding to produce a pseudomycelium. Conidiophores undifferentiated, hyaline, smooth, repent. Conidiogenesis – *events 3, Fig. 24* and *38, Fig. 26*. Conidia produced in acropetal chains from individual (conidiogenous) cells of the mycelium, hyaline, smooth, aseptate, ellipsoid; conidia also formed by fragmentation of hyphae, becoming thick-walled and brown. From Europe and USA, occurring in pickles and vinegar, fruit juices, syrups and sauces.

- *Paecilomyces* (teleomorphs *Byssochlamys*, *Thermoascus*) – conidia in dry chains from conidiogenous cells on solitary, septate, erect conidiophores. Conidiogenous cells (phialides) in groups of two to five, cylindrical at the base, long-necked, on short supporting cells. Conidiogenesis – *event 15, Fig. 25*. Conidia cylindrical, hyaline, aseptate, with flattened ends, yellow. Distributed worldwide, from soil, bottled and tinned fruit, pasteurized food, airtight stored cereals; also produces mycotoxins.

- *Penicillium* (teleomorph *Talaromyces*) – conidia in dry chains from solitary, erect, branched, septate, smooth or rough conidiophores. Conidiogenous cells (phialides) from an apical branching system consisting of branch cells and supporting cells to the phialides (biverticillate), long, lageniform with a short, narrow neck. Conidiogenesis – *event 15, Fig. 25*. Conidia subglobose, ellipsoid or fusiform, smooth or finely spinulose, hyaline, brown, brown-green or pale green. Distributed worldwide, from soil, organic substances, rape, cotton, pears, wheat, barley, milled rice, pecan nuts, bagasse, often in tropical fruit juices, some species heat-resistant. Mycotoxins formed. Used in production of antibiotics and enzymes, manufacture of cheese, sugar, juice and organic acids, and brewing.

- *Penicillium* (teleomorph *Eupenicillium*) – conidia in dry chains from solitary, erect, branched, septate, smooth or rough conidiophores. Conidiogenous cells (phialides) formed directly from conidiophore apices (monoverticillate) or as in species with *Talaromyces* teleomorphs, long or broadly lageniform, with a short, narrow neck. Conidiogenesis – *event 15, Fig. 25*. Conidia smooth, ovoid, subglobose, piriform or ellipsoid, hyaline. Distributed worldwide, from soil, groundnuts, fruit cake, canned fruit, corn, oranges. Produces mycotoxins.

- *Phialophora* (teleomorphs *Mollisia*, *Pyrenopeziza*, *Coniochaeta*) – colonies slow-growing, olivaceous black, sometimes pink or brown. Conidiophores erect, hyaline or pale brown, branched or reduced to simple hyphae. Conidiogenous cells (phialides) clustered or single, lageniform or cylindrical, with a distinct darker collarette. Conidiogenesis – *event 15, Fig. 25*. Conidia formed in slimy heads or in chains, aseptate, globose to ellipsoid or curved, mostly hyaline, smooth. Also linked with *Geaumannomyces*, but several species with no known teleomorph. Distributed worldwide but most common on decaying wood, wood pulp, secondarily soil-borne, from water, fermented corn dough, foodstuffs, butter, wheat.

- *Phoma* – colonies comparatively fast-growing, grey, olivaceous, brown, fluffy. Conidiomata pycnidial, black-brown, ostiolate, sometimes setose. Conidiophores absent. Conidiogenous cells ampulliform to doliiform, hyaline, smooth, phialidic. Conidiogenesis – *event 15, Fig. 25*. Conidia hyaline, smooth, aseptate or sometimes septate, ellipsoid, ovate, cylindrical. Dark-brown unicellular or multicellular chlamydospores sometimes formed. Some teleomorphs in Pleosporaceae (*Pleospora*), but most species with no known teleomorphs. Distributed worldwide, from soil, butter, rice grain, cement, litter, paint, wool, paper. Produces mycotoxins.

- *Rhodotorula* – colonies pink, with carotenoid pigment soluble in organic solvents, mycelium and/or pseudomycelium formed, cells usually small and narrow. Conidiogenesis – *event ?1, Fig. 24*. Conidia spherical, ovate or clavate, with a narrow or rather broad base, budding. Sometimes assimilates nitrate, but fermentation absent. Teliospores absent, but basidium-like structures in some species indicates basidiomycete affinity with *Rhodosporidium* (Sporidiobolaceae). From wood, involved in spoilage of frozen vegetables, dairy products.

- *Scopulariopsis* (teleomorph *Microascus*) – conidia in white to shades of brown, dry powdery masses from erect penicillately or verticillately branched, hyaline to pale brown, smooth, septate, solitary conidiophores. Conidiogenous cells terminal, cylindrical, repeatedly forming basipetal chains of conidia from percurrently proliferating loci giving rise to apical annellations. Conidiogenesis – *event 19, Fig. 25*. Conidia aseptate, hyaline or brown, globose, ovate or mitriform, with a truncate base, smooth, or ornamented. Distributed worldwide, from soil, grain, fruit, soya beans, groundnuts, milled rice, and animal products such as eggs, meat, cheese, milk, butter.

- *Stachybotrys* – colonies black to black-green, powdery. Conidiophores erect, separate, simple or branched, septate, becoming brown and rough at the apex. Conidiogenous cells (phialides) grouped at the conidiophore apex, obovate, ellipsoid, clavate or broadly fusiform, becoming olivaceous, with a small locus and no collarette. Conidiogenesis – *event 15, Fig. 25*. Conidia in large, slimy black heads, ellipsoid, reniform or subglobose, hyaline, grey-green, dark-brown or black, sometimes striate, coarsely rough or warted, aseptate. Distributed worldwide, from soil, paper, cereal seed, textiles. Produces mycotoxins.

- *Trichoderma* (teleomorph *Hypocrea*) – conidia in dry, powdery, green to yellow masses from solitary, repeatedly branched, erect, hyaline, smooth, septate conidiophores which may end in sterile appendages. Conidiogenous cells (phialides) apical and lateral, often irregularly bent, flask-shaped, with a short neck. Conidiogenesis – *event 15, Fig. 25*. Conidia aseptate, hyaline or usually green, smooth or roughened, globose, subglobose, ellipsoid, oblong or piriform. Distributed worldwide, from soil, stored grain, groundnuts, tomatoes, sweet potatoes, citrus fruit, pecan nuts. Produces mycotoxins.

- *Trichosporon* – colonies slow-growing, white to cream, butyrous, smooth or wrinkled. Mycelium repent, hyaline. Conidiophores absent. Conidiogenesis – *events 38, Fig. 26 and 10, Fig. 24*. Conidia of two types: (1) thallic, formed by fragmentation of the mycelium, cylindrical to ellipsoid; (2) blastic, formed in clusters near the ends of the thallic conidia or by budding of the lateral branches of the mycelium, subglobose, with a narrow distinct scar. Distributed worldwide, from humans and animals, saprobic in soil, fresh and sea water, plant material, fermented corn dough.

- *Trichothecium* – colonies powdery, pink. Conidiophores erect, separate, simple, unbranched, septate near the base, rough, apical cell functioning conidiogenously. Conidiogenesis – *event 34, Fig. 26*. Conidia formed in retrogressively delimited basipetal chains, appearance zigzagged, hyaline, smooth, single-septate, ellipsoid or piriform, thick-walled, with an obliquely truncate scar. Distributed worldwide, from soil, water, decaying plant material, leaf litter, cereal seed, pecan nuts, stored apples, fruit juices, foodstuffs especially flour products. A potent producer of mycotoxins.

- *Ulocladium* – colonies black to olivaceous black. Conidiophores erect, separate, simple or branched, septate, smooth, straight, flexuous, often geniculate, geniculations associated with preformed loci (pores). Conidiogenesis – *event 26, Fig. 25*. Conidia dry, solitary or in short chains, obovoid to short ellipsoid, with several transverse and longitudinal or oblique eusepta, medium brown to olivaceous, smooth or verrucose, base conical, apex broadly rounded and becoming conidiogenous.

Not uncommon, widely distributed, from soil, water, dung, paint, grasses, fibres, wood, paper, corn, seeds.

- *Verticillium* – colonies cottony, white to pale yellow, sometimes becoming black due to resting mycelium. Conidiophores erect, separate, septate, smooth, hyaline, simple, unbranched or branched. Conidiogenous cells (phialides) solitary or produced in verticillate divergent whorls, long lageniform to aculeate, hyaline. Conidiogenesis – *event 15, Fig. 25*. Conidia form in droplets at the apices of conidiogenous cells, hyaline, aseptate, smooth, ellipsoid to cylindrical. Hyaline multicellular chlamydospores and microsclerotia sometimes formed. Distributed worldwide, commonly causing plant wilt diseases, from soil, paper, insects, seeds, bakers' yeast, potato tubers, commercially grown fungi. Forms mycotoxins.

- *Wallemia* – colonies xerophilic, restricted, fan-like or stellate, powdery, orange-brown to black-brown. Conidiophores erect, separate, cylindrical, smooth, pale brown. Conidiogenous cells apical, long lageniform to cylindrical, finally verrucose, forming a phialide-like aperture without collarette from which a short chain of four thallic conidia is formed. Conidiogenesis – *events 15/38, Figs 25 and 26*. Conidia initially cuboid, later globose pale brown, finely warted. Distributed worldwide, from dry foodstuffs such as jams, marzipan, dates, bread, cake, salted fish, bacon, salted beans, milk, fruit, soil, air, hay, textiles.

See also: **Alternaria**. **Aspergillus**: Introduction. **Aureobasidium**. **Botrytis**. **Brettanomyces**. **Candida**: Introduction. **Fungi**: Classification of the Peronosporomycetes; Classification of the Zygomycetes; Classification of the Eukaryotic Ascomycetes; Classification of the Hemiascomycetes; Classification of the Deuteromycetes. **Fusarium**. **Geotrichum**. **Penicillium**: Introduction. **Rhodotorula**. **Trichoderma**. **Trichothecium**.

Further Reading

Betina V (1989) *Mycotoxins: Chemical, Biological and Environmental Aspects*. Bioactive Molecules 9. Amsterdam: Elsevier.

Beuchat LR (1987) *Food and Beverage Mycology*, 2nd edn. New York: Van Nostrand Reinhold.

Carmichael JW, Kendrick WB, Conners IL and Sigler L (1980) *Genera of Hyphomycetes*. Edmonton: University of Alberta Press.

Davenport RR (1981) Yeasts and yeast-like organisms. In: Onions AHS, Allsopp D and Eggins HOW (eds) *Smith's Introduction to Industrial Mycology*, 7th edn. London: Edward Arnold.

Domsch KH, Gams W and Anderson TH (1980) *Compendium of Soil Fungi*. London: Academic Press.

Hawksworth DL, Kirk PM, Sutton BC and Pegler DN (1995) *Ainsworth & Bisby's Dictionary of the Fungi*, 8th edn. Wallingford: CAB International.

Hennebert GL and Sutton BC (1994) Unitary parameters in conidiogenesis. In: Hawksworth DL (ed.) *Ascomycete Systematics: Problems and Perspectives in the Nineties*. NATO ASI Series A: Life Sciences 269. New York: Plenum Press.

Kendrick WB (1979) *The Whole Fungus*. Ottawa: National Museums of Canada.

Pitt JI and Hocking AD (1997) *Fungi and Food Spoilage*, 2nd edn. New York: Chapman & Hall.

Samson RA, Hoekstra ES, Frisvad JC and Filtenborg O (1995) *Introduction to Food-borne Fungi*, 4th edn. Baarn: Centraalbureau voor Schimmelcultures.

Sutton BC (1980) *The Coelomycetes*. Kew: Commonwealth Mycological Institute.

Sutton BC (1996) Conidiogenesis, classification and correlation. In: Sutton BC (ed.) *A Century of Mycology*. Cambridge: Cambridge University Press.

Sutton BC and Hennebert GL (1994) Interconnections amongst anamorphs and their possible contribution to ascomycete systematics. In: Hawksworth DL (ed.) *Ascomycete Systematics: Problems and Perspectives in the Nineties*. NATO ASI Series A: Life Sciences 269. New York: Plenum Press.

Classification of the Hemiascomycetes

A K Sarbhoy, Division of Plant Pathology, Indian Agricultural Research Institute, New Delhi, India

Ascomycota has 3255 genera comprising 32 267 species. The presence of lamellate hyphal walls with a thin electron-dense outer layer and a relatively thick electron transparent inner layer of the ascus is one of the diagnostic characters; this enables mitosporic fungi to be recognized as Ascomycetes even in the absence of asci. In the past, much importance has been given to the characteristics of asci (Hemiascomycetes, Plectomycetes, Pyrenomycetes, Discomycetes, Laboulbeniomycetes, Loculoascomycetes). In recent years the development of ascomata and especially the method of discharge of asci has been given paramount importance. The major problem with earlier classifications was that lichen-forming fungi – almost half the Ascomycetes – had often been classified separately. A similar intercalation into a hierarchical system based on variations in non-lichenized fungi was not expected.

There are two schools of thought on the classification of the Hemiascomycetes. Some believe that the presence or absence of abundant mycelium is a fundamental characteristic and that the nutritional characteristic is of secondary importance in delineating the genera of such yeasts. The second school of thought does not recognize the genera of *Endomycopsis* on the basis of presence or absence of mycelium. They refer to cellular as well as filamentous blastoconidial species of *Pichia*. This genus does not assimilate nitrate as the sole source of nitrogen and species may or may not ferment sugars. However, fungi similar to *Pichia* but utilizing nitrate are assigned to *Hansenula*.

There is considerable variation between the systems of higher categories proposed for Ascomycota. In all, 46 orders are accepted, compared with the classification of Eriksson, who proposed 39 orders. Five have already been combined with other accepted orders as more data have been generated; 500 genera could not be assigned with certainty to the 29 accepted families.

Molecular data are adding a new dimension to our understanding of the relationships between the different Ascomycetes orders. In general, certain important criteria have emerged with respect to phylogenies and some groupings are becoming clearer. It is now evident that there is a basal group of Ascomycetes (Archiascomycetes) including Pneumocystidales, Protomycetales, Schizosaccharomycetales and Taphrinales. Saccharomycetales is quite separate from this group. Some authorities argue that the class names Plectomycetes and Pyrenomycetes should be reinstated, but this has not been widely accepted as the circumscriptions differ from those based on the ascomatal stage. Almost half the orders and many families have several members in sequences, and some have been the subject of phylogenetic speculation.

Conidium Ontogeny

In the taxonomy of Ascomycetes and their imperfect states, conidiation (how new cells or conidia are formed) has proved to be a useful criterion. Several kinds of conidiogenous cells and conidia are distinguished. Most are also observed in yeasts and should be used in yeast taxonomy, in addition to fission and budding.

Phialidic (enteroblastic, basipetal) conidiation is observed in the red yeasts, which should be classified into Sporobolomycetales and also in the yeast-like, vegetative state of *Taphrina*, *Symbiotaphrina*, *Protomyces*, *Microstoma* and *Ustilago*. The species belonging to these later genera are mainly plant parasites that cause hypertrophy, swelling or witch's broom growths. They have been classified partly in Ascomycetes and Basidiomycetes or Hyphomycetes,

but may be closely related to each other. The most common feature is the presence of carotenoids.

A practical system of classification has been elucidated which has 10 orders and no higher categories within the group. The main distinguishing features were the appearance of the ascus-containing structure (ascoma), particularly whether it was stromal, more or less absent, disc-like, ostiolate or scaly. Others distinguished two major groups on the basis of the ontogeny of the ascoma: the Ascoloculares in which the asci developed in cavities in a preformed stroma and the Ascohymeniales where the asci develop as a hymenium and not in a preformed stroma.

It was recognized that important fundamental characters such as those of the ascus, or those of ascocarp (ascoma) centrum characters, and major groups were based on the superficial characteristics. It depends mainly whether an ascoma was absent (Hemiascomycetes) or closed (Plectomycetes), whether the opening was a pore or slit (Pyrenomycetes) or whether it was open (Discomycetes).

Saccharomycetales

Saccharomycetales constitute two genera comprising 16 species. There is well-developed mycelium, lacking a polysaccharide sheath. The septa have clusters of minute pores which fragment to produce thallic conidia; asci are formed by the fusion of gametangia from adjacent cells or separate mycelia. These are usually elongated or ellipsoidal; they are rarely ornamented and are covered with a mucous sheath and do not stain blue with iodine.

Many laboratory strains of *Saccharomyces cerevisiae* (yeast form: YF) contain a single mutation that blocks pseudohyphal growth, but wild strains are dimorphic. Mutational studies have shown that the kinase cascade, required for the yeast-mating pathway, is also required for pH growth.

Candida is anamorphic Saccharomycetales. It has pseudomycelium and mycelium. *C. albicans* (candidiasis) is pathogenic for humans and animals. The fungal cell wall contains polymer chitin $(1 \rightarrow 3)$ β $(1 \rightarrow 6)$ β-glucan $(1 \rightarrow 3)$ α-glucan.

Despite early discovery of ballistoconidium-forming yeasts, little is known about this kind of yeast. Recently, however, many new strains of these yeasts have been isolated from natural sources and it has become important to study the classification of these yeasts. For a long time, the morphology of ballistoconidia has been considered an important taxonomic criterion. However, it is well known that the ability to produce ballistoconidia is easily lost. Moreover, recent extensive isolation studies have revealed that three kinds of conidia – ballistoconidia, budding cells and stalked conidia – are produced by the yeasts, although the productivity of these conidia is not stable. Therefore, the conidium ontogeny is not always reliable in the classification of ballistoconidium-forming yeasts. Blastospores with a narrow base and arthrospores are known in yeasts. In this genus the sporulation is multipolar all over the surface. In *Schizosaccharomyces* multiplication takes place by fission.

Blastospores with a broad base followed by the formation of a septum are typical in genera such as *Nadosoma* or *Saccharomycoides*. In this case the formation of the daughter cells takes place only at the end of the mother cells (bipolar) and can repeat itself by meristematic growth of the apical region. A number of daughter cells can be produced in basipetal succession as blown-out ends or by proliferation through the scars of previous daughter cells. In some cases a collarette-like sheath may surround the scar. Currently, cell components such as proteins and nucleic acids are considered to be the important criteria for estimation of relationships among taxa. The phylogenetic tree based on the partial sequences suggested that ballistoconidium-forming yeasts consisted of two branches and these corresponded to the existence of xylose as the cellular monosaccharide component. The phylogenetic tree based on the complete sequences suggested that the branch comprising the genus *Udeniomyces*, which is unique as morphologically from *Sporobolomyces*-like and physiologically *Bullera*-like was distantly aligned. The ubiquinone system and the cellular polysaccharide are considered to be significant at the generic level. The sequencing of 18S ribosomal RNA can provide information for the phylogeny. It is now clear that morphogenic process entails the differential expression of many genes.

Excluding the anamorphs and basidiomyceteous affinity, the genus *Candida* is now limited to the anamorphs of ascosporogenous species. While this action has made the genus more homogeneous, *Candida* still remains a catch-all of unrelated species. The broad definition of the genus permits inclusion of many yeasts with a wide range of characteristics. Currently, about 170 species are listed in *Candida*. Some of these have teleomorphic counterparts. These are represented by 11 teleomorphic names: *Citeromyces*, *Clavispora*, *Issatachenkia*, *Kluyveromyces*, *Metchnikowia*, *Pichia*, *Saccharomyces*, *Stephanoascus*, *Torulaspora*, *Wickerhamiella* and *Yarrowia*.

In the absence of distinguishing characteristics for taxonomic alignment, the use of molecular techniques for strain identification and species evaluation has proved difficult. DNA reassociation studies have demonstrated that lumping physiologically similar strains into a single species is no more plausible than dividing species on the basis of a single or a few physiological traits. Several species have been formed as complexes of unrelated strains and some synonyms have been shown to be valid species. About 4500 DNA sequences from 200 Ascomycetes species are found in the EMBL data library. These are potential phylogenetic markers at different taxonomical levels.

Dipodascales

In a complete revision of the fungus classification, the order Dipodascales is included together with other orders – Protoascales, Protomycetales, Ascolocurales, Spermothorales and Synascales – in the class Periascomycetes, which is not subdivided into subclasses. Only two genera, *Dipodascus* Lagen and *Helicognium* White, are placed in the family Dipodascaceae. The genus *Helicogonium* is characterized by the possession of bifurcate asci formed as a result of the fusion of two gametangia arising from different branches. The asci are eight-spored and do not stain with iodine solution.

Ecological Importance

Some of the unicellular yeasts produce daughter cells either by budding *Saccharomyces* (*S. cerevisiae*) or by binary fission (*Schizosaccharomyces pombe*). The yeasts are a phylogenetically related group of fungi bound together by the characteristics of predominantly unicellular growth and a sexual state not enclosed in a complex fruiting body. The criteria used for species delimitation initially relied on the morphology of vegetative and sexual stages, followed by definition of taxa through biochemical and genetic tests and finally, owing to partial failure of the foregoing to give unambiguous results, reliance on nucleic acid base sequence relatedness and allozyme comparison. At the same time several laboratories have studied the nutritional diversity of yeasts that were phenotypically similar by traditional criteria but shown to be distinct by molecular techniques. Several useful carbon compounds have been found that can reliably distinguish between some species that by traditional criteria would have been considered identical. In a few instances, where the habitat of a species is well defined, its ecology is the only criterion which distinguishes it from phenotypically similar species.

Kloeckera

Kloeckera apiculata (Ress emend. Klocker) Jane is the imperfect stage of *Hanseniaspora valbynesis*. It is found on grapes and associated with the flor of sherry.

Cells are lemon-shaped or oval, 5–8(10) × 2–4.5 μm in liquid media. There is a sediment and a thin surface-ring formation of pseudomycelium is rare. Only glucose, laevulose and mannose are fermented.

Debaryomyces

Debaryomyces Lodder and v. Rij nom conserv. (A de Bary) cells are round or short and oval. They propagate by multipolar budding, producing in liquid media a dry dull pellicle. Ascus formation is almost invariably preceded by conjugation, which is usually heterogamous: a cell conjugates with its own bud. Spores are round, usually with an oil-drop in the middle and sometimes finely warted.

Hansenula

Hansenula H. and P. Sydow grow as dry, wrinkled pellicles on the surface of the liquid. Their metabolism of sugars is more oxidative than fermentative.

Cells are of various shapes – round, oval and elongated, cylindrical with a strong tendency to form pseudomycelium. Ascospores are hemispherical, hat-shaped or angular. Species of *Hansenula* ferment vigorously and can utilize nitrates.

Importance to the Consumer

Ragi and ragi-like starters were obtained from China, India, Sikkim, Indonesia (Java and Bali), Malaysia, Nepal, Philippines and Taiwan. These starters have different names in each country (Murcha, bubood, Chinese yeast). The microorganisms were three genera of Mucorales (*Rhizopus, Mucor, Amylomyces*), yeasts and bacteria. *Amylomyces rouxis* was reported to survive long periods of time – up to 5 years at room temperature – and resulted in retained amylolytic activity. Chlamydospore germination was first observed in *Amylomyces*: these are assumed to be the spores that remain alive when cultures are dried in ragi and similar starters. The remarkable thing about the starters was the consistent occurrence of the filamentous yeast, with the exception of *Amylomyces*

from Nepal. Determining the number of colonies of *Amylomyces* is difficult because they resemble *Rhizopus* and spread rapidly in dilution plates.

See also: **Candida**: Introduction. **Debaromyces**. **Fungi**: Classification of the Eukaryotic Ascomycetes. **Hansenula**. **Saccharomyces**: *Saccharomyces cerevisiae*. **Schizosaccharomyces**. **Yeasts**: Production and Commercial Uses.

Further Reading

Berbee ML and Taylor JW (1992) Two ascomycete classes based on fruiting body characters and ribosomal DNA sequences. *Mol. Biol. Evol.* 9: 278–284.
Bessey EA (1950) *Morphology and Taxonomy of Fungi.* Philadelphia, USA: Blakiston.
Clements FE and Shear CL (1931) *The Genera of Fungi.* P. 227. Minneapolis, USA: H. W. Wilson Co.
Eriksson OE (1981) Bituricate families. *Opera. Bot.* 60: 1–220.
Eriksson OE (1995) DNA and Ascomycete systematics. Presented in V^th Mycological Congress. Abstract, p. 59.
Eriksson OE and Hawksworth DL (1993) Outline of the Ascomycetes. *Syst. Ascomycetum* 12: 51–257.
Fink G and Whitehead R (1995) Institute/MIT, Cambridge MA 02142. V^th Mycological Congress. Abstract, p. 64.
Hawksworth DL (1991) The fungal dimension of biodiversity: magnitude, significance and conservation. *Mycol. Res.* 95: 641–655.
Hawksworth DL (1993) *The Biodiversity of Microorganisms and Invertebrates. Its Role in Sustainable Agriculture.* Wallingford, Oxon: CAB International.
Hesseltine CW, Rogers R and Winario FG (1988) Microbiological studies on amylolytic oriental fermentation starters. *Mycopathologia* 101: 141–155.
Kreger-van Rij NWJ (1970) *Endomycopsis* Dekker, p. 166. Genus 15. *Pichia* Hansen, p. 455. In: Lodder J (ed.) *The Yeasts, A Taxonomic Study.* Amsterdam: North Holland Publishing.
Miller JH (1949) Revision of the classification of the Ascomycetes with special references to the Pyrenomycetes. *Mycologia* 41: 99–127.
von Arx (1979) In: Kendrick (ed.) *The Whole Fungus,* 1: 201.

FUSARIUM

Ulf Thrane, Department of Biotechnology, Technical University of Denmark, Lyngby, Denmark

The genus *Fusarium* is one of the most economically important fungi due to its plant pathogenic capabilities and production of potent mycotoxins which are known to affect animals and humans. In addition some species can also infect humans, especially the immunosuppressed. The genus occurs worldwide; however, not all species are cosmopolitan as some species predominate in cooler temperate regions, and others in tropical and subtropical regions.

During the last 60 years *Fusarium* taxonomy has

been influenced by two major schools: a European school with more than 65 species based on detailed morphological differences and an American school with only nine species. During the 1970s more species were accepted within the American school, which now includes scientists from South Africa and Australia, as some of the original nine species were found to be too broad. Within the last 20 years these schools have moved towards each other and today there is a general agreement among *Fusarium* taxonomists all over the world. Unfortunately, it is difficult to translate information based on one taxonomic system to the other due to differences in nomenclature. This is especially the case with the older literature, and has generated a lot of misunderstanding and misinformation in original as well as in review literature. Many reports on mycotoxin production are based on poorly identified fungal strains so it is often impossible to translate an identification into the present accepted taxonomic systems. This has been highlighted in recommended publications.

A single food product using *Fusarium* is known. Biomass of fermented *F. venenatum* (formerly identified as *F. graminearum*), Quorn™, is produced commercially as an ingredient for pies and similar products in the UK.

Characteristics

The food-borne *Fusarium* species are characterized by fast-growing colonies with a felty or floccose aerial mycelium. Colony pigmentation varies from pale, rose, burgundy to bluish violet depending on species and growth conditions. Conidia are often produced in sporodochia which appear as slimy dots in the culture. In some cultures sporodochia may be so prolific that they merge into a larger slimy layer. The typical *Fusarium* conidium (macroconidium) is fusiform, multi-celled by transverse septa, with a characteristic foot-shaped basal cell and a pointed to whip-like apical cell. In addition, some species may also produce minor conidia (microconidia). These are mostly single celled, in some cases three to five celled, and vary from globose, oval, reniform to fusiform. A few species produce microconidia in chains, the others in slimy heads or solitary.

In general, *Fusarium* species prefer humid conditions, i.e. water activity higher than 0.86, and grow well at temperatures of around 0–37°C, however no *Fusarium* species is thermophilic. Available data for common food-borne species are listed in **Table 1**.

Fusarium avenaceum is probably the most frequent *Fusarium* species on cereal grains in temperate climates. Even though this species is common its effect on the quality of grains has not been reported. The typical mycotoxin problems in grains (trichothecenes) may not be due to *F. avenaceum* as it has not been proved that this species produces these metabolites. A lot of variability in mycelium texture, pigmentation and metabolite profiles has been observed among cultures of *F. avenaceum*. Independent reports also show genetic variability among strains, however at the moment more information is needed before it can be determined whether the present species should be split into more.

Fusarium cerealis is a synonym of *F. crookwellense*. This nomenclature change has not been fully accepted, and is still under debate among *Fusarium* experts, so both names may occur in the literature. Agar cultures of *F. cerealis* are very similar to those of *F. graminearum* and *F. culmorum* in terms of pigmentation, mycelium texture and growth rates. Morphologically, only minor differences separate *F. cerealis* and *F. graminearum*, whereas *F. culmorum* is distinctive. The three species have many metabolites in common, but minor differences in trichothecene derivatives and quantitative differences have been published. However, only a limited number of strains have been investigated. Species-specific DNA primers which can be used for identification of these species have been published.

Fusarium culmorum occurs frequently on cereal plants and grains in temperate climates, and is a potent producer of deoxynivalenol and other trichothecenes. These metabolites have a role in *F. culmorum* rot in cereal plants, and have also been detected in grains and cereal products. Agar cultures of *F. culmorum* appear similar to cultures of *F. cerealis* (q.v.) and *F. graminearum*, however, species-specific DNA primers for identification have been published.

Fusarium equiseti is one of the few *Fusarium* species that never becomes red in culture. At the moment *F. equiseti* is based on a broad species concept where strains from different substrates and temperature regions have pronounced variation in mycotoxin profiles and morphological features. *F. equiseti* is mostly recognized as a secondary invader in agricultural crops after contamination with soil in which it can survive for years as it produces abundant resistant chlamydospores.

Fusarium graminearum is abundant on cereals and its teleomorphic stage, *Gibberella zeae*, is found in nature in warmer to tropical regions. This species can cause severe damage to cereal plants and crops by different types of rots and mycotoxins. *F. graminearum* is believed to cause gushing in beers. In this and many other respects it is similar to *F. culmorum* (q.v.) and *F. cerealis* (q.v.). Species-specific DNA primers for identification of *F. graminearum* have been published. Until recently the producer

Table 1 Important food-borne *Fusarium* species, their habitat (food related), mycotoxins and other biological active metabolites and physiological characteristics

Fusarium species	Habitat	Mycotoxins[a]	Colony diameter[b]	Miscellaneous
F. avenaceum	Worldwide: Cereals, peaches, apples, pears, potatoes, peanuts, peas, asparagus, tomatoes	Antibiotic Y, aureofusarin, chlamydosporol, chrysogine, enniatins, fusarin C, moniliformin	PSA: 30–59 mm TAN: 3–20 mm	Temperature: −3–31°C Min a_w 0.90 at 25°C
F. cerealis	Worldwide: Cereals, potatoes	Aureofusarin, butenolide, chrysogine, culmorin, fusarin C, nivalenol, zearalenone	PSA: 75–90 mm TAN: <2 mm	
F. culmorum	Mainly temperate region: Cereals, potatoes, apples, sugar beet	Aureofusarin, butenolide, chrysogine, culmorin, deoxynivalenol, fusarin C, nivalenol, zearalenone	PSA: 75–90 mm TAN: <2 mm	Temperature: 0–31–(35)°C Min a_w 0.87 at 25°C Tolerates low oxygen tensions
F. equiseti	Worldwide: Cereals and fruits contaminated with soil, vegetables, nuts, spices, UHT-processed juices	Chrysogine, diacetoxyscirpenol, equisetin, fusarochromanone, nivalenol, zearalenone	PSA: 45–69 mm TAN: 3–30 mm	Temperature: −3–35°C Min a_w 0.92 at 25°C Growth at pH 3.3–10.4 Tolerates low oxygen tensions
F. graminearum	Worldwide: Cereals and grasses	Aureofusarin, butenolide, chrysogine, culmorin, deoxynivalenol, fusarin C, nivalenol, zearalenone	PSA: 75–90 mm TAN: <2 mm	Min a_w 0.90 at 25°C Min pH 2.4 at 30°C Max pH 10.2 at <37°C
F. oxysporum	Worldwide: Cereals, peas, beans, nuts, bananas, onions, potatoes, citrus fruits, apples, UHT-processed juices, spices, cheese	Fusaric acid, moniliformin, naphthoquinone pigments	PSA: 30–55 mm TAN: 5–40 mm	Temperature: 5–37°C Can grow anaerobically pH range 2.2–9 Min a_w 0.89 at 25°C
F. poae	Temperate regions: Cereals, soybeans, sugar cane, rice	Butenolide, fusarin C, γ-lactones, nivalenol, T-2 toxin	PSA: 55–85 mm TAN: <2 mm	Temperature: 2.5–33°C Min a_w 0.90 at 25°C
F. proliferatum	Warmer to tropical regions: Corn, rice, figs, fruits	Beauvericin, fumonisins, fusaproliferin, fusaric acid, fusarin C, moniliformin, naphthoquinone pigments	PSA: 35–55 mm TAN: 5–25 mm	Temperature: 2.5–35°C Min a_w 0.92 at 25°C
F. sambucinum	Worldwide: Cereals, potatoes	Aureofusarin, butenolide, diacetoxyscirpenol, enniatins	PSA: 34–59 mm TAN: 5–29 mm	Temperature: 2–32.5°C
F. semitectum	Warmer to tropical regions: Nuts, bananas, citrus, potatoes, melons, tomatoes, spices	Beauvericin, equisetin, fusapyrone, zearalenone	PSA: 45–69 mm TAN: 5–15 mm	Temperature: 3–37°C
F. solani	Worldwide: Fruits and vegetables, spices	Fusaric acid, naphthoquinone pigments	PSA: 25–50 mm TAN: 5–40 mm	Temperature: up to 37°C Can grow anaerobically Min a_w 0.90 at 20°C

Table 1 Important food-borne *Fusarium* species, their habitat (food related), mycotoxins and other biological active metabolites and physiological characteristics (*continued*)

F. sporotrichioides	Worldwide: Cereals, pome fruits	Aureofusarin, butenolide, fusarin C, T-2 toxin	PSA: 65–90 mm TAN: 4–20 mm	Temperature: −2–35°C Min a_w 0.88 at 20°C
F. subglutinans	Worldwide: Corn, pineapple, bananas, spices, sorghum	Beauvericin, fusaproliferin, fusaric acid, moniliformin, naphthoquinone pigments, subglutinols	PSA: 35–55 mm TAN: 5–40 mm	Temperature: 2.5–37°C
F. tricinctum	Worldwide: Cereals	Antibiotic Y, aureofusarin, butenolide, chlamydosporol, chrysogine, fusarin C, visoltricin	PSA: 32–55 mm TAN: 3–20 mm	Temperature: 0–32.5°C
F. venenatum	Temperate region: Cereals, potatoes	Aureofusarin, butenolide, diacetoxyscirpenol	PSA: 45–60 mm TAN: 5–25 mm	Temperature: 2–35°C
F. verticillioides	Subtropical and tropical region: Corn, rice, sugarcane, bananas, asparagus, spices, cheese, garlic	Fumonisins, fusaric acid, fusarin C, moniliformin, naphthoquinone pigments	PSA: 35–55 mm TAN: 5–40 mm	Temperature: 2.5–37.5°C Tolerates > 15% NaCl Min a_w 0.87 at 25°C Can grow anaerobically

[a]Only major components listed, derivatives may also be produced.
[b]Colony diameter on PSA (potato sucrose agar) after 4 days at 25°C; on TAN (tannin–sucrose agar) after 7 days at 25°C.

strain of Quorn™ used in precooked pies was thought to be *F. graminearum*, however, a multidisciplinary study has re-identified the strain as *F. venenatum*.

Fusarium oxysporum is well known as a plant pathogen causing severe damage in many agricultural crops, both in the field and during post-harvest storage. Strains of *F. oxysporum* can grow under very low oxygen tensions, and have often been detected as a re-contaminant in ultra-high temperature (UHT) processed fruit juices.

Fusarium poae is frequently detected in cereals from the temperate region. There have been scattered reports on mycotoxin production by this species, and it is evident that *F. poae* can produce several trichothecenes in cereals. Agar cultures of this species often produce strong aromatic peach-like volatile compounds.

Fusarium proliferatum is characterized by producing microconidial chains borne from polyphialides. Together with the other *Fusarium* species with conidial chains, *F. verticillioides*, this species is very common on corn and other cereals from warm to tropical regions. Many fruits and vegetables can also be damaged by this species, and it is a potent producer of fumonisins and other mycotoxins. In older literature this species was mixed up with *F. moniliforme* which was based on a broad species concept. This is now recognized as outdated and *F. proliferatum*, *F. subglutinans*, *F. verticillioides* and other less important species are now recognized as

individual species. They were all covered by *F. moniliforme*.

Fusarium sambucinum is important to potato breeding and production as this species can cause severe rot. Variation in resistance to thiobendazole fungicides has been observed among strains of this species. Species-specific DNA primers for identification of *F. sambucinum* have been published.

Fusarium semitectum is a species complex covering species of great importance to agricultural crops from tropical and subtropical regions. Other species epithets used in recent reports are *F. incarnatum* and *F. pallidoroseum*, but the taxonomy of this complex has not been settled.

Fusarium solani is a very important soil-borne plant pathogen which can cause damage to many agricultural crops. No known mycotoxins are produced by this species. In agar cultures the teleomorph, *Nectria haematococca*, may be produced but it is very strain dependent.

Fusarium sporotrichioides is a well-known producer of T-2 toxins and other trichothecenes occurring on cereals, especially in temperate regions. It is detected with minor frequency, but due to its potent production of mycotoxins it is still important.

Fusarium subglutinans is frequently found on corn, cereals and fruits from warmer regions. In many respects this species is similar to *F. proliferatum* and *F. verticillioides*, which produce microconidial chains, however, *F. subglutinans* only produces microconidia in slimy heads. In older literature they were all merged into one species, *F. moniliforme*.

Fusarium tricinctum is very common on cereals from temperate regions. Although trichothecene mycotoxins have often been reported from this species, this has never been proved. *F. tricinctum* was formerly used as a collective species epithet for several current species, including *F. poae* and *F. sporotrichioides* which may account for the mycotoxin reports. The effect of *F. tricinctum* occurrence on the quality of the grains is unknown.

Fusarium venenatum is a recently discovered and described species and only a limited amount of information is available. The Quorn™-producing strain has been re-identified as *F. venenatum*. Species-specific DNA primers for identification of *F. venenatum* have been published.

Fusarium verticillioides is the legal synonym for *F. moniliforme* which may also cover *F. proliferatum* and *F. subglutinans*, as well as several species of lesser importance. It has now been agreed among *Fusarium* taxonomists that the name *F. verticillioides* should be used and *F. moniliforme* avoided. *F. verticillioides* which can be recognized by its long microconidial chains borne on monophialides, is a potent producer of fumonisins and other mycotoxins. The teleomorphic *Gibberella* stage can be found on old corn stalks in warmer climates.

Methods of Detection

A number of techniques have been developed for the detection of *Fusarium* in food and other agricultural products, but the focus here is on conventional methods based on agar substrates. The three *Fusarium* selective media mentioned can be used by direct plating of subsamples (e.g. grains) or by dilution plating of homogenates. Plates are incubated at 25°C for 5–7 days under an alternating light regime to enhance sporulation.

Czapek-Dox–iprodione–dichloran agar (CZID), dichloran–chloramphenicol–peptone agar (DCPA), peptone pentachloronitrobenzene (PCNB) agar (PPA) have all been used widely. Differences in selectivity for *Fusarium* between the three media are insignificant. On all media only *Fusarium* cultures are supposed to grow quickly, but care should be taken. If just a small amount of sample is present, or in the case of direct plating, other fungal genera may be able to grow as they will be supported by the nutrition from the sample, thus eliminating the inhibition provided by the *Fusarium* selective medium. Inspection for *Fusarium* conidia is required at this stage or in the following purified culture on a low nutritional medium Spezieller Nährstoffarmer Agar (SNA). For further subculturing low nutritional media are required as *Fusarium* cultures will change irreversibly by prolonged culturing on rich substrates.

Based on a collaborative study using dilution plating CZID is preferred as it is possible to differentiate *Fusarium* species by differences in cultural appearance on this medium, however more data on direct plating are needed before a general recommendation can be made. PCNB is reported to be carcinogenic, and is thus best avoided. CZID is recommended internationally by the Nordic Committee on Food Analysis (Proposal no. 154, 1996) and by the International Commission on Food Mycology.

It should be noted that isolation media for fungi containing rose bengal (e.g. dichloran-rose bengal-chloramphenicol (DRBC) and rose bengal-glucose (RBG) agars) should not be used for *Fusarium* because the light during incubation will convert rose bengal into a potent fungicidal compound which will totally inhibit fungal growth.

Other detection methods based on immuno techniques and molecular biology, which have been shown to be fast and useful, will probably be more widely used in future when problems of cross-reaction and lack of species specificity have been solved. A drawback with these methods is that an array of probes to cover the relevant *Fusarium* species is required, in contrast to conventional isolation methods on agar substrates where purification and culturing is needed prior to identification. The latter is more time consuming but still widely used.

Methods of Identification

A number of identification routines are available of which an operational identification key '*Introduction to food-borne fungi*' by Samson et al (1995) is recommended. For identification of *Fusarium* species the purified SNA cultures should be inoculated on a fresh SNA dish for morphological observations, and on potato sucrose agar (PSA) (or potato dextrose agar (PDA)) for determination of colony diameter after 4 days at 25°C and cultural appearance. As thin layer chromatographic (TLC) analysis of agar plugs from fungal cultures is recommended for identification of food-borne fungi *Fusarium* should also be grown on yeast extract–sucrose agar (YES) for 10–14 days at 25°C. Agar plugs from PSA and YES cultures should then be used for TLC.

Metabolite profiles are not only important for identification of strains, they are also very descriptive characters of an individual strain. For this purpose chromatographic techniques such as high performance liquid chromatography with diode array detection (HPLC-DAD) or gas chromatography with mass spectrometry (GC-MS) may be used in addition to TLC.

DNA probes for species-specific detection (i.e. identification) of *Fusarium* species have been published, however, they have not been used widely. Most reports have been on detection of specific plant pathogenic *Fusarium* species in agricultural crops. Such methods are very promising and without doubt they will be more available in the future.

Regulations

The brewing industry has regulations on *Fusarium* content in malting barley, and most other food industries have regulations on the mycotoxins produced by *Fusarium*. The latter are often identical to governmental (international) regulations. Each country has their own set of regulations on mycotoxins, and some general agreements have been made. Among *Fusarium* mycotoxins, trichothecenes (deoxynivalenol) and fumonisins are of most concern, but there is an increasing understanding that a synergistic effect between several co-existing mycotoxins takes place, and this may be even more important than the individual mycotoxins. Other fungal metabolites, so far not known as mycotoxins, also play an important role in this context.

See also: **Mycotoxins**: Classification. **Spoilage of Plant Products**: Cereals and Cereal Flours. **Spoilage Problems**: Problems caused by Fungi.

Further Reading

Burgess LW, Summerell BA, Bullock S, Gott KP and Backhouse D (1994) *Laboratory Manual for* Fusarium *Research*, 3rd edn. Sydney: University of Sydney.

Marasas WFO, Nelson PE and Toussoun TA (1984) *Toxigenic* Fusarium *Species. Identity and Mycotoxicology*. University Park and London: The Pennsylvania State University Press.

Nelson PE, Toussoun TA and Cook RJ (1981) Fusarium: *Diseases, Biology and Taxonomy*. University Park and London: The Pennsylvania State University Press.

Nelson PE, Toussoun TA and Marasas WFO (1983) Fusarium *Species. An Illustrated Manual for Identification*. University Park and London: The Pennsylvania State University Press.

Pitt JI and Hocking AD (1997) *Fungi and Food Spoilage*, 2nd edn. London: Blackie Academic & Professional.

Samson RA, Hoekstra ES, Frisvad JC and Filtenborg O (eds) (1995) *Introduction to Food-borne Fungi*. 4th edn. Baarn: Centraalbureau voor Schimmelcultures.

Thrane U (1986) The ability of common *Fusarium* species to grow on tannin–sucrose agar. *Letters in Applied Microbiology* 2: 33–35.

Thrane U (1996) *Fusarium*: Determination in foods and feedstuffs. *Nordic Committee on Food Analysis Proposed Method* 154: 6 pp. Helsinki: VTT.

Thrane U (1996) Comparison of three selective media for detecting *Fusarium* species in foods: a collaborative study. *International Journal of Food Microbiology* 29: 149–156.

Yoder WT and Christianson LM (1998) Species-specific primers resolve members of *Fusarium* section *Fusarium* – Taxonomic status of the edible 'Quorn' fungus reevaluated. *Fungal Genetics and Biology* 23: 68–80.

Gastric Ulcers *see Helicobacter.*

GENETIC ENGINEERING

Contents
Modification of Yeasts and Moulds
Modification of Bacteria

Modification of Yeasts and Moulds

R Sandhir, Department of Biochemistry, Dr Ram Manohar Lohia Avadh University, Faizabad, India

S K Garg and **D R Modi**, Department of Microbiology, Dr Ram Manohar Lohia Avadh University, Faizabad, India

The beneficial activities of yeasts and other fungi are of great economic importance. They have long been exploited as food, in food processing, and in brewing. During the 20th century, as the fermentation industry has developed, they have yielded an increasing range of valuable products, including antibiotics, vitamins and various enzymes. However, some fungi can also cause immense economic losses. Recently, with the advent of recombinant DNA technology, fungi have been exploited to produce hormones and proteins which were hitherto only available from mammals. The performance of fungus in an industrial process depends on its genotype under the conditions of the fermentation process. To maximize yield in a process, it is essential that new strains are continuously developed. This involves using genetic techniques combined with selection procedures designed to screen for the desired improvement. This article describes the methods and techniques employed for establishing novel genotypes in various yeasts and moulds.

Techniques for Natural Strain Selection

In nature, there is a great diversity of fungal strains. The successful isolation of a fungal strain from nature is dependent on well-planned and often ingenious screening procedures. The 'seek and discard principle' is frequently called screening and even today represents the beginning of any process of strain improvement. More often the search for strains producing desired products entails making dilute solutions of soil or decaying plant material and carrying out isolations in a way that indicates the desired properties. The organisms that are obtained may be moulds or yeasts but can equally well be actinomycetes. Sometimes what is sought, as in the production of single-cell protein, is an organism capable of utilizing a particular type of substrate, such as petroleum hydrocarbons or starch. Although extensive tests are required to ensure that the protein is of high quality and that the toxic substances are absent, the first step is isolation of organisms from appropriate sites that are able to utilize the selected substrate. The starch-utilizing *Fusarium* that is used to produce mucoprotein, for instance, was isolated from starch-rich effluents.

An enormous range of substances occur in the soil, and the organisms capable of degrading them are also present, but in correspondingly smaller numbers. Their numbers can be accordingly increased prior to attempting isolation by enriching the soil sample with an appropriate substrate and then incubating. The agar medium on which diluted samples are spread will contain a proper carbon source, for example starch, if starch-utilizing organisms are being sought. In this instance, not only is the growth of starch-metabolizing organisms encouraged, but their presence is indicated by a clear circular zone around the colonies, which results from diffusion of amylase from the colonies and the hydrolysis of opaque starch. The *Aspergillus* species used for amylase production is an example of an organism isolated using this technique. Organisms that produce enzymes such as pectinase, cellulase and protease that degrade other macro-molecules have been detected and isolated by similar methods. Production of organic acids by micro-organisms, such as citric acid, can be detected by including calcium carbonate in the medium, which must not, however, contain salts such as ammonium chloride, where the uptake of cation causes the production of mineral acid. Organisms producing an excess of a vitamin can be detected through stimulation of growth of an auxotroph (tester organism) which is unable to synthesize the vitamin; the test organism is seeded into or on to a layer of agar added after the growth of the organism being screened. The production of an antibiotic by a microorganism can be recognized by the presence of a clear zone around the colony, thereby indicating its inhibitory effect. The colony can then be isolated and its ability to inhibit the growth of selected pathogens determined. Although, in the past, this procedure has yielded many useful antibiotics, it now tends to result only in the rediscovery of antibiotics that are already known.

Methods of Genetic Improvement

In biotechnology, as soon as a microorganism is available that forms a certain useful product, the aim exists to increase its yield. This can be achieved, on the one hand, by changing or improving the growth conditions (process optimization) and/or, on the other hand, by genetic manipulation.

Mutagenesis and Selection

Mutagenesis and selection of improved strains have been the most extensively used methods for introducing beneficial changes into industrially important fungi. Mutations are heritable changes which occur in the genome as a result of one or many types of events and may involve individual nucleotide bases or large regions of chromosomes. For a specific gene, the frequency of spontaneous mutation is very low and would be expected to occur 1 in 10^5–10^6 nuclei. Considering the whole genome, such spontaneous mutations are much less rare and provide the basis for the overall genetic variation found in different isolates or strains of fungus. The low frequency of spontaneous mutations of individual genes is clearly not adequate to meet the constant demands of improved strains for a fermentation process. To meet this need, it is necessary to induce mutations using a mutagenic agent which can be either a physical or chemical agent. Under optimal conditions, a mutagen will increase the mutation frequency of a particular gene by at least one order of magnitude. The physical and chemical mutagens used frequently for strain improvement in fungi and their effects on the DNA molecule are summarized in **Table 1**.

The primary effect of a mutagen is to induce a lesion or modification of the base sequence of the DNA molecule. A mutation arises if the alteration to the molecule is not repaired by the host repair mechanisms. Our understanding of these repair processes is based on studies in *Escherichia coli* and *Saccharomyces cerevisiae*. In the former case, it is apparent that the repair of damaged DNA is important in the maintenance of viability because more than 1% of the genome is composed of genes concerned with the function. The mutations that constitute the steps in strain improvement programmes commonly do not affect the morphology, and increase the product yield by not more than 10–15%. Sometimes a mutation may result in a much larger increase in yield, but such a mutation is often accompanied by changes in morphology and behaviour that require extensive modification of the medium and fermentation process. Such infrequent major mutations tend to be less useful than minor mutations, the cumulative effect of which may be more impressive. Interestingly, the haploid status of most fungi ensures that mutations are expressed soon after they are produced.

The synthesis of primary metabolites (such as an amino acid) is controlled in such a way that it is produced in an amount required by the organism. This is because of the inhibition of the enzyme activity and repression of the enzyme synthesis by the end product and such a control is referred to as feedback

Table 1 Mutagens effective against fungi and their mode of action

Mutagen	Action responsible for mutagenesis
Physical mutagens	
X-rays	Single- and double-strand breaks in the phosphodiester backbone. Point mutations because of deamination and dehydroxylation of bases
Ultraviolet light	Pyrimidine dimer formation
Chemical mutagens	
Nitrous acid	Reacts with primary amino group to form hydroxyl group (adenine to hypoxanthine, guanine to xanthine and cytosine to uracil, causing alterations in subsequent replication)
Alkylating agents (dimethylsulphonate, etc.)	Causes alkylation of N and O atoms of bases. Alkylation leads to mispairing of alkylated bases
Acridines (acridine mustard)	Acridine mustard is comprised of an acridine ring and alkylating moiety. The former intercalates in base pairs, causing disruption of base pairing

control. It is obvious that a good commercial mutant should lack the control system and lead to over-production of the product. Mutagenesis has been successful in improving the yield of most of the antibiotics, including penicillin. Mutation frequency has been enhanced in microorganisms by inducing mutations in DNA repair mechanisms.

Strains giving improved performance have to be selected following mutagenesis and this is governed by the type of mutation desired. A variety of selection methods are available for nutritionally defective mutants, resistant mutants, temperature-sensitive and similar types of mutations. The original approach was random screening. Samples of the cells that had been treated by the mutagen were spread on the agar medium and the colonies arising from the single cells were isolated. The isolates were then grown in liquid medium and the culture filtrate assayed for the required product. Such a procedure is tedious and labour-intensive, since only a small percentage will show improvement in yield. A reduction in labour can be achieved where it is possible to detect improved performance on agar media, employing methods similar to those employed for initial detection of products of interest. Alternatively, indirect screening methods are sometimes possible, where what is estimated is not the amount of desired product, but some readily determined attribute likely to be correlated with the amount of desired product. For example, sensitivity to sodium monofluoroacetate is correlated with accumulation of citric acid. This is because the compound inhibits the enzyme aconitate hydratase which converts citric acid to isocitric acid prior to its further metabolism. Hence, a mutant with a low amount of this enzyme will be more sensitive to the sodium monofluoroacetate and will also be one in which citric acid tends to accumulate.

Recombination

Recombination offers an alternative approach for genetic modification and is achieved through either sexual or asexual reproductive processes. Employing genetic recombination, one can obtain the desirable features of two different strains involved in recombination. Recombination between whole genomes or involving small fragments of DNA provides additional approaches to the generation of strains with novel phenotypes that can be included in the strain improvement programmes. Whole genome recombination occurs as a result of sexual and parasexual processes, and recombination involving small fragments is achieved through transformation methods in vivo.

Sexual Hybridization   Fungi exhibit a true sexual process of fusion of two parental cells to form a heterokaryon, containing nuclei of both parental types in a common cytoplasm. These nuclei exist separately, so recombination does not occur at this stage; complementation can, however, be demonstrated at this stage. The duration of the hetero-karyotic stage varies from ephemeral to indefinite (depending on the species) and is succeeded by fusion of two nuclei to give a true diploid cell, usually within a specialized structure. In yeasts the diploid stage may persist, but in filamentous fungi the formation of diploid nucleus is followed by meiosis with the production of haploid spores. The four products of meiosis are retained in a structure called a tetrad. Most fungi of industrial importance reproduce only by asexual sporulation. However, the notable exceptions to the rule are *S. cerevisiae* and *Agaricus bisporus*, the common white mushroom. Other sexually reproducing fungi of industrial interest are *Claviceps purpurea* and *Mucor* species. In both of them, the reproductive process cannot be easily manipulated in the laboratory. Spore germination is poor in *A. bisporus*, and about 70% of the spores are self-fertile, due to the presence of nuclei of both the mating types in a usually binucleate spore. In spite of this, homokaryons of both mating types can be obtained, and mated to yield new dikaryotic strains that give fruiting bodies. However, strains employed in brewing often sporulate poorly or not at all; spores, if obtained, may be difficult to germinate, and mating of the resultant haploids may not be easy. Despite all this, mating has been used in strain improvement. For example, mating among haploids obtained from a diploid lager yeast gives some diploids with better flocculation and lower diacetyl production, both desirable features that occur with the parent strain. Here the beneficial results have been obtained through the elimination of undesirable dominant genes in the original diploid; the introduction of new genetic material requires mating with the haploid progeny of a diploid strain.

Parasexual Hybridization   Many species of *Aspergillus*, *Penicillium* and other fungi of genetic importance lack a sexual cycle. They are, however, able to undergo a parasexual cycle. The first step in this cycle is the formation of a heterokaryon which is accomplished by fusion of protoplasts of the two strains. The next step, vegetative diploidization, is normally a rare event, but the frequency can be greatly increased by exposing to camphor or ultraviolet radiation. Other treatments can be used to increase both the frequency of recombination between homologous chromosomes and of haploidization, the return to haploid strain. The strains are selected by growing the protoplasts in the presence of required nutrients and forcing them to grow under conditions which support the development and growth of the hetero-karyon and selection against the two parental strains. Nutritionally complementing auxotrophic strains are generally used and a minimal medium provides the required selective conditions to promote hetero-karyosis. The parasexual cycle has proved useful in bringing about genetic recombination between closely related industrial strains, for example strains of *Penicillium chrysogenum* with high yields of penicillin and ancestral strains with lower yields but superior growth and sporulation have been obtained. Vegetative compatibility hinders heterokaryon formation between distantly related strains; although protoplast fusion can to some extent by-pass the barrier, there has been rather little success in obtaining recombinants between such strains.

The major differences between the sexual and para-sexual processes are summarized in **Table 2**. The in vivo gene manipulations have been successfully used to increase enzyme production, for overproduction of primary and secondary metabolites.

Recombinant DNA Technology

The more recent technology of DNA recombination in vitro provides a new and more direct method for manipulating the fungal genome. The methods of strain improvement discussed in previous sections are uncertain in their effects. A mutation that has brought about the desired genetic change in a strain may at the same time cause other mutations that are dis-advantageous. Although gene cloning is more time-consuming as well as labour-intensive, its value has been recognized, and now figures in the strategies of strain improvement adopted for many products by most major companies involved in fungal fer-mentation technology. The various steps involved in recombinant DNA technology are described and illus-trated in **Figure 1**.

Obtaining DNA Sequences/Gene of Interest   The first step in gene cloning is isolation of the desired gene. This involves extraction of DNA from the donor animal, plant or microbe, and treatment in various ways with restriction enzymes to obtain the fragment of DNA or the desired gene. If the gene to be cloned in fungi is of mammalian or eukaryotic origin, it may contain one or more introns (non-coding intervening sequences). The nucleotide sequence of such a gene is transcribed into RNA molecules that need to be edited

Table 2   Comparison of sexual and parasexual mechanisms of recombination

Sexual	Parasexual
Specialized cells are involved in hyphal anastomosis (plasmogamy), rarely vegetative cells	Vegetative cells are involved in hyphal anastomosis (plasmogamy)
Heterokaryon formation controlled by compatibility genes in vegetative plasmogamy	Heterokaryon formation controlled by incompatibility genes
Diploid phase is restricted to a single nuclear generation. The diploid nucleus is spatially separated in a specialized cell	Diploid nucleus is a somatic nucleus and persists through many mitotic divisions and can be maintained by employing suitable selection
Chromosomes are involved in multiple cross-over during meiosis	Cross-over during mitotic divisions is rare, and normally involves one arm of a pair of chromosomes
Haploidization is achieved by meiosis, the meiotic products being incorporated into spores	Mitotic non-disjunction results in aneuploidy, ultimately yielding haploid status
Products of meiosis are readily recognized and isolated	Recombinants occur among vegetative cells and are identified by use of suitable markers

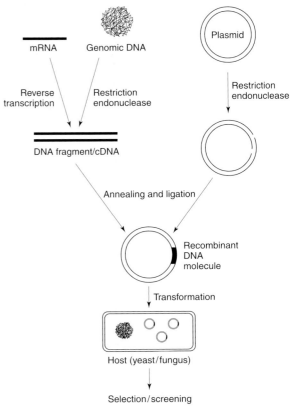

Figure 1   Scheme for gene cloning in yeasts and fungi.

before being passed from nucleus to cytoplasm for synthesis of protein. Prokaryotes do not have the necessary mechanisms for processing eukaryotic RNA and therefore have to be provided with DNA sequences complementary with mRNA. Fungi, being eukaryotes, have the mechanisms required for RNA processing and production of heterologous (mammalian) proteins. Therefore, a gene having introns can be cloned in fungi. However, the fungal genome, having few introns, does not have identical mechanisms for processing DNA as mammals or other eukaryotes and hence complementary DNA (cDNA) is still a better choice. This involves isolation of mRNA for the desired gene and copying to double-stranded cDNA using the enzyme reverse transcriptase. The amount of DNA obtained for cloning can be amplified using the technique of polymerase chain reaction.

Introduction of DNA into Fungal Vectors A prerequisite for gene cloning is the availability of a suitable vector – a DNA molecule that will survive and replicate in the selected host and into which the gene to be cloned can be inserted. Numerous plasmids for fungal transformation have been constructed. The basic components of these plasmids are first, a gene that can be used for selection of fungal transformations, and second, bacterial plasmid sequences that can be used for selection and propagation of the plasmid in *E. coli*. The number of different selectable markers available for fungi is now fairly large and these selective genes can be used in many species, thus providing substantial flexibility in designing plasmids for specific purposes. Selective markers can be divided into at least three functional groups: genes that complement pre-existing mutations and lead to prototrophic growth (*auxotrophic* markers); genes that provide a new function and lead to drug resistance or growth on previously non-utilizable nutrient source (*drug-resistant* or *added-function* markers) and DNA fragments that give rise to selectable mutations when integrated into the genome of the recipient strain at specific locations (*mutagenic* markers). Each type of marker has specific uses. One of the most commonly used marker in *S. cerevisiae* is the *URA3* gene which encodes orotidine-5′-phosphate decarboxylase, an

enzyme of uracil biosynthesis. Acquisition of a functional *URA3* gene by the cells that were previously mutant at this locus allows them to grow in the absence of exogenous uracil. Examples of other frequently used selectable markers in *S. cerevisiae* that can be selected in a similar way are given in **Table 3**. *TUN®* confers resistance to the antibiotic tunicamycin, an inhibitor of glycosylation. Lack of *SUP4* (an ochre chain termination suppressor) can be selected in suitable host backgrounds. One such host has an ochre chain termination mutation in the *CAN1* gene. The wild-type gene encodes a permease that causes the uptake of the toxic arginine analogue canavanine and, consequently, causes cell death in the presence of canavanine. So *CAN1* mutant cells are resistant to canavanine, as they are unable to take it up. However, the chain termination suppressor *SUP4* causes the production of functional permease in the *CAN1* background (because it suppresses the *CAN1* mutation) and, therefore, causes the death of *CAN1* cells in the presence of canavanine. Loss of *SUP4* abolishes canavanine uptake and canavanine resistance, which can be readily selected. Cells containing *SUP4* can also be identified by the use of hosts with a chain termination mutation in the *ADE2* gene for phosphoribosyl amino-imidazole carboxylase, a component of the arginine biosynthesis pathway. The mutation causes the cells to accumulate a red pigment, whereas colonies are of the usual colour if they are wild-type or if the ochre mutation is suppressed by *SUP4*. Cells that have lost *SUP4* therefore acquire the red pigment and are visibly distinguishable from the others. If host strain without *LEU2* is chosen, and a medium lacking leucine is employed, only cells transformed by the vector having *LEU2* marker will grow.

Several selectable markers are available for filamentous fungi and the common examples are given in **Table 4**. The *argB* marker is involved in arginine biosynthesis and needs a host deficient in arginine biosynthesis. The other markers are responsible for conferring resistance to variety of antibiotics.

The yeast *S. cerevisiae* has been extensively used for gene cloning, and a variety of vectors have been employed. Most vectors used in yeast are *shuttle* vectors as they are able to replicate in a second species,

Table 3 Selectable markers used in *Saccharomyces cerevisiae*

Marker	Function	Pathway/phenotype
URA3	Orotidine-5′-phosphate decarboxylase	*de novo* synthesis of pyrimidines
HIS3	Imidazole glycerol phosphate dehydratase	Histidine biosynthesis
LEU2	β-Isopropylmalate dehydrogenase	Leucine biosynthesis
TRP1	N-(5′-phosphoribosyl) anthranilate isomerase	Tryptophan biosynthesis
TUN®	UDP-N-acetylglucosamine-1-P transferase	Glycosylation; confers tunicamycin resistance
SUP4	Tyrosine-tRNA	Ochre suppressor

Table 4 Selectable markers used in filamentous fungi

Marker	Function	Pathway/phenotype
argB	Ornithine transcarbamylase (from *Aspergillus nidulans*)	Synthesis of arginine
hph	Hygromycin phosphotransferase (from *Escherichia coli*)	Hygromycin resistance
benA	β-Tubulin (from *A. nidulans*)	Resistance to fungicide benomyl
amdS	Acetamidase (from *A. nidulans*)	Growth on acetamide as sole carbon or nitrogen source
ble	Bleomycin-binding protein (from *Streptoalloteichus hindustanus*)	Bleomycin resistance (binds to bleomycin and inhibits its DNA-damaging activity)

E. coli. Initial cloning is done in *E. coli* and then the recombinant DNA is transformed into yeast cells. The plasmid vectors most commonly used in yeast can be divided into two categories: *yeast centromeric plasmid* (YCp) and *yeast episomal plasmid* (YEp) vectors. An example of a YCp plasmid is given in **Figure 2**. It contains prokaryotic ampicillin and tetracycline resistance genes, and an origin of replication from *E. coli*, all in a region of the plasmid derived from pBR322. These will allow it to function as a shuttle vector. Many of these vectors are based on pBR322 in this way, but other plasmids have also been used, including pUC and bluescript species. YCp vectors also have a selectable marker and an origin of replication for yeast, *ARS1*. ARS, autonomously replicating sequences, are the sequences that confer on the plasmid the ability to replicate without integration into another replicon. The distinguishing feature of YCp plasmids is the presence of a chromosomal centromeric sequence, *CEN4* in the present example. The

presence of a centromeric sequence allows the plasmid to be partitioned by the spindle apparatus of the cell, in the same way that the endogenous chromosomes are partitioned. This also results in the reduction of copy number per cell, but confers a much greater stability during cell division, allowing the plasmid to be maintained in the absence of selection.

An example of YEp plasmid is shown in **Figure 3**. An episomal plasmid is a plasmid that can replicate independently but can also be integrated into the chromosome. YEps may do this as a result of sequence homologies between the selectable marker gene of the vector and mutant allele of the host, allowing recombination. Most of the features of YEp are similar to those of YCp plasmids, except for the absence of the centromeric sequence and its replacement with a different origin of replication. YEp has been constructed from a naturally occurring episomal plasmid, 2μ plasmid of *S. cerevisiae* using restriction enzyme and DNA ligase. YEps are very effective in bringing out transformation and occur in high copy numbers. Hence, these can have high rates of mRNA transcription and a high output of protein specified by the cloned gene. Apart from a naturally occurring origin of replication, YEps also have a naturally occurring *cis*-acting sequence, *REP3*, which is the site of action of proteins that help in partitioning the plasmid during cell division. This gives YEp plasmids a mitotic stability similar to YCp vector, but with a somewhat high copy number. The 2μ plasmid has the genes essential for DNA replication and maintenance of numerous copies – up to about 100 – in the yeast cell. About half of the plasmid, however, is not concerned with these functions and can be dispensed within vector construction. In the yeast *Kluyveromyces lactis*, vectors equivalent to *S. cerevisiae*

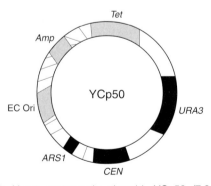

Figure 2 Yeast centromeric plasmid, YCp50 (7.9 kb). The vector has a suitable origin (*Escherichia coli* origin: EC Ori) and selectable markers (*Amp* and *Tet*) that allow it to be used as a shuttle vector in *E. coli*. The vector also contains a yeast selectable marker (*URA3*), a centromeric sequence (*CEN*) and an autonomously replicating sequence (*ARS1*).

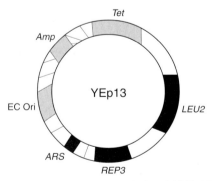

Figure 3 Structure of yeast episomal plasmid, YEp13 (10.8 kb). The YEp is derived from 2 μm plasmid of *Saccharomyces cerevisiae* and *Escherichia coli* plasmid pBR322. The vector has *E. coli* origin of replication (EC Ori) and selectable markers (*Amp* and *Tet*) that allow it to be used as a shuttle vector in *E. coli*. The vector contains a yeast selectable marker (*LEU2*), partitioning sequence (*REP3*) and an autonomously replicating sequence (*ARS*).

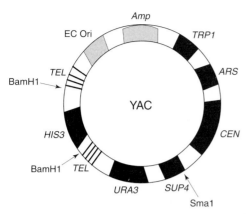

Figure 4 Structure of yeast artificial chromosome (YAC). The vector has two telomeres that constitute the ends of the vector once it is linearized with BamH1 after introducing the desired gene into the cloning site (*SUP4*).

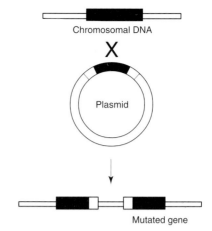

Figure 5 Gene inactivation by homologous plasmid integration. The constructed plasmid contains a DNA fragment from within the gene, shown as a filled rectangle; the deleted regions of the gene are shown by the open area within the rectangle. Chromosomal recombination by a single homologous recombination event (X) leads to duplication of two partially deleted copies of the genes separated by plasmid DNA sequence, leading to gene inactivation.

YEp have been obtained from 1.6μ circular plasmid, and also from linear killer plasmids.

Another vector used is *yeast artificial chromosomes* (YAC), which resemble chromosomes in that they are linear DNA with a centromere and a telomere at each end – a sequence that prevents attack by endonuclease. An example of YAC is shown in **Figure 4**. The vector contains *HIS3*, *URA3*, *TRP1* and *SUP4* markers. *CEN4* is a centromeric sequence and *TEL* is a telomeric sequence derived from the ends of ribosomal RNA encoding molecules from the macronucleus of *Tetrahymena*. In addition, there is a yeast origin of replication, a prokaryotic origin of replication, and an ampicillin resistance gene. The cloning strategy involves cutting the YAC DNA with the restriction enzyme, BamH1, in the present example, to generate linear DNA with telomeres at the two ends, followed by treatment with another restriction enzyme (Sma1) to accommodate the foreign DNA. YACs are large, stable and ideal for cloning large DNA sequences. YACs are around 20 kb in size and can carry inserts up to 500 kb.

Another type of vector used in fungi are *integrating* vectors, which do not have their own origin of replication but rather become integrated into the nuclear DNA by homologous recombination and replicate along with the nuclear genome. As illustrated in **Figure 5**, homologous integration of a circular plasmid carrying an intergenic DNA fragment leads to duplication in which the two partially deleted copies of the genes are obtained separated by plasmid DNA sequences. Such an event most often leads to a loss of gene function that, in some cases, can be used for selection of the integration event and thus for transformation. For example, integration of an internal fragment of *Aspergillus nidulans pyrG* gene, as shown in Figure 5, is expected to lead to loss of orotidine-5′-phosphate decarboxylase, which can be selected by resistance to 5′-fluoroorotic acid.

The transforming DNA is integrated into the chromosomes by the process of homologous and heterologous integration during vegetative growth. Homologous integration of chromosomal DNA resembles that of yeast. The relative frequencies with which homologous and non-homologous recombination events occur depend on the species and the length of the DNA molecule. In *A. nidulans* and *P. chrysogenum*, over half of the integration events are homologous, while in *Neurospora crassa* such events can be less than 5%. Homologous integration is important for manipulation involving gene disruption and deletion. Multiple transformation events are quite common in filamentous fungi and can be a way of increasing gene dosage in commercial applications. Since the integrating vectors give low-frequency transformants, by analogy with yeasts, replicating plasmids are expected to give better results. Naturally replicating vectors are rare in fungi; some species of *Neurospora* contain mitochondrial plasmids of unknown function, but these have not proved useful in the construction of autonomously replicating vectors. A plasmid of *Phanerochaete chrysosporium*, pME, has recently been detected and has been incorporated into a stable, replicating vector with kanamycin resistance marker which is maintained extrachromosomally in the absence of selection. The native circular plasmid exists in low copy number, and is relatively difficult to detect.

In the absence of natural plasmids, and by analogy

with yeasts, it has been possible to construct replicating vectors using chromosomal sequences, carrying putative origins of replication, autonomously replicating sequences. A replicating vector of *Podospora anserina* was constructed by adding telomeric sequences from *Tetrahymena thermophila* ribosomal DNA (a self-replicating element in protozoans). This plasmid becomes linearized in *Podospora* following transfer, and replicates at low copy number without integration. More recently, a sequence isolated from *A. nidulans*, *AMA1*, confers the ability to replicate autonomously to the plasmids that carry this gene, and this leads to an increase in transformation frequency 100-fold or more compared to integrating vectors. Although present in 10–30 copies per cell, these vectors are unstable and are rapidly lost without selection. An *ARS*-based vector that replicates autonomously has been reported for corn smut pathogen *Ustilago maydis*.

After a suitable vector is decided, the vector is opened with the help of restriction enzyme, an enzyme that recognizes a specific nucleotide sequence and cuts the molecule at that point and no other. A vector has to have a cloning site containing sequences that can be cut with the restriction enzymes to permit insertion of DNA to be cloned. Usually a polylinker cloning site is employed: this is a short synthetic double-stranded sequence that consists of a number of sites on which different restriction sites can act, giving experimental versatility. A portion of opened vector is mixed with mammalian DNA or the gene of interest that has to be inserted, and allowed to anneal, followed by treatment with DNA ligase which rejoins the cut DNA vector.

Entry and Expression of Recombinant DNA in Fungal Cells Once the recombinant DNA is constructed, it has to be transformed into a suitable host. The thick cell walls of yeasts and other fungi are a major obstacle to the entry of DNA into the cell. There are a number of ways available for introducing exogenous DNA into fungi. One of the easiest ways is simple treatment of the cells with lithium acetate and polyethylene glycol (PEG), together with transforming DNA and extra nonspecific carrier DNA (for example, from calf thymus) to increase the total amount of DNA present. Although this method is relatively straightforward, the efficiency of transfer can be rather low – approximately 10^3 colonies per microgram of plasmid – although higher frequencies have been claimed and may represent differences between strains.

A second and often more efficient approach of introducing exogenous DNA into *S. cerevisiae* is the transformation of spheroplasts, which are obtained by digestion of cell wall with appropriate enzymes, in the presence of a suitable osmotic buffer to prevent lysis. Calcium chloride and PEG facilitate uptake of transforming DNA by the protoplasts. However, the use of PEG for transformation is disadvantageous as it causes cell fusion and so diploids and polyploids may result. Interestingly, it has been found that lithium ions render fungal walls permeable to DNA, so the use of protoplasts is not always necessary. Recently, electroporation (entry of DNA under the influence of electric charge) has also been used for successful transfer of DNA to protoplasts. A yet more efficient method, the Shotgun approach, involves bombardment of DNA-coated tungsten particles directly into the mycelium.

The filamentous fungi, such as *A. nidulans*, *N. crassa* and *P. chrysogenum*, are also of considerable interest. Transformation can be carried out with protoplasts made from spores or hyphae, by digestion of the cell wall/coat with suitable enzymes. DNA is added in the presence of calcium ions and PEG. The latter causes protoplast fusion and the DNA is taken up at the same time. However, electroporation can also be used for transformation. The protoplasts are then regenerated. Ensuring the stable maintenance of extrachromosomal DNA is difficult, because of the filamentous nature of the fungi. As a result of this filamentous habit, cells that have lost the selectable marker can be maintained by those cells that have retained it. Thus using the system that results in integration of incoming DNA may be more convenient.

After transformation, the cells carrying the desired plasmids are selected/screened using the suitable markers and looked-for expression. The fate of the incoming DNA depends upon whether or not a suitable origin of replication is present. *ARS* elements have been isolated, such as *UARS1* from *U. maydis*, which allows maintenance as a plasmid.

When plasmid has to be inserted, a plasmid-free host has to be employed, so that the plasmids do not compete with the vector. The clones containing the desired vector are selected using markers, since all the treated host cells may not acquire plasmid. *LEU2* and *HIS* are selectable markers and code for enzymes that are essential for leucine and histidine biosynthesis, respectively. In some cases, antibiotic resistance markers have also been used.

Various factors determine the effectiveness with which a gene is expressed, but of special importance is the nature of the promoter. Controlled expression can be directed conveniently from the *alcA* promoter, for alcohol dehydrogenase, in response to carbon substrate. There are several controllable promoters available for expression in yeast. In order to have good expression, the vectors generally have promoter from host cell, e.g. promoters for galactose epimerase, acid phosphatase from *S. cerevisiae*, lactase from *K. lactis*.

Secretion of overexpressed proteins by fusion with a secretion signal may be more useful as it facilitates purification and the proteins produced will be minimally exposed to host proteases. In addition, it may help to bring about correct glycosylation and may also enhance protein folding. A number of secretion signals from yeast and fungi can be used; the most commonly used are those for the mating pheromone alpha-peptide and for invertase. *Aspergillus* spp. have been used to secrete proteins/enzymes into the medium and hence downstream processing is economical in such cases.

Applications of Recombinant DNA Technology

Production of Mammalian Proteins by Fungi The bacterium *E. coli* was not only the first host for gene cloning, but also the first microbe to be used for commercial production of mammalian proteins. In spite of these successes, *E. coli* is not ideal for production of heterologous proteins (production that would not be made by the species in question), especially of mammalian origin. As a polypeptide is synthesized it folds and disulphide bridges are formed to give a protein-precise three-dimensional structure. The folding and disulphide bridge formation can occur in prokaryotic cells giving a protein with the same amino acid sequence, but different properties. In mammalian cells a peptide can be modified in a variety of different ways following its synthesis. Hence the active protein is sometimes a protein in which part of the original polypeptide chain has been removed by proteolytic cleavage. Alternatively, the active product may result from enzymatic addition of groups such as oligosaccharides, phosphates or sulphates to specific amino acids. Bacterial cells are incapable of these post-translational modifications. Furthermore, *E. coli* usually retains the proteins it makes within the cell instead of exporting them into the culture medium. This means that the downstream processing is more difficult, especially since *E. coli* produces endotoxins (toxins associated with the cell) and pyrogens (fever-inducing agents) which must be completely eliminated during purification. These shortcomings have lead to research aimed at finding ways of using other species as hosts for gene cloning.

The first fungus with which success was achieved was *S. cerevisiae*. Another yeast, *K. lactis*, has been found to be an efficient host for synthesizing and exporting mammalian protein chymosin. Bovine chymosin or rennin is used in the cheese industry, and because of its high value and supply problems, it has been the target for microbial expression for some time. As a result of difficulties encountered in earlier attempts at expression in bacterial and yeast systems, *A. awamori* has been successfully developed as a host, using a wide range of genetic manipulation techniques. The preprochymosin gene is expressed using transcription signals derived from host glucoamylase gene to produce recombinant chymosin.

Fungal Enzymes A wide-range of fungal (*A. niger*, *A. oryzae*) enzymes like amylases and cellulases are used in food industry. Most of these are secreted enzymes of low value, made in large quantities by strains already improved by conventional procedures. Nevertheless, many of the genes encoding for these enzymes have been isolated, and attempts have been made to improve their yield by transformation. Many oversecreting recombinant strains of fungi have been produced using recombinant DNA technology, secreting high amounts of enzymes with applications especially in food biotechnology (**Table 5**).

Some of the genes encoding useful enzymes, such as the α-amylase (taka-amylase) of *A. oryzae* and the glucoamylase of *A. niger/A. awamori*, already possess some of the strongest fungal promoters known, particularly when they are derived from commercially improved strains, and these promoters have been exploited for construction of hybrid expression systems. Fungal lipases such as those from *Humicola* and *Rhizopus* species are now used in detergents. Their production has been enhanced by expression in *A. oryzae*, using the α-amylase promoter of this organism.

Strain Improvement Recombinant DNA technology has been widely used to improve strains. For example, in brewing yeast, it has been advantageous in the improvement of fermentation and post-fermentation stages in beer production. Some of the strain improvements achieved in brewing yeasts are shown in **Table 6**. The process can be made more efficient by reducing

Table 5 Enzymes produced by fungi having applications in food biotechnology

Enzyme	Source	Applications
α-Amylase	Aspergillus niger, A. oryzae	Starch conversions
Glucose oxidase	A. niger	Food processing
Rennet	Mucor spp.	Milk coagulation
Lipases	Mucor, Aspergillus, Penicillium spp.	Dairy industry
Hemicellulase	A. niger	Bakery; gum
Pectinases	Aspergillus, Rhizopus spp.	Clarifying fruit juices
Invertase	Yeasts	Confectionery
Proteases (acid, neutral, alkaline)	Aspergillus sp.	Breakdown of proteins (baking, brewing and dairy)

the residence time in beer vessels. Product quality can be improved by inactivating the genes which are responsible for production of off-flavours or reducing susceptibility to contaminants. Yeast capable of secreting enzymes that would otherwise be added to fermentation medium will make the process more efficient and economical. The yield of a particular product can also be improved using in vitro gene manipulation techniques. Employing such approaches, the yield of antibiotic production can be greatly improved by recombinant DNA technology. The commercial β-lactam antibiotic producers *P. chrysogenum* and *Cephalosporium acremonium* have been the main targets for this work. This is achieved by inserting additional copies of such genes.

Table 6   Strain improvement of brewing yeast

Properties	Effect
Amylosis	Low-carbohydrate beer
Proteolytic yeast	Aids colloidal stability
Reduced phenolic compounds	Reduced off-flavours
Reduced diacetyl production	Reduces fermentation time
Catabolite derepression	Improved fermentation
Zymocin production	Protection against contamination

Therapeutics   A number of therapeutic products have been produced in fungal systems, such as cytokines (interleukin-2).

Risks to Consumers and/or Ecology of Non-target Species

Considering the benefits of recombinant DNA technology, any preconceived hazard cannot be ignored. The introduction of modern biotechnology involving genetic manipulation in the last two decades has generated some concern. This highlights the need to ensure that what sometimes seem to be specific hazards from an organism used in biotechnology are adequately assumed and the risks are controlled. The public perception of microorganisms is that they cause disease. The impact of a laboratory infection can stretch beyond the immediate surroundings where industrial processes are employed, and it can have a serious impact on the whole plant, on the marketing of the products, on innocent bystanders, and on the environment. Therefore, a microorganism used in biotechnology should be safe and cause no mortality. A pathogenic organism should not produce an allergic response or endotoxin or a toxic reaction. For example, an *E. coli* strain producing botulism toxin could conceivably be very dangerous; insertion of the penicillinase gene into *Streptococcus pyogenes* (still universally sensitive to penicillin) could make the

treatment of streptococcal infections much more difficult.

It has been observed that single-cell protein consumption caused ill health in consumers. Physical contamination or escape from the laboratory of recombinant organisms may pose a health hazard. Let us take a hypothetical situation; if an interleukin-2-producing microorganism escapes from a laboratory it may produce an immunological imbalance in the host it enters. The modified organisms released should also have no adverse effect on the environment. The growth of a microorganism depends on the substrate present in a particular ecological niche. If the organism shows change in the substrate utilization, it will grow in a different ecological niche from its natural niche and may have a deleterious effect on a given habitat. If an antibiotic-resistant microorganism is used, there is a risk that it may transfer antibiotic resistance genes to other organisms by the processes of conjugation, transformation and transduction. Therefore, a genetically engineered organism should only be released after thorough assessment of the risks.

See also: **Aspergillus**: Introduction. **Ecology of Bacteria and Fungi in Foods**: Influence of Available Water; Influence of Temperature; Influence of Redox Potential and pH. **Escherichia coli**: *Escherichia coli*. **Fusarium**. **Genetic Engineering**: Modification of Bacteria. **Genetics of Microorganisms**: Fungi. **Kluyveromyces**. **Nucleic Acid-based Assays**: Overview. **Penicillium**: Introduction. **Yeasts**: Production and Commercial Uses.

Further Reading

Arora DK, Elander RP and Mukherji KG (1989) In: *Handbook of Applied Mycology*. Vol. 4. New York: Marcel Dekker.

Bennett JW and Lasure LL (1991) In: *Gene Manipulations in Fungi*. San Diego, CA: Academic Press.

Berry DR (1988) In: *Physiology of Industrial Fungi*. Oxford: Blackwell Scientific.

Hambleton P, Melling J and Salusbury TT (1994) In: *Biosafety in Industrial Biotechnology*. London: Blackie Academic & Professional.

Moo-Young M (1985) In: *Comprehensive Biotechnology: The Principles of Biotechnology: Fundamentals*. Vol. 1. Oxford: Pergamon Press.

Peberdy JF, Caten CE, Ogden JE and Bennett JW (1991) In: *Applied Molecular Genetics of Fungi*. Oxford: Cambridge University Press.

Puhler A (1993) In: *Genetic Engineering of Micro-organisms*. Germany: VCH.

Sussman M, Collins CH, Skinner FA and Stewart DE (1988) The release of genetically engineered microorganisms In: *Proceedings of the 1st International Conference on the*

Release of Genetically Engineered Microorganisms. London: Academic Press.

Walker JM and Gingolg EB (1988) In: *Molecular Biology and Biotechnology.* Cambridge: Royal Society of Chemistry.

Wilson M and Lindlow SE (1993) Release of recombinant microorganisms. *Annual Review of Microbiology* 47: 913–944.

Modification of Bacteria

Eric Johansen, Department of Genetics and Microbiology, Chr. Hansen A/S, Hørsholm, Denmark

Genetic engineering refers to the use of recombinant DNA technology for the modification of the properties of various organisms. The resulting organisms are called genetically modified organisms (GMOs). The techniques used allow the modification of bacteria used in food applications as well. Since many foods contain living bacteria, special consideration must be given to the types of modification used. In what follows, food-grade genetic modifications and food-grade GMOs are described with emphasis on lactic acid bacteria from the genus *Lactococcus*.

Definition of 'Food-grade'

Bacteria used in the manufacture of foods often remain viable in the end product. This means they will be eaten, often while still viable, by the consumer. Furthermore, bacteria such as the lactic acid bacteria found in our food are known to pass through the digestive system and to be excreted in a viable form. Thus, they survive and may actually colonize or grow in the human digestive tract. Food-grade genetic modifications are the limited types of modifications that would be considered acceptable for food microorganisms with safety as the chief concern.

A strict definition of food-grade has been useful in initial applications. According to this definition, a food-grade GMO must contain only DNA from the same genus and possibly small stretches of synthetic DNA. These small stretches of synthetic DNA usually range from 2 base pairs to 50 bp and are required for the actual construction of the GMO. They do not code for protein or RNA but are recognition sites for restriction enzymes and are used for the actual construction process. Thus, a *Lactococcus* GMO would only contain DNA from the genus *Lactococcus* plus a small amount of synthetic DNA.

A broader definition is gaining acceptance and would allow the use of DNA from other genera of food microorganisms, provided that the donor belongs to the group of organisms referred to as 'gen-erally recognized as safe' (GRAS). By this definition, a *Lactococcus* GMO could contain DNA from *Streptococcus thermophilus* or other GRAS bacteria.

With either definition, it is acceptable to use DNA from non-GRAS organisms during the construction of the GMO provided that this DNA is removed before the GMO is considered to be food-grade. Sensitive techniques such as DNA–DNA hybridization, polymerase chain reaction and DNA sequencing can be used to confirm that the undesired DNA has been successfully removed.

Most of the food-grade GMOs described here can be used in the USA without further regulatory approval while their use in Europe and other places requires approval from a number of regulatory boards, depending on local laws and international agreements. The relevant regulations are constantly being revised and clarified so researchers must consult their local officials before using GMOs in any food process.

Food-grade Genetic Modifications

A number of genetic modifications can be done in a food-grade manner. Genes can be deleted from a strain. This is especially relevant when a gene has an undesired property, for example the production of an undesirable metabolic end product or the destruction of a desirable component of the food. A gene in a strain can be replaced with the homologous gene from another strain. If a strain has a number of useful properties but a desirable enzyme activity is at too low a level, this can be corrected by introducing a more active copy of the gene from another strain. Genes can be inserted into a strain, for example genes that result in increased resistance to bacteriophage. Finally, the expression of a gene can be increased by increasing the number of copies in the cell, using food-grade cloning vectors.

Homologous Recombination

Most of the food-grade genetic modifications described above are based on the natural ability of bacteria to perform homologous recombination. Recombination is the exchange of genetic material between two DNA molecules or locations on the same molecule. Homologous recombination occurs whenever a bacterial cell has two copies of a DNA sequence which are identical or nearly identical and have a size greater than 500 bp. If a sequence is present on a plasmid and on the chromosome, homologous recombination results in the integration of the plasmid into

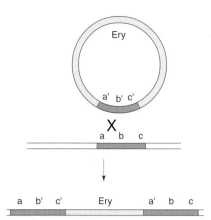

Figure 1 Integration of a plasmid into the chromosome by homologous recombination.

the chromosome (**Fig. 1**). The two copies of the repeated sequence now flank the integrated plasmid and are capable of a second homologous recombination event. This second recombination event results in excision of the plasmid and one copy of the repeated sequence (**Fig. 2**). Since it is unlikely that the second recombination event initiates at exactly the same base pair as the first one, DNA that was originally on the plasmid remains on the chromosome and DNA that was initially on the chromosome is transferred to the plasmid. If the plasmid copy was genetically modified, this modification will be transferred to the chromosome. Simple techniques exist for plasmic curing and the resulting strains are food-grade.

To facilitate the integration process, vectors with temperature-sensitive replication and an antibiotic resistance marker have been used. The plasmid is introduced into the strain, for example by electroporation, and maintained at the permissive temperature. Raising the temperature and selecting for the antibiotic resistance encoded by the vector results in strains in which the desired first recombination

event has occurred, since the plasmid is unable to replicate. The presence of two active replication origins is usually lethal to bacteria, so lowering the temperature again and selecting for survival in the absence of the antibiotic is an effective selection for strains in which the second recombination event has occurred. Screening of a small number of survivors usually results in a strain with the desired modification. Plasmid curing can be done by again raising the temperature, this time in the absence of the antibiotic.

Food-grade Deletions

A food-grade deletion can be made of any dispensable gene of any bacterium using a relatively simple multistep process. Exceptions are strains into which DNA cannot be introduced and those that are incapable of homologous recombination. The first step is to clone the gene to be deleted and flanking regions in a suitable cloning vector. This vector need not be food-grade and *Escherichia coli* vectors are usually used. Deletions are made in vitro either using conveniently located restriction enzyme recognition sites or by polymerase chain reaction. At least 500 bp must remain on either side of the deletion. The plasmid containing the deletion is introduced into the strain of interest. Homologous recombination results in integration of the plasmid into the chromosome producing a duplication of the DNA flanking the deletion. The second recombination event moves the gene of interest to the plasmid and leaves the deletion in the chromosome. In Figures 1 and 2, the deletion is represented by **b'** and the flanking regions by **a'** and **c'**.

Food-grade Gene Replacement

Replacement of a gene with another allele is done by deleting the undesired allele as described above and subsequently inserting the desired allele in a similar multistep process. The desired allele is cloned into a vector so that it is flanked on both sides by at least 500 bp of DNA with homology to the desired integration site. This plasmid is introduced into the deletion strain and is integrated into the chromosome by homologous recombination at one of the flanking sequences. This results in a strain with both the deletion and the desired allele. A second recombination event moves the deletion to the plasmid and leaves the desired allele on the chromosome. This is illustrated in Figures 1 and 2 where the deletion on the chromosome is represented by **b** and the substituting allele on the plasmid by **b'**. Elimination of the plasmid results in a food-grade strain with a gene replacement.

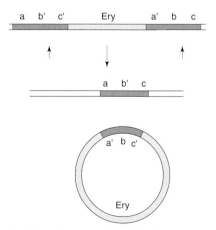

Figure 2 Excision of a plasmid from the chromosome by homologous recombination.

Food-grade Integration into the Chromosome

A simple variation of the gene replacement strategy allows the integration of any gene of interest into the chromosome. The gene to be integrated is inserted into a plasmid and flanked on both sides by at least 500 bp of DNA from the integration site. Integration of the plasmid and subsequent excision from the chromosome can result in a strain containing the gene of interest integrated into the chromosome. In Figures 1 and 2, the gene of interest is represented by **b'**, the flanking regions by **a'** and **c'** on the plasmid and **a** and **c** on the chromosome and the site of integration of the gene into the chromosome by **b**.

Food-grade Cloning Vectors

One simple way to increase the expression of a desirable gene is to increase the number of copies of the gene in the cell. This can usually be done by transferring the gene from the bacterial chromosome (copy number 1 per cell) to a multicopy plasmid. Such plasmids are referred to as cloning vectors and can have copy numbers ranging from 2 or 3 copies per cell to as many as 50 copies per cell.

Plasmids are usually unstable and are eventually lost from the cell unless they provide the cell with a selective advantage. To prevent plasmid loss, a selectable marker is often included in a vector. Cloning vectors for *E. coli* usually contain genes for antibiotic resistance and selection with antibiotics is used to kill plasmid-free cells. This is clearly not acceptable for a food-grade cloning vector.

Many lactic acid bacteria produce bacteriocins such as nisin. These small peptides have antimicrobial activity and yet the producing strains are not affected. This is due to the presence of specific genes which render them resistant to the bacteriocin. These have been used as selectable markers in food-grade cloning vectors. The usefulness of these vectors is, however, limited because selection only occurs in the presence of the bacteriocin and many strains have an intrinsic resistance to the bacteriocin.

A number of other food-grade selectable markers have been developed. In principle, any gene that can be deleted from the chromosome can become a food-grade selectable marker provided that there are growth conditions under which this gene is essential and others under which this gene is dispensable. Inserting the gene onto a plasmid and growing under conditions where the gene is essential selects for the presence of the plasmid. A number of vectors for lactic acid bacteria use the *lac* genes, required for the fermentation of lactose, as selectable marker. The *lac* genes are essential when cells are grown with lactose

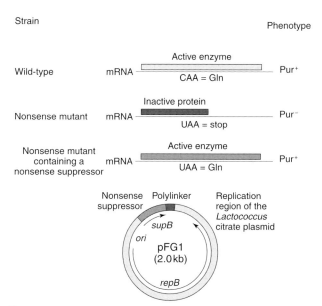

Figure 3 Use of a nonsense suppressor as a selectable marker and the food-grade cloning vector pFG1.

as the sole carbon source but are dispensable when another carbon source is available.

Nonsense suppressor genes have also been used as selectable markers. Mutants of *Lactococcus* with nonsense mutations in the purine biosynthetic pathway have been developed. These strains require purines for growth but will grow in purine-free medium if a nonsense suppressor is present in the cell. This suppressor suppresses the nonsense mutation and allows *de novo* biosynthesis of purines. Thus, the nonsense suppressor is a selectable marker in purine-free medium and has been included in a series of food-grade cloning vectors. The selection principle and one of the cloning vectors is illustrated in **Figure 3**. Nonsense mutations in other biosynthetic pathways such as pyrimidine biosynthesis can be used with the same vectors, giving added versatility. The use of suppressible purine or pyrimidine auxotrophs is particularly useful for lactic acid bacteria because milk is essentially purine- and pyrimidine-free and so acts as a selective medium.

Virtually any gene that can be cloned can be placed on a food-grade cloning vector. The resulting strains are food-grade if the original source of the cloned DNA is from a GRAS organism as discussed earlier. Other sources of DNA might also be declared food-grade after extensive safety evaluations.

Genetically Modified Lactic Acid Bacteria

Lactic acid bacteria are used in the production of a large variety of foods and animal feeds. They produce lactic acid and contribute to the flavour, texture and shelf life of the food.

Food fermentations are often done on an industrial scale following inoculation with appropriate numbers of lactic acid bacteria either produced in the factory or purchased from a commercial culture supply house. It is of utmost importance that the bacteria complete the fermentation process, otherwise large amounts of raw materials may be lost. One major reason for fermentation failure is the presence of bacteriophages, viruses that infect and kill bacteria. A typical bacteriophage of lactic acid bacteria can produce 10^6 progeny in the same time that a single cell can produce eight progeny cells. Clearly, even a small number of infective bacteriophage particles can cause a serious problem for industrial fermentations. Lactic acid bacteria contain a number of mechanisms leading to bacteriophage resistance. The presence of individual mechanisms varies from strain to strain. Combining the best bacteriophage resistance mechanisms into superior industrial strains can be most effectively accomplished by genetic engineering. Since the genes are already contained in GRAS organisms, the resulting strains will be food-grade.

The flavour of fermented food depends on the presence of desirable flavour compounds and the absence of undesirable flavour compounds. Often, the desirable flavour compounds are further degraded by the culture used to manufacture the food. This results in an instability in the quality of the food and a shortening of the shelf life. Genetic engineering can be used to increase the production of the desired compounds and increase stability by prevention of their degradation. Increasing the production of a particular compound can be done by increasing the expression of the genes responsible for its synthesis, elimination of genes responsible for its degradation, or elimination of other biosynthetic pathways that consume a rate-limiting biosynthetic intermediate.

A good example of a flavour compound whose content in food can be changed using food-grade GMOs is diacetyl, a major flavour component of butter and buttermilk. An important intermediate in the production of diacetyl is pyruvate. Inactivating genes coding for enzymes that use pyruvate as a substrate results in a significant increase in the amount of diacetyl produced by a culture. Likewise, increasing the expression of α-acetolactate synthase – an enzyme involved in the conversion of pyruvate to diacetyl – increases diacetyl production. Finally, inactivation of diacetyl reductase, an enzyme that degrades diacetyl to acetoin, results in increased diacetyl levels owing to increased stability. Combining these modifications in one strain has not yet been accomplished but should not be difficult using the food-grade techniques described here.

Another example is the flavour peptides formed during the degradation of casein in cheese production. Some of the peptides produced by the action of the proteolytic enzymes present in the cheese process have a bitter taste and can result in an unpleasant-tasting cheese. The formation of these peptides can be blocked by inactivating the aminopeptidase responsible, or by increasing the expression of an aminopeptidase that degrades the bitter peptide. Food-grade GMOs are currently being studied to determine which aminopeptidases are responsible for the bitter peptides and which can eliminate them. This knowledge will allow the construction of improved industrial strains for cheese manufacturing.

The texture of yoghurt and other fermented milk products results from the interaction of a number of factors. One factor is the extracellular polysaccharide produced by the bacteria in the starter culture used. The relevant genes have been identified from several strains and GMOs with altered texture-producing properties are being constructed.

One major role of lactic acid bacteria is food preservation. This is a consequence of the lactic acid and a variety of bacteriocins produced by the lactic acid bacteria. Modification of the genes involved in lactic acid production can lead to a quicker and more efficient lactic acid production. Likewise, the genes for production and resistance to bacteriocins can be moved between strains in a food-grade manner.

See also: **Bacteriocins**: Nisin. **Cheese**: Microbiology of Cheese-making and Maturation. **Fermented Milks**: Range of Products. **Genetics of Microorganisms**: Bacteria. *Lactococcus*: Introduction. **Milk and Milk Products**: Microbiology of Cream and Butter. **Starter Cultures**: Uses in the Food Industry. *Streptococcus*: *Streptococcus thermophilus*.

Further Reading

Dickely F, Nilsson D, Hansen EB and Johansen E (1995) Isolation of *Lactococcus lactis* nonsense suppressors and construction of a food-grade cloning vector. *Molec. Microbiol.* 15: 839–847.

Gasson M (1997) Genetic manipulation of dairy cultures. *Bulletin of the IDF* 320: 41–44.

Johansen E, Strøman P and Hansen EB (1995) Genetic modification of lactic acid bacteria used in the production of food. In: *Unanswered Safety Questions When Employing GMO's*, pp. 85–88. Coordination Commission Risk Assessment Research, Hoorn, The Netherlands.

MacCormick CA, Griffen HG and Gasson MJ (1995) Construction of a food-grade host/vector system for *Lactococcus lactis* based on the lactose operon. *FEMS Microbiol. Lett.* 127: 105–109.

Mierau I, Haandrikman AJ, Velterop O et al (1994) Tripeptidase gene (*pepT*) of *Lactococcus lactis*: molecular cloning and nucleotide sequencing of *pepT* and construction of a chromosomal deletion mutant. *J. Bacteriol.* 176: 2854–2861.

Venema G, Huis in't Veld J and Hugenholtz J (eds) (1996)

Lactic Acid Bacteria: Genetics, Metabolism and Applications. Dordrecht, The Netherlands: Kluwer Academic Publishers.

GENETICS OF MICROORGANISMS

Contents
Fungi
Bacteria

Fungi

R Sandhir, Department of Biochemistry, Dr Ram Manohar Lohia Avadh University, Faizabad, India

S K Garg and **D R Modi**, Department of Microbiology, Dr Ram Manohar Lohia Avadh University, Faizabad, India

Introduction

Fungi are a major component of the microbial world and typically include filamentous mycelia, but one group, the yeasts, are characteristically unicellular. Yeasts, filamentous fungi or both are important in brewing, in the preparation of many foods, in biodeterioration, and in the fermentation industry. Fungi share the basic characteristics of eukaryotic organisms. This article considers the genetic processes in fungi (eukaryotes). The genetic mechanisms in prokaryotes are simple, whereas eukaryotes have evolved complex genetic systems. Eukaryotic organisms have extensive intracellular compartmentalization, multicellularity and sexuality. Fungi can be differentiated from bacteria by the fact that fungal cells are usually much longer and contain a nucleus, vacuole and mitochondria, typical features of eukaryotic cells.

Structure of the Nucleus

Fungal nuclei have a majority of features common with other eukaryotic organisms. In eukaryotes, most of the DNA is organized into a membrane enclosed structure, the nucleus. This is the defining feature of eukaryotic cells, distinguishing them from prokaryotes. The nucleus is usually a large organelle which is easily visible by light microscopy. The nuclear membrane is a bilayer continuous with the endoplasmic reticulum, and at intervals has pores (9 nm wide) through which nucleoplasmic exchange is believed to occur. Each pore is surrounded by a disc of protein molecules, the nuclear pore complex, that forms a channel wide enough to allow passage of

small molecules (molecular weight < 5000 Da). The nuclear membrane has special transport systems for the exchange of large molecules like RNA and protein. The nuclear envelope functions to separate the cell into nuclear and cytoplasmic compartments. The key genetic processes of DNA replication and RNA synthesis (transcription) occur in the nucleus, whereas, the process of protein synthesis (translation) occurs in the cytoplasm. The nucleus, thus, serves as storehouse and processing factory of genetic information. The complex nuclear organization of eukaryotes and the existence in each nucleus of a number of chromosomes leads to more general mechanisms of gene assortment and segregation in eukaryotes. Within the nucleus is the nucleolus, an area rich in RNA which is the site for synthesis of ribosomal RNA. The small and large subunits of ribosomes are synthesized in the nucleus and exported to the cytoplasm where complete ribosome complex is assembled. **Figure 1** shows the general architecture of nucleus.

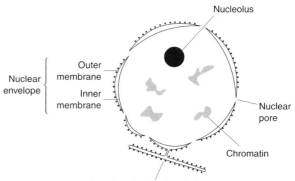

Figure 1 Structure of a typical cell nucleus. The nuclear envelope is a double membrane structure that is continuous with the endoplasmic reticulum. The lipid bilayers of inner and outer membrane fuse, forming nuclear pores. In the resting phase, the chromatin is dispersed and the nucleolus is clearly visible.

The Fungal Nucleus

The nuclei of most fungi (1–2 μm in diameter) are small compared to those of animals and plants and have comparably small chromosomes. The nuclei can either be spherical or oval shaped. The hyphae of most fungi are segmented by septa which separate the mycelium into compartments similar to cells. The fungal nucleus is a plastic structure that can squeeze through pores in the septa. The typical compartments of fungi may contain one or many (multinucleate or coenocytic) nuclei. The apical compartment of *Aspergillus nidulans* can have about 50 nuclei and the smaller compartments about four nuclei each. In some basidiomycetes, the monokaryons have one nucleus per compartment, and the dikaryon two, although in many basidiomycetes there are many nuclei per compartment. Yeasts on the other hand, have a single nucleus per cell. The nuclei of the vegetative phase of most filamentous fungi are haploid, whereas those of oomycetes, which are clearly and distantly related to other fungi, are diploid.

Organization of DNA in Chromosomes

The DNA of prokaryotes is a single free molecule, whereas in eukaryotes, DNA has a more complex structure, the chromosomes (means coloured body), which are structures first observed after staining with certain dyes. Many chromosome stains involve dyes that react strongly with basic (cationic) proteins called histones. Because DNA is negatively charged (due to large number of phosphate groups), there is a strong tendency for various parts of the molecule to repel each other. Histones neutralize some of these negative charges and permit compactness of the chromosomes. The histones are responsible for packing DNA into the nucleus. There are five major types of histone proteins in eukaryotic cell nucleus. These are classified as histone H1, H2A, H2B, H3 and H4. The properties of different histone proteins are given in **Table 1**. The histones are basic proteins of low molecular weight that are readily isolated by salt and acid extraction of chromatin and can be separated on the basis of size and charge. Histones H1, H2A and H2B are lysine rich, whereas, H3 and H4 are arginine rich. The

histone proteins are highly conserved during evolution. They are subject to post-translational modifications that include methylation, acetylation and phosphorylation of specific Arg, His, Lys, Ser and Thr residues. These modifications are all reversible and decrease the positive charge on histones, thereby affecting histone–DNA interaction.

The histone proteins are attached and spaced at regular intervals along the DNA helix, the DNA itself being wound around each histone molecule. The packing forms a discrete structure, or nucleosome, which appears as 'beads on a string' (**Fig. 2**). Each nucleosome has around 200 nucleotide long DNA coiled around histones. A nucleosome consists of a core particle of about 140 bp of DNA wrapped around a disc of four different histones H2A, H2B, H3 and H4 (two subunits each), and a 40–60 bp linker DNA. The H1 histones bind to the linker DNA and participate in nucleosome–nucleosome interaction. In some fungi H1 histones are absent; these fungi lack the ability to condense chromatin. The main difference between fungal and other higher eukaryotes is in the linking of DNA between nucleosomes, which is shorter by 20 bp compared to higher eukaryotes. Nucleosomes aggregate to form a fibrous material called chromatin. Chromatin itself can be compacted by folding and looping to eventually form intact chromosomes. This is a remarkable feat of packing, since the normal nucleus is not more than 2 μm across, whereas if the DNA molecules were put end to end they would stretch for over 3 m. The eukaryotic chromosomes are dynamic structures and their appearance varies dramatically with the stage of the cell cycle. The individual chromosomes only assume their familiar condensed form during cell division (metaphase). During interphase, the remainder of the cell cycle, when the chromosomal DNA is transcribed and replicated, the chromosomes become highly dispersed and can not be individually distinguished.

In general, fungi have DNA content intermediate between prokaryotes and other eukaryotes. The

Table 1 Properties of histone proteins

Type	Mass (kDa)	% Arg	% Lys	Copies per nucleosome
H1	23.0	1	29	1 (bound to linker DNA)
H2A	14.0	9	11	2 (nucleosome octamer)
H2B	13.8	6	16	2 (nucleosome octamer)
H3	15.3	13	10	2 (nucleosome octamer)
H4	11.3	14	11	2 (nucleosome octamer)

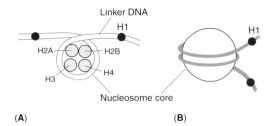

Figure 2 Association of DNA with histones forming nucleosomes. (**A**) DNA helix wrapped around nucleosome core consisting of octamer of four histones H2A, H2B, H3 and H4, and H1 histone bound to linker DNA; (**B**) two turns of DNA around the nucleosome core.

genomes in fungi are smallest of all eukaryotes (12–88 Mbp), usually 3–11 times that of *Escherichia coli* with two notable exceptions, *Saprolegnia* with 200 Mbp and *Entomophaga* with 8200 Mbp. In each chromosome, the DNA is a single linear molecule to which histones are attached. In yeast (and other microorganisms), the length of a DNA molecule is actually shorter than a linearized bacterial chromosome. For instance, the total amount of DNA per yeast cell is only three times that of *E. coli*, but yeast has 17 chromosomes, so the average yeast DNA molecule is shorter than the *E. coli* DNA. The DNA content and nuclear size vary from species to species. The chromosomes of fungi are small and highly condensed. They are difficult to count by conventional microscopy of the stained cells because the nuclear membrane persists during most of the cell cycle. The chromosome number also varies greatly from just a few to many hundred (**Table 2**). The chromosomes of *Aspergillus nidulans* and *Saccharomyces cerevisiae* have been shown to contain a single molecule of double-stranded DNA, and this is probably true of chromosomes from other fungi also.

Table 2 Haploid chromosome number and nuclear genome size in fungi and other organisms

Group and species	Chromosome number	DNA (bp × 10^{-6})
Prokaryotes	1	4
Escherichia coli		
Eukaryotes (fungi)		
Ascomycetes		
Aspergillus nidulans	8	3–5
Saccharomyces cerevisiae	16	0.2–2
Neurospora crassa	7	4–13
Basidiomycetes		
Ustilago maydis	20	0.3–2.0
Deuteromycetes		
Candida albicans	7	1.1–3
Oomycetes		
Phytophthora megaspora	9–14	1.4–4
Eukaryotes (humans)	23	130

In prokaryotes, genes occupy a single uninterrupted sequence of DNA, whereas the majority of eukaryotic genes sequenced so far have been shown to contain non-coding sequences (introns) inserted into the middle of coding sequences (exons). These intervening sequences, or introns, may be small and single or large and multiple. The introns are transcribed into RNA and then removed by internal processing events in the cell nucleus to produce the final messenger RNA product that is translated to the protein product in the cytoplasm (**Fig. 3**). About 60% of the genes sequenced

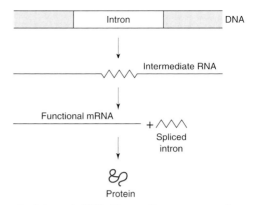

Figure 3 Introns in DNA. The coding sequences (exons) are indicated by shading. The introns are transcribed to an intermediate RNA molecule, followed by splicing to remove intron and forming mRNA that codes for the final protein product.

from filamentous fungi contain introns. These sequences are short, generally less than 100 bp, as compared to those of mammals. In *S. cerevisiae*, only about 10% of the genes sequenced so far have introns, but these introns are, on average, much longer. Like the DNA of other eukaryotes, fungal DNA can consist of a single or multiple copy (repeated) sequences. Little of this DNA usually less than 10% consists of repeated sequences, and it is most likely that these sequences account for genes coding for ribosomal RNA, tRNA and chromosomal proteins, all needed in large quantities. This contrasts with the situation in mammals and in *Physarum polycephalum*, where about one third of the DNA is repetitive.

Extranuclear DNA

In addition to the nuclear DNA of eukaryotes, there is a small amount of DNA present in mitochondria. Yeasts also carry small extra pieces of DNA known as extrachromosomal DNA or plasmids.

Mitochondrial DNA (mtDNA)

Mitochondria are the sites for oxidative phosphorylation in eukaryotic cells and contain DNA molecules. In most fungi, mitochondrial DNA is usually found as circular, double-stranded, supercoiled molecules in the mitochondrial matrix. However, in the yeast *Hansenula marakii*, the oomycete *Pythium* and the slime mould *Physarum*, linear DNA has been reported in mitochondria. mtDNA, unlike nuclear DNA, is not packaged into nucleosomes, which is usually about 15 kb long and codes for 15–20 mitochondrial proteins, tRNA and rRNA. However, some yeast mitochondrial DNAs are 75 kb long. The proteins encoded by mtDNA are the enzymes of the electron transport chain. No mammalian mtDNA contains introns but these are present in yeast mitochondrial genes. In one strain of *S. cerevisiae*, genes account for 16% of the mitochondrial DNA, introns for 22% and the intergenic regions for 62%. *S. cerevisiae*, as a facultative anaerobe, can survive mutations that render the mitochondria non-functional rendering the resulting strains dependent on glycolysis for their energy requirements.

Plasmids

Plasmids are pieces of DNA that can replicate independently of the replication of the chromosome or mtDNA. Although plasmids are common in bacteria, they are infrequent in fungi and other eukaryotes. The plasmids in fungi lack genetic information essential to the life of fungus. An intensively studied fungal plasmid is 2 μm DNA that occurs in the nucleus of *S. cerevisiae*. The 2 μm circular plasmid consists of about 6200 bp. The plasmid can account for almost 3% of the total DNA in a cell and there can be up to 100 copies per cell. Whether this 2 μm DNA has any role in the life of the host is unknown, but it is proving to be a valuable tool in the construction of the vectors for genetic manipulation of yeasts. With the exception of *S. cerevisiae*, most plasmids of fungi are found in mitochondria. Another yeast, *Kluyveromyces fragilis*, has a plasmid in the form of two linear double-stranded DNA molecules of different lengths. Strains with plasmids are 'killers', producing an extracellular protein that kills *Kluyveromyces lactis* strain that lacks the plasmid. In *Podospora anserina*, plasmids that originate from mitochondrial DNA can spread through the mycelium and cause senescence, in which growth rate falls and vegetative propagation finally becomes impossible. The plasmids can be transmitted via maternal parents through ascospores. Senescence in the progeny may be long delayed, allowing continued propagation of the plasmids. It seems that the fungal plasmids are essentially 'selfish-DNA', concerned with their own propagation rather than conferring any desirable property to the species.

Behaviour of Chromosomes During Mitosis and Meiosis

Mitosis

Nuclear division, mitosis, which takes place during the process of vegetative reproduction, ensures the division of genetic material between the two newly formed cells. The classical mitosis has a number of stages.

After the nuclear resting period, the interphase, a preparatory phase, the prophase begins. During this phase, the nuclear matter is arranged into chromosomes that can be readily stained. The chromosomes are at first long and thin, often coiled so that they can not be counted. Later, they become thicker and shorter

and assume a clearly double ribbon-like structure. The doubling of chromosomes gives rise to two ribbon-like structures, the chromatids, which are situated close to each other and are firmly bound by the centromere. The centrioles in the cytoplasm separate and move to opposite poles and the formation of a spindle begins. The nucleolus and nuclear membrane disappear during this stage.

During metaphase the double chromosomes aggregate in the equatorial plane of the spindle. In metaphase of the cell division, chromosomes can be seen with light microscope as compact structures. The DNA is quiescent at this stage serving as a template. The metaphase chromosome is analogous to the quiescent DNA molecule of the infective resting phage particle, packaged for transit to a future host cell. **Figure 4** shows the typical structure of metaphase chromosome. The chromosomes are attached to the spindle by their centromeres.

The anaphase is the third stage of nuclear division during which the pair of chromosomes separate and migrate to opposite poles of the nucleus.

During telophase, the last stage of mitotic division, the nuclei and nuclear membrane reappear, and the two new nuclei are formed.

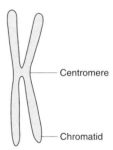

Figure 4 Typical structure of metaphase chromosome. Each chromatid contains two identical DNA molecules created earlier in the cell cycle by the process of DNA replication and the two chromatids are held by a centromere.

In fungi, nonclassical mitosis takes place without the disruption of the nuclear membrane, i.e. endomitosis; the division is also referred to as intranuclear (**Fig. 5**). The nuclear membrane constricts in a dumbbell like fashion and eventually separates into two daughter nuclei. It is believed that the nuclear division without the disruption of the nuclear membrane is advantageous in fungi, which generally have many nuclei in the cytoplasm, in avoiding mixing of chromosomes during division. Similarly, the nucleolus is usually retained during mitosis and may be stretched out and divided between the daughter nuclei. In some fungi, however, the nucleolus is either lost out of the nucleus, disperses and disappears within the nucleus, or disperses within the nucleus but remains detectable during mitosis. In some fungi, centrioles

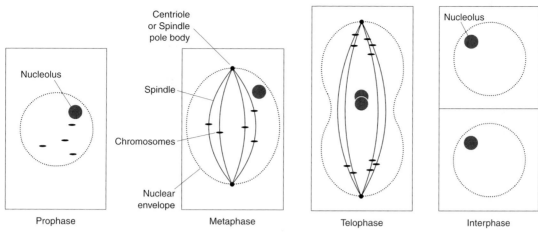

Figure 5 Diagrammatic illustration of the process of mitosis in fungi. Note that the nuclear envelope and nucleolus do not disappear and the chromosomes are randomly distributed at the equatorial plate.

have not been detected and instead there are spindle pole bodies as electron-dense cytoplasmic structures that lie adjacent to the nuclear envelope in most true fungi. Spindle pole bodies are microtubule organizing centres during mitosis and meiosis and have a similar function to that carried out by centrioles in other eukaryotes. In most fungi, the chromosomes are randomly distributed during metaphase and do not define the equatorial plate that is commonly observed in other eukaryotes.

Meiosis

During sexual reproduction, fusion of two haploid nuclei of opposite mating type result in the formation of a zygote. The fusion of nuclei, karyogamy, leads to an increase in nuclear matter and this increase in genetic material is prevented by meiotic division of the nuclei. Meiosis, the process of reduction of chromosome number from diploid to haploid state, requires two nuclear divisions: meiosis I and meiosis II. In meiosis I, pairing of homologous chromosomes at metaphase and the separation of the two homologous chromosomes to opposite spindle poles takes place resulting in two haploid nuclei. Meiosis II is the separation of sister chromatids and is equational as the number of chromosomes remains the same. The majority of the elements of meiosis in fungi are similar to those of eukaryotes.

Meiosis I

Prophase I This is a long and complex phase and differs greatly from mitotic prophase. Prophase is divided in five phases. In the leptotene stage the chromosomes divide longitudinally to form pairs of sister chromatids. Self-duplication of genetic material has already taken place. In the zygotene stage pairing of homologous chromosomes (synapsis) takes place, a requirement for intrachromosomal recombination

process. In the pachytene phase pairing is complete, chromosomes begin to spiralize and appear thick and short. During the diplotene stage the four chromatids of the bivalent are clearly visible as a tetrad. The two sister chromatids remain associated with one another at the undivided centromere. When a pair of chromatids separate, there may frequently be crossing over of non-sister chromatids, called chiasma. Diakinesis, the coiling and resultant shortening of chromosomes reaches its maximum at this stage. The end of prophase is marked by formation of the spindle and dissolution of the nucleolus and nuclear membrane.

Metaphase I The bivalents are arranged at the equatorial plane. The chromosomes attach to spindle fibres through the centromere. The two centromeres of the bivalent are joined to the fibres arising from different poles.

Anaphase I At this stage, the homologous centromeres, which are yet undivided, move to opposite poles of the cell, each carrying with it one of the two chromatid pairs which composed the bivalent. The chiasmata are terminalized and eliminated as homologues separate to the poles.

Telophase I The separation is complete. A daughter nucleus is formed at each pole of the spindle. The chromosomes in the nuclei show loosening of the spirals.

Interphase I Chromosomes are strongly despiralized and appear as thin threads.

Meiosis II

Prophase II Chromosomes again undergo spiralization and become shorter. Centromeres divide longitudinally.

Metaphase II Chromosomes are contracted and the spindle develops from the centromeres. The axis of the spindle is generally at a fixed angle to the direction of meiosis I spindle.

Anaphase II Sister chromatids separate from one another and move to opposite poles. Each centromere has only one chromatid.

Telophase II At the end of this division four nuclei arise from the four chromosome complements originally present in the tetrad.

In fungi, meiosis conforms more closely to the norm for other organisms than does mitosis. In meiosis, both the nucleolus and nuclear envelope degenerate, usually near the end of prophase, although the nucleolus may degenerate later. As in mitosis, the chromosomes are usually irregularly distributed on the spindle at metaphase, but rarely they may be aligned on an equatorial plate. Sometimes the chromosomes are so small and poorly defined even in meiosis that studies of cytology is very difficult. This has led to some confusion in understanding fungal life cycles. The four products of meiosis are usually retained in a single membrane and such meiotic division is referred to as uninuclear meiosis.

Importance with Respect to Compatibility

Sexual reproduction is widespread in nature and allows new genetic combinations to develop and genetic damage to be repaired by recombination. To be effective, it is essential to maintain a high level of out-crossing and in most organisms breeding systems exist which prevent self-fertilization. Heteromorphic systems where sexes are separated and cross-compatible groups are readily distinguishable are rare in fungi. Some species of fungi possess special mechanisms which ensure that only two genetically compatible nuclei are present in each compartment of the hyphae. Genetic barriers to selfing exist and the genes that confer compatibility are known as mating type genes. Extended coexistence of two or more genetically different nuclei in a common cytoplasm (heterokaryosis) is a unique feature of filamentous ascomycetes and basidiomycetes. A heterokaryon can arise as a result of mutation in an originally homokaryotic mycelium, or as a consequence of hyphal fusion followed by intermingling of nuclei. Heterokaryosis confers great potential for variability in fungi. Two kinds of genetic systems affect heterokaryosis:

- heteroallelic compatibility of mating type
- homoallelic compatibility of vegetative type.

Both mating type and vegetative compatibility control self and non-self identification functions, and affect genetic recombination. Mating type identifies a genetically different individual and promotes genetic recombination through the sexual cycle. Vegetative compatibility seems to identify genetically similar individuals, thus limiting heterokaryosis and genetic recombination through a parasexual cycle. The heteroallelic compatibility of mating type requires that the two interacting individuals have different alleles at the *MAT* (mating type) loci to form a sexually functional interaction. The homoallelic compatibility of vegetative systems requires that the two interacting individuals have identical alleles at the *HET* loci to form a vegetatively compatible interaction. The behaviour of heterokaryons, but not their formation is strongly affected by their mating-type loci in most basidiomycetes. Among the ascomycetes, heterokaryon formation in assimilative mycelia is controlled by *HET* genes and rarely by *MAT* genes, *Neurospora crassa* being an exception in that the *MAT* locus also behaves as a *HET* locus.

Mating-type Compatibility

Hyphal fusion (plasmogamy) is unaffected by mating type loci, but the complementation of sexual reproduction requires the two types of nuclei in heterokaryon to have different mating type genes (A and B) at the *MAT* locus, with any two different factors. Several possibilities of heterokaryotic mycelia formation with differing consequences are possible as indicated in **Table 3**. The compatible interaction occurs when the two nuclei have different alleles on mating-type genes A and B and is not possible when either A and B or both have common mating types. The mating-type locus acts as master switch that must be set to the 'on' position for compatible heterokaryotic behaviour. Certain dominant mutations in A and B factors, A_{mut} and B_{mut}, have been obtained that behaved like semi-compatible heterokaryons. The double mutant strain *Coprinus cinereus* has binucleate segments, clamp septa and fertile sporophores. The dominance of these mutations over wild-type counterparts suggests that the mutations allow the formation of a positive genetic regulator comparable to that normally produced by the interaction of different A and B gene products of the dikaryon.

One of the most notable effects of *MAT* loci of eubasidiomycetes is on nuclear migration, which requires the *MATB* locus to be in a compatible configuration. Nuclear migration occurs in sexually compatible dikaryons in common A mating, but not in common B mating, indicating that nuclear migration is controlled by mating-type B. Nuclear migration involves the participation of cytoplasmic

Table 3 Effect of mating type locus on heterokaryotic behaviour in basidiomycetes

Mating type	Compatibility	Phenotype
$A_x/B_r/A_xB_r$	Common AB type; incompatible heterokaryon	No nuclear migration; unstable heterokaryon; sectors to monokaryotic state; simple septa
$A_x/B_r/A_yB_r$	Common B type; hemicompatible heterokaryon	No nuclear migration; unstable heterokaryon; sectors to monokaryotic state; false clamps at septa
$A_x/B_r/A_xB_s$	Common A type; hemicompatible heterokaryon	Nuclear migration; stable heterokaryon; nuclear ratio often strain dependent; simple septa
$A_x/B_r/A_yB_s$	Dikaryon; compatible heterokaryon	Nuclear migration; paired nuclei with complete divisions, clamped septa, sporophores

microtubules, as these have been observed in association with migrating nuclei and the inhibition of microtubule assembly delayed the formation of dikaryon without inhibiting growth. The septa between segments of the hyphae also dissolve during nuclear migration, facilitating nuclear movements. Nuclear migration rates in *Schizophyllum commune* were in the range $500-3000\ \mu m\ h^{-1}$, 5–30 times the hyphal growth rate. Clearly, the nuclear movements could not be accounted for by hyphal growth and by mass protoplasmic flow. Mitochondria do not exchange and migrate during heterokaryon formation either in basidiomycetes or in ascomycetes.

Vegetative Compatibility

Vegetative compatibility is controlled by both allelic and non-allelic systems. Allelic determinants of vegetative compatibility (*HET* locus) prevent compatible interaction between two strains that have different alleles at a particular locus. Only individuals with identical *HET* locus are compatible. *HET* loci, other than the *N. crassa MAT* locus, do not affect mating. Non-allelic vegetative incompatibility occurs when alleles of genes at two separate loci interact as in the R/V incompatibility of *Podospora anserina*. Compatible interaction results in the merging of the two mycelia growing together on agar media. Hyphal fusion occurs and the nuclei may be exchanged between the two mycelia.

Incompatible interactions often result in a barrage reaction with lysis of the hyphae between the two opposed mycelia. Stimulation of asexual sporulation may occur along the borders of the barrage zones. Microinjection studies with *HETC* and *HETD* loci of *N. crassa* demonstrated that heat-labile cytoplasmic constituents are responsible for cell death, suggesting the role of protein factors. Cell death is not the only vegetative incompatibility mechanism. Microinjection studies with *Verticillium dahliae* caused cell death between some, but not all, strains. The *HETI* allelic interaction of *N. crassa* affects heterokaryotic maintenance without causing cell death. Rather, *HETI* nuclei in some way cause the disappearance of *HETi* nuclei, whenever *HETI* nuclei accounted for more than 30% of the nuclei in the heterokaryon. Vegetative compatibility was overcome in *Aspergillus nidulans* and *V. dahliae* by protoplast fusion followed by generation of diploid suggesting an important role of the cell wall in vegetative compatibility.

Importance of Multinucleate States in Hyphae

The hyphae of fungi almost invariably contain large number of nuclei. In aseptate forms, nuclei generally appear to be randomly distributed throughout the cytoplasm of an actively growing hypha. In septate forms, individual hyphal compartments routinely may contain one, two or many nuclei depending on the species and phase of the life cycle. The nuclei of fungi are able to squeeze through the septal pores. Many fungi are haploid, but oomycetes are diploid, and others alternate between haploid and diploid phases. Several nuclei in a common cytoplasm enables the haploid fungi to exploit the advantages of both haploidy and diploidy. This can be understood by considering the relative advantages of both haploidy and diploidy. Haploid organisms have one copy of the genes and express all their genes. They are thus perpetually exposed to selection pressures. Any mutation in these organisms will cause a loss of fitness, and the mutants will be eliminated, or it will lead to an increase in fitness, viz: antibiotic/fungicide resistance genes are acquired. This can be beneficial in the short term, but the corresponding disadvantage is that haploid organisms are not able to accumulate mutations that are not of immediate value. Mutations are often recessive to wild type and so they are not immediately expressed; instead, they accumulate and can recombine in various ways during sexual crossing, so that some of the progeny might have advantageous combinations of mutant genes. The predominance of diploidy in the eukaryotic world indicates that it is a favoured option. Fungi have been able to shield the disadvantages of haploidy by virtue of being multinucleate. If a recessive mutation occurs in one nucleus in a hypha, then the fungus can still behave as a wild type and can store potential variability (mutation) that is the hallmark of diploid organisms. On the other hand mutations can be exposed to selection pressure periodically, when the fungus produces uninucleate spores or when a branch arises with only the mutant nuclear type in it. Diploidy predominates in the rest of the eukaryotic world and in yeasts such as *Candida* spp. and *Saccharomyces* spp., but it seems that the fungal mycelia have remained haploid because they can exploit the advantages of diploid life cycle in any case.

Significance of Parasexual Reproduction

In most organisms, genetic variability results from the sexual cycle, and non-sexual reproduction simply results in procreation of new strains that are genetically identical to the parent. In fungi that lack a sexual cycle, the parasexual cycle mimics sexual reproduction (**Fig. 6**). Parasexuality is the name given to a series of events involving:

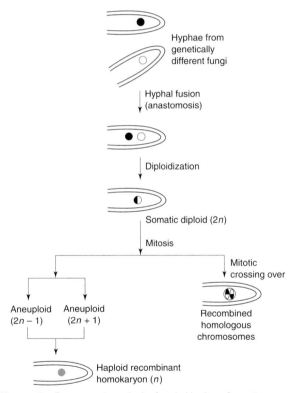

Figure 6 Parasexual cycle in fungi. Hyphae from two genetically different fungi fuse (hyphal anastomosis) giving rise to heterokaryon ($n+n$), followed by nuclear fusion to produce heterozygous diploid nucleus ($2n$, diploidization). The diploid nucleus may undergo recombination between two homologous chromosomes (mitotic crossing over) to produce recombinant fungi. The recombinant can also be produced during mitosis, as a result of mitotic error, resulting in aneuploids with abnormal chromosome number ($2n+1$ and $2n-1$). The aneuploids finally lose chromosomes to restore the haploid chromosome number (haploidization). Since the chromosomes that are lost can come from either of the original homokaryons, a recombinant can result, even if the mitotic crossing over has not occurred.

- heterokaryon formation
- diploidization
- somatic crossing over
- haploidization.

The parasexual cycle is an unusual diploid phase in the asexual cycle of the fungus as a result of fusion of somatic vegetative nuclei which are heterozygous. In many ascomycetes, basidiomycetes and deuteromycetes fusion between vegetative hyphae (hyphal anastomosis) of the same colony is common during colony development. However, hyphal anastomosis is also common when two colonies of the same species come into contact, and is often seen between the hyphae originating from germinating spores. If the hyphae that fuse carry genetically different nuclei, the colony that develops may be a heterokaryon, with nuclei that differ in genotype in a common cytoplasm. If they differ with respect to cytoplasmic genetic elements, e.g. mitochondrial DNA or plasmids, the resulting colony may be a heteroplasmon, with cytoplasmic genetic elements originating from different sources. Heterokaryons can be readily produced by bringing together different mutants that have been derived from the same parental strain. A very effective method is to place two mutants with different nutritional deficiencies on the medium that lacks either nutrient. Under such conditions neither mutant will be able to thrive since each is able to synthesize only one of the missing nutrients. However, heterokaryotic hyphae will have both types of nuclei and hence the enzymes capable of synthesizing both the missing nutrients. They will hence proliferate and a heterokaryotic colony will be established.

Occasionally, two haploid nuclei in a vegetative mycelium as a result of hyphal anastomosis may fuse to give a somatic diploid (diploidization). If the mycelium is heterokaryotic then the fusion may be between genetically unlike nuclei to give a heterozygous diploid nucleus. Such diploidization is rare, occurring perhaps once in a population of a million nuclei. In a growing colony, the number of diploid nuclei (like the number of haploid nuclei) will increase by mitosis. During the division of such nuclei, mitotic crossing over may occur generating variability. This is a rare event compared with meiotic crossing over, occurring perhaps once per five hundred mitoses. Errors in mitosis are quite common in fungi because the chromosomes are randomly distributed and do not lie at the equatorial plane. Sometimes the mitosis of diploid nucleus (chromosome number $2n$) may give rise to one $2n+1$ daughter nucleus (i.e. having three copies of one chromosome) and another $2n-1$ nucleus (i.e. with only one copy of a chromosome). Such abnormalities in chromosome number (aneuploidy) commonly lead to poor growth and any further change in chromosome number that gives improved growth will be favoured. The loss of chromosome present in three copies in a $2n+1$ nucleus will restore the diploid state. In a $2n-1$ nucleus the sequential loss of chromosomes present as two copies can occur to yield a haploid nucleus. Such conversion of diploid into haploid nucleus (haploidization) may occur with about one diploid nucleus in a thousand and take several successive mitoses to accomplish. Not only mitotic crossing over but haploidization can itself result in genetic recombination, since the chromosomes that are lost can come from either of the homokaryons that contributed to heterokaryon formation. The inclusion of recombinant haploid nucleus in a uninucleate spore yields a recombinant heterokaryon that differs from both the original homokaryons. The sequence of events that yields such a recombinant strain has been termed the parasexual cycle.

Although nonsexual reproduction can be an advantage in a static environment, it does not provide the genetic flexibility required if the environmental conditions change. In certain fungi this is achieved by the phenomenon of parasexual reproduction. The parasexual cycle will normally enhance the chances of survival of a fungus because there would be a large genetic reservoir from which to draw the material for adaptations, all accomplished without the delay of passing through a sexual cycle. The parasexual cycle has been of great value in genetic analysis. It has been suggested that it may have been of importance in nature, especially in generating variability in a number of industrially important fungi lacking a sexual cycle, such as *Aspergillus niger* and *Penicillium chrysogenum*. Parasexuality has become a valuable tool for industrial mycologists to produce strains with the desired combination of properties. Now, however, there is considerable doubt as to whether this is the case, since the widespread occurrence of vegetative incompatibility may prevent heterokaryosis except between very closely related strains.

See also: **Aspergillus**: Introduction. **Candida**: Introduction. **Escherichia coli**: Escherichia coli. **Fungi**: The Fungal Hypha. **Kluyveromyces. Saccharomyces**: Introduction; *Saccharomyces cerevisiae*. **Yeasts**: Production and Commercial Uses.

Further Reading

Bennett JW and Lasure LL (1991) *Gene Manipulations in Fungi*. San Diego: Academic Press.
Carlile MJ and Watkins SC (1994) *The Fungi*. London: Academic Press.
Deacon JW (1997) *Modern Mycology*, 3rd edn. Oxford: Blackwell Science Publications.
De Robertis EDP and De Robertis EMF (1987) *Cell and Molecular Biology*. New Delhi: B.I. Waverly.
Fincham JRS, Day PR and Radford A (1979) *Fungal Genetics*, 4th edn. Oxford: Blackwell Science Publishers.
Gow NAR and Gadd GM (1995) *The Growing Fungus*. London: Chapman and Hall.
Griffin DH (1994) *Fungal Physiology*, 2nd edn. New York: Wiley-Liss.
Landecker EM (1996) *Fundamentals of the Fungi*. Englewood Cliffs: Prentice-Hall.
Peberdy JF, Caten CE, Ogdin JE and Bennett JW (1991) *Applied Molecular Genetics of Fungi*. Cambridge: Cambridge University Press.
Pühler A (1993) *Genetic Engineering of Micro-organisms*. Weinhein, Germany: VCH.

Bacteria

S K Garg and **Rajat Sandhir**, Department of Microbiology, Dr Ram Manohar Lohia Avadh University, Faizabad, India

Genetics is the study of mechanisms by which traits are passed from one organism to another. The study of genetics is central to the understanding of an organism. It also provides us with approaches to construct new organisms for potential use in human affairs such as agriculture, medicine and industry. Genetic phenomena include three types of macromolecules: deoxyribonucleic acid (DNA), ribonucleic acid (RNA) and protein.

The DNA molecule is the blueprint of life and carries within its sequence the hereditary information that determines the metabolic and structural nature of an organism. It contains the instructions by which cells grow, divide and differentiate, and has provided a basis for evolutionary process both within and between related species. Expression of genetic information involves synthesis of proteins using information encoded within DNA. Ribonucleic acid is an intermediate molecule that converts the information in DNA into a defined protein sequence. Historically, DNA was first identified as the hereditary material in 1944 by Avery, MacLeod and McCarty who established that DNA was a transforming factor, capable of passing genetic information between cells and responsible for the pathogenicity of the *Pneumococcus* bacterium. This observation was confirmed by Hershey and Chase in 1952 by differentially labelled T2 bacteriophage DNA with ^{32}P and its protein coat with ^{35}S and found that only ^{32}P was transferred to the progeny phage following infection of *E. coli*. Thus

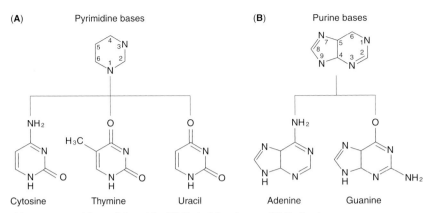

Figure 1 Structure of bases present in nucleic acids. (**A**) Pyrimidine bases. (**B**) Purine bases.

DNA is an information-carrying molecule and is involved with storage of genetic message.

Composition of Nucleic Acids

Structurally, DNA is a polymer of deoxyribonucleotide monomers. Each nucleotide consists of three components: a heterocyclic nitrogen base, a pentose sugar and a phosphate group. The bases are of two types: the *purines* adenine (A) and guanine (G), and the *pyrimidines* cytosine (C) and thymine (T) (**Fig. 1**). The base is connected by a β-glycosidic linkage between carbon 1 of the deoxyribose sugar and nitrogen 1 of pyrimidine or nitrogen 9 of purine base to form a nucleoside (**Fig. 2**). The 5′-hydroxyl group of the sugar is phosphorylated to form a deoxyribonucleotide, the primary building block of the polynucleotide, DNA. Ribonucleic acid, a polymer of ribonucleotide monomers, has uracil (U) instead of thymine and ribose sugar instead of deoxyribose (Fig. 2). The nucleotides of DNA or RNA are linked through phosphodiester bonds between the 3′-OH of the sugar moiety of one nucleotide with the 5′-OH of the other.

Chargaff in 1952 investigated the molar pro-portions of bases in DNA of various organisms and discovered certain regularities in the DNA composition, known as Chargaff's rule: the sum of purines is always equal to the sum of pyrimidines. The knowledge of equivalence of A:T and G:C was extremely important in establishing the concept of the DNA double helix. The physical properties of DNA are strongly influenced by the percentage of G+C content: G+C content as low as 20% and as high as 78% are known in lower eukaryotes and prokaryotes. Since the G+C content of DNA is a characteristic marker of a genus or family, its determination has relevance in taxonomy. Depending on the level of development of an organism, the length of DNA varies from several micrometres in prokaryotes (200 μm in *Mycoplasma*) to approximately 1 m in human cells.

Structure of DNA

The three-dimensional structure of DNA was deduced by Watson and Crick in 1953 from the X-ray diffraction pattern of polycrystalline DNA fibres. They deduced that DNA formed a right-handed double helix containing two polynucleotide chains arranged in an antiparallel manner and coiled around a common axis. The sugar–phosphate formed the helical backbone, and the bases were disposed on the inside of the helix. The two strands of the DNA helix were held together by hydrogen bonding (**Fig. 3**). As a result of specific base pairing (A:T and G:C) the two strands of DNA are complementary. Thus, whereas the sugar and phosphate have structural roles, the bases are responsible for enhancing the stability of the DNA helix and providing specific DNA sequence encoding the genetic information for the production of proteins.

A number of studies showed that DNA can adopt various structures which are dependent on solvent, salt, temperature and DNA sequence. The crystal structure of defined oligonucleotides revealed

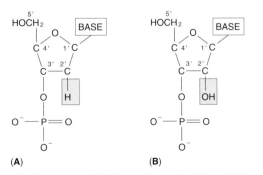

Figure 2 Structure of (**A**) deoxyribonucleotide and (**B**) ribonucleotide. The two nucleotides differ at the 2′ carbon positions of the sugar.

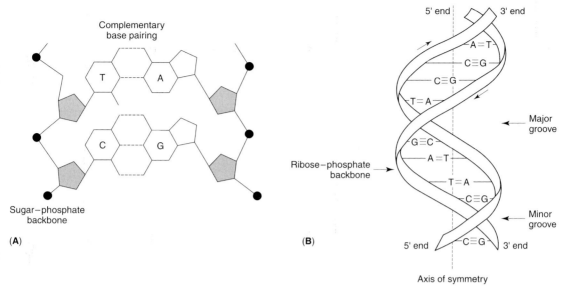

Figure 3 Structure of DNA. (**A**) Base pairing between two strands. (**B**) DNA double helix illustrating major structural features.

systematic sequence-dependent structural modulations in the helices, indicating that DNA is highly polymorphic in nature. The right-handed helices have been classified as type A (A-DNA) or type B (B-DNA), and there is a more recently discovered left-handed helix (Z-DNA). The DNA helix appears to be a dynamic structure comprising many different helical forms which may exist in equilibrium. The helical parameters that determine the overall shape of the double helix for different forms of DNA are given in **Table 1**. According to the structural properties, A-DNA is a more compact helix than the classical B form of DNA which is slim and tall, while Z-DNA is much slimmer and elongated. The exterior of a DNA double helix is discontinuous owing to the base–base interactions displaying two types of grooves – major and minor (Fig. 3). The grooves are important determinants of protein-nucleic acid interactions. The B-DNA has a wider major groove and a narrow minor groove of comparable depth, while Z-DNA has a deep, narrow minor groove, and a major groove which is pushed to the surface so that the groove is

nearly eliminated. The A-DNA structure is favoured under conditions of low relative humidity, and B to A transition can be induced by dehydration.

Polymorphism of DNA in vitro is evident; however, which conformation is present in vivo? Although most of the DNA in the genome is in the classical double-stranded B form, evidence for the presence of A-DNA and Z-DNA is also available. Regions of Z-DNA in the genome may be important for interaction with protein and enzymes. The presence of A-DNA is observed in dormant spores of *Bacillus* species associated with small, acid-soluble, double-stranded DNA-binding protein. It is suggested that presence of A-DNA may result in drastic reduction in RNA synthesis, therefore, decreased cell growth and increased spontaneous mutation and finally cell death. Condensation of a cell's genome to an A-like component in a dormant bacterial spore, in which gene expression as well as metabolic activity is absent, suggests that A-like DNA may be a dormancy adaptation.

Table 1 Helical parameters of DNA types A, B and Z

Parameter	A-DNA	B-DNA	Z-DNA
Helix type	Right-handed	Right-handed	Left-handed
Diameter (approximate)	2.6 nm	2 nm	1.8 nm
Base pairs per helix turn	11	10.4	12
Helix pitch (rise per turn)	2.8 nm	3.4 nm	4.5 nm
Base tilt normal to the axis	20°	6°	7°
Major groove	Narrow, deep	Wide, deep	Flat
Minor groove	Wide, shallow	Narrow, deep	Narrow, deep
Favourable factors	Sodium salt	Sodium salt	Lithium salt
	73% relative humidity	92% relative humidity	66% relative humidity

Arrangement of DNA in Prokaryotes

The molecular mass of typical bacterial DNA is 10^9–10^{10} daltons (e.g. the molecular mass of *Escherichia coli* DNA is 2.7×10^9 Da). Bacterial DNA is not surrounded by a membrane as is typical of the eukaryotes. The bulk of the bacterial DNA is present as a single aggregated molecule termed a nucleoid. Although in the eukaryotes DNA is present in chromosomes, in prokaryotes DNA is present primarily as a naked molecule, arranged as a covalently closed circular molecule which is extensively folded and twisted (supercoiled) to fit in to the bacterial cell (**Fig. 4**). The amount of twisting and coiling can be appreciated when it is considered that the 4.2 million base pairs in the genome of *E. coli*, if opened and linearized, would be about 1 mm in length, yet the *E. coli* cell is only about 2 μm long! The packaging of this DNA into a bacterial cell requires it to be supercoiled, giving more compact shape than the free circularized counterpart. The supercoiling or twisting is achieved with the help of an enzyme, DNA gyrase. In prokaryotes, it is well established that the supercoiled state of DNA plays an important role in various biological processes including replication, transcription and recombination, by altering the accessibility of DNA to protein. The bacterial chromosome lacks the basic proteins, called histones, that are important for coiling of the DNA in the eukaryotic chromosome. Histone-like protein has been reported in only a few archaebacteria and eubacteria. The proteins usually associated with bacterial chromosomes are those involved with replication, transcription and regulation of gene expression.

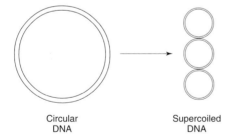

Figure 4 DNA in open circular and supercoiled forms.

Structure of RNA

There are three major types of RNA: ribosomal RNA (rRNA), transfer RNA (tRNA) and messenger RNA (mRNA). Unlike DNA, they are small, single-stranded molecules. There are three distinct species of rRNA in prokaryotes (5S, 16S and 23S) and four species in eukaryotes (5S, 5.8S, 18S and 28S). Ribosomal RNA makes up approximately 80% of the RNA in a cell and is found in association with a number of different proteins as components of the ribosomes – sites for protein synthesis. The second type of RNA, tRNA, is the smallest (4S, 73–93 nucleotides). There is a specific type of tRNA for each amino acid. They constitute 15% of the cell RNA. Transfer RNA contains many unusual bases and has extensive intrachain base pairing. The tRNA acts as an 'adaptor' molecule, carrying the specific amino acid to the site of protein synthesis. Messenger RNA is a carrier of genetic information from the DNA and is used as a template for protein synthesis; it comprises only about 5% of the cell RNA and is heterogeneous in size.

Plasmids

Plasmids are extrachromosomal genetic elements composed of DNA that replicate independently of the host chromosome (**Fig. 5**). Plasmids are found in prokaryotes and some eukaryotes. However, not all bacterial cells possess plasmids. Plasmids are mostly circular, duplex, supercoiled DNA molecules ranging in size from those that encode no more than 10 genes (2.5×10^5 Da) to those that encode for several hundred (1.5×10^8 Da). The large plasmids are present as only 1–2 copies per cell, whereas there may be 20–100 copies per cell of small plasmids. Plasmids can be categorized on the basis of their presence as multiple copies in a cell (*relaxed* plasmids) or as a limited number of copies (*stringent* plasmids).

Plasmids contain a limited number of specific genes that supplement the essential genetic information contained in the bacterial chromosome. The supplemented information can be vital for the cell, establishing mating capabilities (F plasmids), resistance to antibiotics (R plasmids), production of toxins (e.g. enterotoxins), induction of plant tumours (Ti plasmid) and tolerance to toxic metals. Some plasmids contain genes for catabolism of unusual substances, such as aromatic compounds and pesticides; they are referred to as degradative plasmids. **Table 2** lists the properties and phenotypic traits conferred by plasmids on the host cells. Such supplemental genetic capability can permit the survival of the bacterial cell under unfavourable conditions. Pathogenic bacteria containing plasmids encoding for multiple drug resistance are of concern in the treatment of certain

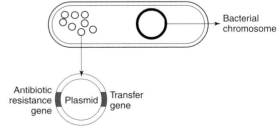

Figure 5 A bacterial cell showing plasmids.

Table 2 Properties of some of the natural plasmids

Phenotypic trait	Host	Plasmid	Relative mol. mass	No. of copies per cell
Resistance plasmids (resistance to antibiotics)	Salmonella paratyphi	R	58×10^6	1–2
Production of antibacterial proteins, e.g. colicins	Escherichia coli	Col E1	4.3×10^6	15
Fertility plasmid (conjugation)	E. coli K12	F	63×10^6	1–2
Transformation plasmids (tumour-induction in plants)	Agrobacterium tumefaciens	Ti	$90–160 \times 10^6$	1–2
Metabolic plasmids (degradation of toluene)	Pseudomonas putida	TOL	78×10^6	?

infectious diseases in humans. Plasmids to which no phenotypic traits remain associated are called *cryptic plasmids*.

Many plasmids can be transferred from one cell to another by means of the conjugation process. Plasmids are categorized as *conjugative* or *non-conjugative* depending on whether or not they carry a set of transfer genes, the *tra* genes, which promote bacterial conjugation. Generally, conjugative plasmids are of relatively high molecular weight and are present as 1–3 copies per cell, whereas non-conjugative plasmids have a low molecular weight and are present as multiple copies per cell. Some plasmids (episomes) also have the ability to integrate into the bacterial chromosomes.

Bacterial cells may contain more than one type of plasmid, but certain pairs of plasmids cannot be stably replicated in the same bacterial cell; these are called incompatibility plasmids.

Plasmids as Vectors

Plasmids are useful in genetic engineering as vehicles for carrying genetic information from different sources. The following properties in plasmids make them vectors of choice for cloning:

1. The small size of the plasmid molecule makes isolation and manipulation easy.
2. The circular form of the DNA molecule makes it more stable during chemical isolation.
3. Independent origin of replication permits replication to proceed without direct cell control.
4. Multiple copies of plasmids in the cell result in amplification of genes.
5. Markers such as antibiotic resistance genes help in detection and selection of plasmid-containing clones.
6. Transformation into the host cell is easy.

An example of a typical cloning plasmid is pBR322

(**Fig. 6**). It is an artificially constructed plasmid vector and has most of the above-mentioned advantageous characteristics. It is small (2.6×10^6 Da), maintained in high copy number (20–30 per cell) and has genes for resistance to ampicillin and tetracyclines. In addition, it has a single cloning site for various restriction enzymes. The advantage of having a single cloning site is that the plasmid DNA will only be linearized to accommodate the foreign DNA, while if it has multiple cloning sites for a restriction enzyme, it will be chopped into pieces.

Reproduction of Bacterial Cells: Binary Fission

Most bacteria reproduce asexually by binary fission, a process in which a cell divides to produce two daughter cells of nearly equal size. Binary fission involves three steps: increase in cell size (elongation); DNA replication; and finally cell division. In a growing culture of a rod-shaped bacterium such as *E. coli*, for example, cells are observed to elongate to approximately twice the length of an average cell and then to form a partition which eventually separates the cell into two daughter cells. The partition (septum) is a result of inward growth of plasma membrane and cell wall from opposite directions until the two daughter cells are produced. Autolysin is believed to be produced by the cells and assists in separation of

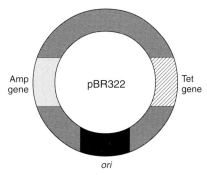

Figure 6 Structure of an artificial plasmid pBR322. The *ori* sequence indicates the origin of replication.

the daughter cells. During the growth cycle all cellular constituents increase in number such that each daughter cell receives a complete chromosome and sufficient copies of all other macromolecules and cellular constituents to exist as an independent cell. Replication of DNA starts at multiple initiation sites in DNA. The gene *dnaA* is associated with initiation of DNA replication, which occurs when the concentration of *dnaA* protein is high. The generation time in *E. coli* is 20 min; a few organisms grow even faster, but many grow more slowly. The mechanisms that control the process are complex.

Mechanisms of Gene Transfer

The ability of organisms to exchange genes within a population is nearly universal. Genetic exchange occurs in bacteria by three different mechanisms: conjugation, transformation and transduction.

Conjugation

Conjugation is a process of 'bacterial mating' in which genetic exchange occurs between two cells in direct contact with one another. In this process plasmid DNA is usually transferred from donor to recipient cell, but sometimes mobilization of chromosomal DNA also occurs.

Conjugation occurs in certain strains designated as F⁺ (fertility plus) acting as donor (male) and others designated as F⁻ (fertility minus) acting as recipient (female). The essential feature of the conjugation process is a plasmid that encodes for the ability to conjugate with other bacterial cells. It is established that all F⁺ strains contain an F plasmid which carries all genes encoding for conjugative genetic transfer. The F plasmid encodes for genes that are responsible for replication of plasmids, synthesis of sex pili and genes for transfer. One or more sex pili are produced by F⁺ cells and help in binding to F⁻ cells. Sex pili form conjugative bridges through which DNA passes. After the formation of a conjugation tube, a nick is formed in the F plasmid, followed by replication employing a rolling circle method in which the intact strand is used as a template and the 3' end of the nicked strand is used as a primer. By this action the 5' end of the nicked strand is displaced and is transferred to the F⁻ cell. The synthesis of complementary strand for newly transferred single-stranded DNA takes place in the recipient cell followed by recircularization (**Fig. 7**).

In certain situations the chromosomal genes are also transferred along with F plasmid cells: such cells are called *high-frequency recombination* (Hfr) strains. The F plasmids are integrated into the bacterial

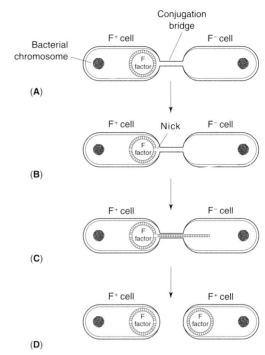

Figure 7 Conjugation process in bacteria. (**A**) The F⁺ cell forms a conjugation tube with an F⁻ cell. (**B**) The F factor nicked by nuclease in one of the strands. (**C**) DNA replication by the 'rolling circle' method, leading to transfer of single-stranded DNA to the recipient cell and simultaneous synthesis of the complementary strand. (**D**) Recircularization of transferred DNA in the F⁻ cell, making it F⁺, followed by cell separation.

chromosome by means of insertion sequences, present in both plasmids and bacterial chromosome. When Hfr cells come in contact with F⁻ cells, replicative transfer begins within the F plasmid region and continues into the chromosomal region leading to transfer of chromosomal genes along with F⁻ DNA. Further, there is also evidence indicating that F plasmids direct the transfer of other non-conjugative plasmids present in the F⁺ cell. The F plasmid-mediated transfer of antibiotic resistance genes on R plasmids is important in the rapid spread of resistance to commonly used antibiotics within the Enterobacteriaceae.

The mechanism of conjugation in Gram-positive bacteria, although not fully understood, differs from Gram-negative bacteria to some extent; conjugation is not mediated by the sex pili, but plasmid-containing cells form clumps with cells that lack the plasmid. The transfer of plasmids takes place within these clumps. Clumping is assisted by interaction between an aggregation substance on the surface of plasmid-containing cells and a binding substance on the surface of plasmid-deficient cells. The recipient cell produces a sex pheromone that induces conjugation by enhancing cell-to-cell contact, through cell aggregation, production of specific surface proteins and functions necessary for physical transfer of DNA.

Transformation

Transformation is the process of genetic modification of bacterial cells by uptake of DNA from the surrounding medium. It was the first mechanism of genetic exchange to be demonstrated in *Pneumococcus* by F. Griffith in 1928. Later in 1944 Avery, Macleod and McCarty identified the pneumococcal transforming factor as DNA. With the increased importance of recombinant DNA technology, transformation has acquired additional importance as the technique for introducing recombinant DNA molecules into the cells.

The critical feature for all bacterial transformations is the ability to take up DNA fragments, a condition known as *competence*. Entry of DNA is encoded by chromosomal genes and signalled by environmental factors; such bacteria are capable of undergoing natural transformation. Certain bacteria do not become competent under ordinary conditions, but can be made competent under artificial conditions such as exposure of the cells to high concentrations of divalent cations and manipulated culture conditions.

The Gram-positive bacterium *Streptococcus pneumoniae* becomes competent during the exponential phase of growth; this process is mediated by proteins called competence factors, which are continuously produced and secreted into the medium. These competence-specific proteins include DNA-binding proteins, a cell wall autolysin and various nucleases. The DNA from the medium binds to the outer surface of the competent cell. One of the strands of the DNA is then digested by the nuclease and the other strand enters the cell while being bound to competence-specific, single-stranded DNA-binding proteins. The cells take up DNA regardless of its source. A DNA molecule that is homologous to the recipient DNA will become integrated and finally alter the recipient cell, whereas non-homologous DNA is degraded. Recombination with homologous DNA is mediated by *RecA* protein. Integration of homologous DNA occurs by the process of strand replacement. A heteroduplex is first formed between the newly transferred DNA fragment and the host DNA. The mismatched regions in the heteroduplex sometimes activate DNA repair mechanisms to remove mismatched bases. This results in removal of bases from the transferred DNA, leading to no heritable change in the recipient cells. If, on the other hand, replication occurs first and two homoduplex copies are made, one of which is identical to the copy taken from outside the cell (exogenote), this results in a genetically altered recipient cell (**Fig. 8**).

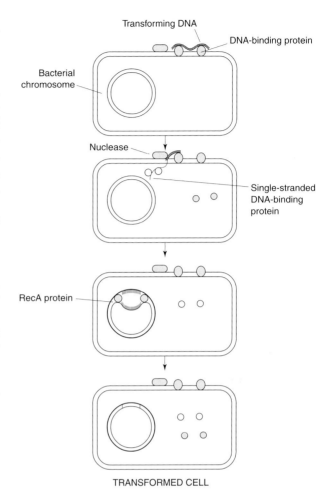

Figure 8 Events involved in transformation in bacteria. The transforming DNA (double-stranded) first binds to the cell surface via DNA-binding proteins followed by degradation of one of the strands by nuclease. The single strand simultaneously moves into the cell, being bound to specific single-stranded DNA-binding proteins. The single-stranded DNA undergoes recombination mediated by RecA protein, finally transforming the recipient cell.

The Gram-negative bacterium *Haemophilus influenzae* is transformed by a mechanism different from that in Gram-positive bacteria in many fundamental aspects (**Table 3**).

1. Competence is not mediated by competence factor (protein); rather, the cells become competent as a result of growth in rich medium.
2. Only homologous DNA (same or closely related species) can bind and is taken up by the cells.
3. Transforming DNA enters in double-stranded form and remains in the same state until the time of integration.
4. The outer membrane of the competent cells on average contains 10 vesicular structures (blebs) with pores at the base. Embedded in this is a protein that specifically binds with an 11 base pair sequence of DNA (5'-AAGTGCGGTCA-3') which

Table 3 Differences in transformation systems in *Streptococcus pneumoniae* and *Haemophilus influenzae*

Characteristics	S. pneumoniae *(Gram-positive)*	H. influenzae *(Gram-negative)*
Competence	Mediated by competence factor (protein)	By growth in rich medium
Form in which DNA enters the cell	Single-stranded	Double-stranded
Source of DNA	Any source	Only homologous
State of DNA in the cell	Protein-bound	Transformasome

occurs at 600 sites on the *Haemophilus* chromosome or about one site per 4000 bp and acts as a signal for uptake. The DNA moves in the cells by forming membranous vesicles called *transformasomes*, which are resistant to the action of DNAase and restriction enzymes.

Transduction

Transduction is a genetic exchange process involving DNA transfer from one prokaryotic cell to another by a viral carrier. Viruses that attack bacterial cells are called bacteriophages or phages. Transduction involves the formation of an aberrant phage virion in which some or all the viral DNA is replaced by bacterial DNA (donor DNA); this 'transducing' phage then attacks and introduces DNA into another susceptible bacterial cell.

The protein capsid of the transducing particles determines the ability of the phage to attach to a sensitive bacterial cell and inject its complement of DNA into the cell. The transducing particle can introduce bacterial DNA derived from the cell in which it is developed into another somatic cell, resulting in transfer of genetic material between these two cells.

Transduction may be *generalized* (nonspecialized) or *restricted* (specialized).

Generalized Transduction Generalized or non-specialized transduction brings about transfer of essentially random regions of the bacterial chromosome. The process results in genetic exchange of homologous alleles. One example of generalized transduction is phage P22 infection of *Salmonella typhimurium*. In the generalized transduction process bacterial DNA fragments are accidentally acquired by developing bacteriophage during their normal replication within the host bacterial cell. During normal replication, the phage invades the host cell, followed by replication of virus within the host cell and finally the lysis of the host cell to release the newly formed phage. During replication, long stretches of phage DNA composed of tandemly repeated phage DNA (concatamers) are synthesized. Before packaging the DNA into phage, concatamers are cleaved by the phage-encoded endonuclease at a specific site called 'pac' (packaging) site. The capacity of virus head to package the DNA is limited. In a lytic infection, the

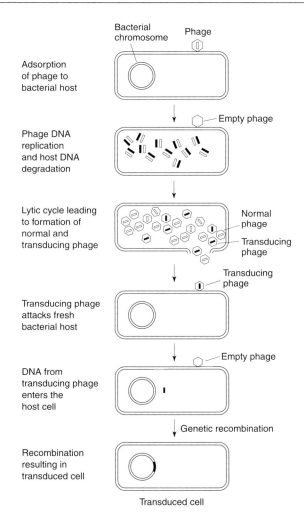

Figure 9 Generalized transduction forming phages with host DNA. The virus particle attacks the bacterial cell initiating a lytic cycle leading to production of normal as well as transducing phage particles (carrying host DNA). The transducing particle then attacks another cell and the transferred DNA undergoes genetic recombination with host DNA to form a transduced cell.

host DNA may also break into virus-sized pieces if it has 'pac'-like sites and some of these pieces become packaged into virus particles. Such phage is called *defective* phage since it does not have viral genes. After lysis of the host cell, the lysate contains mixture of normal and defective virus particles. The defective phage particles can attach and inject DNA into another recipient cell. Once inside the recipient cell, the DNA may be degraded by the nuclease or may

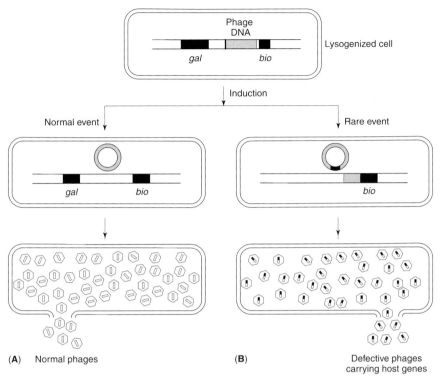

Figure 10 Specialized transduction leading to production of phages carrying host genes. The example shown is λ virus in *E. coli* carrying genes for galactose utilization. Induction of the lysogenized cell is followed either by (**A**) a *normal event* – the phage DNA detaches followed by replication and phage synthesis and finally lysis of the cell releasing normal phages – or by (**B**) a *rare event* – sometimes a portion of host DNA (*gal* in this case) is exchanged for phage DNA, releasing defective phages carrying host genes.

undergo homologous recombination. If recombination occurs, the transduced recipients may possess new combination of genes (**Fig. 9**). Some bacteriophages capable of mediating generalized transduction can transfer up to 90 kb of chromosomal DNA at one time.

Restricted Transduction Restricted or specialized transduction mediates the exchange of only a limited number of specific bacterial genes. The best example involves the transduction of galactose genes by the temperate phage λ of *E. coli* (**Fig. 10**). Bacteriophage λ is capable of establishing lysogeny, in which phage incorporates its DNA into the bacterial genome. Bacteriophage λ DNA becomes integrated into the host DNA at specific sites. The region where λ integrates (λ *att* site) is immediately adjacent to the cluster of host genes that control the enzymes involved in galactose utilization (*gal* genes) and biotin synthesis (*bio* genes). Occasionally during induction of lytic phase, when viral replication occurs, the phage DNA carries within it adjacent bacterial genes and leaves behind some of the viral genes, since the capacity of phage to accommodate DNA is limited. This makes phage defective because some of the viral genes are lost. The

formation of defective phage establishes a viral carrier of bacterial DNA. The defective phage is incapable of development except in a cell that contains another λ phage (helper phage) providing the missing genes. Because there is a specific site of integration of the phage into bacterial chromosome, only genes adjacent to the site of insertion of the viral genes may be transferred by specialized transduction, e.g. λ phage of *E. coli* transfers the genes for galactose utilization or the genes for biotin synthesis by specialized transduction. By manipulating the site of integration of λ into other regions of the *E. coli* chromosome, other genes may be transferred.

Most of the transducing phages are capable of mediating only one of these types of transduction but a limited number are capable of mediating both. The potential role of transduction in natural genetic exchange processes lies in the fact that bacteriophages occur in natural water at much higher titre than had been previously thought.

Applications of Genetic Exchange Processes in Food Technology

The principal aim of the food industry involves construction of strains with desired metabolic properties.

Genetic exchange mechanisms are characterized in these organisms to elucidate the genetic basis of metabolic activity and aid in improving the strains of organisms. The lactic acid bacteria are important in the food industry for a variety of biotechnological reasons. In lactic acid bacteria there are certain plasmid-encoded characteristics such as lactose transport and hydrolysis, proteolytic activity, citrate permease, bacteriocin production and phage resistance. Plasmids capable of conjugal transfer of lactose genes have been extensively studied, and while such genes are usually located on plasmids, there is evidence for their chromosomal integration under some circumstances. *Lactobacillus acidophilus* produces certain antimicrobial proteins or bacteriocins that mediate antagonism towards other microorganisms. The organisms having antibiotic resistances genes may survive conditions of oral antibiotic therapy. Further, a problem in *Lactobacillus* culture is attack by bacteriophages. Therefore, development of bacteriophage-resistant starter cultures is important. The discovery and characterization of insertion sequences in lactic acid bacteria has led to exploitation of systems of genetic exchange in vivo to introduce certain characteristics in the organisms. Genetic exchange mechanisms have been used for lactobacilli to introduce predetermined traits like phage resistance and bacteriocin production.

Genetic Engineering

The terms 'genetic engineering', 'recombinant DNA technology' and 'gene cloning' can be broadly used to include all the techniques designed to manipulate the genetic constitution of organisms. However, in recent years the term has been used in a narrower sense to mean the enzymatic preparation of recombinant DNA molecules and their subsequent cloning in bacterial cells. These techniques made possible the combination in vitro of genes from different sources or organisms as unrelated as humans and bacteria, to produce new combinations that do not occur biologically. The technique essentially involves the steps described below (**Fig. 11**).

Isolation of Genes

The first step in gene cloning is the isolation and identification of a DNA fragment (gene) of interest. Isolation of the gene is carried out by digesting the genomic DNA with a restriction endonuclease and separating the fragment containing the desired gene – the 'shotgun' approach. The other method for obtaining a gene of interest is to synthesize complementary DNA from mRNA for the gene to be isolated using the enzyme reverse transcriptase. This method is more

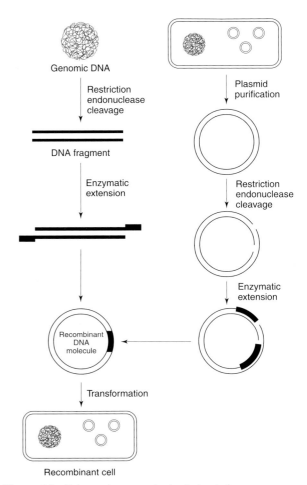

Figure 11 Scheme for gene cloning in bacteria.

suitable in situations where a eukaryotic gene carrying introns has to be expressed.

Combination with the Vector

After isolation of the concerned gene, the next step is to carry the DNA with the help of a suitable vector to the host cells. Plasmids and bacteriophages are the most commonly used vectors. The selected DNA fragment is joined with the vector to obtain a recombinant DNA molecule.

Transformation and Expression of Cloned DNA

After the formation of recombinant DNA molecule, it is transformed to a suitable host cell (recipient cell) for expression. The cells carrying the foreign DNA are selected from the cells not carrying the foreign DNA on the basis of the presence of selectable markers. After selection of the cells, the clones expressing the desired product are cultured for various applications.

Significance of Recombinant DNA Technology

Genetic engineering or recombinant DNA technology has added unprecedented scope to our efforts to modify or improve microorganisms and increase their industrial utility. It has introduced many new possibilities for employing microorganisms to produce economically important substances. The mutation and selection approach is 'hit or miss' compared with the use of recombinant DNA technology, which permits the purposeful manipulation of genetic information to engineer a microorganism that can produce high yields of a variety of products. Until the breakthrough in recombinant DNA technology, a bacterial cell could only produce substances encoded within the bacterial genome.

A number of important products have been made on an industrial scale using microorganisms. Recombinant DNA technology has been used for genetic improvement of the product, to improve the yield of scarce product, to obtain industrially important enzymes and proteins, and to obtain microorganisms with desirable characteristics. The techniques have been successfully used to produce modified antibiotics. Currently microorganisms are used to provide enzyme preparations for use mainly in the chemical and food industries. The major applications of recombinant DNA technology in the food industry are:

1. *Production of enzymes and proteins:* an early example of the application of recombinant DNA technology to the food industry was production by *E. coli* of calf chymosin (also called calf rennin, or rennet), a milk-clotting enzyme used in cheese manufacture. This enzyme was initially isolated from the gut of calves; however, there was a decline in the supply of the enzyme because more calves were being raised to maturity. The deficit was overcome by exploitation of recombinant DNA technology to produce the enzyme in *E. coli*. **Table 4** lists some other enzymes and proteins produced using recombinant DNA technology. Lactase (β-galactosidase) has been commercially prepared to produce lactose-free milk commercially for lactose-intolerant individuals.

2. *Accelerated ripening:* gene manipulation has been employed successfully to obtain setting of cottage cheese curd in 90 min compared with 8–16 h for traditional starters.

3. *Phage-resistant starters:* phage attack is the single major cause of starter culture failure and causes incalculable loss to the industry. Recombinant DNA technology has been used to induce phage resistance in dairy starters.

4. *Construction of strains producing proteinaceous antimicrobial compounds:* certain strains of *Streptococcus*, *Lactobacillus* and *Pediococcus* produce plasmid-encoded proteins such as bacteriocins which have antimicrobial properties. It has been possible using recombinant DNA technology to construct strains producing antibiotic-like substances that help in increasing the shelf-life and therapeutic value of dairy products.

5. *Starters with desirable flavours:* recombinant DNA technology has also been used to improve the flavour and to produce desirable acid during cheese manufacture.

6. *Effluent treatment:* by-products of the food industry, such as whey and buttermilk, are of environmental concern and also lead to economic losses. Several fermented beverages, alcohol, lactic acid, enzymes, vitamins, edible oils, amino acids and single cell proteins have all been produced using recombinant DNA technology.

Table 4 Enzymes produced by recombinant DNA technology having applications in food industry

Enzyme	Applications
Proteases	Improvement in flavour and texture of cheese (rennin)
	Meat tenderizer, baking, etc.
Lipases	Flavour and texture of cheese
Lactase	Production of lactose-free milk
Pectinase	Clarification of wine and fruit juice
Invertase	Confectionery

See also: **Bacteria**: The Bacterial Cell. **Bacteriocins**: Potential in Food Preservation. **Genetic Engineering**: Modification of Yeast and Moulds; Modification of Bacteria. **Genetics of Microorganisms**: Fungi. *Lactobacillus*: Introduction; *Lactobacillus acidophilus*. **Starter Cultures**: Uses in the Food Industry; Cultures Employed in Cheese-making. *Streptococcus*: Introduction.

Further Reading

Adams RLP, Knowler JT and Leader DP (1992) *The Biochemistry of the Nucleic Acids*, 11th edn. London: Chapman & Hall.

Birge EA (1988) *Bacterial and Bacteriophage Genetics – An Introduction*, 2nd edn. New York: Springer.

Cantor CR (1989) *Principles of Nucleic Acid Structure*. Berlin: Springer.

Clewell DB (1993) *Bacterial Conjugation*. New York: Plenum.

Dale JW (1989) *Microbial Genetics of Bacteria*. New York: John Wiley.

Dickerson RE (1992) DNA structure from A to Z. *Methods in Enzymology* 211: 67–111.

Friefelder D (1987) *Microbial Genetics*. Boston: Jones & Bartlet.

Hardy KG (1986) *Bacterial Plasmids*, 2nd edn. Wokingham: Van Nostrand Reinhold.

Miller JH (1991) Bacterial Genetic Systems. *Methods in Enzymology* 204: 18–20.

Murrell JC and Roberts LM (1989) *Understanding Genetic Engineering*. Chichester: Ellis Horwood.

Novik R (1980) Plasmids. *Scientific American* 243 (6): 103–123.

Old RW and Primrose SP (1987) *Principles of Gene Manipulation – An Introduction to Genetic Engineering*, 4th edn. Oxford: Blackwell Scientific.

Streips UN and Yasbin RE (1991) *Modern Microbial Genetics*. New York: Wiley-Liss.

GEOTRICHUM

A Botha, Department of Microbiology, University of Stellenbosch, South Africa

Fungi belonging to the hyphomycetous genus *Geotrichum* Link: Fries are commonly found in nutritionally rich, liquid substrates, such as decaying plant material, industrial effluents, pulp and a wide variety of food types. Previously, genus names like *Odium* and *Oospora* were used for these fungi. However, as more and more related hyphomycetous fungi were isolated and described, a revision of the taxonomy of these fungi became necessary. This was done by De Hoog and his co-workers at the Centraalbureau voor Schimmelcultures in The Netherlands. Today, members of *Geotrichum* Link: Fries are seen as anamorphic or nonsexual fungi of hemiascomycetous relationship. The genus contains anamorphs of two different teleomorphic genera *Dipodascus* de Lagerh. (**Fig. 1**) and *Galactomyces* Redhead and Malloch (**Fig. 2**). Identification of the latter two genera becomes possible when sexual stages (i.e. asci containing ascospores) in the fungal life-cycles are formed. Consequently, *Geotrichum* species may be distantly related, but do have a number of characteristics in common. The sexual stages are absent, either as a result of the presence of only one mating type, or of inadequate culture conditions to induce the sexual stages.

Other characteristics of *Geotrichum* species are the production of white, farinose or hairy colonies on solid substrates. Upon maturation, the hyaline hyphae making up the mycelium tend to brake up into unicellular pieces, called arthric conidia (**Figs 3b** and **4b**). Under adverse conditions, certain species may produce survival structures such as lipid-rich chlamydospores (Fig. 3c) or intracellular endospores (Fig. 3d). The ultrastructure of the cell wall of all *Geotrichum* species appears to consist of two layers. The septum between adjacent cells in a hypha contain micropores. The major ubiquinone system in the electron transport chain of all *Geotrichum* species is coenzyme Q-9.

Eleven species are accepted in *Geotrichum*, which are anamorphs of *Dipodascus* or *Galactomyces* teleomorphs. The main morphological criteria for identification, other than the morphology of the sexual stages of the teleomorphs, are expansion growth of colonies on solid media, branching patterns of marginal hyphae and the prevalent type of conidiogenesis.

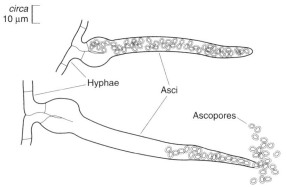

Figure 1 The elongated asci of *Dipodascus albidus* Lagerh., which rupture in the apical region when reaching maturity. The smooth-walled ascospores with mucilaginous sheaths are characteristic of the genus *Dipodascus* Lagerh.

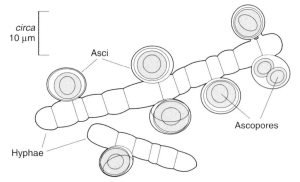

Figure 2 As a typical member of the genus *Galactomyces* Redhead & Malloch, the species *Galactomyces geotrichum* (Butler & Petersen) Redhead & Malloch produces subglubose to ellipsoidal asci, each containing one, or rarely two, verrucose ascopores with median rims.

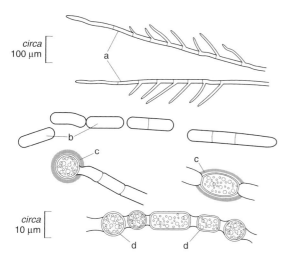

Figure 3 Typical structures formed by *Geotrichum fragrans* during growth on general purpose media. a, expanding hyphae during active growth; b, older hyphae tend to break up into arthric conidia; c, chlamydospores or d, endospores are also produced in older hyphae.

In some cases the shape and dimensions of the hyphae and the presence of chlamydospores are also used to identify species. The physiological criteria that are used to identify *Geotrichum* species include growth responses on D-xylose, cellobiose, salicin, arbutin, sorbitol and D-mannitol.

Generally, only two *Geotrichum* species are commonly associated with foods. The one is *Geotrichum fragrans* (Berkhout) Morenz ex Morenz (Fig. 3), and the other is *Geotrichum candidum* Link: Fries (Fig. 4). The teleomorph of the latter species is known and is named *Galactomyces geotrichum* (Butler and Petersen) Redhead and Malloch (Fig. 2).

The physiological properties of G. *fragrans* and G. *candidum* regarding the ability to ferment a series of carbohydrates, the ability aerobically to utilize a

number of carbon sources and nitrogen sources and to grow in a medium devoid of vitamins, are summarized in **Table 1**. Interestingly, G. *fragrans* has the ability to ferment glucose and sometimes galactose as well, whereas G. *candidum* only shows a very weak

Table 1 The ability of *Geotrichum* species to aerobically assimilate a series of carbon sources, utilize nitrogen sources and grow on media devoid of vitamins

	G. candidum (Galactomyces geotrichum)	G. fragrans
Pentoses		
D-*Arabinose*	–	–
L-*Arabinose*	–	–
D-*Ribose*	V	–
D-*Xylose*	+	–
Hexoses		
D-*Galactose*	+	+
D-*Glucose*	+	+
L-*Sorbose*	+	+
L-*Rhamnose*	–	–
Disaccharides		
Cellobiose	–	–
Lactose	–	–
Maltose	–	–
Melibiose	–	–
Sucrose	–	–
Trehalose	–	–
Trisaccharides		
Melezitose	–	–
Raffinose	–	–
Polysaccharides		
Inulin	–	–
Starch	–	–
Glycoside		
Salicin	–	–
Alcohols		
Erythritol	–	–
Ethanol	+	+
Galactitol	–	–
Glycerol	+	+
Inositol	–	–
D-*Mannitol*	V	V
Ribitol	V	–
Organic acids		
Acetic acid	+	+
Citric acid	V	V
Gluconic acid	–	–
Lactic acid	V	+
Succinic acid	V	+
Nitrogen sources		
Ethylamine	+	+
Nitrate	–	–
Growth without vitamins	+	–

+, assimilated; –, not assimilated; V, variable within species.

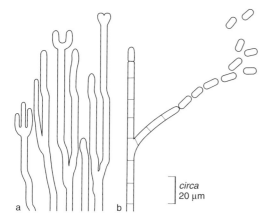

Figure 4 Typical structures formed by *Geotrichum candidum* (*Galactomyces geotrichum*) on general purpose media. a, expanding hyphae during active growth; b, older hyphae tend to brake up into arthric conidia.

ability to ferment sugars (**Table 2**). Both species are able to aerobically utilize the hexoses D-galactose, D-glucose and L-sorbose, but only *G. candidum* can utilize the pentose sugars tested (Table 1). Disaccharides, trisaccharides and polysaccharides, such as starch and inulin are not utilized by these fungi, whereas polyols and organic acids are utilized.

In addition, *G. candidum* is known to produce cellulolytic, lipolytic and proteolytic enzymes. Enzymes, such as diacetyl reductase, glucanase, glycerol dehydrogenase, polygalacturonase and phosphatidase, are also actively produced by strains of this species.

It has been found that although the morphological and physiological properties of strains within the species *G. candidum* are similar, this species seems to be heterogeneous regarding genome composition and habitat. Saprobic as well as pathogenic strains of *G. candidum* have been isolated.

Importance of *Geotrichum* in the Food Industry

Geotrichum fragrans has been isolated from fruit juice, palm wine, milk, figs and mash of *Zea mays*. However, *G. candidum* is by far the most common of the two species and is known to be the causative agent in the spoilage of a number of food types.

G. candidum is a post-harvest pathogen of a wide diversity of fruit and vegetables. This species is known to cause sour rot of citrus fruits. It causes watery soft rot in asparagus, onions, garlic, beans, beetroot, carrots, parsnips, parsley, endives, globe artichoke, lettuce, cabbage, Brussels sprouts, cauliflower, broccoli, radishes, rutabagas, turnips and tomatoes. It also causes post-harvest damage to bananas, mangoes, musk melons and stone fruits. Another similar species, *Geotrichum citri-aurantii* (Ferraris) E.E. Butler, only causes sour rot in citrus. These diseases usually spread when hyphal fragments or arthric conidia from *Geotrichum* are carried by fruit flies or other insects from infected plant material to injuries and lacerations on freshly harvested fruit and vegetables.

G. candidum occurs commonly in raw milk and acts as the spoilage agent of a number of dairy products. It imparts off-flavours on cottage cheese and under certain conditions is able to cause surface defects on hard cheeses. It is able to grow on the surface of butter and has also been isolated from spoilt poultry and from fresh, refrigerated, processed and cured meats.

Geotrichum is known to grow rapidly on equipment in contact with food in processing plants, hence the name 'machinery mould' that is commonly used in the food industry. The fungus usually grows as a slimy layer submerged in liquid on concrete, metal or wood in an unhygienic food processing plant. Dead mycelial fragments then end up in the canned products.

The presence of *Geotrichum* in food is, therefore, an indication of unsanitary conditions during good preparation and/or storage, or the use of inferior raw materials. In addition, it may even constitute a potential health hazard for certain individuals who are delibitated due to an immune deficiency or some other chronic disorder. Geotrichosis may result; this is a bronchial or pulmonary infection caused by certain strains of *G. candidum*. This species has also been implicated in bloodstream infections, as well as in infections of the cornea, ileum, tongue, skin and nails.

However, it is noteworthy that *Geotrichum* not only occurs as a spoilage organism in foods, but also plays a role in the preparation of fermented foods. Mixed cultures of certain *G. candidum* and *Penicillium* strains are used to inoculate curds during the preparation of soft cheeses like Brie and Camembert. Growth of these fungi, as well as growth of certain bacterial strains, is responsible for the characteristic flavour of these cheeses. In West Africa, poisonous cassava roots are peeled and ground and then rendered edible through a fermentation process that involves a number of microorganisms. The characteristic flavour and aroma of this product, called gari, is brought about by the growth of *Geotrichum* strains during the later stages of the fermentation process. The fermented product is fried before it is eaten. High numbers of *G. candidum* are also present during certain stages of the fermentation of cocoa beans. This fermentation process, which involves a wide diversity of microbial species, is essential for the development of the characteristic chocolate aroma after roasting the cocoa beans.

Table 2 Carbohydrates fermented by *Geotrichum* species

	G. candidum (Galactomyces geotrichum)	G. fragrans
Hexoses		
D-*Galactose*	V/W	V
D-*Glucose*	V	+
Disaccharides		
Maltose	–	–
Sucrose	–	–
Lactose	–	–
Trisaccharide		
Raffinose	–	–

+ fermented; –, not fermented; W, weak fermentation; V, variable within species.

Methods of Detection and Enumeration

Viable Counts

Geotrichum and other fungi are enumerated in foods by first homogenizing the food sample, if it is not already in a liquid form. It can then be diluted and plated out onto an appropriate medium. After a sufficient incubation period at an appropriate temperature, the fungal colonies are counted and the number of colony forming units present in the original sample calculated. A typical procedure for estimating *Geotrichum* colonization of foods would therefore be as follows.

Sample Preparation A one-tenth dilution of a food sample, weighing 50–500 g, is prepared using sterile water. The diluted sample is then homogenized. A Colworth Stomacher, applied for 2 min, may be used for this purpose.

Dilution Plates Using 0.1% peptone water, serial 1:9 dilutions of the sample are prepared. Aliquots of 0.1 ml of the appropriate dilutions are spread onto solidified agar medium plates in triplicate.

Media *Geotrichum* species grow well on a number of general purpose media as well as on restrictive media that are used for the enumeration of fungi in foods. General purpose media, two of which are shown in **Table 3**, allow unrestricted growth of a wide diversity of fungal groups. However, antibacterial agents such as chloramphenicol and oxytetracycline are included in some of these media. Other media that are used contain certain antifungal agents such as dichloran or rose bengal. These agents are included in the media to fascilitate enumeration by restricting the spreading of fungal colonies on the plates. Three such media are given in **Table 4**.

Incubation and Identification Inoculated plates are usually incubated at 25°C for 5 days, after which the colonies are counted on those plates containing 25–100 colonies. It must be noted that viable counts may vary, depending on the methodology that was used. For example, the colony forming units may be fragments of mycelium of varying size or single arthric conidia, depending on the degree of homogenization that the food sample was subjected to. Other factors that may induce variable results are changes in the composition of the enumeration medium, the diluent and temperature of incubation. It is therefore critically important, in order to maintain a certain degree of consistency in the results obtained in a specific laboratory, that the protocol for the enumeration of fungal colony forming units must change little over time.

Table 3 General purpose media which allow fungal growth

Oxytetracycline glucose yeast extract agar (OGY)

Glucose	10 g
Yeast extract	5 g
Agar	15 g
Oxytetracycline (0.1% solution, w/v)	100 ml
Water	1000 ml
pH 7.0	

Add all the ingredients to 800 ml water. Steam to dissolve agar and bring solution to 900 ml. Sterilize by autoclaving (15 min, 121°C). Add oxytetracycline solution to sterile medium. Instead of oxytetracycline, 0.1 g chloramphenicol could be added to the above medium before autoclaving.

Potato dextrose agar (PDA)

Potatoes, infusion from	200 g
Glucose	15 g
Agar	20 g
Water	1000 ml
pH 5.6	

Rinse the scrubbed and diced potatoes under running water. Then add the potatoes to 1000 ml water and boil for 1 h. Pass the boiled potatoes and water through a fine sieve or cheesecloth, and squeeze through as much pulp as possible. Add agar to the suspension and boil until it dissolves. Add the glucose and sterilize the medium by autoclaving (15 min, 121°C). The pH can be adjusted to 3.5 by adding approximately 14 ml sterile 10% tartaric acid solution to the cooled (45–50°C) autoclaved medium.

The colour and morphology of *Geotrichum* colonies may vary depending on the medium used. For example, on media containing rose bengal, the colonies may be various shades of pink, whereas on other media the colonies are white. Usually, the colonies are characterized by a flat or creeping fimbriate expanding zone that may contain prostrate to suberect hyphal fasciles. The central portion of each colony contains dry aerial mycelium up to 2 mm high. A typical feature of *Geotrichum* is the fragmentation of the hyphae into arthric conidia (see Figs 3b and 4b,) which can be readily identified by examining a piece of hyphal growth using a compound microscope. Since the ubiquitous basidiomycetous genus *Trichosporon* Behrend is also characterized by the production of arthric conidia, differentiation of the two genera on the basis of the morphology of these structures alone may be problematic for the untrained eye. To confirm the identity of *Geotrichum*, it is advisable to first purify the fungus by repetitive inoculation and incubation on a medium such as yeast extract–malt extract agar (**Table 5**). The diazonium blue B (DBB) colour test can then be performed by dripping the chilled DBB reagent (**Table 6**) directly onto the fungal colony which has grown for three weeks at 25°C on Sabouraud's 4% glucose–0.5% yeast extract agar (**Table 7**). The reagent will turn a reddish colour on *Trichosporon* after 1–2 min, but will remain yellow on

Table 4 Media containing agents which restrict fungal growth and facilitate enumeration

Dichloran 18% glycerol agar (DG18)

Glucose	10 g
Peptone	5 g
KH$_2$PO$_4$	1 g
MgSO$_4$.7H$_2$O	0.5 g
Glycerol	220 g
Agar	15 g
Dichloran (0.2% w/v in ethanol)	1.0 ml
Chloramphenicol	0.1 g
Water	1000 ml
pH 5.6	

Add all the ingredients except the glycerol to 700 ml water. Steam to dissolve agar. Add the glycerol. Bring solution to 1000 ml and sterilize by autoclaving (15 min, 121°C). The a$_w$ of the final medium is 0.955 and it is commonly used to isolate fungi from foods with a low water activity.

Dichloran rose bengal chloramphenicol agar (DRBC)

Glucose	10 g
Peptone	5 g
KH$_2$PO$_4$	1 g
MgSO$_4$.7H$_2$O	0.5 g
Rose bengal (5 % w/v in water)	0.5 ml
Dichloran (0.2 % w/v in ethanol)	1 ml
Chloramphenicol	0.1 g
Agar	15 g
Water	1000 ml
pH 5.6	

Add all the ingredients to 900 ml water. Steam to dissolve agar and bring solution to 1000 ml. Sterilize by autoclaving (15 min, 121°C).

Rose bengal chlortetracycline agar (RBC)

Glucose	10 g
Peptone	5 g
KH$_2$PO$_4$	1 g
MgSO$_4$.7H$_2$O	0.5 g
Rose bengal (0.5 % w/v in water)	10 ml
Chlortetracycline (0.1 % w/v in ethanol)	1 ml
Agar	20 g
Water	1000 ml
pH 7.2	

Add all the ingredients except the rose bengal and chlortetracycline to 900 ml water. Steam to dissolve agar and bring solution to 990 ml and sterilize by autoclaving (15 min, 121°C). Add rose bengal and chlortetracycline solutions to sterile medium.

Table 5 Yeast extract–malt extract agar (YM agar) used for the culture and storage of *Geotrichum* strains

Glucose	10 g
Malt extract	3 g
Yeast extact	3 g
Peptone	5 g
Agar	20 g
Water	1000 ml
pH 5.5	

Add all the ingredients to 1000 ml water. Steam to dissolve agar. Sterilize by autoclaving (15 min, 121°C).

Table 6 Diazonium blue B reagent to distinguish between *Geotrichum* strains and basidiomycetous fungi

Diazonium blue B salt

(o-Dianisidine tetrazotized (Sigma);	
Fast Blue B salt (Hoechst))	15 mg
Chilled 0.25 M Tris buffer, pH 7	15 ml

Dissolve the diazonium blue B salt in the Tris buffer. Keep the solution in an ice bath and use within 30 min.

Table 7 Sabouraud's 4% glucose–0.5% yeast extract agar

Glucose	40 g
Peptone	20 g
Yeast extract	5 g
Agar	20 g
Water	1000 ml
pH 5.6	

Add all the ingredients to 1000 ml water. Steam to dissolve agar and autoclave (15 min, 121°C).

Geotrichum colonies. In order to distinguish between *Geotrichum* species, a number of standardized physiological tests are performed, the results of some are given in Tables 1 and 2 for *G. candidum* and *G. fragrans*. The methodology of these tests is explained in *The Yeasts, a Taxonomic Study*, 4th edn., edited by Kurtzman and Fell.

Non-viable Counts

Viable counts give no indication of the quantity of dead fungal biomass in a particular food product. The dead fungal biomass content is useful for retrospective information concerning the quality of raw materials and hygienic practices used in the production of processed foods. The methodology of determining machinery mould may differ for different food types. However, as an example, the AOAC Method No. 974.34 for determining mould in canned vegetables, fruits and juices, is described.

Obtaining Mycelial Fragments The contents of the can are weighed and then drained on a 2032 mm diameter sieve (2.38 mm aperture or USA standard no. 8). The liquid is collected in a pan. The can and sieve are washed with 300 ml water and the washings and liquid combined. These are then quantitatively transferred onto a 1270 mm diameter sieve (1.18 mm aperture or USA standard no. 16), which is resting on a beaker. The residue on the sieve is washed with about 50 ml water before it is discarded. The combined liquid and washings are quantitatively transferred onto a 1270 mm diameter sieve (63 μm aperture or USA standard no. 230) tilted at an angle of about 30°. The combined liquid and washings are then discarded, but the residue on this sieve is retained. The tissue on the sieve is washed with water to the lower edge of the sieve. Using a spatula and a wash bottle, the residue on the sieve is transferred to a 50 ml graduated centrifuge tube. For a volume of 10–30 ml in the tube, the staining procedure outlined below, is followed. If the volume of the suspension in the tube is less or more than these values, the staining procedure as explained in AOAC Method No. 974.34 (a) or (c) is followed.

Staining of Mycelial Fragments The volume of the suspension in the centrifuge tube is brought to 40 ml, using water. Three drops of a filtered ethanolic solution of crystal violet (10% w/v), are then added and thoroughly mixed with the suspension in the centrifuge tube. A sediment in the tube is obtained by centrifugation (6 min, 528 g) and the supernatant is discarded. Using water, the volume of the sediment in the tube is brought to the nearest 5 ml graduation. An equal volume of stabilizer solution is added, the suspension is thoroughly mixed and the total volume in the centrifuge tube is recorded. The stabilizer solution is prepared by adding 2.5 g sodium carboxymethyl cellulose and 10 ml of approximately 37% (w/w) formaldehyde to 500 ml of boiling water in a high-speed blender, while the blender is running. The stabilizer solution is ready for use after 1 min of blending.

Counting of Mycelial Fragments Using a pipette, 0.5 ml of the well-mixed stabilized suspension is applied as a streak of about 40 mm long to a rot fragment counting slide (**Fig. 5**). Using transmitted diffused bottom illumination, examine the slide at 40× enlargement with a stereoscopic microscope. The number of deep purple mycelial fragments, with three or more hyphal branches, is counted on at least two entire slides. The number of mycelial fragments in 500 g of product is then calculated using the following equation as depicted in AOAC Method 984.30 C:

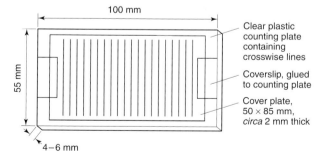

Figure 5 A rot fragment counting plate and cover. The dimensions for a clear plastic plate and glass cover are given. The crosswise parallel lines, 4.5 mm apart with 15 mm spaces at each end, are scribed using a sharp needle. Half of a square cover slip (about 22 mm) is fastened to each end of the counting plate to raise the cover plate about 0.25 mm above the counting plate.

$$N = (S/V(\text{slides})) \times (500/W) \times V(\text{diln})$$

where S is the total number of mycelial fragments counted; $V(\text{slides})$ is the total volume counted on all the slides; W is the total weight of the contents of the can measured in grams; and $V(\text{diln})$ is the volume of the suspension after the final dilution with stabilizer solution.

Immunochemical Detection

Geotrichum and other fungi produce heat-stable extracellular antigens and immunoassays are currently being investigated as rapid alternative methods to highlight the presence of *Geotrichum* in foods. *Geotrichum* antigens may remain in a processed food sample after the fungus has been destroyed during processing. The presence of such antigens may, therefore, also provide information on the quality of raw materials and the standard of hygienic practices used during processing.

Geotrichum candidum antigens consist of protein and polysaccharide moieties. The latter are composed of galactose, glucose and mannose. Enzymatic digestion and competitive inhibition tests using different sugar derivatives have shown that the galactosyl fractions are immunodominant and that they were $\beta(1{\rightarrow}4)$ linked to glucosyl residues within the polysaccharide. The immunoassays that have been developed for *G. candidum* antigens were found to be genus specific. These tests therefore show potential to be developed into a routine diagnostic test for the presence of *Geotrichum* in food.

Regulations

A standard for viable *Geotrichum* mould *per se* does not exist. However, defect action levels for moulds in general do exist for different food classes. An example is the manual published by the American Food and Drug Administration, with the order number PB 88–915400, which can be obtained from National Technical Information, Service, Sales Desk, 5285 Port Royal Road, Springfield, VA 22161, USA. Also, a standard for the numbers of non-viable *Geotrichum* mycelial fragments in processed foods does not exist, but it has been found that the occurrence of mycelial fragments with a distinct feathery appearance is good evidence of unsanitary conditions in the equipment. The numbers of mycelial fragments in the food can usually be drastically reduced by a thorough clean-up of the plant.

See also: **Cheese**: Microbiology of Cheese-making and Maturation; Mould-ripened Varieties. **Fungi**: Overview of the Classification of the Fungi; Food-borne Fungi – Estimation by Classical Culture Techniques. **National**

Legislation, Guidelines & Standards Governing Microbiology: Canada; European Union; Japan. **Spoilage Problems**: Problems Caused by Fungi. **Starter Cultures**: Cultures Employed in Cheese-making. **Total Viable Counts**: Spread Plate Technique. **Cocoa and Coffee Fermentations**. **Fungi**: Classification of the Hemi-Ascomycetes.

Further Reading

De Hoog GS, Smith MTh and Guého E (1986) A revision of the genus *Geotrichum* and its teleomorphs. In: *Studies in Mycology*, no. 29. P. 1. Baarn: Centraalbureau voor Schimmelcultures.

Helrich K (1990) *Official Methods of Analysis of the Association of Official Analytical Chemists*, 15th ed. Arlington: Association of Official Analytical Chemists, Inc.

King AD (1992) Methodology for routine mycological examination of food – a collaborative study. In: Samson RA, Hocking AD, Pitt JI and King AD (eds) *Modern Methods in Food Mycology*. P. 11. Amsterdam: Elsevier.

King AD, Pitt JI, Beuchat LR and Corry JEL (1986) *Methods for the Mycological Examination of Foods*. New York: Plenum Press.

Kurtzman CP and Fell JW (1998) *The Yeasts, A Taxonomic Study*, 4th ed., Amsterdam: Elsevier.

Notermans SHW, Cousin MA, De Ruiter GA and Rombouts FM (1998) Fungal Immunotaxonomy. In: Frisvad JC, Bridge PD and Arora DK (eds) *Chemical Fungal Taxonomy*. P. 121. New York: Marcel Dekker.

GIARDIA

R W A Girdwood and **H V Smith**, Scottish Parasite Diagnostic Laboratory, Stobhill Hospital, Glasgow, UK

Characteristics of the Genus

Giardia have been described traditionally as flagellated protozoans. The genus *Giardia* was created by Kunstler in 1882 and ascribed to the family Hexamitidae. The members of the Hexamitidae are characterized by having bilateral symmetry, two nuclei lying side by side and six or eight flagella. Although a few species of the five genera within the Hexamitidae are free living, the majority are parasites of vertebrates or invertebrates. However, more modern classifications have created a new kingdom, the Archezoa which embraces those eukaryotic organisms which lack mitochondria, plastids, hydrogenosomes, peroxisomes and Golgi bodies. Thus as members of this kingdom, the *Giardia* are not protozoans *sensu stricto*. In these newer classifications *Giardia* are in the order Diplomonadida of the class Trepomonadea and phylum Metamonada. *Giardia* exists in two forms, the parasitic, feeding, multiplying trophozoite and the resistant, transmissive and infective cysts (**Fig. 1**).

The trophozoite is kite or pear shaped *en face* being rounded anteriorly and pointed posteriorly. The dorsal surface is convex and the ventral surface containing the attachment organ, the sucking disc, is concave. There are two similarly sized nuclei within the sucking disc area and four pairs of flagella, two axonemes and two median bodies. The dimensions of the trophozoite are 9–21 μm long, 5–15 μm wide and 2–4 μm thick. The cysts are oval to spherical measuring 8–19 × 7–10 μm (**Table 1**). The immature cysts contain two nuclei but on maturing four nuclei, situated in a group towards one pole become prominent. Longitudinal fibres diagonally bisect the cysts and median bodies transect these.

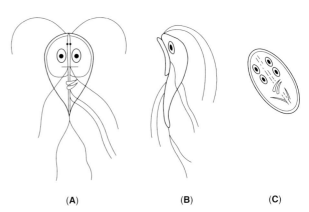

Figure 1 *Giardia duodenalis*: (**A**) trophozoite *en face* view; (**B**) trophozoite lateral view; (**C**) cyst.

Table 1 Characteristic morphological features of *G. duodenalis* cysts by Nomarski differential interference contrast (DIC) microscopy

- Ellipsoid to oval, smooth walled, colourless and refractile
- 8–19 × 7–10 μm (length × width)
- Mature cysts contain four nuclei displaced on one pole of the organism
- Axostyle (flagellar axonemes) lying diagonally across the long axis of the cyst
- Two 'claw-hammer'-shaped median bodies lying transversely in the mid-portion of the organism

Relevant Species

Although first described by Leeuwenhoek in 1681 the speciation and nomenclature of *Giardia* remains confusing and contentious. Various criteria have been used to attempt to discriminate between species. These include morphology and morphometrics, host specificity, isoenzyme and antigenic analysis and molecular methods such as karyotyping.

Based on the appearance of the median bodies of the trophozoite, three type species of *Giardia* have been identified, namely, *G. agilis*, *G. muris* and *G. duodenalis*. Whereas *G. agilis* is a parasite of amphibians and *G. muris* of rodents, birds and reptiles, the *G. duodenalis*-type has been found in a range of birds, reptiles and mammals, including humans. The type species for the *duodenalis* group was isolated from the rabbit; however the parasites which infect humans have similar median bodies and are ascribed to the *duodenalis* group. Those *Giardia* that infect human beings are believed to be a single species, *G. intestinalis* (= *lamblia*). *G. intestinalis* is regarded by some authorities as a race of *G. duodenalis*.

On the basis of constant morphological and/or biological differences, two further *Giardia* species, isolated from birds, have been named: *G. psittaci* isolated from the budgerigar, whose trophozoites lack a marginal groove bordering the adhesive disc, thus differing from all other *G. duodenalis* trophozoites, and *G. ardeae*, isolated from the great blue heron, whose trophozoites possess only one caudal flagellum. Neither of these avian *Giardia* produced infection in gerbils or mice, but both *G. psittaci* and *G. ardeae* possess median bodies typical of the *duodenalis* group.

Life Cycle (Fig. 2)

Infection is by the ingestion of cysts. The cyst wall after passage through the stomach is dissolved in the upper small intestine and two trophozoites are released. The trophozoites which divide by binary fission typically parasitize the duodenum by attaching to the epithelial surface by the sucking disc. There is no invasion locally or systemically. Either by spontaneous detachment or by the natural shedding of the epithelial cells the trophozoites pass down the intestine. Cyst production takes place in the colon. Depending on the speed of transit of trophozoites, trophozoites and cysts, cysts alone or no intact cysts or trophozoites are voided in the faeces. The time from infection to the presence of parasites in stools is about nine days (the prepatent period).

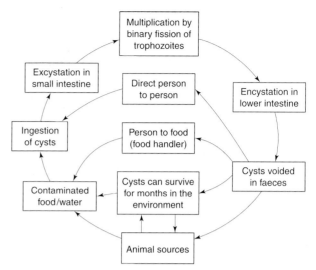

Figure 2 Life cycle(s) of *Giardia duodenalis*.

Detection in Faeces

Diagnosis of giardiasis is dependent on the demonstration of intact parasites (cysts and/or trophozoites) by microscopy, parasite products by immunoassay and/or parasite DNA in faeces or duodenal/jejunal aspirates and serology. Cyst excretion can be intermittent, and some infected individuals may be low cyst excretors. Diagnosis, based on the detection of cysts in faecal concentrates (e.g. formalin–ether method) can be inefficient. Negative microscopy of a single stool sample does not exclude infection, and three stool specimens should be examined before suggesting other diagnostic procedures. During symptomatic infection, up to 10^7 cysts per gram of faeces can be excreted. The erratic nature of cyst excretion, the requirement for experienced staff for microscopic identification and the low detection rate have resulted in the development of alternative methods for the diagnosis of giardiasis. Trophozoites can be demonstrated in duodenal/jejunal aspirates or biopsies when stool microscopy is negative, but it should be noted that a negative aspirate does not definitively exclude the diagnosis.

A number of immunological methods are also available for diagnosis. Epifluorescence microscopy, using fluorescein isothiocyanate-labelled anti-*Giardia* monoclonal antibodies (FITC-G-mAbs), reactive with surface-exposed genus-specific cyst wall epitopes can be used to detect cysts in air-dried smears or formalin–ether (ethyl acetate) concentrates. Detection of antigen in faeces can overcome the problems of erratic cyst excretion, trophozoite and cyst disintegration in vivo, and the low detection rate associated with microscopy. Countercurrent immunoelectrophoresis and subsequently enzyme-linked immunosorbent assays (ELISAs), using a variety of monoclonal and polyclonal antibodies, solid phases and substrates, can also be used to detect *Giardia* antigen (e.g. *Giardia* stool antigen, GSA 65, relative molecular mass = 65 kDa) in faeces. A number of companies market such ELISAs currently.

Total genomic and cDNA probes have been described for the detection of *Giardia* DNA in faeces, although this technology has now been superseded by the more sensitive polymerase chain reaction (PCR)-based technology (see below).

Giardia trophozoites can be isolated and propagated axenically from infected intestinal contents or faeces by culture in vitro using media such as TYI-S-33. Although not used for diagnosis, such in vitro methods can be used to study the organism. Serum antibodies, reactive with cysts and trophozoites have been demonstrated by immunofluorescence and immunodiffusion in patients with giardiasis. Detection of anti-*Giardia* immunoglobulin (Ig)G in serum has not been particularly useful in diagnosis because of the persistence of this isotype after the infection has cleared. IgG antibody is detectable in patients for up to 18 months after drug treatment. Detection of anti-*Giardia* IgM in serum is useful for identifying current infection, although *Giardia*-specific IgM may not be detectable in chronic giardiasis. An inability to detect specific IgM may not necessarily indicate the absence of infection. Further work on the immunochemistry of the parasite, and the human immune responses to infection may resolve *Giardia* serodiagnosis.

The variety of laboratory tests for diagnosing infection indicate that none is totally effective. Fortunately, many laboratory diagnosticians recognize that these tests are complementary. Diagnosis can be augmented by light microscopy of haematoxylin and eosin-stained intestinal mucosal biopsies. Serological methods tend to be used for epidemiological purposes. IgG ELISA-based seroepidemiological surveys indicate prevalence rates from 18% in an inner city area in the USA to 48% in a rural area of Panama.

Sensitivity of Detection

In concentrated faecal smears, the likelihood of detecting cysts, by bright-field microscopy, in a single stool sample excreted by an infected individual is 35–50%, whereas between six and ten stool specimens from such an infected individual must be examined to achieve a detection rate of 70–90%. Although more time consuming than bright-field detection, detection of cysts using FITC-*G*-mAbs has been reported to be more effective. Although the reported sensitivity of

antigen-detection ELISAs is 95–98% of microscopy, they have detected antigen (and hence infection) where standard microscopic techniques have failed (e.g. when trophozoite/cyst antigens, but not cysts, are excreted).

Alternative Detection Procedures

Currently, numerous methods can be used to diagnose infection with both microscopy, either with or without concentration, and antigen detection being most favoured. Although PCR is more sensitive than conventional and immunological assays for the detection of cysts in faeces, it is not used for routine diagnosis. Approaches to developing diagnostic PCR primers for *Giardia* are the same as those identified in the section on *Cryptosporidium* as are some significant inhibitors of PCR in faecal samples. PCR is useful for the identification of asymptomatic carriers and for the detection of low densities of cysts in fomites, food and water. As for *Cryptosporidium*, PCR is more likely to become a tool to assist epidemiological investigations than a routine diagnostic method.

Infectivity and In Vitro Culture

The infectious dose is thought to be small. A human volunteer study indicated that the median infectious dose for *Giardia intestinalis* is between 25 and 100 cysts although as few as 10 cysts initiated infection in two out of two volunteers. Eight of the 22 individuals who received from 10 to 25 cysts became infected. Another volunteer study demonstrated that a human-source isolate can vary in its ability to colonize other humans, suggesting that certain isolates may be less infectious to humans, or cause less clinical signs and symptoms than others.

Both mice and gerbils are effective hosts for *G. muris* and *G. duodenalis*. CF-1 mice are effective hosts for *G. muris* and peak cyst excretion occurs 7–14 days after gavage with 10^3 cysts. Suckling and weaning, but not adult, Swiss mice can be infected with *G. duodenalis* cysts and trophozoites. The Mongolian gerbil (*Meriones unguiculatus*) is highly susceptible to infection with *G. intestinalis* cysts and trophozoites. After infection with 5×10^3 cysts, cyst excretion commences on day 8 post-infection and can persist for 13–39 days. Trophozoites collected from infected gerbils can be used to initiate new cultures in vitro.

G. duodenalis, unlike *Cryptosporidium parvum* is an extracellular parasite, and complete development (from trophozoite to cyst) can be accomplished in vitro. *G. duodenalis* is an aerotolerant anaerobe, and can be grown axenically in a medium containing both thiol-reducing agents, such as cysteine, and mammalian bile. TYI-S-33 supplemented with either 10% (v/v) sterile bovine or equine serum is capable of supporting the growth of trophozoites. Tubes are filled to 90–95% of their capacity to minimize the gas volume above the surface of the medium, and are incubated at 35°C either vertically or at a shallow angle (about 5° from the horizontal). During growth, trophozoites attach to the walls of the culture vessel forming monolayers which become confluent by the end of the logarithmic phase of growth. Trophozoites can be harvested by chilling the culture vessel, which causes trophozoite detachment, and centrifuging (500 g, 5 min) the contents. Trophozoites can encyst in vitro after the addition of 5 mg ml^{-1} bile to TYI-S-33, although the percentage encystation from different isolates is quite variable. Cysts are separated from residual trophozoites by the addition of distilled water which lyses the latter. In vitro-derived cysts are infective for Mongolian gerbils, and trophozoites produced from such infections can multiply and encyst in vitro under the conditions described above.

Determination of Viability

Little can be inferred about the public health significance of cysts detected in food and water samples without knowing whether they are viable or not. Neither animal infectivity nor excystation in vitro is applicable to the small numbers of cysts found in water, food and other environmental concentrates. Viability assessments are also useful for determining whether industrial treatments, processes and procedures used in the food and water industries can kill cysts.

Examination by phase contract or Nomarski differential interference contrast (DIC) microscopy, which can determine gross differences between viable and non-viable organisms, such as the morphological integrity of the parasites within the cyst, is subjective and often compromised by the presence of occluding particulates and other debris. Attempts to develop rapid, objective estimates of organism viability have revolved around the microscopial observation of fluorescence inclusion or exclusion of specific fluorogens as a measure of viability.

The fluorogens fluorescein diacetate (FDA, blue filter block; 480 nm, excitation; 520 nm, emission) and propidium iodide (PI, green filter block; 535 nm, excitation; >610 nm, emission) were used to determine the viability of *Giardia muris* cysts; cysts which included FDA and hydrolysed it to free fluorescein could cause infection in neonatal mice, whereas cysts which included PI were incapable of causing infection in neonatal mice. A positive correlation was demonstrated between the exclusion of 3-[dansylamido-phenyl boronic acid] (FluoroBora 1) and viable *G. muris* cysts, although results using viable *G. duodenalis* cysts were less conclusive.

Two fluorogenic viability assays were developed for *G. muris* cysts (used as a surrogate for *G. duodenalis* although less sensitive to disinfection) and compared to infectivity in neonatal CD-1 mice and in vitro excystation. For ozone and chlorine disinfection studies, SYTO®-9 which stained dead cysts bright yellow (blue filter block; 480 nm, excitation; 520 nm, emission), was the most effective single stain tested for detecting dead cysts, whereas the Live/Dead Bac-Light kit (Molecular Probes, Eugene, Oregon, USA), which stained viable cysts dark green and non-viable cysts light green–orange/yellow, also showed correlation with animal infectivity. Results from both SYTO®-9 and the Live/Dead BacLight kit correlated better with infectivity than with in vitro excystation.

No published correlation between the inclusion of fluorogens and the viability of *G. intestinalis* cysts as determined by infectivity or in vitro excystation has been demonstrated, and a combination of PI inclusion/exclusion and assessment of morphology by DIC remains a currently accepted method. The role of PI in defining cyst death (loss of membrane integrity) has been questioned since cysts exposed to lethal levels of chlorine disinfectant failed to become PI positive. PI is a large planar molecule that will not traverse an intact lipid bilayer. Failure of such cysts to include PI is probably due to the gradual loss of membrane integrity following trophozoite death. Replacing PI with SYTO®-9 might address the comments raised about PI.

The lack of consistency with fluorogenic vital dyes led to the investigation of molecular approaches. One approach measured the differences in the amount of RNA detectable spectrophotometrically (A_{260}) in viable and non-viable cysts, before and after the mild acid induction step of the in vitro excystation protocol. Another approach measured the amount of giardin mRNA by PCR before and after acid induction; viable cysts, having a greater signal than non-viable cysts, could be distinguished from non-viable cysts.

Detection of viable *Giardia* cysts using reverse transcription-PCR (RT-PCR) has also been described. RT-PCR, which amplifies messenger RNA (mRNA) of the *Giardia* heat shock protein, with the GHSP primers, detects viable *G. intestinalis* but has been reported as being inconsistent recently. Previously published primers, other than GHSB, were used to amplify heat shock protein mRNA. An internal positive control was developed to determine the efficiency of mRNA extraction and potential RT-PCR inhibition. Sensitivity was reported in the range of a single viable organism and, in a comparison of the conventional immunofluorescence method for identifying cysts and RT-PCR in water concentrates ($n = 29$), the frequency of detection of viable *Giardia* cysts rose from 24% to 69% with RT-PCR.

Biophysical methods have also been used to determine cyst viability. Electrorotation (ROT) has been used to demonstrate differences between viable and non-viable cysts and *G. muris* and *G. intestinalis* cysts. Before ROT, cysts have to be partially purified and suspended in a low conductivity medium. Currently, for *Giardia*, ROT is still at the experimental stage.

Food-borne Transmission

Documented food-borne outbreaks have been rare in developed countries although some food can be important vehicles of transmission, especially in situations of poor hygiene and endemnicity of infection. In a 1922 outbreak of giardiasis, the food-borne route was suggested when water, vegetables and other foods were found to be contaminated with cysts. Anecdotal evidence from other reported outbreaks has, frequently, implicated food handlers and contaminated fruit and vegetables.

Eight outbreaks of food-borne transmission have been documented. There is only one report of the possibility of food (i.e. tripe) being intrinsically infected *per se*. The seven other outbreaks are associated with contamination by food handlers and include foods such as salmon, fruit salad, raw vegetables, lettuce, onions and tomatoes. Two outbreaks involved contaminated vegetables and a total of 217 individuals were affected in these seven outbreaks between 1979 and 1990. In two outbreaks, the original source of infection was traced to the infected infant of the food handler.

Methods of Detection in Foods

There are no standardized methods for examining food for the presence of *Giardia* cysts. Initially, the methods used were modifications of those developed for detecting cysts in stool samples and concentrates. Up to the 1980s, two methods for isolating cysts from contaminated food had been described namely, emulsification or maceration and immersion in water. In the first approach, soft foods (e.g. some fruit and vegetables) are either homogenized or broken up in saline or water to release contaminating cysts, whereas in the second approach, the contaminating cysts from soft foods (e.g. lettuce) are eluted passively into a suspending medium, such as water, over a 12–24 h period. Where large-sized particulates are generated, the resulting fluid is pre-filtered to remove larger particles, before being concentrated and examined for the presence of cysts by microscopy. Concentration methods include formol–ether sedimentation, sucrose flotation, sucrose discontinuous density gradient flotation and sedimentation followed by zinc sulphate flotation. Some evidence is available to indicate that prolonged exposure of cysts to sucrose can kill cysts.

By using such methods between the 1960s and 1980s, cysts have been isolated from soft foods. In a survey conducted in Rome in 1968, 48 of 64 head of lettuce, collected at random from four markets, were contaminated with cysts. Viable *Giardia* cysts were isolated from strawberries grown in the Poznan area of Poland. Viability was determined by dye exclusion. Thus, viable cysts have been detected as surface contaminants of soft foods.

Current methods involve examination of washings from the surfaces of foods, or of the foods in a liquid phase. Thus the methods used are those or modifications of those which apply to water.

Natural Mineral Waters

In a study with *G. intestinalis* cysts purified from symptomatic individuals, 50–98% cysts were recovered by filtration through 13 mm polycarbonate membranes, from 1 litre volumes of four natural mineral waters (with total dissolved solids of 91–430 mg l^{-1}) seeded with ca. 50 cysts. Cysts were enumerated by epifluorescence microscopy using a commercially available FITC-G-mAb according to the manufacturer's instructions.

Methods of Detection in Water and Other Liquids

Only the general considerations will be discussed below. As *Giardia* cysts occur in low numbers in the aquatic environment and no practicable in vitro culture-enrichment techniques are available for amplifying the very low densities of cysts expected in water concentrates, large volumes are sampled. All methods contain separate sampling, elution, clarification and concentration, and identification elements. 'Standardized' methods, which continually evolve, are available in the UK and USA. Currently, large volume methods are applicable in the USA. In the UK, large volume (depth filter cartridges) and small volume (flat bed membranes, pleated membrane capsule, flocculation) sampling strategies have found favour. Sampling can be performed for various reasons

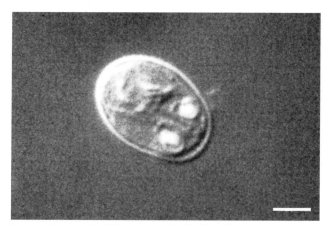

Figure 3 A cyst of *Giardia intestinalis*. Bar = 5 microns.

including monitoring, increased surveillance, outbreak situations, etc. and different strategies apply for different rationales. The eluted retentate is concentrated and analysed for the presence of the organism. In addition to cysts, size-based isolation methods concentrate large amounts of extraneous particulate material which interfere with organism detection and identification. Detergents (0.1% Tween 20, 0.01% Tween 80) are included to prevent cysts and particulates from clumping. flotation media such as sucrose (1.18 sp. gr.) or Percoll–sucrose (1.1 sp. gr.), on which cysts float, are used to concentrate organisms and separate them from extraneous debris which interfere with the microscopic identification of cysts. Both flow cytometry with cell sorting and immunomagnetizable separation (IMS) can also be used to concentrate cysts.

Identification is performed by epifluorescence microscopy and differential interference contrast microscopy, where possible, and putative objects are identified as cysts using a defined series of fluorescence and morphological criteria (see **Table 1**) based on morphometry (the accurate measurement of size and shape) and morphology. Specificity and sensitivity are of paramount importance when attempting to detect small numbers of cysts in concentrates. The majority of commercially available mAbs are genus specific, although one *G. muris*-specific mAb has been reported. As the antibody paratopes bind surface-exposed cyst epitopes, the fluorescence visualized defines the maximum dimensions of the organism, enabling morphometric analyses to be undertaken. The nuclear fluorochrome 4′,6-diamidino-2-phenylindole (DAPI), which binds to trophozoite DNA, is an effective adjunct to FITCG-mAbs for highlighting the nuclei within immature and mature cysts (**Fig. 3**).

Physical changes including distortion, contraction, collapse and rupture of the cyst reduce the number of organisms that conform to accepted criteria (Table 1), resulting in the under-reporting of positives. The 'standardized' methods do not provide information on the species, viability and infectivity of cysts. It must be assumed that each intact cyst detected in the sample is potentially infectious to humans. In order to overcome such difficulties and to present regulators with more definite information regarding the biological status of cysts, the use of more discriminating techniques is necessary.

Nucleic Acid-based Methods

Nucleic acid-based methods can increase both sensitivity and specificity of detection and also address issues including host specificity, infectivity and virulence. Detection of *Giardia* DNA using a genus specific labelled oligonucleotide probe has been reported, but PCR and fluorescence *in situ* hybridization (FISH) can provide more biological information about the parasite.

A genus-specific (265 bp fragment, between base pairs 636 and 900; G+C content ca. 69%) probe was used to detect the presence of *Giardia* nucleic acid, following disruption of cysts with glass beads. Although incapable of distinguishing between *G. duodenalis* and other *Giardia* species which could be present in the aquatic environment, it had a sensitivity of 1–5 cysts per millilitre of surface or waste water concentrate; this is similar to that obtained using a standard immunofluorescence method.

The enhanced sensitivity and specificity of PCR has been used to identify low densities of cysts in water concentrates. IMS can assist in reducing the inhibitors of PCR (humic and fulvic acids, organic compounds, salts and heavy metals, etc.) frequently found in water concentrates. PCR has been used to discriminate *G. duodenalis* from *G. muris* and *G. ardeae*, under ideal laboratory conditions, by amplifying a 218 bp region of the *Giardia* giardin coding gene, and detecting the amplified product with a 28-mer oligonucleotide probe. As few as one cyst could be detected using this method, but PCR analysis with this giardin probe, although able to detect 250 cysts per litre of water, failed to detect species differences when 10^5 cysts were present in a concentrated 400 litre sample.

Concentration of *G. muris* cysts, seeded into turbid river water concentrates, by IMS increased the sensitivity of detection by PCR to 3–30 cysts per millilitre of water. DNA was released from cysts by a freeze–boil Chelex 100 treatment, followed by amplification of a 0.171 kbp segment of the giardin gene by PCR. Formalin-treated *G. muris* cysts yielded no amplifiable DNA.

FISH has also been used to detect cysts in water

Table 2 Potential contributors to cysts detected in raw water. Adapted with permission from Smith et al (1995).

Contribution from infected human beings
- Sewage discharge
- Seepage from septic tanks, pit latrines etc.
- Accidental/deliberate defaecation
- Run-off from night soil

Contribution from human activities
- Disposal of contaminated faeces, and non-controlled effluents from farms
- Accidental spillages from poorly constructed slurry stores and middens
- Slurry spraying and muck spreading
- Intensive husbandry of livestock
- Disposal of faeces from infected animals at abattoirs
- Disposal of sewage sludge to land
- Disposal of contaminated backwash sludge

Contribution from infected animals
- Pasturing of infected livestock
- Infected wild animals
- Watering of infected animals
- Infected domestic/companion animals

Table 3 Factors affecting the potential for transmission via the waterborne route. Adapted with permission from Smith et al (1995).

- Increased chance of contamination of water by sewage (e.g. failure of sewage treatment processes, heavy rain etc.)
- Failure in some water treatment processes
- Faults in operational procedures
- Insufficient barriers in water treatment
- Increased chance of contamination of water by animal faeces (eg. changes in agricultural practices (pasturing, slurry spraying, etc.), increased run-off from land due to weather conditions (heavy rain, snow, thaw etc.)

Table 4 Some documented examples of giardiasis with implication of the food-borne route

Suspected foodstuff	Probable/possible source of infection
Christmas pudding	Rodent faeces
Home-canned salmon	Food handler
Noodle salad	Food handler
Sandwiches	Not known, ? Food handler
Fruit salad	Food handler
Tripe soup	Infected sheep
Ice	Food handler
Raw sliced vegetables	Food handler

concentrates as well as determining the species of individual *Giardia* cysts. Species-specific oligonucleotide rDNA probes to the small subunit rDNA of *G. duodenalis*, *G. muris* and *G. ardeae* were hybridized *in situ* and visualized with laser confocal scanning microscopy. Both species-specific identification by FISH, and morphometry following application of a FITC-G-mAb were undertaken on individual cysts. Using 17–22-mer probes to the 16S-like rRNA of *G. duodenalis*, *G. muris* and *G. ardeae* linked to either FITC or high quantum yield carboxymethylindocyanine dyes (Cy3 or Cy5), together with an FITC-G-mAb, individual *G. duodenalis* cysts present in a sewage lagoon concentrate were identified by laser confocal scanning microscopy.

Importance to the Food and Water Industries

Both surface contamination of fresh produce as well as contaminated water used in food preparation are transmission routes which are significant to the food industry. Surface contamination can occur following contamination by the infected host or transport host (birds, flies), the use of manure or night soil, water for irrigation, fumigation and pesticide application, etc.

Contaminated water is a major source of human infection either by direct consumption or by the use of contaminated water in food processing or preparation. Some of the contributory factors in water contamination are listed in **Table 2** and factors affecting the potential for transmission via the waterborne route are outlined in **Table 3**.

Although the waterborne route is clearly of major importance, the potential for food-borne transmission of giardiasis should not be underestimated and several cases of food-borne giardiasis have been documented (**Table 4**). It is frequently difficult to associate an outbreak of giardiasis with a particular food item and furthermore, if the food-borne route is suspected, to identify how the food implicated became contaminated.

Because of these difficulties, acquisition of *Giardia* infection via the food-borne route is almost certainly under-detected, probably by a factor of ten or more. Data and circumstantial evidence suggest that the most frequent mechanism by which food becomes contaminated with infective cysts is from food handlers who have either been in contact with infected faeces from another source or are themselves infected. Published reports of the prevalence of giardiasis in food handlers in the UK, USA, Korea and Chile describe stool positivity rates of 3.5–15.9%. It has been suggested that food handlers frequently come from lower socioeconomic groups where personal hygiene and knowledge of public health may be poor; whether or not this is so, it will be true that if a food handler has an asymptomatic or minimally symptomatic infection then they may not be aware of their condition and the risk that they pose in transmitting their infection to others.

Other examples of contamination of food are from washing salad vegetables in water containing cysts, from use of excrement (night soil) for fertilizer in the cultivation of food crops or by dissemination of cysts from faeces to food by insects. Vegetables sold in markets have on more than one occasion been found to be contaminated by *Giardia* cysts. It is unclear whether the cysts became associated with the vegetables from cultivation practices, from water, from insects or other animals, or from handling by individual(s) with cysts on their hands. Increased humidity can enhance cyst survival. Food and water can enhance cyst survival by preventing cysts from becoming desiccated. In addition, acquisition of foods from global markets, consumer vogues such as consumption of raw vegetables and undercooking to retain natural taste and preserve heat-labile nutrients can increase the risk of food-borne transmission.

Exposure to a temperature of 55°C for 10 min kills cysts. Storage at −70°C for 1 h kills cysts as does air drying (room temperature) for 4 h. At 4°C, cysts are partially resistant to the concentrations of chlorine disinfection used in water treatment, but susceptible to chlorine dioxide and ozone disinfection. Cysts can remain viable for up to 3 months in cold, raw water sources.

Importance to the Consumer

Human infection is contracted after the ingestion of viable *G. intestinalis* cysts. Transmission is either directly by the faecal oral route or indirectly via faecally contaminated water or food. Person-to-person transmission, is a major route and has been documented between family/household members, sexual partners, health workers and their patients, and children in day-care centres and other institutions. Transmission in day-care centres appears to be particularly common (at least seven documented cases affecting over 200 children), probably due to the lower standards of personal hygiene exhibited by pre-school children. Zoonotic transmission remains controversial. Although the widespread distribution of *Giardia* infection in domestic and wild animals indicates the potential for zoonotic transmission, definitive evidence that this route of transmission occurs and is significant remains elusive.

The most important route of environmental transmission is through the contamination of water by cysts. *Giardia* is the most commonly identified agent of waterborne disease in the USA with over 120 waterborne outbreaks affecting more than 25 000 persons,

since 1965. Up to 60% of all *Giardia* infections in the USA may be acquired from contaminated water, and both human and non-human sources of contamination have been implicated in waterborne outbreaks. *Giardia* cysts occur commonly in both UK and US surface water samples. In the UK, three waterborne outbreaks have been reported. Inadequate water treatment, ineffective filtration or pre-treatment of surface waters and inadequate disinfection when disinfection was the only treatment are identified as the major deficiencies responsible for the majority of these waterborne outbreaks. Although filtration and disinfection is effective in removing *Giardia* cysts, in the USA as many as 21 million individuals have unfiltered drinking water supplies, and in developing countries relatively few individuals will have access to drinking water that has received any treatment.

Worldwide, *Giardia* is one of the most commonly reported protozoal infections of the human intestine and occurs throughout both temperate and tropical locations. Prevalence figures vary and may be based on studies with particular bias, but, in general, the prevalence in developing countries may be 20–30%, and 2–5% in countries of the developed world. Global prevalance has been estimated at 7%; in some developing countries studies have indicated that virtually all children have been infected by three years of age. In the developing countries, *Giardia* can be highly endemic; the possibility of transmission is increased by the coexistence of overcrowding and poverty, poor sanitation and a high level of faecal contamination of the environment. In industrialized nations, the development of water supply systems and sanitation has reduced the availability of transmission routes, but episodic outbreaks occur relatively frequently. It is almost impossible to quantify the cost of giardiasis, although the annual cost of hospital admissions in the USA due to giardiasis has been estimated at over $5 million. Looking at the cost in wider terms, both the primary costs of chemotherapy and direct treatment as well as the secondary costs of lost earnings and, perhaps more importantly, lost education, must also be included.

Ingested trophozoites are destroyed in the stomach. Although theoretically, just one viable cyst can cause infection, it has been demonstrated that as few as ten cysts have produced disease. In most people the disease is self-limiting. The acute phase of giardiasis is usually short-lived and is characterized by flatulence with sometimes sulphurous belching, abdominal bloating and cramps. Diarrhoea is initially frequent and watery but later becomes bulky, sometimes frothy, greasy and offensive and the stool may float on water. Blood and mucus are usually absent and pus cells are not a feature on microscopy. In the immunologically competent, self cure can occur in 6 weeks. Cyst excretion can approach 10^7 per gram of faeces. The incubation period is usually 1–2 weeks. As the prepatent period can exceed the incubation period, initially a patient can have symptoms in the absence of cysts in the faeces.

In the chronic stages of the disease, malaise, weight loss and other features of malabsorption may become prominent. By this time, stools are usually pale or yellow and are frequent and of small volume. Occasionally episodes of constipation intervene with nausea and diarrhoea precipitated by the ingestion of food. Malabsorption of vitamins A and B_{12} and D-xylose can occur. Disaccharidase deficiencies (most commonly lactase) are frequently detected in chronic cases. In young children, 'failure to thrive' is frequently due to giardiasis, and all infants being investigated for causes of malabsorption should have a diagnosis of giardiasis considered.

Trophozoites undergo antigenic variation of their variant-specific surface proteins (VSPs). The patterns of infection in humans and animals fail to show the expected cyclical waves of increasing and decreasing numbers of parasites expressing unique VSPs seen in trypanosomiasis or malaria. Nevertheless, changes in VSP expression occur in vivo within the population due to selection of VSPs both by immune (probably B-cell mediated) and non-immune mechanisms. *Giardia* typing schemes are being developed currently. G. *intestinalis* isolates can be divided into two major genotypes which can be further subtyped by restriction-fragment-length polymorphism (RFLP) analysis of PCR products. These systems have the potential to be used to type cysts from faeces. A sequence-based classification system, based on the gene encoding the metabolic enzyme triose phosphate isomerase (*tim*) has been used to attempt to address the significance of zoonotic transmission. Sequences from a number of G. *duodenalis* isolates of various host origins have been analysed and restriction enzymes identified which can distinguish between isolates without the need for sequencing. Isolates from a previously reported epidemic of giardiasis could be classified accurately using this technique, identifying its possible role in epidemiological investigations.

Treatment

A variety of drugs, including nitroimidazole compounds, quinacrine and furazolidone are available for treatment with cure rates of 75–90% being reported. Tinidazole has the advantage of being effective when given as a single dose. Metronidazole resistance is on the increase. Albendazole and mebendazole, have also been reported to be effective in the treatment of giardiasis. If substantiated by other clinical studies, albendazole and mebendazole will find favour as they are already in use as a broad-spectrum anthelmintic.

Regulations

Although not reportable in England and Wales, giardiasis was made a laboratory reportable disease in Scotland in 1989. The UK Food Safety Act (1990) requires that food, not only for resale but throughout the food chain must not have been rendered injurious to health; be unfit; or be so contaminated – whether by extraneous matter or otherwise – that it would be unreasonable for it to be eaten. A set of horizontal and vertical regulations (which are provisions in food law) covering both foods in general and specific foodstuffs also ensure that food has not been rendered injurious to health. Other than being a potential microbiological contaminant, *Giardia* is not identified in these regulations.

In the USA, the Food and Drug Administration has responsibility for enforcing regulations as detailed by the Federal Food, Drug and Cosmetic Act. As for *Cyclospora*, no regulations specifically address *Giardia*, but products contaminated with the organism are covered by sections of the Act depending on whether the foods are domestic or imported. The reader is referred to the section of *Cyclospora* for a detailed account of regulations in the USA.

Apart from regulations governing agricultural wastes, current regulations in the UK and USA do not require risk-based standards or guidance based on protection from microbial contaminants such as *Giardia*. Good management practices are also suggested for farms and agricultural wastes. The EC Agri-Environment Regulation, under the Common Agricultural Policy, promotes schemes which encourage farmers to undertake positive measures to conserve and enhance the rural environment in Europe.

See also: **Biophysical techniques for enhancing microbiological analysis**: Future Developments. **Enzyme Immunoassays**: Overview. **Food Poisoning Outbreaks**. **Good Manufacturing Practice**. **National Legislation, Guidelines & Standards Governing Microbiology**: European Union. **PCR-based Commercial Tests for Pathogens**. **Water Quality Assessment**: Routine Techniques for Monitoring Bacterial and Viral Contaminants. **Waterborne Parasites**: Detection by Classic and Modern Techniques.

Further Reading

Barnard RJ and Jackson GJ (1984) *Giardia lamblia*. The transfer of human infections by foods. In: Erlandsen SL and Meyer EA (eds) Giardia *and Giardiasis*. P. 365. New York and London: Premium Press.

Craun GF (1990) Waterborne giardiasis. In: Meyer EA (ed.) *Giardiasis. Human Parasitic Diseases*, vol. 3, p. 267. Amsterdam: Elsevier.

Jakubowski W (1990) The control of *Giardia* in water supplies. In: Meyer EA (ed.) *Giardiasis. Human Parasitic Diseases*, vol 3, p. 335. Amsterdam: Elsevier.

Meyer EA (ed.) (1990) *Giardiasis. Human Parasitic Diseases*, vol 3. Amsterdam: Elsevier.

Smith HV, Robertson LJ and Ongerth JE (1995) Crytosporidiosis and giardiasis, the impact of water transmission. *Aqua* 44: 258–274.

Thompson RCS, Reynoldson JA and Lymbery AJ (eds) (1994) Giardia: *From Molecules to Disease*. Wallingford, UK: CAB International.

GLUCONOBACTER

Rolf K Hommel, CellTechnologie Leipzig, Germany

Peter Ahnert, Department of Biochemistry, The Ohio State University, Columbus, USA

Characteristics of the Genus *Gluconobacter*

Gluconobacter constitutes a genus of the acetic acid bacteria, Acetobacteraceae. Former classifications as *Acetomonas* and the inclusion into the family Pseudomonadaceae together with the genus *Acetobacter* were corrected by application of DNA–rRNA hybridization studies. *Acetobacter* and *Gluconobacter* are the sole genera of this family, which belongs to the rRNA superfamily IV (synonymous: alpha group). Both form an rRNA cluster which is also phenotypically well-justified. Each genus consists of a separate rRNA branch and both are linked together at 77°C $T_{m(e)}$, that is the measure of similarities between rRNA cistrons which represents the temperature at which half the DNA–rRNA hybrid is denatured under standard conditions. According to phenotypical classification all strains fall into the species *Gluconobacter oxydans* (formerly *Acetobacter suboxydans*). However, hybridization studies revealed two additional species: *Gluconobacter asaii* and *Gluconobacter cerinus*.

Like most other representatives of the superfamily IV *Gluconobacter* live in the phytosphere as saprophytes, symbionts, or pathogens. The strictly aerobic organisms are well adapted to living in or on plant material and are well adapted to sugary and alcoholic solutions with an acidic pH. *Gluconobacter* is widespread in plants, on fruits, in production facilities, and in different products made from plants or with comparable living conditions.

The strictly aerobic chemoorganotrophic Gram-negative (Gram-variable in a few cases) 0.5–0.8×0.9–$4.2\,\mu m$ rods are motile by polar flagella or non-motile. Endospores are not formed. *Gluconobacter* grows at 4–9°C but not above 38–40°C, with an optimum around 30°C. Acidic pH values are preferred, with an optimum at pH 5–6. Growth is reported down to pH 3 for strains isolated from technical processes. Under production conditions up to 13.5% acetic acid is tolerated. *Gluconobacter* strains are less acid and ethanol tolerant than *Acetobacter*. The metabolism is always respiratory never fermentative. Compared with *Acetobacter* strains the oxidation of ethanol to acetic acid is low whereas high oxidative and ketogenic activities are exhibited by *Gluconobacter* strains. Overoxidation of formed acetic acid or of lactic acid to CO_2 and H_2O does not take place. Single L-amino acids cannot serve as sole source of carbon and nitrogen for growth of *Gluconobacter*. No amino acid is 'essential'. Preferred carbon sources for growth are D-mannitol, sorbitol, glycerol, D-fructose and D-glucose. Ethanol is not a preferred growth substrate but it may be used as an additional carbon source. Whereas *Acetobacter* is able to metabolize only hexoses, *Gluconobacter* metabolizes both pentoses, such as D-xylose, and hexoses. *G. oxydans* is also able to metabolize the trisaccharide raffinose via levansurase. Melobiose, oxidized to melibionic acid, and fructose are formed; the latter is used as a carbon source. Depending on the carbon source used, individual strains require growth factors such as *p*-aminobenzoic acid, biotin, nicotinic acid, thiamine or pantothenic acid.

The phosphorylative breakdown of the carbon source proceeds via the hexose monophosphate pathway. Additionally, the enzymes of the Entner–Doudoroff pathway are present. The glycolysis and tricarboxylic acid cycle are not completely present. *Gluconobacter* displays a high oxidative capacity and ketogenic activities on sugars, alcohols, aldehydes and steroids. Remarkable oxidative capabilities are known for aliphatic monoalcohols, which are converted into the corresponding aldehydes and acids. *G. suboxydans* converts the racemic mixture of 1,2-

propanediol into acetol. The oxidation of diethylene glycol to diglycolic acid, the oxidation of the respective monomethyl ether and the oxidation of other higher aliphatic polyalcohols are other examples of the oxidative capability. Glycerol is oxidized to dihydroxy acetone, a suntanning agent (*G. oxydans*), or to L-glyceric acid (*G. cerinus*). Deoxy sugar alcohols, mannitol, other hexitols, heptitols and octitols, and cyclic polyalcohols, for example, are other substrates oxidized by *Gluconobacter*. Sugar alcohols having a *cis* arrangement of the two secondary alcohol groups in D-configurations contiguous to the primary alcohol group are oxidized to the corresponding ketoses (Bertrand–Hudson rule). This is the basis of the formation of L-sorbose from D-sorbitol, or L-erythrulose from *meso*-erythritol. *G. oxydans* is able to carry out continuously reactions in organic solvents like the formation of isovaleric aldehyde in *iso*-octane.

Gluconobacter (and *Acetobacter*) are the most prominent representatives of oxidative bacteria that can carry out highly effective oxidative fermentations. Oxidative fermentations are incomplete oxidations resulting in the accumulation of huge amounts of the respective oxidation product outside the cell. Substrates converted by *Gluconobacter* in the order of preference are various sugars and sugar alcohols, aliphatic alcohols and aldehydes with glucose and ethanol followed by D-gluconic acid, D-sorbitol and glycerol. The oxidizing system, located in the cytoplasmic membrane, is tightly linked with the respiratory chain which consists of large amounts of cytochrome *c*, ubiquinone and a cytochrome *o* terminal ubiquinol oxidase. In *Gluconobacter*, an alternative cyanide-insensitive terminal oxidase is present. The oxidative power originates in dehydrogenase systems. Their active centres face the periplasm, and form a periplasmatic oxidase system. One group of these enzymes are quinoproteins (with the prosthetic group pyrrolo-quinoline quinone, PQQ) such as alcohol dehydrogenase, aldehyde dehydrogenase, D-glucose dehydrogenase, D-fructose dehydrogenase and glycerol dehydrogenase. The other group covers flavoproteins (with the prosthetic group flavin adenine dinucleotide FAD) such as D-gluconate dehydrogenase, D-sorbitol dehydrogenase, 2-oxo-D-gluconate dehydrogenase and L-sorbose dehydrogenase. Some of these have cytochrome *c* as an additional prosthetic group. Most of these membrane-bound enzymes possess cytoplasmic counterparts, which are NAD(P) dependent with pH optima in the neutral or weak alkaline region. The pH optima of membrane-bound dehydrogenases are mostly acidic. The specific activity of membrane enzymes is up to three orders of magnitude higher than those of the cytoplasmic ones.

Figure 1 demonstrates schematically the organization of the sorbitol oxidizing system.

This membrane-bound oxidase system, a truncated respiratory system, does not require the energy-consuming transport of substrates into the cell, and thus favours the rapid oxidation of large amounts of substrates. It may function as an auxiliary energy-generating system. However, the proton gradient generated by rapid oxidations by the respiratory chain may suppress electron transfer by feedback control. The electron transfer is disturbed by itself. To overcome this contradictory situation, which may be unfavourable for the organism in *Gluconobacter* a cyanide-insensitive bypass (non-energy generating) reduces the energy supply. The little amount of energy formed may not interfere significantly with the electron transfer. Remarkably, a low extracellular pH induces this system. In deficient mutants, the introduction of this bypass enhances oxidation yields.

This system reflects the adaptation to the natural habitats, flowers, fruits and their fermented products like vinegar, saké, wine or beer, in which high concentrations of sugars and alcohol are present. Microbial competitors, yeasts and lactic acid bacteria, favour anaerobic environments. There is no need for *Gluconobacter* to grow rapidly and subsequently no need to permit a high rate of energy generation by the highly aerobic and highly active oxidation system of the respiratory chain.

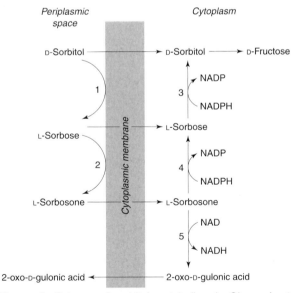

Figure 1 Scheme of sorbitol metabolism in *Gluconobacter*. Numbers indicate participating enzymes: (1) D-sorbitol dehydrogenase; (2) L-sorbose dehydrogenase, both flavoproteins located in the cytoplasmic membrane with the catalytic centre directed to the periplasmic space; (3) NADP⁺-dependent L-sorbose dehydrogenase; (4) NADP⁺-dependent L-sorbosone dehydrogenase; (5) NAD⁺-dependent L-sorbosone reductase, all cytoplasmic enzymes. D-Fructose is metabolized to acetyl-CoA.

Like *Acetobacter* strains, *Gluconobacter* harbour one to eight plasmids. For instance, *G. oxydans* ATCC 9937 has three plasmids of 27.7 kb, 12.3 kb and 18 kb. Host–vector systems and transformation methods have been adapted to *G. oxydans*. At least three host-specific phages for *G. oxydans* are known.

Some strains of *Gluconobacter* excrete inhibitory substances which interfere with the growth of *Saccharomyces cerevisiae* in the early stages of wine making and thus affect fermentation productivity. Polyenic antibiotics, monocyclic β-lactams, against Gram-positive and Gram-negative bacteria have been isolated from *Gluconobacter*. *G. suboxydans* transforms nojirimycin to give a delta-lactam. *G. oxydans* produces brown, water-soluble pigments and pyrones.

The high oxidative capabilities of *Gluconobacter*, and the special organization of the enzymes involved in these reactions, enable the application of whole cells, of isolated enzymes (xylose PQQ-dependent dehydrogenase from *G. oxydans*) or of cytoplasmic membranes in enzyme sensors/electrodes. *Gluconobacter* cells are used for the detection of xylose, (blood) glucose, sucrose and lactose. In one-component media the sensitivity toward ethanol, glucose or glycerol is high, with a range of 10–100 mM.

The natural habitats of *Gluconobacter* are plants and soil. Preferred niches are fruits and flowers, especially nectaries and other floral organs and assorted plant material which are rich in sugars and have an acidic pH. *Gluconobacter* occurs profusely on ripe grapes of the Bordeaux region. Strains are also present on intact, injured, dried and matured grapes in other regions but do not appear during the fermentation process. The presence of *Gluconobacter* and therefore its ability for ketogenesis results in a higher requirement for SO_2 in sulphonation processes. *G. oxydans* may be present in palm wine and is found on the floret of the palm tree, on the tap hole used to collect the juice and in palm sap. *Gluconobacter* strains are also present in cocoa wine. Both beverages are of low alcohol content. Many fruits such as apples, almonds, bananas, mangoes, oranges, plums, strawberries, tomatoes are habitats. *Gluconobacter* strains (and *Acetobacter*) are the causal agents of bacterial brown rot of apples and pears. The oxogluconic acids and (oxo-)phosphogluconates (ketogenic potency) of glucose and fructose metabolism/oxidation are assumed to be causal substances for the pink disease of pineapple, caused by *G. oxydans*, *Acetobacter aceti* and *Erwinia herbicole* or by *Pantoea citrea* or *Acetobacter liquefaciens*. Rot symptoms may be induced by inoculation with 100 cells. The optimal temperature was 25°C, but even at 4°C rotting proceeded. Examples of artificial habitats are soft drinks and beer.

The sugar-loving *Gluconobacter* is one component of the huge microbial population of bees. Like other associated microorganisms, *Gluconobacter* is not present in the larvae. Young bees come into contact with the microbial population via other bees, i.e. in the exchange of food. Depending on the food source, the composition of the microbial flora will vary. However, *Gluconobacter* strains and *Lactobacillus viridescens* constitute the two main bacterial components of the microflora of bees in Spain. A total of 56 *Gluconobacter* strains have been isolated from bees in Belgium. Up to 10^6 *Gluconobacter* cells are spread equally over an individual bee including the intestine and the body surface. In this context, bees may be considered as one important vector in *Gluconobacter* dissimilations.

The aerobic microflora of the nasal, rectal, and preputial or vaginal areas of 37 grizzly and 17 black bears contained *Gluconobacter* and *Acetobacter* at lower frequencies (less than 5%).

Apart from the phytotoxic effects of *Gluconobacter* no pathogenicity toward mammals, including humans, is known. *G. oxydans* and *G. asaii* are on the list of plant pathogenic bacteria.

Methods of Detection

Gluconobacter share their natural habitat with a huge number of other bacteria, such as *Acetobacter*, *Frateuria* and *Zymomonas*. This sometimes makes the isolation and identification to species level more difficult. *Acetobacter* strains are normally co-isolated. Most media used for the detection of *Acetobacter* strains are also applicable to *Gluconobacter*. In general, glucose should be used as the source of carbon and energy (100–400 g l^{-1}). The starting pH should be in the range 6.5–7. Aerobic growth is optimal between 25°C and 30°C. Enrichment becomes necessary when a low viable cell count is expected. In older literature beer is mainly recommended for this purpose. The presence of preserving agents in some beers may reduce the success. Different habitat specific methods have been developed which consider the specific nutrient requirements of adapted strains. Some of those described for *Acetobacter* are also applicable to *Gluconobacter*. One isolation process for *G. oyxdans* includes an enrichment in liquid medium at 30°C and subsequent isolation from the pellicles. In the first step beer is incubated at 30°C. Subsequently the pellicles formed are incubated in beer containing glucose (300 g l^{-1}) and acetic acid (1 g l^{-1}). The third step comprises incubation in yeast water inoculated with a piece of bakers' yeast, that contains glucose and acetic acid at the same concentrations as above. For isolation, differentiation, and purification carbon sources such as glucose ($100\,g\,l^{-1}$), sucrose ($100\,g\,l^{-1}$), ethanol ($30\,g\,l^{-1}$), maltose ($100\,g\,l^{-1}$), or calcium lactate ($20\,g\,l^{-1}$), in the presence of calcium carbonate (20–$30\,g\,l^{-1}$) in yeast water are recommended.

Further differentiation is usually accomplished by morphological consideration and by growth on selective media. The application of modern techniques, such as hybridization and polymerase chain reaction (PCR), still tends to be restricted to some commercially important strains. Typical features of acetic acid bacteria are: Gram-negative (or -variable), strictly aerobic, ellipsoidal to rod-shaped cells with a respiratory type of metabolism, oxidation of ethanol to acetic acid in neutral and acidic media, and oxidation of glucose below pH 4.5. Acetic acid bacteria do not form endospores, are oxidase negative, do not liquefy gelatin, do not form indole or reduce nitrates. Further identification of *Gluconobacter* is based on established features, such as morphology, biochemical activities, etc.

The differentiation between *Acetobacter*, *Frateuria* and *Gluconobacter* tends to be rather simple (**Table 1**). *Gluconobacter* is not able to overoxidize acetic acid or lactic acid, does not form H_2S, is either nonmotile or motile by polar flagella with 3–8 cilia. *Gluconobacter* prefers glucose over ethanol for growth. *G. oxydans* requires sugar to initiate growth. Ethanol may be used as an additional carbon source. All strains produce 2-oxogluconic acid from D-glucose. Starch and lactose are not accepted as carbon sources. The presence of ubiquinone Q_{10} as the major quinone in *Gluconobacter* is a common characteristic of most strains of the superfamily IV.

The saprophytic pseudomonad *Frateuria* settles in habitats normally populated by *Gluconobacter* (nectaries and other floral organs). It shares a number of basic characteristics with *Gluconobacter*, such as polar flagella. However, strains of *Frateuria* do not require organic growth factors, have ubiquinone Q_8 (with a minor component of Q_7) and a different fatty acid profile consisting of iso-branched-chain acid of $C_{15:0}$. Other bacteria also show similarities with *Gluconobacter*.

Zymomonas is genetically (rRNA hybridization studies) and phenotypically related to acetic acid bacteria. Both occur in acidic, sugary and alcohol-containing habitats (tropical plant juice, cider, beer, etc.). It resembles *Gluconobacter* more closely (polar flagella, incomplete tricarboxylic acid cycle, occurrence of the Entner–Doudoroff pathway). *Zymomonas* ferments sugars anaerobically to ethanol and CO_2. It is assumed that both acetic acid bacteria and *Zymomonas* derived from a common aerobic ancestor.

The soil bacterium *Phenylobacterium immobile* showed serologically slight, but significant, crossreactivity with *G. oxydans*. The rod-shaped bacterium *Zoogloea*, with polar flagella, displays some features resembling *Gluconobacter* (and *Acetobacter*) in morphology and oxidative metabolism. Some strains have a greater antigenic relationship to *G. oxydans* subsp. *suboxydans* than to strains of their own genus. *Zoogloea* differs from acetic acid bacteria in its inability to grow at pH 4.5. Additionally, *Pseudomonas diminuta* and *P. vesicularis* resemble strains of *Gluconobacter*; both produce acid from ethanol and aldoses, acetone from isopropanol, and have multiple requirements for growth factors. Some morphological similarities are described, e.g. short flagella (0.6–$1.0\,\mu m$).

The differentiation of the species may be affected by spontaneously occurring mutations. Additionally, acetic acid bacteria are extremely variable, and very potent to adapt functionally and structurally to altered environmental conditions. They achieve this either via activation of cryptic and silent genes, or via selection of mutants in regulatory and structural genes. This is also important for controlling commercial processes. Rapid phenotypical differentiation of *Gluconobacter* species may be achieved by simple assimilation tests as summarized in **Table 2**.

Being isolated and differentiated, *Gluconobacter* may be maintained on various liquid (beer) and solid media. **Table 3** gives a survey on recommended solid media. Agar cultures should be kept at 4°C and transferred monthly. Freeze-dried strains will remain alive for several years.

Table 1 Phenotypic features to differentiate *Gluconobacter*, *Acetobacter* and *Frateuria*

	Gluconobacter	Acetobacter	Frateuria
Overoxidation of acetic acid	–	+	–
Overoxidation of lactic acid	–	+	+
H_2S formation	–	–	+
Main type of ubiquinone	Q_{10}	Q_9 or Q_{10}	Q_8
Type of flagella	Polar	Peritrichous	Polar

Table 2 Differentiation of *Gluconobacter* species (reprinted with permission from Swings 1992)

	Growth on ribitol	Growth on arabitol	Growth on nicotinate
G. asaii	–	–	+
G. cerinus	+	+	+
G. oxydans	–	–	–

Table 3 Common media for maintenance and cultivation of *Gluconobacter*

Medium	Composition		pH at 25°C	Species
Gluconobacter oxydans agar	Glucose	100 g	6.8 +0.2	G. oxydans
	Yeast extract	10 g		
	CaCO₃	20 g		
	Agar	15 g		
	Distilled/deionized water	1 l		
'Acetobacter agar'	Glucose	50 g		G. cerius
	Yeast extract	10 g		G. oxydans
	CaCO₃	30 g		
	Agar	25 g		
	Tap water	1 l		
'Gluconobacter broth'	Glucose	100 g	6.8	G. oxydans
	Yeast extract	10 g		
	CaCO₃	10 g		
	Distilled/deionized water	1 l		
Mannitol agar	Mannitol	25 g		G. asaii
	Yeast extract	5 g		G. cerius
	Peptone	3 g		G. oxydans
	Agar	15 g		
	Distilled/deionized water	1 l		
Yeast glucose agar	Glucose	20 g	7.0 +0.2	G. oxydans
	Yeast extract	10 g		
	CaCO₃	20 g		
	Agar	15 g		
	Distilled/deionized water	1 l		
Yeast glucose agar	Glucose	20 g	7.0 +0.2	G. asaii
	Yeast extract	10 g		G. cerius
	Agar	15 g		G. oxydans
	Distilled/deionized water	1 l		

Importance to the Food Industry

Processes

Gluconobacter strains are involved in a large number of natural fermentations and have large capabilities for other applications in commercial processes.

In cocoa fermentation, beans are naturally fermented by a wild microflora consisting of nearly 50 common species. Acetic acid bacteria constitute this population. Cocoa fermentation is a combination of external microbial processes and autolytic processes in which cocoa bean enzymes are involved. For instance, *G. oxydans* was isolated from cocoa fermentations. Acetic acid bacteria dominate around the second day of fermentation with up to 90% of the whole cell count. It has been shown that the complex natural microflora of this fermentation can be replaced by a defined, mixed starter culture. *Saccharomyces cerevisiae* var. *chevalieri*, *Lactobacillus lactis*, *Lactobacillus plantarum*, *Acetobacter aceti* and *Gluconobacter oxydans* subsp. *suboxydans* are members of this cocktail. *Gluconobacter* strains are members of the microbial consortium of Kombucha/tea fungus.

Gluconobacter strains are involved in palm wine and cocoa wine making. In cider making, *Gluconobacter* strains are present mainly in the early stages of the process, when sugar is plentiful. *Gluconobacter* is not involved in vinegar production.

Gluconic acid formation is most pronounced in *Gluconobacter* (*G. oxydans*). Gluconic acid has some useful properties, and has found a number of applications. Examples are the removal or decomposition, or prevention of milkstone (dairy industry), gentle metal cleaning operations, and the prevention of cloudiness and scaling by calcium compounds in beverages. Applications in the textile and tanning industries and in medicine (gluconates of calcium and iron as carriers) are other examples. According to German law gluconic acid is considered food. Most industrial processes are carried out mainly with moulds such as *Aspergillus niger*. In these, glucose oxidase (β-D-glucose : oxygen 1 oxidoreductase) produces δ-gluconolactone and H_2O_2 and lactone hydrolysing enzymes (lactonase) provide gluconate. Bacterial fermentations are of limited importance due to secondary reactions leading to oxogluconic acids. *Gluconobacter* oxidizes glucose by a membrane-bound PQQ-dependent D-glucose dehydrogenase to gluconate. Cytoplasmic NADP-dependent enzymes are not involved. However, gluconate is frequently further oxidized to 2-oxogluconate (at neutral pH) and to 5-oxogluconate (acidic pH) and 2,5-dioxogluconate; 5-oxogluconate may be useful as the starting material for the production of D(+)-tartaric acid. Oxo-forms of gluconic acid are valuable products with a wide range of applications; 2-oxo-gluconate serves as precursor for *iso*-ascorbic acid which is used, for instance, as an antioxidant.

In aqueous solutions, gluconic acid exists in equilibrium with gluconic acid δ-lactones (1,5 and 1,4, respectively), which may be used in similar ways as gluconic acid; they are preferred for slow acidulation actions (in baking acids, in the production of cured meat products, etc.). In Japan it is used for the coagulation of soybean protein.

Exclusion of oxogluconic acid formation is the basis for gluconic acid production by *Gluconobacter* at an industrial scale. It has been shown, that the pentose phosphate pathway is almost completely repressed if glucose concentrations exceed 5–15 mM and the medium is kept below pH 3.5–4. Maintaining the pH below this range suppresses unwanted oxogluconate formation. Similar to quick vinegar processes, optimal performance of gluconic acid production is achieved by high glucose concentrations, lowered pH and high oxygen concentration.

L-Ascorbic acid (vitamin C) is used as antioxidant in the food industry. The annual demand is up to 40 kt. The production of ascorbic acid by the so-called Reichstein process is a combination of chemical and biotechnological steps:

D-Glucose → D-Sorbitol → L-Sorbose → L-Ascorbic acid

The stereospecific oxidation of D-sorbitol to L-sorbose, is carried out by *G. oxydans* by the action of membrane-bound quinoprotein sorbitol dehydrogenase (Fig. 1). This reaction follows the Bertrand–Hudson rule. The submers fermentation is carried out at 30–35°C. Substrate concentrations are in the range 200–300 g l^{-1} in corn steep liquor medium or yeast extract supplemented with mineral salts under strong aeration. After one to two days the reaction is nearly quantitative. The subsequent reactions are carried out chemically.

As shown in Figure 1, other possible pathways include the formation of 2-oxo-L-gulonate, an other precursor of L-ascorbic acid. Recently, bacteria have been identified that are efficiently able to transform glucose into 2,5-dioxo-D-gluconic acid and this product into 2-oxo-L-idonic acid, a precursor of L-ascorbic acid. When the corresponding strains are used together, it is possible to get 2-oxo-L-idonic acid directly from glucose. Moreover, new strains have been constructed by introducing a gene from a strain responsible for the second step into a strain responsible for the first step. By using one of the new strains,

the transformation can be performed in a single step with only one strain. Even immobilized cells of *G. melanogenus* are capable of converting L-sorbose into L-sorbosone. However, the classical process still remains the most competitive.

Spoiling

Gluconobacter is considered to be a typical spoiler of soft drinks. These constitute highly selective *Gluconobacter* media which contain sugar at low pH, are free of oxygen or growth factors, and contain very little organic nitrogen. Strains of this genus are considered causative agents of a deleterious change in orange juice characterized by a marked staleness of flavour. Alcoholic beverages are good media for *Gluconobacter*; *G. suboxydans* frequently occurs in Dutch beers. *G. oxydans* and *G. industrius* have been identified in beer that is exposed to the atmosphere. *Gluconobacter* has been isolated from palm wine and cocoa wine. In cider making, *Gluconobacter* is present mainly in the early stages of the process, which are rich in sugar. *Gluconobacter* is also present in and on harvested apples, in pressed pomace and in the juice.

Although grapes are one of its natural habitats, *Gluconobacter* is rarely seen in wine. These bacteria affect the wine-making process before it has started. *G. oxydans*, for example, infects almost exclusively sound red grapes. *Acetobacter* appears after the grapes are slightly spoiled and dominates on completely spoiled grapes, which display numbers of acetic acid bacteria equivalent to those of the yeast population (10^5–10^6 cells per millilitre). High humidity during rainy autumn weather supports a varied microflora on grapes. Cell counts of *Botrytis cinerea*, *Penicillium*, *Aspergillus*, *Gluconobacter* and *Acetobacter* may become equal to the cell counts of yeasts. This mouldiness destroys the anthocyanins of grapes and alters their colour. Oxidation reactions lead to oxidative spoilage of wines characterized by mouldy taste and phenolic odour. In both sound and infected (by *B. cinerea*) grapes ketogenic *G. oxydans* (formation of 2-oxogluconic acid and 2,5-dioxogluconic acid) are present. The oxo-compounds are thought to be responsible for enlarged quantities of bound SO_2 (sulphonation) in musts from grapes with enlarged cell counts of acetic acid bacteria. During the initial phase of fermentation oxo-gluconic acids are formed.

In musts from *B. cinerea*-infected grapes, *Gluconobacter* can persist during the first days of fermentation, later being progressively replaced by *Acetobacter*. As already noted, some strains of *Gluconobacter* produce substances affecting the growth of *Saccharomyces cerevisiae*, and therewith fermentation activity, during the first days. The pH minimum tolerated by *Gluconobacter* depends on the ethanol concentration of the wine: with 8.2 vol.% ethanol pH 3 and with 12.5% pH 3.4 are tolerated, respectively. Higher concentrations are normally inhibitory.

Sorbic and benzoic acids, used as food preservatives in the food industry, show minimal inhibitory concentrations (MIC) of 1 and $0.9 \, g \, l^{-1}$, respectively, for *G. oxydans* at pH 3.8. Reduction of the acidity to pH 3.3 reduces the MIC of both to $0.3 \, g \, l^{-1}$. Growth of the bacteria in the presence of sublethal concentrations of both preservatives increased the MIC significantly within 1 h. At extreme temperatures (1°C and 37°C) the preservatives supported the inhibitory temperature effect. The best method to prevent *G. oxydans* infections is the elimination of air from the facility and the addition of sorbic acid ($0.4 \, g \, l^{-1}$).

See also: **Acetobacter. Ecology of Bacteria and Fungi in Foods**: Influence of Redox Potential and pH. **Fermentation (Industrial)**: Production of Xanthan Gum. **Fermented Foods**: Origins and Applications; Fermented Vegetable Products; Fermentations of the Far East. **History of Food Microbiology. Metabolic Pathways**: Release of Energy (Aerobic). **Spoilage Problems**: Problems caused by Bacteria; Problems caused by Fungi. **Wines**: Microbiology of Wine-making.

Further Reading

Adachi T (1968) *Acetic Acid Bacteria. Classification and Biochemical Activities.* Tokyo: University of Tokyo Press.

De Ley J and Swings J (1984) Genus *Gluconobacter.* In: Krieg NR and Holt JG (eds) *Bergey's Manual of Systematic Bacteriology.* Vol. 1, p. 275. Baltimore: Williams and Wilkins.

Fukaya M, Okumura H, Masai H, Uozumi T and Beppu T (1985) Development of a host–vector system for *Gluconobacter suboxydans. Agricultural and Biological Chemistry* 49: 2401–2411.

Matsushita K, Toyama H and Adachi O (1994) Respiratory chains and bioenergetics of acetic acid bacteria. *Advances in Microbial Physiology* 36: 247–301.

Milson PE and Meers JL (1985) Gluconic and itaconic acids. In: Blanch HW, Drews S and Wang DIC (eds) *Comprehensive Biotechnology.* Vol. 3, p. 681. Oxford: Pergamon Press.

Roehr M, Kubicek CP and Kominek J (1996) Gluconic Acid. In: Rehm HJ and Reed G (eds) *Biotechnology.* Vol. 6, p. 348. Weinheim: VCH.

Schwan RF (1998) Cocoa fermentations conducted with a defined microbial cocktail inoculum. *Applied and Environmental Microbiology* 64: 1477–1483.

Swings J (1992) The Genera *Acetobacter* and *Gluconobacter.* In: Balows A, Trüper HG, Dworkin M, Harder W and Schleifer K-H (eds) *The Prokaryotes,* 2nd edn. P. 2269. New York: Springer.

GOOD MANUFACTURING PRACTICE

B Jarvis, Ross Biosciences Ltd, Ross-on-Wye, Herefordshire, UK

Introduction

The term 'Good Manufacturing Practice' (GMP) applies to all the activities associated with manufacturing management and operations. It is especially pertinent to food production, and the principles can be applied at all stages, from the production of raw materials, through intermediate products, to final products. Consumers have the right to expect food to be safe and wholesome, and it must also satisfy their expectations in terms of organoleptic qualities and price. The objective of GMP is to ensure the production of foods which consistently have these characteristics. The application of GMP requires both philosophical and practical adherence, by management and other employees, to a range of principles which have been developed over a number of years.

The senior management of a company must be committed to the need for quality throughout its manufacturing activities, and must adopt systems which ensure GMP. Such systems require the development and application of policies, procedures and practices with regard to:

- personnel, including staff training
- finance and investment
- planning and design of facilities
- purchase of raw materials and ingredients
- manufacturing control
- environmental and personnel hygiene
- quality documentation, monitoring and auditing
- transportation

- new product development
- product information and consumer awareness
- legislative compliance.

Traditionally, food manufacturers relied on quality control (QC) procedures, but modern management recognizes that such procedures are inadequate. QC is predicated on the concept of testing manufactured products for compliance with specification – but this does not ensure that products of adequate quality are manufactured. Positive quality assurance (QA) puts an emphasis on quality from conception and design through to the finished product. However, even QA procedures are inadequate unless they form part of GMP procedures, which in turn comprise part of the operating systems which contribute to Total Quality Management (TQM).

The Fundamentals of GMP

The overriding principle of GMP is the need to provide effective systems for the manufacture and distribution of food products, by appropriately trained and qualified personnel, within an environment which is designed for hygienic manufacture and in accordance with effective quality procedures.

Personnel Policies and Training

Management Structure

Personnel policies must ensure an appropriate management structure. The personnel responsible for quality and those responsible for the purchasing (supply) of raw materials, ingredients and facilities must operate in parallel with those responsible for production, but should not report to them. However, it is essential that these functions interact effectively. The precise organizational structure will differ between organizations, particularly between large, complex organizations and small companies (**Fig. 1**). Yet the principles should be common to all.

Staffing

The need for appropriately qualified and experienced functional managers must be recognized, and each should have an identified deputy, who can assume their role in case of absence. In addition, the level of staffing of key functions must accommodate cover for absences. Temporary staff must not be expected to assume responsibility for key activities without proper training, knowledge and experience of company procedures.

Personnel, particularly key personnel in each of the major functions, should possess appropriate scientific, technical and professional qualifications, together with relevant knowledge, expertise and experience.

What amounts to 'appropriate' varies between companies and countries – in some countries, local or national legislation may prescribe minimum levels of qualifications for functional management.

Training

The policies and procedures associated with training are the key to ensuring that all employees understand and accept responsibility for their actions within their own sphere of operation. Employees should be encouraged to work towards qualifications which reflect their own level of training and experience in a specific operational area. However, it is essential that training is based on best practice and not on inbuilt prejudice and misconception. It should, therefore, reflect internal, national or even international criteria, to ensure that trainees acquire appropriate information and experience. Induction training should be given on recruitment, and should be followed up with further specific training at appropriate intervals. Records of training must be kept for each employee, and from time to time the effectiveness of the training should be assessed.

Responsibility and Authority

Food quality and food safety are dependent on *every* employee within an organization. Each employee should operate within a framework of devolved responsibility and accountability, and must have delegated authority such that they can take decisions appropriate to their areas of responsibility. This is especially important for quality managers, who should be able to perform professionally and without undue interference or pressure from the production personnel.

Quality managers and their staff must have the authority to establish, verify and audit all the parameters that are critical to the quality of manufactured products. Clearly there is a need to ensure that raw materials are of appropriate quality, through the application of quality procedures to suppliers, and a need to assess quality associated with processing, packaging, finished products, storage, distribution and so on. In many organizations, the quality manager has authority to stop production or other processes if there is reason to believe that these are operating outside defined guidelines. However, such action may require ratification by more senior personnel (e.g. a technical or managing director). The quality manager must, therefore, exercise delegated authority with professional judgement in order to ensure that operations are not disrupted unnecessarily and that production and senior management receive prompt and accurate information and advice.

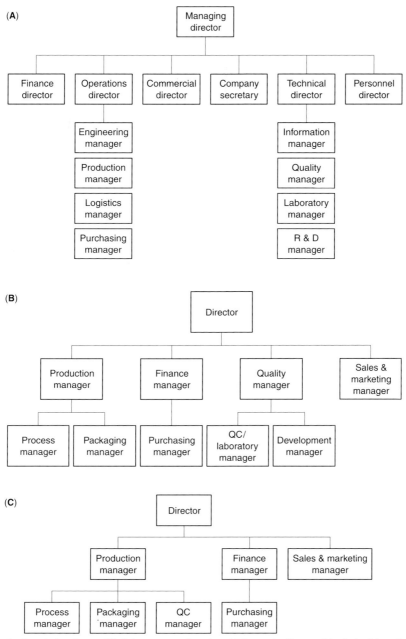

Figure 1 Typical organizational structures. (**A**) A large operation, in which production and technical functions operate in parallel. (**B**) A small operation, in which the quality function operates in parallel with production. (**C**) An unacceptable organizational structure, within a small operation in which the quality function reports to the production manager.

Termination of Employment

When an employee's contract is terminated for any reason, the need for precautionary action must be considered, in order to counteract any threats to food safety or quality which arise due to disaffection or lack of commitment. Disaffected employees may undertake deliberate sabotage, and a lack of commitment may result in failure to comply with GMP.

Finance and Investment

A company's financial policies will impact on its ability to comply with GMP, and so need to take full account of the requirements of GMP. Conversely, inadequate performance in terms of quality will impact on the company's financial performance.

The operating budget must recognize the need for appropriate levels and calibre of personnel in both manufacturing and quality functions. Similarly, budgetary provision for materials, including raw materials, ingredients and packaging materials, should recognize that low contractual purchase prices may be associated with products of low standard. Hence it is essential to operate a 'supplier quality' policy.

Investment proposals must be subjected to careful technical appraisal. The provision of new buildings and of facilities for production, processing, packaging, distribution and laboratories must take into account the GMP requirements applicable to their intended purpose. In drawing up specifications for new facilities and equipment, all the relevant quality and food safety issues must be fully considered.

Facilities

All buildings for food manufacturing should be designed, located, constructed and maintained to suit the purpose intended. Similarly, equipment should be designed, constructed, adapted, located and maintained according to its manufacturing purposes and so as to facilitate GMP. In some countries, food manufacturing facilities must conform to detailed legislative requirements. The layout of production plants should ensure that manufacturing areas do not become general thoroughfares – only designated personnel should be allowed in production areas.

Buildings

Facilities must be situated with regard to the provision of major services, and so as to avoid the risk of contamination from adjacent activities. Proposals for new developments in the vicinity of existing premises must be monitored, with the aim of preventing the introduction of activities which would be inimical to food manufacturing.

The design and construction of premises must provide protection against the entrance and harbouring of vermin, birds and other pests (including feral pets). The design should allow for a logical flow of materials, from the introduction of raw materials to product warehousing and distribution (**Fig. 2**). Areas where high-risk materials are to be stored and/or handled should be segregated from other parts of the manufacturing premises. There should be sufficient space for efficient operation and effective communication and supervision. Building materials must be compatible with food manufacturing activities, for instance to facilitate the maintenance of high hygienic standards. All parts of the premises, including the exterior, should be maintained in a good state of repair and in a clean and tidy condition.

Environmental Issues

Buildings must be efficiently lit and ventilated, with facilities appropriate to the manufacturing activity and the external environment. Working conditions (e.g. temperature, humidity, noise) should be such as to protect the employees and the products, but specific conditions essential for the handling of products (e.g. in cold rooms) must take precedence – operative protection, if necessary, must be considered separately. The supply and extraction of air must not introduce risks of product contamination, e.g. fans must be sited so as to avoid the intake of undesirable materials, such as aromas and gaseous contaminants. Similarly, extractor fans must be situated so as to avoid risks to the external environment. Environmental problems such as noise must be avoided by the appropriate siting and insulation of plant.

Floors and Walls

All floor and wall surfaces within a process plant, and all equipment surfaces must be made of materials which are easy to maintain in a clean condition.

Floors must be level, free from cracks and joints, and laid at levels which allow appropriate drainage. They should be made from materials which are impervious and suited for their purpose – materials suitable in one manufacturing operation may not be suitable in another.

Walls should be sound, with a smooth, impervious surface. Ceilings must be constructed so as to ensure ease of cleaning, and must be maintained, to avoid the risk of contaminating materials falling onto food or process equipment. The junctions of all floors, walls and ceilings must be constructed so as to minimize the build-up of dirt, e.g. through the use of smooth coving.

Doorways and door frames should be of impervious, noncorroding materials. Doors should generally be either opened automatically, or provided with heavy-duty plastic strips which permit easy access by personnel and essential traffic, e.g. fork-lift trucks. Doors also play an important role in pest control. Windows should be made from toughened glass or plastic, should be adequately screened and should have ledges designed to prevent the build-up of dirt and their use for putting down tools and papers, e.g. by incorporating a slope.

Lighting and Overhead Facilities

Adequate levels of natural and artificial light must be provided, and lights must be sited so as to avoid food materials becoming contaminated by glass. The diffusers and shields protecting strip lighting must not become dust traps. All lights and other electrical

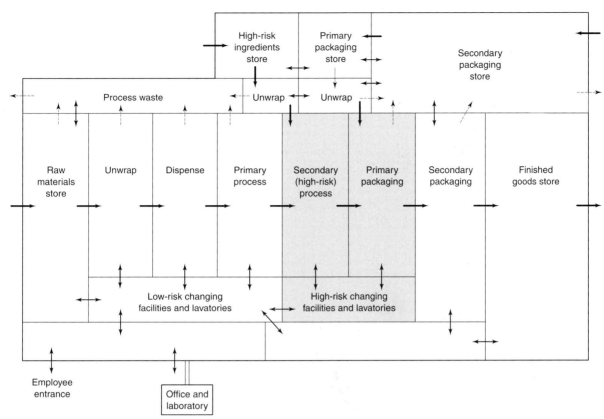

Figure 2 Idealized layout of production plant, separating high-risk areas ▨ and low-risk areas ☐. Flow of materials →. Flow of waste --→. Flow of people ↔.

equipment must be protected against water, and wiring must be protected, e.g. in conduits. Alarms, tannoy systems, overhead pipework and girders must be sited so as to minimize the build-up of dirt. Ideally, services should be run outside processing areas and sealed into the walls and partitions through which they pass.

Drainage and Waste Disposal

Floor drains must be of adequate size and fitted with trapped gullies and ventilation. Open drainage channels must be shallow and any covers must be easy to remove, to facilitate cleaning. Flexible hoses for water or steam must be installed on automatic retraction reels, and should never be left with their ends in gullies. Flexible hoses for product transfer must be fitted with hygienic connectors and screw caps, and the ends should never be allowed to come into contact with floors or drains.

Water services must be installed so as to prevent the backward surge of water into the mains water supply, and there must be no risk of interconnection between effluent drains and mains or indirect water supplies. Water tanks must be sited so that water pressure is adequate, and must be capable of being drained, cleaned and sanitized. An adequate number

of flushable lavatories must be available, and must be connected to a separate drainage system, with no interconnection with the effluent drains from the manufacturing process within the manufacturing environment. Lavatories and associated facilities must be sited without direct access into the areas used for food storage or handling.

Waste from raw materials, including packaging materials, should not be allowed to enter the main production areas. Waste produced within the production and packing areas must not be allowed to accumulate, and must be collected in suitably labelled receptacles of appropriate design, prior to removal to a designated area outside the production buildings. Some waste materials may need to be segregated, to permit effective recycling.

Cleaning

Written cleaning procedures and schedules should be established for all manufacturing and storage areas. Cleaning techniques must reflect the nature of the plant and the process (e.g. dry or wet areas), and any risks to process and product integrity must be minimized. The use of different personnel for cleaning is sometimes advocated, but this removes responsibility from the operatives for maintaining standards in their own areas of operation.

Process Equipment

Process plant should be designed so as to ensure that surfaces in contact with food remain inert under their conditions of use, and that the migration of materials from or into the food cannot occur. Equipment should be easily disassembled for inspection, and for cleaning unless 'in place' cleaning systems (CIP) are used. All surfaces in contact with food should be smooth, non-porous and capable of being effectively cleaned and sanitized. Pipework and equipment should be self-draining and dead ends must be avoided. The food must be protected from contamination either by the equipment, e.g. from leaking glands or lubricants or by the environment. Process equipment should not be inappropriately modified or adapted.

Cleaning and sanitizing agents should be specified for use in particular areas or circumstances. The external, as well as the internal, surfaces of process plant must be capable of being effectively cleaned, particularly those in contact with the floor, walls or supports. Process plant should be serviced, cleaned and sanitized immediately after use. Any faults must be recorded and missing parts (e.g. nuts, bolts, clips) must be reported immediately to the personnel responsible for the quality function. The plant should be checked for cleanliness visually and, ideally, using a hygiene monitoring system such as ATP measurement. The cleaning and sanitizing programme must be repeated immediately if there is any evidence that cleaning has not been effective.

Materials and Ingredients

GMP requires a defined system for ensuring the quality of materials (Supplier Quality Assurance, SQA). Responsibility for this system is vested in either the quality or the purchasing function, but in either case they need to interact effectively. The SQA system should define the specifications of materials, including any agreed tolerances, and also details such as the terms of business and packaging and labelling requirements. Each supplier's compliance with defined quality requirements should be audited, and the supplier should be required to provide definitive information on batch compliance. The audit should take into account the capability and past performance of the supplier.

Negotiations regarding the supply of materials must be predicated on an absolute requirement for the provision of materials within the defined specification. The onus must be placed on the supplier to indicate any deviations from specification and/or any manu-

facturing problems: the documentation should specify that a supplier who attempts to hide problems is in breach of contract. Materials may be delivered on a 'just in time' (JIT) basis, to minimize stock holding, or there may be a predetermined schedule. Although a certificate of compliance should accompany each delivery, the user must demonstrate due diligence by carrying out simple checks to ensure that the materials are fit for purpose. Materials delivered in bulk (e.g. by tanker) should be checked prior to discharge. Any materials that do not conform to specification must be isolated and examined to assess their usability. Clearly, a JIT delivery system that fails to supply materials of the correct standard jeopardizes the manufacturing schedule. Materials should be unloaded under cover, adjacent to the 'goods inward' storage area, and must be protected from environmental contamination.

Manufacturing Control

Products must conform consistently to specifications, and must be protected against environmental contamination and deterioration. As a basis for quality control, the manufacturing procedures must be capable of output which meets the product specifications, taking into account any tolerances acceptable in reflection of tolerances in raw material specifications.

Operating Procedures

Written operating procedures and instructions are essential, forming part of the QA procedures. They should include all ancillary activities and precautions and should define what is to be done, when, how and by whom. They should be clearly written and easy to understand, and should place emphasis on issues which may affect product quality. They must include actions required when defects or other problems are detected, and in the event of stoppages, breakdowns and emergencies. Training must ensure that operatives understand the operating instructions.

All production activities must include a procedure for recording relevant data on ingredients and raw materials, e.g. source, delivery date, lot number and critical process data. These data provide evidence for the demonstration of due diligence by the manufacturer.

Operating instructions should always be treated as definitive, with no opportunity for deviation unless with written authorization from the management of the production and/or quality functions. If the modification of a procedure is justified the reasons must be documented and the change must be authorized formally.

Pre-production Checks

Before processing starts, checks should be made to confirm that the area and plant are clean and that the necessary materials are available. Authorized personnel should check that the information on the pre-printed packages and labels to be used for the product conforms to legal requirements, and that the correct materials are to be used. The accuracy of the settings of the equipment for process monitoring and control (e.g. temperature gauges, metal detectors, check weighers) must be checked, and any final adjustments necessary must be made during the first few minutes of production. Any intermediate or final product prepared prior to the establishment of the correct steady-state conditions must be excluded from the finished product, to ensure that specifications are met.

Intermediate Products

Intermediate products should be labelled and isolated prior to checking and approval by the quality function. Once approved, they should be stored until required for use. Any intermediate product which does not conform to specification should be reworked or rejected, with the recovery of materials if appropriate.

Finished Products

Finished products should be labelled and isolated pending checks for conformity with the product specification. Positive QA procedures should ensure that a defective final product is an exception. Defective products should remain isolated pending a decision on reworking, recovery or disposal, and the reasons for their failure to meet specification should be fully investigated and documented. Final products approved for release should be moved into the relevant warehouse area or, if a separate quarantine area is not used, the quarantine labels should be removed. A system of stock rotation must be used, and products must be protected from environmental contamination. Any products which have been returned or recalled must be clearly identified and segregated from new stock, ideally in a separate store. Warehouse areas should be checked regularly, for cleanliness and good housekeeping (including pest control) and to ensure that conditions such as temperature and humidity conform to those detailed in the storage and/or product specifications.

Process Control and Hygiene

A fundamental principle underlying any effective food manufacturing system is that quality and safety cannot be 'tested into' a product, but must be integral to all stages of manufacture, from design onwards.

The purpose of tests on finished products should be to verify that the products conform to specification.

Effective process control requires a hazard and risk assessment of that process, coupled with the continuous monitoring of the identified critical control points. In food hygiene, this approach is defined as the Hazard Assessment Critical Control Point (HACCP) system. The concept of the HACCP approach is increasingly being applied also to the quality, health and safety, and environmental assessment of processes operating in the context of GMP. Although there is no objection to this, care must be taken not to divert attention from the application of HACCP activities in the context of food safety. Data generated by continuous monitoring of the process conditions and/or inspection and analysis should ideally be analysed using appropriate statistical process control (SPC) procedures, in order to identify and monitor any trends.

Foreign Body Control

Good housekeeping during production is required in response to spillages or breakages, which might result in contamination of the food by foreign materials. The detection of a foreign body, e.g. glass, metal, wood splinter, insect, in a product is *a priori* evidence that the product is unfit for consumption. All production operations must incorporate a facility to detect the presence of foreign bodies and to reject any contaminated food container, filled package or bulk product. Automated equipment for metal detection or for X-ray analysis, linked to automatic rejection systems, provides the most efficient approach. However, such equipment must be tested regularly and the records of both tests and actual rejections must be kept. Contamination may result from a system failure, but may also be deliberate, on the part of disaffected employees or others with access to the products.

Pest Control

All food manufacturers should either use an organization specializing in pest control or provide employees with specialist training. A defined programme of inspection, with action to deter and/or destroy an infestation, is essential and inspections and actions must be documented. The most effective approach to pest control is good housekeeping. Items such as materials, packaging and equipment must always be stored on raised platforms, and never be less than 50 cm from a wall.

Only approved substances may be used for pest control, and extreme care must be taken to avoid the cross contamination of foods and materials likely to come into contact with them. Approved bait boxes should be sited so as not to interfere with manufacturing activities. Electrical insect control devices, fitted with catch trays, should be used unless there is a risk of dust explosion. These must be sited for maximum effect, but never above production or process equipment or packaging lines.

Birds and insects must be excluded from all production and storage areas, by screening. Birds must not be allowed to nest in or around food manufacturing premises, but the legal requirements relating to the protection of birds must be observed. All doorways and windows must have air blowers or strip curtains, which should never be tied back for operator convenience. Domestic and feral pets must be discouraged at all times. Employees should never feed animals, and must ensure that all entrances are kept closed.

Hygiene

Environmental contamination may be biological, chemical or physical and includes: undesirable microorganisms; foreign bodies, such as glass, wood or metal; taints derived from either the environment or cleaning agents; ingredients derived from inadequate cleaning between batches; and the use of wrong ingredients. The avoidance of environmental contamination at all stages of production is critical to GMP.

GMP also requires the avoidance of contamination from personnel and their belongings. The health records of all potential employees should be assessed by experienced occupational nursing or medical staff before their appointment, although a medical examination is not normally necessary. For the manufacture of certain high-risk food products, pre-employment stool (faecal) testing may be carried out, but such testing is not effective at detecting potential carriers of microbial pathogens. Any employee suffering from, or who has been in contact with, any form of enteric disease, including those who have visited an area where specific forms of gastroenteritis may be endemic, must not be allowed to handle food or work in a food processing area until cleared for such work by an occupational nurse or doctor.

All staff should be provided with clean, protective clothing and no-one should be permitted in production areas unless wearing approved protective clothing. Changing rooms and locker facilities must be available for external clothing, and separate lockers are required for clean protective clothing, which often includes overalls, boots, hats, hairnets, beard snoods, ear defenders and protective glasses. Overalls should not have external pockets, from which items such as pens or spectacles could fall into

the process plant or a food product mix. Protective clothing must not be worn outside the production area, including the canteen and office areas, its purpose being both to protect the employee and to protect the product from the employee. In a process plant with one or more high-risk areas, it is essential that distinctively coloured protective clothing is worn in those areas only. Employees should not normally move between high-risk and low-risk areas – if this is unavoidable, great attention to personal hygiene and the changing of clothing is essential, to minimize cross-contamination.

Before entering a process area, employees must be required to remove watches and jewellery, other than a simple marriage band and 'sleeper' earrings (i.e. not stud-type). Employees must scrub and disinfect their hands on each occasion, which necessitates the installation of (preferably automatic) washing facilities adjacent to any entrance to a process area. A major contamination risk is associated with personnel whose job requires them to move between parts of a process plant, e.g. engineering technicians, quality technicians, senior management. Laboratory-based staff should put on fresh overalls before entering a production area.

Any equipment which must be brought into the process area must not be allowed near production lines. Ideally, each production area should be provided with a facility for remote testing, so avoiding the transfer of equipment between the laboratory and production areas. Tools required by engineers should be kept in each process area and cleaned and sanitized appropriately, to avoid the contamination of one production line by another via tools (and engineers' hands).

Laboratory Control

The analytical and microbiological procedures used for food analysis must be fully documented, as should the health and safety risks associated with them, and no-one should be allowed to carry out food analysis without appropriate training. Good laboratory practice requires that technicians must demonstrate their competence in a specific application before their training can be deemed satisfactory. The operational procedures of the laboratory must be documented, and proper records kept of all samples received, analysed and reported on. There must be appropriate systems for the communication of laboratory data, particularly in the case of non-conformity being identified. The analysis of control samples is increasingly considered to be essential in the monitoring of laboratory performance, and interlaboratory tests are considered to be crucial in the demonstration of laboratory competence. The findings may inform audit by a third party, involved in accreditation on the basis of competence.

Documentation

Manufacturers must maintain effective records in order to demonstrate that operations are carried out in conformance with GMP. There are three main types of documentation, covering quality and operational systems, personnel and production programmes and production and quality records and reports. The documentation likely to be included in each category is listed below.

Quality and Operational Systems

- A quality policy statement, approved at board level
- Ingredient specifications
- Purchase contract data
- Plant operating procedures
- Manufacturing procedures and operative instructions
- Intermediate, bulk product and finished product specifications
- QA and QC procedures, including laboratory protocols and reporting systems
- Housekeeping and pest control procedures
- Cleaning procedures and instructions, including health and safety information
- Plant maintenance schedules.

Personnel and Production Programmes

- Training programmes for permanent and temporary employees
- Employment records
- Employee training records
- Production programmes
- Quality and environmental audit reports.

Production and Quality Records and Reports

- Sources of ingredients, raw materials and packaging, lot numbers, receipt dates, etc.
- Process data, including readings from gauges and other instruments, temperature record charts, etc.
- Data relating to weight/volume of packaged product
- Records of cleaning, sanitization and monitoring
- Laboratory records associated with ingredients, intermediate and finished products, process water, in-line cleaning, etc.
- Production and laboratory audit records
- Audit records of suppliers and contract packers
- Records of quarantined products and their eventual fate

- Records of all reused or reworked materials
- Records of product withdrawal or recall, with reports on the cause(s) and remedial actions.

Other Documentation

These lists are not exhaustive. For example, special documentation may be required in relation to specific industrial needs, e.g. customs records or specific customer requirements, especially if 'own brand' products are manufactured.

The quality system should specify the time period for which documentation is to be stored, which will vary according to the particular records, and also the form of storage, e.g. paper, microfiche, computer-based.

Transportation

Raw materials and products may be transported in a variety of vehicles, and it is essential that they are not also used to convey other, incompatible, goods. Vehicles for the transport of chilled or frozen foods must be designed for that purpose, and therefore be capable of maintaining the appropriate temperature. Such vehicles should have double doors or alternative arrangements which ensure that the internal temperature is maintained throughout loading and unloading.

Tankers for the transport of liquid products need to be capable of being cleaned effectively, e.g. by spray systems linked to a recirculation CIP system. Tankers may need to be purged with an inert gas, to minimize the uptake of air during filling, and/or may require heating, e.g. to ensure that sugar syrups remain liquid. Such tankers often require special cleaning before use. Particular attention is needed to ensure the effective cleaning of valves and pipe attachments.

Any vehicle or container for the transport of foods must be inspected prior to use, for cleanliness and to ensure that it is free of contaminants and foreign bodies. If pallets are used, it is essential to ensure that they are in good condition, e.g. without broken slats or protruding nails, and that the stacked products do not overhang, risking damage to the secondary packaging. The vehicle should be loaded evenly, ensuring that neither the permitted gross weight nor the permitted individual axle weights are exceeded. The load should be secured, such that the product is not damaged either during normal driving or in the event of an emergency, e.g. sudden braking. Similarly, containers must be loaded so as to minimize the risks of product movement during transportation. Special arrangements with shipping companies may be needed to ensure that containers are transported in conditions which do not expose the product to extremes of temperature or humidity.

Incoming vehicles must be inspected for evidence of damage to goods, and to ensure that the vehicle is clean and free from pests and other contaminants. Evidence of any defects should be reported to the haulier or supplier. Procedures should be in place to deal with the consequences of damage due to accidents and other incidents during transit.

New Product Development

The development of new products (NPD) poses specific issues in relation to GMP. The technical and marketing staff should operate within defined procedures to ensure that from conception, all considerations relevant to the quality, safety, packaging and labelling of the new product are taken into account. Safety is especially critical when new products are to be assessed organoleptically, either within the company or by external consumer panels. The ingredients required in small quantities for development work must also be available in the quantities and quality required for factory production. This necessitates early attention to the definition of the ingredients, and to the specifications of the processes and the final product. Process safety must be assessed at an early stage, to enable the development of HACCP and other control techniques.

NPD may entail the need to obtain 'change parts' for processing and packaging plants. The identification of this need, and the acquisition of such items must allow time for any necessary trials on production plant. Such trials may require the preparation of specific operator instructions, and must not compromise ongoing production activities. The products from the initial production runs must be isolated pending the detailed evaluation of the product and its packaging needed to ensure compliance with specifications.

Product Information and Consumer Awareness

Information on product labels must comply with all the legal requirements, for example relating to product name and description, packaged weight, ingredients and nutritional information. Responsibility for developing the artwork for packaging is usually vested in the marketing function, but systems must be established to ensure technical and legal approval and to ensure that the proposed labels and packaging materials are compatible with the production plant. Proposed advertising copy must also be evaluated for legal compliance.

Consumer and Official Complaints

The interface between consumers and food manufacturers is critical: effective systems for handling consumer queries and complaints are essential. A documented complaint-handling system should be operated by personnel who liaise closely with technical and legal personnel. A standard letter acknowledging a complaint may be appropriate, but it should give a probable date for a more detailed response. The investigation of complaints is a critical aspect of GMP: responsibility should be vested in the quality function, whose staff must liaise with other relevant personnel. All complaints should be investigated speedily and effectively. Some may merely reflect a consumer's dislike of a product, but very often a complaint is the first indication of a potential problem. If a complaint is justified, the cause must be identified and remedial action taken. Summaries and reports of complaints should be reviewed regularly, for evidence of any trends: an effective complaints database is therefore essential. In recent years, some retailers have operated their own complaint-handling systems and databases, thereby eliminating direct contact between the manufacturer and consumer. However, this can delay remedial action unless the retailer immediately transmits key information regarding the complaint, and ideally the product sample, to the manufacturer.

Official complaints from enforcement authorities must be handled separately from consumer complaints. Ideally, food manufacturers should liaise regularly with enforcement officers so that potential problems can be handled rapidly and effectively through established channels, by professionals operating within an environment of mutual understanding and respect.

Product Withdrawal and Recall

Food manufacturers must have defined systems for the recall of food products from the marketplace. The recall procedures need to be tested regularly, and updated as necessary. Generally, a high-level Crisis Committee is established, which operates within defined criteria and has delegated authority to take any action deemed necessary. Its membership should include representatives of the legal, financial, commercial, production, distribution and technical functions. All companies should establish a 'crisis contact' system, to facilitate communications with customers and with local and central government agencies in the event of an emergency. The police may also need to be involved if deliberate product contamination is suspected.

The nature of the recall should reflect the nature of

the problem (e.g. life-threatening, potentially capable of causing serious risks to public health, capable of causing serious risks to the commercial health of the company). In the case of the less serious issues, the withdrawal of the product from retail and wholesale outlets should be adequate, but actual or potentially life-threatening situations usually justify public recall of the product. The Crisis Committee must evaluate the nature and significance of the issue, and also consider whether sabotage might have been involved.

The first indication of a problem may arise from the evaluation of consumer complaints, or directly from an enforcement or government agency, a retailer or even an individual consumer. Essential information includes the outlet from which the product was obtained and the lot number, which enables information relating to the manufacture and packaging of a particular batch to be traced. The lot number is also important for identifying the lot of product that may need to be recalled. When considering recall it is often desirable to review the data relating to batches of the same product produced within a defined timescale, and/or to batches of other products manufactured around the same date. Notification of a withdrawal or recall must include information on the brand name, pack size, identifying marks (e.g. lot number, sell-by date), the nature of the defect and the action required, and any degree of urgency. In the case of a public recall, telephone numbers must be given so that consumers can obtain further information and advice. The personnel receiving such calls must be fully briefed, using standard written briefing procedures. Arrangements with specialist crisis management advisers, who are accustomed to handling such matters, are essential if a recall is to be achieved promptly and efficiently.

After the crisis, all the documentation must be carefully reviewed and, if appropriate, the quality and production procedures should be modified.

Legislation and GMP

Legislation

In all countries, food manufacturers must operate within the constraints of general or specific legislation with regard to matters such as permitted additives, the avoidance of natural and artificial contaminants and the provision of consumer information on product labels. For many years a decision to operate under conditions of GMP was a matter for individual companies, in accordance with their internal quality regimes. However, increasingly, national and supranational legislation is requiring food manufacturers to demonstrate the adoption of GMP in their day-to-day activities. For instance, European Directives and Regulations, implemented in the UK through legislation such as The Food Safety Act (1990) and its subsidiary regulations, increasingly require the adoption of, and compliance with, GMP principles. The detailed requirements regarding food processing premises and equipment are given, *inter alia*, in The Food Safety (General Food Hygiene) Regulations (1995), The Fresh Meat (Hygiene and Inspection) Regulations (1994), The Meat Products (Hygiene) Regulations (1995), and The Dairy Products (Hygiene) Regulations (1995). The key European legislation on food hygiene is the Directive on Food Hygiene (1993), which includes a requirement to employ the principles of HACCP in food manufacturing and processing.

In the USA, Australia, New Zealand and many other countries the principle requirements of GMP are also embodied in legislation. Some of this is definitive and prescriptive, but most requires at least the demonstration of compliance with the principles of GMP. Internationally, The Codex Alimentarius (1997) has published its Recommended International Code of Practice, which defines general principles of food hygiene and includes guidelines for the application of HACCP systems.

Subsidiary to formal legislation are Codes of Practice dealing with specific GMP issues and hygiene requirements for a range of specific food manufacturing activities. In the UK, such Codes of Practice may be produced in a specific format, defined by the Department of Health. A schedule of Codes of Practice produced by trade associations in the UK has been incorporated in an Institute of Food Science and Technology publication on GMP (1998).

Demonstration of Compliance

Documented evidence of operating a system of GMP, of the implementation of HACCP and of compliance with legal requirements may be considered to demonstrate due diligence in fulfilling the role of a responsible manufacturer. However, it is advisable that any QA or GMP system be vetted by an independent body, which is able to look objectively at the system and the extent to which a company and its personnel comply with the requirements of the system.

There are various approaches to independent appraisal, the most widely used and recognized probably being accreditation to the ISO (International Standards Organization) 9000 series of standards. The disadvantage of this accreditation is that it assesses only the extent to which a company complies with its defined system: in general there is no attempt to define what the system should comprise, although the inclusion of certain basic features is required. It is therefore possible that an ISO-accredited system does not cover all the aspects of GMP. Nevertheless, because it is externally assessed and audited it imparts a degree of credibility to the holder of the accreditation. An alternative approach may be the use of independent third-party audits. Such an approach is currently operated on behalf of the International Bottled Water Association, by the National Sanitation Foundation, and provides an excellent example of the way in which a trade association can regulate the activities of its members in such a way as to protect the reputation of the industrial sector.

Future Developments

The development of GMP systems is never complete: improvements are always possible, although the law of diminishing returns may come into play. Reputable manufacturers operating GMP should also consider the potential benefits of using bench marks. Best Manufacturing Practices (BMP) schemes to facilitate the use of bench marks are operated in the UK by the Department of Trade and Industry and the Confederation of British Industry. In the USA, a BMP Centre of Excellence has been established, with a similar purpose.

See also: **Hazard Appraisal (HACCP)**: Critical Control Points. **Laboratory Design**. **Laboratory Management**:

Accreditation Schemes. **National Legislation, Guidelines & Standards Governing Microbiology**: Canada; European Union; Japan. **Process Hygiene**: Designing for Hygienic Operation; Overall Approach to Hygienic Processing; Involvement of Regulatory Bodies.

Further Reading

Codex Alimentarius Commission (1997) *General Principles of Food Hygiene, Supplement to Volume 1B, Revision 3.* Rome: Food and Agriculture Organization of the United Nations and World Health Organization.

European Council Directive on the Hygiene of Foodstuffs No. 93/43/EEC.

European Council Directive laying down the health rules for the production and placing on the market of raw milk, heat-treated milk and milk-based products. No. 92/46/EEC (subsequently modified on 6 occasions up to 96/23/EEC).

The Food Safety Act (1990). HMSO, London.

IFST (1998) *Food and Drink Good Manufacturing Practice: A Guide to its Responsible Management*, 4th edn. London: Institute of Food Science and Technology.

Tanner AM (1998) Third-party auditing of water bottling facilities. In: Senior DAG and Ashurst PR (eds) *Technology of Bottled Water.* P. 207. Sheffield, UK: Sheffield Academic Press (CRC Press).

Guidelines Governing Microbiology *see* National Legislation, Guidelines & Standards Governing Microbiology: Canada, European Union; Japan.

HAFNIA ALVEI

Jouko Ridell, Department of Food and Environmental Hygiene, University of Helsinki, Finland

Characteristics of the Species

Hafnia alvei is the only species in the *Hafnia* genus. The species has been known by the names *Enterobacter hafnieae*, *Bacterium cadaveris*, *Bacillus asiaticus* and *B. paratyphi alvei*. *Hafnia* is the old name for Copenhagen. The epithet *alvei* is chosen because the bacterium has been isolated from the intestinal tract of honey bees. In spite of the great developments in bacterial taxonomy during recent years, many Enterobacteriaceae species are still genetically diverse. *H. alvei* consists of two distinct genetic groups which are about 50% related at the DNA level. The groups are not reliably distinguishable by the methods used currently in the routine laboratory (**Table 1**). The first genetic group contains a biochemically less active subgroup, *H. alvei* biogroup 1, which is also known as *Obesumbacterium proteus* (genetic group 1). This phenotypically distinct group of *Hafnia* species is found as a contaminant of breweries. *Yokenella regensburgei* is a phenotypically similar species which has been separated from *H. alvei* in 1985 (**Table 2**). *H. alvei* occurs in many natural environments and it is frequently found in food samples. It is known to be an opportunistic human pathogen causing a variety of extraintestinal diseases. True *H. alvei* is almost certainly not a diarrhoeal pathogen. Instead, DNA-based methods showed that the diarrhoeagenic *H. alvei* strains were not true *H. alvei* but were related to the *Escherichia coli–Shigella* group. The inability of the present identification schemes to differentiate true *H. alvei* and these diarrhoeagenic strains explains largely why *H. alvei* has been thought to be an emerging diarrhoeal pathogen. At present the diarrhoeagenic '*H. alvei*' lacks a valid name.

H. alvei is a Gram-negative facultatively anaerobic rod, and it conforms to the general definition of Enterobacteriaceae. It is only about 20% related to the other species of the family. Most strains are motile. *H. alvei* is able to grow at a wide temperature range. Both genetic groups have similar psychrotrophic growth temperature characteristics – a minimum growth temperature of 0.2–3.7°C (mean 2.6°C) and a maximum growth temperature of 41–43°C. In contrast, the diarrhoeagenic '*H. alvei*' strains have much higher mesophilic growth temperatures – minimum growth temperatures of 10.2–11.5°C and maximum growth temperatures of about 46°C.

H. alvei occurs in many natural environments such as surface waters, soil, sewage and vegetation. *H. alvei* has been reported to be one of the major species which contribute to the microbial ecology of silage. It has been isolated from the polluted waters of fish farms as well as from the unpolluted northern rivers of Finland. Based on its ubiquitous distribution in nature it is a frequent inhabitant in the intestinal contents and faeces of fish, insects, birds and mammals. Consequently, *H. alvei* is a common contaminant of meat, dairy products and fish, and all environments where they are handled.

H. alvei is one of the most numerous Enterobacteriaceae species in minced meat at retail level. About two-thirds of minced meat samples collected from retail stores contained *H. alvei*. The initial numbers of *H. alvei* in meat are usually low, but because the refrigeration temperatures used for meat usually exceed the minimum growth temperature of this species, the numbers increase rapidly during storage. The moist anaerobic environment in vacuum packages is favourable for the growth of *H. alvei*. In vacuum-packed beef the numbers of *H. alvei* reach their peak (usually 10^3–10^5 cfu g^{-1}) in about 3 weeks – depending on temperature – and thereafter usually decrease slowly. In the meat of stressed cattle – called DFD-meat (dark, firm, dry) – which has a high pH, *H. alvei* may reach higher levels and even contribute to the spoilage of the meat. Factors affecting the development of the *H. alvei* population include the type of package, composition of the atmosphere, pH and composition of the competing microflora. The inhibition of the growth of *H. alvei* – and other Enterobacteriaceae species as well – is caused by the

Table 1 Phenotypic characteristics of *Hafnia alvei*

Test	%+	Sign
Indole	0	–
Methyl red (35°C)	40	v
Voges–Proskauer (35°C)	59	v
Citrate (35°C)	6	–
Hydrogen sulphide (Triple Sugar Iron)	0	–
Urea hydrolysis	2	–
Phenylalanine deaminase	0	–
Lysine decarboxylase	100	+
Arginine dihydrolase	7	–
Ornithine decarboxylase	100	+
Motility	94	+
Gelatin hydrolysis	0	–
Growth in KCN	97	+
Malonate utilization	83	[+]
D-Glucose acid	100	+
D-Glucose gas	98	+
Acid from:		
Lactose	3[a]	–
Sucrose	12	[–]
D-Mannitol	100	+
Dulcitol	2	–
Salicin	16	[–]
Adonitol	0	–
myo-Inositol	0	–
D-Sorbitol	0	–
L-Arabinose	95	+
Raffinose	4	–
L-Rhamnose	92	+
Maltose	100	+
D-Xylose	99	+
Trehalose	99	+
Cellobiose	67	v
α-Methyl-*d*-glucoside	0	–
Melibiose	0	–
D-Arabitol	0	–
Mucate	0	–
Esculin hydrolysis	5	–
Tartrate (Jordan)	58	v
Acetate utilization	90	+
Lipase (corn oil)	0	–
DNAse	0	–
Oxidase	0	–
ONPG test (O-Nitrophenyl-galactoside)	67	v
Nitrate to nitrite	100	+
2-Ketogluconate util.	100	+
3-Hydroxybenzoate util.	0	–
Minimum growth temperature	2.6°C	
Maximum growth temperature	42°C	

+ = 90–100% positive; [+] = 75–89% positive; v = 26–74% positive; [–] = 11–25% positive; – = 0–10% positive.
[a] 10–15% of the strains ferment lactose slowly.

competing microflora which is usually composed of lactobacilli. The effect is mediated by low pH, bacteriosins and other antagonistic effects of the lactobacilli population which gradually overwhelm the meat surface inside the vacuum package.

Methods of Detection

No specific selective agars are available commercially for the isolation of *H. alvei*. *H. alvei* grows readily on the nonselective and selective isolation media used for enteric bacteria, such as violet red bile (VRB), eosin methylene blue (EMB), MacConkey and xylose-lysine-deoxycholate (XLD) agars. Most of the strains grow even on the highly selective isolation media such as *Salmonella–Shigella* agar. Most of the *H. alvei* strains are lactose-negative and on the above-mentioned isolation media they produce colourless colonies. On less inhibitory agars, such as VRB, the colonies are usually relatively large, translucent, circular, and low convex, with a smooth surface.

For isolation purposes, MacConkey or VRB agar containing sucrose and sorbitol may be used. The colourless colonies may be further tested with biochemical tests. The L-prolineaminopeptidase test is useful to screen the colonies; *H. alvei* and *Serratia* spp. are the only Enterobacteriaceae species to give a positive reaction.

Reliable differentiation of *H. alvei* and most other enteric bacteria can usually be made by routine biochemical tests performed either separately or included in the commercial identification systems. It is noteworthy that the incubation temperature affects the results of some biochemical tests (methyl red, Voges–Proskauer and citrate). *H. alvei* can be differentiated from species of *Enterobacter* and *Serratia* based on positive reaction in lysine, arginine and ornithine decarboxylase tests and a negative reaction in sorbitol and inositol fermentation tests. However, the limited set of tests – 10–20 – which are used in routine clinical laboratories or included in the commercial biochemical panels may not always be sufficient. In particular, the differentiation of *H. alvei* and diarrhoeagenic '*H. alvei*' possessing the attachment effacement gene (*eaeA*+ '*H. alvei*') as well as *H. alvei* biogroup 1 and true *O. proteus* (genetic group 2) is difficult. Other species difficult to distinguish from *H. alvei* include *Y. regensburgei* and inactive (non-lactose-fermenting) *Escherichia coli* (*E. coli* 2). From the diagnostic standpoint, reliable identification of the diarrhoeagenic *eaeA*+ '*H. alvei*' strains is essential. This is not achieved by API20E alone or by the traditional set of biochemical tests. Additional tests such as growth at 4°C, 2-ketogluconate and 3-hydroxybenzoate assimilation tests are needed. DNA-based methods such as polymerase chain reaction detection of the *eaeA* gene offer a reliable alternative for detection.

Table 2 Differential characteristics of *Hafnia alvei* and biochemically similar species

Test	H. alvei	H. alvei *biogroup 1*	eaeA+ 'H. alvei'	Escherichia coli, *inactive*	Yokenella regensburgei
Voges–Proskauer (35°C)	[+]	v	–	–	–
L-Arabinose fermentation	+	–	+	[+]	+
L-Rhamnose fermentation	+	–	–	v	–
Maltose fermentation	+	–	–	[+]	+
ONPG test (O-Nitrophenyl-galactosidase)	+	v	+	v	+
3-Hydroxybenzoate util.	–	–	+	–	NA
2-Ketogluconate util.	+	–	–	NA	–
Growth at 4°C	+	+	–	–	–

+ = 90–100% positive; [+] = 75–89% positive; v = 26–74% positive; [–] = 11–25% positive; – = 0–10% positive; NA = data not available.

Importance to the Food Industry

H. alvei produces biogenic amines, cadaverine and putrescine, and may sometimes contribute to the spoilage of meat. It is probable that only refrigerated meat with an exceptionally high pH (DFD-meat) offers a suitable environment for *H. alvei* to reach population densities which produce sensory changes. *H. alvei* is one of the bacterial species which is responsible for the production of histamine in fish. Tuna fish, in particular, has been associated with histamine food poisoning.

In clinical laboratories *H. alvei* may disturb the detection of *Salmonella*. Many strains grow in the isolation broths used to enrich *Salmonella* and produce colonies resembling those of *Salmonella* on routine isolation media. Moreover, *H. alvei* cultures can sometimes be agglutinated by *Salmonella* O-antisera. Because of their psychrotrophic nature, *H. alvei* organisms may disturb the cold enrichment of *Yersinia enterocolitica* in phosphate-buffered saline. In food laboratories, the occurrence of *H. alvei* in food and water samples may lead to difficulties in interpreting the results of the coliform test – an indicator bacteria analysis which is based on lactose fermentation. Although most *H. alvei* strains are lactose-negative, some strains have a plasmid-mediated ability to ferment lactose with variable speed. This results in colonies of variable size and makes the agar plates difficult to read.

H. alvei can cause fatal extraintestinal infections in farmed fish affected by predisposing factors such as overcrowding. In rainbow trout *H. alvei* has been reported to cause haemorrhagic septicaemic disease, which is associated with high mortality. In sherry salmon *H. alvei* has caused kidney infections. *H. alvei* can infect honey bees affected by a parasitic disease. Thus, it appears that *H. alvei* can cause opportunistic infections in animals in a similar manner as in humans.

Importance to the Consumer

H. alvei is an opportunistic human pathogen which can cause a variety of extraintestinal diseases in susceptible individuals with a severe underlying condition. *H. alvei* has been associated – either in pure or mixed growth – with wound infections, abscesses, septicaemia, pneumonia, etc. Predisposing factors include malignancy, immunodeficiency or any debilitating disease. For healthy persons *H. alvei* apparently does not cause any risk concerning extraintestinal diseases.

H. alvei has been considered to be an emerging human diarrhoeal pathogen. The main reason for the belief is based on studies concerning the diarrhoeagenic *eaeA+* 'H. alvei' strains isolated in Bangladesh. These strains caused diarrhoea with similar mechanisms and comparable symptoms to enteropathogenic *E. coli* – a causative agent of tourist diarrhoea and diarrhoea of children in developing countries. However, taxonomic methods determining the structure of rRNA genes and total DNA homology revealed that these *eaeA+* 'H. alvei' strains were not true *H. alvei*, but closely resembled the *E. coli–Shigella* group, although the taxonomic position of this group remained unclear. At the moment the geographic distribution and clinical importance of the diarrhoeagenic 'H. alvei' is not known.

From time to time *H. alvei* has been reported to be a causative agent of diarrhoea based on isolation from diarrhoeal faeces without further studies concerning causal relationships. However, it is a fact that true *H. alvei* is more often isolated from diarrhoeal than normal human faeces. This epidemiological association of *H. alvei* with diarrhoeal symptoms is carefully documented but has not been explained. No virulence mechanisms have been found in spite of careful studies, and it is possible that the phenomenon has nothing to do with virulence. For instance, the increased intestinal motility during diarrhoea may facilitate the detection of food-originating bacteria from faeces. Thus, it is likely that true *H. alvei* is not a diarrhoeal pathogen, and food-originating *H. alvei* does not cause any risk to the consumer.

See also: **Enterobacteriaceae, Coliforms and *E. coli***: Introduction. **Meat and Poultry**: Spoilage of Meat. ***Salmonella***: Introduction. **Yersinia**: *Yersinia enterocolitica*.

Further Reading

Albert MJ, Faruque S, Ansaruzzaman M et al (1992) Sharing of virulence-associated properties at the phenotypic and genetic levels between enteropathogenic *Escherichia coli* and *Hafnia alvei*. *J. Med. Microbiol.* 37: 310–314.

Brenner DJ (1992) Introduction to the family Enterobacteriaceae. In: Balows A, Tryper HG, Dworkin M, Harder W and Schleifer K (eds) *The Procaryotes*, 2nd edn. P. 2674. New York: Springer-Verlag.

Dainty RH, Edwards RA, Hibbard CM and Ramantanis SV (1986) Bacterial sources of putrescine and cadaverine in chill stored vacuum-packaged beef. *J. Appl. Bacteriol.* 61: 117–123.

Ewing WH (1986) The genus *Hafniae*. In: *Edwards and Ewing's Identification of Enterobacteriaceae*, 4th edn. P. 417. New York: Elsevier.

Fanghänel S, Reissbrodt R and Giesecke H (1991) L-Prolineaminopeptidase activity as a tool for identification and differentiation of *Serratia marcescens*, *Serratia liquefaciens* and *Hafnia alvei* strains. *Zbl. Bakt.* 275: 11–15.

Gelev I, Gelev E, Steigerwalt AG, Carter GP and Brenner DJ (1990) Identification of the bacterium associated with haemorrhagic septicaemia in rainbow trout as *Hafnia alvei. Res. Microbiol. (Inst. Pasteur)* 141: 573–576.

Gunthard H and Pennekamp A (1996) Clinical significance of extraintestinal *Hafnia alvei* isolates from 61 patients and review of the literature. *Clin. Infect. Dis.* 22: 1040–1045.

Hanna MO, Smith GC, Hall LC and Vanderzant C (1979) Role of *Hafnia alvei* and *Lactobacillus* species in the spoilage of vacuum-packaged strip loin steaks. *J. Food Prot.* 42: 569–571.

Prest AG, Hammond JRM and Stewart GSAB (1994) Biochemical and molecular characterization of *Obesumbacterium proteus*, a common contaminant of brewing yeast. *Appl. Environ. Microbiol.* 50: 1635–1640.

Riddell J and Korkeala H (1997) Minimum growth temperatures of *Hafnia alvei* and other Enterobacteriaceae isolated from refrigerated meat determined with a temperature gradient incubator. *Int. J. Food Microbiol.* 35: 287–292.

Ridell J, Siitonen A, Paulin L, Mattila L, Korkeala H and Albert MJ (1994) *Hafnia alvei* in stool specimens of patients with diarrhoea and healthy controls. *J. Clin. Microbiol.* 32: 2335–2337.

Ridell J, Siitonen A, Paulin, L, Lindroos O, Korkeala H and Albert MJ (1995) Characterization of *Hafnia alvei* with biochemical tests, RAPD-PCR and partial sequencing of the 16S rRNA gene. *J. Clin. Microbiol.* 33: 2372–2376.

Sakazaki R and Tamura K (1992) The genus *Hafnia*. In: Balows A, Tryper HG, Dworkin M, Harder W and Schleifer K (eds) *The Procaryotes*, 2nd edn. P. 2817. New York: Springer-Verlag.

HANSENULA

Gerd Gellissen, Rhein Biotech GmbH, Düsseldorf, Germany

Cornelis P Hollenberg, Institut für Mikrobiologie, Heinrich-Heine-Universität Düsseldorf, Germany

Hansenula polymorpha

Hansenula polymorpha belongs to a limited group of yeast species capable of growth on methanol. Isolated from soil, rotten fruits and the gut of insects, it was initially considered to be a good candidate to produce single-cell protein (SCP) in methanol-containing carbon sources. This limited scope was extended and altered after the methodical canon of genetic engineering had been established. The most favourable and most advantageous characteristics of this species have resulted in an increasing number of biotechnological applications. As a consequence, *H. polymorpha* is rapidly becoming a system of choice for heterologous gene expression in yeast. Several production processes for recombinant pharmaceuticals and industrial enzymes have been developed based on gene expression in this species. The system leans on suitable host strains and components derived from genes of the methanol metabolism pathway.

Characteristics of the Species

Morphological, Taxonomic and Physiological Characteristics

Hansenula polymorpha is an ubiquitous microorganism occurring naturally in spoiled orange juice, maize meal, in the gut of various insect species and in soil. It grows as white to cream, butyrous colonies and does not form filaments. Reproduction occurs vegetatively by budding. When cultured on a

Table 1 Characteristics of *Hansenula polymorpha*

Synonyms
> *Hansenula angusta*
> *Torulopsis methanotherma*
> *Pichia angusta*

Related methylotrophic species
> *Pichia pastoris*
> *Pichia methanolica*
> *Candida boidinii*

Taxonomy
Fungi (Kingdom)
  Eumyceta (Division)
    Ascomycotina (Subdivision)
      Hemiascomycetes (Class)
        Endomycetales (Order)
          Saccharomycetaceae (Family)
            Saccharomycetoideae (Subfamily)
              Hansenula polymorpha (Species)

Habitat
> Spoiled orange juice, maize meal, insects, soil

Morphological description
> White to cream, butyrous colonies; vegetative reproduction by budding; no filaments; evanescent asci, containing 1–4 hat-shaped ascospores on 5% Difco malt agar

Growth
> Facultative methylotrophic organism: growth among others on methanol, glucose, glycerol, sucrose and maltose, no growth on lactose, starch, inulin, galactose.
> Growth at 20–45°C, pH 2.5–6.0.

Other characteristics
> Mol % of G+C of nuclear DNA: 48%

Several auxotrophic strains are available as host for heterologous gene expression.
H. polymorpha is classified as S1 organism (no risk for humans and for the environment) for both research and production, according to German gene technology law.

sporulation medium (5% Difco malt agar) evanescent asci form, containing 1–4 hat-shaped ascospores.

H. polymorpha belongs to the fungal family of Saccharomycetaceae, subfamily Saccharomycetoideae. It can be grown on a broad range of carbon sources, including glucose, sucrose, trehalose and maltose, glycerol, erythritol, xylitol and mannitol and methanol and ethanol (**Table 1**).

Methanol Metabolism Pathway

H. polymorpha is a member of a limited group of facultative methylotrophic yeast species that belong to the four genera *Candida*, *Hansenula*, *Pichia* and *Torulopsis* of the Saccharomycetoideae subfamily. They share a specific methanol utilization pathway. The initial reactions take place in specialized microbodies, peroxisomes, followed by subsequent metabolic steps in the cytoplasm. Methanol enters the peroxisomes where it is oxidized by a specific methanol oxidase (MOX) to generate formaldehyde and

hydrogen peroxide. The poor catalytic properties of the enzyme under in vivo conditions are compensated by a massive production of the protein under conditions of enzyme requirement (see below in the description of the *H. polymorpha*-based expression system). The peroxide generated in the oxidase reaction is decomposed to water and molecular oxygen by a peroxisomal catalase. The catalase activity is high during growth on methanol thus being co-regulated to some extent with the MOX activity.

Formaldehyde, generated by the oxidase reaction, enters the cytosolic dissimilatory pathway to yield energy and the assimilatory pathway for the generation of biomass. The initial key reaction of the latter is catalysed by the peroxisomal dihydroxyacetone synthase (DHAS), yielding dihydroxyacetone and glyceraldehyde 3-phosphate in a transketolase reaction between formaldehyde and xylulose 5-phosphate. These C_3 compounds are further assimilated within the cytosol. Dihydroxyacetone is phosphorylated by a dihydroxyacetone kinase. Catalysed by an aldolase it subsequently reacts with glyceraldehyde 3-phosphate to form fructose 1,6-bisphosphate which is then converted into fructose 6-phosphate by a phosphatase. Xylulose 5-phosphate is regenerated by the pentose phosphate cycle via transaldolase, transketolase, pentose phosphate isomerase and epimerase reactions. One-third of the glyceraldehyde 3-phosphate generated is exploited for biomass formation by standard reactions of gluconeogenesis.

Within the dissimilatory pathway the formaldehyde is catabolized in two subsequent dehydrogenase reactions; formaldehyde is oxidized to formate by a formaldehyde dehydrogenase; then formate is oxidized to carbon dioxide by the action of formate dehydrogenase (FMDH).

Growth on different carbon sources results in distinct intracellular protein patterns. In methanol-grown cells key enzymes of the methanol metabolism pathway are present in large amounts, including MOX, FMDH and DHAS. In methanol-grown continuous cell cultures MOX can constitute up to 30% of total cell proteins, and FMDH and DHAS up to 20%. Huge peroxisomes, the cellular compartments for the initial steps of methanol metabolism, can take up to 80% of the cell volume under these conditions. Significant enzyme levels can also be detected in glycerol-grown cells or in glycerol-grown cells subsequently cultured under glucose limitation. They are absent in cells grown in glucose. Thus synthesis of these key enzymes is subject to a repression/derepression/induction mechanism determined by the carbon compound of the culture medium. Peroxisomal proliferation apparently

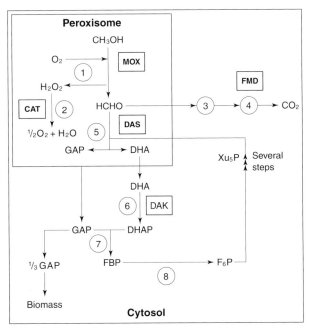

Figure 1 The methanol utilization pathway and its compartmentalization in *H. polymorpha*. (1) Methanol is oxidized by methanol oxidase (MOX) to generate formaldehyde and hydrogen peroxide. (2) The toxic hydrogen peroxide is decomposed to water and oxygen by catalase. (3), (4) Within a dissimilatory pathway the formaldehyde is oxidized by two subsequent dehydrogenase reactions to CO_2. One of the two enzymes involved is formate dehydrogenase (FMD). (5) For assimilation the formaldehyde reacts with xylulose 5-phosphate (Xu_5P) by the action of dihydroxyacetone synthase (DAS) to generate the C_3 compounds glyceraldehyde 3-phosphate (GAP) and dihydroxyacetone (DHA). (6) DHA is phosphorylated (DAK) to dihydroxyacetone phosphate (DHAP). (7) GAP and DHAP yield in an aldolase reaction fructose 1,6-bisphosphate (FBP). (8) In further steps fructose 6-phosphate (F_6P) and finally xylulose 5-phosphate is generated. The genes of the methanol utilization pathway isolated and characterized so far are indicated in the figure at the position of the enzymes they are coding for.

Cloning and Characterization of *H. polymorpha* Genes

In contrast to the situation in *Saccharomyces cerevisiae* in which the entire genome has been sequenced, only about 35 *H. polymorpha* genes have been isolated and characterized to date (**Table 2**). Most of the identified genes belong to a group linked to methanol metabolism and to peroxisomal function. In addition, genes encoding components for phosphate and nitrate utilization and genes encoding heat shock proteins have been cloned. Further isolated sequences include parts of the 26 and 18S rRNAs, autonomously replicating sequences (*ARS*) and telomeric structures. The overall G+C content of the *H. polymorpha* genome was found to be 48%. The G+C content of the genes identified so far ranges between 43% (*URA3*) and 55% observed for *MOX* and other members of the highly expressed methanol utilization

Table 2 Characterized *H. polymorpha* genes (status 1 January 1999)

Gene	Encoded protein
AAO	Amino acid oxidase
ACT	Actin
ALD	Aldehyde dehydrogenase
AMO	Peroxisomal amine oxidase
CAT	Catalase
CPY	Carboxypeptidase Y
DAS	Dihydroxyacetone synthase
DAK	Dihydroxyacetone kinase
EXG1	1,3-β-Glucanase
FAD	Delta 9-fatty acid desaturase
FMD	Formate dehydrogenase
GAP	Glyceraldehyde 3-phosphate dehydrogenase
HARS1	*H. polymorpha*-derived autonomously replicating sequence
HARS48	*H. polymorpha*-derived autonomously replicating sequence
HSA1	Heat shock (HSP70) protein
HSA2	Heat shock protein
LEU2	β-Isopropylmalate dehydrogenase
MOX	Methanol oxidase
PAH2	Peroxisome assembly protein
PEP4	Proteinase A
PER1	Peroxisomal matrix protein
PER3	Peroxisomal matrix import protein
PER8	Peroxisomal integral membrane protein
PER9	Peroxisomal membrane protein
PER10	Peroxisomal membrane protein
PEX4	Peroxin 4p (ubiquitin-conjugating enzyme)
PHO1	Repressible acid phosphatase
PMR1	Ca++-dependent ATPase
URA3	Oritidine 5'-phosphate decarboxylase
YNA1	Transcription factor for genes of the nitrate pathway
YNI1	Nitrite reductase
YNR1	Nitrate reductase
YNT1	Nitrate transport protein
26S rRNA gene (partial)	
18S rRNA gene (partial)	

follows a similar control. The giant organelles present under inductive conditions cannot be detected in cells grown in a repressive carbon source such as glucose and ethanol. Instead, a few small microbodies functioning as glycosomes are observed, from which peroxisomes proliferate after transfer into media with inductive conditions. The enzyme production is controlled on the transcriptional level. The strong and highly regulated promoter structures of the respective genes have been designed as attractive control elements for heterologous gene expression, particularly the *MOX* (methanol oxidase) and *FMD* (formate dehydrogenase)-promoter elements (see below). The enzymatic steps of the methanol utilization pathway and the genes of this pathway isolated so far are summarized in **Figure 1**.

pathway genes. The high G+C content of the methanol utilization pathway genes results in a strong bias
towards codons with G+C in the wobble position.
Thus the nucleotide composition of the *H. polymorpha* genes obeys the rules recognized in other
organisms in that highly expressed genes reveal a bias
to a subset of preferred codons.

Expression Systems Based on *H. polymorpha*

The quality of an expression system depends to a
large extent on the characteristics of the host and the
availability of favourable genetic components that
can be applied for heterologous gene expression. *H.
polymorpha* exhibits several properties beyond those
briefly summarized in the first section that attest to
the suitability of this species as host for recombinant
protein production.

The organism meets safety prerequisites in not harbouring pyrogens, pathogens or viral inclusions. The
eukaryotic yeasts including *H. polymorpha* are
capable of secreting proteins and performing secre-
tion-linked protein modification steps, such as processing and glycosylation. In addition, a vector used
for transformation is stably integrated into the
genome thus providing the base for a reproducible
production process. The cell is robust and requires
inexpensive media for cultivation that can easily be
scaled up to an industrial level (see section on
fermentation). *H. polymorpha* is a thermotolerant
organism and fermentation temperatures up to 45°C
are feasible. Several auxotrophic (*ura⁻* and *leu⁻*)
strains are available as hosts for heterologous gene
expression. The *H. polymorpha* hosts are classified in
the lowest safety class with no risk for humans and
the environment for both research and production,
according to German gene technology law.

The strong and highly regulated promoter structures of the methanol utilization pathway genes
became available after the respective genes had been
identified and isolated. Accordingly, promoter elements derived from the *MOX* and *FMD* genes are
routinely used as components for heterologous gene
expression (see above and Table 2). In an activated
state they can promote an efficient production of
foreign proteins. Activation is achieved under conditions that closely follow the induction/derepression
mechanism observed for the genuine genes.

A standard vector to generate recombinant *H. polymorpha* strains contains the elements described in
Figure 2. The expression cassette consists of the
strong inducible promoter described above, a multiple
cloning site for the insertion of a sequence encoding
a heterologous protein and a terminator sequence

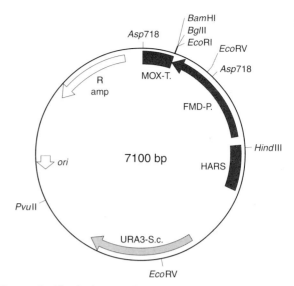

Figure 2 Physical map of a representative *H. polymorpha*
expression vector. The expression cassette consists of the *FMD*
promoter (alternatively the *MOX* promoter), a cloning site for
heterologous sequences and a *MOX* terminator. A segment for
plasmid propagation in bacteria contains an *ori* sequence and
an ampicillin resistance-encoding gene, an *HARS* segment for
propagation in the yeast and a *URA3* gene for complementation
of the host's auxotrophy. A few restriction sites are as indicated.

derived from the *MOX* gene. For propagation and
selection in an *Escherichia coli* host the plasmid is
equipped with an *ori* sequence and a gene that confers
resistance against an antibiotic, in the documented
example against ampicillin. A yeast DNA segment
provides the respective elements for propagation and
selection in the *H. polymorpha* host, namely an *HARS*
(*Hansenula* autonomously replicating) sequence and
a selection marker, in the documented example a *S.
cerevisiae*-derived *URA3* gene that complements the
auxotrophy of the uracil-deficient host. A range of
secretion leaders is available to direct a translation
product into the secretory compartment of the yeast,
among others an element engineered from the *S. cerevisiae*-derived *MFα1* gene or an element engineered
from a *Carcinus maenas* gene encoding a hyper-
glycaemic hormone (*CHH*). Despite the presence of
an *HARS* sequence, plasmids are integrated into the
host chromosome. Transformations result in a variety
of mitotically stable strains containing multiple copies
of the expression cassette in a head-to-tail arrangement; strains with up to a hundred copies have been
identified. Several genes can be co-expressed in *H.
polymorpha* by placing more than one single expression cassette on a transformation vector. Alternatively,
expression cassettes can be transferred in subsequent
transformation steps using different marker genes for
selection. Using this approach strains can be isolated
that harbour different genes in a fixed copy number
ratio. Resulting from these fixed copy numbers, a gene

dosage-dependent production of different recombinant proteins can be envisaged. This provides options to produce complex composite structures or to create new metabolic pathways co-producing suitable recombinant enzymes in an optimal stoichiometric ratio. The *H. polymorpha*-based platform technology, the basic outline of which has been described here, is constantly being extended with new applications, methods and system components. The range of new elements includes alternative auxotrophic strains, alternative *HARS* sequences, new promoter structures and secretion leaders and the development of alternative optimal fermentation schemes.

Fermentation

The commonly used culture media for recombinant *Hansenula polymorpha* strains are based on simple synthetic components. They contain trace metal ions and adequate nitrogen sources which are required for efficient gene expression and cell yield and do not contain proteins. A suitable carbon source has to be added in appropriate concentrations. Different fermentation modes can be applied which vary for the most part in the supplemented carbon source that is selected from glycerol, methanol, glucose and combinations thereof. The optimal fermentation mode depends on the promoter sequence used for heterologous gene expression (*FMD* or *MOX*-promoter) and the inherent control mechanism imposed by this structure (see above).

By product-specific complex fermentation schemes that use different carbon sources and supplements, optimal production of the heterologous compound is obtained. The total fermentation time varies between 80 and 120 h.

The possibility of achieving high yields of a recombinant product without the addition of methanol is a unique feature of the *H. polymorpha* expression system. In contrast, the methanol utilization promoters in related methylotrophic species, namely *Pichia pastoris* and *Candida boidinii*, depend on methanol for induction. Therefore, industrial processes based on these systems require fermentation modes that closely resemble the option described in the following example for the production of the hepatitis B surface antigen (HBsAg).

Case Examples of *H. polymorpha*-based Production Processes

Recombinant *H. polymorpha* strains have been used for a wide range of industrial applications (**Table 3**). Examples include pharmaceuticals that have successfully passed clinical examinations, the high yield production of technical enzymes and the design of

Table 3 Selection of production processes based on recombinant *H. polymorpha* strains

Protein produced	Application	Status (1 January 1999)
Pharmaceuticals		
Hirudin	Therapeutic	Clinical trial phase III
HBsAg	Vaccine	Product launched
Enzymes		
Phytase	Feed additive	Pilot production
Cellulase	Technical enzyme	Feasibility study finished
Metabolic design		
Glycolate oxidase/ catalase T	Biocatalyst for pyruvate/*N*-(phosphonomethyl) glycine	Feasibility study finished

recombinant biocatalysts. The scale of production ranges from five to several thousand litres of fermentation broth. High productivities in the range of several grams per litre of culture have been observed for both intracellular and secreted recombinant proteins.

H. polymorpha-based production systems have been developed for *ad* and *ay* subtypes of the small HBsAg that represent epitopes of the most important serotypes of the hepatitis B virus. A production strain was generated by introducing a plasmid with the gene for a particular viral antigen fused to a *MOX*-promoter element. The identified strain is fermented using glycerol and methanol as carbon sources (see above). In this development a methanol feed is needed since membrane proliferation has to be induced beyond heterologous protein production. Because of intracellular deposition the protein has to be recovered and purified from cellular extracts. The extract is cleared by a sequence of precipitation and centrifugation steps. The HBsAg contained in the cleared supernatant is further purified by a sequence of gel adsorption, ion exchange chromatography, centrifugation and gel filtration. Eventually, highly purified lipoprotein particles are obtained which consist of the heterologous antigen assembled into yeast-derived membranes. The particle structure is essential for stimulating an immunological response. For vaccine formulation the particle is adsorbed to an aluminium hydroxide adjuvant.

A production process for phytase, a secreted protein, follows a production mode that uses growth on glycerol followed by continuous addition of glucose. This process is completely independent from methanol addition.

The possibility of coexpressing different genes that are transferred to the yeast host by subsequent transformation steps provided an option to design recom-

Table 4 Biotechnological applications of *H. polymorpha*

Recombinant strains	Wild-type and mutant strains
Heterologous gene expression	**Protein production**
Recombinant pharmaceuticals	Alcohol oxidase
Recombinant enzymes	Formate dehydrogenase (NAD/NADH recycler)
Metabolic design	Riboflavin kinase
	FMN-adenyltransferase
	Single-cell protein production on methanol and xylose
	Production of other compounds
	Dihydroxyacetone
	Arabitol
	Acetaldehyde
	Amino acids with a branched chain

binant *H. polymorpha* strains as biocatalysts for the bioconversion of specific substrates into valuable industrial compounds. In one particular development, a biocatalyst was generated that was suitable for the conversion of glycolate into glyoxylate and the conversion of lactate into pyruvate. The efficient conversion depends on the presence of a glycolate oxidase and a suitable catalase in optimal stoichiometric amounts. To accomplish this an initial recombinant strain was engineered harbouring 30 copies of a spinach-derived glycolate oxidase (*GO*) gene integrated into the chromosome. This strain served as host for the introduction of 15 copies of the *Saccharomyces cerevisiae*-derived catalase T (*CTT1*) gene. The resulting strain coproduced the enzymes in question in high amounts that were optimal for the anticipated conversion process. For conversion the substrate was added to detergent-treated recombinant cells that had intracellularly accumulated the two enzymes after fermentation under appropriate conditions. Cells can be recycled for at least 25–30 subsequent biocatalytic reactions. A selection of production processes based on genetically engineered *H. polymorpha* strains is provided in Table 3.

Further Biotechnological Applications of *H. polymorpha*

H. polymorpha has become a preferred option among the various yeast expression systems. However, its use is not restricted to this application. Wild-type strains and a variety of mutant strains have been exploited as sources for various compounds. These compounds include enzymes such as the methanol (alcohol) oxidase that can be used as biosensor component for detection of alcohol, formate dehydrogenase, riboflavin kinase and FMN-adenyltransferase.

A range of mutant strains can be used for the production of metabolites such as dihydroxyacetone and acetaldehyde and for the synthesis of amino acids with a branched side chain and for arabitol. The range

of biotechnological applications of *H. polymorpha* is summarized in **Table 4**.

Conclusion

Hansenula polymorpha has become a preferred option as host for heterologous gene expression. The safe and convenient system enables competitive processes for the production of industrial compounds ranging from technical enzymes to pharmaceuticals considered for administration to humans. The technology employing this yeast is essentially mature now and is being complemented with new system components and new applications. A range of wild-type and mutant strains provides the source for industrial enzymes, metabolites, amino acids and arabitol. *H. polymorpha* is, furthermore, a model organism for exploring the function and biogenesis of peroxisomes. A growing list of genes isolated from this organism attest to the increasing recognition of *H. polymorpha* as a target for basic research and industrial application.

See also: **Saccharomyces**: Saccharomyces cerevisiae. **Single Cell Protein**: Yeasts and Bacteria. **Yeasts**: Production and Commercial Uses.

Further Reading

Barnett JA, Payne RW and Yarrow D (1990) *Yeasts: Characterization and Identification*, 2nd edn. Cambridge: Cambridge University Press.

Cleland JL and Craik CS (eds) (1996) *Protein Engineering: Principles and Practice*. New York: Wiley Liss.

Gellissen G and Melber K (1996) Methylotrophic yeast *Hansenula polymorpha* as production organism for recombinant pharmaceuticals. *Drug Research* 9: 943–948.

Hollenberg CP and Gellissen G (1997) Production of recombinant proteins by methylotrophic yeasts. *Current Opinion in Biotechnology* 8: 554–560.

Murooka Y and Imanaka T (eds) (1994) *Recombinant Microorganisms for Industrial and Agricultural Applications*. New York: Marcel Dekker.

Rose AH and Harrison JS (eds) (1987) *The Yeasts*, 2nd edn. Vols 1–3. London: Academic Press.

Smith A (ed.) (1994) *Gene Expression in Recombinant Microbes*. New York: Marcel Dekker.

Wolf K (ed.) (1996) *Nonconventional Yeasts in Biotechnology. A Handbook*. Berlin: Springer.

Hard Cider *see* **Cider (Hard Cider)**.

HAZARD APPRAISAL (HACCP)

Contents

The Overall Concept
Critical Control Points
Establishment of Performance Criteria
Involvement of Regulatory Bodies

The Overall Concept

F Untermann, Institute for Food Safety and Hygiene, University of Zurich, Switzerland

Introduction

Media reports about threats to consumer health from a particular food item or about outbreaks of disease caused by certain products may instantly provide food companies with a great deal of negative publicity. It is to the benefit of commerce and the food industry as well as the food inspection services to take effective precautions against such events. For this purpose the hazard analysis and critical control point (HACCP) concept was conceived. It is founded on a logical and clearly structured system that allows for the early detection and elimination of specific hazards which may cause disease of consumers. However, the correct application of the concept requires comprehensive expert knowledge.

The HACCP concept does not replace traditional hygiene measures. It is rather based on the well-conceived and effective hygiene concept within a food company. Hence, cleaning and disinfection plans, personnel hygiene as well as separation of clean and unclean areas, etc., are the basis and precondition for the establishment of an HACCP system. However, they are not an integral part of a concrete HACCP plan for a particular food item.

The HACCP system is a strictly logical concept aimed at finding out on the basis of scientific facts firstly whether there is need to prevent or eliminate a food safety hazard or reduce it to an acceptable level (hazard analysis principle 1). Then, reliable safety measures have to be taken on the basis of six further principles, which are very strictly defined (preventive management).

The philosophy behind this approach, first to determine the need for action and then to define suitable preventive measures, is not new and naturally applies to all areas where faults should be avoided, including basic hygiene measures. A reflective approach to hygiene is urgently required. However, the application of terms or notions borrowed from the HACCP system, e.g. for basic hygiene measures or in other areas where the seven principles are not wholly applicable, leads to a dilution of the aims and efficacy of the HACCP concept.

The History of the HACCP System

The hazard analysis and critical control point concept was conceived in the USA jointly by the Pillsbury Company, the US Army Natick Research and Development Laboratories and the National Aeronautics and Space Administration as early as 1960 in order to produce safe foods for the space programme. Simply examining finished food products does not ensure a sufficient degree of safety, and for this reason it was necessary to develop and to monitor production processes in such a way that health hazards for astronauts could be precluded safely and reliably.

The concept was adopted during the 1970s and early 1980s by several large food companies, and became known internationally through publications of the International Commission on Microbiological Specifications for Food (ICMSF). The concept has also been adopted by the FAO/WHO *Codex Alimentarius* Commission (ALINORM 97/13A). **Table 1** provides

Table 1 The history of the HACCP system

Concept for US space programme 1959/1960	Simonsen 1987 ICMSF 1988	NACMCF 1989 (National Academy of Science, USA)	FAO/WHO Codex Alimentarius Commission 1996 ALINORM 97/13A
Three HACCP principles 1. Hazard analysis and risk assessment 2. Determination of critical control points 3. Monitoring of CCPs	Six HACCP principles Not included: documentation	Seven HACCP principles	Seven HACCP principles New: definitions
One CCP category	Two CCP categories: CCP I – complete control of potential hazards is effected CCP II – only partial control is effected	One CCP category	One CCP category

a synopsis of the development stages of the HACCP system.

The HACCP System of *Codex Alimentarius*

Definitions and the Seven Principles Exact definitions of HACCP notions assume a central position within the document. The essential definitions are represented in **Table 2**. With the help of these definitions it was possible to give a short version of the seven HACCP principles in the Codex paper:

1. Conduct a hazard analysis.
2. Determine the Critical Control Point (CCP).
3. Establish critical limits.
4. Establish a system to monitor control of the CCP.
5. Establish the corrective actions to be taken when monitoring indicates that a particular CCP is not under control.
6. Establish procedures for verification to confirm that the HACCP system is working effectively.
7. Establish documentations concerning all procedures and records appropriate to these principles and their application.

Guidelines for the Application of the HACCP System

Initially, the procedure for establishing a HACCP plan is presented. Advice is given concerning the necessary preparations, such as assembling the HACCP team, product description, and construction of a flow diagram for the production process. Furthermore, the performing of a hazard analysis and the determination of CCPs are mentioned. A logic sequence for the application of the HACCP is given in **Table 3**.

The HACCP team has to provide the production-specific expertise and experience which are necessary for the development of the HACCP plan. Multidisciplinary expertise is also required. A food safety management which incorporates toxicological, microbiological, medical and epidemiological aspects

necessary for the adequate application of HACCP requires experts with a high degree of scientific training. Alongside scientific and medical knowledge the faculty of structured and systematic thinking is essential in order to apply the elements of quality management intelligently and effectively.

The description of the product is not confined to appearance and structure or the raw materials and additives that were used for its production. Factors that have an influence on the kinetics of microorganisms, e.g. pH and water activity (a_w) values, as well as intended storage conditions (packaging, atmospheric conditions, temperature) and shelf life must also be defined.

The intended use consists of information on whether the product has to be prepared prior to consumption, e.g. by heating, or whether it can be consumed directly. With regard to a possible acceptable risk level for a food safety hazard it has to be stated for which group of the population the food is intended. It is obvious that considerably higher safety requirements are needed for hospitals or old people's homes, for example.

Hazard Analysis

Basic Aspects

In the Codex paper hazard analysis is described as a 'process of collecting and evaluating information on hazards and conditions leading to their presence to decide which are significant for food safety and therefore should be addressed in the HACCP plan'. The term 'hazards and conditions leading to their presence' is a very useful one: the enterotoxin of *Staphylococcus* is an example of a hazard, whereas 'a condition leading to the presence of this hazard' would be the exposure during production or storage of a food product to a temperature at which *Staphylococcus* can grow and produce enterotoxins.

Table 2   Definitions of *Codex Alimentarius*

- **Control** (verb)
  To take all necessary actions to ensure and maintain compliance with criteria established in the HACCP plan
- **Control** (noun)
  The state wherein correct procedures are being followed and criteria are being met
- **Control measure**
  Any action and activity that can be used to prevent or to eliminate a food safety hazard or reduce it to an acceptable level
- **Corrective action**
  Any action to be taken when the results of monitoring at the CCP indicate a loss of control
- **Critical control point**
  A step at which control can be applied and is essential to prevent or eliminate a food safety hazard or reduce it to an acceptable level
- **Critical limit**
  A criterion which separates acceptability from unacceptability
- **HACCP**
  A system that identifies, evaluates and controls hazards which are significant for food safety
- **HACCP plan**
  A document prepared in accordance with the principles of HACCP to ensure control of hazards which are significant for food safety in the segment of the food chain under consideration
- **Hazard**
  A biological, chemical or physical agent in, or condition of, food with the potential to cause an adverse health effect
- **Hazard analysis**
  The process of collecting and evaluating information on hazards and conditions leading to their presence to decide which are significant for food safety and therefore should be addressed in the HACCP plan
- **Monitoring**
  The act of conducting a planned sequence of observations or measurements of control parameters to assess whether a CCP is under control
  In the *Guidelines for the Application of the HACCP System* it is further explained: if the monitoring is not continuous, then the amount or frequency of monitoring must be sufficient to guarantee the CCP is in control
- **Step**
  A point, procedure, operation or stage in the food chain including raw materials, from primary production to final consumption
- **Verification**
  The application of methods, procedures, tests and other evaluations, in addition to monitoring to determine compliance with the HACCP plan

Table 3   Logic sequence for application of HACCP following the FAO/WHO *Codex Alimentarius* Commission principles

Preparation
  Assemble the HACCP team
  Describe product
  Identify intended use
  Construct flow diagram
  On-site verification of flow diagram

Hazard analysis
  List all potential hazards (hazard identification)
  Assess all potential hazards (assessment of risks)
  Determine the need for action

Preventive management
  Examine by which measures the (relevant) hazards can be prevented, eliminated or reduced to an acceptable risk level
  Determine CCPs
  Establish critical limit for each CCP
  Establish a monitoring system for each CCP
  Establish corrective action for deviations that may occur
  Establish verification procedures
  Establish record keeping and documentation analogous to EN ISO 9000

The entire food production process has to be examined to identify potential hazards that might occur during the production or use of a particular food item. Also requiring consideration are the raw materials and ingredients, as well as the type and duration of storage, the method of distribution and the intended use of the final product by the consumer.

Initially, it has to be evaluated whether hazards may be present in raw materials as well as in other ingredients and additives. Next, the possibility of contamination with hazards during individual production steps is assessed. Finally it has to be evaluated whether hazards could develop during the production process, during storage or during the intended utilization of the food product: this could be the growth of pathogenic bacteria or the formation of toxic substances by bacteria or by other chemical reactions (e.g. nitrosamine formation).

This evaluation is followed by an assessment of risk, to estimate the likely occurrence of health hazards and the severity of their adverse health effects.

Within the HACCP system there is a distinction between biological, chemical and physical hazards (**Fig. 1**). It is relatively easy to understand the causality of the occurrence of physical hazards such as splinters of metal, glass or other foreign bodies; it takes logical thinking and the knowledge of the technological production procedures. Here, the expertise lies with the technical staff of the food company. In contrast, the assessment of chemical and biological hazards requires special expertise concerning the pathogenesis of human diseases which are caused by such hazards. The development of effective preventive measures requires comprehensive knowledge of the epidemiological factors which threaten the health of the consumer.

Special Features of the Assessment of Microbiological Hazards

Quantitative scientific assessments of the risks from microorganisms in foods and water on the basis of dose–response relationships and exposure assessment,

Physical hazards

Splinters and other foreign bodies
– stone, metal, glass, wood, bone

Biological hazards
- Macroparasites
  – cysticerci
  – *Trematoda* spp.
  – *Trichinella*
  – *Echinococcus* eggs
  – Anisakidae (in fish)
- Protozoa
  – *Toxoplasma gondii*
  – *Sarcocystis* spp.
  – *Entamoeba histolytica*
  – *Giardia, Cryptosporidium* (water)
  – *Cyclosporum*
- Fungi
  – mycotoxin-producing moulds
- Bacteria
  – species causing infections
    and/or toxi-infections
  – species causing intoxications
- Viruses
- Prions

Chemical hazards
- Toxic substances
  naturally occurring in plants and
  animals (including toxins in fish
  and shellfish originating from
  toxigenic algae)
- Contaminants and residues, e.g.
  – toxic heavy metals
  – PCBs (polychlorinated biphenyl)
  – radionuclides
  – animal drugs
  – feed additives
  – pesticides
- Food additives, e.g.
  – monosodium glutamate
  – nitrate, nitrite

Figure 1 Hazards in foods.

customarily carried out for chemical contaminants, have been developed for some pathogens, especially in drinking water. Two particular difficulties have to be mentioned for the quantification of microbiological hazards associated with the consumption of foods: the determination of the minimal infective dose, and the complex kinetics of bacterial survival, growth and death in foods.

Minimum Infective Dose For most bacterial species the question of the minimum infective dose cannot be answered satisfactorily. Firstly, it must be borne in mind that among consumers there are special risk groups – small children, senior citizens, pregnant women and immunocompromised persons. Furthermore, we are familiar with various physiological factors that influence the minimum infective dose; these include the degree of stomach fluid acidity, the quantity of stomach contents, the intestinal flora, and not least, the immunological status of the person. This status is again influenced by immunity due to previous infections, by the nutritional state, and by stress.

In addition, the fact has to be considered that the quantity of microorganisms in food undergoes continuous change, in contrast to chemical residues. The complex kinetics of dying, survival and growth of bacteria in foods are determined by manifold factors which can be differentiated into intrinsic, extrinsic and process factors. They include the pH, a_w, redox potential and temperature of a food and also the presence of competitive microbial flora. The most important factors influencing microbial growth in

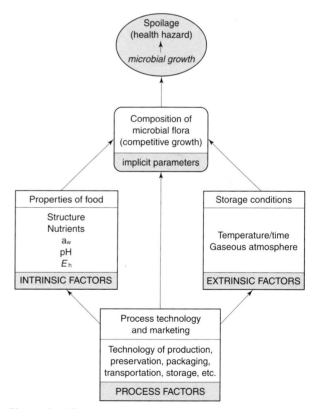

Figure 2 Influences on microbial growth in foods.

foods are shown in **Figure 2**. Hence, the risks resulting from microorganisms – especially from bacteria – vary depending on the composition of the food, on production, processing and preparation procedures, and on the packaging and storage conditions as well.

Predictive Microbiology The synergy of these different factors is a complex one. Computer-based mathematical models have therefore been created that take into account the factors mentioned above. They allow for predictions being made concerning microbial kinetics in foods. This predictive microbiology is based on data obtained from broth cultures under standardized conditions. If such a programme is fed with data on a food's intrinsic factors, on the atmospheric conditions of packaging, and on the projected time and temperature of storage, the prospective behaviour of a microbial species can be computed and depicted in graphs. However, owing to the complex composition of foods, the results from predictive microbiology can only provide a framework for the understanding of the ecology and kinetics of microorganisms in foods. In order to obtain exact data for a particular food, further testing is required. Storage trials are a suitable means of testing the behaviour of certain microorganisms as long as the species in question can be found regularly in the food product under consideration. If this is not the case, challenge tests can be performed where foods or raw materials are spiked with pathogens.

Critical Control Points

In the Codex paper a CCP is defined as 'a step at which control can be applied and is essential to prevent or eliminate a food safety hazard or reduce it to an acceptable level'. There is also a definition for the term 'step' (see Table 2).

The necessity for the determination of a CCP is a consequence of the previously conducted hazard analysis. Moreover, with a CCP it must be possible to reliably prevent or eliminate a health hazard or reduce it to an acceptable level. The 'acceptable level' (acceptable remaining risk level) is also established by the hazard analysis which is carried out beforehand.

Prior to the determination of a CCP, measures have to be defined to bring previously identified hazards under control. In the case of microorganisms these may be heating procedures such as sterilization or pasteurization. In this context, temperatures and inactivation time parameters have to be defined. The decimal reduction time (D value) for the bacteria in question should be indicated. Microbial growth can be influenced by extrinsic factors (e.g. deep freeze or cool storage temperatures) as well as by intrinsic factors (pH and a_w values). Intrinsic factors can be altered by changes in the composition of a food item in such a way that certain bacteria do not multiply even in unrefrigerated foods. A detailed knowledge of the growth kinetics of individual pathogens is indispensable for the definition of such safety measures.

Once the measures are identified that help to prevent or eliminate an identified food safety hazard or to reduce it to an acceptable level, a step in the food production process has to be determined at which the control can be applied – a critical control point (HACCP principle 2). It has to be underlined that CCPs need not be established at every step at which hazards or conditions leading to their presence may occur; this would lead to an HACCP system with too many CCPs.

For the selected preventive measures, critical limits have to be established (HACCP principle 3). Compliance with these conditions has to be guaranteed by a suitable monitoring procedure (HACCP principle 4). The necessary corrective action must also be established (HACCP principle 5).

Monitoring

In the *Codex Alimentarius* document monitoring of CCPs is defined as follows: 'The act of conducting a planned sequence of observations or measurements of control parameters to assess whether a CCP is under control' (i.e. that the control measures function correctly). In the Codex guidelines in the same document it is further explained: 'If the monitoring is not continuous, then the amount or frequency of monitoring must be sufficient to guarantee the CCP is in control'. Hence, in the HACCP system the term 'monitoring' takes on a different meaning from that used in monitoring programmes for chemicals in the environment. The same applies to 'hygiene' or 'bacteriological monitoring', where bacteriological investigations of food establishments are performed with swabs or other samples taken from premises, equipment or foods.

Verification and Documentation

A precondition of hazard analysis for any food item is knowledge of the biological, chemical and physical hazards that may be present in the basic materials. Correspondingly, basic or raw materials must always fulfil the minimal requirements on which the HACCP plan is based. Compliance with these requirements must be ensured by the producers or providers of the basic materials. Controls of incoming materials are a verification measure according to EN ISO 8402 or ALINORM 97/13A, if the supplier operates a corresponding HACCP system. However, such controls of incoming delivery materials cannot function as a CCP.

The term 'verification' summarizes all examinations, measures and information that permit examination of the functioning of the HACCP system for a particular food product. This can be achieved by implementing end product sampling plans, storage trials or microbiological controls during the production process. In addition, regular examination of documentation measures is of equal importance. Regular audits within the company are essential in order to make sure that no changes are implemented in the production process or in food formulations which would call for a revision of the HACCP plan. Most of all, it has to be determined whether the established control measures are effective and meet the requirements of principles 2 to 6 of the HACCP system. For this purpose the preventive measures have to be validated by an expert.

Complete documentation is essential. Documentation should cover the production process (including raw materials) and measures connected with the application of the HACCP system as well as ongoing record taking. Records must be traceable according to EN ISO 8402, otherwise the functioning of the HACCP plan cannot be proved.

The principles of the HACCP concept are illustrated in **Figure 3**, which is intended to underline the connection between controlling and monitoring. Without reliable monitoring the compliance with CCP conditions cannot be guaranteed. A lack of monitoring means also that control measures do not fulfil the requirements for a CCP. A CCP ensures reliable safety which cannot be guaranteed without a monitoring system to indicate deviations from previously established critical limits.

The Role of HACCP

Renaissance of Old Scientific Principles

The principles of the HACCP concept are not new. A good example of their application is the pasteurization of drinking milk (**Fig. 4**), implemented a hundred years ago as an effective measure to protect people against zoonoses such as brucellosis and tuberculosis. Around 1930, this procedure was introduced into the legislation in Germany and elsewhere.

In children the transmission of bovine tuberculosis via drinking milk was an important mechanism for the spread of tuberculosis. The transmission of brucellosis through drinking milk was also a significant hazard. Milk pasteurization was chosen as an effective control measure to inactivate pathogenic bacteria; in contrast to boiling, it hardly affects the taste or vitamin content of milk. Monitoring is assured by continuous temperature control with automatic charting of temperatures. Temperatures and times are defined exactly. Pasteurization devices are constructed so as to ensure that if minimal values are not reached, insufficiently heated milk is directed back into the raw milk storage

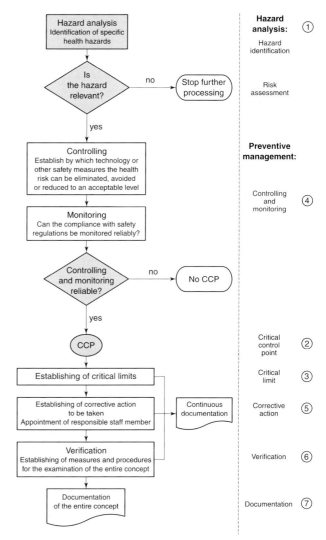

Figure 3 The HACCP concept.

tank through an automatic three-way tap. Proper heating of commercial drinking milk can be validated by the alkaline phosphatase assay: the enzyme, which is normally found in untreated milk, is inactivated by pasteurization. This assay is still in use and constitutes a verification procedure.

Hence, the introduction of the HACCP system to food processing does not constitute a fundamentally new development. It could even be viewed as a renaissance of old scientific principles.

Preventive Measures

The Significance of Basic Hygiene Measures With many types of microorganisms the faeco-oral route of transmission is of great epidemiological significance. This is particularly so with microorganisms that have such a low minimum infective dose that they can induce disease without having to grow in food. In such cases foods act mainly as vectors. The contamination of foods can occur via humans or animal

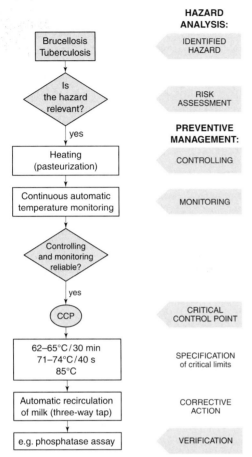

HAZARD ANALYSIS:

Brucellosis
Tuberculosis → IDENTIFIED HAZARD

Is the hazard relevant? → RISK ASSESSMENT

PREVENTIVE MANAGEMENT:

yes

Heating (pasteurization) → CONTROLLING

Continuous automatic temperature monitoring → MONITORING

Controlling and monitoring reliable?

yes

CCP → CRITICAL CONTROL POINT

62–65°C / 30 min
71–74°C / 40 s
85°C → SPECIFICATION of critical limits

Automatic recirculation of milk (three-way tap) → CORRECTIVE ACTION

e.g. phosphatase assay → VERIFICATION

Figure 4 Preventive management in drinking milk production (more than 60 years ago).

species which act as a reservoir for such micro-organisms. Contaminated water is important in this context. Likewise, cross contamination by insects – but more often by tools and equipment – can be found. Effective prevention is ensured by strict compliance with basic hygiene measures, including attitudes towards hygiene as an important factor.

Protozoa, viruses and prions cannot grow in foods. They are either present in raw basic materials, e.g. meat, or they are transmitted into foods as contaminants. In contrast, moulds and bacteria, with the exception of a few individual species, are capable of growing in foods if conditions are suitable.

An overview of bacteria connected with food-borne diseases is shown in **Figure 5**. Pathogens that have other routes of infection besides the oral one are listed separately on the left. In the transmission of (for example) *Mycobacterium bovis* and *Brucella melitensis* to humans, raw milk plays an important role, while the risk of infection with *Coxiella burnetii* in milk is low and therefore other infection pathways are considerably more important.

For *Salmonella typhi*, *S. paratyphi*, *Vibrio cholerae* and *Shigella* ssp. the reservoir is confined to humans

and food mainly figures as a vector. In the case of *Campylobacter jejuni* the decisive pathogen reservoir is poultry. Besides person-to-person transmission, contaminated raw chicken carcasses are likely to play a special role during the infection process, because pathogens can be spread to other foods in kitchens through cross contamination, e.g. by thawing water.

In the USA beef cattle are considered to be the natural reservoir for serotype O157:H7 of verotoxigenic *Escherichia coli* (enterohaemorrhagic *E. coli* (EHEC)). Hence, the consumption of raw or insufficiently heated meat products is seen as an important source of infection. As with all pathogens with a low minimum infective dose, it can be assumed that the faeco-oral route of transmission of verotoxigenic *E. coli* from person to person plays an important part in the spreading of outbreaks and that food products are mainly contaminated by human excretors. Correspondingly, the compliance with basic hygiene measures is a prerequisite for effective prevention.

In contrast, with the remaining infectious agents their growth in food is of major epidemiological significance. The same applies to pathogens which strictly cause intoxications because the formation of toxigenic metabolites in food requires the previous growth of such pathogens. Under those conditions all measures that contribute towards prevention of microbial growth in food constitute an additional safety barrier.

Considering the particular importance of the faeco-oral transmission route in the pathogenesis of many infections caused by foods, it becomes obvious that the HACCP concept under no circumstances replaces common hygiene measures. Rather, it is built on the well-conceived and effective hygiene concept of a food company. This includes personnel hygiene, cleaning, disinfection and pest control. Further components are the temperatures and relative humidity of production and storage sites as well as sufficient separation of production steps and production lines to avoid cross contamination.

These measures are fundamental to the application of an HACCP system. However, they are not part of a concrete HACCP plan for a particular food item. In the *Codex Alimentarius* guideline it is pointed out that 'prior to application of HACCP to any sector of the food chain, that sector should be operating according to the Codex General Principles of Food Hygiene, the appropriate Codex Codes of Practice, and appropriate food safety legislation'. However, it goes without saying that basic hygiene measures need to be applied according to the same logical and scientific criteria on which the HACCP system is based.

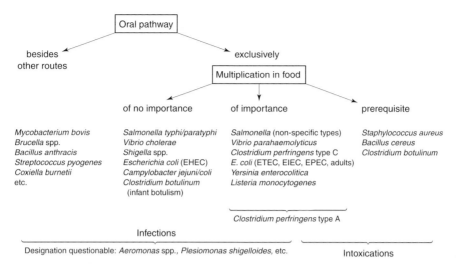

Figure 5 Food-borne bacterial infections and intoxications. (EHEC, EIEC, EPEC, ETEC – enterohaemorrhagic, enteroinvasive, enteropathogenic, enterotoxigenic *Escherichia coli*.)

Differing Interpretations of the HACCP System

As the term 'HACCP' became known throughout the world in the late 1980s, very different views concerning the interpretation and the practical implementation of the concept in food companies developed concurrently. One can gain the impression that 'HACCP' is often used as a synonym for all hygiene and food safety measures. Furthermore, the consistent implementation of principles 2 to 5 during production, processing and preparation of foods is not always possible. In those cases often a purely formal compliance with these principles is sought, and measures and parameters are defined as control and monitoring measures with their respective critical limits which are no longer in agreement with the original meaning of HACCP notions of *Codex Alimentarius*.

It is not uncommon that the logical implementation of basic hygiene measures is declared to be an 'HACCP concept' and that a product- and production-specific hazard analysis is not carried out. Here, the notion of 'hazard analysis and critical control point' is altered into the definition of a 'hygiene analysis and critical check point'.

The 'House of Hygiene'

The HACCP system should be seen as an integral part of an efficient food safety and comprehensive (total) hygiene concept. The 'Zurich House of Food Safety' (**Fig. 6**) may help to elucidate this point. The 'foundations' of the 'house of food safety' are the conditions of premises and equipment. Its 'walls' are the well-known basic hygiene measures. They include cleaning, disinfection and pest control as well as the temperature and relative humidity of production and storage sites. Further components are a sufficient sep-

Product- and production-specific preventive measures based on the HACCP approach and especially on a specific hazard analysis

Basic Hygiene Measures

- temperature and relative humidity of working and storage premises
- sufficent separation of production steps and production lines (avoiding cross contamination)
- personnel hygiene
- cleaning, disinfection and pest control

Conditions of premises and equipment

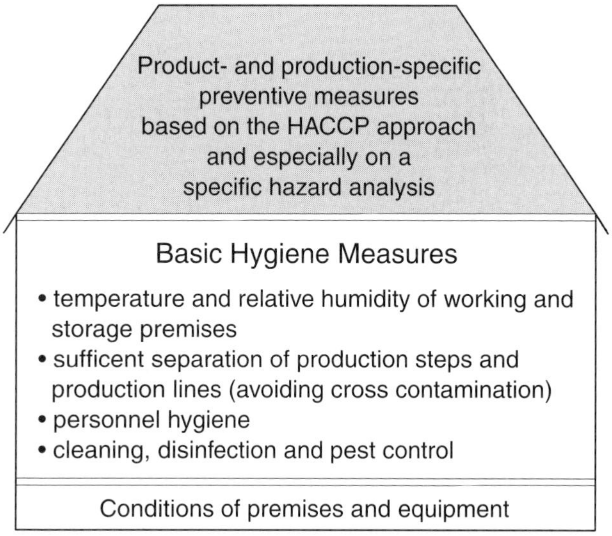

Figure 6 The 'house of hygiene': a representation of a comprehensive food safety concept.

aration of production steps and production lines to avoid cross contamination, and, finally, personnel hygiene. The 'roof' of the house is made up of product- and production-specific preventive measures based on a specific hazard analysis according to the HACCP principles to avoid specific health hazards for the consumer.

See also: **Brucella**: Characteristics. **Escherichia coli O157**: *Escherichia coli* O157:H7. **Hazard Appraisal (HACCP)**: Critical Control Points; Involvement of Regulatory Bodies; Establishment of Performance Criteria. **Heat Treatment of Foods**: Principles of Pasteurization. **Milk and Milk Products**: Microbiology of Liquid Milk. **Molecular Biology – in Microbiological Analysis**. **Mycobacterium**. **Predictive Microbiology & Food Safety**. **Process Hygiene**: Designing for Hygienic Operation; Overall Approach to Hygienic Processing. ***Salmonella***: Introduction; *Salmonella typhi*. ***Shigella***: Introduction and Detection by Classical Cultural Techniques. ***Vibrio***: *Vibrio cholerae*.

Further Reading

Benenson AS (1995) *Control of Communicable Diseases Manual*, 17th edn. Washington: American Public Health Association.

FAO/WHO Codex Alimentarius Commission (1996) *Report of the 29th Session of the Codex Committee on Food Hygiene*. Washington, 21–25 Oct 1996, ALINORM 97/13A, 30–41.

CEN European Committee for Standardization (1994) *Quality Management and Quality Assurance – Vocabulary*. EN ISO 8402.

ICMSF (1988) *Microorganisms in Foods*. Vol. 4, *Application of the Hazard Analysis Critical Control Point (HACCP) System to Ensure Microbiological Safety and Quality*. Oxford: Blackwell Scientific.

ICMSF (1996) *Microorganisms in Foods*. Vol. 5, *Microbiological Specifications of Food Pathogens*. London: Blackie.

Lammerding AM (1997) An overview of microbial food safety risk assessment. *Journal of Food Protection* 60: 1420–1425.

MacNab BW (1997) A literature review linking microbial risk assessment, predictive microbiology and dose-response modeling. *Dairy, Food and Environmental Sanitation* 17: 405–416.

Neumann DA and Jeffery AF (1997) Assessing the risks associated with exposure to waterborne pathogens: an expert panel's report on risk assessment. *Journal of Food Protection* 60: 1426–1431.

Notermans S, in't Veld P, Wijtzes T and Mead GC (1993) A user's guide to microbial challenge testing for ensuring the safety and stability of food products. *Food Microbiology* 10: 145–157.

Teunis PFM, Heijden GO, van der Giessen JWB and Havelaar AH (1996) The dose-response relation in human volunteers for gastrointestinal pathogens. Report 284550002, RIVM, PO Box 1, 3720BA Bilthoven, The Netherlands.

Untermann F (1993) Problems of food hygiene with carriers of microorganisms and permanent excretors. *Zentralblatt für Hygiene* 194: 197–204.

Untermann F (1996) Risk assessment and risk management of food production according to the HACCP concept. *Zentralblatt für Hygiene* 199: 119–130.

Untermann F (1998) Microbial hazards in foods. *Food Control* 9: 119–126.

Vose DJ (1998) The application of quantitative risk assessment to microbial food safety. *Journal of Food Protection* 61: 640–648.

Critical Control Points

S Leaper, Campden & Chorleywood Food Research Association, Chipping Campden, Gloucestershire, UK

A critical control point (CCP) is defined by the *Codex Alimentarius* Commission as a 'step at which control can be applied and is essential to prevent or eliminate a food safety hazard or reduce it to an acceptable level'.

Determination of Critical Control Points

The determination of CCPs is documented as the second of the seven internationally agreed principles of HACCP.

The correct determination of CCPs is essential in the establishment and implementation of an effective food safety assurance system. If CCPs are not correctly identified, the resources of the food business will not be focused appropriately at the steps where control is essential. Too many CCPs in a process can result in an unmanageable system; too few, and safety of the product may not be assured. The determination of CCPs requires a sound knowledge of the process and professional judgment to be exercised. Following the hazard analysis, in which realistic food safety hazards and means for their control have been listed for each process step in a food operation, CCPs can be identified. A sequence of questions (decision tree) can be used to assist in the determination of critical control points. When using a decision tree, each step in the process must be considered in turn for each of the identified hazards. Care is needed in the use of decision trees. A number of decision trees have been published, some of which have been specifically concerned with the raw materials and ingredients, while others focus on the manufacturing or processing steps of the operation. It is not essential to use a decision tree but the use of structured questions can assist in achieving the correct answer consistently.

The example given in the Food and Agriculture Organization of the United Nation's *Hazard Analysis and Critical Control Point (HACCP) System and Guidelines for its Application* (**Fig. 1**) contains the following four questions.

Question 1: Do preventative control measures exist? In answering this question for a new product or a new process development, it is possible that a potential hazard is identified for which no control measure has been established. To determine if control is necessary at this step in the process, a sup-

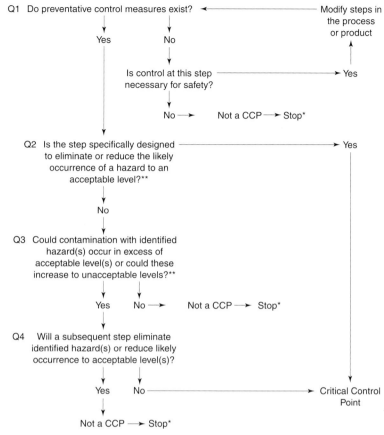

Figure 1 Example of a decision tree to identify critical control points (CCPs). Answer questions in sequence. *Proceed to the next hazard in the process. **Acceptable and unacceptable levels need to be defined within the overall objectives in identifying the CCPs of the HACCP plan. Reproduced from *Food Hygiene Basic Texts* (1997) Joint FAO/WHO Food Standards Programme, Codex Alimentarius Commission, with permission from the Food and Agriculture Organization of the United Nations, Rome.

plementary question can be asked which encourages the consideration of the further process steps and their ability to control the hazard in question. Where no subsequent step exists that has the capability to prevent or eliminate the hazard or reduce the hazard to an acceptable level, then the process must be modified (i.e. a later process step must be introduced that can control the hazard) or the process step must be modified (i.e. a control measure must be introduced at the process step under consideration) – or the product itself must be modified. Whichever modification is appropriate, the change must be introduced before the HACCP system can be implemented.

Question 2: Is the step specifically designed to eliminate or reduce the likely occurrence of a hazard to an acceptable level? Where control measures are in place (i.e. established and implemented), the second question will enable the identification of process steps that are designed to eliminate or reduce the hazard to an acceptable level. If the step under consideration was the pasteurizer, the potential safety hazard would have been identified as survival of bacterial pathogens.

The control measure would be 'correct pasteurization process' (i.e. correct time and temperature). As the pasteurizer is specifically designed to eliminate or reduce the likely occurrence of surviving pathogens the answer to question 2 is 'yes' and the pasteurizer is identified as a critical control point in the process.

Question 3: Could contamination with identified hazard(s) occur in excess of acceptable level(s) or could these increase to unacceptable levels? Where the answer to question 2 is 'no', question 3 should be applied. Consideration is given to the potential for the ingredients to contain the hazard in excess of acceptable levels. This will include consideration of information including epidemiological data and past performance of the supplier. It is also possible that the adjacent processing environment may be a further source of the hazard. Subsequent process steps may also have an effect on the level of hazard in the product. While the hazard may be present at an acceptable level at the step under consideration, the potential to increase at subsequent process steps to unacceptable levels will need to be considered. Where the answer to question 3 is 'no', the conclusion is that the step is not a CCP for the hazard under consideration and the decision tree should be applied to the next hazard. If contamination with the hazard could occur at unacceptable levels or subsequent steps result in an additive effect on the hazard, resulting in unacceptable levels of the hazard, then question 4 must be addressed.

Question 4: Will a subsequent step eliminate identified hazard(s) or reduce likely occurrence to acceptable level(s)? The final question in this decision tree requires the consideration of the subsequent steps in the process and whether or not they will eliminate the hazard or reduce it to an acceptable level. If there are subsequent process step(s) which will eliminate or reduce the hazard under consideration to an acceptable level, then the answer to the question is 'no' and the step being considered is not a CCP. If, however, there are no subsequent steps at which the hazard can be eliminated or reduced to an acceptable level then the step under consideration is a CCP for the identified hazard and the control measures must be implemented effectively. In some instances, additional control measures may be introduced.

At each CCP, critical limits for each of the control measures must be specified, along with procedures that can be used to monitor whether the control measures are operating to the critical limits. Monitoring procedures should be documented and include information on what has to be done, how to carry out the procedure (including details of equipment, materials and records to be used), when the procedure is to be followed and a clearly assigned responsibility. Records made of monitoring procedures are to be signed by the person making the check and countersigned by a senior colleague.

In addition, action will be specified in the HACCP plan which will be taken if the monitoring results show that the critical limit at a CCP has not been met, or that there is a trend towards loss of control. The corrective actions will include a consideration of what to do with the product produced during the 'out-of-control' period, a means of determining the cause of the problem and how to rectify the situation and recommence the production safely. The responsibility for taking any corrective action and release of product held pending any investigation must be clearly assigned. Records of corrective actions taken must be maintained.

See also: **Hazard Appraisal (HACCP)**: The Overall Concept.

Further Reading

Department of Health (1993) *Assured Safe Catering.* London: HMSO.

Dillon M and Griffith C (1996) *How to HACCP: An Illustrated Guide*, 2nd edn. Grimsby: MD Associates.

ILSI Europe (1997) *A Simple Guide to Understanding and Applying the Hazard Analysis Critical Control Point Concept*, 2nd edn. ILSI Concise Monograph Series. Washington: ILSI Press.

Joint FAO/WHO Food Standards Programme, Codex Alimentarius Commission (1997) *Food Hygiene Basic Texts: Hazard Analysis and Critical Control Point (HACCP) System and Guidelines for its Application.* Annex to CAC/RCP 1–1969, Rev. 3 (1997), p. 33–45. Rome: FAO.

Leaper S (ed.) (1997) *HACCP: A Practical Guide*, 2nd edn. Chipping Campden: Campden & Chorleywood Food Research Association.

Mortimore S and Wallace C (1998) *HACCP: A Practical Approach.* 2nd edn. Maryland: Aspen Publications.

MacDonald DJ and Engel DM (1996) *A Guide to HACCP – Hazard Analysis for Small Businesses.* Doncaster: Highfield Publications.

Establishment of Performance Criteria

T Mahmutoğlu, TATKO TAS, Gayrettepe, Istanbul, Turkey

F Bozoğlu, Department of Food Engineering, Middle East Technical University, Ankara, Turkey

Implementation Steps of HACCP

Developing a hazard analysis critical control point (HACCP) plan involves the following steps:

1. Preliminary steps: general information, product description, description of the methods of distribution and storage, identification of the intended use and the expected consumer, and development of a flow diagram.
2. Hazard analysis worksheet: the preparation of this worksheet requires identification of the potential species-related hazards, identification of the potential process-related hazards, understanding the potential hazards and determination of their significance, and identification of the critical control points (CCPs).
3. HACCP plan form: setting the critical limits, estab-

lishing monitoring procedures (what, how, when and who), establishing corrective action procedures, establishing a record-keeping system, and establishing verification procedures.

A crucial step in establishing and maintaining an HACCP system is the setting up of criteria for the parameters that will be monitored. According to Codex Alinorm 93/13A, critical limits are criteria that separate acceptability from unacceptability. Critical limits must be met to ensure that the identified hazards are prevented, eliminated or reduced to acceptable levels. Each control measure has one or more associated critical limits. Thus, some critical limits can be set to reflect national regulatory levels: these may be in the form of action levels, or tolerances for contaminants such as pesticide residues, natural toxins and other contaminants. Established critical limits need to be validated and also verified.

A functioning HACCP system should require little end-product sampling because appropriate safeguards are inherent to the process. Therefore, rather than relying on end-product testing, firms need to conduct frequent reviews of their HACCP records, and ensure that appropriate risk management decisions and product dispositions are made when process deviations occur. Any consumer complaint should be reviewed to determine if the complaint relates to the performance of the HACCP plan or reveals the existence of unidentified CCPs. Although the absence of consumer complaints does not in itself verify the adequacy of a HACCP system, safety-related problems are a guide to the performance of the system.

Microbial Criteria

For every CCP that addresses a microbial hazard, there is an explicit or implicit microbial criterion. This can either be an absolute criterion (e.g. the numbers of indicator organisms) or a performance criterion (e.g. the concentration of a microbial metabolite such as lactic acid). Both types of criteria can be used with respect to a single CCP, a combination of unit operations, or even all the unit operations making up the food production process (i.e. end-product criteria). However, microbiological testing is of limited value for monitoring CCPs, since the time required to obtain results usually does not permit action to be taken while a food is being processed, although there are two exceptions to this:

1. Microbial monitoring of raw materials before their use in processing (e.g. dried milk before its use in chocolate production): in this case the microbiological quality of an ingredient is a CCP and can be assessed by analysis before its use.

2. Microbiological monitoring of finished products prior to their release, where the products are intended for highly sensitive consumers, such as babies.

Microbial analysis is otherwise generally used for the verification and validation of HACCP, rather than for routine monitoring.

A microbial criterion should include the following:

- a statement describing the food or food ingredient
- a statement of the contaminant of concern, i.e. the microorganism or group of microorganisms and/or its toxin
- the analytical method to be used for the detection, enumeration or quantification of the contaminant
- the sampling plan
- the appropriate microbiological limits.

Microbial criteria may be viewed as guidelines for the identification of the minimum microbiological profile that a product should meet. To be meaningful, they should be based on a formal risk analysis relating to the frequency and level of pathogens that are tolerable.

To establish the target the hazard of concern must be identified and the acceptable risk of this hazard determined. A risk profile of the hazard is required as it relates to the product. The performance requirements for each critical processing step for the product should be established, for example, the reduction in bacterial numbers resulting from a heat treatment. The relation between the established microbial criterion, such as the numbers of thermoduric bacteria in a pasteurized food, and the control parameters for a critical operation, such as temperature and duration of heat treatment, should be validated.

Foods or ingredients can be classified into risk categories according to the hazards they possess. Each risk category has a recommended sampling plan to test for the pathogens. Food systems can be classified on the basis of potential health hazards as (i) those containing a 'sensitive' ingredient; (ii) the manufacturing process does not contain a killing step; (iii) those having a potential for microbial abuse during distribution or consumption. Further, the hazard categories are as follows, in order of decreasing risk:

I A special category for non-sterile products designed and intended for consumption by infants, the aged or the infirm.
II Food products subject to all three general hazard characteristics (explained above).
III Food products subject to two general hazard characteristics.
IV Food products subject to one general hazard characteristic.

V Food products subject to none of the general hazard characteristics.

The agents of concern must be identified when developing the HACCP analysis. To do this, the food item has to be considered in terms of ingredients: for example, for eggs the major potential agent of concern is *Salmonella*, and for starchy foods the potential agents are *B. cereus*, pesticides and *Staphylococcus aureus*.

The importance of microbiological criteria to HACCP arises from the fact that while the aim is to control the pathogenic microorganisms, most critical limits for process control refer to chemical or physical attributes, because these can be quick and easy to measure. Among these attributes, time–temperature relations are frequently used to indicate the inactivation of a specific target microorganism; for example, the canning of low-acid foods requires that the product be heated for a specified time at a specified temperature. In fact, this time–temperature limit is a microbiological performance criterion – a reduction in the levels of *Clostridium botulinum* spores by a factor of 10^{12}. This performance criterion, in turn, reflects risk management decisions concerning the known severity of the hazard and 'tolerated' risk associated with the potential for botulism outbreaks.

Besides temperature–time relations, other criteria that may be used include cooling rate, pH, indicator chemicals such as lactic acid, ammonia, trimethylamine, total volatile nitrogen, histamine or pesticide residual levels, organoleptic properties such as flavour, odour, visual appearance, texture, water activity (a_w), packaging material properties and package gas composition.

It is impractical and unnecessary to develop microbiological criteria for every food. Instead, criteria should be developed only for potentially dangerous foods where the danger can be reduced or eliminated by the imposition of microbiological criteria. Experience has indicated that this is the most difficult part of developing an HACCP plan, because it requires the developer to make concrete decisions about the performance of the system, balancing the cost associated with unnecessary stringency against the risk of inadequate control of a microbiological concern. However, the entire purpose of a critical limit is to set a pass/fail criterion that provides the basis for decisions about the operation of a critical control point.

Microorganisms suitable as components of microbial criteria in standards, guidelines or specifications are either pathogens or indicator organisms. Pathogens selected for this purpose are likely to be those found in a food or ingredient, which thereby becomes a potential vehicle for transmission of the organism or its toxin to consumers. Indicator organisms are those whose presence in a food indicates (1) that a pathogen or its toxin of concern may be present, (2) that faulty practices occurred that may adversely affect safety or shelf life of the product, or (3) that the food is unsuited for an intended use. Their presence may be a sign of faecal contamination or post-heat processing contamination.

When selecting pathogens (e.g. *Listeria*, *Salmonella*, *Clostridium*, *Vibrio*), the severity of the hazard should be considered, together with the methods available for their detection and/or enumeration. Indicator organisms for potential faecal contamination or for the possible presence of pathogens include Staphylococcus spp., *Escherichia coli*, other coliforms, enterococci and *Pseudomonas aeruginosa*. Indicators of post-heat processing contamination include coliform bacteria and members of the Enterobacteriaceae such as *Escherichia*, *Salmonella*, *Shigella*, *Enterobacter*, *Serratia*, *Proteus*, *Yersinia* and *Erwinia*.

While most legislated standards are for 'finished foods' (i.e. end products ready for consumption), microbial standards for HACCP monitoring must be established for the different stages of the preparation process. To do this, after establishing the CCPs, the microbiology of the food at different stages of processing must be determined, including product samples and also swabs from the equipment surfaces. Through these studies, the numbers and types of organisms that characterize the flora of a food produced under a given set of conditions can be identified and thus provide a basis for the establishment of a microbiological criterion. Variations in process conditions should be considered, and correlated with the organoleptic quality of the food. An extensive literature survey should also be conducted.

The sampling plan and decision criteria are essential components of a microbiological criterion. They should be based on sound statistical concepts. The choice between sampling portions of foods or rubbing swabs over surfaces (as well as other methods of examining foods) depends on the expected type and quantity of organisms and on their expected location. If serotyping or other definitive typing is done, sources of contamination and the avenues the contaminants followed can sometimes be determined by sampling raw and cooked foods, taking specimens from persons who handle the foods, and swabbing surfaces that come in contact with foods.

The following factors should be considered for the establishment and application of microbial criteria:

- evidence of hazard to health

- microbiology of the raw materials
- effect of processing on the microbiology of the food
- likelihood and consequences of microbiological contamination and/or growth during subsequent handling and storage
- category of consumer at risk
- cost/benefit ratio associated with the application of the criterion.

The terms 'standard', 'guideline' and 'specification' are widely used to describe microbiological criteria for foods.

A *microbiological standard* is a microbial criterion that is a part of a law, ordinance or administrative regulation. A standard is a mandatory criterion, and failure to meet it may result in regulatory action. A Codex microbial criterion is mandatory only when it is contained in a *Codex Alimentarius* standard that refers to an end-product specification. A Codex microbiological standard should contain limits only for pathogenic microorganisms of public health significance in the food concerned, although limits for non-pathogenic microorganisms may be necessary. Microbiological standards may be useful when epidemiological evidence indicates that a food is frequently a vehicle of disease.

A *microbiological guideline* is a criterion that often is used by the food industry or a regulatory agency when monitoring a manufacturing process. Guidelines are helpful in assessing whether microbiological conditions prevailing at CCPs or in the finished product are within the normal range; hence they are used to assess processing efficiency at CCPs and conformity with good manufacturing practices (GMP). They are advisory. When based on GMP, statistically valid data or appropriate experience, such guidelines permit the processor and others to assess the conditions under which certain foods have been processed and stored.

A *microbiological specification* is a microbial criterion that is used as a purchase requirement; conformance with it becomes a condition of purchase between the buyer and the vendor of a food or ingredient. It can be either mandatory or advisory. Microbiological specifications may be used to determine the acceptability of a raw material or finished product in a contractual agreement between two parties (buyer and vendor), and by governmental agencies to assess microbial acceptability of foods purchased.

There is concern in the food industry about some of the proposed applications of mandatory microbiological criteria. It is felt that microbiological standards should be used only for foods with potential hazards that can be eliminated or reduced by the imposition of a microbiological criterion. They should be established and implemented only when there is a need and when the criterion can be shown to be effective and practical. In other words, microbiological standards are only useful when a food is frequently a vehicle of food-borne disease, and microbiological standards should be employed only if they will eliminate or reduce food-borne disease. In addition, the food industry is strongly opposed to microbiological standards based on fixed numbers of non-pathogenic microorganisms. Standards should be adopted only for pathogenic microorganism_, or in particular instances for other organisms when direct tests for pathogens are impractical.

Microbiological quality standards do not relate to safety, and consumers may not perceive any effect on quality despite high microbial counts; however, these standards have been justified on the basis that microbial levels are indicative of (1) the excellence of raw materials and ingredients used, (2) the degree of control during processing, and (3) the conditions of distribution and storage. The assumption is that the quality of the product falls as the numbers of microorganisms rise. However, this assumption is not always true. Also, low numbers of microorganisms do not always reflect good handling and manufacturing practices; the numbers may be low simply because of a lethal step in the process.

According to the Food and Drug Administration (FDA), performance standards are defined in terms of what is to be achieved by compliance with a given regulatory requirement, and represent a shift in focus from 'command and control' regulations because the former specify the ends to be achieved (safe food products), rather than the means of achieving those ends.

Discussions on the tolerance level of risk achieved by requiring interventions that have been validated to achieve a cumulative 'x log reduction' in the target pathogen (the most resistant microorganism of public health significance) or a definite reduction in yearly risk of illness are proceeding. The value of x would depend on the resistance of the target microorganism in the product concerned. The stringency of the performance standard also depends on the target consumer groups: for example, if the consumers are babies, a stringent criterion is expected. The question to be answered is: what is the acceptable risk standard for the related pathogen or hazard in question? In the absence of known specific pathogen-product specifications, some microorganisms (e.g. *L. monocytogenes*, *Salmonella* or *E. coli* O152:H7) could be suggested as the target, based on the number of known outbreaks of disease associated with the microorganism in the food under consideration. In controlling the target organism, other pathogenic organisms are likely to be controlled also. Note that the choice of target microorganism also depends on the chosen process for the x log reduction of the organism: it may be a heat process such as pasteurization, or it may be acidification or a surface treatment. After setting the value of x, the processor must ensure that such a reduction in the target pathogen is occurring. Further, HACCP and safety performance should form the general conceptual framework needed to ensure the safety of foods, and control measures should be based on a thorough hazard analysis.

Industry quality control and assurance departments commonly establish microbiological limits, often based on many years of experience, that should be achievable in foods at CCPs or in the finished product if GMP are observed. Results that exceed these limits serve to signal some divergence from accepted GMP and may suggest remedial measures. Those guidelines may vary from company to company, even for the same product.

The procedures needed for the validation and verification of critical limits will vary according to the nature and specificity of the critical limit under consideration. As a minimum, each critical limit should be based on scientifically established performance criteria that afford an appropriate level of consumer protection.

Validation of critical limits should consist of two activities: (1) confirmation of the appropriateness of the performance criteria and the adequacy of the critical limits for meeting the performance criteria; and (2) assessment of the capability of the system to deliver the product that meets the critical limits or process parameters. Periodic revalidation should also be performed to ensure that the system has not changed such that the critical limits previously identified are no longer adequate. Verification consists of checking compliance with the elements of the HACCP plan, including compliance with prescribed critical limits. The choice of target microorganism also needs to be verified. Verification can include sampling.

Microbial Sampling

There are two prime reasons for microbiological sampling. The first is to enable a decision to be reached on the suitability of a food or ingredient for its intended purpose. The two-class and three-class attributes sampling plans of the International Commission on Microbiological Specifications for Foods (ICMSF) are appropriate for this purpose. The second reason for microbiological sampling is to monitor performance relative to accepted GMP. Related to HACCP, micro-

biological sampling is required for verification and validation purposes.

The purpose of inspection and analysis of food, including microbiological testing, is to obtain information upon which to base a decision to accept or reject the food. The type of plan chosen for this purpose is termed an 'acceptance sampling plan'. The product type, its microbiological history and its intended use will influence the selection of the sampling plan. The sampling plan and the decision criteria should be based on sound statistical concepts.

Acceptance or rejection of a lot can theoretically be based upon 'attributes' or measurement of a variable. When attributes data are used, the decision is based on the number of sample units that are 'positive', i.e. giving results above or below the level specified. Measurements typically involve some continuous variable such as concentration, e.g. the amount in parts per billion of some chemical residue in a sample unit of food. When measurement data are used the decision to accept or reject a lot is commonly based on a summary statistic such as the average (e.g. the average total aerobic count).

Measurement data can be converted to attributes data by referring to the number of sample units above and below a critical level. Transforming a piece of quantitative information, such as the number of colony forming units (cfu) per gram, into an attributive criterion, such as the absence or presence of *Salmonella* in each sample avoids the problem of having to determine the variability.

Attributes sampling is the only way to test ingredients with an unknown history or to determine if they are acceptable for a particular product. Attributes sampling is also used to test held lots where other CCPs may have failed. Although attributes sampling is used for chemical and physical testing, the least understood testing of this type is for hazardous microorganisms.

A sample unit may be regarded as 'defective' if it contains any of certain dangerous microorganisms, or more than some chosen number of other microorganisms. The symbol m (limit value) is used to represent the dividing line separating the sample units into two classes:

- defective (values above m)
- acceptable (values equal to or less than m).

In the case of a dangerous microorganism, such as *Salmonella*, m may be zero. This is called a 'two-class attributes plan'. Where the presence of some microorganisms (e.g. indicator organisms) can be tolerated, it is possible to recognize three classes of quality, where any single sample unit may be:

- wholly acceptable

- marginally acceptable
- defective.

The symbol m is then used to separate acceptable from marginally acceptable quality, while M is used to separate marginally acceptable quality from defective quality. This is a 'three-class attributes plan'. Counts between m and M are undesirable but a few such counts can be accepted.

Besides m and M, two more numbers must be stated in sampling plans: these are n, the number of sample units, and c, the maximum allowable number of sample units yielding unsatisfactory test results.

In a two-class attributes plan, suppose $n = 10$, $c = 0$ and $m = 0$ cfu g^{-1}; 'acceptable' means that all ten measured values are below the m value of 0 cfu g^{-1}. In a three-class attributes plan, suppose $n = 10$, $c = 2$ and $m = 5000$ cfu g^{-1} and $M = 20\,000$ cfu g^{-1}. Here the interpretation is that 'acceptable' means all ten measured values are below the m value of 5000 cfu g^{-1}; 'marginal' means up to two values may lie between m and M, but none exceeds M; 'unacceptable' means one value exceeds M or three or more values exceed m but are less than M.

The ICMSF sampling plans involve either two-class or three-class attributes. Lot acceptance sampling is an established method of assessing the microbiological quality and safety of batches or consignments of food, but the choice between three-class attributes plans and variables plans is not always clear. The three-class plan is easier to understand and apply than the variables plan.

Whenever possible, criteria should not be established at the lower limits of detection of the methods being employed, a concept recommended in relation to any analytical procedure. Instead, the criteria should allow trends to be followed and corrective measures to be taken before 'action levels' are reached. Furthermore, this allows the establishment of criteria that are designed around three-class sampling plans. It should be recognized that this ideal is not always feasible, particularly when dealing with infectious agents (e.g. *Salmonella*) with low minimum infectious doses. In those instances, two-class plans may be more appropriate.

The ICMSF categorized microbiological hazards into 15 cases, according to two factors: severity, and whether the hazard will be reduced, unchanged or increased during the normal conditions of handling between sampling and consumption. The fifteen cases relate to the nature of the concern (e.g. shelf life, low indirect health hazard), the food product (e.g. fresh fish, frozen foods), the bacterial test (e.g. standard plate count) and anticipated conditions of treatment (e.g. subsequent cooking).

The obvious limitation of attribute sampling procedures is that a large number of samples must be analysed. This is time-consuming, expensive and frequently impractical, particularly with microbiological testing. The foregoing recommendations for sampling foods can be used at ports of entry or for foods in domestic commerce. The ICMSF does not recommend the use of these criteria for routine analysis as the method of choice for ensuring safety. Microbiological tests are notoriously difficult to conduct and increased sampling is unrealistic. In fact ICMSF plans involve more sampling than is currently the case in most microbiological testing situations.

No sampling plan can ensure the absence of a pathogen in food. Testing foods at ports of entry or elsewhere in the food chain cannot guarantee food safety. Instead, the ICMSF advocates the use of HACCP from production or harvesting to the consumption of foods. Food processors can use the recommended sampling plans to verify that the HACCP plan is working correctly to assure the control of microorganisms in question. The ICMSF also recommends that highly susceptible individuals be provided with guidelines for the selection and safe handling of foods.

A far more serious consideration is that attributes sampling may not detect a serious contamination problem, particularly by microorganisms, even though a seemingly large number of samples is drawn from a lot. This situation may occur when the level of defective sample units in a lot is under 10%. There may be a very high probability of accepting a lot at a defect rate of 0.1%, 1.0%, 5.0% and 10.0%. This situation improves somewhat for three-class plans. Sometimes, very large numbers of sample units from a lot are required before even one incubated sample will show spoilage present at low levels. In other words, intensive sampling and testing of sensitive ingredients may not prevent hazardous microorganisms contaminating a blended finished product.

Sources of Information

Current sources of performance and validation criteria are experts (consultation), government legislation, review of the literature, epidemiological data on food-borne diseases, suppliers' records, regulatory guidelines, monitoring contaminants in the food manufacturing environment and in the product, data bases, challenge tests and predictive microbiology, and scientific studies based on risk assessment, risk management and agreement on food safety objectives.

Challenge Tests

After the potentially hazardous organisms have been identified and information about their presence in raw food materials has been obtained, then the effects of processing and storage can be tested by microbiological challenge testing (MCT), storage tests, and even by mathematical models for estimation of the number of organisms expected to be present in the product at the times of factory exit, consumption, etc.

Microbiological challenge testing is an established technique within the food industry. It aims to simulate what can happen to a product during processing, distribution and subsequent handling, following inoculation with relevant microorganisms in appropriate numbers. The test product has to be processed correctly and then held under a range of controlled conditions that relate to those used by consumers. It is used to assess risk in a food product.

There are several areas of application of MCT, such as determination of product safety, establishment of shelf life, and formulating products in terms of intrinsic control factors such as pH and a_w. These tests are perceived as a source of experimental documentation of a product's safety.

Challenge tests are carried out if the microorganism of interest is suspected to be present in low numbers or only incidentally, and should be applied if knowledge about the characteristics of a potentially hazardous microorganism is incomplete. The tests involve:

- an appropriate experimental design
- an inoculation procedure
- a test procedure
- interpretation of results.

Times and temperatures used in challenge tests must be selected to reflect local conditions during distribution and retailing, and the conditions of consumer use. Shelf life tests are similarly conducted by shipping samples of the product out to retailers and testing periodically for deterioration. Although these tests provide valuable information, they are time-consuming and costly. One way around this limitation is the use of 'accelerated testing'. Assumptions concerning environmental conditions can be far from reality and can lead to significant errors in estimating shelf life. There must be a balance between the storage temperatures expected and an appropriate level of secondary barriers used to prevent bacterial growth.

When deciding whether microbiological challenge testing should be applied, the first step is to establish the potentially hazardous organisms associated with a particular food product. After producing a list of food-borne disease bacteria, it is necessary to determine whether each microorganism is likely to be present in the raw materials. Only organisms that have never been found can be deleted. Of the remaining organisms, it must be established whether or not they are completely destroyed during processing. Any that are can also be removed from the list.

During (and even after) processing recontamination may occur, and any pathogenic organisms responsible must be included in the list. The next point to consider is whether or not the listed organisms have ever caused a food-borne disease through a similar or related food product. If no case has been reported, the organism can be deleted. The microorganisms remaining on the list are now separated into two groups, according to whether they are infectious or form toxins. All infectious bacteria present are regarded as potentially hazardous. For toxin-forming bacteria, if growth must occur before the toxin is produced the organism is removed from the list; if not, the organism must be regarded as potentially hazardous.

Data Bases

Data bases are being developed that enable the prediction of the survival and growth of selected bacteria in culture media as affected by factors such as pH, water activity, salt content and temperature. Combinations of these and other growth-limiting factors are increasingly being sought to inhibit the growth of microorganisms in food products. The use of data bases is foreseen as a means of allowing access to the vast amount of information necessary to conduct an in-depth risk assessment. Data bases designed to have predictive abilities will provide support for making decisions; more importantly, they should offer aid in formulating more precise questions regarding hazard analysis and risk assessment. The topics of data bases include:

- the prevalence and levels of pathogens on raw materials
- the products associated with food-borne disease
- factors limiting microbial growth and survival
- processing failures that have resulted in food-borne disease related to equipment, sanitation, recontamination, etc.
- the ability of regional distribution and marketing systems to maintain refrigerated storage
- the likelihood that regional consumer practices may lead to thermal abuse of refrigerated products
- cooking before consumption
- information concerning the epidemiological aspects of risk analysis, e.g., minimal infective dose, high-risk consumer groups, etc.

Data bases can also be used to identify target organisms. Microorganisms that can cause spoilage or are potentially hazardous in a product are deduced from knowledge of the ingredients and the process parameters. Data are available on pathogenic and spoilage microorganisms that have posed problems in particular foods. This information is obtained from documented instances of food poisoning and spoilage cases.

Another data set is necessary with the parameters for the calculation of microbial growth and safety. For each relevant microorganism the following are listed:

- minimum and maximum temperature for growth
- minimum and maximum a_w for growth
- minimum and maximum pH for growth
- D value, z value
- growth rate
- parameters needed to calculate the temperature dependence of the maximum specific growth rate and lag phase
- the asymptotic level of the growth rate
- a parameter that determines the transition velocity from growth to inactivation.

These data are not yet available for all relevant microorganisms. In addition to the properties of the food ingredients, the process parameters are important for the choice of target organisms. The complete sequence of unit operations is taken into account. Critical temperatures during heating are chosen based on inactivation data for important food-borne pathogens in cook-chill products with a limited shelf life.

Time–temperature relations can be used as performance criteria for particular processes, especially those of pasteurization and sterilization, if the relevant data are available. In order to do this, the following must be well established for the food product in question:

- suitability of the food product for microbial growth (a_w, pH, etc.)
- the processing steps (time–temperature of the process, hygiene, handling practices)
- storage and distribution conditions
- the intended use.

Correspondingly, the growth conditions for microorganisms (a_w, pH, salt content, temperature, oxygen requirement) should be considered and compared with the conditions of the food product. Data for such purposes are well documented for most foods.

Food Safety Objectives

A food safety standard could be a food safety objective (FSO), which is a statement of the maximum level of a hazard considered acceptable for consumer protection. An FSO should be based on reality and should be established by interactive communication with the parties involved with the product under

consideration, such as consumers, producers, importers/exporters and government representatives. Here, the definition of acceptability of a particular food safety hazard is important; in other words, you need to have a target – an FSO. The targets chosen should be converted into measurable targets on limits. Usually an FSO is a statement of the frequency or concentration of a microbial hazard in a food appropriate for consumer protection, and FSOs provide reference for an effective validation of the HACCP plan and for verification of effectiveness. Food safety objectives have three components: type of food; hazards of concern; and frequency or maximum concentration of the microbiological hazards in the food. The FSOs can be used to establish fresh produce criteria, raw material specifications, process and preparation parameters, critical limits, shelf life conditions and label instructions.

Prerequisite Programmes

Prerequisite programmes may be a basis for the establishment of criteria. According to the FDA, the six prerequisite programmes are:

1. Premises: outside property, building, sanitary facilities and water quality programme.
2. Receiving and storing: receiving of raw materials, ingredients and packaging materials, storage.
3. Equipment performance and maintenance: general equipment design, equipment installation, equipment maintenance.
4. Personnel training: manufacturing controls, hygienic practices, controlled access.
5. Sanitation: sanitation programme, pest control programme.
6. Health and safety recalls: recall system, recall initiation.

Prerequisite programmes need to be effectively monitored and controlled before any attempt to put an HACCP plan in place. They are universal steps or procedures that control the operational conditions within a food establishment to create environmental conditions that are favourable to the production of safe food. Most prerequisite programmes are applicable at time of plant design, rather than used for routine monitoring. However, programmes such as the one related to sanitation may have performance criteria established for routine monitoring. Sanitation programmes must be developed for equipment, personnel, overhead structures, floors, walls, ceilings, drains, lighting devices, refrigeration units and anything else affecting the safety of the food. However, prerequisite programmes and GMP alone are usually not sufficient to ensure a 5 log reduction of target pathogens, but a 3 log reduction may be achieved; it should be validated. Sometimes it may be useful to use a combination of prerequisite programmes (e.g. sanitation of equipment), GMP (e.g. design of the equipment suitable for sanitation) and CCP monitoring (e.g. pasteurization) in HACCP and thus achieve a 5 log reduction. The cumulative 5 log reduction could be measured from the point of the processor's initial treatment of the product. If pathogens are significantly reduced on the raw product through washing or other treatment, and the product is processed under an adequate HACCP programme, the hazard from the presence of pathogens may be controlled.

Suggested operating parameters for sanitation are related to the temperature of application, time of exposure of the sanitizer and the surface, the concentration of the sanitizer and the frequency of cleaning and sanitation. Depending on the contaminant to be removed from the surface, the type of sanitizer (e.g. chlorine or iodophore-based) is also critical for the expected performance. Those variables can be assigned to be criteria for HACCP system. In other words, they indicate the adequacy of the sanitation process for the intended purpose. Each of these variables can be varied independently to adjust the cleaning operation for a particular problem or plant operating practice. The conditions generally are selected so as to obtain the best cleaning at least cost. Some imply that high cleaning temperatures encourage the burning-on of residual soils, while others argue that higher temperatures yield better cleaning. For any food soil containing fat, the minimum effective temperature will be about 3°C higher than the melting point of the fat component. The maximum temperature will depend upon the temperature at which the protein in the system is denatured. For example, 40°C is suggested for cold-wall milk storage tanks, 77°C for plate heat exchangers used for milk processing, and 71°C for milk pipelines. Normally, the longer the solution is in contact with the deposited soil under a given set of conditions, the more soil is removed. Increasing detergent concentration increases detergency up to a limit; higher concentrations have no further effect. Generally, to measure the efficiency of a sanitizer the Chambers test is used: this test requires that sanitizers produce a 99.999% kill of 75–125 million *Escherichia coli* and *Staphylococcus aureus* within 30 s of application at 20°C. The type of sanitizer, the use concentration, exposure time, optimum pH and application temperature should be validated.

See also: **Good Manufacturing Practice**. **Hazard Appraisal (HACCP)**: The Overall Concept; Critical Control Points; Involvement of Regulatory Bodies. **Quantitative Risk Analysis**. **Sampling Regimes & Statistical Evaluation of Microbiological Results**.

Further Reading

Abstracts (1998) *Second International Food Safety HACCP Conference*, June 1998, Nordwijk aan Zee, The Netherlands.

Aldridge JR and Dale BG (1994) Failure mode and effect analysis. In: Dale BD (ed.) *Managing Quality*. P. 451. London: Prentice Hall.

American Society for Quality Control (1987) *Quality Assurance for the Chemical and Process Industries*. Wisconsin: ASQC Quality Press.

Archer DL (1990) The need for flexibility in HACCP. *Food Technology* 5: 174–178.

Baker DA (1995) Application of modeling in HACCP plan development. *International Journal of Food Microbiology* 25: 251–261.

Bauman H (1990) HACCP: concept, development, and application. *Food Technology* 5: 156–158.

Buchanan RL (1995) The role of microbiological criteria and risk assessment in HACCP. *Food Microbiology* 12: 421–424.

Codex Alimentarius Commission: Joint FAO/WHO Food Standards Programme (1993) *Report of the Twenty-sixth Session of the Codex Committee on Food Hygiene*, Alinorm 93/13A. Washington: FAO/WHO.

Giese JH (1991) Sanitation: the key to food safety and public health. *Food Technology* 12: 74–80.

Hamada M, Mackay RJ and Whitney JB (1993) Continuous process improvement. *Journal of Quality Technology* 25(2): 77–84.

ICMSF (1982) *Sampling for Microbiological Analysis: Principles and Specific Applications*. Book 2. Toronto: University of Toronto Press.

ICMSF (1982) Choice of sampling plan and criteria for *Listeria monocytogenes*. *International Journal of Food Microbiology* 22: 89–96.

National Research Council (1985) *An Evaluation of the Role of Microbiological Criteria for Foods and Food Ingredients*. Washington: National Academy Press.

Notermans S and Jouve JL (1995) Quantitative risk analysis and HACCP: some remarks. *Food Microbiology* 12: 425–429.

Notermans S and Veld P (1994) Microbiological challenge testing for ensuring safety of food products. *International Journal of Food Microbiology* 24: 33–39.

Peterson AC and Gunnerson RE (1974) Microbiological critical control points. *Food Technology* 9: 37–44.

Schellekens M, Martens T, Roberts TA et al (1994) Computer aided microbial safety design of food processes. *International Journal of Food Microbiology* 24: 1–9.

Stevenson S (1990) Implementing HACCP in the food industry. *Food Technology* 5: 179–180.

Weingold SE, Guzewich JJ and Fudala JK (1994) Use of foodborne disease data for HACCP risk assessment. *Journal of Food Protection* 57(9): 820–830.

Involvement of Regulatory Bodies

O Peter Snyder, Hospitality Institute of Technology and Management, St Paul, USA

Vijay K Juneja, USDA, Agricultural Research Service, Eastern Regional Research Center, Wyndmoor, USA

Regulatory bodies have a key role to play in the process of hazard appraisal. First, let us review the definition of a hazard. A hazard is a biological, chemical or physical agent that is reasonably likely to cause illness or injury in the absence of its control. The key words to this definition are 'reasonably likely'. Setting standards is not the role of industry. The regulatory body sets standards for what is reasonably likely to cause illness or injury. Does this mean 1 in 20 000 people become ill and 1 in 100 000 die?

The answers to such questions are complex and must be decided by a governmental body. The governmental body must take into account the overall cost to the nation of a standard. This includes the cost to industry for imposing certain levels of control, and the benefits to the consumer of one level of hazard control vs. a second level of control. Therefore, the final decision is based on a balance of each of these stakeholders' – nation, industry and consumer – needs and wants. Industry cannot have safety specifications that are so stringent that it is economically impossible to grow and sell food that meets those standards. The consumer cannot be expected to test food for toxins and pathogens before consuming the food.

Hazards in Food (Table 1)

Pathogenic Microorganisms and their Toxins

These hazards include pathogenic bacteria, yeasts and moulds, viruses and parasites. Bacteria have two basic forms: vegetative cells, which are controlled by pasteurization and usual cooking procedures; and spores, which are not controlled by pasteurization and most cooking procedures, but can be controlled by commercial sterilization. Not all food-borne bacterial pathogens (e.g. *Salmonella*, *Shigella*) can form spores. Such pathogenic bacteria cause illness due to consumption of food contaminated with bacterial cells. The ingested cells penetrate the epithelium and multiply within the epithelial cells of intestine, causing an inflammatory response. The organisms may spread laterally to adjacent cells, producing necrosis, leading to ulcers and there may be septicemia or systemic infection. Other pathogenic bacteria (e.g. *Staphylococcus aureus*, *Clostridium botulinum*) cause illness when food containing toxins produced by the multiplication of these pathogens in the food is ingested.

Most yeasts are non-pathogenic, with the exception of *Candida albicans* (a cause of thrush in infants, elderly people and other individuals with compromised immune systems). Some moulds produce compounds in food that cause allergic reactions in some people. Growth of the mould, *Aspergillus flavus*, in grain produces deadly aflatoxins. Viruses (such as hepatitis A virus and Norwalk virus) may be transmitted through food and water, and in the case of Norwalk virus, air. Parasites (*Cryptosporidium parvum*, *Giardia lamblia*, *Toxoplasma gondii* and *Trichinella spiralis*) are transmitted through food and water and are causes of food-borne illness and disease in the US and other parts of the world.

Pests such as birds, insects and rodents are a nuisance as well as potential carriers of disease and illness. In areas without adequate sewage control, flies are a major source of *Shigella* contamination. Pests must be controlled in any facility through cleanliness, both inside and outside. Debris should never be allowed to accumulate and provide a nesting and/or breeding area for pests. Facilities must have screens and tight-fitting doors and windows to prevent their entry.

It is important to realize that pathogens from pests are a hazard in the retail system. Contamination of food by pests usually occurs during growing and harvesting. Consequently, the retail system must provide control. Controls include washing the raw fruits and vegetables to remove pathogenic microorganisms, chemicals and filth, and cooking raw meat and poultry to temperatures that are adequate to destroy pathogenic microorganisms.

Marine Animals as Sources of Toxic Compounds

Fish and shellfish products may contain some of the most potent toxins known. These toxins are unaffected by cooking, and no antidotes or antitoxins exist to reduce the toxicity of some of these toxins. Poisonings through eating toxic fish and shellfish are significant causes of human illness. Outbreaks are usually due to three types of poisoning: ciguatera poisoning, histamine poisoning and paralytic shellfish poisoning. The best controls are to obtain fish and shellfish that are certified by a supplier with a hazard analysis and critical control point (HACCP) programme to have been taken from safe waters and then

Table 1 Microbiological hazards

Pathogenic microorganisms and their toxins
  Bacteria
    Vegetative cells (Aeromonas hydrophila; Campylobacter jejuni; pathogenic Escherichia coli; Salmonella spp.; Shigella spp.;
    Staphylococcus aureus; Streptococcus, groups A and B; Vibrio spp.; Yersinia enterocolitica)
    Vegetative Cells and Spores (Bacillus cereus; Clostridium botulinum; C. perfringens)
  Moulds (growth of Aspergillus flavus in cereals, legumes, nuts; produces aflatoxin)
  Yeasts (most yeasts are non-pathogenic; however, the presence of Candida albicans in food and beverages is a concern to
    immune-compromised individuals, the elderly and infants)
  Viruses (hepatitis A virus, Norwalk virus, rotaviruses)
  Parasites
    Protozoa (*Cryptospordium parvum; Cyclospora cayetanensis; Entamoeba histolytica; Giardia lamblia*)
    Helminths (*Anisakis; Taenia* spp.; *Toxoplasma gondii; Trichinella spiralis*)
  Pests (as carriers of pathogens)
    Filth from pests and unwanted animal parts or excreta: rodents, insects, birds

Marine animals as sources of toxic compounds
  Fish
    Histamine poisoning from improperly stored scombroid fish, including tuna, snapper, grouper, amberjack and mahi-mahi
    Ciguatera poisoning due to accumulation of toxin within the muscle of some fish (barracuda) when they consume the
      dinoflagellate *Gambierdiscus toxicus*
  Shellfish
    Various paralytic shellfish poisonings due to the ingestion of shellfish that have consumed dinoflagellates

to store these products under conditions that do not allow deterioration.

Physical Hazards (Table 2)

The definitive report on this subject is that of Hyman et al (1993). This report was a careful analysis of 10 923 complaints about food registered with the Food and Drug Administration (FDA) over a 12-month period. Of these complaints, 25% (2726 cases) involved foreign objects in food or drink, and 14% (387 cases) of these involved illness or injury associated with foreign objects ingested in beverages or food. Most of the injuries involved cuts or abrasions

Table 2 Physical hazards

Hard foreign objects
  Glass
  Wood
  Stones, sand and dirt
  Metal
  Packaging materials
  Bones
  Building materials
  Filth from insects, rodents and any other unwanted animal
    parts or excreta
  Personal effects (earrings, jewellery fragments)

Functional hazards
  Particle-size deviation
  Packaging defects
  Sabotage

Choking/food asphyxiation hazards
  Pieces of food

Thermal hazards
  Food so hot that it burns tissue

in the mouth and throat, damage to teeth or dental prostheses or gastrointestinal distress. The foreign objects were ordered from most to least common, as follows: glass, slime or scum, metal, plastic, stones/rocks, crystals/capsules, shells/pits, wood and paper. Foreign-object complaints involving injury and illness were associated most often with soft drinks, baby foods, bakery products, cocoa/chocolate products, fruits, cereals, vegetables and seafood. Injury from hard foreign objects can cause problems if the injury is sufficiently serious to require the attention of a doctor or dentist.

The Federal Food, Drug and Cosmetic (FD&C) Act does not mention the hazards of items such as metal shavings, bits of glass or metal bolts in food. When such items happen to be found in food or drink by regulatory officials, each incident is evaluated and, if deemed actionable under the FD&C Act, it is because the foreign object rendered the food unfit (FD&C Act 404(a) (3)) for human consumption due to the offensive 'mouth feel' of the adulterated food, or has rendered it injurious to health (FD&C Act 404(a) (4)).

Functional hazards occur when particle size deviates from that normally produced or supplied, when there are packaging defects (e.g. improper seals or holes in packaging materials) and when food is sabotaged by employees or consumers. These hazards can be controlled by careful inspection and surveillance techniques by both the food supplier and the consumer. Choking or food asphyxiation hazards include hot dogs, gum drops, nuts, taco chips, steak or any food that is not chewed sufficiently to be swallowed and hence becomes lodged in the pharynx, blocking the opening to the oesophagus and larynx. Even grapes given to very young infants or children have caused a number of fatalities. People may not chew food adequately before swallowing.

Old age, poor dentition and alcohol consumption also contribute to fatal food asphyxiation or choking on food. People should not give large pieces of food to children and the elderly, or to any individuals who are incapable of chewing the food before it is swallowed. It is also beneficial to be trained to perform the Heimlich manoeuvre and cardiopulmonary resuscitation, and to travel with other individuals who are trained in these procedures.

Thermal hazards include serving foods so hot that, when consumed or spilled on people, can cause severe burns or tissue injury. When hot foods are served, people must be warned to handle them properly, or foods should be served at temperatures that will not cause injury or harm. Examples of food causing injury to consumers include pizza, cream soups, chocolate, and coffee served at 190°F (87.8°C) or above; jelly Bismarcks over-heated in a microwave; and baby food or baby formula over-heated in a microwave.

Chemical Hazards (Table 3)

Chemical hazards in food include chemical compounds that, when consumed in sufficient quantities can inhibit absorption and/or destroy nutrients; are

Table 3 Chemical hazards

Poisonous substances
  Toxic plant material
  Intentional (GRAS: generally regarded as safe) food additives
  Chemicals created by the process
  Agricultural chemicals
  Antibiotic and other drug residues in meat, poultry and dairy
    products
  Unintentional additives
  Sabotage
  Equipment material leaching
  Packaging material leaching
  Industrial pollutants
  Heavy metals
Adverse food reactions (food sensitivity)
  Food allergies
  Metabolic disorder-based reactions
  Pharmacological food reactions
  Idiosyncratic reactions to food
  Anaphylactoid reactions
Nutrition
  Excessive addition of nutrients
  Nutritional deficiencies and/or inaccurate formulation of
    synthesized formulas
  Antinutritional factors
  Destruction and unnecessary loss of nutrients during
    processing and storage
  Inaccurate nutritional labelling

carcinogenic, mutagenic or teratogenic; or are toxic and can cause severe illness and possibly death because of their biological effect on the human body.

The following is a summary of poisonous substances that may be present in food.

Toxic plant material includes solanin in potatoes; haemagglutinins and protease inhibitors present in raw beans and peas; cyanogens in fruit kernels; and phytoalexins in sweet potatoes, celery and parsnips. Fortunately, many of these compounds can be eliminated by preparation methods. For example, solanin is eliminated when the green surface portion of potatoes is peeled or trimmed. Fruit seeds and fruit pits containing cyanogens are usually discarded. Haemagglutinins and protease inhibitors in raw plant seeds are altered by cooking with moist heat and thus become harmless.

Intentional food additives include GRAS (generally recognized as safe) compounds that may have inadvertently been added in excessive amounts. Examples include excessive addition of nitrites and nitrates in processed meats, excessive use of monosodium glutamate in prepared foods and excessive use of sulphites in permitted-use items such as dried fruits and wine.

Chemicals created by the process include those created when meat is broiled excessively over hot charcoal and chemical compounds created when fat or oil has been heated excessively or for a long time.

Agricultural chemicals include pesticides and herbicides. It has been noted that with the increased utilization of chemicals in agriculture and animal husbandry, the chances of chemical food contamination are growing throughout the world. Agricultural chemicals have a great impact on water systems. When it rains, these toxic substances are carried into rivers and lakes, affecting fish and aquatic plant life as well as water supplies.

Animal antibiotics and other drug residues are also a problem in terms of food-borne illness hazards. In 1990, the US Department of Agriculture (USDA) sampled 35 561 livestock for drug residues and found unacceptable residual levels in 132 samples. The USDA examined 9132 poultry samples and found violative residues in 12 samples. Drug residues in food can cause violent allergic reactions in sensitized people who consume these products.

Unintentional additives or accidental addition of toxic substances during food handling in the food service and food production operations can also occur. This type of hazard is often traced to storage of caustic or toxic cleaning and sanitizing chemicals in food storage containers.

Equipment material such as copper or lead from

pipes or soldering material can leach into food and water, causing heavy-metal poisoning.

Package material can leach as well. In the US, in the past, there was concern about the leaching of lead from the solder of can seams and polychlorinated biphenols from cardboard packages. These concerns have decreased in the US because these compounds have almost completely been eliminated from packaging systems. However, these types of packaging material may still exist in other regions of the world. There is also concern over the safety of certain plastics, especially those that may be used in the heating or reheating of foods in a microwave.

Heavy metals and radioactive isotopes from the industrial environment can also find their way into food, usually through water sources. An example of this is the level of mercury in fish taken from lakes and rivers.

Sometimes a poisonous substance in food can be controlled (diminished to a minimal risk) if the food is washed or is heated (cooked) sufficiently. However, the best strategy is for the food operator to keep harmful substances out of food by purchasing supplies produced under controlled or known growing, harvesting, processing and storage conditions.

Regarding adverse food reactions, about 1% of the population is allergic to compounds (usually certain proteins) found in food. Allergic reactions may be caused by many types of foods, including milk, eggs, fish, seafood (particularly shrimp), legumes (peanuts), tree nuts and wheat. Other foods, including citrus fruit, melons, bananas, tomatoes, corn, barley, rice and celery, can cause allergic reactions in a few sensitized individuals. Allergic reactions vary with each individual's sensitivity. Some allergic reactions are mild (e.g. watery eyes, nasal discharge, headaches).

However, some people are very sensitive. If they consume an offending food, life-threatening anaphylactic shock can occur within minutes. There must be emphasis on training staff to understand the serious nature of food allergies. Personnel must know, or be able to find, the complete list of all ingredients in food served to customers. Complete disclosure of ingredients used to prepare food should be available to hypersensitive individuals if they request this information. Personnel must recognize that even cross-contact of one food by another can pose a problem for highly sensitive individuals.

In the US, prepared foods must have an ingredient label. Labelling of food and disclosing recipe ingredients enables hypersensitive people to avoid foods with offending components. The use of kitchen chemicals such as monosodium glutamate, food colour (yellow dye no. 5) and aspartame in food items should be disclosed if customers request this information.

Nutrition – or lack of it – is a health problem. The health of people is particularly important in disease prevention and is partially dependent on a properly balanced, nutritious diet. If this is not provided to people, the quality of life and life expectancy are seriously diminished. Both macro- and micronutrients, in required amounts, are necessary to promote and maintain health in humans. In many developing countries, lack of an adequate supply of food contributes to malnutrition and a decreased health status of the general population, particularly infants and children. As a result, a large portion of the population is susceptible to infection and disease.

Nutritional hazards in food products include the following:

1. Nutritional deficiencies and/or inaccurate formulation of synthesized formulas can cause illness complications and possibly death in infants, elderly and critically ill or injured individuals.
2. Antinutritional factors such as phytates in green, leafy vegetables and trypsin-inhibitors in legumes and soybeans must be taken into account in food production and food preparation.
3. Destruction and unnecessary loss of nutrients occur when foods are processed for long periods of time and stored improperly. The nutrient most notably susceptible to destruction is ascorbic acid. Ascorbic acid loss in cooked vegetables is high if these foods are left on steam tables for long periods of time. The B vitamins are also unstable under various process conditions.

The nature of the food system is that there have been, and always will be, hazards in the environment that produces our food. It begins with the very dangerous *C. botulinum*, which is in the soil worldwide and contaminates virtually all food that we purchase. It includes rodents, birds and insects that inhabit farms and carry diseases from the wild into the farm environment. It also includes the water supply coming into this environment used to irrigate and wash fruits and vegetables and to water the animals.

Risk Decision-making

Deciding what should be the allowed consumer risk of illness or death involves three components: risk assessment, risk control specification and risk communication.

Risk Assessment

This includes:

1. Hazard identification: quantitative indication of the hazard that may be associated with the consumption of a particular food product.

2. Exposure assessment – the chance that a hazard will be in the food:
   a. Consumers, what they eat, the consumers' environment and how they handle food.
   b. The amount, frequency and source of the hazard and controls.
   c. Controls associated with the facilities, suppliers, food handlers and training.
3. Dose–response – the probability of consumers becoming ill at various dose intakes.
   a. The level that normally healthy people should consume to stay healthy and maintain the immune system.
   b. Levels that people can consume with no apparent harmful effect.
   c. Levels known to cause a probable illness/injury.
4. Risk characterization: severity and cost of the hazard, as identified from epidemiological evidence.

One of the groups that can do risk assessment, because it determines allowable illness, is the government. Risk assessment is the first step to specifying for the nation the level of risk control for which the food industry must be responsible.

Risk Control Specification (Risk Management)

The second step in the government's role in risk decision-making is the specification of risk controls. If a risk is unacceptable, it is up to the government to decide who should be responsible for controlling that risk and what level of control is necessary.

In the case of toxins and poisons in food, whereby consumers cannot control the risk, because the chemical hazard cannot be washed off or cooked out, the person producing the food must be responsible for keeping toxins and poisons to a tolerable level. What is a tolerable level must, again, be specified by the regulatory body. On the other hand, in the case of an animal such as beef infected with a bacterium such as *Escherichia coli* O157:H7 or chicken with *Salmonella enteritidis* and *Campylobacter jejuni*, the consumer can control the risk by cooking or pasteurizing the food. Producing bacterially contaminated meat, poultry and fish probably reduces the cost of production for producers, provided they are not sued. Again, it falls upon the government to decide that there should be a shared risk between the consumer and the producer. This assures lower-cost food if the consumer is given the responsibility of reducing vegetative pathogens to a tolerable level.

Whatever the outcome of this risk allocation, from the standpoint of management, this must be communicated by the government to the consumer and the producer.

Risk Communication

Risk communication is the third step. Risk is never zero. There is always the possibility of human error. The more the industry can eliminate the risk before a product is available to consumers, the less the risk, because the food supplier will always have better food safety control systems than the home food preparer.

However, there are groups of people who for religious, cultural or health reasons wish to choose high-risk foods such as raw meats, raw milk, forms of fruit and vegetables that contain high levels of plant toxins. It is virtually impossible for a government in a democratic society to ban specific forms of foods or prohibit the availability of the food to the consumer just because it could be hazardous to certain people. The choice is for the government to ensure that the customer understands the risk associated with the food.

There are many examples of human behaviour that puts the consumer at risk for food-borne disease and illness. These include not washing hands and food contact surfaces between handling raw poultry and ready-to-eat food and not using a thermometer to ensure that food is pasteurized. Once the government warns consumers of dangerous food-handling behaviour, there is little it can do if a consumer chooses risky food preparation procedures. In summary, the first step in a national hazard appraisal programme is for the government to determine the hazards and then the degree of control that is economically feasible for that nation without resulting in an unacceptable level of consumer illness and death.

The Government's Role in Specifying Safe Processes

A second major responsibility of the government is to specify safe minimum critical limits for processes to ensure safety. Normally, this takes the form of specifying minimal process parameters. In the case of high-acid foods, this means that, when a pasteurized food has a pH < 4.6, the government declares that food is safe from *C. botulinum* growth at room temperature. Another example is the safety of foods from *S. aureus* toxin production when the food has a water activity < 0.86. These national safe process standards can only be specified by the government.

All food processors should abide by the rules. Therefore, they must be universally applied throughout the industry as minimal critical safety limits. If each operator is required to find critical safety limits, each processor would have to do thousands of dollars worth of research to come up with these limits. If the government determines the critical safety limits for risk, they can be universally applied, because the characteristics of each pathogenic microorganism are the same worldwide. The same is true for safety limits for chemicals – toxins and poisons – and hard foreign objects. For example, if it is decided that a particle approximately 1 mm in diameter does not cause choking and does not break teeth, that is the critical limit for the size of hard foreign objects that can be applied universally, because it represents a tolerable risk to consumers. There would be no need for food processors to carry out redundant research once the safety limit is established.

What are some of the critical procedures and limits?

1. The critical personal hygiene control of correct fingertip washing to reduce transient bacteria to 100 000 to 1.
2. The critical control of cleaning food contact surfaces so that there is a tolerable level of filth on the surface after cleaning, such as 100 organisms per 50 cm.
3. Pasteurizing food to reduce *Salmonella* to < 1 organism per 100 g.
4. Cooling food to prevent the outgrowth of spores of *C. perfringens*.

Each of these processes is governed by universally accepted microorganisms. For example, pasteurization is governed by the inactivation by *Salmonella* spp. Cooling is governed by the control of *C. perfringens* spore outgrowth. Hand-washing is controlled by the removal of *Shigella dysenteriae* from fingertips. Only the regulatory bodies can establish the test validation organism associated with each process and the critical limit for that process.

Having the government establish standards does not preclude the development of new processes. The government should lead the initiative in a country for the development of new processes, such as high-pressure sterilization, irradiation, ozone disinfection and other new ways of controlling the hazards in the food. If the government does not take this role, every food manufacturing company in that country must individually do that research in order to develop a new process. This is economically unfeasible, since it could be done more cost effectively by the government agency working with the industry.

Government Regulation and Communication

The first step in writing the regulations is to develop a firm science base for our process standards in order to optimize commercial food process operations. Whatever is written in the regulations should first undergo the hazard analysis and critical control limit development process. It should be a requirement to have realistic process safety limit data, and limits

controlling the hazards should be founded in correct food process science.

The next major role of the government is to communicate the hazards and the critical process limits to the industry in the form of government regulations. The government is staffed with highly educated and experienced food scientists who work in conjunction with industry food science leaders to approve modifications in hazards, risks and safety limits rapidly.

Implementing and Enforcing Governmentally Developed Regulations

Historically, the preparation of food has been considered more of an art than a science. Therefore, throughout the world, entrepreneurs have been allowed to have licences to prepare food without first demonstrating that they understand the hazards associated with their processes and can establish effective control programmes. Assurance of food safety has relied on inspection after the operator is in business. While existing operators must implement HACCP, the most effective use of HACCP is immediately to require people who want to open a food-processing plant, restaurant, food market, food stand or vending operation – before they are given a licence – to show the government their HACCP plans or operations manual based on their hazards and processes.

In the evolution of the US National Advisory Committee on Microbiological Criteria for Foods' (NACMCF) specification of what constitutes an HACCP plan, we see that the NACMCF is slowly recognizing that an HACCP plan is simply a quality assurance processing plan for a food establishment. The NACMCF requirements fit into the 12 elements of an effective, company HACCP-based total quality management (TQM) plan. These elements are:

1. Company food safety policy.
2. Organization for HACCP-based TQM.
3. System and customer description.
4. Good manufacturing policies, procedures and standards.
5. Supplier HACCP registry.
6. Recipe for HACCP.
7. Cleaning and sanitizing schedule and instructions.
8. Maintenance schedule and instructions.
9. Pest control schedule and instructions.
10. HACCP-TQM employee training programme and record.
11. Company self-inspection and continuous quality improvement.
12. Food safety programme certification.

As operators write their self-control manuals, they simply ensure that they comply with the government hazard lists and limits.

The manual is the easy part of company process control. The major responsibility of the operator is to achieve consistent operations. One cannot have hazard control without stable processes. This means that every day every product is produced exactly the way it was produced the previous day, last week, etc., and will be produced that way until there is a top-management-approved process change within the company. There is no reliable quality control or hazard control unless there are, first and foremost, stable processes.

Enforcing Hazard Control Programmes

The only person who can enforce a zero-defect HACCP programme is the owner of a business. The owner is the one who allocates money for the HACCP pre-control of the processes within the facility. What are these pre-controls?

1. Owner commitment to 'walk' the facility and validate that what has been documented in the operations manual is being done by every employee.
2. Preparation of the procedures that the owner says will be followed in the facility. These procedures are traceable to government rules.
3. Training employees and providing them with the proper tools so that they are capable of doing the processes with zero defects.
4. Requiring employees to monitor their tasks and record actual process values such as times, temperatures, levels of nitrate, weights, volumes and thickness. This way, the owner can show due diligence, if necessary, if there is a claim that a product being produced caused an illness or injury.
5. Review of all monitoring and auditing data by the HACCP team to find gaps in desired performance vs. actual performance and to take corrective action so that the processes stay in control.

In HACCP, the responsibility of regulatory personnel is not to inspect a retail food facility in terms of faults in the facility, but to determine if an owner/operator is operating according to the facility's government-approved HACCP TQM manual and that processes are adequate. The inspector accomplishes this by verifying employee capability. This is done by asking an employee on the line, 'What task are you doing?' The inspector finds that task in the operations manual and asks the employee to describe how he or she has been trained to do that process. Then the inspector can verify that the employee knows and is following the operations manual. In the case of cooking food, the inspector would ask the employee the correct temperatures and then have the employee show that he or she can correctly determine the temperature of the food being cooked. In other words, the employee must demonstrate that he or she has control of the process being performed. The employee knows the hazards and the critical limits and has control of the process.

The key item of government measurement is employee knowledge and performance. In addition to food processing, this performance includes washing pots and pans; cleaning floors, walls and ceilings; maintaining equipment and controls; pest control – in other words, all the components of a good-quality management programme of a reputable food manufacturer.

Summary

Hazards in foods may be microbiological, physical or chemical. The government's role in deciding what should be the consumer risk of illness or death involves three components: risk assessment, which includes hazard identification, exposure assessment, dose–response and risk characterization; risk management; and risk communication. The government's responsibility is to specify safe minimum critical limits for processes to ensure safety. The most effective process is one in which the regulated industry has participated with government in the identification of the hazards as well as of their own process critical control points. Next, the government's role is to communicate the hazards and the critical process limits to the industry in the form of government regulations. HACCP begins with an effective, scientifically valid, regulatory risk analysis and process control development programme, but it is carried out by food processors who understand that they must abide by these mandatory government rules or their licences to operate will be revoked. In HACCP, the responsibility of regulatory personnel is to determine if an owner/operator is operating according to the facility's government-approved HACCP TQM manual.

See also: **Good Manufacturing Practice**. **Hazard Appraisal (HACCP)**: The Overall Concept; Critical Control Points; Establishment of Performance Criteria. **Quantitative Risk Analysis**.

Further Reading

Baker SP and Fisher RS (1980) Childhood asphyxiation by choking or suffocation. *J.A.M.A.* 244: 1343–1346.

Carnevale RA and Sachs S (1991) Counting down on residues. *FSIS Food Safety Rev.* 1: 4–6.

Foster GM and Kaferstein FK (1985) Food safety and the behavioral sciences. *Soc. Sci. Med.* 21: 1273–1277.

Gorham JR (1994) Hard foreign objects in food as a cause of injury and disease: a review. In: Hui YH, Gorham JR, Murrell KD and Cliver DO (eds) *Food-borne Disease Handbook*. Vol. 3, p. 615. New York, NY: Marcel Dekker.

Harris CS, Baker SP, Smith GA and Harris RM (1984) Childhood asphyxiation by food. *J.A.M.A.* 251: 2231–2235.

Hyman FN, Klontz KC and Tollefson L (1993) Food and Drug Administration surveillance of the role of foreign objects in foodborne injuries. *Public Health Rep.* 108: 54–59.

Mittleman RE and Wetli CV (1982) The fatal café coronary. *J.A.M.A.* 247: 1285–1288.

Taylor SL (1985) Food allergies. *Food Technol.* 39: 98–105.

HEAT TREATMENT OF FOODS

Contents

Thermal Processing Required for Canning

Aslan Azizi, Iranian Agricultural Engineering Research Institute, Iran

Introduction

Foods contain microorganisms which may be harmful to the health of consumers unless completely inactivated, or may cause spoilage of canned products. When a product is sterilized, it is completely free from microorganisms. Under conditions of commercial sterility, organisms endangering public health and those likely to cause spoilage of canned foods are destroyed by heat treatment of food for a sufficient time at a particular temperature to produce a condition of commercial sterility.

Thermal process requirements have been determined by trial and error for about a century. The development of a variety of food products, their behaviour under heat penetration and the wide range of microorganisms associated with different food groups demanded a systematic and scientific approach of process evaluation. The main aim of thermal processing of food is to kill the existing microorganisms and their spores which may cause decomposition, food-borne infection, or at worst produce toxins which may result in food poisoning. A sterile product is the aim. In general, all that is attained is commercial sterility. Commercial sterility as often referred to in the literature is a condition sufficient for temperate zones. Many canned products are not 100% sterile. For this reason, the term 'sterilization' is not used. A higher degree of heat treatment is necessary for canned products manufactured in or exported to the tropics or subtropics, for army rations and certain diet foods. In general, a thermal process is evolved by determining:

- the heat resistance of microorganisms likely to cause spoilage
- the rate of heat penetration into the food products
- calculation of thermal process by integration of heat penetration data and thermal resistance data
- confirmation of the calculated process time by microbiological methods (inoculated pack study).

Thermal Resistance of Microorganisms

The thermal resistance of microorganisms, and more so of spores, varies. The relation of the core components of a spore to heat resistance is mainly dependent on pH. As pH has profound effects, foods are classified for purposes of thermal processing on the basis of pH. In the most practical classification, they are divided into three groups:

1. High-acid (pH less than 4.0)
2. Acid (pH between 4.0 and 4.6)
3. Low-acid (pH above 4.6).

The dividing line between acid and low-acid foods is carefully chosen as 4.6, i.e. 0.2 units of a safety factor lower than the minimum pH of 4.8 at which *Clostridium botulinum* is able to grow. High-acid and acid foods need less severe processing in comparison to low-acid foods.

Microorganisms Causing Spoilage of Foods

Microorganisms causing spoilage in acid foods are *Bacillus coagulans*, *B. macerans*, *B. polymyxa*, *C. pasteurianum* and *C. butyricum*. In canning peeled tomatoes of pH 4.2–4.6, spoilage is caused by butyric acid anaerobes such as *C. butyricum* and *C. pasteurianum*. Besides these organisms, spoilage by the flat-sour organism *B. coagulans* cannot be ignored. In high-acid foods, spoilage is caused by non-sporulating bacteria like *Lactobacillus* and *Leuconostoc*, yeasts and moulds. The heat resistance of these organisms is low: processing should ensure their destruction, as otherwise some of the organisms, notably moulds like *Byssochlamys fulva* and *B. nivea*, may grow with a subsequent rise in pH to 4.6 or above which will allow the growth of *C. botulinum*. Peroxidase and pectinesterase enzymes have higher heat resistance than non-spore-forming lactobacilli or yeasts and the sporulating *C. pasteurianum*. Hence, a thermal process schedule for fruits is based on the thermal inactivation of heat-resistant enzymes which has been found to render the canned product microbiologically safe.

Low-acid foods of pH ⩾ 4.6 may support the growth of *C. botulinum*, which produces a potent exotoxin, and several types of this organism (A–G) have been listed. Types A, B and E are associated with cases of botulism in humans. Spores of types A and B are heat-resistant. Types C, D and G are usually not associated with botulism in humans.

Acid foods having pH ⩽ 4.6 – fruit or low-acid products acidified to pH < 4.6 – are not considered potential sources of botulism. However, *C. barati*, *C. perfringens* and *C. butyricum* have been reported to produce toxins in infant food formulations. The production of botulinal toxin by *Clostridium* that does not have the physiological characteristics traditionally associated with *C. botulinum* is of potential significance to the processing of acid foods or acidified low-acid foods.

C. barati and *C. perfringens* can cause spoilage at pH levels as low as 3.7, and *C. butyricum* at pH 4.0. Hence, the growth of these strains may not be prevented by the pH normally used to control *C. botulinum*. The heat resistance of *C. perfringens* (D_{100} = 17.6 min) and *C. butyricum* is low, and can be destroyed by much lower processing than that of a botulinal cook.

Heat Resistance of Bacteria other than *C. botulinum*

Besides *C. botulinum*, mesophilic species of *Clostridum* including *C. sporogenes*, *C. butyricum* and *C. pasteurianum* and the thermophilic species *C. thermosaccharolyticum* can cause putrefactive or saccharolytic spoilage and gas formation, resulting in swelling of cans. Organisms likely to cause flat-sour spoilage of low-acid canned foods are thermophilic and include *B. coagulans* and *B. stearothermophilus*. Of the organisms listed above, *C. pasteurianum* and *C. butyricum* are not likely to cause spoilage in low-acid foods as their heat resistance is low so they cannot withstand the thermal processing given to low-acid foods. *B. cereus*, *B. licheniformis*, *B. megaterium* and *B. polymyxa* behave similarly.

Heat resistance (F value) of spores of *B. coagulans*, the thermophilic aciduric flat-sour spoilage organism of tomato juice does not exceed 0.7 min at 121°C in pH 4.3 tomato juice and hence does not withstand the processing given to low-acid foods.

The spores of thermophiles *C. thermosaccharolyticum*, *B. stearothermophilus* and *Desulfotomaculum nigrificans* are much more heat-resistant than the mesophilic anaerobic spore-formers such as putrefactive anaerobes. The thermophiles produce no toxin and have no consequences on health. Some may remain ungerminated in low-acid canned foods. Commercial sterility allows the presence of some these spores, therefore, canned foods are not completely sterile but may contain dormant spores of thermophilic bacteria. The spores will never germinate if the canned food is properly cooled and held at room temperature. Also, spores of thermophiles will eventually autosterilize when held at temperatures at which they may slowly germinate but not outgrow.

Spoilage in Canned Foods by Thermophiles

C. thermosaccharolyticum has been responsible for spoilage of canned products such as spaghetti with tomato sauce, sweet potato, pumpkin, green beans, mushroom, asparagus, vegetable soup and carrots. It has also caused spoilage of highly acidic products such as fruit and farinaceous ingredient mixtures and tomatoes at pH 4.1–4.5. *C. thermoaceticume* may cause flat-sour spoilage of canned vegetables. The organism has an optimum growth temperature of around 62°C but does not grow above 68°C or below 40°C, and the minimum pH for growth is 5.5. Even F_0 (minutes required to destroy a specified number of spores at 250°F when Z = 18) of 25 min is not

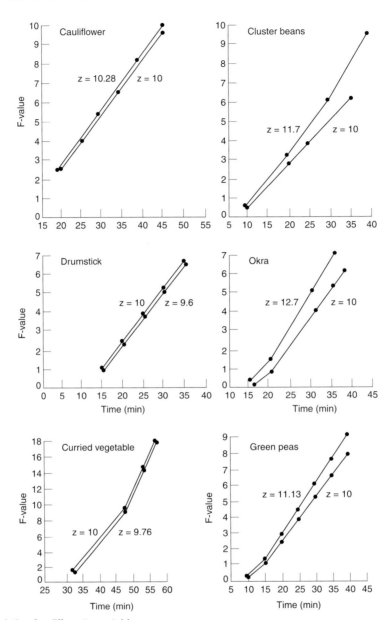

Figure 1 F value calculation for different vegetables.

adequate to render canned products free from spoilage by thermophilic organisms (**Fig. 1**). Ingredients such as vegetables, sugar, dehydrated milk, starch, flour, cereals and rendered meat have been found to be predominant sources of thermophilic anaerobes. This organism occurs widely in soil and raw materials such as mushroom and onion products. It cannot multiply on equipment and handling systems unless there is an anaerobic environment containing nutrients and moisture and elevated temperature. *B. stearothermophilus*, a flat-sour organism, is responsible for spoilage of canned products such as corn, peas and beans. Spores of *B. stearothermophilus* are widely

distributed in the soil. Sugar, starch, vegetables, milk powder and spices may contribute to the contamination.

The temperature ranges for the growth and heat resistance of the spores of different thermophilic bacteria which are important in food spoilage and have been studied by the author are shown in **Table 1**.

An interesting property of flat-sour bacilli is the tendency of spores that survive a heat process to lose viability if stored at temperatures that are too low to support growth. Besides *B. coagulans*, *B. licheniformis* includes strains of facultative thermophiles capable of growth at 30°C as well as 50–55°C, and may cause swelling of cans.

Table 1 Temperature range for growth and spore heat resistance of thermophilic organisms

Organism	Minimum temperature (°C)	Maximum temperature (°C)	Optimum temperature (°C)	D value (min)
Bacillus coagulans	16–21	50–62	35–45	D_{100} 3–9
D. nigrificans	33	55–65	50–55	D_{121} 15–55
Bacillus stearothermophilus	32	75–78	50–62	D_{121} 2.5–14
Clostridium thermosachar-rolyticum	35	65–67	55	D_{121} 65–75

Factors Affecting Heat Resistance of Spores

Effect of Heating Environment

Of particular interest to canners is the behaviour of spores in the food to be canned. pH, acids which are used or which are naturally present, water activity, salt, sugar, carbohydrate and fats affect thermal resistance. Fats, sugars, proteins and acidity present in food have all been shown to be significant in this respect but, unfortunately, their interaction will be different in different products, so it is difficult to predict the resulting D and z values. Products of similar nature and pH may have an entirely different effect on spores on heating.

Concentration

The heat resistance of a microbial suspension increases with an increase in the number of organisms present.

pH

Spores generally manifest lower D values when they are heated under acidic conditions, possibly because of enhanced DNA damage, but pH can also act as a pre-treatment to sensitize (**Fig. 2**). The mechanism by which acid reduces heat resistance is not known. Using various vegetables and salt solutions, the author found only slight variations in the heat resistance of *C. sporogenes* at 121°C but considerable variation in z value (**Table 2**).

The data referred to in Table 2 show that, during heat treatment of a product, the complex situation makes it difficult to predict heat resistance of spores. Hence, heat resistance studies are carried out by suspending the microorganisms in the food in which the thermal process is required to be evolved.

Determination of the Thermal Resistance of Microorganisms

The types of equipment used to determine spore resistance at high temperatures are many and varied:

1. TDT (thermal death time) tube method
2. TDT glass tube method

Table 2 D and z values of *Clostridium sporogenes* in different vegetables

Heating medium	pH	Spore/ tube	D (min) at 250°F min	z (°F)
Cabbage	6.3	6925	1.43	8.94
Cauliflower	6.2	70 000	1.124	10.3
Green beans	5.7	150 000	1.488	11.8
Green peas	5.9	76 167	1.48	11.3
Potato	6.0	76 167	2.11	10.26
Cluster beans	5.8	1 500 000	1.21	11.7
Drumstick	5.0	150 000	1.06	9.8
Okra	5.5	11 200	1.25	13.46
Mushroom (agaricus)	5.8	76 167	2.05	10.53
Mushroom (pluerotus sajor caju)	5.7	761 671	1.566	8.85
Curried vegetables	5.5	761 671	1.805	9.76

3. Unsealed tube method
4. Screw-cap tube technique
5. Serum bottle with a rubber septum crimped with prepunctured aluminium seal
6. TDT pouch method (nylon)
7. TDT cylinder method
8. Flask method
9. Capillary tube method
10. Capillary U-shaped tube method
11. TDT can method
12. Miniature retort time method or tank method
13. Thermoresistometer method.

Production of Spores for TDT Studies

Obligate aerobes or facultative anaerobes from an actively growing culture are inoculated on to a suitable media (**Table 3**).

Obligate anaerobes are best cultivated using screw-capped bottles. The inoculated bottles are incubated at the appropriate temperature until microscopic examination shows sporulation to be virtually complete.

Harvesting the Spores

The spores are harvested using lysozyme (100 mg ml⁻¹, 37°C for 30 min). Adjustment to pH 11.0 destroys vegetative cells almost immediately.

Successive decimal dilutions of the spore suspension

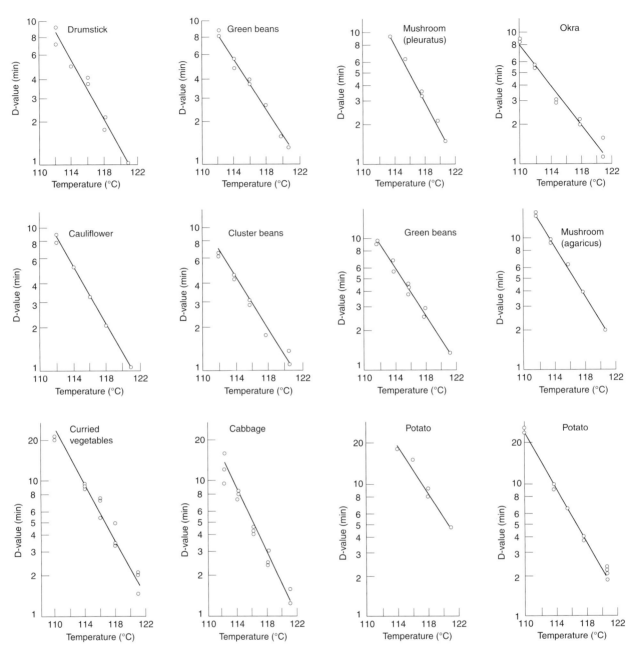

Figure 2 Determination of D values of *C. sporogenes* for different vegetables.

are prepared in sterile water, and used for spore counting.

Determination of Heat Resistance of Spores

A constant-temperature water bath is used for temperatures below 100°C and an oil bath for temperatures above that. Spore suspensions are heated in either of these devices. Heated tubes are subcultured on to a medium and incubated. Gas formation or changes of pH and turbidity production in the tubes are indications of growth.

Death Rate of Bacteria

Bacteria are killed by heat at a rate that is very nearly proportional to the number of bacteria present in the system being heated. This is referred to as the logarithmic order of death. In other words, if a given temperature kills 90% of the population in 1 min, 90% of the remaining population will be killed in the second minute, 90% of what is left will be killed in the third minute, and so on.

Table 3 Media, incubation temperature and time for TDT tubes

Microorganism	Media	Incubation temperature (°C)	Incubation time (days)	Growth sign
Bacillus subtilis	Nutrient broth and agar	28–30	2–4	Turbidity
Bacillus stearothermophilus	Nutrient broth and agar	50	2–4	Turbidity
Bacillus coagulans	Proteose peptone	50–55	2–4	Turbidity
Bacillus betanigrificans	Nutrient agar	37	2–4	Blackening
Clostridium botulinum	Pork, beef liver, heart broth, RCM and DRCM	27–30	7–17	Gas formation
Clostridium sporogenes	Pork, beef liver, heart broth, RCM and DRCM	30	14	Gas formation
Clostridium thermosaccharolyticum	Pea peptone broth with marble or liver broth with iron strip	50–55	4–14	Media digestion

RCM, Reinforce *Clostridium* media; DRCM, differential *Clostridium* media.

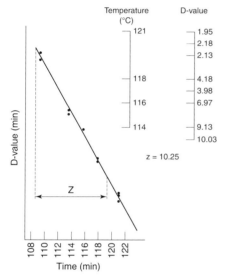

Figure 3 Thermal death time curve (TDT) of *C. sporogenes* for potato.

Heating Time and Temperature

Since the death of organisms follows a logarithmic rate, choose time intervals which represent a definite logarithmic cycle. For example, typical conditions for TDT investigation and suggested time and temperature intervals for potato are given in **Table 4** and **Figure 3**.

Table 4 Calculation of D values in potato with 761 671 spores in 0.1 ml per tube

Temperature (°C)	Time (min)	Number of negative tubes	D value
121	10	4	1.95
	11	3	2.18
	12	5	2.13
118	20	2	4.18
	21	3	3.98
116	32	1	6.97
114	52	6	9.13
	53	2	10.03

Phosphate Buffer as a Reference Medium

A wide variation may occur in the resistance of spore suspensions prepared under apparently identical conditions, and also variations in the composition of spores in phosphate buffer parallel with each determination of mortality. For mixing solutions of M/15 disodium hydrogen phosphate (Na_2HPO_4) and M/15 potassium dihydrogen phosphate (KH_2PO_4) see **Fig. 4**.

Evaluation of Number of Surviving Spores

After each heating, count the number of survivors either by subculturing on a solid medium or by dilutions. Take the number of containers showing survival at the longest time of heating as equivalent to the number of surviving organisms.

Calculate the D values with this formula:

$$D = U/\log a - \log b$$

where D = time required to destroy 90% of cells, a = initial number of cells and b = number of cells surviving after heating.

Heat Penetration into the Can

In canning the food being processed in a hermetically sealed container must be heated thoroughly using an external source of heat. If the food is a liquid (broth) or contains a liquid of low viscosity (peas in brine), the heat is distributed via convection. If it is a solid (meat, fish) or highly viscous (cream-style corn), it can only be heated via conduction; as a result, most of the contents must be severely overprocessed in order to sterilize the small volume occupying the geometric centre.

Transfer of heat from the heating medium to the container is by conduction and from the container to the contents is by conduction or convection or both. The nature of the food and its distribution, the can, its size and shape, the head space, the solid-to-liquid phase, the initial temperature of the product, the location of

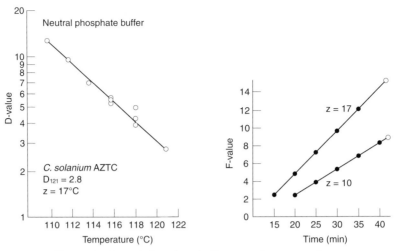

Figure 4   Neutral phosphate buffer as a reference medium.

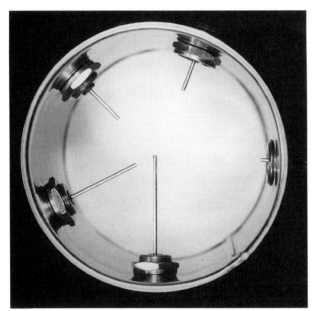

Figure 5   Placement of thermocouples in different positions of the can.

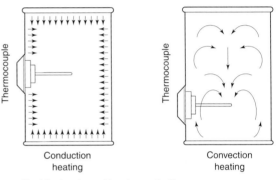

Figure 6   Mechanism of heat penetration.

the can in the heating medium, the processing temperature, the nature of the heating medium and agitation or rotation all affect the rate of heat penetration into a can of food and the lethality of the process.

Temperature changes during heating and cooling must be measured accurately to determine adequate processes for canned foods. The most satisfactory methods involve the use of copper-constantan thermocouples. Heat penetration of each type of vegetable canned in brine or fruit canned in syrup indicates that the mode of heating is by convection, and the slowest heating point inside the can is at about one-tenth of the can height from the bottom on the central axis. In canned curried vegetables and meat pieces, heat transfer is by conduction, and the geometric centre is the cold point. To find the cold point in the can for each different product, thermocouples are placed in different positions in the can (**Fig. 5**). To find the exact heat penetration in each food product, six cans containing thermocouples positioned at the slowest heating point with the tip of each thermocouple embedded in the prepared vegetable material or fruits in syrup were distributed in a crate at different locations, among other cans. When a can of vegetables after exhausting (removal of air), is sealed and placed in a retort, say at 240°F (115.5°C) (10 MPa steam pressure), the retort is at a high energy level and the can is at a low energy level. Heat transfer takes place from the hot body (retort) to the cold body (can). The heat is transferred by conduction from the steam to the can and from the can to the contents by conduction or convection (**Fig. 6**).

Heating by Rapid Convection

This group includes:

● most fruit and vegetable juices
● thin soups
● fruits with large pieces canned in water or syrup (heat penetrates slowly)
● vegetables in brine or water.

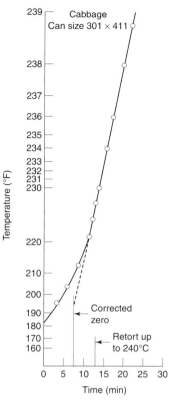

Figure 7   Process calculation by formula method.

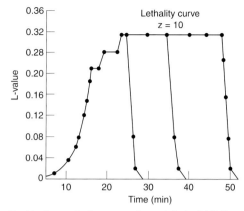

Figure 8   Heat penetration curve for potato in A1 Tall can.

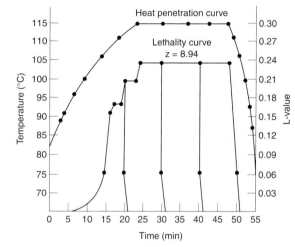

Figure 9   Calculation of lethality for potato in A1 Tall can.

Heating by Slow Convection

In canned fruits and vegetables containing small pieces, or chopped vegetables with low starch content packed in brine, heating takes place by slow convection.

Heating by Conduction

In solid packs with a high moisture content, but with little or no free liquid, e.g. heavy cream-style corn, sweet potato, thick fruit and vegetable purées, thick sauces and soups, fruit juice concentrates and vegetables in sauce, heating is by conduction.

Commercial Sterility

Commercial sterility of equipment and containers used for aseptic processing and packaging of food means the condition achieved by application of heat, chemical sterilants or other appropriate treatment. Such treatment renders equipment and containers free of viable forms of microorganisms having public health significance as well as any microorganisms of non-health significance capable of reproducing in the food under normal non-refrigerated conditions of storage and distribution. Process determination for low acid products:

● Determination of heat penetration rate in product
● Determination of heat resistance of significant spores in product

● Calculations of theoretical process
● Testing of theoretical process by inoculated packs

Inoculated Pack Studies

To check the efficiency of the calculation process (see **Figure 7**), the canned food should be inoculated with heat-resistant organisms, subjected to the normal heat process, and survival of the organisms should be estimated either by subculturing or by incubation. This procedure may also be used for estimating process lethality, where heat penetration data cannot be readily obtained and in the case of agitation retorts (see **Figures 8 and 9**). Do not use C. botulinum for inoculated pack studies. For low-acid foods use PA 3679 and FS 1518 when the contents are subcultured. Test organisms used with acid packs are listed below:

● tomatoes, pears, figs and other fruits with pH above 4: C. pasteurianum
● tomato juice: B. coagulans
● fruit products: Lactobacillus
● citrus juices and pickles: yeasts
● acid fruits: moulds.

The following consideration must be taken into account:

1. The inoculated pack studies should be conducted in the processing equipment to be used for the commercial production of the food
2. If the formulated food contains a dry ingredient, the spores used for the inoculated pack should be incorporated into the dry ingredient

See also: **Clostridium**: *Clostridium botulinum*. **Ecology of Bacteria and Fungi in Foods**: Influence of Temperature; Influence of Redox Potential and pH. **Spoilage Problems**: Problems Caused by Bacteria.

Further Reading

Azizi A and Ranganna S (1993a) Spoilage organisms of canned acidified mango pulp and their relevance to thermal processing of acid foods. *J. Food Sci. Technol.* 30: 241–245.

Azizi A and Ranganna S (1993b) Thermal processing of acid food fermented vegetables. In: *Trends in Food Science and Technology.* Proceedings of the third international food convention (IFCON 93). 1993 at C.F.T.R.I Mysore India.

Azizi A and Ranganna S (1993c) Thermal processing of acidified vegetables. *J. Food Sci. Technol.* 30: 422–428.

Caolari A (1989) Microbial Spoilage of Canned Fruit Juice. In: Cantarelli C and Lanzarini G (eds) *Biotechnology Application in Beverage Production.* London: Elsevier Applied Science.

Hayakawa KI (1977) Mathematical method for estimating proper thermal processes and their computer implementation. *Add. Food Res.* 23: 76–141.

Hayakawa K, De Massaguer P and Trout RJ (1988) Statistical variability of thermal process lethality in conduction heating foods – computerized simulation. *J. Food Sci.* 53: 1887–1894.

Hersom A and Hulland E (1980) *Canned Foods*, 7th edn. Edinburgh: Churchill Livingstone.

Stumbo CR (1973) *Thermobacteriology in Food Processing*, 2nd edn. New York: Academic Press.

Spoilage Problems Associated with Canning

Lahsen Ababouch, Department of Food Microbiology and Quality Control, Institut Agronomique et Vétérinaire Hassan II, Rabat, Morocco

Introduction

Canning is a fairly old method of food preservation which dates back to the beginning of the nineteenth century. Although it started as a process of trial and error, developments in microbiology, heat transfer mechanisms, thermal processing and container closure technology have led gradually to processing techniques that are founded on scientific evidence and outputs.

Nowadays, tens of billions of food cans are produced yearly worldwide, and 20–90% of a harvest is likely to be preserved by canning, depending on the country and the type of food. Despite this large volume of production, the canning industry enjoys good safety and quality records. Indeed, spoilage and poisoning outbreaks involving industrially canned foods are very rare. Unfortunately, these outbreaks can present severe health hazards, as well as damaging the company responsible and undermining the whole industry. Thus, the 1982 outbreak of botulism that caused the death of one person in Belgium following the consumption of contaminated canned salmon led to the examination of the entire 1980 and 1981 output of the Alaskan salmon industry and a series of recalls involving 50 million cans worldwide, the second largest group of food recalls in the history of the US Food and Drug Administration (FDA).

This article reviews the main causes and types of spoilage of canned foods and outlines the preventative measures recommended for the industrial production of safe and good-quality canned food products.

The Canning Process

The preservation of foods by canning involves (1) the use of hermetically sealed containers which are impermeable to liquids, gases and microorganisms, and (2) the application of sufficient heat to inactivate toxins, enzymes and microorganisms capable of proliferating under normal non-refrigerated conditions of storage and distribution. Under these conditions, the food is said to be 'commercially sterile', which should not be taken to mean absolute sterility, i.e. the total absence of viable microorganisms. Indeed, viable microorganisms can be recovered from commercially sterile heat-processed foods under one of three conditions:

- the microorganism is an obligate thermophilic spore-forming bacterium but the normal storage temperature is below the thermophilic range ($< 40°C$)
- the microorganism is acid-tolerant and the food pH is within the high acidity range (pH < 4.6)
- the canning process uses a combination of heat and low water activity to preserve the food.

Food pH is an important parameter which affects the severity of the thermal process used for food preservation. It is well established that a pH of 4.6 or less is sufficient to inhibit the growth of all pathogenic and most spoilage bacteria, of which the most heat-resistant strains are spore-forming species such as *Clostridium botulinum*. Thus, acid foods will need a milder heat treatment to become commercially sterile compared with low-acid foods. **Table 1** presents examples of thermal processes recommended for the preservation of different foods and demonstrates the striking effect of the pH.

Thermal processes of low-acid canned foods are designed to attain a probability of survival of the pathogenic *Clostridium botulinum* sufficiently remote to present no significant health risk to consumers. Experience has shown that such thermal processes should enable the reduction of any population of the most resistant *Clostridium botulinum* spores to 10^{-12} of its initial count. This is known as the 'botulinum cook' or the 12D concept, where D is the decimal reduction time (D value). Canned foods are actually processed beyond the minimum botulinum cook because of the occurrence of non-pathogenic, thermophilic spore-formers of greater heat resistance.

However, the production of a microbially safe and shelf-stable canned food should not unduly impair the food flavour, consistency, colour or nutrient content. Thus, the accepted rate of survival of the most heat-resistant thermophiles is 10^{-2} to 10^{-3} (2D or 3D process). This higher risk of thermophile survival is considered acceptable because thermophiles are not a public health concern and given reasonable storage temperature ($< 35°C$), the survivors will not germinate. **Table 2** presents reference D values for bacteria relevant to low-acid canned foods.

The temperature of storage of the canned foods is a significant factor in the choice of thermal process. **Table 3** groups canned meats into four product classes depending upon the heat treatment and the final storage temperature.

Spoilage of Canned Foods

Spoilage in canned foods is usually indicated by leakage, a swelling of the container, or an abnormal

Table 1 Thermal processing of foods and the effect of pH (Lange, 1983)

Product	pH	Classification	Heat treatment	F_0 value[a]	Cutoff pH	Target organism(s)
Spinach, Peas	7.0					Mesophilic, spore-forming thermophilics and natural enzymes
Milk (evaporated)	6.5	(1) Weakly acid (pH 5.3–7.0)				
Corned beef, Mushrooms, carrots	6.0		Sterilization at temperatures > 100°C	7 to 14		
Asparagus, green beans	5.5				5.3	Lower limit of growth of *Clostridium botulinum*
Tomato soup	5.0	(2) Medium acid (pH 4.5–5.3)		3 to 6		
Tomatoes	4.5				4.5	Acid and spore-forming bacteria, acid-resistant non-spore-formers
Apricots, pears, Peaches	4.0	(3) Acidic (pH 3.7–4.5)		1	4.1	
					3.7	
Orange juice, Candied fruits	3.5		Pasteurization at 69°C to 100°C			Acid-resistant bacteria (non-spore-formers) Yeasts and moulds
Berries, sour vegetables	3.0	(4) Strongly acid (< 3.7)				
Lemon juice	2.5					

[a] F_0 is the sterilization process equivalent time, defined as the number of equivalent minutes at T 121.1°C delivered to a food container calculated using a z value (the temperature increase required for a tenfold decrease in the D value) of 10°C.

Table 2   Decimal reduction times (D values) of bacterial spores of interest in food

Microorganism	Optimal growth temperature (°C)	D value (min)
Bacillus stearothermophilus	55	$D_{121.1°C}$ 4–5
Clostridium thermosaccharolyticum	55	$D_{121.1°C}$ 3–4
Desulfotomaculum nigrificans	55	$D_{121.1°C}$ 2–3
Clostridium botulinum types A and B	37	$D_{121.1°C}$ 0.1–0.23
Clostridium sporogenes PA 3679	37	$D_{121.1°C}$ 0.1–1.5
Bacillus coagulans	37	$D_{121.1°C}$ 0.01–0.07
Clostridium botulinum type E	30–35	$D_{82.2°C}$ 0.3–3

The D value is the decimal reduction time or the time required to reduce, at a given constant temperature, a microbial load to 10% of its initial level.

odour or appearance of its contents. In some cases the presence of microbially induced toxins capable of causing food poisoning is not accompanied by any external or internal visible signs of spoilage.

Four causes can lead to the spoilage of canned foods:

- pre-spoilage or incipient spoilage that takes place before the product is thermally processed
- underprocessing
- thermophilic spoilage
- post-process spoilage.

Pre-spoilage

Pre-spoilage or incipient spoilage takes place before the product or the ingredients are thermally processed. It may be caused by microbial or enzymatic action resulting in gas accumulation, development of off odours and the presence of excessive numbers of dead microbial cells in the end product. If the responsible microorganisms are pathogenic, such as *Staphylococcus aureus*, histamine-producing bacteria or mycotoxin-producing moulds, they produce

thermostable toxins which will not be significantly affected by the thermal process and will cause food poisoning. In fish, meats, fruits or vegetables received at the processing factory, microbial counts of 10^7 per gram or above are not uncommon. Fish habitat and the feed of fish and the soil in which vegetable products are grown are the main sources of organisms found on or in raw products destined for canning. Additional contamination can come from surfaces in contact with the food during harvesting and transportation, from washing water and handling practices, and from added ingredients (e.g. sugar, salt, syrups, starch and spices).

A wide variety of microorganisms can be found on or in raw foods destined for canning. However, bacteria greatly outnumber yeasts and moulds. In fact, bacteria are often the sole concern in the processing of fish and meats. The type and number of these microorganisms will be greatly affected by the different operations that foods will undergo before they are thermally processed. For vegetable materials, the integrity of the product will minimize the risk of pre-spoilage because the microorganisms are unable to invade the food tissues. Unfortunately, many harvesting practices and machinery cause tissue bruising and damage, leading to microbial invasion of the foods and its decay. Likewise, bulk transportation leads to tissue damage and should be avoided. Bulk storage of olives for long periods can lead to mould growth and production of mycotoxins. These toxic metabolites are harmful to humans and very heat-stable. The holding temperature plays a major role in delaying food decay during transportation and storage; refrigeration is very efficient in this respect.

Most foods destined for canning are washed to remove soil, debris, slime and other foreign materials. Washing will often eliminate up to 90% of the food surface microbial load. However, only water with acceptable microbiological quality should be used. Chlorination of the washing water, to levels of 1–4 p.p.m. of residual chlorine, is very useful in this respect.

Preparatory operations such as cutting, slicing,

Table 3   Different classes of canned meats (Lange, 1983)

Type	Heat treatment	Target microorganisms	Food class and storage life
I	65–75°C attained	Vegetative microorganisms	Semipreserves: 6 months at temperature $< 5°C$
II	F_0[a] 0.65–0.80 min	As for I plus spores of mesophilic bacillus	Three-quarter preserves: 6–12 months at $T < 15°C$
III	F_0 5.0–6.0 min	As for II plus mesophilic clostridia	Full preserves: 4 years at $T < 25°C$
IV	F_0 16.0–20.0 min	As for III, plus spores of thermophilic bacilli and clostridia	Full preserves for tropical use: 1 year at $T > 40°C$

[a]F_0 is the sterilization process equivalent time, defined as the number of equivalent minutes at T 121.1°C delivered to a food container calculated using a z value of 10°C.

dicing, beheading or evisceration expose food tissues and accelerate microbial growth and food spoilage if delays occur, especially under high temperatures conducive to bacterial growth. Blanching of vegetables and cooking of fish significantly inactivate heat-sensitive microorganisms such as vegetative bacteria, sensitive spores (e.g. type E *Clostridium botulinum*), yeasts and moulds. However, heat-resistant organisms, especially spores of *Clostridium botulinum* types A and B and thermophilic spore-formers, survive the blanching and cooking processes. These processes soften food tissues, rendering contamination and spoilage easier. Thus, extreme care should be exercised to implement proper hygienic practices and avoid delays, otherwise contamination by hygiene-related organisms such as *Staphylococcus aureus* may lead to spoilage and the accumulation of thermostable toxins which will not be inactivated during thermal processing.

Inappropriate handling practices during harvesting, transportation, storage and preparation of foods for canning can lead to microbial growth, substandard canned food quality (soft texture, repellent odours or colour) and to the accumulation of heat-stable toxic metabolites such as histamine, staphylococcal enterotoxins and mycotoxins. Appropriate handling and hygienic practices, control of temperature and the avoidance of delays during these operations will minimize these problems.

Histamine Spoilage of Canned Seafoods

Histamine is produced in foods by the decarboxylation of histidine (**Fig. 1**). This reaction is catalysed by the enzyme histidine decarboxylase found in some bacterial species. These include various species of Enterobacteriaceae, *Clostridium*, *Vibrio*, *Photobacterium* and *Lactobacillus*. All the strains of certain species such as *Morganella morganii* are capable of histidine decarboxylase activity, whereas the enzyme is present in only few strains of other species such as *Klebsiella pneumoniae* and *Lactobacillus buchneri*. The presence of most of these bacteria is often the result of unacceptable food handling and hygienic practices.

The foods generally incriminated in histamine poisoning are fish, cheese, sauerkraut and sausage. Fish in general, and canned fish in particular, have been involved in an overwhelming majority of the incidents of histamine poisoning. This is because fish

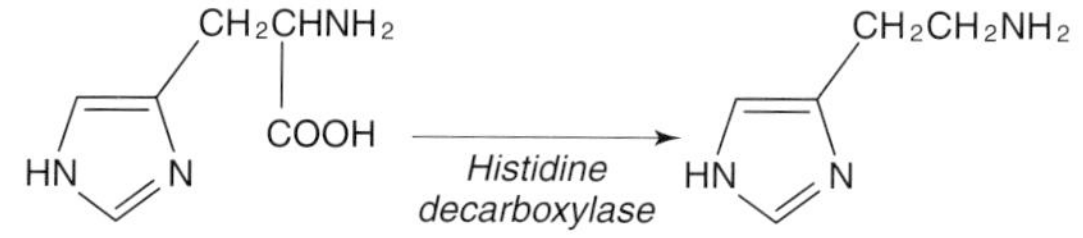

species such as tuna, mackerel, sardines, saury, seerfish and mahi-mahi contain large amounts of free histidine in their muscle tissues, which serves as a substrate for histidine decarboxylase. In addition, proteolysis – autolytic or bacterial – may play a role in the release of histidine from tissue proteins. In the past, histamine poisoning has been referred to as scombrotoxin poisoning because of the frequent association of the illness with the consumption of spoiled scombroid fish such as tuna and mackerel.

Histamine poisoning is usually a mild disorder with a variety of symptoms. The primary symptoms are cutaneous (rash, urticaria, oedema, localized inflammation), gastrointestinal (nausea, vomiting, diarrhoea) and neurological (headache, oral burning and blistering sensation, flushing and perspiration). More serious complications such as cardiac palpitations are rare. Despite its toxicity, histamine is not a substance foreign to the human body. In small physiological doses histamine is a necessary substance involved in the regulation of such critical functions as the release of stomach acid. Furthermore, despite the compelling evidence pointing to histamine as the causative agent in many food poisoning incidents, it has been virtually impossible to reproduce the illness in oral challenge studies with human volunteers. The paradox of the lack of toxicity of pure histamine taken by human volunteers and the apparent toxicity of smaller doses of histamine in spoiled canned fish has been attributed to the possible occurrence of histamine toxicity potentiators in the spoiled fish. These potentiators would act to reduce the threshold dose of histamine needed to precipitate histamine poisoning symptoms in humans challenged orally. Food-borne substances that have been suggested as potentiators of histamine poisoning comprise trimethylamine, trimethylamine oxide, agmatine, putrescine, cadaverine, anserine, spermine and spermidine. Other potentiators include some pharmacological diamine oxidase blockers, heavy intake of alcohol, and certain diseases – liver cirrhosis, upper gut bleeding, bacterial overgrowth in the intestine (**Fig. 2**.

Potentiation of histamine toxicity is likely to result from inhibition of histamine-metabolizing enzymes present in the intestinal tract, namely diamine oxidase (DAO) and) histamine N-methyltransferase (HMT). The latter enzyme is very selective for histamine, whereas DAO also oxidizes other diamines such as putrescine. In the absence of potentiators, these enzymes metabolize histamine and prevent its absorption into the circulation.

The threshold toxic dose for histamine is not precisely known; it has been estimated at 60 mg of histamine per 100 g of food. The major importing countries of fish (including canned fish) have adopted

Figure 1   Formation of histamine.

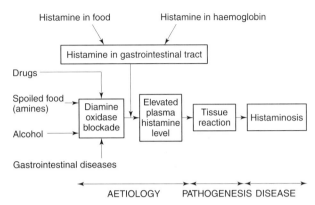

Figure 2 The disease concept of food-induced histaminosis.

regulations limiting the maximum allowable levels of histamine in foods. In the USA the FDA has established a hazard action level (HAL) of 50 mg per 100 g, and a defect action level (DAL), indicative of decomposition, of 20 mg per 100 g. Both levels were based on investigations of previous histamine poisoning episodes and spoilage studies on tuna and mahi-mahi. The DAL has been recently lowered to 5 mg per 100 g for tuna and mahi-mahi. For other fish species such as sardines, mackerel and anchovies, levels of 5 mg per 100 g are indicative of decomposition if supported by sensory evaluation. The 1991 European Union sanitary regulations identify a HAL of 20 mg per 100 g and a DAL of 10 mg per 100 g.

Prevention of histamine accumulation in foods, especially fish destined for canning, relies mostly on cooling the fish as soon as possible after the catch. Because most of the histamine-producing bacteria are mesophiles, fish storage at temperatures below 10°C, coupled with implementation of good hygienic practices, is sufficient to control histamine accumulation. In the case of small pelagic species such as sardines, mackerel and anchovies, which are often caught in large quantities and ice storage is not practical, the fish should be refrigerated quickly using either refrigerated sea-water or chilled sea-water.

Underprocessing

In canning, underprocessing, also referred to as under-sterilization, means that the product did not receive sufficient heat treatment to become commercially sterile. It is often indicated by the survival of bacterial spores exclusively, because these will resist the insufficient heat treatment, while vegetative bacteria, yeasts and moulds will not.

Among the potential survivors, spores of *Clostridium botulinum* are of great concern because of the deadly neurotoxins they produce. The food-borne intoxication they cause, botulism, has been known for over a thousand years, and was first associated

with the consumption of contaminated sausage – hence the name, from the Latin *botulus*, sausage. The bacterium responsible, *Clostridium botulinum*, is a Gram-positive, strictly anaerobic, catalase-negative, spore-forming rod. Its spores are ubiquitous as they occur in soils and sediments. Seven types (A–G) of *Clostridium botulinum* exist based on toxin serology. Strains producing types A, B and E toxins are harmful to humans, while types C and F have been implicated in botulism outbreaks in mammalian and avian species. No type G outbreak has been reported. Strains of *Clostridium botulinum* can also be classified into one of four groups: group I strains are proteolytic and produce toxins A, B and F; group II strains are non-proteolytic and produce toxins B, E and F; group III are weakly proteolytic and produce toxins C and D; and group IV strains are proteolytic but non-saccharolytic and produce toxin G. **Table 4** presents growth conditions for the various types of *Clostridium botulinum*.

Symptoms of botulism typically occur within 18–36 h of the ingestion of the toxin-contaminated foods, but may be delayed up to 8 days depending on the amount of toxin ingested. Symptoms include diplopia, dysarthria and dysphagia resulting from impairment of cranial nerve function. Nausea, vomiting, cramps and diarrhoea often precede neurological symptoms. Early onset of symptoms is associated with greater

Table 4 Conditions for *Clostridium botulinum* growth and toxin production

Minimum temperature	3.3°C (non-proteolytic type B, E and F strains)
	10°C (proteolytic type A, B and F strains)
Optimum temperature	30°C (non-proteolytic type B, E and F strains)
	35°C (proteolytic type A, B and F strains)
Maximum temperature	55°C (type A strains)
Minimum water activity (a_w)	0.97 (non-proteolytic type B, E and F strains)
	0.94 (proteolytic type A, B and F strains)
Minimum pH	5.0 (non-proteolytic type B, E and F strains)
	4.0–4.6 (proteolytic type A, B and F strains)
Maximum salt content (% in aqueous phase)	3–5 (non-proteolytic type B, E and F strains)
	10 (proteolytic type A, B and F strains)
Heat resistance	$D_{121.1°C}$ of spores 0.1–0.25 min (non-proteolytic type B, E and F strains)
	$D_{82.2°C}$ 0.15–2.9 min in broth or $D_{80°C}$ 4.50–10.5 min in products with high proteins and fat content (proteolytic type A, B and F strains)

severity of the disease and a higher probability of fatality.

Understerilization may also be caused by a faulty design of the thermal process, the wrong process being used, or a failure to deliver the designed process; the last can result from a human or mechanical failure, or from deviation of critical factors such as initial product temperature, processing time, type of heat (moist or dry), raw material microbial load, pH of the food, fill of the can or viscosity of the product from the set safe standards. These problems can be overcome by training supervisors and retort operators, using proper retort maintenance schedules, and reliance on a recognized process authority to design and validate thermal processes.

Thermophilic Spoilage

Thermophilic spoilage occurs when the time–temperature conditions are conducive to the growth of thermophilic bacteria. These bacteria are not pathogenic organisms; they occur naturally in soil and their spores are frequently isolated in low numbers from commercially sterile food products. Their number can increase if ingredients such as starch, sugar or spices are used in the product and are excessively loaded with these spores.

Thermophiles can grow in canned foods if cooling of hot retorted cans takes place at ambient temperature, or if finished products are stored at temperatures above 40°C. In the first case, cooling is significantly slow and the temperature of the cans will be in the range 75–40°C for periods long enough to promote growth of thermophiles; this is rare nowadays because most retorts are equipped to provide for rapid cooling of the cans. The most common forms of thermophilic spoilage and their significance in canned foods are presented in **Table 5**.

Prevention of thermophilic spoilage can be achieved by cooling the retorted cans rapidly to a temperature below 40°C and storing finished products at below 35°C to inhibit the growth of any surviving thermophiles. Also, thorough washing of raw vegetables to remove soil and prevention of recontamination with soil or other sources of thermophiles during the preparatory operations is of paramount importance. When ingredients such as sugar, starch and spices are used, processors should exercise great care to ensure that these ingredients do not contain excessive numbers of thermophilic spores. The American National Processors Association has established specifications for these organisms in ingredients used in canning (**Table 6**).

Preheating tanks, blanchers and other flotation washers used in the preparatory steps can provide a breeding area for thermophiles, especially overnight and during periods of shutdown, if the tanks are not properly sanitized. This is particularly of concern in the canning of cream-style sweetcorn and whole kernel corn. As a rule, it is best to hold these tanks overnight empty of foods and full of cold water.

Post-processing Contamination

Post-process contamination or leaker spoilage takes place when microbial contaminants leak into the can after heat sterilization owing to a failure of the container to maintain a hermetic seal. It is undoubtedly of the greatest economic importance as it accounts for 60–80% of the spoilage of canned foods.

The microorganisms involved in leaker spoilage can be any type found on can handling equipment, in cooling water or on the skin of can handlers: they include bacterial cocci, short and long rods, yeasts and moulds, aerobic spore-formers, and are probably a mixture of many of these organisms. Post-process contamination can also result in outbreaks of botulism or *Staphylococcus* enterotoxin poisoning. Leaker spoilage is often associated with the integrity of the can seams, the presence of bacterial contaminants in the cooling water or on wet can runways, and abusive can handling procedures after heat processing. Cooling water can be the primary source of organisms responsible for leaker spoilage.

Although very rare nowadays, defective tin plate, can manufacturing defects and mishandling empty cans will affect the can integrity. Can manufacturing defects of interest are defective side seams, over- or under-flanging which interferes with double seam formation, and defective double seams resulting from faulty seam operation or end compound distribution. The most common type of damage from improper handling of empty cans is bent flanges or cable cuts. The latter occur when cans are held back while the conveyor cable continues to run.

Filling operations can directly affect the quality of the seam. Overfilling, particularly with a cold product, and the subsequent expansion of the product during processing can cause end distortion and seam damage. Also, if the filler leaves product hanging over the can flange, it may interfere with can seaming and result in leakage, especially with fibrous products such as leafy vegetables, or meat and fish products containing pieces of bone.

Double seam deficiencies can be transient or permanent. Transient leaks are reversible in that a leakage path opens but subsequently closes leaving no detectable deformation at the point of the leak. A permanent leak path through the double seam may exist due to an improper seam construction, the use of non-leak-resistant compounds, and improper side-seam soldering and side-seam tightness.

Table 5 Most common thermophilic microorganisms causing spoilage in canned foods

Microorganism	Canned foods affected	Growth conditions				Manifestations
		Opt. temp. (°C)	Temp. range (°C)	pH range	O_2 demand	
Bacillus stearothermophilus	Canned vegetables, meats, UHT milk, products high in starch	55 to 65	45–76	> 4.6	Facultative anaerobic	Flat cans. markedly low pH. Possible loss of vacuum on storage. Food appearance not usually altered. Possible slight abnormal odours or cloudy liquor. Coagulation of UHT milk
Clostridium thermosaccharolyticum	Canned vegetables	55	43–71	4.5–5.0	Strict anaerobic	Can swelling to the point of bursting. Decrease in pH. Production of H_2S and CO_2. Production of fermented, sour, cheesy or butyric odour
Clostridium nigrificans (sulphur stinkers or sulphide spoilage)	Canned vegetables and meats	55	27–70	> 5.8	Strict anaerobic	Production of H_2S which may darken the product because of reaction with container iron. Cans usually remain flat due to solubility of H_2S in food water
Bacillus coagulans	Tomato juice, acidified vegetable foods	55	27–60	> 4.1	Facultative anaerobic	Flat sour, little change in vacuum. Off odours

Table 6 Microbiological specification of the US National Food Processors Association for starch and sugar

Microorganisms	Specifications
Total thermophilic spore count	For the five samples examined, there shall be a maximum of 150 spores and an average of not more than 125 spores per 10 g of sugar or starch
Flat sour spores	For the five samples examined, there shall be a maximum of 75 spores and an average of not more than 50 spores per 10 g of sugar or starch
Thermophilic anaerobic spores	These shall be present in not more than three of the five samples and in any one sample to the extent of not more than four of six tubes inoculated by the standard procedure
Sulphide spoilage spores	These shall be present in not more than two of the five samples and in any one sample to the extent of not more than 5 spores per 10 g

Handling the filled containers can dramatically affect the hermetic seal. Blows due to dropping cans in retort crates without cushioning the fall or to cans rolling and striking solid surfaces (such as another can or the bar of an elevator) can lead to a leak. The effects of repeated blows on the seam are cumulative and may lead to contamination similar to that resulting from a singular violent blow. Seam deformations may also result from mechanical impacts, or from abrupt pressure changes occurring when retorted cans are suddenly exposed to atmospheric pressure for cooling. During thermal processing the can contents expand considerably and may result in permanent distortion of the can ends, unless adequate head-space under vacuum and counterpressure during cooling of large cans is provided.

To minimize post-process contamination, it is necessary to ensure good controls over empty can inspection and handling systems, can seam integrity, adequate chlorination of cooling water and minimal can abuse during in-plant can handling, transportation and distribution. As can integrity is critical, can seams and seaming machines should be inspected as frequently as feasible, at least every 30 min for a visual inspection of a seamed can and every 4 h for a thorough seam tear-down and examination. Cooling water should be chlorinated to 2–5 p.p.m.; this allows for a content of no less than 1 p.p.m. after cooling. Cooled, wet cans should be dried in a restricted access area and must not be handled until they are dry. Care should be exercised thereafter to minimize abuse leading to dents and leakage.

Other Causes of Spoilage

Canned foods can also spoil because of non-microbial causes. Indeed, contamination of foods with metals such as copper or iron before it is placed in the can or a reaction between the food and the container can lead to objectionable food colour defects. These include blue-greying of corn and blackening of peas, corn, shrimps and other fish meats. This is often the result of protein-sulphur compounds breaking up under high temperature during blanching or cooking, and combining with iron to form black iron sulphide. The use of enamel-lined cans for these products eliminates this problem.

Internal can corrosion also leads to the accumulation of hydrogen which relieves the vacuum and swells the can, making it unmarketable. Externally, corrosion often causes pinholes that allow microorganisms to penetrate the can and spoil its contents.

Quality Assurance of Canned Foods

Sampling and examination of end products is of no significant value for assessing the efficacy of a process in canned foods. Indeed, the process failure rate is very low for thermally processed foods, of the order 10^{-6} to 10^{-12}. Consequently, the probability of finding a non-sterile container in a sample of 10 000, for example, based on a Poisson distribution, is very low (< 0.095). Moreover, even in cases where a comprehensive sampling inspection reveals that the process failure rate is exceeded, rational recommendations cannot be given unless the entire line is carefully inspected by often elaborate procedures.

Consequently, the food canning industry must rely on preventative approaches embodied in codes of good manufacturing practices and the implementation of a quality assurance programme based on the hazard analysis critical control point (HACCP) approach. Good manufacturing operations include:

- the use of acceptable quality raw material
- the use of an adequate heat process, designed by a process authority and applied by qualified personnel
- appropriate checking of the integrity of the container closure
- control of post-process hygiene
- control of storage and distribution conditions.

The HACCP approach has been demonstrated to be cost-effective for the assurance of safety and quality of canned foods. This is probably why it has become a mandatory system in the European Union, the USA, Canada and many other countries worldwide.

See also: **Clostridium**: Introduction; *Clostridium botulinum*. **Fish**: Spoilage of Fish. **Food Poisoning Outbreaks**. **Hazard Appraisal (HACCP)**: The Overall Concept; Critical Control Points. **Spoilage Problems**: Problems Caused by Bacteria.

Further Reading

Ababouch LH (1991) Histamine food poisoning. An update. *Fish Tech News* (FAO) 11(1): 3–5, 9.
Ababouch LH (1995) *Assurance de la Qualité dans l'Industrie Halieutique*. Rabat: Actes Editions.
Ababouch LH, Chouguer L and Busta FF (1987) Causes of spoilage of thermally processed fish in Morocco. *Int J Food Sci Technol* 22: 345–354.
Datta AK (1992) Thermal processing: food canning. In: Hui YH (ed.) *Encyclopedia of Food Science and Technology*. Vol. 1, p. 260; vol. 4, p. 2561. New York: Wiley Interscience.
Denny CB and Corlett DA (1992) Canned foods – tests for cause of spoilage. In: *Compendium of Methods for the Microbiological Examination of Foods*, 3rd edn. P. 1051. Washington: American Public Health Association.
Dryer JM and Deibel KE (1992) Canned foods. Test for commercial sterility. In: *Compendium of Methods for the Microbiological Examination of Foods*, 3rd edn. P. 1037. Washington: American Public Health Association.
Gavin A and Weddig LM (1995) *Canned Foods. Principles of Thermal Process Control, Acidification and Container Closure Evaluation*, 6th edn. Washington: Food Processors Institute.
Lange HJ (1983) *Methods of Analysis for the Canning Industry*. Orpington: Food Trade Press.
Smelt JPPM and Mossel DAA (1982) Application of thermal processes in the food industry. In: Russell AD, Hugo WB and Ayliffe GAJ (eds) *Principles and Practice of Disinfection, Preservation and Sterilisation*. Oxford: Blackwell Scientific.
Taylor SL (1986) Histamine food poisoning: toxicology and clinical aspects. *Critical Reviews in Toxicology*. 17: 91–117.
Trique B (1991) Microbiologie des produits végétaux. In: Larousse J (ed.) *La Conserve Appertisée. Aspects Scientifiques, Techniques et Économiques*. P. 136. Paris: Lavoisier.
Warne D (1988) *Manual on Fish Canning*. Technical Paper 285. Rome: FAO.

Ultra-high Temperature (UHT) Treatments

M J Lewis, Department of Food Science and Technology, University of Reading, UK

Introduction

Sterilization of foods in sealed containers has been practised for over 200 years. Practical drawbacks of 'in-container' sterilization processes arise firstly because products heat and cool relatively slowly, and secondly because processing temperatures are limited by the internal pressure generated.

Ultra-high temperature (UHT) processing has been introduced as an alternative sterilization process. This is a process for sterilizing foods, which combines continuous flow thermal processing with aseptic packaging; thus the term 'aseptic processing' is also used. It is possible to use higher temperatures by removing the pressure constraints, and the heating and cooling rates are also potentially faster. Both these factors provide potential for improving product quality. The process can be applied to any foodstuff that can be pumped through a heat exchanger: this ranges from low-viscosity fluids such as cow's milk and soya milk, to fluids of greater viscosity such as creams, ice-cream mixes, soups and starch-based products. It can also be used to process fluids containing discrete particles, up to 25 mm in diameter. A wide variety of UHT products are available; they are commercially sterile and usually have a shelf life of 6 months at ambient temperature. In many cases, shelf life is dictated by chemical and physical changes which continue to take place during storage. Target spoilage rates would be less than 1 in 10^4 containers.

Low-acid foods (pH > 4.5) are rapidly heated to temperatures in excess of 135°C, held for a few seconds and then rapidly cooled. The product is then ideally packed into sterile containers under aseptic conditions. If this cannot be done immediately, the product must be stored in an aseptic tank. For ensuring high quality, the heating and cooling rates should be as fast as possible. It is also possible to treat acidic products, e.g. fruit juices and fermented products, although the heat treatment required is less severe (less than 100°C). Aseptic packaging is still an essential requirement.

A wide variety of heat exchangers are available. For low-viscosity fluids, plate heat exchangers are widely used. As product viscosity increases, tubular heat exchangers become a more suitable choice. Scraped-surface heat exchangers are required for high-viscosity and particulate systems, but there are other options, such as the Ohmic and Jupiter systems. A further option is direct contact of steam with the product, either by injection or infusion. This will result in dilution, so provision needs to be made to remove this added water. After heat treatment it is also crucial to eliminate post-processing contamination.

In continuous processes there is a distribution of residence times. The viscosity and density are two important physical properties, which combined with flow rate and pipe dimensions will also determine whether the flow is streamline or turbulent. This in turn will influence heat transfer rates and the distribution of residence times within the holding tube and also the rest of the plant. For viscous fluids, the flow in the holding tube is likely to be streamline and

there will be a wide distribution of residence times. This can lead to some of the fluid being under-processed, whereas other elements of the fluid may be overprocessed. For Newtonian fluids the minimum residence time will be half the average residence time. Turbulent flow will result in a narrower distribution of residence times, with a minimum residence time of 0.83 times the average residence time. In both cases, the minimum residence time should be greater than the stipulated residence time, to avoid underprocessing.

Review of Kinetic Parameters

The main purpose of heat treatment is to reduce the microbial population. However, when any food is heated many other reactions take place, including enzyme inactivation and other chemical reactions. These may alter the sensory characteristics of the product, i.e. its appearance, colour, flavour and texture, and reduce its nutritional value. The two most important kinetic parameters are the rate of reaction or inactivation at a constant temperature (e.g. D and k values), and the effect of temperature change on reaction rate (z and E values).

The heat resistance of vegetative bacteria and microbial spores at a constant temperature is characterized by their D value; this is the time required to reduce the population by 90% or one log cycle. For vegetative organisms, D values are quoted in the range 60–80°C, and for spores in the range 100–140°C. Generally heat inactivation follows first-order reaction kinetics. The number of decimal reductions, log (N_0/N), can be evaluated from:

$$\log (N_0/N) = \text{heating time/D} \qquad \text{(Equation 1)}$$

where N_0 is the initial population and N is the final population. Two important points follow from this: firstly, it is not possible to achieve 100% reduction of microorganisms; and secondly, for a specified heat treatment, the final population will increase as the initial population increases. Therefore heat treatment is not regarded as an absolute form of sterilization, and the microbial quality of the raw material will have a major effect on the final population and hence the keeping quality.

The temperature dependence of a reaction is measured by the z value, i.e. the temperature change that brings about a tenfold change in the D value. Most heat-resistant spores are found to have a z value of 10°C. Alternatively, one can say that a temperature rise of 10°C will result in a tenfold reduction in the processing time to achieve the same lethality. *Chemical reaction rates are less temperature sensitive than microbial inactivation; using higher temperatures for* shorter times will result in less chemical damage occurring for an equivalent level of microbial inactivation. This is an important principle for UHT processing; it is usually the case that less chemical damage is done at higher temperatures and shorter times. In most instances this will improve product quality – notable exceptions being less inactivation of enzymes and anti-nutritional compounds. **Table 1** gives a summary of heat resistance data for some important spores, enzymes and chemical reactions that occur when milk is heated.

Continuous Processing Options

Ultra-high temperature processes are classified as indirect or direct. For indirect processes, the heat transfer medium does not come into contact with the product. The layout of a typical indirect plant is shown in **Figure 1**. Homogenization may be either upstream or downstream. Energy is conserved by regeneration, where the hot fluid is used to heat the incoming fluid. The most common types of heat exchanger are the plate or tubular types. One of the major practical problems is deposit formation on the surface of the heat exchanger. Such fouling will increase the pressure drop, especially in plate heat exchangers, and must be removed to ensure hygienic processing operations are maintained.

Table 1 Values of D and z for microbial inactivation, enzyme inactivation and chemical reactions. After Burton (1988)

	D_{121} (s)	z (°C)
Bacillus stearothermophilus NCDO 1096, milk	181	9.43
B. stearothermophilus FS 1518, conc. milk	117	9.35
Bacillus stearothermophilus FS 1518, milk	324	6.7
Bacillus stearothermophilus NCDO 1096, milk	372	9.3
B. subtilis 786, milk	20	6.66
B. coagulans 604, milk	60	5.98
B. cereus, milk	3.8	35.9
Clostridium sporogenes PA 3679, conc. milk	43	11.3
C. botulinum NCTC 7272	3.2	36.1
C. botulinum (canning data)	13	10.0
Proteases inactivation	0.5–27 min at 150°C	32.5–28.5
Lipases inactivation	0.5–1.7 min at 150°C	42–25
Browning	–	28.2; 21.3
Total whey protein denaturation, 130–150°C	–	30
Available lysine	–	30.1
Thiamin (B₁) loss	–	31.4–29.4
Lactulose formation	–	27.7–21.0

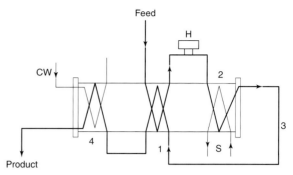

Figure 1 Layout of indirect UHT plant. 1, regeneration section; 2, preheating section; 3, holding tube; 4, cooling section; CW, chilled water; H, homogenizer; S, steam or hot water. (With permission, Cream Processing Manual, Society of Dairy Technology (1989) Huntington, England.)

Direct processes are by injection or infusion, the product being preheated to about 75°C prior to contact with the steam (**Fig. 2**). The condensed steam dilutes the product by about 10–15%. Care has to be taken to avoid contamination by rust, oil droplets or excessive water from the steam. The added water is removed by flash cooling, which also removes some of the more volatile components, which may either improve (in the case of hydrogen sulphide from heated milk) or cause a deterioration in flavour, for example the loss of natural volatiles from fruit juice. It also reduces the dissolved oxygen concentration, which may improve stability to oxidation reactions during storage. Heating and cooling are very rapid, making it a much less harsh process than indirect methods. Both capital costs and running costs are higher.

Liquids containing suspended solids, e.g. baked beans in tomato sauce, are known as 'particulate' systems. These pose a serious problem because it will take longer to sterilize the solid phase than the liquid phase. This becomes even more of a problem as the particle size increases. A second difficulty arises from the fact that the distribution of residence times is different for the solid phase compared with the liquid phase. Factors affecting this are the particle size, the flow regime and the density difference between the solid and liquid phases. The conventional choice would be a scraped-surface heat exchanger; this gives difficulties with agitation. One solution to the heating problem is Ohmic heating system, which is commercially available. The principle of this system is that an electrical current passes through the food, causing it to heat internally in a similar manner to an electric heating element. Factors affecting the degree of heating will be the applied voltage and the electrical resistance of the food. This is particularly useful for particulate systems, as it provides the opportunity of heating the solid phase as quickly as the liquid phase,

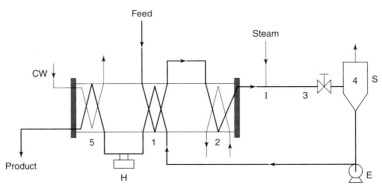

Figure 2 Layout of direct UHT plant. 1, regeneration section; 2, preheating section; 3, holding tube; 4, flash-cooling section; 5, cooling section; CW, chilled water; E, extract pump; H, homogenizer; I, injection point; S, separator. (With permission, Cream Processing Manual, Society of Dairy Technology (1989) Huntington, England.)

thereby reducing the requirement to drastically over-process the liquid stream to obtain adequate particle sterility. Another concept involves sterilizing the solid and liquid phases separately and then recombining them. This is the principle behind the Jupiter system.

Safety and Spoilage Considerations

To some extent, requirements for safety and quality conflict, as a certain amount of chemical change will occur during adequate sterilization of the food, reducing the quality. From a safety standpoint, the main concern is inactivation of the most heat-resistant pathogenic spore, namely *Clostridium botulinum*. The criterion used for UHT processing should be based upon those established for canned and bottled products. A fundamental distinction is made between acidic products (pH < 4.5) and low-acid products (pH > 4.5). For acidic products, yeasts and moulds need to be inactivated. This can be achieved by heating to about 100°C. However, the main concern is with low-acid products, e.g. vegetables, milk, meat and fish, where the minimum criterion should involve 12 decimal reductions for *Clostridium botulinum*. This will involve heating the product at 121°C for 3 min, at its slowest heating point. The microbial severity of a process is traditionally expressed in terms of its F_0 value. This takes into account the contributions of the heating, holding and cooling periods to the total lethality and is expressed in terms of minutes at 121°C. It provides a useful means of comparing processes. The minimum F_0 value for any low-acid food should be 3.

The temperature–time conditions required to achieve the minimum *Clostridium botulinum* cook are given in **Figure 3**, along with conditions for some other well-used criteria. There is experimental evidence to show that the data for *Clostridium botulinum* can be extended up to about 140°C. For UHT

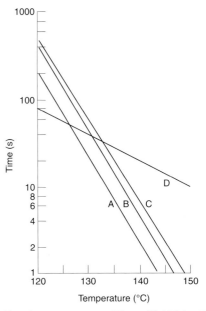

Figure 3 Time/temperature conditions: (A) 12D for *Clostridium botulinum* ($F_0 = 3$). (B) $F_0 = 6$. (C) 9D for thermophilic spores in cream. (D) $C^* = 1$.

products, an approximate value of F_0 can be obtained from the holding temperature (θ) in °C and minimum residence time (t) in seconds (Equation 2). For a process in which F_0 is 3 min, a time of 1.8 s would be required at 141°C. In practice the real value will be higher than this estimated value because of the lethality contributions from the end of the heating period and the beginning of the cooling period, as well as some additional lethality from the distribution of residence times.

$$F_0 = 10^{(\theta-121.1)/10} \, t/60 \qquad \text{(Equation 2)}$$

Therefore, the botulinum cook should be a minimum requirement for all low-acid foods, even those where the microorganism is rarely found in the raw material, e.g. raw milk.

The attainment of commercial sterility is a second important microbiological criterion. The minimum botulinum cook will yield a product that is safe, but not necessarily commercially sterile. The reason for this is the presence of more heat-resistant spores, which may cause spoilage, but which are not pathogenic, such as *Bacillus stearothermophilus*. For foods that may contain such spores, a heat treatment achieving 2 or more decimal reductions is recommended, corresponding to an F_0 value of about 8 min, or 141°C for 4.8 s. Holding times required at other temperatures can be calculated from Equation 2. Such conditions will provide an additional measure of safety as far as the botulinum risk is concerned. Recently, attention has focused upon a heat-resistant mesophilic spore-forming bacteria, which has been causing UHT milks to fail the specified microbiological tests in several European countries. The results from several laboratories are reported in **Table 2**, showing the wide variation between laboratories and some potential for the spores to withstand UHT processing conditions. This microorganism has now been classified as *Bacillus sporothermodurans*. One of its curiosities is that it will grow up to counts of 10^5 per millilitre but will not cause any easily recognized changes in the sensory characteristics of the milk. It is also not considered to be a food poisoning microorganism.

In the UK there are statutory heat treatment regulations, giving the minimum times and temperatures for some UHT products (**Table 3**). Where no guidelines are given, recommended F_0 values for similar canned products would be an appropriate starting point (typically 4–18 min for dairy products).

One important aspect of quality which has already been discussed is the reduction of microbial spoilage. A second important aspect is minimizing the extent of chemical reactions, particularly those adversely affecting the sensory characteristics or nutritional value. In this aspect, UHT processing offers some distinct advantages over in-container sterilization. The different heat exchangers available can heat products at different rates and shear conditions. For a better understanding of the UHT process, it is necessary to know the temperature and time profile for the product. Some examples of such profiles are shown for a number of different UHT process plants in **Figure 4**. In general, direct processes offer the fastest heating and cooling rates. Of the indirect processes, plate systems usually give faster heating and cooling rates than tubular systems and the rates are further reduced as the regeneration efficiency is increased. Because of these differences, similar products processed on different plants may well vary in quality.

Two other parameters introduced for UHT processing of dairy products, which could be more widely used for other UHT products, are B* and C* values. The reference temperature used (135°C) is much closer to UHT processing temperatures than that used for F_0 (121°C) or cooking value (100°C) estimations. The microbial parameter B* is used to measure the total integrated lethal effect of a process. A process in which B* = 1 would be sufficient to produce 9 decimal reductions of mesophilic spores and would be equivalent to 10.1 s at 135°C. The parameter C* measures the amount of chemical damage taking place during

Table 3 Statutory heat treatment regulations in the UK

Product	Minimum temperature (°C)	Minimum time (s)
Milk	135	1
Cream	140	2
Milk-based products	140	2
Ice-cream mix	148.9	2

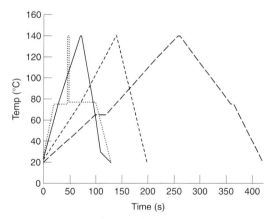

Figure 4 Temperature/time profiles for different UHT plants.

Table 2 Reported heat-resistance data for *Bacillus sporothermodurans* (With permission, Hammer et al, 1996)

Institute	Result
Canning Research Institute, Campden (UK)	$D_{100} = 5.09$ min
Institute for Food Technology, Weihenstephan, Germany	$D_{121} = 8.3$–34 s
Tetra Pak, Lund, Sweden	$F_0 < 10$ min (contaminated milk)
	$F_0 > 68$ min (pilot plant, production plant)
Tetra Pak Research, Stuttgart, Germany	$D_{98} > 60$ min; D_{120} approx 10 min
Netherlands Institute for Dairy Research, Ede, The Netherlands	$D_{126} = 1$ min; $D_{147} = 5$ s

the process. A process in which $C^* = 1$ would cause 3% destruction of thiamin and would be equivalent to 30.5 s at 135°C. Again, the criteria in most cases is to obtain a B^* value greater than 1 and a low C^* value. Calculations of B^* and C^* based on the minimum holding time and temperature are straightforward. Conditions corresponding to B^* and C^* values of 1 are shown in Figure 3.

The effects of increasing, heating and cooling periods on F_0, B^* and C^* are shown in **Table 4**. These results are based on heating the product from 80°C to 140°C, holding it at 140°C for 2 s and then cooling it to 80°C. Heating and cooling periods from instantaneous through to 60 s are shown. Increasing these periods increases both the chemical and the microbial parameters, with the ratio of chemical to microbial values increasing with increasing heating period. At a heating period of about 8 s, the amount of chemical damage done during heating and cooling exceeds that in the holding tube. It is this considerable increase in chemical damage that will be more noticeable in terms of decreasing the quality of the product. However, this may be beneficial where a greater extent of chemical damage may be required, e.g. for inactivating enzymes or for heat inactivation of natural toxic components, such as trypsin inhibitor in soya milk, or for softening of vegetable tissue (cooking).

Chemical damage could be further reduced by using temperatures in excess of 145°C. One problem would be the very short holding times required, and the control of such short holding times. In theory it should be possible to obtain products with very high B^* and low C^* values, at holding times of about 1 s. For indirect processes, the use of higher temperatures may be limited by fouling considerations and it is important to ensure that the heat stability of the formulation is optimized. Generally direct systems give longer processing runs than indirect processes.

Controlling the Process

Strict adherence to these microbiological considerations will ensure that thermal processing is a safe procedure. It is recognized that UHT processing is more complex than conventional thermal processing.

Table 4   Effects of heating and cooling rates on F_0, B^* and C^* values for a holding time of 2 s at 140°C

Heating and cooling period (s)	F_0	B^* (total)	C^*
0.0	2.59	0.59	0.09
0.1	2.61	0.60	0.10
1	2.77	0.64	0.12
10	4.45	1.04	0.31
30	8.20	1.94	0.73
60	13.8	3.29	1.37

The philosophy of UHT processing should be based upon preventing and reducing microbial spoilage by a full understanding of the process, which will lead to procedures for controlling it effectively. One way of achieving this is by using the principles of hazard analysis critical control points (HACCP). The hazards for a UHT process are identified (**Fig. 5**) and procedures are adapted to control them. An acceptable initial target spoilage rate of less than 1 in 10^4 should be aimed for. Such low spoilage rates require large numbers of samples to be taken to verify that the process is being performed and controlled at the desired level; for example, approximately 30 000 samples would need to be analysed and zero defects found. When a new process is being commissioned, the process should be verified by 100% sampling. Once it is established that the process is under control, sampling frequency can be reduced and sampling plans can be designed to detect any spasmodic failures. It is noteworthy that for small sample sizes, detecting one spoiled sample would indicate a gross failure of the process; however, finding no failures would not indicate that the process was being controlled at the desired level.

For milk products, heat treatment regulations require that UHT milk complies with microbiological standards. Milk incubated for 14 days at 30°C should have a microbiological count of less than 100 per millilitre. An alternative incubation period is 55°C for

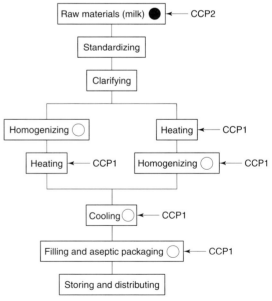

HAZARD ANALYSIS CRITICAL CONTROL POINTS

Figure 5   Identification of critical control points (CCP) for UHT processing. Shaded circle indicates a site of major contamination; open circles indicate sites of minor contamination; CCP1 effective CCP, CCP2 not absolute. From ICMSF (1988). (With permission.)

7 days. This stresses the importance of an incubation period. Without incubation, it is highly unlikely that spoilage would be detected at an early stage. However, where spoilage does occur in UHT products, microbial growth can occur very quickly under suitable incubation conditions. Rapid microbiological test methods are gaining momentum. Whatever testing methods are used, prior incubation is essential. Other indirect indicators of microbial growth can also be used, such as pH and acidity, alcohol stability, clot-on-boiling methods and dissolved oxygen concentration.

More success in detecting spoilage will result from targeting high-risk occurrences such as start-up, shut-down and product changes. Thus holding time and temperature are perhaps the two most critical parameters. Recording thermometers should be checked and calibrated regularly, and accurate flow control is crucial (as for pasteurization). Sufficient pressure must also be applied in order to achieve the required temperature; a working pressure in the holding tube in excess of 1 bar (10^5 Pa) over the saturated vapour pressure corresponding to the UHT temperature has been suggested.

Raw Material and Processing Aspects

A wide variety of raw materials are used, so any UHT product will be potentially complex, containing protein, fat, carbohydrate, minerals, a wide range of nutrients and minor components, plus many active enzymes. It will also have the natural microbial flora of the raw materials. For example, raw milk from healthy animals is almost sterile; however, it will become contaminated with spoilage and perhaps pathogenic organisms. Of particular concern would be high levels of heat-resistant spores and enzymes in the raw materials, as these could lead to increased spoilage and stability problems during storage; dried products such as milk and other dairy powders, cocoa, other functional powders and spices are particularly likely to cause problems. Quality assurance programmes must ensure that contaminated raw materials are not used. It may be worthwhile developing simple tests to assess the heat stability to reduce fouling-related problems; the alcohol stability test has proved useful for milk products. The product formulation is also important, as is the nature of the principal ingredients, the levels of sugar, starch, salt as well as the pH of the mixture, particularly if there are appreciable amounts of protein. Some thought should be given to water quality, particularly the calcium and magnesium content, as these minerals may promote fouling. Reproducibility in metering and weighing ingredients is also important, as

is ensuring that powdered materials are properly dissolved or dispersed and that there are no clumps, which may protect heat-resistant spores. Homogenization conditions may be important; is it necessary to homogenize, and if so, at what pressures? Should the homogenizer be positioned upstream or downstream of the holding tube? Will two-stage homogenization offer any advantages? Homogenization upstream offers the advantage of breaking down any particulate matter to facilitate heat transfer, as well as avoiding the need to keep the homogenizer sterile during processing. All of these aspects will influence both the safety and the quality of the products.

Ultra-high temperature products, like canned products, will be susceptible to post-processing contamination. This will not usually give rise to a public health problem, although contamination with pathogens cannot be ruled out. Contamination may arise from the product being reinfected in the cooling section of the plant, or in the pipelines leading to the aseptic holding or buffer tank or the aseptic fillers. This is avoided by heating all points downstream of the holding tube at 130°C for 30 min.

Sterilization and cleaning procedures are therefore very important: are they adequate and are they properly accomplished? For cleaning, the detergent concentrations and temperatures are important to remove accumulated deposits. Such deposits, if not properly removed, may form breeding grounds for thermoduric and thermophilic spoilage bacteria. Steam barriers should be incorporated if some parts of the equipment are to be kept sterile while other parts are being cleaned.

Recording all the important experimental parameters will help pinpoint when the process plant is deviating from normal behaviour and ensure that any faults are quickly detected. Regular inspection and maintenance of equipment, particularly the elimination of leaks, is essential. Staff education programmes should be implemented. All staff involved with the process should be educated to understand its principles and should be encouraged to be diligent and observant. With experience, further hazards will become apparent and methods for controlling them introduced. The overall aim should be to reduce spoilage rates and to improve the quality of the product.

Aseptic Filling Procedures

A number of aseptic packaging systems are available. They all involve putting a sterile product into a sterile container in an aseptic environment. Pack sizes range from individual portions (14 ml), retail packs (125 ml to 1 l), through to bag-in-the-box systems up to 1000 l. The sterilizing agent is usually hydrogen peroxide (35% at about 75–80°C); the contact time is short and the residual H_2O_2 is decomposed using hot air. The aim is to achieve a 4D process for spores. Superheated steam has been used for sterilization of cans in the Dole process. Irradiation may be used for plastic bags.

Since aseptic packaging systems are complex, there is considerable scope for packaging faults to occur, which will lead to spoiled products. Where faults occur, the spoilage microorganisms would be from the environment and would include microorganisms that would be expected to be inactivated by UHT processing; these often result in blown packages.

Packages should be inspected regularly to ensure that they are airtight, again focusing upon those more critical parts of the process – start-up, shutdown, product changeovers and, for carton systems, reel splices and paper splices. Sterilization procedures should be verified. The seal integrity of the package should be monitored as well as the overall microbial quality of the packaging material itself. Care should be taken to minimize damage during subsequent handing. All these could result in an increase in spoilage rate.

Storage

Usually UHT products are stored at room (ambient) temperature and good-quality products should be microbiologically stable. Nevertheless, chemical reactions and physical changes may take place that will change the quality of the product.

See also: **Bacillus**: Bacillus stearothermophilus. **Clostridium**: Clostridium botulinum. **Hazard Appraisal (HACCP)**: The Overall Concept; Critical Control Points. **Milk and Milk Products**: Microbiology of Liquid Milk. **Packaging of Foods**: Packaging of Solids and Liquids. **Process Hygiene**: Overall Approach to Hygienic Processing; Modern Systems of Plant Cleaning. **Spoilage Problems**: Problems caused by Bacteria; Problems caused by Fungi.

Further Reading

Burton H (1988) *UHT Processing of Milk and Milk Products*. London: Elsevier Applied Science.

Gaze JE and Brown KL (1988) The heat resistance of spores of *Clostridium botulinum* 213B over the temperature range 120 to 140°C. *Int. J. Food Sci. Technol.* 23: 373–378.

Hammer P, Lembke F, Suhren G and Heeschen W (1996) Characterisation of heat resistant mesophilic *Bacillus* species affecting the quality of UHT milk. In: *Heat Treatments and Alternative Methods*. IDF/FIL No 9602.

IFST (1998) *Food and Drink – Good Manufacturing Practice: A Guide to Its Responsible Management*, 4th edn. London: IFST.

ICMSF (1988) *Micro-organisms in Foods.* Vol. 4. *Application of the Hazard Analysis Critical Control Points (HACCP) System to Ensure Microbiological Safety.* Oxford: Blackwell Scientific.

Kessler HG (1981) *Food Engineering and Dairy Technology.* Freising: Kessler.

Kwok KC and Niranjan K (1995) Review: effect of thermal processing on soymilk. *Int. J. Food Sci. Technol.* 30: 263–295.

Petersson B, Lembke F, Hammer P, Stackebrandt E and Priest FG (1996) *Bacillus sporothermodurans*, a new species producing highly heat-resistant endospores. *Int. J. Syst. Bacteriol.* 46 (3): 759–764.

Reznick D (1996) Ohmic heating of fluid foods. *Food Technol.* 50 (5): 250–252.

Statutory Instruments (1995) *Food, Milk and Dairies: The Dairy Products (Hygiene) Regulations.* London: HMSO.

Principles of Pasteurization

R Andrew Wilbey, Department of Food Science, University of Reading, UK

Introduction

Pasteurization is widely accepted as an effective means of destroying vegetative pathogens in food products, with the least possible damage to the sensory qualities of the product. The heat treatment also reduces the general microbial population, so that an increased shelf life is normally obtained. Bacterial spores and some heat-resistant enzymes will survive pasteurization processes and limit the shelf life of the product.

Historical Origins

Cooking is an age-old method of preparing many traditional foodstuffs and for centuries it was generally appreciated that cooked products would normally take longer to putrefy than if they were left raw.

In the latter part of the eighteenth century there

was great interest in understanding the mechanism of putrefaction. Lazzaro Spallanzani demonstrated that putrefaction may not occur in a heated sealed flask of an infusion, but that aerial contamination could result in putrefaction. The presence of microorganisms was demonstrated, and it was recognized that there was a possible division into organisms that could be killed by boiling and those that would survive this heat treatment. Subsequent experiments by Franz Schutze in the early nineteenth century demonstrated that it was not the air itself that caused spontaneous putrefaction, but a contaminant carried in the air. Similar experiments were carried out by Theodor Schwann at the same time.

Methods for the preservation of foodstuffs were developed in parallel with this pioneering work on the basic understanding of why foods spoil. Carl Wilhelm Scheele used heat for the conservation of vinegar, and Nicholas Appert developed the preservation of foods by heating in cans. In 1824 William Dewes recommended that milk for infant feeding be heated to near boiling (but not boiled) and then cooled during its preparation.

The credit for the development of a mild method of processing foods, now particularly associated with milk, has been given to Louis Pasteur, after whom the pasteurization process was named. Pasteur had, amongst his many interests, an interest in fermentations. The poor hygiene conditions associated with the production of food and beverages at that time often led to unwanted fermentations, causing putrefaction and loss of product. His experiments confirmed that fermentations were not spontaneous but were the result of microbial metabolism. While some of his earlier work was with lactic fermentations, most of his work in this field was based on alcoholic fermentations, brought about by yeasts. The conversion of ethanol to acetic acid was demonstrated to be brought about by bacteria, subsequently classified as *Acetobacter*. In an acid medium such as wine, both yeasts and *Acetobacter* could be destroyed by relatively mild heat treatments at about 55°C in closed vessels.

While Pasteur's work on beer, wine and vinegar laid the foundations for hygienic processing, his complementary work on the relation between specific organisms and disease also aided the recognition of the public health implications of hygiene and of heat treatments.

By the late nineteenth century the economic benefits from improving the shelf life of milk and other products were appreciated, though the microbiological and public health implications of pasteurization were not fully understood. Pasteurization of wine was adopted in both France and the USA and is still used for some wines, although filtration, higher alcohol levels and better production hygiene have largely displaced heat treatment.

The heat treatment of milk on a commercial scale did not develop until the end of the nineteenth century, with the production of commercial pasteurizers in Germany and Denmark. The earlier treatment systems were aimed at improving the storage life of milk, often using simple continuous flow techniques to reduce costs. The realization that milk was a potential carrier of diseases such as tuberculosis led to the development of a low temperature long time (LTLT) batch process, the first commercial plant being installed by Charles North in New York in 1907. In-bottle LTLT pasteurization of milk, pioneered by North in 1911, though capable of producing a high-quality product virtually free of post-pasteurization contamination, was too expensive to compete in the marketplace. It was not until 1922 that pasteurization was recognized legally in the UK when the term was defined in the Milk and Dairies (Amendment) Act, using an LTLT process at 62.8–65.5°C for a minimum of 30 min.

Pasteurization by the LTLT process was the first safe method adopted, but the processing of milk was revolutionized by the invention in the UK of the plate heat exchanger which was capable of recovering some of the heat from the hot pasteurized product. The development of a modular heat exchanger that could be relatively easily cleaned, together with a microbiologically effective holding tube system and a flow diversion valve, enabled milk to be heat-treated with safety on a far larger scale than had been possible with the batch-based LTLT system. With better appreciation of the thermal death characteristics of pathogens, this continuous process was able to take advantage of higher process temperatures with a correspondingly shorter hold and became known as the high temperature short time (HTST) process. In the European Community (EC) the minimum heat treatment permitted is 71.7°C for 15 s. It has been suggested that 15 s was originally chosen as the minimum time to allow an adequate safety margin for the response rate of the temperature sensing and control system at that time.

Subsequent developments in the design of process equipment have led to the construction of pasteurization plants that may be cleaned in place, with much greater thermal efficiency and with much more sensitive and responsive instrumentation and control systems. It is now technically possible to pasteurize at higher temperatures with little or no hold, the 'flash' processes.

Aims of Pasteurization

The aim of pasteurization is summarized by the definition adopted by the International Dairy Federation (IDF), in which pasteurization is defined as a process applied to a product with the aim of avoiding public health hazards arising from pathogenic microorganisms associated with milk by heat treatment which is consistent with minimal chemical, physical and organoleptic changes in the product.

This definition would be equally applicable to other commodities if one were to substitute the name of the product for 'milk' in the definition.

To achieve the public health objective of pasteurization in a particular product, it is essential to be aware of the pathogens associated – or potentially associated – with that product. The thermal death characteristics of the organisms in that product must also be known.

In most early work the death characteristics were expressed in terms of a temperature–time combination that would destroy the target pathogen. In the case of milk, tuberculosis was recognized as a major disease associated with milk consumption and *Mycobacterium tuberculosis* was found to be the most heat-resistant pathogen normally associated with milk. Temperature–time combinations needed to destroy *Mycobacterium tuberculosis* were published by North in 1911, North and Park in 1927, Hammer in 1928 and Dahlberg in 1932; these data are included in **Figure 1**. This knowledge enabled safer process conditions to be set up for LTLT and subsequently HTST processes. The methods used are open to the criticism that it is not possible to demonstrate the absence of an organism, only to fail to detect it. However, with better understanding of the kinetics of thermal death rates, more quantitative data may now be obtained.

Thermal Death of Microorganisms

When organisms are subjected to a moist heat above their normal temperature range a number of effects may be noted (**Fig. 2**). With the relatively short heat treatment times normally associated with pasteurization, temperatures just above the normal growth temperature (zone A in Fig. 2) will have little or no effect on the number of survivors although there may be a risk of bacteria becoming more resistant to subsequent treatments. With further increase in temperature (zone B) a small lethal effect will become evident. At higher temperatures (zone C), which are exploited in pasteurization processes, the logarithm of the number of survivors is inversely proportional to the exposure time and to the temperature. Thus for a given temperature within the zone C the time taken for a tenfold reduction in survivors, the D value, may be obtained. These values are expressed in minutes or seconds and must be identified by the temperature, e.g. D_{72} for 72°C. The D values are an approximation for a given strain of a species, the death kinetics of which can include a tail of more temperature-resistant organisms; so, for more accurate work, a more sophisticated model may be appropriate, although for most purposes the D value concept is adequate.

The D value will decrease with increasing temperature. The rate of change is usually given as a z value, being the change in temperature required to give a tenfold change in the D value. Typical z values for mesophiles are 4–8°C in high water activity (a_w) systems; the data in Figure 1 for *Mycobacterium tuberculosis* imply a z value of 6.3°C. By comparison, bacterial spores often have a z value approximating to 10°C at temperatures above 100°C, most spores being able to withstand pasteurization. Care is needed in using z values as they can vary with temperature and, as with the D values, will also vary with the substrate. For example, changes in pH and a_w can produce major changes in the thermal resistance of organisms. Microorgaisms are less susceptible to heat when the a_w is lowered. Reducing the pH will normally increase the susceptibility to LTLT treatments

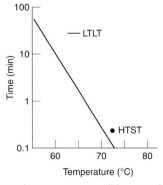

Figure 1 Time and temperature conditions needed for destruction of *Mycobacterium tuberculosis* in milk, together with minimum pasteurization conditions.

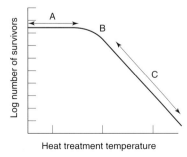

Figure 2 Effect of heat treatment temperature for a fixed time on the survival of organisms in a bacterial culture.

but the effect may not be significant under HTST conditions.

Yeasts and moulds are primarily of interest as spoilage organisms, though moulds may produce mycotoxins. The typical vegetative forms are normally more heat-labile than many spoilage bacteria but the ascospores may be more heat-resistant, though much less so than bacterial spores.

Table 1 gives a range of D and z values for microorganisms; these values must be taken as indicative only, for the reasons outlined above. Though significant numbers of pathogens should not survive the pasteurization process, there is a chance of survival of some heat-resistant organisms if the original numbers are very high or if some protective mechanism is operating. The microbial quality of the raw material and its hygienic handling are thus also important in limiting the challenge to the heat treatment system. In milk pasteurization a treatment at 72°C for 15 s will reduce the total count by approximately 2 orders of magnitude, thus the higher the original numbers then the higher the number of survivors and hence potentially the shorter the shelf life. This gross approximation cannot take into account post-pasteurization contamination or the survival of heat-resistant enzymes originating from the initial microflora. Recently there were concerns about the heat resistance of *Listeria monocytogenes* and *Mycobacterium paratuberculosis*. Under normal conditions the inactivation of *Listeria monocytogenes* appears to be adequate. There is less information available about the thermal destruction of *Mycobacterium paratuberculosis*; however, an IDF expert group concluded in 1997 that the risk can be 'accepted' until more data are available in terms of the thermal death curves in relation to naturally contaminated milks and heat treatments, and also whether there is a link with Crohn's disease.

While the epidemiological evidence supports the safety of pasteurization processes, refinements in the methods for recovering and quantifying specific organisms will lead to a more precise view of thermal death rates and hence the inherent risks.

Modern Pasteurization Processes

Both batch and continuous processes are used in the pasteurization of foods (**Fig. 3**). In-tank batch pasteurization continues to be used for speciality products and small-scale manufacture, e.g. ice cream. The main risks are those of cross contamination from raw materials, the processing environment and from the relatively slow cooling rates which may permit growth of survivors which could contribute to spoilage rates.

The in-container methods may be divided between those applied to unsealed and sealed containers, the former being employed for a few specialized products, e.g. clotted cream and egg-based desserts where there must be zero shear during the cooling process. The heat treatment normally exceeds that required just for pasteurization and the main risks are from contamination during the cooling process. The pasteurization of bottled beer and soft drinks in sealed containers may be carried out on a large scale using tunnel pasteurizers up to 20 m long where the product is conveyed under series of jets spraying progressively hotter then cooler water to effect the heating, holding and cooling parts of the cycle. While the relatively low heat transfer rates require an LTLT approach and can result in some flavour changes, this form of pasteurization has the advantage that the sealed container carries a very low risk of post-process con-

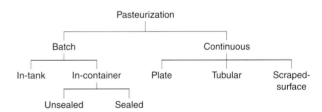

Figure 3 Types of pasteurization process.

Table 1 Examples of the heat resistance of some microorganisms

Organism	Medium	D_{72} value (s)	z value (°C)
Aspergillus niger[a]	Apple juice	0.76	6
Enterococcus faecium	Milk	3	2
Enterococcus faecium	Ham	1.4	7
Escherichia coli	Milk	1	6.5
Lactobacillus fermentum	Orange juice	2	5
Lactobacillus spp.[b]	Beer	10	8.6
Saccharomyces cerevisiae	Orange juice	0.02	4.7
Saccharomyces cerevisiae[b]	Soft drink	0.6	10
Staphylococcus aureus	Milk	0.6	5.1
Streptococcus salivarius subsp. *thermophilus*	Recombined milk	0.04	3.7

[a] Conidiospores.
[b] Heat-resistant strain.

tamination. Risks are further reduced by ensuring that the cooling water is of high microbiological quality, e.g. by hyperchlorination.

The heat treatment of many canned acid fruits (pH < 4.5) may be regarded as pasteurization since the heat treatment is only sufficient to inactivate vegetative spoilage organisms, surviving spores being inhibited by the low pH.

Continuous processes are preferred for large-scale pasteurization, particularly for liquid products. For low-viscosity homogeneous liquids the most commonly used process is based on plate heat exchangers, which are essentially flattened tubes made up from a pair of stainless steel plates separated by elastomeric seals. This creates a large surface area and a relatively thin gap of 2–6 mm so that rapid heat transfer under turbulent flow conditions may be achieved. The plates are assembled into a frame in a mixture of parallel and series flow configurations to give the desired heat transfer characteristics. In many cases a proportion of the heat used in achieving the pasteurization temperature may be recovered and used to preheat the incoming raw product, a process known as regeneration. Regeneration efficiencies of up to 95% are possible, the limits being capital cost and the effects of the increased heat exchange surface area on product quality. The simplest system for regeneration (**Fig. 4**) uses a single pumping system in which the pressure on the 'raw' side of the heat exchanger plate is higher than that of the hot pasteurized product, so that the integrity of the plate is a critical factor in the safety of the process. More sophisticated systems use a second pump and a means of maintaining a higher pressure in the downstream part of the plant. Whatever the pumping system adopted, the flow must be constant,

i.e. not increasing if the pressure drops. Positive displacement pumps are preferred but centrifugal pumps may be employed if accompanied by a flow control valve. Where homogenizers are used these may provide the second pumping system. Tubular heat exchangers have wider clearances than with plates and can cope with viscous and particulate foods more readily. This design will withstand higher pressures but has less effective heat transfer, although this can be improved by adopting an annular design.

Scraped-surface heat exchangers (SSHE) are essentially tubular heat exchangers containing a coaxial rotating shaft carrying blades that rub over the heat exchange surface, ensuring a rapid turnover of material on the surface and preventing the build-up of films that inhibit heat transfer. While SSHEs are effective in terms of heat transfer coefficients, their high capital and running costs limit their use to processes where the other heat exchangers are not appropriate, e.g. cooling of fat spreads and the simultaneous freezing and whipping of ice cream. The use of SSHEs requires rotary seals on the shafts, which can add to cleaning problems.

Estimating the Lethality of a Pasteurization Process

By using the appropriate D and z values it is possible to estimate the risks associated with a temperature–time combination, i.e. the probable level of survivors for a given level of contamination in the raw material. This is easy for a batch process such as LTLT pasteurization, since the hold is easily measured and the contribution of the heating and cooling stages to the overall lethality of the process is small. With HTST processes, however, the temperatures are higher and

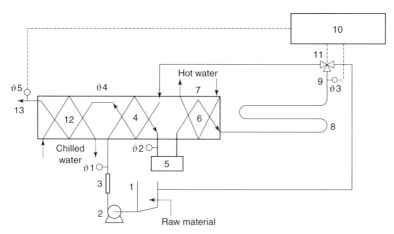

Figure 4 Flow diagram of a simple HTST pasteurizer. 1, balance tank; 2, feed pump; 3, flow controller; 4, preheating section and regenerative cooler; 5 homogenizer (optional); 6, final heating section; 7, hot water set; 8, holding tube; 9, hot product temperature sensor; 10, controller and recorders; 11, flow diversion valve; 12, final cooling; 13, cooled, pasteurized product exit. ϑ 1–5 temperature sensing points.

the heating and cooling stages may make a significant contribution to the overall lethality of the process. It is essential that the process be characterized in terms of temperature and minimum residence time. Minimum time is critical as the microbiological risk (particularly the public health risk) is associated with the minimum heat treatment given to any particle in the product. Since most HTST processes are continuous, the flow characteristics of the system must be taken into account. From a microbiological viewpoint, turbulent flows in the pasteurizer will give the best results as there will be a narrower spread of flow rates and hence of residence times in the equipment. Under turbulent flow conditions the minimum residence time can be up to 0.83 of the average residence time while under streamline flow conditions the minimum residence time is only 0.5 of the average. In practice, slower flow rates may be needed to conserve desirable product characteristics or to avoid excessive pressures.

Once the plant has been characterized it is possible to analyse the process quantitatively in relation to a given risk. One approach has been to define a pasteurization unit (PU) appropriate to the product and its most critical contaminants. A PU of 1 min at 65°C (z = 10°C) has been suggested for acid foods. The brewing industry has used a PU defined as 1 min at 60°C (z about 7°C) with 6–15 PU being used to stabilize bottled beer. A PU for safe HTST treatment of milk, the P*, has been suggested by Kessler, taking 1 P* as equal to 15 s at 72°C (z = 8°C). The implications for each second of a heat treatment are illustrated in **Figure 5**, where a higher z value of 10°C is also included to illustrate the effect of the z value on any estimation. Using a higher z value will lead to an underestimate of the lethality at higher temperatures and vice versa. It can be seen from Figure 2 that the contribution of temperatures below 65°C to overall lethality in an HTST process is so small that it may be ignored. However, at higher temperatures the effect of the temperature during heating and cooling

becomes more important so that by 90°C the total heat treatment is well in excess of the minimum safe treatment even without a hold. Heating and cooling rates are typically 1–3°C per second except in direct steam heating systems.

While the primary concern in pasteurization is to obtain a safe food, this is irrelevant if the sensory quality of the food is noticeably reduced, either by overcooking or by the persistence of other less temperature-labile factors (microbiological or biochemical). Cooked flavours may be acceptable in some foods, e.g. clotted cream, but not in others such as wine.

Heat treatments may also bring about undesirable changes in the stability or functional properties of food products, often related to protein denaturation. In milk the denaturation of agglutinin (an immunoglobulin fraction) reduces the rate at which cream forms in milk on standing, the denaturation being more pronounced when more severe heat treatments are used so that less cream separates from the milk. This has been an important consideration in processing milk for bottling, where the consumer has associated cream separation with milk quality, but is not relevant to the production of homogenized milks where cream separation would indicate a processing failure. Similarly, the protein denaturation associated with pasteurization of liquid egg white reduces the foaming properties of the product slightly, giving less volume and/or a longer whipping time than the raw egg white. Protein denaturation may also be used beneficially to indicate that a satisfactory heat treatment has been given, using an assay of a suitable enzyme occurring naturally in the product. The indicator enzyme should be denatured under conditions slightly more severe than those needed for microbial safety. Ideally the activity of the indicator enzyme should not be subject to wide variation with season or source, or be influenced by varying levels of microbial contaminants.

Alkaline phosphatase is denatured under slightly more severe conditions than are required for destruction of *Mycobacterium tuberculosis*, so the absence of alkaline phosphatase activity is generally used as an indicator for satisfactory pasteurization of milk. European Community milk hygiene regulations specify that pasteurized milk should have a negative reaction in the test for alkaline phosphatase and a positive reaction for lactoperodixase, thus setting minimum and maximum conditions for the heat treatment as lactoperoxidase is deactivated at 78–80°C for 15 s depending on the heat exchanger design.

The pasteurization of liquid egg in the UK (minimum 64.4°C for 2.5 min) is not sufficiently severe to inactivate alkaline phosphatase but will

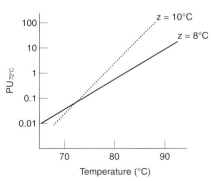

Figure 5 Lethal effect of a 1 s exposure at typical pasteurization temperatures.

denature α-amylase, whereas the milder treatment required in the USA (minimum 60°C for 1.75 min) will leave residual α-amylase activity. The proteins in eggs are more heat-labile than those in milk, which restricts the temperatures that may be used in HTST processes owing to fouling and protein precipitation, conditions favouring the use of tubular heat exchangers.

Bacteria are more resistant to heat treatment when the a_w of the medium is lowered. Thus more severe heat treatments are normally used for pasteurization of sweetened products such as ice cream and dessert products. Minimum heat treatments for ice cream may be 66°C for 30 min, 72°C for 10 min or 80°C for 15 s. Where the ingredients have already been heat-treated, enzyme assays may give misleading results.

In fruit juices the pH is usually below 4.5 so that growth of pathogenic bacteria will not be supported, though some organisms may only die slowly. Yeasts and some lactobacilli may grow and cause spoilage of the juice, and moulds may grow at the surface. Heat treatments to eliminate yeasts and lactobacilli are more severe than the treatments for elimination of vegetative pathogens, e.g. 70°C for 60 s, 85°C for 30 s for citrus juices.

Survival of enzymes can cause problems in the storage of fruit juices. In apple juice extraction polyphenol oxidase will cause rapid browning of cold extracted juices if the juice is not immediately treated with antioxidants such as ascorbic acid or sulphur dioxide. However, HTST treatment at 89°C for 90 s will denature polyphenol oxidase as well as potential spoilage organisms. Thus a check for colour development in the presence of oxygen may be used as a quality criterion.

In citrus juices the presence of pectinase will lead to breakdown of the cloud associated with the fresh juices. The application of HTST treatment at 90°C for 10 s or 85°C for 4 min will denature the pectinase. For many fruit juices, including apple and orange, the juice is extracted in the country of origin, heat-treated and concentrated. The concentrate is then stored and transported in bulk before reconstitution and final heat treatment.

See also: ***Aceterobacter***. **Eggs**: Microbiology of Fresh Eggs. **Heat Treatment of Foods**: Thermal Processing Required for Canning. ***Listeria***: *Listeria monocytogenes*. **Milk and Milk Products**: Microbiology of Liquid Milk. ***Mycobacterium***. **Spoilage Problems**: Problems caused by Bacteria; Problems caused by Fungi. **Wines**: Microbiology of Wine-making.

Further Reading

Cunningham FE (1986) Egg product pasteurisation. In: Stadelman WJ and Cotterill OJ (eds) *Egg Science and Technology*, 3rd edn. Westport: Avi Publishers.

Dubos RJ (1960) *Louis Pasteur: Free Lance of Science*. New York: Da Capo Press.

Green E (1983) *Pasteurising Plant Manual*. Huntingdon: Society of Dairy Technology.

Holdsworth SD (1997) *Thermal Processing of Packaged Foods*. London: Blackie.

International Dairy Federation (1986) *Bulletin 200: Monograph on Pasteurized Milk*. Brussels: IDF.

International Dairy Federation (1998) *Bulletin 330: Significance of* Mycobacterium paratuberculosis *in milk*. Brussels: IDF.

O'Connor-Fox ESC, Yiu PM and Ingledew WM (1991) Pasteurization: thermal death of microbes in brewing. *MBAA Technical Quarterly* 28: 67–77.

Rees JAG and Bettison J (1991) *Processing and Packaging of Heat Preserved Foods*. Glasgow: Blackie.

Action of Microwaves

A Stolle and **Barbara Schalch**, Institute of Hygiene and Technology of Food of Animal Origin, Ludwig-Maximilians-University Munich, Veterinary Faculty, Munich, Germany

Introduction

Nowadays microwave techniques are widely used in industrialized countries. Besides their use for technical, medical and analytical purposes, heating of food can be regarded as the major application. Microwave ovens are used in households and industry (**Fig. 1**) for thawing, heating, drying, pasteurizing and decontamination of food and packaging material. Nevertheless, microwave heating is partially subject to physical laws other than conventional heating methods which must be considered for safe application of microwaves.

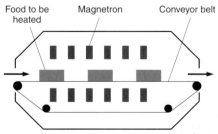

Figure 1   Layout of a continuously working industrial microwave unit.

Physical and Technical Fundamentals

The microwave is an electromagnetic wave with a propagation velocity in a vacuum of approximately $3 \times 10^8 \, \text{m s}^{-1}$ and a frequency range of $10^8–10^{12}$ hertz (Hz). Domestic microwave ovens use a frequency of 2450 MHz.

Hot and cold areas typically occur in material heated by microwaves. There are various explanations for this. The propagation behaviour of the microwaves is quasi-optical, and the occurrence of hot areas in the material to be cooked is the result of the reflection and refraction of the microwaves on the contact areas between the materials being heated and the packaging. **Figure 2** shows, in an idealized manner, field strength distribution in an empty household microwave oven. The distribution of field strengths within the cooking area extends three-dimensionally into the room. The distance between the areas of high energy density is approximately 7 cm. The so-called 'hot spots' occur in areas of high electric field strengths, and the 'cold spots' in areas with lower field strengths. This regular distribution of energy is interfered with by variations in the microwave frequency (2450 ± 25 MHz) as well as by the amount of food to be cooked and the rotating antenna or plate. These factors contribute to a more even temperature distribution in the material to be cooked.

The Effect of Microwaves on Matter

Many foodstuffs contain a considerable amount of water. The water molecule is a dipole as a result of the strongly differing electronegativity of the oxygen and hydrogen atoms. This leads to a partial charge in the water molecule and causes its rotation in the direction of the electric field of the microwave. This form of interaction between electric field and matter is called *orientational polarization*. Put simply, when the electric field changes direction the molecules try to

follow this by rotation and an intermolecular friction occurs, which leads to the heating of the material.

Electrical Properties of the Material to be Heated

In order to assess the heating behaviour of a product in the microwave field, the following parameters have to be taken into account.

The relative dielectric constant (ε_r) of a material indicates the value by which the capacity of a condensator in a vacuum increases when this material is brought between the plates of the condensator. The relative dielectric constant is calculated as the product of the dielectric constant (ε) and the electric field constant (ε_0). The latter is a natural constant and amounts to $8.854 \times 10^{-12} \, \text{F m}^{-1}$ in vacuum. **Table 1** lists the relative dielectric constants of different materials; it shows that the relative dielectric constant is not a fixed value, but varies according to the temperature and phase of the material.

The loss factor is also considered when describing the heating properties of a material in a microwave field: the loss factor ($\tan\delta$) is a measure of the transition of electric energy to heat and is dependent on the temperature and frequency. It is used to calculate the rise in temperature in the material being heated.

Both the electric conductivity of the material being heated and the penetration depth of the microwave have important practical significance. The electric conductivity is in inverse proportion to the penetration depth, which is defined as the distance over which the energy of the microwave rays is reduced to a certain value ($e^{-1} = 0.3678$) during their passage through the dielectric. With an increasing concentration of ions, e.g. as a result of a higher salt content, the conductivity in the material being heated increases and therefore the penetration depth of the microwave decreases. This leads to lower core temperatures. The penetration depth also decreases with increasing frequency of the microwave. This has practical consequences for the Bach process (see below), a pasteurizing method involving microwave heating using two different frequencies.

The required power (P) which expresses itself as a rise in temperature in the product can be calculated with the following formula:

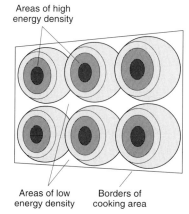

Areas of high energy density

Areas of low energy density

Borders of cooking area

Figure 2 Idealized field strength distribution in an empty microwave oven.

Table 1 Relative dielectric constant of different materials

Material	Relative dielectric constant
Water (0°C)	88
Water (+ 20°C)	81
Ice (−20°C)	16
Ice (0°C)	3
Olive oil	3.1
Air	1.00059

$$P = 2 \pi f E^2 \, \varepsilon_r \varepsilon_0 \tan \delta$$

where P is the absorbed power, $\pi = 3.14$ (rounded), f is the frequency of the respective microwave, E is the electric field strength, ε_r is the relative dielectric constant, ε_0 is the electric field constant and $\tan\delta$ is the loss factor. It can be seen from this formula that the temperature in the material to be heated could be increased, for example, by choosing a higher frequency for the microwave, or a higher relative dielectric constant of the material to be heated, or through a larger loss factor. However, modifications for optimal power absorbance are limited in many cases. Only certain frequencies are permitted for microwave heating in order to prevent interference in radio traffic. Household microwave appliances produce 2450 ± 25 MHz and do not permit frequency modulation. It is also not possible to add a very large amount of water or salt to food, even if this would significantly increase the relative dielectric constant. The formula also shows that air pockets (the ε_r of air is approximately 1) in the foodstuff, which may be inevitable or are necessary for a good sensory quality of the product, reduce the ability of food to be heated in the microwave field.

Thermal Properties of the Material to be Heated

The heating behaviour of a material is further determined by the following properties: the melting and evaporation heat, the thermal conductivity and the specific heat capacity. The melting and evaporation heat is the energy that is required to change phase from ice to water and from water to steam. This has a practical significance during thawing in the microwave. The heat conduction is a material constant and is a measure for the speed of the temperature equalization of two areas of different temperature in a body. This is of great significance for microwave heating, because during the breaks in the microwave treatment (the 'standing' times) the thermal conduction serves to equalize the temperature between cold and hot areas. The heat exchange is greater with larger differences in temperature within a material.

The specific heat capacity of a material is the relation between the amount of heat supplied to the material and the resulting temperature change. Oil heats twice as quickly as water with the same supply of heat; it has a specific heat capacity of $2.9 \, \text{kJ kg}^{-1}$ compared with $4.2 \, \text{kJ kg}^{-1}$ for water. These parameters are relevant to heating with microwaves as well as to conventional heating.

To summarize, it can be stated that heating with microwave energy is partially subject to physical laws other than those relevant to conventional heating

methods. In addition other parameters, such as the frequency of the microwave, the water content, the starting temperature and the dielectric constant of the food, have to be taken into account.

Technical Fundamentals

The elements found in household microwave ovens are also found in continuously working industrial processing units. The microwaves are produced by generators consisting of a cathode and an anode, for example the magnetron, an electron tube which can produce frequencies of 0.2–100 GHz and achieves a high efficiency of up to 70%. The electromagnetic field produced in the magnetron is fed via a waveguide to the cooking area.

Possible Risks of Using Microwaves

Microwaves do not have any ionizing effects. A comparison of the lowest quantum energy of various types of radiation is listed in **Table 2**. The possible dangers of using microwaves include:

- technical malfunctions
- thermal and athermal effects
- changes in the food components.

Technical Malfunctions

One of the technical malfunctions that can be caused by the microwave is interference with telecommunications. In order to avoid this only certain industrial frequencies are permitted. The endangering of people with cardiac pacemakers is another technical malfunction; the stray field that occurs during the production of microwaves can disturb the functioning of the implanted device.

Thermal and Athermal Effects

The effect of microwaves on living beings can be divided into thermal and athermal damage. One example of thermal damage is the cataract, which can occur from radiation densities of $50 \, \text{mW cm}^{-2}$ and over. For this reason the use of microwave devices in households and industry is governed by safety regulations: for example, the German industrial norms indicate that the measured radiation flux density should not be more than $10 \, \text{mW cm}^{-2}$ at a distance of

Table 2 Lowest quantum energy of various types of radiation

Type of radiation	Quantum energy (eV)
X rays	124
Ultraviolet light	4.1
Visible light	2.5
Microwaves	0.12×10^{-4}

5 cm from the empty household device. This is reduced to a maximum of $5\,mW\,cm^{-2}$ for a charged oven.

There is conjecture about possible athermal effects of the microwave. Some authors point to the danger to the population from so-called 'electrosmog', which is defined as electric fields caused by transmitters and industrial and household devices.

Changes in Food Components

Changes in the components of foodstuffs through microwave treatment can be divided as follows:

- destruction of food components
- formation of decomposition products in food.

For an assessment of heating with microwaves from the nutritional physiological point of view, it is important to know how many food components are changed by such treatment (**Table 3**). It can be concluded that the correct use of microwaves for heating foods should have no damaging effects on human health.

Heating Food in the Microwave

The suitability of foodstuffs for preparation in a microwave oven depends very much on the nature of the product. It is necessary to know the heating behaviour and the dielectric properties of the foodstuff to be heated before using microwave technology, whether in the household or in the food industry. This enables calculation of the correct time, power and frequency of the microwave to be used. The popularity of microwave devices in households would not have arisen if the food prepared in this way had sensory defects; there is, however, no browning and no crusts are formed on products that are cooked

Table 3 Influence of microwaves on food and food components

Food/food component	Effects of microwave heating
Milk	Isomerized amino acids content at same level as after conventional heating
Proteins	Negligible changes in amino acid composition. Changes of proteins comparable to conventional heating
Essential fatty acids	Acceleration of lipid oxidation in frozen fish fillet. No effect on fresh fish fillet
Roughage, minerals, iron	Comparable to conventional heating
Ascorbic acid, thiamin	Tendency of decrease after 3 h heating, not different from conventional heating
Pyridoxine	Decrease smaller than with conventional cooking
Riboflavin	Decrease much greater than with conventional cooking

only in the microwave. For this reason combination ovens, which add conventional heat for crust formation during or after microwave cooking, and special ovenware for browning, are on offer. For household purposes it is possible first to heat the food by microwave and then to use a conventional heating method (frying, grilling, etc.) to produce a pleasantly brown or crusty surface and a better sensory quality. The sensory quality of food heated in the microwave depends on the modalities of the microwave. Apart from the absence of crust formation and browning, there is no evidence of a general impairment in taste through treatment in the microwave.

Pasteurizing and Sterilizing in the Microwave

There are many studies concerning the effect of microwaves on microorganisms. Some authors reported the reduction of microorganisms by microwaves with lower time–temperature profiles than with conventional cooking. Oxidation, e.g. of the lipids in the microorganisms' membranes, was supposed to damage their replicability. On the other hand, many reports say that the efficacy of microwaves for the reduction of microorganisms is not different from conventional heating. These results imply the absence of athermal effects. **Table 4** summarizes the reported effects of microwave heating on different food-spoiling and pathogenic microorganisms and their toxins. Microwave heating successfully reduced the numbers of most Gram-negative microorganisms relevant in food hygiene. Nevertheless, it must be kept in mind that the adequate use of microwaves is of immense importance for food hygiene. Outbreaks of salmonellosis have been caused by insufficient microwave treatment.

Among the Gram-positive bacteria, enterococci, lactobacilli, *Listeria* spp. and spores are relatively insensitive to microwave heating. To summarize, there is no evidence for microwaves being more efficient than conventional heating in the destruction of microorganisms. The advantage of pasteurizing with microwaves lies in the lower dependence on heat conduction compared with conventional heating processes. Jam in jars, milk products in plastic tubs and sliced bread are already being pasteurized industrially by microwaves (**Table 5**). The combination of two microwave frequencies is used successfully to extend the shelf life of packed sour milk products – the Bach process, in which the whole product is heated with 10–30 MHz and the surface of the contents with 2450 MHz. A further possible use is to pasteurize ready-cooked meals in the microwave. A shelf life of approximately

Table 4   Effects of microwave heating on microorganisms and toxins

Microorganism or toxin	Reported effects
Staphylococcus aureus	Enzyme and RNA damage; bacteriostatic effect by microwave (30 min, 50°C); increase in sensitivity towards sodium chloride; reduced enterotoxin production; D_{55} value reduced from 17.8 min (conventional heating) to 11.6 min (microwave heating); D_{60} value increased from 3.0 min (conventional heating) to 4.3 min (microwave heating); eradication in artificially contaminated beef patties at 70.6°C
Listeria monocytogenes	Microwave heating up to 70°C caused a 10^6-fold reduction
Salmonella	D_{55} value reduced from 2.4 min (conventional heating) to 2.3 min (microwave heating; D_{60} value 0.2 min with both methods; at 76.6°C (2450 MHz) survival of Salmonella in turkey
Escherichia c. coli	D_{55} value reduced from 3.0 min (conventional heating) to 2.9 min (microwave heating); D_{60} value increased from 1.3 min (conventional heating) to 1.9 min (microwave heating); reduction kinetics of E. coli in phosphate buffer during microwave heating (2450 MHz) proportional to conventional heating
Enterococci	Microwave less effective than conventional heating at identical end point temperature
Psychrotrophic microorganisms	Reduction by microwave heating (60°C): Achromobacter, Aeromonas, Alcaligenes Very resistant to microwave heating: Pseudomonas aeruginosa
Lactobacillus plantarum	Highly resistant to microwave heating
Clostridium perfringens	Elimination of vegetative cells above 70°C; spore inactivation by deep frying (190.5°C)
Bacillus subtilis	Spore reduction at 103°C in phosphate buffer; 137°C in dry heat
Spores	Inactivation possible above 100°C, but indistinguishable from conventional heating
Aflatoxins	Destruction possible with high power and exposure times, but not by microwave heating for domestic purposes

Table 5   Microwave pasteurization of foods

Product	Technical parameters	Methods and results
Dry egg	400 W, 50 MHz	Thickness of layer 23 mm, heating over 1.3 min to 99°C, Salmonella reduction 6 log units
Milk	700 W, 2450 MHz	No reliable pasteurization
Fresh cheese	2.8 and 5 kW	Microwave heating to 48.8°C leads to a shelf-life extension of 1–4 weeks
Fruit preparation	Conveyor	Pasteurization at 85–90°C, 2 min
Cured ham	2450 MHz, 60 MHz	Reduction of juice loss, but higher total count at 60 MHz instead of 2450 MHz Best results with conventional heating
Meat-vegetable-sead patty	Not described	Packing the mass at 70°C surface temperature, post packaging pasteurization at 90–95°C shelf life extension
Sausage for roasting in plastic foil	Conveyor	For micrococci and staphylococci: identical F_0 values[a] provided, conventional and microwave heating were identically effective

[a]Unit of lethality: 1 F_0 refers to 1 min at 121.1°c.

6 weeks after heating the packaged meals to 75–85°C is reported.

To avoid hot and cold spots, which imply sensory and quality defects in foods, microwave long-term treatment was tested with defined standing times. Even various foodstuffs with different container shapes can be heated by computerized processing units with a microwave hybrid system.

See also: **Heat Treatment of Foods**: Principles of Pasteurization.

Further Reading

Bodrero KO, Pearson AM and Magee WT (1980) Optimum cooking times for flavor development and evaluation of flavor quality of beef cooked by microwave and conventional methods. J. Food Sci. 45: 613–616.

Cremer ML (1982) Sensory quality and energy use for scrambled eggs and beef patties heated in institutional microwave and convection ovens. J. Food Sci. 47: 871–874.

Dreyfuss MS and Chipley JR (1980) Comparison of effects of sublethal microwave radiation and conventional heating on the metabolic activity of Staphylococcus aureus. Appl. Environm. Microbiol. 39: 13–16.

Evans MR, Parry SM and Ribeiro CD (1995) Salmonella outbreak from microwave cooked food. Epidemiol. Infect. 115 (2): 227–230.

Harrison DL (1980) Microwave versus conventional cooking methods effects on food quality attributes. J. Food Prot. 43: 633–637.

Jeng DKH, Kaczmarck KA, Woodworth AG and Balasky G (1987) Mechanisms of microwave sterilisation in the dry state. Appl. Environm. Microbiol. 53: 2133–2137.

Jonker D and Til HP (1995) Human diets cooked by microwave or conventionally: comparative sub-chronic (13-

wk) toxicity study in rats. *Food Chem. Toxicol.* 33: 254–256.

Ke PJ, Linke BA and Ackermann RG (1978) Acceleration of lipid oxidation in frozen mackerel fillet by pretreatment with microwave heating. *J. Food Sci.* 43: 38–40.

Khalil H and Villota R (1988) Comparative study on injury and recovery of *Staphylococcus aureus* using microwave and conventional heating. *J. Food Prot.* 51: 181–186.

Mudgett RE (1982) Electrical properties of foods in microwave processing. *Food Technol.* 36 (2): 109–115.

Ockerman HW, Cahill VR, Plimpton RF and Parret NA (1976) Cooking inoculated pork in microwave and conventional ovens. *J. Milk Food Technol.* 39: 771–779.

Patterson JL, Cranston PM and Loh WH (1995) Extending the storage life of chilled beef: microwave processing. *J. Microw. Power Electromagn. Energy* 30 (2): 97–101.

Payre F, Datta A and Seyler C (1997) Influence of the dielectric property on microwave oven heating patterns: applications to food materials. *J. Microw. Power Electromagn. Energy* 32 (1): 3–15.

Rosenberg U (1989) *Einfluss der Hochfrequenzbehandlung auf einige lebensmittelhygienisch bedeutsame mikroorganismen.* Berlin: Diss.-Ing.

Rosenberg U and Bögl W (1987) Microwave thawing, drying and baking in the food industry. *Food Technol.* 41 (6): 85–89.

Schalch B, Eisgruber H and Stolle A (1997) Reduction of numbers of bacteria in vacuum-packed sliced sausage by means of microwave heating. *Dairy, Food Environm. San.* 17: 25–29.

Synergy Between Treatments

E A Murano, Center for Food Safety and Department of Animal Science, Texas A&M University, USA

Introduction

Canning typically involves the heating of low-acid foods at 121°C, with the goal of eliminating all mesophilic microorganisms, as well as spores of *Clostridium botulinum*, leaving the product 'commercially sterile'. In order to accomplish this, the process is applied for a period of time long enough to achieve a 12 $\log_{10}$ reduction in the number of spores of this pathogen (termed '12D processing'). This usually entails heating for at least 2 min, depending on the food composition. Such a process is very effective in maintaining the stability of low-acid foods during room-temperature storage. However, the time/temperature used in canning affects the quality of many food products. It has been suggested that if processing technologies could be used in combination with heat to achieve the 12D process, the heating treatment used in canning would not have to be so severe, i.e. the temperature could be lower and/or the time shorter than that currently being used.

For products not intended to be sold as shelf-stable, the treatment of foods by heat pasteurization is commonly used to eliminate non-spore-forming pathogenic organisms from perishable commodities, as well as to extend their shelf life. This process renders the product free of most disease-causing agents, although refrigeration is still required owing to the presence of spoilage organisms. The temperature of processing varies according to the product being pasteurized, with 100°C commonly used for beverages and fruit products, and 94°C for meats. Products treated by heat pasteurization are no longer considered 'fresh', since texture, flavour and odour become altered. The food industry has tried to develop procedures that will minimally process these foods in order to maintain an almost-fresh quality. However, such technologies as *sous-vide* processing are not as effective as pasteurization in destroying microbial contaminants. It would be advantageous if products could be processed in such a way as to achieve the benefits of pasteurization (i.e. reduction or elimination of microbial pathogens) while maintaining their quality.

Possible Combinations of Treatments

Heat and Ionizing Radiation

Researchers have looked at the possibility of combining canning with technologies such as food irradiation. This process, also termed 'cold pasteurization', is applied to food products after packaging. Ionization of the material is accomplished by exposure to a high-energy source, such as gamma rays from cobalt 60, or to accelerated electrons or X-rays from a linear accelerator. Food components are minimally affected, with only 1 in every 6 million chemical bonds being broken by this procedure. Microorganisms are easily eliminated owing to disruption of bonds between base pairs in their DNA molecule, effectively rendering them unable to replicate or synthesize enzymes needed to carry out essential metabolic reactions. In addition, irradiation results in the formation of free radicals owing to ejection of electrons from their orbitals during ionization. These radicals can also affect the viability of microbial cells since they can disrupt membrane transport systems that rely on ion exchange. **Table 1** depicts the rate of destruction of various foodborne pathogens according to irradiation dose. Given that most of these organisms are found at concentrations no higher than 10^2 cells per gram, irradiation at even 1.0 kGy can reduce their numbers to undetectable levels.

It is important to note that food irradiation by itself is capable of achieving the 12D process required for commercial sterility. However, in order to do this, the food product has to be irradiated at very high doses

Table 1 Log$_{10}$ number of survivors of various food-borne pathogens treated by ionizing radiation in beef

Dose (kGy)	Salmonella[a]	Campylobacter[a]	E. coli O157:H7[a]	Listeria[a]
0	2.0	2.0	2.0	2.0
0.5	1.0	0.7	0.85	1.35
1.0	0.0	−1.15	−0.75	0.20
1.5	−1.0	–	–	−1.0
2.0	−2.0	–	–	−1.7

[a]Average log$_{10}$ number of survivors.

(at least 42 kGy). Such a dose can result in undesirable quality changes due to radical formation leading to off odours and flavours. However, these effects are easily minimized by subjecting the food product to irradiation in the frozen state, thereby significantly lowering the number of radicals produced. The cost associated with using this technology is affected by the dose, since the higher the irradiation dose, the higher the energy requirements. Thus, it would be advantageous if irradiation could be carried out in conjunction with another process, such as heat, in order to minimize the dose required to achieve commercial sterility.

Ionizing radiation can be applied at medium doses (1–10 kGy) in order to 'pasteurize' foods. Such a process could certainly be combined with heat pasteurization in order to achieve these results, without the need to apply as high a temperature, or as high a dose, as would be required if either method were used alone. Such a combination of processes would yield a safer yet high-quality product. In addition, if a product treated by heat as well as radiation pasteurization were to be aseptically packaged, commercial sterility could be achieved at a fraction of the cost of canning or irradiation sterilization alone.

Heat and High Hydrostatic Pressure

Food processing by high hydrostatic pressure (HHP) involves the application of high pressure for a few minutes to food products in order to reduce the number of microbial contaminants. The method is both 'isostatic and instantaneous', meaning that the pressure is transmitted evenly throughout the product immediately upon its application. Membrane permeability of microorganisms is altered, resulting in cell death due to leakage of cytoplasmic contents. Bacterial spores are considerably more resistant than vegetative cells, with endospores being able to survive pressures above 700 MPa.

Depending on the level of treatment, HHP can result in very few changes to the colour, flavour or nutritional content of most foods. Other factors influencing quality of pressurized foods include the duration of the processing event, the rate of depres-

surization, the temperature, and food-related parameters such as pH, water activity and salt concentration. Of course, some of these changes may be desirable, as in the case of tissue tenderization through texture modification due to high pressure processing.

Treatment by HHP is generally accompanied by a temperature increase of about 5–15°C when performed at either refrigerated or room temperature. This effect is reversible, for as soon as the product is decompressed, it returns to its original temperature. Such a moderate rise in temperature does not change the flavour or odour of products treated by this method, resulting in high-quality foods. As an example, vitamin losses due to HHP are usually lower when compared with heat-treated products.

However, detrimental changes can occur if the pressure level is sufficiently high. Using equipment currently available, it is possible to apply high pressure in conjunction with added heat. Such a combination of processes can enhance microbial destruction, further extending the shelf life of treated products. The heat can be applied and controlled easily with a jacketed chamber. Through the use of both heat and HHP, it could be possible to minimize quality changes that would otherwise occur if each treatment were applied by itself. This is especially true regarding commercial sterility, which cannot be accomplished by treatment with HHP alone without ruining the product.

Heat and Organic Acids

The application of organic acid rinsing in the meat industry has taken on new importance in light of recent efforts to decontaminate animal carcasses at the slaughter plant. Lactic, acetic, propionic and citric acids are among those that have been studied and applied successfully in reducing the total number of contaminants, including coliform bacteria, from the surface of beef and pork. The use of these acids in foods has been approved by government agencies in many countries, including the US Food and Drug Administration. The effectiveness of organic acids in reducing microbial contamination, as well as their effect on quality of the product, varies depending on the nature and extent of the contamination, the degree of microbial attachment to the surface being treated, the type of acid, the concentration, the time of contact, and the sensitivity of the microorganism to the specific acid.

Organic acids affect the microorganisms first by forcing the cell to utilize stored ATP in order to maintain ionic balance. Proton transport is affected, ultimately resulting in a lowering of the internal pH of the cell, which impedes most metabolic reactions. Quality of products can also be changed by exposure to organic acids, with flavour, odour and colour changes being most commonly cited. In addition, texture can be altered significantly by changes in the water-holding capacity of ingredients, resulting in loss of water from the product. For these reasons, organic acid concentrations higher than 2% are not recommended if product quality is to be maintained.

In the quest for carcass decontamination strategies, heat has been employed in the form of hot-water rinses and steam pasteurization. These technologies are expensive for the processor owing to their high energy requirements. In addition, the effectiveness of such treatments is limited by the temperatures that can be used, since high temperatures can partially cook the product, a decidedly undesirable result for fresh meats. In contrast, spraying with organic acid solutions has been shown to be an economic way to reduce the microbial load of these products, although its effectiveness is also limited owing to restrictions on concentrations that can be used. Application of both technologies could serve to increase the antimicrobial effect of the individual treatments, while minimizing the expense of their application as well as undesirable quality changes.

Evidence of Combination Effects

Heat and Ionizing Radiation

Heat can be applied to foods prior to, during or after irradiation in order to lower the dose required to achieve a specific $\log_{10}$ reduction in a bacterial population. However, the treatment must be at a temperature high enough to sensitize the cells to the effects of ionizing radiation. As an example, heating of liquid whole eggs at 50°C results in no change to the irradiation D_{10} value of *Salmonella enteritidis*, while heating at 60°C can enhance the bacterial reduction by about $0.5 \log_{10}$ per gram. Interestingly, the temperature used in such combination treatments can vary somewhat without resulting in any further reduction in bacterial populations. However, there is a threshold temperature above which the bacterial reduction is significantly enhanced, with this temperature varying according to the product being treated.

Irradiation and heat act synergistically, as can be seen in studies where irradiation of chicken meat at 2.2 kGy followed by cooking at 65.6°C reduced the number of *Listeria monocytogenes* by about $6.0 \log_{10}$, compared with heat ($0.35 \log_{10}$) or irradiation alone ($2.77 \log_{10}$). The order in which the radiation treatment is applied (before or after heating) has a significant effect on the extent of the microbial reduction. Exposure of chicken meat to heating at 60°C for 3 min followed by irradiation at 0.9 kGy results in a $6.4 \log_{10}$

reduction per gram; however, if irradiation is applied before heating, the resulting reduction is $8.9 \log_{10}$ per gram.

Treatment of foods by thermoradiation – the application of heat during irradiation – is also very effective, with a synergistic effect being evident. Heating of liquid whole eggs at 60°C during irradiation at 0.39 kGy has been shown to be more effective than either treatment alone. In addition, this treatment does not result in changes to food proteins or enzymes such as lysozyme, an important consideration in the treatment of heat-sensitive products.

In applying heat and irradiation for the elimination of moulds, the relative humidity (RH) used in the process must be considered, owing to the ability of surviving spores to produce mycotoxins during subsequent storage. Spores of *Aspergillus flavus* in maize grains heated at 60°C for 30 min at 45% RH, followed by irradiation at 3.5 kGy, produce three times more aflatoxin during storage than spores heated at 85% RH. Thus, the higher the moisture level during heating, the lower the production of mycotoxins. Irradiation at high doses magnifies this effect, with no toxin being produced after irradiation at 4.0 kGy even when the relative humidity is low.

Heat and High Hydrostatic Pressure

Vegetative bacterial cells are easily inactivated by high hydrostatic pressure at levels between 300 MPa and 400 MPa. Spores are considerably more resistant, with some studies showing that endospores of *Bacillus* and *Clostridium* can survive exposure to 200 MPa without any reduction in numbers. **Table 2** depicts the $\log_{10}$ reduction that can be achieved when heat is combined with HHP. Application of 700 MPa has been reported to reduce the spore count of *C. sporogenes* by $4–5 \log_{10}$ per gram. If this same pressure is applied at 80°C for 20 min, the reduction is enhanced by an additional $2 \log_{10}$ per gram. Thus, combining heat and pressure is more effective in spore destruction than pressure alone.

Evidence has been presented in some studies to

Table 2 Log_{10} number of survivors of *Clostridium sporogenes* spores treated by high hydrostatic pressure and heat in whole-muscle chicken

Heating time[a] (min)	Log_{10} no. survivors (g^{-1})
Control[b]	6.8
0	3.5
1	3.4
10	2.7
20	1.2

[a]Heating at 80°C during high-pressure treatment at 700 MPa.
[b]Samples not exposed to pressure or to heating.

suggest that heating can have a counteracting effect on hydrostatic pressure, and vice versa. Death rates of *Lactobacillus casei* during pressurization at 200 MPa decrease with rising temperatures (20–60°C) when compared with non-pressurized cells. The same phenomenon has been observed on denaturation of proteins, with heating under high pressures resulting sometimes in an increase and at other times in a suppression of the denaturation event. At pressures near 760 MPa, proteins are denatured, while at moderate pressures (around 100 MPa) a stabilization effect takes place. This results in an increase in the temperature required for heat denaturation of the proteins.

It is generally recognized that heating causes denaturation of proteins. In contrast, exposure to HHP promotes their coagulation, with covalent bonds being broken at very high pressures. Pressures of the order of 100 MPa applied for about 25 min to post-rigor muscle, along with heat at 60°C, has been shown to improve the tenderness of the meat after subsequent cooking. Juiciness is affected, but the product is still comparable to heat-treated samples.

Studies have shown that exposure of meat to a very high temperature before the application of pressure reduces the tenderization effect. It has been postulated that this may be due to a stabilization of proteins after they have been denatured by the heat treatment. Native myofibrillar proteins (actin and myosin) are considered the main components responsible for tenderization effects of animal tissue. The proteins are easily disaggregated when exposed to high pressures. It is possible that exposure of these to heat, prior to pressurization, results in their denaturation but not their disaggregation. Because they remain associated, treatment by high pressure would have little, if any, effect on these proteins.

Heat and Organic Acids

Studies on specific bacterial pathogens have shown that application of hot lactic or acetic acid can result in a 3–6 $\log_{10}$ reduction per square centimetre of microorganisms on the surface of beef carcasses. The greatest reduction is achieved at a concentration of 3% lactic acid applied at 55°C, with spoilage bacteria being more sensitive than pathogenic organisms.

Organic acids can also be applied in combination with heat in a sequential process to improve the effectiveness of the heat treatment. Spraying with lactic acid after knife trimming and a warm-water wash at 35°C has been shown to reduce the counts of *Salmonella typhimurium* and *Listeria monocytogenes* on the surface of beef carcasses by about 5 $\log_{10}$. Total plate counts on beef carcasses can be reduced by almost 2 $\log_{10}$ per square centimetre if 2% acetic acid

is sprayed after a 74°C water wash. Even when washing at lower temperatures (40°C), subsequent spraying with organic acid can reduce the total microbial counts by an additional 1.5 $\log_{10}$ per square centimetre when compared with water washing alone.

Similarly, steam can be applied to animal carcasses in combination with organic acids to achieve greater bacterial reductions than either treatment applied alone. Sanitation of carcasses with a steam vacuum, followed by treatment with 2% lactic acid at 55°C reduces the number of total coliforms by 4.4 $\log_{10}$ per square centimetre, and of generic *Escherichia coli* (an indicator of faecal contamination) by the same amount. Steam pasteurization systems have been tested for their efficacy in reducing microbial contamination of beef carcasses. Temperatures are typically of the order of 91–95°C, with exposure of the samples to steam lasting about 15 s. Studies have shown that total count reductions achieved by this process are usually about 1.0–1.5 $\log_{10}$ per square centimetre, smaller than reductions after treatment by hot organic acid alone. However, when both technologies are applied sequentially, reductions as large as 3.0 $\log_{10}$ per square centimetre can be obtained.

The data presented in **Table 3** demonstrate that the order in which these treatments is applied can have a significant effect on the microbial reductions. For example, hot water at 70°C, applied after spraying with a 2% lactic acid solution at 55°C, reduces the total microbial counts more than if the hot water is used before spraying with the acid. Similarly, steam pasteurization applied after treatment with lactic acid causes a more pronounced reduction than if applied alone, or before lactic acid.

Effects on Microorganisms

Heat and Ionizing Radiation

Irradiation affects microorganisms in ways that are unique to this technology. The principal target of

Table 3 Reductions in total bacterial counts on the surface of beef carcasses after various decontamination treatments

Treatment[a] (min)	Log_{10} reductions
Hot water (70°C)	4.7–5.7
Lactic acid (2% at 55°C)	5.4–6.1
Hot water + lactic acid	4.1–4.5
Lactic acid + hot water	4.7–5.6
Steam pasteurization (95°C)	2.4–2.7
Steam + lactic acid	2.0–2.5
Lactic acid + steam pasteurization	2.6–3.1

[a]All treatments except for those including steam pasteurization were applied after knife-trimming. Otherwise, they were applied after warm-water wash (35°C) just prior to storing carcasses in a refrigerated chamber.

ionizing radiation is DNA, with membrane effects playing a secondary role. Damage to the DNA of the cell occurs by the breaking of chemical bonds in the molecule. Even though only a few bonds are actually affected, this has a profound influence on the molecule owing to the important role that base pair sequences play on the ability of DNA to replicate and code for proteins crucial to the cell's survival. It has been estimated that a dose as small as 0.1 kGy can damage 2.8% of a bacterium's DNA, while it damages only 0.14% of its enzymes. This difference explains why application of a certain irradiation dose can have a lethal effect on a microorganism without causing much change in the chemical composition of the food.

Combining irradiation with mild heating has been shown to result in an increase in the inactivation of spoilage organisms, compared with application of either treatment. This synergistic effect can result in the sterilization of foods at relatively low doses of radiation. The mechanism by which this combination of treatments affects bacterial cells is rather complex. It is believed that vegetative cells treated first by ionizing radiation will experience damage to their DNA, and that subsequent treatment with heat will inactivate enzymes necessary to effect repairs.

Spore-forming bacteria such as *Clostridium botulinum* are also affected by the combination of heat and irradiation, although the mechanism is slightly different. In this case, ionizing radiation causes the spores to become sensitive to heat, whereas preheating does not significantly alter their radiation sensitivity. Irradiation of *C. sporogenes* in phosphate buffer has been shown to impart long-term sensitivity to heat (at least 2 weeks during room-temperature storage). It is believed that pre-treatment with ionizing radiation stimulates partial spore germination, rendering the spores sensitive to a subsequent heat treatment.

Mycotoxins, like most other molecules, are fairly resistant to the effects of radiation. Inactivation requires a very high dose, one that would not be practical for food applications. Thus, the combination of heat and irradiation has been proposed as an alternative. The sequence of treatments does not matter in regard to destruction of certain moulds (i.e. *Rhizopus stolonifer*), while it is important in others. *Cladosporium herbarum* is more sensitive if heat is applied before irradiation. It is believed that heat causes more damage to mould spore structure than treatment by ionizing radiation, thus its application prior to irradiation sensitizes the cells more to a subsequent process than if irradiation is carried out first.

Heat and High Hydrostatic Pressure

High hydrostatic pressure causes a collapse of gas vacuoles inside microbial cells, as well as an elongation of the organism from the typical 1–2 μm to 100 μm. Motility is also affected, although it is reversible once the cells have been depressurized. Other changes include separation of the cell wall from the cytoplasmic membrane, and a slowing down of cell division.

Most biochemical reactions result in a change in volume for the cell. Electrostatic interactions between molecules are broken by high pressure, thereby exposing more ions to water. This results in electrostriction of water, which decreases the volume. Such a decrease promotes the formation of hydrogen bonds. This alteration of intramolecular structures causes conformational changes to the active site of enzymes, thus inactivating them. Reversible denaturation occurs at pressures between 100 MPa and 300 MPa, whereas pressures above 300 MPa denature bacterial enzymes irreversibly. A reduction in the volume of the membrane lipid bilayer of the cell also occurs, resulting in inhibition of amino acid uptake and in overall damage to the membrane structure.

The higher the pressure to which a microorganism is exposed, and the longer the time of exposure, the more adverse the effects. Cells at the early log phase of growth are more sensitive than those at the stationary phase. In the case of bacterial spores, an interesting phenomenon is observed: pressures of 100–300 MPa are more effective in destroying them than pressures up to 12 000 MPa. This could be because spore germination may be induced at the lower pressures, with the outgrowing cells being more sensitive to environmental conditions to which they are subsequently exposed. This phenomenon is affected by the temperature and duration of compression, and not by the number of compression/decompression steps, nor by the number of spores.

Temperature can affect the sensitivity of microorganisms to HHP. Growth at increased pressures can occur, as long as the temperature is only a few degrees higher than the normal optimum growth temperature of the microorganism at normal pressures. Increasing the pressure has been shown to slow down the lethal effects due to heat that are normally observed at atmospheric pressure. Moreover, it has been reported that the lethal effects due to pressure are slowed down if the temperature is low enough.

Heat and Organic Acids

The mechanism of damage by exposure of microorganisms to weak acids has been extensively studied. It is generally recognized that for the acid to have the highest degree of effectiveness, the pH of the menstruum must be below the pK_a of the acid (typically around 4.0). This is so that transport across the cell membrane and into the cell is not impaired by charge effects. Since the pH of the cytoplasm is near neutrality, the organic acid quickly becomes dissociated once inside the cell. This results in an increase in efflux of protons from the cell, as it attempts to maintain osmotic balance. Such an event requires ATP, which causes its depletion. Eventually, the cell is no longer able to counteract the accumulation of protons in the cytoplasm, resulting in a drop in internal pH. Enzymatic reactions are affected, with cell death being the end result.

Heating applied in addition to, or in conjunction with, organic acids enhances their antimicrobial effect. This is due to the fact that prior heating damages the cell membrane, making it even easier for weak acids to penetrate into the cytoplasm. In addition, a residual effect is observed, with inhibition of growth of survivors during storage. If organic acids are applied before heating, the reduction is not as pronounced. It is believed that damage by a lowering of internal pH is profound enough that when heat is subsequently applied, very little additional damage occurs; this is because enzymes and other proteins are already inactivated before the heat treatment.

Possible Applications

Heat and Ionizing Radiation

Given that irradiation sensitizes microorganisms to heat, and that this effect can last for a few weeks, application of ionizing radiation can be conveniently performed on products that need to be shipped to a different location in order to receive a subsequent thermal process. This system could be applied to fresh meats intended to be further processed in the making of sausage and similar products. In addition, meats that are sold partially or fully cooked (such as hamburger patties destined for institutional kitchens) would benefit from the combination of heat and irradiation, since microbial pathogens would be eliminated from the raw food prior to heating.

Fruit juices are commonly sold as pasteurized products. However, there has been a tendency in recent years toward consumption of 'all natural' products. An outbreak of food poisoning in the USA in 1997 was attributed to apple cider made from apples contaminated with E. coli O157:H7. Even though the pH of the product was low, the organism was able to survive. The manufacturers had not wanted to heat pasteurize the cider because of the negative effect that this can have on the flavour of the juice. Thus, irradiation of the product, in conjunction with a mild heat treatment, would have enhanced the safety of the apple cider without compromising its flavour.

Application of heat and irradiation for the decontamination and disinfestation of fruit has been proposed. Many fruits and vegetables can tolerate doses up to 1.0 kGy, which are necessary to eliminate fruit flies and other arthropod pests. However, some commodities such as mangoes and avocados can be injured by radiation at doses above 0.25 kGy, resulting in skin blemishes and discoloration. Hot-water dips have been used successfully, but these delay ripening to such an extent that the fruit does not mature properly, resulting in a toughening of the flesh. However, heat could be applied together with a very low dose of irradiation to disinfest these fruits without affecting their quality.

Legumes, which can be contaminated with mycotoxin-producing moulds, could be irradiated in order to eliminate these microorganisms. However, this process does not effectively inactivate these moulds if the temperature or moisture level during application is not very high. Thus, treatment of products such as dry cocoa beans and maize at 4.0 kGy could be combined with moist hot air (60°C for 30 min) to inactivate *Aspergillus flavus*, an important cause of chronic liver disease in humans.

Food intended to be sold as commercially sterile can also benefit from the application of both heat and irradiation. It has already been mentioned that applying a medium dose of radiation, along with a heating process less severe than that currently used in canning operations, could aid in producing shelf-stable low-acid foods of high quality. This is also true in reverse: whole muscle products such as beef steaks, made commercially sterile through high-dose irradiation, can spoil owing to enzymatic changes during storage; treatment with heat would inactivate these compounds, preventing quality losses.

Heat and High Hydrostatic Pressure

Most of the applications for which HHP has been considered centre around shelf-life extension of perishable products. Fruits such as peaches can remain commercially sterile for at least 5 years after treatment at 415 MPa. Some of these commodities are negatively affected by the treatment, with toughening of cherry tomatoes being reported. Effects seen in other foods include tenderization of beef, while chicken and fish develop an opaque colour similar to that observed after a slight cook.

Regarding the use of heat together with pressure, pasteurization of fruit can be achieved with minimal quality changes by exposure to 700 MPa for 10 min at temperatures slightly above 25°C. Pressure levels of 550 MPa have been proposed, along with temperatures near 50°C, for meat pasteurization. Under these conditions, beef shelf life can be extended to at least 3 months during refrigerated storage. If the temperature is increased to 60°C, tenderization of meat is enhanced.

Sterilization of low-acid foods can be achieved by HHP, in combination with heat, with minimal changes to quality. Pressures up to 141 MPa, along with heating at 82–103°C, can result in a commercially sterile product, as long as sealed containers are used. The D value of Gram-positive spore-formers in such products is 280 min at 0.34 MPa with heating at 100°C, and 2.2 min at 138 MPa with heating at the same temperature.

Heat and Organic Acids

Currently, the use of organic acid rinsing with heat for the decontamination of animal carcasses is the leading application of these combined technologies. This is because of the need for slaughter facilities and abattoirs to apply a bacterial-reducing treatment that will not alter the appearance, flavour or odour of fresh meats. Such a requirement was brought to the forefront after outbreaks of food-borne illness due to consumption of red meats resulted in the enactment of a 'zero tolerance' policy by US governmental agencies, requiring that no E. coli O157:H7 be permitted on the surface of these products.

In addition to carcass decontamination, acidified products could benefit from the additional application of heat. Fermented sausages, among others, are often prepared without cooking in order to preserve their freshness. It would be possible to apply a heat treatment to these products, while carefully controlling their pH, in order to improve their safety while only minimally affecting their organoleptic quality. Fermented dairy products such as cheese would be a more challenging prospect for combined treatment using heating and organic acid processing. However, a case could be made for the application of a very mild heat treatment in order to reduce the number of spoilage organisms such as lactic acid bacteria. More research is needed to determine the optimal conditions that should be applied to these types of products that would yield the highest quality.

See also: **Clostridium**: Clostridium botulinum. **High Pressure Treatment of Foods**: **Listeria**: Listeria monocytogenes. **Mycotoxins**: Toxicology. **Preservatives**: Traditional Preservatives – Organic Acids. **Salmonella**: Salmonella enteritidis.

Further Reading

Diehl JF (1995) Biological effects of ionizing radiation. In: Diehl DF (ed.) Safety of Irradiated Foods. Pp. 89–132. New York: Marcel Dekker.

Hardin MD, Acuff GR, Lucia LM, Oman JS and Savell JW (1995) Comparison of methods for contamination removal from beef carcass surfaces. Journal of Food Protection 58: 368–374.

Hayashi R (1992) Utilization of pressure in addition to temperature in food science and technology. In: Balny C, Hayashi R, Heremans K and Masson P (eds) High Pressure and Biotechnology. Pp. 185–193. London: John Libbey.

Hoover DG (1993) Pressure effects on biological systems. Food Technology 47: 150–155.

Phebus RK, Nutsch AL, Shafer DE et al (1996) Comparison of steam pasteurization and other methods for reduction of pathogens on surfaces of freshly slaughtered beef. Journal of Food Protection 60: 476–484.

Radomyski T, Murano EA, Olson DG and Murano PS (1994) Elimination of pathogens by low-dose irradiation: a review. Journal of Food Protection 57: 73–86.

Shamsuzzaman K (1988) Effect of combined heat and radiation on the survival of Clostridium sporogenes. Radiation in Physics and Chemistry 31: 187–193.

HELICOBACTER

Irene V Wesley, Enteric Diseases and Food Safety Research, USDA, National Animal Disease Center, Ames, USA

Introduction

Helicobacter pylori was originally designated *Campylobacter pyloridis*. As a member of the RNA superfamily VI, which includes *Campylobacter*, *Helicobacter* and *Arcobacter*, *H. pylori* is distinct from but closely related to these genera. Members of the rRNA superfamily VI of the Proteobacteria are Gram-negative, spiral-shaped, microaerophilic bacteria which are motile by means of flagella. The sheathed flagella are unique to *Helicobacter* and may be an adaptation to survive in gastric juices. The key features of superfamily VI are summarized in **Table 1**.

Characteristics of *Helicobacter* spp.

Members of the genus *Helicobacter* colonize humans and animals and cause enteritis, gastritis and, rarely, abortions in livestock. The host distribution of *Helicobacter* species is summarized in **Table 2**. The species have a narrow host range. To illustrate, natural infections of *H. pylori* occur in humans and non-human primates and infrequently in cats; this may result from close living conditions with the human host.

Antibodies to *H. pylori* reflect infection status in a population and thus may indicate routes of transmission. Field studies have shown that antibodies are common in individuals from rural settings and in populations of low socioeconomic status. In developing countries > 50% of the population is seropositive. In contrast, seroprevalence is less than 50% in industrialized countries. Seropositivity develops early in life when hydrochloric acid secretions, immune competence and gastrointestinal flora have not attained adult levels, and increases with age.

H. pylori (*Campylobacter pylori* or *C. pyloridis*) is the most common human bacterial infection. It is present in 95% of duodenal and in 70–80% of gastric ulcer cases as well as in clinically healthy individuals, including family members of patients.

H. pylori is unique among bacteria in being linked to human gastric carcinoma. Information on the distribution of *H. pylori* in water, food animals, meats and vegetables is critical in determining its potential transmission in foods.

Methods for Enumeration in Foods

Few surveys have been conducted to determine the distribution of *Helicobacter* and, specifically, *H. pylori*, in foods. As a result, isolation methods specifically adapted to its recovery from a food matrix have not been optimized. Isolation protocols which have been used to culture the microbe from human gastric biopsy specimens or faeces may be applicable for foods. Biopsy specimens are homogenized in physiological saline and plated on to selective and nonselective media. Nonselective media include chocolate agar and brain–heart infusion agar supplemented with 5–10% defibrinated blood.

Selective media are commercially available and, in general, are similar to media used for the isolation of *Campylobacter*. Dent's medium, for example, is a

Table 1 Summary of major distinguishing characteristics of members of rRNA superfamily VI

Genus	Growth at 15°C	Oxygen tolerance	Flagella	Urease
Helicobacter	No	Microaerophilic	Multipolar Sheathed[a]	Yes
Campylobacter	No	Microaerophilic	Single polar Unsheathed	No
Arcobacter	Yes	Aerotolerant	Single polar Unsheathed	No

[a]*Helicobacter cinaedi* is unsheathed.

Table 2 *Helicobacter* species and host distribution

Bacterium	Source	Target organ
Helicobacter pylori	Humans	Gastric mucosa
Helicobacter acinomyx	Cheetah	Gastric mucosa
Helicobacter bilis	Mouse	Bile, intestine, liver
Helicobacter bizzozeroni	Canine	Stomach
Helicobacter canis	Dogs, humans	Intestine
Helicobacter colifelis	Kittens	Intestine
Helicobacter cinaedi	Hamsters, humans	Intestine
Helicobacter felis	Dogs, cat	Gastric mucosa
Helicobacter fennelliae	Humans	Intestine
Helicobacter helmannii	Humans, mice, pigs	Gastric mucosa
Helicobacter hepaticus	Mouse	Intestine
Helicobacter muridarum	Rat	Stomach, caecum
Helicobacter mustelae	Ferret	Stomach
Helicobacter nemestrinae	Monkey	Gastric mucosa
Helicobacter pametensis	Pig, tern	Intestine
Helicobacter pullorum	Poultry, humans	Intestine
Helicobacter rappini	Dogs, humans, mice, sheep	Intestine, reproductive tract
Helicobacter rodentium	Laboratory mice	Intestine
Helicobacter salomonis	Dogs	Stomach
Helicobacter trogontum	Rat	Intestine
Helicobacter westmeadii	Humans	Blood culture

modification of Skirrow's medium in which cefsulodin replaces polymyxin B and amphotericin B is added to inhibit *Candida* spp. Freshly poured plates are incubated (35°C, 3–5 days) microaerobically (5–7% O_2, 5–10% CO_2) with the addition of hydrogen (8% H_2) in high humidity. Plates are examined for the presence of small pinpoint translucent non-haemolytic colonies. The presence of Gram-negative, curved or spirally shaped bacilli which grow at 35°C but not at 25°C is indicative of *Helicobacter*. *H. pylori* is positive for urease as well as oxidase and catalase.

Although cultural isolation of *H. pylori* is the gold standard, alternative means for its detection, such as polymerase chain reaction (PCR), have been developed. The PCR method has been shown to be more sensitive in detecting *Helicobacter* than culture, electron microscopy or histological examination. PCR assays have been designed to detect the genes encoding the enzyme urease, which converts urea to CO_2 and ammonia. PCR primers targeting the urease gene have identified *H. pylori* in dental plaque of culture-positive and culture-negative individuals.

Importance of *Helicobacter* to the Food Industry

Route of Transmission

Multiple routes of dissemination are possible for *H. pylori*. Transmission by faecal–oral spread, oral–oral spread via salivary secretions, ingestion of contaminated foods and water, and transmission by flies is supported by epidemiological and laboratory data (**Fig. 1**).

H. pylori has been detected in human faeces either by culture or by PCR. In Gambia, *H. pylori* was isolated from the faeces of 9 of 23 randomly selected children less than 3 years of age. The presence of *H. pylori* in faeces suggests that faecal–oral transmission may occur. Evidence to support a faecal–oral transmission is also derived from the observations that antibodies to *H. pylori* are found in patients with hepatitis A virus. However, not all serosurveys support these observations.

The natural route of infection may be by saliva or gastric juice as a result of vomiting low-pH (achlorhydric) mucus in childhood. *H. pylori* has been isolated from dental plaque and from the surface of the buccal cavity. The higher prevalence of *H. pylori* infections in Chinese immigrants living in Melbourne, Australia, was associated with age, birthplace and the use of chopsticks. These observations indicate transmission by saliva.

Fruits and Vegetables

The possible role of fruits and vegetables in transmission is based on serosurveys and retrospective epidemiological studies. In a study of 1815 Chileans under the age of 35, *H. pylori* antibodies were detected in >60% of lower socioeconomic groups. Seropositivity was correlated with age, low socioeconomic status and consumption of uncooked vegetables. Consumption of uncooked shellfish was also a risk factor that reached marginal significance. Since sewage contamination of irrigation water and subsequent contamination of raw vegetables is a prime route for transmission of enteric pathogens in Chile, *H. pylori* may be similarly disseminated (Fig. 1).

A high intake of fruits and vegetables may actually protect against the risk of gastritis caused by *Helicobacter* spp. Yet no significant differences in *H. pylori* seroprevalence were noted between vegetarians and meat-eaters. Consumption of vitamin C and elevated plasma and gastric levels of vitamin C may eliminate infection. Other dietary risk factors linked to clinical gastritis include poor nutritional status, consumption of salty, smoked or pickled foods, drinking caffeinated beverages and alcoholism.

The presence of *H. pylori* in faeces, epidemiological evidence suggesting a link with consumption of raw vegetables in Chile, and high prevalence of *H. pylori* in developing countries with low hygiene standards indicate widespread exposure to *H. pylori*. When houseflies were exposed to freshly grown *H. pylori* on agar plates, *H. pylori* could be isolated from the flies' surface and excreta for up to 30 h. This suggests that flies may be natural reservoirs and vectors of *H. pylori* and, thus, could easily transfer the microbe from contaminated faeces to foods.

To date, no isolations of *H. pylori* have been reported from vegetables, fruits, shellfish or seafoods.

Methods to Test for *H. pylori*

The urea breath test, which is more sensitive than serological tests, is a non-invasive means of diagnosing *H. pylori* in the stomach (**Fig. 2**).

H. pylori generates high levels of urease, an enzyme which converts urea to CO_2 and ammonia. Following an overnight fast, the patient consumes a specially prepared solution of ^{13}C urea, after which breath samples are collected for up to an hour. If present in the stomach, *H. pylori* converts urea to $^{13}CO_2$, which enters the blood, passes through the lungs, and is exhaled. The amount of $^{13}CO_2$ respired during the interval indicates urease activity and thus *H. pylori* infection.

Figure 1 Possible food-related routes of transmission of *Helicobacter pylori*. (**A**) *H. pylori* has not been detected in cattle or swine. There is no evidence in support of transmission of *Helicobacter* spp. to humans via consumption of beef and pork. However, *H. pullorum* may be transmitted to humans through consumption of contaminated poultry products. (**B**) *H. pylori* has been cultured from human faeces. Contaminated faeces may pollute water which is then used to irrigate vegetables. Recently, *H. pylori* has been shown to survive in flies. Thus, flies may transmit *H. pylori* and thus may disseminate it from faeces by routes similar to other enteric pathogens. (**C**) Since *H. pylori* is present in saliva, transmission by chopsticks has been suggested.

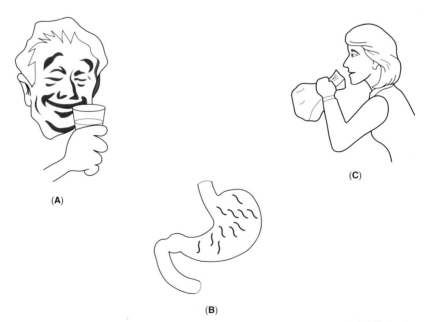

Figure 2 The urea breath test is a non-invasive method of detecting *H. pylori* in the stomach. (**A**) Following an overnight fast, the patient drinks a solution of ^{13}C-labelled urea. (**B**) In the stomach, *H. pylori* utilizes the enzyme urease to convert the urea to ammonia and ^{13}C-labelled CO_2. (**C**) The ^{13}C-labelled CO_2 diffuses into the blood stream and to the lungs where it is exhaled. The amount of ^{13}C–CO_2 exhaled indicates the presence of *H. pylori*.

Evidence for Water as a Vehicle of Transmission

On the basis of the urea breath test, it was concluded that in communities in the city of Lima, Peru, the water source may be a more important risk factor than socioeconomic status in acquiring *H. pylori* infection. In Columbia, drinking stream water, swimming in rivers, eating raw vegetables, contact with sheep, as well as number of children in the household were risk factors correlated with *H. pylori* infection.

H. pylori has been detected via PCR in water, including river water, as well as in sewage samples collected in a midwest American town. It has also been cultured from water in Peru. A viable non-culturable form, induced by refrigeration, has been described for *Campylobacter jejuni*. A similar infective coccoid phase which may revert to a viable replicating form has been described for *H. pylori*. Chlorination studies completed in the US by the Environmental Protection Agency indicate that *H. pylori*, like *C. jejuni* and *Escherichia coli*, is inactivated by standard chlorination regimens.

Evidence for Meat Animals as Sources of Infection

Attempts to isolate *H. pylori* from food animals have been hampered in part due to the fastidious nature of the microbe, its patchy distribution in the stomach and the difficulty of primary isolation. Serosurveys have been used to determine the distribution of *H. pylori* in livestock. Although relatively easy to perform when compared to cultural isolation, the accuracy of a serological test is dependent upon lack of cross-reactivity with other bacteria in the genus *Helicobacter*. Specificity is critical when reagents designed to screen for *H. pylori* in humans are used in livestock serosurveys.

In swine, evidence of infection by *Helicobacter*-like organisms is derived from serosurveys, monoclonal antibody detection of *H. pylori*-like bacteria in stomachs and the presence of *Helicobacter*-like organisms in 53% of pig stomachs. However, *H. heilmannii* (*Gastrospirillum suis*) has been recognized in all stomachs of pigs with ulcers and in 35% of normal stomachs. The phylogenetic proximity of *H. heilmannii* to *H. pylori* suggests the likelihood of antigenic cross-reactivity. Although *H. pametensis* has been found in pigs, in only a single study has *H. pylori* been isolated and verified in a pig by 16S rDNA sequencing. Taken together, the data suggest that pigs are natural carriers of *H. heilmannii*, and are not natural carriers of *H. pylori*. Thus, the risk of pork transmitting *H. pylori* to humans is remote.

For cattle, a case-control study involving 30 unweaned beef calves with fatal perforating or a haemorrhagic ulcer was conducted to determine its association with *H. pylori*. *Helicobacter*-like organisms, including *H. pylori*, were not visualized in or cultured from any of the bovine abomasal tissue samples. In contrast, in another study, spiral shaped organisms were seen in 90% of abomasal samples recovered from clinically healthy cows at slaughter in Germany. A small proportion of these were identified in tissues as indistinguishable in morphology from *H. pylori*. Approximately 60% of biopsy samples from the pylorus region gave a positive urease test. However, although the morphology and a positive urease test are suggestive of a *Helicobacter*-like organism, they are not definitive in identifying *H. pylori*. In another report, *H. pylori* antibodies were detected in 6 of 22 (27%) of calf sera tested. This may be due to possible cross-reactivity of *H. pylori* test reagents with an as yet undescribed bovine species of *Helicobacter*.

H. pylori survives in experimentally inoculated milk refrigerated for up to 6 days and in milk held at room temperature for 3 days. Although milk proteins may buffer gastric acids and would be expected to facilitate colonization, milk consumption actually provides a slight protective effect against human infection.

With respect to poultry, *H. pylori* was not detected in the intestinal tract of chickens. However, a new species, *H. pullorum*, was first cultured from the caeca of healthy broilers, and the livers and intestinal contents of laying hens with hepatic lesions. In a Swiss study of isolates from human gastroenteritis cases, 6 of 387 *Campylobacter* isolates (1.5%) were identified as *H. pullorum*. This study indicates the potential pathogenicity of *H. pullorum* in the human host. Because *H. pullorum* resembles *Campylobacter coli* at the light microscope level, it could be overlooked as a cause of human enteritis. The current availability of PCR primers for the genus *Helicobacter* and *C. coli*, however, would facilitate correct identification.

Indirect evidence of transmission of *H. pylori* from meat animals to humans is derived from serosurveys or blood tests of slaughterhouse workers. Antibodies to *H. pylori* were detected more frequently in slaughterhouse workers exposed to animal carcasses than in clerical workers employed at an Italian plant. The highest titres occurred in female workers who processed rabbits – the only meat animal species described in the study. However, no pre-employment serological titres were included in the study to evaluate pre-existing infection status. Also lacking was information on the country of origin of tested employees. This information is critical in interpreting serological data given the prevalence of *H. pylori* antibodies in citizens of developing countries.

In France, seropositivity for *H. pylori* was greater in slaughterhouse employees exposed to viscera of poultry (24%) and swine (14%) than in age-matched controls who did not work at that abattoir (6%). Interestingly, antibody titres to *H. pylori* were lower in workers with more than 15 years of abattoir experience than in those with less exposure. However, studies in Brazil have shown no correlation between seropositivity and slaughterhouse work.

To summarize, although other *Helicobacter* spp. have been recovered in livestock, *H. pylori* has not been confirmed in cattle or poultry, thus eliminating these meat animals as a source of infection. In developing countries, however, consumption of sewage-contaminated drinking water and vegetables may be a risk factor for *H. pylori* infection. The vector potential of flies links the spread of *H. pylori* from contaminated faeces to foods.

See also: **Campylobacter**: Introduction. **Meat and Poultry**: Spoilage of Cooked Meats and Meat Products. **PCR-based Commercial Tests for Pathogens**.

Further Reading

Axon ATR (1995) Is *Helicobacter pylori* transmitted by the gastro-oral route? *Aliment. Pharmacol. Ther.* 9: 585–588.

Blaser MJ (1997) Ecology of *Helicobacter pylori* in the human stomach. *J. Clin. Invest.* 100: 759–762.

Dunn BE, Cohen H and Blaser MJ (1997) *Helicobacter pylori*. *Clin. Microbiol. Rev.* 10: 720–741.

Fox JG and Lee A (1997) the role of *Helicobacter* species in newly recognized gastrointestinal tract diseases of animals. *Lab. Anim. Sci.* 47: 222–255.

Goodman KHJ, Correa P, Tengana Aux HJKJ et al (1996) *Helicobacter pylori* infection in the Colombian Andes: a population based study of transmission pathways. *Am. J. Epidemiol.* 1244: 290–299.

Goodwin GS and Worsley BW (1993) Helicobacter pylori: *Biology and Clinical Practice*. Boca Raton, Florida: CRC Press.

Grubel P, Hoffman JS, Chong FK, Burstein NA, Mepani C and Cave DR (1997) Vector potential of houseflies (*Musca domestica*) for *Helicobacter pylori*. *J. Clin. Microbiol.* 35: 1300–1303.

Hopkins RJ, Vial PA, Ferreccio C, Ovalle J, Prado P and Sotomayor V (1993) Seroprevalence of *Helicobacter pylori* in Chile: vegetables may serve as one route of transmission. *J. Infect. Dis.* 168: 222–226.

Jerris RC (1995) *Helicobacter*. In: Murray PR, Baron EJ, Pfaller MA, Tenover FC and Yolker RH (eds) *Manual of Clinical Microbiology*. P. 492. Washington, DC: ASM Press.

Johnson CH, Rice EW and Reasoner DJ (1997) Inactivation of *Helicobacter pylori* by chlorination. *Appl. Environ. Microbiol.* 12: 4969–4970.

Hulten K, Han SW, Enroth H et al (1996) *Helicobacter pylori* in the drinking water in Peru. *Gastroenterology* 110: 1031–1035.

Lee A, Fox JG, Otto G, Hegedus DE and Krakowka S (1991) Transmission of *Helicobacter* spp. A challenge to the dogma of faecal–oral spread. *Epidemiol. Infect.* 107: 99–109.

Marshall BJ and Warren JR (1984) Unidentified curved bacilli in the stomach of patients with gastritis and peptic ulceration. *Lancet* 8390: 1311–1314.

Nguyen AH, El-Zaatari AK and Graham DY (1995) *Helicobacter pylori* in the oral cavity: a critical review of the literature. *Oral Surg. Oral Med. Oral Pathol.* 76: 705–709.

Taylor DN and Parsonnet HJ (1995) Epidemiology and Natural History of *Helicobacter pylori* Infection. In: Blaser MJ (ed.) *Infection of the Gastrointestinal Tract*. P. 1. New York: Raven Press.

HELMINTHS AND NEMATODES

K Darwin Murrell, Agricultural Research Service, USDA, Beltsville, USA

Introduction

Food-borne parasitic diseases are an important cause of illness and economic loss worldwide. The public health burden imposed by food-borne parasite zoonoses (FPZ) such as toxoplasmosis, trichinellosis, cysticercosis and trematodosis are substantial even in developed countries. Data, though fragmentary, indicate that these parasites globally cause human illnesses with medical costs and productivity and disability losses totalling billions of dollars annually. For instance, in the USA, congenital toxoplasmosis is estimated to cost up to US$5.3 × 10^9 annually, of which sum perhaps half can be attributed to food sources. The rising concern generally over food safety is causing a reappraisal of the significance of FPZ and the strategies to control them. It is clear that public trust in food production systems will depend on the development of more effective safeguards, which in turn will require much greater understanding of the nature and epidemiology of these zoonoses. The complexities of these parasites' life histories, and the close association of infection risk with entrenched cultural and agricultural practices, make solutions difficult. The application of the hazard analysis critical control point (HACCP) approach will require more information on parasite epidemiology, particularly factors that regulate survival and transmission. Control strategies must address the complete sequence of events encompassed by the food production chain. Especially needed are more effective detection technologies. More concerted efforts to educate consumers, industry, government and public health workers of the hazards of food-borne parasites are also required. This review presents current understanding of the biology and epidemiology of the major FPZ, and recommendations for research and control. Particular attention is paid to species transmitted from fish and meat that are of the greatest public health significance.

Nematodes

Trichinellosis

Life Cycle of *Trichinella spiralis* Globally, the two most important species of the meat-borne parasite *Trichinella* are *T. spiralis* and *T. britovi*, although occasionally other species have been implicated in human infections. *Trichinella spiralis* is the classical agent of human trichinellosis and is nearly always transmitted through pork; *T. britovi* is common in wildlife but is often transmitted to people who consume improperly treated wild game; it is occasionally transmitted from pork (**Table 1**). The life cycle of *Trichinella* is relatively simple, but unusual for a helminth parasite in that all stages of devel-

opment occur within a single host. Epidemiologically, the most important feature of its life cycle is its obligatory transmission through ingestion of meat containing intracellular larvae (trichinae); there is no free-living stage. After ingestion of infected meat, larvae are digested free from the meat in the stomach, pass into the small intestine, and invade the epithelial cells lining the upper small intestine. Here, within 4–6 days, they develop into sexually mature males and females and mate. Their offspring are released from the female worm live (newborn larvae) and migrate via the circulation system throughout the host's body; they invade successfully only striated muscle. Over the next 10–14 days, they develop into fully encapsulated, intracellular larvae, capable of infecting another host. One species, *T. pseudospiralis*, is unique because the larvae reside in the muscle in an unencapsulated state; it too has been reported from humans.

Infection of Humans and Disease Symptoms Human infections result from the ingestion of improperly cooked meat (e.g. pork, game, horsemeat). In developed countries, the epidemiology of human trichinellosis is typified by urban common-source outbreaks. In the USA the largest human outbreaks have occurred among ethnic groups with preferences for raw or only partially cooked pork and game. Infected meat is typically purchased from a local supermarket, butcher shop or other commercial outlet. In recent years, nearly a third or more of human infections in the USA have been derived from wild animal meat.

In Europe, where the safeguard of pork inspection is mandatory, most recent outbreaks have resulted from infected horsemeat or wild boar. The resurgence of trichinellosis in eastern Europe appears to result from increased transmission from both pork and wild game. In Latin America, however, pork appears to be the chief source of infection.

The ingestion of 500 or more larvae by a human carries the risk of clinical disease. There is evidence that most (typically light) infections with *T. spiralis* are unnoticed or confused with some other illness such as influenza. In heavy infections, illness is reflected in gastrointestinal signs such as nausea, abdominal pain and diarrhoea. Coinciding with muscle invasion by the newborn larvae is acute muscular pain, facial oedema, fever and eosinophilia. Cardiomyopathy is not uncommon, and results from unsuccessful invasion by larvae of cardiac muscle. The chief factor in the acute muscle phase of this disease is the host's immune response to invasion; hence, immunosuppressants are often administered in life-threatening cases. The clinical signs and symptoms in humans can differ according to which species is involved. Hence,

Table 1 Examples of recent outbreaks of human trichinellosis

Country	Date	No. of cases	Source
USA	1970–90	1820	Pork, game
Canada	1970–85	924	Polar bear, walrus
Argentina	1992	151	Pork
Lebanon	1981–82	1100	Pork
Egypt	1975	51	Pork
Ethiopia	1990	20	Wild boar
China			
Yunnan Pr.	1983	5558	Pork, bear, other?
Henan Pr.	1992–96	417	Pork, other?
Heilongjiang	1981	147	Mutton?
Thailand	1982–88	3052	Pork
France	1993	200	Horsemeat
France	1985	1073	Horsemeat
Italy	1984–86	582	Wild boar, horsemeat
Lithuania	1992	819	Pork, wild boar
Slovakia	1980	77	Wild boar
Poland	1982–86	1427	Pork, wild boar
Yugoslavia (former)	1983–85	1734	Pork

treatment may be dependent upon proper identification of the infecting species.

Anisakiasis

Life Cycle of *Anisakis simplex* The most important of the nematode diseases of humans acquired from marine fishes is anisakiasis. To date over 12 000 cases have been reported, mostly from Asia. *Anisakis simplex* is the species most often associated with human disease, followed by *Pseudoterranova decipiens*. The diseases are caused by the larval stages; human infection with adult worms has not been documented. The worms are natural parasites of marine mammals such as whales, dolphins, and porpoises (definitive hosts for *A. simplex*), and seals and sea lions (definitive hosts for *P. decipiens*).

Adult anisakids are present in the stomachs of the marine mammalian definitive hosts. Eggs produced by female worms pass in the faeces and embryonate in the ocean waters. Larvae hatch from the eggs, invade small marine microinvertebrates such as euphausid crustaceans, and develop into third-stage larvae. When the crustacean is eaten by a fish or squid paratenic host, the larvae are released and pass through the gastrointestinal tract and enter the mesenteries, viscera or muscle. If the infected fish or squid is eaten by a marine mammal, the larvae are released, become established in the stomach and mature. When humans eat the paratenic hosts (infected fishes) the larvae may enter the tissue of the human gastrointestinal tract and cause disease.

Infection of Humans and Disease Symptoms The foods most commonly associated with anisakiasis are herring, cod, mackerel, salmon or squid which may be raw, inadequately cooked, or poorly salted, pickled or smoked; *P. decipiens* is usually obtained from cod, halibut, flatfish and red snapper. Infection often occurs from traditional fish preparations such as green-herring, lomi lomi salmon, ceviche, sushi and sashimi.

In humans the larvae of *A. simplex* enter the gastric or intestinal mucosa and cause an abscess or eosinophilic granuloma. The worms may also enter the peritoneal cavity and then invade other organs. Some worms may not invade tissue but simply pass out with faeces or vomit, or crawl up the oesophagus. Larvae of *P. decipiens* may also invade tissue but rarely attempt to lodge in the oesophagus; however, infestation may cause 'tickling throat' syndrome in which a tickling sensation occurs and the patient may cough up the larvae.

The symptoms of anisakiasis resemble gastric ulcer or neoplasm. The parasitological diagnosis is made by finding the worms or demonstrating sections of the parasite in biopsied tissue. Serological tests are not conclusive. The treatment of most infections is by removal of the parasites surgically or by the use of forceps through fibreoptic endoscopy. The prognosis is good once the parasite is removed.

Cestodes

Taeniasis, Cysticercosis

Life Cycle of *Taenia* spp. The terms 'cysticercosis' and 'taeniasis' refer to infections with larval and adult tapeworms, respectively. The important features of this particular zoonosis are that the larvae are meat-borne (beef or pork) and that the adult stages develop only in the intestines of humans (obligate host). There are two species: *Taenia saginata* (beef tapeworm) and *T. solium* (pork tapeworm). The latter species, *T. solium*, is of greater clinical importance because humans may also serve as the host for the larval (cysticercus or metacestode) stage if the adult worm's eggs are accidentally ingested; this does not occur with *T. saginata*. The localization of *T. solium* cysticerci in humans may cause severe clinical disease, especially if the larvae develop in the brain (neurocysticercosis). Neurocysticercosis is a major public health problem affecting many in Latin America, Asia and Africa. In Mexico, *T. solium* cysticercosis accounts for 1% of all deaths in general hospitals and 25% of all intracranial tumours. Swine cysticercosis is also a significant economic problem in certain regions owing to condemnation of infected carcasses at slaughter. The bovine form, *T. saginata*, is less severe in humans because it is confined, like the adult stage of *T. solium*, to the human intestine. However, it does represent a considerable economic cost because most countries have instituted costly mandatory meat inspection. In the USA and Europe, the infection rate for *T. saginata* in beef is generally less than 0.3 in 1000; therefore, the inspection cost to find one infected carcass is large. In Africa and Latin America, however, the prevalence of *T. saginata* is relatively high, and the resulting commodity losses from condemnation or treatment to destroy cysticerci are heavy. In Africa as a whole, the cost is about US$1.5–2.0 billion per year. The epizootic nature of bovine cysticercosis (sporadic outbreaks) makes it difficult to assess directly the economic impact of bovine cysticercosis in countries with low prevalences.

The adult tapeworm stages of *T. saginata* and *T. solium* reside in the human small intestine. The adult is composed of a chain of strobila or segments (proglottids) which contain both male and female reproductive systems. As the segments mature and fill with eggs, they become detached and pass out of the anus, either free or in the faecal bolus. The life span of an adult tapeworm may be as long as 30–40 years. The number of eggs shed from a host per day may be very high (500 000–1 million) which results in high environmental contamination. The eggs contain an infective stage (oncosphere) which matures in the environment. It is probably impossible to distinguish the two species on the basis of egg morphology.

When *T. saginata* eggs are ingested by cattle, the oncosphere stage is released in the intestine, and it penetrates the gut and migrates throughout the body via the circulatory system. Oncospheres that invade skeletal muscle or heart muscle develop to the cysticercus stage, a fluid-filled cyst or small bladder. When infected beef that is either raw or improperly cooked is eaten by humans, the larval cyst is freed and it attaches by means of a small head (scolex) with suckers to the intestinal wall. Over the span of a couple of months, the tapeworm develops and begins to shed eggs, completing the life cycle.

The development of *T. solium* is similar in its intermediate host (pigs or humans), except that the cysticerci are distributed throughout the body, especially the liver, brain, central nervous system, skeletal muscle, and myocardium.

Infection rates for the intestinal tapeworm form (taeniasis) are frequently high in developing countries (**Table 2**). The human incidence in Latin America for *T. solium* taeniasis ranges from 0.3% to 1.1%. In some countries the rates of cysticercosis infection in cattle or pigs is as high as 70%.

Table 2 *Taenia solium* and *T. saginata*: incidence in selected countries

Country	Estimated % of population infected	1992 Population (millions)	Estimated cases[a] (no.)
Taenia solium			
USA		255.6	122
Chile	0.3	13.6	40 800
Ecuador	0.9	10.0	90 000
Guatemala	1.1	9.7	106 700
Mexico	1.1	87.7	964 700
Taenia saginata			
USA	0.0002	255.6	519
South America	0.3333	300.0	1 000 000
Cuba	0.1000	10.8	10 800
Guatemala	1.7000	9.7	164 900
Chile	1.9000	13.6	258 400
Asia	0.4677	3 207.0	15 000 000
Europe	2.1526	511.0	11 000 000
Former Yugoslavia	10.0000	10.0	1 000 000
Africa	2.7523	654.0	18 000 000

[a]Except for the USA, these are rough estimates based on computations from incidence estimates, not reported cases.

Infection of Humans and Disease Symptoms
Humans acquire infection by ingestion of improperly cooked pork (*T. solium* larvae) or beef (*T. saginata* larvae). These stages develop to mature tapeworms directly in the intestine. Symptoms vary in their intensity, and some people never realize that they have a tapeworm. Most people, however, experience symptoms which may include nervousness, insomnia, anorexia, loss of weight, abdominal pain and digestive disturbances. Occasionally the appendix, uterus or biliary tract are invaded and serious disorders can occur. Most cases of human cysticercosis (*T. solium*) are asymptomatic and are not recognized by either the individual or a physician. Symptomatic infections may be characterized as either disseminated, ocular or neurological. Disseminated infections may localize in the viscera, muscles, connective tissue and bone; subcutaneous cysticerci may present a nodular appearance. These localizations are often asymptomatic, but may produce pain and muscular weakness. Only about 3% of infections involve the eye. Central nervous system involvement may include the invasion of the cerebral subarachnoid space, ventricles and spinal cord. The larval cysts may persist for years; cysts that die often calcify. Symptoms of infection may include partial paralysis, dementia, encephalitis, headache, meningitis, epileptic seizures and stroke. These manifestations of infection are determined by the numbers and locations of the cysts and the host's inflammatory response against them.

Diphyllobothriasis

Life Cycle of *Diphyllobothrium latum* Cestodes that may be transmitted to humans from marine fish are limited, for the most part, to members of the genus *Diphyllobothrium*. Records of human infection with cestodes associated with fish are generally confined to countries where fish are eaten raw, marinated or undercooked (e.g. Alaska, USA, Canada, Scandinavia, Japan, Chile, Peru and Russia). Six species of *Diphyllobothrium* have been recorded from humans in Alaska, of which *D. latum* was the most common. Worldwide, at least 13 species of *Diphyllobothrium* have been reported from humans, with infections by *D. latum* and *D. dendriticum* being the most prevalent.

The life cycles of these cestodes occur in either marine or freshwater ecosystems, depending on the species. In general, the adult tapeworm, residing in the intestine of the definitive host (either a marine or terrestrial mammal), releases eggs which pass in the host's faeces. If the eggs reach water, they hatch and release a free-swimming stage (coracidium) which may be ingested by a copepod. Within this crustacean intermediate host, the larval tapeworm (procercoid) develops. If the copepod is then ingested by a suitable fish, the larva migrates to the host's body cavity and develops to the plerocercoid stage, which is infective to the definitive host (including humans).

Infection of Humans and Disease Symptoms   The important risk factor for this zoonosis is the consumption of raw or undercooked fish. In Japan since the 1970s diphyllobothriasis (primarily *D. latum*) has increased in incidence; about a hundred cases are recorded annually. Of potential intermediate hosts, the salmonid genus *Oncorhynchus* is the most important. Of the 52 cases of diphyllobothriasis occurring in the west coast states of the USA in 1980, salmon was implicated in 82%. Over 60 cases have been reported from Peru, most considered to be caused by *D. pacificum*. The systematics of this genus is unsettled, and identification is difficult, particularly the larval stage encountered in fish (the plerocercoid). Therefore, post-harvest controls such as proper fish processing and preparation of food are most effective. *Diphyllobothrium latum* has long been of interest because it causes pernicious anaemia, probably due to the worms' competition with the host for vitamin B_{12}.

Trematodes

Liver Parasites

Life Cycles of *Clonorchis sinensis* and *Opisthorchis* spp.   The liver parasites (flukes) *Clonorchis* and *Opisthorchis* are closely related and will be treated together. The most common is *C. sinensis*, which is endemic to China, Korea, Japan, Taiwan, Vietnam and Hong Kong. In China its occurrence in certain provinces is high: 3 million people in Guangdong, 1 million in Guangxi. Pigs, which may serve as reservoir hosts, also have high prevalences of infection judging by the results of various surveys in China (e.g. 11–35%). *Clonorchis sinensis* is important in Korea where the prevalence may exceed 10% in many rural areas. In Hong Kong, the prevalence reported from some villages is 13%. Globally, 290 million people may be at risk and 7 million infected.

The second important human liver fluke disease is opisthorchiasis, caused by *Opisthorchis viverrini* and *O. felineus* (the cat liver fluke). These parasites are found over a wide geographic range; *O. viverrini* occurs chiefly in southeast Asia, and *O. felineus* is generally found in eastern Europe, Poland, Germany and Siberia. There are reports that 46% of Russian territory is endemic for *O. felineus* and that 80 000 to 90 000 new cases occur each year among the 12 million people living in Siberia. The total number of

people infected with *O. felineus* may exceed 4 million. However, the prevalence of *O. viverrini* may be even greater. In Thailand, it is estimated that more than 7 million people are infected. In recent surveys in Laos, the prevalence of *O. viverrini* in a series of villages ranged from 28% to 85%.

Clonorchis sinensis infections are acquired by eating raw or inadequately cooked fish muscle containing the infective larval stage (metacercariae). In the definitive host (human), the trematode matures and produces eggs which pass out into the faeces. If the eggs reach fresh water, hydrobid snails (if present) may ingest them. Within the snail, two asexual proliferative stages ensue (redia and sporocyst). Eventually, a motile or swimming larval stage (cercaria) emerges and seeks out fish, particularly cyprinoid species; over 80 species of fish have been identified as potential hosts in China, for example. The cercariae penetrate beneath the fish's scales and encyst in the muscle, becoming infective in 3–4 weeks.

The epidemiology and life cycle of the opisthorchid flukes are similar to that of *C. sinensis*. The infection is acquired by eating improperly prepared fish harbouring the infective muscle stage (metacercaria). Cats and dogs (reservoir hosts) are commonly infected in endemic areas, complicating efforts to protect snail-inhabiting water sources.

Infections of Humans and Disease Symptoms   The adult *C. sinensis* is a flat, slender trematode that invades the smaller biliary passages of humans. Infected people complain of indigestion, epigastric discomfort and diarrhoea. If the adult worm invades the pancreatic duct, acute pancreatitis may result. Chronic infection may lead to cholangiocarcinoma.

Opisthorchis infections cause clinical symptoms similar to those of *Clonorchis sinensis* infection. However, opisthorchiasis may have greater public health significance because it is more often associated with cholangiocarcinoma than is clonorchiasis, and it is a major cause of death in rural northeast Thailand. The International Agency for Research on Cancer has declared this parasite a Group 1 carcinogen. The estimated direct public health effect in northeast Thailand alone is estimated at US$85 million.

Life Cycle of *Fasciola hepatica*   The liver parasite *Fasciola hepatica* is acquired by eating aquatic plants on which the infective (metacercarial) stage is affixed. In addition to being an occasionally important parasite of humans, it is a cause of serious disease in livestock. The distribution of the parasite is cosmopolitan, being reported from 61 countries. The largest number of human infections occur in Bolivia, Ecuador, Egypt, France, Iran, Peru and Portugal. The life cycle is typical of digenetic trematodes and involves an aquatic snail (generally *Lymnea*). The embryonated egg, on being voided in faeces into water, hatches and releases a ciliated miracidium, which swims about until making contact with a suitable snail. The miracidium penetrates into the snail's viscera and over the next several weeks the parasite develops through several complex asexual stages, before finally producing hundreds of cercariae. These emerge from the snail and swim around until making contact with particular water plants (e.g. watercress) where they affix and encyst and develop into the metacercarial (infective) stage. When the plant is eaten by the mammalian host, the parasite excysts in the digestive tract, penetrates the intestinal wall and migrates to the liver and eventually resides as a mature adult worm in the bile duct. The time required to complete this migration is about 4–6 weeks.

Infection of Humans and Disease Symptoms   The major pathological changes occur during the migration of the immature worm through the liver parenchyma. During this phase, the worm digests liver tissue and causes necrosis. This trauma often yields scars and fibrotic lesions. After arrival in the bile duct, the worms may incite inflammatory gall bladder alterations and fibrosis of the duct. Clinical manifestations are anaemia, jaundice and cholelithiasis.

Intestinal Parasites

Life Cycles of Intestinal Trematodes   Although a large number of intestinal digenetic trematodes have been listed as zoonotic for humans, only a few are important. Chief among these are members of the Heterophyidae family; members of the group are very small trematodes inhabiting the intestine of birds and mammals. The infective stage (metacercaria) can be found in a wide variety of freshwater and marine fishes. Perhaps most important are *Heterophyes heterophyes* and *Metagonimus yokogawai*. These parasites, acquired by eating raw, marinated or improperly cooked fish, are frequently reported from human infections in the Middle East and Asia, especially the Philippines, Indonesia, Thailand, China, Japan and Korea. Large numbers of these parasites in the small intestine may cause inflammation, ulceration and necrosis.

The life cycle is completed when eggs of the intestinal worms are shed in the faeces. If they reach water, they may be ingested by snails in which the asexual stages occur. The cercariae which eventually emerge from the snails seek a suitable fish host to develop to the metacercarial stage. A large number of bird and mammal species may serve as definitive hosts for this parasite species assemblage.

Infection of Humans and Disease Symptoms   As with *Clonorchis* and *Opisthorchis*, the metacercarial infective stage in the fish host's muscle, when consumed raw or improperly cooked, completes its development in the host's intestine. When the intestinal parasite numbers are large, intestinal ulceration and inflammation may occur. Worms deep in the crypts of the intestine release eggs that can become trapped instead of passing completely through the intestine. The trapped eggs may enter the circulatory system and eventually filter out in the spleen, liver, brain, spinal cord and heart; in the heart a foreign body reaction may result, causing fibrosis and calcification of the heart valves.

Prevention and Cure of Infection

Because all of the parasitic infections described above share a common mode of transmission (i.e. foodborne), direct and effective protection for humans rests with proper preparation of food. Specific requirements for preparing parasite-free food are listed in **Table 3**.

Effective control or eradication of food-borne parasites is difficult because of the complexity of the parasites' life histories, and the close association of infection risk with entrenched cultural and agricultural practices. More epidemiological research will be needed for improving control strategies. The broader application of the HACCP approach would be of great value. As with microbial food pathogens, the development of more reliable, sensitive and standardized detection technologies is needed; without them, the magnitude of the problems will be difficult to assess and prevention will be elusive. Although it is important to verify safety through food inspection where possible, an ability to identify the sources of contamination is crucial. These improved diagnostic tools are badly needed for epidemiological surveillance of people and livestock so that infected carriers can be identified and treated, especially at the food production stage.

The education of consumers, industry, governments and public health workers about the hazards of foodborne parasites is also a direct and effective safeguard against infection. Adoption of international codes promulgated by the World Health Organization and the Food and Agriculture Organization for the production of food, especially fish and fishing products, should be encouraged. Piecemeal approaches are a poor substitute for broad holistic approaches to control. An effective strategy must involve all participants, including the beneficiaries, and must permit coordination of all control activities, such as legislation, education and detection. Sustainable control

Table 3 Epidemiology and control of major food-borne helminths

Disease	Parasite	Major food animal sources	Current control methods
Trichinellosis	*Trichinella spiralis* (occasionally other species)	Pigs, horses, game animals	Cook to 60°C Freeze at −23°C for 10 days Curing of meat according to government specifications Meat inspection: in many but not all countries
Taeniasis	*Taenia saginata* (beef) *Taenia solium* (pork)	Meat and organs infected with larvae (cysticercus stage)	Cook to 60°C Freeze at −23°C for 10 days
Cysticercosis	*Taenia solium*	Ingestion of tapeworm eggs may result in larval (cysticercus) invasion of muscle and brain	Avoid contamination of soil, water and food with human faeces which may contain eggs of *T. solium* Meat inspection in most countries
Anisakiasis (USA, Japan, Pacific Islands, northern Europe)	*Anisakis simplex* and *Pseudoterranova decipiens*	Hosts for parasite larvae: herring, mackerel, salmon, cod, whiting, tuna, haddock, smelt, plaice	Cook 60–65°C Freeze at −25°C for 7 days or blast freeze to −35°C for 15 h Salt in 20–30% brine for 10 days
Diphyllobothriasis (northern hemisphere, esp. Baltics, Europe, Russia, USA, Canada)	*Diphyllobothrium latum*	Freshwater fish, especially pike	Cook fish well Freeze to −10°C if to be consumed raw Smoke or pickle well Protect water sources (fishponds, lakes) from human and animal faeces
Clonorchiasis, opisthorchiasis (Asia, Europe)	*Clonorchis sinensis* *Opisthorchis viverrini*, *O. felineus*	Freshwater fish infected with larval stage (metacercaria)	Cook, salt, pickle or smoke fish well Freeze to −10°C Protect fishponds from human and animal faeces
Heterophyiasis (trematodes of the Heterophyidae family) (Europe, Middle East and Asia)	*Heterophyes heterophyes* *Metagonimus yokogawi* Many others	Freshwater or brackish water fish infected with larval stage (metacercaria)	Cook, salt or smoke fish well Freeze to −10°C Protect domestic animals from raw fish Sanitation

demands prevention and intervention at every critical juncture of the production-consumption continuum, especially at the production level.

See also: **Hazard Appraisal (HACCP)**: The Overall Concept.

Further Reading

Dubey JP (1993) *Parasitic Protozoa*. Vol. 6, pp. 1–158. New York: Academic Press.

Dupouy-Camet J, Soule CI and Ancelle T (1994) *Parasite* 1: 99–103.

Hui YH, Gorham JR, Murrell KD and Oliver DO (eds) *Foodborne Diseases Handbook*. Vol. 2, pp. 199–462. New York: Marcel Dekker.

Murrell KD (1995) *International Journal of Environmental Health Research* 5: 63–85.

Pawlowski ZS (1990) *Parasitology Today* 6: 371–373.

Roberts T, Murrell KD and Marks S (1994) *Parasitology Today* 10: 419–423.

Hemiascomycetes – 1 and 2 *see* **Fungi**: Classification of the Hemiascomycetes.

Hepatitis Viruses *see* **Viruses**: Hepatitis Viruses.

HIGH-PRESSURE TREATMENT OF FOODS

Margaret Patterson, Department of Agriculture for Northern Ireland and The Queen's University of Belfast, Agriculture and Food Science Centre, Belfast, UK

The idea of using high hydrostatic pressure (HHP) as a method of food processing is not new. Bert Hite, from West Virginia University, reported in 1899 that high-pressure treatment at ambient temperature could be used to preserve milk. In later studies he also reported that some microorganisms, such as lactic acid bacteria and yeasts, associated with sweet, ripe fruit were more susceptible to pressure than spore-formers associated with vegetables. Research did not progress as the equipment was not available to routinely subject foods to the necessary pressures. However, in recent years there has been renewed scientific and commercial interest in the process. This can be explained by the fact that advances in engineering make it both economically viable and technologically feasible to treat foods at the desired pressures. In addition, consumer demand for high-quality, minimally processed, additive-free and microbiologically safe foods have stimulated research into methods including HHP.

Nature of the Process

Principles of High-pressure Processing

The Système International (SI) unit of pressure is the pascal (Pa) or newton per square metre ($N\,m^{-2}$). This is a very small unit of pressure, but in the metric system prefixes such as 'mega-' (M) equivalent to 10^6 or 'giga-' (G) equivalent to 10^9 pascals are used. Food applications use pressures in the range 100 MPa to 1 GPa. A pressure of 100 MPa is equivalent to 1 kbar, 986.9 atmospheres or $14\,504\,lbf\,in^{-2}$. These pressures are higher than those naturally occurring on earth but are used routinely in industrial processes.

High-pressure is generally a semicontinuous bulk process for liquid foods and a batch process for solid foods. A typical high-pressure system consists of a pressure vessel and a pressure generator. Food packages are loaded into the vessel and the top closed. The pressure transmission fluid, usually water, is pumped into the vessel from the bottom. Once the desired pressure is reached, the pumping is stopped, valves are closed and the pressure is held without the need for further energy input. Processing costs are claimed to be £0.04–0.20 per litre or per kilogram depending on factors such as the pressure applied, process time

and throughput. Pressure vessels capable of processing foods are commercially available. For example, up to 3000 litres of fruit juice can be processed per hour.

Effect of Pressure on Biomolecules

There are two main principles involved in high-pressure processing: the isostatic principle and the Le Chatelier principle. The former states that pressure is transmitted uniformly and instantaneously throughout the sample. This process is independent of the volume or geometry of the sample. This property gives HHP an important advantage over conventional thermal processing. The Le Chatelier principle states that the application of pressure to a system in equilibrium will favour a reduction in volume to minimize the effect of pressure. Thus, reactions that result in a volume decrease are stimulated, while those causing a volume increase are disrupted. Hydrogen bonding tends to be favoured while ionic bonds are broken. Hydrophobic interactions are disrupted below 100 MPa but can be stabilized at higher pressures. Covalent bonds appear to be unaffected by high pressure, so low-molecular-weight molecules – such as those responsible for the sensory and nutritional qualities of food – are left intact. However, the structure of high-molecular-weight molecules can be significantly affected and this can result in altered functionality of proteins and carbohydrates. These changes result in microorganisms being killed, as well as the possibility of producing foods of improved sensory and nutritional quality.

Applications in the Food Industry

The first commercial high-pressure processed product was a high-acid jam, launched in 1990 by the Meidaya Food Factory Co., Japan. Since then several other pressure-treated products have been launched in the Japanese market, including fruit sauces, juices and jellies. A high-pressure processed orange juice is also available in France and a pressure-treated avocado dip is available in the USA. Pressurized fruit products are normally given a treatment of around 400 MPa at 20°C. It is claimed that the colour and flavour of the fresh fruit is maintained and there is only a slight decrease in vitamin C content. Yeasts and moulds, which are the main cause of spoilage in fruit products, are relatively sensitive to pressure, so the microbiological quality of the products can be improved. However, enzymes such as polyphenoloxidase, which causes browning, are resistant to pressure and therefore can limit the shelf life. For these reasons the current commercial applications of HHP as a preservation method have been limited to high-acid, chilled foods where spore-forming bacteria are unlikely to be a problem and enzymatic activity is retarded by refrigerated storage. However, extensive research and development programmes are in progress in Japan, Europe and the USA and it is likely that other foods, including fruit, dairy, meat and fish products, will be launched into the international market (**Table 1**).

Effect on Microbial Cells

The lethal effect of high pressure on microorganisms is thought to be the result of a number of different processes taking place simultaneously, including damage to the cell membrane and inactivation of key enzymes such as those involved in DNA replication and transcription.

The primary site of pressure damage in microorganisms is the cell membrane. Under pressure, a reduction in volume of the membrane bilayers occurs along with a reduction in the cross-sectional area per phospholipid molecule. Protein denaturation also occurs and the activity of membrane-bound enzymes such as Na^+-K^+-ATPase is reduced. These changes disrupt cell membrane function, allowing leakage through the inner and outer membranes. Some enzymes responsible for key biochemical reactions are susceptible to pressure and this can lead to microbial inactivation. The two primary means by which pressure-induced enzyme inactivation occurs are an alteration of intramolecular structures and conformational changes at the active site. Many factors can affect the degree of inactivation including pH, substrate concentration, subunit structure of the enzyme and temperature. Nucleic acids are much more pressure-resistant than proteins, and as the DNA helix is largely a result of hydrogen bond formation which is favoured by pressure, its structure is not affected at least up to 1 GPa. However, despite the stability of

Table 1 Potential applications of high-pressure processing of foods

Application	Typical processing conditions	Examples
Improved quality of fruit products	300–400 MPa, 1–5 min, 30–40°C	Shelf life of citrus juices and fruit jams can exceed 1 month at refrigeration temperatures
Improved shelf life and safety of meat products	400 MPa, 10 min, 50°C	Shelf life of duck foie gras increased to 13 days at 4°C
Restructuring of meat products	100–300 MPa, 30 min	Improved flavour, appearance and elasticity of gels made from minced meat and fish, e.g. surimi
Gelatinization of starch	400–500 MPa, 20–60 min, 40–50°C	Enhanced digestion of wheat, corn and potato starches by α-amylase. Possiblity of cooking rice and pasta with minimal heat
Tenderization of meats	100–150 MPa, 4 min, 35°C	Suitable for pre-rigor meat
	100–150 MPa, 60 min, 60°C	Suitable for post-rigor meat
Smoked 'raw ham'	Smoking at 65°C for 90 min followed by 250 MPa, 3 h, 20–40°C	New type of cured meat product
Improved shelf life and safety of dairy products	450 MPa, 15 min, 2–10°C	Increased shelf life of goat's cheese stored at 4°C
Improvement in rate of cheese maturation	10–250 MPa	Ripening time reduced to 3 days
Prevention of over-acidification of yoghurt	200–300 MPa, 10 min, 10–20°C	Shelf life of > 2 weeks at 10°C
Improved texture of dairy products	200–600 MPa, up to 120 min, 20°C	Increased strength of acid-set gels made from pressurized milk
Specific hydrolysis of β-lactoglobulin in bovine milk by thermolysin	200 MPa	Milk allergenicity reduced
Tempering of chocolate	150 MPa, 32°C, 5 min twice	Improved crystal structure in cocoa butter
Pressure-shift freezing	200 MPa, −20°C, rapid pressure release and storage at −20°C	Only small ice crystals formed during freezing so improved product quality
Pressure-shift thawing	50 MPa, 20°C, 30 min	Rapid thawing of beef with reduced drip loss

DNA, the enzyme-mediated steps of DNA replication and transcription are disrupted by pressure. As the pressure treatment is isostatic, microbial cells are not 'burst' open by the treatment and there are usually no obvious changes to the external structure of vegetative cells. However, transmission electron micrographs of bacteria such as *Listeria monocytogenes* show changes in the internal cell organization. In some cases clear areas, devoid of ribosomes, are found adjacent to the cytoplasm. The cause of the clear regions is not known but may be due to phase or other conformational changes to the membrane or to the localized destruction of some of the ribosomes.

Microorganisms vary in their response to HHP and the kinetics of high-pressure inactivation are different from those observed with other food processing methods. Typical inactivation curves are shown in **Figure 1**. It is obvious that pressure inactivation is not always first-order (a straight line relationship) and death curves often show pronounced survivor tails. Several theories have been proposed to explain the curve shape. The tail may be independent of the mechanisms of inactivation or survival and be the result of population heterogeneity, for example due to cell age, clumping, genetic variation or experimental error. Tailing may also be a normal feature associated with resistance. There are reports that when pressure 'resistant' tail populations of *Salmonella* spp. were isolated, grown and again exposed to pressure there was no significant difference in the pressure resistance between the re-treated and the original cultures. Other reports suggest that repeated pressurizing (up to 18 cycles) can select for pressure-resistant mutants of *Escherichia coli*, although some of these mutants are more sensitive to heat. Temperature during pressure treatment may also affect the shape of the inactivation curve. First-order inactivation is more common at temperatures above 30°C while second-order inactivation is often found below 30°C. This may be explained by a less pressure-sensitive sub-population which has an altered membrane composition below 30°C as a consequence of liquid–gel transformations in the membrane.

The surviving tail populations are of concern in food processing and have to be taken into account when choosing appropriate treatments to ensure microbiological safety. The traditional method of calculating D values (decimal reduction time) and Z values (the temperature increase required for a tenfold decrease in D value) used in thermal processing cannot be applied successfully to pressure treatments. Various approaches are being taken to try to model the complex inactivation kinetics of pressure-treated microorganisms, taking into account other process variables such as temperature, water activity and ionic strength. These models will have to be tested rigorously in different foodstuffs to ensure their reliability and should eventually be available as computer-based predictive models for use in process calculations.

Intrinsic and Extrinsic Factors Affecting Sensitivity

Species and Strain Variation

Microorganisms vary in their resistance to HHP (**Table 2**). Eukaryotic cells tend to be more pressure-sensitive than prokaryotes. Yeasts are among the most sensitive of the microorganisms and treatments of 300–400 MPa for a few minutes at 20°C can result in more than a 6 log reduction in numbers. Vegetative forms of moulds are also relatively sensitive but mould spores are more resistant.

It was initially thought that Gram-negative bacteria were more pressure-sensitive than Gram-positive ones. A suggested explanation was that the cell membrane structure is more complex in Gram-negative bacteria so it was likely to be more susceptible to environmental changes caused by pressure. More recent studies have shown that vegetative bacteria do vary greatly in their resistance to pressure and some Gram-negative bacteria, such as certain strains of *E. coli* O157:H7, are surprisingly resistant (**Fig. 2**). To date, the reason for such strain variation is not understood.

The heat resistance of some vegetative organisms is correlated with pressure resistance, but there are many exceptions. Gram-positive cocci such as enterococci and *Staphylococcus aureus* are more resistant to both heat and pressure than Gram-negative rods such as

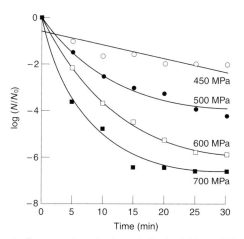

Figure 1 Pressure inactivation of *Escherichia coli* O157:H7 (NCTC 12079) in phosphate-buffered saline (pH 7.0) at 20°C. N_0, original number of bacteria; N, number of surviving bacteria. Reprinted with permission from *Journal of Food Protection*, © IAMFES.

Table 2 The sensitivity of vegetative pathogens to high pressure in various foods

Microorganism	Substrate	Treatment conditions	Inactivation (log$_{10}$ units of reduction)
Vegetative bacteria			
Campylobacter jejuni	Pork slurry	300 MPa, 10 min, 25°C	6
	Poultry meat		
Citrobacter freundii	Minced meat	300 MPa, 20 min, 20°C	> 5
	Strained chicken baby food	340 MPa, 10 min, 23°C	< 2
Escherichia coli (non-pathogenic)	Goat's cheese	400 MPa, 10 min, 25°C	> 7
E. coli O157:H7	UHT milk	800 MPa, 10 min, 20°C	< 2
	Poultry meat	700 MPa, 30 min, 20°C	5
	Poultry meat		
Lactobacillus spp.	Moscato wine, pH 3.0	400 MPa, 2 min, 20°C	6
Listeria monocytogenes (Scott A)	UHT milk	340 MPa, 80 min, 23°C	6
	Raw milk	340 MPa, 60 min, 23°C	6
Listeria innocua	Ewe's milk	350 MPa, 10 min, 2°C	4
		350 MPa, 10 min, 25°C	1
Pseudomonas aeruginosa	Pork slurry	300 MPa, 10 min, 25°C	6
Pseudomonas fluorescens	Minced meat	200 MPa, 20 min, 20°C	> 5
Staphylococcus aureus	Pork slurry	600 MPa, 10 min, 25°C	6
	Poultry meat	600 MPa, 30 min, 20°C	4
	UHT milk	600 MPa, 30 min, 20°C	2
Streptococcus faecalis	Pork slurry	300 MPa, 10 min, 25°C	< 1
Vibrio parahaemolyticus	Canned clam juice	170 MPa, 10 min, 23°C	> 5
Yersinia enterocolitica	Pork slurry	300 MPa, 10 min, 25°C	6
Spore-forming bacteria			
Bacillus coagulans spores	PBS (100 mmol l^{-1}) pH 8.0	400 MPa, 30 min, 45°C plating out immediately after pressurizing	2
		pressurized spores heat treated (70°C for 30 min) before enumeration	7
B. coagulans spores	McIlvaine's citrate-phosphate buffer	400 MPa, 30 min, 45°C	
	pH 7.0		2
	pH 4.0		4
B. stearothermophilus spores	Not given	800 MPa, 60 min, 60°C	4
		400 MPa, 6 × 5 min, 70°C	4
Clostridium botulinum type A spores	'Rich medium'	250 MPa, 60 min, 70°C	6
Clostridium sporogenes spores	Meat broth, pH 7.0	800 MPa, 5 min, 80–90°C	> 5
		1500 MPa, 5 min, 20°C	No inactivation
Yeasts and moulds			
Byssochlamys nives ascospores	Bilberry jam, a$_w$ 0.84	700 MPa, 30 min, 70°C	< 1
	Grape juice, a$_w$ 0.97		4
Candida utilis	Pork slurry	300 MPa, 10 min, 25°C	2
Rhodotorula rubra	Sucrose solution	400 MPa, 15 min, 25°C	
	a$_w$ 0.92		< 1
	a$_w$ 0.96		> 7
Saccharomyces cerevisiae	Mosca wine, pH 3.0	400 MPa, 2 min, 20°C	6
	Satsuma mandarin juice, pH 3.1	100 MPa, 5 min, 47°C	3
	Pork slurry	300 MPa, 10 min, 25°C	2
Zygosaccharomyces bailii	Mosca wine, pH 3.0	400 MPa, 2 min, 20°C	6
Viruses			
HIV-1	Laboratory culture	550 MPa, 10 min, 25°C	Infectivity titre reduced by 4 log units
Bacteriophage T4	0.01 mol l^{-1} TRIS buffer	400 MPa, 10 min, 4°C	> 6

a$_w$, activity of water; PBS, phosphate-buffered saline; TRIS, tris(hydroxymethyl)aminomethane; UHT, ultra-high temperature.

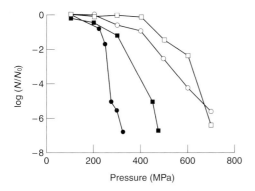

Figure 2 Pressure inactivation of pathogens after 15 min treatment in phosphate-buffered saline at 20°C. Solid circles, *Yersinia enterocolitica*; solid squares, *Salmonella enteritidis*; open circles, *Escherichia coli* O157:H7; open squares, *Staphylococcus aureus*; N_0, original number of bacteria; N, number of surviving bacteria.

Campylobacter jejuni and *Pseudomonas aeruginosa*. However, a heat-resistant strain of *Sal. senftenberg* was found to be less pressure-resistant than a heat-sensitive strain of *Sal. typhimurium*.

Bacterial spores are extremely resistant to the pressures normally applied to foods, although there can be significant variation between different species and strains (Table 2). Spores of *Clostridium* spp. tend to be more pressure-resistant than those of *Bacillus*, although there is relatively little information on the effect of pressure on *C. botulinum*. The extreme resistance of bacterial spores means that it is likely that HHP will have to be used in combination with other preservation technologies to give an acceptable level of inactivation in foods.

Although extremely high pressures are needed to inactivate spores directly, lower pressures, usually less than 400 MPa, have been found to trigger spores to germinate. This effect has still not been explained at the molecular level but it can be promoted by common germinants for the particular type of spore, such as L-alanine, adenosine and inosine. If a subsequent pressure and/or temperature treatment is high enough, it may inactivate the now more pressure-sensitive germinated spores. The germinative effect of pressure is greatly enhanced at raised temperatures so such combinations are likely to be most effective at giving adequate inactivations, particularly in low-acid foods.

Pressure cycling between low and high pressures, with or without heating, has also been proposed as an effective method of inactivating spores. In theory, a relatively low pressure can be used to cause germination and this can be followed by a higher-pressure treatment to kill the germinated cells. The process can be repeated on a number of occasions to increase the overall inactivation. This approach may reduce the severity of the heat treatment needed. Repeated

treatment at one pressure can also be effective. For example, cycling treatment of 400 MPa at 70°C (5 min per cycle for 6 cycles) gave the same level of inactivation (a log 4 reduction) of *B. stearothermophilus* spores as 60 min continuous treatment of 600 MPa at 70°C.

A 12 log inactivation of *C. botulinum* is required by the thermal processing industry for shelf-stable foods of high pH and high water activity. It is clear that much more research is required before the equivalent pressure-processed products will be available.

There is little published research on the effect of high pressure on viruses, although it has been suggested that enveloped viruses, such as influenza and herpesviruses, lose their infectivity when treated with HHP. It is reported that pressures of 400 MPa or higher can inactivate human immunodeficiency virus type 1, although pressure sensitivity differed between strains (Table 2).

There is also little information on the destruction of bacterial toxins by high pressure. There is some evidence that botulinum toxin may be partially inactivated at pressures greater than 600 MPa. Pressure-treating the mycotoxin patulin does cause some decrease in toxin levels. However, relatively long treatment times are needed and toxin inactivation is incomplete. Pressurizing at 500 MPa for 1 h at room temperature decreased the patulin content to 80% and 47% of the original values in apple concentrate and apple juice respectively.

Stage of Growth

Cells in the stationary phase of growth are generally more resistant to pressure than those in the exponential phase. When bacteria enter the stationary phase they are known to synthesize new proteins that protect cells against a variety of adverse conditions including elevated temperatures, oxidative stress and high salt concentration. It is not known if similar mechanisms also protect bacteria against high pressure.

Substrate

The chemical composition of the substrate can influence the response of bacteria to pressure (Table 2). Proteins, carbohydrates and lipids can confer protection. Rich media are thought to be more protective since they provide essential amino acids and vitamins to stressed cells. Studies in buffer solutions have generally shown greater microbial inactivation than that achieved in foods, and some foods appear to give more protection than others. For example, a treatment of 375 MPa for 30 min at 20°C in phosphate buffer (pH 7.0) resulted in a 6 log inactivation of *E. coli* O157:H7; however, the same treatment gave a 2.5 log reduction in poultry meat and a 1.75 log reduction in milk. Similarly, when *Sal. typhimurium* was pressurized for 10 min in pork slurry at 300 MPa, a 6 log reduction in numbers was observed compared with less than a log 2 reduction when baby food containing chicken was pressurized at 340 MPa for 10 min. These differences in results may be partly due to different experimental conditions but it is likely that the substrate has a very significant effect.

A reduction in water activity (a_w) can confer protection on cells during pressurization. In one study, *Rhodotorula rubra* was suspended in solutions of sucrose, glucose, fructose or sodium chloride, to give a range of water activities, and then treated for 15 min at 25°C and pressures up to 400 MPa. A protective effect was observed when the a_w fell below 0.92 and there was no inactivation of the yeast. However, at an a_w of 0.96, cell counts were reduced by log 7. The nature of the solute can have a significant effect on pressure resistance of spores. Ionic solutes such as NaCl and $CaCl_2$ conferred more protection on *B. coagulans* than non-ionic solutes such as sucrose and glycerol. This effect was especially pronounced above a_w 0.96 and suggests that spores are protected more by high concentrations of ions rather than by low a_w. From a practical viewpoint, dry products such as spices cannot be successfully decontaminated by pressure alone, but if the a_w is increased, as in spice pastes, microbial inactivation can be increased.

The pH of aqueous solutions decreases with increasing pressure because of electrostriction, although the extent of the change is extremely difficult to measure during pressure treatment. For example, from theoretical calculations fruit juices, which tend to be acidic, given a treatment of 500 MPa should undergo a pH shift of 1 towards the acid side. When the pressure is released, the pH increases to the original value. It is not known if this sudden drop in pH affects microbial survival in addition to the effect of pressure. Studies on the effect of initial substrate pH of the substrate on pressure resistance of microbial cells have given mixed results. *Salmonella* spp. in brain–heart infusion broth were more pressure-resistant at pH 4.5 than at pH 7.0. The pressure resistance of *Saccharomyces cerevisiae* inoculated into various fruit juices, however, was not affected by the type of juice, the pH (2.5–4.5) or the kinds of organic acids. Reducing the pH also caused a progressive increase in the sensitivity of *L. monocytogenes* to pressure at 23–24°C. At pH 7.1 there was complete survival after 10 min at 300 MPa, whereas at pH 5.3 the same treatment reduced viable numbers by log 1.8.

Spores of *B. coagulans* are more sensitive to pressure when treated at lower pH and increased temperatures. Samples treated at 400 MPa for 30 min

at 70°C in pH 7.0 buffer showed a 4 log decrease compared with a log 6 reduction at pH 4.0. No reduction was observed when the treatments were carried out at 25°C.

Food additives can have a variety of effects on pressure resistance. Potassium sorbate and the antioxidant butylated hydroxyanisole (BHA) were found to increase the pressure inactivation of *L. monocytogenes* while other antioxidants (sodium ascorbate and butylated hydroxytoluene) had no effect. None of these compounds affected the viability of non-pressurized cells exposed for the same length of time. The mode of action is not thought to simply involve the sensitization of cells to BHA by pressure or vice versa, because sequential exposure to pressure and BHA did not produce the synergistic inactivation of the pathogen observed with simultaneous exposure to these treatment.

It is clear that many factors in the substrate can affect the response of microorganisms to pressure. Thus, it is important to evaluate the process conditions in the food of interest rather than extrapolating data from other substrates.

Temperature

Temperature during pressurization is an important factor affecting survival of vegetative microorganisms. Increased inactivation is usually observed at temperatures either below or above the ambient temperature. There are several reports of enhanced lethal effects when pressurizing at −20°C compared with +20°C. It has been postulated that the enhanced lethal effects may be related to the fact that different proteins can be denatured at low temperatures compared with ambient temperature.

Refrigeration temperatures can also enhance pressure inactivation. Ewe's and goat's milk pressurized at 2°C or 10°C resulted in lower microbial numbers than the milks treated at 25°C.

Pressure applied with mild heating is also effective, particularly at inactivating the more pressure-resistant vegetative pathogens such as *Staph. aureus* and certain strains of *E. coli* O157:H7 (**Fig. 3**). Simple inactivation models, based on the combined effects of pressure and temperature, are being developed and may be useful in predicting appropriate processing conditions to ensure the microbiological safety of certain pressure-treated foods.

As discussed above, increased temperature can encourage spores to germinate, resulting in them being more susceptible to pressure treatments. A preheat treatment followed by pressurization is generally more effective at inactivating spores than heating during pressurization. In a study of *C. sporogenes* spores, no inactivation was observed at 600 MPa at

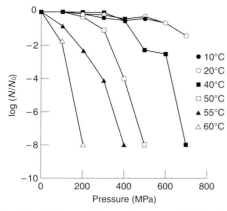

Figure 3 Effect of temperature and pressure (15 min treatment) on the inactivation of *Escherichia coli* O157:H7 (NCTC 12079) in UHT milk. N_0, original number of bacteria; N, number of surviving bacteria. Reprinted with permission from *Journal of Food Protection*, © IAMFES.

60°C for 60 min. However, preheating followed by a milder pressure treatment (80°C for 10 min followed by 400 MPA at 60°C for 30 min) resulted in a 2 log reduction.

Combination Treatments

A number of high-pressure processing combinations have been proposed. The benefits of combining pressure and heat have been discussed above. Other combinations being investigated include pressure with ultrasound, electrical resistance heating or supercritical carbon dioxide. All show some enhanced antimicrobial activity but the commercial benefits must be clearly identified, such as producing commercially sterile products, before they will be adopted by the food industry.

Significance of Sublethally Injured Cells

As with all physical preservation techniques, high pressure can cause sublethal injury in microorganisms. The possibility exists that injured cells may not be detected by plate methods because of their failure to initiate growth when plated out immediately after pressurization. However, given the right conditions such as prolonged storage in an appropriate substrate, they may be able to repair. The potential for recovery, particularly of pathogens, is of importance in all food processing; high-pressure processing is not unique in this respect. There are many reports comparing recoveries of pressure-treated bacteria on nonselective and selective agars. In general, there is a lower survival rate on the latter as they contain ingredients that are inhibitory to injured cells. For example, there was a lower recovery of pressurized *B. coagulans* spores on nutrient agar containing 0.18 IU ml^{-1} of nisin compared with control plates

without nisin. The selective agar eosin–methylene blue plus 2% sodium chloride (EMBS) gave a lower recovery of pressurized *Sal. typhimurium* than on tryptic soy agar (TSA). Cells treated in phosphate buffer were found to be more susceptible than those treated in strained chicken. Injured cells appeared to recover within 4 h (no significant difference in counts obtained on EMBS and TSA). However, a similar recovery did not occur in the phosphate buffer, suggesting that lack of nutrients may hinder recovery of pressure-injured cells.

In conclusion, HHP alone or in combination with other preservation techniques has many potential advantages and is in keeping with demands for minimally processed foods. There is currently great interest in the process, but compared with other techniques such as heating, knowledge is still limited. In particular, more targeted research towards commercial products is needed to satisfy regulatory bodies and consumers that the technology can produce safe, nutritious, high-quality foods.

See also: **Bacillus**: Bacillus stearothermophilus. **Bacteria**: The Bacterial Cell. **Clostridium**: Clostridium botulinum. **Ecology of Bacteria and Fungi in Foods**: Influence of Temperature; Influence of Redox Potential and pH. **Escherichia coli**: Escherichia coli. **Heat Treatment of Foods**: Synergy Between Treatments. **Saccharomyces**: Saccharomyces sake.

Further Reading

Cheftel JC (1992) Effects of high hydrostatic pressure on food constitutents: an overview. In: Balny C, Hayashi R, Heremans K and Masson P (eds) *High Pressure and Biotechnology*. P. 195. London: INSERM/John Libbey.

Heremans K (ed.) (1997) *High Pressure Research in the Biosciences and Biotechnology*. Leuven University Press.

Hoover DG, Metrick C, Papineau AM, Farkas DF and Knorr D (1989) Biological effects of high hydrostatic pressure of microorganisms. *Food Technology* 43: 99–107.

Knorr D, Böttcher A, Dörnenburg H et al (1992) High pressure effects on micro-organisms, enzyme activity and food functionality. In: Balny C, Hayashi R, Heremans K and Masson P (eds) *High Pressure and Biotechnology*. P. 211. London: INSERM/John Libbey.

Leadley CE and Williams A (1997) *High-pressure Processing of Food and Drink – An Overview of Recent Developments and Future Potential*. New Technologies Bulletin 14. Campden and Chorleywood Food Research Association.

Ledward DA, Johnston DE, Earnshaw RG and Hastings APM (eds) (1995) *High Pressure Processing of Foods*. Nottingham University Press.

HISTORY OF FOOD MICROBIOLOGY

N D Cowell, Elstead, Godalming, Surrey, UK

Introduction

In 1663 Robert Boyle commented that 'much hath been already performed, as to the preservation of Aliments, even by those that have not troubled themselves to make Philosophical enquiries after Causes and Remedies of Putrefaction in Bodies, but only have been taught by obvious and daily Observations'. He continues by concentrating his attention on the exclusion of air, a technique that is considered further below, but his remark is pertinent and of general applicability, since examination of traditional indigenous practices the world over shows communities without any concept or understanding of microbiology itself making extensive use of physical and chemical methods of inhibiting the microbial decay of foods and, additionally, carrying out food conversion by the use of various fermentations. This can be most easily seen in the records that have come down to us from early literate civilizations, of which those of the Mediterranean basin are a good example. These writings also provide us with the earliest known speculations concerning the existence and actions of bodies that cannot be resolved by the unaided human eye.

The Classical Period

The books on agriculture written by Cato, Varro and Columella, the *Geography* of Stabo, Pliny's *Natural History* and the philological and culinary ramblings of Athenaeus, provide considerable evidence of the food technology of their times, particularly with reference to methods of food preservation. It is clear that a full range of techniques was being exploited to defeat microbial decay.

Considerable attention was given to the manufacture of wine in the writings referred to above. One sees evidence of the problems encountered in controlling a fermentation to produce a stable, palatable product. Rudimentary sanitary procedures are suggested, as is the use of herb-based preservatives which, if the texts are to be believed, were sufficiently powerful to arrest fermentation.

Concentration by evaporation was used to reduce the water activity of grape juice sufficiently to inhibit fermentation – a process mentioned in Virgil's *Georgics*. Salting and smoking were used to preserve meat and fish. Pickling recipes abound. A range of processes are mentioned for preserving vegetables, combining in different ways partial dehydration by wilting the material in sunlight, dry-salting and pressing and pickling in various vinegar/brine mixtures with or without the addition of herbs. Both tree and vine fruits were preserved by solar drying. Fruit was also stored immersed in honey, wine or concentrated grape juice. Mention is made of weather-frozen fish and the use of ice for cooling – so use was being made of the limited opportunities for exploiting refrigerated storage.

Two techniques are of particular importance with respect to later developments. The first is preservation by heat. Columella recommends constructing a chamber, the 'fumarium', into which should issue the fumes from the bathhouse fire. This was to be used for seasoning timber and treating amphorae of new wine. Though Columella regards the smoke of the fire as the agent of preservation, Galen in his *De Antidotis* gives a very full description of a similar process and specifically points to heat preserving the wine 'so that it never became acidic'.

The second technique, to which frequent reference is made in classical texts, is hermetic storage. One version of this was to coat fruit in wax or plaster before storage. Another was to store fruit in sealed, pitch-lined vessels.

Philosophers were already turning their attention to the causes of the decay which the above processes were seeking to remedy. Lucretius's poem *De Rerum Natura* (first century BC), among many other things, presents a view of disease and putrefaction. Arguing from the random motion of dust particles in the air, the existence of disease-causing entities below visible size is postulated. These, it is said, could be generated in damp ground, which could become putrescent in hot weather, and then rise into the air. When circumstances generated a high concentration of such entities, the air is thought of as becoming diseased. This dangerous air then, like a cloud or mist, creeps over the landscape falling on water, on crops or, through inhalation, affecting human beings.

Other writers in the same century reflect similar views. Vitruvius says that towns should not be situated by marshes as clouds arising from these could infect the inhabitants with the poisonous breath of marsh creatures. Varro explicitly expresses a microbial origin for disease, saying that precautions had to be taken in the neighbourhood of swamps because they were the breeding ground of creatures below visible size which could float in the air and be inhaled to cause serious disease.

These could be no more than speculations at this period, but the conceptual stage was already being set for developments that would take place once microscopical observation became possible.

To the Invention of the Microscope

The works that have been discussed above were recycled in various forms with little or no development or extension for something like fifteen centuries after they were written. At the end of this period, in 1546, Girolamo Fracastoro published a book, *De Contagione*, which set forth a germ theory both of contagious diseases and putrefaction which, though still essentially speculative, did refine the concepts of Lucretius. Disease could spread from person to person by transfer of an infecting agent by direct contact, by transfer on some intermediate object, or through the air. Similarly, putrefaction could be spread from item to item through the same mechanisms.

The principal developments over this period, however, were in the study of optics. There are scattered references in classical literature to the use of a transparent sphere (e.g. a water-filled bulb) as a burning glass or to the apparent enlargement of objects seen through such a device, but there is no indication that the nature of a lens was understood. It was the Arabic writer Alhazen, who lived at the end of the tenth and the beginning of the eleventh centuries, who first described this.

Roger Bacon (1214–1292) speaks of experiments with lenses and the use of a magnifying glass as a reading aid. He also wrote a passage suggesting that he had at least some concept of a telescope, though it is not certain that he ever constructed one. By the beginning of the fourteenth century spectacles for the correction of defective vision were being made and the consequent increase in lens manufacture led to further developments. The telescope and the compound microscope appeared at around the beginning of the seventeenth century. Credit for the latter is probably due to Hans Jansen and his son Zacharias.

At this time a change of attitude to knowledge becomes obvious. Giambattista della Porta (1534/5–1615), while still recycling the beliefs and practices of the classical writers referred to above, gives an indication that in some instances at least he had also put them to the test experimentally, and Francis Bacon (who, ironically, died in 1626 from a chill contracted from the experiment of stuffing a chicken with snow to find the effect of low temperatures in delaying putrefaction) shows an even stronger bias towards practical experimentation in his *Sylva Sylvarum*, where a number of experiments on food preservation are recorded. Thus with the development of the microscope and a basic scientific methodology, the ground was prepared for the advances of the seventeenth century.

The Seventeenth Century

Because of imperfections in their lenses, the early compound microscopes offered little advantage over a well-made simple microscope of one lens. It was with the latter that Anton van Leeuwenhoek made his notable discoveries in the second half of the seventeenth century. Leeuwenhoek showed great skill in grinding small biconvex lenses which he mounted between two metal sheets. Fluid preparations were held in a vertical glass tube mounted in a metal cage. The lens assembly was fixed to the cage by a flexible metal strip. Focusing was achieved by a thumbscrew which worked against the spring-force of this strip. While Leeuwenhoek did produce lenses magnifying 300 diameters, he found that lenses of half this power (with a focal length of just over 1 mm) were more convenient.

With this simple arrangement Leeuwenhoek investigated blood, milk, muscle tissue, infusions of ground pepper and many other materials and, as far as is known, became the first person to observe yeasts, protozoa and bacteria under the microscope. As an amateur with no formal education in medicine or science, Leeuwenhoek was generally content to describe his observations and refrain from theorizing. The channels of scientific communication were at that time in their infancy and it was fortunate that he had been introduced to Henry Oldenburgh, secretary of the Royal Society of London. As a consequence of this he reported his observations in a series of letters to the Society, some of which the Society published. These communications stimulated others, in particular Robert Hooke.

Hooke's microscope consisted of a double-convex objective fitted with a pinhole diaphragm and an eyepiece consisting of a field lens (to increase the field of view) followed by an eye lens and an eyepiece cup to control the distance of the observer's eye from the eye lens. It is a sign of the quality of lenses in those days that Hooke records that he removed the field lens when he wished to observe fine structure. Hooke replicated and confirmed many of Leeuwenhoek's observations. He observed that the 'living worms' from pepper-water survived freezing but that the 'worms' he had discovered in vinegar died 'as if they had been stifled' when sealed away from ambient air in a vial 'stopped very close'. He records in his diary for 1 January 1675/6 that he had confided, at a meeting with his scientific friends, his view that 'there was upon every great plant a herball of smaller plants microscopicall growing on them and a heard of various animalls microscopicall ranging among them'. The microscope was substantiating the existence of the microflora postulated by the ancient writers, but several factors worked against an early understanding of the role that they played in the scheme of things.

One factor was the doctrinc of spontaneous generation. This concept, which appears in the writings of Aristotle, had been largely discredited for larger animals, but it was by no means clear how microorganisms were generated. Did putrefaction result from microbes, or did microbes result from putrefaction?

Another factor was the contemporary developments in the understanding of respiration. The traditional, Galenical, view that the purpose of breathing was to cool the body was, in the seventeenth century, being abandoned in favour of the belief that something necessary for the continuation of life was being taken from the air during respiration. In 1670 Boyle reported a series of experiments in which living creatures, held in chambers evacuated by the newly invented mechanical vacuum pump, either died or became dormant. If air were necessary for all life, microorganisms placed under vacuum would suffer this fate also.

As early as 1663 Boyle had pointed to exclusion of air as a method of food preservation and from about 1668 to 1679 he carried through a series of experimental investigations of food preservation under vacuum. His early experiments showed that while holding foods under vacuum had some effect, preservation was not perfect. Preliminary heat treatment of the foodstuffs was then found to improve preservation – the security of preservation increasing with the severity of the heat treatment. These investigations were continued by Denis Papin, who in 1687 published a method of preservation in which foodstuffs, vacuum packed in glass containers, were subsequently given a heat treatment – a process that seems startlingly modern.

From 1700 to Pasteur

Hermann Boerhaave (1668–1738), the illustrious professor of Leyden University and foremost medical teacher of his time, was influenced by Boyle and recommended his students to study Boyle's writings. In his major work, *Elementa Chemiae*, Boerhaave distinguishes three stages of fermentation (vinous, acetous and putrefactive) and mentions the action of heat in inhibiting fermentation. He also states that the most accurate modern experiments had shown that no creatures – even the minutest insects – or plants would breed or produce an embryo in the absence of air.

The classification of fermentation processes used by Boerhaave, which was also followed by later writers, did little to advance understanding since so many disparate processes, usually involving some sort of gas production, came to be grouped under this heading. Stephen Hales (1677–1761) even spoke of purely inorganic chemical reactions involving gas release as 'fermentations'.

John Pringle, one of Boerhaave's many pupils, undertook an extensive study of putrefaction which he presented in a series of papers read to the Royal Society in 1750–1. He defined putrefaction as a deteriorative change producing a bad smell, and demonstrated that, contrary to the belief of the time, alkaline substances acted as inhibitors rather than promoters of putrefaction. He established a procedure for testing the efficacy of substances as inhibitors of putrefaction and reported tests on a wide range of materials – inorganic compounds and extracts of herbs, spices and other vegetable preparations. Pringle's table of results was reprinted 75 years later as still being the only one available. Next, Pringle started to seek promoters of putrefaction. He showed that inoculation of egg yolk with putrid egg yolk hastened its putrefaction, and following experiments in which farina/flesh/water combinations were incubated, he concluded that 'animal substances have a power of exciting fermentation in the farinacea in proportion to their degree of corruption'.

David MacBride (1726–1778) repeated and extended Pringle's experiments and reached the more general conclusion 'that bodies in a state of putrefaction are exciting ferments to such as are sweet'.

This concept, that fermentation was caused by a ferment, in the form of decayed matter, coming into contact with and catalysing the decay of other material was to be championed by Justus von Liebig (1803–1873) a century later. The nineteenth century saw the development of an alternative concept, that fermentations were caused by the metabolic processes of living microorganisms. In 1838 Charles Cagniard-Latour, using a new microscope that allowed ×300–400 magnification, examined yeast and concluded that it was composed of small globules which were able to reproduce, belonged to the vegetable kingdom, and could decompose sugar only when alive.

The eighteenth century saw the start of the attack on the problem of the origin of microbes. The initial experiments, conducted by John Turbevill Needham between 1745 and 1750, favoured the concept of spontaneous generation. Needham boiled infusions of meat and other substances with the intention of sterilizing them, and then held them in corked phials which, after a period of storage, were found to teem with microbial life. Needham's results were challenged by the more highly organized experiments of Lazaro Spallanzani, who worked with vegetable infusions sealed into glass phials which he heated for varying periods. He identified two classes of organism present, the larger of which (probably protozoa) were destroyed by a very short heat treatment, while the smaller (bacteria) required considerably longer heat treatment before a situation arose where there was no development of organisms on storage – no 'spontaneous generation'. Spallanzani was aware from earlier experiments that absence of air hindered microbial growth, so he provided generous air space in his flasks and took precautions in sealing them. Nevertheless, he found that the air 'was almost always condensed' on opening. Spallanzani was therefore open to the criticism, made by Needham, that he had degraded the air left in his phials and it was for this reason that no microbes developed. Support was given to this view in 1810 when Gay-Lussac investigated the atmospheres in Nicholas Appert's bottles of heat-preserved foods and found that they contained no oxygen – which brought him to the conclusion that the absence of this gas was a necessary condition for the preservation of animal and vegetable substances.

Spallanzani's experiments were refined by other workers to try to overcome these objections. In 1836, Schultze admitted air to sterilized food preparations by first passing it through strong chemical reagents, and in 1837 Schwann carried out a similar experiment in which the air passed through a heated tube. Twenty years later Schroeder and Dusch showed that a cotton-wool filter would serve the same purpose, and removed the objection that the air had been altered by the processes previously used. The success of this technique led to the practice of plugging culture vessels with cotton wool. It was left to Pasteur to demonstrate that an open flask with a long, thin, downward-pointing neck could be heated to destroy the microorganisms in any infusion within it and then be left simply open to the air without any recurrence of microbial growth. This was a body blow to the theory of spontaneous generation.

Louis Pasteur (1822–1895)

Pasteur's early research lay in the field of optical activity and crystallography. In 1856, while Dean of Science at the University of Lille, he was asked by a local businessman to investigate some problems encountered in the manufacture of alcohol from beets.

This changed the direction of his studies which were from that time devoted to microbiology – a science of which he was the principal originator. A paper he published in 1858, in which he shows that yeast could ferment the dextro-rotatory form of ammonium tartrate, but not the laevo-rotatory form, links his earlier and later interests. In his first paper on fermentation, published in 1857, he claims that the fermentation of sugar to lactic acid is effected by the growth of organized bodies. By adopting the position that fermentation was the result of microbial growth, he brought himself into direct confrontation with Liebig.

In 1860 Pasteur published his *Memoir on Alcoholic Fermentation*, which reviewed both the history of the subject and his own researches. The principal features of Pasteur's own work, which effectively undermined Liebig's position, were, first, the observation that alcoholic fermentation never occurred without the presence of live yeast cells, and correspondingly yeast would not grow without the fermentation of a carbohydrate material; and second, the demonstration that yeast could be grown on a medium free of proteinaceous material, consisting of a sucrose solution to which small quantities of ammonium tartrate and yeast ash had been added, the yeast increasing in weight, taking in nitrogen and carbon from the medium and at the same time producing alcohol.

A year later, as a consequence of his work on lactic acid fermentations, Pasteur announced the discovery of an organism that produced butyric acid but could live only in the complete absence of air – the first obligate anaerobe to be identified. The same year he announced the discovery that yeast could grow both aerobically and anaerobically, and also reported his experiments with narrow-necked flasks. The latter report also included his demonstration of the presence of organisms in the air by filtering the air through guncotton, dissolving the guncotton in solvents and examining the resulting solution microscopically. These experiments did not immediately destroy belief in spontaneous generation, but it withered in a few years as the objections of the remaining believers were answered.

Liebig remained reluctant to abandon his views on fermentation. After Pasteur had published his work on vinegar manufacture, Liebig responded with a paper attacking Pasteur's ideas and claiming that vinegar, at least in German factories, was made without Pasteur's *Mycoderma* (an organism for which he had searched one of the production barrels in vain), nothing being required to ferment alcohol to vinegar but air and the surfaces of wood and charcoal. Pasteur replied that in that case if one of the barrels were treated in boiling water for half an hour and the process restarted, the barrel should operate without change, whereas in fact it would not produce vinegar until a new population of *Mycoderma* had grown again. The Academy offered to pay the expenses of this experiment, but Liebig did not accept the challenge.

From vinegar Pasteur moved on to wine, showing the microbiological origins of wine diseases and developing a refined and controlled heating technique, now known as pasteurization, which could destroy undesirable ferments while retaining the sensory quality of the wine. Heating wine for preservation, as we have seen, was practised in antiquity, but it was the measure of control involved that distinguished Pasteur's method even from that which Appert had practised earlier in the century. Later still Pasteur extended his studies to the brewing of beer.

Other Nineteenth Century Developments

As the century progressed, Pasteur's interests moved away from what might be regarded as food microbiology. In the late 1860s he was studying the diseases of the silkworm, and subsequently infectious diseases in both animals and humans, and developing vaccination. More and more workers were now taking up Pasteur's ideas and in the last decades of the nineteenth century the infective agents of disease after disease – human, animal and plant – were identified. While these matters are strictly outside the direct scope of this history, they are nevertheless important in that insights, concepts and techniques were being developed that were of wide applicability, welding microbiology into a coherent science.

Koch's postulates (or rules) were intended as a scheme for proving that a specific organism was the cause of a given disease, but the framework that they provided was applicable to identifying the connection between an organism and any specific phenomenon. The emphasis Koch placed on the isolation of organisms in pure culture led to his development of gelatin and agar gel culture plates, a technique which was subsequently refined by R. J. Petri in 1887. Prior to this only the more cumbersome and uncertain dilution technique of Lister was available. The introduction of staining methods by Weigert, Erlich, Koch, Gram and others aided the identification and classification of bacteria (the latter having already been attempted by Cohn), and led to photomicrography. Later, Chr. Hansen grew pure cultures of yeasts, starting from a single cell, and devised a classification of them based on their observed properties. The discovery of bacterial spores and their enhanced heat resistance by Cohn in 1876, during his study of hay infusions, and by Koch in 1877 in respect of anthrax, shed light both on the variable results that had bedevilled the

controversy on spontaneous generation and on the prolonged heating times that had been found necessary by food canners to produce a stable product. The insights and techniques developed in the late nineteenth century, largely from the study of human and animal diseases, now provided powerful weapons for attacking problems of food processing and preservation.

Food canning had developed from the publication of Appert's *L'Art de Conserver* in 1810 on a purely empirical basis. Though Appert had stressed that his glass containers should be hermetically closed when heated, as canning in tinplate containers developed it became standard practice to provide a channel from the can to the atmosphere at least during the first stages of heating. Even up to the beginning of the twentieth century descriptions of the process appear which attribute its success to the elimination of air from the pack, and canners seem to have held to the traditional belief that air was responsible for putrefaction.

In 1895 Russell published the results of an investigation he had been asked to conduct by American pea-canners – the first microbiological approach to a practical canning problem. Cans that experience suggested should have been adequately processed were suffering 'swells'. Russell discovered an infection by a thermoduric anaerobe in these cans and showed that processing at a higher temperature dramatically reduced the spoilage rate. Other investigations, e.g. on canned lobsters and sweetcorn by Prescott and Underwood, followed in the USA and – at first reluctantly – packers abandoned their old beliefs in the requirement to remove all air from the pack and accepted the microbiological basis of canned food spoilage and the time/temperature basis of a satisfactory process. Nevertheless it was not until the 1920s, with the work of Bigelow and Esty, that heat processing requirements were given theoretical basis and their calculation a priori became possible.

It seems highly probable that the processing failures examined by Russell, Prescott and Underwood arose from the build-up of an infection of thermoduric organisms brought about by failures in factory hygiene. When Lister championed the antiseptic principle in surgery in the 1860s he initially met resistance from the medical profession, until the clear results of his procedures carried the day and strengthened the case for the germ theory of disease. The food industry also had to learn that what happened in the operating theatre had a relevance to them, and as the nineteenth century moved into the twentieth increasing attention was given to sanitary procedures, factory hygiene and the cleanliness and dress of operatives.

See also: **Fermentation (Industrial)**: Basic Considerations. **Fermented Foods**: Origins and Applications. **Heat Treatment of Foods**: Thermal Processing Required for Canning. **Microscopy**: Light Microscopy. **Preservatives**: Traditional Preservatives – Oils and Spices; Traditional Preservatives – Sodium Chloride; Traditional Preservatives – Organic Acids. **Vinegar. Wines**: Microbiology of Wine-making. **Yeasts**: Production and commercial uses.

Further Reading

Appert N (1810) *L'Art de Conserver Pendant Plusiers Années toutes les Substances Animales et Végétales.* Paris: Patris.

Bitting AW (ed.) (1937) *Appertizing or the Art of Canning.* San Francisco: Trade Pressroom.

Boyle R (1744) *The Works of the Honorable Robert Boyle in Five Volumes to which is prefaced the Life of the Author.* London: A. Miller.

Brock TH (1961) *Milestones in Microbiology.* London: Prentice-Hall.

Clay RS and Court TH (1932) *The History of the Microscope.* London: Charles Griffin.

Cowell ND (1996) Our founding fathers – a view of Roman food technology. *Food Science and Technology Today* 10 (3): 143–146.

Lechevalier HA and Solotorovsky M (1965) *Three Centuries of Microbiology.* New York: McGraw-Hill.

Papin D (1687) *Continuation of the New Digester of Bones.* London: Joseph Streater.

Thorne S (1986) *The History of Food Preservation.* Kirkby Lonsdale: Parthenon Publishing.

HURDLE TECHNOLOGY

Leon G M Gorris, Unit Microbiology and Preservation, Unilever Research Vlaardingen, The Netherlands

Introduction

Hurdle technology has existed for many years as a concept for the production of safe, stable and nutritious foods using basic food preservation technologies. The concept was recently revived because it fits well with the current consumer trend for minimally processed foods. Whereas a single preservation factor ('hurdle') such as sterilization can be used for maximum safety, the trend in food production is towards maximum quality, without a reduction in food safety. Hurdle technology involves the educated use of a combination of preservation factors that together provide a mild preservation effect while maximizing food quality. In order to use the concept adequately and safely, detailed knowledge of the impact of both single hurdles and combinations of hurdles on target microorganisms at the cellular level is necessary.

The Concept of Hurdle Technology

The stability and safety of a food depend on a range of physical, chemical and microbiological reactions taking place within it. Spoilage or health risks due to microorganisms are a problem only when the food matrix and the environmental conditions support growth and proliferation of these microorganisms. In cases where, for instance, the water activity (a_w) or pH in the food matrix is below a critical limit microorganisms will be unable to survive and cause spoilage

problems. When high-temperature treatments are applied, microorganisms in the food undergoing processing may be killed or sufficiently inactivated not to compromise food stability and safety. Excluding oxygen from the atmosphere the food is stored in will hamper the growth of all spoilage microorganisms that rely on oxygen for energy production. Thus, by manipulating the food matrix or its surroundings to conditions that are not supportive to microbial growth, foods can be efficiently preserved. These basic rules of food preservation have been known for centuries, as is apparent from the widespread use of elementary preservation technologies such as drying or salting (to reduce a_w), fermentation (to reduce pH), cooking (for heat treatment) and storage of food in containers covered with grass (to reduce O_2 in the head-space).

Modern food industry has many different types of processes and preservative techniques in its toolbox that help to produce stable, safe foods. Many of these act by preventing or inhibiting microbial growth (e.g. chilling, freezing, drying, curing, vacuum packing, modified-atmosphere packing, acidifying, fermenting and adding preservatives). Other methods inactivate microorganisms (e.g. pasteurization, sterilization and irradiation) or restrict the access of microorganisms to products (e.g. aseptic processing and packaging). Several new preservation techniques that act by inactivation (e.g. ultra-high pressure, electroporation, manothermosonication and addition of lytic enzymes) are under development. The food industry has long been engaged in a quest to find the 'golden bullet' of preservative technology. None of the techniques mentioned above has yet proved to be that technology. Sterilization, for example, is perfect for assuring food stability and safety but where food is available in relative abundance consumers want foods of better quality and nutritional value than sterilization can provide. Consumers prefer a processed food to be very much like the raw product because they associate this with naturalness and healthiness. Classical combinations of curing and the use of chemical preservatives also generally provide safe food but are again less appreciated by consumers who dislike the altered taste and find the use of preservatives suspect.

Indeed, consumers seem to be dictating food preservative practice in industry today. To a large extent this trend has been initiated by major retailers and supermarket chains, but it is strongly supported by consumer organizations. Their opinion is that food production should use preservation techniques that deliver products that are less heavily preserved, have a higher quality, are more natural, are essentially free of additives, and are nutritionally healthier or functional compared with the current standards. In many cases, optimization of the use of a preservative technique in practice can help to avoid loss of food quality to a certain extent. Convection heating of a food product in practice often results in inhomogeneous heating throughout the product, resulting in too cold or too hot (burned) spots. The heating process can be improved by using microwave-assisted convection heating, which enables the food to be heated from the inside and the outside simultaneously.

The problem of achieving preservation procedures that have less of an impact on product quality while assuring food safety has led to the rediscovery of an old concept: hurdle technology.

Hurdle technology (also called combined processes, combination processing, combination preservation, combination techniques, or barrier technology) advocates the use of different preservation techniques in combination. Individually, the strength or intensity of the individual hurdles would not be sufficient to preserve the food, but in concert they have the desired level of effect. Individual hurdles used are, for instance, temperature, water activity (a_w), pH, redox potential (E_h), and preservatives (**Table 1**). It requires a certain amount of effort from a microorganism to overcome each hurdle. The higher a hurdle, the greater this effort is. Some hurdles, such as pasteurization, can be high for a large number of different types of microorganisms, whereas others, such as salt content, have a less strong effect or the effect is limited in the range of types of microorganisms it affects.

In employing this technology, food manufacturers choose a set of hurdles that is specific for a particular food and the processing applied to produce it in terms of the nature and strength of its effect. They should base their choice on knowledge of the impact of the individual hurdles available to them on different types of microorganisms, as it has become clear that using hurdles at mild impact levels (hurdles at low strength or intensity) will narrow their antimicrobial spectrum considerably. Together, these hurdles stabilize the food and assure its safety by keeping the growth of spoilage or pathogenic microorganisms under control as these are not able to 'jump over' all of the hurdles used. **Figure 1** illustrates a set of food-processing hurdles consisting of chilling during storage, low water activity, acidity, low redox potential and preservatives. Some of the target microorganisms present can overcome a number of hurdles, but none can jump over the last hurdle. Thus, the food is sufficiently stable and safe. However, the selection of the individual hurdles included in a particular set and the strength or intensity at which they are used is very much dependent on the composition of the raw materials and the expected variation in the level of microbial contamination. When food ingredients are generally

Table 1 Potential hurdles for the preservation of foods. Reproduced from Leistner and Gorris (1994) with permission

Physical hurdles

High temperature (sterilization, pasteurization, blanching), low temperature (chilling, freezing); ultraviolet radiation, ionizing radiation; electromagnetic energy (microwave energy, radiofrequency energy, oscillating magnetic field pulses, high electric field pulses); photodynamic inactivation; ultra-high pressure; ultrasonication; packaging film (plastic, multilayer, active coatings, edible coatings); modified-atmosphere packaging (gas packaging, vacuum, moderate vacuum, active packaging); aseptic packaging; food microstructure

Physicochemical hurdles

Water activity (a_w), pH, redox potential (E_h), salt (NaCl), nitrite, nitrate, carbon dioxide, oxygen, ozone, organic acids, lactic acid, lactate, acetic acid, acetate, ascorbic acid, sulphite, smoking, phosphates, gluconolactone, phenols, chelators, surface treatment agents, ethanol, propylene glycol, Maillard reaction products, spices, herbs, lactoperoxidase, lysozyme

Other hurdles

Competitive flora, protective cultures, bacteriocins, antibiotics, monolaurin, free fatty acids, chitosan, chlorine

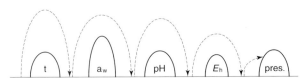

Figure 1 An example of the hurdle technology effect. t, chilling during storage; a_w, low water activity; pH, acidity; E_h, low redox potential; pres., preservatives.

low in contamination but high levels do occur occasionally, the set of hurdles used should be able to control the high contamination levels; often, this would still result in overprocessing and unnecessary loss of food quality. Optimization of the selection of individual hurdles may be a solution, but it should be realized that knowledge of the food product composition is as vital to the successful use of the hurdle technology concept as is knowledge of the impact of individual hurdles on different types of microorganisms. Variation in the composition of a food product needs also to be considered. The availability of carbon and energy sources may differ between apparently similar batches or types of products, and there may also be trace amounts of certain compounds (e.g. osmoprotectants, essential amino acids, vitamins) present in certain food ingredients that enable or strongly stimulate growth of specific microorganisms. Nevertheless, when used with the appropriate care and knowledge base, the hurdle technology concept is an adequate working hypothesis for optimization of mild food preservation systems.

Multi-target Preservation Using Hurdle Technology

Many of the existing and emerging preservation techniques act by interfering with the homeostasis mechanisms that microorganisms have evolved in order to survive environmental stresses. Homeostasis is the constant tendency of microorganisms to keep their internal environment stable and balanced. For instance, although the pH values in different foods may be variable, the microorganisms living in them spend considerable effort keeping their internal pH values within very narrow limits. In an acid food, they will actively expel protons against the pressure of a passive proton influx. Another important homeostasis mechanism regulates the internal osmotic pressure (osmohomeostasis). The osmotic strength (which is inversely related to the a_w) of a food is a crucial physical parameter, greatly determining the ability of microorganisms to proliferate. Cells have to maintain a positive turgor by keeping the osmolarity of the cytoplasm higher than that of the environment, and they often achieve this using osmoprotective compounds such as proline and betaine.

When the homeostasis or internal equilibrium of a microorganism is disturbed by a preservative factor, the microorganisms will not multiply but remain in the lag phase or even die before homeostasis is re-established. The most effective procedures in food preservation, such as sterilization, freezing and irradiation, will disturb most of the homeostasis mechanisms available to undesired microorganisms

simultaneously. Hurdle technology preservation should therefore be most efficient when the various hurdles relate to different systems, such as the cell wall, membrane, genetic material, or enzyme systems. This multi-targeted approach is fundamental to hurdle technology: it is more effective than a single-targeted approach and enables the use of hurdles of lower intensity, thereby minimizing the reduction in product quality. Also, the possibility exists that different hurdles in a food will not just have an additive effect on stability, but act synergistically. Employing different hurdles in the preservation of a particular food, microbial stability and safety can be achieved with a combination of less severe treatments. In practical terms this means that it may be more effective to use a combination of different preservative factors with low intensities when these hit different targets or act synergistically, rather than a single preservative factor with a high intensity. One example of synergism is the combined use of low water activity or low pH with modified-atmosphere or vacuum packaging. The latter conditions restrict the oxygen available to the stressed microorganisms for energy production and thus for operating the osmotic or pH homeostasis reaction. Furthermore, using multi-target hurdle technology may reduce the chances of the microorganism developing tolerance or resistance to the preservative treatment.

Applications of Hurdle Technology

By definition, hurdle technology preserved foods are products whose shelf life and microbiological quality are extended by use of several preservative factors in concert, none of which individually would be totally lethal towards the target spoilage or pathogenic microorganisms. Hurdle techniques of food preservation were developed empirically many centuries ago, and consequently many different types of food preserved in this way are commonly produced and marketed. For instance, fermented food products (sausages, cheeses, vegetables) are actually made safe and shelf-stable for long periods through a sequence of hurdles that arise at different stages of the ripening/fermentation process. With salami-type fermented sausages, salt and nitrite inhibit many microorganisms in the batter and are thus important hurdles in the early stage of the ripening process. Other bacteria multiply and use up oxygen. This reduces the redox potential of the product and enhances the E_h hurdle. This reduces growth of aerobic microorganisms and favours the selection of lactic acid bacteria. They are a competitive flora and cause acidification of the product, thus increasing the pH hurdle. Also with non-fermented foods, such as ready-to-eat products that are composed of different types of raw or minimally processed (washed, trimmed, sliced) vegetables, hurdle technology has been used to assure proper safety. Refrigeration and modified-atmosphere packaging are the main hurdles used to stabilize these perishable foods. Because it is difficult to maintain sufficiently low temperatures throughout the chain of production and processing to consumption in practice, and also because a number of cold-tolerant pathogens (e.g. *Listeria monocytogenes*, *Aeromonas hydrophila*) occurring on produce can proliferate under modified atmospheres, additional hurdles are required for optimal safety. One such hurdle can be biopreservation, using specific lactic acid bacteria that produce low amounts of acids (in order to minimize the influence of the hurdle on the food quality) but sufficient quantities of antilisterial bacteriocins. Bacteriocins are small proteinaceous compounds that generally have an antimicrobial activity spectrum restricted to Gram-positive microorganisms such as – fortunately – *L. monocytogenes*. When a broader or a complementary spectrum of activity is desirable, natural antimicrobial compounds present in aromatic plants, herbs and spices can be used; a number of them have been found to be successful in controlling different pathogenic bacteria (e.g. *L. monocytogenes*, *Salmonella enteritidis*, *Aeromonas hydrophila*, *Clostridium botulinum*, *Staphylococcus aureus*) and spoilage bacteria, yeasts or fungi. A preservative treatment that would be very compatible with the use of natural antimicrobial compounds such as bacteriocins in certain food products would be the use of edible coatings. Edible coatings are prepared from natural biopolymers (carbohydrates, proteins, fats) and are applied directly to the surface of a food product. They act as a physical protection against food contamination. When antimicrobial compounds are added to the coating, these are immobilized at the product surface from which they slowly migrate to have effect. When food contamination at the product surface is of concern, edible coatings may allow the delivery of antimicrobials to the specific site where their presence and activity are required. Sometimes, less of the antimicrobial compound needs to be used in a coating system compared with dipping or spraying the food, which could further minimize the impact of this antimicrobial hurdle on food quality.

An overview of the combinations of hurdles studied or already employed for specific food applications is given in **Table 2**. In a number of recently developed food products, an almost infinite shelf life can be obtained. An example of this is the use of the heat-stable bacteriocin nisin as an extra hurdle in the canning of peas. Normally, heating and reduction of pH are the only two hurdles employed, but these do not suppress growth of surviving acid-tolerant, spore-forming clostridia, which are completely inhibited by nisin.

Limitations to Hurdle Technology

Because hurdle technology – unlike sterilization – does not eliminate all the microorganisms present in food, but only (temporarily) inactivates them, it introduces a level of uncertainty with respect to safety. It is therefore of the utmost importance for food manufacturers applying hurdle technology to be able to assess or estimate this uncertainty and to be aware of the possible limits to the use of combinations of hurdles as a preservation system. Microorganisms have developed many different mechanisms to overcome unfavourable conditions, such as the homeostasis systems and stress reactions. One prominent limit to mild preservation systems and a very good example of microbial adaptation is the formation of spores by certain microorganisms. Spore formation limits the impact of heating, and also blocks the access of acids or preservatives to the vegetative cell structure. More transient types of adaptations are triggered once a stress has been detected by a cell and involve structural or biochemical changes on different cellular levels: membrane transport systems, receptor functioning, signal transduction, control of gene expression, bioenergetics, reserve material status, etc. Detailed knowledge on the physiological level of how stress adaptation processes are triggered and proceed may give new leads that are essential for improving mild preservation technology.

An elementary homeostasis mechanism is related to the intracellular pH of microorganisms. A tightly regulated pH is essential for continued growth and viability by maintaining the functionality of key cell components. Food preservatives such as acetic, lactic, propionic and benzoic acids all affect the pH homeostasis mechanism. Membrane-associated transport systems for hydrogen ions are continuously operational to balance the internal pH with that of the environment. As the external pH is reduced, hydrogen ions are actively imported into the cytoplasm; the reverse occurs as the external pH rises. Adaptation to low pH stress is a phenomenon that has been frequently observed in food-borne pathogens. For instance, cells of *Salmonella typhimurium* exposed to hydrochloric acid at pH 5.8 for one or two doublings were more resistant than non-exposed cells to inactivation by lactic, propionic and acetic acids. With *Escherichia coli* O157:H7, cells exposed to lactic acid (pH 5.0) survived better than non-adapted cells in shredded dry salami (pH 5.0) and apple cider (pH 3.4). With *Listeria monocytogenes*, cells adapted

Table 2 An overview of different types of hurdle technology-preserved food products. Reproduced from Leistner and Gorris (1995) with kind permission from Elsevier Science Ltd, The Boulevard, Langford Lane, Kidlington OX5 1GB, UK

	Cottage cheese	Potato crisps	Ham	Meida-Ya	jam	Modified atmosphere-packaged salad	Peas, canned using nisin	Cake, packaged using ethanol vapour	Bread, packaged using a flush of CO$_2$ gas	Cold smoked salmon	Pasta sauce	Modified atmosphere-packaged fresh pasta	Acidified, pasteurized vegetable
Main cause of spoilage													
Microbiological	X		X			X				X			
Biochemical		X	X	X		X		X	X	X	X	X	X
Physical	X			X									
Type of hurdle													
High temperature		X	X				X				X	X	X
Low temperature	X		X			X					X	X	X
Increased acidity – low pH	X			X			X				X		X
Reduced water activity (a$_w$)		X	X					X				X	
Reduced redox potential (E$_h$)						X			X			X	
Preservative(s)	X		X				X	X		X			
Competitive flora			X							X			
Modified gas atmosphere						X		X	X			X	
Packaging film												X	
Ultra-high pressure				X									
Product origin													
Traditional	X	X	X							X			
Recently developed				X		X	X	X	X		X	X	X
Country in which developed or marketed	UK, USA		USA	Japan		France, UK, USA	UK	Japan		UK, USA	UK	UK, USA	Europe
Shelf life (weeks)	2	>25	>2	12		1	∞	12	12	8		4	>4

to sublethal levels of lactic acid showed increased survival in milk products acidified with lactic acid, including cottage cheese (pH 4.71), natural yoghurt (pH 3.9) and full-fat Cheddar cheese (pH 5.16). Additionally, these cells survived better in low-pH foods, e.g. orange juice (pH 3.76) and salad dressing (pH 3.0), containing acids other than lactic acid. Acid adaptation is an important phenomenon in food preservation. On the one hand, it is a powerful counteractive tool helping microorganisms to survive one particular stress. On the other hand, it may turn out to be the Achilles' heel of microorganisms, because a mild preservation factor that successfully influences pH homeostasis could be extremely attractive in its own right, and even more so in combination processing. One mechanism by which an effective preservation factor could function might be that of an ionophore. The antibiotic gramicidin, for instance, makes membranes permeable to hydrogen ions so that the internal pH cannot be maintained above the minimum level required. In fact, growth ceases at pH values well above those that still allow the growth of control cells when the pH of the environment is gradually decreased in the presence of gramicidin. While it is not possible to use gramicidin for food applications, food-grade alternatives would be ideally suited for mild preservation use.

Adaptation or stress reactions of target microorganisms should not be studied only in relation to well-known or obvious facts. Stress reactions may have a non-specific effect, which means that exposing bacteria to one sublethal stress may significantly improve their response to other, apparently different, types of stress at a later stage. One such example of 'cross-tolerance' occurs in *L. monocytogenes*, which can overcome osmotic stress by taking up proline from the food matrix. When osmotic stress is one of the preservative hurdles chosen to control the pathogen, the presence of proline and other food ingredients that act as osmoprotectants (carnitine, betaine, etc.) should be evaluated as they may diminish the effectiveness of this hurdle. However, it was recently discovered that proline and other osmoprotectants may help the pathogen to grow better at low temperature even in the absence of osmotic stress. Conceivably, temperature stress and osmotic stress (partly) share a common stress signalling process that activates the uptake of the osmoprotectant from the environment to counteract the unfavourable condition. The aspecificity of this signalling process and the cellular response it initiates may be important for the survival of the pathogen. The establishment of adequate food preservation systems based on hurdle technology requires cross-tolerance reactions to be investigated much more systematically.

Much to the relief of all involved in food preservation, however, physiological and genetic changes put a high toll on the energy and material resources of microorganisms. Thus, successful adaptation of microorganisms is very much dependent on optimal growth conditions. In foods, microorganisms often grow at much less than optimal rates and do not have the opportunity to accumulate sufficient amounts of reserves that will be necessary under (multiple) stress conditions.

The Future for Hurdle Technology

With the increasing popularity of minimally processed foods, which are highly convenient for the consumer but preserved only by relatively mild techniques, the environmental conditions in foods as a habitat for microorganisms have changed dramatically, giving many new options for their survival and growth compared with traditionally preserved foods. In order to control food poisoning and spoilage microorganisms in these new food habitats, while keeping loss of product quality to a minimum, a hurdle technology approach is advocated which involves the educated selection and use of a set of preservative factors that can adequately ensure product stability and safety. Minimal processing, however, should not lead to minimum food safety. It is appreciated that combined processes used today do not eliminate microorganisms but rather inactivate them. Any process or factor that influences the efficacy of inactivation may have an impact on the overall hurdle technology preservation effect. The proper use of hurdle technology needs to include sound information on factors affecting the survival and growth of target microorganisms at both the macroscopic and the microscopic level. Perfecting hurdle technology will enable food manufacturers to improve food quality further without compromising on food safety, at some stage achieving processing and preservation treatments that are undetectable by the consumer.

See also: **Bacteriocins**: Nisin. **Chilled Storage of Foods**: Use of Modified Atmosphere Packaging; Packaging with Antimicrobial Properties. **Ecology of Bacteria and Fungi in Foods**: Influence of Available Water; Influence of Temperature; Influence of Redox Potential and pH. **Heat Treatment of Foods**: Synergy Between Treatments. *Listeria*: Listeria monocytogenes. **Packaging of Foods**: Packaging of Solids and Liquids. **Ultra-violet Light**.

Further Reading

Barbosa-Cánovas GV and Welti-Chanes J (eds) (1995) *Food Preservation by Moisture Control, Fundamentals and Applications*. Lancaster: Technomic.

Gorris LGM and Peck MW (1998) Microbiological safety considerations when using hurdle technology with refrigerated processed foods of extended durability. In: Ghazala S (ed.) *Sous-vide and Cook Chill Processing for the Food Industry*. Gaithersberg: Aspen Publishers Inc., pp. 206–233.

Gould GW (1992) Ecosystem approach to food preservation. *J Appl Bacteriol* 73 (supplement): 58S–68S.

Gould GW (ed) (1995) *New Methods of Food Preservation*. London: Blackie.

Leistner L and Gorris LGM (eds) (1994) *Food Preservation by Combined Processes*. European Commission Publication EUR 15776, ISBN 90-900-7303-5. Brussels: EC.

Leistner L and Gorris LGM (1995) Food preservation by hurdle technology. *Trends Food Sci Technol* 6 (2): 41–46.

Whittenbury R, Gould GW, Banks JG and Board RG (eds) (1988) *Homeostatic Mechanisms in Micro-organisms*. Bath: Bath University Press.

HYDROPHOBIC GRID MEMBRANE FILTER TECHNIQUES (HGMF)

Phyllis Entis, QA Life Sciences, Inc., San Diego, USA

Introduction

Since its introduction 25 years ago, the hydrophobic grid membrane filter (HGMF) has found numerous uses both as an analytical tool in food and water microbiology and as a research tool in molecular biology laboratories. The properties of HGMF confer several benefits to the user, notably: adaptability to enumerating microorganisms over a wide counting range; possibility of inoculating with larger volumes of sample than agar plates; differential detection of low numbers of a specific target microorganism in the presence of as much as a three $\log_{10}$ excess of competitors; repair of injured cells prior to selective enumeration without compromising the reliability of the resultant counts; sequential determination of two or more differential biochemical characteristics of all colonies present on a filter without disturbing growth; and development of enzyme-labelled antibody or DNA colony hybridization reactions simultaneously on all colonies on the HGMF. This article explores

Figure 1 Hydrophobic grid membrane filter containing a mixture of *Salmonella* colonies (green) and non-salmonellae (yellow) after 18 h incubation on EF-18 Agar at 42°C.

the principles and selected applications of HGMF technology.

Principles of HGMF

The hydrophobic grid membrane filter consists of a microporous membrane filter (most commonly, 0.45 μm pore size) on which has been imposed a grid pattern. When a conventional membrane filter is placed on the surface of an agar medium, water containing dissolved nutrients from the medium is carried by capillary action up through the filter's pores to feed the microorganisms which are retained on the top surface during filtration. This produces a microscopic film of moisture on the top surface of the membrane filter through which motile organisms can travel. The hydrophobic grid pattern of an HGMF interrupts this film of moisture, preventing motile organisms from travelling beyond the boundaries of the grid squares in which they have landed. Also, wherever the hydrophobic material has been deposited on the filter, the pores are blocked and water is prevented from bringing nutrients to any stray microorganisms which might be pushed onto a grid line as a colony increases in size and fills its square. Thus, as illustrated in **Figure 1**, colonies tend to remain confined to the squares in which they originated.

Counting Colonies on HGMF

Most Probable Number (MPN) Although the hydrophobic grid can prevent colonies from spreading beyond their initial squares and fusing with adjacent colonies, it cannot prevent more than one viable organism from landing in any one square. The probability of this occurrence increases in logarithmic proportion to the total number of viable microorganisms in the sample being filtered. Since it is often impossible

to tell whether an occupied square (i.e. one containing a colony) originated from just one or from multiple microorganisms, counting individual colonies tends to produce an underestimate of the true population. In addition, since multiple individual colonies might be discernible inside some squares but not inside others, the error produced by counting individual colonies is neither consistent nor predictable. Therefore, it is necessary to determine the most probable number (MPN) of microorganisms present in the sample by deducing it from the total number of occupied squares. (It should be noted that this same risk of underestimation also applies to conventional pour plate or spread plate procedures. In those cases, the analyst has no choice but to operate on the assumption that each colony originated from just a single viable microorganism.)

Ensuring Validity of MPN Certain conditions regarding distribution of the microorganisms in the sample homogenate and on the HGMF must be met for the MPN determination to be applied with confidence. First, the microorganisms targeted for enumeration must be distributed randomly throughout the entire sample portion to be filtered. Secondly, each individual microorganism in the filtered sample portion must have an equal chance of landing in any one of the individual grid squares. Finally, each square must be equally capable of supporting growth of the target microorganisms.

The first and second conditions depend on the correct preparation and mixing of the sample homogenate, use of the correct filtration equipment and handling techniques, careful placement of the filtration apparatus on its vacuum flask or manifold and correct placement of the membrane filter on the filtration unit, as shown in **Figure 2**. If, for example, the filtration unit is not positioned vertically, the surface on which the HGMF rests will be tilted, resulting in unequal volumes of liquid filtering through each of the individual squares. This situation invalidates the condition of equal probability of distribution of microorganisms amongst all the grid squares. The third condition for validity depends on complete contact between the surface of the culture medium and the underside of the HGMF and on the equal ability of all the grid squares to transfer nutrients by capillary action from the culture medium to the top surface of the square. For example, an air bubble trapped between the agar surface and the underside of the HGMF would prevent nutrients from being transferred to the top surface of the filter, thus inhibiting growth of any target microorganisms that might be present in those squares directly over the air bubble.

Figure 2 Hydrophobic grid membrane filter being positioned onto the base of a filtration apparatus. Following placement of the HGMF, the filtration funnel is rotated into a vertical position and clamped using a stainless-steel jaw clamp (not shown). The stainless-steel pre-filter (5 μm pore size) located at the bottom of the cylindrical portion of the funnel screens out food particles during the filtration process.

Calculating MPN per gram If all the above conditions have been met, the MPN can be calculated for any given number of positive squares using the formula:

$$MPN = N \times \log_e[N/(N-X)]$$

where N = total number of grid squares on the filter; and X = number of positive grid squares.

Calculating the MPN per gram for any quantitative analysis must always be done in the following sequence.

1. Determine the score by counting the number of squares containing target colonies (positive squares). If the score has been determined over only a portion of the HGMF surface (e.g. 20%), multiply by the appropriate factor to estimate the score over the entire HGMF.
2. Convert the score for the entire HGMF to the corresponding MPN by using the formula given above.
3. Multiply by the dilution factor of the sample portion that had been filtered to determine the MPN per gram.

Precision of MPN Traditionally, MPN determinations in food and water microbiology have been carried out by inoculating a replicate series of three or five tubes at each of three dilutions. Although this procedure is usually accurate, it is not very precise. Note that accuracy refers to how closely the mean result of a large number of replicate analyses of the same sample reflects the true content of the sample, whereas precision is a measure of the amount of scatter of the individual results about the mean.

Precision is usually reported as a 95% confidence interval.

In the past, the imprecision of multiple-tube/multiple-dilution MPN tests has cast a shadow of doubt over other MPN measurement systems. In reality, the lack of precision of these tests is entirely a function of the very low number of replicates (either three or five) typically run at each dilution. As with any statistical sampling method, the larger the number of replicates, the greater the precision of the measuring system. The HGMF provides the microbiologist with a number of replicates per sample that would be totally impractical in a conventional tube MPN system. The commercial ISO-GRID® HGMF comprises a 40×40 matrix, providing 1600 individual growth compartments of equal size. Carrying out a single filtration using this HGMF is equivalent in precision to inoculating 1600 separate test tubes, all at the same dilution.

The impact of the number of replicates on MPN precision can be illustrated by calculating theoretical 95% confidence intervals for HGMF's with varying numbers of squares. The contents of **Table 1** and the graphic representation in **Figure 3** illustrate the impact of the number of replicates on precision when 50% of the squares are positive, a level of saturation at which precision should be optimum. The upper and lower 95% confidence intervals shown in this table were calculated using the following pairs of equations:

Lower 95% confidence interval:

$$Q_2 = (X/N) + 1.96 \sqrt{\{[X/N][1-(X/N)]/N\}}$$
$$MPN_U = -N \times \log_e(1-Q_2)$$

where: X = number of positive squares ('Score');

Table 1 Effect of number of replicate growth compartments on MPN precision, calculated at 50% saturation of the grid matrix

N^a	X^b	*MPN*	$MPN_L{}^c$ (% of MPN)		$MPN_U{}^d$ (% of MPN)	
10	5	7	2	(29)	17	(243)
20	10	14	7	(50)	25	(179)
40	20	28	17	(61)	43	(154)
80	40	55	40	(73)	75	(136)
100	50	69	51	(74)	91	(132)
200	100	139	113	(81)	168	(121)
400	200	277	240	(87)	319	(115)
800	400	555	501	(90)	612	(110)
1000	500	693	633	(91)	757	(109)
1200	600	832	766	(92)	902	(108)
1400	700	970	899	(93)	1046	(108)
1600	800	1109	1032	(93)	1189	(107)

[a] Number of replicate growth compartments.
[b] Number of positive growth compartments.
[c] Lower 95% confidence limit of MPN.
[d] Upper 95% confidence limit of MPN.

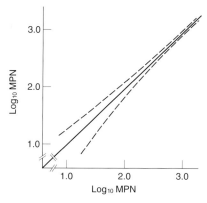

Figure 3 95% confidence intervals relative to MPN as influenced by the number of replicate growth compartments. MPN at 50% saturation relative to numbers of replicate growth compartments (——); upper and lower 95% confidence interval of MPN at 50% saturation relative to numbers of replicate growth compartments (– – –).

N = total number of squares; MPN_L = MPN index of the lower extreme of the 95% confidence interval.

Upper 95% confidence interval:

$$Q_2 = (X/N) + 1.96 \sqrt{\left\{ [X/N][1-(X/N)]/N \right\}}$$
$$MPN_U = -N \times \log_e (1-Q_2)$$

where: X = number of positive squares ('Score'); N = total number of squares; MPN_U = MPN index of the upper extreme of the 95% confidence interval.

History of HGMF

The hydrophobic grid membrane filter concept was first described in 1974. HGMF was viewed as a convenient research counting tool and the basis for developing the first truly reliable automatic colony counter. It was reasoned that conventional agar pour plates presented too many random patterns and sources of interference (for example from food particles or from highly coloured foods) to enable an automatic counter to produce a consistent and reliable result. By forcing the colonies to develop in an ordered matrix of rows and columns, it was believed that it would be far simpler to develop automated counting devices.

Initially, the HGMF was used for enumerating microorganisms from aqueous suspensions of pure cultures and from water samples. In 1978, a major study determining the ability of a range of food homogenates to pass through a 0.45 μm membrane filter was published. This led to further investigation into the use of pre-filter screens to remove particles from a homogenate prior to or during filtration and to the development of a series of enzyme digestion procedures to enable filtration of proteinaceous foods or foods containing significant concentrations of starch, cellulose or gum.

Applications of HGMF to Food Microbiology

Several characteristics of membrane filters in general and HGMF in particular have made them very attractive to food microbiologists. Membrane filtration, especially combined with a pre-filtration step to remove particles, is an effective means of separating target microorganisms from a food matrix, thus eliminating interactions between the organisms and the food or between the food and the culture medium. Also, once the organisms have been captured on the surface of a membrane filter, they can be transported from culture medium to culture medium, or from culture medium to one or more biochemical reagents, all without disturbing the organisms or affecting the reliability of the enumeration result. By subdividing the surface of a membrane filter into a large number of squares, the hydrophobic grid pattern greatly increases the counting range of a single membrane filter, eliminating the need to test multiple dilutions of a sample in most cases. Also, the physical barrier provided by the grid lines prevents highly motile or rapidly growing organisms from overgrowing other organisms on the filter. This results in a more accurate count as well as making it easier to obtain a pure subculture from an individual colony without repeated purification steps. The membrane filter material used in the commercial ISO-GRID® hydrophobic grid membrane filter is virtually non-reactive, allowing dyes to be incorporated into culture media and enhancing the contrast between the colonies and the membrane filter.

Culture-based Applications

The hydrophobic grid membrane filter has formed the basis for a wide range of quantitative food microbiology applications including total bacterial counts, coliform and *Escherichia coli*, fluorescent pseudomonads, lactic acid bacteria, yeasts and moulds, *Aeromonas, E. coli* O157:H7, *Vibrio parahaemolyticus*, faecal streptococci, *Staphylococcus aureus* and *Zygosaccharomyces bailii*. It has also been used for rapid detection of *Salmonella* and *Yersinia*, as well as for disinfectant efficacy tests. Several of these applications have been validated and are recognized as Official Methods by AOAC International. These applications include aerobic plate count (AOAC Method no. 986.32), yeast and mould count (AOAC Method no. 995.21), coliform and *E. coli* count (AOAC Method no. 990.11), *E. coli* O157:H7 count (AOAC Method no. 997.11) and *Salmonella* detection (AOAC Method no. 991.12).

Coliform and *E. coli* Count   This application, outlined in **Figure 4**, takes advantage of several of the characteristics of the HGMF. As with virtually all HGMF-based applications, only a single filtration is required. Carbohydrate contained in the sample is eliminated during filtration, avoiding the potential for false-positive fermentation reactions due to the introduction of a carbohydrate other than lactose into the culture medium. Therefore, no subculture or confirming step is needed to verify the coliform result. Also, the coliform and *E. coli* method makes use of the ability to transfer colonies from medium to medium without disturbing their growth or numbers.

Coliforms produce blue colonies on lactose monensin glucuronate (LMG) due to fermentation of lactose, which is detected by the aniline blue dye incorporated into the medium. *E. coli* produce β-glucuronidase which splits the 4-methylumbelliferyl-β-D-glucuronide (MUG) reagent contained in buffered MUG (BMA) agar to release 4-methylumbelliferone, a fluorescent molecule. There is incompatibility between these two reactions which makes it difficult to combine both into one test and obtain consistent and timely results. When lactose is fermented, acid is produced, resulting in a pH drop in the medium. The higher the colony density, the more acidity is produced. However, the β-glucuronidase enzyme of *E. coli* requires a near-neutral pH for optimum per-

formance, and 4-methylumbelliferone fluoresces poorly or not at all at acid pH. This method takes advantage of the portability of the HGMF, allowing all of the growth to be transferred undisturbed from LMG agar to BMA agar which quickly raises the pH to 7.4 ± 0.2 enabling the *E. coli* colonies to develop fluorescence quickly and consistently.

Yeast and Mould Count   The 2-day yeast and mould procedure is shown in **Figure 5**. As can be seen by comparing this flow diagram with Figure 4, the initial steps performed for all HGMF-based quantitative tests are very similar. The methods differ in the choice of culture medium, incubation time and incubation temperature. The yeast and mould count procedure illustrated in Figure 5 is a one-stage test; no secondary culture medium or incubation step is required.

This method benefits from several characteristics of the commercial HGMF to produce a reliable yeast and mould count in just 2 days. A prefilter built into the ISO-GRID® filtration unit removes food particles that might otherwise interfere with the visibility of small colonies (or, conversely, be counted erroneously as colonies). The medium, YM-11 Agar, contains trypan blue dye, which stains the fungi during their growth without causing the membrane filter to change in colour, thus enhancing the visibility of small colonies. The grid lines help to contain the spread of fast-growing moulds, thus enabling use of a culture medium designed to stimulate more rapid growth. Any constituents of the food sample (such as preservatives, acids, or natural microbial inhibitors present in some foods) are separated during filtration and are thus prevented from interfering with colony development.

Salmonella   Due to the ability of even low levels of *Salmonella* to cause disease, it is almost always necessary to test for the presence of *Salmonella* in at least a 25 g sample of finished product. Often, much

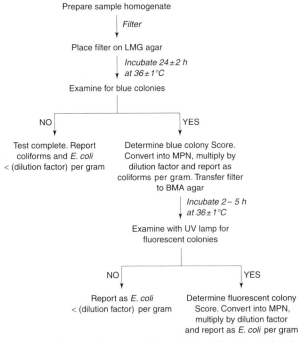

Figure 4 Total coliform and *Escherichia coli* enumeration by HGMF-based AOAC Official Method no. 990.11.

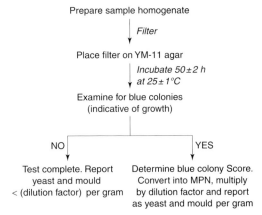

Figure 5 Yeast and mould enumeration by HGMF-based AOAC Official Method no. 995.21.

larger sample sizes are used. The initial step of virtually all *Salmonella* methods comprises an overnight nonselective enrichment broth culture to enable any injured *Salmonella* to repair and to allow the concentration of *Salmonella* to reach easily detectable levels. This is followed by some type of selective enrichment process to improve the ratio of *Salmonella* to other organisms that might be present in the sample. The HGMF-based *Salmonella* method follows this same enrichment approach, but takes advantage of the colony separation properties of the hydrophobic grid pattern to shorten the enrichment steps.

The AOAC Official HGMF *Salmonella* method is illustrated in **Figure 6**. Since the hydrophobic grid permits detection of target colonies in the presence of a thousandfold excess of competitors, use of this technique enables the pre-enrichment incubation to be shortened to as little as 18 h and the selective enrichment incubation time from 24 h to 6 h without affecting reliability. EF-18 agar, the medium designed for use with the HGMF *Salmonella* method, requires only 18–24 h incubation as compared to up to 48 h for bismuth sulphite agar, one of the three plating media used with the conventional method. Overall, the HGMF *Salmonella* method can be completed to the negative screen stage in as little as 42 h.

The colony separation properties of the hydrophobic grid also facilitate confirmation of presumptive positive samples. Since even highly motile organisms such as *Proteus* spp. are contained by the grid, pure isolates of the presumptive positive colonies can usually be obtained on initial subculture. This enables direct inoculation of biochemical screening media from the isolated colonies on the HGMF, resulting in confirmation of presumptive positive results in only an additional 24 h.

Listeria Although not yet approved by AOAC, a 24-h presumptive enumeration method has been developed for *Listeria* spp. and *L. monocytogenes*. The procedure comprises filtration of a portion of sample homogenate, incubation of the HGMF on a selective agar medium (LM-137 agar), enumeration of presumptive positive colonies, and subculture of a selected number of presumptive positive colonies for confirmation. It is presently proceeding through the AOAC approval process, having completed the pre-collaborative study stage.

Immunological and Colony Hybridization Applications

The hydrophobic grid membrane filter has found application in both enzyme-labelled antibody (ELA) and DNA colony hybridization techniques for confirming the identity of specific target organisms. Researchers have taken advantage of the ordered growth matrix to replicate the primary HGMF (for example, from a presumptive-positive *Salmonella* test) to one or more secondary filters. This enables the primary filter to be saved as a viable 'back-up' for any additional detailed characterization of positive isolates while the secondary filter is probed using either an enzyme-labelled antibody or chromogenic DNA probe hybridization assay to detect the presence and location of the target colonies. A positive 'spot' at a specific location on the probed filter can be correlated directly to the same row and column coordinates on the original HGMF.

Other Possible Applications

Any procedure that can benefit from the ordering of colonies into a two-dimensional array, from the prevention of colony overgrowth, or from any of the other characteristics of the hydrophobic grid membrane filter is a candidate for application of HGMF techniques. Several possible HGMF applications have not yet been fully exploited. These include, for example, screening for antibiotic-resistant cultures, antimicrobial and disinfectant efficacy testing, screening large numbers of transformed cultures for specific target nucleic acid sequences, or screening environmental samples for the presence of microorganisms that exhibit a very specific characteristic such as the ability to metabolize and render harmless an environmental pollutant.

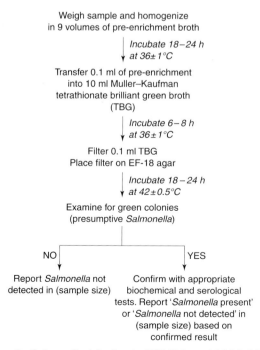

Weigh sample and homogenize
in 9 volumes of pre-enrichment broth

*Incubate 18–24 h
at 36±1°C*

Transfer 0.1 ml of pre-enrichment
into 10 ml Muller–Kaufman
tetrathionate brilliant green broth
(TBG)

*Incubate 6–8 h
at 36±1°C*

Filter 0.1 ml TBG
Place filter on EF-18 agar

*Incubate 18–24 h
at 42±0.5°C*

Examine for green colonies
(presumptive *Salmonella*)

NO — YES

Report *Salmonella* not
detected in (sample size)

Confirm with appropriate
biochemical and serological
tests. Report '*Salmonella* present'
or '*Salmonella* not detected' in
(sample size) based on
confirmed result

Figure 6 *Salmonella* detection by HGMF-based AOAC Official Method no. 991.12.

Advantages and Limitations of HGMF

HGMF technology offers several advantages, most of which are apparent from reading the above material. HGMF is versatile, being readily applicable to a wide range of food and environmental microbiology analyses. The filtration procedure efficiently separates the sample matrix from the microorganisms of interest, eliminating interfering components of the sample that might on the one hand produce false positive results or, in other cases, underestimate the concentration of the target microorganisms. The counting range of a single HGMF is so broad as to eliminate the need to prepare and analyse multiple dilutions in most cases. The ordered grid array lends itself to automated counting and also facilitates manual counting. Finally, the portability of the HGMF from medium to medium or from medium to reagent without disturbing the colonies allows for repair of injured cells on a recovery medium, sequential development of biochemical reactions, and the use of confirmation techniques involving enzyme-labelled antibody or nucleic acid colony hybridization procedures.

Limitations of HGMF technology are few, the most important of which is the need, on occasion, to digest a portion of sample homogenate using sterile enzyme in order to render the sample capable of passing through the filtration system. Rarely, it is not possible to establish a fully satisfactory digestion protocol, resulting in some loss of test sensitivity. The other significant limitation is the need to wash and sterilize the filtration apparatus. One approach to avoiding this requirement for certain types of samples was the development of the 'spread-filter' device. Another is the potential development of a fully disposable filtration system incorporating the HGMF. At the moment, neither of these devices is commercially available.

See also: **Enterobacteriaceae, Coliforms and *E. coli*:** Classical and Modern Methods for Detection/Enumeration. **Escherichia coli:** *Escherichia coli*. **Listeria:** Detection by Classical Cultural Techniques. **Physical Removal of Microfloras:** Filtration. **Salmonella:** Detection by Classical Cultural Techniques. **Wines:** Microbiology of Wine-making; Specific Aspects of Oenology.

Further Reading

Brock TD (1983) *Membrane Filtration: A User's Guide and Reference Manual.* Madison: Science Tech.

Cunniff P (ed.) (1997) *Official Methods of Analysis of AOAC International*, 16th edn, 3rd rev. Gaithersburg, MD: AOAC International.

Entis P, Brodsky MH and Sharpe AN (1982) Effect of prefiltration and enzyme treatment on membrane filtration of foods. *Journal of Food Protection* 45: 8–11.

Peterkin PI, Idziak ES and Sharpe AN (1992) Use of a hydrophobic grid-membrane filter DNA probe method to detect *Listeria monocytogenes* in artificially-contaminated foods. *Food Microbiology* 9: 155–160.

Sharpe AN and Michaud GL (1974) Hydrophobic grid-membrane filters: new approach to microbiological enumeration. *Applied Microbiology* 28: 223–225.

Sharpe AN and Peterkin PI (1988) *Membrane Filter Food Microbiology.* Innovation in Microbiology Research Studies Series. Chichester: Wiley.

Todd ECD, Szabo RA, Peterkin P et al (1988) Rapid hydrophobic grid-membrane filter-enzyme-labeled antibody procedure for identification and enumeration of *Escherichia coli* O157 in foods. *Applied and Environmental Microbiology* 54: 2536–2540.

Hydroxybenzoic Acid *see* **Preservatives:** Permitted preservatives – Hydroxybenzoic Acid.

Hygiene Monitoring *see* **ATP Bioluminescence:** Application in Hygiene Monitoring.

Hygienic Processing *see* **Process Hygiene:** Overall Approach to Hygienic Processing.

I

ICE CREAM

A Kambamanoli-Dimou, Department of Animal Production, Technological Education Institute, Larissa, Greece

Ice cream is made from a liquid mix which is based on milk, cream, water, milk solids-not-fat, milk fat or other fat as may be legally required, sugar, emulsifying and stabilizing agents, flavours and colours. This wide variety of ingredients, the possible variations in their microbiological standard and quality, as well as the conditions and methods used to prepare the final product, greatly affect the quality of the ice cream. Consequently, the microbiological quality of ice cream depends on many factors, the most important of which are described here.

Raw Materials (Major Components and Additives)

Any of the wide variety of ingredients that may be used to produce ice cream may contribute microorganisms to the product and affect its quality. The heat treatment process gives only an adequate reduction in bacterial numbers as well as the destruction of pathogenic organisms. It cannot entirely correct the hygienic quality of poor ingredients. So carefully selected ingredients are essential for the manufacture of ice cream of the highest quality.

Liquid milk, cream, skimmed milk and concentrated skimmed milk may contain appreciable numbers of bacteria, including some that are pathogenic (e.g. *Mycobacterium* spp., *Streptococcus* spp.). Adequate heat treatment by the supplier, together with handling and storage under sanitary conditions (kept under refrigeration and used promptly) leads to raw materials of a satisfactory quality. The main organisms present in these dairy materials are a few spore-forming bacilli, micrococci, and psychrotrophic and thermoduric microorganisms, which may sometimes spoil the mix but are not a major health hazard. Milk powders may contain large numbers of spore-forming bacilli or may be contaminated by salmonellae. Numerous outbreaks of food poisoning attributed to salmonellae or staphylococci from milk powder provide evidence that these pathogens do on occasions survive in the final product. A special hazard of staphylococcal enterotoxin in ice cream may be present if whey powder is used as a source of milk solids. Careful control and storage of these powders under dry and cool conditions are necessary.

Dry sugars used as ingredients should be almost sterile if properly prepared, processed and stored. Sugar syrups also are used as sweetening ingredients. The contamination of sugars or sugar syrups is limited, mainly consisting of small numbers of osmophilic microorganisms. Certain yeasts and moulds would be the principal flora. Some species of bacteria have also been suggested as possible spoilage problems, including species of *Bacillus* and *Leuconostoc*. Osmophilic yeasts may be able to grow in these syrups and moulds may grow on the surface if contamination should occur, so it is suggested that tests for yeasts should be carried out on sugars and sugar syrups.

Butter and butter oil (anhydrous milk fat) are made from pasteurized cream, in which pathogenic and most spoilage organisms have been destroyed. Relatively small numbers of mesophilic bacteria, coliforms and lipolytic organisms, particularly the *Pseudomonas* sp. responsible for butter spoilage, as well as moulds and yeasts, can be found. Butter commonly is kept refrigerated, and during commercial storage is kept at about –20°C at which temperature no microbial growth can take place. For these reasons bacteria usually do not grow in butter, and when they do, their growth is not extensive. The flavour of good butter is so delicate, however, that small amounts of growth may cause appreciable damage to the flavour. When satisfactory hygienic conditions are applied during the production of the above ingredients and they are properly handled (storage temperatures of no more than –20°C for butter, and dry, refrigerated conditions for butter oil), combined with tests for the above-mentioned microorganisms, a high microbiological quality will be ensured.

Fats other than milk – commonly vegetable fats – may also be used. The high temperature used during their processing and their low moisture content give raw materials containing very few microorganisms.

Dry, refrigerated conditions should be used for their storage.

Stabilizers do not constitute an important source of bacteria if they have been packaged under hygienic conditions, as they are usually produced by methods involving high temperatures. Gelatin as an animal product may be a hazard if produced under insanitary conditions, so it should be obtained from a reputable supplier and kept under cool and dry conditions. Emulsifiers should not present any problem except for eggs, which may be contaminated by *Salmonella* spp., and pasteurization is required to avoid any hazard arising from their use.

Many other foodstuffs are added to ice cream, either mixed with it or as coatings. These may be flavouring materials such as vanilla, chocolate and cocoa, fruit (canned, fresh or frozen) and nuts, as well as colours. All of them are potential sources of hazard, particularly if they are added after the heat treatment of the mix. Yeasts and moulds will predominate in fresh fruits, while canned fruits, because of their heat treatment, should be of a satisfactory microbiological standard. Nuts may be contaminated by moulds and may possibly contain mycotoxins. Coconuts can also contain salmonellae. For these reasons, it is much better for all these materials to be used after their heat treatment (roasted nuts, for example, or pasteurized chocolate), especially if they are added to the mix after its pasteurization. Also, they should be stored in cool, dry conditions. The examination of these materials should include a visual inspection and the enumeration of mesophilic bacteria, coliforms, yeasts and moulds. Colours manufactured and handled carelessly may cause microbiological problems, but this can be avoided if they are obtained from a good supplier and stored properly.

All these potential hazards highlight the necessity of using high-quality raw materials, purchased from a reputable supplier, and carefully stored under conditions that will not allow the proliferation of microorganisms. In addition, it is suggested that appropriate microbiological tests should be carried out on raw materials, and the use of strict stock rotation is essential.

Hygiene During Production

After the high-quality raw materials have been chosen, the mix is ready for processing. This begins with combining the ingredients into a homogeneous suspension that can be pasteurized, homogenized, cooled, aged, flavoured and frozen. This is a complex operation comprising a series of steps which all have some effect on the microbiological quality of the final product, so they must be carefully controlled to produce a product that will safeguard the consumer's health.

After the ingredients have been weighed or measured, they are blended to make a liquid mix. This mixture is then subjected to a heat treatment process, which is specified by legal requirements in most countries. This pasteurization renders the mix substantially free of vegetative microorganisms, killing all of the pathogens likely to be present. The ice cream mix is always homogenized, often as a step in the pasteurization process. The homogenizer is a complex piece of equipment and must be carefully cleaned and disinfected each time it is used, or the mix may be seriously contaminated. It is therefore suggested that homogenization of the ice cream mix is carried out before it is finally heat-treated, where this is possible. Cooling of the mix to about 4°C is followed by an ageing period; then the mix is passed to the freezer where it is subjected to considerable agitation and reduction in temperature, as well as incorporation of air. On leaving the freezer the ice cream will normally be packaged (in family packs, individual retail packs, or any other forms), frozen hard in wind tunnels at −40°C or in hardening rooms, and then kept at a temperature of about −30°C until and during distribution. Some ice cream is sold direct from a dispensing freezer as 'soft-serve' ice cream in cones, or in various types of made-up desserts in restaurants and cafés, or from vehicles complete with their own electricity generation equipment.

Ice cream mixtures must not be kept for more than 1 h at a high temperature (more than 7.2°C) before being pasteurized, in order to avoid the proliferation of organisms in the ingredients. During pasteurization, time as well as temperatures should be strictly observed in order to avoid on the one hand excessive heat treatment, which may lead to undesirable flavour changes, and on the other hand, to ensure the destruction of pathogenic organisms and the adequate reduction in bacterial numbers. The cooling of the ice cream mix to about 4°C must be rapid and the mix must be kept at that low temperature until it is frozen, otherwise the proliferation of any viable organisms may occur. This can lead to a product with a high microbial count and possibly to a disease outbreak. The same danger exists if the cooling system stops during the ageing of the mix. In this case, the mix must be discarded because, although a new pasteurization will kill the organisms, it will not destroy possible toxins already present. Normally the mix should be frozen within 24 h of heat treatment, as undue prolongation of storage may lead to proliferation of psychrotrophic organisms, with a serious risk of spoilage of the mix.

The incorporation of air into the ice cream mix

during the freezing process may cause air-borne contamination, so air filters are required to prevent the ingress of organisms. For some products the ice cream is blended with water ice on leaving the freezer, and may be frozen on a stick or in a cone (single portions) and covered with a chocolate or flavoured couverture, together with broken biscuits or nuts. All this handling has to be done under sanitary conditions in order to minimize the microbiological contamination of ice cream.

In addition to the careful operation of the equipment, the proper cleaning and sanitizing of the plant and equipment is of great importance. Any equipment that comes into contact with the ice cream or ice cream ingredients must be carefully and effectively cleaned and sanitized immediately after use. This is usually done at the end of each day's operation; however, if scale builds up during the operation, it may be necessary to clean the plant thoroughly before further processing the same day. Not only is the processing equipment important, but ancillary equipment, in particular at the final sales point, must be kept in a satisfactory hygienic condition. Poor cleaning and sanitizing of the plant and equipment may lead to pockets of ice cream residues where intense proliferation occurs, which results in recontamination of the pasteurized mix with a large number of bacteria. For soft-serve freezers manufacturers typically provide specific instructions for cleaning and sanitizing, and these should be followed closely.

Clean surroundings are essential if equipment is to be kept in hygienic condition. All rooms, especially toilets and locker rooms, must be kept as clean and sanitary as the area immediately surrounding the packaging equipment. Surrounding activities (e.g. sewerage plants, rubbish tips) often represent potential sources of contamination, and birds, rodents and insects are all important vectors of such contamination. In addition to preventing access of pests to the process area, it is important that the factory yard is kept free of food waste, rubbish and spilled material that might attract birds, rodents and insects. All such material should be kept in lidded containers and removed on a daily basis. Pet animals such as dogs and cats similarly have no place in a food production factory.

Operations must be segregated to minimize the chances of pathogenic microorganisms being carried from raw materials to finished products. Persons handling raw milk or cream must not be allowed access to rooms where pasteurized products are exposed unless those persons have first changed their clothes completely and have disinfected themselves. Room air pressures should be maintained at successively higher levels from the mix room to the processing room, to the freezing operation and to the packaging operation. Thus, the flow of air will be away from the most critical area, the packaging room. The supply of hot and cold water must be unrestricted and facilities for disposal of both liquid and solid wastes must be adequate. All water that is used in food formulation, or will be used on (or could gain access to) food contact surfaces, should be of potable quality and be stored in enclosed tanks and distributed in piping that is completely segregated from other pipe systems. Potable water may be derived from public mains supplies or other sources such as boreholes which must be protected against contamination by surface water or underground contamination from drains, seeping from farm or industrial tips and similar potentially hazardous areas. Whatever the origin of water it should be routinely examined microbiologically at the point of entry to the site and at the point of use, particularly if there is on-site storage.

The hygienic standards of the workforce are crucial to the ice cream manufacturer. No worker should be allowed to perform tasks in the plant who has not been adequately taught the necessities of personal hygiene and approved practices within the plant. Every employee must dress in a clean uniform, wear a hair restraint, wash and sanitize hands, disinfect footwear on entry to the process area, and refrain from touching any product contact surface without properly sanitizing the hands, or gloves if worn. Proper sanitary practices are essential to the ice cream plant. No one should be allowed to enter the processing environment who is not familiar with the required sanitary procedures or who does not conform to the required dress and personal hygiene measures. Freedom from chronic contagious diseases should be confirmed yearly by medical examination. Provided that preparation of ice cream is conducted in a closed processing cycle with modern industrial equipment of high hygiene standards, the opportunities for contamination by human contact are few.

The packaging materials may occasionally cause trouble, but there should be no problem if they have been handled and stored under hygienic conditions.

Tests carried out during the production process should be indicative of the standard of hygiene.

Microbial Changes During Storage

Ice cream is a perfect substratum for the proliferation of microorganisms because of its composition. It contains sugar, proteins and oxygen, all the essential components that microbes need, as well as a suitably high pH. The only factor not available is high temperature. If a rise in temperature does occur, all the conditions are in place for the development and proliferation of microbes that may have survived the heat treatment, or come from post-pasteurization contamination or insanitary processing and packaging.

Ice cream may be sold direct from the freezer as a 'soft-serve' product, or it may be further reduced in temperature and frozen in wind tunnels at –40°C or in hardening rooms, to produce 'hard' ice cream, which will be stored at a temperature of about –30°C until it is sold. Deep-freezing stabilizes the microbial content of ice cream: microorganisms found in it no longer proliferate. Some sensitive species (Gram-negative) die and their populations reduce. Even if the period between freezing and final sale is several months, there will be little change, if any, in the microbial content of the ice cream. Extensive research has shown that both *Mycobacterium* and *Salmonella*, as well as many other less harmful but often more resistant types, can survive at the low temperature of storage for very long periods. They do not multiply, provided that the temperature is low enough for the ice cream to remain hard; in effect, the microbial quality of ice cream is 'locked in' by the hardening process. It is, therefore, essential that the bacteriological content of the ice cream from the freezer is as low as possible, as neither the final hardening process nor the low temperature storage can be relied upon to reduce the numbers appreciably, and pathogenic organisms should be absent.

If the mix is frozen promptly spoilage is impossible, for microorganisms cannot grow in the frozen product. However, if there is a delay between pasteurizing and freezing, spoilage can occur, as well as in cases of melting and refreezing of the product resulting from temperature variations or failure of the freezing systems. Such delays are unusual in the manufacture of hardened ice cream. Special care is needed for mix with soft-serve ice cream that has to be transported, often for long distances, by trucks to retail soft-serve stores or stands where it is kept soft-frozen and dispensed to consumers. Both contamination and temperature abuse of the mix may easily occur. Furthermore, refrigeration space is usually limited, and adequate facilities for cleaning and sanitizing the freezer and the associated equipment are often lacking or are, at best, marginal.

Under these conditions, especially if wrong practices had occurred previously, the ice cream is overloaded with microbes which can lead to quality deterioration or even to cases of food poisoning. Food poisoning is known to have followed the consumption of ice cream contaminated by microbes, such as *Staphylococcus*, *Salmonella*, *Shigella*, *Listeria* and *Streptococcus* group A organisms. With few exceptions, outbreaks that have occurred in recent years have been caused by ice cream made not in commercial establishments but at home, where a combination of faulty practices may occur. The use of raw milk or cream, the addition of raw eggs containing *Salmonella* to the mix and the use of contaminated equipment, in addition to inadequate heat treatment and contamination from infected persons, give rise to products with high microbial loads, especially of pathogenic bacteria, which, if they are present, may survive for many months in ice cream.

Problems at Point of Sale

Even when the greatest care has been taken to produce an ice cream of the highest quality, it is still liable to contamination at the point of sale. The largest proportion of microbiological problems, in general, are due to poor techniques of selling and serving at the final point of distribution.

Although much ice cream is retailed in its final packaging, a significant quantity is portioned from bulk packs at point of sale. The method of sale has a major bearing on the amount of contamination to which the product is subjected. Ice cream sold prepacked as a single retail portion, and which has only to be handled by the consumer in its wrapping, should have the least contamination of all. Greater degrees of contamination may occur in ice creams served in cones or other individual portions scooped from bulk ice cream in restaurants or coffee shops, or from vehicles complete with their own electricity generation equipment. Here there is a possibility of considerable contamination, unless all the equipment used (servers, etc.), the method of dispensing and the personal hygiene of the operator are all of a very high standard. The equipment (servers, wafer holders and so on) has to be kept free of all residues of ice cream, which might otherwise melt and allow the growth of bacteria to recommence. Wherever possible these items of equipment should be kept in running cold water. If they have to be kept in a jug of water, this water must be regularly changed to avoid it becoming a source of contaminating bacteria. The personal hygiene of the server is also of the utmost importance. This applies even if pre-wrapped ice cream is being sold. Cleanliness of hands, clothing and habits must be above reproach, and the operators must be trained in the best ways of maintaining this, and in the distribution of the individual portions of ice cream.

Soft-serve ice cream sold directly from a dispensing freezer can become contaminated very easily unless stringent precautions are taken. The product is usually manufactured at the point of sale, which may be a specialist outlet or café, a non-food outlet such as a filling station, or a mobile outlet or kiosk. Soft-serve ice cream may be manufactured from a conventional mix produced on the premises, from an ultra-high temperature processed, aseptically packaged mix, or from a spray-dried powder mix. Powder mixes may be formulated for reconstitution in either hot or cold water. Hot-water mixes are preferable with respect to hygiene, but cold-water mixes are often considered more convenient. Attention must be paid to the reconstitution of the mixes; this must be done under satisfactory hygienic conditions to prohibit the proliferation of *Salmonella*, which may survive if mixes are not prepared with care.

Generally, special dispensing freezers should be kept in constant operation and placed inside the shop with the taps facing the interior; they must not be directly exposed to the sun, dust or flies. A strict cleaning and disinfecting regimen for the freezer must be instituted, adhered to and performed thoroughly on a daily basis. In particular, ice cream must not be allowed to remain in the freezer overnight, and the personal hygiene of the operator must be of a very high standard. It must be recognized that maintenance of the necessary hygiene standards can be more difficult in an environment primarily concerned with retailing than in one wholly concerned with manufacturing. Particular difficulties may be encountered in outlets that are predominantly non-food, such as filling stations, and those with inherently limited facilities, such as kiosks.

Special self-pasteurizing dispensing freezers are now available which have a heat treatment process as part of their operation, and provided the process is used each and every day, they do not normally need daily cleaning. At the end of each day's operation (or other convenient time) the freezer is switched to the 'self-pasteurize' mode, and all the mix, and every part of the freezer that can come into contact with ice cream or mix, are raised to a temperature above that required for pasteurization of the mix, and held at that temperature for the legally required time. The freezer and its contents are then cooled rapidly to about 4°C and held at that temperature. Tests have shown that there is little or no increase in the bacterial content of the product over a period of more than a week. It is to be recommended, however, that these freezers are cleaned out and disinfected at regular intervals of, say, a week. It must be emphasized that self-pasteurizing freezers are not intended to process unpasteurized mix.

Probably the most serious and dangerous sources of contamination are operation and serving. Many major food poisoning outbreaks have been caused by human contamination. Cases of typhoid fever, including deaths, have been reported to be caused by ice cream contaminated by the manufacturer who was a urinary excretor of *Salmonella typhi*. There has been a case of *Shigella* dysentery caused by an ice cream that was accidentally touched by a monkey. Also, outbreaks involving salmonellae and staphylococci have been reported. The personal hygiene and habits of vendors at sale points are important. Education, in addition to medical inspection, is absolutely necessary and no employee must be allowed to work without full medical clearance.

Finally, birds, rodents, insects and pet animals have no place at the retail selling point.

To some extent ice cream retains a reputation as a high-risk food. This is unjustified for commercially produced ice cream in developed countries which have legislated microbiological standards for ice cream, as well as for the conditions and methods to be used for heat treatment and subsequent storage and sale.

See also: **Eggs**: Microbiology of Fresh Eggs. **Food Poisoning Outbreaks**. **Freezing of Foods**: Growth and Survival of Microorganisms. **Heat Treatment of Foods**: Principles of Pasteurization. **Milk and Milk Products**: Microbiology of Liquid Milk; Microbiology of Cream and Butter. **Salmonella**: Introduction. **Process Hygiene**: Overall Approach to Hygienic Processing; Hygiene in the Catering Industry.

Further Reading

Frazier WC and Westhoff DC (1988) *Food Microbiology*, 4th edn. New York: McGraw-Hill.

International Commission on Microbiological Specifications for Food (1980) *Microbial Ecology of Foods*. Vol. 2. *Food Commodities*. London: ICMSF/Academic Press.

Jervis DI (1992) Hygiene in milk product manufacture. In: Early R (ed.) *The Technology of Dairy Products*. P. 272. Chichester: John Wiley.

Mantis AI (1993) *Hygiene and the Technology of Milk and its Products*. Thessaloniki: Kiriakidis.

Marshall R and Arbuckle WS (1996) *Ice Cream*, 5th edn. New York: Chapman & Hall.

Rothwell J (1985) Microbiology of frozen dairy products. In: Robinson RK (ed.) *Microbiology of Frozen Foods*. P. 209. London: Elsevier.

Rothwell J (1990) Microbiology of the icecream and related products. In: Robinson RK (ed.) *Dairy Microbiology*. Vol. 2, p. 1. London: Elsevier.

Varnam A and Sutherland J (1993) *Milk and Milk Products*. Food and Commodities Series no. 1. London: Chapman & Hall.

Immunological Techniques *see* **Mycotoxins**. Immunological Techniques for Detection and Analysis.

Immunomagnetic Particle-Based Assays *see* *Escherichia coli* **O157**: Detection by Commercial Immunomagnetic Particle-based Assays; *Listeria*: Detection by Commercial Immunomagnetic Particle-based Assays; *Salmonella*: Detection by Immunomagnetic Particle-based Assays.

IMMUNOMAGNETIC PARTICLE-BASED TECHNIQUES: OVERVIEW

Kofitsyo S Cudjoe, Department of Pharmacology, Norwegian College of Veterinary Medicine, Oslo, Norway

Copyright © 1999 Academic Press

Introduction

Methods for the isolation, detection and identification of pathogenic bacteria in foods have previously been based on lengthy conventional cultural techniques. This is so because in foods, specific organisms must be detected among a mixed population of dominating background flora. Modifications to the different stages of these culture methods tend to focus on detection. Thus, rapid detection, based on molecular biology and immunoassay techniques and/or a combination of both, have diverted attention from the time-consuming enrichment steps needed to increase and select the target numbers needed for detection. Although these new detection methods are sensitive and specific, they often fail to combine the two key factors of simplicity and speed.

Immunomagnetic separation (IMS), represents a time-saving replacement and/or supplement for the current lengthy selective enrichment protocols and allows the flexibility of applying different end-point detection methods. IMS employs magnetizable particles coated with specific antibodies to concentrate target bacterial cells selectively from a sample matrix. The technique represents a new generation of concentration and physical separation techniques like centrifugation and filtration. It is rapid, inherently specific due to the coated antibodies having been raised against target specific surface markers and requires simple and relatively inexpensive equipment. The technique has been applied in a wide range of fields including food, environmental and clinical microbiology, waste water treatment, plant pathogen

Table 1 Some applications of immunomagnetic separation in food microbiology

Target organism	Detection method
Clostridium perfringens enterotoxin A	IMS-ELISA
Staphylococcus aureus enterotoxin B	Magnetic enzyme immunoassay (MEIA)
Salmonella	Plating
	Plating and serology
	IMP-ELISA
	Electrochemiluminescence
	Conductance
	PCR and DNA hybridization
E. coli O157	Plating
	Electrochemiluminescence
Vibrio parahaemolyticus	Plating
Listeria	Plating
Yersinia enterocolitica	Plating, PCR and DIANA
Campylobacter jejuni	PCR
Shigella dysenteriae	PCR
Cryptosporidium parvum	Fluorescence microscopy
Erwinia carotovora	Plating and PCR

detection, biotechnology production processes and downstream applications such as isolation and sequencing of nucleic acids from various microorganisms (**Table 1**).

Magnetic separation (MS) employs other proteins and/or lectins as ligand molecules rather than antibodies to remove target proteins or organisms from sample matrices. This article describes the principles and types of immunomagnetic systems and illustrates how IMS can be used in combination with other techniques for the rapid detection of food-borne pathogens and spoilage flora.

Table 2 Some magnetizable particles available commercially (Adapted and reproduced with permission from Kroll et al (1993) and Cudjoe et al *Immunomagnetic Separation Techniques for the Detection of Pathogenic Bacteria in Foods*. Oxford: Blackwell Scientific Publications)

Particles	Size (μm)	Manufacturer
Biomag 4300: silanized magnetic iron oxide	0.5–1.5	Metachem Diagnostics Ltd, Northampton, UK
Polystyrene paramagnetic microparticles	1–2	Polysciences Ltd, Northampton, UK
Magnetic Polyacrolein/iron oxide particles	1–10	Scipac, Sittingbourne, UK
Polystyrene/divinyl-benzene	0.7	Sepadyn, Indianapolis, USA
Dynabeads (Polystyrene)	2.8 or 4.5	Dynal A/S, Oslo, Norway

Table 3 Commercial immunomagnetic separation-based products for the detection of food- and waterborne pathogens

Organism	Product name	Manufacturer
Salmonella	Dynabeads anti-*Salmonella*	Dynal
	Salmonella Screen/Salmonella Verify	Vicam
Escherichia coli O157	Dynabeads anti-*E. coli* O157	Dynal
	Captivate O157	Lab M
Listeria	Dynabeads anti-*Listeria*	Dynal
	ListerTest	Vicam
Cryptosporidium	Dynabeads anti-*Cryptosporidium*	Dynal
	CryptoScan	Clear Water Diagnostics

Principles of Immunomagnetic Separation Systems

Although several companies produce magnetizable particles, also referred to as beads (**Table 2**), only two commercial companies (Dynal, Norway; Vicam, USA) actively produce and market antibody-coated magnetizable particles for the specific isolation and detection of food- and waterborne pathogenic microorganisms (**Table 3**). Common to all IMS systems is the principle that bacteria or cells bound to magnetizable beads by specific antibodies can be isolated from heterogeneous suspensions by the application of a magnetic field. Bacteria immunologically bound to beads can subsequently multiply when nutritional requirements are provided.

The procedure involves mixing the antibody-coated particles with the prepared sample, and after incubation, the bead–bacteria mixture is extracted with a magnet. The isolated bead–bacteria complexes are washed before transfer to suitable growth media or used with other detection systems (**Fig. 1**).

Two fundamentally different approaches exist for the detection of food-borne pathogenic bacteria, and have independently been validated for the detection of *Salmonella*, *Escherichia coli* O157 and *Listeria* spp. from food and environmental samples. The approach gaining most favour due to its simplicity and flexibility involves the maceration, blending or mixing of samples in a stomacher bag with a filter before pre-enrichment. IMS of the target organisms is then performed on 1 ml aliquots of the filtered pre-enriched samples. By this approach, low numbers and/or sub-lethally injured target organisms are able to multiply during the pre-enrichment phase to detectable levels before being specifically and selectively concentrated by IMS within 10–30 min. The resulting particle–bacteria complexes are washed before being returned to the conventional detection system by plating onto standard selective plating media or other current rapid detection methods.

The alternative approach is novel as the initial stages of sample preparation vary according to individual sample and involves no pre-enrichment. In this approach, homogenates of weighed samples are usually filtered and/or centrifuged. After centrifugation, the supernatant is decanted and the pellet resuspended in buffer. IMS is then performed on 2 ml aliquots of the filtrate or resuspended pellets for 2 h before washing and plating onto selective plating media. This alternative sample preparation procedure is considered time consuming and labour intensive. Furthermore, the prolonged 2 h incubation might cause the target organisms to multiply by at least two generations although the procedure is assumed to enable quantitation of the initial contamination level of the target organism.

For both approaches, however, the washing regimes are perceived to involve too many hand manipulations and are therefore labour intensive.

Types of Magnetizable Particles

A variety of magnetizable particles are commercially available for use in IMS techniques (Table 1). Most of these particles are collected readily on permanent magnets, are paramagnetic or superparamagnetic, that is, the particles have no magnetic remanence or memory and are therefore not attracted to each other. The paramagnetic property allows the particles to be readily suspended as homogeneous mixtures and enhances their further use in food homogenates. All the particles listed in Table 1, except Dynabeads, are non-uniform in shape and content of magnetic oxides and range in size from typically 50 nm to 1 μm. Dyna-

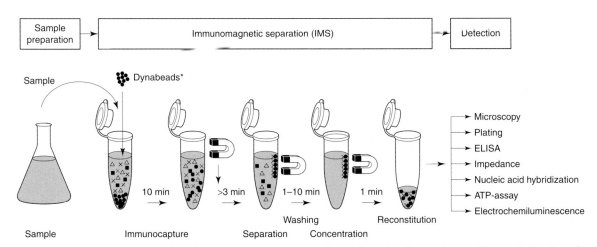

Figure 1 Diagram to show the principles of immunomagnetic separation. Adapted and reproduced with permission from Cudjoe KS, Patel PD, Olsen E, Skjerve E and Olsvik Ø (1993) *Immunomagnetic Separation for the Detection of Pathogenic Bacteria in Foods.* In: Kroll RG, Gilmour A and Sussman M. Oxford: Blackwell Scientific Publications.

beads, the most frequently cited for IMS, are of uniform size (either 2.8 or 4.5 µm), shape and magnetic oxide content due to the unique production process of activated swelling. The chemical composition of magnetizable particle surfaces is also critically important for their successful use in bioseparation. Since it is desirable to have a covalent binding between the particle surface and the protein ligand, the particle surfaces carry functional groups, such as acid, carboxyl, amino, epoxy, thiol, hydroxyl, to enhance protein uptake.

Immobilization of Antibodies and Types of IMS

The target specific monoclonal or polyclonal antibodies have often been covalently linked directly to particles or indirectly to particles pre-coated with anti-antibodies. Both particle types can be used in direct or indirect IMS techniques for the specific isolation of target organisms. In direct IMS, anti-target antibodies are either bound directly or indirectly to the particles and incubated with the sample containing the target or analyte. The particle–target complexes thus formed are separated from the suspension with the use of magnetic particle concentrator (MPC). In indirect IMS, the target sample is pre-treated with target specific antibodies so that all binding sites are blocked. The antibody–target complexes are then collected by centrifugation, washed to remove excess antibody and resuspended into an appropriate volume. Subsequently, the addition and incubation of particles already immobilized with antibodies against the anti-target antibodies will result in specific antibody–antibody interactions and hence the retrieval of the target organisms. The indirect IMS approach is a less user-friendly method because of the additional

handling by centrifugation. However, for samples with low numbers of target organisms and which are devoid of particulate matter, isolation may be significantly improved.

Blocking Proteins and Antibody Immobilization

All antibody-coated particles must be blocked with an inert or irrelevant protein, for example, bovine serum albumin or casein. The use of a blocking protein during immobilization of antibodies reduces the potential for methodological artefacts and maximizes the sensitivity, specificity and reproducibility of the binding ability of coated particles to target organisms. Blocking proteins serve the primary purpose of covering uncoated portions on beads and thus prevent bead–bacteria interactions. A blocking protein is not intended to prevent direct contact between antibodies and the uncoated polystyrene beads, and therefore the type of protein used is not important. This way of using blocking proteins contrasts with their use in enzyme immunoassays carried out in microtitre wells. For the latter purpose, the type of protein competitor is essential to achieve a full blocking effect and so reduce and/or prevent non-specific binding of non-targets and enzyme conjugates.

Magnetic Particle Concentrators

Magnetic particle concentrators (MPC®s) are important complementary accessories in immunomagnetic separation and are used to separate bead–bacteria complexes from the unwanted debris in test samples. The MPCs are made from strong permanent magnets based on rare earth metals. MPCs used in IMS must be of uniform strength and capable of concentrating the particles from all types of prepared food samples. In the future, with completely automated IMS systems, electromagnets might serve the same purpose as the current MPCs.

Factors Affecting the Performance of Immunomagnetic Separation Systems

The performance of antibody-coated beads during IMS is solely dependent on the extent to which particles are recovered from different sample matrices. In this regard, the proper use of MPCs is paramount. Failure to recover the bead–bacteria complexes could result in failure to detect the presence of target organisms in a positive sample. In extremely fatty, viscous and/or particulate samples, a two- to tenfold dilution of the 24 h pre-enriched sample must be made prior to IMS analysis. Such a dilution will not limit detection of the target organism but rather ensure that beads are recovered (Table 6). The user must, however, practice care not to aspirate and discard the isolated bead–bacteria complexes. Some samples, such as cocoa powder, bacteriostatic spices and other materials containing inhibitory substances should not be diluted. To prevent the loss of bead–bacteria complexes from such samples, approximately 10% of the original sample must be left in the tube to be diluted further with wash buffer.

In practice, the performance of antibody-coated beads directed against bacterial surface antigens depends on a number of factors. The type of antibody coated, its functional affinity, the concentration of coated antibody per milligram particle, the number of particles per test, the incubation time, mode and temperature of incubation, type of sample and the washing regimes employed determine to a large extent the performance of the IMS procedure (**Tables 4–8**). If the antibodies coated are too restricted in specificity as may be the case of monoclonal antibodies, specificity becomes restricted to only bacteria with the common antigenic epitopes. Target bacterial epitopes may only be loosely associated with the bacterial cells and, in old cultures, free antigen from dead or lysed cells can interfere with the binding abilities of the particle-bound antibodies, particularly if the recovery of viable cells is intended. Some capsulate bacterial strains are 'sticky' and attach nonspecifically to the beads. This may interfere with the binding of the antibodies to the specific targets and may result in isolation of mixed bacterial populations. High coating antibody concentrations may lead to aggregation of bound targets. Use of too many particles and too long an incubation time normally increases nonspecific

Table 4 The effect of mixing during incubation and sample particulate matter on the recovery of seeded *Salmonella* Give (400 cells per ml) as determined by IMS followed by plating[a] (Reproduced with permission from Cadjoe KS, Patel PD, Olsen E, Skjerve E and Olsvik Ø (1993). *Immunomagnetic Separation for the Detection of Pathogenic Bacteria in Foods*. In: Kroll RG, Gilmour A and Sussman M. Oxford, Blackwell Scientific Publications

	% Recovery		
	Broth	Whey milk powder	
Type of mixing		Whole suspension	Clear supernatant
Continuous mixing	33.0	4.0	12.0
Intermittent mixing	20.0	3.0	5.0
No mixing (Static)	11.0	2.0	2.0

[a]Continuous sample mixing during incubation improves binding kinetics and results in increased recovery of target organisms. The presence of particulate matter and/or fat micelles results in loss of beads during washing with concomitant reduced recovery of target organisms. The recovery of bead–bacteria complexes is therefore improved if samples are devoid of particulate matter.

Table 5 The effect of washing during IMS on the recovery of non-*Salmonella* organisms in a 20 h enriched whey milk powder and seeded *Salmonella enteritidis* in broth culture[a]

No. of washings during IMS	% Recovery	
	Non-Salmonella organisms (10^8 cfu ml^{-1})	S. enteritidis (10^6 cfu ml^{-1})
0	35	26
1	4	17
2	2	10

[a]Washing during IMS is important in eliminating nonspecifically bound background flora and thereby shifting the binding ratio in favour of the specifically bound target organisms.

Table 6 Level of iodine-labelled particles recovered from dilutions of blended egg after IMS[a]

Dilutions	Particle counts (cpm)		% Recovered
	Initial	Final	
1 : 1	25 458	5 579	22.0
1 : 3	25 600	13 648	53.0
1 : 5	25 814	21 754	84.0
1 : 10	26 360	24 126	92.0

[a]Dilution of samples ensures that particles are recovered with a concomitant improved recovery of target organisms.

binding, whereas the use of sub-optimum particle concentrations reduces sensitivity.

Incubation During IMS

During incubation, intermittent mixing enhances binding kinetics, resulting in the isolation and separation of higher levels of target organisms than when

Table 7 The correlation between particle number and the level of E. coli O157 recovered when using IMS[a]

Particle number (×10⁶)	% Recovery
1	12.3
2	31.0
3	33.0
4	54.8

[a]Increasing particle number per test improves the level of target organisms recovered (see Table 8).

Table 8 The correlation between extended incubation time and the level of target bacteria recovered when using IMS[a]

	E. coli O157 (300 cells per ml)			
Incubation time (min)	10	15	20	30
%Recovery	20	34	50	49

[a]Although the actual number of cfu recovered increases with prolonged incubation and increased particle number per test, the efficiency of recovery decreases with increasing levels of initial cells per millilitre.

static incubation is used. However, continuous mixing during incubation gives even better binding kinetics resulting in increased recovery of target organisms. The fear of cross-contamination from the tube's cap has been expressed as a basis for static incubation without closing tubes. Static incubation could be used for those IMS protocols that involve less than 3 min incubation, but the beads must be thoroughly mixed with the sample before static incubation commences otherwise they will sediment. A major source of contamination, however, is the use of contaminated wash buffer, improper handling of pipettes and lack of good laboratory practice. Care should be taken to distinguish between wash buffers meant for first and second washes and from those meant for reconstitution of particles after the final wash. Main batch wash buffers should be aliquoted into smaller volumes and used appropriately. Care must be taken to avoid the development of aerosols while aspirating sample supernatants, wash buffer supernatants or when adding wash buffers. When opening Eppendorf tubes, the index finger should be placed over the cap and the thumb used gently to lift it upwards, or an opening device for Eppendorf tubes should be used.

Washing During IMS

Washing is an important step in the IMS technique. Like any immunoassay procedure, washing is essential to remove unbound target and non-target organisms, remove sample matrix, some of which could contain bacteriostatic agents, and minimize non-specific binding. Too much washing may result in the loss of target bacteria specifically bound, whereas under-washing may not remove non-target bacteria nonspecifically bound to the particles. The latter

may mask the isolated target organisms and so hinder detection, particularly if growth is used for detection. The proper use of MPCs to concentrate bead–bacteria complexes during the washing procedures determines to a large extent if the IMS procedure will be successful. Normally, the entire IMS procedure is performed on a bench top at a room temperature of 18–28°C.

Principles of Combination Techniques for Food-borne Pathogens

Primarily, IMS was developed in conjunction with plating so that target organisms could be isolated and further characterized. Bead–bacteria complexes are therefore resuspended into a small volume to achieve the desired concentration and to enable plating all the suspension. Resulting colony-forming units do not reflect an absolute quantitative measure as to the actual level of organisms recovered. This is because more than one organism can bind to a bead and because of the formation of aggregates among the bead–bacteria complexes during the concentration process. If alternative methods are desired, for example, measuring the metabolic activity as in the ATP assay, then bead–bacteria complexes must be resuspended into a larger volume (1 ml) to break up the aggregate. If for example a polymerase chain reaction (PCR) detection is desired, the bead–bacteria complexes can be resuspended directly into lysis buffer to an optimized bead concentration of less than $100 \,\mu g \, ml^{-1}$ (or approximately 10^6 beads per millilitre) so that the beads do not interfere with the reaction. No matter which detection combination is adopted, sample preparation prior to IMS is critical and the appropriately resuspended bead–bacteria complexes then serve as a template for the detection method in question.

Food Sample Preparation

Some food types require special enrichment broths because of the presence of substances that may be toxic for the target bacteria or for other reasons connected with the composition of the food. For most food samples, however, maceration by stomaching for 1 min in a stomacher bag with a filter is important. The process releases the target bacteria into suspension which then passes easily through the pore. After incubation, the filtrate, now devoid of particulate matter is then used for IMS. For IMS procedures based on samples not pre-enriched, filtration of the stomachate, followed by centrifugation is recommended. Sample preparation involving centrifugation permits the detection of lower levels of contamination when a larger gram-equivalent of sample is used in the examination.

Water Sample Preparation

Examination of potable water supplies for the presence of water-borne parasites, for example *Cryptosporidium parvum* oocysts, involves the filtration of large volumes (10–1000 litres) of water. The suspension of particulate matter and parasites eluted from the filter is then concentrated by centrifugation. The sediments (retenate) are then resuspended in a reagent grade water and used for IMS which currently involves two recognized protocols. In one protocol, a pre-clear procedure involving beads coated with irrelevant antibodies is used to remove interfering material, before performing the main IMS capture procedure, followed by staining and detection. The second recognized protocol involves addition of controlled buffer formulations to the sample during the oocysts capture by IMS to inhibit interfering substances. Thereafter, the captured oocysts are dissociated from the beads and detected after a staining procedure. The scientific merits of both methods are subject to intense disagreements, but currently, it appears that the second recognized protocol is gaining favour because the pre-clear treatment of the retenate appears to deplete the oocysts as well, often resulting in false negative results. Furthermore, in turbid water the recovery efficiency of the former method is both low and variable (**Table 9**). Nevertheless, the only accepted method for detecting *Cryptosporidium* oocysts after IMS is immunofluorescence assay using a fluorescein isothiocyanate (FITC) monoclonal antibody. If viability of the oocysts is to be determined, the immunofluorescence assay can be combined with vital fluorogenic dyes.

Detection Methods

Immunomagnetically isolated target organisms from sample matrices allows the use of any end-point detec-

tion method. For purposes of epidemiological tracing, bead–bacteria complexes of most pathogenic bacteria, for example, *Salmonella*, *E. coli* O157 and *Listeria* have been plated onto respective selective plating media. However, for the purposes of screening large numbers of samples, rapid detection methods are preferred. Bacteria or oocysts attached to the beads have been stained with acridine orange or reacted with FITC-labelled antibodies that recognize specifically the bound organisms for rapid presumptive identification under a fluorescence microscope. Bead–bacteria complexes have been post-selectively enriched to augment bacteria numbers and to enhance antigen production prior to detection by enzyme immunoassays. The concept of PCR and other DNA hybridization-based approaches have tended to dominate the more recent literature in view of the specific detection of the target organisms it confers. The magnetic immuno PCR assay (MIPA) uses PCR as the detection method after IMS. IMS enhances the sensitivity of MIPA or similar type assays by removing PCR inhibitory compounds from the sample. Furthermore, amplified fragments of nucleic acids prepared by MIPA have been detected by a method that involved a marriage between DNA technology and enzyme immunoassay known as DIANA (detection of immobilized amplified nucleic acids). In DIANA, the amplified biotinylated DNA is selectively bound by streptavidin-coated beads.

IMS has been combined with a commercial sensor for electrochemiluminescence (ECL) detection. The concept of ECL appears promising for rapid detection of enteric pathogenic bacteria from food and environmental samples. The total processing and assay time is estimated to be less than one hour, with a detection limit of approximately 10^3 cells per millilitre.

Advantages and Limitations

IMS has several advantages over other separation and concentration methods.

1. The target bacteria are separated from the source and from other bacteria not bound to the particles and may be concentrated by resuspension in a small volume suitable for subsequent cultivation or detection by other means.
2. Possible growth inhibitory substances in the sample are removed, thus facilitating cultivation. IMS is not a detection system, but may provide a significant time-saving for all existing methods based on selective enrichment.
3. It enables the flexibility of applying different end-point detection techniques. Depending on requirements, detection techniques including microscopy,

Table 9 Comparison of two IMS systems for concentration of *Cryptosporidium* oocysts from water (Data presented by Bukhari Z, McCuis RM, Fricker CR and Clancy JL (1997) AWWA. Water Quality Technology Conference. November 9–12. Denver, CO)

| Sample | % Recovery by IMS test system produced by: | | | |
| | Dynal | | Clearwater | |
	Replicate 1	Replicate 2	Replicate 1	Replicate 2
Deionized water	75	81	13	52
Surface water (500 NTU)	74	85	0	0.23

NTU, Nephelometric Turbidity Units.

plating, enzyme-linked immunosorbent assay (ELISA), polymerase chain reaction (PCR), and impedance methods can be used.

4. The technique is, however, limited by the requirement of antibodies directed against surface antigens of the organism.

5. The presence of high concentrations of dead or lysed cells in the sample, which would include free surface antigens would negatively affect retrieval of viable bacteria.

6. Although IMS is accepted as a reliable alternative to conventional selective enrichment protocols, perceived fears of cross-contamination exist due to the manual nature of the technique. The fact that antibodies are not absolutely specific, mean that some cross-reacting organisms may bind to the beads. Bacteria frequently attach non-specifically to surfaces and the problems of non-specific binding often arise.

7. Furthermore, inherent in the IMS procedure itself, is the entrapment of microorganisms within the aggregates formed during the concentrating step. Therefore, when not too specific detection methods are used after IMS, for example, plating onto agar media, the problem can become accentuated.

Future Perspectives

IMS is now recognized as an essential technique which despite its manual nature improves the detection of food-borne pathogens in a relatively short time. In the future, automation of the whole concept will resolve many of the perceived problems of cross-contamination and enhance the user-friendliness of the system. IMS would be fully integrated into all conceivable detection systems and adapted for continuous on-line screening and monitoring of processing plants designed to assure consumer food safety.

See also: **Bacteriophage-based Techniques for Detection of Foodborne Pathogens**. **Biophysical Techniques for Enhancing Microbiological Analysis**: Future Developments. **Biosensors**: Scope in Microbiological Analysis. ***Campylobacter***: Detection by Cultural and Modern Techniques; Detection by Latex Agglutination Techniques. ***Clostridium***: Detection of Enterotoxins of *C. perfringens*; Detection of Neurotoxins of *C. botulinum*. **Direct (and Indirect) Conductimetric/Impedimetric Techniques**: Food-borne pathogens. ***Escherichia coli* O157**: Detection by Latex Agglutination Techniques; Detection by Commercial Immunomagnetic Particle-based Assays. **Flow Cytometry**. **Fungi**: Food-borne Fungi – Estimation by Classical Culture Techniques. **Hydrophobic Grid Membrane Filter Techniques (HGMF)**. *Listeria*: Detection by Commercial Enzyme Immunoassays; Detection by Colorimetric DNA Hybridization; *Listeria monocytogenes* –

Detection by Chemiluminescent DNA Hybridization; *Listeria monocytogenes* – Detection using NASBA (an Isothermal Nucleic Acid Amplification System). **Molecular Biology – in Microbiological Analysis**. **Petrifilm – An Enhanced Cultural Technique**. **Polymer Technologies for Control of Bacterial Adhesion**. ***Salmonella***: Detection by Latex Agglutination Techniques; Detection by Enzyme Immunoassays; Detection by Colorimetric DNA Hybridization; Detection by Immunomagnetic Particle-based Assays. **Sampling Regimes & Statistical Evaluation of Microbiological Results**. ***Staphylococcus***: Detection of Staphylococcal Enterotoxins. **Verotoxigenic *E. coli***: Detection by Commercial Enzyme Immunoassays. ***Vibrio***: Detection by Cultural and Modern Techniques.

Further Reading

Avoyne C, Butin M, Delaval J and Bind J-L (1997) Detection of *Listeria* spp. in food samples by immunomagnetic capture: ListerScreen Method. *Journal of Food Protection* 60: 377–384.

Bolton FJ, Crozier L and Williamson JK (1996) Isolation of *Escherichia coli* O157 from raw meat products. *Letters in Applied Microbiology* 23: 317–321.

Bukhari Z, McCuin RM, Fricker CR and Clancy JL (1998) Immunomagnetic separation of *Cryptosporidium parvum* from source water samples of various turbidities. *Applied and Environmental Microbiology* 64: 4495–4499.

Campbell AT, Grøn B and Johnsen SE (1997) Immunomagnetic separation of *Cryptosporidium* oocysts from high turbidity water sample concentrates. In: Fricker CR, Clancy JL and Rochelle PA (eds) *Proceedings of the 1997 AWWA International Symposium on Waterborne Cryptosporidium*, p. 91–96.

Cudjoe KS, Thorsen LI, Sørensen T et al (1991) Detection of *Clostridium perfringens* type A enterotoxin in faecal and food samples using immunomagnetic separation (IMS)-ELISA. *International Journal of Food Microbiology* 12: 313–322.

Cudjoe KS, Hagtvedt T and Dainty R (1995) Immunomagnetic separation of *Salmonella* from foods and their detection using immunomagnetic particle (IMP)-ELISA. *International Journal of Food Microbiology* 27: 11–25.

Cudjoe KS and Krona R (1997) Detection of salmonella from raw food samples using Dynabeads anti-Salmonella and a conventional reference method. *International Journal of Food Microbiology* 37: 55–62.

Docherty L, Adams MR, Patel P and McFadden J (1996) The magnetic immunopolymerase chain reaction assay for the detection of *Campylobacter* in milk and poultry. *Letters in Applied Microbiology* 22: 288–292.

Fierens H, Eyers M, Bastyns K, Huyghebaert A and De Wachter R (1993) Rapid and sensitive isolation and detection of *Campylobacter* species in poultry products by using an immunomagnetic separation technique combined with the polymerase chain reaction (PCR). *Med. Fac. Landbouww. Univ. Gent* 58/4b: 1879–1884.

Hafeli U, Schütt W, Teller J and Zborowski M (eds) (1997) *Scientific and Clinical Applications of Magnetic Carriers*. New York: Plenum.

Heuvelink AE, Zwartkruis-Narhuis JTM and De Boer E (1997) Evaluation of media and test kits for the detection and isolation of *Escherichia coli* O157 from minced beef. *Journal of Food Protection* 60: 817–824.

Islam D and Lindberg AA (1992) Detection of *Shigella dysenteriae* type I and *Shigella flexneri* in feces by immunomagnetic isolation and polymerase chain reaction. *Journal of Clinical Microbiology* 30: 2801–2806.

Kapperud G, Vardun T, Skjerve E, Hornes E and Michaelsen TE (1993) Detection of pathogenic *Yersinia enterocolitica* in food and water by immunomagnetic separation, nested polymerase chain reactions, and colorimetric detection of amplified DNA. *Applied and Environmental Microbiology* 59: 2938–2944.

Kemshead JT (ed.) (1991) Magnetic separation techniques applied to cellular and molecular biology. *Proceedings of the First John Ugelstad Conference*. Christ Church, Oxford; Somerset, UK: Wordsmith's Conference Publications.

Kroll RG, Gilmour A and Sussman M (ed.) (1993) *New Techniques in Food and Beverage Microbiology*, p. 1. *The Society for Applied Bacteriology Technical Series* 31. Oxford: Blackwell Scientific.

Olsvik Ø, Popovic T, Skjerve E et al (1993) Magnetic separation techniques in diagnostic microbiology. *Clinical Microbiology Reviews* 7: 43–54.

Patel PD (1991) Advances in methods for rapid microbiological food analysis. *Food Technology Europe* 235–239.

Safarik M, Safaríková and Forsythe SJ (1995) The application of magnetic separation in applied microbiology. *Journal of Applied Bacteriology* 78: 575–585.

Skjerve E, Rørvik LM and Olsvik Ø (1990) Detection of *Listeria monocytogenes* in food by immunomagnetic separation. *Applied and Environmental Microbiology* 56: 3478–3481.

Stannard CJ, Patel PD, Haines SD and Gibs PA (1987) Magnetic Enzyme Immunoassay (MEIA) for Staphyloccocal Enterotoxin B. In: Grange JM, Fox A and Morgan NL (eds) *Immunological Techniques in Microbiology*, p. 59. *The Society for Applied Bacteriology Technical Series* no. 24. London: Blackwell Scientific Publications.

Toyomasu T (1992) Development of the immunomagnetic enrichment method selective for *Vibrio parahaemolyticus* serotype K and its application to food poisoning study. *Applied and Environmental Microbiology* 58: 2679–2682.

Uhlén M, Hornes E and Olsvik Ø (ed.) (1994) *Advances in Biomagnetic Separation*. Nartick, MA: Eaton Publishing.

Van der Wolf JM, van Beckhoven JRCM, de Vries PhM and van Vuurde JWL (1994) Verification of ELISA results by immunomagnetic isolation of antigens from extracts and analysis with SDS-PAGE and Western blotting, demonstrated by *Erwinia* spp. in potatoes. *Journal of Applied Bacteriology* 77: 160–168.

Wai-Ching L and Phang STW (1997) A comparison of conventional and Vicam immunomagnetic capture enrichment methods for the detection of *Salmonella* spp. in chicken litter. *Asian Pacific Journal of Molecular Biology and Biotechnology* 5: 94–97.

Widjoatmodjo MN, Fluit AC, Torensma R, Keller BHI and Verhoef J (1991) Evaluation of magnetic immuno PCR assay for rapid detection of *Salmonella*. *European Journal of Clinical Microbiology of Infectious Disease* 10: 935–938.

Yu H and Bruno JG (1996) Immunomagnetic–electrochemiluminescent detection of *Escherichia coli* O157 and *Salmonella typhimurium* in foods and environmental water samples. *Applied and Environmental Microbiology* 62: 587–592.

Inactivation Techniques *see* **Lasers:** Inactivation Techniques.

Indirect Conductimetric/Impedimetric Techniques *see* **Direct (and Indirect) Conductimetric/Impedimetric Techniques:** Food-borne Pathogens; Enterobacteriaceae, Coliforms and *E. coli*.

Industrial Fermentation *see* **Fermentation (Industrial):** Basic Considerations; Media for Industrial Fermentations; Control of Fermentation Conditions; Recovery of Metabolites; Production of Xanthan Gum; Production of Organic Acids; Production of Oils and Fatty Acids; Colours/Flavours Derived by Fermentation.

INTERMEDIATE MOISTURE FOODS

K Prabhakar, Department of Meat Science and Technology, College of Veterinary Science, Tirupati, India

Foods that can be preserved in a simple way at ambient temperatures and which are moist enough to be consumed without rehydration have traditionally come to be known as intermediate moisture (IM) foods. Their shelf-stability at ambient temperatures is mainly attributed to adjustment and control of water activity. Moisture content of IM foods varies between dried foods with levels of less than 7% which can be stored at room temperatures, and fresh foods with levels of 60–80% and above which need to be preserved by freezing, canning or other similar mechanisms. Usual ranges of water levels are between 10% and 50% and water activities (a_w) vary from 0.65 to 0.90. So IM foods can be classified as partially dehydrated foods with suitable concentrations of dissolved solids to inhibit the growth of bacteria, moulds and yeasts and to control undesirable enzymatic activity. Fruits, vegetables, fish and meats are successfully processed into 'intermediate moisture' range products.

In the preparation of IM foods, some water is removed from the fresh food and the availability of the rest of water to microbial growth is reduced by the addition of suitable solutes. Compounds added to foods for this purpose are termed 'humectants'. Humectants keep the food products moist because they allow adsorption of water and pass it on to the product, compensating for natural drying. Ideally, humectants should be harmless to the consumer, should not alter the normal character of the food product and must be highly soluble in water at ambient temperatures. They should preferably be chemically inert but be capable of being metabolized as a source of energy. Humectants commonly used in food manufacture are glycerol, sugars, propylene glycol, polyethylene glycol, polyhydric alcohols such as sorbitol, and salts such as sodium chloride and potassium chloride. Permissible chemical preservatives and antimycotic agents can be incorporated for enhancing stability of IM foods. However, all the humectants so far known to us fall short of these ideal requirements.

Principles Behind Formulation of IM Foods

For a proper understanding of the principles and problems of IM food manufacture, an insight is necessary into the following spoilage-controlling factors which delicately balance the two opposing requirements of IM food products – consumer safety and eating quality:

- water content
- water activity
- pH
- chemical preservatives and additives
- oxygen availability
- temperature of storage.

With proper control over these factors, growth of most of the spoilage and pathogenic microorganisms can be prevented. A judicious combination of factors has to be achieved to produce IM foods with optimum safety, higher yield and satisfactory eating quality.

Water Content

Water is needed for growth of microorganisms. Many preservation methods attempt to decrease the water content of foods to enhance their keeping quality. Dried meats are dehydrated to moisture levels of 7.0% and below to arrest growth of bacteria, moulds and yeasts, but with adverse effects on texture and palatability. The most tolerant bacteria require at least 18% of water to grow and moulds require about 13%.

Water is available in two forms in muscle tissue: bound water and free water. Out of about 75% of water in meats, bound water which is an integral component of the structure is reported to be present at levels of around 5–7%. The rest is free water held loosely within the muscle protein network and electrostatic forces between peptide chains. When the protein network is tightened, as in severe heating or cooking, free water is released as proteins become denatured and lose their ability to hold water. Processing losses will be higher in such situations and cooked meat products may taste dry. These adverse changes are dependent on the severity of temperature increases. Hence, it is preferable to withdraw only moderate amounts of water from foods during processing and to stabilize such foods with additives which will make the remaining water unavailable to the growth of microbes.

Water Activity

Although it is known that foods with higher moisture contents generally spoil more quickly microbiologically than those with lower moisture levels, the moisture content alone does not give a clear picture of microbiological stability. The chemical potential of water present in the food in relation to that of pure water at the same temperature and pressure, known as water activity (a_w), is generally accepted as a better indicator in this context. It is regarded as a measure of the availability of water for microbial growth, and plays an important role in the fabrication of IM foods.

Through manipulation of water activity, the osmoregulatory capacity of microorganisms and the osmotic stress in the food can be interfered with and the growth of microbes and the activity of most of the important enzymes of muscle can be arrested. Humectants when added to meats can bring about these changes. Lowered water activity is reported to make water unavailable for solubilization and transport of nutrients required for the growth of microorganisms. Denaturation due to increased osmotic concentration and movement of water towards the concentrated solution makes growth and multiplication of bacteria difficult. Each genus of microorganism has a specific water activity at which its growth is maximum. For reduction of water activity levels through humectants alone from 0.99 to 0.85, a safe water activity level desired in IM meats, higher levels (above 15–20%) have to be incorporated which adversely affect the palatability of meat products.

Several investigators observed correlations of variations in organoleptic quality parameters, enzymatic and microbial spoilage indicators with specific changes in water activity levels in meat products. Water activity values range between 0 and 1, and are around 1 (0.99) in fresh meats and 0.3–0.4 in dried meats. Though a_w is an important criterion in IM foods, it is realized that the nature of the water activity-controlling solute has also a role to play. Variations were observed in the effects of different solutes.

Acidity (pH)

The degree of acidity or alkalinity of any aqueous solution is expressed as units of pH. The muscle pH in vivo is around 7.0 (neutral), which is required for optimum enzymatic and metabolic activities. After slaughter of an animal, muscle glycogen content (which amounts to around 0.8–1.0%) is broken down by muscle enzymes leading to the accumulation of lactic acid, which reduces the muscle pH from 7.0 to around 6.4–6.0 in 6 h and to 5.6–5.7 in about 12 h at room temperature in tropical conditions. Subsequently protein breakdown starts due to the muscle's own cathepsin enzymes and bacterial enzymes which multiply and bring about spoilage. The pH gradually rises again from 5.6 to 6.4–6.8 with perceptible changes in colour, texture and odour, indicating the onset of spoilage.

A direct relationship is observed between pH (acidity) and keeping qualities of meats. Meats with higher pH values (6.0 and above) tend to spoil more quickly than those with lower pH values (5.5 to 5.6). Most bacteria have optimum pH values for growth at around 7.0 and minimum values at around 5.0. In the development of fermented meat products, lactic acid-producing bacteria are allowed to multiply under controlled conditions, so that an acidic pH (4.8 to 5.2) is produced which helps shelf-stability during subsequent storage at higher temperatures. A 'tangy' flavour is also produced in such foods, which are popular in Europe and other places. However, a low pH is not conducive to optimum conditions of other quality parameters such as water-holding capacity and juiciness. Lower pH values tend to denature muscle proteins and lead to a decrease in the water-binding properties. Processing losses will be higher in such conditions and meat products after cooking are likely to have less juiciness.

Chemical Preservatives and Additives

Chemical preservatives are substances which when incorporated into foods at effective levels inhibit specific groups of spoilage and pathogenic microorganisms through direct action, thus extending storage life. Some of the commonly used preservatives such as sodium chloride, sodium or potassium nitrite and weak organic acids were accidentally discovered by ancient civilizations and effectively used in combinations with processing procedures for preservation of surplus foods during periods of plenty. Their uses are among the cheapest of the food preservation methods. Growing urbanization, steadily increasing affluence and a current fad for convenience foods have necessitated mass production, transportation and marketing of ready-to-eat manufactured food products, even in developing countries. Market competition is compelling the creation of products which are appealing and superior in physical, nutritional and organoleptic qualities with enhanced shelf-life, and are within the reach of the vast majority of people. To improve functional properties and stability of these foods, additives such as antioxidants, flavour enhancers and chelating agents are being used in addition to preservatives.

In the manufacture of IM foods, approved chemical preservatives and additives in permitted doses are incorporated to enhance stability or improve quality. A present trend is to utilize the synergistic effects of sub-inhibitory doses of permitted preservatives so that inhibitory effects are maximum and any residual toxicity problems are minimized. The potential of proven, harmless and beneficial preservatives should be properly and judiciously explored, keeping the public health requirements and the needs of the processed food industries in mind. For IM foods, there is need to incorporate agents that give protection against fungi and yeasts.

Oxygen Availability

Oxygen is necessary for growth of aerobic bacteria. If meats are wrapped in vacuum packages, bacterial multiplication is significantly reduced. Residual oxygen within the package is utilized by microbes and carbon dioxide is produced which in turn inhibits growth of obligate aerobes, while encouraging facultative anaerobes such as *Lactobacillus*. The shelf life of chilled meat products in such packages may be around 20–30 days at 2–4°C. Similar advantages accrue if meats are packaged with 20% carbon dioxide and 80% nitrogen in gas-impermeable packages.

Though IM foods are supposed to be shelf-stable in simple packages, vacuum packaging as a hurdle against microbial growth can significantly reduce the levels of other inhibitory agents and enable optimum exploitation of synergistic effects. However, the cost-benefit effects have to be carefully analysed before a decision is taken.

Cooked, cured and sliced meats are marketed in vacuum packages to prevent oxidation of cured meat pigments and to improve keeping quality.

Temperature of Storage

Low temperatures retard microbial growth and enzymatic activity in foods and prolong keeping quality. In temperate regions IM meat products can be stored for long periods at ambient temperatures, which are normally 10–20°C or less, whereas in tropical areas shelf life is lower because ambient temperatures are higher (30–45°C). This has resulted in the historical development of succulent IM foods like cured and smoked meats in regions with cold climates. In tropical areas, IM meat products are more dry in nature as the demands on stability requirements are higher. There is a considerable energy saving if foods preserved by deep freezing (–18°C to –20°C) can be processed as IM foods which can be kept at refrigeration temperatures (2–4°C) with the same efficiency.

Traditional and Commercial IM Foods

Non-comminuted traditional IM meat delicacies of Europe include Bundnerfleisch, Rohschinken, Coppa and Speck. The spicy, dehydrated Dendeng is a traditional IM beef product with a characteristic flavour. It is prepared by curing with salt, sugar, spices and other ingredients and dried by heating or cooking over charcoal, or deep fat fried to a moisture level of 15–20% and an a_w of around 0.70. Sun-dried shrimp processed and sold as a snack item in the USA is claimed as an IM product because of its high salt content (25–27%) with an a_w of 0.70 to 0.75. Pastarma, an IM product made from beef or buffalo meat, is highly esteemed in Balkan countries. Thick slices cut from the big muscle groups are salted for 1–2 days, washed and compressed. A thick paste of garlic and other spices is applied to the slices and they are dried in hot air or sun-dried for a few days until 40–50% of the original weight is obtained. After ripening, the meat is cut into thin slices and eaten raw or lightly roasted. Cecina is an IM meat product of Mexico. Thin strips of muscles are brine-cured for a few hours and then sun-dried for several days to a_w levels of around 0.85.

Jerky is a popular IM product in the USA. Meat is cured in hot brine and later subjected to smoking over a low fire. It is mass-produced in humidity-controlled smokehouses to a_w levels of around 0.70 to 0.75. Country-cured hams are also popular in parts of the USA; the hams are processed by dry curing, and are stable at higher temperatures because of their high salt content and a_w values of 0.80 to 0.85. Precooked sliced bacon is another IM product popular in the USA.

Processing of fermented sausages in dry or semidry condition for shelf-stability at higher temperatures dates back to prehistoric times. Salami is processed to dry at 10–15°C and 75% relative humidity for about 2–3 months for a semidry variety, or 4–6 months for a dry variety. Cervelats are popular semidry sausages with a_w values around 0.85. Mortadella is an Italian sausage made with cured pork and beef with added cubes of back fat. It is fermented, smoked and then air-dried to a_w values of 0.85 to 0.88. Pepperoni is a dry sausage of Italian origin made from cured pork and beef. Pepper is used along with other spices, and the meat is dried to moisture levels of 20–22% and a_w values of 0.60 to 0.65. Low-acid sausages are usually dried without fermentation to reduce water activity. Some casserole-type products like chicken à la king and sweet and sour pork are processed to a_w values of 0.82 to 0.85. In spite of having a slightly sweetish taste, these items (which are usually consumed with specially prepared sauces) are popular as delicacies.

Some examples of commercial IM products, along with storage times claimed by the manufacturers, are:

1. Cold sausage with a moisture content of 40%, protein content of 16% and salt content of 3.8% which can be stored in a cool place for up to 14–21 days.
2. Ring sausage of Cracow with moisture content of 58%, fat 45%, protein 14% and salt 2.8% – storable in a cool place for up to 14–21 days.
3. Tourist cold sausage made of pork and beef with moisture content of 40%, protein 16% and salt 3.8% – storable in a cool place up to 14–21 days.
4. Sausage of Alps with moisture content of 40%, fat 42%, protein 17% and salt 3.8%. Storage period is 30 days in a cool place.
5. Trussed smoked ham or chopped smoked ham with a salt content of 7.0% which can be kept at 20°C for 21–28 days.
6. Cooked ham in foil with a salt content of 4% which can be stored at 2–4°C for about 6 months.
7. Sausage in vacuum foil which can be stored at 0°C for 50 days.

Fruits, vegetables, jams and some bakery products are also processed to intermediate moisture ranges by withdrawal of water and addition of sugar, salt, etc. Processing and marketing of such products do not pose serious problems as they are not highly perishable, unlike meat and fish. Besides, sweetness or saltiness is a natural taste to these products. Fruit bars are prepared from a combination of figs, dates, pears, cherries, raisins, etc. in a compressed state. They are meant for direct consumption from the package without rehydration.

Advantages of IM Foods

1. *Storage at ambient temperatures in simple packages:* if highly perishable foods can be processed for storage at ambient temperatures, they can revolutionize mass production and distribution of foods through marketing.
2. *Economical food preservation method:* IM foods can be fabricated with low-cost technology utilizing local climatic conditions.
3. *Energy conservation during storage:* once IM foods are fabricated for shelf-stability at ambient temperature, no energy is needed for subsequent storage.
4. *Feasibility in developing countries:* IM food processing is the obvious choice for developing regions which experience severe energy limitations with frequent supply interruptions.
5. *Convenience:* handling IM foods is easier as they can be stored in open shelves in rooms with good ventilation.
6. *Safety and effectiveness:* if foods are processed to the safe water activity level with suitable antimicrobial and antimycotic agents, they are safe for consumption. This technology needs refinement to utilize more natural antimicrobial systems and restrict incorporation of chemical preservatives and other additives.
7. *Efficiency in packaging:* IM foods do not need absolute protection as in the case of canned meats. They can be moulded or compressed into convenient shapes for maximum efficiency in packaging.
8. *Suitability for special situations and applications:* IM foods are convenient foods for space exploration, high-altitude military operations and other similar situations as they are of low bulk and weight and offer concentrated sources of nutrients. This is facilitated by removal of more than half of the total moisture present in the foods.
9. *Acceptability:* most IM foods have problems with palatability which is affecting their globalization. With improvements in processing methods they have the capability to dominate food markets in all parts of the world.
10. *Versatile processing-cum-preservation method:* food processors can innovate and use their creative zeal to produce new products with variations in appeal, texture and flavour.

Processing Technologies of IM Foods

Basic IM food processing involves addition of humectants and partial dehydration through mild heating and surface-drying to achieve the desired water activity and moisture levels in the end product. Successful IM food manufacture depends on the art of manipulating several hurdles to microbial and enzymatic degradations. Important points to consider in this regard are:

1. Rapid microbial spoilage is not likely to occur in foods with water activity at or below 0.85. Toxin production by *Staphylococcus aureus* stops at water activity of 0.86. Some yeasts and moulds may grow slowly at water activities just above 0.6.
2. In foods with water activity values above 0.85, pH plays an important role in the control of spoilage organisms. At pH 5.0 or below, their growth – except for desirable strains such as *Lactobacillus* and *Leuconostoc* – is suppressed. Many moulds and yeasts may grow at pH values of 3.0 or below to 8.5.

Depending on the importance given to processing steps, the resulting IM products can be grouped into three categories:

- glycerol (or similar humectant) and salt-based IM products
- cured and smoked meats
- fermented meats such as dry and semidry sausages.

Glycerol and Salt-based IM Products

Processing of glycerol–salt desorbed IM meats involves equilibration of meat cubes in infusion solutions at 1:1.5 ratio for 20–24 h at ambient temperatures followed by heating in ovens and drying under circulating air. The infusion solutions are prepared with lower levels of humectants and synergistic combinations of permitted chemical preservatives and antifungal agents to enable processing of safe and stable IM products. This also results in prevention or minimizing of undesirable cross-linking in proteins, non-enzymic browning, reduced palatability, etc., attributed to higher levels of humectants like glycerol. Processed samples are packaged for storage at ambient temperatures (25–45°C). They are usually cooked before consumption if heating schedules are shorter during processing.

Cured and Smoked Meats

Since cured and smoked meats can be classified as salt-based IM products, a brief description is included. Curing of meat chunks is done in brine solutions containing sodium chloride (10–20%), sodium or potassium nitrate (0.1%), sodium or potassium nitrite (0.01%), ascorbate (0.05%), etc. for 5–6 days at 4–5°C. Cured meats are held for maturation for 5–7 days at 4–5°C to enable even distribution of curing salts. Later, they are subjected to smoking-cum-heating in controlled smokehouses to reduce moisture levels and to impart a smoky flavour, besides fixing the desirable bright pink colour of cured meats. Processing schedules and brine compositions vary depending on the type of product being processed.

Nitrite has a specific inhibitory action on the spores of *Clostridium botulinum*, which is a potentially important pathogen in this class of meat products. Common salt, in addition to its direct inhibitory action, lowers water activity and contributes to the safety of cured and smoked meat products. There is a trend to decrease salt levels in cured products because of consumer demand for mild-cured products without compromise on safety and stability. Sliced cured and smoked meat products which are ready to eat can be kept under refrigerated conditions for 2–3 weeks but at higher ambient temperatures spoilage is noticed within 3–5 days. Their keeping quality can be enhanced by a few days if they are subjected to further air-drying or vacuum packaging. Though the commercial production of cured and smoked meats is a success story, it has not made a big impact in tropical countries because of cost and food habits.

Fermented Meat Products

Processing of fermented sausages is still an art. Semidry sausages are processed to ultimate moisture levels of around 50% and they can be kept at refrigeration temperatures (2–4°C) for several months. Dry sausages are processed to final moisture levels of 30–35%, They can be stored at higher ambient temperatures for 4–6 months. Salt contents usually range between 3.5% and 5.5% in commercial varieties. Though traditional processing procedures were refined with technological advances and automated process control equipment for uniform quality in the finished products, processing schedules vary depending on the intuition and imagination of the processors.

Meat is first subjected to coarse grinding. Curing salts at permitted levels, sugars (usually 2–10% level or higher) and combinations of spices carefully chosen for product-specific aroma and flavour are added to the meat and thoroughly mixed, before being stuffed into casings. The stuffed sausages are hung on sticks and held in the 'green-room' for 5–10 days at 20–25°C with a relative humidity of 75–80% for development of cured colour and initiation of fermentation. Sausage casings are punctured with small holes over the entire surface to allow entrapped gases (released due to fermentation) to escape. In the traditional methods, chance contaminants within the processing room were utilized for fermentation, which resulted in variable quality. Now, frozen-stored starter cultures are incorporated to achieve uniform, desired levels of acidity. Fermentation periods have been considerably reduced. The fermented sausages are next moved to drying rooms for further moisture removal under controlled conditions. Control of the drying step is very important because slow drying results in a soft surface of the sausages, whereas rapid drying results in hardening at the surface which hinders further evaporation of moisture from the interior and the product acquires a shrivelled appearance. The optimum drying rate should be sufficient to remove the moisture as it moves from the interior to the surface. A combination of 12–18°C room temperature, 65–75% relative humidity and 15–20 air changes per hour is reported to provide desirable drying rates. Individual processors arrive at effective drying rates suitable to specific product requirements through various combinations of the above three parameters. Uniform air distribution in the drying room is necessary to avoid formation of hollow spots which encourage growth of bacteria and fungi. Semidry sausages are usually dried for up to 25–50 days and dried products for up to 60–90 days.

Though semidry and dry sausages offer the convenience of higher-temperature storage and consumption without cooking, marketing of this class of products is reported to be limited even in affluent countries. Products are expensive because of high processing costs, due to prolonged processing times, the requirement for large processing spaces as the sausages are air-dried for several weeks to months, higher energy costs, and the higher risks involved. However, this form of preservation is an attractive proposition in developing countries and tropical regions, where low-cost processing with simple equipment to produce safe products is desirable. Non-uniformity in the composition of the finished products does not matter much, but the food products should be developed to the tastes favoured by people of tropical regions.

The Future

Studies are needed to develop IM meat foods with sodium chloride levels in the final products not exceeding 5–6%, through incorporation of additional hurdles to microbial and enzymic spoilage such as physical barriers, naturally occurring antimicrobial agents, or atmosphere and direct and indirect effects of microbial competition. The IM food processing technology has to be exploited to the fullest extent for food preservation and 'readily consumable product' development suitable for ambient temperature storage.

See also: **Confectionary Products**: Cakes and pastries. **Ecology of Bacteria and Fungi in Foods**: Influence of Available Water; Influence of Temperature; Influence of Redox Potential and pH. **Fermented Foods**: Fermented Meat Products. **Meat and Poultry**: Spoilage of Meat; Curing of Meat; Spoilage of Cooked Meats and Meat Products. **Packaging of Foods**: Packaging of Solids and Liquids. **Preservatives**: Classification and Properties; Traditional Preservatives – Sodium Chloride; Permitted Preservatives – Nitrate and Nitrite.

Further Reading

Burke CS (1980) International Legislation. In: Tilbury RH (ed.) *Developments in Food Preservatives.* P.25. London: Applied Science Publishers.

Campbell-Platt G (1995) Fermented meats – a world perspective. In: Campbell-Platt G and Cook PE (eds) *Fermented Meats* P. 39. Glasgow: Blackie.

Chang SF, Huang TC and Pearson AM (1996) Control of the dehydration process in production of intermediate moisture meat products. a review. *Adv Food Nutr Res* 39: 71–160.

Heidelbaugh ND and Karel M (1975) Intermediate moisture food technology. In: Goldblith SA, Rey L and Rothmayer WW (eds) *Freeze Drying and Advanced Food Technology.* P. 619. New York: Academic Press.

Karel M (1976) Technology and application of new intermediate moisture foods. In: Davies R, Birch GG and Parker KJ (eds) *Intermediate Moisture Foods.* P. 4. London: Applied Science Publishers.

Labuza TP (1980) Water activity: physical and chemical properties. In: Linko P, Melkki Y, Olkku J and Larinkari J (eds) *Food Process Engineering.* P. 320. London: Applied Science Publishers.

Ledward DA (1981) Intermediate moisture meats. In: Ralston Lawrie R (ed.) *Developments in Meat Science.* Vol. 2, p. 159. Oxford: Pergamon Press.

Leistner L (1995) Principles and applications of hurdle technology. In: Gould GW (ed.) *New Methods of Food Preservation.* P. 1. Glasgow: Blackie.

Leistner L (1995) Stable and safe fermented sausages worldwide. In: Campbell-Platt G and Cook PE (eds) *Fermented Meats.* P. 160. Glasgow: Blackie.

Leistner L and Rodel W (1976) The stability of intermediate moisture foods with respect to micro-organisms. In: Davies R, Birch GG and Parker KJ (eds) *Intermediate Moisture Foods.* P. 120. London: Applied Science Publishers.

Smith JL and Pintauro ND (1980) New preservatives and future trends. In: Tilbury RH (ed) *Developments in Food Preservatives.* P. 137. London: Applied Science Publishers.

Taylor RJ (1980) *Food Additives.* Chichester: John Wiley.

Tilbury RH (1980) Introduction. In: Tilbury RH (ed.) *Developments in Food Preservatives.* P. 1. London: Applied Science Publishers.

INTERNATIONAL CONTROL OF MICROBIOLOGY

B Pourkomailian, Department of Food Safety and Preservation, Leatherhead Food RA, Surrey, UK

Introduction

The control, regulation and inspection of food has one primary objective: to ensure that the consumer receives foods and food products that are of the quality claimed, healthful and pure. With this aim, enforcement and control agencies (ranging from international to private) have been set up to regulate food industry processes. In order to be able to achieve this clearly and efficiently, microbiological standards have been established by the regulatory and other agencies; these standards are set according to the microbiological criterion associated with the specific food under review.

Enforcement and Control Agencies and Their Role

Enforcement and control agencies have an important role in international, regional and governmental control of microbiology in the food sector. The food industry joins the regulatory bodies in assuring the safety and quality of product which reaches the consumer. These two bodies are the major groups interested in establishing controlling criteria for the production of safe and good-quality foods. Although both the regulatory agencies and the food industry have the same final goal, they approach the problem with different incentives. The regulatory agencies have to assure food safety and quality in order to fulfil their statutory responsibility to protect the public from hazardous or inferior-quality foods. It must be noted that the regulatory authorities can only operate within the food laws of that country. Commercial food companies have different responsibilities, and those are to their shareholders. By producing safe, high-quality products, they ensure an increase in their market size and enhance their good reputation.

There are many national, regional and international

enforcement and control agencies; the major agencies involved in food microbiology control are listed below.

International Agencies

The *Association of Official Analytical Chemists* (AOAC) has been responsible for studies on sampling plans and laboratory methodology.

The *Food and Agricultural Organization* (FAO) is primarily concerned with food production through improved methods of production, processing, preservation and distribution of foods.

The *International Children's Emergency Fund* (UNICEF) is concerned with international trade and safety of food.

The *International Commission on Microbiological Specifications for Food* (ICMSF) is a voluntary body which has as its primary objective the establishment of international sampling plans and methods for analysis.

The *International Dairy Federation* (IDF) has been responsible for studies on sampling plans and laboratory methodology.

The *World Health Organization* (WHO) is concerned with the health of the consumer and the maintenance of food wholesomeness.

Federal Agencies

One of the most important national agencies in the USA is the *Food and Drug Administration* (FDA); this organization is responsible for ensuring that food is safe and wholesome and that all foods are labelled informatively and honestly.

The *National Marine Fisheries Service* (NMFS) is a free service fisheries products inspection programme which is linked to the USDA in establishing microbiological criteria for food.

The *United States Department of Agriculture* (USDA) has the legal authority to promote the marketing of safe and high-quality agricultural products.

The *United States Army Natick Research and Development Center* is involved in establishing microbiological criteria and solving microbiological problems in military ration development.

Others

Commercial Some food associations or institutes make their own recommendations for their own industries, e.g. the National Food Processors Association specifies microbiological standards for sugar and starch for the canning industry.

Cooperative Programmes Certain food industries voluntarily cooperate with regulatory agencies for the standardization of regulations, e.g. the National Shellfish Sanitation Program.

Private Private agencies approve and list tested foods, e.g. the Good Housekeeping Institute.

Processing Industry Food processing companies may establish their own microbiological standards.

Professional Societies Recommendations on methods for the microbiological examination of foods have been published by societies such as the American Public Health Association (APHA).

State Some states in the USA have their own microbiological standards or guidelines for foods. These are often enforced through a state department of public health, agriculture or sanitary engineering.

Standardization

Standardization can be defined as the activity of establishing specifications for common and repeated use that aim to achieve the optimum degree of order in a given context, relating to actual or potential problems. Activities that require particular attention are the processes of formulating, issuing and implementing standards. The importance of standardization lies in the benefits it brings for the improvement in suitability of products, processes and services for their intended purposes. Standardization will bring down the barriers to trade and facilitate technological cooperation.

The Aims of International Standardization

International standardization aims to achieve various targets. Primarily it is implemented in order to promote the quality of product, processes and services. This can be achieved by defining those features and characteristics that control their ability to satisfy given needs. These needs are those that illustrate their fitness for purpose. Safety and health are a major concern of the food regulatory agencies and hence another aim of standardization is to promote improvement in the safety and quality of food. In all forms of processes, it has been noted that clear communication between interested parties must be established for progress to begin, so a third aim of standardization is to promote clear and unambiguous communication between all interested parties. This must be in a form suitable for reference or quotation in legally binding documents. Different countries, regions and continents have their own particular operational practices; in some cases these may overlap, but mostly they differ and result in trading barriers. Setting international standards would promote international trade by the removal of barriers caused by these different practices. Also, when systematic operations are not in place and tests and procedures vary from day to day, inefficiency will increase. The control of variety will improve industrial efficiency. The final aim of standardization is to ensure the economic use of materials and human resources in the production of foods.

Principles of Standardization

Standardization involves both the preparation and the use of standards. Therefore, several key issues, principles, must be considered before commencing with setting standards. Primarily, the standards must be wanted by all interested parties. All parties involved must be willing and agree voluntarily among themselves for one or more stated purposes. Secondly, all parties involved with the standardization must be certain that the standard will be used; it is of little value if the standard is published but not exploited. The setting and use of the standard are solely dependent upon the voluntary commitment of those setting the standards. When setting standards, the intended use and application must be clearly understood throughout the preparation. It must be noted that different types of standards are written in different ways for particular purposes. This can be illustrated when writing standards for specifications for products, materials, processes or services. These are in many ways different in layout from written standards for codes of practice (recommendations governing actions). Different still is the manner in which various kinds of methods and glossaries of terms are written. The written standards must allow the efficient and easy retrieval of information; the text should be clear, concise and unambiguous as well as being well arranged, indexed and referenced. It should always be possible to verify consistency with specific requirements within a realistic time and at a reasonable cost. The next principle of standardization is planning. Planning of standards involves many factors. The benefits, both economic and social, should be compared with the total costs that would arise from the preparation, publication and monitoring of the standards. The time scale for the completion of the proposed standard is also a matter for concern. The selection committee must consider whether it would be possible to complete the standard, to an acceptable level both technically and commercially, in time to be of use. Standards must be planned with foresight, otherwise by the time the standard is ready for implementation the technology may have developed further, rendering the proposal irrelevant. Therefore, the selection committee must be careful that their selected criteria will allow the standard to be useful at the time of its intended use. A standard will, in any event, contain only those criteria that the selection committee agreed at the time of proposal to include.

An important issue is pinpointed in areas of rapid development, where setting standards too late may be a cause of increased costs of any subsequent re-standardization. It is therefore necessary to have a clear understanding of the suitable timing for the application of the proposed standard. In addition, standards must be regularly reviewed and any necessary changes and modifications made to the original specifications. Failure to do so would lead to the standard becoming irrelevant and/or inhibit progress.

Another principle of standardization is the assurance that standards are not duplicated. A common occurrence in many establishments is the duplication of work. Since standards are proposed at various levels (individuals, firms, associations, countries, regions and worldwide) it is inevitable for some duplication to occur. Ideally, standards should be proposed and set at the widest level – logically worldwide – and should be consistent with satisfying the needs of all interested parties in an acceptable time scale. The simultaneous preparation of standards at different levels for the same subject often occurs. It is therefore essential to research previously established standards for the specific area where a standard proposal is to be made. This has to be investigated at all levels, in one's own and other authorities. It will be the aim of international standardization to adopt as international standards harmonized documents that are ideally identical or at least are technically equivalent, in each country.

Finally, the last principle of standardization is the application of unbiased standards. Since standards are to be established for the harmonization of whole communities, they should not be targeted towards a specific group. Standards should ideally be suited to use by worldwide suppliers and not exclusive to an individual.

International Organizations Concerned with Standards

The *International Organization for Standardization* (ISO) was founded in 1947. It comprises national standards bodies of 75 countries and 16 correspondent members, and is made up of over 160 technical committees, 650 subcommittees and 1500 working groups. More than 6000 ISO standards have been published. The ISO also provides the secretariat of the International Federation for the Application of Standards (IFAN), comprising official standards user bodies recognized by their national standards bodies.

Other organizations include the *Codex Alimentarius Commission*, which was created to implement the joint FAO–WHO Food Standards Programme; the *European Economic Community* (EEC), the 'Common Market' founded under the Treaty of Rome, 1957; and the *FAO–WHO Food Standards Commission*, which is a forum for the cooperation among nations for the development of or agreement on the various international standards for the food industry.

Microbiological Criteria (Specifications)

Improving the safety and quality of foods requires a greater understanding of microbiological criteria. These are the necessary microbiological standards which are the basis for sound judgments on the safety and quality of the food. Setting up such criteria will inevitably affect certain sectors in the food industry. The implementation of microbiological limits has and always will have a great impact upon the processes required for the production of the food in the food industry. It is therefore important to address the issues that concern microbiological criteria and understand their meaning. The need for implementation and use of a microbiological criterion can clearly be demonstrated in all food sectors. It is clear that food safety and quality are issues that are directly associated with the presence of microorganisms in the food. The presence and possible growth of pathogens and the presence of their toxins may lead to food poisoning, which is a safety issue. Also related to safety is the extent to which one can achieve the control or destruction of the food's microflora. Food quality issues, however, are associated with the presence of non-pathogenic microorganisms and their growth. The control and destruction is also related to food quality as is the lack of good hygienic practices.

It is, therefore, by the application of a microbiological criterion that the acceptability or unacceptability of a food can be established from the handling of the product to its processing and through to the final product. In setting microbiological specifications, contradictory advice is often inevitable.

Contradictions in Specifications   The production of foods and food products that are both safe and of good quality requires the assignment of microbiological limits. In the early part of the twentieth century microbial limits for some foods were suggested – one of the widely accepted limits was set for pasteurized milk – and ever since researchers have reviewed safety and quality issues of food with reference to microbial limits. Different research bodies have issued different specifications for the safe microbial load of foods; this arose as a result of the ever-increasing number of regulatory agencies which grew with the number of food companies. As a result, so did the numbers of unnecessary and badly chosen criteria. This became a matter of growing concern to those knowledgeable in setting microbiological specifications. In some cases the limits were unnecessarily stringent; in others the processing regimes implemented were inadequate for the product. Also there has been direct conflict between the requirements of different national food legislations. These differing requirements led to the demonstration of serious non-tariff obstacles to trade in the world food market. As a direct response to this problem, the FAO/WHO Expert Consultation, which was appointed at the request of Codex Committee on Food Hygiene, set out to establish a modified version of the document originally issued by the *Codex Alimentarius* Committee, voicing their concern on this problem. The *Codex Alimentarius* Committee later issued a modified version of the document which set out to achieve six targets:

- to identify the components of a microbiological criterion
- to establish the definitions of a microbiological criterion
- to identify the purpose of a microbiological criterion
- to identify the fulfilment factors necessary for the purpose of a microbiological criterion
- to establish the reasoning behind the selection of the components of a microbiological criterion
- to state the actions that are necessary to be implemented, based on the results acquired after the application of the criterion

In order to be able to proceed with setting a microbiological criterion, the statements and terms to be used must be defined.

Definitions   In order to clarify, harmonize and create an international consensus, the *Codex Alimentarius* Commission established definitions, which were modified by the ICMSF and endorsed. From the definitions established by the *Codex Alimentarius* Commission and ICMSF the following list has been derived:

Mandatory Criterion (microbiological standards)   These are specific microbiological standards, which the ICMSF regards as part of a law or regulation to be enforced by a control agency, for limits of pathogenic microorganisms which may be of concern to public health; however, non-pathogenic microorganism limits may also be set under this category.

Advisory Criterion   This can be sectioned into two possible categories contained in the code of practice:

- a microbiological end-product specification intended to increase assurance that hygienic significance has been met
- a microbiological guideline that is applied in a food

establishment at a point during or after processing to monitor hygiene.

The ICMSF and the Codex have also included other microbiological criterion statements which further expand the definitions:

- a statement specifying the type of food under review
- a statement about the microorganisms of concern and/or their toxins
- a statement detailing the analytical methods to be used to identify the microorganisms and toxins that may be present
- details of the plans in place for sampling (this would include the time and the location of sample removal)
- a statement considering the microbiological limits which would be appropriate for the food of concern and the number of samples required to conform with these limits.

Microbiological criterion use necessitates the incorporation of these definitions in the sampling plan. However, before the sampling plan can be established some areas of importance have to be considered as discussed below.

Considerations In order to give recommendations on microbiological criteria, one must have available all relevant information. Primarily, it is necessary to define the exact constituents of the food. Foods and food products vary in composition dramatically and this in itself plays a major part in establishing the correct criterion. The natural or contaminating microorganisms and their behaviour are to a great extent influenced by the surrounding conditions, i.e. those of the food product. Information on the microorganisms that may be present and could proliferate in such systems are a key issue. Another area where information is required is the process that a product endures. It may be that all safety and quality issues that may have arisen owing to the presence of microorganisms have been eliminated as a result of final product processing.

The selection of standards, product specifications and guidelines is not an easy task and therefore it is essential to consider all information before assigning microbiological criteria. It is generally accepted that in order to be able to assign correct microbiological criteria, it is necessary to (1) identify any possible evidence of hazard to health; (2) gather all information on the microbiology of the raw materials used; (3) clearly understand the effects on the microbiology of the food as a result of processing regimes that are in place; (4) establish the possibilities of microbial contamination and the subsequent consequences of their presence and/or possible growth in the product during its handling and storage; (5) be able to identify

consumers who may be at risk; and (6) establish the costs of implementing a criterion in relation to its benefits.

Another area to consider is the categorization of microorganisms in relation to the potential risk. This risk category may vary depending on the product in question. Particular microorganisms will behave significantly differently in different foods. Hence, the significance of a pathogen varies depending on the food; its presence in a particular food may or may not be significant. In a microbial criterion, only pathogens that are significant to the specific food are included. To assist in the selection and assigning significance, one must turn to annual summaries, periodical reports, reviews and textbooks dealing with food-borne diseases.

It is also necessary to be knowledgeable about the current methodologies used for the detection and enumeration of microorganisms. The assurance of the correct methodology can be ascertained from standard methods that have been validated by several established and accepted international standards organizations. It is essential that the methods used be reviewed regularly and modified when necessary. Failure to do so may lead to the method becoming irrelevant or inhibitory to progress.

The above definitions and considerations are all captured within the sampling plan.

Sampling Plan The sampling plan is a procedure for the appropriate examination, that should be carried out on the product, using the required number of samples and using a specific method. This can proceed under two different plans, a 'two-class' plan or a 'three-class' plan. A two-class plan is one that categorizes a product as acceptable or unacceptable. A three-class plan distinguishes between acceptable, marginally acceptable and unacceptable product. The specifications used by both plans are symbolized by n, m and c, plus M for the three-class plan:

n the total number of samples to be taken from the product lot

m a maximum count of microorganisms per gram; often given the value 0 for a two-class plan (presence or absence), and a non-zero value for a three-class plan; it is used to distinguish between acceptable and marginally acceptable product in a three-class plan

c the maximum number of samples that may show the microbiological specification designated by m;

M a maximum count of microorganisms that if exceeded in any of the samples (sample numbers specified by n) would lead to the rejection of product lot; this is only used in a three-class plan to distinguish between an acceptable product and a marginally acceptable product.

When applying a two-class plan a product may be assigned as either acceptable or unacceptable (see examples 1 and 2 below).

Example 1

$n = 5$; $c = 0$; $m = 0$

If of the five samples tested none shows the presence of microorganisms tested for, the product is acceptable.

Example 2

$n = 5$; $c = 2$; $m = 10^3$

If of the five samples tested two show the presence of microorganisms below 10^3 per gram of product, the product is acceptable. However, if more than two samples show the presence of microorganisms or one of the samples exceeds 10^3 per gram of product, then the product lot is rejected.

In a three-class plan M is also used (example 3).

Example 3

$n = 5$; $c = 3$; $m = 10^5$

The two-class plan criteria apply; in addition, if of the five samples tested any one sample shows a count higher than M, then the product lot is rejected. If three (c) or fewer of the samples show counts between m and M, the product lot is accepted (marginally).

A two-class plan is used to establish the acceptability or unacceptability of a product. A three-class plan is used for the enumeration of microorganisms present in the product and for assigning acceptability, marginal acceptability and unacceptability to the product. The plans can be used to assign probabilities to acceptability by using the numbers of n and c.

Future Development

To ensure more successful control of microbiology internationally, it is best to use microbiological criteria as part of a comprehensive programme, such as hazard analysis and critical control point (HACCP). It has been pointed out by various researchers that such a combination leads to increased numbers of safe and acceptable-quality products.

See also: **Hazard Appraisal (HACCP)**: The Overall Concept. **National Legislation, Guidleines & Standards Governing Microbiology**: Canada; European Union; Japan. **Sampling Regimes & Statistical Evaluation of Microbiological Results**.

Further Reading

Adams RM and Moss OM (1997) *Food Microbiology* P. 323. Cambridge: Royal Society of Chemistry.

Codex Alimentarius Commission (1981) Codex Alimentarius Commission, Fourteenth Session, 1981: *Report of 17th Session of the Codex Committee on Food Hygiene, 17–21 November 1980*. ALINORM 81/13. Washington: Codex.

Elliot PH and Michener DH (1961) Microbiological standards and handling codes for chilled and frozen foods: a review. *Applied Microbiology* 9: 452–468.

Frazier CW and Westhoff CD (1988) *Food Microbiology*, 4th edn. P. 495. New York: McGraw-Hill.

ICMSF (1986) *Micro-organisms in Foods*. Vol. 2. *Sampling for Microbiological Analysis: Principals and Specific Applications*, 2nd edn. Toronto: University of Toronto Press.

International Standards Organization (1983) *Catalogue*. Geneva: ISO.

Jay MJ (1996) *Modern Food Microbiology*, 5th edn. P. 417. London: Chapman & Hall.

Miskimin KD, Berkowitz AK, Solberg M et al (1976) Relationship between indicator organisms and specific pathogens in potentially hazardous foods. *Journal of Food Science* 41: 1001–1006.

Sanders BRT (ed.) (1972) *The Aims and Principles of Standardisation*. Geneva: ISO.

Solberg M, Miskimin DK, Martin BA, Page G, Goldner S and Libfeld M (1977) Indicator organisms, food borne pathogens and food safety. *Association for Food and Drugs Official Quarterly Bulletin* 41(1): 9–21.

KLEBSIELLA

P T Vanhooren, **S De Baets**, **G Bruggeman** and **E J Vandamme**, Department of Biochemical and Microbial Technology, University of Gent, Belgium

Introduction

In the ninth edition of *Bergey's Manual of Determinative Bacteriology*, the genus *Klebsiella* is taxonomically classified in subgroup 1: family *Enterobacteriaceae*, which belongs to group 5, describing the facultatively anaerobic Gram-negative rods. *Klebsiella* bacteria occur worldwide and are commonly found in soil, carbohydrate-rich waste water, effluents from paper, pulp and textile mills, cooling water, aerosols, woodland, on fruits, sugar cane, grains and on vegetables such as radish, lettuce, carrots and tomatoes. Drinking water reservoirs, made from redwood (Californian sequoia), also harbour *Klebsiella* sp. and it is a frequent contaminant of water supplies in general. These bacteria also occur in the human respiratory tract, in faeces and in clinical specimens. As such, they can contaminate food and contribute to disease and spoilage. Certain species (*K. pneumoniae*, *K. oxytoca* and occasionally others) are opportunistic pathogens. They are frequently the cause of hospital acquired (nosocomial) infections in urological, neonatal, intensive care and geriatric patients. Often, they display multiple antibiotic resistance and resistance to biocides. Some strains could be exploited in industrial fermentation processes for the synthesis of (fine) chemicals.

Characteristics of the Genus *Klebsiella* and Relevant Species

General Physiological Properties of *Klebsiella* Strains

Klebsiella is a non-spore-forming, non-motile, facultative anaerobic Gram-negative straight rod, 0.3–1.0 µm in diameter and 0.6–6.0 µm in length. The rods are arranged singly, in pairs or in short chains. The cells are capsulated. The optimal temperature for growth is 37°C.

Klebsiella sp. are chemoorganotrophic, having both a respiratory and a fermentative type of metabolism. Glucose is fermented with the production of acid and gas (more CO_2 is produced than H_2). Most strains produce 2,3-butanediol as a major end product of glucose fermentation, whereas lactic acid, acetic acid and formic acid are formed in smaller amounts and ethanol in larger amounts than in typical mixed acid bacterial fermentations. No special growth factor requirements are known. Some strains have the ability to fix molecular nitrogen under anaerobic conditions.

Klebsiella strains may be lysogenic, but phages have been isolated from stools and sewage and used in phage typing. Many *Klebsiella* strains produce bacteriocins, klebecin being an example.

Klebsiella species are cytochrome oxidase negative and catalase positive. Other typical Enterobacteriaceae taxonomical tests vary among the species: indole, methyl red, Voges–Proskauer and Simmons citrate. They are usually lysine decarboxylase positive and ornithine decarboxylase negative. Several species hydrolyse urea and most reduce nitrates (except *K. pneumoniae* subsp. *ozaenae.*) They are arginine dihydrolase negative and H_2S is not produced. Most species ferment all commonly tested carbohydrates, except dulcitol and erythritol; they also grow in the presence of KCN. The mol% G+C of the DNA is 53–58 (T_m).

Successful genetic recombinations have been reported in *Klebsiella* and *K. pneumoniae* has been used by several workers for detailed genetic analysis of the genes involved in N_2-fixation (the *Nif*-genes). These genes are clustered near the His-region in the genome, but can be mobilized and transferred to other Gram-negative bacteria. Strains from clinical origin or those from nosocomial infections harbour R-plasmids, which determine resistance to a variety of antibiotics, such as penicillins, cephalosporins, aminoglycosides, tetracyclines, chloramphenicol, sulphonamides and trimethoprim. They are good recipients of R-plasmids, making them a culprit in serious nosocomial epidemic diseases. The cell wall of *Klebsiella* consists of the typical Gram-negative bacterial type with two distinct layers; the peptidoglycan layer and the 'outer membrane layer'. The latter consists of a complex of lipopolysaccharide (LPS), phospholipid and protein. The LPS structure determines the O-antigen type, of which a few examples are given in **Figure 1**.

The *Klebsiella* Polysaccharide Capsule

A polysaccharide capsule surrounds the cell wall structure of *Klebsiella* species and makes up the K-antigen. Serological typing is based on the examination of the K-antigens, since the number of O-antigen types is lower than that of the K-antigen types and also because O-antigen determination can be masked by the heat-stable K-antigens.

Especially, *K. pneumoniae* and *K. oxytoca* have a thick polysaccharide capsule, which gives rise to large mucoid colonies, especially when grown on carbohydrate rich agar media.

The K-antigens of *Klebsiella* bacteria have been divided into 82 different types and most of these capsular polysaccharides have been characterized. The K-antigens are constituted of repeating units. Most K-antigens contain only one charged monosaccharide, either D-glucuronic or D-galacturonic acid, and two to four of the following monosaccharides: D-glucose, D-mannose, D-galactose, L-rhamnose, L-fucose. In addition, non-carbohydrate constituents such as acetate, pyruvate or succinate can occur in the capsules. For example, types 1, 6, 16, 54, 58 and 63 contain L-fucose; their structure is represented in **Figure 2**.

Klebsiella strains with the capsular (K-antigen) types 1, 2 and 3 are a major cause of respiratory tract infections (pneumonia), whereas those with the capsular (K-antigen) types 8, 9, 10 and 24 are involved in infections of the urinary tract.

Klebsiella Strain Differentiation Methodology

Because of the ubiquitous distribution and the (opportunistic) pathogenicity of *Klebsiella* sp., scientists and field workers have become increasingly aware of the potential health hazard and the need for monitoring the organism in environmental, food and clinical sources. Its clinical significance has been well established, but its potential as a public health problem in natural, industrial and food environments needs more attention. In this respect, the need for a quick, cheap and reliable methodology to screen,

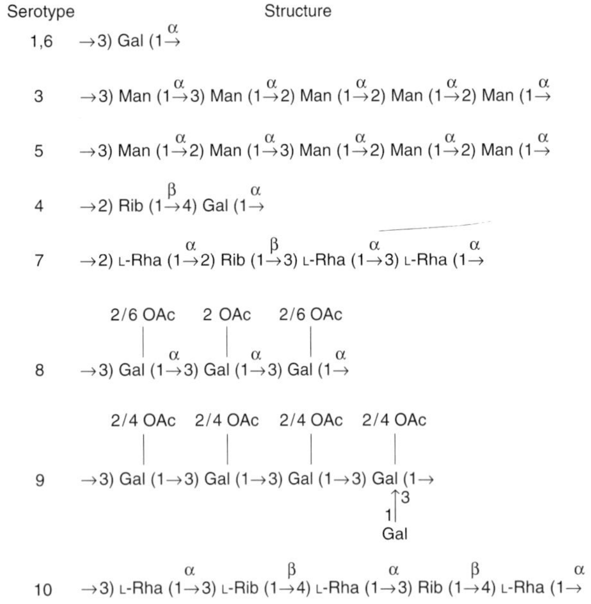

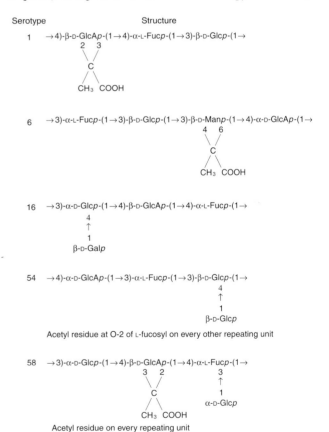

Figure 1 Repetitive units (O-antigen) of the O-specific polysaccharide part of the lipopolysaccharide of *Klebsiella* species. Redrawn from Ørskov and Ørskov (1984).

Figure 2 Structure of the L-fucose-containing tri- and tetrasaccharide repeating units of *Klebsiella* types 1, 6, 16, 54, 58 and 63.

culture and identify *Klebsiella* to the species and sub-species level remains a goal to be achieved.

Within the genus *Klebsiella*, the following species are currently distinguished: *K. oxytoca*; *K. planticola*; *K. pneumoniae* subsp. *pneumoniae*; *K. pneumoniae* subsp. *ozaenae*; *K. pneumoniae* subsp. *rhinoscleromatis*; *K. terrigena*.

K. pneumoniae subsp. *pneumoniae* is considered as the type species. As to the differentiation of the family Enterobacteriaceae from other families and genera, classified in Bergey's group 5, the reader is referred to the ninth edition of *Bergey's Manual*.

As to the differentiation of about 30 genera (> 115 species and subspecies) in the family Enterobacteriaceae, about 50 differential biochemical characteristics can be verified; these can also be found in *Bergey's Manual*. Due to the large number of data to be compared in such an identification attempt, one tries to identify members of this family directly to the genus or species level using the above battery of about 50 tests. Additional tests are then available to differentiate between certain species and subspecies; for *Klebsiella* species/subspecies, these are summarized in **Table 1**.

The usefulness of the pyrrolidonyl-arylamidase activity test to differentiate among Enterobacteriaceae and non-fermentative Gram-negative rods has also been studied. Positive results were uniformly obtained with *Citrobacter*, *Klebsiella*, *Enterobacter* and *Serratia* species. Negative results were displayed by *Escherichia coli*, *Proteus*, *Salmonella*, *Shigella*, *Pseudomonas* and *Flavobacterium* species, indicating its value as a complementary differentiation test.

The Biolog System has been evaluated for the identification of 55 Gram-negative taxa (789 strains), likely to be encountered commonly in clinical laboratories. It performed best with oxidase-positive fermenters, but although for 39 of the 55 taxa an identification rate of 70% was achieved, problems were encountered, particularly with the identification of capsulated strains of *Klebsiella*.

The new BBL CrystalEnteric/Non Fermenter (Crystal, Becton Dickinson Microbiology Systems) identification system for Gram-negative rods has been compared with the well-known API 20 E or API 20 NE (Bio-Merieux) system. More than 100 clinical isolates were studied, including six *K. oxytoca* strains and 12 *K. pneumoniae* strains; it was concluded that the Becton Dickinson Crystal test allowed a quicker, easier and more accurate identification of Gram-negative clinical isolates compared to the API system.

A remaining problem is to distinguish *K. pneumoniae* strains from non-motile *Enterobacter aerogenes* strains; however, the latter liquefy gelatin very slowly and are urease negative.

Serological tests based on the K-antigen can be used to confirm the identification results. A total of 62 serotypes among 72 different serological *Klebsiella* strains could be distinguished, based on a unique agglutination pattern with plant lectins.

Based on an extensive survey of about 160 strains of *Bacillus*, lactic acid bacteria, Enterobacteriaceae and *Staphylococcus*, it can be concluded that gas chromatographic analysis of cellular fatty acid composition was not sufficiently specific to the species level in several cases; characterization of food-borne bacteria, including *Klebsiella* sp. by the analysis of their cellular fatty acids should thus only be used to complement other taxonomical methods.

Recently, several molecular identification techniques have been proposed for a wide range of medically and food-related bacteria, including *Klebsiella* sp.

A fluorescence-based polymerase chain reaction–single strand conformation polymorphism (PCR–SSCP) analysis of the 16S rRNA gene has been described to identify a broad range of Gram-positive and Gram-negative bacteria: 178 bacterial strains, representing 51 species in 21 genera were examined. All strains gave species-specific patterns, except *Shigella* which resembled *E. coli*. This sensitive technique can be applied on very low numbers of bacteria, i.e. 10 colony forming units (cfu).

Two 16S DNA targeted oligonucleotides were used as PCR primers for the specific detection of *Salmonella* serotypes in food. Some of the primers (16S III) also hybridized with *Klebsiella* and *Serratia* sp., however.

Recommended Methods for Detection and Enumeration of *Klebsiella*

Detection and Enumeration

Conventionally, eventual resuscitation (2 h at 17–25°C) in tryptic soy broth and (subsequent) plating on violet red bile glucose agar (VRBG) allows an efficient presumptive enumeration of Enterobacteriaceae, in foods or in other substrates. Other isolation media commonly used in this respect are: Simmons citrate agar, MacConkey agar and eosin methylene blue agar. Incubation is at 35°C (clinical samples) and 10°C (environmental samples) for 24–48 h. Raised mucoid colonies are selected and further differentiation and confirmation of *Klebsiella* is then based on the battery of tests mentioned above.

The detection and isolation from sources such as faeces, soil, water and food can be facilitated by use of the standard selective media (see Table 2). Specific *Klebsiella* enumeration is also of great importance to environmental microbiologists investigating the

Table 1 Differential biochemical characterization of *Klebsiella* spp.

Test	Klebsiella oxytoca	Klebsiella planticola	Klebsiella pneumoniae subsp. ozaenae	Klebsiella pneumoniae subsp. pneumoniae	Klebsiella pneumoniae subsp. rhinoscleromatis	Klebsiella terrigena
Gram stain (24 h)	–	–	–	–	–	–
Oxidase (24 h)	–	–	–	–	–	–
Indole production	–	[–]	–	–	–	–
Methyl red	[+]	+	+	[–]	+	d
Voges–Proskauer	+	+	–	+	–	+
Citrate (Simmons)	+	+	d	+	–	d
Hydrogen sulphide production	–	–	–	–	–	–
Urea hydrolysis	–	+	–	+	–	–
Phenylalanine deaminase	–	–	–	–	–	–
Lysine decarboxylase	–	+	d	+	–	+
Arginine dihydrolase	–	–	–	–	–	–
Ornithine decarboxylase	–	–	–	–	–	[–]
Motility	d	–	–	–	–	–
Gelatin hydrolysis, 22°C	–	–	–	–	–	–
KCN, growth	–	+	[+]	+	[+]	+
Acid production						
D-Adonitol	–	+	+	+	+	+
L-Arabinose	–	+	+	+	+	+
Cellobiose	–	+	+	+	+	+
Dulcitol	–	[–]	–	d	–	[–]
β-Gentibiose	+	+	+	+	–	+
D-Glucose	+	+	+	+	+	+
Glycerol	[–]	+	d	+	d	+
myo-Inositol	–	+	d	+	+	[+]
Lactose	d	+	d	+	–	+
Maltose	[–]	+	+	+	+	+
D-Mannitol	+	+	+	+	+	+
D-Mannose	+	+	+	+	+	+
D-Melezitose	d	–	–	–	–	+
Melibiose	–	+	+	+	+	+
α-Methyl-D-glucoside	–	+	d	+	–	+
Mucate	+	+	[–]	+	–	+
Raffinose	–	+	+	+	+	+
L-Rhamnose	[–]	+	d	+	+	+
Salicin	[+]	+	+	+	+	+
D-Sorbitol	–	+	d	+	+	+
Sucrose	–	+	[–]	+	[+]	+
Trehalose	+	+	+	+	+	+
D-Xylose	[–]	+	+	+	+	+
Tartrate, Jordans	d	+	d	+	d	+
Aesculin hydrolysis	d	+	[+]	+	d	+
Pectate hydrolysis	+	–	–	–	–	–
Utilization						
Acetate	–	d	–	[+]	–	[–]
Gentisate	+	–	–	–	–	+
m-Hydroxybenzoate	+	–	–	–	–	+
Malonate	–	+	–	+	+	+
D-Glucose, gas production	–	+	d	+	–	[+]
Growth 10°C	+	+	–	–	–	+
Lactose, gas production 44°C	–	–	+	+	+	–
Nitrate reduction	+	+	[+]	+	+	+
Deoxyribonuclease, 25°C	–	–	–	–	–	–
Lipase	–	–	–	–	–	–
ONPG[a]	[+]	+	[+]	+	–	+
Pigment	–	–	–	–	–	–

Table 1 contd

Test	Klebsiella oxytoca	Klebsiella planticola	Klebsiella pneumoniae subsp. ozaenae	Klebsiella pneumoniae subsp. pneumoniae	Klebsiella pneumoniae subsp. rhinoscleromatis	Klebsiella terrigena
Flagella arrangement[b]	P	−	−	−	−	−
Catalase production (24 h)	+	+	+	+	+	+
Oxidation-fermentation[c]	F	F	F	F	F	F

−, 0–10% positive; [−], 11–25% positive; d, 26–75% positive; [+], 76–89% positive; +, 90–100% positive. Results are for 48 h incubation.
[a]ONPG, *o*-nitrophenyl-β-D-galactopyranoside.
[b]P, peritrichous.
[c]F, fermentative.

effects of pulp and paper mill and cannery effluents in receiving waters.

The search for improved selective media and diagnostic tests goes on and is outlined below. A synthetic medium, based on *myo*-inositol as the sole carbon source has also been proposed for selection of *Klebsiella* (and *Serratia*). As a further elaboration, a Mac-Conkey–inositol–carbenicillin agar medium (MCIC) was proposed, the selectivity of this medium being based on the high resistance of the capsulated *Klebsiella* cells towards carbenicillin. Based on its oligotrophic characteristics, the development of a synthetic medium was claimed for the detection of *K. pneumoniae*; it only contains, apart from agar, $1 \, g \, l^{-1}$ KNO_3, $2 \, g \, l^{-1}$ KH_2PO_4, $20 \, g \, l^{-1}$ sucrose, but it is supplemented with $10 \, \mu g \, l^{-1}$ carbenicillin.

A highly selective, differential medium for the enumeration and isolation of *Klebsiella* species has been devised: MacConkey–inositol–potassium tellurite (MCIK) agar. With pure cultures, 100% recovery of *Klebsiella* was observed, and with environmental samples recovery of *Klebsiella* was as good as or better than on MCIC agar. MCIK agar was subsequently field tested for its ability to selectively enumerate *Klebsiella* species from the cold waters (1–6°C) of the Saint John River Basin (New Brunswick, Canada) which include fresh and marine waters. Results of this study indicate that 77% of the typical colonies on MCIK agar were *Klebsiella* species, but the total *Klebsiella* population enumerated was greatly underestimated; the MICK medium seems to be more specific for its target organisms but appears to lack sensitivity.

Various selective media have been assessed as to their ability to detect and differentiate *K. oxytoca* and *E. coli* in drinking-water samples. Only two media, membrane lauryl sulphate agar and deoxycholate agar allowed differentiation, with *K. oxytoca* only able to grow at 37°C and not at 44°C.

The CPS ID2 medium (Bio-Merieux) enabled the presumptive identification of urinary tract bacterial isolates, including *Klebsiella*, in specimens from a rehabilitation centre. Recently, a new chromogenic plate medium (CHROMagar Orientation) for the visual differentiation and presumptive identification of Gram-negative bacterial species and enterococcal isolates was evaluated. Similarity in colour resulted in failure to discriminate between *Klebsiella*, *Enterobacter* and *Citrobacter* species, but these species could be readily differentiated from other members of the Enterobacteriaceae. These data indicate an urgent need for the development of a simple, reliable and specific detection and enumeration methodology for *Klebsiella* sp.

Media Composition Suited for Cultivation of *Klebsiella* Strains

The composition of media suitable for cultivation of *Klebsiella* strains is listed in **Table 2**.

The media are routinely sterilized by autoclaving (20 min at 121°C and 2.1 atm). The carbon source is

Table 2 Composition of media for cultivation of *Klebsiella* strains

	Concn. $(g \, l^{-1})$
Nutrient broth (Oxoid) – pH 7.4	
'Lab-Lemco' beef broth	1.0
Yeast extract	2.0
Peptone	5.0
NaCl	5.0
***Klebsiella* medium** – pH 7.0	
Glucose	100.0
Soya peptone	10.0
Yeast extract	0.5
$MgSO_4.7H_2O$	0.5
K_2HPO_4	0.7
NaH_2PO_4	0.7
Ørskov and Ørskov medium – pH 7.0	
Glucose	15.0
Bacteriological peptone	7.0
NaCl	5.0

thereby separated from the nitrogen source to prevent Maillard reactions. Solidified media are obtained by adding $20\,g\,l^{-1}$ of agar before sterilization.

Culture Maintenance

Klebsiella strains can be easily maintained in meat extract agar stabs at room temperature. They can also be preserved either by storage in broth, containing 10% glycerol at −80°C or by lyophilization.

Procedures Specified in National/International Regulations or Guidelines

No official guidelines/regulations seem to exist at national or international level with regard to specific *Klebsiella* detection, enumeration or threshold numbers. The reader is referred to water and food quality guidelines, related to Enterobacteriaceae or coliforms.

Medical Aspects of *Klebsiella* Bacteria

As indicated above, *Klebsiella* bacteria are present in the respiratory tract and faeces of about 5% of normal individuals. They cause a minor proportion (about 3%) of bacterial pneumonias and can cause extensive haemorrhagic necrotizing consolidation of the lung. They occasionally produce urinary tract infection, septicaemia, bacteraemia with focal lesions and meningitis in debilitated patients. *K. pneumoniae* and *K. oxytoca*, especially, cause hospital-acquired infections. Two other *Klebsiella* subspecies are associated with inflammatory conditions of the upper respiratory tract:

- *K. pneumoniae* subsp. *ozaenae* has been isolated from the nasal mucosa in ozena, a fetid, progressive atrophy of mucous membranes;
- *K. pneumoniae* subsp. *rhinoscleromatis* has been isolated from rhinoscleroma, a destructive granuloma of the nose and pharynx.

K. pneumoniae is resistant to penicillin and ampicillin; resistant strains usually produce R-plasmid encoded β-lactamase. Broad-spectrum third generation cephalosporins such as cefotaxime, or aminoglycosides are used to combat normal strains (community-acquired), whereas hospital-acquired strains are multiple antibiotic resistant. Often, *K. pneumoniae* infections commonly occur following antibiotic treatment.

Environmental Relevance of *Klebsiella* Bacteria

Since *Klebsiella* species are widely distributed in the environment and in water systems, and since often little or no differences can be detected between environmental and clinical strains, there is increasing concern about the potential health hazard related to *Klebsiella*. There is, therefore, a growing need to monitor these organisms in the environment, especially in (drinking) water systems, soils, aerosols, cooling waters, biofilms and industrial effluents.

Relevance of *Klebsiella* in the Food Sector

General Aspects

The relevance of *Klebsiella* in foods as a contaminant or spoilage organism is only recently being addressed. Even in standard texts on food microbiology, there is little or no mention of problems related to *Klebsiella* food spoilage, contamination or transmission to humans. This is in sharp contrast with its ubiquitous presence in the daily human environment, and the pathogenic character of the clinical *Klebsiella* isolates, which are taxonomically very similar to the environmental strains.

As indicated already, *Klebsiella* species are opportunistic pathogens, that can give rise to bacteraemia, pneumonia, urinary tract and several other types of human infection. The origin and transfer of the infection is not always clear, since *Klebsiella* spp. are widely distributed in nature, occurring in soil, water, grain, fruits, vegetables, biofilm etc. Many of these environmental strains belong to the species *K. terrigena* and *K. planticola*. Since they also occur, albeit in low numbers, in the human respiratory and intestinal tracts, these seem to form the main reservoir for human-to-human infection (via food or otherwise), particularly in hospitals, where the hands of personnel and aerosol formation are the main factors in transmission. Outbreaks particularly occur in urological patients and in neonatal and intensive care units. Enterotoxin-producing *Klebsiella* strains have also been described.

K. pneumoniae strains are mainly isolated in association with several pathological processes in humans (respiratory and urinary tract infections) and in animals (metritis in mares and mastitis in cows) and as such can also enter the food chain.

Equally, the importance of the occurrence of *Klebsiella* species in the food sector is difficult to judge. *Klebsiella* species are usually not selectively cultured from foodstuffs, but are present when total counts or the presence of Enterobacteriaceae are tested. It is assumed that *K. terrigena*, isolated mainly from aquatic and soil environments, and *K. planticola*, mainly isolated from botanical, aquatic and soil environments, are saprophytic strains, that are easily transmitted to food. *K. oxytoca* is present in the intestinal tract of humans and animals, and has also been isolated from botanical and aquatic environments and as such it can also contaminate food and infect humans, indicating again an urgent need for specific *Klebsiella* detection. In this respect, a bioluminescence-based detection of lux⁺ *K. oxytoca* strains was developed and their survival in the barley rhizosphere was studied; it was found that *K. oxytoca* specifically survived on the plant roots during the whole vegetative period and that it could not be isolated from the soil.

Drinking Water Quality

Although it has been claimed that the attachment of bacteria, i.e. *Klebsiella* sp., to surfaces via capsule formation is a cause of increased bacterial disinfection resistance (for instance survival in chlorinated water supplies), it has also been found that resistance to chlorine was not related to the presence of polysaccharide, but to the formation of cell aggregates. Indeed, *K. pneumoniae* and *K. oxytoca* grown in low-nutrient media were found to be more resistant to chloramine than cells grown in rich media, and to form large aggregates/flocs of 10–10^3 cells per millilitre. This formation of flocs and aggregates allows the cells to survive chlorination and to enter the water distribution system. Indeed, *Klebsiella* sp. is one of the principal microorganisms involved in bacterial regrowth within chlorinated drinking water systems. The regrowth of this organism is of particular importance since, as a coliform, it will make compliance to water quality guidelines difficult, and it may be involved in opportunistic infections.

The growth kinetics of coliform bacteria, including *K. pneumoniae*, have been studied under conditions relevant to drinking water distribution systems. It was found that most of them – and *Klebsiella* in particular – could develop in unsupplemented mineral salts medium and in the unsupplemented distribution water. This proves that environmental coliforms, and equally *K. pneumoniae*, can develop under the conditions found in operating municipal drinking water systems. In this context, the ability of *K. pneumoniae*, *Enterobacter aerogenes*, *Agrobacterium tumefaciens*, *Bacillus subtilis* and *Pseudomonas* strains to grow and maintain motility and viability in drinking water has been studied. Plate counts dropped below the detection limit within 7 days for all strains mentioned, except for *Bacillus* and *Pseudomonas* strains.

The drinking water quality in a major South African metropolitan area, in collecting water samples from private houses, apartments and public places was assessed over a period of two years. Enterobacteriaceae bacteria were found in 33% of the samples, as well as *Bacillus* sp. *Klebsiella* was also frequently found. The age of the plumbing system was clearly correlated with poorer microbiological quality of the potable water. Among 62 trademarks of bottled drinking water, a sampling of 158 bottles revealed the presence of *K. oxytoca*, along with other coliforms, in three bottles.

The quality of packaged ice, sold in retail establishments in Iowa, USA has also been studied. A total of 18 samples were analysed in relation to the drinking water standards of the US-EPA. Only one sample exceeded the health standard and contained *K. pneumoniae*. Several samples had heterotrophic plate counts, which exceeded the recommendation ($< 500\,\mathrm{cfu\,ml^{-1}}$) of the Packed Ice Association. Although such ice consumption does not represent an immediate threat to personal or public health, the potential for disease transmission exists in a sector, which is in this respect self-regulated.

It is clear that *Klebsiella* species comprise a large part of the coliforms, usually detected as indicators of water quality; in most instances however, they are not differentiated.

Food Quality

As to their occurrence in foods, the microbiology of contaminated foods in health-care facilities in the USA was surveyed and the importance of microbial surveillance, quality assurance and employee education was stressed. *K. pneumoniae* is mentioned as one of the encountered contaminants, together with *E. coli*, *Yersinia intermedia*, *Aeromonas hydrophila*, *Enterobacter agglomerans*, *E. cloaca*, *Campylobacter jejuni*, *Acinetobacter anitratum*, *Streptococcus viridans*, *Serratia liquefaciens*, *Staphylococcus* sp., *Salmonella* sp., *Corynebacterium* sp., *Lactobacillus* sp., *Listeria* sp. and others.

High numbers of *Klebsiella* species were found in samples of the local food, pap 'akamu', prepared in Nigeria from cereals (maize, guinea corn and millet); *Klebsiella* sp., *Enterobacter* sp. and *Staphylococcus* sp. were the most common bacteria found. These data indicate the widespread occurrence of *Klebsiella* sp. in these indigenous foods, in combination with other opportunistic pathogens.

K. pneumoniae subsp. *pneumoniae* (using the API 20 E system) was also found to be present in the industrial fermentation process of 'Saccharina' production (fermented fodder) from sugar cane.

'Coliforms' were enumerated in fresh and processed

mangoes (puree and cheeks) in order to establish the source of the organisms in the production chain, to determine whether they have any public health significance, and to devise methods for their control. Products from four processors were tested on two occasions. The retail packs of cheeks-in-puree having the highest coliform counts were those in which raw puree was added to the cheeks. Coliform counts in these samples ranged between 1.4×10^3 and 5.4×10^4 cfu g^{-1}. Pasteurization reduced the coliform count of raw puree to 5 cfu g^{-1}. Around 47% of the 73 colonies, isolated as coliforms on the basis of their colony morphology on violet red bile agar, were identified as *K. pneumoniae*, using the ATB 23 E Identification System. *Klebsiella* strains were tested for growth at 10°C, faecal coliform response and fermentation of D-melezitose; these tests are used commonly to differentiate the three phenotypically similar strains *K. pneumoniae*, *K. terrigena* and *K. planticola*. Results indicated that 41% of the isolates gave reactions typical of *K. pneumoniae*. A further 44% of strains gave an atypical reaction pattern and were designated 'psychrotrophic' *K. pneumoniae*. *K. pneumoniae* counts of 2.1×10^3–4.9×10^4 cfu g^{-1} were predicted to occur in the retail packs of mango cheeks-in-puree produced by the processors, who constituted this product with raw puree. In view of the opportunistic pathogenic nature of *K. pneumoniae*, its presence in these products is considered undesirable and steps, such as pasteurization of puree, should be taken in order to inactivate it.

Recently attempts have been made to correlate the presence of selected pathogens (*Campylobacter jejuni*, *C. coli*, *Salmonella*, *K. pneumoniae* and *E. coli* O157:H7) in fresh hand-picked blue crab meat and general microbial quality to sanitation practices by the processors (Chesapeake Bay region, USA). *K. pneumoniae* was isolated from 51 samples out of the 240 (21%) (0.3–4.3 most probable number (MPN) g^{-1}), followed by *C. jejuni* (36 out of 240), *C. coli* (14 out of 240). *Salmonella* and *E. coli* O157:H7 were not detected in any of the 240 samples analysed.

The foregoing data indicate again that *Klebsiella* sp. is frequently present as a contaminant in water and food, often in high numbers. They are commonly lumped within the group of the Enterobacteriaceae or coliforms, with most attention always focused on the well known food pathogen members of the group. It is only recently that *Klebsiella* is being selectively searched for and 'looked after' as a genus/species, relevant to food microbiologists too.

Industrial Aspects of *Klebsiella* Bacteria

Production of 2,3-butanediol

Most *Klebsiella* species are saprophytic, some are pathogenic and only a few are of industrial use. Under controlled fermentation conditions, *K. oxytoca* strains produce high levels of 2,3-butanediol, an interesting chemical feedstock or liquid fuel, from sugary substrates such as glucose, xylose and whey permeate. Due to its toxicity to the producer cells, only moderate concentrations (approximately 100 g l^{-1}) can be obtained in even optimized fermentation processes. This, together with the high boiling point and hygroscopicity of 2,3-butanediol, makes recovery costs high. 2,3-Butanediol can be chemically converted into butadiene, the raw feedstock for synthetic rubber, or into other derivatives such as ethylmethylketone (used as a solvent, fuel additive) and tetramethylether (antifreeze) or into polyester plastics.

Biofilm Formation by *Klebsiella* sp.

As a result of its capacity to form capsules, *Klebsiella* species are often a main cause of (undesirable) biofilm formation and fouling in cooling water systems, piping and other industrial equipment. The biofilm-forming capacity of several *Klebsiella* species, isolated from pulp and paper mill water, and of *Klebsiella terrigena* BCCM strains has been studied in vitro by the authors in 2 litre lab fermenters. The capsular polysaccharide from one isolate was recovered (up to 4.6 g l^{-1}), its rheological properties were identified as pseudoplastic and its sugar composition was identified as: L-fucose, L-rhamnose, D-galactose, D-glucose, D-mannose and D-glucuronic acid. Enzymes which can efficiently hydrolyse and remove biofilm have been looked for.

The *Klebsiella* Capsule as a Source of Unusual Sugars

The *Klebsiella* capsule, as described above, often contains unusual sugar moieties such as L-fucose and L-rhamnose, and the authors have cultivated such capsular bacteria on a large scale, as a source of these specialty sugars, which are otherwise difficult to obtain.

Klebsiella as a Vitamin Producer in Fermented Foods

Recently, the formation of vitamin B_{12} was demonstrated by strains of *K. pneumoniae*, isolated from Indonesian tempeh samples, during controlled solid-state tempeh fermentation. The absence of enterotoxins in these strains was confirmed by using PCR techniques, and it was even suggested that these safe strains should be used in Indonesian tempeh fermentation to enrich it in vitamins.

Acknowledgements

The authors are grateful to F. Callewaert, M. Naessens and S. Vandedrinck for their help with the preparation and critical reading of the manuscript and for providing pertinent practical and theoretical background information.

See also: **Biofilms. Enterobacteriaceae, Coliforms and *E. coli*:** Introduction; Classical and Modern Methods for Detection/Enumeration. **Fermented Foods:** Fermentations of the Far East. **Water Quality Assessment:** Routine Techniques for Monitoring Bacterial and Viral Contaminants; Modern Microbiological Techniques. **Waterborne Parasites:** Detection by Classic and Modern Techniques.

Further Reading

Augoustinos MT, Venter SN and Kfir R (1995) Assessment of water quality problems due to microbial growth in drinking water distribution systems. *Environ. Toxicol. Water Qual.* 10: 295–299.

Brooks S, Malcolm S, Lamont P, Khan A, Madahar B and Hameed R (1995) Contaminated foods in healthcare facilities. *Infect. Control Hosp. Epidemiol.* 16: 675.

Bruggeman G, Van Speybroeck M, De Moor L and Vandamme EJ (1995) Exopolysaccharide production by *Klebsiella* strains and characterisation of its sugar composition. *Med. Fac. Landbouww. Univ. Gent*, 60(4a): 2017–2020.

Byrd JJ, Xu HS and Colwell RR (1991) Viable but non-culturable bacteria in drinking water. *Appl. Environ. Microbiol.* 57: 875–878.

Doyle MP, Beuchat LR and Montville TJ (eds) (1997) *Food Microbiology: Fundamentals and Frontiers.* Herndon: ASM Press.

Holt JG, Krieg NR, Sneath PHA, Staley JT and Williams ST (eds) (1994) *Bergey's Manual of Determinative Microbiology,* 9th edn. London: Williams and Wilkins.

Jones K and Bradshaw SB (1996) Biofilm formation by the *Enterobacteriaceae*: a comparison between *Salmonella enteritidis*, *Escherichia coli* and a nitrogen-fixing strain of *Klebsiella pneumoniae. J. Appl. Bacteriol.* 80: 458–464.

Keuth S and Bisping B (1994) Vitamin B_{12} production by *Citrobacter freundii* or *Klebsiella pneumoniae* during Tempeh fermentation and proof of enterotoxin absence by PCR. *Appl. Environ. Microbiol.* 60: 1495–1499.

Magee RJ and Kosaric N (1987) The microbial production of 2,3-butanediol. *Adv. Appl. Microbiol.* 32: 89–161.

Ørskov I and Ørskov F (1984) Serotyping of *Klebsiella. Methods Microbiol.* 14: 143–164.

Power KN, Schneider RP and Marshall KC (1997) The effect of growth conditions on survival and recovery of *Klebsiella oxytoca* after exposure to chlorine. *Water Res.* 31: 135–139.

Stewart MH and Olson BH (1992) Impact of growth conditions on resistance of *Klebsiella pneumoniae* to chloramines. *Appl. Environ. Microbiol* 58: 2649–2653.

Tortorello ML and Gendel SM (eds) (1997) *Food Microbiological Analysis: New Technologies.* New York: Marcel Dekker.

Vandamme EJ, Bruggeman G, De Baets S and Vanhooren PT (1996) Useful polymers of microbial origin. *Agro-Food-Industry Hi-Tech* 7: 21–25.

Van Speybroeck M, Bruggeman G, Van Poele J, Vandamme E and Van Pee K (1995) Exopolysaccharide degrading enzyme and method of use. US-patent No. 08/527905, WR Grace, USA.

KLUYVEROMYCES

Carl A Batt, Department of Food Science, Cornell University, USA

Characteristics of the Species

Members of the *Kluyveromyces* genus belong to the Class Ascomycetes, and families within this class form endospores, and mycelia without clamp connections. In general, members of the Ascomycetes do not produce urease, can ferment certain saccharides, and have G+C contents of 30–52%. Nitrate is not assimilated and these yeasts require the addition of vitamins for growth. *Kluyveromyces* cells are oval to spherical and reach approximately 3–5 μm in size.

Kluyveromyces, similar to *Saccharomyces*, is a budding yeast and this is the most common mode of vegetative reproduction. Budding yeasts reproduce by developing daughter cells as compared to fission yeast (e.g. *Schizosaccharomyces*) where there is a relatively equal division of one cell to form two cells. The buds arise at any site on the mother cell, a process that is regulated by a number of genes that are responsible for the production of enzymes that give rise to new cell wall material. The cells may appear to be branched chains when the buds fail to separate. Like other yeasts, *Kluyveromyces* can also form hyphae and are therefore dimorphic, able to grow either as single cells or in filaments. The shift between yeast and filamentous growth can be affected by oxygen and other growth conditions. The filamentous growth phase of *Kluyveromyces* may be advantageous for certain industrial applications. For example, the hyphae have a much greater surface area as compared to the yeast cell and therefore may be easier to immobilize. In addition the larger surface area may increase the yield of secreted enzymes.

Historically some of the *Kluyveromyces* were classified as *Saccharomyces*, another member of the Class Ascomycetes, Family Saccharomycetaceae, and differ principally in spore shape and appearance as well as their ability to agglutinate. Within the *Kluyveromyces* genus, other nomenclature changes include renaming *K. fragilis* to *K. marxianus*.

There are approximately 17 species within the *Kluyveromyces* genus. The species are subdivided based on their ability to form single or multispore asci. Among the *Kluyveromyces in sensu stricto* are *K. polysporus* which ferments sucrose and *K. africanus* which does not. Spore shape is also diagnostic. In *Kluyveromyces* species both reniform (subgenus *Fabospora*) and round spores (subgenus *Globospora*) are found. Sugar fermentation is also used to speciate the multispore asci forming *Kluyveromyces*. The ability to utilize a particular sugar is a function of the presence or absence of a critical enzyme. For example, the ability to assimilate maltose requires the enzyme α-glucosidase and is diagnostic of *K. drosophilarum*. Notable is the ability of certain species to ferment lactose, including *K. marxianus* and *K. lactis*. Other sugars whose assimilation is variant among the *Kluyveromyces* genus are xylose, maltose and inulin.

K. lactis and *K. marxianus* (formerly *K. fragilis*) are among the most well-known members of the *Kluyveromyces* genus. They are morphologically identical with the exception of their spore shape and other characteristics, most notably the ability to ferment lactose, which suggest that *K. lactis* is similar to *K. fragilis* (now *K. marxianus*). Some references cite them as subspecies, *K. marxianus* var. *lactis* and var. *marxianus* respectively. Their DNA homology is, however, only 15–20% and as a consequence they are most likely distinct species.

The ability of *Kluyveromyces* spp. to utilize lactose is one of the major reasons for its being pursued as a host for the biotechnological processes. Although *S. cerevisiae* cannot metabolize lactose, *Kluyveromyces* spp. transport lactose using an inducible lactose permease. The *K. lactis* and *K. marxianus* have lactose permeases that are proton symports. These can also transport galactose. The lactose permeases are induced by lactose or galactose. Introduction of this lactose transporter (*LAC12*) along with a β-gal-actosidase into *S. cerevisiae* renders this yeast able to utilize lactose.

Biotechnological Importance of *Kluyveromyces*

Members of the *Kluyveromyces* genus are important to the biotechnology industry by virtue of the enzymes and other biomolecules that they produce and their ability to serve as hosts for the expression of recombinant enzymes. As a host for the production of proteins *K. lactis* has a number of advantages over other yeasts. First it is classified as generally recognized as safe (GRAS). GRAS status is given to microorganisms or other ingredients that have a historical record of usage in foods without any incidence of toxicity. As such, GRAS status is a very powerful inducement for its usage in food applications as it obviates the need for it to clear any regulatory hurdles. Second, it can be grown to relatively high cell densities which translates into significant increases in product yield. Third, it is able to secrete a number of enzymes and has the capacity to secrete heterologous proteins. It has been reported that *Kluyveromyces* may have a greater capacity to secrete proteins as compared to *Saccharomyces*. Finally there is a robust set of recombinant DNA tools including both integrative and replicating vectors.

Members of the *Kluyveromyces* genus are the source of a number of different hydrolytic enzymes. Either whole cells or purified (and semi-purified) enzymes are used as biocatalysts (**Table 1**). Among the most cited examples of enzymes from *Kluyveromyces* are lactases, β-glucosidases, cellulase and inulinases.

Production of lactase from whey is an important application for *Kluyveromyces*. Lactase (β-D-galactosidase), the enzyme which degrades lactose, is used to reduce the lactose content of milk and dairy products. Individuals that are lactose-intolerant can drink lactose-reduced milk. Further the characteristics of a number of food products benefit from the use of lactose-reduced milk including the stability of frozen concentrated milk and the colour of fried foods.

Table 1 Selected products produced by members of the *Kluyveromyces* species

Class	Product	Species
Vitamins	D-Erythro-ascorbic acid	*Kluyveromyces* spp.
Enzymes	Lactase	*K. marxianus*
	Cellulase	*K. cellobiovorans*
	Inulinase	*Kluyveromyces*
Whole-cell	Biosorbents (heavy metals)	*K. marxianus*
	Animal feeds	*K. lactis*,
		K. marxianus

Plate 14 Quorn myco-protein obtained by growing a selected mycelial fungus in a carbohydrate-rich medium, and then filtering-off the mycelium. (With permission from Marlow Foods.) See also entry **Fungi: The Fungal Hypha**.

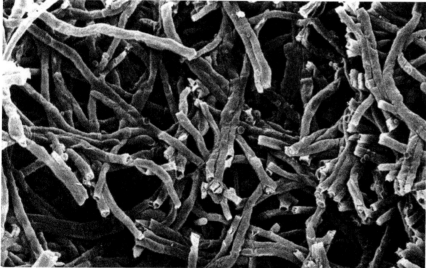

Plate 15 (*left*) Micrograph of myco-protein, showing its composition of individual fungal hyphae. (With permission from Marlow Foods.) See also entry **Fungi: The Fungal Hypha**.

Plate 16 (*right*) A mould growing on Rose Bengal Agar. Note that the use of Rose Bengal has restricted the growth of the individual colonies. (With permission from The Water Quality Centre, UK. www.wqc.co.uk.) See also entry **Fungi: The Fungal Hypha**.

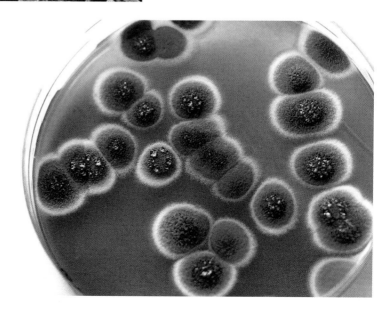

Plate 17 *Amanita muscaria* ('fly agaric'). An ectomycorrhizal toadstool, typically found with birch. Poisonous and hallucinogenic, used by shamans visiting spirit worlds. (With permission from P. Roberts, Kew, Surrey, UK.) See also entry **Fungi: The Fungal Hypha**.

Plate 18 *Amanita phalloides* ('death cap'). An ectomycorrhizal toadstool, found with oaks and other deciduous trees. As the English name suggests, this species is lethally poisonous. (With permission from P. Roberts, Kew, Surrey, UK.) See also entry **Fungi: The Fungal Hypha**.

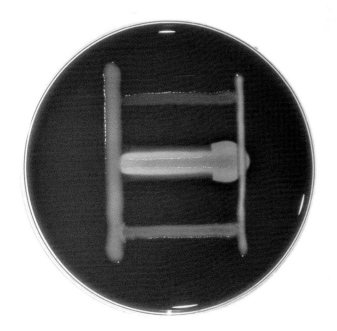

Plate 19 A typical pattern in the CAMP reaction that is employed to confirm the identity of species of *Listeria*. Horizontal streaks: top, *Listeria monocytogenes* NCTC 11994; middle, *Listeria ivanovii* ATCC 19119; bottom, *Listeria seeligeri* ATCC 35967. Vertical streaks: left, *Rhodococcus equi* ATCC 25972; right, *Staphylococcus aureus* ATCC 25923. (With permission from OXOID Ltd.) See also **Listeria: Detection by Classical Cultural Techniques**.

Plate 20 Fresh pork sausages stored for **(a)** 0h and **(b)** 72h, at 4°C. Colour deterioration is obvious due to proliferation of bacteria. (Courtesy of I. Guerrero and L. P. Chabela.) See also **Meat and Poultry: Spoilage of Cooked Meats and Meat Products.**

Plate 21 (*right*) pH distribution in electrolyte solution induced by electrolysis (a) before electrolysis, (b–o) after the first electrolysis and (p–z) after the second electrolysis. See also entry **Microscopy: Sensing Microscopy.**

Plate 22 (*below*) pH distribution in agar medium induced by an *E. coli* colony. Incubation for (a) 10h, (b) 13h and (c) 16h. See also entry **Microscopy: Sensing Microscopy.**

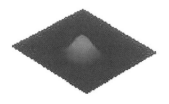

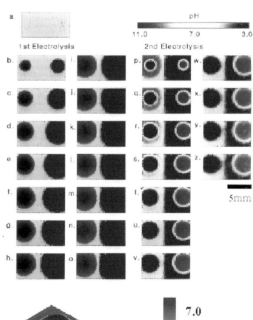

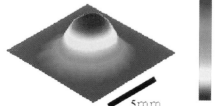

(a) 10 hours incubation   (b) 13 hours incubation   (c) 16 hours incubation

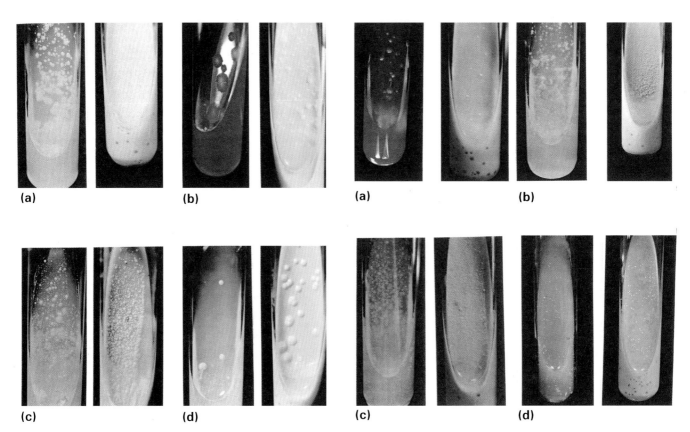

Plate 23 Colonial morphology of **(a)** *Mycobacterium avium*, **(b)** *M. bovis*, **(c)** *M. chelonae* and **(d)** *M. fortuitum*. (Courtesy of Jerrald Jarnagin, USDA.) See also entry **Mycobacterium**.

Plate 24 Colonial morphology of **(a)** *Mycobacterium intracellulare*, **(b)** *M. kansasii*, **(c)** *M. marinum* and **(d)** *M. paratuberculosis*. (Courtesy of Jerrald Jarnagin, USDA.) See also entry **Mycobacterium**.

Plate 25 Colonial morphology of **(a)** *Mycobacterium scrofulaceum*, **(b)** *M. smegmatis*, **(c)** *M. tuberculosis* and **(d)** *M. xenobi*. (Courtesy of Jerrald Jarnagin, USDA.) See also entry **Mycobacterium**.

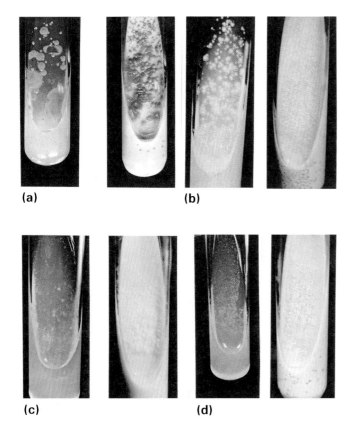

Kluyveromyces was one of the first suggested sources of lactase to reduce the lactose content of milk. A process that dates back to the 1950s for growing *K. marxianus* var. *lactis* on whey has been developed that results in a dried product which retains good activity. The lactase has a pH range of 6–7 and optimum hydrolytic activity is observed at 37°C. In practice, however, hydrolysis of milk is carried out at 4–6°C to preserve milk quality and to prevent the growth of bacteria.

In a number of biotechnological applications the catalyst is immobilized to render the system continuous and/or to assist in the recovery of the product. *K. lactis* and *K. bulgaricus* have been used to hydrolyse lactose in whey and immobilization of whole cells in glass wool and alginate, respectively, have been reported. In a similar approach, *K. marxianus* has been immobilized in alginate and used to hydrolyse inulin. In contrast, a purified form of β-galactosidase has been immobilized on to solid supports including cellulose triacetate fibres to produce lactose-reduced milk. Although widely used in the industry, the *Kluyveromyces* lactase has largely been replaced by other fungal lactases. The *Aspergillus* enzyme has a lower pH optimum although it is more thermostable and therefore more difficult to inactivate and stop the process when the desired lactose content is reached.

Kluyveromyces has also been used to produce single-cell protein (SCP) from whey and whey fractions. A number of SCP processes were developed using whey to help reduce the problems of waste disposal from cheese-making processes. Now a number of uses of whey proteins as well as lactose obviates the need to utilize it for SCP production. SCP has a relatively good balance of amino acids and is not limited in its lysine content. As dried yeast, *K. marxianus* var. *lactis* has been used to fortify foods and to serve as a vitamin supplement in both human and animal feeds.

The fermentation process is largely exothermic and heat generated from these fermentations must be removed. Up to 12% of the energy obtained from the metabolism of lactose is lost to heat. Capturing the heat for other purposes has been proposed and most of it is lost through the venting process. Mathematical models that can be used to describe the growth of *Kluyveromyces* in whey have been developed. A series of these models have been reported to account for a number of the parameters in the fermentation process and can be employed to optimize the process.

The most well-characterized yeast is *Saccharomyces cerevisiae* and there is a wealth of recombinant expression vectors, protocols and host strains for this organism. In comparison the repertoire for *Kluyveromyces* is much less extensive but nevertheless suitable for

developing systems for the expression of a wide variety of products (**Table 2**). Among these products a number of pharmaceutical targets have been reported including interleukin, hepatitis B surface antigen, human serum albumin and granulocyte colony stimulating factor. Production of these targets in *K. lactis* as compared to *S. cerevisiae* is largely driven by the potentially lower cost of substrate for the fermentation. Lactose is a lower cost substrate as compared to glucose and these processes can be easily coupled to other processes that generate whey (i.e. cheese-making).

Vectors for *Kluyveromyces* have been developed and both self-replicating and integrative plasmids are available. Three types of plasmid vectors have been reported that can be used to develop recombinant strains of *Kluyveromyces*. One is based on the cytoplasmic linear double-stranded killer plasmids. Killer plasmids are 8–14 kb in length and can be maintained at a copy number of 100–200 per cell. Unfortunately given their non-circular nature and the presence of proteins covalently linked to their 5′ ends, killer plasmid-based vectors cannot easily be manipulated using *Escherichia coli* as an intermediate host. Vectors have also been constructed based on autonomously replicating sequences (ARS). However, these vectors suffer from mitotic instability and therefore cannot be used in industrial fermentations. Even under selective pressure, only 35% of the cells harboured the plasmid after 20 generations. The more well-studied vectors are derived from pKD1 which is a circular plasmid that was found in a *Kluyveromyces drosophilarum*. It is a 2µ-like plasmid with an origin of replication that appears to function in a number of different yeast strains. The original pKD1 was 4.8 kb in size and a number of open reading frames were identified which appear essential for its replicative function. Vectors based on pKD1 are stably maintained at 70–100 copies per cell. The addition of an antibiotic-resistance marker changed it into a rudimentary vector from which a number of second generation vectors have been constructed.

Copy number is an important feature of a recom-

Table 2 Selected recombinant products produced in *Kluyveromyces* species

Class	Product	Species
Enzymes	Chymosin	*K. lactis*
	α-Galactosidase	*K. lactis*
Pharmaceutical	Interleukin-1β	*K. lactis*
	Hepatitis B surface antigen	*K. lactis*
	Granulocyte colony stimulating factor	*K. lactis*
	Human serum albumin	*K. lactis*
	β-Lactoglobulin	*K. lactis*

binant vector. Although high copy number is desirable in terms of delivering high level expression due to the presence of multiple copies of the targeted genes, these vectors are also sometimes unstable. In contrast, low-copy number vectors are typically more stable but also generate lower product yields. Low-copy vectors for *Kluyveromyces* have been constructed using the *Klori*, which is an origin of replication that has been isolated from the plasmid pKD1. In contrast, a high copy number vector has been engineered by combining the *Klori* with an ARS and the centromeric sequences from *S. cerevisiae*.

Selection of *Kluyveromyces* transformants can be accomplished by complementation of mutations to genes involved in amino acid, nucleotide or other essential biosynthetic pathways. The genes from *Kluyveromyces* including *LEU2*, *TRP1*, *HIS4*, *ARG8*, *URA3* have been cloned and are suitable markers. In these systems, the wild-type gene is carried on a plasmid and when introduced into the respective auxotrophic strain, it complements the mutation allowing the transformed strain to grow in the absence of that nutrient.

A number of factors determine the yield of product from recombinant expression hosts. The growth rate of the organism, the level of expression of the heterologous gene and most importantly the maximum cell density all contribute to product yield. Expression yields are typically expressed as the amount of product obtained, per unit time, per fermentation volume. *K. lactis* and *K. marxianus* can grow to very high cell densities of the order of > 100 g dry cell weight per litre.

Production of heterologous proteins in *Kluyveromyces* is best accomplished using a promoter that directs expression at a relatively high level in comparison to other genes. One example is the phosphoglycerate kinase gene (PGK) which is highly expressed in *Kluyveromyces*. Efficient expression of recombinant proteins, however, usually requires a promoter that can be regulated. This is especially important when trying to express proteins that are either toxic or impede the growth of the host. With a regulated promoter, cell biomass can be increased by growing the *Kluyveromyces* under conditions where the target protein is not expressed. Once the maximum biomass is obtained, then the regulated promoter can be turned on by adding an inducer or removing a repressor. Regulated promoters have been identified including *LAC4* which encodes lactase, *KlPHO5* which encodes an acid phosphatase and *KlADH4* which encodes a mitochondrial alcohol dehydrogenase (**Table 3**).

The host strain itself can have a dramatic impact

Table 3 Regulated promoters from *Kluyveromyces* species

Gene	Enzyme	Regulated by:
LAC4	Lactase	Added lactose
KlPHO5	Acid phosphatase	Low phosphate
KlADH4	Alcohol dehydrogenase	Added ethanol
KlDLD	D-Lactate ferricytochrome *c* oxidoreductase	Added lactate Repressed by glucose

on product yields, yet the factors that are responsible for these strain differences are not fully understood. Dramatic differences in the stability of certain vectors including those based on pKD1 has been observed. This mitotic instability can lead to a significant reduction in product yield especially in large-scale fermentations where a greater number of generations between seed stock and final fermentation is needed. For example the production of human serum albumin was tested in a total of 50 wild-type strains of *Kluyveromyces* with yield variations from barely detectable levels to over 300 mg l^{-1}.

The major characteristic of interest, as mentioned, is that both *K. lactis* and *K. marxianus* grow on lactose. Lactose is a relatively inexpensive carbon source since it is a by product of cheese-making. Whey is an abundant source of lactose and although major applications have been developed for whey proteins, the lactose component is still an underutilized commodity. Progress made in furthering the development of *Kluyveromyces* as a recombinant host will likely have a significant impact on its utility to the biotechnology industry especially as substrate costs become a rate-determining factor in production.

See also: **Milk and Milk Products**: Microbiology of Liquid Milk. **Saccharomyces**: *Saccharomyces cerevisiae*. **Single Cell Protein**: Yeasts and Bacteria. **Yeasts**: Production and Commercial Uses. **Fungi**: Classification of the Eukaryotic Ascomycetes.

Further Reading

Buckholz RG and Gleeson MA (1991) Yeast systems for the commercial production of heterologous proteins. *Biotechnology* 9: 1067–1072.

Fleer R (1992) Engineering yeast for high level expression. *Current Opinions in Biotechnology* 3: 486–496.

Gellissen G, Melber K, Janowicz ZA et al (1992) Heterologous protein production in yeast. *Antonie Van Leeuwenhoek* 62: 79–93.

Gellissen G and Hollenberg CP (1997) Application of yeasts in gene expression studies: a comparison of *Saccharomyces cerevisiae*, *Hansenula polymorpha* and *Kluyveromyces lactis* – a review. *Gene* 190: 87–97.

LABORATORY DESIGN

M Ahmed, Food Control Laboratory, Dubai, United Arab Emirates

Introduction

Food microbiology laboratories play an important role in the control of food hygiene, quality and safety. The potential hazards associated with pathogenic microorganisms in these laboratories together with the development of strict legislation to promote health and safety at work have led to higher standards of laboratory design. Contamination of samples within the laboratory through air and other sources has been a major problem associated with microbiological analysis. The laboratory design must meet the requirements to avoid contamination. Cleanliness, ventilation, accessibility, storage, waste disposal, security, fire protection and emergency precautions must all be considered at the initial stage of design.

Even though the final design of the laboratory is made by architects and engineers, involvement of microbiologists is essential when taking important decisions that affect the working environment and conditions. Microbiologists should work in close association with the architect and explain all the technical and safety requirements of each room. They should also follow the building through the different stages of construction to ensure that all the requirements included in the design are fulfilled.

Study Report

Microbiological laboratories can be broadly classified into three categories:

1. Hygiene control laboratories performing limited microbiological tests to evaluate sanitation and hygiene procedures followed in food production plants, restaurants and catering establishments.
2. Quality control laboratories involved in the testing of imported and locally produced foods as well as hygiene control, which perform a wide range of analyses and carry out work-related research and investigations.
3. Research laboratories involved in carrying out research and development (R&D) but not involved in quality control.

Before designing a laboratory, a study report should be prepared by a consultant with good knowledge and experience of designing laboratories. The technical experts of the consultant should meet the management, microbiologists and other technical staff and discuss in detail their requirements. Due consideration should be given to their views and recommendations while preparing the study report, which should consist of the following:

- scope and objectives of the laboratory
- organization chart indicating the various functions of the laboratory
- number of technical, administrative and support staff
- expected number of samples to be analysed
- details of technical facilities required
- service requirements
- the interrelationships, if any, between the functions of the laboratory and other disciplines (chemistry, biochemistry, nutrition, etc.)
- budget requirements.

The study report should also address the scientific and technical developments in the area and make provisions for future expansion of the laboratory.

The recommended organization of a food microbiology laboratory suitable for routine quality control analysis is shown in **Figure 1**. The laboratory consists of a general microbiology unit, with culture techniques and media preparation sub-units, and an advanced microbiology unit, with rapid diagnostic techniques and instrumental techniques sub-units. The administration sample management, quality management, R&D and training management, and calibration and maintenance constitute other functions. These may be common to a laboratory consisting of multiple disciplines such as chemistry, biochemistry and nutrition. The major activities of different functions in the laboratory are listed in **Table 1**.

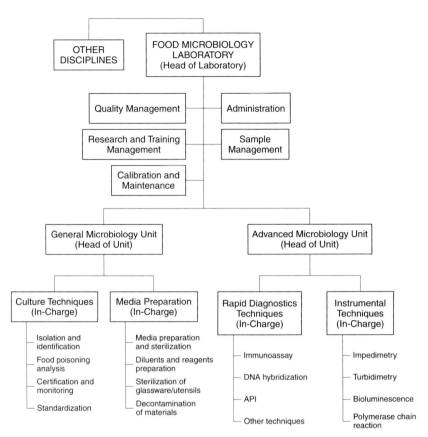

Figure 1 Food laboratory organization chart.

Building Layout

Many types of laboratory layouts are possible, depending on scope of work, space and budget. The building layout for a food microbiology laboratory carrying out routine quality control analysis of a wide range of samples in addition to conducting a limited number of applied research projects is shown in **Figure 2**.

Table 1 Major activities of a food microbiology laboratory

Unit/function	Sub-unit	Major activity
General microbiology	Culture techniques	Certification and monitoring programmes, food poisoning –emergency analysis, standardization
	Media preparation and sterilization	Preparation and sterilization of media, glassware, sample utensils, decontamination and washing of used materials
Advanced microbiology	Rapid diagnostic techniques	Application development and implementation of immunoassay, DNA hybridization, API, etc.
	Instrumental techniques	Application development and implementation of impedimetry, turbidimetry, bioluminescence, PCR, etc.
Quality management	Quality management	Implementation of quality assurance system (ISO 9002/ISO Guide 25), internal quality control, proficiency testing, audits, etc.
Calibration and maintenance	Calibration and maintenance	Equipment and building maintenance, calibration of equipment, maintenance of services
R&D, training management	R&, training management	Planning, budgeting of R&D work, coordination with different units, training requirements and their planning and scheduling, management of external training programmes
Sample management	Sample management	Receipt, identification, registration, preparation of composite samples, assigning code numbers, distribution of samples to different functions
Administration	Administration	Secretariat, personnel management, budget/accounts, purchase/stores, library and housekeeping

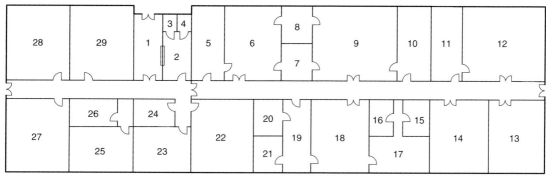

Figure 2 Building layout for a food microbiology laboratory. 1, Entrance and reception; 2, sample management; 3, freezer room; 4, cold room; 5, pretreatment (high contaminants); 6, culture techniques I; 7, walk-in incubator; 8, anaerobic work station; 9, culture techniques II; 10, pretreatment (low contaminants); 11, preparation room; 12, advanced microbiological techniques; 13, rapid diagnostic techniques; 14, isolation and identification techniques; 15, dry media store; 16, autoclave decontamination of materials; 17, washing room; 18, media preparation; 19, aseptic filling room; 20, autoclave sterilization of materials; 21, chilled room; 22, library; 23, head of microbiology laboratory; 24, secretary; 25, R&D, training management; 26, quality management; 27, toilet and changing rooms (female); 28, toilet and changing rooms (male); 29, staff room.

General Considerations

The microbiology laboratories have a unique contamination problem and should have a central air conditioning system if possible. This system should be divided into zones depending on the type of work carried out in different rooms to facilitate exchange of fresh air and to take necessary precautions against environmental contamination. The incoming air is filtered through 0.2 μm filters to reduce the risk of environmental contamination of the laboratory. The humidity must be kept low to reduce problems with hygroscopic materials such as media and chemicals, and to avoid growth of moulds on laboratory surfaces. Air conditioning also stabilizes room temperatures, enabling incubators to function more efficiently. Temperatures and relative humidity should be comfortable for workers and suitable to the requirements of the laboratory equipment. Normally an ambient temperature of 21–23°C and a relative humidity of 40–50% are recommended.

Lighting It is recommended that laboratory lighting be maintained at an average intensity of 0.5–1 klx (50–100 footcandles). Dependence on natural sunlight during the day is discouraged: direct exposure to sunlight is known to alter the performance of media, chemicals and reagents. Likewise, analytical work must not be performed in direct sunlight since final results are affected.

Storage Sufficient storage space should be provided for equipment, materials and samples. The laboratory wall space should be utilized for additional shelving, protected by glass-enclosed cabinets to provide a dust-free environment for storage of media, chemicals and other materials. Samples should be stored in refrigerators, freezers or at room temperature according to the procedures outlined in the operational manual.

Future Expansion Future expansion of activities, increases in workload and staff should be considered when designing a laboratory. The design should include provision for a minimum of 25% of expansion. The design should also be flexible to allow room functional changes and allocation of new activities.

Allocation of Space The design should allow maximum utilization of laboratory space. The subunit of media preparation and filling, decontamination of used material and cleaning of glassware should be separated from the analytical area. Within the analytical area, isolation and identification of pathogenic materials should be carried out in a separate room if possible.

Safety The safety of the laboratory personnel should be taken care of in the design. The laboratory should be equipped with fire extinguishers and alarms, a sprinkler system, eye-wash stations and safety showers. Fire and smoke detectors are also recommended. A comprehensive safety programme should be a vital part of all laboratory procedures.

Access Two exits should be provided for the building for prompt exit in the event of fire or other emergencies. Entrances should be designed to minimize pedestrian traffic.

Security A security system must be provided to restrict entry into the laboratory building. Laboratory rooms should be separated from offices by another security system, apart from the general security system, to restrict unauthorized entry into the laboratory rooms, to avoid contamination and for effective operation.

The Building Programme

The building programme is a written document that describes and quantifies the design goals for a building. The goal of a good programme is to define a building that will have ample space, meet the technical requirements of the user, function safely and meet the owners' budget. The design team with the assistance of the laboratory management and technical staff will develop the building programme from the analysis of data collected on the following:

- the range of analyses to be carried out
- the number and types of personnel who will occupy the building
- the interrelationships of functions and personnel
- the expected workload.

The programme should describe the architectural, mechanical, electrical, plumbing and fire protection criteria for the functions to be accommodated. It should also identify areas of special concern for safety such as high hazard areas containing flammable, toxic or pathogenic materials, and should also address the problem of waste removal.

The main tasks and sequence of a building programme are as follows:

- analyse the study report
- interview management, technical staff and other users
- establish space standards
- list various activities and room types required for such activities
- draw a layout diagram for different room types
- determine the number and area of each room type
- develop room data sheets specifying details of functions
- establish building net and gross areas
- describe basic mechanical, electrical, electromechanical and plumbing systems
- describe the services
- estimate the cost of construction.

Supply of Services

Proper supply of services such as electrical connections, gases, hot water, demineralized or distilled water, compressed air, vacuum, telephone and data networks, fire protection systems, smoke detection system and alarms, emergency showers, sprinklers, eye-wash stations, etc. is essential for efficient running of a laboratory. The services should be installed in appropriate places in each room (**Figs 3, 4**). Centralized services for gases, deionized or distilled water, etc., are preferred.

Electrical Connections

It is essential to determine the total electrical load of each room. In order to achieve this, the equipment to be placed in each room must be decided upon and its power requirements (voltage, current rating, etc.) listed and supplied to the consultant or contractor. Equipment such as autoclaves and washing machines may require three-phase connections. These have to be identified and separated from equipment of low-energy consumption. It is recommended that items of high electrical rating are placed in different rooms to balance the power consumption. Proper earth

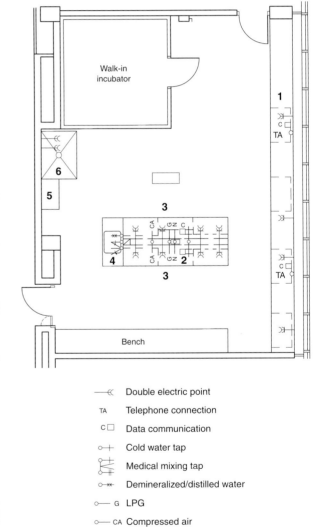

—«	Double electric point
TA	Telephone connection
c☐	Data communication
o—┼	Cold water tap
	Medical mixing tap
o—✳✳	Demineralized/distilled water
o— G	LPG
o— CA	Compressed air
o— N	Nitrogen

Figure 3 Room data sheet – culture techniques. General laboratory furniture: 1, window bench, 75 cm high; 2, island bench, 90 cm high; 3, stool; 4, slab sink; 5, safety storage cabinet; 6, biohazard safety cupboard.

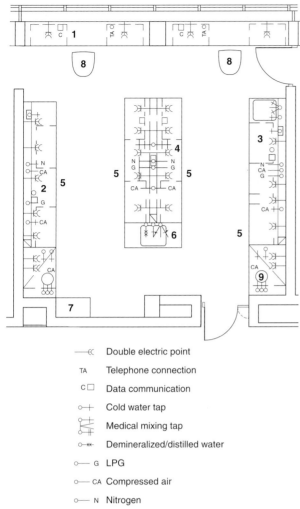

—《	Double electric point
TA	Telephone connection
C□	Data communication
o—+	Cold water tap
	Medical mixing tap
o—××	Demineralized/distilled water
o— G	LPG
o— CA	Compressed air
o— N	Nitrogen

Figure 4 Room data sheet – instrumental techniques. General laboratory furniture: 1, window bench, 75 cm high; 2, wall bench, 90 cm high; 3, wall bench with sink, 90 cm high; 4, island bench, 90 cm high; 5, stool; 6, slab sink; 7, safety storage cabinet; 8, adjustable chair.

connections must be provided with bonding resistance per earth of less than 1 ohm. The minimum resistance of the earthing net should be 1.2 ohm. Circuit breakers must be installed at each workbench.

Gases

Most microbiological laboratories require the following gases:

- liquefied petroleum gas (LPG)
- nitrogen
- carbon dioxide
- oxygen.

The rooms and the locations in each room requiring supply of different gases should be identified and listed. It is possible to provide all laboratory rooms with a supply of piped gases; the gases are supplied in bulk cylinders and stored in an outhouse built for the purpose. The piped supply runs along the corridors, with branches into the laboratory rooms. Each branch should be equipped with a valve enabling the supply to be shut off in an emergency. Stainless steel tubing with Swagelock® fittings is recommended for piping the gases. Welding should be avoided. Pressure checks and certification from the contractor duly approved by the consultant and approval from the civil defence authority are required prior to the actual supply of gas. All the lines must be accessible for future leak checks. Gases such as nitrogen and CO_2 may be required only for anaerobic work stations. If the use of such gases is limited to one or two rooms, the cylinders may be housed in a purpose-built cabinet near to the point of use or within the laboratory, as they are not inflammable or hazardous.

Compressed Air and Vacuum

Laboratories requiring compressed air may be supplied from a centrally located compressor connected to the laboratory by a system of copper or high-pressure plastic pipes. The air should be dried to a dewpoint of – 15°C and freed from oil droplets with the aid of filters. The pressure in the system as far as the branches into the laboratories should be 7 bar, which in the laboratories should be reduced to a working pressure of 3 bar. Vacuum may be supplied through a central system if it is required in many rooms; otherwise, small vacuum pumps may be used.

Hot and Cold Water

The building should be provided with a supply of process water and also drinking water if necessary. The process water should be equipped, downstream of the meter, with a break installation. The pressure measured at the highest tap should be 2.5 bar. The pipes should be laid in such a way that water nowhere becomes stagnant. Wherever necessary, hot water should be provided from closed-in boilers. The minimum temperature of the water should be 60°C. Medical mixing taps with a lever should be provided for mixing cold and hot water, in order to avoid contamination from the hands of the microbiologist.

Demineralized and Distilled Water

A supply of demineralized or distilled water should be available in all the laboratories. Demineralized water can be prepared with the aid of an automatic double-column demineralization system housed in a centrally located room. Distilled water can be prepared with the aid of electrically operated distillation equipment. In both cases, the water should be transported through plastic tubing to the laboratories. The demineralized water should have specific conductance less than $5\,\mu\Omega\,cm^{-1}$.

Telephone and Data Network Connections

Rooms and laboratories requiring telephones must be identified and the appropriate connections provided. Most modern laboratories use client–server technology to manage sample information and analytical data. Several laboratory information management systems (LIMS) are commercially available, and could be customized. Data network connections are provided in the laboratories to facilitate data entry and necessary approvals. They should be located preferably on the side workbenches at a height of 75 cm or near the desks, slightly away from the working area. Network connections are also required on island benches where analytical instruments with data stations are located, to hook up with LIMS for direct transport of instrumental data.

Fire Protection, Smoke Detectors and Alarms

For the purpose of preventing fires, and quelling any fires that break out, it is necessary to draw up plans for fire prevention and firefighting systems. The laboratory building should be divided into compartments separated by fireproof walls and ceilings. The floor surface area of a compartment may vary from 500 m² to 750 m², or as necessary to achieve a logical arrangement of the compartments. In accordance with internationally accepted test methods, the fire retardance of floors and ceilings should not be less than 1 h. All electric and other cables should be passed through fireproof blocks. All spots within the building should be within reach of the jet of a fire hose connected to the process water mains. The reels should be hung on the walls of the corridors, and the hoses should not exceed 30 m in length. In addition to the fire hoses, fire-extinguishers should be distributed throughout the building. Their filling should be in accordance with the type of fires they are likely to be used against. The building should be equipped with an automatic fire alarm system. Ionization detectors should be mounted in all rooms or spaces where fires may start, and are mandatory in rooms where people are at work. A (repeater) fire alarm should accompany each fire-hose reel. The system should be combined with an acoustic alarm system (hooters or sirens) and should be fully independent of the building control system. An emergency power supply system is needed to illuminate and mark escape routes, enabling people to leave the building in the shortest possible time in emergencies. It should be equipped with a no-break unit. The emergency power supply must serve all the electrical equipment that must be kept in operation in emergencies.

Eye-washes and Emergency Showers

All laboratory rooms should be provided with eye-wash stations if possible. Emergency showers are required in laboratories where hazardous chemicals or other materials are being used, and must be easily accessible.

Design of Furniture and Choice of Finishes

Laboratory furniture normally consists of work-benches, cupboards, wall units, desks and drawers. Prefabricated furniture units are available.

Benches

Workbenches can be wall-mounted or island type. The framework should comprise a mild steel tubular framework based on a modular construction with an epoxy-based plastic coating, and should incorporate adjustable levelling jacks, pipe clips and cableways. The bench top should be set at a height of 90–95 cm for normal work in a standing position. The desk tops or 'sit down' benches can be at a height of 65–79 cm as needed to accommodate microscopes, plate counting, computer usage or paperwork. The low-level benches should be mounted on the window side walls to accommodate microscopes, network computers, etc. It is also necessary to keep instruments on low-level island benches for easy access to the reverse of the instrument. Services such as electrical sockets and gas connections on island benches meant for installing instruments should run at the side of the bench for optimum utilization of the bench space. The storage cabinets and drawers should be suspended from the bench connections, and there should be a combination of cabinets and drawers on each bench. Cabinets may be built with WBP grade plywood with an inert and corrosion-resistant finish with minimum seams (e.g. seamless melamine). Drawers may be constructed with corrosion-resistant faced plywood. The cabinets and drawers on the workbench should be fixed in such a way that adequate legroom remains. Ample space should be allowed for refrigerators and writing desks when installing wall-mounted workbenches.

Bench Tops

The bench tops may be constructed from solid melamine, WBP plywood with a seamless melamine finish, WBP plywood with stainless steel top and edges, or solid hardwood with a laminate finish. The bench tops should have a smooth surface and be easily disinfected. Cracks and crevices should be minimal as they provide an opportunity for the build-up of debris which may contribute to cross contamination of samples. Stainless steel tops must be provided on the benches in the washing room.

Sinks

At either end of the benches (apart from benches meant for installing instruments) stainless steel sinks should be mounted with 60 cm side adjoining them, a 50 cm side jutting out, and a depth of 25 cm. Medical mixing taps (cold and hot water) and a deionized or distilled water tap should be mounted above the sinks, as required.

Seating

Laboratory stools and chairs of adjustable or fixed heights should be provided. Stools should be used at the workbenches, and chairs may be used at computer desks.

Wall-mounted Cupboards

Cupboards with sliding glass doors may be mounted on the walls for storing reagents, media, etc. Other cupboards may be used for books, catalogues and instrument files.

Laminar Flow and Biohazard Safety Cabinets

Safety cabinets should comply with standards set by organizations such as the British Standards Institution, the Standards Association of Australia and the National Sanitation Foundation of the USA. Care must be taken in siting equipment that might generate air currents, e.g. fans and heaters. The safety cabinets should be installed in proper sites in the laboratory. Safety cabinets are intended to protect the worker from airborne infection. Work should be done in the middle to the rear of the cabinet, not near the front and workers should not remove their arms from the cabinet until the procedure is completed. After each set of manipulations, aerosols should be swept into the filters. The operator's hands and arms may be contaminated and should be washed immediately after ceasing work. Bunsen burners and micro-incinerators should not be used as they disturb airflow.

Facilities for Incubation and Refrigeration

Incubators

Incubators and incubator rooms must be properly constructed and controlled. It is best to obtain the largest possible models to prevent crowding of the interior. Small incubators suffer wider temperature fluctuations when their doors are open than do larger models. Incubator rooms, if used, must be well insulated, equipped with properly distributed heating units and have appropriate air circulation. They should be installed by specialist suppliers. The rooms should be supplied with stainless steel shelves suitable for holding Petri dishes, flasks, etc. Wooden shelves are not recommended because of the problem of mould growth in a humid atmosphere. The recommended temperatures for incubators in food laboratories are 15–20°C, 30–37°C and 55°C. Cooled incubators must be fitted with a refrigeration system and heating and cooling controls, which must be correctly balanced.

Incubators should be kept in rooms where temperatures are within the range 16–27°C. The incubator temperature must not vary by more than ± 1°C. Chamber temperature must be checked twice daily (morning and afternoon). The thermometer bulbs and stem must be submerged in water or glycerol to the stem mark. For best results use a recording thermometer.

Water Baths

Water baths should be of an appropriate size for the required workload with a suitable water level maintained. When the level of water in the bath is at half to two-thirds the level of the column of liquid in the tube, convection currents keep the constituents of the tube well mixed and hasten reactions such as agglutination. Water baths should be equipped with electrical stirrers to prevent temperature stratification. They must also be lagged to prevent heat loss, although the walls are fitted with sloping lids to prevent heat loss and dripping of condensed water on materials. To avoid choke deposits on tubes and internal surfaces, distilled or deionized water should be used. Only racks made with stainless steel, heat-resistant rubber, plastic, plastic-coated substances or corrosion-proof materials should be used. The temperature of the water bath must be monitored and recorded daily using a certified thermometer.

Refrigerators

A refrigerator maintained at 0–4°C for storing untested food samples is required. Another refrigerator to cool and maintain the temperature of media and reagents may also be used. The temperature of the refrigerator should be checked and recorded daily, and it should be cleaned monthly or more often when required. Refrigerated rooms, if used, must be well insulated and equipped with a distributed cooling system. A continuous temperature monitoring and recording system equipped with an alarm must be used. The temperature at different points should be recorded daily. Stainless steel shelves should be installed for storing samples. Stored materials should be identified and dated, and stored in such a way that cross contamination does not occur. Expired materials should be discarded at regular intervals, e.g. quarterly.

Freezers

A freezer or a freezer room to maintain the temperature of frozen food items at – 18°C is required. The temperature should be recorded daily. A recording thermometer with an alarm system is highly desirable. The freezer should be defrosted and cleaned twice a year. Materials should be identified and dated, and outdated materials should be discarded quarterly. A separate freezing space should be identified for storing freeze-dried bacterial cultures.

Clean/Dirty Sterilization Facilities

Sterilization facilities are required for sterilizing prepared media, diluents, etc., and used glassware, Petri dishes, flasks, tubes, etc. prior to washing or disposal.

The use of heat, particularly moist heat, is the most desirable and widely used method of sterilization in the microbiology laboratory. When using heat sterilizing techniques, it is necessary to know the difference between dry and moist heat and the limitations of each. Moist heat leads to the destruction of microorganisms through the irreversible denaturation of enzymes and structured proteins. The temperature at which denaturation occurs varies with the latent heat of steam. With dry heating, the primary lethal process is considered to be oxidation of cell constituents. Thus, sterilization methods involving dry heat require higher temperatures and longer exposure time than are required with moist heat.

Hot-air Oven

Sterilization by hot-air oven is achieved by the slow penetration of heat into the materials. The efficiency of this process can be increased by the use of circulating fans. Modern equipment has electronic controls which can be set to raise the temperature to the required level, heat for a specified time and switch off automatically. These ovens are fitted with solenoid locks to prevent the oven being opened before the cycle is completed. This protects the staff from accidental burns and safeguards the sterility of the materials. The load should be packed in the oven chamber in such a way that sufficient space remains between the articles for circulation of hot air. The high temperature needed to achieve dry heat sterilization has a damaging effect on many materials. This method should therefore be used only for thermostable materials that cannot be sterilized by steam owing to deleterious effects or failure to penetrate. Materials that can be sterilized by this method include heat-resistant articles such as glass Petri dishes, flasks, pipettes, metallic objects and coated materials.

The performance of a hot-air oven should be tested quarterly with commercially available spore strips or spore suspension. The temperature should be monitored with a certified thermometer, accurate in the temperature range of 160–180°C.

Autoclaves

The minimum recommended standard for sterilization by autoclaves is the exposure to steam at approximately 1 bar pressure, equivalent to 121°C, for 15 min. Saturated steam is a much more efficient means of destroying microorganisms than either boiling water or dry heat. Air has an important influence on the efficiency of autoclaving. If about 50% of the air remains in the autoclave, the temperature of the steam-air mixture at 1 bar is only 112°C. As successful autoclaving depends on the removal of all the air from the chamber, the materials to be sterilized should be packed loosely. There are two types of laboratory autoclaves:

- pressure cooker models
- gravity displacement models.

The pressure cooker is a simple benchtop autoclave consisting of a vertical metal chamber with a strong metal lid which can be fastened down and sealed with a suitable gasket. The lid is fitted with an air/steam discharge trap, a pressure gauge and a safety valve (**Fig. 5**). Steam is generated from the water in the bottom of the autoclave by an external immersion heater or a steam coil.

The gravity displacement autoclave, widely used in microbiological laboratories, consists of a chamber surrounded with a jacket containing steam under pressure, which heats the chamber wall. The steam enters the jacket from the main supply which is at high pressure, thus forcing the air and condensate to flow out of the drain by gravity displacement (**Fig. 6**). In modern autoclaves, air and steam are removed by

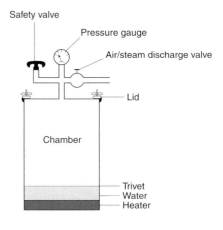

Figure 5 Pressure cooker autoclave.

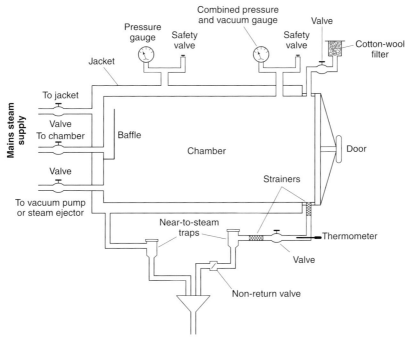

Figure 6   Gravity displacement autoclave.

vacuum pumps and flexible thermocouple probes are fitted in the chamber so that the temperatures at various parts of the load may be recorded.

The performance of the autoclave should be checked monthly using spore strips or suspension. Log books and other records should be maintained for each run, listing the items sterilized, temperature, pressure and time.

Washing Machines

A washing machine may be used for cleaning and drying glassware and other heat-resistant articles. The machine should be capable of washing, rinsing and drying cycles. A log book should be maintained with the details of the programmes used and the materials washed.

Personnel Requirements

Lockers

Lockers are needed to hold the personal belongings of the staff. They should be spacious enough to hold laboratory coats, etc. They may be kept in a staff room.

Laboratory Coats

Laboratory coats must be composed of 100% cotton materials. Polyester or polyester blends must not be used as they easily catch fire. Coats should be long-sleeved and knee-length. They should be washed and decontaminated at least once a week.

Safety Glasses or Goggles

Safety goggles are essential for viewing ultraviolet cabinets and other equipment that may emit UV radiation.

Masks

Face masks with various filters are available for use in laboratories. Appropriate filters are required for working with pathogenic microorganisms and spores, acid fumes, solvent vapours, etc.

Clothing for Entering Freezers or Cold Rooms

Special clothing is available to protect staff entering freezers or cold rooms, and must be worn if staff intend to work for long periods in such rooms.

Gloves

Latex, rubber, leather and heat-resistant gloves are available for use. Gloves, *Hot Hand*® Protector pads must be used when handling hot beakers, conical flasks, etc.

First Aid

A first-aid box and fireproof blankets must be kept in a conspicuous place near the door for use in an emergency.

See also: **Good Manufacturing Practice**. **Laboratory Management**:   Accreditation   Schemes.   **Process Hygiene**: Designing for Hygienic Operation.

Further Reading

Ashbrook P and Renfrew M (eds) (1991) *Safe Laboratories: Principles and Practices for Design and Remodelling.* Lewis Publishers.

Barker JH, Blank CH and Steere NV (eds) (1989) *Designing a Laboratory.* Washington: American Public Health Association.

Collins CH, Lyne PM and Grange JM (1998) *Microbiological Methods*, 7th edn. Oxford: Butterworth Heinemann.

Diberardinis LJ, Baum JS and First MW (1993) *Guidelines for Laboratory Design: Health and Safety Consideration*, 2nd edn. Chichester: John Wiley.

FAO (1986) *Manual of Food Quality Control. 1. The Food Control Laboratory.* FAO Food and Nutrition Paper 14/1 Rev. 1. Rome: FAO.

Laboratory Design Handbook (1994) Lewis Publishers.

Speck ML (ed.) (1984) *Compendium of Methods for the Microbiological Examination of Foods.* Washington: American Public Health Association.

LABORATORY MANAGEMENT – ACCREDITATION SCHEMES

Catherine Bowles, Leatherhead Food Research Association, Surrey, UK

Introduction

Accreditation of laboratories carrying out analysis in the food industry is an important aspect of the management of such laboratories. Accreditation requires constant monitoring of the quality system, enabling management to keep a tight control of the laboratory's activities and to have confidence in the results produced.

Accreditation revolves around the concept of *traceability*. The external formal accreditation of testing laboratories establishes a recognized standard of providing accurate, reliable and repeatable test results under controlled conditions. Accreditation is a tool enabling both those working within a laboratory and those relying upon its services to be sure of the validity of results. All aspects of a laboratory's activities can be traced, from receipt of samples through analysis to test report generation, including the staff involved in analysis, the media used and the results recorded. Formal accreditation of laboratories performing microbiological testing in the food industry is increasingly demanded by retailers, manufacturers and government representatives.

Accreditation Standards

Laboratory accreditation is, at present, not required by law in the UK, except for laboratories involved in the production of data for licensing substances (Good Laboratory Practice (GLP)) assurance is required). However, guidelines do exist to determine good manufacturing practice through due diligence and the use hazard analysis critical control point (HACCP) systems. Much of the UK food manufacturing industry works to British Standard BS 5750 (ISO 9000, see below) to enable recognition of a standard of operation. The Control of Substances Hazardous to Health regulations (COSHH) and all Health and Safety at Work acts apply to food microbiology laboratories. Legislation also exists governing the qualifications required by public analysts or food examiners from the Food Safety Act.

The recognized international standard for management quality systems established by the International Standards Organization (ISO) is ISO 9000. The British Standard BS 5750 has been incorporated with the ISO requirements; BS 5750 became BS EN ISO 9000 in 1994. The ISO 9000 Standards take on a generic role in standards. The ISO Guide 25 defines the application of ISO 9000 in laboratories. The United Kingdom Accreditation Service (UKAS) uses the NAMAS (National Measurement Accreditation Service) standard (which is ISO Guide 25 compliant) to assess laboratories and grant certification. Accreditation by UKAS has government recognition and can only be applied to commercial testing and calibration laboratories. Although UKAS is mainly a UK certification body, it has increasing recognition worldwide.

There is considerable movement towards consolidating assurance standards on a worldwide scale. The Organization for Economic Cooperation and Development (OECD) within the EEC has achieved considerable progress in the move towards international standards of laboratory practice. Along with several guidelines originally published between 1986 and 1989 concerning GLP, Council Directive 93/99/EEC was published in 1993 to set out conditions for results from accredited laboratories to be accepted throughout the EEC. This move towards international cooperation will benefit the manufacturer, the consumer and related concerns.

In 1997, the European cooperation for Accreditation (EA) was formed from a merger of the European Accreditation of Certification (EAC) and the European cooperation for the Accreditation of Laboratories (EAL). This indicates a further progression of cooperation between countries to consolidate the accreditation of calibration and testing laboratories to one internationally recognized standard. At the time of writing, the members of the EA are Austria, Belgium, Denmark, Finland, France, Germany, Greece, Iceland, Ireland, Italy, Netherlands, Norway, Portugal, Spain, Sweden, Switzerland and the UK; associate member countries are the Czech Republic, Hungary and Slovakia.

Advantages and Disadvantages of Accreditation Schemes

An accreditation scheme must be relevant to the work of the laboratory, which may be pharmaceutical or medical, or related to manufacturing of food or drinks, or to the commercial testing of food, water and environmental samples. If the laboratory is sited within a manufacturing company that already has accreditation, it may become part of the assurance scheme operating throughout the organization. The ISO 9000 Standard is currently applied to manufacturing organizations and can include any laboratories only carrying out analysis in-house. The main advantage of maintaining an accreditation scheme is the national and international recognition of the quality of results produced from the testing laboratory. Formal accreditation recognizes the effectiveness of the quality scheme operating in the laboratory. It enables management and staff to operate to a documented quality system. This secures work from customers requiring national and international recognition of the test results produced by the laboratory in the ever-increasing competitive markets.

Through the laboratory's internal audit and quality review systems, the compliance of the system with documented procedures must be constantly monitored. This enables anomalies either from external complaints or from an internal source to be dealt with efficiently to a specific protocol. The requirement for traceability throughout the laboratory's daily activities should facilitate prompt identification of any major problems in the quality system before the report is compiled and sent to the customer. Results should be consistent, providing a reliable service to customers and industry alike from the operation of fully validated and documented methodology. Staff training must reflect the scope of accreditation, ensuring that all testing is carried out by competent personnel familiar with the analysis that they are performing. Equipment performance must be constantly monitored with respect to operating parameters enabling testing to be carried out under controlled conditions. This also should rapidly identify any problems with test conditions so that the appropriate action can be taken.

Customers may request that some products are analysed by an accredited laboratory, despite the fact that the particular test employed may be beyond the scope of the laboratory's accreditation. The fact that the analysis is carried out in accredited conditions using accredited equipment, controlled media and reagents, competent staff working to recognized standards and results reported using traceability, may satisfy the customer. Thus, an accredited laboratory could attract more business than an unaccredited laboratory.

The main disadvantage of formal accreditation schemes is the cost of operating a quality system, both in terms of revenue and staff time. An accreditation scheme also restricts the flexibility of the laboratory to operating a testing regimen compliant with the accredited scope. A large scope of accreditation enables a large range of services to be offered to customers, but can be difficult to maintain because of staff training, equipment monitoring and documentation commitments. In the competitive commercial market of microbiological analysis of food, test prices should reflect the accreditation status of a laboratory.

Laboratories dedicated to carrying out research work are restricted as to the type of formal accreditation applicable to the nature of research. Normal strict control on methodology to recognized protocols and documentation of procedures to be adhered to is more difficult to apply to the uncertain routes that research must take to advance technology in a particular area. It is possible to apply ISO 9000 on the basis that the systems used in research can be documented and appropriate records maintained.

Accreditation schemes require a high degree of commitment from the staff as the time involved in maintaining a quality system is considerable. This time is non-profitable in a commercial laboratory which has to be justified to the budget controllers. For microbiology there is an added problem with samples having to be tested as soon as possible on receipt, which can lead to sample analysis taking priority over maintaining an accredited system. A compromise must be reached, which should be possible with a well-documented and well-run quality system. If staff are well trained time spent on maintaining the system can be reduced by keeping all accreditation-related activities as up-to-date as possible. Once a system is allowed to fall behind on areas such as record keeping, auditing and document amending, it takes a considerable extra effort to bring things back into line. Staff may be opposed to working within a quality system owing to the increase in records which is thought of as 'a waste of time' in relation to the 'real' work. This attitude must be overcome. In the long term, formal accreditation should increase the confidence of laboratory results and bring a recognized central standard of working which is known worldwide.

Accreditation Bodies

The main accreditation body granting accreditation to testing and calibration laboratories within the UK is UKAS. Other accreditation can be granted under the authority of the LABCRED, CLAS, BS 5750 and GLP schemes. Each scheme has its own merits depending upon the requirements of the laboratory seeking accreditation.

GLP

GLP is an assurance scheme specifically for laboratories providing toxicological analysis for the licensing of substances. It was set up to avoid duplicate testing of pharmaceutical products and chemicals and is an older scheme than the ISO 9000 Standard, originating from an investigation in the USA by the Food and Drug Administration into allegations of data validity in the 1970s. The GLP regulations were proposed in 1976, becoming effective in 1979. The OECD developed the principles of GLP in order to enable European countries to accept each other's work. These principles were adopted in 1981 with implementation from sources of the individual governments concerned (for example, the Department of Health in the UK).

CLAS and LABCRED

The CLAS and LABCRED accreditation schemes were set up in the UK following the perceived need for a 'practical' standard of accreditation for microbiology and chemistry testing laboratories within the food and drink industry. Both schemes have recognition within the industry but no recognition by the government.

The CLAS accreditation scheme is run by the Campden and Chorleywood Food and Drink Research Association (CCFDRA). The scheme is based on the European standard Euronorm (EN) 45001 (general criteria for the operation of testing laboratories). Assessors are trained to NAMAS or BS 5750 standards and are taken from a pool within CCFDRA.

The LABCRED accreditation scheme is operated by Law Laboratories Ltd in Birmingham. It was set up in 1993 and sets its own standard based on EN 45001, BS 5750 and other recognized standards. It is maintained by assessors from Law Laboratories Ltd assuring the granting of certification to laboratories operating to its required standard.

BS 5750

The British Standard BS 5750 is identical to the ISO 9000 Standard in Europe; part 1 is the specification for design/development, production, installation and servicing; part 2 is the specification for production and installation; and part 3 is the specification for final inspection and test. Published in 1994, BS 5750/ISO 9000 is a standard for Quality Management Systems and is applied to the manufacturing industries. It is a standard less concerned with quality of the product (or test result), but can be used as a generic scheme for laboratories. Certification of compliance with BS 5750/ISO 9000 is granted by bodies such as the British Standards Institution with a licence to certify from the National Accreditation Council for Certification Bodies (NACCB).

UKAS

The NAMAS Standard accreditation is awarded through the UK Accreditation Service. The National Measurement Accreditation Service originated in 1985 from an amalgamation between the British Calibration Service (BCS) and the National Testing Laboratory Accreditation Service (NATLAS). The NAMAS Standard M10 was published in 1989, superseding the original NATLAS N1 document. In 1995 UKAS was formed through a merger of NAMAS and NACCB and continues to grant accreditation on compliance to the NAMAS Standard.

The NAMAS Standard originated in the engineering sector and has diversified and adapted (since 1990) to incorporate assessment of food testing laboratories. The UK Accreditation Service has government recognition for the awarding of accreditation certification for commercial testing and calibration laboratories against the NAMAS Standard (ISO Guide 25 compliant). As previously stated, ISO 9000 was written for production and manufacturing industries and software development, whereas NAMAS is the recognized application of ISO 9000 Standard in laboratories. Certification is granted by UKAS following assessment of a laboratory against the NAMAS Standard by permanently employed assessment managers in conjunction with a team of independent technical assessors.

Implementation of an Accreditation Scheme

Before implementing an accreditation scheme, the laboratory's commitment to the introduction and maintenance of the scheme must be established. Any accreditation scheme is a considerable drain on finances and staff time; therefore, once committed, a laboratory must be dedicated to maintaining the scheme. The resources involved in the initial achievement of an accreditation scheme can be reduced if a quality system is already operating within the laboratory which is adaptable to the scheme chosen.

Once commitment has been assured, the scheme to implement must be decided upon. The scheme must be appropriate for the laboratory's activities. The main consideration is the purpose of the work carried out by the laboratory (for example commercial testing, internal testing within a factory, or government trials work). Other considerations include the type of client base the laboratory has and any quality systems already operating within the laboratory or associated business. It is far easier to adapt a quality system with which staff are already familiar than to introduce a completely new scheme (provided the existing scheme is workable). The principle of keeping the documentation and the system itself as simple as possible (with no loss of purpose) must be kept in mind. An unnecessarily involved system will become too cumbersome to maintain effectively. Once the scheme has been decided upon, the body granting accreditation must be contacted. Starter documentation with guidelines as to the areas and standards to be met to achieve formal certification must be purchased from the accreditation body.

Independent consultants will provide help in implementing an accreditation scheme – at a price. They have experience in writing documentation to comply with the requirements of accreditation. They also have experience of various schemes and the best approach to be taken to fulfil the requirements. Alternatively, training courses are run by several organizations (often with the cooperation of the accreditation bodies) to make the task of attaining accreditation less daunting. The courses cover most aspects of accreditation from quality management and how to prepare a quality manual to auditing a quality system (again, cost is involved). Laboratories that are already certified by an accreditation body are another valuable source of help and advice in the best approach to particular areas. The actual system may differ, but the requirements of a standard still have to be met.

The Quality Manual

The main document to prepare for most accreditation schemes is the Quality Manual. This exists to lay down the policy and procedure of the operation of the laboratory. The manual is set out in sections which include any or all of the following, depending on the accreditation body:

- management and staff
- the quality system
- documentation and records
- quality assurance
- audit and review
- test methods
- sample handling
- housekeeping
- equipment
- calibration of equipment
- administration procedures
- test reports
- the environment
- security and confidentiality
- subcontracting of testing.

The detail contained in the manual must meet the requirements of the accreditation standard. The manual must be operated under authorization by a specific member of staff (usually the technical manager, laboratory manager or quality manager) with documented commitment from the highest level of management. The style of layout must be compliant with the requirements of the accreditation body. Further associated documents within the quality system must be referred to including any of the following:

- standard operating procedures (equipment operation and/or calibration procedures; laboratory practices on daily operations)
- methods manuals
- files or records specific to the daily operation of the laboratory
- equipment records
- staff training records.

The Quality Manager

A quality system usually demands the appointment of a quality manager (or somebody within the quality assurance area) to maintain the implementation of the quality system. This role will involve organizing audits, liaising with the accreditation body and maintaining the controlled documentation governing the quality system. Regular reviews by the management of the whole system are required to monitor performance.

Documentation

The laboratory should prepare as much documentation as possible within the guidelines set out by the accreditation body before the initial assessment visit will be set up. Documentation preparation is a daunting task and is not to be undertaken lightly. Under a central coordinator, the involvement of all laboratory staff in preparing documentation for their quality system can be beneficial; the introduction of a quality system may meet with opposition, and staff involved in the implementation of their scheme are more likely to operate it effectively. This approach also spreads the load of documentation preparation if time will permit. The following list is a guide to the areas to be addressed for accreditation purposes. The specific requirements will depend upon the accreditation body:

1. Test methodology must be based on recognized standard methods (for example British Standards or the International Dairy Federation) and written appropriately.
2. Test methods usually require validation by the laboratory operating them.
3. Documentation of staff training in operation of equipment and carrying out analysis is required.
4. Appropriate equipment calibration (externally or internally) must be documented. The equipment will include balances, autoclaves, pH meters, incubators, thermometers, water baths and freezers, for example.
5. Reference equipment must be kept only for the purpose of internal calibrations if appropriate (e.g. balance weights and reference thermometers).
6. The requirement for the laboratory to incorporate external and internal quality assurance schemes must be addressed (in the UK the Public Health Laboratory Service and the Ministry of Agriculture, Fisheries and Food both run external quality assurance schemes designed for laboratories).
7. The audit procedure must be documented adequately with all aspects of laboratory activities covered in a specified time and non-compliances identified to be addressed in a given time period.
8. Suppliers of reagents, media and consumables are usually required to be accredited (ISO 9000 is sufficient) to ensure traceability.
9. Media prepared are required to have adequate quality control records with respect to preparation (autoclave load) and performance (positive and negative control organisms).
10. Control cultures must be traceable to national standards such as the National Collection of Type Cultures (NCTC) or the National Collection of Food Bacteria (NCFB). Storage and use of the reference cultures must be subject to control.
11. Computer systems (for example the use of a Laboratory Information Management System (LIMS) system in report generation).
12. Environmental control (swabbing of laboratory areas for microbiology laboratories).
13. Correct disposal of samples and waste.

Auditing

Internal auditing is an integral part of the quality system and is a useful tool to be started before the initial assessment visit from representatives of the accreditation body. The accreditation body cannot usually audit all aspects of a laboratory's activities during one visit, therefore internal auditing is an excellent way of policing a quality system between visits. Audits must be carried out by staff independent of the area under audit and must be trained in the process. Many bodies provide training courses for auditors to achieve certified auditor/assessor status (ISO 9000 assessor for example).

Assessment

The preparations for assessment can take up to 2 years to complete. The timescale is not the same for any two laboratories: it is dependent upon the existing systems, the scope of accreditation sought and the staff involved in the implementing of a scheme. Once the laboratory feels that it is ready for assessment, the actual accreditation procedure is started by the accreditation body. Usually this involves an assessment visit to audit all aspects of the laboratory's quality system to ensure that the required standards are being met. There are documented standards to be followed by assessors conducting the visit. The laboratory is entitled to question points and ask for guidance on any issue.

There is no 'right' or 'wrong' implementation of a quality system provided the standards of the accreditation body are met. Each laboratory will have its own system tailored to the work undertaken, customer requirements and laboratory resources. As long as the testing is carried out to the documented system which in turn is compliant with the required standard, then accreditation should be granted. Prior to formal assessment, forms are completed by the laboratory seeking accreditation, detailing information such as:

- the staff involved in the laboratory with respect to position and qualifications held
- management structure of the laboratory
- work undertaken by the laboratory and scope of accreditation sought

- equipment calibration procedures and policy
- external quality assurance schemes participated in.

Some accreditation bodies may require access to the documentation prepared to control the quality system before an assessment visit. This will mainly involve the quality manual, and documented methods for which accreditation is sought. The formal assessment is carried out to a documented procedure by assessors appointed by the accreditation body. The date is mutually agreed between the assessment team and the laboratory once the laboratory feels it is ready for a preliminary visit and the accreditation body's application requirements have been met. The time involved for assessment is dependent upon the size of the scope of the accreditation for which the laboratory is seeking. Usually it is not possible to audit all aspects of a quality system at one visit. The initial assessment visit, however, must cover sufficient areas to an adequate depth to satisfy the accreditation body that the laboratory is operating to the required standards.

An initial meeting between the assessment team and the laboratory management will begin the visit. Formal introductions will precede the lead assessor's explanation of the structure and purpose of the assessment visit. The assessment team will investigate all aspects of the quality system in operation with cooperation of all laboratory staff. Laboratory staff are expected to be familiar with the policy and procedure for the area in which they are working (or to know where to look for documentation governing their operation). A written report will detail the findings of the visit, specifying any non-compliances and the level of their severity. Actions are agreed between assessors and laboratory staff and a timescale is set for clearance of them.

Attainment of an accreditation scheme entitles the accredited laboratory to display a logotype relevant to the accreditation body. The use of a logotype is governed by the accreditation body and can only usually be used in association with the accredited activities of the laboratory.

Maintenance of an Accreditation Scheme

Once accreditation is granted, the hard work cannot stop. Formal accreditation requires much commitment, time and work to set up. It takes an equal amount of commitment to maintain – maybe even more, as once accreditation is achieved the system must be adhered to even though the pressure of a goal of attainment no longer applies.

A good relationship between the laboratory and the accreditation body is important in maintaining an accreditation scheme. The accreditation body will have personnel available for help and advice at any time between assessment visits. They also need to be kept informed of any major changes in key management staff or practices within the operation of the quality system. A laboratory may voluntarily suspend accreditation if it cannot meet the standard required at any time for any reason.

Documentation and Auditing

Internal audit and review must be kept up according to the schedule laid out in the laboratory's quality manual. It is the mark of an efficient system to determine its own instances of non-compliance before the external assessors can locate them. Documentation controlling the quality system must be up-to-date, with amendments made and authorized following recorded procedure. Amended copies are required to be retained for a stated period along with all records involved in the accredited system. For UKAS accreditation records must be retained for a minimum of 6 years in a secure place. Access to records must be restricted to ensure confidentiality. All documented checks and calibrations, external quality assurance and internal quality assurance must be maintained.

Staff Training

The initial training of staff when employed is important to the maintenance of a quality system. A thorough understanding of the laboratory's operation within documented parameters must be emphasized to ensure adherence to the system. Many non-compliances exposed at assessment visits (and from internal auditing) are due to 'human error' in not completing the required documentation or carrying out calibration checks. The daily operation of a quality system must become second nature to those involved.

Continued Monitoring

Assessment visits will highlight areas where the standard has slipped or not been adhered to. The regularity of assessment visits is dependent upon the accreditation body (annually for UKAS accreditation). Reports on the visits will take the documented form of the accreditation body, allowing a period of time for the laboratory to put right any problems found during the visit before accreditation is suspended or revoked (if of a level not requiring immediate suspension).

The Future

Finally, an important aspect of the maintenance of a quality system must be the continuous growth and development of the system. A quality system must not remain static, but must be able to develop with the laboratory as practices change to incorporate customer, manufacturer and technological advances.

See also: **Hazard Appraisal (HACCP)**: The Overall Concept; Critical Control Points; Involvement of Regulatory Bodies; Establishment of Performance Criteria. **National Legislation, Guidelines & Standards Governing Microbiology**: European Union.

Further Reading

Campden and Chorleywood Food and Drink Research Association (1993) CLAS – The Campden Laboratory Accreditation Scheme.

Department of Health (1989) *Good laboratory practice – The United Kingdom Compliance Programme*. London: DOH.

Institute of Food Science and Technology (1991) *Food and Drink – Good Manufacturing Practice: A Guide to its Responsible Management*, 3rd edn.

International Standards Organization (1990) ISO/IEC Guide 25. *General requirements for the competence of calibration and testing laboratories*. 3rd edn.

Law Laboratories Ltd (1993) LABCRED – Laboratory Accreditation Scheme. Issue No 1.

United Kingdom Accreditation Service (1989) NAMAS Standard M10. Edition 1.

Wilson S and Weir G (1995) *Food and Drink Laboratory Accreditation*. London: Chapman & Hall.

Lactic Acid Bacteria *see* **Lactobacillus**: Introduction; *Lactobacillus bulgaricus; Lactobacillus brevis; Lactobacillus acidophilus; Lactobacillus casei;* **Lactococcus**: Introduction; *Lactococcus lactis* Sub-species *lactis* and *cremoris;* **Pediococcus**

LACTOBACILLUS

Contents
Introduction
Lactobacillus bulgaricus
Lactobacillus brevis
Lactobacillus acidophilus
Lactobacillus casei

Introduction

Carl A Batt, Department of Food Science, Cornell University, USA

Characteristics of the Species

The genus *Lactobacillus* is quite diverse and consists of a number of different species with little commonality. A measure of their diversity can be estimated by the range in the G+C% content among the lactobacilli. Members of the species have G+C% of 32–53%, which is a much wider range than is encountered with other lactic acid bacteria. Their common taxonomical features are restricted to their rod shape and their ability to produce lactic acid either as an exclusive or at least a major end product. In addition, they are Gram-positive and do not form spores. *Lactobacillus* cells are typically rod-shaped with a size range of $0.5–1.2 \times 1–10\,\mu m$. Under certain growth conditions they can look almost coccoid-like hence this characteristic is not absolutely diagnostic. In fact the former *Lactobacillus xylosus* has been reclassified as *Lactococcus lactis* subsp. *lactis*, although its historical designation as a lactobacillus must have been on the basis of its rod-shape coupled to its ability to ferment xylose. It is the latter phenotype which before the 1980s probably excluded it from the typical xylose non-fermenting *L. lactis* subsp. *lactis* (itself formerly known as *Streptococcus lactis*). The lactobacilli are facultative anaerobes that, in general, grow poorly in air, but their growth is sometimes enhanced by 5% carbon dioxide. Because they are auxotrophic for a number of different nutrients, they grow best in rich complex media. It is auxotrophies in some strains that have been exploited to develop bioassays for a number of vitamins and other micronutrients. Their optimum growth temperature is 30–40°C, but they can grow over a range of 5–53°C. They are also aciduric with an optimum growth pH of 5.5–5.8 but in general they can grow at a pH < 5.

A number of traits distinguish members of the *Lactobacillus* genus from other lactic acid bacteria. The characteristics are presented in **Table 1**.

In addition to growth characteristics at different temperatures, pH values and salt concentrations, other methods to distinguish the lactobacilli include carbohydrate fermentation patterns, hydrolysis of arginine, peptidoglycan content and DNA–DNA homology. As with many taxonomic schemes the usage of 16S rRNA sequence data is increasingly viewed as definitive. Some assemblage can be made using a subset of characteristics as shown in **Table 2**. Concordance between these various classification schemes can never be realized, and different clas-

Table 1 Characteristics for discrimination of lactic acid bacteria

Character	Carno	Lactob	Aeroc	Enteroc	Lacto/ Vagno	Leuco/ Oenoc.	Pedio	Strepto	Tetragen	Weissella
Tetrad formation	–	–	+	–	–	–	+	–	+	–
CO_2 from glucose	–	+/–	–	–	–	+	–	–	–	+
Growth at 10°C	+	+/–	+	+	+	+	+/–	–	+	+
Growth at 45°C	–	+/–	–	+	–	–	+/–	+/–	–	–
Growth at 6.5% NaCl	ND	+/–	+	+	–	+/–	+/–	–	+	+/–
Growth at 18% NaCl	–	–	–	–	–	–	–	–	+	–
Growth at pH 4.4	Ns	+/–	–	+	+/–	+/–	+	–	–	+/–
Growth at pH 9.6	–	–	+	+	–	–	–	–	+	–
Lactic acid	L	D,L,DL	L	L	L	D	L,DL	L	L	D, DL

Table 2 Group classification of lactobacilli (from Salminen and von Wright 1998)

Character	Group I obligate homofermenters	Group II facultative heterofermenters	Group III obligate heterofermenters
Pentose fermentation	–	+	+
CO_2 from glucose	–	–	+
CO_2 from gluconate	–	+	+
FDP aldolase	+	+	–
Phosphoketolase	–	+	+
	L. acidophilus	L. casei	L. brevis
	L. delbrueckii	L. curvatus	L. buchneri
	L. helveticus	L. plantarum	L. fermentum
	L. salivarius	L. sake	L. reuteri

sification schemes, tailored to the particular problem, should probably be tolerated.

The lactobacilli include over 25 unique species, and the first level of differentiation is based on end-product composition; some are homofermentative whereas others are heterofermentative. The former are classified as organisms that produce > 85% lactic acid as their end product from glucose. The latter include organisms that produce approximately 50% lactic acid as an end product, with considerable amounts of carbon dioxide, acetate and ethanol. Notable among the homofermenters are *L. delbrueckii*, *L. leichmannii* and *L. acidophilus*. Heterofermenters include *L. fermentum*, *L. brevis*, *L. casei* and *L. buchneri*. Although they all produce lactic acid as a major end product they differ in the isomeric composition. Some produce exclusively L(+) lactic acid and these include *L. salivarius* and *L. casei*. Others, for example *L. bulgaricus* and *L. jensenii* produce just D(–), and finally *L. acidophilus* and *L. helveticus* produce a mixture of D(+) and L(–) lactic acid. The next major criterion for distinguishing among the lactobacilli is the production of gas from carbon sources including glucose and gluconate. In addition there is a great degree of diversity in the ability of various *Lactobacillus* spp. to ferment pentose sugars including ribose and xylose.

Importance to the Food Industry

Given the diversity of metabolic properties exhibited by members of the *Lactobacillus* genus they are found in a number of fermented food products. In these products the lactobacilli contribute to their preservation, nutrition availability and flavour. Lactobacilli are added as deliberate starters or take part in the fermentation as a result of their being natural contaminants of the starting substrates.

A number of dairy products are produced using *Lactobacillus* either alone or in combination with other lactic acid bacteria. Acidophilus milk is an example of a fermented dairy product and *L. acidophilus* is the organism used to produce it. *L. bulgaricus* in combination with *Streptococcus thermophilus* is used to produce yoghurt and a balance between these two starters can affect product quality.

Vegetables are fermented with lactobacilli to produce products including pickles, olives and sauerkraut. Members of the *Lactobacillus* genus are natural contaminants of vegetables and take their place in the fermentation process along with a number of other microorganisms. The lactobacilli produce modest amounts of acid and are usually a transient flora in the process.

Lactobacillus species play an essential role in bread-making and a number of unique strains have been identified in products most notably sourdough bread. Typical species of lactobacilli identified in sourdough bread include *L. acidophilus*, *L. farciminis*, *L. delbrueckii* subsp. *delbrueckii*, *L. casei*, *L. plantarum*, *L. rhamnosus*, *L. brevis*, *L. sanfrancisco* and *L. fermentum*. The exact composition of most sourdough breads is not known and attempts to blend starters to mimic a particular product are sometimes less than satisfactory. Traditional sourdough fermentations are carried out by 'back-slopping', a process where a small fraction of a prior batch is used to start the next batch. The indigenous lactobacilli are able to overcome other contaminating microflora largely by thriving under the fermentation conditions. During the fermentation process lactic acid builds up levels approaching 1% and a small amount of acetic acid is also produced. The number of lactic acid bacteria can reach $10^7 \, \text{cfu g}^{-1}$.

Another property of lactobacilli that has become more appreciated is their ability to produce bacteriocins. The bacteriocins produced by lactobacilli are presented in **Table 3**. Bacteriocins probably evolved to provide the producing organism with a selective advantage in a complex microbial niche. Incorporation of *Lactobacillus* spp. as starters or the inclusion of a purified or semi-purified bacteriocin preparation as an ingredient in a food product may provide a margin of safety in preventing pathogen growth.

Importance to the Consumer

Lactobacillus species are important either as deliberate or accidental ingredients in many food products. A great deal of attention has been directed toward their potential role as probiotics. Strains which have been examined for their probiotic effects include *L. acidophilus* LA1, *L. acidophilus* NCFB 1748, *Lactobacillus* GG, *L. casei* Shirota, *L. gasseri* ADH and *L. reuteri*. Reported clinical effects attributed to the consumption of *Lactobacillus* consist of immune enhancement, lowering faecal enzyme activity, preventing intestinal disorders, and reducing viral diarrhoea. Most probiotic strains are believed to have an ability to colonize the intestinal tract and thereby positively affecting the microflora and perhaps excluding colonization by pathogens.

Although the potential benefits from the consumption of probiotics is significant, documentation of their efficacy is limited. Efficacy trials are difficult to perform especially in humans where compliance with experimental protocols and normalizing for other genetic and environmental factors are difficult. Another unappreciated problem is the lack of strain identification. There are for example no strict classification schemes that for example would denote *Lactobacillus* GG from other *Lactobacillus* species.

See also: **Bacteriocins**: Potential in Food Preservation. **Bread**: Bread from Wheat Flour; Sourdough bread. **Fermented Foods**: Fermented Vegetable Products. **Fermented Milks**: Range of Products; Yoghurt. ***Lactobacillus***: *Lactobacillus bulgaricus*; *Lactobacillus brevis*; *Lactobacillus acidophilus*; *Lactobacillus casei*. **Lactococcus**: *Lactococcus lactis* Sub-species *lactis* and *cremoris*. **Probiotic Bacteria**: Detection and Estimation in Fermented and Non-fermented Dairy Products. **Starter Cultures**: Uses in the Food Industry.

Further Reading

Ludwig W, Seewaldt E, Kilpper BR et al (1985) The phylogenetic position of *Streptococcus* and *Enterococcus*. *Journal of General Microbiology* 131: 543–551.

Salminen S and von Wright A (1998) *Lactic Acid Bacteria*. New York: Marcel Dekker.

Stiles ME and Holzapfel WH (1997) Lactic acid bacteria of foods and their current taxonomy. *International Journal of Food Microbiology* 36: 1–29.

Lactobacillus bulgaricus

Paula C M Teixeira, Escola Superior de Biotecnologia, Porto, Portugal

Characteristics

Lactobacillus delbrueckii subsp. *bulgaricus* (termed *L. bulgaricus* hereafter) is an aerobic to anaerobic homofermentative bacterium (converts hexoses into lactic acid via the Emden–Meyerhof pathway) normally isolated from yoghurt and cheese. Carbohydrates fermented by *L. bulgaricus* (90% or more strains) are fructose, glucose and lactose. Lactic acid is the major end product of fermentation; however

Table 3 Selected bacteriocins produced by members of the *Lactobacillus* species

Bacteriocin	Organism	Sensitive strains
Lactacin B	*L. acidophilus*	*L. delbrueckii*, *L. helveticus*
Lactacin F	*L. acidophilus*	*L. fermentum*, *S. aureus*, *E. faecalis*
Brevicin 37	*L. brevis*	*P. damnosus*, *Leu. oenos*
Lacticin A	*L. delbrueckii*	*L. delbrueckii* subsp. *lactis*
Helveticin J	*L. helveticus*	*L. helveticus*, *L. delbrueckii* subsp. *bulgaricus*
Sakacin A	*L. sake*	*Carno. Piscicola*, *L. monocytogenes*
Plantaricin A	*L. plantarum*	*Lactococcus lactis*, *E. faecalis*
Gassericin A	*L. gasseri*	*L. acidophilus*, *L. brevis*

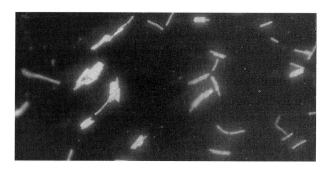

Figure 1 Acridine orange staining of *Lactobacillus bulgaricus* at early stationary phase of growth (×100).

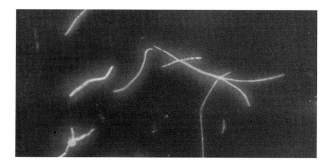

Figure 2 Acridine orange staining of *Lactobacillus bulgaricus* at late stationary phase of growth (×100).

secondary end products such as acetaldehyde, acetone, acetoin and diacetyl can also be produced in very low concentrations.

In lactic acid bacteria that do not possess superoxide dismutase, the dismutation of superoxide is normally catalysed by internally accumulated manganese. *L. bulgaricus*, however, has a low capacity to scavenge O_2^- since it has neither superoxide dismutase nor high levels of Mn (II) and is very sensitive to O_2 (the ability to grow aerobically must be distinguished from the ability to survive exposure to O_2).

Cells are rod shaped with rounded ends, 0.5–0.8 by about 2–9 μm. They are usually separate or in short chains (**Fig. 1**) but long chains can also be observed in late stationary phase cells (**Fig. 2**). They are generally short but some times long, straight and often arranged in palisades. Internal granulations are observed with the Gram reaction or methylene blue stain especially when cells become older. In addition to age, variability of *L. bulgaricus* cellular morphology is dependent on the composition of the growth medium and oxygen tension. Additional physiological and bio-

chemical characteristics are presented in **Table 1** and **Table 2**.

Fatty acid composition has been used to group and classify these microorganisms. As shown in Table 2, lipid compositions are different in the different strains. However, hexadecanoic (16 : 0), hexadecenoic (16 : 1), octadecenoic (18 : 1), and lactobacillic (19 : 0) acids are the major fatty acids present which are common to the three *L. bulgaricus* strains. In addition to the strain, variability can occur as a result of different growth conditions (medium composition, temperature), phase of growth and even the methodology used for lipid extraction.

Methods of Detection and Enumeration of *Lactobacillus bulgaricus* in Foods

Plating Methodologies

Rogosa agar (RA) is used for the isolation and enumeration of lactobacilli from milk, cheese and other fermented milk products (**Table 3**). With the exception of yoghurt and bio-yoghurts the usual purpose of a lactobacillus count is to ensure that numbers are high in products to which they are added. Since RA may not be optimum for isolation of some thermophilic lactobacilli from dairy sources, it is recommended that it is supplemented with 0.5% (w/v) meat extract (leuconostocs and pediococci are not inhibited and colonies may have to be further identified). Typical colony appearance of *L. bulgaricus* on RA is 0.5–2.5 mm diameter, greyish-white, flat or raised and smooth, rough or intermediate. Several media have been developed for selective isolation and enumeration of *L. bulgaricus* from yoghurt (**Table 4**). Enumeration can be performed using two different types of media:

1. Media formulated to isolate *L. bulgaricus* selectively such as acidified De Man Rogosa and Sharp agar (MRS) and acidified Reinforced Clostridial agar (RCA);
2. Differentiating media that permit the enumeration of all the organisms as separate, and visually identifiable colonies on the same plate such as Lee's medium and L-S differential medium (L-S).

Acidified MRS (Table 4) is the medium of choice of the International Dairy Federation (IDF), for differential enumeration of *L. bulgaricus* in yoghurt. RCA at pH 5.5, L-S agar and Lee's agar have also been used (Table 4). The typical colony appearances of *L. bulgaricus* on different media used for isolation from yoghurt is presented in **Table 5**.

Table 1 Physiological and biochemical characteristics of *Lactobacillus bulgaricus*

GC content (mol %)	DNA melting temperature (°C)[a]	Chromosome size (Mbp)	Peptidoglycan type	Teichoic acid	Antigenic group	Lactic acid isomer	Electophoretic motility D-LDH[b]	Optimum growth (°C)	Minimum growth (°C)	Maximum growth (°C)	Optimum pH	NH₃ from arginine
49–51	91.7	2.3	Lys-D-Asp	Glycerol	E	D	1.70	40–50	22	62	5.5–5.8	No

[a] Approximate value determined by differential scanning calorimetry; individual strains vary.
[b] LDH, lactic acid dehydrogenase.

Table 2 Fatty acid composition of lipids from different *Lactobacillus bulgaricus* strains determined by gas–liquid chromatography

Strain origin	Fatty acid content (%)								
	12:0	14:0	15:0	16:0	16:1	17:1	18:0	18:1	19:0
Commercial (NCS1)	tr[a]	3.2	0.7	18.8	13.8	1.0	1.1	35.8	25.5
NCFB-1489	ND[b]	ND	ND	6.28	30.44	ND	0.72	35.39	27.17
State Univ, Ultrecht-9LB	2.9	10.4	ND	23.0	13.8	ND	0.6	27.1	22.2

[a] The presence of a compound in an amount less than 0.5% is denoted tr.
[b] ND, not detected.

Table 3 Composition of Rogosa media used for isolation of lactobacilli from milk, cheese and fermented milks

Component	RA[a]
Tryptone (g)	10.0
Yeast extract (g)	5.00
Glucose (g)	20.0
KH₂PO₄ (g)	6.00
Na acetate .3H₂O (g)	25.0
NH₄ citrate (g)	2.00
MgSO₄.7H₂O (g)	0.58
MnSO₄.4H₂O (g)	0.15
FeSO₄.7H₂O (g)	0.03
Tween 80 (ml)	1.00
Agar (g)	15.0
Water (l)	1.00
Incubation	42°C/48 h

[a]Dissolve the agar in 500 ml of boiling water. Dissolve all other ingredients in 500 ml water, adjust pH to 5.4 with acetic acid (100%, glacial) and mix with the melted agar. Boil for a further 5 min. No further sterilization is given.

Table 4 Composition of some media used for enumeration of *Lactobacillus bulgaricus* in yoghurt

Component	MRS[a]	RCA[b]	L-S[c]	Lee's[d]
Peptone (g)	10.0	10.0	5.00	–
Tryptone (g)	–	–	10.0	10.0
Meat extract (g)	10.0	10.0	5.00	–
Yeast extract (g)	5.00	3.00	5.00	10
Glucose (g)	20.0	5.00	20.0	–
Lactose (g)	–	–	–	5.00
Sucrose (g)	–	–	–	5.00
K₂HPO₄ (g)	2.00	–	–	0.50
Na acetate.3H₂O (g)	5.00	3.00	–	–
NH₄ citrate (g)	2.00	–	–	–
MgSO₄.7H₂O (g)	0.20	–	–	–
MnSO₄.4H₂O (g)	0.05	–	–	–
Tween 80 (ml)	1.00	–	–	–
Soluble starch (g)	–	1.00	–	–
NaCl (g)	–	5.00	5.00	–
L-Cysteine.HCl (g)	–	0.50	0.3	–
Bromocresol purple (g)	–	–	–	0.02
Agar (g)	15.0	12.0	13.0	18.0
Water (l)	1.00	1.00	1.00	1.00
Incubation	Anaerobic 37°C/72 h	Anaerobic 45°C/72 h	Aerobic 43°C/48 h	Anaerobic 37°C/48 h

[a] Adjust pH to 5.4 with acetic acid (100%, glacial). Sterilize at 121°C for 15 min.
[b] Adjust pH to 5.5 with 1.0 M HCl. Sterlize at 121°C for 15 min.
[c] Sterilize at 121°C for 15 min. Cool to 50°C and just prior to use add: (1) 100 ml of a 10% w/v antibiotic-free skim-milk powder, sterilized by autoclaving at 121°C for 5 min; (2) 10 ml of a 2% w/v triphenyltetrazolium chloride solution, sterilized by filtration (these solutions should be warmed to 50°C before adding them to the base medium).
[d] Adjust pH to 7.0. Sterilize at 121°C for 15 min. Bromocresol purple is added in the form of 1 ml of sterile 0.2% solution (autoclaved at 121°C for 15 min) per 100 ml of sterile agar just before pouring into Petri plates.

Bifidobacterium spp. and *Lactobacillus acidophilus* are increasingly to be found in yoghurts as probiotic organisms. Media traditionally used for the isolation of *L. bulgaricus* can no longer be used since these media support the growth of some of these species. Differentiating media that permit the enumeration of all the organisms as separate, and visually identifiable colonies on the same plate were also developed for these products. Examples include tryptose–proteose–peptone–yeast extract–eriochrome T (TPPY), TPPY agar with added prussian blue (TPPYPB) and Reinforced Clostridial prussian blue (RCPB) (**Table 6**). **Table 7** presents the typical colony appearance of *L. bulgaricus* on media used for its enumeration in probiotic yoghurts.

Table 5 Colony appearance of *Lactobacillus bulgaricus* on various media used for its enumeration in yoghurt

Medium	Appearance of colonies
Acidified MRS	Lenticular often sharp-shaped, 1–3 mm diameter
RCA pH 5.5	Lenticular, rough
L-S	Irregular red, rhizoidal, 1.0–1.5 mm diameter, surrounded by a white opaque zone
Lee's	White

Table 6 Composition of some media used for enumeration of *Lactobacillus bulgaricus* in yoghurt containing probiotic organisms

Component	TPPY[a]	RCPB[b]	TPPYPB[b]
Tryptone (g)	7.00	–	7.00
Peptone (g)	7.00	10.0	7.00
Yeast extract (g)	2.00	3.00	2.00
Meat extract (g)	–	10.0	–
Glucose (g)	10.0	5.00	10.0
Lactose (g)	10.0	–	10.0
Na acetate (g)	–	3.00	–
Soluble starch (g)	–	1.00	–
Tween 80 (ml)	1.00	–	1.00
L-Cysteine.HCl (g)	–	0.50	–
Prussian blue (g)	–	0.30	0.30
Agar (g)	15.0	12.0	15.0
Water (l)	1.00	1.00	1.00
Incubation	Anaerobic 42°C 24 h	Anaerobic 37°C 48 h	Anaerobic 37°C 48 h

[a] Adjust pH to 6.8 with NaOH and sterilize at 120°C for 20 min. Before pouring plates add 1% (v/v) of a 0.4% (w/v) sterile solution of Eriochrome T in distilled water.
[b] Sterilize at 121°C for 15 min. Cool to 50°C and then add the Prussian blue.

Table 7 Colony appearance of *Lactobacillus bulgaricus* on various media used for enumeration in yoghurt containing probiotic organisms

Media	Appearance of colonies
TPPY	Flat, transparent, diffuse, 4–6 mm in diameter, undefined shape, irregular edge
RCPB	Small, discrete light blue with white centres, about 1 mm in diameter, surrounded by wide clear blue zones
TPPYPB	Small, shiny, white, surrounded by a wide royal blue zone

Oligonucleotide Probes for Detection and Identification

Molecular biological identification is based on the constitutive composition of nucleic acids and is therefore considered more reliable for identification purposes than conventional microbiological identification based on morphological, physiological and biochemical characteristics which only reflect that portion of the genome expressed under a particular set of conditions.

A molecular probe designated pY85 was isolated from genomic library DNA of *L. bulgaricus*. This probe was capable of distinguishing the closely related species, *Lactobacillus helveticus*, which may also be used in yoghurt manufacture.

Enantioselective Analysis

Since *L. bulgaricus* and *Streptococcus thermophilus* produce different enantiomers of lactic acid, D and L, respectively, it is possible to follow the growth of these microorganisms in yoghurt by measuring the L/D lactic acid ratio. A reliable correlation has been made between microbial counts and the D/(D + L)% ratio measured by HPLC.

Impedimetric Analysis

The principle of this technique is the measurement of changes occurring in a substrate as evidence of bacterial metabolism. A good correlation was found between values obtained using the Bactometer (Impedimetric measurement) or the Malthus instrument and those obtained by standard plate counts in the selective count of the two microorganisms specific for yoghurt: *L. bulgaricus* and *S. thermophilus*. Results can be determined within 12 h.

Culture Maintenance and Conservation

Isolated cells can be cultured in MRS broth and kept at 4°C and periodically transferred to fresh media or maintained for periods no longer than 1–2 months in MRS agar stabs. The major disadvantages of serial transfer techniques are the risks of contamination, selection of mutants, loss of culture and transposition of strain numbers or designations.

For long-term maintenance freeze drying is one of the most economical and effective methods. When freeze-drying facilities are not available cryogenic storage of the cells (with added glycerol) at –70°C on glass beads is also a good method.

Importance in the Food Industry

Fermented Products

Together with drying and smoking, fermentation is one of the oldest known forms of food preservation. *L. bulgaricus* and many other lactic acid bacteria play an important role in food fermentations, causing the characteristic flavour changes and having a preservative effect on the fermented product. *L. bulgaricus* is used in a large number of food product fermentations worldwide (**Table 8**). In those parts of the world where fermented products are still made on the farm, the inoculum may be a naturally soured milk of acceptable taste whereas in countries with a more advanced industry, the cultures are specially selected and developed for their ability to confer the correct properties on the final fermented product. Sophisticated starter strains which arrive at the factory in frozen or freeze-dried form are available and commercialized by specialized culture-producing companies. Selected starters containing one or several strains of *L. bulgaricus* in addition to other species are used particularly in the industrial manufacture of yoghurt and different types of cheeses (which require elevated temperatures during the process of curd preparation) such as Emmental, Gruyère, Gorgonzola, mozzarella, Cacciocavallo and Provolone.

Adjunct Cultures

L. bulgaricus has been used as an adjunct strain added with mesophilic starters to the cheese vat to decrease bitter flavours in Cheddar cheese and other cheeses. Increased dipeptidase activity in the thermophile may degrade bitter peptides in cheese produced by mesophilic strains. It was demonstrated that exopolysaccharide (EPS)-producing strains of *L. bulgaricus* can be useful to increase moisture retention in low fat mozzarella cheese.

Table 8 Application of *Lactobacillus bulgaricus* in European and indigenous fermented foods

Product	Raw material	Origin
Bulgarian buttermilk	Milk	Bulgaria
Cheese	Milk	Unknown (Southwest Asia)
Dahi	Cow or buffalo milk	India, Pakistan, Bangladesh, Sri Lanka
Ginseng whey	Whey, ginseng, honey, apricot, sweetener	Japan, Korea, China
Kefir	Milk	Russia
Kisra	Sorghum flour	Sudan
Koumiss	Mare's milk	USSR
Laban Zeer	Sour milk	Egypt
Mahewu	Maize	South Africa
Rice masa	Rice	Nigeria
Siljo	Horsebean flour, safflower	Ethiopia
Skyr	Milk	Iceland
Trahanas	Wheat, sheep (or cow) milk	Greece, Turkey, Cyprus
Yoghurt	Milk	Unknown (Southwest Asia)

Lactic Acid Production

Lactic acid was the first organic acid produced with microbes, carried out in 1880. Today, synthetic processes for the production of lactic acid (from, e.g. lactonitrile) are very competitive at the same costs as biological processes; lactic acid production is divided about equally between the two processes. The major part of lactic acid in Europe is produced by fermentation using strains of *L. bulgaricus* when whey is used as the substrate and other lactobacilli when different substrates are used.

Lactic acid is included in the FDA GRAS additives for miscellaneous and/or general purpose uses. It was one of the earliest organic acids used in foods. Lactic acid is used by the food industry in a number of ways: it is used in packing Spanish olives, where it inhibits spoilage and further fermentation; it aids in the stabilization of dried-egg powder; it improves the taste of certain pickles when added to vinegar; it is used to acidify the fruit juice in the wine-making; in frozen confections it imparts a 'milk' tart taste and does not mask other natural flavours. Lactic acid is also used in the production of the emulsifiers calcium and sodium stearoyl lactylates which function as dough conditioners. The sodium and potassium salts of lactic acid have significant antimicrobial properties: in meat products against toxin production by *Clostridium botulinum*, and against *Listeria monocytogenes* in chicken, beef and smoked salmon.

Important Characteristics of *Lactobacillus bulgaricus* as Food Producers

Proteolytic Activity The proteolytic systems of lactic acid bacteria are important as a means of making protein and peptide available for growth and as part of the curing or maturation processes which give foods their characteristic rheological and organoleptic properties. Proteolytic systems of lactobacilli are complex and are composed of proteinases and peptidases with different subcellular locations. Proteinases of *L. bulgaricus* are associated with the cell wall and are regulated by temperature and growth phase.

L. bulgaricus is auxotrophic for a number of amino acids and relies on caseins as its major source of amino acids during growth in milk which contains very low amounts of free amino acids and peptides. As *S. thermophilus* exhibits very little proteolysis (compared with that of lactobacilli), the high proteolytic activity of *L. bulgaricus* is an important characteristic in yoghurt production since the peptides which it releases from milk proteins act as stimuli to the growth of *S. thermophilus*. Additionally, the release of threonine by peptidases is important in flavour development since much of the acetaldehyde is derived from this amino acid via the threonine aldolase of *S. thermophilus*. Considerable variation in the proteolytic ability among *L. bulgaricus* strains has been demonstrated. Such pronounced proteolytic diversity concerns commercial manufacturers because culture rotation practices could result in significant variability of the final product.

Exopolysaccharide Production EPS produced by slime-producing strains of *L. bulgaricus* are thought to play a role in the viscosity and texture of fermented milks by binding free water and in prevention of gel fracture and wheying-off, which are common problems in the manufacturing of these products. For this reason there is now a tendency to incorporate these strains into the traditional starter cultures. For example, in yoghurt with low or no fat content, use of EPS-producing *L. bulgaricus* for the fermentation process increases the thickening properties of the yoghurt which are necessary to compensate for the lower fat content.

The sugar composition of the EPS changes during the fermentation cycle. It is also dependent on the growth conditions, such as temperature, medium composition and incubation time, as well as on the carbon source.

pH Homeostasis pH homeostasis is very important for *L. bulgaricus* since it has to cope with low pH during growth and fermentation. Cytoplasmic pH in *L. bulgaricus* decreases as the pH of the medium decreases. It is thought that contrary to bacteria that maintain intracellular pH at near neutral value (pH gradient increases as extracellular pH declines), bacteria (e.g. *L. bulgaricus*) in which the cytoplasmic pH declines as a function of extracellular pH are more resistant to the toxic effects of fermentation acids. It has been proposed that by maintaining a relatively constant pH gradient across the cell membrane such bacteria do not accumulate high and potentially toxic concentrations of fermentation acid anions at low pH.

Factors Causing Inhibition of *Lactobacillus bulgaricus* during Fermentation

Starters must be able to multiply rapidly in order to produce enough lactic acid to complete the conversion of milk into an acid curd. There are, however, some factors such as bacteriophage infections, presence of antibiotics and other inhibitory compounds that can result in failure or 'slow' acid production. In addition to economic losses, end products with a high pH value can support the growth of pathogenic organisms present in the raw materials.

Bacteriophages Bacteriophages are bacterial viruses which infect the bacterial cell, multiply within it, eventually causing the cell to lyse. Phage infection is the major cause of slow acid production in dairy fermentations.

Virulent phages specific for *L. bulgaricus* have been reported. In the dairy plant, phages can be isolated from raw milk, the cheese whey, the air and milk residues due to inadequate sanitation. Another important source of phages in the dairy industry is thought to be the starter culture organisms themselves which carry within them lysogenic phages that can be induced into a virulent state. Lysogeny has been demonstrated in *L. bulgaricus*. Problems occur particularly when starters contain a single strain or only a few strains and the same culture is re-used over an extended period.

The industry tries to prevent phage problems by practising aseptic techniques, rotating the starter cultures, propagating the starters in phage-inhibitory media (containing phosphate salts to chelate Ca^{++} and Mg^{++} required for successful phage adsorption to the bacterial cell) and by selecting for bacterial cultures that are phage-resistant.

Antibiotics Many antibiotics are used for mastitis therapy in milk-producing cows, and this leads to their excretion in the milk if adequate precautions are not taken. In addition to the serious consequences that can occur in sensitive or allergic consumers, antibiotics in milk can result in inhibition of acid production by starter organisms leading to poor quality fermented milk. *L. bulgaricus* has shown high sensitivities towards many antibiotics (**Table 9**).

The values presented in Table 9, however, must be interpreted with care and used as indicators only since antibiotic resistance may differ depending, for example, on the strain, the methodology used, the basal medium used and mutation.

Detergent and Disinfectant Residues Residues of detergents and disinfectants used for cleaning and disinfection of dairy equipment can reduce *L. bulgaricus* activity. The sensitivities of *L. bulgaricus* to some compounds normally used are presented in **Table 10**.

Miscellaneous Inhibitors Growth of starter organisms can be inhibited by natural compounds such as lactins and agglutinins present in milk. These compounds, however, are heat-sensitive and therefore destroyed by pasteurization of milk.

Phagocytosis of starter organisms can occur as a result of the presence of leucocytes in mastitic milk.

Starter organisms can also be inhibited by environmental pollutants such as insecticides.

Table 9 Sensitivity of *Lactobacillus bulgaricus* strains to antibiotics

Antibiotic	Concentration
Ampicillin	0.5 µg
Aureomycin	10 µg
Bacitracin	5 U
Carbenicillin	50 µg
Cephaloridin	30 µg
Cephalothin	30 µg
Chloromycetin	5 µg
Clindamycin	2 µg
Cloxacillin	1 µg
Dicloxacillin	1 µg
Dihydrostreptomycin	10 µg
Doxycycline	30 µg
Erythromycin	2 µg
Mandelamine	3 µg
Methicillin	5 µg
Novobiocin	10 µg
Oleandomycin	15 µg
Oxacillin	1 µg
Penicillin G	0.5 U
Rifampicin	5 µg
Streptomycin	2 µg
Terramycin	10 µg
Tetracycline	1 µg

Table 10 Inhibitory levels of some compounds normally used as detergents and disinfectants in the dairy industry towards *Lactobacillus bulgaricus*

Chlorine compounds (mg l^{-1})	Quaternary ammonium compounds (mg l^{-1})	Iodophors (mg l^{-1})
100–500	60	60

Importance for the Consumer

Natural Preservation

Lactic acid production during fermentation lowers the pH and creates an environment that is unfavourable to pathogens and spoilage organisms. In addition, the low pH potentiates the antimicrobial effects of organic acids which show greater lethality to bacteria than the inorganic acids.

Hydrogen peroxide is another antagonistic metabolite produced by *L. bulgaricus* in the presence of air. The antimicrobial action of hydrogen peroxide has been attributed to its ability to produce toxic compounds, such as the superoxide anion and other free radicals. Lactic acid bacteria, however, show higher resistance to the effects of hydrogen peroxide than do many other organisms. This higher resistance has been attributed to an inducible oxidative stress response when cells are exposed to sublethal concentrations of H_2O_2. This response protects them against subsequent exposure at lethal concentrations of H_2O_2.

Other compounds produced by *L. bulgaricus* and showing antimicrobial activity have been reported. Bulgarican, which showed maximum activity and stability at pH 2.2 and was not affected by autoclaving at 120°C for 1 h, was inhibitory towards both Gram-positive and Gram-negative bacteria but had no apparent antifungal activity. Inhibitory compounds against *Staphylococcus* and *Clostridium* have been found, which were insensitive to proteolytic enzymes, resistant to heat and active over a wide range of pH.

Many consumers are concerned about the nutrient content of processed foods, and there is enthusiasm for the use of 'natural' methods of improving shelf life and safety. Fermented foods have a positive image in all these respects.

Improvement in Nutritional Value and Health Benefits

Suggestions of probiotic properties associated with *L. bulgaricus* were first made by Metchnikoff in 1907 and supported by Louden Douglas and Kopeloff, both in the 1920s. Their claims, however, were later disproved when it was demonstrated that most strains of *L. bulgaricus* are highly sensitive to gastric acid and bile acids, and show poor survival during transit through the gastrointestinal tract to the colon.

Milk is nutritious and provides high-quality protein in the diet. However, the major carbohydrate is lactose and a large proportion of the world's adult population is intolerant to lactose due to deficiency in the lactose-hydrolysing enzyme, lactase. Intolerance of lactose can also occur due to reduction in lactase activity during intestinal disorders. By fermenting the milk, in yoghurt and cheesemaking, the lactose is converted largely into lactate, which is more readily digestible. It is noteworthy that when appreciable quantities of lactate are present, people are more tolerant to lactose, so that even when yoghurts contain added skim-milk powders lactose-intolerant individuals can usually consume yoghurt without problems. This appears to be due to the presence of the lactose-hydrolysing enzyme β-galactosidase in the yoghurt organisms. The enzyme is partially protected by the bacterial cell envelope during its passage through the stomach but, in the presence of bile in the small intestine, it leaks from the cell and assists in the breakdown of lactose.

Zinc and selenium are essential elements for humans. It has been suggested that fermentation of milk (for example in yoghurt) increases the availability of zinc and selenium.

Increased Variety in Diet

In order to obtain a well-balanced diet with the maximum possible range of nutrients, a varied diet is desirable. *L. bulgaricus* is involved in the production of various fermented foods (Table 8) providing a wide range of ingredients, flavours and textures. Cheese and yoghurt are major sources of proteins and fat in the diet. They also provide calcium, sodium, potassium, magnesium and phosphorus as well as useful amounts of vitamins B, A and D. It is important to point out the number of different types of cheeses and yoghurts, which creates interest in the diet and makes eating more pleasurable, all made with the same raw material.

Pathogenicity

Lactobacilli have usually been considered to be non-pathogenic but there is now increasing evidence that they can act as opportunistic pathogens especially in people with an underlying disease or immuno-suppression. There was, however, only one clinical case reported involving *L. bulgaricus* since 1938; it was the only organism isolated in blood culture from a case of leukaemia.

Concluding Comments

Despite the economic importance of *L. bulgaricus* in the food industry many possibilities are still being investigated for the use of this species.

- It is thought that growing EPS-producing strains in whey (cheese whey is produced in high amounts) may provide polymers that could be used as food stabilisers.
- Techniques for converting lactose present in whey permeate into lactic acid are being improved.
- Different techniques are being tested to avoid post-acidification in yoghurt. For example, *L. bulgaricus* starter strains have been screened for the presence of spontaneous mutants with no or reduced residual β-galactosidase activity.
- Different substrates are being investigated for the production of yoghurt, e.g. soya milk and grape must.
- Reduction of microbial populations on carcasses has been achieved by spraying solutions of lactic acid on the meat surface. However, lactic acid is very expensive. Production of lactic acid *in situ* seems to be a good alternative. Inoculation of *L. bulgaricus* on meat surfaces has been shown to reduce the growth rate of *Pseudomonas*.
- *L. bulgaricus* strains have been screened for amino-peptidase activity to evaluate the possibility of their use for accelerated cheese ripening. Results seemed to be promising.

Other interesting possible applications of *L. bulgaricus* in the food industry may have not been mentioned here. In conclusion, an optimistic view of the future indicates that progress in microbiology, biochemistry, genetics, immunology and cellular biology will increase the availability of *L. bulgaricus* strains with important characteristics to the food industry and may allow the improvement of existing products and the development of new ones.

See also: **Biochemical and Modern Identification Techniques**: Microfloras of Fermented Foods. **Cheese**: Microbiology of Cheese-making and Maturation; Role of Specific Groups of Bacteria. **Electrical Techniques**: Lactics and Other Bacteria. **Fermented Foods**: Origins and Applications; Fermentations of the Far East. **Fermented Milks**: Range of Products; Yoghurt; Products from Northern Europe. ***Lactobacillus***: Introduction. **Starter Cultures**: Uses in the Food Industry; Importance of Selected Genera; Cultures Employed in Cheese-making. ***Streptococcus***: *Streptococcus thermophilus*. **Preservatives**: Traditional Preservatives – Organic Acids.

Further Reading

Auclair J and Accolas JP (1983) Use of thermophilic lactic starters in the dairy industry. *Antonie van Leeuwenhoek* 49: 313–326.

Campbell-Platt GC (1990) Fermented foods. In: Birch GG, Campbell-Platt GC and Lindly MG (eds) *Foods for the 90s*. P. 39. London: Elsevier.

Charteris WP, Kelly PM, Morelli L and Collins JK (1997) Selective detection and identification of potentially probiotic *Lactobacillus* and *Bifidobacterium* species in mixed bacterial populations (Review article). *International Journal of Food Microbiology* 35: 1–24.

Crueger W and Crueger A (1989) In: Brock TD (ed.) *Biotechnology: A Text book of Industrial Microbiology*, 2nd edn. Sunderland, USA: Sinauer.

Frank HK (1992) *Dictionary of Food Microbiology*. Lancaster, USA: Technomic.

Gasser F (1994) Safety of lactic acid bacteria and their occurrence in human clinical infections. *Bulletin Institute Pasteur* 92: 45–67.

Hammes WP, Weiss N and Holzapfel W (1992) The Genera *Lactobacillus* and *Carnobacterium*. In: Balows A, Truper HG, Dworkin M, Harder W and Schleifer KH (eds) *The Prokaryotes*, 2nd edn. P. 1535. New York: Springer-Verlarg.

Kandler O and Weiss N (1986) Regular, Nonsporing Gram-positive Rods. In: Sneath PHA, Mair NS, Sharp ME and Holt JG (eds) *Bergey's Manual of Determinative Bacteriology*. Vol. 2, p. 1209. Baltimore: William and Wilkins.

Kunji E, Mierau I, Hagting A, Pooman B and Konings WN (1986) The proteolytic systems of lactic acid bacteria. *Antonie van Leeuwenhoek* 70: 187–221.

Porubcan RS and Sellars RL (1979) Lactic Starter Culture Concentrates. In: Peppler HJ and Perlman D (eds) *Microbial Technology. Microbial Processes*. P. 59. New York: Academic Press.

Prescott LM, Harley JP and Klein DA (1990) *Microbiology*, 2nd edn. DuBuque, USA: Wm. C. Brown Communications.

Steinkraus KH, Cullen RE and Gavitt BK (1983) *Handbook of Indigenous Fermented Foods*. New York: Marcel Dekker.

Tamine AY (1981) Microbiology of Starter Cultures. In: Robinson RK (ed.) *Dairy Microbiology*. P. 113. New York: Pergamon Press.

Lactobacillus brevis

Paula C M Teixeira, Escola Superior de Biotecnologia, Porto, Portugal

Characteristics of *Lactobacillus brevis*

Lactobacillus brevis is a microaerophilic, obligately heterofermentative lactic acid bacterium (uses the phosphoketolase pathway to produce a mixture of lactic acid, ethanol, acetic acid and CO_2 as products of hexose fermentation) which has been reported to lack phosphotransferase systems specific for glucose, fructose and lactose. Recent studies, however, demonstrated that anaerobic growth of *L. brevis* in the presence of fructose induces the synthesis of a phosphotransferase system and glycolytic enzymes that allow fructose to be metabolized via the Embden–Meyerhof pathway.

L. brevis is included in the second phylogenetic group of the lactobacilli, the *L. casei–Pediococcus* group.

L. brevis is normally isolated from milk, cheese, plants, sewage, cereal products, silage, fermented vegetables, fermented meats, cow manure, faeces, mouth and intestinal tract of humans and rats.

Carbohydrates fermented by *L. brevis* (90% or more strains) are arabinose, fructose, glucose, gluconate, maltose, melibiose and ribose. Esculin, galactose, lactose, raffinose, sucrose and xylose are fermented by 11–89% of the strains. *L. brevis* is unable to grow in chemically defined media having pentoses as a sole source of fermentable sugar. Identification of *L. brevis* strains by carbohydrate fermentation reactions or additional simple phenotypic tests has proved to be insufficient. Some strains earlier assigned to *L. brevis* have been assigned to new species on the basis of nucleic acid and biochemical data. In addition to DNA studies, electrophoretic mobility of lactic acid dehydrogenases is recommended to clearly distinguish *L. brevis* from *L. buchneri*, *L. hilgardii*, *L. collinoides* or *L. kefir* since some DNA of *L. brevis* strains hybridizes with that of some of these other lactobacilli.

Calcium pantothenate, niacin, thiamin and folic acid are essential growth factors but riboflavin, pyridoxal, and vitamin B_{12} are not required.

L. brevis is considered to be a weakly proteolytic species.

Haem-independent nitrite reductases and haematin-requiring catalase activity have been found in *L. brevis*.

Cells are rod shaped with rounded ends, generally short and straight ($0.7–1.0 \times 2.0–4.0\,\mu m$). Long rods, however, are always present. They are usually separate or in short chains. Bipolar or other internal granulations are observed with the Gram reaction or methylene blue stain especially when cells become older. Most *L. brevis* strains possess immunologically heterogeneous S-layer proteins with molecular weights in the range 38–55 kDa.

Colonies are generally rough or intermediate, flat, and they may be nearly translucent. Although some strains are pigmented orange to red, they are generally non-pigmented. Additional physiological and biochemical characteristics are presented in **Table 1**.

Fatty acid composition has been used in the grouping and classification of microorganisms. As shown in **Table 2**, hexadecanoic acid ($16:0$), octadecenoic acid ($18:1$) and lactobacillic acid ($19:0$) are the major fatty acids present in *L. brevis*. However, there is variability at different stages of growth, between strains, as a result of different growth conditions (medium composition, temperature), and if different methodologies are used for lipids extraction.

Antibiotic resistance in lactic acid bacteria has been studied as a potential means of identification. However, no definite patterns of resistance have emerged to allow for use in a classification scheme.

Table 1 Physiological and biochemical characteristics of *Lactobacillus brevis*

G+C content (mol%)	44–47
Peptidoglycan type	Lys-D-Asp
Techoic acid	Glycerol
Antigenic group	E
Lactic acid isomer	DL
Electrophoretic mobility	
D-*LDH*[a]	1.62
L-*LDH*	1.40
Optimum pH	4.0–5.0
Growth temp.	
Optimum (°C)	30
Minimum (°C)	2–4
NH_3 from arginine	+

[a]LDH, lactate dehydrogenase.

Table 2 Major fatty acid components of lipids from *Lactobacillus brevis* strains determined by gas–liquid chromatography

L. brevis *strain*	*14:0*	*15:0*	*16:0*	*16:1*	*17:0*	*18:0*	*18:1*	*19:0*
CIP 7135	2.58	0.13	37.43	4.20	0.60	1.15	33.50	20.37
NCIB 4617	2.74	0.15	36.60	4.15	0.38	3.08	33.32	16.38

Table 3 Sensitivity of *Lactobacillus brevis* strains to antibiotics

Antibiotic	Concentration
Ampicillin	10 µg
Bacitracin	10 U
Cephaloridin	30 µg
Chloramphenicol	30 µg
Colistin	10 µg[a]
Erythromycin	15 µg
Kanamycin	30 µg[a]
Methicillin	5 µg[a]
Neomycin	30 µg
Novobiocin	5 µg
Penicillin	10 U
Polymyxin B	300 U[a]
Rifampicin	5 µg
Streptomycin	10 µg[a]
Tetracycline	30 µg

[a]Some strains are resistant.

Susceptibility of *L. brevis* to some antibiotics is presented in **Table 3**.

Methods of Detection and Enumeration of *Lactobacillus brevis* in Foods

Many media have been described over the years for the isolation of lactobacilli from various foods. Semi-selective de Man, Rogosa and Sharp (MRS), all purpose Tween 80 (APT) and modified Homohiochii media (mHom) (**Table 4**) have been shown to be suitable as general culture media for isolating lactobacilli and other lactic acid bacteria.

Rogosa agar (RA) is commonly used when a selective medium is necessary for the detection of fastidious lactobacilli such as *L. brevis*. This medium, however, allows the growth of some pediococci, leuconostocs and yeasts (cycloheximide, $10 \, \text{mg} \, \text{l}^{-1}$ can be added to inhibit yeasts). The use of RA is recommended for isolation from a wide variety of foods including milk and fermented milks, meat products, fermented vegetables, and salad dressings.

In some cases, it is difficult to detect *L. brevis*, as well as other microorganisms in foods. They are often present in low numbers, sublethally damaged due to the environmental conditions or may be adapted to specialized environments (fruit juices, wine, beer, etc.) and become very reluctant to multiply in other environments such as highly nutritious laboratory media. The addition of the natural substrate is often necessary to supply any unknown but essential growth factors.

Table 4 Composition of some general media used for isolation of *Lactobacillus brevis* and other lactobacilli

Component	Medium		
	MRS	APT	mHom[a]
Peptone (g)	10.0	–	–
Tryptone (g)	–	10.0	10.0
Meat extract (g)	10.0	–	2.0
Yeast extract (g)	5.0	5.0	7.0
Glucose (g)	20.0	10.0	5.0
Fructose (g)	–	–	5.0
Maltose (g)	–	–	2.0
Na acetate.3H$_2$O (g)	5.0	–	5.0
Na citrate (g)	–	5.0	–
Na gluconate (g)	–	–	2.0
NH$_4$ citrate (g)	2.0	–	2.0
K$_2$HPO$_4$ (g)	2.0	5.0	–
MgSO$_4$.7H$_2$O (g)	0.2	0.8	0.2
MnSO$_4$.4H$_2$O (g)	0.05	–	0.05
MnCl$_2$.4H$_2$O (g)	–	0.14	–
FeSO$_4$.7H$_2$O (g)	–	0.04	0.01
NaCl (g)	–	5.0	–
Mevalonic acid lactone (g)	–	–	0.03
Tween 80 (ml)	1.0	1.0	1.0
Cysteine HCl (g)	–	–	0.5
Agar (g)	15.0	15.0	15.0
Water (l)	1.0	1.0	1.0
pH	6.2–6.5	6.7–7.0	5.4

[a]After sterilization, add 40 ml ethanol per litre.

Table 5 Composition of orange serum agar for isolation and enumeration of spoilage organisms of citrus products

Component	Orange serum agar
Tryptone (g)	10.0
Yeast extract (g)	3.0
Glucose (g)	4.0
K$_2$HPO$_4$ (g)	3.0
Orange extract (g)	5.0
Agar (g)	17.0
Water (l)	1.0
pH	5.5

Media prepared with orange juice have been used for the control of the processing of citrus products. An orange serum agar (**Table 5**) has been developed for the isolation of microorganisms responsible for spoilage of citrus products. Due to their strongly stimulatory effects, tomato juice and yeast extract are normally included in media for the isolation of lactobacilli from wine. Addition of ethanol to all media is also recommended.

Many media have been used for isolation of beer

Table 6 Composition of standard media used for enumeration of lactobacilli

Component	Raka-Ray agar[a]	Sucrose agar[b]
Casein peptone (g)	–	10.0
Yeast extract (g)	5.0	5.0
Liver concentrate (g)	1.0	–
Tryptone (g)	20.0	–
Tween 80	10.0 ml	0.1 g
Glucose (g)	5.0	–
Fructose (g)	5.0	–
Maltose (g)	10.0	–
Sucrose (g)	–	50.0
NH_4 citrate (g)	2.0	–
Potassium aspartate (g)	2.5	–
Potassium glutamate (g)	2.5	–
Potassium phosphate (g)	2.0	–
NaCl (g)	–	5.0
$MnSO_4.4H_2O$ (g)	0.7	0.5
$MgSO_4.7H_2O$ (g)	2.0	0.5
$CaCO_3$ (g)	–	3.0
N-Acetylglucosamine (g)	0.5	–
Betaine HCl (g)	2.0	–
Bromocresol green (mg)	–	20.0
Cycloheximide (mg)	7.0	–
Agar (g)	17.0	20.0
Distilled water (l)	1.0	1.0
pH	5.4	6.2

[a]After autoclaving, just before pouring the plates add 3.0 g 2-phenyl ethanol.
[b]After autoclaving, just before pouring the plates add filter sterilized cycloheximide (final concentration 10.0 mg ml^{-1}) and 3.0 g 2-phenyl ethanol.

Table 7 Composition of some media developed for enumeration of lactobacilli in beer

Component	Medium KOT	Medium M NBB	Medium UBA
Casein peptone (g)	–	5.0	–
Peptonized milk (g)	–	–	15.0
Yeast extract (g)	2.5	5.0	6.1
Liver concentrate (g)	1.0	–	–
Trypticase peptone (g)	5.0	–	–
Malt extract (g)	2.5	–	–
Meat extract (g)	–	2.0	–
Glucose (g)	5.0	15.0	16.1
Maltose (g)	5.0	15.0	–
Tween 80 (ml)	1.0	0.5	–
Ferrous sulphate (mg)	–	–	6.0
Potassium acetate (g)	–	6.0	–
K_2HPO_4 (g)	0.5	2.0	0.3
KH_2PO_4 (g)	–	–	0.3
NaCl (mg)	–	–	6.0
$MnSO_4.4H_2O$ (mg)	25.0	–	6.0
$MgSO_4.7H_2O$ (g)	0.12	–	0.12
NaN_3 (mg)	50.0	–	–
Cysteine HCl (g)	0.5	0.2	–
Chlorophenol red (mg)	–	70.0	–
L-Malic acid (g)	0.5	0.5	–
Tomato supplement (g)	–	–	12.2
Cytidine (g)	0.2	–	–
Thymidine (g)	0.2	–	–
Cycloheximide (g)	0.1	–	–
Agar (g)	20.0	15.0	12.0
Beer (l)	0.8	0.5	0.25
Distilled water (l)	1.0	0.5	0.75
pH	6.3	5.8	6.1

spoiling lactobacilli. Some of these media e.g. MRS medium (Table 4), MRS medium supplemented with maltose, Raka-Ray medium (**Table 6**) and sucrose medium (Table 6), are standard media for the detection of these organisms. Other media have been developed especially for the brewing industry and normally include wort or beer in their formulation (composition of some of these media is presented in **Table 7**), e.g. Nachweismedium fur biererchadliche bacterien (NBB) and modified NBB medium (m NBB), Universal Beer Agar (UBA), VLB-S7-Agar, KOT medium, etc. A double-concentrated MRS medium adjusted with beer to normal concentration before autoclaving is also often used. Avoparcin (20 µg ml^{-1}) and vancomycin (30 µg ml^{-1}) are added to media used for isolation of beer spoilage organisms as these antibiotics inhibit the growth of most Gram-positive bacteria and have no effect on growth of heterofermentative lactobacilli, *L. salivarius* and bacteria of the genera *Leuconostoc* and *Pediococcus* which constitute the main beer spoilage lactic acid bacteria.

Although several comparisons between different media have been done the optimal medium for detection of lactobacilli in beer has not yet been identified.

The use of Raka-Ray agar is recommended by the American Society of Brewing Chemists and the European Brewing Convention.

Plating methods take a long time (incubation time is normally 5–7 days at 28–30°C) to detect beer spoilage organisms and have generally a poor selectivity, so that more rapid alternative methods have been developed, e.g. bioluminescence techniques, direct epifluorescence filter techniques, immunoassays, conductimetric analysis, flow cytometry, polymerase chain reaction (PCR), DNA hybridization techniques, etc. As *L. brevis* is a common brewery contaminant that rapidly proliferates in beer, it is considered to be a suitable indicator for monitoring spoilage but most of the existing rapid methods are not efficient enough for its early detection or include the use of advanced, expensive equipment and reagents.

Bioluminescence methods, based on the measurement of light produced when ATP reacts with the firefly luciferin/luciferase enzyme system, can detect about 100 bacteria cells per millilitre. With this technique, however, detected microorganisms are not identified and it is difficult to determine numbers in beer which contains both yeast and bacteria. It is also difficult to use bioluminescence as a routine analytical tool in breweries as the reagents are rather unstable. The basis of the direct epifluorescence filter techniques is the concentration of the cells on a membrane filter and staining them with acridine orange, a fluorescent dye, which binds to the nucleic acids. Viable cells fluoresce orange, non-viable cells fluoresce green. This technique has been used with success for milk but with heat-treated beverages differentiation between viable and non-viable cells is unreliable. Conductance measurements for the rapid detection of lactobacilli in beer has been investigated and has shown promise. Samples containing less than about 50 cfu ml^{-1} were not detected within 50 h but higher levels were detected in 30 h or less. At present, however, these methods only indicate the presence or absence of contaminants and cannot be used when actual counts are required.

PCR assays involving amplification of DNA and RNA fragments were developed for the rapid detection of *L. brevis*. PCR assays are generally highly specific and sensitive, but the procedure is complicated and not suitable for use in the brewery as it involves time-consuming and delicate steps such as DNA extraction, PCR amplification, and electrophoresis. Additionally, the presence of PCR inhibitors in beer decreases the sensitivity of PCR assays which seem to be limited to laboratory use at present.

Recently, a new method has been developed in which imaging of single cells and microcolonies without a microscope by an ultrasensitive chemiluminescence enzyme immunoassay with a photon-counting television camera allows the rapid detection and quantification of *L. brevis* contaminants in beer and pitching yeast, i.e. the MicroStar Rapid Micro Detection System. It is claimed that optimization of membrane filtration, bioluminescent chemistry and advanced image analysis, enables the user to rapidly (within minutes or hours rather than days) enumerate 0–200 cfu per sample in variable sample volumes. This method compares well with PCR in terms of sensitivity but it is less labour-intensive and more rapid.

Culture Maintenance and Conservation

As for most other species of lactobacilli, *L. brevis* can be cultured in MRS broth or yeast glucose chalk litmus milk medium, kept at 4°C, and periodically transferred to fresh medium or maintained for periods no longer than 1–2 months in MRS or tomato juice agar stabs. Addition of glycerol to the cultures (1:1) allows storage at −20°C for at last one year without significant loss of viability and reduces the risks of contamination, selection of mutants, loss of culture

and transposition of strain numbers or designations associated with serial transfer techniques. Freezing in liquid nitrogen and freeze-drying are the recommended methods for long-term preservation. If these facilities are not available, cryogenic storage of the cells (with added glycerol) at −70°C on glass beads is also a good method.

Importance in the Food Industry

Fermented Products

Fermentation is considered to be one of the most economical methods of producing and preserving foods for human consumption. It is extensively used for these purposes in the underdeveloped world where the low levels of disposable income and limited infrastructure available in the food-processing industry greatly restrict the use of more advanced technologies.

L. brevis is involved in the production of a wide variety of fermented products (**Table 8**), reflecting the different diets and needs in various parts of the world. Some of these fermented foods have developed from natural fermentations into the selection and use of specific starter strains; however, even in Europe, several industrial lactic acid food fermentations are still 'spontaneous' processes.

In contrast to most vegetable fermentations, which are still produced on a small scale, sauerkraut, pickles and olives are fermented vegetables of significant commercial importance in the western world. Sauerkraut is made from salted shredded cabbage. Fermentation starts with *Leuconostoc mesenteroides*, present in high numbers in fresh cabbage, producing lactic acid, acetic acid and CO_2. Then *L. brevis* grows, producing more acid, and finally *L. plantarum* lowers the pH to below 4.0. The early dominance of heterofermenters

Table 8 Application of *Lactobacillus brevis* in fermented foods

Product	Raw material	Area
Burong mustala	Mustard	Philippines
Busaa	Maize, finger millet, sorghum	Kenya
Cheese	Milk	Worldwide
Fufu	Cassava tubers	Nigeria
Kefir	Milk	Caucasus
Kimchi	Korean cabbage, Korean radish root	Korea
Kishk	Milk, wheat	Egypt, Iraq
Laban zeer	Sour milk	Egypt
Mesu	Bamboo shoot	India
Nham	Pork, rice	Thailand
Olives	Green olives	Worldwide
Pickles	Vegetables, cucumbers	Worldwide
Pulque	Agave juice	Mexico
Sauerkraut	White cabbage	Worldwide
Sausages	Pork, beef	Worldwide
Sourdoughs	Wheat, rye	Worldwide

in the fermentation is important in the inhibition of undesirable organisms ensuring the stability and consistency of the natural fermentation process. Although they produce less total acidity, acetic acid, with a higher pKa, is a more potent antimicrobial than lactic acid. Various studies have been performed to develop starter cultures to sauerkraut fermentation but industrial production is still based on natural fermentation processes.

For the production of pickled cucumbers, whole vegetables are washed and covered with brine. Aerobic microorganisms develop first and create favourable conditions for the growth of lactic acid bacteria (LAB) which are responsible for the main fermentation process. The succession of LAB in cucumber fermentation is similar to that of sauerkraut: the heterofermentative LAB, initially *Leuconostoc* spp. followed by *L. brevis*, are soon overgrown by homofermentative species such as *L. plantarum* and *Pediococcus pentosaceus*.

A sourdough for leavening bread doughs, is one of the oldest biotechnological processes in food production. Although nowadays breads from wheat may be leavened with yeasts exclusively, sourdoughs containing *Lactobacillus* (*L. brevis*, *L. delbrueckii*, *L. fermentum*, *L. plantarum*, *L. sanfrancisco* and others) are still used mainly in the production of rye and rye–mixed grain breads, cake leavened baked products (e.g. Panettone) and wheat bread. The use of 'spontaneous' fermentation to produce a sourdough results in small deviations between fermentations because the composition of the microflora is not critically controlled. It is known that heterofermentative strains of lactic acid bacteria are needed to obtain the sensory properties characteristically associated with sourdough breads. Although the application of starter cultures for the production of fermented foods of plant origin has still not been very successful in practice, some lactic acid bacteria strains including *L. brevis* are now being industrially produced in highly concentrated freeze-dried form. It is convenient and quick to use these cultures to make sourdough bread.

Nitrite reduction is a rare property of lactic acid bacteria. Two types of nitrite reductases are known, those depending on the presence of haematin (ammonia is produced from nitrite reduction) and haem-independent enzymes (NO and N_2O are produced from nitrite reduction). *L. brevis* possesses haem-independent nitrite reductases. This is an important characteristic for technological or toxicological purposes with regard to potential applications as starter cultures in food fermentations. The production of NO is desirable in meat technology since this intermediate is required in the reddening reaction. It is usually produced from nitrite in chem-

ical reactions and provides the substrate for the formation of nitrosomyoglobin. The production of N_2O might be an advantage over ammonia since it is more effective and it requires less reduction equivalents to remove nitrite from the substrate.

Kefir grains, which are necessary to inoculate milk to produce Kefir (Table 8), are conglomerates of lactic acid bacteria and yeasts held together by a polysaccharide gum. This polysaccharide, kefiran, is produced by the predominating bacterial species, including *L. brevis*.

The secondary flora of many hard and semihard cheeses, such as Cheddar, Gouda and Edam, is dominated by mesophilic lactobacilli such as *L. brevis*. Their exact role in the cheese is not fully understood, but it is considered that they have an important function in flavour development. The substrates used as energy sources within the ripening cheese are not well known. Since only residual amounts of lactose are present, insufficient to support significant growth of lactobacilli, other sources of metabolites/nutrients must be considered (e.g. galactose, citrate, lactate, starter cell autolysate material, free amino acids, peptides and glycerol from lipolysis).

The importance of silage in the diet of ruminants is well established. The increasing practice of conserving fodder as silage, rather than hay, for feeding cattle in winter has been one of the most important developments in farming in the past several decades. Silage is made from various raw materials, of which grass, hay and maize play the major role. *L. brevis* and some other lactobacilli are very important in the acidification of silage made from green forage. Silage acidification is normally initiated by homofermentative lactic acid bacteria. As fermentation proceeds, *L. brevis* and other heterofermentatives, become dominant.

Food Spoilage

L. brevis in some circumstances can be a cause of spoilage of various food products. The organism is commonly isolated from grapes and wines worldwide. It is possible that some strains or species could contribute desirable characteristics to wine quality, although excessive growth could be undesirable. When present in trace amounts, diacetyl enhances wine flavour. However, excessive production of diacetyl from citric acid by lactobacilli causes spoilage. *L. brevis* produces mannitol from glucose. The production of ethanol and glycerol decreases as fructose is reduced to mannitol. Furthermore, the excess production of mannitol may result in mannite spoilage of wine. Mannite spoilage is accompanied by formation of excess acetic acid. The decomposition of tartaric acid is normally associated with severe spoil-

Table 9 Some characteristics of *Lactobacillus brevis* important for beer spoilage

Hop tolerance[a]	25–35
Maltose	+
$O_2 < 0.4$ (mg l^{-1})	+
pH	< 4.2
Alcohol tolerance	> 14%
Minimum temperature	2–4°C

+, Growth.
[a]EBC bitterness unit.

age of wine. Tartarate decomposition by *L. brevis* results in the formation of CO_2, lactate, acetate and succinate.

Sorbate in wines, generally around 200 p.p.m., although being inhibitory to most yeasts and some lactic acid bacteria, is not inhibitory for *L. brevis* which shows almost no inhibition by sorbate levels up to 1000 p.p.m. Addition of SO_2 to crushed grapes (minimum 30 ml l^{-1}) has proved useful to delay the growth of *L. brevis*.

L. brevis is potentially one of the most undesirable beer spoilage microorganisms because of its micro-aerophilic nature, its resistance to hop-derived compounds, such as isohumulone, to ethanol and to low pH (**Table 9**). Beer spoilage by lactobacilli is characterized by 'silky' turbidity accompanied by acid, 'dirty' (acetoin) or 'buttery' (diacetyl) off-flavours.

One of the major bacterial spoilage agents in citrus juices is *L. brevis*. This organism can multiply at a pH of < 3.5 and at a temperature of 10°C and is responsible for the production of diacetyl which imparts an undesirable 'buttery' flavour to juice, and fermented off-flavours due to ethanol, carbon dioxide and higher-molecular-weight alcohols.

Heterofermentative isolates from cider are usually *L. brevis*. They metabolize fructose actively to produce acetate which is detrimental to flavour.

L. brevis has also been responsible for gas production in salad dressings, vigorous fermentation in canned tomato resulting in can swelling and acid odour in canned tomato ketchup, Worcestershire sauce and similar products, milk stringiness produced by the growth of cord like strains, production of carbon dioxide in marinated herring, and coloured spots in cheese as a consequence of growth of orange-pigmented strains. If present in excessive numbers (> 10^8 cfu ml^{-1}), *L. brevis* can be responsible for certain cheese defects: undesirable gas pockets and blowing of packaged cheeses due to excessive production of CO_2, formation of biogenic amines, 'green spot' development, excessive build up of calcium lactate crystals, unclean flavours, and acidic texture and flavour.

Importance for the Consumer

As previously noted, *L. brevis* occurs naturally in different food materials or can be deliberately introduced in order to produce different fermented foods (Table 8). In addition to the preservation effect, fermentation is a means of improving sensory quality and acceptability of many raw materials, improving the nutritional value and providing the consumer with a wide variety of flavours, aromas and textures to enrich the human diet. Additionally, preparation of lactic acid-fermented products has low, if any, energy requirements and can be consumed without (or with little) cooking (e.g. pickled vegetables, fermented cabbages, olives). Energy saving is very important in countries where housewives spend many hours collecting leaves, twigs, wood and dried dung with which to cook every day.

Natural Preservation

The ability of lactic acid bacteria to inhibit growth and survival of spoilage microflora and pathogens has been used as a means of improving safety and keeping quality of foods. Whereas in the Western world foods are preserved by refrigeration, freezing, canning or modified atmosphere packaging, in developing countries these techniques are prohibitively expensive and fermentation and drying are the methods available. The contribution of lactic acid fermentations to food safety is very important in developing countries.

The use of 'natural' methods of preservation has increased during recent years when the nutrient content of processed foods became a concern among consumers. Fermentation is a useful natural preservation system that meets consumer concerns.

The decrease in pH during fermentation creates an environment that is unfavourable to pathogens and spoilage organisms. Additionally, *L. brevis* produces significant quantities of acetic acid which is a more effective antimicrobial agent than lactic acid. Although acidity is the most important antimicrobial factor, other inhibitory agents produced by *L. brevis* should not be ignored, e.g. bacteriocins, CO_2, hydrogen peroxide, ethanol and diacetyl.

Bacteriocins are antimicrobial compounds, containing a biologically active protein/peptide moiety, produced by bacteria which are inhibitory to a limited range of organisms, normally very closely related bacteria. Various *L. brevis* strains produce bacteriocins (**Table 10**).

The antimicrobial activity of CO_2, hydrogen peroxide, ethanol and diacetyl is well established. The individual contribution of each of these agents, however, is relatively minor, particularly compared with the acid production that occurs at the same time

Table 10 Bacteriocins produced by *Lactobacillus brevis* strains

Bacteriocin producer	Bacteriocin	Activity against
Isolated from kimchi	Unnamed	*Enterococococcus faecalis, E. faecium, L. brevis, L. sake, Leuc. mesenteroides, P. pentosaceus*
L. brevis B37 (isolated from plant and fermenting material)	Brevicin 37	*L. brevis, Leuc. oenos, Nocardia corallina, P. damnosus*
L. brevis SB27 (isolated from sausages)	Brevicin 27	*L. brevis, L. buchneri, L. plantarum, P. pentosaceus,* some *Bacillus* spp.
L. brevis VB286 (isolated from vacuum packaged meat)	Brevicin 286	*E. faecalis, E. faecium, L. curvatus, Listeria* spp.

(reduced pH and presence of undissociated organic acids).

Nutritional Value and Health Considerations

Improvement in Nutritional Value and Health Benefits Increased nutritional quality of fermented foods has been attributed to improvement in the nutrient density, increase in the amount and bioavailability of nutrients, detoxification of food raw materials, improvement of functional properties, and improvement in digestibility. The role of individual microorganisms in the increased nutritional value of these products is sometimes unclear since investigations on the microorganisms involved in most of the fermentation processes appear to terminate at the isolation and identification stages.

Fermentations which involve yeasts tend to be enriched in the B vitamins. Pulque (Table 8), produced by the fermentation of juices of agave, continues to be an important source of nutrition of peasants and other low-income people in the poorest semi-arid areas of Mexico (agave is the only plant that can grow on the very poor soil under the extremely low water availability). Pulque is rich in thiamin, riboflavin, niacin, pantothenic acid, *p*-aminobenzoic acid, pyridoxine and biotin. Additionally, ethanol present in pulque is an important source of calories.

Microbial activity during the production of fufu (Table 8) softens cassava root tissues allowing linamarase to breakdown linamarin, a cyanogenic glycoside responsible for severe intoxications following the consumption of raw cassava. *L. brevis* isolated from fermented cassava products possessed considerable linamarase activity but did not possess tissue-degrading enzymes.

In legumes, carbohydrates are often present as oligosaccharides, such as raffinose, stachyose and verbascose, which are not readily digestible, and can cause flatus, diarrhoea and indigestion when broken down by bacteria in the large intestine. These oligosaccharides possess α-D-galactosidic bonds which are hydrolysed by α-galactosidases. α-Galactosidase production is a constitutive property of *L. brevis*.

Surpluses of vegetables can be safely preserved by farmers using lactic acid fermentation (e.g. pickled vegetables). This improves the supply and availability of vegetable foods throughout the year and improves the nutrition of the population. For example, kimchi (Table 8) is an important source of vitamins and minerals in Korea during the wintertime when fresh vegetables are not available.

Various *L. brevis* strains produce γ-aminobutyric acid, reported as having antihypertensive and diuretic effects.

D-Lactate D-Lactate is a non-physiological isomer in mammals, which is poorly metabolized, and accumulates in the blood, especially if there is thiamin deficiency, causing acidosis (disturbance of the acid–alkali balance in the blood) and mineral mismanagement. The FAO/WHO Joint Committee on Food Additives reviewed the toxicological evidence available and concluded that there was evidence that babies in their first 3 months of life have difficulties in utilizing small amounts of DL or D-lactate and that neither should be used for infant foods.

Biogenic Amines The formation of biogenic amines of toxicological significance occurs in foods and *L. brevis* has been identified as the causative agent (e.g. tyramine in Gouda cheese). γ-Aminobutyric acid, cadaverine and histamine are formed during spoilage of food products by lactobacilli including *L. brevis*.

Pathogenicity *L. brevis* is generally considered to be non-pathogenic. However, it is associated with lung infections, complicated by lung cancer, indicating an opportunistic behaviour.

See also: **ATP Bioluminescence**: Application in Beverage Microbiology. **Bacteria**: Classification of the Bacteria – Phylogenetic Approach. **Bacteriocins**: Potential in Food Preservation. **Biochemical and Modern Identification Techniques**: Microfloras of Fermented Foods. **Bread**: Sourdough Bread. **Cheese**: Microbiology of Cheese-making and Maturation; Role of Specific Groups of Bacteria. **Direct Epifluorescent Filter Techniques (DEFT)**. **Electrical Techniques**: Food Spoilage Flora and Total Viable Count (TVC); Lactics and other Bacteria.

Fermented Foods: Origins and Applications; Fermented Vegetable Products; Fermented Meat Products; Fermentations of the Far East. **Fermented Milks**: Range of Products; Yoghurt; Products from Northern Europe. *Lactobacillus*: Introduction. **Starter Cultures**: Uses in the Food Industry; Importance of Selected Genera. **Wines**: Microbiology of Wine-making. **Preservatives**: Traditional Preservatives – Organic acids.

Further Reading

Coventry MJ, Wan J, Gordon JB, Mawson RF and Hickey MW (1996) Production of brevicin 286 by *Lactobacillus brevis* VB286 and partial characterization. *Journal of Applied Bacteriology* 80: 91–98.

Gasser F (1994) Safety of lactic acid bacteria and their occurrence in human clinical infections. *Bulletin Institute Pasteur* 92: 45–67.

Hammes WP, Weiss N and Holzapfel W (1992) The genera *Lactobacillus* and *Carnobacterium*. In: Balows A, Truper HG, Dworkin M, Harder W and Schleifer KH (eds) *The Prokaryotes*, 2nd edn. Pp. 1535–1594. New York: Springer-Verlag.

Hammes WP and Vogel RF (1995) The genus *Lactobacillus*. In: Wood BJB and Holzapfel WH (eds) *The Lactic Acid Bacteria. The Genera of Lactic Acid Bacteria*, pp. 19–54. London: Blackie Academic.

Jespersen L and Jakobsen M (1996) Specific spoilage organisms in breweries and laboratory media for their detection. *International Journal of Food Microbiology* 33: 139–155.

Kandler O and Weiss N (1986) Regular, nonsporing Gram-positive rods. In: Sneath PHA, Mair NS, Sharp ME and Holt JG (eds) *Bergey's Manual of Determinative Bacteriology*. Vol. 2, p. 1209. Baltimore: William and Wilkins.

Kyriades AL and Thurston PA (1989) Conductance techniques for the detection of contaminants in beer. In: Stannard CJ, Petitt SB and Skinner FA (eds) *Rapid Microbiological Methods for Foods, Beverages and Pharmaceuticals*. Pp. 101–117. London: Blackwell Scientific Publications.

Saier Jr. MH, Ye JJ, Klinke S and Nino E (1996) Identification of an anaerobically induced phosphoenolpyruvate-dependent fructose-specific phosphotransferase system and evidence for the Embden–Meyerhof glycolytic pathway in the heterofermentative bacterium *Lactobacillus brevis*. *Journal of Bacteriology* 178: 314–316.

Steinkraus KH, Cullen RE and Gavitt BK (1983) *Handbook of Indigenous Fermented Foods*. New York: Marcel Dekker.

Vandamme P, Bot B, Gillis M et al (1996) Polyphasic taxonomy, a consensus approach to bacterial systematics. *Microbiological Reviews* 60: 407–438.

Wood BJB (1998) *Microbiology of Fermented Foods*, 2nd edn. London: Blackie Academic and Professional.

Yasui T and Yoda K (1997) Imaging of *Lactobacillus brevis* single cells and microcolonies without a microscope by an ultrasensitive chemiluminescent enzyme immunoassay with a photon-counting television camera. *Applied Environmental Microbiology* 63: 4528–4533.

Lactobacillus acidophilus

Todd R Klaenhammer and **W Michael Russell**, Department of Food Science and Microbiology, North Carolina State University, Raleigh, USA

Introduction

Lactobacillus acidophilus, first isolated by Moro (1900) from infant faeces, has undergone many transformations in the description of its metabolic, taxonomic and functional characteristics. The *acidophilus* (meaning 'acid-loving') bacterium is isolated from the intestinal tract of humans and animals and is also reported in the faeces of milk-fed infants and older persons consuming high milk-, lactose- or dextrin diets. Historically, *L. acidophilus* is the *Lactobacillus* species most often implicated as an intestinal probiotic capable of eliciting beneficial effects on the microflora of the gastrointestinal tract (GIT). Metchnikoff's 1906 publication *The Prolongation of Life: Optimistic Studies*, implicated a *lactic acid bacillus* in Bulgarian yoghurts as the agent responsible for preventing intestinal putrefaction and ageing. Later, it was discovered that Metchnikoff's bulgarian strain did not survive passage through the gastrointestinal tract, prompting substitution of *Lactobacillus acidophilus* as the most likely candidate to fulfil the primary criteria expected of an intestinal probiotic. It has since been discovered that a variety of homofermentative and heterofermentative lactobacilli inhabit the GIT, mouth and vagina and each may elicit a variety of benefits as constituents of the normal microflora. The most predominant among these are six species of homofermentative lactobacilli that now constitute the group known as the *L. acidophilus* complex.

The six species shown in **Table 1** collectively demonstrate the metabolic and functional properties that have typically been assigned to the bacteria called *L. acidophilus* over the last century. Members of the *L. acidophilus* complex are generally considered to facilitate the establishment of the normal gastrointestinal microflora, represented by a complex population of microorganisms that are known to exert beneficial influences on the host. Probiotic lactobacilli

Table 1 Species of the *Lactobacillus acidophilus* complex

Species	DNA homology groups		G+C %	Type strains
	Johnson et al	Lauer et al		
Lactobacillus acidophilus	A1	Ia	34–37	ATCC 4356
Lactobacillus crispatus	A2	Ic	35–38	ATCC 33820
Lactobacillus amylovorus	A3	Ib	40–41	ATCC 33620
Lactobacillus gallinarum	A4	Id	36–37	ATCC 33199
Lactobacillus gasseri	B1	IIa	33–35	ATCC 33323
Lactobacillus johnsonii	B2	IIb	33–35	ATCC 33200

have further been implicated in a variety of beneficial roles that include:

- maintenance of the normal microflora
- pathogen interference, exclusion and antagonism
- immunostimulation and immunomodulation
- antimutagenic and anticarcinogenic activity
- deconjugation of bile acids
- presentation of lactase in vivo
- lowering of serum cholesterol.

Although these proposed benefits are widely touted, there is insufficient and often conflicting information about the actual roles, ecology and benefits of probiotic *Lactobacillus* species in the gastrointestinal tract. With the diversity that exists in the phenotypic and genotypic characteristics of the six *L. acidophilus*-type species, establishing their potential probiotic roles, individually and in combination, will be a significant challenge to future researchers in this field.

Taxonomy

The *L. acidophilus* species are Gram-positive rods (dimensions are in the range 0.5–1 × 2–10 μm), with rounded ends, occurring in pairs or short chains. The group was initially categorized in the thermobacteria classification of lactic acid bacteria based on their homofermentative metabolism and ability to grow at 45°C. In 1980, *L. acidophilus* was recognized as a heterogeneous group by DNA hybridization studies and separated into two main DNA homology groups (A and B or I and II), eventually forming six distinct species composed of *L. acidophilus*, *L. amylovorus*, *L. crispatus*, *L. gallinarum*, *L. gasseri* and *L. johnsonii*.

Figure 1 shows the phylogenetic relatedness of the *L. acidophilus* group based on analysis of their 16S ribosomal RNA sequences. *L. acidophilus* is most closely related to *L. helveticus*, a milk fermenting *Lactobacillus*, and the other members of the A-homology group, *L. crispatus* and *L. amylovorus*. *L. gasseri* and *L. johnsonii* are related in the acidophilus phylogenetic group, but are found more distant from *L. acidophilus* than either of the fermentative strains of *L. helveticus* or *L. delbrueckii*. The genetic relationship between the gastrointestinal lactobacilli of

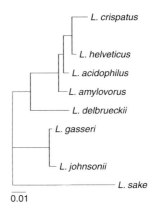

Figure 1 Phylogenetic relationships between members of the *L. acidophilus* group based on an analysis of aligned 16S rRNA gene sequences. The tree was rooted with *L. sake* and created by applying the neighbour-joining method, of Saitou and Nei (1987), to a matrix of pairwise distances. The bar indicates 0.01 fixed mutations per nucleotide position. Phylogenetic analysis and tree construction by M Kullen.

the *L. acidophilus* complex and the milk-fermenting species is curious. Genetic analysis of the probiotic versus fermentation roles of these species is likely to reveal some important similarities and differences in this regard.

Metabolism and End Products

Currently, members of the *L. acidophilus* complex are classified as obligate homofermenters. Hexoses are fermented primarily to lactic acid (> 85%) by the Embden–Meyerhof–Parnas (EMP) pathway. They possess aldolase and lack phosphoketolase. Neither gluconate nor pentoses are fermented. All species produce both D and L isomers of lactic acid. *L. acidophilus*, *L. amylovorus*, *L. crispatus*, *L. gallinarum*, and *L. johnsonii* all possess β-galactosidase, whereas *L. gasseri* lacks β-galactosidase, but has phospho-β-galactosidase. Fermentation patterns and key distinguishing characteristics of species within the *L. acidophilus* complex are listed in **Table 2**. All species possess a Lys-D-Asp type peptidoglycan. Analysis of cell wall components and genetic analysis for *slp*-related sequences have generally shown that members

Table 2 Distinguishing characteristics and fermentation patterns of species in the *Lactobacillus acidophilus* complex

	Species					
	acidophilus	crispatus	amylovorus	gallinarum	gasseri	johnsonii
Growth at 30°C	−	−	−	−	−	−
Growth at 45°C	+	+	+	+	+	+
Lactic acid isomers	DL	DL	DL	DL	DL	DL
S-layer	+	+	+	+	−	−
Aesculin	+	+	+	+	+	+
Amygdalin	+	+	+	+	+	+
Cellobiose	+	+	+	+	+	+
Glycogen	+	+	−	+	−	−
Galactose	+	+	+	+	+	+
Lactose	+	+	−	v	v	v
Maltose	+	+	+	+	v	+
Mannitol	−	−	−	−	−	−
Mannose	+	+	+	+	+	+
Melibiose	v	−	−	+	v	v
Raffinose	−	−	−	+	v	v
Rhamnose	−	−	−	−	−	−
Salicin	+	+	+	+	+	+
Sucrose	+	+	+	+	+	+
Trehalose	v	−	+	−	v	v
Xylose	−	−	−	−	−	−

+, >90% of strains are positive.
−, >90% of strains are negative.
v, variable.

of the A homology group possess an S-layer, whereas members of the B group do not.

In 1995, Hammes and Vogel grouped lactobacilli and related genera based on both their fermentation patterns and phylogenetic relatedness. The *L. acidophilus*-complex species belong to group Aa, being defined as obligately homofermentative organisms closely related to *L. delbrueckii*. In addition to the six species in the *L. acidophilus* complex, there are five other recognized species in the Aa group: *L. delbrueckii*, *L. jensenii*, *L. helveticus*, *L. amylophilus* and *L. kefiranofaciens*. Based on the difference in G+C content between *L. delbrueckii* (49–51 mol%) and the rest of the species in the group (33–41 mol%), the 'L. delbrueckii group' was renamed the 'L. acidophilus group' because of the more typical G+C content (34–37 mol%) for the majority of species defined in this group.

The *L. acidophilus* group also produces a variety of antimicrobial compounds, including lactic acid (>85%), hydrogen peroxide, and a variety of bacteriocins. Early reports of 'antibiotic-like substances' have not been substantiated by purification or characterization of the active compounds, their properties often resemble well-known antimicrobials like lactic acid and hydrogen peroxide. In assays screening for antimicrobial activities, the impacts of acid and hydrogen peroxide must be eliminated by neutralization of the culture supernatant to pH 7.0 or addition of catalase (3%), respectively. *L. acidophilus*

Table 3 Bacteriocin production by probiotic lactobacilli

Bacteriocin	Producer
Lactacin B	*L. acidophilus*
Acidophilucin A	
Acidocin 8912	
Acidocin A	
Acidocin B	
Acidocin JCM1132	
Acidocin J1229	
Lactobin A	*L. amylovorus*
Lactacin F	*L. johnsonii*
Gassericin A	*L. gasseri*
Reutericin 6	*L. reuteri*

Klaenhammer, 1998

cultures are known to produce hydrogen peroxide, but the amount produced varies considerably between strains. A list of peptide bacteriocins produced by members of the *L. acidophilus* complex is shown in **Table 3**. Most of the bacteriocins are class II peptide bacteriocins, such as lactacin F, lactacin B and acidocin A, which are active against other lactobacilli and enterococci. Recently, an unusual *L. acidophilus* bacteriocin, acidocin B, was isolated which showed activity against *Listeria monocytogenes*, *Clostridium sporogenes* and *Brochothrix thermosphacta*, but not other lactobacilli. Bacteriocin production is assumed to enhance the competitiveness of lactobacilli against closely related lactic acid bacteria in the gastrointestinal tract, and potentially antagonize undesirable intestinal pathogens. However, the impacts of bacteriocin production by probiotic lactobacilli, in vivo, remain to be investigated.

Identification and Differentiation

It is critically important to have rapid and accurate methods available to identify and differentiate species within the L. acidophilus complex from each other as well as from closely related lactobacilli. Identification based on traditional criteria has been unreliable due to the physiological and biochemical diversity of this genus. Molecular techniques targeting highly conserved rRNA sequences are rapid, accurate and reliable. Differentiation of species within the L. acidophilus group has been reported through the use of specific rRNA oligonucleotide probes for hybridization, generation of Random Amplified Polymorphic DNA (RAPD) patterns, and the use of species-specific oligonucleotide primers in polymerase chain reactions. A list of probes and primers compiled for the species of the L. acidophilus complex is given in **Table 4**. These protocols are highly sensitive, require very specific temperatures and conditions, and

can be difficult to reproduce between laboratories. Therefore, the identities of unknown organisms are often difficult to resolve. In these cases, it is recommended that unknown lactobacilli be identified by sequencing the 16S or 23S rRNA regions that define the species.

Becoming more common is the need to differentiate a strain within a given species. Traditional methods of strain typing include plasmid profiles and total soluble protein patterns. Recent efforts have resulted in two methods that generate simple and reproducible DNA restriction fragment patterns that allow differentiation of strains of lactobacilli. Ribotypes are the patterns of an rRNA-oligonucleotide probe following its Southern hybridization to the DNA restriction digests of the strain being characterized. Pulsed field gel electrophoresis (PFGE) uses alternating currents to separate large restriction fragments (> 50 kb) which are generated with restriction enzymes that cut infrequently in the genome. For lactobacilli, the preferred PFGE methods use mutanolysin, to carry out in-block cell lysis, followed by lengthy digestions with enzymes such as SmaI (5' CCCGGG 3') or ApaI (5' GGGCCC 3'). The Sma I banding patterns for the ATCC neotype strains of the six species of the L. acidophilus complex are shown in **Figure 2**. Analysis

Table 4 Oligonucleotide probes and primers reported for the differentiation of species of the L. acidophilus complex

Probes	Target (position)	Reference
TCTTTCGATGCATCCACA	23S – Lactobacillus acidophilus (1159–1180)	Pot et al. (1993) J. Gen. Microbiol 139: 513–517
CAATCTCTTGGCTAGCAC	23S – L. crispatus	Ehrmann et al (1992) Syst. Appl Microbiol 15: 453–455
GTAAATCTGTTGGTTCCGC	16S – L. amylovorus	Ehrmann et al (1994) FEMS Microbiol Lett 117: 143–149
none	L. gallinarum	
TCCTTTGATATGCATCCA	23S – L. gasseri (1160–1178)	Pot et al (1993) J. Gen. Microbiol 139: 513–517
ATAATATATGCATCCACAG	23S – L. johnsonii (1158–1179)	Pot et al. (1993) J. Gen. Microbiol 139: 513–517
Specific primers		
Aci I – TCTAAGGAAGCGAAGGAT and Aci II – CTCTTCTCGGTCGCTCTA	16S–23S intergenic spacer region, L. acidophilus ATCC 4356	Tilsala-Timisjarvi and Alatossava (1997) Inr J. Food Microbiol 35: 49–56
HE1 – AGCAGATCGCATGATCAGCT and SS2 – CACGGATCCTACGGGTACC TTGTTACGACTT	16S – L. acidophilus[a]	Drake et al (1996) J. Food Protection 59: 1031–1036
RAPD primers		
OPL-1 GGCATGACCT	All L. acidophilus group	Du Plessis and Dicks (1995) Curr Microbiol 31: 114–188
OPL-4 GACTGCACAC	All L. acidophilus group	
AGCAGCGTGG	Differentiates L. acidophilus from other lactobacilli	Cocconcelli et al (1995) Lett. Appl Microbiol 21: 376–376

[a]Also generates an amplicon in L. helveticus which can be differentiated by restriction fragment length polymorphism analysis.

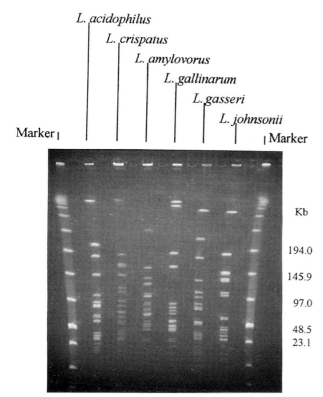

Figure 2 PFGE. *Sma*I fragmentation patterns for the type strains comprising the *L. acidophilus* group. Samples of 18 h digestions were run in 1.1% agarose at 200 V for 22 h at 14°C. Lanes 1 and 8, molecular weight marker; lane 2, *L. acidophilus*; lane 3, *L. crispatus*; lane 4, *L. amylovorus*; lane 5, *L. gallinarum*; lane 6, *L. gasseri*; lane 7, *L. johnsonii*.

of large DNA fragmentation patterns can discriminate effectively between strains on the basis of overall genomic organization. However, PFGE cannot detect minor genetic differences within the large fragments (e.g. point mutations, minor additions, deletions and rearrangements) that can occur between clonal variants. These small genetic changes can impact the phenotypic characteristics, activities and behaviour of strains that may appear identical by PFGE. Therefore, descriptions of any member of the lactobacilli should include both genetic and phenotypic information.

Genetics

The genomes of *L. gasseri* and *L. acidophilus* have been estimated at 1.85 and 2.0 Mb, respectively. This is a small size relative to other prokaryotes such as *Escherichia coli* (4.7 Mb) and *Bacillus cereus* (5.7 Mb). The smaller genome of the *L. acidophilus* group may reflect their long-term evolution to the bounteous nutritional environment of the small intestine, where biosynthetic pathways may not be necessary. Indeed, it has been shown that *L. acidophilus* carries mutations that silence a number of amino acid biosynthetic pathways found active in other lacto-

bacilli. Work has been initiated to determine the sequence of the *Lactobacillus acidophilus* genome. The strain to be sequenced has been deposited with the American Type Culture Collection under the accession number ATCC 700396. A description of the organism deposited to ATCC is as follows:

Lactobacillus acidophilus ATCC 700396. A Gram-positive, catalase negative, rod. Growth at 37C, variable growth at 45C and no growth at 15C. Carbohydrate fermentation patterns are identical to those of *L. acidophilus*. ATCC 700396 was isolated as a derivative of the human *L. acidophilus* strain RL8K, which is tolerant to >0.8% oxgall when grown on MRS-Lactobacillus agar (Difco Laboratories). Varying media components can alter colonial morphology between rough and smooth. The progenitor strain, RL8K, was isolated from a human and later named NCFM (North Carolina Food Microbiology) by Dr Marvin Speck. Bile-tolerant derivatives of RL8K, such as NCFM/N2, have been widely used in scientific studies and human clinical trials. The progenitor strain RL8K hybridizes to the *L. acidophilus*-specific oligonucleotide probe Lba 5'-TCTTTCGATGCATCCACA-3'. The *Sma*I genomic fingerprints of RL8K, NCFM, NCFM/N2 and ATCC 700396 are identical to each other and highly similar, but not identical, to the neotype *L. acidophilus* strain ATCC 4356.

Growth and Culture Conditions

Lactobacillus acidophilus cultures are micro-aerophilic and capable of aerobic growth in static cultures without shaking. Anaerobic conditions are preferable and growth is stimulated in broth or agar under a standard anaerobic gas mixture of 5% carbon dioxide, 10% hydrogen, and 85% nitrogen. The nutritional requirements of *L. acidophilus* reflect the fastidious nature of these bacteria. Media for standard propagation are rich in amino acids and vitamins (peptones, tryptones, yeast/beef extracts), and usually contain sorbitan monooleate (Tween 80), sodium acetate and magnesium salts which stimulate growth. The primary propagation medium is MRS (de Man, Rogosa and Sharpe) broth. The formulation sold commercially (Difco Laboratories, Detroit, MI) is as shown in **Table 5**.

Derived from the original selection medium, using high levels of acetate, a number of media are now available for selection of the *L. acidophilus* complex species from food products and biological samples (intestinal, vaginal, mouth, faeces) where mixed bacterial populations exist. Three primary selective agents can be used, individually or in combinations: sodium acetate ($15–25\,g\,l^{-1}$), tomato juice and bile (ranging from 0.15% to 1.0% oxgall). The formulations for Rogosa SL media (Difco Laboratories, Detroit, MI) and Lactobacillus Selection Agar (LBS, Becton Dickenson, Cockeysville, MD) are given in **Tables 6 and 7**.

For isolation of *L. acidophilus* from biological or food samples, our laboratory uses LBS + tomato juice. Plates are incubated for 48–72 h at 37°C under anaerobic gas. Either MRS (Difco) or LBS (BBL) agars containing 0.15% oxgall (BBL) have been used to assess injury of *L. acidophilus* populations suspended in dried or frozen cultures.

Selection Criteria

Selection criteria for comparison of potential probiotic lactobacilli have been widely reported. The collective list of desired characteristics grows continuously, but no consensus among researchers has so far been reached. A list of selection criteria is shown in **Table 8**. The question marks identify proposed criteria that remain subject to debate as to their in vivo significance.

Conclusions

L. acidophilus type strains under consideration for use as probiotics are often expected to serve in two very contrasting roles: survival or fermentation in a delivery vehicle or food, followed by survival and in vivo activity during passage into the gastrointestinal tract. In this regard, a number of major challenges continue to face the production of *Lactobacillus* cultures targeted for use as probiotics. First, the stability of cultures in dried, frozen, or suspended states remains a critical issue and efforts to improve the stress tolerance of *L. acidophilus* type species is needed. Second, the potential impact of culture propagation and storage conditions on the in vivo conditioning of the probiotic must be considered. Third, understanding the genetic components and controls that direct in vivo performance will be the key to uncovering the in vivo activities of probiotic lactobacilli and will ensure expression of vital characteristics during production, storage and, ultimately, in the gastrointestinal tract. Therefore, as molecular biology has now elucidated the taxonomic rela-

Table 5 Composition of commercially available *Lactobacillus* MRS broth

Ingredient	Concentration (g l^{-1})
Bacto proteose peptone no. 3	10
Bacto beef extract	10
Bacto yeast extract	5
Dextrose	20
Sorbitan monooleate complex	1
Ammonium citrate	2
Sodium acetate	5
Magnesium sulphate	0.1
Manganese sulphate	0.05
Disodium phosphate	2

Final pH 6.5 at 25°C
One pound will make 8.25 l of medium

Table 6 Bacto Rogosa SL Broth, dehydrated

Ingredient	g l^{-1}
Bacto tryptone	10
Bacto yeast extract	5
Bacto dextrose	10
Bacto arabinose	5
Bacto saccharose	5
Sodium acetate	15
Ammonium citrate	2
Monopotassium phosphate	6
Magnesium sulphate	0.57
Manganese sulphate	0.12
Ferrous sulphate	0.03
Sorbitan monooleate	1

Final pH 5.4 at 25°C

Table 7 LBS agar (BBL)

Ingredient	g l^{-1}
Pancreatic digest of casein	10.0
Yeast extract	5.0
Monopotassium phosphate	6.0
Ammonium citrate	2.0
Dextrose	20.0
Polysorbate 80	1.0
Sodium acetate hydrate	25.0
Magnesium sulphate	0.575
Manganese sulphate	0.12
Ferrous sulphate	0.034
Agar	15.0

Final pH 5.5

Table 8 Selection criteria for intestinal probiotics

- Species specific origin: human isolates for human probiotics
- GRAS
- Acid tolerant
- Bile tolerance?
- Bile-salt hydrolase activity?
- Adherence – cell type in vivo or in vitro
- Antimicrobial
- Immunogenic
- Non-invasive
- Non-pathogenic
- Coaggregation?
- Reduces cholesterol?
- Fermentation compatible
- Genetically amenable

tionships of the species defined within the *L. acidophilus* complex, genetic information will ultimately reveal the potential activities and interactions of these bacteria in the microecology of the gastrointestinal tract.

See also: **Enzyme Immunoassays**: Overview. **Fermented Milks**: Range of Products; Yoghurt. **Microflora of the Intestine**: Biology of *Lactobacillus acidophilus*. **Probiotic Bacteria**: Detection and Estimation in Fermented and Non-fermented Dairy Products.

Further Reading

Bibel DJ (1988) Elie Mechnikoff's bacillus of long life. *ASM News* 54: 661–665.

De Man JC, Rogosa M and Sharpe ME (1960) A medium for the cultivation of *Lactobacilli*. *J. Bacteriol.* 23: 130–135.

Fujisawa T, Benno Y, Yaeshima T and Mitsuoka T (1992) Taxonomic study of the *Lactobacillus acidophilus* group, with recognition of *Lactobacillus gallinarum* sp. nov. and *Lactobacillus johnsonii* sp. nov. and synonymy of *Lactobacillus acidophilus* group A3 (Johnson et al 1980) with the type strain of *Lactobacillus amylovorus* (Nakamura, 1981). *Int. J. Syst. Bacteriol.* 32: 487–491.

Hammes WP and Vogel RF (1995) The genus *Lactobacillus*. In: Wood BJ and Holzapfel WH (eds) *The Genera of Lactic Acid Bacteria*. Ch. 3, p. 19. London: Chapman & Hall.

Johnson JL, Phelps CF, Cummins CS, London J and Gasser F (1980) Taxonomy of the *Lactobacillus acidophilus* group. *Int. J. Syst. Bacteriol.* 30: 53–68.

Klaenhammer TR (1993) Genetics of bacteriocins produced by lactic acid bacteria. *FEMS Microbial. Rev.* 12: 39–86.

Klaenhammer TR (1995) Genetics of intestinal lactobacilli. *Int. Dairy J.* 5: 1019–1058.

Klaenhammer TR (1998) Functional activities of *Lactobacillus* probiotics: genetic mandate. *Int. Dairy J.* 8: 497–506.

Lauer E, Helming C and Kandler O (1980) Heterogeneity of the species *Lactobacillus acidophilus* (Moro) Hansen & Moquot as revealed by biochemical characteristics and DNA–DNA hybridization. *Zentralbl. Bakteriol. Mikrobiol. Hyg.* 1. Abt. Orig. C1: 150–168.

Leer RJ, van der Vossen JM, van Giezen M, van Noort JM and Pouwels PH (1995) Genetic analysis of acidocin B, a novel bacteriocin produced by *Lactobacillus acidophilus*. *Microbiology* 141: 1629–1635.

Rogosa M, Mitchell JA and Wiseman RF (1951) A selective medium for the isolation and enumeration of oral and fecal *Lactobacilli*. *J. Bacteriol.* 62: 132–133.

Saitou N and Nei M (1987) The neighbor-joining method: a new method for reconstructing phylogenetic trees. *Mol. Biol. Evol.* 4: 406–425.

Schleifer KH and Ludwig W (1995) Phylogeny of the genus *Lactobacillus* and related genera. *Syst. Appl. Microbiol.* 18: 461–467.

Schleifer KH, Ehrmann M, Beimfohr C, Brockman E, Ludwig W, Aman R (1995) Application of molecular methods for the classification and identification of lactic acid bacteria. *Int. Dairy J.* 5: 1081–1094.

Tannock GW (1995) *Normal Microflora: An Introduction to Microbes Inhabiting the Human Body*. London: Chapman & Hall.

Teuber M (1993) *Lactic Acid Bacteria*. In *Biotechnology*, 2nd edn. Vol. I, ch. 10, p. 325. Weinheim: VCH Publishers.

(1981) *Bergey's Manual of Systematic Bacteriology*, 8th edn. P. 1208. Baltimore: Williams & Wilkins.

(1957) *Bergey's Manual of Systematic Bacteriology*, 7th edn. P. 546. Baltimore: Williams & Wilkins.

Lactobacillus casei

Marco Gobbetti, Instituto di Produtionie Paparationi Alimentari, Agriculture Faculty of Foggia, Italy

Lactobacillus casei consists of phenotypically and genetically heterogeneous strains which colonize various food ecosystems. *L. casei* is favoured or deliberately added to enhance the quality of foodstuffs and to promote human and animal health.

This article describes the taxonomy, metabolism, genetics and probiotics, as well as the food applications of *L. casei* and related taxa.

General Characteristics of the Species

Lactobacillus casei is a Gram-positive, non-motile, non-sporulating and catalase-negative bacterium. Cells are rods of $0.7–1.1 \times 2.0–4.0\,\mu m$, often with square ends, which tend to form chains (**Fig. 1**). The cell wall contains L-Lys-D-Asp peptidoglycan and

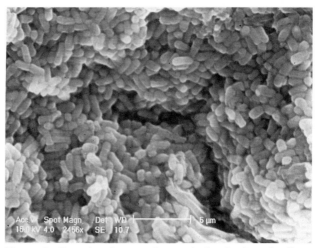

Figure 1 Scanning electron micrograph showing *Lactobacillus casei* cells.

polysaccharides which determine the serological specificity (B or C) based on the content of rhamnose or glucose–galactose. Teichoic acids are not present. The G+C content of the DNA is 45–47%. Phylogenetic relatedness of lactic acid bacteria based on sequence analysis of 16S rRNA includes *L. casei* in the *L. casei–Pediococcus* group.

L. casei is a typical cheese bacterium isolated mainly from milk and dairy products but also from silage, sourdough, cow dung, human intestinal tract and mouth and vagina. De Man–Rogosa–Sharpe (MRS) and Lactobacillus selection agar (LBS) media are recommended for enumeration and isolation from various habitats. *L. casei* can be distinguished by its capacity to grow on substrates (gluconate, malate and pentitols) rarely used by lactic acid bacteria. A modified Rogosa SL medium, with melezitose as the only sugar, is suitable for isolation from human saliva. Riboflavin, folic acid, calcium pantothenate and niacin are required for growth. Pyridoxal or pyridoxamine is essential or stimulatory. Growth which occurs at 15°C but not at 45°C, is optimal at 30°C. *Bergey's Manual of Systematic Bacteriology* recognizes four subspecies: *L. casei* subsp. *casei*, *pseudoplantarum* (non-fermenting raffinose and melibiose), *rhamnosus* (rhamnose-fermenting) and *tolerans* (tolerates heating at 72°C for 40 min). However, the taxonomy of *L. casei* is continuously changing.

DNA–DNA hybridizations have shown high levels of relatedness among strains of *L. casei* subspp. *casei*, *pseudoplantarum* and *tolerans*, which all differed from the type strain *L. casei* subsp. *casei* ATCC 393. Hence, members of *L. casei* subsp. *pseudoplantarum*, *tolerans* and the majority of *L. casei* subsp. *casei* strains were included in *L. paracasei* subspp. *paracasei* and *tolerans*. Strains of *L. casei* subsp. *rhamnosus* which form a genomically homogeneous group, were elevated to the status of species as *L. rhamnosus*. Afterwards, it was shown that *L. casei* subsp. *casei* ATCC 393 has the highest level of similarity with *L. rhamnosus* ATCC 15820, the original type strain of *Lactobacterium zeae*, whereas *L. casei* ATCC 334 is genetically related to *L. casei* subsp. *casei* and *L. paracasei* strains. The subsequent proposal was that *L. casei* subsp. *casei* ATCC 393 and *L. rhamnosus* ATCC 15820 should be included as members of *Lactobacillus zeae* nom. rev., with strain ATCC 334 designated as the neotype strain of *L. casei* subsp. *casei* and species *L. paracasei* should be rejected. Recently, the structures of the 16S rRNA genes have shown that *L. casei* and related taxa should be classified in three species: *L. zeae* which includes *L. zeae* and *L. casei*, a species that includes the strains of *L. paracasei* and *L. casei* ATCC 334, and *L. rhamnosus*.

This article considers *L. casei* and related taxa.

Metabolism and Enzymes

The latest grouping of lactobacilli based on biochemical-physiological criteria includes *L. casei* in the facultatively heterofermentative group. Hexoses are almost entirely converted into lactic acid via the Embden–Meyerhof pathway and pentoses are used by the induced phosphoketolase, to produce lactic acid and acetic acid. The enzymes of the phosphogluconate pathway are repressed by glucose.

The phosphoenolpyruvate-dependent phosphotransferase transport system (PEP-PTS) is the most frequently used by *L. casei*. However, both β-galactosidase and β-phosphogalactosidase are found. H⁺ATPase, responsible for ATP hydrolysis which pumps protons out of the cells thereby establishing a proton motive force (pmf) is also found. Oxygen is used as an external electron acceptor: pyruvate oxidase and acetate kinase, respectively convert pyruvate into acetyl phosphate and then to acetate thus increasing ATP yield.

When fermentable carbohydrates are absent, *L. casei* uses citrate as the energy source. Adding sodium citrate to the broth medium increases the growth yield. Lactic, formic or acetic acids, CO_2 and traces of acetoin, diacetyl and ethyl alcohol are produced from citrate. The uptake of citrate is inhibited by low concentrations of glucose.

L. casei strains possess threonine aldolase and may decarboxylate α-ketoisocaproic or α-ketoisovaleric acids.

Strains of *L. casei* isolated from various habitats produce exopolysaccharides (EPS) resulting in mucoid, slimy colonies. The production and composition of EPS are markedly influenced by the available carbon sources. Glucose, rhamnose and traces of galactose and arabinose generally compose EPS. By adding glucose or sucrose to milk, EPS production increases and the monosaccharide composition changes with glucose becoming dominant.

Proteolytic enzymes from *L. casei* have been purified and characterized with respect also to their involvement in cheese ripening (**Table 1**). Features common to these enzymes are: optimum activity at 30–40°C and pH 6.5–7.5 with considerable activity at low temperatures and pH values.

A cell wall-bound serine proteinase with a molecular mass of ca. 180 kDa has been purified from *L. casei* subsp. *casei* and *L. paracasei* subsp. *paracasei* strains. The chromosomally located gene has been cloned and sequenced. The unprocessed proteinase comprises 1902 amino acid residues and the primary sequence shares more than 95% homology with lactococcal proteinases. Differences occur in only two positions which are important for substrate specificity

Table 1 Proteolytic enzymes purified and characterized from *Lactobacillus casei* and related taxa

Strain	Enzyme	Molecular mass (kDa)	Type	pH	Localization
L. paracasei subsp. *paracasei* HN1	Proteinase	n.d.	Serine-enzyme	n.d.	Cell wall
L. casei subsp. *casei* NCDO151	Proteinase	181	Serine-enzyme	6.5	Cell wall
L. casei subsp. *casei* LLG	Aminopeptidase	87	Metalloenzyme	7.0	Intracellular
L. casei subsp. *casei* LLG	X-prolyl dipeptidyl peptidase	79	Serine-enzyme	7.0	Intracellular
L. casei subsp. *casei* UL21	X-prolyl dipeptidyl peptidase	n.d.	Serine-enzyme	7.0	Intracellular
L. casei subsp. *rhamnosus* UL26	X-prolyl dipeptidyl peptidase	n.d.	Serine-enzyme	7.0	Intracellular
L. casei subsp. *casei* LLG	Proline iminopeptidase	46	Cysteine-enzyme	7.5	Intracellular

n.d., not determined.

and binding. Comparing the N-terminal sequence of the mature proteinase with that deduced from the nucleotide sequence reveals a typical signal sequence for Sec-dependent translocation and a prosequence, both of which are removed by post-translational processing. A proteolytic enzyme is involved in proteinase maturation. The N-terminal part of the mature enzyme constitutes the catalytic domain. The outermost part of the C-terminal carries a sorting signal, followed by a putative membrane spanning α-helix and a small charged tail. The substrate specificity of proteinases from *L. casei* strains differs: enzymes which only hydrolyse β-casein or with a broad specificity have been characterized.

The proteinase activity is generally weaker than peptidase activity. *L. casei* subspecies generally have high amino- and dipeptidase activities and relatively weak tripeptidase activity. Differences have been found between the subspecies: *L. casei* subsp. *casei* has the highest di- and tripeptidase activities, whereas *L. casei* subsp. *rhamnosus* has the highest activities for Leu-, Lys-, Ala-, Val- and Met-aminopeptidase and iminopeptidase. An 87 kDa general metalloaminopeptidase (PepN) has been purified and characterized from *L. casei* subsp. *casei*. It seems to account for all the aminopeptidase and some of the dipeptidase activities found in the cell-free extracts. A serine-dependent X-prolyl dipeptidyl peptidase (PepX) and a sulphydryl proline iminopeptidase (PepP) have also been purified. There is some controversy about the role attributed to enzymes which hydrolyse proline-containing peptides in preventing bitter cheese taste.

Esterase and lipase activities of *L. casei* are strain specific. In general, esterase activity is greatest on β-naphthyl butyrate but limited on β-naphthyl derivatives of C_{10}–C_{14} fatty acids. *L. casei* is among the strongest esterolytic species of the mesophilic lactobacilli. Strains which have high esterase activity do not necessarily have high lipolytic activity. Lipase activity decreases with increasing fatty acid chain length from tributyrin to tripalmitin and is almost absent on triolein. The lipases have an optimum activity close to

neutrality and at temperatures of 30–40°C. A 320 kDa butyric esterase has been reported in *L. casei* subsp. *casei* which has a broad pH range of activity, is stable at room temperature and resistant to moderate heating. Because of the large population reached by non-starter lactobacilli, such as *L. casei*, in cheese curd and because of the long ripening time of several cheeses, the role of esterases and lipases from mesophilic lactobacilli must be carefully considered.

Genetics and Bacteriophages

Most of the gene cloning results concerning members of the genus *Lactobacillus* refer to *L. casei*.

ISL1 from *L. casei* was the first insertion sequence (IS) element to be characterized from a lactic acid bacterium. It has the industrially important property of converting a temperate phage into a lytic phage. ISL1 may be present in both chromosomal and plasmid DNA of *L. casei* but is not widely distributed among strains.

The first evidence of plasmids in lactobacilli was found in *L. casei* subsp. *casei* (39 kb) and *L. casei* subsp. *rhamnosus* (28.7 kb). Plasmids are largely distributed in *L. casei* strains where one to five plasmids of ca. 1.5–68 kb are generally harboured. The *lac* operon consisting of genes for lactose transport (*lac*EF), lactose-6-P hydrolysis (*lac*G) and tagatose-6-P pathway (*lac*ABCD) is plasmid located. The operon shares similarities with a lactose phosphoenolpyruvate-dependent phosphotransferase system (PEP-PTS) of *Staphylococcus aureus* and *Lactococcus lactis* subsp. *lactis* which consist of a soluble IIA domain and a membrane-bound IICB complex. IIB contains a highly conserved and essential cysteine residue. Some strains harbour a 62.8 kb plasmid which only encodes phospho-β-galactosidase activity. Therefore lac⁻ mutants grow in a galactose medium because they retain the complete set of genes for the (PEP-PTS) and the predominant galactose metabolic pathway. EPS production is also encoded by plasmid DNA. However, most of the plasmids harboured by *L. casei* strains are still cryptic.

Table 2 Heterologous gene expression in *Lactobacillus casei* and related taxa

Protein	Origin	Species	Promoter
Xylose isomerase	*L. pentosus*	*L. casei*	*xyl*A
Catalase	*L. sake*	*L. casei*	pGKV210 derived
Lysostaphin	*S. staphylolyticus*	*L. casei*	
FDMV-β-gal	FMDV-*E. coli*	*L. casei*	*xyl*R
FDMV-α-amylase	FMDV-*L. amylovorous*	*L. casei*	*amy* A, L-*ldh*
Chloramphenicol acetyltransferase	pC194	*L. casei*	*xyl*A, *xyl*R
Pyruvate decarboxylase	*Z. mobilis*	*L. casei*	pGKV432 derived
Alcohol dehydrogenase	*Z. mobilis*	*L. casei*	pGKV432 derived

Genera: *L.*, *Lactobacillus*; *S.*, *Staphylococcus*; *E.*, *Escherichia*, *Z. Zymomonas*.

The first report on a successful application of the electroporation technique in lactobacilli concerned *L. casei*. Transformants were obtained at efficiencies of ca. 10^3–10^5 per µg of DNA, which were later improved to ca. 10^7 per mg. The 28 kb plasmid pLZ15, coding for β-galactosidase activity, the streptococcal shuttle vector pSA3 for erythromycin resistance, pNCO937, which carries a *Streptomyces* sp. cholesterol gene, and plasmids pLP825 and pNZ12 for chloramphenicol resistance are some examples of plasmid vectors used to transform *L. casei* strains. Transfection has also been successfully used to demonstrate that large molecules can be introduced into *L. casei* by electroporation. Conjugal self-transmission of lactose plasmids in *L. casei* ATCC 4646 as well as transmission of the conjugal plasmid pAMβ1 from *L. lactis* subsp. *lactis* into *L. casei* have been reported. The β-galactosidase-encoding plasmid pLZ15 was transformed into protoplasts of *L. casei* ATCC 393 at a high frequency by liposome-mediated polyethyleneglycol fusion. The same technique was also used for transfection of phage J1 DNA to *L. casei* sphaeroplasts.

Heterologous genes have often been expressed in *L. casei* (**Table 2**). In particular, the L-*ldh* promoter of *L. casei* drove the expression of fused genes such as *cat* gene (encoding chloramphenicol acetyl transferase). Expression and secretion of extracellular proteins have been reported for α-amylase from *L. amylovorus* in *L. casei* under control of its own promoter.

The first *L. casei* lytic phage, J1, was isolated during fermentation mishaps in the manufacture of yakult. A second phage PL1 was later isolated under similar circumstances. The main characteristics of these two phages are shown in **Table 3**.

The first characterization of a *Lactobacillus* lysin involved the purification of bacteriophage PL1 lysin from lysates of *L. casei*. The enzyme is an *N*-acetyl-muramidase with a molecular mass of ca. 37 kDa and with a C-terminal amino acid like Tyr. The lysin has a narrow spectrum of activity which is very effective against the host strain *L. casei* ATCC 27092 but is inactive or weakly active against several other *L. casei* strains. The lysin of bacteriophage PL1 has been used to produce protoplasts and to demonstrate the transfection of PL1 phage DNA in *L. casei*.

A temperate phage ΦFSW of *L. casei* ATCC 27139 is the source of a number of virulent phages called ΦFSV which have been isolated from different Japanese dairy plants. The lysogenic state may be altered by inserting ISL1.

Applications in Food Fermentations

Lactobacillus casei is frequently encountered in food ecosystems such as raw milk, cheese, fermented milk, fresh and fermented vegetables, raw sausage and meat. Because of the positive metabolic activity, *L. casei* is favoured or deliberately added to enhance the quality of foodstuffs and to promote human and animal health.

Cheeses

Although generally not used as starter cultures in cheesemaking, *L. casei* and its subspecies can be isolated from a variety of cheeses.

L. casei is probably the most important adventitious non-starter lactic acid bacterium (NSLAB) which populates cheeses such as Cheddar, Grana and Gouda during ripening. NSLAB withstand pasteurization and/or contaminate the post-pasteurized cheese milk. *L. casei* is one of the most heat-resistant NSLAB. In contrast to the starter lactic acid bacteria, NSLAB grow from small numbers ($\approx 10^2$ cfu g^{-1}) in fresh curd to dominate the microbiota of the mature cheese ($\approx 10^8$ cfu g^{-1}) without a decline during ripening. They tolerate the hostile environment of the curd-cheese (low moisture content, pH and temperature, and high salt concentration) and use residual lactose, lactate, citrate, peptides, amino acids and nutrients released from autolysing starter bacteria. NSLAB generally perform a secondary fermentation, after the primary one conducted by starters.

The NSLAB positively influence the proteolysis, lipolysis and flavour development of Cheddar cheese

Table 3 Main characteristics of *Lactobacillus casei* virulent phages

Characteristics	Phage J1	Phage PL1
Host	*L. casei* Shirota ATCC 27092	*L. casei* Shirota ATCC 27092
		L. casei Shirota mutant[a]
Nucleic acid type	Double-stranded DNA, linear cohesive ends	Double-stranded DNA, mostly linear cohesive ends
Molecular weight	24.4×10^6	25.3×10^6
G+C%	45–48	$46–47 \pm 2$
Isometric head diameter	55 nm	56 nm and 69 nm
Non-contractile tail	290×10 nm	$276 \times 10–12$ nm
		282×12 nm
Fibre	–	50 nm and 60 nm
Bacterial receptor	Membrane bound D-galactosamine, cell-wall rhamnose	Cell-wall L-rhamnose and D-glucose
Adsorption inhibitors	L-Rhamnose, D-Galactose	L-Rhamnose

[a] Derived strain resistant to J1 phage.
Reproduced by permission of Vescovo et al (1995) (*Ann. Microniol. Enzimol.* 45: 51–83).

mainly due to peptidase, lipase and esterase activities. Peptidases of *L. casei* are active for a long time during Cheddar ripening. Debittering properties are also attributed to *L. casei* strains which possess a broad aminopeptidase activity.

L. casei is used as a starter in cheeses from several countries: Egyptian pickled Brinza and Domiati, Romanian Brine and Telema, Italian Silter, Argentinian Tafi, Bulgarian Kachkaval and Ras. Economic and possibly technological incentives to accelerate cheese ripening draw attention to the use of *L. casei*. Cell-free extracts of *L. casei*, cheese slurries which contain NSLAB cells and enzymes, and especially mesophilic NSLAB as adjunct starters are used because they are effective and economic. Cultures of *L. casei* are now available from commercial starter suppliers. NSLAB cultures do not interfere with the growth and acid production of starter bacteria, although a more rapid decrease of starter lactococci during ripening has been observed with *L. casei* in cheese. Addition of *L. casei* subsp. *casei* as live and heat-shocked cells increases the flavour intensity and acceptability of Cheddar cheese due to high levels of total free amino acids. A good equilibrium between proteinases and peptidases supplied mainly by *L. casei* is important for flavour development and bitterness prevention. Differences in improved flavour acceptability are found between the subspecies of *L. casei*. Adding *L. casei* or related mesophilic lactobacilli influences proteolysis in cheese ripening to a greater extent in the absence of lactococci starters. A balance between lactic acid bacteria starters and *L. casei* adjunct must be carefully considered since an increase in proteolysis may not be as apparent when both are at elevated concentration, possibly due to the lack of available substrate for the wide range of enzymes supplied. Although the *L. casei* strains used are generally isolated from highly flavoured raw milk cheese,

using these strains has much less impact than the more heterogeneous indigenous microflora of raw milk.

Other seemingly negative biochemical activities attributed to NSLAB are production of flavour defects, slits and the formation of Ca-lactate crystals. These negative effects are attributed mainly to heterofermentative NSLAB. Furthermore, the addition of selected *L. casei* cultures limits the defects caused by the endogenous microflora.

Mesophilic lactobacilli, and *L. casei* in particular, can potentially modify cheese flavour and accelerate flavour development but further research is needed to select the best strains.

Fermented Milks and Beverages

L. casei has also been isolated from traditional cultured milks produced in different countries: Indonesian dadih, Indian dahi, Ethiopian ititu and Kenyan maziwa lala. *L. casei* Shirota ATCC 27092 is used in the industrial production of yakult, celebrated for its therapeutic effects.

In soya milk, *L. casei* is a suitable starter for its proteolytic activity. When associated with *Kluyveromyces fragilis* it produces more acid in a short time.

Cereal-cultured milks prepared with whole wheat, rice and maize flours, and yoghurt starter cultures show improved organoleptic quality when *L. casei* is included.

Vegetables

L. casei occurs at low cell numbers as an epiphyte on several vegetable crops suitable for silage preparation. *L. casei* in association with other homofermentative lactic acid bacteria is occasionally used as silage inoculant to preserve the nutritional and healthy value of the forage.

L. casei strains are effective in rapid acidification,

inhibit the loss of dry matter and gas production, and result in extended storage of silage produced from Graminaceae and Leguminosae (when the latter are used for silage by addition of carbohydrates). A high level of lactic acid has the beneficial effect of improving silage quality but cases of lactic acidosis in animals may be related to high contents of the D(−)-lactic acid isomer. *L. casei* is selected as the strain producing the L(+)-isomer. *L. casei* as inoculant in grass silage may improve the feed intake as well as the efficiency of utilization of wheat silage as feed. Strains of *L. casei* possess fructan hydrolase activity which permit the ensiling of herbage from the late cut with high proportions of fructans and low sugar contents.

L. casei together with *L. acidophilus* is used as an additive in maize-based diets for hens. It contributes positively to the daily egg production, feed conservation ratio, egg weight and albumen quality.

The fermentation of vegetable juices with *L. casei* as starter influences the flavour and degrades malic acid to L(+)-lactic acid. The malolactic fermentation is sometimes required to improve the stability and flavour of wine. *L. casei* can be added directly to the wine due to its tolerance of acid and alcohol.

L. casei contributes to the fermentation of soya beans into 'Tempeh', together with *Rhizopus oligosporus*, yeasts and other bacteria, being the species responsible for acidification. It is also isolated from different types of pickles produced in oriental countries. Soya bean and peanut milks with added glucose are fermented with a mixed population of endogenous lactic acid bacteria including *L. casei*.

L. casei and *L. casei* subsp. *rhamnosus* are occasionally isolated from rye and wheat sourdoughs used in breadmaking. Facultatively heterofermentative lactic acid bacteria are fundamental because lactic acid influences flavour, dough elasticity and microbial shelf life of sourdough baked products.

Ready-to-use raw fresh vegetables which are washed, sliced, chopped or grated and usually packaged in semipermeable polyethylene films and stored at low temperatures may suffer from contamination by spoilage or pathogenic bacteria. *L. casei* can be used as inoculant to produce hygienically safe and sensorially acceptable ready-to-eat mixed salads excluding contamination by mesophilic psychrotrophic pathogens.

Meat

Although *L. casei* is not a typical inhabitant of meat products, it has sometimes been isolated from or deliberately used in refrigerated meat stored under vacuum or modified atmospheres and in fermented sausages.

Strains of *L. casei* isolated from dry sausages contribute to the acidification and proteolytic activities during ripening. They may also protect dry cured sausages and ground beef against contamination by *Listeria monocytogenes*, *Staphylococcus aureus* and a wide range of Gram-negative bacteria.

Increased levels of CO_2 in the atmosphere cause a selective pressure on the endogenous microflora of raw meat which favours lactic acid bacteria, occasionally including *L. casei* strains which contribute to the product safety.

Food Biopreservation

Lactobacillus casei produces substances with antimicrobial activity against a variety of Gram-positive and Gram-negative bacteria.

A heat-labile protein of large molecular mass (ca. 40 kDa) with a pI of 4.5, included in the third class of *Lactobacillus* bacteriocins, has been purified from *L. casei* and named Caseicin 80. It has a very narrow antimicrobial spectrum, being active only against strains of the same species. Another bacteriocin-like inhibitory substance, Lactocin 705, produced by a different strain of *L. casei*, inhibits the growth of *Listeria monocytogenes* in ground beef.

Pyroglutamic acid produced by *L. casei* subspp. *casei* and *pseudoplantarum* inhibits spoilage bacteria, such as *Bacillus subtilis*, *Enterobacter cloacae* and *Pseudomonas putida*.

Inhibitory compounds resistant to proteolytic enzymes and heat treatment, with a molecular mass lower than 1000 Da which are presumably antibiotics, peptides or short-chain fatty acids, are also responsible for the antagonistic activity of *L. casei* strains against Gram-positive and Gram-negative bacteria including *Staphylococcus aureus*, *Clostridium tyrobutyricum* and *Escherichia coli*. The inhibitory activity of these substances is synergistic with lactic and acetic acids.

L. casei is one of the lesser nisin-sensitive lactic acid bacteria, since it can tolerate concentrations up to $1000 \, IU \, ml^{-1}$.

Food Spoilage

Depending on the circumstances, *Lactobacillus casei* can also play a role in the spoilage of processed foods.

In the case of underprocessing or leakage of food preserves, juices and juices containing beverages, *L. casei* occasionally grows and causes formation of slime, gas, off flavours, turbidity and changes in acidity. Elevated cell concentrations of slime-producing strains of *L. casei* subsp. *tolerans* which withstand pasteurization can produce ropiness in liquid milk. The fermentation of citrate by *L. casei* may cause excessive CO_2 giving rise to unwanted gas pockets and blowing of packaged cheeses. A potential for formation of tyramine and histamine is also reported during cheese ripening. The occasional growth of *L. casei* in beer causes a silky turbidity accompanied by acidity and off flavours due to the presence of diacetyl.

Probiotics

The therapeutic role of *Lactobacillus casei* administration has been largely and positively evaluated mainly in animals. Nowadays, a number of lactic acid bacteria including *L. casei* contained in special or food preparations are used as probiotics. Responses in probiotic activity are strictly strain specific.

L. casei is able to establish itself in the porcine intestine, by adhering to the duodenal cells; it is resistant to low pH and to the environment of the duodenum. In the small intestine of rats, *L. casei* becomes attached to specific glycosphingolipids that have a short sugar chain and a galactosyl residue on the non-reducing terminal. By preventing colonization by pathogenic species through competition for the glycosphingolipid receptors it maintains the host's health.

L. casei enhances host resistance against various types of microbial infections, in particular those induced by a number of opportunistic pathogens (*Pseudomonas aeruginosa*, *Listeria monocytogenes* and *Escherichia coli*). The protective and therapeutic activity is mainly based on the activation of the host macrophages in either a T cell-dependent or -independent manner, causing a remarkable enhancement of mobilization of blood monocytes to the sites of infections. In vivo antagonistic activity of *L. casei* strains in the gut against *Salmonella typhimurium* infection significantly delays the death of germ-free mice. *L. casei* is one of the more effective lactic acid bacteria on the secretory immune system of mice with a dose-dependent activity.

L. casei strains are effective in the treatment of rotavirus diarrhoea in infants and in antibiotic-associated or traveller's diarrhoea. *L. casei* causes an increase in immunoglobulin-secreting cells, specific rotavirus antibody secreting cells and serum IgA antibody. It probably promotes a systemic and local immune response to rotavirus which may be important for protective immunity against reinfection.

There is evidence that *L. casei* strains significantly decrease faecal and urinary mutagenicity as well as the levels of faecal microbial enzymes, such as β-glucuronidase, β-glucosidase, nitroreductase and urease which produce carcinogenic derivatives. It has been shown that parental administration of *L. casei*

Shirota has antitumour and immunostimulating activities on chemically induced tumours in animals. *L. casei* has been shown to reduce the size and number of murine tumours by stimulating the lymphocytes and natural killer cells which destroy tumour cells.

Extracts of autologous *L. casei* cell lysates, orally administered to hypertensive rats decrease systolic blood pressure. The antihypertensive compound is linked to polysaccharide glycopeptide complexes found in the cell wall.

A strain of *L. casei* subsp. *paracasei*, a species commonly found in the microbiota of the human intestine and vagina, which is plasmid free and highly transformable, has been proposed as a vaccine delivery vehicle with an intrinsic capacity to stimulate the immune system in an aspecific manner.

Uncertain Diseases

Although lactic acid bacteria have a good safety record supported by their commensal status and by their use in food fermentation, biopreservation and as probiotics, the adjective 'pathogenic' or better 'opportunistic pathogenic' has sometimes been applied to *L. casei* and related species, due to their uncertain involvement in clinical cases of endocarditis and infections of bloodstream.

Lactobacilli developing in the mouths of 2- to 5-year-old children consisted mostly of *L. casei* and *L. rhamnosus*. These species are also dominant in deep dental plaque and in softened and hard caries of deciduous molars. Nevertheless, neither *L. casei* nor *L. rhamnosus* is able to form plaque alone and without the participation of *Streptococcus mutans* or *S. sanguis*. Also, the low pH found in carious cavities favours lactobacilli and the high cell numbers may, at least in part, be the result of caries and not the cause.

Recent detailed reviews exclude lactic acid bacteria as determinants of clinical diseases in healthy humans because currently there is insufficient evidence to suggest that their use poses any danger.

See also: **Bacteriocins**: Potential in Food Preservation. **Cheese**: Microbiology of Cheese-making and Maturation; Role of Specific Groups of Bacteria. **Fermented Foods**: Fermented Vegetable Products. **Fermented Milks**: Range of Products; Yoghurt; Products from Northern Europe; Products of Eastern Europe and Asia. *Lactobacillus*: Introduction. **Probiotic Bacteria**: Detection and Estimation in Fermented and Non-fermented Dairy Products. **Spoilage Problems**: Problems caused by Bacteria.

Further Reading

Axelsson LT (1993) In: Salminen S and von Wright A (eds) *Lactic Acid Bacteria.* P. 1. New York: Marcel Dekker.

Davidson BE, Kordias N, Dobos M and Hillier AJ (1996) Genomic organization of lactic acid bacteria. *Antonie van Leeuwenhoek* 70: 161–183.

Fox PF, Wallace JM, Morgan S et al (1996) Acceleration of cheese ripening. *Antonie van Leeuwenhoek* 70: 271–297.

Gouet P (1994) In: de Roissart H and Luquet FM (eds) *Bactéries Lactiques.* P. 257. Chemin de Saint Georges, France: Lorica.

Hammes WP and Vogel RF (1995) In: Wood BJB and Holzapfel WH (eds) *The Genera of Lactic Acid Bacteria.* P. 19. London: Chapman & Hall.

Kaudler O and Weiss N (1986) In: Smeath PHA, Maic NS, Sharpe ME and Hoer SG (eds) Bergey's Manual of Systematic Bacteriology, Vol. 2, p. 1208. Williams and Williams.

Kok J and de Vos WM (1994) In: Gasson, MJ and de Vos WM (eds) *Genetics and Biotechnology of Lactic Acid Bacteria.* P. 169. London: Chapman & Hall.

Kunji RS, Mierau I, Hagting A, Poolman B and Konings WN (1996) The proteolytic systems of lactic acid bacteria. *Antonie van Leeuwenhoek* 70: 187–221.

Pouwels PH and Leer RJ (1993) Genetics of lactobacilli: plasmids and gene expression. *Antonie van Leeuwenhoek* 64: 85–107.

Stiles ME and Holtapfel WH (1997) Lactic acid bacteria of foods and their current taxonomy. *Int. J. Food Microbiol.* 36: 1–29.

Tomioka H and Saito H (1992) In: Wood BJB (ed.) *The Lactic Acid Bacteria in Health and Disease.* P. 263. London: Elsevier Applied Science.

de Vos WM (1996) Metabolic engineering of sugar catabolism in lactic acid bacteria. *Antonie van Leeuwenhoek* 70: 223–242.

LACTOCOCCUS

Contents

Introduction

Lactococcus lactis* Sub-species *lactis* and *cremoris

Introduction

Carl A Batt, Department of Food Science, Cornell University, USA

Copyright © 1999 Academic Press

Characteristics

Members of the *Lactococcus* genus are Gram-positive cocci that can, depending on growth conditions, appear ovoid and are typically 0.5–1.5 µm in size. They do not form spores and they are not motile. *Lactococcus* species grow in pairs or in short chains and unlike many members of the *Streptococcus* genus, these organisms do not grow in long chains. They have a fermentative metabolism and as expected for lactic acid bacteria, they produce copious amounts of lactic acid. They have complicated nutritional requirements and are auxotrophic for a number of amino acids and vitamins. Their optimum growth temperature is 30°C and they can grow at temperatures as low as 10°C but not at 45°C. They also cannot grow in 0.5% NaCl. Both their maximum growth temperature and their failure to tolerate salt are somewhat diagnostic of this genus as compared to closely related members of the *Streptococcus* genus, most notably *S. thermophilus*. The lactococci are usually members of the Lancefield serological Group N. Although serotyping was historically used for taxonomic purposes, it is now less important except to identify potentially pathogenic streptococci. Lactic acid bacteria have as a common feature: the ability to produce lactic acid as a major end product of their fermentation of hexoses. Beyond that lactic acid bacteria diverge into a wide array of microorganisms that have few other common features. Initial classification schemes as proposed by Orla-Jensen have proven relatively accurate even in the face of challenges raised by the advent of molecular classification methods.

The genus *Lactococcus* is relatively new and most members of this genus previously belonged to the genera *Streptococcus* and *Lactobacillus*. In 1985, Schleifer and colleagues proposed the genus *Lactococcus* and included the species formerly known as *Streptococcus lactis*, *Streptococcus raffinolactis*, *Lactobacillus hordniae* and *Lactobacillus xylosus*. The fact that the latter two species were formerly classified within the genus *Lactobacillus* is curious and suggests that the pleomorphic nature of these organisms can confound classification based on cell shape. The former *Lactobacillus xylosus* has in fact been reclassified to *Lactococcus lactis* subsp. *lactis*, with the single distinguishing characteristic being the ability of the former *Lactobacillus xylosus* to metabolize xylose, whereas most strains of *Lactococcus lactis* subsp. *lactis* fail to metabolize that sugar. Reclassification was on the basis of cell wall structure,

Table 1 Characteristics for discrimination of lactic acid bacteria

Character	Carno-bacterium	Lacto-bacillus	Aero-coccus	Entero-coccus	Lactoco-coccus/ Vagno-coccus	Leuconostoc/ Oenococcus	Pedio-coccus	Strepto-coccus	Tetrageno-coccus	Weissella
Tetrad formation	–	–	+	–	–	–	+	–	+	–
CO$_2$ from glucose	–	±	–	–	–	+	–	–	–	+
Growth at 10°C	+	±	+	+	+	+	±	–	+	+
Growth at 45°C	–	±	–	+	–	–	±	±	–	–
Growth at 6.5% NaCl	ND	±	+	+	–	±	±	–	+	±
Growth at 18% NaCl	–	–	–	–	–	–	–	–	+	–
Growth at pH 4.4	Ns	±	–	+	±	±	+	–	–	±
Growth at pH 9.6	–	–	+	+	–	–	–	–	+	–
Lactic acid	L	D, L, DL	L	L	L	D	L, DL	L	L	D, DL

Table 2 Speciation tests for *Lactococcus*

Test	L. garvieae	L. lactis subsp. cremoris	L. lactis subsp. hord-niae	L. lactis subsp. lactis	L. piscium	L. plan-tarum	L. raffi-nolactis
Growth at 40°C	+	–	–	(+)	–	–	–
Growth with 4% NaCl	+	–	–	+	ND	+	–
Arginine hydrolysis	+	–	+	+	–	–	(–)
Acid from lactose	+	+	–	+	+	–	+
Acid from mannitol	(+)	–	–	(–)	+	+	ND
Acid from raffinose	–	–	–	–	+	–	+
Pyrrolidonyarylamidase	+	–	–	(–)	ND	–	–

fatty acid composition and menaquinone composition. In addition to the lactococci, other 'new' genera including *Vagococcus* and *Enterococcus* were established.

The characteristics that distinguish the lactococci from other lactic acid bacteria including former members of the *Streptococcus* genus include pH, salt and temperature tolerances for growth. Whereas some lactic acid bacteria produce D-lactic acid, L-lactic acid and/or a combination of D and L, the lactococci produce only L-lactic acid. The other characteristics are presented in **Table 1**.

There are five major species within the genus *Lactococcus*. They can be distinguished by their ability to grow above 40°C and at > 4% sodium chloride. In addition they differ in their ability to produce acid from sugars including lactose, mannitol and raffinose. The ability to ferment lactose is an important characteristic especially for those species used to produce dairy products. Specific tests to distinguish among the various *Lactococcus* species are shown in **Table 2**.

These various tests can help to distinguish between the various lactococcal species but they do not give unequivocal results. Therefore, the use of 16S and 23S rRNA sequences is becoming the method of choice for taxonomic purposes. In fact rRNA and immunological analyses of superoxide dismutase have been used to justify the establishment of the *Lactococcus* species. Below the species level other nucleic acid-based methods including ribotyping and random amplified polymorphic DNA (RAPD) have been employed. The latter, although potentially more discriminatory in terms of being able to separate out individual strains, is difficult to perform on a routine basis. The former is less discriminatory; however, automated instrumentation-based methods are available and databases are being established.

Importance to the Food Industry

Lactococci are typically used for the production of dairy products. Within the species *Lactococcus lactis*, two subspecies *L. lactis* subsp. *lactis* and *cremoris* are the most widely encountered being used for dairy fermentations. In contrast, a third *L. lactis* subsp. *hordniae* is not commonly employed for industrial fermentations. *L. lactis* subsp. *lactis* and subsp. *cremoris* are mesophilic starters and their role in the fermentation is primarily to produce lactic acid. They utilize less than 0.5% of the lactose in milk during a typical fermentation and are not as acid resistant as other lactic acid bacteria (e.g. *Lactobacillus*). The lactococci all produce L(+)-lactic acid as a primary end product although one *L. lactis* subsp. *lactis* biovar *diacetylactis* produces a mixture of lactic acids as well as acetaldehyde, diacetyl and acetoin. The latter strain ferments citrate.

Lactococci are employed either as single strain starters or as part of a multiple-strain starter mix. The latter mixture can consist of different strains from a single species, or multiple strains of different species. The lactococci are frequently used as starters in combination with other lactic acid bacteria, including *Lactobacillus* and *Streptococcus*. Starter cultures ferment sugars to produce lactic acid which serves to acidify the product and preserve it as well as impart flavour. They also hydrolyse proteins altering the texture of the product and again affecting the taste. To a limited extent, *Lactococcus* species with the exception of *L. lactis* subsp. *lactis* biovar *diacetylactis*, impart flavour due to the production of certain organic acids. Finally, starters can help to preserve the product by the production of bacteriocins in addition to lactic acid. Of these bacteriocins, the most studied is nisin, whose genetics of biosynthesis and host resistance have been extensively studied.

See also: **Lactobacillus**: Introduction. **Lactococcus**: *Lactococcus lactis* Sub-species *lactis* and *cremoris*. **Milk and Milk Products**: Microbiology of Liquid Milk. **Streptococcus**: Introduction; *Streptococcus thermophilus*.

Further Reading

Ludwig W, Seewaldt E, Kilpper BR et al (1985) The phylogenetic position of *Streptococcus* and *Enterococcus*. *Journal of General Microbiology* 131: 543–551.

Salminen S and von Wright A (1998) *Lactic Acid Bacteria*. New York: Marcel Dekker.

Schleifer KH, Kraus J, Dvorak C et al (1985) Transfer of *Streptococcus lactis* and related streptococci to the genus *Lactococcus* gen. nov. *Systematic and Applied Microbiology* 6: 183–195.

Stiles ME and Holzapfel WH (1997) Lactic acid bacteria of foods and their current taxonomy. *International Journal of Food Microbiology* 36: 1–29.

Lactococcus lactis Subspecies *lactis* and *cremoris*

Polly D Courtney, Department of Food Science and Technology, Ohio State University, Columbus, USA

Bacteria in the species *Lactococcus lactis* are used extensively as starter cultures in dairy fermentations throughout the world. They were first isolated in the late 1800s from spontaneous milk fermentations. The realization that these organisms are fully or partly responsible for some fermented dairy products was the beginning of what has become a sophisticated starter culture industry. Rather than relying on spontaneous fermentations, most dairy fermentations are now initiated by the addition of defined or undefined starter cultures to pasteurized milk. *L. lactis* strains are used to produce foods such as cheeses, sour cream and buttermilk. Not only do they contribute to the characteristic flavour, aroma and texture of these products, but they also help preserve the product by producing organic acids, bacteriocins and hydrogen peroxide. The significance of these organisms to the food industry has fuelled research on their microbiology, biochemistry and molecular biology.

Characteristics of the Species

General Characteristics

Members of the species *L. lactis* are Gram-positive, mesophilic, facultative anaerobes. They are non-spore-forming and non-motile. They have a spherical or ovoid morphology and occur in pairs or chains. Their carbohydrate metabolism is homofermentative, producing primarily L(+)-lactic acid from glucose. Phylogenetically, they are classified within the *Clostridium* subdivision of the Gram-positive eubacteria. Originally, these organisms were classified in the genus *Streptococcus* (Lancefield group N). In 1985, using nucleic acid hybridization and immunological data, the genus *Lactococcus* was created, separating the mesophilic lactic streptococci from the genera *Streptococcus* and *Enterococcus*.

Two subspecies are recognized, *L. lactis* subspp. *lactis* and *cremoris*. Traditionally, the subspecies have been differentiated based on phenotypic characteristics. *L. lactis* subsp. *lactis* grows at 40°C, grows in 4% NaCl, and produces ammonia from arginine, whereas *L. lactis* subsp. *cremoris* does none of the above. Within the subspecies *lactis*, is a biovariant, *L. lactis* subsp. *lactis* biovar. *diacetylactis*, that is distinguished by its ability to metabolize citrate. Nucleic acid-based techniques have strengthened researchers' ability to categorize members of the *L. lactis* species. Restriction enzyme fragmentation patterns, DNA–DNA homology, pulsed-field gel electrophoresis, and 16S rRNA gene sequencing have confirmed the phenotypic classification for most strains. A few strains are phenotypically *L. lactis* subsp. *lactis* and genetically subsp. *cremoris*. These include the common laboratory strains MG1363 and LM0230 that are both derived from the industrial strain 712.

Growth Requirements

The lactococci are nutritionally fastidious organisms. In nature, they most commonly inhabit environments rich in carbon and nitrogen sources, such as raw milk and plant material. During dairy fermentations, the growth medium for *L. lactis* is milk. In the laboratory, an undefined, buffered medium, such as M17, is generally used for *L. lactis* propagation. Chemically defined media that support growth have also been reported. *L. lactis* strains are auxotrophic for many amino acids. The specific amino acids that they are unable to synthesize differ from strain to strain. Common amino acids that must be added to defined media for *L. lactis* growth are isoleucine, valine, leucine, histidine, methionine, arginine, proline, glutamate, serine and threonine. Vitamins necessary for optimum growth include biotin, pyridoxal, folic acid, riboflavin, niacin, thiamin and pantothenic acid. Glucose, buffer and various minerals are also components of the defined media.

Lactococci are mesophilic organisms with an optimum growth temperature of 25–30°C. Their maximum generation time is 60–70 min in milk or 35–40 min in undefined synthetic medium. Growth becomes inhibited at pH 4.5 or lower; therefore, the highest cell densities may be obtained in buffer media. *L. lactis* does not require oxygen for growth, but can tolerate the presence of oxygen due to oxygen-metabolizing enzymes.

To protect company interests, many of the details of commercial starter culture preparations are not public knowledge. Generally, starter culture manufacturers prepare high cell density cultures in large fermenters. Cells are then concentrated and preserved by freezing or freeze-drying. Certain commercial culture preparations are designed for direct inoculation into the fermentation vessel (DVS, or direct-to-vat set). Others are intended for propagation in the dairy plant, with subsequent inoculation of the bulk starter culture into the fermentation vessel.

The *L. lactis* Genome

L. lactis strains have one, circular chromosome. Chromosome size estimates from various strains range from 2.0–3.1 Mbp. Physical and genetic maps have been constructed for two *L. lactis* subsp. *lactis* strains and two *L. lactis* subsp. *cremoris* strains. Comparisons of maps indicate a somewhat flexible nature of the genome with several large-scale rearrangements and inversions. The most detailed map is for *L. lactis* subsp. *cremoris* MG1363 on which at least 115 genetic loci have been mapped. Currently, the sequencing of the entire *L. lactis* subsp. *lactis* IL1403 chromosome is underway. The results from this effort will vastly increase our understanding of the *L. lactis* chromosome.

Plasmids are common components of *L. lactis* genomes. In fact, most strains carry multiple plasmids. Lactococcal plasmids are diverse in size, copy number and distribution. They often carry determinants for industrially significant characteristics, such as lactose metabolism, proteolysis, bacteriophage resistance, citrate utilization, polysaccharide production and bacteriocin production and immunity. Thus, plasmid loss can result in a poor fermentation. Due to their industrial importance, many plasmids have received detailed research attention. Some plasmids also encode the determinants necessary for conjugation. Transfer of conjugal plasmids may be a significant factor in strain evolution by horizontal gene transfer during mixed culture fermentations. Culture manufacturers may also use plasmid conjugation to introduce desirable traits into specific industrial strains.

Transposable elements have been identified in the genomes of many *L. lactis* strains. Transposable elements are mobile genetic entities capable of moving from one location to another within a piece of DNA or between two pieces of DNA. In lactococci, transposable elements are often associated with genes encoding essential properties for growth in milk (lactose metabolism, proteolysis and phage resistance). At least five distinct classes of insertion sequences have been identified and are commonly located in lactococcal chromosomes and plasmids. When associated with genes encoding conjugal ability, these elements can also promote genetic exchange between organisms. As with plasmids, transposable elements may also play a fundamental role in strain evolution in the milk environment. The identification of lactococcal transposable elements was also instrumental in the development of various integration vectors used for random mutagenesis of the lactococcal genome and for integration of genes into the chromosomes.

The large-scale, worldwide growth of multiple-strain starter cultures provides ample opportunity for genetic exchange between *L. lactis* strains and for mutations or rearrangements within one cell. The prevalence of plasmids and transposable elements in this species undoubtedly contributes to some of the changes observed.

Properties of Significance to the Food Industry

Industrially relevant characteristics of *L. lactis* have been investigated extensively. The metabolism of milk components and production of various compounds

Table 1 Contributions of *L. lactis* metabolism to fermented dairy product sensory and microbiological quality

Metabolic trait	Contribution to flavour, aroma and texture	Contribution to preservation
Carbohydrate metabolism	Production of lactic acid, ethanol, acetate, diacetyl, exopolysaccharides, and other compounds Consumption of lactose	Production of lactic acid, other organic acids, and diacetyl
Protein metabolism	Production of peptides, amino acids, and amino acid derivatives Consumption of bitter peptides	
Lipid metabolism	May produce volatile fatty acids	
Oxygen metabolism		Hydrogen peroxide
Bacteriocin production		Bacteriocin

are of particular interest due to their impact on product flavour, texture and preservation (**Table 1**).

Carbohydrate Metabolism

Rapid metabolism of lactose to lactic acid is a desirable trait in dairy starter cultures. Under conditions of carbon excess, *Lactococcus* strains ferment sugars by homolactic fermentation. Thus, mainly L(+)-lactic acid is produced from the lactose in milk. Lactic acid production is the major contributor to the preservation of fermented dairy products. The resulting low pH inhibits the growth of many spoilage and pathogenic bacteria. Under carbon-limiting conditions, homolactic fermentation may shift to a mixed acid fermentation yielding acetate, ethanol and CO_2, in addition to lactate.

In *L. lactis*, lactose is imported into the cell and concomitantly phosphorylated via a lactose-specific phosphoenolpyruvate-dependent phosphotransferase system (PEP-PTS). The resulting lactose-phosphate is cleaved by phospho-β-galactosidase into glucose and phosphogalactose. These monosaccharides are subsequently metabolized via the glycolytic and tagatose 6-phosphate pathways, respectively. Both pathways lead to the formation of L(+)-lactic acid which is excreted into the medium.

Milk contains citrate at a concentration of 8–9 mM. Citrate metabolism by *L. lactis* subsp. *lactis* biovar. *diacetylactis* produces diacetyl, a butter-flavoured compound. Other end products include CO_2, acetate, lactate, formate, acetoin and butanediol. Citrate-metabolising *L. lactis* strains possess unique citrate permease and citrate lyase enzymes not found in non-citrate metabolizing strains. Diacetyl production is especially important in the production of buttermilk.

Some *Lactococcus* strains can assimilate simple sugars into complex polysaccharides that are exported and remain associated with the cell surface (exopolysaccharides). These compounds provide a thick texture to some products and can act as stabilizers. Exopolysaccharides may also increase moisture retention and yield in some cheeses.

The understanding of carbohydrate metabolic pathways and their regulation has helped to assure fast acid production and consistent flavour and texture in products fermented with *L. lactis*. The cloning and characterization of many of the genes involved in carbohydrate metabolism have allowed researchers to design or select *L. lactis* strains that produce more or less of specific compounds.

Protein and Amino Acid Metabolism

The proteolytic system of *L. lactis* is among the best characterized in the bacterial world. *L. lactis* cells are auxotrophic for many amino acids; therefore, they must acquire these amino acids from their environment. The specific amino acids required are strain dependent. Milk is high in protein content, but is not a good source of free amino acids. An elaborate proteolytic system exists in *L. lactis* that allows these organisms to acquire the necessary amino acids from milk proteins and peptides. Effective proteolysis of milk proteins is essential for cell growth, and therefore, is essential for acid production during the fermentation.

Not only does proteolysis impact the overall cell growth, but it is also required for proper cheese flavour development. Some peptides, amino acids and amino acid derivatives produced by starter culture enzymes are thought to contribute positively to cheese flavour and aroma. On the other hand, some peptides are perceived as bitter and must be degraded by bacterial enzymes to prevent defective cheese flavour.

The proteolytic system of *L. lactis* consists of a cell-wall-associated protease, a variety of peptidases, and peptide and amino acid transport proteins. During cheese-making, the cell-wall-associated protease generates peptides and amino acids from κ-, β- and α$_{s1}$-caseins. The amino acids can be transported into the cell via specific transport proteins and then used for cellular protein synthesis. Peptides are also imported via specialized transport proteins. Once inside the cell, peptides are degraded to amino acids by a battery of intracellular peptidases with different specificities

Table 2 Peptidases characterized in *Lactococcus lactis*

Enzyme	Activity	Substrate specificity
PepN	Aminopeptidase	Broad; di-, tri-, oligopeptides
PepC	Aminopeptidase	Broad; di-, tri-, oligopeptides
PepT	Tripeptidase	Broad; tripeptides only
PepV	Dipeptidase	Broad; dipeptides only
PepX	X-prolyl-dipeptidyl aminopeptidase	X-Pro*(X)n
PepP	Aminopeptidase	X*Pro-(X)n
PepI	Proline iminopeptidase	Pro*X-(X)n
PepQ	Prolidase	X*Pro
PepR	Prolidase	Pro*X
PepA	Glutamyl aminopeptidase	Di-, tri-, or oligopeptides with N-terminal Glu or Asp
PepO	Endopeptidase	Oligopeptides
PepF	Endopeptidase	Oligopeptides

*, indicates where enzyme cleaves the peptide. Pro, proline; Glu, glutamic acid; Asp, aspartic acid; X, any amino acid.

(**Table 2**). In the later stages of cheese ripening, lysis of the *L. lactis* cells releases the intracellular peptidases into the cheese matrix. The released peptidases continue to break down peptides, producing free amino acids and smaller peptides that may contribute to cheese flavour.

Lipid Metabolism

Limited lipolysis of the triacylglycerol component of milk may also contribute to flavour in fermented dairy products. Lactococci possess low levels of lipase or esterase activities whose primary role is likely to be maintaining the physiological functions of the cell. Milk triacylglycerol is not a good substrate for these enzymes. However, only slight lipolysis may be required for a significant flavour contribution since short-chain fatty acids are potent flavour compounds at low levels. The contribution of lactococcal lipolytic enzymes to flavour of fermented products is not clear, though current research into several esterases may soon shed light on this issue.

Oxygen Metabolism

L. lactis are anaerobic organisms, lacking the tricarboxylic acid cycle and an electron transport system. However, they tolerate oxygen due to several oxygen-metabolizing enzymes. With NADH oxidases and superoxide dismutase, *L. lactis* can convert O_2 or oxygen radicals (O_2^-) into hydrogen peroxide or water (**Table 3**). The hydrogen peroxide generated by superoxide dismutase and $NADH:H_2O_2$ oxidase accumulates in the growth medium because *L. lactis* does not have catalase or NADH peroxidase activities to remove it. The H_2O_2 produced may have an inhibitory effect on other bacteria providing an additional pre-

Table 3 Enzymes involved in oxygen metabolism in *Lactococcus lactis*

Enzyme	Reaction catalysed
$NADH:H_2O_2$ oxidase	$NADH + H^+ + O_2 \rightarrow NAD^+ + H_2O_2$
$NADH:H_2O$ oxidase	$2\,NADH + 2H^+ + O_2 \rightarrow 2\,NAD^+ + H_2O$
Superoxide dismutase	$2\,O_2^- + 2H^+ \rightarrow H_2O_2 + O_2$

servative effect to the dairy product. When high cell masses of lactococci are desired, such as during preparation of starter cultures, catalase may be added to the growth medium to prevent inhibition of the *L. lactis* by H_2O_2.

Bacteriocins

Some *L. lactis* strains produce bacteriocins. Bacteriocins are peptides or proteins produced by one bacterium that inhibit closely related bacteria. Bacteriocins produced by *L. lactis* include those in two categories: the lantibiotics and the non-lantibiotic, small heat-stable bacteriocins. Nisin, a lantiobiotic, is the best characterized of the lactic acid bacteriocins and is the only bacteriocin approved as a food additive worldwide. Nisin inhibits vegetative cells of *Listeria monocytogenes*, *Bacillus*, *Clostridium* and some lactic acid bacteria, and prevents outgrowth of *Clostridium* and *Bacillus* spores. An *L. lactis* strain is used as the commercial source for nisin. Other well-characterized bacteriocins of *L. lactis* are the small, heat-stable bacteriocins, lactococcins A, B and M. The lactococcins have a narrower inhibitory spectrum than nisin, inhibiting only other lactococcal strains. Foods fermented with bacteriocin-producing *L. lactis* strains naturally contain bacteriocins that may provide an additional preservation to the product.

Bacteriophages and Bacteriophage Resistance in *L. lactis*

Bacteriophages, or phages, are viruses that infect bacteria, usually causing cell death. Bacteriophage infection of *L. lactis* cultures are the main cause of slowed or stopped dairy fermentations. Cultures in dairy fermentations are particularly susceptible to phage infection for the following reasons: pasteurized milk is not sterile and may contain phages; the fluid nature of milk allows rapid dispersion of phage; repeated use of pure cultures provides constant hosts for phage proliferation. *L. lactis* phages and phage resistance have received much research attention due to their potential impact on productivity and product quality.

Numerous bacteriophages that infect *L. lactis* have been isolated and studied. They are classified into species based on morphology, DNA hybridization, protein composition, and serology. Both lysogenic and lytic phages infect *Lactococcus*. Lytic phages are responsible for fermentation disruptions. Upon infection, a lytic phage rapidly takes over the cellular functions to replicate phage DNA and synthesize phage proteins. When newly synthesized phages are assembled, specific phage proteins direct the lysis of the bacterial cell, resulting in dispersal of the progeny phages. Infection of an entire culture with a lytic phage can proceed rapidly since each infected cell may generate 10–400 progeny phage. The progeny phage can then infect neighbouring cells, resulting in an exponential increase in phage numbers and cell death.

Lysogenic phages have a different life cycle than their lytic counterparts. Upon infection, lysogenic phage usually integrate their genomes into the bacterial host genome and lie dormant until an environmental trigger induces them to excise, replicate and lyse the cell. Most lactococcal strains have lysogenic phages in their genomes that are undetected until induced in the laboratory. No major fermentation disruption has been attributed to a lactococcal lysogenic phage. However, these phages can not be discounted as potential sources of lytic phage or of genetic exchange between phage genomes during a lytic infection.

Many lytic and lysogenic lactococcal phages have been well characterized, with several phage genomes completely or partially sequenced. Many of the genes necessary for phage replication, assembly and host cell lysis have been identified. This information can be exploited to develop novel methods of controlling phages in fermentations.

Precautions should be taken to control bacteriophage contamination and propagation in dairies. Utilization of bacteriophage-resistant *L. lactis* strains in fermentations is one method to reduce the threat of phage infection. The isolation of bacteriophage-resistant mutants (BIMs) of infected strains is a common practice, though phage-resistant mutants do not always possess the same fermentation characteristics as the original strain. Many naturally occurring phage resistance mechanisms have been studied and are described in **Table 4**. The genes encoding most of these mechanisms are found on plasmids. Though individual phage resistance mechanisms can effectively inhibit phages for a short time, phages evolve quickly. Frequently, phage mutants appear that are no longer inhibited by the mechanism. In highly phage-resistant industrial strains, researchers have found a combination of several different phage-resistance mechanisms in one strain. Thus, if a phage becomes insensitive to one phage-resistance mechanism, other mechanisms are present to prevent phage proliferation.

Table 4 Natural bacteriophage resistance mechanisms in *L. lactis*

Type of mechanism	Description
Adsorption interference	Prevents phage from attaching to the cell surface by alteration or obstruction of surface receptor
DNA injection inhibition	Phage adsorption proceeds normally, but phage DNA never enters the cell
Restriction and modification systems	Restriction endonuclease cleaves unmethylated DNA, including phage DNA. Host DNA is protected from degradation by methylation
Abortive bacteriophage infection	Phage infection is inhibited at some point after adsorption, DNA entry, and early phage gene expression

Biotechnology

The tools of molecular biology have dramatically advanced our knowledge about *Lactococcus*. To date, the most valuable contribution of these technologies has been the detailed exploration of industrially important characteristics. Genetic tools can also be used to design strains for specific applications by stabilizing or overexpressing desirable traits, or removing undesirable traits.

The availability of various types of cloning vectors has allowed in-depth studies of lactococcal gene function and regulation. Basic cloning vectors have been in use for many years. These plasmids are constructed to replicate in *Lactococcus* and in *E. coli* (to facilitate cloning) with selectable markers for each. Plasmids with high, medium or low copy numbers and different modes of replication (theta or rolling circle) are now available.

The study of lactococcal promoter sequences has provided the tools for constitutive and inducible gene expression in *Lactococcus*. Expression plasmids are available that have promoters followed by a multiple cloning site where the gene of interest may be cloned. Inducible promoters characterized from *L. lactis* include those controlled by lactose, chloride, phage proteins and nisin. Plasmids containing promoterless reporter genes have been designed for promoter cloning. Signals for secretion or cell surface anchoring have also been studied and used successfully to promote secretion or anchoring of specific proteins.

Various systems for integrating DNA into the *L. lactis* genome have been used in studying these organisms. Reported integration plasmids are based on transposable elements, bacteriophage attachment sites and recombination between homologous sequences. Transposable element-based plasmids have been used for random mutagenesis of the lactococcal chromosome. Integration plasmids based on homologous recombination can inactivate specific chromosomal genes. All types of integration plasmids may be used to integrate plasmid-borne genes into the chromosome where they may be inherited more consistently.

Plasmid DNA can be introduced to *L. lactis* via the natural processes of conjugation and transduction. Protoplast transformation and electroporation are also effective methods used in the laboratory. Naturally occurring conjugative plasmids carrying genes encoding desirable traits, such as phage resistance, have been introduced to industrial strains to improve their performance. Since these plasmids are found naturally and are introduced to the cell by a natural phenomenon, there has not been a regulatory barrier to this method of strain improvement. On the other hand, live organisms containing recombinant plasmids are not currently used in dairy fermentations. Several food-grade cloning vectors have been developed using DNA derived only from *L. lactis*. The acceptability of food-grade recombinant plasmids by regulatory agencies may be tested in the future.

The common use of *L. lactis* in foods makes it an appealing host for producing food ingredients or pharmaceuticals. With the expression and secretion cloning vectors currently available, many heterologous proteins can be expressed and/or secreted. Examples of heterologous proteins expressed in *L. lactis* include lysozyme, bacteriocins, and proteases. Because *L. lactis* is non-pathogenic and non-colonizing, it has been proposed as a vaccine delivery vehicle for immunization via the intestinal mucosa. Mice were fed lactococcal cells in which an antigen (tetanus toxin fragment C) was expressed. These mice subsequently exhibited an antigen-specific immune response.

Conclusion

Lactococci are essential for the production of many popular dairy products. Their metabolic activities provide preservative and sensory attributes to the fermented product. Substantial research into the industrially important characteristics of *L. lactis* has allowed more control over fermentations such that large-scale fermentations may be concluded with consistent product quality and production times.

See also: **Bacteriocins**: Potential in Food Preservation; Nisin. **Biochemical and Modern Identification Techniques**: Microfloras of Fermented Foods. **Cheese**: Microbiology of Cheese-making and Maturation. **Fermented Milks**: Range of Products. **Metabolic Pathways**: Release of Energy (Anaerobic); Nitrogen Metabolism. **Preservatives**: Traditional Preservatives – Organic Acids. **Starter Cultures**: Uses in the Food Industry; Cultures Employed in Cheese-making.

Further Reading

Chassy BM and Murphy CM (1993) *Lactococcus* and *Lactobacillus*. In: Sonenshein AL, Hoch JA, and Losick R (eds) Bacillus subtilis *and other Gram-positive Bacteria: Biochemistry, Physiology and Molecular Genetics*. P. 65. Washington, DC: American Society for Microbiology Press.

Cocaign-Bousquet M, Garrigues C, Loubiere P, and Lindley ND (1996) Physiology of pyruvate metabolism in *Lactococcus lactis*. *Antonie van Leeuwenhoek* 70: 253–267.

Davidson BE, Kordias N, Dobos M and Hillier AJ (1996) Genomic organization of lactic acid bacteria. *Antonie van Leeuwenhoek* 70: 161–183.

Delves-Broughton J, Blackburn P, Evans RJ and Hugenholtz J (1996) Applications of the bacteriocin, nisin. *Antonie van Leeuwenhoek* 69: 193–202.

deVos WM (1996) Metabolic engineering of sugar catabolism in lactic acid bacteria. *Antonie van Leeuwenhoek* 70: 223–242.

Djordjevic GM and Klaenhammer TR (1998) Inducible gene expression systems in *Lactococcus lactis*. *Molecular Biotechnology* 9: 127–139.

Dodd HM and Gasson MJ (1994) Bacteriocins of lactic acid bacteria. In: Gasson MJ and deVos WM (eds) *Genetics and Biotechnology of Lactic Acid Bacteria*. P. 211. London: Blackie Academic & Professional.

Klaenhammer TR and Fitzgerald GF (1994) Bacteriophages and bacteriophage resistance. In: Gasson MJ and deVos WM (eds) *Genetics and Biotechnology of Lactic Acid Bacteria*. P. 106. London: Blackie Academic & Professional.

Klijn N, Weerkamp AH and deVos WM (1995) Biosafety assessment of the application of genetically modified *Lactococcus lactis* spp. in the production of fermented milk products. *Systematic and Applied Microbiology* 18: 486–492.

Kunji ERS, Mierau I, Hagting A, Poolman B and Konings WN (1996) The proteolytic systems of lactic acid bacteria. *Antonie van Leeuwenhoek* 70: 187–221.

Romero DA and Klaenhammer TR (1993) Transposable elements in lactococci: a review. *Journal of Dairy Science* 76: 1–19.

Teuber M (1992) The genus *Lactococcus*. In: Wood BJB and Holzapfel WH (eds) *The Lactic Acid Bacteria*, Vol. 2: *The Genera of Lactic Acid Bacteria*. P. 173. London: Blackie Academic & Professional.

Wells JM, Robinson K, Chamberlain LM, Schofield KM and Le Page RW (1996) Lactic acid bacteria as vaccine delivery vehicles. *Antonie van Leeuwenhoek* 70: 317–330.

Lactoferrin *see* **Natural Anti-Microbial Systems**: Lactoperoxidase and Lactoferrin.

Lactoperoxidase *see* **Natural Anti-Microbial Systems**: Lactoperoxidase and Lactoferrin.

LAGER

Iain Campbell, International Centre for Brewing and Distilling, Heriot-Watt University, Edinburgh, Scotland

Copyright © 1999 Academic Press

Introduction

The name lager, derived from the German word for storage, was introduced to the British vocabulary to denote that such beers had a lengthy period of post-fermentation storage or maturation. Originally such beer was produced in only a few Bavarian monastery-breweries. After acquisition, early in the 19th century, of the special yeast culture by the brewers of Pilsen (now Plzen in the Czech Republic), demand for Pilsener beer soon spread, initially to the German states and Denmark, and by the beginning of the 20th century that type of beer had achieved worldwide market dominance.

However, not all lager-type beers are the light-coloured Pilsener product with its characteristic balance of flavour compounds. In Bavaria in particular there are speciality dark beers with an important local share of the market, predating the development of lager beers in Pilsen, but brewed and matured in a similar way. Belgium and Germany continued a moderate production of the older traditional beers, but only in Britain (which included Ireland at that time) did the established ales and stouts retain a majority market share. In other countries, even most English-speaking countries, the word beer would be assumed to mean pils. Indeed, a request for lager would probably not be understood.

Therefore it is only in Britain that the question, 'What is the difference between ale and lager?' has any real significance. Until about 1960 there would have been no difficulty in explaining to a British beer-drinker the differences between ale and lager, but by 1980 brands of the two types of beer overlapped in most characteristics. Although the word lager refers specifically to storage, the original differences also extended to the types of malt, hops and yeast, and the mashing and fermentation procedures. Now, although it is unusual for a brewer to use the same malt and hops for ale and lager, there are no other consistent differences between the two types of beer over the British brewing industry as a whole. Most British beer-drinkers profess the ability to distinguish ale from lager, and are usually right, but would be unable to explain exactly what they perceive as the difference.

Brewing: Preparation of the Wort

In most respects the production of ale, stout and lager beers is the same. The outline of the brewing process shown in **Table 1** is equally applicable to all three types. There are various detail differences which are discussed in the following pages.

Malt

Of the characteristics of Pilsener beer, two in particular are attributable to the malt: light colour and a trace aroma of dimethyl sulphide (DMS). Both are due to the low temperature of kilning at the end of the malting process. Although originally discovered by trial and error, now with our modern knowledge we would attribute the low-temperature drying to a desire first, to maximize enzyme activity in the malt, for maximal starch conversion during mashing, and second, to cause least darkening of the malt, in order to produce a light-coloured beer.

Therefore the steeping and germination stages of the malting process are the same as for ale or whisky; only the drying process is different. Typically, kilning of lager malts begins at 45–50°C, with only a brief final period at 80°, by which time the malt is already almost dried, usually to a specified value about 5% moisture. Carrying out most of the drying process at relatively low temperature restricts the Maillard browning reactions, which occur most readily at high temperature and moisture content, although not only temperature and dryness affect these reactions. *S*-Methyl methionine (SMM) is synthesized during the germination stage of malting and is the main precursor of DMS, although dimethyl sulphoxide

Table 1 The brewing process

Malting

Development of hydrolytic enzymes for conversion of barley starch to fermentable sugars (glucose, maltose, maltotriose) and protein to free amino nitrogen

Milling, mashing

Extraction of sugars, amino acids, other yeast nutrients and enzymes with hot water: sweet wort

Wort boiling

Boiling with hops to extract aroma and bittering compounds; also sterilization: hopped wort

Fermentation

Conversion by yeast (*Saccharomyces cerevisiae*) of fermentable sugars to ethanol + CO_2; also production of flavour volatiles

Post-fermentation treatments

Maturation (improvement of flavour), clarification, packaging, pasteurization

(DMSO) is also present in higher amounts in lager malts. SMM and DMSO are unaffected by the low-temperature kilning of lager malts, whereas the higher kiln temperatures used for darker-coloured ale malts destroy SMM and DMSO.

Water

The quality of the local water supply, usually from wells, has been an important influence on the development of the brewing industry in Britain and continental Europe. When Pilsener beers became the dominant form of lager, in order to produce similar beers elsewhere, the water supply for mashing (brewing liquor) had, logically, to be of similar composition. Pilsen water is soft, with about $7\,mg\,l^{-1}\,Ca^{++}$ and $14\,mg\,l^{-1}\,HCO_3^-$. Although the raw materials and yeast are more important for flavour of the beer, flavour is certainly influenced by the concentrations of ions such as Ca^{--}, Cl^- and SO_4^{--}. Hardness such as at Burton-on-Trent (about $270\,mg\,l^{-1}\,Ca^{++}$; $640\,mg\,l^{-1}$ SO_4^{--}), although ideal for the local ales, is certainly incompatible with the requirements for lager brewing, or the flavour of the lager beer, and requires artificial softening.

Milling and Mashing

Milling of lager malts is the same as the process for ales and stouts, although the mashing process and subsequent clarification of the wort may be different. The traditional multi-roll mills in the brewery produce a fine malt flour but with minimal damage to the husks, which are required as a filter medium to clarify the wort after mashing. Recently, some breweries have changed to hammer mills, which grind all of the grain, including husk material, even more finely. This allows more efficient and more rapid extraction of the nutrients from the wort, but requires a mash filter for acceptable clarification of the wort.

Certainly it was true in the past, and is to some extent true even now, that lager malts are under-modified in comparison with ale malts. Infusion mashing for ale brewing requires a well-modified malt, since there is no opportunity to control temperature to the various optima for the different hydrolytic enzymes for conversion of starch and protein to fermentable sugars and amino acids during mashing. The traditional lager brewhouse can use, and in the past had to use, poorly modified malts, i.e. with lower hydrolytic activity, and increase the mashing temperature in reproducible steps.

Beers for sale in Germany, and some lagers brewed in other countries, must be produced from a 100% malt grist. Otherwise, flavour and aroma, and also colour, can be modified by incorporation of unmalted cereal or simple fermentable sugar in the recipe. Whole cereal grains, or grits which are prepared without heat treatment, must first be heated to boiling in a cooker vessel in order to gelatinize the starch. Other adjuncts, e.g. flaked cereals that have been crushed between heated rollers, have already been gelatinized and are milled and mashed directly. Glucose, maltose, sucrose or other already-fermentable sugar powders or syrups are added later, during boiling with hops in the copper. All of these ingredients affect flavour first, by their direct contribution; second, by the effect on yeast growth of the altered ratios of individual sugars to each other; and third, by the effect of the different fermentable sugar/amino nitrogen ratio of the wort on yeast metabolism, and therefore on production of flavour compounds.

Breweries which pride themselves on traditional brewing may still use decoction mashing. In this process a known proportion of the mash is removed, heated to boiling and returned to the original mash to raise its temperature by a known amount. This system became common in the past not only as a versatile method for utilizing malts of variable quality, but also for consistent control of temperature before the introduction of reliable thermometers. Poorly modified malts require an initial low-temperature stage in mashing – about 45°C – to achieve the neces-

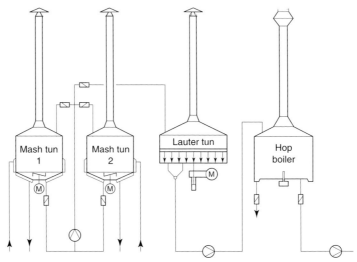

Figure 1 Typical lager brewhouse, showing two mashing vessels, lauter tun and hop boiler.

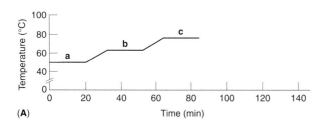

(A)

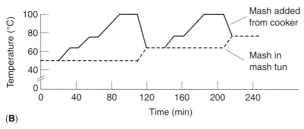

(B)

Figure 2 Typical temperature profiles (**A**) in mash tun; (**B**) in double-decoction mash (two vessels). Optimum temperatures of enzymes: **a**, protease; **b**, β-amylase; **c**, x-amylase.

sary enzyme activity. **Figure 1** shows a typical brewhouse suitable for decoction brewing for the required mashing temperature profile (**Fig. 2**): a measured amount of mash is transferred from the mash tun to cooker, heated to boiling, and returned to the mash tun to raise the mash to the next temperature step of the process. Alternatively, a modern temperature-controlled mash mixer, preceded by a cereal cooker if necessary to handle a proportion of unmalted cereal in the recipe, can follow the profile shown in Figure 2. Most mashing programmes are arranged to hold the temperature at about 50°C for proteolysis, 62°C for β-amylase and 70–75°C for α-amylase.

Lautering

Spent grains must be removed before hop boiling. In some breweries great care is taken to have a perfectly bright (haze-free) wort, and certainly it is important to produce a wort as clear as possible. The false bottom of the lauter tun has numerous thin slits about 1 mm wide; filtration is accomplished by the bed of 60–80 cm husk and grist debris supported on the slotted plates. In infusion mashing for traditional English ales, the grist with entrapped air floats clear of the false bottom, and the flow rate must be restricted to maintain such flotation. Run-off is fast in comparison with infusion mashing, since flotation is irrelevant, but also because of the large filtration area of the false bottom (Fig. 1 shows a typical broad lauter tun) and the knives carried on a rotating arm, cutting through the bed to prevent blockage. After draining the initial strong wort, residual sugar is extracted by spraying with hot water. Disturbance of the bed during sparging can also be remedied by rotation of the knives. Often the shallow space below the false bottom is divided into sections with individual control of run-off, at the fastest rate for that section compatible with clarity. Certainly, to prevent development of off flavour it is important to avoid any substantial carry-over of grain solids to the following stage of hop boiling.

Although faster than infusion mashing, the 3-h operating cycle of the lauter tun restricts the brewery to a maximum of eight brews per day. In recent years mash filters have been introduced as an alternative to the traditional lauter tun. The equipment is more compact and is claimed to be cheaper to install and operate, and is more efficient in several respects. The finer grind of the grist allows more efficient extraction of wort nutrients; operation with a thin bed under pressure gives more efficient filtration. The mash filter uses less sparge water and can therefore produce high-gravity worts, even with an all-malt grist. It is possible to operate to a cycle time of 2 h, increasing brewhouse production to 12 brews per day. The mash filter consists of a horizontal frame carrying a set of robust polypropylene filter sheets; at the end of the mashing programme the mash is run into the filter units and compressed to expel the wort. After sparging, and draining the weaker wort, the compressed spent grains in the filter are in the form of easily removed sheets of low moisture content.

Hops and Wort Boiling

Hops can be used as the natural cones, or as pellets or extracts. Liquid CO_2 is now normally used as an acceptable food-grade solvent for the preparation of hop extracts. There are no important technical differences between ale and lager breweries in the technology for handling hops or hop products, or for boiling and cooling worts. However, certain aromatic hop varieties, e.g. Hallertau, Saaz, Styrian and Tettnang, are especially desirable for lager brewing; other varieties are used in particular brands of beer. A whirlpool separator for spent hop material and the hot-break precipitate from hop boiling is now considered to be more cost-effective than centrifugation or filtration for clarification of hopped worts before cooling, aeration and fermentation. Most brewers insist on a perfectly bright wort: residual hot break is believed to affect yeast growth, and therefore flavour.

Fermentation

Pilsener beers are fermented by bottom yeast. In traditional ale fermentations, the yeast forms a substantial head which is collected during fermentation and which provides the inoculum of pitching yeast for the next fermentation. One of the fortuitious incidents in the development of lager was the introduction of a different yeast, perhaps a contaminant from neighbouring vineyards or wineries, which did not form such a head but was collected from the bottom of the fermentation vessel while draining off the beer. This operation, which is fundamentally different from top-cropping as for traditional ales, requires the yeast to remain viable even after 2 weeks in the high ethanol concentration of the fermentation.

Lager yeast was first isolated in pure culture by E.C. Hansen of the Carlsberg Laboratories in Denmark and named as a distinct species, *Saccharomyces carlsbergensis*. Subsequently, *S. carlsbergensis* was shown to be identical in all taxonomic

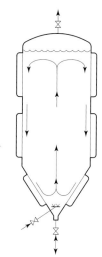

Figure 3 Cylindroconical fermentation vessel, showing circulation of fermenting beer.

features to the wine yeast *S. uvarum*, but the differences between *S. uvarum*, *S. carlsbergensis* and the traditional ale yeast, originally named *S. cerevisiae*, are so slight that taxonomists now include all three in a single species, *S. cerevisiae*. Although technically wrong, it is convenient to refer in this article to lager strains of *S. cerevisiae* as *S. carlsbergensis*. Traditionally the primary fermentation was carried out at 8–10°C, with an inoculum of $1–2 \times 10^7$ cells per millilitre, over 7–10 days, reducing the concentration of fermentable sugar to about 1%.

As with wort production, traditional procedures have been largely superseded by modern methods. Some smaller breweries continue to use the traditional rectangular vessels, but most breweries now use cylindroconical fermentation vessels (CCFVs; **Fig. 3**). Fermentation in the cone generates a central upward flow of CO_2, and the circulation is enhanced by the downward flow over the cooling jackets on the vertical walls. Therefore, even at 8–10°C, the mixing in CCFVs gives a much faster fermentation, and with some increase of temperature, to 12°C or even 15°C, the primary fermentation could be complete within 4 days.

Maturation

Traditionally, with about $1–5 \times 10^5$ yeast cells per millilitre remaining in suspension, the beer after primary fermentation was transferred to chilled storage vessels, where a slow secondary fermentation – maturation – continued over several months. This is the storage from which the term lager was derived. Unlike *S. cerevisiae*, *S. carlsbergensis* is capable of growth at 0°C. The tradition of long low-temperature secondary fermentation arose from necessity: *S. carlsbergensis*, unlike most production ale yeasts, is relatively non-flocculent, i.e. does not form clumps which would settle quickly from the beer, and a long period of still conditions is required for yeast and other hazard-forming material to sediment out of the beer. Finings, as used in British cask-conditioned ales, were forbidden in Bavaria and the tradition of avoiding such additives continued in most countries until recent years. Evolution of CO_2 during the slow secondary fermentation purged H_2S and other volatile sulphur compounds from the beer, and the yeast itself removed diacetyl, its precursor α-acetolactate and other strongly flavoured diketones, from the beer.

These effects are now achieved more economically by short-term storage in contact with the yeast. With modern understanding of the process, the prolonged secondary fermentation is no longer required for maturation, and diacetyl, acetaldehyde and other undesirable flavours can be removed by maturation of less than 1 week. For part of the time, but only after yeast multiplication has ceased, the temperature could be raised to 18–20°C for faster removal of diacetyl and its precursors and homologues (diacetyl rest). After removal of the yeast by centrifugation or filtration, injection of CO_2 purges remaining volatiles and replaces carbonation lost during clarification. Even the diacetyl rest as a separate maturation process could be omitted provided the primary fermentation itself, with a sufficiently high yeast population, continued long enough for yeast to remove the relevant compounds. This could require additional yeast, obtained from a later fermentation at the active stage of log-phase growth and development of the frothy head (German: *Krausen*). Krausen, added at late fermentation or to post-fermentation maturation as necessary, sufficiently increases the yeast population that maturation processes are achieved in 1 week or less. Actively fermenting beer itself must be transferred since, unlike yeast head of an ale fermentation, krausen head is largely foam; hence the impossibility of using such head alone as pitching yeast.

A modern brewery with CCFVs could use the same vessel for both primary fermentation and secondary maturation. In theory, there are advantages in avoiding the turbulence of transfer between vessels, which could cause entrapment of air, perhaps creating a problem of oxidized off flavour. In practice, this is seldom done, so the complex and expensive fermentation vessel can be reused as soon as possible.

Post-maturation Processing

Pasteurization, although important for the shelf-stability of beer, is an aggressive treatment. Even low levels of dissolved oxygen can cause oxidation reactions of flavour spoilage at the high temperatures of pasteurization – typically, 67–70°C. With their more delicate flavour, lager beers are more sensitive than robust ales to such stress. Otherwise, there are no important differences between ale and lager in clarification or packaging. Since Pilsener-type beers are consumed chilled and are expected to form a substantial head of good retention time, additives (when permitted) to avoid chill-haze and to improve head may be more important than in ales. However, modern preference often requires similar serving of ales, so that such additives or processing aids are not exclusive to lager beers.

Lager in Britain

In Britain as a whole, the consumption of lager beer has increased, particularly over the years 1970–1990, to the point where sales of ale and lager are now approximately equal. This largely reflects the trend in England. In Scotland, three large lager breweries have operated throughout most of this century, and lager had overtaken ale sales by 1960. Also there are short-term variations; lager consumption is substantially increased in relation to ale by prolonged spells of hot weather. For various reasons, Pilsener-type lager beer now seems to be firmly established in British culture.

See also: **Heat Treatment of Foods**: Principles of Pasteurization. *Saccharomyces*: *Saccharomyces cerevisiae*; *Saccharomyces*: Brewer's Yeast. **Yeasts**: Production and Commercial Uses.

Further Reading

Lewis MJ and Young TW (1995) *Brewing*. London: Chapman & Hall.
Priest FG and Campbell I (eds) (1996) *Brewing Microbiology*, 2nd edn. London: Chapman & Hall.
Palmer GH (ed.) (1988) *Cereal Science and Technology*. Aberdeen: Aberdeen University Press.

LASERS: INACTIVATION TECHNIQUES

Ian A Watson and **Duncan E S Stewart-Tull**, University of Glasgow, UK

Introduction

There is increasing pressure to ensure the effective decontamination of foods, for two main reasons: firstly, to reduce the likelihood of contamination by pathogenic microorganisms and hence food-borne infection, and secondly, to retard food spoilage by indigenous microorganisms. Consumers expect products that are risk-free, so decontamination must be achieved by processes which are reliable and do not affect the nutritive value of the food, reduce its quality or leave potentially toxic residues.

The chemical treatment of foods, beverages and water is becoming less popular with consumers, and there is an increase in legislation to ban the use of chemicals or reduce their permitted concentrations, making them less effective. Chemical or bacterial treatments may leave residues in foods and produce effluents which can contaminate water supplies, and may result in the formation of toxic by-products. For example, fruit or vegetable wash water may contain chlorine, at concentrations up to 100 p.p.m., and may be disposed of into the water drainage system regardless of its environmental impact. Furthermore the working environment may be hazardous – for example, chlorine-based washing systems can affect the workers and corrode the process equipment. The enforcement of legislation intended to reduce concentrations of chemicals in wash water and to limit the dumping of waste is expensive and difficult. Therefore, alternative methods of decontaminating foods are needed.

There is public concern over the safety of some decontamination processes, including the use of gamma irradiation. The heat treatment of foods at high temperatures, although acceptable to the public, can adversely affect the appearance, flavour and texture of the product, and less severe thermal treatments may not decontaminate adequately. Consequently, minimal processing methods have become attractive to consumers and industry. Among these methods are a number of emerging processes, including pulsed electric fields, pulsed light, ohmic heating and high hydrostatic pressure. Little has been published on the use of laser beams for the decontamination of foods. The major advantage of such optical decontamination – whether photochemical, photomechanical or photothermal – is the absence of additives. Laser technology offers major advantages compared with other forms of optical decontamination, including high intensities and a range of wavelengths. In addition, the energy density can be precisely and easily controlled, so that minimal processing can be achieved.

Electromagnetic Radiation

Various systems use electromagnetic radiation to inactivate bacteria, for example cobalt 60 sources, which generate gamma radiation, light sources and generators of microwaves and waves of radio frequency. Their applications range from the sterilization of food to that of medical equipment.

The difference between the various portions of the electromagnetic spectrum is the energy level or wavelength of the constituent photons. Gamma radiation has a wavelength of about 10^{-12} m, that of visible light is about 10^{-7} m, microwave radiation typically has a wavelength of the order of 10^{-3} m, and waves of radio frequency are about 1 m or over in wavelength. The effect of each wavelength on a bacterium depends on the energy of the photons and the optical characteristics of the bacterium. The reactions that occur can be described in terms of the cross-section or the probability of that reaction occurring. For example, ultraviolet radiation with a wavelength around 260 nm is most likely to be absorbed by thymine dimers, so this reaction would have a relatively high cross-section. As the wavelength of the light is increased the probability of this reaction occurring falls. Ionizing radiation such as ultraviolet light, X-rays and gamma rays has a mutagenic effect on microorganisms.

The methods of generating and delivering electromagnetic radiation must be considered in the context of the optimization of decontamination systems. Lasers are one of the most efficient ways of generating extreme intensities of light. Also, photosensitizers can increase the biocidal effect of electromagnetic radiation, so that sources of lower power can be used.

Infrared Radiation

Infrared radiation can be generated by lamps or lasers. Commercially available lamps allow fine control over the rate of delivery of thermal energy into food, so minimizing the loss of quality of the food.

Ultraviolet Radiation

Ultraviolet (UV) radiation can be generated by excimer lasers or conventional UV lamps, which differ in terms of their output spectrum and method of generation of the light. UV irradiation can kill bacteria extremely efficiently, requiring low energy densities compared with some other forms of radiation. However some forms of bacteria, spores and oocysts are difficult to kill using conventional, continuous-wave UV – for example, *Cryptosporidium* and its oocysts, which can contaminate water. These micro-organisms cannot be filtered out efficiently during water treatment.

A procedure that prevents bacterial growth in a small tank of water has been described, which relies on a light-emitting diode and a system encompassing visible light and infrared and ultraviolet radiation.

High-intensity Flash Lamps

A process that uses intense flashes of broad-spectrum white light to kill microorganisms on food and packaging has been developed by Dunn et al (1995). The wavelength of the light ranges from 200 nm (UV) to 1.0 μm (near infrared). The spectral distribution is similar to that of sunlight, but the pulsed light is about 20 000 times stronger. A single flash at $0.5–1.0 \, J \, cm^{-2}$ killed *Listeria monocytogenes*, *Escherichia coli* O157:H7, *Bacillus pumilus* and *Aspergillus niger* at concentrations of about $1 \times 10^5 \, cm^{-2}$, with 7–9 log reduction values. The oocysts of *Cryptosporodium parvum* at concentrations of $1.0 \times 10^6 – 1.0 \times 10^7$ oocysts per cm^{-2} were destroyed by a single flash at $1.0 \, J \, cm^{-2}$: this method could be useful for treating supplies of drinking water. The irradiation process requires a transparent medium, which will allow the white light to penetrate. Dunn et al described the use of short-duration pulsed light flashes with 90% of their energy distributed between 300 nm and 2500 nm and a flash duration of about 0.001 ms to 100 ms at an energy density of $0.01–20 \, J \, cm^{-2}$. It was suggested that the heat produced on the surface would be conducted into the interior of the product, but even if a series of pulses are delivered, the heat produced at the surface does not seem to raise the temperature inside the product substantially. The inactivation of *E. coli*, *Bacillus subtilis*, *Saccharomyces cerevisiae*, *A. niger* and *Staphylococcus aureus* on surfaces, using intermittent pulses of light of visible and near-visible wavelengths, has also been described by Dunn et al.

Lasers

Many lasers have demonstrated potential for laser inactivation. In a recent study investigating the performance of a range of laser wavelengths from 118 μm

to 355 nm, bactericidal activity was observed in the case of lasers generating wavelengths of 1.06 μm, 10.6 μm and 355 nm. There is particular interest in lasers that emit radiation at the UV end of the spectrum, for example the krypton fluoride laser, which emits a wavelength of 248 nm. A few reports have mentioned the inactivation by lasers of bacteria on surfaces, including the sterilization of reamers and scalpel blades in a CO_2 laser beam. However these investigations were limited and no detailed results were presented.

The rapid sterilization of inert surfaces using lasers has been demonstrated, and the efficiency and dependency of the process quantified for a range of variables. The optical and thermal properties of the surface material are important. Discs of metal, glass (borosilicate) and plastic (nylon) were fabricated, inoculated with bacteria, either dried or left wet, and then exposed to light from the neodymium:yttrium-aluminium-garnet (Nd:YAG) and carbon dioxide lasers. The Nd:YAG laser was most effective at decontaminating the discs in the order metal > plastic > glass. In contrast, the CO_2 laser was most effective at decontamination in the order plastic > glass > metal. This work was extended to include multifactorially designed experiments to identify the effect of the variables associated with the lasers and the growth of the bacteria. The range of variables included pulse energy, pulse repetition frequency (PRF), exposure time, pH, NaCl, and either wet or dry conditions during exposure.

How Lasers Work

Since the introduction of the first laser in early 1960s, laster technology has progressed in leaps and bounds. Nowadays lasers are available that can produce radiation of practically any wavelength, from microwaves to X-rays. Commercially available lasers can produce peak powers measured in terawatts, or mean powers in excess of 100 kW. Their applications are diverse, from initiating nuclear fusion to manipulating single cells. The word 'laser' is derived from 'Light Amplification by Stimulated Emission of Radiation'. Amplification produces a beam of photons with identical scalar and vector properties, i.e. frequency, phase, direction and polarization.

Lasers comprise an active medium, a pumping or power source and, generally, a resonator. The active medium may be gas, liquid or solid, and the pumping source may be electrical, optical or chemical. The resonator is an optical system which provides feedback and achieves amplification. The components are often used as descriptors of the type of laser – for example CO_2 gas laser, solid-state ruby laser and Nd:YAG laser.

Laser Radiation

Laser radiation has the unique characteristics of being highly intense, monochromatic, coherent and unidirectional. 'Coherent' means 'in phase', and this property underpins many applications of lasers, including interferometry; the measurement of planarity, physical dimensions and distances between objects; and holography. 'Monochromatic' means 'of one wavelength' – although in practice physical effects usually cause very slight wavelength broadening, for example Doppler effects and pressure. Lasers are the best monochromatic sources. Laser radiation has high directionality and low divergence, so that a beam reaches a distant object without significant loss of intensity, provided that it is passing through a medium which does not absorb radiation of that wavelength. Laser beams do diverge, but this can be limited to fractions of a milliradian. The high brightness of laser beams is very useful in the drilling or cutting of metals and in surgery.

The inactivation of bacteria and spores is well within the capability of many lasers. In this context, it is essential that the substrate is not damaged – this can be achieved because the output of the laser can be controlled precisely.

Laser Irradiation of Liquids

In selecting a laser for a particular decontamination function, the interaction between a beam of the particular wavelength delivered and the particular microorganism is of vital importance. For example, pulses of 7 ns from an Nd:YAG laser emitting radiation of wavelength 1.06 μm had no effect on *Escherichia coli* lawned on agar plates, whereas pulses of 8 ms produced significant killing – perhaps indicating that the killing mechanism is partly thermal. Radiation of this wavelength is transmitted through water with about 8% absorption per centimetre, and is transmitted through plastic with little attenuation. Therefore this laser could conceivably be used to sterilize liquids through plastic. In contrast, CO_2 or krypton fluoride lasers emit radiation of wavelengths 10.6 μm and 248 nm respectively, which would not penetrate plastic or water – these lasers are, however, proving to be useful in the decontamination of surfaces.

Figure 1 shows a laser system for the sterilization of water. The water is passed from the reservoir or tank through the laser beam, and the treated water can be either recycled or safely discharged. Commercially, systems utilizing lasers to produce ultra-pure water have been developed for the electronics industry.

Important considerations in the design of sterilization systems for water are the absorption characteristics of the liquid at the relevant wavelength of radiation; the energy densities needed to decontaminate the liquid; and the power requirements of the laser given the rate of flow of the liquid.

Effect of Laser Radiation on *Escherichia coli*

The decontamination of various dilutions of *E. coli* using an Nd:YAG laser is shown in **Figure 2**. The laser parameters were 10 J, 10 Hz, pulse length 8 ms, beam diameter 1.35 cm. For each concentration of *E. coli*, the transient temperature profile of the liquid was measured. As the bacterial concentration was increased from about 1×10^4 to about 1×10^8 cfu ml^{-1}, the temperature profiles were similar, with reduction in viability of log 2.2–4.5. When the concentration of the suspension was increased to 2×10^9 cfu ml^{-1}, the reduction in viability decreased. The lower reduction in viability with increased concentrations reflected less penetration of the laser radiation through the denser suspension.

These results have important implications for the limitations of the optical decontamination of opaque materials or turbid liquids. Water decontamination systems have been developed in which turbulence

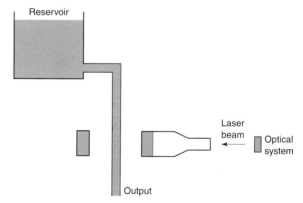

Figure 1 System for decontaminating water by laser beam.

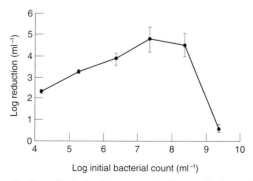

Figure 2 The effect of culture concentration on the bactericidal action of Nd:YAG laser light. The logarithmic cell reduction after laser exposure is plotted for six different *Escherichia coli* concentrations.

is introduced, so that the radiation interacts more effectively with the water.

Modelling the Effects of Laser Radiation

Computer simulation of the rise in temperature induced by exposure to laser radiation has been used extensively in medicine. A mathematical model has been developed to predict the temperature field of a target irradiated by laser, taking into account system variables including the wavelength, pulse shape, energy density and bacterial concentration. The temperature dependency of the absorption coefficient, thermal conductivity and thermal diffusivity were considered in the development of the model. The results helped in the identification of the optimum operating variables of the decontamination system, which can maximize the efficacy of decontamination by laser radiation.

Figure 3 shows a grey-level image of the temperature profile at the instant the laser was turned off, where the black level represents a temperature value of 40°C and the white level a temperature of 100°C. The dark regions are of lower temperature, and bacteria in this area are less likely to be killed or inactivated; whereas in the bright region, cell death is highly likely within the exposure time. The sterilized zone in the liquid suspension can be identified if the lethal temperature–time threshold is known. Moreover, energy density grey-level images could be produced to indicate likely areas where sterilization or inactivation may not be sufficiently effective. In an attempt to show the sterilization boundaries, 65°C after a 21.6 s exposure was taken to be the lethal temperature–time threshold for the *E. coli* under the laser treatment. **Figure 4** illustrates the sterilization zones in the liquid suspension for this case, where the white area represents the sterilized zone and the black region where bacteria may survive.

Scanning electron microscopy of *E. coli* exposed to different temperatures induced by laser radiation

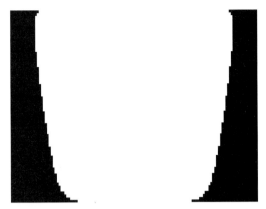

Figure 4 Sterilization zone in the liquid suspension, where the white area represents the sterilized zone, and the black region is where bacteria may survive.

indicated that the cells developed blebs on the surface at about 60°C, and ruptured at about 70°C. In a water bath, exposure to 100°C for 5 min caused neither blebs nor rupture. In addition, analysis of the RNA released from the laser-irradiated samples indicated that its value was six times greater than that of the untreated samples. Differential scanning calorimetry analysis showed that significantly more extensive killing was achieved by laser radiation than by using a water bath with intracellular temperatures below 60°C.

Laser Irradiation of Foods and Surfaces

Surfaces can be decontaminated by laser radiation by either scanning the laser beam across the surface, moving the surface under a stationary laser beam, or using a hybrid configuration in which both the laser beam and surface are moved (**Fig. 5**). Stationary laser beams could also be used to decontaminate containers, and perhaps their contents, on a production

Figure 3 Grey-level image of the temperature profile when the laser was turned off (see text). Height 3 cm, radius 0.6 cm.

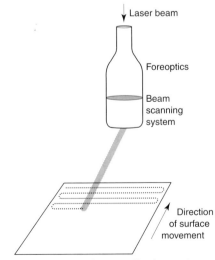

Figure 5 Hybrid laser surface sterilization system.

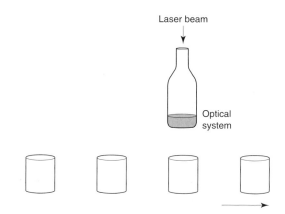

Figure 6 Production line decontamination system.

line (**Fig. 6**). To estimate the power requirements of the laser system and the typical throughput that could be expected, the relationship between the bioburden and the energy density needed to achieve a given level of decontamination needs to be identified. Further work is needed for the evaluation of the optical decontamination of rough organic surfaces.

Sterilization by laser radiation has been studied in dentistry and medicine, but not to any significant extent in the context of the food industry. Most studies have used photosensitizers and low-power lasers, producing conditions in which bactericidal action is generally dependent on the generation of free radicals. Further, few studies have systematically investigated and quantified the effects of laser-related variables on the efficacy of sterilization. However, some studies have demonstrated the potential of CO_2 lasers for the decontamination of foods, e.g. fruits, vegetables and meats, as well as that of solid surfaces, including metals and some plastics.

Results in Carrots and Potatoes

Figure 7 shows the loss in viability of *E. coli* on carrot and potato as a function of time of exposure to laser radiation. Discs of carrot and potato were sterilized, dried or partially dried and inoculated with *E. coli* to a concentration of about 1×10^8 cfu per disc. The discs

were then exposed to a CO_2 laser beam of 1 kW for pulse durations of 2–10 ms. Figure 7 shows the loss in viability as a function of exposure time. With an exposure time of 10 ms, a 5 log reduction was observed with the partially dried samples and a 3 log reduction for the dry samples. However, this level of contamination is extremely high compared with levels likely in practice, and so the experiments were repeated with partially dried samples and an inoculum of about 1×10^5 cfu per disc. **Figure 8** shows that after an exposure of 8 ms, both carrots and potatoes were completely decontaminated.

It is informative to scale up these results to predict the potential performance of an optical decontamination system. The samples were approximately 0.78 cm^2, and consequently the rate of inactivation of the microorganisms on the carrot and potato was about 98 cm^2 s^{-1}. With an average carrot of about 2.0 cm in diameter at the base and about 10 cm long, the area would be about 31 cm^2. Consequently, it would be possible to decontaminate three carrots per second, or 260 000 carrots per day, with a 5 log reduction in the numbers of bacteria. If each carrot weighed 100 g, then this would represent a throughput of about 26 tonnes per day, assuming favourable geometry of exposure.

Results in Ham, Bacon and Fish

Serratia marcescens, *Staphylococcus aureus* and *Pseudomonas aeruginosa* were inoculated onto ham, bacon and fish (whiting and herring) and a low-power laser beam was scanned across the surface of the samples. On ham and bacon the appearance of *Serratia marcescens* growth was pink, while *Staphylococcus aureus* and *P. aeruginosa* appeared colourless. On fish, whiting and herring, all three species caused distinct colour changes. A distinct zone of clearance was evident on the bacon (**Figure 9**). A zone of clear-

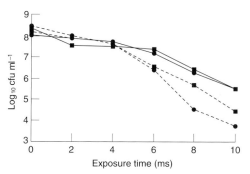

Figure 7 Effect of exposure of *Escherichia coli* to CO_2 laser light. The bacteria were dried on the surfaces of potato (circles) or carrots (squares); the dotted lines indicate partially dried samples.

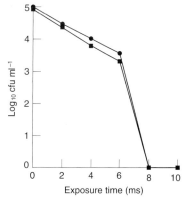

Figure 8 Effect of exposure of *Escherichia coli* to CO_2 laser light at lower inoculum concentrations (10^5 cfu ml^{-1} compared with 10^8 cfu ml^{-1}) than those in Figure 7. Samples partially dried on potato (circles) or carrot (squares).

ance was also apparent on the herring skin. A faint zone of clearance was seen on the wafer-thin ham, but no zones of clearing were observed on the fish flesh.

Linear and Rotary Scan on Agar, Contaminated Ham, Bacon and Fish   The effect of speed (m s⁻¹) and laser power (W) were investigated with a scanning laser beam and a rotary system. Nutrient agar plates seeded with *Serratia marcescens* were exposed to the laser light and examined after overnight incubation. The effect of low-power laser exposure is clearly seen in **Figure 10** where growth alternates with clearance at a fast rotating speed; here the system was operating too close to the sterilization threshold for consistent clearing to be observed. Distinct clearance occurred when the rotating speed was reduced to half the higher value.

Effect of Laser Radiation on Food Quality

In order to satisfy the bodies responsible for food safety and regulation, as well as consumers, the treatment of foods must not cause unacceptable changes. Preliminary studies on the nutrient content and lipid oxidation effect of laser-treated ham suggest that exposure has no significant effect.

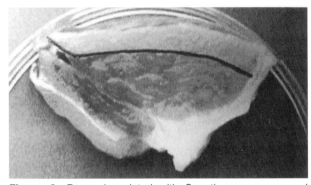

Figure 9 Bacon inoculated with *Serratia marcescens* and exposed to CO$_2$ laser irradiation. A clearance zone can be seen above the line at the top of the picture.

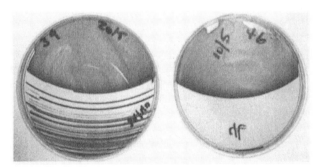

Figure 10 Low-power laser scan of agar lawned with *Serratia marcescens*. The plate on the left was scanned at twice the speed of the right-hand plate, which shows complete clearing.

Future Developments

Approval has been given by the US Food and Drug Administration for the use of flash lamps for the decontamination of foods, and optical decontamination seems to be a potentially valuable approach to minimal processing in the case of some applications and is likely to be a growth area in food technology in the twenty-first century.

See also: **Cryptosporidium**. **Escherichia coli**: *Escherichia coli*. **Process Hygiene**: Designing for Hygienic Operation; Modern Systems of Plant Cleaning. **Ultraviolet Light**.

Further Reading

Bruhn C and Schutz H (1989) Consumer awareness and outlook for acceptance of food irradiation. *Food Technology* 43: 93–97.

Dunn J, Ott T and Clark W (1995) Pulsed light treatment of food and packaging. *Food Technology* 49(9): 95–98.

Gunes G, Splittstoesser DF and Lee CY (1998) Microbial quality of fresh potatoes: effect of minimal processing. *Journal of Food Protection* 60(7): 863–866.

Hocking AD, Arnold G, Jenson I, Newton K and Sutherland P (1997) *Foodborne Microorganisms of Public Health Significance*. Sydney: CSIRO Food Science and Technology.

Mertens B and Knorr D (1992) Development of nonthermal processes for food preservation. *Food Technology* 46(5): 124–133.

Monk J, Beuchat L and Doyal M (1995) Irradiation inactivation of food-borne microorganisms. *Journal of Food Protection* 58(2): 197–208.

Sapers GM and Simmons GF (1998) Hydrogen peroxide disinfection of minimally processed fruits and vegetables. *Food Technology* 52(2): 48–52.

Seigman AE (1986) *Lasers* University Science Books. Oxford: OUP.

Schultz R, Harvey G, Fernandez-Beros M et al (1986) Bactericidal effects of the neodymium:YAG laser. An *in vitro* study. *Lasers in Surgery and Medicine* 6: 445–448.

Wang RK, Watson IA, Ward GD, Stewart-Tull DE and Wardlaw AC (1997) Temperature distribution in *Escherichia coli* liquid suspensions during irradiation by a high-power Nd:YAG laser for sterilization applications. *Journal of Biomedical Optics* 2: 295–303.

Ward G, Watson IA, Stewart-Tull D, Wardlaw A and Chatwin C (1996) Inactivation of bacteria and yeasts on agar surfaces with high power Nd:YAG laser light. *Letters in Applied Bacteriology* 23: 136–140.

Watson IA, Ward G, Wang R et al (1996) Comparative bactericidal activities of lasers operating at seven different wavelengths. *Journal of Biomedical Optics* 1(4): 1–7.

Witteman WJ (1987) *The CO$_2$ Laser*. Berlin: Springer.

Latex Agglutination Techniques *see* **Campylobacter**: Detection by Latex Agglutination Techniques; **Escherichia coli O157**: Detection by Latex Agglutination Techniques; **Salmonella**: Detection by Latex Agglutination Techniques.

Legislation *see* **National Legislation, Guidelines & Standards Governing Microbiology**: Canada; European Union; Japan.

LEUCONOSTOC

Aline Lonvaud-Funel, Faculty of Œnology, University Victor Segalen, Bordeaux, France

Introduction

The genus *Leuconostoc* belongs to the 'sensu stricto' group of lactic acid bacteria (LAB), defined as bacteria producing lactic acid exclusively or not, from carbohydrate fermentation. They are Gram-positive, non-sporulating bacteria, exhibiting a G+C DNA content less than 50%. They develop in anaerobiosis or aerobiosis, and no catalase is generally present. Like the other groups of LAB, they need complex media due to their multiple demands for amino acids, peptides, carbohydrates, vitamins and metallic ions.

About 12% of isolated LAB from many ecosystems, mostly in plant material, are *Leuconostoc* species. Some are isolated from the surface of a wide range of vegetables and fruits, and others from refrigerated meats and dairy products. They are regarded as non-pathogenic except for a very few species from clinical sources which are vancomycin-resistant.

Leuconostoc ferments glucose by the pentose phosphate pathway, producing D-lactate ethanol and CO_2. In addition, acetate is also produced when the reduced coenzyme NADH is reoxidized by electron acceptors, such as fructose or O_2, via NADH oxidase. Until recently the genus *Leuconostoc* included heterofermentative cocci, which exclusively produce D-lactate from glucose. DNA analyses together with phenotypic methods have provided new data for phylogenetic studies. The relationships established among LAB have led to a new classification comprising new genera and new species.

Like other LAB species, *Leuconostoc* spp. are of technological interest in the food and beverage industry. In general, lactic fermentations are traditional biotechnological processes, but most remain uncontrolled. However, interest in selected *Leuconostoc* strains has been shown in some food industries (dairy, wine and meat) and will probably concern others in the future. Indeed, some species of the *Leuconostoc* group have special commercial importance due to their ability to produce aroma compounds, valuable polysaccharides and malolactic fermentation, for example. On the other hand, some other species can induce spoilage by producing undesirable compounds in food or dextran in sugar processes.

Description of the Genus *Leuconostoc* and Species Related to the Food Industry

Classification

In earlier definitions, *Leuconostoc* species were, according to their physiology, LAB very close to the heterofermentative lactobacilli. They were differentiated from lactobacilli by their morphology and the exclusive production of D-lactate from glucose. Today, molecular characters, especially those based on nucleic acid analyses, are considered more and more both for classification and identification. The new molecular tools have led to significant changes for the *Leuconostoc* genus. The sequences of 16S rRNA, 23S rRNA and of the *rpo*C gene (encoding the β-subunit of DNA-dependent RNA polymerase) of *Leuconostoc* and other heterofermentative bacteria show that they are phylogenetically grouped in three distinct lineages; the *Leuconostoc* genus 'sensu stricto' and the new genera *Weissella* and *Œnococcus* (**Table 1**). The *Leuconostoc* genus comprises *L. mesenteroides*, with three subspecies *mesenteroides*, *dextranicum* and *cremoris* and seven other species. *Lactobacillus fructosus* is also classified in the *Leuconostoc* group. They are well separated from the *Weissella* genus which includes the former *Leuconostoc paramesenteroides* and four heterofermentative lactobacilli and from *Œnococcus*.

The *Leuconostoc oenos* species was formerly differentiated from the other *Leuconostoc* members by its acidophilic nature. Its originality was highly confirmed by the analysis of 16S rRNA and 23S rRNA

Table 1 The *Leuconostoc* group and its subdivisions

Genus	Species
Leuconostoc (Type species: *Leuconostoc* *mesenteroides*)	*mesenteroides* subsp. *mesenteroides* *mesenteroides* subsp. *dextranicum* *mesenteroides* subsp. *cremoris* *carnosum* *gelidum* *citreum* (*amelibiosum*) *lactis* *pseudomesenteroides* *fallax* *argentinum* *paramesenteroides*
Weissella (Type species: *Weissella viridescens*)	*confusus* *kandleri* *minor* *viridescens* *hellenica*
(Type species: *Œnococcus œni*)	

which have a very low level of homology with the other *Leuconostoc* sp. It represents such a distant phylogenetic group that a new *Œnococcus* genus was created, with only one species *Œnococcus œni*. The restriction profile of ribosomal genes and the intergenic 16S–23S rDNA sequence of several strains of *Œ. œni* are highly conserved and demonstrate the very high degree of homogeneity of this species.

Identification

The identification of bacteria in a fermented food or in the raw material first necessitates strain isolation. This is performed by inoculating suitable nutritive media onto plates. Colonies are isolated and grown to obtain sufficient biomass for all the identification tests.

Morphology is usually the first determination done with Gram staining. *Leuconostoc* and *Œnococcus* are almost spherical cells, sometimes rather lenticular, and resemble very short bacilli with rounded ends (**Fig. 1**). Their size is approximately 0.5–0.7 × 0.7–1.2 μm. The cells are arranged in pairs or chains. In nutrient media during the active growth phase, they are often in short chains, whereas in their natural environment and in more stressful conditions the chains are longer. Most of these bacteria grow at between 20°C and 30°C, the optimal temperature zone. The initial pH of the growth medium of 6.5 drops to 4.4–5.0 during culture due to acid production. The acidophilic *Œ. œni* grows better at an initial pH of 4.8.

According to *Bergey's Manual*, one of the distinctive traits of *Leuconostoc* species is their inability to hydrolyse arginine. However, as at least some strains of *Œ. œni* and perhaps some *Leuconostoc* possess the arginine deiminase pathway, this feature no longer seems convenient for identification. An interesting property of *Leuconostoc* and *Œnococcus* is the exclusive production of D-lactate from glucose. This is very pertinent since most other LAB produce DL-lactate. However, the confusing species *Weisella viridescens* and *Lactobacillus fructosus* also predominantly form the D-isomer. The fermentation pattern of carbohydrates gives other keys for strain identification at the species level. The miniaturized API tests (Bio-Mérieux) are composed of small tubes, each containing one carbohydrate as a possible fermentable substrate. Cells suspended in a medium containing all the nutrients except the main energy source and bromocresol purple are poured into each microtube and overlayed by paraffin oil to manage anaerobiosis. Positive tests are visualized by the change of the medium from blue to yellow. The fermentation pattern (**Fig. 2**) allows the isolate to be identified by comparing the pattern to the characteristics of the type strains (**Table 2**).

This traditional method of identification is still very useful and routinely used. However, it is admitted that some results may be ambiguous, especially when determining the fermentable carbohydrates. Even in optimized conditions, the change of colour may sometimes take so long for a given strain that it is difficult to conclude whether there is a positive or negative reaction. On the other hand, LAB often contain plasmids coding for key enzymes involved in biochemical pathways. Due to their instability especially in the absence of selective pressure, some tests which are usually positive can turn negative. Moreover, phage-mediated characters may also induce the same problems. In case of difficulties, other investigations can clarify the situation. These are based on the electrophoretic mobility of enzymes, the amino acid sequence of the interpeptide bridge of the peptidoglycan, fatty acid composition, pyrolysis mass spectrometry and infrared spectroscopy. In spite of their interest, these methods cannot be routinely considered and feature only in specialized laboratories.

The genomic definition of a species is a group of strains with more than 70% DNA homology. In consequence, another approach for bacterial identification at the species level is based on DNA sequence comparison. Nucleic acid hybridization, polymerase chain reaction (PCR) amplification or nucleic acid sequencing may be applied to *Leuconostoc* identification. DNA–DNA hybridization is the most useful and has confirmed the phenotypic identification of LAB in most cases. Hybridization can be targeted against specific sequences or whole

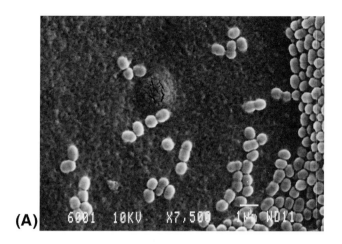

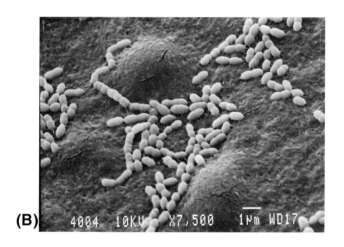

Figure 1 Scanning electron micrograph of (**A**) *Leuconostoc mesenteroides* and (**B**) *Œnococcus œni*. Scale bar = 1 μm.

genomic DNA. It uses DNA probes which are single-strand DNA fragments labelled either by isotopic or non-isotopic nucleotide derivatives. In defined conditions of temperature and ionic strength, the DNA probe binds specifically to complementary target DNA. Binding occurs if probe and target DNAs are homologous. According to the probe, the identification may be done at different levels, genus, species or group of strain sharing in common a gene or specific sequence.

Leuconostoc whole-cell DNA probes have shown their specificity, reproducibility and reliability at the species level. The method is very well adapted to such strains, which grow very slowly. Another interest of the method is that hybridization can be performed directly on bacterial colonies which are transferred onto the nylon membrane. Successive de-hybridization and re-hybridization with different DNA probes allows the identification of several species in a mixture as is often the case in natural fermenting foods. The first whole-cell DNA probe was prepared for *Œ œni*. It allows its enumeration among the mixed microflora of wine.

In addition to whole-cell DNA probes, oligonucleotides have been designed. The 16S rDNA gene, or part of it has been frequently chosen. Partial

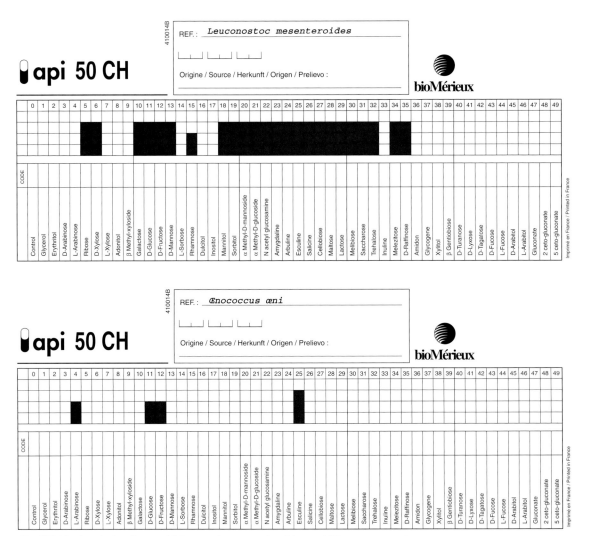

Figure 2 API-biochemical patterns of *Leuconostoc mesenteroides* and *Œnococcus œni*.

sequences based on a variable region of 16S rDNA have been successfully used to identify *Leuconostoc lactis* and *Leuconostoc mesenteroides*. However, such specific oligonucleotides cannot be found for every species, and it is recommended to use 16S rDNA probes at the genus level.

Variability exists within a species. Strains classified in the same species share a common phenotype and may differ for few characters. For example, all Œ. *œni* strains have the same phenotype characterized by a small number of fermentable carbohydrates and acidophily. However, their growth either in laboratory conditions or in wine can vary greatly from one strain to the other. Moreover in practice, wine flavour is different according to the strain. Several molecular tools may be used for intraspecific characterization. Total genomic DNA restriction followed by electrophoresis separation and Southern hybridization with a 16S + 23S rDNA probe (ribotyping) generate a pattern which is possibly strain-specific. The same

results are obtained by PCR amplification with random primers (RAPD) or restriction analysis of PCR products such as the 16S/23S ribosomal intergenic sequence. Finally, pulse field gel electrophoresis (PFGE) analysis of genomic DNA fragments generated by rare cutting enzymes is also a very efficient method to distinguish one strain from another. The methods are chosen according to their reliability and reproducibility, and according to the homogeneity of the species. PFGE of genomic DNA seems to be the most widely applicable. One practical example is the characterization of Œ. *œni* strains although they belong to a very phylogenetically homogenous species. Restriction by one or two enzymes is often sufficient. The variability of the profiles makes the recognition of each strain possible. This is particularly interesting for characterizing starter strains, and to follow their establishment in a fermentation process where they are opposed to the natural microflora.

Table 2 Distinctive fermentative characters for specific identification of *Leuconostoc* and *Œnococcus*

| | Leuconostoc species | | | | | | | | | | Œnococcus |
	argentinum	carnosum	citreum	fallax	gelidum	lactis	mesenteroides subsp. cremoris	mesenteroides subsp. dextranicum	mesenteroides subsp. mesenteroides	pseudo-mesenteroides	œni
Amygdaline	–	–	+	–	–	–	–	d	d	–	–
L-Arabinose	d	–	+	–	–	–	–	–	+	+	d
Arbutin	–	–	+	–	–	–	–	–	d	–	–
Cellobiose	d	d	+	–	d	–	–	d	d	+	d
Fructose	d	+	+	+	+	+	–	+	+	+	+
Galactose	+	–	–	–	–	+	d	d	+	+	d
Lactose	+	–	–	–	–	+	+	+	d	+	–
Maltose	+	d	+	+	d	+	d	+	+	+	–
Mannitol	d	–	+	+	–	–	–	d	d	–	–
Mannose	+	d	+	+	d	d	–	d	+	+	d
Melibiose	+	d	–	–	d	d	–	d	d	+	d
Raffinose	+	–	–	–	–	d	–	d	d	+	–
Ribose	–	d	–	+	d	–	+	–	+	+	+
Salicin	–	d	+	–	d	d	–	d	d	d	d
Sucrose	+	+	+	+	+	+	–	+	+	+	–
Trehalose	d	+	+	+	+	–	–	+	+	+	+
D-Xylose	d	–	–	–	–	–	–	d	d	d	d
Aesculin	–	d	+	+	+	–	–	d	d	+	+
Dextran	–	+	+	+	d	–	–	+	+	d	–

+, 90% or more strains positive; –, 90% or more strains negative; d, 11–89% positive.

Leuconostoc species in Their Natural Environment

Plants are the natural habitat for many LAB. Even if *Lactobacillus* seems to be the predominant genus, *Leuconostoc* is frequently isolated on the plant surface together with other LAB, acetic acid bacteria, yeasts and moulds. Their population depends on meteorological conditions, humidity, ultraviolet light, temperature and nutritional conditions. Whatever the plant, as a general rule, the population increases during maturation and much more rapidly at harvest since nutrients from sap or fruits are more available. At that time, the distribution of the microflora changes and some genera and species progressively dominate according to the new environment caused by the fermenting process. Growth of *Leuconostoc* strains induces the process or takes part in the fermentation together with yeasts, for example. Generally, sugars are fermented to lactic acid and acidification guarantees against spoilage of raw material by moulds or other undesirable organisms. The succession of microorganisms results in several types of interactions. Acidity is one of the factors of natural selection; not only does the pH decrease, but also the organic acids (lactic and acetic acid) produced are toxic or inhibitory. Another mechanism involved is H_2O_2 production in certain conditions and the effect of bacteriocins is now being increasingly studied.

Leuconostoc mesenteroides

Of the three subspecies, *L. mesenteroides* subsp. *mesenteroides* is the most frequently isolated on plants and fruits. Generally *L. mesenteroides* initiates the fermentation process and increases the total acidity. Mannitol is also produced when fructose is used as an electron acceptor. Dextrans are synthesized from sucrose. As the pH increases, *L. mesenteroides* which is less acid tolerant disappears giving way to other LAB species. The two subspecies *mesenteroides* and *dextranicum* are present on the surface and in the soil of sugar cane fields; there is twice the amount of subspecies *mesenteroides* as *dextranicum*. Burning of the cane does not eliminate the bacteria. After the harvest, they grow and are then spontaneously eliminated by acidity, as more tolerant lactobacilli develop.

Leuconostoc species are also important in the beverage industry. They participate in the fermentation of wine, cider and more traditional products such as kefir or other very localized products. In a Turkish traditional beverage from cereals, LAB ferment together with yeasts; *L. mesenteroides* subsp. *mesenteroides* represents about 20% of the bacteria and subsp. *dextranicum* about 50%.

Œnococcus œni

Œ. œni is isolated in fermenting fruit juices. It is essential in wine and cider production and it can be a spoiling organism in other media, as in canned mango juice. *Œ. œni* is normally present at a very low population at the early stage of wine-making or cider-making. It finally becomes the predominant and sometimes nearly exclusive species at the end of production after alcoholic fermentation which is conducted by yeasts. It is very well adapted to acidic media, relatively tolerant to ethanol, and is responsible for malolactic fermentation.

Leuconostoc gelidum and Leuconostoc carnosum

These *Leuconostoc* species occur among the microbial population of meats during storage in vacuum packs or CO_2 atmosphere. They are associated with several other LAB genera and spoil meat by producing off-flavours, off-odours and colour alteration. Such bacteria were formerly classified in the *L. mesenteroides* group. However, the numerical taxonomic techniques have shown that these strains can be classified in two new species. Both produce dextran from sucrose, but *L. carnosum* ferments a lower number of sugars compared to *L. gelidum*.

Leuconostoc fallax

Some *Leuconostoc* strains first isolated from sauerkraut proved to be atypical according to a phylogenetic study based on 16S rDNA sequences. They were genetically distinct from other *Leuconostoc* species but very close to this group with 94–95% DNA homology. They are now considered as a new species in the genus even though peripheral. A strain of *L. fallax* was further isolated from *Gerbera* sap. This isolate is acid and ethanol tolerant (up to 9% v/v), properties it shares with *Œ. œni*, although it does not degrade either malic acid or citric acid.

Leuconostoc citreum (Leuconostoc amelibiosum)

The *L. citreum* species was initially included in *L. mesenteroides* and was differentiated on the basis of DNA sequence analysis. Strains previously considered as *L. amelibiosum* proved to belong to *L. citreum* after chemotaxonomic and phylogenetic studies. *L. citreum* was identified among the LAB isolated from a fermented rice cake consumed in the Philippines, together with *L. mesenteroides* and *L. fallax*. Such strains are also responsible for dextrans in sugar solutions.

Leuconostoc argentinum

L. argentinum strains were isolated from Argentinian raw milk. They grouped in one cluster exhibiting a low level DNA homology with other *Leuconostoc* species. In particular, on this basis they differ from *L. lactis* and *L. mesenteroides* which are frequently isolated in milk.

However, these strains cannot be distinguished from other *Leuconostoc* species by their carbohydrate fermentation profile.

Leuconostoc lactis This species has only been isolated from milk products. One characteristic is its higher heat resistance compared to other *Leuconostoc* species. It has mainly been studied in the dairy industry for its ability to produce aroma compounds together with *L. mesenteroides*.

Detection and Enumeration of *Leuconostoc* in Beverages and Foods

In fermented food processes, the number and species of microorganisms must be controlled. Fermentation often occurs during the growth phase. In this case, the total population is close to the viable population which can be determined by microscope counting. However, although this method is applicable to large microorganisms, e.g. yeasts, it is more difficult to count small *Leuconostoc* cocci. On the other hand, a viable population is indispensable for microbial analyses after production to prevent or to evaluate the bacterial spoilage risk. Moreover, the determination of the species of dead microorganisms particularly after spoilage is valuable in terms of diagnosis, and further prevention or risk evaluation.

Culture Media

Normally, viable cell counts are performed. The primary definition of viability is the ability of the microorganism to multiply. Optimal culture media and conditions are required. *Leuconostoc* are fastidious chemoorganotrophic bacteria. Therefore, culture media must be rich in nutrients such as carbohydrate as energy providers, indispensable amino acids as nitrogen compounds, salts and vitamins. Other ingredients may be added, such as Tween 80 which usually increases growth of *Leuconostoc* by providing oleic acid incorporated into the cell membrane. Tomato juice is also recognized as a growth factor supplier; it contains a glucoside derivative of pantothenic acid.

Several culture media have been described for *Leuconostoc*. They all have essential compounds in common such as glucose, peptone, meat and/or yeast extract, potassium (or sodium) phosphate, manganese and magnesium sulphate. The most widely used medium is MRS medium. It is often called 'Lactobaccilli medium' broth but it also meets the *Leuco-*

nostoc requirements (**Table 3**). A decrease in pH is observed during growth to 5.0 or less because of the lactic acid production. But an initial pH near 4.8–5.0 is also convenient for most *Leuconostoc* and particularly for the acidophilic *Œnococcus*.

The addition of sterile wine or cider to MRS or other similar media is beneficial to the LAB recovery of the former. Colonies are larger and generally more numerous than on MRS alone. This demonstrates the difficulty of obtaining a colony number corresponding to the actual viable population. *Œnococcus* also actively grows in medium added with grape juice instead of tomato juice, as both contain the same growth factor. Moreover fructose is a better energy source than glucose. More ATP is produced by fructose fermentation because of its capacity of electron acceptor for coenzyme reoxidation; this finally leads to more acetic acid production than from glucose alone.

Techniques for Counting

Colony Count Enumeration of *Leuconostoc* strains in growing pure cultures is easy. The correct nutritive medium is added to $20 \, g \, l^{-1}$ agar to obtain a solid medium. This is inoculated with the sample or suitable decimal dilutions corresponding to the development of 30–300 colonies per plate. Dilutions are ideally performed in $(g \, l^{-1})$ peptone 0.1, KH_2PO_4 0.3, Na_2HPO_4O6, NaCl 8.5. However, for most *Leuconostoc* species results are the same when dilutions are done in sterile water or physiological saline solution. Two methods may be used for inoculation. First in poured plates, the molten agar medium (45°C) is poured onto the sample dilution and the Petri dish is swirled to mix the sample and medium. Second, in spread plates the solid agar medium is first poured and left to harden in the dish, then the sample (0.1 ml of dilution) is spread on the surface. Plates are incubated at 25°C, preferably under an atmosphere of $CO_2 + N_2$ either in jars (Gas pack) or in a specially equipped incubator. Some *Leuconostoc* strains are particularly slow-growing bacteria, and need at least six days to form visible colonies. If the population is too low, so that no colonies develop even with the undiluted sample, it must be filtered through a sterile membrane (0.45 μm porosity) in order to concentrate the bacteria. The membrane is then laid onto the surface of the nutritive agar. Colonies grow at the surface of the membrane in the same conditions as for diluted samples.

Leuconostoc strains frequently have the same habitats as other types of microorganisms. They can be separated on plates from moulds, yeasts and acetic acid bacteria. Anaerobiosis eliminates strictly aerobic microorganisms such as moulds, part of yeasts and

Table 3 Composition of several culture media for *Leuconostoc*

	MRS	ATB	TJB	104	Lafon–Lafourcade	DSMZ
Glucose	1.0	1.0	0.5		2.0	1.0
Fructose						0.5
Tryptone			0.2			
Peptone	1.0	1.0	0.5	0.5	1.5	1.0
Meat extract	0.8				1.0	
Yeast extract	0.5	0.5	0.5	0.5	0.5	0.5
K_2HPO_4	0.2					
$MgSO_4, 7H_2O$	0.2	0.2		0.005	0.02	0.2
$MnSO_4, 7H_2O$	0.005	0.005		0.005	0.005	0.005
Ammonium citrate	0.2					0.35
Citric acid				0.2		
DL-Malic acid			0.6			
Sodium acetate	0.5			0.5		
Tween 80	0.1		0.1	0.1		0.1
Tomato juice		25	40	25		10
pH	6.8	4.8	5.5	4.8	5.4	4.8

Values are given as percent (w/v) of each compound.
To improve reductive conditions 0.05% cysteine–HCl, H_2O sterilized by filtration may be added before use.
MRS, de Man Rogosa and Sharpe
ATB, acidic tomato broth
DSMZ, Deutsche Sammlung von Mikroorganismen und Zellkulturen GmbH
TJB, tomato juice broth
104, Dubois medium

all acetic acid bacteria. All yeasts are inhibited by addition of sorbate (0.2%) or more efficiently by pimaricine (0.1%). There are no selective conditions that will discriminate *Leuconostoc* species from other LAB. From food samples, all LAB are counted since the complex nutrient requirements and optimal conditions for growth are roughly comparable for *Lactobacillus*, *Pediococcus* and *Leuconostoc*. Tentative selective media have been described but none has as yet proven to be effective. Antibiotics have been tried. *Lactococcus* spp. and some homofermentative *Lactobacillus* are inhibited by vancomycin. However, resistance to this antibiotic seems to be a general feature of *Leuconostoc* spp. although *Pediococcus* and *Lactobacillus* have also been shown to be resistant. The vancomycin (30 mg ml^{-1}) may be added to the medium for *Leuconostoc* enrichment. No other variability in antibiotic resistance has allowed specific growth of *Leuconostoc* spp. Addition of 10% solution thallium acetate (1% v/v) has given satisfactory results for *Leuconostoc* isolation from plant materials and meat, but *Carnobacterium* at least also proved resistant.

Providing there are pronounced differences, other physiological features may lead to the detection of partial specificity. For example, the acidophilic nature of Œ. *œni*, compared to other *Leuconostoc* species allows growth at pH 4.8. However, adjustment to this pH does not prevent growth of *Pediococcus* and *Lactobacillus* species in acidic fruit juices, cider or wine. Indeed, in their natural environment, bacteria

are adapted to low pH and easily multiply even if acidophily is not recognized as a specific trait of the species. Similarly, the psychrophily of *Leuconostoc gelidium* and *L. carnosum*, spoilage bacteria from chilled meat, may be used. Overall, no selective *Leuconostoc* count is really possible.

The only means to specifically enumerate *Leuconostoc* associates colony counts and colony hybridization with suitable DNA probes. This method was first applied to the specific count of wine LAB. It is especially well adapted for Œ. *œni* and *Leuconostoc mesenteroides*, both present in grape juice and wines. DNA probes are prepared by labelling genomic DNA extracted from reference strains. The easier protocol is digoxygenin labelling (Boehringer Mannheim) applicable in any laboratory. After incubation of spread plates, colonies are transferred onto the nylon membrane. Both species (and possibly other LAB) can be identified on the same plate. After the first hybridization with the Œ. *œni* probe the membrane can be dehybridized, then re-hybridized with the *L. mesenteroides* probe. Each colony is identified at the species level. The method is obviously applicable to specific enumeration of the other *Leuconostoc* species provided that genomic DNA probes do not cross-hybridize with other species.

Rapid Methods Methods that are more rapid than plate counts are sometimes indispensable (**Table 4**). For example, sterility control after food production

Table 4 Rapid methods for enumeration of *Leuconostoc*

	Principle	Performance	Delay	Limit of detection (cells per ml)
Bioluminescence[a]	ATP measurement	Viable cells	15–20 min	10^4
	Enzymatic determination of intra-cellular ATP	No specificity		
Direct epifluorescent filter technique[b]	Fluorescence Acridine (RNA/DNA)	Viable (red) Dead cells (green) No specificity	30 min	10^5
	Fluorochrome (enzymatic activity of cell)	Viable No specificity	30 min	10^5
	Immunofluorescence	Viable + dead cells Species-specific	2 h	10^5
	In situ hybridization DNA probe	Viable + dead cells Species-specific	10 h	10^5

[a]Sample sterility can be checked after filtration, incubation of the membrane and ATP measurement.
[b]Sensitivity is easily increased if a picture analyser is coupled to the microscope.

and packaging or monitoring of the process requires rapid results. Bioluminescence is based on ATP measurement. By definition viable cells contain ATP and as soon as they die, the ATP reserve decreases to zero. Therefore, ATP measurement is similar to viability measurement. It consists of the enzymatic determination of ATP. The substrates are ATP, luciferin and O_2 and the products are $AMP + PP_i +$ luminescence which is measured by photometry. The enzyme is firefly luciferase. The protocol comprises two main steps: (1) determination of free and intracellular ATP after permeabilization of bacteria; (2) determination of free ATP after sterile filtration of the sample to retain bacteria. As a bacterial cell such as *Leuconostoc* contains 1 fg ATP, the results can be converted and expressed as viable cells per millilitre. However, the method is not specific and not adapted to the control of fermenting foods which often contain yeasts (100 fg ATP per cell) but, it can be adapted to sterility control and it is very useful in the control of pure cultures.

The direct epifluorescent filter technique (DEFT) is also very rapid. The most frequently used fluorochrome is acridine orange which intercalates into nucleic acids. Another non-fluorescent reagent is a precursor that enters the viable cell, is hydrolysed inside and releases a fluorochrome in the cell. This method is applicable to *Leuconostoc* species but it is limited by the sensitivity. The threshold level is 10^5 cells per millilitre, due to the small size of bacteria which implies large magnification. Its performance can be increased if a picture analyser is coupled to the microscope. However, it is not specific for bacterial species, but bacteria can be distinguished from yeasts. Variants of DEFT, immunofluorescence and *in situ* hybridization, are both rapid and specific. The first uses fluorescent-labelled antibodies that recognize the antigens on the surface of bacteria Anti-Œ. œni polyclonal sera were proved to be specific after exhaustion

with yeasts and other bacteria to detect Œ œni cells. The method was applied to the control of malolactic starters and wines. The second method uses digoxigenin (DIG)-labelled DNA probes that are detected by fluorescent anti-DIG-Fab fragments. Hybridization is directly performed on the slide where bacteria have been fixed with paraformaldehyde and permeabilized with lysozyme. The same whole-cell DNA probe as for colony hybridization was used and has been shown to directly detect Œ. œni in wine samples. These two latter methods determine dead and living cells.

Colony and *in situ* hybridization are the only methods which allow species-specific counting. On the other hand, polymerase chain reaction (PCR) amplification is suitable for detecting the presence of *Leuconostoc* even at a low population, provided specific primers are known. As examples, *Leuconostoc* in mixed dairy starter culture and histamine-producing *Œnococcus* have been detected using 16S rDNA oligonucleotide and sequences of the histidine-decarboxylase gene as primers, respectively.

Importance of *Leuconostoc* in the Food Industry

Leuconostoc strains are used in several industrial fermenting processes for the production of food and beverages, but they are highly undesirable in some particular industries. In some cases they can improve or decrease quality according to the strain and to the conditions.

Leuconostoc spp. as Fermenting Agents

Dairy Industry *Leuconostoc* species are used in the production of fermented milks, butter and cheese together with *Lactococcus*, *Streptococcus* and *Lactobacillus* spp. They have a poor acidifying power and are mainly chosen for their capacity to produce typical aroma compounds such as ethanal, acetoin and diacetyl. The balance between diacetyl, which is the most aromatic, and the other products is very dependent on the pH of the medium, temperature and redox potential, probably much more than on the strain itself. The sensory quality of fermented milk also depends on the viscosity. Slime results from the synthesis of polysaccharides. Besides other ropy strains, *L. mesenteroides* subsp. *mesenteroides* and *dextranicum* strains synthesize dextrans from saccharose. This inducible and unstable property must be controlled when preparing starters. Excessive ropiness may also depreciate the quality of yoghurts.

Acidified Vegetables Lactic acid-fermented fruit juices and vegetables are receiving increasing interest. More than 20 different fermented vegetables, mixtures of vegetables and fruit juices have been reported. Until now, only three have real commercial importance: sauerkraut, olives and cucumbers. Vegetables are fermented by LAB leading to foods naturally stabilized against spoilage microorganisms due to the dramatic decrease in pH and the production of inhibitors. Sauerkraut is an example of such foods. After washing, cabbage still retains Gram-negative and Gram-positive microflora together with yeasts and moulds. Here LAB comprising *Lactobacillus* spp., *Pediococcus* spp. and *Leuconostoc mesenteroides* are not numerous. Environmental factors such as temperature, anaerobiosis, pH and salt concentration are judiciously adjusted to direct the interactions and optimize LAB growth. After cabbage salting, the liquid phase formed by plasmolysis with the intracellular water-carrying vitamins and other growth factors, serves as nutrient medium and as an antioxygen. The aerobic microflora is discarded while the LAB multiply. Generally *Leuconostoc mesenteroides* initiates the fermentation, then it disappears gradually to be replaced by *Lactobacillus plantarum*. Of the LAB species associated with sauerkraut production, *Leuconostoc mesenteroides* is the most sensitive to decreasing pH and undissociated forms of lactic and acetic acids. In addition to acidification, *Leuconostoc mesenteroides* is also responsible for flavour compounds but also in some cases for spoilage by dextran production.

Olives can be prepared only by salting and pasteurization. However, many are traditionally stored in salt brine before processing. As for sauerkraut, potential spoilage microorganisms are eliminated by decreasing pH due to the growing population of LAB. During the initial phase *Leuconostoc*, *Pediococcus* and *Streptococcus* are present. Then *Leuconostoc* first predominates before the higher-acid tolerant *Lactobacillus* invades the medium. The fermentation of green cucumber is roughly the same as with olives. Cucumber are brined in tanks. LAB initially in low numbers quickly grow and compete with yeasts. *Leuconostoc mesenteroides* is most present in natural fermentation although the preferable starters are *Lactobacillus plantarum* and *P. damnosus*.

Wine Industry After grape berries have been crushed and directed into fermentation vats, the natural microflora (moulds, yeasts and bacteria) rapidly evolves as a result of alterations in environmental conditions. First, yeasts well-adapted to grow in acidic and very sweet media develop. Of the LAB population, *Lactobacillus* spp. are predominant, with *Leuconostoc mesenteroides* at a low concentration and Œ. *œni* almost undetectable. As alcoholic fermentation progresses, the LAB population is redistributed. *Leuconostoc mesenteroides* disappears followed by *Lactobacillus* and *Pediococcus* spp.; eventually by the end of alcoholic fermentation, the LAB population is roughly restricted to Œ. *œni*. The spontaneous selection of this species is explained by its best tolerance to ethanol, acidic pH and other inhibitors resulting from yeast metabolism. Œ. *œni* induces the second stage of winemaking, malolactic fermentation. The main factors that control the process are pH, temperature, ethanol content and sulphur dioxide. The most significant change is the transformation of L-malic acid to L-lactic acid and CO_2. Citric acid is metabolized and produces aroma compounds (diacetyl, acetoin, and others). When bacterial growth is slow, due to pH and temperature, more diacetyl and less acetate are formed. Many other alterations of wine composition occur and finally lead to more complex and improved sensory quality. Besides its effect on taste, malolactic fermentation increases the biological stability of wine because most of the nutrients available after alcoholic fermentation are used by Œ. *œni*. Malolactic fermentation is also encountered in cider-making, mostly with the same bacteria.

Leuconostoc spp as Spoilage Agents

The useful property of *Leuconostoc mesenteroides* subsp. *mesenteroides* or *dextranicum* to produce dextran in some cases, becomes a real spoilage factor in others. Sugar solutions appear to be adequate growth media for both subspecies. The biodeterioration of sugar cane includes souring and dextran formation. Both lead to loss of sucrose estimated as 4–9% of recoverable sugar. High viscosity also induces significant processing problems such as retardation of crystallization and reduced yields. Similar problems occur in the sugar beet industry. Contamination in refineries is propagated by the circulation of sweet water, and slime can be produced in preferential locations. In the rum industry, *Leuconostoc* also forms dextrans from milling to fermentation. Moreover, if their population is high enough, they inhibit yeasts and delay or even stop alcoholic fermentation.

In cured meat products, slime-forming *Leuconostoc* strains are also identified among the discolouring, off-odour and off-flavour producing LAB. In addition, ropy strains are isolated from vacuum-packed cooked meat products. *Leuconostoc amelibiosum* and *Leuconostoc mesenteroides* produce polysaccharides, acetic acid, acetoinic compounds and ethanol. A little different is the spoilage of wine and cider called 'piqûre lactique'. It is not the development of Œ *eni*, that is desirable for malolactic fermentation, but the time when it occurs which induces the spoilage. If Œ. *œni* grows before the end of alcoholic fermentation, it ferments sugar to lactic acid and above all to acetic acid which depreciates wine quality.

Production of biogenic amines is another concern in the fermented food industry. This is important for hygiene as these compounds have physiological functions in humans with possible disturbances. The main amines found are histamine, tyramine, diaminobutane (putrescine) and diaminopentane (cadaverine). Amino acid decarboxylase-positive strains are responsible for their production. As shown by studies in wine, this property is not linked to a species but to strains. The histidine-decarboxylase DNA probe prepared by labelling the corresponding gene has revealed that a significant percentage of Œ *œni* is likely to produce histamine. This was also the case for a collection of *Leuconostoc mesenteroides* from the dairy industry.

Leuconostoc Starters

Bacterial starter cultures are applied in the beverage and food industry to take advantage of their metabolism in the transformation process. *Leuconostoc* spp. participate in starters in combination with other LAB and other microorganisms in the dairy industry, the production of sauerkraut, wine and bread-making. If the raw material is sterilized as in the dairy industry, it is necessary to add a starter. However, most of the time, *Leuconostoc* starters used in fermented vegetables and wine must compete with the indigenous microflora. Such starters may be single-strain or multiple-strain cultures. Since the objective is to control fermentation to produce high quality products, the strains must be carefully selected. As a general rule, they are isolated from naturally fermenting foods. Selection itself involves criteria directly related to the product in terms of sensorial and hygienic quality and stability. They also include the capacity of the strain

to multiply easily on an industrial scale and its ability to be freeze-dried or dried. Finally, selected strains must adapt rapidly to technological conditions in order to outgrow indigenous microflora.

Until now the wine industry has probably been the field in which most experimentation on *Leuconostoc* (Œnococcus) starters has taken place. The first studies on malolactic starters began about 1970 since malolactic fermentation was recognized as indispensable but very unpredictable. Quality and economic constraints require that malolactic fermentation should be finished as soon as possible. Therefore winemakers currently use starters if malolactic fermentation has not started within 2–3 weeks after the end of alcoholic fermentation. Œ. *œni* strains are selected among isolates from wine, mainly on their malolactic activity, and their tolerance to harsh conditions (low pH, ethanol). Research on membrane composition has finally led to a better understanding of their adaptation to wine and to the preparation of malolactic starters for direct inoculation. From now on, they will be more frequently used. Therefore new selection criteria have been added to the basic ones, such as influence on wine aroma components and inability to produce biogenic amines.

Leuconostoc Strains as Food Preservatives

Like other LAB, *Leuconostoc* strains preserve food by producing antagonistic compounds, or when competing with the indigenous microflora by exhausting most of the available nutrients. They exhibit antagonistic activities against close-related bacteria and potential pathogenic microorganisms. Food can be preserved either by adding such bacteria that inhibit undesirable microorganisms when growing or by adding purified inhibitory products such as bacteriocins. Bacteriocinogenic strains and their bacteriocins including *Leuconostoc* strains are receiving much attention. Such strains have so far been isolated from a number of habitats such as meat, fish and dairy products. They usually have a narrow inhibition spectrum and some are active against *Listeria monocytogenes*.

In chilled beef stored under vacuum, off-flavours and discoloration by *Lactobacillus sake* and *Carnobacterium maltaromicus* are prevented by seeding with an antagonistic strain of *Leuconostoc gelidium*. This strain produces the bacteriocin leucocin A. Similarly two strains of *Leuconostoc carnosum* and *Leuconostoc mesenteroides* subsp. *dextranicum* isolated from meat produce bacteriocin active against LAB and *Listeria*. The bacteriocin-coding genes are homologous to the corresponding N-terminal coding region of leucocin A. Moreover, mesentericin (Y 105) from *Leuconostoc mesenteroides*, although from a different source, differs from leucocin A by only two amino acids and inhibits *Lactobacillus*, *Carnobacterium* and *Listeria*. This suggests that bacteriocins close to leucocin A may occur in several other *Leuconostoc* species. Basic research has already led to the characterization of the structural and putative immunity genes of the bacteriocin of *Leuconostoc* spp. Selection of *Leuconostoc* strains for some of their various applications in the food industry may also include their potential antagonistic properties. Co-inoculation with such inhibitory *Leuconostoc* strains (bacteriocinogenic or not) is the future pathway for obtaining the natural microbial stabilization of fermented beverages and foods.

See also: **ATP Bioluminescence**: Application in Dairy Industry; Application in Beverage Microbiology. **Bacteria**: Classification of the Bacteria – Traditional; Classification of the Bacteria – Phylogenetic Approach. **Bacteriocins**: Potential in Food Preservation. **Biochemical and Modern Identification Techniques**: Microfloras of Fermented Foods. **Cheese**: Role of Specific Groups of Bacteria. **Cider (Hard Cider). Direct Epifluorescent Filter Techniques (DEFT). Fermented Foods**: Origins and Applications; Fermented Vegetable Products. **Fermented Milks**: Range of Products. **Meat and Poultry**: Spoilage of Meat. **Nucleic Acid-based Assays**: Overview. **Spoilage Problems**: Problems caused by Bacteria. **Starter Cultures**: Uses in the Food Industry; Importance of Selected Genera. **Total Counts**: Microscopy. **Total Viable Counts**: Pour Plate Technique; Spread Plate Technique. **Wines**: The Malolactic Fermentation.

Further Reading

Daeschel MA, Andersson RE and Fleming HP (1987) Microbial ecology of fermenting plant materials. *FEMS Microbiol. Rev.* 46 357–367.

Damelin LH, Dykes A and von Holy A (1995) Biodiversity of lactic acid bacteria from food-related ecosystems. *Microbios* 83: 13–22.

De Roissart H and Luquet FM (1991) *Bactéries Lactiques*. Uriage, France: Lorica.

Garvie EI (1986) Genus *Leuconostoc* In: Sneath PHA, Mair NS, Sharpe ME and Holt JC (eds) *Bergey's Manual of Systematic Bacteriology*. P. 1071–1075. Baltimore: Williams and Wilkins.

Goodfellow M and O'Donnell AG (1993) *Handbook of New Bacteral Systematics*. London: Academic Press.

Holzapfel WH and Schillinger U (1992) The genus *Leuconostoc*. In: Balows A, Trüper HG, Dworkin M, Harder W and Schleifer KH (eds) *The Procaryotes*, 2nd edn. Vol. II, p. 1508. New York: Springer Verlag.

Stiles ME and Holzapfel WH (1997) Lactic acid bacteria for foods and their current taxonomy. Review article. *Int. J. Food Microbiol.* 36: 1–29.

Vandamme P, Pot B, Gillis M et al (1996) Polyphasic taxonomy, a consensus approach to bacterial systematics. *Microbiol. Rev.* 60: 407–438.

Yang D and Woese CR (1989) Phylogenetic structure of the 'Leuconostocs': an interesting case of a rapidly evolving organism. *System. Appl. Microbiol.* 12: 145–149.

Light Microscopy *see* **Microscopy**: Light microscopy.

Lipid Metabolism *see* **Metabolic Pathways**: Lipid metabolism.

LISTERIA

Contents

Introduction
Detection by Classical Cultural Techniques
Detection by Commercial Enzyme Immunoassays
Detection by Colorimetric DNA Hybridization
Detection by Commercial Immunomagnetic Particle-based Assays
Listeria monocytogenes
Listeria monocytogenes **– Detection by Chemiluminescent DNA Hybridization**
Listeria monocytogenes **– Detection using NASBA (an Isothermal Nucleic Acid Amplification System)**

Introduction

Carl A Batt, Department of Food Science, Cornell University, USA

Characteristics of the Species

Listeria is a Gram-positive rod that is typically of 0.5–2 µm in length. It is non-spore-forming, and is not encapsulated. *Listeria* can appear coccoid and motile depending upon the growth temperature. They have an optimum growth temperature of 30–37°C, and some species, most notably *L. monocytogenes* can grow at temperatures as low as 4°C. As such these species are a particular food-borne hazard because of their ability to replicate, albeit slowly, at refrigerated temperatures. At 20–25°C, they form flagella (and other antigens as well as virulence factors) and are therefore motile, whereas at 37°C they are not. *Listeria* is a facultative anaerobe and grows vigorously on a variety of complex media.

The genus *Listeria* is characterized by its catalase activity, its lack of hydrogen sulphide production and its production of acid from glucose. It has a positive methyl red reaction and a positive Vogues–Proskauer reaction. It does not produce indole, utilize citrate or possess urease activity. At one time, there was only a single species, *L. monocytogenes* in the genus *Listeria*. Subsequently, *L. denitrificans*, *L. grayi*, *L. murrayi*, *L. innocua*, *L. ivanovii*, *L. welshimeri* and finally *L. seeligeri* were added. *L. denitrificans* was subsequently reclassified as *Jonesia denitrificans*. Finally, it has been suggested based on rRNA sequences, that *L. murrayi* and *L. grayi* are a single species. Multilocus enzyme electrophoresis reveals that *L. monocytogenes*, *L. ivanovii*, *L. welshimeri* and *L. seeligeri* all form distinct clusters with no overlap. 16S rRNA sequences help to form two groups; one consists of *L. grayi* and the other consists of *L. monocytogenes*, *L. ivanovii*, *L. innocua*, *L. welshimeri* and *L. seeligeri*. From this latter group a further division which clusters *L. monocytogenes* and *L. innocua* appears distinct from *L. ivanovii*, *L. seeligeri* and *L. welshimeri*. This next stage of distinction is curious as only *L. monocytogenes* and to a lesser extent *L. ivanovii* are considered to be virulent. Among the various *Listeria* species, the most studied is *L. monocytogenes*. *L. monocytogenes* is covered in detail elsewhere. Among the other *Listeria* species, none are considered to be highly virulent and apart from *L. monocytogenes* only *L. ivanovii* has been associated with disease in animals. There are rare reports of human disease caused by *L. ivanovii* but these may be compromised by difficulty in accurately identifying the organism to the species level.

Virulence in *Listeria* is mediated by a number of factors, some of which are unique to *L. monocytogenes* whereas a number are also shared by the non-*L. monocytogenes* species including *L. ivanovii* and *L. welshimeri*. **Table 1** presents a list of a selected group of virulence factors. Most of these virulence genes including *prfA*, *plcA*, *hlyA* and *actaA* are clustered into a single operon.

Table 1 Selected virulence genes found in *Listeria*

Protein	Gene(s)	Comments
PrfA	*prfA*	Regulatory protein for operon
PI-PLC	*plcA*	Phospholipase
LLO	*hylA*	Haemolysis
ActA	*actA*	Actin polymerization
InlA	*InlA*	Internalin needed for cell entry

Listeria may be an indictor of *L. monocytogenes*. Surveys of foods, processing plants and other environments document that non-*L. monocytogenes* are often found in samples that contain *L. monocytogenes*. For example, whereas a total of 12.5% of fresh chicken wings tested positive for *L. monocytogenes* over 42% tested positive for all *Listeria* species. Although the ecology of *Listeria* species is not completely understood nor is the overlap in the ecology of non-*L. monocytogenes* vs. *L. monocytogenes* known, surveys for all *Listeria* might be a useful indicator of the presence of *L. monocytogenes*.

Methods of Detection

Listeria was difficult to isolate and because of its emergence during the early 1980s, it was the target of a large number of efforts to develop both cultural and rapid methods for its detection. Although initial efforts have focused on the isolation of *Listeria* and its presumed utility as an indicator for the human pathogen *L. monocytogenes*, the focus has now shifted to detection of *L. monocytogenes*. Generic *Listeria* detection is still valuable since its absence is a strong presumptive negative for *L. monocytogenes*.

Early efforts focused on its ability to grow at refrigerated temperatures, and the use of cold enrichments was part of the isolation practice although it could take up to a few months to complete the procedure. A number of enrichment and recovery methods have been developed and each is specific for a given food product (**Table 2**). Sublethal injury of *Listeria* is a significant issue and injury can occur as a result of processing and/or acid stress in fermented foods.

There are a number of enrichment media for *Lis-*

Table 2 Enrichment and selective plating media used for the detection of *Listeria*

Enrichment	Selective plating
Listeria enrichment broth (LEB)	Modified McBride agar
Buffered *Listeria* enrichment broth (BLEB)	Lithium chloride phenylethanol moxalactam (LPM)
University of Vermont medium (UVM)	Modified Oxford agar
Fraser broth	Polymyxin acriflavin lithium chloride
	Ceftazidime aesculin mannitol (PALCAM)

teria most of which are based around the inclusion of two or more antibiotics (i.e. acriflavin, polymyxin, ceftazidime). These enrichments are typically carried out for 24 h after which a second round of enrichment may be used. The type of food being tested is a factor in the choice of a protocol for isolation including media and growth conditions. For example, *L. monocytogenes* in foods that are dried during processing would be injured and a pre-enrichment is buffered peptone water might be recommended. To help with recovery, antibiotics are sometimes withheld during a period of incubation to allow the injured cells to recover. Secondary enrichments are recommended for all foods except dairy products. The use of a secondary enrichment depends on the endogenous microflora and the anticipated levels of *Listeria* in the product.

After selective enrichment, the next stage in culture-based detection of *Listeria* typically employs plating onto a selective media. The most common ingredient is again an antibiotic to promote the selective growth of *Listeria* (e.g. lithium chloride) and aesculin to detect hydrolysis activity. The plates are incubated for 24–48 h and on medium containing aesculin, the colonies appear black, although they are actually clear and the black aesculin hydrolysis product is seen beneath the colony.

A number of *Listeria* species appear similar on selective media, i.e. PALCAM. Although this result can be taken as a presumptive positive, further confirmation is needed to speciate the organism and to confirm its identity. Different schemes for confirmation have been established which use a number of morphological and biochemical tests. One of the most widely accepted tests is the CAMP (named for the inventors, Christie, Atkins and Munch-Peterson) that involves cross-streaking the test organism on a blood agar plate perpendicular to streaks of *Staphylococcus aureus* and *Rhodococcus equi*. Although *L. ivanovii* and to a lesser extent *L. monocytogenes* and *L. seeligeri* are all β-haemolytic, their haemolysis patterns are greatly altered by *S. aureus* and *R. equi*. Therefore, a CAMP result is presented in terms of the haemolytic reaction scored around the intersection of these cross-streaks. Since haemolytic activity in *L. monocytogenes* is relatively weak, difficulties in scoring CAMP have been reported. Specific tests to distinguish between the various *Listeria* species are shown in **Table 3**.

A number of biochemical tests can be used to speciate *Listeria* and acid production from carbohydrates including xylose, L-rhamnose, mannitol, α-methyl-D-mannoside and soluble starch can help distinguish between the species. These and others have been incorporated into a variety of 'miniaturized' identification tests that include API *Listeria*, API Coryne, API 50CH and Mast-ID.

In addition to the biochemical tests a number of typing methods have been established that allow species and subspecies discrimination. Serotyping reveals at least 16 different serovars some of which overlap more than one species. The serotyping is dependent on variation in the O- (somatic) and H- (flagellar) antigenic factors. The *L. monocytogenes/L. seeligeri* group has the most serovars, with a total of 13 among these two species.

In addition to serotyping, the *Listeria* species can be typed by phage, although not all strains are typable with the current array of phages. Multilocus enzyme electrophoresis (MEE) is based on polymorphisms in the electrophoretic mobility of 10–25 different enzymes which can be histochemically stained. There are far more MEE types than serovars and in one sampling a total of 82 distinct types were found among 390 isolates. At least three nucleic acid-based methods have been established, random amplified polymorphic DNA (RAPD), pulsed-field gel electrophoresis (PFGE) and ribotyping. The last is based on sequence polymorphisms that can be scored by probing the rRNA operon. To date, well over 50 different ribotypes have been discovered in the *L. monocytogenes* species using the restriction enzyme *Eco*RI. RAPDs can potentially be more discriminating; they are based on the patterns resolved on polymerase chain reaction (PCR) amplification with a small oligonucleotide primer which has a 'random' sequence. However, RAPDs are notorious for their lack of reproducibility and transferability of the protocols between laboratories. PFGE can be effective and has been used to subtype a single serovar.

A number of commercial suppliers have produced rapid methods to detect either *Listeria* or more specifically *L. monocytogenes*. For the most part these assays are extensions of formats used to detect other microorganisms (or their toxins) but their components have been altered to specifically detect *Lis-*

Table 3 Speciation tests for *Listeria*

Test	L. grayii	L. innocua	L. ivanovii	L. mono-cytogenes	L. murrayi	L. seeligeri	L. welshimeri
β-Haemolysis	−	−	+	+	−	+	−
CAMP (*S. aureus*)	−	−	−	+	−	+	−
CAMP (*R. equi*)	−	−	+	−	−	−	−
Acid production from							
Mannitol	+	−	−	−	+	−	−
α-Methyl-D-mannoside	ND	+	−	−	ND	−	+
L-Rhamnose	−	+/−	−	+	+/−	−	+/−
Soluble starch	+	−	−	−	+	+/−	+/−
D-Xylose	−	−	+	−	−	+	+
Hippurate hydrolysis	−	+	+	+	−	+/−	+/−
Nitrate reduction	−	−	−	−	+	+/−	+/−
Mouse lethality	−	−	+	+	−	−	−

ND, not determined

teria. Assays based on nucleic acid hybridization or antibody–antigen interactions are available as well as one that employs nucleic acid amplification. For all of these assays some prior enrichment is necessary to selectively increase the target population to a level at which they can be detected. At best, these assays can be completed in 24 h although this depends on the food source and obviously the intended level of sensitivity. In general these assays attempt to integrate into a standard *Listeria* detection method. After an initial one or two rounds of culture enrichment and selection, the culture is then subjected to the rapid assay. Any positive samples can then be carried through the standard microbiological detection for eventual confirmation. This approach is critical especially in advance of regulatory acceptance of a rapid method. Therefore these rapid methods can be used as a screening tool to quickly assay a large number of samples.

Enzyme-linked immunosorbent assays (ELISAs) for *Listeria* have been developed by Organon-Teknika. This ELISA is formatted for a 96-well microtitre plate and the readout is colorimetric. To obviate the liquid handling normally associated with microtitre plates, Tecra has developed an antibody-coated dipstick and incorporated this into their Unique system. After an initial capture step, an interim culture replication step helps to increase cell number. The readout is colorimetric.

In general most immunological assays for *Listeria* have not been successfully further specified for *L. monocytogenes*. Despite apparent successes with the development of antibodies specific for *L. monocytogenes*, incorporation of these into commercial assay formats has not followed. A single exception is the development of an immunomagnetic bead-colony immunoassay by Vicam.

A nucleic acid hybridization assay is available from GeneTrak which is based on their sandwich capture format. Two assays have been developed, one which detects the genus *Listeria* in general and the other which is specific for *L. monocytogenes*. Both assays target 16S rRNA and have a colorimetric readout. In each assay a capture probe binds to a region of the 16S rRNA and a second probe binds to a spatially separated region of the 16S rRNA molecule. The complex is then removed from solution using a poly(A) tail which hybridizes to a poly(T)-coated solid support.

Finally a PCR-based amplification method for the detection of *L. monocytogenes* has been developed by Qualicon (a subsidiary of DuPont). The nature of the amplicon has not been detailed but it is reported to be specific for *L. monocytogenes*. All of the assay components are contained in a single tablet which is added to the sample after processing. This assay requires gel electrophoresis to confirm the presence of the appropriate amplification product.

Regulations

With the advent of methodology that can detect *L. monocytogenes* as compared to just *Listeria* species, most regulations are moving toward the former as a target for establishing numerical thresholds. In many countries including the United States, Australia and New Zealand, a zero-tolerance has been established for ready-to-eat foods. In Canada and Europe, quantitative limits greater than zero (< 100) have been established and these are further specified depending on the food product and in effect the targeted consuming population. Therefore, foods destined for consumption by infants and other people who might be more susceptible to listeriosis typically have a more stringent specification. A second factor is the anticipated storage life of the food. Foods that are highly perishable even under refrigerated storage have less stringent specifications since the opportunity for *Listeria* to multiply is shorter.

Importance to the Food Industry

Contamination of food products with *L. monocytogenes* can lead to a recall of the product even in advance of any reported food-borne illness. It can be endemic in a processing plant and reservoirs have been shown to be the source of continuous contamination of the product even in plants that have a 'good' hygiene plan in place. Floor drains are notorious as reservoirs for *Listeria* and the need to keep production areas dry is essential. Since *Listeria* can be found in a wide variety of food products, most food industries need to be diligent in their process as well as in their plant design and operation. As with other food pathogens the design and implementation of a hazard analysis critical control point (HACCP) can be the first line of defence in preventing *Listeria* contamination of a food product. Within an HACCP and a good manufacturing program (GMP) are all the elements to significantly reduce the problems associated with *Listeria*. The standard concerns of separation of raw and finished product, the use of effective cleaning and sanitation procedures and the maintenance of a dry environment are all effective at reducing contamination by *Listeria*. Effective cleaning of processing equipment is essential, and the recent concern for detecting *Listeria* reservoirs in the plant environment has resulted in increased attention to biofilm formation. Factors that promote the formation of biofilms and their recalcitrance to removal by standard cleaning have been studied. These factors and methods to reduce biofilm formation may have a positive impact on decreasing environmental loads of *Listeria*. Collectively therefore, attention to these matters by the food processor should lead to reduced incidence of *Listeria*-contaminated food products.

Importance to the Consumer

Listeriosis in humans can take any one of three clinical forms: encephalitis, septicaemia and abortion. Symptoms can range from flu-like to more substantial clinical manifestations. For the most part, only *L. monocytogenes* appears to be pathogenic in humans, whereas other species, including *L. ivanovii*, have been reported to cause ovine abortions and retarded growth in lambs. Listeriosis appears to have a higher mortality rate than other food-borne bacterial pathogens. The mortality rate has been reported to be 20% but this may be misleading as probably few cases of listeriosis that do not progress past the flu-like symptoms are diagnosed correctly. A sensitive population including the young, the elderly and persons with compromised immune systems need to be diligent in not only avoiding foods that may contain *Listeria* but also in seeking medical attention when flu-like symptoms do not quickly dissipate. Persons with acquired immunodeficiency syndrome (AIDS) are among the sensitive population and fetal abortions are a prevalent outcome in both outbreak and sporadic cases of listeriosis.

There are a number of well-documented outbreaks of listeriosis in humans, all of which involve *L. monocytogenes*. The most common serotype of *L. monocytogenes* associated with food-borne illness is type 4B. A variety of foods have been implicated including, animal-, dairy- and vegetable-based products. The most common food is soft cheeses, which, due to the nature of their manufacture, are particularly susceptible to contamination by *L. monocytogenes*. Proper in-home preparation of foods especially hot dogs can help to reduce the likelihood of contracting food-borne listeriosis. In 1998, a large outbreak of listeriosis that resulted in a number of deaths and fetal abortions appeared to involve hot dogs and perhaps other ready-to-eat meat products. No conclusive source of contamination has been identified to date.

See also: **Enzyme Immunoassays**: Overview. **Food Poisoning Outbreaks. Good Manufacturing Practice. Hazard Appraisal (HACCP)**: The Overall Concept. *Listeria*: Detection by Classical Cultural Techniques; Detection by Commercial Enzyme Immunoassays; Detection by Colorimetric DNA Hybridization; Detection by Commercial Immunomagnetic Particle-based Assays; *Listeria monocytogenes*; *Listeria monocytogenes* – Detection by Chemiluminescent DNA Hybridization; *Listeria monocytogenes* – Detection using NASBA (an Isothermic Nucleic Acid Amplification System). **Process Hygiene**: Hygiene in the Catering Industry. **Spoilage Problems**: Problems caused by Bacteria.

Further Reading

Farber JM and Peterkin PI (1991) *Listeria monocytogenes, a food-borne pathogen. Microbiological Reviews* 55: 476–511.

Reyser ET and Marth EH (1999) *Listeria, Listeriosis and Food Safety.* New York: Marcel Dekker.

Sutherland PS and Porritt RJ (1997) *Listeria monocytogenes.* In: Hocking AD (ed.) *Foodborne Microorganisms of Public Health Significance.* Sydney: Australian Institute of Food Science and Technology Inc. NSW Branch, Food Microbiology Group.

Detection by Classical Cultural Techniques

G D W Curtis, Bacteriology Department, John Radcliffe Hospital, Oxford, UK

Introduction

Until the rapid increase of interest in listeria in the mid-1980s food microbiologists had little choice but to use culture media and methods for listeria isolation developed by clinical bacteriologists. Since most clinical isolates were made from normally sterile sites or after prolonged cold enrichment these media were of limited selectivity. In recent years many formulations of selective media designed for food microbiology have appeared which incorporate antibiotics, dyes and other inhibitors (see **Tables 1** and **2**). Nutrient requirements of listeria are simple and can be provided by any basic nutrient broth or agar although most formulae include yeast extract. More recently sodium pyruvate has been used to supplement some enrichment broths to aid the resuscitation of sublethally injured organisms.

Since the minimum infective dose of *Listeria monocytogenes* has not been established, some authorities have adopted a zero tolerance for this organism in ready-to-eat foods. In consequence, most methods of examination relate to presence or absence in a defined quantity of the product, usually 25 g. Thus the methods developed have followed the sequence of pre-enrichment, selective enrichment and selective plating established as the norm for pathogens such as salmonellae. Methods for enumeration exist but are generally less well developed.

Culture Media for Detection

In 1989 the Working Party on Culture Media of the International Committee on Food Microbiology and Hygiene published monographs on 16 media for listeria. In the following sections only those media which have proved of ongoing value either by general consensus or by their incorporation into regulatory or standard methods are described. Some commercial sources of dehydrated media are listed in **Table 2**.

Some manufacturers and regulatory bodies refer to media by descriptive names, e.g. Listeria enrichment broth. Such names are confusing in that they describe a number of media which fulfil that function as well as the specific medium which it is the intention to describe, in this case the enrichment broth (M52) recommended by the Food and Drug Administration (FDA). With this medium the problem is further confounded by changes in specification of the broth in each successive edition from the supplement to the sixth edition of the *Bacteriological Analytical Manual* (1987) to the eighth edition (1995). Throughout this chapter reference is made to the original formulation of 1987 as FDA (Lovett) and the current as FDA-1995.

Pre-enrichment Broths

Buffered Peptone Water (BPW), 'Universal' Enrichment Broth, University of Vermont (UVM) I, Half Fraser Broth, FDA-1995 Enrichment Broth (FDA-1995), Listeria Repair Broth (LRB) These fall into two categories: those without any selective agent and those where the selective agents have been reduced in quantity or are added after a preliminary period of incubation. BPW and 'Universal' pre-enrichment broth have the advantage that the same pre-enrichment can be used for subsequent examination for salmonellae as well as listeria. Both media are low in carbohydrates and buffered to prevent a rapid drop in pH which might further injure sublethally damaged cells.

Selective enrichment broths modified for pre-enrichment contain reduced amounts of the DNA intercalating agent acriflavine which inhibits the RNA synthesis necessary for the resuscitation of injured cells (e.g. UVM I). Other selective agents may also be reduced in concentration (e.g. Half Fraser). Half Fraser broth also contains aesculin and ferric iron which may provide presumptive evidence of the presence of listeria by the blackening of the broth after incubation but this cannot be relied upon as both false positive and false negative results may occur. FDA-1995 is prepared without selective ingredients and sodium pyruvate is added before inoculation. The selective ingredients are then added after incubation for four hours. The oxygen-scavenging pyruvate has been shown to have a beneficial effect on the repair of heat-injured cells but the additional manipulations required and the four hour delay before adding the selective supplement may restrict the usefulness of this broth. The addition of phosphate buffers to the original formulation of FDA broth has brought it into line with the University of Vermont broths in this respect. Similar to the FDA-1995 broth is the specially formulated Listeria repair broth (LRB) which is buffered with MOPS (3-morpholinopropanesulphonic acid) instead of phosphates and contains glucose, magnesium sulphate, ferrous sulphate and sodium pyruvate, all of which have been shown to enhance the recovery of heat-injured listeria. As with FDA-1995 selective agents are added after a preliminary incubation period, in this case five hours.

Table 1 Constituents of selective enrichment broths (g l^{-1} unless otherwise stated)

	Nutrient base		Acriflavine	Nalidixic acid	NaCl	Buffer		Indicators		Other	
FDA (Lovett)	Tryptone	17.0	0.015	0.04	5.0	K$_2$HPO$_4$	2.5			Cycloheximide	0.05
	Yeast extract	6.0									
	Soya peptone	3.0									
	Glucose	2.5									
FDA-1995	Tryptone	17.0	0.01	0.04	5.0	K$_2$HPO$_4$	2.5			Cycloheximide	0.05
	Yeast extract	6.0				KH$_2$PO$_4$	1.35			Sodium pyruvate[a]	1.0
	Soya peptone	3.0				Na$_2$HPO$_4$	9.6				
	Glucose	2.5									
Half Fraser	Meat peptone	5.0	0.0125	0.01	20.0	KH$_2$PO$_4$	1.35	Aesculin	1.0	LiCl	3.0
	Tryptone	5.0				Na$_2$HPO$_4$	9.6	Iron (III) ammonium citrate	0.5		
	Beef extract	5.0									
	Yeast extract	5.0									
Fraser	Meat peptone	5.0	0.025	0.02	20.0	KH$_2$PO$_4$	1.35	Aesculin	1.0	LiCl	3.0
	Tryptone	5.0				Na$_2$HPO$_4$	9.6	Iron (III) ammonium citrate	0.5		
	Beef extract	5.0									
	Yeast extract	5.0									
LRB	Tryptone	17.0	0.015	0.04	5.0	K$_2$HPO$_4$	2.5			Cycloheximide	0.05
	Yeast extract	6.0				MOPS (free acid)	8.5			MgSO$_4$	4.94
										Sodium pyruvate	10.0
	Soya peptone	3.0				MOPS (Na salt)	13.7			FeSO$_4$	0.3
	Glucose	7.5									
L-PALCAMY	Special peptone	23.0	0.005	nil	5.0	nil		Mannitol	5.0	LiCl	10.0
	Yeast extract	5.0						Phenol red	0.08	Polymyxin B	0.01
	Beef extract	5.0						Aesculin	0.8	Ceftazidime	0.03
	Peptonized milk	5.0						Iron (III) ammonium citrate	0.5		
								Egg yolk emulsion	25 ml		
UVM I & II	Proteose peptone	5.0	0.012 (I)	0.02	20.0	KH$_2$PO$_4$	1.35			Aesculin	1.0
	Tryptone	5.0	or			Na$_2$HPO$_4$	9.6				
	Yeast extract	5.0	0.025 (II)								
	Beef extract	5.0									

[a]Optional ingredient.

In summary, buffering is essential for pre-enrichment broths and selective agents should be absent or reduced in concentration. Pyruvate and divalent cations may have a beneficial effect in creating conditions favourable to repair of injured organisms. All pre-enrichment broths should be incubated at 30°C and subcultured after 18–24 h.

Comparisons of the performance of pre-enrichment broths have failed to show significant differences, many authors concluding that the use of more than one primary enrichment is necessary to achieve optimal sensitivity. The use of a broth specifically supplemented to resuscitate injured cells together with one of greater selectivity may be the best choice; the one allows repair of sublethally injured listeria whereas the other reduces interference from non-target organisms.

Selective Enrichment Broths

UVM II, Fraser Broth, FDA (Lovett) Listeria Enrichment Broth, FDA-1995 Enrichment Broth (FDA-1995), Polymyxin Acriflavine Lithium Chloride

Table 2 Code numbers of dehydrated media available from some commercial sources. This list is not exhaustive since media manufacturers' ranges are updated in accordance with perceived demands for new media. Some formulations may not correspond exactly with the descriptions which are sometimes imprecise, e.g. 'listeria enrichment broth' (see text)

	B-D	Lab M	Difco	Merck	Oxoid
Broths					
Buffered Peptone Water	4312367	LAB 046	1810	7228	CM 509
FDA-1995	N/A	LAB 139	N/A	N/A	N/A
FDA (Lovett)	4312333	LAB 138	0222	10549	CM 862 + SR 141
Half Fraser	N/A	LAB 164 + X 164	0653 + 0211	10398 + 10399	N/A
Fraser	4312395 + 4312401	LAB 164 + X 165	0219 + 0211	10398 + 10399	CM 895 + SR 156
Listeria recovery broth	N/A	N/A	N/A	N/A	N/A
L-PALCAMY	N/A	LAB 144 + X 144	N/A	10823 + 12122	N/A
Universal enrichment broth	N/A	N/A	0235	N/A	N/A
UVM I	4312348	LAB 155 + X 155	0223	10824	CM 863 + SR 142
UVM II	N/A	LAB 155 + X 156	N/A	10824 + 4039	CM 863 + SR 143
Agars					
EHA (Cox)	N/A	N/A	N/A	N/A	N/A
Heisick 1	N/A	N/A	N/A	N/A	N/A
Heisick 2	N/A	N/A	N/A	N/A	N/A
LPM	4312336 + 4312402	N/A	0221 + 0216	N/A	N/A
MOX	4312397 + 4312402	N/A	0225 + 0218	N/A	CM 856 + SR 157
Oxford	N/A	LAB 122 + X 122	0225 + 0214	7004 + 7006	CM 856 + SR 140
PALCAM	N/A	LAB 148 + X 144	N/A	11755 + 12122	CM 877 + SR 150

Table 3 Constituents of diagnostic selective agars (g l^{-1} unless otherwise stated)

	Nutrient base		Acriflavine	LiCl	Polymyxin (P) or Colistin (C)	Ceftaz- idime	Indicators		Other	
EHA	Columbia	39.0	0.005	10.0	C	0.02	nil	MUG[a]	0.05	Sphingomyelinase	10 U
								Sheep blood	50 ml	Cycloheximide	0.4
										Cefotetan	0.002
										Fosfomycin	0.01
Heisick 1	Columbia-EH	39.0	0.005	7.5	C	0.02	nil	MUG	0.05	Sphingomyelinase	10 U
								Sheep blood	50 ml	Cycloheximide	0.4
										Cefotetan	0.002
										Fosfomycin	0.01
Heisick 2	Columbia-EH	39.0	0.01	7.5	P	0.01	0.03	MUG	0.05	Sphingomyelinase	10 U
								Sheep blood	50 ml	Amphotericin B	0.0025
LPM	Tryptone	5.0	nil	5.0	nil		nil	nil		NaCl	5.0
	Peptone	5.0								Glycine anhydride	10.0
	Beef extract	3.0								2-Phenylethanol	2.5
										Moxalactam	0.02
MOX	Columbia	39.0	nil	15.0	C	0.01	nil	Aesculin	1.0	Moxalactam	0.015
								Iron (III) ammonium citrate	0.5		
Oxford	Columbia	39.0	0.005	15.0	C	0.02	nil	Aesculin	1.0	Cycloheximide	0.4
								Iron (III) ammonium citrate	0.5	Cefotetan	0.002
										Fosfomycin	0.01
PALCAM	Columbia	39.0	0.005	15.0	P	0.01	0.02	Aesculin	0.8	Glucose	0.5
								Iron (III) ammonium citrate	0.5		
								Mannitol	10.0		
								Phenol red	0.08		

[a]4-Methylumbelliferyl β-D-glucoside.

Aesculin Mannitol Egg Yolk broth (L-PAL-CAMY) There are fewer variations in secondary enrichment media than in those for pre-enrichment. Since resuscitation has either been accomplished or is not an issue it is sufficient to supplement a basal broth with selective agents and buffers with or without an indicator system. Concentrations of inhibitors, particularly acriflavine, can be greater than in pre-enrichment broths and, whereas primary enrichment requires the examination of 25 g quantities of sample and hence 225 ml of broth to provide a 1/10 dilution, quantities for secondary enrichment are usually much less. Most authorities recommend the addition of 1 or 10 ml of primary enrichment to 9 or 90 ml, respectively, of secondary enrichment.

UVM II differs from UVM I only in containing twice the amount of acriflavine. Fraser broth is a modification of UVM II containing in addition lithium chloride and ferric ammonium citrate to provide extra selectivity and an indicator system. As with Half Fraser broth, the indicator system cannot be relied upon. Small inocula of the target organism may fail to produce blackening of the medium and some unwanted flora, notably *Enterococcus* spp., may grow and produce a false positive result in the absence of *Listeria* spp. One estimate of the predictive value of a positive test with Fraser broth was a disappointing 35% and the modification is best viewed as simply rendering the broth more selective. Lovett's listeria enrichment broth was originally adopted by the FDA and later modified to allow the incorporation of differing amounts of acriflavine for non-dairy foods (15 mg l^{-1}) and milk and dairy products (10 mg l^{-1}). It has since been further adapted to provide a combined repair and selective enrichment broth as detailed above (FDA-1995). L-PALCAMY is the most complex of selective enrichment broths. It contains ceftazidime in addition to lithium chloride and acriflavine. Two indicator systems are also incorporated, aesculin and ferric iron and mannitol and phenol red though, as with Fraser broth, the value of these is doubtful and all 'negative' broths should be subcultured.

Selective enrichment broths require subculturing to diagnostic selective agar for the presumptive identification of *Listeria* spp. and most authorities agree that this should be done after incubation at the chosen temperature for 24 and 48 h. Some workers claim better results if incubation is prolonged with subculture at 7 days but this is not generally thought to be necessary. Incubation temperatures of 30, 35 and 37°C have all been recommended.

A particular problem with enrichment broths used for the isolation of listeria is that the non-pathogenic *L. innocua* has a shorter generation time than *L. monocytogenes* and this effect is exaggerated in media containing acriflavine. The effect is lessened by protein contained in the food sample which binds to acriflavine and renders it less active. This binding increases as the pH drops below 5.8, being maximal at values < 4.8. At this pH, however, growth of *L. monocytogenes* is affected. This problem has not yet been overcome, although use of a selective agar that differentiates *L. monocytogenes* from *L. innocua* may increase the overall chance of success in isolating the former from mixed cultures. Recent work on the substitution of cycloheximide with natamycin and acriflavine with moxalactam and bacitracin has shown promise but the comparative effect on *L. monocytogenes* and *L. innocua* has not been tested.

Diagnostic Selective Agars

Lithium Chloride Phenylethanol Moxalactam Agar (LPM), Oxford Agar, Modified Oxford Agar (MOX), Polymyxin Acriflavine Lithium Chloride Ceftazidime Aesculin Mannitol Agar (PALCAM), Haemolysis Agars Since the selectivity and specificity of enrichment broths is not absolute it is necessary to subculture to agar media to establish the presumptive presence of *Listeria* spp. These agars contain selective agents (**Table 3**) and most include an indicator system, usually aesculin and ferric iron.

LPM is the oldest of these agars. It contains no indicator but, when colonies are viewed with the 45° transillumination method of Henry, listeria appear as whitish piles of crushed glass with blue–grey iridescent areas. A modification of this medium (LPM–aesculin Fe^{+++}) incorporates aesculin and ferric iron and renders this special viewing technique unnecessary. Early attempts at producing a selective agar for listeria were frustrated by the growth of enterococci. This problem is overcome in Oxford agar by increasing the concentration of lithium chloride to 15 g l^{-1}. Colistin is incorporated to inhibit Gram-negative bacteria and a small amount of cefotetan prevents the growth of colistin-resistant *Proteus* spp. Acriflavine and fosfomycin provide additional selectivity with regard to unwanted Gram-positive organisms and moulds are inhibited by cycloheximide. In MOX agar ceftazidime is substituted for acriflavine, cefotetan and fosfomycin and cycloheximide is omitted. Another development of the basic principle of Oxford agar is PALCAM where polymyxin B is substituted for colistin and ceftazidime for cefotetan and fosfomycin. Mannitol and phenol red are included to provide a second indicator system. None of the above agars is capable of differentiating between *L. monocytogenes* and other species of *Listeria*. Attempts to make this distinction have resulted in various 'haemolysis agars' in which blood is incorporated to distinguish *L. mono-*

cytogenes, *L. ivanovii* and *L. seeligeri* from the non-haemolytic species. The earliest of these agars were used as overlays on plates where presumptive listeria colonies had already been identified. This was a tedious procedure and prolonged the examination by a further 24 h. It also required colonies to be dug out of the agar for further confirmation or subtyping. Cox's enhanced haemolysis agar (EHA) has emerged as a convenient medium which can be inoculated direct from enrichment broths without the need for overlaying. It is similar to Oxford agar but with a reduced lithium chloride content, the substitution of 4-methylumbelliferyl-β-D-glucoside for the aesculin–ferric iron indicator system, the incorporation of sheep blood to demonstrate haemolysis and sphingomyelinase to enhance this feature. Later versions of this medium (Heisick 1 and 2) use Oxford (Heisick 1) or PALCAM (Heisick 2) supplements to provide different levels of selectivity with Columbia base-EH to enhance haemolysis.

An incubation temperature of 30°C is recommended for LPM, MOX and PALCAM and 37°C for EHA. Oxford agar may be incubated at 30 or 37°C but if *L. seeligeri* or *L. ivanovii* are sought 30°C is mandatory since these species (but not *L. monocytogenes*) have been shown to have increased sensitivity to cefotetan and fosfomycin at the higher temperature. A similar phenomenon has been demonstrated with ceftazidime which is more active at 37°C against strains of *L. monocytogenes* and *L. seeligeri* and moxalactam which shows greater activity against *L. seeligeri* at 37°C. In general therefore, it is preferable to incubate at 30°C. Presumptive positives can usually be detected on those agars containing aesculin and iron after 18–24 h but, since the reaction develops after the formation of the colony, negative plates should be returned to the incubator for a further 20–24 h to ensure the detection of slow-growing or slow-reacting strains. Originally it was considered necessary to incubate PALCAM microaerobically but this is no longer the case.

These agars show various degrees of selectivity and the choice of which medium to use should have regard to the amount of interference from unwanted organisms that is likely to be encountered. Those media with high contents of β-lactam antibiotics tend to be more selective than Oxford agar where the cefotetan content has been deliberately restricted (but can be increased to 10 mg l^{-1} to increase the selectivity subject to some loss in productivity). Where comparisons have been made between agars it has been found that very selective agars perform well in very 'dirty' situations but that less selective media may grow strains which are not isolated on more inhibitory agars. The best counsel is to use two isolation media of differing degrees of selectivity, e.g. Oxford and PALCAM.

Culture Methods

Methods proposed by some standards organizations and other national bodies are shown in **Table 4**. Where pre-enrichment is entirely omitted (ISO 10560: 1993) or where inhibitors are added after a few hours of the commencement of incubation (FDA; 1995) a negative result may be obtained in four days whereas with other methods five days are required before the result can be declared negative. Presumptive positive results, as demonstrated by typical colonies on selective agars, may be obtained after two days with the FDA (1995) method and three days with the others but the need for confirmatory tests will delay a final report.

Mention has been made above of the desirability of using more than one plating medium in any detection method and this argument has also been extended to the use of two enrichment broths. This latter idea has not been generally accepted. The use of broth and agar media with the same selective spectrum, e.g. L-PALCAMY broth and PALCAM agar, should be avoided as interfering organisms which will grow in the broth may also be expected to grow on the agar.

Table 4 Media used in regulatory and other methods for the detection of *L. monocytogenes*

Authority	Primary or pre-enrichment	Secondary or selective enrichment	Diagnostic plating[a]
Canadian (Warburton et al, 1992)	UVM I	Fraser	**Oxford + either** LPM or MOX or PALCAM
FDA (BAM, 1995)	Single FDA-1995 enrichment broth with selective agents added after 4 h incubation		**Oxford + either** LPM or LPM-aesculin-Fe^{+++} or PALCAM
ISO 11290-1: 1996	Half Fraser	Fraser	**Oxford + PALCAM**
ISO 10560: 1993 (= IDF 143a: 1995)	(none)	'modified FDA'[b]	**Oxford** ± PALCAM
UK PHLS	BPW	FDA (Lovett)	Oxford or PALCAM
USDA (draft, 1997)	UVM I	Fraser	**MOX**

[a] Mandatory agars are printed in **bold** type.
[b] This modification of the original Lovett's broth has a reduced acriflavine content (0.01 g l^{-1}).

Where different selective systems are used in broth and agar the effect may, to some extent, be complementary. Simpler systems were once proposed for milk and dairy products whereas more complex procedures were recommended for other samples. In more recent times there has been some standardization and methods now tend to serve for all types of sample (ISO 11290–1: 1996) (**Fig. 1**) though at the expense of becoming more complex.

Of the many comparisons of methods that have been made, the series performed by Warburton and co-workers is the most comprehensive. The three reports cover various combinations of broths and agars, a wide range of foods and environmental samples, some naturally contaminated and some spiked. Target organisms were examined after stressing by heat and freezing and low numbers of unstressed cells were mixed with interfering organisms to test the selective ability of each system. The pH of enrichment broths at the time of streaking was also determined. Some of the trials involved up to

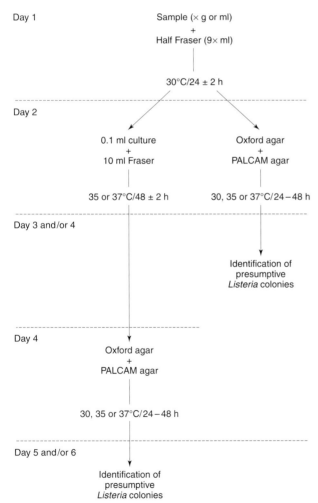

Figure 1 Diagrammatic representation of the ISO 11290-1; 1996 method.

19 laboratories across Canada in a series of tests originally conceived as a comparison of FDA and USDA methods. As a result of modifications made to the original methods during the course of the study and conclusions formed during the assessment of early results, the third report (1992) provides an excellent and comprehensive survey of the state of the art at that time. Since the preparation of that report the FDA method has been revised (1995), the ISO standard has been published (1996) and a new draft USDA method prepared (1997). Any attempt to choose a suitable method should take account of all these publications.

Factors extrinsic to the test system which need to be considered when a method is selected or devised include the nature of the sample, the time available for completion of the test, whether the search is for *L. monocytogenes* only or for all *Listeria* spp. and whether a regulatory test is necessary. Samples such as raw milk should not require pre-enrichment whereas samples processed for safety certainly will. Although in the past some media or methods were found to be more appropriate for certain types of sample, with the increased use of buffering in enrichment media it now seems unlikely that particular types of sample require individual consideration. If a product is to be quarantined until the results are available it may prove more economical to follow pre-enrichment with a rapid non-cultural method rather than compromise on sensitivity by curtailing the time taken for conventional cultural examination. If all *Listeria* spp. are to be sought then a lower (30°C) incubation temperature is to be preferred to 35 or 37°C for selective agars containing β-lactam antibiotics. If the product is being examined for regulatory purposes then, of course, the appropriate method of the relevant regulatory body must be used and strictly followed.

Since there are several regulatory or standard methods (Table 3) which provide a satisfactory range of choices, further systems should not be devised without good cause. For unstressed samples, particularly those of dairy origin, the ISO 10560:1993 method, prepared in cooperation with the International Dairy Federation (IDF), is the simplest and will give a negative result in four days, requiring only a single subculture of the enrichment broth. However, the standard suggests adjusting samples with low pH to 7.0 ± 0.5 before examination. To improve the sensitivity, the same standard also recommends subculturing the enrichment broth at 24 h as well as 48 h when examining raw milk and extending the incubation to seven days when examining red smear and blue cheeses. There is evidence that, even with dairy samples, Fraser broth provides more reliable enrichment by virtue of its buffering. The UK PHLS method has the advantage of a nonspecific buffered pre-enrichment broth followed by a simple subculture procedure. This was originally designed for milk and dairy products but has been much used in the UK for other foods such as paté, shellfish and bakery products. Unfortunately no comparisons have been made with other methods. The desirability of a buffered secondary enrichment and plating on at least two selective differential agars means that neither of these methods can be recommended.

Of the remaining methods the Canadian is, in effect, an expansion of ISO 11290-1 1996, using the less selective UVM I for pre-enrichment in place of Half Fraser broth and making subcultures of UVM I at 48 h as well as 24 h. Also the pattern of subculture of the Fraser broth is more complex in the Canadian scheme. The FDA (1995) method provides a 4 h nonspecific pre-enrichment in buffered broth followed by the addition of selective agents before continuing incubation for a total of 48 h with subculture to two selective media at 24 and 48 h. This shortens the pre-enrichment period to produce a negative result in four days. The price paid for this is in the extra manipulation of the initial broth with the requirement to make aseptic additions prior to and at 4 h after inoculation with the sample. This may prove difficult to fit into the working schedule in some laboratories unless all samples are received at the beginning of the day. The choice of method lies then between the ISO 11290-1: 1996, FDA (1995) and Canadian systems. The ISO method is too recent for comparisons with the other methods to have been published and the Canadian method has been compared with a previous version of FDA. Those workers who can accommodate the handling programme of the FDA (1995) on the first day may feel that its shorter time to a final result commends it. Others will look to ISO or the more extensive Canadian method. Where, as in FDA (1995) and the Canadian method, there is a choice of a second plating medium, LPM is contraindicated due to its reliance on the Henry illumination technique to identify presumptive listeria colonies. Workers who have the opportunity might include EHA or Heisick 1 or 2 in the system to determine their usefulness in comparison to the established agars.

Methods for Enumeration

Although regulatory authorities may specify presence/absence testing for *Listeria monocytogenes* there is sometimes a need to count listeria in food samples. Counting methods are poorly developed compared to presence/absence testing and are based either on direct plating or most probable number (MPN) techniques, the latter being more sensitive but less accurate. A major problem in enumeration is posed by sublethally injured cells and methods for the inclusion of these in counts are not well established. Those that have been described rely on seeding the sample dilutions to a nonselective agar and overlaying with a selective agar after a period of incubation to allow resuscitation. On agars used for counting there is a need to distinguish presumptive listeria colonies from those of interfering organisms. This is made more difficult by the fact that the inoculum has not been cultured previously in a selective broth as with techniques for presence/absence testing. Increasing the selective properties of the diagnostic agar used for enumeration may reduce the count of interfering organisms but is likely to render the medium unsuitable for resuscitation of sublethally injured organisms.

Early attempts at enumeration found LPM, Oxford and MOX agars to be of use in the examination of various foods. FDA (1995) recommend direct plating on Oxford agar, USDA (draft, 1997) on MOX. Heisick 1 and 2 agars, designed for small and large populations of background microflora respectively, have also been used successfully for direct plating of naturally contaminated and spiked foods with better recovery of repaired stressed cells than on Oxford agar. A system incorporating direct plating on tryptone soya agar (TSA) followed by incubation at 30°C for 5 h and subsequent overlay with Oxford agar produced counts up to $\log_{10} 2.5\,cfu\,g^{-1}$ higher than standard recovery methods with meat and fish products.

MPN methods are not as yet fully developed. FDA (1995) have published an optional method using a five tube MPN culture procedure on 10, 1, 0.1 and 0.01 g samples in FDA-1995 enrichment broth followed by plating on Oxford agar. *L. monocytogenes* is detected by pooling five colonies and performing the Accuprobe test to confirm the presumptive presence of *L. monocytogenes* in each tube yielding black colonies on the agar. USDA (draft, 1997) outlines a 3×3 method which uses UVM I as the primary broth and then follows the presence/absence procedure for secondary enrichment and diagnostic plating.

There is little published material on quantitative methods that compares results using the current selection of culture media. An ISO standard is in course of preparation but until that is available for consideration workers requiring to enumerate listeria in foods should use the FDA (1995) method if they need the sensitivity of an MPN method or the Heisick agars for direct plating. The two-stage plating on TSA and overlay with Oxford agar may prove to be the best method if sublethally injured cells are likely to be present in large numbers.

Media for Identification

Identification of presumptive *Listeria* spp. is essential in view of the common colonial appearance of all species on agar media which do not contain blood whilst even on blood-containing agars it is necessary to differentiate *L. monocytogenes* from the other haemolytic species. It is quite common to find more than one *Listeria* spp. in a sample of food and, to increase the chances of detection of *L. monocytogenes*, it is necessary to examine several colonies (usually at least five) from each agar plate. Several biochemical tests serve to differentiate the various *Listeria* species (**Table 5**) and with one exception these are performed in the conventional way. Peculiar to the speciation of *Listeria* is the CAMP (Christie, Atkins and Munch-Petersen) test which demonstrates enhancement of haemolysis of sheep red blood cells by certain *Listeria* spp. in conjunction with cultures of *Staphylococcus aureus* or *Rhodococcus equi*. Cultures are streaked on a specially prepared sheep blood agar plate as shown in **Fig. 2**. The plate consists of a layer of nutrient agar overpoured with a thin layer of the same medium containing 3–4% washed sheep cells. It is essential to use sheep cells and washing will remove any trace of listeria antibodies which may be present in the serum. The choice of *Rhodococcus* strain does not seem critical but some *S. aureus* strains perform better than others; *R. equi* (ATCC 6939; NCTC 1621) and *S. aureus* (ATCC 49444; CIP 5710; NCTC 7428 or NCTC 1803) have been found satisfactory. In streaking the listeria strains it is important that they are placed close to the vertical streaks but not touching them. This is to prevent the carrying of the vertically streaked organism across the plate with the listeria which would mask the synergistic haemolysis which occurs where the two lysins interact. The

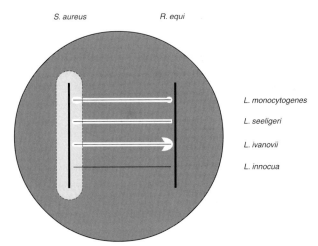

Figure 2 Method of inoculation and typical results of CAMP test. A zone of partial lysis extends around the *Staphylococcus aureus* streak. Haemolytic species growing within this zone are surrounded by a narrow band of complete haemolysis which may be more easily seen if the listeria growth is wiped from the plate with a swab. The spade-shaped zone of complete haemolysis at the junction of the *Listeria ivanovii* and *Rhodococcus equi* growths is characteristic; a small area of enhancement at the *L. monocytogenes*/*R. equi* junction is often seen but should be read as a negative result.

plates should be incubated at 35 or 37°C for 18–24 h. Interpretation of the results is straightforward except that the enhancement of haemolysis by *L. ivanovii* near to the *R. equi* streak is very much clearer and spread over a wider area than the enhancement seen between *L. monocytogenes* or *L. seeligeri* and *S. aureus*. Slight enhancement is often seen between *L. monocytogenes* and *R. equi* but this may be ignored. As other *Listeria* spp. do not react in this test, it is a useful confirmation test for *L. innocua* which is indistinguishable from *L. monocytogenes* by biochemical reactions (see **Table 5**).

Table 5 Identification of *Listeria* spp. All *Listeria* spp. are Gram-positive rods which are catalase and Voges-Proskauer test positive, oxidase and urease negative. They produce acid but not gas from D-glucose and D-salicin and exhibit tumbling motility at room temperature (adapted from McLauchlin, 1987)

| Species | Haemolysis on horse blood | Nitrates reduced to nitrites | CAMP test | | Production of acid from | | | |
			Staphylococcus aureus	Rhodococcus equi	D-mannitol	L-rhamnose	D-xylose	α-methyl D-mannoside
L. monocytogenes	+	−	+	−	−	+	−	+
L. ivanovii	++	−	−	+	−	−	+	−
L. innocua	−	−	−	−	−	V	−	+
L. welshimeri	−	−	−	−	−	V	+	+
L. seeligeri	+	−	+	−	−	−	+	V
L. grayi	−	V	−	−	+	V	−	V

V = variable reaction.

Subspeciation of *L. ivanovii* can be achieved by testing the capability of an isolate to ferment ribose and to produce acid from *N*-acetyl-β-D-mannosamine within 24 h. *L. ivanovii* ssp. *ivanovii* produces acid from ribose but not from *N*-acetyl-β-D-mannosamine; with *L. ivanovii* ssp. *londoniensis* these reactions are reversed.

L. grayi ssp. *murrayi* reduces nitrates to nitrites which differentiates it from *L. grayi* ssp. *grayi*.

See color Plate 19.

See also: **ATP Bioluminescence**: Application in Dairy Industry. **Bacteriophage-based Techniques for Detection of Foodborne Pathogens**. **Biochemical and Modern Identification Techniques**: Food-poisoning Organisms. **Food Poisoning Outbreaks**. *Listeria*: Detection by Commercial Enzyme Immunoassays; *Listeria monocytogenes* – Detection by Chemiluminescent DNA Hybridization; *Listeria monocytogenes* – Detection using NASBA (an Isothermic Nucleic Acid Amplification System). **Nucleic Acid-based Assays**: Overview. **PCR-based Commercial Tests for Pathogens**. **Polymer Technologies for Control of Bacterial Adhesion**. **Reference Materials**.

Further Reading

Anon. (1993) ISO 10560:1993 Microbiological examination for dairy purposes Part 3. Methods for detection and/or enumeration of specific groups of organisms. Sect. 3.15 Detection of *Listeria monocytogenes*. Geneva: International Standards Organization.

Anon. (1995) Pharmacopoeia of Culture Media. In: Corry JEL, Curtis GDW and Baird RM (eds) *Culture Media for Food Microbiology*, Part 2. Amsterdam: Elsevier.

Anon. (1996) ISO 11290–1:1996 Microbiology of food and animal feeding stuffs. Horizontal method for the detection and enumeration of *Listeria monocytogenes* Part 1. Detection method. Geneva: International Standards Organization.

Curtis GDW and Lee WH (1995) Culture media and methods for the isolation of *Listeria monocytogenes*. In: Corry JEL, Curtis GDW and Baird RM (eds) *Culture Media for Food Microbiology*. Part 1, p. 63. Amsterdam: Elsevier.

Farber JM and Peterkin PI (1991) *Listeria monocytogenes*, a food-borne pathogen. *Microbiology Reviews* 55: 476–511.

Hitchins AD (1995) *FDA Bacteriological Analytical Manual*, 8th edn. Ch. 10 *Listeria monocytogenes*. Gaithersburg, MD: AOAC International.

Johnson JL (1997) *USDA/FSIS Microbiology Laboratory Guidebook*, 3rd edn (draft). Ch. 8 Isolation and identification of *Listeria monocytogenes* from meat, poultry and egg products. Washington, DC: USDA/FSIS.

McLauchlin J (1987) A review: *Listeria monocytogenes*, recent advances in the taxonomy and epidemiology of listeriosis in humans. *Journal of Applied Bacteriology* 63: 1–12.

Mossel DAA, Corry JEL, Struijk CB and Baird RM (1995) *Essentials of the Microbiology of Foods*. Chichester: Wiley.

Roberts D, Hooper W and Greenwood M (eds) (1995) *Practical Food Microbiology*, 2nd edn. London: Public Health Laboratory Service.

Warburton DW, Farber JM, Powell C et al (1992) Comparison of methods for optimum detection of stressed and low levels of *Listeria monocytogenes*. *Food Microbiology* 9: 127–145.

Detection by Commercial Enzyme Immunoassays

Martin Wagner, Institute for Milk Hygiene, University for Veterinary Medicine, Vienna, Austria

Andreas Bubert, Department for MIcrobiology, University of Würzburg, Germany

Introduction

Listeria spp.

Listeria are Gram-positive microorganisms which are widespread in nature, including water, plant and mammalian habitats. The genus *Listeria* comprises pathogenic species such as *L. monocytogenes* and *L. ivanovii* as well as the non-pathogenic species like *L. innocua*, *L. seeligeri*, *L. welshimeri* and *L. grayi*. Although known almost since the beginning of this century, the pathogenic species were thought predominantly to cause disease in animals such as ruminants and rodents. The role of *L. monocytogenes* as a major food-borne pathogen was first described by tracking large *Listeria* epidemics to the consumption of foodstuffs, particularly brands of soft cheese. Hence, modern food production has to ensure safety of products in respect to listerial contamination thus protecting the consumer's health.

Microbiological versus Proprietary Techniques

Stringent national regulations require zero tolerance of the presence of *L. monocytogenes* in a defined amount of foodstuff in most countries. To monitor food safety in respect to this pathogen, microbiological methods have been the methods of choice. However, microbiological detection and enumeration methods as published by the International Dairy Federation (IDF) or the International Standardization Organization (ISO) lack online control of food products, leading to unsatisfactory surveillance as products might be consumed before microbiological investigation is finished. This is particularly the case where suspected colonies are detected after selective enrichment of pathogen: confirmation of colonies by biotyping takes a couple of days. Additionally, *L. monocytogenes* strains were isolated being different major biochemical features such as catalase reaction and haemolysis. Although haemolysis-negative *L. monocytogenes* clones are thought to be non-pathogenic, it is possible that a weak haemolysis reaction may lead to false-negative results. This point is worth being considered as this reaction is the most important classification marker for *Listeria monocytogenes* and hence used to differentiate pathogenic from non-pathogenic listerial species.

To overcome the disadvantages of microbiological *Listeria* detection, a range of rapid detection methods was developed using automated biochemical typing, antigen–antibody-based detection or detection of listerial chromosomal DNA or similar genetic products.

Criteria for Evaluation of Rapid Detection Principles in Food Hygiene

The important parameters describing the usefulness of detection procedures in food hygiene are:

1. Intrinsic and extrinsic sensitivity.
2. Specificity.
3. Detection of viable organisms.
4. Time span required for detection.
5. Test configuration and design.
6. Cost of commercially available test kits.

Regarding occurrence of pathogens, an optimized test should combine 100% specificity with maximum sensitivity, especially when zero tolerance is required. Further, it should detect viable organisms in a short time. Cheap and easy-to-handle test kits are generally preferable for mass investigations in the field.

Immunological Methods

General Principles

Enzyme immunoassays comprise an antibody–antigen reaction essential for recognition of specific targets and detection of immunological reaction by enzyme-mediated colour or fluorescence reaction (**Fig. 1**). Cell-specific targets for antibody recognition are cell-surface-bound substructures (e.g. teichoic acids), flagellar antigens or proteins secreted by the organism being detected (e.g. the p60 protein). For *Listeria* and *L. monocytogenes*, numerous proteins such as flagella, listeriolysin, actin polymerization protein ActA, internalins, p60, and various cell wall-associated proteins of unknown function have been used as targets for polyclonal and monoclonal antibody production. Some are highly specific for *Listeria* and *L. monocytogenes*, and some display cross-reactivity between pathogenic and some non-pathogenic *Listeria* spp. However, only a few of these antibodies are suitable for diagnostic purposes because expression of the corresponding genes is relaxed enough to reach satisfactory detection levels in immunoassays. In particular the virulence genes such as *hly, actA* and *inlA* of *L. monocytogenes* are controlled by the transcriptional activator protein PrfA; most are efficiently activated during infection in mammalian cells but to a much lower extent in nutrient-rich bacterial culture media.

The specificity of an immunoassay is largely determined by the intrinsic specificity of either the poly-

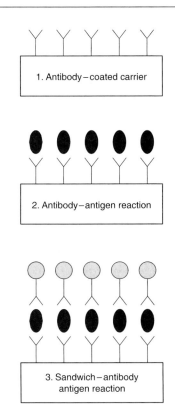

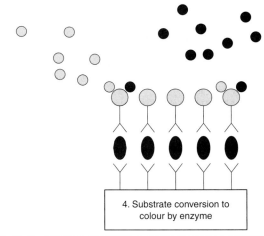

Figure 1 Scheme depicting a commonly used configuration for enzyme immunoassays.

clonal or monoclonal antibody to the target. The detection limit of an immunoassay depends on the number of organisms necessary to show a positive signal after enzyme-mediated detection of antigen–antibody complexes. Numbers of organisms available for detection are strongly dependent on the numbers originally contaminating the foodstuff and on the efficiency of extraction of organisms from food matrix and the subsequent enrichment of small numbers of organisms to a detectable level. Using immunoassays in food hygiene usually requires two steps, including enrichment of targets (**Fig. 2**). As enrichment efficiency may vary according to food matrix effects,

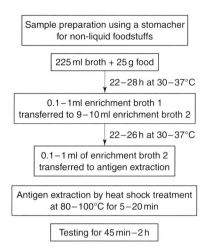

Figure 2 Scheme depicting a standard enrichment and antigen extraction procedure before testing.

leading to preinjured bacteria, different enrichment protocols which allow resuscitation are recommended by manufacturers. The enrichment step should foster production of antigenic epitopes, e.g. as flagellar antigens. Although enrichment ensures that cells are viable, it greatly increases the detection time needed.

Available test kits are either single-use format, mainly dipsticks (e.g. the *Listeria* Unique assay®, Tecra) or miniplates (e.g. Visual immunoprecipitate assay® – VIP, Biocontrol Systems; Clearview *Listeria*®, Oxoid), microplate formats (Pathalert *Listeria*® and Pathalert *L. monocytogenes*®, Merck Germany; Tecra *Listeria* Visual Immunoassay®, VIA; Transia *Listeria*®, Diffchamb; *Listeria* Tek®, Organon Teknika; *Listeria* Assurance®, Biocontrol Systems) or special formats (e.g. the Solid Phase Receptacle®, SPR, bioMérieux; EIA Foss®, Foss Electric). Although numerous test kits and formats are available, only test principles and a few widely distributed test kits can be reviewed in this article (**Table 1**).

Single-use Assays

These test kits are generally used as screening tests. Both environmental and food samples may be screened. Only a test unit and standard microbiological laboratory equipment are needed.

Dipsticks
Principles   The test principle is an immunocapture-based assay using a dipstick format. In respect to the *Listeria* Unique assay *Listeria*-specific antibodies are coated to the surface of the dipstick. All reagents are ready-to-use and packed in a self-contained module: 24-h incubated enrichment broth is added to tube 1 of this test in a first step. Additionally, tube 1 contains the moveable dipstick device. *Listeria* are captured and transferred to another tube to allow replication of cells to detectable numbers. Enzyme-labelled antibodies recognize targets bound to the dipstick. After a brief washing step to remove excess conjugate the dipstick is freshly transferred to tube 3, containing substrate for the enzyme.

Enrichment and Extraction of Antigens   Twenty-four hour enrichment is recommended by manufacturers. Extraction of antigens by an individual step is not required.

Interpretation of Results and Controls   *Listeria*-positive samples are indicated if a purple colour appears on the lower half of the dipstick. The upper half of the dipstick which remains white serves as the negative control. Additionally, a positive control is included in the test. It appears as a coloured cross at the dipstick base. Testing is finished after 32 h.

Detection Limit and Specificity   Detection limit ranged from 3×10^4 to 4×10^5 in six different *Listeria* spp. tested. Specificity for *Listeria* spp. was tested against 28 non-*Listeria* organisms and found to be 100%.

Evaluation/Validation   Comparative trials demonstrated excellent correlation between Unique *Listeria* and standard cultural methods in a variety of naturally and artificially contaminated foods. There were no false positives in either group.

Table 1   Manufacturers' addresses

Assay	Company	Address
Unique *Listeria*	Tecra	28 Barcoo Street, Roseville, NSW 2069, Australia
VIP *Listeria*	Biocontrol Systems	19805 N. Creek Parkway, Bothell, WA 98011, US
Clearview *Listeria*	Oxoid	Wade Rd, Basingstoke, Hampshire, RG24 8PW, UK
Listeria Tek	Organon Teknika	100 Akzo Avenue, Durham, NC 27712, US
Listeria VIA	Tecra	28 Barcoo Street, Roseville, NSW 2069, Australia
Assurance *Listeria* EIA	Biocontrol Systems	19805 N. Creek Parkway, Bothell, WA 98011, US
Transia *Listeria*	Diffchamb	Backa Bergögota 5, S-42246 Hisings, Backa, Sweden
Pathalert *Listeria* spp. + *L. monocytogenes*	Merck	Frankfurterstrasse 250, D-64293 Darmstadt, Germany
Vidas	bioMérieux	69 280 Marcy l'Etoile, France
EIA Foss	Foss Electric	69 Siangerupgade, DK 3400, Hillerød, Denmark

Miniplate formals

Principles Test units are usually offered in miniplate formats made from plastic. A sample inoculation zone, a detection zone and a verification zone can be identified at the upper side of the device. Between the sample inoculation zone and the detection zone an interface zone usually contains the specific antibody as main reagent. More complex test kits harbour the *Listeria*-specific antibody in the inoculation zone, predominantly bound to coloured latex particles or similar antibody-coated carriers. As a result of capillary attraction the latex–antibody–antigen complex is moved to the detection zone where the complex is captured by a second antibody which recognizes the antigen. In the control zone, antibody-coupled latex particles not associated with antigens are bound to an antibody (Clearview *Listeria*®, Oxoid). The coloured latex beads group together closely.

Enrichment and Extraction of Antigens A two-step enrichment procedure is usually performed before testing and interpreting results. Each step normally lasts 21–24 h. Broths recommended by the manufacturers are mainly half Frazer broth (first-step enrichment) and buffer *Listeria* enrichment broth (bLEB) or Frazer broth for second-step enrichment (**Table 2**). Lower concentrated half Frazer broth is appropriate to allow resuscitation for heavily injured bacteria. Extraction of antigens is achieved by heating an aliquot of secondary enrichment to 80°C for almost 20 min.

Interpretation of Results and Controls A positive reaction is seen as a coloured line in the detection field. A similar reaction indicates presence of particle-bound antibody in any test, and this is a positive control. This reaction is visible in the control window. Enzyme immunoassays in single-use format are usually screening tests capable of detecting *Listeria* spp. (Clearview *Listeria*®).

Detection Limit and Specificity In-house investigations and comparative studies confirmed an overall correlation between standard cultural methods and the Clearview *Listeria* assay® of > 99%. The *Listeria* assay was tested against a panel of closely related organisms and further food contaminants, thus indicating specificity of 100% for *Listeria* spp.

Evaluation/Validation The VIP® assay was extensively evaluated by the Association of Official Analytical Chemists International (AOAC). As a result it is an approved AOAC official method number 997.03.

Conclusion Single-use test kits in dipstick or miniplate formals are easy-to-handle, rapid *Listeria* detection assays which have particular benefits for screening *Listeria* within the frame of facility-specific quality-control programmes. Standard laboratory equipment is sufficient: test-specific hardware and specifically trained personnel are not required. The *Listeria* Unique assay® provides results within a short time. However, confirmation of presumed positive samples, including differentiation of pathogenic from non-pathogenic species considerably lengthens investigation time. This aspect applies to all immunoassay systems described in this article which contain *Listeria* genus-specific antibodies.

Microplate Formals

Principle The test configuration is a sandwich ELISA applied in microtitre plates. It is either performed manually (Tecra, Merck, Diffchamb, Organon Teknika, Biocontrol) or is fully automated (Diffchamb). The wells are precoated with antibodies which remain immobilized throughout the procedure and recognize and capture whole bacteria (Tecra, Diffchamb, Organon Teknika, Biocontrol) or proteins secreted into the culture supernatant (Merck). After the samples are added to the wells, the bound targets are recognized by antibodies added to the wells. These form an immobilized antibody–target–antibody (sandwich) complex (Fig. 1). The latter antibody is either directly conjugated with an enzyme (e.g. Tecra, Diffchamb) or recognized by an antibody conjugated with an enzyme (e.g. Merck). The enzyme converts a chemical substrate to colour. If *Listeria* spp. are absent, there is no change in colour.

Assays in microplates usually detect the genus *Listeria* by recognizing whole bacteria via flagella antigens, while both Merck Pathalert tests specifically identify both the genus *Listeria* and *L. monocytogenes* by recognizing the secreted p60 protein. This protein represents an essential component of *Listeria* for cell propagation; it is involved in septum separation of the daughter cells. Thus, the detection is focused on the viable status of *Listeria*. The p60 protein, which is produced by all *Listeria* over a broad temperature range (10–40°C) at all stages of growth, shares conserved and species-specific variable protein portions. The molecular characterization of the whole listerial p60 protein family led to the development of *L. monocytogenes*-specific antibodies which are included in the Pathalert *L. monocytogenes* test.

Enrichment and Extraction of Antigens Two-step enrichment is performed before testing and interpreting results. Each enrichment normally lasts approximately 24 h at 30 or 37°C depending on

Table 2 Review of enzyme immunoassays, including characteristics of enrichment and testing

Assay	Format	Application	Enrichment/ time (approx.)	First-step enrichment broth	Second-step enrichment broth	Extraction of antigens	Detection limit	Specificity	Approval
Unique Listeria	Dipstick	Manually	Yes/24 h	bLEB		No	3×10^4– 4×10^5	L. spp	
VIP Listeria	Miniplates	Manually	Yes/48 h	Frazer	bLEB	Yes		L. spp	AOAC
Clearview Listeria	Miniplates	Manually	Yes/48 h	Frazer	bLEB	Yes		L. spp	
Listeria Tek	Microplates	Manually	Yes/48 h	mFrazer	bLEB	Yes		L. spp	AOAC
Listeria VIA	Microplates	Manually	Yes/48 h	bLEB	bLEB or Frazer	Yes	10^3– 5×10^5	L. spp	AOAC and others
Assurance Listeria EIA	Microplates	Manually	Yes/48 h			Yes		L. spp	AOAC
Transia Listeria	Microplates	Manually/ automated	Yes/48 h	Frazer	Frazer	Yes	10^5–10^6	L. spp	AFNOR and others
Pathalert Listeria spp. + L. monocytogenes	Microplates	Manually automated	Yes/48 h	Frazer	Frazer or Palcam	Yes	10^5–10^6	L. spp L. m.	
VIDAS	Special	Automated	Yes/48 h	Frazer	Frazer	Yes	10^5–10^6	L. spp. L. m	AFNOR
EIA Foss	Special	Automated	Yes/48 h	Depending on foodstuff	Depending on foodstuff	Yes	10^5	L. spp	

AOAC, Association of Analytical Chemists International; AFNOR, Association Française de Normalisation; bLEB, buffered Listeria enrichment broth; Frazer, modified Frazer.

expression of the target protein. When flagella are used as antigens, enrichments must be performed at 30°C because expression of flagella is inhibited at temperatures higher than 30°C. As mentioned above, for the detection of p60, enrichments can be performed at the optimal growth temperature of 37°C. Broths and procedures for antigen extraction as recommended by manufacturers are listed in Table 1.

Interpretation of Results A positive reaction is indicated by colour development of the solution in the microtitre plate well. This may be analysed either by eye (e.g. Tecra) or automatically by measuring absorbance in a microtitre plate reader (e.g. Diffchamb, Merck).

Detection Limit and Specificity The ELISA produced by Tecra reaches a detection limit of 10^3 to 5×10^5 cells per millilitre of broth. The Pathalert tests as well as the Transia assay have a detection limit of 10^5 to 10^6 per millilitre.

Evaluation The Tecra assay was evaluated with 360 samples from dairy foods and compared with cultural techniques, and showed excellent results. According to the manufacturers' brochure the test has several approvals in Australia, New Zealand, and a first-action approval of the AOAC. The Listeria Tek assay was also validated in a collaborative study by the AOAC. It is an approved AOAC method, as is the Assurance Listeria EIA. Both Pathalert tests, which do not have approval, were evaluated with more than 150 Listeria strains and demonstrated excellent correlation with standard identification techniques for Listeria. Moreover, since the antibodies used in this test recognize the essential p60 protein which displays a low genetic molecular drift in most protein portions, Pathalert® L. monocytogenes has also been shown to identify unusual L. monocytogenes strains such as haemolysis-, catalase- and flagella-negative isolates.

Automated Assays

Vitek Immuno Diagnostic Assay®
Principle The Vidas is an automated ELISA which

can be used for extended screening trials in the food industry. Looking like a pipette tip, the solid phase receptacle (SPR) is coated with *Listeria*-specific monoclonal antibodies, capturing targets from enriched food samples. It simultaneously serves as a pipetting device. In addition a sealed reagent strip is provided, containing all further reagents such as washing buffers and conjugates. After enrichment of targets and extraction of antigens, 500 µl of liquid culture is pipetted into the reagent strip. Then the sample volume is taken up by the solid phase receptacle. Listerial cells are specifically bound to the capture antibody-coated solid phase. Further sample components are washed away. Alkaline phosphatase-labelled secondary antibodies are subsequently bound to the capture antibody–antigen complex. Unbound conjugate is removed by an additional washing step. A fluorescent substrate, whose conversion to a fluorescent product is catalysed by the enzyme, is finally introduced. Fluorescence is measured at 450 nm wavelength by an optical scanner and test values are printed on data sheets. All analytical steps are automatically carried out by the Vidas analyser. Test strips for detection of both *Listeria* spp. and *L. monocytogenes* are available.

Enrichment and Extraction of Antigens Two-step enrichment of food samples before testing is essential. Enrichment in Frazer broth containing ferric ammonium citrate for 22 h followed by 22 h enrichment in Frazer broth without ferric ammonium citrate is recommended by the manufacturers. Heating broth aliquots to 100°C for 15 min deliberates antigenic targets. Regarding detection of *Listeria* from raw milk cheeses, a special protocol was recently introduced by the manufacturers. The second enrichment step, usually performed in Frazer broth, was replaced by streaking sample to selective plate. Suspected colonies are harvested using peptone water and loops for scraping. Suspensions are directly subjected to testing. Testing requires an additional 45 min.

Interpretation of Results and Controls The result of an individual investigation is displayed by relative fluorescence units (RFU). Unspecific background fluorescence is considered at the point of analysis. Then the ratio of the sample-specific fluorescence to a standard specific fluorescence is calculated, resulting in a test value. Samples are interpreted as positive when the test value exceeds a defined limit of 0.10. Positive control and *Listeria* standard solutions, both containing purified inactivated *Listeria* antigen as well as a negative control solution, are provided.

Detection Limit and Specificity The detection limit of the Vidas assay® was revealed using different *Listeria* spp., including *L. monocytogenes*. Several *L. monocytogenes* serovars were also tested. The detection limit ranged from 4×10^5 to 10^6 cfu depending on the *L. monocytogenes* serovar used. Specificity was examined both against 217 *Listeria* and against 35 non-listerial strains. Specificity was sufficiently validated.

Evaluation/Validation The use of the Vidas assay® was extensively evaluated by a panel of scientific institutions. For use in *Listeria* detection in foods, it is a validated rapid screening method approved by the Association Française de Normalisation (AFNOR). Specific protocols for environmental samples are also provided.

Further Automated Assays

Principle Both the Transia plate *Listeria* – Elisamatic II® and the EIA Foss assay® are fully automated *Listeria* screening assays based on a standard sandwich ELISA working in microtitre plates. The principle and, for Transia Plate *Listeria*, plates and antibodies are described in greater detail in the section microtitre plate formals. Both the Transia Plate *Listeria* – Elisamatic II assay® and the EIA Foss assay® are specific for detection of *Listeria* spp. from preenriched samples and reveal results within 2 h of analysis.

Enrichment and Extraction of Antigens This step is almost identical to that described above in the section microplate assays. Two-step enrichment of food samples is necessary before testing. The enrichment scheme is mainly based on the ISO standard method. Inoculation of food samples in half Frazer broth for 24–26 h at 30°C is followed by secondary enrichment in Frazer broth for 24–26 h at 26°C. Flagellar antigens are extracted by heat shock treatment at 100°C for 20 min (Transia Plate *Listeria* – Elisamatic II assay®). The EIA Foss assay® requires slightly different enrichment procedures depending on whether dairy products (modified *Listeria* enrichment broth, Frazer broth) or alternative products are tested (ISO-based protocols).

Interpretation of Results and Controls Microtitre plates contain wells for either positive or negative controls as well as for samples (Transia Plate *Listeria* – Elisamatic II assay®); 100 µl of either sample and controls (one positive, two negative controls) are pipetted into the wells followed by equal amounts of antibody–peroxidase conjugate. Plates are incubated at room temperature for 1 h. After repeated washing steps (five times), a substrate–chromogen mixture is added and mixtures are incubated for 30 min. Testing is stopped and results are read at 450 nm wavelength. Interpretation of results is carried out according to calculation of a positive threshold (PT) value. The PT relies on the optical density of two negative controls plus a correction factor. Positive samples must show optical density values exceeding the PT; negative samples may not.

Detection Limit and Specificity   The detection limit of the Transia plate assay® was calculated as 10^5–10^6 cfu. The EIA Foss assay® specific detection limit is 10^5 cells per millilitre according to the manufacturers product information. Specificity of Transia plate assay® was tested against 108 *Listeria* spp. strains: 107 out of 108 tested revealed a positive signal. None of 50 heterologous strains tested were detected. Reproducibility data were revealed by an extensive collaborative study among European laboratories. Agreement of data varied according to levels of cfu *Listeria* inoculated. For a sample containing 6 to 252 cells per 25 g food, 98% of experiments detected *Listeria*.

Evaluation/Validation   For use per *Listeria* detection in foods, the Transia plate assay® is a validated rapid screening method approved by AFNOR.

Conclusion   The automated enzyme immunoassays are useful screening procedures for *Listeria* spp. and *L. monocytogenes*. The procedure prevents possible testing errors which may occur from variance in chemicals or handling discrepancies during testing. Cross contamination during testing is almost impossible. The test procedures are usually harmonized with the enrichment protocols, as recommended by the manufacturers. As these fully automated *Listeria* screening assays reveal results in a short time, they are suitable as opportune testing procedures in hazard analysis of critical control point (HACCP) schemes. Furthermore, most test designs can be applied to a range of additional food-borne pathogens and their toxins. However, an analysing device must be bought for some test designs in order to perform an analysis. This might be a considerable expense, especially for small laboratories.

Immunological versus Alternative Rapid Techniques for Detection of *Listeria*

Both immunogenic and DNA-based detection techniques have revolutionized the detection of food-borne pathogens in recent years. In addition to their detection limit and specificity, these assays are normally easy to handle, time-saving and either single-use formats for application in food-processing routine or automated for mass screening purposes. Most assays are approved by internationally well-reputed validation authorities. Due to the test principle and test design, the use of enzyme immunogenic assays is generally scarcely susceptible to handling errors. The reliability of an assay mainly depends on the efficiency of enrichment, which must ensure that the number of pathogens exceeds the individual detection limit of the assay. Cross contamination and decreased amplification or hydridization efficacy, which may be caused by food and enrichment matrix components in DNA-based technology, are of minor importance when using enzyme immunogenic assays. Although immunogenic techniques show benefits in comparison to alternative test principles, they are usually validated as screening procedures. Confirmation of positive samples must be performed using conventional microbiology as these methods still serve as gold standard procedures.

Outlook

Most tests described here are only specific for the genus *Listeria*. Although non-pathogenic *Listeria* species such as *L. innocua* can be considered as indicator bacteria for the presence of *L. monocytogenes*, the crucial question for many users of these tests is whether *L. monocytogenes* is detectable or not. To get an answer to this question, extra time-consuming biochemical methods must be applied. There is a further need to characterize surface-associated and secreted substructures or proteins on the molecular level as this may lead to the development of more different antibodies for the specific identification of *L. monocytogenes* and for highly virulent *L. monocytogenes* isolates. These antibodies will probably lead to more specific and more sensitive immunological assays in convenient test formats.

See also: **Bacteriophage-based Techniques for Detection of Foodborne Pathogens. Biochemical and Modern Identification Techniques**: Food-poisoning Organisms. **Enzyme Immunoassays**: Overview. **Food Poisoning Outbreaks. Hydrophobic Grid Membrane Filter Techniques (HGMF). Immunomagnetic Particle-based Techniques**: Overview. *Listeria*: Detection by Classical Cultural Techniques; Detection by Colorimetric DNA Hybridization; *Listeria monocytogenes* – Detection by Chemiluminescent DNA Hybridization; *Listeria monocytogenes* – Detection using NASBA (an Isothermic Nucleic Acid Amplification System). **National Legislation, Guidelines & Standards Governing Microbiology**: European Union; Japan. **Nucleic Acid-based Assays**: Overview. **PCR-based**

Commercial Tests for Pathogens. Sampling Regimes & Statistical Evaluation of Microbiological Results.

Further Reading

Allerberger F, Dierich M, Petranyi G, Lalic M and Bubert A (1997) Nonhemolytic strains of *Listeria monocytogenes* detected in milk products using VIDAS immunoassay kit. *Zentralblatt für Hygiene und Umweltmedizin* 200: 189–195.

Bhunia AK, Ball PH, Fuad AT, Kurz BW, Emerson JW and Johnson MG (1991) Development and characterization of a monoclonal antibody specific for *Listeria monocytogenes* and *Listeria innocua*. *Infection and Immunity* 59: 3176–3184.

Bohne J, Sokolovic Z and Goebel W (1994) Transcriptional regulation of *PrfA* and *PrfA*-regulated virulence genes in *Listeria monocytogenes*. *Molecular Microbiology* 11: 1141–1150.

Bubert A, Kuhn M, Goebel W and Köhler S (1992) Structural and functional properties of the p60 proteins from different *Listeria* species. *Journal of Bacteriology* 174: 8166–8171.

Bubert A, Schubert P, Köhler S, Frank R and Goebel W (1994) Synthetic peptides derived from the *Listeria monocytogenes* p60 protein as antigens for the generation of polyclonal antibodies specific for secreted cell-free *L. monocytogenes* p60 proteins. *Applied and Environmental Microbiology* 60: 3120–3127.

Bubert A, Riebe J, Schnitzler N, Schönberg A, Goebel W and Schubert P (1997a) Isolation of catalase-negative *Listeria monocytogenes* strains from listeriosis patients and their rapid identification by anti-p60 antibodies and/or PCR. *Journal of Clinical Microbiology* 35: 179–183.

Bubert A, Kestler H, Götz M, Böckmann R and Goebel W (1997b) The *Listeria monocytogenes iap* gene as an indicator gene for the study of *PrfA*-dependent regulation. *Molecular and General Genetics* 256: 54–62.

Bubert A, Sokolovic Z, Chun S-K, Papatheodorou L, Simm A and Goebel W (1999) Differential virulence gene expression by *Listeria monocytogenes* growing within host cells. *Molecular and General Genetics* 261: 323–336.

Butman BT, Plank MC, Durham RJ and Mattingly JA (1988) Monoclonal antibodies which identify a genus-specific *Listeria* antigen. *Applied and Environmental Microbiology* 54: 1564–1569.

Curiale MS and Lepper W (1994) Enzyme-linked immunoassay for detection of *Listeria monocytogenes* in dairy products, seafoods and meats: collaborative study. *Journal of AOAC International* 77: 1472–1489.

Farber JM and Peterkin PI (1991) *L. monocytogenes*, a foodborne pathogen. *Microbiological Review* 55: 476–511.

Feldsine PT, Lienau AH, Forgey RL and Calhoon RD (1997) Visual immunoprecipitate assay (VIP) for *Listeria monocytogenes* and related *Listeria* species detection in selected foods: collaborative study. *Journal of AOAC International* 80: 791–805.

Mattingly JA, Butman BT, Plank MC and Durham RJ (1988) Rapid monoclonal antibody-based enzyme-linked immunosorbent assay for detection of *Listeria monocytogenes* in food products. *Journal of AOAC International* 71: 679–681.

Niebuhr K, Chakraborty T, Rohde M et al (1993) Localization of the ActA polypeptide of *Listeria monocytogenes* in infected tissue culture cell lines: ActA is not associated with actin 'comets'. *Infection and Immunity* 61: 2793–2802.

Portnoy, DA, Chakraborty T, Goebel W and Cossart P (1992) Molecular determinants of *Listeria monocytogenes* pathogenesis. *Infect. Immun.* 60: 1263–1267.

Swaminathan B, Dewitt WE, Carlone GM et al (1989) A monoclonal antibody specific for certain serotypes of *Listeria monocytogenes*. Abstract B-255, p 73. Abstr. 89th Annual Meeting of the American Society of Microbiology.

Torensma R, Visser MJ, Aarsman CJ, Poppelier MJ, Fluit AC and Verhoef J (1993) Monoclonal antibodies that react with live *Listeria* spp. *Applied Environmental Microbiology* 59: 2713–2716.

Detection by Colorimetric DNA Hybridization

A D Hitchins, Center for Food Safety and Applied Nutrition, Food and Drug Administration, Washington, DC, US

Colorimetric DNA hybridization (DNAH) methodology can be used for rapid detection of members of the genus *Listeria*, including the human pathogen *Listeria monocytogenes*, in contaminated foods after selective cultural enrichment. The key feature of this methodology is the detection of specific nucleic acid probe-target hybrids by indirect linkage to a colour-developing chemical reaction. Currently two commercial *Listeria* colorimetric DNAH test kits are available, both marketed by the same corporation (Gene-Trak® Systems, Hopkinton, MA, US). The first is the Gene-Trak *Listeria* assay for detection of all species of the genus *Listeria*. The second is the Gene-Trak *L. monocytogenes* assay for detection of only one species, *L. monocytogenes*, the causative agent of human and animal listeriosis.

Principle of Gene-Trak DNAH Assays

The two Gene-Trak DNAH assays for the genus *Listeria* and for *L. monocytogenes* are based on the same common principle. The assay systems utilize synthetic oligonucleotide DNA probes directed against ribosomal ribonucleic acids (rRNAs) of the target *Listeria*. Each assay employs two probes, the hybrid capture and hybrid detector probes, which are homologous to unique regions of the rRNAs that are specific to the target microbes and do not occur in other bacteria. The names of the two probes reflect their different roles.

In both assays, cells of *Listeria* spp. are lysed enzymatically with lysozyme and mutanolysin to expose their ribosomes. The rRNAs are released from the ribosomes by denaturation with a detergent. Next the two specific probes are added. If the target rRNA sequence is present, the capture and detector probes will hybridize, in a temperature-dependent way, to different specific regions of the same target molecule. Thus, a triplex hybrid is formed (**Fig. 1**). The capture probe contains a polydeoxyadenylic acid (poly dA) 3′ tail region. The detector probe is fluorescein-labelled at both the 3′ and 5′ ends of its deoxynucleotide backbone.

The triplex hybrids are captured on the polystyrene region of a dipstick. Triplex capture results from polydeoxythymidylic acid (poly dT) molecules attached to the polystyrene hybridizing with the poly dA region of the capture probe component of the triplex. The captured hybrids can now be moved manually between the various reagent and washing solutions involved in the method. Thus, the captured hybrids are purified for the subsequent steps without using cumbersome chemical separation techniques.

The twin fluorescein haptens of the detector probe in the captured triplex are affinity-complexed with polyclonal antifluorescein antibody conjugated to the

enzyme horseradish peroxidase. The enzyme conjugate is detected by adding the substrates, hydrogen peroxide and a chromogen, tetramethylbenzene. A blue colour develops. The colour intensity is proportional to the amount of enzyme which in turn is proportional to the amount of target rRNA in the triplex hybrid that was captured by the dipstick. The enzyme reaction is stopped by acidification which also turns the blue colour to yellow. The yellow intensity is measured photometrically and, if it is in excess of the stated cutoff value, the presence of the *Listeria* target in the test sample is indicated. It can then be inferred that viable *Listeria* was present in the sample but this must be confirmed by cultural isolation and identification of the suspected *Listeria*.

Protocols for DNAH Detection

The kits are recommended by the manufacturer for use with specifically suggested *Listeria*-selective enrichment culture protocols. They are primarily intended for screening secondary enrichment cultures at 48 h and not for identifying colony isolates, although that is possible. The point of application of either of the two DNAH detection methods is at the end of the obligatory preliminary cultural enrichment steps (**Fig. 2**). The primary fluid enrichment differs depending on the nature of the food or environmental sample matrices. The secondary solid-phase enrichment step is the same for all matrices.

All procedures should only be performed by suitably trained personnel using precautions necessary with pathogenic organisms, such as no mouth pip-

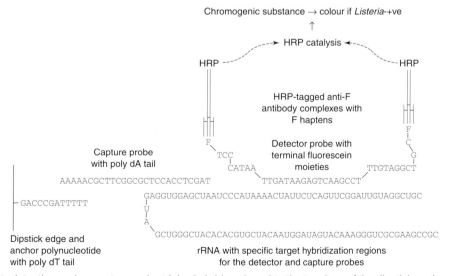

Figure 1 Target : detection probe : capture probe triplex hybrid anchored to the terminus of the dipstick anchor poly dT nucleotide. F = fluorescein moiety; HRP = horseradish peroxidase; A, T, G, C and U signify the nucleotide bases adenine, thymine, guanosine, cytosine and uracil respectively. The nucleotides are deoxyribonucleotides in the DNA probes and ribonucleotides in the RNA target. Specific hybridization regions among the various nucleic acids are indicated by the stretches composed of closely parallel juxtapositions of bases representing the hybridizing interstrand basepairs AT, AU and GC.

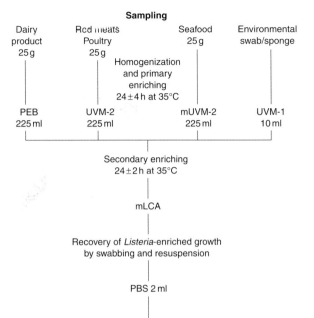

Figure 2 Selective culturing steps for preparing samples for the *Listeria* genus and *L. monocytogenes* DNA hybridization (DNAH) assays. PEB = phosphate-buffered *Listeria* enrichment broth; UVM-1, UVM-2 = University of Vermont media 1 and 2; mUVM-2 = modified UVM-2; mLCA = modified lithium chloride-ceftazidime agar; PBS = phosphate-buffered saline.

etting. This laboratory protocol is for analysing test and control samples by either the colorimetric DNAH test for *Listeria* or that for *L. monocytogenes*. The kits may be used separately or in parallel on the same sample. The amounts suggested in the following protocol for assaying the culture samples are suitable for up to 25 assays. The major equipment, reagents and other items needed to perform the Gene-Trak assay procedure for both *Listeria* and *L. monocytogenes* test kits are shown in **Tables 1** and **2**.

Table 1 Culturing requisites for the DNA hybridization *Listeria* and *L. monocytogenes* assays

Equipment	Media
Incubator (35°C)	Phosphate-buffered *Listeria* enrichment broth (PEB)
Blender or homogenizer	Phosphate-buffered saline (PBS)
Culture bottles (250 ml liquid capacity)	University of Vermont medium 1 (UVM-1)
Culture bottles (110 ml liquid capacity)	University of Vermont medium 2 (UVM-2)
Culture tubes (10 ml liquid capacity)	Modified University of Vermont broth 1 (mUVM-1)
Sterile cotton swabs	Modified lithium chloride-ceftazidime agar (mLCA)

Assay Set-up

The equipment, reagents and accessories needed for assaying the culture samples, obtained by the culturing procedure shown in Figure 2, are listed in **Table 3**.

1. Prewarm water baths to operating temperatures, 37 ± 1 and 65 ± 1°C. Water needs to be about 4 cm deep.
2. To the pretreatment reagent concentrate (bottle 1A) add 12 ml solution 1B (pretreatment reagent buffer). Mix gently to dissolve and store on ice.
3. To reagent bottle 2A (lysis reagent concentrate) add 6 ml solution 2B (lysis reagent buffer). Dissolve and store on ice.
4. Prepare 1.3 l of 1×-wash solution by diluting 65 ml of wash solution 20×-concentrate with 1235 ml deionized (or distilled) water at room temperature.
5. Fill a wash basin with 300 ± 10 ml 1×-wash solution (using level-mark on basin); cover and place in 65°C water bath.
6. Prepare three additional covered basins of wash solution in the same manner but keep them at room temperature.
7. Label a 12×75 mm glass tube for each specimen and the positive and negative controls.

Running the Assay

1. Gently agitate positive and negative control solutions *thoroughly* to mix and add 0.5 ml to appropriately prelabelled tubes. Add 0.5 ml amounts of controls and 0.5 ml amounts of samples from the Gene-Trak secondary enrichment cultures to appropriately prelabelled tubes.
2. Add 0.10 ml reconstituted pretreatment reagent (bottle 1A) to control and test sample tubes. Shake rack manually for 5 s. The mixture will be purple. Incubate in the 37°C water bath for 15 min.
3. With the rack still in the water bath, add 0.1 ml reconstituted lysis reagent (bottle 2A). Manually shake rack of tubes for 5 s. Tube contents will be green at this stage. Replace rack of tubes in water bath at 37°C for 15 min more.
4. Meanwhile, place appropriate number of dipsticks (two for the controls and one each for unknown) into the dipstick holders. Care must be taken to use the handles and not to touch the dipstick fin portions.
5. Using the dipstick holder rinse dipstick fins in 1×-wash solution for 2–3 min at room temperature. Remove excess liquid by blotting tip of fin on to absorbent paper.
6. Add 0.1 ml of solution 3 (*Listeria* genus or *L. monocytogenes* probe solution) to each tube.

Table 2 Formulation and preparation of culture media needed for DNA hybridization assays

Note: Commercial products may be substituted as components or as media *in toto* as consistent with these media formulations

Phosphate-buffered saline (PBS) – 10× stock solution
Sodium phosphate, dibasic, anhydrous 12 g; sodium phosphate, monobasic, monohydrate 2.2 g; sodium chloride 85 g. Dissolve components in distilled water to 1 l final volume. Autoclave at 121°C for 15 min. For use, dilute stock 1 : 9 with distilled water, mix well and, as necessary, adjust pH to 7.5 with 0.1 N HCl or 0.1 N NaOH

Phosphate-buffered *Listeria* enrichment broth (PEB)
Trypticase soy broth powder 30 g; yeast extract 6.0 g; potassium phosphate, monobasic, monohydrate 1.35 g; sodium phosphate, dibasic, anhydrous 9.6 g; distilled water 1 l; acriflavine monohydrochloride (1.2% w/v) 1.2 ml; sodium nalidixate acid (10 mg ml^{-1}) 4.0 ml; cycloheximide (50 mg ml^{-1}) 1.0 ml. Autoclave at 121°C for 15 min

University of Vermont media 1, 2 and modified 2 (UVM-1, UVM-2, mUVM-2)
UVM basal medium: proteose peptone 5.0 g; tryptone 5.0 g; Lab Lemco powder 5.0 g; yeast extract 5.0 g; sodium chloride 20.0 g; potassium phosphate, monobasic, monohydrate, 1.35 g; sodium phosphate, dibasic, anhydrous 12.0 g; aesculin 1.0 g; nalidixic acid (2% w/v in 0.1 N NaOH) 1.0 ml; distilled water 1 l. Dissolve solid ingredients in the water. (*Note:* this basal medium requires addition of 5.0 g lithium chloride before autoclaving, when it is to be modified for use as mUVM-2 medium; see later). Autoclave at 121°C for 15 min. Store at 5 ± 3°C
    Just before use as UVM-1 add 1.0 ml filter-sterile 1.2% w/v acriflavine per litre of medium base
    Just before use as UVM-2 add 1.0 ml filter-sterile 2.5% acriflavine per litre of medium base
    Just before use as mUVM-2 add 1.0 ml filter-sterile 2.5% acriflavine per litre of modified medium base containing 5.0 g lithium chloride per litre (see base medium preparation)

Modified lithium chloride-ceftazidime agar (mLCA)
Brain heart infusion agar 52 g; lithium chloride 10 g: glycine anhydride 10 g; distilled water 1 l; ceftazidine, aqueous solution (10 mg ml^{-1}) 5.0 ml. Combine first four ingredients. Autoclave at 121°C for 15 min. Cool to 45–50°C. Add filter-sterile ceftazidine solution, mix and pour 20 ml amounts into 100 mm diameter Petri dishes. Store moist at 5 ± 3°C in plastic bags for up to 1 month

Table 3 Equipment, reagents and supplies required for the Gene-Trak DNA hybridization *Listeria* genus and *L. monocytogenes* assays

Apparatus and supplies	Reagents
1. Heating water bath at 37 ± 1°C, water depth ca. 4 cm	1A. Pretreatment reagent concentrate
2. Heating water bath at 65 ± 1°C, water depth ca. 4 cm	1A. Pretreatment reagent buffer
3. Photometer (450 nm)	2A. Lysis reagent concentrate
4. Test tube racks, 72-tube (3)	2B. Lysis reagent buffer
5. Dipstick holders (6)	3. Probe solution (*Listeria* or *L. monocytogenes*) (care: sodium azide)
6. Wash basins (4)	4. Wash solution ×20 concentrate
7. Test tubes, borosilicate glass, 12 × 75 mm	5. Enzyme concentrate
8. Graduate cylinder (1 l)	6. Substrate concentrate
9. Minute timer	7. Chromogen
10. Repeater-pipette and tips for 0.10, 0.25 and 0.75 ml volumes (optional)	8. Stop solution (4 N sulphuric acid)
11. Micropipette and tips, 100–1000 µl range (optional)	9. Dipsticks (do not touch fins)
12. Absorbent paper	10. Positive control (care: sodium azide)
	11. Negative control

Generally store reagents at 5 ± 3°C (some may be stored at ambient temperatures). Number corresponds to manufacturer's labelling system for the two assays.

(Solution will turn red.) Using dipstick holders, immerse fins into sample and control tubes in the 37°C waterbath. Rinse and lower dipsticks five times.

7. Transfer rack of tubes from 37°C water bath to 65°C water bath and incubate for 1 h.

8. Meanwhile, label a second set of 12 × 75 mm tubes in another rack. Put 0.75 ml 1×-enzyme conjugate in each empty tube. To make the 1×-enzyme conjugate mix one part of solution 5 (100×-enzyme conjugate concentrate) with 100 parts 1×-wash solution. Use dilute enzyme conjugate within 1 h.

9. After 1 h at 65°C, remove the dipsticks from the tubes, rinse by gentle agitation for 1 min in the 65°C wash solution and then in the room temperature wash solution.

10. Blot the dipsticks dry on clean absorbent paper, and place dipsticks into the second set of tubes containing the enzyme conjugate (step 8). Incubate at room temperature for 20 min.

11. Meanwhile, set up a third rack of appropriately

labelled 12 × 75 mm tubes including an additional tube labelled reagent blank. Add 0.75 ml substrate–chromogen mixture to each empty tube. To prepare the substrate–chromogen reagent, mix solution 6 (substrate solution) and solution 7 (chromogen solution) in a 2:1 proportion.

12. When the 20 min incubation is ended, remove the dipsticks from the enzyme conjugate tubes and wash them by gently shaking each for 1 min in one of the remaining basins of fresh 1×-wash solution. Repeat the 1 min wash in the last basin of fresh 1×-wash solution.

13. Blot dipsticks dry on clean absorbent paper, and then place them in the third set of tubes containing the substrate–chromogen reagent (step 1). Incubate at ambient temperature for 30 min.

14. Remove the dipsticks from their tubes. Discard the dipsticks; add 0.25 ml solution 8 (stop solution) to each tube containing substrate–chromogen and to a reagent blank tube. Manually shake rack of tubes to mix solutions.

15. Measure absorbance of stopped substrate chromogen reaction mixture photometrically at 450 μm using the Gene-Trak photometer. Determine the absorbance values of the positive and negative control reaction mixtures against the reference blank. Determine the absorbance of each test sample reaction mixture against the negative control reaction mixture.

16. The absorbance value for the negative control must be $\leqslant 0.15$ and that for the positive control must be $\geqslant 1.0$ for the assay to be valid. Otherwise, the DNAH assay must be repeated.

17. If the absorbance of the test sample reaction mixture is $\leqslant 0.10$, the sample is inferred to be free of either any *Listeria* spp. or else free of just *L. monocytogenes*, according to which of the two assays is used. Similarly, if the absorbance of the test sample reaction mixture is $\geqslant 0.10$ the sample is inferred to contain the particular target microorganism assayed.

18. If the target microorganism is indicated, its viability should be confirmed by culturally isolating it from the phosphate-buffered saline-suspended LCA growth. Identify the isolate. Recognized methods, such as those of the US Food and Drug Administration or the US Food Safety and Inspection Service, may be used for selective agar isolation and speciation.

Advantages and Limitations of the Colorimetric DNA Probe

The main advantage is the relative rapidity of the test. It takes about 2.5 h minimum, starting with a *Listeria* broth culture or isolated colony. In this respect it compares favourably with other rapid kits for *Listeria* detection. It is inconvenient that frequent manipulations are needed throughout this time period. As with most rapid kits nowadays, no radioactive material is involved, although the originally marketed *Listeria* genus kit did require a radioactive probe. The threshold of detection limit of the method is reported to be about 5×10^5 to 1×10^6 cells per millilitre which compares favourably with the limits of other kinds of rapid detection methods. The observed limit depends on the fact that there are thousands of ribosomes and therefore rRNA targets per cell. Like other DNA–rRNA hybridization probes, these will detect biochemical and morphological variants of *Listeria* and *L. monocytogenes*. The photometer is supplied by the manufacturer, but with proper allowances, other brands of photometer could be used.

Results and Reported Data on Collaborative Evaluations and Validations

A number of institutions worldwide are involved in validating food microbe detection methods. Of these institutions, the Association of Official Analytical Chemists (AOAC) International has one of the most formalized procedures for validating microbial and chemical detection methods. The Gene-Trak *Listeria* genus detection method was validated by the AOAC procedure. Validation of detection methods is conventionally divided into two parts: specificity and sensitivity. Method specificity is preliminarily assessed when the microbial analyte is in excess and under nonselective conditions. Additionally, it is also measured in assay sample conditions. In the recent past, when the Gene-Trak detection method was evaluated, the assumption was made that specificity was not affected by selective conditions in assay samples. Since then this assumption has not been taken for granted but there is no evidence that it is not valid, at least to a first approximation.

Method Specificity

Specificity is measured by the percentage of *bona fide* analyte strains correctly positively identified (the method's inclusivity) and by the percentage of *bona fide* non-analyte strains correctly negatively identified (the method's exclusivity) by the test detection method.

Listeria **Genus Assay**   The strain exclusivity (**Table 4**) and inclusivity (**Table 5**) of the Gene-Trak *Listeria* genus method have been properly validated. Many inclusive strains representative of the number of species and serotypes in the genus were tested. Although the number of exclusive strains examined

Table 4 Generic exclusivity of the Gene-Trak DNA hybridization assays[a]

Listeria genus test kit
   No cross-reactivity with non-*Listeria* genera[b]
L. monocytogenes test kit[c]
   Bacillus (10), *Brocothrix*, *Corynebacterium* (4), *Enterococcus* (22), *Micrococcus* (2), *Staphylococcus* (44), *Streptococcus* (2)

[a] Company data sheet. Species not indicated. One strain per genus was studied, except where stated otherwise in parentheses.
[b] Details not published.
[c] For data on non-reacting species of *Listeria*, see Table 5.

has not been publicly documented, no cross-reactivity was observed. It is interesting that the numbers of exclusive and inclusive strains necessary to validate microbial detection methods have never been formalized. In practice, the inclusivity strain numbers examined vary with the microbial analyte, depending mainly on the number of species and serotypes, their availability and occurrence in food-borne disease outbreaks. The numbers of exclusive strains examined, as is typical in such validations, were more restricted, being strains representative of the exclusivity genera. Emphasis was placed on strains of the genera more likely to be encountered in the analytical and assay samples, in this case on Gram-positive genera.

***Listeria monocytogenes* Assay**   The exclusivity and inclusivity validation strains tested with this assay are also shown in Tables 4 and 5. Again, reliance is placed mainly on manufacturer's data. The remarks made for the generic assay method also apply to this species method. However, in this case greater emphasis was appropriately placed in the exclusivity study on non-analyte *Listeria* spp. since the purpose of the method is confidently to exclude the possibility of confusing other *Listeria* spp. with the target microbe, *L. mono-cytogenes*. The author's laboratory has also examined

exclusivity paying particular attention to *Listeria* strains isolated from foods.

Method Sensitivity

Sensitivity is determined on the method's assay samples worked up from food samples that are naturally or artificially (spiked) contaminated. Either way, the contaminating concentration is determined quantitatively. One problem associated with sensitivity determinations at limiting concentrations of the microbial analyte is the uncertainty associated with microbial enumeration techniques, as this author has pointed out elsewhere. Given a common analyte strain, differences in sensitivity observed between samples will be due to differences in contamination concentrations and differences in the food matrix composition, e.g. the microflora (microbiota) or adverse physico-chemical factors. The latter, in contrast to the former, can often be controlled. Thus, food acidity (low pH) can be neutralized. The microbiota, which can vary in both concentration and composition strain from lot to lot of a food, is supposed to be controlled by the use of appropriate selective factors. But no selective system is perfect and sometimes selective factor-resistant microbiota will decrease the assay's sensitivity.

***Listeria* Genus Assay**   This assay has been subjected to comprehensive intra- and interlaboratory validation tests. The intralaboratory testing was performed by one laboratory with the following food samples: 2% fat milk, chocolate milk, dried non-fat milk, ice cream, cottage cheese, Cheddar cheese, raw shrimp, fish fillet, crabmeat, roast beef, ground raw turkey, ground raw pork, frankfurters, cured ham and fermented sausage. Each food was homogeneously and quantitatively spiked with single various strains representative of relevant *Listeria* spp. and serotypes.

Table 5 Intrageneric strain inclusivities and exclusivities of the Gene-Trak *Listeria* genus and *L. monocytogenes* DNA hybridization assays[a]

	Number of strains tested					
	Listeria *genus test kit*			L. monocytogenes *test kit*		
Species	Total	Positive	Negative	Total	Positive	Negative
L. monocytogenes	188	188	0	143	143	0
L. grayi[b]	6	6	0			
L. innocua	40	40	0	142	0	142
L. ivanovii	11	11	0	2	0	2
L. seeligeri	22	22	0	16	0	16
Listeria spp.				10	0	10
L. welshimeri	19	19	0	7	0	7

[a] For intergeneric exclusivity, see Table 4.
[b] Includes three strains of *L. graysi* subsp. *murrayi*.
[c] Unspecified non-*L. monocytogenes* strains.

Spiking was at low (about $0.04\,\mathrm{cfu\,g}^{-1}$) and high (about $0.4\,\mathrm{cfu\,g}^{-1}$) levels. The uninoculated controls contained $0.04\,\mathrm{cfu\,g}^{-1}$. Sometimes, however, samples with a determined level of natural *Listeria* contamination were used, so there was not a negative control. Any *Listeria* in the analytical samples, 25 g, were sequentially cultured for 48 h total incubation time in two different selective enrichment media, as described earlier. Samples of the secondary cultures were assayed by the colorimetric hybridization method according to the technique already described. Twenty samples per inoculation level per food were analysed. The observed specificities, ranging from 91 to 100% (mainly 100%), were acceptable (note: % false positive rate = 100 − % specificity). These are the specificities under culture sample work-up conditions as opposed to the pure culture specificities described earlier.

Observed sensitivities ranging from 30 to 100%, depending on the actual spiking level, were judged acceptable by AOAC in comparison to conventional culture method controls (note: % false negative rate = 100 − % sensitivity). Theoretically, conventional culture detection methods, and culturally based detection methods like this one, are at best capable of detecting a mean of 1.0 cfu per analytical sample: that is, $0.04\,\mathrm{cfu\,g}^{-1}$ with a 25 g analytical sample size. But, due to the Poisson distribution effect applicable at these low spiking levels, the associated observed positive detection rates can be appreciably below 100%, e.g. about 60%. Therefore, the corresponding negative rates will be about 40%. These true or pseudo false-negative rates are due to unavoidable technical failure to spike the sample homogeneously. When the published data were amenable, the author calculated the 50% positive endpoint levels. They ranged from 0.01–$0.12\,\mathrm{cfu\,g}^{-1}$ or 0.25–3.0 cfu per 25 g analytical sample, with seafoods and cottage cheese being on the high end. Compared with conventional culture methods and given the relative inaccuracies in enumerating the spiked levels, these values are acceptable validations.

Based on these intralaboratory results a representative number of these foods (six) were chosen as matrices for more intensive validation of the detection method: 2% milk, Brie cheese, crab meat, cooked frankfurter sausage, roast beef and raw ground pork. The experimental protocol was similar to the intralaboratory study. An important difference was that the number of laboratories involved was increased to 12–20, varying arbitrarily with the particular food sample. There were five analytical sample replicates per level per food matrix. Also, a critical point was that the laboratories validated spiked or naturally contaminated samples centrally prepared by and distributed from one laboratory. These crucial characteristics of an AOAC interlaboratory collaborative study place emphasis on measuring the reproducibility of the method between different laboratories and their analysts whilst attempting to keep sample homogeneity constant within the limitations of the Poisson distribution at low spiking levels. The results are shown in **Table 6**. The sample culture specificities varied from 92 to 100%. The sensitivities ranged from 60 to 100%, depending on the spiking level. Allowing for the spiking variability, calculated 50% end point values were 0.01–$0.08\,\mathrm{cfu\,g}^{-1}$, with Brie and crab being at the high end of this range. These end points are acceptably variable about the theoretical mean value of $0.04\,\mathrm{cfu\,g}^{-1}$. There was good agreement between the method and the cultural control method.

The method has been approved for use with dairy products, meats and seafoods and has been designated AOAC Official Method 993.09. In addition, the method has been studied with foods by several other laboratories (see Further Reading). The original method, which used a different selective agar from the current modified Oxford medium (MOX), performed slightly less well than the International Dairy Federation (IDF) cultural method for samples of dairy products and related production wastes. The slight superiority disappeared when the IDF selective medium, Oxford agar, was substituted into the Gene-Trak *Listeria* method. This improvement may be due to the Oxford agar, like the current MOX agar, supporting better growth of *Listeria* while still suppressing growth of interfering microflora. In another study with raw milk, the method performed better than both the cultural control method and the Listeria-Tek colorimetric monoclonal ELISA method. In a French study, the method did not perform as well as the control cultural method with vegetable products and pastries, though performance with raw-milk cheeses and shrimp was comparable. It should be noted that these are results from single laboratory studies and not from multilaboratory studies using common samples.

***Listeria monocytogenes* Assay**   This assay has not been officially validated but the manufacturer has done an in-house validation. The results obtained are presented in **Table 7**. Sensitivities calculated from the artificially inoculated dairy product sample results ranged from 67 to 83% at a spiking level of $0.4\,\mathrm{cfu\,g}^{-1}$. These were at least comparable to the corresponding values obtained with the control culture method. However, values closer to 100% would be expected at this level (10 cfu per 25 g analytical sample). A possible explanation for the deficit is the fact that the spike deliberately contained a numerically equivalent

Table 6 Comparison of Gene-Trak *Listeria* and conventional culture *Listeria* detections in selective enrichments of artificially and naturally contaminated food samples[a]

Food	Spike (cfu g^{-1})	Sensitivity rate[b]		Specificity rate[c]	
		DNAH	Culture	DNAH	Culture
Milk, 2%	0.24	84.4	95.3		
	0.24	83.6	91.8		
	< 0.003			98.4	100.0
Cheese, Brie	11.0	98.3	98.3		
	1.5	96.6	89.8		
	< 0.003			98.3	93.3
Crab	0.24	85.9	65.6		
	0.24	71.4	47.6		
	<0.003			100.0	79.4
Frankfurters	0.043	69.5	96.6		
	0.015	60.3	91.4		
	< 0.003			92.3	92.3
Beef, roast[d]	0.46	100.0	98.7		
	0.093	97.3	91.9		
	0.023	94.2	85.5		
Pork, ground[d]	0.24	100.0	89.2		
	0.24	98.4	89.1		
	0.024	83.1	91.5		

[a]Subsamples (25 g) from a centrally prepared homogeneous sample analysed by US Food and Drug Administration (dairy foods) or US Food Safety and Inspection Service (meats, seafood and sausages) methods by 15–17 laboratories.
[b] Incidence of positive samples among total positive samples by test and control methods.
False-negative percentage rate = 100 – sensitivity.
[c] Incidence of negative samples among total negative samples by test and control methods.
False-positive percentage rate = 100 – specificity.
[d] Naturally contaminated samples.

Table 7 Comparison of Gene-Trak *Listeria monocytogenes* and conventional culture *L. monocytogenes* detection in selective enrichments of artificially and naturally contaminated food samples[a]

Food	Number of observations				
	Total assays	Test + culture +	Test + culture −	Test + culture +	Test − culture −
Artificially contaminated dairy products[b]					
Cheese, Brie	10	7	1	0	2
Cheese, Cheddar	12	7	3	0	2
Cheese, cottage	15	10	0	0	5
Cheese, soft blue	12	7	3	0	2
Cream, half and half	25	18	0	0	7
Milk, chocolate	12	9	1	0	2
Milk, non-fat dry	12	8	2	0	2
Naturally contaminated seafoods[b]					
Crabmeat	10	5	0	0	5
Scallops	10	1	3	0	6
Shrimp	10	0	1	0	9
Naturally contaminated meats[c]					
Beef, ground	10	0	10	0	0
Pork, ground	7	1	6	0	0
Beef and lamb, ground	8	0	7	0	1
Sirloin, ground	8	7	1	0	0
Chicken, ground	8	2	1	0	5
Turkey, ground	10	0	3	0	7

[a] Foods inoculated with 10 cfu each of *L. monocytogenes* and *L. innocua* per 25 g.
[b] Compared with the US Food and Drug Administration *Bacteriological Analytical Manual* method.
[c] Compared with the US Food Safety and Inspection Service method.

level of *L. innocua*. This species is known to be capable of competing effectively with *L. monocytogenes*, since at least some strains of the former are known to grow faster than some strains of the latter species. If so, the target analyte strain may not always have been able to achieve the method's required threshold level. Also, the control cultural method's performance would be expected to be adversely affected, since it depended on being able to find the *L. monocytogenes* colonies among the, probably preponderant, look-alike *L. innocua* colonies also present on the selective agar medium. Thus, these methods were evaluated under deliberately stringent conditions.

Also shown in Table 7 are results with naturally contaminated seafoods and meats. Firm comparisons are difficult to make when the level of contamination is unspecified but it appears that the method was as good as the control culture method and even better in the case of some of the meats. This incompletely validated method is probably most economically and appropriately used in conjunction with the fully validated *Listeria* genus method. When *Listeria*-positive enrichments or isolated *Listeria* colonies are obtained with the genus kit they can be tested further to see if they contain, or are, *L. monocytogenes*.

See also: **Bacteriophage-based Techniques for Detection of Foodborne Pathogens**. **Biochemical and Modern Identification Techniques**: Food-poisoning organisms. **Enzyme Immunoassays**: Overview. **Food Poisoning Outbreaks**. **Hydrophobic Grid Membrane Filter Techniques (HGMF)**. **Immunomagnetic Particle-based Techniques**: Overview. **Listeria**: Detection by Classical Cultural Techniques; Detection by Commercial Enzyme Immunoassays; *Listeria monocytogenes* – Detection by Chemiluminescent DNA Hybridization; *Listeria monocytogenes* – Detection using NASBA (an Isothermic Nucleic Acid Amplification System). **National Legislation, Guidelines & Standards Governing Microbiology**: European Union; Japan. **Nucleic Acid-based Assays**: Overview. **PCR-based Commercial Tests for Pathogens**. **Polymer Technologies for Control of Bacterial Adhesion**. **Sampling Regimes & Statistical Evaluation of Microbiological Results**.

Further Reading

Since only two colorimetric DNAH assays for *Listeria* are currently marketed and they are closely related, being produced by the same company, this article has relied heavily on the package inserts and product information sheets to provide an example of the detail needed for the technical performance of such assays.

AOAC Official Method 993.09 (1996) *Listeria* in dairy products, seafoods, and meats. Colorimetric deoxyribonucleic acid hybridization method (Gene-Trak *Listeria* assay). Official Methods of Analysis, Supplement. Gaithersburg, MD, US: AOAC International.

Curiale MS, Sons T, Fanning L et al (1994) Deoxyribonucleic acid hybridization method for the detection of *Listeria* in dairy products, seafoods, and meats: collaborative study. *J. AOAC Int.* 77: 602–617.

Duvall RE and Hitchins AD (1997) Pooling of non-collaborative multilaboratory data for evaluation of the use of DNA probe test kits in identifying *Listeria monocytogenes* strains. *J. Food Protect.* 60: 995–997.

Food and Drug Administration (1995) *Bacteriological Analytical Manual*, 8th edn. Gaithersburg, MD, US: Association of Official Analytical Chemists International.

Hitchins AD (1998) Retrospective interpretation of qualitative collaborative study results: *Listeria*. 112th AOAC Interntl. Ann. Mtg. Final Program. Abstract J-702, p. 102.

Lachica VR (1990) Selective plating medium for quantitative recovery of food-borne *Listeria monocytogenes*. *Appl. Environ. Microbiol.* 56: 167–169.

Ravomanana D, Richard N and Rosec J (1993) *Listeria* spp. dans des produits alimentaires – étude comparative de différents protocoles de recherche et d'une méthode rapide par hybridation nucléique. *Microbiol. Aliment. Nutr.* 11: 57–70.

Rodriguez JL, Gaya P, Medina M and Nuñez M (1993) A comparative study of the Gene-Trak *Listeria* assay, the *Listeria*-Tek ELISA test and the FDA method for the detection of *Listeria* species in raw milk. *Lett. Appl. Microbiol.* 17: 178–181.

Url B, Heitzer A and Brandl E (1993) Determination of *Listeria* in dairy and environmental samples: comparison of a cultural method and a colorimetric nucleic acid hybridization assay. *J. Food Protect.* 56: 581–592.

US Department of Agriculture (1989) FSIS method for the isolation and identification of *Listeria monocytogenes* from processed meat and poultry products. Laboratory Communication no. 57. Washington, DC, US: USDA.

Detection by Commercial Immunomagnetic Particle-based Assays

Barb Kohn, VICAM LP, Watertown, USA

Introduction

Immunomagnetic separation (IMS) for isolation of *Listeria* serves as the basis for assays that are ideally suited to food industry applications. These include hazard analysis and critical control point (HACCP) programmes as well as in-process and finished product testing. IMS can be integrated into such testing in two ways. There are IMS methodologies that have been specifically developed for these applications. Alternatively, IMS can be performed during execution of standard cultural methods to increase sensitivity and reduce test time.

Several attributes of IMS account for its useful properties of speed and sensitivity. Primary among these is the concentrating effect of IMS beads. Whereas enrichment increases bacterial number per unit volume as bacteria multiply over time, IMS immediately concentrates bacteria from a larger volume into a smaller volume. This is because IMS beads can be dispersed into a significant sample volume to bind pathogens, and can then be magnetically concentrated and resuspended in a smaller volume for further analysis. Enrichment can therefore be eliminated or minimized. Because of this, one IMS method for *Listeria* detection produces results – positive and negative – in 24 h. Eliminating 2–6 days of test time allows earlier shipping of finished product and faster decontamination when pathogens are detected in the processing environment.

Omission of enrichment renders the combination of IMS with an appropriate detection method capable of enumerating pathogens and detecting injured cells. Enumeration of cells subsequent to IMS provides more information than enrichment-based methods which only report pathogens as present or absent. For example, during sanitizing to eliminate contamination, pathogen numbers may be decreasing, but may still be sufficient to result in continued positive results in tests that give either a positive or a negative result only. This could obscure the effectiveness of the chosen sanitation protocols during the decontamination process. Injured *Listeria* cells are often present in environmental and food samples. These injured organisms are more readily detected with IMS-based methods than with enrichment-based methods. This is because the separating and purifying properties of IMS minimize the presence of background flora which must otherwise be controlled, as in enrichment-based methods. Background flora are traditionally suppressed by use of selective agents which are lethal to injured cells and which can be omitted from IMS-based methods.

IMS can be performed without culturing organisms, facilitating the detection of the only *Listeria* species whose presence is regulated – *Listeria monocytogenes*. When *Listeria monocytogenes* is co-present with other *Listeria* species, it is often overgrown and rendered undetectable. Accurate detection becomes possible with IMS even under conditions where other *Listeria* species would overgrow, obscuring detection

by other methods. These characteristics make IMS valuable for *Listeria* control by both food and environmental monitoring throughout the food industry.

Development and Use of IMS Products for Food Testing

Historical Origins of Magnetic Beads

IMS has only recently been extended to food, having been used historically in the separation of mammalian cell types. One example is the removal of white blood cells from bone marrow aspirates that would otherwise cause graft-versus-host disease when used to reconstitute individuals with lymphoid cancers who had received radiation therapy. Beads derived from mammalian applications, but coated with antibodies to bacteria, were applied to the challenging task of isolating target pathogens from foodstuffs and processing environment samples. These beads, perhaps due to size or surface chemistry, could not purify the targeted bacteria efficiently, and exhibited extensive nonspecific binding.

Later, beads designed specifically for food applications were developed. This development was important because the requirements for beads used with foods are unique. Separation of mammalian cell types usually presents many large targets in a defined mixture. In foods, smaller pathogens are present at vanishingly low levels in very complex matrices, against a background of other flora. This is especially so if the food is raw or microbiologically cultured during production, as are some sausages, cheese, yoghurt, and many other staples.

Commercial IMS Products for Isolation of *Listeria*

Although IMS has been applied to isolation of many target pathogens, two companies have developed applications for *Listeria*. Dynal markets a superparamagnetic polystyrene bead of 2.8 μm diameter (Dynal AS, Oslo, Norway). This is used to capture *Listeria* from any food, feed or environmental sample that has been pre-enriched for 18–24 h in UVM I or half Fraser broth. According to Dynal representatives, scientific publications for the *Listeria* bead product are not available; however, there are many references for the use of Dynabead IMS products for *Salmonella* and *Escherichia coli* O157:H7. Dynal *Listeria* beads do not have any formal approvals.

VICAM (Watertown, MA, USA) manufactures a complete test kit for IMS-based detection of *Listeria* in foods, feeds and environmental samples. Developed specifically for the food industry, the patented beads are superparamagnetic iron-containing complexes several hundred nanometres in size. VICAM's *Listeria* kit is approved by the French Association of Normalization (AFNOR) in France and by the Association of Official Analytical Chemists' Research Institute (AOAC RI) in the United States.

Sample analysis depends on the bead being used. All of the applications for Dynal beads begin with pre-enrichment for 18–24 h in selective broth, dictated by the cultural method in use. Then 1 ml pre-enriched sample is treated with beads, which are washed and plated onto selective medium, usually Oxford or PALCAM. Isolated colonies are then analysed by standard serological or biochemical tests of the user's choice.

The _Listeria_ IMS kit by VICAM is sold worldwide under the trade name Listertest, except in France where a variant is called Listerscreen. The IMS kit consists of specially developed magnetic beads and a detection system for analysing colonies on plates (**Fig. 1**). IMS can be done with or without enrichment, as the beads efficiently bind _Listeria_ (**Fig. 2**). Various media can be used for plating beads to allow growth of _Listeria_. For use of IMS with cultural methods, 2 ml of either the pre-enrichment or the enrichment can be analysed. After IMS separation, the beads are plated on the medium described in the cultural method protocol. If high bacterial levels are expected, it is possible to streak for single colonies.

Detection of IMS-isolated _Listeria_ depends on the user's chosen method. Detection steps in the cultural method are followed as described once colonies are of sufficient size for processing. The detection step in the VICAM IMS method is carried out by creating an imprint of the master plate on a plastic membrane. Bacteria are killed and exposed to anti-_Listeria_ antibodies. Enzyme-conjugated second antibody is bound to _Listeria_-specific antibody on the membrane. After this a substrate for the enzyme is added. Sites on the membrane that form dark purple spots or rings thus correspond to _Listeria_ colonies on the master plate (**Fig. 3**). Speciation is accomplished by picking isolated _Listeria_ colonies onto rhamnose agar and blood agar. Those colonies which are positive on both are _Listeria monocytogenes_.

Results of IMS Testing of Food and Food Processing Environment Samples

Dynal Publications

The use of beads that were the precursor to those sold by Dynal to isolate _Listeria_ from food samples was reported by Skjerve and colleagues. The beads were coated with an antibody to flagella which recognized a subset of _Listeria_ species. Using these beads, Skjerve and colleagues isolated _Listeria_ from samples spiked at high levels with the organism. Maximal efficiency of capture was only about 10% from a defined buffer solution. This low capture efficiency was cor-

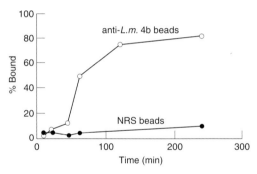

Figure 2 Graph of IMS-based capture of _Listeria_ from buffer. Several hundred cfu of _Listeria_ were added to buffer and subjected to IMS isolation. The magnetic beads used in IMS were either coated with anti-_Listeria_ antibody (anti-_L.m._) or with antibodies from rabbit serum obtained from rabbits not exposed to _Listeria_ (NRS). The number of cfu recovered was evaluated with various capture times and expressed as a percentage of the original cfu added to buffer.

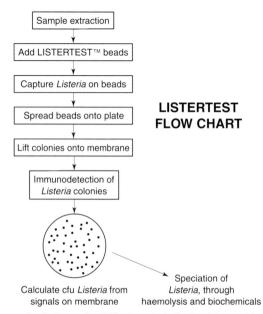

Figure 1 Pathway for IMS of analysis for food-borne _Listeria_.

ListerTest™ Results

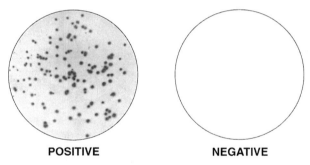

Figure 3 Detection of IMS-isolated _Listeria_ on plastic membranes, processed as described in the Listertest manual.

roborated by a VICAM study of commercially available Dynal beads (**Table 1**). Capture of *Listeria* was below 10%, versus capture rates in excess of 80% for VICAM beads.

VICAM Publications

Jackson (now Kohn) and coworkers presented studies on a 24 h (total elapsed time) quantitative assay of *Listeria* in milk, meats and environmental samples. Their work included some basic information about bead performance. They studied capture kinetics for 0.1 mg beads (by iron weight) in 2 ml of buffer containing 260 colony-forming units (cfu) of *Listeria*. Capture of *Listeria* was almost half maximal at 50 min. Capture was almost complete within 200 min, with binding of the majority of the input organisms. This high capture efficiency was specific, as it did not occur with beads coated with antibody not directed against *Listeria* (Fig. 2). Capture was linear over the entire range tested, from less than 10 to 300 000 cfu. Interference by other organisms was negligible, with capture efficiency towards 200 cfu of *Listeria* remaining above 79% in the presence of a 2–5-fold excess of *Enterococcus faecium*, *Bacillus subtilis* or *Escherichia coli*.

Assay of food samples revealed that as few as 0.05–0.13 cfu g^{-1} were recovered from milk and meat, respectively, without enrichment. For environmental samples and sausage which were both naturally contaminated, Listertest detected low levels of *Listeria* which were missed by standard cultural methods. In analysis of desludge, a waste product from milk which is produced during cheese-making, Listertest captured

74% of the 95 cfu of *Listeria* spiked into desludge containing about 10^8 mixed flora. In their marketing literature, Dynal also cites the ability to efficiently capture *Listeria*, when as few as 100 cfu are present in mixed enteric flora of 10^6 organisms.

It is also possible to recover injured cells with Listertest, in contrast to methods based on enrichment, due to the type and concentration of selective agents used. Enrichment-based methods would include standard cultural methods and some IMS methods such as those of Dynal, which require enrichment prior to IMS. Mitchell and her colleagues injured over 93% of the *Listeria* in a culture by drying the cells onto environmental surfaces. Degree of injury was assessed by differential plating on agar with and without salt. Healthy cells plate with equal efficiency on both media whereas injured cells fail to grow under the added osmotic stress of salty medium. Under conditions where the extreme degree of cell injury resulted in 64% false negatives by the cultural method, Listertest successfully detected *Listeria* in 100% of the spiked environmental samples analysed by Mitchell and her colleagues.

Other Publications

The French variant of Listertest is called Listerscreen and was described in detail by Avoyne and her colleagues, who evaluated the test at the Laboratoire de Touraine and in a multi-laboratory collaborative study in France. They found IMS capture of less than 5% of input cfu for most Gram-positive bacteria from the genera of *Bacillus*, *Brochothrix*, *Enterococcus*, *Erysipelothrix*, *Jonesia*, *Kurthia*, *Lactobacillus*, *Micrococcus*, *Nocardia*, *Sarcine*, *Staphylococcus*, *Streptococcus* and *Streptomyces*. Capture of *Corynebacterium* strains was noted, but at levels less than half that of *Listeria monocytogenes* serogroups 1/2 or 4. None of these strains except *Listeria* grew on the Listerscreen medium except one strain of *Streptomyces* which was not captured, so all non-*Listeria* produced negative results.

Listerscreen was compared to a cultural method for analysis of meat, dairy products, vegetables, seafood, and ready-to-eat products. Each method produced the same number of positive and negative results. One false negative was produced by each method, out of 211 samples.

In a multi-laboratory collaborative study using milk spiked with *Listeria monocytogenes*, nine laboratories agreed on 143 out of 144 samples. Spike levels ranged from 12 to 117 organisms per 25 grams. These results were among those resulting in approval by AFNOR of the Listerscreen method in 1995.

The United States Food and Drug Administration (US-FDA) performed a study comparing Listertest to

Table 1 Comparison of VICAM and Dynal beads for IMS isolation of *Listeria*

Listeria added (cfu)	Listeria captured (cfu)	
	Dynal	VICAM
28	2	26
244	5	245
2440	52	1552
24 400	625	15 520
244 000	4270	TNTC[a]

Listeria monocytogenes were spiked into 2 ml of sterile phosphate-buffered saline at the indicated numbers of cfu. Either Dynal or VICAM immunomagnetic beads were added at the concentration recommended by the manufacturers, and incubation was carried out for one hour with gentle rotation. Subsequently, beads were isolated on a magnetic rack containing rare earth magnets to generate high magnetic field strength. After two washes with 2 ml buffer, beads were resuspended in buffer, diluted as necessary and plated onto brain heart infusion medium. After overnight incubation at 37°C, the number of colonies on the plates was counted and used to calculate total number of *Listeria* captured.

[a]TNTC, too numerous to count.

Table 2 IMS permits detection of *Listeria monocytogenes* in the presence of *Listeria* spp. under conditions where cultural methods fail

Ratio L.m./L. spp	USDA method	VICAM
L.m.: *L. innocua*		
2 : 1	L. spp.	L.m.
1 : 1	L. spp.	L.m.
1 : 10	L. spp.	L. spp.
L.m.: *L. welshimeri*		
1 : 2	L. spp.	L.m.
1 : 4	L. spp.	L.m.

Mixtures of *L. monocytogenes* (*L.m.*) and *L. innocua* or *L. welshimeri* were made in buffer and subjected to analysis by the USDA cultural method or by VICAM's Listertest.

Table 3 Detection of *Listeria* at 1 cfu per 29.4 g milk by IMS

Cfu added	Cfu recovered	Detection limit (cfu g^{-1})
9	9	0.034
17	14	0.068
170	120	0.68

Unenriched whole milk (250 g) was spiked with *Listeria monocytogenes* and centrifuged for 30 min at 7000 r.p.m. in a JA-14 rotor in a Beckman J2-21 centrifuge. The pellet was resuspended in 10 ml of buffer to which was added 0.5 ml of beads (0.5 mg iron). After 1 h for capture, beads were washed twice with buffer, resuspended in a small volume, and plated onto nonselective medium. After overnight incubation, *Listeria* were visualized using a colony blot procedure similar to that described in the Listertest user's manual.

the most probable number (MPN) method of analysis for *Listeria monocytogenes* in cooked crabmeat and cold smoked salmon. In this study, McCarthy found that the results for both methods were statistically similar for salmon, whereas Listertest was superior in recovery efficiency from crabmeat. It was concluded that Listertest was both faster and less cumbersome than the MPN method, and was suitable for enumeration of *Listeria* from cooked crabmeat and smoked salmon.

Listertest was deemed effective for recovery of *Listeria* from raw and pasteurized milk in a Canadian study in which it was noted that centrifugation was necessary to remove fat from raw milk, in order to improve the detection step. It is highly likely that centrifugation was necessary for IMS to be effective as well.

The United States Army performed a study of Listertest in foods, by comparing it to the standard cultural method devised by the United States Department of Agriculture. In this work, Metcalf and Lachica analysed ham, hot dog, shrimp and roast beef, by spiking them with *Listeria* at 0.18–7.3 cfu g^{-1} and evaluating the percentage of organisms recovered. They found variable recoveries, with poorer recovery at very low spike levels or in fatty samples. They also showed that *Listeria monocytogenes* could be recovered more effectively from ground beef using Listertest than by the cultural method. This was true even when *Listeria innocua* was co-present, a finding corroborated by a similar study at VICAM (**Table 2**) on recovery of *Listeria monocytogenes* in the presence of other *Listeria* species. It was reported that the cultural method gave presumptive positive results in all cases, requiring days of biochemical testing to confirm or exclude *Listeria*. Upon completion, the cultural method gave uniformly negative results, even though *Listeria monocytogenes* had been spiked into all samples. In contrast, Listertest detected *Listeria monocytogenes* in seven of the ten spiked samples.

A study in the Republic of South Africa focused on a comparison of rapid methods, including two enzyme-linked immunosorbent assay (ELISA) methods and the IMS method of Listertest. In a study of these methods on spiked hake mince carried out at the Fishing Industry Research Institute, Listertest was shown to detect *Listeria* with equal efficiency regardless of the enrichment medium used, whereas detection in the two ELISA tests was dependent on enrichment. The Listertest IMS method and one of the ELISA methods were shown to be similar in sensitivity to the cultural method used by the Fishing Industry Research Institute. Listertest was more complex and costly, but the ability to quantitate and speed of result generation were praised.

Performance Limits of IMS

The concentrating effect of beads developed specifically for food is striking indeed. By using conditions different from those recommended for the kit, capture of *Listeria monocytogenes* spiked at 0.034 cfu ml^{-1} into milk was achieved (**Table 3**). This was done without enrichment. It simply required a larger sample to be analysed, using proportionately more beads. This performance breaks the 1 cfu per 25 grams sensitivity limit associated even with enrichment, since it represents a sensitivity of 1 cfu in 29.4 grams of non-enriched food sample.

Advantages and Limitations of IMS

IMS for *Listeria* has many attractive features. It is 50–80% shorter than cultural methods. The results are also sensitive, and permit recovery of even injured *Listeria* in the case of the VICAM IMS system. IMS without enrichment is more capable than other methods of identifying *Listeria monocytogenes* when these are mixed with other *Listeria* spp. This is important because food industry regulations generally address the presence of *L. monocytogenes* only. Test selectivity is good enough to minimize false positives, because IMS can select and isolate the target organism from a background of competing microflora. As the minimization or elimination of enrichment vastly reduces exposure of the operator to pathogens, IMS is an extremely safe analytical method.

IMS is, however, considered in some cases more tedious due to washing steps. This concern has been addressed in the case of the VICAM IMS by introduction of a semi-automated multi-channel pipette. Both Dynal and VICAM beads can also be used with automated systems.

Sensitivity of IMS-based tests can be compromised by factors that reduce efficiency of bead capture of *Listeria* or bead binding to magnets. These factors include fat content and viscosity of samples. Enrichment minimizes these effects. In the absence of enrichment, interference with bead capture can be reduced by centrifugation or addition of non-toxic amounts of surfactants or celite (a form of diatomaceous earth).

Cost has been described as higher than for cultural methods. The additional cost can be offset by savings in other areas. Processors can ship finished goods more quickly, especially important for products with short shelf lives. The costs of carrying inventory while waiting for test results are also reduced, and recalls are minimized.

The final analysis of IMS advantages and disadvantages must address the purpose of IMS use by the food industry. Pathogen detection programmes reduce risk and liability. They improve product quality and marketability. For these reasons, the quality advantage and superior risk management IMS affords testers and purchasers of IMS-tested goods are the greatest benefits of this powerful technology.

Case Study of Decontamination of a Processing Environment

The ability to recover injured cells was put to use in a case study reporting decontamination of processing lines. In this report, Listertest was used to rank sites by contamination levels, in order to target sanitizing to sites containing *Listeria*. The effectiveness of sanitizing was then measured, by enumeration of declining numbers of *Listeria* as cleaning progressed. Turnaround time for each assay was 24 hours total, which significantly decreased the time required to act on reported positives by implementing additional sanitizing steps.

Several processing lines were studied, with special emphasis on machines near the end of the lines, which could contact finished product. One line ended in a manually operated slicing machine which was shown to be slightly contaminated with *Listeria* spp. on the handle and product contact regions, at levels of 20–

$30 \, cfu \, ml^{-1}$ of environmental sample. The rest of the line appeared negative at less than $0.5 \, cfu \, Listeria$ spp. per environmental sample of $4 \, cm^2$. Low levels of *Listeria* were detected on incoming raw product. After the normal sanitation procedure which included disassembly, the slicer was more extensively re-tested. Although the handle and product contact surface did not show the presence of any *Listeria*, the hinges, screws and knobs holding the blade in place were contaminated at levels too numerous to count (in excess of $250 \, cfu \, Listeria$ spp. per environmental sample of a $4 \, cm^2$ area). A more intensive sanitizing which included a heat treatment rendered the machine negative for *Listeria* upon further environmental sampling.

A second processing line had an automated slicer at the end. Raw product entering the line was contaminated at low levels with *Listeria* spp., and the slicer was contaminated to different degrees depending on the area sampled. The area between the array of slicing blades was weakly contaminated at $2 \, cfu \, ml^{-1}$, the product contact region was contaminated at $200 \, cfu \, ml^{-1}$, and the blade contact region was highly contaminated at $1000 \, cfu \, ml^{-1}$, measurable only by sample dilution. The standard sanitizing method recommended by the manufacturer was only partially successful in decontaminating the machine, a fact revealed by Listertest that would not have been noticed in a standard enrichment assay which is not quantitative. When the machine was further sanitized, no further positive samples were detected.

A similar situation existed with a processing line mixer bowl. Low levels of *Listeria* were detected on the lid. The locking rings and shaft used to connect the bottom of the bowl to the mixer were too highly contaminated to quantitate. The first cycle of sanitizing reduced contamination to $22–410 \, cfu \, ml^{-1}$ on these areas, whereas a final cycle of sanitizing eliminated all positive samples.

Use of IMS for *Listeria* in this processing application produced several benefits. The quantitative nature of the test revealed sources of high levels of *Listeria* which could be eliminated to minimize the risk of cross contamination. It also provided a measure of sanitizing effectiveness, as numbers of *Listeria* isolated from the slicers and bowl began decreasing. The rapidity of testing allowed lines to be down for the minimal amount of time. As of this writing, three years have elapsed without further isolation of *Listeria*-positive samples from these processing environments. This is the strongest endorsement of the use of IMS for *Listeria* isolation that could exist in the food industry.

See also: **Escherichia coli O157**: Detection by Commercial Immunomagnetic Particle-based Assays. **Immunomagnetic Particle-based Techniques**: Overview. **Listeria**: Detection by Classical Cultural Techniques; Detection by Colorimetric DNA Hybridization; *Listeria monocytogenes* – Detection by Chemiluminescent DNA Hybridization; *Listeria monocytogenes* – Detection using NASBA (an Isothermic Nucleic Acid Amplification System. **National Legislation, Guidelines & Standards Governing Microbiology**: Japan. **Nucleic Acid-based Assays**: Overview. **PCR-based Commercial Tests for Pathogens. Polymer Technologies for Control of Bacterial Adhesion. Reference Materials. Salmonella**: Detection by Immunomagnetic Particle-based Assays. **Sampling Regimes & Statistical Evaluation of Microbiological Results**.

Further Reading

Avoyne C, Butin M, Delaval J and Bind J-L (1997) Detection of *Listeria* spp. in food samples by immunomagnetic capture: ListerScreen method. *Journal of Food Protection* 60: 377–384.

Farber JM (1993) Current research on *Listeria monocytogenes* in foods: an overview. *Journal of Food Protection* 56: 640–643.

Jackson B, Brookins A, Tetreault D and Costello K (1993) Detection of *Listeria* in food and environmental samples by immunomagnetic bead capture and by cultural methods. *Journal of Rapid Methods and Automation in Microbiology* 2: 39–54.

Kohn B, Costello K and Phillips A (1997) HACCP verification procedures made easier by quantitative *Listeria* testing. *Dairy, Food, Environment and Sanitation* 17: 76–80.

McCarthy S (1997) Evaluation of the Listertest method for quantitation of *Listeria monocytogenes* in seafoods. *Journal of Food Protection* 60: 424–425.

Mitchell B, Milbury J, Brooking A and Jackson B (1994) Use of immunomagnetic capture on beads to recover *Listeria* from environmental samples. *Journal of Food Protection* 57: 743–745.

Skjerve E, Rorvik L and Olsvik O (1990) Detection of *Listeria monocytogenes* in foods by immunomagnetic separation. *Applied and Environmental Microbiology* 56: 3478–3481.

Listeria monocytogenes

Scott E Martin and **Christopher W Fisher**, Department of Food Science and Human Nutrition, University of Illinois, Urbana, USA

Listeria monocytogenes is the causative agent of the disease listeriosis and was discovered almost 90 years ago. In 1926 *L. monocytogenes* infection in rabbits and guinea pigs was reported. A typical mononucleosis-like infection was observed in these animals and the organism was called *Bacterium monocytogenes*. The organism can be found intracellularly within monocytes and neutrophils, and the name is derived from the observation that increased numbers of monocytes are found in the peripheral blood of infected animals. In 1929 the organism was isolated for the first time from the blood of a human with a mononucleosis-like illness. The genus *Listerella* was given to the organism to honour the surgeon Lord Lister. The genus was changed to *Listeria* in 1940 for taxonomic reasons. In 1981 it was recognized as an important food-borne pathogen. *L. monocytogenes* causes a high percentage of fatalities in food-borne disease, exceeding even *Clostridium botulinum* and *Salmonella*. It has been suggested that listeriosis may be the leading fatal food-borne infection in the US. Consumption of foods contaminated with *Listeria* can cause both sporadic illness as well as food-borne disease epidemics. The annual rate of listeriosis incidence is 0.7 cases per 100 000. The rate is three times higher for persons over 70 and is 17 times higher for pregnant women. The overall fatality rate for recent outbreaks has been 33%.

The number of recognized listerial species varies with the source of reference. Some authors have suggested that there are five species recognized in the *Listeria* genus: *L. monocytogenes*, *L. ivanovii*, *L. seeligeri*, *L. innocua* and *L. welshimeri*. In addition to these five species, the species *L. grayi* has been included, as well as a seventh species, *L. murrayi*. *L. monocytogenes* is pathogenic for humans and animals, *L. ivanovii* is an animal pathogen, and the remaining species are generally considered to be nonpathogenic. However, both *L. innocua* and *L. seeligeri* have rarely been reported to cause human illness. *L. monocytogenes* is generally regarded as the most important species causing disease in humans. Biochemical characteristics of the six species are found in **Table 1**.

L. monocytogenes are small (0.5–2.0 μm in length × 0.4–0.5 μm in diameter), Gram-positive, facultatively anaerobic, non-spore-forming coccoid rods. Older cultures have the tendency to vary between Gram-positive and Gram-negative. In addition, they may also occur as elongated rods and long filaments in older cultures or when grown under adverse conditions (**Fig. 1**). They are motile by means of peritrichous flagella and exhibit end-over-end tumbling motility when grown below 25°C, but not at 37°C. When the bacterium is inoculated into motility agar and incubated at 25°C an umbrella-like zone of growth occurs 3–5 mm below the surface. *L. monocytogenes* can grow over the temperature range of < 1°C to < 50°C (optimum between 30°C and 37°C)

Table 1 Characteristics of *Listeria* species

Characteristic	L. monocytogenes	L. ivanovii	L. seeligeri	L. innocua	L. welshimeri	L. grayi
β-Haemolysis	+	+	+	−	−	−
CAMP-*Staphylococcus aureus*	+	−	+	−	−	−
CAMP-*Rhodococcus equi*	−	+	−	−	−	−
Fermentation of:						
Mannitol	−	−	−	−	−	+
Xylose	−	+	+	−	+	−
Rhamnose	+	−	−	+/−	+/−	−
Virulent (mouse)	+	+	−	−	−	−

CAMP, Christie–Atkins–Munch–Peterson.

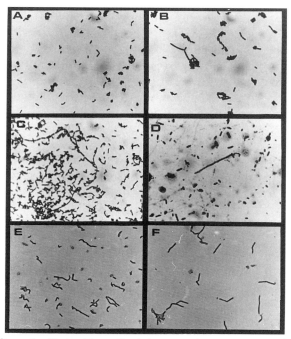

Figure 1 Filament formation in *L. monocytogenes* grown under environmental stress. Cells were photographed (magnification 412.5×) after growth in tryptic soy broth: (**A**) without acid; (**B**) pH 6.0, citric acid; (**C**) pH 9.2, NaOH; (**D**) pH 9.4, NaOH; (**E**) 1200 mmol l⁻¹ NaCl; (**F**) 1500 mmol l⁻¹ NaCl.

and survives freezing. It grows over a pH range of 4.0–9.5, and in medium containing up to 10% NaCl. The minimum water activity for growth ranges from 0.90 to 0.97. They do not survive heating at 60°C for 30 min. It has been calculated that the probability was less than 2 in 100 that one viable *L. monocytogenes* cell occurs in 3.8×10^{10} gallons of bovine milk processed at 74.4°C for 20 s.

The biochemistry of the cell surface of *L. monocytogenes* was examined: it was typical of other Gram-positive bacteria. The cell wall had a thick homogeneous layer surrounding the cytoplasmic membrane. The peptidoglycan layer accounted for about 35% of the dry cell wall. The glycosamine content was high. Teichoic acids are covalently linked to the glycan layers and constitute about 60–75% of the cell wall dry weight. Lipoteichoic acids are also present and resemble lipopolysaccharides of Gram-negative bacteria in both structure and biological function. The antigenic composition of *L. monocytogenes* consists of O (somatic, designated with numbers and lower-case letters) and H (flagellar, designated with upper-case letters) antigens. Serovars exhibit the following antigenic composition: 1/2a, 1/2b, 1/2c, 3a, 3b, 3c, 4a, 4ab, 4b, 4c, 4d, 4e and serovar 7. Types 1/2a, 1/2b and 4b predominate (92%) among isolates from humans and animals. All non-pathogenic species, except *L. welshimeri*, share one or more somatic and flagellar antigens with *L. monocytogenes*. Studies have indicated that immunological cross-reactions occur between strains of *Streptococcus*, *Staphylococcus*, *Escherichia* and some *Corynebacterium* species. No correlation has been reported between the various serotypes and any particular clinical syndrome or specific host. There is, however, a significant geographic distribution of the various serotypes with 4b being predominant in the US, Canada and Europe.

L. monocytogenes requires at least four B vitamins (biotin, riboflavin, thiamin, thioctic acid) as well as several amino acids (cysteine, glutamine, isoleucine, leucine, valine). Carbohydrates are essential for growth, with glucose being the optimum. Under anaerobic conditions only hexoses and pentoses support growth; aerobically, maltose and lactose, but not sucrose, also supported growth. The organism grows well in nutrient-rich media such as tryptic soy broth and brain-heart infusion broth. Colonies of *L. monocytogenes* on nutrient agar are small (0.5–1.5 mm/24–48 h), round and translucent. The colonies appear bluish grey under normal illumination but have a blue-green sheen when viewed using obliquely transmitted light. Good growth occurs on blood agar, upon which colonies are surrounded by a narrow band of β-haemolysis. However, strains isolated from faecal matter and environmental sources are frequently non-haemolytic on 5% sheep blood agar.

Sources

Listeria monocytogenes is an ubiquitous organism that has been isolated from a variety of sources. A primary source is soil and decaying vegetation. It has been found associated with numerous species of mammals (at least 37, including sheep, cattle, swine, dogs), poultry, sewage, milk, cheese, coleslaw, lettuce and meat products. Silage is an important source of *L. monocytogenes* due to the correlation of silage feeding and outbreaks of listeriosis in ruminants. The presence of *L. monocytogenes* in silage has been found to be strongly influenced by pH: silage with a pH 5.0–6.0 and above is far more likely to contain *L. monocytogenes* than silage below pH 5.0. The organism has been observed in cattle and to cause udder infections. *L. monocytogenes* has been isolated from raw milk, pasteurized milk, naturally contaminated cheese, hard salami, raw meat and poultry, and ground beef. Healthy human intestinal carriers also occur at a rate of 1–5% of the total population.

The microorganism can be isolated from a number of sites in food-processing plants. In dairies it has been found in raw milk, on cooler floors, freezers, processing rooms, cases and mats. It has been isolated from various sites in meat-processing facilities and from seafood.

Plasmids

The presence of plasmids in *Listeria* was first reported in 1982 with the discovery of a large cryptic plasmid in 7 of 32 strains investigated. A total of 139 listerial isolates were screened and 107 were found to carry a single plasmid and one strain with two plasmids. It was suggested that the occurrence of a single plasmid was prevalent among most food isolates. There is little information regarding the function of listerial plasmids. Plasmids from *L. monocytogenes* have been shown to confer resistance to cadmium and antibiotics. It was reported that 35.8% of listerial strains are cadmium-resistant. Some authors reported that the presence of a plasmid influenced the specific activities of listerial catalase and superoxide dismutase.

Growth and pH

The influence of pH has been studied both in terms of absolute value and acid type, with special emphasis on dairy products. When the pH was lowered to 5.0, survival time of *L. monocytogenes* was significantly reduced. *Listeria* would grow at pH 5.0 at temperatures > 13°C. It was established that acetic acid was most detrimental to *L. monocytogenes* growth, followed by lactic and citric acids, coinciding with their degree of undissociation. The growth-inhibitory pH was found to be 5.0 for propionic acid, 4.5 for acetic and lactic acids, and 4.0 for citric and hydrochloric acids at 4°C. It appears that pH, acid type and temperature all influence the growth and survival of *L. monocytogenes*.

Temperature

Listeria monocytogenes is a psychrotrophic bacterium able to grow at refrigeration temperatures, although optimal growth occurs between 30 and 37°C. The influence of temperature on the survival and growth of *L. monocytogenes* has been examined, frequently in conjunction with other environmental variables. Some authors studied temperature and its interactions with initial pH, atmosphere, sodium chloride content and sodium nitrite concentration. They found that growth kinetics were dependent on the interaction of the five variables. It was reported that the mean minimum growth temperature of *L. monocytogenes* was $+1.1 \pm 0.3$°C. It was also found that haemolytic listerial isolates grew better than non-haemolytic strains under cold conditions. Generation times in various milk products ranged from 1.2 to 1.7 days at 4°C, 5.0 to 7.2 h at 13°C, and 0.65 to 0.69 h at 35°C. It was established that the listerial minimum growth temperatures ranged from -0.1 to -0.4°C in chicken broth, with generation times ranging from 13 to 24 h at 5°C to 62 to 131 h at 0°C.

Growth temperature has also been shown to influence the virulence of *L. monocytogenes*. Growth at 4°C significantly increased the virulence of three clinical listerial isolates in mice when compared to cells grown at 37°C for intravenously, but not intragastrically, inoculated mice. It was also shown that growth at 4°C significantly decreased the killing of test strains by human neutrophils. *L. monocytogenes* strain 10403S did not increase in virulence when grown at low temperature or in increased levels of NaCl.

Salt

Sodium chloride is frequently added to food to improve flavour and as an agent to reduce water activity. It was found that *L. monocytogenes* is tolerant to NaCl, being capable of growth in 10% NaCl, and surviving for 1 year in 16% NaCl. As with pH and temperature, the influence of salt is often examined with other variables. Thermal inactivation of *L. monocytogenes* was not significantly influenced by the presence of up to 2% NaCl, but heat-stressed cells had increased sensitivity to the salt. Low salt concentrations (4–6%) improved survival while higher concentrations reduced survival of listerial cells at limiting pH values.

Modified Atmosphere

Modified atmosphere and its effects on *L. monocytogenes* has been examined by fewer investigators than pH, temperature or NaCl. Growth of *L. monocytogenes* was enhanced under decreased oxygen concentrations and when supplemented with carbon dioxide. Some authors examined the survival and growth of *L. monocytogenes* on three fresh vegetables under controlled atmosphere storage (asparagus: 15% O_2, 6% CO_2, 79% N_2; broccoli: 11% O_2, 10% CO_2 79% N_2; cauliflower: 18% O_2, 3% CO_2, 79% N_2). It was found that the listerial cells increased in number during storage and the type of atmosphere did not influence the rate of growth. Strictly anaerobic conditions significantly increased the recovery of heat-injured *Listeria* when compared with aerobically incubated controls.

Flagella

Listeria monocytogenes cells are able to survive throughout the environment and undergo morphological changes to adapt to new surroundings. It is known that temperature and osmolarity of the growth medium have a role in flagellar expression. Therefore flagella of *L. monocytogenes* may play a role in survival of the organism.

The degree of listerial flagellation and extent of motility at 37°C remains controversial. Several reports indicate that the production of flagella in *L. monocytogenes* is markedly reduced at 37°C, with repressed motility. Other reports indicate the presence of flagella and flagella-based motility at this temperature. When grown between 20 and 25°C, *L. monocytogenes* are peritrichously flagellated, vigorously motile and display a tumbling motility interspersed with smooth swimming. Listerial strains are vigorously motile via a peritrichous flagellar arrangement that causes the cell to exhibit a tumbling motility with bursts of rapid motion occurring in one direction for 10–20 cell lengths in liquid medium. One test for motility in *L. monocytogenes* utilizes motility agar and involves stabbing a semisolid motility agar tube (about 1 cm deep) and incubation at 20–25°C. Studies using soft agar in Petri dishes have also been conducted to correlate swarming with the degree of motility. The growth environment has been found to influence listerial flagellation and motility.

There are conflicting reports on the extent of flagellation and degree of motility at different temperatures. There was one report of a characteristic temperature-dependent motility, where *L. monocytogenes* was only motile between 20 and 25°C. Studies on chemotaxis showed that *L. monocytogenes* exhibited directional motility at 37°C, and it has been hypothesized that directional motility of *L. monocytogenes* may facilitate penetration of the intestinal epithelium. The small quantity of flagellin produced at 37°C was thought to be highly immunogenic upon infection; the largest single antigenic component of *L. monocytogenes* grown at 20°C is flagellin.

In addition to the effects of the growth environment on the activities of virulence enzymes (see below), it also influences listerial morphology and flagella formation. *L. monocytogenes* was grown in media containing increasing concentrations of NaCl from 0 to 1500 mmol l^{-1} or at pH values ranging from 3.5 to 9.5. Filament formation occurred at NaCl concentrations above 1000 mmol l^{-1} with an increase in filament length as the NaCl concentration increased (see Fig. 1). The same phenomenon was observed at pH values of 5.0–6.0 (adjusted with citric acid) and at pH > 9.0 (adjusted with NaOH). The length of the filaments increased as the growth environment became more challenging. Flagella formation is influenced by the growth environment. It was found that the optimum temperature for flagella formation was 25°C. Cells were flagellated at 4°C but were not motile in soft agar. Cells grown at 37°C were motile but with significantly reduced flagella production. Both NaCl concentration and pH were found to influence motility: increasing the NaCl concentration from 0 mmol l^{-1} caused a decrease in motility; increasing the pH from 5.5 to 9.5 resulted in a linear increase in motility.

Listeriosis

In 1983, 49 pregnant women and immunocompromised adults in Massachusetts contracted listeriosis caused by *L. monocytogenes*. Fourteen people died as a result of this food-borne disease outbreak, including two unborn infants. The epidemiologically implicated source of the organisms was pasteurized whole and 2% low-fat milk produced by a single dairy plant. *L. monocytogenes* was the cause of one of the most deadly outbreaks of food poisoning in the history of the US: the Jalisco cheese episode in southern California. Following consumption of this soft Mexican-style cheese, 142 cases of listeriosis were recorded. Of 93 pregnant women infected, 29 lost their babies; of 49 immunocompromised adults, 18 died. The overall fatality rate was 33%. An outbreak of listeriosis involving gastrointestinal symptoms occurred in San Giorgio di Piano, Italy, in 1993. Eighteen young, previously healthy, non-pregnant adults developed symptoms (mostly gastrointestinal) after ingestion of rice salad. **Table 2** lists several representative outbreaks of *L. monocytogenes*.

Most cases of human listeriosis occur as sporadic illnesses. Retrospective epidemiological studies have

Table 2 Outbreaks of listeriosis associated with food

Year	Location	Serotype	Outbreak and associated food
1979	Boston, MA	4b	Salads or pasteurized milk implicated in nosocomial outbreak among immunocompromised patients
1979	Nova Scotia, Canada	4b	66 cases in pregnant women, 18 deaths (infants); cabbage thought to be contaminated with sheep manure
1981	Auckland, New Zealand		Consumption of raw fish and oysters, epidemiologically linked to perinatal listeriosis
1983	Massachusetts	4b	Pasteurized milk linked to 49 cases with 14 deaths; mainly immunocompromised adults
1985	Los Angeles, California	4b	Mexican-style soft cheese; more than 100 cases and 30 deaths, mostly infants
1983–1987	Vaub, Switzerland	4b	Vacherin Mont d'Or cheese over a period of years; 31 deaths
1993	San Giorgio di Piano, Italy	1/2b	Rice salad, 18 victims with gastrointestinal symptoms; no deaths

identified a number of small food-borne outbreaks of listeriosis. In a 25-month study, the Centers for Disease Control identified 301 cases of listeriosis. Underlying patient conditions included pregnancy, steroid therapy, cancer, renal or liver disease, diabetes, HIV infection and organ transplant patients. Of the 301 cases, 32% were attributed to delicatessen foods.

Listeriosis can occur in persons of all ages, but occurs most frequently in individuals who have an underlying condition leading to suppression of T-cell-mediated immunity. It is clinically defined when the organism is isolated from a normally sterile site such as blood or cerebrospinal fluid. The disease may take a number of complications, with the most common being meningitis in the elderly and in immunocompromised patients. In the average, healthy individual infections are usually asymptomatic or cause mild influenza-like symptoms (fever, fatigue, malaise, nausea, cramps, vomiting and diarrhoea). The estimated infective dose varies with the strain of *L. monocytogenes* and the host, but ingestion of fewer than 1000 cells is thought sufficient to cause disease. Time of onset of gastrointestinal symptoms is unknown but is probably more than 12 h. Gastrointestinal symptoms have been epidemiologically associated with consumption of antacids or cimetidine.

Serious complications of human listeriosis include septicaemia, infectious mononucleosis-like syndrome, pneumonia, endocarditis, aortic aneurysm, localized abscesses, cutaneous lesions, conjunctivitis, hepatitis and urethritis. Time of onset for the more serious forms of listeriosis may range from a few days to 3 weeks. The mortality rate among untreated patients is approximately 70%. The disease is most common in hosts whose immune systems are compromised because of pre-existing disease or administration of immunosuppressive agents (AIDS, cancer, diabetes, age, alcoholism). *L. monocytogenes* can survive and multiply within macrophages. Selecting plating, monoclonal antibodies, polymerase chain reaction and DNA probe techniques have been developed for detecting *Listeria* in foods. Clinical diagnosis is frequently initially accomplished by microscopic observation of infected fluids such as blood, cerebrospinal fluid, amniotic fluid and genital tract secretions. Most clinical isolates are made on conventional media such as heart infusion agar containing 5% blood (from sheep, horse or rabbit). Treatment with the antibiotics ampicillin or penicillin is usually effective in treating listeriosis. Trimethoprim sulphamethoxazole may be used for patients allergic to penicillin.

A serious manifestation of *Listeria* infection is an intrauterine infection in pregnant women that can lead to abortion, stillbirth or to delivery of a healthy child who develops meningitis after birth. This occurs most frequently during the third trimester, although it may occur any time during pregnancy. The mother may have had a low-grade uterine infection and few or no symptoms. **Table 3** lists recommendations to prevent listeriosis in pregnant women and in other susceptible individuals.

Transplacental transmission causes a second category of infection unique to *L. monocytogenes*: granperinatal listeric septicaemia (also known as granulomatosis infantiseptica). The immune system of the pregnant woman is altered so that the fetus is able to escape immunological attack by the maternal host defences. Such changes in immune competence affect the defence of the pregnant host leaving the woman and her fetus at high risk of infection. Two clinical forms of neonatal listeriosis are recognized: early- and late-onset. In the early-onset form symptoms occur in 1.5 days. *L. monocytogenes* is found in numerous foci of necrosis throughout the infant body, especially in the liver, spleen, lungs, kidney and brain. The poorest prognosis appears to occur in the early-onset group. In late-onset neonatal listeriosis the mean onset of symptoms is 14.3 days with meningitis as the predominant form of the disease.

Symptoms of late-onset neonatal listeriosis include

Table 3 Recommendations to protect against listerial infection

- Avoid raw/unpasteurized milk
- Keep raw and cooked foods separate at all times
- Wash hands, knives and all food contact surfaces after handling uncooked foods
- Wash raw vegetables thoroughly before eating
- Thoroughly cook all food of animal origin, including eggs; cook raw meat and fish to an internal temperature of 160°F, raw poultry to 180°F; reheat leftovers thoroughly
- Read and follow label instructions to 'keep refrigerated' and 'use by' a certain date
- Keep hot food hot (above 140°F); do not keep foods at room temperature for longer than 2 h
- Keep cold food cold (at or below 40°F); do not keep foods at room temperature for longer than 2 h
- Refrigerate small portions of food so that they chill rapidly and evenly
- Keep refrigerator clean and the temperature at 34–40°C

High-risk individuals (pregnant women, the elderly, the immunosuppressed):
- Avoid soft cheeses such as Mexican-style, feta, Brie, Camembert and blue cheeses
- If soft cheeses are eaten during pregnancy, cook them until they boil
- Do not eat hard cheese made from unpasteurized milk; use only hard cheeses aged at least 60 days
- Reheat leftovers and ready-to-eat foods throroughly until steaming hot before eating
- Reheat all meats purchased at deli counters

respiratory distress syndrome, rash, conjunctivitis, pneumonia, hyperexcitability, vomiting, cramps, shock, haematological abnormalities and either hyper- or hypothermia. Most neonatal deaths are from pneumonia and respiratory failure. Prompt antibiotic therapy significantly lowers mortality, although the mortality rate is still about 36%. Because listeriosis has no distinctive clinical manifestations, isolation and identification of *Listeria* is an important aid to diagnosis.

The most common syndrome of listeriosis in ruminants is encephalitis, leading to observations of nervous system involvement in cattle and sheep. These animals become disoriented, causing them to circle in one direction. Listeriosis in ruminants is often referred to as circling disease.

Listeria monocytogenes Infection

Listeria monocytogenes is acquired by ingestion of contaminated food and attaches to the intestinal mucosa. In mouse and guinea pig models, M cells or Peyer's patches may be the primary intestinal invasion site. It is thought that α-D-galactose residues on the bacterial surface bind the α-D-galactose receptors on intestinal cells. Flagella may be an important virulence factor in the attachment in that they may aid the bacteria in moving to the intestinal target site of infection. Following translocation across the intestinal barrier, *L. monocytogenes* cells can be seen in phagocytic cells in the underlying lamina propria. In mice, the bacteria become disseminated to the liver and spleen. Although most listerial cells are killed by phagocytes in the liver, the survivors multiply in hepatocytes. In cultured epithelial cells host cell entry requires the presence of surface proteins InlA (internalin) and InlB. The receptor for InlA is the cell adhesion molecule E-cadherin. InlB-mediated entry

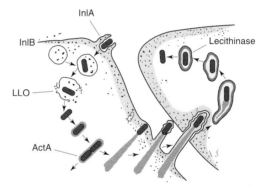

Figure 2 The steps of *L. monocytogenes* infectious process. Derived from Ireton and Cossart (1997).

requires activation of host cell protein phosphoinositide 3-kinase. Once inside the host cell, the *L. monocytogenes* cell is encased in a membrane vesicle (**Fig. 2**). Escape from this vesicle is mediated by listeriolysin O and phospholipases C. A zinc metalloprotease is directly involved in the maturation of one of the phospholipases C. Once free from the vesicle the listerial cells are free in the host cell cytoplasm where rapid multiplication occurs. The final steps of infection involve intra- and intercellular movement, utilizing host cell actin. Actin polymerization is catalysed by a listerial surface protein of the *actA* gene ActA. Intracytoplasmic *L. monocytogenes* appear to be covered by a cloud of host cell actin filaments, which rearrange into a polymerized comet tail. As the comet tail grows the listerial cell is propelled to the host cell membrane, forming protrusions that extend into adjacent cells. These protrusions are internalized by the neighbouring host cell. Again, with the aid of listeriolysin O and phospholipases C, the double membrane is lysed and the cell cycle begins again. Molecular genetic techniques have been used to analyse the factors responsible for the various steps of the infectious process. Virulence

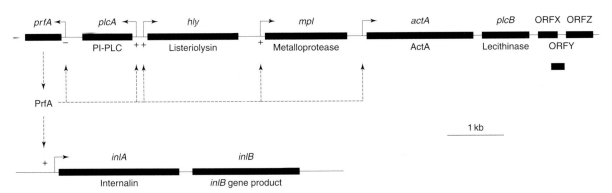

Figure 3 Organization and regulatory control of some *L. monocytogenes* virulence genes that are clustered on the chromosome. *prfA*, regulatory protein (activator); genes activated by PrfA are indicated by arrows: *plcA*, PI-PLC; *plcB*, PC-PLC (lecithinase); *hly*, listeriolysin O; *actA*, protein involved in actin polymerization (?); *orfX, Y, Z* orfs of unknown function; probable locations of promoters are indicated by P; dotted lines indicate mRNA transcripts. Derived from Sheehan et al. (1994).

genes identified include those involved in entry of *L. monocytogenes* in epithelial cells (*inlA* and *inlB*); escape from the phagosome (*hly* and *plcA*), intra- and intercellular movement (*actA*), and lysis of the two-membrane vacuoles (*plcB*). All of these genes are co-regulated and, except for the *inl* genes, are clustered on the same region of the chromosome (**Fig. 3**). In addition, the enzymes catalase (CAT) and superoxide dismutase (SOD) may function as secondary virulence factors. These two enzymes act by degrading toxic oxygen free radicals produced during the respiratory burst following macrophage phagocytosis.

Active Oxygen and the Phagocyte

Cell-mediated immunity is critical in resistance to infection by *L. monocytogenes* and includes mono-nuclear phagocytes as early response to infection and specific T cells as the secondary response. When *L. monocytogenes* cells penetrate the structural barriers of the host, phagocytes (neutrophils and mononuclear phagocytes) are important in the destruction of these bacteria, with neutrophils constituting the first line of defence. A neutrophil is a short-lived cell, produced in the bone marrow, which circulates in the blood for only 6 or 7 hours. They then penetrate blood vessel walls, where they reside in the tissues, surviving for a few days. It is estimated that in the human adult there are at least 60×10^{11} neutrophils circulating at any given time. Phagocytosis consists of three activities: attraction of the neutrophil by the microorganism, ingestion and digestion. Attraction occurs by chemotaxis, in which the neutrophil senses the presence of the bacterium, and migrates towards it. Serum-derived proteins (opsonins), which coat the bacteria, serve to identify the microorganism as a target. The neutrophil establishes intimate contact with the invading cell by engulfing the organism through phagocytosis. Two mechanisms are then used to kill the

bacteria. The first involves the discharge of the lysosomal contents on to the surface of the invading organism. The second involves the manufacture of cytotoxic oxidants by the neutrophil.

Cytotoxic oxidants are produced during a period of neutrophil metabolism known collectively as the respiratory burst. At the onset of phagocytosis, neutrophils and macrophages both show a marked increase in O_2 uptake, which is not sensitive to cyanide and is therefore not related to mitochondrial electron transport. During the respiratory burst, phagocytes produce the superoxide radical ($\cdot O_2^-$), hydrogen peroxide (H_2O_2) and the hydroxyl radical (OH$\cdot$). The production of these oxygen metabolites is due to the activation of a membrane-bound $\cdot O_2^-$-generating reduced pyridine nucleotide oxidase, which catalyses the reduction of oxygen to $\cdot O_2^-$ at the expense of a reduced pyridine nucleotide. A second microbicidal mechanism, which may be related to the production of oxidizing radicals, may also be involved in bacterial death. A haem enzyme, myeloperoxidase, is present in large quantities in azurophil granules, one of two types of lysosomes present in neutrophils. During phagocytosis, H_2O_2, produced during the respiratory burst, plus myeloperoxidase, in the presence of a halide ion (probably Cl$^-$) constitutes an exceedingly potent microbicidal system.

Listeria-sensitized T cells function to attract, focus and activate macrophages at infection sites by liberating macrophage-inhibiting factors. These factors cause blood monocytes to localize at the infection site, and by liberating macrophage-activating factor (interferon-γ), enhance macrophage phagocytic activity. Activation of T lymphocytes leads to acquired cellular immunity. See **Figure 4** for a representation of some of the events taking place during *Listeria* infection.

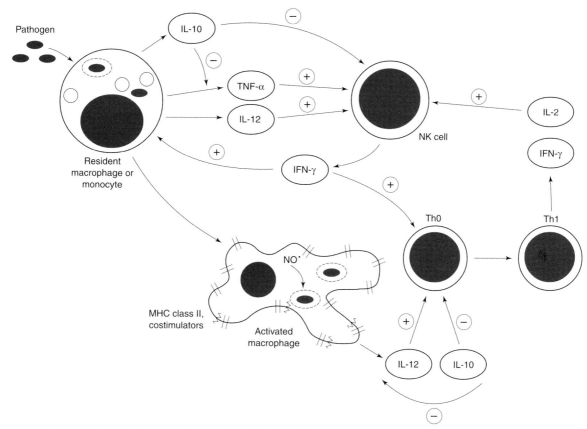

Figure 4 Events during *Listeria monocytogenes* infection. A resident macrophage that takes up *Listeria* produces a number of cytokines, some of which induce natural killer (NK) cells to produce interferon-γ (IFN-γ: induction is represented by +). Other cytokines, like interleukin-10 (IL-10), inhibit this process (inhibition is represented by –). The IFN-γ drives the macrophage to an activated state. The activated macrophages express high levels of major histocompatibility complex (MHC) class II molecules (represented by two parallel lines) and costimulators, like intercellular adhesion molecule (ICAM)-1 and B7 molecules (represented by a wavy line). The activated macrophage is active: reducing the growth of *Listeria* in vacuoles (closed circles) or in the cytosol following vesicle lysis (broken circle); in presenting peptides to *Listeria*-specific T cells (marked as Th0); and in inducing the differentiation of the Th0 cell to Th 1 (by its release of IL-12). NO˙ represents nitric oxide. Derived from Unanue (1997).

Listeriolysin O

All virulent strains of *L. monocytogenes* produce an extracellular SH-activated haemolysin. Listeriolysin O is an extracellular protein of 60 kDa, and, as with other bacterial sulphydryl-activated toxins is inhibited by low amounts of cholesterol; activated by reducing agents and suppressed by the lytic activity of oxidation; and has antigenic cross-reactivity with streptolysin O. Listeriolysin O differs from other toxins in that its cytolytic activity towards erythrocytes is maximum at low pH (5.5) and is undetectable at pH 7.0. The protein is able to lyse not only erythrocytes, but also eukaryotic membranes in general. Loss of haemolysin production is associated with loss of virulence. Non-haemolytic bacteria were unable to grow in mouse tissues and were rapidly eliminated from infected animals. The loss of haemolysin significantly reduced the survival of the bacteria in mouse peritoneal macrophages but did not reduce their uptake.

Phospholipases C

Bacterial phospholipases C are structurally related enzymes whose function appears to be securing a supply of phosphate for bacterial cells. This hypothesis has been supported by regulating phospholipase C genes in the presence of exogenous phosphate levels in some bacterial species. Phospholipases C are characterized by the site at which phospholipids are cleaved. Phospholipases have been isolated from both Gram-positive and Gram-negative bacteria, and all of these enzymes have been identified as a single polypeptide chain.

L. monocytogenes contains two genes which encode phospholipase C enzymes: phosphatidylinositol-specific phospholipase C (PI-PLC) is encoded by the *plcA* gene and phosphatidylcholine-specific phospholipase C (PC-PLC) is encoded by the *plcB* gene. The phospholipases C disrupt host cell membranes by hydrolysing membrane phospholipids such as phosphatidylinositol and phosphatidylcholine.

Both PI-PLC and PC-PLC may have a significant role in the pathogenesis of listeriosis. Studies have suggested that phospholipases C are important determinants in this organism's intracellular spread.

The gene *plcA* is adjacent to the haemolysin gene (*hlyA*). PI-PLC appears to be important in the progression of infection. The role of PI-PLC seems to be most crucial after ingestion by professional phagocytes. PI-PLC mutants were unable to invade liver hepatocytes, replicate and cause progressive infections. Other research in the role of PI-PLC showed that a PI-PLC mutant was able to invade, but was unable to replicate in mouse peritoneal macrophages. This result may be due to reduced ability of the organism to escape from the host cell phagolysosome. It was suggested that PI-PLC removes glycosyl phosphatidylinositol-anchored host cell membrane proteins which potentiate the cell membrane to the damaging effects of listeriolysin O and PC-PLC.

The *plcB* gene encodes a zinc metallophospholipase C, PC-PLC or lecithinase and is located downstream of the previous identified gene *mpl*. In one *L. monocytogenes* strain producing a high amount of lecithinase activity, the protein was secreted as a single polypeptide between 29 and 32 kDa. This polypeptide was, additionally, post-translationally activated by the removal of 14 or 26 N-terminal amino acids. The enzyme has been suggested to be a polypeptide of 289 amino acids with sequence similarity to the PC-PLC of *Bacillus cereus* and *Clostridium perfringens* (α-toxin). *L. monocytogenes* with this type of activity produces a dense zone of opacity surrounding the colony on egg yolk agar. PC-PLC activation has been suggested to occur after export from the cell. The function of PC-PLC may be partially to disrupt the cell–cell fusion vacuole membrane and later complete disruption may occur because of the action of listeriolysin O or PI-PLC.

The specific activities of five virulence enzymes (listeriolysin O, PC-PLC, PI-PLC, catalase and superoxide dismutase) have been shown to be influenced by the growth temperature, salt concentration and pH of the growth medium. In general, maximum activities were observed when *L. monocytogenes* strains were grown in medium containing 428 mmol l^{-1} NaCl at 37°C at pH 7.5

Isolation and Detection

In the US, both the Food and Drug Administration (FDA) and the US Department of Agriculture Food Safety and Inspection Service (USDA-FSIS) have developed methods to detect *L. monocytogenes*. The FDA method for the isolation and detection of *L. monocytogenes* utilizes an enrichment procedure. A 1 : 10 dilution is made in enrichment broth (tryptic soy broth with added acriflavin, nalidixic acid, cycloheximide and pyruvate), homogenized and incubated at 30°C for 48 h (see the *Bacteriological Analytical Manual*, 7th edition (1992), pp. 141–152, for specific details). At 24 and 48 h samples are removed and streaked for isolation on oxoid medium (OM) and LiCl-phenylethanol-moxalactam agar (LPM) and incubated for 24–48 h at 35°C(OX) or 30°C (LPM). *L. monocytogenes* colonies on OM are black with a black halo while, using indirect illumination, colonies on LPM appear sparkling blue or white. Purified suspect isolates are identified using the characteristics described in **Table 4**.

The (Christie–Atkins–Munch–Peterson (CAMP) test is a haemolytic test useful in the detection of *L. monocytogenes*. A sheep red blood agar plate is streaked in parallel and diametrically opposite each other with β-haemolytic cultures of *Staphylococcus aureus* and *Rhodococcus equi*. Suspect cultures of *L. monocytogenes* are streaked parallel to one another, but at right angles to and between the *S. aureus* and *R. equi* cultures. After incubation at 35°C for 24–48 h, the β-haemolytic activity of *L. monocytogenes* is enhanced in the zone influenced by the streak of *S. aureus*.

The USDA-FSIS method differs from the FDA procedure primarily in the selective enrichment and plating medium used, along with the size of the sample. The Netherlands Government Food Inspection Service utilizes a detection procedure which is said to be more sensitive in detecting low levels of listerial contamination (< 10 cfu g^{-1}). This procedure utilizes L-PALCAMY enrichment broth at 30°C for 24 and 48 h, following which growth is streaked on PALCAM agar.

As with other food-borne microbial pathogens, a number of new and more rapid methods have been developed for listerial detection. Some of these procedures employ enzyme-linked immunosorbent assays (ELISA), DNA probes, polymerase chain reaction and ligase chain reaction.

Table 4 Characteristics of *Listeria monocytogenes*

- Blue-grey to blue colonies upon indirect illumination
- Wet mount – motile, slim, short rods
- Catalase-positive
- Gram-positive
- β-Haemolytic on sheep blood agar
- Nitrite reduction negative
- Motility agar – umbrella-like growth
- Carbohydrate utilization – glucose (+), maltose (+), rhamnose (+), xylose (–) and mannitol (–)
- Serology
- CAMP test

CAMP, Christie–Atkins–Munch–Peterson.

Cells of *L. monocytogenes* may be sublethally injected by a number of treatments (heating, freezing, drying, or exposure to chemical – acids, sanitizers, preservatives). Selective agents used in isolation media may inhibit the growth of these sublethally injured cells. A period of recovery in non-inhibitory recovery medium may be required to detect this injured population.

Strain Typing

Serotyping is of limited epidemiological usefulness because human listeriosis is caused by so few serotypes (4b, 1/2a and 1/2b). Several other methods to type *L. monocytogenes* strains have been employed.

Phage Typing

Phage typing is a useful epidemiological tool for outbreaks of listeriosis. A common set of 26 bacteriophages (International Phage Set) has been found to be highly discriminatory and capable of analysing a large number of isolates.

Isoenzyme Typing

Isoenzyme typing allows species differentiation by the variation in the electrophoretic mobility of any of a large number of metabolic enzymes. Examination of many strains permits the construction of distinctive allele profiles or electrophoretic types (ETs). Two electrophoretic typing analyses found 45 ETs (175 listerial isolates) and 56 ETs (310 isolates). All of the major food-borne disease outbreaks appeared to have been caused by strains of the same or similar ETs. This technique may be useful in confirming or eliminating a strain as a source of food-borne listeriosis.

DNA Fingerprinting

DNA fingerprinting using restriction enzyme analysis has been used to examine isolates from outbreaks of listeriosis. Repetitive element sequence-based polymerase chain reaction of strain and species discrimination was used in the *Listeria* genus. Specific band profiles enabled identification of listerial species as well as the ability to distinguish between *L. monocytogenes* serotypes.

Plasmid and Monocine Typing

Plasmid typing and monocine typing have also been employed as epidemiological tools. Both techniques have been found to have limited usefulness.

Summary

Listeria monocytogenes has been recognized as a human pathogen for over 70 years. In 1981 it was identified as an important and dangerous food-borne disease-causing bacterium, having a fatality rate of 33%. *L. monocytogenes* is resistant to many environmental challenges and is one of the few food-borne pathogens able to grow at refrigeration temperatures. The bacterium is ubiquitous and can be isolated from a wide variety of sources, both the environment and food. Listeriosis is a potentially fatal illness for immunosuppressed victims. Serious complications of human listeriosis include meningitis and granperinatal listeric septicaemia. *L. monocytogenes* cells are able to survive phagocytosis and multiply within host cells. Isolation and detection of listerial cells from foods requires the use of enrichment and plating on antibiotic-containing media.

See also: **Bacteriophage-based Techniques for Detection of Food-borne Pathogens**. **Biochemical and Modern Identification Techniques**: Food Poisoning Organisms. **Food Poisoning Outbreaks**. ***Listeria***: Introduction; Detection by Classical Cultural Techniques; Detection by Commercial Enzyme Immunoassays; Detection by Colorimetric DNA Hybridization; Detection by Commercial Immunomagnetic Particle-based Assays; *Listeria monocytogenes* – Detection by Chemiluminescent DNA Hybridization; *Listeria monocytogenes* – Detection using NASBA (an Isothermic Nucleic Acid Amplification System). **Nucleic Acid-based Assays**: Overview.

Further Reading

Cossart P (1994) *Listeria monocytogenes*: Strategies for entry and survival in cells and tissues. *Baillere's Clin. Infect. Dis.* ch. 7, pp. 285–304.

Donnelly CW (1994) *Listeria monocytogenes*. In: Hui YH, Gorham JR, Murrell KD and Cliver DO (eds), Food-borne Disease Handbook, vol. 1, pp. 215–252. Marcel Dekker, NY.

Farber JM and Peterkin PI (1991) *Listeria monocytogenes*, a Food-Borne Pathogen. *Microbiol. Rev.* 55: 476–511.

Gray ML and Killinger AH (1966) *Listeria monocytogenes* and listeric infections. *Bacteriol. Rev.* 30: 309–382.

Ireton K and Cossart P (1997) Host-Pathogen Interactions During Entry and Actin-Based Movement of *Listeria monocytogenes*. *Ann. Rev. Genet.* 31: 113–138.

McLauchlin J (1987) *Listeria monocytogenes*, recent advances in the taxonomy and epidemiology of listeriosis in humans. *J. Appl. Bacteriol.* 63: 1–11.

Salyers AA and Whitt DD (1994) *Listeria monocytogenes*. Bacterial Pathogenesis: A Molecular Approach, ch. 15, pp. 182–189. Univ. of Ill. Press.

Seeliger H and Höhne K (1979) Serotyping of *Listeria*

monocytogenes and related species. *Meth. in Microbiol.*, ch. 11. 13: 31–49.

Sheehan B, Kocks C, Dramsi S, Gouin E, Klarsfeld AD, Mengaud J and Cossart P (1994) Molecular and Genetic Determinants of *Listeria monocytogenes* Infectious Process. *Curr. Top. Microbiol. Immunol.* 192: 187–216.

Unanue ER (1997) Inter-relationship Among Macrophages, Natural Killer Cells and Neutrophils in Early Stages of *Listeria* Resistance. *Curr. Opin. Immunol.* 9: 35–43.

Listeria monocytogenes – Detection by Chemiluminescent DNA Hybridization

A D Hitchins, Center for Food Safety and Applied Nutrition, Food and Drug Administration, Washington DC, USA

Chemiluminescent DNA hybridization methodology can be used for rapid and specific identification of isolates of the human pathogen *Listeria monocytogenes* obtained from contaminated foods by conventional selective cultural enrichment and isolation methods, such as that described in the *Bacteriological Analytical Manual* (BAM) of the US Food and Drug Administration. The reagents necessary to perform this kind of hybridization can, in principle, be synthesized in any laboratory with chemical and molecular biological capabilities. In practice, as with other kinds of hybridizations, it is more convenient to rely on commercially prepared reagents available as a test kit package. In this case, there is currently only one commercially available *Listeria* chemiluminescence test kit, the AccuProbe *Listeria monocytogenes* Culture Identification Test (Gen-Probe Inc., San Diego, California, USA). This test kit was designed primarily for clinical microbiology laboratories; nevertheless, it has been successfully applied to identifying *L. monocytogenes* isolated from food, and developments in sample processing have improved the kit's application to the screening of food sample selective culture enrichments for this pathogen. These applications are described here, together with as much detail as possible about how the assay works, given that some crucial information is proprietary and not available.

The Gen-Probe Assay for *L. monocytogenes*

Principle of the Assay

In this assay, cells of *Listeria monocytogenes* are lysed enzymatically and chemically to expose their three macromolecular ribosomal ribonucleic acids (rRNAs). At least one of these rRNAs contains nucleotide sequences that are specific to *L. monocytogenes* and so do not occur in the rRNAs of other species of *Listeria* or those of other bacteria. One of these sequences, presumably in the rRNA of intermediate centrifugal sedimentation size, 16S, is the hybridization target of the proprietary DNA probe. The kit reagents provide the optimal conditions for the DNA probe to hybridize, by complementary nucleotide base pairing involving hydrogen bonding, with its rRNA target. Complementarity is due to the fact that the DNA probe's nucleotide sequence corresponds to the region of the transcribing strand of the gene which includes the code for the rRNA target sequence.

The probe is pre-tagged with a chemiluminescinogenic chemical group (**Table 1**). The resulting hybrid hetero-duplex molecules (DNA:RNA) can be detected by adding a chemical reagent which causes the tagging moiety on the DNA strand to release chemical energy as photons, that is to chemiluminesce. The particular tag used in the AccuProbe kit is an acridinium ester moiety which releases light when treated with hydrogen peroxide and alkali. The reactions involved are shown in **Table 2**. If there are sufficient tagged hybrid DNA:RNA duplexes, enough photons are rapidly released (within 2 s) as a pulse of light at 430 nm. The photon pulse can then be detected and amplified by a photomultiplier tube to give a luminometric value indicating the presence or absence of the pathogen in the sample.

A crucial feature of the AccuProbe system is that

Table 1 Definition and examples of chemiluminescent reactions

- Definition: chemiluminescence, including biochemiluminescence, is the emission of radiation by a chemical or biochemical reaction product which is produced in an electronic excited state. Reversion to the electronic ground state occurs at a rate characteristic of the particular product, from less than 1 s to more than 1 day.
- Generalized reaction:[a]

$$A + B \rightarrow C^* + D$$

$$C^* \rightarrow C + photon$$

- Example 1:

$$luminol + H_2O_2 \rightarrow 3\text{-aminophthalate} + N_2 + photon$$

(reaction requires alkaline conditions or enzymic catalysis by horseradish peroxidase)

- Example 2: firefly lantern luciferase system
- Example 3: the AccuProbe chemiluminescent system (see Table 2)

[a]Excited chemical intermediate is denoted by an asterisk.

Table 2   Chemiluminescence reaction steps of the acridinium derivative used to label the AccuProbe *Listeria monocytogenes* DNA probe[a]

1. Acridinium-*N*-hydroxysuccinylamide ester   →   peroxylated acridinium ester
   $+ H_2O_2$

2. Peroxylated acridinium ester   →   succinylamide moiety + acridinium cyclooxetane product
   $+ NaOH$

3. Acridinium cyclooxetane product   →   excited state acridone $+ CO_2$

4. Excited state acridone   →   ground state acridone + light (430 nm)

[a] The acridinium derivative is covalently linked to the synthetic DNA probe by a residue. An alkylamine reagent is used in the linkage reaction.

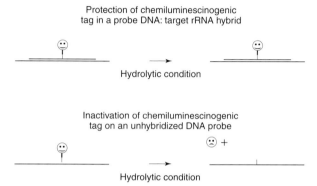

Figure 1   Schematization of the relative degree of inactivation of the chemiluminescinogenic tags on hybridized and unhybridized probe molecules in the AccuProbe hybridization protection system. Symbols: ☺, DNA probe tag; ☹, inactivated DNA probe tag.

the ester bonds of acridinium tags on probe molecules that are hybridized to rRNA targets are more resistant to hydrolysis than the ones on unhybridized probe molecules (**Fig. 1**). Since the tag does not chemiluminesce when de-esterified, the differential hydrolysis system ensures the tags on unhybridized probe molecules are preferentially inactivated. Thus, the chemiluminescence observed is due solely to tagged probe that is hybridized to target rRNA and thus indicates that the target's host bacterium, *Listeria monocytogenes*, is present in the sample.

Protocol for the Detection of *L. monocytogenes*

The major equipment, reagents and other items needed to perform the AccuProbe test are listed in **Table 3**. Reagents are kept refrigerated or stored at temperatures not exceeding 25°C, according to the manufacturer's package insert instructions. Unused probe reagent tubes should be stored with desiccant at 5°C. Refer to the instrument manual for the preparation, care and calibration of the luminometer (**Fig. 2**).

The protocol for analysing test and control samples is outlined in Table 4. When broth cultures are being analysed, uninoculated medium controls should be run the first time a new medium is tried in order to check for possible interference by certain components.

Table 3   Equipment, reagents and accessory requisites for the AccuProbe *Listeria monocytogenes* kit

Equipment
1. Luminometer – for example:
   Leader; AccuLDR, formerly PAL (Gen-Probe)[a]
   Optocomp I; Optocomp II (MGM)[b]
   If other models desired, confirm their compatibility with
      AccuProbe assay system
2. Incubator or water bath, 35–37°C
3. Heating block with 12 mm diameter holes or water bath, 60 ± 1°C

Reagents
1. Probe reagent tubes (20 assay tubes in box of four packages) made of low-phosphorescence plastic, containing desiccated DNA probe
2. Identification reagent kit (200 assays per kit):
   reagent 1 – lysis reagent for use with solid culture medium growth
   reagent 2 – hybridization buffer (promotes the desired specific hybridization)
   reagent 3 – selection reagent (inactivates unhybridized probe tag)
3. Detection reagent kit (1200 assays per kit) – induces chemiluminescence of tag
   component 1 – hydrogen peroxide solution stabilized with 0.001 mol/l nitric acid
   component 2 – sodium hydroxide (1 mol/l)

Accessories
1. Plastic sterile 1 µl loops (or wire loops, plastic sterile needles, applicator sticks)
2. Positive (*L. monocytogenes*) and negative (e.g. *L. innocua*) control cultures
3. Micropipettes (50 µl, 300 µl)
4. Vortex mixer

[a] Gen-Probe Inc., San Diego, California, USA.
[b] MGM Instruments, Hamden, Connecticut, USA.

It may be necessary to analyse sedimented cell pellets instead.

The luminometer result read-out usually includes the background reading, the sample reading and a positive or negative statement for presence of *L. monocytogenes*. Readings are given as numbers of photometric light units (PLU) or relative light units (RLU) depending on the instrument model; 33.3 RLU are equivalent to 1 PLU. Readings at or above the manufacturer's designated cutoff value (1500 PLU)

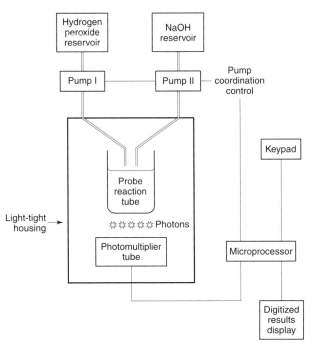

Figure 2 Essential features of the chemiluminometer (not to scale). Reagent tubing is denoted by double lines, component outlines and connections by single lines.

Table 4 Laboratory protocol for the chemiluminescinogenic DNA probe for *Listeria monocytogenes* (use appropriate microbiological safety precautions when working with this pathogen)

1. Take one probe reagent tube (PRT) and add 50 μl reagent 1 (lysis reagent) if solid medium growth or sedimented cells from a broth culture are to be tested. For enrichment cultures or extracts proceed directly to step 3
2. Take a loopful of an isolated 1 mm diameter colony (or combine several smaller colonies) or of confluent bacterial growth from solid culture medium, or sedimented growth from liquid culture with a plastic loop, emulsify in the prepared PRT, and proceed directly to step 4
3. Micropipette 50 μl of pure or selective enrichment cultures or RNA extracts into the prepared PRT. Recap the PRT
4. Incubate PRT at 35–37°C for 10 min (5 min in water bath)
5. Micropipette 50 μl of reagent 2, the hybridization buffer, into the PRT and recap the tube
6. Incubate PRT at 60 ± 1°C for 15 min [**Critical Control Point**]
7. Add 300 μl of reagent 3 (selection agent) to PRT and recap the tube
8. Mix very thoroughly (vortex if possible); incubate the PRT for at least 5 min at 60 ± 1°C [**Critical Control Point**]
9. Incubate the PRT for at least 5 min but no more than 60 min at ambient temperature
10. Remove cap from the PRT and conduct away any static electricity on the exterior of the PRT by wiping it with a damp tissue. Immediately insert the PRT in the prepared luminometer. Follow manufacturer's instructions for preparing, calibrating and using the instrument. Proper care and flushing of the luminometer detection reagent lines is essential [**Critical Control Point**]
11. Observe digital display messages. Record readings by hand or print-out

are interpreted as positive for *L. monocyogenes*. The suggested repeat range for marginally positive sample readings starts at 80% of the cutoff value.

There are three feasible points of application of the AccuProbe *Listeria monocytogenes* culture confirmation test kit in the common methodologies used to isolate the pathogen from food samples. These are the purification agar stage, the selective agar stage and the enrichment stage (Fig. 3).

Identification of colonies isolated from selective or purification agar media as *L. monocytogenes*, or not, takes about 45 min. This is a considerable time saving as identification by conventional tests takes 2 days minimum and up to 7 days in some cases. Also, use of the kit can save the time, effort and expense of media preparation and storage. The method's application at these stages requires one isolated colony of about 1 mm diameter (or several smaller colonies). Suspensions of confluent growth harvested from the Oxford selective agar plate can also be used if no isolated colonies are present, but the confluent growth tends to darken owing to esculin hydrolysis. In such cases the presence of listeria should be subsequently confirmed by restreaking the confluent growth to selective agar.

Although not the original purpose for which it was designed, the kit can be used at the selective enrichment stage (Fig. 3). This application is still being developed. The chances of false negative results with direct testing at 48 h are appreciable when the initial

level of *L. monocytogenes* is low and there is interference by any of the food's microflora that are selective agent resistant. Usually food sample contamination levels are low, i.e. a few listerial cells per 25 g. The problem can be largely alleviated by concentrating the culture sample.

One possible strategy involves first directly testing 50 μl of 48 h enrichment culture. If this yields a negative response, the test is repeated by sampling a larger volume and concentrating it by centrifugation. If this test is also negative, the remainder of the culture is used to extract the RNA fraction. The volume to be sedimented will be limited to about 0.5 ml because usually 10% of the volume is solid food sample and the AccuProbe test sample size is only 50 μl. Sampling somewhat larger culture volumes (5–10 ml) may be feasible if solid food particles can be centrifugally sedimented at about 500 g for 10 min and then the listerial cells recovered from the decanted supernatant by sedimentation at 8000 g for 15 min. The maximum feasible sample volume obtainable by differential cen-

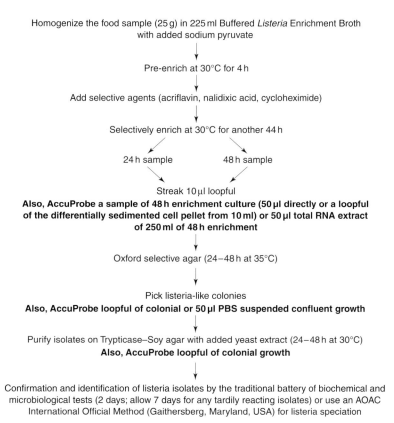

Homogenize the food sample (25 g) in 225 ml Buffered *Listeria* Enrichment Broth
with added sodium pyruvate

Pre-enrich at 30°C for 4 h

Add selective agents (acriflavin, nalidixic acid, cycloheximide)

Selectively enrich at 30°C for another 44 h

24 h sample   48 h sample

Streak 10 µl loopful
**Also, AccuProbe a sample of 48 h enrichment culture (50 µl directly or a loopful
of the differentially sedimented cell pellet from 10 ml) or 50 µl total RNA extract
of 250 ml of 48 h enrichment**

Oxford selective agar (24–48 h at 35°C)

Pick listeria-like colonies
Also, AccuProbe loopful of colonial or 50 µl PBS suspended confluent growth

Purify isolates on Trypticase–Soy agar with added yeast extract (24–48 h at 30°C)
Also, AccuProbe loopful of colonial growth

Confirmation and identification of listeria isolates by the traditional battery of biochemical and
microbiological tests (2 days; allow 7 days for any tardily reacting isolates) or use an AOAC
International Official Method (Gaithersberg, Maryland, USA) for listeria speciation

Figure 3 Possible sampling points for application of the AccuProbe test kit to foods contaminated at levels requiring a selective enrichment as exemplified by the BAM procedure for the isolation of *Listeria monocytogenes*. PBS, phosphate-buffered saline.

trifugation will still be limited by the volume of fine particles of food unavoidably remaining in the sediment.

Maximal concentration of the enrichment cultures can be achieved by chemical fractionation (**Fig. 3**). The sample is processed to produce the bacterial and food RNA fraction. Enrichment cultures can be reduced in this way from 250 ml down to 0.50–0.05 ml, i.e. a 500- to 5000-fold concentration depending on the volume of the food RNA precipitate. Routine optimization of the concentration factor may be achievable by reduction of the food RNA yield by differentially sedimenting away as much food as possible from the enrichment culture before extraction. Some food RNA is beneficial as it can act as a carrier for the minute amounts of bacterial RNA expected.

The points considered for 48 h enrichment samples also apply to 24 h samples. However, they are potentially exacerbated by the shorter time available for the pathogen to attain measurable levels. One pathogen cell per 25 g food sample or per 250 ml enrichment broth, growing unimpeded for 24 h, can produce a population of the order of 10^4 cells per ml of the 250 ml enrichment culture. Thus, extracting the entire 24 h enrichment culture could provide marginally detectable amounts of RNA as long as microflora interference is weak. With 10 cells or more per 25 g of food, the situation would be more favourable. If a fast positive indication is urgently required, it may be worthwhile setting up two 250 ml enrichments of the sample, one for RNA extraction at 24 h and the other as a 48 h backup RNA extraction. The latter would not be needed if the 24 h Oxford plates showed listeria growth, as it could be tested directly with the Accu-Probe kit. It is important to note that when kits are used in regulatory analyses the positive results have to be backed up by isolation of the target microorganism. Hence the use of Oxford medium in **Figure 4**.

Advantages and Limitations of the Chemiluminescent Probe

The main advantage of the AccuProbe assay is the rapidity of the test. It takes less than 45 min starting with a listeria broth culture or colonies. Conveniently, very few manipulations such as pipettings are needed. In these respects it compares very favourably with other methods of *Listeria* detection. As with most rapid kits nowadays, no radioactive material is involved, except a sealed vial containing a tritiated standard for checking the performance of the lum-

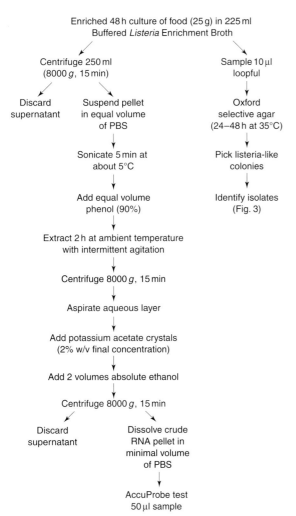

Figure 4 RNA extraction from selective enrichment culture samples for the AccuProbe test. Cautions: use ultra-sound-proof earmuffs; wear disposable gloves and eyewear; phenol is corrosive. PBS, phosphate-buffered saline.

Table 5 Genera and number of species of microorganisms used by various laboratories to evaluate the AccuProbe DNA probe test kit (one species per genus was studied except where stated otherwise in parentheses)

Acinetobacter (2), *Actinomyces*, *Aerococcus*, *Aeromonas*, *Alcaligenes*, *Arcanobacterium*, *Bacillus* (3), *Bacteroides*, *Bordetella*, *Branhamella*, *Brevibacterium*, *Brochothrix*, *Campylobacter*, *Candida*, *Capnocytophaga*, *Chromobacterium*, *Clostridium* (2), *Corynebacterium* (7), *Cryptococcus*, *Deinococcus*, *Derxia*, *Enterobacter* (2), *Enterococcus* (4), *Erysipelothrix*, *Escherichia*, *Flavobacterium*, *Gemella*, *Haemophilus*, *Jonesia*, *Klebsiella*, *Kurthia*, *Lactococcus*, *Lactobacillus* (4), *Legionella*, *Listeria* (6), *Leuconostoc* (2), *Micrococcus* (2), *Mycobacterium* (2), *Mycoplasma* (2), *Neisseria*, *Nocardia*, *Oerskovia* (2), *Paracoccus*, *Pediococcus*, *Peptostreptococcus* (2), *Propionibacterium*, *Proteus*, *Pseudomonas*, *Rahnella*, *Rhodococcus*, *Rhodospirillum*, *Staphylococcus* (3), *Streptococcus*, *Streptomyces*, *Vibrio* (11), *Yersinia*

ization probes, can detect non-motile variants of *L. monocytogenes*. It will also detect non-haemolytic and rhamnose-negative variants.

The only major disadvantage of the kit is the expense of the required luminometer. One-sample models without the optional printer are the cheapest. Leasing arrangements are generally available for large-volume purchasers of the test kit, such as clinical laboratories.

Results of Collaborative Evaluations

The test kit performance characteristics, as coefficients of variation, are stated by the manufacturer to be 4.3–9.3% for the within-run and 10.8–13.5% for the between-run method precisions. This test kit has not been formally validated but it has been tested with about 1560 strains of microorganisms representing over 50 genera (**Table 5**) by ten independent laboratory studies. Only 2 false negatives and 1 false positive were found (**Table 6**) but these are justifiably discountable. The ten laboratories did not collaboratively examine a common set of strains. However, a much larger number of strains than the 50 or so studied in real interlaboratory collaborations were collectively examined. Notwithstanding that the collaboration was only a virtual one, the false positive and negative rates, which are the indicators of interlaboratory reproducibility, could still be estimated. Pooling the results obtained by the ten laboratories, the false negative rate was estimated as 0.38% or less and the false positive rate as 0.34% or less. Thus, the estimated sensitivity and specificity rates were at least 99.62 and 99.66%, respectively. These validation results provide the primary requisite for application of the method to detecting *Listeria monocytogenes* in food matrices, namely highly acceptable strain inclusivity and exclusivity values.

inometer's optical system at recommended intervals. The published detection limits of the kit vary from about 10^5 to 10^6 cells per analytical sample (50 µl). This compares favourably with the limits of other kinds of rapid detection kits.

The observed limit depends on the fact that there are thousands of ribosomes and therefore rRNA targets per cell. It is rarely possible to compare kits on a number of targets basis. Because of their proprietary nature, kit targets are often unspecified and so generally their number per cell cannot be estimated.

Since ribosomes provide a stringently conserved cell function – protein synthesis – there will be a negligible chance of variants arising with altered rRNA targets, which would not hybridize with the synthetic probe. Non-hybridization kits depending on targets that may not have obligately conserved functions, e.g. flagellar components, will not detect any targetless variants. The AccuProbe test, like other DNA:rRNA hybrid-

Table 6   Comparison of the AccuProbe *Listeria monocytogenes* DNA probe test kit with conventional methods for determining the identities of strains of a wide variety of microorganisms[a]

Listeria monocytogenes *strains*		Strains of other species and genera	
True positives (AccuProbe +ve, conventional +ve)	False negatives (AccuProbe −ve, conventional +ve)	False positives, (AccuProbe −ve, conventional +ve)	True negatives (AccuProbe −ve, conventional −ve)
732	2[b]	1[c]	828

[a]For list of genera and numbers of species see Table 5.
[b]Ascribable to technical artefact.
[c]Not confirmed upon test repetition.

Table 7   Conventional culture detections of *Listeria monocytogenes* compared with AccuProbe test kit detections in enrichments of artificially and naturally contaminated food samples[a]

Time (h)[b]	Inoculum type[c]	No. of tests	Concentration[d]	Percentage of number tested			
				True positives (AccuProbe +ve, conventional +ve)	False positives (AccuProbe +ve, conventional −ve)	False negatives (AccuProbe −ve, conventional +ve)	True negatives (AccuProbe −ve, conventional −ve)
24	N	24	−	0	0	25	75
24	N	33	+	3	6	27	64
48	N	54	−	9	7	24	59
48	N	76	+	16	1	20	63
24	A	90	+	31	0	69	0
48	A	90	+	76	0	24	0

[a]Condensed and recalculated results based on studies by Niederhauser et al. (1993) and by Bobbitt and Betts (1992); the overall trends of the separate parts of the studies were unaffected by pooling.
[b]Enrichment times. Samples were mainly from soft cheeses. A few samples were from pâté and raw chicken. Several standard enrichments were used.
[c]N, naturally contaminated at unspecified level; A, artificially contaminated at 1–10 cfu per 25 g.
[d]Concentration: +, sample concentrated by centrifugal sedimentation; −, sample not concentrated.

Two different laboratories have studied the application of the AccuProbe method to the detection of *L. monocytogenes* in selective enrichment cultures of artificially and naturally contaminated foods. They tried several standard selective enrichment broths and tested mainly soft cheeses contaminated with low levels of *L. monocytogenes*. A representative compilation of their results (**Table 7**) shows the method detects the pathogen best at 48 h of enrichment. Regular and differential centrifugation improved detection somewhat but was more or less limited by the volume of solid particle food in the samples. There were some unexplained false positives but not enough to be a nuisance. The false negative rates were disappointingly high but, considering the foods tested are among the most refractory to analysis by selective enrichment, were not unexpected. The method described earlier for increasing sample size by extracting the RNA of the 48 h enrichment culture was designed to ameliorate the problems apparent in Table 7. That method works well with 48 h selective enrichments but its possible limitations with 24 h samples have not yet been defined, and it also needs testing with a greater variety of foods.

The AccuProbe test is somewhat affected by the medium used to suspend the test microorganism. This particularly applies to selective enrichment media and appears to be due to interference with the chemiluminescence by medium components. The FDA BAM medium was marginally inhibitory and the least inhibitory of the common selective enrichment media used with *Listeria*. Any inhibitory effects can be avoided by centrifuging the sample and resuspending the pellet in phosphate-buffered saline or by using the RNA extraction method.

See also: **ATP Bioluminescence**: Application in Meat Industry; Application in Dairy Industry; Application in Hygiene Monitoring; Application in Beverage Microbiology. **Bacteriophage-based Techniques for Detection of Food-borne Pathogens**. **Biophysical Techniques for Enhancing Microbiological Analysis**: Future Developments. **Biosensors**: Scope in Microbiological Analysis. **Immunomagnetic Particle-based Techniques**: Overview. *Listeria*: Detection by Classical Cultural Techniques; Detection by Commercial Enzyme Immunoassays; Detection by Colorimetric DNA Hybridization; *Listeria monocytogenes* – Detection using NASBA (an Isothermal Nucleic Acid Amplification System). **Molecular Biology – in Microbiological Analysis**.

National Legislation, Guidelines & Standards Governing Microbiology: European Union. Nucleic Acid-based Assays: Overview. PCR-based Commercial Tests for Pathogens. Reference Materials.

Further Reading

Arnold LJ, Hammon PW, Wiese WA and Nelson NC (1989) Assay formats involving acridinium-ester-labelled DNA probes. *Clinical Chemistry* 35: 1588–1594.

Bobbitt JA and Betts RP (1992) Confirmation of *Listeria monocytogenes* using a commercially available nucleic acid probe. *Food Microbiology* 9: 311–317.

Collins MD, Wallbanks S, Lane DJ et al (1991) Phylogenetic analysis of the genus *Listeria* based on reverse transcriptase sequencing of 16S RNA. *International Journal of Systematic Bacteriology* 41: 240–246.

Duvall RE and Hitchins AD (1997) Pooling of non-collaborative multilaboratory data for evaluation of the use of DNA probe test kits in identifying *Listeria monocytogenes* strains. *Journal of Food Protection* 60: 995–997.

Food and Drug Administration (1995) *Bacteriological Analytical Manual*, 8th edn. Gaithersburg, Maryland: Association of Official Analytical Chemists International.

Niederhauser C, Hofelein C, Luthy J, Kaufmann U, Buhler HP and Candrian U (1993) Comparison of 'Gen-Probe' DNA probe and PCR for detection of *Listeria monocytogenes* in naturally contaminated soft cheese and semi-soft cheese. *Research Microbiology* 144: 147–154.

Okwumabua O, Swaminathan B, Edmonds P, Wenger J, Hogan J and Alden M (1992) Evaluation of a chemiluminescent DNA probe assay for the rapid confirmation of *Listeria monocytogenes*. *Research Microbiology* 143: 183–189.

Septak M (1989) Acridinium-ester-labelled DNA oligonucleotide probes. *Journal of Bioluminescence and Chemiluminescence* 4: 351–356.

Sutherland IW and Wilkinson JF (1971) Chemical extraction of microbial cells. In: Norris JR and Ribbons DW (eds) *Methods in Microbiology*. Vol. 5B, p. 346. New York: Academic Press.

Listeria monocytogenes–Detection using NASBA (an Isothermal Nucleic Acid Amplification System)

M R Uyttendaele and **J M Debevere**, Faculty of Agricultural and Applied Biological Sciences, University of Ghent, Ghent, Belgium

Introduction

Nucleic acid amplification techniques show great promise as highly sensitive and specific tools for the detection of food-borne pathogens. Characterization of microorganisms based on the sequence of DNA or RNA allows a more reliable and stable identification than the use of biochemical and physiological tests. Expression of a phenotypic characteristic often depends upon environmental changes. The nucleotide sequence of DNA contains all the information that specifies the characteristics and potential activities of an organism. The in vitro amplification of a preselected region of nucleic acid improves the sensitivity of these methods. It eliminates the need for time-consuming preliminary enrichment procedures to attain high numbers (10^4–10^5 cfu ml^{-1}) as is required for immunological or hybridization techniques. Numerous polymerase chain reaction (PCR) tests using specific primer sets for *Listeria monocytogenes* have been described. Most of these sets target the *hlyA* gene coding for listeriolysin O, although some other genes have also been used as a template. As a consequence of the success and opportunities of the PCR technique for research and diagnostic purposes, various other amplification methods were developed, including the nucleic acid sequence-based amplification technique (NASBA).

NASBA is an isothermal nucleic acid amplification system with the ability to selectively amplify RNA. The presence of RNA sequences is related to biological activity, thus NASBA can be applied to check for gene expression and cell viability. The isothermal nature of a NASBA reaction eliminates the need for a thermal cycler. This article deals with the history, principle and (dis)advantages of the NASBA amplification system, its development for detection of both *L. monocytogenes*' ribosomal and messenger RNA and the application of the amplification method for detection of the pathogen in foods.

The NASBA nucleic acid amplification system

History

The NASBA amplification system as an alternative to PCR amplification was first reported in clinical diagnostics in 1991. The NASBA amplification system was optimized for the detection of HIV-1 sequences. Detection levels were 10 molecules of HIV-1 RNA in a model system with in vitro-generated HIV-1 RNA as input. The NASBA method can also be used as a diagnostic tool for other RNA viruses.

NASBA has the possibility for identification of bacteria based on the highly variable ribosomal RNA (rRNA) sequences. The NASBA technique was developed for identification of mycobacteria on the basis of a highly conserved region of the 16S rRNA sequence sandwiching a variable sequence. Species-specific probes were shown to hybridize specifically to the amplified single-stranded RNA of the different mycobacteria species.

Apart from clinical microbiology, the amplification system can also serve as a diagnostic tool in food microbiology. The NASBA system was developed to enable rapid detection of pathogens such as *L. monocytogenes* and pathogenic campylobacters in food samples.

The rights for the NASBA amplification system are held by Organon Teknika (Boxtel, the Netherlands).

Principle

Isothermal nucleic acid amplification of RNA is achieved through the concerted action of reverse transcriptase (RT), T7 RNA polymerase and RNase H (**Fig. 1**). The reaction starts with a non-cyclic phase, in which a downstream primer containing a tail sequence of the T7 promoter sequence (P1) anneals to the single-stranded target sequence. Through the action of RT, P1 is extended and DNA is synthesized. The RNase H hydrolyses the RNA from the RNA-DNA hybrid, which results in a single strand of DNA

to which the upstream primer (P2) can anneal. The RT synthesizes, through its DNA polymerase activity, a second DNA strand, producing a double-stranded DNA intermediate (cDNA) with a transcriptionally active T7 promoter sequence. The T7 RNA polymerase generates from this intermediate single-stranded antisense RNA copies (10–1000) which serve as a template in the now cyclic phase of NASBA. In the cyclic phase the events are the same, but P1 and P2 are incorporated in reverse order.

While the steps of cDNA synthesis and transcription have been outlined separately for clarity, all the described reactions occur concomitantly and continuously during isothermal incubation at 41°C. During the NASBA reaction there is a major accumulation of specific single-stranded antisense RNA. Incubation of the reaction tubes at 41°C for 1.5–2 h achieves an amplification of about 10^6–10^9-fold of the original target sequence. The fidelity of the amplification reaction was confirmed by direct sequencing of amplified RNA; $< 0.3\%$ error frequency was reported. When DNA is used as input two heat-denaturing steps are compulsory to obtain single-stranded DNA intermediates, available for primer annealing (P1 and P2).

Advantages/Disadvantages

NASBA allows selective amplification of RNA while PCR was originally designed for amplification of DNA, although RNA can also be amplified by PCR using an additional RT step. However, the isothermal nature of the NASBA reaction avoids amplification of homologous double-stranded DNA, avoiding the need of DNA-free RNA as an input for the NASBA reaction.

The use of RNA as target can be beneficial. Ribosomal RNA possesses a large diagnostic potential because of its structure of genus, species and intra-species-specific sequences. Specific gene products can be detected by NASBA amplification of the corresponding messenger RNA (mRNA). In addition, a higher number of copies of RNA are present in the cell than DNA. RNA, however, is chemically and biologically more labile than DNA and susceptible to RNase degradation. Hence, precautions should be taken for storage and manipulation of RNA. On the other hand, RNA is related to cell viability.

Major differences between PCR and NASBA are indicated in **Table 1**. NASBA is performed isothermally. This means that NASBA reactions can generally be performed in a common laboratory incubator while PCR requires a thermocycling device. The constant temperature maintained throughout the amplification reaction allows each step of the reaction to proceed as soon as an amplification intermediate

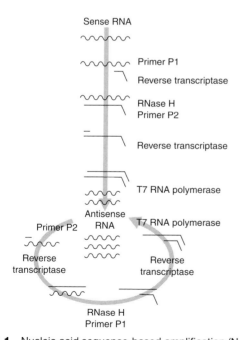

Figure 1 Nucleic acid sequence-based amplification (NASBA) amplification: the method is based on an extension of primer 1 (containing a T7 promotor) by reverse transcriptase (RT), degradation of the RNA strand by RNase H, synthesis of the second strand of DNA by RT and RNA synthesis by T7 RNA polymerase. Courtesy of P. Oudshoorn, Organon Teknika, Boxtel, The Netherlands.

Table 1 Major differences between the nucleic acid sequence-based amplification (NASBA) and polymerase chain reaction (PCR) amplification systems

NASBA	PCR
Target:	*Target:*
RNA, ssDNA	dsDNA
dsDNA by prior heat denaturation	RNA by reverse transcriptase step
Two phases:	*Three phases:*
Conversion of RNA to dsDNA	Denaturation (92–97°C, 1–2 min)
Transcription	Hybridization (40–60°C, 1–2 min)
	Extension (72°C, 1–2 min)
Isothermal reaction: 41°C, 90 min	*Thermocycling* required: 25–45 cycles
Three enzymes:	*One* thermostable *enzyme:*
AMV-reverse transcriptase	*Taq* DNA polymerase
RNase H	
T7-RNA polymerase	
Two specific primers with primer 1 (antisense, complementary) holding a T7-RNA polymerase promotor site	*Two specific primers*
Amplification factor: x^n	*Amplification factor:* 2^n
n = number of hypothetic cycles	n = number of cycles
x = RNA copies generated during transcription (10–1000)	
Amplification product:	*Amplification product:*
antisense (3'–5'), complementary ssRNA	ds DNA

becomes available. Also, since amplification in the NASBA system is based in part on the transcription of RNA, multiple copies of RNA are transcribed in each (theoretical) cycle. Thus, the augmented kinetics of the NASBA process is intrinsically more efficient than PCR amplification that is limited to binary increases per cycle.

NASBA does have some disadvantages. First, to make it an isothermal reaction thermolabile enzymes must be used. This means that the reaction must be effected at a lower temperature than the PCR process using thermocycling, sometimes resulting in an increase in nonspecific primer interactions. Including dimethyl sulphoxide (DMSO) and a 65°C priming step for the primer and template RNA in a NASBA reaction has solved many of the mispriming problems. Nevertheless, the identity of the amplification product should be confirmed by a subsequent hybridization step. Because the amplification product is single-stranded, no denaturation before hybridization is necessary.

Like the other amplification methods, NASBA is subject to contamination by extraneous nucleic acids, especially own amplification product carryover.

Detection of the Amplification Product

Detection of amplification products uses agarose gel electrophoresis and staining of the nucleic acid with ethidium bromide. The identity of the amplification products is determined by hybridization. Initially, a rapid and simple non-radioactive hybridization assay (enzyme-linked gel assay, ELGA) was used to identify the NASBA products. First, a liquid hybridization with specific oligonucleotide probes (ELGA probes)

labelled with horseradish peroxidase (HRP) is performed. After hybridization, excess of non-hybridized ELGA probes is separated from the homologous hybridized product by vertical gel electrophoresis. The free ELGA probes and the hybridized products can be directly visualized in the acrylamide gel by incubating the gel in a substrate solution for HRP. Due to its lower mobility the homologous hybridized product migrates in the gel above the free ELGA probe.

Subsequently, a bead-based enzymatic assay was developed. Polystyrene paramagnetic beads are coated with streptavidin and incubated with a biotinylated capture probe. Washed beads are then incubated with NASBA reaction mixture. Next, the beads are washed twice and a HRP-labelled detection oligonucleotide probe is added. The bead-capture oligonucleotide-NASBA amplificate–detection probe complex is detected by colour development upon the addition of substrate for the enzyme. Similarly, an electrochemiluminescent (ECL) bead-based detection was described using an ECL-labelled oligonucleotide detection probe and the appropriate substrate. This bead-based assay enabled automation of the detection procedure of the hybridized product.

In addition, an immunoassay for detection of amplification product was reported. This assay uses an antibody recognizing the RNA–DNA hybrids formed upon hybridization of the amplification product (RNA) of the NASBA reaction with probe DNA immobilized in a microtitre plate. The microtitre plate format facilitates the simultaneous assay of multiple samples.

Detection of *L. monocytogenes* on the Basis of Ribosomal RNA Using NASBA

Ribosomal RNA

A large number of rRNA sequences are readily available in computer data banks, which can be used for setting up diagnostic amplification tests. The abundance of ribosomes in the bacterial cell (each cell contains over 10 000 copies) can further improve sensitivity of the assay. RNA is also suggested as a valuable indicator of viability. The possibility of NASBA for assessment of bacterial viability was evaluated using *Mycobacterium smegmatis* 16S rRNA. During exposure of the bacterium to various concentrations of antibiotics, the lack of the NASBA signal coincided with a loss of viability, expressed by absence of cfu on agar plating. In contrast, the NASBA amplification system could not be used to distinguish viable and dead campylobacters. It was shown that 16S rRNA of a killed campylobacter culture was fairly stable and took a long period (> 10 days) for complete degradation, especially if high inocula were used.

The 16S rRNA sequences of *Listeria* spp. exhibit very high levels of sequence relatedness. *L. monocytogenes* and *L. innocua* possess 99.2% similarity, corresponding to only 11 base differences. Only two regions are useful in distinguishing *Listeria* spp.: the V2 region (nt 171–213) and the V9 region (nt 1244–1287; **Fig. 2**).

Detection on the Basis of the V2 Region of Ribosomal RNA

The NASBA system allowed amplification of a *L. monocytogenes*-specific region situated in the 16S rRNA. An amplification product was clearly visible for all of the *Listeria* spp. as the NASBA primers are not unique for *L. monocytogenes* but fit to 16S rRNA of all *Listeria* spp. The identity of the amplimers was determined using the ELGA assay. No hybridization with the ELGA probe occurred for these other species, confirming the specificity of the probe (**Fig. 3**). Also the negative controls showed an amplification product, although this was not always noticed when

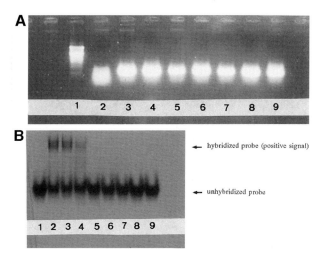

Figure 3 (A) Ethidium bromide-stained 2% agarose gel showing the nucleic acid sequence-based amplification (NASBA) products of *Listeria monocytogenes* LMG 10470 (lane 2), *L. monocytogenes* LMG 13304 (lane 3), *L. monocytogenes* LMG 13305 (lane 4), *L. seeligeri* LMG 11386 (lane 5), *L. innocua* LMG 11387 (lane 6), *L. ivanovii* LMG 11388 (lane 7) and *L. welshimeri* LMG 11389 (lane 8) with primer set OT1683/OT1686. Input was in lane 1, marker (2 µg *Escherichia coli* 16S rRNA) and for lane 9, amplification product of the negative control sample. (B) Specifity of NASBA in combination with the enzyme-linked gel assay (ELGA) probe. After in-solution hybridization (ELGA) of the NASBA product, the hybridization reaction mixtures were run in a 7% acrylamide gel, to separate the free ELGA probe (fast mobility) from the homologous hybridized NASBA product (slow mobility). Lane 1, Free ELGA probe; lane 2, *L. monocytogenes* LMG 13304; lane 3, *L. monocytogenes* LMG 13305; lane 4, *L. monocytogenes* LMG 10470; lane 5, *L. seeligeri* LMG 11386; lane 6, *L. innocua* LMG 11387; lane 7, *L. ivanovii* LMG 11388; lane 8, *L. welshimeri* LMG 11389; lane 9, negative control sample. Reprinted from the *International Journal of Food Microbiology* 1995; 27: 77–89.

repeating NASBA reactions. It was shown to be a nonspecific amplification product that does not hybridize with the probe.

The latter system required at least 2×10^5 cfu ml^{-1} or 4×10^3 cfu of *L. monocytogenes* per NASBA reaction to produce a positive result. The poor detection capability of the NASBA-ELGA detection system could in part have been caused by self-annealing of

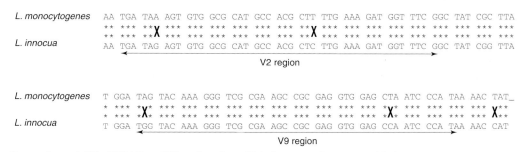

Figure 2 Comparison of 16S rDNA V2 and V9 regions from *Listeria monocytogenes* and *L. innocua*.

the detector probe. The 5 end (AAAG) and the 3 end (CTTT) of the ELGA probe are complementary to one another but these final nucleotide sequences also include the mutations differentiating *L. monocytogenes* from *L. innocuca*. The NASBA system was applied for detection of *L. monocytogenes* in food samples after an enrichment step.

Detection on the Basis of the V9 Region of Ribosomal RNA

Multiple primer sets and probes were evaluated for NASBA amplification and ELGA detection of the *L. monocytogenes*-specific V9 region. It was noted that the choice of the primer set and the probe influences sensitivity of the NASBA-ELGA assay. It was also found that *L. monocytogenes*-specific probe, containing a single point mutation towards the other *Listeria* spp., under the applied hybridization conditions lacked specificity to identify *L. monocytogenes* solely. To overcome this problem, hybridization was performed in the presence of a competitive probe, containing the exact complementary sequence to nonpathogenic *Listeria* spp. amplification product. A ratio of 1000 : 1 of competitive probe : specific probe sustained specific identification of *L. monocytogenes*. The sensitivity of this latter ELGA assay, however, decreased about 2 log units if compared to the ELGA assay using a single detector probe. Addition of a competitive flora to a *L. monocytogenes* culture before lysis and amplification also decreased the test's sensitivity. This can be explained by the universal occurrence of ribosomal RNA in bacterial cells and a possible nonspecific priming during the NASBA reaction. The NASBA-ELGA assay for detection of *L. monocytogenes* on the basis of the 16S rRNA V9 region was not applied to food samples.

Detection of *L. monocytogenes* on the Basis of mRNA Using NASBA

mRNA

Although ribosomal RNA offers highly discriminatory sequences, this may not always be sufficient to distinguish to subspecies levels or to identify pathogenic strains within a species. Characteristic gene products or virulence factors can be detected using NASBA by amplification of the corresponding mRNA. mRNA has a short half-life and is more associated with metabolic activity than rRNA. Detection of mRNA of a virulence gene might be more relevant to health issues than detection of the gene itself, since only cells capable of virulence gene expression would constitute a potential health hazard. But detection on the basis of mRNA is only possible if the corresponding genes are expressed. Some genes, such as those encoding for flagella or outer membrane proteins, are often constitutively expressed while others such as some genes coding for metabolic enzymes or virulence factors are only expressed in a particular environment.

Detection on the Basis of the Listeriolysin O gene (*hlyA* Gene)

A NASBA system targeting the mRNA sequence from the *L. monocytogenes hlyA* gene was developed. Synthesis of *hlyA* mRNA in broth cultures of *L. monocytogenes* was induced by incubating the culture for 30 min at 37°C in the presence of 0.1% sodium azide. After extraction of total RNA, a 133-mer region of *hlyA* mRNA was amplified using the NASBA reaction. The identity of the biotin-labelled amplimers was confirmed by hybridization with an immobilized capture probe. Treatment of the total RNA extracts with RNase-free DNase prior to performing the NASBA assay confirmed that this amplification method is highly specific for mRNA. The NASBA hybridization assay was shown to enable specific detection of *L. monocytogenes*. The assay detected the pathogen in samples for which a minimum of 500 cfu were introduced into the NASBA reaction mixture, corresponding to 5000 cfu ml^{-1} broth culture. This suggests that if the NASBA assay system is to be applied for detection of this pathogen in foods, it will first be necessary to subject the sample to pre-enrichment in a nutrient-rich broth in order to allow multiplication of the cells to a detectable level.

The same research group undertook the development of the NASBA system targeting the DNA sequence of the *L. monocytogenes hlyA* gene. Two heat-denaturing steps (5 min, 100°C) needed to be performed prior to the NASBA isothermal amplification procedure. The identity of the amplimers was confirmed using an antibody recognizing the RNA–DNA hybrids formed upon hybridization of the amplimer (RNA) with the probe DNA. Specific detection of *L. monocytogenes* was obtained. The system permitted detection of ca. 0.05 pg of genomic DNA, or about 10 cell equivalents, per NASBA reaction. Sensitivity, however, was not determined using a dilution serial of *L. monocytogenes* cells. The NASBA assay using *hlyA* gene DNA as a target detected another part of the gene than the NASBA assay targeting the *hlyA* gene mRNA. Also, the detection format of NASBA amplimers differed. This makes it impossible to compare the efficiency of both NASBA reactions, based either on mRNA or DNA.

NASBA Detection of *L. monocytogenes* in Foods

The NASBA amplification system based either on detection of 16S ribosomal RNA or on *hlyA* mRNA sequences makes a specific assay for detection of *L. monocytogenes* in foods. Sensitivity of the isothermal amplification method is determined at 500–4000 cfu/NASBA reaction dependent upon the choice of RNA used as a target for amplification. The actual sensitivity of the method, when used as a detection technique for the pathogen in foods, will depend upon the accessibility of the target organisms, the presence of other organisms and the presence of inhibitory substances in the food sample.

Most often pathogen screening tests are zero tolerance, meaning that the required sensitivity is one pathogen cell in a test sample (generally 25 g). This necessitates the optimization of methods to isolate and harvest bacteria from the food such as filtration, centrifugation, enzymatic digestion or addition of chemical compounds. These procedures are time-consuming, cumbersome and lack efficiency. Although, it is theoretically possible using nucleic acid amplification techniques to detect pathogens in the food sample directly (without prior enrichment), this has been done so far on very few occasions. Pre-enrichment for several hours sustains the increase of the target cell concentration (typically greater than 10^4-fold) to the test's threshold detection level. In addition, non-target organisms and inhibitory substances in the food matrix are diluted out. Inhibitors in foods are a common problem in amplification assays which require well-defined conditions for functioning of the enzymatic reactions. The inhibiting substances can be fats, proteinases which degrade the required assay enzymes, haemoglobin degradation products and selective components of the broths. Some foods, especially cheeses and minced meat, need apart from dilution supplementary treatments such as immunomagnetic separation, buoyant density centrifugation or nucleic acid extraction procedures to enable the amplification assay to function well.

A typical detailed protocol for detection/identification of *L. monocytogenes* starting from raw or processed foodstuff is shown in **Figure 4**. The NASBA assay was performed after a 48 h enrichment period in a selective enrichment broth and subsequent extraction of total RNA using a guanidinium thiocyanate (GuSCN) lysis and silica purification protocol. The identity of the NASBA amplification product was confirmed by the ELGA assay. After pre-enrichment the molecular procedure took about 6 h to complete.

The NASBA amplification technique based on the

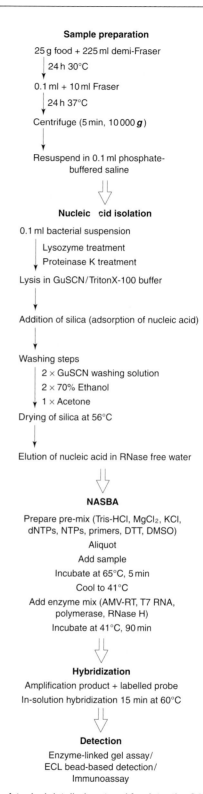

Sample preparation

25 g food + 225 ml demi-Fraser
 ↓ 24 h 30°C
0.1 ml + 10 ml Fraser
 ↓ 24 h 37°C
Centrifuge (5 min, 10 000 ***g***)

Resuspend in 0.1 ml phosphate-buffered saline

Nucleic cid isolation

0.1 ml bacterial suspension
 ↓ Lysozyme treatment
 ↓ Proteinase K treatment
Lysis in GuSCN/TritonX-100 buffer

Addition of silica (adsorption of nucleic acid)

Washing steps
 ↓ 2 × GuSCN washing solution
 ↓ 2 × 70% Ethanol
 ↓ 1 × Acetone
Drying of silica at 56°C

Elution of nucleic acid in RNase free water

NASBA

Prepare pre-mix (Tris-HCl, MgCl₂, KCl, dNTPs, NTPs, primers, DTT, DMSO)
Aliquot
Add sample
Incubate at 65°C, 5 min
Cool to 41°C
Add enzyme mix (AMV-RT, T7 RNA, polymerase, RNase H)
Incubate at 41°C, 90 min

Hybridization
Amplification product + labelled probe
In-solution hybridization 15 min at 60°C

Detection
Enzyme-linked gel assay/
ECL bead-based detection/
Immunoassay

Figure 4 A typical detailed protocol for detection/identification of *Listeria monocytogenes* starting from raw or processed foodstuff.

detection of the *L. monocytogenes* V2 region of 16S ribosomal RNA was compared to an ELISA-based method (Listeria-Tek, Organon Teknika) and to a

Table 2 Detection of *Listeria monocytogenes* in different food products using a modified FDA method, ELISA (Listeria-Tek, Organon Teknika) and the NASBA-ELGA assay

Food sample (CFU of L. monocytogenes *per 25 g*)	Detection of L. monocytogenes with the modified FDA method	Detection of L. monocytogenes *with ELISA*	Detection of L. monocytogenes using the NASBA technique
Chicken breast meat			
Control	–	–	–
7	+	+	+
70	+	+	+
700	+	+	+
Minced meat			
Control	+	+	+
1	+	+	+
10	+	+	+
100	+	+	+
Shrimps			
Control	–	–	–
1	–	+	+
10	+	+	+
100	+	+	+
Raw milk			
Control	–	–	–
10	+	+	+
100	+	+	+
1000	+	+	+
Soft cheese			
Control	–	–	–
2	+	+	+
20	+	+	+
200	+	+	+
Dry sausage			
Control	–	–	–
2	–	+	+
20	+	+	+
200	+	+	+
Mushrooms			
Control	–	–	–
6	+	+	+
60	+	+	+
600	+	+	+
Radish			
Control	–	–	–
10	+	+	+
100	+	+	+
1000	+	+	+

+ Detection of *L. monocytogenes*, – no detection of *L. monocytogenes*
FDA, Food and Drug Administration
ELISA, enzyme-linked immunosorbent assay; NASBA-ELGA, nucleic acid sequence-based amplification-enzyme-linked gel assay.
From *International Journal of Food Microbiology* 1995; 25: 77–89.

modification of the cultural method proposed by the US Food and Drug Administration (FDA). The three methods were applied to detect the pathogen in a variety of artificially inoculated foods. The NASBA amplification assay allowed detection of low numbers (1–100 cfu 25 g^{-1}) of *L. monocytogenes* in these foods (**Table 2**). The identification by NASBA of *L. monocytogenes* in the different food products shows the potential general applicability of the method.

The NASBA system amplifying the *L. monocytogenes hlyA* gene was applied for detection of low numbers of the pathogen (1–45 cfu 5 g^{-1}) into a number of dairy products (five sorts of cheese, milk and milk powder, ice cream and egg powder). The NASBA assay system, combined with a hybridization assay using an immobilized capture probe, could reliably detect *L. monocytogenes* in the range of 1–2 cfu g^{-1} of food after a 48 h enrichment period. One positive reaction for three replicates was obtained with uninoculated samples of egg powder and mozzarella cheese. These results were believed to be false-positives. The reason for these false-positives may be cross contamination of the NASBA reactions with amplimers from previous amplifications carried out at the same site. Another possibility is cross contamination of the samples during RNA extraction. The amplification power of the NASBA reaction results in the detection of even trace amounts of any contaminants. Proper negative controls should be included during each of the manipulations of the detection procedure.

Future Perspective

More research is necessary to reduce analysis time by increasing sensitivity of the NASBA assay, thus shortening the previous required enrichment time and to evaluate whether this amplification method can be used to detect *L. monocytogenes* in naturally contaminated food products. Further developments, especially with regard to the sample preparation protocol and the detection format of amplimers to reduce the complexity of the procedures enabling rapid handling of multiple samples, are required before considering applicability of the NASBA assay in routine examinations.

Acknowledgements

Mieke Uyttendaele is indebted to the Fund of Scientific Research – Flanders (Belgium) (FWO) for a position as research assistant.

See also: **Nucleic Acid-based Assays**: Overview. **PCR-based Commercial Tests for Pathogens**.

Further Reading

Barbour WM and Rice G (1997) Genetic and immunologic techniques for detecting foodborne pathogens and toxins. In: Doyle MP, Beuchat LR, Montville TJ (eds) *Food Microbiology, Fundamental and Frontiers*. p. 710. Washington DC: ASM Press.

Blais BW, Turner G, Sooknanan T, Malek LT and Phillippe LM (1996) A nucleic acid sequence-based amplification (NASBA) system for *Listeria monocytogenes* and simple method for detection of amplimers. *Biotechnol. Techn.* 10: 189–194.

Blais BW, Turner G, Sooknanan T and Malek LT (1997) A nucleic acid sequence-based amplification system for detection of *Listeria monocytogenes hlyA* sequences. *Appl. Environm. Microbiol.* 63: 310–313.

Collins MD, Wallbanks S, Lane DJ et al (1991) Phylogenetic analysis of the genus *Listeria* based on reverse transcriptase sequencing of 16S rRNA. *Int. J. Syst. Bacteriol.* 41: 240–246.

Kievits T, van Gemen B, van Strijp D et al (1991) NASBA isothermal enzymatic in vitro nucle · acid amplification optimized for the diagnosis of HIV-1 infection. *J. Virol. Methods* 35: 273–286.

Sooknanan R and Malek TL (1995) NASBA, a detection and amplification system uniquely suited for RNA. *Biotechnology* 13: 563–564.

Uyttendaele M, Schukkink R, van Gemen B and Debevere J (1995a) Detection of *Campylobacter jejuni* added to foods by using a combined selective enrichment and nucleic acid sequence-based amplification (NASBA). *Appl. Environm. Microbiol.* 61: 1341–1347.

Uyttendaele M, Schukkink R, van Gemen B and Debevere JM (1995b) Development of NASBA, a nucleic acid amplification system, for identification of *Listeria monocytogenes* and comparison to ELISA and a modified FDA method. *Int. J. Food Microbiol.* 27: 77–89.

Uyttendaele M, Bastiaansen A and Debevere J (1997) Evaluation of the NASBA nucleic acid amplification system for assessment of the viability of *Campylobacter jejuni*. *Int. J. Food Microbiol.* 37: 13–20.

Van der Vliet GME, Schukkink RAF, van Gemen B, Schepers P and Klatser PR (1993) Nucleic acid sequence based amplification (NASBA) for the identification of mycobacteria. *J. Gen. Microbiol.* 139: 2423–2429.

Van der Vliet GME, Schepers P, Schukkink RAF, van Gemen B and Klatser PR (1994) Assessment of mycobacterial viability by RNA amplification. *Antimicrob Agents Chemother.* 38: 1959–1965.

Van Gemen B, van Beuningen R, Nabbe A et al (1994) A one-tube HIV-RNA NASBA nucleic acid amplification assay using electrochemiluminescent (ECL) labeled probes. *J. Virol. Methods* 49: 157–168.

Lysins *see* **Minimal Methods of Processing**: Potential Use of Phages and/or Lysins.

Lysozyme *see* **Natural Antimicrobial Systems**: Lysozyme and other Proteins in Eggs.

Malolactic Fermentation *see* **Wines:** The Malolactic Fermentation.

Manothermosonication *see* **Minimal Methods of Processing:** Manothermosonication.

Manufacturing Practice *see* **Good Manufacturing Practice**.

Mathematical Modelling *see* **Predictive Microbiology and Food Safety**.

MEAT AND POULTRY

Contents
Spoilage of Meat
Curing of Meat
Spoilage of Cooked Meats and Meat Products

Spoilage of Meat

George-John E Nychas and **Eleftherios H Drosinos**, Department of Food Science and Technology, Laboratory of Microbiology and Biotechnology of Foods, Agricultural University of Athens, Greece

Copyright © 1999 Academic Press

Introduction

Spoilage of meat is an ecological phenomenon, encompassing the changes of the meat ecosystem during the development of its microbial association. The establishment of a particular microbial association of meat depends on the ecological factors that persist during processing, storage, transportation and in the market. In meat, five categories of ecological determinants influence the development of the initial and transient microbial associations and determine the rate of attainment of a climax population by the ephemeral spoilage microorganisms (those that fill the niche available by adopting R-ecological strategy as a result of enrichment disturbance of an ecosystem).

These are (1) the intrinsic factors associated with the physico-chemical attributes and structure of meat (e.g. pH, water activity, buffering power, the presence of naturally occurring or added antimicrobial components, E_h and redox poising capacity, and nutrient composition – carbohydrate content and, in particular, the concentration of glucose); (2) the processing factors; (3) extrinsic parameters that have selective influences, such as temperature, relative humidity and the composition of the gaseous atmosphere obtaining during distribution and storage; (4) the implicit factors (intrinsic biotic parameters) that play an important role in the genesis of spoilage associations; and (5) the emergent effects due to those factors that interact to produce effects greater than would be expected from their action in isolation. In essence all of the above determinants constitute the dimensions of a particular ecological niche – an n-dimensional hypervolume. Indeed, the ecosystem approach is pertinent to an analysis of changes occurring in meat or meat products. In practice, therefore, meat technologists attempt to modify some or all of the dimensions noted above in order to extend the shelf life of meat or to create new products.

Typical Microflora of Fresh or Frozen Meat

Contamination and Spoilage

The microbiology of carcass meats is greatly dependent on the conditions under which animals are reared, slaughtered and processed. Thus the physiological status of the animal at slaughter, the spread of contamination during slaughter and processing, the temperature, and other conditions of storage and distribution are the most important factors that determine the microbiological quality of meat. The characteristic microbial associations developing on meat and in meat products are the result of the determinants noted above on the growth of microbes initially present in the fresh meat or, more generally, introduced during processing. As the inherent antimicrobial defence mechanisms of the live animal are destroyed at slaughter, the resultant meat is liable to rapid microbial decay. Unless effectively controlled, the slaughtering process may cause extensive contamination of the cut face of muscle tissue with a vast range of both Gram-negative and Gram-positive bacteria as well as yeasts (**Table 1**). Some of these microorganisms will be derived from the animal's intestinal tract, and others from the environment with which the animal had contact at some time before or during slaughter. Studies on the origin of the contaminants have shown that the source of Enterobacteriaceae on meats is associated with work surfaces and not with direct faecal contamination. Moreover, psychrotrophic bacteria are recovered from hides and work surfaces within an abattoir as well as from carcasses and butchered meat at all stages of processing.

Microorganisms of the Spoilage Association

Although a range of microbial taxa are found in meat (Table 1), its spoilage in developed countries is caused by the selection of relatively few of these organisms (**Table 2**). It is evident from **Table 3** that chill storage and the gaseous composition around meat packed in vacuum or in modified atmospheres exert strong selectivity on its microflora. Selective factors favour the growth of particular organisms and, as a consequence, a characteristic microbial association is present at the time of spoilage and it will manifest characteristic spoilage features. For example, with the advent of supermarkets in the late 1950s, storage of meat aerobically at chill temperatures and high relative humidity became a major selective factor and

Table 1. The genera of bacteria and yeasts most frequently found on meats and poultry

Genus	Fresh meat	Processed meat	VP/MAP	Poultry
Bacteria				
Acinetobacter	××	×	×	××
Aeromonas	××		×	×
Alcaligenes	×			×
Alteromonas	×	×	×	×
Arthrobacter	×	×	×	×
Bacillus	×	×		×
Bacteroides	×			
Brochothrix	×	×	××	×
Campylobacter				××
Carnobacterium	×	×	××	×
Chromobacterium				×
Citrobacter	×			×
Clostridium	×		×	×
Corynebacterium	×	×	×	×
Enterobacter	×	×	×	×
Enterococcus			××	
Escherichia	×			×
Flavobacterium	×			××
Hafnia	×		×	
Janthinobacterium			×	
Klebsiella			××	
Kluyvera			××	
Kurthia	×		×	
Lactobacillus	×	××	××	
Leuconostoc	×	×	×	
Listeria			×	
Micrococcus	×	×	×	×
Moraxella	××			×
Neisseria	×	×		×
Pantoea			×	
Pediococcus	×	×	×	
Planococcus				×
Plesiomonas				×
Providencia			×	
Proteus	×			×
Pseudomonas	××	×	×	××
Psychrobacter	×			
Serratia	×	×	×	×
Shewanella			×	
Streptococcus	×	××	×	×
Streptomyces	×			
Staphylococcus		×		
Vibrio		×		
Weissella		×	×	
Yeasts				
Candida	××	××		×
Debaryomyces	×			×
Rhodotorula	×	××		
Saccharomyces		×		
Torulaspora		××		
Trichosporon		×		×

Key: x, known to occur; xx, most frequently reported.
VP/MAP, meat stored under vacuum or modified-atmosphere packaging.

Table 2 Psychrotrophic bacteria associated with chilled meats and meat products

Gram-negative bacteria	Gram-positive bacteria
Aerobes	**Catalase reaction weak**
Pseudomonas spp.	Brochothrix
rRNA homology, group I:	thermosphacta
P. fluorescens	
biovars I, II, III, IV, V	
(includes 7 clusters)	
P. lundensis, P. fragi	
Shewanella putrefaciens	
Alteromonas spp.	
Alcaligenes spp.,	
Achromobacter spp.	
Flavobacterium spp.	
Moraxella spp.	
Psychrobacter spp.	
P. immobilis,	
P. phenylpyruvica	
Acinetobacter spp.	
A. lwoffi, A. johnsonii	
Facultative anaerobes	**Catalase reaction**
Photobacterium spp.	**negative**
Vibrio spp.	Lactobacillus spp.
Aeromonas spp.	L. sake
Plesiomonas spp.	L. curvatus
Serratia spp.	L. bavaricus
S. liquefaciens	Carnobacterium spp.
S. marcescens	C. divergens
Citrobacter spp.	C. piscicola
C. freundii, C. koseri	Leuconostoc spp.
Providencia spp.	L. carnosum
P. alcalifaciens, P. stuartii,	L. gelidum
P. rettgeri	L. amelibiosum
Hafnia spp.	L. mesenteroides subsp.
Hafnia alvei	mesenteroides
Pantoea agglomerans	Weissella spp.
Enterobacter spp.	W. hellenica
E. cloacae,	W. paramesenteroides
E. aerogenes	Lactococcus raffinolactis
E. agglomerans/	Clostridium estertheticum
Erwinia herbicola	
complex	
Klebsiella spp.	
K. pneumoniae	
Kluyvera spp.	
Proteus spp.	
P. vulgaris, P. mirabilis	

Pseudomonas spp. are considered to be the main spoilage organisms. Gram-positive bacteria (lactic acid bacteria and *Brochothrix thermosphacta*) are the main spoilage organisms in chill meat stored in a modified atmosphere. To date studies on the contribution of yeasts to the spoilage of meat, whole or minced, has attracted little attention even though they are common contaminants. Yeasts do not outgrow bacteria on meat or meat products unless a bacteriostatic agent is included, such as sulphite in British fresh sausages, or the water activity is reduced.

Table 3 Specific spoilage flora on fresh meat stored at 0–4°C in different gas atmospheres

Gas composition	Specific spoilage flora
Air	Pseudomonas spp.
>50% CO_2 mixed with O_2	Brochothrix thermosphacta
>50% CO_2	Enterobacteriaceae
<50% CO_2 mixed with O_2	Brochothrix thermosphacta, lactic acid bacteria
>50% CO_2	Lactic acid bacteria
100% CO_2	Lactic acid bacteria
Vacuum pack	Lactic acid bacteria, Brochothrix thermosphacta

Under Aerobic Conditions Although the Gram-negative aerobic psychrotrophic bacteria of meat include a number of well-defined species (see Table 2), it is now well established that under aerobic storage three species of *Pseudomonas* – *P. fragi*, *P. fluorescens* and *P. lundensis* – are the most important. Off odours are present when the population of pseudomonads is of the order of 10^7 per square centimetre and slime when these organisms reach 10^8 per square centimetre. In practice off odours become evident when the pseudomonads have exhausted the glucose and lactate present in meat and begin to metabolize the amino acids.

Although rarely, if ever, contributing significantly to the spoilage flora on meat and meat products, the Enterobacteriaceae have been considered as indicators of food safety. With ground beef, *Pantoea agglomerans*, *Escherichia coli* and *Serratia liquefaciens* were the major representatives of this family (see Table 2).

Brochothrix thermosphacta has been detected in the aerobic spoilage flora of chilled meat but it is not considered to be important in spoilage except possibly of lamb. This organism has been isolated from beef carcasses during boning, dressing and chilling. Moreover, lairage slurry, cattle hair, rumen contents, soil from the walls of slaughterhouses, the hands of workers, the air in the chill room, the neck and skin of the animal as well as the cut muscle surfaces have been shown to be contaminated with this organism. *Brochothrix thermosphacta* is one of the main – if not the most important – cause of spoilage which can be recognized as souring rather than putrefaction. This type of spoilage is commonly associated with meat packed under modified atmospheres.

Under Vacuum or Modified-atmosphere Packaging Conditions The atmosphere may be modified by vacuum packaging or storage of meat in atmospheres containing a mixture of gases (N_2, CO_2 and O_2). Meat in a vacuum pack or modified atmosphere (protective atmosphere) has an extensive shelf life when compared with meat stored aerobically. Shelf life is determined by the choice of atmosphere, storage temperature and meat type. As the bacterial population of meat (particularly the aerobes, e.g. pseudomonads) is restricted by the relative high concentration of CO_2 and the oxygen limitation, the spoilage of meat stored under vacuum or modified atmosphere occurs later than that of meat stored aerobically. In meat samples stored under vacuum or modified-atmosphere packaging lactic acid bacteria are recognized as important members of the spoilage association (Table 3). Many of the isolates could not be identified with existing species of *Lactobacillus* (see Table 2). It is now recognized that many of these isolates belong to a recently defined genus, *Carnobacterium*.

It needs to be stressed that each of the atmospheres in Table 3 selects a microbial flora dominated by Gram-positive bacteria (principally *Brochothrix thermosphacta* and lactic acid bacteria) rather than the Gram-negative ones that develop on meat stored aerobically at chill temperatures. As the former grow much more slowly than the latter, the shelf life of meat is extended. It needs to be stressed also that there are differences in the metabolic attributes of these two groups of spoilage organisms. These are manifested at different times and in different ways as judged by odours coming from the meat.

Another cold-tolerant microorganism, *Clostridium estertheticum*, causes distension or explosion of packs of vacuum-packaged meat. The optimum growth temperature of these organisms is 20°C. It is tempting to speculate that the production of a spore protects this organism from those factors in meat processing that kills psychrophiles lacking this means of protection.

Spoilage in Frozen Meat Studies of microbial growth at subfreezing temperatures clearly indicate that microbial growth does not occur in meat ecosystems with a temperature below −8°C. Thus, the main determinant for the storage period of a properly frozen meat ecosystem is the physical, chemical or biochemical changes which are unrelated to microbiological proliferation. Therefore, frozen storage life is limited by changes in other qualities such as appearance or taste which are unrelated to microbiological activity.

The key problem with frozen meat is the enumeration of the microbial populations of such ecosystems. Microorganisms are injured by exposure to reduced temperatures leading to sublethal damage, the effects of which in microbial populations include (1) increased lag times and (2) the inability to develop quantitatively on selective media that do not exert any inhibitory effect on undamaged populations of the same taxon. These effects – especially the prolonged lag phase – are less noticeable when the meat ecosystem is refrozen and analysed again after a short period of storage. The appropriate resuscitation of frozen meat flora prior to its enumeration is crucial; resuscitation of the injured flora may take place in the meat ecosystem during thawing, or in nonselective culture media. Studies on the effect of different environmental stresses on the enumeration and the recovery of microorganisms are focused on pathogenic microorganisms; in which case the important feature is to ascertain the presence or absence of the pathogenic bacterium. The results obtained have a cardinal role in the evaluation of microbiological hazards.

Roles of Microbes and Enzymes in Spoilage

The role of the microbial flora is cardinal for the spoilage of meat. The metabolic activity of the organisms selected in a meat ecosystem leads to the manifestation of changes or spoilage of meat. This manifestation is related to the level of (1) the population and (2) the substrates in meat. Under both aerobic and vacuum or modified-atmosphere packaging the corresponding flora catabolize glucose for growth. By the end of this phase changes and subsequently overt spoilage are due to catabolism of nitrogenous compounds and amino acids as well as secondary metabolic reactions. The contribution of indigenous meat enzymes to spoilage is negligible compared with the action of the microbial flora. Postmortem glycolysis ceases after the death of the animal when ultimate pH reaches a value of 5.4–5.5. During storage, however, there is a proteolytic activity by indigenous enzymes. The activity of these enzymes has a role in the conditioning (ageing) of meat. Added enzymes in meat may be used to artificially ameliorate its organoleptic properties. Enzymes used for their tenderizing effects are proteolytic and of bacterial, fungal or plant origin.

Chemistry of Spoilage

The critical physico-chemical changes occurring during spoilage take place in the aqueous phase of meat. This phase contains glucose, lactic acid, certain amino acids, nucleotides, urea and water-soluble proteins which are utilized by most of the bacteria of the meat microflora. The concentration of these low-molecular-mass compounds is sufficient to support massive microbial growth. Glucose is the prime nutrient in a meat ecosystem and it is catabolized initially during microbial growth. This substrate is attacked by almost all groups of spoilage bacteria, under aerobic and anaerobic conditions (**Table 4**). Until spoilage is evident organoleptically, the major detectable

Table 4 Substrates used for growth of major meat spoilage microorganisms

| Microorganism | Substrates used for growth | |
	Aerobic	Anaerobic
Pseudomonas spp.	Glucose, glucose 6-phosphate, lactic acid, pyruvate, gluconate, 6-phosphogluconate, amino acids, creatine, creatinine, citrate, aspartate, glutamate	Glucose, lactic acid, pyruvate, gluconate, amino acids
Acinetobacter/Moraxella	Amino acids, lactic acid, glucose	Glucose, amino acids
Shewanella putrefaciens	Glucose, lactic acid, pyruvate, gluconate, propionic acid, ethanol, acetate, amino acids	Formate
Brochothrix thermosphacta	Glucose, amino acids, ribose, glycerol	Glucose
Enterobacter spp.	Glucose, glucose 6-phosphate, amino acids, lactic acid	Glucose, glucose 6-phosphate, amino acids
Lactobacillus spp.	Glucose	Glucose, lactic acid, amino acids

effect of bacterial growth is a reduction of the glucose concentration. This does not alter the organoleptic properties of meat. When glucose or its oxidative products are reduced to non-substrate levels, lactic acid is catabolized. It needs to be stressed that when this second major carbon and energy source is exhausted the microbial association is at its climax stage.

Under Aerobic Conditions The relative potential of bacteria depends upon which species predominate, and upon their ability to form malodorous compounds such as H_2S, volatile amines, esters and acetoin. *Pseudomonas* spp. are important because of their dominance in the aerobic climax associations at chill temperatures. The key chemical changes associated with the metabolic attributes of pseudomonads have been studied extensively in broth and in model systems such as meat juice (**Table 5**). Among these major attributes are (1) the sequential catabolism of D-glucose and L- and D-lactic acid with D-glucose used in preference to lactate, and (2) the oxidization of glucose and glucose 6-phosphate via the extracellular pathway causing a transient accumulation of D-glu-

conate and an increase in the concentration of 6-phosphogluconate. The increase in the concentration of D-gluconate led investigators to propose a method for controlling the microbial activity in meat by the addition of glucose to meat and its transformation to gluconate. The rationale for this suggestion is the fall in pH due to the accumulation of oxidative products. The transient pool of gluconate and its inability to be catabolized by all the taxa of the association may offer a selective determinant on the meat ecosystem. Another important feature is the catabolism of creatine and creatinine by *Pseudomonas fragi* under aerobic conditions. The phenomenal release of ammonia and the increase in pH are inextricably linked with the catabolism of these substrates. Ammonia, which is the major cause of the increase of pH, can be produced by many microbes, including pseudomonads during their amino acid metabolism. A list of other volatile compounds found in spoiled meat is given in **Table 6**. Pseudomonad species growing on the surface of meat will preferentially consume glucose until the rate of diffusion of glucose from underlying tissues becomes inadequate to meet

Table 5 Metabolic activity of pseudomonads in meat juice at 0–4°C

Substrate	Pseudomonas fragi	Pseudomonas lundensis	Pseudomonas fluorescens
D-Glucose	c	c	c
D-Glucose 6-phosphate	c	c	−
D-Gluconate	f	f	f
6-Phospho-D-gluconate	f	f	−
L-lactic acid	c	c	c
D-lactic acid	c	c	c
Pyruvate	f/c	f/c	f/c
Acetic acid	c	nd	nd
Amino acids	c	c	c
Creatine	c	−	−
Creatinine	c	−	−
Proteolysis	+	nd	+
Ammonia	f	f	f

Key: The substrate was catabolized (c) or formed (f) during growth; − neither catabolized nor formed; +, positive; nd, no available data.

Table 6 End product formation of Gram-negative bacteria (e.g. *Pseudomonas* spp., *Shewanella putrefaciens*, *Moraxella* etc) when grown in broth, sterile meat model system and in naturally spoiled meat

Sulphur compounds:
sulphides, dimethylsulphide, dimethyldisulphite, methylmercaptan, methanethiol, hydrogen sulphide, dimethyltrisulphide
Esters:
methyl esters (acetate), ethyl esters (acetate)
Ketones:
acetone, 2-butanone, acetoin/diacetyl
Aromatic hydrocarbons:
diethylbenzene, trimethylbenzene, toluene
Aliphatic hydrocarbons:
hexane, 2,4-dimethylhexane, methyl heptone
Aldehydes:
2-methylbutanal
Alcohols:
methanol, ethanol, 2-methylpropanol, 2-methylbutanol, 3-methylbutanol
Biogenic amines – other compounds:
cadaverine, ammonia, putrescine, methylamine, trimethylamine

their demand; when high numbers (10^8 per cm^2) are reached and glucose becomes depleted at the meat surface, pseudomonads start proteolysis and use nitrogenous compounds and free amino acids as their growth substrate with production of malodorous sulphides and esters (Table 6).

The Enterobacteriaceae can be important in spoilage if the meat ecosystem favours their growth. This group utilize mainly glucose and glucose 6-phosphate as the main carbon sources; the exhaustion of these substances allows amino acid degradation. Moreover, some members of this family produce ammonia, volatile sulphides including H_2S and malodorous amines from amino acid metabolism (Table 6).

Acinetobacter and *Moraxella* constitute a major part of the aerobic spoilage population. These organisms are of low spoilage potential. They utilize amino acids as their growth substrate but do not form malodorous by-products from amino acid degradation; they rather enhance the spoilage activities of pseudomonads and *Shewanella putrefaciens* by restricting the availability of O_2 to these organisms. When O_2 limits growth, pseudomonads attack amino acids, even when glucose is present, with the subsequent production of malodorous substances. Under anaerobic conditions *S. putrefaciens* will generate H_2S, resulting in the greening of meat due to sulphmyoglobin formation.

Under Vacuum or Modified-atmosphere Packaging Conditions A shift from a diverse initial flora to one dominated by Gram-positive facultative anaerobic microflora (lactic acid bacteria and *Brochothrix*

thermosphacta) usually occurs in meat during its storage under modified atmosphere packaging. Among these, the physiological attributes of the lactic acid bacteria and *B. thermosphacta* have been studied extensively. Environmental determinants such as the oxygen tension, glucose concentration and the initial pH have a major influence on the physiology of these organisms, and hence on the end products formed. *Brochothrix thermosphacta* has a much greater spoilage potential than the lactobacilli and can be important in both aerobic and anaerobic spoilage of meat. This organism utilizes glucose and glutamate but no other amino acid during aerobic incubation. It produces a mixture of end products including acetoin, acetic, *iso*-butyric and *iso*-valeric acids, 2,3-butanediol, diacetyl, 3-methylbutanal, 2-methylpropanol and 3-methylbutanol during its aerobic metabolism in media containing glucose, ribose or glycerol as the main carbon and energy source (**Table 7**). The precise proportion of these end products is affected by the concentration of glucose, pH and temperature.

Lactobacillus spp. constitute only a small proportion of the initial spoilage bacterial population of meat. When oxygen is in low concentration, as in vacuum-packed meats, the developing microflora is usually dominated by *Lactobacillus* spp. These fermentative organisms probably grow faster than would-be competitors because they are unaffected by pH and antimicrobial products such as lactic acid, H_2O_2 and antibiotics. These organisms utilize glucose for growth and produce lactic acid. When carbohydrates are exhausted, amino acids are utilized with

Table 7 End products of homofermentative lactic acid bacteria (HO), heterofermentative lactic acid bacteria (HT) and *Brochothrix thermosphacta* (BT) when grown in broth, sterile meat model system and in naturally spoiled meat

Aerobic	In different gaseous atmospheres
Acetoin – HO, HT, BT	Acetoin – HO
Acetic acid – HO, HT, BT	Acetic acid – HO, HT, BT
L-Lactic acid – HO, HT, BT	L-Lactic acid – HO, HT, BT
D-Lactic acid – HO, HT	D-Lactic acid – HO, HT
Formic acid – HO, HT, BT	Formic acid – HO, HT, BT
Ethanol – HO, HT, BT	Ethanol – HO, HT, BT
CO_2 – HO, HT, BT	
H_2O_2 – HO, HT	
iso-Butyric acid – BT	
iso-Valeric acid – BT	
2-Methylbutyric acid – BT	
3-Methylbutanol – BT	
2-Methylbutanol – BT	
2,3-Butanediol – BT	
Diacetyl – HO, HT, BT	
2-Methylpropanol – BT	
2-Methylpropanal –BT	
Free fatty acids – BT	

the consequent production of volatile fatty acids which impart a 'dairy' or 'cheesy' odour to the vacuum-packaged meat. Because with meat stored under modified atmosphere increased concentration of CO_2 inhibits growth of aerobic flora – and glucose assimilation by pseudomonads – the cheesy odours are found mainly in samples stored in gas mixtures containing CO_2, where they are probably produced by *Brochothrix thermosphacta* and lactic acid bacteria. These also form diacetyl, acetoin and alcohols from glucose under aerobic conditions or low partial pressure of oxygen (Po_2). In addition, alcohols (ethanol and propanol) are present at only trace levels at the beginning of storage but their concentrations increase significantly before the onset of spoilage, making them promising compounds as indicators of spoilage (Table 7).

Evaluation of Spoilage

Enumeration of bacterial population by culture techniques (agar media) and rapid methods (malthusian) in food is used as indicator of its hygiene. As the spoilage of meat is caused by specific spoilage bacteria, different selective media should be applied. Because the correlation between the population of specific spoilage bacteria and the sensorial manifestation of spoilage is imprecise, it is difficult to use bacterial levels as an estimate of spoilage.

The time-consuming microbiological analyses can be replaced by assessment of the chemical, enzymatic and physico-chemical changes associated with microbial growth on meat. For this reason a number of chemical and physical methods have been proposed for the estimation of bacterial spoilage in meats. However, there is as yet no single test available to assess meat quality. Spoilage is a subjective evaluation and therefore a sound definition is required to develop a suitable method. The lack of a general agreement on the signs of incipient spoilage in meat and the changes in the technology of meat preservation (e.g. vacuum or modified-atmosphere packaging) make it difficult to identify spoilage indicators.

The spoilage indicators or microbial metabolite should meet the following criteria:

1. The indicator should be absent or initially at low levels in meat.
2. It should increase proportionally with the storage period.
3. It should be produced by the dominant flora and have a good correlation with sensory evaluation.

As noted earlier, physico-chemical analyses of meat can be used instead of microbiological ones for the evaluation of spoilage. For this reason, numerous attempts have been made since the 1970s to associate given metabolites with the microbial spoilage of meat. The idea for these methods is that as the bacteria grow on meat they utilize nutrients and create by-products. The quantitative determination of these metabolites could provide information about the degree of spoilage. The identification of the ideal metabolite, fulfilling the criteria noted above, has proved a difficult task for the following reasons:

1. Most metabolites are specific to certain organisms (e.g. gluconate to pseudomonads).
2. Although the metabolites are the product of the metabolism of a specific substrate, the absence of the given substrate or its presence in low quantities does not preclude spoilage.
3. The rate of microbial metabolite production and the metabolic pathways of spoilage bacteria are affected by the environmental conditions (e.g. pH, oxygen tension, temperature).
4. The accurate detection and measurement of metabolites require sophisticated procedures.
5. Many of them provide retrospective information.

Role of Cooking in Susceptibility to Spoilage

Cooking raw meat results in the death of its microbial association. Subsequent recontamination of the cooked meat and temperature abuse lead to the development of a new spoilage association. As the antagonists belonging to the initial microflora of raw meat are absent, pathogens that contaminate cooked meat have a rich substratum for their proliferation. The microbiological stability of cooked meat products depends on extrinsic factors, mainly the packaging method and storage temperature, as well as on intrinsic ones, e.g. product composition.

Special Problems Associated with Meat

Production of biogenic amines by microbial flora is a problem in stored meat. Amines have been detected in fresh meat stored under aerobic or vacuum/modified-atmosphere packaging conditions. Among them, putrescine and cadaverine show a constant increase during storage. Concentrations of spermine, spermidine and tryptamine remain steady, and a small increase in tyramine concentration is usually observed after long storage periods. As lactic acid bacteria and *Brochothrix thermosphacta* do not produce amines, the formation of these compounds has been attributed to Enterobacteriaceae. However, tyramine could also be formed by some strains of the genus *Lactobacillus*.

The limiting factors of meat stored under modified-atmosphere packaging is another issue. Concerns have been expressed by regulatory authorities, meat industries and meat research institutes that this practice may represent a safety hazard. Indeed, despite increasing commercial interest in the storage of meat under modified atmospheres to extend its shelf life, the potential growth of pathogenic bacteria which could survive and grow even at refrigeration temperatures remains the limiting factor to further expansion of this packaging method. Studies on the effect of modified-atmosphere packaging on the growth and survival of the food-borne psychrotrophic pathogens are mentioned in the following section.

Consumer Risks from Meat Products

Risks to the consumer from consumption of meat and meat products are associated mainly with the presence of pathogens on meat. As a meat ecosystem is intended to be stored for protracted periods, the potential incubation period for the species of the microbial associations is prolonged. Although pathogenic genera do not constitute a part of the spoilage association per se, their occurrence is possible owing to their presence in the raw meat or transfer during unhygienic processing of a product.

It is generally recognized that the new ecosystems of meat stored under modified atmosphere do not add any more direct microbial hazards and that their safety aspects are equivalent to meat stored in an aerobic ecosystem under refrigeration. The safety of packaged meat is governed mainly by two basic but indirect influences: first by suppression of the spoilage flora with a further effect, a reduction in its potential to suppress the growth of pathogens; and second, by delaying or suppressing the normal and well-documented course of spoilage. Another possible direct impact is, however, stimulation of germination and toxin production by *Clostridium botulinum*. Even so, the direct impact from the stimulation of pathogenic species is of minor importance.

Studies with psychrotrophic pathogenic bacteria have shown that vacuum and modified-atmosphere packaging with nitrogen create atmospheres that may readily support the growth of *Yersinia enterocolitica*, *Aeromonas hydrophila* and *Listeria monocytogenes*. It also seems to be the case that the growth of these organisms is inhibited by the enrichment of atmosphere with carbon dioxide, and in general terms the higher the CO_2 concentration and the lower the temperature and pH, the greater is the growth inhibition. At normal meat pH (5.5) and at a low temperature (1°C) the growth of psychrotrophic pathogens is stopped when the CO_2 concentration is 40% (v/v). However, when the meat pH is high (6.0 or more) or an abuse of storage temperature occurs, these organisms, like other non-psychrotrophic pathogens, may grow resulting in microbial hazards (toxigenesis or growth of pathogenic species to infective doses) without detectable organoleptic degradation owing to inhibition of the 'prophylactic' action of spoilage flora. With unpacked meat, spoilage precedes the creation of microbial hazards and thus alerts the consumer to its hygienic status.

It is impossible to estimate or to predict accurately (model) the safety of a meat system. Studies of these topics have demonstrated that the 'history effects' and the complexity of an ecosystem enable the emanation of general principles only about the prediction of safety. The most important factors for any prediction are the initial hygienic status and the adherence to the selected ecological determinants, mainly temperature.

See also: **Acinetobacter**. **Aeromonas**: Introduction. **Clostridium**: Introduction; *Clostridium botulinum*. **Enterobacteriaceae, Coliforms and E. coli**: Introduction. **Freezing of Foods**: Growth and Survival of Microorganisms. **Lactobacillus**: Introduction. **Listeria**: *Listeria monocytogenes*. **Meat and Poultry**: Spoilage of Cooked Meats and Meat Products. **Moraxella**. **Pseudomonas**: Introduction. **Spoilage Problems**: Problems caused by Bacteria. **Yeasts**: Production and Commercial Uses. **Yersinia**: *Yersinia enterocolitica*.

Further Reading

Dainty RH and Mackey BM (1992) The relationship between the phenotypic properties of bacteria from chill-stored meat and spoilage processes. *Journal of Applied Bacteriology Symposium* (Supplement) 73: 103S–114S.

Gill CO (1986) The control of microbial spoilage in fresh meats. In: Pearson AM and Dutson TR (eds) *Advances in Meat Research: Meat and Poultry Microbiology.* P. 49. New York: Macmillan.

Davies AR and Board RG (eds) (1998) *The Microbiology of Meat and Poultry.* London: Blackie.

Huis in't Veld JHJ (ed.) (1996) Specific spoilage organisms. *International Journal of Food Microbiology* 33: 1–156.

Mossel DAA, Corry JEL, Struijk CB and Baird RM (1995) *Essentials of the Microbiology of Foods.* Chichester: John Wiley.

Curing of Meat

K Prabhakar, Department of Meat Science and Technology, Veterinary College Tirupathi, India

Cured meats are those in which the haem pigment of the muscle, myoglobin is converted into red-coloured

Table 1 Curing salts and their levels of incorporation

S. no.	Curing salts	Level of incorporation (%)	Usual permitted level in cured products
Curing agents			
1.	Sodium chloride	10.0–25.0	≤7.5%
2.	Sugar	2.0–4.0	
3	Sodium nitrate/potassium nitrate	0.10–1.50	Not > 100 p.p.m. as nitrite
4.	Sodium nitrite/potassium nitrite	0.01–0.05	Not > 100 p.p.m. as nitrite
Curing adjuncts			
1.	Sodium ascorbate	0.2–1.0	
2.	Polyphosphates	2.0–4.0	0.2–0.3%

nitrosomyoglobin through addition of common salt. Salting and smoking of meats for storage without refrigeration is an age-old practice. Application of common salt and sugar to meats withdraws water by osmosis from the microorganisms and inactivates them. Sodium nitrate (saltpetre), an impurity of common salt, on reduction to nitrite and nitric oxide combines with myoglobin to produce the cured meat colour. This attractive colour and the peculiar flavour developed as a result of curing and smoking-cum-heating (exposure to wood smoke) have made cured meats popular. The method originally conceived for preservation is now used for the colour and flavour alterations. The latter part of the nineteenth century marked the beginning of the curing and smoking of meats on a scientific basis. Mostly pork and beef and to a lesser extent poultry meat and sheep meat are cured and smoked worldwide.

The stability of cured meats depends on salt content and moisture content. High salt (8–15%) cured meats with moisture contents of 25–30% are reported to keep for up to six months. However, due to consumer health considerations and public demand, cured meats are now processed with lower salt levels (4–8%). As lower salt levels and higher moisture levels (45–55%) lead to decreased stability, such cured meats when stored under refrigeration may keep for up to 4–6 weeks.

Curing Reaction in Meats

Curing agents are the substances which are directly involved in the curing reaction and curing adjuncts are those which are not very essential but help to quicken the curing reaction and stabilize the cured meat colour. Levels of incorporation of some of them vary with the curing methods like dry addition or immersion in solution with water (brine).

Important curing agents and adjuncts and their levels of incorporation during processing are shown in **Table 1**.

Figure 1 shows the important steps of meat curing reaction and processing.

Nitrate-reducing bacteria effect reduction of nitrate to nitrite. Nitrite is highly reactive and acts as both a reducing agent and an oxidizing agent. In acid medium, it ionizes to yield nitrous acid which further decomposes to yield nitric oxide. Nitric oxide produces the desired red colour in cured meats. The various transformations of myoglobin in fresh meats and cured meats are indicated in **Figure 2**.

Other Functions of Curing Salts

Sodium Chloride   This is the major microbial inhibitory factor in cured meats. It inhibits the growth of bacteria by lowering water activity and causes plasmolysis of cells through increased osmotic pressure. It ionizes to yield chlorine ion which is inhibitory to microorganisms. Though it encourages the growth of halophilic organisms, their growth is comparatively slow compared to non-halophilic bacteria. Sodium chloride also contributes to flavour.

Sugar   Sugar counteracts the astringent quality of sodium chloride and enhances the flavour of cured products. It helps in lowering the pH of cure through bacterial fermentation so that nitrite is more effective. However, the importance of sugar is negligible under the current commercial conditions of controlled processing.

Nitrates   Nitrates serve as a source for nitrites. They are also bacteriostatic in acid medium especially against anaerobes. In prolonged curing methods, nitrates serve as a reservoir from which nitrite can be formed. However, most commercial processors add nitrites directly to avoid dependence on reduction by bacteria.

Nitrites   In addition to colour fixation, nitrite has bacteriostatic and bactericidal properties. It offers protection against botulism food poisoning and is also claimed to inhibit the growth of other food poisoning

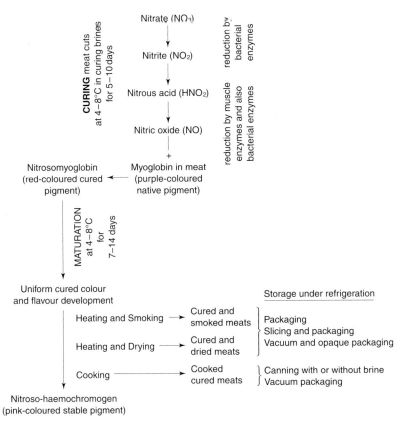

Figure 1   Curing reaction and processing of cured meats.

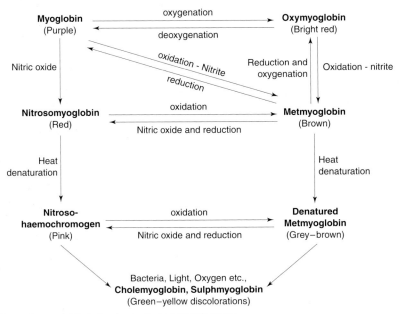

Figure 2   Transformations of myoglobin in fresh meats and cured meats.

microbes such as *Clostridium perfringens*, *Bacillus cereus*, *Staphylococcus aureus*, *Salmonella* etc. at a concentration of 100–200 p.p.m. The inhibitory effect of nitrite on bacterial spores is attributed to its ability to enhance destruction of spores by heat, to stimulate spore germination during heating which is inactivated during subsequent heating and to inhibit spore germination after heating. Inhibition is more effective in acidic pH levels (5.0–6.0) and in the presence of undissociated nitrous acid. Nitrite in combination with sodium chloride has been found to have greater antimicrobial properties.

Ascorbates Reducing or chelating agents like sodium ascorbate enhance the anti-botulinal effect of nitrites. They quicken colour formation by ensuring rapid conversion of nitrite into nitric oxide. Ascorbates hasten the development of the cure colour by converting metmyoglobin back to myoglobin which is more reactive to become nitrosomyoglobin. They give stability to cured meat colour through inhibition of oxidative changes initiated by exposure to light and the presence of oxygen.

Polyphosphates Polyphosphates contribute to the increased water-holding capacity of muscle proteins and decreased processing losses. They chelate trace metal irons and retard development of rancidity in meat products.

Processes Available for Curing Meats

Curing salts are applied to meats by five methods:

1. Dry cure method: Dry ingredients are rubbed into meat cuts such as pork legs (ham) or pork bellies (bacon) and stacked in curing rooms or kept in layers in barrels. Each layer is covered with curing salts.
2. Pickling: Meat cuts, chunks or slices are immersed in a solution of the curing salts in water.
3. Stitch pumping: Brine solution containing cure ingredients is injected under pressure through long, perforated multiple needles into meat cuts or chunks. This enables the brine to be distributed to all parts of meat quickly and ensures antibacterial conditions in the deeper portions of musculature immediately. Brine equivalent to 10.0% of weight of meat cuts is pumped. The solution normally contains sodium chloride up to 25–30%. This method enables shorter curing schedules to be used. In commercial processing, carcass sides pass along a conveyor and two to four rows of needles inject brine into fleshy parts.
4. Arterial pumping: Brine is injected under pressure into the main artery of major cuts like leg and shoulder of carcasses to ensure even distribution through the vascular network.
5. Direct addition method: Curing salts are added directly to comminuted meats or sausage meats. Curing reaction is quicker and lower levels of salts are sufficient.

In commercial practice, combinations of the above methods are used depending on the size of meat cuts, expected colour and flavour development and shelf-life. Curing is usually done at 3–7°C. Higher temperatures ensure quicker penetration of brine but microbial growth is encouraged leading to quicker onset of spoilage. Curing times vary with the methods used and the meats to be cured. Dry curing or brine curing and bigger cuts of meat require several weeks of curing but methods like multineedle machine injection (stitch pumping) require less than 1–2 weeks. Tumbling meat chunks in rotating drums where they massage against one another in the presence of salt shortens the curing periods to 1–2 days. Alternatively, large muscle groups of leg are stitch pumped with brine and tumbled to increase the rate of diffusion of curing salts throughout the meat and to solubilize the surface proteins of meat pieces in the brine. During subsequent heat processing, the pieces are bound together by a protein gel. This facilitates moulding of ham into desired dimensions and ensures better slicing and packaging.

Curing by dry salting or pickling of entire sides of trimmed pork carcasses with stitch pumping into the fleshy parts is described as Wiltshire cure. Cured comminuted beef is known as corned beef. In earlier days, grains of salt were used to cure comminuted beef; hence the product is popular by that name.

Stabilization of Cured Meats

Cured meat stability is further enhanced by heating-cum-smoking, drying, cooking or canning as indicated in Figure 1.

Cured and Smoked Meats

Heating-cum-smoking of cured meats is done in controlled smoke houses to enhance stability, to improve and stabilize cured meat colour, to give better finish and gloss to cured meat surface and to impart the desired flavour. Surface tissues coagulate and form a physical barrier for entry of microorganisms. Smoke is generated by burning hard wood, saw dust, wood shavings, corn cobs etc., and supplied to a smoke chamber where cured meats are held. The temperature and relative humidity are controlled. The temperature of smoking varies from 45 to 75°C and the smoke period from a few hours to several days. Wood smoke contains a large number of volatile compounds that have bacteriostatic and fungistatic effects. Formaldehyde is considered the most effective followed by phenols, cresols, aliphatic acids, and primary and secondary alcohols. The effects of drying, concentration of curing salts and smoke constituents reduce the number of microorganisms on the surface and create conditions where the subsequent growth rate of survivors is considerably reduced.

Cured and Dried Meats

Curing alone cannot prevent eventual spoilage even under refrigeration. Cured meats are, therefore, exposed to moderate heat (30–60°C) for several hours so that the moisture content is lowered and a dry atmosphere is created at the surface. This enables stability even at higher temperatures of storage (10–15°C).

Cooked Cured Meats

Cured hams are cooked at 85°C for 3–4 h, usually to internal temperatures of 65–75°C and then canned or sliced and vacuum packaged. This results in vegetative organisms of *Bacillus* and *Clostridium* being eliminated but their spores survive. These products have a shelf life of about six months in cans at refrigeration temperature. Low bacterial loads, nitrites and anaerobiasis ensure safety of such products.

Role of Packaging

Cured and smoked meats are packaged for the sake of retail sale convenience and attractive presentation to consumers. Keeping quality is improved as post-processing contamination is avoided. For better consumer appeal, cured meats require to be packaged in transparent films but exposure to light and the presence of salt, initiate oxidation of fat and destabilization of cured pigment. Hence the usual practice is to ensure that one side (exposed in the display cabinet) of the package is of opaque material and the non-exposed transparent side is shown to the buyer at the time of sale only. Vacuum packaging is a subsequent development. It protects the bright red colour from oxidation due to absence of oxygen and enhances keeping quality by at least two weeks at refrigeration temperatures. Cooked cured hams which are exposed to pasteurization temperatures only are canned in brine for enhanced stability.

Microflora and Spoilage of Cured Meats

Carefully processed and low temperature-stored cured and smoked meats are usually low in bacteria and comparatively free from moulds and yeasts. The microflora of cured meats depends on the initial contamination of meats, the microbial load of the curing brines, further processing, type of packaging and temperature and relative humidity of storage.

Curing brines are usually recirculated after filtration and reconstitution. Ideal brines are prepared to contain total counts of less than $50–100 \times 10^3$, *Pseudomonas* counts of less than $0.1–1.0 \times 10^3$, *Vibrio* counts of less than $1.0–10.0 \times 10^3$ and *Escherichia coli* of less than 10 per millilitre. *Vibrio* counts indicate spoilage potential of cured pork and their presence indicates contamination from earlier batches of cured meats.

Curing salts make meat more favourable to the growth of Gram-positive bacteria, moulds and yeasts than to the Gram-negative bacteria. Higher salt concentrations and processing or storage at refrigeration temperatures permit growth of psychrophilic/psychrotrophic and salt-requiring organisms. In vacuum packaged cured meats, facultative aerobes and anaerobic bacteria multiply and cause spoilage.

Surface spoilage (slime, stickiness and taints) in bacon is usually caused by *Streptococcus*, *Vibrio*, *Micrococcus*, *Lactobacillus*, *Staphylococcus* and *Acinetobacter*. Others which may also be seen are *Aeromonas*, *Alcaligenes* and *Achromobacter*. *Vibrio* spp. usually grow in bacons with pH values above 5.9–6.0. *Acinetobacter* may be commonly seen in high salt bacons (8–12%) with pH values above 6.5. *Corynebacterium* multiplies mostly in low salt bacons (4–7%) with pH values above 6.5. Higher fat content and lower water activity values discourage bacterial growth. Consequently, mould growth is seen on the surface of cured meats. Most common among them are *Aspergillus*, *Penicillium*, *Cladosporium*, *Alternaria*, *Monilia*, *Fusarium*, *Mucor*, *Rhizopus* and *Botrytis*. Yeasts of the genera *Torulopsis*, *Candida* and *Rhodotorula* may also be noticed on spoiled Wiltshire bacon, but they are reported to be a major problem in dried and smoked bacons.

Bacteria which produce souring in hams are *Acinetobacter*, *Bacillus*, *Pseudomonas*, *Lactobacillus*, *Proteus*, *Micrococcus* and *Clostridium*. When improperly chilled meats are used for curing, internal spoilage (ham taint or bone taint) is noticed due to growth of *Vibrio*, *Proteus*, clostridia and halotolerant micrococci.

In vacuum packaged sliced bacon stored at refrigeration temperatures, lactobacilli, pediococci, *Leuconostoc*, group D streptococci and micrococci are usually noticed. At slightly higher storage temperatures, *Vibrio* and *Proteus* multiply and produce putrefactive spoilage.

In canned cured cooked ham, heat resistant and non-sporing genera like *Streptococcus*, *Corynebacterium* and *Microbacterium* and spores of *Bacillus* and *Clostridium* are usually present. Enterococci and lactobacilli are responsible for greening in cooked hams. In the presence of air, they produce hydrogen peroxide which oxidizes cooked cured pigment to green-coloured cholemyoglobin. When H_2S-producing strains predominate, the green-coloured pigment sulphmyoglobin is produced. Gas production in cans is due to growth of *Bacillus* species, such as *Bacillus cereus* or *Bacillus coagulans*. Putrefactive spoilage is caused by mesophilic putrefactive anaerobes, such as *Clostridium* species. If processing is not done correctly, heterofermentative lactic acid bacteria such as *Leuconostoc* spp. predominate. *Bacillus* spp. and homofermentative lactic acid bacteria such as *Enterococcus faecium* and *Enterococcus faecalis* produce souring and abnormal colour and texture.

In vacuum-packaged cooked cured meats, spoilage is mainly due to growth of *Lactobacillus* and *Brochothrix thermosphacta*. Other genera which are also seen are *Micrococcus*, *Streptococcus*, and *Leuconostoc*. Lactobacilli produce sourness spoilage whereas Enterobacteriaceae and *Vibrio* spp. cause green discolorations due to sulphide spoilage.

Hygienic and careful processing followed by refrigeration-temperature storage eliminate most of the spoilage problems in cured meats.

Risk to the Consumer

Nitrite is highly reactive in muscle/meat and its concentration decreases gradually. Therefore, microbial growth increases and causes spoilage in stored cured meats. Fresh cured meats with bacterial loads of less than 10^4–10^5 per square centimetre are known to keep satisfactorily for 20–30 days at 4–5°C. High-salt cured meats or dry cured hams can be stored for a few months without any problem. If cured meats are consumed before the onset of spoilage, manifested by surface slime, off-odours and discoloration, risks to the consumers on account of microbial spoilage are few. Curing and smoking and low temperature storage permit mostly psychrophilic and halotolerant strains whose growth effectively checks the survival and multiplication of pathogenic bacteria such as *Salmonella*, *Staphylococcus aureus* and *Clostridium*. Growth of the latter is thought to be limited even at 25°C.

If the spoilage is in the initial stages and confined only to the surface, it can be eliminated by washing with curing brine and subjecting to quick drying or smoking after sprinkling salt. Mould growth can also be removed by scraping and it can be prevented by spraying potassium sorbate (0.1–0.2%) on cured meats. It is clearly established that even inoculated common pathogenic bacteria and viruses are inactivated during commercial meat curing. However, there is a build-up of international opinion to reduce the levels of nitrite incorporated or to replace it totally with alternatives because of the perceived risks to consumers. Nitrite can react with secondary and tertiary amines in cured meats to form nitrosamines which are known to be carcinogenic. Several studies have been undertaken to reduce the levels of nitrite

without compromise on safety. It has been suggested that alternatives to nitrites or combinations such as lower levels of nitrite (50 p.p.m.) with sorbate (80 p.p.m.), should be used. However, these proposals are not likely to eliminate nitrite from meat curing in the near future because of the proven protection offered by nitrites especially against botulism. The risk to the consumers from exposure to other sources of nitrate/nitrite, such as vegetables, water and endogenous synthesis, has been found to be several times higher than that posed from cured meats. So far, no direct evidence has been established to suggest consumer risk with cured meats. Only frying or severe heating of bacon has been reported to result in significant levels of nitrosamines. The usual nitrosamine content of cured meats is below 1–2 parts per billion.

Another risk to consumers is from polycyclic hydrocarbons present in wood smoke which are reported to be carcinogenic. Liquid smoke is now produced which is free from such compounds. It can be incorporated directly into products or applied to the surface.

See also: **Meat and Poultry**: Spoilage of Cooked Meats and Meat Products. **Preservatives**: Traditional Preservatives – Sodium Chloride; Traditional Preservatives – Wood Smoke; Permitted Preservatives – Nitrate and Nitrite.

Further Reading

Cassens RG (1997) Composition and safety of cured meats in the USA. Mediterranean aspects of meat quality as related to muscle biochemistry. *Food Chemistry* 59: 561–566.

Frazier WC and Westhoff DC (eds) (1995) Contamination, preservation and spoilage of meats and meat products. In: *Food Microbiology*, 4th edn. P. 218. New Delhi: Tata McGraw-Hill.

Gardner GA (1982) Microbiology of processing bacon and ham. In: Brown HM (ed.) *Meat Microbiology*. P. 129. London: Applied Science Publishers.

Gardner GA (1983) Microbial spoilage of cured meats. In: Roberts TA and Skinner FA (eds) *Food Microbiology: Advances and Prospects*. P. 179. London: Academic Press.

Gray TJB (1980) Toxicology. In: Tilbury RH (ed.) *Developments in Food Preservatives*. P. 53. London: Applied Science Publishers.

Lawrie R (1995) The structure, composition and preservation of meat. In: Campbell-Platt G. and Cook PE (eds) *Fermented Meats*. P. 1. Glasgow, Blackie Academic & Professional.

Pearson AM and Gillett TA (eds) (1997) Curing. In: *Processed Meats*, 3rd edn. P. 53. New Delhi: CBS Publishers and Distributors.

Shahidi F, Rubin LJ and Wood DF (1988) Stabilisation of

meat lipids with nitrite – free curing mixtures. *Meat Science* 22: 73–80.

Sen NP, Baddoo PA and Seaman SW (1993) Nitrosamines in cured pork products packaged in elastic rubber nettings. An update. *Food Chemistry* 47: 387–390.

Sinskey AJ (1980) Mode of action and effective application. In: Tilbury RH (ed.) *Developments in Food Preservatives*. P. 111. London: Applied Science Publishers.

Spoilage of Cooked Meats and Meat Products

Isabel Guerrero and **Lourdes Pérez Chabela**, Universidad Autónoma Metropolitana-Iztapalapa, Mexico

A combination of intrinsic, extrinsic and processing factors determines the composition of the microflora found on raw meat. Spoilage processes will therefore vary according to the metabolic pathways of the predominant microorganisms. When the physicochemical characteristics of the meat are changed the microflora changes in parallel, with increased prevalence of microorganisms that can tolerate the new environment or are able to grow faster under the new conditions. At a later stage when meat products are stored, the composition of the microflora may change again, according to conditions such as temperature, gas atmosphere and length of storage. Therefore, the type and rapidity of spoilage of processed meats depend on (a) the microflora on the raw meat, (b) the type of preservation process applied and (c) the storage conditions.

Spoilage Microflora in Meat Products

The surface of carcasses of healthy animals may become contaminated by microorganisms from tools, walls, water or handling. During the carcass operations of dressing, cooling and cutting in the processing rooms, microorganisms come into contact with sectioned muscles, increasing their growth rate owing to the rich substrate and adequate humidity. Even in operations where good sanitation standards are observed, microorganisms proliferate during these stages. Most of the spoilage microorganisms grow well at low temperatures (3–10°C) and preferentially utilize glucose. Only when it is not available do they utilize other substrates, releasing odoriferous compounds.

In addition to microbial contamination during dressing operations, external contamination by faecal microorganisms – especially enterococci – occurs in poor sanitary conditions. These microorganisms are found in many habitats, mainly in the intestinal tracts of humans and animals. They are resistant to adverse conditions such as high temperature, pH and high ionic strengths. Once introduced into a meat processing plant, enterococci can become established, and subsequent contamination of a food product does not necessarily indicate faecal pollution.

Microorganisms surviving at low temperatures are mostly Gram-negative aerobic bacteria, but after long-term storage fungi can also be found. The most important bacterium found in this stage is *Pseudomonas*, which is also of economic importance owing to the active proteases it produces. An important Gram-positive spoilage coccus also found in fresh meat is *Streptococcus*. *Brochothrix thermosphacta* is present in considerable numbers when meat is stored aerobically or vacuum packaged at chill temperatures; this microorganism produces mostly lactic acid under anaerobic conditions, but if small amounts of oxygen are present it can produce highly odoriferous compounds such as acetoin and acetic, isobutyric and valeric acids.

Spoilage occurs when bacterial numbers reach 10^7 per square centimetre or per gram; it is caused not only by the microorganisms themselves but also by the metabolites produced from meat constituents as substrates. Many of these metabolites are volatile and specific for any given microorganism.

Strains of psychrotrophic bacteria most often found in raw and processed meat, in addition to *Pseudomonas* and *Brochothrix thermosphacta*, are *Moraxella*, *Acinetobacter*, *Shewanella putrefaciens*, *Lactobacillus*, *Aeromonas* and certain genera of Enterobacteriaceae. The spoilage potential of *Moraxella*, *Acinetobacter*, *Brochothrix thermosphacta* and *Shewanella putrefaciens* is only evident when the pseudomonad population is reduced. *Aeromonas* spp., *Pseudomonas* spp., *Enterobacter* spp. and *B. thermosphacta* generate odour-related volatile compounds, whereas *Pseudomonas* spp. and *Aeromonas* also alter the texture. At temperatures higher than 10°C, e.g. due to refrigeration failure, Enterobacteriaceae strains overgrow other species, and spoilage is evident owing to the production of biogenic amines, sulphur compounds and H_2S.

Biogenic amines are aliphatic, aromatic or heterocyclic organic bases of low molecular weight. They are psychoactive or vasoactive, and are therefore involved in several human physiological functions. In the majority of foods they are formed by decarboxylation of the corresponding amino acids through the action of substrate-specific enzymes derived from microorganisms of the Enterobacteriaceae and Bacillaceae, species of *Lactobacillus*, *Pediococcus* and *Streptococcus*. *Pseudomonas* found in meat and *B.*

thermosphacta show no evidence of producing biogenic amines.

Besides changes in texture due to protease production by some spoilage microflora such as *Pseudomonas* and *Aeromonas*, and in aroma due to production of odoriferous compounds (for example by *Brochothrix thermosphacta*, Enterobacteriaceae and some lactic acid bacteria), water-holding capacity also changes owing to an increase in the formation of intermolecular bridges establishing a gel-like structure, the consequences of texture changes due to proteolysis. The looser structure allows a faster attack by bacteria. Faster spoilage in products with a high water-holding capacity is more evident in low-fat meats where total counts are higher during chilled storage than in similar products traditionally formulated.

The most common processing factors affecting the type of prevalent microflora in a meat product are:

- pH changes
- addition of curing salts
- changes in water activity (a_w)
- thermal processing
- other, such as competitive microflora, bactericides, etc.

Storage conditions affecting microflora are:

- chilling
- freezing
- gas atmospheres.

Generally, the factors are applied in combination, therefore creating more than one hurdle to prevent microbial growth.

Spoilage in Processed Meats

Acid-treated and Dry Sausages

Application of organic acids, of either chemical or microbial origin, has a bacteriostatic effect. The mechanisms involved in bacterial inhibition are mainly related to the amount of undissociated acid. The cytoplasm is acidified by the dissociation in the inner part of the cell of acid that has previously diffused through the cell wall in an undissociated form. The result is the alteration of several cell metabolic pathways by disruption of the proton motive force. Lactic, acetic and propionic acids have been the most widely applied. Lactic acid is a stronger acid than acetic acid (pK_a 3.86 and 4.75, respectively) in buffer foods; at a moderate pH of 4–6 acetate is expected to be the more potent antimicrobial since a greater proportion would be dissociated. However, in processed meats

pH reduction is not always desirable (with the exception of fermented or acidified products).

Reduction in pH can be also achieved by the action of acid-producing strains. The growth of the original Gram-negative microflora in raw meat is inhibited by the presence of salt, reduction of water activity and limited oxygen availability. A predominantly Gram-positive microflora dominated by *Lactobacillus* is then established. As a microaerophilic bacterium, *Lactobacillus* becomes the dominant microorganism when availability of oxygen in the interior of the sausage is limited by the fermentation processes. The reduction of water activity and lack of oxygen limit the growth of aerobic spoilage microflora present in raw meat; at the same time, salt inhibits the growth of *Leuconostoc* and *Carnobacterium*.

Spoilage in acid-added or fermented meat products is only evident when maximum cell densities, determined by the availability of substrate for fermentation, are above 10^8 colony forming units (cfu) per square centimetre. Meat lactic acid bacteria do not produce off flavours or off odours in vacuum-packaged meat stored at 0–5°C as rapidly as spoilage microflora. Spoilage by meat lactic acid bacteria is generally secondary to microbial growth, unless H_2S-producing strains predominate.

Lipolysis can be due to release of free fatty acids by chemical means, but microbial metabolism could be involved in the breakdown of fatty acids. Lipolysis has been related to bacterial lipase activity, whereas oxidative changes of unsaturated fatty acids resulting in peroxide and carbonyl production have been related to both chemical and microbial reactions. Added or native lactic acid strains can produce lipases which in some cases promote flavour development. Lipids form the major fraction in fermented sausages and are considered, together with proteins, to be the main substrate for the production of flavour compounds, including alcohols, aldehydes, ketones and short-chain fatty acids. During ripening the lipid fraction undergoes hydrolytic and oxidative changes involving the liberation of fatty acids and the oxidation of unsaturated fatty acids, with the concomitant production of carbonyls. Nonetheless, when excessive lipolysis occurs, it is associated with spoilage. This produces highly odoriferous compounds such as non-branched aliphatic compounds, alcohols and furans. Lipase production is repressed by readily metabolized carbohydrates.

Originally dry sausages were made without adding starter cultures, relying instead on inoculation with microorganisms contaminating the ripening rooms, but the resulting products were variable in quality. Since 1961 dry sausage starters have been available. A number of metabolic products with antibacterial

Table 1 Spoilage microorganisms in meat products and alteration of sensory characteristics

Meat product	Microorganism	Alteration
Acid-treated and fermented sausages	Lactic acid bacteria *Lactobacillus, Streptococcus*	Off odour
	Leuconostoc, Pediococcus	Greening
Cooked cured meat	*Lactobacillus, Streptococcus, Leuconostoc*	Greening
	Enterobacteriaceae	Off odour
	Pseudomonas, Aeromonas	Texture alteration and off odour
	Alcaligenes, Acinetobacter, Flavobacterium	Off odour
Smoked products	Heterofermentative lactic acid bacteria, *Micrococcus*, yeasts, moulds	Off odour
with liquid smoke	*Lactobacillus plantarum, Leuconostoc mesenteroides, Pediococcus damnosus*	Souring
Canned hams	*Enterococcus faecalis*	Discoloration
Cooked ham	Enterococci, lactobacilli	Discoloration
Vacuum-packaged cooked meats	*Leuconostoc, Lactobacillus, Carnobacterium*	Souring, off odour
Processed meats in 100% CO_2 (pâté, ripened sausages, fermented meats)	*Brochothrix thermosphacta*	Souring
Vacuum-packaged processed meats	*Lactobacillus carnosus, L. gelidium, Leuconostoc mesenteroides, L. paramesenteroides*	Souring, off odour

action are found by the mixed microflora used in dry sausages, such as antibiotics, bacteriocins, diacetyl, acetic acid, CO_2, lactic acid and hydrogen peroxide. Processing of dry sausages, owing to their reduced a_w, promotes the prevalence of some lactic acid bacteria, mainly of *Lactobacillus* and *Pediococcus* spp., and catalase-positive cocci, such as *Staphylococcus carnosus, S. xylosus* and *Micrococcus varians*. In some dry sausages yeasts and moulds such as *Debaryomyces hansenii* and *Penicillium nalgiovense* are also used to cover the sausage surface. This microbial cover further promotes anaerobic conditions in the inner part of the sausage, encouraging the growth of microaerophilic lactic acid bacteria.

Spoilage in dry sausage is due to excessive lactic acid production, promoting acid flavour. In addition, greening can occur owing to the action of H_2O_2-producing lactobacilli, such as *Lactobacillus fructovorans, L. jensenii* and *L. viridescens*. Other types of bacteria that are responsible for greening are enterococci such as *Enterococcus faecium* and *E. faecalis*, *Leuconostoc* and *Pediococcus*.

Cooked Cured Meat

The term 'cooked cured meat products' describes those in which only nitrite is used as the curing salt, except in the case of dietetic meat products. These foods develop a pink-red colour due to the reaction of nitrite with myoglobin, producing a red pigment, haemochromogen, which is stabilized by thermal treatment. Cooked cured meat products are generally made of pork, free of bones, rind, jelly and any visible fat.

The curing mixture contains several bacteriostatic compounds, such as sodium chloride, sodium nitrite and spices such as onion, garlic and pepper. Usually, these components of the meat curing mixture are dissolved in water to form a brine which is kept at about 4°C and injected into whole joints of meat, such as hams or bacons. After filtration to eliminate meat debris, the brine is returned to the reservoir of the injection machine and recycled. The microbial quality of the brine is therefore inferior after recycling. A predominance of *Micrococcus, Acinetobacter, Flavobacterium* and coryneform bacteria, common in pork carcasses, is found. Pathogens such as *Salmonella, E. coli* and other faecal species cannot grow in brines, but they survive. As *Pseudomonas* does not grow in brines but is able to survive, it can be taken as an indicator of manufacturing practices.

In cured products the microflora can develop from several sources, including pork carcass and curing brine; its growth is encouraged by external factors such as relative humidity and temperature at maturation. Micrococci seem to originate in the pig skin, and later are present in the brine. Gram-negative bacteria, mainly *Vibrio, Acinetobacter, Alcaligenes, Aeromonas, Pseudomonas*, and Enterobacteriaceae, also constitute an important microflora in cured bacon. However, sodium nitrate inhibits the growth of *Brochothrix thermosphacta* and Enterobacteriaceae at low temperatures.

Brochothrix thermosphacta does not possess a nitrite reductase enzyme system which reduces nitrates under anaerobic conditions, unlike many lactic acid bacteria. The result is that sodium nitrite inhibits the growth of *B. thermosphacta* at concentrations as low as $25 \, mg \, kg^{-1}$. Gram-positive cocci are not affected by the addition of nitrite but are not competitive. Because lactic acid bacteria are not affected by nitrite or microaerophilic conditions they become a larger proportion of the total microflora.

Greening is also an important spoilage reaction in cured meat products. It is caused by some catalase-negative bacteria. Peroxides are formed by exposure of oxygen to air and as catalase in meat is destroyed by heat treatment and nitrites, peroxides react with the cured pigment, nitrosohaemochrome. The iron of the haem ring is eliminated, producing oxidized porphyrins responsible for the green colour. For this reason, when catalase-negative bacteria are growing under anaerobic conditions they are able to multiply but are unable to produce greening.

The main microorganisms involved in cured meat greening are lactic acid bacteria, in particular the heterofermentative species of *Lactobacillus*, *Streptococcus* and *Leuconostoc*. Most of these bacteria are tolerant of NaCl and NaNO$_2$, heat and smoke resistant, and capable of growing at low temperatures in vacuum-packed cured meats. Lactic acid bacteria are also competitive, suppressing other microflora by antagonism. The presence of one of these greening organisms in the meat processing plant, especially *Lactobacillus viridescens* can cause problems. *L. viridescens* grows well in the presence of 4% salt concentration and more slowly at 6%, although no growth is detected at 7%, the normal salt concentration in cured meat of 2.5–3% is not enough to inhibit the growth of greening bacteria. Bacterial greening is often a direct reflection of malpractice in sanitation prior to or after processing.

Smoked Products

Processing of smoked meat products usually is a secondary operation to cooking, although some sausages are cold-smoked. Smoking implies the incorporation into the meat product of some of the numerous chemical compounds derived from the combustion of sawdust, extending the product's shelf life by microbial and enzyme inhibition, and promoting desirable new flavours and aromas. The main compounds originated by the pyrolysis of wood are phenols, acids, alcohols, carbonyls and hydrocarbons.

Smoking of cured meats has a similar effect to the use of salt and nitrite on the microflora of the finished product. In general lactobacilli are resistant to smoke; the volatile acids of the smoke reduce the pH considerably to levels at which lactobacilli are able to survive, but not Gram-negative rods, which are also sensitive to the dry environment and excessive NaCl during the last stages of smoking. The low a$_w$ also accounts for the low proportion of Gram-negative bacteria found.

The surface of smoked meat products has a different microenvironment to the inner part; the microflora is usually formed of micrococci, yeasts and moulds, but Gram-negative bacteria are rarely found owing to high NaCl concentrations and dry conditions. In general, smoking promotes an environment low in spoilage microflora, with the exception of heterofermentative lactic acid bacteria, yeasts and moulds. Liquid smoke inhibits the growth of motile lactobacilli as well as staphylococci and micrococci. However, some strains of *Lactobacillus plantarum*, *Leuconostoc mesenteroides* and *Pediococcus damnosus* are not affected. As a result, spoilage in smoked or liquid smoke-treated products is mainly caused by souring, or is located in circumscribed areas.

Thermally Processed Meats

Temperature is the most important environmental parameter involved in the growth of microorganisms in meat. Preservation is achieved if all microorganisms are killed or severely damaged. The main objectives of heat treatment are enzyme inactivation to avoid decomposition reactions, and destruction of pathogens and spoilage microorganisms. Heating acts on at least one key enzyme of the bacterial metabolism; as a result microbial population reduction is a first-order kinetic reaction.

Heat treatment can be divided into:

1. Scalding, applied to inactivate enzymes in products that will be further processed; the processing temperature is about 65°C, non-sporulated microorganisms are destroyed but cells of *Enterococcus faecium* and *E. faecalis* can survive, as well as spores of *Bacillus* and *Clostridium*.
2. Cooking is applied at around 85°C, and improves the food's sensory characteristics; in addition to the microorganisms destroyed by scalding, spore-forming psychrophiles are destroyed.
3. Pasteurization is a process involving a time–temperature combination (145–160°C for 1–45 s or 71–72°C for 15–20 s) destroying vegetative cells, heat-resistant, spore-forming genera *Streptococcus* and *Corynebacterium*, as well as spores of *Bacillus* and *Clostridium*.
4. Sterilization destroys vegetative microorganisms and spores; the time–temperature ratio depends on a particular microorganism and product liability.

Canning is the most widely applied thermal process, and it ensures the destruction of vegetative cells and spores of *Clostridium botulinum*. It is a process producing safe foods, where sanitary problems are present only when processing failures occur. In this situation the most important genera present are

Micrococcus, *Enterococcus*, *Bacillus*, *Clostridium*, Enterobacteriaceae, *Flavobacterium* and *Pseudomonas*. Properly processed canned hams may include genera such as *Enterococcus* associated with the gut of the animal, but not causing extensive spoilage. However, other changes can occur. Enterococci, particularly *Enterococcus faecalis* var. *liquefaciens*, can cause discoloration in the centre of canned hams, or liquefaction of the gel surrounding the product.

Being a mildly processed product from the thermal point of view, cooked ham presents spoilage problems owing to its low salt content (around 2%), pH above 6.0 and high a_w (0.94). After cooking, the normal microflora of the product, consisting of lactic acid bacteria, is inadequate to protect the product against the growth of Gram-negative, saprophytic microorganisms. In this situation a number of deterioration changes can occur. Discoloration is caused by enterococci and lactobacilli, in particular if there is not enough nitrite reacting with myoglobin to produce nitrosomyoglobin, or when the nitrite is unevenly distributed. A major failure is the presence of areas of grey, cooked, non-cured meat. Good massaging of the cured meat can prevent this problem. As the effect of nitrite on bacteria is pH-dependent, a combination of low pH and low temperature increases the bacteriostatic effect of nitrite. Sodium nitrate depletion rate doubles for every 12°C increase in cooking temperature or every 0.86 decrease in pH. Therefore, its effect is considerably reduced after thermal processing.

Vacuum packaging is the most widely used packaging technique for cooked meat products. The shelf life of sliced, vacuum-packed cooked ham is up to 20 days at 4°C. However, in low-heat-processed, cured, vacuum-packed chilled meat a number of microorganisms can grow, including *Leuconostoc*, *Lactobacillus*, *Carnobacterium* and *Brochothrix*, resulting in gas accumulation, cloudy fluid in the packages, slime in the meat surface and sour off odours.

Gas Atmosphere Packaging

In modified-atmosphere packaging (MAP) of meats a gas mixture is injected into a gas-impermeable package in which meat or a meat product is stored. Vacuum packaging is a modification of MAP, in which air has been evacuated, and highly gas-impermeable package material is used. Gases used in meat and meat products packaging are CO_2, O_2 and N_2. Carbon dioxide has a selective inhibitory effect upon moulds and aerobic bacteria such as *Pseudomonas*, whereas it has a limited effect on yeasts, anaerobes, lactic acid bacteria, Enterobacteriaceae and *Brochothrix thermosphacta*. Inhibition by CO_2 follows various mechanisms including the inhibition of enzymes catalysing

pigment oxidation such as malate and citrate dehydrogenases. Carbon dioxide also alters the pH of the medium by the production of carbonic acid. Nitrogen is an inert gas used as a filler; however, when meat is stored in 100% N_2, decoloration occurs. Oxygen is used for its ability to promote 'blooming' of raw meat, although it also promotes growth of aerobic microflora, lipid oxidation and formation of metmyoglobin.

Anaerobic conditions due to vacuum packaging or 100% CO_2 atmospheres, and chilled storage favour the growth of psychrotrophic lactic acid bacteria, mainly *Leuconostoc* spp., *Lactobacillus* spp. and *Carnobacterium* spp. Spoilage is evident by souring, as lactic acid bacteria only attack amino acids to a minor extent under anaerobic conditions and formation of biogenic amines, putrescine and cadaverine is not important if Enterobacteriaceae such as *Hafnia alvei* or *Serratia liquefaciens* are not present. Souring is due to the growth and metabolism of Lactobacillaceae, although sometimes sulphide spoilage can also occur, the flora dominating in this case being Enterobacteriaceae and *Vibrio*, or sulphide-producing lactobacilli. The bacteria originate in operations after cooking and before packaging.

Heterofermentative facultative anaerobic *Leuconostoc* spp. occur naturally in unprocessed and processed meat. *Lactobacillus carnosum* and *L. gelidium* occur predominantly together with *Leuconostoc mesenteroides* subsp. *mesenteroides* and *L. paramesenteroides*. The isolates are psychrotrophic and dominate vacuum-packaged meat. Other predominant bacteria include homofermentative *Lactobacillus sake*, *L. curvatus*, *Carnobacterium* spp. and *Brochothrix thermosphacta* as well as *L. viridescens*.

Under anaerobic conditions spoilage by *Brochothrix thermosphacta* is minimal, but the presence of small quantities of oxygen leads to the formation of end products of aerobic metabolism, highly odoriferous products such as acetoin, acetic, isobutyric, isovaleric and formic acids, diacetyl and 3-methylbutanal. *Brochothrix thermosphacta* is able to grow in atmospheres containing 100% CO_2 or SO_2, usually inhibiting normal Gram-negative spoilage microorganisms. Therefore, it is an important microorganism in modified-atmosphere-packaged meat products such as pâté, ripened sausages, fermented meat and meat products.

In oxygen-rich atmospheres, pseudomonads are the main contaminants responsible for putrid odours, the volatile compounds produced only appearing when the substrate metabolized consists of amino acids. In addition, *Pseudomonas* is an important protease-producing bacterium. However, the dominance of pseudomonads decreases the concentration of diacetyl and acetoin because closely related compounds, such as propylene and butylene glycols, can serve as a carbon source for some strains of *Pseudomonas* spp.

Under conditions of poor refrigeration or refrigeration failure, Enterobacteriaceae predominate and cause off odours, generally when the number of microorganisms reaches 10^7 cfu cm^{-2} or g^{-1}. If there is a high concentration of CO_2 in the package, growth of aerobic flora is discouraged and the slower-growing lactic acid bacteria such as *Lactobacillus* are encouraged.

In addition to *Lactobacillus* and *Brochothrix thermosphacta*, other bacteria identified as important in vacuum-packed, cooked, cured meat are micrococci and *Corynebacterium*, streptococci, *Leuconostoc* and streptobacteria. Gram-negative bacteria are also important, such as Enterobacteriaceae, *Acinetobacter* and *Vibrio*. The difference between the microfloras of particular products can be due to factors such as pH, NaCl and NaNO$_2$ concentration, degree of vacuum in the package, and storage conditions. Some species can even inhibit another. For example, *Vibrio* inhibits *Leuconostoc*, *Lactobacillus viridescens* and some streptobacteria.

Meat Preservation by Microbially Produced Bacteriostatic Compounds

Although spoilage processes are almost always microbially initiated, some strains also produce antibacterial compounds, providing a 'natural' way of preserving meat. Unfortunately, the application of these natural bacteriostatic compounds alone is not enough to assure wholesomeness in meat products.

Probably the most widely studied antibacterial compounds are the antibacterial proteins or bacteriocins; however, these are only active against other phylogenetically related organisms. Bacteriocin-producing microorganisms can also have unwanted effects, for example some strains of *Leuconostoc* spp. produce bacteriocins along with other antimicrobial compounds such as H_2O_2, diacetyl, and lactic and acetic acids: besides the discoloration effect of H_2O_2, being an heterofermentative organism, *Leuconostoc* also produces gas and other changes associated with spoilage. *Lactobacillus reuterii*, a heterofermentative lactic acid bacterium, produces reuterin, a low-molecular-weight compound which inhibits the growth of species of *Salmonella*, *Shigella*, *Clostridium*, *Staphylococcus* and *Listeria*; however, the use of *Lactobacillus reuterii* in sausages is undesirable because it also produces CO_2. Other heterofermentative lactic acid bacteria can produce CO_2, inhibiting the growth of Gram-negative aerobes such as *Pseudomonas* spp., but CO_2 is undesirable in meat

products, producing texture failures such as pores and fractures, and promoting a fizzy taste.

When catalase activity is insufficient, some aerobes and anaerobes produce H_2O_2, an oxidative agent that inhibits or destroys bacterial enzymes of other bacterial genera such as *Pseudomonas*. Peroxides also promote lipid oxidation, producing a rancid flavour, and have also a bleaching action causing colour faults.

Some lactic acid bacteria produce acetoin and diacetyl, both bacteriostatic compounds. Diacetyl is produced from citrate via pyruvate; it inhibits yeasts and Gram-positive and Gram-negative bacteria.

Bacterial preparations used as protective cultures have been shown to extend the shelf life of certain meat products. Protective cultures are strains that can suppress the growth of undesirable microorganisms such as pathogens or spoilage microflora. These are usually mixtures of lactic acid bacteria able to grow fast and become the predominant microflora; in addition they can reduce pH and produce other bacteriostatic compounds. The main microorganisms commercially produced as bioprotectors are *Lactobacillus alimentarius*, *Staphylococcus xylosus*, *S. carnosus* and *Pediococcus pentosaceus*.

See color Plate 20.

coccus: Introduction; *Streptococcus thermophilus*. *Vibrio*: Introduction, including *Vibrio vulnificus*, and *Vibrio parahaemolyticus*; *Vibrio cholerae*.

Further Reading

Bailey ME, Intarapichet KO, Gerhardt KO, Gutheil RA and Noland LA (1993) Bacterial shelf-life of meat and volatile compounds produced by selected meat spoilage microorganisms. In: Charalambous G (ed.) *Shelf Life Studies of Foods and Beverages. Chemical, Biological and Nutritional Aspects*. P. 63. Amsterdam: Elsevier.

Gill CO (1979) A review: intrinsic bacteria in meat. *Journal of Applied Bacteriology* 47: 367–378.

Grant GF, McCurdy AR and Osborne AD (1988) Bacterial greening in cured meats: a review. *Canadian Institute of Food Science and Technology Journal* 21: 50–56.

Greer GG (1989) Red meats, poultry and fish. In: McKellar RC (ed.) *Enzymes of Psychrotrophs in Raw Foods*. P. 267. Boca Raton: CRC Press.

Knudtson L and Hartman PA (1993) Enterococci in pork processing. *Journal of Food Protection* 56: 6–9.

Nychas GJ, Dillon VM and Board RG (1988) Glucose: the key substrate in the microbiological changes occurring in meat and certain meat products. *Biotechnology and Applied Biochemistry* 10: 203–231.

Thumel H (1995) Preserving meat and meat products: possibilities and methods. *Fleischwirtschaft International* 3: 3–8.

Weber H (1994) Dry sausage manufacture. The importance of protective cultures and their metabolic products. *Fleischwirtschaft International* 3: 23–28.

Metabolic Activity Tests *see* **Total Viable Counts**: Metabolic Activity Tests.

METABOLIC PATHWAYS

Contents
Release of Energy (Aerobic)
Release of Energy (Anaerobic)
Nitrogen Metabolism
Lipid Metabolism
Metabolism of Minerals and Vitamins
Production of Secondary Metabolites – Fungi
Production of Secondary Metabolites – Bacteria

Release of Energy (Aerobic)

Astrid Brandis-Heep, Philipps Universität, Fachbereich Biologie, Laboratorium für Mikrobiologie, Marburg, Germany

Microbial cells consist of a wide variety of chemical substances which have to be synthesized or taken up from outside the cell. These processes require a lot of energy. Each cell has to provide the necessary energy and different possibilities for its supply are developed: a number of organisms can use light energy; most microorganisms, of course, obtain it from the oxidation of chemical compounds. Chemicals used as sources for energy are metabolized and energy is conserved either by substrate-level phosphorylation or by building up an electrochemical gradient across the cytoplasmic membrane.

The different processes by which energy is converted, especially from glucose to ATP, under aerobic conditions will be referred to in this article. At the end of this article it is outlined how substrates other than glucose can be made available for the pathways described.

The major carbohydrate-metabolizing pathways are:

- Embden–Meyerhof–Parnas (EMP) pathway, also called glycolysis
- Entner–Doudoroff (ED) pathway
- pentose phosphate (PP) pathway.

The three pathways differ in many ways, but two generalizations can be made:

1. All three pathways convert glucose to glyceraldehyde 3-phosphate (GAP) by different routes.
2. The GAP is converted to pyruvate via reactions that are the same in all three pathways.

Transport into the Cell

The cytoplasmic membranes of the cell are not simply permeable for substrates being metabolized. There are three different transport systems which control their entry into the cell (**Table 1**):

1. Passive transport: facilitated diffusion (e.g. *Zymomonas mobilis* and erythrocytes). As facilitated diffusion is only found in a few organisms or cell types, it will not be discussed further here.
2. Active transport: symport with a proton, protons or sodium ions. In this process of active transport the substrate can accumulate to a high concentration in the cytoplasm in a chemically unaltered form. Active transport requires energy and is linked to energy available from ion gradients or ATP hydrolysis.
3. Group translocation: phosphotransferase system (PTS). Group translocation is the process whereby a substance is transported while simultaneously being chemically modified, generally by phosphorylation.

Active Transport: Symport of Sugars

Active transport is an energy-dependent system. The substance being transported combines with a membrane-bound carrier, which then releases the chemically unchanged substance inside the cell. Substances transported by active transport are sugars, most amino acids, organic acids and a number of inorganic ions such as sulphate, phosphate and potassium.

In bacteria the driving force of the active transport comes from ATP hydrolysis or, more commonly, from the electrochemical H^+ gradient ($\Delta\mu H^+$) across the membrane, called the proton motive force.

The proton concentration outside the cell (+) is higher than inside (−) and there is a potential of about 200 mV across the membrane. It is this electrochemical potential that drives the uptake of cationic nutrients by active transport. For neutral or anionic nutrients, the transport must be driven by a cation,

H^+, or in some cases by Na^+. The transport of lactose is driven by a proton, and this process is called symport. Each carrier has two specific sites: one for the substrate (for example, glucose or lactose) and one for a proton (or protons). As the substrate is taken up, protons move across the membrane and the proton motive force is diminished.

Group Translocation: Phosphotransferase System

In this transport process, the substance transported is chemically altered in the course of its passage across the membrane and so no actual concentration gradient of the external solute is produced. PTS is the best-studied group translocation system which involves transport of the sugars glucose, mannose, fructose, α-acetylglucosamine, and β-glucosides, which are phosphorylated during transport.

The PTS in *Escherichia coli* is composed of three reactions with 24 proteins; only four are necessary to transport a given sugar. The proteins are themselves alternately phosphorylated and dephosphorylated in a cascading fashion until the transmembrane transport protein, called E III, receives the phosphate group and phosphorylates the sugar (**Fig. 1**). The high-

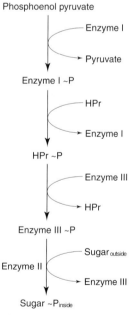

Figure 1 Flow of a phosphoryl group from phosphoenol pyruvate to the sugar, which is translocated by the phosphotransferase system. HPr, heat-stable protein.

Table 1 Transport systems for sugars in aerobic, facultative and anaerobic organisms

Sugar	Aerobic organisms	Facultative organisms	Anaerobic organisms
Monosaccharide	H^+ symport	PTS	PTS
Disaccharide	H^+ symport	H^+ symport	PTS
	e.g. *Bacillus*	e.g. *Escherichia coli*	e.g. *Lactobacillus*

PTS = Phosphotransferase system.

energy phosphate bond that supplies the necessary energy for the PTS comes from the key metabolic intermediate phosphoenol pyruvate (PEP). A small heat stable protein, called HPr, is phosphorylated by PEP.

$$\text{Glucose} + \text{PEP} \xrightarrow[\text{Mg}^{++}]{\text{PTS}} \text{Glucose 6-phosphate} + \text{Pyruvate}$$

This reaction is not specific to the glucose transport system but in general is involved in the sugar transport. Enzyme I and HPr are soluble cytoplasmic enzymes, whereas the enzymes II and III are membrane-bound and specific for the uptake of each individual sugar. For example, there are different enzymes II and III for the transport of glucose, lactose and fructose. Mutants defective in HPr or enzyme I are unable to transport many different sugars, whereas mutants defective in enzyme II or III are unable to transport a particular sugar.

Embden–Meyerhof–Parnas Pathway

The process of sugar breakdown is called glycolysis. The enzymatic reactions of a glycolytic pathway will form pyruvate coupled to ATP synthesis by substrate-level phosphorylation.

In the EMP pathway glucose (C_6) is first converted in a series of reactions to form fructose 1,6-bisphosphate, which is cleaved to form two interconvertible C_3 sugars. They enter a common set of catabolic reactions to form two pyruvates (**Fig. 2**). The breakdown of one glucose to two pyruvates releases sufficient free energy to permit the synthesis of four ATP from ADP and P_i. The conversion of glucose is also accompanied by the formation of two reduced coenzymes (NADH).

Cells initiate the EMP pathway by activation of glucose by phosphorylation with ATP (hexokinase) to glucose 6-phosphate or using the PTS (see above). Glucose 6-phosphate is isomerized to fructose 6-phosphate (isomerase) and then converted to fructose 1,6-bisphosphate with ATP (phosphofructokinase). The phosphofructokinase is the key enzyme regulating the rate of glycolysis. It is allosterically regulated by the effector molecules AMP and ADP. When both of these are high, the ATP concentration in the cell is low and glycolysis must be stimulated to regenerate ATP. Besides phosphofructokinase is feedback inhibited by PEP and fructose 6-phosphate (end product inhibition).

Fructose 1,6-bisphosphate can be broken down into two phosphorylated 3-carbon units, glycer-

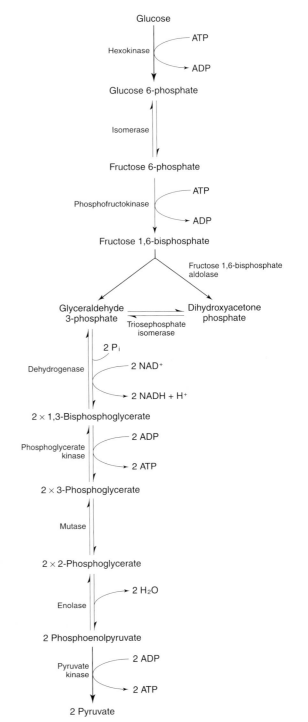

Figure 2 Embden–Meyerhof–Parnas pathway.

aldehyde 3-phosphate and dihydroxy-acetone phosphate, by an aldolase reaction (fructose 1,6-bisphosphate aldolase). Both compounds are in equilibrium (triosephosphate isomerase) with each other. However the constant removal of GAP by the glycolytic pathway shifts the balance so that for each glucose molecule two GAPs are formed.

The oxidation of two GAPs to pyruvate starts with

the exergonic GAP dehydrogenase reaction in which inorganic phosphate is incorporated to form the two 1,3-bisphosphoglycerates and two reduced coenzymes NADH. The mixed anhydride 1-phosphate of 1,3-bisphosphoglycerate is coupled with the synthesis of ATP (3-phosphoglycerate kinase) by substrate-level phosphorylation. The final reactions (mutase, enolase and pyruvate kinase) form two pyruvates and two ATP from ADP.

The overall reaction is:

$$Glucose + 2\ ADP^{3-} + 2\ Pi^{2-} + 2\ NAD^{+} \longrightarrow$$
$$2\ Pyruvate^{-} + 2\ ATP^{4-} + 2\ NADH + 2\ H^{+} + 2\ H_2O$$

Although the initial series of reactions of the EMP pathway require the input of two ATP, the overall reaction is exergonic, so that finally a net gain of two ATP will result from one glucose degraded in glycolysis.

Entner–Doudoroff Pathway

There is a second important pathway for the breakdown of carbohydrates which is only found in prokaryotes. It was first discovered in 1952 by Entner and Doudoroff in *Pseudomonas saccharophila*. It is widespread, especially among aerobic Gram-negative bacteria, and can operate in different modes: in linear and catabolic modes, in cyclic mode, in a modified mode involving non-phosphorylated intermediates, or in alternative modes involving C_1 metabolism and anabolism. The ED pathway will here be viewed as an alternative to the EMP pathway (**Fig. 3**).

Prokaryotes, which carry out the ED pathway, lack the key enzyme phosphofructokinase of the EMP pathway. Therefore glucose is oxidized to 6-phosphogluconate and then via a lactonase and a dehydratase reaction converted to 2-dehydro-3-deoxy-6-phosphogluconate (erroneous KDPG). KDPG is cleaved directly to pyruvate and glyceraldehyde 3-phosphate. Because of this direct formation of pyruvate, some of the ATP-generating steps are lost. So the breakdown of glucose via the ED pathway results in the net production of only one ATP, one NADPH and one NADH.

The overall reaction is:

$$Glucose + NADP^{+} + NAD^{+} + ADP^{3-} + P_i^{2-} \rightarrow$$
$$2\ Pyruvate^{-} + NADPH + NADH + 3\ H^{+} + ATP^{4-}$$

Because of the net production of only 1 mole ATP per mole glucose fermented, this pathway is usually found in aerobic bacteria. Anaerobes do not have a respiratory chain and are restricted to ATP synthesis

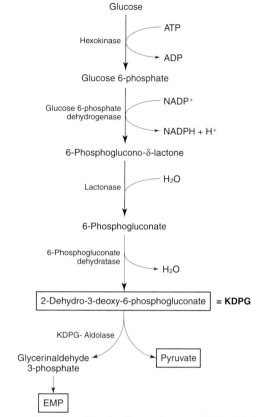

Figure 3 Entner–Doudoroff pathway. EMP, Embden–Meyerhof–Parnas.

by substrate-level phosphorylation. So they are dependent on the twice-as-efficient EMP pathway in the degradation of glucose to two pyruvate.

The ED pathway is important when substrates such as gluconate (or other aldonic acids) serve as nutrients, for example, when *E. coli* is transferred from a glucose-containing medium to one in which gluconate is the source of carbon, three new enzymes will be induced: a gluconokinase, 6-phosphogluconate dehydratase and KDPG-aldolase. Hence, *E. coli* switches from the EMP to the ED pathway.

The overall reaction is:

$$Gluconate^{-} + NAD^{+} + ADP^{3-} + P_i^{2-} \rightarrow 2\ Pyruvate^{-} + NADH + H^{+} + ATP^{4-}$$

In addition, the ED pathway provides an important role for the synthesis of NADPH + H$^+$, which is often essential for anabolic (biosynthetic) reactions.

Pentose Phosphate Pathway

In bacteria, about 80% of the glucose is degraded aerobically via the EMP and ED pathway, and about 20% enters the PP pathway for generation of ATP, regeneration of NADPH and the synthesis of precursors for nucleotide and aromatic amino acid biosynthesis. So many variations of the PP pathway are possible, depending on the need of the growing cell.

The reactions of the PP pathway are divided into three stages:

- oxidation and decarboxylation reactions
- isomerization reactions
- transaldolase and transketolase reactions in the rearrangement of sugars.

In one version of the PP, glucose is converted into ribulose 5-phosphate and CO_2, a process that requires one ATP and generates two NADPH (**Fig. 4**). First, glucose 6-phosphate is oxidized to 6-phospho-gluconolactone by an $NADP^+$-specific glucose 6-phosphate dehydrogenase. The lactone is then hydrolysed to 6-phosphogluconate by gluconolactonase. In this oxidation process energy is lost as heat but cannot be used for substrate-level phosphorylation. 6-Phosphogluconate dehydrogenase reduces a second NADPH under the formation of ribulose 5-phosphate and CO_2. Ribulose 5-phosphate can be isomerized to ribose 5-phosphate and epimerized to xylulose 5-phosphate.

The overall reaction is:

$$\text{Glucose} + \text{ATP}^{4-} + 2\,\text{NADP}^+ + \text{H}_2\text{O} \rightarrow \text{Pentose 5-P}^{2-}$$
$$+\,\text{CO}_2 + \text{ADP}^{3-} + 2\,\text{NADPH} + 3\,\text{H}^+$$

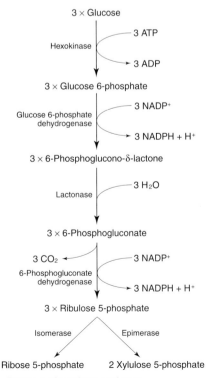

Figure 4 Pentose phosphate pathway.

When there is a further need for more NADPH, the excess PP has to be removed and the enzymes transketolase (TK) and transaldolase (TA) can convert PPs back into hexose phosphates (**Fig. 5**). TK transfers C_2 units with thiamine diphosphate as the prosthetic group, whereas TA transfers C_3 units.

$$2\ \text{Xylulose-5-P} + \text{Ribose-5-P} \rightleftharpoons$$
$$2\ \text{C}_5 + \text{C}_5$$

$$2\ \text{Fructose-6-P} + \text{Glyceraldehyde-5-P}$$
$$2\ \text{C}_6 + \text{C}_3$$

Because of the reversibility of the TK- and TA reaction the synthesis of PP from hexose phosphates is also possible under conditions where NADPH is not needed (**Table 2**).

Oxidation of Pyruvate to Acetyl Coenzyme A

In all the major carbohydrate catabolic pathways, pyruvate is a common product, oxidized during aerobic growth by the pyruvate–dehydrogenase complex:

$$\text{Pyruvate}^- + \text{NAD}^+ + \text{CoASH} \rightarrow \text{Acetyl CoA}$$
$$+\text{CO}_2 + \text{NADH}$$

This enzyme complex is located in the cytosol of prokaryotes and in the mitochondrial matrix of eukaryotic cells and consists of three enzymes: pyruvate dehydrogenase with thiamine pyrophosphate (TPP) as cofactor, dihydrolipoate transacetylase and the dihydrolipoate dehydrogenase with flavin adenine dinucleotide (FAD), lipoic acid, NAD^+ and coenzyme A. The complex catalyses a short metabolic pathway rather than a simple reaction. Under physiological conditions it is irreversible and under the control of several allosteric effectors, e.g. the *E. coli* pyruvate dehydrogenase is feedback-inhibited by the products of the reaction, acetyl coenzyme A and NADH. Only as much acetyl coenzyme A and NADH are produced as can be used. The enzyme is also stimulated by PEP. High AMP levels also influence the enzyme, signalling low ATP concentrations.

The Tricarboxylic Citric Acid Cycle

To finish the respiratory metabolism of glucose, acetyl coenzyme A enters the tricarboxylic acid (TCA) cycle to produce carbon dioxide, water, reduced coenzymes and ATP (**Fig. 6**).

The overall reaction is:

$$\text{Acetyl CoA} + \text{ADP}^{3-} + \text{P}_i^{2-} + \text{Q} + 2\,\text{H}_2\text{O} + \text{NADP}^+ +$$
$$2\,\text{NAD}^+ \rightarrow 2\,\text{CO}_2 + \text{ATP}^{4-} + \text{QH}_2 + \text{NADPH} +$$
$$2\,\text{NADH}^+ + \text{H}^+ + 3\,\text{H}^+ + \text{CoASH}$$

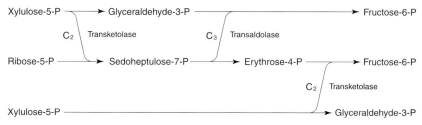

Figure 5 Conversion of pentose-P into hexose-P. P, Phosphate.

Table 2 Distribution of Embden–Meyerhof–Parnas (EMP), Entner–Doudoroff (ED), and pentose phosphate (PP) pathway in bacteria and some eukarya

Organism	EMP	ED	PP
Bacteria			
Acetobacter aceti subsp. xylinum	–	+i	N
Azotobacter chroococcum	+	–	N
Alcaligenes eutrophus	–	+i	N
Bacillus larvae	+	+c	+
Bacillus subtilis	+	–	–
Escherichia coli	+	+i	+
Gluconobacter oxidans	N	+	N
Methylococcus capsulatus	–	+c	–
Pseudomonas aeruginosa	–	+i	–
Pseudomonas fluorescens	–	+i	+
Rhizobium leguminosarum	–	+i	+
Bradyrhizobium japonicum	+	+	–
Enterococcus (Streptococcus) faecalis	+	+i	+
Salmonella typhimurium	+	+i	+
Thiobacillus sp.	–	+i	+
Yersinia pseudotuberculosis	+	+i	–
Xanthomonas campestris	–	+c	+
Zymomonas mobilis	–	+c	–
Eukarya			
Aspergillus nidulans	+	+i*	+
Penicillium notatum	+	+c*	+

+ = Present; – = not present.
i = Inducible; c = constitutive; * = modified; N = not known.

There are four oxidation steps per acetyl coenzyme A, producing two NADH (2-oxoglutarate dehydrogenase, malate dehydrogenase), one NADPH (isocitrate dehydrogenase, in bacteria mostly NADP-dependent), and one QH_2 (succinate dehydrogenase). One ATP (succinate thiokinase) is formed via substrate-level phosphorylation. The cycle usually operates in conjunction with the respiratory chain which reoxidizes NAD(P)H+H$^+$ and QH_2 using the proton motive force (in mitochondria, intramitochondrial NAD and extramitochondrial NADP is used) to generate ATP in the ATP-synthase reaction. Finally, the respiratory chain uses the reduced cofactors produced in several oxidation steps, described previously. The respiratory chain with its components will not be referred to here.

Other Substrates as Sources for Metabolic Activity

Living organisms can use a variety of substrates for growth: almost every naturally occurring organic compound can serve as a source for cell carbon or energy. These can be low-molecular-mass compounds or polymers, such as glycogen, starch, cellulose, polysaccharides, lipids, fatty acids and proteins. Polymers cannot enter the cell; they must be cleaved outside into monomers and dimers which are small enough to be transported into the cell and then enter the metabolic pathways.

Carbohydrates

Glucose is not the only carbohydrate that can be converted to pyruvate by glycolysis. Many other mono-, di- and polysaccharides are substrates for ATP synthesis (**Fig. 7**; **Table 3**).

Lipids and Fatty Acids

Lipids can also serve as substrates for the production of ATP. The fatty acids are cleaved from the glycerol backbone of a triglyceride lipid molecule by the action of lipases (**Fig. 8**).

Glycerol is converted to dihydroxyacetone phosphate and then to glyceraldehyde 3-phosphate, entering the EMP pathway. Fatty acids are broken down by a process called β-oxidation. The fatty acid chain is first converted to the corresponding coenzyme A ester by an acyl coenzyme A synthetase. The coenzyme A ester is then oxidized in the β-position and cleaved into acetyl coenzyme A and the acyl coenzyme A (coenzyme A ester of the fatty acid shortened by two carbon atoms) which enters a new degrading cycle. Every reaction sequence forms QH_2 and NADH. Acetyl coenzyme A is passed into the TCA cycle.

Amino Acids

Amino acids and short-chain peptides are actively taken up and metabolized. The degradation starts with the removal of the amino group by an oxidase to the corresponding keto acid, by a dehydrogenase coupled with a transaminase, or by a deaminase. All 20 amino acids are degraded to the following inter-

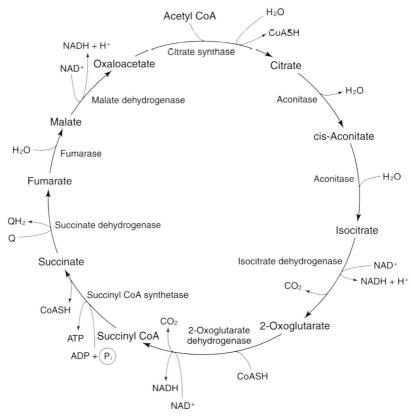

Figure 6 Tricarboxylic acid cycle.

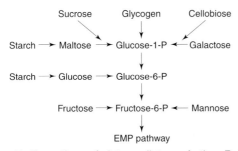

Figure 7 Formation of intermediates of the Embden–Meyerhof–Parnas (EMP) and Entner–Doudoroff pathway from carbohydrates other than glucose. P, phosphate.

mediates entering the TCA cycle: pyruvate, acetyl coenzyme A, acetoacetyl coenzyme A, 2-oxoglutarate, succinyl coenzyme A, fumarate and oxaloacetate. Acetoacetyl coenzyme A itself is a precursor to acetyl coenzyme A.

Aromatic Substrates

Plants produce many substrates with aromatic ring systems which become available when organic material is decomposed. In particular bacteria and fungi can degrade the aromatic rings. Under aerobic conditions aromatic compounds are transformed by mono-

Table 3 Conversion of carbohydrates to intermediates of the Embden–Meyerhof–Parnas pathway

Carbohydrate	Enzyme	End products
Maltose	Maltase	Glucose
Maltose	Maltose phosphorylase	Glucose + β-D-glucose 1-phosphate
Cellobiose	Cellobiose phosphorylase	Glucose + α-D-glucose 1-phosphate
Sucrose	Sucrase	Glucose + fructose
Sucrose	Sucrose phosphorylase	Fructose + α-D-glucose 1-phosphate
Lactose	β-Galactosidase	Glucose + galactose
Fructose	Fructokinase	Fructose 6-phosphate
Fructose	Fructokinase	Fructose 1-phosphate
Galactose	Galactokinase, glucose : galactose 1-phosphate uridylyltransferase; UDP-glucose epimerase	Glucose 1-phosphate

UDP, uridyl diphosphate

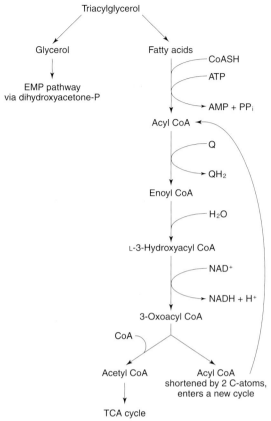

Figure 8 Formation of intermediates of the Embden–Meyerhof–Parnas (EMP) pathway and tricarboxylic acid (TCA) cycle from lipids.

oxygenases and dioxygenases into a few central intermediates such as catechol, protocatechuate or gentisate and homogentisate. These dihydroxylated compounds are suitable for an oxidative ortho- or meta-cleavage via a dioxygenase before other reaction sequences follow to form intermediates of the TCA cycle.

See also: **Escherichia coli**: Escherichia coli. **Metabolic Pathways**: Release of Energy (Anaerobic); Lipid Metabolism; Metabolism of Minerals and Vitamins; Production of Secondary Metabolites – Fungi; Production of Secondary Metabolites – Bacteria. **Pseudomonas**: Introduction.

Acknowledgement

I thank Prof Wolfgang Buckel for his scientific advice and for proof-reading the manuscript and Michael Liesert for preparing the figures.

Further Reading

Atlas RM (1995) *Principles of Microbiology.* Dubuque, IA: Wm. C. Brown.
Conway T (1992) The Entner–Doudoroff pathway: history, physiology and molecular biology. *FEMS Microbiological Reviews* 103: 1–28.
DiMarco AA and Romano AH (1985) D-Glucose transport system of *Zymomonas mobilis. Applied Environmental Microbiology* 49: 151–157.
Gottschalk G (1986) *Bacterial Metabolism.* New York/Berlin: Springer Verlag.
Lengeler JW, Drews G and Schlegel HG (eds) (1999) *Biology of Prokaryotes.* Stuttgart: Thieme Verlag.
Lissie TG and Phibbs PV (1984) Alternative pathways of carbohydrate utilization in *pseudomonads. Annual Review of Microbiology* 38: 359–387.
Stryer L (1988) *Biochemistry.* New York: WH Freeman.
White D (1995) *The Physiology and Biochemistry of Prokaryotes.* New York/Oxford: Oxford University Press.

Release of Energy (Anaerobic)

M D Alur, Food Technology Division, Bhabha Atomic Research Centre, Mumbai, India

This article describes the pathways involved in the anaerobic catabolism of carbohydrates by different bacteria, leading to the release of energy as well as the formation of various important fermentation products such as organic acids and alcohols. The mechanism of uptake of substrates, the site of action and other less prominent ways of energy metabolism are briefly discussed.

Metabolism (from the Greek *metabole*, meaning change) is the totality of an organism's chemical processes. A cell's metabolism is an elaborate road map of numerous reactions that occur in the cell. These reactions are arranged in an intricately branched metabolic pathway along which molecules are transformed by a series of steps. The cell routes matter through the metabolic pathways by means of enzymes which selectively accelerate each of the steps in the labyrinth of reactions. Thus, metabolism is concerned with managing the material and energy resources of the cell. Some metabolic pathways release energy by breaking down complex molecules to simpler compounds. These degradative processes are called catabolic pathways. Certain chemical transformations are involved in the synthesis of macromolecules. This part of metabolism is termed biosynthesis.

An organism obtains two resources for synthesizing organic compounds: energy and a source of carbon. Species that use light energy are termed phototrophs. Chemotrophs obtain their energy from chemicals taken from the environment. If an organism needs only the inorganic compound CO_2 as a carbon source, it is an autotroph. Heterotrophs require at least one organic nutrient – glucose, for instance – as a source of carbon for making other organic compounds. Depending on how bacteria obtain energy and carbon, they can be divided into four categories:

- chemo-organotrophs (chemoheterotrophs)
- chemolithotrophs
- photo-organotrophs (photoheterotrophs)
- photolithotrophs.

Substrates Utilized by Bacteria and Fungi

Fungi, being osmotrophic chemoheterotrophs, utilize substrates ranging from simple sugars to cellulose, hydrocarbons, lignin, pectins and xylans. Energy-yielding metabolism may involve respiration or fermentation. Heterotrophic bacteria can use a variety of organic compounds as energy sources. These compounds include carbohydrates, fatty acids and amino acids. For many microorganisms, the six-carbon sugar, glucose, is preferred. The lower fungi and the moulds are endowed with a rich enzymatic make-up which attacks carbohydrate and complex ones such as cellulose as well as protein and fats. For example, cellulose can be attacked by several species of fungi belonging to the genera *Aspergillus*, *Penicillium*, *Fusarium*, *Cladosporium* and *Trichoderma* which act simultaneously on the pectin, as well as fats and proteins present in decomposing vegetable matter. Yeasts and moulds can grow in a substrate or medium containing concentrations of sugars that inhibit most bacteria. Thus, jams and jellies are spoiled by moulds but not by bacteria. The simple carbohydrates, such as sugars and starches and their derivatives, are attacked by many microorganisms which ferment them and turn them into alcohols and organic acids, such as lactic, acetic, formic and butyric acid.

Mechanisms of Uptake

Small-molecular-weight substances, such as nucleosides, fatty acids and carbohydrates from monosaccharides to oligosaccharides, can be transported in the bacterial cell. Usually, the nutrient is bound stereospecifically by a carrier protein present in the cytoplasmic membrane and transported against a concentration gradient through the expenditure of energy. This process is referred to as active transport and operates in the accumulation of nucleosides and the disaccharides maltose, melibiose and lactose. The carbohydrate glycerol appears to enter the cell by facilitated diffusion, a process characterized by the participation of a stereospecific membrane-associated transport protein carrier, but without the participation of energy. Thus, glycerol cannot be transported into the cell against a concentration gradient. Transport by both facilitated diffusion and active transport results in the presence of unmodified nutrients within the cell. However, in group translocation, this energy-dependent transport mechanism involves modification of some sugars during their passage through the membrane. Sugars that are transported by this mechanism become phosphorylated during passage and appear within the cell as sugar phosphates. The mechanism is referred to as the phosphotransferase system (PTS) and transports mannitol, sorbitol, lactose, glucose, fructose and *N*-acetyl-glucosamine.

In bacteria, active transport is often associated with group translocation, a process in which a molecule is actively transported into the cell and chemically modified at the same time. Thus, glucose transport occurs via the carbohydrate PTS found in prokaryotes. Glucose in the cell is found in the form of glucose 6-phosphate while glucose outside the cell occurs as free glucose. The phosphate group is supplied by phosphoenolpyruvate (PEP), a compound with energy equivalent to an ATP molecule. The energetic phosphate is transferred along protein molecules until it is attached to the glucose cell membrane (**Fig. 1**).

The phosphorylated glucose carries a net negative charge, and consequently is less likely to be able to cross the cell membrane and escape from the cytoplasm than is an unchanged free glucose molecule. However, the Gram-positive bacterium, *Enterococcus faecalis*, has an anion–anion antipolar system that can catalyse the exchange of a cytoplasmic phosphorylated glucose for an external inorganic phosphate molecule. The exchange reaction system prevents excessive accumulation of a substance in the cytoplasm.

In the basic PTS pathway described above, there are four proteins involved. PEP reacts with enzyme I in the cytoplasm to form a phosphorylated enzyme I molecule. This phosphorylated protein then reacts with heat-stable protein (Hpr), another cytoplasmic protein, and transfers the phosphate group to it. Enzyme II, located in the cell membrane, carries out the actual transport of the sugar. It receives a phosphate from either Hpr or from enzyme III and uses the energy to transport the sugar molecule and phosphorylate it.

Cell membranes are normally impermeable to phosphate esters, but *Escherichia coli*, *Salmonella typhimurium*, *Staphylococcus aureus* and some other bacteria form an inducible active transport system catalysing uptake of glucose 6-phosphate, 2-deoxy-

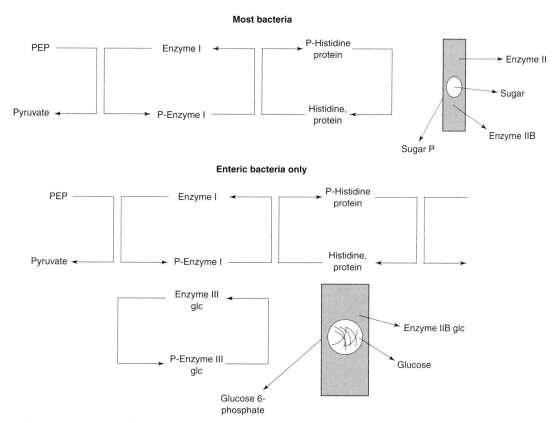

Figure 1 Glucose phosphotransferase systems (PTS) in enteric and most other bacteria. PEP, phosphoenolpyruvate.

glucose 6-phosphate, mannose 6-phosphate, fructose 6-phosphate, glucose 1-phosphate and fructose 6-phosphate.

Pathways Involved in Catabolism

Anaerobic Breakdown of Carbohydrates

Glucose occupies an important position in the metabolism of most biological forms and its anerobic dissimilation provides a metabolic pathway common to most forms of life. The terms glycolysis and fermentation have been applied to the anaerobic decomposition of carbohydrate to the level of lactic acid. The dissimilation of carbohydrates involves a complicated series of catalysed reactions including oxido-reduction and phosphorylation. The final product in some organisms is lactic acid; in others, the lactic acid is further metabolized anaerobically to butyric acid, butyl alcohol, acetone and propionic acid. The two most common forms of fermentation are lactic and alcoholic. These two fermentations proceed along the same path to fermentation of pyruvic acid, which is the key substance in fermentation reactions.

Carbohydrates are usually phosphorylated at one or two positions, and the energy utilized in phosphorylation is derived from the change of ATP to ADP. Diphosphorylated hexoses are characteristically broken down into two triose units. Each triose produces two high-energy bonds in its conversion to pyruvate. One high-energy bond results from phosphate esterified into a carbon with an oxygen double-bond attachment, and the other results from a carbon that is double-bonded to another carbon. ADP forms ATP. This energy change is termed substrate-level phosphorylation and it produces four high-energy bonds per mole of hexose. Since two bonds were utilized in hexose phosphorylation, only two are gained. Two hydrogen are given off by hexose breakage into two trioses, and the hydrogens thus produced are picked up by NADP. Pyruvate formed in this process may be broken down into acetaldehydride and CO_2, as in alcoholic fermentation, and the acetaldehyde is reduced to ethanol by hydrogen given off when the hexose molecule was split.

Yeasts and other microorganisms degrade hexoses according to the Embden–Meyerhof–Parnas (EMP) pathway. Hexoses may be converted by other pathways, however, and pentose is involved after primary decarboxylation. Only one high-energy bond is derived per mole of hexose in this system. Pentose as well as hexose may be broken down by the pentose shunt system.

The Embden–Meyerhof Scheme

The basic concepts of glycolysis are incorporated in the Embden–Meyerhof scheme, which provides the major pathway for glucose breakdown in many organisms. Glucose must be phosphorylated before fermentation. The initial reaction is between glucose and ATP, forming glucose 6-phosphate. This reaction is catalysed by the enzyme hexokinase. Glucose 6-phosphate is in equilibrium with glucose 1-phosphate but the latter is concerned with the formation of polysaccharides. The next step is the formation of fructose 6-phosphate from glucose 6-phosphate, which reacts with ATP to form fructose 1,6-diphosphate. The cleavage of the diphosphate in two three-carbon fragments results in the formation of glyceraldehyde 3-phosphate, which is then oxidized in the presence of inorganic phosphate to 1,3-diphosphoglyceric acid. This intermediate reacts with ADP to form ATP plus 3-phosphoglyceric acid. The next reaction involves an internal shift of the phosphate group of 3-phosphoglyceric acid to form 2-phosphoglyceric acid. The 2-phosphoglyceric acid undergoes dehydration to yield the enol form of phosphopyruvic acid, which transfers its phosphate to ADP to enolpyruvic acid which is in equilibrium with the keto form of pyruvic acid (**Fig. 2**).

The Monophosphate Shunt

In an alternative pathway, termed the monophosphate shunt, glucose 6-phosphate is oxidized to phosphogluconic acid which is decarboxylated to yield ribose 5-phosphate and other pentose phosphates. A split into two and three-carbon fragments then occurs. This scheme provides a means for the metabolism of the pentoses, ribose and deoxyribose, into constituents of the nucleic acids and would permit an entrance of the pentoses into the EMP pathway for the organisms. Fructose, galactose and other monosaccharides are converted into their corresponding phosphates by reacting with ATP and are converted to glucose 6-phosphate which gains entrance to the main metabolic pathway (**Fig. 3**).

The hexose monophosphate shunt (HMS) pentose phosphate pathway operates in conjunction with glycolysis in many bacteria. This set of reactions generates ribose for nucleic acid synthesis and produces NADPH for other synthetic reactions. The phosphorylation of glucose to glucose 6-phosphate is the same as that found in glycolysis and the Entner–Doudoroff (ED) pathway. The next reaction converts glucose 6-phosphate to 6-phosphogluconic acid, as in the ED pathway, followed by the conversion to ribulose 5-phosphate and then to ribose 5-phosphate. Ribulose 5-phosphate also combines with CO_2 in the

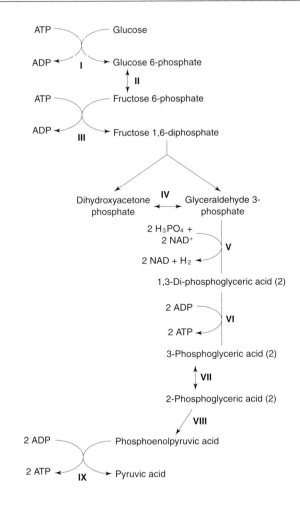

Figure 2 The Embden–Meyerhof pathway of glucose catabolism (glycolysis).

dark reaction of photosynthesis. Here $NADP^+$ is reduced at two reaction sites, rather than the commonly encountered NAD^+. The pentose units can subsequently be converted to two different intermediates in glycolysis and into pyruvic acid. Thus, the HMS serves as the loop in glycolysis for the production of pentose units and NADPH (**Fig. 4**).

Entner–Doudoroff Pathway

In the ED pathway, glucose is converted to pyruvic acid in fewer steps than it is in the pathway of glycolysis. In HMS, glucose is converted to five-carbon carbohydrates (pentose units).

The ED pathway involves an initial phosphorylation as in glycolysis, but is then followed by

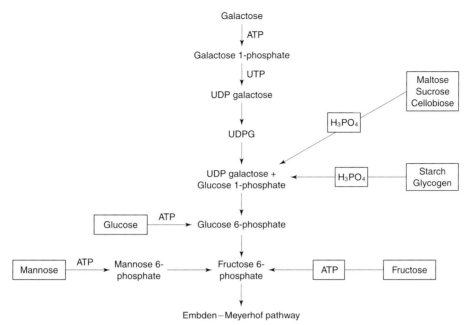

Figure 3 Entrance of different carbohydrates to the Embden–Meyerhof pathway.
UDP, uridine diphosphate
UTP, uridine triphosphate
UDPG, uridine diphosphogalactose

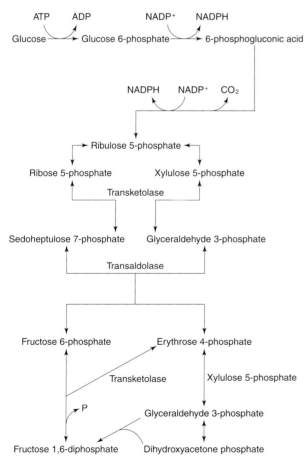

Figure 4 The pentose phosphate pathway of glucose catabolism.

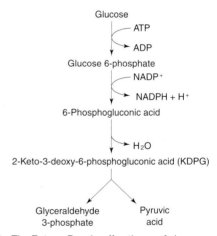

Figure 5 The Entner–Doudoroff pathway of glucose catabolism in aerobic and anaerobic Gram-negative bacteria.

an oxidative step of the compound to an acid (phosphogluconic acid). Subsequently, dehydration occurs, with the formation of keto-deoxy-phosphogluconic acid. The last reaction produces pyruvic acid and glyceraldehyde phosphate, which can be converted to pyruvic acid (**Fig. 5**).

The ED pathway is found in some Gram-negative bacteria such as *Pseudomonas*, *Rhizobium* and *Agrobacterium*. It is generally not found in Gram-positive bacteria.

Fermentation

In glycolysis, the same reactions occur whether oxygen is present or not. The products are primarily pyruvic acid, NADH and ATP. The essential difference between aerobic and anaerobic processes occurs with pyruvic acid and NADH. In the case of fermentation reactions, pyruvic acid is converted to a variety of organic compounds. These reactions involve the transfer of electrons and hydrogen from NADH to organic compounds.

Fermentation is a major source of energy for those organisms that can only survive in the absence of air (obligate anaerobes). Other fermentative organisms that can grow in the presence or absence of air (facultative anerobes) use fermentation as a source of energy only when oxygen is absent. In fermentation, energy gain is very low and occurs as a result of substrate-level phosphorylation. The synthesis of ATP in fermentation is restricted to the amount formed during glycolysis.

During glycolysis, glucose is oxidized to pyruvic acid, which is the physiologically important first intermediate product in the aerobic or anaerobic dissimilation of glucose. The genus *Pseudomonas* follows the ED cycle in the dissimilation of glucose. In this cycle, the pathway to pyruvate progresses via glucose 6-phosphate, 6-phosphogluconic acid, 2-keto-3-deoxy-6-phosphogluconic acid to pyruvate plus 3-phosphoglyceraldehyde.

Pyruvate may also be reached via the metabolism of sugars other than glucose or the metabolism of fatty acids and amino acids. The monosaccharides, fatty acids and amino acids are mainly derived from the hydrolysis of starch, glycogen, cellulose, fats, poly-β-hydroxybutyric acid, chitin or proteins, depending on the organism.

Pyruvate is a sort of Grand Central Station, in that it is the point of arrival and departure of a wide variety of metabolic substrates and products. Pyruvate is reduced to lactic acid. It may also be decarboxylated and reduced to ethyl alcohol. Conversely, it may serve as the source of amino acids, fatty acids and aldehydes.

The anaerobic system of biological oxidations that does not use oxygen as the final acceptor of electrons is called anaerobic respiration. In anaerobic respiration, compounds such as carbonates, nitrates and sulphates are ultimately reduced. Many facultative anaerobic bacteria can reduce nitrate to nitrite under anaerobic conditions. This type of reaction permits continued growth when free oxygen is absent, but the accumulation of nitrite which is produced by the reduction of nitrate is eventually toxic to the organisms. Certain species of *Bacillus* and *Pseudomonas* are able to reduce nitrite to gaseous nitrogen. This process occurs when aerobic organisms are grown under anaerobic conditions. The organisms which reduce sulphate and carbonate are strictly anaerobic. *Desulfovibrio desulfuricans* reduces sulphate to hydrogen sulphide as it oxidizes carbohydrate to acetic acid. *Methanobacterium bryanti* is able to couple the reduction of CO_2 to methane with the oxidation of carbohydrate to acetic acid. Some anaerobes do not have a functional glycolytic system. They may have carbohydrate fermentation pathways that use the pentose phosphate pathway and the ED pathway. The pentose phosphate pathway of glucose catabolism yields ribose 5-phosphate and $NADPH + H^+$.

Some organisms (strict aerobes) are enzymically equipped to use only free oxygen as the final hydrogen (e^-) acceptor, but others (facultative aerobes) are equipped to use as the final hydrogen (e^-) acceptor either free oxygen or some reducible inorganic substrate, commonly a nitrate. In fermentations, usually only NAD or NADP functions as the hydrogen (e^-) carrier. Flavine adenine dinucleotide (FAD) and cytochrome systems are not required since the final hydrogen (e^-) acceptor is not oxygen but an organic substance – commonly pyruvic acid. Fermentation may be caused by facultative organisms under anaerobic conditions, e.g. *Saccharomyces cerevisiae*, by strictly anaerobic organisms, e.g. *Clostridium* or by organisms that do not utilize free oxygen, e.g. species of *Lactobacillus*.

Depending on the conditions of growth, the substrate and the organisms involved, the end products of fermentation vary greatly. *Clostridium* species normally ferment glucose to yield butyl and other alcohols and certain acids.

Lactic Fermentation

The products of glucose fermentation by all species of *Streptococcus*, many species of *Lactobacillus* and several other species of bacteria are mainly lactic acid with minor amounts of acetic acid, formic acid and ethanol. Several species of *Streptococcus* produce more than 90% of lactic acid based on the sugar used, and hence this type of fermentation is referred to as homolactic fermentation. This is the simplest fermentation – a step reaction catalysed by NAD-linked lactic dehydrogenase, which reduces pyruvate to lactate. Since two ATP molecules are consumed in the formation of hexose diphosphate from glucose and four ATP molecules are subsequently produced, the net yield is two ATP per hexose. This fermentation is the first stage in cheese manufacture.

The homolactic fermentation which forms only lactate is the characteristic of many of the lactic acid bacteria, e.g. *Lactobacillus casei*, *Streptococcus crem-*

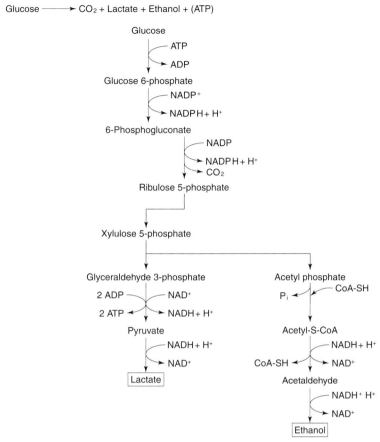

Figure 6 The heterofermentative metabolic sequence in *Leuconostoc* and some species of *Lactobacillus*.

oris and pathogenic streptococci, a heterolactic fermentation converts only half of each glucose molecule to lactate. Both these fermentations are responsible for the souring of milk and pickles. The heterofermentative metabolic sequence found in *Leuconostoc* and some species of *Lactobacillus* ferments glucose according to the equation:

$$Glucose \rightarrow Lactate + Ethanol + CO_2$$

Lactic acid bacteria include the genera *Lactobacillus*, *Sporolactobacillus*, *Streptococcus*, *Leuconostoc*, *Pediococcus* and *Bifidobacterium*, which produce lactic acid as a major fermentation product (**Fig. 6**).

Alcoholic Fermentation

The major substrates yielding ethanol are the sugars which in yeasts are degraded to pyruvate by the EMP or glycolytic pathway. There is a net yield of one ATP for each pyruvate formed from glucose. The identical metabolic route for ethanol formation is found in the bacterial species *Sarcina ventriculi*, *Erwinia amylovora* and *Zymomonas mobilis*, which also possess the enzymes pyruvate decarboxylase and alcohol dehydrogenase. Yeasts ferment glucose to pyruvate, which is converted to CO_2 plus acetaldehyde. Acet-

aldehyde is then reduced to ethanol by an NAD-linked reaction.

Butyric Fermentation

The butyric fermentation is initiated by a conversion of sugars to pyruvate through the EMP pathway. Pyruvate undergoes a thiolytic cleavage to acetyl-coenzyme A (CoA), CO_2 and H_2. Acetate is derived from acetyl-CoA via acetyl phosphate accompanied by the synthesis of ATP. Butyrate synthesis starts through an initial condensation of 2 mol of acetyl-CoA to form acetato-acetyl-CoA, which is then reduced to butyryl-CoA. Butyrate is then formed by CoA transfer to acetate. The ATP yield per mole of glucose fermented is 2.5 mol (**Fig. 7**).

Mixed Acid (Formic) Fermentation

This is a characteristic of most Enterobacteriaceae. These organisms dispose of their substrate in part by lactic fermentation but mostly through pyruvate breaking down into formate and acetyl-CoA, which in turn generates an ATP. The formic fermentation yields three ATP per mole of glucose fermented (compared with two in lactic fermentation; **Fig. 8**).

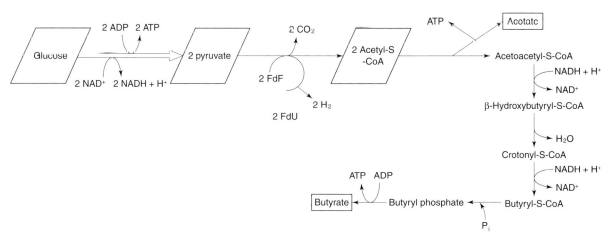

Figure 7 Fermentation of glucose to butyrate by *Clostridium butyricum, C. kluyveri* and *C. pasteurianum*. FdF, Ferredoxin F

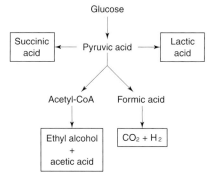

Figure 8 Glucose fermentation by *Escherichia coli*.

Propionic Fermentation

This pathway extracts additional energy from the substrate. Pyruvate is carboxylated to yield oxaloacetate, which is reduced to yield succinate and is then decarboxylated to yield propionate.

The lactate is first oxidized to pyruvate; part is then reduced to propionate and the rest is oxidized to acetate and CO_2:

$$3\,CH_3 \cdot CHOH \cdot COOH \xrightarrow{-6H} 3\,CH_3 \cdot CO \cdot COOH$$
$$\text{Pyruvate}$$
$$\xrightarrow{-6(H^+)} 2\,CH_3CH_2COOH + CH_3COOH +$$
$$\text{Propionate} \qquad \text{Acetate}$$
$$CO_2 + H_2O$$

This process of extracting energy from lactate yields only one ATP per nine carbon atoms fermented. Hence, propionic acid bacteria grow slowly (**Fig. 9**).

Butyric-Butanol Fermentation

This pattern of pyruvate reduction is found in certain strict anaerobes (*Clostridium*) spp.). The initial scission yields H_2, CO_2 and 2-C fragments at the acetate level of oxidation. Two such fragments are then con-

densed. The resulting aceto-acetyl-CoA undergoes decarboxylation to acetate and reduction by H_2 (activated by ferredoxin) to yield isopropanol, butyric acid and *n*-butanol (**Fig. 10**). The general view of fermentation products formed by different bacteria is depicted in **Figure 11**.

Sites of Activity (Mitochondria and Membrane)

Mitochondria possess a folded inner membrane (forming crests), with a large functional surface. The major oxidation–reduction processes take place in the inner membrane or on the inner face (matrix side) of the inner membrane. The lipoic acid dehydrogenases of pyruvate and α-ketoglutarate and the succino-dehydrogenase enzymes of the Kreb's cycle and β-oxidation of fatty acids, the carriers participating in the respiration chain and the system of synthesis of ATP coupled with the respiratory chain, are found in the inner membrane.

The electron carries of the respiratory chain drain the maximum number of protons and electrons originating from the dehydrogenations taking place in the mitochondria (notably in the inner membrane) or in the cytosol. The inner membrane has selective permeability, and nucleotides of the NADP type cannot pass through it, so that the electrons and protons from the cytosol are used to reduce substrates like oxaloacetate into malate, which can cross the inner membrane. The carrier can then be reoxidized (oxaloacetate), yielding NADPH and H^+, which will enter the respiratory chain.

In oxidative phosphorylation, the synthesis of ATP is coupled to the flow of electrons from NADH or $FADH_2$ (reduced adenine dinucleotide) to O_2 by a proton gradient across the inner mitochondrial membrane. Electron flow through three asymmetrically

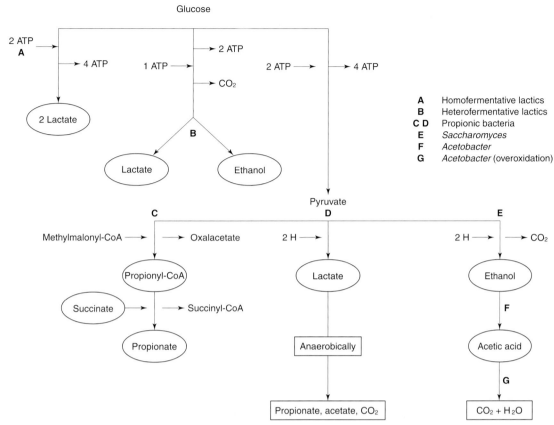

Figure 9 General pathways for the formation of fermentation products from glucose by various organisms.

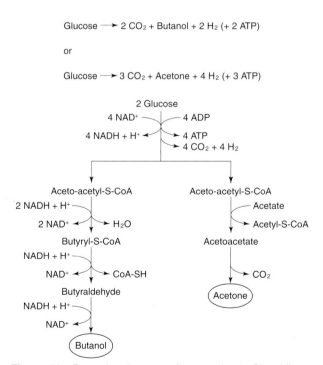

Figure 10 Butanol and acetone fermentation in *Clostridium acetobutylicum*.

oriented transmembrane complexes results in the pumping of protons out of the mitochondrial matrix and the generation of a membrane potential. ATP is synthesized where the protons flow back to the matrix through a channel in an ATP-synthesizing complex, called the ATP synthetase system.

The membrane plays an essential role in the synthesis of ATP in mitochondria (inner membrane) as well as in chloroplasts (thylakoids) and bacteria (plasmic membrane, respiration or photosynthesis). A general mechanism was proposed for the synthesis of ATP linked with the transducer membranes (mitochondria, bacteria, chloroplasts). The chain of e$^-$ carriers implanted in the membrane has an anisotropic function and, because of the orientation of its sites, causes the translocation of protons and their ejection at a specific face of the membrane. Since the membrane is impermeable to ions, the functioning of the chain creates a gradient of protons on either side of the membrane and a promotive force which is utilized for the synthesis of ATP. ATP is thus synthesized at the cost of the gradient protons (**Fig. 12**).

See also: **Clostridium**: Introduction; *Clostridium aceto-butyricum*. **Escherichia coli**: *Escherichia coli*. **Lactobacillus**: Introduction. **Leuconostoc. Streptococcus**: Introduction.

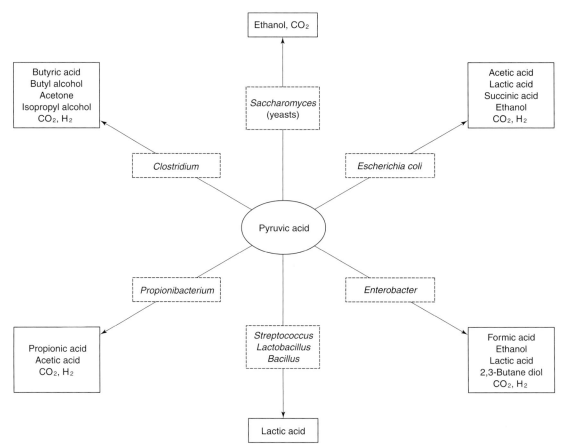

Figure 11 Overview of fermentation products formed from pyruvic acid by different bacteria.

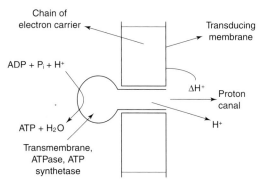

Figure 12 Involvement of protons in the synthesis of ATP in energy-transducing mechanisms.

Further Reading

Builock J and Kristiansenceds B (1987) *Basic Biotechnology*. Ch. 2. *Biochemistry of Growth and Metabolism*. New York: WH Freeman.

Davis BD, Dulbecco R, Eisen HN and Ginsberg HS (eds) (1980) *Microbiology*. New York: Harper.

Hamilton WA (1988) Energy transduction in anaerobic bacteria. In: Antony C (ed.) *Bacterial Energy Transduction*. London: Academic Press.

Ketchum PA (1984) *Microbiology Introduction for Health Professional*. New York: John Wiley.

Nester EW, Roberts CE and Nestler MT (1995) *Microbiology: A Human Perspective*. Oxford, UK: Brown Publishers.

Pelczar MG Jr, Chan ECG and Krieg NR (1986) *Microbiology*, Vth edn. New York: McGraw-Hill.

Schlegel HG (1993) *General Microbiology*. Cambridge: Cambridge University Press.

Starr MP, Stolp H, Truper HF, Balows A and Schlegel HF (eds) (1986) *The Prokaryotes*. Vol. II. Berlin: Springer-Verlag.

Stryer L (1995) *Biochemistry. Enzymes: Basic Concepts and Kinetics*, IVth edn. New York: W. H. Freeman.

Nitrogen Metabolism

M D Alur, Food Technology Division, Bhabha Atomic Research Centre, Mumbai, India

This article gives an account of nitrogen metabolism in bacteria, particularly focusing on the different ways in which amino acids are catabolized by various bacteria. The synthesis of cellular components, e.g. proteins and nucleic acids and the mechanism of uptake of substrate and site of activity are briefly dealt with.

Bacteria vary widely in their ability to utilize various sources of nitrogen for synthesis of proteins.

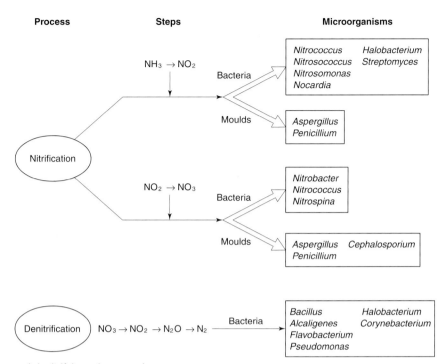

Figure 1 Nitrifying and denitrifying microorganisms.

In nitrogen metabolism ammonia forms the key intermediate and most organisms can use ammonia as the principal source of protein nitrogen. For most heterotrophic organisms, amino acids and peptones provide the principal nitrogen sources. Since the native protein molecule is too large to enter the bacterial cell in order to be utilized, it must be broken down to smaller units by extracellular proteolytic enzymes excreted into the medium. Certain species of *Clostridium*, such as *C. histolyticum* and *C. perfringens*, produce proteolytic enzymes which attack native proteins to be hydrolysed to peptones. Most heterotrophic organisms complete the breakdown of proteins to their constituent amino acids in favourable growth conditions.

Chemoautotrophic bacteria such as *Nitrosomonas* sp. convert ammonia to nitrites. Facultatively anaerobic bacteria such as *Pseudomonas* and *Bacillus* denitrify nitrate or nitrite to atmospheric nitrogen gas. The reduction of nitrates to molecular nitrogen via nitrite as an intermediate is termed denitrification. When the process proceeds all the way from nitrate to nitrite to molecular nitrogen or ammonia, more energy is available than in the formation of nitrites. This type of denitrification is exhibited by *P. aeruginosa* under anaerobic conditions (**Fig. 1**).

Cyanobacteria are able to use atmospheric nitrogen as a source of nitrogen. In this process of nitrogen fixation, bacteria convert atmospheric nitrogen to ammonia. Nitrogen-fixing cyanobacteria only require

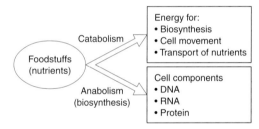

Figure 2 General view of metabolism.

light energy, CO_2, N_2, H_2O and some minerals for growth.

Metabolism is the sum total of all the degradative (catabolic) and biosynthetic (anabolic) reactions performed by the cell. Growth of the organism requires the synthesis of all cell components. This synthesis involves the breakdown of foodstuffs, both to provide compounds that are converted to cell material and to supply energy and reducing equivalents. The energy and reducing equivalents are used to drive the large numbers of biosynthetic reactions by which the cell converts relatively simple molecules to the supramolecular structures that make up cells (**Fig. 2**).

Sources of Nitrogen for Fungi and Bacteria

Molecular nitrogen is fixed by nitrogenase in *Azotobacter vivelandii*. Most fungi utilize nitrate as a nitrogen source due to the presence of nitrate reductase. *Aspergillus niger* can use nitrite as a nitrogen source for its accelerated growth. With the exception of ammonium sulphate, all ammonium compounds serve as substrate for *Saccharomyces cerevisiae*, *A. niger* and *Alternaria tenui*. *S. cerevisiae* can also utilize urea as nitrogen substrate for growth.

Microorganisms can utilize a number of macromolecules (e.g. DNA, RNA, peptides, chitin, etc.) and highly charged molecules (e.g. nucleotides) as substrates for growth that cannot pass the cell membrane. These substrates are enzymatically hydrolysed (degraded) in the external medium by enzymes such as chitinase, proteinase (peptidase), nucleases (deoxyribonuclease and ribonuclease) secreted by bacteria, namely *Staphylococcus aureus*, *Bacillus megaterium*, *Streptococcus haemolyticus* and *B. subtilis*, respectively. The hydrolytic products of these substrates then enter the cell by specific transport systems.

The solutes are taken up by microorganisms by four distinct processes:

- passive diffusion
- facilitated diffusion
- active transport
- group translocation.

A solute passes through a membrane by passive diffusion as a result of random molecular movement. Diffusion occurs down a concentration gradient; the rate of diffusion is proportional to the difference in solute concentration on the two sides of the membrane (Fick's first law).

Facilitated diffusion displays high stereospecificity because of competition between a solute and its structural analogues. The rate of translocation is independent of the concentration difference across the membrane, but exhibits saturation kinetics, as in the case of an enzyme reaction. Although facilitated diffusion does not require metabolic energy, starved *Escherichia coli* is unable to effect facilitated diffusion.

Active transport resembles facilitated diffusion, with one important difference. Systems promoting active transport consist of a solute-specific carrier which acts catalytically and supplies the metabolic energy, thus allowing net movement of solute against an electrochemical gradient.

Group translocations are mechanisms that chemically alter substrates to impermeable derivatives as they cross the cytoplasmic membrane. These mechanisms are not active transport, because a concentration gradient is not established, but the same metabolic function is accomplished. The concentration of the substrate derivative inside the cell is greater than the concentration of the free substrate outside the cell. Entry of a substrate by group trans-

location constitutes savings of metabolic energy. Although the transport-related reaction requires expenditure of a high-energy phosphate bond, the derivative is formed as it is identical to the product of the first intracellular metabolic reaction. In view of the energy savings associated with group translocation, such mechanisms are encountered in strict and facultative anaerobes.

The well-known group translocation is the phosphotransferase system (PTS) by which certain carbohydrates are brought into some bacteria. Proteins (enzymes I and II (EI and EII) and heat-stable protein (HPr)) are needed to transport and phosphorylate PTS carbohydrates. These proteins form a chain of carriers that transfer a high-energy phosphate group from phosphoenolpyruvate to the incoming carbohydrate. EI and HPr are the first carriers in all PTS chains. Substrate specificity residues in membrane-bound EII, which recognizes a series of structurally related carbohydrates (**Fig. 3**).

Since PTS transport is energy-conserving, it is predominantly associated with fermentative organisms. The facultative anaerobes possessing PTS include *Escherichia*, *Salmonella*, *Staphylococcus* and *Vibrio*. The strict anaerobes include *Clostridium* and *Fusobacterium*. Similarly, the class of reactions:

$$\text{Purine or pyrimidine} + \xrightarrow[\text{transferase}]{\text{Phosphoribosyl}} \text{Nucleoside monophosphate} + \text{Pyrophosphate}$$

constitutes a group of translocation mechanisms for the uptake of adenine, guanine, hypoxanthine, xanthine and uracil.

Amino acids are transported by shock-sensitive or secondary active transport systems. In *Escherichia coli*, 14 different transport systems are involved in bringing amino acids into the cell. Some of the systems transport a group of amino acids with similar structures. In addition to amino acid transport systems, *E. coli* has a variety of transport systems for cofactors, various ions and metabolic intermediates.

Uptake and accumulation of amino acids in *E. coli* appear to be by active transport. The kinetics of amino acid transport are basically similar to active transport systems, with regard to β-galactosidase transport. Periplasmic binding proteins are involved in the uptake of glutamate, glutamine, cystine, basic amino acids and the branched chain amino acids (leucine, isoleucine and valine). Most amino acid transport systems are constitutive, although those involving periplasmic binding proteins are repressible. *E. coli* forms an inducible system for both glutamate and tryptophan. Similar inducible systems have been reported in *Pseudomonas aeruginosa* and *B. subtilis*

Group translocation

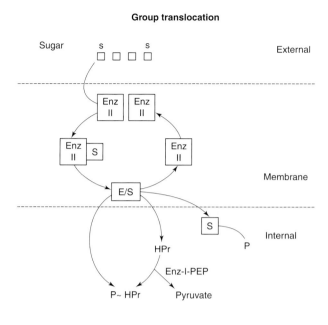

Active transport

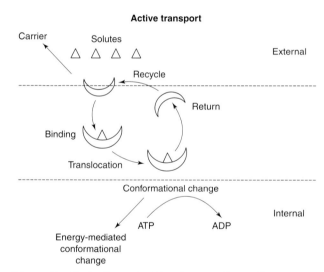

Figure 3 Mechanisms of nutrient transport into cells.

for glutamate and proline as well as arginine, respectively. Transport of aromatic amino acids, viz., phenyl alanine, tyrosine and tryptophan, occurs in *E. coli* and *Salmonella typhimurium* by multiple transport systems.

Peptides are transported by systems, which are quite distinct from those responsible for uptake of amino acids. *E. coli* possesses a transport system for active uptake of dipeptides of L–L configuration. The presence of N-terminal α-amino group and C-terminal carboxyl groups is essential for transport of dipeptides. An oligopeptide for transport must possess a free N-terminal α-amino group, though the presence of C-terminal carboxyl group is not essential. Due to the sieving effect of the cell wall, only small oligo-

peptides are transported, and this is limited by Stokes radius. Stokes law states that at low velocities, the frictional force on a spherical body moving through a fluid at constant velocity is equal to 6π times the product of the velocity, the fluid viscosity and the radius of the sphere.

Pathways Involved in Utilization

Evolution of Degradative Pathways

Heterotrophic microorganisms utilize organic compounds as carbon and energy sources for growth. The glycolytic pathway has often been cited as a primitive catabolic pathway since it was thought that most primitive organisms were anaerobic heterotrophs. Thus, substrate-level phosphorylation which permits anaerobic growth on glucose is more primitive than the highly efficient oxidative phosphorylation of aerobes. The primitive environment was rich in amino acids. Thus, amino acid fermentation is common among anaerobic *Clostridium*, and some species obtain all their energy from the Stickland reaction in which one amino acid is oxidized at the expense of the reduction of another amino acid. Even in the aerobic Pseudomonads the arginine deaminase pathway enables ATP to be synthesized by substrate-level phosphorylation and thus *P. aeruginosa* can live anaerobically in the absence of nitrate. The arginine deaminase pathway can fulfil the major functions of catabolic pathway.

Amino Acids

The amino acids liberated by the action of the proteinases and peptidases on proteins may undergo a variety of metabolic transformations. Certain metabolic pathways for amino acids involve an initial removal of the α-amino group of the molecule. This may be accomplished by first, the process of deamination, which involves the initial separation of the nitrogen from the carbon chain of one amino acid and the utilization of the ammonia so formed for the synthesis of other amino acids, and second, transamination, in which free ammonia is not formed, but the nitrogen is transferred directly.

Oxidative Deamination

In oxidative deamination, amino groups are removed from amino acids, resulting in the formation of corresponding keto acids and ammonia. The reaction catalysed by a flavo protein may be represented as follows:

$$\underset{\text{L-Alanine}}{\overset{\displaystyle CH_3}{\underset{\displaystyle COOH}{\overset{|}{\underset{|}{H-C-NH_2}}}}} + \tfrac{1}{2}\,O_2 \longrightarrow \underset{\text{Pyruvic acid}}{\overset{\displaystyle CH_3}{\underset{\displaystyle COOH}{\overset{|}{\underset{|}{C=O}}}}} + NH_3$$

The Enterobacteriaceae possess active deaminases for oxidative deamination of the dicarboxylic acids such as aspartic acid and glutamic acid, forming the corresponding α-keto acids. Thus, the oxidative deamination of glutamic acid, which yields α-keto glutaric acid, involves the participation of L-glutamic acid dehydrogenase. This reversible conversion of L-glutamic acid to α-keto glutaric acid, an intermediate of the citric acid cycle, serves as a link between the metabolism of this amino acid and that of the carbohydrate. Some organisms metabolize aspartic acid to acetic acid, ammonia and CO_2.

$$\underset{\text{Aspartic acid}}{\overset{\displaystyle COOH}{\underset{\displaystyle COOH}{\overset{|}{\underset{|}{\overset{\displaystyle CH_2}{\overset{|}{CH-NH_2}}}}}}} + O_2 \longrightarrow \underset{\text{Acetic acid}}{CH_3COOH} + NH_3 + 2CO_2$$

Reductive Deamination

Reductive deamination results in the formation of saturated acids, generally produced by strict or facultative anaerobes grown under anaerobic conditions.

Reductive deamination takes place anaerobically by the action of *Clostridium*. Thus, glycine is reduced to acetic acid.

$$\underset{\text{Glycine}}{\overset{\displaystyle CH_2-NH_2}{\underset{\displaystyle COOH}{\overset{|}{\underset{|}{}}}}} \xrightarrow{\text{Clostridia}} \underset{\text{Acetic acid}}{\overset{\displaystyle CH_3}{\underset{\displaystyle COOH}{\overset{|}{\underset{|}{}}}}} + NH_3$$

Both alanine and serine are deaminated to propionic acid by *C. putrefium*. *E. coli* deaminates reductively tryptophan and histidine to β-indole propionic acid and β-imidazone propionic acid, respectively. Both *E. coli* and *Proteus vulgaris* reduce aspartic acid to succinic acid.

Alanine can be hydrolytically deaminated to form lactic acid.

$$\underset{\text{L-Alanine}}{\overset{\displaystyle CH_3}{\underset{\displaystyle COOH}{\overset{|}{\underset{|}{H-C-NH_2}}}}} + H_2O \longrightarrow \underset{\text{Lactic acid}}{\overset{\displaystyle CH_3}{\underset{\displaystyle COOH}{\overset{|}{\underset{|}{H-C-OH}}}}} + NH_3$$

Hydrolytic deamination, coupled with decarboxylation, yields a primary alcohol with one less carbon atom.

$$\underset{\text{L-Alanine}}{\overset{\displaystyle CH_3}{\underset{\displaystyle COOH}{\overset{|}{\underset{|}{H-C-NH_2}}}}} + H_2O \longrightarrow \underset{\text{Ethyl alcohol}}{\overset{\displaystyle CH_3}{\underset{}{\overset{|}{CH_2-OH}}}} + CO_2 + NH_3$$

Decarboxylation

Bacteria possess enzymes that attack amino acids at the carboxyl groups and catalyse the decarboxylation of amino acids, to yield CO_2 and an amine with one less carbon atom. Thus,

$$\underset{\text{L-Alanine}}{\overset{\displaystyle CH_3}{\underset{\displaystyle COOH}{\overset{|}{\underset{|}{H-C-NH_2}}}}} \longrightarrow \underset{\text{Ethylamine}}{\overset{\displaystyle CH_3}{\underset{\displaystyle CH_2-NH_2}{\overset{|}{\underset{|}{}}}}} + CO_2$$

The decarboxylases are highly specific and pyridoxal phosphate is required as the coenzyme for these reactions. Decarboxylases for lysine, arginine, histidine, ornithine, phenylalanine, tyrosine, glutamic acid and aspartic acid have been isolated and purified from several bacteria. Some of the amines such as histamine and tyramine have pharmacological activity in animals.

Transamination Reactions

The chief mechanism by which amino acids are formed is by transamination reactions. In these reactions, the amino group of an α-amino acid is transferred to the α-position of an α-keto acid under the influence of the transaminase, of which pyridoxal phosphate is the cofactor. The amino group from either aspartic or glutamic acid is transferred to the α-position of an α-keto acid.

$$\underset{\substack{\text{Oxaloacetic}\\\text{acid}}}{\overset{\displaystyle COOH}{\underset{\displaystyle COOH}{\overset{|}{\underset{|}{\overset{\displaystyle CH_2}{\overset{|}{C=O}}}}}}} + \underset{\substack{\text{Glutamic}\\\text{acid}}}{\overset{\displaystyle COOH}{\underset{\displaystyle COOH}{\overset{|}{\underset{|}{\overset{\displaystyle CH_2}{\overset{|}{\overset{\displaystyle CH_2}{\overset{|}{CH-NH_2}}}}}}}}} \longrightarrow \underset{\substack{\text{Aspartic}\\\text{acid}}}{\overset{\displaystyle COOH}{\underset{\displaystyle COOH}{\overset{|}{\underset{|}{\overset{\displaystyle CH_2}{\overset{|}{CH-NH_2}}}}}}} + \underset{\substack{\text{α-Keto-glutaric}\\\text{acid}}}{\overset{\displaystyle COOH}{\underset{\displaystyle COOH}{\overset{|}{\underset{|}{\overset{\displaystyle CH_2}{\overset{|}{\overset{\displaystyle CH_2}{\overset{|}{C=O}}}}}}}}}$$

Other amino acids which participate in such reactions involving the transfer of amino group to and from glutamic acid are alanine, valine, leucine, tyrosine, phenylalanine, tryptophan and methionine.

In mutase reaction, one molecule of an amino acid is oxidized while another is reduced:

$$3CH_3-CH-COOH \quad \xrightarrow{+2H} \quad 2CH_3$$
$$\underset{\text{NH}_2}{|} \qquad\qquad \underset{\text{CH}_2-COOH}{|}$$

L-Alanine   Propionic acid   +

$$\begin{array}{l} CH_3 \\ | \\ COOH \end{array} + CO_2 + 3NH_3$$

Acetic acid

This reaction is employed by some strictly anaerobic microorganisms such as *C. sporogenes* and *C. botulinum*.

The oxidation of aromatic compounds such as tyrosine by anaerobic species appears to be restricted to their aliphatic side chain because of the need for direct participation of oxygen in hydroxylation and rupture of benzene ring. Aliphatic compounds are oxidized to fatty acids. Thus, *C. sporogenes* converts proline to δ-amino valeric acid and tryptophan to indole propionic acid. Similarly, purine fermenting *Clostridium* converts uric acid to xanthine. *C. propionicum* converts serine to propionate via pyruvate. *C. tetanomorphum* converts glutamate to butyrate via pyruvate and acetyl coenzyme A.

Deamination may also lead to the formation of the unsaturated acids. Thus, histidine is converted to urocanic acid by *E. coli*.

$$CH=CH-CH_2-CH-COOH$$
$$\underset{HN}{|} \quad \underset{N}{|} \qquad \underset{NH_2}{|}$$
$$\diagdown\; C\; \diagup$$
$$|$$
$$H$$

Histidine

$$\xrightarrow[\text{Histidase}]{E.\ coli}$$

$$CH=CH-CH=CH-COOH$$
$$\underset{HN}{|} \quad \underset{N}{|} \qquad\qquad + NH_3$$
$$\diagdown\; C\; \diagup$$
$$|$$
$$H$$

Urocanic acid

Similarly, aspartic acid is converted to fumaric acid by desaturative deamination in *Pseudomonas fluorescens* and *E. coli* by aspartase found in these bacteria.

$$\begin{array}{l} COOH \\ | \\ CH-NH_2 \\ | \\ CH_2 \\ | \\ COOH \end{array} \xrightarrow[\text{Aspartase}]{P.\ fluorescens} \begin{array}{l} CH-COOH \\ \| \\ CH-COOH \end{array} + NH_3$$

Aspartic acid   Fumaric acid

Deamination of hydroxy amino acids results in the removal of H_2O, forming an unsaturated acid. On tautomerization to imino acid, it is further hydrolysed to keto acid. Thus, L-serine yields pyruvate.

$$\begin{array}{l} CH_2-OH \\ | \\ CH-NH_2 \\ | \\ COOH \end{array} \xrightarrow[-H_2O]{\text{Coliform bacteria}} \begin{array}{l} CH_2 \\ \| \\ C-NH_2 \\ | \\ COOH \end{array} +$$

L-Serine

$$\begin{array}{l} CH_3 \\ | \\ C=NH \\ | \\ COOH \end{array} \xrightarrow{+\ H_2O} \begin{array}{l} CH_3 \\ | \\ C=O \\ | \\ COOH \end{array} +NH_3$$

Pyruvate

Threonine yields α-keto butyric acid. This reaction is carried out by coliform bacteria containing the enzyme serine deaminase. A parallel reaction is the removal of H_2S from cysteine. *Proteus vulgaris* liberates H_2S from both cysteine and homocysteine.

$$\begin{array}{l} CH_2-SH \\ | \\ CH-NH_2 \\ | \\ COOH \end{array} \xrightarrow[P.\ vulgaris]{H_2S} \begin{array}{l} CH_2 \\ \| \\ C-NH_2 \\ | \\ COOH \end{array} \longrightarrow$$

Cysteine

$$\begin{array}{l} CH_2 \\ | \\ C=NH \\ | \\ COOH \end{array} \xrightarrow{H_2O} \begin{array}{l} CH_3 \\ | \\ C=O \\ | \\ COOH \end{array} +NH_3$$

Pyruvic acid

Arginine metabolism by bacteria leading to the formation of ornithine and citrulline is depicted in **Figure 4**.

Nitrification

Although molecular nitrogen is the major constituent of the earth's atmosphere, the nitrogen molecule is chemically inert and is not a suitable source for most

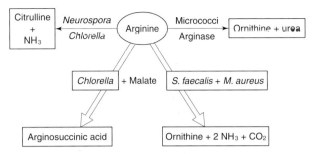

Figure 4 Arginine metabolism by several bacteria. *S. faecalis*, *Streptococcus faecalis*; *M. aureus*; *Staphylococcus aureus*.

living forms. Combined nitrogen in the form of ammonia, nitrate and organic compounds is scarce in soil and water. For this reason, cyclic transformation of nitrogenous compounds is important in the biological sphere.

The conversion of ammonia to nitrate (nitrification) is brought about by obligate aerobic chemolithotrophic bacteria. Nitrification occurs in two steps: first, ammonia is oxidized to nitrite by the *Nitrosomonas* group, and second, nitrite is oxidized to nitrate by *Nitrobacter*.

Denitrification

Many types of bacteria use nitrates in place of oxygen as a final hydrogen acceptor. Whenever organic matter is decomposed in the soil and conditions become anaerobic, the nitrate present tends to become reduced. In the process of denitrification molecular nitrogen is the principal end product. Nitrogen-fixing organisms compensate for the loss of combined nitrogen due to denitrification. Two principal types of nitrogen fixation are encountered in nature: symbiotic and non-symbiotic nitrogen fixation.

Sites of Activity

The active site of an enzyme is the region that binds the substrates (and the prosthetic group) and contains the residues that directly participate in the making and breaking of bonds. These residues are called catalytic groups. Although enzymes differ widely in structure, specificity and mode of catalysis, a number of generalizations can be made regarding their active sites:

1. The active site takes up a relatively small part of the total volume of an enzyme.
2. The active site is a three-dimensional entity formed by groups of different parts of the linear amino acid sequence – residues. Amino acid residues situated far apart in the linear sequence may interact more strongly than adjacent residues in the amino acid sequence. This is true for haemoglobin and myoglobin molecules. In lysozyme also the important groups in the active sites have been shown to

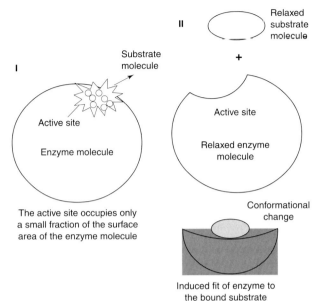

Figure 5 Induced-fit model of enzyme–substrate complex formation.

be formed of residues numbered 35, 52, 62, 63 and 101 in the linear sequence of 129 amino acids.

3. Substrates are bound to enzymes in multiple weak attractions.
4. Active sites are clefts or crevices. In all enzymes of known structure, substrate molecules are bound to the cleft or crevice. Water is normally excluded. The non-polar character of the cleft enhances the binding of substrate.
5. The specificity of binding depends on the arrangement of atoms in an active site. To fit into the site a substrate must have a matching shape. The lock-and-key mechanism is appropriate. However, it has been proposed that the active sites of many enzymes are modified by the binding of substrate. After the substrate is bound, the active sites of enzymes assume suitable shapes complementary to those of the substrates. This process of dynamic recognition is known as induced fit (**Fig. 5**). Prokaryotic cells, including bacteria, are very small and simple, with only a single membrane. These cells contain no compartment separated by internal membranes, yet there is some degree of segregation of certain enzyme systems in bacteria. Most enzymes participating in the biosynthesis of proteins are located in the ribosomes and some enzymes participating in the biosynthesis of phospholipids are located in the bacterial cell membrane.

Synthesis of Cellular Components

A few bacteria synthesize their amino acids via reductive amination and transamination but do not obtain their nitrogen in the form of ammonium ions. Instead, they carry out nitrogen fixation, reducing nitrogen gas from the atmosphere. The reaction is catalysed by the enzyme nitrogenase.

N_2 + 3 reduced cofactor + $3H^+$ + ~ 15 ATP $2NH_3$ + 3 oxidized cofactor + 15 ADP + 15 P_i

The nitrogenase enzyme is very sensitive to the presence of oxygen. Therefore, nitrogen-fixing organisms have evolved mechanisms to protect the enzyme from atmospheric oxygen. Species of the genus *Rhizobium* have a symbiotic relationship with certain plants to eliminate oxygen. Other organisms such as *Clostridium* only grow anaerobically. Organisms like *Azotobacter* that only grow aerobically seem to have enzymes located in the cell membrane while the nitrogenase resides in the cytoplasm. The cell can protect its nitrogen-fixing ability by maintaining a rate of aerobic electron transport that is high enough to reduce all oxygen in the air into water molecules before it can diffuse through the cell.

Synthesis of Amino Acids

Although a number of different pathways are involved in synthesizing 20 amino acids, certain amino acids such as glutamic acid, aspartic acid and alanine are derived from precursor metabolites of the glycolytic pathway and tricarboxylic acid (TCA) cycle.

The 20 amino acids can be grouped into families. Thus, aromatic acids mainly consist of three amino acids, tyrosine, phenylalanine and tryptophan, all of which have a benzene ring as part of their structure.

In the first step of synthesis of these aromatic amino acids, the 4-carbon compounds produced by the pentose phosphate pathway combine with a 3-carbon compound, which comes from the glycolytic pathway to form a 7-carbon compound. The 7-carbon compound is then metabolized, resulting in the formation of three aromatic amino acids. A number of enzymes are common in the formation of all three amino acids. If one of these enzymes is defective, none of the amino acids can be synthesized. Energy must be supplied at several steps in the pathway. The high-energy compounds are formed in the glycolytic pathway and TCA cycle. If the energy sources of the cell are limited, the cell will metabolize the energy-rich compound, phosphoenolpyruvic acid, through the glycolytic pathway and generate ATP rather than use it for biosynthesis. In amino acid biosynthesis, nitrogen is introduced to form the amino (-NH₂) group. A number of enzymatic reactions catalyse the incorp-

oration of ammonia into an organic molecule. A key reaction links energy metabolism to biosynthesis. In this reaction, α-ketoglutaric acid reacts with NH_4^+ and is converted to the amino acid, glutamic acid.

α-Ketoglutaric acid

Since the reaction is reversible, glutamic acid can be readily converted to α-ketoglutaric acid, which enters the TCA cycle, producing CO_2 and H_2O as well as energy.

Amino acids are synthesized using metabolic intermediates and ammonia (NH_3). The synthesis of glutamic acid and aspartic acid is a useful example. The enzyme glutamic acid dehydrogenase takes ammonia and incorporates it into α-ketoglutarate, which is an organic intermediate of the TCA cycle. The energy is expended in this reaction through utilization of NAD red. Once glutamic acid is formed, it can be used to form other amino acids. The formation of aspartic acid employs another TCA cycle intermediate. Aspartic acid is formed by a transaminase that catalyses the transfer of the amino group (-NH₂) from glutamic acid to oxaloacetic acid.

Oxaloacetic acid   Glutamic acid   Aspartic acid   α-Ketoglutaric acid

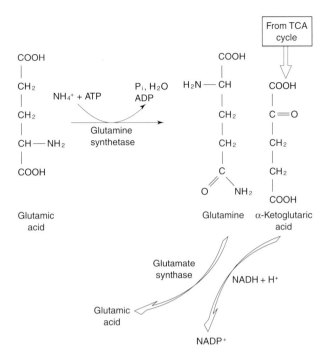

Figure 6 Conversion of ammonium (NH_4^+) ion to organic nitrogen. TCA, tricarboxylic acid.

Inorganic nitrogen in the form of ammonia (NH_4^+) is used to form the amino acid glutamic acid or glutamine. These amino acids are then used to donate an amino group to enzymatic reactions that form other amino acids. No energy is required for transamination because there is a simple exchange of functional groups (**Fig. 6**).

Two families of amino acids cannot be derived by transamination of intermediates from energy metabolism. These are histidine and aromatic amino acids (phenylalanine, tyrosine and tryptophan). The aromatic amino acids have a ring structure with carbon–carbon double bonds and one derived from the oxidative pentose phosphate cycle, which begins with erythrose-4-phosphate and phosphoenolpyruvate from the Embden–Meyerhof pathway. The histidine pathway begins with ribose-5-phosphate, which can be derived from glucose in the hexose monophosphate pathway. The biosynthesis of the aspartate family (lysine, methionine, threonine and isoleucine) is shown in **Figure 7**.

Biosynthesis of Purines and Pyrimidines

Purine and pyrimidine nucleotides are the building blocks of nucleic acids. They are also contained in several coenzymes and function in the activation and transfer of amino groups, sugars, cell wall components and lipids. The synthesis of purine nucleotides follows a common path which only diverges at the level of inosine monophosphate into adenylate and guanylate. Pyrimidine nucleotides are also syn-

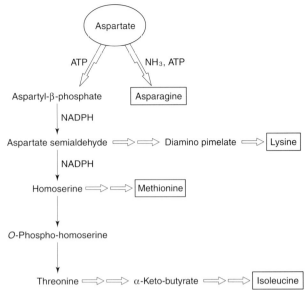

Figure 7 Biosynthesis of the aspartate family.

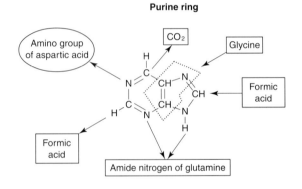

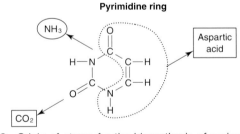

Figure 8 Origin of atoms for the biosynthesis of purine and pyrimidine molecules.

thesized via a single pathway which branches out from uridylic acid.

The nitrogenous bases, adenine, cytosine, guanine, thymine and uracil, comprising the nucleic acids are derived from certain amino acids and their precursors (**Fig. 8**). When the appropriate ribosyl sugar is added they are called nucleosides (adenosine, cytidine, guanosine, thymidine and uridine). As successive phosphates are added the molecules are designated as mono-, di- or trinucleotides. The purine pathway produces inosinic acid first, and then adenosine and

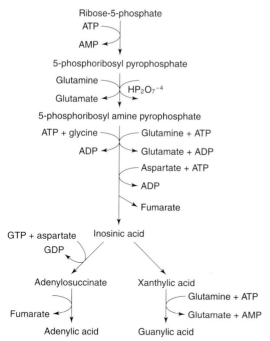

Figure 9 Pathways of formation of adenylic and guanylic acids.

from ribose-5-phosphate. This can be formed in two ways: oxidatively from glucose-6-phosphate via the pentose phosphate pathway, and non-oxidatively from fructose-6-phosphate and glyceraldehyde-3-phosphate via the transaldolase–transketolase reaction. Ribose-5-phosphate is used in the energy-rich form of phosphoribosyl diphosphate for the synthesis of purine and pyrimidine nucleotides. The reduction of ribose to deoxyribose occurs at the level of the ribonucleotide.

The sugars for the nucleosides are derived from the same phosphoribosylpyrophosphate (PRPP) used in the synthesis of histidine and the aromatic amino acids. They are added at an early stage in the biosynthetic pathway so that the final products are mononucleotides. Nucleotides of higher order are produced by the addition of phosphate groups sequentially, and ATP is used as a donor. ATP is then regenerated by any of the energy-transducing pathways. A summary of the interrelationship between anabolic and catabolic reactions in the cell is shown in **Figure 10**.

Nitrogen Metabolism in Fungi

Three inorganic forms of nitrogen, ammonium salts, nitrites and nitrates, are utilized by fungi. Nitrates are excellent sources of nitrogen for many fungi, e.g. actinomycetes. Preferential utilization of ammonia from ammonium nitrate is influenced by pH in

guanosine are derived from it. The pyrimidine pathway produces uridine first and thymidine and cytosine triphosphate are derived from uridine (**Fig. 9**).

The pentose moiety of the nucleotide is derived

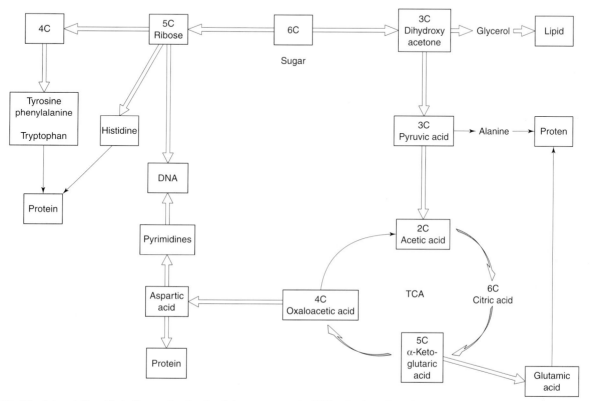

Figure 10 Interrelationship in the synthesis of cellular components. TCA, tricarboxylic acid.

Aspergillus niger. Nitrate reduction results in the incorporation of nitrate nitrogen into the cell. Nitrites serve as sole sources of nitrogen for *Fusarium niveum.* However, *Aspergillus* sp. cannot utilize nitrite due to its toxicity. Cyanide, though a respiratory poison, acting on iron enzyme, has been reported to provide assimilable nitrogen to preformed mycelium of *A. niger* and to cause an apparent increase in growth of *Fusarium lini.*

The fungi and Actinomycetes as a group are active in the decomposition of protein in soil and other materials. In pure culture, proteins such as gelatin, casein and egg albumin can serve as sources of nitrogen for the common saprophytic fungi and Actinomycetes. Dipeptides and tripeptides also support the growth of *Aspergillus* sp. *Streptomyces* sp. and *Rhizopus* sp. grow on and decompose casein, zein, gliadin, fibrin, gelatin and egg albumin.

The D- and L-amino acid oxidases of *Neurospora crassa* act on a number of amino acids, producing the corresponding keto acids with liberation of ammonia. Glutamic dehydrogenase found in *N. crassa* mycelium requires pyridine nucleotides for its activity. Glutamic dehydrogenase is the key enzyme for the synthesis of amino acids and to the entrance of ammonia into organic combination. Non-oxidative deamination of L-serine and L-threonine is also effected by serine dehydratase of *N. crassa.* The final products produced in this reaction are pyruvate and α-ketoglutarate respectively. Aspartase, which converts aspartate non-oxidatively to fumaric acid and ammonia, is also found in *N. crassa* and *Penicillium notatum.* Threonine is cleaved to glycine and acetaldehyde non-oxidatively by *N. crassa.* Arginine is broken down in the urea cycle by *Streptomyces griseus* containing an inducible enzyme which converts arginine to guanidinobutyramide. Glutamic acid is the principal donor of amino groups in transamination. In this process it is converted to α-ketoglutaric acid.

Fungi, like bacteria, attack nucleic acids first by nucleodepolymerase action, yielding mononucleotides. The second step in the pathway of nucleic acid degradation is the dephosphorylation of nucleotides yielding inorganic phosphate. Deamination of nucleotides and nucleosides is usually effected by adenyl deaminase of *A. oryzae.* Thus, there is great similarity in the metabolism of nitrogen by fungi and bacteria.

See also: ***Aspergillus***: Introduction. ***Clostridium***: *Clostridium perfringens.* ***Pseudomonas***: *Pseudomonas aeruginosa.* ***Staphylococcus***: *Staphylococcus aureus.*

Further Reading

Campbell NA (1993) *Biology,* 3rd edn. New York: Benjamin/Cummings.

Cochrane VW (1958) *Physiology of Fungi.* New York: John Wiley/Chapman & Hall.

Greiberg DM (ed.) (1969) *Metabolic Pathway, Nitrogen Metabolism of Amino Acids.* New York: Academic Press.

Joklik WK, Willett HP, Amos DB and Wilfert CM (1988) *Zinsser Microbiology,* 19th edn. Prentice-Hall International.

Lyles ST (1969) *Biology of Microorganisms.* St Louis, MO: CV Mosby.

Neidhardt FC, Ingraham JL and Schaechter M (1990) *Physiology of the Bacterial Cell: A Molecular Approach.* Sunderland, MA: Sinauer.

Nester EW, Roberts CE and Nester MT (1995) *Microbiology: A Human Perspective.* Wim C Brown.

Purves GH, Orians GH and Craig GH (1995) *Life. The Science of Biology.* USA: Sinauer/W.H. Freeman.

Rose AHC (1976) *Chemical Microbiology: An Introduction to Microbial Physiology,* III edn. London: Butterworths.

Stainer RY, Ingraham JL, Wheelis ML and Painter PR (1993) *General Microbiology,* 5th edn. London: Macmillan Press.

Virge EA (1992) *Modern Microbiology: Principles and Applications.* Wim C Brown.

Weil JH (1996) *General Biochemistry,* 6th edn. Mumbai, India: New Age International.

Lipid Metabolism

Rajat Sandhir, Department of Biochemistry, Dr Ram Manohar Lohia Avadh University, Faizabad, India

Lipids are a heterogeneous group of compounds that are soluble in non-polar solvents such as ether, chloroform and benzene, and are completely immiscible in water. They are a major group of organic compounds found in living matter. Lipids are essential to the structure and function of membranes, that separate living cells from their environment. Lipids also function as energy reserves, which can be mobilized as sources of carbon, water and energy. Microorganisms like all living cells, contain lipids. In yeasts and fungi they are important in the assembly of membranes of various intracellular organelles. As in plants and animals, there is a wide diversity of lipid types in microorganisms, and can be broadly divided as having fatty acids (non-terpenoid) or terpenoid lipids. The former include triacylglycerols and phospholipids and the latter include the sterols and carotenoids. Some lipid types are confined to individual groups of microorganisms, whereas, others have almost ubiquitous distribution. Lipids may also be combined in nature with other compounds such as proteins, amino acids and polysaccharides. The variety of types of lipids and even of individual fatty acids, is so great that it is possible to use lipids as an important aid in the taxonomy of bacteria. With the advent of genetic engineering we can envisage increasing use being made of unicellular algae, bacteria and yeasts as providers of economically important lipid molecules in areas such as food and the pharmaceutical industry. This article covers the major lipids and their metabolism in bacteria, yeasts and fungi.

Lipids in Fungi and Bacteria

Lipids can serve as sources of energy and carbon for bacteria and fungi. Most of the microorganisms are capable of synthesizing lipids and do not require exogenous supplementation. The lipid content of bacteria is in the range of 2–5% of dry weight, for yeast 7–15% dry weight, and for fungi 6–9%. However, in certain strains of microorganisms lipid content of 40–50% of dry weight has been reported and in some fungi lipid content as high as 90% has been reported. Such strains may be used for commercial production of fats and are referred to as 'oleaginous organisms'.

Major Lipid Classes

Non-terpenoid Lipids

This class of lipids includes simple and complex lipids essentially having fatty acids.

Fatty Acids Fatty acids are the main constituents of the non-terpenoid class of lipids. Over 500 different types of fatty acids have been reported in microorganisms, but only few are important. Fatty acids do not occur in a free form but are normally found as esters, usually esterified with glycerol, alcohol or sterol. In addition, fatty acids are constituents of more complex structures like lipopolysaccharides. Fatty acids have interesting chemical properties because of highly hydrophobic (water repelling) and highly hydrophilic (water soluble) regions. Palmitate for example is a 16-carbon fatty acid composed of a chain of 15 saturated (fully hydrogenated) carbon atoms and a single carboxyl group (hydrophilic region). The fatty acid composition in microorganisms is quite variable and depends on the substrate and also on the growth conditions. The fatty acids of bacteria are generally 10–20 carbons in length. Odd-chain fatty acids have also been reported in trace quantities.

There are mainly four types of fatty acids: straight chain saturated; straight chain monounsaturated, branched chain and cyclopropane fatty acids (**Table 1**). The common fatty acids are stearic acid (C_{16} saturated) and oleic acid (C_{18} monounsaturated). The monounsaturated fatty acids are of mainly two types depending on the route of synthesis: *cis*-vaccenic acid type, which is the most common, or the oleic acid type. The double bond present in fatty acids can give rise to two stereoisomers: *cis* and *trans*. In most of the physiologically occurring fatty acids the double bond is in *cis* position. Another type of fatty acid, cyclopropane fatty acids, such as lactobacillic acid, have a three-member cyclic structure and are present in *Lactobacillus* spp. In certain Gram-negative bacteria, such as *Escherichia coli*, hydroxy fatty acids like β-hydroxylauric acid and β-hydroxymyristic acids are present. In *Saccharomyces cerevisiae*, very long-chain hydroxy fatty acids, such as 2-hydroxyhexacosanoic acid, are present as important components of sphingolipids. In mushrooms hydroxy fatty acids may represent up to a quarter of total fatty acids. In addition some yeasts or yeast-like fungi produce extracellular hydroxy fatty acids.

With few exceptions bacteria do not contain polyunsaturated fatty acids. The longer chain polyunsaturated fatty acids are usually absent. Yeasts, however, produce polyunsaturated fatty acids, though *S. cerevisiae* is unable to produce either *cis*,*cis*-linoleic acid [18:2, (9,12)] and α-linolenic, acid [18:3 (9,12,15)], because of the absence of the enzyme Δ^{12} desaturase, although these are present in other yeasts. However, in some fungi, arachidonic acid, a 20-carbon polyunsaturated fatty acid, has been found as a minor component.

Other fatty acids that are considered unique to bacteria include branched-chain fatty acids, where a methyl group is located at the ω-1 or ω-2 position, and tuberculostearic acid, 10-methylstearic acid (br-19:10 Me), where the methyl group is in the middle of the molecule. Some bacteria of the *Corynebacterium* and *Mycobacterium* group synthesize multi-branched very-long-chain fatty acids, called mycolic acids, that are up to C_{90} in length (**Fig. 1**). Mycolic acids have a 2-branch, 3-hydroxy long chain of 20–90 carbon atoms. These are associated with the construction of the lipophilic cell envelope of these bacteria, which include tubercle and leprosy bacilli (*Mycobacterium tuberculosis* and *M. leprae*). In *S. cerevisiae*, branched-chain fatty acids are unusual.

Simple Lipids Simple lipids (fats) consist of fatty acids esterified to a C_3 alcohol, glycerol (**Fig. 2**). Depending on the number of fatty acyl residues, they can be monoacylglycerol, diacylglycerol or triacylglycerol (triglycerides). The predominant type of simple lipid is triacylglycerol. The three carbons of glycerol are stereochemically different and are named

Table 1 Structures and nomenclature of the fatty acids in *Escherichia coli*

Type	Structure	Systemic name	Trivial name
Saturated	$CH_3\text{-}(CH_2)_n COOH$	Dodecanoic acid ($n = 10$)	Lauric acid
		Tetradecanoic acid ($n = 12$)	Myristic acid
		Hexadecanoic acid ($n = 14$)	Palmitic acid
		Octadecanoic acid ($n = 16$)	Stearic acid
Unsaturated	$CH_3\text{-}(CH_2)_5 CH=CH\text{-}(CH_2)_n COOH$	*cis*-9-hexadecanoic acid ($n = 7$)	Palmitoleic acid
		cis-11-octadecanoic acid ($n = 9$)	*cis*-Vaccenic acid
Cyclopropane	$CH_3\text{-}(CH_2)_5\text{-}C \overset{\overset{\displaystyle CH_2}{\diagup \ \ \diagdown}}{\underset{H}{\mid}} C\underset{H}{\overset{\mid}{}}\text{-}(CH_2)_n COOH$	*cis*-9,10 methylene hexadecanoic acid ($n = 7$)	None
		cis-11,12 methylene octadecanoic acid ($n = 9$)	Lactobacillic acid
Hydroxy	$CH_3\text{-}(CH_2)_n\text{-}CH(OH)\text{-}CH_2\text{-}COOH$	D (−)-3-hydroxy tetradecanoic acid ($n = 10$)	β-Hydroxy myristic acid

$$CH_3(CH_2)_n CO\!-\!O\!-\!CH_2(CH_2)_n CH_3$$

Figure 1 Structure of mycolic acid found in mycobacteria.

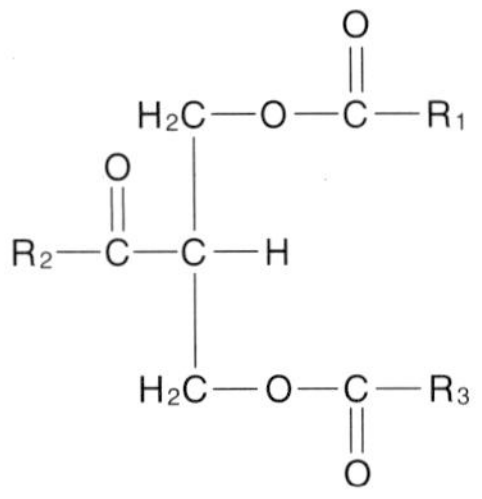

Figure 2 Structural formula of triacylglycerol (simple lipid). Fatty acids are linked to glycerol by ester linkage. R_1, R_2 and R_3 represent different acyl groups of fatty acids. The three acyl groups in a fatty acid may be similar or different.

as *sn*-1, *sn*-2 and *sn*-3. The three fatty acids that are esterified with the three carbon atoms can be the same or different (mixed triacylglycerol). Bacteria usually do not have triacylglycerides as a major storage lipid. However, triacylglycerol is the major storage lipid in yeasts and fungi and may constitute up to 90% of the total lipids. Such microorganisms are referred to as oleaginous. In oleaginous microorganisms, the enzyme ATP:citrate lyase plays a key role as the principal provider of acetyl-CoA. Commercial interest in the use of yeasts as potential sources of edible oils, known as single cell oils, have chiefly centred on the ability of some species, such as *Candida curvata*, to produce facsimile of cocoa butter in which the Δ^9 desaturase gene which converts stearic acid ($18:0$) into oleic acid ($18:1$) has been deleted by genetic manipulation. Such cells contain up to 50% stearic acid in an overall content of 35–40%. Diacylglycerol serves as an intermediate in the biosynthesis of phospholipids.

Waxes Waxes are esters of fatty acids with monohy-

dric fatty alcohols (**Fig. 3**). Waxes have been found in few genera of bacteria: *Acinetobacter* sp., *Micrococcus cryophilus* and *Clostridium*, where they may constitute 13% of the total lipid. Waxes are also important components of the cell wall of acid fast bacteria.

Polyhydroxyalkaonates Many bacteria accumulate polyhydroxyalkonates (PHA) as a major source of carbon and energy. These are peculiar compounds as the molecular formula is similar to that of carbohydrates, but the solubility characteristics are that of lipids. The structures of the various types of PHA are shown in **Figure 4**. The number of repeating units in the polymer can be as high as 20 000 with a molecular weight of 2×10^6 Da. PHA may constitute 40–60% of the dry weight in bacteria and have been found in both Gram-positive and Gram-negative organisms as well as actinomycetes. In some species such as

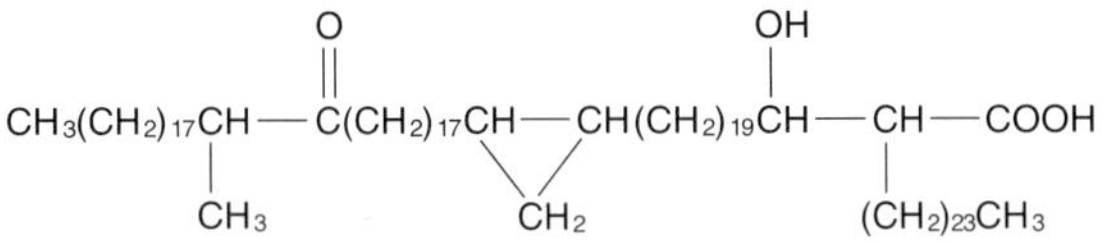

Figure 3 General structure of wax. Wax is an ester of fatty acid with monohydric alcohol.

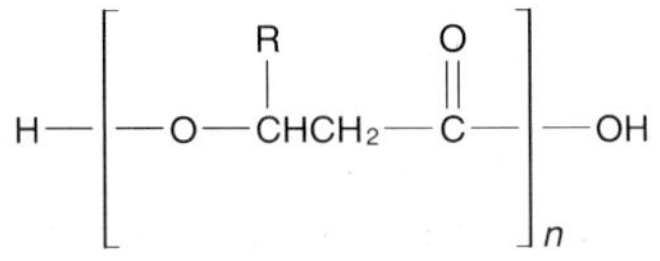

R=	– CH_3	Poly β-hydroxybutyrate
	– CH_2CH_3	Poly- β-hydroxyvalerate
	– $(CH_2)_7$ – CH_3	Poly β-hydroxyoctanoate

Figure 4 Structure of polyhydroxyalkaonates from bacteria. The different types of polyhydroxyalkaonates have different R groups. The value of n for different PHA varies between 10 000 and 20 000 (molecular weight 2×10^6 Da).

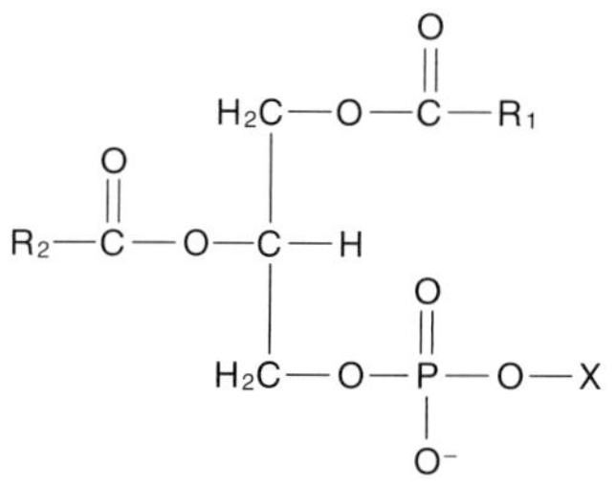

$X =$
- $-H$ Phosphatidic acid (PA)
- $-CH_2CH_2NH_3$ Phosphatidylethanolamine (PE)
- $-CH_2CH_2N^+H_2CH_3$ Phosphatidyl monomethyl ethanolamine (PME)
- $-CH_2CH_2N^+H(CH_3)_2$ Phosphatidyl dimethyl ethanolamine (PDE)
- $-CH_2CH_2N^+(CH_3)_3$ Phosphatidylcholine (PC)
- $-CH_2CH(COOH)NH^+_3$ Phosphatidylserine (PS)

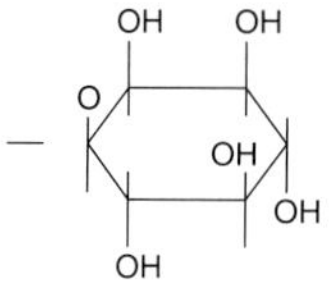

Phosphatidylinositol (PI)

Figure 5 Structural formula of phospholipids (complex lipids). The phospholipids differ from triacylglycerol in having a phosphate group linked with a substituent base (X) at *sn*-3 position instead of fatty acyl group. X is different in the various phospholipid types.

Alcaligenes PHA may constitute up to 85% of the cell biomass. There is commercial interest in the production of PHA as biodegradable plastics.

Complex Lipids Complex lipids are simple lipids which contain additional elements such as phosphate, nitrogen or sulphur, or small hydrophilic carbon compounds such as sugars, ethanolamine, serine or choline.

Phospholipids These are an important class of complex lipids as they have a major structural role in cytoplasmic membranes. In phospholipids, the *sn*-1 and *sn*-2 carbon is esterified to hydrophobic fatty acids and the *sn*-3 position of glycerol is linked to a phosphate residue and a substituted base (**Fig. 5**). The ionizable 'head-group' of the phospholipids has a negative charge from the phosphate. The negative charge is balanced by positive charge on the substituent group (X). The chemical properties of phospholipids make them ideal structural components of membranes because of their dual properties of hydrophobicity and hydrophilicity, lipids aggregate in membranes with hydrophobic portion towards the external or internal (cytoplasmic) environment. Such structures are ideal permeability barriers because of the inability of water-soluble substances to flow through the hydrophobic portion of lipids. The major classes of phospholipids are phosphatidylinositol (PI), phosphatidylethanolamine (PE), phosphatidic acid (PA), phosphatidylserine (PS), phosphatidylglycerol (PG) and phosphatidylcholine (PC, lecithin). Phospholipids are the major lipids of bacterial and fungal membranes. In bacteria PG is widely distributed in

all types except actinomycetes (e.g. mycobacteria). Diphosphatidylglycerol also occurs together with phosphatidylglycerol. Diphosphatidylglycerol (cardiolipin) is a major lipid of bacteria, in contrast to eukaryotic microbes where it is present in inner mitochondrial membrane. PE is generally the major phospholipid in Gram-negative bacteria and is a major component of Gram-positive bacteria such as *Bacillus*. In contrast to higher organisms PC is rarely a major lipid in bacteria, although monomethyl- and dimethylethanolamine-containing lipids are reported. PI is uncommon in bacteria and is confined to few Gram-positive genera only. In certain cases mannosides of phosphatidylinositol may be present, e.g. actinomycetes. The major phospholipids of fungi are phosphatidylcholine, phosphatidylglycerol and phosphatidylethanolamine with small amounts of phosphatidylinositol. In yeast, 3–7% of the total lipids are phospholipids. The order of predominance is PC>PE>PI>PS, whereas PA and PG are minor components.

Plasmalogens Certain anaerobic bacteria contain an ether-linked residue at the *sn*-1 position of glycerol in a regular phospholipid structure. These are known as plasmalogens and have mainly been reported from rumen bacteria (**Fig. 6**).

Sphingolipids Lipids containing sphingosine (**Fig. 7**) or related amino alcohols are of minor importance in microorganisms. In bacteria, sphingolipids are rare except in certain Gram-negative anaerobic bacteria such as *Bacteroides levii*. Most fungi seem to contain small amounts of the usual sphingolipids. In *Amanita*

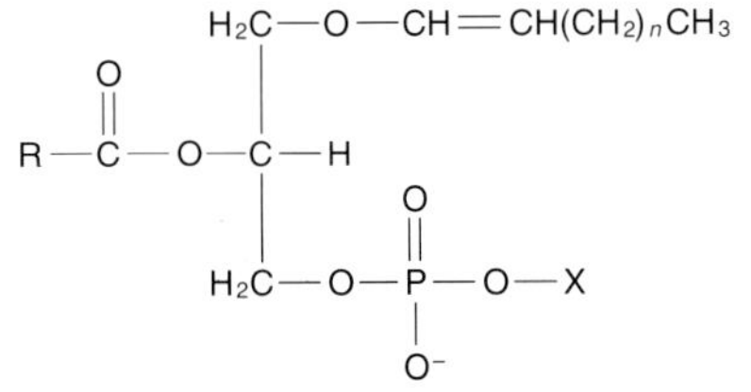

Figure 6 Structure of plasmalogens. Plasmalogens have an ether linkage at the sn-1 position and an ester at sn-2 position. The ether-linked residue may be saturated or unsaturated at the carbon adjacent to the ether link.

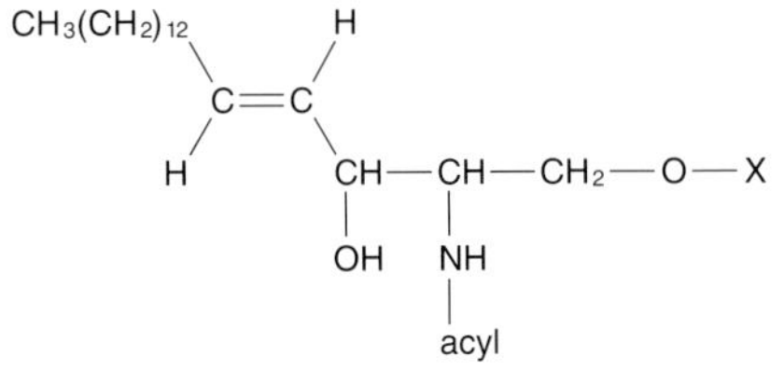

X = H	Ceramide (acylated sphingosine)
X = sugar	Cerebroside
X = phosphocholine	Sphingomyelin
X = phosphoinositol	Ceramide phosphoinositol (present in yeast)

Figure 7 Structure of sphingosine-based lipids. The various substituents in different types of sphingolipids are shown by X.

muscaria, a filamentous fungi, ceramides and cerebrosides represent 1% of the mycelial dry weight. A number of unusual sphingolipids have been detected in different species, including several inositol-containing phosphorylceramides.

Glycolipids Glycolipids are a class of lipids containing carbohydrate residues and are usually the major lipids of bacterial and fungal walls (**Fig. 8**). Phosphatidylinositol and sphingolipids have been excluded from this category. Some glycerides contain a carbohydrate moiety attached to the sn-3 position of the glycerol. Although these glycosylglycerides are present in small quantities in fungal and bacterial species, they are really characteristic of photosynthetic

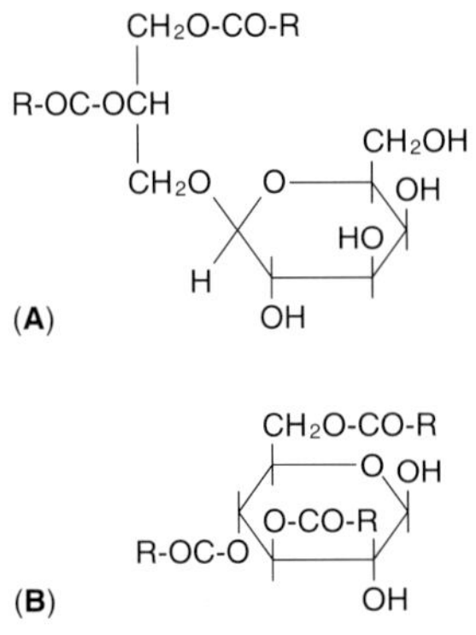

Figure 8 Structure of some of the glycolipids. (**A**) Diacyl-galactosylglycerol; (**B**) acylated glycerol.

membranes of algae and cyanobacteria. The two galactose-containing lipids that are common are diacylgalactosylglycerol and diacyldigalactosylglycerol. These compounds represent 40% of the dry weight of photosynthetic membranes and are consequently the most prevalent membrane lipids in the world. The two galactose molecules are linked by $\alpha(1 \rightarrow 6)$ linkage in diacyldigalactosylglycerol and the two glucose molecules are linked by $\alpha(1 \rightarrow 2)$ or $\beta(1 \rightarrow 6)$ linkage in diacyldiglucosylglycerol. The commonest bacterial glycolipids are diacyldiglucosylglycerols. They are found more frequently in Gram-positive organisms. Another important group of bacterial glycolipids is the phosphoglycolipids found in a number of Gram-positive organisms. There are four types of phosphatidylglycolipids confined to N-group streptococci; the sn-glycero-3-phosphoglycolipids of mycoplasma; the sn-glycerol-1-phosphoglycolipids; and sn-glycero-1-phosphatidyl glycolipids of several Gram-positive organisms.

The other glycolipids include acylated sugars, such as acylglucose of mycobacterial 'cord factor', the rhamnolipid of *Pseudomonas aeruginosa* and the diacylated glucose attached to D-glyceric acid in the cell envelope of *Nocardia otidis-caviarum*. Further variants are mycobacterial mannosides which are glycosides of p-phenol with normal and branched-chain fatty acids esterified to oligosaccharide. These glycolipids are involved in the pathogenicity of mycobacteria. In *Saccharomyces cerevisiae* two categories, glycosyldiglyceride (1,2-acylglycerol and carbohydrate residue) and acylated sugar derivatives have been reported. Fungi, particularly yeasts and yeast-like organisms produce a variety of unusual lipids, mostly extracellular, that include glycolipids. The glycolipids usually contain a fatty acid linked glycosidically or via an ester bond to a carbohydrate moiety. Some of these glycolipids, such as ustilagic acid, are antibiotics and serve as important survival factors when the fungi are competing with bacteria for nutrients.

A number of sulpholipids have been detected in species of algae and bacteria. Usually they have sparse distribution and most are sulphate esters of the carbohydrate moiety in a glycolipid, e.g. trehalose mycolates of mycobactria. An exception is the thermoacidophile *Bacillus acidocaldarius* which contain 10% of its lipids as the sulphonic acid derivative of diacylgalactosylglycerol.

Terpenoid Lipids

Terpenoids are a class of compounds based on five carbon building blocks (isoprene units). **Figure 9** shows the structure of some of the terpenoid lipids.

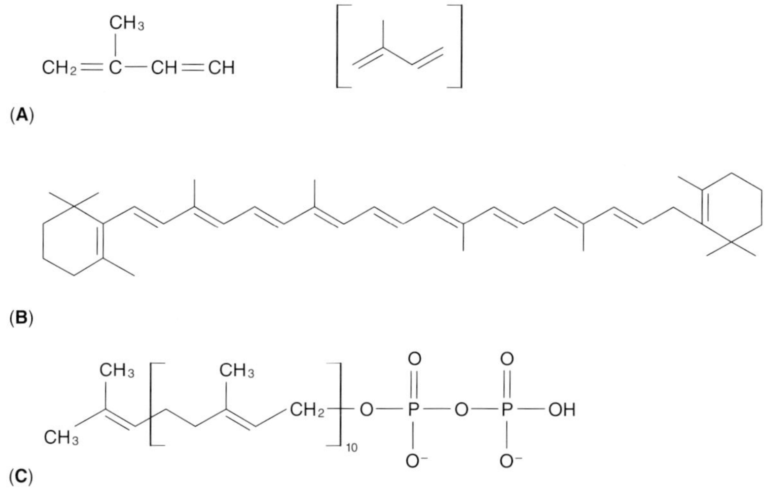

Figure 9 Structure of commonly occurring terpenoids. **(A)** Isoprene unit; **(B)** β-carotene; **(C)** undecaprenol pyrophosphate.

Some of these are primary metabolites for compounds such as sterols and carotenoids. Others may contribute to important structures, as in the side chain of chlorophyll or in enzyme prosthetic groups. Numerous fungal antibiotics are terpenoids. Gibberellins are examples of diterpenes (C_{20}). These were originally discovered as phytotoxins produced by plant pathogenic fungi. These substances are produced by fungi to manipulate the physiology of the plant host. The electron transport chain components such as quinones and plastoquinones are isoprene derivatives containing prenyl side chains. A family of *cis*-poly prenols, called dolichols, are important as carriers of sugar residues in cell-wall synthesis and protein glycosylation, facilitating the movement of hydrophilic molecules across the hydrophobic membrane barrier. The bacterial derivative is a C_{55} isoprenoid alcohol called undecaprenol or bactoprenol. Its pyrophosphate derivative serves a similar function to a lipophilic sugar carrier in the biosynthesis of peptidoglycan, lipopolysaccharide O-antigens, techoic acid and several other extracellular polysaccharides.

Carotenoids These are pigmented terpenoids which usually contain eight isoprene units (C_{40}), comprising two 20-carbon halves joined 'head' to 'head' and nine conjugated double bonds, which makes them brightly coloured. Carotenoids containing oxygen are called xanthophils, often as carboxylic acid esters e.g. torularhodin. There are many different types of carotenoids present in bacteria and fungi. Carotenoids have been reported from almost all yeasts and fungi. γ-Carotene is particularly common, whereas α-carotene has not been detected. Yeasts, particularly *Rhodotorula*, *Crytococcus* and *Sporobolomyces* produce a variety of colours; yellow (β-carotene), red (torularhodin) giving a coloured appearance to the

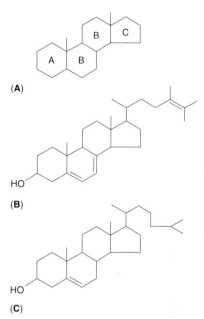

Figure 10 Structure of sterols. **(A)** The four-member cyclopentanoperhydrophenanthrene ring; **(B)** ergosterol; **(C)** cholesterol.

organism/medium. Carotenoids are reported to be absent from *Candida* spp.

Sterols The distinction between terpenoid and sterols is not clear cut. The structures based on a tetracyclic cyclopentanoperhydrophenanthrene ring structure are called sterols (**Fig. 10**). Sterols are derived from squalene. They can be present in free form or esterified to fatty acids. The free sterol is associated with membrane function, whereas, esterified sterols are biosynthetic intermediates or fulfil storage or pool functions. Bacteria characteristically lack sterol in the membrane except methylotrophs. The major forms of sterols are zymosterol, ergosterol and cholesterol. Yeasts accumulate large amounts of sterols, up to 10% of the dry weight. In yeasts and mushrooms ergosterol is the most abundant sterol. In *Saccharomyces cerevisiae* under anaerobic conditions lanosterol is the predominant sterol. Cholesterol has been reported from *S. cerevisiae* and *Candida krusei*. In *Mucor* spp. 90% of the sterol is ergosterol.

Lipopolysaccharide   Gram-negative bacteria have a cell envelope containing two membranes, the outer membrane is characterized by the presence of lipopolysaccharide in the outer leaflet of the bilayer structure. The lipopolysaccharide is involved in several aspects of pathogenicity. It serves as the hydrophobic anchor of Gram-negative bacteria. Lipopolysaccharide is a complex polymer of four parts. Outside of the cell there is a polysaccharide of variable structure known as O-antigen which carries several antigenic determinants. This is attached to a core polysaccharide of two parts, an outer core and a backbone. The cores vary between different bacteria. The backbone is connected to a glycolipid called lipid A, through a short link composed of 3-deoxy-D-mannooctulosonic acid. Lipid A consists of disaccharides of glucosamines that are highly substituted with phosphate, fatty acid and 3-deoxy-D-mannooctulosonic acid. The amino groups are substituted exclusively by 3-hydroxymyristate whereas the remaining hydroxyl groups are acylated with C_{12}, C_{14} and C_{16} saturated fatty acids and 3-myristoxymyristate. There is microheterogeneity in bacteria with respect to fatty acids that are present in lipopolysaccharides.

Biochemical Mechanisms of Uptake

Direct utilization of fatty acids can occur within many microorganisms. The lipids in the external medium can be hydrolysed by the action of extracellular lipases to yield fatty acids and glycerol. It is common knowledge that certain substances diffuse spontaneously across various membranes. Such passive diffusion processes are driven by differences in the chemical potential of the solute bathing the two sides of the membrane and the chemical nature of the solute molecule. Most biological membranes are known to allow the transverse diffusion of hydrophobic compounds. Before β-oxidation, fatty acids must enter the cell via uptake mechanisms that translocate them across the membranes. It was originally thought that as the fatty acids are hydrophobic they can diffuse through the membrane without requiring carrier protein. However, it has been observed that the uptake of fatty acids by bacteria involves active transport process. The 'fad regulon' of bacteria encodes for proteins

necessary for fatty acid translocation. The induction of *fad* regulon (fatty acid regulator gene) requires uptake of exogenous fatty acids. Genetic and biochemical studies indicate that two proteins are involved in the uptake of fatty acids encoded by genes *fadC* and *fadD*. The *fadD* locus encodes for membrane-bound acyl-CoA synthetase with broad substrate specificity for C_7–C_{18} fatty acids. The *fadC* gene encodes for 33 000 Da intrinsic membrane protein (FLP) which appears to be essential for long-chain fatty acid transport. FLP is the first membrane protein shown to be involved in fatty acid uptake. The *fadC* structural gene is localized as a 2.8 kb EcoRV fragment of *Escherichia coli* genome and has been cloned, mapped and analysed for gene expression. Plasmids, that contain this gene, complement *fad* C mutants to increase long-chain fatty acid uptake by two- to threefold and direct the synthesis of specific membrane protein of 33 000 Da. A model is proposed for fatty acid uptake in *E. coli* K-12 whereby long-chain fatty acids and medium-chain fatty acids adsorb to FLP present in the outer membrane. These fatty acids pass unidirectionally through the outer membrane via FLP and cross the cytoplasmic membrane to become activated by acyl-CoA synthetase. When FLP is inactive, medium-chain fatty acids diffuse freely through the membrane, become activated and are finally metabolized by β-oxidation enzymes.

In yeasts, as typified by *Saccharomyces cerevisiae*, the uptake of fatty acids is facilitated coupled with passive diffusion for lauric (C_{12}) and oleic ($C_{18:1}$) acid. At lower concentrations it has been shown to be carrier-mediated without the involvement of energy whereas at higher concentrations passive diffusion predominates. In all these organisms fatty acids once inside the cell are readily converted into their acyl-CoA esters to minimize their inhibitory effects on the cell.

Transformations Within the Cell

In a dynamic biological system there is a fine balance between the activities of synthetic and degradative enzymes. The levels of these activities control the accumulation of individual products, the turnover of most cell components and the utilization of the stored reserves. Lipids can serve as substrate for the cellular production of energy (ATP).

Degradation of Triacylglycerols

The degradation (hydrolysis) of triacylglycerols is achieved with the help of the enzyme lipase. Lipases are the enzymes that can cleave triacylglycerols to release fatty acids and glycerol. Lipases can be non-specific cleaving fatty acid either at *sn*-1,3 position or are fatty acid specific. Fatty acids released are oxidized and glycerol can be metabolized to form dihydroxyacetone phosphate, and then glyceraldedyde 2-phosphate thereby entering the glycolytic pathway. Lipases are widespread in nature and found in all phyla. Bacteria do not store triacylglycerol, and the major function of bacterial lipases is the breakdown of exogenous triacylglycerol as a food source. The same is true for most yeasts and fungi, although a few species do store triacylglycerol. The lipases can be intracellular or extracellular, being released into the medium. The extracellular lipases are inducible enzymes. The majority of the microbial lipases are secretory enzymes. Utilization of oils and fats has been reported in several bacteria, yeasts and fungi. *Yarrowia lipolytica* is a frequently studied yeast. High activities are found in microorganisms growing on lipid substrates. Many kinds of microorganisms including bacteria, fungi and yeasts, produce lipases, and some are isolated commercially on a large scale for medical and industrial use, for example those with activity similar to that of pancreatic lipase can be used as a substitute digestive aid in cases of pancreatic insufficiency.

Degradation of Phospholipids

In the case of a phospholipid, the enzymes that carry out hydrolysis are called phospholipases. Phospholipases are a heterogeneous group of hydrolytic enzymes involved in the catabolism of phospholipids. Historically these enzymes have been called by letters A–D (**Fig. 11**). Phospholipase (PL) A_1 and A_2 are carboxylic ester hydrolases cleaving at *sn*-1 and -2 position of diacylglycerol phospholipids releasing fatty acids. PLA_1 is widely distributed in bacteria. In Gram-negative bacteria it is present in outer membrane. Fungal phospholipase A_1 is poorly documented, although activity has been reported in many fungal systems. Phospholipase B is a monoacyl phospholipid hydrolase that attacks the product of PLA action. Phospholipase C, hydrolyses the phosphoester bond between glycerol and phosphate forming diacylglycerol and phosphoryl-X (X = substituent base).

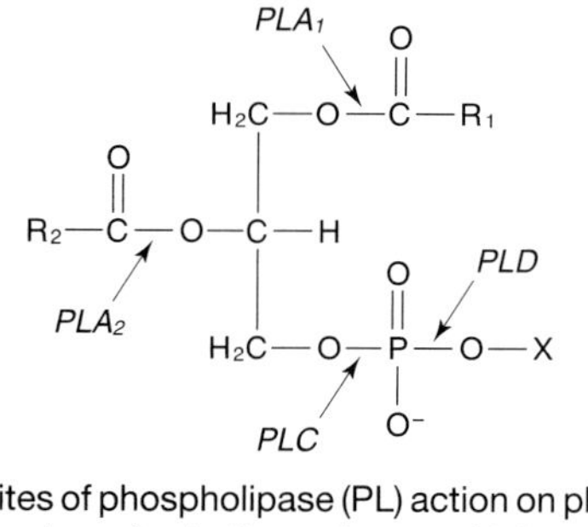

Phospholipase D hydrolyses the phosphate ester bond between the phosphate and the substituent base releasing the substituent base.

Phospholipases have been detected in a wide range of bacterial genera, sometimes being associated with pathogenicity (virulence). Phospholipases are components of toxins, e.g. *Clostridium perfringens*, the causative organism for gas gangrene, secretes phospholipase C that attacks the host phospholipids.

Fatty Acid Degradation

The free fatty acid released by the action of lipases can be oxidized by a number of enzyme systems which are named after the position of their attack on the acyl chain as is shown in **Figure 12** and detailed below. Of the three systems of oxidation of fatty acids, β-oxidation is predominant pathway.

α-**Oxidation** Oxidative decarboxylation (α-oxidation) of fatty acids has been studied in plant and animal systems with infrequent reports occurring for prokaryotic microorganisms. This pathway is of minor importance resulting in sequential decrease of one carbon atom from the fatty acid. This type of oxidation is important when the β-oxidation pathway is blocked by the presence of a methyl branch and α-oxidation releases the side chain as CO_2. In the α-oxidation pathway, a non-esterified fatty acid is attacked by molecular oxygen to generate an unstable 2-hydroperoxy intermediate. This intermediate releases CO_2 to yield a fatty aldehyde which is oxidized to yield an odd-chain fatty acid that is shorter by one carbon atom. Odd-chain fatty acid will be formed as a result of α-oxidation of even-chain fatty acids and has been studied in *Candida utilis* and *Arthrobacter simplex*. The cofactor requirement for this type of reaction is O_2 and NADH.

$$R(CH_2)_n COOH \rightarrow R(CH_2)_{n-1} COOH + CO_2$$

In certain bacteria the 2-hydroperoxy intermediate may be reduced by bacterial peroxidase to yield D-2 hydroxy fatty acid that cannot be metabolized.

β-**Oxidation** β-Oxidation is a major pathway for oxidation of fatty acids to small 2-carbon acetyl coenzyme A (CoA) units. In bacteria, fatty acid β-oxidation is cytoplasmic, whereas in yeast and fungi it takes place predominantly in peroxisomes

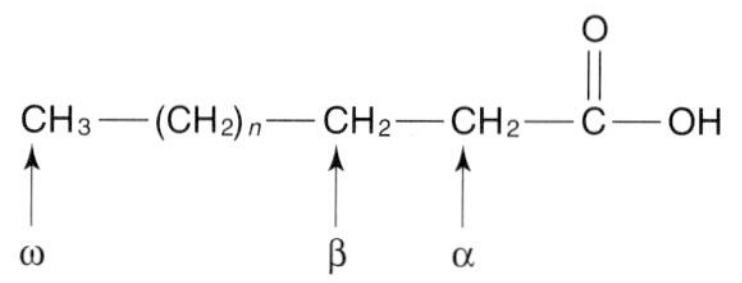

Figure 11 Sites of phospholipase (PL) action on phospholipids. X refers to a number of substituent bases which may be present in different phospholipids.

Figure 12 Different sites of fatty acid oxidation.

(microbodies) and in small amounts in mitochondria.

The growth of bacteria on fatty acids requires coordinated induction of β-oxidation enzymes plus a fatty acid transport system. The enzymes of fatty acid β-oxidation in bacteria are under the control of *fad* regulon. The regulation of *fad* regulon is similar to *lac* operon. Before oxidation the fatty acid molecule is activated. Activation involves thiol esterification of fatty acid with coenzyme A (CoASH) to form activated fatty acid (fatty acyl CoA). The activation of fatty acids to its CoA-derivative is coupled with utilization of energy from ATP and requires the enzyme acyl-CoA synthetase (*fad*D gene) and has a broad substrate specificity. The activated fatty acid undergoes dehydrogenation to form fatty enoyl-CoA requiring the enzyme acyl-CoA dehydrogenase encoded by *fad*E gene. Further metabolism involves enzymes enoyl-CoA hydratase (*fad*B), 3-hydroxy acyl-CoA dehydrogenase (*fad*B) catalysing dehydrogenation to form 3-ketoacyl-CoA, and finally thiolase releases acetyl-CoA and a fatty acyl-CoA molecule that is two carbons shorter than the parent fatty acid molecule (**Fig. 13**). The same reactions are repeated leading to shortening of the fatty acid molecule and release of acetyl-CoA units till the fatty acyl-CoA is completely degraded. The intermediates formed during fatty acid oxidation remain enzyme bound. The long-chain fatty acids (C_{12}–C_{18}) induce *fad* regulon, the medium-chain fatty acids (C_7–C_{11}) can not induce the *fad* regulon genes but are substrates for oxidation. The enzymes acyl-CoA synthetase and acyl-CoA dehydrogenase are at different positions, not linked to the genes of enoyl-CoA hydratase, 3-hydroxy acyl-CoA dehydrogenase, 3-ketoacyl-CoA thiolase, 3-hydroxy-CoA epimerase and *cis*-Δ^3 CoA epimerase. The enzymes enoyl-CoA hydratase, 3-hydroxy acyl-CoA dehydrogenase, 3-ketoacyl-CoA thiolase, 3-hydroxy-CoA epimerase and *cis*-Δ^3 CoA epimerase are present as a multifunctional complex of 260 000 Da consisting of $\alpha_2\beta_2$. The α-subunit and β-subunits have molecular weights of 78 000 and 42 000 Da respectively. They are encoded by *fad*B (enoyl-CoA hydratase, 3-hydroxyacyl-CoA dehydrogenase, 3-,3-hydroxy-CoA epimerase and *cis*-Δ^3 CoA epimerase) and *fad*A (ketoacyl-CoA thiolase) genes, respectively. The position of the different structural genes of *fad* regulon are indicated in **Figure 14**.

The release of acetyl-CoA is coupled with the formation of reduced coenzyme: one molecule of reduced flavin adenine dinucleotide (FADH$_2$) and one molecule of NADH. The acetyl-CoA produced during β-oxidation of fatty acids is passed to the tricarboxylic acid (TCA) cycle to be oxidized accompanied by synthesis of ATP.

The oxidation of unsaturated fatty acids is similar to that of saturated fatty acids. In bacteria C_{12}–C_{18} *cis*-monounsaturated fatty acids are present which have a double bond between carbons 9 and 10, and 11 and 12. However, in eukaryotic microbes there are multiple double bonds. Monounsaturated fatty acid is oxidized by normal β-oxidation to give rise to Δ^3 *cis*-enoyl-CoA or Δ^2 *cis*-enoyl-CoA that cannot be a substrate for the fatty acyl-CoA dehydrogenase. Δ^3 *cis*-enoyl-CoA is therefore isomerized to Δ^2 *trans*-enoyl-CoA by Δ^3-*trans* Δ^2 enoyl-CoA isomerase, a normal substrate for enoyl-CoA hydratase to form L(+)-β-hydroxyacyl-CoA that can be metabolized by the β-oxidation enzymes. In the case of Δ^2 *cis*-enoyl-CoA, it is first hydrated by Δ^2-enoyl-CoA hydratase to D(−)-β-hydroxyacyl-CoA derivative that is epimerized to L(+)-β-hydroxyacyl-CoA involving the enzyme 3-hydoxyacyl-CoA epimerase.

Peroxisomal Oxidation An alternative to β-oxidation in yeast and fungi involves the subcellular organelle termed peroxisomes. In these organisms peroxisomes play a key role in β-oxidation of fatty acids. Peroxisomes are unit membrane limited organelles containing catalase, long-chain fatty alcohol and fatty acid dehydrogenase activities, β-oxidation enzymes plus other enzyme activities. Peroxisomal β-oxidation enzymes are induced when cells are grown in the presence of fatty acids or alkanes. The first step involving dehydrogenation is catalysed by the enzyme fatty acyl-CoA oxidase (instead of dehydrogenase) to form 2-unsaturated fatty acyl-CoA ester (**Fig. 15**). The enzyme requires O_2 and FADH$_2$ as cofactors and forms H_2O_2 that is detoxified by catalase. The other enzymes enoyl-CoA hydratase, 3-hydroxyacyl CoA dehydrogenase, 3-hydroxyacyl-CoA epimerase and 3-ketoacyl-CoA thiolase are present as a multienzyme complex, acting in sequential manner to release acetyl-CoA. A limited amount of fatty acid β-oxidation in yeast occurs in the mitochondria which also involves acyl-CoA dehydrogenase rather than oxidase. Transport of fatty acids into the mitochondria requires carnitine for the transport of acyl-CoA derivatives.

ω-Oxidation ω-Oxidation is a pathway of fatty acid oxidation present in both bacteria and fungi. Generally speaking ω-oxidation plays only a minor role in the oxidation of fatty acids or related compounds compared with α- and β-oxidation. This type of oxidation is important when carboxyl end is unavailable or for the formation of ω-hydroxy fatty acids. It is particularly useful in microorganisms capable of utilizing alkanes as sole energy and carbon source. It requires the O_2-dependent ω-hydroxylase similar to one described in *Pseudomonas putida*. The ω-

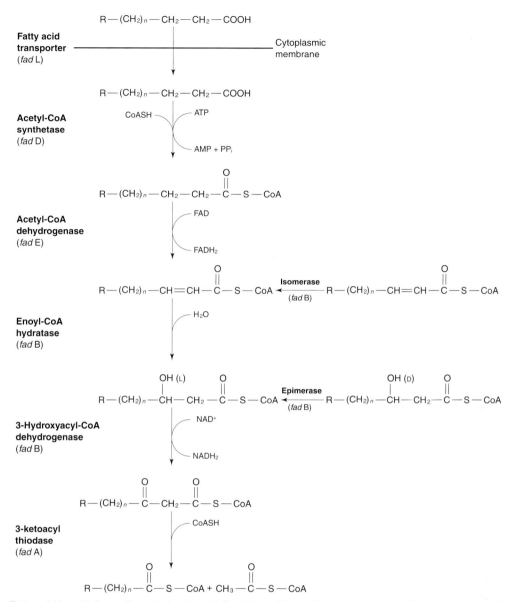

Figure 13 Fatty acid β-oxidation pathway in bacteria. β-Oxidation of fatty acids releases acetyl-CoA and fatty acyl-CoA that is two carbons shorter than the parent fatty acid.

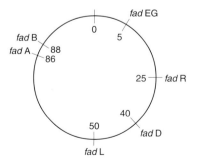

Figure 14 Fatty acid β-oxidation genes on bacterial genome.

hydroxylation requires O_2 and NAD(P)H as reducing equivalent donor. The ω-hydroxyl fatty acid is then converted into ω-aldehyde and finally to α, ω-dicarboxylic acid.

$$CH_3(CH_2)_n COOH \rightarrow HOCH_2(CH_2)_n COOH \rightarrow$$
$$OHC(CH_2)_n COOH \rightarrow HOOC(CH_2)_n COOH$$

Fatty Acid Biosynthesis

Most naturally occurring fatty acids have an even number of carbon atoms. The biosynthesis of fatty acids proceeds by the sequential addition of 2-carbon units derived from acetyl-CoA and is cytoplasmic in bacteria and fungi (**Fig. 16**). The first step in the synthesis of fatty acids is carboxylation of acetyl-CoA to form malonyl-CoA, a key intermediate in the synthesis of fatty acids, catalysed by the enzyme acetyl-CoA carboxylase. The formation of malonyl-CoA requires ATP and biotin and CO_2 as cofactor.

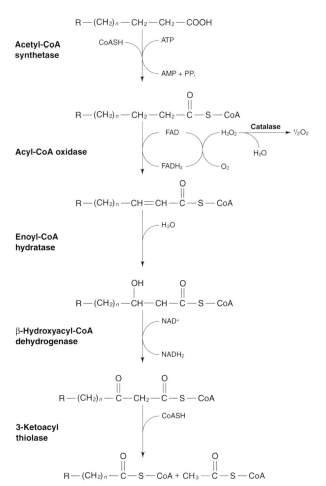

Figure 15 Peroxisomal fatty acid β-oxidation in fungi and yeast.

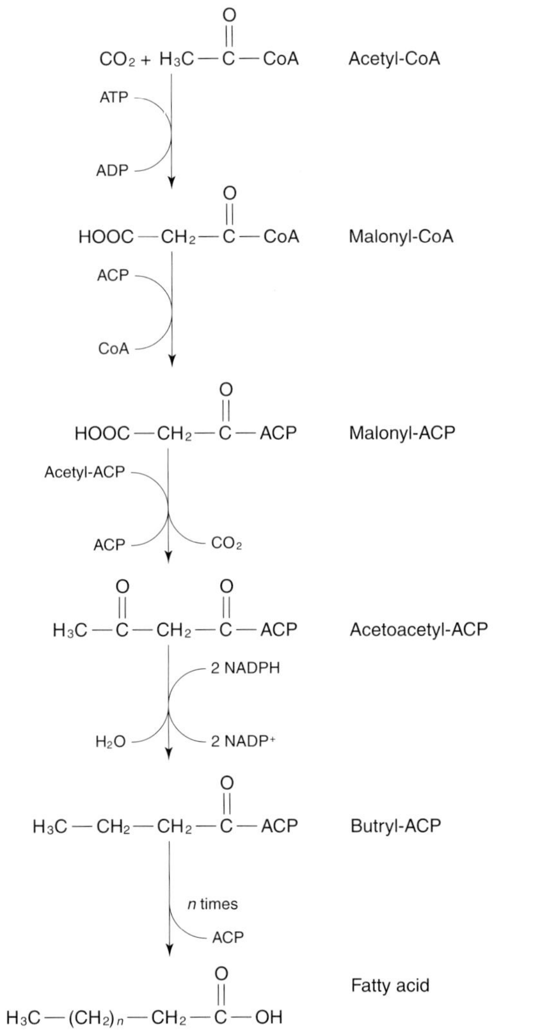

Figure 16 Fatty acid biosynthesis. The fatty acid biosynthesis involves 2-carbon additions on acyl carrier protein (ACP).

The requirement of biotin in this reaction is one reason why many organisms require biotin in trace quantities as a growth factor. In *Escherichia coli*, acetyl-CoA carboxylase consists of three disociable components. One protein contains biotin, has a molecular weight of 22.5 kDa and is the biotin carboxyl carrier protein (BCCP). A second, 102 kDa component consists of two subunits and catalyses the biotin carboxylase reaction. The final component (130 kDa) contains two pairs of non-identical subunits of 30 and 35 kDa and catalyses the carboxyltransferase reaction. Evidence from various studies indicates that acetyl-CoA carboxylase from yeasts and fungi consists of a single multifunctional protein. The molecular mass of these proteins is in the range of 189–230 kDa for yeasts, *S. cerevisiae* and *Candida lipolytica*. The yeast enzyme can be prepared in a form that is activated by citrate, but this activation is not accompanied by polymerization of the enzyme as in higher eukaryotes. However, the bacterial acetyl-CoA carboxylase is not regulated by citrate and is also not under phosphorylation control; it is, instead

regulated by the nucleosides guanosine-3'-diphosphate, 5'-diphosphate (ppGpp) and guanosine-3'diphosphate, 5'-triphosphate (pppGpp), which reduce the enzyme activity by inhibiting carboxyltransferase. These guanosine nucleosides, that are unique to bacteria, are formed by phosphoribosyl transfer from ATP to GDP or GTP on the ribosome in response to amino acid starvation, or when other conditions of reduced growth rate lead to ribosome 'idling'.

The malonyl-CoA generated by acetyl-CoA carboxylase forms the source of all the carbons of the fatty acyl chain. The group of enzymes catalysing the synthesis of fatty acids is known collectively as fatty acid synthase (FAS) and involves seven separate enzyme activities. Initially, a priming molecule of acetyl-CoA is transferred to -SH group of acyl carrier protein (ACP) on FAS involving the enzyme transacylase, followed by transfer of malonyl-CoA to 4'-

phosphopantothein of ACP releasing CoA catalysed by the enzyme malonyl transacylase. The acetyl group attacks the methylene group of malonyl residue, catalysed by β-ketoacyl synthase, forming acetoacetyl-ACP with release of CO_2. This frees the -SH group of ACP that was occupied by acetyl group. The aceto-acetyl-ACP is then reduced, dehydrated and again reduced to form butryl-ACP catalysed by β-ketoacyl reductase, hydratase and enoyl-reductase, respectively. The reductases require NADPH as cofactor. Further, elongation is by malonyl-CoA contributing to the successive 2-carbon units to growing acyl chain and the cycle is repeated finally giving C_{16} fatty acid (palmitic acid). The palmitic acid formed is liberated by the seventh enzyme thioesterase. The overall reaction of palmitic acid is as follows:

$$8 \text{ Acetyl-CoA} + 7 \text{ ATP} + 14 \text{ NADPH} \rightarrow$$
$$\text{Palmitic acid} + 14 \text{ NADP}^+ + 8 \text{ CoA} + 6 \text{ H}_2\text{O} + 7 \text{ P}_i$$

Fatty acid synthases can be divided mainly into type I and type II enzymes. Type I synthases are multifunctional proteins in which the proteins catalyse the individual partial reactions in discrete domains and the acyl carrier protein is covalently linked to protein. This type include synthases from higher bacteria (*Mycobacterium smegmatis* and *Cornybacterium* spp.) and yeasts (*S. cerevisiae*). Type I synthases are characteristically high-molecular-weight proteins (0.4×10^6–2.5×10^6) comprising two or more large multifunctional polypeptide chains (molecular weight 1.8×10^5–2.7×10^5). The type I of *M. smegmatis* is unusual in several respects; the two reductases have different reduced pyridine nucleotide specificities, β-ketoacyl-ACP reductase requires NADPH and enoyl-ACP reductase requires NADH (other type I enzymes use only NADPH). In animals it is generally accepted that the two chains are identical. Recent genetic analysis of fatty acid synthase mutants of yeast have demonstrated that there are two unrelated polycistronic genes designated as *fas*1 and *fas*2. The *fas*1 gene encodes for the acetyl transferase, malonyl (palmitoyl) transacylase, dehydratase and enoyl reductase enzymes, whereas *fas* 2 encodes for the phospho-pantothein-binding region and the β-ketoacyl synthase and reductase enzymes. The yeast synthase is probably A_6B_6 complex made of two different multifunctional proteins (A and B). Type I of yeast is inhibited by palmitoyl-CoA and that of *Aspergillus fumigatus* is inhibited by malonyl-CoA.

Type II synthases contain enzymes that can be separated, purified and studied individually and are present in most of the lower bacteria (*E. coli*) and the acyl carrier protein readily dissociate from the

enzyme. Type II synthase has been most studied from *E. coli*, seven proteins having been isolated. It is assumed that inside the cell individual enzymes associate to form a loosely bound multienzyme complex. The site of association may be the cell membrane because in *E. coli*, ACP is localized in the membrane. During these reactions, the substrates are bound to ACP. The type II enzyme synthesizes both saturated and unsaturated fatty acids. The reason for this is the presence of β-hydroxydecanoyl-ACP-β,γ-dehydrase which releases *cis*-3-decenoyl-ACP (precursor of unsaturated fatty acids) instead of *trans*-2-decenoyl-ACP that is converted into saturated fatty acid.

In the case of yeasts and fungi, the saturated fatty acids, palmitic (C_{16}) and stearic (C_{18}) serve as precursors of the monounsaturated fatty acids palmitoleic acid and oleic acid, respectively. The double bond is formed as a result of the action of the enzyme fatty acyl-CoA oxygenase in an oxidation reaction. In the reaction NADPH is oxidized to $NADP^+$.

Incorporation of Fatty Acids into Triglycerides and other Intracellular Structures

Fatty acids are essential precursors in many lipid components found in bacteria and fungi, including mono-, di- and triacylglycerols, phospholipids and a variety of other lipids.

Biosynthesis of triacylglycerols

Triacylglycerols are present as storage lipids in fungi and yeasts, but not in bacteria. The biosynthetic pathway involved in the synthesis of triacylglycerol is described in **Figure 17**. Glycerol and fatty acyl-CoA

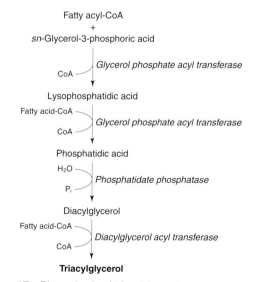

Figure 17 Biosynthesis of triacylglycerol.

are precursors for the synthesis of triacylglycerol. Glycerol is first converted into glycerol-3-phosphate by the action of glycerokinase, followed by transfer of acyl group from acyl-CoA resulting in the formation of phosphatidic acid that involves the enzyme glycerol phosphate acyl transferase. The phosphate from phosphatidic acid is removed by the action of the enzyme phosphatidic acid phosphatase to form diacylglycerol. The final product triacylglycerol is formed by transfer of a third molecule of acyl group catalysed by the enzyme diacylglycerol acyl transferase.

Apart from *de novo* synthesis as described above, triacylglycerol can also be formed from phosphoglycerides via diacylglycerol (**Fig. 18**). Major phosphoglycerides can be converted into diacylglycerol by the action of the enzyme phospholipase C or alternatively by a reversal of CDP-choline: diacylglycerolcholine phosphotransferase.

Biosynthesis of Phospholipids

The biosynthesis of phospholipids, which are synthesized exclusively for use in the biogenesis of membranes is shown in Figure 18. The steps in the biosynthesis of major phospholipid classes of *E. coli* have been established. All the enzymes of phospholipid biosynthesis in bacteria are in the cytoplasmic membrane, apart from phosphatidylserine synthetase, the location of which still remains an enigma. In the biosynthesis of phospholipids the major difference is acyl-ACP is the donor instead of acyl-CoA. The phospholipid biosynthesis involves the addition of fatty acids to glycerol-3-phosphate. The glycerol-3-phophate reacts with acylated-ACP to form phosphatidic acid, a common intermediary metabolite in the synthesis of phospholipids and triglycerides. The phosphatidic acid is activated by cytosine triphosphate (CTP) to form CDP-diacylglycerol, and the CDP is finally displaced by serine, alcohols or glycerol to produce the completed phospholipids, phosphatidylserine, phosphatidylethanolamine or phosphatidylglycerol, respectively. The enzymes catalysing the different steps are shown in Figure 18. In bacteria, phosphatidylcholine is not synthesized except in a few bacteria where the transfer of a methyl group to phosphatidylethanolamine is mediated by N-methyltransferase (*S*-adenosylmethionine is the methyl donor).

In *S. cerevisiae*, the biosynthesis of phospholipid is similar to that of *E. coli*. However, there are a few differences. Firstly, the fatty acyl group is transferred as acyl-CoA and not as acyl-ACP. Secondly phosphatidylinositol is synthesized in addition to phosphatidylserine, phosphatidylethanolamine, phosphatidylglycerol and disphosphatidylglycerol by

a mechanism that involves the exchange of CDP from CDP-diacylglycerol with inositol. In addition, phosphatidylcholine, a major membrane phospholipid in yeast and fungi, is synthesized by exchange of choline from CDP-choline to diacylglycerol, whereas in bacteria phosphatidylcholine is not usually synthesized and in the few bacteria where it is synthesized it involves methyl transfer to phosphatidylcholine. Although, phosphatidylethanol amine is made by decarboxylation of phosphatidylserine in yeast, the CDP-base exchange pathway is also operative for synthesis of phosphatidylethanolamine.

Biosynthesis of Terpenoids and Sterols

The membranes of many eukaryotic cells contain sterols, such as cholesterol, that are made up of repeating units of the unsaturated hydrocarbon isoprene (isoprenoid hydrocarbons). The early steps of steroid and terpenoid synthesis are the same. The first intermediate is 3-hydroxy-3-methyl-glutaryl-CoA (HMG-CoA), formed by condensation of three molecules of acetyl-CoA, followed by reduction of HMG-CoA to give mevalonic acid. This reaction is catalysed by HMG-CoA reductase, an important regulatory enzyme that controls the rate of sterol biosynthesis (**Fig. 19**). Mevalonic acid then gives rise to isopentenyl pyrophosphate, a 5-carbon structure characteristic of terpenoids, through a series of phosphorylations. Isoprenoid hydrocarbons are synthesized from acetyl-CoA molecules in a reaction requiring ATP. Isopentenyl pyrophosphate and its isomer dimethylallyl pyrophosphate condense to form farnesyl pyrophosphate that gives rise to a series of cyclic and noncyclic terpenoids and finally to carotenoids, dolichol, etc. In the biosynthesis of sterols, farnesyl pyrophosphate forms squalene, a non-cyclic intermediate that after a series of transformations gives rise to sterols. The biosynthesis of cholesterol exemplifies the fundamental mechanism of long-chain carbon skeleton formation from 5-carbon isoprenoid units.

Biosynthesis of Poly-β-hydroxybutyric Acid

The pathway for synthesis of poly-β-hydroxybutyric acid (PHB), a common storage product of bacteria, is similar to the pathway for the synthesis of fatty acids (**Fig. 20**). The enzymes involved in polyhydroxybutyrate synthesis are 3-ketothiolase, acetoacetyl reductase and PHB synthase. PHB synthesis involves condensation of two molecules of acetyl-CoA to form acetoacetyl-CoA (C_4 derivative of CoA), followed by reduction of acetoacetyl-CoA to form β-hydroxybutyryl-CoA and finally repetitive sequential addition of acetyl-CoA resulting in chain length elongation, and subsequent removal of the CoA portion of the molecule forming poly-β-hydroxybutyric acid, which

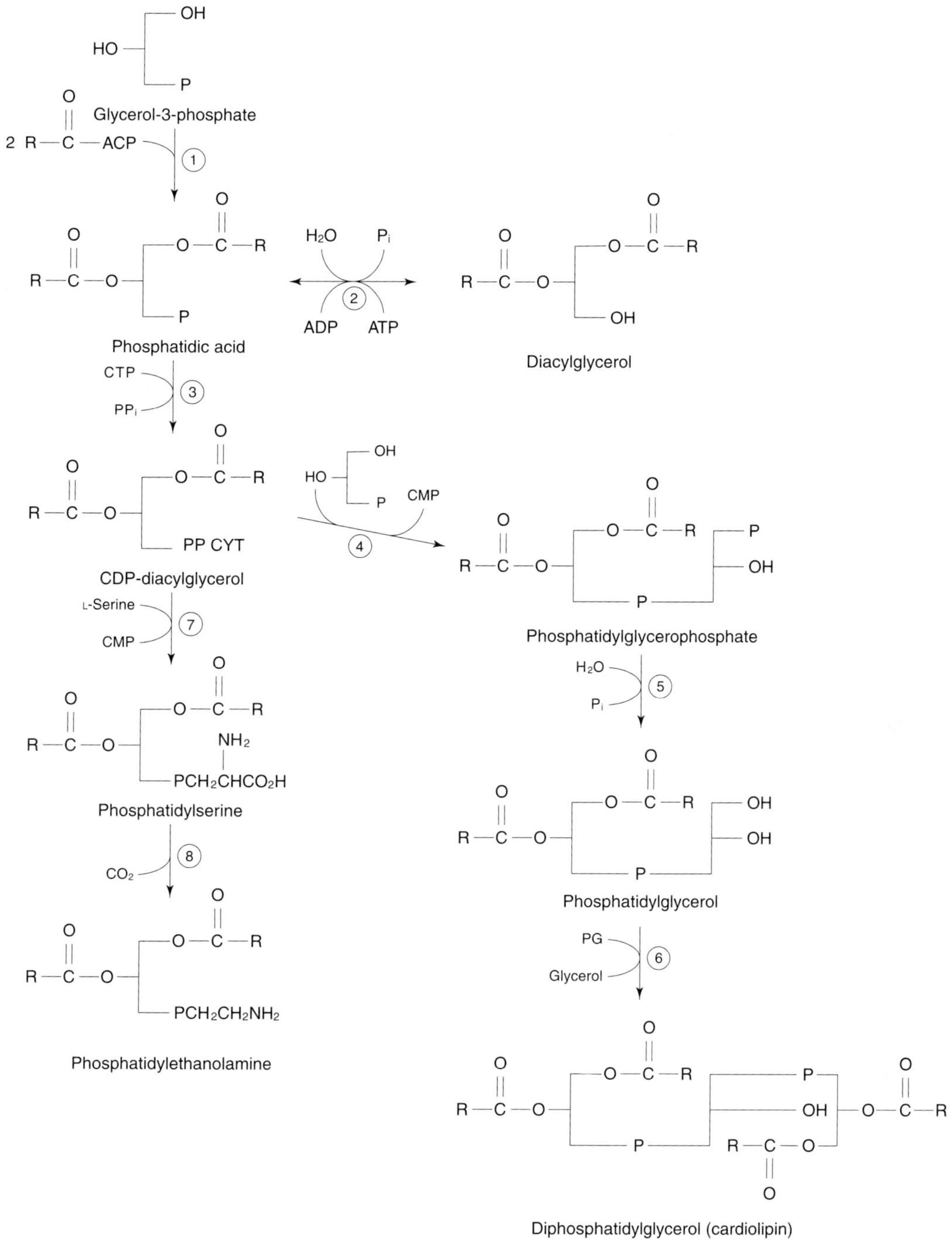

Figure 18 Biosynthesis of phospholipids. The formation of phospholipids starts with glycerol-3-phosphate. Step (1) is catalysed by the enzyme glycerol phosphate acyl transferase (*pls*A, *pls*B); (2) the forward reaction involving the formation of diacylglycerol by phosphatidic acid phosphatase and the reverse reaction involving the formation of phosphatidic acid from diacylglycerol involves diacylglycerol kinase (*dgk*); (3) CDP-diacylglycerol synthetase (*cds*); (4) phosphatidylglycerolphosphate synthetase (*pgs* A, *pgs* B); (5) phosphatidylglycerolphosphatate phosphatase (*pgp* AB); (6) diphosphatidylglycerol (cardiolipin) synthetase (*cls*); (7) phosphatidylserine synthetase (*pss*); (8) phosphatidylserine decarboxylase (*psd*). The names of genes encoding the respective enzymes are given in parentheses.

can accumulate in large amounts in bacteria and serve as energy and carbon source. Interestingly, unlike other biosynthetic reactions, the formation of poly-β-hydroxybutyrate uses the coenzyme NADH rather than NADPH.

See also: **Escherichia coli**: Escherichia coli. **Preservatives**: Traditional Preservatives – Vegetable Oils. **Saccharomyces**: Saccharomyces cerevisiae.

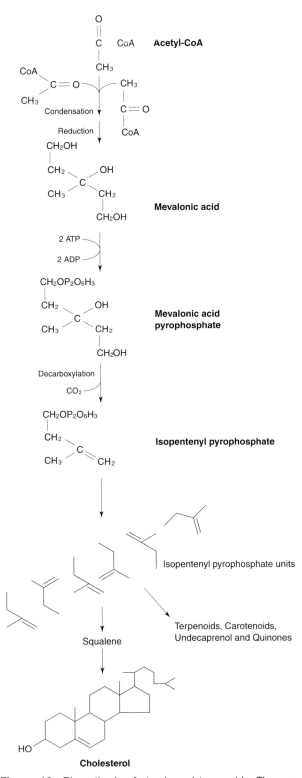

Figure 19 Biosynthesis of sterols and terpenoids. The synthesis of cholesterol and terpenoids involves the formation and subsequent condensation of isopentenyl pyrophosphate units from acetyl-CoA.

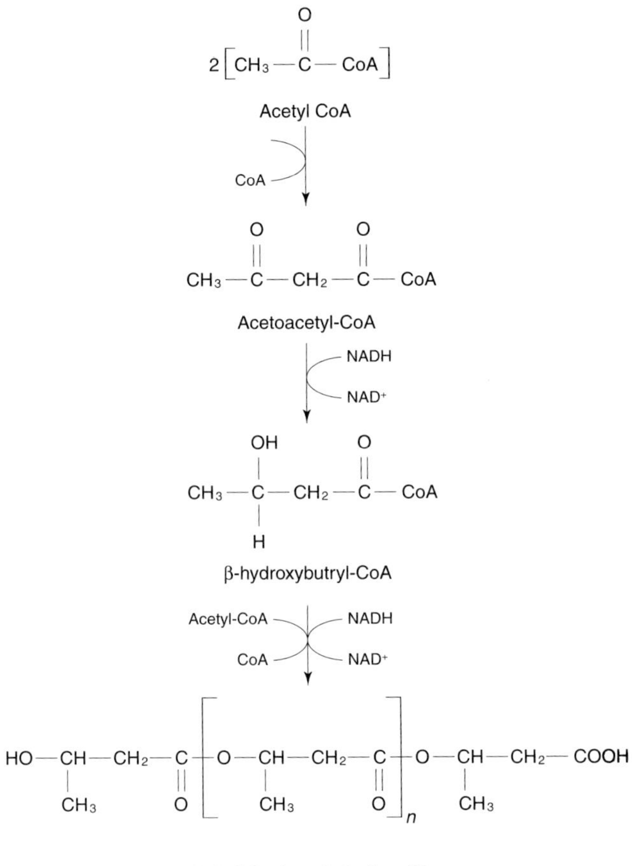

Figure 20 The synthesis of polyhydroxybutyrate in bacteria.

Further Reading

Boom TV and Cronan JE (1989) Genetics and regulation of bacterial lipid metabolism. *Annual Reviews of Microbiology* 43: 317–343.

Brenan PJ and Lasel DM (1978) Physiology of fungal lipids: selected topics. *Advances in Microbial Physiology* 17: 47–179.

Carman GM and Henry SA (1989) Phospholipid biosynthesis in yeast. *Annual Reviews of Biochemistry* 58: 635–669.

Goldf H (1972) Comparative aspects of bacterial lipid metabolism. *Advances in Microbial Physiology* 8: 1–98.

Gurr MI and Harwood JL (1991) *Lipid Biochemistry*, 4th edn. London: Chapman & Hall.

Harwood JL and Russell NJ (1984) *Lipids in Plants and Microbes*. London: George Allen & Unwin.

Nunn WD (1986) A molecular view of fatty acid catabolism in *Escherichia coli*. *Microbiological Reviews* 50: 179–192.

Ratledge C and Wilkinson SG (1989) *Microbial Lipids*, vols 1 and 2. London: Academic Press.

Vance DE and Vance JE (1988) *Biochemistry of Lipids and Membranes*. Menlo Park, CA: Benjamin/Cummings.

Metabolism of Minerals and Vitamins

C Umezawa and **M Shin**, Faculty of Pharmaceutical Sciences, Kobe Gakuin University, Japan

Copyright © 1999 Academic Press

Vitamins and minerals are essential for the function of most forms of life, including microorganisms. Some enzymes require non-protein components called cofactors. The cofactor may be a metal ion or an organic molecule called a coenzyme. Many vitamins are precursors of the coenzymes. Structures of vitamins are complex and diverse. Most vitamins are heterocyclic compounds; pyrimidine and thiazole rings of thiamin, isoalloxazine ring of riboflavin, pyridine ring of pyridoxine and niacin, thiophene ring of biotin, corrin ring of cobalamin, and so on. Many microorganisms readily produce the vitamins *de novo*. Genes participating in biosynthesis of vitamin are often clustered as an operon. The transcription of operon genes is generally regulated strictly. Vitamins and minerals may be present in extremely low quantities in the environment and microorganisms are equipped with the transport systems to scavenge the vitamins and minerals.

Uptake of Minerals by Bacteria

Copper

Copper is an essential prosthetic group of proteins such as superoxide dismutase and cytochrome oxidase. On the other hand, excess copper is toxic. Homeostatic control of copper uptake is, therefore, needed to solve the problem of acquiring sufficient copper and at the same time avoiding toxic copper excess.

In *Saccharomyces cerevisiae*, reduction of copper by the surface reductase, the *fre1* gene product facilitates cellular uptake. In most laboratory strains of *Saccharomyces cerevisiae*, cellular copper acquisition requires Ctr1 protein, which exists in the plasma membrane as a multimer. The Ctr1 protein contains the methionine-rich domain homologous with copper-binding proteins, which suggests that it will form a pocket with affinity for Cu^+.

The regulation of *fre1* and *ctr1* genes is mediated at the level of copper-dependent transcription. Copper deprivation induces, and copper loading represses, transcription of both genes. Mac1 protein is a cellular component which functions as the copper sensor–regulator. Mac1 protein controls the expression of surface reductase and copper uptake activity in yeast

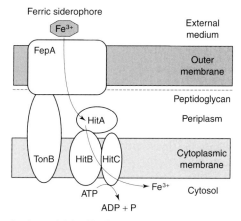

Figure 1 A model for TonB-dependent ferric ion transport in Gram-negative bacteria. The ferric siderophore binds to the cell surface receptor (FepA) and transports the iron across the outer membrane in a process requiring TonB. HitA in the periplasm binds the iron and passes it to the HitB and HitC proteins in the inner membrane. Passage across the inner membrane requires ATP through HitC protein.

and thus provides homeostatic control of copper acquisition.

As in *Saccharomyces cerevisiae*, bacterial copper transport is the cooperation of several proteins: the CurA and CutB proteins both play a role in *Escherichia coli* copper import. Bacteria and *Saccharomyces cerevisiae* rely on the sequestration of copper to handle excessive copper. *Pseudomonas syringae* retains copper in the periplasm, while in yeast vacuoles play a role in storage or detoxification of copper.

Iron

Microorganisms growing under aerobic conditions need iron for formation of haem to reduce oxygen for synthesis of ATP. Iron is the second most abundant metal in the earth's crust, but usable iron is in short supply because of its extreme insolubility at physiological pH. To scavenge iron from the environment, most aerobic and facultative anaerobic microorganisms synthesize at least one ferric iron-specific chelating protein (siderophore) under low iron stress.

Siderophores differ substantially in structure. Enterobactin, the siderophore indigenous to *E. coli*, is a cyclic trimer of 2,3-dihydroxybenzoylserine. Its synthesis and uptake across the membranes are encoded by the *ent-fep* gene cluster of 14 genes. In the outer membrane of Gram-negative bacteria there is a ferric enterobactin receptor called FepA. FepA is a gated porin, which transports iron in the form of ferric siderophores in a TonB-dependent manner (**Fig. 1**). Fur, the product of *fur* gene, responds to the intracellular iron level and acts as a negative repressor of transcription of the siderophore. Fe^{2+} is a co-repressor to organize Fur to bind the operator.

Magnesium

Magnesium is transported by three distinct systems in *Salmonella typhimurium*. The CorA Mg^{2+} transport system is the dominant constitutive uptake mechanism and is ubiquitous within Gram-negative bacteria. Under conditions of Mg^{2+} depletion, expression of MgtA and MgtB transport systems (which belong to the P type ATPase superfamily) greatly increases. At concentrations of extracellular Mg^{2+} below $10\,\mu mol\,l^{-1}$, MgtB becomes the dominant transporter. The P type ATPases are so termed because the mechanism of transport involves direct phosphorylation of a conserved aspartyl residue by ATP and the subsequent hydrolysis of this aspartylphosphate as a necessary part of the transport cycle. The CorA Mg^{2+} transport system has a broad specificity, and Zn^{2+} and Co^{2+} are also transported via this system.

Sodium

Most animal cells maintain intracellular K^+ at a relatively high and constant concentration, whereas the intracellular Na^+ concentration is usually much lower. In order to generate a solute gradient, the membrane must be equipped with a primary pump. The pumps such as the ATP-driven importers (or exporters of poisonous chemicals) – Na^+-translocating (F_1F_0)-ATPase, Na^+-ATPases of the V and P type – serve the cells directly and have no function in energy transduction. The second type of pump creates an energized state of the membrane in the form of an electrochemical ion gradient, which can be used by other membrane-linked systems to perform work. Decarboxylase reactions catalysed by oxaloacetate decarboxylase of *Klebsiella pneumoniae* (**Fig. 2**), methylmalonyl-CoA decarboxylase of *Propionigenium modestum* and others are found to couple to the vectorial movement of Na^+ ions. Methyltetrahydromethanopterin:coenzyme M methyltransferase of *Methanobacterium thermoautotrophicum* is also an integral membrane protein functioning as a Na^+ pump. Na^+-translocating NADH:ubiquinone oxidoreductase is structurally not related to its H^+-translocating counterparts. The K^+-transporting ATPases of *Escherichia coli* and *Streptococcus faecalis* are also of the P type.

Mechanisms for Incorporation of Minerals into Enzymes

Copper Chaperone for Superoxide Dismutase

The insertion of copper into copper/zinc superoxide dismutase (SOD1) involves a specific metal carrier protein, identified as Lys7 in *Saccharomyces cerevisiae*. This 'copper chaperone' is specific for SOD1 and does not deliver copper to other proteins. A yeast mutant for the SOD1 copper chaperone has normal level of SOD1 protein but fails to incorporate copper into SOD1, which is therefore devoid of superoxide scavenging activity.

Iron Storage: Ferritin as a Source of Iron

The ferritin molecule is a protein (M_r about 500 000) composed of 24 polypeptide chains and capable of storing up to 4500 Fe (III) atoms as an inorganic complex. Ferritins distribute ubiquitously among living species including fungi, yeast and bacteria. Iron is normally in the Fe^{3+} state, but when reduced, it remains within the ferritin unless removed by an Fe^{2+} acceptor. Since ferritin itself acts as a ferroxidase during iron uptake, no external oxidase is needed if O_2 is available.

Haem, the prosthetic group of proteins such as catalase, peroxidase and cytochrome C, is made by the insertion of the ferrous form of iron by ferrochelatase.

Heavy Metal Efflux Systems and Metallothionein

In contrast to most cation transport systems, which are determined by chromosomal genes, toxic metal ion resistance systems are encoded by plasmid genes. Plasmid-based transport systems are toxic ion efflux mechanisms and the ions effluxed by the systems are generally not essential; Co^{2+}, Zn^{2+} and Cu^{2+} are exceptions.

The major group of metal resistance systems are efflux pumps. Some are ATPases (the cadmium and copper ATPases of the Gram-positive bacteria and the arsenite ATPase of the Gram-negative bacteria) and others are chemiosmotic cation/proton antiporters (the divalent cation efflux systems of *Alcaligenes* and the arsenite efflux system of the Gram-positive and the Gram-negative bacteria).

Studies of copper resistance in the Gram-positive bacterium *Enterococcus hirae* led to the discovery of two copper-transporting ATPases. An operon involved in copper homeostasis contains *copY*, *copZ*, *copA* and *copB*. Genes *copA* and *copB* encode P type ATPases: CopA serves in the uptake of copper and CopB in its extrusion. Genes *copY* and *copZ* located upstream of the *copAB* region control the

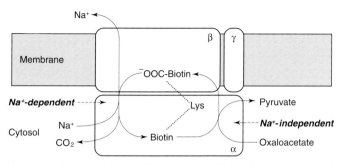

Figure 2 A model for the oxaloacetate decarboxylase sodium pump of *Klebsiella pneumoniae*. The α subunit (63 kDa) has the carboxyl-transferase activity on the N-terminal domain and the biotin-binding Lys residue on the C-terminal domain. The β (45 kDa; with the decarboxylase site) and the γ (9 kDa) sub-units are membrane-spanning proteins and the α subunit is attached to them. The carboxyl group is transferred from oxaloacetate to the prosthetic biotin of the enzyme and the carboxybiotin intermediate is decarboxylated sodium ion-dependently. As a result, sodium ion is transported across the membrane.

expression of the two ATPases: CopY acts as a repressor and CopZ as an activator (**Fig. 3**).

Metallothioneins, approximately 60 amino acids long with one-third of these as cysteines, are widely reported in eukaryotes including yeast and *Neurospora*. *Synechococcus* is the only genus with well-studied metallothionein in bacteria. Metallothionein from *Synechococcus* is a 56 amino acid product of the *smtA* gene and contains nine cysteine residues. The preference of cation binding for *Synechococcus* metallothionein is $Zn^{2+} > Cd^{2+} >> Cu^{2+}$. Cells that lack metallothionein have reduced uptake of Zn^{2+} and reduced tolerance to high Zn^{2+} concentrations.

Biosynthesis and Uptake of Vitamins

Biotin

Biotin functions only as a protein-bound cofactor which carries active CO_2 groups in metabolism. The biosynthetic pathway of biotin has been studied mainly by using strains of *Escherichia coli*. The biotin biosynthetic genes are organized in an operon in *Escherichia coli* but in two different clusters in *Bacillus sphaericus*. The enzymes of the pathway of dethiobiotin synthesis from pimelic acid are pimelyl-CoA synthetase (*bioW* gene product), 7-keto-8-amino-pelargonic acid synthetase (*bioF* gene product), 7,8-diaminopelargonic acid aminotransferase (*bioA* gene product) and dethiobiotin synthase (*bioD* gene product). The final step in biotin synthesis, the conversion of dethiobiotin to biotin, remains to be elucidated.

The biotin operon of *Escherichia coli* is a striking example of regulation, where the transcriptional regulatory protein (BirA) is also the enzyme, biotin protein ligase, that catalyses the covalent attachment of biotin to apoenzymes. The enzymes of biotin biosynthesis in *Escherichia coli* are encoded by a cluster of genes.

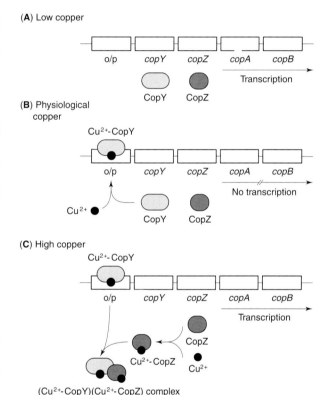

Figure 3 A model for the regulation of the *cop* operon in *Enterococcus hirae*. Expression of the *cop* operon of *Enterococcus hirae* is induced at both low and high copper concentrations. (**A**) Under copper-limiting conditions, CopY and CopZ are free so that the *cop* operon is expressed. (**B**) When the copper concentration is in the physiological range, CopY binds copper and represses transcription of the *cop* operon. (**C**) Under toxic copper concentration, CopZ also binds copper and the copper-CopZ complex binds to CopY and releases it from the operator.

Transcription of this cluster of genes is blocked in vivo upon addition of high levels of exogenous biotin in the medium of growing *Escherichia coli* cultures, whereas biotin starvation results in greatly increased transcription. The biotin operon regulation is unique in that the repressor is also the biotin protein ligase

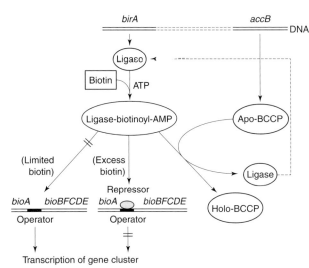

Figure 4 Regulation of biotin synthesis in *Escherichia coli*. Ligase-biotinoyl-AMP, repressor of the *bio* operon; BCCP, biotin carboxyl carrier protein. Transcription of the *bio* operon is blocked upon addition of excessive biotin, while biotin starvation results in increased transcription.

and the co-repressor is not biotin but biotinoyl-AMP, the product of the first half of the ligase reaction.

When the biotin supply is severely limited, the biotinoyl-AMP synthesized is rapidly consumed in biotination of apoenzyme molecules. Thus, the *bio* operator is seldom occupied and transcription is maximal. Repression of *bio* operon transcription occurs when the supply of biotin is in excess. Under these conditions the ligase-biotinoyl-AMP complex accumulates, binds to the *bio* operator and represses transcription (**Fig. 4**). The rate of biotin operon transcription is therefore sensitive not only to the intracellular concentration of biotin, but also to the supply of the protein to which the biotin must be attached in order to fulfil its metabolic role. Thus, accumulation of the apoenzyme increases the rate of biotin needed for its modification to holoenzyme.

Biotin is covalently attached to the enzyme through an amide bond that links the carboxyl group of the valeric acid side chain of biotin to the ε-amino group of a lysine residue in the enzyme. The covalent attachment of biotin to a specific lysine in apoenzyme is catalysed by biotin protein ligase in a two-step reaction:

$$ATP + biotin \rightleftharpoons biotinoyl\text{-}AMP + PPi$$
$$biotinoyl\text{-}AMP + apoenzyme \rightarrow holoenzyme + AMP$$

Escherichia coli biotin protein ligase, a protein called BirA, is the repressor protein that regulates transcription of the biotin operon as described earlier.

Cobalamin (Vitamin B$_{12}$)

Cobalamin (vitamin B$_{12}$) has a large molecular size and a complex chemical structure. Under anaerobic conditions, *Escherichia coli* and *Salmonella typhimurium* synthesize cobalamin *de novo*. About 30 enzymes are involved in the synthesis of cobalamin. Genes for biosynthesis and for transport of cobalamin in *Salmonella typhimurium* were located in a single, 20-gene operon (the *cob* operon) in this organism's genetic map.

In constructing B$_{12}$, multiple components are synthesized individually and then assembled. The largest component of B$_{12}$ is the corrinoid ring, synthesized from uroporphyrinogen III, a precursor common to haem. The nucleotide loop is assembled by first activating the aminopropanol side chain of cobinamide to form GDP-cobinamide. The Cobalt's lower (Co α) axial ligand, dimethylbenzimidazole is synthesized separately and converted to a nucleotide by addition of ribose derived from nicotinic acid mononucleotide. Ultimately, dimethylbenzimidazole nucleotide is added to the end of the activated isopropanol side chain to form the nucleotide pool and complete the synthesis of 5'-deoxyadenosylcobalamin. The *cob* operon encoding B$_{12}$ synthetic enzymes is induced by propanediol, the first step of which degradative pathway is the B$_{12}$-dependent diol dehydratase, using a regulatory protein (the PocR protein). Transcription of the *cob* operon is reduced in the presence of cobalamin.

Passage of B$_{12}$ through the outer membrane requires a specific transport system. Studies using *Escherichia coli* revealed one system for transport across the outer membrane and another for transport across the inner membrane. Transport through the outer membrane requires the BtuB protein acting with TonB. This system has a high affinity for vitamin B$_{12}$. Without the BtuB/TonB system, B$_{12}$ penetrates the outer membrane with extremely low efficiency. Once bound to the BtuB protein, B$_{12}$ is moved into the periplasm in an energy-dependent process which requires the TonB protein. Co-transport of calcium is required for successful passage of B$_{12}$ through the outer membrane.

The synthesis of these outer membrane transport proteins are regulated by the availability of cobalamin, which controls the expression of the genes. Transport across the inner membrane is provided by the *Escherichia coli* BtuC and BtuD proteins encoded by the *btuCED* operon. The BtuD protein has an ATP-binding site.

Folic Acid, Methanopterin

Folic acid derivatives are essential cofactors for a variety of enzymes involved in one-carbon transfer reactions in the biosynthesis of purines, pyrimidines and amino acids in living cells. Most bacterial species synthesize their own folate.

The folate biosynthetic pathway was elucidated by studies on bacterial species including *Streptococcus pneumoniae*. All five enzymatic activities that function in folate biosynthesis are coded for by a single gene cluster in *Streptococcus pneumoniae*. It begins with GTP, which is converted by GTP cyclohydrolase (encoded by *sulC* gene) to dihydroneopterin triphosphate. After phosphate residues are removed from dihydroneopterin triphosphate dihydroneopterin aldolase (*sulD* gene product) converts 7,8-dihydroneopterin to 6-hydroxymethyl-7,8-dihydropterin. The latter compound is converted to 6-hydroxymethyl-7,8-dihydropterin pyrophosphate by 6-hydroxymethyl-7,8-dihydropterin pyrophosphokinase (*sulD* gene product; a bifunctional enzyme) and ATP. Dihydropteroate synthase (*sulA* gene product) catalyses the linkage to p-aminobenzoate to give 7,8-dihydropteroate. The final addition of glutamate to dihydropteroate to produce dihydrofolate is catalysed by dihydrofolate synthetase (*sulB* gene product). Certain intestinal microflora are capable of *de novo* synthesis of folate, some of which is incorporated into the tissue folate of the host.

In the methanogenic bacteria such as *Methanococcus volta* and *Methanobacterium formicicum*, methanopterin, a structurally modified folic acid, functions in the same way folic acid does in other cells. The pterin portion of methanopterin is biosynthetically derived from 7,8-dihydro-6-(hydroxymethyl)pterin, which is coupled to methaniline by a pathway analogous to the biosynthesis of folic acid.

Niacin, NAD

Nicotinamide adenine dinucleotide (NAD) is synthesized from quinolinic acid in a similar pathway in all organisms and functions as a cofactor in numerous oxidation-reduction reactions. Quinolinic acid is synthesized from L-aspartate and dihydroxyacetone phosphate by quinolinic acid synthetase complex in *Escherichia coli* and *Salmonella typhimurium*. This complex consists of L-aspartate oxidase and quinolinate synthetase A, encoded by the genes *nadB* and *nadA*, respectively. Quinolinic acid is then converted to nicotinic acid mononucleotide (NaMN) by quinolinate phosphoribosyltransferase, which is coded by the *nadC* gene. NaMN then reacts with ATP in a reaction catalysed by the *nadD* gene product to give nicotinic acid adenine dinucleotide, which is then amidated by the *nadE* gene product to NAD.

Cells can also convert exogenous nicotinic acid to NaMN by nicotinic acid phosphoribosyltransferase. When *nadA*, *nadB* or *nadC* is inactivated, cells become dependent on nicotinic acid for growth.

In *Salmonella typhimurium*, the initial steps of the *de novo* biosynthetic pathway, L-aspartate oxidase (*nadB*) and quinolinic acid synthetase (*nadA*) and one of the components of the scavenging system, nicotinic acid phosphoribosyltransferase (*pncB*) are negatively controlled at the transcription level by a repressor encoded by the *nadR* gene (referred to as *nadI* by other investigators). A gene involved in nicotinamide mononucleotide (NMN) transport, *pnuA*, and *nadI* gene are believed to be a single bifunctional gene. Both functions of this protein appear to exert their control in response to internal NAD levels.

Pantothenic Acid, Coenzyme A

Pantothenic acid is the precursor of coenzyme A. Coenzyme A (CoA) and 4′-phosphopantetheine, the cofactor forms of pantothenic acid, participate in the enzyme-catalysed reactions of several metabolic pathways by formation of thioester bonds as the predominant acyl group carriers. Nearly a hundred enzymes require CoA. *Escherichia coli* and *Salmonella typhimurium* produce this vitamin from the condensation of β-alanine and pantoic acid (synthesized from α-ketoisovaleric acid via ketopantoic acid) by pantothenate synthetase encoded by *panC* gene.

Coenzyme A is synthesized by a sequence of five reactions from pantothenate. The phosphorylation of pantothenate by pantothenate kinase (coded by *coaA*) is the first reaction in the pathway. Next, 4′-phosphopantothenate is converted to 4′-phosphopantothenoyl-cysteine and then decarboxylated to 4′-phosphopantetheine. This is then converted to dephospho-CoA by phosphopantetheine adenylyltransferase. Biosynthesis of CoA is regulated at pantothenate kinase level by feedback inhibition by nonesterified CoA.

A saturable pantothenate uptake system is present in *Escherichia coli* which is sensitive to uncouplers of oxidative phosphorylation. Sodium ions stimulate pantothenate uptake.

Pyridoxine (Vitamin B₆)

Pyridoxine is the pyridine ring-containing precursor of essential coenzyme pyridoxal phosphate, which is utilized by enzymes in all phases of amino acid metabolism. Pyridoxine is synthesized by numerous bacteria and fungi.

In *Escherichia coli* K12, erythrose 4-phosphate is converted to 3-hydroxy-4-phosphohydroxy-α-ketobutyrate by two successive dehydrogenase reactions. The latter compound is transaminated to 4-phosphohydroxy-L-threonine by *serC* gene product;

4-phosphohydroxy-L-threonine and 1-deoxy-D-xylulose 5-phosphate are then joined to form pyridoxine 5'-phosphate by *pdxJ*- and *pdxA*-encoded enzymes. Pyridoxine 5'-phosphate is oxidized to pyridoxal 5'-phosphate by PNP/PMP oxidase (encoded by *pdxH*).

Pyridoxal/pyridoxamine/pyridoxine kinase (*pdxK* gene product) does not catalyse an obligatory step in *de novo* pyridoxine 5'-phosphate biosynthesis. Instead, it acts solely in a salvage pathway.

Riboflavin (Vitamin B₂)

Riboflavin-derived FMN and FAD are the versatile redox cofactors of a large number of flavoenzymes. The first step in the biosynthesis of riboflavin is the conversion of GTP to 2,5-diamino-6-ribosylamino-4-pyrimidinone 5'-phosphate. The subsequent steps are different in bacteria and fungi, though they ultimately lead to the same intermediate, 5-amino-6-ribitylamino-2,4-pyrimidinedione 5'-phosphate. In yeast, the ribosyl side chain of 2,5-diamino-6-ribosylamino-4-pyrimidinone 5'-phosphate is reduced and then deaminated to yield 5-amino-6-ribitylamino-2,4-pyrimidinedione 5'-phosphate. On the other hand, in *Escherichia coli* the deamination of the pyrimidine ring is followed by reduction of the side chain. The final step in the biosynthesis of riboflavin is dismutation of two molecules of 6,7-dimethyl-8-ribityllumazine by the enzyme riboflavin synthase to yield riboflavin and 5-amino-6-ribitylamino-2,4-pyrimidinedione.

The genes encoding the riboflavin biosynthetic enzymes of *Bacillus subtilis* were found to be clustered in a single operon (*rib* operon). The gene products of the *rib* operon (RibG, RibB, RibA and RibH) catalyse the conversion of GTP and ribulose 5-phosphate to riboflavin. Flavokinases catalyse the conversion of riboflavin to FMN and FAD-synthetase converts FMN to FAD. In *Bacillus subtilis*, *ribC* gene encodes a bifunctional flavokinase/FAD-synthetase. In *Bacillus subtilis*, flavin nucleotides (FMN and/or FAD), but not riboflavin, act as effector molecules controlling riboflavin biosynthesis.

Both FAD and FMN are covalently bound to a histidine residue of the apoflavoprotein polypeptide by an 8α-(N^3-histidyl)-riboflavin linkage.

Thiamin (Vitamin B₁)

Thiamin pyrophosphate is a cofactor for a number of enzymes such as transketolase, pyruvate dehydrogenase and α-ketoglutarate dehydrogenase. Thiamin monophosphate is formed in *Salmonella typhimurium* and *Escherichia coli* by coupling 4-methyl-5-(β-hydroxyethyl)thiazole monophosphate and 4-amino-5-hydroxymethyl-2-methylpyrimidine pyrophosphate. The thiazole moiety is thought to be derived from cysteine, tyrosine and 1-deoxy-D-threo-2-pentulose, while the pyrimidine moiety is thought to be derived from aminoimidazole ribotide, an intermediate in purine biosythesis.

The *thi* cluster of *Escherichia coli* contains five genes involved in thiamin synthesis, designated *thiCEFGH*. The *thi* cluster forms an operon. In *Salmonella typhimurium*, the *thi* cluster encodes an operon whose transcription is regulated by thiamin. The thiamin pyrophosphate is the effector molecule and exogenously added thiamin is converted to thiamin pyrophosphate to exert repression of *thiCEFGH* transcription.

In *Saccharomyces cerevisiae*, two regulatory genes, *thi2* and *thi3*, are involved in the expression of the thiamin-sensitive genes; *thi2* controls expression of the thiamin-sensitive acid phosphatase and the thiamin biosynthetic genes, while *thi3* controls thiamin transport in addition to the phosphatase and the biosynthetic genes.

Ubiquinone (Coenzyme Q)

Ubiquinone is a lipid consisting of a quinone head group and a polyprenyl tail which varies in length depending on the organism. Ubiquinone is a component of the membrane-bound electron transport chains and serves as a redox mediator in aerobic respiration via reversible redox cycling between ubiquinol, the reduced form of coenzyme Q, and ubiquinone. Ubiquinol possesses significant antioxidant properties and may play an important role in the protection of lipids in the cell.

The quinone ring of ubiquinone derives from tyrosine, the isoprenoid side chain from mevalonic acid and methyl and methoxyl groups attached to the quinone ring derive from *S*-adenosyl-L-methionine.

See also: **Bacillus**: Bacillus subtilis. **Escherichia coli**: Escherichia coli. **Metabolic Pathways**: Release of energy (aerobic); Release of energy (anaerobic); Nitrogen metabolism. **Saccharomyces**: Saccharomyces cerevisiae. **Salmonella**: Introduction. **Streptococcus**: Introduction.

Further Reading

Brandsch R (1994) Regulation of gene expression by cofactors derived from vitamins. *Journal of Nutritional Science and Vitaminology* 40: 371–399.

Cronan JE (1989) The *E. coli* bio operon: transcriptional repression by an essential protein modification enzyme. *Cell* 58: 427–429.

Dimroth P (1997) Primary sodium ion translocating enzymes. *Biochimica et Biophysica Acta* 1318: 11–51.

Harrison PM and Arosio P (1996) The ferritins: molecular properties, iron storage function and cellular regulation. *Biochimica et Biophysica Acta* 1275: 161–203.

Nikaido H and Saier MH (1992) Transport proteins in bacteria: common themes in their design. *Science* 258: 936–942.

Roth JR, Lawrence JG and Bobik TA (1996) Cobalamin (coenzyme B_{12}): synthesis and biological significance. *Annual Review of Microbiology* 50: 137–181.

Silver S and Walderhaug M (1992) Gene regulation of plasmid- and chromosome-determined inorganic ion transport in Bacteria. *Microbiological Reviews* 56: 195–228.

Silver S and Phung LeT (1996) Bacterial heavy metal resistance: new surprises. *Annual Review of Microbiology* 50: 753–789.

Umezawa C and Kishi T (1989) Vitamin metabolism. In: Rose AH and Harrison JS (eds) *The Yeast*, 2nd edn. Vol. 3, p. 457. London: Academic Press.

Vulpe CD and Packman S (1995) Cellular copper transport. *Annual Review of Nutrition* 15: 293–322.

Winzerling JJ and Law JH (1997) Comparative nutrition of iron and copper. *Annual Review of Nutrition* 17: 501–526.

Production of Secondary Metabolites – Fungi

Poonam Nigam, Biotechnology Research Group, University of Ulster, Coleraine, UK

Dalel Singh, Microbiology Department, CCS Haryana Agricultural University, Hisar, India

Copyright © 1999 Academic Press

Introduction

Secondary metabolites usually accumulate during the later stage of fermentation, known as the idiophase, which follows the active growth phase called the trophophase. Compounds produced in the idiophase have no direct relationship to the synthesis of cell material and normal growth of the microorganisms. Secondary metabolites are formed in a fermentation medium after the microbial growth is completed. The most common secondary metabolites are antibiotics; others include mycotoxins, ergot alkaloids, the widely used immunosuppressant cyclosporin, and fumagillin, an inhibitor of angiogenesis and a suppressor of tumour growth.

Characteristics of Secondary Metabolites

In secondary metabolism, the desired product is usually not derived from the primary growth substrate but a product formed from the primary growth substrate acts as a substrate for the production of a secondary metabolite. Secondary metabolites have the following characteristics:

- secondary metabolites can be produced only by a few microorganisms
- these compounds are not essential for the organisms's own growth and reproduction
- growth conditions, especially the composition of medium within a fermentation system, control the formation of secondary metabolites
- these compounds are produced as a group of closely related structures
- secondary metabolic compounds can be overproduced.

Metabolic Pathway for Secondary Metabolites

For the production of a desired secondary metabolite, it is essential to ensure that appropriate conditions for metabolic pathways are provided during the trophophase to maximize growth of the microbial species; it is also important that the conditions are properly altered at the appropriate time of fermentation to obtain the best product yield.

Microorganisms cultured under ideal conditions for primary metabolism without environmental limitations attempt to maximize the microbial biomass formation. Under conditions of balanced growth, however, the microbial cell minimizes the accumulation of any particular cellular building blocks in amounts beyond those required for growth. Hence, the metabolic pathway of a particular microorganism can be manipulated for the production of a large excess of the desired metabolite.

Most secondary metabolites are complex organic molecules which require a large number of specific enzymatic reactions for synthesis. One characteristic of a secondary metabolite is that the enzymes involved in the production of the secondary metabolite are regulated separately from the enzymes of primary metabolism. In some cases, specific inducers of secondary metabolite production have been identified. The metabolic pathways of these secondary metabolites start from primary metabolism, because the starting materials for the secondary metabolism come from the major biosynthetic pathways. Many structurally complex secondary metabolites originate from structurally quite similar precursors. Thus, the secondary metabolite is generally produced from several intermediate products that accumulate in the fermentation medium and or in microbial cells, during primary metabolism.

Table 1 Antibiotics produced by fungal cultures

Antibiotic group	Produced by	Spectrum of action	Cell target
Cephalosporin	Cephalosporium acremonium	Broad spectrum	Cell wall
Fumagillin	Aspergillus fumigatus	Amoebae	
Griseofulvin	Penicillium griseofulvum	Fungi	Microtubules in fungi
	P. nigricans		
	P. urticae		
Penicillins	Penicillium chrysogenum	Gram-positive bacteria	Cell wall
	Aspergillus nidulans		
	Cephalosporium acremonium		
Pleuromutilin	Pleurotus mutilus	Gram-positive bacteria	Mycoplasms
	Pleurotus passeckerianus		
Fusidic acid	Fusidium coccinium	Prokaryotes and eukaryotes	Ribosomal translocation
	Acremonium fusidioides		

Transformation Within Cells

There are several hypotheses about the role of secondary metabolites. Besides the five phases of the cell's own metabolism – intermediary metabolism, regulation, transport, differentiation and morphogenesis – secondary metabolism is the activity centre for the evolution of further biochemical development. This development can proceed without damaging primary metabolite production. Genetic changes leading to the modification of secondary metabolites would not be expected to have any major effect on normal cell function. If a genetic change leads to the formation of a compound that may be beneficial, then this genetic change would be fixed in the cell's genome and becomes essential, with the result that this secondary metabolite would be converted into a primary metabolite.

Antibiotics

Antibiotics are chemical substances produced by certain microorganisms as products of secondary metabolism. These substances possess activity to inhibit growth processes or kill other microorganisms, even used at low concentrations. Growth inhibition of one organism by another organism in mixed culture has been known for a long time. The most famous example is the growth inhibition observed by Alexander Fleming in 1929. He noticed that staphylococcal growth on plate culture was inhibited by a contaminant, *Penicillium notatum*, which produced the antibiotic penicillin. Medicinally useful antibiotics have shown their impact on the treatment of infectious diseases. Some less effective antibiotics work after a chemical modification, making them semi-synthetic antibiotics. The sensitivity of microorganisms and other chemotherapeutic agents varies. Gram-positive bacteria are more sensitive to antibiotics than are Gram-negative bacteria. Broad-spectrum antibiotics act on Gram-positive as well as

Gram-negative bacteria and are therefore used more widely in medicine than narrow-spectrum antibiotics, which are effective for only a single group of microorganisms.

Antibiotics are produced by bacteria (about 950 types of antibiotics), actinomycetes (about 4600 types) and fungi (about 1600 types). This article deals only with secondary metabolites that are produced by fungal cultures (**Table 1**).

The antibiotics produced by the Aspergillaceae and Moniliales are of practical importance. Only 10 of the known fungal antibiotics are produced commercially and only the penicillins, cephalosporin C, griseofulvin, and fusidic acid are clinically important. Penicillins, cephalosporins and cephamycins belong to the β-lactam group of antibiotics, so called because their structure consists of a β-lactam ring system (**Fig. 1**). All of these are medically useful antibiotics produced by fungi.

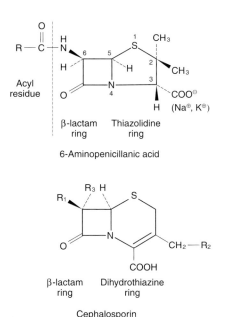

Figure 1 Beta-lactam antibiotics from fungi.

Table 2 Three main types of penicillins

Natural	Biosynthetic	Semi-synthetic
Benzylpenicillin (penicillin G)	Benzylpenicillin (penicillin G): acid labile low activity against Gram-negative bacteria β-lactamese sensitive	Propicillin: acid stable β-lactamase sensitive
2-Pentenylpenicillin (penicillin F)		Methicillin: acid stable β-lactamase resistant
n-Amylpenicillin (penicillin-DihydroF)	Phenoxymethylpenicillin (penicillin V): acid-stable β-lactamase sensitive Low activity against Gram-negative bacteria	Oxacillin acid stable β-lactamase resistant
Methylpenicillin n-Heptylpenicillin (penicillin K) p-Hydroxybenzylpenicillin (penicillin X)	Allylmercaptomethyl penicillin (penicillin O): reduced allergenic properties	Ampicillin: broadened spectrum of activity (against Gram-negative bacteria) acid stable β-lactamase sensitive Carbenicillin: broadened spectrum of activity (against *Pseudomonas aeruginosa*) acid stable ineffective orally β-lactamase sensitive
Penicillin N (synnematin B) (D-4-amino-4-carboxy-n-butylpenicillin)		
Isopenicillin N (L-4-amino-4-carboxy-n-butylpenicillin)		

Penicillins

The basic structure of the penicillins is 6-amino-penicillanic acid which consists of a thiazolidine ring with a condensed β-lactam ring. The types of penicillins are presented in **Table 2**.

- natural penicillins: the fermentation is carried out without addition of side-chain precursors
- biosynthetic penicillins: out of over a hundred biosynthetic types, only benzylpenicillin, phenoxy-methylpenicillin and allylmercaptomethylpenicillin (penicillins G, V, and O) are commercially produced
- semi-synthetic penicillins: benzylpenicillin and phenoxymethylpenicillin (penicillins G and V) are used in their synthesis; these penicillins have broadened action spectrum and improved characteristics such as acid stability, resistance to plasmid or chromosomally coded β-lactamases and expanded antimicrobial effectiveness, and are therefore used extensively in therapy.

Synthetic Pathway and Regulation of Penicillin Formation The β-lactam–thiazolidine ring of penicillin is constructed from L-cysteine and L-valine in a non-ribosomal process by means of a dipeptide composed of L-α-aminoadipic acid (L-α-AAA) and L-cysteine. Subsequently, L-valine is connected by an epi-merization reaction resulting in formation of the tripeptide. The first product of the cyclization of tripeptide is isopenicillin N. Benzylpenicillin is produced in the exchange of L-α-AAA with activated

phenylacetic acid (**Fig. 2**). Penicillin biosynthesis is affected by phosphate concentration, shows a distinct catabolite repression by glucose, and is regulated by ammonium ion concentration.

Industrial Production of Penicillin Benzylpenicillin and phenoxymethylpenicillin (penicillins G and V) are produced in a submerged process (**Fig. 3**) in fermenters from 40 000 litres to 200 000 litres in size. The process is highly aerobic with a volumetric oxygen absorption rate of 0.4–0.8 mmol l^{-1} min^{-1}, an aeration rate of 0.5–1.0 vvm, and an optimal temperature range 25–27°C. A typical penicillin fermentation medium consists of corn-steep liquor, an additional nitrogen source such as yeast extract, whey or soya meal, and a carbon source such as lactose; the pH is maintained at 6.5 and phenylacetic acid or phenoxyacetic acid is fed continuously as a precursor.

Cephalosporins

Cephalosporins are β-lactam antibiotics containing a dihydrothiazine ring with D-α-aminoadipic acid. Cephalosporins are produced by *Cephalosporium acremonium* (*Acremonium chrysogenum*), *Emericellopsis* and *Paecilomyces* spp. Cephalosporins are less toxic and have a broader spectrum of action than ampicillin. Thirteen therapeutically important semi-synthetic cephalosporins are commercially produced.

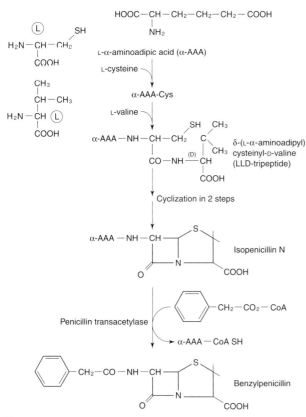

Figure 2   Biosynthesis of penicillin in *Penicillium chrysogenum*.

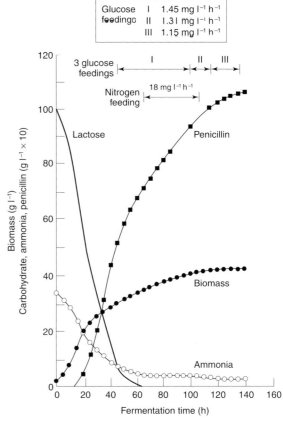

Figure 3   Penicillin fermentation with *Penicillium chrysogenum*.

Synthetic Pathway of Cephalosporins in Fungi

Cephalosporin biosynthesis (**Fig. 4**) proceeds from δ-(α-aminoadipyl-L-cysteinyl-D-valine to isopenicillin N. In the next stage, penicillin N is produced by transformation of the L-α-AAA side chain into the D-form, by the action of a very labile racemase. After ring expansion to deacetoxycephalosporin C by the expandase reaction, hydroxylation via a dioxygenase to deacetylcephalosporin C occurs. The acetylation of cephalosporin C by an acetyl-CoA dependent transferase is the end point of the pathway in fungi.

Industrial Production of Cephalosporins

Fermentations are carried out as fed-batch processes with semicontinuous addition of nutrients at pH 6.0–7.0, temperature 24–28°C in complex media with corn-steep liquor, meat meal, sucrose, glucose and ammonium acetate. The biosynthesis is affected by phosphate, nitrogen and carbohydrate catabolite regulation. Rapidly metabolizable carbon sources, such as glucose, maltose or glycerol, reduce the production. The repression of expandase is the most significant effect. Lysine in low concentrations and methionine stimulate the synthesis.

Fusidic Acid

The antibiotic fusidic acid was first isolated in 1960 from fermentations of the imperfect fungus *Fusidium coccineum* (Moniliaceae) or *Acremonium fusidioides*. In addition, the production of fusidic acid by strains of *Cephalosporium*, various dermatophytes and *Isaria kogane* has been reported. Ramycin, an antibiotic mixture, has been isolated from the culture fluid of the zygomycete *Mucor ramannianus* and the identity of one of the components with fusidic acid was proved later.

Fusidic acid belongs chemically to the group of tetracyclic triterpenoids with a fusidane skeleton. This type of hydrocarbon skeleton is also present in many natural steroids and triterpenes and contains a cyclopentanoper-hydrophenanthrene ring connected at C17 with an α,β-unsaturated carboxylic acid side chain and a β-(16,21) *cis*-oriented acetoxy group on C16. Naturally occurring antibiotics related to fusidic acid comprise helvolic acid from cultures of *Aspergillus fumigatus* and *Cephalosporium caerulens*; cephalosporin P1 and related derivatives from *Cephalosporium acremonium*; and the viridominic acids A, B and C from a *Cladosporium* species.

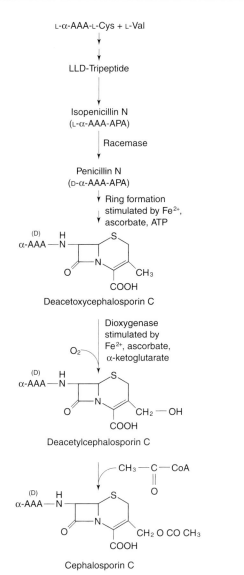

Figure 4 Biosynthesis of cephalosporin C by *Cephalosporium acremonium*.

Biosynthetic Pathway of Fusidic Acid   The biosynthesis of fusidic acid follows the general pathway for the formation of sterols and polycyclic triterpenes. The isolation of common intermediates such as several protosterols in the biosynthesis of fusidic acid and helvolic acid from the mycelium of *Fusidium coccineum* and *Cephalosporium caerulens* indicates that the biogenetic pathways leading to these antibiotics are identical.

Besides the total syntheses of fusidic acid, several hundred semi-synthetic derivatives have been synthesized by chemical or microbial modification in order to achieve a broader antibacterial spectrum, increased potency, modified pharmacokinetics or better stability in solution. One derivative, 16-deacetoxy-16β-acetylthiofusidic acid, is more stable and twice as active as fusidic acid against several Gram-positive bacteria.

Large-scale production of fusidic acid is carried out in batch fermentations using a complex medium containing sucrose, glycerol or glucose as the carbon source, soya bean meal, corn-steep liquor or milk powder as the nitrogen source, vitamins (biotin) and inorganic salts. The fermentation is carried out at 27–28°C for 180–200 h with efficient aeration and vigorous agitation with a high-producing mutant of *Fusidium coccineum*.

Applications of Fusidic Acid   Fusidic acid is used for the treatment of multiply resistant staphylococcal infections or in combination with other antibiotics. Systemic application includes treatment of septicaemias, endocarditis, staphylococcal pneumonia, osteomyelitis and wound infections. Topically applied, fusidic acid is effective in the treatment of staphylococcal and streptococcal skin infections, wounds, burns and ulcers. Fusidic acid is available in various forms of pharmaceutical presentations (Fucidin), either in the form of tablets containing sodium fusidate, as an aqueous solution for parenterally or intravenous infusion in a sterile buffer, and as an ointment for topical applications. Immuno-suppressive activities have been reported for fusidic acid on activated blood mononuclear cells. Fusidic acid inhibits protein biosynthesis in both prokaryotes and eukaryotes. The antibiotic binds to the translocation factors in prokaryotic and eukaryotic cell-free systems. Resistance to fusidic acid has been mainly studied in *Staphylococcus aureus* and *Escherichia coli*.

Griseofulvin

The systemic antifungal antibiotic griseofulvin was first isolated from the mycelium of *Penicillium griseofulvum* Dierckx. In 1946 a compound named 'curling factor' was isolated from the mycelium and the culture filtrate of *Penicillium janczewskii* Zal. It caused abnormal curling of the hyphae of *Botrytis alii* and was later identified as griseofulvin. Many other fungi were shown to produce griseofulvin with most of these species belonging to the genus *Penicillium* (e.g. *P. urticae*, *P. raistrickii*, *P. raciborskii*, *P. kapuscinskii*, *P. albidium*, *P. melinii*, *P. brefeldianum* and some mutant strains of *P. patullum*). In addition, *Aspergillus versicolor* and *Nematospora coryli* have been shown to produce the antibiotic. The carbon skeleton of griseofulvin is a tricyclic-spiro system based on grisan and consists of a chlorine-substituted coumaranone and an enone containing a cyclohexane ring adjacent to the asymmetric spirane centre.

Biosynthesis of Griseofulvin Griseofulvin is formed by linear combination of acetate units with the benzophenone as a possible intermediate. Oxidative coupling followed by saturation of one of the double bonds in the resulting dienone may form griseofulvin, as demonstrated by labelling with [2-^{3}H]- and [^{14}C]acetate. The double bond saturation in the intermediate dienone occurs via *trans* addition of hydrogen. Labelling with [1-^{13}C, ^{18}O$_2$]acetate and analysis by ^{13}C nuclear magnetic resonance spectroscopy proved that all oxygen atoms derive from acetate. The occurrence of dechlorogriseofulvin in some producing fungi (e.g. *Aspergillus versicolor*) indicates that the chlorination must occur as a late step in the biosynthesis of griseofulvin, although the exact mechanism of this reaction has not yet been elucidated.

Commercial Production Griseofulvin is commercially produced in submerged culture with mutant strains of *Penicillium patulum*, *P. raistrikii* or *P. urticae* which have been obtained by mutating the spores with ultraviolet light, chemical mutagens or sulphur isotopes, in a corn-steep medium in which the factors of pH (through intermittent addition of glucose), aeration and the concentration of chloride and nitrogen are carefully controlled. A typical griseofulvin titre of 6–8 g l^{-1} is achieved after 10 days of fermentation. Higher yields (up to 12–15.5 g l^{-1}) can be obtained by the addition of various methyl donors (choline salts, methyl xanthate and folic acid) to the medium.

Applications of Griseofulvin Griseofulvin is used for the treatment of infections caused by species of certain dermatophytic fungi (*Epidermophyton*, *Trichophyton* and *Microsporum*) which cannot be cured by topical therapy with other antifungal drugs. In vitro, minimal inhibitory concentrations ranging from 0.18 μg ml^{-1} to 0.42 μg ml^{-1} against various dermatophytes have been reported. The drug has no effect on bacteria, other pathogenic fungi and yeasts. Griseofulvin is effective in vivo in cutaneous mycoses because, when administered orally, it concentrates in the deep cutaneous layer and the keratin cells. The uptake of griseofulvin into the susceptible fungal cells is an energy-requiring process dependent on concentration, temperature, pH and an energy source such as glucose. It has been suggested that insensitive fungi and yeasts do not bind sufficient amounts of the antibiotic.

Pleuromutilin (Tiamulin)

The only commercial antibiotic produced by a basidiomycete is the diterpene pleuromutilin. Pleuromutilin was first isolated from *Pleurotus mutilus*

and *Pleurotus passeckerianus* in a screening for antibacterial compounds. Pleuromutilin is active against Gram-positive bacteria but its most interesting biological activity is its effectiveness against various forms of mycoplasms. The preparation of more than 66 derivatives of pleuromutilin resulted in the development of tiamulin, which exceeds the activity of the parent compound against Gram-positive bacteria and mycoplasms by a factor of 10–50. The minimal inhibitory concentrations against different strains of *Mycoplasma* were in the range 0.0039–6.25 μg ml^{-1}.

Production of Pleuromutilin Pleuromutilin can be produced by fermentation in a medium composed of glucose 50 g, autolysed brewer's yeast 50 g, KH$_2$PO$_4$ 50 g, MgSO$_4$·7H$_2$O 0.5 g, Ca(NO$_3$)$_2$ 0.5 g, NaCl 0.1 g, FeSO$_4$·7H$_2$O 0.5 g, water to 1 litre, pH 6.0. The yield after six days of growth in a 1000 litre fermenter was reported as 2.2 g l^{-1}. It could be demonstrated that during fermentation of pleuromutilin, derivatives differing in the acetyl portions attached to the 14-OH group of mutilin were formed. The biosynthesis of these derivatives was stimulated strongly by addition of corn oil as the carbon source during fermentation. Pleuromutilin overproducers have been obtained by conventional mutagenesis and selection programmes as well as by protoplast fusion and genetic studies.

Applications of Pleuromutilin Studies on the mode of action revealed that pleuromutilin and its derivatives act as inhibitors of prokaryotic protein synthesis by interfering with the activities of the 70S ribosomal sub-unit. The ribosome-bound antibiotics lead to the formation of inactive initiation complexes which are unable to enter the peptide chain elongation cycle. In various bacteria, resistance to the drug develops in a stepwise fashion. Because of its outstanding properties, pleuromutilin is used for the treatment of mycoplasma infections in animals.

Other Secondary Metabolites

Mycotoxins

Mycotoxins are natural products produced by filamentous fungi and their biological activity is toxicity against vertebrates. Although both chemically and biologically diverse, they are all fungal secondary metabolites. As such, the principles of their biosynthesis, physiology and evolution are similar to those of antibiotics and other pharmacologically active secondary metabolites.

Table 3 Important mycotoxins and their producer fungi

Class	Chemical taxonomy	Major producing species
Aflatoxins	Polyketides	*Aspergillus flavus, A. parasiticus, A. nomius*
Citrinin	Polyketides	*Penicillium citrinum, P. verrucosum,* numerous *Aspergillus* and *Penicillium* spp.
Ergot	Amino acid-derived	Numerous *Claviceps* spp.
Fumonisin	Amino acid-derived	*Fusarium moniliforme, F. proliferatum, F. napiforme, F. nygamai,* other *Fusarium* spp.
Ochratoxin	Polyketides	*Aspergillus ochraceus, Penicillium verrucosum,* numerous *Aspergillus* and *Penicillium* spp.
Patulin	Polyketides	*Penicillium expansum, P. griseofulvum* and *Aspergillus* spp.
Rubratoxin		*Penicillium rubrum*
Sterigmatocystin	Polyketides	*Aspergillus versicolor,* numerous *Aspergillus* spp.
Tremorgens		*Penicillium cyclopium, Alternaria tenuis, Phoma sorghina, Pyricularia oryzae*
Trichothecenes	Sesquiterpenoids	*Fusarium roseum, F. nivale,* several *Myrothecium roridum, M. verrucaria,* several *Fusarium* spp. *Trichothecium roseum*
Zearalenone	Polyketide	*Fusarium graminarium* and numerous *Fusarium* spp.

Impact of Mycotoxins Mycotoxins cause adverse health effects in human and livestock populations which range from acute toxicity and death to milder chronic conditions and impairment of reproductive efficiency. In addition, some mycotoxins show insecticidal, antimicrobial and phytotoxic effects. Mycotoxins cause huge economic losses in agriculture because they contaminate crops in the field, after harvest or during storage.

A few companies produce and sell mycotoxins as analytical standards. Otherwise the economic impact of mycotoxins is largely negative, and the major biotechnological emphasis in mycotoxin research is on prevention rather than production. These compounds play a major role in agricultural ecosystems. Elimination or minimization of mycotoxin contamination in the raw materials for industrial fermentations is a continuing biotechnological challenge. In addition, several classes of mycotoxins have emerged as models for research in the biosynthesis and molecular biology of fungal secondary metabolism. The genetic engineering of existing mycotoxin biosynthetic pathways ultimately could yield novel products for medicinal use.

Major Classes of Mycotoxins The extreme toxicity and carcinogenicity of the aflatoxins and the common occurrence of their producer *Aspergillus* species means that moulds are more than mere agents of deterioration. Reports that even trace levels of aflatoxin in feeds had disastrous consequences for young poultry led to the awareness that other mould metabolites may also have serious consequences for human and veterinary health.

Mycotoxins have been discovered in various ways. Aflatoxins were identified after outbreaks of turkey X disease. The symptoms caused by ingestion of ergot alkaloids – gangrenous necrosis, neurological disturbances and the human disease called St Anthony's

fire – have been known for centuries. Trichothecenes have also been implicated in several natural intoxications, e.g. alimentary toxic aleukia in human beings and a variety of mouldy corn toxicoses of domesticated animals. Ochratoxins, on the other hand, were discovered by laboratory screening targeted specifically at finding toxigenic fungi. Patulin and trichothecin were originally discovered as part of screens for new antibacterial compounds from fungi. During the 1960s they were reclassified from 'antibiotics too toxic for drug use' to 'mycotoxins'. A list of major classes of mycotoxins and producing fungal species is presented in **Table 3**.

Aflatoxins are the most biologically potent, economically important and scientifically understood of the mycotoxins. In addition to the acute toxicity leading to such conditions as turkey X disease, in laboratory tests aflatoxin B_1 is one of the most potent carcinogens known. There is strong epidemiological evidence linking aflatoxin to human liver cancer.

Under appropriate environmental conditions, aflatoxins are produced by toxigenic strains of *Aspergillus flavus* and *A. parasiticus*. The crops at greatest risk for aflatoxin contamination are corn, peanuts and cottonseed, but rice, nuts and spices are also susceptible. When animals consume aflatoxin-contaminated feeds, the toxic factor may be transferred to animal products such as meat and milk. After the aflatoxins, trichothecenes are the next most important group of mycotoxins.

Health Impact of Mycotoxins The toxic effects of mycotoxins can be divided into two broad categories:

- acute effects which cause rapid, often fatal diseases
- chronic effects which may cause weight loss, immunosuppression, cancer, reduced milk yields and other sublethal changes.

The wide range of pathological effects are listed in

Table 4 Range of pathological effects of mycotoxins

Mycotoxin group	Pathological effect
Aflatoxin	Hepatotoxicity, haematopoiesis, carcinogenicity
Ergot alkaloid	Vasoconstriction, neurotoxicity, reproductive irregularities
Fumonisins	Neurotoxicity
Ochratoxin	Haematopoiesis, carcinogenicity
Ochratoxin A	Nephrotoxicity
Sterigmatocystin	Carcinogenicity
Tremorgens	Neurotoxicity
Trichothecenes	Dermal toxicity
Zearalenone	Reproductive irregularities

Table 4. Diseases caused by mycotoxins – mycotoxicoses – are not only clinically diverse, but often extremely difficult to diagnose owing to the numerous pharmacological effects of mycotoxins. Human diseases associated with mycotoxin ingestion include St Anthony's fire (ergot alkaloids), alimentary toxic aleukia (T-2 toxin) and yellow rice disease (citrinin and citreoviridin). Most mycotoxicoses are known as veterinary syndromes. Some of the best-known include zearalenone as the cause of an oestrogenic syndrome in swine, fumonisins as the cause of a brain encephalopathy in horses and ochratoxins as the cause of a porcine nephropathy. Examples of specific human and veterinary mycotoxicoses are listed in **Table 5**.

There are virtually no effective treatments for any of these mycotoxicoses. Therefore, prevention is of the utmost importance.

Fungal Secondary Metabolites in Fermented Foods

Most of the mould-fermented foods are considered to be safe, even when they are produced using species of *Aspergillus* and *Penicillium* that include strains capable of producing mycotoxins. The inability of *Aspergillus oryzae* and *A. sojae* to produce aflatoxins is not understood; presumably aflatoxin production offers no selective advantage in the koji environment. Some species used for food fermentations are even able to reduce the mycotoxin concentration in the substrate. For example *Rhizopus oligosporus*, used

for tempeh fermentation, reduces aflatoxin present in the substrate to 40% and is able to inhibit growth, sporulation and aflatoxin production of *A. flavus*. Species of *Neurospora* used to prepare oncom (from peanuts) inhibit aflatoxin-producing strains of *A. flavus* and *A. parasiticus* by competition or antagonism. Nevertheless, the application of defined starter organisms will improve the quality and consistency of the products without the production of undesired secondary metabolites.

Meat Products The spontaneous mycoflora of mould-ripened salami and ham mainly consists of *Penicillium* spp. In mould-ripened sausages in Europe 50% of the *Penicillium* population was identified as *P. nalgiovense*; *P. verrucosum*, *P. oxalicum* and *P. commune* were only minor components of the mycoflora. The *Aspergillus* spp. *A. candidus*, *A. flavus*, *A. fumigatus*, *A. caespitosus*, *A. niger*, *A. sulphureus* and *A. wentii* were isolated from Italian hams. Mycelium of the mould penetrates the product, causing some biochemical changes by its metabolism.

Production of Secondary Metabolites in Meat Products Not only products of the fungal primary metabolism are formed but also secondary metabolites such as mycotoxins and antibiotics. For example, the antibiotic penicillin may be produced from *Penicillium chrysogenum*, *P. nalgiovense* and further species of the genus growing on fermented meat. The production of penicillin in meat products is not desirable, as it may cause allergic reactions in sensitive people. A continual ingestion of low doses of penicillin or other antibiotics may lead to the development of resistant bacteria in the human digestive tract. This antibiotic-resistant flora is able to transfer its genetic information to pathogenic bacteria and prevent therapy with this antibiotic. Also for technological reasons the presence of penicillin is undesirable; although pathogenic bacteria can be suppressed, it may also inhibit the bacterial starter organisms. The production of penicillin is a consistent characteristic of *P. nalgiovense* when grown on a medium optimal for penicillin production. It seems highly probable

Table 5 Selected human and veterinary diseases associated with mycotoxins

Causative toxin	Disease	Affected species	Food/Feed
T-2 and other *Fusarium* toxins	Alimentary toxic aleukia	Humans	Over-wintered wheat
Fumonisins	Encephalopathy	Horses	Grains
Sporidesmin	Facial eczema	Sheep	Pasture grass
Zearalenone	Hypoestrogenism	Swine	Corn
Ochratoxin	Nephropathy	Pigs, poultry	Barley, oats
Satratoxin H, roriden, verrucarin	Stachybotryotoxicosis	Horses, cattle	Hay, straw
Ergot alkaloids	St Anthony's fire	Humans	Rye bread
Citreoviridin, citrinin	Yellow rice disease	Humans	Rice
Aflatoxins	Turkey X disease	Turkeys, other poultry	Peanut meal, grain

that *P. nalgiovense* cannot produce penicillin on meat-based substrates, but selection of non-penicillin-producing strains is advised.

The problem of mycotoxin production in mould-ripened sausages and ham is often discussed. About 70–80% of the *Penicillium* species of the spontaneous flora of salami are potential producers of mycotoxins such as ochratoxin A and cyclopiazonic acid. From country-cured ham stored under dry conditions, aspergilli from the species *A. flavus* and *A. parasiticus* were rarely identified. No case has been reported of aflatoxin detection in fermented meat products from the market. The same can be said about the presence of sterigmatocystin. Ochratoxin can be produced on ham by *Aspergillus ochraceus* and *Penicillium verrucosum* under experimental conditions but no reports are available about occurrence of ochratoxin in market products of mould-ripened ham and sausages. Cyclopiazonic acid has frequently been isolated from *Penicillium* strains grown on mould-ripened sausages. Penicillic acid could not be detected after experimental inoculations of sausages with producer strains. It is suggested that this toxin is inactivated by reactions with amino acids in the meat. Although moulds isolated from fermented meat products have the potential to produce mycotoxins under appropriate conditions in laboratory media – and in some cases even on the fermented product – scant evidence exists that the market-ready products contain dangerous concentrations of mycotoxins; there is usually little carry-over of mycotoxins to the muscle tissues of animal.

Cheeses   Mould-ripened cheeses include the blue-veined cheeses, e.g. Roquefort, Blue (France), Gorgonzola (Italy), Brick, Muenster and Monterey (USA), Limburger (Belgium) and Stilton (UK), and the surface-ripened Camembert and Brie (France). Blue-veined cheeses are produced by inoculation of the curd with cultures of *Penicillium roqueforti*, which produces blue-green spores. Proteolytic enzymes of the mould contribute to the ripening of the cheese and influence texture and aroma; concomitantly, water-soluble lipolytic enzymes produce free fatty acids and mono-and diacylglycerols from milk fat. For the production of Camembert and Brie, white strains of *P. camemberti* form the surface crust.

Secondary Metabolites in Cheese *Penicillium camemberti* is able to produce the mycotoxin cyclopiazonic acid. From 61 strains tested, all synthesized cyclopiazonic acid. This mycotoxin is isolated from both laboratory media and commercial cheeses. In cheeses it is mainly produced in the rind and after storage at too-high temperatures. No risk to human health exists according to toxicological data and consumption habits. Mutants of *P. camemberti* that cannot produce cyclopiazonic acid were isolated. This may be a first step to improve the starter organisms by the methods of genetic engineering. For *P. roqueforti* the production of isofumigaclavine A and B, marfortines, mycophenolic acid, PR toxin and roquefortine C were described for chemotype I and botryodiploidin, mycophenolic acid, patulin, penicillic acid and roquefortine C for chemotype II.

In samples of commercial blue-veined cheeses, roquefortine was observed in all samples, and isofumigaclavine A and traces of isofumigaclavine B were observed in several samples; PR toxin was not detected. Mycophenolic acid is produced by some starter cultures in laboratory media and in cheese. Today starter cultures are available which do not have the ability to produce patulin, PR toxin, penicillic acid and mycophenolic acid. The toxicity of roquefortine and isofumigaclavines is relatively low. Adequate handling of the cheese during ripening and storage, and screening of strains with low potential for the production of roquefortine and isofumigaclavines and/or a modification with genetic methods, will improve the production of Roquefort.

Secondary Metabolites in Soy Sauce   The koji moulds are yellow-green aspergilli morphologically characterized as *Aspergillus oryzae* and *A. sojae*. A clear separation of these strains from the aflatoxin-producing *A. flavus* and *A. parasiticus* is difficult, because of the occurrence of intermediate forms. The conidia of the domesticated *A. oryzae* are larger and germinate faster than those of the wild *A. flavus*. Apparently the domesticated strains of *A. oryzae* and *A. sojae* have lost the ability to produce aflatoxins. Because of the relatedness of the koji strains to the aflatoxin-producing strains of *A. flavus*, there is a fundamental interest in the mycotoxin-producing abilities of the koji strains. No aflatoxin production has been demonstrated in *A. oryzae*, *A. sojae* and *A. tamarii*. Other mycotoxins are reported to be produced by these strains under special conditions. *Aspergillus oryzae* produces cyclopiazonic acid, kojic acid, 3-nitropropionic acid and maltoryzine; *A. sojae* produces aspergillic acid and kojic acid; and *A. tamarii* produces cyclopiazonic acid and kojic acid. Nevertheless *A. oryzae* has 'generally recognized as safe' status and is used for the production of enzymes. There is only scant evidence that these mycotoxins exist in industrial products. Generally the koji fermentation lasts 48–72 h, whereas toxin production needs a longer incubation (5–8 days). In addition the soya bean may be an unsuitable substrate for the production of mycotoxins, and the subsequent fermentation by bacteria and yeasts may inactivate any mycotoxins. The large industries use well-defined, non-toxigenic koji moulds as starters, but some small-scale manufacturers continue to use the house flora, where the risk exists of contamination by aflatoxin-producing strains of *A. flavus* and *A. parasiticus*.

See also: **Aspergillus**: Introduction; *Aspergillus oryzae*; *Aspergillus flavus*. **Cheese**: Mould-ripened Varieties. **Fermented Foods**: Origins and Applications; Fermented Meat Products; Fermentations of the Far East. **Fermented Milks**: Range of Products. **Mycotoxins**: Classification; Occurrence; Detection and Analysis by Classical Techniques; Immunological Techniques for Detection and Analysis; Toxicology. **Penicillium**: Introduction; *Penicillium in Food Production*.

Further Reading

Bery J (1986) Further antibiotics with practical applications. In: Rehm HJ and Reed G (eds) *Biotechnology*. Vol. 4, p. 465. Weinheim: VCH Verlagsgessellschaft.

Bhatnagar D, Lillehoj EB and Arora A (1992) *Handbook of Applied Mycology*. Vol. 5: *Mycotoxins in Ecological Systems*. New York: Marcel Dekker.

Buckland BC, Omstead DR and Santamaria V (1985) Novel β-lactam antibiotics. In: Moo-Young M (ed.) *Comprehensive Biotechnology*. Vol. 3, p. 49. Oxford: Pergamon.

Cole RJ (1986) *Modern Methods in the Analysis and Structural Elucidation of Mycotoxins*. San Diego: Academic Press.

Crueger W and Crueger A (1989) Antibiotics. In: Brock TD (ed.) *Biotechnology: A Textbook of Industrial Microbiology*, 2nd edn. P. 229. MA: Sinauer Associates Inc., USA.

Ellis WO, Smith JP, Simpson BK et al (1991) Aflatoxins in food: occurrence, biosynthesis, effects on organisms detection, and methods of control. *Critical Reviews on Food Science and Nutrition* 30: 403–439.

Hesseltine CW (1986) Global significance of mycotoxins. In: Steyn PS and Vleggaar R (eds) *Mycotoxins and Phycotoxins*. P. 1. Amsterdam: Elsevier.

Page MI (ed.) (1992) *The Chemistry of β-Lactams*. London: Chapman & Hall.

Tim A (ed.) (1997) *Fungal Biotechnology*. London: Chapman & Hall.

Vining LC and Stuttard C (eds) (1994) *Genetics and Biochemistry of Antibiotic Production*. Oxford: Butterworth-Heinemann.

Production of Secondary Metabolites – Bacteria

M D Alur, Food Technology Division, Bhabha Atomic Research Centre, Mumbai, India

This article briefly discusses the production of bacterial secondary metabolites such as antibiotics, alkaloids, steroids, and polyketides as well as their applications in food and pharmaceutical industries. Regulation of secondary metabolites, site of action of antibiotics, production of toxin and the impact of recombinant DNA technology on the production of useful products by bacteria are described.

Metabolism which is not essential for and plays no part in growth is referred to as secondary metabolism. It commonly occurs maximally under conditions of no restricted growth or absence of growth, e.g. at the end of log-phase or trophophase growth in batch cultures. Secondary metabolites include substances such as antibiotics and mycotoxins and are produced from substrate provided by primary metabolism, particularly shikimic acid (precursor of many aromatic compounds), amino acids (precursor of many alkaloids and antibiotics) and acetate (precursor of isoprenoids and many toxins).

Secondary Metabolites

Substances such as pigments, alkaloids, antibiotics and terpenes which only occur in certain organisms are the products of secondary metabolism. They are thus distinct from primary metabolites (products of primary metabolism) involved in energy metabolism and growth. Primary metabolites consist of a large group of compounds such as glycolysis and tricarboxylic acid (TCA) cycle intermediates, amino acids, proteins, purines, pyrimidine bases, polysaccharides and fatty acids.

In certain microorganisms, secondary metabolites are usually synthesized at the end of the exponential growth (trophophase) or at the beginning of the stationary phase (idiophase) and their formation is usually repressed during rapid growth. The risk of metabolic suicide diminishes when cell division is complete before toxic secondary products (e.g. antibiotics) are produced.

The operation of secondary metabolism is advantageous by keeping metabolism running at a low rate, rather than closing it down completely after growth ceases. The unbalanced growth hypothesis suggests that mechanisms controlling primary metabolism are

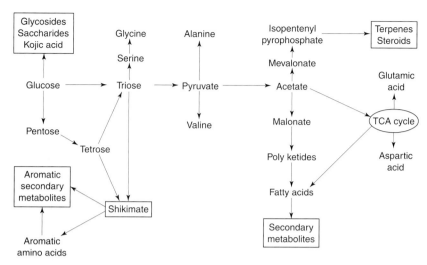

Figure 1   Interdependent pathways for primary and secondary metabolism. TCA, tricarboxylic acid.

not adequate to prevent overproduction of some compounds when balanced growth ceases. Since these compounds may be toxic to the cell, secondary metabolism diverts synthesis to the production of harmless products which are excreted. According to this theory, secondary metabolism should increase long-term viability. There is some evidence that *Pseudomonas aeruginosa* loses viability when grown under conditions that prevent secondary metabolism.

Secondary metabolites are distinguished from primary metabolites since they have a restricted distribution and are often characteristic of individual genera, species or strains. They are formed along specialized pathways from primary metabolites. Secondary metabolites are not essential to life, although they are important to the organism that produces them. Thus, the sporulation antibiotic produced by *Bacillus subtilis* is implicated in spore formation. Furthermore, they may help the survival of the organism in nature, e.g. as effectors for differentiation and weapons against competition. Thus, secondary metabolites may be redefined as compounds which are not essential for exponential growth.

Secondary metabolites are biosynthesized from a handful of primary metabolites, α-amino acids, acetyl coenzyme A (CoA), mevalonic acid and intermediates of shikimic acid pathway. Typically, secondary metabolites are synthesized as families, i.e. mixtures of chemically related components. Formation of secondary metabolites may be ascribed to the low specificity of some enzymes involved in secondary metabolite synthesis and action of isozymes on slightly modified substrates (**Fig. 1**).

Secondary metabolites of a specific type are usually produced by some strains of a species. A pathway may evolve in one organism and then be transferred

as a whole or in part to another organism in the same habitat. This may be speculated with regard to cephalosporin synthesis, which may have evolved in Streptomycetes or other soil bacteria and parts of which might have been transferred to various filamentous fungi in soil. Also, reduced stringency on structure or biosynthesis may be a reflection of the fact that secondary metabolism may be a playground of evolution.

All microbial taxa do not undergo secondary metabolism. It is a common feature of the filamentous fungi, bacteria and the sporing bacteria, but it is not a feature of the Enterobacteriaceae. Thus, although taxonomic distribution of secondary metabolism is limited, the range of secondary products produced is enormous. Many secondary metabolites may be involved in competition in the natural environments (**Fig. 2**).

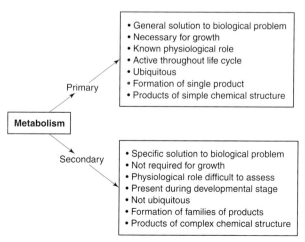

Figure 2   Primary and secondary metabolism of bacteria.

Production of Antibiotics

'Antibiotic' refers to any substance produced by a microorganism which kills certain other microorganisms. Antibiotics were the first industrially produced microbial metabolites which were not major metabolite end products. The synthesis of antibiotics normally begins after growth of the organism that produces them has virtually ceased. No antibiotic is active against all bacteria. Some are active against a narrow range of species, while others are active against a broad spectrum, including both Gram-positive and Gram-negative bacteria. Among the antibiotics, penicillin is produced in the greatest quantity (10^7 kg per year). Other important antibiotics produced by fermentation are the aminoglycosides, cephalosporins, polyenes, non-polyenes, macrolides and the tetracyclines.

Many microorganisms excrete products that are not related to the basic metabolism of the producing organism. Bacteria produce a host of substances which play significant roles as therapeutics and stimulants and consequently gain greater economic importance. Antibiotics are generally produced by special synthetic pathway, using enzymes that are not essential for the growth and maintenance of the cell. Hence, the genetic apparatus for the synthesis of antibiotics would be a burden to the organism and such a burden would be eliminated during evolution. During evolution, some genetic material with unknown function is carried along, and the concept of a burden to the organism has been accepted.

One of the more important antibiotics, streptomycin, was isolated from the culture medium of *Streptomyces griseus*. The antibiotic consists of three groups: N-methyl-L-2-glucosamine, a methyl pentose and an inositol derivative with two guanidyl residues. Streptomycin is effective against acid-fast and Gram-negative bacteria that are resistant to penicillin, but it produces marked allergic side reactions in humans.

Chloromycetin (chloramphenicol), another antibiotic recovered from cultures of *S. venezuelae*, is effective against Gram-negative bacteria, spirochaetes, *Rickettsia*, actinomycetes and viruses.

The tetracyclines, which are related chemically to a naphthacene skeleton, are excreted by *S. aureofaciens*. The carbon skeleton of tetracyclines is derived from acetate units and methionine. Chlortetracycline (aureomycin), oxytetracycline (terramycin) and tetracyclines are effective broad-spectrum antibiotics which are well tolerated by humans.

Of the polypeptide antibiotics (gramicidin S, polymyxins, bacitracin and ristocetin), polymyxin B consists of a ring of seven amino acids with a side chain attached by a peptide bond. The polypeptide antibiotics have a high affinity for the cytoplasmic membrane and are toxic for bacteria and for eukaryotes, which make their clinical usefulness limited.

Aminoglycosides are bases containing amino sugar, for example, streptomycin, produced by *S. griseus*. Streptomycin production takes place immediately after the exponential phase. The process requires high levels of oxygen transfer and optimal phosphate uptake for growth and antibiotic synthesis. The macrolide antibiotic erythromycin is produced by *S. erythreus* from propionyl CoA and methyl malonyl-CoA precursors in a metabolic pathway, which is analogous to that of fatty acid biosynthesis (**Fig. 3**).

Polyketide

This compound (commonly a secondary metabolite) is, or may be regarded as, a condensation product of acetate. Polyketides may be classified according to the number of C2 units they contain (e.g. a tetraketide is a C8 compound). Metabolites derived from polyketides include many antibiotics (e.g. tetracyclines) and mycotoxins (e.g. patulin, a tetraketide, aflatoxins and cytochalasins). Such compounds are synthesized from precursors such as acetyl CoA and malonyl CoA; in aflatoxin biosynthesis the malonyl CoA subunits condense with an acetyl CoA 'starter', with concomitant decarboxylation, in a process analogous to fatty acid biosynthesis but without the dehydration and reduction steps.

Site of Activity

An antibiotic characteristically acts at a precise site in the bacterial cell. Depending on the antibiotic, the site of action may be the cell wall, the cell membrane, the protein-synthesizing machinery or an enzyme involved in nucleic acid synthesis. Thus, polymyxins (produced by *Bacillus polymyxa*) increase the permeability of the bacterial cell membrane and are mainly effective against Gram-negative bacteria. Chloramphenicol binds to the 50S subunit of bacterial ribosome and prevents the formation of a peptide bond during protein synthesis. The tetracyclines (produced by species of *Streptomyces*) exert an antibacterial action by binding to the 30S subunit of ribosome, preventing aminoacyl t-RNA molecules from binding to the A-side of the ribosome. Novobiocin (obtained by *Streptomyces* sp.) specifically inhibits bacterial DNA polymerase, thus preventing DNA replication.

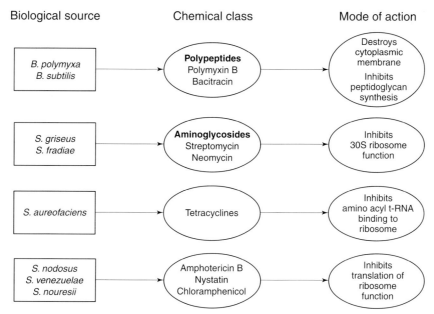

Figure 3 Action of certain antibiotics produced by different bacteria.

The Production of Chemicals by Microorganisms

The widespread use of microorganisms in the chemical and pharmaceutical industries is partly due to the cheaper cost of production through biosynthesis than through chemical synthesis. Microbial syntheses also have distinguished advantages in the preparation of optically active compounds since chemical synthesis leads to a racemic mixture which must subsequently be resolved. Some pigments produced by microorganisms are secondary products of metabolism. The fungus *Monaseus anka* synthesizes a group of red pigments which are used to colour foodstuffs.

Microorganisms also produce a variety of compounds that have the ability to stimulate growth of other living organisms. Thus, gibberellins, the secondary products of metabolism of fungus *Gibberella fujikuroi* (the perfect stage of *Fusarium moniliforme*) stimulated the growth of rice seedlings. Now gibberellins are extensively used in horticulture and agriculture.

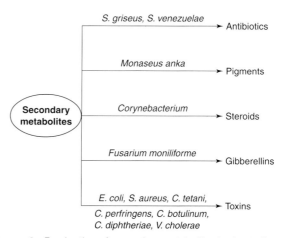

Figure 4 Production of secondary metabolites by bacteria at a glance.

tial in the synthesis of a cortisone derivative, prednisolone (**Fig. 4**).

Microbial Transformation of Steroids

Many microorganisms are capable of performing limited oxidation of steroids which cause small and highly specific structural changes. The positions and nature of these changes are often characteristic of the microbial species so that, by selecting an appropriate microorganism as an agent, it is possible to bring about any one of a large number of different modifications of the steroid molecule. The introduction of a double bond by dehydrogenation between position one and two mediated by a *Corynebacterium* is essen-

Enzymes

Enzyme proteins of secondary metabolism are of three different groups. The first group of enzymes of secondary metabolism are only found in certain organisms, the second group catalysing special catabolic pathways operates in the metabolism of foreign substances, e.g. dietary compounds, drugs and exoenzymes including bacterial exotoxins, while the third group exhibits special functions such as luciferase system and special transport system.

Regulation of Secondary Metabolite Fermentation

Secondary metabolites are molecules produced by a narrow spectrum of organisms. They are usually produced like mixtures of members of a closely related chemical family – there are at least three neomycins, five mitomycins, 10 bacitracins, eight aflatoxins, 10 polymyxins, 10 natural penicillins and more than 20 actinomycins. The ability of an organism to produce secondary products is easily lost by mutation. Another characteristic of secondary metabolites is at the end of trophophase marked changes occur in the enzymatic composition of the cell and enzymes specifically related to formation of secondary products suddenly appear. Thus, many antibiotics appear to be assembled using an alternative mechanism to the normal ribosomal system. Analogues to t- and m-RNA exist in the form of antibiotic-committed enzymes. The control parameter for antibiotic-committed enzyme formation may provide the basis for regulatory factors for secondary metabolite formation and include induction, carbon catabolite repression and energy charge regulation. Induction may result from the addition or accumulation of a metabolite intermediate.

The biosynthesis of gramicidin synthetase (GS) by *Bacillus brevis* does not occur until late in the cell cycle because the GS are not formed until the latter part of the logarithmic growth phase. During the synthesis of gramicidin the antibiotic-committed enzyme GSII initiates the biosynthetic sequence by catalysing the activation of L-phenyl alanine by ATP to form the amino-acyl derivative. This process is analogous to the formation of amino-acyl t-RNA in primary metabolism where amino acids are activated prior to stepwise biosynthesis of protein molecule. GSII also affects the conversion of L-phenyl alanine to the D-form, which then becomes bound to GSI. This latter antibiotic-committed enzyme catalyses the adenylation of other precursor amino acids which then bind on to GSI as thiol esters. The polymerization of all the amino acids results in the formation of linear gramicidin (a linear pentapeptide) which is then released from the enzyme complex. Linear gramicidin molecules then combine as dimers to form the final cyclic product, gramicidin (a cyclic decapeptide; **Fig. 5**).

Carbon Catabolite Repression

Glucose, an excellent carbon energy source for microbial growth, exerts a repressive effect on secondary metabolism. It will suppress the formation of β-lactam antibiotics, tetracyclines, bacitracin and the alkaloids. Glucose has been shown to inhibit the synthesis of antibiotic-committed enzymes, e.g. phenoxazinone

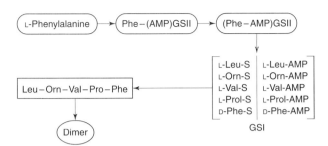

Figure 5 Involvement of antibiotic-committed enzymes in biosynthesis of gramicidins in *Bacillus subtilis*. GS, gramicidin synthetase.

synthetase. This enzyme is involved in the production of actinomycin antibiotic by *S. antibioticus*. There may exist a control mechanism in secondary metabolism similar to the catabolite repression process observed in *Escherichia coli* lac operon system. Similarly, phosphate and nitrogen sources have also been reported to repress secondary metabolite formation.

Inducible Effects

The synthesis of a number of antibiotics is stimulated by the addition of inducer compounds to the culture medium. Thus, ergoline alkaloid production by the mould *Claviceps purpurea* is stimulated by the addition of L-tryptophan to the culture medium during the exponential phase. This happens even though tryptophan is not a precursor of the product; it is produced after the exponential phase. The reason may be that antibiotic-committed enzyme in the sequence responsible for alkaloid production is tryptophan-inducible. Methionine has been shown to induce cephalosporin biosynthesis.

Bacterial Toxins

Pathogenic bacteria more commonly cause illness by producing toxins. A toxin is a substance synthesized by a pathogen which damages or kills the host by interfering with host functions. These poisons are of two types: exotoxins and endotoxins. Exotoxins are proteins secreted by the bacterial cell. The exotoxins can produce symptoms even without the bacteria actually being present. For example, when *Clostridium botulinum* grows anaerobically in poorly canned foods, one of the by-products of fermentation is an exotoxin that causes botulism. Another exotoxin-producing species is the enteric bacterium *Vibrio cholerae* which can infect the lower intestine of humans and cause cholera. *E. coli* is an exotoxin-releasing culprit.

In contrast to exotoxins, endotoxins are not secreted by the pathogen, but are instead components of the outer membrane of certain Gram-negative bacteria. Endotoxins of Gram-negative bacteria are lipopolysaccharides which form part of the outer membrane of the pathogen cell wall. Some of the toxins secreted by pathogenic bacteria are briefly outlined here.

Tetanus

Tetanus involves uncontrollable contraction of the skeletal muscles, often leading to death by exhaustion. The disease develops when deep anaerobic wounds are contaminated with *Clostridium tetani*. This organism produces a powerful neurotoxin – a toxin which specifically acts against nerve tissue. The protein neurotoxin (tetanospasmin) acts mainly on the central nervous system.

Cholera

This disease is caused by certain strains of *Vibro cholerae*; following ingestion, *V. cholerae* multiplies in the lower intestines, liberating a powerful enterotoxin – a toxin active against intestinal mucosa.

Botulism

This disease develops following the ingestion of foodstuffs contaminated with protein neurotoxin of *C. botulinum*. Food commonly implicated in cases of botulism include cold meats and imperfectly canned non-acid vegetables. The toxins of *C. botulinum* act on the peripheral nervous system.

Staphylococcus aureus produces toxins and other pharmacologically active proteins. Factors such as lipase, protease, staphylokinase, hyaluronidase and enterotoxin contribute to pathogenesis. Most likely candidates include α, β and δ haemolysins, leucocidins and coagulase.

In some cases a toxin can only be synthesized when the cells of the pathogen contain a particular plasmid or a bacteriophage. In these cases, the genes coding for toxin production are present in the nucleic acid of the plasmid or phage and not in the chromosome of the pathogen. Thus, a cell of the same species which lacks the particular plasmid or phage cannot produce toxins. An example of a phage-specified toxin is the diphtheria toxin of *Corynebacterium diphtheriae*. Some strains of *E. coli* produce a plasmid-specified enterotoxin and such enteropathogenic strains cause intestinal disorder in humans. *Clostridium perfringens*, a strict anaerobe, is responsible for gas gangrene in humans (**Fig. 6**).

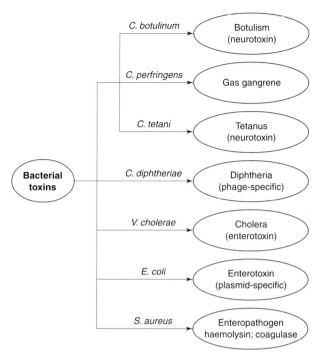

Figure 6 Diseases caused by exo- and endotoxins elaborated by pathogenic bacteria.

The Impact of DNA Recombinant Technology

The advent of DNA recombinant technology opened new possibilities for the production of useful products by microorganisms. Up until that time, a microorganism could only produce material of which the synthesis was encoded in the cell's own genome. Genetic selection might improve the level of production or modify the chemical nature of the product but production of new protein was totally out of the question. Using techniques of recombinant DNA technology, it is possible to introduce any gene or set of genes into a microorganism and thereby produce the immediate gene products whose synthesis is catalysed by their action.

In the last few years, it has been demonstrated that for secondary metabolites such as antibiotics, the yields can be increased by overcoming rate-limiting steps in the biosynthetic pathways using gene amplification techniques. For example, the production of cephalosporin C by *Cephalosporium acremonium* could be increased by overexpressing the *cef EF* gene. This gene codes for a bifunctional protein that exhibits two sequentially acting cephalosporin biosynthetic enzyme activities: deacetoxy cephalosporin C synthetase and deacetyl cephalosporin C synthetase. The parent strain of *C. acremonium* excreted a substantial quantity of penicillin N, a precursor of cephalosporin C, into the culture medium. The recombinant strain with a twofold increase in the specific activity of deacetoxy cephalosprin C synthetase was able to convert penicillin N completely into cephalosporin C.

There will be great industrial benefits since the genes for the biosynthesis of antibiotics are generally clustered together with genes for regulation. Thus, by disrupting the regulatory region, it is possible to increase the production of methyl enomycin as, in this instance, a negative control system is involved. However, positive control genes are normally found in the biosynthetic gene clusters for various secondary metabolites. Therefore, overexpression of these regulatory genes caused overproduction of streptomycin and undecylprodigiosin of actinorhodin, respectively, by the wild-type strains.

Using a cosmid clone containing the penicillin biosynthetic gene cluster from *Penicillium chrysogenum* it is possible to transform filamentous fungi *Neurospora crassa* and *Aspergillus niger* which do not produce β-lactam antibiotics. Both of these transformed hosts produce penicillin. Thus, it is possible that *P. chrysogenum* genes encoding for all enzymes necessary for the biosynthesis of penicillin form a gene cluster which can be expressed in heterologous fungal hosts.

Recombinant DNA techniques can be used for the production of hybrid or even novel antibiotics. Genes for biosynthetic steps in different organisms may be combined in the same organism, leading to the production of novel antibiotics.

A still greater challenge for getting novel antibiotics is to alter the backbone structure of a metabolite. *Streptomyces galilaecus* normally produces actacinomycin A and B. After transformation with the genes for polyketide synthase involved in the synthesis of actinorhodin, clones were obtained which produced an anthraquinone. This is an interesting observation in the use of this approach to produce antibiotics with novel structures.

See also: **Bacillus**: *Bacillus subtilis.* **Clostridium**: Introduction; *Clostridium botulinum*; Detection of neurotoxins of *C. botulinum.* **Escherichia coli**: *Escherichia coli.* **Mycotoxins**: Classification. **Nucleic Acid-based Assays**: Overview. **Pseudomonas**: *Pseudomonas aeruginosa.* **Staphylococcus**: *Staphylococcus aureus.* **Streptomyces**.

Further Reading

Herbert RB (1981) *The Biosynthesis of Secondary Metabolites.* London: Chapman & Hall.

Jay JM (1992) *Modern Food Microbiology*, 4th edn. New York: Chapman & Hall.

Rehm HJ and Read G (eds) (1981) *Microbial Fundamentals.* Vol. 1. Florida: Verlag Chemie.

Rehm HJ and Reed G (eds) (1986) *Biotechnology.* Vol 4, *Microbial Products* 11. Verlagchemie VCH: Weinheim.

Rehm HJ and Reed G (eds) (1993) *Biotechnology: Biological Fundamentals.* Vol. 1. New York: VCH Weinheim.

Rose AH (ed.) (1979) *Secondary Products of Metabolism.* London: Academic Press.

Schlegel HJ (1993) *General Microbiology*, 8th edn. Cambridge: Cambridge University Press.

Singleton P and Sainsbury D (1981) *Introduction to Bacteria.* New York: John Wiley.

Stanier RY, Ingraham JL, Wheelis ML and Painter PR (1987) *General Microbiology*, 5th edn. London: Macmillan.

Walker JM and Gingold EB (eds) (1998) *Molecular Biology and Biotechnology*, 3nd edn. London: Royal Society of Chemistry.

Wiseman A (ed.) (1983) *Principles of Biotechnology.* New York: Surrey University Press. New York: Chapman & Hall.

Metabolite Recovery *see* **Fermentation (Industrial):** Recovery of Metabolites.

METHANOGENS

W Kim and **W B Whitman**, Department of Microbiology, University of Georgia, USA

Methanogens are prokaryotes that produce methane gas as an essential component of their energy metabolism. Although they are responsible for copious methane production by the gastrointestinal tract of humans and many domestic animals, none are known to be responsible for food-borne disease. For the food industry, the primary interest has been in their importance in anaerobic bioreactors for wastewater treatment and animal nutrition.

Methanogens are obligate anaerobes and are common in many environments with a low redox potential, such as the gastrointestinal tracts of animals and termites, sediments of freshwater lakes, rice paddies, sewage, landfills and anaerobic digestors. In these habitats, methanogens catalyse the terminal step in an anaerobic food chain that converts organic matter into methane and CO_2. Anaerobic bacteria and anaerobic eukaryotes perform the initial transformations of the biopolymers in organic matter to the substrates for methanogenesis. Thus, consortia of microorganisms are required to produce methane in most habitats, a striking example of which are the symbiotic methanogens of many anaerobic protozoa.

Methanogens are the major source of atmospheric methane, an important greenhouse gas that has been increasing in concentration for the last two hundred years. Given that about half of the methane produced is oxidized by methane-oxidizing bacteria, about 1–2% of all the organic matter produced by plants each year passes through the methanogenic food chain. Thus, this process is a significant component of the carbon cycle.

The substrate range of the methanogens themselves is quite limited, and only three types are utilized (**Fig. 1**). In the first type, CO_2 is reduced to methane by a fairly narrow range of electron donors. H_2 and formate are utilized by most methanogens. Some methanogens can also slowly oxidize a few alcohols, especially ethanol, isopropanol, isobutanol and cyclopentanol. The alcohols are incompletely oxidized to form ketones or carboxylic acids. In the second type of methanogenesis, acetate is fermented to methane and CO_2. Even though this aceticlastic reaction accounts for most of the methane generated in many habitats, it is catalysed by only a few genera of methanogens. The last type of methanogenesis is the reduction of the methyl groups of C_1 compounds to methane. C_1 compounds utilized include methanol, monomethylamine, dimethylamine, trimethylamine, methanethiol and dimethylsulphide. Usually, a portion of the C_1 compound is oxidized to provide electrons for the reduction. Thus, about one quarter of the methyl groups are oxidized to CO_2 and three-quarters are reduced to methane. Some methylotrophic methanogens can not oxidize methyl groups, and H_2 is the electron donor.

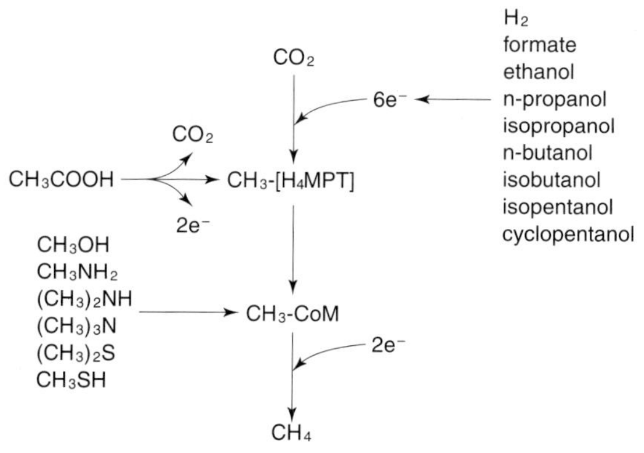

Diversity and Taxonomy

All known methanogens are archaea, a diverse phylogenetic group that also includes many thermophilic and halophilic prokaryotes. Although first proposed by Carl Woese and his collaborators in 1977 on the basis of their unusual 16S rRNA sequences, other features also characterize the archaea, including unique structures of their lipids, novel cell walls, abundance in extreme habitats, and eukaryotic type of DNA-dependent RNA polymerase and promoter structure. Recent genomic sequencing has confirmed many of the profound differences between archaea and the more familiar eubacteria.

Lipids are important chemotaxonomic markers in methanogens and other archaea. There are four major differences from the lipids of eubacteria and eukaryotes. First, the hydrocarbon side chains are linked to glycerol with ether instead of ester bonds. Second, the hydrocarbon side chains are based on the C_5 isoprenoid unit instead of the C_2 acetyl moiety. Third, archaeal lipids contain sn-2,3-glycerol, the opposite stereoisomer of the glycerol in eubacterial and eukaryotic lipids. Fourth, archaeal lipids are frequently tetraethers that span the membrane and are formed by a condensation of the isoprenoid side chains. Although ether-linked and branched lipids are found in some eubacteria, the combination of unusual features in archaeal lipids argues strongly for an entirely different biosynthetic pathway. Within the methanogens, the major core lipids are archaeol (2,3-di-O-phytanyl sn-glycerol diether), caldarchaeol (ditetraterpenediyl glycerol tetraether), sn-2-hydroxyarchaeol, and sn-3-hydroxyarchaeol. Glycolipids also contain glucose, galactose, N-acetylglucosamine and mannose. Phospholipids also contain inositol,

Figure 1 Overview of the three types of methanogenesis. Type I: CO_2 is reduced to methane via the C_1 carriers methanofuran (not shown) and tetrahydromethanopterin (H_4MPT), a folate analogue that is virtually unique to methanogens. The electrons for the reduction of CO_2 to methane are obtained from H_2, formate, or a few secondary and primary alcohols. Type II: in aceticlastic methanogenesis, the methyl group of acetate is transferred to H_4MPT. The electrons for reduction of the methyl group to methane are obtained from the oxidation of the carboxy group of acetate. Type III: in methylotrophic methanogenesis, the methyl group is transferred to coenzyme M (CoM, 2-mercaptoethanesulphonic acid), the terminal C_1 carrier in the pathway of methanogenesis from CO_2 and acetate. CoM, is unique to methanogens. Methyl groups are also oxidized by a reversal of the pathway of CO_2 reduction. The last step in the pathway, the reduction of methyl-CoM to methane, is common to all types of methanogenesis and requires two additional unique coenzymes (not shown). The first is coenzyme F_{430}, a nickel tetrapyrrole that is tightly bound to the methylreductase enzyme. The second is 7-mercaptoheptanoylthreonine phosphate, which serves as the proximal electron donor to the methylreductase.

ethanolamine, serine, aminopentanetetrols and glycerol.

Several different types of cell envelopes are present in methanogens. In the simplest type, the cell envelope is composed solely of a protein surface layer or S-layer. The S-layer contains hexagonally arranged protein subunits, which vary in molecular weight and antigenicity between species. Frequently, the S-layer protein is glycosylated. In other methanogens, the wall contains additional polymers, such as methanochondroitin. This compound is similar in structure to chondroitin, which is found in the connective tissue of animals, and plays a vital role in cellular aggregation in some genera. In other methanogens, the cell envelope also contains a protein sheath that is strongly resistant to detergents and proteases. Lastly, methanogens that stain Gram-positive contain pseudomurein, a peptidoglycan that superficially resembles the common murein of eubacteria. In pseudomurein, the sugar backbone is composed of L-N-acetyltalosaminuronic acid and D-N-acetylglucosamine. The interpeptide bridge contains only L-amino acids.

Among the archaea, methanogens are unusual in that they are common in moderate as well as extreme habitats. Thus, growth temperatures of methanogens span 100°C, from psychrotrophic to hyperthermophilic. Optimal salinities for growth vary from freshwater to saturated brine. In contrast, most methanogens prefer neutral to moderately alkaline pH values and few acidophiles have been described.

Recent taxonomic proposals place the methanogens in 25 genera, representing 12 families and five orders (**Table 1**). This taxonomy is consistent with the high degree of phenotypic and genotypic diversity found within this group. The wide diversity within the group suggests that methanogenesis is an ancient life style. Because modern methanogens are monophyletic, it is also likely that methanogenesis evolved only once and that all modern methanogens share a single ancestor.

Methanogenic Bioreactors

In a typical anaerobic bioreactor, oxidation of organic matter rapidly exhausts good electron acceptors such as O_2, nitrate, and sulphate. When CO_2 is the only abundant electron acceptor remaining, the conditions become favourable for methanogenesis. Because CO_2 is such a poor electron acceptor, only about 5% of the total energy of combustion in the organic matter is available to support microbial growth, and the formation of microbial biomass or sludge is correspondingly less than that found in a typical aerobic bioreactor. In addition, the fuel requirements and reactor volumes are much smaller for a methanogenic bioreactor, and the process may be significantly less expensive than for typical aerobic waste treatment. For instance, a 223 000 gallon (843 000 l) anaerobic fluidized bed reactor removing 9700 lbs (4400 kg) biochemical oxygen demand (BOD) per day would produce 72 000 ft^3 (2 040 000 l^3) of CH_4 per day and 180 tons (180 000 kg) of sludge per year. A comparable aerobic treatment would require a 1 700 000 gallon (6 400 000 l) reactor and produce 1800 tons (1 800 000 kg) of sludge per year.

During anaerobic digestion, the initial fermentation of organic matter generates H_2, formate, and a wide variety of organic acids such as acetate, propionate, butyrate and lactate as well as alcohols (**Fig. 2**). Consumption of the H_2 and formate by the methanogens enables syntrophic bacteria to oxidize the organic acids and alcohols to additional H_2 and acetate. This interaction between the syntrophic bacteria and the methanogens, called *interspecies H_2 transfer*, is necessary because the fermentation of propionate and butyrate to acetate and H_2 is only feasible when the H_2 concentration is extremely low, or below 10^{-4} atm. Similarly, fermentation of lactate and ethanol is also greatly stimulated when the H_2 concentration is low. Thus, the anaerobic food chain of fermentative and syntrophic bacteria converts the organic matter into the major substrates for methanogenesis: CO_2, H_2, formate and acetate.

For the fermentation of sugars, typically about two-thirds of the methane is derived from acetate, whereas the remainder is formed by CO_2 reduction. These ratios are in close agreement with that expected based on the biochemistry of glycolysis. In this pathway, one molecule of hexose is oxidized to two molecules of pyruvate with the reduction of two molecules of NAD^+. Pyruvate is then oxidized to acetate and CO_2 with the reduction of two more molecules of NAD^+ or NAD^+ equivalents. The four molecules of NADH (or NADH equivalents) are used to reduce one molecule of CO_2 to methane. The two molecules of acetate are utilized to form two additional molecules of methane and CO_2. Thus, one molecule of hexose is converted into three molecules of methane and CO_2. Even though aceticlastic methanogenesis is more abundant, CO_2 reduction to methane is essential to generate most of the acetate, and both processes are interrelated.

The generation times of the aceticlastic methanogens and the syntrophic bacteria frequently exceed 24 h, which limits the turnover time of the anaerobic bioreactor. This problem is overcome by recycling or trapping the anaerobic microbes within the bioreactor, and high volumetric loading rates can be achieved. Nevertheless, the slow growth of these microbes may partially explain the long start-up times for anaerobic bioreactors.

Table 1 Genera of methanogenic archaea

Genus	Morphology	Substrates[a]	Optimal temperature (°C)	Habitats
Methanobacterium	Rod	H_2, (for, iP, iB)	35–65	Sewage, bioreactor, marshy soil, alkaline lake sediment, oil reservoir waters
Methanothermobacter	Rod	H_2, (for)	55–70	Sewage, river sediment
Methanobrevibacter	Coccobacillus	H_2, for	33–40	Rumen, sludge, human and animal faeces, wet wood of trees
Methanosphaera	Coccus	H_2 + m	35–40	Human faeces, rabbit colon
Methanothermus	Rod	H_2	77–88	Solfataric water and mud
Methanopyrus	Rod	H_2	98	Heated marine sediment
Methanococcus	Coccus	H_2, for	35–40	Marine sediments, salt marsh
Methanothermococcus	Coccus	H_2, for	65	Heated marine sediments
Methanocaldococcus	Coccus	H_2, for	85	Marine hydrothermal vents
Methanoignis	Coccus	H_2	88	Marine hydrothermal vents
Methanomicrobium	Short rod	H_2, for	40	Rumen
Methanolacinia	Irregular rod	H_2, iP, iB, cPe	40	Marine sediment
Methanogenium	Irregular coccus	H_2, for, (E, iP, P, iB, B, cPe, iPe)	30–57	Marine and fresh sediments, bioreactors
Methanoplanus	Irregular disk, plate	H_2, for	32–40	Swamp, marine sediment, oil reservoir waters
Methanoculleus	Irregular coccus	H_2, for, (iP, iB)	37–60	Marine and river sediment, bioreactor
Methanofollis	Irregular coccus	H_2, for	37–40	Solfataric pool mud
Methanospirillum	Spirillum	H_2, for, (iP, iB)	35–40	Bioreactor, freshwater sediments
Methanocorpusculum	Irregular coccus	H_2, for, (iP, iB)	30–37	Bioreactor, lake sediment
Methanosarcina	Aggregates, coccus, macrocyst	m, MeN, ac, (H_2)	35–50	Freshwater and marine sediments, bioreactor, rumen, soil, sewage
Methanolobus	Irregular coccus	m, MeN, (MeS)	35–40	Marine sediment
Methanococcoides	Irregular coccus	m, MeN	20–35	Marine water and sediment
Methanohalophilus	Irregular coccus	m, MeN	26–36	Saline lake sediment, stromatolite associated mat
Methanohalobium	Flat polygon	MeN	50	Salt lagoons
Methanosalsus	Irregular coccus	m, MeN, MeS	45	Saline lake sediment
Methanosaeta	Sheathed rod	ac	35–60	Sewage, bioreactor, landfill

[a]ac: acetate, B: butanol, cP: cyclopentanol, E: ethanol, for: formate, iB: isobutanol, iP: isopropanol, iPe: isopentanol, m: methanol, MeN: methylamines, MeS: dimethylsulphide or methanethiol, P: l-propanol.
Parentheses indicate that the substrate is utilized by only some species or strains.

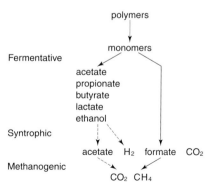

Figure 2 Methanogenic food chains in a bioreactor and rumen. Organic polymers are degraded to monomers and then organic acids and alcohols, H_2 and CO_2 by fermentative organisms. Syntrophic organisms convert the organic acids and alcohols into acetate, H_2, formate and CO_2, which are the substrates for the methanogenic archaea. Broken lines indicate reactions that do not occur in the rumen and colon habitats.

Numerous species of CO_2-reducing methanogens have been isolated from bioreactors, including species of *Methanobacterium*, *Methanothermobacter*, *Methanobrevibacter*, *Methanogenium*, *Methanocorpusculum* and *Methanospirillum*. For the aceticlastic methanogens, species of both *Methanosarcina* and *Methanosaeta* are usually present. Although *Methanosarcina* grows more rapidly, it is unable to utilize the low concentrations of acetate taken up by *Methanosaeta*.

Methanogenesis in the Gastrointestinal Tract of Animals

Methanogenesis in the rumen of cattle and other ruminants is a major source of atmospheric methane as well as of considerable importance in the nutrition of these animals. The anaerobic food chain is similar to that found in bioreactors except that CO_2 reduction is the major source of methanogenesis (Fig. 2).

Because the residence time of the rumen contents is less than one day, aceticlastic methanogens and syntrophic bacteria are washed out, preventing the metabolism of acetate and other organic acids. The organic acids can then accumulate to concentrations that can be absorbed by the animal. Although species of *Methanobrevibacter* are commonly isolated from the rumen, species of *Methanomicrobium* and methylotrophic *Methanosarcina* are also present.

The bovine rumen produces 200–400 l of methane per day, which represents a significant loss in energy for the animal. Two common feed additives, monensin (or rumensin) and lasalocid, derive their effectiveness from their ability to inhibit H_2 production by Gram-positive eubacteria in the rumen. The lower availability of H_2 limits methanogenesis, and propionate production is stimulated. The net result is more efficient utilization of low fibre feeds by the animal.

In humans, a similar process occurs in the colon. However, the methanogenic fermentation of only about 10% of people on a Western diet is significant enough to produce more than a litre of methane per day. At least superficially, this fermentation resembles that found in the rumen, where organic acids are absorbed by the intestines and CO_2 reduction is the source of most of the methane. People who do not produce large amounts of methane probably contain homoacetogenic bacteria instead. These eubacteria are strict anaerobes that oxidize H_2 to reduce CO_2 to acetate. Because this fermentation is slightly less favourable energetically than methanogenesis, CO_2 reduction to acetate is not a major process in many methanogenic habitats, and the factors which enable acetogenesis to dominate in the colon are not understood. In the colon, the most abundant methanogen is *Methanobrevibacter smithii*. However, low numbers of the methylotroph *Methanosphaera* have also been detected.

See also: **Microflora of the Intestine**: The Natural Microflora of Humans.

Further Reading

Ferry JG (ed.) (1993) *Methanogenesis: Ecology, Physiology, Biochemistry and Genetics*. New York: Chapman & Hall.

Koga Y, Nishihara M, Morii H and Akagawa-Matsushita M (1993) Ether polar lipids of methanogenic bacteria: structures, comparative aspects, and biosynthesis. *Microbiol. Rev.* 57: 164–182.

Rogers JE and Whitman WB (eds) (1991) *Microbial Production and Consumption of Greenhouse Gases: Methane, Nitrogen Oxides and Halomethanes*. Washington DC: American Society of Microbiology Press.

Speece RE (1996) *Anaerobic Biotechnology for Industrial Wastewaters*. Nashville, Tennessee: Archae Press.

Whitman WB, Bowen TL and Boone DR (1992) The methanogenic bacteria. In: Balows A, Truper HG, Dworkin M, Harder W and Schleifer K-H (eds) *The Prokaryotes*. P. 719. New York: Springer-Verlag.

MICROBIOLOGY OF SOUS-VIDE PRODUCTS

Frédéric Carlin, Institut National de la Recherche Agronomique, Unité de Technologie des Produits Végétaux, Avignon, France

Introduction

The literal meaning of the French term 'sous-vide' is 'under vacuum'. Sous-vide foods, and other cooked and chilled foods, are becoming increasingly popular. They meet increasing consumer demand for convenient, fresh foods of high organoleptic quality, reflecting a reduction in the time devoted to food preparation by a growing number of single-parent families, senior citizens and households with microwave ovens. To maintain the appearance and taste of freshly home-prepared foods, sous-vide foods are given only a mild heat treatment. This allows the survival of microorganisms, mainly bacteria: these foods are therefore non-sterile by design, and require refrigeration during a shelf life which usually lasts for several weeks. The survival of bacteria and the extended shelf life together have caused some concern about the microbial safety of sous-vide foods for the consumer.

Foods that are vacuum-packed and either receive no heat treatment, or receive a high-temperature treatment similar to that applied to canned foods, are not considered in this article.

Processing of Sous-vide Foods

Sous-vide technology began in the early 1960s. There have been many different applications worldwide, in restaurants, catering and industrial products (particularly meat and ham), and, particularly in the UK and France, it has been applied to retail products.

The processing of sous-vide foods is depicted in **Figure 1**. The ingredients can be either raw or pre-

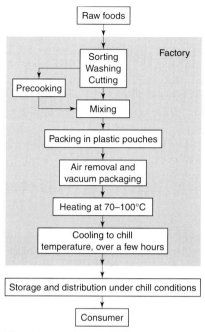

Figure 1   Manufacture of sous-vide foods.

cooked. After mixing, they are packaged in plastic pouches, usually without additives or preservatives, in keeping with their image of freshness and minimal processing. Air is extracted from the package mechanically, immediately prior to sealing (vacuum packaging). The pack is then heat-processed ('sous-vide cooking'). The final product appears to be draped by a plastic film, which assumes the same shape as that of the product.

The application of sous-vide technology to foods has some advantages, as follows:

- no contamination of the foods after packing
- heat transfer and the cooking of foods in their own juices are facilitated by the absence of an air layer
- losses of food flavours, aromas and nutrients are low
- oxidation of the foods is prevented by the removal of air (and therefore O_2): this effect is reinforced by the use of packaging films which act as barriers to O_2
- inhibition of the growth of undesirable aerobic contaminants.

Heating is usually achieved by hot air or steam, or by immersion in large tanks of water (see below). Heating is followed by rapid cooling. The sous-vide products are then stored, distributed and retailed under chill conditions (i.e. at > 0°C). Shelf lives are very variable and range from 1 week to 3 months, depending on the food, the particular process, and national recommendations and regulations.

Heat Treatment

The main objective of heat treatment in the processing of sous-vide foods is to obtain products with optimal organoleptic qualities. This contrasts with the objectives of the canning and frozen food industries, which focus on microbial stability and safety.

The heat treatment of sous-vide foods depends on the time and temperature needed to cook them. For example, the time needed to achieve the optimal texture of root vegetables ranges from 4 to 13 min at 100°C, that needed for potato ranges from 35 to 50 min at 90°C, and from 4 to 12 min at 100°C. The optimal taste and/or texture of cooked red meat, poultry and fish are obtained when heated at relatively low temperatures (50–75°C) for as long as several hours – heating to temperatures approaching 100°C is usually detrimental to quality. Sous-vide foods are usually processed at temperatures in the range 70–100°C. Temperatures of < 70°C are sometimes used, but temperatures of > 100°C are rare.

Physico-chemical Characteristics of Sous-vide Foods

Sous-vide foods are usually prepared without the additives and preservatives used in traditional food processing. The pH and water activity of sous-vide foods are therefore close to those of the raw materials: the pH of raw meat, fish, milk and most vegetables is between 5.0 and 7.0, and often between 6.0 and 7.0, the only exceptions being some fruits and vegetables, such as the tomato. Salt is added to improve the taste, but in concentrations which will not significantly affect water activity – in the final product, this is generally > 0.98 and often > 0.99.

There has been some debate about the O_2 concentration in and the related redox potential of sous-vide foods. Sous-vide foods were considered either to be strictly anaerobic, because of the exclusion of air (and hence O_2) by vacuum packaging, or alternatively to contain, despite vacuum packaging, a small amount of residual O_2 that might prevent the growth of strictly anaerobic bacteria. Experimental data indicates that:

- The redox potential of foods is naturally low, and suitable for the growth of anaerobic bacteria such as clostridia.
- Concentrations of O_2 of up to 2% around foods do not fully prevent the growth of *Clostridium botulinum*, a strictly anaerobic bacterium.
- Some outbreaks of food-borne botulism have been caused by foods which were stored in aerobic conditions, but in which anaerobic conditions developed locally.
- After experimental inoculation, *C. botulinum* can grow and produce botulinum neurotoxin in sous-

Table 1 Characteristics of food-borne pathogenic bacteria

Bacterium	Need for O₂	Spore production	Minimal growth conditions			Heat resistance (as decimal reduction time or D value)[c]
			Temperature	pH	Water activity	
Bacillus cereus	Aerobic, facultatively anaerobic	+	4°C	5.0	0.93	$D_{85} > 30$ min; $D_{95} = 2$–35 min; $D_{121} \leqslant 2$ min
Clostridium botulinum Group I (mesophilic and proteolytic)	Strictly anaerobic	+	10°C	4.6	0.94	$D_{100} = 20$ min; $D_{121} = 0.15$ min
Clostridium botulinum Group II (psychrotrophic and non-proteolytic)	Strictly anaerobic	+	3°C	5.0	0.97	$D_{82.2} = 1$–80 min
Clostridium perfringens	Strictly anaerobic	+	12°C	5.5–5.8	0.93	$D_{100} = 1$–30 min
Listeria monocytogenes	Aerobic, facultatively anaerobic	–	0°C	4.4	0.92	$D_{70} = 0.01$–0.3 min
Salmonella	Aerobic, facultatively anaerobic	–	7°C[a]	3.8	0.94	$D_{66} = 0.1$–0.25 min[b]
Escherichia coli, including serotype O157:H7	Aerobic, facultatively anaerobic	–	7–8°C	4.4	0.95	$D_{60} = 0.2$–3 min

[a] Some strains are able to grow at 5.2°C
[b] For most strains: D_{66} of some rare strains = 1 min.
[c] For a given species, D values show variations between different works. Hence D values are given as ranges.

vide foods. This can occur at refrigeration temperatures in the case of the psychrotrophic strains of Group II.

The physico-chemical characteristics of sous-vide foods are therefore favourable to the growth of a wide range of bacteria.

Bacteria in Sous-vide Foods

Effects of Heat Treatment

The heat treatment applied during the processing of sous-vide foods determines the nature and the number of microorganisms in the final product. The resistance of bacteria to heat is expressed as the 'decimal reduction time' or 'D value', which is the time required, at a given temperature, for a tenfold reduction of a bacterial population (i.e. a 90% reduction, a reduction to 10%, or a 1 log reduction). The temperature (in °C) at which the D value is determined may be indicated by a subscript, e.g. D_{70}. For a given microorganism, the D value decreases as the temperature increases. The effect of temperature on the D value is expressed as the 'z value', which is the increase in temperature needed to produce a tenfold reduction in the D value.

The D values of vegetative cells are lower than those of bacterial spores (Table 1). The practical consequences of this are that at 70–100°C, vegetative bacteria will virtually disappear but spores will survive prolonged heat treatment. This elimination of vegetative cells has led to the heat treatment applied in sous-vide processing being compared to a pasteurization process.

The effect of heat treatment on vegetative cells can be quantified as the pasteurization value, P. The reference bacterium is *Streptococcus faecalis*, one of the most heat-resistant bacteria among non-spore-formers. Its D value at 70°C is close to 3 min, and its z value is 10°C. The pasteurization value can be defined as the period of heating at 70°C that would cause the same reduction in a population of *Streptococcus faecalis* as heat treatment at a different temperature for a different length of time. The pasteurization value can be calculated using the formula:

$$P = \int 10^{(T - T_{ref})/z} \cdot dt$$

T_{ref} is the reference temperature (i.e. 70°C); z = 10°C; and t is the heating time in minutes.

Using this formula, it can be shown that if a heat treatment of 40 min at 70°C has a pasteurization value of 40, approximately a 13 log reduction in *Streptococcus faecalis* will be achieved.

Heat treatments of 100 min at 70°C; 10 min at 80°C; or 1 min at 90°C have the same pasteurization value, $P = 100$. Their effects on the reduction in the number of *Streptococcus faecalis* are, therefore, similar.

The pasteurization value is particularly applicable in the estimation of the number of surviving cells of non-spore-formers, including spoilage bacteria (e.g. pseudomonads, enterobacteria, and lactic acid bacteria) and pathogenic bacteria (e.g. *Salmonella*, *Listeria monocytogenes* and *Escherichia coli*). Therefore, the pasteurization value is widely used in the sous-vide industry. Similar calculations, using the concept of the sterilization value' (F_0) used in the canning industry, and referring to the inactivation of *Clostridium botulinum* spores, would show that the sterilization values involved in sous-vide processing are very low.

The diversity of heating times and temperatures in sous-vide processing is a reflection of the diversity of the ingredients and recipes. From the microbial point of view, sous-vide foods can be divided into two categories: pasteurized foods, in which only spores survive; and non-pasteurized sous-vide foods, in which the vegetative cells of non-spore-forming bacteria survive, along with the spores of other species.

Bacteria of Concern

The bacteria of concern regarding the spoilage or pathogenicity of sous-vide foods have the following general characteristics:

- They are known contaminants of traditionally unprocessed and processed foods.
- They can resist heat processing to some extent, along the lines described above.
- They are able to grow in the conditions present in sous-vide foods. These foods should be kept at refrigeration temperatures and so the bacteria of concern grow at $< 10°C$.
- They are able to grow in microaerophilic/anaerobic environments.

Lactic acid bacteria, *Bacillus* species and *Clostridium* species fulfil these conditions and have been associated with the spoilage of sous-vide foods. This may involve the swelling of packs and/or the development of off flavours and off odours.

The pathogenic bacteria of main concern are *Listeria monocytogenes*, *Bacillus cereus* and *Clostridium botulinum*. *L. monocytogenes* is a frequent contaminant of foods, and combines a relatively high heat resistance with the ability to grow at low temperatures and a greater tolerance to low water activity and high acidity than associated with other pathogenic bacteria. *B. cereus* and *C. botulinum* are also considered to be major pathogens associated with heat-processed foods, because of the production of spores

and toxins. Some strains of *B. cereus* and of *C. botulinum* Group II are able to grow at low temperatures (see Table 1), but other pathogenic bacteria are less well-adapted to the conditions in sous-vide foods, and/or less tolerant to extreme conditions. The control of *C. botulinum*, *B. cereus* and *L. monocytogenes* will therefore have the effect of controlling the other pathogens.

Bacterial Growth and Concentration

Outbreaks of food poisoning result from the consumption of food in which: pathogenic bacteria are present; they survive heat treatment(s); and the surviving pathogenic bacteria multiply to a critical level. The food must be consumed: therefore the critical level must be reached before both the end of the shelf life of the food and spoilage – assuming that most consumers will not purchase or eat either products after their sell-by date, or sous-vide foods in swollen packs and with an aggressive odour.

Little is known about the contamination of sous-vide foods by pathogenic bacteria, but in contrast, the contamination of raw, unprocessed food is well-documented. Such contamination is very variable, the proportion in a range of foods (raw meat, sea foods and fish, fresh vegetables, milk) of samples positive for *Listeria monocytogenes* or *Clostridium botulinum* ranging from 0 to 100% in different surveys. The levels of contamination are generally low, usually being $< 10^2$ cfu g^{-1} of food in the case of *L. monocytogenes* or *B. cereus*, and < 1 cfu g^{-1} of food in the case of *C. botulinum*.

The heat resistance of these bacteria is well-documented. The D and z values may vary according to the genetic heterogeneity of the bacteria, the composition of the food matrix, and the nature of the medium used for recovery after heat treatment. For instance, in the case of *B. cereus*, D values of approximately 2 min have been reported at 90–95°C as well as at 121°C, and the D values at 90°C among a collection of strains of different origins varied from 4 min to > 200 min. Industrial equipment enables the precise monitoring of the temperature inside and outside a product, but the temperature varies during processing, and in different parts of the product. These variations must be taken into account when estimating the number of bacteria likely to survive heat treatment.

After heat treatment, sous-vide foods are cooled down to refrigeration temperatures. During the cooling period, the temperature of the food varies from 50°C to 15°C, a range suitable for the growth of mesophilic bacteria. The spore-former *Clostridium perfringens* has been identified as the main hazard in such circumstances, because of its ability to survive heat processing and its rapid growth in favourable conditions.

For the bacteria surviving heat processing, sous-vide foods offer a favourable environment for growth. The experimental inoculation of sous-vide foods with *L. monocytogenes* and *C. botulinum* (known as a 'challenge-testing' experiment) shows that the time needed to multiply to a critical level at refrigeration temperatures may be shorter than the shelf lives of some retail sous-vide foods. In the case of *B. cereus* and *L. monocytogenes*, critical concentrations may be reached before visible spoilage of the product. Critical concentrations may be defined as, for instance, the minimum infective dose of *L. monocytogenes* ($> 10^2$ cfu g^{-1}) or *B. cereus* (10^4–10^5 cfu g^{-1}) or the detection of the botulinum neurotoxin. The growth of these pathogens may be not as rapid in practice as in challenge testing, for several reasons. Natural inocula can be expected to be significantly smaller than those used in challenge testing, and the bacterial cells and spores will be markedly affected by heat processing. The growth of heat-stressed cells is either delayed or inhibited compared to that of non-heated cells as has been shown with psychrotrophic strains of *C. botulinum* (**Fig. 2**).

Recommendations and Regulations

Potential hazards in the sous-vide industry result from extended shelf lives, the possible survival and growth of pathogenic bacteria, retail and domestic storage in inappropriate conditions and insufficient experience of food microbiology on the part of manufacturers. There is an obvious need for recommendations and regulations, to protect both the consumer and the future expansion of the industry.

Recommendations were first laid down by the countries in which sous-vide foods were first introduced, including France, the UK, the US and Canada. However, they are now being established throughout Europe through the efforts of professional associations such as the European Chilled Food Federation (ECFF) and intergovernmental organizations. At the time of writing, a 'Code of hygienic practice for refrigerated packaged foods with extended shelf-life' was being prepared by the Codex Alimentarius, for worldwide application. However, no country at the time of writing has specific regulations for sous-vide foods.

The recommendations cover, *inter alia*, general hygienic practices: the microbial quality of raw foods, personal hygiene of employees, hygiene of food handling and storage areas, cleaning of equipment and so on. These considerations are not specific to the sous-vide industry.

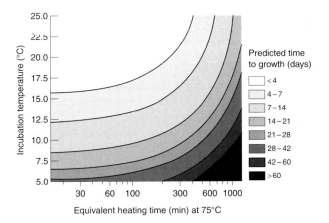

Figure 2 Effect of combined heat-treatment and temperature incubation on the growth of psychrotrophic *Clostridium botulinum*. Temperatures of heating ranged from 70°C to 80°C. Heating times were converted to equivalent heating time at 75°C assuming a linear z value of 6°C, 1 min at 75°C being for instance equivalent to 10 min at 69°C or to 0.1 min at 81°C. Inoculum contained 10^6 spores of psychrotrophic *C. botulinum*. Reprinted with permission from Journal of Food Protection (1997), 60, 1064–1071. Copyright held by the International Association of Milk, Food and Environmental Sanitarians, Inc. Authors are PS Fernandez, Universidad Miguel Hernandez, Orihuela, Spain and MW Peck, Institute of Food Research, Norwich, UK.

The recommendations also address heat treatment, the target microorganisms being *Listeria monocytogenes* and non-proteolytic and psychrotrophic strains of *Clostridium botulinum*. For example, a 10^6-fold reduction of these bacteria is considered as a safe treatment by the ECFF. Information on the lethal effect of a heat treatment at a given temperature, derived from the scientific literature, is summarized in tables available to manufacturers (see **Table 2** and **Table 3**). For instance, treatment for 2 min at 70°C will reduce contamination in a food initially containing 100 viable *L. monocytogenes* cells per gram to 0.0001 cells per gram. The same result would be achieved by treating for 5 s at 80°C. Taking into account the variability in the heat resistance of bacteria and the uneven distribution of temperature

Table 2 Heating time to produce a 10^6-fold reduction in *Listeria monocytogenes* (Adapted from ECFF (1996). The values above 70°C have been obtained by extrapolation, assuming a D_{70} of 0.3 min on a linear z value of 7.5°C. These values must be used only as an indication of lethal effect.)

Temperature (°C)	Time to produce 10^6-fold reduction (min)
60	43.5
65	9.3
70	2.0
75	0.43
80	0.083
85	0.017

Table 3 Heating time to produce a 10^6-fold reduction in psychrotrophic *Clostridium botulinum* (Adapted from ECFF (1996). The values have been calculated assuming a D_{90} of 1.5 min, and a z value of 7°C for temperatures < 90°C and a z value of 10°C for temperatures > 90°C.)

Temperature (°C)	Time to produce 10^6-fold reduction (min)
80	270
85	52
90	10.0
95	3.2
100	1.0

within a product, a substantial safety margin must be built into any predictions. Manufacturers are encouraged to monitor the temperature carefully during heat treatment.

Sous-vide foods must be cooled after heat treatment, to prevent bacterial growth. Recommendations suggest cooling to 4°C in < 2 h, as well as cooling to 20°C in < 5 h, followed by cooling to 7°C in < 3 h. Despite these variations, all national guidelines and codes of practice agree that cooling should not extend beyond a few hours.

The possible growth of *C. botulinum* during storage is universally recognized as the main hazard associated with sous-vide foods. Refrigeration substantially reduces its growth, and storage below 3°C totally prevents growth. Consequently, sous-vide foods must be kept refrigerated constantly from the time of packaging to the time of consumption, and the consumer must be so informed, e.g. on the packaging. The importance of the storage temperature of sous-vide foods is reflected in some national regulations on refrigerated foods, e.g. in France, where refrigerated foods have to be kept at 3°C by manufacturers, distributors and carriers.

Surveys involving supermarkets and domestic refrigerators show that refrigerated foods are almost always subjected to variations in temperature during storage. In addition, handling by consumers often involves temperature abuse for relatively short times, e.g. between the supermarket and home. In most countries, therefore, there are recommendations which introduce additional safety factors.

Shelf life, defined as the period during which the product maintains its microbiological safety and organoleptic qualities, must take into account storage temperature. Shelf lives may be prolonged by acidification of the product or by adjustment of its water activity. In countries where the main hazard is considered to be the mild temperature abuse of sous-vide foods (up to 10°C), either increasing the acidity to pH < 5.0, or reducing the water activity to 0.97, is recommended in order to prevent the growth of

psychrotrophic *C. botulinum*. For instance, such a recommendation appears in the ECFF Guidelines for the hygienic manufacture of chilled foods. If prolonged exposure at room temperature is considered a hazard, increasing the acidity to pH < 4.6 or reducing the water activity to 0.93 is recommended in order to prevent the growth of mesophilic proteolytic *C. botulinum*. This is the case, for instance, in North America, where several cases of food-borne botulism were caused by refrigerated foods exposed to room temperature for a prolonged period. However, although such modifications make the product safer, they also significantly alter its organoleptic qualities.

In general, it is possible to apply any combination of heat treatment, adjustment of pH, application of modified atmosphere, adjustment of water activity and control of storage temperature that has been shown to be effective against the target bacteria (mainly *C. botulinum*) during the shelf life proposed by the manufacturer. This represents an application of hurdle technology, in which none of the barriers to bacterial growth is independently high enough to be preventive, but in combination they show inhibitory effects. Suitable combinations of conditions which are inhibitory towards *C. botulinum*, *L. monocytogenes* and other pathogens are predicted by models, such as the Food Micromodel proposed by the Leatherhead Food Research Association and the Pathogen Modeling Program proposed by the US Department of Agriculture. However, these models are often too conservative, growth in actual foods usually being slower than predicted. Nevertheless, they are useful for predicting the effects of modifications of the composition and formulation of a food on bacterial growth. These effects may also be demonstrated by challenge testing, and guidelines for a protocol to evaluate the risk of *C. botulinum* toxin production in refrigerated foods with extended shelf lives have been proposed. However, challenge testing is generally time-consuming (only one product can be tested at a time) and expensive, and experimentation, particularly when pathogenic organisms are involved, requires safety precautions, including suitable facilities and trained personnel. Hence challenge testing is not universally applicable.

Future Developments

Sous-vide products now represent many thousands of tonnes and millions of individual portions of foods, and have had an excellent safety record in Europe. This could be explained by either low initial contamination with pathogenic bacteria, or by the fact that a combination of sublethal heat treatment and sub-optimal growth and storage conditions inhibit bacterial growth. However, a consistent safety record cannot verify these empirical observations. Research is needed in several areas to investigate the shelf life of sous-vide foods on a scientific basis. These areas include: time-to-toxicity versus time-to-spoilage; the effects of intrinsic food factors, in addition to pH and water activity; and the effects of temperature variations during storage. Sous-vide products must remain chilled throughout the distribution chain. Therefore the development of adapted time–temperature indicators, to help to ensure that manufacturers, distributors and consumers store products in appropriate conditions, would be useful, e.g. in the case of prolonged exposure to temperatures > 10°C, which significantly increase the risk of *C. botulinum* poisoning.

The continued success of sous-vide foods, and other chilled foods, will depend on an understanding of their specific microbiology.

See also: **Bacillus**: Introduction; *Bacillus cereus*. **Bacteria**: The Bacterial Cell; Bacterial Endospores. **Chilled Storage of Foods**: Principles; Use of Modified Atmosphere Packaging; Packaging with Antimicrobial Properties. **Clostridium**: *Clostridium botulinum*. **Ecology of Bacteria and Fungi in Foods**: Influence of Available Water; Influence of Temperature; Influence of Redox Potential and pH. **Heat Treatment of Foods**: Principles of Pasteurization. **Listeria**: *Listeria monocytogenes*. **Packaging of Foods**: Packaging of Solids and Liquids. **Predictive Microbiology & Food Safety. Process Hygiene**: Designing for Hygienic Operation; Overall Approach to Hygienic Processing; Modern Systems of Plant Cleaning; Involvement of Regulatory Bodies.

Further Reading

Advisory Committee on the Microbiological Safety of Food (1992) *Report on Vacuum Packaging and Associated Processes*. London: HMSO.

Baird B (1990) Sous vide: What's all the excitement about? *Food Technology* 44(11): 92, 94–95.

Codex Alimentarius (1998) Draft code of hygienic practice for refrigerated packaged foods with extended shelf-life. ALINORM 99/13, appendix III.

Doyle MP (ed.) (1989) *Foodborne Bacterial Pathogens*. New York: Marcel Dekker.

Doyle MP (1991) Evaluating the potential risk from extended-shelf-life refrigerated foods by *Clostridium botulinum* inoculation studies. *Food Technology* 45(4): 154–156.

Doyle MP, Beuchat LR and Montville TJ (eds) (1997) *Food Microbiology. Fundamentals and Frontiers*. Washington: ASM Press.

European Chilled Food Federation (ECFF) (1996) *Guidelines for the Hygienic Manufacture of Chilled Foods*. London–Paris: Chilled Food Association and SYNAFAP.

Farber JM and Dodds KL (eds) (1995) *Principles of Modified Atmosphere and Sous-vide Product Packaging*. Lancaster: Technomic Publishing.

Harada T and Paulus K (1987) Effects of cooking treatment on the texture of root vegetables. *Agricultural Biology and Chemistry* 51: 837–844.

Harada T, Tirtohusodo H and Paulus K (1985) Influence of temperature and time on cooking kinetics of potatoes. *Journal of Food Science* 50: 459–462, 472.

Hauschild AHW and Dodds KL (eds) (1993) *Clostridium botulinum. Ecology and Control in Foods*. New York: Marcel Dekker.

International Commission of Microbiological Specifications of Foods (ICMSF) (1980) *Microbial Ecology of Foods*. Vol. 1. San Diego: Academic Press.

International Commission of Microbiological Specifications of Foods (ICMSF) (1996) *Micro-organisms in Foods. 5. Characteristics of Microbial Pathogens*. London: Blackie Academic & Professional.

Priestley RJ (ed.) (1979) *Effects of Heating on Foodstuffs*. London: Applied Science Publishers.

Schellekens W and Martens T (1993) *Sous-vide Cooking. FLAIR-study Report No 2*. Brussels: Commission of the European Communities.

MICROCOCCUS

Maria-Luisa García-López, Jesús-Ángel Santos and **Andrés Otero**, Department of Food Hygiene and Food Technology, University of León, Spain

Introduction/Current Taxonomic Status

According to *Bergey's Manual of Determinative Bacteriology*, the family Micrococcaceae, which includes aerobic and facultatively anaerobic Gram-positive cocci, giving a positive reaction in the catalase test, consists of four genera: *Micrococcus, Staphylococcus,* *Stomatococcus* and *Planococcus*. Recent phylogenetic studies have showed that *Staphylococcus caseolyticus* and other related Gram-positive, catalase-positive cocci should be placed into a separate genus, which has been given the name *Macrococcus*. In addition to *M. caseolyticus*, three new species of macrococci have been described, *M. bovicus, M. equipercicus* and *M.*

carouselicus. The second edition of the *Prokaryotes* described *Micrococcus* spp. as Gram-positive non-spore-forming spherical cells occurring in pairs, tetrads or irregular clusters that are usually non-motile and aerobic with strictly respiratory metabolism. The G+C content of the DNA ranges from 65 to 75 mol%. Both Bergey's Manual and the *Prokaryotes* recognized nine species of *Micrococcus*: *M. agilis*, *M. halobius*, *M. kristinae* (or *M. kristiniae*), *M. luteus*, *M. lylae*, *M. nishinomiyaensis*, *M. roseus*, *M. varians* and *M. sedentarius*. More recently, data on G+C content in the DNA and studies of fatty and mycolic acid patterns, peptydoglycan type and 16S rDNA sequence – analysis have shown that the genus *Micrococcus* is significantly heterogeneous and isolates previously identified as *Micrococcus* spp. have been allocated to six different genera: *Arthrobacter*, *Dermacoccus*, *Kocuria*, *Kytococcus*, *Micrococcus* and *Nesterenkonia*. Thus, *M. agilis* is now *Arthrobacter agilis*; *M. halobius* has been transferred to the genus *Nesterenkonia* as *N. halobia*; *M. roseus*, *M. varians* and *M. kristinae* have been moved to the genus *Kocuria* as *K. rosea*, *K. varians* and *K. kristinae*, respectively; *M. nishinomiyaensis* is now in the genus *Dermacoccus* and *M. sedentarius* is *Kytococcus sedentarius*. The genus *Micrococcus* currently comprises only two species, *M. luteus* and *M. lylae*.

Habitats and Pathogenicity for Humans

The skin of humans and warm-blooded animals is the main reservoir for most of the former micrococcal species; meat and dairy products are considered the secondary source. They are also commonly found in a variety of terrestrial and aquatic ecosystems, including soil, fresh and marine water, sand and vegetation. The initial microflora of raw foods of animal origin such as meat and poultry, milk, freshwater fish and seafood commonly includes these bacteria but their role as spoilage organisms is not significant. By contrast, their tolerance to reduced water activity (a_w) and their thermoduric properties means that they are of interest in some processed foods, e.g. cured meats, sweetened condensed milk and intermediate-moisture products.

Although the pathogenicity of 'micrococci' is controversial, they are considered to be opportunistic pathogens in extraintestinal infections, particularly in immunocompromised patients.

Strains of *M. luteus* and *M. lylae* have been isolated from clinical specimens of patients suffering from pneumonia and septicaemia. Also *Kytococcus sedentarius* and *Kocuria kristinae* have been involved in several types of infections.

Characteristics of *Micrococcus* and Related Genera

The main characteristics of taxonomic interest for the differentiation of species of *Micrococcus* and related genera are presented in **Table 1** and **Table 2**.

Arthrobacter agilis

It has been suggested that the transfer of *M. agilis* to the genus *Arthrobacter* implies that the description of this genus must be emended to include a typical coccoid form.

In addition to the features given in Table 2, this species has the following characteristics. Spheres (0.8–1.2 µm in diameter) occur in pairs and tetrads. The dark rose-red pigment is water-insoluble. The species is motile by means of one to three flagella, though non-motile strains can occur. The polar lipids are phosphatidylglycerol, diphosphatidylglycerol, phosphatidylinositol and an unknown glycolipid. The optimal growth temperature range is 20–30°C. *A. agilis* is saprophytic, isolated from water, soil and human skin and is not common.

Dermacoccus nishinomiyaensis

In addition to the characteristics presented in Table 2, this monospecific genus has the following properties. Spheres (0.9–1.6 µm in diameter) occur in pairs, tetrads or irregular clusters. Some strains produce a water-soluble orange pigment. The polar lipids are phosphatidylglycerol, diphosphatidylglycerol and phosphatidylinositol. The optimal growth temperature range is 25–37°C. If present, the major aliphatic hydrocarbons (br-δ-C) are C_{22} and C_{23} with minor amounts of C_{25}, C_{26} and C_{27} hydrocarbons. The cytochromes are aa_3, c_{549}, c_{555}, b_{559}, b_{564} and d_{626}. *D. nishinomiyaensis* is saprophytic and its habitat is mammalian skin and water.

Kocuria

Like the other genera included in this article, cells of *Kocuria* are Gram-positive cocci (0.7–1.5 µm in diameter, depending upon the species), mostly arranged in tetrads and irregular clusters; *K. rosea* cells also occur in pairs. *Kocuria* is non-encapsulated, non-spore-forming and catalase-positive. They are aerobic, but two species (Table 2) are slightly facultatively anaerobic. Data on mol% G+C, amino sugars in cell wall polysaccharide, menaquinone composition and fatty acid composition are given in Table 2. Mycolic acids and teichoic acids are absent. The long chain aliphatic hydrocarbons (br-Δ-C) are C_{24} to C_{29} hydrocarbons. The polar lipids include diphosphatidylglycerol and phosphatidylglycerol. The type species is *K. rosea*.

Table 1 Differential characteristics of *Arthrobacter agilis* and the current genera of the family Micrococcaceae

	Tetrads[a]	Capsule	Motility	Anaerobic growth[b]	Oxidase	Lysostaphin resistance[c]	Furazolidone[c]	Lysozyme[c]	Teichoic acid in cell wall[d]	Peptidoglycan type	G–C content
A. agilis	+	–	+	–	+	R	R	R	–	Lys-Thr-L-Ala$_3$	67–69
Dermacoccus	+	–	–	–	+	R	R	SR	–	L-Lys-L-Ser$_{1,2}$-L-Ala-D-Glu	66–71
Kocuria	+	–	–	–/±	v[e]	R	R	R/SR	–	L-Lys-L-Ala$_{3-4}$	66–75
Kytococcus	+	–	–	–	–	R	R	SR	–	L-Lys-D-Glu$_2$	68–69
Micrococcus	+	–	–	–	+	R	R	S/SR	–	L-Lys-peptide subunit or L-Lys-D-Asp	69–76
Nesterenkonia	–/+	–	–	–	+	R	R	R	–	L-Lys-Gly-L-Glu	70–72
Staphylococcus	–	–	–	+	–	S	S	R	+	L-Lys-MCA var	30–39
Macrococcus	–/+	–	–	±	+	S/R	S	R	–/+	L-Lys-Gly$_{3-4}$, L-Ser	38–45
Planococcus	–	–	+	–	–	R	R	S	–	L-Lys-D-Glu	39–52
Stomatococcus	–/+	+	–	+	–	R	R	R	–	L-Lys-L-Ala	56–61

[a] –/+, Arranged mostly in clusters and occasionally in pairs or tetrads.

[b] –/±, Two species of *Kocuria* may grow anaerobically and macrococci (*M. caseolyticus*) are marginally anaerobic.

[c] S, Sensitive; R, resistant; SR, slightly resistant; S/R, only *M. caseolyticus* is resistant to lysostaphin.

[d] –/+, Only *M. caseolyticus* appears to contain cell-wall teichoic acid.

[e] v, variable within species (see Table 2).

Table 2 Scheme for the differentiation of *Arthrobacter agilis* and species of the genera *Dermacoccus*, *Kocuria*, *Kytococcus*, *Micrococcus* and *Nesterenkonia*

	A. agilis	D. nishinomiyaensis	Kocuria			Kytococcus sedentarius	Micrococcus		N. halobia
			K. kristinae	K. rosea	K. varians		M. luteus	M. lylae	
Pigment[a]	R	O	PC/PO	P/R	Y	CW/BY	Y/CW	CW/U	U
Facultatively anaerobic	–	–	±	–	±	–	–	–	–
Oxidase	+	+	+	–	±	–	+	+	+
Growth at 37°C	–	+	+	+	±	–	+	+	+
Salt tolerance	<5%	<7%	10%	7.5%	7.5%	10%	10%	10%	10%
Motility	+	–	–	–	+	–	–	–	–
Growth on Simmons citrate agar	–	–	–	–	–	–	–	–	–
Growth on inorganic nitrogen agar[b]		w/–	–/w	–	–	–	+	–	–
Gelatin	+	+	–	–	–	+	+	+	–
Arginine dihydrolase	–	–	–	+	–	+	–	–	–
Nitrate reduction	–	±	±	+	+	–	±	±	–
β-Galactosidase	+	–	±	–	–	–	–	–	+
Acid from glucose	–	±	+	±	+	–	–	–	+
Acid from glycerol	–	–	+	–	–	–	–	–	+
Lysozyme[c]	R	SR	R	SR	R	SR	S	SR	R
Main fatty acids	ai-C15:0	i-C17:1, i-C17:0, aiC17:0	ai-C15:0	ai-C15:0	ai-C15:0	ai-C17:0, i-C17:1	i- ai-C15:0, i-C15:0	i-C15:0, ai-C15:0	ai-C15:0, ai-C17:0
Major menaquinones	MK-9(H₂)	MK-8(H₂)	MK-7(H₂), MK-8(H₂)	MK-8(H₂)	MK-7(H₂)	MK-8, MK-9, MK-10	MK-8, MK-8(H₂)	MK-8(H₂)	MK-8, MK-9
Amino sugar in cell wall polysaccharide	Glucosamine	Galactosamine	Glucosamine	Galactosamine	Galactosamine		Mannosamine, uronic acid	Galactosamine	Galactosamine
G+C content	67–69	66–71	67	66–75	66–72	68–69	70–75.5	68.6	70–72

[a]R, Dark rose-red; O, orange; PC/PO, pale cream or pale orange; P/R, pink or red; Y, yellow; CW/BY, cream white or buttercup yellow; Y/CW, yellow or cream white; CW/U, cream white or unpigmented; U, unpigmented; [b] w, weak reaction; [c] R, resistant (minimum inhibitory concentration (MIC) above 100 µg ml^{-1}); SR, slightly resistant (MIC 5–50 µg ml^{-1}); S, susceptible (MIC below 5 µg ml^{-1}).

1. *K. rosea* does not produce water-soluble exopigment. The optimal growth temperature range is 25–37°C. The cytochromes are aa_3, c_{550}, c_{557}, b_{564} and d_{626}. *K. rosea* is distributed in soil and water.

2. *K. varians* does not produce water-soluble exopigment. The growth temperature range is 22–37°C. The cytochromes are aa_3, c_{549}, c_{557}, b_{564} and d_{626}. *K. varians* is saprophytic and is isolated from mammalian skin, soil and water.

3. *K. kristinae* does not produce water-soluble exopigment. The colony colour becomes more distinctive with colony age. The optimal growth temperature range is 25–37°C. The cytochromes are aa_3, c_{548}, c_{557}, b_{561} and d_{626}. In addition to the genus-specific phospholipids, nynhidrin-negative phospholipids and glycolipids are present. Human skin is the primary habitat. *K. kristinae* is occasionally involved in human infections.

Kytococcus sedentarius

This monospecific genus has the following additional characteristics. Spheres (0.8–1.1 μm in diameter) occur predominantly in tetrads or in tetrads in cubical packets. Some strains produce a brownish pigment. The optimum temperature range is 28–37°C. The cytochromes are aa_3, c_{626}, c_{550}, b_{557}, b_{561} and b_{564}. The polar lipids are diphosphatidylglycerol, phosphatidylglycerol and phosphatidylinositol. The major aliphatic hydrocarbons (br-δ-C) are C_{30}–C_{33} hydrocarbons. The primary habitat is human skin. *K. sedentarius* is occasionally involved in human infection.

Micrococcus

The emended genus includes spherical cells (0.8–1.8 μm in diameter) occurring predominantly in tetrads and in irregular tetrads or clusters. They are non-motile and non-spore-forming, Gram-positive and aerobic. *Micrococcus* is chemo-organotrophic with strictly respiratory metabolism, and is catalase- and oxidase-positive. Optimal growth temperature is 25–37°C. *Micrococcus* is non-halophilic. The cytochromes are aa_3, b_{557}, b_{567} and d_{626}; cytochromes c_{550}, c_{551}, b_{563}, b_{564} and b_{567} may be present. Mycolic acids and teichoic acids are absent. The polar lipids are phophatidylglycerol, diphosphatidylglycerol and unknown ninhydrin-negative phospholipids and glycolipids; phosphatidylinositol may be present. The major aliphatic hydrocarbons (br-Δ-C) and C_{27} to C_{29} hydrocarbons. Other characteristics are listed in Table 2. The primary habitat is mammalian skin. The type species is *M. lylae*. Some strains of this species produce a violet pigment which diffuses into the medium. *M. luteus* is the other species included in the emended genus *Micrococcus*. Both species are occasionally isolated from clinical specimens.

Nesterenkonia halobia

In addition to the features listed in Table 2, this monospecific genus has the following characteristics. Spheres (0.8–1.8 μm in diameter) occur singly, in pairs and occasionally in tetrads or clusters. *N. halobia* is moderately halophilic. Best growth occurs on media containing 1–2 mol l^{-1} NaCl; moderate growth occurs in media containing 4 mol l^{-1} NaCl and the organism is unable to grow on media without KCl or NaCl. Mycolic acids are absent. The phospholipids are diphosphatidylglycerol, phosphatidylglycerol and phosphatidylinositol. The growth temperature range is 20–40°C. *N. halobia* is presumably distributed in saline habitats.

Applications

'Micrococci' have been involved in the ripening of different types of cheese made from cow's, goat's and/or ewe's milk. Salt-tolerant 'micrococci' are a major component of the surface microflora of smear-ripened cheeses and certain blue-veined cheeses. Furthermore, 'micrococci' have been related to the development of the desired body and texture of several soft cheeses, as well as to their flavour. Selected strains of 'micrococci' have been patented for their inclusion as starters in the production of different varieties of cheeses.

The contribution of 'micrococci' to the ripening of cheeses has been related to their proteolytic, lipolytic and esterolytic activities, as well as to their ability to produce methanethiol. Extracellular proteinases are produced by several species of 'micrococci', although their activity is optimal in an alkaline pH range, being almost negligible at pH values below 5.5. Commercial preparations of micrococcal proteinases are available for accelerated cheese-ripening. Intracellular proteinase, endopeptidase, aminopeptidase and dipeptidase activities have been demonstrated in cells of different strains of 'micrococci'.

Several intracellular and extracellular lipolytic, as well as esterolytic activities have been described in strains of 'micrococci' isolated from cheeses. However, their role in the ripening of cheeses is controversial. Their optimal pH is in the alkaline range, and most are very sensitive to low pH and salt.

Micrococcaceae become one of the dominant microbial groups throughout curing of different kinds of dry cured hams, as well as during ripening of dry fermented sausages. Isolates from these and other similar dry cured meats are ascribed to the genera *Staphylococcus* and *Micrococcus*. Their positive role in the organoleptic properties of such meat products is linked to some enzymatic activities: nitrate reductase, catalase and lipolytic and proteolytic activities.

Kocuria varians produces a nitrate reductase which is active even at 10°C. The contribution of nitrate reductase and catalase of micrococcal origin to the development and maintenance of the colour in cured meats is well established. Both enzymes can also contribute to the flavour by limiting fatty acid oxidation and aldehyde production. However, the real contribution of lipolytic and proteolytic enzymes of microbial origin to the ripening of dry cured meats does not seem to be of great significance. Strains of *K. varians* are included in several meat starter cultures.

Some strains of *Micrococcus* of food origin can degrade in vitro histamine and tyramine, and one strain of *K. varians* showed high tyramine oxidase activity while it exhibits no decarboxylase activity towards histidine, tyrosine, lysine, ornithine or phenylalanine. However, other isolates of *K. varians* have been shown to produce in vitro biogenic amines.

Although some strains of *K. varians* isolated from fermented meat products produce a bacteriocin (variacin) antagonistic towards several spoilage microorganisms and *Listeria monocytogenes*, the growth of several beneficial microorganisms (*Lactobacillus sake*, etc.) is also inhibited.

Enzymes from *Macrococcus caseolyticus* have been patented for synthesizing aspartame. Strains of *K. varians* have been proposed for the industrial production of trehalose. Salt-tolerant glutaminases produced by *Micrococcus luteus* of marine origin are of interest for the production of soy sauces. Selected strains of *M. luteus* and other micrococci can be used in biological assays, for the detection of antibiotic residues in foods, as well as the vitamin biotin and the enzyme lysozyme.

Recommended Methods of Detection and Enumeration in Foods

Isolation and Preservation

Plating media used for enumeration and isolation of the former micrococcal species are of four types: a non-selective medium devised for direct isolation from skin (P agar), selective nitrofuran-containing media (FTO agar) for distinguishing micrococci from staphylococci in human skin samples and clinical specimens, chemically defined media especially devised for *M. luteus* and *K. rosea*, and mannitol-salt agar, a selective medium for *Staphylococcus aureus*, which is widely used for enumeration of 'micrococci', mainly *Kocuria*, from cured meats, fermented sausages and cheese.

P agar has the following composition: peptone (Difco) $10\,g\,l^{-1}$; yeast extract (Difco) $5\,g\,l^{-1}$; sodium chloride $5\,g\,l^{-1}$; glucose $1\,g\,l^{-1}$; agar (Difco) $15\,g\,l^{-1}$.

Yeast extract-free P agar and P agar supplemented with the mould inhibitor cycloheximide ($50\,\mu g\,ml^{-1}$) can also be used. FTO agar is conveniently prepared in 700 ml amounts in 1 l Erlenmeyer flasks. Tryptic soy agar (Difco) or trypticase soy agar (BBL) fortified with 0.1% yeast extract and 0.5% Tween 80 is the basal medium. After autoclaving and cooling to 48°C, 0.1% of a 0.5% acetone stock of the dye oil red O and 10% of a 0.5% acetone stock of nitrofuran (Furoxone) are added. To prevent flocculation, the latter is added slowly from a 100 ml graduate into swirling agar. Flasks are then left open or loosely covered in a water bath for 3–5 min to allow acetone to evaporate. The modified FTO agar has the same basal composition as P agar. After autoclaving and cooling, 100 ml of a 0.02% acetone solution of furazolidone is mixed as described above. FTO agar supports the growth of 'micrococci' and prevents the growth of staphylococci.

Mannitol salt agar is a commercially available medium containing 7.5% NaCl. For the isolation of strains from marine origin, sea-water agar is recommended.

Inoculated agar media are incubated under aerobic conditions at 30–37°C for 4 days. It should be noted that *A. agilis*, which does not grow in media containing 5% NaCl, grows best at 20–25°C, and that media for the isolation of *N. halobia* should include 5% NaCl.

For short-term preservation, cultures may be kept on nutrient agar slopes at 5°C. They may also be stored on the same medium under liquid paraffin at refrigeration temperatures (5°C) for 1–2 years. Lyophilization or storage in liquid nitrogen are recommended for longer periods (5 years or more).

Identification

Separation of the Former Micrococcal Species from Related Genera

Gram stain, cell morphology and catalase production are the first screening tests for Micrococcaceae. For routine purposes, distinction of 'micrococci' from staphylococci can be made by demonstrating the susceptibility of staphylococci to $200\,\mu g\,ml^{-1}$ of lysostaphin and their ability to produce acid from glycerol aerobically in the presence of $0.4\,\mu g\,ml^{-1}$ of erythromycin. Sensitivity of staphylococci to furazolidone (100 μg disc) and bacitracin (0.04 U disc) is an alternative procedure. Furthermore, staphylococci are facultative anaerobic (thioglycolate medium) and, with the exception of a few species, such as *S. sciuri* and *S. lentus*, give a negative reaction in the modified oxidase test of Faller and Schleifer. The remaining genera may be differentiated by the tests listed in

Table 1. Analysis of the cellular fatty acids, isoprenoid quinones, polar lipids and mycolic acids, as well as peptydoglycan composition and genetic methods such as determination of the DNA base composition, 16S rDNA sequencing, etc. permit an accurate separation of the microorganisms included in this article. However, these characters cannot be easily determined in the routine laboratory.

Differentiation of Species

Species may be differentiated by means of tests listed in Table 2. In addition to identification based on conventional methodology, there are commercial kits that allow the identification of some of the former micrococci, though none of these kits reflects the current taxonomy of these bacteria. Thus, API Staph-Ident database includes an entry 'Micrococcus spp.', with an additional test table to differentiate M. luteus, M. lylae, Kocuria varians, K. kristinae and Kytococcus sedentarius. Also, the database of Microscan conventional and Microscan 2-h includes Micrococcus entries. The list of taxa of the BBL Crystak Gram-positive ID panel includes the true micrococci and some of the related species (K. kristinae and Kytococcus sedentarius). Finally, the Vitek II system, a new fluorogenic automated system, can identify species of Micrococcus and Kocuria.

See also: **Arthrobacter. Bacteria**: Classification of the Bacteria – Phylogenetic Approach. **Cheese**: Mould-ripened Varieties. **Fermented Foods**: Fermented Meat Products. **Meat and Poultry**: Curing of Meat. **Micrococcus. Milk and Milk Products**: Microbiology of Liquid Milk. **Process Hygiene**: Designing for Hygienic Operation. **Staphylococcus**: Introduction.

Further Reading

Bascomb S and Manafi M (1998) Use of enzyme tests in characterization and identification of aerobic and facultatively anaerobic Gram-positive cocci. Clin. Microbiol. Rev. 11: 318–340.

Bhowmik T and Marth EH (1990) Role of Micrococcus and Pediococcus species in cheese ripening: a review. J. Dairy Sci. 73: 859–866.

Hammes WP and Hertel C (1998) New developments in meat starter cultures. Meat Sci. 49: S125–S138.

Kloos WE, Tornabene TG and Schleifer KH (1974) Isolation and characterization of micrococci from human skin, including two new species: Micrococcus lylae and Micrococcus kristinae. Int. J. Syst. Bacteriol. 24: 79–101.

Kloos WE, Ballard DN, George CG et al (1998) Delimiting the genus Staphylococcus through description of Macrococcoccus caseolyticus gen. nov., com. nov. and Macrococus equipercicus sp. nov., Macrococcus bovicus sp. nov. and Macrococcus carouselicus sp. nov. Int. J. Syst. Bacteriol. 48: 859–877.

Koch C, Schumann P and Stackebrandt E (1995) Reclassification of Micrococcus agilis (Ali-Cohen 1889) to the genus Arthrobacter as Arthrobacter agilis comb. nov. and emendation of the genus Arthrobacter. Int. J. Syst. Bacteriol. 45: 837–839.

Kocur M, Kloos WE and Schleifer K-H (1992) The genus Micrococcus. In: Balows A, Trüper HG, Dworkin M, Harder W and Schleifer K-H (eds) The Prokaryotes. A Handbook on Habitat, Isolation and Identification of Bacteria, 2nd edn. P. 1300. New York: Springer-Verlag.

Stackebrandt E, Koch C, Gvozdiak O and Schumann P (1995) Taxonomic dissection of the genus Micrococcus: Kocuria gen. nov., Nesterenkonia gen. nov., Dermacoccus gen. nov., and Micrococcus Cohn 1872 gen. emend. Int. J. Syst. Bacteriol. 45: 682–692.

MICROFLORA OF THE INTESTINE

Contents

The Natural Microflora of Humans

Caroline L Willis, Public Health Laboratory Service, Southampton, UK

Glenn R Gibson, Microbiology Department, Institute of Food Research, Reading, UK

Copyright © 1999 Academic Press

Introduction

This article overviews the composition and activities of the microbiota that resides in the human gastrointestinal tract. Profiles of the resident flora show differences depending on the host organ. Gastric contents are essentially sterile, with at the most 10^3 microorganisms per millilitre. *Helicobacter pylori* is one important colonizer of the stomach. In the small intestine, the rapid transit of contents, as well as the input of secretions such as bile, maintain populations at around 10^3–10^6 bacteria per millilitre. In the colon, however, the picture changes enormously. Bacterial populations may be as high as 10^{12} per gram. This anaerobic microflora makes the large intestine by far the most heavily colonized organ in humans, possibly representing over 90% of total cells present in the body. The bacteria carry out fermentation whereby dietary and endogenous growth substrates are metabolized to give a variety of end products. Some of the products are benign and may even have health-promoting values, whereas others are toxic. Principal groups that maintain colonic function are *Bacteroides* spp., bifidobacteria, clostridia, *Eubacterium* spp., lactobacilli, coliforms and Gram-positive cocci.

The Stomach

Mainly because of the low luminal pH, relatively few microorganisms have the ability to reside in the human stomach. One important exception is *Helicobacter pylori*. Along with certain acid-tolerant lactobacilli and streptococci, this is the main bacterium able to survive and colonize the gastric environment. To overcome the hostile effects of stomach acid, *H. pylori* uses its intense ureolytic activity to generate ammonia, which offers increased protection by helping to neutralize an acidic microenvironment surrounding the bacterium. Moreover, the bacterium is highly motile, and it is thought that *H. pylori* uses this motility to invade the semisolid mucus layer of the stomach and thereafter adhere to epithelial cells. As a result, *H. pylori* compromises the mucosal barrier and may then secrete a cytotoxin. It is believed that *H. pylori* infection can cause peptic ulceration, type B gastritis and perhaps even stomach cancer. As a result of these epidemiological links, the organism has attracted recent intensive research interest. Future studies are likely to be directed towards the pathogenic mechanisms elaborated by *H. pylori*.

The Small Intestine

The transit time of gut contents in the small bowel, as well as intestinal secretions and physicochemical variables such as pH and redox potential (E_h), all contribute towards microflora development. The upper small gut is dominated by facultatively anaerobic and aerotolerant bacteria such as streptococci, staphylococci and lactobacilli, with bacterial numbers showing a progressive increase both in terms of numbers and degree of anaerobiosis. Although lactobacilli and streptococci tend to predominate in the terminal regions of the small intestine, there is a higher proportion of *Bacteroides* spp. and enterobacteria. In the small gut, host immune defences may also be a significant factor limiting bacterial establishment.

The Large Intestine

The human colon extends from the ileocaecal junction to the anus, with a total length of approximately 150 cm, and has about 200 g of contents. It can be divided into a number of regions, namely the caecum, ascending colon, transverse colon, descending colon and sigmoid rectum.

Contrary to the original belief that the main role of the large intestine was simply to absorb water and salt, and store and dispose of waste materials, it is now recognized that the colon is an extremely complex organ with a variety of important biological functions. It actively transports sodium and chloride ions, probably by means of sodium-hydrogen and chloride-bicarbonate exchanges, and also absorbs many products of bacterial fermentation such as short-chain fatty acids (SCFA), ammonia and other metabolites. The colonic mucosa can also secrete fluid and is a potential source of hormones or neuropeptides which may influence colonic absorption by acting as local messengers.

In addition, the large intestine and its indigenous microflora make a major contribution to the digestion of dietary components which escape breakdown and absorption in the small intestine. Dietary starches may be resistant to digestion in the small intestine for a number of reasons. In whole or partly milled grains and seeds, starch may be physically inaccessible, while ungelatinized starch granules found in bananas and other plants exist in a closely packed, partially crystalline form which resists degradation. Potatoes, rice and maize undergo retrogradation, a process by which cooked starchy foods become less soluble as they cool, and are thus less susceptible to enzymatic hydrolysis. Plant polysaccharides such as cellulose, hemicellulose and pectin, which are not digested by host enzymes, contribute a further fermentable substrate, as do sugars such as lactose, raffinose and stachyose which also escape digestion in the small intestine. In addition, a number of endogenously produced substances are present in the large intestine, for example mucus, pancreatic enzymes and sloughed epithelial cells, which may serve as effective bacterial growth substrates.

Water, undigested food and endogenous sources of nutrients enter the caecum from the small intestine. As the nutrient supply is rich at this point, bacteria living in the proximal colon are active and their growth is rapid. Short-chain fatty acids, produced from bacterial fermentation, lower the pH of gut contents. As digesta progress through the large intestine, nutrients (especially carbohydrates) become depleted. Thus, in the distal colon there is reduced SCFA production and the pH approaches neutrality. Since water is progressively absorbed as material moves through the colon, the moisture content of the digesta decreases.

Transit times through the gut are variable. While food residues may remain in the colon for as much as a week after the intake of a meal, most is evacuated within 20–140 h, with the average transit time being around 60 h. Faecal composition also varies greatly. The water content remains constant at about 70% of total stool weight, while in individuals living on a typical Western diet, bacteria constitute around 55% of faecal solids.

Bacteriology of the Colon

Because of the heterogeneous nature of the large intestine and the wide variety of nutrients available, many different types of microorganism can flourish, making the large bowel the most densely colonized region of the human digestive tract. There are, however, many difficulties involved in the study of the colonic microflora, predominantly the problem of obtaining samples of gut contents. For this reason, most work has focused on the bacteriology of faecal specimens.

A further issue is that studies of faeces and luminal contents give little information about the microflora associated with the intestinal mucosa. In general, little is known about such populations in humans. However, it is likely that biofilm formation plays a significant biological role in the colonic microecology.

Many species in the human large intestine are unculturable using currently available bacteriological techniques, and therefore viable counts are likely to underestimate numbers of organisms present in gut contents. This problem may be at least partially overcome by the use of molecular biology. For example, total DNA can be extracted from faecal samples, 'shotgun' cloned, and the cloned 16S ribosomal RNA genes subjected to sequence analysis. By comparison of sequences with existing reference collections of rRNA sequences, the probable identity of the original organism can be established. Alternatively, labelled genomic DNA from a particular bacterium can be used to probe for the presence of the organism and other bacteria with highly homologous genomes in total DNA obtained from samples of gut contents. It is likely that modern molecular methodologies will shed new light on the gut microbial diversity and allow the development of powerful genetic tools (e.g. genotypic probes) to elucidate the microbiological role of the human colon.

There are in the region of 10^{12} colony forming units per gram (dry weight) of contents in the colon, a figure that approaches the theoretical limit that can be accommodated by this mass. These belong to at least 50 different bacterial genera, comprising well over 400 species. The most numerous genera are *Bacteroides* (which may account for up to 30% of culturable bacteria in the large bowel), *Eubacterium* and *Bifidobacterium*.

Before birth the gut of a normal fetus is sterile, but bacteria are acquired from the mother's vaginal and faecal flora, and the environment, during birth. Initially, facultative anaerobes such as *Escherichia coli* and *Enterococcus faecalis* predominate, although the pattern of colonization is influenced by exogenous

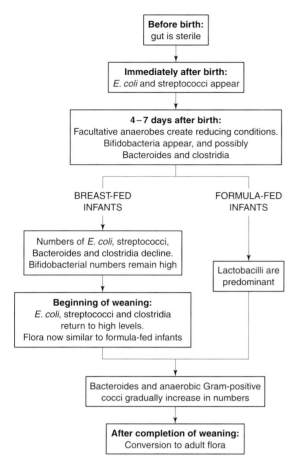

Figure 1 Acquisition and development of the human large intestinal microflora.

factors such as diet, type of delivery and gestational age. As these initial facultative species reduce the redox potential, obligate anaerobes become established (**Fig. 1**). Soon after birth, bifidobacteria are thought to predominate in breast-fed infants, while in formula-fed babies lactobacilli are prevalent. After weaning, conversion to an adult-type flora pattern occurs, and by the end of the second year of life the predominant gut populations are similar to those observed in adults.

Once a climax community has been established, the composition of the colonic microflora remains relatively constant, being largely controlled by competition for limiting nutrients. Other factors affecting colonization by different organisms are pH, redox potential and mucosal immune responses exerted by the host, as well as complex interactions, both beneficial and antagonistic, between bacterial species.

Since the host diet provides a large proportion of fermentable substrates for colonic bacteria, it is likely that dietary composition has a profound effect on the microflora composition. This has led to the development of strategies that can improve the microbial 'balance' of the gut. Probiotics, prebiotics and syn-

biotics are all examples of such mechanisms. The premise is that the colonic microbiota contains beneficial bacterial genera. In this context, bifidobacteria and lactobacilli are usual target organisms. A probiotic approach advocates the use of live microbial additions to appropriate food vehicles. For example, many fermented milk products contain viable species of lactic acid-producing bacteria. A prebiotic targets 'advantageous' organisms that are indigenous to the gut. Finally, a synbiotic combines both the probiotic and prebiotic concepts.

Major disturbances in the colonic ecosystem can occur through the use of antibiotics. The importance of the normal colonic flora to host health can be illustrated by observing the effects of antibiotic treatment on certain patients. One example is the association of pseudomembranous colitis with the use of broad-spectrum antibiotics, especially clindamycin and ampicillin. This disease is caused by *Clostridium difficile*, a minor commensal in the normal gut. When antibiotic treatment disrupts the homeostasis of bacteria in the bowel, *C. difficile* is able to proliferate, resulting in severe colitis and toxaemia. Similarly, staphylococcal enterocolitis can result from overgrowth of *Staphylococcus aureus* in patients receiving wide-spectrum antibiotic therapy. The infective dose of gut pathogens required to cause disease is known to be much lower in germ-free animals or those that have been pre-treated with antibiotics.

Bacterial Fermentation in the Colon

Fermentation is the term used to describe the anaerobic breakdown of organic matter by bacteria. In the colon, various bacterial groups follow numerous nutritional pathways, reflecting the wide variety of potential substrates available. There is a large amount of cross-feeding in this complex ecosystem, with one species utilizing another's metabolic end products. The degradation of some complex molecules therefore requires the concerted action of several species.

The majority of bacteria in the human colon are saccharolytic, obtaining their energy from the fermentation of carbohydrates. The most numerous and versatile polysaccharide utilizers belong to the genus *Bacteroides*. Other common saccharolytic species belong to the genera *Bifidobacterium*, *Ruminococcus*, *Eubacterium*, *Lactobacillus* and *Clostridium*. Complex polysaccharides are broken down by a wide range of bacterial polysaccharidases and glycosidases, in a series of hydrolytic steps, to produce simple sugars. The majority of saccharolytic anaerobes use the Embden–Meyerhof–Parnas pathway to metabolize these sugars, resulting in the formation of pyruvate. This is further metabolized by a number of pathways. The main products of carbohydrate fer-

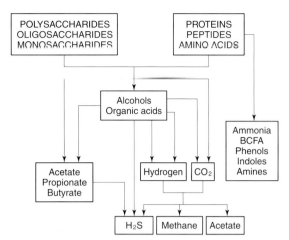

Figure 2 Summary scheme of bacterial fermentation in the large intestine. BCFA, branched-chain fatty acids.

mentation are SCFA (mostly acetate, propionate and butyrate), hydrogen and carbon dioxide, but electron sink products such as lactate, succinate and ethanol may also accumulate, although to a lesser extent (**Fig. 2**).

Asaccharolytic organisms such as species of *Peptococcus*, *Peptostreptococcus*, *Eubacterium* and *Clostridium* are dependent on proteolytic bacteria to hydrolyse proteins and thus provide amino acids for their fermentation. Organisms reported to be proteolytic include species of *Clostridium*, *Lactobacillus*, *Fusobacterium*, *Streptococcus*, *Propionibacterium* and members of the *Bacteroides fragilis* group. Bacteria belonging to the *B. fragilis* group constitute the numerically predominant type of proteolytic bacteria. The enzymes are particularly active against casein, azocasein, trypsin and chymotrypsin, but not against elastin, collagen, gelatin or globular proteins such as bovine serum albumin. Of the clostridia, *Clostridium perfringens*, *C. bifermentans* and *C. sporogenes* are strongly proteolytic. In addition to providing a substrate for asaccharolytic bacteria, proteolysis is important in supplying nitrogen.

Protein fermentation produces similar metabolites to those obtained from carbohydrate fermentation, but also ammonia, phenols, indoles, amines and branched-chain fatty acids (BCFA). Ammonia is formed mainly by the deamination of amino acids. Deamination of the aromatic amino acids, tyrosine, phenylalanine and tryptophan, results in the production of phenolic and indolic compounds. Amines are formed by the decarboxylation of amino acids.

The various fermentation products have a number of possible fates, some being beneficial to the host and some harmful. The electron sink products may be absorbed or further metabolized by bacteria to produce SCFA and gases. Similarly, hydrogen and carbon dioxide may be metabolized or excreted in

breath or flatus. Of the large amounts of SCFA produced, most are absorbed by the colonic mucosa. The precise mechanism of absorption is not clear, although the process is associated with an appearance of bicarbonate ions and stimulation of sodium absorption. One route is through passive diffusion of acids across the colonocyte cell membrane, providing a powerful driving force for movement of water out of the colonic lumen.

After absorption from the gut, acetate is mainly metabolized in the brain, muscle and peripheral tissues, while propionate has been shown to be a major precursor of gluconeogenesis in ruminants. It has also been speculated that propionate-supplemented diets have a cholesterol-lowering effect. Butyrate is largely metabolized by the colonic epithelial cells, acting as an important source of energy. Butyrate also appears to have a potential protective role against colon carcinoma. It reversibly inhibits histone deacetylase enzymes in colonocytes, resulting in hyperacetylation of the histone proteins and is a powerful stimulator of apoptosis.

Ammonia, in contrast, may select for neoplastic growth in the colon, owing to its greater toxicity to healthy cells than to those that have been transformed. It has been found to alter the morphology and metabolism of intestinal epithelial cells, reduce their life span, and also increase the rate of colonic cell turnover. Ammonia can, however, be detoxified by the formation of urea in the liver. Phenols and indoles are usually absorbed from the gut and detoxified in the liver by conjugation with glucuronide or sulphate. They are believed to have the ability to act as cocarcinogens and have been implicated in bladder and bowel cancer. They are also found in higher concentrations in some disease states, such as diarrhoea and schizophrenia. Amines are also rapidly absorbed and transported to the liver where they may be detoxified by deamination. Like phenolic compounds, amines have been associated with a number of diseases such as migraine, hypertension, heart failure, schizophrenia and cancer. The condensation of secondary amines with nitrite results in the formation of *N*-nitrosamines and there is some evidence to suggest that this group of compounds are carcinogenic.

Conclusion

It is clear that the human large intestine contains a vast array of microbial diversity. Much of the resident flora may not yet have been characterized. New methodological developments will more appropriately address this issue. Probably more importantly, the functional aspects of gut microbiology are now being more fully defined. It has long been recognized that the microbiota contains pathogenic species. However, it is also apparent that certain intestinal bacteria are essential for host wellbeing, through their nutritional aspects. For example, the provision of SCFA as significant end products of metabolism has a variety of biological consequences. Moreover, the normal gut flora has powerful 'colonization resistance' aspects. Transient pathogens may be suppressed through competition for substrates and colonization sites, as well as the elaboration of antagonistic substances like antibiotics and acids. Steps that fortify the benign or 'beneficial' genera hold much promise and will further elucidate the biological relevance of the gastrointestinal microflora.

See also: **Bacteriodes**. **Biofilms**. **Clostridium**: Introduction. **Enterobacteriacea, Coliforms and E. coli**: Introduction. **Eschericia coli**: *Escherichia coli*. **Lactobacillus**: Introduction. **Microflora of the Intestine**: Biology of Bifidobacteria; Biology of *Lactobacillus acidophilus*; Biology of the *Enterococcus spp.*; Detection and enumeration of Probiotic Cultures. **Helicobacter**.

Further Reading

Cummings JH, Rombeau JL and Sakata T (eds) (1995) *Physiological and Clinical Aspects of Short Chain Fatty Acid Metabolism.* Cambridge: Cambridge University Press.

Cummings JH and Macfarlane GT (1991) The control and consequences of bacterial fermentation in the human colon. *Journal of Applied Bacteriology* 70: 443–459.

Dixon MF (1992) *Helicobacter pylori* and chronic gastritis. In: Rathbone BJ and Heatley RV (eds) Helicobacter pylori *and Gastroduodenal Disease*, 2nd edn. P. 124. Oxford: Blackwell Scientific.

Fuller R (ed.) (1992) *Probiotics: The Scientific Basis.* London: Chapman & Hall.

Fuller R (ed.) (1997) *Probiotics 2: Applications and Practical Aspects.* London: Chapman & Hall.

Gibson GR and Beaumont A (1996) An overview of human colonic bacteriology in health and disease. In: Leeds AR and Rowland IR (eds) *Gut Flora and Health – Past, Present and Future.* P. 3. London: RSM Press.

Gibson GR and Macfarlane GT (eds) (1995) *Human Colonic Bacteria: Role in Nutrition, Physiology and Pathology.* Boca Raton: CRC Press.

Gibson GR and McCartney AL (1998) Modification of gut flora by dietary means. *Biochemical Society Transactions* 26: 222–228.

Gibson GR and Roberfroid MB (1995) Dietary modulation of the human colonic microbiota: introducing the concept of prebiotics. *Journal of Nutrition* 125: 1401–1412.

Gibson GR, Willems A, Reading S and Collins MD (1996) Fermentation of non-digestible oligosaccharides by human colonic bacteria. *Proceedings of the Nutrition Society* 55: 899–912.

Gilliland SE (1990) Health and nutritional benefits from lactic acid bacteria. *FEMS Microbiology Reviews* 87: 175–188.

Macfarlane GT and Gibson GR (1994) Metabolic activities of the normal colonic flora. In: Gibson SAW (ed.) *Human Health – The Contribution of Microorganisms.* P. 17. London: Springer.

Tannock GW (1995) *Normal Flora.* London: Chapman & Hall.

Whitehead R (ed.) (1995) *Gastrointestinal and Oesophageal Physiology*, 2nd edn. Edinburgh: Churchill Livingstone.

Biology of Bifidobacteria

A Y Tamime, Scottish Agricultural College, Auchincruive, Ayr, UK

The human intestinal microflora has not been fully characterized, but several hundred different species of bacteria have been detected in the faeces of humans. Anaerobic bacteria are present in abundance and outnumber other types of microorganisms; predominant organisms are Gram-negative rods belonging to the genera *Bacteroides*, *Enterococcus* and *Fusobacterium*. Also, in a healthy adult colon anaerobic Gram-positive bacteria such as bifidobacteria, lactobacilli and streptococci share the habitat with other anaerobic rods such as *Clostridium* and *Eubacterium*. Although probiotic lactic acid bacteria, which are present in the gastrointestinal tract, constitute only a minor component, *Bifidobacterium* species may comprise up to 25% of the cultivable human gut microflora.

Role of Bifidobacteria in the Intestine

The discovery of a rod-shaped anaerobic bacterium in the faeces of newly born babies by Tisser dates back to the beginning of the twentieth century. This microorganism with its distinctive bifid morphology was named *Bacillus bifidus* (currently known as *Bifidobacterium bifidum*). Some bifidobacterial species are among the 25 species of bacteria most frequently isolated from healthy adult humans. Although the habitat of some species of this genus is the human colon, *Bifidobacterium* spp. have been isolated from different ecological niches in the environment: the human intestine, vagina and oral cavity; the animal intestine; the intestine of the honey bee; and sewage. At present, 30 different species of bifidobacteria have been isolated from these sources, and six species of *Bifidobacterium* are used in the dairy industry for the manufacture of therapeutic/probiotic fermented milk products. However, only the indigenous microflora of the human are discussed here.

Bifidobacterial species found in humans are *Bifidobacterium adolescentis*, *B. bifidum*, *B. breve*, *B. infantis*, *B. lactis* and *B. longum*. *Bifidobacterium lactis* strains LW 420 and URI, which have been isolated from fermented milk products, have different physiological characteristics from other known species of bifidobacteria, and their 16S rRNA sequencing homology reflects a different phylogenic position; *B. lactis* is therefore currently regarded as a new species. The origin of the organism is not well established, and its efficacy as a probiotic culture merits further investigation to establish if it is able to colonize the human intestinal tract. Some human bifidobacteria have a distinctive cell morphology, such as amphora-like, specific epithet, thin, short, elongated or non-specific (**Fig. 1**).

Colonization of the Intestine, Causes of Depletion and Possible Effects

Factors that can affect bacterial colonization of the human intestine may include age, drug therapy, diet, peristalsis, host physiology, local immunity and microflora interactions. It is evident that the fractions of bifidobacteria in the gut flora of infants are high compared with the fractions in elderly people. A number of mechanisms have been identified whereby the components of the gut microflora can interact during colonization. For example, facultative anaerobic bacteria (e.g. coliforms, streptococci and/or staphylococci) rapidly utilize the traces of oxygen that diffuse into the intestinal lumen. As a consequence, a low redox potential (E_h) is maintained which allows microorganisms such as *Bifidobacterium*, *Bacteroides* and *Eubacterium* to colonize the intestine.

Other factors that may influence bifidobacterial colonization of the intestine include the following:

1. Infant prematurity makes it difficult for the implantation of *Bifidobacterium* species in the gut epithelium because of the lack of receptors or endogenous substrates.
2. The method of feeding (breast or bottle) can affect the proportions of bifidobacteria and other microbial species.
3. Endogenous substrates present in the digestive tract without a dietary source (blood group antigens or mucin oligosaccharides) are degraded by enzymes produced by *Bifidobacterium* species.
4. The microflora of the environment – in particular that of hospitals – can influence the rapidity of bifidobacterial colonization.

The end products of fermentation (e.g. organic acids) by the colonic microflora are inhibitory to some

invasive bacteria. Acid production by *Bifidobacterium* species lowers the pH in the intestine, while maintenance of a low E_h, the ability to compete for available nutrients and adhesion sites at the colonic mucosa or on food particles are important factors in determining the colonization of the intestine by bifidobacteria. Bacterial species that are unable to compete are rapidly eliminated from the intestine. However, it is still unclear whether bacterial interaction determines whether certain microorganisms are indigenous to the intestine or transient in the luminal contents. Nevertheless, organic acids such as acetic and lactic acids are produced by bifidobacteria, and the theoretical ratio of fermentation is 2 hexose → 3 acetate and 2 lactate. In addition, a slight increase in uric and formic acids was observed when milk was fermented with bifidobacteria. The inhibitory activity of these organic acids is governed by their dissociation constants (pK_a), and the acid concentration at a given pH. Therefore, an organic acid of high pK_a value is more acid in the undissociated form, and has a stronger antimicrobial activity. Thus, the activity and pK_a values of some organic acids are: lactic 3.85 < acetic 4.74 < propionic 4.87 < benzoic 4.91.

Other proteinaceous metabolites, bacteriocins, can manifest antimicrobial activities against closely related bacteria and pathogenic microorganisms. The bacteriocins produced by bifidobacteria have not been as well characterized as those from some other lactic acid bacteria. Current studies of bifidobacterial bacteriocins have shown activity against species of *Lactococcus*, *Bifidobacterium*, *Lactobacillus* (*L. acidophilus*), *Clostridium* (*C. perfringens* and *C. tyrobutyricum*) and *Streptococcus* (*S. thermophilus*).

Growth Promoting Factors

In vitro studies on the ability of bifidobacteria to grow and produce large numbers of viable cells in milk and synthetic media have resulted in the description of many growth promoting factors (**Table 1**). Three main groups of bifidogenic factors, which differ depending on the species of human origin, are:

- the BB factors (BF1, BF2 and glycoproteins)
- the BI factor
- the BL factors.

The BB factors are characterized as the elements found in human and animal milks, including colostrum, and in yeast and liver extracts, while the BI and BL factors are abundant in many plant extracts as well as in liver and milk extracts.

Other compounds that have a promoting effect on the growth of certain *Bifidobacterium* spp. are: (a) lactoferrin and its metal complexes with Fe, Cu or

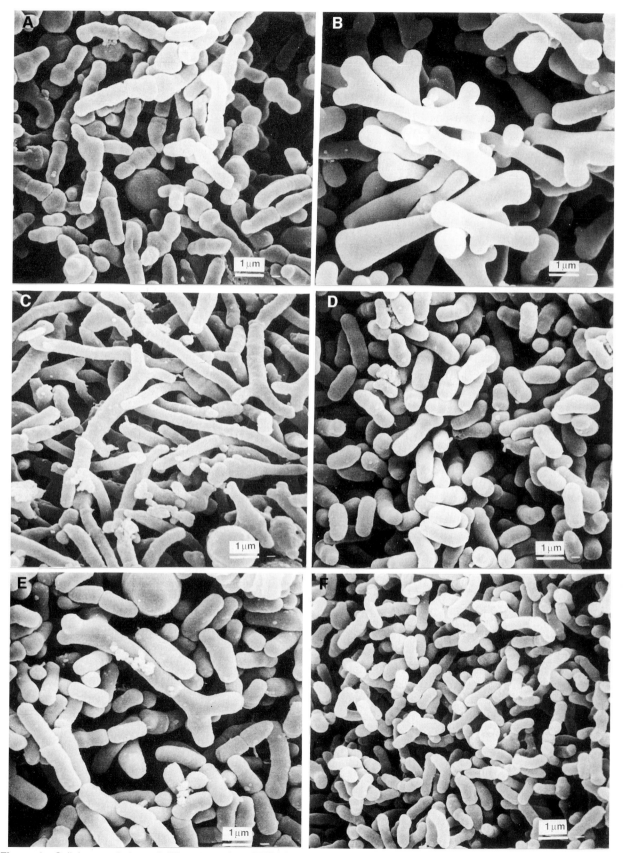

Figure 1 Cellular morphology of different species of *Bifidobacterium*. (**A**) *B. infantis*. (**B**) Bifidobacteria isolated from a commercial dairy product. (**C**) *B. longum*. (**D**) *B. pseudolongum*. (**E**) *B. animalis*. (**F**) *Bifidobacterium* sp. Reproduced by courtesy of Professor V. Bottazzi-Bianchi, UCSC, Institute of Microbiology, Piacenza, Italy.

Table 1 Growth factors of bifidobacteria. After Tamime (1997).

Source	Component
Human milk, milk from some animals and cow's colostrum	Glycoside of N-acetylneuraminosyllactose
Human blood and hog gastric mucin	As above
Human milk	Oligosaccharides containing: N-acetylglucosamine, glucose, galactose, fructose or lactose
Human κ-casein or its trypsin derivative	Consists of low molecular weight and high carbohydrate content (about 75%) or glycomacropeptide
Bovine casein digest	Known as casein bifidus factor (CBF) containing: lactose, N-acetylgalactosamine, or amino acids
Mucus secreted by the salivary glands, the small intestine and the colon	Glycoprotein with N-acetylgalactosamine, N-acetylgalactosamine and sialic acid
Milk and whey after hydrolysis using proteinase	Non-glycosylated peptide
Naturally occurring carbohydrates	Raffinose and stachyose, fructo-oligosaccharides, lactulose, lactitol, lactohionic acid
Synthetic products	Transgalactosylated oligosaccharide, galsucrose, lactosucrose
Associative growth between lactobacilli and bifidobacteria	Due to proteolytic enzymes and aminopeptidase of lactobacilli
Soya bean oligosaccharide extracts	Stachyose and raffinose
Biological materials	Yeast extract and β-lactoglobulin
Hydrolysed lactose	Galacto-oligosaccharides
Chicory	Oligofructose, inulin

Zn; (b) lactulose and lactitol, (c) oligoholosides and polyholosides (but not amylose and cellulose); and (d) fructo-oligosaccharides.

Health-promoting Activities

Bifidobacteria are considered as probiotic microorganisms which, in general, are helpful in maintaining appropriate balances between the various floras in different sections of the human intestine. The established health-promoting properties associated with the ingestion of *Bifidobacterium* spp. are:

- enhanced lactose digestion
- increase in faecal bifidobacteria
- decease in faecal enzyme activity
- colonization of the intestinal tract
- prevention and/or treatment of acute diarrhoea caused by food-borne infection
- prevention and/or treatment of rotavirus diarrhoea
- prevention of antibiotic-induced diarrhoea.

Other health benefits attributed to bifidobacteria include: (a) activity against *Helicobacter pylori*; (b) stimulation of intestinal immunity; (c) stabilization of intestinal peristalsis; (d) reduced carriage time for *Salmonella* spp.; (e) improved immunity to various diseases; (f) suppression of some cancers; (g) reduction in serum cholesterol levels; and (h) reduction in hypertension. Some of these benefits have been proposed on the basis of studies in vitro or in animals, and insufficient information is available to determine whether such findings are applicable to humans. It is evident that appropriate studies in vivo are required to confirm the claims listed above.

In general, the sensitivity of bifidobacteria to antibiotics is not well established. From the published data, it seems that most bifidobacterial strains are resistant to streptomycin, polymyxin B, neomycin, nalidixic acid, gentamicin, kanamycin and metronidazole. In general, the sensitivity to antibiotics varies from $10 \, \mu g \, ml^{-1}$ to $500 \, \mu g \, ml^{-1}$ or more, but some antibiotics, e.g. benzylpenicillin (penicillin G), erythromycin, ampicillin and chloramphenicol, strongly inhibit the growth of most *Bifidobacterium* species.

Some *Bifidobacterium* strains of human origin are producers of certain vitamins. For example, thiamin, folic acid, biotin and nicotinic acid are synthesized in some quantity by *Bifidobacterium bifidum* and *B. infantis*, whereas *B. breve* and *B. longum* release only small quantities of these vitamins; the latter species are recognized producers of riboflavin, pyridoxine, cobalamin and ascorbic acid.

Intestinal Bifidobacterial Counts and Dietary Supplementation

Within a few days of birth, *Bifidobacterium longum*, *B. infantis*, *B. breve*, *B. bifidum* and/or *B. adolescentis* colonize the intestinal tract, and the average total count of bifidobacteria is then at least $10 \log_{10}$ colony forming units (cfu) per gram of stools. The distribution of the strains depends on whether the infant is breast- or bottle-fed. In contrast, *B. longum* and *B. adolescentis* always predominate in the colon of adult and elderly people. The count averages $8 \log_{10}$ cfu g^{-1} of faeces in adults, and gradually reduces with age.

Dietary supplementation with products containing bifidobacteria helps to either maintain or increase the counts of these organisms in the intestine. There are now many food products which contain bifidobacteria; examples include infant formulae, nondigestible oligosaccharides, pharmaceutical preparations and fermented milks (see below). The survival of ingested bifidobacteria during passage through the digestive tract is dependent on the strains being of human origin, and on their resistance to acidic conditions, proteolytic enzymes and bile salts.

The count of bifidobacteria in the colon increases after ingestion of a food supplement containing these organisms. In general, the increase is at least 2 $\log_{10}$ cfu g^{-1} of colon content; however, when the supplementation regimen is stopped, the bifidobacterial count drops to its original level. Such an effect is difficult to explain, but it is most likely that the bacterial ecosystem in the intestine is self-regulatory and the *Bifidobacterium* count is maintained at about 10 $\log_{10}$ cfu g^{-1} under normal conditions.

Available Products and Survival of Bifidobacteria until Consumption

Some dietary adjuncts, which are used to increase the numbers of bifidobacteria in the intestine, are:

Infant Feed Formulae These are made from cow's milk modified to resemble human milk and are supplemented with essential minerals and vitamins. Products of this type, such as Lactana-B®, Eledon® and Eugalan Forte®, contain *B. bifidum*, alone or in combination with *L. acidophilus* and *Pediococcus acidilactici*. Another milk product containing *B. bifidum*, Pelargon®, is made from milk fermented using *Lactococcus lactis* subsp. *lactis*.

Pharmaceutical Preparations Freeze-dried tablets, which contain viable bifidobacteria, include Bifider®, Bifidogène®, Lyobifidus®, Liobif®, Life Start Two® or Original®, Euga-Lein® and Lactopriv®. Other preparations that may contain *B. bifidum* or *B. longum* and other desirable organisms are Synerlac®, with *L. acidophilus* and *L. delbrueckii* subsp. *bulgaricus*; Infloran Berna®, with *L. acidophilus*; and Omniflora®, with *L. acidophilus* and *Escherichia coli*.

Dairy Foods Products such as soft-serve or hard ice-cream, cheese such as fresh type, Gouda or Cottage cheese, ultrafiltered (UF) milk, milk powders such as formula feeds for infants, strained yoghurt and fermented milks are used as vehicles for implantation of bifidobacteria in the human intestinal tract. Of these, fermented milks including 'bio-yoghurts' are apparently the most popular dairy products for bifidobacterial supplementation of the human intestine.

The species that are common in fermented milks are *B. bifidum*, *B. longum* and *B. infantis*, in combination with other lactic acid bacteria. These therapeutic organisms should be present at the time of consumption as viable cell counts of at least 10^6 cfu g^{-1} or ml^{-1} of the product. Isolation and characterization of bifidobacteria in some commercial yoghurts sold in Europe have, an occasion, identified species other than those stated on the labels. In many such instances *Bifidobacterium animalis* was the only species present, while in other products, the viable counts of bifidobacteria have been found to be less than the recommended level at the time of consumption.

Characterization of Culture Stock and Isolates

In fermented milk in India, strains of *B. bifidum* that originated from infants were 10^8 cfu g^{-1} at the end of 3 weeks. However, in flavoured yoghurts, the viable counts of *B. bifidum* dropped from 10^7 cfu ml^{-1} to 10^2–10^5 cfu ml^{-1} after 21 days at pH 4, or to 10^3–10^6 cfu ml^{-1} during the same period at pH 4.5. In a Spanish study, the counts of bifidobacteria in yoghurt were reduced to between 78% and 60% of the initial numbers after 10 days and 30 days of storage, respectively (**Table 2**). Studies at the Scottish Agricultural College suggested that survival of bifidobacteria in fermented milks made with commercial starter cultures varied after 20 days' storage (**Fig. 2**). The co-culture of bifidobacteria species with yoghurt starter cultures suppressed the growth of the former, but the count did not decline significantly during storage.

It is evident from such studies that strains of bifidobacteria that are tolerant to acidic conditions in fermented milks must be used with such products if they are to have high viable cell counts of bifidobacteria at the time of consumption.

B. bifidum and *L. acidophilus* have also been used in ice-cream making: after storage for 16 weeks at $-20°C$, the counts of each organism were approximately 1.0×10^7 cfu ml^{-1}; however, a slight drop in the count occurred before and after freezing owing to the incorporation of air at the whipping stage and to the actual freezing stage. Other workers observed no survival of *B. bifidum* in ice-cream mixes of pH 3.9–4.6. Recently, Zabady (an Egyptian fermented milk) was made by replacing 33% and 50% of the yoghurt starter culture with *B. bifidum* DI and BB12 respectively, and the resulting product was used to make frozen Zadady. The numbers of bifidobacteria that survived after 5 weeks' storage averaged 10^7 cfu ml^{-1}.

Supplementation of non-fermented milk with 5×16^8 cfu ml^{-1} of *B. longum* gave a product which was a good source of β-galactosidase. Other researchers have recommended a strain of *B. breve* for cheese-making. A cream dressing fermented with *B. infantis*

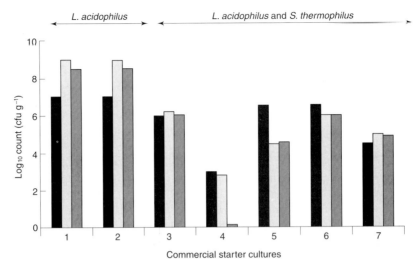

Figure 2 Numbers of bifidobacteria in milk (black bars) and fermented milks (fresh, white bars; stored, shaded bars), using different commercial starter cultures: 1, *B. lactis*; 2, *B. longum* and *B. infantis*; 3, *B. bifidum*; 4, *B. longum*; 5–7, *B. lactis*. Results are averages of three trials, and counts of *L. acidophilus* and *S. thermophilus* are not shown; incubation periods ranged between 6 h and 22 h; pH of stored products ranged between 4.15 and 4.64.

Table 2 Numbers of bifidobacteria in commercially produced fermented milk during storage in Spain. Adapted from Medina and Jordano (1994)

Storage time (days)	pH	Bifidobacteria ($\times 10^6$ cfu g^{-1})[a]	Percentage[b]
0	4.57	7.4	100.0
10	4.34	5.8	78.3
31	4.19	4.4	59.4
42	4.23	4.0	54.0
51	4.24	1.0	13.5
84	3.81	< 0.0001	–

[a] Mean values from five samples.
[b] Percentage of surviving bifidobacteria.

was used successfully to blend Cottage Cheese, but a loss of activity of bifidobacteria was evident at pH 4.5.

See also: **Bacteriocins**: Potential in Food Preservation. **Bacteriodes**. **Bifidobacterium**. **Clostridium**: Introduction. **Enterococcus**. **Fermented Milks**: Range of Products; Yoghurt; Products from Northern Europe; Products of Eastern Europe and Asia. **Ice Cream**. **Milk and Milk Products**: Microbiology of Liquid Milk.

Further Reading

Charteris WP, Kelly PM, Morelli L and Collins JK (1998) Ingredient selection criteria for probiotic micro-organisms in functional diary foods. *International Journal of Dairy Technology* 51: 123–136.

Fuller R (ed.) (1992) *Probiotics: The Scientific Basis.* London: Chapman & Hall.

Fuller R (ed.) (1997) *Probiotics 2: Applications and Practical Aspects.* London: Chapman & Hall.

Kok RG, de Waal A, Schut F, Welling GW, Weenk G and Hellingwerf KJ (1996) Specific detection and analysis of a probiotic *Bifidobacterium* strain in infant faeces. *Applied and Environmental Microbiology* 62: 3668–3672.

Marshall VME and Tamime AY (1997) Physiology and biochemistry of fermented milks. In: Law BA (ed.) *Microbiology and Biochemistry of Cheese and Fermented Milk*, 2nd edn. P. 153. London: Blackie.

Medina LM and Jordano R (1994) Survival of constitutive microflora in commercially fermented milk containing bifidobacteria during refrigerated storage. *Journal of Food Protection* 56: 731–733.

Meile L, Ludwig W, Rueger U et al. (1997) *Bifidobacterium lactis* sp. nov., a moderately oxygen tolerant species isolated from fermented milk. *Systematic and Applied Microbiology* 20: 57–64.

Salminen S and von Wright A (eds) (1998) *Lactic Acid Bacteria – Microbiology and Functional Aspects*, 2nd edn. New York: Marcel Dekker.

Tamime AY (1997) Bifidobacteria – an overview of physiological, biochemical and technological aspects. In: Hartemink R (ed.) *Non-Digestible Oligosaccharides: Healthy Food for the Colon.* P. 9. Bennekom: Drukkerij Modern.

Tamime AY and Marshall VME (1997) Microbiology and technology of fermented milks. In: Law BA (ed.) *Microbiology and Biochemistry of Cheese and Fermented Milk.* P. 57. London: Blackie.

Tamime AY, Marshall VME and Robinson RK (1995) Microbiological and technological aspects of milks fermented by bifidobacteria. *Journal of Dairy Research* 62: 151–187.

Tisser H (1900) *Rechercher sur la Flora Intestinale Normale et Pathologique des Nourrissons.* PhD Thesis, University of Paris.

Biology of *Lactobacillus acidophilus*

William R Aimutis, Land O' Lakes Inc., St Paul, Minnesota, USA

Introduction

The human gastrointestinal tract is colonized at birth by a simple microbial population that begins undergoing dynamic transformation when the infant first consumes food. Functions of the intestinal microflora can be divided into beneficial and harmful activities. Beneficial bacteria normally predominate during a host's healthy state. If the microbial balance is upset and harmful bacteria predominate the human can experience an increase in various intestinal diseases. The goal is to maintain a healthy balance of bacteria in the intestinal tract with maximum numbers of beneficial bacteria and a minimum of harmful, or potentially pathogenic bacteria. One group of beneficial intestinal bacteria is species in the genus *Lactobacillus*.

Lactobacilli Colonization in the Gastrointestinal Tract

Colonization of the human gastrointestinal tract occurs as the infant is born. The course of colonization will be determined by gestational age, type of delivery, hospital environment and dietary constituents. The influence of various factors of host and microbial origin are interconnected in the establishment of the individual gastrointestinal microbial ecosystem. Commensal bacteria derived from the mother's vagina, intestine and skin contaminate infants vaginally delivered. Many of these bacteria are unable to survive in the intestinal tract of the newborn and will disappear shortly after the infant begins eating. Caesarean section infants are colonized by environmental contamination, which changes after the infant begins feeding.

Although *Lactobacillus acidophilus* was first isolated from infant faeces early in the century, the frequency of lactobacilli occurrence in the intestine during the first days of life is variable (15–100%). However, it is widely believed that the first organisms to colonize the newborn large intestine are species of the genera *Lactobacillus* and *Bifidobacterium* in formula-fed and breast-fed infants, respectively. Both groups of organisms must compete against other organisms that are attempting to establish in the infant's intestinal tract. *Lactobacillus acidophilus* remains a resident in the intestinal tract of humans until death, overcoming many obstacles along the way.

Continued *L. acidophilus* colonization is achieved by an ability to survive in the environment and secretions of the host, resistance against antagonistic activity of other microorganisms and an ability to depress or inhibit their growth, adherence to host intestinal epithelial cells or mucin secretions, and degradation of the host endogenous nutrients. All of these factors must work simultaneously to allow any microorganism to remain a part of the intestinal microflora. Although the gut flora remains relatively stable, it can be disturbed by exogenous factors such as antibiotic treatment, hormone (oestrogen) therapy, intestinal diseases (bacterial overgrowth) and radiation exposure. Normal diet seems to have little if any effect on the continued establishment of *L. acidophilus* in the human intestinal tract. Several researchers have reported the presence of this species in the intestinal tract of subjects consuming vegetarian, Far Eastern, or Western diets.

Role of *L. acidophilus* in the Intestine

The human intestinal tract is inhabited by more than 400 different species of microorganisms, among them lactobacilli. Sufficient data have been accumulated about the species and numbers of lactobacilli in the proximal and distal parts of the intestine, but information concerning lactobacilli in the ileum, caecum and colon of healthy humans is inadequate. Interest in the occurrence, numbers and role of lactobacilli in the human intestine has received considerable attention since Metchnikoff first proposed in 1908 that lactobacilli are responsible for increased longevity.

Lactobacilli are Gram-positive, rod-shaped facultatively anaerobic, non-sporulating, acid-tolerant, catalase-negative bacteria with a DNA base composition of less than G+C 53%. The most frequent lactobacilli isolated from the human intestinal tract belong to *Lactobacillus acidophilus*, *L. salivarius*, *L. casei*, *L. plantarum*, *L. fermentum*, *L. reuteri* and *L. brevis*. The population numbers vary according to intestinal location (proximal to distal, and lumen to mucosa). The human stomach, once thought to be sterile, contains approximately 10^3 colony forming units (cfu) per gram lactobacilli at the mucosal surface and in the lumen. The population increases as the intestinal tract is transversed to the faeces (**Table 1**). Faeces contain an average of 10^6–10^{10} cfu g^{-1} lactobacilli.

The metabolic capacity of *L. acidophilus* is extremely diverse, but has not been extensively studied in vivo. Much of the published data were

Table 1 Lactobacilli numbers in a healthy human gastrointestinal tract

	Lumen (log$_{10}$ cfu g^{-1})	Mucosa (log$_{10}$ cfu g^{-1})
Stomach	< 3.0	> 3.0
Duodenum	< 3.0	
Jejunum	2.1–4.3	3.1–4.6
Ileum		
Colon	3.0–7.0	
Rectum	4.0–10.0	0–6.3
Faeces	6.0–10.0	

generated by in vitro or animal experimentation, and the information must be accepted as conjecture. None the less it is worth examining.

Lactose and glucose are metabolized by *L. acidophilus* to form large amounts of lactic acid in the intestinal tract. In an infant this may be essential for pH regulation and formation of an acid barrier. Mucin and other endogenous secretions hydrolysed by *L. acidophilus* form organic acids and supply other intestinal bacteria with metabolizable substrates. Although *L. acidophilus* is not proteolytic, some strains produce peptidases to further digest peptides formed by other organisms with proteolytic activity.

Bile acid conjugates are hydrolysed by *L. acidophilus*. Taurocholate and glycocholate are hydrolysed by most strains of *L. acidophilus*. Deconjugated primary bile acids may be less soluble and less absorbable. This may have importance to the host in prevention of colon cancer.

A number of substrates have been tested to determine the type of reactions *Lactobacillus* catalyses. It has been reported that *L. acidophilus* cleaves bonds in a number of antimicrobials. The azo bond of sulphasalazine, a drug used to treat ulcerative colitis, is cleaved by *L. acidophilus*. It is also reported that *L. acidophilus* partially degrades the antimicrobials phthalylsulphathiazole and chloramphenicol. The physiological significance of these reactions is elusive. However, it is speculated that bond cleavage accentuates the antimicrobial activity in vivo.

Causes of Depletion and Possible Effects

Healthy individuals have a stable intestinal microflora, but it can be altered to an abnormal flora by many endogenous and exogenous factors such as cancer, peristalsis disorders, surgery, liver diseases, radiation therapy, emotional stress, and administration of antibiotics. Diet can influence the composition of the human intestinal microflora. Subjects on Japanese and Seventh-Day Adventist 'low-risk' diets (characterized by low protein and high fibre levels) had higher lactobacilli populations than subjects on a Western 'high-risk' diet (characterized by high levels of protein, fat and sodium). Changing from

a low-risk to a high-risk diet causes a decrease in the intestinal lactobacilli population.

Antimicrobial therapy causes reduction or elimination of lactobacilli in the intestinal tract. Clindamycin, chloramphenicol and cephalosporins reduce or eliminate intestinal lactobacilli. The use of neomycin or a combination of neomycin and metronidazole significantly decreases the number of lactobacilli. Suppression of the normal microflora can lead to establishment of potentially pathogenic microorganisms in the intestinal tract. Colitis or pseudomembranous colitis is caused by an overgrowth of the toxin-producing *Clostridium difficile*.

Stress causes the intestinal lactobacilli population to decrease. Patients with severe diarrhoeal diseases will shed lactobacilli in their faeces owing to denuding of the intestinal microvilli and the attached lactobacilli. Eventually all the lactobacilli will be washed from the intestine, and the organisms will need to be re-established. Astronauts before and during space flight excrete high numbers of lactobacilli. Workers at the Chernobyl atomic power plant exposed to radiation excreted high numbers of lactobacilli in their faeces after the accident at that site. Starvation reduces the number of lactobacilli in the intestinal tract, and favours a predominance of coliforms. It is believed lactobacilli are affected by starvation because they derive their nutrients from the diet of the host. Starvation predisposes the host to intestinal infections caused by opportunistic pathogens such as *Escherichia coli* or *Salmonella typhimurium*.

Advantages of Dietary Supplementation

Fermentation by lactic acid bacteria is used as an effective food preservation method. Cultured dairy products have been consumed by every civilization in the world. The establishment of certain lactobacilli in the intestinal tract is believed to exercise beneficial effects in the health of the consumer.

Products containing *L. acidophilus* have attracted interest in recent years from commercial entities, researchers and consumers. The 'functional food' concept is of worldwide interest as the population explores better ways to prevent diseases and increase life expectancy. Several companies have begun marketing viable lactic acid bacterial products to promote a healthy intestinal environment. Such products have been termed 'probiotics'. The functional properties of many products are inferred from published data. Although research support is still sketchy, researchers continue to study the effects of lactic acid bacteria, including *L. acidophilus* to support the claims of many commercial companies.

Any potential new colonizer or probiotic must overcome the chemical and physical defence mechanisms of the gastrointestinal tract, and compete with the established flora for suitable nutrients, atmospheric conditions and attachment sites in the gut mucosa. Many lactobacilli attach to the gut mucosa, but attachment is a highly specific interaction between bacterial protein adhesions and complementary host cell receptors. Therefore, colonization or attachment of a probiotic strain in the intestinal tract of humans is best accomplished by strains of human origin.

Diarrhoea

Several clinical studies have examined the efficacy of lactic acid bacteria in decreasing the symptoms or duration of diarrhoeal diseases. In one study, the incidence of traveller's diarrhoea was reduced when subjects ingested 3×10^9 cfu daily of *L. acidophilus*, *Bifidobacterium bifidum*, *Lactobacillus bulgaricus* and *Streptococcus thermophilus*. Travellers began consuming capsules containing these organisms 2 days prior to departure and continued until the last day of travel.

People infected with *Salmonella* remain as carriers of the organism long after clinical symptoms have disappeared. This creates a major problem for food service workers who could potentially handle food and contaminate it unknowingly. The period of carriage for patients with *Salmonella* can be reduced by daily consumption of *L. acidophilus*. Patients needed to consume 500 ml of milk containing 6×10^9 cfu ml^{-1} of *L. acidophilus*.

Several studies have shown that infants recovering from gastroenteritis will benefit from consumption of lactic acid bacteria including *L. acidophilus*. The results suggest viable lactic acid bacteria are able to colonize the gut and shorten the duration of acute diarrhoea.

Antibiotic-associated diarrhoea is caused by *Clostridium difficile* overgrowth in the large intestine. Patients consuming *L. acidophilus* yoghurt while receiving antibiotics had less diarrhoea than control groups. Other side effects such as abdominal distress, stomach cramps and flatulence were also less common. Several studies have also shown continued use of *L. acidophilus* yoghurt after antibiotic treatment reduced the incidence of recurrent *C. difficile* colitis.

Lactose Digestion

A large segment of the population is lactose intolerant because of a deficiency of the enzyme lactase. Failure to hydrolyse lactose leads to its fermentation in the large intestine and causes intestinal distress in the consumer. Symptoms include flatulence, abdominal pain, bloating and diarrhoea. There is good evidence

that lactose-intolerant subjects can consume significant amounts of lactose from milk or milk products if lactic acid bacteria are present.

Lactose-intolerant subjects consuming yoghurt showed lower levels of hydrogen breath excretion. Yoghurt made with just *L. acidophilus* showed less encouraging results. Unfermented *acidophilus* milk made with sonicated cells achieved better results in hydrolysing lactose. Heat-treated yoghurt is not as effective as unheated, or traditional, yoghurt at lowering breath hydrogen levels. Lactase activity in yoghurt can be increased by using a combination of traditional yoghurt cultures (*Lactobacillus delbrueckii* subspp. *bulgaricus* and *Streptococcus thermophilus*) and higher levels of *L. acidophilus* (4×10^6 cfu ml^{-1}).

Cholesterol Reduction

Some intestinal bacteria have enzymes capable of modifying cholesterol. Although lactic acid bacteria have not been shown to possess this activity, some have been shown to metabolize bile. Bacteria modifying cholesterol would need to accomplish the conversion in the small intestine. Presumably most probiotic cultures would be most active in the small intestine.

A strain of *Lactobacillus acidophilus* was selected for its ability to grow in the presence of bile and assimilate cholesterol. In an animal trial, feeding this strain to pigs for 10 days reduced diet-induced elevation of cholesterol. Other studies have been less conclusive about the role of *L. acidophilus* in reducing serum cholesterol levels. In a study of healthy adults consuming sweet *acidophilus* milk no significant differences were observed in levels of triacylglycerols, high-density lipoproteins or low-density lipoproteins compared with controls. In another study involving over 300 subjects fed a capsule containing *L. acidophilus*, no difference was observed in serum concentrations of triacylglycerols, high-density lipoproteins and low-density lipoproteins.

More research is needed in this area to determine if lactic acid cultures can reduce serum cholesterol levels in humans, and if the decrease will provide a substantial health benefit to the consumer. Altered cholesterol metabolism by intestinal bacteria may place different populations at a higher risk of developing the disorder.

Cancer Suppression

Human intestinal cancer can be caused by factors in the environment and diet. Colon cancer has been theorized to be caused by carcinogens in the food or by the ability of intestinal bacteria to alter the human intestinal chemical environment. Several experimental studies have shown that fermented dairy products can alter the activity of some faecal enzymes thought to be important in the development of colon cancer.

Faecal mutagenicity has been detected in humans after ingestion of fried meat. Heterocyclic aromatic amines are promutagens formed during cooking of meats and fish, and the carcinogenic activity of these compounds has been confirmed in animal studies. Studies with human faecal microflora have demonstrated that these compounds can be metabolized.

Lower faecal mutagen activity was noted in a study where healthy volunteers were fed a fried meat diet with *L. acidophilus* fermented milk. Faecal mutagen activity was 33% lower in the experimental group compared with controls by day 2 after supplementation was started. The response among subjects in the experiment were highly variable and may be explained by individual variations in intestinal microflora and environment (bile flow, gastric acidity, etc.). An increase in the faecal lactobacilli population was noted especially in subjects with lower mutagen excretion.

Another study suggested that *L. acidophilus* is capable of suppressing colonic microflora metabolic activity and subsequently reducing the formation of carcinogens in the intestine. Humans fed *L. acidophilus* had a significant decline in the faecal levels of bacterial β-glucuronidase, azoreductase and nitroreductase. These enzymes are thought to transform procarcinogens to proximate carcinogens and cause colon cancer. In another study, faecal nitroreductase was reduced in subjects ingesting a fermented *L. acidophilus* dairy product.

Colonic carcinogenesis may be caused by a cytotoxic effect on colonic epithelium by bile acids in the faeces followed by an increased intestinal cell proliferation. Total soluble faecal bile acids were reduced in colon cancer patients after consumption of *L. acidophilus* for 6 weeks. Soluble deoxycholic acid was reduced by 20% after prolonged ingestion of fermented dairy products. The decrease in soluble bile acids may be explained by several factors. Lower concentrations of soluble faecal bile acids were not caused by lower fat ingestion, as fat levels were constant before and during the study. Faecal bile acid concentrations may have decreased because of the larger amount of calcium ingested by patients in this study: increased levels of dietary calcium in the colon decrease the levels of bile acids by binding them to form calcium soaps. The researchers also reported a change in the faecal microflora of patients ingesting *L. acidophilus*. The altered microflora may have controlled the formation of secondary bile acids (more soluble) from primary bile acids.

Constipation

Fermented *acidophilus* milk has been shown to relieve constipation in elderly people. The majority of elderly people suffer from constipation and seek relief by daily consumption of laxatives. Hospitalized elderly patients were given fermented *acidophilus* milk ad libitum at breakfast for 56 days. The experiment was designed as a crossover trial, and therefore patients served as their own controls. Hospitalized elderly patients who consumed 200–300 ml per day had less need of laxatives to relieve constipation. Fermented *acidophilus* milk was more effective than buttermilk. This study also examined defecation rates, but was flawed because laxatives were given as needed throughout the study. Nonetheless, the study did demonstrate that elderly constipated patients could use less laxatives.

Immune System Stimulation

The increased incidence of immune system deficiency diseases worldwide has created an increased interest in studying the interaction of intestinal bacteria and the immune system. Conventional animals with a complete indigenous gut flora have higher immunoglobulin levels and phagocytic activity than their germ-free counterparts. Probiotic bacteria should enhance immunity both locally on the mucosal surfaces and at the systemic level. Probiotic research on immune system stimulation has centred on: (1) the response of mice to oral ingestion of lactic cultures or fermented milk; (2) the response of mice to intraperitoneal injection of lactic acid bacteria or cellular extracts of the same; (3) the response of cell cultures to exposure of lactic acid bacteria or their cellular components; and (4) human feeding studies.

Several studies have shown that *L. acidophilus* fed to mice is capable of inducing release of lysosomal enzymes from macrophages, activating the cell population of the phagocytic mononuclear system, and stimulating the lymphocytes. This leads to the speculation that there is an enhancement of the systemic immune response in hosts that ingest *L. acidophilus*. Furthermore, in vivo macrophage activation may be important in suppressing tumour growth.

The role of *L. acidophilus* in allergy treatment is beginning to be examined. Yoghurt fermented with *L. acidophilus* and fed to adult patients with asthma increased the levels of interferon gamma and decreased eosinophilia. However, there were no changes in clinical parameters in asthma patients. Subsequent studies will need to be done with longer administration of yoghurt fermented with *L. acidophilus*, to understand the mechanism of how interferon gamma was increased.

Available Products and Survival until Consumption

Numerous lactic acid bacteria have been used as probiotic cultures to manage intestinal disorders including lactose intolerance, diarrhoeal diseases, constipation, food allergy and inflammatory bowel disease. Most of these diseases are associated with an imbalance of the intestinal microflora and various degrees of inflammation in the intestinal tract mucosa. To treat these conditions successfully a probiotic strain should be able to survive gastric acidity, adhere to the intestinal mucosa or antagonize pathogens by antimicrobial activity, and temporarily colonize the intestinal tract. Other criteria are important for commercial application of probiotic strains (**Table 2**).

Probiotic microorganisms have been used by the food industry for many years in dairy products without causing any major health problems for consumers. There have been rare cases of lactobacilli being implicated in a clinical infection, but the origin of the strains has been speculative. Three approaches are used to assess the safety of probiotic strains: (1) studies on the intrinsic properties of the strain; (2) studies on the pharmacokinetics of the strain; and (3) studies searching for interactions between the host and strain. Intrinsic properties include enzymatic activity of the strains. For example, does the strain excessively deconjugate bile acids or degrade the mucin that lines the intestinal tract? Do the strains possess platelet aggregating properties that may cause problems in heart valve replacement patients? Pharmacokinetic properties include the ability of the strain to translocate and colonize in the intestine. Also of importance is the fate of any active compounds the strain may produce. Interaction studies will examine that a proposed probiotic strain does not have invasion potential in the host.

A study conducted in Finland compared lactobacilli isolates from dairy foods with lactobacilli isolated from blood cultures. None of the clinical isolates was identical to isolates from dairy products. The authors concluded *Lactobacillus* strains used by the food industry are safe to the consumer.

Common commercial products that contain *L.*

acidophilus are yoghurts, yoghurt drinks, sweet *acidophilus* milks, fermented *acidophilus* milk, and fruit juices that contain live cultures. Several other forms of viable cells are marketed as capsules, suppositories and pastes. There is considerable interest in extending probiotic cultures beyond dairy foods to infant formulae, baby foods, cereal products and pharmaceuticals.

Storage conditions are critical for maintenance of viability. Products that contain viable cultures should be maintained under refrigerated or frozen conditions. Products held at ambient temperature in a dried form must be used shortly after manufacture unless they have been manufactured by patented technology.

See also: **Lactobacillus**: Lactobacillus acidophilus. **Microflora of the Intestine**: The Natural Microflora of Humans; Detection and enumeration of Probiotic Cultures. **Probiotic Bacteria**: Detection and Estimation in Fermented and Non-fermented Dairy Products.

Further Reading

Gilliland SE and Walker DK (1990) Factors to consider when selecting a culture of *Lactobacillus acidophilus* as a dietary adjunct to produce a hypercholesterolemic effect in humans. *Journal of Dairy Science* 73: 905–911.

Gorbach SL (1986) Function of the normal human microflora. *Scandinavian Journal of Infectious Diseases* 49: 17–30.

Lidbeck A, Nord CE, Gustafsson JA and Rafter J (1992) Lactobacilli, anticarcinogenic activities and human intestinal microflora. *European Journal of Cancer Prevention* 1: 341–353.

Robinson RK (ed.) (1991) *Therapeutic Properties of Fermented Milks*. London: Chapman & Hall.

Salminen S and Deighton M (1992) Lactic acid bacteria in the gut in normal and disordered states. *Digestive Diseases* 10: 227–238.

Salminen S and von Wright A (eds) (1998) *Lactic Acid Bacteria: Microbiology and Functional Aspects*. New York: Marcel Dekker.

Saxelin M, Rautelin S, Salminen S and Makela PH (1996) Safety of commercial products with viable *Lactobacillus* strains. *Infectious Diseases in Clinical Practice* 5: 331–335.

Table 2 Criteria for commercial strain selection: these criteria are essential for an efficacious probiotic product or product line

- Ability to commercially prepare and maintain culture
- Maintain guaranteed level of viability through product shelf life
- Accurate strain identification and genetic stability of desired traits (intrinsic properties)
- Statistically significant clinical evidence of health claim (host and strain interaction)
- Verification of important parameters for functionality (pharmacokinetics)

Biology of the *Enterococcus* spp.

Nezihe Tunail, Department of Food Engineering, University of Ankara, Turkey

The following discussion of the properties and functions of the enterococci is limited to microorganisms belonging to the genus *Enterococcus*. These organisms are sometimes grouped as D streptococci or enterococci and sometimes as faecal streptococci. These terms are used synonymously. For convenience a brief taxonomic background is given, and the differences between these groups are also discussed below.

Prior to 1984 when they were assigned to the genus *Enterococcus*, the enterococci were grouped in the *Streptococcaceae* family, genus *Streptococcus*. In 1937, the *Streptococcus* spp. were classified according to their physiological properties by Sherman into four groups, namely pyogenic, viridans, lactic streptococci and enterococci. The classical enterococci consisted of the species *Streptococcus faecalis*, *Streptococcus faecium* and their subspecies *Streptococcus faecalis* subsp. *liquefaciens* and *Streptococcus faecalis* subsp. *zymogenes*. Moreover, *S. durans* was sometimes regarded as a synonym for *S. faecium* and sometimes as the subspecies of this bacterium. These organisms, plus *S. bovis* and *S. equinus*, were grouped as D streptococci based on Lancefield's antigenic classification of the *streptococci*. The species *S. suis* and *S. avium* were also placed in this group later by Buchanan and Gibbons. Serologically, *S. avium* is included in group Q, but contains also the group D antigen. Faecal streptococci, being the largest group, include *S. mitis* and *S. salivarius* along with the species mentioned above. The classical enterococci (*S. faecalis*, *S. faecium*, *S. durans*) were transferred to the new genus *Enterococcus* as *E. faecalis*, *E. faecium* and *E. durans*. Subsequently *S. avium* and *S. gallinarum* also were transferred to this genus as *E. avium* and *E. gallinarum*. Along with the above-mentioned species 11 new species (*E. casseliflavus*, *E. cecorum*, *E. dispar*, *E. hirae*, *E. malodoratus*, *E. mundtii*, *E. pseudoavium*, *E. raffinosus*, *E. saccharolyticus*, *E. seriolicida*, and *E. solitarius*) were recognized in the new, separately assigned *Enterococcus* genus.

The genus *Enterococcus* was not included in *Bergey's Manual of Systematic Bacteriology*, published in 1986. In *Bergey's Manual of Determinative Bacteriology*, 9th edition, published in 1994, *E. faecalis*, *E. faecium* (type species), *E. gallinarum*, *E. avium*, *E. durans*, *E. casseliflavus* (*Streptococcus faecium* subsp. *casseliflavus*) and *E. malodoratus* species were included in the genus *Enterococcus*. Recent literature reports the inclusion of a total of 18 species in this genus.

Role in the Intestine

The enterococci are part of the natural microflora of the small and large intestine of humans and animals. In the fetal period the intestine is sterile. During birth the vaginal and faecal flora of the mother contaminate the intestine. Even though a diverse population in the newborn colon could occur as a result of environmental contamination, in the end the human digestive system establishes its own beneficial balance of flora. One day after birth the faecal population containing *Escherichia coli* and the enterococci can reach numbers as high as 10^{10} cells per gram and remain at this level for a couple of days. Depending on the intensity of breast-feeding of the baby, the numbers of lactobacilli and bifidobacteria can reach levels as high as 10^7–$10^{10}\,g^{-1}$. In contrast to the increase in the numbers of representatives of these two genera, a decrease in the numbers of the colon's first residents is observed.

The diversity of genera in the microbial flora of the intestine does not seem to differ much between babies who are breast-fed, formula-fed, breast-fed with supplementary food, or (later) fed solely on solid food. However, the populations of the genera do show a great variation depending on the dietary regimen. While *Lactobacillus* and *Bifidobacterium* are found to be the dominant species in breast-fed babies, *E. coli*, *Enterococcus*, *Clostridium* and *Bacteroides* predominate in formula-fed babies. As soon as the babies are introduced to supplementary foods *Bifidobacterium* preserve their level while *Escherichia coli*, *Enterococcus*, *Clostridium* and *Bacteroides* tend to increase in number. In babies solely fed on solid foods, *Bacteroides* and *Bifidobacterium* become dominant in the microflora while the other species tend to decrease. The predominant microflora in the intestinal system of adult humans gradually becomes established in children beginning from their second year.

Of the 400–500 species that are found in the intestinal flora of adults, only 30–40 species are predominant. Obligate anaerobic bacteria are found more commonly than facultative anaerobic bacteria and comprise 95% of the total intestinal flora, which possesses a wide range of metabolic activities. The total flora is reported to weigh around 1 kg, reaching 10^{16} in number. However the distribution, numerical densities and genera of these microorganisms in the small intestine and colon vary greatly. Anatomically, the small intestine is divided into three regions. The first region, the duodenum, is influenced by the secretions of the stomach, pancreas and bile. The intestinal bacterial population is generally localized in the jejunum and ileum and ranges between 10^6 and 10^7 cells per gram. Together with species of *Bacteroides*, *Clostridium*, *Mycobacterium* and Enterobacteriaceae, species of *Lactobacillus* and *Enterococcus* are the dominant flora here. In the colon the population reaches the level of 10^{10}–10^{12} cells per gram and is mainly composed of species of *Fusobacterium* and Enterobacteriaceae, along with *Bacteroides*,

Eubacterium, Enterococcus, Clostridium, Bifidobacterium and *Lactobacillus.*

The microbiota that decompose in the small intestine and colon are excreted in the form of faeces as result of physiological processes. In an adult 10^{14} microorganisms are excreted daily with the faeces as a result of peristaltic movements in the colon. Faeces is essentially a desquamate of surface (epithelial) cells of the colon on which the microorganisms attach and then colonize, with a constant flow of mucus also hosting the microorganisms. This loss is constantly compensated for in the intestine and the microflora exists in harmony with the host. The quantity of enterococci and faecal streptococci in human faeces has been reported to be 10^4–10^9 cells per gram wet weight. **Table 1** indicates the major bacterial groups and shows the high proportion of *Enterococcus* species in the intestinal contents of healthy individuals.

The handling of iron in the intestinal system of mammals is known to play a role as a defence mechanism against excessive microorganism growth. This mechanism involves the depletion of iron in such a way that the host has adequate iron, but microorganisms have difficulty in obtaining sufficient iron and thus their growth is limited. Possessing a perfect iron-independent metabolism, the enterococci grow in larger numbers as the absence of iron is not a growth-limiting factor for enterococci. Their iron-independent metabolism enables these organisms to grow rapidly and thus quickly colonize the intestinal system of newly born and breast-fed mammals. In the colon of adults they reach their peak numbers.

Some species of *Lactobacillus* and *Bifidobacterium* found among the indigenous microflora of the gastrointestinal system are known to have a positive effect on the health of their host. Many studies have been carried out on the predominant species of the intestinal system, namely *Lactobacillus acidophilus*, *L. reuteri*, *L. lactis*, *L. casei* and *L. fermentum*. However it has not yet been determined whether all or only a few of them are indigenous. *Lactobacillus bulgaricus*, which is unable to survive for long in the intestinal system, has also been well studied. Although there have been few studies of classical enterococci, there are hints regarding the probiotic potential of enterococci in the intestinal system.

The capability of *Enterococcus faecalis* and *E. faecium* to produce bacteriocin has been known for many years. However it is interesting to note that these bacteriocins are active against pathogenic bacteria such as *Listeria* and *Clostridium* spp. The *E. faecium* 68 (SF68) strain in particular has proved to be capable of curing certain diseases. Three bacteriocins (BC-48, enterocin 1146, enterocin 222 NWC) and one peptide-antibiotic have been extracted from different strains of *E. faecalis*, and more recently a new antilisterial bacteriocin (enterococcin EF S2) has been isolated, purified, and partially characterized from *E. faecalis*. Another bacteriocin, enterocin CRL35, has also been purified from *E. faecium* CRL35. The potential of *Enterococcus* species to produce antibacterial substances could be regarded as an indication of their probiotic value. Relevant studies in this field have made the functions of the enterococci more significant and meaningful. The value of these organisms as dietary adjuncts is discussed below.

Enterococci as Indicator Organisms

Microbial contamination in food is determined by detecting indicator organisms. Detection of the coliform group, *Escherichia coli*, members of *Enterobacteriaceae*, and two members of *Enterococcus* (*E. faecalis* and *E. faecium*) in food and water are important signs of faecal contamination. The presence of these indicator microorganisms in food or water indicates the possible coexistence of pathogens threatening human health and inadequate hygienic conditions in food or water processing and distribution chains. Even though the enterococci are regarded as important sanitation indicators, they have never been able to reach the status held by *Escherichia coli* and other members of the coliform group. Although maximum allowable levels have been established for *E. coli* and coliforms as sanitation and faecal indicators, no such limits have yet been determined for the enterococci.

Upon leaving the intestinal system, the enterococci spread and contaminate foods and water either directly or indirectly. The enterococci have a better survival rate than other members of the coliform group outside the intestinal system owing to their higher resistance to adverse environmental conditions. Cool, moist conditions prolong the survival of *Enterococcus faecalis* in soil. Enterococci are able to grow within a wide pH range. They can survive for long time in acidic foods and at high salinity. They can also survive

Table 1 Average population ($\log_{10}$) of major bacterial groups in the intestinal contents of normal people according to Drasar and Hill (Ray, 1996)

Bacterial groups	Jejunum	Ileum	Colon	Faeces
Lactobacillus	3	5	6	6
Gram-positive non-sporing anaerobes[a]	2	2	5	6
Enterococcus	3	5	7	7
Bacteroides	3	3	7	9
Enterobacteriaceae	3	4	6	6

[a]Including *Bifidobacterium* species.

heat processes owing to their thermoduric characteristics. *Enterococcus faecalis* remains viable longer than *Escherichia coli* in acidic foods, and the enterococci have proved to have a better performance than *E. coli* and other members of the coliform group in studies of the shelf life of spray-dried egg powder. Moreover, the enterococci have proved to be a more effective indicator of contamination than the coliform group in frozen foods. For example, the enterococci remain viable for a long period and survive better than the coliform group both short-term and long-term in food samples frozen at −20°C. In different frozen vegetables and seafoods, the enterococci have been observed to be numerically greater than the coliform group. However, these promising findings have not ended the dispute regarding their use as indicators, because typical and atypical *Enterococcus faecalis* and *E. faecium* as well as other enterococci similar to the classical enterococci have also been detected in plants, soil, insects and wild animals. The majority of plant and insect isolates of *E. faecalis* are capable of digesting starch and produce stratiform digestion of a soft, rennet-like curd in litmus milk. These properties have not been observed in human isolates.

Most of the findings regarding the natural occurrence of 16 *Enterococcus* species relate to the classical enterococci. Although *E. faecalis* and *E. faecium* are usually the indigenous intestinal flora of human and animals, nevertheless they are also commonly found on plants, insects and wild animals. Further studies are needed to determine the natural occurrence of new species and especially their faecal and environmental origins. While *E. faecalis* has been found in the faeces of all mammalian animals, it is more specifically observed in the intestinal system of humans. *Enterococcus dispar* has also been found in human gut. In contrast, *E. faecium* and *E. durans* are more common in the faeces of pigs. Although *E. durans* and *E. hirae* have been isolated from six mammalian species, they have been reported to be more prevalent in poultry; *E. gallinarum* also appears to be more specific to poultry. While *E. cecorum* has been isolated from chicken caecae, *E. avium* has been found in chicken faeces and in mammalian intestinal systems. In contrast to these species, *E. casseliflavus* has been observed in silage, soils and on plants. *Enterococcus mundtii* has been found in cattle, on the hands of milkers, and in soils and plants. Since the habitats of different species show a wide variation, it is difficult to distinguish the origins of these species employing the methods used for classical enterococci. Consequently this handicap makes the definite diagnosis of faecal contamination of foods impossible. Currently the faecal and non-faecal strains could be dis-

tinguished by evaluating the resultant fermentation in litmus milk, melezitose and melibiose broths, but a diagnostic agar medium is not yet available for this purpose. Moreover, the close similarity between 2334 strains of *E. faecalis* isolated from dried and frozen foods with vegetation-resident types has decreased the significance of their use as sanitary indicators.

Classical enterococci are resistant to heating, drying and freezing treatments employed in food processing. They are also quite resistant to cleaning and disinfectant substances. Therefore their use as a sanitation or faecal indicator in processed and frozen foods is favoured by many researchers. Similarly, the use of the enterococci as sanitation or faecal indicators in frozen foods is favoured more than the coliform group. Generally, however, enterococci can be considered as indicator organisms only in a broad sense.

The enterococci are not used as indicator organisms in dairy products, but are usually counted in order to measure the level of good manufacturing practices (GMPs) of the different production lines in an enterprise. The interpretation of results would vary according to the local conditions and the types of factories and products. There is no international consensus regarding the assessment of GMPs.

The enterococci are also used as indicator organisms in microbial analysis of water samples. The pertinent feature of the classical enterococci leading to their use as a pollution indicator for water is that they will only grow in water rich in organic matter. Generally, the ratio of faecal coliforms to enterococci in human faeces is $4:1$. In addition to the reasons mentioned above, the enterococci die off at a slower rate than coliforms in water and thus would normally outlive the pathogens. Therefore tests for enterococci are believed to reflect the existence of pathogens better than tests for faecal coliforms. The enterococci are widely chosen as indicator organisms for both tap water and sea water.

Classical enterococci are Gram-positive, catalase-negative, facultative anaerobes. They are capable of growth at pH 9.6 and develop in 6.5% NaCl. In cooked ham, the decimal reduction time (D value) for *E. faecalis* at 66°C is 1.39 min and the z value is 6.85°C; for *E. faecium* the values are D 66°C; 29.04 min, z 7.46°C. *Enterococcus faecalis* rapidly reduces triphenyltetrazolium chloride (TTC) to the formazan derivative and forms deep-red colonies; *E. faecium* cannot – or can only weakly – reduce TTC and forms pink colonies. The value of minimal water activity (a_w) for these bacteria is 0.93 and their minimal temperature requirement is 6–10°C.

For isolation and enumeration of the enterococci, m-enterococcus agar, KF streptococcus agar, citrate-azide Tween carbonate (CATC) agar or kanamycin

Table 2 Major biochemical tests of classical enterococci. From Baumgart (1994)

Biochemical tests	E. faecalis	E. faecium
Reduction of potassium tellurite	+	V
Reduction of tritetrazolium chloride (TTC)[a]	+	V
Ammonia from arginine	+	+
Acid from melibiose	–	+
Acid from sorbose	–	+

V, variable reaction (+ or –).

[a] On m-enterococcus agar or CATC agar.

esculin azide (KEA) agar are used. Dilutions of samples can be plated by spread or dropping techniques. The plates are incubated at 37°C for 28–72 h. On the first three media, pink colonies are counted; on KEA agar black colonies are counted. For confirmation five to ten colonies are picked and transferred to brain–heart infusion broth and incubated for 18–24 h at 37°C. Gram staining and catalase tests are performed on samples taken from these tubes. The key tests used in differentiating *E. faecalis* and *E. faecium* are given in **Table 2**. It is also possible to differentiate the two species by the Rapid Strep. System (API).

Use in Cheeses and Fermented Products

The use of enterococci as a starter in the fermentation of milk, cultured butter, different kinds of cheese, meat products and vegetable-originated foods is not a common practice. However, enterococci are reported to have been used as a starter in the production of Cheddar and some soft cheeses.

The enterococci are part of the natural flora in cheeses manufactured with raw or pasteurized milk and in marinated food products, and are believed to contribute to the ripening process and the development of flavour and aroma in these foods. Enterococci are therefore unsuitable faecal or sanitation indicators in such products. They are found in high numbers in many kinds of cheese. Classical enterococci are predominantly found in the intestinal system, and being part of the natural flora, are consumed with the above-mentioned products. Therefore the use of selected strains of classical enterococci as starters or their addition to lactic starter combinations has proved to be possible. Studies carried out in the 1960s and 1970s in the USA indicated that enterococci are intensively found in Cheddar cheese and contribute to the ripening process. Furthermore, it has been observed that the enterococci contribute to the development of flavour and aroma in some hard cheeses. Researchers have frequently recommended the use of the enterococci as starters in the production of cheese, e.g. soft cheese in Italy, Feta and Teleme

cheese in Greece, white-pickled cheese in Turkey and Bulgaria, Domiati cheese in Egypt and various kinds of cheese in Argentina.

The production and sale in the 1970s of lyophilized cultures of *Streptococcus durans* CH and *S. faecalis* CH as commercial preparations, and the use of these organisms in the production of Cheddar cheese and certain types of Italian soft cheese are also a matter of record.

The enterococci are believed to be more efficient than *Lactococcus* spp. as starter strains in the ripening process of white-pickled cheese, owing to their high level of tolerance for acidic conditions, high salinity and low temperatures during the ripening process in the cold storage.

Depending on the technology employed, raw milk, pasteurized milk or natural lactic starter cultures are the main sources of the enterococci in many kinds of cheese. These cultures are obtained by pasteurization of good-quality milk, followed by incubation at 42–44°C for 12–15 h. The enterococci become dominant in many kinds of cheese, especially during the ripening period, because of their thermoresistant characteristics and high adaptability.

Although the enterococci possess many useful properties as cheese starters, caution is necessary because cases of food poisoning in association with enterococci have been reported. These have been found to be similar to food poisonings caused by *Bacillus cereus* and *Clostridium perfringens*, in terms of both symptoms and incubation period. Turkey dressing, cured ham, barbecued beef, Vienna sausage, cheese and evaporated milk have been reported as foods typically causing these poisonings. The actual role of these bacteria categorized by nonspecific food poisoning is in dispute because attempts to confirm the role of enterococci by reproducing these symptoms in human volunteers have been unsuccessful.

In food items produced by starter culture or subjected to natural fermentation, the free amino acids formed by proteolytic activity are converted to biogenic amines by decarboxylating bacteria. The enterococci are generally decarboxylase-positive strains and are therefore held responsible for poisonings caused by consumption of food rich in biogenic amines (e.g. histamine, tyramine) such as cheese and fish. The formation of amines by *Enterococcus faecalis* and *E. faecium* has also been reported. It has been demonstrated that enterococci isolated from dairy products possessed only tyrosine decarboxylase. In a more recent study it was reported that 11 of 29 *E. faecium* strains, 4 of 7 *E. faecalis* strains and 5 of 18 *E. durans* strains isolated from white-pickled cheeses had the ability to form biogenic amines. These strains had a significant ability to produce tyramine. Three of 13

Lactococcus lactis strains were also found to produce tyramine and phenylethylamine. It should also be noted that species of many different genera including *Lactococcus* found in foods could also form amines. Detoxification of biogenic amines by acetylation and oxidation is achieved through the activities of monoamine oxidase (MAO) and diamine oxidase (DAO) enzymes present in the gastrointestinal system and liver. Activities of MAO and DAO may be inhibited by painkillers and other drugs used for the treatment of stress and depression. These enzymes may also be insufficient in preventing histamine poisoning when foods with high biogenic content are consumed.

The control of enzyme activities (e.g. histidine and tyrosine decarboxylase) in potential starter strains and the selection of mutant or genetically redesigned strains deficient in such enzyme activities may be a good approach to the assessment of starter cultures. This approach would probably contribute to the wider use of the enterococci for this purpose; *E. faecalis* INIA 4-07 mutant has been reported to inhibit tyramine formation when used as a starter in Manchego cheese (Spain) owing to the bacteriocin activity present in the strain.

The disadvantages of enterococci are not only associated with biogenic amines. Their pathogenetic and toxicogenic activities are also in dispute. Lactic acid bacteria and *E. faecalis* have been reported to occur in clinical infections, and in particular to be the cause of endocarditis, although generally accepted as being nonpathogenic. Haemolytic activity is known to be associated with the virulence of the enterococci. Beta-haemolytic strains of *E. faecalis* have caused anxiety. Thermonuclease-producing strains of *E. faecalis* and *E. faecium* have been reported, although there are contradictory published data. Enterotoxin production has been observed in the potentially toxigenic, thermonuclease-positive *E. faecium* IF-100 strain isolated from dried baby foods, but attempts to produce enterotoxin in 33 strains of *E. faecalis*, *E. faecium* and *E. durans* have been unsuccessful. Toxigenic strains of enterococci are believed to be very rare.

Although the abandonment of production of lyophilized cultures containing *E. faecalis* and *E. durans* by starter manufacturing companies could be considered as general disapproval of the use of the enterococci as starters, *E. faecalis* is successfully used as an inoculant in silage processes along with *Lactobacillus plantarum*. Promising results have been also obtained in Scandinavian countries with the same combination in fish silage. *Enterococcus faecalis* and *Lactobacillus plantarum* are inoculated following the addition of a fermentable carbohydrate to the minced

biomass. The development of acidic conditions prevents microbial deterioration in the final liquid product.

Use as a Dietary Adjunct

The normal intestinal flora of humans performs protective, defensive and supportive functions. As shown in Table 1, *Lactobacillus*, *Enterococcus* and *Bifidobacterium* are the most common microorganisms in the human intestinal system. Several species and strains belonging to these genera are known to possess important probiotic functions.

The term 'probiotics' generally refers to viable bacteria found in cultured dairy products or to food supplements containing viable lactic acid bacteria. Many species and strains belonging to different genera of lactic acid bacteria (LAB) have the potential of producing different types of antibacterial substances, namely *Streptococcus thermophilus*, *Lactobacillus lactis*, *Leuconostoc mesenteroides*, *Lactobacillus bulgaricus*, *Lactobacillus casei*, *Lactobacillus acidophilus*, *Lactobacillus reuteri* and some species of *Bifidobacterium*. These are used in the preparation of dietary adjuncts. Generally the benefits of LABs to human health can be summarized as follows:

1. They help re-establish the normal flora of the intestine.
2. They control enteric pathogens in the intestine and prevent infection in the urogenital system.
3. They help lactose-intolerant individuals.
4. They inhibit the formation of tumours in the colon and other organs.
5. They reduce the risk of cardiac disorders.
6. They reduce the serum cholesterol level.
7. They stimulate intestinal movements and the immune system.

These beneficial probiotic properties differ in each strain. A good probiotic strain should:

- be stabile to gastric and bile acids
- be derived from a human host
- have the potential to adhere to human intestinal cells
- be able to produce antimicrobial substances
- have the potential to colonize the human intestinal tract.

Adherence is a highly specific process. It involves the attachment of the bacterial surface to complementary host receptors found on epithelial cells. In the case of *Lactobacillus acidophilus*, the surface layer protein (SLP) and/or lipoteichoic acid are believed to play an important role in host-specific adherence. However such detailed studies are lacking for commercial probiotics and for the enterococci tried as probiotics. In addition to adherence potential, the probiotic strain should form several antimicrobial metabolites, e.g. lactic acid, a number of organic acids, hydrogen peroxide and bacteriocins and other similar inhibitors. An antagonistic effect against both pathogenic and cariogenic bacteria is also desirable. The probiotic strain should also grow well in vitro. Obviously, safety for human use is required as well.

Enterococcus faecalis and some other *Enterococcus* species are suspected of being opportunistic pathogens and are generally not desirable in foods because of an enterococci–amine interaction. Nevertheless, along with mixed or single commercial preparations of *Lactobacillus acidophilus*, *L. bulgaricus* and *L. caucasicus* in the form of tablets and capsules, *E. faecium* and *E. faecalis* preparations are used as probiotics. Since human and animal intestines are the natural habitats of the enterococci, their use as probiotics seems rational. *Enterococcus* species are capable of producing enterococcins like nisin and pediocin which are effective bacteriocins. Moreover, because they meet certain criteria as probiotics, their use as a dietary adjunct is favoured.

Strain characteristics must be well defined before the microorganism can be considered as a probiotic product. Furthermore, the benefits to the host must be proved by statistically well-designed, randomized, double-blind clinical studies. Limited studies have been carried out on the enterococci, but more data are available for *Lactobacillus* species. Preparations containing *E. faecium* 68 strain (*E. faecium* SF68) have been used as treatments for acute enteritis and other intestinal disorders. Having a short generation time (20 min), being more stable at low pH than *L. acidophilus* and possessing antibacterial activity against some enteropathogens have made this strain an attractive probiotic. Other *E. faecium* strains have been used for the same purpose.

Enterococcus faecium has been proved to provide a faster recovery than inert control preparations used to treat young children with diarrhoea. In a double-blind clinical trial, children with diarrhoea were treated with either a LAB mixture (*Lactobacillus acidophilus* 5×10^8 cfu, *L. bulgaricus* 5×10^8 cfu, *L. lactis* 4×10^9 cfu) in lyophilized form, or *E. faecium* (3.75×10^7 cfu). The speed of recovery in the second treatment was found to be better than that of the first treatment. In one study *E. faecium* SF68 was not found to provide an efficient protection against traveller's diarrhoea among Austrian tourists. The therapeutic efficiency of the same bacteria has been also tested against the watery diarrhoea caused by *Vibrio cholerae* and enterotoxigenic *Escherichia coli*. The study included 41 *E. coli* patients and 114 *V. cholerae*

patients who were treated either with placebo or *E. faecium* preparation (1×10^9 cfu) three times a day. No antidiarrhoeal effect was reported to be observed. However, *E. faecium* preparations have been found to be effective in reducing diarrhoeal intensity in patients treated with antitubercular drugs. The *E. faecium* SF68 strain has also been tested on patients with hepatic encephalopathy and was found to be as effective as *L. acidophilus* in lowering blood ammonia levels and in providing clinical improvements. The effect of the *Enterococcus* preparation was found to be more persistent than that of lactulose after the completion of the treatment. No side effects were reported.

It is not known clearly whether *E. faecium* strains used as probiotics are the same as those present in the faeces of healthy adults. It has been observed that strains identified as *E. faecium* and used as probiotics are not actually *E. faecium*, but some other unidentified enterococcus. Strains used as probiotics must be derived from the human intestinal system, be tested for their relevancy in terms of set-up criteria, and must have been positively identified.

The dispute over the probable role of classical enterococci in food poisoning, associable interaction with *Staphylococcus aureus* toxins and other biogenic amines, evaluation of certain species and strains as opportunistic pathogens and inadequate data regarding their associable role in nonspecific gastrointestinal disorders or participant role in clinical infections could be regarded as the disadvantages of using enterococci as dietary adjuncts.

Many factors, inherent or environmental, can cause the elimination or reduction of intestinal microorganisms used as dietary adjuncts. One of these factors is the intake of antibiotics. The susceptibility of enterococci to antibiotics could be regarded as a criterion for evaluating their potential use as a probiotic or dietary adjunct. As opportunistic pathogens, some enterococcal strains from clinical sources are naturally tolerant of antibiotics (streptomycin, kanamycin, gentamicin, penicillin, ampicillin, chloramphenicol, erythromycin and tetracycline), and generally multiple antibiotic resistance is the case.

The enterococcal strains isolated from dairy products have been found to be more sensitive to certain antibiotics than strains isolated from environmental or clinical sources. The susceptibility to vancomycin of 11 strains of *Enterococcus faecium* and *E. faecalis* isolated from dairy products was investigated in a study in 1977; a 399 bp intragenic fragment of the *vanA* gene responsible for resistance to vancomycin, found in clinical isolates of *E. faecium* BM 4147 and transconjugant *E. faecalis* 206, was not observed in these susceptible strains.

As with their pathogenic potential, resistance to antibiotics seems to be strain-specific and changes with the habitat. In nonpathogenic enterococcal strains used as probiotics, resistance to antibiotics should be regarded as an advantage, in that the probiotic strains could inhabit the intestinal system and preserve their viability and probiotic function despite antibiotic therapy. This could be achieved by selecting antibiotic-resistant strains or by transferring resistance genes to the probiotic strain.

Conclusion

The functions and use of classical enterococci as starters and dietary adjuncts in the food industry and their potential pathogenicity are still issues of disputes. Even their use in frozen foods as indicator organisms has not yet gained wide acceptability. The properties of the enterococci in dispute need to be clarified. The species diversification, origins and identification of the enterococci grouped into a new genus by a taxonomic revision also need further study. New criteria for identification of all species should be developed.

In addition to their potential probiotic value, the use of enterococci as a dietary adjunct, or as a natural or commercial starter for cheese, creates the need for an investigation of their potential pathogenic character. If their association with biogenic amine poisoning was the only problem, the cheese industry could compensate by using tyrosine decarboxylase-negative mutants or natural varieties not showing this enzyme activity. Recombinant DNA technology could also be used to produce bacteriocin-positive, tyrosine decarboxylase-negative recombinant strains. For the present, priority must be given to studies of β-haemolytic and thermonuclease activities and enterotoxin production in *Enterococcus* species with a potential use as starter or probiotic. The origins and the frequencies of *Enterococcus* strains should be determined and their enterotoxicity confirmed by animal tests. Such an approach will contribute to a better understanding of the rare and strain-specific pathogenicity and toxicity.

Constituting the normal intestinal microflora in humans and animals and being consumed in large amounts with the daily food intake, some enterococci seem to have a use in treating diarrhoeal diseases and in food technology. Research into the isolation, purification and characterization of enterococcins such as bacteriocins and the value of enterococci as food additives will enhance the importance of this newly revised genus.

See also: ***Bifidobacterium***. **Cheese:** Microbiology of Cheese-making and Maturation. ***Enterococcus***. ***Lactobacillus***: Introduction; *Lactobacillus bulgaricus*; *Lacto-* *bacillus casei*. **Lactococcus**: *Lactococcus lactis* Subspecies *lactis* and *cremoris*. **Probiotic Bacteria**: Detection and Estimation in Fermented and Non-fermented Dairy Products. **Staphylococcus**: *Staphylococcus aureus*. **Starter Cultures**: Cultures Employed in Cheesemaking.

Further Reading

Banwart GJ (1989) *Basic Food Microbiology*, 2nd edn. New York: Chapman & Hall.

Baumgart J (1994) Markerorganismen (III. 2. 1–12) (Ed. J Baumgarten) Microbiologische Untersuchung von Lebensmitteln. 7. Aktualisierungs-Lieferung 1999. Hamburg: Behr.

Durlu-Özkaya F, Alichanidis E, Litopoulou-Tzanetaki E and Tunail N (1999) Determination of biogenic amine contents of Beyaz cheese and biogenic amine production capacity of some lactic acid bacteria by reversed phase high performance liquid chromatography (RP-HPLC).

Giraffa G and Sisto F (1997) Susceptibility to vancomycin of enterococci isolated from dairy products. *Letters in Applied Microbiology* 25: 335–338.

Giraffa G, Carminati D and Neviani E (1997) Enterococci isolated from dairy products: a review of risks and potential technological use. *Journal of Food Protection* 60: 732–738.

Holt JG, Krieg NR, Sneath PHA, Stanley JT and Williams ST (1994) *Bergey's Manual of Determinative Bacteriology*, 9th edn. Baltimore: Williams & Wilkins.

Jay JM (1992) *Modern Food Microbiology*, 4th edn. New York: Chapman & Hall.

Joosten HMLJ, Gaya P and Nunez M (1995) Isolation of tyrosine decarboxylase-less mutants of a bacteriocin-producing *Enterococcus faecalis* strain and their application in cheese. *Journal of Food Protection* 58: 1222–1226.

Koch AL (1996) The gram-positive coccus: *Enterococcus hirae*. In: Koch AL (ed.) Bacterial Growth and Form. P. 197. New York: Chapman & Hall.

Martin AM (1996) Role of lactic acid fermentation in bioconversion of wastes. In: Bozoğlu TF and Ray B (eds) *Lactic Acid Bacteria: Current Advances in Metabolism, Genetics and Applications.* P. 219. Nato ASI Series. Series H Cell Biology, Vol. 98. Berlin: Springer.

Mendosa MC, Scarinci HE, Garat MH and Simonetta AC (1992) Technological properties of enterococci in lactic starters: acidifying and lipolytic activities. *Microbiology Aliments Nutrition* 10: 289–193.

Ray B (1996) Probiotics of lactic acid bacteria: science or myth? In: Bozoğlu TF and Ray B (eds). *Lactic Acid Bacteria: Current Advances in Metabolism, Genetics and Applications.* P. 101. Nato ASI Series. Series H Cell Biology, Vol. 98. Berlin: Springer.

Richard JA (1996) Use of bacteriocin producing starters advantageously in dairy industry. In: Bozoğlu TF and Ray B (eds) *Lactic Acid Bacteria: Current Advances in*

Metabolism, Genetics and Applications P. 137. Nato ASI Series. Series H Cell Biology, Vol. 98. Berlin: Springer.

Salminen S, Deighton M and Gorbach S (1993) Lactic acid bacteria in health and disease. In: Salminen S and von Wright A (eds) *Lactic Acid Bacteria*. P. 199. New York: Marcel-Dekker.

Detection and Enumeration of Probiotic Cultures

G Kalantzopoulos, Department of Food Science and Technology, Agricultural University of Athens, Greece

Introduction

The utilization of selected intestinal bacteria, which are capable of growth on acid production milk, for the industrial manufacture of fermented milk products as well as in the pharmaceutical preparation is relatively recent. However, it is believed that fermented milk products, prepared with selected intestinal bacterial cultures, will be readily available by the year 2000. In Japan, these products already account for approximately 25% of cultured milks sold. The most important point for these products is the survival of probiotic microorganisms before consumption. The detection and enumeration of probiotic bacteria in fermented products is the subject of this article.

Application of Cultures in the Food Industry

A probiotic strain is a live microbial culture used in the production of cultured dairy products, which beneficially influences the health and nutrition of the host. In addition to the general desirable properties of probiotic bacteria the strains used must also possess characteristics identified in carefully conducted clinical studies. Some guidelines are stated below:

- Each strain must be documented and tested independently, on its own merit
- Extrapolation of data from closely related strains is not acceptable
- They must be well-defined probiotic strains, from well-defined study preparations
- Double-blind, placebo-controlled human studies must be performed
- Randomized human studies must be performed
- Results must be cross-validated by several independent research groups

- Results must be published in peer-reviewed journals.

Probiotic bacteria with these properties include:

- *Lactobacillus acidophilus* (NCFB 1478)
- *Lactobacillus casei* Shirota strain
- Lactobacillus GG (ATCC 53103)
- *Lactobacillus acidophilus* LAI

There are a large number of published studies for each strain which document their health effects. All of these strains are currently being studied further for possible intestinal disorders.

New strains which are likely to be included in our diet are, among others: *Lactobacillus reuteri, Bifidobacterium bifidum, B. animalis, B. adolescentis, B. longum, B. infantis, Lactobacillus casei* Danone 001, and strains such as *Enterococcus faecium*.

Lactic acid bacteria (LAB) strains, whether used in combination with the bifidobacteria or alone, should be studied to determine their adaptability to technological factors, whether the cells retain their optimal physiological state and to establish what physiological characteristics could be used as indices of behaviour in the gastrointestinal tract.

Various technologies have been applied to the production of fermented milks. Selected viable intestinal strains are used either as bulk starter cultures or as highly concentrated and standardized frozen or freeze-dried cultures, which are inoculated directly into milk.

For product manufacture, a 3–7% bulk starter is inoculated into the heat-treated milk and incubated at 37–42°C to a pH value lower than 4.75 for set milk products and a pH of 4.45–4.50 for stirred milk products.

Tables 1, 2 and 3, presents the species combination and product quality of culture types mixed with selected intestinal microorganisms (viscosity is related to the strains used).

In the case of direct inoculation, a significant number of probiotics, 1.0–1.5% inoculum, can be directly added. A concentrated frozen culture contains at least 10×10^{11} cfu m^{-1}, thus the addition of 1 kg per 10 000 l of milk is sufficient to give an initial count of 10×10^{7} cfu ml^{-1}.

Properties

The technological properties necessary for a probiotic strain to be used in the production of fermented products depend on the selection and characterization of human probiotic strains, the process of incorporation into fermented foods and their evaluation in selected animal models. Some of the more important factors are listed below:

Table 1 Species combination and product quality

Product quality	Species composition
Mild acid taste, low to high viscosity	Streptococcus thermophilus, Lactobacillus acidophilus
Medium acid taste, weak yoghurt flavour, low to medium viscosity	S. thermophilus, L. acidophilus, L. delbrueckii subsp. bulgaricus
Medium acid taste, weak yoghurt flavour, high viscosity	S. thermophilus, L. acidophilus, L. delbrueckii subsp. bulgaricus
Mild acid taste with a mild acetic acid flavour, low viscosity	L. acidophilus, S. thermophilus, Bifidobacterium sp.

Table 2 Fermented milk products containing *Lactobacillus acidophilus*

Product	Microorganisms involved
Acidophilus milk	L. acidophilus
Acidophilus paste	L. acidophilus
Acidophilus buttermilk	L. acidophilus, mesophilic starter
Acidophilus natural buttermilk	L. acidophilus, mesophilic starter
Acidophilus yoghurt	L. acidophilus, Streptococcus thermophilus, L. delbrueckii subsp. bulgaricus
Biogurt	L. acidophilus, S. thermophilus
Acidophilus-yeast milk	L. acidophilus, lactose-fermenting yeasts
Acidophilin	L. acidophilus, Lactococcus lactis, kefir culture
LC1	L. acidophilus La 1
ABC Ferment	L. acidophilus, Bifidobacterium, L. casei
LA-7 plus	L. acidophilus, B. bifidum
Vita	L. acidophilus LA-H3, B. bifidum LB-H1, L. casei LC-H2

Table 3 Fermented milk products containing bifidobacteria

Product	Microorganisms involved
Bifidus milk	Bifidobacterium bifidum or B. longum
Bifigurt	B. bifidum, Streptococcus thermophilus
Biogarde	B. bifidum, Lactobacillus acidophilus, S. thermophilus
Biokys	B. bifidum, L. acidophilus, Pediococcus acidilactici
Special yoghurt	B. bifidum, (B. longum), S. thermophilus, L. delbrueckii subsp. bulgaricus
Special yoghurt	B. bifidum (B. longum), L. acidophilus, S. thermophilus, L. delbrueckii subsp. bulgaricus
Cultura	B. bifidum, L. acidophilus
Cultura	B. bifidum, L. acidophilus
Mi-Mil	B. bifidum, B. breve, L. acidophilus
Progurt	Lactococcus lactis biovar diacetilactis, Lactococcus lactis subsp. cremoris, L. acidophilus and/or B. bifidum

- The ability to grow in milk or in fortified milk with or without the normal lactic starter cultures and according to the thermal cycle production;

- The nutritional requirements of probiotic strains in milk as basal medium;
- Their ecological relationship with commercially available yoghurt starter cultures, in order to detect the presence of antagonistic or synergistic reactions among enteric and dairy bacteria;
- Resistance to the acidity of fermented milk; guaranteed viability of the cells during shelf life of the product. Thus, LAB cells will reach the gastrointestinal tract in their best physiological state;
- Good organoleptic properties of the strains cofermented or incorporated into fermented milk;
- The maintenance of good flavour and aroma profiles after fermentation;
- The ability to maintain viability;
- The maintenance of the mild acidity throughout storage time with a good acidity profile;
- The maintenance of colonizing properties throughout processing and storage;
- The development of good storage stability in fermented products;
- Stability after freeze-drying and other drying methods;
- Accurate strain identification at the genus and species level;
- Dose–response data for required effects.

Also, the nitrogen source is important, the oxygen content is crucial for cell reproduction and the redox potential has a strong influence on the growth of bifidobacteria.

Another current area of investigation is the effect of 'bifidogenic growth factors' or the 'Bifidus factor' on the survival of *B. longum*, *B. infantis* and *B. adolescentis* in various dairy products. Many factors which stimulate growth of bifidobacteria in vivo have been reported in the literature. These include raftilose, a fructo-oligosaccharide (FOS), transgalactosyloligosaccharide (TOS), lactulose and the enzyme, oxidase. The advantage of these carbohydrates is that they are not hydrolysed by the human intestinal galactosidase and pass to the lower intestine, promoting bifidobacteria growth. The data with the use of the 'bifidus factor' in fermented products are interesting and emphasis is now centred on incorp-

Table 4 Enzymatic characteristics of LAB and bifidobacteria species used in fermented milks

Enzymes	% of frequency					
	Lact. API 20 Z	S. therm. LRA ZYM	L. bulg. LRA ZYM	L. casei API 20 Z	L. acid. API 20 Z	Bifidobacteria API 20 ZYM
In mucin degradation						
α-D-galactosidase	0	40	60	20	100	100
α-D-glucosidase	80	60	60	100	30	100
α-L-arabinosidase	nd	100	90	nd	nd	[23] 0 B. breve [23] 96 B. longum [23] 96 B. infantis
α-D-xylosidase	nd	90	90	nd	nd	nd
α-L-fucosidase	0	20	10	0	0	0
β-D-fucosidase	nd	100	100	nd	nd	[23] 100 B. breve [23] 90 B. longum [23] 90 B. infantis
β-L-fucosidase	nd	0	11	nd	nd	nd
N-Ac-α & β-D- glucosaminidase	10	0 or week	0 or week	10	100	10 B. bifidum B. infantis
Versus other substrata						
β-D-galactosidase	100	100	100	100	100	100
β-D-glucosidase	30	0	0	90	90	100
α-maltosidase	nd	93	83	nd	nd	nd
β-D-mannosidase	nd	43	35	nd	nd	nd
α-D-mannosidase	nd	7	4	nd	nd	nd
Proteases	Lact.	S. therm.	L. bulg.	L. casei	L. acid.	Bifidobacteria
Trypsin	0	0	0	0	0	0
Chymotrypsin	0	13	0	0	0	0

Lact. = Lactococcus; *S. therm.* = *Streptococcus thermophilus*; *L. bulg.* = *Lactobacillus bulgaricus*; *L. casei* = *Lactobacillus casei*; *L. acid.* = *Lactobacillus acidophilus*.

orating both bifidobacteria (*B. longum*) and bifidogenic factors (FOS, TOS and lactulose) into cottage and Edam cheeses.

Products in which the aforementioned strains have been used are as follows.

- Pasteurized milk: the milk is inoculated at a temperature of 4–5°C.
- Fermented milk products: a large number of products are produced from heat-treated and homogenized milk, with an increased protein content and fat content over the whole range. These are subdivided into three main groups, according to their microflora:
  Group A: yoghurt, mesophilic fermented milks prepared with specific dairy bacteria having euprobiotic properties, i.e. *Streptococcus thermophilus*, *Lactobacillus delbrueckii* subsp. *bulgaricus* and lactococci.
  Group B: fermented milks prepared with intestinal LAB possessing properties suitable for dairy preparation, i.e. *Lactobacillus casei* and *Lactobacillus acidophilus*.
  Group C: bacteria exclusively of intestinal origin, with highly acclaimed probiotic properties but with little adaptability to dairy preparation, i.e. bifidobacteria (*B. bifidum*, *B. longum*, *B. infantis*). The

minimum count of bifidobacteria should not be lower than 10^6 cfu ml^{-1} at the time of sale.

- Cheese: mainly cottage and Maasdamer cheeses. The most appropriate method for the addition to cottage cheese is during the creaming mixture and for the Edam cheese, immediately after cooking, just prior to hopping. Added probiotic LAB, *L. acidophilus* La-H3, *B. bifidum* LB-H1, *L. casei* LC-H3.
- Ice cream and other dairy desserts, with the addition of *B. bifidum*, *L. acidophilus* and/or *casei*, mainly in the yoghurt ice cream
- Butter milk, with the addition of *L. casei*
- Sour cream, with the addition of *B. longum*.

Stability of Probiotic Bacteria During Storage

The most important problem is survival of the probiotic bacteria in the products during storage at 4°C. This is related to the kind and strains of microorganisms used, the technology, the final pH of the product and the length of storage. For example, pasteurized milk with the addition of probiotics has a shelf life corresponding to normal milk stored at 4°C. The shelf life of the different kinds of fermented products varies from four days to three weeks.

Table 5 Characteristics of *Lactobacillus* spp. possibly used in the production of new fermented milks

	L. delbrueckii subsp. bulgaricus (old name L. bulgaricus)	L. acidophilus	L. casei subsp. casei
Type strain	ATCC 11842	ATCC 4356	ATCC 393[a]
Ecology	Milk and milk products	Human and animal intestinal tracts, human mouth and vagina	Milk, cheese, sour dough, silage, human intestinal tract, mouth and vagina, sewage
Phenotypic characteristics			
Morphology	Rods with round ends, variable with age, singly and in short chains, metachromatic granules in cells	Rods with round ends singly, in pairs or short chains, no metachromatic granules	Rods, often with square ends, tending to form chains
Catalase reaction	–	–	–
Growth at:	– (15°C), + (45°C)	– (15°C), + (45°C)	+ (15°C), + (45°C)
Reaction in milk	+ (coagulation)	+ (slow coagulation)	+ (slow coagulation)
Lactic acid isomers	D(–)	DL	L(+)
Technological characteristics			
Activity (organic acid production)	Obligately homofermentative and citrate negative; lactic acid (1.7% in milk)	Obligately homofermentative and citrate negative; lactic acid (0.6–0.9% in milk)	Facultatively heterofermentative and citrate negative; lactic acid (1.2–1.5% in milk), acetic acid in small quantities
Proteolytic activity	+ (free amino acids)	Weak	+
Lipolytic activity	Weak	Weak	Weak
Aroma/flavour production	+ (acetaldehyde, diacetyl, acetoin, acetone)	±/+ weak (acetaldehyde)	Weak
Gas production	–	–	–
Production of polysaccharides	+strong/variable	Weak	Weak
Alcohol production	± (ethanol: trace)	± (ethanol: trace)	± (ethanol: trace)
H_2O_2 production	+	+	+
Vitamin production	+ (niacin, folic acid, B_6, B_{12})	Unknown	Unknown
Other	Benzoic acid production		

[a] There is a request to reject this type strain because of its low DNA homology with other strains of *L. casei*.

The probiotic properties of desirable bacteria are largely dependent on their ability to remain viable and to colonize the surface of human intestinal cells. Therefore, it is necessary that sufficient numbers of viable bacteria are present at the time of consumption. A concentration of 1×10^5 cfu ml^{-1} has been suggested as the 'therapeutic minimum'. Most probiotic bacteria, for example *L. acidophilus* and *B. bifidum*, show a short stationary growth phase, which is followed by rapid loss of cell viability. Their short shelf life represents a logistic problem for both manufacturers and retailers and a technical challenge for researchers.

Physiological and Technological Characteristics of the Most Important Probiotic Bacteria

In the new generation of fermented products lactic acid bacteria with specific and, in some cases, very diverse physiological and metabolic traits are combined, and sometimes different from those of traditional yoghurt cultures. With the increasing demand for these products, there are a number of technical issues to be addressed, such as:

- the development of suitable production processes
- monitoring of the number of each species during the process
- checking the interaction between different species
- establishing selective procedures for their differentiation.

Table 4 gives the enzymatic characteristics of LAB and bifidobacteria species used in fermented products. This information is very important because, at the intestinal level, the enzymes produced both extra- and endocellularly by LAB and bifidobacteria may have activities against the mucin or the diet, i.e. meat and vegetables, and therefore the availability of different substrates, and thus affect the level of nutrition. Consequently, LAB are important from the probiotic point of view, not only in comparison with other intestinal

Table 6 Characteristics of *Bifidobacterium* spp. possibly used in the production of new fermented milks

	B. bifidum (old name B. bifidum var. pennsylvanicus)	B. infantis	B. longum
Type strain	ATCC 29521	ATCC 15697	ATCC 15707
Ecology	Predominant in the faeces of neonates	Predominant in the faeces of breast-fed infants	Biovar *a* more frequent in human adults; biovar *b* more frequent in neonates
Phenotypic characteristics			
Morphology	Cells are highly variable; groupings of 'amphora like' cells are characteristic	Does not present specific traits, similar to that of many species of the genus	Elongated cells with irregular contours, branching rare
Catalase reaction	–	–	–
Growth at:	25–45°C	37°C (anaerobic)	24–43°C (anaerobic)
Reaction in milk	+ (weak coagulum)	+ (weak coagulum)	+ (weak–medium coagulum)
Lactic acid isomers	L(+)	L(+)	L(+)
Technological characteristics			
Activity (organic acid production)	Lactic acid (0.4–0.8%), acetic acid, small amounts of formic and succinic acids	Lactic acid (0.4–0.8%), acetic acid, small quantities of formic and succinic acids	Lactic acid (0.9–0.9%), acetic acid, formic acid and succinic acid in small amounts
Proteolytic activity	+ (free amino acids, esosamine)	+ (free amino acids, esosamine)	+ (free amino acids, esosamine)
Lipolytic activity	Unknown	Unknown	Unknown
Aroma/flavour production	+ (acetaldehyde, acetone, acetoin, diacetyl)	+ (acetaldehyde, acetone, acetoin, diacetyl)	+ (acetaldehyde, acetone, acetoin, diacetyl)
Gas production	–	–	–
Production of polysaccharides	±	±	+
Alcohol production	+ (ethanol: trace)	± (ethanol: trace)	+ (ethanol: trace)
H_2O_2 production	–	–	–
Vitamin production	+ (B_6, B_{12}, folic acid)	+ (B_6, B_{12}, folic acid)	+ (B_6, B_{12}, folic acid)

Table 7 Characteristics of *Streptococcus* sp. possibly used in the production of new fermented milks

	Streptococcus thermophilus (old name 'Streptococcus salivarius *subsp.* thermophilus
Type strain:	ATCC 19258
Ecology	Milk and milk products
Phenotypic characteristics	
Morphology	Spherical to ovoid cells, in pairs or long chains, strong polymorphism in old cells
Catalase reaction	–
Growth at:	– (10°C); + (45°C)
Reaction in milk	+ (coagulation)
Lactic acid isomers	L(+)
Technological characteristics	
Activity (organic acid production):	Obligately homofermentative, citrate negative, lactic acid (0.7–0.8% in milk)
Proteolytic activity:	Weak
Lipolytic activity:	Weak
Aroma/flavour production:	+ (acetaldehyde, acetone, acetoin, diacetyl)
Gas production:	–
Production of polysaccharides:	±
Alcohol production:	± (ethanol: trace)
H_2O_2 production:	±
Vitamin production:	+ (B_6, B_{12})

microbial species, but also for their effect on the host physiology and dietary contents. Thus strains that produce glycosidase, which attack fucosides, mannosides or xylosides, can digest fibre, and leave carbohydrates that are available as an energy source for their growth or that of other species. **Tables 5, 6 and**

7 present the characteristics of probiotic bacteria used in the production of fermented products.

Methods and Media for Detection of Probiotic Cultures

A variety of fermented dairy products containing probiotic bacteria have been developed as a result of increasing evidence for the beneficial functions of these bacteria in human health. Varying numbers of live probiotic bacteria have been reported in different products indicating the effect of manufacturing technology on the growth of these organisms, as well as of the microecosystem of the products.

For the detection and enumeration of bifidobacteria, two factors are important: an adequate culture medium and anaerobic conditions. Generally, media containing many nutrients and special metal salts are required. They are usually supplemented with substances to lower the redox potential, for example, cysteine, cystine, ascorbic acid, sodium sulphite, liver extract containing cysteine in addition to certain growth substances, etc. and frequently fortified with horse or sheep blood.

Incubation is carried out under anaerobic conditions, in an atmosphere of either CO_2 and H_2 ($10:90$) or CO_2 for at least 48 h. The culture medium for enumerating bifidobacteria in fermented dairy products depends, first of all, on the type of product under investigation. If the product contains bifidobacteria alone, a rich medium can be used: BL agar, modified Rogosa agar and others. The following conditions are recommended: MRS-agar + NNL, anaerobic incubation, 37°C for 3 days.

In products containing bifidobacteria together with lactic acid, bacteria selective media should be used. NPNL agar may be a medium of choice for plate counts using anaerobic incubation. For the enumeration of bifidobacteria and *L. acidophilus* Hansen recommends MRS agar under anaerobic incubation at 37°C, 3 days.

Culture Media for Detection and Enumeration of *Lactobacillus acidophilus*

Concerning the detection and enumeration of *Lactobacillus acidophilus* in fermented and non-fermented milk products, the joint IDF/ISO/AOAC group of experts for lactic acid bacteria, proposes a large number of media. The choice of medium depends principally on the type of product to be investigated. In products containing *L. acidophilus* only, detection and enumeration are rather simple; complications arise when *L. acidophilus* is combined with other types of lactic acid bacteria, as these generally grow on the same type of media. The temperature of incu-

bation and the atmosphere must be specifically adjusted to obtain accurate counts.

According to Nighswonger et al, all cultures of *L. acidophilus* and *L. casei* formed equal numbers of colonies on CLBS (cellobiose *Lactobacillus* agar) and CLBSO (CLBS supplemented with 0.15% oxgall agar). Thus agars were considered to be suitable for the enumeration of total and bile-tolerant *L. acidophilus* and *L. casei* in cultured buttermilk and yoghurt prepared using the indicated starter cultures.

Genetic methods, for example plasmid analysis or DNA restriction mapping, may prove useful for unequivocal identification. By this technique, DNA molecules are cut with different restriction enzymes and the DNA fragments are separated by electrophoresis. The pattern is specific for each strain and in some cases different strains can be identified. On the other hand, different strains within a species may have the same DNA pattern and still behave differently from a physiological point of view.

See also: **Bifidobacterium**. **Enterococcus**. **Lactobacillus**: Introduction; *Lactobacillus acidophilus*; *Lactobacillus casei*. **Probiotic Bacteria**: Detection and Estimation in Fermented and Non-fermented Dairy Products.

Further Reading

Arroyo L, Cotton LN and Martin JH (1994) Evaluation of media for enumeration of *Bifidobacterium adolescentis*, *B. infantis* and *B. longum* from pure culture. *Cultured Dairy Products J*. May: 20–24.

Dellaglio F, Torriani S, Vlaeminck G and Cornet R (1992) Specific characteristics of microorganisms used for new fermented milks. *IDF Bull*. 277: 4–16.

Flair Research Program (1994) Final Technical Report. The selection and characterization of human probiotic strains, the process technology necessary for their incorporation into fermented foods and evaluation in selected animal models. Contract No AGRF-CT91-0053.

Hansen C (1996) Bulletin. Nutrich cultures. Methods of analysis for the identification of *L. acidophilus* and bifidobacteria. P. 27–30.

International Dairy Federation/ISO/AOAC Group of Experts (1995) Detection and enumeration of *Lactobacillus acidophilus*. *IDF Bull*. 306: 23–33.

Martin JH (1996) Technical considerations for incorporating bifidobacteria and bifidogenic factors into dairy products. *IDF Bull*. 313: 49–50.

Nighswonger BD, Brashears MM and Gilliland ES (1996) Viability of *Lactobacillus acidophilus* and *Lactobacillus casei*. *J. Dairy Sci*. 79: 212–219.

O'Sullivan MG (1996) Metabolism of bifidogenic factors by gut flora – an overview. *IDF Bull*. 313: 23–30.

Rasic JLj (1990) Culture media for detection and enumeration of bifidobacteria in fermented milk products. *IDF Bull.* 252: 24–30.

Reute G (1997) Present and future of probiotics in Germany and in Central Europe. *Bioscience Microflora* 16: 43–51.

Salminen S (1996) Uniqueness of probiotic strains. *IDF Nutrition Newsletter* 5: 18–19.

Salminen S and Tanaka R (1996) Annual review on cultured milks and probiotics. *IDF Nutrition Newsletter* 4: 47–50.

MICROSCOPY

Contents

Light Microscopy

Robert W Lovitt and **Chris J Wright**, Department of Chemical and Biological Process Engineering, University of Wales Swansea, UK

Introduction

Ever since the role of microorganisms in fermentation was realized light microscopy has become an essential technology for the food microbiologist. Historically the techniques of light microscopy developed from the use of single convex lens as magnifying glasses, the invention of the first practical microscope being accredited to Antony van Leeuwenhoek in 1668. Leeuwenhoek's microscope had a magnification ratio of 270 to 1 and consisted of a single lens that was moved up and down by a screw mechanism. Today there is a vast variety of microscopes available of different design and quality employing a number of light microscopy techniques to study microbiological structures and their interactions with the environment.

Light microscopy is routinely used in microbiological laboratories often without consideration for the finer details of the instrumentation. The full potential of light microscopy can be realized when a light microscope is used by an operator knowledgeable of the underlying principles of the instrument and of microbiological staining methods. The specimen preparation and correct staining procedures are paramount to effective light microscopic study of food microbiological samples. Direct examination by the experienced eye is a rapid cost-effective diagnostic aid to estimating the quality of food, the nature of an infection or the purity of the culture. Despite many limitations, microscopic observation of microbes is normally necessary for identification. There are many methods for observing and characterizing microbes and many date back to the last century. The primary aim is to identify and enumerate the microorganisms present. It may also be used to determine the strategy for further microbiological examination and investigation.

Principles of Light Microscopy

A basic knowledge of the theory of light microscopy is useful to understand how the general instrument works, optimization procedures, instrumentation specifications and technological innovations. A good physics textbook should be consulted if a more detailed explanation of image production by lenses is required. A light microscope essentially consists of three components, the eyepiece, the objective and an illumination source (condenser). The last two have greatest influence on image quality. Both objective and condenser are constructed from lenses. Thus their performance is governed by the efficiency of their component lenses light transmission.

Convex lenses are used in light microscopy because they converge incident light into a principal focus (**Fig. 1A**). Figure 1(B) shows how a convex lens forms an image. The human eye contains a convex lens that produces images of distant objects on the retina at the back of the eyeball. The ciliary muscles change the shape of the lens to change the principal focus point and thus allow the eye to focus objects at different distances. The eye can see an object with greatest clarity when it is positioned at the near point. A distance away from the eye termed the least distance of distinct vision (D)(Fig. 1C). When using a magnifying glass the observer moves the lens until the image is situated at the near point (Fig. 1D). The

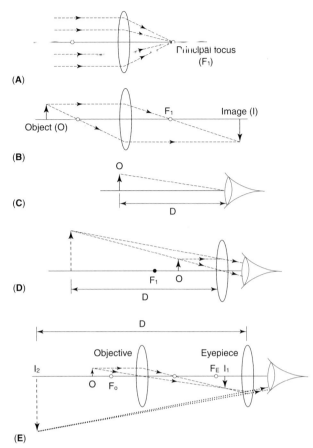

Figure 1 Ray diagrams illustrating how light travels through the lenses of optical instruments: (**A**) and (**B**) convex lens; (**C**) human eye; (**D**) magnifying glass; (**E**) compound microscope.

object is located within the focal distance. For high magnification a lens of short focal length is required. The fabrication of lens with focal lengths below a certain limit is impracticable, to increase higher magnification two separate convex lenses are used. This is the basis of a compound microscope (Fig. 1E). The lens close to the object is termed an objective. The lens that is used to observe the final image is called the eyepiece.

The power of an objective is shown by its resolution. The resolution (R) of the objective lens is the smallest distance between two points that can be differentiated in the generated image. The resolution is found to equal $0.61\lambda/A$ where λ is the wavelength of the light and A is the numerical aperture. This equation defines the limits of light microscopy. Small values of λ improve resolution. The numerical aperture represents the ability of the objective to gather the rays coming from each illuminated point on the specimen and it is a function of the refractive index of the material between the lens and the sample. The theoretical maximum value for the numerical aperture is equal to the refractive index of the material between the objective and the sample. When the numerical

aperture is large then the resolution of the system is increased. Thus the addition of immersion oil increases the magnification of a sample, the refractive index of oil is about 1.52 compared to 1.00 that of air.

The efficiency of lenses and thus objectives and condensers are subject to aberrations. Spherical aberration occurs with all single lenses and results in a halo of unfocused light around the image that is then blurred. It is due to the uneven bending of transmitted light rays and forms images at various distances along the axis. Chromatic aberration is due to the differential bending of light of different wavelengths. The shorter the wavelength the larger the bending. In this case, different coloured images are formed at different points along the principal axis, red furthest away from the lens, violet nearest the lens. On examination of these images they are found to have a rainbow-like fringe due to the unfocused light of the other colours. Spherical aberration and chromatic aberration of the system can be corrected simultaneously by joining lenses, of equal and opposite error, to form doublet or triplet lenses.

The choice of objective is paramount in light microscopy and there are many types available each designed to give best results for particular purposes. Objectives are multiple lens systems constructed to minimize aberrations and to enhance differentiation of sample features. To this end objective lens systems have different numerical apertures, and different working distances and are constructed from different lens combinations. The working distance is an important specification of an objective and it is the distance between the object and the objectives front lens, when the system is correctly focused. The objective will be redundant if the depth of the sample and cover slip is greater than the working distance of the objective. Achromatic objectives are economic and the most widely used objectives, they have moderate numerical apertures and working distances. Apochromatic objectives are more complex lens systems than achromatic lenses and produce very high quality images. Combinations of doublets, triplets, specially fabricated and shaped lenses ensure that apochromatic objectives have high numerical apertures and perfect colour rendering.

Fluorite objectives are examples of objectives that contain lenses fabricated from a different material to glass or quartz. Fluorite objectives characteristically increase the contrast between objects and their surroundings; they also act to darken some colours. Hence these lenses are useful when used in conjunction with staining methods. Fluorite objectives have performance close to that of apochromatic

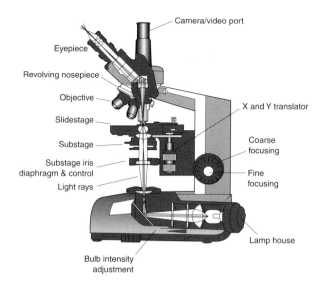

Figure 2 The general design of a typical microbiological laboratory light microscope.

lenses. The construction of objectives for phase contrast is discussed below.

An important consideration in light microscopy that is often overlooked is sample illumination. It is imperative that the plane at which the objective is focused is uniformly illuminated with light that passes through the object and forms a cone that is large enough to fill the front of the objective. To meet the illumination requirements of the different types of objectives there are equally as many types of condenser. A compromise is made with the condenser and in most instruments a general-purpose condenser is used possibly with the addition of a specialist unit if more advanced techniques such as phase contrast are routine. If the specimen is mounted on an opaque substrate then reflected light as opposed to transmitted light can be used to view the surface. Epi-illumination is used to direct light actually through the objective focusing light and image simultaneously.

General Design and Operation

Figure 2 shows the general design of the simple light microscope used in microbiological laboratories; the following steps describe its routine operation. A suitable slide is placed onto the microscope stage with the lowest power lens in position. Before viewing the sample through the eyepiece the objective should be brought close to the slide surface using the coarse adjustment. The sample is then focused by retracting the objective whilst viewing through the eyepiece. This procedure will prevent damage to the objective and the slide. The illumination should then be optimized by adjusting the position of the condenser and the size of the aperture until a homogeneous bright illumination is achieved without glare. Final focus

correction can be achieved by the intuitive movement of the fine adjustment.

If required, the sample can next be viewed under a higher magnification. Many microscopes hold parfocal objectives that have the same focal positions. Thus only small adjustments will be needed to focus the higher-powered lens. Care should be taken when changing objectives as adjacent lenses may not be parfocal and oil immersion lenses have short working distances. In addition sample and cover slip depth can vary. If the microscope has an achromatic oil-immersion lens, a drop of oil can be placed on top of the cover slip so that when the objective is positioned a meniscus of oil forms between the two surfaces increasing resolution. If after focusing and illumination optimization procedures, the image remains poor then the meniscus of oil may be compromised by bubbles or lens surface grime. The solution to this problem is to raise the objective and clean the surface oil and grime from the lens surface before reforming the meniscus. This is modern light microscopy in its simplest form and is often adequate when used alongside effective staining techniques (see below). If however staining is not a viable option or greater image differentiation is required further light microscope techniques should be considered.

Dark-field Illumination

In normal light microscopy there are many objects that cannot be seen because they are transparent. To improve the image an increase in contrast between the object and its surroundings is required. Dark-field illumination functions to improve contrast. In normal light microscopy the condenser functions to ensure the sample is uniformly illuminated and that light enters the objective correctly. In dark-field illumination the condenser functions to produce a hollow cone of light which is focused on the specimen. Only light that is scattered by the specimen will enter the objective. Thus the specimen will be seen illuminated against a dark background. In low-power dark-field illumination the long working distance of the objective means that it is relatively easy to reduce the entry of light into the objective. A disc or a diaphragm with moveable leaves stops the central rays of the cone of light. An annular cone of light can be focused on the specimen without any direct light entering the objective. High power dark-field illumination is used regularly alongside oil immersion to view living organisms. However, the short working distance of the required objectives means that a special dark-field condenser is needed. The modern design of these condensers is effective for objectives with numerical apertures between 1 and 1.15, for greater apertures a further means, such as a in-built iris diaphragm within the objective, is required to reduce the passage of direct rays. The microbiological sample to be viewed should be well spaced and within a single plane, thus samples should be dilute. The spaces between the organisms provide the dark background that contrast with the illuminated specimen. If the objects are in multiple planes then the scattering of light by all the planes will mask the dark field of the individual plane. Both specimen and immersion oil should be free of bubbles or dust that act on the passage of light compromising dark-field illumination.

Phase-contrast Microscopy

Phase-contrast microscopy is an extremely useful technique for observing specimens that have not been stained and are in their natural state. Objects, that under normal light microscopy cannot be seen, are observed in sharp outline and in good contrast to their surroundings. Light travels in the form of waves of energy, and so has wave length and amplitude parameters. When two waves meet, the resultant light ray will have wave parameters that will depend on whether the two original waves were in phase, for example peak arriving with peak, or out of phase, valley arriving with peak. In a light microscope the component parts ensure that direct and diffracted rays meet at the eyepiece producing brightness and shade according to the phase relationship of the incident rays. The image observed is in reality a complex interference pattern. The greatest contrast is achieved when the direct and diffracted rays meet at the point of interference out of phase by about half a wavelength. The specimen has regions of different thickness and different refractive indices, thus rays are diffracted to different degrees. If the specimen is transparent with a density and thickness insufficient to create the required phase difference then contrast will be poor.

Phase-contrast objectives contain a phase plate, or a lens face with phase rings, that retard the diffracted rays so that their phase is altered. The phase-contrast condenser ensures that both diffracted and direct rays are incident on the phase changing device. The annular grooves of the phase plate allow a differential retardation of the diffracted rays with a phase change of about a quarter of a wavelength. This is added to the phase change caused by the passage through the sample. The total change is adequate to produce an image of excellent contrast when the rays interfere at the eyepiece.

Fluorescence Microscopy

Fluorescence microscopy has become an important imaging technique in microbiology. It is used in conjunction with staining techniques (see below) to visualize a whole range of intracellular structures. When certain substances are excited by illumination of short wavelength, for example ultraviolet, the emergent rays are converted into longer-wavelength light. Thus blue, green, yellow or red light is emitted depending on the composition of the substance. If this emission of long-wavelength light only occurs while the excitation illumination is present it is called fluorescence. If the emission continues when excitation has been removed, this is termed phosphorescence. Many natural substances display fluorescence in specific colours when excited by short-wavelength light. This is termed primary fluorescence. Many dyes exhibit primary fluorescence and can be used in dilute aqueous solutions to stain tissues and cells, which in turn fluoresce. This is termed secondary fluorescence and is exploited by fluorescence microscopy. The fluorescence stains are used in very dilute concentrations so living cells are exposed to minimal damage. There are many light sources available, which excite samples not only with light within the UV range but with other short wavelengths such as the blue short-wave component of the visible spectrum. The light sources are used in conjunction with filters to remove unwanted wavelengths of light. The objectives of fluorescence microscopy are constructed from quartz when short-wave UV light is used, as glass is opaque to these wavelengths. When longer UV wavelength light is used optical glass will allow the passage of these wavelengths, negating the need for quartz lenses and slides. A further requirement for the objectives of fluorescence microscopy is that none of their optical components fluoresce themselves. A further addition to the microscope equipment is a filter that removes UV light once it has induced the fluorescence of the specimen. This filter is essential to protect the eyes of the operator. Advanced forms of fluorescence microscopy use combinations of other light microscopy techniques to improve performance.

Micrometry

Micrometry is the term used to describe the technique for measuring detail in a microscopic sample using a light microscope. For optimum results the specimen is viewed with a superimposed grid at the highest magnification with the greatest resolution possible. The grid is superimposed by mounting near the focal point of the eyepiece a glass disc with an engraved or photographically reproduced rule. This glass disc is called an eyepiece micrometer. The micrometer must be calibrated for each objective fitted to the microscope. To calibrate, a grid of known dimensions, termed a graticule, is imaged. A typical graticule will have a 100 μm line split into 100 divisions. The number of divisions on the eyepiece micrometer per division on the graticule is recorded. Features in subsequent images can be measured by counting how many divisions in the eyepiece micrometer they occupy and calculating the length using the previously recorded calibration constant.

Another routinely used method in food microbiology is counting cells in a known volume of solution. To do this the cell solution is placed on a microscope slide with a chamber of known volume. This slide is called a haemocytometer. The chamber of known volume is normally split into smaller sections, which simplifies counting. All micrometry methods are aided by computer techniques. There are numerous computer packages that can be used instead of the eyepiece micrometer; calibration of the objective is in terms of pixels. In addition imaging software can calculate the area of an image that is within a contrast range, this facilitates cell counting.

Cytological Light Microscopy

Histological Basis of Staining

The main problem with visualizing microbes under the microscope is that organisms are virtually transparent. The refractive index of vegetative cells is very similar to that of water. One approach to visualizing organisms is to use phase contrast, dark-field or fluorescence as discussed above. However, from the earliest investigations, stains have been used to distinguish microbes from their surroundings. The main advantages over the more sophisticated methods are the speed and simplicity of this approach. Thus the main objective of staining is to increase the contrast of the cells. Although natural dyes were used initially, artificial dyes derived from coal tar and other aromatic materials have now replaced them. These dyes either act directly as chromophores or become chromogenic when they react with specific chemicals within the cell. The chemistry of the chromaticity of these compounds is based on the delocalization of the electronic structure as they contain many functional groups (carbon–carbon double bonds, carbon=oxygen carbon=nitrogen, nitrogen=nitrogen) that cause absorption spectra in the light region. The structure of crystal violet is shown in **Figure 3** and typifies the properties of a chromogenic compound.

For a stain to bind effectively to the cell it must be able to react or associate with cellular materials. Dyes are normally charged, either as cations or anions, and so bind to the many charged sites within cellular materials (proteins, polysaccharides and nucleic acids). Binding may also be aided by the addition of a mordant that enhances the interaction or acts as a bridging compound. Stains may be used individually, termed simple stains, or in combinations, termed dif-

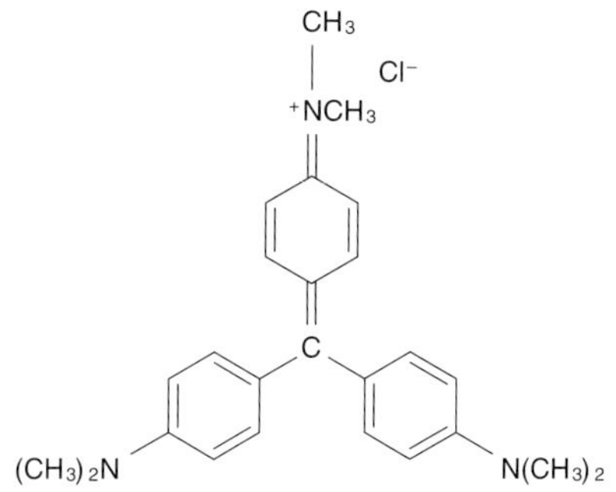

Figure 3 Crystal violet.

ferential staining, e.g. the Gram stain or the acid fast stain.

Obtaining Samples

Appropriate sampling techniques are important if bacteria and other microbes are to be observed. It must be remembered that to observe a good number of bacteria in a single field at least 10^7 organisms per millilitre are required in a typical wet mount under a coverslip. As a consequence sampling from air usually requires a filtration technique or a culturing process such as growth on an exposed agar plate. Water can be examined directly, but in many cases the microbes in fluids are concentrated by filtration. Soil generally has large numbers of organisms present and can be sampled by the addition of water. There are also a number of specialized methods to observe the growth and activity of microbes in soil. Samples from food usually involve techniques that enrich the microbes prior to observation.

Wet Mounts

One of the most common methods of examining microbes is to make a wet mount of the sample. Typically this is made by placing a small drop of the microbial suspension on a glass slide and positioning a coverslip over the drop, the slide can then be placed under the microscope. The hanging drop method can also be used to ensure better conditions for observing motility. The main advantages of wet mounts are that the samples remain viable and it avoids drying or fixation that could alter cell morphology. Therefore, size, shape and motility are normally observed using wet mounts. After appropriate simple staining of the sample, cells may be counted directly to give total cell counts. Counts typically involve the use of a counting chamber (see micrometry section above). Viable counts can be made if an appropriate vital stain is used. For example, fluorescein diacetate can be used if cells contain esterases. The enzymes will hydrolyse the fluorescein diacetate to release fluorescence.

Permanent Mounts

For extended preservation and permanent mounting of stained specimens, dry samples should be cleaned in xylene prior to mounting in Canada balsam. Other mounting materials include plastics dissolved in solvents, e.g. 'Permount', that may be painted directly on the stained film.

Negative Stains

Another wet mount procedure involves negative staining. This provides the simplest and often the quickest means of gaining information about cell shape and size, refractive inclusions and spores. This staining technique relies on colloidal carbon in the form of nigrosin or Indian ink. These stains reveal unstained bacteria against a dark sepia background. These methods are especially useful to observe capsular layers. **Table 1** details some typical methods.

Fixation of Smears and Suspensions

For more complex differential staining methods, cells are fixed to surfaces so that they do not wash off between the stages of staining. Fixation also preserves structure avoiding digestion or changes in the shape of cells. Two types of fixation methods achieve this: heat or chemical means.

Heat Fixation Heat fixation is the most common method used for stabilizing microbial smears. Typically the smear of a bacterial aqueous suspension is allowed to dry on a glass slide. The slide is then passed over a flame several times taking care not to overheat the sample. The smear is now ready for staining procedures.

Chemical Fixation Chemical fixation is generally better than heat fixation at preservation of morphology. The procedures are, however, more time consuming. Some of the most useful procedures are fixation with aldehydes such as glutaraldehyde or formalin. For example, a sample is mixed with formaldehyde to give a final concentration of 1.5–2% formaldehyde. After a few minutes the cells are smeared and dried onto a slide. Osmium tetroxide can be used to fix wet films by exposing the wet smear to the fumes of a 1% (w/v) solution in a closed vessel for 2 min.

Simple Staining

Morphological studies of bacteria in culture or from natural specimens are generally heat fixed and then stained with basic dyes. Where a single stain is used the process is referred to as simple staining. Slides are typically bathed in a 2% (w/v) crystal violet/ammonium oxalate (for 10 s) or 0.8% methylene blue in alkaline aqueous solution (for 30 s).

Other simple stains include alkaline lactophenol cotton blue that is good at enhancing the visibility of fungal cell walls. Nigrosin or Indian ink has also been used for negative staining wet mounts when capsular structures are present.

Fixed differential staining methods

Cells There are a number of important differential staining methods that are important in the characterization of microbes. Stains for different cell wall and membrane structures and intracellular bodies have been developed. Some of the most common are discussed here.

The Gram stain is one of the most important differential staining techniques applied to bacteria and was first developed by Christian Gram in 1884. In theory it should be possible to divide bacteria into two groups: Gram-positive and Gram-negative. However, in practice it is common to observe Gram-variable organisms. This is not surprising considering that physiological conditions of the cell can affect wall structure.

In the Gram stain, the cells are treated with crystal violet and then with an iodine solution that acts as a mordant binding the dye to the cellular materials. The cells are then washed with ethanol. Gram-positive cells retain the crystal violet whereas Gram-negative cells do not. To make the difference obvious, a red counter stain is used such a safranin or fuchsin. The basis of the stain is the differential permeability to the iodine–crystal violet complex. The complex can cross the murein layer of Gram-negative bacteria whereas it is unable to cross the cell wall layer of Gram-positive bacteria. A typical procedure is outlined in Table 1.

The stain requires practice and good technique to obtain reproducible results. For example Gram-negative bacteria can appear to be Gram-positive if the film is too thick or if decolorization washes, with ethanol, are too short. Gram-positive organisms can appear Gram-negative if over washed with the ethanol.

The acid-fast stain exploits the presence of waxy fatty acid compounds in the cell wall. Certain groups of bacteria contain long-chain fatty acids (50–90 carbons long) called mycolic acids that give them a waxy coat which is impervious to basic dyes such as crystal violet. Normally a detergent is required to allow the dye to penetrate the cell. Once inside the cell the normal acidic alcoholic solvent is unable to decolorize the cell. The acid-fast stain is particularly useful to identify specific groups of bacteria: *Nocardia* and *Mycobacterium* and the spores of *Crypto-*

Table 1 Microscopic stains and staining procedures common in food microbiology

Type	Solution	Method	Comments
Simple stains			
Crystal violet	Crystal violet staining reagent: mix 20 ml 10% (w/v) crystal violet in ethanol with 80 ml 1% ammonium oxalate		
Methylene blue	Loeffers methylene blue reagent: 30 ml of 1.6% (w/v) methylene blue ethanol solution is mixed with 100 ml of 0.01% KOH aqueous solution	1. Flood heat-fixed smear with crystal violet or methylene blue for 10 or 30 s, respectively 2. Wash with water then blot dry	Cells should take the stain to which they are exposed.
Carbol fuchsin	Carbol–fuchsin stain: basic fuchsin 0.3%, phenol 5%, ethanol 10%, water 85%	If the cells do not stain with either of the above then Carbol fuchsin can be used	
Lactophenol cotton blue	Lactophenol cotton blue: A solution 0.5% cotton blue in phenol (20%), lactic acid (20%), glycerol 40%, and water (10%)	This stains colorizes fungal cell walls	
Negative stain			
Nigrosin	Nigrosin stain: 7% nigrosin	1. Place a droplet of nigrosin on a slide 2. Mix with small sample containing microbe 3. Take another slide and smear out the mixture to produce a gradient of film thickness 4. Allow to dry completely	These preparations reveal bacteria unstained and standing out brightly against a sepia background
Differential stains			
Gram stain	Crystal violet staining reagent Mix 20 ml 10% (w/v) crystal violet in ethanol with 80 ml 1% ammonium oxalate Mordant A solution of 0.33% (w/v) iodine 0.66% KI Decolorizing reagent Ethanol 95% solution Counterstain: 0.25% (w/v) safranin in 10% ethanol water solution	1. Flood an air-dried and heat-fixed smear with crystal violet stain for 1 min 2. Wash with water 3. Flood the smear with mordant for 1 min 4. Wash smear with water 5. Decolorise with ethanol for 30 s 6. Wash with water 7. Flood smear with counter stain for 30 s 8. Wash smear and blot dry	Gram-positive organisms appear blue/black whereas Gram-negative organisms appear pink/red
Acid fast staining: Ziehl–Neelsen stain	Carbol–fuchsin stain: Basic fuchsin 0.3%, phenol 5%, ethanol 10%, water 85% Decolorizing solvent: 3 ml conc. HCl in 95% ethanol Counter stain: 0.3% methylene blue solution	1. Place an air-dried and heat-fixed smear, cover the slide with basic fuchsin dye and place in steam for 3–5 min to allow the dye to penetrate 2. After washing decolorize with the decolorizing solvent 3. Flood smear with counter stain for 20–30 s and wash	Acid fast bacteria appear red whereas non-acid-fast appear blue.

Table 1 Microscopic stains and staining procedures common in food microbiology—continued

Type	Solution	Method	Comments
Endospores: Malachite green	Malachite green: 0.5% (w/v) malchite green Counterstain: 0.25% (w/v) of safranin in 10% ethanol water solution	1. Take a slide with an air-dried and heat-fixed smear and flood with malachite green. 2. Place slide in steam for 5 min 3. Wash in water 4. Flood slide with counter stain 5. Then wash and blot dry	Endospores appear bright green and vegetative cells appear brownish red
Cytoplasmic inclusions: Poly-β-hydroxybutyrate (PHB)	Sudan Black III: 0.3% (w/v) in ethylene glycol Washing agent Xylene Counterstain: 0.5% (w/v) aqueous safranin	1. Flood heat-fixed film with Sudan Black for 5–15 min 2. Drain and air dry 3. Wash in xylene 4. Blot dry 5. Counterstain with safranin 6. Rinse with water and blot dry	PHB appears as a black droplet while the cytoplasm appears pink.
Polyphosphate	Loeffers methylene blue reagent: 30 ml of 1.6% (w/v) methylene blue ethanol solution is mixed with 100 ml of 0.01% KOH aqueous solution or 1% toluidine blue	1. Stain a heat-fixed smear for 10–30 s with either Loffers methylene blue or toluidine blue 2. Rinse and blot dry	Under the methylene blue stain polyphosphate appears as deep blue/ violet spheres. Using toluidine, polyphosphate appears as red spheres while the cytoplasm appears blue
Glycogen-like polysaccharides	Periodic acid solution: 1% periodic acid in solution of 20 mM sodium acetate in 70% ethanol 70% ethanol washing solution Reducing solution: 0.05% (w/v) sodium thiosulphate, 0.1% (w/v), 20 mM HCl in 60% aqueous ethanol Schiff reagent: 0.25% basic fuchsin in about 100 mM HCl and 1% potassium metabisulphite Metabisulphite washing solution: 4% potassium metabisulphite and 1% HCl in aqueous solution Malachite green counter stain 0.002% aqueous solution	1. Flood heat-fixed slide with periodic acid reagent for 5 min 2. Wash with ethanol solution 3. Flood with reducing reagent for 5 min 4. Wash with ethanol solution 5. Stain with reducing solution for 14–45 min 6. Wash several times with metabisulphite solution 7. Counterstain with malachite green for 2–3 s 8. Wash with water and blot dry	Polysaccharides appear red; other cytoplasmic components appear green
Fat	Sudan black III reagent: 0.01% (w/v) Sudan III in a solution of 50% ethanol and glycerol Counterstain: 0.3% (w/v) methylene blue solution	Flood heat-fixed Sudan black III Wash Counterstain with methylene blue	Fat is stained red. This will work for bacteria, yeasts or fungi

Table 1 Microscopic stains and staining procedures common in food microbiology—continued

Type	Solution	Method	Comments
Fluorescent stains			
Acridine orange	Acridine orange reagent: 0.1% acridine orange in 0.2 M acetate buffer (pH 4.5) Washing reagent: 0.2 M sodium acetate buffer pH 4.5	Methanol is used to fix the smear Immerse in sodium acetate buffer Stain with 0.1% acridine orange 20–120 min	In yeast, cytoplasm is normally orange while the DNA is green or green/yellow fluorescence
Calcofluor white Primulin	Calcofluor white stain: 0.1 mg l⁻¹ calcofluor, 0.1 mg l⁻¹ Primulin	Add 1 drop of calcofluor white to fixed preparation Add 1 drop of KOH Add coverslip	Fungal elements appear bright green or blue depending on UV filter used. Primulin, yeast cell walls fluoresce green yellow. Will allow the observation of budding scars. In some cases can be used to distinguish between live and dead cells, as the dyes are not membrane permeable so if the cytoplasm stains then membranes are damaged.
DAPI (4′,6′-diamidino-2-phenylindol)	DAPI staining solution DAPI 2 mg l⁻¹	Mix DAPI solution with cell suspension Mount the stained suspension on the slide Apply coverslip and view	This stain is used extensively for mapping nuclei, clear view of well spread chromosomes is possible In yeast will stain chromosomes blue/white
Conjugate fluorescent antibodies	Antibodies (preferable monoclonal) are raised to specific organisms and then conjugated with fluorescein isothiocyanate (FITC)	Cover the air-dried and heat-fixed or formalin-fixed samples with the conjugate serum for 20–30 min taking care not to allow the preparation to dry out. Then wet mount using (10%) glycerol solution which is phosphate buffered (pH 7.0) under a coverslip and examine fluorescence	These techniques of rapid identification are now being replaced by DNA binding/probe methodologies

sporidium. The Ziehl–Neelsen staining procedure is shown in Table 1.

Subcellular Structures

Endospores The position of the developing spores in vegetative cells can also be an important piece of diagnostic information. Spores can easily be observed under phase-contrast microscope and are strongly refractile shining brightly slightly above the true focus. Spores can also be negatively stained using nigrosin. Spores by their nature are very resistant to simple staining and it requires quite harsh techniques to achieve this, however, once stained they are also difficult to decolorize. The malachite green stain is one such staining method and is outlined in Table 1.

Capsules Slime layers and capsules are produced by many bacteria and are best demonstrated by wet preparations because their highly hydrated polymers shrink when fixed and dried. Dulguid Indian ink is the simplest capsule stain.

Cytoplasmic Inclusions Many bacteria will produce, as a result of metabolism, inclusion bodies within the cell. These include fat droplets, poly-hydroxybutyrate, polyphosphate and polysaccharides, sulphur and protein crystals. Many of these can be visualized using appropriate staining methods. Polyhydroxybutyrate is observed using negative staining or by dyes which bind specifically to the polymer. Polyphosphates can also be stored by cells in the form of metachromatic granules and are detected by treating the fixed cells with methylene blue or toluidine blue. Glycogen-like polymers can be stained with periodic Schiff stain. Bacterial nuclear bodies are not easy to stain directly with basic dyes, however, fluorescence staining is particularly useful for detecting nuclear materials.

Fluorescent Staining Procedures

There are a number of procedures that use fluorescent stains. Their main advantages are that they allow observation at lower magnification and detection in complex backgrounds such as blood or animal tissues. The main disadvantage is the requirement for a specialized microscope.

There are several popular fluorescent dyes. Acridine orange, a fluorochrome that intercalates into nucleic acids, in both native and denatured forms, is able to distinguish between fungal and bacterial DNAs. Another nuclear fluorochrome is DAPI (4',6'-diamidino-2-phenylindol) which is widely used for visualization of chromosomes in eukaryotic microbes.

Calcofluor white nonspecifically binds to polysaccharides, such as cellulose and chitin; it is thus useful in detecting fungi and yeast. Similarly primulin will bind to chitin and, for example, will enhance the visibility of yeast bud scars.

Antibodies conjugated with fluorochrome have also been developed for the rapid identification of specific pathogenic organisms such as *Legionella*. This is a very good method for the detection of specific organisms. However, the modern molecular biology techniques using polymerase chain reaction (PCR) technology, DNA fingerprinting and DNA probes are beginning to replace this technology.

Conclusions

Light microscopy is an established and vital tool for the food microbiologist. The instrumentation continues to be improved and exciting techniques such as confocal light microscopy are providing new insights into the relationship between microorganisms and their environment. With the advent of modern computation methods substantial advances in image analysis are proving extremely useful in the enumeration and morphological measurement of microbes. The correct use of instrumentation and optimized staining procedures will ensure that the food microbiologist can exploit the full potential of light microscopy.

See also: **Total Viable Counts**: Microscopy.

Further Reading

Gerhardt P, Murray RGE, Wood WA and Krieg NR (eds) (1994) *Methods for General and Molecular Bacteriology.* Washington, DC: ASM Press.

Murray PR, Baron EJ, Pfaller MA, Tenover FC and Yolken RH (eds) (1995) *Manual of Clinical Microbiology*, 6th edn. Washington DC: ASM Press.

Richardson JH (1991) *Handbook for the Light Microscope – A Users Guide.* New Jersey: Noyes Publications.

Shotton D (ed.) (1993) *Electronic Light Microscopy – Techniques in Modern Biomedical Microscopy* New York: Wiley-Liss.

Confocal Laser Scanning Microscopy

D F Lewis, Food Systems Division, SAC, Auchincruive, Ayr, Scotland

Description

The art of all microscopy is to balance the achieving of contrast with the retention of resolution. Thus, in bright-field microscopy the microscopist will adjust the condenser diaphragm in accordance with the inherent contrast of the specimen, closing the diaphragm increases contrast but at the expense of resolution and vice versa. The object of increasing contrast whilst maintaining resolution has led to the development of numerous optical contrast methods. Examples of microscopy contrast techniques include dark-field, phase-contrast, polarized light, differential interference contrast, Hoffman modulation contrast and the use of ultraviolet light and fluorescent techniques. The development of electron microscopy techniques used the shorter wavelength of the electron to improve resolution well beyond that achievable by light microscopy and atomic force microscopes improve on electron microscopy. So where does confocal laser scanning microscopy fit into this array of techniques and what are its main areas of application?

The main advantages of confocal laser scanning microscopy (CLSM) over 'normal' light microscopy techniques are improved contrast at high resolution, improved resolution in fluorescent specimens and improved axial (depth) resolution which allows optical sectioning of specimens. The main disadvantages over 'normal' light microscopy are the cost and the loss of true colour. Compared with the transmission electron microscope the CLSM has much poorer resolution but requires considerably less specimen preparation and can give three-dimensional information on internal structures much more conveniently.

In biological applications, CLSM has been mostly developed with fluorescent staining and especially for localization with fluorescent markers and for deriving three-dimensional representations of biological structure.

Principle of Operation

The principle of the confocal microscope is illustrated in **Figure 1**.

A point source of illumination is produced by placing a pinhole in front of an illumination source and the pinhole is focused onto the specimen by a lens (lens 1); the light emitted by the specimen is focused

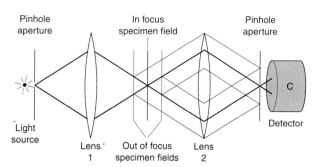

Figure 1   The principle of the confocal microscope.

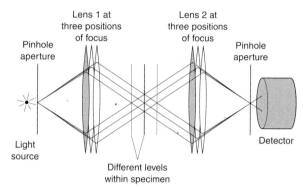

Figure 2   Diagram to show how the confocal arrangement can be used to collect sequential optical sections from a sample.

by lens 2 onto another pinhole placed in front of a detector. The effect is that the light collected by the detector is derived mostly from the in-focus specimen field. Light from out-of-focus material (shown in Fig. 1 in grey) is diffuse when it reaches the pinhole and hence relatively little light from these fields reaches the detector. The size of the pinhole defines the depth of the in-focus field in the specimen; the smaller the pinhole the narrower the depth of the in-focus field. As the pinhole size is reduced the level of illumination also decreases and in practice the pinhole size is adjusted to give the best balance of axial resolution and illumination for a particular specimen.

Figure 2 illustrates how the confocal arrangement can be used to collect sequential optical sections from a sample. This figure shows how focusing the lenses to bring three fields within the specimen into focus can allow optical sections to be obtained. These optical sections can be stored on computer data storage media and can be reconstructed to give either rotational 3-D presentations of the specimen or to give stereo pairs. Stereo pairs can be presented as side by side images to be merged by either special viewers or by manipulation of the viewer's eyes. **Figure 3** shows stereo pairs laid side by side.

The alternative method of presenting stereo images is as combined red/green images which are viewed with red/green glasses to separate the images to the

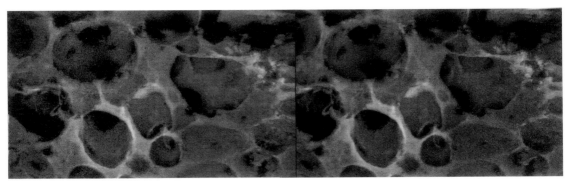

Figure 3 Stereo pair of freeze substituted ice cream. CLSM.

left and right eyes. An example of a red/green stereo pair is shown in **Figure 4**.

In most applications of the confocal principle in microscopy the light used for imaging is either reflected or fluoresced from the specimen and lens 1 and lens 2 are the same lens with a beam splitting device such as a dichroic mirror separating the emitted light to the second pinhole and the detector. In some cases the emitted light may also be further divided to pass to separate detectors. This arrangement is illustrated in **Figure 5** which illustrates a common arrangement for commercial CLSMs.

In Figure 5 the light from the laser, in this case shown as green and blue lines, passes through an adjustable pinhole and is formed into a raster by two mirrors vibrating at 90° to each other. The light is now passed to a dichroic mirror which reflects the blue and green light through the objective lens of the microscope onto the specimen. The specimen is illuminated with a fine beam making a raster similar to that of a television screen. Some of the light is reflected back into the objective and in suitable specimens some light is re-emitted at a longer wavelength (shown here as red and yellow). Some of the reflected light will be reflected back towards the light source but most of the fluoresced light will pass through the dichroic mirror towards the detectors. The light beam

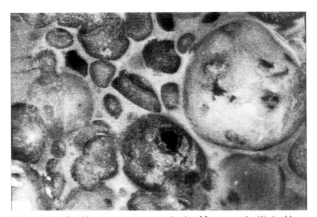

Figure 4 Red/green stereo anaglyph of freeze substituted ice cream. CLSM.

can be split again and different filters can be inserted to give complementary signals from differently fluorescing regions.

History

The concept of the confocal microscope can be traced back to 1884 when Nipkow developed a spinning disc with fine holes arranged in a spiral to produce a 'raster' of illumination spots on an image and allow information to be transmitted from a two-dimensional image as a single stream of data. This 'spinning disc' arrangement was developed by Petránň in the late 1960s to produce the tandem scanning disc confocal microscope. In 1957 Minsky recognized the potential of confocal microscopy and patented a microscope where a confocal system of illumination and imaging was combined with a scanning stage. This was in effect the principle now used in CD players. The development of lasers and the growth of computing capabilities from the 1970s allowed both the development of beam scanning instruments and the capability of capturing and manipulating the image data generated to allow realistic commercial instruments to be produced.

Features

Lasers

Although the confocal principle is applicable with normal, non-coherent light sources the realization of their potential has come about with the advances in laser technology. Dye-based lasers are less commonly used in confocal microscopy but can be used to give a wide range of wavelengths by tuning to particular frequencies.

The laser light source with its greater penetration and low divergence allows imaging in thicker specimens than is possible with normal light sources. The disadvantages of lasers have been their cost and the restricted wavelengths available. The restriction in wavelengths has in turn restricted the number of fluorescent stains that can be used.

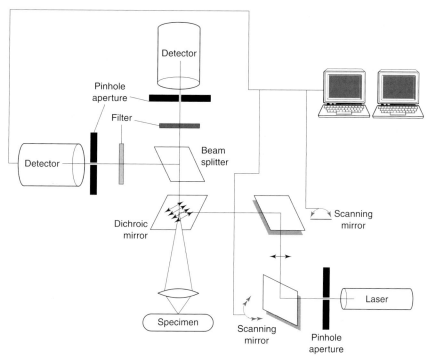

Figure 5 Schematic of CLSM set-up.

Lasers can be divided into two broad cateogires, continuous wave lasers and pulsed lasers. Both types can be used in confocal microscopy but most commonly continuous wave type lasers are preferred for routine examinations. The classes of continuous wave lasers include gas lasers, dye lasers and solid-state lasers. To date the most popular of these have been gas lasers and the basic confocal laser is the argon-ion laser; these are most commonly tuned to give spectral lines at 488 and 514 nm wavelengths.

Increasingly, microscopes are being fitted with dual and triple laser sources to extend the wavelengths available for excitation; the attainment of additional wavelengths is often achieved at the expense of robustness and some lasers have a relatively short lifetime. Krypton-based lasers are used to extend the range of argon lasers into the red region of the spectrum whereas mixed argon–krypton lasers will give lines at 649, 567 and 488 nm. Mixed helium–neon lasers will give lines at 534, 594, 612 and 632 nm. Finally in the gas laser class are the helium–cadmium lasers which give lines in the violet and UV region (325, 354 and 442 nm). With a mixture of gas lasers it is possible to cover a wide range of excitation wavelengths, though at some expense.

More recently solid-state lasers have become more available and these are likely to represent the light source for future generations of confocal microscopes. Gallium-based semiconductor lasers can be used to give lines in the red and near IR region. Titanium–sapphire lasers can be tuned to give a broad range of

wavelengths in the red and near IR region and other combinations promise to provide sources from the visible region well into the IR part of the spectrum.

Pulsed lasers give intense bursts of laser emission and can consequently be used to study time domain phenomena and in the rather specialized derivative of CLSM, two-photon microscopy.

Two- or Multi-photon Microscopy

Normal confocal fluorescence microscopy can give problems, particularly in observing living biological materials where changes over short periods of time need to be observed. One problem is photobleaching of the fluorescent label (chromophore). Because the small confocal aperture blocks most of the light emitted by the tissue the exciting laser must be very bright to give an adequate signal-to-noise ratio. This bright light causes rapid fading of fluorescent dyes. Thus, the fluorescence signal weakens as subsequent scans are made, either to produce a three-dimensional image, or to observe a single slice at several time points. Also phototoxicity can be a problem either from the blue light itself or from the formation of toxic free radicals from the dyes. Thus, the scanning time or light intensity must be limited, if one hopes to keep the specimen alive.

A relatively new type of microscope that greatly reduces both of these problems has been developed. This device depends on the two-photon effect, by which a chromophore is excited not by a single photon of visible light, but by two lower-energy (infrared) photons that are absorbed simultaneously (within 1 fs). Fluorescently labelled specimens are illuminated by a titanium – sapphire laser that produces very short pulses of infrared light – with a large peak amplitude.

Fluorescence from the two-photon effect depends on the square of the incident light intensity, which in turn decreases approximately as the square of the distance from the focus. Hence only dye molecules near the focus of the beam are excited. The tissue above and below the plane of focus is merely subjected to infrared light that causes neither photobleaching nor phototoxicity. Although the peak amplitude of the IR pulses is large, the mean power of the beam is small and does not cause significant heating in the sample.

The two-photon laser-scanning microscope promises to greatly expand the application of confocal techniques. Its usefulness with a variety of living specimens and a variety of chromophores has been demonstrated with a variety of specimens and has recently become available in commercial form from a number of manufacturers. The use of an IR laser allows better penetration of the specimen and hence extended 3D images compared with conventional CLSM and the reduction in interference from out-of-focus regions of the specimen allows better resolution and contrast to be achieved.

Objectives

As with conventional light microscopes choice of objectives is critical to the performance of the instrument. To make best use of the CLSM's ability to optically section into thick specimens long working distance objectives are to be preferred. This coupled with high numerical aperture and good transmission in the visible and UV range are the primary considerations for a working confocal objective. In biological CLSM, fluorescent stains are generally used and the objectives should ideally be corrected at the excitation and emission wavelengths, in general this will not be the case with conventional optics.

As a general rule a microscope with a good set of optics for conventional light microscopy will give a reasonable performance with a CLSM, although there may be some distortions in 3D reconstructions. When purchasing an objective specifically for confocal use the best strategy is to take the lens on loan and evaluate it in the particular CLSM before deciding to purchase.

Image Recording

When CLSM first became a commercially available technique the image recording options were relatively limited. Optical disc storage was expensive and the hard copy output was likely to be through a video printer or through a camera photographing a specially devised high resolution VDU screen which would show sequentially red, green and blue exposures. Currently the output of even relatively inexpensive colour printers is adequate for most confocal images. This factor coupled with the dramatic increase in availability of computer data storage media allows the storage of large numbers of images in an easy and rapidly accessible form.

Filters

Several types of filters are used in CLSM. Neutral density or grey filters may be used to reduce the intensity of the incident light without changing the colour of the light. Polarizing filters are used to reduce reflections and the attendant 'Newton's Rings' interference patterns and dichroic mirrors used in fluorescence microscopy allow separation of reflected incident light from light that has been absorbed and re-emitted at a longer wavelength. Since the confocal image is essentially monochromatic there is a need to use optical filters to distinguish between different fluorochromes and hence allow meaningful colours to be introduced into the image. These filters can be classified as three main types: longpass (shortpass) filters, bandpass filters and rejection band filters.

Longpass filters allow all light above a certain wavelength (or below in the case of shortpass filters) to pass through the filter while light of shorter wavelength is rejected. These filters are often coded ***LP to signify the wavelength at which the cut off occurs.

Bandpass filters allow the transmission of a defined spectral band. The bands may be from less than one to several hundred nanometres wide. These filters are often coded NB for narrow band (transmission band less than 4 nm), BP bandpass (transmission range up to 100 nm.), DF discriminating filter (filters with a sharp cut-off and cut-on and a particularly even transmission region) and WB wideband (transmission range up to several hundred nanometres.

Rejection filters (RF) will reject a defined wavelength band or bands and will allow other wavelengths to pass through. These filters can be designed to filter out specific laser lines to eliminate reflection in confocal microscopy.

In designing the optical system to discriminate on the colour of light emitted from the specimen the selection of emission filters is critical. Take the example of two stains both excited in the 500 nm region of the spectrum but one giving an emission at about 520 nm and the other at 600 nm. In order to distinguish between the two stains the beam can be split and directed towards two detectors as shown in Figure 5. If a bandpass filter with maximum transmission at 520 nm is placed in front of detector 1 and a longpass filter with a cut-off wavelength of 580 nm is placed in front of detector 2 then the signals from the two fluorochromes will be seen separately by the two detectors. Colouring the image from detector 1 in shades of green and the image from detector in shades of red and superimposing the two images will produce an image that reflects the distribution of the two stains and bears some similarity to the 'real' colours from the specimen. In the case given above there is a large separation between the emission spectra of the two stains and separation is relatively straightforward and some tolerance can be given on the ranges of the filters. When the differences are smaller it may be necessary to select filters with narrower tolerances to avoid colour bleed between the two images.

Presentation

One of the features of all microscopy is that 'seeing is believing' and the production of a convincing picture is often as important in delivering a message as the technical accuracy of the methodology. In the raw state confocal images are often relatively unattractive monochrome images. The digital format of the images, however, allows manipulation to make them more attractive and to highlight features that are masked in the raw data. False colour images can quite easily be made with the use of look-up tables (LUTs) that convert a grey intensity scale into specific colours. Similarly with dual or triple detector systems it is possible to ascribe each detection channel with a particular colour coding and consequently show the location of different fluorochrome binding sites as different colours in the final image. A wide range of image manipulation can also be undertaken to improve contrast and reduce background noise in the image. These techniques should always be treated with caution, however, since they can quite easily give an appealing image that does not accurately reflect the specimen structure. The presentational benefits of CLSM really come in 3D and to make best use of those generally requires colour images or the capacity to present moving images. CLSM images in publications are often disappointing; not least to the author who has seen the 'on-screen' representation of the image. The explosive development of computer graphic and imaging technologies will allow the appreciation of CLSM to a wider audience.

Figure 6 shows a CLSM of yeast presented in different ways. All are stained with acridine orange and excited with 488 and 514 lines from an argon ion laser. Figure 6(A) shows the image obtained as a monochrome, Figure 6(B) shows the image with false colour using one LUT set and Figure 6(C) shows the same image with a second LUT set. Figure 6(B) shows

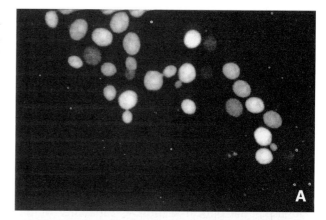

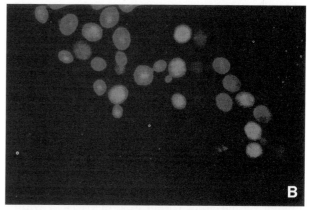

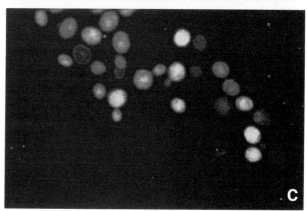

Figure 6   CLSM of yeast presented in different ways. (**A**) Image obtained as a monochrome; (**B**) image with false colour using one look-up table (LUT) set; (**C**) the same image with a second LUT set. All were stained with acridine orange and excited with 488 and 514 lines from an argon ion laser.

enhanced edges to the yeast cells and the red edges to the cells now appears to be a cell wall which is not portrayed in the raw image. Figure 6(C) further modifies the view of the cell walls to apparently reveal inhomogeneities seen as orange/red speckles. The coloured images undoubtedly have more appeal but whether the highlighted cell walls and other structural detail is a genuine reflection of cellular detail is doubtful.

Specimen Preparation

Foods

For *in situ* studies of microorganisms in a food matrix the normal specimen preparation procedures can be used. Normally foods will need to be sectioned either from frozen specimens or specimens embedded in wax or plastic resin. Prepared sections can be beneficially much thicker than those for normal light microscopy. Some foods are best examined in the confocal as hand cut cubes that are then stained en-bloc and mounted in a glycerol solution or a similar mountant and examined using the optical sectioning capability of the CLSM. Special slides can be fabricated to avoid squashing the sample with the coverslip when covering the preparation.

Isolated Cultures

Microbiological slides can be prepared as heat-fixed smears in the normal way although the CLSM offers the opportunity to examine thicker samples such as a colony lifted directly from an agar plate and stained to show the 3D interactions between the organisms. Vital stains (i.e. those that do not kill the cells) can be used to follow the growth or movement of microbiological cells either within food matrices or within model cultures.

Staining

Basis for Selection of Stains

CLSM specimens must be labelled with fluorescent stains. These stains or probes are available for a wide range of applications allowing the CLSM to image molecular structures (such as proteins, lipids, carbohydrates and nucleic acids) and physiological ions (such as calcium and pH). Several factors must be considered when selecting fluorescent probes for use with a CLSM. The major considerations are the excitation/emission wavelengths of the probe to be used, the laser lines available and the filter sets used.

Major factors that influence fluorochrome selection are the emission spectrum and the absorption spectrum. Both the emission and absorption spectrum can be greatly affected by environmental factors (e.g. pH, mineral ions, polarity). They can also change their properties after they are conjugated to antibodies or other molecules, and after they are bound to their target.

Generally the excitation wavelengths available for CLSM will be limited. Most CLSM instruments will be capable of using 488 nm and 514 nm lines and there are a number of fluorochromes that absorb around these wavelengths. Other possible lines for CLSM include 325, 354, 442, 534, 567, 594, 612, 632 and 649 nm.

Solubility is an important consideration in choosing stains; simple aqueous solutions are to be preferred although for some fat stains a less polar solvent (ethanol or propan-2-ol) may be more effective. Generally stains will be applied to samples on the microscope slide although in some cases the specimen may be stained en-bloc before further processing.

Confocal microscopy allows the use of thicker specimens and for this reason conventional staining procedures may need to be adapted to suit the CLSM. Generally more concentrated stains and longer staining times should be tried where the specimen thickness significantly exceeds that of a conventional histology preparation.

A list of useful stains for CLSM is given in **Table 1**.

Proteins

For general morphology, proteins can be stained with eosin which images well with an argon ion laser. Similarly unconjugated fluorescein isothiocyanate (FITC) will serve as an acceptable protein stain though neither stain is absolutely specific. Both stains are simple to use with aqueous solutions, eosin can be made as an ethanolic stain. Specific proteins can be stained using antibodies labelled with fluorescein, BODIPY (a range of stains produced by Molecular Probes Inc. having the basic fluorophore structure – 4,4-difluoro-4-bara-3a,4a-diaza-s-indacene), Oregon green, Texas red or tetramethylrhodamine isothiocyanate (TRITC).

Fats

The majority of fat stains are based on preferential solubility where the dye partitions itself largely into the low polarity lipid phase rather than the more polar staining solution. Nile red is the stain of choice for CLSM having both good spectral properties and a suitable partition coefficient to allow stain to concentrate in the lipid phase.

Carbohydrates

Starches can be stained with Congo red or Nile blue/Nile red; in both these cases the staining mechanism is probably similar to that with lipids in that the stain prefers to be in a less polar environment. However, with Congo red there is probably also some hydrogen bonding. With both stains penetration of intact starch grains is poor and the stains can be used as a diagnosis of starch damage. Congo red is also a stain for amyloid deposits in animal tissues. Periodic acid Schiff staining is the established method for polysaccharides. The method is based on the cleaving of 1 : 2 glycol groups by periodic acid and the subsequent

Table 1 Useful stains for CLSM

Stain	Absorption (nm)	Emission (nm)	Applications
Acridine orange	460–500	520–530	DNA/RNA binding; pH dependent protein and carbohydrate staining, fluorescent replacement for Schiff's reagent in PAS method
BCECF	505	531	Fluoresces at alkaline pH
BODIPY range of stains (Molecular Probes)	500–581	510–591	Mostly as fluorophores for labelling proteins and oligonucleides
Congo red	488–500		Starch, glucans, amyloid, pH dependent (blue at acid pH)
Eosin	517	537	General morphology, stains proteins
Evans blue	611		Quenching agent for autofluorescence, used in viability and damage tests
Fluo 3	506	526	Fluorescence dependent on calcium concentration, used in cell viability tests
Fluoroscein isothiocyanate (FITC)	490	520	Mostly as a fluorophore for labelling proteins and oligonucleides
Green fluorescent protein	395,470	509	Genetically expressed marker
Neutral red	540	610	Viability assays, as part of Gram stain for tissue sections
Nile blue	600–690	670+	Damaged starch and with Nile red for lipids
Nile red	485	525	Lipids
Oregon green range of stains (Molecular Probes)	496–511	522–530	Mostly as a fluorophore for labelling proteins and oligonucleides
Periodic acid Schiff	540–555		Polysaccharide stain
Propidium iodide	536	617	DNA binding
Rhodamine 123	505–511	534	Vital stain
Saffranin			General morphology, part of Gram stain
SYTO 25 (Molecular Probes)	521	556	Live cell stain, nucleic acid stain
SYTO 59 (Molecular Probes)	630	649	Live cell stain, nucleic acid stain
SYTOX Green nucleic acid stain (Molecular Probes)	504	523	Dead cell stain, permeability based
Tetramethylrhodamine isothiocyanate (TRITC)	547	572	Mostly as a fluorophore for labelling proteins and oligonucleides
Texas red	596	620	Mostly as a fluorophore for labelling proteins and oligonucleides

BCECF = 2′,7′-bis-(2-carboxyethyl)-5-(and-6)-carboxyfluorescein.

recolouring of sulphite decolorized magenta solution by the resultant aldehyde groups. Acridine orange can be used instead of Schiff's reagent as the final stain. Lectin-coupled dyes can stain specific sugar residues in fluorescent mode, alternatively lectin-coated colloidal gold can be used with the CLSM in reflectance rather than fluorescence mode.

Mineral Elements and pH

The fluorescent characteristics of some dyes are markedly altered by changes in the mineral ion concentration or by changes in pH. Fluo 3 is an example of a stain that varies according to calcium concentration; generally the fluorescence is greater as the calcium concentration increases. Many dyes are pH sensitive and either change the intensity of fluorescence or change the absorption or emission spectra according to the pH of the media in which they are located. These dyes allow mapping of changes in pH and calcium at a cellular level in microbiological and food systems. Congo red and acridine orange are examples of pH-sensitive dyes, whereas BCECF (2′,7′-bis-(2-carboxyethyl)-5-(and-6)-carboxyfluorescein)

and similar dyes are purpose designed as pH probes.

Immunological Techniques

The CLSM is particularly well suited to immunological labelling techniques since the vertical resolution is improved over conventional fluorescence microscopy. The reagents and fluorophores used for fluorescent microscopy are in the main suitable for CLSM although there has naturally been a number of fluorophores that absorb in the 488–514 nm region developed for linking to antibodies or lectins. FITC and Texas red are common fluorophores for conventional fluorescent microscopy and these can be used for CLSM although the Texas red signal is rather weak. Purpose-designed stains (Oregon green and BODIPY) have been developed to allow dual labelling with CLSM.

Fluorescent *In Situ* Hybridization (FISH)

This technique identifies genetic markers in cells and uses oligonucleotide probes which specifically complement target rRNA in bacterial cells. The probes are mostly 16–20 bases long and can be targeted

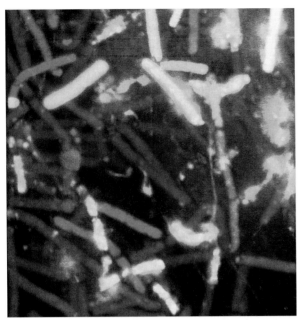

Figure 7 A mixture of live and heat-killed *Bacillus cereus* cells simultaneously stained with the cell-permeant SYTO® 59 red fluorescent nucleic acid stain (*S-11341*) and the cell-impermeant SYTOX® green nucleic acid stain (*S-7020*), each at a concentration of 5 μM. Bacteria with intact cell membranes stain fluorescent red, whereas bacteria with damaged membranes stain fluorescent green. This maximum-projection image was generated from a series of 10 images taken at 0.2 μm increments through the specimen with a confocal laser scanning microscope. Photo courtesy of Molecular Probes, Inc.

against a bacterial group, species or genus. The fluorescently labelled organisms can then be detected by CLSM. The development of polymerase chain reaction (PCR) techniques has allowed the relatively easy production of RNA probes and this in turn has led to methods for detecting non-culturable cells as well as culturable ones. Potentially this approach could allow all the bacterial cells on a microscope preparation to be identified.

Live/Dead Stains

Staining bacteria and other cellular material to discriminate between live and dead cells has been applied to CLSM. The methods are based on nucleic acid stains for live cells and differential membrane penetration stains for dead cells. A variety of stains with different spectral properties are available in each category. An example of the application of the technique is shown in **Figure 7**.

Benefits

Confocal microscopes are considerably more expensive than light microscopes and are often more expensive than electron microscopes so why are they considered so useful? In resolution terms they are still limited by the wavelength of light and so will never rival electron microscopes in that regard. In fluorescent mode the achievable resolution of a CLSM can be better than for a conventional microscope and this means that detail can be more comfortably seen in CLSM than in normal light microscopy. The axial resolution of the CLSM is considerably improved on the conventional light microscope and this is the basis of the real attractiveness of confocal microscopy. On a technical level this allows more accurate localization studies than are possible with normal fluorescence microscopy. By rejecting out of focus information CLSM is able to reduce some of the rigours of specimen preparation and to reveal 3D information in a way which no other microscopy technique can. The reduction in specimen preparation and in particular the ability to examine thick preparations allows specimens to be examined closer to their natural state and live samples can be studied *in situ*. The ability to collect images under computer control allows time-lapse imaging in a much simpler and more convenient way than was previously possible. The future development of lasers and of two- and multi-photon techniques will extend the CLSM benefits even further.

In short the CLSM has extended the scope of microscopy into three and four dimensions with improved resolution and simpler specimen preparation. The wise scientist, however, will always use as many approaches to obtain structural information as possible and the CLSM will be complementary to other microscopy techniques.

See also: **Immunomagnetic Particle-based Techniques**: Overview. **Nucleic Acid-based Assays**: Overview. **PCR-based Commercial Tests for Pathogens**.

Further Reading

Haugland RP (1996) In: Spence MTZ (ed.) *Handbook of Fluorescent Probes and Research Chemicals*, 6th edn. Eugene, USA: Molecular Probes.

Mason WT (1993) *Fluorescent and Luminescent Probes for Biological Activity*. London: Academic Press.

Pawley JB (1998) *Handbook of Biological Confocal Microscopy*, 2nd edn. New York: Plenum Press.

Scanning Electron Microscopy

U J Potter and **G Love**, Centre for Electron Optical Studies, University of Bath, UK

Introduction

The scanning electron microscope (SEM) is a powerful tool for imaging the surfaces of a wide range of materials, including small biological organisms. It encompasses the magnification range of the conventional optical microscope but possesses also the resolution necessary to provide meaningful information at magnifications of up to 100 000. A further attribute of the instrument is its large depth of field, which enables all parts of a specimen's surface to be sharply focused even when there are relatively large height variations on some of the features imaged.

The SEM

The scanning electron microscope employs a finely focused electron beam to scan across the specimen's surface. The beam is generated in what is known as the electron gun; here a thin, pointed filament of wire is electrically heated causing electrons to be emitted from its tip. The electrons are given energy and direction by applying a voltage between the filament (forming the cathode) and an annular metal plate (the anode) placed beneath the filament. The diameter of the electron beam at its point of impact on the specimen crucially affects the microscope's performance because the detail that is resolvable in the image will not be finer than this. Although some initial focusing of the beam is performed in the electron gun itself, the diameter (approximately $50 \mu m$) of the emergent electron beam is much too large for most SEM work. Reduction can be accomplished using localized electric or magnetic fields to focus the electron beam. In the SEM the magnetic fields are created by electromagnetic lenses, the strength being controlled by adjusting the electric current passing through each of the lens coils. There are typically three lenses in the instrument, the final one being used to focus the beam onto the surface of the specimen. To minimize interaction of the electron beam with air molecules and to prevent electrical breakdown in the gun, the electron microscope column and specimen chamber are evacuated to a pressure of $\sim 1 \times 10^{-3}$ Pa. (**Fig. 1**).

When the energetic electrons bombard the surface some are backscattered and others cause low-energy (secondary) electrons to be emitted from the specimen close to the point of impact of the electron beam. Both types of emergent electron may be collected using an appropriate detector, although the

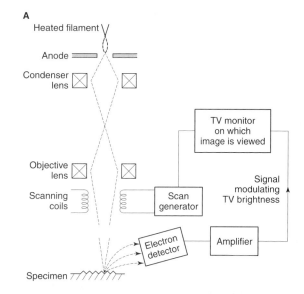

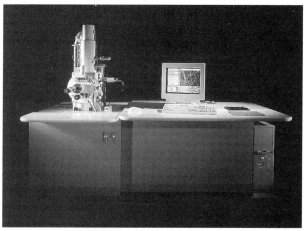

Figure 1 (**A**) Simplified schematic diagram of an SEM. Dashed lines show the approximate path of electron beam. Note that the second condenser lens is omitted for clarity. (**B**) Scanning electron microscope JSM6310, courtesy of JEOL UK Ltd.

scintillator/photomultiplier device most commonly used records mainly the low-energy secondary electrons. These electrons give topographical information because the number of them detected is influenced strongly by the local orientation of the specimen surface and the specimen/detector geometry, a higher proportion of electrons being collected from specimen features angled towards the detector.

To create an image, information from more than one point on the specimen is required. Therefore sets of current-carrying coils in the electron column are used to generate rapidly changing magnetic fields, causing the electron beam to scan a raster on the specimen's surface. The secondary electron signal, detected from each of the points within the scanned area, is then electronically amplified and used to modulate the intensity of a spot scanning across the face of a television monitor. Correspondence between

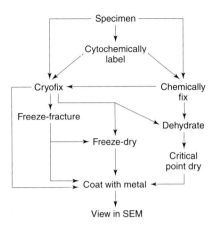

Figure 2 The preparation of samples for scanning electron microscopy.

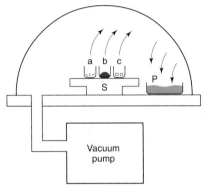

Figure 3 Freeze-drying. Cryofixed samples in metal carriers (a, b and c) are placed on the cold-stage (S) at between –60°C and –100°C. The chamber is evacuated with a vacuum pump. Drying takes place over several hours under vacuum as the water vapour (arrows) is absorbed by phosphorus pentoxide (P).

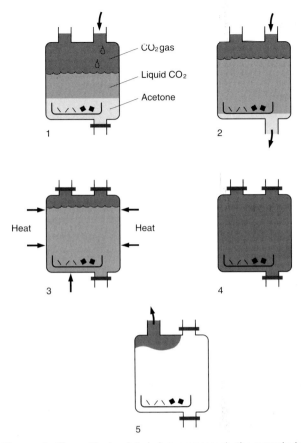

Figure 4 The critical point drying process: 1, the sample in acetone is introduced into the chamber, and liquid CO_2 is added; 2, as acetone is drained from the apparatus more liquid CO_2 enters until the acetone is completely removed from the sample (repeated draining and filling of the chamber with CO_2 is necessary, keeping the sample under the liquid surface); 3, the outlets are sealed and the apparatus slowly heated to just above the critical temperature and pressure; 4, as these points are reached all the liquid becomes a gas; 5, while maintaining the temperature the gas is bled away slowly and the sample removed.

the movement of the beam inside the electron microscope column and the spot on the monitor is maintained by using the same scan generator to deflect both, although the size of the deflection in the former case will be smaller than in the latter. Indeed, in the SEM the magnification is controlled by the relative size of the scans on monitor and specimen, i.e. the magnification is increased by reducing the area scanned on the specimen surface while maintaining a fixed sized image on the monitor. Thus, as the electron beam scans across the sample surface, the variation in brightness of image features observed on the monitor (a function of the number of electrons collected) directly relates to changes in the surface topography of the corresponding structures in the sample.

Specialized Techniques

Backscattered Electron Imaging

Topography is the usual imaging mode in SEM but sometimes one needs to distinguish between a sample and its electron-dense marker as in techniques of gold labelling (see below). This may be accomplished by using a solid state detector sensitive to high-energy backscattered electrons rather than secondaries. Because the backscattered electron signal increases with the mean atomic number of the sample, localized agglomerations of heavy elements (for example) are clearly identifiable as bright regions in the image.

X-ray Analysis

The technique of X-ray analysis requires an energy dispersive or wavelength dispersive spectrometer to be fitted as an attachment to the SEM. It enables chemical elements in the sample to be identified by studying the X-ray emission spectra excited by the electron beam. A stationary electron beam typically interacts with a few cubic micrometres of material in the sample, and examination of the spectrum (see Fig. 16B) generally allows elements to be identified,

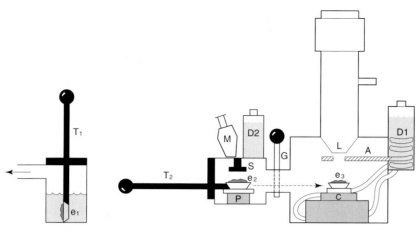

Figure 5 Low-temperature SEM. The equipment consists of a cold-stage (C) fitted inside the SEM specimen chamber. The cold-stage is cooled with nitrogen gas which circulates within a Dewar flask (D1) filled with liquid nitrogen. A metal anticontaminator (A) is situated above the sample (e) and below the SEM final lens (L). The cryopreparation chamber, attached to the side of the SEM and evacuated by a vacuum pump, is fitted with a second cold-stage (P) thermally cooled from a Dewar flask (D2). A sputter head for gold coating (S) is positioned above the sample, and a lens (M) allows the sample to be viewed for accurate fracturing. A gate valve (G) separates the chambers. A sample fixed to a carrier (e_1) and attached to a transfer rod (T_1) is cryofixed in liquid nitrogen slush within a container linked to a vacuum pump (arrow). The transfer rod (T_2) carries the frozen sample (e_2) to the preparation chamber where fracturing and gold coating occur. The sample is pushed via the open gate valve (broken arrow) onto the SEM cold-stage for sublimation of frost or final examination (e_3).

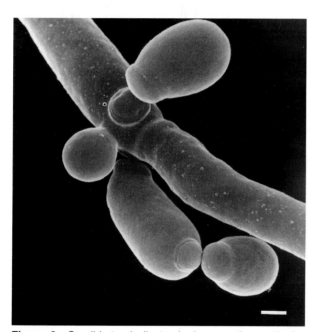

Figure 6 *Candida tropicali*, standard preparation method. Scale bar 1 μm. Courtesy of John Forsdyke.

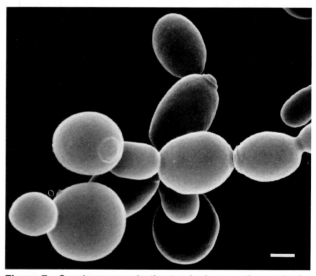

Figure 7 *Saccharomyces lactis*, standard preparation method. Scale bar 1 μm. Courtesy of John Forsdyke.

provided that their localized concentration exceeds about 0.1 wt%. Alternatively, elemental X-ray maps may be produced by scanning the surface of the sample with the electron beam in the usual way, and using the X-ray detector to record points on the sample where a particular X-ray line is detected. If an area is scanned for a few minutes a concentration map is produced in which the brightest regions correspond to high levels of the measured element (see Fig. 16C).

Low-voltage SEM

Most SEM investigations tend to use gun voltages of between 10 kV and 15 kV. These voltages can disrupt certain biological materials and lower values (< 5 kV) may reduce damage produced by electron beams in sensitive specimens while also providing improved specimen contrast of fine surface features. The disadvantages are that viewing the sample at fast (television) scan rates can be difficult because of low signal levels, and resolution of the SEM is significantly worse than at higher voltages. Therefore, high-resolution studies using low-voltage SEM will generally

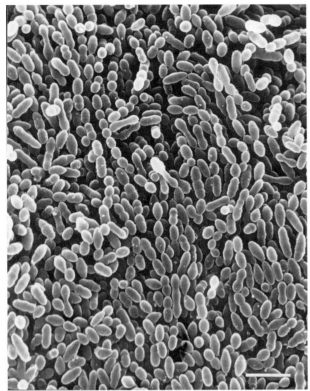

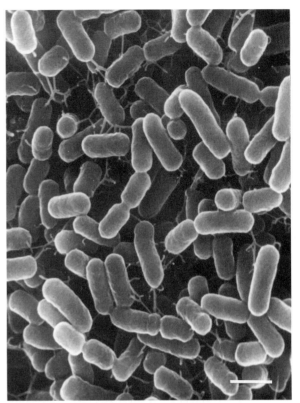

Figure 8 *Lactobacillus bulgaricus*, standard preparation method. Scale bar 5 μm. Courtesy of John Forsdyke.

Figure 9 *Escherichia coli*, standard preparation method. Scale bar 1 μm. Courtesy of John Forsdyke.

require that the instrument be fitted with a special design of electron gun – a field emission gun (FEGSEM) which is capable of delivering a large current into a small diameter electron beam. This overcomes, to a large extent, the limitations of low-voltage operation referred to above.

Environmental SEM

In the environmental SEM a small aperture separates the electron column (under high vacuum) from the specimen stage which may be at 6 MPa. Thus, hydrated specimens may be examined directly without any time-consuming pre-treatment. However, a major disadvantage of the instrument is that the poorer the specimen chamber vacuum, the greater is the broadening of the incident electron beam from collisions with air molecules. This factor limits the resolution of the environmental microscope, and its usefulness in microbiological research is yet to be proved.

Preparation of Microbiological Samples

Conditions within the conventional SEM and the process by which the image is formed determine the type of sample suitable for examination. Principal sample requirements are that it should be:

- dry – to tolerate the high vacuum in the SEM
- mechanically and thermally stable – to withstand

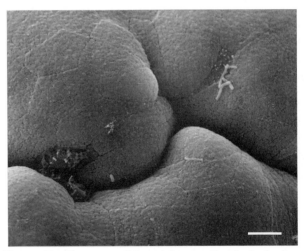

Figure 10 Human small bowel tissue with attached micro-organisms. The OTOTO process is used following standard fixation to enable a greater amount of osmium tetroxide to be incorporated into the tissue. Scale bar 5 μm.

the heat generated on interaction of the electron beam with the sample
- electrically conducting on its surface – to prevent electron charge build-up on the sample.

Biological samples do not meet these requirements, and therefore must undergo a series of preparatory steps to render them suitable. At the same time, the preparation must aim to preserve biological material

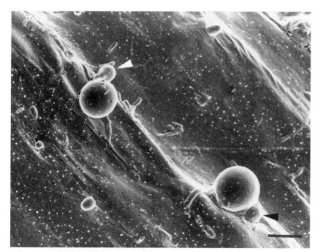

Figure 11 Encysted zoospores of *Spongospora subterranea* f. sp. *nasturtii*, the causal agent of crook root disease in watercress. Zoospores are seen on the surface of a ridge of the watercress root; adhesoria (arrowed) are clearly visible. Scale bar 5 μm. From Claxton et al (1996) *Mycol Res* 100 (12): 1431–1439. Reproduced with permission from the *British Mycological Society*.

in as lifelike a state as possible. The complexity of sample preparation for scanning electron microscopy (SEM) will depend on the nature of the sample and the type of information required from the inves-tigation. An outline of the steps, essential in following a particular pathway through preparation to specific sample information, is shown in **Figure 2**.

Pre-fixation Labelling

The gold labelling technique, one of the most exciting developments in biological microscopy, long used in light microscopy and transmission electron micro-scopy (TEM), is now increasingly employed with SEM. The labelling of macromolecules (proteins, carbohydrates or enzymes) with an electron-dense marker, such as colloidal gold or lead (an enzyme reaction product), is carried out before fixation on the surface of samples destined for SEM examination.

Immunocytochemistry, the technique of labelling proteins, makes use of the very specific binding which occurs in animals between an antibody and a foreign protein, termed the antigen. The sample is exposed to antibodies, raised in a laboratory animal, against the protein of interest. Colloidal gold particles (made from the controlled reduction of tetrachloroauric acid by a reducing agent) are complexed to the antibodies in order to mark the sites on the sample surface, of the protein under study.

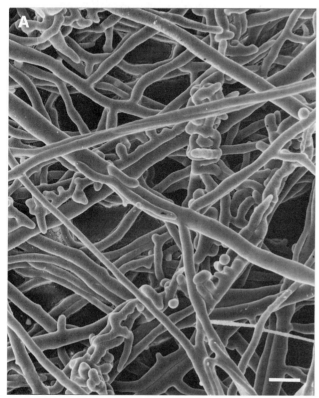

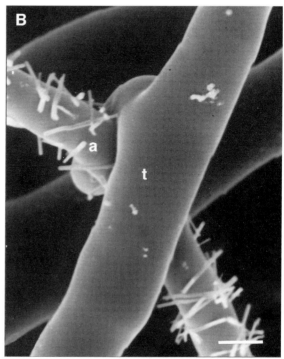

Figure 12 *Trichoderma larzianum* and *Agaricus bisporus* cryofixation. (**A**) Fungal hyphae of *Trichoderma larzianum*. Scale bar 10 μm. (**B**) Fungal hyphae of *Trichoderma larzianum* (t) and *Agaricus bisporus* (a) showing *Trichoderma* aggressive towards *Agaricus*. Scale bar 2 μm.

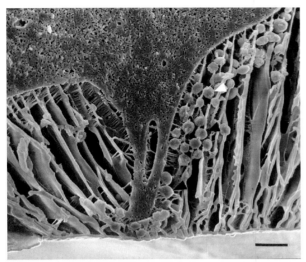

Figure 13 *Saccharomyces cerevisiae* cells (arrows) immobilized in a cellulose-based matrix (BPS Separations Ltd., Sperrymoor, U.K. Productiv™) for use during primary beer fermentation. The matrix is freeze-fractured and partially freeze-dried to reveal the location of the yeast within the matrix. Scale bar 10 µm. Micrograph courtesy of Ann O'Reilly.

Fixation

Fixation by immersion in chemical solutions or by freezing is the first step in preparing samples for SEM examination. The aim of fixation is to stabilize and immobilize cell components, protecting them from subsequent chemical treatment while preserving the whole cell in a form as true to life as possible. Double fixation is still the most widely used method of chemical fixation. Primary fixation is carried out with one or more aldehydes to cross-link proteins, while secondary fixation in osmium tetroxide stabilizes the unsaturated lipids of cell membranes.

Chemical vapours are widely used in the preparation of samples, and delicate fungal structures, mucilages and bacterial biofilms may be fixed in osmium tetroxide vapour. Techniques exist to improve the impregnation of a sample with osmium tetroxide thereby also providing extra rigidity and conductivity. The OTOTO process – alternate treatments with osmium tetroxide (O) and thiocarbohydrazide (T) – uses thiocarbohydrazide as a ligand to incorporate a greater amount of osmium into the sample, and the increased conductivity imparted may obviate the need for a coating step. Although the OTOTO method is used mainly with animal cells, it is also useful in studies of microorganisms in association with animal cells, as in the human gut (see Fig. 10).

The double fixation method does not adequately fix polysaccharides, and a technique termed the PATOTO (periodic acid, thiocarbohydrazide, osmium tetroxide, etc.) process may be used to

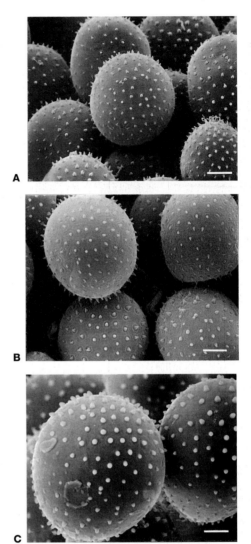

Figure 14 Urediospores of *Uromyces viciae-fabae* showing the influence of different methods of preparation on cell volume. (**A**) Standard preparation and CPD. Scale bar 3 µm. (**B**) Osmium tetroxide vapour fixation, cryofixation and freeze-drying. Scale bar 3 µm. (**C**) Cryofixation and LTSEM in fully hydrated state. Scale bar 3 µm. Micrographs courtesy of Alan Beckett, University of Bristol.

improve the preservation of fungal cell walls and extracellular mucilage while also increasing cell wall rigidity and conductivity. This process, while similar to the OTOTO method, involves periodic acid as an extra oxidizing step to create aldehyde groups. Thiocarbohydrazide is complexed to these groups forming the ligand which binds extra osmium into the sample. Preservation of extracellular carbohydrates, fungal mucilage and bacterial biofilms is also improved by the addition of ruthenium red or ruthenium hexamine trichloride to the primary fixative.

Cryofixation, a physical method of fixation that does not involve the use of chemicals, will preserve the natural composition of a cell, and stabilize and arrest cellular processes almost instantaneously. It

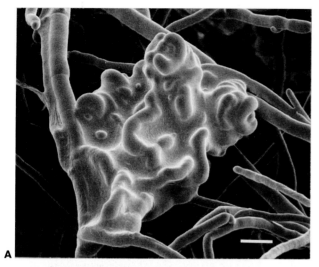

Figure 15 *Ascodesmis sphaerospora*, ascogonial coils. **(A)** Ascogonial coils surrounded by a well-preserved extracellular matrix. Cryofixation, LTSEM and partial freeze-drying. Scale bar 10 μm. **(B)** Conventional preparation consisting of primary fixation in glutaraldehyde, secondary fixation in osmium tetroxide followed by dehydration in ethanol and CPD. Note that most of the extracellular material seen in (A) is extracted during preparation. Only insoluble components remain on the ascogonial coils. Also note that this specimen is considerably shrunken relative to that in (A). Scale bar 10 μm. From Read ND (1991) Low temperature scanning electron microscopy of fungi and fungus-plant interactions. In: Mendgen K and Lesemann DE (eds) *Electron Microscopy of Plant Pathogens*. P. 17. Berlin: Springer. Copyright © 1991 Springer-Verlag. Reproduced with permission.

involves the rapid extraction of heat by plunging the sample into a cryogen, typically liquid nitrogen slush. Ideally, cryofixation should produce a layer of amorphous ice in the sample and avoid the formation of ice crystals. While this is possible in microorganisms, the relatively large size of the majority of SEM samples results in ice crystal formation. Fortunately, in the study of internal structure at relatively low magnification (< 5000 ×) and sample surfaces, ice crystal artefacts can be tolerated. However, the use

of high-resolution FEGSEM, in studies of membrane surfaces and internal cell structures for example, will require the use of superior cryofixation methods.

Samples preserved by cryofixation are suitable for a number of subsequent preparation techniques such as freeze-drying, or freeze-substitution and critical point drying. Alternatively, cryofixed samples can be imaged in a fully hydrated frozen state by low-temperature SEM (LTSEM – see over).

Dehydration and Drying

Biological samples contain a large amount of water so that dehydration and drying of samples is necessary to remove water from cells before they are introduced into the microscope vacuum (except for LTSEM). The different methods of dehydration inevitably result in some shrinkage and distortion. Air-drying, the simplest means of removing water, causes complete disruption of samples by surface tension forces which tear through the tissue at the air–liquid interface. However, the use of solvents with low surface tensions and high evaporation rates followed by air-drying will reduce this damage. Bacteria and fungi have been successfully prepared by fixation, dehydration in a solvent, and air-drying from tetramethylsilane or hexamethyldisilazane.

An alternative method of removing water from biological samples is by freeze-drying (**Fig. 3**). Frozen material is freeze-dried by the sublimation of ice under a vacuum using temperatures of –60°C to –100°C for a few hours or days, depending on the size of the sample. The main advantage of freeze-drying is that samples avoid exposure to solvents. For example, fungal mucilage and bacterial biofilms (not retained after exposure to solvents) are preserved by freeze-drying and artefacts are minimized by chemical vapour fixation prior to freezing. The composition of a sample is preserved for X-ray microanalysis by freeze-drying without chemical fixation (see Fig. 16), and immunoreactivity of proteins may be retained for correlative TEM immunocytochemistry.

A third technique exists to remove water from biological samples. Critical point drying (CPD) makes use of the critical temperature and pressure of a liquid. At the critical point a liquid turns into a gas without the presence of a meniscus and the associated disruptive surface tension forces. Unfortunately water has an impractical critical temperature (374°C) and pressure (about 22 MPa), too high to be safely used. Therefore, samples are dried from a transitional fluid – usually carbon dioxide (critical temperature 31°C and pressure 7.4 MPa) or Freon (29°C and 3.8 MPa). The CPD technique (**Fig. 4**) is performed on samples dehydrated with an intermediate fluid such as acetone miscible with both water and the transitional fluid.

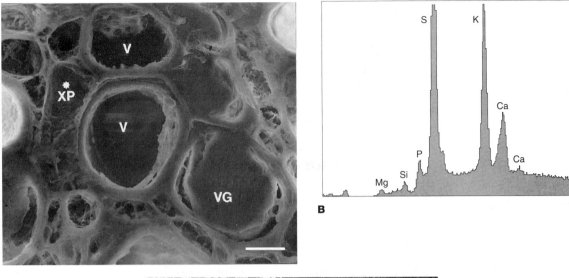

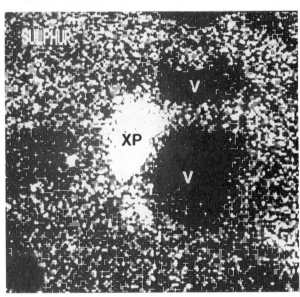

Figure 16 Freeze-drying and X-ray analysis of *Theobroma cacao*. (**A**) A section through a stem of a resistant genotype of *Theobroma cacao* inoculated with *Verticillium dahliae* and prepared by cryofixation and freeze-drying. X-ray microanalysis reveals a high accumulation of sulphur only in the cells and structures in potential contact with the vascular pathogen. XP, xylem parenchyma cell; V, xylem vessel; VG, vascular gel; scale bar 10 μm. (**B**) X-ray analysis taken from the region marked with an asterisk in (A): the spectrum shows a sulphur peak greater than potassium. (**C**) X-ray map of sulphur taken from equivalent area in (A), showing a parenchyma cell with a high accumulation of sulphur. Reprinted with permission from *Nature*: Cooper et al (1996) *Nature* 379: 159–162. Copyright (1996) Macmillan Magazines Ltd.

To minimize shrinkage and distortion samples should be kept as small as possible. Complete dehydration with the intermediate fluid must be carried out, and a subsequent slow temperature rise during CPD is necessary to achieve the best results.

Samples may be prepared in a variety of ways prior to CPD, the standard preparation method being that of chemical fixation and subsequent dehydration (see Fig. 2). However, they may also be cryofixed and freeze-substituted by replacing the ice with a solvent-fixative. Although freeze-substitution is mostly used for TEM preparation of samples, the superior results of this technique (compared with the standard

preparation) allows dynamic cell processes to be studied. Excellent results have been obtained by treating fungal material in this way. Yeast fimbriae have been preserved by freeze-substitution, and naturally dehydrated samples such as fungal spores may be studied without the use of aqueous solutions. Correlative TEM studies may also be undertaken by embedding and sectioning a sample prepared by CPD after SEM examination.

Conductive Coating

Biological material is a poor electrical conductor, and primary electrons induce a build-up of charge in the

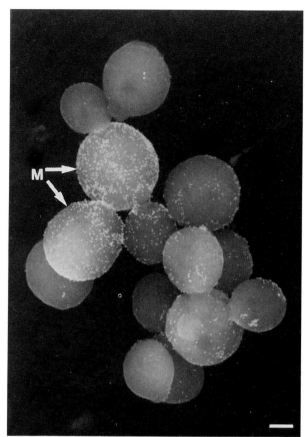

Fig. 17 Gold labelling of *Saccharomyces cerevisiae* (mnn9 mutant). Localization of surface-exposed chitin using wheat-germ agglutinin-gold (71 nm diameter). Mother cell surfaces (M) are labelled. Other cells are almost free of label. Bud and birth scars are unmarked. The gold is successfully imaged on an uncoated sample with secondary electrons (at 30 kV). Scale bar 1 μm. From Horisberger and Clerc (1987) *Eur J Cell Biol* 45: 62–71. Reproduced with permission from the *European Journal of Cell Biology*.

sample which does not dissipate from the interaction site. This results in distortion, overbrightness and poor focus of the SEM image. It is therefore generally necessary to sputter coat (or evaporate) a thin (about 10 nm) conductive layer onto the surface of a sample to remove charge and improve the quality of the SEM image. The coating is the main source of secondary electrons, and provides the sample with some protection against heat generated on exposure to the electron beam. Gold is the most commonly used coating as its efficiency of secondary electron production is high. However, the gold is not totally structureless and in high-resolution FEGSEM some fine detail from the coating itself may obscure features of interest. In such studies sputtering materials producing less structure such as chromium or tungsten can be used to replace gold, which improves matters to some extent. For X-ray microanalysis carbon is often used as the conductive coating because carbon

does not produce X-ray emissions which interfere with X-ray lines from higher atomic number elements present in the sample (see Fig. 16).

Low-temperature SEM

The standard preparation procedure of chemical fixation and dehydration/drying, which may produce shrinkage and distortion of the sample, is now no longer the preferred procedure for most SEM studies involving fungi and labile materials (biofilms, mucilage and foodstuffs). The most suitable pathway for examination of these samples is by the relatively new and very versatile technique of low-temperature SEM (LTSEM). This technique is designed for studying biological samples in a fully hydrated and lifelike state, at temperatures between –100°C and –175°C. Preparation of a biological sample is as follows:

1. Cryofixation in liquid nitrogen slush at –210°C.
2. Transfer to the SEM cold-stage for sublimation (at about –80°C) of contaminating surface frost (formed from condensation of water during the transfer from liquid nitrogen to the sample freezing apparatus).
3. Withdrawal into the cryopreparation chamber where optional fracturing of the sample and gold coating is carried out.
4. Transfer to the SEM cold-stage for examination at a temperature between –160°C and –180°C. **Figure 5** illustrates the LTSEM apparatus.

Freeze-fracture of samples reveals internal cells with a very smooth fracture face, and sublimation of surface ice is required to expose structural detail. When sublimation is carried out the sample is termed 'partially freeze-dried' although beneath the surface it is frozen-hydrated.

Examples of SEM Techniques

Figures 6 to **16** illustrate the wide variety of techniques available for the SEM examination of bacteria and fungi. Figures 6–9 illustrate the standard preparation method of chemical fixation, dehydration in a solvent and critical point drying. Figures 11–13 show samples prepared by cryofixation, partial freeze-drying and LTSEM.

Conclusion

The advantages of using SEM in the study of bacteria and fungi are many and are highlighted in the illustrations. The ability to freeze a sample and view it in the frozen state is of particular benefit in the study of these organisms. Extracellular matrices (ECMs) associated with fungal material and bacteria are often completely extracted with solvents (see Fig. 15), and the shrinkage and distortion associated with drying techniques are overcome by examining samples in their naturally hydrated state (see Figs 14 and 15). Low-temperature SEM may be essential in studies of fungus–plant interactions and bacterial spread. The ECMs at fungal adhesion and penetration sites are of prime importance, and bacterial biofilms are significant in the spread, growth and colonization of bacteria on plants and foodstuffs. Low-temperature techniques are also advantageous for the examination of a whole infection process since tissue may be harvested and cryofixed at specific time intervals and stored in liquid nitrogen for subsequent LTSEM study. If necessary samples may be subjected to an immunocytochemical or cytochemical technique prior to cryofixation and examined in the SEM using secondary or backscattered electrons or X-rays. Optimum cryofixation enables a sample examined under LTSEM to be freeze-substituted, resin embedded and sectioned for correlative TEM. The use of LTSEM coupled with FEGSEM offers high resolution and magnifications that overlap those of TEM, allowing surfaces to be studied without the necessity of preparing TEM replicas. In some isolated instances, uncoated biological samples may be examined by LTSEM and FEGSEM.

A less critical approach to SEM sample preparation, compared with TEM, has usually been the case because of the lower resolution of the former. High-resolution SEM, however, will require specialized and better-quality methods of specimen preparation, which will only evolve from a more detailed investigation into the mechanisms of sample preparation and further adaptation of TEM techniques.

Scanning electron microscopes have improved dramatically since the first instrument became commercially available in the 1960s. Field emission guns, backscattered electron detectors, energy-dispersive X-ray detectors, cryostages and environmental instruments are some of the major developments since then. Although many of these are pertinent only to specialized investigations, advances in computing have affected all aspects of SEM. Modern computer control with Windows simulation makes the instrument user-friendly, and microprocessor control of the functions permits easy column alignment and optimum imaging conditions to be achieved even by a novice operator. Digital imaging and storage together with the ability to transfer images across networks also facilitates off-line processing such as image analysis. However, the SEM is one technique among many, and to solve a problem it must always be remembered that other related techniques such as light, confocal, transmission electron and scanning probe microscopy may have a significant role to play.

See also: **Bacteria**: Classification of the Bacteria – Traditional. **Biofilms**. **Candida**: Introduction. **Escherichia coli**: *Escherichia Coli*; Detection of Enterotoxins of *E. coli*. **Fungi**: Classification of the Zygomycetes; Classification of the Eukaryotic Ascomycetes (Ascomycotina); Classification of the Hemi-Ascomycetes; Classification of the Deuteromycetes. **Lactobacillus**: *Lactobacillus bulgaricus*. **Microflora of the Intestine**: Detection and enumeration of probiotic cultures. **Microscopy**: Light Microscopy; Confocal Laser Scanning Microscopy; Transmission Electron Microscopy. **Saccharomyces**: *Saccharomyces cerevisiae*. **Salmonella**: Introduction. **Trichoderma**. **Viruses**: Detection. **Yeasts**: Production and Commercial Uses.

Further Reading

Aldrich HC and Todd WJ (eds) (1986) *Ultrastructural Techniques for Microorganisms*. New York: Plenum Press.

Crang RFE and Klomparens KL (eds) (1990) *Artifacts in Biological Electron Microscopy*. New York: Plenum Press.

Echlin P (1992) *Low-temperature Microscopy and Analysis*. New York: Plenum Press.

Goldstein JI, Newbury DE, Echlin P et al (1992) *Scanning Electron Microscopy and X-Ray Microanalysis*, 2nd edn. New York: Plenum Press.

Hall JL and Hawes C (1991) *Electron Microscopy of Plant Cells*. London: Academic Press.

Hayat MA (ed.) (1985) *Principles and Techniques of Scanning Electron Microscopy: Biological Applications*. New York: Van Nostrand Reinhold.

Howitt DG (ed.) (1991) *Microbeam Analysis*. San Francisco: San Francisco Press.

Kochler JK (ed.) (1986) *Advanced Techniques in Biological Electron Microscopy*. Berlin: Springer.

Mendgen K and Lesemann DE (eds) (1991) *Electron Microscopy of Plant Pathogens*. Berlin: Springer.

Muller R, Becker, P, Boyde A and Wolosewick JJ (eds) (1985) *The Science of Biological Specimen Preparation*. Chicago: SEM.

Robards AW and Wilson AJ (eds) (1993) *Procedures in Electron Microscopy*. Chichester: John Wiley.

Steinbrecht RA and Zierold K (eds) (1987) *Cryotechniques in Biological Electron Microscopy*. Berlin: Springer.

Verkleij AJ and Leunissen JLM (eds) (1989) *Immuno-gold Labelling in Cell Biology*. Boca Raton: CRC Press.

Transmission Electron Microscopy

U J Potter and **G Love**, Centre for Electron Optical Studies, University of Bath, Claverton Down, Bath, UK

Introduction

In microbiological research examination of small organisms cannot be accomplished using a conventional optical microscope because its resolution is limited, by diffraction effects, to approximately half the wavelength of the light illuminating the specimen. Consequently, structures below a few micrometres in size are difficult, or impossible, to image with clarity. High-energy electrons however have desirable properties as an illumination source in a high-resolution microscope since they possess a very short wavelength and may be readily focused using electric or magnetic fields. The first transmission electron microscope (TEM) was developed in the 1930s but only following improvements in electromagnetic lens design during the 1940s and 1950s was the true potential of the technique realized and high-resolution images of ultrastructure obtained.

The TEM

The transmission electron microscope (**Figs 1 and 2**) consists of:

- an electron gun, to generate electrons of the required energy
- a series of condenser lenses, to focus the electron beam onto the specimen area of interest
- a goniometer stage, to manipulate the specimen under the electron beam
- a further series of lenses, to produce a magnified image of the specimen
- a phosphor screen, on to which an image of the specimen is projected.

In the electron gun a thin filament, usually of tungsten wire (the cathode), is heated to several thousand degrees Celsius causing electrons to be emitted from its tip. The electrons are accelerated towards the anode, by a voltage applied between the anode and cathode (typically 60 000–100 000 volts), to produce a short-wavelength (about 0.005 nm) electron beam. This passes through the central hole of the anode and the goniometer stage is then used to position the specimen in its path. Between anode and specimen the magnetic fields, produced by the electrical current passing through the coils of the condenser lenses, are adjusted to vary the beam diameter and cause it to illuminate the required area on the specimen. Thus, if a high-magnification image is required the beam diameter will be small (a few hundreds of nanometres in diameter) whereas at lower magnifications the beam diameter may be some micrometres in size. Electrons that pass through the specimen without scattering through angles larger than 1–2° enter the first magnifying lens (the objective) placed immediately beneath the specimen. These electrons are then used to form a magnified (about ×50) image of the object. Intermediate lenses magnify this image further and the final lens projects the image onto a phosphor-coated screen. Electron impact causes light emission from the phosphor and enables an image of the specimen to be viewed by the operator through a leaded glass screen. When a permanent recording is required the phosphor screen may be raised to allow the electrons to fall upon a special photographic film or onto a charge-coupled detector (CCD) camera.

Imaging the Specimen

Because of lens aberrations the resolution of the transmission electron microscope does not approach the theoretical limit of about half the electron wavelength, but 0.3 nm is achievable on certain specimens and this permits well-focused images to be recorded at magnifications of up to a million. However, there are some drawbacks with electrons which place particular constraints upon the specimen:

1. The specimen must be stable in a vacuum because the whole of the TEM column, including the spe-

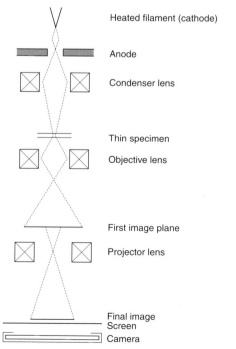

Figure 1 Simplified schematic diagram of a transmission electron microscope. Dashed lines show approximate path of electron beam. Intermediate lenses are omitted for clarity.

Heated filament (cathode)

Anode

Condenser lens

Thin specimen

Objective lens

First image plane

Projector lens

Final image
Screen
Camera

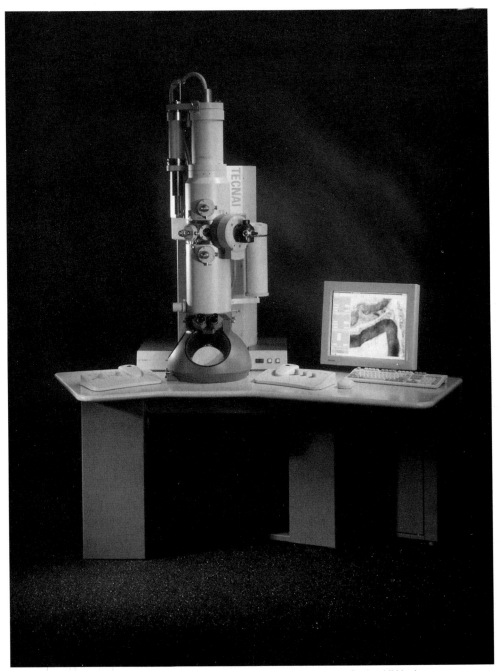

Figure 2 Transmission electron microscope TECNAI 12, courtesy of Philips Electron Optics UK Ltd.

cimen chamber, is evacuated to allow the free passage of electrons.

2. The specimen must be very thin to ensure that sufficient electrons pass through it to form an image (thicker specimens scatter electrons more and therefore many will fall outside the acceptance angle of the objective lens).

3. In order to create contrast in the image different parts of the specimen must scatter electrons more than others. If it is uniformly thin and non-crystalline any contrast arising will be from density or atomic number differences (heavier atoms scatter electrons more than lighter ones). In biological material the composition and density of different features is often very similar and it is necessary to improve contrast by selectively staining the specimen using heavy metal salts or by gold labelling, for example. Alternatively a specialized technique known as scanning transmission electron microscopy may be employed to provide contrast enhancement without the need for staining. Here, instead of using a static electron beam as in con-

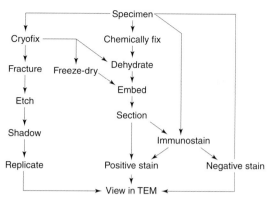

Figure 3 Steps involved in the preparation of samples for transmission electron microscopy.

ventional TEM, a finely focused beam scans an area of the specimen sequentially. An electron detector placed beneath the specimen records the changes in signal level as the beam passes through different parts of the specimen and electronic amplification of the signal is used to optimize contrast.

4. The specimen must be able to withstand the damaging effects produced by the high current density of the electron beam. If damage is observed, as in high-resolution studies of viruses for example, it may be eliminated or minimized by (a) sample freezing and subsequent examination in a cryogenic specimen stage, or (b) using a low electron beam current. This can produce a dim image, and observation of the sample by means of a camera designed for low light levels may prove necessary. Alternatively, a low-dose technique can be used whereby focusing is performed on one area of the sample and the beam is displaced electronically so that the region of interest is only irradiated for the time necessary to acquire the micrograph (see Figure 12).

Preparation of Microbiological Samples

The type of sample suitable for investigation with a TEM is determined by the microscope itself, as outlined above. Microorganisms are not generally dry, thin enough or stable under the electron beam and therefore must undergo a series of preparatory steps to render them suitable for examination. At the same time, the preparation must aim to preserve biological material as true to life as possible. The complexity of the preparative procedure for TEM will depend on the nature of the sample and the type of information required from the investigation. An outline of the steps, essential in following a particular pathway through preparation to specific sample information, is shown in **Fig. 3**.

Organisms such as bacteria, fungi and their spores

are difficult to prepare for TEM. When following the standard preparative pathway (**Fig. 4**), problems arise owing to insufficient fixative and resin penetration of cells, caused by the dense, impermeable walls of these organisms. However, modifications to preparative methods and recent developments in preparation techniques help to overcome these difficulties, as described below.

Fixation

The aim of the fixation steps is to stabilize and immobilize cell components (glycoproteins, nucleoproteins and phospholipids), protecting them from further chemical treatment while preserving the whole cell structure in as lifelike a state as possible. The quality of cellular ultrastructure revealed in the electron image will depend on fixation conditions (osmolarity, pH, temperature, duration) and will differ with the preparative pathway chosen.

> **WARNING**
> Chemicals (fixatives, buffers, stains, drying agents, solvents and cryogens) used in sample preparation for electron microscopy are mostly toxic and hazards to health. No work should be attempted without reference to an appropriate safety manual and safety data sheets for individual chemicals.

Primary fixation with glutaraldehyde (a dialdehyde) forms protein cross-links, while osmium tetroxide, used as the secondary fixative, stabilizes unsaturated lipids. Modifications at the fixation stage have improved the preservation of bacteria and fungi. For example, bacteria have been successfully fixed by microwaves while immersed in the primary fixative, thereby reducing fixation time and resulting in improved preservation. Monoaldehydes (para-formaldehyde, acrolein) added to the primary fixative penetrate biological material rapidly, improving the preservation of microorganisms and fungi. The choice of buffer, in which the fixative is dissolved, will depend on the investigation and the ultrastructural features of importance, for instance, the use of piperazine-N,N'-bis[z-ethanesulphonic acid (PIPES) buffer reveals extra detail in cell walls. The inclusion of ruthenium red or ruthenium hexamine trichloride to the primary fixative allows carbohydrates and proteoglycans to be retained, resulting in improved preservation of bacterial biofilms and fungal mucilage (see Fig. 14).

In the last two decades investigations with the TEM have moved rapidly from descriptive morphological analyses to studies of dynamic cellular processes, requiring the identification, localization and quantification of macromolecules. The techniques of

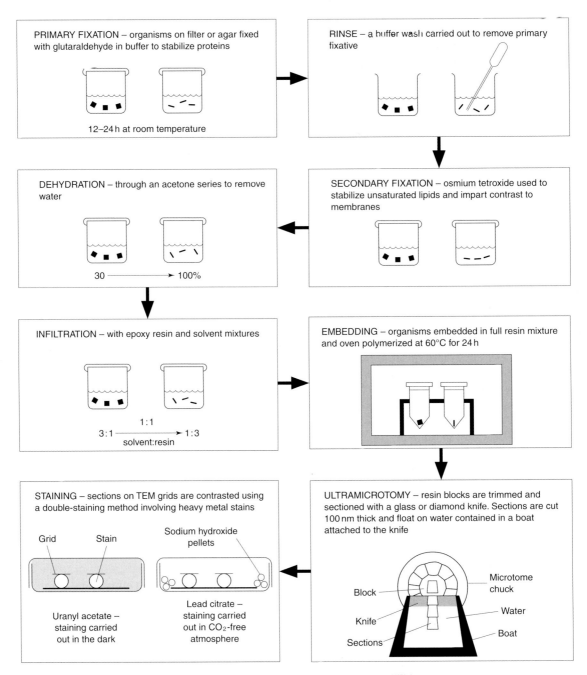

Figure 4   Steps involved in the standard method of preparing biological material for TEM.

enzyme and lectin cytochemistry, immuno-cytochemistry and *in situ* hybridization are needed to pinpoint specific macromolecules in cells and organisms. These specialized techniques often require a very different preparative pathway to that outlined in Figure 4. A procedure involving light aldehyde fixation only, and low-temperature embedding is often necessary to stabilize cell components, while at the same time preserving their particular activity within a cell, which may be destroyed by high aldehyde concentrations, osmium tetroxide and epoxy resins.

Using specialized techniques may necessitate a compromise at the fixation stage between good ultrastructural preservation and retention of cellular activity.

An alternative to chemical fixation, proving very successful with bacteria and fungi, is the physical method of cryofixation. Rapid extraction of heat from the sample, by freezing in a cryogen, preserves cells and arrests cellular processes and activities almost instantaneously. Because of its speed – 4 to 5 orders of magnitude faster than chemical fixation – cryo-

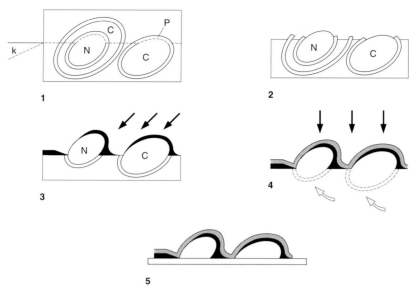

Figure 5 The freeze-fracture technique: the steps involved in the preparation of a replica of the sample surface for TEM. 1. A frozen drop of yeast cells is fractured using a cold knife (K). The fracture plane (dashed line) passes through the double membranes of the nucleus (N) and the plasma membrane (P). C is the cytoplasm. 2. The fractured sample is etched by the sublimation of a small amount of ice from the surface to reveal membrane structure. 3. A heavy metal such as platinum is evaporated onto the sample surface from the direction of the arrows, and 'shadows' the surface structure. 4. A thin film of carbon is evaporated onto the sample surface from the direction of the solid arrows to give strength to the replica. The replica is floated onto a chemical which dissolves away biological material (open arrows). 5. The final well-washed replica is picked up on a grid (a fine metal mesh) and examined in the TEM.

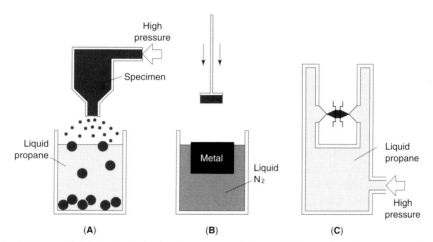

Figure 6 Freezing techniques suitable for bacteria, fungi, spores and viruses. (**A**) Spray freezing: a suspension of bacteria, spores or viruses is sprayed as microdroplets of 10–50 μm into liquid propane. The small size of the droplets ensures extremely quick cooling. (**B**) Slam freezing: bacteria or fungi grown on agar are frozen by slamming a portion of the agar plus organisms against a very cold copper block or metal mirror. Heat is transferred very quickly from the sample to the cold metal. (**C**) Jet freezing: small specimens, such as viruses or macromolecules in suspension, are sandwiched between two thin supporting sheets. The sandwich is kept static and the cryogen is squirted over it. Adapted with permission from Robards (1991) Rapid-freezing methods and their application. In: Hall JL and Hawes C (eds) *Electron Microscopy of Plant Cells*. London: Academic Press.

fixation is suited to studies of cell processes, particularly those involving membranes. Also, internal membrane surfaces and transmembrane particles (especially in yeast) have been examined using cryofixation and freeze-fracture followed by the production of a replica of the fractured sample surface (**Fig. 5**). Cryofixation is an alternative technique in

situations where fixation chemicals compromise immunoreactivity, as in the localization of a protein.

Ideally, cryofixation should be as rapid as possible, at a temperature as low as practically achievable to produce a layer of amorphous ice (vitrification), and to avoid the formation of ice crystals which damage ultrastructure. However, complete vitrification is dif-

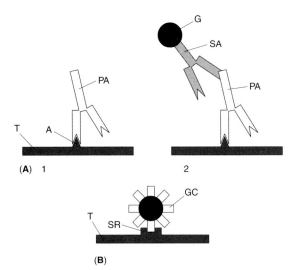

Figure 7 Protein and carbohydrate labelling in thin sections using antibodies, lectins and colloidal gold particles. (**A**) The doubling immunolabelling procedure. 1, a protein in the tissue section (T) termed the antigen (A) is labelled with a primary antibody (PA) originally raised, for example, in a rabbit. 2, a secondary anti-rabbit antibody (SA) is complexed with an electron-dense gold particle (G) that makes visible the site of the protein of interest in the TEM. (**B**) One-step lectin labelling. The sugar residue (SR) in the tissue section (T) is labelled directly with the lectin-complexed gold particle (GC).

ficult to achieve and is not always necessary. Provided any microcrystallites of ice produced, and the segregation artefacts associated with them, are very small and below the TEM resolution required for the investigation, then no deleterious effects will be observed. High-resolution TEM studies of viruses will require complete vitrification but, fortunately, the small size of viruses permits rapid cooling and the attainment of an amorphous ice layer. Three methods of freezing small samples are outlined in **Figure 6** (see Fig. 12 for an image of cryofixed rotavirus). Larger specimens do present a problem, however, because many freezing techniques will vitrify to less than 20 µm in depth. New instruments and techniques for cryofixation are now being developed with the aim of achieving vitrification of ice to a greater depth in such samples.

Dehydration and Embedding

Dehydration, to remove water from cells, is required before infiltration of a hydrophobic epoxy resin into the specimen. Standard epoxy mixtures produce excellent clarity of ultrastructure in the TEM, but their high viscosity may cause infiltration problems owing to the dense nature of the walls of bacteria and spores. Modifications to standard infiltration procedures at this step include a vacuum treatment, to draw resin into the sample, or the use of low-viscosity

resins. The final step of infiltration is polymerization, to produce a resin block suitable for sectioning.

Freeze-substitution, a dehydration step following cryofixation, involves the replacement of ice in the sample with a solvent-fixative mixture, and infiltration with a resin, at low temperature (–75°C to –100°C). Acrylic resins for use at low temperatures retain protein antigenicity to a greater extent than conventional epoxy resins, and may be used in immunocytochemical work. However, bacteria and fungi exhibit resin infiltration difficulties at low temperatures. Consequently, the best procedure is to infiltrate these at 0°C or at room temperature and polymerize by ultraviolet light or heat, depending on the resin in use and the type of investigation. Studies of bacteria and fungi have shown that the freeze-substitution technique is superior to conventional preparation methods for preserving internal morphology and immunoreactivity in these organisms (see Fig. 18).

Freeze-drying, an alternative to freeze-substitution, involves the sublimation of ice in the sample into water vapour using a cooled stage under a vacuum; resin impregnation of the sample follows. Samples are not exposed to solvents (and chemical fixation steps may be omitted), therefore protein antigenicity is preserved and elemental composition is unaltered. Molecular distillation drying (MDD) is a new technique developed from freeze-drying. It involves optimizing the freeze-drying process by using very low temperatures (–177°C) and ultrahigh vacuum pumping. With all the advantages of conventional freeze-drying plus the ability to preserve ultrastructure, MDD is now beginning to produce excellent results, and may supersede other methods.

Sectioning

An ultramicrotome allows the automatic slicing of sections 20–200 nm thick from specimens embedded in blocks of cured resin (see Fig. 4). Glass is routinely used as the ultramicrotome knife; however, bacteria and fungi benefit from sectioning with a diamond knife to avoid cutting artefacts.

Labelling

Labelling of ultrastructure is a step usually carried out on sections, and is designed to give electron opacity, and contrast, to particular cell components in the TEM. Specialized techniques allow specific macromolecules to be targeted and visualized. High-resolution TEM requires that any label used to mark macromolecules must be electron-dense, and small enough not to obscure the ultrastructure beneath it.

The ease of producing colloidal gold particles has allowed the development of specialized techniques for

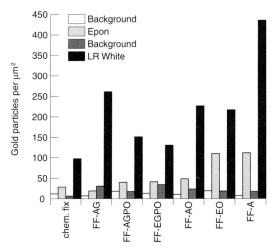

Figure 8 Differences in immunolabelling intensity with varying conditions of fixation, freeze-substitution and embedding. The enzyme luciferase found in the bioluminescent bacteria *Vibrio harveyi* was detected by immunogold labelling using an anti-luciferase antibody. The number of gold particles per μm^2 measures the labelling intensity in the cytoplasmic area. The intensity depends on the type of fixation (chem. fix., chemical fixation versus FF, fast freeze-fixation), the embedding resin (LR white versus Epon), the solvent used during cryosubstitution (A, acetone versus E, ethanol) and the fixatives added to the solvent (G, glutaraldehyde; O, osmium tetroxide; P, paraformaldehyde). The best compromise between preservation of ultrastructure and antigenicity was, in this case, fast freeze-fixation and freeze-substitution in acetone alone and embedding in Epon (see Fig. 18). From Nicolas MT and Bassot JM (1993) Immunocytochemical techniques for TEM and SEM. In: Robards AW and Wilson AJ (eds) *Procedures in Electron Microscopy*. Copyright John Wiley & Sons Ltd. Reproduced with permission.

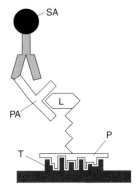

Figure 9 Labelling of specific nucleic acid sequences in tissue sections. The nucleic acid sequence of interest (T) in the tissue section is hybridized to a complementary sequence which forms the probe (P). A label (L), biotin or digoxygenin, attached to the probe is visualized using the double immunolabelling procedure. A primary antibody (PA) is raised against the label, and a gold-complexed secondary antibody (SA) makes visible the site of the nucleic acid sequence of interest in the TEM.

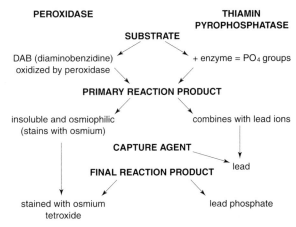

Figure 10 Examples of enzyme localization methods.

labelling macromolecules. Colloidal gold particles, made from the controlled reduction of tetrachloroauric acid by a reducing agent, are electron-opaque and easily distinguished in the TEM. The size of gold particles can be varied from 1 nm to 40 nm, and they may be complexed with macromolecules such as antibodies and lectins.

Protein Labelling Immunocytochemistry, the most widely used technique for labelling proteins in the TEM, makes use of the very specific binding that takes place in animals between an antibody and a foreign protein, termed the antigen. The ability to complex gold particles with antibodies, the high affinity of an antibody for a particular antigen and the strength of the binding between them, allows specific and reliable localization of proteins within cells. Post-embedding immunolabelling (the most common method) is carried out on sections of material (**Fig. 7**), and the intensity of the immunolabelling seen in the TEM, will vary with the sample preparation technique used (**Fig. 8**) (see also Figs 17 and 18).

DNA/RNA Labelling Another specialized technique used with the TEM is *in situ* hybridization, performed to locate and label specific DNA or RNA sequences in sectioned material. This technique is founded on the base-pairing (termed 'hybridization') that takes place between two complementary nucleic acids. A probe, in the form of a labelled polynucleotide of known base sequence, is hybridized to a target of complementary polynucleotides in the section of biological material. To visualize the DNA or RNA sequence of interest, a hapten label (biotin or digoxygenin) is attached to the probe and exposed by immunocytochemistry first to an antihapten antibody and then to a colloidal gold conjugated antibody (**Fig. 9**). Exposure of a section to two differently labelled nucleic acid probes, followed by immunocytochemistry using gold labels of varying size, allows

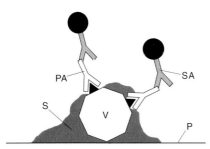

Figure 11 The combined technique of immunocytochemistry and negative staining carried out *in situ* on a plastic-coated TEM grid. A virus (V) is dried onto the plastic film (P), and immunolabelled with primary antibody (PA), raised against a viral coat protein, and a secondary antibody–gold complex (SA). The preparation is negatively stained with a heavy metal stain that surrounds the virus (S). This diagnostic method may be used to label a particular virus in a preparation containing different viruses.

the double labelling of different nucleic acid sequences on the same section.

The combined technique of immunocytochemistry and *in situ* hybridization is an exciting tool, allowing localization not only of a protein but also of the genes and transcripts coding for that protein. Virus replication studies have used antibodies to detect viral proteins, and now the spread of viral genetic material in a cell may also be detected by the localization of specific viral RNA or DNA, using the *in situ* hybridization technique.

Carbohydrate Labelling   Sugar residues of complex carbohydrates can be visualized on sections in the TEM, using carbohydrate-binding glycoproteins called lectins that are bound to colloidal gold particles. Lectins differ in their sugar-binding specificity, and most commonly are complementary to between two and six sugar residues. Fungal cells, for example, have been labelled with wheatgerm agglutinin-gold, specific for *N*-acetylglucosamine residues found in chitin and other related compounds (see Fig. 7B).

Enzyme Labelling   Enzyme cytochemistry (EC) is a technique used for the localization of specific enzyme activity in cells. It is based on the strong affinity that an enzyme possesses for its substrate, and localization is performed by incubation of a lightly fixed sample with an artificial substrate. Reaction between substrate and enzyme yields either an insoluble electron-dense reaction product, or a soluble electron-translucent one that requires a further 'capture' step to produce a final electron-dense precipitate visible in the TEM. Standard sample preparation (see Fig. 4) is performed after enzyme localization. **Figure 10** outlines the steps in the localization of peroxidase and thiamin pyrophosphatase.

Enzyme cytochemistry has been eclipsed by immu-

nocytochemistry in recent years. Actual enzyme activity is localized in cells by EC, while immunodetection of enzymes may label inactive proteins as well. However, for some enzymes cytochemical methods may be very involved, or not exist, and immunocytochemistry, which is able to localize almost any protein, provides an alternative technique (see Fig. 19).

Positive Staining

Staining of thin sections to improve the overall contrast of biological material in the TEM uses heavy metal salts that scatter electrons. Positive staining uses uranyl acetate to contrast nucleic acids and proteins, followed by lead citrate to reveal a wide range of cell and tissue components, particularly lipid-containing structures fixed by osmium tetroxide. This double staining method is used purely for increasing the contrast in sections of biological material, and is generally not specific. However, there are many staining methods, often involving chemical reactions, which may be carried out on sections to produce a more specific staining of a cell component. Polysaccharides are detected using a chemical reaction involving periodic acid and thiocarbohydrazide with silver proteinate staining (see Fig. 15). Phosphotungstic acid, used at a low pH, will selectively stain the plasma membrane. At the fixation step, techniques exist to enhance the uptake of osmium by cell components, for example, zinc iodide with osmium tetroxide, used as a secondary fixative, will impregnate and increase the contrast of double membranes in cells (see Fig. 16).

Negative Staining

Particles such as viruses, macromolecules and bacteria may be visualized whole in the TEM. The particles, on a plastic-coated TEM grid, are stained with a solution of heavy metal, forming a surrounding layer which scatters electrons. The background appears dark with the particle outlined against it (see Figs 12, 20). High-resolution morphology and identification may be carried out on viruses and bacterial pili using combined techniques of negative staining and immunocytochemistry (**Fig. 11**). (See also Figs 12(C), 21.)

Examples of TEM Techniques

Figures 12 to 21 illustrate the wide variety of applications involved in the study of bacteria, fungi and viruses with the TEM.

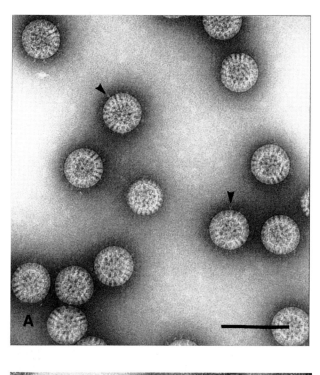

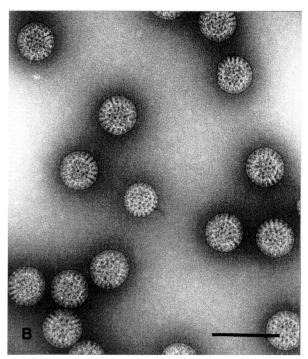

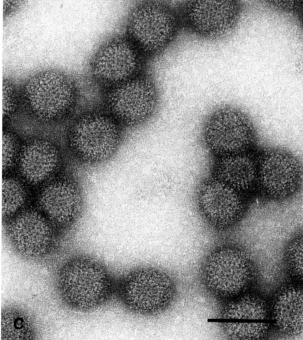

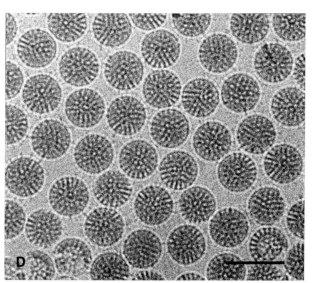

Figure 12 Techniques for examination of viruses (micrographs courtesy of John Berriman). (**A**) Negatively stained rotavirus particles are imaged using a low-dose method (between 10 and 20 electrons per Å^2). Surface projections or spikes (arrows) are retained. Scale bar 100 nm. (**B**) The same particles as in (A) imaged after the doses used in conventional microscopy, at least an order of magnitude higher. Surface spikes are lost owing to high-dose damage. Scale bar 100 nm. (**C**) Rotavirus particles (outer layer removed) treated with an antibody (αSG1) to an inner exposed protein (VP6). When the particles are immobilized on the carbon surface of the grid and treated in this way the technique is a form of the ELISA method for electron microscopy. Scale bar 100 nm. (**D**) Cryofixed rotavirus particles viewed in their frozen state by cryo-TEM. Improved preservation is apparent compared with the negatively stained particles in (A) and (B), which suffer flattening distortion caused by drying from the stain solution. The spherical shape of the cryofixed virus allows three-dimensional reconstruction using image analysis. The spikes can be seen projecting from the surface of each particle. Scale bar 100 nm.

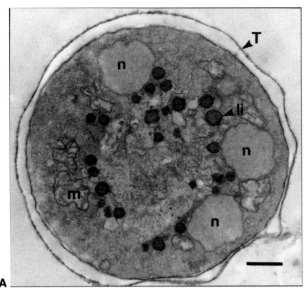
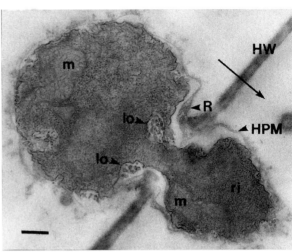

Figure 13 Modifications to the standard sample preparation method were necessary to image stages in the life cycle of *Spongospora subterranea* f. sp. *nasturtii*, the causal agent of crook root disease in watercress. Simultaneous glutaraldehyde and osmium tetroxide primary fixation was carried out with secondary osmium fixation, and low viscosity resin embedding. From Claxton et al (1996) *Mycol. Res.* 100 (12): 1431–1439, reproduced with permission of the British Mycological Society. (**A**) A myxamoeba surrounded by the tonoplast (T) of the watercress root cell, nuclei (n), lipid drops (li) and mitochondrion (m). Scale bar 1 μm. (**B**) Cell-to-cell movement (in direction of arrow) of a myxamoeba through the host cell wall (HW) which has formed a rim (R) adjacent to the site of penetration. Host plasma membrane (HPM), lomasomes (lo), free ribosomes (ri) and mitochondria (m). Scale bar 0.2 μm.

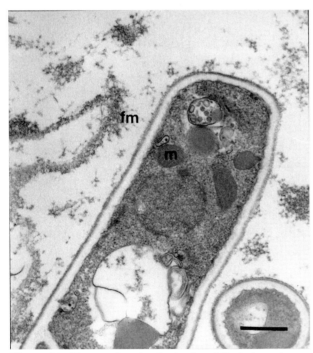

Figure 14 The growing tip of a hypha of *Septoria nodorum*. Fungal mucilage (fm) is preserved by the addition of ruthenium hexamine trichloride to the primary fixative. Mitochondrion (m). Scale bar 0.5 μm.

Conclusion

Viewing, recording and collecting data from a sample using the TEM requires an understanding of the different modes of TEM operation, such as bright-field

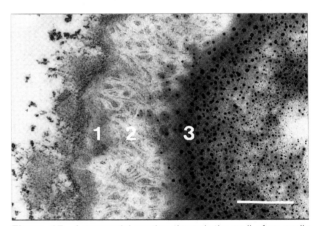

Figure 15 A tangential section through the wall of an undischarged ascospore of *Sordaria humana*. The multilayered secondary wall (1, 2, 3) was treated with periodic acid–thiocarbohydrazide–silver proteinate to demonstrate polysaccharides (black dots in 3). Scale bar 0.5 μm. From Read N and Beckett A (1996) *Mycol. Res.* 100 (11): 1281–1314, reproduced with permission of the British Mycological Society.

(normal mode), dark-field and low-dose microscopy, in order to extract as much information as possible from a particular sample. In addition to the variety of operational modes, a TEM may be fitted with attachments to perform specialist functions, such as X-ray analysis (identification of chemical elements in a sample) or cryo-TEM (viewing a frozen sample).

The advantages of using a TEM in studies of bacteria, fungi and viruses are highlighted by Figures 12–21. A minute organism, like a virus, may be identified

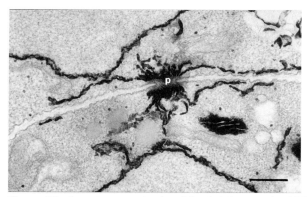

Figure 16 An ascogenous hypha of *Sordaria humana* in which double membranes have been selectively stained with zinc iodide–osmium tetroxide to show the continuity between the membranes of the pore structure (p), the endoplasmic reticulum and the nuclear envelope. Scale bar 0.5 μm. From Read N and Beckett A (1996) *Mycol. Res.* 100 (11): 1281–1314, reproduced with permission of the British Mycological Society.

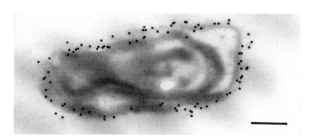

Figure 17 Section through a spore of *Bacillus stearothermophilus*. A light fixation in glutaraldehyde alone, and embedding in an acrylic resin, was performed. Immunolocalization of α-glucosidase shows good specific labelling of the spore coat with gold particles. However, spore ultrastructure and resin infiltration are poor. Scale bar 0.2 μm. From Albert H (1995) PhD Thesis, University of Bath, reproduced with permission.

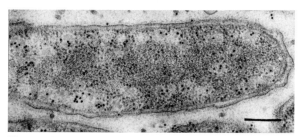

Figure 18 A thin section through the bioluminescent bacterium *Vibrio harveyi*. Preparation of the organism was by fast freeze-fixation, freeze-substitution in acetone alone and epoxy resin embedding. The labelling with antiluciferase antibodies, is concentrated over enigmatic patches in the cytoplasmic area. Ultrastructural preservation and enzyme labelling are excellent. Reproduced with permission, from Marie-Thérèse Nicolas, Jean-Marie Bassot, and Gisèle Nicolas: Immunogold Labelling of Luciferase in the Luminous Bacterium *Vibrio harveyi* After Fast-freeze Fixation and Different Freeze-substitution and Embedding Procedures. *Journal of Histochemistry and Cytochemistry* 37: 663–674, 1989.

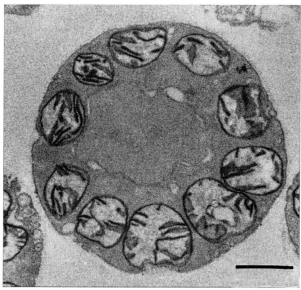

Figure 19 Cytochrome *c* has been localized by enzyme cytochemistry in yeast cells using diaminobenzidine as the enzyme substrate. The reaction product produces osmium black on exposure to osmium tetroxide, during preparation of the yeast. A strong positive reaction appears in the mitochondrial membranes. Scale bar 0.5 μm. From Keyhani E (1972) *Journal de Microscopie* 15: 343–352, reproduced with permission of the Société Française des Microscopies Paris.

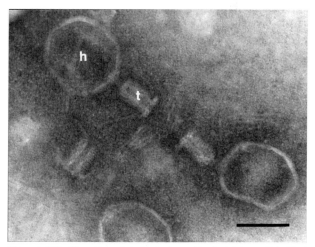

Figure 20 Negative stain technique. Bacteriophage T4 dried onto a carbon-coated TEM grid and stained *in situ*. This virus infects the *Escherichia coli* bacterium and injects DNA from the head (h) through the cylindrical tail (t) into the bacterium. Scale bar 50 nm.

and examined at high resolution and its three-dimensional structure determined. Specific cell components and structures such as proteins, enzymes, nucleic acids, polysaccharides, the plasma membrane, the Golgi apparatus and other double-membraned structures may be studied by selective staining, or labelled and traced during cell processes using specialized labelling techniques. The number of preparation methods suitable for the TEM study of micro-

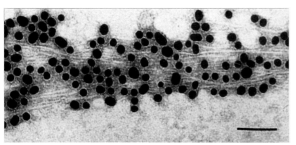

Figure 21 Immunonegative stain technique. *Bacteroides nodosus* pili dried onto a TEM coated grid and immunolabelled *in situ* with specific antibody–gold. Gold particles of 20 nm are used for low-magnification examination of bacterial populations to determine whether all the bacteria present on the grid are of one serotype. Scale bar 100 nm. From Beesley J (1985) *Proceedings of the RMS* 20: 187–196, reproduced with permission of the Royal Microscopical Society.

organisms and fungi is expanding with new freezing techniques and MDD is becoming more widely used.

Progress in TEM instrumentation is advancing every year, and the TEM is becoming easier to use, with computers and microprocessors allowing automatic alignment and reproducibility of particular settings. Digital images may now be produced for computer storage, image processing, transfer over a computer network or for hard copy output rather than using conventional photographic methods.

Different types of microscopy (light, confocal, scanning electron, transmission electron and scanning probe microscopy) also provide complementary information for many scientific applications. Therefore, a critical approach to biological investigations of microorganisms is necessary, possibly using more than one type of microscopy to achieve the desired result.

See also: **Bacillus**: Bacillus stearothermophilus. **Bacteria**: The Bacterial Cell. **Bacteriophage-based Techniques for Detection of Food-borne Pathogens**. **Enzyme Immunoassays**: Overview. **Fungi**: Classification of the Eukaryotic Ascomycetes (Ascamycotina); Classification of the Hemi-ascomycetes. **Microscopy**: Light Microscopy; Confocal Laser Scanning Microscopy; Scanning Electron Microscopy; Atomic Force Microscopy. **Saccharomyces**: Saccharomyces cerevisiae. **Vibrio**: Introduction, including *Vibrio vulnificus*, and *Vibrio parahaemolyticus*. **Viruses**: Introduction.

Further Reading

Beesley JE (1989) *Colloidal Gold: A New Perspective for Cytochemical Marking*. Royal Microscopical Society Microscopy Handbook 17. Oxford: Oxford University Press.

Crang RFE and Klomparens KL (eds) (1990) *Artifacts in Biological Electron Microscopy*. New York: Plenum Press.

Hall JL and Hawes C (eds) (1991) *Electron Microscopy of Plant Cells*. London: Academic Press

Harris N and Wilkinson D (1990) *In Situ Hybridization*. SEB Seminar Series. Cambridge: Cambridge University Press.

Hayat MA (ed.) (1974) *Electron Microscopy of Enzymes: Principles and Methods*. Vols 1–3. New York: Van Nostrand Reinhold.

Hayat MA (1989) *Principles and Techniques of Electron Microscopy – Biological Applications*, 3rd edn. Boca Raton: CRC Press.

Hayat MA (ed.) (1989) *Colloidal Gold: Principles, Methods and Applications*. New York: Academic Press.

Lewis PR and Knight DP (1977) Staining methods for sectioned material. In: AM Glauert (ed.) *Practical Methods in Electron Microscopy*. Vol. 5. Amsterdam: Elsevier.

Peachey LD and Williams DB (eds) (1990) *Electron Microscopy*. San Francisco: San Francisco Press.

Polak JM and Varndell IM (1984) *Immunolabelling for Electron Microscopy*. Amsterdam: Elsevier.

Read ND and Beckett B (1996) Centenary review: ascus and ascospore morphogenesis. *Mycol. Res.* 100 (11): 1281–1314.

Robards AW and Sleytr UB (1985) Low temperature methods in biological electron microscopy. In: Glauert AM (ed.) *Practical Methods in Electron Microscopy*. Vol. 10. Amsterdam: Elsevier.

Robards AW and Wilson AJ (eds) (1993) *Procedures In Electron Microscopy*. Chichester: John Wiley.

Steinbrecht RA and Zierold K (1987) *Cryotechniques in Biological Electron Microscopy*. Berlin: Springer.

Zierold K and Hagler HK (eds) *Electron Probe Microanalysis*. Berlin: Springer.

Atomic Force Microscopy

W Richard Bowen, **Nidal Hilal**, **Robert W Lovitt** and **Chris J Wright**, Department of Chemical and Biological Process Engineering, University of Wales Swansea, Singleton Park, Swansea, UK

Introduction

Atomic force microscopy (AFM) is a new technique that offers great potential to food microbiologists. Since its emergence from the physics laboratory in 1986 great excitement has followed this revolutionary microscopy technique, which holds the promise of imaging with subnanometre resolution living biological samples within their own aqueous environments. This technology, however, is still maturing. Initially AFM was used as a comparison method with established technologies such as scanning electron microscopy (SEM). However, AFM is now an essential imaging technique. Not only can it image non-conducting surfaces in liquid, but it can also probe spatial variations in surface properties, such as adhe-

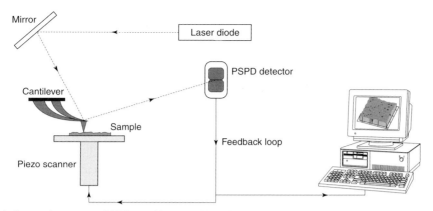

Figure 1 The atomic force microscope. PSPD, position-sensitive photodetector.

siveness and elasticity. Thus AFM has evolved into an essential surface probing technique.

Principles of Atomic Force Microscopy

The atomic force microscope (**Fig. 1**) is made up of the following components. A very small, sharp tip held at the free end of a cantilever systematically scans a surface of interest in order to generate a topographical image. The cantilever is 100–200 µm long, with a tip that is only a few micrometres long and about 10 nm in diameter at its apex. As the tip tracks the surface the forces between the tip and the surface cause the cantilever to bend. The deflection of the cantilever is measured by a device such as an optical lever and used to generate a map of surface topography.

The optical lever consists of a laser beam that is focused on the reflective gold-plated back of the cantilever and a position-sensitive photodetector (PSPD) which registers the position of the reflected beam. As the cantilever bends the PSPD measures the change in the position of the incident laser beam. The PSPD can measure displacements of the incident beam as small as 1 nm. The ratio of the path length between the cantilever and the detector to the length of the cantilever itself produces a mechanical amplification. Thus, the system can detect subnanometre vertical movements of the cantilever tip.

As the AFM tip and cantilever are rastered across a surface, several forces contribute to the system's deflection. **Figure 2** shows the dependence of the total interatomic force in air upon tip-to-sample separation distance. The AFM exploits two distinct regions of this curve. In contact mode the tip is held less than a nanometre from the surface, within the repulsive region of the interatomic force curve. The dominant force is Born repulsion. In the non-contact mode the tip is held several nanometres from the surface. The interatomic force between the tip and the surface is

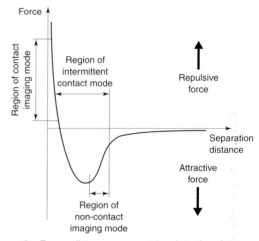

Figure 2 Force–distance curve showing the tip-to-sample separation of different AFM operating modes.

dominated by long-range attractive van der Waals interactions.

In the initial stages of imaging a new surface a systematic procedure should be adopted using the different imaging modes to optimize the image production. In addition, the choice of AFM instrument for a particular application will be governed by software considerations. The accompanying software of an instrument controls its operation and the access to the data set. Hence, the software defines the limits of the AFM experimentation.

Contact Mode

In contact mode the equilibrium between the spring force of the deflected cantilever and the incident force changes as the sample is systematically scanned (Fig. 2). Once this equilibrium shifts the AFM operates in either of two ways. In constant force mode, the total force between the tip and the sample is kept constant by means of a feedback loop. The piezo scanner moves up and down to maintain the equilibrium between the deflected cantilever spring force and the incident force as the topography changes under the cantilever.

The feedback signal is used to generate the image data set. In constant height mode, this equilibrium is not maintained, so that changes in the cantilever deflection are used directly to generate topographic images. Constant force mode is the preferred mode of operation because it allows the total force of the tip on the sample to be kept within controlled limits.

The force exerted on a sample can also be controlled by the choice of cantilever (typically about 10^{-8} N). A soft cantilever that is sensitive to changes in applied force resolves greater surface detail than a stiffer cantilever. A soft cantilever, however, is more likely to crash into the surface, possibly damaging the sample surface and cantilever tip and thus reducing the image quality. A stiffer cantilever will reduce the danger of tip crashing but the image resolution is reduced. A compromise must be reached.

Non-contact Mode

Non-contact mode is a vibrating cantilever technique that relies on the fact that an incident force serves to change the vibrational amplitude and resonant frequency of a vibrating cantilever. The cantilever in this case is held 5–10 nm away from the surface (Fig. 2). As the sample is rastered underneath the vibrating cantilever the non-contact mode AFM measures the change in the vibrational parameters of the cantilever. A feedback system keeps the monitored vibration constant by moving the sample up or down as the topography changes. The motion of the scanner is used to generate the data set.

As the name of this imaging mode suggests there is very little contact, if any, of the tip with the sample surface. The detection method procedure must be sensitive enough to measure the small change in the vibrational parameters of stiff cantilevers. Soft cantilevers are not used because they are too easily pulled into the sample surface. The total force between the tip and the sample is about 10^{-12} N. Samples are not contaminated or damaged by the act of imaging. In general, non-contact AFM is more effective than contact mode at imaging soft biological samples.

Intermittent Contact Mode

In intermittent contact or tapping mode a vibrating cantilever is held at a tip-to-sample distance close to the region of the force–distance curve exploited by contact mode. The lowest extreme of the cantilever's vibrational movement just touches or taps the surface. As the sample is rastered beneath the tip, the changes in the cantilever's vibrational parameters are monitored and the feedback parameters correction used to keep these parameters constant is processed to produce a topographical image.

The intermittent contact reduces the degree of friction or drag on a sample compared with imaging in contact mode. Also, the method allows penetration of covering layers such as water which may compromise the non-contact AFM operation. These factors combine to make intermittent contact extremely useful when imaging soft biological samples.

Imaging in Liquids – Double Layer Mode

In liquid, an electrical double layer is formed as ions are attracted to the sample surface charge. The thickness of this layer depends on the ionic strength of the solution. This phenomenon can be exploited by the AFM operator to image soft samples using double layer imaging mode. A scanning cantilever pressed onto the surface with a certain force will be held at a distance from the surface dependent on the thickness of the double layer and the magnitude of sample and tip charges. The closer the tip is held to the surface the greater the image resolution, but the greater the risk of sample damage. A compromise must be reached between image clarity and applied force when food microbiological samples are imaged under liquid.

Tip Geometry

An AFM image is a composite of the surface topography and the geometry of the scanning tip. When imaging at the nanometre scale the geometry of the tip becomes a critical parameter. There may be information missing from an image if the tip is unable to interact with the small structures. To reduce this problem tips of higher aspect ratio can be used. Electron deposition within an SEM can be used to generate very fine tips. If the geometry of the tip is known, by direct measurement or interaction with known sample geometry, then algorithms can be written to remove the tip shape contribution from the image data set.

Image Analysis

Image analysis is essential for the correct identification of landmarks and the differentiation of image or preparation artefacts. Most commercially available AFMs are accompanied by sophisticated image analysis software which generates surface statistics such as surface roughness, average height and maximum peak-to-valley distance allowing quantitative interpretation of the three-dimensional image data set.

Force–Distance Curves

The AFM can measure the forces of interactions between surfaces, which has obvious implications in any science that needs to study interfacial phenomena in order to understand and control a process. In the

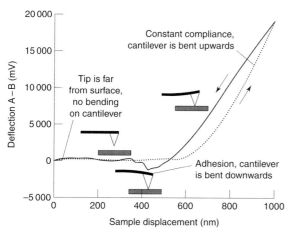

Figure 3 Annotated raw data plot of cantilever deflection versus piezo displacement measured between an AFM silica tip and silica surface in electrolyte solution.

past, the surface force apparatus has been used to study surface forces. The AFM has the advantage of allowing the imaging and identification of points of interest on a surface prior to the measurement of the forces of interaction. In addition, surface forces are measured over very small contact areas, minimizing contamination problems. Moreover, AFM can measure the forces experienced by small particles such as cells.

To generate a force–distance curve, the deflection is recorded as a function of tip-to-sample separation as the piezo scanner of the AFM raises the sample towards the tip. **Figure 3** shows the raw experimental data measured between an AFM tip and a silica surface in an electrolyte solution with schematic annotation of the tip and surface positioning.

Force–distance curves are characteristic of the system under study. They have features that reflect chemical and physical attributes of the surface that are interacting. For example, the retraction curve of Figure 3 has a distinctive jump back to zero force position. This feature is due to adhesion between tip and sample. As the scanner retracts, the tip and the sample adhere until a threshold is reached where the adhesive force is equalled by the spring forces of the bent cantilever, and contact is broken.

In order to convert the raw data to a force versus separation distance curve, it is necessary to know the spring constant of the cantilever and to define zeros of both force and separation distance. A number of different methods for the determination of the cantilever spring constant have been reported. The zero of force is defined when the cantilever is undeflected and the tip and the sample are far apart. Zero distance is chosen when the tip and the sample move in unison, the onset of the 'constant compliance' region (Fig. 3).

Colloid Probe Technique

To compare AFM force measurements with those made using the surface force apparatus and theoretical predictions, the geometry of the AFM tip must be known. This is often not the case. To enable comparison, the AFM tip is replaced by a small sphere to produce a colloid probe. Prior to experimentation the sphere may be sized using SEM or AFM. The colloid probe technique can be adapted by adsorbing molecules such as proteins to the sphere to produce a coated colloid probe. Similarly, living cells can be immobilized at the apex of a tipless cantilever to produce a cell probe (**Fig. 4**).

Atomic Force Microscopic Imaging in Food Microbiology

Atomic force microscopy has proved to be a useful extension to the imaging techniques available to the food microbiologist. In the past the microbiologist has been restricted by the resolution of the light microscope or by the sample preparation and vacuum requirements of the SEM. However, AFM offers the possibility of molecular level imaging of samples in a suitable aqueous environment. Such detailed examination of surfaces has and will give useful insights of structure–function relationships within food microbiology. The physical and chemical relationships that exist between surfaces and bacteria have been the focus of much scientific endeavour. The AFM is unique in that it allows the study of these relationships within relevant environments using living cells.

Surfaces in Food Microbiology

Figure 5 shows an AFM image and a line profile of a stainless steel surface with an SMS code Super Bright No. 7 finish, BS1449, a standard highly polished steel surface used in the construction of processing equipment. This is the most effective way of viewing flat uniform surfaces. The AFM imaging of food preparation and processing surfaces is relatively straightforward. Contact mode in air is the first choice of

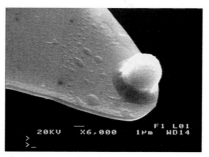

Figure 4 Scanning electron micrograph of cell probe – *Saccharomyces cerevisiae*.

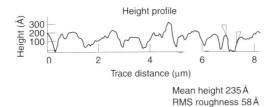

Height profile

Mean height 235 Å
RMS roughness 58 Å

Figure 5 AFM image and line profile of stainless steel (SMS code Super Bright No. 7, BS1449, Sillavan Metal Services, Wolverhampton, UK). The two arrows identify a surface scratch that is 404 nm wide. 1 Å = 10^{-1} nm. RMS = root mean square.

imaging mode for inorganic hard surfaces. If the surface roughness is greater than the piezo z movement capability a poor image will be generated. Aluminium or Teflon surfaces can be very rough. The image will then be restricted to a local scale, and features of interest may not be located. The optical microscope usually accompanying the AFM may well serve to position the cantilever in the correct area. Instruments are available that allow the imaging of a large surface area. Instruments that have a scanning tip, as opposed to a moving sample, are not limited in sample size. Such instruments have the advantage that they can be used to study surfaces *in situ*, for example within a food processing plant.

Figure 6 is an AFM image of a tomato skin, in air, in which the lamella of the cell walls can be clearly seen. There are small features upon the plant cell surface that may be bacterial cells. The waxy cuticle forms a layer covering the entire cell. The waxy cuticle

and its deposition patterns dominate the high-resolution images of a large range of vegetable and fruit surfaces, imaged both in electrolyte and tapping mode in air. The plant epidermis cells are large, the typical limit of the x–y scanning area is $100\,\mu\text{m}^2$. Features larger than this limit should be examined using light microscopy.

There are no published AFM images of meat surfaces. However, there are a large number of AFM studies of mammalian cells immobilized on surfaces. The surface relationships that exist between bacteria and animal tissue surfaces can be investigated using AFM.

Bacterial, Yeast and Animal Cells

The real power of the AFM lies in its ability to image soft samples at a nanometre level within solution. While a population of bacterial cells can be studied using a light microscope, AFM allows the real-time study of individual living microbiological surfaces within an aqueous environment. The technique can simultaneously image and probe some of the mechanical properties of living cells.

Fixed cells in general can withstand the high imaging forces in contact mode. The methods of sample preparation can be quick and simple, such as air-drying of the washed bacterial population on glass coverslips or the rapid dehydration of cells in ethanol. If the study of specific features is required then established sample preparations similar to those used in SEM should be followed. The cell morphology that

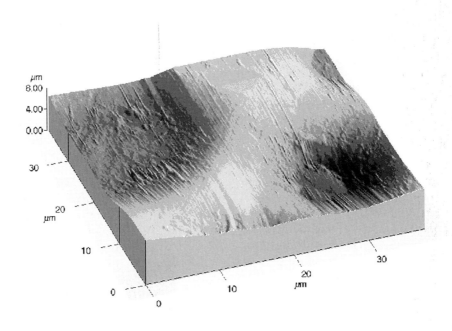

Figure 6 An AFM image of the surface of a tomato imaged in electrolyte.

can be viewed includes the cytoskeleton, yeast budding scars and nuclei. The fixing process tends to compromise the integrity of the cell surface, however, and there have been few reports of AFM images of intact fixed plasma membranes and their components.

The capability of imaging living cells with such high resolution sets AFM apart from other techniques. To ensure that cells are alive, sample preparation and imaging must be within an aqueous environment. To stop the cells moving with the action of a scanning tip, the cells must be attached to a suitable substrate. Cell adhesion to a surface can be promoted by the use of surface treatments such as polylysine coating. Cells may also be held in place by vacuum and orifices such as membrane filtration pores.

Yeast cells have a rigid cell wall, which means they can withstand higher imaging forces. When mammalian cell surfaces have been studied the cytoskeletal elements were clearly visible. This suggests that the plasma membrane is shaped by the underlying structures or that the tip actually penetrates the surface. There are few reports of images of living bacterial cells, perhaps a reflection of present limits of AFM resolution of soft samples, the existence of adequate SEM morphological studies and that the surface components, such as the bacterial surface layer (S layer), can be isolated and imaged with high resolution (see below).

Viruses

The action of viruses is governed by their relationship with surfaces. As a surface analytical technique, AFM is well suited to the study of viruses. Viruses are relevant to the food industry not only as pathogens or spoilage entities but also because of their dominance in genetic research methods.

Intact viruses can be studied by simply depositing the viruses from solution onto atomically flat surfaces and allowing the solution to dry. With this method T4 bacteriophages have been studied. Bundles of DNA strands were seen emerging from the heads of lysed viral particles.

As with all biological samples the softness of the viral surface has limited AFM study. Most methods of study have imaged the crystal structure of the component viral proteins. Atomic force microscopy has added to the knowledge of viral assembly. Subnanometre resolution of the φ29 bacteriophage head-tail connector has been achieved, providing structural evidence that the connector and its movement play an important role in the packing of the viral DNA.

Once a living biological specimen is under the AFM tip and in a favourable environment the system can be maintained and studied for hours. This has been exploited to produce a set of real-time images of viral interactions with an animal cell membrane. Living monkey kidney cells were reproducibly imaged with a resolution on the 10 nm scale and then a solution of poxviruses was added to the medium. Three sets of surface structure changes were observed and related to the action of the virus. The cell membrane was seen to soften for a short period as the viral infection of the cell took place. Exocytosis events were observed as viral proteins were expelled from the cell. Finally, the emergence of the progeny viruses was witnessed when large temporary protrusions of 200–300 nm cross section appeared in the membrane, leaving scars.

Macromolecule Components of Cells

There are numerous reported examples of microbiological macromolecules that have been imaged using the AFM. These include proteins, lipids, DNA, RNA and glycoproteins. These molecules are normally fixed to a substrate such as freshly cleaved mica, covalently linked to a surface or immobilized in self-assembled monolayers (SAMs). Imaging in multivalent cations serves to promote the adhesion of the macromolecules. These procedures serve to anchor the soft sample to a harder surface so that the action of the rastering tip does not dislodge them. Any movement of the molecule will appear as image artefacts.

The AFM has been routinely used to study the structure and biological interactions of DNA. Applications include the splicing of DNA in selected locations, the estimation of base pair number and DNA length within a nucleosome, and the study of DNA tertiary structures. In addition, using the real-time and liquid capabilities of the AFM, the interactions of DNA and DNA enzymes such as RNA polymerase have been observed.

The bacterial S layer component structure and its symmetries have been differentiated by AFM. In particular, remarkable resolution has been achieved of OmpF porin protein, which is found in Gram-negative bacterial envelopes. OmpF porin is a major outer membrane surface which functions as a molecular sieve allowing the diffusion of solutes into the cell. When a large number of the OmpF porin ring structures have been simultaneously imaged, to high resolution, variation in their configuration has suggested that the protein structure essentially opens and closes.

Non-membrane proteins are harder to image than those bound to a membrane. To improve resolution several methods have been developed, such as crosslinking the protein to the surface, cooling the sample and crystallization. The best images of proteins have featured large molecules such as fibrinogen, RNA polymerase, actin, collagen and protein arrays. Other proteins imaged include immunoglobulins.

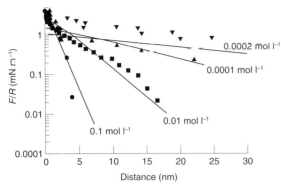

Figure 7 The AFM colloid probe technique: measurement of protein–protein interactions in different NaCl solutions. The lines represent theoretical predictions. F = Force mN; R = Radius of probe (m)

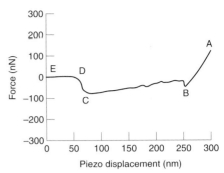

Figure 8 AFM cell probe measurement of the adhesion of a *Saccharomyces cerevisiae* cell to a mica surface after momentary contact in 10^{-2} mol l^{-1} NaCl, pH 4. A–B, cell probe and surface moving together; B–C, cell stretching; C–D is a quantitative measure of adhesive force; D–E, cell and surface move apart.

Measuring Forces of Interaction in Food Microbiology

An important potential application of AFM to the food microbiologist is its force measurement capability within a relevant aqueous environment. Many processes pertinent to food microbiology are governed by the interactions of surfaces. Surface forces are important to the understanding and control of processes such as the flocculation of brewers' yeast at the end of a fermentation, the adhesion of microbes to food or preparation surfaces, and the initial protein coating of surfaces prior to microbial mat formation in processing equipment. The interaction of whole cells can be studied or the interaction of purified membrane proteins can be investigated to quantify their role in the cell–surface relationship.

The first reported surface force AFM studies were based on inorganic systems used to investigate the interaction of the AFM cantilever tips with surfaces. The materials that could be studied and the comparisons that could be drawn were limited. The colloid probe technique meant that these limits were removed. **Figure 7** shows the forces of interaction measured by an AFM between bovine serum albumin (BSA) layers adsorbed onto silica surfaces. The forces were measured at different electrolyte concentrations and pH values and were in good quantitative agreement with predictions based on the DLVO theory using zeta potentials (outer Helmholtz plane potential) calculated for BSA from an independently validated site binding site dissociation model. This work looked at the approach of proteins towards a surface already coated in proteins. Further useful information can be gained by examining the retraction of surfaces after contact. The coated colloid probe technique has been used to study the adhesion of BSA to a filtration membrane. The knowledge of adhesive forces is useful in the development of adhesion prevention regimes such as surface treatments or equipment washing and the assessment of immobilization techniques.

A further example of the use of AFM as a force sensing technique is its ability to estimate the bond strength of different biological ligand and receptor molecule pairs. In this procedure one type of molecule is attached to the probe and the other fixed to the surface. The adhesive force region of the force–distance curve gives a direct measurement of the binding strength. Systems that have been successfully studied include antigen–antibody, biotin–streptavidin, complementary DNA strands and cell adhesion proteoglycans. The stretching of proteins has also been studied in this manner with the measurement of tertiary structure domain strengths from the 'saw-tooth' multiple peaks of the measured force curve within its adhesive region. The number of peaks in the adhesion region changed with the length of protein and number of domains.

Figure 8 shows the retraction force curve of a living yeast cell being pulled from an inorganic surface (freshly cleaved mica) after momentary contact in an electrolyte solution. The adhesion of a yeast cell to the surface shows a number of important features compared with the adhesion of inorganic oxide particles at similar surfaces. Firstly, the adhesion of cells tends to be greater than that of inorganic particles. Secondly, the detachment of inorganic particles takes place over a very narrow range of the force curve (less than 5 nm), whereas the cells show a much more complex behaviour. This includes a staggered snap back to zero force indicative of the breaking of multiple bonds formed in the area of cell–surface contact. In addition the detachment of the cell from the surface suggested that the cell was being stretched over tens of nanometres. The degree of cell stretching and adhesive force increased when the cell was left in contact with the surface, demonstrating that the cell was responding to the presence of the surface. This work has shown the tremendous potential that the AFM offers for the study of cell–surface interactions. Adhesion of cells such as *Escherichia coli* to food surfaces or *Lactobacillus* to food processing equipment can be quantified with the aim of reducing cell adhesion.

Other micromechanical properties of cells that can be measured with AFM include compressibility, viscoelasticity and shear stress effects.

Future Prospects

Atomic force microscopy is an essential technique for the study of surfaces and their interactions. As more scientists realize the importance of the surface region in prediction and control of microbiological phenomena the applications of AFM will grow. The technology is still improving. The study of biological samples is advancing simply because more biologists are becoming AFM operators. Biologists with prior knowledge of structure and function are using the AFM to explore biological applications.

For the food microbiologist there are a large number of surfaces that have yet to be imaged using the AFM. In a number of cases the technology must advance for it to surpass existing SEM images. However, there are reports of subnanometre resolution of biological samples within an aqueous environment. The imaging potential of AFM is well on the way to being realized.

The surface force sensor capability of AFM will also be exploited by the food microbiologist. Direct quantification of the interactions of microorganisms with each other or with relevant surfaces will give useful insights, not available before, that will help food microbiologists in the control of pathogenic or industrial microorganisms. These insights promise to become more specific. Rather than examining the interaction of protein layers or cell surfaces, individual ligands and bond strengths will be studied.

As an addition to the family of microscopic techniques AFM has earned its place alongside other techniques in new hybrid instruments that combine different devices for the study of surfaces. An AFM unit may well soon replace a lens objective on many optical microscopes found in microbiological laboratories. The AFM has become an essential tool for the microbiologist.

See also: **Bacteria**: The Bacterial Cell. **Microscopy**: Scanning Electron Microscopy. **Total Counts**: Microscopy. **Total Viable Counts**: Microscopy. **Viruses**: Detection. **Yeasts**: Production and Commercial Uses.

Further Reading

Bowen WR, Hilal N, Lovitt RW and Wright CJ (1999) Atomic force microscope studies of membranes. In: *Surface Chemistry and Electrochemistry of Membrane Surfaces*. Surface Science Series. New York: Marcel Dekker.

Bowen WR, Hilal N, Lovitt RW and Wright CJ (1998) Direct measurement of the force of adhesion of a single cell using an atomic force microscope. *Colloids and Surfaces A: Physicochemical and Engineering Aspects* 136: 231–234.

Bowen WR, Hilal N, Lovitt RW and Wright CJ (1998) Direct measurement of interactions between adsorbed protein layers using an atomic force microscope. *Journal of Colloid and Interface Science* 139: 269–274.

Firtel M and Beveridge TJ (1995) Scanning probe microscopy in microbiology. *Micron* 26: 347–362.

Haberle W, Horber JKH, Ohnesorge F, Smith DPE and Binnig G (1992) In situ investigation of single living cells infected by viruses. *Ultramicroscopy* 42–44: 1161–1167.

Hansma HG and Hoh JH (1994) Biomolecular imaging with the atomic force microscope. *Annual Review of Biophysical and Biomolecular Structure* 23: 115–139.

Henderson E (1994) Imaging of living cells by atomic force microscopy. *Progress in Surface Science* 46: 39–60

Muller DJ, Engel A, Carrascosa JL and Velez M (1997) The bacteriophage φ29 head-tail connector imaged at high resolution with the atomic force microscope in buffer solution. *EMBO Journal* 16: 2547–2553.

Sarid D (1994) *Scanning Force Microscopy*. Oxford: Oxford University Press.

Sensing Microscopy

Motoi Nakao, Horiba Ltd, Miyanohigashimachi, Kisshoin, Minami-ku, Kyoto, Japan

Introduction

In a wide variety field such as food hygiene and water quality assessment, it is becoming more and more important to be able to measure the number of microorganisms in a short period of time. The plate-count method is the standard for determining the number of microorganisms due to its high reliability, despite the requirement for a long incubation period lasting for 12 hours to a few days. A number of other methods, which measure the number of microorganisms in shorter periods, have been developed and commercialized. For example, optical methods using light scattering or penetration, chemical analysis such as chromatography, impedance methods and ATP methods are attractive alternatives to the conventional plate-count method. The ATP method can detect microorganisms in a particularly short time period, typically a few minutes. The new methods are, however, not very reliable and/or count dead cells as well as living cells.

For example, visualization of the intracellular calcium ion and pH distribution has been widely carried out in the field of cell biology. Progress in this imaging technology has greatly contributed to understanding the role calcium ions play in cell functions. However, the method for measurement of extracellular calcium has not been widely used.

In order to count only the living microorganisms with high reliability, a scanning-laser-beam chemical imaging sensor has been developed which allows the detection of changes in the pH value in agar medium acidified by microorganisms. Microorganisms consume nutrients, such as glucose, and excrete carbon dioxide produced by aerobic respiration and lactic acid by glycolysis.

The chemical imaging sensor is based on the Light Addressable Potentiometric Sensor (LAPS). LAPS was introduced by Hafeman et al in 1988. This is similar to the conventional semiconductor pH sensor, ion sensitive field effect transistor (ISFET), as the LAPS detects Si surface potential change in order to detect pH in solution. However, the ISFET uses aluminium electrodes on the sensor surface as source and drain electrodes, with an epoxy resin to prevent an electric short. This is disadvantageous for measuring the solution. Although it is easy to make the single sensor of the ISFET, it is difficult to attain the integration of the sensor. Since the LAPS locally addresses the sensor surface by a light, unlike ISFET, aluminium electrodes are not needed at the surface. Therefore, the LAPS is a very attractive method for the integration and array of a chemical sensor.

Chemical Imaging Sensor (pH-sensing microscope)

Principle

Figure 1 shows the energy band diagram of an electrolyte insulator semiconductor (EIS) structure. When light with more energy than the band gap of Si (1.1 eV) illuminates the back of the Si substrate, electrons in the valence band are excited into the conduction band. Positive holes in the valence band as well as electrons in the conduction band are generated as photocarriers. If the carriers diffuse into the depletion layer between the Si and the insulator, charge separation is induced due to the potential gradient of the depletion layer, and a transient photocurrent flows through the EIS structure. The alternating photocurrent flows by illuminating a modulated light. The photocurrent can be expressed by **Equation 1**.

$$I = q\,\varphi\,\eta\,(1-\Theta)\exp\left(-\frac{d}{L_p}\right)\frac{C_i}{C_i + C_d} \qquad \text{(Equation 1)}$$

where q is an elementary charge, φ the number of illuminated photons, η the quantum efficiency, Θ the reflective index, d the Si substrate thickness, L_p the diffusion length, C_i the insulator film capacitance, and C_d the depletion layer capacitance. The diffusion length L_p can be expressed with **Equation 2**.

$$L_p = \sqrt{D_\tau} \qquad \text{(Equation 2)}$$

where D and τ are the diffusion coefficient and the life time of minority carriers, respectively. The depletion layer capacity C_d is:

$$C_d = \frac{\varepsilon_s}{W} = \frac{\varepsilon_s}{L_D}\frac{\left(1 - e^{\beta\psi}s + \dfrac{p_{no}}{n_{no}}\left(e^{-\beta\psi}s - 1\right)\right)}{\sqrt{\left(e^{\beta\psi}s - \beta\psi_s - 1\right) + \dfrac{p_{no}}{n_{no}}\left(e^{-\beta\psi}s - \beta\psi_s - 1\right)}}$$

$$\text{(Equation 3)}$$

where ε_s is the permittivity of Si, W the width of the depletion layer, n_{no} and p_{no} the electron and the hole concentration of the Si substrate, respectively. Ψ is the surface potential of Si, and L_D the Debye length. $\beta = q/kT$, where k is the Boltzmann constant, T the absolute temperature.

Figures 2 and **3** show the photocurrent–surface potential or –bias voltage (I–V) characteristics. Figure 2 is the simulated result, and Figure 3 is the experimental result. The capacitance–voltage (C–V) characteristics are also represented in each figure. The vertical axis corresponds to the amplitude of ac photocurrent. Since n-type Si is used, the negative side of the surface potential or bias voltage corresponds to the inversion condition for the Si region between the semiconductor and the insulator, and the positive side corresponds to the accumulation condition. No

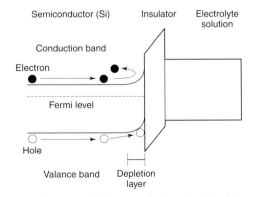

Figure 1 Energy band diagram of electrolyte insulator semiconductor (EIS) structure.

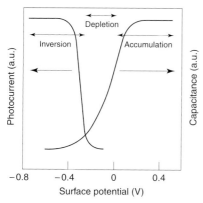

Figure 2 Dependence of photocurrent and capacitance on Si surface potential. (Simulated result.)

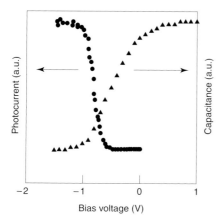

Figure 3 Dependence of photocurrent and capacitance on bias voltage. (Experimental result.)

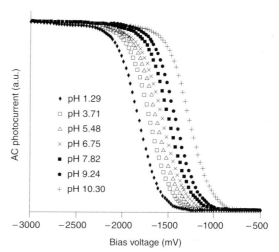

Figure 4 Photocurrent versus bias voltage for various pH values.

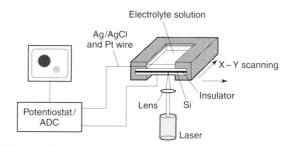

Figure 5 Block diagram of chemical imaging sensor.

photocurrent flows through the EIS structure for the accumulation. As surface potential or bias voltage is negative, the photocurrent increases steeply. The photocurrent is saturated for the inversion. The detectable current flowing through the extra circuit depends on the depletion capacitance, as shown in Equation 1. For inversion, the width of the depletion layer is saturated with the maximum value, and the capacitance of depletion layer is constant with a minimum value. Therefore, the current flowing through the extra circuit is large. On the other hand, for depletion and weak inversion, the width of the depletion layer changes as the bias voltage varies. The photocurrent strongly depends on bias voltage or surface potential. For accumulation, no photocurrent flows because the charge separation does not occur. The capacitance changes in the depletion, and the photocurrent changes in the weak inversion, as shown in Figures 2 and 3. The region of transition for the I–V curve is narrower than that for the C–V curve. This is due to the conductance of the majority carriers. The value of conductance is large for the depletion, so the current for the region is suppressed.

Figure 4 shows the I–V characteristics for various electrolyte solution pH values. As the pH of the electrolyte solution changes, the I–V curve shifts along the bias voltage direction. This can be explained by the site-binding model. When the pH of the electrolyte solution is low, or acidic, the proton in the electrolyte solution binds with the silanol site (SiOH) and the amino base (NH_2), existing on an Si_3N_4 surface. As a result, $SiOH_2^+$ and NH_3^+ form, and the sensor surface has a positive charge. Conversely, for the electrolyte solution of an alkali, the silanol and amino sites become SiO^- and NH^-, respectively, and the sensor surface has a negative charge. This surface charge induces a potential change in semiconductor Si, and the potential leads to the I–V curve shift. In the present system, the shift value is approximately 56 mV/pH, even though the theoretical value is 59 mV/pH at room temperature. In order to shorten the measurement time, the pH value is determined by measuring the photocurrent at a fixed bias voltage instead of directly measuring the shift value.

Figure 5 shows a block diagram of the chemical imaging sensor. The sensor is made using the following process. The 10–20Ω·cm n-type Si wafer is thermally oxidized, and then the Si_3N_4 film is deposited on the SiO_2/Si using low pressure chemical vapour deposition. The thickness of the SiO_2 and Si_3N_4 films are 50 nm and 100 nm, respectively. After removal of the backside of the insulator film (Si_3N_4/SiO_2) on the Si substrate, a gold film containing 0.5% antimony is deposited on the backside of the Si substrate to form an ohmic contact with the Si substrate. An electrolyte solution is in contact with the insulator on the Si substrate, which forms the EIS structure. A potentiostat is employed for applying bias voltage to the semiconductor Si with respect to the electrolyte solution. Silver/silver chloride and platinum wires work as reference and counter electrode, respectively. The photocurrent converts the voltage by I–V converter, and then the value is determined using an analog-digital converter (ADC) and a personal computer.

Spatial Resolution

As a focused laser beam illuminates the backside of the Si substrate, the photocarriers induced at this side must diffuse across the Si substrate to the depletion layer between the Si and the insulator to produce photocurrent. Therefore, the lateral diffusion of photocarriers restricts the spatial resolution of this sensor. Equation 1 can be converted to **Equation 4** by considering the diffusion of carriers in the Si substrate so as that the photocurrent flows through the EIS structure.

$$I = \iiint q \, \phi \, \eta \, (1-\Theta) f\big(\mathrm{pH}(x,y)\big) I_p \alpha \exp(-\alpha z)$$

$$\times \quad \frac{d-z}{4\pi\left(x^2 + y^2 + (d-z)^2\right)^{\frac{3}{2}}}$$

$$\times \quad \exp\left(-\frac{\sqrt{x^2 + y^2 + (d-z)^2}}{L_p}\right) dx\,dy\,dz$$

(Equation 4)

where (x, y, z) shows the coordinate, the illumination point is $(0, 0, 0)$, pH (x, y) is the pH value at the point (x, y) on the surface, f (pH) is a response function ranging from 0 (for lower pH) to 1 (for higher pH), and α is the absorption coefficient for Si. For simplicity, provided that all photocarriers are generated at the backside of the Si substrate, **Equation 5** can be used for the simulation.

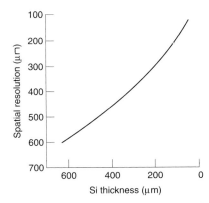

Figure 6 Relation between spatial resolution and Si thickness.

$$I = \iint q \, \phi \, \eta \, (1-\Theta) f\big(\mathrm{pH}(x,y)\big)$$

$$\times \quad \frac{d}{4\pi\left(x^2 + y^2 + d^2\right)^{\frac{3}{2}}}$$

$$\times \quad \exp\left(-\frac{\sqrt{x^2 + y^2 + d^2}}{L_p}\right) dx\,dy$$

(Equation 5)

Figure 6 is the simulated result of spatial resolution as a function of the Si substrate thickness using Equation 5. The thinner the Si substrate, the better the spatial resolution. Previously, we performed the following experiment. The photoresist with a various line and space (L&S) pattern is formed on the sensor, and then a metal film is deposited on the whole surface. In this way, the microscopic distribution of the Si surface potential is produced instead of the pH distribution in order to perform the spatial resolution experiment. The images obtained experimentally agreed well with the simulated image using Equation 5 for Si thicknesses of 630, 300 and 100 μm. These results indicate that the spatial resolution can be estimated using the carrier diffusion model, and that the thinner Si layer of the sensor results in improvement of the spatial resolution.

A mechanical polish and etching technique is used to make the sensor with Si thickness of 100 μm. However, in order to make the Si substrate thinner than 100 μm, it is difficult to thin the whole Si substrate due to the mechanical intensity of the sensor structure. We attained a sensor with high spatial resolution and strong mechanical intensity by thinning a part of the Si substrate by etching. We have been able to reduce the Si substrate to 20 μm and to obtain spatial resolution better than 10 μm. The results showed that the Si layer thickness was not uniform after etching. This nonuniform Si substrate thickness

10 μm L&S

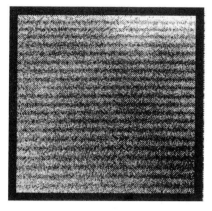

5 μm L&S

Figure 7 Experimental result of spatial resolution using SOI wafer.

created an unevenness in the two-dimensional image, so it was difficult to observe the pH distribution practically.

The sensor was constructed with an Si layer of uniform thickness using bonded silicon-on-insulator (SOI) wafers instead of Si wafers. Anisotropic chemical etching was used for removing the Si bulk layer of SOI wafer. In order to protect the sensor surface completely, the one-sided etching technique was employed, and the sensor surface did not contact the etching solution. The etching solution was 20% potassium hydroxide (KOH) solution at 80°C.

The image of the spatial resolution from the sensor with a 20 μm Si layer is shown in **Figure 7**. In this figure, the 5 μm and 10 μm L&S patterns are shown clearly.

The uniformity of the Si layer thickness is thus improved. In addition, this result indicates that a practical sensor with a spatial resolution better than 5 μm is fabricated. The two-dimensional chemical information in a solution can be microscopically measured using a sensor with such high spatial resolution. For example, we would be able to observe the pH distribution induced by the metabolism in a single cell, and estimate the metabolic activity of each microorganism.

pH Resolution

In addition to spatial resolution, pH resolution has to be considered as a factor influencing the sensor characteristic. The pH resolution is mainly governed by the signal to noise ratio (S/N) of the measuring signal. It is possible to improve the S/N by lengthening the measurement time or integrating the signal. However, a pH gradient in a solution decreases with time because of diffusion of protons and finally, the pH distribution becomes uniform. Therefore, a shorter measurement time for one image is necessary.

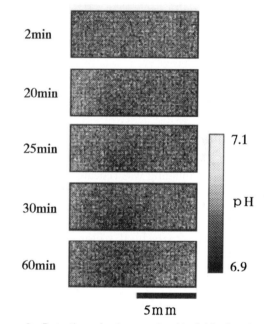

Figure 8 Detection of microscopic pH distribution by ion-exchange resin.

Since the X–Y stage is presently used for the two-dimensional scanning, it dominates the measuring time for one image, and for example, it takes about 30 s to measure one $64 \times 64\,mm^2$ image (64×64 pixels). The evaluation of measuring time on pH resolution was performed experimentally. The result is shown in **Figure 8**. To form a very small pH gradient, a very small and unpurified ion-exchanger was placed on a thin agar film. The pH distribution of the agar was then measured. The line profile of the acidified region in Figure 8 is shown in **Figure 9**. Although the line profile of the pH value is almost constant for 2 min, the central value for 30 and 60 min is smaller than the surrounding region, by 0.045 and 0.03 pH units, respectively. This result indicates that the pH resolution is 0.01.

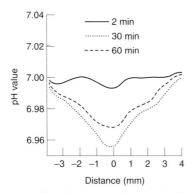

Figure 9 Line profile of the results illustrated in Figure 8.

Application

Ion Exchange Resin

Figure 10 shows an example observation of two-dimensional pH distribution using the chemical imaging sensor. The pH distribution was measured when the ion-exchange resin was placed on the agar film. We used a cation-exchange resin (Amberlite IR-120B, sulphonated type, Organo, Japan), which received K^+ and Na^+ and released the protons. The gel film was prepared with a solution containing 1.5% agar and 0.1 M KCl. The pH was adjusted to 7.4 with NaOH. The thickness of gel film is 0.5 mm. Two-dimensional pH imaging was repeated using a single cation ion-exchange resin as the source for transient microscopic pH distribution. We found that, after placing a resin particle on agar film, the acidified area, which corresponds to the black region, becomes larger over time. Also, as shown in Figure 10 the resin purified with HCl produces a larger acidified region and makes the pH of the region lower than with the unpurified resin. From these results, we concluded that the resin performance was reflected in the pH image.

The amount of protons released from the particles of ion-exchange resin can be calculated by spatial integration of proton concentration from the pH image using **Equation 6**.

$$N = \iint 10^{-pH(x,y)} - 10^{-pH} xy \qquad \text{(Equation 6)}$$

where pH (x,y) and pH are the pH values at the point (x,y) and background pH value of the agar film (pH = 7.4), respectively. The pH distribution perpendicular to the sensor was assumed to be zero. **Figure 11** shows the calculated result. The total amount of the released protons was confirmed to be 100 times larger with the purified resin than with the unpurified resin. Moreover, the amount of released protons was maximal by around 9 min and 5 min with the purified and the unpurified resin, respectively. From this result, the activity of the ion-exchange resin was evaluated as a two-dimensional pH image using this sensor. The pH distribution can also be used for the accurate evaluation of the proton diffusion in agar because the pH can be quantitatively measured, unlike evaluation with a fluorescence microscope.

Electrolysis

Two-dimensional pH measurements of electrolyte were carried out at the bottom of an electrolysis cell. The measured area was 9.6 × 5.0 mm and the number of measuring points was 48 × 25. The measured area included the points directly under both the anode and

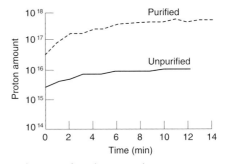

Figure 11 Amount of total proton release.

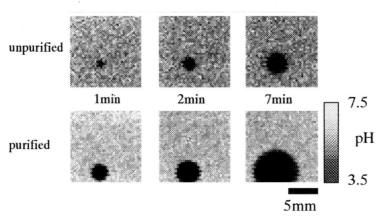

Figure 10 pH distribution in gel induced by ion-exchange resin.

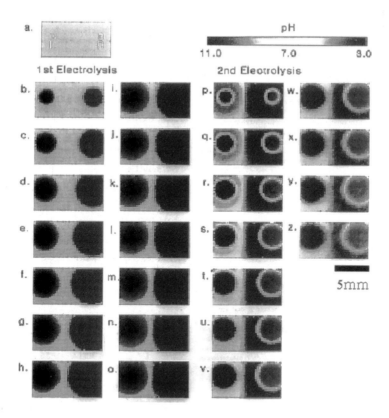

Figure 12 pH distribution in electrolyte solution induced by electrolysis. (a) Before electrolysis; (b–o) after the first electrolysis; (p–z) after the second electrolysis. (See also color **Plate 21**.)

the cathode. The measuring time for each image was 25 s. After the 15th measurement, another electrolysis was carried out under the same conditions except for the polarity of the current.

Figure 12 shows the change of pH distribution in electrolyte solution before and after electrolysis. Generation of pH distribution was clearly shown in the pH images (Fig. 12b), even though there was no pH distribution before electrolysis (Fig. 12a). After the second electrolysis, generation of an opposite pH distribution was observed inside the pH distribution generated by the first electrolysis (Fig. 12p).

On repetition of pH imaging, expansion of the pH-distributed region was observed (Fig. 12c–o). After the first electrolysis, expansion of the lower pH region seemed to occur faster than that of the higher pH region. As the two regions became closer, the expansion became slower and then became distorted. After the second electrolysis, expansion of the newly generated pH distribution was also observed in the pH distributions already present (Fig. 12p–z). There were neutral pH regions surrounding both the lower and the higher pH region generated after the second electrolysis. The expansions of both lower and higher pH regions generated by the second electrolysis were apparently slower than those of the regions generated

by the first electrolysis. In addition, as both the regions generated by the second electrolysis expanded, those already generated by the first electrolysis became diffused. This result shows that the expansion of proton and hydroxide depends on the background pH, and that diffusion involving acid–base neutralization could also be visualized on pH images.

It was confirmed that the electrogenerated pH distribution and its expansion could be imaged by two-dimensional potentiometric pH imaging. Using the pH values represented in pH images, the pH distribution was studied quantitatively. Acid and base neutralization in a very small region was also observed. Such measurement and imaging is rarely possible when a conventional method of potentiometric pH measurement is used.

It is important in electrochemistry that the distributed pH values around the electrodes are obtained separately from those in the bulk region because most electrode processes involve pH change around the electrodes. In particular, recent advanced technology requires electrodeposition or electrochemical etching on a microscopic scale. The preliminary results show the applicability of this chemical imaging sensor to such processes.

Observation of Microorganisms

Yeast

When microorganisms are grown on agar medium, the two-dimensional pH distribution of agar film can be visualized. When microorganisms are incubated in culture medium, they consume nutrients, and generally excrete acidic products, such as carbon dioxide and/or lactic acid. Therefore, the surrounding area becomes acidified. It should be possible to direct the metabolic activity of microorganisms by observing the atmosphere in which the microorganisms are cultivated.

Saccharomyces cerevisiae IFO203 was incubated on standard agarose plates (0.25% yeast extract, 0.5% tryptone, 0.1% glucose and 1.5% agar) containing 0.1 M KCl. KCl was added to reduce the impedance of agar and increase the signal intensity. The thickness of the agar film was about 1–2 mm. A surface smear technique was adopted for inoculation.

Figure 13 shows the two-dimensional pH image of yeast colonies; the colonies were incubated on agar for 24 h at room temperature before being placed upside-down onto the sensing surface. Since the dark area in the image corresponds to the lower pH value, these areas show the existence of colonies. This result indicated that the chemical imaging sensor enables colonies of microorganisms on agar medium to be observed. Furthermore, automatic colony-counting may be possible by enlarging the measuring region and combining this technique with image-processing software.

In terms of further applications, the incubation of microorganisms at the interface between the sensing surface and agar is very attractive because *in situ* observation of microorganisms during incubation can be carried out. *In situ* observation provides detailed information about the growth of branches and diffusion of various materials on agar. To perform *in situ* observations, the agar is brought into contact with the sensor Si$_3$N$_4$ surface without air bubbles after the microorganisms are placed on the agar. Air bubbles cause an undesirable reduction of photocurrents due to an air gap. The incubation is performed in order to detect the two-dimensional distribution of pH value due to metabolism of the microorganism. Aseptic incubation can be performed by treating the Si$_3$N$_4$ surface with alcohol. The growth rates of microorganisms between the agar and the sensing surface were about half that on an agar surface.

Figure 14 shows the *in situ* observation of yeast colony growth. The incubation time was 48 h at room temperature. However, further incubation resulted in the generation of fermentation gas which prevented further observations. It is noteworthy that colonies of different sizes could be seen, even though the visual sizes of the colonies were almost the same. This is related to the initial state or activity of the microorganisms before incubation, which would result in different lag periods. The growth process of the microorganisms can be studied by this technique.

Escherichia coli

The *E. coli* (JM109) used in this study was stored at < 0°C before use. The microorganism was incubated on standard agarose plates (0.25% yeast extract, 0.5% tryptone, 0.1% glucose and 1.5% agar) containing 0.1 M KCl.

The pH distribution around an *E. coli* colony that originated from a single cell was observed. Three agar plates (2–3 mm in thickness) were prepared in sterilized Petri dishes (9 cm in diameter). *E. coli* was planted on the plate by the surface smear technique. After incubation at 36°C, a 2 × 2 cm piece of the agar film was cut out from one of the agar plates. The piece was placed on the sensor so that the *E. coli* colony would come into contact with the sensor.

Figure 15 shows that the acidified region appears after cultivation for 10 h, and then the pH of the region decreases and the acidified area expands with the cultivation time. After cultivation for 10 h, the number of cells in and size of the colony is approximately 10 000 and 0.3 mm², respectively. A colony of

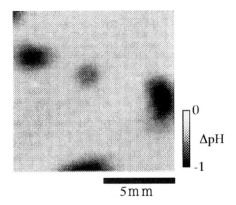

ΔpH

5 mm

Figure 14 *In situ* observation of yeast colonies incubated between the sensor and agar medium.

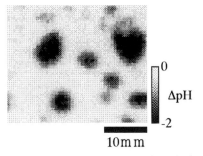

ΔpH

10 mm

Figure 13 pH distribution of agar induced by colonies of yeast.

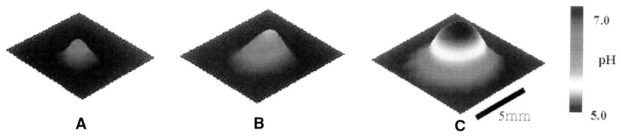

Figure 15 pH distribution in agar medium induced by an *E. coli* colony. Incubation for (**A**) 10 h, (**B**) 13 h and (**C**) 16 h. (See also color **Plate 22**.)

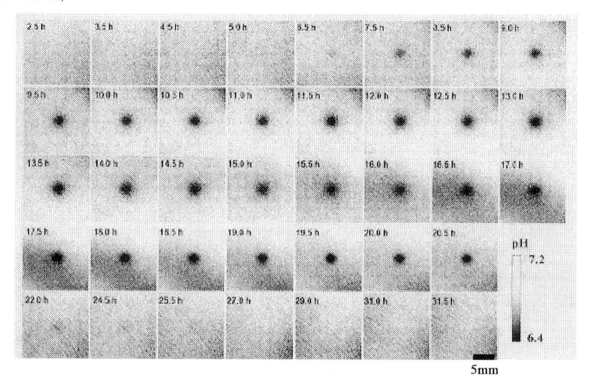

Figure 16 Change of pH distribution around an *E. coli* colony (JM 109).

such size is difficult to count by human eye. Surely, this technique can be applied to the screening of microorganisms and counting cells.

In another experiment, *E. coli* colonies were incubated on a thin agar film (1–2 mm in thickness) prepared on the sensor. A small portion of an *E. coli* colony was picked up with a sterilized platinum wire (500 µm diam.) and placed on the agar film. The size of this starting colony was about 100 µm in diameter. The pH change in the agar film was measured every 0.5–2.5 h until 36 h after the start of the experiment. The colony was kept at 36°C by keeping the sensor unit in the incubator except during the pH measurement. During the night time (between 8.5 h and 22 h after the start of the experiment), the sensor unit was kept on the equipment for the imaging. The colony was, therefore, incubated at room temperature (10–25°C) during this period.

The pH started to fall after 6.5 h, and the region kept expanding until 14 h (**Fig. 16**). This indicates

that *E. coli* cells in the colony became active after incubation for around 6.5 h and excreted acidic products until 14 h. After 14 h, the region of lower pH became smaller and pH increased until 18 h, at which time the pH decreased again. This change seems to correspond to the change of incubation temperature; the metabolic activity was low during the night time. After 22 h, the lower pH region disappeared indicating that *E. coli* cells in the colony were no longer active. After even longer incubation, however, pH_L increased slowly, suggesting a different metabolic path of *E. coli*.

E. coli IFO 3301 was also used. Both microorganisms were incubated on standard agarose plate (0.25% yeast extract, 0.5% tryptone, 0.1% glucose, and 1.5% agar) containing 0.1 M KCl.

Figure 17 (A) and (B) shows the results for *E. coli* colonies incubated on agar for 8 and 12 h, respectively. In Figure 17(A), the pH value of the colony region and that of the surrounding area differs by

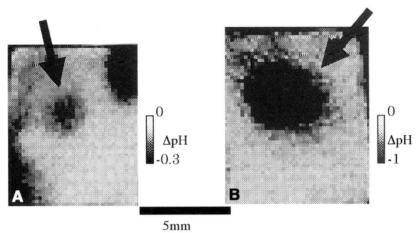

Figure 17 pH distribution of agar medium by an *E. coli* colony (IFO 3301). Incubation for (**A**) 8 h and (**B**) 12 h.

only 0.3. Incubation for 12 h is sufficient for detection of a colony of *E. coli* using the chemical imaging sensor (Fig. 17B). After 8 h incubation (Fig. 17A), the smaller dark region at the centre in the image is easily recognized as a colony. The other dark region at the right is due to the uneven thickness of the sensor substrate. The dark region due to the colony can be distinguished from that due to thickness fluctuation of substrate by measuring blank agar in advance, or measuring the I–V characteristics at each point.

It is surprising that *E. coli* incubated on agar for only 8 h can be seen. Normally, it is difficult to detect colonies visually at 12 h, and impossible at 8 h. The visual sizes of *E. coli* colonies incubated for 8 and 12 h were 0.2 and 0.7 mm^2, respectively, and they contained 800 and 10^6 living cells in the single colonies, respectively. The acidic regions produced by the colonies were much larger after both 8 and 12 h incubation than the visual sizes of the respective colonies. These results indicate that the lateral diffusion of acidic ions produced by *E. coli* on agar is not negligible.

The pH distribution around the colonies due to the generation and diffusion of metabolic products, namely, acidic ions, was examined, i.e. the total generation rate of the number of microorganisms. The number of living cells present in a single colony on agar at a certain incubation time was measured using the plate-count method. The result showed that the number of microorganisms, $N(t)$, at a time, t (s), in the exponential growth phase (4.5–12 h) is given by **Equation 7**.

$$N(t) = 2^{(t-t1)/td}$$

(Equation 7)

where the lag period, $t1$, is 4.5 h and the doubling time, td, is 22 min. The generation rate of acidic ions per cell, G, was determined by measuring the pH change in the culture liquid using a titration method.

The time-dependence of G during incubation is neglected here for simplicity. The experimental result yields $G = 1.0 \times 10^{-18}$ (mol s^{-1} cell) assuming that the generation rate per cell is constant in the course of multiplication. The effective diffusion coefficient of acidic ions, D, was estimated to be 2.0×10^{-5} cm^2 s^{-1} by fitting with the experimental result. This value is comparable to that for the sulphur ion in agar. For simplicity, the diffusion of acidic ions along a thin surface region is only considered since surface diffusion may be faster than bulk diffusion in agar. The size of the colony was also neglected. When the generation point of acidic ions was set at (0,0) and the starting time of incubation as $t = 0$, the surface density of acidic ions at a point (x,y) on the plate $Z = 0$ and at incubation time, t, $A(x,y,t)$ can be expressed as

$$A(x,y,t) = \int_0^t \frac{G(\tau)}{4\pi D(t-\tau)} \exp\left(-\frac{x^2 + y^2}{4D(t-\tau)}\right) d\tau$$

(Equation 8)

The exponential factor represents the in-plate diffusion on the surface of the agar.

Figure 18 presents the simulation results using Equation 2. The experimental results of the distribution for the 8 h incubation is also shown. The simulation curve for 8 h incubation agrees well with the experimental results, except for the slight difference at the tail. The difference is believed to be due to vertical diffusion into the agar during the diffusion of acidic ions from the colony to a distant surface point, a portion of the acidic ions diffuse into the agar medium and cannot reach the surface point. The results in Figure 18 indicate that the pH distribution on the agar can be adequately explained by the generation and diffusion model of the metabolic products. Since the chemical imaging sensor measures the two-dimensional pH distribution, background fluctuation must be removed. Therefore, the thickness of

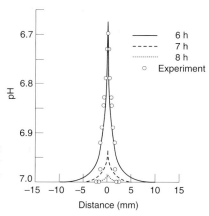

Figure 18 Simulated and experimental line profile of pH in acidified area by *E. coli* colony (IFO 3301).

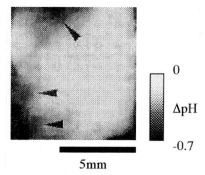

Figure 19 pH distribution of agar medium induced by *P. diminuta* (ATCC 19146).

the Si substrate and the Si$_3$N$_4$ film must be homogeneous in the measuring region. Further improvement of the signal to noise ratio would enable the detection of microorganisms with a shorter incubation period, and furthermore, would enable the observation of detailed information about the microorganisms such as cell membrane potential which is typically of the order of nanovolts to microvolts.

In addition, we could make a different type of pH imaging from those already done by existing pH imaging methods such as fluorescence microscopy and fluorescence confocal microscopy, which mainly deal with the pH distributions in a single cell.

Regarding the spatial resolution and time resolution, the chemical imaging sensor cannot compete with the fluorescence microscope and fluorescence confocal microscope. However, the performance of the sensor was good enough for observation of the metabolic activity of the microorganism. Since this sensor does not require the use of a fluorescent dye, it would be a better and easier method of pH imaging in certain applications. Another attractive feature of the sensor is that pH is measured by potentiometric principle, which is defined as the standard pH measurement method. Therefore, for the numerical evaluation of the pH distribution, the sensor is thought to be a more reliable pH imaging technique than present ones, such as the colorimetric method and the fluorescence method.

Pseudomonas diminuta

Colonies of *P. diminuta* (ATCC19146), which is a typical bacterium existing in ultra-pure water, could also be detected after 3 days incubation as shown in

Figure 19. The acidified area (arrowhead) induced by *P. diminuta* can be seen. Ultra-pure water plays an important part in the semiconductor process because of contact with the Si water surface. A protein of the microorganism itself and metabolic products from the microorganisms reduce the purity of ultra-pure water. It is very important to monitor the number of *P. diminuta* in ultra-pure water. A plate counting method is used at present. However, the growth rate for incubation of *P. diminuta* is very slow. Typically, incubation for about one week is required in order to count the number of *P. diminuta* colonies. Also there are very few living cells (less than five cells per 100 ml). A shorter incubation period is required to detect *P. diminuta* cells using the chemical imaging sensor.

See also: **Escherichia coli**: *Escherichia coli.* **Pseudomonas**: Introduction. **Saccharomyces**: *Saccharomyces cerevisiae.* **Total Counts**: Microscopy. **Total Viable Counts**: Pour Plate Technique; Spread Plate Technique; Specific Techniques; Microscopy. **Yeasts**: Production and Commercial Uses.

Further Reading

Grove AS (1967) *Physics and technology of semi-conductor devices.* New York: John Wiley and Sons.

Hafeman DG, Parce JW and McConnell HM (1988) Light addressable potentiometric sensor for biochemical systems. *Science* 240: 1182.

Nakao M, Yoshinobu T and Iwasaki H (1994) Scanning-laser-beam semiconductor pH-imaging sensor. *Sensors and Actuators* B20: 119.

Sze SM (1981) *Physics of semi-conductor devices.* New York: Wiley Interscience.

Microwaves *see* **Heat Treatment of Foods:** Action of Microwaves.

MILK AND MILK PRODUCTS

Contents
Microbiology of Liquid Milk
Microbiology of Dried Milk Products
Microbiology of Cream and Butter

Microbiology of Liquid Milk

Barbaros H Özer, Department of Food Science and Technology, University of Harran, Şanlıurfa, Turkey

Introduction

In general, cows are milked twice a day on farms worldwide. The collection of milk varies from primitive hand milking to the use of complex machines for milking herds of thousands of cows. The ambient temperature where milk is produced varies from subzero to 30°C or higher. In the former conditions milk must be protected from freezing, while in the hotter climates refrigeration is essential to keep milk for processing without any microbiological deterioration. Furthermore, the temperature and period of milk storage on the farm can vary widely, so that the numbers and types of microorganisms present when the milk leaves the farm differ, often unpredictably, even under apparently similar conditions.

It is unfortunate that milk is an excellent culture medium for certain microorganisms, in particular certain pathogens and lactic acid bacteria, whose growth and activity depend on temperature and the absence of competing microorganisms and their metabolic products. Milk drawn aseptically from the healthy udder is not sterile, but often contains low numbers of microorganisms, the so-called 'udder commensals'. These microorganisms are predominantly micrococci and streptococci, although coryneform bacteria including *Corynebacterium bovis* are also fairly common. The numbers and types of microorganisms in milk can be useful indicators of the origin of microbial contamination.

The Initial Microflora of Raw Milk

Under proper hygienic conditions the level of contaminants is minimal – less than 1000 colony forming units (cfu) per millilitre of milk. However, heavily contaminated milk may contain more than 10^6 cfu ml^{-1}. According to the International Dairy Federation (IDF), a total colony count of more than 10^5 cfu ml^{-1} of milk indicates a serious fault in production hygiene, whereas lower figures ($< 20\,000$ cfu ml^{-1}) indicate that milk has been harvested under good hygienic conditions.

The main groups of aerobic and mesophilic, microorganisms present in raw milk are listed in **Table 1**.

Sources of Contamination

The main sources of contamination are the cow itself, the human handler and the environment, including air, water supplies and dairy equipment. Temperature is of paramount importance in controlling the growth of pathogenic and non-pathogenic microorganisms in raw milk.

The numbers of bacteria in milk from an unhealthy udder are normally high, although the level varies depending on the type of disease and external conditions. Of the cattle diseases that are of major economic importance to dairy farmers, mastitis is accepted to be the most stubborn. Trauma or physiological disturbance can be considered as potential causes of mastitis, but udder infections with microorganisms are the main cause. The main group of microorganisms responsible for mastitis is the bacteria, but fungi, mycoplasma, algae and even multicellular parasites can also be implicated. *Staphylococcus aureus*, *Streptococcus agalactiae*, *Streptococcus dysagalactiae*, *Streptococcus uberis*, coliforms (*Escherichia coli*), *Pseudomonas aeruginosa*, *Mycoplasma bovis* and *Corynebacteria* spp. are the most frequently present bacteria in infected animals.

Apart from contamination transmitting through the udder directly into the milk or originating from the skin and mucous membranes of the animal, extraneous sources of infection play an important role in determining the keeping quality of raw milk and processed dairy products. It is quite possible that the udder is contaminated with dung, mud and bedding materials, and unless the udder is washed properly, contaminants can pass into milk during milking. Bedding material may contain 10^8–10^{10} cfu g^{-1}. Any equipment in a dairy plant including bulk tanks, churns, separators, heating and cooling units is a potential source of contamination. Strains of *Yersinia enterocolitica*, *Salmonella typhimurium* and *Listeria monocytogenes* are likely to be present in the pasteurization and cooling systems, and on the surface of stainless steel equipment.

Table 1 The main groups of aerobic mesophilic microorganisms in raw milk

Spore-formers	Micrococci	Gram-positive rods	Streptococci	Gram-negative rods
Bacillus spp.	Micrococcus	Microbacterium	Enterococcus	Pseudomonas
	Staphylococcus	Corynebacterium	Streptococcus	Acinetobacter
		Arthrobacter	S. agalactiae	Flavobacterium
		Kurthia	S. dysagalactiae	Enterobacter
			S. uberis	Klebsiella
				Aerobacter
				Escherichia
				Serratia
				Alcaligenes

Water supply is another source of contamination, especially of *Cryptosporidium parvum* which is a protozoan parasite and causes cryptosporidiosis, is associated with water and has been found in dairy plants and products. Cryptosporidiosis is spread through the faecal-oral route. *Cryptosporidium* oocysts can remain viable for at least 12 months at 4°C. Heat treatments under conditions normally used for pasteurization of milk are unlikely to render oocysts non-viable, and the usual high temperature short time pasteurization protocol (72°C for 15 s) is not sufficient to destroy the infectivity of *C. parvum* oocysts in milk. Therefore, untreated or improperly heated milks and some types of cheeses produced from raw milk and low-heat-treated milk, especially cottage cheese in which the curd is washed, carry a potential risk of cryptosporidiosis.

Air is not a particularly important source of contamination; however, microorganisms including micrococci, coryneforms, *Bacillus* spores, streptococci and Gram-negative rods may be present in the air in cowsheds or milking parlours.

Insects, rodents, dirt and manure can play an important role introducing pathogens into the milk.

Collection and Storage of Raw Milk

The frequency at which milk is collected depends on the farm's storage capacity and the refrigeration temperatures which can be achieved. Because it is more costly than collecting the milk every 2–3 days, daily collection is becoming less common throughout the world. During the storage period on the farm, milk should be kept below 10°C as such temperatures have an inhibitory effect on most pathogens. Milk can be collected in churns or cans or by tankers; churns or cans must be cleaned properly and kept away from direct sunlight, and filled churns should be transferred to the milk collection point as soon as possible. The main drawback with bulk collection is the risk that an undetected faulty consignment from one farm may spoil a whole load of milk.

The period of storage and the temperature are determinative factors with respect to the micro-

biological quality of raw milk, as well as the type and number of bacteria present. Unless proper hygienic and storage conditions are provided, the spoilage bacteria multiply in the milk, and at 25–30°C streptococci and coliforms – both of which increase the acidity of the milk – become predominant. Until inhibited by developed acidity, Gram-negative rods and micrococci (including staphylococci) also multiply at moderate temperatures. Some measures which should be taken in order to prevent contamination and the growth of pathogens and non-pathogens in milk are:

- general hygiene
- disinfection of the udder, utensils and equipment in contact with milk
- cooling below 10°C immediately after milking
- separation of abnormal milks (unusual smell or colour)
- the 'cold chain' should not be broken between milking and processing
- if necessary, thermization should be applied.

All the above precautions can have a positive effect on the microbiological quality of raw milk. However, the storage of raw milk under cold conditions may cause some quality problems depending on the time elapsed between milking and processing.

Cold Storage and Growth of Psychrotrophs

Storage of milk for long periods at low temperature has brought about new quality problems for the dairy industry. These problems are related to growth and metabolic activities of microorganisms at low temperatures. These microorganisms, which are termed 'psychrotrophs', are ubiquitous in nature and common contaminants of milk. Although conventional psychrotrophic bacteria are heat-labile, the bacterial metabolites or enzymes (lipases or proteinases) released by psychrotrophic bacteria may be heat resistant, and such enzymes can cause major flavour and textural defects in processed dairy products, e.g. bitterness in cheese and a rancid aroma in high-fat items like butter and cheese. Psychrotrophs may include long or short rods, cocci and vibrios; Gram-positive or Gram-negative bacteria; spore-

Table 2 Thermoduric and psychrotrophic microorganisms in raw milk

Thermoduric microorganisms		Psychrotrophic microorganisms	
Gram-positive	Gram-negative	Gram-positive	Gram-negative
Microbacterium	Alcaligenes	Micrococcus	Pseudomonas
Micrococcus		Bacillus	Achromobacter[a]
Bacillus spores		Arthrobacter	Alcaligenes
Clostridium spores		Clostridium	Chromobacterium
		Corynebacterium	Enterobacter
		Lactobacillus	Citrobacter
		Microbacterium	Escherichia coli
		Sarcina	Flavobacterium
		Staphylococcus	Klebsiella
		Streptococcus	Serratia

[a] Genus is no longer recognized in Bergey's *Manual of Determinative Bacteriology*, 8th edn, and most have been reclassified as *Alcaligenes* or *Pseudomonas* species.

formers and non-spore-formers; and aerobic, facultative anaerobic or anaerobic microorganisms. Bacteria in the genus *Pseudomonas* are the most frequently encountered Gram-negative group in raw milk; other Gram-negative and Gram-positive bacteria present in raw milk are listed in **Table 2**.

Most psychrotrophic microorganisms in milk and dairy products come from soil and vegetation; coryneforms and *Arthrobacter* spp. constitute the majority. Water is also a source of psychrotrophic contamination in milk and dairy products. *Pseudomonas*, *Alcaligenes* and *Flavobacterium* dominate the psychrotrophic flora in water with *Chromobacterium*, *Bacillus* and coliforms in lesser numbers. Air and faecal contamination may contribute psychrotrophs to milk, since both fresh and dry manure contain millions of such bacteria per gram.

Post-pasteurization contamination with psychrophilic bacteria can also be a problem unless extensive care is taken thoroughly to clean the lines and filters leading out of the pasteurization chamber.

Biochemical Changes Caused by Psychrotrophs
Psychrotrophic bacteria are able to spoil milk by biochemically altering the compounds present in milk. Psychrotrophs can cause the decomposition of urea, hydrolysation of starch, reduction of nitrate to nitrite and hydrolysis of proteins and lipids at temperatures as low as subzero. The growth of psychrotrophic bacteria is primarily responsible for the loss of keeping quality of milk and dairy products held at temperatures below 7°C. During the early stage of growth of psychrotrophic microorganisms, biochemical changes occur at a low level resulting in a lack of freshness or a stale taste. At the later stages, biochemical transformations gain velocity and aroma and flavour defects become apparent. Development of these off flavours and odours is usually a result of proteolysis and/or lipolysis, and both are of major concern to the dairy industry.

Lipolysis A lipase is defined as an enzyme that is able to hydrolyse the esters from emulsified acylglycerols at an oil-water interface. The production of heat-stable, exocellular lipases by psychrotrophic bacteria, including the genera *Pseudomonas*, *Acinetobacter*, *Serratia*, *Aerobacter*, *Alcaligenes* and *Moraxella*, is well known and, of the above genera, species of *Pseudomonas*, *Acinetobacter* and *Moraxella* produce lipases capable of hydrolysing tributyrin and milk fat at both 6°C and 20°C. Fluorescent *Pseudomonas* species and *Flavobacterium* and *Alcaligenes* species are recognized as the most active lipolytic bacteria. The behaviour of microbial lipases in relation to heat treatment depends on the species of microorganism. In general, lipases of *Pseudomonas* and *Serratia* spp. are heat-resistant, but not those of *Alcaligenes* and *Flavobacterium* species. In any case, the total thermal destruction of microbial lipase requires a higher temperature than that needed to kill the microorganisms that synthesize the lipases. Microbial lipases are able to remain active over a wide range of temperatures. Thus, while *Pseudomonas fluorescens*, *Pseudomonas mucidolens* and some strains of *Pseudomonas fragi* produce lipases that are stable at temperatures as high as 100°C, some strains of *Pseudomonas fragi*, *Staphylococcus aureus*, *Geotrichum candidum*, *Candida lipolytica*, *Penicillium roqueforti* and other *Penicillium* spp. have been reported to produce lipases that are active at −7°C, −18°C and even −29°C.

Proteolysis Although most psychrotrophs (excluding *Bacillus* spp.) in raw milk are killed by pasteurization, most of them produce extracellular proteinases which are extremely thermostable and can withstand high temperature short time (HTST) (72°C for 15 s) and ultra-high temperature (UHT) (138°C for 2 s) treatments. Of the bacteria that can secrete exocellular proteinases, the genus *Pseudomonas* is highly proteolytic and, therefore, most of the studies on thermal stability of proteinases have concerned the pseudomonads. Most proteinases of *Pseudomonas* can survive heat treatment at 149°C for 10 s and, for example, one proteinase from *Pseudomonas* is about 4000 and 400 times more heat-resistant than spores of *Bacillus stearothermophilus* and *Clostridium sporogenes*, respectively. This heat resistance and the ability to hydrolyse casein at temperatures as low as 2°C are among the main characteristics of these proteinases. *Pseudomonas aeruginosa* is able to produce an exocellular proteinase which can remain active at 2°C for up to 1 month and can hydrolyse casein at this temperature; most of the proteinases show optimum activity at pH 6.5–8.0.

As far as milk and dairy products are concerned, proteolysis leads to gelation, unclean and bitter off-flavours, but the level of population that causes these defects varies from one species to another. It is a well-established fact that heat-treated milk is more readily subject to proteolysis than raw milk, probably owing to the existence of natural proteinase inhibitors in raw milk.

For many years, the quality of refrigerated raw milk for cheese-making has been of concern, in particular with regard to bacterial numbers and types, rennetability, starter activity and, eventually, quality of the finished cheese. The growth of psychrotrophs in cheese milk has been reported to result in a reduced rennet coagulation time (RCT), and psychrotrophic *Pseudomonas* and *Flavobacterium* species can cause excess bitterness in Cheddar cheese. Similar observations can be adapted to other types of cheeses and, to a lesser extent, yoghurt.

Microbiology of Pasteurized Milk

Pasteurization is intended to make the milk and milk products safe by destroying all the vegetative pathogenic microorganisms. With pasteurization, not only are pathogenic microorganisms killed, but also a wide range of spoilage organisms. Pasteurization is the most frequently applied heat treatment to liquid milk. Typical pasteurization conditions should be as follows:

- not less than 62.8°C or more than 65.6°C for at least 30 min (Holder method)
- not less than 71.7°C for at least 15 s (HTST).

Raw milk often contains microorganisms at levels of 10^4–10^5 cfu ml^{-1}, and the extent to which the number of microorganisms can be reduced by pasteurization depends not only on the number present initially, but also on the types of organisms. The spoilage microflora of pasteurized milk is of two types: post-pasteurization contaminants, which have entered the milk after heating; and heat-resistant bacteria, which have survived heating. In general, almost all Gram-negative organisms in milk are destroyed with pasteurization at 63°C for 30 min, and while some thermophilic and mesophilic bacteria, e.g. micrococci and *Streptococcus* spp., which are thermoduric, may survive pasteurization, they grow very slowly once the pasteurized milk is chilled to 4°C; coryneform bacteria are another group often present in pasteurized milk, but they grow very slowly in cooled milk and rarely cause defects.

Table 3 Criteria for the acceptability of pasteurized milk

Criteria	Limits
Newly pasteurized milk	
Total plate count (30°C for 72 h)	$<3 \times 10^4$ cfu ml^{-1}
After storage	
Total plate count (21°C for 25 h)	$<1 \times 10^5$ cfu ml^{-1}
Thermophilic count	$<1 \times 10^4$ cfu ml^{-1}
Psychrotrophic count	$<1 \times 10^5$ cfu ml^{-1}
Coliform	<1 ml^{-1}
Methylene blue test	Decolorization >30 min
Threshold for bitterness and off flavour	$<1 \times 10^7$ cfu ml^{-1}
Minimum shelf life	>4 days
Freeze point depression	>0.53°C

The endospore-forming genera such as *Bacillus* and – to lesser extent – *Clostridium* can be important in terms of spoilage of products made from contaminated milk. Although the anaerobic spore-formers may survive in pasteurized milk they are usually unable to reproduce owing to high redox potential; the genus *Bacillus* is, in contrast, capable of remaining active after pasteurization, and its spores may cause spoilage of heat-treated milk. Some criteria for the acceptability of pasteurized milk are shown in **Table 3**.

The principal microorganisms growing and causing spoilage of refrigerated pasteurized milk are psychrotrophic microorganisms, and as these are heat-labile, the most common origin of psychrotrophs is post-pasteurization contamination. After heating, certain members of the Enterobacteriaceae, including *Serratia*, *Enterobacter*, *Citrobacter* and *Hafnia*, may be numerically dominant, but nevertheless the ultimate spoilage microflora consists of psychrotrophic Gram-negative rods, e.g. *Pseudomonas*, *Alcaligenes* and *Flavobacterium*.

Pasteurized milk is required to satisfy a phosphatase test. Phosphatase is an enzyme that is present in raw milk indigenously and is destroyed at a temperature only slightly higher than that used to destroy *Mycobacterium tuberculosis*.

Pathogenic Microorganisms in Milk and Milk Products

Raw milk may contain microorganisms that are human pathogens. The sources of pathogenic microorganisms may be within or outside the udder, and some of the pathogenic microorganisms that are likely to be present in milk and dairy products are discussed below.

Staphylococcus aureus

Staphylococci, and in particular *S. aureus* – some strains of which produce enterotoxins – are among the leading causes of food-borne disease outbreaks throughout the world. It has long been known that *S. aureus* may cause mastitis or skin diseases in dairy cattle or lead to food-borne intoxications in consumers of milk and milk products.

In terms of product quality and the hygienic acceptability of dairy foods, *S. aureus* is of paramount importance, because heat-treated milk is more suitable for its growth and enterotoxin production than raw milk. This may be due to the nutritional status of treated milk, but water activity, pH value and oxidation–reduction potential have also been cited. An improper heating regimen or post-pasteurization contamination may allow the multiplication of *S. aureus*.

Campylobacter jejuni

Campylobacter jejuni is now recognized as a leading cause of acute bacterial gastroenteritis in humans and raw milk has been implicated as a vehicle responsible for transmitting campylobacters to susceptible individuals. *Campylobacter jejuni* is thermally inactivated at minimal pasteurization temperatures. Its optimum growth temperature is between 30°C and 45°C, and it is killed at pH 3.0–4.5; it is also sensitive to NaCl. Since it is not particularly resistant to heat, pH, NaCl or the other usual treatments of milk, the major cause of disease outbreaks due to *C. jejuni* is the consumption of raw milk or cheese made from raw or improperly heated milk.

Yersinia enterocolitica

Yersinia enterocolitica has received considerable attention as a causative agent of human gastroenteritis. Raw and pasteurized milk have been shown to be possible vehicles for the transfer of the bacteria to human beings. As this microorganism shows little tolerance of low pH and is easily killed under pasteurization conditions, the consumption of raw milk, in particular, carries a potential health risk. The main reason for disease outbreak caused by *Y. enterocolitica*

is the post-pasteurization contamination of milk in containers and especially in holding tanks.

Yersinia enterocolitica is most likely to be found in whole pasteurized milk, cheeses such as Brie, Camembert and blue-veined cheeses, whipped cream and ice cream. It becomes inactive and finally dies in yoghurt. However, recontamination of milk with *Y. enterocolitica* after heat treatment may be hazardous for the consumer, as in the course of yoghurt preparation the inactivation rate of the bacterium may not be satisfactory.

Salmonella spp.

Salmonellae are Gram-negative, non-spore-forming rods which grow at 35–37°C and pH 6.5–7.5; at lower pH levels these microorganisms are killed off. *Salmonella* spp. are widespread in the environment and appear in a broad range of dairy products, thus posing a great problem for the industry.

Almost any milk or milk product may be contaminated with salmonellae as a consequence of mishandling or improper hygiene. In general, salmonellae are rarely indigenously present in raw milk.

Escherichia coli

Escherichia coli is a member of the family Enterobacteriaceae. The bacteria are Gram-negative, non-spore-forming, straight rods. Some strains can produce enterotoxins which may be thermolabile or thermostable; enteropathogenic strains cause diarrhoea. Some strains are able to grow at low temperature (e.g. 5°C), but most isolates grow only slowly under refrigeration conditions.

This bacterium can spoil milk and dairy products, usually with the production of gas. Early blowing in white-brined cheeses due to the metabolic activity of *E. coli* is common, and some encapsulated strains may cause ropiness in milk and cheese-brine. Only a few strains isolated from Camembert and French Brie cheeses, yoghurt, milk and other milk products have been found to cause disease.

Listeria monocytogenes

Listeria monocytogenes is the most notorious pathogen associated with cheese-related outbreaks of disease. It is a small, Gram-positive, non-spore-forming, aerobic, rod-shaped bacterium. It may also be able to survive pasteurization and grow during refrigerated storage, but the main source of *L. monocytogenes* is post-pasteurization contamination. The optimum pH range for growth is pH 6–9, and it can grow in the presence of 10% NaCl solution and at 1–5°C.

All over the world, one can expect between 3% and 7% of raw milk samples to be positive for *L.*

monocytogenes, but according to a worldwide survey carried out by the International Dairy Federation, sour cream and goat's milk or sheep's milk cheeses are more likely to be the cause of listeriosis.

Mycobacteria

Mycobacterium tuberculosis produces tuberculosis in humans, while *M. bovis* causes tuberculosis in cattle, both domestic and wild ruminants, humans and other mammals. Both species are slow-growing, and do not grow outside the range 25°C to 45°C; the optimum growth temperature is 37°C.

The incidence of *M. tuberculosis* and *M. bovis* in raw milk is quite high. Of all dairy products, cheese invites the most attention regarding the presence of these bacteria: *M. tuberculosis* may remain active in Camembert, Cheddar and Tilsit cheeses for up to 180 days, 220 days and 300 days, respectively. White cheese varieties and butter produced from raw milk may also be contaminated with mycobacteria.

Brucella abortus

Brucella abortus is a rod-shaped, non-motile, catalase-positive and oxidase-positive organism. The optimum growth pH varies from 6.6 to 7.4. The optimum growth temperature is 36–38°C, but this species can grow between 20°C and 40°C. It is a causative agent of brucellosis, a zoonosis of worldwide importance, also called 'Malta fever', 'Mediterranean fever' or 'undulant fever'.

Although *B. abortus* is killed off by pasteurization, the consumption of raw milk and dairy products derived from raw milk are among the potential sources of brucellosis. This microorganism may remain active in cheese for up to 60 days. In general, full-cream dairy products are good media for *B. abortus*. The survival times of *B. abortus* in cream, ice cream, butter and whey are approximately 6 weeks, 4 weeks, 20 weeks and 1 week, respectively.

See also: **Bacillus**: Introduction. **Bovine Spongiform Encephalopathy (BSE)**. *Campylobacter*: Introduction. **Cheese**: Microbiology of Cheese-making and Maturation. **Chilled Storage of Foods**: Principles. *Cryptosporidium*. *Escherichia coli*: *Escherichia coli*. **Heat Treatment of Foods**: Ultra-high Temperature (UHT) Treatments; Principles of Pasteurization. **Listeria**: Introduction; *Listeria monocytogenes*. *Mycobacterium*. *Psychrobacter*. *Salmonella*: Introduction. **Waterborne Parasites**: Detection by Conventional and Developing Techniques. *Yersinia*: *Yersinia enterocolitica*.

Further Reading

Celestino EL, Iyer M and Roginski H (1996) The effects of refrigerated storage on the quality of raw milk. *Australian Journal of Dairy Technology* 51: 59–63.

Cousin MA (1982) Presence and activity of psychrotrophic micro-organisms in milk and dairy products: a review. *Journal of Food Protection* 45: 172–207.

Donnelly JK and Steniford EL (1997) The Cryptosporidium problem in water and food supplies. *Lebensm-Wiss. U.-Technol.* 30: 111–120.

Fairbairn DJ and Law B (1986) Proteinases of psychrotrophic bacteria: their production, properties, effects and control. *Journal of Dairy Research* 53: 139–177.

International Dairy Federation Bulletin (1980) *Behaviour of Pathogens in Cheese.* Doc. 122: 1–23.

International Dairy Federation (1994) *Monograph on the Significance of Pathogenic Micro-organisms in Raw Milk*, p. 125. IDF.

International Dairy Federation (1996) *Proceedings of a Symposium on Bacteriological Qı :lity of Raw Milk*, p. 178, 13–15 March, Wolfpassing, Austria. IDF.

Muir DD, Kelly ME and Philiphs JD (1978) The effect of storage temperature on bacterial growth and lipolysis in raw milk. *Journal of the Society of Dairy Technology* 31: 203–208.

Robinson RK (1990) *Dairy Microbiology.* Vol. 1, p. 301. London: Elsevier.

Microbiology of Dried Milk Products

Donald Muir, Hannah Research Institute, Ayr, UK

The microbiology of dried milk products is predicated by the quality of the raw material, the conditions employed during manufacture of the product and on post-processing contamination. Most dried milk is now produced by spray drying and, for this reason, other drying methods (e.g. roller drying or freeze-drying) are excluded from consideration. Because of the complexity of the manufacturing process each step is considered in turn.

Manufacturing Processes

A schematic of the production sequence for dried whole milk is shown in **Table 1**. Milk is essentially heat treated, concentrated in an evaporator to remove most of the water then spray dried. The process is expensive both in terms of capital outlay and in terms of running and cleaning costs. Thus every effort is made to improve cost efficiency without prejudicing quality.

Table 1   Schematic of manufacturing process for dried whole milk

Material	Process	Comments
Raw milk ex. farm	Transport to creamery	
	Optional treatment: deep cooling to 2°C or thermization	Allows safe storage for 2–3 days in an insulated silo
Stored bulk milk	Clarification	Removes particulate solids and reduces bacterial count
	Standardization of fat content	
Standardized milk	Heat treatment	Primary aim to reduce level of microbiological contamination; secondary purpose is to enhance functionality of product
Heat-treated milk	Concentrate in an evaporator, usually of the falling-film type, to a solids content of 45–52%	
Concentrated milk	Homogenized at 200/20 bar at 50–70°C	Ensures stability of fat dispersion
Homogenized concentrate	Spray drying	Reduces water content to ca. 2% for dried whole milk; 4% for skim milk powder
Primary powder	Agglomeration and spray drying	Produces a cold water-soluble product
Instant powder	Packed into multilayer bags or filled into lacquered tinned steel cans, gas flushed and sealed	Improves shelf stability
Final product		

Table 2   Pathogens and potential pathogens found in raw milk

Organism	Growth at $<6°C$	Survives pasteurization[a]
Bacillus cereus	Yes[b]	Yes (spores)
Campylobacter jejuni	No	No
Clostridium spp.	(No)[c]	Yes (spores)
Escherichia coli	?	No
Listeria monocytogenes	Yes	No
Mycobacterium paratuberculosis	?	Yes (limited)
Salmonella spp.	No	No
Staphylococcus aureus	No	No
Yersinia enterocolitica	Yes	No

[a] Heat-treatment at 72°C for 15 s.
[b] Some species only.
[c] Some proteolytic species can grow at low temperature.

Table 3   Spoilage bacteria in raw milk and associated extracellular enzyme activity

	Pseudomonas		Other Gram-negative flora[a]
	Fluorescing	Non-fluorescing	
Proportion of population (%)			
Creamery silo	33.5	44.1	22.4
Farm bulk tank	50.5	31.5	18.0
Proportion of isolates with stated activity (%)			
Lipase only	5	32	0–25
Protease only	2	1	0–9
Lipase and protease	71	11	24–92

[a] Includes bacteria classified as: Enterobacteriaceae, Aeromonas, Pasturella or Vibrio; Acinetobacter, Moraxella or Brucella; Flavobacterium; Chromobacterium; Alcaligines.

Raw Milk

Most dried milk is produced in the UK from cows' milk, although small volumes of caprine milk are converted into powder. In the case of bovine milk, there are tightly focused schemes in which payment is related to milk quality. Among the quality indices, a measure of total viable count of bacteria is used as a basis for payment. A clear distinction must be made between two classes of organisms found in raw milk – pathogens and potential pathogens, and spoilage bacteria. In refrigerated, bulk milk spoilage bacteria are prevalent and pathogens are seldom present at high counts.

Nevertheless, the presence of pathogens in dried milk must be avoided. Therefore, processing conditions must assure that this is achieved. The pertinent properties of pathogens found in milk are summarized in **Table 2**. Only three of these organisms survive

pasteurization: *Bacillus cereus*, *Clostridia* spp. and, to a limited extent, *Mycobacterium paratuberculosis*. None of these bacteria present a major risk in dried milk. The other pathogens listed in Table 2 do not survive heat treatment and thus can only find their way into dried milk by post-processing contamination from the environment.

In contrast, the main spoilage organisms in refrigerated, raw bulk milk are Gram-negative, psychrotrophic bacteria (**Table 3**). *Pseudomonas* spp. of both fluorescing and non-fluorescing types predominate. The Gram-negative organisms are readily killed by pasteurization and, *per se*, pose no threat to the quality of milk or products manufactured from it.

However, many of the Gram-negative psychrotrophs produce extracellular enzymes with the potential to degrade the constituents of milk. Lipase, protease and combined lipase and protease activity are found in a substantial proportion of bacteria isolated from refrigerated, raw bulk milk (Table 3). Moreover, the enzymes are noted for their heat tolerance. Substantial proportions of activity remain after pasteurization (**Table 4**) and, surprisingly, after ultra-high temperature (UHT) at 140°C for 5 s (**Table 5**). The corollary to this observation is that, to avoid breakdown of milk constituents in products, psychrotroph numbers must not be allowed to reach the critical level at which there is sufficient activity to initiate degradation.

Several studies have sought to define this critical level. Rancidity, caused by lipase breaking down milk fat to liberate free fatty acid, was detected in Cheddar cheese made from milk in which the count of psychrotophic bacteria exceeded 7×10^6 colony forming units (cfu) per millilitre. Lipase activity has also been implicated as the cause of soapy character in chocolate. The offending ingredient was dried milk made from raw milk of poor quality. Parallel research has determined that protease activity, expressed by gelation in UHT milk, can be exacerbated when the psychrotroph count in the raw material exceeds 3×10^6 cfu ml^{-1}. Product quality must, therefore, be safeguarded by not using milk in which the count of psychrotroph bacteria exceeds 10^6 cfu ml^{-1}.

The psychrotrophic bacteria in milk grow remarkably fast in refrigerated milk with typical generation times in the range 4–12 h. In addition, growth is sensitive to small (1–2°C) differences in storage temperature. Typically, raw silo milk with an initial psychrotroph count of 5×10^4 cfu ml^{-1} has an expected 'safe' shelf life of 36 h during storage at 6°C. If the milk was deep cooled to 2°C on reception at the factory an extension of the 'safe' storage period by 24 h might be anticipated. When further extensions in 'safe' storage time are required, more drastic treatment of the raw milk is necessary. The most useful technique is thermization. Thermization is the generic description of a range of subpasteurization heat treatments that kill most spoilage bacteria found in raw milk (but not all pathogens) with minimum collateral heat damage. Thermization of good quality raw milk at 65°C for 15 s, followed by prompt cooling to 2°C offers a 'safe' shelf life of 72 h.

Milk Processing

Clarification and Fat Standardization

Early in the processing sequence, the fat content of the raw milk is adjusted. Cream is separated from skim milk using a high-speed centrifugal separator. This device separates the milk on the basis of the density difference between the 'light' milk fat globules and the relatively dense serum. The two liquid streams are then recombined to yield product with the required fat content. The overall effect of fat standardization on the microbial population of milk is modest. However, another subtly different separation process – clarification, sometimes called bactofugation – may also be applied. Clarifiers or bactofuges are special separators, which remove microorganisms from milk based on the density difference between the bacterium and the serum phase of the milk. This density difference is greatest in the case of the spores of spore-forming bacteria found in raw milk (e.g. *Bacillus* spp. or *Clostridium* spp.). A modern clarifier can achieve a 90% reduction in spore count in a single pass. Clarification is particularly valuable because, although the vegetative cells of spore-forming bacteria are inactivated by modest heat treatment, the spores can resist fairly severe heating (see below). Thus, clarification offers an alternate method of controlling the spore count of the finished product.

Heat Treatment

Heat treatment during production of milk powder serves two distinct purposes. It not only controls microbial quality, but also influences functionality. For example, it is usual to apply a severe heat treatment to milk destined for manufacture into whole milk powder. Such heating results in denaturation of whey protein. The presence of denatured protein

Table 4 Residual activity of extracellular enzymes after pasteurization at 72°C for 15 s

Enzyme activity	Residual activity (%)
Lipase	59
Protease	66
Phospholipase C	30

Table 5 Residual enzyme activity after heat treatment of cell-free supernatant at 140°C for 5 s

Bacterial type	Enzyme activity (%)		Phospholipase C
	Protease	Lipase	
Pseudomonas			
Fluorescent	17–50	14–51	31–57
Non-fluorescent	5–48	0–73	5
Other Gram negative[a]	0–57	0–82	0–40
Bacillus cereus	<5	–	–
Bacillus firmus	<5	–	–

[a] Includes bacteria classified as: Enterobacteriaceae, *Aeromonas*, *Pasturella* or *Vibrio*; *Acinetobacter*, *Moraxella* or *Brucella*; *Flavobacterium*; *Chromobacterium*; *Alcaligines*.

reduces the rate of lipid oxidation during subsequent storage. Pasteurization (63°C/30 min or 72°C/15 s) kills most pathogens (Table 2) and all Gram-negative, psychrotrophic spoilage bacteria. However, a residual population of heat-resistant bacteria remains. These bacteria are called thermoduric and comprise members of the coryneform group, heat-resistant streptococci, micrococci and spore-forming bacteria. The predominant spore-forming organisms found in heated milk are *Bacillus* spp., which survive heat treatment in the spore form (**Table 6**). *Bacillus* spp. degrade milk readily and are noted for their phospholipase activity. Spoilage due to these organisms is often associated with damage to the milk fat globule membrane and is characterized by the defect known as 'bitty cream'. Two species of *Bacillus*, *B. stearothermophilus* and *B. thermodurans*, pose particular threats because of their extreme heat resistance. As described above, the population of spores in milk can be reduced by clarification. If very low spore counts are required in the product, then severe heat treatments must be applied to the milk: typically 110–120°C for 30 s.

Concentration and Homogenization

Milk is concentrated to 45–55% solids in an evaporator before spray drying. Apart from the expected increase in count caused by the concentration process, there is an additional potential hazard. In multistage evaporators, typical of modern dairy plant, concentrate may be held for extended periods at the temperature range 45–55°C. Some heat-resistant bacteria can grow under these conditions and, as a result, the bacterial content of the concentrate can increase disproportionately. After concentration, the product is homogenized to reduce fat globule size and inhibit creaming. Homogenization may cause an increase in bacterial count as a result of the disaggregation of bacterial clusters.

Table 6 Heat-resistant bacteria in milk and their associated enzyme activity

	Bacillus *spp.*	Coryneform group
Proportion of isolates (%)		
63°C/30 min	54	46
80°C/10 min	61	37
Isolates with enzyme activity (%)		
Protease + lipase	37	10
Protease only	34	3
Phospholipase	80	0
Inactive	12	67

Spray Drying, Agglomeration and Coating

Powder is formed by atomizing the concentrated milk in a stream of very hot air (typically 180–200°C). However, as a result of evaporative cooling, the temperature of the droplet remains low during the drying process. As a result, the bacterial load of the concentrate largely determines that of the powder. Provided the ultimate moisture content of the powder is below 4%, bacteriostasis is assured. (In the case of dried whole milk the moisture content may be as low as 2%, to inhibit fat oxidation during extended storage.) Operations on the powder downstream of the drier have little further effect *per se* on bacterial load. Nevertheless, serious deterioration of powder quality can occur from environmental contamination.

Environmental Contamination

When powder comes into contact with contaminated air surfaces, serious problems can arise. For example, a crack in a spray drier wall can result in a reservoir of active bacteria within the material insulating the spray drier. Such reservoirs are resistant to normal cleaning and disinfection procedures and can harbour pathogenic or spoilage bacteria. In addition, the air used for conveying powder must be sterile. The modern strategy to prevent powder recontamination involves careful separation of raw from heated product, tight control of environmental hazards and scrupulous attention to cleaning and disinfection of surfaces that come into contact with the dried milk.

Process Monitoring

It is apparent that limited information on the microbiological status of a spray-drying plant can be deduced from examination of the quality of the powder alone. Multipoint sampling is more effective especially if the bacterial load of (a) the raw milk, (b) stored milk from the balance tank of the heat exchanger, (c) concentrate ex. evaporator, (d) powder ex. primary cyclone and (e) packed product are monitored. It is prudent to include routine swabs from drains, walls etc. in the monitoring operation because these can be valuable indicators of potential hazards.

Suggested Standards

There is no single standard for the microbial status of dried milk products. Specifications vary from country to country and from customer to customer within countries. Nevertheless, there is an overall measure of agreement and this is reflected in the suggested values proposed in **Table 7**. It should be noted that no

Table 7 Suggested microbiological standards for dried milk products

Contaminant	Skim/whole milk powder	Casein/caseinates	Whey powder
Total viable count	3×10^4 cfu g^{-1}	3×10^4 cfu g^{-1}	5×10^4 cfu g^{-1}
Coliforms	< 10 cfu g^{-1}	Absent in 0.1 g	< 10 cfu g^{-1}
Escherichia coli	Absent in 25 g	Absent in 25 g	Absent in 25 g
Staphylococcus aureus	Absent in 25 g	Absent in 25 g	Absent in 25 g
Salmonella spp.	Absent in 200 g	Absent in 200 g	Absent in 200 g
Yeasts and moulds	< 10 per g	< 10 per g	< 10 per g
Thermophilic spores	< 30 per 2 g		

account is generally taken of the potential threat of residual enzyme activity derived from psychrotrophic bacteria in the raw material from which the powder was made. Protection from this undesirable occurrence could be assured by the specification that, at the point of manufacture, the total viable count should not exceed 1×10^6 cfu ml^{-1}.

See also: **Bacillus**: Introduction; *Bacillus cereus*. **Cheese**: Microbiology of Cheese-making and Maturation. **Clostridium**: Introduction. **Dried Foods**. **Heat Treatment of Foods**: Ultra-high Temperature (UHT) Treatments; Principles of Pasteurization. **Milk and Milk Products**: Microbiology of Liquid Milk; Microbiology of Cream and Butter. **Mycobacterium**. **Process Hygiene**: Designing for Hygienic Operation; Overall Approach to Hygienic Processing; Modern Systems of Plant Cleaning; Risk and Control of Aerial Contamination. **Pseudomonas**: Introduction.

Further Reading

Early R (1998) *The Technology of Dairy Products*, 2nd edn. London: Blackie Academic and Professional.

Harding F (1995) *Milk Quality*. London: Blackie Academic and Professional.

Masters K (1988) *Spray Drying Handbook*, 4th edn. London: George Goodwin.

Shapton DA and Shapton NF (1991) *Principles and Practice for the Safe Handling of Food*. Oxford: Butterworth Heinemann.

MICROBIOLOGY OF CREAM AND BUTTER

Rekha S Singhal and **Pushpa R Kulkarni**, University Department of Chemical Technology, University of Mumbai, India

Cream and butter are valuable ingredients as well as products useful to the food processing industry. Owing to their high fat content, these products pose unique spoilage problems of a chemical and microbial nature. This article summarizes the information on various types of creams and butters with special reference to the manufacturing processes, risks of contamination during processing, typical microflora associated with spoilage and public health risks, recommended microbial standards and possible problems during storage.

Butter was one of the first dairy products manufactured by humans and has been traded internationally since the fourteenth century. Its manufacture relies on cream. Cream was obtained by gravity separation of milk until the 1850s when factories began producing butter on a small scale. Large-scale manufacture became possible only after development of the mechanical cream separator in 1877. The world consumption of butter and butterfat products was estimated at 2 420 000 tonnes in 1993.

Cream

Definition and Types

Cream products are dairy products that are enriched to a varying degree with milk fat. Creams may be acidified or non-acidified, whipped, and may or may not have additives. Classification of cream is mainly on the basis of fat content, application and manufacture. The cream types available in the European market are slightly different from those available in the USA. **Table 1** lists some of the different types of commercial creams. The Food and Agriculture Organization (FAO) classification of cream is given in **Table 2**. In Germany, coffee cream and whipping cream with 10% and 35% minimum fat respectively are also available.

Manufacture

The quality of cream depends on the chemical, physical and microbiological properties of the milk, and hence milk for cream production should be properly specified. In addition, milk should be carefully handled to prevent damage to the fat globules during pumping and agitation, since this may result in free fat which may coalesce or 'churn', making the separation difficult. The general steps involved in the industrial manufacture of cream are as follows.

Table 1 Commercially available creams

Cream type	Fat content (% by weight)	Applications
Half cream or single cream	10–18	As pouring cream for use in desserts and beverages; as breakfast cream poured over fruit and cereals; used industrially as an ingredient of canned soups and sauces
Coffee cream	Up to 25	To give an attractive appearance to coffee with appropriate modification in flavour
Cultured or sour cream	< 25 normally Occasionally up to 40	In confectionery, and in meat and vegetable dishes
Whipping cream	30–40	For toppings and fillings for baked goods
Double cream (marketed in Europe)	> 48	Used in desserts and whipped in gateaux
Clotted cream	> 55	Used as spread on scones in conjunction with fruit preserves
High-fat creams (plastic cream)	70–80	For ice cream manufacture

Table 2 Food and Agriculture Organization classification of cream

Product type	Fat content (%)
Cream	18–26
Light cream (or cream with additional terms such as coffee cream)	> 10
Whipping cream	> 28
Heavy cream	> 35
Double cream	> 45

Production of Milk on the Farm　This should be done in as hygienic a manner as possible, since although vegetative cells may be killed by subsequent heat treatment, spores and organisms such as *Bacillus cereus* can survive and cause subsequent spoilage of the milk.

Transport to the Dairy　Milk should be transported at 5°C. It is a common practice to hold the milk at 5°C for up to 48 h in creameries.

Storage of Milk in the Dairy　Milk should be stored at less than 5°C in silos or suitable tanks until cream manufacture.

Separation and Standardization of Cream　Milk is heated at 44–55°C for separation of the cream. Temperatures less than 40°C yield a product with high viscosity and the potential for development of off flavours due to lipolytic activity, whereas temperatures above 55°C cause the cream to thicken rapidly and excessively during storage. The cream separation is carried out continuously in centrifugal separators which have separate ports for skimmed milk and cream. The mechanics of keeping cream separated from skimmed milk depends on the type of centrifugal separator used. The fat content in the cream is controlled by restricting the cream outlet by means of a throttling valve or 'cream screw'.

Homogenization of Cream　Homogenization is an extra treatment and hence increases the risk of contamination of cream. Homogenization increases the viscosity, which is preferred by consumers, but also increases the potential for light-induced rancidity (manifested as oxidized flavour) owing to the increased surface area resulting from homogenization. It is used only for some types of creams, such as half cream or single cream, to prevent fat separation. Double cream may also be lightly homogenized. Whipping creams in general are not homogenized since inhibition in the formation of a stable foam is encountered. Homogenization is carried out after standardization at 65°C and 17 MPa.

Heat Treatment of Cream　Cream is a high-moisture product and has a short shelf life. Heat treatment extends the shelf life by inhibiting the growth of pathogenic and spoilage organisms and denaturing the indigenous lipases which may promote rancidity. Heat treatments must conform to one of the following minima:

- pasteurization at 63°C for at least 30 min or not less than 72°C for 15 s; temperatures up to 80°C are also used
- sterilization at 108°C for 45 min
- ultra-high temperature (UHT) treatment at 140°C for 2 s.

Pasteurization causes reduction in the viscosity of the cream, and also produces some sulphurous notes which disappear on storage. Higher temperatures result in cooked flavours, and may impair cream quality by possibly activating bacterial spores. A major defect of pasteurized cream that has not been homogenized is the formation of 'cream plugs'. This is attributed to the free fat which welds the globules together, and in extreme cases may even solidify the cream into gel. The fat composition and rate of cooling of the cream are other factors that affect plug formation.

The efficacy of the heat treatment must be checked by testing for *Escherichia coli* and phosphatase. Application of the phosphatase test to pasteurized creams presents special problems owing to reactivation of phosphatase on storage.

Packaging Pasteurized cream for domestic consumption is packed in plastic pots or cardboard cartons. Polystyrene containers should be avoided because this material can cause taints; polypropylene pots are generally preferred. These plastic packaging materials are generally used for holding about 5–10 l of cream. Sterilized cream is mostly produced in cans. Cans are sterilized with superheated steam, while aerosol cans are sterilized by hydrogen peroxide. Bulk quantities of cream (2000–15 000 l) are transported in stainless steel tankers.

Storage and Distribution of Cartoned Cream A temperature of not more than 10°C during storage and distribution is recommended, with 5°C being preferred. Cream should not be stored in places that house odoriferous materials such as disinfectants, paints, varnishes, scents or strong-smelling foods, since the cream may be rendered inedible.

Sale – Possibly a Multistage Operation Cream presents more problems than milk owing to distribution methods and the requirements for longer keeping quality. Sales are erratic, depending on the weather, holiday seasons, local activities and other such factors. Cream should be dispatched from the manufacturing dairies in chilled distribution vehicles to larger retailers who may then supply the product under identical conditions to smaller retailers or their shops via wholesalers.

A typical flow sheet for the manufacture of sterilized cream is shown in **Figure 1**. Apart from clotted cream, most creams are produced by means of mechanical separators. Clotted cream has a very high viscosity, a golden creamy colour and granular texture. **Figure 2** is a flow sheet for the manufacture of clotted cream.

In whipping creams, air is incorporated at the air–water interface and there is a disruption of the milk fat globule membrane. Some important factors for whipping cream are:

- the amount of beating required to form a stable aerated structure
- overrun, expressed as the percentage volume increase of the cream due to air incorporation
- the stiffness of the whipped cream
- serum leakage from whipped cream which is due to overwhipping and leads to sogginess if used in cakes.

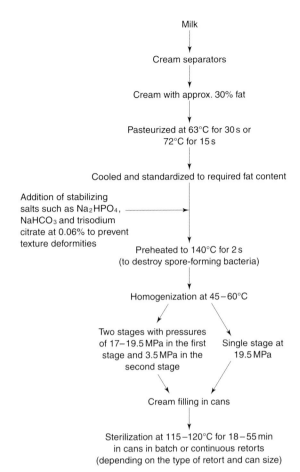

Figure 1 Flow sheet of typical manufacturing process for sterilized cream.

Factors that affect whipping properties of cream are the fat content, temperature (should be below 10°C) and the distribution of the size of fat globules and membrane structure. Whipping creams can also be foamed by aerosols. In this process, cream is filled into hermetically sealed cans that are prefilled with an inert gas such as nitrogen. Although the foam stability in aerosol-foamed creams is low, it can be compensated for with stabilizers; this also prevents microbial spoilage.

Sour cream is made by inoculating cream with cultures of lactic acid-producing bacteria such as *Lactococcus lactis* subspp. *lactis* and *cremoris*, and flavour-producing bacteria such as *Leuconostoc mesenteroides* subspp. *cremoris* and *dextranicum*. Souring takes place at 20°C and avoids spoilage by thermophilic organisms.

The creams are processed in different ways and sold accordingly. For example, sterilized cream has a distinct caramelized flavour due to the in-can sterilization process, is usually sold in cans and has a shelf life of about 2 years. The temperatures employed are 110–120°C for 10–20 min. This severe heating brings

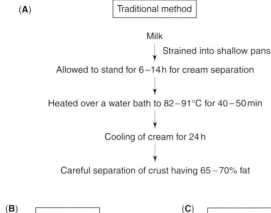

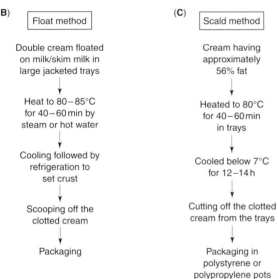

Figure 2 Flow sheet of typical manufacturing processes for clotted cream. (**A**) Traditional process. (**B, C**) Commercial processes.

about protein denaturation, Maillard browning and fat agglomeration which collectively modify the texture and flavour of the cream. A process for rapid sterilization of cream, known as the ATAD friction process, consists of preheating the cream to about 70°C and then heating to 140°C for 0.54 s. This process can be applied successfully to creams ranging in fat content from 12% to 33%. Double, whipping, single and half cream may be UHT treated, or frozen after adequate pasteurization. Ultra-high temperature sterilization at 135–150°C for 3–5 s followed by aseptic packaging does not induce chemical changes in the same way as in-container sterilization, but creaming and fat agglomeration do take place on storage. In this process the shelf-life is limited by biochemical rather than microbiological considerations. Since all forms of microorganisms are destroyed, the cream can be stored indefinitely without refrigeration. Calcium–casein interactions destabilize the emulsion, and any proteases surviving the heat treatment may bring about gelation. The

development of a stale or 'cardboardy' flavour generally limits the shelf life to 3–6 months. Problems arise in controlling the UHT method for high-fat creams.

Bulk storage of surplus cream may be done by freezing after pasteurization. The temperatures employed are –18°C to –26°C. A shelf life of 2–18 months (average 6 months) is achieved. The cream is frozen in rotary drum freezers: a rotating drum containing recirculating refrigerant is immersed in a vat of cream to form a frozen film. The process can be made continuous and ensures rapid freezing with less damage to the cream than plate freezing. In plate freezing, plates containing the refrigerant are arranged vertically in parallel. The cream is poured into gaps between the plates, and this causes instantaneous surface freezing which progresses towards the centre. Cryogenic freezing using liquid nitrogen coming in contact with cream in a countercurrent manner in an insulated tunnel has also been shown to be a useful technique. However, for a good freeze-thaw stability of frozen creams, care must be taken to preserve the natural milk fat globule membrane. It is for this reason that homogenization should be avoided for frozen creams.

Additives for stabilization and improvement of whipping properties of cream are permitted in many countries. Gelatin and carboxymethylcellulose serve mainly to increase the viscosity, while the alginates and carrageenan interact with calcium–casein–phosphate complex to enhance whipping properties. Emulsifiers and stabilizers improve the freeze-thaw stability of cream. Low-molecular-weight sugars such as glucose and sucrose also impart freeze-thaw stability. Nutritive sweeteners and characteristic flavouring and colouring ingredients are also sometimes used. Cream powders, and imitation creams produced by emulsifying edible oils and fats in water, are other products available for industrial use.

The keeping quality of creams can be enhanced by following Good Manufacturing Practices which take into account the hygienic conditions of the plant and quality control of the products. Some steps that can ensure this quality assurance to the manufacturer and the consumer are as follows:

- sanitizing all items that come into contact with cream at any stage by removing any dirt and killing residual organisms by heat or chemical disinfectants such as chlorine compounds
- in a manually operated plant, ensuring good supervision by managerial staff
- controlling the contamination of the air around the fillers – this is often neglected

- packaging creams in rooms that are separate from processing activities
- using water of acceptable bacteriological quality (5 p.p.m. available chlorine is generally used)
- in-line testing of cream equipment.

Microflora of Retail Cream

The main source of microorganisms in butter made under excellent sanitary conditions is cream. Fat globules in the raw milk carry microbial pathogens as well as spoilage organisms which originate from the udder of the cow or from the hides and milking lines used in the processing. Bacteria found in milk may be placed in three groups from the point of view of the keeping quality of milk:

- thermolabile organisms such as *Pseudomonas* spp. and coliforms which are destroyed during pasteurization
- thermoduric non-spore-formers such as *Enterococcus faecalis*, which survive pasteurization
- aerobic spore-formers such as *Bacillus cereus*.

Generally, Gram-negative organisms, yeasts and moulds are destroyed during the pasteurization conditions used in processing cream. Psychrotrophic *Bacillus* and *Clostridium* spp. survive cream pasteurization and so do heat-resistant microbes such as some strains of *Lactobacillus*, *Enterococcus* and spores of *Bacillus* and *Clostridium*. *Bacillus* spp., particularly *B. cereus*, have been shown to be more important contributors to product failure in late summer and early autumn than at other times of the year. *Bacillus cereus* is also of particular importance in creams, since it can reduce methylene blue and hence lead to failure in the official Public Health Laboratory Services (PHLS) test. *Bacillus licheniformis* and *B. subtilis* are other species that produce spores and cause thinning and bitterness in cream. Yeasts are occasionally involved in the spoilage of pasteurized whipping cream containing added sucrose: *Candida lipolyticum* and *Geotrichum candidum* are of greatest importance. Controversy exists about the survival of *Mycobacterium paratuberculosis* during cream pasteurization. In unpasteurized cream, *Streptococcus agalactiae*, *S. pyogenes*, *Staphylococcus aureus* and *Brucella abortus* survive for varying periods of time and find their way into butter. Pasteurization at 62.8°C for 30 min of butter made from contaminated cream can eliminate these organisms. Spoilage of UHT cream is normally due to failure of packaging systems and entry of post-processing contaminants. Endospores of *Bacillus* species may survive both UHT and in-container sterilization. In pasteurized clotted cream, the microflora depends on the nature of the process, the degree of

control and the standard of hygiene. In most cases, *Bacillus* spp. are dominant, although non-endospore-forming thermoduric species such as *Enterococcus* are present where lower cooking temperatures are used.

Development of flavour defects in cream is associated with high numbers ($> 10^7$ per millilitre) of psychrotrophs due to post-processing contamination of milk or cream. These produce lipolytic enzymes which cause hydrolytic rancidity. The rancid flavour is caused mainly by fatty acids of C_4 to C_{12} while long-chain acids of C_{14} to C_{18} make little contribution. This is due to the lower threshold of short-chain fatty acids in fat than in water compared with long-chain fatty acids. Processing steps such as agitation of milk at 5–10°C or 37°C results in a severalfold increase in lipase activity associated with the cream. The transferred enzyme is bound to the milk fat globule membrane wherein it has enhanced heat stability. This redistribution is of relevance in butter manufacture. Homogenization of the cream at high pressure, slow cooling and subsequent storage at higher temperatures, slow freezing and repeated freeze-thawing also promote lipolysis. Difficulties have been experienced in churning cream made from rancid milk. The cream foams excessively and may take up to five times longer than normal cream to churn. Lipolysed milk and products prepared therefrom may slow down the manufacture of fermented products (in this case, sour cream) owing to the inhibitory effects of free fatty acids developed during the hydrolytic rancidity. Proteolytic enzymes can originate from germination of heat-resistant spores of *Bacillus subtilis* in sterilized cream. This may produce bitterness and also cause coagulation. The lipolytic enzymes in these products cause thinning of the cream. These enzymes produce a 'bitter' taint, which reflects the use of poor-quality raw milk rich in spores or the use of dirty equipment. Presence of non-spore-forming organisms in sterilized cream indicates contamination after sterilization, and in canned cream indicates a defective or leaking can. *Proteus* may cause bitterness and thinning due to attack on proteins to produce peptones and polypeptides. Associated growth of organisms may also be responsible for bitterness. This is exemplified by the production of bitterness by *Rhodotorula mucilaginosa* after souring by lactic organisms such as *Lactococcus lactis*. Coliforms may produce gas, and cocci may give rise to acid curdling. All these are common water-borne organisms. Yeasts such as *Torula cremoris*, *Candida pseudotropicalis* and *Torulopsis sphaerica* may produce fruity flavours in whipped cream containing added sugar.

Herbage-derived substances that are fat-soluble may also produce flavour taints in cream. This can be readily removed by the steam distillation process developed in New Zealand, which pasteurizes the cream and also removes the undesirable volatiles. Other types of chemical taints have been reported to be caused by cows eating certain plants such as garlic and decaying fruits.

Butter

Manufacture and Typical Microflora of Fresh Butter

Butter is a water-in-oil emulsion with fat as the continuous phase, obtained by the phase reversal of cream during the churning process in its manufacture. The manufacture of butter is shown in **Figure 3** and the steps involved along with their process conditions and

significance are outlined in **Table 3**. Cream for butter manufacture should have at least undergone a heat treatment of 74–76°C for 15 s, should have a fat content of about 40% so as to be amenable for continuous butter-making, and should be cooled to 4–5°C for at least 4 h (preferably overnight) to enable sufficient crystallization of the liquid fat. This is followed first by ageing to bring about an equilibration of the α and β forms of fat crystals, and then by agitation in tanks, washing, salting, packaging, storage and then repackaging.

Butter may be either sweet cream butter which may or may not be salted, or ripened cream butter in which lactic acid bacteria ferment the citrate in the cream to flavour-imparting compounds such as acetoin and diacetyl. Sweet cream butter is bland in taste but has

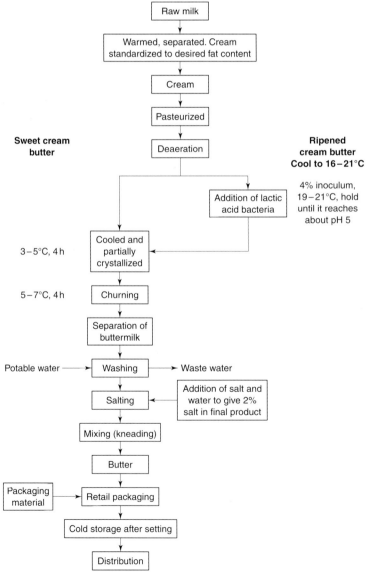

Figure 3 Flow sheet for manufacture of butter.

Table 3 Steps in butter manufacture

Step	Process	Significance
1. Concentration of fat phase in the milk	Using cream separators	For separation and standardization of resultant cream to the desired fat content
2. Crystallization of the fat phase	(a) For sweet cream – cooling at 5°C for at least 4 h after pasteurization of cream at 66°C for 30 min (b) for ripened cream – addition of lactic acid bacteria to pasteurized cream after cooling to 16–21°C until a pH of 5.0 is reached and then followed by cooling to 3–5°C	Develops an extensive network of stable fat crystals
3. Phase separation and formation of water in oil emulsion	Churning and working a proper blend of solid and liquid fat, usually at 5–7°C	(a) Disrupts membranes on milk fat globules, followed by effective clumping that further causes butterfat to harden (b) Enhances diacetyl production in ripened cream butter
4. Washing	Rinsing with water	Removes excess buttermilk
5. Salting	Using finely ground salt or brine containing 26% w/w salt, or slurries of salt in saturated brine containing 70% sodium chloride	Inhibits microbial growth
6. Packaging	Cardboard boxes lined with vegetable parchment, aluminium foil or plastic films for bulk packaging	Protection from air, workers, plant environment and temperatures that may promote spoilage
7. Storage	−15°C to −30°C	
8. Repackaging for retail outlets		

a nutty or boiled milk flavour, this is preferred in America, Australia and New Zealand. In Europe, Latin America and Asia the preference is for intense flavour which can be developed by use of milk cultures. In ripened cream butter, diacetyl formation can be enhanced by incorporation of air by intensive stirring of the cream, using ripening temperatures below 15°C, maintaining an optimal pH below 5.2 and adding 0.15% citric acid to cream. The level of diacetyl in ripened cream butter is $0.5–2.0\,\mathrm{mg\,kg^{-1}}$ butter. Diacetyl also inhibits Gram-negative bacteria and fungi. Another type of butter is whey cream butter, which is processed from whey cream and is indistinguishable from sweet cream butter. Whey cream in turn is obtained from milk fat recovered from the whey produced during cheese-making. In the manufacture of ripened cream butter, pasteurized cream is cooled to 6–8°C for 2 h or more to initiate fat crystallization followed by warming to 19–21°C and then inoculation with pure or mixed strains of *Lactococcus lactis* subspp. *lactis*, *cremoris* and *diacetylactis*, and *Leuconostoc mesenteroides*. In some areas of Europe *Candida krussi* has been tried in mixed cultures. Ripening occurs for 4–6 h until a pH of 4.6–4.7 is achieved, and the product is then cooled. Spoilage microorganisms are primarily controlled through the bacteriostatic effect of lactic acid produced in the fermentation.

During the churning and working of the cream, most of the starter culture is retained in the buttermilk; however, about 0.5–2.0% of the cultures remain in the butter.

The NIZO method for manufacture of cultured butter is used in many factories in western Europe. In this method, starter culture is not added to the cream, but instead a mixture of diacetyl-rich permeate and starter cultures is worked into butter. The permeate is itself produced by the fermentation of delactolized whey or other suitable medium. The method has advantages of greater control over the manufacturing process, lower risks of oxidative defects, lower chances of hydrolytic rancidity, less need of starter cultures, better quality of butter even after 3 years of cold storage, and elimination of pumping problems often encountered with viscous ripened cream. The pH of the butter made with this process is also easier to adjust to the desired range of 4.8–5.3.

Salt is added to butter after removal of the excess buttermilk. It should distribute evenly in the moisture phase of the product and can inhibit microbial growth. Salt can be used in finely ground form or as a slurry in saturated brine solutions containing up to 70% sodium chloride. The microbial quality of water used for washing or for preparation of brine is of great importance: *Listeria* is known to survive in a saturated brine solution at 4°C for 132 days, hence water used to prepare brine must be free of *Listeria*. Psychrotrophic organisms multiply in salted butter stored at temperatures as low as −6°C, owing to the fact that salt lowers the freezing point of water and allows the growth of these organisms.

Addition of herbs for herb butters is also practised in certain countries to increase the variety of foods.

Table 4 Most common spoilage and defects encountered in butter

Nature of spoilage or defect	Causative factors
Spoilage	
Bacterial spoilage	Contaminated water supplies
	Improper distribution of salt
	Temperature abuse in sweet cream butter
Putridity or 'surface taint' and hydrolytic rancidity	*Pseudomonas* spp. such as *P. fragi*, *Shewanella putrefaciens* and *P. fluorescens* which grow on butter surfaces at 4–7°C and produce proteases and lipases
Mould growth producing musty flavour	Growth of *Rhizopus*, *Geotrichum*, *Penicillium* and *Cladosporium* which may cause hydrolytic rancidity
	Humidity above 70%
	Improper personal hygiene of workers in the manufacturing plant
Malty flavour, skunk-like odour and black discoloration	Growth of *Lactococcus lactis* var. *maltigenes*, *Pseudomonas mephitica* and *Alteromonas nigrifaciens*
Colour changes	Surface growth of various fungi which produce coloured spores
Acid production	Growth of yeasts such as *Saccharomyces*, *Candida mycoderma*, *Torulopsis holmii*
Defects	
Metallic taste and smell	Over-acidification of cream, high level of metallic ions in wash water, defects of tinned utensils
Soapy taste and smell	Contamination with cleansing agent residues
Short, brittle structure	Butterfat too hard, improper cooling during ripening
Salve-like, greasy structure	Too much liquid fat in fat globules, defects in ripening, too high buttering and kneading temperature
Streaky, marbled appearance	Uneven salt distribution, blending of butter
Flat or insipid flavour in freshly made butter	Excessive washing of butter grains during manufacture, dilution of cream with water, initial stages of bacterial deterioration
Medicinal flavour in butter	Use of medicaments for treating cows, presence of chlorine compounds in milk or cream

Soft butters that are spreadable at 5°C can be produced by the following ways:

1. Using cream from the summer period, which is the softest.
2. Subjecting the butter to extra working.
3. Incorporating about 10% soft vegetable fat with milk fat to give a normal butter composition (80% fat minimum, 16% moisture).
4. Reducing the percentage fat below the traditional 80% to as little as 50% fat. This low-fat dairy spread contains about 11–15% milk solids and emulsifiers in order to maintain a stable water-in-fat emulsion.

However, traditionally and legally, butter must contain not less than 80% of only milk fat. The products described in (3) and (4) above cannot be called butter, but are dairy-type spreads.

At least six different continuous processes of butter manufacture are available, each of which utilizes specialized processing equipment and technology. Continuous processes for butter manufacture are generally economically advantageous and are finding increasing use.

Possible Problems During Storage All commercial butter is produced from pasteurized cream. If properly performed, pasteurization destroys all pathogens and more than 99.9% of organisms present in milk or cream. The only avenue for infection and spoilage during storage is post-pasteurization carelessness. Microbiologically induced flavours developed prior to pasteurization may be carried over into butter. The introduction of stainless steel equipment has eliminated many flavour problems, particularly those due to yeasts and moulds.

Spoilage of Butter

The low temperatures (below –10°C) employed for bulk storage of butter are inhibitory to the growth of most microorganisms, and a general decline in numbers is expected. The lethal effect of temperature is, however, selective. Survival of *Micrococcus* spp. and yeasts is generally greater than that of the Enterobacteriaceae. Microorganisms that enter during reworking and packaging and those that survive low temperatures are capable of growing during retail and domestic storage at temperatures greater than 0°C.

Various types of spoilages and defects have been encountered in butter (**Table 4**). All of the *Pseudomonas* groups found in butter are psychrophiles, and have been traced to wash water. They grow well at refrigerator temperatures and produce putrid or lipolytic flavours in 5–10 days. These psychrophiles also produce extracellular phospholipases, which are important, since they have the ability to degrade the phospholipids of the milk fat globule membrane,

Table 5 Suggested microbiological standards for cream and butter

Product	Test	Count or results[a]
Raw cream for direct consumption	Total bacterial count	<30 000 (10 000) per ml
	Total coliform count	<30 (10) per ml
	E. coli (faecal type)	1 (10) per ml
	Methylene blue reduction time (at 36°C)	Not less than 7 h
	3 h resazurin test (at 36°C, Lovibond disc No. 4/9)	Not less than 4
	Staphylococcus aureus (coagulase-positive)	<10 (1) per ml
	Somatic cell count	<500 000 (250 000) per ml
Pasteurized cream	Total bacterial count	<30 000 (5000) per ml
	Total coliforms	<1 (0.1) per ml
	E. coli (faecal type)	Absent in 1 ml
Butter	Contaminating organisms (non-lactic-acid bacteria)	<10 000 (5000) per g
	Total bacterial count (non-cultured butter only)	<50 000 per g
	Total coliforms	<10 (1) per g
	E. coli (faecal type)	Absent in 1 g
	Staphylococcus aureus (coagulase-positive)	Absent in 1 g
	Yeasts and moulds	<10 per g
	Proteolytic organisms	<100 per g
	Lipolytic organisms	<50 per g

[a] Figures in parentheses are target values.

thereby increasing the susceptibility of milk fat to lipolytic attack. Most lactic starters used in the manufacture of fermented milk products have a weak lipolytic activity. While it is the natural milk lipase that accounts for hydrolytic rancidity in milk and cream, microbial lipases assume greater significance in stored products. The off flavours developed as a result of hydrolytic rancidity are described variously as 'bitter', 'unclean', wintry', 'butyric' or 'rancid'. These defects are sometimes evident during manufacture, but may also develop during storage. Butters made from creams that have undergone hydrolytic rancidity may not show this defect, since the rancid flavours arising due to short-chain fatty acids (C_4 and C_6) are water-soluble and are readily lost in buttermilk. However, if lipolysis occurs after the manufacture of cream, off flavours associated with butters with a low degree of hydrolytic rancidity may also appear to be intense. This discussion is valid only for sweet cream butter. Ripened cream butter is believed to be less prone to hydrolytic rancidity. However, yeasts such as *Candida lipolyticum*, *Torulopsis*, *Cryptococcus* and *Rhodotorula* are capable of growth and lipolysis at low temperatures. These are particularly favoured at the low pH of some cultured cream butter.

Microbiological Standards for Cream and Butter

Microbiological standards for cream are not favoured by many because of the complexity of the factors involved. However, based on total colony count, methylene blue reduction test and coliforms, a distinction can be made between satisfactory, doubtful and unsatisfactory types. Suggested standards for cream satisfactory for buttermaking are counts of less than 1 cfu ml^{-1} for yeasts, moulds and coliforms, and a total colony count of less than 1000 cfu ml^{-1}. **Table 5** gives the limits that have been proposed as microbiological standards for cream and butter. Complying with these standards may be a useful tool to give good-quality products. Bacteriological standards for cream in some countries have been laid down as follows:

● Northern Ireland:
  untreated – bacterial count < 50 000 per gram
  pasteurized – no coliforms in 1 g

● Canada:
  count less than 50 000 per gram
  no coliforms in 1 g
  phosphatase negative

● Sweden:
  count < 10 000 per gram
  coliforms < 10 per gram
  aerobic spores < 100 per gram.

Public Health Concerns

The incidence of documented food poisoning associated with butter consumption was low even before the widespread pasteurization of cream for manufacture of butter. Early outbreaks of diphtheria (caused by *Corynebacterium diphtheriae*) and tuberculosis (caused by *Mycobacterium tuberculosis* or *M. bovis* in naturally contaminated cream) in the USA and Europe, and of typhoid fever (caused by *Salmonella typhi*) in the USA from 1915 to 1927, have been reported to be caused by butter. Butter contaminated by a convalescent carrier of *S. typhi* was responsible for 35–40 cases (including 6 deaths) of typhoid fever in Minnesota in 1913. *Salmonella typhi* can survive for 2–4 weeks in butter prepared from contaminated cream and hence acts as a vehicle of infection in outbreaks. Some major public health concerns with respect to cream and butter are outlined below.

Aflatoxins

Aflatoxins are secondary metabolites produced by certain moulds, namely *Aspergillus flavus*, *A. parasiticus* and *A. nomius*, and are recognized as extremely potent liver carcinogens for both animals and humans. Four types of aflatoxin designated AFB1, AFB2, AFG1 and AFG2 are currently recognized, of which AFB1 is the most potent and comes from contaminated feeds. Ingestion of aflatoxin-contaminated animal feed leads to the excretion of the less toxigenic AFM1 in milk within 12–24 h. While many countries have legislation regarding aflatoxin limits in animal feed, the USA and many European countries also have legislations for maximum levels of AFM1 in milk and other dairy products. Present evidence indicates the level of AFM1 in milk and dairy products to be relatively unaffected by pasteurization, sterilization, fermentation, cold storage, freezing, concentration or drying. However, treatment with hydrogen peroxide, benzoyl peroxide, ultraviolet light, bisulphites, riboflavin or lactoperoxidase have been shown to be effective in reducing the levels of AFM1 in experimental trials. Only about 0.4–2.2% of the ingested AFB1 appears in milk as AFM1. Furthermore, since AFM1 is water-soluble it partitions naturally during manufacture of cream and butter. Typically, about 10% of the AFM1 in milk appears in cream and about 2% appears in butter. Rigid monitoring of animal feed for AFB1 can control the AFM1 levels in dairy products including cream and butter.

Brucellosis

Brucellosis remains one of the most widespread and costly diseases afflicting humans and animals and is acquired by direct or indirect contact with infected animals harbouring any of three of the six bacterial species belonging to the genus *Brucella*. Human brucellosis ranges from a mild, flu-like illness to a severe disease. The severity depends on the species involved, with *B. melitensis* being the most pathogenic for humans, followed by *B. suis* and *B. abortus*. Osteomyelitis is the most common complication of *B. melitensis* infection, followed by skeletal, genitourinary, cardiovascular and neurological complaints. Cream and butter are unusual sources for *Brucella* spp. with only 5 out of 916 cream samples being positive in one outbreak-related survey. Both these products can

extend survival of *B. melitensis* and *B. abortus* for 4–6 weeks stored at 4°C. These microorganisms can survive even longer in refrigerated butter, persisting for 6 months and 13 months in salted and unsalted butter respectively. Dairy-related brucellosis outbreaks have been virtually eliminated as a result of immunization of livestock, slaughtering of infected animals and mandatory pasteurization of milk.

Listeriosis

Listeria monocytogenes, the causative agent of this disease, has recently emerged as a serious food-borne pathogen which can cause abortion in pregnant women, and meningitis, encephalitis and septicaemia in newborn infants and immunocompromised adults. The organism is a Gram-positive, non-spore-forming, facultatively anaerobic, short, diphtheroid-like rod-shaped bacterium that occurs singly or in short chains. *Listeria* infections are devastating, with a mortality rate of 20–30%. Dairy-related outbreaks of listeriosis, two in Switzerland and one in the USA in the 1980s have been linked to consumption of various products including pasteurized milk. Cream has been implicated in a major outbreak of listeriosis in Halle, East Germany, during the period 1949–1957. *Listeria monocytogenes* can attain populations of 10^6 cfu ml^{-1} in whipping cream after 8 days of storage at 8°C. It has occasionally been recovered from commercially produced butter, with survival up to 70 days also being reported in butter prepared from inoculated cream.

Salmonellosis

The causative agent is *Salmonella typhi*. It produces infections ranging from a mild, self-limiting form of gastroenteritis to septicaemia and life-threatening typhoid fever. Salmonellae are short, Gram-negative, facultatively anaerobic, rod-shaped bacteria and can grow at 5–45°C. Inadequate pasteurization and post-processing contamination have occasionally resulted in milk and cream that test positive for *Salmonella*, evidenced by various outbreaks of salmonellosis. The numbers of salmonellae decrease in fluid milk products and butter prepared from inoculated cream during extended storage at 7°C or below.

Staphylococcal Poisoning

Staphylococcal poisoning results from ingesting a preformed, heat-stable toxin (enterotoxin) which is produced by the bacterium *Staphylococcus aureus*. The bacteria are facultatively anaerobic, non-motile, small Gram-positive cocci growing at temperatures of 10–45°C and a pH of 4.2–9.3. Ten serologically distinct enterotoxigenic proteins known as enterotoxin types A, B, C_1, C_2, C_3, D, E, F, G and H are recognized in

S. aureus. The severe intoxication is of short duration and develops 1–6 h after ingestion of the enterotoxin. The common symptoms are nausea, vomiting, diarrhoea, abdominal cramps and mild leg cramps. Between 1951 and 1970 cream has been implicated in six outbreaks of staphylococcal poisoning in the USA involving 131 cases. Large numbers of *S. aureus* are seldom found in butter since the product composition and storage conditions severely limit its growth. However, when cream was inoculated with *S. aureus*, incubated for 24 h at 37°C and then churned to butter, the finished product contained at least 1 µg of enterotoxin per 100 g, or approximately 10% of the enterotoxin present originally in the cream. Since 0.1 µg of enterotoxin can induce symptoms of staphylococcal poisoning, ingesting such a dose poses a potential health hazard. This has been demonstrated by butter-related outbreaks.

See also: **Aspergillus**: Introduction; *Aspergillus flavus*. **Bacillus**: *Bacillus cereus*. **Food Poisoning Outbreaks**. **Heat Treatment of Foods**: Ultra-high Temperature (UHT) Treatments; Principles of Pasteurization. **Lactococcus**: *Lactococcus lactis* Sub-species *lactis* and *cremoris*. **Leuconostoc**. **Listeria**: *Listeria monocytogenes*. **Milk and Milk Products**: Microbiology of Liquid Milk; Microbiology of Dried Milk Products. **Packaging of Foods**. **Pseudomonas**: Introduction. *Salmonella*: *Salmonella typhi*. **Spoilage Problems**: Problems Caused by Bacteria. **Staphylococcus**: Introduction; Detection of Staphylococcal Enterotoxins. **Ultrasonic Imaging**.

Further Reading

Early R (ed.) (1992) *The Technology of Dairy Products*. Glasgow: Blackie.

Fox PF (ed.) (1983) *Developments in Dairy Chemistry 2: Lipids*. London: Applied Science.

Lapides DN (ed.) (1977) *McGraw-Hill Encyclopedia of Food, Agriculture and Nutrition*. New York: McGraw-Hill.

Marth EH and Steele JL (eds) (1998) *Applied Dairy Microbiology*. New York: Marcel Dekker.

Robinson RK (ed.) (1990) *Dairy Microbiology*. Vol. 1: *The Microbiology of Milk*, 2nd edn. London: Elsevier.

Robinson RK (ed.) (1990) *Dairy Microbiology*. Vol 2: *The Microbiology of Milk Products*, 2nd edn. London: Elsevier.

Robinson RK (ed.) (1994) *Modern Dairy Technology*. Vol 1: *Advances in Milk Processing*, 2nd edn. London: Chapman & Hall.

Spreer E (1998) *Milk and Dairy Product Technology*. New York: Marcel Dekker.

Varnam AH and Sutherland JP (1994) *Milk and Milk Products Technology, Chemistry and Microbiology*. London: Chapman & Hall.

Millet *see* **Fermented Foods**: Beverages from Sorghum and Millet.

Mineral Metabolism *see* **Metabolic Pathways**: Metabolism of Minerals and Vitamins.

MINIMAL METHODS OF PROCESSING

Contents

Electroporation – Pulsed Electric Fields

María Luisa Calderón-Miranda, Javier Raso, M Marcela Góngora-Nieto and **Gustavo V Barbosa-Cánovas**, Biological Systems Engineering, Washington State University, USA

Barry G Swanson, Food Science and Human Nutrition, Washington State University, USA

Copyright © 1999 Academic Press

Introduction

Electroporation is a physical phenomenon that consists of the formation of pores in the membranes of cells exposed to high-intensity pulsed electric fields. The formation of these pores can be either reversible or irreversible, depending on the intensity of the electric field applied. Reversible electroporation is of practical importance in genetic engineering and biotechnology, because the change in membrane permeability allows the introduction of DNA and other material into cells, as well as cell fusion. Irreversible electroporation of the cell membrane is the basis of the pasteurization of food by means of pulsed electric fields.

The non-thermal inactivation of microorganisms using pulsed electric fields (PEF) was demonstrated in the 1960s by Doevenspeck. The preservation of foods by exposure to high-intensity PEF is now an emerging technology with a promising future. While the inactivation of enzymes and of pathogenic and spoilage microorganisms is accomplished, the colour, flavour, texture and nutrients of the food are not affected. PEF provides an excellent alternative to conventional thermal methods of preservation such as pasteurization, in which the inactivation of the microorganisms results in the loss of valuable nutrients and organoleptic attributes. In addition, in most cases PEF is more efficient in energy usage, making the processing system more economic in the long run. PEF can be used either as a single technology, or as a hurdle in a combination of various preservation methods.

Fundamental Aspects of PEF

PEF treatment is based on the application of a high-intensity electric field to a food product. The PEF processing system consists of a high-voltage power supply, a capacitor for energy storage and a treatment chamber (**Fig. 1**).

High Voltage Generation

The power supply converts normal utility voltage into high-intensity voltage, and can supply either direct current (d.c.) or capacitor-charging power. The alternating current (a.c.) from the mains supply (60 Hz) is transformed into high-voltage a.c., and then rectified to produce high-voltage d.c. A capacitor bank is used to store large amounts of energy. When the capacitor is charged, high voltage is applied to the treatment chamber by using high-voltage switches. The switches are controlled by a command signal generator, and operate at high repetition rates (0.1–5000 Hz). Commonly used types of discharge switches include mercury-ignitrion, gas spark-gap, vacuum spark-gap, thyratron or magnetic, and mechanical rotary switches.

Treatment Chamber

The treatment chamber is used to expose the food to a high-intensity PEF ($20–80\,kV\,cm^{-1}$) by means of a short pulse, with a duration of $0.14–5\,\mu s$. The chamber consists of two electrodes, a spacer and two lids. One electrode is earthed and the other is connected to the high-voltage source. The electrodes are made of stainless steel or other conductive materials, and have a smooth surface to avoid electric field enhancement. The lids and spacer are made of insu-

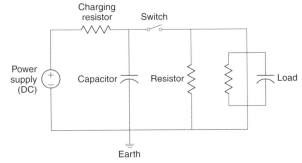

Figure 1 Pulsed electric field circuit.

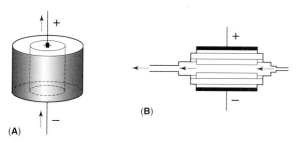

Figure 2 Pulsed electric field treatment chambers. (**A**) Coaxial cylinder chamber. (**B**) Parallel plate chamber.

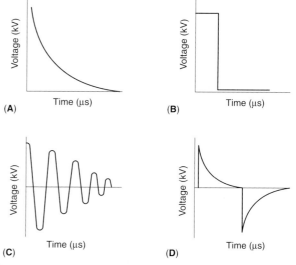

Figure 3 Pulsed electric field waveforms. (**A**) Exponentially decaying pulse. (**B**) Square-waveform pulse. (**C**) Oscillatory decaying pulse. (**D**) Bipolar exponentially decaying pulse.

lating materials (e.g. polysulphones). There are several types of treatment chambers, including batch parallel plate, continuous-flow parallel plate, and continuous coaxial cylinder (**Fig. 2**).

The batch parallel plate chamber contains two disc-shaped electrodes. The continuous-flow parallel plate chamber is a modification of the batch chamber, with the addition of baffled channels to facilitate the pumping of liquid foods. In the coaxial chamber, concentric cylinders allow the food to flow continuously through a uniform electric field. The gap between the electrodes can be modified by changing the diameter of the inner electrode. Turbulent flow within the cross section of the chamber causes food to flow towards the centre of the treatment chamber, resulting in the food being subjected to a uniform electric field throughout the chamber.

Dielectric Breakdown of Food

The major technical drawback of PEF technology is the dielectric breakdown of food. This occurs when the strength of the applied electric field is equal to the dielectric strength of the food system. The possible consequences include a luminous spark, the generation of bubbles, a pressure increase in the system, and the formation of pits on the electrodes. To avoid dielectric breakdown, the degassing and cooling of the food is recommended. Cooling food before placing it in the treatment chamber avoids a large increase in temperature during processing. Electrodes with an inner cooling zone are recommended: cooling both the electrode and the food product enables PEF treatment at low temperatures (e.g. 15°C) and minimizes the generation of air bubbles.

Parameters

The key variables involved in PEF are electric field strength (E), pulse duration or pulse width (τ), treatment time (t), pulse repetition rate (f), the waveform of the pulse, and temperature (T).

The intensity of the electric field defining an infinite parallel plate capacitor is the potential difference between two given points divided by the distance between the points (**Equation 1**):

$$E = V / d \qquad \text{(Equation 1)}$$

where V is the voltage discharged into the chamber (in kV) and d is the distance or gap between the two electrodes (in cm).

The pulse duration τ is defined by **Equation 2** and is expressed in µs:

$$\tau = RC \qquad \text{(Equation 2)}$$

where R (in Ω) is the resistance of the discharge circuit and C (in µF) is the capacitance of the energy-storage capacitor.

The treatment time t (in s) is given by multiplying the pulse duration by the number of pulses, n (**Equation 3**):

$$t = n\tau \qquad \text{(Equation 3)}$$

The pulse repetition rate or frequency (f) is the number of pulses discharged in a second.

Waveform of the Electric Field

High-intensity PEF can be generated as waveforms of bipolar, exponentially decaying, square or oscillatory decaying types.

Exponentially Decaying Pulse   This comprises a unidirectional voltage that rises rapidly to maximum intensity and decays exponentially to zero (**Fig. 3**). The power rating P_e (in W) for an exponentially decaying pulse is described by **Equation 4**:

$$P_e = (V^2 f\tau) / 2R \qquad \text{(Equation 4)}$$

where V (in kV) is the peak voltage. The energy input Q_e (in J cm^{-3}) for an exponentially decaying pulse is given by **Equation 5**:

$$Q_e = (V_0^2 t)/(Rv) \qquad \text{(Equation 5)}$$

where V_o (in kV) is the initial charge voltage, and v (in cm^3) is the volume of the treatment chamber.

Square-waveform Pulse This maintains a peak voltage for a longer time (Fig. 3 (B)). To generate a square-waveform pulse, a pulse-forming network (PFN) is necessary. The PFN and the treatment chamber must match in terms of impedance. The power rating P_s (in W) of a square-waveform pulse can be expressed as **Equation 6**:

$$P_s = (V^2 f\tau)/R \qquad \text{(Equation 6)}$$

where V (in kV) is the peak voltage. The energy input Q_s (in J cm^3) for square-waveform pulses is approximated by **Equation 7**:

$$Q_s = (V^2 tI)/(Rv) \qquad \text{(Equation 7)}$$

where I (in kA) is the current and V (in kV) is the voltage.

Oscillatory Decaying Pulses The frequency is determined by the capacitance, inductance, and resistance of the PEF system (see Fig. 3 (C)).

Bipolar Pulses Using this type of waveform, the food is subjected to a positive pulsed electric field followed by a negative pulsed electric field (Fig. 3 (D)). Bipolar pulses can be applied to foods as exponentially decaying or square-waveform pulses.

Mechanism of Microbial Inactivation

The mechanism of inactivation of microorganisms by high-intensity PEF is not entirely understood. One of the most accepted theories is the dielectric rupture theory. This explains electroporation by considering the cell membrane as a capacitor with weak dielectric properties. When an external electric field (E) is applied to the cell, electrical charges accumulate so as to increase the transmembrane potential (**Fig. 4**). The accumulated opposite charges are attracted to each other, with the result that the membrane surface is compressed and the membrane thickness is reduced. An elastic or viscoelastic force opposes the compression, but as the distance between the charges decreases, the transmembrane potential increases rapidly, giving rise to local breakdown of the cell membrane or the formation of pores. The proteins and lipids of the cell membrane are sensitive to applied PEF, and may reorient, forming pores, channels or gates. When cells are exposed to an electric field of 1–10 kV cm^{-1} for 2–20 μs, reversible poration is observed and the permeability of the membrane increases. When the electric field is removed, the original permeability of the cell membrane is recovered.

The electroporation of the cell membrane results in the destruction of the cell when the size and number of the pores is large compared to the cell size. Destruction occurs when a specific threshold electric field intensity (E_c) is exceeded. If E is equal to or slightly higher than E_c, reversible pore formation results. If an electric field of 2–20 kV cm^{-1} is applied here, poration of the cell is irreversible because E_c will be exceeded and the threshold transmembrane potential of 1 V will also be exceeded.

E_c and the threshold transmembrane potential depend on the microbial cell size and shape, its growth conditions and the electrical characteristics of the food being subjected to PEF. For a spherical cell of radius a, the transmembrane potential ΔV is expressed by **Equation 8**:

$$\Delta V = 1.5 f_m a E_0 \cos\theta \left\{ 1 - \exp\left(-t/\tau\right) \right\} \qquad \text{(Equation 8)}$$

where f_m is a constant, depending on membrane properties; E_0 is the electric field applied; θ is the angle between the radius vector and the direction of the electric field; t is the duration of application of the electric field; and τ is the relaxation time. τ is given by **Equation 9**:

$$\tau = f_m a C_m \left(r_i + r_e/2 \right) \qquad \text{(Equation 9)}$$

where C_m is the membrane capacitance per unit area, and r_i and r_e are the specific resistances of the internal and external media, respectively. The factor f_m is given by **Equation 10**:

$$f_m = 1/\left\{ 1 + a G_m \left(r_i + r_e \right) \right\} \qquad \text{(Equation 10)}$$

where G_m is the membrane conductance per unit area.

Transmission electron microscopy (TEM) of ultrastructural changes in microorganisms subjected to PEF provides evidence of the mechanical breakdown

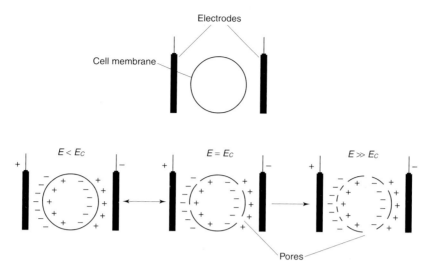

Figure 4 Microbial inactivation, explained by the dielectric rupture theory. E: electric field applied to the cell; E_c: threshold electric field intensity.

of cell membranes, supporting the dielectric rupture theory. In *Staphylococcus aureus* treated with PEF of $60\,kV\,cm^{-1}$ in the form of 64 exponentially decaying pulses with a total duration of $2\,\mu s$, the cell membranes were disrupted and the leaking of cytoplasm was observed. In *Saccharomyces cerevisiae* treated with a PEF of $40\,kV\,cm^{-1}$ in the form of 64 exponentially decaying pulses with a total duration of $4\,\mu s$, the cell membranes ruptured, cellular organelles were disrupted and the absence of ribosomes was observed.

Survival of Microorganisms after PEF

Microbial inactivation by PEF exhibits a linear correlation between the logarithm of the survival fraction (s) and the electric field applied (**Equation 11**):

$$\ln s = -b_E\left(E - E_c\right) \tag{Equation 11}$$

where b_E is the linear regression coefficient, and E_c is the extrapolated critical value of the electric field E for the initial microbial concentration.

The survival fraction can be expressed as a function of the treatment time (**Equation 12**):

$$1n\ s = -b_t \ln\left(t/t_c\right) \tag{Equation 12}$$

where b_t is the linear regression coefficient and t_c is the critical treatment time, which can be derived by extrapolating the value of t for the initial microbial concentration.

If Equations 11 and 12 are combined, a parameter k, which is specific to the microorganism, can be obtained (**Equation 13**):

$$k = (E - E_c)/b_t = \ln(t/t_c)/b_E \tag{Equation 13}$$

The survival fraction can be expressed as a function

of both the treatment time and the electric field by **Equation 14**:

$$s = (t/t_c)^{(E-E_c)/k} \tag{Equation 14}$$

where $k < 0$, and E_c, t_c and k are obtained from empirical data for each microorganism.

Parameters affecting Microbial Inactivation

There are reports of the inactivation by PEF of the vegetative cells of *Escherichia coli*, *Salmonella typhimurium*, *Salmonella dublin*, *Streptococcus thermophilus*, *Lactobacillus brevis*, *Lactobacillus delbrueckii* subsp. *bulgaricus*, *Micrococcus lysodeikticus*, *Pseudomonas fragi*, *Pseudomonas fluorescens*, *Klebsiella pneumoniae*, *Staphylococcus aureus*, *Listeria monocytogenes*, *Bacillus subtilis*, *Clostridium perfringens*, *Candida albicans*, *Candida utilis*, and *Saccharomyces cerevisiae*. The inactivation of *Bacillus cereus* and *Bacillus subtilis* spores inoculated into foods or model systems has also been reported.

Parameters affecting the inactivation of microorganisms by PEF can be divided into three groups, dependent on the microorganisms, the treatment conditions and the treatment media.

Parameters Dependent on Microorganisms

Type of Microorganism

In general, bacteria are less sensitive to PEF treatment than yeasts, and Gram-positive bacteria are more resistant than Gram-negative bacteria. For example, treatment applying an electric field of $16\,kV\,cm^{-1}$

reduces the population of *Escherichia coli* (Gram-negative) by 4–5 log cycles, but reduces the population of *Staphylococcus aureus* (Gram-positive) by only 3–4 log cycles. The inactivation of the spores of *Bacillus subtilis* and *B. cereus* has been reported, but most efforts to inactivate bacterial spores by PEF are unsuccessful. *Clostridium tyrobutyricum* spores are not inactivated by PEF, and neither are the ascospores of the fungus *Byssochlamys nivea*.

Growth Stage of the Cells

Inactivation by PEF varies according to the microbial growth stage. Cells in the logarithmic phase of growth are highly sensitive to electric field treatment compared to cells in the stationary and lag phases. After exposure to four pulses of a $36\,kV\,cm^{-1}$ electric field, a 2 log cycle reduction of a population of *Escherichia coli* in the logarithmic phase, but a 0.5 log cycle reduction of a population of *E. coli* in the stationary or lag phase, were observed.

Cell Size

A given PEF treatment induces a transmembrane potential which is proportional to the size of the cell. As cell size increases, the critical external electric field for the cell decreases, and hence the cell is more susceptible to PEF.

Parameters Dependent on Treatment Conditions

Electric Field Intensity

The intensity of the applied electric field is one of the most important factors influencing microbial inactivation by PEF. Once the applied E exceeds E_c, microbial inactivation increases with E. There is an exponential correlation between the survival fraction and the electric field intensity for individual microorganisms.

Treatment Time

Microbial inactivation by PEF is governed by the treatment time (number of pulses × duration of the pulse). With a constant pulse duration, microbial inactivation increases with an increase in the number of pulses. Similarly, for a constant number of pulses, inactivation increases with an increase in pulse duration. However, a large increase in the pulse duration may result in an increase in food temperature.

There is a linear relationship between the logarithm of the survival fraction and the logarithm of the treatment time when the field intensity is fixed. For microorganisms inactivated by PEF, a threshold value of the treatment time, t_c can be calculated.

Pulse Waveform

For a given peak value of field intensity and amount of electric energy input, the inactivation of microorganisms by PEF is closely related to the waveform of the applied pulses. Microbial inactivation by PEF has been studied using different pulse waveforms: exponentially decaying, oscillatory decaying, square and bipolar. Comparisons between pulses involving equivalent amounts of energy show that square-waveform pulses are more lethal than exponentially decaying pulses. This is because the peak voltage of square-waveform pulses is maintained throughout the duration of the pulse. Exponentially decaying pulses inactivate microorganisms more effectively than do oscillatory decaying pulses, and bipolar pulses are more lethal than monopolar pulses. In addition, bipolar pulses offer the advantages of reduced electrolysis of liquid foods, minimum energy utilization, and reduced fouling of electrode surfaces.

Treatment Temperature

There is a synergistic effect between the use of PEF and of moderate temperatures for microbial inactivation. Even at nonlethal temperatures, the rate of microbial inactivation by PEF increases with increasing temperature. The inactivation of *Escherichia coli* by PEF increased from 2 to 3 log cycles when the temperature was increased from 7°C to 40°C.

Parameters Dependent on Treatment Medium

Conductivity

Conductivity can be defined as the ability of a material to conduct electric current. The conductivity of a liquid medium is dependent on the ionic strength of that medium. An increase in ionic strength increases the conductivity of the liquid. The greater the conductivity of the food, the smaller the intensity of the electric field generated. Consequently, an increase in the ionic strength of a food results in a decrease in the inactivation rate of microorganisms contained in it. In a solution of potassium chloride of concentration $0.028\,mol\,l^{-1}$, the population of *Escherichia coli* decreased by 2.5 log cycles after eight pulses at $40\,kV\,cm^{-1}$, whereas no inactivation was obtained with the same treatment conditions in a solution containing $0.168\,mol\,l^{-1}$ of potassium chloride.

A high conductivity results in an increase in temperature, thus electrical breakdown of the treatment medium is favoured.

The conductivity of a liquid can be increased by the addition of electrolytes. When monovalent cations (e.g. Na^+ and K^+) are added to a food, microbial

Table 1 PEF treatment conditions for liquid foods

Food	Peak electric field (kV cm^{-1})	Pulse duration (µs)	Pulse number	Treatment temperature (°C)
Fresh apple juice	50	2	16	45
Raw skim milk	40	2	20	50
Green-pea soup	35	2	32	53

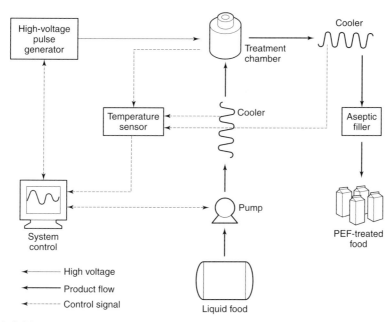

Figure 5 Pulsed electric field treatment system.

inactivation by PEF is not affected. However, with bivalent cations (e.g. Mg^{++} and Ca^{++}), microbial inactivation by PEF is less effective.

pH

Adjustment of the pH of the treatment medium with phosphate buffer in the pH range 5–9 does not affect microbial inactivation. However, adjusting the pH of the treatment medium by adding hydrochloric acid or sodium hydroxide resulted in a slightly greater inactivation of *E. coli* at pH 5.7 than at pH 6.8.

Antimicrobials

The inactivation of microorganisms by PEF can be enhanced by the use of antimicrobials. The presence of pediocin AcH or nisin in the treatment medium increases the lethality of PEF on *Listeria monocytogenes*, *E. coli*, and *Salmonella typhimurium*. When a population of *E. coli* was treated with one pulse of an electric field of 12.5 kV cm^{-1} in a solution containing 1000 p.p.m. of benzoic acid with a pH of 3.4, the microbial population was reduced by 10^4, whereas the same treatment in a solution containing 1000 p.p.m. of sorbic acid with a pH of 3.4 reduced the population of microorganisms by 10^3.

Pasteurization of Food by PEF

A variety of liquid foods have been treated with PEF. These include orange, apple and cranberry juices, pea soup, skim milk, yoghurt, liquid whole eggs and tap water. Treatment extended the shelf life of the food, and no differences in colour or flavour compared to those of fresh food were observed.

Two processes based on PEF have been developed by Krupp Maschinentechnik GmbH (Hamburg, Germany). Elcrack® involves the cracking of vegetable and animal cells, and Elsteril® is a pasteurization process for foods which can be pumped and for electrically conductive treatment media. Pure-Pulse Technologies (San Diego, US) have also developed a PEF process, CoolPure®, for the treatment of liquids and foods which can be pumped. The first commercial-scale CoolPure® system was used to extend the shelf life of refrigerated liquid whole eggs, at a capacity 1000 l h^{-1}.

A pilot-scale laboratory system has been used at Washington State University to treat several liquid food products. The shelf life at ambient temperature of fresh apple juice treated with PEF increased to 28 days. The shelf lives of raw milk and green-pea soup treated with PEF and stored at 4–6°C are 14 and 10 days respectively (**Table 1**).

Interest within the food industry in non-thermal processing with high-intensity PEF is increasing, with the possibility of it being used commercially to treat acid fruit juices. A possible treatment system is depicted in **Figure 5**.

See also: ***Bacillus***: *Bacillus cereus*; *Bacillus subtilis.* **Bacteriocins**: Potential in Food Preservation. ***Byssochlamys***. ***Candida***: Introduction. ***Clostridium***: Introduction. **Eggs**: Microbiology of Egg Products. ***Escherichia coli***: *Escherichia coli.* **Heat Treatment of Foods**: Principles of Pasteurization. **Hurdle Technology**. ***Klebsiella***. ***Lactobacillus***: *Lactobacillus bulgaricus*; *Lactobacillus brevis.* ***Listeria***: *Listeria monocytogenes.* ***Micrococcus***. **Microscopy**: Transmission Electron Microscopy. **Milk and Milk Products**: Microbiology of Liquid Milk. ***Pseudomonas***: Introduction. ***Saccharomyces***: *Saccharomyces cerevisiae.* ***Salmonella***: Introduction. ***Staphylococcus***: *Staphylococcus aureus.* ***Streptococcus***: *Streptococcus thermophilus.*

Further Reading

Barbosa-Cánovas GV, Pothakamury UR, Palou E and Swanson BG (1997) *Nonthermal Preservation of Foods.* New York: Marcel Dekker.

Castro AJ, Barbosa-Cánovas GV and Swanson BG (1993) Microbial inactivation of foods by pulsed electric fields. *J. Food Proc. Pres.* 17: 47–73.

Doevenspeck H (1961) Influencing cells and cell walls by electrostatic impulses. *Fleischwirtschaft* 13: 986–987.

Grahl T and Markl H (1996) Killing of microorganisms by pulsed electric fields. *App. Microbiol. Biotechnol.* 45: 148–157.

Harrison SL, Barbosa-Cánovas GV and Swanson BG (1997) *Saccharomyces cerevisiae* structural changes induced by pulsed electric field treatment. *Lebensm-Wiss. U-Technol.* 30(3): 236–240.

Hulsheger H, Potel J and Niemann EG (1981) Killing of bacteria with electric pulses of high field strength. *Radiat. Environ. Biophys.* 20: 53–65.

Mertens B and Knorr D (1992) Developments of non-thermal processes of food preservation: novel non-thermal preservation methods are being developed to produce high-quality foods. *Food Technol.* 4: 124–133.

Qin B, Pothakamury UR, Vega UH et al (1995) Food pasteurization using high intensity pulsed electric fields. *Food Technol.* 12: 55–60.

Qin B, Zhang Q, Barbosa-Cánovas GV, Swanson BG and Pedrow PD (1995) Pulsed electric field treatment chamber design for liquid food pasteurization using a finite element method. *ASAE* 38(2): 557–565.

Qin B, Pothakamury UR, Barbosa-Cánovas GV and Swanson BG (1996) Nonthermal pasteurization of liquid foods using high intensity pulsed electric fields. *Crit. Rev. Food Sci. Nutrition* 36(6): 603–627.

Shoenbach KH, Peterkin FE, Alden (III) RW and Beebe SJ (1997) The effect of pulsed electric fields on biological cells: experiments and applications. *IEEE Trans. Plasma Sci.* 25(2): 284–292.

Sitzmann W (1994) High-voltage pulse techniques for food preservation. In: Gould GW (ed.) *New Methods of Food Preservation.* P. 236. Glasgow: Blackie.

Vega-Mercado H, Martín-Belloso O, Quin BL et al (1997) Nonthermal food preservation: pulsed electric fields. *Trends Food Sci. Technol.* 8: 151–156.

Zhang Q, Barbosa-Cánovas GV and Swanson BG (1995) Engineering aspects of pulsed electric field pasteurization. *J. Food. Eng.* 25: 261–281.

Zimmermann U, Pilwat G and Riemann F (1974) Dielectric breakdown on cell membranes. *Biophys. J.* 14: 881–899.

Manothermosonication

Justino Burgos, Department of Animal Production and Food Science, University of Zaragoza, Spain

Introduction

The preservation of food requires the destruction of microbes and the inactivation of enzymes, the most frequently used method being heating. However the heat treatment of food may undesirably affect its organoleptic properties, or its nutrient content or availability. There is, therefore, growing interest in alternative procedures, able to destroy microorganisms and inactivate enzymes with little or no heat. One approach involves using physical or chemical agents to add to or potentiate the effects of heat, so that the intensity of the heat treatment necessary can be reduced. Synergistic combinations of agents can be identified, and are obviously the most desirable.

Among the physical agents able to destroy microbial cells and inactivate enzymes are ultrasonic waves. Manothermosonication (MTS) attempts to combine the effects of heat and ultrasonic waves. As a method of food preservation, it is still in its infancy. MTS consists of irradiation with ultrasonic waves, under moderate static pressure (in the range 200–700 kPa), at temperatures reflecting the resistance of the prevalent spoilage agents. Its main potential seems to be related to problems caused by thermoresistant enzymes.

Basic Principles

Ultrasonic waves are acoustic waves not perceived by the human sense of hearing. They are defined by their frequency, 10 kHz–10 MHz. Ultrasonic waves have various applications in food technology, involving the use of different frequencies. The ultrasonic waves used in MTS are the so-called 'power ultrasounds' – frequencies in the range of 20 kHz–100 KHz which produce chemical and mechanical effects involving cavitation.

Ultrasonic waves are propagated through materials as rapidly alternating cycles of pressure, oscillating sinusoidally around the equilibrium value (the static pressure). Particles in the path of the wave also oscillate, around their equilibrium position at a frequency equal to that of the ultrasonic wave. The maximum deviation of the pressure from its equilibrium value defines the wave amplitude, which can be expressed in terms of either pressure amplitude (in Pa) or the maximum displacement of particles from their equilibrium position (in μm).

When ultrasonic waves are applied to an aqueous solution, during the rarefaction stage bubbles of either previously dissolved gases or water vapour can be generated. Driven by the acoustic field, the bubbles can undergo relatively stable, low-energy oscillations. This phenomenon is termed 'stable' or 'non-inertial' cavitation. Under some circumstances the bubbles so produced coalesce, creating microcurrents – a phenomenon known as 'microstreaming'.

Bubbles may suffer transient ('inertial') cavitation. This involves very rapid growth, in which the radius can multiply several times in a few microseconds, until the bubble reaches a critical size. Before this, a pressure equilibrium is maintained across the bubble boundary. Beyond the critical size, the bubble cannot absorb enough energy to maintain the equilibrium, resulting in negative pressure inside the bubble. This causes the collapse of the bubble, known as 'implosion'. Implosion is associated with local pressures of about 1000 bar and local temperatures estimated at no less 5000°C (micro hot spots). Under these conditions, water is decomposed and free radicals ($H^{\cdot}$ and $OH^{\cdot}$) are generated. Owing to their high reactivity, free radicals can induce a number of chemical reactions, according to the composition of the medium. Implosion also produces very intense shock waves towards the centre of the bubbles, with the result that molecules of solute may be submitted to shearing forces.

In order to obtain chemical and biological effects as a result of cavitation, inertial cavitation is needed. This is a threshold phenomenon, depending critically on a number of parameters including frequency of the ultrasonic waves, pressure amplitude and initial bubble size. The frequency depends on the nature of the transducer, and hence is usually constant. Inertial cavitation is commonly produced using frequencies in the range 20–100 kHz. At frequencies in this range, the key factors determining the number of implosions are amplitude and initial bubble size. At a given amplitude, bubbles with an initial radius within a specific range will undergo inertial cavitation, and for a given initial bubble radius, a minimum pressure amplitude of the wave is required.

The amount of energy flowing during the insonation of a medium, per unit area per unit of time, is known as the 'field intensity'. Field intensity is proportional to the square of the pressure amplitude. Many cavitation effects increase with the field intensity, because of the increased number of bubbles undergoing cavitation. However, increases in field intensity beyond a certain value do not provoke more cavitations. This has been explained in terms of expanding bubbles reaching a size that impairs their complete collapse. The cavitation threshold decreases with increases in temperature, due mainly to the increase in water vapour pressure inside the bubble, and reaches zero at the boiling point of water.

The 'amount of inertial cavitation', i.e. the magnitude of the effects resulting from inertial cavitation, depends on the number of bubbles and the intensity of their collapse. This intensity is dependent mainly on static pressure. Increasing the static pressure affects the collapse of a single bubble in two ways. Firstly, the expansion of the bubble becomes more difficult – hence it is less likely to reach the critical radius and there is a decrease in the number of collapses. Secondly, however, an increased static pressure also increases the compressive forces that cause the bubble to collapse, and hence the energy associated with each individual collapse. As a result of these effects, the plots of some effects of sonication under applied static pressure versus static pressure show a maximum at moderate pressures (around 5 bar).

Effects of Ultrasonic Waves on Microbial Cells

The bactericidal effects of ultrasound have been known since about 1920. They were initially attributed to compression generated in the liquid medium, and later to the intense current provoked by insonation. However, now it is generally believed that most of the destructive effects on microbes are due to inertial cavitation. The lethality seems to be due mainly to cell disruption by shearing forces associated with the extreme changes of pressure caused by the collapse of the bubbles, the eddies created by the vibration of the bubbles and the flow of liquid towards the centre of the imploding bubbles. These mechanical forces damage the cell membrane and cause leakage.

The contribution to microbial lethality of the highly reactive free radicals produced by sonolysis, and of the H_2O_2 generated by the combination of two hydroxyl free radicals, seems to be of minimal importance, despite their known bactericidal effects. The contribution of the extremely high temperatures reached in hot spots is thought to be even less important, because of the very low proportion of the total volume of medium which reaches these temperatures and because a temperature equilibrium with the surroundings is reached extremely quickly – in a fraction of a microsecond.

A number of similarities between microbial destruction by heating and by insonation are evident. In both cases, sensitivity varies between species and development stages. Larger cells are usually more sensitive than smaller ones; rod-shaped bacteria are more sensitive than the coccal forms; Gram-positive bacteria are more sensitive than Gram-negative; anaerobic organisms are more sensitive than aerobes; and younger cells are more sensitive than older cells. The greatest difference in sensitivity to ultrasonic waves is shown by vegetative forms and spores, spores being much more resistant. Microbial sensitivity to both heat and ultrasonic waves is affected by the composition of the medium, sensitivity decreasing with the presence of fat and protein.

Microbial destruction follows first-order kinetics. Inactivation rates may therefore be expressed in terms of the decimal reduction time at a given temperature, also known as the D value. The temperature (T, in °C) at which the D value applies may be indicated by a subscript, e.g. D_T. The D value is the time needed, at the insonation temperature, to reduce the number of viable cells – or the enzyme activity –to one tenth of its original value. It is usually quoted in minutes. Alternatively, the inactivation rate can be expressed in terms of the inactivation rate constant (s^{-1}).

Ultrasonic waves are practically unable to kill spores, but may reduce their heat resistance. This effect has been attributed to the liberation of substances such as dipicolinic acid and some glycopeptides, which may be involved in the heat resistance of spores.

Enzyme Inactivation

Ultrasonic waves have various chemical effects on proteins. Polymeric globular proteins are split into subunits. Lipoproteins undergo delipidation, and haemoglobin dissociates, producing haem groups and globin. If the exposure to ultrasonic waves is suf-

ficiently long, polypeptides may be split into fragments. Cyclic amino acids can be split off and the aromatic residues can be oxidized.

Many of these effects on proteins can result in enzyme inactivation. The rate of enzyme inactivation depends on the characteristics of the acoustic field and of the insonation medium, as well as on the molecular structure of the enzyme. Some enzymes, including catalase, yeast invertase (saccharase), ribonuclease and pepsin, are very resistant to ultrasonic waves at low (or room) temperatures. Others, including alcohol dehydrogenase, malate dehydrogenase, polyphenoloxidase, tomato endopolygalacturonase and *Pseudomonas fluorescens* lipase, are much more sensitive. Generally, enzyme inactivation by ultrasonic waves at low (or room) temperatures and atmospheric pressure requires long periods of exposure.

The inactivation of enzymes by ultrasonic waves is thought to be due to either the shearing forces associated with the eddies and currents provoked in the medium by the oscillation and implosion of bubbles, or to the chemical reactions induced by the free radicals generated.

Principles Underlying Manothermosonication

The combined application of ultrasonic waves and heat, at atmospheric pressure, is termed 'thermosonication'. When either vegetative cells or spores are subjected to thermosonication, the rate of inactivation is greater than that predicted by adding together the rate of inactivation due to heat treatment and that due to ultrasonic waves at room temperature. Therefore, synergism has been postulated. The efficiency of microbial inactivation by thermosonication decreases as the temperature increases, and it is lost when the temperature approaches the boiling point of the medium. This is due mainly to decreased intensity of the collapse of bubbles when the vapour pressure of the liquid medium approaches static (atmospheric) pressure.

Manothermosonication (MTS) attempts to overcome this decreased intensity of bubble collapse at high temperatures, by increasing the static pressure. The energy of collapse is a function of the pressure differential (static pressure minus vapour pressure), assuming a direct effect. The efficiency of manothermosonication can be expressed as the ratio of the rate of microbial or enzyme inactivation by MTS to the rate of microbial or enzyme inactivation by heat treatment at the same temperature. The rate of inactivation increases with applied pressure, until a critical pressure, at which the increase in the energy of collapse is counterbalanced by a decrease in the number of bubbles undergoing cavitation. The same efficiency can be achieved at different temperatures by adjusting the pressure so that the differential pressure is constant. Thus for every temperature of treatment, an optimum applied pressure can be defined.

Theoretically, MTS could be a good method of enzyme inactivation and microbial destruction. As with heat treatment, the resistance of microorganisms and enzymes to MTS is expressed in terms of either the D value or the inactivation rate constant. The temperature dependence of an inactivating reaction may be expressed as the 'z value', i.e. the temperature increase in Celsius degrees needed to reduce the D value to one tenth of its original value. Alternatively, temperature dependence may be expressed in terms of activation energy E_a, which can be estimated from Arrhenius plots. Investigations are complicated by the fact that the geometry of the reaction chamber also has a marked influence on the extent of the effects of ultrasonic waves. Since published results will not have been obtained using the same experimental equipment, the specific experimental conditions under which data were obtained must be stated.

Effects of Manothermosonication

Microbial Destruction

In the treatment of spores with MTS, the heat and ultrasonic waves seem to combine synergistically. However, in the manothermosonication of vegetative cells, heat and ultrasonic waves seem to combine additively and not synergistically. This effect is seen with *Yersinia enterocolitica*, *Listeria monocytogenes* and *Aeromonas hydrophila*. The exceptions are organisms previously subjected to treatments such as thermal shock, which substantially increases thermoresistance. In contrast to simple heat treatment, MTS does not result in cells which are injured but cells capable of recovery – all cells which are injured are killed. Microbial destruction by manothermosonication increases with pressure and temperature at a given ultrasonic wave amplitude, up to at least 150 µm.

The contribution of ultrasonic waves to microbial destruction by MTS does not seem to be related to the generation of free radicals – free radical scavengers do not affect the rate of inactivation.

Production of Free Radicals

The production of free radicals, measured with a thereftalic acid dosimeter, correspond well to the-oretical predictions (**Fig. 1**). Production increases with pressure, at constant temperature. At 70°C and an amplitude of 117 µm, a maximum rate of production is reached at about 250 kPa. At constant pressure and amplitude, the rate of production decreases linearly with temperature, and at constant pressure and temperature it increases exponentially with the square of the amplitude.

Hydroxyl free radicals, and other free radicals, are constantly being formed in vivo and are thought to be involved in several human diseases. Consequently there is some concern about the possible effects on human health of the production of free radicals in foods as a result of MTS. However, free hydroxyl and other radicals are also formed in foods, including meat and fish, processed using more conventional methods. The free radicals react with food components and cause oxidative deterioration by mechanisms similar to those operating in human tissues. Their high reactivity makes their absorption in the gut extremely unlikely, but they do reduce the antioxidant content of food and cause oxidation, particularly of lipids.

Enzyme Inactivation

MTS is particularly effective in the inactivation of enzymes. The first to be investigated were peroxidase, lipoxygenase and polyphenoloxidase. The effects of ultrasonic waves and heat on all three seem to combine synergistically – the inactivation rate caused by MTS is higher than that predicted by adding the inactivation rate caused by ultrasonic waves at 37°C to that caused by simple heating.

There are, however, some interesting differences between the behaviour of these enzymes. In the inactivation of peroxidase by MTS, the synergy between the ultrasonic waves and the heat is less marked than in the case of the other two enzymes. Also, the relationships between temperature and pH and the inactivation rate are the same in the case of both MTS and simple heat treatment. Because of this and because identical reactivation phenomena after marginal treatments are observed, a common inactivation mechanism within the two treatments has been proposed.

The rate of inactivation of lipoxygenase by MTS is much less temperature-dependent and pH-dependent than is simple heat inactivation (i.e. in the case of MTS, the z value is higher and E_a is lower). Consequently treatment by MTS is particularly effective in the acidic conditions in fruit (and many vegetable) juices (**Fig. 2**). It has therefore been proposed that ultrasonic waves and heating within MTS cause the inactivation of lipoxygenase by different mechanisms. The ultrasonic waves may cause a 'caged Fenton reaction'.

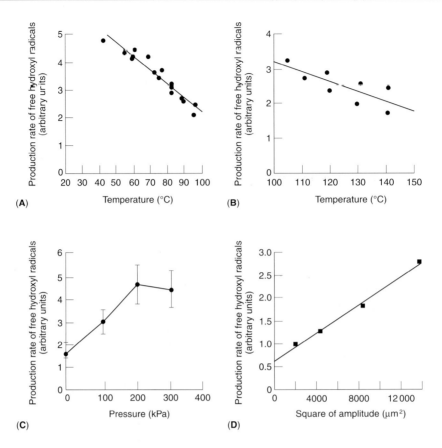

Figure 1 Effects of temperature, pressure and the amplitude of ultrasonic waves on the production of free radicals by mano-thermosonication. (**A**) Gauge pressure 200 kPa, amplitude 117 µm. (**B**) Constant differential pressure (gauge 200 kPa) and amplitude (117 µm). (**C**) Constant temperature (70°C) and amplitude (117 µm). (**D**) Constant temperature (130°C) and pressure (gauge 500 kPa). (Reprinted from A Vercet, P López and J Burgos (1998) with permission from Elsevier Science.)

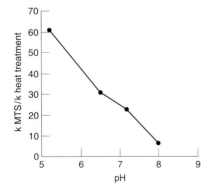

Figure 2 Dependence on pH of the efficiency of the inactivation of lipoxygenase by manothermosonication.

Treatment of Foods with Manothermosonication

The intensity of the heat treatment applied to foods is usually determined on the bases of microbial load and microbial heat resistance, enzymes generally being more heat-sensitive than are microorganisms. However, in a few cases, thermoresistant enzymes become a major problem in the preservation of foods by heat treatment. These include lipases and proteases, secreted into milk by psychrotrophic bacteria during storage at refrigeration temperatures; pectin methylesterase (PME), from oranges and other citric fruits used for juice production; and endo-polygalacturonase (PG) from tomatoes. MTS appears to be a potential alternative to simple heat treatment for achieving the inactivation of these undesirable enzymes.

Lipases and Proteases in Milk

Lipases and proteases from psychrotrophic bacteria withstand sterilization treatments and are particularly resistant to UHT (ultra-high temperature) treatment (D_{140} value is around 1.5 min), due to their high z values (around 35°C). Consequently, they limit the shelf life of UHT milk and often cause deterioration of the organoleptic properties of dairy products made from pasteurized milk. The exocellular lipase from *Pseudomonas fluorescens* is highly sensitive to ultrasonic waves and is thermoresistant, and hence is more easily inactivated by MTS at 30°C, 450 kPa and ultrasonic amplitudes of around 120 μm than by simple heating at 115°C. Exocellular protease from the same organism is much less sensitive to ultrasonic waves at room temperature, and to achieve reasonable inactivation rates by MTS, the temperature has to be raised. Even so, MTS at 100°C, 450 kPa and amplitudes of around 120 μm inactivates the enzyme as effectively as simple heating at temperatures of about 120°C. The inactivation of both enzymes by MTS is less dependent on temperature than is inactivation by heat. Therefore the efficiency of MTS decreases with increasing temperatures (**Fig. 3**).

However, the rate of lipase inactivation by MTS does not show uniform temperature dependence. The temperature dependence of the inactivation at 450 kPa and 117 μm of amplitude changes at around 110°C. Below this temperature, the inactivation rate is lower ($E_a = 19.4$ kcal mol^{-1} K^{-1}; z = 107°C) than in the range 110–140°C ($E_a = 54.4$ kJ mol^{-1} K^{-1}; z = 58.5°C). Even in the high temperature range, the temperature dependence of inactivation by MTS is less than that of inactivation by simple heating. This cannot be entirely explained in terms of changes in the differential pressure with increases in temperature. For protease inactivation at the same pressure and amplitude in the range 100–140°C, $E_a = 62.7$ kJ mol^{-1} K^{-1} and z = 47.6°C. The efficiency of the inactivation of protease by MTS increases linearly with the square of amplitude between 60 μm and 150 μm, but that of lipase is much more amplitude-dependent above 117 μm than at amplitudes below this value (**Fig. 4**). In spite of the decreased efficiency of MTS with increasing temperature, the rate of lipase inactivation achieved at UHT temperatures (about 140°C) by applying a pressure of 500 kPa and ultrasonic waves of 150 μm is 8-times that of simple heating at the same temperature. The efficiency of the inactivation of protease by MTS under the same

conditions is lower – 2–2.5 times that of simple heating.

The potential advantages of MTS as a technique for stabilizing dairy products thus seem to be greater in connection with processes in which the milk is heat-treated at temperatures below the UHT range. In UHT milk and dairy products derived from it, MTS could delay the appearance of problems caused by enzymes secreted by psychrotrophic bacteria by a factor of between two and eight.

Pectin Methylesterase in Citric Juices

The PME of citrus fruits is the cause of the loss of turbidity in orange juice. If the enzyme is not inactivated or inhibited immediately after juice extraction, the pectin which contributes to the turbidity (about 30%) is de-esterified, to form calcium pectate. On standing, the juice separates into a clear supernatant and a sediment. In concentrated juice, a gel forms, from which single-strength juice cannot be reconstituted.

There are several PME isozymes in citric fruits. Some of them, for example PME of high molecular weight from oranges, are active at low pH and low temperatures and show rather high heat resistance. The percentage of the total PME activity due to thermoresistant isozymes, and the D and z values of the isozymes, vary with species and cultivars. In oranges, z values of between 6.5°C and 10.8°C, and D_{90} values of between 0.35 min and 33 min, have been reported. The high heat resistance of this enzyme seems to be related to its association with pectin in orange juice – its D values can be multiplied by 50 by adding 0.1% of pectin to a solution containing the main soluble components of orange juice, at standard juice concentrations and pH.

The full inactivation of the thermoresistant fraction of the PME citric fruits by heat requires heat treat-

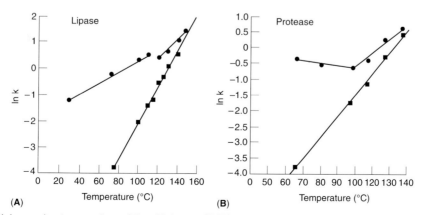

Figure 3 Decay with increasing temperature of the efficiency of MTS in the inactivation of exocellular enzymes from *Pseudomonas fluorescens*. Squares, simple heating; circles, MTS.

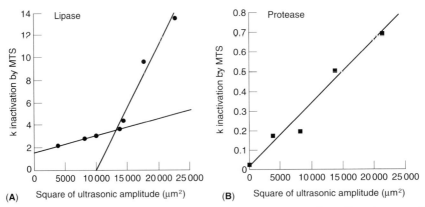

Figure 4 Effect of amplitude on the rate of inactivation of exocellular enzymes from *Pseudomonas fluorescens* by MTS. (**A**) Inactivation of lipase at 500 kPa and 140°C. (**B**) Inactivation of protease at constant differential pressure (200 kPa) and 76°C.

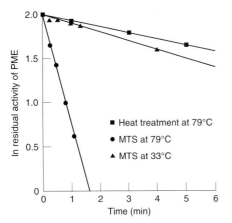

Figure 5 Inactivation of the thermoresistant fraction of PME from orange by MTS in orange juice. Amplitude 117 μm, gauge pressure 200 kPa.

ments which have undesirable effects on the flavour of orange juice, which is very heat-sensitive. None of the available alternatives is satisfactory, but MTS may be useful for this purpose. The typical microbial flora of orange juice is very heat-sensitive, and is also easily destroyed by ultrasonic waves. The thermoresistant fraction of PME is also very sensitive to ultrasonic waves under applied pressure at comparatively low temperatures, and the presence of pectin does not increase the resistance of PME to MTS. In orange juice, the rate of inactivation of PME by MTS at 35°C, 200 kPa of applied pressure and abut 110 μm amplitude is equal to that attained by simple heating at 80°C (**Fig. 5**). Unfortunately, the temperature dependence of the inactivation is about 7 times less with MTS than with simple heating. Because of this, the efficiency of MTS decreases rapidly with increasing temperature. Even so, at 80°C the rate of inactivation of PME is about 10 times higher using MTS than by simple heating, and matches that achieved by heat treatment at 90°C. Therefore MTS could substantially reduce either the temperature or the time needed for treatment.

Pectin Methylesterase and Endopolygalacturonase in Tomato-juice Products

Two basic attributes of tomato paste and other tomato-juice products are flow behaviour (mainly controlled by pectins) and colour (mainly due to lycopene). The combined action of two of the pectic enzymes in tomato, PME and PG, can degrade pectins and provoke a loss of consistency (lower apparent viscosity). In order to protect the characteristic flow behaviour of tomato extracts, these enzymes must be inactivated rapidly, immediately after the extract is prepared. This is frequently accomplished by means of a so-called 'hot break', in which temperatures in the range of 85–102°C are applied. These temperatures damage the colour of the extract, but are demanded by the heat resistance of PG. The PG of tomatoes comprises two species: PGII and PGI. PGI is fairly heat-resistant ($D_{82} = 200$ min; $z = 5.6$°C). At 200 kPa of applied pressure and about 120 μm of amplitude, the rate of inactivation of PME by MTS at 60–65°C is about 50 times higher than that reached by simple heating. The rate of inactivation of the most thermoresistant form of PG, PGI, is multiplied by 85 by MTS at 85°C. Therefore, to protect the flow behaviour of tomato juice products, MTS treatment either at low temperatures or for shorter times than those used in hot breaks should suffice.

Industrial Applications of Manothermosonication

Much work remains to be done before MTS can be used commercially. For example, more experimental data, obtained using treatment chambers with different geometries, would facilitate process modelling.

Very little is known about the effects of MTS on nutrient content. MTS treatments which inactivate the PME of orange juice do not seem to affect its ascorbic acid (one of the nutrients more susceptible to oxidation) any more than simple heat treatment.

However, the effects of MTS on other labile nutrients, including thiamin, riboflavin and folic acid, at different pH values and in the context of differently composed food, are unknown.

The effects of MTS on the organoleptic and functional properties of foods are practically unexplored, although it is known that MTS changes the renettability of milk, in a way that could be advantageous for some dairy products but disadvantageous for others. MTS treatments able to inactivate the PGI of tomato juice and the PME of orange juice do not seem to have negative effects on the rheological properties of these juices, but nothing is known about their effects on volatiles.

There are a number of designs for machines which can apply continuous ultrasonic waves – for various purposes and to Newtonian and non-Newtonian solutions, as well as to emulsions and slurries. The basic characteristics of these designs could be valid for the development of industrial equipment for MTS. The geometric characteristics of some of the designs are excellent in terms of ensuring the maximum efficiency of the ultrasonic waves. However, these instruments operate with a range of amplitudes lower than those used for MTS so far. The efficient control of pressure and temperature would also be necessary for industrial applications.

See also: **Heat Treatment of Foods**: Ultra-high Temperature (UHT) Treatments; Principles of Pasteurization. **Milk and Milk Products**: Microbiology of Liquid Milk. *Psychrobacter.* **Ultrasonic Standing Waves**.

Further Reading

Earnshaw RG (1998) Ultrasound: a new opportunity for food preservation. In: Povey MSW and Mason T (eds) *Ultrasound in Food Processing*. London: Blackie.

El' Piner I (1964) *Ultrasounds: Physical, Chemical and Biological Effects.* New York: Consultants Bureau.

García ML, Burgos J, Sanz B and Ordóñez JA Effect of heat and ultrasonic waves on the survival of two strains of *Bacillus subtilis*. *Journal of Applied Bacteriology* 67: 619–628.

Lopez P and Burgos J (1995) Lipoxygenase inactivation by manothermosonication. *Journal of Agricultural and Food Chemistry* 43: 620–625.

López P and Burgos J (1995) Peroxidase stability and reactivation after heat treatment and manothermosonication. *Journal of Food Science* 60: 451–455.

López P, Vercet A, Sanchez AG and Burgos J (1998) Inactivation of tomato pectic enzymes by manothermosonication. *Zeitschrift für Lebensmittel Untersuchung und Forschung* 207: 249–252.

Mason TJ and Lorimer JP (1988) *Sonochemistry, Theory, Application and Uses of Ultrasonics in Chemistry*. New York: Ellis Horwood.

Raso J, Pagan R, Condon S and Sala FJ (1998) Influence of temperature and pressure on the lethality of ultrasounds. *Applied and Environmental Microbiology* 64: 466–471.

Sala F, Burgos J, Condón S, López P and Raso J (1995) Effect of heat and ultrasound on microorganisms and enzymes. In: Gould GW (ed.) *New Methods of Food Preservation*. London: Blackie.

Vercet A, López P and Burgos J (1996) Inactivation of heat-resistant lipase and protease from *Pseudomonas fluorescens* by manothermosonication. *Journal of Dairy Science* 29–36.

Vercet A, López P and Burgos J (1998) Free radical production by manothermosonication. *Ultrasonics* 36: 617–620.

Potential Use of Phages and/or Lysins

Juan Jofre and **Maite Muniesa**, Department of Microbiology, University of Barcelona, Spain

Introduction

This article describes the potential uses of bacteriophages and phage lysins in the control of bacterial pathogens in food. Phage lysins are enzymes which hydrolyse cell walls. They are synthesized during late gene expression in the lytic cycle of multiplication of most phages. A consideration of the nature and main features of phages and their lysins together with the main traits of their interactions with host bacteria indicates the chances of there being useful applications in the control of pathogens in food. A few examples, which are discussed in the article, show that lysins have a good chance of being applicable, whereas the potential uses of phages are very limited.

Phages

Bacteriophages, also known as phages or bacterial viruses, are viruses that infect bacteria. They were discovered in the middle of the 1910s by Twort and d'Herelle, and since then phages which infect the majority of bacteria have been described, amounting to a few thousand phages to date.

Phages have been recovered from many sorts of foods, either infecting bacteria that constitute the microflora of the foods, or as exogenous contaminants. They can influence bacterial populations in foods in different ways, including destruction by virulent phages; lysogenization of bacteria and the conferment of new phenotypic characteristics by temperate phages; and the transduction of genes to bacteria. These interactions in food may lead to both beneficial and harmful effects. Potentially valuable applications of phage–host interactions in food include the inhibition of spoilage bacteria in refrigerated or perishable foods; the elimination of pathogens in foods; and the development of phage typing schemes for the precise identification of spoilage and pathogenic bacteria in foods. Harmful effects could include the destruction of bacteria used for the fermentation of foods (e.g. cheese-making, malolactic fermentation in wine-making) and the transfer of virulence factors between related bacteria via lysogenic conversion (e.g. the acquisition of virulence genes by *Escherichia coli* O157:H7).

Phages which infect food pathogens have been described for both psychrotrophic pathogens, able to grow in different sorts of food (e.g. *Aeromonas*, *Pseudomonas* and *Listeria*) and pathogens that contaminate food (e.g. *E. coli*, including *E. coli* O157:H7, *Salmonella*, *Shigella*, *Campylobacter*, *Staphylococcus*, *Vibrio* and *Clostridium*). Phages infecting pathogenic bacteria do not differ from those that infect non-pathogenic species. For example, **Figure 1** shows phages infecting *E. coli* O157:H7 isolated from sewage, which are similar to those infecting non-pathogenic stains of *E. coli*.

Characteristics

Phages basically consist of one nucleic acid molecule – the genome – surrounded by a protein coat – the capsid – which is made up of morphological sub-units called capsomers. Many phages contain additional structures such as tails and spikes, and some may also contain lipids.

There is great diversity in terms of the nature and characteristics of the nucleic acid, the structure and composition of the viral particles, and size. Phages have been classified into 11 families by the International Committee on Taxonomy of Viruses. The characteristics of the phages most frequently isolated from water and foods are summarized in **Figure 2**. Phage structure may be as simple as that of Leviviridae (e.g. f2, MS2) which consists of a molecule of RNA and an associated RNA polymerase, both surrounded by an icosahedral capsid. However, phage morphology may also be complex, such as that of Myoviridae (e.g. T2, T4) which have a head, which contains a double-stranded DNA molecule, connected through a collar to a contractile tail, at the end of which there is a base-plate with pins and fibres. Phages with a tail are the most common. Amongst these, the Siphoviridae account for half of all the phages described to date. The size of phages ranges from the 20 nm of the Leviviridae to the 110 × 20 nm of the Myoviridae, which have an elongated head plus > 100 nm of tail.

Viruses, including phages, are not mobile. Their

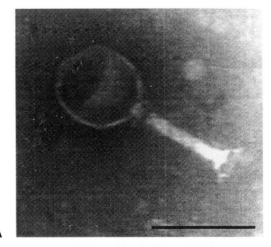

A

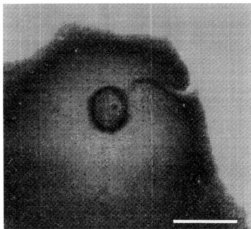

B

Figure 1 Transmission electron micrographs of two bacteriophages infecting *Escherichia coli* O157:H7, representing the morphotypes most frequently isolated from sewage. (**A**) Myoviridae family. (**B**) Siphoviridae family. Bar = 100 nm.

movement in a given environment occurs only through diffusion and Brownian or random motion.

Phages can only multiply inside appropriate host cells. However they can persist outside the host cell under a great variety of conditions, and usually persist much better than their bacterial host under adverse conditions. Most phages are far more resistant to heat, freezing, radiation, chemical disinfection and natural inactivation than their host bacteria.

Replication

Phages can only replicate within a bacterial cell. All known phages use the ribosomes, protein-synthesizing factors, amino acids, enzymes, nucleotides and energy-generating systems of the host cell to replicate – hence a phage can grow only in a metabolizing bacterium.

The bacteria in which a given phage can replicate determine its specificity. In general, this corresponds to bacterial groups. However, phages differ in terms of the range of hosts which they can infect – some

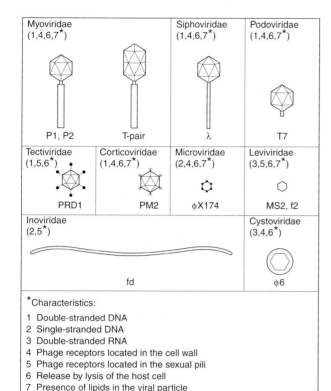

Figure 2 Classification and characteristics of the main groups of phages isolated from water and foods. Characteristics are shown in parentheses.

have broad host ranges, including more than one species or even more than one genera (e.g. the polyvalent phages of *Listeria* or those able to infect *Escherichia coli* and *Shigella*), and others have very narrow host ranges, infecting only a few strains of one species (e.g. some phages of *Listeria* infect only some serovars). The latter may be used for the phage typing of bacteria, and are frequently used for typing food pathogens.

Host specificity is due mainly to the nature of the receptors located on the surface of the cell. These may be outer membrane proteins, lipopolysaccharides, capsules, or appendixes such as flagella and pili. These receptors have more than one function, for example some may be related to virulence, as in the case of the *E. coli* K antigens, which some phages use as receptors. In the case of virulent phages, the receptors of some of the host strains show high genetic stability – in this case, the great majority of the bacterial cells are sensitive and may be infected. More frequently, the rate of appearance of resistant mutants is high enough to allow the survival or persistence of a given bacterial population, even in the presence of high densities of phages.

In the context of multiplication, the nature of the relationships between phages and their host cells is diverse. Virulent and temperate phages follow different patterns of replication. Most relevant with regard to foods are the virulent phages, which kill the cells that they infect. The basic sequence of events during phage replication, named the lytic cycle, is similar for most phages and occurs only in metabolizing host cells. The steps of phage replication are as follows, attachment and host lysis being the steps most important for the topics discussed in these articles:

1. Adsorption or attachment of the phage onto the surface of the host cell.
2. Introduction of the phage nucleic acid into the host cell.
3. Synthesis of phage nucleic acid and other molecules required for the reproduction of complete viral particles. Cessation of synthesis of host cell components.
4. Assembly of the new phage particles.
5. Release of the new phage particles, usually by sudden lysis of the host cell.

The first step in phage replication is the adsorption or attachment of phage particles to the bacterial host. Essential for this is that the phages encounter bacteria with specific phage receptors on their surface. The efficiency of adsorption depends on a number of environmental conditions. At the end of the lytic cycle, the newly formed virus particles are usually released by sudden lysis of the bacterial host. Lysis is caused by one or more enzymes coded by the phages, generically known as phage lysins (see below). The lytic cycle of phages is usually short – sometimes as short as the 20 min of the phage T4 infective cycle. The number of phages released by one infected bacterium, known as the 'burst size', ranges from a few hundred in the case of the Myoviridae to many thousands in the case of the Leviviridae.

Temperate phages may follow a lysogenic cycle, in which the phage genome becomes part of that of the host bacterium. As the bacterium reproduces, the genome remains integrated and is replicated along with the host genome. The lysogenic cycle plays an important role in the maintenance of the genetic variability of bacteria, through the phenomenon known as 'lysogenic conversion' that is the name given to the changes in the bacterial host phenotype caused by the temperate bacteriophages.

The phage genome will produce new phages through a lytic cycle only after induction by various agents and conditions, such as ultraviolet light, some chemicals and host stress. Temperate phages may influence food safety and quality in an indirect manner, for example by the lysogenic conversion of a non-virulent strain to a virulent strain.

Use of Phages for the Control of Pathogens in Foods

The discovery in the second half of the 1910s of phages that attack pathogenic bacteria in vitro led to an intensive period of inquiry into their use for treating infectious diseases. The early claims of success were not, however, substantiated by subsequent research and the introduction of antibiotics resulted in investigations into the use of phages for treating infectious diseases being largely abandoned.

The main reasons postulated for the failure of phages to cure infections have been their low degree of activity in vivo compared with that in vitro; the appearance of antibodies against phages; and primarily, the rapid emergence of phage-resistant bacterial mutants. Experiments in the 1980s have demonstrated the successful control of *Escherichia coli* infections by using phages whose receptor is the K antigen, which is a virulence factor. This prevents the rapid emergence of virulent phage-resistant mutants, because phage-resistant mutants lack the virulence factor. One problem with the practical application of this approach, however, might be the narrow range of the specificity of phages, and the high numbers of phages needed to control the infection.

In theory, phages could also be used for the elimination of bacterial pathogens in food. This has been seen as an innovative approach to the problem of microbial food contamination, because as a consequence of their specificity, phages should have a minimal impact on the microbial ecology of foods, while eliminating their target. However, the influence that virulent phages exert on a susceptible bacterial population depends on the probability of phage–host contact and the suitability of the environment for the infection to succeed. The net effect of phage–host interactions may also be influenced by the rate of mutation of the bacteria to produce phage-resistant strains. The densities of bacteria likely in contaminated food suggest that mutation would probably be a minor problem. The main foreseeable problems in the use of phages for the control of bacterial pathogens in food are related to the threshold densities of phages and host bacteria needed to guarantee bacterial destruction, and to the occurrence in food of conditions suitable for the successful infection of the target bacteria.

Sufficiently high concentrations of both phages and hosts, as well as an appropriate medium for the movement and contact of phages and hosts, are very important factors in phage–bacteria interactions in foods. It is assumed that the first stage of the infective process, the initial contact between a phage and its host, occurs by chance and that the phage particles act as passive entities, contact between freely moving cells in a liquid medium occurring due to the mobility of the bacteria and the diffusion and Brownian motion of the phages. At low densities of phages and bacteria, the probability of an encounter is low and productive infection may never occur. Most of the available information about the concentrations of phages and host bacteria needed to guarantee infection concerns water. Different authors have concluded, for different phage–host systems, that in water, phage replication does not occur successfully unless the concentration of host bacteria is at least approximately 10^4 per millilitre, and that the lower the number of host cells, the higher the number of phages needed to kill them. For example, in suspensions containing as few as 1 or 2 bacteria, 10^8 phages were needed to kill the bacteria. The threshold host density could reflect the time needed for the phage particles to come into contact with the host cells. The time required for one phage to contact a host can be calculated from a first-order equation describing the adsorption of phages to host cells. Using measured constants for bacteriophage T4 and an *E. coli* density of 100 cfu (colony forming units) ml^{-1}, which was below the measured threshold (10^4 cfu ml^{-1}), it can be calculated that an average of 4000 min would be required for one phage particle in an initial density of 1000 pfu (plaque-forming units – phage infection on a lawn of bacteria is observed as cleared areas called plaques, each produced by a single bacteriophage) ml^{-1} to contact a host. At a host density of 10^5 cfu ml^{-1}, which is above the threshold, it would take 4 min.

The data available on the concentrations of phages and host cells needed to guarantee contact in foods are very limited. However, the data available indicate that the concentrations needed to guarantee encounters in solid foods seem to be higher than those applicable in water, probably because of reduced mobility of both the phages and the bacteria. Minimum phage concentrations of 10^5–10^6 pfu ml^{-1}, and approximately the same concentrations of host bacteria, have been described as the thresholds necessary to guarantee the replication of phages of *Pseudomonas* in skim milk or beef juice, or on the surface of beefsteaks or pigskin maintained in vitro. It is very likely that the minimum concentrations of phages and host bacteria are even higher inside solid foods, resulting in the need for the inoculation of large numbers of phages in order to guarantee contact with bacteria. This should not be a problem in terms of phage production, because high-titre (10^{10}–10^{11} pfu ml^{-1}) phage suspensions are attainable. However, the introduction of high numbers of phages into foods could present problems from the point of view of approval by legislators and acceptability by consumers.

If the probability of encounters between phages and bacteria depends on time, the stability of phages in food may also be an important factor. However, it does not seem to be a limiting factor because phages are relatively resistant to adverse environmental conditions, and it can be assumed that they will survive successfully in many foods.

When a potentially successful encounter between a phage and a host bacterium occurs, the subsequent development of the infective process depends on environmental factors, which may affect the phage during adsorption and penetration or the activity of the host after infection. Conditions affecting phage development include the temperature, the ionic environment, pH, nutrient concentration and the growth phase of the host cell. Much evidence indicates that infection only occurs when the conditions are suitable for the proliferation of the host. In the case of a few host–phage systems, e.g. *Pseudomonas*, *Aeromonas* and *Listeria* and their phages, the conditions will be suitable for replication in some foods. However, in the case of most pathogen–phage systems, the conditions for successful infection and replication are very unlikely to occur in foods.

It can be concluded that the use of phages for the control of pathogens in foods is not a suitable method in the case of most pathogens, for example *E. coli* and *Salmonella*, which are found in low concentrations in foods and which do not replicate in food in conditions suitable for food preservation. However, psychrotrophic pathogens, e.g. *Aeromonas* and *Listeria monocytogenes*, and food spoilage bacteria, e.g. *Pseudomonas*, may replicate to high densities (up to 10^7 cfu g^{-1}) – well over the threshold host density for phage replication – in many different foods (e.g. dairy foods, meat, vegetables and seafood). Moreover, the conditions for phage replication may be favourable, because the hosts replicate actively. Consequently, phages have been proposed as both disinfectants of surfaces in food processing plants and as a means of controlling these pathogens in foods. For example, listeriophage suspensions at concentrations of up to 3.5×10^8 pfu ml^{-1} were at least as efficient as a $20 \, \mu g \, ml^{-1}$ solution of a quaternary ammonium compound in reducing *L. monocytogenes* populations from artificially contaminated surfaces. In addition, phages (at $> 10^6$ pfu ml^{-1}) have been used successfully for the control of the growth of *Pseudomonas* and the spoilage of beef.

Lysins

Phage lysins, or endolysins, are enzymes which hydrolyse cell walls. They are synthesized during late gene expression in the lytic cycle of multiplication of most phages, thereby enabling the release of progeny phages.

The mechanisms of lysis are not the same in the case of all phages, and may even differ in the same bacterial host, depending on the phage. There are at least two different mechanisms. Some small phages, exemplified by the *Escherichia coli* phages φX174 and MS2, have developed a single gene for a lysin which cannot degrade murein (peptidoglycan). These phages simply cause cells to empty, leaving non-refractile, rod-shaped cell ghosts. However, the great majority of known phages has a more complex mechanism of lysis, involving two different kinds of enzymes, known as lysins and holins. Only the joint activity of the two enzymes leads to lysis of the host cell. Lysins, also known as endolysins, are highly efficient and specific peptidoglycan-hydrolysing enzymes, which are expressed as soluble cytoplasmic proteins. However lysins can only reach their peptidoglycan substrate in the cell wall through the action of a second group of phage-encoded proteins, known as holins. These produce holes in the cytoplasmic membrane, through which the lysin molecules move into the periplasm, where they come into contact with the peptidoglycan. This is the dominant strategy for lysis by phages of both Gram-positive bacteria (e.g. *Lactococcus*, *Streptococcus*, *Listeria* and *Bacillus*) and Gram-negative bacteria (e.g. *E. coli*, *Salmonella* and *Haemophilus*).

Phage-encoded endolysins can be of any one of several unrelated types of enzymes (e.g. lysozyme, amidase, transglycosylase). These attack either glycosidic bonds (lysozyme and transglycosylase) or peptide bonds (amidase), which in combination confer mechanical rigidity on peptidoglycan.

The lysins of the phages of Gram-positive bacteria were recognized early on as being strongly active against the cell walls of their host bacteria when added exogenously, i.e. the cells can be lysed from the outside by the specific lysin. In Gram-negative bacteria, the situation seems more complex. When these bacteria are infected with phages at a high multiplicity of infection, i.e. with many phages per host cell, the bacteria lyse before phage replication. This phenomenon is known as 'lysis from without', and is due partially to the phage lysins. However, at least one other molecule is needed, probably because the outer membrane of the bacterium hinders contact between the lysins and their peptidoglycan substrate. Thus it appears that independently, only the lysins of Gram-positive bacteria have clear exogenous activity.

Even when applied exogenously, lysins retain a certain degree of specificity, the causes of which are not yet well-understood. For example, when the lysins of listeriophages are applied exogenously they induce rapid lysis of *Listeria* strains from all species, but generally do not affect other bacteria. Even the lysins coded by phages with limited host ranges are exogenously active against all listerial cell walls, regardless of serovars and species. However, they do not affect other Gram-positive bacteria with the same peptidoglycan type (A1γ-variation of directly cross-linked *meso*-diaminopimelic acid peptidoglycan). See Loessner et al (1995) *Molecular Biology* 16(6): 1231–1241. The molecular basis of this substrate specificity remains to be elucidated. A similar pattern can be observed in the case of phages infecting a number of different lactic acid bacteria. Certain lysins, e.g. some pneumococcal lysins, may have a broader substrate specificity than others, and even have some activity in taxonomically unrelated bacteria. However, the specificity shown by the lysins of the phages of lactic acid bacteria and listeriophages is a general phenomenon, common to the lysins of the phages of Gram-positive bacteria. Thus phage lysins can be described as very specific compared to other antimicrobial agents, although less specific than phages. The differences in the structure of the cells walls of Gram-positive and Gram-negative bacteria suggest that the exogenous activity of lysins will be more efficient in Gram-positive than in Gram-negative bacteria, in which the outer membrane is likely to somehow obstruct contact between the phage lysins and their peptidoglycan substrate.

Commercial Use of Lysins

Both the endogenous and the exogenous activities of phage lysins have potential commercial applications in food processing.

The endogenous activity of phage lysins may be applied through the construction of bacterial strains with an 'autolytic' phenotype, for example, lactococci possessing the gene encoding phage lysin for *Lactococcus* species. Such strains may release their intracellular enzymes (e.g. proteases and lipases) more efficiently in the curd, thus providing accelerated ripening of cheese. Such tailored strains of lactococci have been constructed successfully, but difficulties related to their survival and the control of the expression of the 'autolytic' genes in the food environment remain to be resolved before this approach can be applied successfully.

The exogenous activity of phage lysins may also be exploited. Firstly, because of their specificity and potency, phage lysins may underlie an effective method of eliminating contaminating food pathogens without affecting other organisms. Second, they may have applications in food manufacturing. For example, in the cheese industry the use of lysin to accelerate the release of intracellular enzymes from starter cultures at the end of the primary milk fermentation accelerates ripening.

Two ways of introducing phage lysins into food can be envisaged. Firstly, engineered bacteria that release phage lysins could be introduced into the food. However this approach is not yet feasible, because the survival of the tailored bacterial strains in the food environment and the control of the expression of their genes are problematic. Secondly, the enzyme could be produced by the highly efficient fermentation techniques which already exist, and subsequently be added to the food. This approach seems more feasible than the former, although limitations, similar to those of the many enzymes already used in the food industry (e.g. α-amylases, proteases, lipases) must be anticipated. If necessary, however, solutions similar to those already used may be applied.

In theory, there are two ways of producing phage lysins. The most direct way involves phage-infected specific host cells. The optimum conditions in terms of pH, temperature and multiplicity of phage infection for the production of phage-associated lysins can be determined. They have, for example, been established for the production of group C *Streptococcus* phage-associated lysin. The other method involves recombinant DNA technology. Phage lysin genes can be cloned and overexpressed to very high levels in bacteria unrelated to the original host, and engineered so as to facilitate their purification, increase their stability, or improve their performance in other ways. This has already been achieved with *Listeria* phage lysin genes, which have been successfully cloned in *Escherichia coli*. One of the cloned lysins, an L-alanoyl-D-glutamate peptidase, is extremely active on all listerial cells when added exogenously, and can be overexpressed in *E. coli* to very high levels. The corresponding gene has been modified by gene fusion, in order to facilitate the purification of the lysin, a modification of the amino terminal allowing highly efficient purification without loss of specific activity. Promising preliminary results are claimed regarding the experimental use of this recombinant phage lysin to combat *L. monocytogenes* in soft cheeses and other dairy products.

Thus it seems that the application of lysins in the control of food pathogens is more feasible than that of phages. Problems related to the industrial production and commercial use of phage lysins are likely to be similar to those encountered and solved in connection with the many enzymes already used in the food industry, and so their use is likely to be influenced more by economic factors than by technical feasibility.

See also: **Bacteriophage-based Techniques for Detection of Foodborne Pathogens**. *Escherichia coli*: Escherichia coli. **Genetic Engineering**: Modification of Bacteria. *Listeria*: Introduction. **Starter Cultures**: Uses in the Food Industry; Cultures Employed in Cheesemaking.

Further Reading

Greer GG (1986) Homologous Bacteriophage Control of *Pseudomonas* Growth and Beef Spoilage. *J. Food Prot.* 49: 104–109.

Kennedy JR and Bitton G (1987) Bacteriophages in Foods. In: Goyal SM, Gerba CP and Bitton G (eds) *Phage Ecology*. New York: John Wiley.

Loessner MJ, Schneider A and Scherer S (1996) Modified *Listeria* Lysin Genes (*pyl*) Allow Overexpression and One-Step Purification of Biochemically Active Fusion Proteins. *Appl. Environ. Microbiol.* 62: 3057–3060.

Sable S and Lortal S (1995) The lysins of bacteriophages infecting lactic acid bacteria. *Appl. Microbiol. Biotechnol* 43: 1–6.

Young RY (1992) Bacteriophage Lysis: Mechanisms and Regulation. *Microbiol. Rev.* 56: 430–481.

Modified Atmosphere Packaging see **Chilled Storage of Foods:** Use of Modified Atmosphere Packaging.

MOLECULAR BIOLOGY IN MICROBIOLOGICAL ANALYSIS – DNA-BASED METHODS FOR THE DETECTION OF FOOD-BORNE PATHOGENS

Louise O'Connor and **Majella Maher**, National Diagnostics Centre, National University of Ireland, Ireland

Introduction

Current consumer demand for high-quality foods is putting pressure on the industry to ensure that its food products are pleasing to the palate and safe for consumption. To date, quality testing of foods has relied almost entirely on conventional microbiological methods, isolating and enumerating bacteria from food on specialized microbiological media, yielding results only after several days and repeated culture enrichment steps. Developments in the fields of immunology and molecular biology offer the potential to develop rapid, high-throughput tests that will allow the food industry to make timely assessments on the microbiological safety of its food products. This article outlines recent developments in molecular biology and their current and future applications for detecting and identifying existing and emerging food-borne pathogens.

Microbiological Analysis of Foods

The traditional method for isolating and identifying bacteria from a food sample involves the homogenization of the food in a buffer, its inoculation into selective media and incubation for a predetermined period. This primary enrichment is followed by another period where the broth is streaked on to a solid media to yield isolated colonies that can be identified definitely on the basis of their biochemical and immunological characteristics. There have been

a number of developments to speed up or automate these traditional procedures, including the addition of colorimetric/fluorimetric substrates in media for enumeration of coliforms, the availability of rehydratable nutrients eliminating the requirement for pour plates, the compilation and comprehensive packaging of biochemical and morphological identification kits such as API, BBL-Crystal, Biolog and Enterotube, the correlation of simple and rapid indicators of microbial viability such as ATP-bioluminescence with hygiene standards and the invention of sophisticated machinery, such as Bactometer, Malthus and Rabit for early detection of bacterial growth based on measurement of impedence or conductance by the growing culture in liquid media. Immunological enzyme-linked immunosorbent assay (ELISA) test kits are available for a large number of food-borne pathogens but, with detection limits of 10^4–10^6 cfu ml^{-1}, culture enrichment is required before application of the tests.

Nucleic Acid-based Tests

The genome of every microorganism contains stretches of DNA sequence that are homologous to all members of that family or genus. They also have unique sequences that are particular to a species which can be exploited to determine the presence of that bacterial species in a sample. Genes associated with virulence are commonly used for identifying pathogenic microorganisms – listeriolysin in *Listeria monocytogenes*, cytotoxin gene in *Vibrio cholerae*, neurotoxin genes in *Clostridium botulinum*, and verotoxin-encoding genes in *Escherichia coli*. The ribosomal RNA represents an attractive target for DNA probe design as it is present in multiple copies in most organisms, with the exception of slow-growing mycobacteria, which have one copy per cell. This region contains stretches of conserved sequence interspersed with variable sequence regions, providing the scope to design single-stranded DNA probes providing the broad-ranging or specific target detection an assay requires. Other targets include genes encoding for flagellar proteins and outer membrane antigenic proteins. These DNA probes can be employed directly in hybridization assays or may be combined with an in vitro amplification step for pathogen detection.

Direct Hybridization Assays

DNA probe direct hybridization assays generally follow two basic formats. In the first format, direct colony hybridization involves impression transfer of bacterial cells as primary colony isolates on to a nylon or nitrocellulose membrane. Alkaline treatment of the membrane releases the organism's DNA as single-stranded molecules which are fixed on to the membrane by baking or by brief exposure to ultraviolet (UV) light. The DNA probe is labelled, formerly by radiolabelling but more recently and more appropriately for wide-scale use, by non-isotopic labels such as biotin or digoxigenin. The labelled probe is hybridized to the membrane containing the bound DNA using conditions dictated by the DNA probe length and nucleotide composition, which, combined with temperature and salt concentration, determine the assay stringency to allow the probe to bind specifically to its appropriate target.

In the second format the DNA probe is linked to the membrane or a microtitre solid phase as a capture probe and the DNA released from the bacteria and hybridized using reverse hybridization kinetics to this capture probe. This probe/DNA hybrid can be detected by the addition of a second tagged reporter probe, increasing the sensitivity and specificity of the assay. These assays do not require sophisticated equipment, are simple to perform but since they have a detection limit of 10^4–10^5 bacterial cells they require selective enrichment of the target organism from food for up to 48 h before probe hybridization.

A number of hybridization assays for food-borne pathogens are described in the literature. Gen Probe Inc (San Diego, CA) are marketing a chemiluminescent solution hybridization method which can rapidly identify colonies on primary isolation, with

assays available for the detection of *L. monocytogenes*, *Salmonella*, *E. coli* and *Campylobacter*.

Amplification-based Methods

The application of a test or assay for food-borne pathogen identification and detection that includes an in vitro amplification step has the potential to increase the speed and sensitivity of food quality testing, facilitating more expedient release of products from the industry. In vitro amplification technologies are designed to amplify either a target nucleic acid or a detection signal. While the polymerase chain reaction (PCR) has been most widely adapted for these rapid tests, a number of other amplification technologies are being developed, adapted or incorporated into tests for food-borne pathogens. These include the following.

Strand Displacement Amplification

Strand displacement amplification is an isothermal reaction exploiting the ability of DNA polymerase to initiate polymerization at a single-stranded nick following restriction enzyme digestion of the duplex DNA leading to the generation of a new strand yielding two templates for second round synthesis with a capability of producing 10^7–10^8 copies of the original target in two hours.

Transcription-mediated Amplification

Transcription-mediated amplification involves the isothermal amplification of rRNA by reverse transcription and subsequent generation of numerous transcripts by RNA polymerase (**Figures 1–4**). Following amplification, these RNA copies are hybridized with a complementary oligonucleotide probe for detection via a chemiluminescent tag.

Ligase Chain Reaction

The ligase chain reaction covalently ligates two selected probes with 3′ and 5′ ends which are immediately adjacent following homologous binding to the target DNA. Ligation of the two probes generates a new target for second-round covalent ligation, leading to geometric amplification of the target of interest.

Qβ Replicase

Qβ replicase, the RNA-directed RNA polymerase from Qβ bacteriophage, increases the amount of target RNA present in a sample. A specialized detection probe containing stretches of DNA complementary to the target of interest and to Qβ is constructed for hybridization to the target nucleic acid while the addition of Qβ replicase increases the

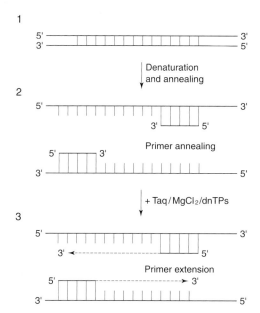

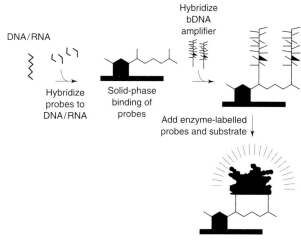

Figure 3 Branched DNA (bDNA) signal amplification is a technology for increasing sensitivity of direct detection using specialized branched DNA probes.

Figure 1 The polymerase chain reaction. A specific sequence of DNA is chosen for amplification. The strands of DNA are separated by heating and oligonucleotide primers anneal to their complementary sequence on the separated strand. New strands of DNA are synthesized, using the original strands as templates, by the enzymatic polymerization of DNA by a thermostable enzyme in the presence of $MgCl_2$ and excess deoxyribonucleotide triphosphates.

detection probe signal, indicating the presence of target RNA in a sample.

Nucleic Acid Sequence-based Amplification (NASBA)

NASBA or 3SR is an isothermal amplification system combining RNA polymerase with reverse transcriptase to generate a cDNA intermediate containing

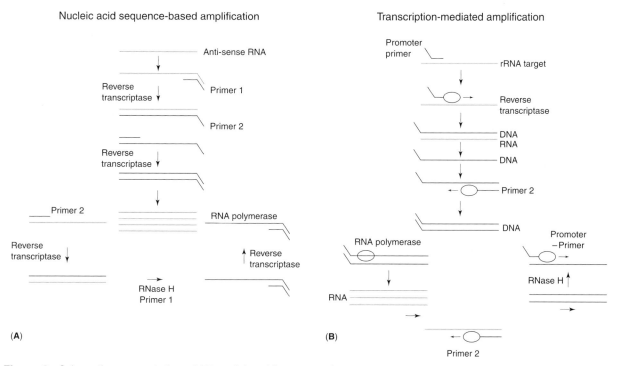

Figure 2 Schematic representation of (**A**) nucleic acid sequence-based amplification (NASBA) and (**B**) transcription-mediated amplification (TMA) technologies. Both isothermal methods combine the activities of reverse transcriptase and RNA polymerase to generate cDNA intermediates for RNA polymerase, leading to billion-fold target amplification.

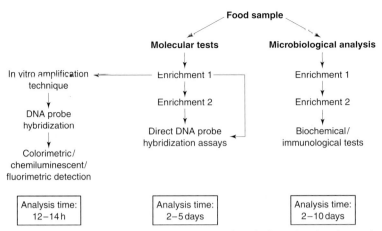

Figure 4 Steps and time frame involved in conventional microbiological analysis and molecular methods for the detection and identification of food-borne pathogens.

a T_7 bacteriophage promoter as a target for T_7 RNA polymerase. Increases of greater than 10^8 copies of the target in 30 min have been reported (Fig. 2).

Branched DNA Technology (bDNA)

The sensitivity of direct detection can be increased by using specialized branched DNA (bDNA) probes. The bDNA molecule is essentially a tree which binds to the target complex. Each tree contains 15 branches in a staggered array, allowing maximum binding of the detection substrates in the last step of the assay, and thereby significantly increasing the sensitivity over single-probe binding.

Polymerase Chain Reaction

PCR is currently one of the most widely employed techniques to complement classical microbiological methods for the detection of pathogenic microorganisms in foods. PCR is an in vitro technique used for amplifying a specific segment of DNA. Each reaction cycle consists of three steps. The first involves the separation of the double-stranded DNA to be used as a template by heat denaturation. The second is an annealing step, where the temperature is lowered to allow a pair of specific oligonucleotide primers to bind to the template DNA. The third and final step is the extension of the primers with a thermostable enzyme, DNA polymerase (Fig. 1). Subsequent cycles involve further denaturation and extension steps, during which the original target region is amplified in addition to the amplification product. Because newly synthesized copies also serve as templates for subsequent rounds of synthesis, the amount of DNA generated increases exponentially.

Detection of amplification products can be carried out using gel electrophoresis, ethidium bromide staining and visual examination of the gel using UV light. Southern blotting and hybridization with a specific DNA probe can follow electrophoresis, allowing confirmation of the identity of the PCR product. Colorimetric or fluorimetric hybridization of amplification products can also be carried out using specific DNA probes bound to solid phases, such as microtitre plates allowing rapid and simplified detection.

For the detection of microorganisms in food using PCR assays, the two important criteria are sensitivity and specificity. Specificity is determined by the sequence of the oligonucleotide primers used and the annealing temperature. Primers must be carefully designed to ensure amplification of a single specific region. If the annealing temperature is too low the primers can bind non-specifically to the template and several regions may be amplified. Specificity of PCR assays can be increased by DNA probes which only recognize the correctly amplified target DNA. The sensitivity of PCR depends on the reaction conditions used, the food matrix and the post-PCR detection method employed. Sensitivity is considerably influenced by the food matrix, and detection limits using pure cultures of microorganisms can be significantly greater than those achieved in food samples.

Application of PCR in Food Samples

Several detailed reviews of PCR assays for the detection of microorganisms in foods have been published. **Table 1** shows some of the more recent assays. One of the limitations of PCR assays for detection of food pathogens is that PCR will detect both viable and nonviable microorganisms if the correct target sequence is present in the food samples. The inclusion of a pre-enrichment step prior to PCR increases the number of target cells in the media, dilutes the inhibitory effects of the food matrix and confines detection to viable and culturable cells. The use of intact RNA as target has also been suggested to distinguish between dead and living cells; however, the isolation of RNA

Table 1 Recently developed polymerase chain reaction assays for the detection of microorganisms in foods

Organism	Food type	Reference
Escherichia coli O157	Milk	Desmarchelier et al 1998
Shigella-like toxin producing E. coli	Milk	Lehmacher et al 1998
Shigella flexneri	Vegetables/salads	Rafii and Lunsfird 1997
Camplyobacter jejuni	Skim milk/poultry	Ng et al 1997
Listeria monocytogenes	Milk/cheese	Manzano et al 1997
Listeria monocytogenes	Seafoods	Agersberg et al 1997
Listeria monocytogenes	Cooked ground beef	Klein and Juneja 1997
Listeria monocytogenes	Rainbow trout	Ericsson and Stalhandske 1997
13 food-borne pathogens	Seafood/soft cheese	Wang et al 1997

is technically more difficult than isolation of DNA as RNA is considerably less stable, which makes it more difficult to manipulate.

One of the main problems associated with the use of PCR assays for food samples is the presence of PCR inhibitors in the food. False-negative results may occur for a variety of reasons, including first, nuclease degradation of target nucleic acid sequences and/or primers; second the presence of substances which chelate divalent magnesium ions necessary for the PCR; and, last, inhibition of the DNA polymerase. The degree of inhibition varies greatly with food type. Studies have shown that high levels of oil, salt, carbohydrate and amino acids have no inhibitory effect, while casein hydrolysate, calcium ions and certain components of some enrichment broths are inhibitory for PCR. The removal of inhibitory substances from DNA to be amplified is an important prerequisite to successful PCR amplification.

Several methods of sample preparation have been reported, including filtration, centrifugation, use of detergents and organic solvents, enzyme treatment, immunomagnetic capture and sample dilution. Inhibition of PCR by substances present at low concentrations can be overcome by diluting the food sample before amplification. However, dilution of the sample results in a corresponding decrease in cell numbers, with consequent reduction in PCR sensitivity.

Centrifugation is another technique which has been used to remove inhibitors from food samples. A disadvantage of this technique is that large particles in the food may trap bacteria as they settle. Buoyant density centrifugation (BDC) has recently been used to overcome the problem of inhibitory substances in

food. This is achieved by layering food homogenates on top of Percoll media; following centrifugation, the food remains in the upper part of the tube while the organisms of interest are concentrated below the light Percoll layer. Separation of target organisms and PCR inhibitors using filtration can be based on differences in solubility or molecular weight. However, large particles may clog filters and inhibitory substances may be concentrated together with the bacteria which are being isolated. A commonly used sample preparation method is lysis of bacterial or viral cells to release nucleic acids, making them available for PCR. Methods of lysis include heating, the use of detergents such as sodium dodecyl sulphate (SDS) and Triton X-100, and proteases such as proteinase K. Purification of DNA is sometimes carried out following lysis. The DNA is purified and concentrated by organic solvent extraction and then precipitated using ethanol. While such a step may remove PCR inhibitors found in food samples, organic solvents are generally not considered suitable for routine use.

A user-friendly method for the separation of organisms from inhibitors is immunomagnetic separation (IMS). Magnetic particles coated with antibodies to the organism of interest can be used to capture organisms from the food sample for inclusion directly in the PCR. It should be noted that specificity will be determined by the antibody used for coating the magnetic particles. A potential problem is that certain food components can interfere with the antibody–organism interaction.

The inclusion of an isothermal amplification procedure in place of PCR may prove useful in terms of less stringent requirements for preparation of the sample and removal of inhibiting substances prior to amplification. In the clinical sector, for example, it has been demonstrated that these isothermal methodologies seem to be less affected by substances that are inhibitory to PCR.

The type of sample preparation method which needs to be used depends on the food being analysed. Certain foods are more problematic than others. Soft cheeses can completely inhibit PCR assays. Calcium ions in milk have also been identified as a source of PCR inhibition. It is obvious therefore that no one method of sample preparation can be applied to all food types. Sample preparation methods for routine use must be rapid, safe and user-friendly, particularly if they are to be used for analysis of a large number of samples. Most of the PCR assays currently applied to food samples include a pre-enrichment step of 18 h or more to increase cell numbers while diluting potential PCR inhibitors present in the food matrix: researchers using this strategy have reported the successful application of PCR tests on a broad range of food matrices.

Detection Technologies

Another important aspect of PCR assays is the method to be used for detection of amplified products. Like the sample preparation method, the detection method should be rapid and simple to use. The purpose of the detection method is to confirm the identity of the amplified product as well as to increase the sensitivity of the assay.

Hybridization of the amplified product to a specific DNA probe is the commonest detection method. Various formats have been used for this hybridization, including Southern blots and dot blots, where nucleic acids are immobilized on a membrane following separation on a gel or from solution. Alternatively, specific capture probes can be immobilized on to a nylon membrane or on to the wells of a microtitre plate. Once the hybrid is formed various detection techniques can be used. Many of these are similar to the techniques used in immunoassays for the detection of antibody–antigen complexes. The method of detection used depends how the probe has been labelled. Radioactive and fluorescently labelled probes allow the direct detection of product–probe hybrids. Indirect detection methods involve the binding of pairs such as biotin–streptavidin, digoxigenin–antidigoxigenin and fluorescein–antifluorescein. Biotinylated probes can be generated either chemically or enzymatically and operate on the principle that biotin has a high affinity for streptavidin or avidin. Following hybridization of the probe to the PCR product, a streptavidin conjugate is added, allowing detection of activity.

An alternative format for detection is a reverse hybridization assay. This method incorporates biotin into the PCR product by including a biotinylated primer in the PCR. This biotinylated product is hybridized to a capture probe immobilized on to a chemically modified nylon membrane or on to specially treated microtitre wells. The captured product can then be detected colorimetrically using a conjugate such as streptavidin alkaline phosphatase and a chromogenic substrate (**Fig. 5**). This technology is currently being applied in our laboratory to compile rapid qualitative tests for a range of pathogenic bacteria and fungi. A panel of capture probes to different pathogenic bacteria can be immobilized on a single membrane, allowing multi-pathogen detection. There are also a broad range of microtitre plates available with various surface chemistries (Nunc A/S, Costar) to enable capture probes, the amplified product, or, indeed, the PCR primer to be immobilized on the well

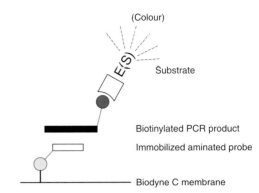

Figure 5 Reverse hybridization detection of a biotinylated polymerase chain reaction (PCR) product. The probe on the membrane specifically captures the biotinylated PCR product. This probe/product hybrid is then detected via a streptavidin alkaline phosphatase conjugate and conversion of nitroblue tetrazolium chloride/5-bromo-4-chloro-3-indolyl-phosphate 4 toluidine salt to insoluble product, yielding a purple colour on the membrane.

surface for subsequent colorimetric fluorescent and quantitative detection of a target molecule. These detection methods and technologies are rapid, reliable, reproducible and economical and, particularly in the case of microtitre formats, can be automated, making them useful for routine analysis of large numbers of samples.

Future Developments

The increasing numbers of innovations in nucleic acid detection technologies are overcoming the technical challenges which to date have hampered the adaptation of these techniques for wide-scale, high-throughput routine diagnostic use. Currently the detection methods employed to analyse PCR products run the risk of carry-over contamination, leading to false-positive results being determined. The ideal test would incorporate the dual processes of amplification and detection in an automated system. A number of companies are investing significant resources in the development of such instrumentation and also, where possible, automating the process of sample preparation such that a single instrument includes modules for sample preparation, PCR amplification and PCR amplicon detection and the test sample is moved through these steps by robotic arms. The operator is simply required to ensure that the instrument is supplied with reservoirs of reagents to allow it to process up to 50 samples in a single run and to interpret the test results on completion. The microbiology laboratory of the future may also feature high-technology devices such as DNA chips. DNA chips or array's are ordered arrays of oligonucleotides or DNA samples synthesized on glass or other suitable substrates which offer a powerful alternative to gels and blots for high-throughput hybridization analyses. The emergence of this next generation of highly informative diagnostic tools will bring speed and simplicity to the analysis of nucleic acids and the applications for such invaluable technologies will be innumerable.

See also: **ATP Bioluminescence**: Application in Dairy Industry; Application in Hygiene Monitoring; Application in Beverage Microbiology. **Enzyme Immunoassays**: Overview. **PCR-based Commercial Tests for Pathogens**. **Immunomagnetic Particle-based Techniques**: Overview. **Nucleic Acid-based Assays**: Overview.

Further Reading

Agersborg A, Dahl R and Martinez I (1997) Sample preparation and DNA extraction procedures for polymerase chain reaction identification of Listeria monocytogenes in seafoods. Int. J. Food Microbiol. 15: 275–280.

Barbour WM and Tice G (1997) Genetic and immunologic techniques for detecting foodborne pathogens and toxins. In: Doyle MP, Beuchat LR and Montville TJ (eds) Food Microbiology, Fundamentals and Frontiers. P. 710. Washington DC: ASM Press.

Cahill P, Foster K and Mahan DE (1991) Polymerase chain reaction and QB replicase amplification. J. Virol. 37: 1482.

de Boer E and Beumer R (1998) Developments in the microbiological analysis of foods. De Ware(n)- Chemicus 28: 3–8.

Desmarchelier PM, Bilge SS, Fegan N, Mills L, Vary JC Jr and Tarr PI (1998) A PCR specific for Escherichia coli O157 based on the rfb locus encoding O157 lipopolysaccharide. J. Clin. Microbiol. 36: 1801–1804.

Doyle MP, Beuchat LR and Montville TJ (1997) Food Microbiology Fundamentals and Frontiers. Washington DC: ASM Press.

Hill WE (1996) The polymerase chain reaction: applications for detection of foodborne pathogens. Crit. Rev. Food Sci. Nutr. 36: 123–173.

Klein PG and Juneja VK (1997) Sensitive detection of viable Listeria monocytogenes by reverse transcription-PCR. Appl. Environ. Microbiol. 63: 4441–4448.

Lehmacher A, Meier H, Aleksic S and Bockemuhl J (1998) Detection of hemolysin variants of Shiga toxin-producing Escherichia coli PCR and culture on vancomycin-cefixime-cefsulodin blood agar. Appl. Environ. Microbiol. 64: 2449–2453.

Meng J and Doyle MP (1997) Emerging issues in microbiological food safety. Annu. Rev. Nutr. 17: 255–275.

Olsen JE, Aabo S, Hill W et al (1995) Probes and polymerase chain reaction for detection of foodborne pathogens. Int. J. Food. Microbiol. 1: 1–78.

Panaccio M, Good RT and Reed MB (1994) A road map for PCR from clinical material. J. Clin. Lab. Anal. 8: 315–322.

Scheu PM, Berghof K and Stahl U (1998) Detection of pathogenic microorganisms in food with the polymerase chain reaction. Food Microbiol. 15: 13–31.

Swaminathan B and Feng P (1994) Rapid detection of foodborne bacteria. Annu. Rev. Microbiol. 48: 401–426.

Uyttendaele MR and Debevere JM (1998) Nucleic acid based methods for detection and typing of foodborne pathogens. De Ware(n)-Chemicus 28: 19–23.

Walker GT, Fraiser MS, Schram JL, Little MC, Nadeau JD and Malinowski DP (1992) Strand displacement amplification – an isothermal in vitro DNA amplification technique. Nucleic Acids Res. 20: 1691–1696.

Wang RF, Cao WW and Cerniglia CE (1997) A universal protocol for PCR detection of 13 species of foodborne pathogens in foods. J. Appl. Microbiol. 83: 727–736.

Molluscs see **Shellfish (Molluscs and Crustacea)**: Characteristics of the Groups; Contamination and Spoilage.

MONASCUS

L Martínková, Institute of Microbiology, Academy of Sciences of the Czech Republic, Prague, Czech Republic

P Patáková, Institute of Chemical Technology, Prague, Czech Republic

The fungus Monascus has been traditionally used for manufacturing food colorants (red rice) and fermented foods and beverages in southern China, Taiwan, Japan, Thailand, Indonesia and the Philippines.

The typical pigmentation of Monascus is caused by the formation of structurally related yellow, orange–red and purple oligoketides (**Fig. 1**) and their derivatives originating from reactions with amino compounds. Some of the Monascus pigments, as well as other secondary metabolites, are probably responsible for its disinfectant effects known in the Chinese folk medicine for centuries.

In past decades, with growing demands for natural dyes Monascus has attracted the attention of food producers all over the world. Particularly, extracts

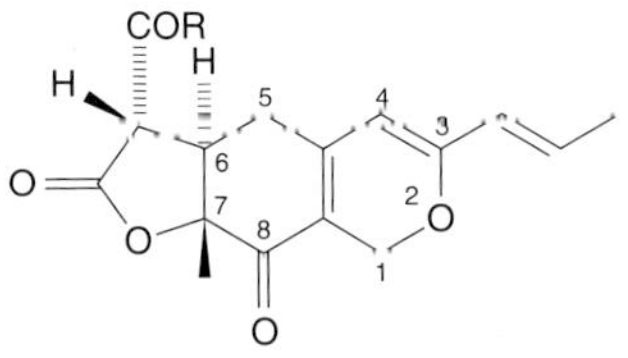

R = C$_5$H$_{11}$ Monascin; yellow, mol. wt 358.4, m.p. 143–145°C

R = C$_7$H$_{15}$ Ankaflavin; yellow, mol. wt 386.5, m.p. 120–121°C

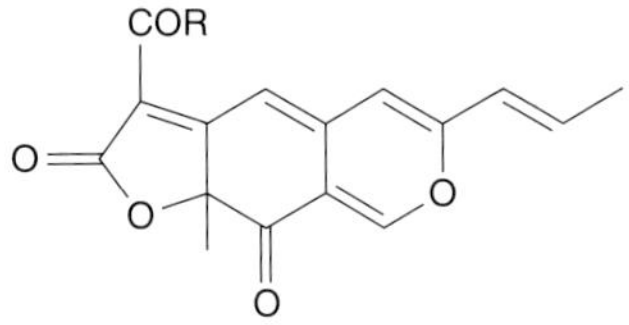

R = C$_5$H$_{11}$ Rubropunctatin; orange–red, mol. wt 354.4, m.p. 156–157°C

R = C$_7$H$_{15}$ Monascorubrin; orange–red, mol. wt 382.4, m.p. 142–143°C

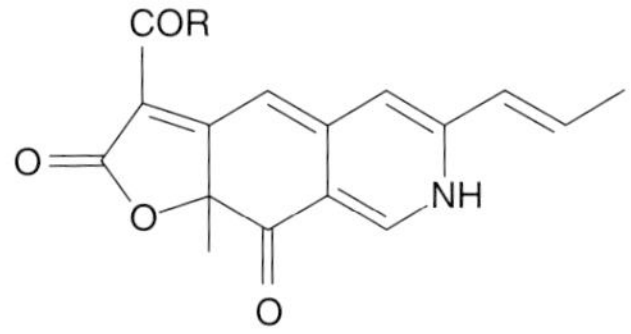

R = C$_5$H$_{11}$ Rubropunctamine; purple, mol. wt 352.4, m.p. 217–218°C

R = C$_7$H$_{15}$ Monascorubramine; purple, mol. wt 380.4, m.p. 207–208°C

Figure 1 Major pigments of *Monascus*. For spectral data see Martínková et al (1999) and Sweeny et al (1981) and citations therein.

from *Monascus* cultures have been examined as potential substitutes for nitrite curing salts in cured meat products. Although a number of patents dealing with the solid-state or submerged production and applications of *Monascus* pigments have been issued, not only in Asia but also in Europe and Northern America, the use of *Monascus* pigments as a natural colorant for human food has not been approved either in the EC or in the USA. In Japan, on the other hand, all natural colours, i.e. the substances which are obtained from natural sources by natural means, are permitted as food additives.

Addition of *Monascus* colorants to foods can be detected by spectrophotometric and chromatographic methods (**Table 1**). Methods for separation of the pigment components on a preparative scale are given in **Table 2**.

Physiology

The genus *Monascus* (placed in the phylum Eumycota, subphylum Ascomycotina, class Plectomycetes, order Eurotiales family Monascaceae) is divided into six species: *M. pilosus*, *M. purpureus*, *M. ruber*, *M. floridanus*, *M. pallens* and *M. sanguineus*. *M. pallens* can easily be distinguished from other *Monascus* species by its lack of pigmentation. The species with the

Table 1 Outline of selected analytical methods for detection and quantitation of *Monascus* pigments

Method	Application	Reference
TLC on silica gel	Pigment detection in meat products	Henning et al (1992)
Spectrophotometry	Semi-quantitative pigment detection in meat products	Henning at al (1992)
Reversed-phase HPLC	Analysis of natural pigments	Chen and Johns (1993)
Reversed-phase HPLC	Analysis of natural and semisynthetic pigments	Lin et al (1992)

Table 2 Outline of selected methods for isolation and purification of *Monascus* pigments

Method	Application	Reference
Crystallization	Preparation of natural pigments	Sweeny et al (1981), and citations therein
Silica gel column chromatography	Preparation of natural pigments	Wong and Koehler (1983)
Reversed-phase HPLC	Preparation of semisynthetic pigments	Lin et al (1992)

greatest significance in food industry are *M. purpureus* (**Fig. 2**), also known as *M. anka* and *M. pilosus*. Mutants that do not produce pigment (albino) can be obtained from pigment-forming strains by u.v. irradiation. In nature, *Monascus* occurs as a contaminant in silage, cereals, starch and cow's milk.

The reproduction of the homothallic fungus *Monascus* can be either asexual or sexual. In the asexual cycle, conidia give rise to filaments that branch and form mycelium. Aleurioconidia (single or in short chains) are created on the tips of specialized hyphae, named conidiophores. Lateral or sympodial aleurioconidia, intercalar chlamydoconidia or arthroconidia are produced by some species.

Briefly, the sexual reproduction involves migration of male nuclei from an antheridium through the trichogyne and into an ascogonium containing female nuclei; this is followed by karyogamia and the formation of asci with eight haploid ascospores each. The ascospores are closed in a cleistothecium. They are released after rupture of both the ascus wall and the cleistothecium wall. *Monascus* ascospores are about half the size of the conidia.

The strictly aerobic, saprophytic and chemoorganotrophic fungus *Monascus* grows on various substrates due to its broad spectrum of lytic enzymes with amylolytic, proteolytic and nucleolytic activities. *M. purpureus* and *M. ruber* differ from other *Monascus*

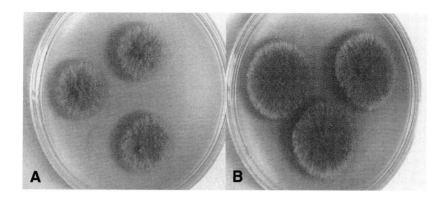

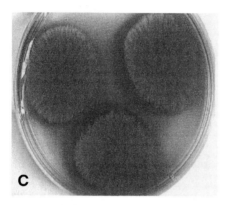

Figure 2 Colonies of *Monascus purpureus* CCM 8152 grown on starch agar, containing (g l⁻) soluble starch 20, peptone 5, casamino acids 2, yeast extract 1, agar 20, at 30°C for (**A**) 5, (**B**) 7 and (**C**) 10 days.

species in having a strong polypectase activity at pH 6 and a cellulolytic activity, respectively.

Secondary Metabolites

Major pigments of *Monascus* include three pairs of homologues (see Fig. 1): monascin–ankaflavin (yellow), rubropunctatin–monascorubrin (orange–red) and rubropunctamine–monascorubramine (purple). These secondary metabolites belong to the group of oligoketides. The molecules are synthesized from acetyl coenzyme A subunits except for a methyl group (attached to C-7) which originates from *S*-adenosylmethionine. Monascin and ankaflavin are reduced forms of rubropunctatin and monascorubrin, respectively.

In addition, some *Monascus* strains produce other secondary metabolites, i.e. related minor pigments (xanthomonasin A, xanthomonasin B, 'yellow II'), citrinin, monankarins (coumarin derivatives with monoamine oxidase inhibitory activity), an αβ-unsaturated γ-lactone, named ankalactone, and mevinolin (monacolin K, lovastatin) which has found (together with its derivatives) broad use as a hypocholesterolaemic drug. Mevinolin is probably also responsible for the changes in lipoprotein blood levels in laboratory animals fed with *Monascus* colorants.

Pigments

Physico-chemical Properties

Rubropunctatin and monascorubrin belong to the azaphilones, that is compounds that readily react with ammonia and methylamine. The conversion proceeds via Schiff base formation and dehydration leading to substitution of the oxygen atom at position 2 by nitrogen. By reaction with ammonia, rubropunctamine and monascorubramine are formed. Furthermore, various compounds containing one or more primary amino group such as amino acids, peptides, proteins, amino sugars, amino alcohols, chitosan, nucleotides and nucleic acids can enter such reactions and afford derivatives of rubropunctamine and monascorubramine having a side chain attached to the nitrogen atom.

The major pigments are nearly insoluble in water and soluble in organic solvents. However, the attachment of a side chain to the nitrogen atom of rubropunctamine or monascorubramine results in a markedly greater water solubility, e.g. by 500 times in derivatives containing a glutamic acid side chain. Moreover, some of these water-soluble pigments are more stable to heat, light and pH changes and display increased absorption coefficients at their absorption maxima. Therefore, such compounds are promising as new chemically defined food or cosmetic colours.

Methods of Manufacture

Red rice, i.e. rice fermented with *Monascus*, has been used for more than 1500 years in China. In different countries of the Far East this product is known as ang-kak, anka, ankak, angkhak, angquac, Chinese red rice, beni-koji or aga-koji.

A red rice cultivation method used in the Philippines involves the following steps: rice is washed thoroughly, soaked overnight in water, drained, sterilized at 121°C for 30 min, inoculated with actively growing *M. purpureus* and incubated in woven bamboo trays (bilao) lined and covered with banana leaves at ambient temperature for 10 days. In the Taiwanese traditional manufacture, the substrate (prepared by washing rice, steaming for 60 min, spraying with water, steaming for a further 30 min and cooling to 36°C) is mixed with a *M. purpureus* inoculum (chu chong tsaw) and heaped in a bamboo chamber until its temperature rises to 42°C. Then the rice is spread on plates, shelved and incubated using repeated moistening.

An optimal substrate humidity (25–50% depending on the strain) should be maintained during the entire cultivation time in order to prevent the grains sticking together and to keep the fungal glucoamylase activity at a low level in favour of pigment production. Ang-kak should be deep purplish-red and the pigments should be distributed evenly across a broken kernel. The quality and colour of ang-kak can differ according to the rice variety.

Another red rice product (chu chong) is made in a similar manner as ang-kak using an inoculum (chu kong tsaw) containing not only *M. purpureus* but also *Saccharomyces formosensis*.

The final step in red rice manufacture is drying. Typically, temperatures of 40°, 45° or 60°C are applied for 1 day. Before use as a colouring agent, the red rice can be powdered.

To reduce the risk of contamination associated with the traditional way of manufacture, cultivation can be carried out in autoclavable plastic bags (**Fig. 3**). After inoculation with a spore suspension of *M. purpureus*, incubation proceeds for about 10 days at 30°C with occasional shaking of the bags. Alternatively, a moving bed fermenter can be used. In addition to rice, steamed bread (mantou), corn, oat or wheat grains are suitable substrates.

Figure 3 *Monascus purpureus* CCM 8152 grown on rice at 30°C for 10 days. The cultivation vessel is a plastic bag plugged with a cotton stopper.

A more efficient approach to the manufacture of *Monascus* colorants is submerged cultivation. The demands for carbon source are strain dependent. The following substrates were evaluated as suitable for pigment biosynthesis in some *Monascus* strains: glucose, fructose, maltose, sucrose, galactose, ethanol, maltitol and starch. The nitrogen source influences growth, sporulation and the type of pigments produced. In culture media with ammonium chloride medium acidification impairs the reactions of rubropunctatin and monascorubrin with amines and therefore these orange–red oligoketides (together with the yellow pigments) are formed preferentially. Under these conditions, most of the pigments are located intracellularly. Ammonium ions also suppress conidia and ascospore formation. Purple pigments (rubropunctamine, monascorubramine and water-soluble pigments) can be obtained from cultures grown at pH values above 5. A buffered medium (at an initial pH 5.5) with monosodium glutamate, for example, is suitable for preparation of derivatives containing a glutamic acid side chain. Excretion of the pigments is favoured so that they can be isolated from culture broth. In general, organic nitrogen sources stimulate growth and conidiation and suppress sexual reproduction and pigment yield.

Food Colouring and Other Applications

In the Orient, *Monascus* is used as red rice to prepare coloured and flavoured traditional foods, e.g., fish food, red soybean cheese (red sufu, hung-lu chiu, hon-fan), pickled vegetables and salted meats, or alcoholic beverages (rice wine: fu chiu, hung-chu). *Monascus* pigments are also considered as suitable colorants of tomato ketchups, imitated crab meat, soya bean meat, jellies, syrups, milk and ice cream. In the manufacture of fermented foods such as red rice wine, kaoliang brandy and anka mash pork (hong chao zou) the potential of *Monascus* fermentation starters (usually inocula prepared on rice) as sources of hydrolytic enzymes is exploited.

Examination of *Monascus* extracts as potential substitutes for nitrite curing salts yielded some promising results. The dyes tested were dried and water-resuspended methanol or methanol–chloroform extracts from *M. purpureus* mycelium grown in a rice medium of from filtrate of a *M. ruber* submerged culture grown on monosodium glutamate. The meat products were Bologna-type, frankfurter-type, Toulosan, dry and Strasbourgan sausages or liver pâté. Addition of a *Monascus* extract (4000 p.p.m.) to a Frankfurter-type sausage enabled nitrite reduction by 90%, i.e. to 7.2 p.p.m., without affecting the intensity of the red colour of the product. The colour showed a good stability towards daylight and heat (120°C).

Monascus-coloured sausages without nitrite had an attractive red–brown colour but they lacked the typical aroma and colour of cured products. Moreover, nitrite curing salts are also added to meat products to inhibit growth of undesirable microorganisms. Although *Monascus* extracts show some antibacterial activity, their effects cannot be compared to those of nitrite. Therefore, *Monascus* dye cannot be considered as an equivalent of nitrite curing salt but its use could meet the demands for cured meat products with less or no nitrite. The addition of *Monascus* colorant should be labelled as it may otherwise lead to a fallacious estimation of the product freshness or its meat and fat content.

Some strains of *Monascus* are used, together with acetic acid bacteria and acid-tolerant yeast in the production of an orange fruit vinegar containing vitamin C and monacolin.

In addition to food colouring, *Monascus* pigments can also be used for marking food products and in cosmetics, e.g. in the production of hair dyes.

Antibiotic Properties

In a book about Chinese medicine (Pen Chaw Kang Mu, Peking, 1590) Li Shih-Chun wrote that red rice was also utilized for the treatment of diseases such as indigestion, muscle bruises, dysentery and anthrax.

The effects of *Monascus* as a disinfectant and a food preservative are at least partly caused by rubropunctatin and monascorubrin. These compounds inhibit growth of both bacteria (*Bacillus subtilis* and *Escherichia coli*) and fungi (*Candida pseudotropicalis*). Tentatively, the mechanism of their action is postulated to involve aminophilic reactions with

amino compounds in the organism. This view is supported by the observation that pigments lacking the reactive oxygen atom (rubropunctamine, monascorubramine and their derivatives containing an amino acid side chain) show minor or no antibiotic effects.

An extract from the fungal mycelium was found to be bacteriostatic towards *Bacillus* sp., *Listeria monocytogenes* and *Staphylococcus aureus* in model systems and in meat products. Some fractions active against *S. aureus* were not pigmented. This indicates that compounds other than the pigments are also involved in the antibiotic effects. An antibiotic substance named monascidin A that inhibited growth of bacteria belonging to the genera *Bacillus*, *Streptococcus* and *Pseudomonas* has been identified as citrinin. Ankalactone was also found to be inhibitory against *B. subtilis* and *E. coli* but to a lesser extent than rubropunctatin and monascorubrin.

Toxicological and Dietetic Properties

Tests of a *M. purpureus* culture extract in mice in vivo indicated its nontoxicity (mean lethal dose (LD_{50}) > 10 g kg^{-1} body weight) on oral administration. Similarly, oral doses of up to 18 g of red rice per kilogram body weight caused no toxic effects in mice. Examination of intraperitoneal toxicity in the same animal model showed LD_{50} values of 7 g of red rice per kilogram body weight.

An in vitro genotoxicity study showed that the genotoxic potential of a *Monascus* extract was much lower than that of nitrosamines which can occur in cured meat products. None of 40 examined mycotoxins was found in an extract from *M. purpureus* DSM 1379. However, this conclusion cannot be extrapolated to other *Monascus* strains as some of them produce, for example, the mycotoxin citrinin. Finally, purified rubropunctatin and monascorubrin themselves displayed effects comparable to medium potent mycotoxins in tests using chick embryos as sensitive models. However, these compounds are unlikely to occur in red rice in significant concentrations as they undergo reactions with amino compounds. An extract from red rice did not induce any adverse effects in the chick embryos.

The most severe biological effect demonstrated with *Monascus* pigments was a marked increase in extracellular activities of γ-glutamyltranspeptidase and glutathione *S*-transferase in cultured fetal rat hepatocytes.

In rats with dietary-induced hyperlipoproteinaemia oral administration of *Monascus* extracts induced a decrease in cholesterol and triglyceride plasma levels. However, the high density cholesterol (HDL) fraction was also lowered so that the total cholesterol/HDL cholesterol (TC/HDL-C) ratio (a measure of the atherogenic potential of increased lipoprotein levels) was not changed favourably. On the other hand, in rabbits with induced hyperlipidaemia treatment with red rice lowered the TC/HDL-C ratio and suppressed atherosclerotic changes.

Monascus yellow (zanthomonasin A and B) and red pigments (glycyl-rubropunctatin and glycyl-monascorubrin) showed inhibitory effects on mutagenicity of the direct-acting heterocyclic amine Trp-P-2(NHOH) (3-hydroxyamino-1-methyl-5*H*-pyrido[4,3-*b*]indole) present in cooked meat and fish.

Detection and Isolation

A qualitative assay of *Monascus* pigment in meat products involves fat removal with petroleum ether, extraction of the pigments with methanol/concentrated ammonium hydroxide (98 : 2), extraction on reversed phase, thin layer chromatography on silica gel plates, using, for example, a mobile phase consisting of ethanol/methanol (1 : 1, v/v), and detection of the spots at 366 nm. Alternatively, addition of red rice is detectable by a characteristic u.v./visible spectrum displaying a maximum at approx. 400 nm in the extracts from meat products. This spectrophotometric method is also suitable for a semi-quantitative assay of the red rice concentrate in the product.

For the quantitative analysis of individual components of *Monascus* pigment there are a number of HPLC methods using, typically, a reversed-phase column and a mobile phase consisting of acetonitrile or methanol and water or buffer (phosphate, acetate). Such methods are also applicable for the analysis of the semisynthetic rubropunctamine and monascorubramine derivatives.

Selected methods of the qualitative and quantitative analysis of *Monascus* pigments are given in Table 1.

As standards of the oligoketides of *Monascus* have not been available commercially till now, demands for their quantitative analysis may require preparation of the standards in the analyst's laboratory. A typical procedure includes extraction of a freeze-dried *Monascus* mycelium with petroleum ether or hexane and flash chromatography of the evaporation residue on silica gel to elute monascin and ankaflavin with, for example, petroleum ether/ethyl acetate (20 : 1, v/v) and a mixture of rubropunctatin and monascorubrin with ethyl acetate. Rubropunctatin and monascorubrin can be separated from each other by flash chromatography on reversed phase using acetonitrile/water (55 : 45, v/v).

Alternatively, the hexane extract can be worked up by crystallization of the mixture of yellow and orange pigments from ethanol. Subsequent crystallization

from ether gives the yellow pigments (monascin/ankaflavin). After concentration, the crystallized ether-soluble fraction affords orange–red pigments (rubropunctatin/monascorubrin) by recrystallization from ethanol. The resulting homologue mixtures can be separated on a reversed phase.

Purple pigments (rubropunctamine/monascorubramine) can be isolated from benzene extract of red rice by successive washing with acetone and ether and recrystallization from ethanol.

An outline of selected isolation and purification methods is given in Table 2.

See also: **Fermented Foods**: Origins and Applications; Fermented Vegetable Products; Fermented Meat Products; Fermentations of the Far East. **Fungi**: Overview of the Classification of the Fungi. **Preservatives**: Permitted Preservatives – Nitrate and Nitrite.

Acknowledgement

The work done in our laboratory was supported by grant A6020509 from the Grant Agency of the Academy of Sciences of the Czech Republic.

Further Reading

Bridge PD and Hawksworth DL (1985) Biochemical tests as an aid to the identification of *Monascus* species. *Letters in Applied Microbiology* 1: 25–29.

Chen MH and Johns MR (1993) Effect of pH and nitrogen source on pigment production by *Monascus purpureus*. *Applied Microbiology and Biotechnology* 40: 132–138.

Hawksworth DL and Pitt JI (1983) A new taxonomy for *Monascus* species based on cultural and microscopical characters. *Australian Journal of Botany* 31: 51–61.

Henning W, Wedekind R and Bertling L (1992) Nachweis von Farbstoffen des '*Monascus*' in Fleischerzeugnissen. *Fleischwirtschaft* 72: 1428–1431.

Hesseltine CW (1965) A millenium of fungi, food and fermentation. *Mycologia* 57: 149–197.

Hesseltine CW (1983) Microbiology of oriental fermented foods. *Annual Review of Microbiology* 37: 575–601.

Jůzlová P, Martínková L and Křen V (1996) Secondary metabolites of the fungus *Monascus*: a review. *Journal of Industrial Microbiology* 16: 163–170.

Leistner L, Fink-Gremmels J and Dresel J (1991) *Monascus* extract – a possible alternative to nitrite in meats. *Proceedings of the 37th International Congress of Meat Science and Technology* 3: 1252–1256.

Lin TF, Yakushijin K, Büchi GH and Demain AL (1992) Formation of water-soluble *Monascus* red pigments by biological and semi-synthetic processes. *Journal of Industrial Microbiology* 9: 173–179.

Martínková L, Patáková-Jůzlová P, Křen V, Kučerová Z, Havlíček V, Olšovský P, Hovorka O, Říhová B, Veselý D, Veselá D, Ulrichová J and Přikrylová V (1999) Biological activities of oligoketide pigments of *Monascus purpureus*. *Food Additives and Contaminants* 16: 15–24.

Steinkraus KH (1983) Chinese Red Rice: Anka (Ang-kak). In: Steinkraus KH, Cullen RE, Pederson CS, Nellis LF and Gavitt BK (eds) *Handbook of Indigenous Fermented Food.* P. 547. New York and Basel: Marcel Dekker.

Sweeny JG, Estrada-Valdes MC, Iacobucci GA, Sato H and Sakamura S (1981) Photoprotection of the red pigments of *Monascus anka* in aqueous media by 1,4,6-tri-hydroxynaphthalene. *Journal of Agricultural and Food Chemistry* 29: 1189–1193.

Wong HC and Koehler PE (1983) Production of red water-soluble *Monascus* pigments. *Journal of Food Science* 48: 1200–1203.

MORAXELLA

Jesús-Angel Santos, **María-Luisa García-López** and **Andrés Otero**, Department of Food Hygiene and Food Technology, University of León, Spain

Introduction

Bacteria of the genus *Moraxella*, now placed within the family Moraxellaceae, have received a great deal of attention from taxonomists because of their former unsatisfactory status (classification and nomenclature) and from food microbiologists because of their association with spoilage development in foods, especially in raw proteinaceous products held in air at refrigeration temperatures. Several *Moraxella* species are involved in a number of human diseases, usually as opportunistic pathogens and others are capable of producing infections in a variety of animals. In addition, there is evidence that at least some strains could be exploited in industrial process.

Based on our experience together with data compiled from the literature, this article reviews the taxonomy, isolation and prevalence of moraxellae in foods. Information on their spoilage capacity, resistance to several treatments, pathogenicity and possible applications is also given.

Moraxella and Related Genera

The genus *Moraxella* comprises Gram-negative microorganisms grouped in two subgenera, *Moraxella* (rod-shaped bacteria) and *Branhamella* (cocci).

Until recently, the genus was included in the family Neisseraceae, but in 1991 a proposal was made to accommodate *Moraxella*, together with *Acinetobacter*, *Psychrobacter* and related organisms, in a new family, Moraxellaceae. Moreover, it has also been shown that the subgenus *Branhamella* does not constitute a genotypic entity and that its species form a separate group within the genus *Moraxella*.

Several taxonomic studies of the family Moraxellaceae have demonstrated that it contains four different groups.

- genus *Moraxella sensu stricto*
- *Moraxella osloensis* and *Moraxella atlantae*
- genus *Psychrobacter*, including *P. phenylpyruvicus* (former [*Moraxella*] *phenylpyruvica*)
- genus *Acinetobacter sensu stricto*.

All of these are Gram-negative bacteria, but with a tendency to resist decolorization, rod-shaped or coccobacilli (with the exception of the subgenus *Branhamella*), non-flagellated and non-motile in liquid media, aerobic and oxidase positive (except the organisms belonging to the genus *Acinetobacter*). A table with the main differential characteristics of these genera is given in the article dedicated to the genus *Psychrobacter* and, in general, it is relatively easy to allocate isolates into one of these genera because the majority of the strains show natural competence and therefore transformation assays may be performed.

Moraxella Species

The taxonomy of the genus *Moraxella* is complex, with two clearly defined groups. The first group contains the classical moraxellae (rod-shaped organisms of the subgenus *Moraxella* and coccal organisms of the subgenus *Branhamella*), and the second group is constituted by the species *M. atlantae* and *M. osloensis*.

Bacteria within this genus are Gram-negative, aerobic, chemoorganotrophic, positive to the oxidase test using either the tetra- or dimethyl-*p*-phenylenediamine reagents, and unable to produce acid from carbohydrates. They are non-flagellated and non-motile in liquid media though they can be fimbriated and, in some species, fimbriation may be associated with pathogenicity, natural competence and colony-type variation (the fimbriated type of colony spreads and corrodes the agar in solid media and expresses twitching motility).

To date, there are 13 recognized *Moraxella* species: *M. boevrei*, *M. bovis*, *M. caprae*, *M. lacunata*, *M. lincolnii*, *M. nonliquefaciens*, *M. [B.] catarrhalis*, *M. canis*, *M. [B.] caviae*, *M. [B.] [cuniculi]*, *M. [B.] ovis*, *M. atlantae* and *M. osloensis*.

Methods of Detection and Enumeration in Foods

The species of *Moraxella* can grow in standard media at 33–35°C. Blood agar or chocolate agar are usually utilized in clinical microbiology, but, in our experience, *Moraxella* strains from a variety of foods (meat, fish, dairy products) are readily isolated by using standard count media such as tryptone soy agar (TSA) or plate count agar (PCA). Selective procedures and media have also been used but are generally not well developed. Media containing antibiotics have been devised for the selection of *M. osloensis*, *M. nonliquefaciens* and *M. lacunata* and there are several described media for the recovery of *Moraxella* [*B.*] *catarrhalis*, based on the DNase activity of this species; however, none of these media have been used in food microbiology.

After prolonged incubation, the colonies may present variation of the morphology, including spreading and corrosion of the agar, usually associated with fimbriation and twitching motility.

Gram-negative, non-pigmented, non-motile (in liquid media), non-sporeforming, aerobic and oxidase-positive rods or coccobacilli are ascribed to the *Moraxella–Psychrobacter* groups. Strains can be preserved as recommended for *Psychrobacter*.

A polymerase chain reaction (PCR) assay is also available for the generic detection of several Gram-negative bacteria involved in meat spoilage (including moraxellae). Universal primers based on conserved 23S rDNA sequence of *Pseudomonas aeruginosa*, as well as gel electrophoresis detection of PCR products are used.

Characterization and Identification

Although the isolates are not usually ascribed to a defined species of *Moraxella* in the microbiological analysis of foods, phenotypic, chemical and genotypic approaches are useful for their identification.

Species within the subgenus *Branhamella* can be easily differentiated from the remaining *Moraxella* species by observation of the cellular shape.

Table 1 shows some of the phenotypic tests useful for differentiating the species of *Moraxella*, but it is important to be aware that due to their relative inactivity, the biochemical differentiation of the species is not always easy to achieve.

The genetic relationships of new isolates of suspected *Moraxella* with some of the defined species can be determined by performing transformation assays. For the identification of *M. bovis*, *M. lacunata*, *M. nonliquefaciens*, *M. osloensis* and *M. atlantae* quantitative assays with streptomycin resistance marker are available. In these assays, high ratios of interstrain to intrastrain transformation (values of 10^{-2} or higher) indicate very close affinity. In addition, simple qualitative transformation assays with mutant auxotrophic recipient strains can be used for rapid identification of *Moraxella osloensis* and *Moraxella bovis*.

Other approaches, such as DNA–DNA or DNA–rRNA hybridizations, analysis of polyacrylamide gel electrophoresis (PAGE) of whole-cell proteins or analysis of fatty acid profiles have also been used to differentiate the species of moraxellae. Oleic acid ($C_{18:1}$ *cis*-9) is one of the major fatty acid components of *Moraxella*. The cellular lipid composition is rather similar among the *Moraxella* species, but *M. osloensis* gives a distinct and reproducible pattern that may be of use for the identification of this species.

Studies on 16S rDNA sequences showed that diversity at this level is higher among the species of *Moraxella* than could be expected by other methods, probably due to the natural competence of these bacteria, which allows for recombination with DNA of different species. This finding limits the use of this methodology for species differentiation.

Moraxella in Foods: Prevalence and Spoilage Potential

It is widely accepted that representatives of the genus *Moraxella* form a significant portion of the psychrotrophic aerobic flora of many fresh and spoiled proteinaceous foods. The taxonomy of the genus *Moraxella* and allied bacteria has been extensively studied. However, most of the published information on the incidence, significance and spoilage activity of non-motile, non-pigmented, aerobic, oxidase-positive Gram-negative rods and coccobacilli obtained from foodstuffs, identified the isolates as *Moraxella–Acinetobacter* group or type, oxidase-positive *Moraxella*, *Moraxella*-like or *Moraxella* spp. Since several researchers have claimed that, with a few exceptions, strains formerly identified as *Moraxella* or *Moraxella*-like are almost certainly *Psychrobacter* spp., the real importance of the moraxellae in food is not known with certainty. In spite of this, the 'recognized status' of *Moraxella* as a food contaminant justifies the sections discussed below.

Fish and Shellfish and Their Products

Presumptive moraxellae appear to be the predominant bacterial flora on the surface of whole finfish of marine origin caught in cold water regions (pollock, cod, rock sole, halibut, hake, Cape hake, blue grenadier, sardines, mackerel, etc.) and also constitute a considerable proportion of the psychrotrophic flora of tropical sea water fish and both wild and farmed

Table 1 Phenotypic characteristics of currently recognized species of *Moraxella*

	M. boevrei	M. bovis	M. caprae	M. lacunata	M. lincolnii	M. non-lique-faciens	M. [B.] catar-rhalis	M. canis	M. [B.] caviae	M. [B.] cuniculi	M. [B.] ovis	M. atlantae	M. oslo-ensis
Morphology	SR[a]	R	R	R	C, R	C	C	C	C	C	C	R	R
Gelatin liquefaction	+	+	−	+	−	−	−	−	−	−	−	−	−
Proteolysis	+	+	−	+	−	−	−	−	−	−	−	−	−
Haemolysis	+	(+)[c]	+	−	−	−	−	+	w	−	(+)	−	−
Nitrate reduction	+	(−)	+	+	−	+	(+)	+	+	−	+	−	(−)
Nitrite reduction	−	−	−	−	(+)	−	(−)	−	(+)	−	−	−	−
MM[b]	−	−	−	−	−	−	−	(+)	−	−	−	−	+
Hydrolysis of Tween 80	(+)	(+)	(+)	(+)	−	−	−	−	−	(−)	−	−	−
Alkaline phosphatase activity[d]	−	−	−	+	−	−	+	+	+	(+)	+	+	+
Esterase activity[d]	+	+	−	+	+	+	+	+	(+)	(+)	+	+	+

[a] Symbols: +, positive; −, negative; (+), positive for most strains; (−), negative for most strains; w, weak reaction; C, coccus; R, rod; SR, short rod.
[b] MM, growth in a minimal medium containing ammonium and acetate.
[c] Non-haemolytic strains of '*M. equi*'.
[d] Tests can be done with miniaturized kits (API ZYM).

freshwater fish (tilapia, striped bass, Victorian Nile perch, white amur and salmon). The ubiquity of these organisms on fresh fish means that their presence is inevitable in a variety of fish products such as dressed fish, frozen whole fish, fillets (frozen and chilled) and minced fish.

Bivalve molluscs (oysters and scallops) consistently carry presumptive moraxellae which probably are part of the resident bacterial population. As in shellfish, the microflora of raw crustaceans largely reflects the quality of the water from which the animals were taken; however, shrimps and crabs harvested from different geographical areas usually have significant numbers of these bacteria.

There is considerable confusion over the role of moraxellae in spoilage of seafood and freshwater fish. Specific bacteria commonly associated with spoilage of fresh fish kept in air at temperatures below 5°C are *Pseudomonas* species and *Shewanella putrefaciens* whereas storage at 5°C or above favours the growth of *Aeromonas*. Deterioration of packed fish stored under vacuum or modified gas atmospheres is mainly attributed to the latter genus and *Photobacterium phosphoreum*. However, *Moraxella* spp. and *Moraxella*-like isolates have been found to be predominant among the spoilage flora of aerobically stored tropical sea water fish, herring fillets, striped bass, sardines, mackerel and thawed blue grenadier fillets. Their presence has also been reported in vacuum-packed fillets after storage. The spoilage flora of molluscan shellfish are both proteolytic (*Pseudomonas* and *Vibrio*) and saccharolytic (*Lacto-*

bacillus) bacteria, but the *Moraxella–Acinetobacter* group may also occur in spoiled bivalve molluscs such as scallops. Studies on the microbiological deterioration of raw crustaceans (shrimps and crabs) show that, more often than not, members of the *Moraxella–Acinetobacter* group can be a cause of spoilage. Presumptive moraxellae are also likely to play an important role in deteriorating packed crab meat and cooked shrimps stored chilled. Pure culture studies have shown that one strain of *Moraxella nonliquefaciens* produces phenethyl alcohol (one of the major volatile compounds generated in haddock fillets during storage at 2°C) from L-phenylalanine. In our experience, *Moraxella* (mainly [M.] *phenylpyruvica*, now reclassified as *Psychrobacter phenylpyruvicus*) occurs throughout the storage life of wild and farmed freshwater fish (brown trout, rainbow trout and pike) but rarely accounts for more than a small portion of the total flora.

Meat and Poultry and Their Products

Moraxellae isolates of doubtful identity have been readily recovered from red meat carcasses (beef, pork and lamb), retail cuts (joints and steaks) and fresh and frozen ground meat. In spite of their strict aerobic character, these bacteria are also capable of growth on fresh and cured meats stored under modified gas atmospheres. Although it is assumed that *Moraxella* forms a significant portion of any spoilage flora on aerobically stored meat, it has also been reported that their importance is often overstated as they constitute a minor part of the spoilage microflora (<1%) and have low spoilage potential. Moraxellae strains associated with abattoirs or red meat spoilage are reported as *Moraxella* spp., [*M. phenylpyruvica*] or *M. osloensis*. Growth of moraxellae on meat, particularly on fat surfaces, has been reported. Furthermore, growth of pure cultures on sterile pieces of fat resulted in acidic or fruity odours after 10 days. By contrast, other studies have shown that *Moraxella* spp., at least *M. osloensis*, do not appear to be suited for attachment and adherence to the meat surface. Attachment may be considered a first step in the microbial spoilage of meat. The few works dedicated to investigating the spoilage activity of presumptive moraxellae in red meat report that isolates from ground beef were responsible for the unpleasant rotten vegetable odour detected. In meat drip, one strain referred to as *Moraxella*-like (probably a *Psychrobacter* isolate) was able to metabolize glucose and glucose 6-phosphate simultaneously. During the storage period this strain also used a group of drip meat amino acids which were not attacked by other spoilers (*Pseudomonas* and *Serratia*). The amino acids and other related substances concerned were aspartic acid, glutamic acid, phenylalanine, ornithine and urea.

The Gram-negative flora on finished poultry carcasses include presumptive moraxellae which enter poultry processing plants on birds' feathers and feet and in water and ice supplies. Although they do not appear to constitute a major component of the spoilage flora of chilled raw poultry meat held in air, pure cultures of the *Acinetobacter–Moraxella* group have been described as imparting fruity odours to chicken leg muscle at 2°C. Oxidase-positive moraxellae have also been shown to produce volatile organic compounds (ethyl acetate, methyl isopropyl sulphide, 2-propanethiol) involved in the odour of spoiled chicken.

As in carcasses and cut-up parts stored aerobically, *Pseudomonas* spp. and *Shewanella putrefaciens* are the bacteria that most commonly spoil poultry meat packed in oxygen-permeable films, but growth of *Moraxella* is thought to be of importance during the storage life of these products.

Irradiated Meat and Poultry Products

Extensive research has been conducted on the effects of non-sterilizing ionizing irradiation treatments (< 10 kGy) on the microflora of meat and poultry (mainly ground beef, chicken, pork and Vienna sausages); most of the work was carried out between late 1960s and 1980s. In irradiated meats, the composition of the surviving flora depends on several factors (e.g. composition of the initial flora, radiation dose, etc.); however, certain trends, such as the radiation resistance of non-spore-forming termed *Moraxella* sp., intermediate type *Moraxella–Acinetobacter*, *Moraxella*-like, *M. osloensis* and [*M. phenylpyruvica*], were always observed. Subsequent storage of irradiated meats and poultry kept in aerobic conditions results, particularly in poultry, in spoilage due to overgrowth of these radiation-resistant organisms. Concern has been expressed that irradiation might change the properties of food-associated bacteria so dramatically that they might be more pathogenic and/or competitive than their wild-type counterparts. With *Moraxella* such fears appear to be unfounded because radiation-resistant strains of [*M. phenylpyruvica*] and *M. osloensis* are more demanding in their growth requirements (higher limiting temperature and water activity for growth, lower tolerance to sodium chloride and polyphosphates, higher sensitivity to modified atmospheres, etc.). Furthermore, other changed properties (i.e. morphological characteristics in *M. osloensis* strains) are usually temporary and can be reversed after a few subcultures. Factors affecting the survival of *Moraxella* in irradiated meats have also been investigated. Radiation resistance is modified by a number of factors, including pre- and post-irradiation temperature, temperature during irradiation (increased resistance in beef and ground beef at subfreezing and freezing temperatures), age of cells (also greater during the log phase) and atmosphere composition (more sensitive in poultry packed in modified atmospheres). It has also been found that a highly resistant strain of *Moraxella* from beef had a mechanism of resistance to UV light which was not associated with induced mutagenesis although it was influenced by other factors such as the age of the cells (log-phase cultures are much more resistant).

Reptiles

Because of their presence in aquatic environments, *Moraxella* spp. may be recovered from meat of reptiles dedicated to human consumption, their behaviour during storage life being similar to that reported for beef.

Milk and Dairy Products

Moraxellae represent a minor constituent of the psychrotrophic flora found in milk and certain dairy products. Thus, *Moraxella* spp. have been isolated from tank milk (5.6% of the isolates), pasteurized double cream at spoilage (0.1–0.5%), salted butter and Camembert cheese made from raw milk. Occasionally *M. osloensis* have been detected in fresh cheeses, butter and ewes' milk. Although these bacteria are relatively rare in milk and dairy products and their presence is not particularly troublesome, the lipolytic and proteolytic activities of isolates from tank milk and the ability of some strains to multiply in salted butter at 18°C have been demonstrated.

Eggs and Egg Products

Spoilage of eggs is most often associated with fluorescent pseudomonads which are able to penetrate through the shells and grow into the interior more quickly than any other group of bacteria, but *Moraxella*, which may follow these primary invaders (presumably due to their capacity to utilize metal ions sequestered by the siderophores produced by the former organisms), are implicated in colourless rots that can enter the liquid egg undetected. Pure cultures obtained from this source and then inoculated into sterile whole egg did not cause organoleptic changes after 72 h.

The laying environment can be a potential source of moraxellae for the external surface of eggs. In egg-grading plants, considerable numbers of these bacteria are found on equipment surfaces and washing water, their incidence being much lower on shells (<2% of the total flora), especially on unwashed eggs. These organisms have also been isolated (1.5% of the total flora) from unpasteurized liquid whole egg after freezing.

Water

Depending on the source, raw water may contain a variety of heterotrophic organisms, including *Moraxella*. These bacteria which are readily inactivated by chlorine concentrations of $1.0\,\mathrm{mg\,l^{-1}}$ or less, are commonly isolated from tap water and drinking water produced by domestic water filtration units. Growth of moraxellae in distribution systems is associated with the presence of metabolizable organic compounds, absence or reduced levels of disinfectants, attachment and biofilm development and secondary recontamination. *Moraxella* spp. are also readily found in water used in food factories for a number of purposes. Since they need only small amounts of organic compounds and are characterized by their low nitrogen requirements, they have consistently been detected among the main flora of non-carbonated and even carbonated mineral waters bottled in glass and plastic.

The reported high incidence of multiply antibiotic-resistant (MAR) moraxellae in rivers and distribution water has been considered of concern because their proliferation may hamper the efficacy of antibiotics in chemotherapy. This concern is particularly relevant because MAR isolates, which may carry transmissible R-plasmids, appear to be selected by treatment of raw water.

Tolerance or Susceptibility to Preservation Methods

As stated above, moraxellae strains in foods are psychrotrophic, sensitive to heat treatments, able to survive during frozen storage and, overall, resistant to non-sterilizing doses of ionizing radiation. Although the number of studies is low, these bacteria appear to be susceptible to modified atmospheres, with CO_2 being the most inhibitory, resistant to UV light and high pressures (200 MPa), sensitive to sodium chloride (though they have been detected in brine), potassium sorbate (0.53%), $NaNO_2$ (200 p.p.m.), organic acids, such as acetic and citric acids, and perhaps to polyphosphates. They are also effectively killed by microwave treatments and certain disinfectants (i.e. chlorine).

Applications

The potential usefulness of cold-active enzymes produced by psychrophilic bacteria created considerable interest in both fundamental and industrial fields. Lipases excreted by some Antarctic *Moraxella* strains have been shown to be highly active at 3°C; however, as a consequence of their cold-dependence in enzyme excretion, which in practice implies considerable technical problems, these wild isolates have not been considered appropriate for large-scale lipase production. This was circumvented by cloning the lipase genes of one strain, designated TA 144, into *Escherichia coli*. Recently, [*M. phenylpyruvica*] strain O100 has been proposed as a potential meat starter for ensuring superior meat flavour development in cured meats. This is because of its ability to metabolize the amino acids L-leucine and L-phenylalanine into methyl-branched aldehydes which participate in the formation of a characteristic aroma in high-quality cured meat products. One strain, obtained from fermenting corn meal and designated *Moraxella* sp., produced vitamin B_6. This is of interest because humans consuming corn as the main component of their diet may need additional vitamin B_6. Old published information showed that one strain belonging to the *Moraxella–Acinetobacter* group produced a tea-like odour in wild rice, comparable to that of the naturally fermented product. Finally, it has been reported that a considerable percentage of *Moraxella* strains obtained from drinking water distribution systems produced bacteriocin-like substances specific for at least one of the following Enterobacteriaceae: *E. coli*, *Klebsiella* spp. and *Enterobacter aerogenes*.

Pathogenicity

Species of the genus *Moraxella* found in association with mucosal surfaces of humans have been identified as causes of human disease. For example, *M. lacunata* may cause eye infections and conjunctivitis, *M. nonliquefaciens* acute tracheobronchitis and even septicaemia in leukaemic patients, *M. atlantae*, *M. osloensis* and [*M. phenylpyruvica*] opportunistic systemic infections, especially in immunocompromised individuals, and *M. [Branhamella] catarrhalis* ear, eye and sinus infections in children, as well as chronic bronchitis. Furthermore, in animals, *M. bovis* is a common cause of bovine keratoconjunctivitis and *M. ovis*, '*M. equi*' and *M. [B.] cuniculi* are involved in infectious processes in sheep (usually respiratory tract), horses (conjunctivitis), rabbits (oral mucosa and pharynx) and guinea pigs (pharynx). To date, none of the above species have been implicated as agents of food-borne disease.

See also: **Acinetobacter**. **Cheese**: In the Market Place. **Eggs**: Microbiology of Fresh Eggs. **Fish**: Spoilage of Fish. **Meat and Poultry**: Spoilage of Meat. **Milk and Milk Products**: Microbiology of Liquid Milk. **Psychrobacter**. **Shellfish (Molluscs and Crustacea)**: Contamination and Spoilage. **Spoilage Problems**: Problems caused by Bacteria.

Further Reading

Bovre K and Hagen N (1981) In: Starr MP, Stolp H, Trüper HG, Balows A and Schlegel HG (eds) *The Prokaryotes. A Handbook on Habitats, Isolation and Identification of Bacteria.* P. 1506. New York: Springer-Verlag.

Doern GV (1992) In: Balows A, Trüper HG, Dworkin M, Harder W and Schleifer K-H (eds) *The Prokaryotes. A Handbook on Habitats, Isolation and Identification of Bacteria,* 2nd edn. P. 3276. New York: Springer-Verlag.

Eribo BE, Lall SD and Jay JM (1985) Incidence of Moraxella and other Gram-negative oxidase-positive bacteria in fresh and spoiled ground beef. *Food Microbiol.* 2: 237–240.

Feller G, Thiry M, Arpigny JL, Mergeay M and Gerday C (1990) Lipases from psychrotrophic Antarctic bacteria. *FEMS Microbiol. Lett.* 66: 239–244.

ICMSF (1998) *Microorganisms in Foods 6. Microbial Ecology of Food Commodities.* London: Blackie Academic & Professional.

Ito H, Sato T and Iizuka H (1976) Study of the intermediate type of *Moraxella* and *Acinetobacter* occurring in radurized Vienna sausages. *Agric. Biol. Chem.* 40: 867–873.

Moss CW, Wallace PL, Hollis DG and Weaver RE (1988) Cultural and chemical characterization of CDC groups EO-2, M-5, and M-6, *Moraxella (Moraxella)* species, *Oligella urethralis*, *Acinetobacter* species, and *Psychobacter immobilis*. *J. Clin. Microbiol.* 26: 484–492.

Petterson B, Kodjo A, Ronaghi M, Uhlén M and Tonjum T (1998) Phylogeny of the family *Moraxellaceae* by 16S rDNA sequence analysis, with special emphasis on differentiation of *Moraxella* species. *Int. J. Syst. Bacteriol.* 48: 75–89.

Prieto M, García-Armesto MR, García-López ML, Otero A and Moreno B (1992) Numerical taxonomy of Gram-negative, nonmotile nonfermentative bacteria isolated during chilled storage of lamb carcasses. *Appl. Environ. Microbiol.* 58: 2245–2249.

Rossau R, Van Landschoot A, Gillis M and De Ley J (1991) Taxonomy of *Moraxellaceae* fam. nov., a new bacterial family to accommodate the genera *Moraxella*, *Acinetobacter*, and *Psychrobacter* and related organisms. *Int. J. Syst. Bacteriol.* 41: 310–319.

Moulds *see* **Biochemical and Modern Identification Techniques**: Food Spoilage Flora (Yeasts and Moulds); **Fungi**; **Genetic Engineering**: Modification of Yeast and Moulds; **Starter Cultures**: Moulds Employed in Food Processing.

MPN *see* **Total Viable Counts**: MPN.

MUCOR

A Botha and **J C du Preez**, Department of Microbiology and Biochemistry, University of the Orange Free State, Bloemfontein, South Africa

Characteristics of the Genus *Mucor*

Morphological Properties

The genus *Mucor* belongs to the zygomycotan order Mucorales. These fungi are characterized by eucarpic, mostly coenocytic thalli containing haploid nuclei. Asexual reproduction is characterized by the formation of one to many sporangiospores in a mitosporangium. Sexual reproduction occurs when two similar gametangia conjugate to produce a zygospore. Generally, the families and other taxa in Mucorales can be distinguished from one another by the morphology of the asexual reproductive structures, specifically the characteristic features of the sporangiophores, sporangia, columellae and sporangiospores.

The family Mucoraceae, which includes *Mucor*, is characterized by columellate multi-spored sporangia. In addition, rhizoids and stolons are either much reduced or completely absent in members of this family. Arthrospores are formed under unfavourable nutritional or environmental conditions through septation of normally coenocytic hyphae, followed by hyphal fragmentation. The typical morphological characteristics of the genus *Mucor* are illustrated in **Figure 1**. Various *Mucor* species or strains of species exhibit dimorphism, for example *M. rouxii*, *M. racemosus*, *M. genevensis*, *M. bacilliformis*, *M. subtilissimus* and *M. circinelloides*, which grow as spherical multipolar budding yeasts under certain conditions. Growth on a fermentable hexose is always required for growth in the yeast-like form and although, in general, anaerobiosis favours the transition to a yeast-like morphology, this is not a prerequisite for all species. Furthermore, the hexose concentration, partial pressure of CO_2 and the nitrogen source can also be important effectors of dimorphism in certain species (Fig. 1).

Physiological Properties

Mucor, as well as many other mucoralean fungi, are generally the first saprophytic colonizers on dead or decaying plant material. They are able rapidly to utilize the limited number of simple carbohydrate molecules available before other fungi that are able to utilize complex carbohydrates such as cellulose and lignin dominate the decomposition process.

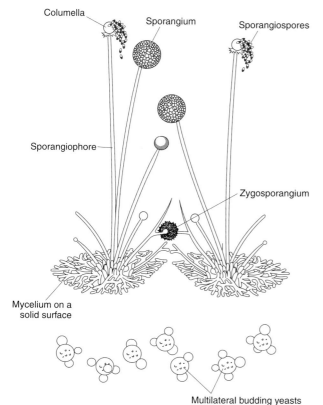

Figure 1 Typical morphological characteristics of fungi belonging to the genus *Mucor*.

Columella

Sporangium

Sporangiospores

Sporangiophore

Zygosporangium

Mycelium on a solid surface

Multilateral budding yeasts

Table 1 Enzymes produced by mucoralean fungal strains

Species and strain	Enzymes produced
Mucor circinelloides van Tieghem f. *circinelloides* Schipper	
ATCC 12166	β-Glucosidase
CCRC 31544	α-Glucosidase
CBS 119.08	Lipase
Mucor circinelloides van Tieghem f. *lusitanicus* (Bruderlein) Schipper	
CBS 108.17	Lipase
CBS 242.33	Lipase
CBS 277.49	Lipase/β-glucosidase
Mucor indicus Lendner	
CBS 120.08	α-Amylase
Mucor mucedo (Linnaeus) Fresenius	
ATCC 38694	Proteolytic and lipolytic activity
Mucor piriformis Fischer	
ATCC 42556	Pectolytic enzymes

ATCC = American Type Culture Collection, Rockville, MD, USA; CCRC = Culture Collection and Research Center, Food Industry Research and Development Institution, Hsinchu, Taiwan; CBS = Centraalbureau voor Schimmelcultures, Baarn, the Netherlands.

Table 2 Carbon compounds aerobically assimilated as the sole carbon source by *Mucor* strains in synthetic liquid media

	M. circillenoides			M. rouxii	M. flavus	M. mucedo
	CBS 119.08	CBS 108.16	CBS 203.28	CBS 416.77	CBS 234.35	CBS 109.16
Pentoses						
D-Arabinose	–	–	–	–	–	–
L-Arabinose	+	+	+	+	+	+
D-Ribose	+	+	+	+	–	–
D-Xylose	+	+	+	+	+	+
Hexoses						
D-Galactose	+	+	+	+	+	+
D-Glucose	+	+	+	+	+	+
D-Mannose	+	+	+	+	+	+
D-Fructose	+	+	+	+	+	+
L-Sorbose	–	–	–	–	–	–
L-Rhamnose	+	–	–	–	+	–
Disaccharides						
Cellobiose	+	+	+	+	+	+
Lactose	–	–	–	–	–	–
Maltose	+	+	+	+	+	+
Melibiose	+	–	–	–	–	+
Sucrose	+	–	–	–	–	–
Trehalose	+	+	+	+	+	+
Trisaccharides						
Melezitose	+	+	+	+	+	+
Raffinose	–	–	–	–	+	–
Polysaccharides						
Inulin	+	+	+	+	+	+
Starch	+	+	+	+	+	+
Glycoside						
Salicin	+	+	+	+	+	+
Alcohols						
Erythritol	–	–	–	–	+	–
Ethanol	+	+	+	+	+	–
Galactitol	+	–	–	–	–	–
Glycerol	–	–	–	–	+	–
Inositol	–	–	–	–	+	–
D-Mannitol	+	+	+	+	+	+
Methanol	–	–	–	–	–	–
Ribitol	+	+	+	+	–	–
Sorbitol	+	+	+	+	+	+
Organic acids						
Acetic acid	+	+	+	+	+	+
Butanoic acid	+	+	+	+	–	–
Citric acid	–	–	–	–	–	–
Formic acid	–	–	–	–	–	–
Gluconic acid	+	+	+	+	+	–
Lactic acid	+	+	+	+	+	+
Succinic acid	+	+	+	+	+	+
Propionic acid	–	–	–	–	–	–

+ Assimilated; – not assimilated

Investigations into the physiological properties of *Mucor* have indicated the biotechnological potential of this group of fungi. **Table 1** lists enzymes that are known to be produced by strains identified to species level and maintained in internationally recognized culture collections. The lipid metabolism of species such as *M. circinelloides* and *M. rouxii* are particularly well studied. Representatives of these species are known to be oleaginous (accumulating at least 20% lipids, on dry weight) and produce substantial quantities of high-value fatty acids such as γ-linolenic acid, which have applications in medicine as lipid constituents.

Members of *Mucor* are able to utilize a wide range of carbon sources. **Table 2** shows that strains of *Mucor* can aerobically utilize pentoses, hexoses, disaccharides, trisaccharides, polysaccharides, glucosides, alcohols and organic acids. Hexose sugars are also fermented (**Table 3**).

Whereas several *Mucor* species have been isolated

Table 3 Carbohydrates fermented by *Mucor circinelloides* f. *circinelloides* CBS 108.16

Pentoses	
D-Arabinose	–
L-Arabinose	–
D-Ribose	–
D-Xylose	–
Hexoses	
D-Galactose	+
D-Glucose	+
Disaccharides	
Maltose	+
Sucrose	–
Trisaccharide	
Raffinose	–

+ Fermented; – not fermented

from fruits with pH values of between 4 and 5, growth can also occur at pH 8. The limiting water activity value for *Mucor* seems to be between 0.92 and 0.93 and most *Mucor* species are able to grow and sporulate at temperatures from 20°C to 30°C. Strains of *M. recurvus* can grow and sporulate at temperatures of up to 40°C, but strains of *M. flavus*, *M. piriformis*, *M. plasmaticus* and *M. racemosus* grow at temperatures as low as 0°C. *Mucor* spp. can grow over a wide range of temperatures, aerobically use many carbon sources, ferment carbohydrates and use ammonia or organic nitrogen; therefore, the genus is ubiquitous in many habitats, including various foods.

Occurrence in Foods

Data from various laboratories indicate that mucoralean fungi such as *Rhizopus* or *Mucor* are ubiquitous in unspoiled foods, especially spices, flour and nuts. Mucoralean fungi also occur in low numbers in fresh and dried fruit, fresh vegetables and protein-rich foods such as milk products, dog food and meats. However, when storage conditions permit mould growth, *Mucor* strains as well as other fungi are known to cause spoilage of these foodstuffs. For example, mouldiness of bacon, the formation of 'whiskers' on beef, egg spoilage, mould growth on butter and cereals, as well as pickle softening, may all be caused by members of the genus *Mucor*. **Table 4** lists *Mucor* strains isolated from different foods and that have been identified up to species level and are maintained in internationally recognized culture collections. It is noteworthy that these moulds not only occur as spoilage organisms in foods, but that some species are used in the preparation of fermented foods such as meju, ragi and sufu.

Although some oriental fermented foods are prepared using *Mucor* species, the presence of *Mucor* in other foods may have two negative implications. First, the contaminating fungus may constitute a limited potential health hazard. No specific mycotoxin has been isolated and characterized in *Mucor*; however, the results of bioassays did indicate that toxins are present in extracts from certain *Mucor* species. Some authors found that aqueous fungal extracts of *M. mucedo* were weakly toxic to brine shrimp. However, others found that although ethanol-chloroform extracts of this species were only moderately toxic to brine shrimp, these were highly toxic to chicken embryos. Similarly, mycotoxin production was demonstrated in *M. indicus* and *M. circinelloides* in tests where ducklings were used. The genus *Mucor* is generally accepted to be non-toxic to humans. Cases of mucormycosis were, however, reported to be caused by representatives of this genus. For example, people debilitated due to diabetes, leukaemia or an immune deficiency are susceptible to infection by *Mucor*.

The second negative implication for the presence of *Mucor* in food is an indication of unsanitary conditions during food preparation and/or storage, or of the use of inferior raw materials. An example of such a situation is the spoilage of soft cheeses. *Mucor* strains present in the air and on equipment in cheese factories cause surface growth on cheese. This results in a variety of defects in the cheese, making it commercially unacceptable. Another example where *Mucor* in particular causes defects in foods occurs during the cold storage of fruit. Strains of *M. piriformis*, characterized by the ability to grow at low temperatures, caused post-harvest decay of fruit such as pears, peaches, strawberries and tomatoes. This, of course, prohibits the preparation of high-quality foods and beverages from such damaged fruits.

Detection and Enumeration in Foods

Enumeration of Colony-forming Units

To enumerate fungi belonging to the genus *Mucor* in foods, the food sample is first homogenized, if it is not already in liquid form. It can then be diluted and plated out on to an appropriate medium. Either general-purpose, relatively nonselective media or more selective media can be used for this. After an appropriate incubation period the fungal colonies are counted and the number of colony-forming fungal units present in the original sample calculated. A typical procedure would be as follows.

Sample Preparation A one-tenth dilution of the food sample, weighing 50–500 g, is homogenized in sterile 0.1% peptone water. A Colworth Stomacher, applied for 2 min, may be used for this purpose.

Table 4 Mucoralean fungal strains and the food types from which they were isolated

Species and strain	Type of food
Mucor circinelloides van Tieghem	
ATCC 48558	Tomatoes
MUCL 18550	Maize
Mucor circinelloides van Tieghem f. circinelloides Schipper	
ATCC 38313	Tomatoes
Mucor circinelloides f. *griseocyanus* (Hagem) Schipper	
CBS 698.68	Maize
CBS 907.69	Maize
CBS 366.70	Canned strawberries
CBS 541.78	Maize
Mucor circinelloides f. *janssonii* (Lendner) Schipper	
CBS 762.74	Milk powder
Mucor circinelloides van Tieghem f. *lusitanicus* (Bruderlein) Schipper	
CBS 633.65	Maize
Mucor falcatus Schipper	
CBS 251.35	Honey comb
Mucor genevensis Lendner	
CBS 564.75	Apples
Mucor hiemalis Wehmer	
ATCC 32469	Guava fruit
ATCC 46126	Production of sufu
ATCC 46128	Production of sufu
Mucor hiemalis f. *corticola*	
MUCL 15858	Dust from bakery
Mucor hiemalis Wehmer f. *hiemalis*	
MUCL 15859	Dust from bakery
MUCL 15870	Dust from bakery
MUCL 18551	Maize
Mucor hiemalis f. *silvaticus* (Hagem) Schipper	
MUCL 15868	Dust from bakery, flour
MUCL 15869	Dust from bakery, flour
Mucor inaequisporus Dade	
CBS 255.36	Spanish plums
CBS 351.50	Bananas
CBS 496.66	Japanese persimmons
Mucor indicus Lendner	
CBS 670.79	Fermenting rice/cassava
CBS 671.79	Fermenting rice/cassava
CBS 535.80	Sorghum malt
CBS 545.80	Sorghum malt
Mucor mucedo (Linnaeus) Fresenius	
ATCC 36628	Grapes
NCAIM F.00840	Red pepper
ATCC 48559	Decaying tomatoes
MUCL 18552	Maize
MUCL 18553	Maize
ATCC 36628	Grapes
Mucor piriformis Fischer	
CBS 255.85	Decaying pears
CBS 256.85	Decaying pears
ATCC 38314	Peaches
ATCC 42556	Decaying strawberries
ATCC 52553	Apricots
ATCC 52554	Nectarines

Table 4 (contd)

Species and strain	Type of food
ATCC 52555	Peaches
ATCC 60988	Decaying pears
Mucor plumbeus Bonorden	
MUCL 941	Lemons
MUCL 14187	Dairy contaminant
MUCL 16154	Meat meal, cattle feed
MUCL 18842	French Brie cheese
ATCC 8771	Pea seed
ATCC 8773	Pea seed
JCM 3900	Fermented soya beans, meju
Mucor racemosus Fresenius	
NCAIM F.00841	Red pepper
ATCC 46129	Production of sufu
ATCC 46130	Production of sufu
Mucor racemosus Fresenius f. *racemosus*	
CBS 632.65	Maize
CBS 657.68	Contaminated cheese
CBS 906.69	Spices
CBS 222.81	Nut of *Juglas regia*
Mucor racemosus f. *sphaerosporus* (Hagem) Schipper	
CBS 574.70	Steamed sweet potato
MUCL 9130	Rotting cheese
Mucor recurvus Butler	
MUCL 28170	Fermented cassava
Mucor rouxii (Calmette) Wehmer sensu Bartnicki-Garcia cf. *Mucor indicus*	
CBS 416.77	Production of fermented rice
Mucor sinensis Milko & Beljakova	
CBS 204.74	Production of soy cheese

ATCC = American Type Culture Collection, Rockville, MD, USA; MUCL = Mycothèque de l'Université Catholique de Louvain, Louvain-la-Neuve, Belgium; CBS = Centraalbureau voor Schimmelcultures, Baarn, the Netherlands; NCAIM = National Collection of Agricultural and Industrial Microorganisms, Budapest, Hungary; JCM = Japan Collection of Microorganisms, Saitama, Japan

Dilution Plates Using 0.1% peptone water, serial 1:9 dilutions of the sample are made. Aliquots of 0.1 ml of the appropriate dilutions are spread on to solidified agar medium plates in triplicate.

Media The media used for the isolation or enumeration of *Mucor* species can be divided into two categories. The first category comprises general-purpose media that are not very selective among members of the fungal domain, allowing growth of a wide diversity of fungal groups. These media, three of which are shown in **Table 5**, are most commonly used to enumerate or isolate members of *Mucor*. Antibacterial agents such as chloramphenicol are included in some of these media. Rose Bengal and dichloran,

Table 5 General-purpose media for the enumeration of mucoralean fungi

Dichloran 18% glycerol agar (DG18)

Glucose	10 g
Peptone	5 g
KH_2PO_4	1 g
$MgSO_4.7H_2O$	0.5 g
Glycerol	220 g
Agar	15 g
Dichloran (0.2% w/v in ethanol)	1.0 ml
Chloramphenicol	0.1 g
Water	1000 ml
pH 5.6	

Add all the ingredients except the glycerol to 700 ml water. Steam to dissolve agar. Add the glycerol. Bring solution to 1000 ml and sterilize by autoclaving (15 min, 121°C). The water activity of the final medium is 0.955 and it is commonly used to isolate fungi from foods with a low water activity

Dichloran Rose Bengal chloramphenicol agar (DRBC)

Glucose	10 g
Peptone	5 g
KH_2PO_4	1 g
$MgSO_4.7H_2O$	0.5 g
Rose Bengal (5% w/v in water)	0.5 ml
Dichloran (0.2% w/v in ethanol)	1 ml
Chloramphenicol	0.1 g
Agar	15 g
Water	1000 ml
pH 5.6	

Add all the ingredients to 900 ml water. Steam to dissolve agar and bring solution to 1000 ml. Sterilize by autoclaving (15 min, 121°C)

Malt extract agar (MEA)

Malt extract	20 g
Agar	16 g
Water	1000 ml

Add the ingredients to 800 ml water. Steam to dissolve agar. Bring solution to 1000 ml and autoclave (15 min, 121°C)

Table 6 Selective media for fungi belonging to the genus *Mucor*

Ketoconazole medium

Malt extract	20 g
Yeast extract	2 g
Agar	15 g
Chloramphenicol	0.5 g
Ketoconazole (1% w/v in ethanol)	5 ml
Water	1000 ml
pH 5.6	

Add all the ingredients to 900 ml water. Steam to dissolve agar. Bring solution to 995 ml. Sterilize by autoclaving (15 min, 121°C). Filter-sterilize the ketoconazole and add to cooled molten medium at 50°C, just before pouring into Petri dishes

Benomyl-containing medium

Malt extract	20 g
Benomyl	0.02 g
Agar	15 g
Water	1000 ml
pH 5–6	

Add the ingredients to 900 ml water. Steam to dissolve agar. Bring solution to 1000 ml. Sterilize by autoclaving (15 min, 121°C)

Table 7. Media used in the identification of *Mucor* species

Synthetic *Mucor* agar (SMA)

Glucose	40 g
Asparagine	2 g
KH_2PO_4	0.5 g
$MgSO_4.7H_2O$	0.5 g
Thiamine chloride	0.005 g
Agar	15 g
Water	1000 ml

Add all the ingredients to 900 ml water. Steam to dissolve agar. Bring solution to 980 ml and autoclave (15 min, 121°). Dissolve thiamin chloride in 20 ml water. Filter-sterilize and add to cooled molten medium before pouring into Petri dishes

Malt extract agar (MEA)

Malt extract	20 g
Agar	16 g
Water	1000 ml

Add the ingredients to 800 ml water. Steam to dissolve agar. Bring solution to 1000 ml and autoclave (15 min, 121°)

an antifungal agent, are used to restrict fungal growth and thereby facilitate enumeration of colonies on the plate. The other category includes media used to enumerate selectively or isolate members of *Mucor* from habitats containing, in most cases, predominantly other fungal groups. Two media in this category are given in **Table 6**. The antifungal agents included in these media, benomyl and ketoconazole, allow growth of *Mucor* species, but inhibit growth of ascomycetous and hyphomycetous fungi.

Incubation and Identification Usually, inoculated plates are incubated at 25°C for 5 days, whereafter the colonies are counted on those plates containing 15–150 colonies. It must be borne in mind that *Mucor* colonies may develop from various morphologically and physiologically different structures. Germinating chlamydospores, sporangiospores and zygospores or even hyphal fragments may result in colony formation. Growth from these developing colonies can be transferred to media appropriate for identification (**Table 7**). Most of the known species of *Mucor* can be identified using the keys of Schipper (see Further Reading), a former employee of the Centraalbureau voor Schimmel Cultures in the Netherlands. The species within this genus differ from one another mainly in the diameter of the sporangia and sporangiophores, as well as in the morphology and measurements of the sporangiospores. In addition, the morphology of the columellae, the presence and

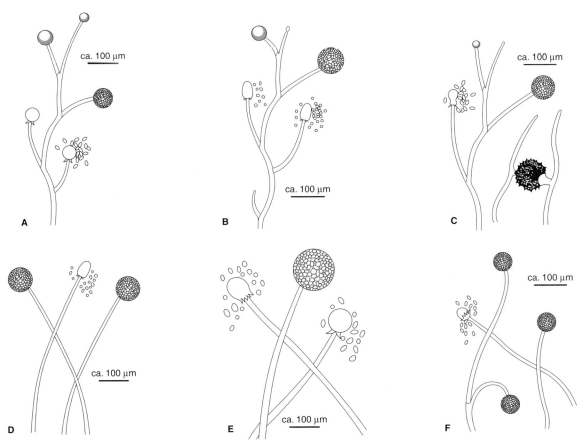

Figure 2 A summary of the characteristic features of some of the more frequently encountered *Mucor* species in foods. (**A**) *Mucor circinelloides* van Tieghem. Sporangia, brown to black, 40–80 μm in diameter, rarely 100 μm. Sympodially branched sporangiophores 20–30 mm in length and up to 17 μm in diameter. Sporangiospores broadly ellipsoidal, 4–9 μm in length and 3–5 μm in width. Colony up to 20 mm, rarely 30 mm in height. Growth and sporulation between 5°C and 30°C. No growth at 40°C. (**B**) *Mucor falcatus* Schipper. Sporangia, yellow to brown, up to 100 μm in diameter, rarely reaching 130 μm. Sympodially branched sporangiophores up to 18 μm in diameter. Sporangiospores globose; some are ellipsoidal, 6–10 μm in diameter. Columellae conical or cylindrical up to 60 × 55 μm. Colony up to 8 mm in height. Growth and sporulation between 10°C and 25°C. No growth at or above 37°C. (**C**) *Mucor genevensis* Lendler. Sporangia yellow to brown, up to 70 μm in diameter. Sympodially branched sporangiophores up to 10 μm in diameter. Sporangiospores are ellipsoidal, 4–8 μm in length and 2–4 μm in width. Columellae piriform-ellipsoidal up to 40 × 32 μm. Homothallic species forming dark brown spiny zygosporangia with a diameter of 80 μm. Colony up to 5 mm in height. Growth and sporulation between 5°C and 25°C. No growth at or above 37°C. (**D**) *Mucor hiemalis* Wehmer. Sporangia yellow to brown, up to 70 μm in diameter. Sporangiophores unbranched or slightly sympodically branched, up to 15 μm in diameter. Sporangiospores oblong to ellipsoidal, up to 10 μm in length and 5 μm in width. Columellae globose or oval. Colony up to 20 mm in height. Growth and sporulation between 5°C and 25°C. No growth at or above 37°C. (**E**) *Mucor inaequisporus* Dade. Sporangia yellow to brown, up to 150 μm in diameter, rarely reaching 175 μm in diameter. Sporangiophores up to 30 μm in diameter, mostly unbranched, but sympodial branches may occur. Sporangiospores variable in shape and size, 5–30 μm in length and 3–23 μm in width. Columellae are up to 83 × 75 μm and are subglobose, conical to applanate in shape. Colony up to 30 mm in height. Growth and sporulation between 10°C and 25°C. No growth at or above 30°C. (**F**) *Mucor indicus* Lendner. Sporangia yellow to brown, 40–50 μm in diameter. Sporangiophores are branched sympodially up to 14 μm in diameter. Sporangiospores are subglobose to ellipsoidal, 5–6 μm in length and up to 4 μm in width. Colony up to 10 mm in height. Growth and sporulation between 20°C and 37°C. Optimal growth at 30°C. At 40°C growth without sporulation occurs. (**G**) *Mucor mucedo* (Linnaeus) Fresenius. Sporangia grey, up to 250 μm in diameter. Sporangiophores are unbranched or branched, up to 40 μm in diameter. Sporangiospores are subglobose or ellipsoidal, 11–14 μm in length and 6–8 μm in width, or 8–9 μm in diameter. Columellae ovoid to ellipsoidal up to 160 × 125 μm. Colony up to 25 mm in height. Growth and sporulation between 5°C and 25°C. No growth at 30°C or higher. (**H**) *Mucor piriformis* Fischer. Sporangia black, up to 350 μm in diameter. Sporangiophores are unbranched or branched, up to 40 μm in diameter. Sporangiospores mostly ellipsoidal, 7–10 μm in length and 4–7 μm in width. Columellae ellipsoidal, pyriform or subglobose up to 190 × 175 μm. Growth and sporulation between 5°C and 25°C. Optimal growth between 10°C and 15°C. No growth at 30°C or higher. (**I**) *Mucor plumbeus* Bonorden. Sporangia grey, up to 80 μm in diameter, rarely 100 μm. Sporangiophores branch sympodially or monopodially, up to 21 μm in diameter. Sporangiospores mostly globose, 7–8 μm in diameter. Columellae pyriform, ovoid-ellipsoidal to cylindrical or conical, 49 × 25 μm. Some collumellae contain one or more projections. Colony up to 20 mm in height. Growth and sporulation at 5–28°C. No growth at or above 37°C. (**J**) *Mucor racemosus* Fresenius. Sporangia grey to brown, up to 80 μm in diameter, rarely 90 μm. Sporangiophores branch sympodially and monopodially, up to 18 μm in diameter. Sporangiospores broadly ellipsoidal to subglobose, up to 10 μm in length and 7 μm in width, or up to 8 μm in diameter. Columellae ovoid, ellipsoidal, cylindrical, subglobose or pyriform up to 55 × 37 μm. Chlamydospores frequently occur in cultures. Colony up to 45 mm in height.

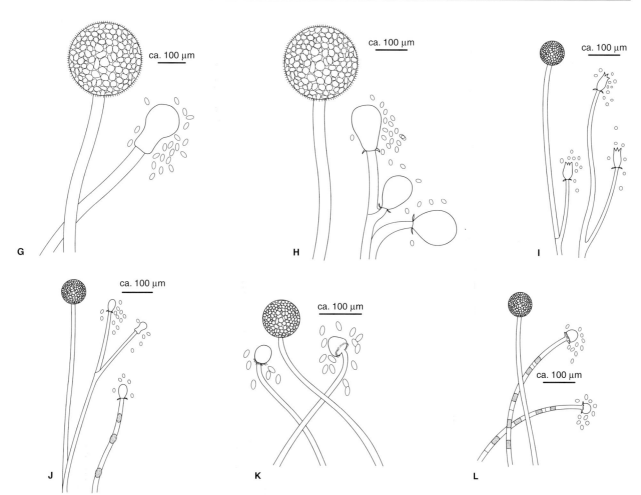

Figure 2 continued —Growth and sporulation at 5–30°C. Optimal growth and sporulation 20–25°C. No growth at or above 37°C. (**K**) *Mucor recurvus* Butler. Sporangia yellow, up to 125 µm in diameter. Sporangiophores are unbranched or sympodially branched, transitorily recurved, up to 18 µm in diameter. Sporangiospores are ellipsoidal, up to 27 µm in length and 11 µm in width. Columellae applanate, conical or cylindrical, up to 70 × 60 µm. Colony up to 40 mm in height. Growth and sporulation at 10–40°C. Optimal growth and sporulation 20–30°C. (**L**) *Mucor sinensis* Milko & Beljakova. Sporangia pale yellow to brown, up to 70 µm in diameter, rarely 100 µm in diameter. Sporangiophores mostly unbranched, up to 14 µm in diameter. Numerous chlamydospores in sporangiophores. Sporangiospores globose to irregular in shape, up to 12 µm in length and 11 µm in width. Globose spores are up to 16 µm in diameter. Columellae cylindrical, ellipsoidal or conical, up to 42 × 35 µm. Growth and sporulation at 5–25°C. No growth at or above 37°C.

type of branching of the sporangiophores and the maximum growth temperatures are determined to enable identification of some species. Some of the characteristic features of the *Mucor* species that have been found in foods are depicted in **Figure 2**.

Immunochemical Detection

Conventional detection methods, such as counting the number of colony-forming units in a food sample, have their limitations. This is especially true for processed foods in which the fungi were killed or removed during preparation. It is for this reason that an immunological detection method was developed in order to detect the immunologically active extracellular polysaccharides (EPS) of mucoralean fungi present in foods.

These EPS, which are water-extractable and heat-resistant, are mainly composed of fucose, galactose, glucuronic acid, mannose and small amounts of protein. Mannosyl residues with α-linkages are the immunodominant sugars in the EPS of members of *Mucor*. Enzyme-linked immunosorbent assays (ELISAs) have been developed for the detection of the EPS of *M. circinelloides*, *M. hiemalis* and *M. racemosus*. Although polyclonal antibodies raised against EPS of *M. racemosus* showed cross-reactions with the EPS obtained from members of other mucoralean genera in an ELISA, as well as with the yeast *Pichia membranaefaciens*, no cross-reactivity occurred with the major species of *Penicillium* and *Aspergillus*. The testing of ELISAs prepared from antibodies raised against the EPS of *M. hiemalis* showed that this type of assay is a sensitive, rapid and reliable method for detecting mucoralean fungi in a number of food products. However, false-positive reactions do occur, especially in foods containing jams or walnuts. In an attempt to make the tests more specific, a monoclonal antibody was raised against the EPS from *M. racemosus* after intrasplenic immunization of mice. The IgG antibody was specific for mucoralean fungi representing various genera, including *Mucor*, *Rhizopus*, *Rhizomucor* and *Absidia*. No cross-reactivity occurred with the EPS from representatives of various species of *Aspergillus*, *Penicillium* and *Fusarium*. In addition, no cross-reactivity could be detected with the EPS from representatives of *Pichia membranaefaciens*. Commercial ELISA kits using monoclonal antibodies may in the future be used for the rapid and sensitive detection of mucoralean fungi in foods.

See also: **Biochemical and Modern Identification Techniques**: Food Spoilage Flora (i.e. Yeasts and Moulds). **Fungi**: Food-borne Fungi – Estimation by Classical Culture Techniques. **Spoilage Problems**: Problems caused by Fungi.

Further Reading

Bärtschi C, Berthier J, Guiguettez C and Valla G (1991) A selective medium for the isolation and enumeration of *Mucor* species. *Mycological Research* 95: 373–374.

Benjamin RK (1979) Zygomycetes and their spores. In: Kendrick B (ed.) *The Whole Fungus, The Sexual–Asexual Synthesis*. P. 573. Alberta, Canada: National Museum of Natural Science, National Museums of Canada and the Kananaskis Foundation.

Botha A, Strauss T, Kock JLF, Pohl CH and Coetzee DJ (1997) Carbon source utilization and γ linolenic acid production by mucoralean fungi. *Systematic and Applied Microbiology* 20: 165–170.

De Ruiter GA, Hoopman T, Van der Lugt AW, Notermans SHW and Nout MJR (1992) Immunochemical detection of *Mucorales* species in foods. In: Samson RA, Hocking AD, Pitt JI and King AD (eds) *Modern Methods in Food Mycology*. P. 221. Amsterdam: Elsevier.

Hesseltine CW and Ellis JJ (1977) Mucorales. In: Ainsworth GC, Sparrow FK and Sussman AS (eds) *The Fungi, An Advanced Treatise*. P. 187. New York: Academic Press.

King AD (1992) Methodology for routine mycological examination of food – a collaborative study. In: Samsom RA, Hocking AD, Pitt JI and King AD (eds) *Modern Methods in Food Mycology*. P. 11. Amsterdam: Elsevier.

King AD, Pitt JI, Beuchat LR and Corry JEL (1986) *Methods for the Mycological Examination of Foods*. P. 63. New York: Plenum Press.

Orlowski M (1991) *Mucor* dimorphism. *Microbial Reviews* 55: 234–258.

Pitt JI and Hocking AD (1997) *Fungi and Food Spoilage*, 2nd edn. New York: Chapman & Hall.

Reiss J (1993) Biotoxic activity in the *Mucorales*. *Mycopathologia* 121: 123–127.

Samson RA, Hoekstra ES, Frisvad JC and Filtenborg O (1995) *Introduction to Food-borne Fungi*, 4th edn. Baarn, the Netherlands: Centraalbureau voor Schimmelcultures.

Schipper MAA (1978) On certain species of *Mucor* with a key to all accepted species. In: *Studies in Mycology No. 17*. P. 48. Baarn, the Netherlands: Centraalbureau voor Schimmelcultures.

Mycelial Fungi *see* **Single-cell Protein**: Mycelial Fungi.

MYCOBACTERIUM

Janet B Payeur, National Veterinary Services Laboratories, Animal and Plant Health Inspection Service, US Department of Agriculture, Ames, USA

Characteristics of the Genus

Mycobacteria are members of the order Actinomycetales, and the only genus in the family Mycobacteriaceae. Over 50 species are recognized in *Bergey's Manual of Systematic Bacteriology*, including numerous pathogens and saprophytic organisms of warm-blooded animals. The distinguishing characteristics of this genus include acid-fastness and the presence of mycolic acids. Mycobacteria are slender, non-spore-forming, rod-shaped, aerobic, slow-growing and free-living in soil and water. These bacteria have a generation time of about 20 h, thus isolation and identification may take up to 6 weeks (although a few species may grow in only 5–7 days).

These bacteria are acid-alcohol-fast, which means that after staining they resist decolorization with acidified alcohol as well as strong mineral acids. The property of acid-fastness, due to waxy materials in the cell walls, is particularly important for recognizing mycobacteria. The staining procedures must be carefully performed because other Gram-positive bacteria (e.g. *Nocardia*, *Corynebacterium* and *Rhodococcus*) are often partially acid-fast.

Mycobacterium tuberculosis, *M. africanum*, *M. bovis*, *M. bovis* BCG and *M. microti* are collectively referred to as the *M. tuberculosis* complex because these organisms cause tuberculosis (TB), a disease characterized by the formation of tubercles and caseous necrosis in tissues. The source of tubercle bacilli is tuberculous individuals. Humans perpetuate *M. tuberculosis*; cattle, bison and deer perpetuate *M. bovis*; and chickens perpetuate *M. avium*. *M. bovis* and *M. avium* can infect wild mammals and birds respectively, and these animals occasionally become sources of infection for domestic animals. In contrast, most non-tuberculous mycobacteria are saprophytes, and some are normal commensal bacteria of animals – diseased individuals are not significant sources of infection.

Tuberculosis in Birds

Tuberculosis is a chronic disease of birds manifested by progressive weight loss. It is usually found in older chickens that have been kept beyond the laying season. The causative agent is *M. avium*. Clinical signs of tuberculosis include dull, ruffled feathers; pale skin of the face, wattles, and comb; diarrhoea; and progressive emaciation. If the bone marrow of the leg bones is involved, a jerky, hopping gait is observed. Common gross lesions include greyish-white lesions in the liver, spleen, intestinal serosa and, in advanced cases, the bone marrow. Tubercle formation stops short of calcification. Although the organism has been isolated from eggs, transovarian infection of chicks is rare. *M. avium* affects many species of birds, but psittacines and canaries are resistant – they are more susceptible to *M. tuberculosis* rather than *M. avium*.

Tuberculosis in Ruminants, Swine and Horses

Clinical TB in ruminants is typically a debilitating disease characterized by progressive emaciation, erratic appetite, irregular low-grade fever and occasionally, by localizing signs such as enlarged lymph nodes, cough and diarrhoea.

Cattle, sheep and goats are the species most often infected with *M. bovis*. Infection is centred in the respiratory tract and adjacent lymph nodes and serous cavities. The disease commonly progresses via air spaces and passages, but haematogenous dissemination involving liver and kidney also occurs. Fetuses may be infected *in utero* and surviving offspring commonly develop liver and spleen lesions.

Udder infections are rare ($< 2\%$ of cases) but have obvious public health implications because *M. bovis* may be secreted in the milk. *M. tuberculosis* causes minor, non-progressive lesions in cattle, sheep and goats. Infection, with *M. avium* is generally subclinical.

In the US and Canada, mycobacterial infections in swine are usually caused by *M. avium* and are associated with the gastrointestinal system. It does not produce classic tubercles (e.g. caseation, calcification or liquefaction) but it may disseminate to viscera, bone and meninges. Swine may get avian TB from either bird droppings or from eating dead birds. In swine, *M. bovis* causes progressive disease with classical lesions. *M. tuberculosis* infections do not advance past regional lymph nodes.

Horses are rarely infected, but relatively more often with *M. avium* than with *M. bovis*. Infection is usually by the oral route, with primary lesions in the pharynx and intestine. Secondary lesions may be in lung, liver, spleen and serous membranes. Gross lesions are tumour-like but lack the caseation and gross calcification of classical tubercles.

Government Regulations

In Canada and the US, poultry carcasses affected with TB (*M. avium*) are condemned on postmortem examination. According to the US Department of Agriculture (USDA) Poultry Inspection Regulations, suspected birds are segregated from the other poultry and held for separate slaughter, evisceration and postmortem inspection.

The USDA Meat Inspection Regulations state that when lesions similar to those caused by *M. bovis* are detected in carcasses of animals from premises being depopulated because of TB, the carcasses shall be condemned regardless of extent of infection. If carcasses are presented as part of the regular kill, the disposition of affected carcasses reflects both the location and extent of lesions detected. Affected carcasses are condemned if lesions are detected in one or more primary sites and one or more body lymph nodes; or lesions are detected in any other organ, e.g. lungs, liver or spleen. When carcasses are affected to a lesser extent, the affected lymph node and the corresponding portion of the carcass is condemned, e.g. head and tongue, lungs or intestine and stomach. In swine, if a carcass has two or more isolated lesions of mycobacteriosis, it must be cooked at 76°C (170°F) for 30 min. Carcasses that are 'passed for cooking' lose most of their commercial value, and the additional labour in cooking is an added expense. Many processing plants have no facilities for cooking, so the carcasses are condemned.

Although TB (*M. bovis*) in ruminants is now relatively infrequent in the US and Canada, it is still an important public health threat in many parts of the world where it has not been eradicated. In addition to meat, milk may be contaminated by mycobacteria. Although milk is a potentially significant vehicle for transmission of infection to humans, its importance has been drastically reduced in the US and Canada by careful sanitation and by pasteurization. One still risks acquiring TB by consuming dairy products when in foreign countries that do not have a TB eradication and control programme in place and where pasteurization of milk and milk products is not mandatory.

M. bovis occurs in some cervid populations and is a problem in Canada and the US. There is a risk in handling deer carcasses and in eating meat which has not been inspected and passed for human consumption by federal and state meat inspectors.

Mycobacteria Species of Public Health Importance

Robert Koch was the first to establish the causal relationship between the tubercle bacillus (*M. tuberculosis*) and the disease TB. It causes TB in humans and may infect domestic and wild animals, usually directly or indirectly from humans. Simians are particularly likely to become infected.

Many species of mycobacteria that normally exist as environmental saprophytes occasionally cause disease in humans and animals. Such infections may be caused by most of the slowly growing mycobacteria such as *M. avium*, *M. intracellulare*, *M. scrofulaceum*, *M. kansasii*, *M. marinum*, *M. simiae*, *M. ulcerans* and *M. xenopi*. The only rapidly growing pathogenic species are *M. chelonae* and *M. fortuitum* (**Table 1**). Unlike TB, these mycobacterial infections are acquired from the environment and rarely, if ever, transmitted from person to person. The principal source of these infections seems to be water. Contact with waterborne mycobacteria by drinking, washing or inhalation of aerosols is common, yet the incidence of overt disease is very low.

M. bovis typically causes TB in cattle and bison but may also infect other animals, including dogs, cats, swine, rabbits, cervids, badgers, coyotes, raccoons and brush-tailed possums. This is the classic bovine tubercle bacillus, common in dairy cows before eradication schemes were introduced. It is still occasionally associated with human disease, both pulmonary and extrapulmonary, but infections are now rarely associated with the consumption of contaminated milk. Most occurrences of disease are reactivation of infections acquired much earlier.

M. marinum was originally isolated from diseased fish and is the causative agent of a superficial granulomatous skin disease of humans known as swimming-pool granuloma, fish-tank granuloma or fish-fancier's finger. It is found in sea-bathing pools and in tanks where tropical fish are kept and is a pathogen of some fish, amphibians and aquatic mammals.

M. kansasii causes pulmonary lesions and was originally isolated from an infected human lung. It is one of the most frequent causes of opportunist mycobacterial disease. It has been isolated on several occasions from piped water supplies but is rarely encountered in the natural environment. It has also been found in noduloulcerative skin lesions of cats and tuberculous lesions in lymph nodes of the alimentary tract in pigs.

M. scrofulaceum is most commonly associated with cervical lymphadenitis in young children. Uncommon extranodal manifestations include pulmonary disease, disseminated disease and rare cases of conjunctivitis, osteomyelitis, meningitis and granulomatous hepatitis in humans. Tuberculous lesions in cervical and intestinal lymph nodes are seen in domestic and wild pigs, cattle and bison. It has been isolated from soil, water (including tap water), raw milk and other dairy products and oysters.

M. avium–intracellulare complex is widely distributed in water, soil, plants, dust, mammals, poultry and other environmental sources. It has been isolated from meat, milk and eggs. They are opportunistic pathogens of humans, associated with cervical adenitis, especially in young children, but pulmonary infections also occur. They frequently cause opportunistic disease in patients with AIDS. Such disease is often disseminated, and the organisms may be isolated from many sites, including blood, bone marrow and faeces.

M. xenopi was first isolated from an African toad. Hot- and cold-water taps, including water storage tanks and hot-water generators of hospitals, are potential sources for nosocomial infections. It is an opportunist pathogen in human lung disease and is rarely significant in other sites. It is a frequent contaminant of pathological material, especially urine. This organism is very common in the UK, France, Denmark, Australia and the US.

M. paratuberculosis causes chronic enteritis in cattle, goats, sheep and certain captive wild ruminants in many countries. It has also been isolated from several cases of humans with Crohn's disease, but its role in that disease is uncertain.

M. chelonae–fortuitum complex are opportunist pathogens usually occurring in superficial infections (e.g. needle injection abscesses, post surgical wound infections, accidental trauma) and occasionally as secondary agents in pulmonary disease. They have also

Table 1 Environmental sources and clinical significance of selective mycobacteria

Species	Environmental source	Clinical significance
Mycobacterium avium–intracellulare complex	Soil, water, birds and other animals (especially chickens, swine, cattle), foods such as meat, milk and eggs	Chronic pulmonary disease, local lymphadenitis and joint disease, disseminated disease in patients with AIDS, skin and soft-tissue infections including abscesses and corneal infections, mycobacterial diseases in animals
M. bovis	Cattle, bison, cervids, sheep, goats, possums, non-human primates, badgers, swine, dogs, cats, milk and dairy products, meat	Bovine and human tuberculosis
M. chelonae–fortuitum complex	Water, soil, dust	Disseminated disease, cutaneous lesions, pulmonary disease, soft-tissue infections, postoperative wound infections, keratitis
M. kansasii	Tap water, tissues from cattle, deer, swine	Chronic pulmonary disease, bone and joint disease, disseminated disease, cervical lymphadenitis, cutaneous disease
M. marinum	Fresh- and salt-water fish, amphibians, aquatic mammals, swimming pools and aquariums	Cutaneous disease
M. scrofulaceum	Soil, water, raw milk, other dairy products, oysters	Cervical lymphadenitis in children, chronic pulmonary disease in adults, disseminated disease in children
M. smegmatis	Tissues from cattle, raw milk	Mastitis in cattle, skin or soft-tissue infections
M. tuberculosis	Tissues from cattle, other animals, primates, raw milk	Human tuberculosis
M. xenopi	Swine tissues, water, birds, hot-water systems	Chronic pulmonary disease

Adapted from Howard et al (1994) and Nolte and Metchock (1995).

been associated with a wide variety of infections involving the lungs, skin, bone, central nervous system, prosthetic heart valves and with disseminated disease. They are common in the environment and frequently appear as laboratory contaminants.

M. smegmatis has been associated with granulomatous mastitis in cattle and ulcerative skin lesions in cats. Generally, it is considered non-pathogenic in humans and animals.

Isolation of the Agent

From Tissue: NaOH method

All procedures should be performed in a biological safety cabinet since mycobacteria are classified as Class II and III organisms by the Centers for Disease Control and Prevention.

Tissue samples are treated with sodium hydroxide (NaOH) to eliminate contaminating organisms prior to culture on selective media.

1. Tissue samples are homogenized in a blender jar with 50 ml of phenol red broth for 1–2 min.
2. To 7.0 ml of macerated tissue suspension, 5.0 ml of 0.5 N NaOH is added into a 50 ml screw-cap test tube. Do not allow exposure of tissue suspension to NaOH to exceed 10 min.
3. The remaining macerated tissue is added to a second screw-cap test tube containing no NaOH and is used to inoculate selective media, i.e. Middlebrook 7H10 and Middlebrook 7H11. The

untreated suspension is then frozen at – 70°C for future reference.

4. To the NaOH-treated tissue suspension approximately 10–15 drops at 6.0 N HCl are added until the mixture turns yellow. The suspension is brought back from yellow to pale pink with 1.0 N NaOH.
5. The NaOH-treated tubes with neutralized tissue suspension are centrifuged for 20 min at 1650 RCF (relative centrifugal force). The centrifuge should have sealed dome carriers to contain contents if a tube breaks during centrifugation.
6. The pellicle is removed and 85% of the overlying fluid is decanted.
7. Selective media are inoculated with treated sediment, i.e. Middlebrook 7H10, Middlebrook 7H11, Stonebrink, Herrold egg yolk with malachite green and mycobactin, Lowenstein–Jensen and BACTEC®12B.
8. The inoculated media are incubated at 37 ± 2°C and examined every week for 8 weeks for the presence of mycobacterial colonies. If bacterial colonies resembling those of mycobacteria are found, a smear is made from each type of colony, stained by the Ziehl–Neelsen technique and observed for the presence of acid-fast bacilli.

From Milk

Samples of milk may be collected where tuberculous mastitis is suspected. About 25–30 ml is drawn from each quarter under aseptic conditions towards the end of milking.

1. At least 100 ml of milk from each animal is centrifuged for 20 min at 1450 RCF.
2. The supernatant is decanted.
3. The cream and sediment are treated separately by the NaOH method (see tissue procedure, above).
4. A variety of media is inoculated with treated and untreated sediment.

From Water

1. Up to 2 l water is passed from cold and hot taps through membrane filters.
2. The membranes are drained and placed in 3% sulphuric acid (H_2SO_4) for 3 min and then in sterile water for 5 min.
3. The membranes are cut into strips and placed on the surface of the culture medium in screw-capped bottles.
4. Lowenstein–Jensen medium and Middlebrook 7H11 agar containing antibiotics is inoculated.

From Cold and Hot Water Pipes

1. The insides of cold- and hot-water taps are swabbed.
2. The swab is placed in a tube containing 1 N NaOH solution for 5 min.
3. The swab is removed and placed in another tube containing 14% potassium dihydrogen orthophosphate (KH_2PO_4) solution for 5 min.
4. The swab is removed and used to inoculate a variety of culture media.

Identification of the Agent – Methods of Detection

Microscopy to Demonstrate Acid-fast Bacilli

Acid-fast bacilli are straight or slightly curved rods, $0.2–0.7 \times 1.0–10 \, \mu m$, sometimes branching.

They are acid-alcohol-fast at some stage of growth; they are not readily stained by Gram's method; they are usually weakly Gram-positive. No aerial hyphae are grossly visible. They are non-motile, non-spore-forming, without conidia or capsules.

Ziehl–Neelsen Stain

Mycobacterial cells are difficult to stain with common aniline dyes; however, they will stain with basic fuchsin. Once stained, they retain the dye despite treatment with strong mineral acids, such as HCl. The mechanism responsible for the retention of basic dyes is not clearly understood. It has been postulated that acid-fastness is due to absorption of dye by the mycolic acid residues that are linked to the arabinogalactan-peptidoglycan layer of the cell wall skeleton (**Figs 1–3**):

- Positive test: organisms retain carbol fuchsin and stain red (*M. tuberculosis* ATCC 25177)
- Negative test: organisms stain blue with the methylene blue counterstain (*Corynebacterium*).

Colonial Morphology

Growth is slow or very slow; visible colonies appear in 2–60 days at optimum temperature. Colonies are often buff, pink, orange or yellow, especially when exposed to light. Pigment is not diffusing; the surface is commonly dull or rough (**Figs 4–6**). Some species are fastidious, requiring special supplements (e.g. *M. paratuberculosis*) or non-culturable (*M. leprae*).

Growth Rate and Pigment Production

Mycobacteria may be separated into two groups based on growth rate. Those that form visible colonies within 7 days are called rapid-growers, and those that require longer periods are the slow-growers. The rapid-growers encompass the Runyon group IV mycobacteria, e.g. *M. fortuitum*; the slow-growers include the *M. tuberculosis* complex and groups I–III. The photochromogens (group I) are slow-growing, photoreactive mycobacteria. Some mycobacteria tolerate higher temperatures, e.g. *M. xenopi*, with an optimum growth rate at 35–45°C; others may be inhibited at higher temperatures, e.g. *M. marinum*, whose optimum growth is at 30–32°C and may not grow at 37°C. Some species of mycobacteria possess carotenoid pigments in the presence or absence of light and others are dramatically induced to form yellow-orange β-carotene crystals only by photoactivation. Those producing pigment either in the presence or absence of light are described as scotochromogenic and those whose pigment is induced only by photoactivation are described as photochromogenic. Some species of mycobacteria lack β-carotene and are non-chromogenic. Colonies that are white, cream or buff are described as non-pigmented or non-chromogenic. Colonies that are lemon-yellow, orange or red are described as pigmented or chromogenic.

Intermediate colorations such as pink, pale yellow or tan sometimes occur, and these are recorded as observed. Such cultures generally are regarded as non-photochromogens unless the pigment becomes more intense on exposure to light (**Table 2**).

Differential Characteristics of Commonly Isolated Mycobacteria

Niacin

Certain mycobacteria, e.g. most isolates of *M. tuber-*

Table 2 Growth characteristics of commonly isolated mycobacteria

Organism	Growth rate (days)	Optimal temperature	Pigment light	Production dark	Colonial morphology on Middlebrook 7H10 agar	Colonial morphology on Lowenstein–Jensen (LJ) medium
Group I: Photochromogens						
Mycobacterium kansasii	10–21	37°C	Yellow	Buff	Raised and smooth; some are rough and wrinkled; numerous carotene crystals after exposure to light	Smooth or rough; pigmentation same as on 7H10
M. marinum	5–14	30°C	Yellow	Buff	Round, smooth or intermediate in roughness; some may be wrinkled	Same as 7H10
Group II: Scotochromogens						
M. scrofulaceum	10–14	37°C	Yellow	Yellow	Smooth, moist, yellow and round	Same as 7H10
M. xenopi	28–42	42°C	Yellow	Yellow	Small, yellow colonies with compact centres surrounded by fringe of branching filaments; at 45°C, resemble a miniature bird's nest	Small, smooth, dysgonic, dome-shaped, non-pigmented colonies that become yellow on ageing
M. avium–intracellulare	10–21	37°C	Buff to yellow	Buff to yellow	Thin, transparent, glistening or matte, smooth, circular, pyramid-shaped; some colonies rough and wrinkled	Smooth, dome-shaped, buff-coloured; rough, wrinkled colonies are sometimes seen
M. bovis	25–90	37°C	Colourless to buff	Colourless to buff	Small, thin, often non-pigmented, raised, rough, later wrinkled and dry; some colonies inhibited on this medium	Low, smooth, colourless, pyramid-shaped
M. paratuber-culosis	42–112	37°C	Buff	Buff	Smooth, yellow colonies; appear domed with entire margin or flattened irregular periphery; rough colonies are rarely seen. Needs mycobactin in media to grow	Initial growth is smooth, but with continued incubation becomes rough, dry, umbonated and heaped; needs mycobactin in media
M. tuberculosis	12–28	37°C	Buff	Buff	Non-pigmented, flat, dry, rough, and corded	Non-pigmented, dry, rough, with nodular surface and irregular, thin periphery
Group IV: Rapid growers						
M. chelonae	3–7	28°C	Buff	Buff	Rounded, smooth, matte, periphery entire or scalloped, no branching filaments; some colonies are rough and wrinkled	Rounded, smooth, colourless and hemispheric, rough colonies are occasionally seen on prolonged incubation
M. fortuitum	3–7	28°C	Buff	Buff	Circular, convex, wrinkled or matte; smooth or rough branching filaments on periphery	Soft, butyrous, hemispheric and multilobate or rough with heaped centres; although non-pigmented, they may appear green owing to absorption of malachite green
M. smegmatis	3–7	28°C	Buff to yellow	Buff to yellow	Raised, rough, wrinkled and with scalloped edges	Same as 7H10

Adapted from Musial and Roberts (1987), Nolte and Metchock (1995) and Hall and Howard (1994).

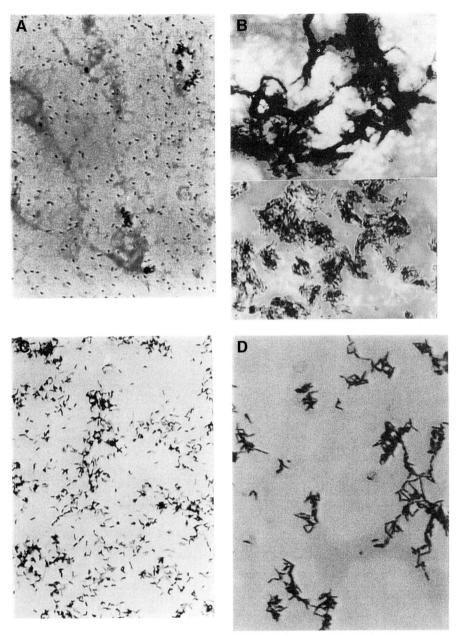

Figure 1 Microscopic morphology of (**A**) *M. avium*; (**B**) *M. bovis*; (**C**) *M. chelonae*; and (**D**) *M. fortuitum*. ×1000.

culosis and *M. simiae*, accumulate niacin and excrete it into the culture media. Commercially available reagent-impregnated filter-paper strips are incubated with the test medium, and a yellow colour is indicative of niacin accumulation and a positive test.

Nitrate Reduction

Only a few species of mycobacteria produce nitro-reductase which catalyses the reduction of inorganic nitrate to nitrite. The development of a red colour on addition of sulphanilic acid and N-naphthyl-ethylenediamine to an extract of the unknown culture is indicative of the presence of nitrite and a positive test. Species that reduce nitrate include *M. tuber-*

culosis, *M. kansasii*, *M. szulgai*, *M. terrae* complex and *M. flavescens*.

Tween-80 Hydrolysis

Tween-80 is the trade name of a detergent that can be useful in identifying those mycobacteria that possess a lipase that splits the compound into oleic acid and polyoxyethylated sorbitol. The released oleic acid changes the optical characteristics of the substrate so that the neutral red indicator changes from an original amber colour to pink. This test is helpful in identifying *M. kansasii* which is positive in 3–6 h and differentiating *M. gordonae* (positive) from *M. scrof-ulaceum* (negative).

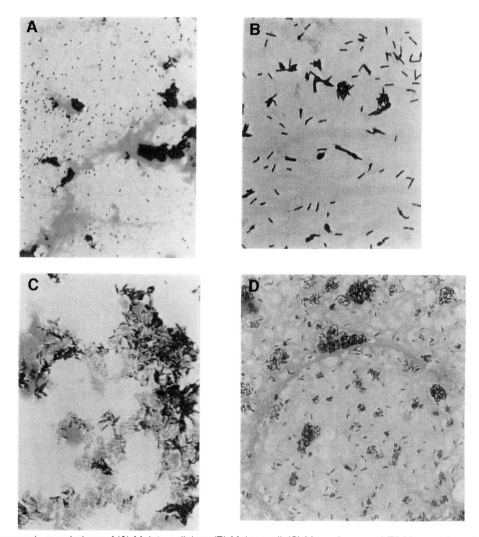

Figure 2 Microscopic morphology of (**A**) *M. intracellulare*; (**B**) *M. kansasii*; (**C**) *M. marinum*; and (**D**) *M. paratuberculosis*. ×1000.

Catalase

The enzyme catalase splits hydrogen peroxide (H_2O_2) into water and oxygen, which appears as bubbles. The semi-quantitative catalase test detects differences among certain mycobacteria in their production of catalase by measuring the height of the column of bubbles produced after the addition of H_2O_2, i.e. those species producing < 45 mm of bubbles and those producing > 45 mm of bubbles. Most mycobacteria produce catalase, with the exception of *M. gastri*, INH Isoniazid-resistant *M. tuberculosis* and *M. bovis*. In addition, certain mycobacteria produce a catalase that is heat-labile and can be detected by heating the culture to 68°C before adding H_2O_2.

Arylsulphatase

The enzyme arylsulphatase, which is primarily produced by rapidly growing mycobacteria, is detected by its degradation of the sulphate molecules of a tripotassium phenolphthalein disulphate salt into free phenolphthalein and the remaining salts. The addition of a base, sodium carbonate, reacts with the phenolphthalein and produces a red diazo reaction that is easily visible. The 3-day test is used to identify and distinguish some rapid-growers (*M. fortuitum*, *M. chelonae*) which give a positive reaction, from other rapid-growers. The 14-day test identifies slower-growing species (*M. marinum*, *M. xenopi*) and some rapid-growers (*M. smegmatis*).

Urease

Urease is an enzyme possessed by many *Mycobacterium* spp. that can hydrolyse urea to form ammonia and carbon dioxide. The ammonia reacts in solution to form ammonium carbonate, resulting in alkalinization and an increase in the pH of the medium. A colour change from amber to pink or red is a positive reaction.

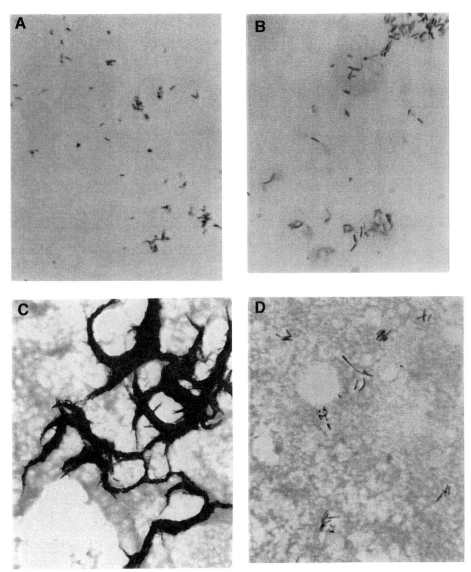

Figure 3   Microscopic morphology of (**A**) *M. scrofulaceum*; (**B**) *M. smegmatis*; (**C**) *M. tuberculosis*; and (**D**) *M. xenopi*. ×1000.

Pyrazinamidase

The enzyme pyrazinamidase hydrolyses pyrazinamide to pyrazinoic acid. Pyrazinoic acid is detected by the addition of ferrous ammonium sulphate to the culture medium. The formation of a pink ferrous–pyrazinoic acid complex indicates a positive test. This test is most useful in separating *M. marinum* from *M. kansasii* and *M. bovis* from *M. tuberculosis*. *M. bovis* is negative, even at 7 days, whereas *M. tuberculosis* is positive within 4 days.

Iron Uptake

M. fortuitum and a few other rapid- and slow-growers are capable of converting ferric ammonium citrate to iron oxide. The iron oxide is visible as a rust colour in the colonies when grown in the presence of ferric ammonium citrate. *M. chelonae* lacks this property.

Triophene-2-carboxylic Acid Hydrazide (TCH) Tolerance Susceptibility

TCH selectively inhibits the growth of *M. bovis*; however, *M. tuberculosis* and most other slowly growing mycobacteria are resistant to TCH at levels of $10\,\mu g\,ml^{-1}$ in the medium.

Growth on 5% NaCl

Few mycobacteria are able to grow in culture media containing 5% sodium chloride. The exceptions include *M. triviale* and most of the rapid-growers except the *M. chelonae* complex.

Growth on MacConkey Agar without Crystal Violet

Most isolates of the *M. fortuitum* and *M. chelonae* complexes will grow on MacConkey agar without crystal violet, whereas most other rapid-growers will not.

Table 3 Differential characteristics of commonly isolated mycobacteria

	Niacin	Nitrate reduction	Tween hydrolysis	Catalase semi-quantitative	Catalase 68°C	Aryl-sulphatase 3 days	Urease	Growth on PZA 4 days	Iron update	Growth on TCH	Growth on 5% NaCl	Growth on MacConkey	DNA probes available
Mycobacterium avium–intracellulare	–	–	–	<45	+	–	–	+	–	+	–	–/+	Yes
M. bovis	–	–	–	<45	–	–	+	–	–	–	–	–	Yes
M. chelonae	–/+	–	v	>45	+/–	+	+	+	–	+	v	+	
M. fortuitum	–/+	+	v	>45	+	+	+	+	+	+	+	+	
M. kansasii	–	+	+	>45	+	–	+	–	–	+	–	–	Yes
M. marinum	–/+	–	+	<45	–	–/+	+	+	–	+	–	–	
M. paratuberculosis	–	–	+	<45	+	–	–		–	+	–		Yes
M. scrofulaceum	–	–	–	>45	+	–	v	v	–	+	–	–	
M. smegmatis	–	+	+	<45	–	–	+	v	+	+	–	–	
M. tuberculosis	+	+	+/–	<45	–	–	+	+	–	+	–	–	Yes
M. xenopi	–	–	–	<45	+/–	+	–		–	+	–	–	

Adapted from Lutz (1992), Holt (1994) and Nolte and Metchock (1995).
PZA = Pyrazinamidase; TCH = triophene-2-carboxylic acid hydrazide; V = variable.

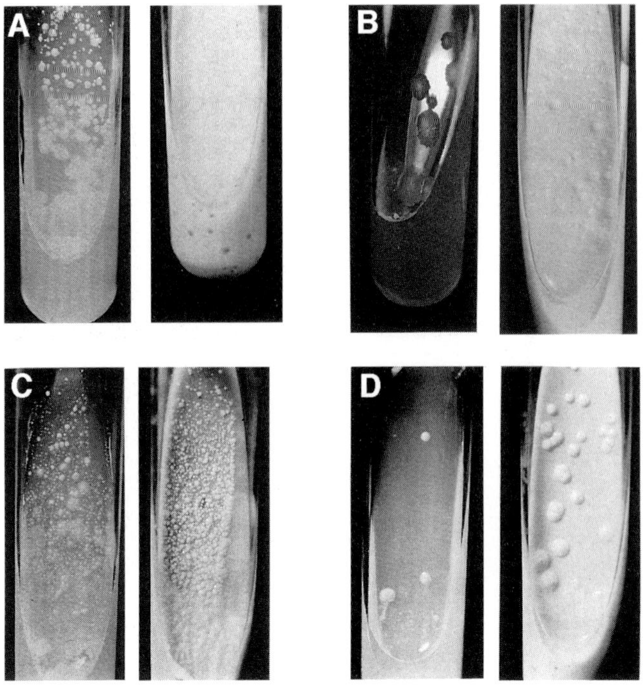

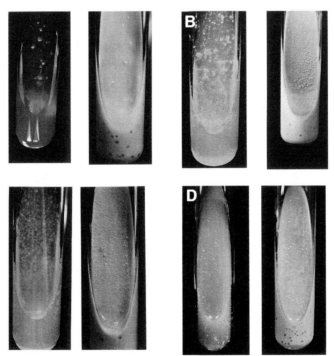

Figure 4 Colonial morphology of (**A**) *M. avium*; (**B**) *M. bovis*; (**C**) *M. chelonae* and (**D**) *M. fortuitum*. ×1000. (See also color **Plate 23**.)

Figure 5 Colonial morphology of (**A**) *M. intracellulare*; (**B**) *M. kansasii*; (**C**) *M. marinum*; and (**D**) *M. paratuberculosis*. (See also color **Plate 24**.)

DNA Probes for Culture Confirmation

DNA probes complementary to species-specific sequences of rRNA are available for the identification of the *M. tuberculosis* complex, *M. avium* complex, *M. avium*, *M. intracellulare*, *M. gordonae* and *M. kansasii* (Accuprobe®, Gen-Probe, San Diego, CA). These probes can be used to identify isolates that arise on solid culture media or from broth culture (**Table 3**).

See also: **Fish**: Spoilage of Fish. **Heat Treatment of Foods**: Principles of Pasteurization. **Meat and Poultry**: Spoilage of Meat. **Milk and Milk Products**: Microbiology

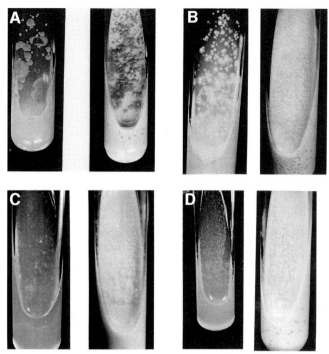

Figure 6 Colonial morphology of (**A**) *M. scrofulaceum*; (**B**) *M. smegmatis*; (**C**) *M. tuberculosis*; and (**D**) *M. xenopi*. (See also color **Plate 25**.)

of Liquid Milk. **Nucleic Acid-based Assays**: Overview.

Further Reading

Collins CH, Lyne PM and Grange JM (eds) (1995) Mycobacterium. In: *Collins and Lyne's Microbiological Methods*, 7th edn. Pp. 410–431. Oxford: Butterworth Heinemann.

Della-Latta P and Weitzman I (1998) Mycobacteriology. In: Isenberg, HD (ed.) *Essential Procedures for Clinical Microbiology*. Pp. 169–203. Washington, DC: American Society for Microbiology.

Hall GS and Howard BJ (1994) Mycobacterium. In: Howard BJ (ed.) *Clinical and Pathogenic Microbiology*, pp. 503–28. St. Louis: Mosby.

Holt JG (ed.) (1986) The Mycobacteria. In: *Bergey's Manual of Systemic Bacteriology*. Vol. 1, pp. 1435–57. Baltimore: Williams & Wilkins.

Holt JG, Krieg NR, Sneath PH, Staley JT and Williams ST (eds) (1994) The Mycobacteria. In: *Bergey's Manual of Determinative Bacteriology*, 9th edn. Pp. 597–603. Baltimore: Williams & Wilkins.

Howard BJ, Keiser JF, Smith TF, Weissfeld AS and Tilton RC (eds) (1994) Mycobacterium. In: *Clinical and Pathogenic Microbiology*. P. 503. St Louis: Mosby.

Koneman EW, Allen SD, Janda WM, Schreckenberger PC and Winn WC (eds) (1997) Mycobacteria. In: *Color Atlas and Textbook of Diagnostic Microbiology*, 5th edn. Pp. 893–952. Philadelphia: Lippincott JB.

Kubica GP (1984) Clinical Microbiology. In Kubica GP and Wayne LG (eds): *The Mycobacteria, A Sourcebook*. Pp. 133–75. New York: Marcel Dekker.

Lutz B (1992) Identification tests for mycobacteria. In: Isenberg HD (ed.) *Clinical Microbiology Procedures Handbook*. Vol. 1, pp. 3.11.6–3.11.7. Washington, DC: American Society for Microbiology.

Musial CE and Roberts GD (1987) Tuberculosis and other Mycobacterioses. In: Wentworth BB (ed.) *Diagnostic Prcoedures for Bacterial Infections*, 7th edn. Pp. 562–63. Washington, DC: American Public Health Association.

Nolte FS and Metchock B (1995) Mycobacterium. In: Murray PR (ed.) *Manual of Clinical Microbiology*, 6th edn. Pp. 400–36. Washington, DC: ASM Press.

Roberts GD (1985) Mycobacteria and Nocardia. In: Washington JA (ed.) *Laboratory Procedures in Clinical Microbiology*. Pp. 379–411. New York: Springer-Verlag.

Wayne LG (1984) Mycobacterial Speciation. In: Kubica GP and Wayne LG (eds) *The Mycobacteria, A Sourcebook*. Pp. 25–65. New York: Marcel Dekker.

MYCOTOXINS

Contents

Classification

Lloyd B Bullerman, Department of Food Science and Technology, University of Nebraska, Lincoln, USA

Copyright © 1999 Academic Press

What is a Mycotoxin?

Mycotoxins are a group of structurally diverse, naturally occurring chemical substances produced by a wide range of filamentous microfungi or moulds. The term mycotoxin is derived from the Greek word *mykes*, which means fungus, and the Latin word *toxicum*, which means toxin or poison. The term mycotoxin literally means fungus toxin or fungus poison, i.e. a toxic substance produced by a mould or fungus, as opposed to a substance that is toxic to the mould or fungus as the term phytotoxin implies toxicity to a plant, zootoxin toxicity to an animal, etc. The term is further restricted to mean the metabolites of microfungi, or moulds, as opposed to the toxic principles of certain macrofungi, i.e. mushrooms. Mycotoxins first became recognized as potential dangers to human and animal health in 1960 with the outbreak of the so-called turkey X disease in the UK, that led to the discovery of the aflatoxins. In this disease outbreak more than 100 000 turkey poults and other young farm animals were lost as a result of a toxic substance in a feed ingredient – peanut meal – from Brazil. The peanut meal, also called groundnut meal, was heavily contaminated with a common storage mould, *Aspergillus flavus (parasiticus)* that had produced the toxic substance. The toxic compound was dubbed aflatoxin, which was an acronym for *A. flavus* toxin. This was the major event that led to the realization that mould metabolites could pose hazards to human and animal health and stimulated extensive and intensive research on mycotoxins.

The Toxins

Since the discovery of aflatoxins, numerous moulds have been tested in the laboratory for the production of toxic metabolites. Of the hundreds of mycotoxins produced under laboratory conditions, only about 20 are known to occur naturally in foods and feeds with sufficient frequency and in potentially toxic amounts to be of concern to food safety. The moulds that produce the mycotoxins of most potential concern can be found in five taxonomic genera – *Aspergillus, Penicillium, Fusarium, Alternaria* and *Claviceps*.

Aspergillus species produce aflatoxins B_1, B_2, G_1, G_2, M_1 and M_2, ochratoxin A, sterigmatocystin and cyclopiazonic acid. *Penicillium* species produce patulin, citrinin, penitrem A, rubratoxin and a number of other toxic substances as well as ochratoxin A and cyclopiazonic acid. *Fusarium* species produce the trichothecenes: deoxynivalenol (DON, vomitoxin), 3-acetyl deoxynivalenol, 15-acetyldeoxynivalenol, nivalenol, diacetoxyscirpenol and T-2 toxin; zearalenone, fumonisins and moniliformin as well as other potentially toxic and possibly unknown toxic substances. *Alternaria* species produce a number of biologically active compounds of questionable mammalian toxicity, including tenuazonic acid, alternariol and alternariol methyl ether. *Claviceps* toxins are primarily the ergot alkaloids that can be found in ergot-parasitized grasses and small grains.

Of the 20 or so naturally occurring mycotoxins mentioned above, there are five toxins, or groups of related compounds, that are of greatest concern. These are the aflatoxins, ochratoxin, zearalenone, deoxynivalenol (or nivalenol in some regions) and fumonisins. Toxins of growing concern that can be added to that list are patulin, cyclopiazonic acid and moniliformin.

These mycotoxins of greatest concern are produced by mould species found in three main genera – *Aspergillus, Penicillium* and *Fusarium*. Aflatoxins are produced by *A. flavus, A. parasiticus* and *A. nomius*. *A. flavus* can also produce cyclopiazonic acid. Ochratoxin is produced by *A. ochraceus* and *Penicillium verrucosum*. *P. expansum*, as well as other *Penicillium* species and some *Aspergillus* species, can produce patulin. Zearalenone is produced by *Fusarium graminearum, F. culmorum* and *F. crookwellense*; deoxynivalenol or nivalenol are produced by the same three species, depending on the geographic origin of the producing strain. These same species may also

Table 1 Mycotoxins of greatest concern and the moulds that produce them

Mycotoxin	Producing mould
Aflatoxins B_1 and B_2	Aspergillus flavus
Aflatoxins B_1, B_2, G_1 and G_2	Aspergillus parasiticus
	Aspergillus nomius
Ochratoxin	Aspergillus ochraceus
	Penicillium verrucosum
Patulin	Penicillium expansum, other
	Penicillium species, some
	Aspergillus species
Cyclopiazonic acid	Aspergillus flavus
Zearalenone	Fusarium graminearum,
	F. culmorum, F. crookwellense
Deoxynivalenol	Fusarium graminearum,
3-acetyldeoxynivalenol	F. culmorum, F. crookwellense
15-acetyldeoxynivalenol	
nivalenol, trichothecenes	
Fumonisins	Fusarium moniliforme,
	F. proliferatum, F. subglutinans
Moniliformin	Fusarium proliferatum,
	F. subglutinans

produce the related toxins 3-acetyldeoxynivalenol and 15-acetyldeoxynivalenol, again depending on the geographic origin of the organism. Fumonisins are produced by *F. moniliforme*, *F. proliferatum* and *F. subglutinans*. *F. proliferatum* and *F. subglutinans* are also capable of producing moniliformin. The mycotoxins of greatest concern and the producing organisms are summarized in **Table 1**.

The Mycotoxin-producing Moulds

The moulds that produce the mycotoxins of greatest concern are species of *Aspergillus*, *Penicillium* and *Fusarium*. *Aspergillus* and *Penicillium* species tend to be saprophytic and often attack commodities such as cereal grains and nuts while in storage, though some aspergilli can also invade in the field. *Fusarium* species may be plant pathogenic as well as saprophytic types. A number of factors affect mycotoxin production by moulds, including moisture, relative humidity, temperature, substrate, pH, competitive and associative growth of other fungi and microorganisms, and stress on plants, such as drought and damage to seed coats from hail, insects and mechanical harvesting equipment. The major commodities that are susceptible to contamination with mycotoxins include corn (maize), peanuts, cottonseed and some tree nuts. Wheat and barley are also susceptible to contamination, primarily with deoxynivalenol, but also in some ochratoxin in some regions.

Contamination of Foods by Mycotoxins

Mycotoxins can enter the food supply in two ways: by direct contamination or by indirect contamination. Direct contamination occurs when there is mould growth and mycotoxin production directly on the food or commodity itself. Most foods are susceptible to mould growth during some stage of production, processing, storage or transport, and therefore have the potential for direct contamination. Contamination of grains, peanuts, tree nuts, etc. in the field prior to harvest is one of the main avenues for mycotoxins to enter the food supply. Mould growth on foods in storage such as grains, pulses, aged cheeses and cured and smoked meats can also result in direct contamination. Indirect contamination of foods can occur when a contaminated ingredient is used in the manufacture of a food, such as when peanuts are made into peanut butter, corn (maize) into corn meal, or wheat into flour. Then these products may be further used as ingredients in other foods containing many other ingredients such as baked goods, batters and breadings. Processed and prepared foods are most likely to be involved in indirect contamination and exposure to mycotoxins. Indirect exposure to mycotoxins can also result from consumption of animal products, such as milk and organ meats, that contain mycotoxin residues. Consumption of mouldy feed by food-producing animals can result in mycotoxin residues in these and other animal tissues. Dairy products, primarily milk for example, can become contaminated with aflatoxins M_1 and M_2 as a result of feeding contaminated feed to dairy animals. Cheese made from contaminated milk will also be contaminated.

Exposure to mycotoxins as a result of direct contamination of foods appears to be the greatest problem in tropical areas and regions where food preservation systems are inadequate and shortages exist. Indirect contamination, on the other hand, is more of a problem in those areas of the world where food is more highly processed, such as Canada, Europe, Japan and the US.

Toxicity and Biological Effects of Mycotoxins

In general, mycotoxins produce a number of adverse effects in a range of biological systems, including microorganisms, plants, animals and humans. The toxic effects of mycotoxins in humans and animals, depending upon dose, may include:

- acute toxicity and death as a result of exposure to high amounts of a mycotoxin
- reduced milk and egg production and lack of weight

gain in food-producing animals from subchronic exposure

- impairment or suppression of immune functions and reduced resistance to infections from chronic exposures to low levels of toxins
- tumour formation, cancers and other chronic diseases from prolonged exposure to very low levels of a toxin.

In addition, mycotoxins may be mutagenic, capable of inducing mutations in susceptible cells and organisms, and teratogenic, capable of causing deformities in developing embryos. Other manifestations of mycotoxins that can affect the food supply in economic terms are reduced growth rates and increased reproductive problems in food-producing animals and livestock.

Specific Mycotoxins of Greatest Concern

Aflatoxins

Aflatoxins are primarily produced by some strains of *Aspergillus flavus* and most, if not all, strains of *A. parasiticus*. Aflatoxins are also produced by *A. nomius* which so far has only been found in soils of the western US. There are four main aflatoxins, B_1, B_2, G_1 and G_2, plus two additional ones that are of significance, M_1, and M_2 (**Fig. 1**). The M toxins were first isolated from the milk of lactating animals fed aflatoxin preparations, hence the M designation, though some mould strains may also produce low amounts of these toxins.

Aflatoxins are potent liver toxins in all animals in which they have been tested and carcinogens in some species. Aflatoxin B_1 is the most toxic and most carcinogenic of the group. Effects of aflatoxins in animal tests vary with dose, length of exposure, species, breed and diet or nutritional status. These toxins may be lethal when consumed in large doses; sublethal doses produce chronic toxicity and low levels of chronic exposure result in cancers, primarily liver cancer in a number of animal species. In general, young animals of any species are more susceptible to the acute toxic effects of aflatoxins than are older animals of the same species. Susceptibility also varies between species. Of all the mycotoxins, the aflatoxins are of greatest concern because they are highly toxic and potently carcinogenic. Aflatoxins have been found in the diets of people in regions in which there are high incidences of human liver cancer. The International Agency for Research on Cancer (IARC) classifies aflatoxin B_1 as a human carcinogen. Mould growth and aflatoxin production are favoured by warm temperatures and high humidity, typical of tropical and subtropical regions, primarily on corn, peanuts, cottonseed and tree nuts.

Ochratoxins

Ochratoxins are a group of related compounds that are produced by *A. ochraceus* and related species, as well as *Penicillium verrucosum*. The main toxin in this group, ochratoxin A, is a potent mycotoxin that causes kidney damage in rats, dogs and swine (**Fig. 2**). Ochratoxin is thought to be involved in a disease of swine in Denmark known as porcine nephropathy, which has been associated with the feeding of mouldy barley. Ochratoxin is teratogenic to mice, rats and chicken embryos, and has been suggested as a possible causative factor, though never proven, in a human disease known as Balkan endemic nephropathy, which occurs in the Balkan countries of eastern Europe. Ochratoxin is also now thought to be a carcinogen.

Patulin

Patulin is toxic to many biological systems, including bacteria, mammalian cell cultures, higher plants and animals, but its role in causing animal and human disease is unclear. Patulin has a lactone structure and is carcinogenic when injected intradermally into mice (**Fig. 3**). Patulin is produced by numerous *Penicillium* and *Aspergillus* species and *Byssochlamys nivea*. *Penicillium expansum*, which commonly occurs in rotting apples, is the most common producer of patulin. Patulin is of some public health concern because of its potential carcinogenic properties, and because it has frequently been found in commercial apple juice. Patulin appears to be unstable in grains, cured meats and cheese, reacting with sulphydryl-containing compounds and becoming nontoxic. When administered orally to rats, patulin showed no toxicity or carcinogenicity.

Cyclopiazonic Acid

Cyclopiazonic acid (CPA) was originally isolated from *P. cyclopium* (**Fig. 4**). It now appears that CPA is produced by several moulds which commonly occur on agricultural commodities or which are used in certain food fermentations. Besides *P. cyclopium*, CPA has been reported to be produced by *A. flavus*, *A. versicolor* and *A. tamarii* as well as several other *Penicillium* species, some of which are used in the production of fermented sausages in Europe. Other moulds used in food fermentations that produce CPA are *P. camemberti*, used to produce Camembert cheese, and *A. oryzae*, used to produce fermented soy sauces. CPA occurs naturally in corn and peanuts, and a type of millet (kodo) that reportedly caused human intoxication in India. It is also possible that CPA was involved along with aflatoxins in the turkey

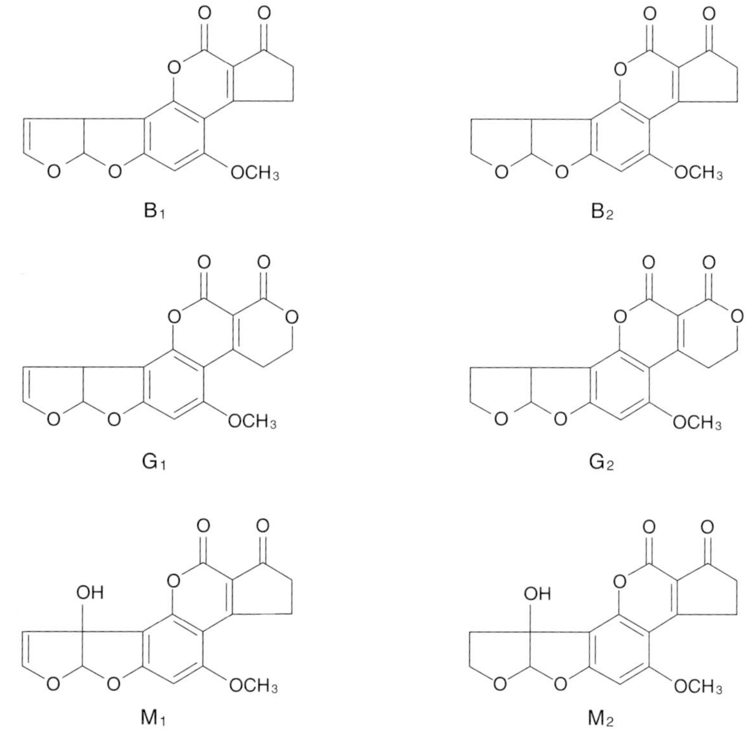

Figure 1 The chemical structures of aflatoxins B_1, B_2, G_1, G_2, M_1 and M_2.

X disease in the UK in 1960, since some isolates of *A. flavus* produced both aflatoxins and CPA.

Cyclopiazonic acid affects rats, dogs, pigs and chickens. Clinical signs of intoxication include anorexia, diarrhoea, pyrexia, dehydration, weight loss, ataxia, immobility and extensor spasm at the time of death. Histopathological changes in CPA-exposed

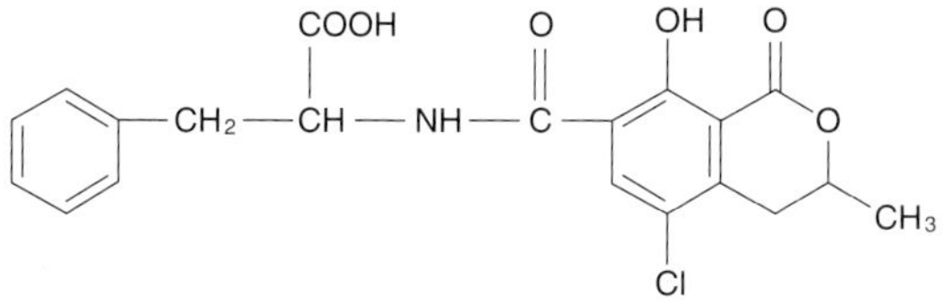

Figure 2 The chemical structure of ochratoxin A.

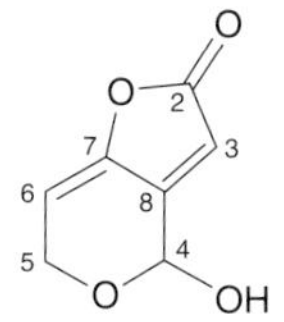

Figure 3 The chemical structure of patulin.

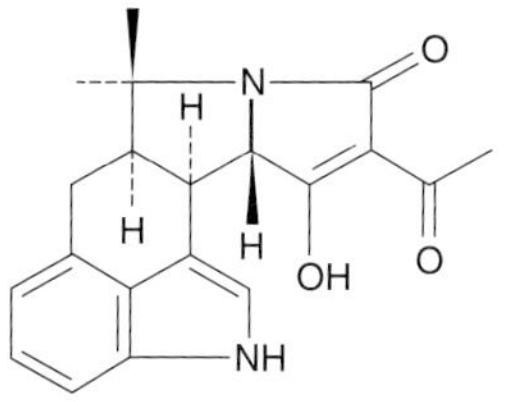

Figure 4 The chemical structure of cyclopiazonic acid.

animals include alimentary tract hyperaemia, haemorrhage and focal ulceration. Focal necrosis can be found in the liver, spleen, kidneys, pancreas and myocardium. In broiler chicks given CPA, skeletal muscle degeneration characterized by myofibular swelling and fragmentation has been observed. About 50% of a dose of CPA given orally or intraperitoneally to rats or chickens is distributed to skeletal muscle within 3 h. Cyclopiazonic acid has the ability to chelate metal cations. Chelation of such cations as calcium, magnesium and iron may be an important mechanism of toxicity of CPA.

Zearalenone

Zearalenone (**Fig. 5**), an oestrogenic compound also known as F-2 toxin, causes vulvovaginitis and oestrogenic responses in swine. Zearalenone is produced by several *Fusarium* species, but in particular *F. graminearum*, *F. culmorum* and *F. crookwellense*. Zearalenone occurs naturally in high-moisture corn in late autumn and winter, primarily from the growth of *F. graminearum* in North America and *F. culmorum*

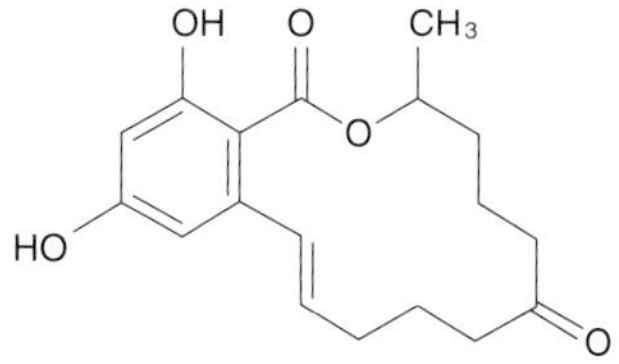

Figure 5 The chemical structure of zearalenone.

in northern Europe. Although the compound is not especially toxic, 1–5 p.p.m. is sufficient to cause physiological responses in swine. Zearalenone can be transmitted to piglets in sows' milk and cause oestrogenism in the young pigs. Zearalenone has been found in mouldy hay, high-moisture corn, corn infected before harvest and pelleted feed rations. The involvement of zearalenone in human toxicoses has not been confirmed, but it is classified as an endocrine disrupter and is considered a potential hazard. The formation of zearalenone and other *Fusarium* toxins is favoured by high humidity and temperatures fluctuating between moderate and low values. These conditions often occur in temperate regions during autumn harvest.

Trichothecenes/T-2 Toxin and Deoxynivalenol

The trichothecenes are a family of closely related compounds produced by several *Fusarium* species. There are more than 20 naturally occurring compounds produced by *Fusarium* species which have similar structures, including T-2 toxin, diacetoxyscirpenal, neosolaniol, nivalenol, diacetylnivalenol, deoxynivalenol (DON, vomitoxin), HT-2 toxin, and fusarenon X.

T-2 toxin (**Fig. 6**) has been implicated in a disease known as mouldy corn toxicosis of swine, symptoms of which include refusal to eat (refusal factor), lack of weight gain, digestive disorders and diarrhoea, ultimately leading to death. T-2 toxin, while rarely occurring, is quite toxic to rats, trout and calves. Large doses of T-2 toxin fed to chickens result in severe oedema of the body cavity and haemorrhage of the large intestine, along with neurotoxic effects, oral lesions and, finally, death. T-2 toxin is also thought to be one of the toxins involved in a human disease, alimentary toxic aleukia, that occurred in Russia during World War II and in the early 20th century. T-2 toxin causes several dermal responses in rabbits, rats and other animals, including humans, when applied to the skin. However, it is not thought to be carcinogenic. Although T-2 toxin is very toxic, as mentioned earlier, it occurs more rarely and its natural contamination in foods and feeds appears to be low.

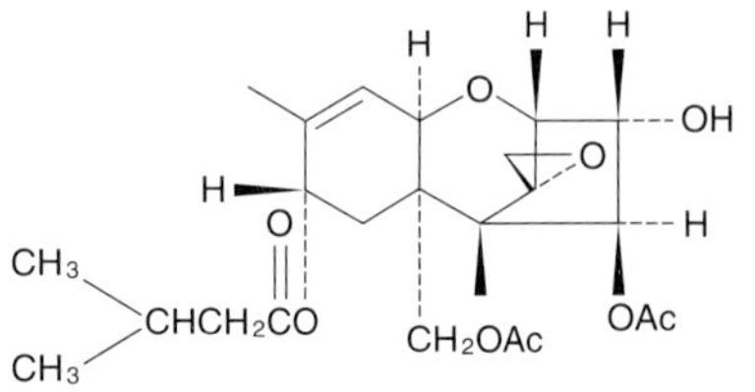

Deoxynivalenol or DON (**Fig. 7**) is the most commonly occurring trichothecene. Deoxynivalenol causes vomiting in animals. Short-term feeding trials with animals suggest low acute toxicity, but other evidence indicates that deoxynivalenol may have teratogenic potential. Unlike T-2 toxin, deoxynivalenol contamination in commodities, especially wheat and corn, is significant in certain years, and has been found in processed-food products. The trichothecenes are also of concern because chronic exposure to low levels of these compounds may cause immunosuppression-like illnesses in humans and animals.

Fumonisins

Fumonisins (**Fig. 8**) are a group of compounds produced primarily by *Fusarium moniliforme* and *F. proliferatum*. In animals, fumonisins have been shown to cause several diseases, such as equine leuko-encephalomalacia, porcine pulmonary oedema and liver cancer in rats. While not as potent a liver carcinogen as aflatoxins, it has been suggested that the lifetime carcinogenic potential of fumonisin B_1 in the rat falls somewhere between carbon tetrachloride and dimethylnitrosamine. In terms of human disease, fumonisins are of concern because they have been linked to oesophageal cancer. There is a significant correlation between the consumption of corn contaminated with *F. moniliforme* and fumonisins and oesophageal cancer in humans in the Transkei region of South Africa. *F. moniliforme* and fumonisins have also been linked to corn consumption and a high incidence of oesophageal cancer in certain provinces of northern China and north-eastern Italy. *F. moniliforme* commonly occurs in corn-growing regions of

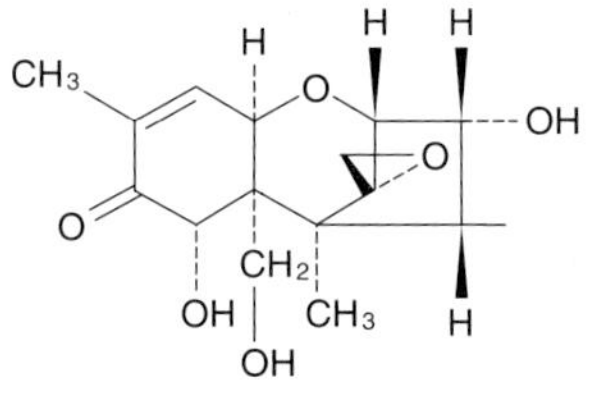

Figure 7 The chemical structure of deoxynivalenol.

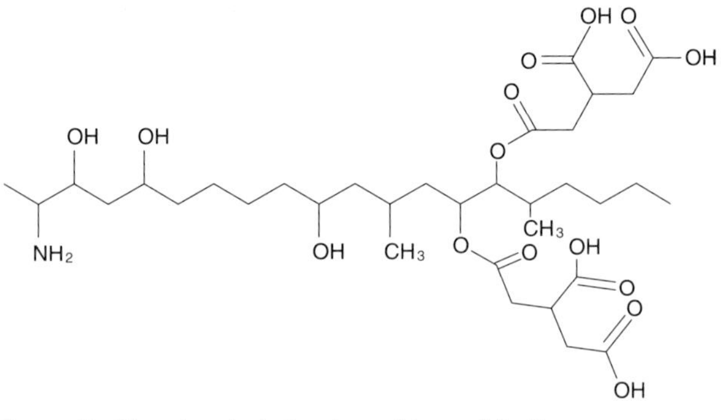

Figure 6 The chemical structure of T-2 toxin.

Figure 8 The chemical structure of fumonisin B_1.

the world and is found virtually everywhere that corn is grown. *F. moniliforme* invades the corn plant and can exist there as an endophyte, and may not be evident as a surface contaminant or infestation. Infection of the corn kernel by *F. moniliforme* may occur by invasion through the silk, fissures in the kernel pericarp and/or through systemic infection of the plant.

Moniliformin

Moniliformin (**Fig. 9**) was first reported to be produced by *F. moniliforme* isolated from corn. While the toxin was apparently named after this organism, the name has turned out to be a misnomer, since subsequent work has shown that most strains of *F. moniliforme* do not produce moniliformin, or are only weak producers. The toxin is produced by *F. proliferatum* and *F. subglutinans*, as well as other *Fusarium* species. Moniliformin is highly toxic when given orally to experimental animals and causes rapid death without severe cellular damage. One-day-old cockerels and mice have been shown to be quite sensitive to the toxin. Clinical lesions observed include acute degenerative lesions in the myocardium and other tissues.

Mycotoxins of Lesser Concern

Sterigmatocystin

Sterigmatocystin is produced by several species of *Aspergillus*, *Penicillium luteum* and a *Bipolaris* species. Chemically, sterigmatocystin resembles the aflatoxins and is thought to be a precursor in the biosynthesis of aflatoxin. The acute toxicity of a sterigmatocystin is low, and the main concern is that it is carcinogenic; it is about one-tenth as potent a carcinogen as aflatoxin B_1. Sterigmatocystin has been detected at low levels in green coffee, mouldy wheat and in the rind of hard Dutch cheese.

Citrinin

Citrinin is a yellow-coloured compound which is produced by several *Penicillium* species as well as *Aspergillus* species. Like ochratoxin A, citrinin causes kidney damage in laboratory animals, similar to swine nephropathy. Citrinin may be involved with ochra-

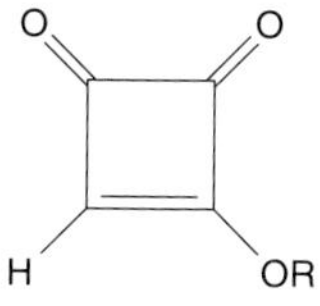

R = Na$^+$ or K$^+$

Figure 9 The chemical structure of moniliformin.

toxin A in cases of swine nephropathy in Denmark. However, the toxicity of citrinin is low compared to ochratoxin, although possible synergistic activity with ochratoxin A cannot be ruled out.

Penicillic Acid

Penicillic acid has low oral toxicity. The concern about penicillic acid in foods is because the compound bears structural similarity to known carcinogens such as patulin, and is carcinogenic to rats when injected subcutaneously. However, the potencies of penicillic acid and patulin as carcinogens are much lower than aflatoxins. When given in lethal doses, penicillic acid caused fatty liver degeneration in quail and liver cell necrosis in mice. Mixtures of penicillic acid with ochratoxin A are synergistic and cause death in mice. Pharmacologically, penicillic acid dilates blood vessels and has antidiuretic effects. However, penicillic acid is similar to patulin in its rapid reaction with sulphydryl-containing compounds in foods to form non-toxic products.

Penicillic acid is produced by strains of *A. ochraceus* and related species and several *Penicillium* species. Some strains of *A. ochraceus* are capable of producing penicillic acid along with ochratoxin A. Penicillic acid has been found in large quantities in high-moisture corn stored at low temperatures.

Mycophenolic Acid, β-Nitropropionic Acid, Tremorgens (Penitrem) and Rubratoxin

Many toxic compounds have been obtained from mould cultures; however, not all cause disease in humans or animals. Other mycotoxins, such as mycophenolic acid, β-nitroprionic acid and tremorgens, have not been studied extensively. Tremorgenic mycotoxins called penitrems have been reported to have caused poisoning of dogs which consumed mouldy cream cheese, mouldy walnuts and other mouldy debris. The toxins caused severe muscle tremors, uncoordinated movements and generalized seizures and weakness in dogs. The disease can also occur in cattle, where it is called staggers.

Tremorgenic mycotoxins can be produced by fungi in the genera *Aspergillus*, *Penicillium*, *Claviceps* and *Acremonium*. Mycophenolic acid and β-nitropropionic acid have been associated with cheeses produced in Europe and are believed to be antibiotic substances of low oral toxicity. Rubratoxin B produced hepatic degeneration, centrilobular necrosis and haemorrhage of the liver and intestine when given to experimental animals. However, natural occurrence of disease caused by this toxin has not been documented, although it is suspected of causing a hepatotoxic, haemorrhagic disease of cattle and pigs fed mouldy corn. Rubratoxin is produced by *Pen-*

icillium rubrum and may exert a synergistic effect with aflatoxins.

Potential Toxicity of *Penicillium roqueforti*

Penicillium roqueforti produces several toxic compounds, including roquefortine, PR toxin and festuclavine. Toxicities of PR toxin and roquefortine are low. Roquefortine is a neurotoxin that reportedly causes convulsive seizures, liver damage and haemorrhage in the digestive tract in mice. However, repeated studies have failed to reproduce these results. Roquefortine has been recovered from blue cheese and was associated with the mould mycelia rather than the non-mouldy areas of the cheese. A toxic factor in the fat of Roquefort cheese that caused severe injury to the liver and other organs of rats has been reported. Atypical wild strains of *P. roqueforti* have been shown to produce patulin and penicillic acid simultaneously, patulin alone, patulin plus citrinin and mycophenolic acid. The significance of the various toxins produced by *P. roqueforti* to public health is not clear.

Patulin, penicillic acid and citrinin have only been observed in wild-type isolates of the organism and not in commercial strains, nor in any cheese produced by commercial strains. As such, the wild isolates represent no greater significance than any other toxigenic isolates of other species. The significance of PR toxin, mycophenolic acid, the roquefortines and related alkaloids to human health is likewise unclear, particularly in view of the limited toxicological information available on these compounds. PR toxin apparently reacts with cheese components and is neutralized. The fact that blue-veined cheeses have been consumed for centuries without apparent ill effect suggests that the hazard to human health is minimal or non-existent.

Alternaria Toxins

Alternaria species, including *A. alternata*, *A. citri*, *A. tenuis*, *A. tenuissima* and others, produce several toxic compounds, alternariols, altenuene, tentoxin and tenuazonic acid. These organisms are common in many foods and grains. *Alternaria* require high moisture conditions and tend to be found in foods that are high in moisture such as grains prior to harvest and fruits and vegetables. The *Alternaria* toxins can be found in grains when drying in the field and harvest are delayed by rain, high humidity or early frost. *Alternaria* are most common in sorghum grain, but the toxins have only been found in heavily weathered sorghum. Post-harvest occurrence of *Alternaria* in

fruits and vegetables is more common because the moisture content remains high after harvest.

Alternaria infection of fruits and vegetables has been observed in apples, oranges, tomatoes and bell peppers. *Alternaria* toxins have been detected in oranges, tomatoes, tomato paste and commercial apple products. *Alternaria* toxins are toxic to *Bacillus mycoides* and HeLa cells. The toxins, however, are only weakly toxic to mice and do not appear to be toxic to rats or chicks when administered as single purified compounds. There is some evidence that mixtures of the compounds may be more toxic. The *Alternaria* toxins are also phytotoxins which affect various plants. A race or strain of *Alternaria*, *A. alternata* f. sp. *lycopersici*, that is pathogenic to tomatoes, produces a toxin known as AAL toxin. This toxin is structurally and toxicologically similar to the fumonisins, and is also a phytotoxin in tomatoes.

Ergot

Ergot is a disease of plants, particularly small grains such as rye and barley and other grasses, which is caused by species of *Claviceps*, in particular *C. purpurea*, *C. paspalli* and *C. fusiformis*. These fungi invade the female sex organs of the host plant and replace the ovary with a mass of fungal tissue known as a sclerotium. The sclerotia, also called ergots, are about the same size and density as the grain kernels and tend to go with the grain when harvested. The sclerotia contain alkaloids that are produced by the fungus. The alkaloids, ergotamine, ergosine and others, are derivatives of lysergic acid, and cause disease in animals and humans. The disease is manifested by a sensation of cold hands and feet followed by an intense burning sensation. As the disease progresses, the extremities may become gangrenous and necrotic, and in animals sometimes the extremities are sloughed. In severe cases death may ensue. Ergotism, also known as St Anthony's fire, reached epidemic proportions during the Middle Ages. At that time the cause of the disease was not known, but it was probably associated with bread made from flours of rye and other grains that were infested with ergot sclerotia. In recent times outbreaks involving humans have occurred in Africa and India. Outbreaks of animal poisonings still occur in areas where rye, barley and other susceptible small grains and grasses are grown.

Effects of Mycotoxins on Animal and Human Health

Mycotoxins can cause a broad range of harmful toxicological effects in animals and would probably cause similar effects in humans if exposure of humans to mycotoxins occurs. Mycotoxins can cause acute poisoning and death in animals and humans if sufficiently high amounts of a toxin are consumed. In animals consumption of subacute amounts of mycotoxins result in economic impacts on productivity, increased disease incidence because of immunosuppressive effects, chronic damage to tissues and organs and reproductive problems.

In acute doses, aflatoxins cause severe liver damage and death in animals. Swine, young calves and poultry are quite susceptible, whereas mature ruminants and chickens are more resistant. Mature sheep seem to be particularly resistant. Subacute and chronic exposures to aflatoxin cause liver damage, decreased milk production, decreased egg production, lack of weight gain and immune suppression. The young of all species are more susceptible, but older individuals will also be affected. Clinical signs of subacute or chronic exposures of animals to aflatoxins include gastrointestinal problems, decreased feed intake and efficiency, reproductive problems, anaemia and jaundice.

Ochratoxin A causes kidney damage in many animals and is most commonly associated with mycotoxic nephropathy of swine. In addition, ochratoxin may be immunosuppressive and is now classified as a carcinogen. In high doses ochratoxin can also cause liver damage, intestinal necrosis and haemorrhage. While swine are very susceptible to ochratoxin, ruminants are more resistant, presumably due to degradation in the rumen.

The trichothecenes include deoxynivalenol. These toxins generally cause gastroenteritis, feed refusal, necrosis and haemorrhage in the digestive tract, destruction of bone marrow and suppression of blood cell formation and suppression of the immune system. Clinically, animals show signs of gastrointestinal problems, vomiting, loss of appetite, poor feed utilization and efficiency, bloody diarrhoea, reproductive problems, abortions and death. Poultry frequently develop mouth lesions and extensive haemorrhaging in the intestines.

Zearalenone causes reproductive problems in animals, especially swine, where it disrupts oestrus cycles and causes vulvovaginitis in females and feminization of males. In high concentrations it can interfere with conception, ovulation, implantation, fetal development and viability of newborn animals. Ruminants are more resistant to zearalenone than monogastric animals, presumably due again to degradation in the rumen.

The range of adverse effects caused by mycotoxins in animals includes embryonic death, inhibition of fetal development, abortions and teratogenicity (deformities) in developing embryos. Nervous system dysfunctions are also observed, including signs such as tremors, weakness of limbs, uncoordinated movement, staggering, sudden muscular collapse and loss of comprehension, due to brain tissue destruction. Other symptoms include seizures, profuse salivation and gangrene of limbs, ears and tails. Several mycotoxins also cause cancers in liver, kidney, urinary tract, digestive tract and lungs.

Human exposure to acute dosages of aflatoxins has resulted in oedema, liver damage and death. Aflatoxins have also been associated, along with hepatitis B virus, with liver cancer in regions where liver cancer is endemic. Ochratoxin A has been associated with the human kidney disease known as Balkan endemic nephropathy. Zearalenone was implicated in an outbreak of precocious pubertal changes in thousands of young children in Puerto Rico. Zearalenone is now considered to be an endocrine disrupter and speculation has suggested that it could play a role in human breast cancer. T-2 toxin is believed to be the cause of alimentary toxic aleukia, a severe human disease that occurred in Russia during World War II as a result of food shortages that forced people to eat overwintered mouldy cereal grains. The disease was manifested by destruction of bone marrow, damage to the haematopoietic system and loss of blood-making capacity, severe haemorrhaging, anaemia and death. Deoxynivalenol is believed to be the cause of a number of gastrointestinal syndromes reported in different parts of the world, including the former Soviet Union, China, Korea, Japan and India. *Fusarium moniliforme* and possible fumonisins have been linked with high rates of oesophogeal cancer in the Transkei region of South Africa, north-eastern Italy and northern China. Moniliformin has been suggested by Chinese scientists as a possible cause of a degenerative heart disease known as Keshan disease occurring in regions of China where corn contaminated with moniliform is eaten. The disease is a human myocardiopathy involving myocardial necrosis. Immunotoxicity in humans may be another effect of mycotoxins, particularly *Fusarium* mycotoxins. T-2 toxin is known to be highly immunosuppressive and there are reports of so-called sick houses where individuals have contracted diseases such as leukaemia and where *Fusarium* species have been found in the houses. Deoxynivalenol can cause elevated immunoglobulin A levels in mice, resulting in kidney damage that is very similar to a human kidney disease known as glomerulonephritis or immunoglobulin A nephropathy. The involvement of mycotoxins in human disease is less clear than their involvement in animal diseases, but there is growing evidence that these toxins are also causative factors in human diseases.

See also: **Alternaria**. **Aspergillus**: Introduction; *Aspergillus flavus*. **Byssochlamys**. **Fusarium**. **Mycotoxins**: Occurrence; Detection and Analysis by Classical Techniques; Immunological Techniques for Detection and Analysis; Toxicology. **Penicillium**: Introduction. **Spoilage Problems**: Problems Caused by Fungi.

Further Reading

Allcroft R, Carnaghan RBA, Sargeant K and O'Kelly J (1961) A toxic factor in Brazilian groundnut meal. *Veterinary Record* 73: 428–429.

Betina V (ed.) (1984) *Mycotoxins: Production, Isolation, Separation and Purification*. New York: Elsevier Scientific.

Betina V (ed.) (1989) *Mycotoxins: Chemical, Biological and Environmental Aspects*. New York: Elsevier Scientific.

Blount WP (1961) Turkey 'X' disease. *Turkeys (Journal of the British Turkey Federation)* 9(2): 52, 55–58, 61–71, 77.

Bullerman LB (1979) Significance of mycotoxins to food safety and human health. *Journal of Food Protection* 42: 65–86.

Bullerman LB (1986) *Mycotoxins and Food Safety. A Scientific Status Summary by the Institute of Food Technologists' Expert Panel on Food Safety and Nutrition*. Chicago, IL: Institute of Food Technologists.

CAST (1989) *Mycotoxins. Economic and Health Risks*. Task force report no. 116. Ames, IA: Council for Agricultural Science and Technology.

Eaton DL and Groopman JD (eds) (1994) *The Toxicology of Aflatoxins, Human Health, Veterinary and Agricultural Significance*. New York: Academic Press.

International Agency for Research on Cancer (1987) Aflatoxins. In: *IARC Monograph on the Evaluation of Carcinogenic Risks to Humans*. Suppl. 7, p. 83. Lyon, France: IARC.

Miller JD and Trenholm HL (eds) (1994) *Mycotoxins in Grain. Compounds other than Aflatoxins*. St Paul, MN: Eagan Press.

Sharma RP and Salunkhe DK (eds) (1991) *Mycotoxins and Phytoalexins*. Boca Raton, FL: CRC Press.

Sinha KK and Bhatnagar D (eds) (1998) *Mycotoxins in Agriculture and Food Safety*. New York: Marcel Dekker.

Smith JE and Henderson RS (eds) (1991) *Mycotoxins and Animal Foods*. Boca Raton, FL: CRC Press.

Occurrence

M de Nijs and **S H W Notermans**, TNO Nutrition and Food Research Institute, Division of Microbiology and Quality Management, AJ Zeist, The Netherlands

Introduction

Moulds are ubiquitous in nature and in domestic environments. As a result, they can easily contaminate food crops during the growing period in the field and after harvest. The negative effects of fungal infection in cereal crops are most often related to loss of grain yields and decrease of processing quality, but attention is now focusing on food safety owing to the possible effects of toxigenic fungi. The toxic metabolites excreted by these fungi can adversely influence the health of consumers of infected crops. The worldwide transportation of agricultural produce and the changing trends in the consumer food market require insight into mycotoxin contamination of foods both before and after processing.

Routes of Contamination

Several fungal genera are associated with mycotoxin contamination of raw and processed foods; these include *Aspergillus, Penicillium, Byssochlamys, Alternaria* and *Fusarium*. Mycotoxin production depends both on fungal characteristics and environmental factors. **Figure 1** describes the possible routes of contamination of foods with mycotoxins.

The fungal genera *Aspergillus* and *Penicillium* rarely infect cereal crops in the temperate climate zone. The carcinogenic mycotoxins, aflatoxins and ochratoxin A, are produced by these fungi during growth. These fungi pose a risk when the harvested cereals are stored under sub-optimal conditions. The most significant factor is rise in water activity. A moist atmosphere during harvest, increased environmental humidity, leaking water in storehouses, or condensation of moisture during storage or transportation cause these fungi to grow. Mycotoxins produced by these fungi are most often concentrated at certain spots and are not always uniformly distributed throughout the batch of produce. This affects sampling procedures. These fungi can infect crops in the field in the tropics and subtropics. Most reports are of aflatoxin contamination of (ground-) nuts and figs. Ochratoxin A contamination of coffee is reported but it is not yet clear if this contamination takes place before harvest or during the fermentation of the beans. Aflatoxins B_1, B_2, G_1 and G_2 and ochratoxin A can frequently be detected in grains and various nuts and figs. Aflatoxin B_1 is metabolized in the rumen of cattle into another toxic derivative, aflatoxin M_1, which is excreted by cows in the milk. Ochratoxin A is stored in the fat layer of pigs after ingestion, through which it enters the human food chain. Humans are exposed to the mycotoxins mentioned in this paragraph through cereals, nuts, pork meat and milk.

Fusarium is generally regarded as the most important fungus invading cereals in the field the world over. The distribution of *Fusarium* mycotoxins is, therefore, assumed to be uniform in the crop from a lot and possibly in all lots from a certain region. Species of

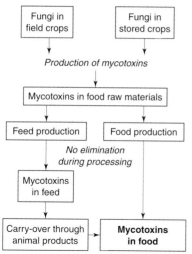

Figure 1 Routes of contamination of foods by mycotoxins.

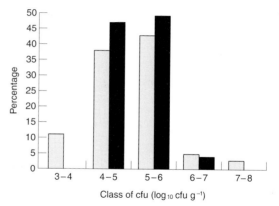

Figure 2 Frequency distribution of fungal infection of cereals, sampled in The Netherlands at harvest in 1991 (shaded bars) and 1993 (solid bars); cfu, colony forming unit.

Table 1 Frequency distribution of different fungal genera isolated from cereals in The Netherlands in 1991 and 1993

Genus	Contaminated samples (%)	
	1991 (n = 65)	1993 (n = 69)
Aspergillus	29	3
Penicillium	65	20
Mucor/Rhizopus	35	30
Cladosporium	89	100
Fusarium	34	83
Other fungi	92	97

n, total number of cereal samples.

this genus are able to secrete toxic secondary metabolites in the plant during the field period of the cereal crops. The most frequently mentioned *Fusarium* mycotoxins are:

- deoxynivalenol (DON), causing growth depletion in pigs and immunosuppression in laboratory animals
- zearalenone (ZEA), causing hyperoestrogenism in female pigs and possible precocious pubertal changes in children
- fumonisin B_1, related to human oesophageal cancer and kidney failure in both animals and humans.

There is no evidence for carry-over of *Fusarium* mycotoxins or altered metabolites from feed through animals to humans. Human exposure is mainly through cereals and foods produced from cereals.

The mycotoxin patulin is produced by several fungi, such as *Byssochlamys*. These fungi and patulin are most often found in fruit and fruit products. The ascospores produced by these species can survive heat treatments up to 95°C. The ascospores can germinate and grow at the low pH present in apple juice, grape juice or strawberry fillings, for example. The patulin produced during the fungal growth causes deterioration in the taste of the product.

Mycotoxin Production

Mycotoxin contamination of food crops depends on the infecting fungal species and on the environmental conditions. Changing agricultural practices can affect the incidence of fungal infection and the fungal genera present in crops over the years. Counts of fungal colony forming units (cfu) present at the moment of

harvest show little fluctuation over the years (**Fig. 2**); however, the genera involved can differ widely between years (**Table 1**). The *Fusarium* species present on the crops is influenced by agricultural practices and climatic conditions (**Table 2**).

The quality of seed plays a key role in infection and spreading of the fungi. Fungal mycelium can be present internally in seed kernels without visible symptoms of infection. Owing to the varying amount and seasonal distribution of rainfall, fungal infection of cereals can fluctuate from year to year and even during the growth season. High rainfall creates conditions favourable for *Fusarium* growth and spread within the crop by splash dispersal of conidia. The increase in the use of fertilizers and the acquired resistance of moulds to fungicides are presumably the major causes for the increase of *Fusarium* infection in wheat. Other environmental factors influencing fungal infection and diversity of the species are temperature, crop rotation, insect epidemics, use of fungicides, susceptibility of plant cultivar, and geographic region. Dry storage conditions after harvest can limit the growth of *Fusarium* mycelium, but mycotoxins may be produced by the mycelium present in and on the grain kernels at the time of harvest. Most of the mycotoxins in the cereals are stable under storage and

Table 2 Frequency distribution of *Fusarium* species isolated from cereals in The Netherlands

| Fusarium *section* | Fusarium *species* | Occurring species (%) | |
		1991 (n = 45)	1993 (n = 56)
Discolor	F. crookwellense/cerealis	0	11
	F. culmorum	40	25
	F. graminearum	2	5
	F. sambucinum	2	11
Sporotrichiella	F. poae	13	39
	F. tricinctum	0	2
Roseum	F. avenaceum	38	7
Gibbosum	F. acuminatum	4	0

n, total number of *Fusarium* isolates identified.

Table 3 Mycotoxins produced by *Fusarium* species isolated from cereals in The Netherlands

Fusarium *species*	Mycotoxins produced
F. acuminatum	Zearalenone, neosolaniol, diacetoxyscirpenol, T-2 toxin, deoxynivalenol, chlamydosporol, moniliformin
F. avenaceum	Zearalenone, nivalenol, deoxynivalenol, fusarin C, moniliformin
F. crookwellense/cerealis	Zearalenone, trichodermin derivatives, sambucinol, diacetoxyscirpenol, T-2 toxin, nivalenol derivatives, butenolide, culmorin, fusarin C, sambucoin
F. culmorum	Zearalenone, calonectrin, sambucinol, diacetoxyscirpenol, deoxynivalenol, butenolide, culmorin, sambucoin
F. graminearum	Zearalenone, sambucinol, diacetoxyscirpenol, T-2 toxin, nivalenol, deoxynivalenol, butenolide, culmorin
F. poae	Diacetoxyscirpenol, T-2 toxin, nivalenol, fusarin C
F. sambucinum	Zearalenone, neosolaniol, diacetoxyscirpenol, T-2 toxin, nivalenol, deoxynivalenol, butenolide
F. tricinctum	Neosolaniol, diacetoxyscirpenol, T-2 tetraol, T-2 toxin

process conditions and may enter final food and feed products.

Fruits are contaminated with ascospore-producing fungi through airborne infections (apples) or by contamination with soil (strawberries). It is known that patulin can be produced during storage.

The genetic characteristics of the fungal species and strains present in the crop, as well as environmental factors such as humidity and temperature, influence mycotoxin production. The ability to produce a certain mycotoxin can vary not only between species (**Table 3**) but also between isolates of the same species. Carbon and nitrogen substrates may influence myco-toxin production. Rice may support trichothecene production by *Fusarium* better than barley and wheat. Rice may support the production of deoxynivalenol better than corn. Both the amount and type of myco-toxins produced are influenced by these parameters. Several authors report increased mycotoxin production by *Fusarium* due to the use of fungicides. The application of fertilizers and the use of susceptible cereal cultivars can increase fungal infection and thus increase the risk of mycotoxin contamination of cereals.

The number of mycotoxins produced by *Fusarium* species has been extensively studied. At least 61 *Fusarium* species have been isolated from food crops, and of these, 35 were reported to produce a total of 137 secondary metabolites under laboratory conditions. All these secondary metabolites should be considered as potentially present in food crops and related foods.

Methods of Analysis

Various methods have been developed to determine the mycotoxin content of food materials. Most methods start with extraction of the mycotoxin from the matrix, followed by clean-up. This clean-up step may include the use of immunoaffinity columns. The cleaned extract is further separated and analysed. The aflatoxins, ochratoxin A, deoxynivalenol and zearalenone can be determined by fluorescence detection after chromatographic separation. Fumonisin B_1 can be detected after derivatization with a fluorescent group. Trichothecene contamination can be determined by gas chromatography with mass spectrometry detection. Thin-layer chromatography is suitable for confirmation of positive samples.

Immunodetection allows quick scanning of samples. The enzyme-linked immunosorbent assay (ELISA) method and dipstick techniques are fast techniques and well developed for food materials. These techniques have, in general, a higher level of detection which need not be a drawback in screening programmes.

Table 4 Aflatoxins in foods other than (ground-) nuts

Country marketed	Product	Year of sampling	Number pos./total[a]	Conc (ng g^{-1}) Range	Average
Bolivia	Rice, wheat, maize	1986–97	1273/2460	7–144	34
Brazil	Maize	1997	142/321	LOD–2440	–
		1997	2546[NR]	–	35
	Rice	–	40/401	–	2
	Wheat	–	45/237	–	2
	Malt	–	30[NR]	–	30
	Sorghum	–	20/59	–	3
	Soya beans	–	24/143	–	1
China	Maize	1997	117/486	5–251	–
	Wheat	1997	9/597	–	–
	Sorghum	1997	1/58	–	–
	Rice	1997	7/747	–	–
	Soya beans	1997	43/388	51–100	–
	Soy sauce	1997	14/308	–	–
	Soya oil	1997	10/379	LOD–251	–
Colombia	Sorghum	1995–96	11/45	1–43	11
	Maize	1996	4/33	4–66	21
	Soya beans	1996	0/25	–	–
	Rice	1996	8/22	1–53	21
Costa Rica	Maize	–	1/49	–	–
Cuba	Maize	–	924/4620	–	–
	Rice	–	54/340	–	–
	Frijoles	–	54/413	–	–
France	Fats, oils, oilseed	–	56[NR]	LOD–25	4.5
Japan	Maize	1995	16/371	–	–
Mexico	Maize	1992–96	265/1710	–	–
USA	Maize	1991	27/28	0–321	73

LOD, level of determination.
[a] Number of positive samples/total number of samples.
[NR] not reported.

Table 5 Fumonisin B$_1$ contamination of maize from various geographic regions

Continent marketed	Year of sampling	Number pos./total[a]	Concentration of fumonisin B^{-1} (ng g^{-1}) Range	Av. pos[b]	Av. all[c]	Median (all) (ng g^{-1})	LOD (ng g^{-1})
Africa	1994	34/37	nd–1910	236	217	105	20
	1989–92	29/31	nd–2630	282	263	80	10
Asia	1992	8/9	nd–1450	794	706	740	50
Europe	1989–92	51/67	nd–2330	382	291	30	10
	1992–93	8/8	100–5310	2899	2899	2920	10
	1994–96	53/54	nd–3353	676	663	615	25
North America	1990–96	70/70	36–2940	703	703	542	20
South America	1991	17/17	1110–6695	2876	2876	2385	50
	1990–91	47/48	nd–18 520	5491	5376	5065	nr
	nr	8/8	85–8791	2131	2131	410	20
Overall		325/349	nd–18 520	1459	1359	420	

[a] Number of positive samples/total number of samples.
[b] Average concentration of positive samples.
[c] Average concentration of all samples, with 0 ng g^{-1} taken for the samples below the limit of determination.
LOD, limit of determination; nd, below limit of determination; nr, not reported.

On-line detection of aflatoxin contamination has been developed for groundnuts. This system is based on the fluorescent aflatoxin excreted by the fungus on the surface of the groundnuts.

Levels of Contamination

Levels of contamination have been determined for several mycotoxins. Mass fractions of aflatoxins, ochratoxin A and fumonisin B$_1$ in several commodities

Table 6 Fumonisin B$_1$ contamination of maize-based foods

Country marketed	Product	Year of sampling	Number pos./total[a]	Concentration (ng g^{-1})	
				Range	Average
Canada	Semolina	1991–94	18/53	100–3500	530
	Corn breakfast cereal	1991–94	11/52	100–320	140
Germany	Corn semolina/polenta	1992–93	4/5	10–30	20
	Corn semolina/polenta	1992–93	6/6	50–1230	590
	Sweetcorn	1993–94	11/40	10–190	–
	Grit meal and semolina	1992–94	38/49	10–130	–
	Grit meal and semolina	1992–94	22/22	180–16 000	–
	Popcorn	1992–93	4/6	10–110	70
	Popcorn	1992–94	13/29	10–160	ND
Italy	Polenta	1992–93	6/6	420–3730	2150
	Polenta	–	20/20	150–3760	1380
	Puffed corn	1992–93	6/6	790–6100	3150
	Sweetcorn	1992–93	5/5	60–790	175
	Cornflakes	1992–93	1/2	0–10	10
Japan	Sweetcorn	1988–92	0/8	–	–
The Netherlands	Polenta	1991–94	2/3	ND–40	–
	Tostada	1991–94	0/1	ND	–
	Canned maize	1991–94	0/6	ND	–
	Maize starch	1991–94	0/5	ND	–
	Maize bread	1991–94	1/2	ND–80	–
	Popped maize	1991–94	3/5	ND–300	–
	Flour mixes	1991–94	2/6	ND–trace	–
	Maize chips	1991–94	3/9	ND–160	–
	Cornflakes	1991–94	1/5	ND–1430	–
South Africa	Cornflakes	1991–94	0/3	–	–
Spain	Corn grits	1993	3/15	0–90	60
	Cornflakes	1993	2/12	0–100	60
	Snacks	1993	2/11	0–200	130
	Toasted corn	1993	0/9	ND	ND
	Cornflour	1993	1/3	0–70	70
Switzerland	Corn grits	1991	34/55	0–790	260
	Cornflakes	1991	1/12	0–55	55
	Cornmeal	1991	2/7	0–110	85
	Sweetcorn	1991	1/7	0–70	70
UK	Polenta	1994–95	16/20	16–2124	529
	Breakfast cereals	1994–95	12/50	11–194	22
	Popcorn (ready made)	1994–95	0/9	ND	ND
	Popcorn	1994–95	6/13	14–784	76
	Corn snacks	1994–95	31/40	11–220	42
	Tortilla, taco, enchilada	1994–95	6/20	10–31	6
	Corn-based thickeners	1994–95	4/21	14–110	15
	Corn-on-the-cob	1994–95	0/20	ND	ND
	Sweetcorn	1994–95	1/22	11	1
USA	Breakfast cereals	1990–91	9/17	10–330	130
	Popcorn	1990–91	2/2	10–60	40
	Tortillas	1990–91	1/3	50–60	60

[a]Number of positive samples/total number of samples.
ND, not detected.

are presented in **Tables 4–6. Table 7** gives examples of the occurrence of several mycotoxins in one commodity. From reports of contamination of many different food commodities it can be concluded that the range of contamination level is generally wide. However, the data may be influenced by the method of determination: for example, if the method's lower limit of determination (LOD) is high the percentage of positive samples may be low. The co-occurrence of several secondary metabolites in foods gives reason for concern as these metabolites may alter the toxic effects of mycotoxins. The large variation in the available data makes the assessment of human exposure difficult. Furthermore, it is difficult to trace the geographic regions that are highly infected because the country of production may not be known.

Table 7   Combinations of mycotoxins in food commodities

Country marketed	Commodity	Year	No.[a]	OA	FB1	FB2	DON	NIV	T-2	HT-2	T2ol	ZEA	AFB1	AFB2	AFG1	AFG2	FA	FCH	MON
Canada	Yellow seedcorn (feed)	1995	1	–	–	–	–	–	–	–	–	–	–	–	–	–	1	–	–
	Corn screenings (feed)	1995	1	–	–	–	–	–	–	–	–	–	–	–	–	–	1	–	–
	Corn screenings (feed)	1995	1	–	–	–	–	–	–	–	–	–	–	–	–	–	1	–	–
	'Bee wings' (feed)	1995	1	–	–	–	–	–	–	–	–	–	–	–	–	–	1	–	–
	Duck feed A	1995	1	–	1	–	1	–	0	–	–	–	1	1	–	–	1	–	–
	Duck feed B	1995	1	–	1	–	1	–	0	–	–	–	1	1	–	–	1	–	–
	Duck feed C	1995	1	–	1	–	1	–	0	–	–	–	1	1	–	–	1	–	–
	Ostrich feed D	1995	1	–	1	–	1	–	0	–	–	–	1	1	–	–	1	–	–
	Ostrich feed E	1995	1	–	1	–	1	–	0	–	–	–	1	1	–	–	1	–	–
	Ostrich feed F	1995	1	–	1	–	1	–	0	–	–	–	1	1	–	–	1	–	–
	Bird feed	1995	1	–	–	–	–	–	–	–	–	–	–	–	–	–	1	–	–
	Rodent chow A	1995	1	–	–	–	–	–	–	–	–	–	–	–	–	–	1	–	–
	Rodent chow B	1995	1	–	–	–	–	–	–	–	–	–	–	–	–	–	1	–	–
	Mixed feed	1995	1	–	–	–	–	–	–	–	–	–	–	–	–	–	1	–	–
Denmark	Feed	1989	12	–	–	–	–	–	–	–	–	–	–	–	–	–	–	12	–
Indonesia	Maize	–	12	–	7	3	–	0	–	–	–	0	10	8	0	0	–	–	–
Korea	Rice	1989	8	–	–	–	0	0	0	–	–	–	–	–	–	–	–	–	–
	Barley	1989	10	–	–	–	3	2	0	–	–	–	–	–	–	–	–	–	–
	Corn	1989	3	–	–	–	1	1	0	–	–	–	–	–	–	–	–	–	–
	Millet	1989	2	–	–	–	0	0	0	–	–	–	–	–	–	–	–	–	–
	Indian Millet	1989	4	–	–	–	1	1	0	–	–	–	–	–	–	–	–	–	–
The Netherlands	Barley	1989	8	2	–	–	0	–	–	–	–	1	–	–	–	–	–	–	–
	Wheat	1989	7	0	–	–	2	–	–	–	–	1	–	–	–	–	–	–	–
	Wheat bran	1989	4	3	–	–	0	–	–	–	–	1	–	–	–	–	–	–	–
	Triticale	1989	1	0	–	–	0	–	–	–	–	0	–	–	–	–	–	–	–
	Oats	1989	1	0	–	–	0	–	–	–	–	0	–	–	–	–	–	–	–
	Rye	1989	1	0	–	–	0	–	–	–	–	0	–	–	–	–	–	–	–
	Maize	1989	6	1	–	–	3	–	–	–	–	2	–	–	–	–	–	–	–
	Maize gluten/feed	1989	6	0	–	–	6	–	–	–	–	6	–	–	–	–	–	–	–
	Maize feed/meal	1989	6	1	–	–	5	–	–	–	–	5	–	–	–	–	–	–	–
	Milo	1989	2	2	–	–	0	–	–	–	–	0	–	–	–	–	–	–	–
	Rice bran	1989	2	0	–	–	0	–	–	–	–	0	–	–	–	–	–	–	–
	Soya bean	1989	2	0	–	–	1	–	–	–	–	0	–	–	–	–	–	–	–
	Peas	1989	6	1	–	–	0	–	–	–	–	0	–	–	–	–	–	–	–
	Vicia faba	1989	6	1	–	–	0	–	–	–	–	0	–	–	–	–	–	–	–
	Palm kernel expeller	1989	3	0	–	–	0	–	–	–	–	0	–	–	–	–	–	–	–
	Coconut expeller	1989	4	3	–	–	1	–	–	–	–	0	–	–	–	–	–	–	–
	Rapeseed (extracted)	1989	3	0	–	–	0	–	–	–	–	0	–	–	–	–	–	–	–
	Cottonseed (extracted)	1989	1	0	–	–	0	–	–	–	–	0	–	–	–	–	–	–	–
	Sunflower seed (extracted)	1989	3	3	–	–	0	–	–	–	–	0	–	–	–	–	–	–	–
	Soya beans (extracted)	1989	7	0	–	–	0	–	–	–	–	1	–	–	–	–	–	–	–
	Tapioca	1989	6	3	–	–	0	–	–	–	–	6	–	–	–	–	–	–	–
	Citrus pulp	1989	4	1	–	–	0	–	–	–	–	0	–	–	–	–	–	–	–
New Zealand	Maize	1986	29	–	–	–	–	–	–	–	–	–	–	–	–	–	–	–	15
	Wheat	1986–89	151	–	–	–	–	–	–	–	–	48	–	–	–	–	–	–	–
	Barley	1986–89	85	–	–	–	–	–	–	–	–	15	–	–	–	–	–	–	–
	Oats	1986–89	29	–	–	–	–	–	–	–	–	10	–	–	–	–	–	–	–
	Maize	1986–89	91	–	–	–	–	–	–	–	–	69	–	–	–	–	–	–	–
	Wheat	1986–89	90	–	–	–	45	36	–	–	–	–	–	–	–	–	–	–	–
	Barley	1986–89	44	–	–	–	11	11	–	–	–	–	–	–	–	–	–	–	–
	Oats	1986–89	3	–	–	–	0	1	–	–	–	–	–	–	–	–	–	–	–
	Maize	1986–89	91	–	–	–	45	59	–	–	–	–	–	–	–	–	–	–	–
Norway	Barley	1988–92	131	–	–	–	90	–	–	–	–	–	–	–	–	–	–	–	–
	Oats	1988–92	209	–	–	–	203	–	–	–	–	–	–	–	–	–	–	–	–
	Wheat	1988–92	531	–	–	–	352	–	–	–	–	–	–	–	–	–	–	–	–
Philippines	Maize	–	50	–	26	6	–	7	–	–	–	2	44	34	2	2	–	–	–
Poland	Barley	1988–92	24	–	–	–	–	–	12	5	2	–	–	–	–	–	–	–	–
Thailand	Maize	–	27	–	19	12	–	0	–	–	–	1	17	11	3	0	–	–	–
UK	Maize	1990	36	–	–	–	–	–	–	–	–	–	–	–	–	–	–	–	36
USA	Maize	1991	49	–	24	–	–	–	–	–	–	–	27	–	–	–	–	–	–
	Wheat	1993	116	–	–	–	23	–	–	–	–	1	7	–	–	–	–	–	–

[a] Number of samples analysed.

OA, ochratoxin A; FB1, fumonisin B_1; FB2, fumonisin B_2; DON, deoxynivalenol; NIV, nivalenol; T-2, T-2 toxin; HT-2, HT-2 toxin; T2ol, T-2 tetraol; ZEA, zearalenone; AFB1, aflatoxin B_1; AFB2, aflatoxin B_2; AFG1, aflatoxin G_1; AFG2, aflatoxin G_2; FA, fusaric acid; FCH, fusarochromanone; MON, moniliformin.

Conclusion

Aflatoxins, ochratoxin A and fumonisins are regarded as the most important mycotoxins owing to their carcinogenic properties. Aflatoxins B_1, B_2, G_1 and G_2 and ochratoxin A can frequently be detected in grains, various nuts and figs. Aflatoxin B_1 is metabolized in ruminants into the toxic derivative aflatoxin M_1 and is excreted in cow's milk. Ochratoxin A is stored in the fat layer of pigs after ingestion. Most reports on fumonisins concern maize but this mycotoxin is also reported in rice and pulses. The oestrogen zearalenone and the immunotoxic mycotoxin deoxynivalenol are routinely detected in cereals. Patulin is mainly found in apples and is regarded as an indicator for quality. Most mycotoxins are chemically stable during processing and can thus be detected in many foods. Many of the mycotoxins discussed above occur together in food crops and consequently in food products and animal feed. This will have consequences for toxic effects.

See also: **Alternaria**. **Aspergillus**: Introduction. **Biochemical and Modern Identification Techniques**: Food Spoilage Flora (i.e. Yeasts and Moulds). **Byssochlamys**. **Fungi**: Overview of Classification of the Fungi. **Fusarium**. **Genetics of Microorganisms**: Fungi. **Mycotoxins**: Classification; Detection and Analysis by Classical Techniques; Immunological Techniques for Detection and Analysis; Toxicology. **Penicillium**: Introduction. **Spoilage Problems**: Problems caused by Fungi.

Further Reading

Bhat RV, Shetty PH, Amruth RP and Sudershan RV (1997) A foodborne disease outbreak due to the consumption of moldy sorghum and maize containing fumonisin mycotoxins. *Clin. Toxicol.* 35: 249–255.

Lacey J (1990) Mycotoxins in UK cereals and their control. *Asp. Appl. Biol. Cereal Qual.* 25: 395–405.

Langseth W, Stenwig H, Sogn L and Mo E (1993) Growth of moulds and production of mycotoxins in wheat during drying and storage. *Acta Agric. Scand. B: Soil Plant Sci.* 43: 32–37.

Lauren DR, Agnew MP, Smith WA and Sayer ST (1991) A survey of the natural occurrence of *Fusarium* mycotoxins in cereals grown in New Zealand in 1986–1989. *Food Addit. Contam.* 8: 599–605.

de Nijs M, van Egmond HP, Nauta M, Rombouts FM and Notermans SHW (1998) Human exposure assessment to fumonisin B_1. *J. Food Protect.* 61: 879–884.

Patel S, Hazel CM, Winterton AGM and Gleadl AE (1997) Surveillance of fumonisins in UK maize-based foods and other cereals. *Food Addit. Contam.* 14: 187–191.

Sydenham EW, Thiel PG, Marasas WFP, Shephard GS, van Schalkwyk DJ and Koch KR (1990) Natural occurrence of some *Fusarium* mycotoxins in corn from low and high esophageal cancer prevalence areas of the Transkei, Southern Africa. *J. Agric. Food Chem.* 38: 1900–1903.

Yamashita A, Yoshizawa T, Aiura Y et al. (1995) *Fusarium* mycotoxins (Fumonisins, nivalenol, and zearalenone) and aflatoxins in corn from Southeast Asia. *Biosci. Biotech. Biochem.* 59: 1804–1807.

Zhi-gang W, Jian-nan F and Zhe T (1993) Human toxicosis caused by moldy rice contaminated with *Fusarium* and T-2 toxin. *Biomed. Environment. Sci.* 6: 65–70.

Detection and Analysis by Classical Techniques

Imad Ali Ahmed, Central Food Control Laboratory, Ajman Municipality, UAE

Introduction

An awareness of the hazards posed by mycotoxins to both animals and humans drew attention, quite early, to the crucial importance of assay methods in any attempt to control the problem. Early methods of detection depended on biological assays which used, for example, one day-old ducklings, chick embryos, microorganisms or tissue cultures. However, biological assays are now used less frequently, and physico-chemical assays, which are characterized by being quicker, more precise and less tedious, have come into favour.

Certain physico-chemical characteristics of mycotoxins that were observed by early workers in this field, namely that they are more soluble in chloroform than in hydrophilic solvents and fluoresce brightly when exposed to long-wave UV light, have remained the basis for all analytical procedures. However, the widespread distribution of mycotoxins among very diverse agricultural products, of different components and ingredients has tended to complicate the analytical procedures which need to be fully validated by collaborative studies before being officially adopted. This task has been carried out by specialized organizations, including for example, the Association of Official Analytical Chemists (AOAC), the American Oil Chemists' Society (AOCS), the American Association of Cereal Chemists (AACS), the International Union of Pure and Applied Chemistry (IUPAC) and the European Community (EC).

General Description of Chemical Assays

All analytical procedures consist of the same basic steps which include: (a) sampling and sample preparation, (b) extraction, (c) purification and clean-up,

(d) separation and quantification and (e) confirmation of identity.

Sampling and Sample Preparation

Due to the ubiquitous nature of toxigenic fungi, mycotoxin contamination of a consignment of food can occur at numerous sites of invasion, so giving rise to small, local batches of high concentration. Even among these batches, contamination may not be uniform, and this situation often contributes great variability to the analytical results. Generally, the distribution of any mycotoxin in a food product is affected by the type of processing to which the food has been subjected after contamination. In checking all the factors associated with variability of mycotoxin results, it was found that sampling contributes the greatest single source of error. Consequently, much effort has been expended to minimize the effect of sampling on the results of analyses for mycotoxins.

Variations in mycotoxin concentration among samples of different agricultural crops could be minimized by size reduction and mixing to achieve effective distribution of any contaminated portion. Many methods recommend that the lot sample should be divided and subdivided, using any random dividing procedure until its size is close to the weight of the sample required for analysis. More than twelve sample preparation devices have been used to provide more representative samples e.g. Hobart Vertical Mixer (HVM), Waring blender, food cutter, Wiley mill, hammer mill, disk mill and meat chopper. Also, the use of water slurries have been used for many food products as an effective approach to minimizing the variability associated with the sampling steps. It is true, therefore, that the larger the particle size of the food product the greater the problem associates with sampling. Liquids and fine powdered products need only immediate mixing before sampling. Sampling procedures used by the US Food and Drug Administration (FDA) for many products recommend sample sizes ranging from 10 lb (4.54 kg) for tree nuts to 60 lb (27.3 kg) for cottonseeds.

Extraction

Due to differences in the physico-chemical properties of mycotoxins and the diverse nature of the products affected, there is no single extraction method that can be used for all mycotoxins and commodities. The selection of the method of extraction, therefore, depends on the nature of the sample and the interfering materials that might be co-extracted with the toxin under question.

Most mycotoxins are extracted using chloroform or methanol mixed with water and their combinations. A few mycotoxins, such as patulin, are not readily soluble in chloroform, and other systems containing ethyl acetate, acetone or acetonitrile in water are used (**Table 1**).

As early as 1961, workers succeeded in isolating the toxic agent (aflatoxin) responsible for turkey X-disease by means of exhaustive extraction methods. These early methods involve a laborious Soxhlet procedure with methanol and chloroform as the most frequently used solvents. Besides being time-consuming, the main disadvantage of this approach is the large amount of lipids, pigments and carbohydrates co-extracted with the aflatoxins. Also, the prolonged contact with warm methanol leads to a loss of aflatoxins and the formation of new spots of methanol adduct.

Progress in reducing the analysis time and achieving simplicity came with the recognition that aqueous solvents might allow improved penetration of the substrate. Quicker extraction procedures, i.e. 1–3 min using high-shear blenders or 30 min using mechanical shakers, were employed; both techniques are satisfactory with many standard methods, whereas some methods suggest the use of one over the other. In fact, chloroform is considered to be the best solvent for mycotoxins, but in a Waring blender, it causes considerable splash and expulsion of the extract mixture from the blender jar as it fails to wet the substrate. This problem can be solved by the use of just enough water (a poor solvent for aflatoxin) to wet the substrate and improve the penetration of the chloroform. The chloroform–water system was later adopted by the AOAC in the widely used Contaminants Branch (CB) method for aflatoxin detection. However, due to the carcinogenicity of chloroform, methylene chloride has been recommended in its place.

Table 1 Solvents used for extraction of different mycotoxins

Mycotoxin	Extraction solvent(s)
Aflatoxins	Chloroform : water (250 : 100, v/v); methanol : water (55 : 45, v/v); acetone : water (85 : 15, v/v)
Ochratoxins	Chlorform : phosphoric acid (10 : 1, v/v); ethyl acetate : phosphoric acid (10 : 1, v/v)
Sterigmatocystin	Acetonitrile : 4% potassium chloride (9 : 1, v/v)
Trichothecenes	Acetonitrile (twice). Methanol and acetonitrile and their mixture with water
Rubratoxin	Methyl acetate, acetonitrile, ethanol acetone
Patulin	Ethyl acetate (150 ml)
Zearalenone	Ethyl acetate; acetonitrile : 4% KCl (9 : 1, v/v); chloroform : water (10 : 1, v/v)
Citrinin	Acetonitrile : 8% KCl : 10% NaEDTA : 20% H_2SO_4 (18 : 1 : 1 : 0.2, v/v/v/v)
Penicillic acid	Chloroform : methanol (9 : 1, v/v); methylene : chloride methanol (1 : 1, v/v); acetonitrile : 4% KCl (9 : 1, v/v)

Aqueous hydrophilic solvents, mainly methanol, ethyl acetate, acetone and acetonitrile, are used for mycotoxin extraction in connection with the quick extraction methods. A methanol–water (55 : 45 v/v) solvent system with high-speed blending which was used in the early 1960s, formed the basis for a method adopted by the AOAC for aflatoxin extraction and named the Best Food (BF) method. A higher ratio of methanol to water (80 : 20 v/v) was also used later in an attempt to improve extraction, but a methanol concentration of 55% proved to be effective without lowering the amount of aflatoxin extracted.

Aqueous acetone was used early for aflatoxin extraction using a high-speed blender or mechanical shaker. Equilibrium extraction with acetone : water (85 : 15 v/v) was achieved, and this approach formed the basis for the AOAC method known as the Romer Minicolumn (RMC) method. One of the problems associated with acetone is that it extracts more pigments and interfering materials than chloroform, and hence the extract requires rigorous cleaning and purification. Alternatively, aqueous acetonitrile can be used as an effective solvent, as it is sufficiently selective to simplify the required clean-up procedures. Ethyl acetate is used for extraction of some mycotoxins like ochratoxin, patulin and rubratoxin.

With all the above solvent systems, hydrophilic or chlorinated, the ratio of organic solvent to water has to be adjusted to achieve the best balance between the amounts of toxin and the interfering substances that are extracted. The volume (in millilitres) of extracting solvents should be at least three times the weight (in grams) of the sample, and most of the officially adopted methods use a volume of organic solvent equivalent to five times the weight of the sample. Usually, after filtration, a representative sample of the filtrate is taken, and the result of the analysis adjusted to compensate for the toxin retained in the substrate. In this way, a short analysis time is achieved and there are less interfering substances to be dealt with during the subsequent steps of analysis.

For some food products, additives have been used to facilitate mycotoxin extraction, e.g. saturation of the solvent with sodium or potassium chloride to achieve an ionic effect during the extraction of some mycotoxins. Citric acid, ammonium sulphate or orthophosphoric acid are added to compete with aflatoxin for adsorption sites on proteins and nucleic acids during the extraction of liver.

Purification and Clean-up

Almost all organic solvents used for mycotoxin extraction are good solvents for lipids, plant pigments and other food components which, unless otherwise removed, will interfere with mycotoxin detection and quantification. Lipids are the major food components co-extracted with mycotoxins, but fortunately, they are the most easily removed. Thus, as mycotoxins are immiscible in non-polar solvents like petroleum ether, hexane and iso-octane, these solvents can be used to remove the lipid contaminants; the final choice depends on the solvent used for the original extraction.

In some of the early methods, lipids were removed before aflatoxin extraction by defatting the sample in a Soxhlet extractor. Alternatively, the lipids were removed after extraction by first developing the 'spotted' TLC plates with petroleum ether, which moves the lipids with the solvent front leaving the toxin at the origin. Current methods employ extraction of the mycotoxins and lipids together, followed by a number of clean-up steps.

The first method adopted by the AOAC for cleaning a chloroform extract used a Celite column; later, activated silica gel was found to be more effective. Mycotoxins, which are normally adsorbed onto the silica gel layer, are washed, successively, with hexane and petroleum ether, and then eluted with a chloroform–methanol mixture. For the best removal of interfering substances, the elution power and polarity of the solvent should be adjusted just short of the conditions necessary to elute the toxins. In an attempt to reduce analysis time, a system was developed in which a small portion of the sample extract was transferred to a small glass column filled with silica gel. However, this column system was found to be inadequate for cleaning, and a multilayer column of Florisil, sand, alumina and silica gel is now used to obtain cleaner extracts.

When hydrophilic solvents are used for extraction, the liquid–liquid partition between the extract and the petroleum solvent, such as hexane and iso-octane, has been effectively used for defatting and cleaning. Separating funnels have been mainly used for this purpose, and their operation has proved to be simple and fast. Some methods involve the addition of the fat solvent and extraction solvent(s) together, so that extraction and defatting are carried out in one step.

Non-lipid interfering substances, mainly pigments, can be precipitated by heavy metals, e.g. lead in the form of lead acetate, copper as cupric carbonate, iron as freshly prepared ferric hydroxide or chloride and zinc as zinc acetate. Silver nitrate has been used to remove the natural component of cocoa, theobromine, in the method of analysis for aflatoxin described by the AOAC (1995 – 16th edn. Washington, DC).

Generally, it has been concluded that the transfer of mycotoxins to chloroform, regardless of the primary extraction system, helps in the clean-up stage, and also facilitates the final concentration by evaporation.

There has been widespread use of cartridges pre-packed with silica gel for the purification and clean-up steps, and many different designs of support materials are commercially available. Bonded-phase columns for sample clean-up have been used to obtain extracts clean enough for HPLC analysis. Recently, disposable, monoclonal antibody affinity columns have gained acceptance for the complete and rapid removal of interfering materials from biologically diverse samples.

Isolation and Quantification

Mycotoxins can be separated by means of chromatography depending on their inherent fluorescence properties under long-wave ultraviolet (UV) illumination, and this provides the basis for their detection and quantification.

Paper chromatography was initially employed for aflatoxin isolation, but the technique suffers from poor resolution, limited quantification and difficulty in confirmation of identity. As the discovery of aflatoxin coincided with the emergence of thin layer chromatography (TLC) as a new tool in chromatographic separation, investigators turned to TLC as a means of obtaining better resolution. Glass plates covered with alumina were used initially, but the alumina was then replaced by silica gel which has been found to give more reproducible R_f values for many mycotoxins The use of TLC plates, coated with silica gels of different properties, developed rapidly, and it has become the most widely used technique for detection, quantification and confirmation of identity. In addition, it has been shown to be very sensitive, and can detect levels as low as 0.5 ng of the toxin per TLC spot.

The successful application of TLC for mycotoxin analysis requires a delicate balance between the adsorbent properties of the gel and the elution properties of the developing solvents. The initial system employed chloroform, which still remains the major component of most of the developing systems, to which hydrophilic solvents are added to impart polarity. Originally, methanol was added to chloroform but, later, acetone was substituted for methanol as it was found to be more sensitive and gave better resolution. Numerous combinations of adsorbent and mobile phases have been employed with different food commodities, the choice depending on the interfering substances co-extracted with mycotoxins (**Table 2**). The use of unequilibrated and unlined developing chambers is favoured for better separation of closely related mycotoxins. Two-dimensional TLC is important for certain products, to overcome the problem of interfering substances which are otherwise difficult to

Table 2 Developing solvents used for separation of mycotoxins by TLC

Mycotoxin	Developing solvents
Aflatoxins	Chloroform : acetone (9 : 1, v/v); benzene : methanol : acetic acid (90 : 5 : 5, v/v/v); benzene : ethanol : water (46 : 35 : 19, v/v/v); toluene : ethyl acetate : acetone : acetic acid (50 : 35 : 15 : 2, v/v/v/v) and many more
Ochratoxins	Benzene : acetic acid (3 : 1 or 4 : 1, v/v); toluene : ethyl acetate : acetic acid (5 : 4 : 1, v/v/v); benzene : methanol : acetic acid (18 : 1 : 1, v/v/v); hexane : acetone : acetic acid (18 : 1 : 2, v/v/v)
Sterigmatocystin	Benzene : methanol : acetic acid (90 : 5 : 5, v/v/v).
Trichothecenes	Chloroform : methanol (98 : 2 or 97 : 3, v/v); benzene : tetrahydrofuran (85 : 15, v/v)
Rubratoxin	Chloroform : methanol : glacial acetic acid : water (80 : 20 : 1 : 1, v/v/v/v)
Patulin	Toluene : ethyl acetate : 90% formic acid (5 : 4 : 1, v/v/v); chloroform : acetone (9 : 1, v/v); chloroform : methanol (95 : 5, v/v); benzene : methanol : acetic acid (24 : 2 : 1, v/v/v); pentane : ethyl acetate (96 : 4, v/v)
Zearalenone	Alcohol : chloroform (5 : 95 or 3.5 : 96.5, v/v); benzene : acetic acid (95 : 5 or 95 : 10, v/v)
Citrinin	Benzene : acetic acid (95 : 5, v/v); toluene : acetic acid : formic acid : dimethyl sulphoxide (70 : 20 : 7 : 3, v/v/v/v); chloroform : methanol : hexane (64 : 1 : 35, v/v/v)
Penicillic acid	Chloroform : methanol : water : formic acid (250 : 24 : 25 : 1, v/v/v/v); chloroform : 90% ethyl acetate : formic acid (60 : 40 : 10 or 50 : 40 : 10, v/v/v); chloroform : methanol (95 : 5 or 90 : 10, v/v)

separate. For example, it is used successfully in the analysis of animal tissues, dairy products, spices and human fluids.

Generally, the most significant advantage of the TLC technique is the low cost of each analysis. However, its major disadvantage is the quantification step. Visual detection of the developed spots on a TLC plate is considered as a 'semi-quantitative' method and suffers from a high level of variation; a 20% change in the amount of the toxin in the spot is required to provide a difference to the eye. A possible error of ± 30–50% can happen when a standard and an unknown are matched, whereas ± 15–20% can arise when the unknown is spotted between two standards. Measurement of the fluorescence intensity of the TLC spots using fluorodensitometers has proved to be accurate, precise and superior to visual estimation, but the system still suffers from the intrinsic weakness(es) associated with the TLC procedure.

Due to the poor fluorescence intensity of some mycotoxins, it is essential to spray the TLC plates to detect the toxin spots. Methanolic boron trifluoride is used for detection of ochratoxin and aluminium chloride for sterigmatocystin. In order to detect penicillic acid under long-wave UV, the TLC plate should be sprayed with p-anisaldehyde solution.

In order to improve the quantitative accuracy and precision of mycotoxin analysis, a recent trend has been towards the use of HPLC to replace the existing TLC-based methods. With HPLC, reproducible results can be achieved with minimum handling of the toxin and better protection of the extract from light. Also the attachment of sophisticated data retrieval equipment, automatic sample injectors and an electronic integrator makes automation possible. Considerable effort has been targeted on the achievement of better separation and quantification through the optimization of the column and detector performance.

Silica gel-packed columns have been used in a normal-phase (NP) system of water-saturated chloroform or dichloromethane to separate the four major aflatoxins. This system is characterized by an extremely long retention time (30 min), which can be reduced by the addition of a small amount of methanol (0.3%) to the mobile phase. Generally, reproducibility is poor with the NP system due to the difficulty of maintaining the saturated condition of the solvents which is temperature dependent. The chlorinated solvents, associated with NP, also markedly quench the fluorescence of aflatoxins B_1 and B_2 during UV detection. In order to continue with the use of the NP system, some attempts were made to overcome this quenching problem; thus, a silica gel-packed flow cell with fluorometric determination has been used to improve the sensitivity of detection, and a toluene-based mobile phase, which does not quench the fluorescence of the B-group aflatoxins, was also introduced. This latter system did not gain popularity due to the corrosive action of the formic acid which was part of the mobile phase.

Reverse-phase (RP) columns were developed to achieve excellent separation, practical retention times and improved reproducibility. Methanol and acetonitrile formed the basis for the RP mobile phases, together with water. Continuous adjustment of the solvent ratios in the mobile phase is important to accommodate changes in column properties. The RP system is favoured over the NP system due to the economy and safety of the solvents used, and the introduction of microparticulate RP columns provided excellent baseline resolution. The RP system allows the separation of many mycotoxins in one HPLC run. However, a major problem associated with the RP system is the weak fluorescence of some mycotoxins in the solvent used. This lack of intensity

can be solved either by pre-column derivatization using trifluoroacetic acid (TFA), or a post-column derivatization (PCD) reaction with iodine.

The TFA-catalysed addition of water is to give the more fluorescent derivatives, and the TFA solution can be added to the final concentrated extract to induce the reaction just before injection. This rapid procedure is harmless to the column, and the TFA-treated solution is stable for more than one week when refrigerated.

After being collaboratively studied in many laboratories using artificially contaminated samples, the TFA method of derivatization has been widely adopted for the determination of aflatoxin, trichothecenes, sterigmatocystin and penicillic acid. However, pre-column derivatization is found to suffer from reduced accuracy and precision due to the additional handling steps which lead to losses of aflatoxin.

As an alternative to the TFA reaction, another method which uses iodine at elevated temperature (100°C) was introduced; this approach greatly improves the intensity of fluorescence of aflatoxins B_1 and G_1. Although this method succeeds in freeing the aflatoxins from potential interfering compounds, the method is characterized by the relatively long retention times of the eluted toxins. This problem was solved successfully by inducing the iodine reaction after separation in the column, so that elution is not affected, but before the detector; the fluorescence intensity of aflatoxins B_1 and G_1 increases by a factor of 50. This derivatization procedure formed the basis of the PCD (post-column derivation) procedure which has been optimized and used in many laboratories. Furthermore, automation becomes possible with the use of the PCD method of analysis, and it can also be applied to other mycotoxins.

Quantification of mycotoxins by HPLC analysis, which is done through electronic measurements of integrated detector responses, has improved the precision and reproducibility of testing, and mycotoxins can be quantified either by UV or fluorescence detectors.

UV detectors with variable wavelengths allow: (a) the selection of the most suitable wavelength for mycotoxin determination; (b) maintenance of a steady baseline; and (c) tolerance of a variety of injection solvents. Unfortunately, the UV detection suffers from sensitivity to the interfering substances normally present in an extract, and a rigorous clean-up is necessary.

The fluorescence detection system has proved to be more sensitive, more selective and less liable to background interference than UV. This tolerance means that the normal clean-up procedures adopted for TLC procedures are also applicable to HPLC

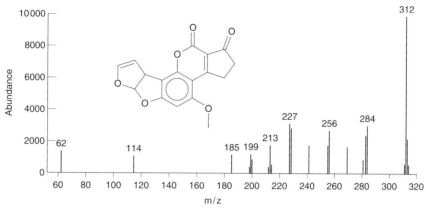

Figure 1 Mass spectrum of aflatoxin B_1.

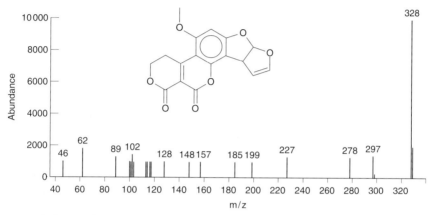

Figure 2 Mass spectrum of aflatoxin G_1.

determinations. However, fluorescence detection suffers, as mentioned earlier, from the weak fluorescence property of certain mycotoxins, e.g. aflatoxins B_1 and G_1 in the solvents used in the RP system. Eventually, with the help of one or other of the derivatization procedures mentioned earlier (TFA and PCD), fluorescence detection exhibited sufficient sensitivity to achieve measurement of picogram amounts

In certain circumstances, fluorescence of aflatoxin-related compounds can prove useful, and the bright green yellow fluorescence (BGYF) test, originally designated for cottonseeds and corn but now extended to other products, is based on a reaction between kojic acid and the plant enzyme, peroxidase. Kojic acid, a secondary metabolite usually excreted with aflatoxins, produces fluorescent compounds when it reacts naturally with peroxidase. These fluorescent compounds can be detected under long UV light, and their presence indicates the possible presence of aflatoxins. The method is considered fast and simple, but it is relatively inaccurate and the results do not always correlate well with quantitative analyses.

Gas chromatography (GC) and mass spectrometry (MS) have been successfully used also for quantification of many mycotoxins after the formation of

volatile derivatives. Trichothecenes can be quantified by GC or GC-MS after the formation of trimethyl silyl (TMS) derivatives. The formation of TFA derivatives of mycotoxins helps in the detection and quantification of penicillic acid using GC with an electron capture detector.

Rapid screening methods based on immunological techniques, are commercially available in many forms, including, for example, microtitre plates (such as ELISA), the card test for rapid visual detection, or the radioimmunoassay (RIA) method.

Confirmation of Identity

Unfortunately, some interfering substances, which can fluoresce or absorb UV light and show the same retention time as mycotoxins, can lead to false 'positives'. Consequently, it has been widely recommended that, in any sort of chemical analysis, the identity of the suspected mycotoxins should be confirmed.

Early methods used bioassays for confirmatory purposes, but these methods are rarely used today because they are time-consuming and some of the test factors are difficult to control. Running the unknown samples alongside internal standards under different chromatographic conditions has been used to confirm the presence of mycotoxins when using the TLC method. Alternatively, spraying the TLC plates with certain chemicals could be used for the identification and confirmation of mycotoxins. For example, spraying aflatoxins with 25% methanolic sulphuric acid can help to identify interfering substances which give yellow fluorescence under UV light; no change in aflatoxin fluorescence takes place. Zearalenone identity can be confirmed by spraying the TLC plates with silver chloride solution and then heating the plate for 5 min at 130°C before examination under long UV light.

The formation of derivatives can also be used for the confirmation of mycotoxins. The production of hemiacetals using TFA can be performed directly on TLC plates or in the HPLC injection vials for the confirmation of aflatoxins, sterigmatocystin, penicillic acid and trichothecenes. Methyl esters can be formed using methanolic boron trifluoride and 3-methyl-2-benzothiazolinone hydrazone hydrochloride solution to confirm the identity of ochratoxins and patulin, respectively. Direct acetylation on TLC plates is also used as a chemical confirmation method for aflatoxins B_1 and G_1. The change in fluorescence intensity, due to the iodine reaction in PCD, can be used as a confirmatory tool by sequential injections into the HPLC of material with and without PCD.

The use of MS spectra for confirming the chemical identity of mycotoxins is now widely used as a quick and accurate tool (**Figs 1** and **2**).

Lastly, it is important to note that due to the toxicological nature of mycotoxins, a separate room with restricted entry should be assigned for the analytical work. Staff involved in mycotoxin work should be technically prepared and equipped with enough knowledge of the safety and precaution measures required for such work. Procedures for the decontamination of the laboratory wastes, glassware and working area should be followed carefully and documented in the laboratory records.

See also: **Aspergillus**: *Aspergillus flavus*. **Mycotoxins**: Classification; Occurrence; Immunological Techniques for Detection and Analysis; Toxicology.

Further Reading

Betina V (1985) Thin layer chromatography of mycotoxins. *Journal of Chromatography* 334: 211–276.

Betina V (ed.) (1984) *Mycotoxins Production, Isolation, Separation and Purification*. Amsterdam: Elsevier Science.

Castegnaro M, Hunt SC, Sansone EB et al (1980) *Laboratory Decontamination and Destruction of Aflatoxin B_1, B_2, G_1 and G_2 in Laboratory Wastes*, IARC Publications no. 37 Lyon, France: International Agency for Research on Cancer.

Cole RJ (ed.) (1986) *Modern Methods in the Analysis and Structure Elucidation of Mycotoxins*. New York: Academic Press.

Egan H (ed.) (1982) *Environmental Carcinogens Selected Methods of Analysis*. Vol. 5. IRAC Scientific Publication No. 44. Lyon, France: International Agency for Research on Cancer.

Ellis WO, Smith JP and Simpson BK (1991) Aflatoxin in food: occurrence, biosynthesis, effect of organism, detection and methods of control. *Critical Reviews in Food Science and Nutrition* 30: 403–439.

Heathcote JG and Hibbert JR (1978) *Development in Food Science 1. Aflatoxin: Chemical and Biological Aspects*. Amsterdam, Oxford, New York: Elsevier Scientific.

Macrae R, Robinson RK and Sadler M (eds) (1992) Mycotoxins. In: *Encyclopedia of Food Science, Food Technology and Nutrition*. P. 3196. New York: Academic Press.

Scott PM (1995) Natural toxins. In: *Official Methods of Analysis of the Association of Official Analytical Chemists*, 16th edn. Washington, DC: AOAC.

Immunological Techniques for Detection and Analysis

Arun Sharma, Food Technology Division, Bhabha Atomic Research Centre, Mumbai, India

M R A Pillai, Isotope Division, Bhabha Atomic Research Centre, Mumbai, India

Immunological methods of analysis, commonly referred to as immunoassays, which were developed in the early 1960s by Solomon Berson and Rasalyn Yalow, are now routinely used in the management of food quality. Due to their high specificity, sensitivity, speed and ease of application, these methods are used for the analysis of pathogens and their toxins, contaminants, adulterants and even some of the constituents of foods. Analysis of mycotoxins and their metabolites in food, feed, serum and milk is one of the important areas, which has greatly benefited from the development of immunoassays.

Mycotoxins

Certain groups of fungi are known to produce families of closely related low-molecular-weight compounds during their growth on agricultural commodities and other foods. These metabolic products have not been assigned any function in the growth and physiology of these organisms. Moreover, in the laboratory these substances have been found to appear in the medium after a lag towards the late growth phase and continue to be produced even after the cessation of active growth of the organism. Therefore, these compounds represent a class of substances termed secondary metabolites. Many of these compounds have been found to have adverse effects on human health and are therefore commonly referred to as mycotoxins.

Mycotoxins can induce both acute and chronic effects on human and animal health. Many of these diseases, collectively known as mycotoxicoses, may be responsible for some of the human diseases commonly attributed to unknown origin. As many of these diseases surface much after the consumption of contaminated food, mycotoxin-contaminated food remains as an unsuspected health hazard. With the advent of newer methods of toxicological testing and improved isolation and detection techniques it is now possible to learn more about the role of mycotoxins in human health. With newer mycotoxins being added to the list of dangerous substances, it has become extremely important to regulate mycotoxin contamination in food for human and animal consumption. Many countries around the world have invoked stringent regulations to check mycotoxin contamination and have imposed extremely low tolerance limits on the presence of mycotoxins in food.

Important Mycotoxins in Food

There have been a number of mycotoxins identified and isolated from food. **Table 1** gives a list of some of the widely occurring mycotoxins. The three major genera of mycotoxin-producing fungi are *Aspergillus*, *Fusarium* and *Penicillium*. Aflatoxin B_1 is the most commonly encountered mycotoxin and this has been identified as one of the most potent naturally occurring carcinogens known. The International Agency for Research on Cancer rated aflatoxin in Category 1 (**Table 2**) which indicates that there is sufficient evidence of its carcinogenicity in humans and animals. Aflatoxin B_1 and B_2 are hydroxylated in ruminants to form aflatoxin M_1 and M_2, respectively, and appear in milk of the animals raised on aflatoxin-contaminated diet. Aflatoxin M_1 is also identified to be a potent carcinogen as aflatoxin B_1. Food commodities most commonly affected by aflatoxin contamination include peanut, corn, wheat, rice, cottonseed, copra, nuts, milk, cheese and many other commonly consumed food articles. Of all the mycotoxins identified, only aflatoxin contamination is legally regulated in foods at present. Tolerance levels for aflatoxin

Table 1 Important mycotoxins identified in infested food

Mycotoxin	Fungi	Toxicity
Aflatoxin	*Aspergillos flavus*	Carcinogenic
	A. parasiticus	Mutagenic
		Teratogenic
		Toxic
Trichothecine	*Fusarium sporotrichoides*	Skin irritant
	F. culmorum	Emetic
	Trichothecium roseum	
Ochratoxin	*A. ochraceous*	Nephrotoxic
Fumonisin	*Fusarium moniliformae*	Oesophagal Cancer
Citrinin	*Penicillium viridicatum*	Nephrotoxin
	P. citrinum	
Patulin	*P. patulum*	Sarcoma
	P. urticae	
Zearalenone	*Fusarium roseum*	Oestrogenic
	F. nivale	
Ergot alkaloids	*Claviceps* sp.	Ergotism

Table 2 Carcinogenic potential of mycotoxins as given by the International Agency for Research on Cancer, WHO (1993)

Mycotoxin	Degree of evidence[a]		IARC rating[b]
	Human	Animal	
Aflatoxin	S	S	1
Ochratoxin	I	S	2B
Fumonisin	I	S	2B
Trichothecenes (T-2)	I	L	3

[a]S, sufficient evidence; I, inadequate evidence; L, limited evidence.
[b]1, The toxin is carcinogenic; 2B, the toxin is possibly carcinogenic to humans; 3, the toxin is not classifiable as carcinogenic to humans.

Table 3 Tolerance limit for aflatoxin as imposed by different countries

Country	Product	Limit (ng g⁻¹ or ppb)
Australia	Peanut products	15
Belgium	All foods	5
Canada	Nut and nut products	15
China	Rice and other cereals	50
France	All foods	10
	Infant foods	5
UK	Nuts and products for direct use	4
	For further process	10
US	All foods	20
	In dairy feed	20
	Beef cattle feed	300–400
India	All foods	30

approved in some of the countries are given in **Table 3**.

Ochratoxin has an IARC rating of 2B that is the

toxin is possibly carcinogenic to humans. The commodities contaminated with this mycotoxin could be wheat, barley, oats, corn, dry beans, peanuts, cheese and coffee. Fumonisin is also probably carcinogenic to humans and hence IARC has given it a rating 2B based on sufficient evidence of its carcinogenic potential demonstrated in animals. Presence of fumonisin has been detected in corn produced in several regions of the world.

Trichothecenes have an IARC rating 3, which means the toxins are not classifiable as carcinogenic to humans. There is only limited evidence of its carcinogenicity in animals. T2 is one of the important trichothecine mycotoxins. Trichothecene toxins have been detected in corn wheat and cattle feed.

Among the other mycotoxins, the oestrogenic toxin, zearalenone may be found in corn, feed and hay. Patulin is mainly encountered in apple and apple juice, other mouldy fruit juices and feed. Citrinin is found on cereal grains, wheat, barley, corn and rice. As the mould spores are ubiquitous in nature and agricultural commodities provide good substrate for the mould growth, the chances of mycotoxin contamination in foods are high, especially under storage conditions which promote mould growth. The food items likely to support mould growth therefore need to be screened for mycotoxin contamination to avoid risk to human health.

Mycotoxin Assay

Development of suitable analytical techniques for the accurate estimation of a variety of mycotoxins present in food commodities is a challenging task. Appropriate sampling plans and sample preparation procedures have to be devised keeping in view the nature and the bulk of the commodity. Extraction and isolation procedures may be complicated due to the presence of major food components such as starch, protein and fat. Interfering substances and pigments need to be removed by special clean-up procedures before the extracted samples can be taken for analysis.

Detection and estimation of a mycotoxin in a given food thus needs standardization of the sampling plan, preparation of sample, extraction, clean up and finally analysis by a suitable method which offers sensitivity and specificity. Whereas sampling, sample preparation, extraction and clean-up procedures may be standardized and kept constant, the assay procedures could vary depending on the analytical method followed. Conventional techniques of quantification such as fluorometry could give the necessary sensitivity. However, these methods suffer from lack of specificity.

Among all the analytical techniques, immunoassays offer higher sensitivity and specificity. With the immunoassay techniques, simple clean-up procedures such as filtration and centrifugation of the extracted sample are sufficient for sample preparation. However, the efficiency of immunoassays could be greatly improved with a reasonably clean sample.

Use of Immunoaffinity Columns

In addition to their use in the detection and quantification of mycotoxins, immunological methods have also been used in sample clean-up and concentration procedures. Antibodies, either monoclonal or polyclonal, against a particular mycotoxin which is to be isolated are used for this purpose. An immunoaffinity column is made by either physically adsorbing or more preferably chemically coupling the antibodies to the solid supports. The solid support used could be polystyrene beads, biopolymers such as cellulose or other similar polymeric substances. These solid-phase supports immobilized with the specific antibodies are packed into small glass or plastic columns. The extract from food containing the mycotoxin is then passed through the column. The mycotoxin present in the extract will bind with the antibody molecules immobilized on the solid support. After washing the columns with an appropriate solvent to remove interfering substances, the toxin could be eluted by desorption of the antigen–antibody complex with the help of a suitable eluent. Immunoaffinity clean-up and concentration procedure can provide highly purified mycotoxins that can be quantified using any of the available assay techniques such as flourometry or HPLC. The major advantage of immunoaffinity purification is that it is highly specific and capable of removing all other contaminating molecules from the test sample. Immunoaffinity columns are now available commercially for all major mycotoxins. In liquid foods such as milk and fruit juices, clean-up and extraction of mycotoxins could be simultaneously accomplished by using immunoaffinity columns.

Immunological Methods for Detection and Analysis of Mycotoxins

Immunoassays are a class of analytical techniques in which the reaction between an antigen and its antibody is utilized for the quantification of the antigen in an unknown sample. Immunoassays offer several advantages over conventional methods used for mycotoxin detection and quantification. In immunoassays, a large number of samples can be analysed at a time and the turn around time of the assay is relatively low. Due to the high specificity of the immunoassay procedures, elaborate sample preparation is not necessary.

Immunoassays for mycotoxin could be divided into two broad categories, radioisotopic and non-isotopic immunoassays. In radioisotopic immunoassays, a radioactive isotope is used for detection, whereas in non-isotopic immunoassays detection could be achieved by using an enzyme, a chemiluminescent or a fluorescent label.

Radioimmunoassays (RIA)

The principle of the radioimmunoassay technique is as follows

$$
\begin{array}{ccc}
Ag^* & & Ag^*Ab \\
& + \quad Ab \rightleftharpoons & \\
Ag & & AgAb
\end{array}
$$

     Ag – Antigen
     Ag* – Labelled antigen
     Ab – Antibody

Radioimmunoassay is a competitive assay technique in which the reagent, the antibody (Ab) is used in a limited amount as compared to the amount of analyte (antigen, Ag). In order to perform a RIA, it is essential to have a radioactively labelled antigen (Ag^*) which is also referred to as the tracer. In RIA, the tracer antigen and the antibody are used in fixed concentration. The amount of antibody used in these assays is always less than the labelled antigen. Hence, these assays are also called 'limited reagent assays'.

In RIA, the unlabelled and labelled antigens compete for a limited amount of antibody molecules to form antigen–antibody complex molecules. At the end of the reaction, the antibody-bound and free antigens are separated by a suitable separation method. The amount of labelled antigen–antibody complex formed is inversely related to the concentration of the unlabelled antigen. A standard curve could be set up with known concentrations of unlabelled antigen as standards. The concentration of the antigen in the unknown sample could be estimated from the standard curve.

One of the critical steps in RIA is the separation of the antibody-bound and free antigen. In modern assays, the separation of immune-complex is easily achieved by immobilizing the antibodies on a solid support. Microtitre plates are first coated with an appropriate amount of antibody specific to the mycotoxin (antigen). An aliquot of the sample (containing mycotoxin) and labelled mycotoxin are added to this. The labelled mycotoxin and the unlabelled mycotoxin from the sample compete for the limited number of antibody molecules. The principle of the competitive solid-phase RIA is illustrated below.

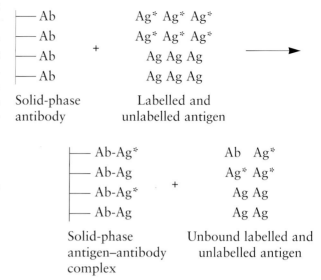

In the absence of unlabelled antigen, the labelled antigen will occupy all the binding sites of the antibody molecules. As the concentration of the Ag in the sample increases, fewer Ag^* molecules will be able to bind to the immobilized antibody molecules. Hence, the amount of activity bound to the antibody on the solid phase will be inversely related to the concentration of the analyte mycotoxin in the sample. By using different concentrations of the authentic mycotoxin in the assay as standard, a dose–response curve could be prepared and the concentration of the mycotoxin in a given sample can be read from this dose–response curve.

RIA offers several advantages over conventional mycotoxin assays. Due to the exquisite sensitivity of the RIA technique, only very small quantities of the samples are required to carry out the assay. By careful experimental conditions, it is possible to prepare specific antibodies and by using these specific antibodies highly specific RIAs can be developed. A major advantage of immunoassays are that they do not need elaborate purification of the sample from other mycotoxins. This is because of the remarkable specificity and sensitivity of the assay. In the RIA procedure, sample clean-up is significantly simplified allowing more samples to be analysed in a given time.

The success of a good radioimmunoassay procedure depends on the availability of antibody and labelled antigen against mycotoxins. Almost all the mycotoxins without exception are small molecular weight antigens. Being haptens they do not induce antibody production. These molecules are synthetically modified and coupled to a large molecular weight protein such as bovine serum albumin to make them immunogenic. A scheme for the preparation of a bovine serum albumin–aflatoxin B_1 conjugate is given in **Figure 1**. These conjugates are injected into labora-

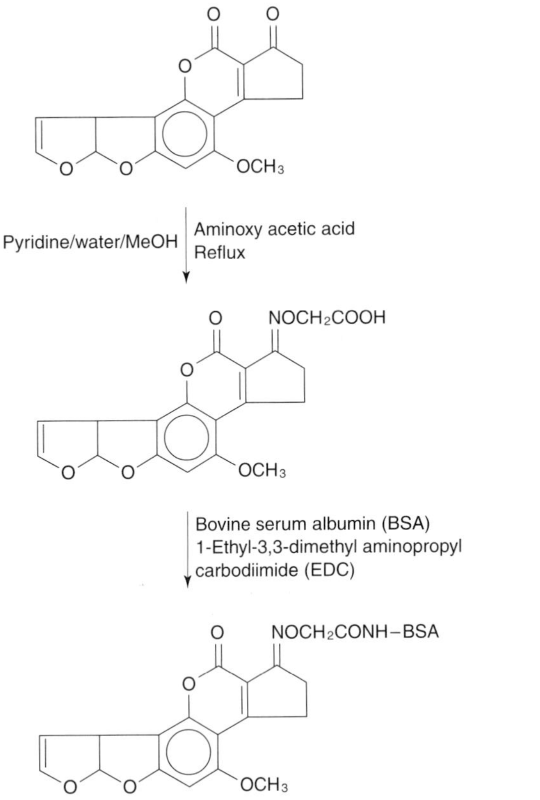

Figure 1 Preparation of aflatoxin B$_1$–bovine serum albumin conjugate.

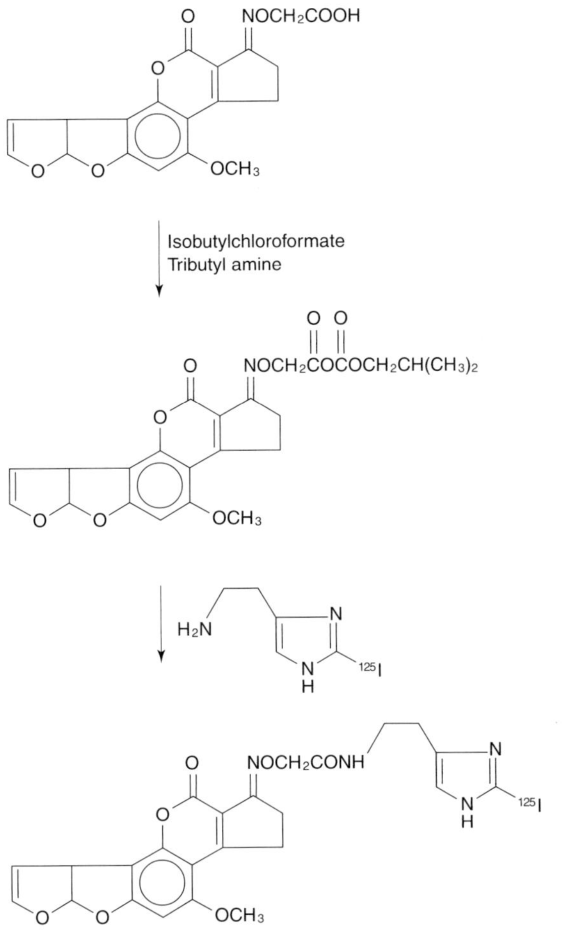

Figure 2 Preparation of aflatoxin B$_1$ tracer.

tory animals after emulsification in Freund's adjuvant. The serum of immunized animals, which contain the antibodies, is used in the assays after appropriate dilution.

The second most important reagent for RIA of mycotoxin is radioactively labelled mycotoxin. The most commonly used radioisotope for RIA is an isotope of iodine called iodine-125 (^{125}I). Iodine-125 offers several advantages, as it can be prepared with very high specific radioactivity and with almost 100% isotopic abundance. It has a convenient half-life of 60 days and hence the tracer could have a long shelf-life. Iodine can be easily introduced into many molecules. As ^{125}I decays by electron capture emitting low energy (35 keV) gamma photons, it will not damage the molecule. These gamma photons can be detected by using a simple solid scintillation counter having a NaI(TI) crystal. In the case of small molecules such as mycotoxins, ^{125}I cannot be easily incorporated into the molecules. Incorporating a pendant group, which can be labelled with ^{125}I, synthetically modifies the mycotoxins and these modified mycotoxins can be used as tracer in RIA. **Figure 2** illustrates the preparation of a radiotracer for aflatoxin B$_1$ by conjugating it with an ^{125}I-labelled histamine. An alternative method, used earlier, was to label the mycotoxins with tritium (^{3}H). A liquid scintillation counter is essential to measure the radioactivity emitted by this isotope. However, the use of tritiated tracers is now not in practice.

Immunoradiometric Assays

An alternate technique called the immunoradiometric assay (IRMA) which is capable of providing higher sensitivity than RIA was developed in the late 1960s. In IRMA, the antibody is labelled instead of the antigen. IRMA is essentially an excess reagent assay in which an excess concentration of a radiolabelled antibody is used as the reagent. An excess concentration of a labelled antibody and the antigen (either from the standard or sample) are allowed to react. At the end of the assay, the antigen-bound and free antibody are separated and the antigen-bound fraction is assayed for radioactivity. The activity associated with this fraction is directly proportional to the concentration of the antigen (analyte).

After the availability of monoclonal antibodies, a new type of IRMA called the two-site IRMA or sandwich IRMA was developed. In this technique, two antibodies both specific for the same antigen, but binding to two different epitopes, are used. One of the antibodies is coated on to the solid phase and used an immuno-extractant for the antigen. A second antibody labelled with ^{125}I is added to the solid phase. This labelled antibody binds with the antigen, which is already bound to the first antibody. At the end of the reaction unreacted-labelled antibody is aspirated out and the solid phase washed with a wash solution. The radioactivity in the solid phase is directly proportional to the concentration of analyte present.

Being an excess reagent technique, IRMA offers higher sensitivity than RIA. The use of two antibodies makes the two-site IRMA more specific than RIA. The IRMA being an excess reagent technique, the assay can be performed in a very short time.

The IRMA technique has not found much application in the detection of mycotoxins. This is mainly due to the fact that mycotoxins are small molecular weight antigens and hence cannot bind to two different antibodies simultaneously.

Enzyme-linked Immunoassays for Mycotoxins

Although radioimmunoassays were initially developed for the detection of mycotoxins such as aflatoxin, they were soon replaced with the non-isotopic assays such as enzyme immunoassays. This was due to the fact that radioisotopes are generally not desirable in food environments, they pose special handling and disposal problems and also the mycotoxins labelled with radioisotopes have limited shelf life and usability. Currently, the most commonly used assay techniques for mycotoxins are based on the enzymes as marker. The preparation of an enzyme-labelled mycotoxin is similar to the preparation of mycotoxin–bovine serum albumin conjugate (Fig. 1). In the reaction, the enzyme is used instead of the bovine serum albumin molecule. The most commonly used enzyme is horseradish peroxidase.

As all the enzyme immunoassays are based on solid-phase antibody techniques, the enzyme immunoassays are commonly referred to as enzyme-linked immunosorbent assays (ELISA). Two types of ELISAs have been developed for mycotoxins.

Direct Competitive ELISA The principle of this assay is similar to that of the RIA technique. The antibody is coated in the wells of microtitre plates to which fixed amounts of enzyme-labelled toxin and the sample containing the toxin are added. Toxin in the sample and the enzyme-labelled toxin compete for the binding sites on the coated antibody. After incubation the unreacted reagents are washed away. The enzyme activity of the labelled mycotoxin bound to the coated antibody is estimated by adding a substrate of the enzyme. H_2O_2 in combination with tetramethylbenzidine is used for the colour development when horseradish peroxidase is used as the enzyme

marker. The colour formed is generally measured in an ELISA reader. A darker colour indicates less concentration of mycotoxins from the test samples, whereas, a lighter colour indicates the presence of a higher concentration of mycotoxins. The ELISA procedure has been used in different formats for the development of simple and rapid qualitative and quantitative methods.

Indirect Competitive ELISA In this form of ELISA, a mycotoxin–protein conjugate is coated on the solid support. To these coated microtitre plates, a mycotoxin antibody labelled with an enzyme and the sample containing the mycotoxin are added and incubated. The toxin in the test sample and the toxin on the coated plate compete for the antibody present in the solution. At the end of the reaction a wash step is performed and the enzyme activity associated with the microtitre plate is determined by using the substrate of the enzyme. The enzyme activity associated with the solid phase is inversely related to the mycotoxin in the test sample. A darker colour indicates the presence of less mycotoxin and vice versa. The indirect competitive ELISA could also be done using an enzyme-labelled second antibody. In this method, the first antibody is used without any modification. After the reaction and wash step a second antibody labelled with the enzyme is added to the plates. The amount of second antibody bound to the microtitre plates can be estimated by measuring the enzyme activity associated with the solid phase as before. The darker the colour the less the concentration of mycotoxin present. The advantage of this technique is that much less specific antibody is required for the assay. Moreover, the second antibody (anti-antibody), can be used as a universal reagent in all mycotoxin assays. Such enzyme-labelled second antibodies are available from commercial sources.

ELISA Protocols for Mycotoxins

Direct Competitive ELISA A typical protocol for direct competitive ELISA for aflatoxin is as follows:

1. Pipette 100 µl of an appropriate dilution of horseradish peroxidase-conjugated aflatoxin into a non-coated microtitre plate well.
2. Add 100 µl test sample or standard solution (10–50 p.p.b.) and mix thoroughly using pipette.
3. Transfer 100 µl of the solution from the above well to a microtitre plate well coated with the antibody and incubate for 10 min at ambient temperature.
4. Shake the solution out of the well and wash 5–10 times with distilled water.
5. Invert the microtitre plate on a filter paper and let dry.
6. Add 100 µl enzyme substrate (1 ml of 0.5 mg ml^{-1}

2,2′ azino bis (3-ethylbenzthiazonline-6-sulphonate) in citrate buffer, pH 4.0 + 1 ml of 30% H_2O_2 in citrate buffer pH 4.0) and incubate for 10 min at ambient temperature.
7. Add 100 µl enzyme stopping solution (3.5 mg HF, 10.5 g Na citrate, 6 ml 1N NaOH, and 400 mg sodium EDTA per litre of H_2O), mix gently.
8. Read absorbency at 405 nm.

Indirect double antibody ELISA Indirect competitive ELISA has also been developed for the estimation of aflatoxin. A typical indirect double antibody ELISA protocol for aflatoxin is as follows.

1. Coat microtitre plate wells by adding aflatoxin–carrier protein conjugate (0.5 µg ml⁻¹) and incubating for 3 h at 37°C.
2. Wash with phosphate-buffered saline containing 0.05% Tween 20 (PBST) and store dry at 4°C.
3. Add 100 µl of the test sample to the wells of the above microtitre plate in triplicate.
4. Add aflatoxin B_1 mouse or rabbit antibody (diluted 1 : 5000 in PBST containing 0.1% BSA).

5. Incubate for 90 min at 30°C and wash 3 times with PBST.
6. Add 200 µl goat anti-mouse/rabbit IgG-alkaline phosphatase conjugate (1 : 1000 diluted in PBST containing 0.1% BSA).
7. Incubate for 90 min at 37°C and wash 3 times with PBST.
8. Add 200 µl *p*-nitrophenyl phosphate (1 mg ml⁻¹ in a suitable buffer, pH 9.6) and incubate for 30 min at 37°C.
9. Record absorbency at 405 nm.

Commercial Immunoassay Kits Immunoassay kits for detection and quantification of most of the major mycotoxins are available from commercial sources. Many of these kits have been evaluated in collaborative studies between different laboratories and adopted as official method by the Association of Official Analytical Chemists (AOAC). The AOAC Research Institute also validates and certifies ELISA test kits.

A good immunoassay kit should have

Table 4 Commercial immunoassay kits for aflatoxin available in the USA

Trade name	Detection	Sensitivity (ng g⁻¹ or ppb)	Supplier	Commodity
AflaCup	Visual	5, 10, 20	Romer Labs Inc., MO	Peanuts, peanut butter, corn, cotton seed
Afla 5, 10, 20 Cup	Visual	5, 10, 20	International Diagnostics, MI	Peanuts, peanut butter, corn, cotton seed
Aflatest	Fluorometry	10	Vicam, MA	Corn, peanuts, peanut butter, milk
Agri-screen	Visual/ELISA reader	5	Neogen Corp., MI	Corn, peanuts, cotton seed, feed
One-Step ELISA	ELISA reader	5	International Diagnostics, MI	Corn, peanuts, peanut butter, cotton seed, feed, milk
Veratox	ELISA reader	5	Neogen Corp., MI	Corn, peanuts, cotton seed, feed, milk
Cite Probe	Visual	5, 20	Idexx, Labs, ME	Corn, cotton seed
EZ-screen	Visual	5, 20	Medtox Diagnostics Inc., NC	Corn, peanuts

Table 5 Commercial immunoassay kits for mycotoxins

Trade name	Mycotoxin	Detection	Sensitivity (ng g⁻¹ or ppb)	Supplier
EZ SCREEN	Aflatoxin M1	Visual (Pass/Fail)	0.5	DiAgnostix Inc. NC
	Ochratoxin	Visual (Pass/Fail)	5	DiAgnostix Inc. NC
	T-2 toxin	Visual (Pass/Fail)	12.5	DiAgnostix Inc. NC
	Zearalenone	Visual (Pass/Fail)	50	DiAgnostix Inc. NC
One Step ELISA	Aflatoxin M1	Quantitative	0.5	International Diagnostics, MI
	Zearalenone	Quantitative	200	International Diagnostics, MI
AgriScreen	DON	Visual (Pass/Fail)	1000	Neogen Corp, MI
	Fumonisin	Visual (Pass/Fail)	500	Neogen Corp, MI
	Ochratoxin	Visual (Pass/Fail)	20	Neogen Corp, MI
	T-2 toxin	Visual (Pass/Fail)	500	Neogen Corp, MI
	Zearalenone	Visual (Pass/Fail)	250	Neogen Corp, MI
Veratox	Aflatoxin M1	Quantitative	0.25	Neogen Corp, MI
	DON	Quantitative	300	Neogen Corp, MI
	T-2 toxin	Quantitative	50	Neogen Corp, MI
	Zearalenone	Quantitative	250	Neogen Corp, MI

- the desired limits of detection and sensitivity;
- high specificity for the mycotoxin;
- short screening/quantification time;
- adaptability to a wide range of foods;
- adaptability to recommended sample extraction procedures;
- inter- and intra-assay reproducibility.

Table 4 lists some of the commercial immunoassay kits available for detection and quantification of aflatoxin.

A typical assay using an ELISA kit is as follows. A sample of corn or peanut is blended at high speed with 80:20 methanol:water (2 ml of solvent per gram of food product) for 3 min. The mixture is then filtered using a coarse and a fine filter paper to remove particulate matter. The filtrate is used as sample in any one of the commercial ELISA kits. In the AflaCup™ ELISA kit (Romer Labs Inc., MO, USA) an aliquot of the filtrate is added to the microtitre plates, followed by the enzyme conjugate and the substrate solution. The colour change can be visually observed. The presence of colour indicates that the aflatoxin concentration is less than the detection limit of the kit. Test kits for 5, 10 or 20 p.p.b. detection limits are available. ELISA kits for the actual quantification of the aflatoxin levels are also available. The enzyme activity is measured in a microtitre plate reader (spectrophotometer) and dose–response curve can be set up. It is also possible to program the ELISA reader to indicate a Yes/No type of answer.

Immunoassay techniques have also been used for the detection of aflatoxin metabolites (aflatoxin M_1 and M_2) in milk, serum and urine. Detection of aflatoxin–albumin adducts in blood and alfatoxin–DNA adducts in body tissues have also been made possible with the help of immunoassays.

Ochratoxin is yet another toxin for which immunoassays were developed in the early stage. Both RIA and ELISA have been described for ochratoxin. Polyclonal as well as monoclonal antibodies have been used in these assays. Commercial immunoassay kits are also available for some other mycotoxins as shown in **Table 5**. Although several commercial kits are available the testing and validation of many new kits needs validation. The Official Methods Board – Task Force on Test Kits and Proprietary Methods of AOAC provides the necessary guidelines for the approval of test kits. Immunoassay kits are less cumbersome for operators in the field and are less expensive and because of the widespread prevalence of mycotoxins in food and increasing regulatory controls they will have a major role in strengthening quality control in the food and feed industry.

See also: **Enzyme Immunoassays**: Overview. **Fungi**: Food-borne Fungi – Estimation by Classical Culture Techniques. **Immunomagnetic Particle-based Techniques**: Overview. **Mycotoxins**: Detection and Analysis by Classical Techniques. **National Legislation, Guidelines & Standards Governing Microbiology**: European Union; Japan. **Reference Materials**. **Sampling Regimes & Statistical Evaluation of Microbiological Results**.

Further Reading

Butt WR (ed.) (1984) *Practical Immunoassays*. New York: Marcel Dekker.

Chard T (1990) *An Introduction to Radioimmunoassays and Related Techniques*. Amsterdam: Elsevier.

Chu FS and Ueno I (1977) Production of antibody against aflatoxin B1. *Applied and Environmental Microbiology* 33: 1125–1128.

Collins WP (ed.) (1985) *Alternative Immunoassays*. New York: John Wiley.

Pestka JJ (1988) Enhanced surveillance of foodborne mycotoxins by immunochemical assays. *Journal of the Association of Official Analytical Chemists* 71: 1075–1081.

Pestka JJ, Abouzied MN and Sutikno (1995) Immunological assays for mycotoxin detection. *Food Technology* 49: 120–128.

Ram BP and Hart LP (1986) Enzyme-linked immunosorbent assay of aflatoxin B₁ in naturally contaminated corn and cottonseed. *Journal of the Association of Official Analytical Chemists* 69: 904–907.

Samarajeeva U, Wei CI, Huang TS and Marshal MR (1991) Application of radioimmunoassay in food industry. *CRC Critical Reviews in Food Science and Nutrition* 29: 403–434.

Sinha KK and Bhatnagar DD (eds) (1998) *Mycotoxins in Agriculture and Food Safety*. New York: Marcel Dekker.

Wyatt GM, Lee HA and Morgan MRA (1995) *Immunoassays for Food Poisoning Bacteria and Bacterial Toxins*. London: Chapman & Hall.

Toxicology

D Abramson, Agriculture & Agri-Food Canada, Cereal Research Centre, Winnipeg, Canada

Introduction

Research efforts since the beginning of the twentieth century have identified a number of mycotoxins that affect human and livestock health. Worldwide studies have established that human exposure is frequent as a result of ingesting mycotoxin-contaminated plant products, meat, milk or eggs. Mycotoxins have been implicated in a wide variety of human maladies. Furthermore, economic losses have been documented worldwide as a result of deterioration of foods and feeds by moulds. Overviews of detection methods, and summaries of commodity surveys for major mycotoxins, are updated frequently by the Association of Official Analytical Chemists International, which publishes an annual synopsis in the January issue of its journal.

To maintain the focus on food-related issues, this discussion is limited to mycotoxins satisfying two criteria: proven oral toxicity, and natural occurrence in agricultural produce. The toxins are arranged in three groups to reflect their health significance. Toxins in the first group are implicated in human illness, while those in the second group are implicated in livestock toxicosis. Toxins associated with human illness often appear at first in epidemics of livestock poisoning. For example, aflatoxin B_1, associated with human liver cancer, first appeared as a toxic contaminant of peanut meal fed to turkeys in the UK. Another example is fumonisin B_1, now associated with human oesophageal cancer, which first appeared in the feed of horses suffering encephalomalacia in the USA. Toxins of the third group have significant oral toxicity, are found in foods and in animal feeds, and therefore have the potential to cause both human and livestock health problems.

The fungal metabolite zearalenone is considered by many researchers to be mycotoxin, even though it exhibits no oral toxicity. Zearalenone is produced by *Fusarium* species, and is commonly found in maize and maize-based feeds. Although zearalenone has negligible oral toxicity in test animals, it exhibits oestrogenic activity in swine, and to a lesser extent in ruminants. In swine, the most conspicuous abnormality is vulvovaginitis, and interference with reproduction is the major problem. Levels of 1000–3000 $ng\,g^{-1}$ of zearalenone in feed reportedly cause continuous oestrus, pseudopregnancy and ovarian abnormalities, and on a biochemical level, elevated serum progesterone levels, and decreased prolactin and luteinizing hormone levels. There is little evidence implicating the fungal oestrogen zearalenone in any human health problems.

Mycotoxins Implicated in Human Illness

Aflatoxins

The aflatoxins were first isolated from *Aspergillus flavus* and were named after the fungus. The *A. flavus* group contains several closely related species, but *A. flavus*, *A. parasiticus* and *A. nomius* are the only species that are aflatoxigenic. There are four naturally occurring aflatoxins produced by these aspergilli: aflatoxins B_1, B_2, G_1 and G_2. In culture, toxigenic *A. flavus* strains produce only aflatoxins B_1 and B_2 while *A. parasiticus* usually produces all four.

Aflatoxins occur in many foods. In fact, the *A. flavus* group can grow and produce aflatoxins in almost any commodity stored under conditions of high moisture and temperature. Consequently, aflatoxins occur widely in foods, feed ingredients and agricultural commodities. Plant products with a high risk of aflatoxin contamination include maize, peanuts, brazil nuts, cottonseed, pistachio nuts and copra. Commodities with a slightly lower risk of aflatoxin contamination include figs, almonds, pecans, walnuts, raisins and spices. Soya beans, beans, pulses, cassava, grain sorghum, millet, wheat, oats, barley and rice show minimal susceptibility to aflatoxin contamination.

Aflatoxin B_1 is the biologically most active of the structurally related aflatoxin group, and is a hepatocarcinogen. The oral median (LD_{50}) lethal dose in rats is 7.2 $mg\,kg^{-1}$. In test animals, aflatoxin B_1 induces a dose-related increase in unscheduled DNA synthesis in primary hepatocytes, and also induces DNA adduct formation. The aflatoxins are potent hepatotoxins, and are subject to official regulation in many countries. In terms of animal health, aflatoxin toxicity can be acute or chronic and can affect poultry, swine and other livestock. In mammals, some aflatoxin B_1 is converted to aflatoxin M_1, and these toxic residues are transmissible through milk. Contamination of milk by aflatoxin M_1, which is also carcinogenic, is of sufficient concern for aflatoxin M_1 content also to be officially regulated.

In human populations, high aflatoxin intake is positively correlated with the high incidence of primary liver cancer in some developing countries, but there is also a high prevalence of hepatitis B in many of these countries. Debate continues about the relative importance and contribution of aflatoxin intake and hepatitis B in the aetiology of primary liver cancer, and so the impact of aflatoxin contamination of food on human health in developed countries is at present somewhat controversial. Reports of acute human toxicosis arising from aflatoxin-contaminated foods occasionally emerge from developing countries. The structures of aflatoxin B_1 and other mycotoxins are shown in **Figure 1**.

Ochratoxin A

Ochratoxin A is the most abundant and most toxic member of the ochratoxin group. In tropical and semitropical zones the main producers are aspergilli, particularly *Aspergillus ochraceus*, while in temperate zones, *Penicillium* spp. are the prime sources. Microbiologists have considered several *Penicillium* species as ochratoxin-producing, but in the current taxonomic scheme, only *P. verrucosum* and its chemotypes are regarded as producers.

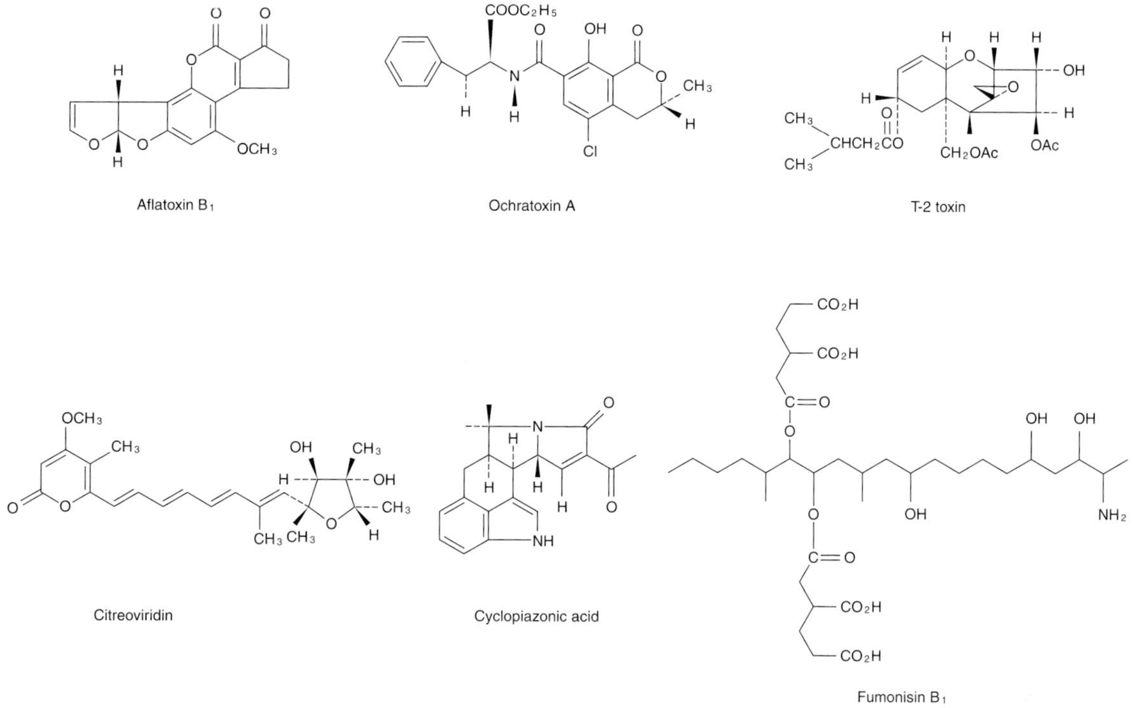

Fig. 1 Structures of mycotoxins implicated in human illness.

The ochratoxins act primarily as nephrotoxins, and ochratoxin A is also a potent teratogen in rats, hamsters and mice. This mycotoxin acts as an inhibitor of protein synthesis, and adducts of ochratoxin A and DNA have been found in the kidneys and livers of orally dosed test animals. The oral LD_{50} for ochratoxin A in rats is $22\ mg\ kg^{-1}$. In day-old chicks, the oral LD_{50} is $3.6\ mg\ kg^{-1}$ for ochratoxin A and $54\ mg\ kg^{-1}$ for ochratoxin B, with acute kidney damage observed; ochratoxin C shows similar effects. The carcinogenicity of ochratoxin A in mice has been documented. A study at the National Institutes of Health in the USA also showed ochratoxin A to be a carcinogen in rats.

Ochratoxin A contaminates a variety of foods, and is particularly likely to appear in cereal grains and their products. Although ochratoxin A is a worldwide problem, its significance is greatest in temperate climates where much of the world's grain is produced and stored. In commodity surveys, ochratoxin A has been found in cereals from Australia, Japan, North America and Europe. Cheese is evidently also a good substrate for ochratoxin production, and this toxin has been detected in mouldy cheese from Britain. Ochratoxin A is found in meat products from monogastric animals, and has been detected in pork sausages in Germany and Switzerland. In poultry, this toxin can be carried over from contaminated feed into the muscle tissue and eggs.

In terms of livestock operations, ochratoxin A is

the most significant of the *Penicillium* mycotoxins. Ochratoxin contamination of feed grains is of concern to the swine and poultry industries because, unlike ruminants, monogastric animals lack the ability to rapidly hydrolyse and excrete ochratoxin A. As a result, monogastric livestock are far more susceptible to the nephrotoxic effects of ochratoxin A than ruminants. Furthermore, ochratoxin A residues can enter the human food supply through swine and poultry products. In ruminants, the main adverse effect of ochratoxin A appears to be suppression of milk production, which is transient.

Ochratoxin A is a major concern to livestock producers, especially in Europe. Ochratoxin A in the feed of monogastric livestock can contaminate eggs, organs, fat, muscle tissue and biological fluids. Although limited data are available for poultry, swine are known to be quite susceptible owing to a long serum half-life of 72–120 h. Surveys have detected ochratoxin A as a contaminant of swine blood in Canada and in European countries. Ochratoxin A has also been found in swine kidneys in the USA and in several European countries.

Ochratoxin A is often found in the blood of humans, and has been associated with a high incidence of nephropathic illness in eastern Europe. This disease is encountered in Bulgaria and in the former state of Yugoslavia, and is known as Balkan endemic nephropathy. Ochratoxin A residues have been compared in food and blood samples from Balkan regions showing high and low incidence of nephropathic disease. Ochratoxin A has been found in human blood samples from other parts of Europe, including Germany, Poland and the Czech Republic and in Canada and Japan. Inhalation of ochratoxin A in grain dust has recently been implicated in a case of acute renal failure, but this toxin was not isolated from the dust or from the patient's blood.

T-2 Toxin

Fusarium sporotrichioides and *F. poae* generally develop in cereal grains under the relatively cool conditions of a late wet harvest. These fungi are commonly found on grains overwintered under snow cover. Both species are weakly pathogenic to cereal plants, but are capable of producing comparatively toxic compounds having a characteristic trichothecene structure. One of the most potent is T-2 toxin, which produces necrotic degeneration of the gastrointestinal tract, and also inhibits leucocyte production, thus acting as an immunosuppressor. The oral LD_{50} for T-2 toxin in rats is $5.2 \, mg \, kg^{-1}$. In chicks, the oral LD_{50} is $5.0 \, mg \, kg^{-1}$ for T-2 toxin, and $7.2 \, mg \, kg^{-1}$ for the related HT-2 toxin. Residues from these toxins are non-transmissible.

Overwintered or wet-harvested grain, contaminated with *Fusarium* and trichothecenes, is often conspicuously mouldy and is usually rejected as a foodstuff. It is therefore rare to see T-2 or HT-2 toxins mentioned in any commodity survey. Grain contaminated with *Fusarium* trichothecenes of the T-2 type is occasionally fed to livestock, and mycotoxicosis occurs. In times of dire scarcity, overwintered grain has been used as a foodstuff, with disastrous consequences. During 1942–1946 in the USSR wartime scarcity prompted the consumption of contaminated grain, resulting in a condition called alimentary toxic aleukia (ATA), which affected tens of thousands of people. In a typical occurrence of ATA, consumption of contaminated food initially results in oral and upper gastrointestinal tract inflammation, nausea, vomiting, and diarrhoea. In the second stage, destruction of the bone marrow commences, leucocyte counts decrease, and subcutaneous haemorrhagic spots appear. In the third stage, destruction of the bone marrow continues, with continuing decrease in levels of leucocytes, erythrocytes and granulocytes, and establishment of various infections. In the USSR outbreak death was often the outcome and mortality rates in some districts were as high as 60%.

The evidence for T-2 toxin involvement in the outbreak of ATA during 1942–1946 in the USSR is retrospective but valid. Original and authentic isolates of *F. sporotrichioides* and *F. poae*, collected from overwintered grain in the hard-hit Orenburg district at the time of the outbreak, were re-cultured three decades later. Cultures were grown on wheat at 12°C, and produced T-2 toxin at concentrations up to $2.1 \, g \, kg^{-1}$ of substrate. The toxin was assayed by gas chromatography and confirmed by thin-layer and bioassay techniques.

Human toxicosis arising from T-2 toxin still occurs occasionally. A wet harvest and contamination of rice by *F. heterosporum* in China resulted in an episode affecting 97 individuals. Victims showed typical first-stage symptoms: upper gastrointestinal tract inflammation, nausea, vomiting and diarrhoea. Timely intervention and supportive medical treatment ensured recovery in all cases. The T-2 toxin was identified in rice samples by enzyme immunoassay.

Fumonisins

In maize crops, *Fusarium moniliforme* and the related fungi *F. subglutinans* and *F. proliferatum* cause kernel rot, and contamination by a recently characterized group of toxins known as the fumonisins. Fusarium kernel rot occurs during warm weather, and is facilitated by insect damage. *Fusarium moniliforme* and *F. subglutinans* are very common in maize kernels in Canada, the USA and Europe. In southern USA and tropical locales, *F. moniliforme* is the main cause of fusarium kernel rot. Generally, insect damage appears to promote *F. moniliforme* occurrence regardless of moisture conditions.

Fusarium moniliforme and *F. subglutinans* produce the fumonisin group of mycotoxins, although this group is also produced by other *Fusarium* species. Four main structurally related fumonisins are known: fumonisins B_1 and B_2 are reported to cause leucoencephalomalacia in horses, pulmonary oedema in pigs and hepatic cancer in rats. The mechanisms of action have not yet been fully elucidated. Fumonisins occur naturally in maize in the USA, Poland, South Africa and other countries. Fumonisin-producing isolates of *F. moniliforme* are also widely distributed among various crops in Taiwan. High fumonisin levels in maize have been reported in areas with high incidence of human oesophageal cancer. This has prompted a guideline of $1000 \, ng \, g^{-1}$ as a preliminary tolerance level for foods in Switzerland.

In maize, *F. moniliforme*, *F. subglutinans*, *F. proliferatum* and other species also produce the mycotoxin moniliformin. This toxin has been found in naturally infected maize from North America, Europe and Africa. Moniliformin is an oral toxicant (see Table 1), but has not been strongly implicated in any episodes of human or livestock illness.

Citreoviridin

In the early twentieth century in Japan, citreoviridin was associated with acute cardiac disease arising from consumption of rice infected with *Penicillium citreoviride*. Investigators later discovered that rice is an optimal substrate for *P. citreoviride*, and in 1958 this fungus was reported in 7.4% of rice surveyed in Italy, Spain, Thailand, Burma and other countries. Citreoviridin is also produced by *P. pulvullorum*, *P. ochrasalmoneum*, and *P. charlesii*.

In mice, citreoviridin causes progressive paralysis in the limbs, vomiting, convulsions, cardiovascular damage and respiratory arrest. In cats and dogs, the symptoms are the same, with vomiting preceding the progressive paralysis. The toxin appears to attack motor neurons and interneurons of the spinal cord, medulla oblongata and central nervous system. This is followed by dilation of the right side of the heart and paralysis of the diaphragm. Biochemical studies indicate that citreoviridin inhibits ATPase activity and mitochondrial energy-linked reactions such as ADP-stimulated respiration.

Modern analysis methods and reference materials for citreoviridin have been available since 1970. This mycotoxin has been found in the USA in pecans and maize. Identification of citreoviridin as the aetiological agent in mouldy rice associated with the human cardiac disease epidemics in Japan is based on fairly conclusive evidence. The documentation includes published accounts of the epidemic, isolation of fungus, toxicology studies with fungal extracts, and structural elucidation of the toxin.

The cardiac disease associated with citreoviridin was first thought to be identical with the thiamin deficiency syndrome from eating polished rice. Between 1890 and 1925, Japanese investigators showed that a distinct type of beriberi, due to toxicosis rather than avitaminosis, was on the increase in Japan, principally in urban centres such as Tokyo. From 1909 onwards, increasing government pressure to inspect rice for mould contamination reduced the Tokyo death rate from more than 11 per 10 000 to approximately 2 per 10 000. By 1929 the epidemic was effectively controlled.

During the course of this illness, a typically afflicted patient initially undergoes cardiac distress and tachypnoea, followed by nausea, vomiting and painful seizures. With falling blood pressure, the patient experiences tachycardia, dyspnoea, pupillary dilation and cold cyanotic extremities. Increasing paralysis of the respiratory muscles impairs pulmonary circulation, overloading the right ventricle of the heart, and finally causing cardiac failure.

Cyclopiazonic Acid

Cyclopiazonic acid was isolated from *Penicillium cyclopium* in 1968. In the current taxonomic scheme, this mycotoxin appears to be produced mainly by *P. camemberti*, *P. commune*, *P. griseofulvum*, and several aspergilli.

Oral administration to male and female rats produces LD_{50} values of $36\,mg\,kg^{-1}$ and $63\,mg\,kg^{-1}$ respectively, with death ensuing after 1–5 days. In rats, cyclopiazonic acid produces degenerative changes and necrosis in the liver, pancreas, spleen, kidney, salivary glands, myocardium and skeletal muscle. Although the teratogenic potential is low in rats, retardation in embryonic skeletal development was evident after administration of 5–10 mg of cyclopiazonic acid during pregnancy. When feed containing $100\,mg\,kg^{-1}$ cyclopiazonic acid was provided to chickens over a 7-week period, the test group experienced decreased weight gain, poor feed conversion and a sixfold increase in mortality, compared with control chickens receiving toxin-free feed. Post-mortem examination revealed proventricular lesions characterized by mucosal erosion and hyperaemia, mucosal necrosis in the gizzard, and hepatic and splenetic necrosis and inflammation.

Cyclopiazonic acid and aflatoxins are often produced together by *Aspergillus flavus* in maize and peanuts in the USA. In a 1990 survey of maize and peanut crops, 51% of the maize samples contained this mycotoxin at levels up to $2800\,ng\,g^{-1}$ (average $470\,ng\,g^{-1}$); 90% of the peanut samples contained up to $2900\,ng\,g^{-1}$ (average $460\,ng\,g^{-1}$). Cyclopiazonic acid may also be produced by *P. camemberti* in improperly ripened or stored Camembert-type cheeses. A study showed that this mycotoxin did not appear during the 9-day ripening period at 14–18°C, nor during 12 days' storage at 8°C, but cheese kept at 25°C for 5 days accumulated this toxin at levels up to $4000\,ng\,g^{-1}$. In chickens, there is carry-over of cyclopiazonic acid from feed into muscle tissue, and a potential for contamination of human diets. Analysis of chicken meat following oral dosing with 10 mg cyclopiazonic acid per kg body weight indicated that 15% of the dose was in the meat 48 h later.

In India, cyclopiazonic acid has been found in millet associated with human toxicosis. Toxicosis from mouldy millet, known as 'kodua poisoning', is a non-lethal illness in humans; it is characterized by fatigue, nausea, tremors and slurring of speech. Although the disease has been known since antiquity, some recent episodes of human toxicosis arising from millet contaminated with cyclopiazonic acid have been reported. In these cases, typical symptoms include fatigue and nausea. Cyclopiazonic acid was isolated from the suspect millet by preparative thin layer chromatography, and identified from infrared and mass spectrum comparisons.

Toxins Implicated in Animal Illness

Deoxynivalenol and Related Compounds

Fusarium graminearum and *F. culmorum* develop in the field during midsummer to produce fusarium head blight (known as 'scab' in the USA) in wheat and barley, along with trichothecene mycotoxins. An epidemic affecting less than a quarter of the heads in a field may still produce a crop that exceeds accepted tolerances for mycotoxins in food or feed. The main toxin associated with fusarium head blight is deoxynivalenol, although toxic acetyl-deoxynivalenol homologues may also be present. Compared with T-2 and HT-2 toxins, deoxynivalenol is one of the less toxic substances produced by *Fusarium* fungi. Current Canadian guidelines recommend that feed intended for swine should contain no more than $1000 \, \text{ng} \, \text{g}^{-1}$ deoxynivalenol in order to avoid feed refusal problems, but other livestock such as poultry and cattle appear to be far more tolerant. Like T-2 toxin and the other trichothecenes, deoxynivalenol

inhibits protein synthesis. Deoxynivalenol also acts primarily on the gastrointestinal tract, but the effects are far less severe than those of T-2 toxin, and necrotic lesions are usually absent.

Fusarium graminearum predominates in blighted wheat from Canada, the USA and China. It is highly pathogenic to both wheat and maize. North American isolates of this species produce deoxynivalenol along with the related 8-keto-trichothecenes 15-acetyl-deoxynivalenol and nivalenol. Isolates from Australia, New Zealand, China and northern Europe tend to produce 3-acetyl-deoxynivalenol instead of 15-acetyl-deoxynivalenol. *Fusarium graminearum* isolates from western Canadian grains can also produce significant levels of fusarenone-X (4-acetyl-nivalenol) in culture. The symptoms arising from infection by *F. graminearum* in wheat are generally quite characteristic. For example, fusarium head blight in western Canadian hard red spring wheat produces characteristic 'tombstone' kernels which are typically smaller than normal, shrivelled in appearance, and white or pale pink in colour.

In maize, *F. graminearum* and *F. culmorum* cause gibberella ear rot, also called pink ear rot. This disease is prevalent in northern temperate climates, especially

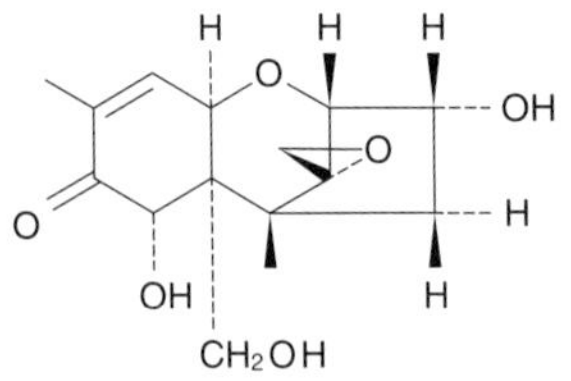

Deoxynivalenol

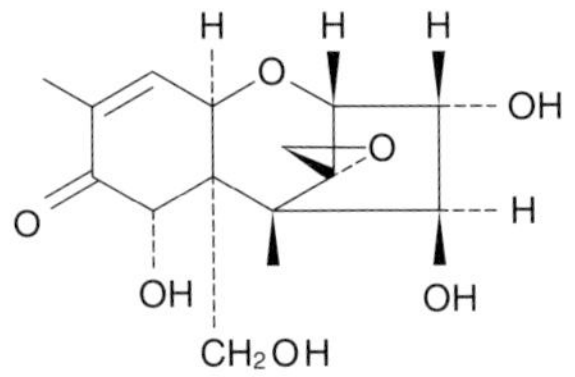

Nivalenol

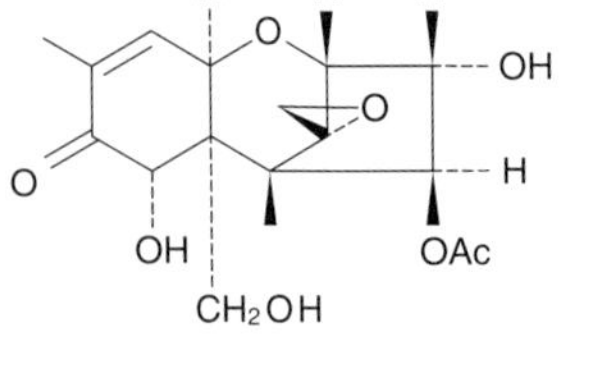

Fusarenone -X

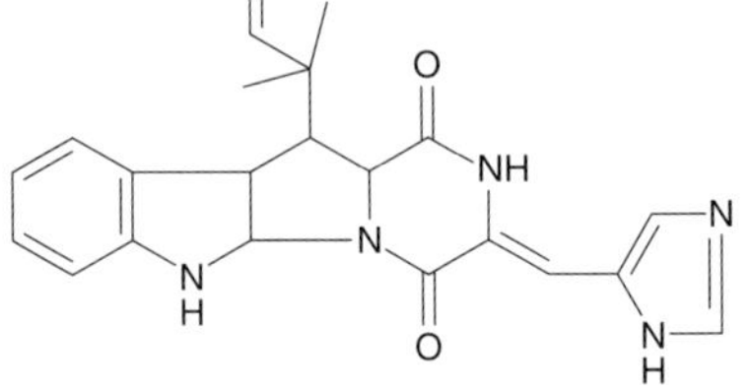

Roquefortine

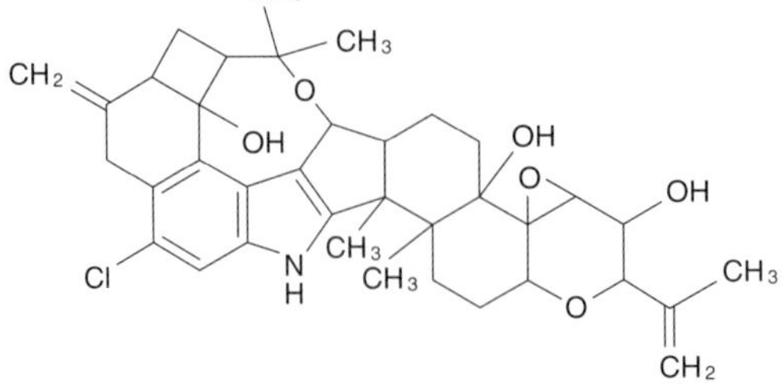

Penitrem A

Fig. 2 Structures of mycotoxins implicated in animal illness.

in wet years. In outbreaks of this disease, *F. graminearum* is predominant in Canada and northern USA, while *F. culmorum* tends to be the dominant species in north-central and eastern Europe. In maize, North American isolates *F. graminearum* can produce high levels of deoxynivalenol and 15-acetyl-deoxynivalenol; 20 000 ng g^{-1} and 16 000 ng g^{-1} respectively were reported in a sample of maize involved in feed refusal by swine. Structures of some of the mycotoxins are shown in **Figure 2**.

Roquefortine C

The toxin roquefortine C is produced by several species of *Penicillium* including *P. chrysogenum*, *P. crustosum*, *P. expansum*, *P. griseofulvum*, *P. hirsutum*, *P. hordei* and *P. melanoconidium*, but most notably by *P. roqueforti*, which is used in cheese production. Attempts to find completely non-toxigenic strains for use in cheese production have so far been unsuccessful. Although oral LD$_{50}$ figures have not been established for roquefortine C, day-old chicks, orally dosed via intubation, have been used as a bioassay for this toxin. With increasing dosage, the chicks initially lose their balance, and remain in sitting and leaning postures, and die in characteristic postures with head and neck extended backwards, and legs and feet extended outwards from the body.

Roquefortine C has been found mainly in blue cheese and blue cheese products. In a survey of 16 blue cheese samples from Denmark, Finland, Germany, France, Britain, Italy and Canada, all samples contained this mycotoxin at levels up to 6800 ng g^{-1} (average 950 ng kg^{-1}). In a later survey, 12 American blue cheese samples all contained roquefortine C at average levels of 420 ng g^{-1}; for American blue cheese salad dressing, 2 samples contained an average of 45 ng g^{-1}.

In a cattle poisoning incident, roquefortine C was found in barley-based feed at 25 000 ng g^{-1} on a Swedish farm. The grain was heavily infected by *P. roqueforti*, with a large accumulation of mycelium. Cattle developed extensive paralysis which did not respond to calcium treatment. The disease symptoms disappeared when the cattle were fed sound grain.

Roquefortine C was recently found in the stomach contents of dogs showing strychnine-like symptoms during episodes of fatal food poisoning. These particular animals had evidently eaten garbage which had become mouldy during warm weather. Fatal canine poisoning due to ingestion of roquefortine from mouldy blue cheese has also been reported. Because of their scavenging nature, many canine species are likely to seek out and consume mouldy foods from garbage, and thus become affected by food-borne mycotoxins.

Penitrem A

Penitrem A, also called tremortin A, is a tremorgenic toxin produced by *Penicillium crustosum*, *P. melanoconidium* and *P. commune*. Owing to the high incidence of *P. crustosum* isolates producing significant levels of this toxin, the identification of this species in food or feed is often regarded as a warning sign of penitrem contamination.

Of the tremorgenic mycotoxins produced by *Penicillum* spp., penitrem A is the most potent. Like citreoviridin, penitrem A affects the central nervous system, but this neurotoxin also causes sustained tremors in animals which are otherwise able to feed and function normally. Penitrem A acts primarily on the spinal cord, and is thought to affect the cerebro-cortical synapses by stimulating release of neuro-transmitters. Fivefold to twentyfold increases in dosage are rapidly fatal with characteristic pathological changes, particularly to the hepatocytes. The amount of toxin producing a given response varies among animal species. The oral LD$_{50}$ for penitrem A in mice is 10 mg kg^{-1}; in chickens, it is 42 mg kg^{-1}.

Penitrem A has been implicated in toxicosis of dogs suffering tremors after eating mouldy cream cheese or mouldy walnuts. In the former instance, isolation of *P. crustosum* from mouldy cream cheese and inoculation onto normal cream cheese resulted in production of penitrem A by the fungus. In the latter case, a chloroform extract of the walnuts yielded a lipid residue which produced tremors in mice after oral administration in olive oil. A case of canine poisoning by penitrem A from mouldy bread indicated that levels of 35 000 ng g^{-1} in bread could be achieved by *P. crustosum*, and that approximately 0.175 mg penitrem A per kilogram body weight was sufficient to produce severe muscle tremors in dogs.

Other Oral Toxicants Occurring in Foodstuffs

The mycotoxins in **Table 1** have well-documented oral toxicity, and are found in foodstuffs. Some, such as citrinin and sterigmatocystin, have been tentatively associated with livestock health problems. Others, such as fusarenone-X and nivalenol, have been found in mouldy rice associated with human health problems, and may act synergistically with other mycotoxins. Although patulin has not been associated with human or livestock illness, it is subject to regulatory control in 10 countries (**Table 2**).

There has been a significant spread of *Fusarium* disease in North America and Europe, while economic considerations and transportation difficulties often lengthen crop storage times and exposure to *Penicillium* and *Aspergillus* moulds. These factors will

Table 1 Oral toxicity of mycotoxins produced in foodstuffs by *Aspergillus*, *Penicillium*, *Fusarium* and *Alternaria* species

Mycotoxin	Producers	Foodstuff contamination	Oral LD$_{50}$ (mg kg^{-1} body weight)
Sterigmatocystin	*A. versicolor*	Cereals	120 (rat)
Citrinin	*P. citrinum* *P. expansum* *P. verrucosum*	Cereals Oilseeds Oilseed products	134 (rabbit) 43 (guinea pig) 105 (mouse) 95 (chicken) 57 (duck)
Patulin	*P. expansum* *P. griseofulvum*	Apples	35 (mouse)
Penicillic acid	*P. aurantiogriseum* *P. aurantiovirens* *P. cyclopium* *P. freii* *P. viridicatum*	Maize Beans	90 (chicken) 600 (mouse)
Rubratoxin B	*P. purpurogenum* *P. rubrum*	Maize	63 (duckling) 83 (chick) 120 (mouse)
Secalonic acid D	*P. oxalicum*	Maize (dust)	25 (rat)
3-Acetyl-DON	*F. graminearum* *F. culmorum*	Cereals	34 (mouse)
15-Acetyl-DON	*F. graminearum*	Cereals	34 (mouse)
Nivalenol	*F. graminearum* *F. nivale*	Cereals	19.5 (rat)
Fusarenone-X	*F. graminearum*	Cereals	4.5 (mouse) 33.8 (chick)
Moniliformin	*F. moniliforme* *F. proliferatum*	Maize	4 (chick)
Diacetoxyscirpenol	*F. poae* *F. sporotrichioides*	Cereals	7.3 (rat) 3.8 (chick)
Tenuazonic acid	*Alternaria* spp.	Wheat Tomatoes Olives Oranges	37.5 (chick) 174 (rat)

result in future mycotoxin problems in foodstuffs, as the incidence of toxigenic fungi increases, along with favourable conditions for toxin development and longer storage times.

Regulatory Aspects

For aflatoxins, concern about long-term health effects has resulted in widespread regulations. At least 50 countries today legislate the aflatoxin limits for animal feeds and human foodstuffs. The actual amount allowed varies from country to country and some-

times the guidelines are tightened or relaxed depending on the quality and availability of the regulated commodity. Aflatoxin contamination frequently disrupts orderly marketing and affects the price of maize, especially in the southeastern USA.

Regulations concerning mycotoxins are also changing in certain countries owing to public perception of risk. For example, although the European Community has specified 5 ng g^{-1} as the limit for ochratoxin A in foodstuffs, the German Department of Health announced its intention in late 1997 to enforce a 3 ng g^{-1} limit for foodstuffs and a 0.3 ng g^{-1} limit for infant foods.

Table 2 Current regulatory limits for some mycotoxins

	Country/organization	Commodity	Limit (ng g^{-1})
Aflatoxins (B$_1$, B$_2$, G$_1$, G$_2$)	USA	Foodstuffs	20
		Feed maize	20–300
		Other feedstuffs	
	UN-WHO-FAO	Foodstuffs	10
Aflatoxin B$_1$	European Community	Dairy cattle feed	5
Aflatoxin M$_1$	USA	Milk	0.5
Ochratoxin A	European Community	Foodstuffs	5
T-2 toxin	Two countries	Grains	100
Deoxynivalenol	USA	Wheat	1000–2000
	Canada	Wheat	600–2000
Patulin	Ten countries	Apple juice	20–50
Diacetoxyscirpenol	Israel	Grains	100

Chemical detoxification procedures for mycotoxin-contaminated commodities are generally ineffectual, since the quality of the commodity is often degraded by the treatment, and this offsets the benefit arising from reduction of mycotoxin levels. The continuing decline in international grain prices make most detoxification procedures cost-ineffective. A major exception has been the use of ammonia in reducing aflatoxin B$_1$ in feed ingredients. In food and feed grains, the most common method for bringing mycotoxin levels within regulatory limits has been blending. This procedure requires large stocks of uncontaminated grain, and very accurate sampling and analysis procedures. The best detoxification strategy is to minimize mycotoxin contamination through safe storage practices, and through development of cultivars resistant to field fungi such as *Fusarium*.

See also: **Aspergillus**: Introduction; *Aspergillus flavus*. **Cheese**: Microbiology of Cheese-making and Maturation; Mould-ripened Varieties. **Fusarium**. **Mycotoxins**: Classification; Occurrence; Detection and Analysis by Classical Techniques; Immunological Techniques for Detection and Analysis. **Penicillium**: Introduction.

Further Reading

Abramson D (1998) Mycotoxin formation and environmental factors. In: Sinha KK and Bhatnagar D (eds) *Mycotoxins in Agriculture and Food Safety.* p. 255. New York: Marcel Dekker.

Cole RJ and Cox RH (1981) *Handbook of Toxic Fungal Metabolites.* New York: Academic Press.

Denning DW (1987) Aflatoxin and human disease. *Adverse Drug Reactions and Acute Poisoning Reviews* 4: 175–209.

Joffe AZ (1978) *Fusarium poae* and *F. sporotrichioides* as principal causal agents of alimentary toxic aleukia. In: Wyllie TD and Morehouse LG (eds) *Mycotoxic Fungi, Mycotoxins, Mycotoxicoses.* Vol. 3, p. 21. New York: John Wiley.

Samson RA and Frisvad JC (1991) Current taxonomic concepts in *Penicillium* and *Aspergillus*. *Developments on Food Science.* Vol. 26, p. 405. Amsterdam: Elsevier.

Scott PM (1994) *Penicillium* and *Aspergillus* toxins. In: Miller JD and Trenholm HL (eds) *Mycotoxins in Grain: Compounds other than Aflatoxin.* P. 261. St Paul: Eagan Press.

Smith JE and Henderson RS (eds) (1991) *Mycotoxins and Animal Foods.* Boca Raton: CRC Press.

Ueno Y (1985) The toxicology of mycotoxins. *Critical Reviews in Toxicology* 14: 99–132.

Uraguchi K (1969) Mycotoxic origin of cardiac beriberi. *Journal of Stored Product Research* 5: 227–236.

Van Egmond HP (1989) Current situation on regulations for mycotoxins. Overview of tolerances and status of standard methods of sampling and analysis. *Food Additives and Contaminants* 6: 139–188.

APPENDIX I: BACTERIA AND FUNGI

The genera listed here are those associated with food, agricultural products and environments in which food is prepared or handled.

Abiotrophia
Acinetobacter
Actinobacillus
Actinomyces
Aerococcus
Aeromonas
Agrobacterium
Alcaligenes
Alloiococcus
Anaerobiospirillum
Arcanobacterium
Arcobacter
Arthrobacter
Aureobacterium
Bacillus
Bacteroides
Bergeyella
Bifidobacterium
Blastoschizomyces
Bordetella
Branhamella
Brevibacillus
Brevibacterium
Brevundimonas
Brochothrix
Brucella
Budvicia
Burkholderia
Buttiauxella
Campylobacter
Candida
Capnocytophaga
Cardiobacterium
Carnobacterium
CDC
Cedecea
Cellulomonas

Chromobacterium
Chryseobacterium
Chryseomonas
Citrobacter
Clostridium
Comamonas
Corynebacterium
Cryptococcus
Debaryomyces
Dermabacter
Dermacoccus
Dietzia
Edwardsiella
Eikenella
Empedobacter
Enterobacter
Enterococcus
Erwinia
Erysipelothrix
Escherichia
Eubacterium
Ewingella
Flavimonas
Flavobacterium
Fusobacterium
Gardnerella
Gemella
Geotrichum
Gordona
Haemophilus
Hafnia
Hansenula
Helicobacter
Kingella
Klebsiella
Kloeckera
Kluyvera

Kocuria
Kytococcus
Lactobacillus
Lactococcus
Leclercia
Leptotrichia
Leuconostoc
Listeria
Malassezia
Methylobacterium
Microbacterium
Micrococcus
Mobiluncus
Moellerella
Moraxella
Morganella
Myroides
Neisseria
Nocardia
Ochrobactrum
Oerskovia
Oligella
Paenibacillus
Pantoea
Pasteurella
Pediococcus
Peptococcus
Peptostreptococcus
Photobacterium
Pichia
Plesiomonas
Porphyromonas
Prevotella
Propionibacterium
Proteus
Prototheca
Providencia

Pseudomonas
Psychrobacter
Rahnella
Ralstonia
Rhodococcus
Rhodotorula
Rothia
Saccharomyces
Salmonella
Serratia
Shewanella

Shigella
Sphingobacterium
Sphingomonas
Sporobolomyces
Staphylococcus
Stenotrophomonas
Stomatococcus
Streptococcus
Suttonella
Tatumella
Tetragenococcus

Trichosporon
Turicella
Veillonella
Vibrio
Weeksella
Weissella
Xanthomonas
Yarrowia
Yersinia
Yokenella
Zygosaccharomyces

APPENDIX II: LIST OF SUPPLIERS

The suppliers below are mentioned in the text as main sources of specialist equipment, culture media or diagnostic materials. This list is not intended to be comprehensive.

3M Microbiology Products
3M Center
Building 260–6B-01
St Paul
MN 55144–1000
USA

ABC Research Corporation
3437 SW 24th Avenue
Gainesville
FL 32607
USA

Adgen Ltd
Nellies Gate
Auchincruive
Ayr KA6 5HW
UK

Agi-Diagnostics Associates
Cinnaminson
New Jersey
USA

ANI Biotech OY
Temppelikatu 3–5, 00100
Helsinki
Finland

Applied Biosystems
The Perkin-Elmer Corporation
12855 Flushing Meadow Drive
St Louis
MO 63131 1824
USA

Becton Dickinson Microbiology Systems
7 Loveton Circle
Sparks
MD 21152–0999
USA

bio resources
9304 Canterbury
Leawood
KS 66206
USA

BioControl Systems
19805 North Creek Parkway
Bothwell
WA 98011
USA

BioControl Systems, Inc
12822 SE
32nd Street
Bellevue
WA 98005
USA

Bioenterprises Pty Ltd
28 Barcoo Street
PO Box 20 Roseville
NSW 2069
Australia

Biolog, Inc
3938 Trust way
Hayward
CA 94545
USA

Bioman Products, Inc
400 Matheson Blvd
Unit 4
Mississauga
Ontario
LAZ 1N8
Canada

bioMérieux
Chemin de l'Orme
69280 Marcy L'Étoile
France

bioMérieux (UK)
Grafton House
Grafton Way
Basingstoke
Hants RG22 6HY
UK

bioMérieux Vitek, Inc
595 Anglum Drive
Hazelwood
MO 63042 2320
USA

Bioscience International
11607 Mcgruder Lane
Rockville
MD 20852 4365
USA

Biosynth AG
PO Box 125
9422 Staad
Switzerland

Biotecon
Hermannswerder haus 17
14473 Potsdam
Germany

Biotrace
666 Plainsboro Road
Suite 1116
Plainsboro
NJ 08536
USA

Celsis
2948 Old Britain Circle
Chattanooga
TN 37421
USA

Celsis International plc
Cambridge Science Park
Milton Road
Cambridge
CB4 4FX
UK

Celsis-Lumac Ltd
Cambridge Science Park
Milton Road
Cambridge
CB4 4FX
UK

Charm Sciences Inc
36 Franklin Street
Malden
MA 02148 3141
USA

Chemunex Corporation
St John's Innovation Centre
Cowley Road
Cambridge
CB4 4WS
UK

Crescent Chemical Co, Inc
1324 Motor Parkway
Hauppauge
NY 11788
USA

diAgnostix, Inc
1238 Anthony Road
Burlington
NC 27215
USA

DIFCO
PO Box 331058
Detroit
MI 48232
USA

Diffchamb (UK)
1 Unit 12 Block 2/3
Old Mill Trading Estate
Mansfield Woodhouse
Nottingham NG19 9BG
UK

Diffchamb SA
8 Rue St Jean de Dieu
69007 Lyons
France

Digen Ltd
65 High Street
Wheatley
Oxford OX33 1UL
UK

DiverseyLever
Weston Favell Centre
Northampton
NN3 8PD
UK

Diversy Ltd
Technical Lane
Greenhill Lane
Riddings
DE55 4BA
UK

Don Whitley Scientific Ltd
14 Otley Road
Shipley
West Yorkshire
BD17 7SE
UK

DuPont/Qualicon
E357/1001A
Rouote 141 & Henry Clay Road
PO Box 80357
Wilmington
DE 19880 0357
USA

Dynal
PO Box 158 Skoyen
0212 Oslo
Norway

Dynal (UK) Ltd
Station House
26 Grove Street
New Ferry
Wirral
Merseyside L62 5AZ
UK

Dynal (USA)
5 Delaware Drive
Lake Success
NY 11042
USA

Dynatech Laboratories Inc
14340 Sulleyfield Circle
Chantilly
VA 22021
USA

Ecolab Ltd
David Murray John Building
Swindon
Wiltshire
SN1 1NH
UK

Envirotrace (BioProbe)
675 Potomac River Road
McLean
VA 22100
USA

Foss Electric (UK)
Parkway House
Station Road
Didcot
Oxon OX11 7NN
UK

Fluorochem Ltd
Wesley Street
Old Glossop
Derbyshire
SK13 9RY
UK

Foss Electric A/S
69 Slangerupgade
PO Box 260
DK-3400 Hillerod
Denmark

GENE-TRAK Systems
94 South Street
Hopkinton
MA 01748
USA

Gist-Brocades Australia
PO Box 83
Moorebank
NSW 2170
Australia

Gist-Brocades BV
PO Box 1345
2600 M A Delft
The Netherlands

I.U.L.
1670 Dolwick Drive
Suite 8
Erlanger
KY 41018

IDEXX Laboratories, Inc
One IDEXX Drive
Westbrook
ME 04092
USA

**Industrial Municipal Equipment
Inc (ime, Inc)**
1430 Progress Way
Suite 105
Ridersburg
MD 21784
USA

Innovative Diagnostic Systems
2797 Peterson Place
Norcross
GA 30071
USA

**International BioProducts Tecra
Diagnostics**
14780 NE 95th Street
Redmond
WA 98052
USA

Lab M Ltd
Topley House
52 Wash Lane
Bury
Lancashire
BL9 6AU
UK

Launch Diagnostics Ltd
Ash House
Ash Road
New Ash Green
Longfield
Kent DA3 8JD
UK

**Lionheart Diagnostics Bio-Tek
Instruments, Inc**
Highland Park
Box 998
Winooski
VT 05404 0998
USA

M. I. Biol
BioPharma Technology Ltd
BioPharma House
Winnall Valley Road
Winchester SO23 0LD
UK

Malthus Instruments Ltd
Topley House
52 Wash Lane
Bury
Lancashire
BL9 6AU
UK

Merck (UK) Ltd
Merck House
Poole
Dorset BH15 1TD
UK

Meridian Diagnostics Inc
3741 River Hills Drive
Cincinnati
OH 45244
USA

MicroBioLogics
217 Osseo Ave N
St Cloud
MN 56303 4455
USA

Microbiology International
10242 Little Rock Lane
Fredrick
MD 21702
USA

Microgen Bioproducts
1 Admiralty Way
Camberley
Surrey GU15 3DT
UK

MicroSys, Inc
2210 Brockman
Ann Arbor
MI 48104
USA

Minitek-BBL
BD Microbiology Systems
7 Loveton Circle
Sparks
MD 21152
USA

**Mitsubishi Gas Chemical
America, Inc**
520 Madison Avenue
25th Floor
New York
NY 10022
USA

M-Tech Diagnostics
49 Barley Road
Thelwall
Warrington
Cheshire WA4 2EZ
UK

National Food Processors Assoc
1401 New York
NW
Washington
DC 20005
USA

Neogen Corporation
620 Lesher Place
Lansing
MI 48912
USA

New Horizons Diagnostic Corp
9110 Red Branch Road
Suite B
Columbis
MD 21045 2014
USA

Olympus Precision Instruments Division
10551 Barkley
Suite 140
Overland Park
KS 66212
USA

Organon Teknika AKZO NOBEL
100 Akzo Avenue
Durham
NC 27712
USA

Oxoid, Inc
217 Colonnade Road
Nepean
Ontario
K2E 7K3
Canada

Oxyrase Inc
PO Box 1345
Mansfield
OH 44901
USA

Perkin Elmer Corporation
50 Tanbury Road
Mail Station 251
Wilton
CT 06897 0251
USA

Pharmacia Biotech
800 Centennial Avenue
PO Box 1327
Piscataway
NJ 08855 1327
USA

Prolab Diagnostics
Unit 7 Westwood Court
Clayhill Industrial Estate
Neston
Cheshire L64 3UJ
UK

QA Life Sciences Inc
6645 Nancy Ridge Drive
San Diego
CA 92121
USA

Radiometer Ltd
Manor Court
Manor Royal
Crawley
West Sussex
RH10 2PY
UK

R-Biopharm GmbH
Dolivostr 10
D-64293
Darmstadt
Germany

RCR Scientific Inc
206 West Lincoln
PO Box 340
Goshen
IN 46526 0340
USA

Remel
12076 Santa Fe Drive
Lenexa
KS 66215
USA

Rhone-Poulenc Diagnostics Ltd
3.06 Kelvin Campus
West of Scotland Science Park
Maryhill Road
Glasgow G20 0SP
UK

SciLog, Inc
14 Ellis Potter Ct
Madison
WI 53711–2478
USA

Silliker Laboratory Inc
1304 Halstead Street
Chicago Heights
IL 60411
USA

Spiral Biotech
7830 Old Georgetown Road
Bethesda
MD 20814
USA

Tecra Diagnostics
28 Barcoo Street
PO Box 20
Roseville
NSW
Australia

Tecra Diagnostics (UK)
Batley Business Centre
Technology Drive
Batley
W Yorkshire WF17 6ER
UK

Unipath, Oxoid Division
Wade Road
Basingstoke
Hampshire
RG24 8PW
UK

Unipath, Oxoid Division (USA)
800 Proctor Avenue
Ogdensburg
NY 13669
USA

Vicam
29 Mystic Avenue
Somerville
MA 02145
USA

Wescor, Inc
1220 E
1220 N
Logan
UT 84321
USA

INDEX

NOTE

Page numbers in **bold** refer to major discussions. Page numbers suffixed by T refer to Tables; page numbers suffixed by F refer to Figures. *vs* denotes comparisons.

This index is in letter-by-letter order, whereby hyphens and spaces within index headings are ignored in the alphabetization. Terms in parentheses are excluded from the initial alphabetization.

Cross-reference terms in *italics* are general cross-references, or refer to subentry terms within the same main entry (the main entry is not repeated to save space).

Readers are also advised to refer to the end of each article for additional cross-references – not all of these cross-references have been included in the index cross-references.

bacteria *(continued)*
envelope *(continued)*
S-layer 1336
structure 161–164
F⁺ and F⁻ strains 934
fat and lipid composition 718, 720T
in fermented foods 249T
fish 807
flagella *see* flagella
flavours produced by 733T
food poisoning due to 835
see also food poisoning
food spoilage by *see* spoilage of food
G+C values *see* DNA, G+C content
generation time, calibration of
impedimetric technique 587
genetic engineering *see* genetic
engineering
genetics **929–940**
gene transfer 934–938
genus 175
glycocalyx, fermented milk products 792
growth 205
after rehydration 536
environment-dependence 1709–1710
factors influencing 542–543
freezing effect 846–847
as function of environment and models
1708–1710
lag phase 543–544, 550, 665
limits 550–551
low pH foods 561–562
at low temperatures 845, 845T
minimum/maximum pH 558T, 1729F
minimum temperatures 846T
minimum water activity 841T
modified atmosphere packaging effect
414–415
normal profiles 665F
optimization 548
phases and effect of freezing 842
reaction rates and 551–552
redox dependence 557F
requirements 542–545
sous-vide products 1341–1342
submerged fermentations 665–666,
665F
suppression by salt 1724–1725
temperature control 548
temperature effect 575, 845
temperature interaction with other
factors 550
temperatures for 840
tolerance of low water activity 542,
543T
water activity levels 542, 542T
growth curve 548, 548F, 665F, 1709F
maximum carrying capacity 548, 548F
growth limit models 1706, 1706F
growth rate 1723
absolute and specific 1709F
Arrhenius plot 552, 552F
carbon dioxide effect 559
comparison of food-borne bacteria
550F
effect of freezing 848
effect of water activity 543, 543F
environmental factors effect 543, 544F
fastest-growing organisms 549–550
in food 548
interactions of factors 551, 551F
modelling 1704, 1709F
models 1707T
optimum temperature 549
solute tolerances 174F
temperature effect 549–550, 552–553,
549F
see also temperature
harvesting, centrifugation application
1685–1686
heat resistance 1340, 1340, 1341
higher taxa 175

bacteria *(continued)*
high-frequency recombination (hfr)
strains 934
identification 174, 175
electrical techniques 582
inactivation *see* bacteria, destruction;
microbial inactivation
inhibition of undesirable microbes
by fermentation 1726
by salt 1725, 1726–1727
in intestine *see* gastrointestinal flora
intracellular structures, water activity
effect 545–546
lipids 1299
lysis
by bacteriophage 203–204
PCR sample preparation 1479
by phage lysins 1473
marine, tetrodotoxin production 1674
metabolism, temperature effect 553–555
mineral uptake 1313–1314
morphology 159–160, 159F
environmental influences 160
variations and flow cytometry 828
nomenclature 174, 174
nucleoid 166
organelles 165
origin of term 173
outer membrane in Gram-negative cells
163
pathogen detection by phage-based
techniques *see* bacteriophage-based
techniques
periplasm 163
phage adherence 1471
phage as viability indicator 205
phage interactions *see* bacteriophage
phage-resistant mutants 1472
phage typing *see* phage typing
pili 164, 934
polysomes 165
preservatives active against 1712T
protective cultures, meat preservation
1271
replication 205, 933–934
replication rate, phage rate comparison
208, 208F
reproduction, binary fission 933–934
riboflavin production 730
secondary metabolites **1328–1334**
see also secondary metabolites
selective adsorption 1696–1699
single-cell protein *see* single-cell protein
(SCP)
size 160T, 1676
S-layer 163–164
atomic force microscopy 1423
slime 792
composition 793T
determinants affecting production 793
production process 793–794
production rate 794
sourdough bread 300
spore-forming 168
see also endospores; spore-formers
spores *see* endospores, bacterial
sporulation 543–544
starter cultures *see* starter cultures
storage granules 166
structure 158, 159T
sublethal injury due to freezing 844
recovery 844
surface-ripening, as starter culture 2085T
survival at low temperatures 847–848
see also psychrophiles
taxonomy 174
chemical 176–177
classical and numerical 175
genetic methods 175
major taxa 177–178
serology 176
toxin production 543–544

bacteria *(continued)*
transduction *see* transduction
transformation 935–936, 935F
competence 935
'type strains' 174
viability
freezing effect 847, 847F
freezing rates and 841
staining for 830–832
viable cell counts
alternative methods 219–220
method 219
virulence factors 1472
viruses *see* bacteriophage
water activity
inhibitory levels 1724T
requirements 542, 542T, 1723
tolerance of low levels 542, 543T
yield, water activity effect 544
bacterial adhesion
conditioning layer of organic materials
1693
control by polymer technologies **1692–
1699**
free energy change 1693
importance 1692
inhibition 1692–1696
hydrophilic surface polymers 1693–
1695
low surface energy polymers 1695
mobile surface polymers 1695–1696
polymers for retardation 1692–1693
polymer structures 1694T
thermodynamic treatments 1693
selective adsorption of bacteria 1696–
1699
see also adherence; biofilms; polymers
bacteria-specific adsorbents 1696–1698
bactericidal barriers, bacteriocins as 190
bactericides, ultrasound 1463–1464
bacteriocins **183–191**, 1711
advantages/disadvantages 188–189
applications 185
as bactericidal barriers 190
Bacteroides 202
Brevibacterium linens 309
cost-effectiveness issues 190
definition and description 184–186
detection 184, 184F
effect in fermented meat products 746
effect on *Bifidobacterium* 1356
enterococci producing 623–624
Enterococcus faecalis 1367
Enterococcus faecium 1367
fast- and slow-acting 190
future prospects 190
genetics 189
GRAS status 184, 190
harvesting 189–190
hurdle technology 1074
hydrophobicity 189
lactic acid bacteria 2103–2104
lactobacilli 1136, 1136T
Lactobacillus acidophilus 1153–1154,
1153T
Lactobacillus brevis 1149, 1150T
Lactococcus lactis 1169
Leuconostoc 1193–1194
as markers for food-grade cloning vectors
919
meat preservation 1271
Moraxella 1491
mutants resistant to 189
natural antimicrobials 1573–1574
as natural food preservatives 185
as natural products 188
origin of term 184
pediocin-like 187–188
potential uses 188
production 189–190
Proteus 1863
safety aspects 189

D

Related Journal

Food Microbiology

Editor
C.A. Batt
Cornell University, Ithaca
New York, U.S.A.

Contributing Editors
R. L. Buchanan, USDA-ARS-ERRC, Philadelphia, U.S.A.

C.O. Gill, Agriculture Canada Research Station, Alberta, Canada

S. Notermans, National Institute of Public Health and Environmental Hygiene, The Netherlands

J. I. Pitt, CSIRO Division of Food Research, North Ryde, Australia

A.N. Sharpe, Health and Welfare Canada, Tunney's Pasture, Canada

P. Teufel, Federal Health Office, Berlin, Germany

M. L. Tortorello, FDA/NCFST Summit, U.S.A.

Food Microbiology publishes primary research papers, short communications, reviews, reports of meetings, book reviews, and news items dealing with all aspects of the microbiology of foods. The editors aim to publish manuscripts of the highest quality which are both relevant and applicable to the broad field covered by the journal. Although all manuscripts will be considered, authors are encouraged to submit manuscripts dealing with the novel methods of detecting microorganisms in foods, especially pathogens of emerging importance, for example, *Listeria* and *Aeromonas*. Papers relating to the genetics and biochemistry of microorganisms that are either used to make foods or that represent safety problems are also welcomed, as are studies on preservatives, packaging systems, and evaluations of potential hazards of new food formulations. The editors make every effort to ensure rapid and fair reviews, resulting in timely publication of accepted manuscripts.

Research Areas Include

Food spoilage and safety

Predictive microbiology

Rapid methodology

Application of chemical and physical approaches to food microbiology

Biotechnological aspects of established processes

Production of fermented foods

Use of novel microbial processes to produce flavors and food-related enzymes

Database coverage includes Biological Abstracts (BIOSIS); Chemical Abstracts; Current Contents/Agriculture, Biology, and Environmental Science; Dairy Science Abstracts; Food Science and Technology Abstracts; Maize Abstracts; and Research Alerts

For further information: www.academicpress.com/foodmicro

Related Journal

LWT/Food Science and Technology

Published for the Swiss Society of Food Science and Technology by Academic Press
Official Publication of the International Union of Food Science and Technology

Editor-in-Chief
M. W. Rüegg
Federal Research Institute,
Liebefeld-Bern, Switzerland

Editor
A. D. Clarke
University of Missouri, Columbia, U.S.A.

Associate Editor
A. T. Temperli
Federal Research Institute
Wadenswil, Switzerland

Food Science and Technology/Lebensmittel-Wissenschaft und Technologie **is an established international bimonthly journal pertaining to all aspects of food science. Papers are published in the fields of chemistry, microbiology, biotechnology, food processing, and nutrition, and are written in English, German, or French. However, since English is the predominant language of scientific exchange, manuscripts submitted in English will be given preference with regard to speed of publication. Contributions are welcomed in the form of review articles, research papers, and research notes.**

For further information
www.academicpress.com/lwt

Research Areas Include

▲ **Biochemistry: food constituents, enzyme chemistry, industrial enzyme applications, analytical methods, carbohydrate and protein metabolism, and food pigments and natural colorants**

▲ **Food processing: food conservation, engineering problems, influence of processing methods on product quality, and physical properties**

▲ **Microbiology: spoilage of food by microorganisms, food fermentations, methods for detection and determination of microorganisms, and microbial toxins**

▲ **Nutrition: effects of food processing, dietary constituents, metabolism of nitrogenous biomolecules, and dietary fibers in food**

Database Coverage includes AGRICOLA, Biological Abstracts, Chemical Abstracts, and Dairy Science Abstracts

Journal of Food Composition and Analysis

Editor
Barbara Burlingame
FAO, Rome, Italy

An International Journal

An Official Publication of The United Nations University International Network of Food Data Systems (INFOODS)

The **Journal of Food Composition and Analysis** is devoted to all scientific aspects of the chemical composition of human foods. Emphasis is placed on new methods of analysis; data on composition of foods; studies on the manipulation, storage, distribution, and use of food composition data; and studies on the statistics and distribution of such data and data systems. The journal plans to expand its strong base in nutrient composition and to place increasing emphasis on other food components such as anti-carcinogens, natural toxicants, flavors, colors, functional additives, pesticides, agricultural chemicals, heavy metals, general environmental contaminants, and chemical and biochemical toxicants of microbiological origin.

RESEARCH AREAS INCLUDE

- Computer technology and information systems theory directly relating to food composition database development, management, and utilization

- Effects of processing, genetics, storage conditions, growing conditions, and other factors on the levels of chemical and biochemical components of foods

- Processes of development and selection of single-value entries for food composition tables

- Quality control procedures and standard reference materials for use in the assay of food components

- Statistical and mathematical manipulations involved with the preparation and utilization of food composition data

FEATURES

- Data and methods for natural and/or normal chemical and biochemical components of human foods—including nutrients, toxicants, flavors, colors, and functional additives

- Methods for determination of inadvertent materials in foods—including pesticides, agricultural chemicals, heavy metals, general environmental contaminants, and chemical and biochemical toxicants of microbiological origin

For further information
www.academicpress.com/jfca

Journal of
CEREAL SCIENCE

Editor-in-Chief
J. D. Schofield
The University of Reading, U.K.

Regional Editors

R. J. Hamer
Wageningen Centre for Food Sciences
The Netherlands

D. Lafiandra
University of Tuscia, Viterbo, Italy

B. A. Stone
La Trob University, Bundoora, Australia

The *Journal of Cereal Science* was established in 1983 to provide an international forum for the publication of original research papers of high standing covering all aspects of cereal science related to the functional and nutritional quality of cereal grains and their products.

The journal also publishes concise and critical review articles appraising the status and future directions of specific areas of cereal science and short rapid communications that present news of important advances in research. The journal aims at topicality and at providing comprehensive coverage of progress in the field.

Fo further information
www.academicpress.com/jcs

Research Areas Include

■ Composition and analysis of cereal grains in relation to quality in end use

■ Morphology, biochemistry, and biophysics of cereal grains relevant to functional and nutritional characteristics

■ Structure and physicochemical properties of functionally and nutritionally important components of cereal grains such as polysaccharides, proteins, oils, enzymes, vitamins, and minerals

■ Storage of cereal grains and derivatives and effects on nutritional and functional quality

■ Genetics, agronomy, and pathology of cereal crops in relation to end-use properties of cereal grains

■ Functional and nutritional aspects of cereal-based foods and beverages, whether baked, fermented, or extruded

■ Technology of human food and animal foodstuffs production

■ Industrial products (e.g., starch derivatives, syrups, protein concentrates, and isolates) from cereal grains, and their technology

Database coverage includes AGRICOLA, Biological Abstracts (BIOSIS), Chemical Abstracts, Current Contents, Food Science and Technology Abstracts, Maize Abstracts, Research Abstracts, and Science Citation Index

V500 M 80 B 25 DP 8 71521:5 P 5

EFM

ISBN 0-12-227070-3